NOBEL PRIZES AWARDED FOR RESEARCH IN MICROBIOLOGY OR IMMUNOLOGY

Year	Winners	Research in Microbiology	Research in Immunology
1969	M. **Delbruck** A. D. **Hershey** S. E. **Luria**	Discovered genetic structure and replication of bacteriophage	
1971	E. W. **Sutherland**, Jr.	Made biochemical discoveries pertinent to pathogenesis	
1972	Gerald M. **Edelman** Rodney R. **Porter**		Determined the structure of immunoglobulins
1975	David **Baltimore** Howard **Temin** Renato **Dulbecco**	Discovered reverse transcriptase and made fundamental discoveries on the interaction between tumor viruses and the host nucleic acid	
1976	B. S. **Blumberg** D. C. **Gajdusek**	Discovered antigen important in diagnosing serum hepatitis	
1977	Rosalyn S. **Yalow** R. C. I. **Guillemin** A. V. **Schally**	Demonstrated that the neurodegenerative disease kuru is caused by an "unconventional virus" now called a prion	Developed radioimmunossay (RIA) techniques; used RIA to analyze peptide hormones in the brain
1978	Daniel **Nathans** H. O. **Smith** Werner **Arber**	Used restriction enzymes to map viral genomes	
1980	Baruj **Benacerraf** George **Snell** Jean **Dausse**		Studied genetically determined structures in the cell surface that regulate immunological reactions
	Paul **Berg**	Made fundamental contributions to the biochemistry of recombinant DNA	
	Walter **Gilbert** Frederick **Sanger**	Made fundamental contributions on the sequencing of DNA	
1983	Barbara **McClintock**	Discovered mobile genetic elements (transposons)	
1984	Cesar **Milstein** Georges J. F. **Koehler** Niels **Jerne**		Developed a method to produce large quantities of monoclonal antibodies
1987	Susumu **Tonegawa**		Discovered the genetic basis of antibody diversity
1988	J. **Deisenhofer** R. **Huber** H. **Michel**	Described the structure of photosynthetic reaction center in bacteria	
1989	J. M. **Bishop** H. E. **Varmus**	Discovered the cellular origin of retroviral oncogenes, the proto-oncogenes	
1990	Joseph E. **Murray** E. Donnall **Thomas**		Advances in organ and cell transplantation in the treatment of human disease
1993	Kary **Mullis**	Developed the polymerdse chain reaction	
	Michael **Smith**	Developed a technique for generating site specific mutants	
1996	Peter C. **Doherty** Rolf M. **Zinkernagel**		Discovered how the immune system recognizes virus-infected cells
1997	Stanley B. **Prusiner**	Discovered and characterized prions as a new biological infectious agent containing only protein and no nucleic acid	

Microbiology

A HUMAN PERSPECTIVE

third edition

Eugene W. Nester
University of Washington

Denise G. Anderson
University of Washington

C. Evans Roberts, Jr.
University of Washington

Nancy N. Pearsall

Martha T. Nester

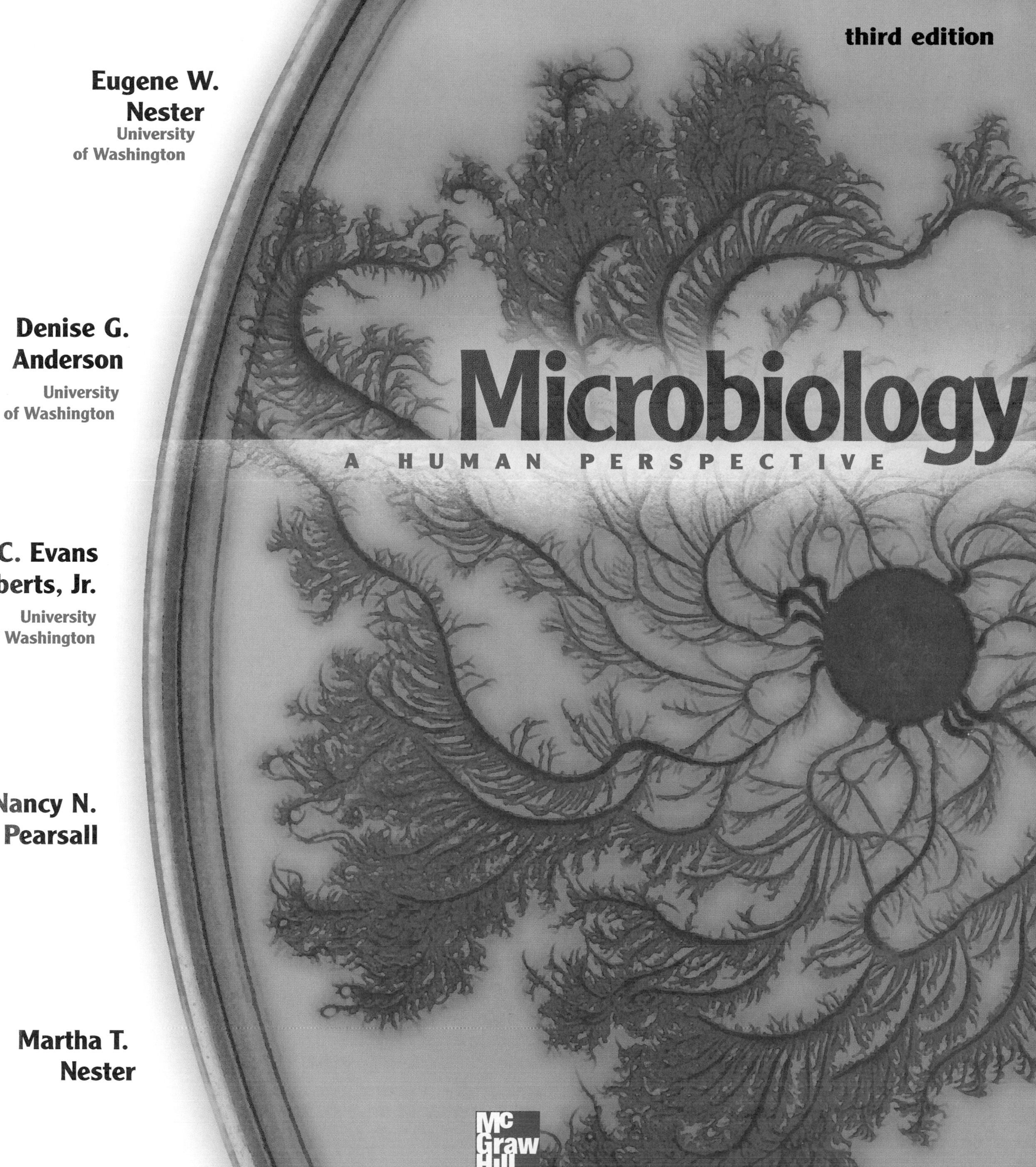

Microbiology

A HUMAN PERSPECTIVE

McGraw Hill

Boston Burr Ridge, IL Dubuque, IA Madison, WI New York San Francisco St. Louis
Bangkok Bogotá Caracas Kuala Lumpur Lisbon London Madrid Mexico City
Milan Montreal New Delhi Santiago Seoul Singapore Sydney Taipei Toronto

MICROBIOLOGY: A HUMAN PERSPECTIVE, THIRD EDITION

Published by McGraw-Hill, an imprint of The McGraw-Hill Companies, Inc., 1221 Avenue of the Americas, New York, NY 10020.

2 3 4 5 6 7 8 9 0 VNH/VNH 0 9 8 7 6 5 4 3 2 1

ISBN 0-07-231878-3
ISBN 0-07-118083-4 (ISE)

Vice president and editor-in-chief: *Kevin T. Kane*
Publisher: *James M. Smith*
Senior developmental editor: *Deborah Allen*
Marketing manager: *Thomas D. Timp*
Senior project manager: *Marilyn Rothenberger*
Media technology producer: *Lori A. Welsh*
Production supervisor: *Laura Fuller*
Coordinator of freelance design: *Rick D. Noel*
Cover/interior designer: *Christopher Reese*
Photo research coordinator: *John C. Leland*
Photo research: *Connie Mueller*
Senior supplement producer: *Stacy A. Patch*
Compositor: *Electronic Publishing Services Inc., NYC*
Typeface: *10/12 Galliard*
Printer: *Von Hoffmann Press, Inc.*

The credit section for this book begins on page C-1 and is considered an extension of the copyright page.

About the front cover: Complex patterns shown during growth of bacterial colonies result from cooperative effects and chemical communication among the bacteria. On the front cover, you see a chiral pattern developed by a colony of *Paenibacillus dendritiformis* grown on a soft thin agar substrate.
Photo provided by E. Ben Jacob, grown by E. Braines, photography by S. Avikam.

About the back cover: Multicellular organization of bacteria revealed by the regular appearance of differentiated cell groups in colonies. The three colonies at the top right contain *Escherichia coli* cells that turn blue when they produce the enzyme beta-galactosidase. The colony on the left contains *Chromobacterium violaceum* cells that can produce a purple pigment.
Photos provided by James A. Shapiro, University of Chicago.

Library of Congress Cataloging-in-Publication Data

Microbiology : a human perspective / Eugene W. Nester . . . [et al.]. 3rd ed.
p. cm.
Includes index.
ISBN 0-07-231878-3
1. Microbiology. I. Nester, Eugene W.

QR41.2 .N47 2001
579—dc21 00-055449
CIP

INTERNATIONAL EDITION ISBN 0-07-118083-4

www.mhhe.com

We dedicate this book to our students; we hope it helps to enrich their lives and to make them better informed citizens,

to our families whose patience and endurance made completion of this project a reality,

to Anne Nongthanat Panarak Roberts in recognition of her invaluable help, patience, and understanding,

to our colleagues for continuing encouragement and advice.

TABLE OF CONTENTS

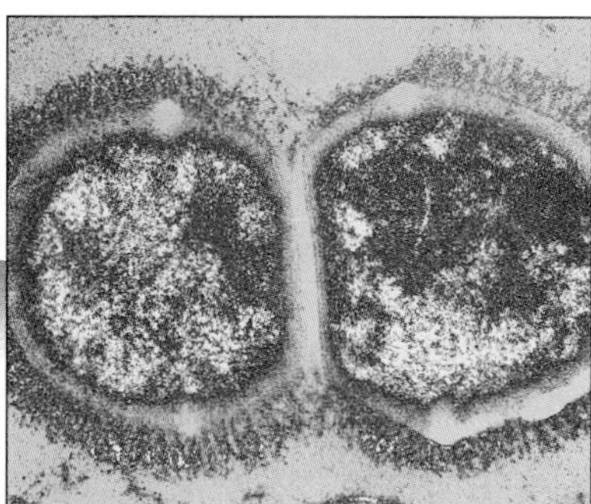

PART ONE

LIFE AND DEATH OF MICROORGANISMS

CHAPTER 1

Humans and the Microbial World 1

CHAPTER 2

The Molecules of Life 19

C H A P T E R 3

Microscopy and Cell Structure 43

C H A P T E R 4

Dynamics of Prokaryotic Growth 87

CHAPTER 9

Biotechnology and Recombinant DNA 221

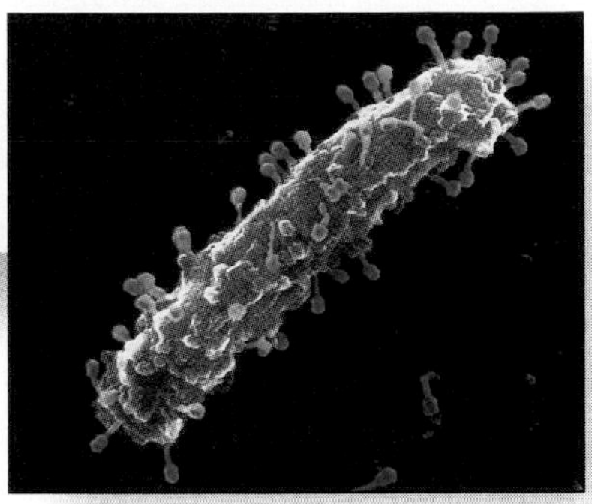

PART TWO

THE MICROBIAL WORLD

CHAPTER 10

Classification and Identification of Prokaryotes 245

CHAPTER 13

Viruses of Bacteria 323

CHAPTER 14

Viruses of Animals and Plants 341

PART THREE

MICROORGANISMS AND HUMANS

CHAPTER 15

Nonspecific Immunity 369

CHAPTER 21

Antimicrobial Medications 495

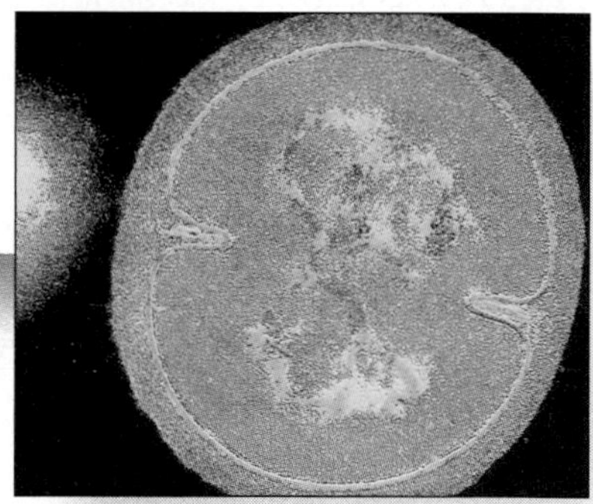

PART FOUR

INFECTIOUS DISEASES

CHAPTER 22

Skin Infections 521

CHAPTER 27

Wound Infections 691

CHAPTER 28

Blood and Lymphatic Infections 715

CHAPTER 29

HIV Disease and Complications of Immunodeficiency 741

PART FIVE

APPLIED MICROBIOLOGY

CHAPTER 30

Environmental Microbiology 767

CHAPTER 31

Water and Waste Treatment 787

CHAPTER 32

Food Microbiology 803

Eugene Nester

Eugene (Gene) Nester did his undergraduate work at Cornell University and received his Ph.D. in Microbiology from Case Western University. He then did postdoctoral work in the Department of Genetics at Stanford University with Joshua Lederberg. Since 1962, Gene has been a faculty member in the Department of Microbiology at the University of Washington. Gene's research has focused on gene transfer systems in bacteria. His laboratory demonstrated that *Agrobacterium* transfers DNA into plant cells, the basis for the disease, crown gall. He continues to study this unique system of gene transfer which has become a corner stone of plant biotechnology.

In 1990, Gene Nester was awarded the inaugural Australia Prize along with an Australian and a German scientist for their work on *Agrobacterium* transformation of plants. In 1991, he was awarded the Cetus Prize in Biotechnology by the American Society of Microbiology. He has been elected to Fellowship in the National Academy of Sciences, the American Academy for the Advancement of Science, the American Academy of Microbiology, and the National Academy of Sciences in India. Throughout his career, Gene has been actively involved with the American Society for Microbiology and currently serves as Chair of the Board of Governors of the American Academy of Microbiology.

In addition to his research activities, Gene has taught an introductory microbiology course for students in the allied health sciences for many years. He wrote the original version of the present text, *Microbiology: Molecules, Microbes and Man*, with Evans Roberts and Nancy Pearsall more than 25 years ago because they felt no suitable text was available for this group of students. The original text pioneered the organ system approach to the study of infectious disease.

Gene enjoys traveling, museum hopping, and the study and collecting of Northwest Coast Indian Art. He and his wife, Martha, live on Lake Washington with a seldom used sailboat and their dog, Otis. Their two children and four grandchildren live in the Seattle area.

Denise Anderson

Denise Anderson is a Senior Lecturer in the Department of Microbiology at the University of Washington, where she teaches a variety of courses including general microbiology, recombinant DNA techniques, medical bacteriology laboratory, and medical mycology/parasitology laboratory. Equipped with a diverse educational background, including undergraduate work in nutrition and graduate work in food science and in microbiology, she first discovered a passion for teaching when she taught microbiology laboratory courses as part of her graduate training. Her enthusiastic teaching style, fueled by regular doses of Seattle's famous caffeine, receives high reviews by her students.

Outside of academic life, Denise relaxes in the Phinney Ridge neighborhood of Seattle, where she lives with her husband, Richard Moore, two dogs, and two cats, none of which are very well trained. When not planning lectures, grading papers, or writing textbook chapters, she can usually be found chatting with the neighbors, fighting the weeds in her garden, or enjoying a fermented beverage at the local pub.

Evans Roberts, Jr.

Evans Roberts was a marginally motivated mathematics student at Haverford College when a chance encounter landed him a summer job at the Marine Biological Laboratory in Woods Hole, Massachusetts. There, interaction with leading scientists awakened his interest in biology and medicine. After completing his undergraduate work in mathematics, he studied for his M.D. at Columbia University, completed an internship at the University of Rochester School of Medicine and Dentistry, and held a Residency in Medicine at the University of Washington. Further, he received a fellowship in Infectious Disease with Dr. William M. M. Kirby and fulfilled a traineeship in Diagnostic Microbiology with Dr. John Sherris.

Subsequently, Dr. Roberts has taught microbiology, directed diagnostic microbiology laboratories, worked on hospital infection control committees, and helped in a refugee camp for Karen people in northern Thailand. He has had extensive experience in the practice of medicine as it relates to infectious diseases. He is certified both by the American Board of Microbiology and the Academy of Family Physicians.

Evans Roberts worked with Gene Nester in the early development of *Microbiology: A Human Perspective.* His professional publications concern susceptibility testing as a guide to treatment of infectious diseases, Whipple's disease, Group A streptococcal epidemiology, use of fluorescent antibody in diagnosis, bacteriocin typing, antimicrobial resistance of tuberculosis and gonorrhea, viral encephalitis, and rabies. Dr. Roberts has traveled extensively around the world. For relaxation he enjoys hiking and gardening, especially the cultivation of flowers and exotic tropical fruits.

Nancy Pearsall

Nancy Neville Pearsall attended the College of William and Mary, the University of Virginia, and the University of Washington School of Medicine, where she earned M.S. and Ph.D degrees in the areas of immunology and medical microbiology. Her research has included work on cell-mediated immunity, immunity to candidiasis, and immune responses to urinary tract infections.

Nancy's love of teaching led to writing a number of textbooks in immunology and also medical microbiology for medical students, and in microbiology for college students. The affiliation of Nester, Roberts, and Pearsall in teaching microbiology courses and writing textbooks for the courses spans more than two decades. She has also coauthored a monograph, *The Macrophage*, and studied the role of macrophages in the response to tissue transplants, as well as to infectious agents. She has served on the editorial board of several scientific journals and as a consultant to the National Institutes of Health.

Dr. Pearsall was a faculty member in the Department of Microbiology and Immunology at the University of Washington, and served for 10 years as Professor and Head of the Department of Pathology and Microbiology in the School of Medicine at the University of Zambia, part of the time as a Fulbright Professor. While in Zambia, she also helped establish a program allowing postgraduate doctors to be trained within the country of Zambia, rather than having to go overseas to specialize. She has two sons, two daughters-in-law, four grandchildren, two cats, and a springer spaniel. Huge bald eagle neighbors that fish and soar nearby and Stellar blue jays that come for breakfast every day help make her northwestern United States home a perfect substitute for the excellent animal-viewing safaris of Zambia, Zimbabwe, Botswana, Kenya, and other countries of southern and central Africa.

Martha Nester

Martha Nester received an undergraduate degree in biology from Oberlin College and a Master's degree in education from Stanford University. She has worked in university research laboratories and has taught elementary school. She currently works in an environmental education program at the Seattle Audubon Society. Martha has worked with her husband, Gene, for more than 35 years on microbiology textbook projects, at first informally as an editor and sounding board, and then in the last 22 years as one of the authors of *Microbiology: A Human Perspective.* Martha's favorite activities include spending time with their four grandchildren, all of whom live in the Seattle area. She also enjoys playing the cello with a number of musical groups in the Seattle area.

This is an exciting, enjoyable yet challenging time to be teaching and learning about microbiology. Almost every day a newspaper article describes the discovery of microbes in an ecological niche thought to be inhospitable to life, the sequencing of another microbial genome, or the death of an individual from a rare infectious disease. Anyone even glancing at the front page can't help but realize that microorganisms are very important in our daily lives. With the announcements of the many scientific advances being made about the microbial world, there also come many vehement arguments that also are played out in the popular press. Are plants that contain genes of microorganisms safe to eat? Is it wise to put antimicrobial agents in soaps and animal feed? What is the likelihood of getting AIDS from a person infected with HIV? What are the chances of finding life on Mars? This book presents what we to believe are the most important facts and concepts about the microbial world and the important role its members play in our daily lives.

An important consideration in revising this textbook was the diversity of interests among students who take an introductory microbiology course today. As always, many students are taking microbiology as a prerequisite for nursing, pharmacy, and dental programs. A suitable textbook must provide a solid foundation in health-related aspects of microbiology, including coverage of medically important bacteria, antimicrobial medications, and immunization. There is also an increasing number of students who take microbiology as a step in the pursuit of other fields, including biotechnology, food science and ecology. For these students, topics such as recombinant DNA techniques, fermentation processes and microbial diversity are essential. Microbiology is also becoming more popular as an elective for biology students, who are particularly interested in topics that highlight the relevance of microorganisms to shaping the biological world. Because of the wide range of career goals and interests of students, we have made a particular effort to broaden the scope of the previous editions, providing a more balanced approach, yet retaining our strength in medical microbiology.

Diversity in the student population is manifested not only in the array of career goals, but also in educational backgrounds. For some, microbiology may be their first college-level science course; for others microbiology builds on an already strong background in biology and chemistry. To address this broad range of student backgrounds, we have incorporated learning aids that will facilitate review for some advanced students, and will be a tremendous support to those who are seeing this material for the first time. We recognize that the number of terms in microbiology is almost enough to constitute a new language for beginning students. Therefore, we have defined key terms when they are introduced and include a comprehensive glossary. A pronunciation key is provided for names of microorganisms. We make frequent use of cross references that direct students to sections with a more thorough explanation of a concept. We recognize that a textbook, no matter how exciting the subject material, is not a novel. Few students will read the text from cover to cover and few instructors will include all of the topics covered in their course. Accordingly, we have used judicious redundancy to help present each major topic as a complete unit. We have avoided the chatty, superficial style of writing in favor of clarity and conciseness. The text is not "watered down" but rather provides students the depth of coverage needed to fully understand and appreciate the role of microorganisms in the biological sciences and human affairs.

Preparing a textbook that satisfies the needs of such a broad range of needs and interests is a daunting task, but also extremely rewarding. We hope you will find that the approach and structure of this edition presents a modern and balanced view of microbiology in our world today, acknowledging the profound and essential impact that microbes have on our lives.

What's New in This Edition?

No element of the book was left unexamined in this revision. Although previous and current users will recognize familiar features such as the Glimpse of History and the unparalleled coverage of medical microbiology, there are many new elements to explore. Some reorganization has occurred and each chapter contains material that appears for the first time in this edition. New learning aids have been incorporated such as critical thinking questions and illustrated tables. Many new topics such as quorum sensing and edible vaccines are included, and many familiar topics such as prokaryotic classification and horizontal gene transfer are presented in a more modern context.

Organization

Major changes in organization include:

1. The Table of Contents is now presented in 5 parts, with separate parts devoted to Microorganisms and Humans (Chapters 15–21) and Infectious Diseases (Chapters 22–29).
2. Control of Microbial Growth (Chapter 5) now immediately follows Dynamics of Prokaryotic Growth (Chapter 4).

3. Due to their importance in global health issues, coverage of arthropods and parasitic worms has been moved from an appendix and integrated into the text coverage where appropriate. These appear within distinct sections so that they still may be skipped in the interest of time.
4. Coverage of Respiratory System Infections now appears in a single chapter (Chapter 23) and coverage of Alimentary System Infections also appears in a single chapter (Chapter 24) so that common elements within the systems may be consolidated.
5. A "super" structure has been incorporated into lengthy chapters with more than one major topic so that students can identify discrete topics and instructors can make more directed assignments. For example in Chapter 3, Microscopy and Cell Structure, there are super-sections covering Microscopy and Cell Morphology, The Structure of the Prokaryotic Cell, and The Eukaryotic Cell.
6. Coverage of infectious diseases is indicated with yellow highlighting on the corners of pages. We hope this will facilitate its use in courses that use this material as a reference section.

Updates

With such rapid and sometimes surprising growth in the field of microbiology, new topics are found throughout the text. Moreover, familiar topics have been expanded with the addition of new information and insights as well as presented in new contexts made possible by advances in molecular biology. Recognizing the newly emerging significance of microbiology in the scientific community, each chapter ends with an essay (Future Challenges) presenting an issue of human concern facing us now or in the near future that will be tackled and possibly solved by the microbiologists of tomorrow. Epidemiological statistics have been updated. A partial list of new and updated topics follows.

- Antimicrobial resistance and gene transfer
- Overuse and misuse of germicidal chemicals
- Genomics
- Nucleotide array technology, ribotyping and other modern molecular methods
- Development of new vaccines
- New products of genetic engineering
- Quorum sensing
- Biofilms in ecology, medicine and applied microbiology
- Molecular methods to study microorganisms in nature
- Horizontal gene transfer and implications in microbial evolution
- Microbial life in the universe
- Ribosomal RNA and classification of microorganisms
- Role of viruses in cancer
- Nosocomial infections
- Emerging and re-emerging diseases
- Eradification of diseases such as measles and polio
- Updates on HIV infections and AIDS worldwide

Learning Aids

In order for students to succeed in their study of microbiology, they must be able to understand the material presented, utilize the text as a tool for learning, and enjoy reading the text. Therefore, we have continued use of many of the learning aids of previous editions and added some new ones to this edition to make the study of microbiology efficient and enjoyable. Many of these are shown in the Visual Preview included in this preface.

1. The art program has been completely revised to facilitate learning. New summary figures allow efficient review of complex systems and processes such as metabolic pathways or immune responses. More explanation appears within figures so that students can do not have to track events in a lengthy figure legend. Figures are more closely integrated with the text through boldface text references to parts and references to steps clearly identified within the text and the figure. Finally, all figures now benefit from attention to strict color coding and use of consistent icons to aid student comprehension.
2. Brief essays of human interest and contemporary relevance appear throughout the text. Each chapter opens with a Glimpse of History and closes with Future Challenges. Within a chapter, a human perspective is provided in the Perspective essays. Chapters on infectious disease contain a realistic Case Presentation.
3. New Microchecks provide opportunity to review each major section and test factual knowledge through review questions as well as offer practice in valuable critical thinking skills (through the blue questions).
4. Cross references with page numbers are provided at the ends of paragraphs mentioning concepts that are presented in more depth elsewhere in the text.
5. Figure and table references appear in boldface type throughout the text so that the context for figures and tables is easy to locate on the page.
6. Many new tables that summarize terms and concepts have been added to this edition.

   Tables outlining the major parameters of each disease appear in the chapters on infectious disease (Chapters 22–29); major diseases are also supported by a Disease Summary table that includes a visual presentation of the course of the infection.
7. Chapter summaries clearly indicate the key points under each major heading. Key figure references and table references appear in blue type. Key terms appear in boldface.
8. Each chapter still ends with Review Questions that encourage students to recap chapter content. New to this edition are Multiple Choice Questions, Applications Questions and Critical Thinking Questions. Multiple Choice questions provide self-testing; answers are available in Appendix 3. Students can apply their knowledge to real-life issues in the Applications. Critical Thinking questions, written by Robert Allen, a leading expert in critical thinking, provide practice in analyzing and using information in ways that will benefit students in any discipline. See the essay by Robert Allen in this preface.
9. A full glossary appears at the back of the book along with a pronunciation guide for names of microorganisms (Appendix 5) and appendices that provide help with

To the Student:

Included at the end of each chapter are critical thinking exercises that provide practice for applying the concepts and information included in the chapter. These exercises extend well beyond the correct recall of information or "looking up the answer." They require you to utilize critical thinking skills such as interpreting data and experimental results, predicting outcomes when conditions are changed, proposing and evaluating experimental designs, and establishing sound arguments and lines of reasoning. In other words, the emphasis is on skill development and application.

In each exercise, you should develop a logical argument that is based on information in the chapter and sound reasoning, not on opinion. The strength of your argument and reasoning will depend on appropriate concept application and structuring your argument so that it clearly leads to your conclusion and/or interpretation.

A few general guidelines for completing the exercises will be helpful:

1. Know what the question is asking. Does the question ask for a prediction? an interpretation? an experimental analysis?
2. What information is needed to solve the exercise? Now is the time for you to go back through the chapter, if necessary, to review concepts and information. Be sure to apply appropriate information to the problem.
3. What new information is provided in the exercise? How does this exercise differ from examples included in the chapter? Decide how this new information fits with and extends information in the chapter.
4. Draw a diagram or outline of the exercise. This will help organize your thoughts and insure understanding of the relationships in the exercise. It will also indicate the structure of your logic leading to your solution.

For most students, these exercises are difficult, especially at first. But as with any skill, ability and skill will improve with practice. Do not be discouraged with early difficulties; consistent effort and practice will lead to significant improvement.

Robert Allen

mathematics (Appendix 1), metabolic pathways (Appendix 2), and classification of microorganisms (Appendix 4).

Chapter Highlights

Chapter 1

Humans and the Microbial World Presents a more balanced view of the importance of microorganisms as well as the present and future challenges facing microbiologists. Includes multicellular parasites as members of the microbial world.

Chapter 2

The Molecules of Life Incorporates more biological relevance in coverage of chemistry. New sections on water chemistry and pH.

Chapter 3

Microscopy and Cell Structure Divided into three major sections - Microscopy and Cell Morphology, The Structure of the Prokaryotic Cell, and The Eukaryotic Cell. New emphasis on practical applications of various types of microscopes and staining techniques, updated coverage of transport mechanisms, expanded coverage of eukaryotic anatomy relevant to microbial infection processes. Many stunning new micrographs.

Chapter 4

Dynamics of Prokaryotic Growth New or expanded coverage of stock cultures, most probable number method, nutritional diversity, colony growth, and growth of biofilms in nature.

Chapter 5

Control of Microbial Growth Organization reflects the extent of use of a method rather than the traditional division into physical and chemical methods. Covers the factors included in choosing a control method, including associated risks and benefits.

Chapter 6

Metabolism: Fueling Cell Growth Completely revised chapter with new integrated art program and supportive summary tables. Initial overview of the fundamental principles of metabolism is followed by a deeper discussion of metabolic pathways; organization allows even beginning students to see the forest through the trees.

Chapter 7

The Blueprint of Life, From DNA to Protein The impact of genomics demands that today's students understand how cells extract and utilize information encoded in their DNA. Expansion of discussions on mechanisms make the processes of replication, transcription and translation easier to understand. Completely new art program reviewed for clarity and accuracy. Regulation now includes signal transduction, quorum sensing, and phase variation. New coverage of genomics.

Chapter 8

Bacterial Genetics Updated to include relevance of gene transfer in the spread of antibiotic resistance. Completely new art program continues clarity and style of Chapter 7. Divided into two major sections—Gene Mutation and Mechanisms of Gene Transfer.

Chapter 9

Biotechnology and Recombinant DNA Students become equipped to truly understand today's science with 1) a new discussion of DNA sequencing methods and 2) an expanded discussion of PCR that clarifies how it generates a fragment of a discrete size. Updated and relevant discussion of modern techniques including nucleotide array technology. Art program continues color and icon coding established in Chapters 7 and 8.

Chapter 10

Classification and Identification of Prokaryotes Chapter opens with identification which is the aspect of taxonomy that students will most likely experience. Routine phenotypic methods are covered, and building on the foundations laid in Chapter 9, current genotypic methods, including 16S rDNA sequencing and ribotyping are described. Phenotypic and genotypic methods of classification include the impact of 16S rDNA sequencing.

Chapter 11

The Diversity of Prokaryotic Organisms Organization and focus has been radically revised to highlight the extraordinary diversity of prokaryotes rather than their classification. Major sections presenting more than 75 genera include Metabolic Diversity and Ecophysiology. For those incorporating classification schemes into their course, Appendix 4 includes the index of the latest edition of *Bergey's Manual of Determinative Bacteriology* which is now based on phylogeny rather than upon phenotype. Many stunning new micrographs portray microbial diversity.

Chapter 12

The Eukaryotic Members of the Microbial World New groupings of algae, protozoa, and fungi are presented based upon rRNA data. New discussion of arthropod vectors and the helminth parasites.

Chapter 13

Viruses of Bacteria Reorganization presents the general principles of virology followed by biology of bacteriophage, including several new groups.

Chapter 14

Viruses of Animals and Plants Now includes methods for studying viruses. Many new tables and summary figures, including comparisons of phage and animal viruses. Updated coverage on the role of viruses in cancer. Expanded coverage of prions and viroids.

Chapter 15

Nonspecific Immunity New overview of cells and tissues involved host defense supported by completely new art program. Updated coverage of cytokines. Phagocytosis now covered before inflammation.

Chapter 16

Specific Immunity Completely revised art program helps student to follow elements from Chapter 15 through discussion of specific immunity and to a final integrative overview figure. New summary tables. Revised and updated presentation of T-cell dependent and T-cell independent antigens, antibody diversity, and immunological tolerance.

Chapter 17

Applications of Immune Responses Antibody-mediated and cell-mediated immune responses are now included in Chapter 16. New section on quantifying antigen-antibody reactions, including principle of serial dilutions. New section on tests used in cellular immunology. Updated and expanded information on vaccines and immunization procedures, including current progress in the development of new vaccines.

Chapter 18

Immunologic Disorders New art program in continuity with preceding chapters on immunology. New section on transplantation immunity, expanded treatment of autoimmune diseases.

Chapter 19

Host-Microbe Interactions New overview with supporting summary tables introduces terms and concepts used in the study of infectious diseases. Updated coverage of A-B toxins and pathogenicity islands. New sections on 1) establishing the cause of an infectious disease, including discussion of Molecular Postulates as well as Koch's Postulates, 2) modes of transmission of infectious agents, and 3) pathogenic effects of viruses and other nonbacterial agents.

Chapter 20

Epidemiology New overview section on the principles of epidemiology. Coverage of nosocomial diseases has been expanded. New section on trends in disease, including emerging diseases. New heading structure adds clarity.

Chapter 21

Antimicrobial Medications New overview introduces the important concepts of antimicrobial medications. Discussion of penicillin family expanded to provide one in-depth example. Expanded coverage of mechanisms and transfer of antimicrobial resistance.

Chapter 22

Skin Infections Updated discussion and figures on Lyme disease, expanded material on acne. Common names of diseases are emphasized as well as clinical terminology. Illustrated disease summaries are presented for Lyme disease, chickenpox, measles and rubella.

Chapter 23

Respiratory System Infections Major sections include Infections of the Upper Respiratory System and Infections of the Lower Respiratory System. Updated revision of figures showing action of diphtheria and pertussis toxins, updated coverage of hantavirus pulmonary syndrome. Illustrated disease summaries are presented for strep throat, diphtheria, tuberculosis, and influenza.

Chapter 24

Alimentary System Infections Major sections include Infections of the Upper Alimentary System and Infections of the Lower Alimentary System. Improved illustration of system anatomy. New figures showing mode of action of cholera toxin, pathogenesis of shigellosis, and hepatitis A distribution. Revised figure on hepatitis B replication. Diseases caused by parasitic worms are included, with illustrations of life cycles and course of disease. Illustrated disease summaries are presented for cholera and *E. coli* disease.

Chapter 25

Genitourinary Infections Updates and new discussion on controlling sexually transmitted diseases. New photos showing symptoms of syphilis. New sections on non-venereal genital tract infections, pubic lice, and scabies. Illustrated disease summaries are presented for leptospirosis, gonorrhea, and syphilis.

Chapter 26

Nervous System Infections New figures on the causes of meningitis, viral encephalitis, polio, and the natural cycle of the Lacrosse virus. New discussion on prospects for the eradication of polio. Illustrated disease summaries are provided for meningococcal meningitis and listeriosis.

Chapter 27

Wound Infections New figures on "flesh-eating" streptococcal disease, cat scratch disease, and the mode of action of tetanus toxin. Illustrated disease summaries include lockjaw and gas gangrene.

Chapter 28

Blood and Lymphatic Infections New coverage of the blood fluke (*Schistosoma*), and the infection hypothesis as a contributing cause of arteriosclerosis. New or revised figures on Gram : negative sepsis, distribution of tularemia, plague-infected fleas, infectious mononucleosis, and the malaria life cycle. Illustrated disease summaries include brucellosis and plague.

Chapter 29

HIV Disease and Complications of Immunodeficiency Extensively updated coverage throughout including cellular targets of HIV, global epidemiology of AIDS, effect of new treatments such as HAART, and the relationship between viral load and mortality.

Chapter 30

Environmental Microbiology Revised introductory section on the principles of microbial ecology. New section on bacteria in low-nutrient environments. Section on aquatic environments now immediately follows a completely revised section on terrestrial environments, including an overview discussion of soil.

Chapter 31

Water and Waste Treatment Updated coverage of health effects associated with waterborne organisms. Revised discussion of large-scale sewage treatment methods with new overview illustration.

Chapter 32

Food Microbiology Presentation of fermented foods has been reorganized to emphasize the microbial processes used to make food products—lactic acid fermentation by bacteria, alcohol fermentation by yeast, and changes due to mold growth. Although at the end of the text, this chapter can be a perfect follow-up to coverage of growth and metabolism.

Supplements

McGraw-Hill has developed an extensive teaching and learning package to accompany *Microbiology: A Human Perspective, third edition* that will provide unparalleled support for both students and instructors. Supplementary material appears in print versions, on CD-ROM, and on the Web. Check out some of the items available.

For the Student

Microbes in Motion Interactive CD-ROM This interactive CD-ROM for both Windows and Mac brings microbiology to life through interactive screens, video, audio, animations, and hyperlinking. This easy to use tutorial can go from the classroom to the resource center to the student's own personal computer. Ideal for self-quizzing, class preparation, or review of microbiological concepts.

Hyperclinic CD-ROM Students will have fun with this interactive CD-ROM while learning valuable concepts and gaining practical experience in clinical microbiology. Packed with over 100 case studies and over 200 pathogens supported with audio, video, and interactive screens, students will gain confidence as they take on the role of the professionals. Available Fall 2001.

Student Study Guide This valuable student resource written by Rick Corbett and Michael Lema of Midlands Technical College goes beyond the standard multiple choice and true-false self-quizzing. The authors have provided a wealth of study assets to help students truly master the material. In addition to unique learning activities, it includes key concepts, vocabulary review, self-tests, and more.

Laboratory Manual The third edition of *Microbiology Experiments: A Health Science Perspective* by John Kleyn and Mary Bicknell has been prepared to directly support the text (although it may be used with other microbiology textbooks). The laboratory manual features health-oriented experiments and endeavors also to reflect the goals and safety regulation guidelines of the American Society for Microbiologists. The class-tested exercises are modular and do not require a great deal of time either in lab or in preparation. Equipment and materials for the labs, including the laboratory manual itself, are inexpensive. Finally, all experiments are safe—they do not call for use of pathogens or human samples.

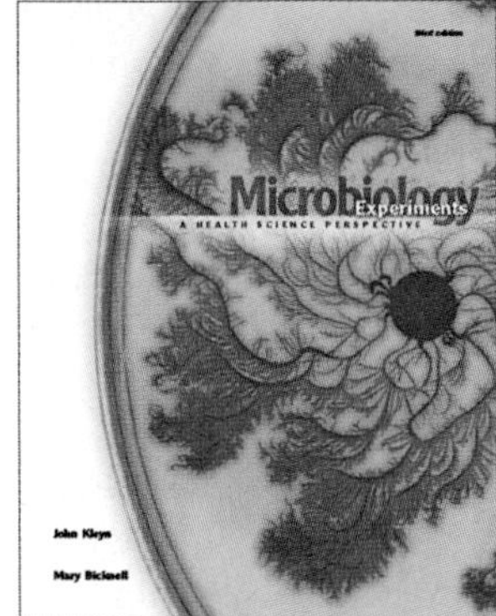

New to this edition is a series of engaging student projects that introduce some more intriguing members of the microbial world and expand the breadth of the manual beyond health-related topics. New experiments introduce modern techniques in biotechnology such as use of restriction enzymes and use of a computer database to identify sequence information. Five new appendices have been added to provide background in basic technique and practice problems. A new *Preparator's Manual* including answers to exercises, tips for successful experiments, lists of microbial cultures with sources and storage information, formulae and sources for stains and reagents, directions and recipes for preparing culture media, and sources of supplies is available to instructors for this edition.

For the Instructor

Transparencies A set of 200 images from the textbook are enhanced for classroom projection and available to adopters of the third edition of *Microbiology: A Human Perspective.*

Visual Resource Library This valuable CD-ROM contains all of the images from the textbook as well as the tables that appear in the textbook. This presentation software allows you to create your own multimedia presentations or export images into other programs. Images may be sorted by a number of criteria, and may be viewed in groups using the Small Gallery view.

Projection Slides Slide sets are available that show clinical examples of diseases or examples of microbial specimens.

Instructor's Manual and Test Item File Prepared by Michael Lema and Rick Corbett of Midlands Technical College, this valuable resource provides approximately 2,000 test items including multiple choice questions coded for level of difficulty, plus matching, true-false, and critical thinking questions. The manual also includes Learning Objectives keyed to the Student Study Guide, correlations to the multimedia and Web resources available with the text, answers to questions in the text, and support for teaching using a critical thinking approach.

Computerized Test Bank This helpful computerized testing and classroom management software from Microtest provides a sortable database of objective questions from the Test Item File for preparing exams. Available for Windows or Mac, it also includes an easy-to-use grade-recording program.

PageOut, PageOut Lite, McGraw-Hill Course Solutions Designed specifically to help you with your individual course needs, these products and services will assist you in integrating your syllabus with *Microbiology: A Human Perspective, third edition* state-of-the-art media tools. Create your own course-specific web page supported by McGraw-Hill's extensive electronic resources, set up a class message board or chat room online, provide online testing opportunities for your students, and more! For further information on these features, visit the Nester Web site at www.mhhe.com/nester.

Multimedia Resources

Online Learning Center Through the Nester 2001 Online Learning Center, everything you need for effective, interactive teaching and learning is at your fingertips. Moreover, this vast McGraw-Hill resource is easily loaded into course management systems such as Web CT or Blackboard.

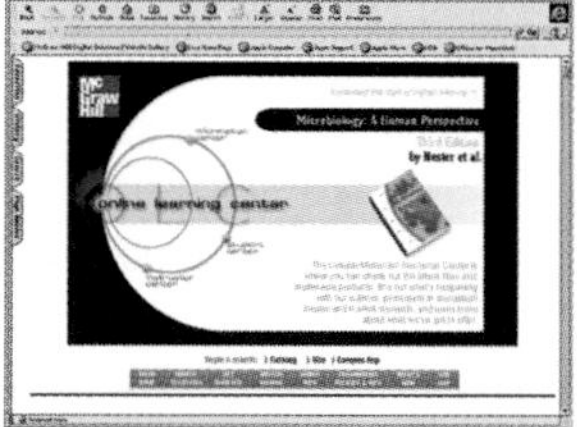

Some of the online features you will find to support your use of *Microbiology: A Human Perspective, third edition* include:

For the Student:

- Additional multiple choice questions in a self-quizzing interactive format
- Electronic flash cards to review key vocabulary
- Study Outlines
- Tips for Solving Critical Thinking exercises
- Student Tutorial Service

For the Instructor:

- All of the images and tables from the text in an uploadable format for classroom presentation
- Correlation guides for use of all resources available with the text and to the ASM Guidelines
- Answers to text questions
- Course Consultant service to answer your specific questions about using McGraw-Hill resources with your syllabus
- Tips for teaching using Critical Thinking exercises

Plus

- An interactive Time Line detailing events and highlighting personalities critical to the development of the science of microbiology
- A supplementary section on clinical and diagnostic microbiology including additional Case Presentations and other resources
- Web exercises encouraging practice of use of the Web to gather and evaluate information
- Further Reading and Web Links to explore topics in microbiology more extensively

www.mhhe.com/nester

Interactive E-Source

The Interactive E-Source is an exciting student resource that combines McGraw-Hill print, media, study and web-based materials into one easy-to-use CD-ROM. This CD-ROM provides cutting edge technology that accommodates all learning styles, and complements the printed text. It provides a truly non-linear experience by using video, art, web-based, and other course materials to help students organize their study. Features of the Interactive E-Source include:

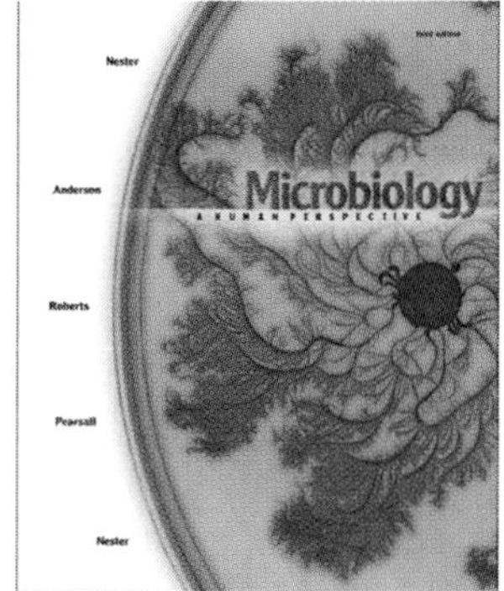

All narrative, art and photos, plus expertly crafted animations and video are included in the interlinked HTML files of the full textbook and study guide. Students can perform non-linear searches as well as retrieve any media content relevant to a topic.

Targeted web links supported by a powerful search engine encourage focused web research.

An annotation function allows students to enhance their study by personalizing e-Source material with their own notes and comments. A customized binder allows students to extract information and customize materials for future reference. The highlighting function allows students to quickly identify important concepts as they review their materials.

A special Read feature converts written text to audio, enabling students to listen to important concepts and hear the proper pronunciation of key terms and names.

Available Fall, 2001.

Reviewers of the Third Edition

We offer our sincere appreciation to the many gracious and expert professionals who helped us with this revision by offering helpful suggestions. In addition to thanking those individuals listed here who carefully reviewed revised chapters, we also thank those who responded to our informal surveys, those who viewed illustrations as they were rendered and revised, those who solicited feedback for us from their students, those who participated in regional focus groups, and those participants who chose not to be identified. All of you have contributed significantly to this work and we thank you.

Jameel Al-Dujali, Louisiana State University at Eunice
Barry Anderson, Portland Community College
Delia Anderson, University of Southern Mississippi
Rao Ayyagari, Lindenwood College
Al Brown, Auburn University
Dan Brown, Sante Fe Community College
Anne Camper, Montana State University
Daniel Caprioglio, University of Southern Colorado
Elizabeth Carrington, Tarrant County Junior College
Bret Clark, Newberry College
John Clausz, Carroll College
William Coleman, University of Hartford
Rick Corbett, Midlands Technical College
Donna Daugherty, Floyd College
Michael Davis, Central Connecticut State University
Ted Drouin, University of Alberta
David Filmer, Purdue University
S. Marvin Friedman, Hunter College, City University of New York
Juliet Fuhrman, Tufts University
Joseph Gauthier, University of Alabama at Birmingham
David Giron, Wright State University
Terry Giugni, Chaffey College
Diane Godin, Richland Community College
Steve Greenwald, Gordon College
Dana Haldeman, Community College of Southern Nevada
James Helliger, Cancer Research Institute of New England
Dawn Holsapple, Schenectady County Community College
David Hurley, South Dakota State University
Suzanne Huth, Louisiana Tech University
Suzanne Kelly, Scottsdale Community College
Harry Kestler, Lorain County Community College
Christopher Kirk, University of Michigan Medical Center
Ed Leadbetter, University of Connecticut
Michael Lockhart, Truman State University
Andrea Mastro, Pennsylvania State University
Trudy McKee, Thomas Jefferson Medical College
Blair McMillan, Madison Area Technical College
Catherine McVay, Texas Tech University Health Sciences Center
Brian Merkel, University of Wisconsin at Green Bay
Robert Moldenhauer, St. Clair County Community College
Thomas Montie, University of Tennessee
Douglas Oba, Brigham Young University at Hawaii
Mark Peppler, University of Alberta
Bobbie Pettriess, Wichita State University
Barbara Poole, Bossier Parish Community College
Laraine Powers, East Tennessee State University
Fred Rosenberg, California Lutheran University
Harry Rowen, University of Nebraska at Omaha
Doug Schelhaas, University of Mary, North Dakota
Wendy Schlucter, University of New Orleans
Thomas Schmidt, Michigan State University
Brian Shmaefsky, Kingwood College
Sara Silverstone, State University of New York at Brockport
Ann Smith, University of Maryland
Jim Smith, Emory University
Kathy Smith, Emory University
Cynthia Sommer, University of Wisconsin at Milwaukee
Angela Spence, Southwest Missouri State College
Christine Tachibana, University of Washington
Marcelo Tolmasky, California State University at Fullerton
Thomas Matthew Walker, University of Central Arkansas
Paul Wanda, Southern Illinois University at Edwardsville
Terry Werner, Harris Stowe State College
Luman Wing, San Diego State University
Chris Woolverton, Kent State University

Acknowledgments

We thank our colleagues in the Department of Microbiology at the University of Washington who have lent their support of this project over many years. Our special thanks go to: John Leigh for advice on the coverage of microbial diversity, James Staley for advice on the classification of prokaryotes, and Mary Bicknell, Mark Chandler, Jimmie Lara, and Sharon Shultz for their general suggestions and encouragement.

We would also like to thank Richard Moore, who inadvertently learned a great deal about microbiology while he critiqued and proofed many of the chapters. As a person with no formal training in science, he gave many helpful suggestions for making the fundamental chapters "reader-friendly."

Above all we would like to thank our extraordinary developmental editor Deborah Allen, who redefined the concept of dogged determination. Surprisingly, she still has hair left to pull, but her constant suggestions for improvement proved to be worth the ordeal. Her abilities and persistence are truly remarkable.

Additionally, we would like to thank Kathy Naylor who developed, created, and sketched the new and revised figures that appear in this edition. Her care and talent have transformed our acclaimed art program into one that we believe is truly exceptional. We thank Robert Allen and Brian Shmaefsky for their valuable contributions of critical thinking questions and applications. We are also grateful for the skillful assistance of the McGraw-Hill staff, including Jim Smith, our editor, Connie Mueller and John Leland, our photo editors, Rick Noel, our designer, and Stacy Patch and Lori Welsh, our print and multimedia supplements coordinators. Special appreciation goes to Marilyn Rothenberger, our project manager, who directed this project through the complexities of the publishing process while maintaining good humor along with the highest standards for accuracy and quality.

We hope very much that this text will be interesting and educational for students and a help to their instructors. We would appreciate any comments and suggestions from our readers.

Eugene Nester
Denise Anderson
C. Evans Roberts, Jr.
Nancy Pearsall
Martha Nester

Visual Preview

The next few pages show you the tools found throughout the text to help you in your study of microbiology.

CHAPTER 11

The Diversity of Prokaryotic Organisms

In his native country, the Netherlands, Cornelis B. van Niel (1897–1985) earned a degree in chemical engineering from the Technological University at Delft. At the Delft School, as it is often called, an outstanding general and applied microbiology program within the Department of Chemical Technology was chaired in succession by two prominent microbiologists—Martinus Beijerinck and Albert Kluyver.

After earning his degree in 1923, van Niel accepted a position as assistant to Kluyver, caring for an extensive culture collection and helping prepare demonstrations for lecture courses. Kluyver was relatively new to the school, but he had a vast knowledge of microbiology and biochemistry. Although little was known at the time about metabolic pathways, Kluyver believed that biochemical processes were fundamentally the same in all cells and that microorganisms, which can be grown in pure culture, could be an important research tool, serving as a model to study biochemical processes. Thirty years later, Kluyver and van Niel would present lectures that would be published in a book entitled The Microbe's Contribution to Biology. *Under Kluyver's direction, van Niel began studying the photosynthetic activities of vividly colored purple bacteria such as* Chromatium *species, a subject for which he developed a lifelong interest.*

Shortly after earning his Ph.D. in 1928, van Niel moved to the United States, bringing with him the intense appreciation for general microbiology that had been fostered at the Delft School. Settling at the Hopkins Marine Station in California, he continued his work on purple photosynthetic bacteria. Using systematic methods, van Niel conclusively showed that the growth of these organisms is light dependent, yet they do not evolve O_2. Furthermore, his experiments showed that in order to incorporate CO_2 into cellular material, these anoxygenic phototrophs oxidize hydrogen sulfide. He noted that the reaction stoichiometry of this process was remarkably similar to that of the photosynthesis of green plants and algae, except hydrogen sulfide was used in place of water, and oxidized sulfur compounds were produced instead of O_2. This finding raised the possibility that O_2 generated by plants did not come from carbon dioxide, as was believed at the time, but rather from water.

In addition to his scientific contributions, van Niel was recognized as an outstanding teacher. During the summers at Hopkins Marine Station, he taught a bacteriology course, inspiring many microbiologists with his enthusiasm for the diversity of microorganisms and their importance in nature. His keen memory and knowledge of the literature, along with his appreciation for the remarkable abilities of microorganisms, enabled him to successfully impart the awe and wonder of the microbial world to his students.

—A Glimpse of History

SCIENTISTS ARE ONLY BEGINNING TO UNDERSTAND the vast diversity of microbial life. Although a million species of prokaryotes are thought to exist, only approximately 6,000 of these, grouped into 850 genera, have been actually described and classified. Traditional culture and isolation techniques have not supported the growth, and subsequent study, of the vast majority. Not surprisingly, most effort has been put into the study of microbes intimately associated with the human population, especially those causing disease, and these have been most extensively described. This situation is changing as new molecular techniques aid in the discovery and characterization of previously unrecognized species. The sheer volume of the rapidly accruing information made possible by this modern technology, however, can be daunting for scientists and students alike.

The phylogenetic relationships being elucidated by the ribosomal RNA studies discussed in the previous chapter are causing significant upheaval in prokaryotic classification schemes. Some organisms, once grouped together based on their phenotypic similarities, have now been split into different taxonomic units based on their ribosomal RNA differences.

267

Glimpse of History

Each chapter opens with an engaging story about the men and women who pioneered the field of microbiology.

Learn of the heartbreaks, triumphs, and strokes of luck that produced the subject you study today.

294 Chapter 11 The Diversity of Prokaryotic Organisms

Figure 11.33 **Typical Habitat of *Sulfolobus*** Sulfur hot spring in Yellowstone National Park.

Thermophilic Extreme Acidophiles

Members of two genera, *Thermoplasma* and *Picrophilus*, are notable for their preference of growing in extremely acidic, hot environments. *Thermoplasma* species grow optimally at pH 2; in fact, *T. acidophilum* lyses at neutral pH. It was originally isolated from coal refuse piles. *Picrophilus* species tolerate conditions that are even more acidic, growing optimally at a pH below 1. Two species have been isolated in Japan from acidic areas in regions that exude sulfurous gases.

MICROCHECK 11.9

Archaea typically inhabit extreme environments that are otherwise devoid of life. These include conditions of high salinity, heat, acidity, and alkalinity.

- Why do seawater ponds sometimes turn pink as the water evaporates?
- At which relative depth in a sulfur hot spring would a sulfur reducer likely be found? How about a sulfur oxidizer?
- What characteristic of the methanogens makes it logical to discuss them with the Bacteria rather than the other Archaea?

FUTURE CHALLENGES

Astrobiology: The Search for Life on Other Planets

If life as we know it exists on other planets, one form it will take will likely be microbial. The task, then, is to figure out how to find and detect such extraterrestrial microorganisms. Considering that we still know relatively little about the microbial life on our own planet, coupled with the extreme difficulty of obtaining or testing extraterrestrial samples, this is a daunting challenge with many as yet unanswered questions. For example, what is the most likely source of life on other planets? What is the best way to preserve specimens for study on earth? What will be the culture requirements to grow such organisms? ***Astrobiology****, the study of life in the universe, is a new field that is bringing together scientists from a wide range of disciplines, including microbiology, geology, astronomy, biology, and chemistry, to begin answering some of these questions. The goal is to determine the origin, evolution, distribution, and destiny of life in the universe. These Astrobiologists are also given the task of developing lightweight, dependable, and meaningful testing devices to be used in future space missions.*

Astrobiologists believe that within our solar system, life would most likely be found either on Europa, a moon of Jupiter, or on Mars. This is because Europa and Mars appear to have, or have had, water, which is crucial for all known forms of life. Europa has an icy crust, beneath which may be liquid water or even a liquid ocean. Mars is the planet that is closest to Earth, and it has the most similar environment. Photographs suggest that flowing water once existed there. Besides missions to both these bodies, NASA also has future plans to return material from both a comet and an asteroid.

To prepare for researching life on other planets, microbiologists have turned to some of the most extreme environments here on Earth. These include glaciers and ice shelves, hot springs, deserts, volcanoes, deep ocean hydrothermal vents, and subterranean features such as caves. Because select microorganisms can survive in these environs, which are analogous to conditions expected on other planets, they are good testing grounds for the technology to be used on future missions.

Future Challenges

Each chapter ends with a pending challenge facing microbiologists and future microbiologists.

See how learning to share this planet more effectively with the multitude of astonishing microorganisms will shape our lifestyles and our world of tomorrow.

786 Chapter 30 Environmental Microbiology

S U M M A R Y

Principles of Microbial Ecology

1. Microorganisms are found throughout the biosphere. The best adapted organism takes over its environment.
2. Within the **biosphere, ecosystems** vary in their **biodiversity** and **biomass**.
3. The **microenvironment** immediately surrounding a microorganism is most relevant to its survival and growth.
4. Microorganisms can change their environments and can adapt to environmental change.

Bacteria in Low-Nutrient Environments

1. Low-nutrient environments are common in nature, and most organisms in such environments grow in biofilms. (Figure 30.1)
2. Organisms in low-nutrient environments transport nutrients into their cells very efficiently.

Microbial Competition and Antagonism

1. Microbial competition demands rapid reproduction and efficient nutrient use. (Figure 30.2)
2. Antagonism helps determine the make-up of a community.

Microorganisms and Environmental Changes

1. Microbial populations both cause and adapt to environmental changes. (Figure 30.4)
2. Mutants may be selected or enzymes may be induced to allow microorganisms to adapt to a new environment. (Figure 30.3)
3. Growth and metabolism of organisms may change environmental conditions significantly.

Terrestrial Environments

1. The soil teems with a broad diversity of organisms that are essential for modifying, degrading, and producing biologically important substances.
2. Environmental influences such as moisture, pH, temperature, and nutrient supply affect the numbers and kinds of organisms in soil.
3. Some soil microorganisms are pathogens of plants, animals, or people.

Microorganisms and Soil

1. Bacteria called **actinomycetes** are the most common soil bacteria. They can produce antibiotics and **geosmins**, and are essential in biogeochemical cycling.
2. Soil fungi may be free-living or they may be symbiotic in **mycorrhizae** or **lichens**. They are important in decomposing plant matter.
3. The algae in soil serve as nutrients for fungi, protozoa, and worms.
4. Protozoa are consumers of soil bacteria and algae. Together with termites, they decompose wood.

Environmental Influences in Soil

1. Important environmental influences in soil include moisture, acidity, temperature, and nutrient availability.

Aquatic Environments

1. Virtually all water contains living organisms, which vary greatly with different aquatic environments, from fresh water to seawater and some extreme habitats, such as salt lakes.
2. **Primary producers** in aquatic environments convert inorganic substances such as CO_2 to organic matter by photosynthesis.

Energy Sources for Ecosystems

1. Energy for ecosystems can come from sunlight via **photosynthesis** or from chemical synthesis of inorganic and organic materials by **chemoautotrophic** microorganisms.

Biogeochemical Cycling

1. Microorganisms are essential in biogeochemical cycling of biologically important elements such as oxygen, carbon, nitrogen, sulfur, and phosphorus, among others. (Figure 30.7)
2. Recycling processes require the activities of **producers, consumers**, and **decomposers**.

Oxygen Cycle (Figure 30.8)

1. Oxygen is cycled by the processes of photosynthesis and respiration.

Carbon Cycle (Figure 30.9)

1. The carbon cycle revolves around CO_2, its fixation into organic compounds by primary producers, and its regeneration, mostly by microorganisms.

Nitrogen Cycle (Figure 30.11)

1. Processes fueling the nitrogen cycle include **ammonification, nitrification, denitrification**, and **nitrogen fixation** by free-living and symbiotic nitrogen fixers. (Table 30.1)
2. A vital facet of the nitrogen cycle is the ability of microorganisms to convert atmospheric nitrogen to biologically useful forms.

Sulfur Cycle (Figure 30.13)

1. The sulfur cycle bears many resemblances to the nitrogen cycle.

Phosphorus Cycle and Other Cycles

1. Bacteria and fungi are largely responsible for making organic phosphorus available through the action of the enzyme phosphatase.

Bioremediation: The Biological Cleanup of Pollutants

1. Bioremediation is the biological cleanup of pollutants. It may involve the use of specially selected organisms introduced into the polluted habitat, or it may use organisms already present, perhaps with added nutrients to encourage their activities.

Pollutants

1. Biodegradable pollutants are removed within a relatively short time, but many synthetic compounds remain in the environment for long periods of time, or indefinitely. (Figure 30.15)

Chapter Summary

Important points are listed under each major heading.

Key figure references and table references are highlighted in blue.

Key terms appear in boldface type.

End-of-Chapter Review

Short Answer Questions review major chapter concepts.

Multiple Choice Questions allow self-testing; answers are provided in Appendix 3.

Applications provide an opportunity to use knowledge of microbiology to solve real-world problems.

Critical Thinking Questions encourage practice in analysis and problem-solving that can be used in the study of any subject.

2. Biological magnification occurs when compounds are taken up by sequential members of the food chain, thereby concentrating large amounts of the polluting compound in higher levels in the food chain. (Table 30.2)

Means of Bioremediation

1. Organisms already in the environment or specially selected organisms are used to degrade and/or recycle materials in pollutants. Sometimes, nutrients are added to encourage growth of the organisms. (Figure 30.16)

R E V I E W   Q U E S T I O N S

Short Answer

1. Why are microorganisms well suited to recycle elements?
2. How is the decomposition of organic matter achieved?
3. List several functions of fungi in soil.
4. How are proteins decomposed in natural environments?
5. What is the importance of nitrogen fixation?
6. List at least four genera of soil microorganisms that are pathogenic for humans and note whether each genus is bacterial, fungal, or something else.
7. Why is there a high concentration of microbes in the rhizosphere of plants?
8. Contrast ecosystems supported by photosynthesis with those that depend on chemoautotrophy.
9. What are some differences between warm sea vents and black smokers, and how do these differences affect the microbial flora found in each location?
10. How can apparently barren basalt rocks deep under the surface of the earth support microbial growth?
11. Give examples of free-living and symbiotic nitrogen-fixing microorganisms. Are these prokaryotic or eukaryotic?
12. Outline the symbiotic relationship between rhizobia and leguminous plants.
13. Describe the use of bioremediation in the cleanup of oil spills.

Multiple Choice Questions

1. Atmospheric nitrogen can be used…
   A. directly by all living organisms.
   B. only by aerobic bacteria.
   C. only by anaerobic bacteria.
   D. in symbiotic relationships between rhizobia and plants.
   E. in photosynthesis.
2. Extremophiles…
   A. are found in the polar ice caps.
   B. exist under great atmospheric pressure in the ocean floor.
   C. live in salt lakes.
   D. live within basalt rocks.
   E. All of the above
3. Mycorrhizae represent associations between plant roots and microorganisms that…
   A. are antagonistic.
   B. help plants take up phosphorus and other nutrients from soil.
   C. involve algae in the association with plant roots.
   D. form nodules on the plant's leaves.
   E. lead to the production of antibiotics.
4. The decomposition of organic matter…
   A. is carried out by only a few bacterial species.
   B. causes the production of oxygen.
   C. involves all the biogeochemical cycles discussed.
   D. involves photosynthesis.
   E. is largely symbiotic.
5. In symbiotic nitrogen fixation by rhizobia and legumes…
   A. the amount of nitrogen fixed is much greater than by nonsymbiotic organisms.
   B. neither the bacteria nor the legume can exist independently.
   C. the bacteria enter the leaves of the legumes.
   D. bacteroids are found in the leaves of the legumes.
   E. the bacteria operate independently of the legume.
6. To compete successfully, microorganisms must…
   A. produce endospores.
   B. reproduce more rapidly than their competitors.
   C. be aerobic.
   D. be anaerobic.
   E. be symbiotic.
7. Individual microorganisms are most affected by…
   A. the amount of oxygen in the region.
   B. sunlight.
   C. their microenvironment.
   D. the temperature.
   E. the pH.
8. Energy for ecosystems can come from…
   A. sunlight via photosynthesis.
   B. chemical synthesis by chemoautotrophs.
   C. Both A and B
   D. Only A
   E. Only B
9. Chemically synthesized compounds are most likely to be biodegradable if they…
   A. are totally different from anything found in nature.
   B. have three chlorine atoms per molecule.
   C. are plastics.
   D. are present in very large amounts.
   E. are chemically similar to naturally occurring substances.
10. Numbers of living organisms in an environment can be estimated by all of the following techniques, *except*…
    A. measurement of ATP in the sample.
    B. nucleic acid probes.
    C. Gram staining.
    D. staining with dyes that only stain living cells.
    E. A and D

Applications

1. A farmer who was growing soybeans, a type of legume, saw an Internet site advertising an agricultural product for safely killing soil bacteria. The ad claimed that soil bacteria were responsible for most crop losses. The farmer called the agricultural extension office at a local university for advice. Explain what the extension office crop adviser most likely told the farmer about the usefulness of the product.
2. Recent reports suggest that human activities, such as the generous use of nitrogen fertilizers, have doubled the rate at which elemental nitrogen is fixed, raising concerns of environmental overload of nitrogen. What problems could arise from too much fixed nitrogen, and what could be done about this situation?
3. Nearly every summer, a huge area of oxygen-depleted, almost lifeless ocean spreads off the coast of Louisiana into the Gulf of Mexico. Scientists claim that this "dead zone" is the result of nitrogen- and phosphorous-containing fertilizer used in farmland along the Mississippi River. Fertilizer washes down the river, into the Gulf of Mexico, and nourishes the increased growth of algae and tiny organisms that feed on the algae. Two questions can be asked about this effect:
   a. How can increased algae and other organisms deplete oxygen from the surface layers of the water?
   b. Algae live only near the surface, where sunlight is available. Why would oxygen be depleted far below the surface?

Critical Thinking

1. A student argued that if soil particles were all the same size and all the same composition, then only one kind of microorganism would be found in the soil. What was the student's probable reasoning? How would you criticize the student's argument?
2. Large populations of bacteria are found living almost 3 km underground. Most of these are autotrophs and derive their nutrients and energy from inorganic chemicals in the immediate environment. Surprisingly, many other bacterial species are found that require organic material as a nutrient and energy source. Since organic material will not be carried from the surface to such depths, where can these bacteria obtain the necessary organic material?
3. Each colony of microorganisms growing on an agar plate arises from a single cell (see photo). Colonies growing close together are much smaller than those that are well separated. Why would this be so?

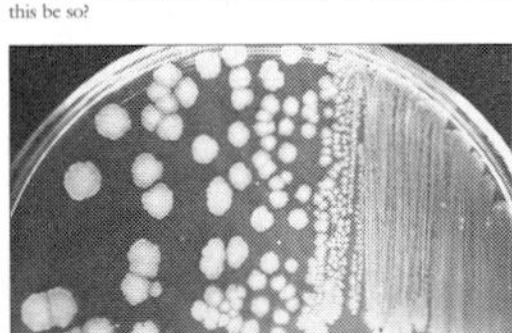

4. An entrepreneur found an economically feasible way of collecting large amounts of sulfur from underwater hot vents in the Pacific Ocean. The sulfur will be harvested from the microorganisms found in the vent areas. A group of ecologists argued that the project would destroy the fragile ecosystem by depleting it of usable sulfur. The entrepreneur argued that the environment would not be harmed because the vents produce an unlimited source of sulfur for the clams and tube worms in the area. Explain who is correct in their assessment.
5. Describe the kind of life forms one would expect to find on Mars or the moon, given what we know about conditions there.

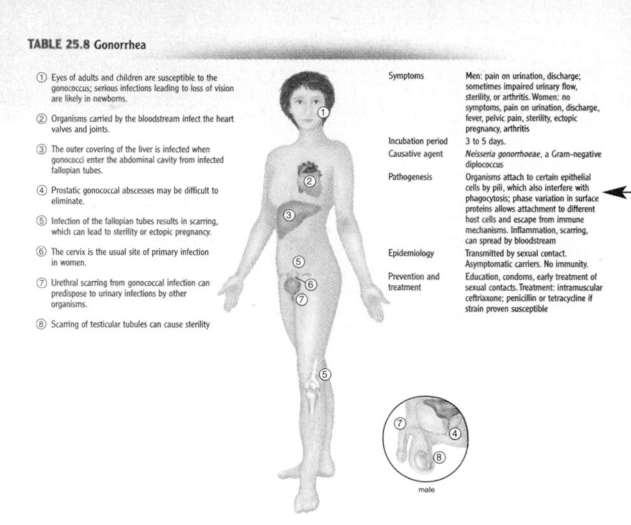

TABLE 25.8 Gonorrhea

① Eyes of adults and children are susceptible to the gonococcus; serious infections leading to loss of vision are likely in newborns.

② Organisms carried by the bloodstream infect the heart valves and joints.

③ The outer covering of the liver is infected when gonococci enter the abdominal cavity from infected fallopian tubes.

④ Prostatic gonococcal abscesses may be difficult to eliminate.

⑤ Infection of the fallopian tubes results in scarring, which can lead to sterility or ectopic pregnancy.

⑥ The cervix is the usual site of primary infection in women.

⑦ Urethral scarring from gonococcal infection can predispose to urinary infections by other organisms.

⑧ Scarring of testicular tubules can cause sterility

Symptoms	Men: pain on urination, discharge; sometimes impaired urinary flow, sterility, or arthritis. Women: no symptoms, pain on urination, discharge, fever, pelvic pain, sterility, ectopic pregnancy, arthritis
Incubation period	3 to 5 days.
Causative agent	*Neisseria gonorrhoeae*, a Gram-negative diplococcus
Pathogenesis	Organisms attach to certain epithelial cells by pili, which also interfere with phagocytosis; phase variation in surface proteins allows attachment to different host cells and escape from immune mechanisms. Inflammation, scarring, can spread by bloodstream
Epidemiology	Transmitted by sexual contact. Asymptomatic carriers. No immunity.
Prevention and treatment	Education, condoms, early treatment of sexual contacts. Treatment: intramuscular ceftriaxone; penicillin or tetracycline if strain proven susceptible

Disease Summaries

Major diseases are represented with a summary table and an outline of pathogenesis keyed to a human figure showing the entry and exit of the pathogen as well as the course of the infection.

Infectious Disease Coverage

Diseases are organized by human body system with background anatomy and physiology.

Each disease is presented systematically and predictably including Symptoms, Causative Agents, Pathogenesis, Epidemiology, and Prevention and Treatment.

Each disease is summarized in a table.

Each chapter includes a Case Presentation of a realistic clinical situation.

Chapters on infectious diseases are highlighted with yellow shading for easy reference.

Additional Case Presentations and Clinical Reference material is available on the Nester Web Site.

708 | Chapter 27 Wound Infections

CASE PRESENTATION

The patient was a 24-year-old woman, a surgical nurse, seen in the clinic for evaluation of a needle puncture wound to the hand. Earlier in the day, while assisting in a frantic attempt to revive a man with cardiac arrest, she sustained a deep puncture wound to her right palm from a needle that had accidentally dropped into the bedclothes. The needle was visibly contaminated with blood. She immediately washed her hand thoroughly with soap and water, applied an antiseptic, and dressed the puncture site with a loose adhesive bandage.

She was married, with one 14-month-old child. There was no history of blood transfusion or injected-drug abuse. She had donated blood the previous month, and it was not rejected. Her tetanus immunization was up to date, but she had not been immunized against hepatitis B.

Two days after the clinic visit, tests for antibody in the cardiac arrest patient's blood revealed that he had a chronic viral infection.

1. What were the main diagnostic considerations?
2. What risk of infection did the patient face?
3. What measures could be taken to reduce the risk? How much time could expire before preventive measures became ineffective?
4. What was the nurse's prognosis?

Discussion

1. The viruses of concern are hepatitis B virus (HBV), human immunodeficiency virus (HIV), and hepatitis C virus (HCV). Each of these could be transmitted to the nurse by a needle stick and cause serious illness.
2. There are an estimated 750,000 to 1 million carriers of HBV in the United States. They typically have large amounts of circulating infectious virus, so that even a tiny amount of their blood can transmit the disease. The risk of infection from a needle puncture wound when the blood originates from a hepatitis B virus carrier is estimated to be 10% to 35%. The AIDS-causing human immunodeficiency virus (HIV) infects approximately 1 million Americans. The blood of these persons is also potentially infectious, but the risk of transmission by a needle stick is considerably lower than the risk for hepatitis B, averaging about 0.4%. The lower risk results from smaller amounts of circulating infectious virus in HIV-infected individuals. The risk is probably higher early in HIV disease, during the acute infection, and later, when AIDS develops, because much higher levels of circulating infectious virus are then present.

   Hepatitis C virus transmission by blood accounts for most cases of post-transfusion hepatitis. Transmission from surgeon to patient has been documented, presumably by the multiple pricks from surgical needles that often penetrate the surgeons' gloves during major surgery. The risk of transmission by needle stick from an HCV-positive individual is about 1.8%. The number of new hepatitis C virus infections in the United States each year has been estimated at between 150,000 and 170,000, but the mode of transmission is unknown in most cases.

   Other viruses, such as cytomegalovirus (CMV) and Epstein-Barr virus (EBV), can be transmitted by blood. The risk from a needle stick injury is unknown but is probably much lower than from the viruses already mentioned. Obviously, all blood should be considered potentially infectious.
3. In the case of needle puncture wounds that expose a person to HBV, hepatitis B immune globulin (HBIG) is given as soon as possible after the wound occurs. HBIG is gamma globulin obtained from individuals that have a high titer of antibody against HBV. At the same time, active immunization is started with hepatitis B vaccine. These measures must be initiated within 7 days of the injury to be effective. This nurse, as with all persons at high risk of blood exposure, should have already been immunized with hepatitis B vaccine; then, no other preventive measures would need to be taken.

   Those exposed to HIV by needle punctures should be given zidovudine (AZT), plus one or more other anti-HIV medications, immediately and for 4 weeks. There is probably little protective effect if therapy is delayed beyond 2 hours.

   There is no proven preventive measure for HCV exposure. Approaches similar to those for HBV may become available in the future.
4. Preventive measures for hepatitis B exposure are highly effective, reducing the risk of infection by 75% or more. Also, the already relatively low risk from needle puncture wound for HIV exposure can probably be reduced by 75% to 80% with preventive medication. The patient's prognosis for remaining free of infection was good. The small chance of becoming infected and the long incubation period of these diseases, however, add up to considerable worry. Every effort should be made to avoid needle puncture wounds in the first place.

Human Bites

Wounds caused from human bites, striking the teeth of another person, or resulting from objects that have been in a person's mouth are common and can result in very serious infections. Rarely, diseases such as syphilis, tuberculosis, and hepatitis B are transmitted this way. Much more commonly, it is the normal mouth flora that cause trouble.

Symptoms

The wound may appear insignificant at first but then becomes painful and swells massively. Discharged pus often has a foul smell. Most of the wounds are on the extensor surface on the hand, and here the swelling may soon involve the palm also, and movement of some or all of the fingers becomes difficult or impossible.

Causative Agents

Stains and cultures usually show members of the normal mouth flora, including anaerobic streptococci, fusiforms, spirochetes, and *Bacteroides* sp., often in association with *Staphylococcus aureus*.

Pathogenesis

The crushing nature of bite wounds provides suitable conditions for anaerobic bacteria to establish infection. Although most members of the mouth flora are harmless alone, together they produce an impressive number of toxins and destructive enzymes. These include leukocidin, collagenase, hyaluronidase, ribonuclease, various proteinases, neuraminidase, and enzymes that destroy complement and antibody. Capsules of some species inhibit phagocytosis. Facultatively anaerobic organisms reduce available oxygen and thus encourage the growth of anaerobes. The result of all these factors is a **synergistic infection**, meaning that the sum effect of all the organisms acting together is greater than sum of their individual effects. Irreversible destruction of tissues such as tendons and permanent loss of function can be the result. ■ facultative anaerobes, p. XXX

Epidemiology

Most of the serious human bite infections occur in association with violent confrontations related to alcohol ingestion, or during forcible restraint, as in law enforcement and in mental institutions. The risk is greatly increased when the biting individual has poor mouth care and extensive dental disease. Bites by little children are usually inconsequential.

Prevention and Treatment

Prevention involves avoiding situations that lead to uncivilized behavior such as biting and hitting. Prompt cleansing of wounds followed by application of an antiseptic are advised, and

Definitions

Key terms appear in boldface type, are defined when introduced and may be found in the glossary.

Cross References

Page numbers direct students to sections elsewhere in the text with additional background to support the concepts mentioned within a paragraph.

TABLE 27.10 Human Bite Wound Infections

Symptoms	Rapid onset, pain, massive swelling, drainage of foul-smelling pus
Incubation period	Usually 6 to 24 hours
Causative agent	Mixed mouth flora: anaerobic streptococci, fusiforms, spirochetes, anaerobic Gram-negative rods; sometimes *Staphylococcus aureus*
Pathogenesis	Various mouth bacteria act synergistically to destroy tissue
Epidemiology	Alcohol-related violence; forcible restraint; poor mouth care and extensive dental disease
Prevention and treatment	No proven preventive measures except to avoid altercations. Prompt cleansing of wound and application of antiseptic is advised. Treatment is usually surgical

most important is immediate medical attention if there is any suspicion of developing infection. Treatment of infected wounds consists of opening the infected area widely with a scalpel, washing the wound thoroughly with sterile fluid, and removing dirt and dead tissue. The choice of antibacterial medication includes one effective against anaerobes.

The main features of human bite wound infections are presented in **table 27.10**.

MICROCHECK 27.4

A single species of Gram-negative, encapsulated, facultatively anaerobic rods, *Pasteurella multocida*, can infect bite wounds caused by a number of different animals, notably cats. Cat bites and scratches can also transmit *Bartonella henselae*, cause of cat scratch disease, characterized typically by local lymph node enlargement, but the disease may involve other parts of the body. Streptobacillary rat bite fever, acquired from bites of rats and mice, and animals that prey on them, is marked by fever that comes and goes and a rash. The causative bacterium, *Streptobacillus moniliformis*, spontaneously develops L-forms. Human bite infections can be dangerous because certain members of the mouth flora with little invasive ability when growing alone can invade and destroy tissue when growing together.

- What Gram-negative organism commonly infects wounds caused by animal bites?
- What is the most common cause of chronic localized lymph node enlargement in young children?
- What unusual kind of variant occurs spontaneously in *Streptobacillus moniliformis* cultures?
- Why would normal mouth flora be a more common cause of serious human bite infections than the causative agents of syphilis, tuberculosis, and hepatitis B, which can also be transmitted by human bites?

Fungal Wound Infections | 709

Fungal Wound Infections

Fungal infections of wounds are unusual in economically developed countries, except that the yeast *Candida albicans* can be troublesome in severe burns and in those with wounds and underlying diseases such as diabetes and cancer. This yeast, commonly present among the normal flora and kept in check by it, becomes pathogenic when the competing microorganisms are eliminated, as in individuals receiving antibacterial therapy. Other fungal wound infections are much more common in impoverished people around the world. For example, Madura foot, a condition caused by various species of fungi, occurs in areas of the world where foot injuries are common, resulting from lack of shoes. Named after the city in India where it was first described, Madura foot is characterized by swellings and draining passageways that spit out yellow or black granules of fungal material. Only a minority of those with foot injuries contract the disease despite exposure to the same fungi, suggesting that other factors such as malnutrition may play a role. Sporotrichosis, another kind of fungal wound infection, occurs worldwide and is not poverty-related.

"Rose Gardener's Disease" (Sporotrichosis)

Sporotrichosis, also known as "rose gardener's disease," is widely distributed around the world, associated with activities that lead to puncture wounds from vegetation. Although many cases are sporadic, the disease can occur in groups of people engaged in the same occupation. Thousands of workers in the warm humid mines of South Africa have contracted the disease from splinters on mine timbers. Epidemics have occurred in the United States among handlers of spagnum moss from Wisconsin.

Symptoms

In most cases, a hand or arm is involved, but the trunk, legs, and face can also be sites of infection. Typically, a chronic ulcer forms at the wound site, followed by a slowly progressing series of ulcerating nodules that develop sequentially toward the center of the body (**figure 27.16**). Lymph nodes in the region of the wound enlarge, but patients generally do not become ill. If they have AIDS or other immunodeficiency, however, the disease can spread throughout the body, threatening life.

Causative Agent

Sporotrichosis is caused by the dimorphic fungus *Sporothrix schenckii* (**figure 27.17**), which lives in soil and on vegetation. ■ dimorphic fungus, p. XXX

Pathogenesis

Sporothrix schenckii spores are usually introduced with an injury caused by plant material. After an incubation period that usually ranges from 1 to 3 weeks but can be much longer, the multiplying fungi cause formation of a small nodule or pimple at the site of the injury. This lesion slowly enlarges and ulcerates, producing a red, easily bleeding skin defect. Unless the ulcer becomes secondarily infected with bacteria, there is little or no pus, and the lesion is pain-free. After a week or

Perspective Boxes

Perspective boxes introduce a "human" perspective by showing how microorganisms and their products influence our lives in a myriad of different ways.

PERSPECTIVE 8.2 Bacteria Can Conjugate with Plants: A Natural Case of Genetic Engineering

For more than 50 years, scientists have known that DNA can be transferred between bacteria. Twenty-five years ago, it was shown that a bacterium can even transfer its genes into plant cells, such as tobacco, carrots, and cedar trees, through a process analogous to conjugation. What led to this discovery started almost 100 years ago in the laboratory of the plant pathologist, Dr. Erwin Smith. He showed that the causative agent of a common plant disease, termed **crown gall** is a bacterium, *Agrobacterium tumefaciens*. This disease is characterized by large galls or swellings that occur on the plant at the site of infection, usually near the soil line, the crown of the plant. When other investigators cultured the diseased plant tissue on agar plates containing nutrients necessary for the growth of plant tissue, it displayed properties that differed from normal plant tissue. Whereas normal tissue requires several plant hormones for growth, crown gall tissue grows in the absence of these added hormones. In addition, crown gall tissue synthesizes large amounts of a compound termed an **opine,** which neither normal plant tissue nor *Agrobacterium* synthesizes. The most surprising observation was that the plant cells maintained their altered nutritional requirements and the ability to synthesize opine even after the bacteria were killed by penicillin. Investigators concluded that the crown gall plant cells are permanently transformed. Although *Agrobacterium* is required to start the infection, they are not necessary to maintain the altered nutritional requirements and biosynthetic capabilities of the plant cells.

The explanation of the process by which *Agrobacterium* causes crown gall tumors and transforms plant cells was established in 1977 following a report from a team of Belgian investigators that all strains of *Agrobacterium* capable of causing crown gall tumors contained a large plasmid termed the **tumor inducing** or **Ti plasmid.** A group of microbiologists at the University of Washington showed that a specific piece of the Ti plasmid, termed the **transferred DNA**, or **T-DNA**, is transferred from the bacterial cell to the plant cell, where it becomes incorporated into the plant chromsome **(figure 1)**. Like conjugation between bacteria, a pilus is required for DNA transfer.

The studies of investigators from around the world yielded many surprising observations. This bacterial DNA acts like plant DNA because the promoter regions of the transferred genes resemble those of plants rather than those of bacteria. Therefore, the genetic information in the T-DNA is expressed in the plants but not in *Agrobacterium*. This DNA encodes enzymes for the synthesis of the plant hormones as well as for the opine. The expression of these genes supplies the plant cells with the plant hormones, explaining why the transformed plant cells can grow in the absence of added hormones and are able to synthesize opine. Thus once incorporated into the plant chromosome, the DNA provides the transformed cell with additional genetic information that confers new properties on the plant cell.

Why does *Agrobacterium* transform plants? This bacterium has the ability to use the opine as a source of carbon and energy, whereas most other bacteria in the soil, as well as plants, cannot. Therefore, *Agrobacterium* subverts the metabolism of the plant to produce food that only *Agrobacterium* can use. Thus, *Agrobacterium* is a natural genetic engineer of plants.

The *Agrobacterium*–crown gall system is of great interest for several reasons. First, it shows that DNA can be transferred from prokaryotes to eukaryotes. Many people believed that such transfer would be impossible in nature and could only occur in the laboratory. Second, this system has spawned an industry of plant biotechnology dedicated to improving the quality of higher plants. Thus, it is possible to replace the genes of hormone synthesis in the Ti plasmid with any other genes, which will then be transferred and incorporated into the plant. With this technology, genes conferring resistance to bacteria, viruses, insects, and different herbicides have been incorporated into a wide variety of plants. Rice has been transformed to synthesize high levels of β-carotene, the precursor of vitamin A. Edible vaccines are being synthesized in bananas following transformation by *Agrobacterium*.

Genetic engineering of plants became a reality once scientists learned how a common soil bacterium caused a well-recognized and serious plant disease. This system serves as a beautiful example of how solving a riddle in basic science can lead to major industrial applications.

Chromosome
T-DNA
Ti plasmid
Chloroplast
T-DNA
Plant DNA
Plant cell nucleus
Agar medium

Agrobacterium tumefaciens cell

Agrobacterium cells inoculated into wounded plant.

Tumor develops at wound site.

Bacteria-free tumor tissue is hormone independent and synthesizes opine.

Plant tumor cell containing T-DNA integrated into plant chromosome

Figure 1 ***Agrobacterium* sp. Causes Crown Gall and Transforms Plant Cells**

people can serve as a reservoir for R plasmids, which then can be transferred to disease-causing organisms.

MICROCHECK 8.8

All members of the microbial world contain plasmids, which in most cases code for

Plasmids vary in size, copy nu

genetic composition, and the

to other cells. One of the most important plasmids is the R plasmid, which codes for resistance to various antimicrobial medications and heavy metals.

- What functions must a plasmid code for in order to be self-transmissible?

Microchecks

Major sections end with a short "Microcheck" that summarizes the major concepts in that section.

Microchecks also offer several review questions to assess your understanding of the preceding material. Finally there is an opportunity to sharpen your critical thinking skills with the questions in blue.

MICROCHECK 4.5

Organisms require a source of major and trace elements. Heterotrophs use an organic carbon source, and autotrophs use CO_2. Bacteria that lack the ability to synthesize certain small molecules require these for growth. Phototrophs harvest the energy of sunlight, and chemotrophs obtain energy by degrading chemicals.

- List the major elements other than carbon required for growth of bacteria.
- What is the carbon source in a photoautotroph? In a chemoautotroph?
- Why would human-made materials (such as plastics) be degraded only slowly or not at all?

Cultivating Prokaryotes in the Laboratory

By knowing the environmental and nutritional factors that influence the growth of specific prokaryotes, it is often possible to provide the appropriate conditions for their cultivation. These include a medium on which to grow the organisms and a suitable atmosphere.

General Categories of Culture Media

Considering the diversity of bacteria, it is not surprising that a wide variety of media is used to culture them. For routine purposes, one of the many types of **complex media** is used. In contrast, **chemically defined media** are generally used for specific research experiments when nutrients must be precisely controlled. **Table 4.6** summarizes the characteristics of various types of media.

Complex Media

A **complex medium** contains a variety of ingredients such as meat juices and digested proteins, making what might be viewed as a tasty soup for microbes. Although a specific amount of each ingredient is in the medium, the exact chemical composition of these ingredients can be highly variable. One common ingredient is **peptone**. This is protein taken from any of a variety of sources that has been hydrolyzed to amino acids and short peptides by treatment with enzymes, acids, or alkali. **Extracts**, which are the water-soluble components of a substance, are also used. For example, beef extract is a water extract of lean meat and provides vitamins, minerals, and other nutrients. A commonly used complex medium, **nutrient broth,** consists of only 5 grams of peptone and 3 grams of beef extract per liter of distilled water. If agar is added, then **nutrient agar** results.

Many medically important bacteria are fastidious, requiring a medium that is even richer than nutrient agar. One rich medium commonly used in clinical laboratories is **blood agar**.

Table 4.6 Characteristics of Media Used to Cultivate Bacteria

Medium	Characteristic
Categories	
Complex	Composed of ingredients such as peptones and extracts, which may vary in their chemical composition.
Chemically defined	Composed of precise mixtures of pure chemicals such as ammonium sulfate.
Selective	Medium to which additional ingredients have been added that inhibit the growth of many organisms other than the one being sought.
Differential	Medium that contains an ingredient that can be changed by certain bacteria in a recognizable way.
Representative Types of Agar Media	
Blood agar	Complex medium used routinely in clinical labs. Not selective. Differential because colonies of hemolytic organisms are surrounded by a zone of clearing of the red blood cells.
Chocolate agar	Complex medium used to culture fastidious bacteria, particularly those found in clinical specimens. Not selective or differential.
Glucose-salts	Chemically defined medium. Used in laboratory experiments to study nutritional requirements of bacteria. Not selective or differential.
MacConkey agar	Complex medium used to isolate Gram-negative rods that typically reside in the intestine. Selective because bile salts and dyes inhibit Gram-positive organisms and Gram-negative cocci. Differential because the pH indicator turns red when the sugar in the medium, lactose, is fermented.
Nutrient agar	Complex medium used for routine laboratory work. Supports the growth of a variety of nonfastidious bacteria.
Thayer-Martin	Complex medium used to isolate *Neisseria* species, which are fastidious. Selective—contains antibiotics that inhibit most organisms except *Neisseria* species.

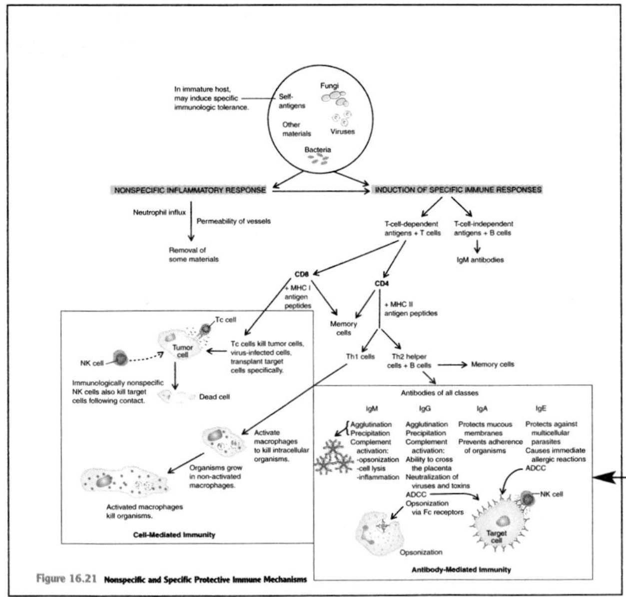

Figure 16.21 Nonspecific and Specific Protective Immune Mechanisms

Summary Figures and Tables

Many new figures and tables have been added to this edition that summarize complex information in a concise presentation.

All figure and table references appear in bold type within the text for easy correlation between text and visual support elements.

CHAPTER 1

Humans and the Microbial World

Microbiology as a science was born in 1674 when Antony van Leeuwenhoek (1632–1723), an inquisitive Dutch drapery merchant, peered at a drop of lake water through a glass lens that he had carefully ground. It was known for several centuries previous that curved glass would magnify objects, but it took the skillful hands of a craftsman coupled with the questioning mind of an amateur scientist to revolutionize the understanding of the world in which we live. What he observed through this simple magnifying glass was undoubtedly one of the most startling and amazing sights that humans have ever beheld—the first glimpse of the world of microbes. As van Leeuwenhoek wrote in a letter to the Royal Society of London, he saw

> *"Very many little animalcules, whereof some were roundish, while others a bit bigger consisted of an oval. On these last, I saw two little legs near the head, and two little fins at the hind most end of the body. Others were somewhat longer than an oval, and these were very slow a-moving, and few in number. These animalcules had diverse colours, some being whitish and transparent; others with green and very glittering little scales, others again were green in the middle, and before and behind white; others yet were ashed grey. And the motion of most of these animalcules in the water was so swift, and so various, upwards, dowwards, and round about, that 'twas wonderful to see."*

It is quite surprising that not only was van Leeuwenhoek able to see bacteria, but he even described their numerous shapes (chapter opening figure). His simple microscope increased the size of the object he was viewing only about 300 times. Microscopes commonly used in the laboratory today magnify objects over 1,000 times. Seeing small objects through the lens of a microscope depends not only on magnifying the size of the specimen being viewed, but also on illuminating it. Van Leeuwenhoek must have developed an unusually good method for lighting the specimens to see what he reported.

—A Glimpse of History

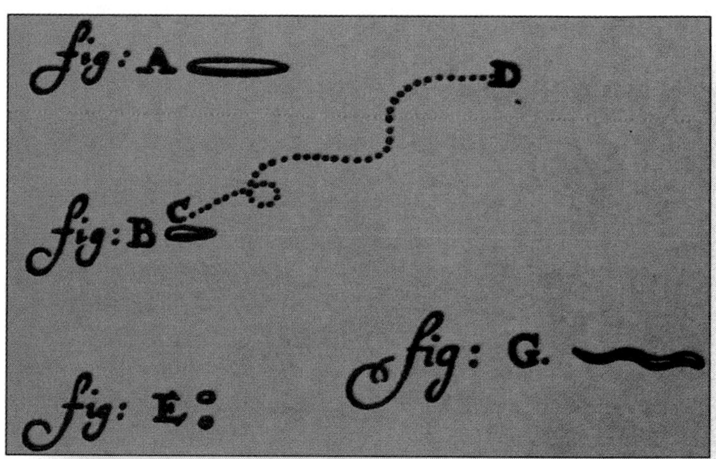

MICROORGANISMS ARE THE FOUNDATION FOR all life on earth. It has been said that the twentieth century was the age of physics. Now we can say that the twenty-first century will be the age of biology and biotechnology, with microbiology as the most important branch. From the fossil record, microorganisms have existed on earth for 3.5 to 3.8 billion years, and over this time, plants and animals have evolved from these microscopic forms (**figure 1.1**). Microorganisms themselves have also evolved into a very diverse group. They vary in their appearance, in their ability to carry out different biochemical transformations, in their remarkable ability to grow in a wide variety of different environments, and in their interactions with other microorganisms and all other forms of life, in particular humans. One of their most important characteristics is their ability to quickly change their properties and readily adapt to a changing environment. And yet, for all of these amazing abilities, we cannot really assess their true importance and capabilities because less than one in a hundred can be grown in the laboratory! As the exploration of the microbial world continues there are certain to be many new surprises.

The Origin of Microorganisms

The discovery of microorganisms raised an intriguing question: "Where did these microscopic forms originate?" The theory of spontaneous generation suggested that organisms, such as tiny worms, can arise spontaneously from nonliving material. It was completely debunked by Francesco Redi, an Italian biologist and physician, at the end of the seventeenth century. By a simple

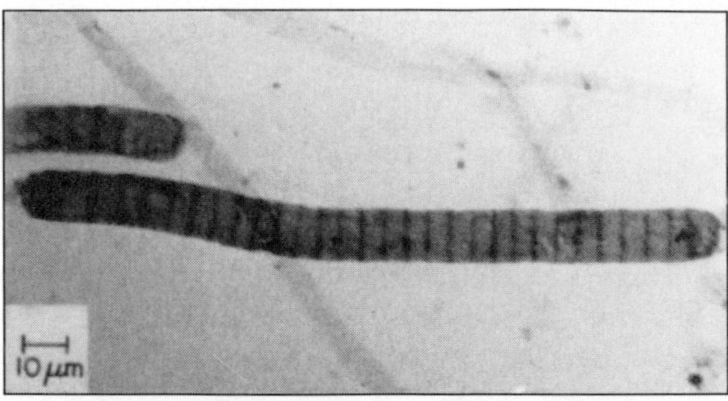
10 μm

Figure 1.1 Microfossil Resembling Cyanobacteria The microfossil *Palaeolyngbya* isolated from shale in eastern Siberia, is 950 million years old.

experiment, he demonstrated conclusively that worms found on rotting meat originated from the eggs of flies, not directly from the decaying meat as proponents of spontaneous generation believed. To prove this, he simply covered the meat with gauze fine enough to prevent flies from depositing their eggs. No worms appeared.

Theory of Spontaneous Generation Revisited

Despite Redi's findings, the theory of spontaneous generation was difficult to disprove, and it took about 200 years more to refute this idea. Because the gauze used by Redi could not prevent the development of microscopic organisms, new experiments were needed to completely refute the theory. The typical experiment designed to determine whether microbes could arise from nonliving material consisted of boiling organic material in a vessel to kill all forms of life (sterilize) and then sealing the vessel to prevent any air from entering. If the solution became cloudy after standing, then one could conclude that microorganisms must have arisen from the organic material in the vessel, thus supporting the theory of spontaneous generation. Unfortunately, this experiment did not consider several alternative possibilities: that the flask might be improperly sealed, that microorganisms might be present in the air, or that boiling might not kill all forms of life. Therefore, it was not surprising that different investigators obtained different results when they performed this experiment.

Experiments of Pasteur

One giant in science who did much to disprove the theory of spontaneous generation was the French chemist Louis Pasteur, considered by many to be the father of modern microbiology. In 1861, Pasteur published a refutation of spontaneous generation that was a masterpiece of logic. First, he demonstrated that air is filled with microorganisms. He did this by filtering air through a cotton plug, trapping organisms that he then examined with a microscope. Many of these trapped organisms looked identical microscopically to those that had previously been observed by others in many **infusions**. Infusions are liquids that contain nutrients in which microorganisms can grow. Pasteur further showed that if the cotton plug was then dropped into a sterilized infusion, it became cloudy because the organisms quickly multiplied.

Most importantly, Pasteur's experiment demonstrated that sterile infusions would remain sterile in specially constructed flasks even when they were left open to the air. Organisms from the air settled in the bends and sides of these swan-necked flasks, never reaching the fluid in the bottom of the flask (**figure 1.2**). Only when the flasks were tipped would bacteria be able to enter the broth and grow. These simple and elegant experiments should have ended claims put forth by some individuals, that unheated air or the infusions themselves contained a "vital force" necessary for spontaneous generation.

Experiments of Tyndall

Although most people in the scientific community, especially in France, were convinced by Pasteur's experiments, other respected scientists refused to give up the concept of spontaneous generation. This persistence undoubtedly stemmed in part from the inability of some scientists to verify Pasteur's results. One of these was an English physicist, John Tyndall. It was Tyndall who finally explained the differences in experimental results obtained in different laboratories and proved Pasteur correct. Tyndall concluded that different infusions required different boiling times to be sterilized. Thus, boiling for 5 minutes would sterilize some materials, whereas others, most notably hay infusions, could be boiled for

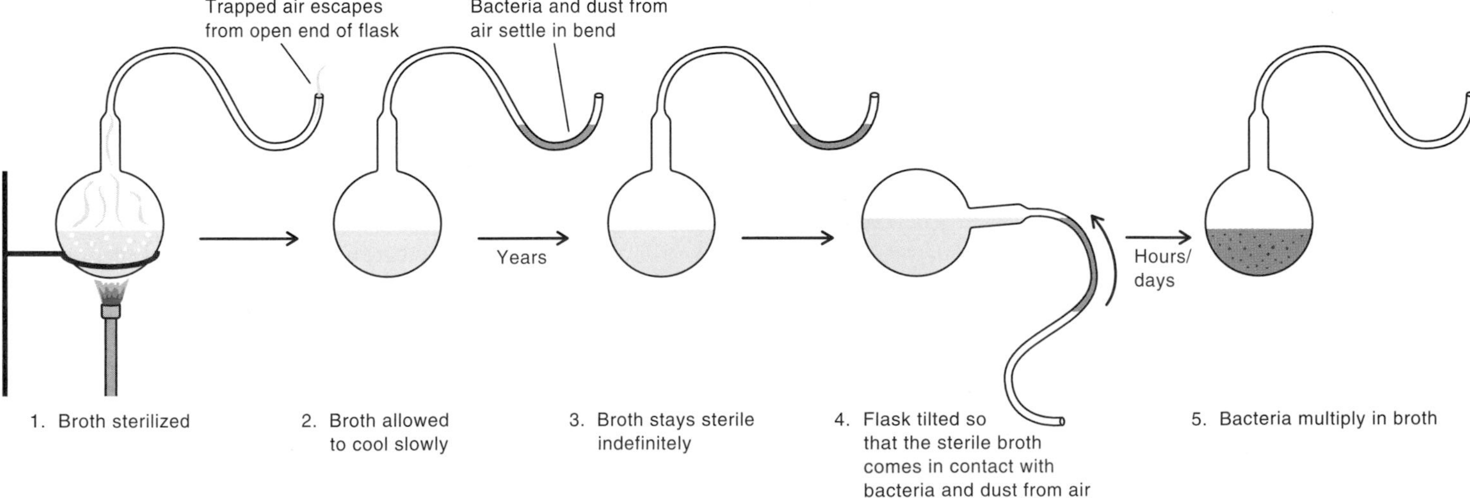
Trapped air escapes from open end of flask
Bacteria and dust from air settle in bend
Years
Hours/ days
1. Broth sterilized
2. Broth allowed to cool slowly
3. Broth stays sterile indefinitely
4. Flask tilted so that the sterile broth comes in contact with bacteria and dust from air
5. Bacteria multiply in broth

Figure 1.2 Pasteur's Experiment with the Swan-Necked Flask If the flask remains upright, no microbial growth occurs. **(1–3)** If the flask is tipped, microorganisms trapped in the neck reach the sterile liquid and grow. **(4, 5)** Why did bacteria grow in the flask only after the flask was tipped?

5 hours and they still contained living organisms! Furthermore, if hay was present in the laboratory, it became almost impossible to sterilize the infusions that had previously been sterilized by boiling for 5 minutes. What did hay contain that caused this effect? Tyndall finally realized that heat-resistant forms of life were being brought into his laboratory on the hay. These heat-resistant life forms must then have been transferred to all other infusions in his laboratory on dust particles, thereby making everything difficult to sterilize. Tyndall concluded that some microorganisms could exist in two forms: a cell that is readily killed by boiling, and one that is heat resistant. In the same year (1876), a German botanist, Ferdinand Cohn, also discovered the heat-resistant forms of bacteria, now termed **endospores.** ■ endospores, p. 72

The extreme heat resistance of endospores explains the differences between Pasteur's results and those of other investigators. Organisms that produce endospores are commonly found in the soil and most likely were present in hay infusions. Because Pasteur used only infusions prepared from sugar or yeast extract, his broth most likely did not contain endospores. At the time these experiments on spontaneous generation were performed, the importance of the source of the infusion was not appreciated. In hindsight, the infusion source was critical to the results observed and conclusions drawn.

These experiments on spontaneous generation point out an important lesson for all scientists. In repeating an experiment and comparing results with previous experiments, it is absolutely essential to reproduce all conditions of an experiment as closely as possible. It may seem surprising that the concept of spontaneous generation was disproved less than a century and a half ago. **Table 1.1** lists some of the other important advances in microbiology that have been made in the course of history. We will return to many of these major milestones as our study of microbiology continues. How far the science of microbiology and all biological sciences have advanced over the last 140 years!

The First Microorganisms

Where did microorganisms really come from? No one knows, but many theories abound. The progenitors of microorganisms were the first biological entities, and they no longer exist. Which microorganisms first evolved from these progenitors can only be speculated upon. One current idea is that the first microorganisms must have grown at very high temperatures in the absence of oxygen because the early earth was likely very hot with active volcanoes and no oxygen. Bacteria that can grow at temperatures higher than boiling water in the absence of oxygen have been isolated from ocean vents. Perhaps some suggestion as to how life arose on earth will be provided by finding life on other planets, the field of **astrobiology**.

MICROCHECK 1.1

Antony van Leeuwenhoek first observed microorganisms about 300 years ago. Pasteur and Tyndall refuted the theory of spontaneous generation only 140 years ago.

- Give two reasons why it took so long to disprove the theory of spontaneous generation.
- What experiment disproved the notion that a "vital force" in air was responsible for spontaneous generation?
- If Pasteur's flasks had contained endospores, what results would have been observed?

Microbiology: A Human Perspective

Microoganisms have had, and continue to have, an enormous impact on the lives of all living things. On the one hand, microorganisms and other infectious agents, the viruses, have killed far more people than have ever been killed in war. On the other hand, without microorganisms, life as we know it could not exist on the earth. They are responsible for continually providing the oxygen and nitrogen that all living beings require.

Which organisms are included in the microbial world? Microbiology encompasses the study of many diverse organisms, which include bacteria, viruses, protozoa, algae, fungi, and some multicellular parasites. The overwhelming majority of these organisms are too small to be seen without the aid of a microscope. Indeed, the main feature of most members of the microbial world is their small size. Microorganisms are also very diverse in all their aspects: appearance, metabolism, physiology, and genetics. They are far more diverse than the organisms that are known as plants and animals. Even though they seem to be relatively simple, some members of the microbial world can be extraordinarily complex.

Let us now consider some of the roles that microorganisms play in our lives, both beneficial and harmful. In large part this section and the remainder of this chapter will introduce you to what will be covered in more detail in later chapters.

Vital Activities of Microorganisms

The activities of microorganisms are responsible for the survival of all other organisms, including humans. A few examples can be cited. Nitrogen is an essential part of most of the important molecules in our bodies, such as nucleic acids and proteins. Nitrogen is also the most common gas in the atmosphere. Neither plants nor animals, however, can use nitrogen gas. Without certain bacteria that are able to convert the nitrogen in air into a chemical form that plants can use, life as we know it would not exist on earth.

Moreover, all animals including humans require oxygen to breathe. The supply of oxygen in the atmosphere, however, would be depleted in about 20 years, were it not replenished. On land, plants are important producers of oxygen, but when all land and aquatic environments are considered, microorganisms are primarily responsible for continually replenishing the supply of oxygen.

Microorganisms can also break down a wide variety of materials that no other forms of life can degrade. For example, the bulk of the carbohydrate in terrestrial (land) plants is in the form of cellulose, which humans and most animals cannot digest. Certain microorganisms can, however. As a result, leaves and downed trees do not pile up in the environment. Cellulose is also degraded by billions of microorganisms in the stomachs of cattle, sheep, deer, and other ruminants. The digestion products are used by the cattle for energy. Without these bacteria, ruminants would not be a readily available

TABLE 1.1 Some Major Milestones in Microbiology

Date	Event
1500 B.C.	Egyptians ferment cereal grains to make beer.
A.D. 1546	Italian physician Girolamo Fracastoro suggests that invisible organisms may cause disease.
1665	Robert Hooke publishes his discovery of cells in cork.
1676	Antony van Leeuwenhoek observes bacteria and protozoa using his homemade microscope.
1796	Edward Jenner introduces a vaccination procedure for smallpox.
1838–1839	Mathias Schleiden and Theodor Schwann independently propose that all organisms are composed of cells, the basic unit of life.
1840	J. Henle presents a clear exposition of the germ theory of disease.
1847–1850	Ignaz Semmelweis demonstrates that puerperal or childbed fever is a contagious disease transmitted by physicians to their patients during childbirth.
1853–1854	John Snow demonstrates the epidemic spread of cholera through a water supply contaminated with human sewage.
1857	Louis Pasteur demonstrates that yeast can degrade sugar to ethanol and carbon dioxide as they multiply.
1861	Louis Pasteur publishes experiments that refute the theory of spontaneous generation.
1864	Louis Pasteur develops pasteurization as a method to destroy unwanted organisms in wine.
1867	Joseph Lister publishes the first work on antiseptic surgery, beginning the trend toward modern aseptic techniques in medicine.
1876	Robert Koch demonstrates that anthrax is caused by a bacterium.
1881	Robert Koch introduces the use of pure culture techniques for handling bacteria in the laboratory.
	Walter and Fanny Hesse introduce agar-agar as a solidifying gel for culture media.
1882	Koch identifies the causative agent of tuberculosis.
1884	Koch states Koch's postulates.
	Elie Metchnikoff discovers phagocytic cells and their role in engulfing bacteria.
	Christian Gram publishes a paper describing the Gram stain.
1892	Dmitri Ivanowski discovers that tobacco-mosaic disease is caused by a filterable agent-a virus.
1908	Paul Ehrlich develops the drug salvarsan to treat syphilis, thereby starting the use of chemotherapy to treat diseases.
1911	F. Peyton Rous discovers that a virus can cause cancer in chickens.
1928	Frederick Griffith discovers genetic transformation in bacteria, thereby raising a key question in genetics: What chemical caused the transformation?
1929	Alexander Fleming discovers and describes the properties of the first antibiotic, penicillin.
1944	Oswald Avery, Colin MacLeod, and Maclyn McCarty demonstrate that Griffith's transforming principle is DNA.
	Joshua Lederberg and Edward Tatum demonstrate that DNA can be transferred from one bacterium to another.
1948	Barbara McClintock demonstrates transposable elements in maize, and almost two decades later they are discovered in bacteria.
1953	James Watson, Francis Crick, Rosalind Franklin, and Maurice Wilkins determine the structure of DNA.
1957	D. Carlton Gajdusek demonstrates the slow infectious nature of the disease kuru, which is later shown to be caused by a prion.
1970	Hamilton Smith reports the discovery of the first restriction enzyme.
1971	Theodor Diener demonstrates the fundamental differences between viroids and viruses.
1973	Herbert Boyer and Stanley Cohen, using plasmids, are the first to clone DNA.
1975	Cesar Milstein, Georges Kohler, and Niels Kai Jerne develop the technique for making monoclonal antibodies.
1976	Michael Bishop and Harold Varmus discover the cancer-causing genes, called oncogenes, and find that such genes are in normal tissues.
1977	Carl Woese classifies all organisms into three domains.
1980	A rare cancer in humans is shown to be caused by a retrovirus.
	World Health Organization declares eradication of smallpox in the world.
1982	Stanley Prusiner isolates a protein from a slow disease infection and suggests that it might direct its own replication. He suggests the agent be termed a prion.
	Barry Marshall demonstrates that a bacterium, *Helicobacter pylori* causes ulcers.
1983	Luc Montagnier of France and Robert Gallo of the United States independently isolate and characterize the human immunodeficiency virus (HIV), the cause of AIDS.
	Kary Mullis invents the polymerase chain reaction.
1994	The Food and Drug Administration approves the first genetically engineered food for human consumption, a slower ripening tomato.
1995	The Food and Drug Administration approves the first protease inhibitor, a major weapon against the progression of AIDS.
	The first complete nucleotide sequence of a chromosome of a bacterium, (*Haemophilus influenzae*) is reported.
1997	The first complete nucleotide sequence of all of the chromosomes of a eukaryote (yeast) is reported.
1999	Ford Doolittle proposes that evolution proceeded through horizontal gene transfer between the three domains.
2000	The first new antibiotic in 35 years, Zyvox or linezolid, is approved by the Food and Drug Administration.

source of protein. Microorganisms also play an indispensable role in degrading a wide variety of materials in sewage and wastewater treatment.

Economic Applications of Microbiology

In addition to the crucial roles that microorganisms play in maintaining all life on earth, they also have made life more comfortable for humans over the centuries. **Biotechnology** is the application of biology to solve practical problems and produce useful products economically.

Food Production

By taking advantage of what microorganisms do naturally, Egyptian bakers as early as 2100 B.C. used yeast to make bread. Modern bakeries use essentially the same technology. ■ **bread making, p. 812**

The excavation of early tombs in Egypt revealed that by 1500 B.C., Egyptians employed a highly complex and sophisticated procedure for fermenting cereal grains to produce beer. Today, brewers use the same fundamental techniques to make beer and other fermented drinks. ■ **beer, p. 810**

Virtually every human culture that has domesticated milk-producing animals such as cows and goats also has developed the technology to ferment milk to produce foods such as yogurt, cheeses, and buttermilk. ■ **milk products, p. 807**

Bioremediation

The use of living organisms to degrade environmental pollutants is termed **bioremediation**. Bacteria are being used to destroy such dangerous chemical pollutants as polychlorinated biphenyls (PCBs),dichlorodiphenyltrichloroethane (DDT), and trichloroethylene, a highly toxic solvent used in dry cleaning. All three organic compounds and many more have been detected in soil and water. Bacteria are also being used to degrade oil and assist in the cleanup of oil spills as well as in the treatment of radioactive wastes. A bacterium was discovered recently that can live on trinitrotoluene (TNT). ■ **bioremediation, p. 781**

Useful Products from Bacteria

Bacteria can synthesize a wide variety of different products in the course of their metabolism. Many of these products have great commercial value. Although these same products can be synthesized in factories, bacteria often can do it faster and cheaper. For example, different bacteria produce:

- Cellulose used in stereo headsets
- Hydroxybutyric acid used in the manufacture of disposable diapers and plastics
- Ethanol, which is added to gasoline to make it burn cleaner
- Chemicals poisonous to insects
- Antibiotics used in the treatment of disease
- Amino acids, which are used as dietary supplements

Genetic Engineering

It is now possible to introduce genes from one organism into a related or an unrelated organism and confer new properties on that organism. This is the process of **genetic engineering.** Genetically engineered microorganisms often appear in the popular press because they are being used to solve many problems associated with an industrial society. Genetic engineering has expanded the power of biotechnology enormously. Examples of the roles that microorganisms play in this new biotechnology are the following:

- Microorganisms can be genetically engineered to produce a variety of medically important products. These include interferon, insulin, human growth hormone, blood clotting factors, and enzymes that dissolve blood clots. ■ **genetic engineering, p. 222**
- Microorganisms are being modified so that they will produce vaccines against rabies, gonorrhea, herpes, leprosy, malaria, and hepatitis. ■ **vaccines, p. 427**
- A bacterium can be used to genetically engineer plants so they become resistant to insect attacks and viral diseases, and produce large amounts of β-carotene. ■ **plant transformation, p. 230**
- A bacterium can transfer genes into bananas so that vaccines against certain diarrheal diseases are produced in the fruit. ■ **vaccines, p. 425**
- Viruses are being studied as a means of delivering genes into humans to correct genetic defects such as cystic fibrosis. This is the process of **gene therapy.** ■ **gene therapy, p. 365**

These examples represent only a few of the ways that microorganisms and viruses are being used to promote human welfare. In the past, microorganisms were considered only as dangerous organisms because they caused disease. The current and future use of microorganisms to increase the quality of human life, however, will receive increasing attention in scientific laboratories.

Medical Microbiology

In addition to the important roles that microorganisms play in our daily lives, some also play a sinister role. For example, more Americans died of influenza in 1918–1919 than were killed in World War I, World War II, the Korean War, and the Vietnam War combined. Modern sanitation, vaccination, and effective antibiotic treatments have reduced the incidence of some of the worst diseases, such as smallpox, bubonic plague, and influenza, to a small fraction of their former numbers. Another disease, acquired immunodeficiency syndrome (AIDS), however, has risen as a modern-day plague.

Past Triumphs

About the time that spontaneous generation was finally disproved to everyone's satisfaction, the Golden Age of medical microbiology was born. Between the years 1875 and 1918, most disease-causing bacteria were identified, and early work on viruses had begun. Once people realized that some of these invisible agents could cause disease, they tried to prevent their spread from sick to healthy people. The great successes in the area of human health in the last 100 years have resulted from the prevention of infectious diseases with vaccines and treatment of these diseases with antibiotics. The results have been astounding!

The viral disease smallpox was one of greatest killers the world has ever known. Approximately 10 million people have died from this disease over the past 4,000 years. It was brought to the New World by the Spaniards and made it possible for Hernando Cortez, with fewer than 600 soldiers, to conquer the Aztec Empire, whose subjects numbered in the millions. During a crucial battle in Mexico City, an epidemic of smallpox raged, killing only the Aztecs who had never been exposed to the disease before. In recent times, an active worldwide vaccination program has resulted in no cases being reported since 1977. The disease will probably never reappear.

Bubonic plague has been another great killer. One-third of the entire population of Europe, approximately 25 million people, died of this bacterial disease between 1346 and 1350. Now, generally less than 100 people in the entire world die each year from bubonic plague. In large part, this dramatic decrease is a result of eliminating the black rats that harbor the bacterium. Further, the discovery of antibiotics in the early twentieth century made the isolated outbreaks treatable and the disease no longer the scourge it once was.

Present and Future Challenges

Although progress has been very impressive against bacterial diseases, a great deal still remains to be done, especially in the treatment of viral diseases and diseases that are prevalent in developing countries. Even in wealthy developed countries with their sophisticated health care systems, however, infectious diseases remain a serious threat. For example, about 750 million cases of infectious diseases of all types occur in the United States each year. Every year these diseases lead to 200,000 deaths and cost tens of billions of health care dollars. Respiratory infections and diarrheal diseases cause most illness and deaths in the world today.

Emerging Diseases In addition to the well-recognized diseases, seemingly "new" **emerging** diseases continue to arise. In the last several decades, they have included:

- Legionnaire's disease, p. 571
- Toxic shock syndrome, p. 641
- Lyme disease, p. 531
- Acquired immunodeficiency syndrome (AIDS), p. 655
- Hantavirus pulmonary syndrome, p. 577
- Hemolytic-uremic syndrome, p. 603
- Cryptosporidiosis, p. 614

None of these diseases are really new, but an increased occurrence and wider distribution have brought them to the attention of the scientific community. Using the latest techniques, biomedical scientists have isolated, characterized, and identified these agents of disease. Now, methods need to be developed to prevent them.

A number of factors account for these emerging diseases arising even in industrially advanced countries. One reason is that changing lifestyles bring new opportunities for infectious agents to cause disease. For example, the vaginal tampons used by women provide an environment in which the organism causing toxic shock syndrome can grow and produce a toxin. In another example, the suburbs of cities are expanding into rural areas, bringing people into closer contact with animals previously isolated from humans. Consequently, people become exposed to viruses and infectious organisms that had been far removed from their environment. A good example is the hantavirus. This virus infects rodents, usually without causing disease. The infected animals, however, shed virus in urine, feces, and saliva; from there, it can be inhaled by humans as an aerosol. This disease, as well as Lyme disease, are only two of many emerging human diseases associated with small-animal reservoirs.

Some emerging diseases arise because the infectious agents change abruptly and gain the ability to infect new hosts. It is possible that HIV (human immunodeficiency virus), the cause of AIDS, arose from a virus that once could infect only monkeys. Some bacterial pathogens differ from their nonpathogenic close relatives in that the disease-causing organisms, the **pathogens**, contain large pieces of DNA that code for the ability of the organism to cause disease—that is, to be pathogenic. These pieces of DNA, which are often transferred from unrelated organisms, are termed **pathogenicity islands.** ■ **pathogenicity islands, p. 465**

Figure 1.3 shows the countries in the world where, since 1976, new infectious diseases of humans and animals have first appeared. Are there other agents out there that may cause "new" diseases in the future? The answer is undoubtedly yes!

Resurgence of Old Diseases Not only are "new" diseases emerging, but many infectious diseases once on the wane in the United States have begun to increase again. Further, many of these diseases are more serious today because the causative agents are resistant to the antibiotics once used to treat them. One reason for this resurgence is that thousands of foreign visitors and U.S. citizens returning from travel abroad enter this country daily. About one in five comes from a country where such diseases as malaria, cholera, plague, and yellow fever still exist. In developed countries these diseases have been eliminated largely through sanitation, vaccination, and quarantine. An international traveler incubating a disease in his or her body, however, could theoretically circle the globe, touch down in several countries, and expose many people before he or she became ill. As a result these diseases are recurring in countries where they had been virtually eliminated.

A second reason that certain diseases are on the rise is that in both developed and developing countries many childhood diseases have been so effectively controlled by childhood vaccinations that some parents have become lax about having their children vaccinated. The unvaccinated children are highly susceptible, and the number of those infected has increased dramatically. These diseases include measles, polio, mumps, whooping cough, and diphtheria.

A third reason for the rise in infectious diseases is that the medications and treatments being used increasingly to prolong the life of the elderly generally lower their disease resistance by impairing their immune system. Further, the population contains an increasing proportion of elderly people, who have weakened immune systems and are susceptible to diseases that

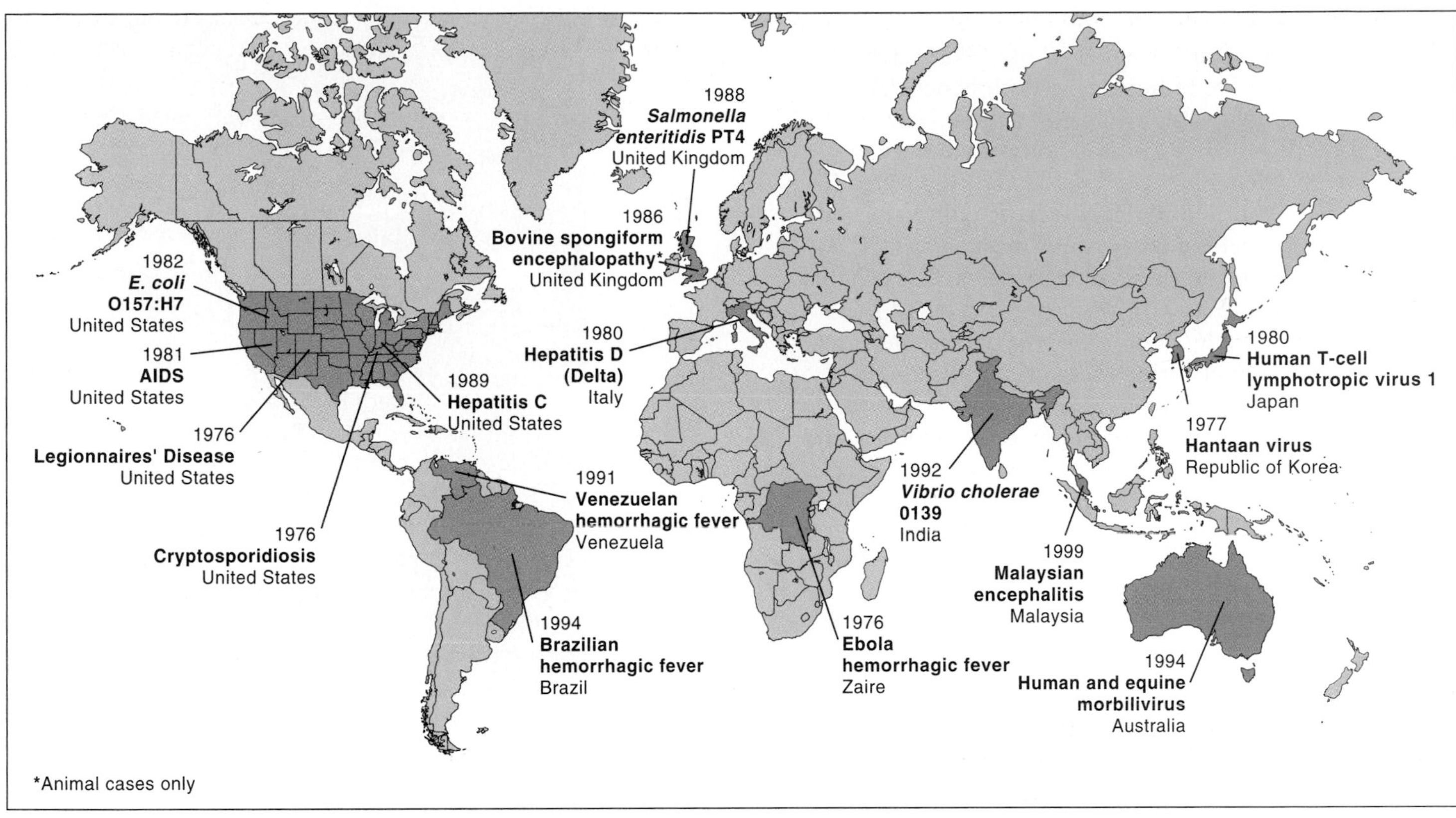

Figure 1.3 **"New" Infectious Diseases in Humans and Animals Since 1976** Countries where cases first appeared or were identified appear in a darker shade.

younger people readily resist. Likewise, individuals infected with HIV are especially susceptible to a wide variety of diseases, such as tuberculosis and Kaposi's sarcoma.

Chronic Diseases Caused by Bacteria In addition to the diseases long recognized as being caused by microorganisms or viruses, some illnesses once attributed to other causes may in fact be caused by bacteria. The best known example is peptic ulcers. This common affliction has recently been shown to be caused by a bacterium, *Helicobacter pylori*, and is treatable with antibiotics. Chronic indigestion, which affects 25% to 40% of the people in the Western world, may also be caused by the same bacterium.

Some scientists have also suggested that a bacterium is involved in cardiovascular disease. Studies are now under way to see if this disease improves when antibiotics are administered. It is likely that bacteria and viruses will be shown to play a role in a number of chronic diseases that are currently of unknown origin.

Host-Pathogen Interactions

Only a very small minority of bacteria cause disease. All surfaces of the human body both inside and outside are populated with bacteria, most of which are highly protective. They successfully compete with the occasional disease-causing bacteria and keep them from breaching host defenses. The body is an ecological niche for a vast number of different bacteria interacting with one another and with the surfaces of the body. Pathogenic bacteria have the ability to invade the cells in the body where they can escape competition from other bacteria, and the defenses of the host, and find a source of nutrients. These **host-pathogen interactions** are very complex, involving combat between host defenses and the disease-causing ability of the invading organisms.

Microorganisms As Subjects for Study

Microorganisms are wonderful model organisms to study because they display the same fundamental metabolic and genetic properties found in higher forms of life. For example, all cells synthesize protein from the same amino acids by the same mechanism. They all duplicate their DNA by similar processes, and they degrade food materials to generate energy via the same metabolic pathways. To paraphrase a Nobel Prize–winning microbiologist, Dr. Jacques Monod, what is true of an elephant is also true of bacteria. Bacteria are easy to study and results can be obtained very quickly because they grow rapidly and form billions of cells per milliliter on simple inexpensive media. Thus, most of the major advances that have been made in the last century toward understanding life have come through the study of microorganisms. The number of Nobel Prizes that have been awarded to microbiologists proves this point (see inside cover). Such studies constitute basic research, and they continue today.

MICROCHECK 1.2

Microorganisms are essential to all life on earth and affect the life of humans in both beneficial and harmful ways. Microorganisms have been used for food production for thousands of years using essentially the same techniques as are used today. They are now being used to degrade toxic pollutants and produce a variety of compounds more cheaply than can be done in the chemical laboratory. The genetic engineering of microorganisms has expanded these capabilities greatly. Enormous progress has been made in preventing and curing most infectious diseases, but new ones continue to arise around the world. Microbes represent wonderful model organisms, and many principles of biochemistry and genetics have been discovered from studying bacteria.

- Discuss activities that microbes carry out that are essential to life on earth.
- Discuss several reasons for the reemergence of old diseases.
- Why would it seem logical, even inevitable, that at least some bacteria would attack the human body and be disease-causing agents?

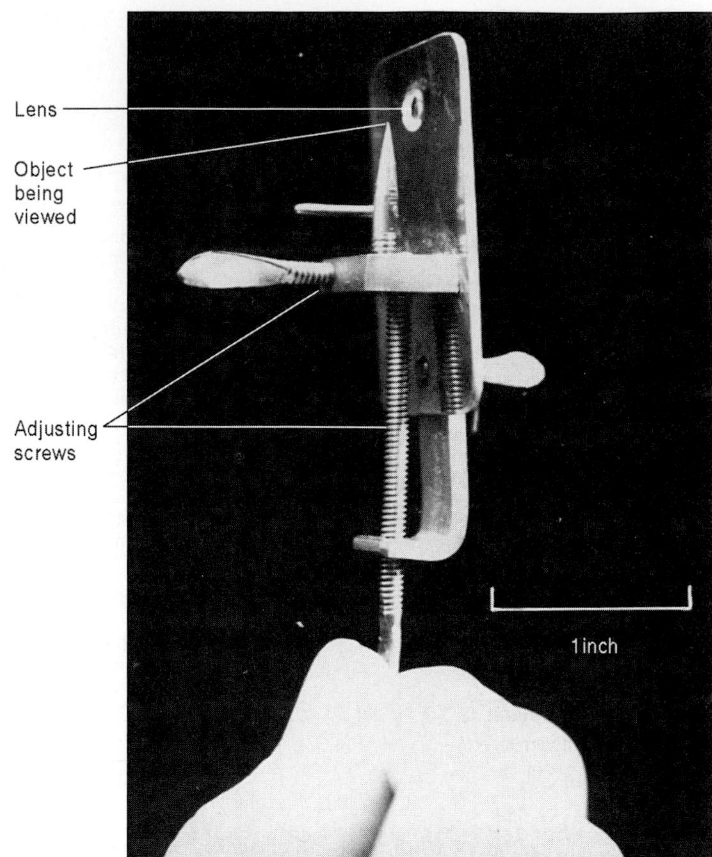

Figure 1.4 Model of Van Leeuwenhoek's Microscope The original made in 1673 could magnify the object being viewed almost 300 times. The object being viewed is brought into focus with the adjusting screws. This replica was made according to directions given in the *American Biology Teacher* 30:537, 1958. Note the small size.

The Microbial World

The microbial world includes the kinds of cells that van Leeuwenhoek observed looking through his simple microscope **(figure 1.4)**. Although he could not realize it at the time, the microbial world, in fact all living organisms can be classified into one of three major groups called **domains**. Organisms in each domain share properties of their cells that distinguish them from members of the other domains. Many properties, however, are shared among members of different domains because genes were transferred between domains billions of years ago. The three domains are the **Bacteria** (formerly called Eubacteria), the **Archaea** (meaning ancient), and the **Eukarya**. Microscopically, members of the Bacteria and Archaea look identical **(figure 1.5)**. Both are single-celled organisms that do not contain a membrane-bound nucleus nor any other intracellular lipid-bound **organelles**. Their genetic information is stored in fibrils composed of deoxyribonucleic acid (DNA) in a region called the **nucleoid**. These simple cell types have their cytoplasm surrounded by a rigid cell wall and are termed **prokaryotes**, which means "prenucleus." All bacteria and archaea are prokaryotes. These two groups of prokaryotes, however, differ significantly in their chemical composition.

Members of the Eukarya, termed **eukaryotes,** which means "true nucleus," are distinctly different from members of the Bacteria and Archaea. Eukaryotes may be single-celled or multicellular, but they always contain a true membrane-bound nucleus and other internal cell organelles, making them far more complex than the simple prokaryotes **(figure 1.6)**. These structures include **mitochondria**, organelles for generating energy, and **chloroplasts,** which harvest light energy in plants.

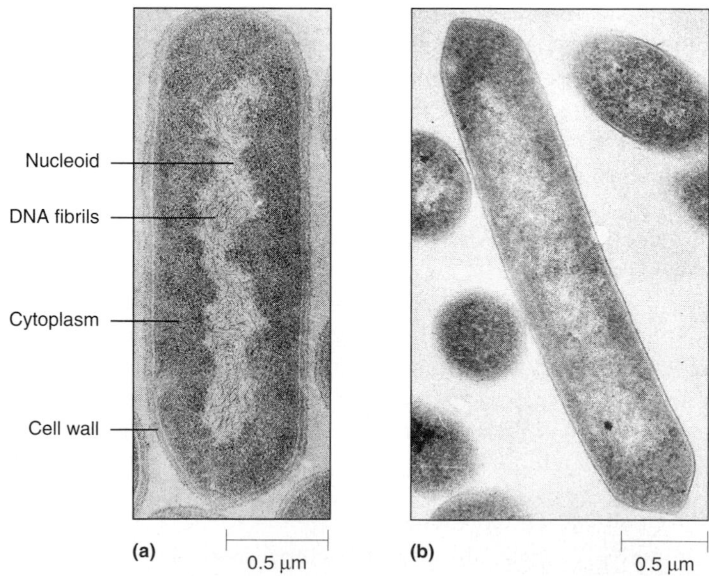

Figure 1.5 Electron Photomicrograph of (a) a Bacterium, *E. coli* and (b) a Member of the Archaea, *Methanobacterium foricum* Note that there are no major differences between the two organisms that can be seen using an electron microscope. The differences are at the chemical level.

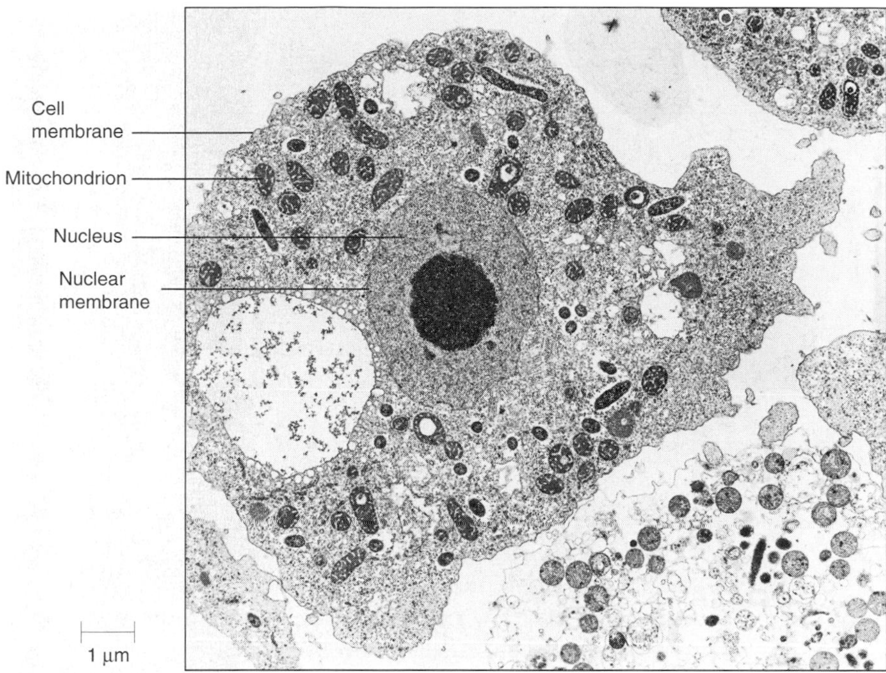

Figure 1.6 Eukaryotic Cell Note that this cell contains several internal structures surrounded by a membrane, including the mitochondria (sing: mitochondrion) and a nucleus. The cell is considerably larger than the prokaryotic cell.

Eukaryotes also have an internal scaffolding, the **cytoskeleton**, which gives the cells their shape. All algae, fungi, protozoa, and multicellular parasites considered in this book are eukaryotes.

The prokaryotes and eukaryotes are compared in **table 1.2**.

The Bacteria

Most of the prokaryotes covered in this text are members of the Domain Bacteria. Even within this group, much diversity is seen in the shape and properties of the organisms. At this time we will only describe the properties of the typical bacteria that most frequently cause disease and carry out the biogeochemical reactions necessary to support life on earth. Their most prominent features are:

- They are all single-celled prokaryotes.
- They have specific shapes, most commonly cylindrical (rod-shaped), spherical (round), or spiral (**figure 1.7**).
- They have rigid cell walls, which are responsible for the shape of the organism (see figure 1.5). The walls contain an unusual chemical compound called **peptidoglycan,** which is not found in organisms in the other domains.
- They multiply by **binary fission** in which one cell divides into two cells, each identical to the original cell (see figure 1.7a).
- Many can move using appendages extending from the cell, called **flagella** (sing: **flagellum**) (see figure 1.7a).

The Archaea

The Archaea have the same shape, size, and appearance as the Bacteria (see figure 1.5). Like the Bacteria, the Archaea multiply by binary fission and move primarily by means of flagella. They also have rigid cell walls. The chemical composition of their cell wall, however, differs from that in the Bacteria. The Archaea do not have peptidoglycan as part of their cell walls. Other chemical differences also exist between these two groups.

Perhaps the most interesting and distinguishing feature of the Archaea as a group is their ability to grow in environments in which most organisms, including the Bacteria, cannot survive. For example, some archaea can grow in salt concentrations 10 times as high as that found in seawater. These organisms grow in such habitats as the Great Salt Lake and the Dead Sea. Other archaea grow best at extremely high temperatures. One member grows best at temperatures above 105°C (100°C is the temperature at which water boils at sea level). Some archaea can be found in the boiling hot springs at Yellowstone National Park. Members of the Archaea, however, are spread far beyond extreme environments. They are widely distributed in the oceans, and they are found in the cold surface waters of Antarctica and Alaska.

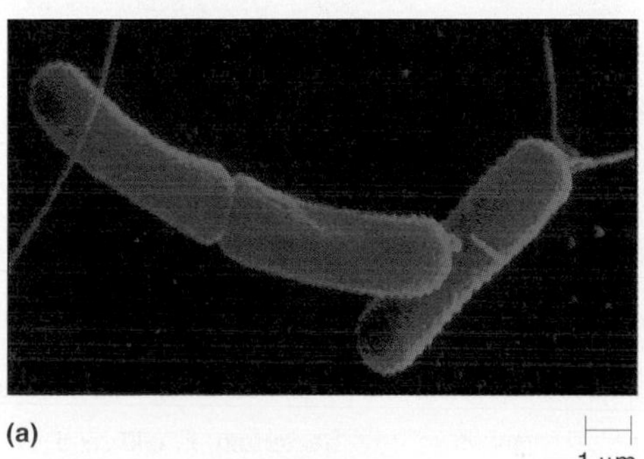

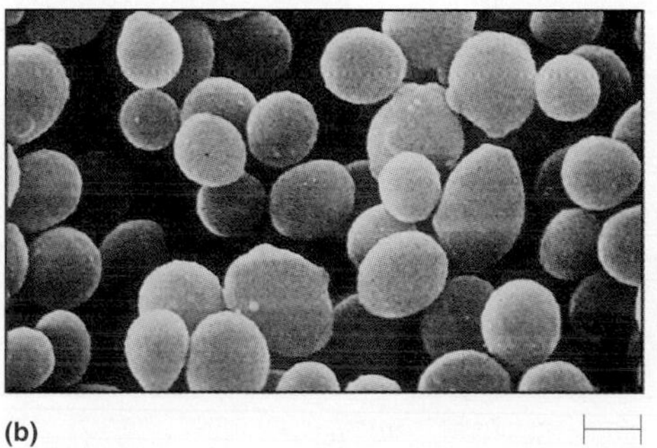

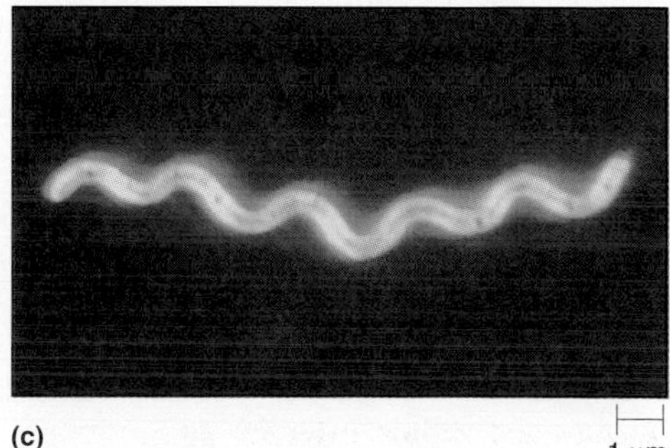

Figure 1.7 The Three Most Common Shapes of Bacteria As Viewed Through a Scanning Electron Microscope **(a)** Cylindrical or rod-shaped; **(b)** spherical; **(c)** spiral-shaped.

Table 1.2 Comparison of Prokaryotic and Eukaryotic Cells

Features of Cells	Prokaryotic	Eukaryotic
Size	0.3–2 μm	2–20 μm
Nuclear Membrane	No	Yes
Cell Wall	Unique chemical components; peptidoglycan	Not always present; no peptidoglycan
Cytoplasmic Structures		
Mitochondria	No	Yes
Chloroplasts	No	Yes, in plant and algal cells
Cytoskeleton	No	Yes

The Eukarya

All members of the living world except the prokaryotes are in the Domain Eukarya, and all members of this domain consist of eukaryotic cells. The microbial world is composed of single-celled members of the Eukarya as well as their close multicellular relatives. These members include **algae** (sing: **alga**), **fungi** (sing: **fungus**), and **protozoa** (sing: **protozoan**). Algae and protozoa are also referred to as **protists**. In addition, the eukaryotic multicellular parasites are considered in this text because of their importance in causing disease.

The Bacteria, Archaea, and Eukarya are compared in **table 1.3**.

Algae

The algae are a diverse group of eukaryotes; some are single-celled and others are multicellular. Many different shapes and sizes are represented, but they all share some fundamental characteristics **(figure 1.8)**. They all contain a green pigment, **chlorophyll**, and some also contain other pigments that give them characteristic colors. The pigments absorb light, which algae use as a source of energy. Algae are usually found near the surface of either salt or fresh water. Like bacteria, their cell walls are rigid, but their chemical composition is quite distinct from that of the Bacteria and the Archaea. Like bacteria and archaea, many algae move by means of flagella, which are structurally more complex than flagella in prokaryotes.

Fungi

Fungi are also a diverse group of eukaryotes. Some are single-celled yeasts, but many are large multicellular organisms including molds and mushrooms **(figure 1.9)**. In contrast to algae, which derive energy from sunlight, fungi gain their energy from organic materials. Interestingly, fungi are found wherever organic materials are present. Unlike algae, which live primarily in water, fungi live mostly on land.

Protozoa

Protozoa are a diverse group of microscopic, single-celled organisms that live in both aquatic and terrestrial environments. Although microscopic, they are very complex organisms and much larger than prokaryotes **(figure 1.10)**. Unlike algae and fungi, protozoa do not have a rigid cell wall. However, most of them do have a specific shape based on a rigid covering just beneath the outer membrane of the cell. Most protozoa require

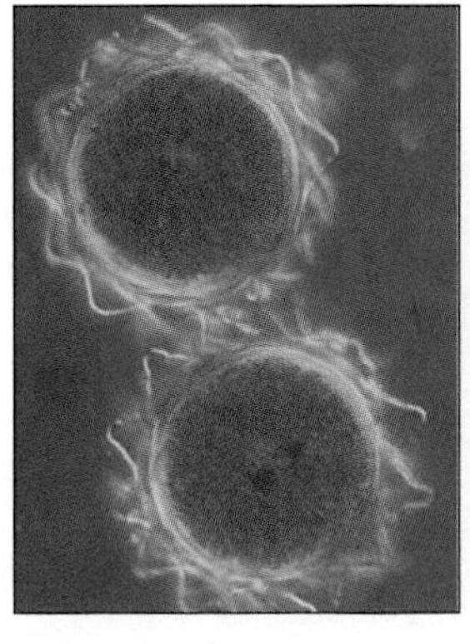

(a)

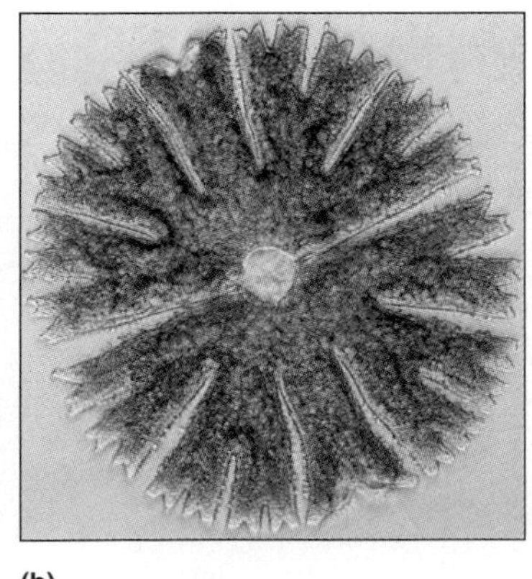

(b)

(c)

Figure 1.8 Algae **(a)** *Volvox* (400×), a motile alga that grows in colonies. Each colony contains up to 50,000 cells; **(b)** *Micrasterias,* a green alga composed of two symmetrical halves (100×); **(c)** Diatoms—organisms that serve as food for aquatic animals, as viewed through a scanning electron microscope (2,400×).

Table 1.3 Comparison of Bacteria, Archaea, and Eukarya

	Bacteria	Archaea	Eukarya
Size	0.3–2 μm	0.3–2 μm	2–20 μm
Nuclear Membrane	No	No	Yes
Cell Wall	Peptidoglycan present	No peptidoglycan	No peptidoglycan
Cytoplasmic Structures			
Mitochondria	No	No	Yes
Chloroplasts	No	No	In plant and algal cells
Cytoskeleton	No	No	Yes
Where Found	In all environments that are not extreme	Extreme environments	In environments that are not extreme

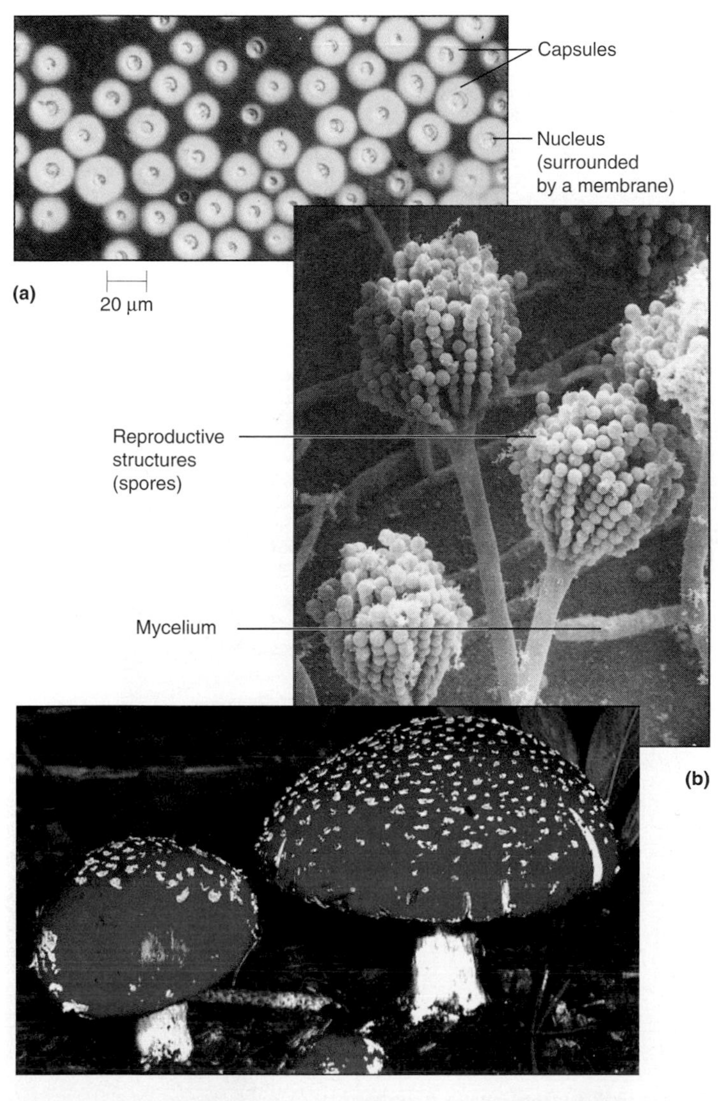

Figure 1.9 Three Forms of Fungi (a) Living cells of yeast form *Cryptococcus neoformans* stained with India ink to reveal the large capsules that surround the cell. **(b)** *Aspergillus*, a typical mold form whose dark reproductive structures rise above the mycelium. **(c)** A mushroom, *Amanita muscaria*.

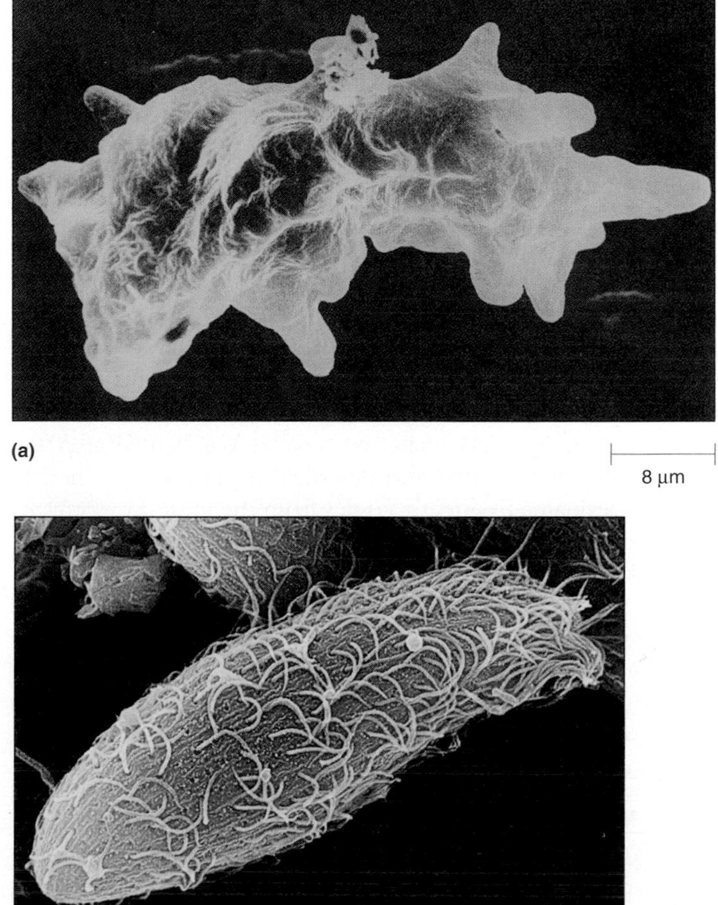

Figure 1.10 Protozoa (a) A typical ameba changes shape as it moves; **(b)** a paramecium moves with the aid of cilia on the cell surface.

Table 1.4 Comparison of Eukaryotic Members of the Microbial World

	Algae	Fungi	Protozoa	Helminths
Cell Organization	Single- or multicellular	Single- or multicellular	Single-celled	Multicellular
Source of Energy	Sunlight	Organic compounds	Organic compounds	Organic compounds derived from host
Size	Microscopic or macroscopic	Microscopic or macroscopic	Microscopic	Macroscopic

organic compounds as sources of food, which they ingest as particles. Most groups of protozoa are motile, and a major feature of their classification is their means of locomotion.

Multicellular Parasites

Several disease-causing multicellular organisms considered in this text are called **parasites** because they live on or within a host organism and derive nutrients from it. These parasites fall into three groups: **roundworms**, **tapeworms**, and **flukes** (**figure 1.11**). Collectively, they are known as **helminths.** These multicellular parasites have been controlled in the developed nations, but they continue to kill many millions in underdeveloped parts of the world.

Helminths enter the body in a number of ways. They may be eaten in contaminated food, be passed through insect bites, or directly penetrate the skin. They cause disease by invading the host tissues or robbing the host of nutrients.

The eukaryotic members of the microbial world are compared in **table 1.4**.

Nomenclature

In biology, the Binomial System of Nomenclature refers to a two-word naming system. The first word in the name is the **genus**, with the first letter always capitalized. The second word is the **species** name, which is not capitalized. Both words are always italicized. For example, *Escherichia coli* is a member of the genus *Escherichia*. Like many bacteria, it was named after the person who first isolated and described it, Theodor Escherich. The species name frequently comes from the location where the organism is most commonly found, in this case the colon. The genus name is commonly abbreviated, with the first letter capitalized: that is, *E. coli*. A number of different species are included in the same genus. Members of the same species may vary from one another in minor ways, but not enough to give the organisms different species names. These differences, however, may result in the organism being given different **strain** designations, for example, *E. coli* strain B or *E. coli* strain K12.

MICROCHECK 1.3

All organisms fall into one of three large groups based on their cell structure and chemical composition: the Bacteria, the Archaea, or the Eukarya. The Bacteria and the Archaea have a simple cell type and are termed prokaryotes. Both are identical in appearance but distinctly different in many aspects of their chemical composition. The Eukarya have a complex cell structure and are termed eukaryotes. The algae, fungi, protozoa, and multicellular parasites belong to this group. Bacteria, like all organisms, are classified according to the Binomial System of Nomenclature.

- Name one feature that distinguishes the Bacteria from the Archaea.

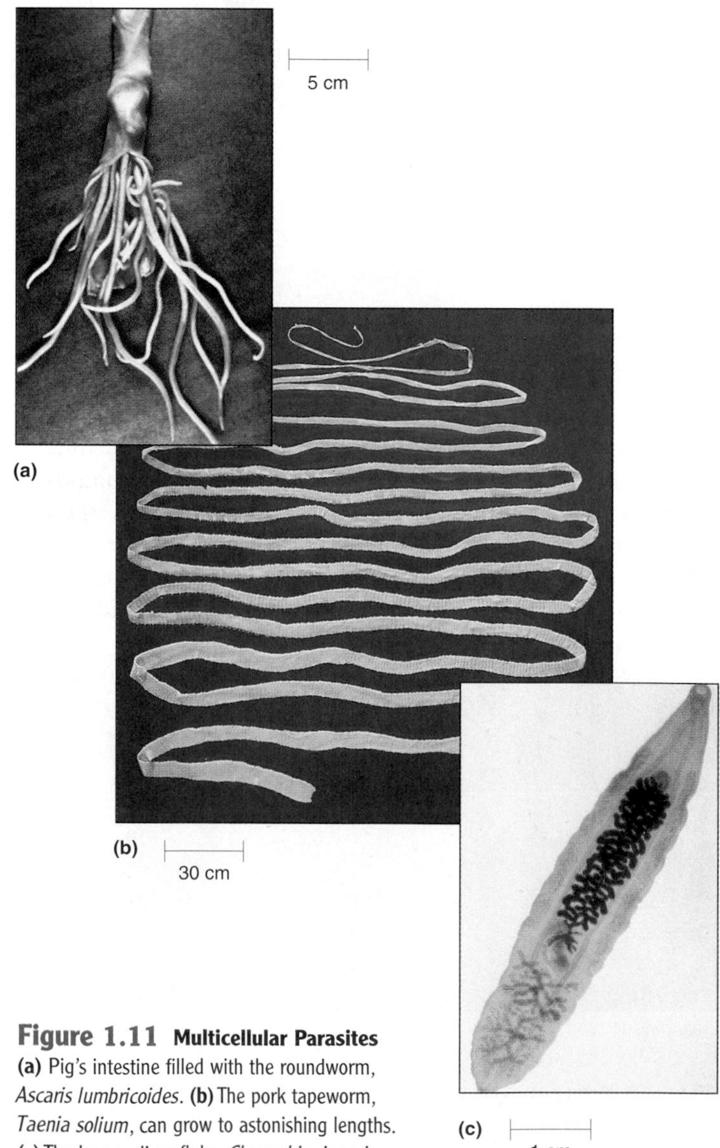

Figure 1.11 Multicellular Parasites **(a)** Pig's intestine filled with the roundworm, *Ascaris lumbricoides*. **(b)** The pork tapeworm, *Taenia solium*, can grow to astonishing lengths. **(c)** The human liver fluke, *Clonorchis sinensis*.

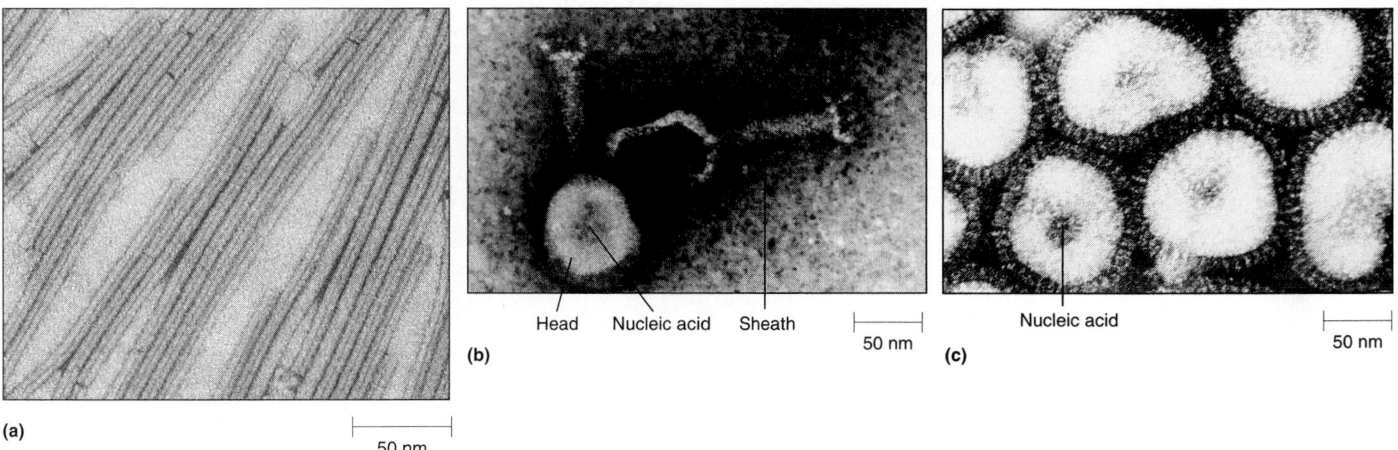

Figure 1.12 Viruses that Infect Three Kinds of Organisms **(a)** Tobacco mosaic virus that infects tobacco plants. A long hollow protein coat surrounds a molecule of RNA. **(b)** A bacterial virus (bacteriophage), which invades bacteria. Nucleic acid is surrounded by a protein coat. **(c)** Influenza virus, thin section.

- List four groups of organisms in the eukaryotic world.
- List two features that distinguish prokaryotes from eukaryotes.
- The binomial system of classification uses both a genus and a species name. Why bother with two names? Wouldn't it be easier to use a single, unique name for each different kind of microorganism?

Viruses, Viroids, and Prions

The organisms discussed so far are living members of the microbial world. In order to be alive, an organism must be composed of one or more cells. Organisms are **cellular** whereas **agents**, which are acellular. Many infectious agents consist of only a few of the molecules typically found in cells. These nonliving agents of the microbial world are the **viruses, viroids, and prions.** ■ viroids, p. 364 ■ prions, p. 364

Viruses consist of a piece of nucleic acid surrounded by a protective protein coat. They come in a variety of shapes, conferred by the shape of the coat (**figure 1.12**). Viruses share with all organisms the need to reproduce copies of themselves. Viruses can only multiply inside living host cells, whose machinery and nutrients they must borrow for reproduction. Outside the hosts, they are inactive. Thus, viruses may be considered **obligate intracellular parasites.** All forms of life including members of the Bacteria, Archaea, and Eukarya can be infected by viruses. Although viruses frequently kill the cells in which they multiply, some viruses live harmoniously within the host cell without causing obvious ill effects.

Viroids are simpler than viruses, consisting of a single, short piece of nucleic acid, specifically ribonucleic acid (RNA), without a protective coat. They are much smaller than viruses (**figure 1.13**), and like viruses they can reproduce only inside cells. Viroids cause a number of plant diseases, and some scientists speculate that they may cause diseases in humans.

Prions are very unusual agents that are responsible for at least six neurodegenerative diseases in humans and animals;

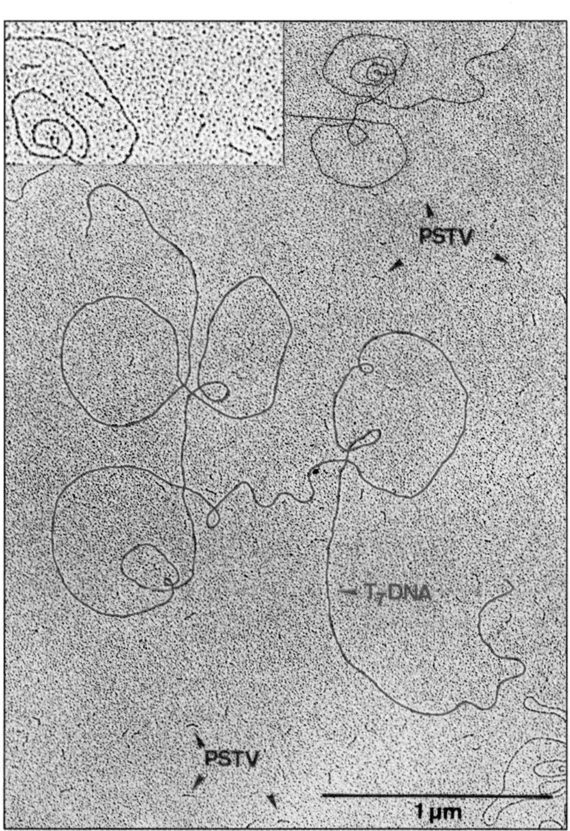

Figure 1.13 The Size of a Viroid Compared with a Molecule of DNA from the Bacteriophage T_7 The red arrows point to the potato spindle tuber viroids (PSTV); the other arrow points to T_7 DNA. The PSTV consists of RNA of about 350 nucleotides.

these are always fatal. They appear to be only protein, without any nucleic acid (**figure 1.14**). Although unlikely, it is possible that another agent that is very difficult to isolate might also be involved in causing the neurodegenerative disease.

The distinguishing features of the nonliving members of the microbial world are given in **table 1.5**. The relationships of the major groups of the microbial world to one another are presented in **figure 1.15**.

MICROCHECK 1.4

The acellular agents are viruses, viroids, and prions, all of which can be considered to be obligate intracellular parasites.

- Compare the chemical composition of viruses, viroids, and prions.
- What groups of organisms are infected by each of the following: viruses, viroids, prions?
- How might one argue that viruses are actually living organisms?

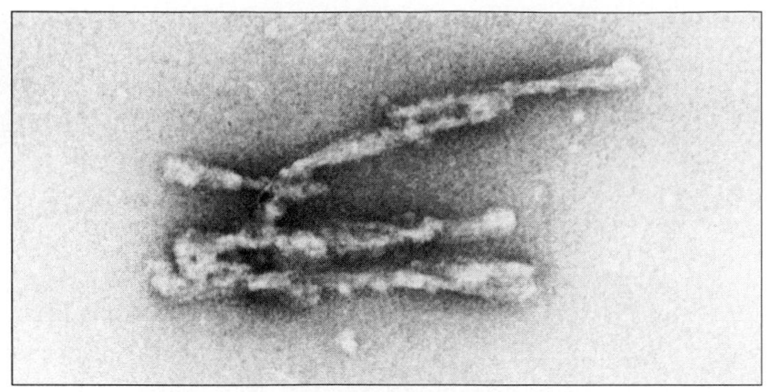

Figure 1.14 **Prion**—Prions from a human patient with Creutzfeldt-Jakob disease. This neurodegenerative disease is caused by a prion.

Figure 1.15
The Microbial World

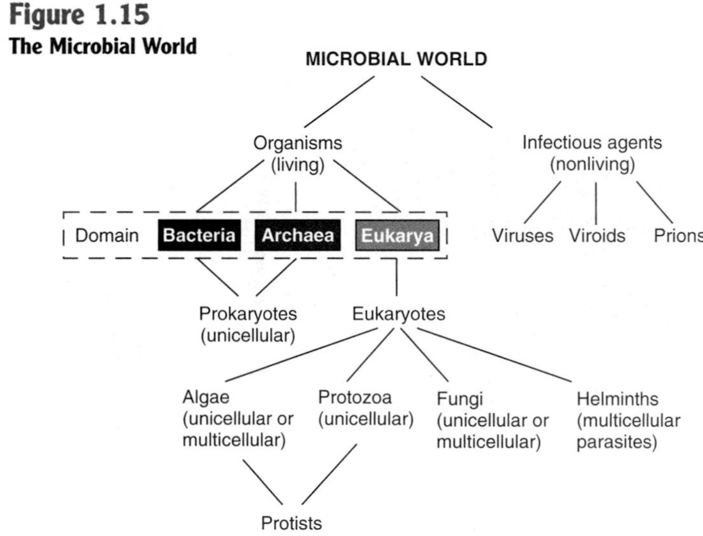

Size in the Microbial World

Members of the microbial world cover a tremendous range in their sizes, as is seen in **figure 1.16**. The smallest viruses are about 1 million times smaller than the largest eukaryotic cells. Even within a single group, wide variations exist. For example, *Bacillus megaterium* and *Mycoplasma* are both bacteria, but they differ enormously in size. (see figure 1.16). The variation in size of bacteria was recently expanded when a bacterium that is longer than 0.5 mm was discovered. (see **Perspective 1.1**, p. 15). In fact, it is so big that it is visible to the naked eye. More recently, an even

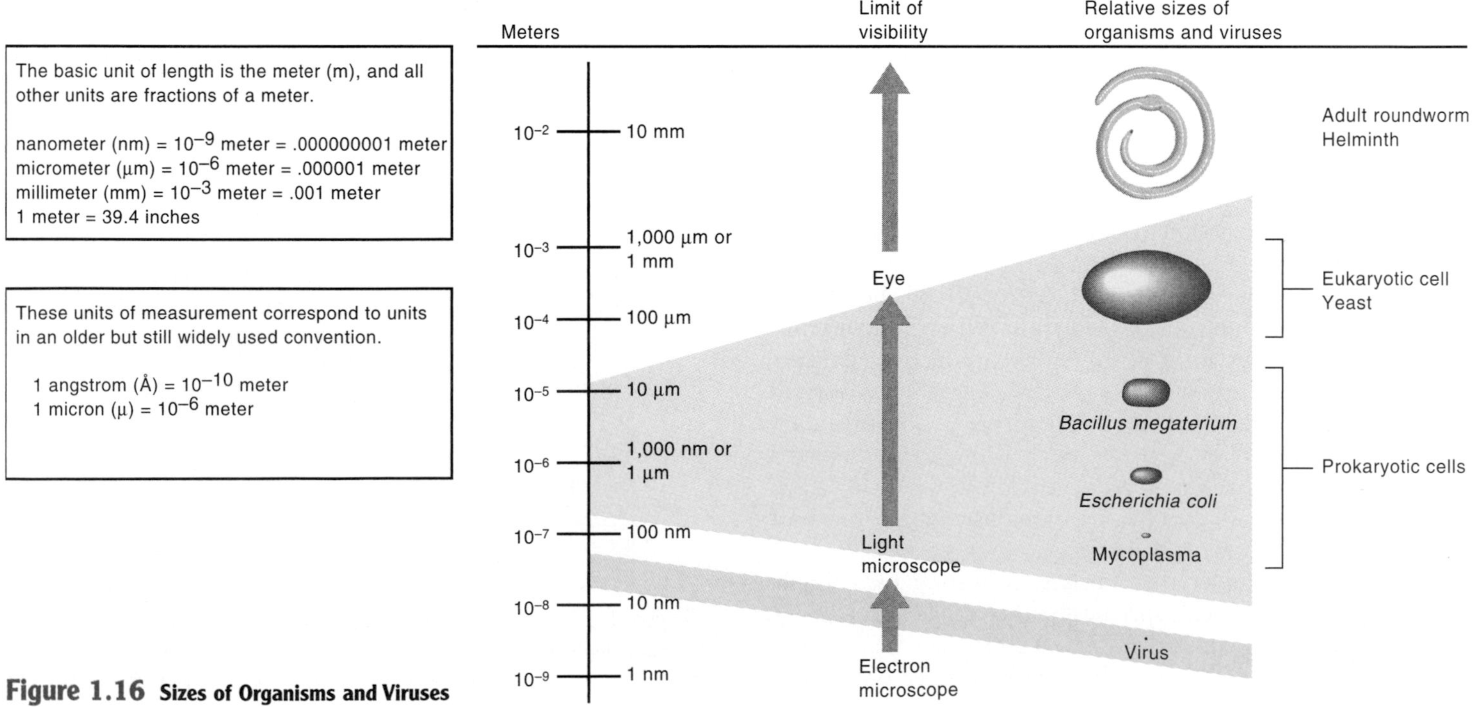

Figure 1.16 **Sizes of Organisms and Viruses**

Perspective 1.1 The Long and the Short of It

We might assume that because prokaryotes have been so intensively studied over the past hundred years, no major surprises are left to be discovered. This, however, is far from the truth. In the mid-1990s, a large, peculiar-looking organism was seen when the intestinal tracts of sturgeon fish from both the Red Sea in the Middle East and the Great Barrier Reef in Australia were examined **(figure 1)**. This organism, named *Epulopiscium*, cannot be cultured in the laboratory. Its large size, 600 μm long and 80 μm wide, which makes it clearly visible without any magnification, suggested that this organism was a eukaryote. It did not, however, have a nuclear membrane. A chemical analysis of the cell confirmed that it was a prokaryote and a member of the Domain Bacteria. This very long, slender organism is an exception to the rule that prokaryotes are always smaller than eukaryotes.

In 1999, an even larger prokaryote in volume was isolated from the sulfurous muck of the ocean floor off the coast of Namibia in Africa. It is a spherical organism 70 times as big in volume as *Epulopiscium* **(figure 2)**. Since it grows on sulfur compounds and contains glistening globules of sulfur, it was named *Thiomargarita namibiensis*, which means "sulfur pearl of Namibia." Although scientists were initially skeptical that prokaryotes could be so large, there is no question in their minds now. In contrast to these large bacteria, a cell was isolated in the Mediterranean Sea that is 1 μm in width. It is a eukaryote because it contains a nucleus and a mitochondrion, even though it is about the size of a typical bacterium.

On the other side of the coin, investigators are asking how small can a living organism be. Investigators from Finland claim that they have discovered a new form of life, the nanobacteria, which are roughly the size of viruses, about 50 to 500 nanometers (nm) in diameter (see figure 1.16). These nanobacteria are coated with thick shells of minerals, which render them very resistant to heat and chemicals. There is no clear evidence that the nanobacteria are living, however, and their apparent multiplication could result from crystal formation.

More recently, Australian investigators have observed fuzzy tangles of filaments, 20 to 150 nm in length, projecting from sandstone retrieved from an oil-drilling site in the ocean off western Australia **(figure 3)**. They also appear to be another form of nanobacteria. The Australian researchers reported that the filaments appear to reproduce quickly and form colonies, suggesting that they are alive.

Most scientists are very skeptical that "organisms" of such a small size can live independent lives. The molecules required for life such as nucleic acids and the protein-synthesizing machinery cannot be packed into "organisms" of this small size. If they are not free-living, might they be part of a larger biological entity? Might they have a biological function? Although there is great skepticism about the biological significance of these extraordinarily small "organisms," most microbiologists are convinced that many unusual microbes exist in nature that have not yet been discovered. Are nanobacteria an example?

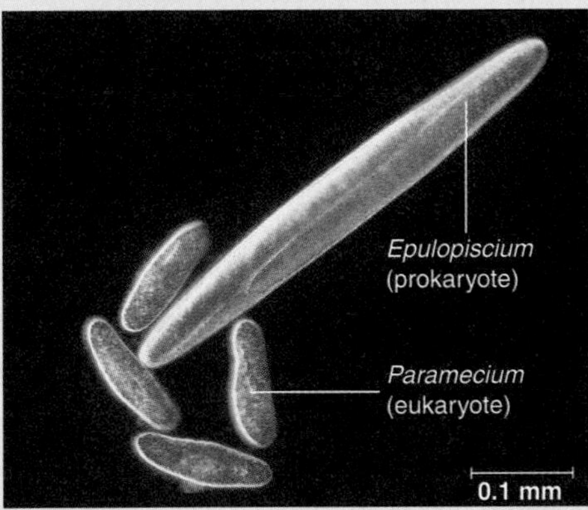

Figure 1 Photomicrograph of the longest known bacterium, *Epulopiscium* mixed with paramecia. Note how large this prokaryote is compared with the four eukaryotic paramecia.

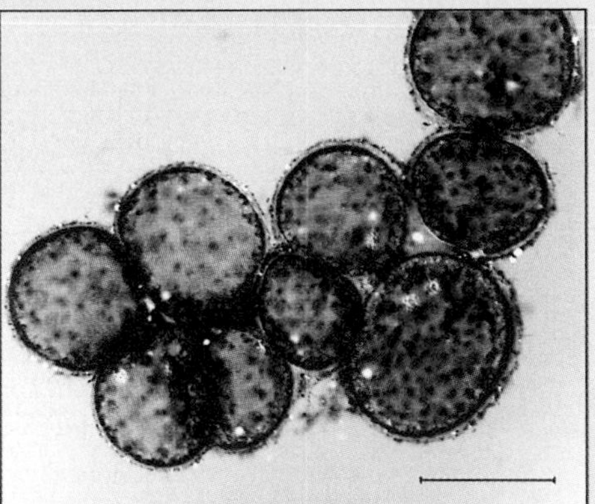

Figure 2 The average *Thiomargarita namibiensis* is two tenths of a millimeter, but some reach three times that size.

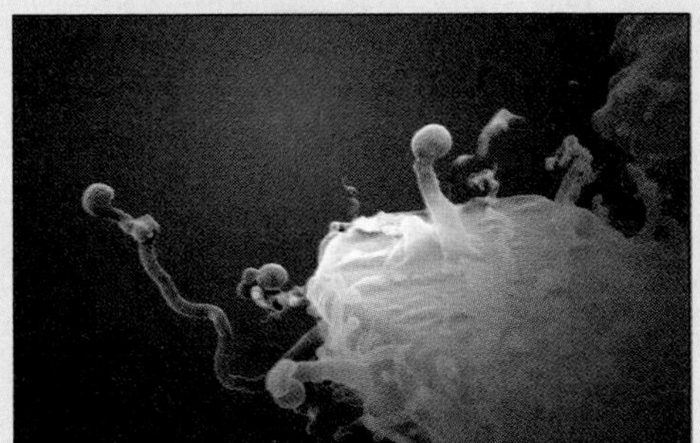

Figure 3 An electron micrograph shows filaments projecting from sandstone. The filaments, 20 to 150 nanometers in size, are smaller than cells.

larger bacterium, round in shape, was discovered. Its volume is 70 times larger than the previous record holder. Likewise, a eukaryotic cell was recently discovered that is not much larger than a typical bacterium. (see Perspective 1.1). These, however, are rare exceptions to the rule that eukaryotes are larger than prokaryotes, which in turn are larger than viruses. Helminths, being multicellular eukaryotes, are also very large and display a wide range of sizes.

As you might expect, the small size and broad size range of some members of the microbial world have required the use of measurements not commonly used in everyday life. The use of logarithms has proved to be enormously helpful, especially in designating the sizes of prokaryotes and viruses. A brief discussion of measurements and logarithms is given in appendix I.

Table 1.5 Distinguishing Characteristics of Viruses, Viroids, and Prions

Viruses	Viroids	Prions
Obligate intracellular agents	Obligate intracellular agents	Obligate intracellular agents
Consist of either DNA or RNA, surrounded by a protein coat	Consist only of RNA; no protein coat	Consist only of protein; no DNA or RNA

MICROCHECK 1.5

The range in sizes of the members of the microbial world is tremendous. As a general rule, the obligate intracellular parasites are the smallest and the eukaryotes the largest.

- Why do eukaryotic cells need to be larger than prokaryotic cells?
- What is the limiting factor in the size of free-living cells?

FUTURE CHALLENGES

Exploring the Unknown

For all the information that has been gathered about the microbial world, it is remarkable how little we know about its prokaryotic members. This is not surprising in view of the fact that less than 1% of the prokaryotes have ever been studied. In large part, this is because only one in a hundred of the prokaryotes in the environment can be cultured in the laboratory. Part of the current revolution in microbiology, however, will allow us to inventory the millions of species that are out there waiting to be discovered. Exploring the unknown of the microbial world is a major challenge and should answer many intriguing questions fundamental to understanding the biological world. What are the extremes of temperature, salt, pH, radioactivity, and pressure in which prokaryotes can live? Who would have thought that some organisms, members of the Archaea, could live at temperatures above boiling water and a pH of sulfuric acid? Are there organisms out there growing in even more extreme environments? If life can exist on this planet under such extreme conditions, what does this mean about the possibility of finding living organisms on other planets?

Although considered highly unlikely, is it possible that living organisms exist whose chemical structure is not based on the carbon atom? Will living organisms be found whose genetic information is coded in a chemical other than deoxyribonucleic acid? What new metabolic pathways remain to be discovered? As extreme environments are mined for their living biological diversity, there seems little doubt that many surprises will be found. In many cases these surprises will be translated into new biotechnology products on this planet, and they will help shape the way we look for life on other planets.

SUMMARY

The Origin of Microorganisms

Theory of Spontaneous Generation Revisited

1. The experiments of Pasteur refuted the theory of spontaneous generation. (Figure 1.2)
2. The experiments of Tyndall and Cohn demonstrated the existence of heat-resistant forms of bacteria that could account for the growth of bacteria in infusions that had been heated.

The First Microorganisms (Figure 1.1)

1. The progenitors of microorganisms no longer exist, but it seems likely that the first microorganisms probably grew in the absence of air and at very high temperatures. These were the conditions of the early earth.

Microbiology: A Human Perspective

Vital Activities of Microorganisms

1. The activities of microorganisms are vital for the survival of all other organisms, including humans.
2. Bacteria are necessary to convert the nitrogen gas in air into a form that plants and other organisms can use.
3. Microorganisms replenish the oxygen on earth.
4. Microorganisms degrade organic waste materials.

Economic Applications of Microbiology

1. For thousands of years, bread, wine, beer, and cheeses have been made by using technology still applied today.
2. Bacteria are being used to degrade dangerous toxic pollutants.
3. Bacteria are used to synthesize a variety of different products, such as cellulose, hydroxybutyric acid, ethanol, antibiotics, and amino acids.

Genetic Engineering

1. **Genetic engineering** is the process in which genes from one organism are introduced into related or unrelated organisms resulting in new properties.
2. Genetically engineered organisms have expanded the capabilities of microorganisms enormously.
3. Microorganisms produce medically important products and can produce vaccines against a variety of diseases.
4. A bacterium can transfer genes into plants and modify their properties.

Medical Microbiology

1. Many devastating diseases such as smallpox, bubonic plague, and influenza have determined the course of history.
2. "New" emerging diseases are arising in developed countries. Partly, this is because people are engaging in different lifestyles and living in regions where formerly only animals lived. (Figure 1.3)
3. "Old" diseases that were on the wane have begun to reemerge. Many are brought to this country by people visiting foreign lands.
4. Several chronic diseases such as ulcers and perhaps even heart disease may be caused by bacteria.
5. Bacteria use the body as an ecological niche and interact with other bacteria on its surface. Pathogens gain entrance to the body and find a protected niche inside host cells.

Microoganisms As Subjects for Study

1. Microorganisms are excellent model organisms to study because they grow rapidly on simple, inexpensive media, but follow the same genetic, metabolic, and biochemical principles as higher organisms.

The Microbial World (Figure 1.15)

1. Members of the microbial world consist of two major cell types: the simple prokaryotic and the complex eukaryotic. (Figures 1.5 and 1.6, Table 1.2)
2. All organisms fall into one of three **domains**, based on the chemical composition and cell structure. These are the **Bacteria**, the **Archaea**, and the **Eukarya**. (Table 1.3)

The Bacteria

1. Bacteria are single-celled **prokaryotes** that have **peptidoglycan** in their cell wall.

The Archaea

1. Archaea are single-celled prokaryotes that are identical in appearance to the Bacteria. They do not have peptidoglycan in their cell wall.
2. Archaea tend to grow in extreme environments such as hot springs and salt flats.

The Eukarya

1. Eukarya have eukaryotic cell structures and may be single-celled or multicellular.
2. Microbial members of the Eukarya are the **algae**, **fungi**, **protozoa**, and **helminths**.
3. Algae can be single-celled or multicellular, and they can use sunlight as a source of energy. (Figure 1.8 and Table 1.4)
4. Fungi are either single-celled yeasts or multicellular molds and mushrooms. They use organic compounds as food. (Figure 1.9 and Table 1.4)
5. Protozoa are single-celled organisms that are motile by a variety of means. They use organic compounds as food. (Figure 1.10 and Table 1.4)
6. Multicellular **parasites** called the helminths derive food from the host organism on which they live. They include; **tapeworms**, **roundworms**, and **flukes**. (Figure 1.11 and Table 1.4)

Nomenclature

1. Organisms are named according to a binomial system.
2. Each organism has a **genus** and a **species** name, written in italics.

Viruses, Viroids, and Prions

1. The nonliving members of the microbial world are not composed of cells. They are considered **obligate intracellular parasites** and include **viruses**, **viroids**, and **prions.**
2. Viruses are a piece of nucleic acid surrounded by a protein coat. They can infect members of all three domains. (Figure 1.12 and Table 1.5)
3. Viroids are composed of a single, short RNA molecule. Thus far, they are only known to cause diseases in plants. (Figure 1.13 and Table 1.5)
4. Prions consist only of protein, without any nucleic acid. They cause several different neurodegenerative diseases of humans and animals. (Figure 1.14 and Table 1.5)

Size in the Microbial World

1. Sizes of members of the microbial world vary enormously from the proteins of prions to the multicellular helminths. (Figure 1.16)

REVIEW QUESTIONS

Short Answer

1. Name the prokaryotic groups in the microbial world.
2. List five beneficial applications of bacteria.
3. Name three nonliving groups in the microbial world and describe their major properties.
4. In the designation *Escherichia coli* O157:H7, what is the genus? What is the species? What is the strain?
5. Where would you go to isolate members of the Archaea?
6. How might you distinguish a prokaryotic cell from a eukaryotic cell?
7. Give three reasons that life could not exist on earth without the activities of microorganisms.
8. Differentiate between biotechnology and genetic engineering.
9. Name two diseases that have been especially destructive in the past. What is the status of those diseases today?
10. State three reasons that there is a resurgence of infectious diseases today.

Multiple Choice

1. The prokaryotic members of the microbial world include
   1. algae.
   2. fungi.
   3. prions.
   4. bacteria.
   5. archaea.

   A. 1,2 B. 2,3 C. 3,4 D. 4,5 E. 1,5
2. The Archaea
   1. are microscopic.
   2. are commonly found in extreme environments.
   3. contain peptidoglycan.
   4. contain mitochondria.
   5. are most commonly found in the soil.

   A. 1,2 B. 2,3 C. 3,4 D. 4,5 E. 1,5
3. The most fundamental division of cell types is between
   A. algae, fungi, and protozoa.
   B. eukaryotes and prokaryotes.
   C. viruses and viroids.
   D. typical bacteria and archaea.
   E. eukarya, bacteria, and archaea.
4. Astrobiology involves the study of
   1. life on other planets.
   2. viroids.
   3. emerging infectious diseases.

4. biotechnology.
5. life under extreme conditions.

A. 1,2 B. 2,3 C. 3,4 D. 4,5 E. 1,5

5. An organism isolated from a hot spring in an acidic environment is most likely a member of the
   A. Bacteria.
   B. Archaea.
   C. Eukarya.
   D. virus family.
   E. Fungi.
6. The agent that contains no nucleic acid is a
   A. virus.
   B. prion.
   C. viroid.
   D. bacterium.
   E. fungus.
7. Prokaryotes do not have
   A. cell walls.
   B. flagella.
   C. a nuclear membrane.
   D. specific shapes.
   E. genetic information.
8. Nucleoids are associated with
   1. genetic information.
   2. prokaryotes.
   3. eukaryotes.
   4. viruses.
   5. prions.

   A. 1,2 B. 2,3 C. 3,4 D. 4,5 E. 1,5
9. Pathogenicity islands
   1. are composed of DNA.
   2. are found in viruses.
   3. are found only in eukaryotes.
   4. are found mainly in archaea.
   5. are found in some bacteria.

   A. 1,2 B. 2,3 C. 3,4 D. 4,5 E. 1.5
10. The person best known for his microscopy of microorganisms is
    A. Antony van Leeuwenhoek.
    B. Louis Pasteur.
    C. John Tyndall.
    D. Ferdinand Cohn.

Applications

1. The American Society of Microbiology is preparing a "Microbe-Free" banquet to emphasize the importance of microorganisms in the diet. What foods would not be on the menu if microorganisms were not available for our use?
2. If you were asked to nominate one of the individuals mentioned in this chapter for the Nobel Prize, who would it be? Make a statement supporting your choice.

Critical Thinking

1. An early microbiologist, who was inclined to accept the theory of spontaneous generation, criticized Pasteur's experiments (see figure 1.2). He claimed that a few spontaneously generated bacteria would be present in the boiled and cooled broth but shaking the broth was necessary to stimulate these few to start dividing. Pasteur had not included this step, and bacteria had not started dividing until the flask was tipped and the broth agitated. Does this microbiologist have a valid criticism? What kind of experiment would check this possibility?
2. A microbiologist obtained two pure isolated biological samples: one of a virus, and the other of a viroid. Unfortunately, the labels had been lost from the two samples. The microbiologist felt she could distinguish the two by analyzing for the presence or absence of a single chemical element. What element would she search for and why?
3. Chlamydias and rickettsias were once classified as viruses because of their small size and obligate intracellular growth requirement. What techniques can be used to show that they are really bacteria?

CHAPTER 2

The Molecules of Life

Louis Pasteur (1822–1895) is often considered the father of bacteriology. His contributions to this science, especially in its early formative years, were enormous and are discussed in many of the succeeding chapters. Pasteur started his scientific career as a chemist, initially working in the science of crystallography. It was not until he was on the Faculty of Science at Lille, France, a town with important brewing industries, that he became interested in the biological aspects of chemical problems. All of his later studies—including his work on spontaneous generation, infectious disease, and protection against infectious diseases through vaccination—employed the analytical experimental methods and thinking of a trained chemist.

His first chemical studies were performed on two compounds, tartaric and paratartaric acids, which formed thick crusts within wine barrels. These two substances form crystals that have the same number and arrangement of atoms, yet they twist (rotate) a plane of light differently when that light passes through the crystal. Tartaric acid twists the light; paratartaric acid does not. Therefore, the two molecules must differ in some way, even though they are chemically identical. Pasteur was intrigued by these observations and set about to understand how the crystals differed. Looking at them under a microscope, he saw that the crystals of tartaric acid all looked identical but paratartaric acid consisted of two different kinds of crystals. Using tweezers, he carefully separated the two kinds into two piles and dissolved each kind in a separate flask of water. When he shone polarized light through each solution, one solution twisted the light to the left and the other twisted it to the right. When he mixed equal numbers of each kind of crystal into water and shone polarized light through the solutions, the light was not twisted. Apparently, the two components of the mixture neutralized each other, and as a result, the mixture did not rotate the light. Pasteur concluded that paratartaric acid is a mixture of two compounds, each being the mirror image, or **optical isomer**, of the other. Optical isomers are often called **stereoisomers**. This mixture of two optical isomers can be conceived of as a mixture of right- and left-handed molecules, represented as a right and left hand facing each other (see figure 2.15). They cannot be superimposed on each other, much as a right-handed glove cannot fit the left hand.

Pasteur thus solved a long-standing mystery in chemistry. At age 25 he had established the fundamental concept of **stereochemistry**. These studies proved to be far more significant in biology than Pasteur could have imagined. In later studies, he showed that molecules synthesized by living organisms have a preferred

handedness, whereas molecules synthesized by chemical means are always a mixture of right- and left-handed molecules. Further, living organisms can only use one form. Indeed, an organism would starve to death if it consumed only the wrong-handed sugars, because such sugars are indigestible.

Stereoisomers of the same molecule have greatly different properties. For example, the amino acid phenylalanine, one of the key ingredients in the artificial sweetener aspartame, makes aspartame sweet when it is in one optical form but bitter when in the other form. As another example, researchers believe that the birth defects caused by the drug thalidomide, once used to prevent nausea in pregnant women, resulted from one, but not the other optical form in a mixture of the two forms. Unfortunately, it is difficult to separate the two forms on a large scale.

Thus, what Pasteur studied as a straightforward problem in chemistry has implications far beyond what he ever imagined. It is often difficult to predict where research will lead or the significance of interesting but seemingly unimportant observations.

—***A Glimpse of History***

TO UNDERSTAND HOW CELLS LIVE AND HOW THEY interact with one another and with their environment, we must be familiar with certain facts related to the molecules that compose all living matter. The molecules of life make up the structure of cells and in so doing interact with one another with great specificity. Certain properties of molecules allow cells to carry out the myriad of functions necessary for life. What are these properties that allow such specific interactions? Of what are the molecules of life composed to give them such properties? In addition to providing answers to such questions, this chapter will also introduce some of the vocabulary necessary to understand the text.

For some, this information may serve as a review of material already encountered. For others, it may be a first encounter with the chemistry of biological molecules. In this case, you likely will return to this chapter frequently. The discussion proceeds from the lowest level of organization, the atoms and elements, to the highly complex associations between molecules to form large molecules, the macromolecules.

Atoms and Elements

Atoms, the basic units of all matter, are are made up of three major components: the negatively charged **electrons**; positively charged **protons**; and uncharged **neutrons (figure 2.1)**. The protons and neutrons, the heaviest components, are found in the heaviest part of the atom, the nucleus. The very light electrons orbit the nucleus. The number of protons normally equals the number of electrons, and so the atom as a whole is uncharged. The relative sizes and motion of the parts of an atom can be illustrated by the following analogy. If a single atom were enlarged to the size of a football stadium, the nucleus would be the size of a marble and it would be positioned somewhere above the 50-yard line. The electrons would resemble fruit flies zipping around the stands. Their orbits would be mostly inside the stadium, but on occasion they would travel outside it.

An **element** is a pure substance that consists of a single type of atom. Although 92 naturally occurring elements exist, four elements make up over 98% of all living material by weight. These elements are carbon (abbreviated C), hydrogen (H), oxygen (O), and nitrogen (N). Two other elements, phosphorus (P), and sulfur (S), together make up an additional 0.5% of the elements in living systems **(table 2.1)**. All of the remaining elements together account for less than 0.5% of living material. In general, the basic chemical composition of all living cells is remarkably similar.

Each element is identified by two numbers: its **atomic number** and its **atomic weight** or **mass**. The atomic number is the number of protons, which equals the number of electrons. For example, hydrogen has 1 proton, and thus its atomic number is 1; oxygen has 8 protons, and its atomic number is 8. The atomic weight is the sum of the number of protons and neutrons, since electrons are too light to contribute to the weight. The atomic weight of hydrogen is 1, the atom is abbreviated ^{1}H, reflecting 1 proton and no neutrons. It is the lightest element known. The atomic weight of oxygen is approximately 16 and the atom is abbreviated ^{16}O, reflecting 8 protons and 8 neutrons.

The electrons are arranged in **shells** that orbit the nucleus (see figure 2.1). Each shell can contain only a certain number of

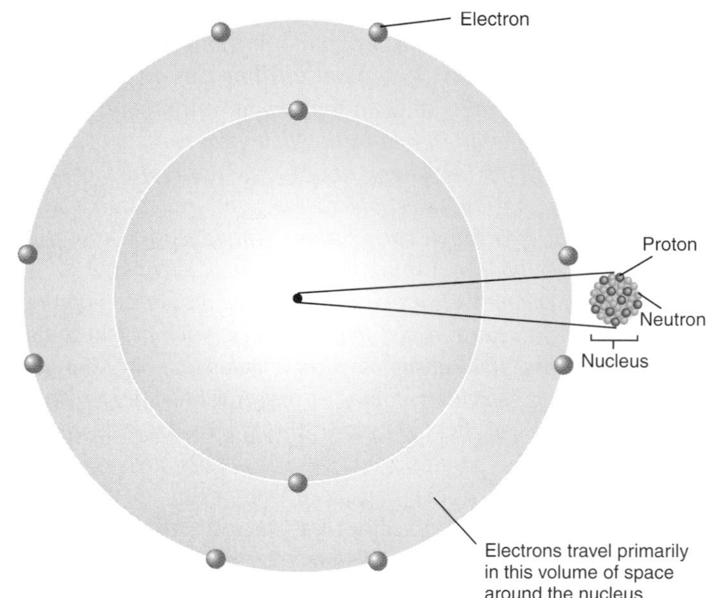

Electron
Proton
Neutron
Nucleus
Electrons travel primarily in this volume of space around the nucleus.

Figure 2.1 Atom The proton has a positive charge, the neutron has a neutral charge, and the electron has a negative charge. The electrons that orbit the nucleus are arranged in shells. For stability, the shells of the atom are filled in order. The shell closest to the nucleus is filled first, then the next shell is filled, and so on.

TABLE 2.1 Atomic Structure of Elements Commonly Found in the Living World

Element	Symbol	Atomic Number (Total Number of Protons)	Atomic Weight (Protons and Neutrons)	Number of Possible Covalent Bonds*	% of Atoms in Cells
Hydrogen	H	1	1	1	49
Carbon	C	6	12	4	25
Nitrogen	N	7	14	3	0.5
Oxygen	O	8	16	2	25
Phosphorus	P	15	31	3	0.1
Sulfur	S	16	32	2	0.4

*The number of electrons required to fill the outer shell equals the number of possible covalent bonds. The number of electrons in a completed outer shell varies depending on the distance of the shell from the nucleus.

electrons. The first shell closest to the nucleus contains a maximum of 2 electrons, the next 8, and the next also 8. Other atoms, which have little biological importance, have additional electrons. Each shell must be filled, starting with the one closest to the nucleus, before electrons can occupy the next outer shell.

MICROCHECK 2.1

All living organisms contain the same elements. The four most important are carbon, hydrogen, oxygen, and nitrogen. The basic unit of all matter, the atom, is composed of protons, electrons, and neutrons.

- Of all the elements found in cells, which element is found most frequently?
- Give the three major components of atoms. Describe their properties.

Chemical Bonds and the Formation of Molecules

For maximum stability, the outer shell of an atom must contain the maximum number of electrons. If an atom does not have its outer shell full, then it tends to fill its outer shell by bonding with other atoms.

Most atoms do not have their outer shells filled with the maximum number of electrons. To fill their outer shells, atoms can either gain electrons from other atoms or lose electrons to other atoms. To gain or lose electrons, atoms bond with other atoms to form **molecules**. A molecule consists of two or more atoms held together by **chemical bonds**. The atoms that make up a molecule may be of the same or different elements. For example, H_2 is a molecule of hydrogen gas formed from two atoms of hydrogen; water (H_2O) is an association of two hydrogen atoms with one oxygen atom. A compound consists of two or more different elements. The molecular weight of a molecule is the sum of the atomic weights of the component atoms. Thus, the molecular weight of water is 18 (1 + 1 + 16).

The chemical bonds that hold atoms together to form molecules are of various types, which differ in strength. These include covalent bonds, ionic bonds, and hydrogen bonds.

Covalent Bonds

Atoms often achieve stability by sharing electrons with other atoms, thereby filling the outer shells of both atoms simultaneously. This sharing of electrons creates strong bonds, called **covalent bonds**. Carbon, the most important single atom in biology, is frequently involved in covalent bonding. Carbon has four electrons but requires a total of eight to fill its outer shell. It can do this by sharing four electrons with four hydrogen atoms to form CH_4, which is the gas methane (**figure 2.2,** and see table 2.1). Each H atom requires one additional electron to fill its outer shell, which it gains by sharing an electron of the carbon atom. Thus, the formation of methane involves one carbon atom sharing eight electrons with four hydrogen atoms. The outer shells of both carbon and hydrogen are then filled with shared electrons. This molecule is highly stable and cannot form additional covalent bonds. A single covalent bond is designated by a dash or colon between the two atoms sharing the electrons and is written as C—H or C : H. Sometimes two pairs of electrons are shared between atoms in order for their outer shells to be filled. This results in the formation of a double covalent bond indicated by two lines between the atoms—for example O=C=O (CO_2).

Different elements require different numbers of electrons to fill their outer shells, and these numbers determine the number of covalent bonds that each element can make. This information is given in table 2.1 for the elements most important in the living world.

All covalent bonds are strong. The stronger the bond, the more difficult it is to break. Consequently, covalent bonds do not break unless they are exposed to strong chemicals or are supplied with large amounts of energy, generally as heat. Molecules

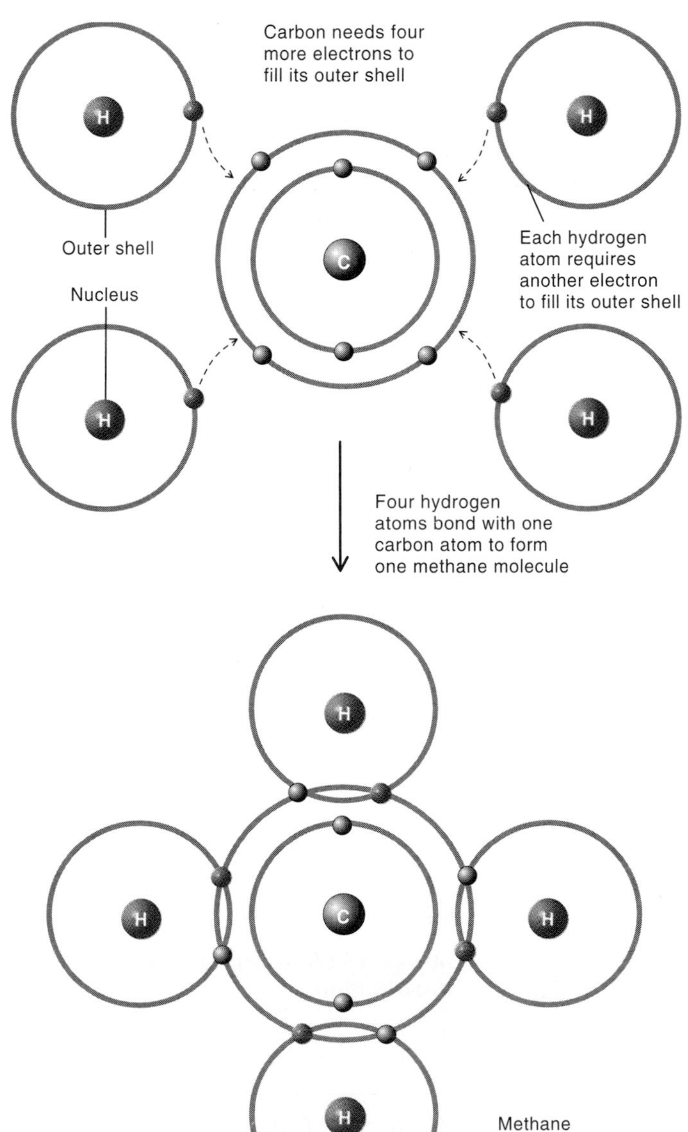

Figure 2.2 Covalent Bonds The carbon atom fills its outer electron shell by sharing a total of eight electrons with four H atoms. Each hydrogen atom shares two electrons with a C atom, thereby filling its outer shell.

formed by covalent bonds never break apart at temperatures compatible with life. Since most biological systems function only within a narrow temperature range (approximately 5°C–100°C) and cannot tolerate the high temperatures required to break these covalent bonds, cells utilize protein catalysts called **enzymes**, which can break these covalent bonds at the lower temperatures found in living systems. We will consider the function of enzymes in chapter 6.

Nonpolar and Polar Covalent Bonds

Two atoms connected by a covalent bond may exert the same or different attractions for the shared electrons of the bond. In covalent bonds between identical atoms, such as H—H, the electrons are shared equally. Such equal bonds also exist between different atoms, such as C—H, if both atoms have a similar attraction for electrons **(table 2.2)**. The bond is termed a **nonpolar covalent bond**. If, however, one atom has a much greater attraction for electrons than the other, the electrons are shared unequally, and **polar covalent bonds** are formed. One part of the molecule has a slightly positive charge and another part a slightly negative charge. An example of a polar molecule is water, in which the oxygen atom has a greater attraction for the shared electrons than do the hydrogen atoms **(figure 2.3)**. Consequently, the oxygen atom has a slight negative charge and the hydrogen atoms a slight positive charge. Polar covalent bonds play a key role in biological systems because they allow weak bonds, termed **hydrogen bonds,** to be formed. These weak bonds are responsible for different molecules to recognize one another and join together to carry out biological functions. These bonds will be discussed shortly.

Ionic Bonds

As we have already mentioned, an atom can fill its outer shell by either gaining or losing electrons. If electrons from one atom are attracted very strongly by another nearby atom, the electrons completely leave the first atom and become a part of the outer electron shell of the second, without any sharing. The attraction of the positively charged atom to the negatively charged atom forms the bond. Such charged atoms are termed **ions**, and this bond is called an **ionic bond (figure 2.4)**. The atom that gains the electrons becomes negatively charged, while the atom that gives up the electrons becomes positively charged. The difference between the number of protons and electrons in the ion is indicated by a superscript number. If only a + or – is indicated, then the charge is 1. For example, Na^+ indicates a Na ion with one positive charge. Note that the charge is much greater when electrons are transferred from one atom to another than when polar covalent bonds are formed. Ionic bonds are an example of extreme polarity.

Ionic bonds are important in biology because they are common among the weak forces holding ions, atoms, and molecules together. In water (aqueous solutions), ionic bonds are about 100 times weaker than covalent bonds, because water molecules tend to move between the ions and thereby greatly reduce their attraction for one another. Thus, in aqueous solution, which is common in all biological systems, ionic bonds are readily broken at room temperature. In all biological systems, recognition between molecules is very important. This recognition depends on large numbers of atoms on the surfaces of

Table 2.2 Nonpolar and Polar Covalent Bonds

Type of Covalent Bond	Atoms Involved and Charge Distribution	
Nonpolar	C—C C—H H—H	C and H have equal attractions for electrons, so there is an equivalent charge on each atom.
Polar	O—H N—H O—C N—C	The O and N atoms have a stronger attraction for electrons than do C and H, so the O and N have a slight negative charge; the C and H have a slight positive charge.

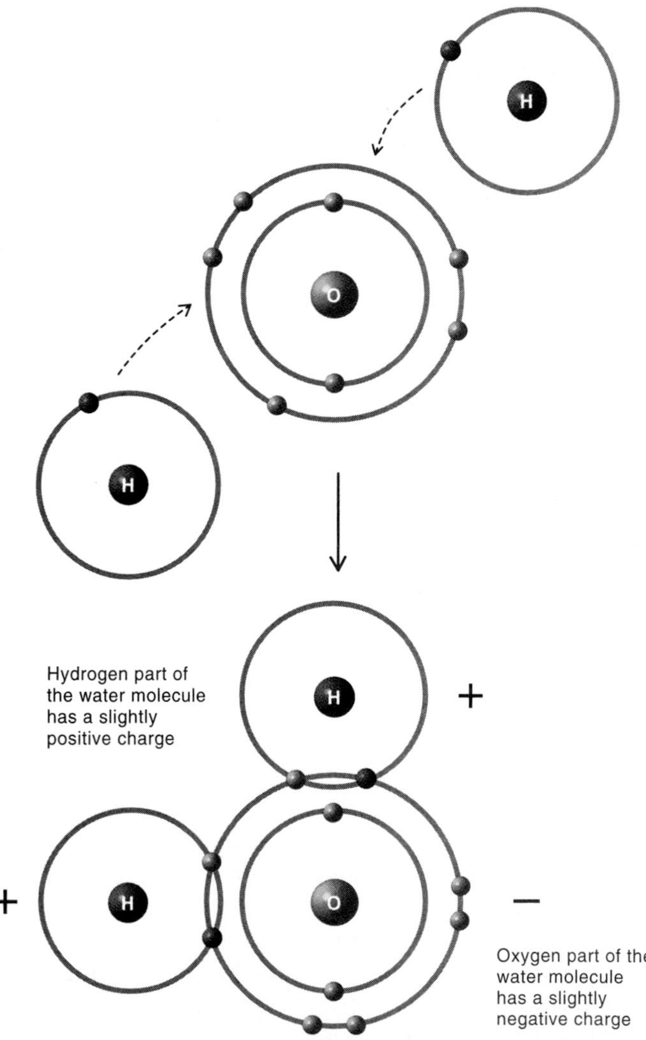
H
O
H
Hydrogen part of the water molecule has a slightly positive charge
H
+
+
H
O
–
Oxygen part of the water molecule has a slightly negative charge

Figure 2.3 Formation of Polar Covalent Bonds in a Water Molecule The oxygen atom has a greater attraction for the shared electrons than do the hydrogen atoms. Hence, the electron is closer to the oxygen and confers a negative charge on a portion of this atom. Each of the hydrogen has a positive charge on a portion of the atom. Because of these charges, water is a polar molecule.

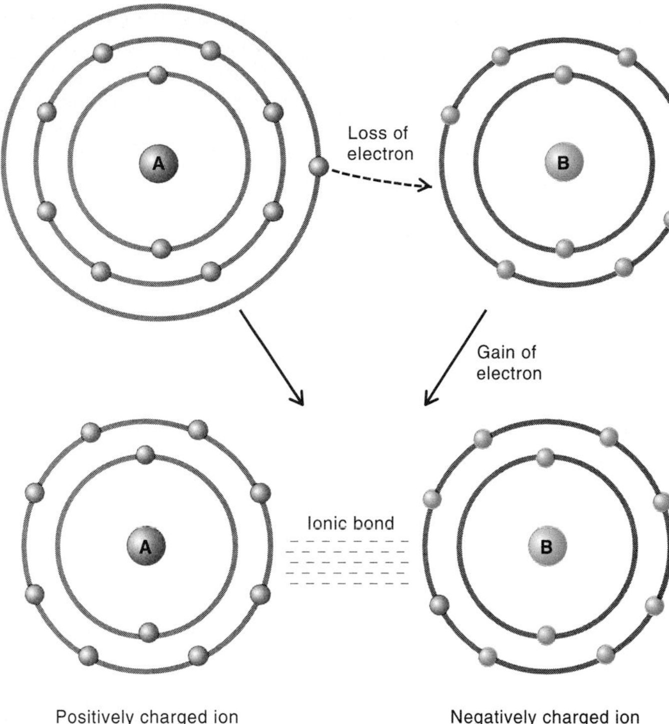

Figure 2.4 Ionic Bond Atom A gives up an electron to atom B; hence, atom A acquires a positive charge and atom B a negative charge. Both atoms then have their outer shells filled with the maximum number of electrons leading to maximum stability. The attraction of the positively charged ion to the negatively charged ion forms the bond.

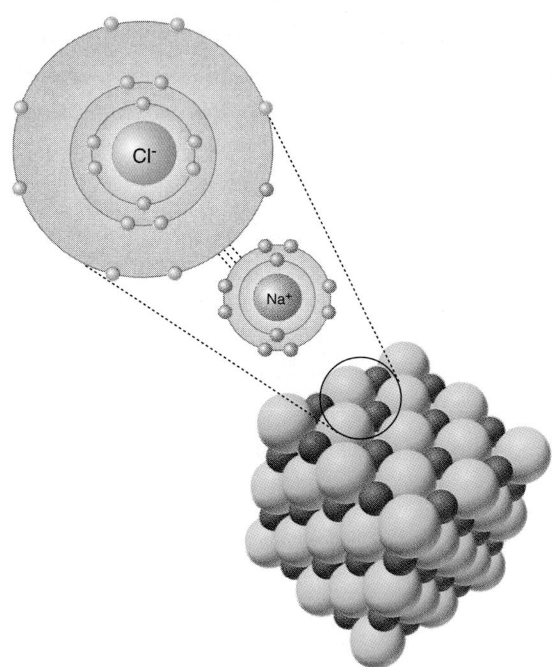

Figure 2.6 Crystal of NaCl (Salt) Resulting from Ionic Bonds Between Na^+ and Cl^- ions This crystal structure is maintained only as long as water is not present to pull apart and keep apart the Na^+ and Cl^- ions.

molecules matching each other precisely **(figure 2.5)**. Large numbers of weak bonds and hydrogen bonds such as ionic bonds are necessary to hold the molecules together.

In the absence of water, a condition that rarely occurs in biological systems, ionic bonds are strong and account for crystal formation. Such bonds account for the strength of minerals like marble and agate as well as crystals of salt (sodium chloride) **(figure 2.6)**.

Hydrogen Bonds

Hydrogen bonds are weak bonds that result from the attraction of a positively charged hydrogen atom in a polar molecule to a negatively charged atom, frequently oxygen (O) or nitrogen (N) in another polar molecule (**figure 2.7**, and see table 2.2). Oxygen and nitrogen attract electrons within a molecule, and therefore they often have a slightly negative charge. Note that hydrogen bonds form between molecules such as water molecules or within molecules such as DNA, whereas covalent bonds occur between the atoms that make up these molecules. Hydrogen bonds are important in biological systems. Living organisms are composed of many molecules that contain hydrogen atoms bonded to nitrogen or oxygen atoms, thereby creating the possibility for many hydrogen bonds. These weak bonds are important in recognizing matching surfaces and holding the molecules on these surfaces together (see figure 2.5).

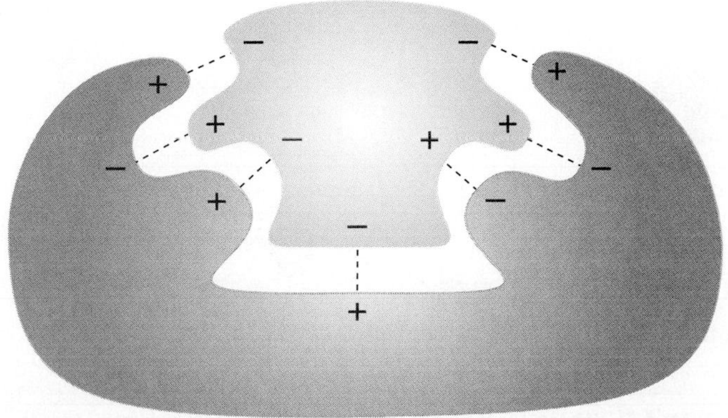

Figure 2.5 Weak Ionic Bonds and Molecular Recognition Weak bonds, such as ionic and hydrogen bonds, are important for molecules to recognize each other. Many weak bonds are required to hold the two molecules together.

In contrast to covalent bonds, but like ionic bonds in an aqueous environment, hydrogen bonds are constantly being formed and broken at room temperature because the energy produced by the movement of water is enough to break these bonds. The average lifetime of a single hydrogen bond is only a fraction of a second at room temperature, so that enzymes, which speed up reactions under physiological conditions, are not necessary to form or break hydrogen bonds.

Although a single hydrogen bond is too weak to bind molecules together, a large number can hold molecules together firmly. A good example is the double-stranded DNA molecule. The two strands of this molecule are held together by many hydrogen bonds up and down the length of the molecule. The two strands can be unraveled if energy is supplied, usually in the form of heat approaching temperatures of 100°C.

MICROCHECK 2.2

Molecules are formed by bonding between atoms. Bonds are formed when electrons from one atom interact with another atom. The bonds between the

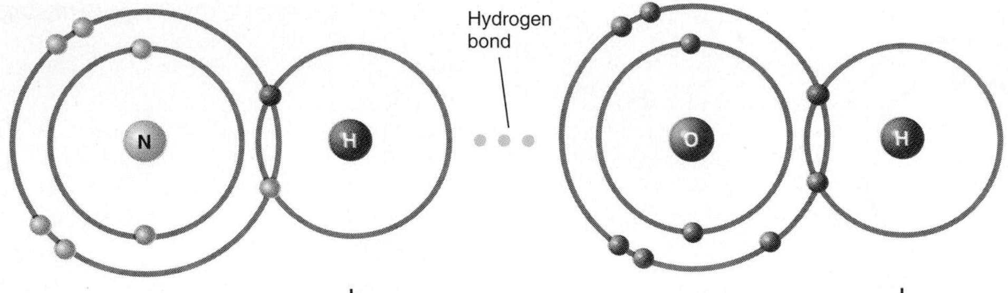

Figure 2.7 Hydrogen Bond Formation The N atom has a greater attraction for electrons than the H atom, thereby conferring a slight positive charge on the H atom. The O atom has a greater attraction for electrons than the H atom to which it is covalently bonded, thereby gaining a slight negative charge. The positively charged H atom hydrogen bonds to the negatively charged O atom.

atoms which make up a molecule are strong covalent bonds; bonds between molecules are generally weak bonds such as ionic and hydrogen bonds.

- Compare the relative strengths of covalent, hydrogen, and ionic bonds.
- Which of these bonds requires an enzyme to break it?
- Why does an atom that gives up electrons become positively charged? What causes the positive charge?

Chemical Components of the Cell

The most important molecule in the cell is water; all organisms have been built on its special properties.

Water

Unquestionably, water is the most important molecule in the world. Water makes up over 70% of all living organisms by weight. Most of the compounds of living beings are in solution inside cells. The importance of water to life on earth in large part depends on its unusual properties.

Bonding Properties of Water

Hydrogen bonding plays a very important role in the properties of water. Since water is a polar molecule, the positive H portion of the molecule is attracted to the negative O portion of other water molecules, thereby creating hydrogen bonds. The number of hydrogen bonds that are formed between water molecules depends on the temperature. At room temperature, when water is in a liquid state, the weak bonds are continually forming and breaking in a shifting pattern **(figure 2.8a)**. As the temperature is lowered, the breakage and formation decreases, and in ice, a crystalline structure is formed. Each water molecule bonds to four other molecules to form a rigid lattice structure (figure 2.8b). When ice melts, the water molecules can move closer together. Consequently, liquid water is more dense than ice, which explains why ice floats. This explains how fish and bacteria can apparently live in frozen bodies of water. They actually live in the water, which remains liquid below the ice.

The polar nature of water also accounts for its ability to dissolve a large number of compounds. Water has been referred to as *the universal solvent of life* because it dissolves so many compounds placed into it. To dissolve in water, compounds must contain atoms with positive or negative charges. When placed in water, they ionize or split into their component charged atoms. For example, NaCl dissolves in water to form Na^+ ions, termed **cations** because they move toward the negatively charged cathode in an electrical field, and Cl^- ions, termed **anions** because they move toward the positively charged anode. In solution, ions such as Na^+ and Cl^- tend to be surrounded by water molecules in such a way that the OH^- of HOH forms weak bonds with Na^+, and the H^+ forms weak bonds with Cl^- **(figure 2.9)**. The Na^+ and the Cl^- cannot come together, and this accounts for the solubility of NaCl in water.

Water containing dissolved substances freezes at a lower temperature than pure water. Because the water molecules are bonded

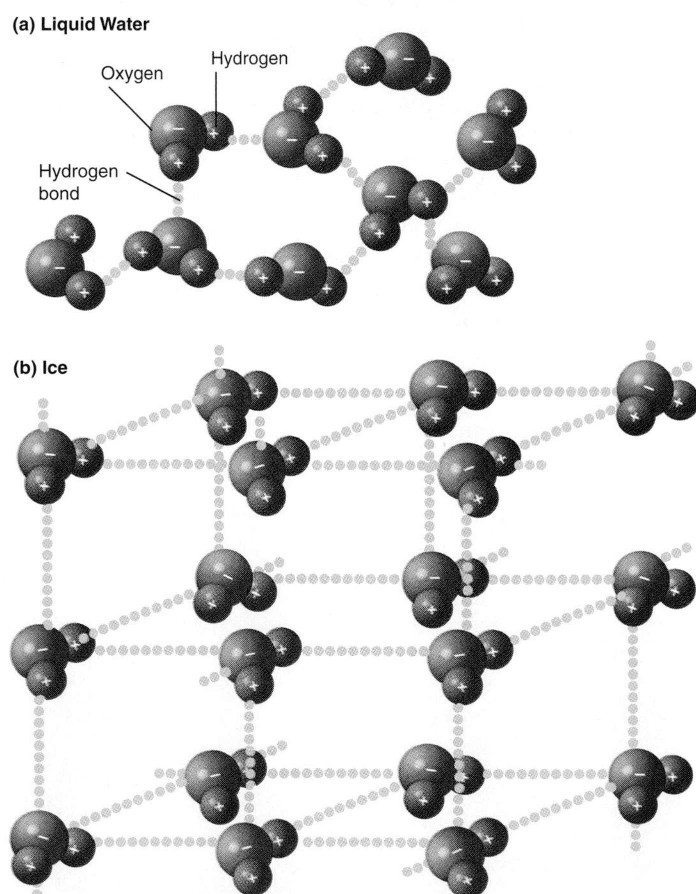

Figure 2.8 Water (a) In liquid water, each H_2O molecule hydrogen bonds to one or more other H_2O molecules. These bonds continually break and re-form. **(b)** In ice, each H_2O molecule is hydrogen bonded to four other H_2O molecules, forming a rigid crystalline structure. The bonds do not break continuously.

Perspective 2.1 Isotopes: Valuable Tools for the Study of Biological Systems

One important tool in the analysis of living cells is the use of **isotopes,** variant forms of the same element that have different atomic weights. The nuclei of certain elements can have greater or fewer neutrons than usual and thereby be heavier or lighter than is typical. For example, the most common form of the hydrogen atom contains 1 proton and 0 neutrons and has an atomic weight of 1 (^1H). Another form, however, also exists in nature in very low amounts. This isotope, ^2H (deuterium), contains 1 neutron. A third, even heavier isotope, ^3H (tritium), is not found in nature but can be made by a nuclear reaction in which stable atoms are bombarded with high-energy particles. This latter isotope is unstable and gives off radiation (decays) in the form of rays or electrons, which can be very sensitively measured by a radioactivity counter. Once the atom has finished disintegrating, it no longer gives off radiation and is stable.

An important feature of radioactive isotopes is that their other properties are very similar to their nonradioactive counterparts. For example, tritium combines with oxygen to form water and with carbon to form hydrocarbons, and both molecules have biological properties very similar to their nonradioactive counterparts. The only difference is that the molecules containing tritium can be detected by the radiation they emit.

Isotopes are used in numerous ways in biological research. They are frequently added to growing cells in order to label particular molecules in cells. For example, when tritiated thymidine (a component of DNA) is added to growing bacteria, the thymidine will become a part of the DNA of the cells. The DNA molecules can then be detected easily by their radioactivity. The shape and size of the radioactive DNA molecule can actually be seen by placing the DNA on photographic film and storing the material in the dark for several days to allow the radioactivity to make its marks on the photographic emulsion. The location of the radioactivity in the cell can be determined from the position of the developed silver grains produced by the radioactivity, making it possible to locate the molecules in the cell. For example, DNA is in the nucleus only **(figure 1a)**, whereas RNA, synthesized from ^3H-uridine (tritiated uridine, a component of RNA) is found throughout the cytoplasm (figure 1b). This technique involving radioisotopes and photography is termed **autoradiography**.

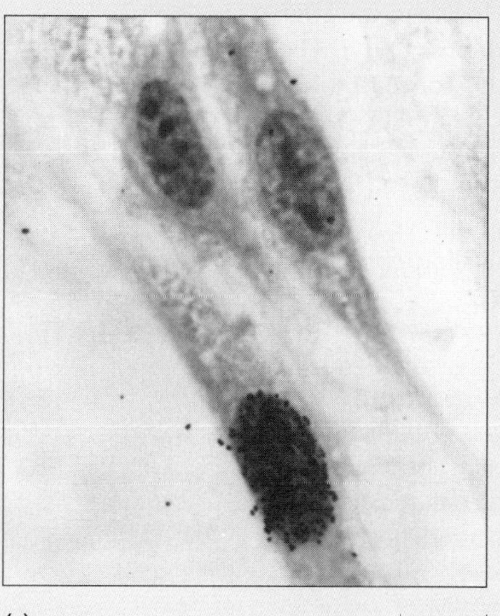

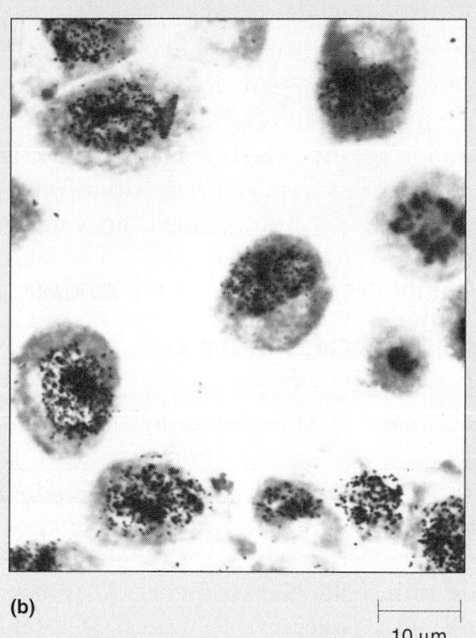

Figure 1 Localization of Macromolecules Using Radioactive Isotopes (a) Cells labeled with tritiated thymine, which is incorporated into DNA only, and **(b)** cells labeled with tritiated uridine, which is incorporated into RNA..

to the dissolved ions, a much lower temperature is required for the water molecules to assume the rigid lattice structure of ice. Thus, in nature, most water does not freeze unless the temperature drops below 0°C. Consequently, microorganisms can usually grow in liquids at 0°C, the freezing temperature of pure water.

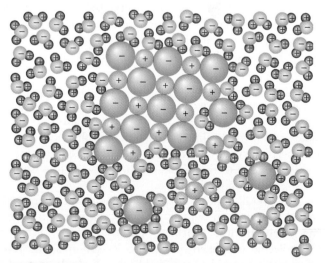

Figure 2.9 Salt (NaCl) Dissolving in Water In the absence of water, the salt is highly structured because of ionic bond formation between Na^+ and Cl^-. In water, the Na^+ and Cl^- are separated by H_2O molecules. The Na^+ hydrogen bonds to the slightly negatively charged O^- and the Cl^- hydrogen bonds to the slightly positively charged H^+ portion of the water molecules.

pH

An important property of every aqueous solution is its degree of acidity. This property is measured as the **pH** of the solution (an abbreviation for **p**otential **H**ydrogen), defined to be the concentration of H^+ in moles per liter. pH is measured on a logarithmic scale of 0 to 14 in which the lower numbers represent more acidic solutions. The acidity of a solution is based on several properties of water. Water itself has a slight tendency to split (ionize) into hydrogen ions H^+ (protons), which are acidic, and OH^- ions, which are basic or alkaline. In pure water, only one molecule in 10 million molecules (1 in 10^7) undergoes this breakdown spontaneously:

$$HOH \longrightarrow H^+ + OH^-$$

As this reaction indicates, when water splits into its component parts, the number of H^+ and OH^- ions is equal, and in pure water, the concentration of each is 10^{-7} molar (10^{-7} M). The product of the concentration of H^+ and OH^- must always be 10^{-14} M ($10^{-7} \times 10^{-7}$) (Recall that exponents are added when numbers are multiplied.)

The concentration of H^+ in water can be increased by adding a compound, termed an **acid**, that contains hydrogen atoms and ionizes to give H^+ but not OH^- ions. Conversely, a **base** increases the concentration of OH^- ions. HCl is an example of an acid; it ionizes completely to H^+ and Cl^-. Each increase in H^+ concentration must be balanced by a corresponding decrease in OH^- concentration since the product of H^+ and OH^- must equal 10^{-14} M. If the concentration of H^+ increases 10-fold to 10^{-6} M, then the concentration of OH^- must decrease by a factor of 10 (to 10^{-8} M).

The pH scale simplifies the relationship between the concentration of hydrogen ions (H^+) and hydroxyl ions (OH^-). The scale ranges from zero to fourteen because the concentrations of H^+ and OH^- ions varies within these limits. The pH of some common solutions is shown in **figure 2.10**.

Most bacteria can live within only a narrow pH range, near neutrality. Some, however, can live under very acidic conditions (**acidophiles**) and a few under alkaline conditions (**alkalophiles**). ■ acidophiles, p. 99 ■ alkalophiles, p. 99

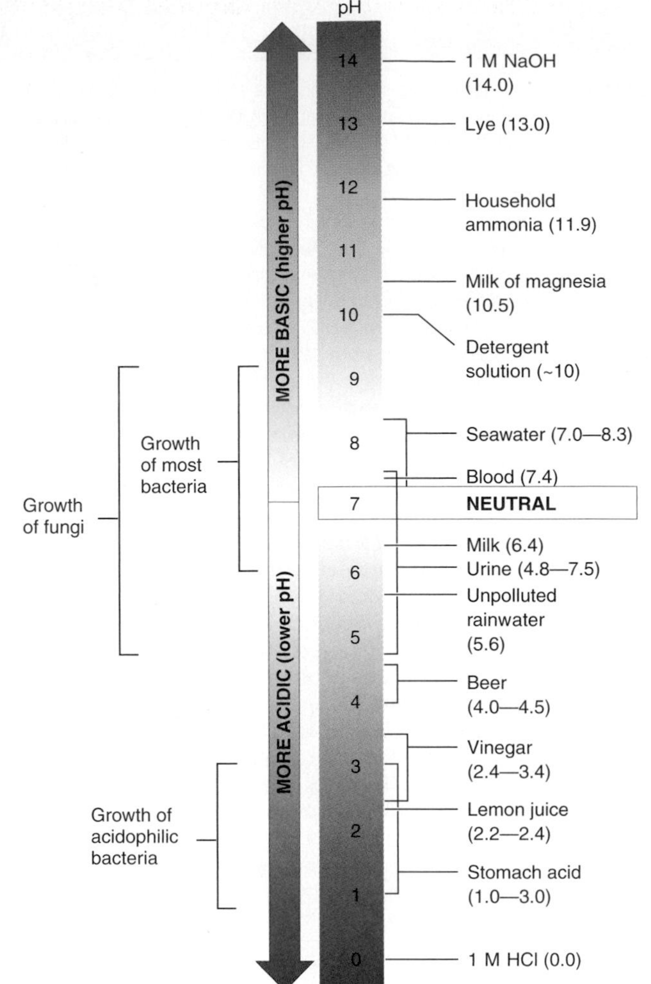

Figure 2.10 **pH Values of Various Solutions**

Small Molecules in the Cell

All cells contain a variety of small organic and inorganic molecules, many of which occur in the form of ions. Organic molecules contain carbon atoms bonded to each other or to hydrogen atoms; inorganic molecules do not contain carbon to carbon bonds. About 1% of the weight of a bacterial cell, once the water is removed (dry weight), is composed of inorganic ions, principally Na^+ (sodium), K^+ (potassium), Mg^{2+} (magnesium), Ca^{2+} (calcium), Fe^{2+} (iron), Cl^- (chloride), PO_4^{3-} (phosphate), and SO_4^{2-} (sulfate). Certain enzymes require positively charged ions in minute amounts in order to function. The negatively charged phosphate ion is especially important, because it plays a key role in energy metabolism. This will be discussed in chapter 6.

The key atom in all cells is carbon. This atom is extremely versatile; it can form strong covalent bonds with four other atoms, including carbon, and thereby build up a wide variety of organic molecules. The biologically important small organic molecules are mainly compounds that are being metabolized or have accumulated as a result of metabolism. Many of these molecules can serve as the building blocks of much larger molecules. Small molecules can be grouped into categories based on the distinctive groups of atoms they contain **(table 2.3)**.

As table 2.3 indicates, the most important covalent bonds in these molecules are between carbon atoms (C—C), carbon and hydrogen (C—H), carbon and oxygen (C═O), carbon and nitrogen (C—N), oxygen and hydrogen (O—H), and nitrogen and hydrogen (N—H). Some of the distinctive groups are the hydroxyl (—OH), amino ($—NH_2$), and carboxyl (—COOH) groups.

Macromolecules and Their Component Parts

Macromolecules are very large molecules (*macro* means "large") consisting of several thousand atoms each. The four major classes of biologically important macromolecules are **proteins**, **polysaccharides**, **nucleic acids,** and **lipids**. Although macromolecules are very large, their structure and mechanism of synthesis have several features in common. Understanding these common features makes the structure of these large molecules surprisingly simple to learn and understand.

All macromolecules are **polymers** (*poly* means "many"), large molecules formed by joining together small molecules, the **subunits**. Each different class of macromolecule is composed of different subunits. The subunits of the same macromolecule, however, have a similar structure.

The synthesis of macromolecules involves two steps: first, the subunits are synthesized; and then, they are joined together, one by one. The synthesis of the various subunits is very complex and involves almost 100 different chemical reactions. These specific reactions, which are carried out by enzymes, are beyond the scope of this book. An overview of how some are synthesized is given in chapter 6.

The overall process of joining two subunits involves a chemical reaction in which H_2O is removed, a reaction termed **dehydration synthesis (figure 2.11a)**. When the macromolecule is broken down into its subunits, the reverse reaction occurs, and H_2O is added back, a **hydrolytic reaction,** or **hydrolysis** (figure 2.11b). This type of reversible reaction, involving the removal and addition of H_2O molecules, is common to the synthesis and degradation of a large number of molecules and requires the action of specific enzymes.

Table 2.3 Classification of Small Molecules by Functional Groups

Name of Group	Example of Group in a Biologically Important Molecule*	Name of Compound	Biological Significance	Class of Molecule Found in
Methyl	H H—C—H H	Methane	Contributes to greenhouse warming	
Hydroxyl	H H H—C—C—O—H H H	Ethanol (ethyl alcohol)	Beverage	Alcohols
Carboxyl	H O H—C—C—O—H H	Acetic acid	Part of energy pathway	Acids
Amino	H H N O H—C—C—O—H	Glycine	Component of proteins	Amines and amino acids
Keto	H O H H—C—C—C—H H H	Acetone	Commercial solvent	Ketones
Aldehyde	H O H—C—C—H H	Acetaldehyde	On pathway to ethanol	Aldehydes
Sulfhydryl	H H H N O H—S—C—C—C—O—H H	Cysteine	Component of proteins	A few amino acids
Ketones and Carboxyl	H O O H—C—C—C—OH H	Pyruvic acid	Part of energy pathway	Keto acids
Hydroxyl and Carboxyl	H H O O H—C—C—C—OH H H	Lactic acid	Sours milk	Hydroxy acids

*Key functional groups are shaded.

MICROCHECK 2.3

The weak polar bonds of water molecules are responsible for the many properties of water required for life on earth. The degree of acidity of water is also an important property in biological systems. Macromolecules consist of many repeating subunits, each subunit being similar or identical to the other subunits.

- Why is water a polar molecule? Give three examples of why this property is important in microbiology.

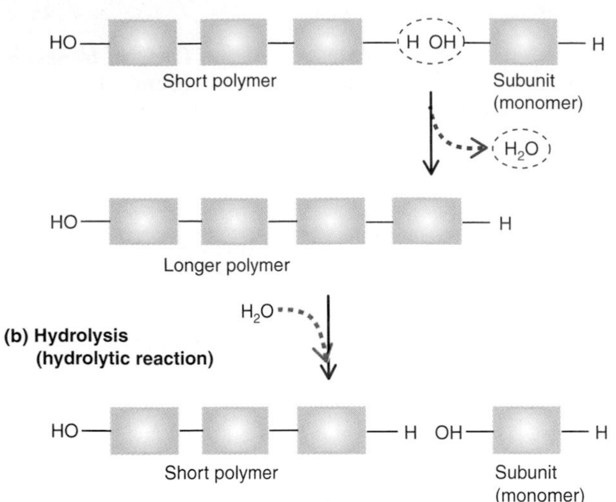

Figure 2.11 The Synthesis and Breakdown of Polymers (a) Subunits are joined (polymerized) by removal of water, a dehydration reaction. **(b)** In the reverse reaction, hydrolysis, the addition of water molecules breaks bonds between the subunits. These reactions take place in the formation of many different polymers (macromolecules).

- Name the four important classes of large molecules in cells.
- In pure water, what must be done to decrease the OH^- concentration? To decrease the H^+ concentration?

Proteins and Their Functions

Proteins constitute more than 50% of the dry weight of cells. Of all the macromolecules, they are the most versatile in what they do in cells.

In the microbial world, proteins are responsible for:

- Catalyzing all reactions of the cell required for life. ■ **enzymes, p. 142**
- The structure and shape of certain structures such as ribosomes, the protein-building machinery in all cells. ■ **ribosomes, p. 70**
- Cell movement by flagella. ■ **flagella, p. 67**
- Taking nutrients into the cell. ■ **binding proteins, p. 61**
- Turning genes on and off. ■ **gene regulation, p. 179**
- The structure of various membranes in the cell. ■ **inner membrane, p. 63** ■ **outer membrane, p. 63**

Amino Acid Subunits

Proteins are composed of numerous combinations of twenty major amino acids. The properties of a protein depend mainly on its shape. Its shape depends on the arrangement of the amino acids that make up the protein.

All amino acids share a carboxyl group (—COOH) at one end and an amino group (—NH_2) bonded to the same carbon atom to which the —COOH group is bonded **(figure 2.12)**. This carbon atom, termed the alpha (α) carbon, also is bonded to a side chain or backbone (labeled R), which gives each amino acid its characteristic properties. The amino acids are subdivided into several different groups based on similarities in their side chains **(figure 2.13)**. One important property of the side chains is whether they are polar or nonpolar; this feature determines the solubility properties of the protein, its shape, and how it interacts with other proteins inside the cell. Amino acids that contain many methyl (CH_3) groups are nonpolar and therefore do not interact with water molecules. Thus, they are weakly soluble in water and are termed **hydrophobic** (means "water-fearing"). Side chains with carboxyl or amino groups readily form ions, which give negative (COO^-) or positive (NH_3^+) electrical charges to the amino acid **(figure 2.14)**. Such amino acids are termed **acidic** or **basic amino acids**, respectively. They are readily soluble in water because the charged water molecules can hydrogen bond to the positive or negative charges of the amino acids. Thus, they are termed **hydrophilic** (means "water-loving"). Other amino acids have distinctive features that confer important biological functions. For example, the sulfur-containing amino acid cysteine can form covalent bonds with other cysteine molecules in the same protein. The unusual structure of proline allows it to form kinks in the proteins and helps to determine how the protein folds.

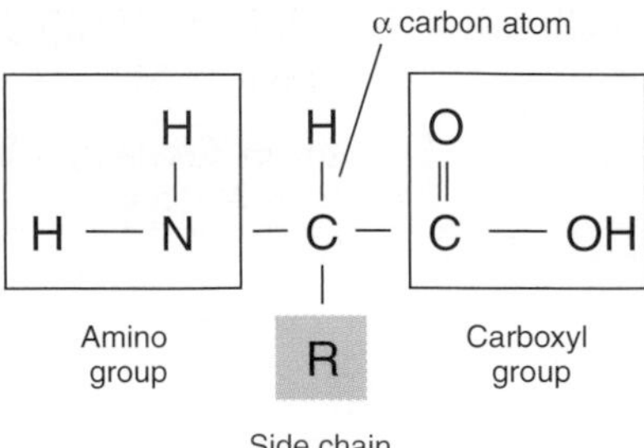

Figure 2.12 Generalized Amino Acid This figure illustrates the three groups that all amino acids possess. The R side chain differs with each amino acid and determines the properties of the amino acid.

All amino acids except glycine can exist in two stereoisomeric forms, a D (right-handed) or L (left-handed) form. Each is a mirror image of the other (**figure 2.15**; see **Glimpse of History**). Only L-amino acids occur in proteins, and accordingly, they are designated the **natural amino acids**. D-amino acids, the **unnatural amino acids**, are rare in nature and are found in only a few compounds mostly associated with bacteria—where they are found primarily in the cell walls and antibiotics that bacteria produce. Interestingly, the bacterium *Bacillus anthracis*, which causes the disease anthrax and is a prime candidate as a biological warfare agent, has an outer coat of D-glutamic acid.

Peptide Bonds and Their Synthesis

Amino acids making up proteins are held together by **peptide bonds**, a unique type of covalent linkage formed when the carboxyl group of one amino acid reacts with the amino group of another amino acid, with the release of water (dehydration synthesis) **(figure 2.16)**.

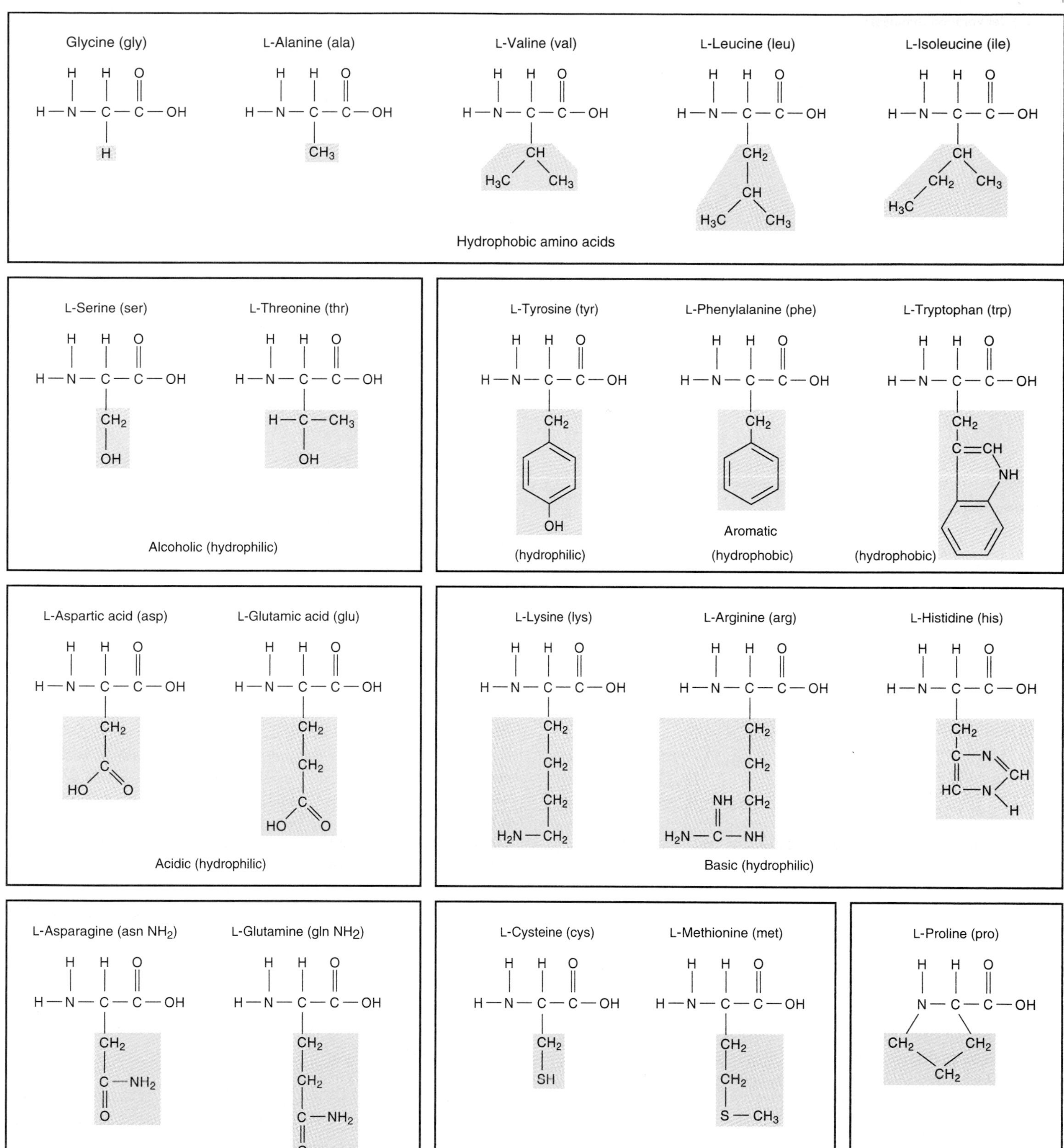

Figure 2.13 **Amino Acids** All amino acids have one feature in common—a carboxyl group and an amino group bonded to the same carbon atom. This carbon atom is also bonded to a side chain (shaded). The structure of the R group gives the amino acid its distinctive features. The three letter code name for each amino acid is given.

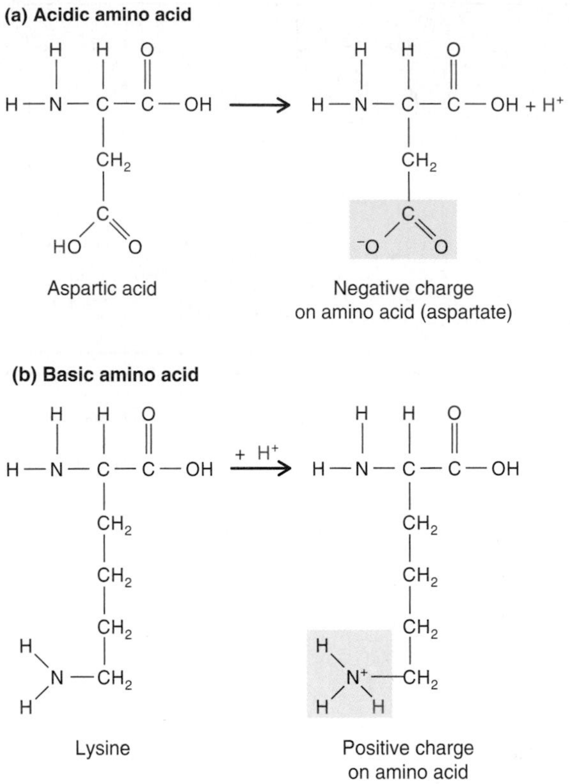

(a) Acidic amino acid

Aspartic acid: $H_2N-CH(CH_2COOH)-COOH \longrightarrow$ Negative charge on amino acid (aspartate): $H_2N-CH(CH_2COO^-)-COOH + H^+$

(b) Basic amino acid

Lysine: $H_2N-CH((CH_2)_4NH_2)-COOH \xrightarrow{+H^+}$ Positive charge on amino acid: $H_2N-CH((CH_2)_4NH_3^+)-COOH$

Figure 2.14 Acidic (Aspartic Acid) and Basic (Lysine) Amino Acids At a neutral pH (7), aspartic acid **(a)** is negatively charged and lysine **(b)** is positively charged because it has one less or one more proton, respectively.

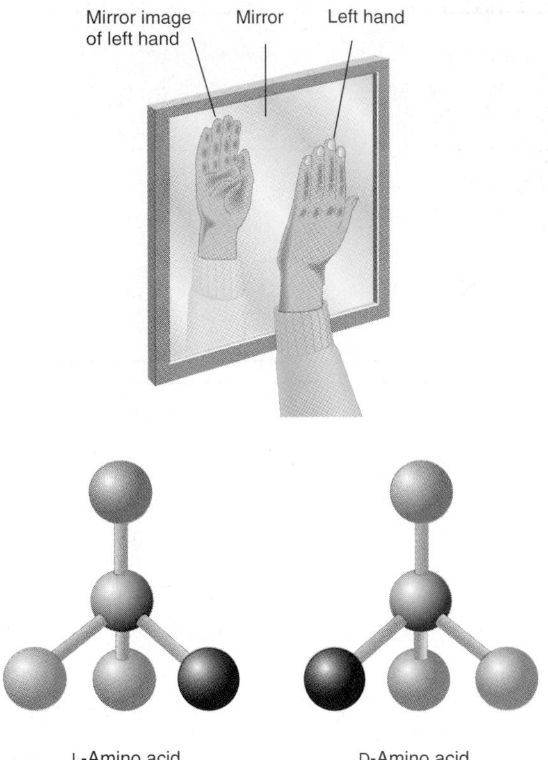

Figure 2.15 Mirror Images (Stereoisomers) of an Amino Acid The joining of a carbon atom to four different groups leads to asymmetry in the molecule. The molecule can exist in either the L- or D- form, each being the mirror image of the other. There is no way that the two molecules can be rotated in space to give two identical molecules.

The chain of amino acids formed when a large number of amino acids are joined by peptide bonds is called a **polypeptide chain (figure 2.17)**. A protein is a long polypeptide chain. One end of the chain has a free amino ($-NH_2$) group, which is termed the **N terminal,** or **amino terminal**, end. The other end has a free carboxyl ($-COOH$) group, which is termed the **C terminal,** or **carboxy terminal**, end. Proteins are always synthesized in cells starting from the N terminal end. Some proteins consist of a single polypeptide chain, whereas others consist of one or more chains joined together by weak bonds. Sometimes, the chains are identical; in other cases, they are different. The total molecule comprising one or more polypeptide chains is the protein. In proteins that consist of several chains, the individual polypeptide chains generally do not have biological activity by themselves. Proteins vary greatly in size, but an average-size protein consists of a single polypeptide chain of about 400 amino acids. ■ **protein synthesis, p. 175**

Protein Structure

The structure of proteins can be described at four levels: primary, secondary, tertiary, and quaternary. The number and arrangement or sequence of amino acids in a protein determines its **primary structure (figure 2.18a)**. The primary structure in large part determines the other features of the protein. Some amino acids are especially important to functioning of the protein and the substitution of such a critical amino acid with another usually destroys the ability of an enzyme to carry out its catalytic function. This will be discussed in greater detail in chapter 8.

Once the specific amino acids are joined together to form the primary structure, the amino acids will form several different arrangements within the protein. This is the protein's **secondary structure** (figure 2.18b). Certain sequences of amino acids will arrange themselves into a helical structure termed an **alpha** (α) **helix**. Others will form a **pleated structure** termed a **beta** (β) **sheet** (figure 2.18b). These structures are formed when the amino acids form weak bonds, such as hydrogen bonds, between other amino acids. This is why certain sequences of amino acids lead to distinctive secondary structures in various parts of the molecule.

The protein next folds into its distinctive three-dimensional shape, its **tertiary structure** (figure 2.18c). Two major shapes

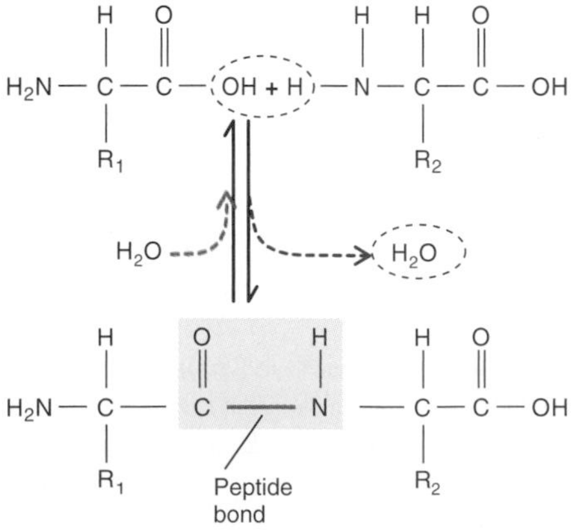

$H_2N-CH(R_1)-CO-OH + H-NH-CH(R_2)-COOH \rightleftharpoons H_2N-CH(R_1)-CO-NH-CH(R_2)-COOH + H_2O$

Peptide bond

Figure 2.16 Peptide Bond Formation by Dehydration Synthesis

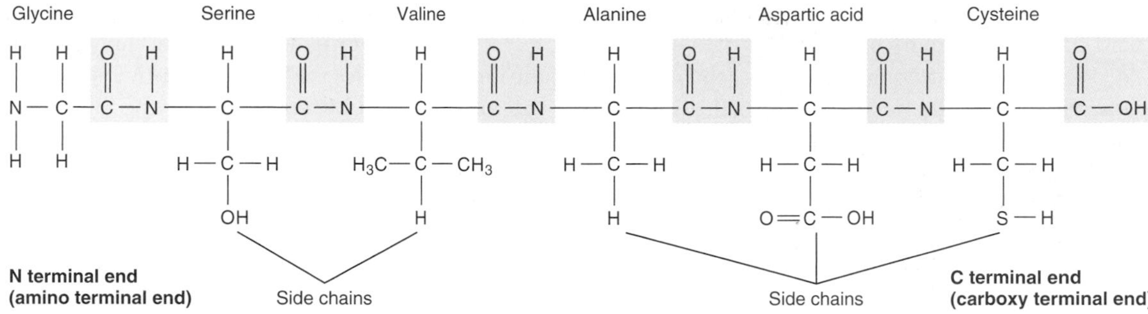

Figure 2.17 Short Polypeptide Chain The shaded areas represent the atoms involved in forming the peptide bonds. This chain is much shorter than would be found in enzymes.

exist: **globular**, which tends to be spherical; and **fibrous** which has an elongated structure (figure 2.18c). The shape is determined in large part by the sequence of the amino acids and whether or not they interact with water. Amino acids that have charged polar groups are found on the outside of the protein molecule, where they can interact with the charged polar water molecules. Amino acids that are composed largely of nonpolar methyl ($—CH_3$) groups are pushed together and cluster inside the molecule to avoid water molecules. This phenomenon also explains why nonpolar molecules of fat form droplets in an aqueous environment. The nonpolar amino acids form weak interactions with each other, termed **hydrophobic interactions**. In addition to these weak bonds, some amino acids can form strong covalent bonds with other amino acids. One example is the formation of bonds between sulfur atoms (S—S bonds) in different cysteine molecules.

The combination of strong and weak bonds between the various amino acids results in the tertiary structure of the protein. Proteins that contain extensive regions of sheets generally form globular proteins. Extensive helices are found in fibrous proteins. Globular proteins can contain short stretches of helices.

Proteins often consist of more than one polypeptide chain held together by many weak bonds. The chains also assume a specific shape, termed the **quaternary structure** of the protein (figure 2.18d). These chains may be identical or different. Of

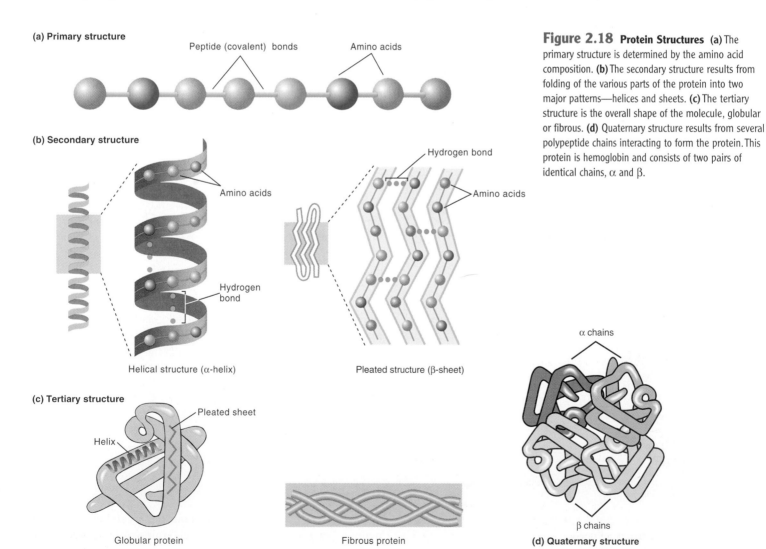

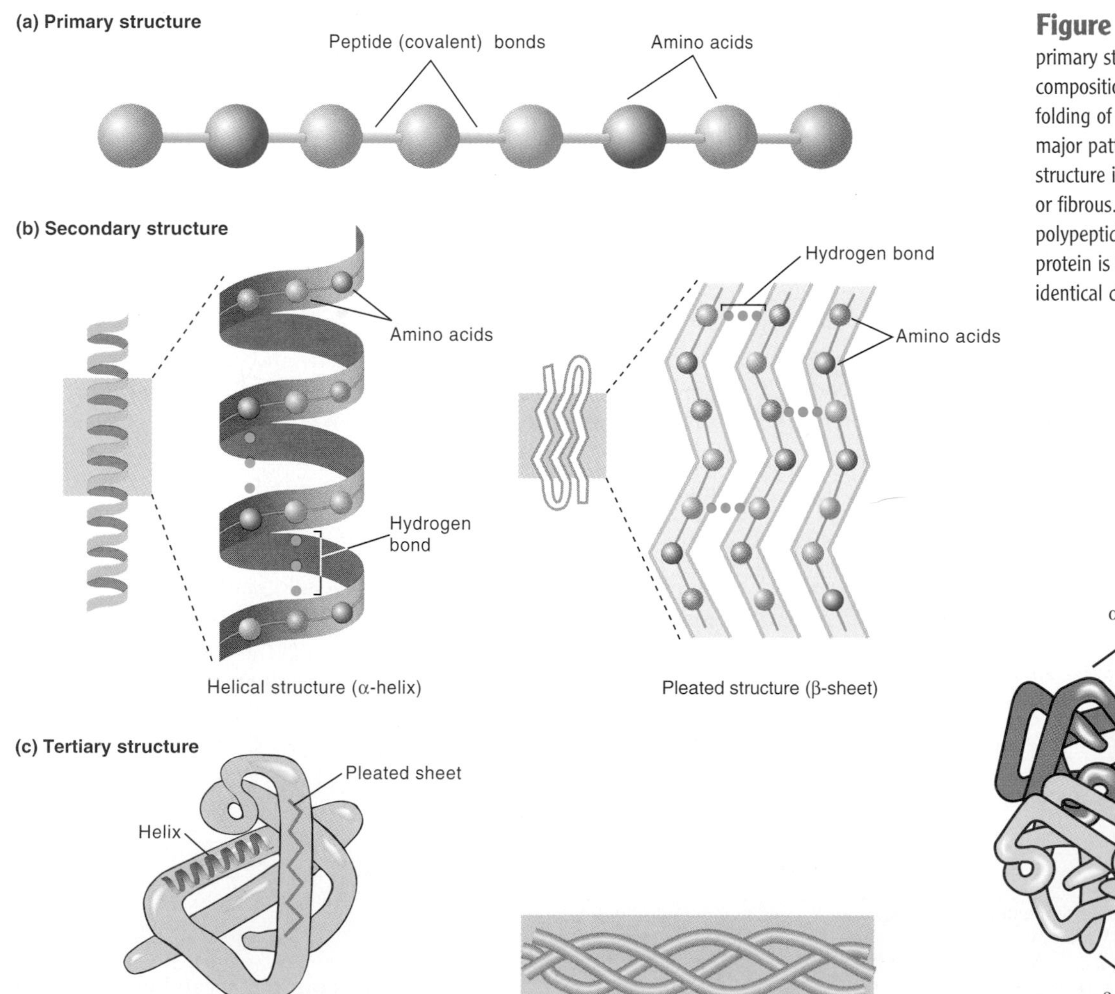

Figure 2.18 Protein Structures **(a)** The primary structure is determined by the amino acid composition. **(b)** The secondary structure results from folding of the various parts of the protein into two major patterns—helices and sheets. **(c)** The tertiary structure is the overall shape of the molecule, globular or fibrous. **(d)** Quaternary structure results from several polypeptide chains interacting to form the protein. This protein is hemoglobin and consists of two pairs of identical chains, α and β.

course, only proteins that consist of more than one polypeptide chain have a quaternary structure.

Sometimes different proteins, each having different functions, associate with one another to make even larger structures termed **multiprotein complexes**. For example, sometimes several enzymes involved in the synthesis pathway of the same amino acid are joined in a **multi-enzyme complex**. A similar situation can occur for several enzymes involved in the degradation of a particular compound.

The formation of proteins occurs extremely rapidly. In seconds, amino acids are joined together to yield a polypeptide chain. The process by which this occurs will be discussed in chapter 7. The polypeptide chain then folds into its correct shape. Although there are many possible shapes that a protein can assume, only one is functional. Most proteins will fold spontaneously into their most stable state correctly. To help some proteins assume the proper shape, however, all cells have proteins called **chaperones** that help proteins fold. The cell detects incorrectly folded proteins and degrades them into amino acids that can then be used to make more proteins.

Protein Denaturation

A protein must have its proper shape to function. When proteins encounter different conditions such as high temperature, high or low pH, or certain solvents, bonds within the protein are broken and its shape changes **(figure 2.19)**. The protein becomes **denatured** and no longer functions. Most bacteria cannot grow at very high temperatures because their enzymes are denatured by the heat. Denaturation may be reversible in some cases; in other cases, it is irreversible. For example, boiling an egg denatures the egg white protein, an irreversible process since cooling the egg does not restore the protein to its original appearance. If a denaturing solvent is removed, however, the protein may refold spontaneously into its original shape.

Substituted Proteins

The proteins that play important roles in certain structures of the cell often consist of other molecules covalently bonded to the side chains of amino acids. These are called **substituted** or **conjugated** proteins. The proteins are named after the molecules that are covalently joined to the amino acids. If sugar molecules are bonded, the protein is termed a **glycoprotein**. If lipids are bonded, then the protein is termed a **lipoprotein**. Sugars and lipids are covered later in this chapter.

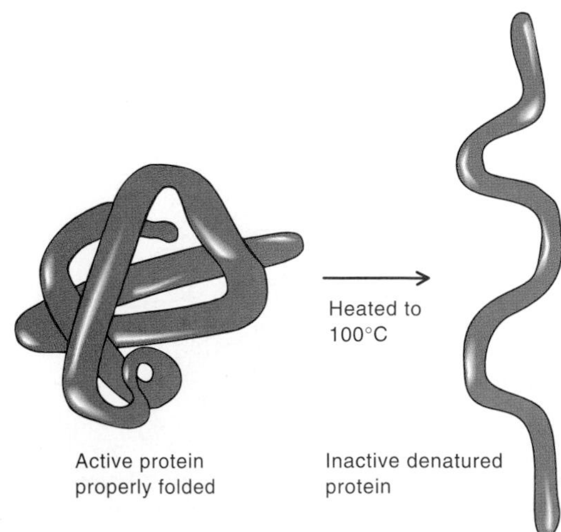

Figure 2.19 Denaturation of a Protein

MICROCHECK 2.4

The side chains of amino acids are responsible for their properties. The sequence of amino acids in a protein determines how the protein folds to assume its three-dimensional shape.

- What type of bond joins amino acids to form proteins?
- Name two groups of amino acids that are hydrophilic.
- What elements must all amino acids contain? What elements will only some amino acids contain?

Carbohydrates

Carbohydrates comprise a heterogeneous group of compounds of various sizes that play important roles in the life of bacteria. These include the following:

- Carbohydrates are a common food source from which bacteria and other organisms can obtain energy and make cellular material. ■ metabolism, chapter 6
- Two sugars form a part of the nucleic acids, DNA and RNA. ■ nucleic acids, p. 35
- Certain carbohydrates serve as a reserve source of food in bacteria. ■ storage granules, p. 71
- Sugars form a part of the bacterial cell wall. ■ cell wall structure, p. 61

The one feature common to all diverse carbohydrates is that they contain carbon, hydrogen, and oxygen atoms in an approximate ratio of 1:2:1. This is because they contain a large number of alcohol groups (—OH) in which the C is also bonded to an H atom to form H—C—OH. **Polysaccharides** are high molecular weight compounds and are linear or branched polymers of their subunits. **Oligosaccharides** are short chains. The term **sugar** is often applied to **monosaccharides** (*mono* means "one"), a single molecule, and **disaccharides** (*di* means "two"), which are two monosaccharides joined together by covalent bonds.

Carbohydrates also usually have an aldehyde group

$$\begin{array}{c} O \\ \| \\ -C-H \end{array}$$

and less commonly, a keto group

$$\begin{array}{c} O \\ \| \\ -C- \end{array}$$

(see table 2.3). The —OH groups on sugars can be replaced by many other groups such as carboxyl, amino, and acetyl groups to form molecules that are important in the structures of the cell. For example, acetyl glucosamine is an important component of the cell wall of bacteria. ■ bacterial cell wall, p. 61

Monosaccharides

Monosaccharides are classified by the number of carbon atoms they contain. The most common monosaccharides are those with 5- or 6-carbon atoms. The 5-carbon sugars, **ribose** and **deoxyribose**, are the sugars in nucleic acids **(figure 2.20)**. Note that these monosaccharides are identical except that deoxyribose has one less molecule of oxygen than does ribose (*de* means "away from"). Thus, deoxyribose is ribose "away from" oxygen. Common 6-carbon sugars include glucose, galactose, and fructose. It is convenient to number the carbon atoms with carbon atom 1 being closest to the aldehyde or keto group.

Sugars can be drawn in two forms, a linear and a ring form. The linear form of ribose, the ring form, and the relationship between the two representations are shown in figure 2.20. Both forms naturally occur in the cell, but most molecules are in the ring form. The forms are interconvertible.

Two different kinds of isomers occur in sugars. These are **stereoisomers** (mirror images) and **structural isomers**. Two stereoisomers can result as the ring structure is formed from the linear form, because the —OH group on the C atom involved in the formation of the ring can be above or below the plane of the ring **(figure 2.21)**. These two different positions yield molecules with different properties when this —OH is linked to another monosaccharide to form larger molecules. The two different forms are termed alpha (α) and beta (β). Every sugar can also exist in two different forms, D and L, which also are mirror images of one another **(figure 2.22)**. Most monosaccharides in living organisms are of the D-configuration, which is opposite to the situation observed with amino acids. The sugars of different isometric forms still have the same name.

Sugars also form **structural** isomers**.** These are molecules that contain the same elements but in different arrangements that are not mirror images. They are different sugars and have different names. For example, common hexoses of biological importance include glucose, galactose, and mannose **(figure 2.23)**. They all contain the same atoms but differ in the arrangements of the —H and —OH groups relative to the carbon atoms. Glucose and galactose are identical except for the arrangement of the —H and —OH groups attached to carbon 4. Mannose and glucose differ in the arrangement of the —H and —OH groups joined to carbon 2. Structural isomers result in three distinct sugars with

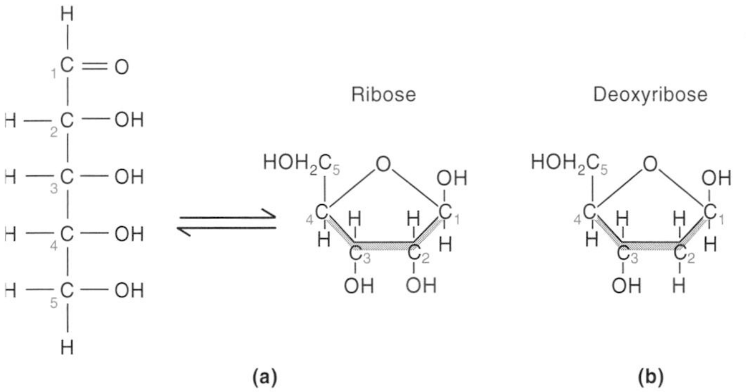

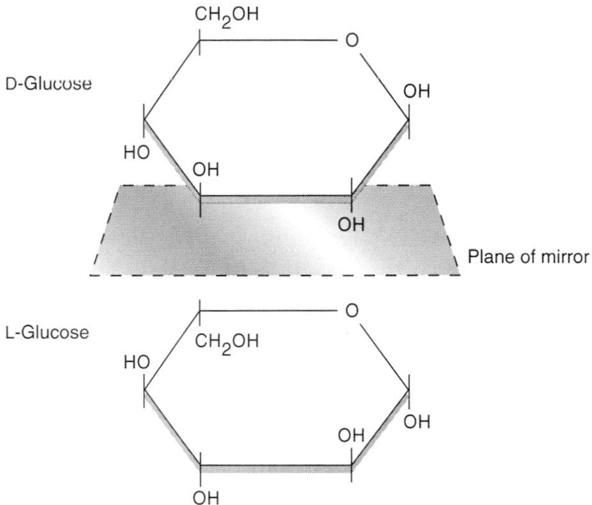

Figure 2.22 The Stereoisomers of Glucose By convention, the carbon atoms at the intersections of the lines in the ring structure are understood to be present and are not labeled.

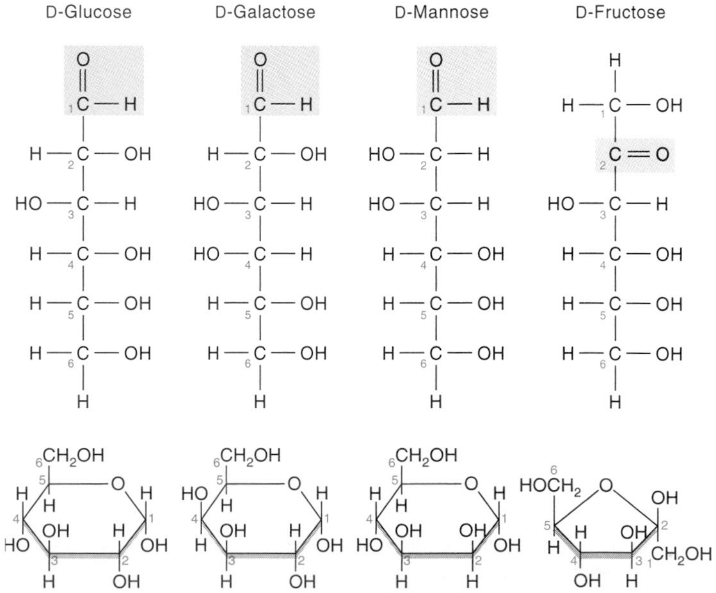

Figure 2.20 Ribose and Deoxyribose (a) Ribose in linear and ring form. **(b)** Deoxyribose in ring form.

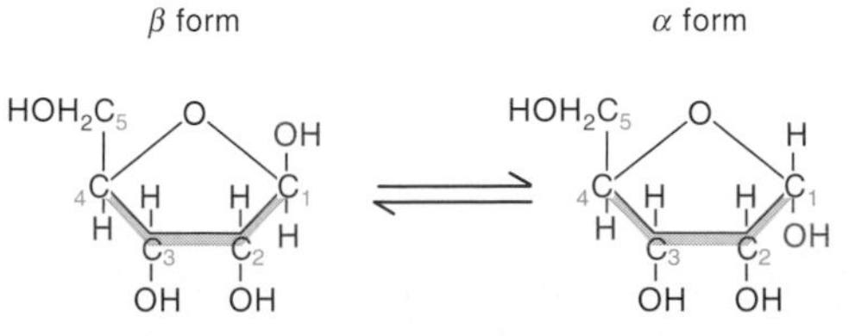

Figure 2.21 Stereoisomers of Ribose The α and β forms of ribose are interconvertible and only differ in whether the OH group on carbon 1 is above or below the plane of the ring.

Figure 2.23 Formulas of Some Common Sugars Represented in Their Linear and Ring Forms Note that glucose, galactose, and mannose all have an aldehyde group, involving C atom 1 (shaded), whereas fructose has a keto group, involving C atom 2 (shaded).

different properties and different names. For example, glucose has a sweet taste as does mannose, but mannose has a bitter aftertaste.

Disaccharides

The two most common disaccharides in nature are the milk sugar, **lactose**, and the common table sugar, **sucrose**. Lactose consists of glucose and galactose, while sucrose, which comes from sugar cane or sugar beets, is composed of glucose and fructose. The monosaccharides are joined together by a dehydration reaction between hydroxyl groups of two monosaccharides, with the loss of a molecule of water **(figure 2.24)**. Note that this reaction is similar to the joining of two amino acids. The bond is called a **glycosidic bond.** The reaction is reversible, so that the addition of a water molecule, the process of hydrolysis, yields the two original molecules (see figure 2.24). Great diversity is possible in molecules formed by joining monosaccharides. The carbon atoms involved in the joining together of the monosaccharides may differ and the position of the —OH groups, α and β, involved in the bonding may also differ.

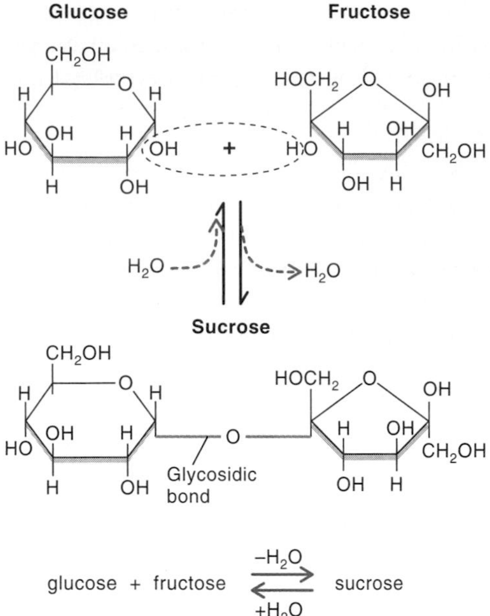

Figure 2.24 Formation of Sucrose from Glucose and Fructose by Dehydration Synthesis and the Formation of a Glycosidic Bond The reaction is reversible by the addition of water (hydrolysis).

Polysaccharides

Polysaccharides, which are found in many different places in nature, serve different functions. **Cellulose**, the most abundant organic molecule on earth, is a polymer of glucose subunits and is the principal constituent of plant cell walls. Some bacteria synthesize cellulose in the form of fibrils that attach the bacteria to various surfaces. **Glycogen**, a carbohydrate storage product of animals and some bacteria, and **dextran**, which is also synthesized by bacteria as a storage product for carbon and energy, resemble cellulose in some ways.

All of the above three polysaccharides are composed of glucose subunits, but they differ from one another in many important ways. These include (1) the size of the polymer; (2) the degree of chain branching, since the side chains of monosaccharides can branch from the main chain; (3) the particular carbon atoms of the two sugar molecules involved in covalent bond formation, such as a 1, 4 linkage when the carbon atom number 1 of one sugar is joined to the number 4 carbon atom of the adjacent sugar; and (4) the orientation of the covalent bond between the sugar molecules, whether α or β. Thus, these subunits can yield a large variety of polysaccharides that have different properties but are composed of the same subunits. How these various aspects of the structure of a polysaccharide fit into the structures of cellulose, glycogen, and dextran is shown in **figure 2.25**.

Polysaccharides and oligosaccharides can also contain different monosaccharide subunits in the same molecule. For

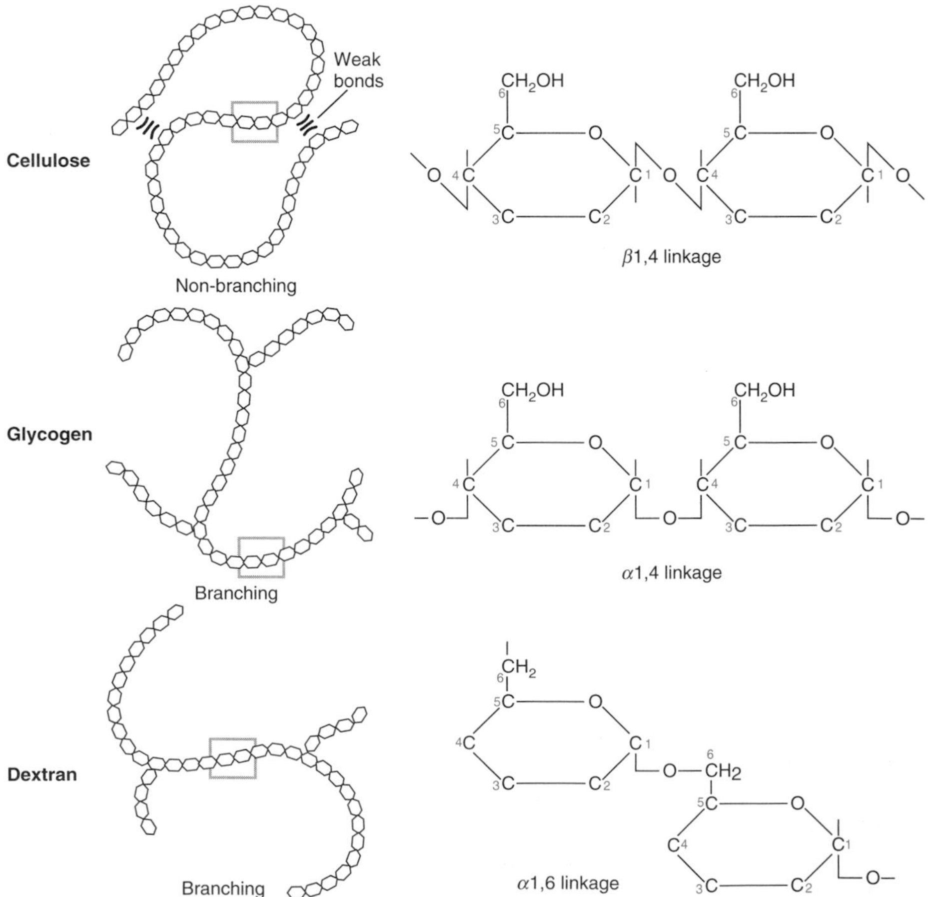

Figure 2.25 Structure of Three Important Polysaccharides The three molecules shown consist of the same subunit, D-glucose, yet they are distinctly different molecules because of the differences in linkage that join the molecules (α and β; 1,4 or 1,6), the degree of branching, and the bonds involved in branching (not shown). Hydrogen bonds are also involved.

example, the cell walls of bacteria but not archaea contain a polysaccharide consisting of alternating subunit molecules of two different amino sugars. A polysaccharide consisting of different subunits is termed a **complex polysaccharide**.

MICROCHECK 2.5

Carbohydrates perform a variety of functions in cells, including serving as a source of energy and forming part of the cells' structures. Carbohydrates with the same atomic composition can have distinct properties because of different arrangements of the atoms in the molecules.

- Distinguish between structural isomers and stereoisomers.
- What is the general name given to a single sugar?
- How could you distinguish sucrose and lactose from protein by analyzing the elements in the molecules?

Nucleic Acids

Nucleic acids are macromolecules whose subunits are **nucleotides**. There are two types of nucleic acids: **deoxyribonucleic acid (DNA)** and **ribonucleic acid (RNA)**.

DNA

DNA is the master molecule of the cell—all of the cell's properties are determined by its DNA. This information is coded in the sequence of DNA's subunits, the **nucleotides**. The code is then converted into a specific arrangement of amino acids that make up the protein molecules of the cell. The details of this process are covered in chapter 7.

In addition to their role in the structure of DNA, nucleotides play additional roles in the cell. These include the following:

- They carry chemical energy in their bonds. ■ **adenosine triphosphate, p. 137**
- They are part of certain enzymes. ■ **CoA, p. 143**
- They serve as specific signaling molecules. ■ **cyclic AMP, p. 182**

The nucleotides of DNA are composed of three units: a nitrogen-containing ring compound, called a **base**; which is covalently bonded to a 5-carbon sugar molecule, **deoxyribose**; which in turn is bonded to a **phosphate** molecule **(figure 2.26)**. The four different nitrogen-containing bases found in DNA can be divided into two groups according to their ring structures: two **purines**, adenine and guanine, which consist of two rings; and two **pyrimidines**, cytosine and thymine, which consist of a single ring **(figure 2.27)**. Each one of these bases is linked to the same carbon atom (C_1) as occurs in sugar-sugar bonds.

These four bases are named after the original source from which the base was isolated. Adenine was isolated from the pancreas and is derived from the Greek word for gland. Guanine was isolated from bird guano (excrement). Thymine was isolated from the thymus gland, and cytosine was isolated from cells.

The nucleotide subunits are joined by a covalent bond, called a **phosphodiester** bond, between the phosphate of one nucleotide of one subunit to the sugar of the adjacent nucleotide **(figure 2.28a)**. Thus, the phosphate is a bridge that joins the number 3 carbon atom (termed 3′) of one sugar to the number 5 carbon atom (termed 5′) of the other. Water is removed in the reaction that joins the sugar and phosphate molecules. This yields a molecule with a backbone of alternating sugar and phosphate molecules. The two ends of the molecule are different. The 5′ end has a phosphate molecule attached to

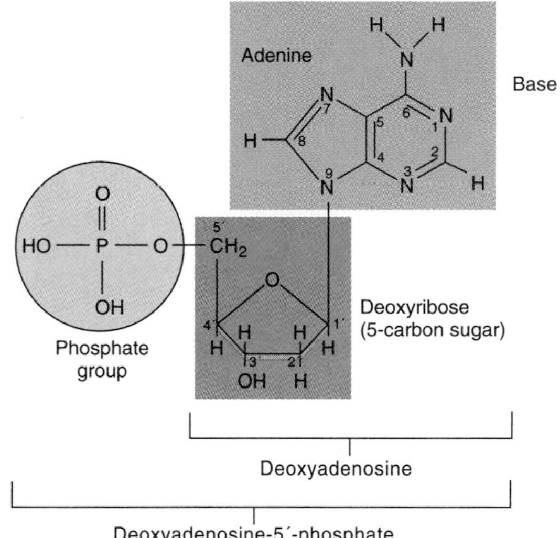

Figure 2.26 A Nucleotide This is one subunit of DNA. This subunit is called adenylic acid or deoxyadenosine-5′-phosphate because the base is adenine. If the base is thymine, the nucleotide is thymidylic acid; if guanine, guanylic acid; and if cytosine, cytidylic acid. If the nucleotide lacks the phosphate molecule, it is called a nucleoside, in this case, deoxyadenosine.

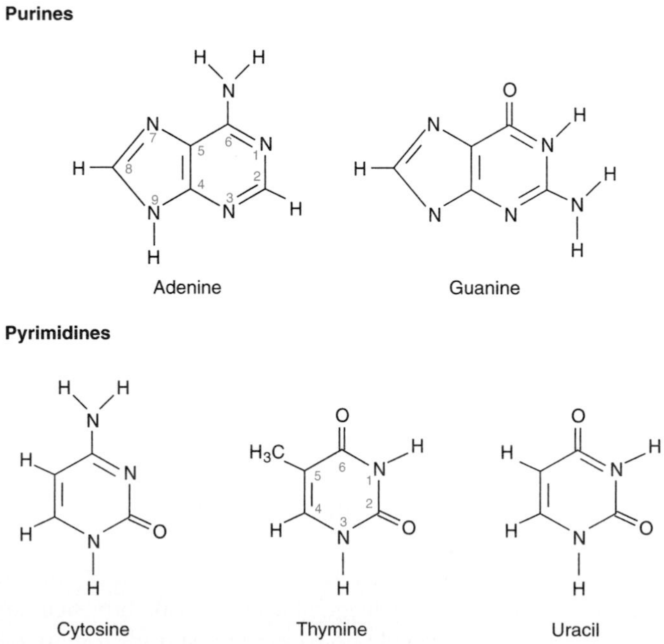

Purines

Adenine

Guanine

Pyrimidines

Cytosine

Thymine

Uracil

Figure 2.27 Formulas of Purines and Pyrimidines DNA contains the pyrimidine thymine. RNA contains uracil in place of thymine.

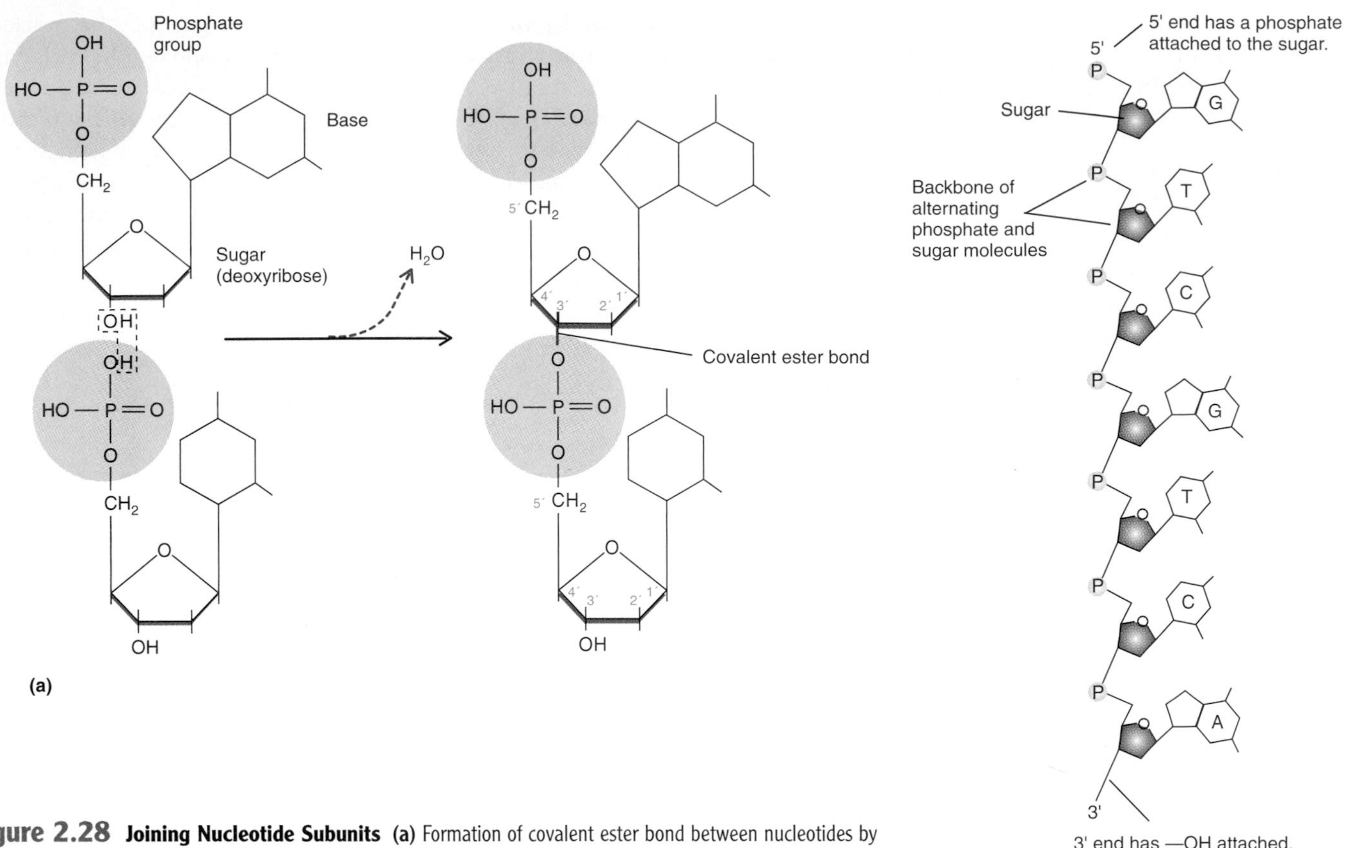

Phosphate group
Base
Sugar (deoxyribose)
H2O
Covalent ester bond
(a)
5' end has a phosphate attached to the sugar.
Sugar
Backbone of alternating phosphate and sugar molecules
3' end has —OH attached. The DNA molecule grows by adding more nucleotides to this end.
(b)

Figure 2.28 Joining Nucleotide Subunits **(a)** Formation of covalent ester bond between nucleotides by dehydration synthesis. **(b)** Chain of nucleotides showing the differences between 5′ end and 3′ end. The chain always is extended at the 3′ end, which has the free —OH group.

the sugar; the 3′ end has a hydroxyl group (figure 2.28b). Accordingly, the end of the chain that grows by adding more nucleotides through dehydration synthesis is always the 3′ end. The synthesis of DNA is covered in chapter 7.

The DNA of a typical bacterium is a single molecule composed of nucleotides joined together and arranged in a double-stranded helix, with about 4 million nucleotides in each strand **(figure 2.29)**. This double-stranded helical molecule can be pictured as a spiral staircase with two railings and stairs split in half. The railings represent the sugar-phosphate backbone of the molecule, and the stairs attached to the railings are the bases. One half of each stair is strongly attached to one railing, and the other half is strongly attached to the other railing. Each pair of stairs (bases) is held together by weak hydrogen bonds. A specificity exists in the bonding between bases, however, in that adenine (A) can only hydrogen bond to thymine (T), and guanine (G) to cytosine (C). The pair of bases that bond are **complementary** to each other. Thus, G is complementary to C, and A to T. As a result, one entire strand of DNA is complementary to the other strand. This explains why in all DNA molecules, the total number of adenine molecules is equal to the number of thymine molecules and the number of guanines equals the number of cytosines.

Three hydrogen bonds join each G to C, but only two join A to T. Each of the hydrogen bonds is weak, but their large number in a DNA molecule holds the two strands together. In addition to the differences in their sequence of bases, the two complementary strands differ from each other in their orientation. From top to bottom, the two strands are arranged in opposite directions. One goes in the 3′ to the 5′ direction; the other in the 5′ to 3′ direction. Consequently, two ends of the helix opposite each other differ; one is a 5′ end, the other, a 3′ end (see figure 2.29).

RNA

RNA is involved in decoding the information in the DNA into a sequence of amino acids in protein molecules. This is a complex multi-step process that will be examined in chapter 7.

The structure of RNA is similar to that of DNA, but it differs in several ways. First, RNA contains the pyrimidine **uracil** in place of thymine and the sugar **ribose** in place of deoxyribose (see figures 2.21 and 2.27). Also, whereas DNA is a long, double-stranded helix, RNA is considerably shorter and exists as a single chain of nucleotides that may form short double-stranded stretches as a result of intramolecular hydrogen bonding between complementary bases.

MICROCHECK 2.6

DNA carries the genetic code in the sequence of purine and pyrimidine bases in its double helical structure. The information is transferred to RNA and then into a sequence of amino acid in proteins.

- What are the two types of nucleic acids?
- If the DNA molecule were placed in boiling water, how would the molecule change?

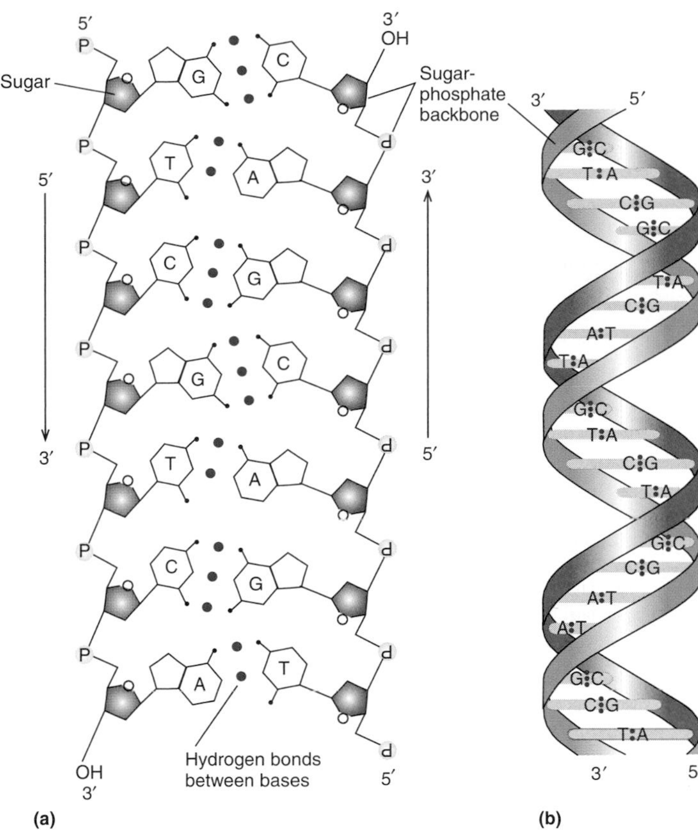

Figure 2.29 DNA Double-Stranded Helix (a) The sugar-phosphate backbone and the hydrogen bonding between bases. **(b)** The spiral staircase of the sugar-phosphate backbone with the bases on the inside. The railings go in opposite directions.

Lipids

Lipids play an indispensible role in all living cells: They are critically important in

- the structure of all membranes in cells. ■ **cytoplasmic membranes, p. 57**
- these membranes act as the gatekeepers of the cells. ■ **transport mechanisms, p. 59**
- membranes keep a cell's internal contents inside the cell and enable the cell to rapidly take up nutrients and other molecules required for growth.

Lipids are a very heterogeneous group of molecules. Their defining feature is slight solubility in water contrasted with great solubility in most organic solvents such as ether, benzene, and chloroform. These solubility properties result from their nonpolar, hydrophobic nature. Lipids have molecular weights of no more than a few thousand and so are the smallest of the macromolecules we have discussed. Further, unlike the other macromolecules, they are not composed of similar subunits; rather, they consist of a wide variety of substances that differ in their chemical structure. Lipids can be divided into two general classes: the **simple** and the **compound lipids**, which differ in important aspects of their chemical composition.

Simple Lipids

Simple lipids contain only carbon, hydrogen, and oxygen. The most common are the **fats**, a combination of fatty acids and glycerol that are solid at room temperature **(figure 2.30)**. A molecule that contains only carbon and hydrogen is a **hydrocarbon**. Fatty acids are molecules with long chains of C atoms bonded to H atoms with an acidic group (—COOH) on one end. Since glycerol has three hydroxyl groups, a maximum of three fatty acid molecules, either the same or different, can be linked through **ester bonds** formed between the —OH group of glycerol and the —COOH group of the fatty acid. Note that water is eliminated in the formation of this ester bond, another example of dehydration synthesis. If only one fatty acid is bound to glycerol, the fat is called a **monoglyceride**; when two fatty acids are joined, it is a **diglyceride**; when three fatty acids are bound, a **triglyceride** is formed. Fatty acids are stored in the body as an energy reserve by forming triglycerides.

Unlike carbohydrates, fatty acids do not have a ratio of approximately 1 : 2 : 1 (C : H : O)—since they have few oxygen atoms. Although hundreds of different fatty acids exist, they can be divided into two groups based on whether or not any double bonds are present in the hydrocarbon portion. If there are no double bonds, the fatty acid is termed **saturated** with H atoms. If it contains one or more double bonds, it is **unsaturated**. Unsaturated fats tend to be liquid and are then called **oils**. Oils are liquid because these unsaturated fatty acids develop kinks in their long tails that prevent tight packing. The saturated fats can pack their straight long tails tightly together and therefore are solid **(figure 2.31a)**. Two of the most common saturated fatty acids in nature are **palmitic acid** (16 carbon atoms), and **stearic acid** (18 carbon atoms). A common unsaturated fatty acid is oleic acid (18 carbon atoms and one double bond in the molecule). Because oleic acid has only one double bond, it is a **monounsaturated** fatty acid (figure 2.31b). Other fatty acids contain numerous double bonds that make them **polyunsaturated**. Different lipids are called highly saturated or highly unsaturated when they contain highly saturated or unsaturated fatty acids.

Another very important group of simple lipids is the **steroids**. All members of this group have the four-membered ring structure shown in **figure 2.32a**. These compounds differ from the fats in chemical structure, but both are classified as lipids because they are both insoluble in water. If a hydroxyl

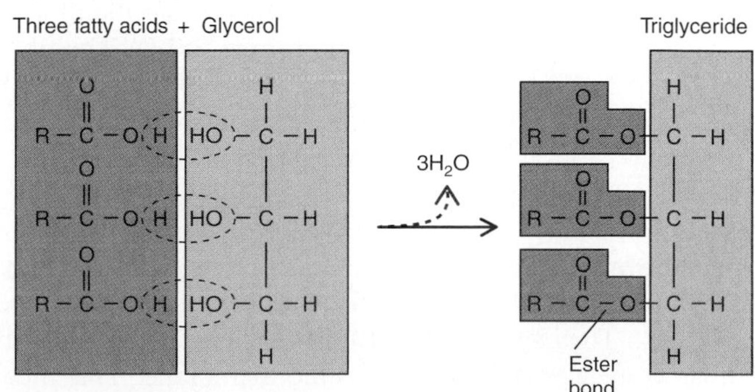

Figure 2.30 Formation of a Fat The R group of the fatty acids commonly contains 16 or 18 carbon atoms.

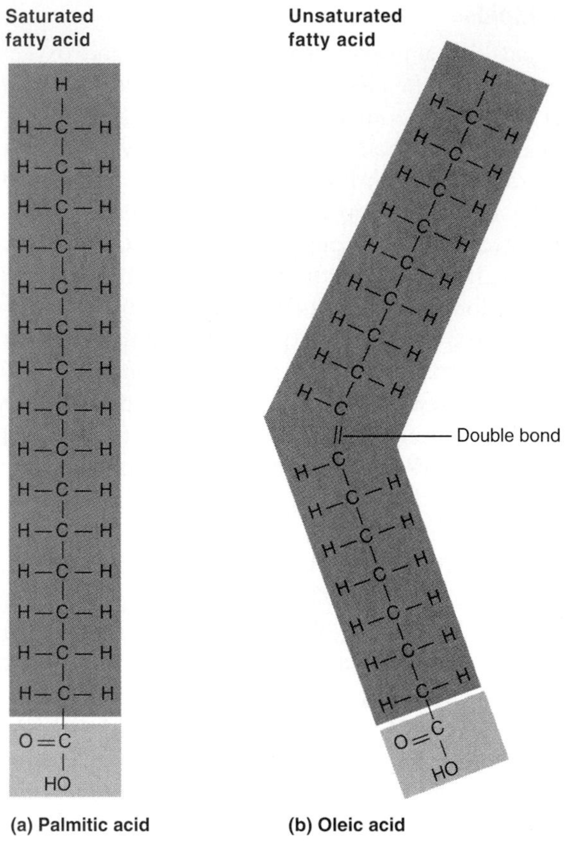

Figure 2.31 **Fatty Acids** The saturated fatty acids **(a)** are solids, and the unsaturated fatty acids **(b)**, are liquids.

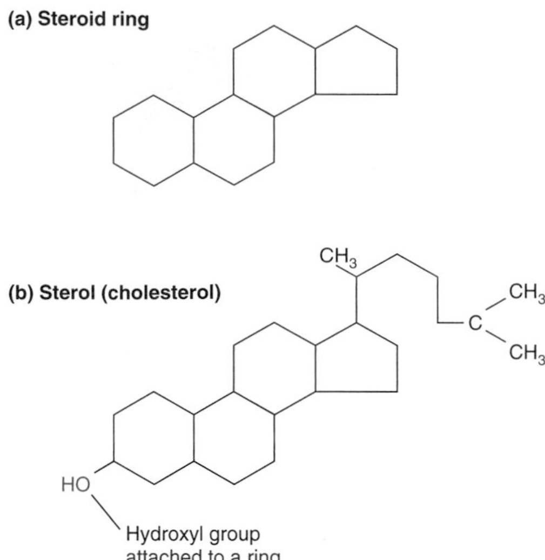

Figure 2.32 **Steroid** **(a)** General formula showing the four-membered ring and **(b)** the OH group that make the molecule a sterol. The sterol shown here is cholesterol. The carbon atoms in the ring structures and the attached hydrogen atoms are not shown.

group is attached to one of the rings, the steroid is called a **sterol**; in the case shown, it is **cholesterol** (figure 2.32b). Sterols are commonly found in the cytoplasmic membrane of eukaryotic cells, but only one group of prokaryotic cells; the mycoplasmas, contain sterols (see chapter 3). Other important compounds in this group of lipids are certain hormones such as cortisone, progesterone, and testosterone.

Compound Lipids

Compound lipids contain fatty acids and glycerol as well as elements other than carbon, hydrogen, and oxygen. Some of the most important members of this group in biology are the **phospholipids**, which contain a phosphate molecule in addition to the fatty acids and glycerol **(figure 2.33)**. The phosphate is further linked to a variety of other polar molecules, such as an alcohol, a sugar, or one of certain amino acids. This entire group is often referred to as a **polar head group** and is soluble in water.

Phospholipids occur as a double layer, termed a **bilayer** or **unit membrane**, in the cytoplasmic membrane, a structure that separates the outside of the cell from its internal contents (see figure 2.33). The chemical structure of the bilayer produces the essential properties of the cytoplasmic membrane. This membrane acts as the major barrier to the entry and exit of substances from the cell. The structure of the phospholipids is responsible for this barrier. A phospholipid consists of two parts, each with unique properties. The end with the phosphate bonded to the polar head group is hydrophilic and therefore soluble in water. The long fatty acid chains consisting of only C and H atoms are hydrophobic and therefore water insoluble. The hydrophilic regions orient themselves toward the external or internal (cytoplasmic) environment, with its high concentration of water. The long-chain fatty acids orient themselves away from the aqueous areas and inward interacting hydrophobically with the long hydrocarbon chains of other phospholipid molecules (see figure 2.33). Water-soluble substances, which are the most common and important in the cell's environment, cannot pass through the hydrophobic portion. Therefore, the cell has special mechanisms to bring these molecules into the cell. These will be discussed in chapter 3.

Other compound lipids are found in the outer covering of bacterial cells and will be discussed in chapter 3. These include the **lipoproteins**, covalent associations of proteins and lipids, and the **lipopolysaccharides**, molecules of lipid linked with polysaccharides through covalent bonds.

Some of the most important properties of macromolecules of biological importance are summarized in **table 2.4**.

MICROCHECK 2.7

Phospholipids, with one end polar and the other nonpolar form a major part of cell membranes. They limit the entry and exit of molecules into and out of cells.

- What are the two main types of lipids and how do they differ from one another?
- What are the main functions of lipids in cells?
- Some molecules such as many alcohols are soluble in water and in hydrophobic liquids such as oils. How easily do you think these molecules would cross the cell membrane?

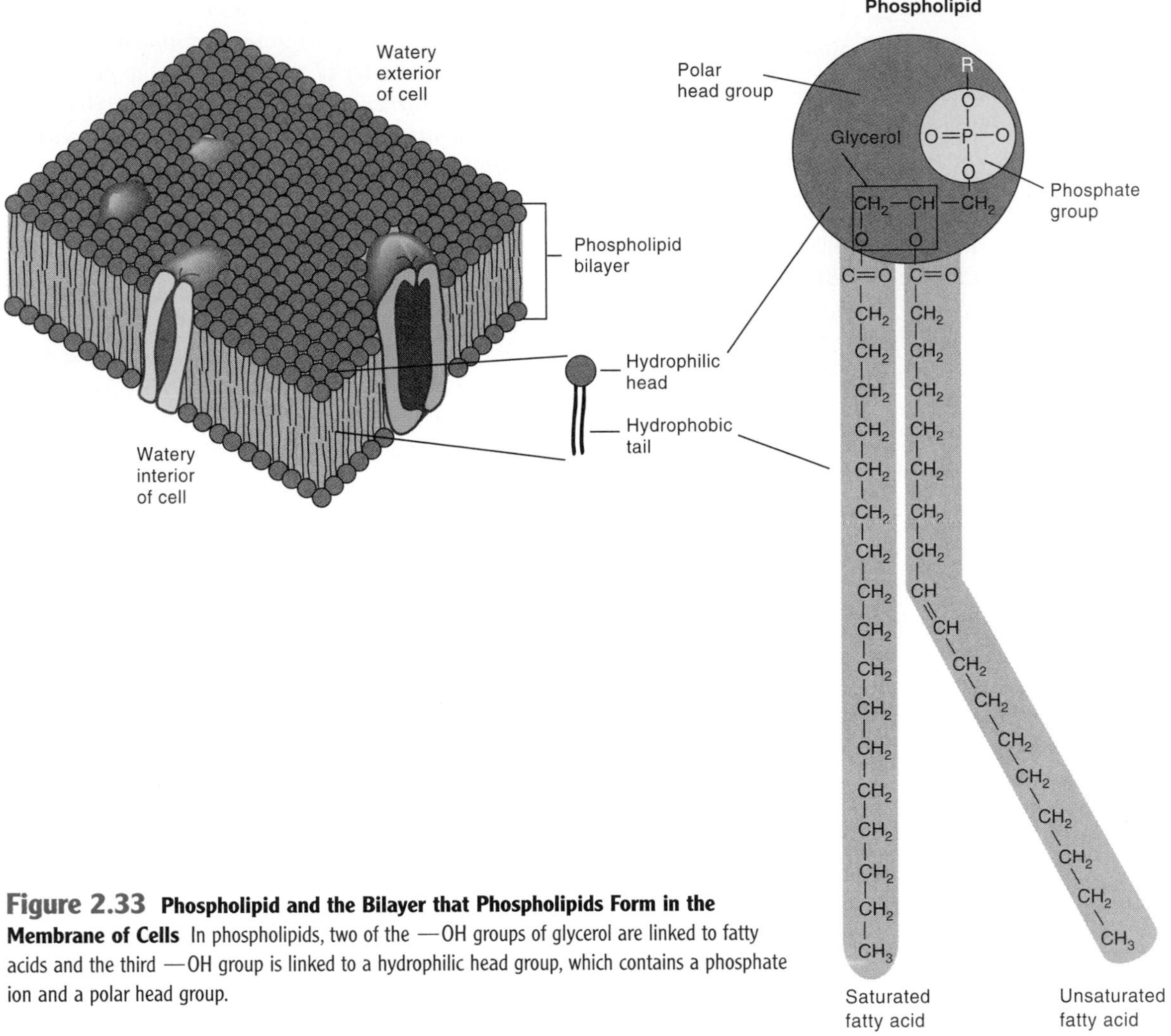

Figure 2.33 Phospholipid and the Bilayer that Phospholipids Form in the Membrane of Cells In phospholipids, two of the —OH groups of glycerol are linked to fatty acids and the third —OH group is linked to a hydrophilic head group, which contains a phosphate ion and a polar head group.

Table 2.4 Structure and Function of Macromolecules

Name	Subunit	Bond Joining Subunits	Atoms in Bond	Some Functions of Macromolecules
Protein	Amino acid	Peptide	$-\overset{\overset{\displaystyle O}{\Vert}}{C}-\underset{\underset{\displaystyle H}{\vert}}{N}-$	Catalysts; structural portion of cell organelles
Nucleic acids	Nucleotide	Phospho-Diester	$-\overset{\overset{\displaystyle H}{\vert}}{\underset{\underset{\displaystyle H}{\vert}}{C}}-O-\overset{\overset{\displaystyle O}{\Vert}}{\underset{\underset{\displaystyle OH}{\vert}}{P}}-O-\overset{\overset{\displaystyle H}{\vert}}{\underset{\underset{\displaystyle H}{\vert}}{C}}-$	RNA—Various roles in protein synthesis DNA—Carrier of genetic information
Polysaccharide	Monosaccharide	Glycosidic	$-\overset{\vert}{\underset{\vert}{C}}-O-\overset{\vert}{\underset{\vert}{C}}-$	Structural component of plant cell wall; storage products
Lipids	Varies—No similar subunits	Ester Bond Common	$R-\overset{\overset{\displaystyle O}{\Vert}}{C}-O-\overset{\overset{\displaystyle H}{\vert}}{\underset{\underset{\displaystyle H}{\vert}}{C}}-H$	Important in structure of cell membranes

FUTURE CHALLENGES

The Protein Folding Problem

The properties of all organisms depend on the proteins they contain. These include the structural proteins as well as enzymes. Even though a cell may be able to synthesize a protein, unless that protein is folded correctly and achieves its correct shape, it will not function properly. A major challenge associated with the molecules of life is to understand how proteins fold correctly. This is called the **protein folding problem**. *This is not only an important problem from a purely scientific point of view, but a number of serious neuro-degenerative diseases appear to be a result of protein misfolding. These include Alzheimer's disease and the neurodegenerative diseases caused by prions. If we could understand how proteins fold correctly, we might be able to prevent such diseases.*

The information that determines how a protein folds into its 3-dimensional configuration is contained in the sequence of its amino acids. It is not yet possible, however, to predict accurately how a protein will fold from its amino acid sequence. The folding occurs in a matter of seconds after the protein is synthesized. It appears that the protein folds rapidly into its secondary structure and then more slowly into its tertiary structure. In these slower reactions, many different adjustments of the amino acid side chains are tried out before the protein assumes the correct tertiary structure. It is in this slow stage that the proteins that help other proteins to fold correctly, the chaperones, play their role. The cell is able to dispose of improperly folded proteins. Enzymes, called **proteases**, *recognize improperly folded proteins and degrade them. The protein folding problem has such important implications for medicine and is so challenging a scientific question, that a super computer is being built to help solve the problem of predicting the 3-dimensional structure of a protein from its amino acid sequence.* ■ **prions, p. 13-14.**

SUMMARY

Atoms and Elements

1. Atoms are composed of **electrons**, **protons**, and **neutrons**. (Figure 2.1)
2. An **element** is a pure substance.

Chemical Bonds and the Formation of Molecules

1. For maximum stability, the outer **shell** of electrons of an atom must be filled. **Bonds** form between atoms to fill their outer shells.

Covalent Bonds

1. **Covalent bonds** are strong bonds formed by atoms sharing electrons. (Figure 2.2)

Nonpolar and Polar Covalent Bonds

1. When atoms have an equal attraction for electrons, a **nonpolar covalent bond** is formed between them. (Table 2.2)
2. When one atom has a greater attraction for electrons than another atom, **polar covalent bonds** are formed between them. (Figure 2.3)

Ionic Bonds

1. When electrons leave the shell of one atom and enter the shell of another atom, an **ionic bond** forms between the atoms. (Figure 2.4)

Hydrogen Bonds

1. **Hydrogen bonds** are weak bonds that result from the attraction of a positively charged hydrogen atom in a polar molecule to a negatively charged atom in another polar molecule. (Figure 2.7)

Chemical Components of the Cell

Water

1. Water is the most important molecule in the cell.
2. Water makes up over 70% of all living organisms by weight.
3. Hydrogen bonding plays a very important role in the properties of water. (Figure 2.8)
4. **pH** is the degree of acidity of a solution; it is measured on a scale of 0 to 14. (Figure 2.10)

Small Molecules in the Cell

1. All cells contain a variety of small organic and inorganic molecules.
2. A key element in all cells is carbon; it occurs in all organic molecules.

Macromolecules and Their Component Parts

1. **Macromolecules** are large molecules composed of **subunits** with similar properties.
2. Synthesis of macromolecules occurs by **dehydration synthesis**, and their degradation occurs by **hydrolysis**.

Proteins and Their Functions

1. **Proteins** are the most versatile of the macromolecules in what they do.
2. Activities of proteins include catalyzing reactions, being a component of cell structures, moving cells, taking nutrients into the cell, turning genes on and off, and being a part of cell membranes.

Amino Acid Subunits

1. Proteins are composed of 20 major **amino acids**. (Figure 2.13)
2. All amino acids consist of a **carboxyl group** at one end and an **amino group** bonded to the same carbon atom as the carboxyl group. (Figure 2.12)
3. The side chain of the amino acid confers unique properties on the amino acid.

Peptide Bonds and Their Synthesis

1. Amino acids are joined through **peptide bonds,** joining amino with carboxyl groups and splitting out water. (Figure 2.16)

Protein Structure (Figure 2.18)

1. The **primary structure** of a protein is its amino acid sequence.

2. The **secondary structure** of a protein is determined by intramolecular bonding to form **helices** and **sheets**.
3. The **tertiary structure** of a protein describes the three-dimensional shape of the protein, either **globular** or **fibrous**.
4. The **quaternary structure** describes the structure resulting from the interaction of several **polypeptide** chains.
5. When the intramolecular bonds within the protein are broken, the protein changes shape and no longer functions; the proteins are **denatured**. (Figure 2.19)

Substituted Proteins

1. **Substituted proteins** contain other molecules such as **sugars** and **lipids**, bonded to the side chains of amino acids in the protein.

Carbohydrates

1. **Carbohydrates** comprise a heterogeneous group of compounds that perform a variety of functions in the cell.
2. Carbohydrates have carbon, hydrogen, and oxygen atoms in a ratio of approximately 1 : 2 : 1.

Monosaccharides

1. **Monosaccharides** are classified by the number of carbon atoms they contain, most commonly 5 or 6.
2. Two different isomers can exist in sugars: **stereoisomers** and **structural isomers**. (Figures 2.21, 2.22, and 2.23)

Disaccharides

1. **Disaccharides** consist of two monosaccharides joined by dehydration synthesis between their hydroxyl groups. (Figure 2.24)

Polysaccharides

1. **Polysaccharides** are macromolecules consisting of monosaccharide subunits, sometimes identical, other times not.

Nucleic Acids

1. **Nucleic acids** are macromolecules whose subunits are **nucleotides**.
2. There are two types of nucleic acid: **deoxyribonucleic acid (DNA)** and **ribonucleic acid (RNA).**

DNA

1. DNA is the master molecule of the cell and carries all of the cell's genetic information in its sequence of nucleotides.
2. DNA is a double-stranded helical molecule with a backbone composed of covalently bonded sugar and phosphate groups. The **purine** and **pyrimidine** bases extend into the center of the helix. (Figure 2.28)
3. The two strands of DNA are **complementary** and are held together by hydrogen bonds between the **bases**. (Figure 2.29)

RNA

1. RNA is involved in decoding the genetic information contained in DNA.
2. RNA is a single-stranded molecule and contains **uracil** in place of the thymine in DNA. (Figure 2.27)

Lipids

1. **Lipids** are a heterogeneous group of molecules that are slightly soluble in water and very soluble in most organic solvents.
2. They comprise two groups: **simple** and **compound lipids.**

Simple Lipids

1. **Simple lipids** contain carbon, hydrogen, and oxygen and may be liquid or solid at room temperature.
2. **Fats** are common simple lipids and consist of **glycerol** bound to **fatty acids**. (Figure 2.30)
3. Lipids may be **saturated**, in which the fatty acid contains no double bonds between carbon atoms, or **unsaturated**, in which one or more double bonds exist. (Figure 2.31)
4. Some simple lipids consist of a four-membered ring, and include **steroids** and **sterols**. (Figure 2.32)

Compound Lipids

1. **Compound lipids** contain elements other than carbon, hydrogen, and oxygen.
2. **Phospholipids** are common and important examples of compound lipids. They are essential components of bilayer membranes in cells. (Figure 2.33)

REVIEW QUESTIONS

Short Answer

1. Differentiate between an atom, an element, an ion, and a molecule.
2. Which solution is more acidic, one with a pH of 4 or a pH of 5? What is the concentration of H^+ ions in each? The concentration of OH^- ions?
3. How do the two types of nucleic acids differ from one another in (a) composition, (b) size, and (c) function?
4. Name the subunits of proteins, polysaccharides, and nucleic acids.
5. What are the two major groups of lipids? Give an example of each group. What feature is common to all lipids?
6. How does the primary structure of a protein determine its overall structure?
7. Why is water a good solvent?
8. Give an example of a dehydration synthesis reaction. Give an example of a hydrolysis reaction. How are these types of reactions related?
9. List four functions of proteins.
10. What is a steroid?

Multiple Choice

1. Choose the list that goes from the lightest to the heaviest:
   A. Proton, atom, molecule, compound, electron
   B. Atom, proton, compound, molecule, electron
   C. Electron, proton, atom, molecule, compound
   D. Atom, electron, proton, molecule, compound
   E. Proton, atom, electron, molecule, compound

2. The strongest chemical bonds between two atoms in solution are
   A. covalent.
   B. ionic.
   C. hydrogen bonds.
   D. hydrophobic interactions.
3. Dehydration synthesis is involved in the synthesis of all of the following, *except*
   A. DNA.
   B. proteins.
   C. polysaccharides.
   D. lipids.
   E. monosaccharides.
4. The primary structure of a protein relates to its
   A. sequence of amino acids.
   B. length.
   C. shape.
   D. solubility.
   E. bonds between amino acids.
5. Pure water has all of the following properties, *except*
   A. polarity.
   B. ability to dissolve lipids.
   C. pH of 7.
   D. covalent joining of its atoms.
   E. ability to form hydrogen bonds.
6. The macromolecules that are composed of carbon, hydrogen, and oxygen in an approximate ratio of 1 : 2 : 1 are
   A. proteins.
   B. lipids.
   C. polysaccharides.
   D. DNA.
   E. RNA.
7. In proteins, α helices and β pleated structures are associated with the
   A. primary structure.
   B. secondary structure.
   C. tertiary structure.
   D. quaternary structure.
   E. multiprotein complexes.
8. Complementarity plays a major role in the structure of
   A. proteins.
   B. lipids.
   C. polysaccharides.
   D. DNA.
   E. RNA.
9. A bilayer is associated with
   A. proteins.
   B. DNA.
   C. RNA.
   D. complex polysaccharides.
   E. phospholipids.
10. Isomers are associated with
    1. carbohydrates.
    2. amino acids.
    3. nucleotides.
    4. RNA.
    5. fatty acids.

    A. 1, 2 B. 2, 3 C. 3, 4 D. 4, 5 E. 1, 5

Applications

1. A group of bacteria known as thermophiles thrive at high temperatures that would normally destroy other bacteria. Yet these thermophiles cannot survive well at the lower temperatures normally found on the earth. Propose a plausible explanation for this observation.
2. Microorganisms use hydrogen bonds to attach themselves to the surfaces that they live upon. Many of them lose hold of the surface and end up dying because of the weak nature of these bonds. Contrast the benefits and disadvantages of using covalent bonds as a means of attaching to surfaces.

Critical Thinking

1. What properties of the carbon atom make it ideal as the key atom for all molecules in organisms?
2. A biologist determined the amounts of several amino acids in two separate samples of pure protein. His data are shown below:

Amino Acid	Leucine	Alanine	Histidine	Cysteine	Glycine
Protein A	7%	12%	4%	2%	5%
Protein B	7%	12%	4%	2%	5%

   He concluded that protein A and protein B were the same protein. Do you agree with this conclusion? Justify your answer.

3. The table below indicates the freezing and boiling points of several molecules:

Molecule	Freezing Point (°C)	Boiling Point (°C)
Water	0	100
Carbon tetrachloride (CCl_4)	–23	77
Methane (CH_4)	–182	–164

   Carbon tetrachloride and methane are nonpolar molecules. How does the polarity and nonpolarity of these molecules explain why the freezing and boiling points for methane and carbon tetrachloride are so much lower than those for water?

CHAPTER 3

Microscopy and Cell Structure

Hans Christian Joachim Gram (1853–1938) was a Danish physician working in a laboratory at the morgue of the City Hospital in Berlin, microscopically examining the lungs of patients who had died of pneumonia. He was working under the direction of Dr. Carl Friedlander, who was trying to identify the cause of pneumonia by studying patients who had died of it. Gram's task was to stain the infected lung tissue to make the bacteria easier to see under the microscope. Strangely, one of the methods he developed did not stain all bacteria equally; some types retained the first dye applied in this multistep procedure, whereas others did not. Gram's staining method revealed that two different kinds of bacteria were causing pneumonia, and that these types retained the dye differently. We now recognize that this important staining method, called the ***Gram stain****, efficiently identifies two large, distinct groups of bacteria: Gram-positive and Gram-negative. The variation in the staining outcome of these two groups reflects a fundamental difference in the structure and chemistry of their cell walls.*

For a long time, historians thought that Gram did not appreciate the significance of his discovery. In more recent years, however, several letters show that Gram did not want to offend the famous Dr. Friedlander under whom he worked; therefore, he played down the importance of his staining method. In fact, the Gram stain has been used as a key test in the initial identification of bacterial species ever since the late 1880s.

—A Glimpse of History

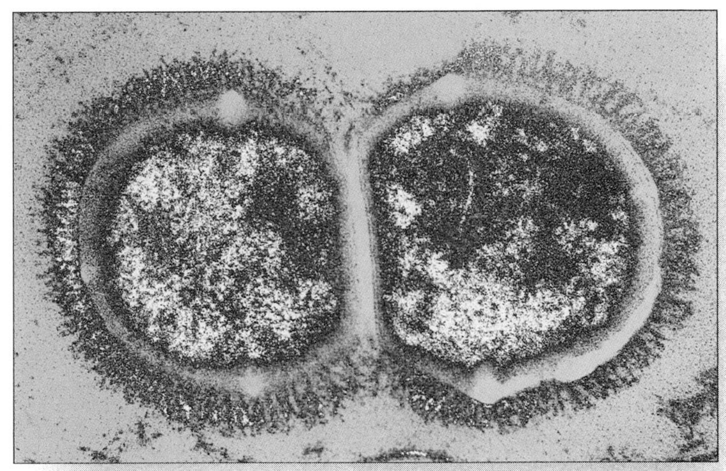

IMAGINE THE ASTONISHMENT ANTONY VAN Leeuwenhoek must have felt in the 1700s when he first observed microorganisms with his handcrafted microscope, an instrument that could magnify an image 300-fold (300×). Even today, observing the activities of diverse microbes interacting in a sample of stagnant pond water can provide enormous education and entertainment.

Microscopic study of cells has revealed that there are two fundamental types: prokaryotic and eukaryotic. The cells of all members of the Domains Bacteria and Archaea are prokaryotic. In contrast, cells of all animals, plants, protozoa, fungi, and algae are eukaryotic. The similarities and differences between these two basic cell types are important from a scientific standpoint and also have significant consequences to human health. For example, chemicals that interfere with processes unique to prokaryotic cells can be used to selectively destroy bacteria without harming humans.

Prokaryotic cells are much smaller than most eukaryotic cells—a trait that carries with it certain advantages as well as disadvantages. On one hand, their high surface area relative to their low volume makes it easier for these cells to take in nutrients and excrete waste products. Because of this, they can multiply much more rapidly than their eukaryotic counterparts. On the other hand, their small size makes them vulnerable to an array of threats. Predators, parasites, and fierce competition constantly surround them. Prokaryotic cells, while generally simple in their structure, have developed many unique attributes, which enhance their evolutionary success.

Eukaryotic cells are considerably more complex than prokaryotic cells. Not only are they larger, but many of their cellular processes take place within membrane-bound compartments. Eukaryotic cells are defined by the presence of a membrane-bound nucleus, which harbors the chromosomes. Although eukaryotic cells share many of the same characteristics as prokaryotic cells, many of their structures and cellular processes are fundamentally different.

MICROSCOPY AND CELL MORPHOLOGY

The study of microbiology has progressed concurrently with the development of microscopes. Today, common microscopes can magnify images 1,000× and more specialized instruments have much higher powers. Special staining procedures can enhance the use of the microscope by highlighting certain structures. Microscopes and accompanying staining techniques enable researchers to study minute aspects of cells.

Microscopic Techniques: The Instruments

Even today, one of the most important tools for studying microorganisms is the **light microscope**, which, like van Leeuwenhoek's instrument, uses visible light for observing objects. These instruments can magnify images approximately 1,000×, making it relatively easy to observe the size, shape, and motility of prokaryotic cells. The **electron microscope**, introduced in 1931, can magnify images in excess of 100,000×. This microscope was instrumental in elucidating many of the fine details of cell structure. The most recent major advancement came in 1981 with the introduction of the first **scanning probe microscope**. In turn, this has led to even more sophisticated technologies, now enabling scientists to view individual atoms. The types of microscopes are summarized in **table 3.1**.

Principles of Light Microscopy: The Bright-Field Microscope

In light microscopy, light typically passes through a specimen and then through a series of magnifying lenses. The most common type of light microscope, and the easiest to use, is the **bright-field microscope**, which illuminates the field of view evenly.

Magnification

The modern light microscope has two magnifying lenses—an **objective lens** and an **ocular lens**—and is called a **compound**

TABLE 3.1 A Summary of Microscopic Instruments and Their Characteristics

Instrument	Mechanism	Uses/Comment
Light Microscopes	Visible light passes through through a series of lenses to produce a magnified image.	Relatively easy to use. Considerably lessexpensive than confocal and electron microscopes.
Bright-field	Illuminates the field of view evenly.	Most common type of microscope.
Phase-contrast	Amplifies differences in refractive index to create contrast.	Makes unstained cells more readily visible.
Interference	Two light beams pass through the specimen and then recombine.	Causes the specimen to appear as a three-dimensional image.
Dark-field	Light is directed toward the specimen at an angle.	Makes unstained cells more readily visible; organisms stand out as bright objects against a dark background.
Fluorescence	Projects ultraviolet light, causing fluorescent molecules in the specimen to emit longer wavelength light.	Used to observe cells that have been stained or tagged with a fluorescent dye.
Confocal	Mirrors scan a laser beam across successive regions and planes of a specimen. From that data, a computer constructs an image.	Used to construct a three-dimensional image structure; provides detailed sectional views of intact cells.
Electron Microscopes	Uses electron beams in place of visible light to produce the magnified image.	Can clearly magnify images 100,000×.
Transmission	Transmits a beam of electrons through a specimen.	Elaborate specimin preparation, which may introduce artifacts, is required.
Scanning	A beam of electrons scans back and forth over the surface of a specimen.	Used for observing surface details; produces a three-dimensional effect.
Scanning Probe Microscopes	Makes it possible to view images on an atomic scale.	Produces a map showing the bumps and valleys of the atoms on that surface.
Scanning tunneling	A sharp metallic probe causes electrons to tunnel between the probe and a conductive surface.	First scanning probe microscope developed.
Atomic force	A tip bends in response to even the slightest force between it and the sample.	Can operate in air and in liquids.

microscope (**figure 3.1**). These lenses in combination visually enlarge an object by a factor equal to the product of each lens' magnification. For example, an object is magnified 1,000-fold when it is viewed through a 10× ocular lens and an objective lens with a power of 100×. Most compound microscopes have a selection of objective lenses that are of different powers—typically 4×, 10×, 40×, and 100×. This makes a choice of different magnifications possible with the same instrument.

The **condenser lens** does not affect the magnification but, positioned between the light source and the specimen, is used to focus the illumination.

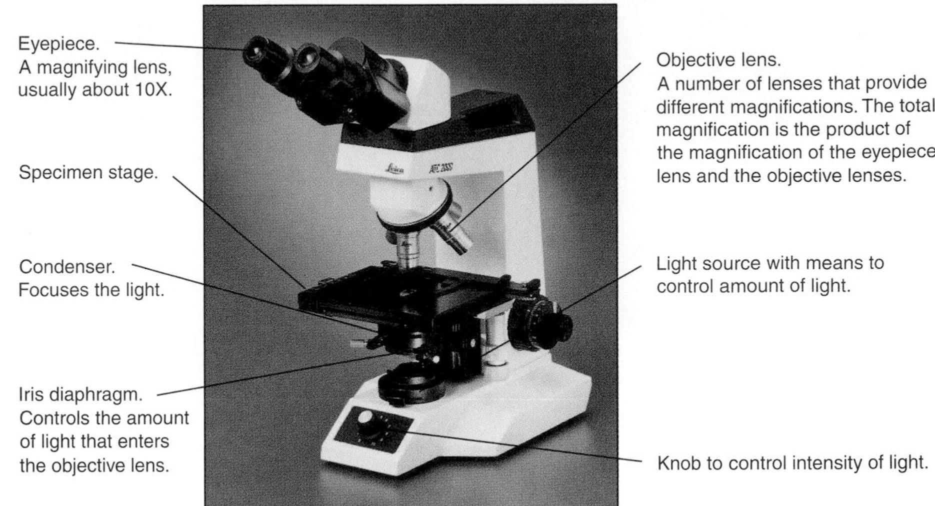

Eyepiece.
A magnifying lens, usually about 10X.
Specimen stage.
Condenser.
Focuses the light.
Iris diaphragm.
Controls the amount of light that enters the objective lens.
Objective lens.
A number of lenses that provide different magnifications. The total magnification is the product of the magnification of the eyepiece lens and the objective lenses.
Light source with means to control amount of light.
Knob to control intensity of light.

Figure 3.1 A Modern Light Microscope The compound microscope employs a series of magnifying lenses.

Resolution

The usefulness of a microscope depends not so much on its degree of magnification, but on its ability to clearly separate, or **resolve**, two objects that are very close together. The **resolving power** is defined as the minimum distance that can exist between two objects when those objects can still be observed as separate entities. The resolving power therefore determines how much detail actually can be seen (**figure 3.2**). The importance of resolving power was demonstrated in Seventeenth-century Europe by microscopists Robert Hooke and van Leeuwenhoek, who employed very different instruments. Although Hooke's microscope had greater magnification, his lenses had serious optical defects, resulting in blurred images. The simpler instrument utilized by van Leeuwenhoek had less magnification but greater resolving power and proved to be the better for observing the details of microorganisms.

The resolving power of a microscope depends on the quality and type of lens, the magnification, and how the specimen under observation has been prepared. At greater magnifications it can also be limited by the wavelength of the light used for illumination. The shorter the wavelength, the greater the resolution. The maximum resolving power of the best light microscope is 0.2 μm. This is sufficient to observe the general morphology of a prokaryotic cell but too low to distinguish a particle the size of a virus.

To obtain maximum resolution when using certain high-power objectives such as the 100× lens, an oil is employed to displace the air between the lens and the specimen. This is to avoid the bending of light rays, or **refraction**, that occurs when light passes from glass to air (**figure 3.3**). Refraction can prevent those rays from entering the relatively small openings of higher power objective lenses. The oil used has nearly the same **refractive index** as glass. Refractive index is a measure of the relative velocity of light as it passes through a medium. As light travels from a medium of one refractive index to another, those rays are bent. When the oil is used to displace the air at the interface of the glass slide and the glass lens, light rays pass with little refraction occurring.

Contrast

Contrast reflects the number of visible shades in a specimen—high contrast being just the two shades, black and white. Different specimens require various degrees of contrast to reveal the most information. One example is bacteria, which are essentially transparent against a bright colorless background. The lack of contrast presents a problem when viewing objects. One way to overcome this is to stain the bacteria with any one of a number of dyes. The types and characteristics of these stains will be discussed shortly.

Light Microscopes that Increase Contrast

Special light microscopes that increase the contrast between microorganisms and their surroundings overcome some of the difficulties of observing unstained bacteria. Staining kills

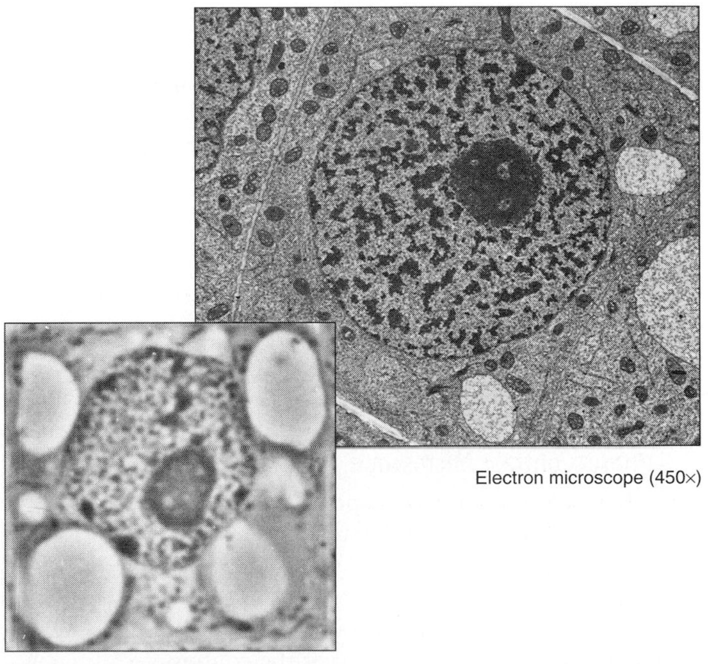

Figure 3.2 Comparison of the Resolving Power of the Light Microscope and an Electron Microscope In this case, an onion root tip was magnified 450×. Note the difference in the degree of detail that can be seen at the same magnification.

(a)

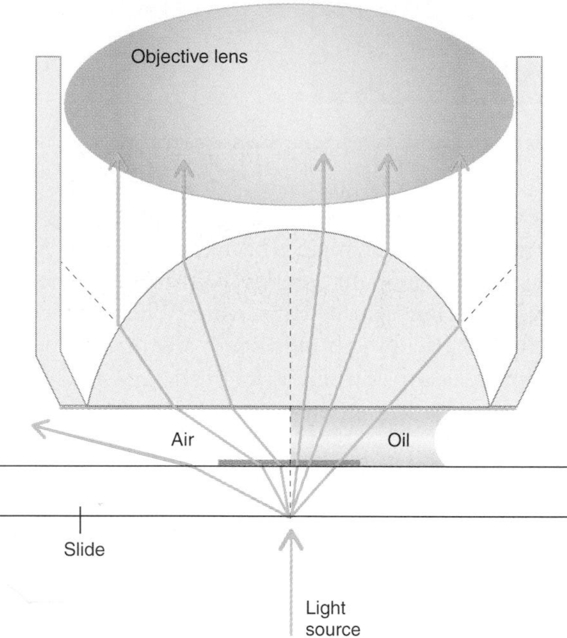

(b)

Figure 3.3 Refraction As light passes from one medium to another, the light rays may bend, depending on the refractive index of the two media. **(a)** The pencil in water appears bent because the refractive index of water is different from that of air. **(b)** Light rays bend as they pass from air to glass because of the different refractive indexes of these two media; some rays are lost to the objective. Oil and glass have the same refractive index, and therefore the light rays are not bent.

microbes; therefore, some of these microscopes are invaluable when the goal is to examine characteristics of living organisms such as motility and behavior.

The Phase-Contrast Microscope

The **phase-contrast microscope** exploits the difference in the refractive index of cells and the surrounding medium and results in a darker appearance of the denser material **(figure 3.4)**. As light passes through cells, it is refracted slightly differently from when it passes through the surroundings. Special optical devices amplify those differences, increasing the contrast.

The Interference Microscope

The **interference microscope** causes the specimen to appear as a three-dimensional image **(figure 3.5)**. This microscope, like the phase-contrast microscope, depends on differences in refractive index as light passes through different materials. The most frequently employed microscope of this type is the **Nomarski differential interference contrast microscope**, which has a device for separating light into two beams that pass through the specimen and then recombine. The light waves are out of phase when they recombine, yielding the three-dimensional appearance of the specimen.

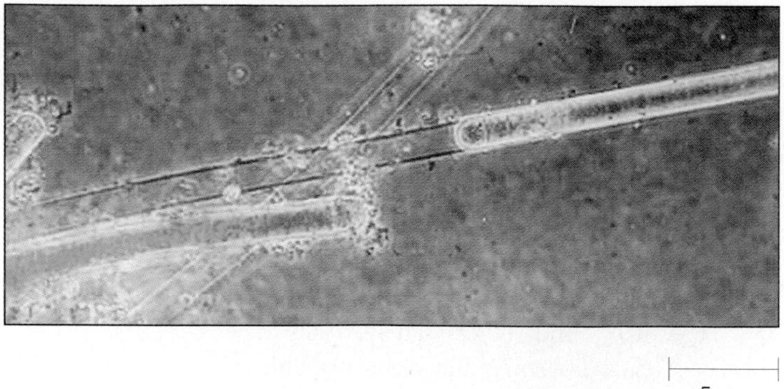

Figure 3.4 Phase-Contrast Photomicrograph Sheathed filaments of a species of the cyanobacterium *Lyngbya*.

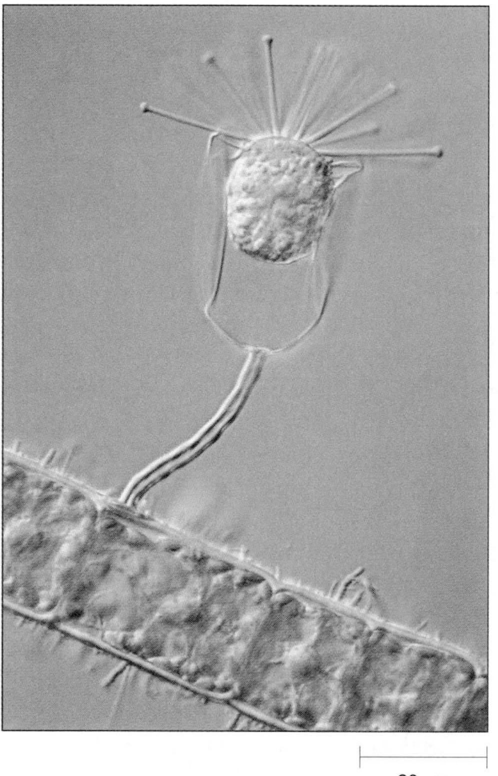

Figure 3.5 Nomarski Differential Interference Contrast Microscopy Protozoan (*Paracineta*) attached to green algae (*Spongomorpha*) (320×).

The Dark-Field Microscope

Organisms viewed through a **dark-field microscope** stand out as bright objects against a dark background **(figure 3.6)**. The

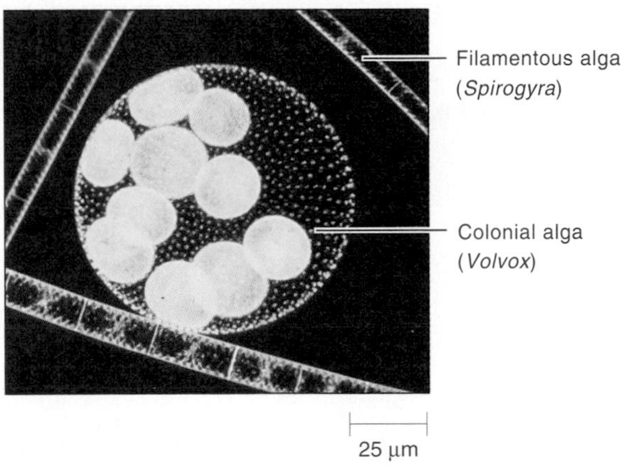

Figure 3.6 Dark-Field Photomicrograph *Volvox* (sphere) and *Spirogyra* (filaments), both of which are eukaryotes.

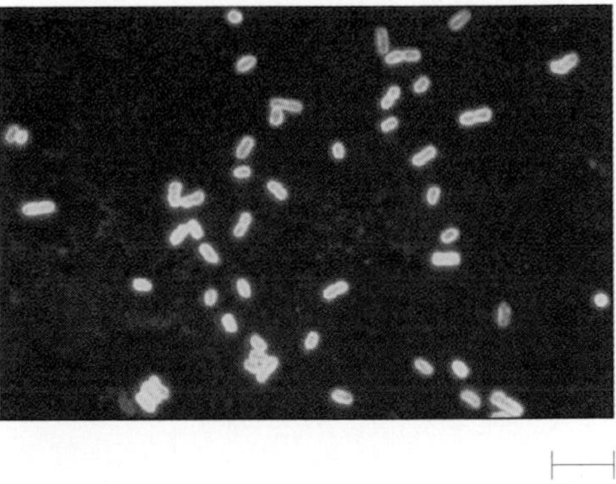

Figure 3.7 Fluorescence Photomicrograph A rod-shaped bacterium tagged with a fluorescent marker.

microscope operates on the same principle that makes dust visible when a beam of bright light shines into a dark room. A special mechanism directs light toward the specimen at an angle, so that only light scattered by the specimen enters the objective lens. Dark-field microscopy is used to detect members of the genus *Treponema*. These very thin, spiral-shaped organisms stain poorly and are very difficult to see via bright-field microscopy (see figure 11.28).

The Fluorescence Microscope

The **fluorescence microscope** is used to observe cells that have been stained or tagged with fluorescent dyes, enabling the cells to glow against a dark background **(figure 3.7)**. Because the stained cells stand out so clearly, fluorescent microscopy is a more sensitive method of detecting microorganisms than is bright-field microscopy.

The fluorescence microscope can project ultraviolet light through the objective lens and onto the specimen. The light is absorbed by the fluorescent dyes, which then release that energy in the form of longer wavelength light. Because the light can be projected onto a specimen, rather than transmitted through it, even cells attached to soil particles or other opaque materials can be observed. The color that the cells will appear depends on the type of dye and the light filters used. The types and characteristics of fluorescent stains will be discussed shortly.

The Confocal Scanning Laser Microscope

The **confocal scanning laser microscope** is used to construct a three-dimensional image of a thick structure such as a natural community of microorganisms. The instrument can also provide detailed sectional views of the interior of an intact cell. In confocal microscopy, the lenses focus a laser beam to illuminate a given point on one vertical plane of a specimen. Mirrors then scan the laser beam across the specimen, illuminating successive regions and planes until the entire specimen has been scanned. Each plane corresponds to an image of one fine slice of the specimen. A computer then assembles the data and constructs a three-dimensional image, which is displayed on a screen. In effect, this microscope is a miniature CAT scan for cells.

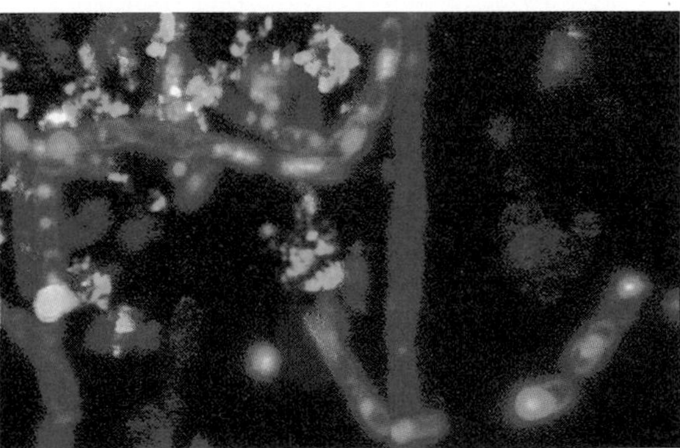

Figure 3.8 Confocal Photomicrograph A biofilm of *Pseudomonas oleovorans* stained with fluorescent dyes; the reddish-stained material within cells is a type of storage granule. The biofilm is approximately 6 μm thick.

Frequently, the specimens are first stained or tagged with a fluorescent dye. By employing certain fluorescent tags that bind specifically to a given protein or other compound, the precise cellular location of that compound can be determined **(figure 3.8)**. In some cases, multiple different tags that bind to specific molecules are used, each a distinct color.

Electron Microscopes

Electron microscopy is in some ways comparable to light microscopy. Rather than using glass lenses, visible light, and the eye to observe the specimen, the electron microscope uses electromagnetic lenses, electrons, and a fluorescent screen to produce the magnified image **(figure 3.9)**. That image can be captured on photographic film to create an **electron photomicrograph**. Sometimes, the black and white images are artificially enhanced with color to add visual clarity.

Since the electrons have a wavelength about 1,000 times shorter than visible light, the resolving power increases about 1,000-fold, to about 0.3 nanometers (nm)or 0.3×10^{-3} μm. Consequently, considerably greater detail can be observed due to the much higher resolution. These instruments can clearly magnify an image 100,000×. One of the biggest drawbacks is

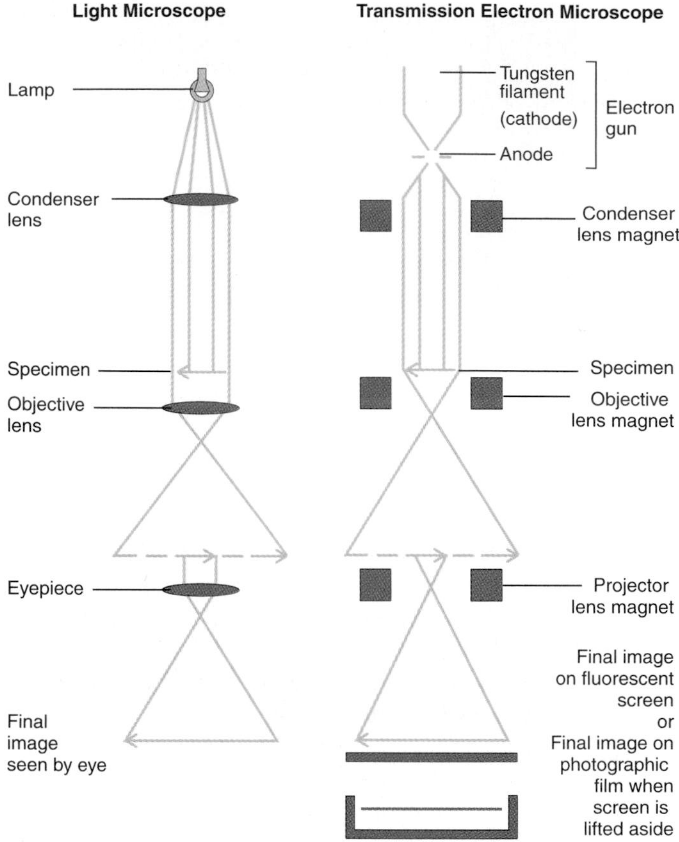

Figure 3.9 Comparison of the Principles of Light and the Electron Microscopy For the sake of comparison, the light source for the light microscope has been inverted (the light is shown at the top and the eyepiece at the bottom).

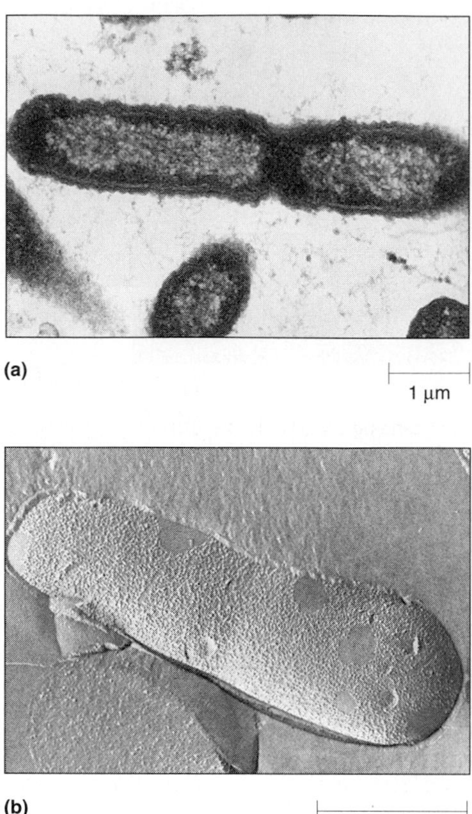

Figure 3.10 Transmission Electron Photomicrograph A rod-shaped bacterium prepared by **(a)** thin section; **(b)** freeze etching.

that the microscope requires a vacuum. Otherwise, the molecules composing air would interfere with the path of the electrons. This results in an expensive and bulky unit and also requires substantial and complex preparation of the specimen.

The Transmission Electron Microscope

The **transmission electron microscope (TEM)** is used to observe fine details of cell structure, such as the number of layers that envelop a cell. The instrument directs a beam of electrons at a specimen. Depending on the density of a particular region in the specimen, electrons will either pass through or be scattered to varying degrees. The darker areas of the resulting image correspond to the denser portions of the specimen (**figure 3.10**).

Transmission electron microscopy requires elaborate and painstaking specimen preparation. To view details of internal structure, a process called **thin sectioning** is used. Cells are carefully treated with a preservative and dehydrated in an organic solvent before embedding them in a plastic resin. Once embedded, they can be cut into exceptionally thin slices with a diamond or glass knife and then stained with heavy metals. Even a single bacterial cell must be cut into slices this way to be viewed via TEM. Unfortunately, the procedure can severely distort the cells. Consequently, a major concern in using TEM is distinguishing actual cell components from artifacts that occur as a result of specimen preparation.

A process called **freeze fracturing** is used to observe the shape of structures within the cell. The specimen is rapidly frozen and then fractured by striking it with a knife blade. The cells break open, usually along the middle of internal membranes. The surface of the section is then coated with a thin layer of carbon to create a replica of the surface. This replica is then examined in the electron microscope. A variation of freeze fracturing is **freeze etching**. In this process, the frozen surface exposed by fracturing is desiccated slightly under vacuum, which allows underlying regions to be exposed.

The Scanning Electron Microscope

The **scanning electron microscope (SEM)** is used for observing surface details, but not internal structures of cells. The SEM does not transmit electrons through the specimen; instead, a beam of electrons scans back and forth over the surface of a specimen that has been coated with a thin film of metal. As those beams move, electrons are released from the specimen and are reflected back into the viewing chamber. It is this reflected radiation that is observed with the microscope. Relatively large specimens can be viewed, and a dramatic three-dimensional effect is observed with the SEM (**figure 3.11**).

Scanning Probe Microscopes

Scanning probe microscopes make it possible to view images at an atomic scale. Their resolving power is much greater than

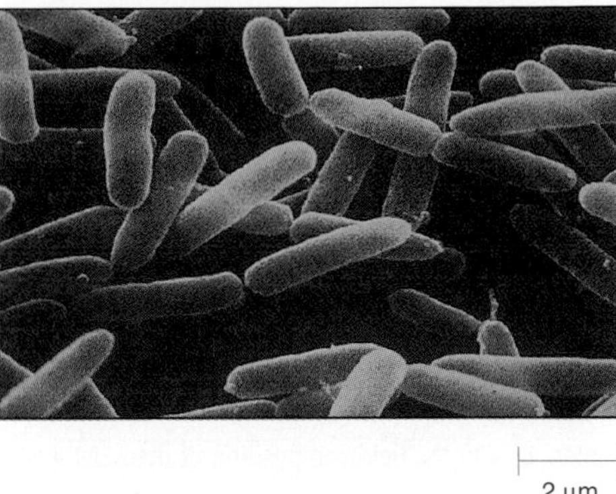

Figure 3.11 Scanning Electron Photomicrograph A rod-shaped bacterium.

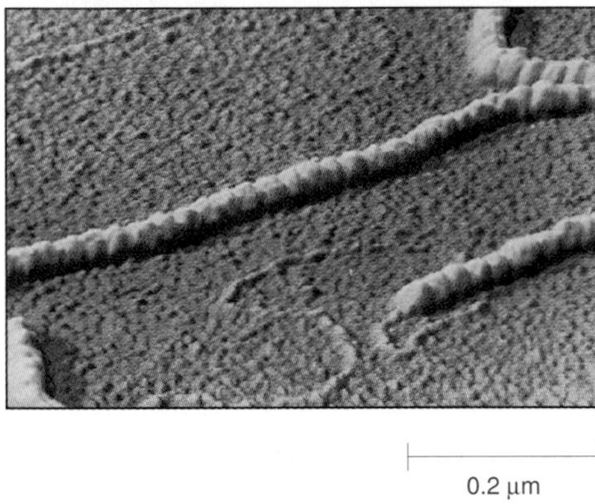

Figure 3.12 Scanning Probe Photomicrograph DNA-protein complex.

the electron microscope, and the samples do not have to be prepared in a special way as they do for electron microscopy. One such microscope, developed in the early 1980s, is the **scanning tunneling microscope**. This instrument has a sharp metallic probe, the tip of which is the size of a single atom. Electrons tunnel between the probe and a conductive surface. By scanning across a surface, this instrument produces a map, which shows the bumps and valleys of the atoms on that surface (**figure 3.12**). A more recent instrument is the **atomic force microscope**, which has a tip that is mounted so it can bend in response to even the slightest force between the tip and the sample. By monitoring the motion of the tip as it scans across a surface, a map of the surface is produced. Unlike the scanning tunneling microscope, the atomic force microscope does not rely on a vacuum and thus can operate in air and in liquids.

MICROCHECK 3.1

The usefulness of a microscope depends on its resolving power. The most common type of microscope is the bright-field microscope. Variations of the light microscopes are designed to increase the contrast between a bacterium and its surroundings. The fluorescence microscope is used to observe microbes that have been stained with special dyes. The confocal scanning laser microscope is used to construct a three-dimensional image of a thick structure. Electron microscopes can magnify images 100,000×. Scanning probe microscopes map images on an atomic scale.

- Why might oil be employed when using the 100× lens?
- Why are microscopes that enhance contrast used to view live rather than stained specimens?
- If an object being viewed under the phase-contrast microscope has the same refractive index as the background material, how would it appear?

Microscopic Techniques: Dyes and Staining

It can be difficult with the bright-field microscope to observe living microorganisms. Most are nearly transparent and often move rapidly about the slide. Consequently, cells are frequently immobilized and stained with dyes. These dyes may be attracted to one or more cellular components. As a result, an entire organism, or its specific parts, can be made to contrast with the unstained background. Many different dyes and staining procedures can be employed; each has specific applications (**table 3.2**).

Basic dyes, which carry a positive charge, are more commonly used for staining than are negatively charged **acidic dyes**. Because opposite charges attract, the basic dyes are drawn to the many negatively charged components of cells, including nucleic acid and many proteins, whereas acidic dyes would be repelled. Common basic dyes include methylene blue, crystal violet, safranin, and malachite green. **Simple staining** employs one of these basic dyes to impart color to the cells. Acidic dyes are sometimes used to stain backgrounds against which colorless cells can be seen, a technique called **negative staining**.

To stain microorganisms, a drop of a liquid containing the microbe is placed on a glass microscope slide and allowed to air dry. The resulting specimen appears as a film and is called a **smear**. The organisms are then attached, or **fixed**, to the slide, usually by passing the slide over a flame (**figure 3.13**). Dye is then applied and washed off with water. The heat fixing and subsequent staining steps kill the microorganisms and may distort their shape.

Differential Stains

Differential staining techniques are used to distinguish one group of bacteria from another. They take advantage of the fact that certain bacteria have distinctly different chemical structures in some of their components. These lead to differences in their staining properties. The two most frequently used differential staining techniques are the **Gram stain** and the **acid-fast stain**.

Gram Stain

The **Gram stain** is by far the most widely used procedure for staining bacteria. The basis for it was developed over a century ago by Dr. Hans Christian Gram, when he noticed that in lung specimens, certain bacteria that cause pneumonia retained a stain while

Table 3.2 A Summary of Stains and Their Characteristics

Stain	Characteristics
Simple Stains	Employs a basic dye to impart a color to a cell. Easy way to increase the contrast between otherwise colorless cells and a transparent background.
Differential Stains	Distinguishes one group of microorganisms from another.
Gram stain	Used to separate bacteria into two major groups, Gram-positive and Gram-negative. The staining characteristics of these groups reflect a fundamental difference in the chemical structure of their cell walls. This is by far the most widely used staining procedure.
Acid-fast stain	Used to detect members of the genus *Mycobacterium* in a specimen. Due to the lipid composition of their cell walls, these organisms do not readily take up stains.
Special Stains	Stains specific structures inside or outside of a cell.
Capsule stain	Because the viscous capsule does not readily take up stains, it stands out against a stained background. This is an example of a negative stain.
Endospore stain	Stains endospores, a type of dormant cell that does not readily take up stains. These are produced by the genera *Bacillus* and *Clostridium*.
Flagella stain	The staining agent adheres to and coats the otherwise thin flagella, enabling them to be seen with the light microscope.
Fluorescent Dyes and Tags	Fluorescent dyes and tags absorb ultraviolet light and then emit light of a longer wavelength. They are used in conjunction with a fluorescence microscope.
Fluorescent dyes	Some fluorescent dyes bind to compounds found in all cells; others bind to compounds specific to only certain types of cells.
Fluorescent tags	Antibodies to which a fluorescent molecule has been attached are used to tag specific molecules.

surrounding tissues did not. His observations led to procedures by which bacteria can be separated into two major groups: **Gram-positive** and **Gram-negative**. We now know that the difference in the staining attributes of these two groups reflects a fundamental difference in the chemical structure of their cell walls.

The Gram stain procedure involves four basic steps (**figure 3.14**):

1. The smear is first flooded with the **primary stain**, crystal violet in this case. The primary stain is the first dye applied in multistep staining procedures and generally stains all the cells.
2. The smear is rinsed to remove excess crystal violet and then is flooded with a dilute solution of iodine, called **Gram's iodine**. Iodine is a **mordant**, a substance that increases the affinity of cellular components for a dye. The iodine combines with the crystal violet to form a dye-iodine complex, thereby decreasing the solubility of the dye within the cell.
3. The stained smear is rinsed again, and then 95% alcohol or a mixture of alcohol and acetone is briefly added. These solvents act as a **decolorizing agent** and readily remove the dye-iodine complex from Gram-negative, but not Gram-positive, bacteria.
4. A **counterstain** is then applied to impart a contrasting color to the now colorless Gram-negative bacteria. For this purpose, the red dye safranin is used. This dye stains Gram-negative as well as Gram-positive bacteria, but because the latter are already stained purple, it imparts little difference.

To obtain reliable results, it is essential that the Gram stain be done properly. One of the most common mistakes is to decolorize a smear for too long a time period. Even Gram-positive cells

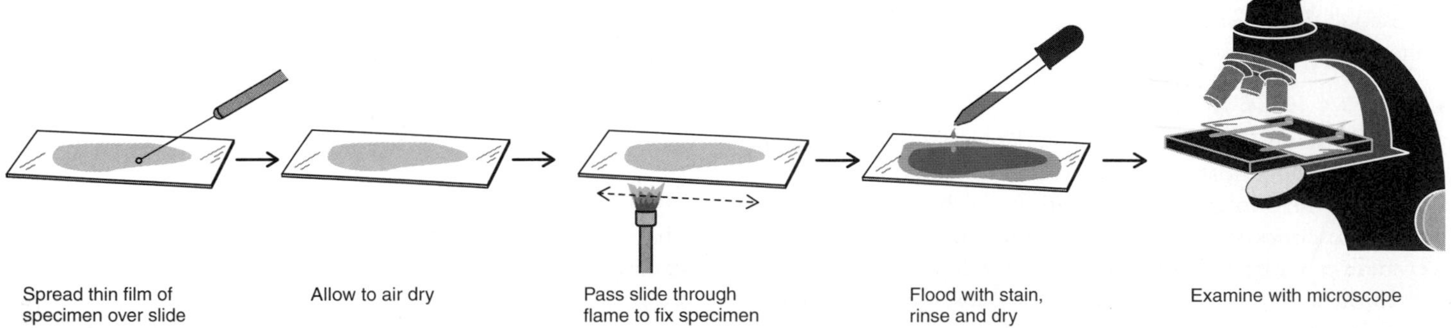
Spread thin film of specimen over slide
Allow to air dry
Pass slide through flame to fix specimen
Flood with stain, rinse and dry
Examine with microscope

Figure 3.13 Staining Bacteria for Microscopic Observation

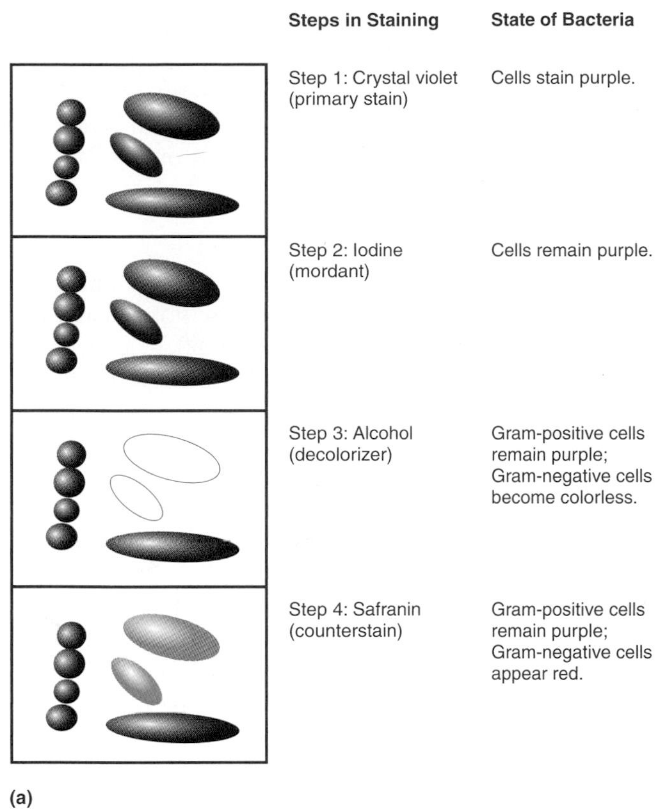
Steps in Staining
State of Bacteria
Step 1: Crystal violet (primary stain)
Cells stain purple.
Step 2: Iodine (mordant)
Cells remain purple.
Step 3: Alcohol (decolorizer)
Gram-positive cells remain purple; Gram-negative cells become colorless.
Step 4: Safranin (counterstain)
Gram-positive cells remain purple; Gram-negative cells appear red.
(a)

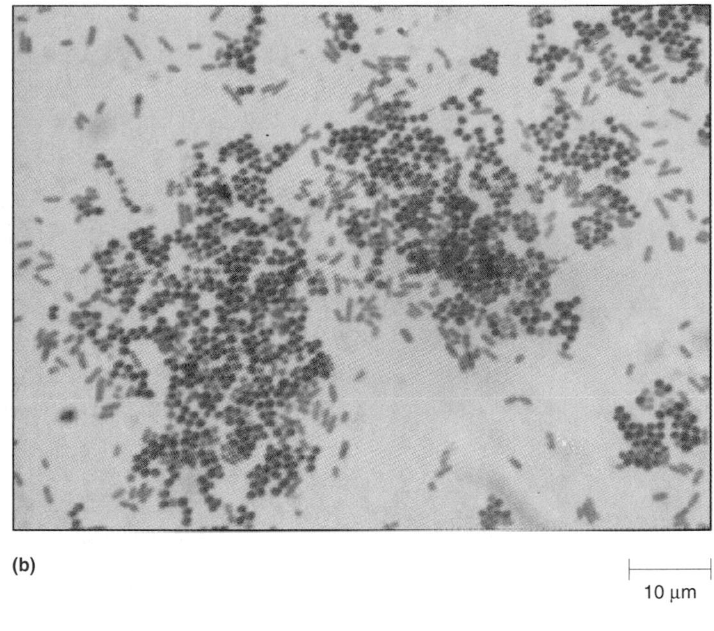
(b)
10 μm

Figure 3.14 Gram Stain (a) Steps in the Gram stain procedure. **(b)** Results of a Gram stain. The Gram-positive cells (purple) are *Staphylococcus aureus*; the Gram-negative cells (reddish-pink) are *Escherichia coli*.

can lose the crystal violet–iodine complex during prolonged decolorization. An over-decolorized Gram-positive cell will appear pink after counterstaining. Another important consideration is the age of the culture. As bacterial cells age, they lose their ability to retain the crystal violet–iodine dye complex, presumably because of changes in their cell wall. Like the effect of over-decolorization, this will cause the cells to appear pink. In the case of certain types of Gram-positive bacteria, this aging may occur within 24 hours. Thus, the Gram stain results of fresh cultures are more reliable.

Acid-Fast Stain

The **acid-fast stain** is a procedure used to stain a small group of organisms that do not readily take up stains. Among these are members of the genus *Mycobacterium*, which includes a species that causes tuberculosis and one that causes Hansen's disease (leprosy). The cell wall of these acid-fast bacteria contains high concentrations of a waxy lipid that prevents the uptake of dyes, including those used in the Gram stain. Therefore, harsh methods are needed to stain these organisms. Once stained, however, these same cells are very resistant to decolorization. Because mycobacteria are among the few organisms that retain the stain in this procedure, the acid-fast stain can be used to presumptively identify them in clinical specimens that might contain a variety of different bacteria. ■ **tuberculosis, p. 569** ■ **Hansen's disease, p. 672**

The acid-fast stain, like the Gram stain, requires multiple steps. The primary stain in this procedure is carbol fuchsin, a red dye. In the classic procedure, the stain-flooded slide is heated over boiling water, which facilitates the staining. A current variation does not use heat, instead employing a prolonged application of a more concentrated solution of dye. The slide is then rinsed briefly to remove the residual stain before being flooded with acid-alcohol, a potent decolorizing agent. This step removes the carbol fuchsin from nearly all cells, including tissue cells and most bacteria. Those few exceptional organisms that retain the dye are called **acid-fast**. Methylene blue is then used as a counterstain to make the non-acid-fast cells visible. Acid-fast organisms, which do not take up the methylene blue, appear a bright reddish-pink. They stand out against a background of the other cells, which are stained blue **(figure 3.15)**.

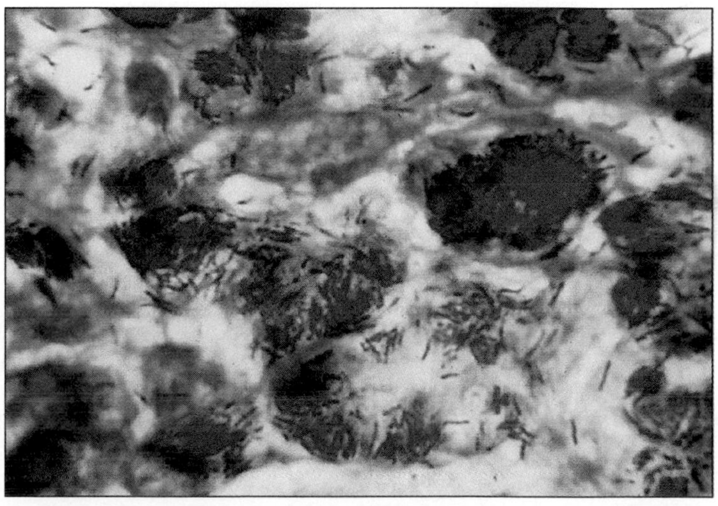
10 μm

Figure 3.15 Acid Fast Stain *Mycobacterium* species retain the red primary stain, carbol fuchsin. Counterstaining with methylene blue imparts a blue color to cells that are not acid-fast.

Special Stains to Observe Cell Structures

Dyes can also be used to stain specific structures inside or outside the cell. The staining procedure for each component of the cell is different, being geared to the chemical composition and properties of that structure. The function of each of these structures will be discussed in more depth later in the chapter.

Capsule Stain

A **capsule** is a viscous layer that envelops a cell and is sometimes correlated with an organism's ability to cause disease. Capsules stain poorly, a characteristic that is exploited with a **capsule stain**, an example of a negative stain. It colors the background, allowing the capsule to stand out as a halo around an organism **(figure 3.16)**.

To observe capsules, a liquid specimen is placed on a slide next to a drop of India ink. A thin glass coverslip is then placed over the two drops, causing them to flow together. This creates a gradient of India ink concentration across the specimen. Unlike the stains discussed previously, the capsule stain is done as a **wet mount**—a drop of liquid on which a coverslip has been placed—rather than as a smear. At the optimum concentration of India ink, the fine dark particles of the stain color the background enough to allow the capsule to be visible.

Endospore Stain

Members of certain Gram-positive genera including *Bacillus* and *Clostridium* form a special type of dormant cell, an **endospore**, that is resistant to destruction and to staining. Although these structures do not stain with a Gram stain, they can often be seen as a clear smooth object within an otherwise purple-stained cell. To make endospores more readily noticeable, a **spore stain** is used. This stain, like the classic acid-fast staining procedure, uses heat to facilitate staining.

The endospore stain, like the differential stains discussed previously, is a multistep procedure that employs a primary stain as well as a counterstain. Generally, malachite green is used as a primary stain. Its uptake by the endospore is facilitated by gentle heat. When water is then used to rinse the smear, only endospores retain the malachite green. The smear is then counterstained, most often with the red dye safranin. When viewed with a microscope, the spores appear green amid a background of pink cells **(figure 3.17)**.

Flagella Stain

Flagella are the appendages that provide the most common mechanism of motility for bacterial cells. Because not all bacteria have flagella, and those that do can have them in different arrangements around a cell, the presence and location of these structures can be used to classify and identify bacteria. However, the diameter of a single flagellum is generally 30 nm or less and, ordinarily, is too thin to be seen with a light microscope. The flagella stain overcomes this limitation by employing a mordant that enables the staining agent to adhere to and coat these thin structures, effectively increasing their diameter **(figure 3.18)**. Unfortunately, this staining procedure is difficult and requires patience and expertise.

Fluorescent Dyes and Tags

Depending on the procedure employed, fluorescence can be used to observe total cells, a subset of cells, or cells that have certain proteins on their surface **(figure 3.19)**.

Fluorescent Dyes

Some fluorescent dyes bind to compounds found in all cells. For example, **acridine orange** binds DNA and is often used to determine the number of total microorganisms in a sample. However, dyes that bind DNA do not discriminate between cells that are alive, or **viable**, and those that are dead. To detect only viable cells, other dyes can be employed. For example, the dye CTC is made fluorescent by cellular proteins involved in the process of respiration. Consequently, CTC only fluoresces when it is bound to viable cells. ■ respiration, p. 140

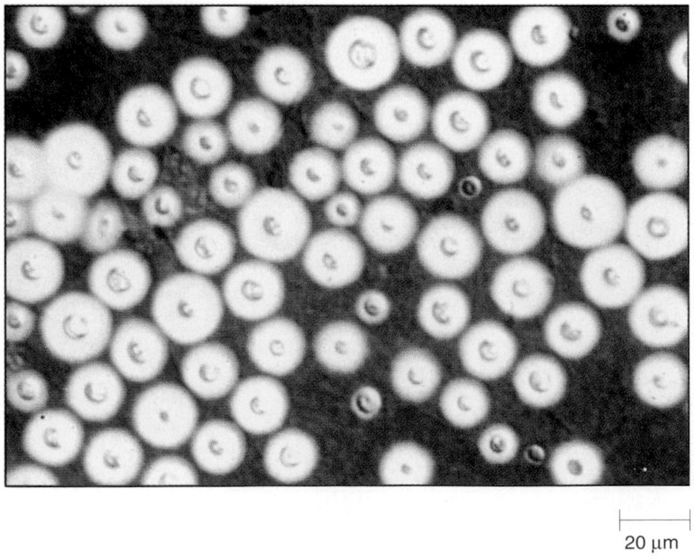

20 μm

Figure 3.16 Capsule Stain Capsules stain poorly, and so they stand out against the India ink-stained background as a halo around the organism. This photomicrograph shows an encapsulated yeast.

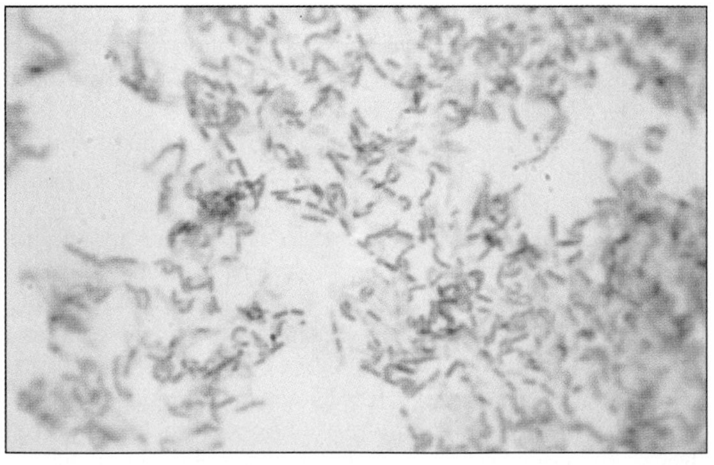

Figure 3.17 Endospore Stain Endospores retain the green primary stain, malachite green. Counterstaining with safranin imparts a red color to other cells.

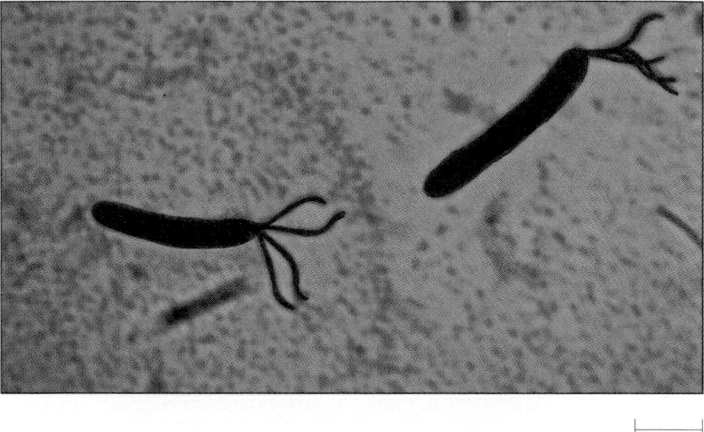

Figure 3.18 Flagella Stain The staining agent adheres to and coats the flagella. This increases their diameter so they can be seen with the light microscope.

Other fluorescent dyes bind to compounds specific to only certain types of cells. For example, **calcofluor white** binds to a component that characterizes the cell walls of fungi, causing those cells to fluoresce bright blue. The fluorescent dyes **auramine** and **rhodamine** bind to a compound found in the cell walls of members of the genus *Mycobacterium*. These two dyes can be used in a staining procedure analogous to the acid-fast stain; cells of *Mycobacterium* will emit a bright yellow or orange fluorescence.

Immunofluorescence

Immunofluorescence is a technique used to tag specific proteins of interest with a fluorescent compound. By tagging a protein that is unique to a given organism, immunofluorescence can be used to detect that specific microbe in a sample containing numerous different bacteria. Immunofluorescence employs an **antibody** to deliver the fluorescent tag. An antibody is produced by the immune system in response to a foreign compound, usually a protein; it binds with exquisite specificity to that compound. Immunofluorescence exploits the natural function of antibodies. This important part of the body's immune response will be discussed in chapter 16.

MICROCHECK 3.2

A number of different dyes can be used to stain cells so they stand out against the unstained background. The Gram stain is by far the most widely used differential stain. The acid-fast stain detects species of *Mycobacterium*. Specific dyes and techniques can be used to observe cell structures such as capsules, endospores, and flagella. Fluorescent dyes and tags can be used to observe total cells, a subset of cells, or cells that have certain proteins on their surface.

- What are the functions of a primary stain and a counterstain?
- Describe one error in the staining procedure that would result in a Gram-positive bacterium appearing pink.
- What color would a Gram-negative bacterium be in an acid-fast stain?

Morphology of Prokaryotic Cells

Prokaryotes come in a variety of simple shapes and often group together, forming chains or clusters. These characteristics, particularly the shape, are used to describe, classify, and identify these microorganisms.

Shapes

Most common bacteria are one of two shapes: spherical, called a **coccus** (plural: cocci); and cylindrical, called a **rod (figure 3.20)**. A rod-shaped bacterium is sometimes called a **bacillus** (plural:

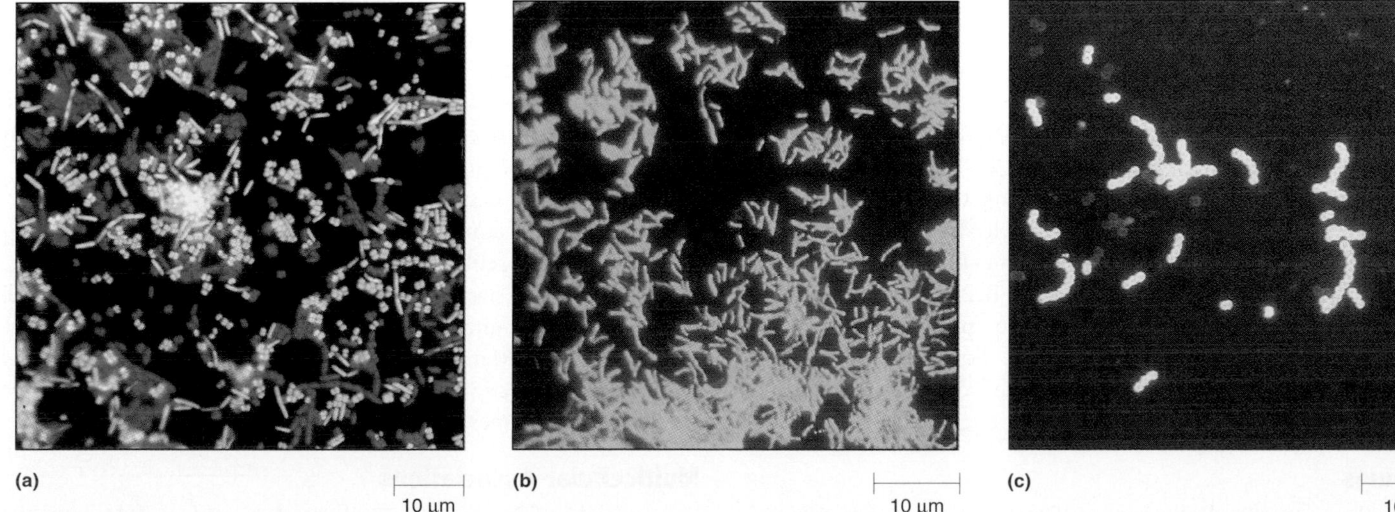

Figure 3.19 Fluorescent Dyes and Tags (a) Dyes that cause live cells to fluoresce green and dead ones red; **(b)** Auramine is used to stain *Mycobacterium* species in a modification of the acid-fast technique; **(c)** Fluorescent antibodies tag specific molecules. In this case, the antibody binds to a molecule that is unique to *Streptococcus pyogenes*.

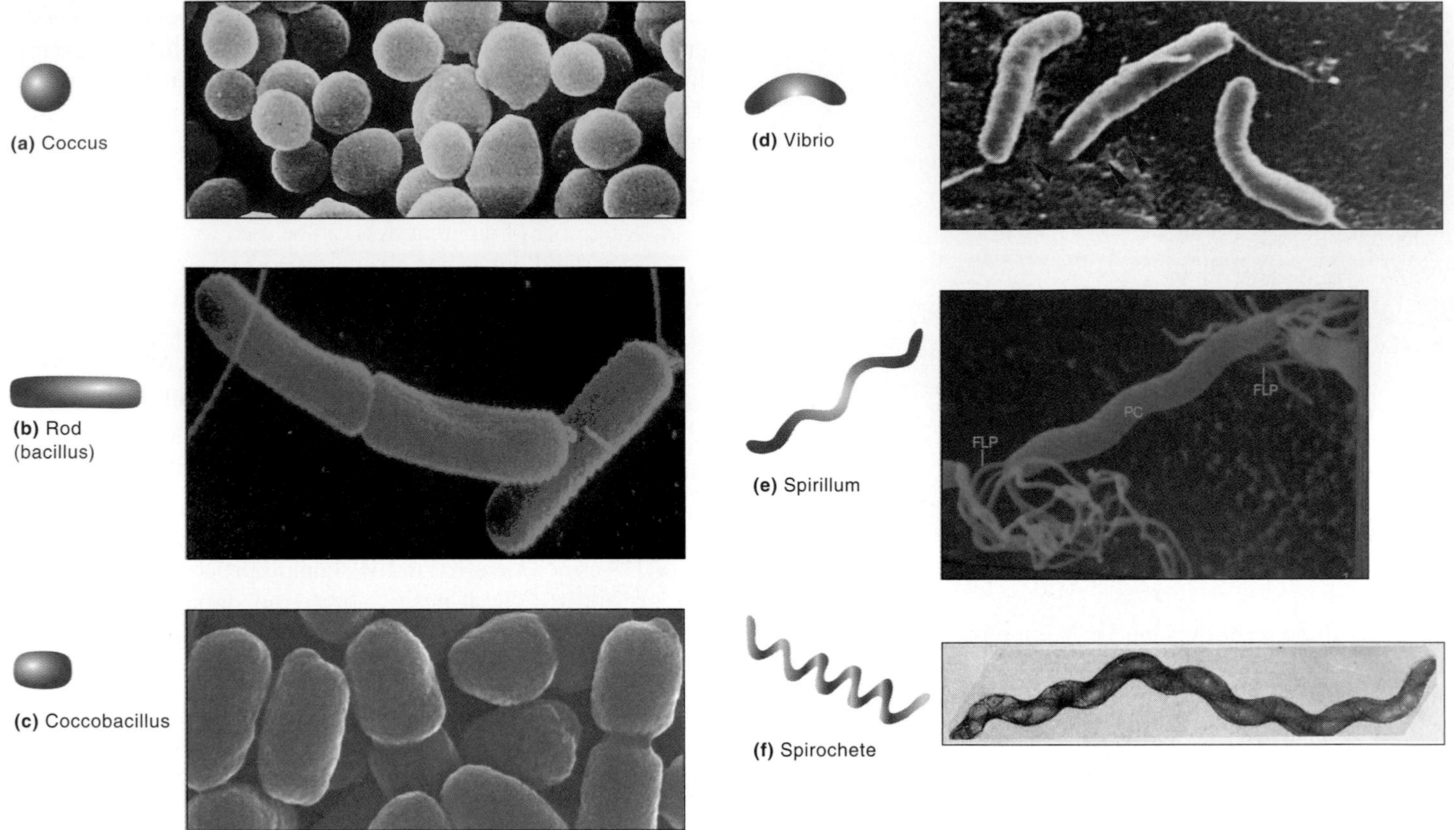

Figure 3.20 **Typical Shapes of Common Bacteria** **(a)** Coccus; **(b)** rod; **(c)** coccobacillus; **(d)** vibrio; **(e)** spirillum; **(f)** spirochete. Electron micrographs.

bacilli). Note, however, that the descriptive term bacillus is easily confused with *Bacillus*, the name of a genus. While members of the genus *Bacillus* are rod-shaped, so are many other bacteria, including *Escherichia coli*.

A variety of other cell shapes also occur. A rod-shaped bacterium that is so short that it can easily be mistaken for a coccus is often called a **coccobacillus**. A short curved rod is called a **vibrio** (plural: vibrios), whereas a curved rod that is long enough to form spirals is called a **spirillum** (plural: spirilla). A long helical cell with a flexible cell wall and a unique mechanism of motility is a **spirochete**. Bacteria that characteristically vary in their shape are called **pleomorphic** (*pleo* meaning many and *morphic* referring to shape).

Perhaps the greatest diversity in cell shapes is found in aquatic environments, where maximizing their surface area helps microbes absorb nutrients (**figure 3.21**). Some aquatic bacteria have extensions on their surface, **prosthecae**, which give the organisms a starlike appearance. Square, tilelike cells have been found in the salty pools of the Sinai Peninsula in Egypt. ■ **prosthecate bacteria, p. 286**

Groupings

Most prokaryotes divide by **binary fission**, a process in which one cell divides into two identical cells. Cells adhering to one another following division form a characteristic arrangement that depends on the planes in which the organisms divide. This is seen especially in the cocci because they may divide in more than one plane (**figure 3.22**). Cells that divide in one plane may form chains of varying length. Cocci that typically occur in pairs are routinely called **diplococci.** An important clue in the identification of *Neisseria gonorrhoeae* is its characteristic diplococcus arrangement. Some cocci form long chains; this characteristic is typical of some, but not all, members of the genus *Streptococcus.*

Cocci that divide in two or three planes perpendicular to one another form cubical packets. Members of the genus *Sarcina* form such packets. Cocci that divide in several planes at random may form clusters. Species of *Staphylococcus* typically form characteristic grapelike clusters.

It should be noted that the various groupings are sometimes described with a Latin word. For example, chains of cocci may be called streptococci, cubical packets may be called sarcinae, and clusters may be called staphylococci. Unfortunately, these same words have also been assigned as genus names, which are a formal taxonomic category. Many microbiologists intentionally avoid using these words as descriptive terms because they are so easily confused with the genus names *Streptococcus*, *Sarcina*, and *Staphylococcus.*

Multicellular Associations

Some types of bacteria typically live as multicellular associations. For example, members of a group of bacteria called myxobacteria glide over moist surfaces together, forming a swarm of cells that moves as a pack. These cells release extracellular enzymes, which enables the pack to degrade organic material, including other bacterial cells. When water or nutrients are depleted, cells aggregate to form a structure called a **fruiting body**, which is visible to the naked eye (see figure 11.9). Within this structure, cells of myxobacteria form a dormant resting stage. ■ **myxobacteria, p. 284**

Many types of bacteria live on surfaces in associations called biofilms. Cells within these aggregates alter their activities when a critical number of cells are present. They sense the density of cells within their population by a mechanism called quorum sensing. ■ **biofilms, p. 108** ■ **quorum sensing, p. 184**

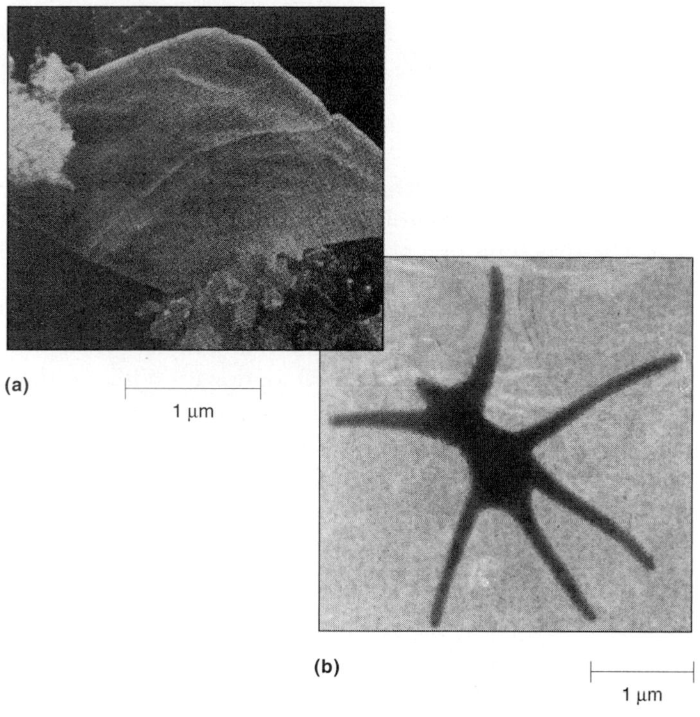

Figure 3.21 Diverse Shapes of Aquatic Bacteria **(a)** Square, tilelike bacterial cell. **(b)** *Ancalomicrobium*, an example of a prosthecate bacterium. Note that the cytoplasmic membrane and the cytoplasm are a part of each arm.

MICROCHECK 3.3

Most common prokaryotes are cocci or rods, but as a group, they come in a variety of shapes and sizes. Cells may form characteristic arrangements such as chains or clusters. Some types of bacteria form multicellular associations.

- Which environmental habitat has the greatest diversity of bacterial shapes?
- What causes some bacteria to form characteristic cell arrangements?

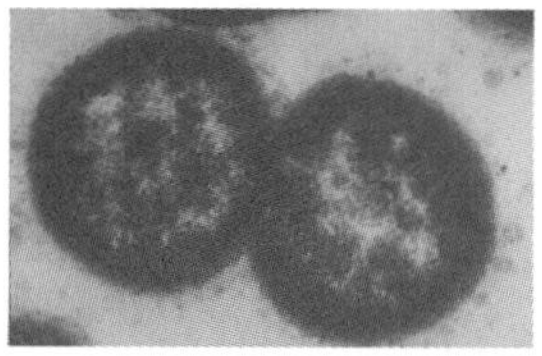

(a) Chains

Cell divides in one plane

Diplococcus

Chain of cocci

(b) Packets

Cell divides in two or more planes perpendicular to one another

Packet

(c) Clusters

Cell divides in several planes at random

Cluster

Figure 3.22 Typical Cell Groupings The planes in which cells divide determine the arrangement of the cells. These characteristic arrangements can provide important clues in the identification of certain bacteria: **(a)** chains; **(b)** packets; **(c)** clusters.

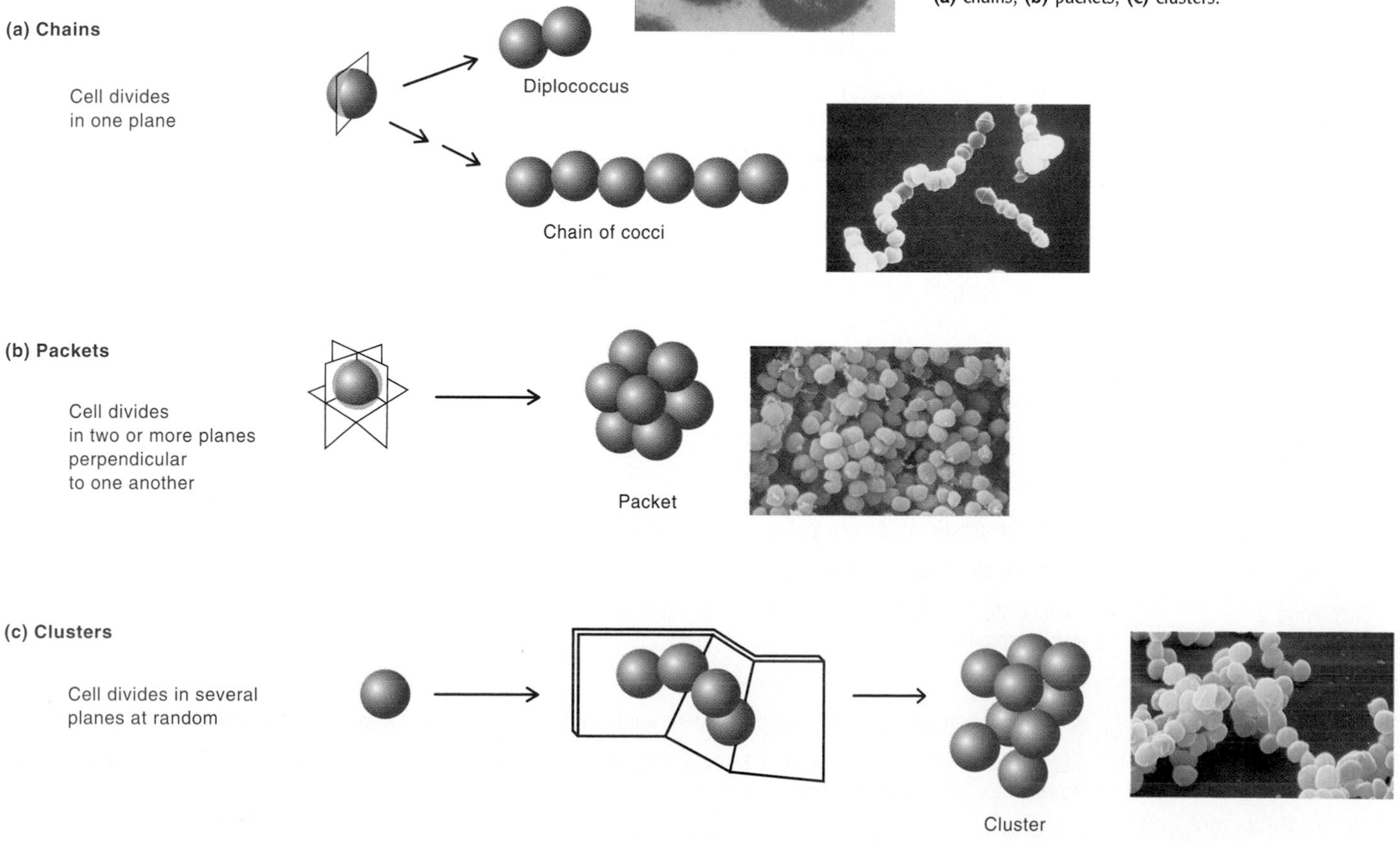

THE STRUCTURE OF THE PROKARYOTIC CELL

The overall structure of the prokaryotic cell is deceptively simple (**figure 3.23**). The **cytoplasmic membrane** surrounds the cell, acting as a barrier between the external environment and the cell contents, the **cytoplasm**. This membrane permits the passage of only certain molecules into and out of the cell. Enclosing the cytoplasmic membrane is the **cell wall**, a rigid barrier that functions as a tight corset to keep the cell contents from bursting out. Cloaking the wall may be additional layers, some of which serve to protect the cell from predators and environmental assaults. The cell may also be adorned with appendages giving it useful traits, including motility and the ability to adhere to select surfaces. Within all of these layers are the internal structures such as DNA, ribosomes, and storage granules, all of which contribute to essential cell processes.

The structures of a prokaryote, together, enable the cell to survive and multiply in a given environment—the very definition of evolutionary success. Some structures are essential for survival and, as such, are common among all prokaryotic cells; others might be considered optional. Without these "optional" structures, the cell might be able to exist in the protected confines of a laboratory, but it may not be suited for the competitive surroundings of the outside world. The characteristics of typical structures of prokaryotic cells are summarized in **table 3.3**.

Table 3.3 A Summary of Prokaryotic Cell Structures

Structure	Characteristics
Extracellular	
Filamentous appendages	Composed of subunits of proteins that form a helical chain.
Flagella	Provides the most common mechanism of motility.
Pili	Different types of pili have different functions. The common types, often called fimbriae, enable cells to adhere to surfaces. A few types mediate twitching or gliding motility. Sex pili are involved in a mechanism of DNA transfer.
Surface layers	
Glycocalyx	Layer outside the cell wall, usually made of polysaccharide.
Capsule	Distinct and gelatinous. Enables bacteria to adhere to specific surfaces; allows some organisms to thwart innate defense systems and thus cause disease.
Slime layer	Diffuse and irregular. Enables bacteria to adhere to specific surfaces.
Sheath	Protect aquatic bacteria from disruption by turbulent waters.
Cell wall	Peptidoglycan is the molecule common to all bacterial cell walls. Provides rigidity to prevent the cell from lysing.
Gram-positive	Thick layer of peptidoglycan that contains teichoic acids and lipoteichoic acids.
Gram-negative	Thin layer of peptidoglycan surrounded by an outer membrane. The outer leaflet of the outer membrane is lipopolysaccharide.
Cell Boundary	
Cytoplasmic membrane	Phospholipid bilayer embedded with proteins. A barrier between the cytoplasm and the outside environment. Also functions as a discriminating conduit between the cell and its surroundings.
Intracellular	
DNA	Contains the genetic information of the cell.
Chromosomal	Carries the genetic information that is essential to a cell. Typically a single, circular, double-stranded DNA molecule.
Plasmid	Carries genetic information that may be advantageous to a cell in certain situations.
Endospores	A type of dormant cell that is extraordinarily resistant to damaging conditions including heat, desiccation, ultraviolet light, and toxic chemicals.
Gas vesicles	Small, rigid structure that provides buoyancy to a cell.
Granules	Accumulations of high molecular weight polymers, which are synthesized from a nutrient that a cell has in relative excess.
Ribosomes	Intimately involved in protein synthesis. Two subunits, 30S and 50S, join to form the 70S ribosome, which serves as the structure that facilitates the joining of amino acids.

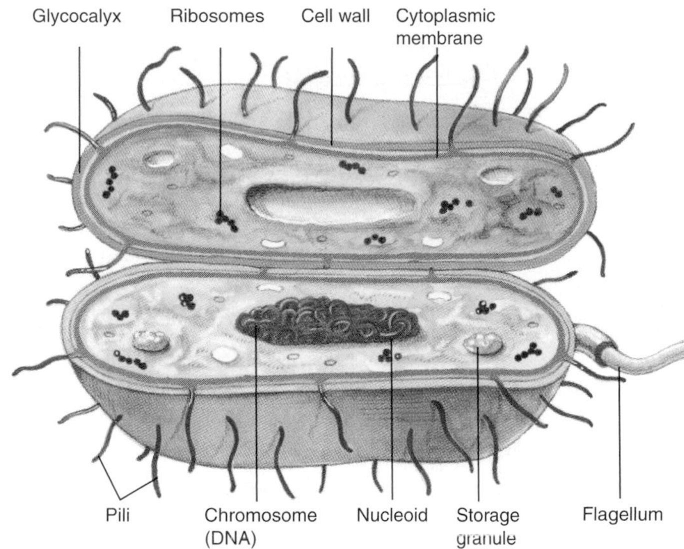

Figure 3.23 **Typical Prokaryotic Cell** A representation of typical structures within and outside a bacterial cell.

The components of the prokaryotic cell have important consequences for human health. Recall that structures and processes unique to bacteria are potential targets for selective toxicity. By interfering with these attributes, bacteria can be killed or their growth inhibited without harming the human host. More importantly, the defense systems of our bodies have evolved to recognize specific bacterial surface components as the sign of an "invader." For example, certain cell wall components shared by nearly all bacteria are found nowhere else in nature. Our innate immune defenses—such as our phagocytic cells—are poised to mount an attack against them. When these molecules enter our body, our defense systems respond rapidly, generally eliminating the invader before it has a chance to multiply. It is because of this swift response that the vast majority of bacteria cannot cause disease in a healthy person. Those that do are able to disguise their surface components or otherwise interfere with our immune defenses. This complex interaction between our immune defenses and microbes will be discussed in detail in later chapters. ■ phagocytes, p. 372

Cell structures are also important in identifying bacteria. Certain components are characteristic of select groups of microbes and can be used as identifying markers.

The Cytoplasmic Membrane

The **cytoplasmic membrane** is a delicate thin fluid structure that surrounds the cytoplasm and defines the boundary of the cell. It serves as an important semipermeable barrier between the cell and its external environment. Although the membrane's chemical composition primarily allows only water, gases, and some small hydrophobic molecules to pass through freely, there are specific proteins embedded within the membrane that act as selective gates. These permit desirable nutrients to enter the cell, and waste products to exit. Other proteins within the membrane serve as sensors of environmental conditions. Thus, while the cytoplasmic membrane acts as a barrier, it also functions as an effective and highly discriminating conduit between the cell and its surroundings.

Structure and Chemistry of the Cytoplasmic Membrane

The structure of the cytoplasmic membrane is typical of other biological membranes—a lipid bilayer embedded with proteins **(figure 3.24)**. The bilayer consists of two opposing **leaflets** composed of phospholipids. At one end of each phospholipid molecule are two fatty acids chains, which act as **hydrophobic tails**. At the other end, a combination of glycerol, a phosphate group, and other polar molecules functions as a **hydrophilic**

Figure 3.24 **The Structure of the Cytoplasmic Membrane** Two opposing leaflets make up the phospholipid bilayer. Embedded within it are a variety of different proteins, some of which span the membrane.

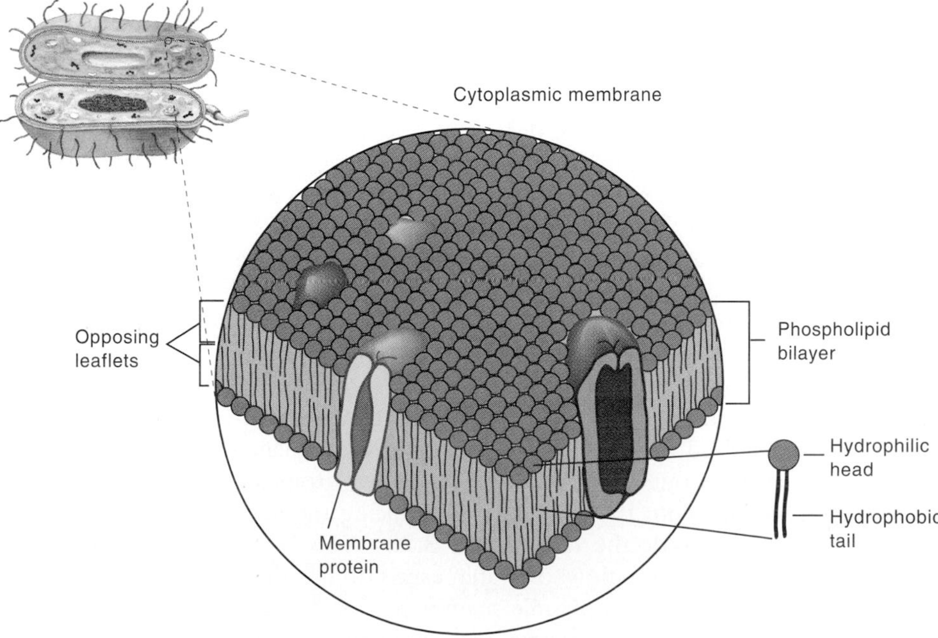

head. The phospholipid molecules are arranged in each leaflet of the bilayer so that their hydrophobic tails face in, toward the other leaflet. Their hydrophilic heads face outward. As a consequence, the inside of the bilayer is water insoluble whereas the two surfaces, which interface the internal and external environments, interact freely with aqueous solutions. ■ **phospholipids, p. 38** ■ **hydrophobic, p 28** ■ **hydrophilic, p. 28**

The proteins embedded in the bilayer are quite diverse; more than 200 different **membrane proteins** have been found in *E. coli*. Many function as **receptors**, binding to highly specific molecules in the environment. In turn, this provides a mechanism for the cell to sense and adjust to its surroundings. Proteins are not static within the fluid bilayer; rather, they are constantly changing position. Such movement is necessary for the important functions that the membrane performs. This structure, with its resulting dynamic nature, is called the **fluid mosaic model**.

Members of the Bacteria and Archaea have the same general structure of their cytoplasmic membranes, but the lipid composition is distinctly different. The side chains of the membrane lipids of Archaea are connected to glycerol through an ether linkage instead of an ester linkage. In addition, the side chains are hydrocarbons rather than fatty acids. These differences represent important distinguishing characteristics between these two domains of prokaryotes. ■ **ester bond, p. 37** ■ **hydrocarbon, p. 37**

Permeability of the Cytoplasmic Membrane

The cytoplasmic membrane is **selectively permeable**; relatively few types of molecules can pass through freely. These move through the membrane by a process called **simple diffusion**, whereas other molecules must be transported across it by very specific mechanisms. The transport mechanisms, which generally require an expenditure of energy, will be discussed in detail later.

Simple Diffusion

Simple diffusion is the process by which some molecules move freely into and out of the cell. Water, small hydrophobic molecules, and gases such as oxygen and carbon dioxide are among the few compounds that move through the cytoplasmic membrane by simple diffusion. The speed and direction of diffusion depend on the relative concentration of molecules on each side of the membrane. The greater the difference in concentration, the higher the rate of diffusion. The molecules continue to pass through at a diminishing rate until their concentration is balanced on both sides of the membrane.

The ability of water to move freely through the membrane has important biological consequences. The cytoplasm of a cell is a very concentrated solution of inorganic salts, sugars, amino acids, and various other molecules. However, the environments in which prokaryotes normally grow contain only small amounts of some salts and other small molecules. Since the concentration of dissolved molecules, or **solute**, tends to equalize on the inside and the outside of the cell, water flows from the surrounding medium into the cell, thereby reducing the concentration of solute inside the cell (**figure 3.25**). This is the process of **osmosis**. This inflow of water exerts tremendous **osmotic pressure** on the cytoplasmic membrane, much more than it generally can resist. However, the rigid cell wall that envelops the membrane generally withstands such high pressure. The cytoplasmic membrane is forced up against the wall but cannot balloon further. Damage to the cell wall weakens the structure, and consequently, cells may burst or **lyse**.

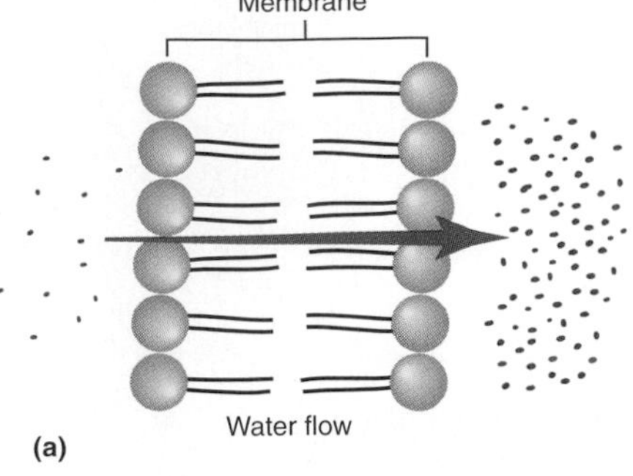

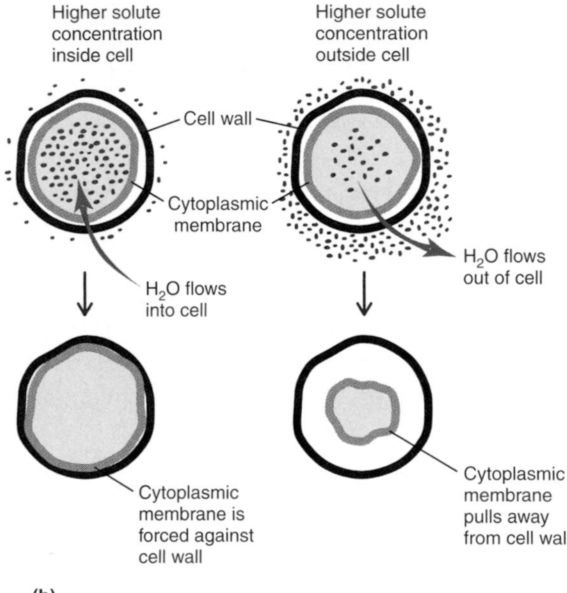

Figure 3.25 Osmosis (a) Water flows across a membrane toward the side that has the highest concentration of molecules and ions, thereby equalizing the concentrations on both sides. **(b)** The effect of osmosis on cells.

The Role of the Cytoplasmic Membrane in Energy Transformation

The cytoplasmic membrane of a prokaryotic cell plays an indispensable role in its ability to convert energy into a usable form. This is an important distinction between prokaryotic and eukaryotic cells; in eukaryotic cells energy is transformed in membrane-bound organelles, which will be discussed later in this chapter.

Prokaryotes that transform energy by either respiration or photosynthesis have a series of compounds, the **electron transport chain,** embedded in their membrane. These sequentially transfer electrons and, in the process, eject protons from the cell. The details of these processes will be explained in chapter 6. The expulsion of protons by the electron transport chain results in the formation of a proton gradient across the cell membrane. Positively charged protons are concentrated immediately outside the membrane, whereas negatively charged hydroxyl ions accumulate directly inside the membrane (**figure 3.26**). This separation of

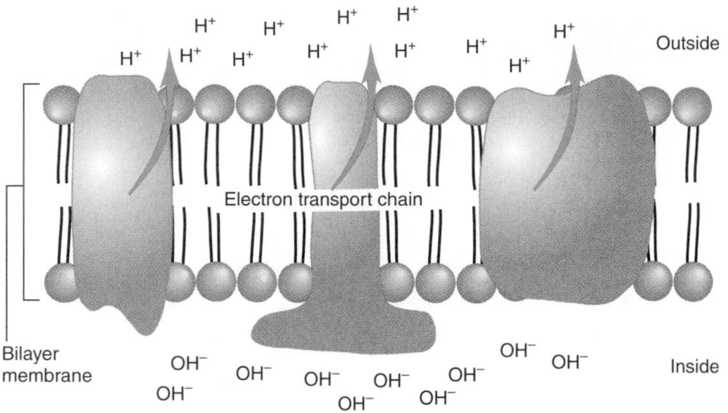

Figure 3.26 Proton Motive Force The series of compounds within the membrane called the electron transport chain ejects protons from the cell. This creates an electrochemical gradient, which is a form of energy called proton motive force. The controlled flow of protons back into the cell releases energy. This is used to drive flagella and certain transport systems. It can also be used to synthesize ATP.

charged ions creates an **electrochemical gradient** across the membrane; inherent in it is a form of energy, called **proton motive force**. This is analogous to the energy stored in a battery. ■ respiration, p. 140 ■ photosynthesis, p. 158 ■ electron transport chain, p. 150

The energy of a proton motive force can be harvested by mechanisms that allow protons to move back across the membrane and into the cell. This is used directly to drive certain cellular processes, including some transport mechanisms that carry small molecules across the membrane and some forms of motility. It is also used to synthesize ATP.

MICROCHECK 3.4

The cytoplasmic membrane is a phospholipid bilayer embedded with a variety of different proteins. It serves as a barrier between the cell and the surrounding environment, allowing relatively few molecules to pass through freely. The electron transport chain within the membrane expels protons, generating proton motive force.

- Explain the fluid mosaic model.
- Name three molecules that can pass freely through the lipid bilayer.
- Why is the word fluid in fluid mosaic model an appropriate term?

Directed Movement of Molecules Across the Cytoplasmic Membrane

Nearly all molecules that enter or exit a cell must cross the otherwise impermeable cytoplasmic membrane through proteins that function as selective gates. Mechanisms that allow nutrients and other small molecules to enter the cell are called **transport** systems. It is now recognized that some of these same mechanisms are used to expel wastes and compounds that are otherwise deleterious to the cell. This function of transport systems is sometimes called **efflux.** Efflux is of considerable medical importance, because it allows certain bacteria to oust antimicrobial medications and disinfectants that are meant to destroy them. By doing so, these bacteria can resist those antimicrobial chemicals. ■ antimicrobial drugs, p. 497 ■ disinfectants, p. 114

Cells actively move certain proteins they synthesize out of the cell—a process called **secretion**. Some of these secreted proteins make up structures such as the flagella, which are the appendages used for motility. Others are enzymes secreted to break down substances such as oligosaccharides and peptides, which would otherwise be too large to transport into the cell.

Transport Systems

The mechanisms used to transport molecules across the membrane employ highly specific proteins. These proteins, sometimes called **transport proteins**, **permeases**, or **carriers**, span the membrane. One end projects into the surrounding environment and the other into the cell. The interaction between the transport protein and the molecule it carries is highly specific. Consequently, a single carrier generally transports only a specific type of molecule. As a carrier transports a molecule, its shape changes, facilitating the passage of the molecule **(figure 3.27)**.

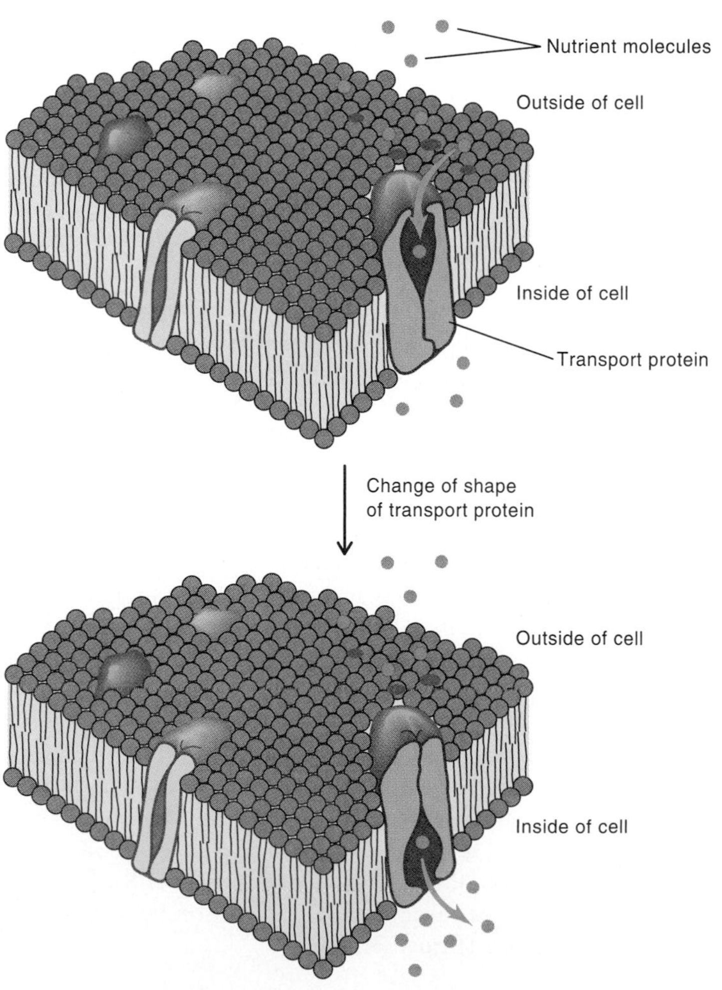

Figure 3.27 Transport Protein A transport protein changes its shape to facilitate passage of a compound through the cytoplasmic membrane.

Table 3.4 A Summary of Transport Mechanisms Used by Prokaryotic Cells

Transport Mechanism	Characteristics
Facilitated Diffusion	Rarely used by bacteria. Exploits a concentration gradient to move molecules; can only eliminate a gradient, not create one. No energy is expended.
Active Transport	Energy is expended to accumulate molecules against a concentration gradient.
Major facilitator superfamily	In bacteria, the proton motive force drives these transporters.
Symporter	Two different substances are transported at the same time and in the same direction.
Antiporter	One substance is transported in as a different one exits.
Uniporter	One substance is transported.
ABC transporters	ATP is used during transport. Extracellular binding proteins deliver a molecule to the transporter.
Group Translocation	The transported molecule is chemically altered as it passes into the cell. Consequently, uptake of that molecule does not affect its concentration gradient.

Cells do not produce all of the transport proteins all of the time. Instead, sophisticated regulatory mechanisms sense the presence of specific compounds and, based on the concentration of those compounds and the availability of other molecules, determine which of the various transport proteins are synthesized. The mechanisms of transport are summarized in **table 3.4**. ■ regulation of gene expression, p. 179

Facilitated Diffusion

Facilitated diffusion moves impermeable compounds from one side of the membrane to the other by exploiting a concentration gradient. The molecules are transported across until their concentration is balanced on both sides of the membrane. This mechanism can only eliminate a difference in concentration; it cannot create one. As cells typically grow in relatively nutrient-poor environments, they generally cannot rely on this **passive transport** mechanism to acquire nutrients. Glycerol is one of the few nutrients known to enter *E. coli* cells this way.

Active Transport

Active transport mechanisms can accumulate compounds against a concentration gradient, so that it may be much higher on one side of the membrane than the other. This attribute requires an expenditure of energy. There are two primary mechanisms of active transport, each utilizing a different form of energy.

Transport Systems that Use Proton Motive Force Many bacterial transport systems can accumulate or extrude small molecules and ions using the energy of a proton motive force. These systems are part of a large group of transporters, collectively known as the **major facilitator superfamily (MFS)**, found in prokaryotes as well as eukaryotes. They function by one of three general mechanisms (**figure 3.28**):

- **Symporters** transport two different substances at the same time and in the same direction. The passage of one facilitates the transport of the other. In bacteria, a proton usually facilitates the transfer of another substance. For example, a lactose molecule enters a cell with one proton.
- **Antiporters** exchange the location of one substance for another; in other words, one is transported in while a different one exits. As an example, sodium is transported out of the cell as a proton passes in. Efflux systems used by bacteria to expel antimicrobial drugs are often antiporters.
- **Uniporters** transport a single substance. In bacteria, this is either a cation coming in or an anion going out. Although a proton is not directly involved in the uniport function, in bacteria the ionic gradient of proton motive force plays an integral role. Potassium enters the cell via a uniporter. ■ cation, p. 24 ■ anion, p. 24

Transport Systems that Use ATP **ABC transport systems** require ATP as an energy source (ABC stands for ATP Binding-Cassette).

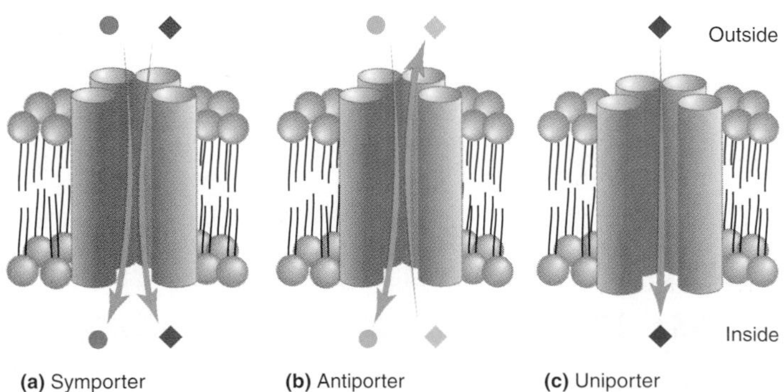

Figure 3.28 Active Transport by the Major Facilitator Superfamily Each of these transporters requires an electrochemical gradient (proton motive force) and can accumulate a compound within a cell. **(a)** A symporter transports two different substances at the same time and in the same direction. **(b)** An antiporter exchanges the location of one substance for another. **(c)** A uniporter transports a single substance.

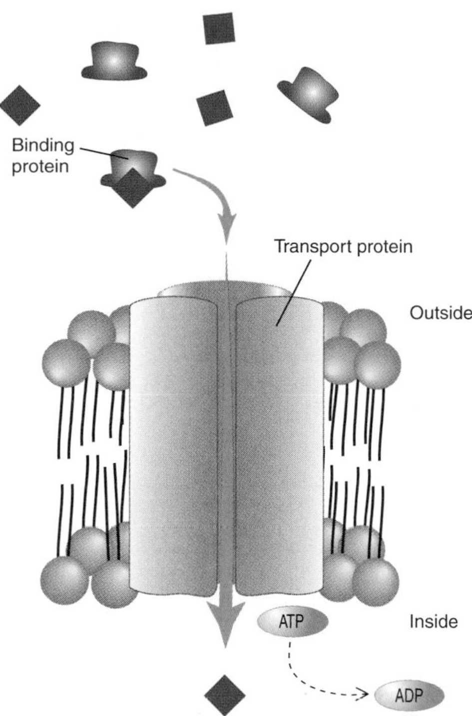

Figure 3.29 Active Transport by an ABC Transporter These transporters require energy in the form of ATP and can accumulate a compound within a cell. A binding protein that resides outside of the cytoplasmic membrane delivers a given molecule to a specific transport protein.

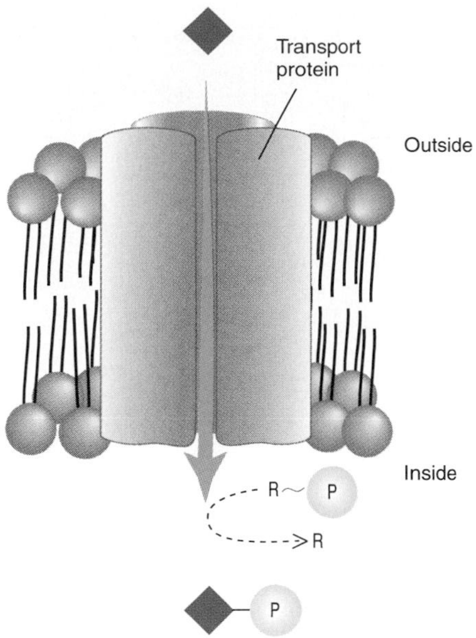

Figure 3.30 Group Translocation This process chemically alters a molecule during its passage through the cytoplasmic membrane. Consequently, uptake of that molecule does not alter the concentration gradient.

These systems are relatively elaborate, involving multiple protein components (**figure 3.29**). The ABC transport system utilizes a **binding protein** that resides immediately outside of the cytoplasmic membrane to deliver a given molecule to a specific transport complex within the membrane. Numerous different binding proteins exist for the various sugars and amino acids. Each binds tightly to its respective molecule and, therefore, the binding protein can effectively scavenge even low concentrations of a molecule and bring them into the cell. The sugar maltose is an example of a molecule that is transported by an ABC transport system.

Group Translocation

Group translocation is a transport process that chemically alters a molecule during its passage through the cytoplasmic membrane (**figure 3.30**). Consequently, uptake of that molecule does not alter the concentration gradient. As an example, glucose and several other sugars are phosphorylated during their transport into the cell by the **phosphotransferase system**. The energy expended to phosphorylate the sugar can be regained when that sugar is later broken down to provide energy.

Secretion

The **general secretory pathway** is the primary mechanism used to secrete proteins. It recognizes appropriate proteins by their characteristic sequence of amino acids that make up the amino terminal end. This **signal sequence** consists of 20 or so hydrophobic amino acids and functions as a tag, which is specifically removed by an enzyme during the process of secretion.

The general secretory pathway requires at least 11 different proteins and utilizes ATP to drive the process. However, the precise mechanism by which the proteins are translocated across the membrane is still poorly understood.

MICROCHECK 3.5

Facilitated transport does not require energy but is rarely used by bacteria. Active transport via the major facilitator superfamily uses proton motive force. Active transport via an ABC transporter uses ATP as an energy source. Group translocation chemically modifies a molecule as it enters the cell. Proteins that have a leader sequence are secreted from a cell by the general secretory pathway.

- Explain why antiporters are medically important.
- Describe the role of binding proteins in an ABC transport system.
- How could it be argued that bacterial uniporters are a form of facilitated diffusion?

Cell Wall

The cell wall of most common prokaryotes is a rigid structure, which determines the shape of the organism. A primary function of the wall is to hold the cell together and prevent it from bursting. If the cell wall is somehow breached, undamaged parts maintain their original shape (**figure 3.31**). The cell wall is composed of unique structures and molecules, some of which are recognized by our immune system as the sign of an invader. Antimicrobial medications target some of the unique structures.

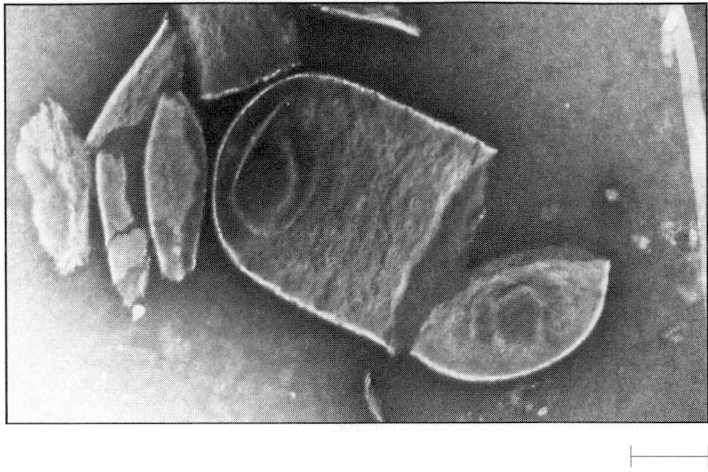

Figure 3.31 The Rigid Cell Wall Determines the Shape of the Bacterium Even though the cell has split apart, the cell wall maintains its original shape.

Peptidoglycan

Although the structure varies in Gram-positive and Gram-negative cells,the rigidity of bacterial cell walls is due to a layer of **peptidoglycan**, a macromolecule found only in bacteria. The basic structure of peptidoglycan is an alternating series of two major subunits, ***N*-acetylmuramic acid** (**NAM**) and ***N*-acetylglucosamine** (**NAG**). These subunits, which are related to glucose in their structure, are covalently joined to one another to form a **glycan chain (figure 3.32)**. This high molecular weight linear polymer serves as the backbone of the peptidoglycan molecule.

Attached to each of the NAM molecules is a string of four amino acids, a **tetrapeptide chain**, that plays an integral role in the structure of the peptidoglycan molecule. Cross-linkages can form between tetrapeptide chains, thus joining adjacent glycan chains to form a single, very large three-dimensional molecule. In Gram-negative bacteria, the tetrapeptides are joined directly. In Gram-positive bacteria, they are usually joined indirectly by a **peptide interbridge,** the composition of which may vary among species.

An assortment of only a few different amino acids make up the tetrapeptide chain. One of these, diaminopimelic acid, which is related to the amino acid lysine, is not found in any other place in nature. Some of the others are D-isomers, a form not found in proteins.

The Gram-Positive Cell Wall

A relatively thick layer of peptidoglycan characterizes the cell wall of Gram-positive bacteria **(figure 3.33)**. Depending on the species, this single component makes up 40% to 80% of the dry weight of the wall. As many as 30 layers, or sheets, of interconnected glycan chains make up the polymer. Regardless of its thickness, peptidoglycan is fully permeable to many substances including sugars, amino acids, and ions.

Teichoic Acids

A prominent component of the Gram-positive cell wall is a group of molecules called **teichoic acids** (from the Greek word *teichos,* meaning wall). These are chains of a common subunit, either ribitol-phosphate or glycerol-phosphate, to which various sugars and D-alanine are usually attached. They are attached to the peptidoglycan molecule through covalent bonds to *N*-acetylmuramic acid. Some, which are called **lipoteichoic acids**, are linked to the cytoplasmic membrane. Teichoic acids and lipoteichoic acids both stick out above the peptidoglycan layer and, because they are negatively charged, give the cell its negative polarity.

Lipoteichoic acids play an important role in our body's ability to recognize invading bacteria. Phagocytic cells within the body have molecules on their surface, which bind lipoteichoic acid. If Gram-positive bacteria gain access to the body, the phagocytic cells will recognize their presence and not only engulf them but also send chemical signals to alert other cells of the immune system.

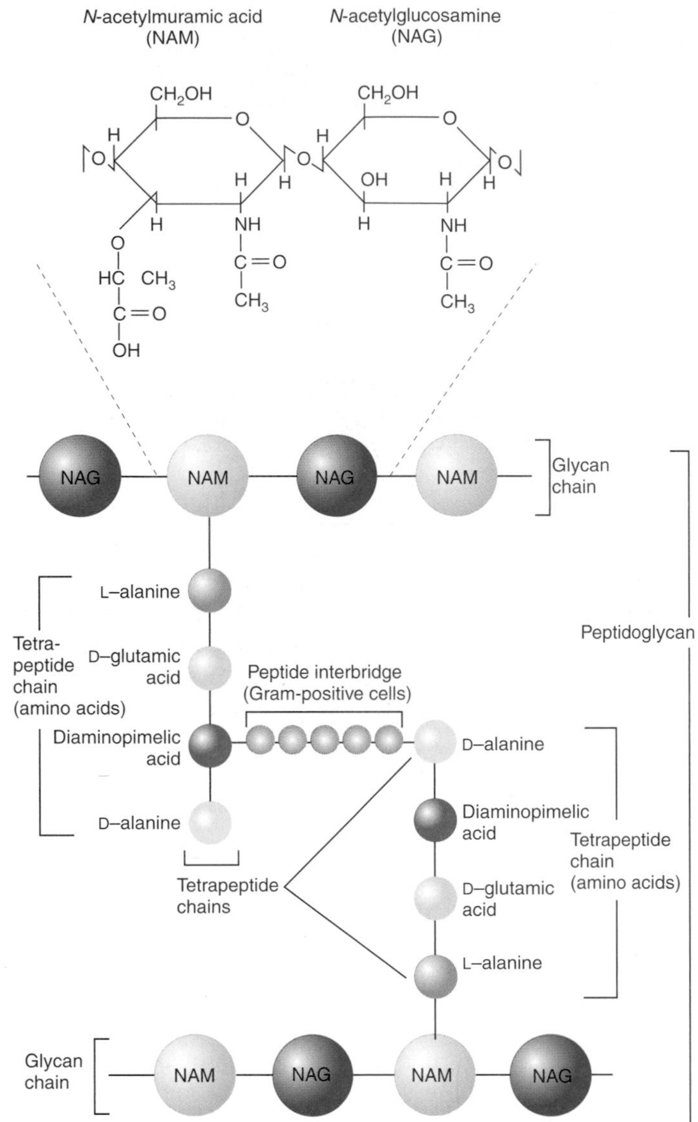

Figure 3.32 Components and Structure of Peptidoglycan Chemical structure of *N*-acetylglucosamine (NAG) and *N*-acetylmuramic acid (NAM); the ring structures of the two molecules are glucose. Glycan chains are composed of alternating subunits of NAG and NAM joined by covalent bonds. Adjacent glycan chains are cross-linked via their tetrapeptide chains to create peptidoglycan.

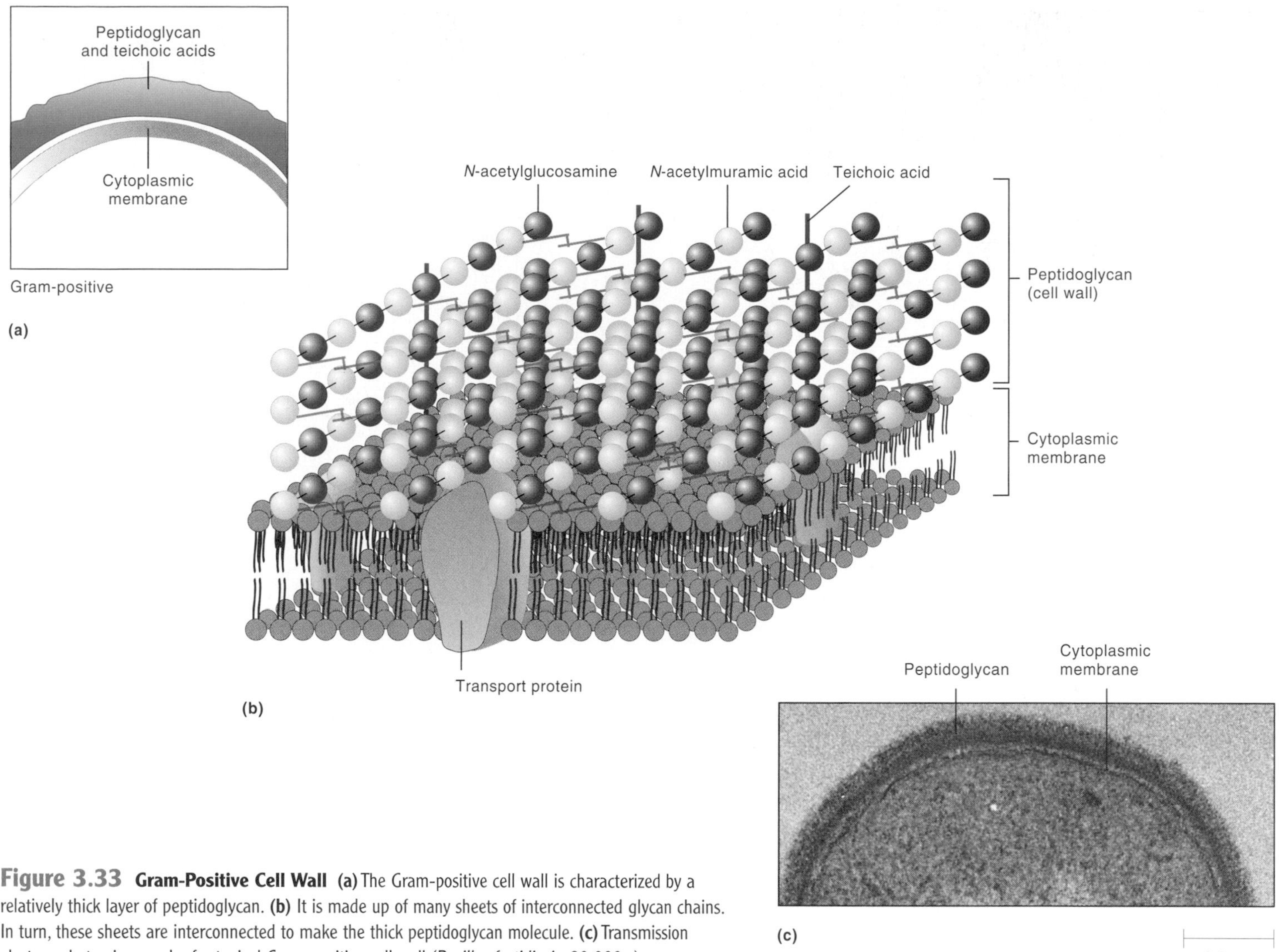

Figure 3.33 Gram-Positive Cell Wall **(a)** The Gram-positive cell wall is characterized by a relatively thick layer of peptidoglycan. **(b)** It is made up of many sheets of interconnected glycan chains. In turn, these sheets are interconnected to make the thick peptidoglycan molecule. **(c)** Transmission electron photomicrograph of a typical Gram-positive cell wall (*Bacillus fastidiosis*, 90,000×)

The Gram-Negative Cell Wall

The cell wall of Gram-negative bacteria is far more complex than that of Gram-positive organisms **(figure 3.34)**. It contains only a thin layer of peptidoglycan. Outside of that layer is the **outer membrane**, a unique lipid bilayer embedded with proteins. The peptidoglycan layer is sandwiched between the cytoplasmic membrane and the outer membrane.

Periplasm

The region between the cytoplasmic membrane and the outer membrane is filled with a gel-like fluid called **periplasm.** In Gram-negative bacteria, all secreted proteins are contained within the periplasm unless they are specifically translocated across the outer membrane as well. Thus, the periplasm is filled with proteins involved in a variety of cellular activities, including nutrient degradation and transport. For example, the enzymes that cells secrete to break down peptides and other molecules are found in the periplasm. Similarly, the binding proteins of the ABC transport systems are found there.

The Outer Membrane

The outer membrane is unlike any other membrane in nature. Its lipid bilayer structure is typical of other membranes, but the outside leaflet is made up of **lipopolysaccharides** rather than phospholipids. For this reason, the outer membrane is also called the **lipopolysaccharide layer** or **LPS**. The outer membrane is joined to peptidoglycan by means of lipoprotein molecules. ■ **lipoproteins, p. 38**

Like the cytoplasmic membrane, which in Gram-negative bacteria is sometimes called the **inner membrane**, the outer membrane serves as a barrier to the passage of most molecules. However, specialized proteins, **porins**, span the outer membrane. These proteins have a small channel that runs through them, allowing passage of certain small molecules and ions either into or out of the periplasm. Some porins are specific for the molecules that pass through them; others allow many different molecules to pass. The size of the porin channel partially determines the size of the molecule that can pass through it. In addition to the general secretory pathway already discussed,

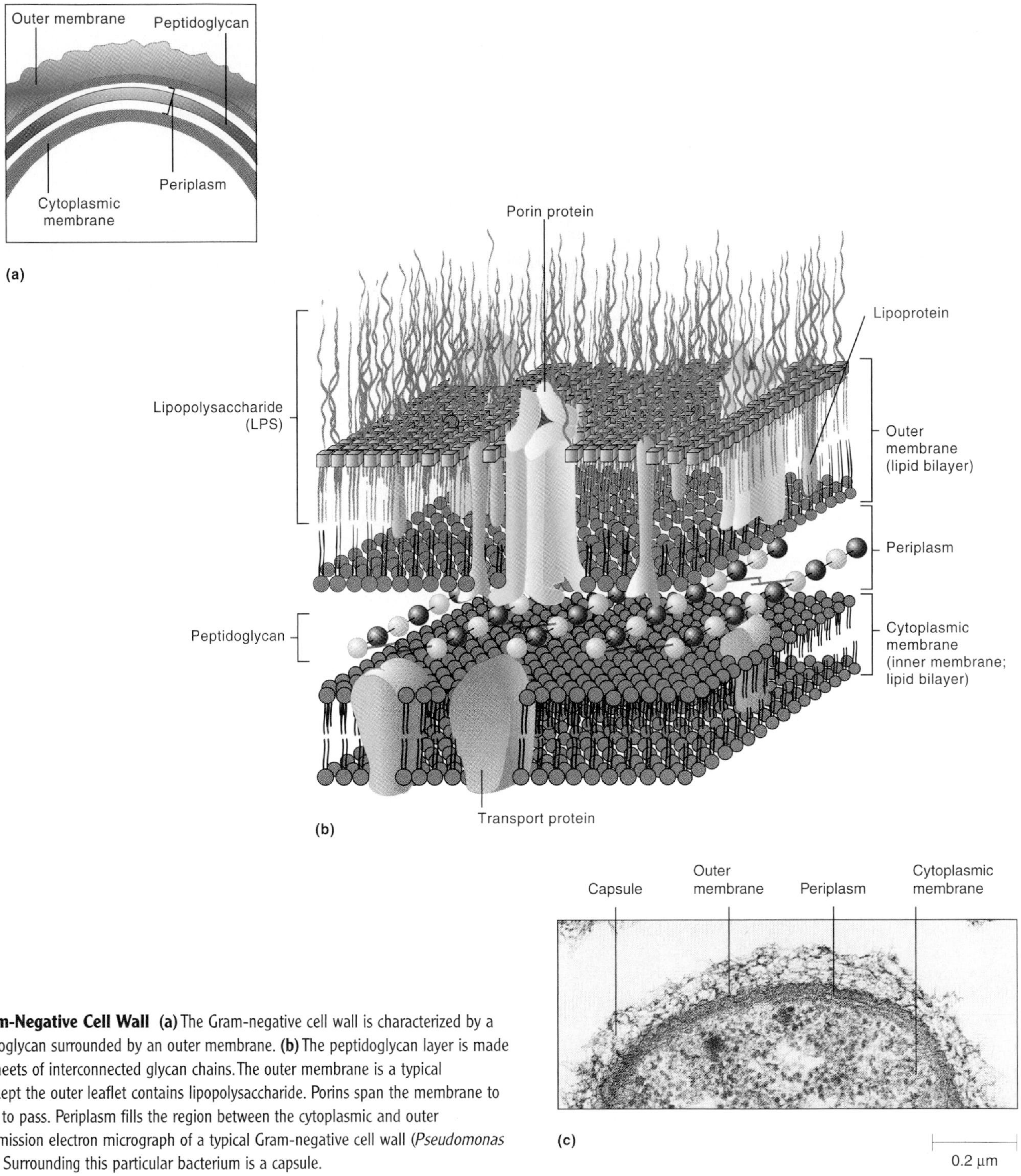

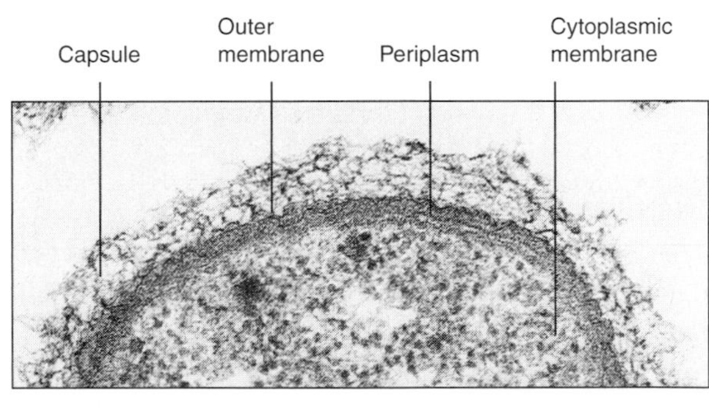

Figure 3.34 Gram-Negative Cell Wall (a) The Gram-negative cell wall is characterized by a very thin layer of peptidoglycan surrounded by an outer membrane. **(b)** The peptidoglycan layer is made up of only one or two sheets of interconnected glycan chains. The outer membrane is a typical phospholipid bilayer, except the outer leaflet contains lipopolysaccharide. Porins span the membrane to allow specific molecules to pass. Periplasm fills the region between the cytoplasmic and outer membranes. **(c)** A transmission electron micrograph of a typical Gram-negative cell wall (*Pseudomonas aeruginosa*, 120,000×). Surrounding this particular bacterium is a capsule.

Gram-negative bacteria have additional secretion systems to translocate proteins to the outside of the outer membrane. Some disease-causing bacteria use these secretion systems to transfer proteins not only outside of the bacterial cell, but also into mammalian cells.

The outer membrane functions as a protective barrier and excludes many toxic compounds. These include certain antimicrobial medications. This is one reason that Gram-negative bacteria are generally less sensitive to many such medications.

The Lipopolysaccharide Molecule

The lipopolysaccharide molecule is extremely important from a medical standpoint. When purified lipopolysaccharide is injected into an animal, it elicits symptoms characteristic of infections caused by live bacteria. The same symptoms occur regardless of the bacterial species. To reflect the fact that the toxic activity is an inherent part of the cell wall, it is called **endotoxin**. Because endotoxin cannot be destroyed with heat, even though such treatment kills the bacteria, manufacturers of intravenous

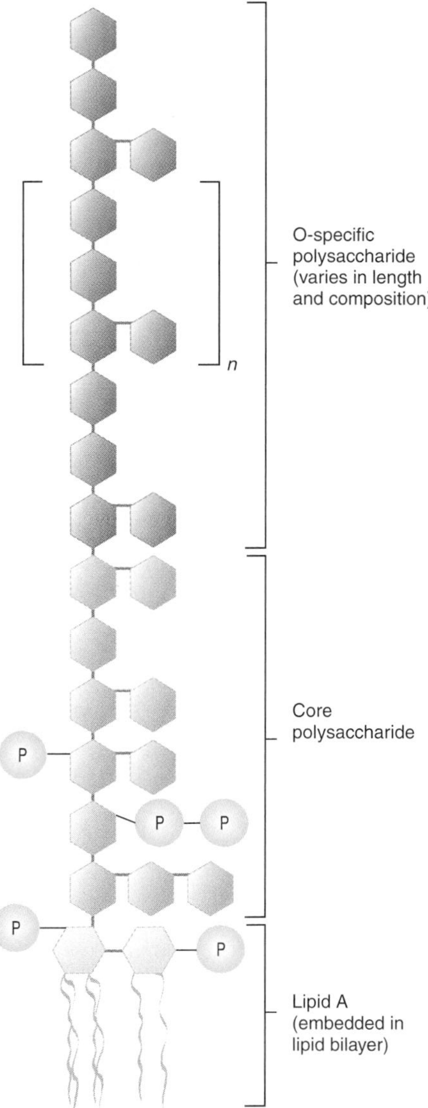

Figure 3.35 Chemical Structure of Lipopolysaccharide The Lipid A portion, which anchors the LPS molecule in the lipid bilayer, is responsible for the symptoms associated with endotoxin. The composition and length of the O-specific polysaccharide side chain varies among different species of bacteria.

solutions must take scrupulous precautions to ensure that their solutions do not become contaminated with bacteria, or even cell wall fragments. ■ endotoxin, p. 461

Two parts of the LPS molecule are notable for their medical significance **(figure 3.35)**. **Lipid A** is the portion that anchors the LPS molecule in the lipid bilayer. Its chemical make-up plays a significant role in our body's ability to recognize the presence of invading bacteria. When Lipid A is introduced into the body in small amounts, such as when microbes contaminate a small lesion, the defense system responds at an appropriate level to effectively eliminate the invader. If, however, large amounts of Lipid A are present, such as when Gram-negative bacteria are actively growing in the bloodstream, the magnitude of the response damages even our own cells. It is actually this response to Lipid A that is responsible for the symptoms associated with endotoxin. The chemical structure of Lipid A varies between species of bacteria, and some variations do not elicit as strong of a response as others.

The **O-specific polysaccharide side chain** is the portion of LPS that is directed away from the membrane, at the end opposite that of Lipid A. It is made up of a chain of sugar molecules, the number and composition of which varies among different species of bacteria. This characteristic can be exploited to identify certain species or strains. For example, the "O157" in *E. coli* O157:H7 refers to that particular strain's characteristic O-side chain. The structure of the O-side chain of some bacteria protects them from the otherwise destructive effects of certain host defenses.

Antibacterial Compounds that Target Peptidoglycan

Compounds that interfere with the synthesis of peptidoglycan or alter its structural integrity weaken the rigid molecule to a point where it is not strong enough to prevent the cell from bursting. These compounds include the antibiotic penicillin and the enzyme **lysozyme**, which is found in many body fluids including tears and saliva.

Penicillin

Penicillin is the most thoroughly studied of a group of antibiotics that interfere with peptidoglycan synthesis. Penicillin binds to proteins involved in cell wall synthesis and, subsequently, prevents the cross-linking of adjacent glycan chains. These proteins are called **penicillin-binding proteins**, a name that reflects their medical importance rather than their role in peptidoglycan synthesis. ■ penicillin, p. 502

Generally, but with notable exceptions, penicillin is far more effective against Gram-positive cells than Gram-negative cells. This is because the outer membrane of Gram-negative cells prevents the medication from reaching its site of action, the peptidoglycan layer. However, the structure of penicillin can be modified to create penicillin derivatives that can pass through porin channels. These drugs are effective against a range of Gram-negative bacteria. ■ penicillin-derivatives, p. 502

Lysozyme

Lysozyme breaks the bond that links the alternating *N*-acetylglucosamine and *N*-acetylmuramic acid molecules and thus destroys the structural integrity of the glycan chain, the backbone of the peptidoglycan molecule.

Lysozyme is sometimes used in the laboratory to remove the peptidoglycan layer from bacteria for experimental purposes. Removing that layer from a Gram-positive bacterium creates a **protoplast** that lacks a cell wall. In contrast, removing the peptidoglycan layer from a Gram-negative bacterium creates a **spheroplast**. Spheroplasts retain some portions of the outer membrane. Because they lack their rigid cell wall, protoplasts and spheroplasts both become spherical regardless of the original cell shape. Due to osmosis, they will burst unless maintained in a solution that has the same relative concentration of ions and small molecules as the cytoplasm.

Differences in Cell Wall Composition and the Gram Stain

Differences in the cell wall composition of Gram-positive and Gram-negative bacteria account for their staining characteristics. It is not the cell wall, however, but the inside of the cell that is stained by the crystal violet–iodine complex. The Gram-positive cell wall somehow retains the crystal violet–iodine complex even when subjected to the trauma of acetone-alcohol treatment, whereas the Gram-negative cell wall cannot. The precise mechanism that accounts for the differential aspect of the Gram stain is not entirely understood. Presumably, the decolorizing agent dehydrates the thick layer of peptidoglycan and in this dehydrated state the wall acts as a permeability barrier, retaining the dye. In contrast, the solvent action of acetone-alcohol easily damages the outer membrane of Gram-negative bacteria; their relatively thin layer of peptidoglycan cannot retain the dye complex. These bacteria lose the dye complex more readily than their Gram-positive counterparts. Also, as Gram-positive cells age, they often lose their ability to retain the dye. This probably results from damage to their peptidoglycan layer that occurs as a consequence of aging.

Characteristics of Bacteria that Lack a Cell Wall

Some bacteria naturally lack a cell wall. Species of *Mycoplasma*, one of which causes a mild form of pneumonia, have an extremely variable shape because they lack a rigid cell wall (**figure 3.36**). As expected, neither penicillin nor lysozyme affects these organisms. *Mycoplasma* and related bacteria can survive without a cell wall because their cytoplasmic membrane is stronger than that of most other bacteria. They have **sterols** in their membrane; these rigid, planar molecules stabilize membranes, making them stronger.

Cell Walls of the Domain Archaea

As a group, members of the Archaea inhabit a wide range of extreme environments, and so it is not surprising they contain a greater variety of cell wall types than do members of the Bacteria. However, because most of these organisms have not been studied as extensively as the Bacteria, less is known about the structure of their walls. None contain peptidoglycan, but some do have a similar molecule, **pseudopeptidoglycan**.

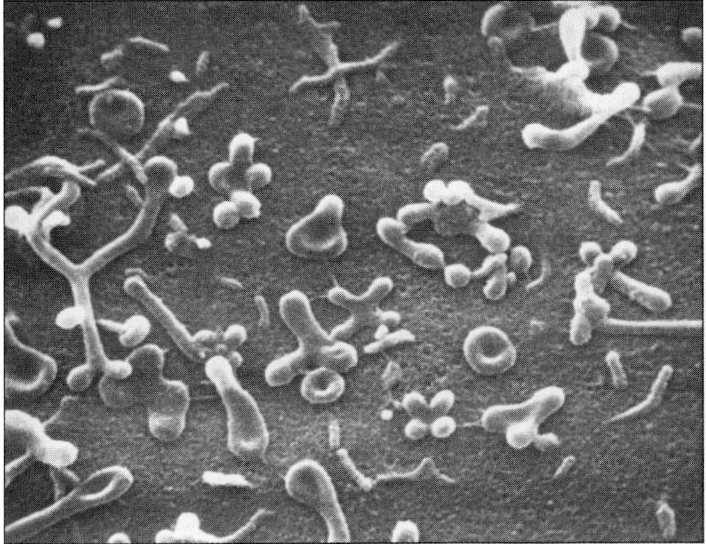

Figure 3.36 **The Plastic Shape of *Mycoplasma pneumoniae*, Which Lacks a Rigid Cell Wall**

MICROCHECK 3.6

Peptidoglycan is a molecule unique to bacteria that provides rigidity to the cell wall. The Gram-positive cell wall is composed of a relatively thick layer of peptidoglycan. The Gram-negative cell wall has a thin layer of peptidoglycan and an outer membrane, which contains lipopolysaccharide. The outer membrane excludes molecules with the exception of those that pass through porins; proteins are secreted via special mechanisms. Penicillin and lysozyme interfere with the structural integrity of peptidoglycan. *Mycoplasma* species lack a cell wall. Members of the Archaea have a variety of cell wall types.

- What is the significance of Lipid A?
- How does the action of penicillin differ from that of lysozyme?
- Explain why penicillin will kill only actively multiplying cells, whereas lysozyme will kill cells in any stage of growth.

Surface Layers External to the Cell Wall

Bacteria may have one or more layers outside of the cell wall. The functions of some of these are well established, but that of others are unknown.

Glycocalyx

Many bacteria envelop themselves with a gel-like layer called a **glycocalyx** that generally functions as a mechanism of either protection or attachment (**figure 3.37**). If the layer is distinct and gelatinous, it is called a **capsule**; recall that capsules can be seen microscopically using a capsule stain. If, instead, the layer is diffuse and irregular, it is called a **slime layer**. Colonies of bacteria that form either of these extracellular layers often appear moist and glistening.

Capsules and slime layers vary in their chemical composition depending on the species of bacteria. Most are composed of polysaccharides such as dextrans and glucans. These take the form of tiny, short, hairlike structures or fibrils, which form a network on the outside of the cell wall. A few capsules consist of polypeptides made up of repeating subunits of only one or two amino acids. Interestingly, the amino acids are generally of the D-stereoisomeric form, one of the few places D-amino acids are found in nature. ■ **polysaccharide, p. 34** ■ **D-amino acid, p. 28**

Some types of capsules and slime layers enable bacteria to adhere to specific surfaces, including teeth, rocks, and other bacteria. These often enable microorganisms to grow as a biofilm, a mass of bacteria coating a surface. One example is **dental plaque**, a biofilm on teeth. *Streptococcus mutans* uses sucrose to synthesize a capsule, which enables it to adhere to and grow in the crevices

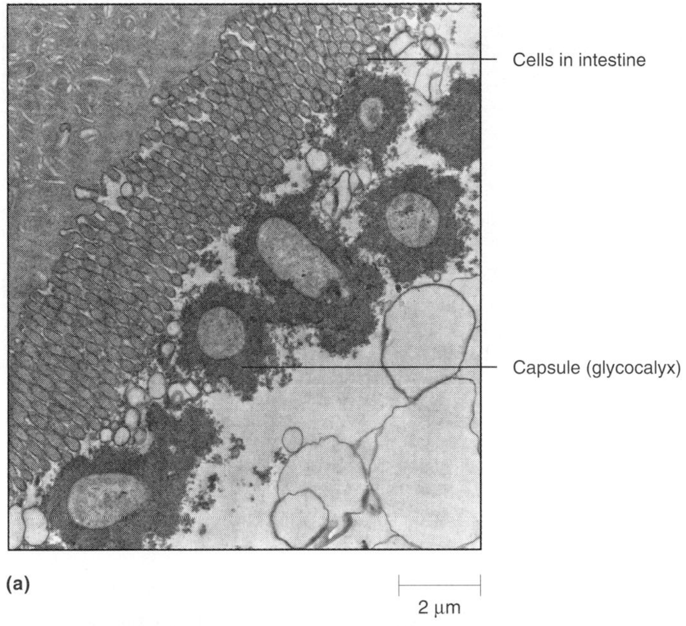

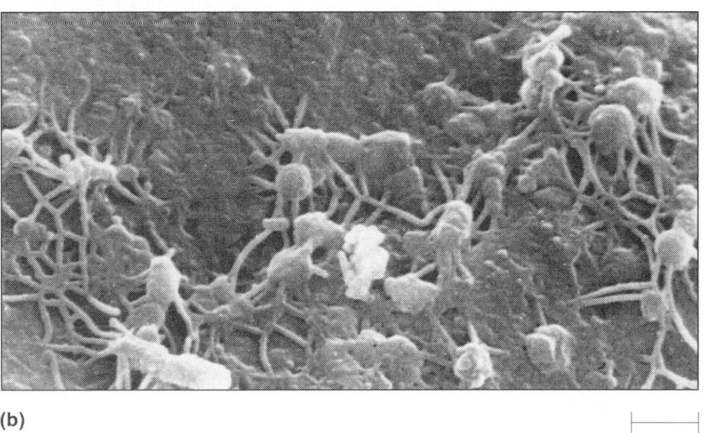

Figure 3.37 Glycocalyx This layer enables bacteria to attach to specific surfaces. **(a)** Capsules facilitating the attachment of bacteria to cells in the intestine (EM). **(b)** Masses of cells of *Eikenella corrodens* adhering in a layer of slime (SEM).

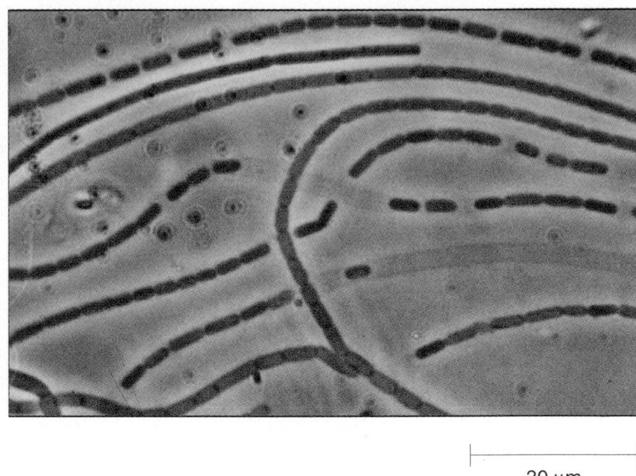

Figure 3.38 Sheath A sheath surrounds chains of some aquatic bacteria, protecting the enclosed community from being disrupted by turbulence.

of the tooth. Other bacteria can then adhere to the layer created by the growth of *S. mutans*. Acid production by bacteria in the biofilm damages the tooth surface. ■ ***Streptococcus mutans*, p. 590** ■ **sucrose, p. 34** ■ **dental caries, p. 589**

Some capsules enable bacteria to thwart innate defense systems that otherwise protect against infection. This is well illustrated in the case of the organism that causes bacterial pneumonia, *Streptococcus pneumoniae*. This organism can only cause disease if it has a capsule. Unencapsulated cells are quickly engulfed and killed by phagocytes, an important cell of our innate defense system. ■ ***Streptococcus pneumoniae*, p. 564** ■ **phagocytes, p. 372**

Sheaths

Some bacteria that live in aquatic habitats have a **sheath**, which is a tube that surrounds and holds a linear chain of cells **(figure 3.38)**. These are thought to protect the enclosed community from being disrupted by the turbulence of a flowing stream, which might otherwise disperse cells to a less suitable habitat. The filamentous growth of sheathed bacteria is easily seen with a light microscope. ■ **sheathed bacteria, p. 286**

MICROCHECK 3.7

Capsules and slime layers enable organisms to adhere to surfaces and sometimes protect bacteria from our innate defense system. Sheaths protect an aquatic community of bacteria from disruption by turbulence.

- How do capsules differ from slime layers?
- What is dental plaque?
- Explain why a sugary diet can lead to tooth decay.

Filamentous Protein Appendages

Many bacteria have protein appendages that are anchored in the membrane and protrude out from the surface. These structures are not essential to the life of the cell, but they do allow some bacteria to exist in certain environments in which they otherwise might not survive.

Flagella

The **flagellum** (plural: flagella) is a long protein structure that is responsible for most types of bacterial motility **(figure 3.39)**. By spinning like a propeller, using proton motive force as energy, the flagellum pushes the bacterium through liquid much as a ship is driven through water. Flagella must work very hard to move a cell, since water has the same relative viscosity to bacteria as molasses has to humans. Nevertheless, their speed is quite phenomenal; flagella can rotate more than a thousand revolutions per second, propelling *E. coli* at a rate of 20 body lengths per second. This is the equivalent of a 6-foot man running 82 miles per hour!

In some cases, flagella are important in the ability of an organism to cause disease. For example, *Helicobacter pylori*, the

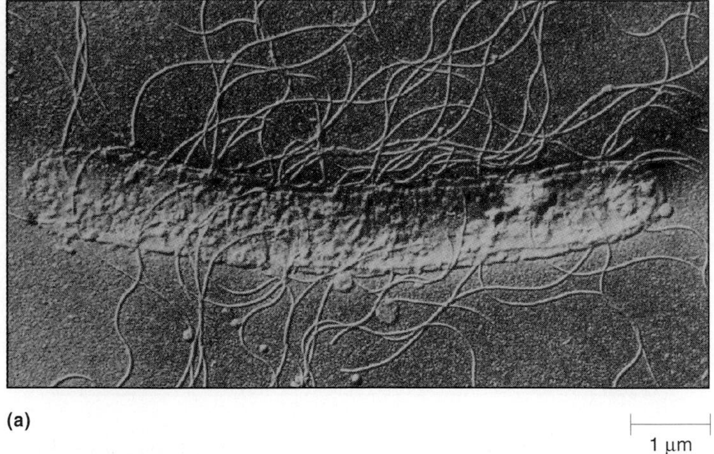

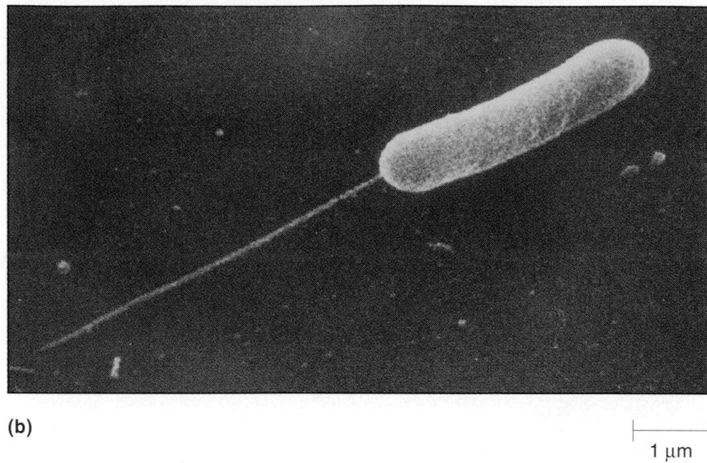

Figure 3.39 **Flagella** **(a)** Peritrichous flagella (SEM); **(b)** polar flagellum (SEM).

bacterium that causes gastric ulcers, has powerful multiple flagella at one end of its spiral-shaped cell. These flagella allow *H. pylori* to penetrate the viscous mucous gel that coats the stomach epithelium. ■ *Helicobacter pylori*, p. 593

Structure and Arrangement of Flagella

Flagella are composed of three basic parts **(figure 3.40)**. The **filament** is the portion that extends into the exterior environment. It is composed of identical subunits of a protein called **flagellin**. These subunits form a chain that twists into a helical structure with a hollow core. Connecting the filament to the cell surface is a curved structure, the **hook**. The **basal body** anchors the flagellum to the cell wall and cytoplasmic membrane

The numbers and arrangement of flagella can be used to characterize flagellated bacteria. For example, *E. coli* have flagella distributed over the entire surface, an arrangement called **peritrichous** (*peri* means "around"). Other common bacteria have a **polar flagellum**, a single flagellum at one end of the cell. Less commonly, bacteria may have a tuft of flagella at one or both ends of a cell.

Chemotaxis

Motile bacteria sense the presence of chemicals and respond by moving in a certain direction—a phenomenon called **chemotaxis**. If a compound is a nutrient, it may serve as an **attractant**, enticing cells to move toward it. On the other hand, if the compound is toxic, it may act as a **repellent**, causing cells to move away.

The movement of a bacterium toward an attractant is anything but direct **(figure 3.41)**. When *E. coli* travels, it progresses in a given direction for a short time, then stops and

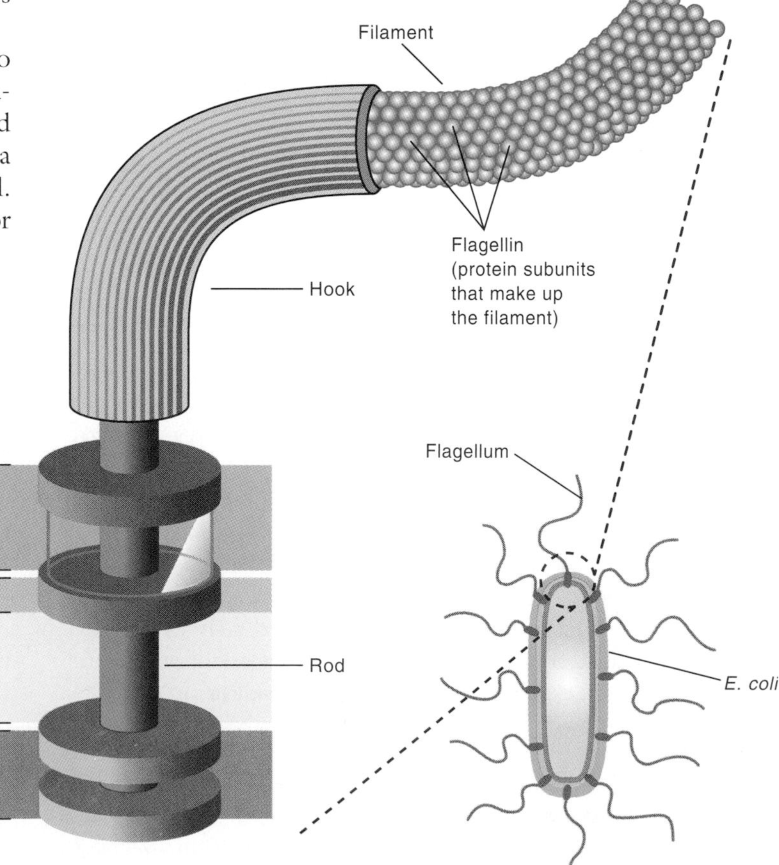

Figure 3.40 **The Structure of a Flagellum in a Gram-Negative Bacterium** The flagellum is composed of a filament, a hook, and a basal body.

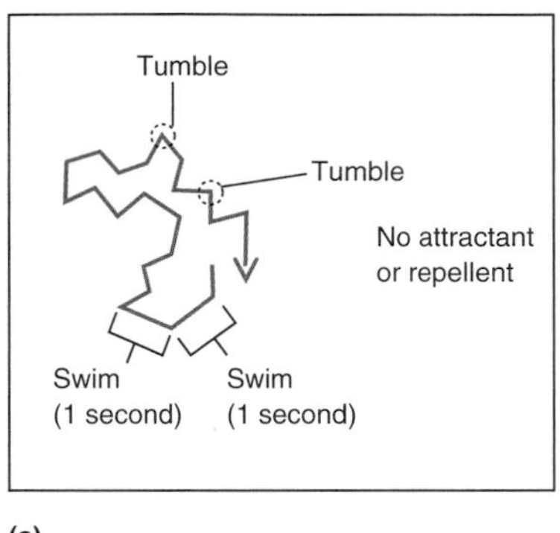

(a)

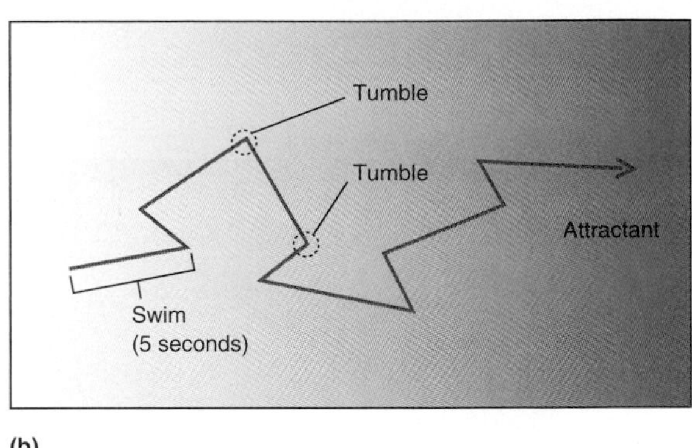

(b)

Figure 3.41 Chemotaxis (a) Cells move via a random series of runs and tumbles when the attractant or repellent is uniformly distributed. **(b)** The cell tumbles less frequently when it senses that it is moving closer to the attractant.

tumbles for a fraction of a second, and then moves again in a relatively straight line. But after rolling around, the cell is often oriented in a completely different direction. The seemingly odd pattern of movement is due to the rotation of the flagella. When the flagella rotate counterclockwise, the bacterium is propelled in a forward movement called a **run**. The flagella of *E. coli* and other peritrichously flagellated bacteria rotate in a coordinated fashion, forming a tight propelling bundle. After a brief period of time, the direction of rotation of the flagella is reversed. This abrupt change causes the cell to stop and roll, called a **tumble**. Movement toward an attractant is due to runs of longer duration that occur when cells are going in the right direction.

In addition to reacting to chemicals, some bacteria can respond to variations in light, **phototaxis**. Other bacteria can respond to the concentration of oxygen, **aerotaxis**. Organisms that require oxygen for growth will move toward it, whereas bacteria that grow only in its absence tend to be repelled by it. Certain motile bacteria can react to the earth's magnetic field by the process of **magnetotaxis**. They actually contain a row of magnetic particles that cause the cells to line up in a north-to-south direction much as a compass does **(figure 3.42)**. The magnetic forces of the earth attract the organisms so that they move downward and into sediments where the concentration of oxygen is low, which is the environment best suited for their growth.

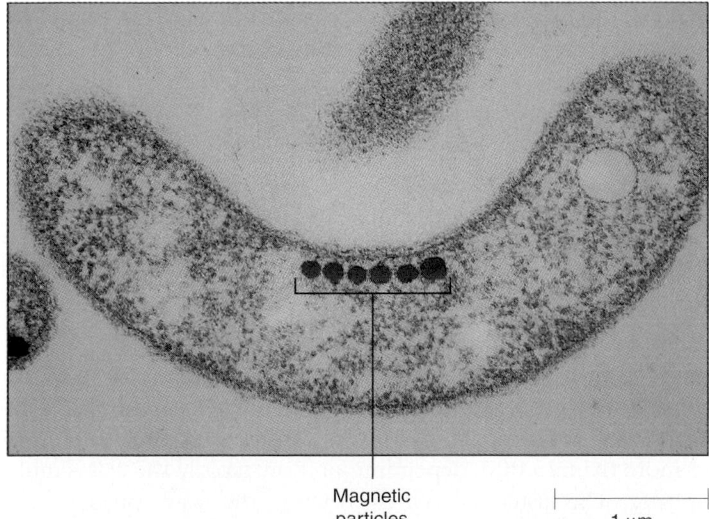

Figure 3.42 Magnetic Particles Within a Magnetotactic Bacterium The chain of particles of magnetite (Fe_3O_4) within the spirillum *Magnetospirillum* (*Aquaspirillum*) *magnetotacticum* serve to align the cell along geomagnetic lines (TEM).

Pili

Pili are considerably shorter and thinner than flagella, but they have a similar structural theme to the filament of flagella—a string of protein subunits arranged helically to form a long cylindrical molecule with a hollow core **(figure 3.43)**. The functions of pili, however, are distinctly different from those of flagella.

Many types of pili enable attachment of cells to specific surfaces; these pili are also called **fimbriae**. At the tip or along the length of the molecule is located another protein, an **adhesin**, that adheres by binding to a very specific molecule. For example, certain strains of *E. coli* that cause a severe watery diarrhea can attach to the cells that line the small intestine. They do this through specific interactions between adhesins on their pili and the intestinal cell surface. Without the ability to attach, these cells would simply be propelled through the small intestine along with the other intestinal contents. ■ **enterotoxigenic *E. coli*, p. 603**

Pili also appear to play a role in the movement of populations of cells on solid media. Some types of bacteria show a short, jerking movement called **twitching**, whereas other bacteria move smoothly, called **gliding**. Both of these forms of motility occur on solid media and appear to require some form of cell-to-cell contact. They do not, however, involve flagella.

Another type of pilus is involved in **conjugation**, a mechanism of DNA transfer from one bacterial cell to another. **Sex pili** are used to join those two cells. An example is the **F pili** of *E. coli*. Typically, sex pili are somewhat longer than the types of pili that mediate other characteristics. ■ **conjugation, p. 208**

MICROCHECK 3.8

Flagella are the most common mechanism for bacterial motility. Chemotaxis is the directed movement of cells toward an attractant or away from a repellent. Pili provide a mechanism for attachment to specific surfaces and, in some cases, a type of motility. Sex pili are involved in the transfer of DNA from one bacterium to another.

- What role does a series of runs and tumbles play in chemotaxis?
- What energy source is used to rotate flagella?

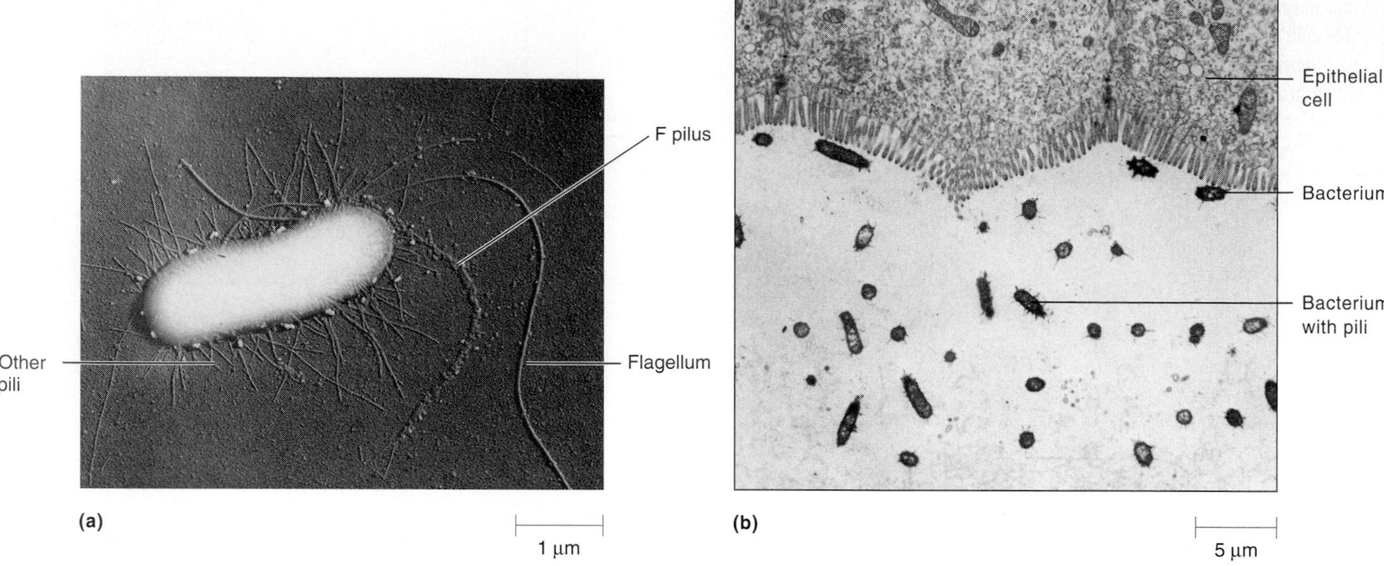

Figure 3.43 **Pili** **(a)** Pili on an *Escherichia coli* cell. The short pili (fimbriae) mediate adherence; the F pilus is involved in DNA transfer. (11,980×). **(b)** *Escherichia coli* attaching to epithelial cells in the small intestine of a pig.

Internal Structures

Prokaryotic cells have a variety of structures within the cell. Some, such as the chromosome and ribosomes, are essential for the life of all cells. Others, such as plasmids, are optional but confer certain selective advantages. Storage granules, vesicles, and endospores are characteristic of only certain types of bacteria.

The Chromosome

The chromosome of prokaryotes is not contained within a membrane-bound nucleus. Instead, it resides as an irregular mass within the cytoplasm, forming a gel-like region called the **nucleoid**. Typically, it is a single, circular, double-stranded DNA molecule that contains all the genetic information required by a cell. If that circular molecule is cut to form a linear piece and extended to its full length, it is about 1 mm long, approximately 1,000 times as long as the cell itself!

Chromosomal DNA is tightly packed into about 10% of the total volume of the cell. Rather than being a loose circle, it is typically in a twisted form called **supercoiled**, which appears to be stabilized by the binding of positively charged proteins (**figure 3.44**). Supercoiling can be visualized by cutting a rubber band and twisting one end several times before rejoining the ends. The resulting circle will twist and coil in response. The replication of DNA and the utilization of the genetic information encoded by it will be discussed in detail in chapter 7.

Plasmids

Plasmids are circular double-stranded DNA molecules that are present in many bacteria. They are generally 0.1% to 10% of the size of the chromosome and carry from a few to several hundred genes. A single cell can carry multiple types of plasmids.

A cell does not typically require the genetic information carried by a plasmid. However, the encoded genetic characteristics may be advantageous in certain situations. For example, many plasmids code for the production of one or more enzymes that destroy certain antibiotics, enabling the organism to resist the otherwise lethal effect of these medications. Because bacteria can sometimes transfer a copy of a plasmid to another bacterial cell, this accessory genetic information can spread, which accounts in large part for the increasing frequency of antibiotic-resistant organisms worldwide. At the same time, excess genetic information can be disadvantageous to a cell, slowing its multiplication. Occasionally, a cell will divide without evenly distributing its plasmids, giving rise to one cell that lacks a plasmid. If that cell can multiply faster as a consequence of losing that plasmid, and that plasmid was not essential to viability in that environment, then progeny of the cell will eventually predominate. Thus, populations can gain and lose plasmids. In the laboratory, cells can be grown under conditions that increase the likelihood that a plasmid will be lost. The resulting population has been **cured** of the plasmid.

Ribosomes

Ribosomes are intimately involved in protein synthesis, where they serve as the structures that facilitate the joining of amino acids. They are not passive workbenches; rather, they play an active role in the complex process of protein synthesis. The number of ribosomes in *E. coli* varies from approximately 7,000 to more than 25,000, depending on how rapidly the cell is multiplying. The faster the cell is growing, the faster proteins are being synthesized and the greater the number of ribosomes needed. Each ribosome is composed of a large and a small subunit, which are made up of **ribosomal proteins** and **ribosomal RNAs**. ■ **function of ribosomes, p. 175**

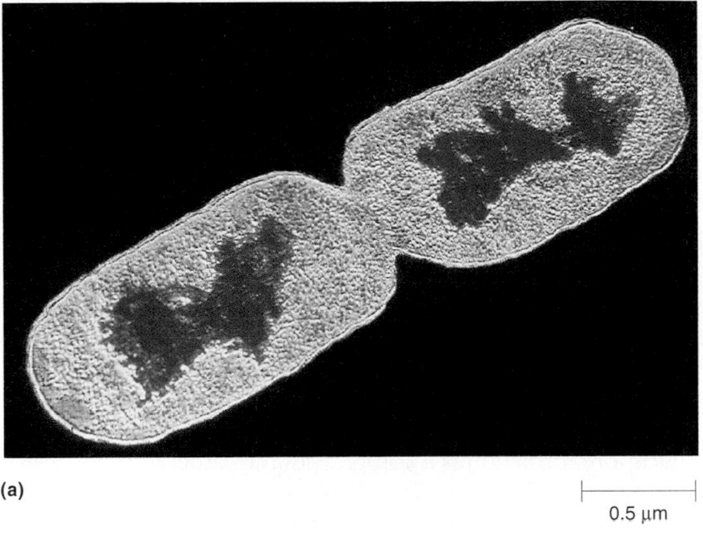

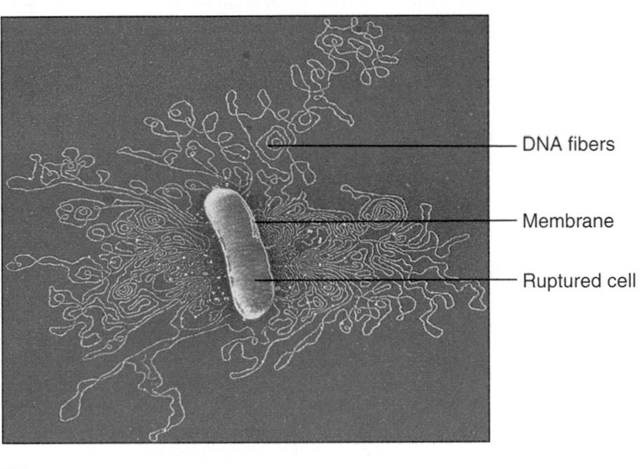

Figure 3.44 The Chromosome **(a)** Color-enhanced transmission electron micrograph of a thin section of *Escherichia coli,* with the DNA shown in red. **(b)** Chromosome released from a gently lysed cell of *E. coli*. Note how tightly packed the DNA must be inside the bacterium.

The relative size and density of ribosomes and their subunits is expressed as a distinct unit, **S** (for Svedberg), that reflects how fast they move when they are spun at very high speeds in an **ultracentrifuge**. The faster they move toward the bottom, the higher the S value and the greater the density. Prokaryotic ribosomes are 70S ribosomes. Note that S units are not strictly arithmetic; the 70S ribosome is composed of a 30S and a 50S subunit **(figure 3.45)**.

Prokaryotic ribosomes differ from the eukaryotic ribo somes, which are 80S. This variance serves as a target for antibiotics, which preferentially bind to the 70S ribosome and thus inhibit protein synthesis in bacteria.

Storage Granules

Storage granules are accumulations of high molecular weight polymers, which are synthesized from a nutrient that a cell has in relative excess. For example, if nitrogen and/or phosphorus are lacking, *E. coli* cannot multiply even if a carbon and energy source such as glucose is plentiful. Rather than waste the carbon/energy source, cells use it to produce

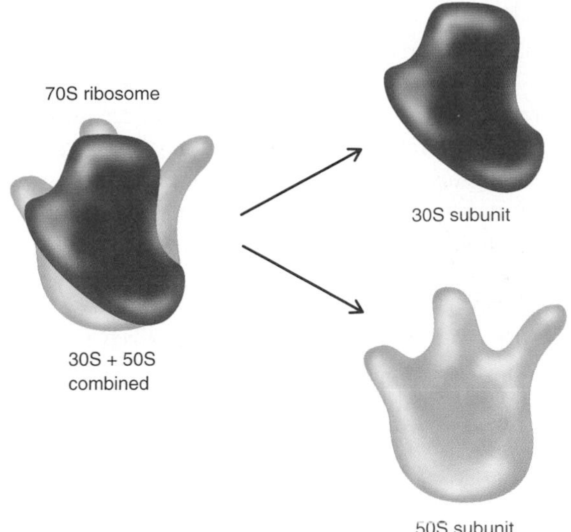

Figure 3.45 The Ribosome The 70S ribosome is composed of 50S and 30S subunits.

glycogen, a glucose polymer. A single large molecule such as glycogen has little osmotic effect on the cell. Later, when conditions are appropriate, cells degrade and use the glycogen granule. Other bacterial species store carbon and energy as **poly-β-hydroxybutyrate (figure 3.46)**. This microbial compound is now being employed to produce a biodegradable polymer, which can be used in place of petroleum-based plastics. As a general rule, only one type of storage granule is produced by a given organism.

Some types of granules can be readily detected by light microscopy. **Volutin** granules, a storage form of phosphate, stain red with blue dyes such as methylene blue, whereas the surrounding cellular material stains blue. Because of this, they are often called **metachromatic granules** (*meta* means "change" and *chromatic* means "color"). Bacteria that store volutin are beneficial in wastewater treatment because they scavenge phosphate, which is an environmental pollutant.

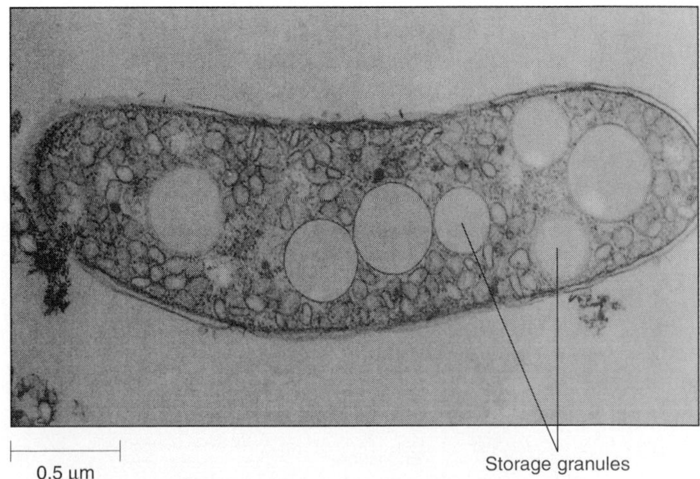

Figure 3.46 Storage Granules The large unstained areas in the photosynthetic bacterium *Rhodospirillum rubrum* are granules of poly-β-hydroxybutyrate.

Gas Vesicles

Some aquatic bacteria produce **gas vesicles**, small rigid compartments that provide buoyancy to the cell. Gases, but not water, flow freely into the vesicles, thereby decreasing the density of the cell. By regulating the number of gas vesicles within the cell, an organism can float or sink to its ideal position in the water column. For example, bacteria that use sunlight as a source of energy float closer to the surface, where light is available.

Endospores

An **endospore** is a unique type of dormant cell type produced by a process called **sporulation** within cells of certain bacterial species, primarily members of the genera *Bacillus* and *Clostridium* (**figure 3.47**). The structures may remain dormant for perhaps 100 years, or even longer, and are extraordinarily resistant to damaging conditions including heat, desiccation, toxic chemicals, and ultraviolet irradiation. Immersion in boiling water for hours may not kill them. Endospores that survive these treatments can **germinate**, or exit the dormant stage, to become a typical, actively multiplying cell, called a **vegetative cell**.

The consequences of these resistant dormant forms are far-reaching. Because endospores can survive so long in a variety of conditions, they can be found virtually anywhere. They are quite common in soil, which can make its way into environments such as laboratories and hospitals and onto products such as food, media used to cultivate microbes, and medical devices. Because the exclusion of microbes in these environments and on these products is of paramount importance, special precautions must be taken to destroy these resistant structures.

Endospores are sometimes called **spores**. However, this latter term is also used to refer to the structures produced by unrelated microorganisms such as fungi. Bacterial endospores are much more resistant to environmental conditions than are other types of spores.

Pathogenic Endospore-Formers

Several species of endospore-formers can cause disease. Botulism results from the ingestion of a deadly toxin produced by vegetative cells of *Clostridium botulinum*. If endospores of this bacterium contaminate vegetables or other foods that are then canned without adequate heat treatment, those endospores can survive. They can then germinate to form vegetative cells, which would be killed by oxygen, but thrive within the oxygen-free, or **anaerobic**, conditions provided by the canned food. There, they will produce **botulinum toxin**, which can cause a fatal paralysis in people who consume it. Other disease-causing species of endospore-formers include *Clostridium tetani*, which causes tetanus; *Clostridium perfringens*, which causes gas gangrene; and *Bacillus anthracis*, which causes anthrax. ■ **botulism, p. 675** ■ **canning, p. 119** ■ **tetanus, p. 698** ■ **gas gangrene, p. 701** ■ **anthrax, p. 491**

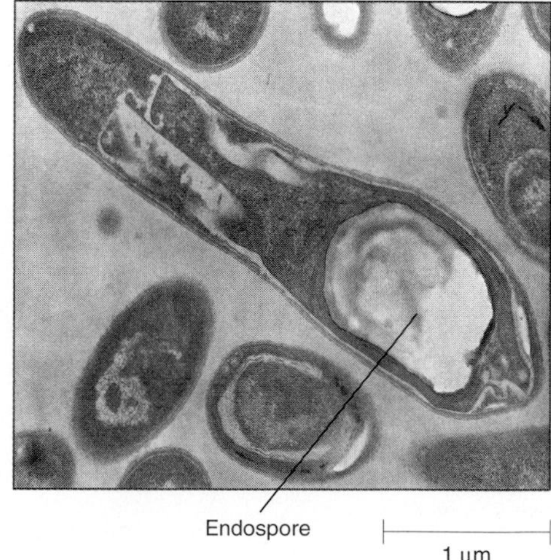
Endospore
1 μm

Figure 3.47 Endospores Endospore inside a vegetative cell of a *Clostridium* species (TEM).

Sporulation

Endospore formation is a complex, highly ordered sequence of changes that initiates when cells are grown in low amounts of carbon or nitrogen (**figure 3.48**). Apparently the bacteria sense that they are facing starvation conditions and therefore begin the 8-hour process that prepares them for rough times ahead.

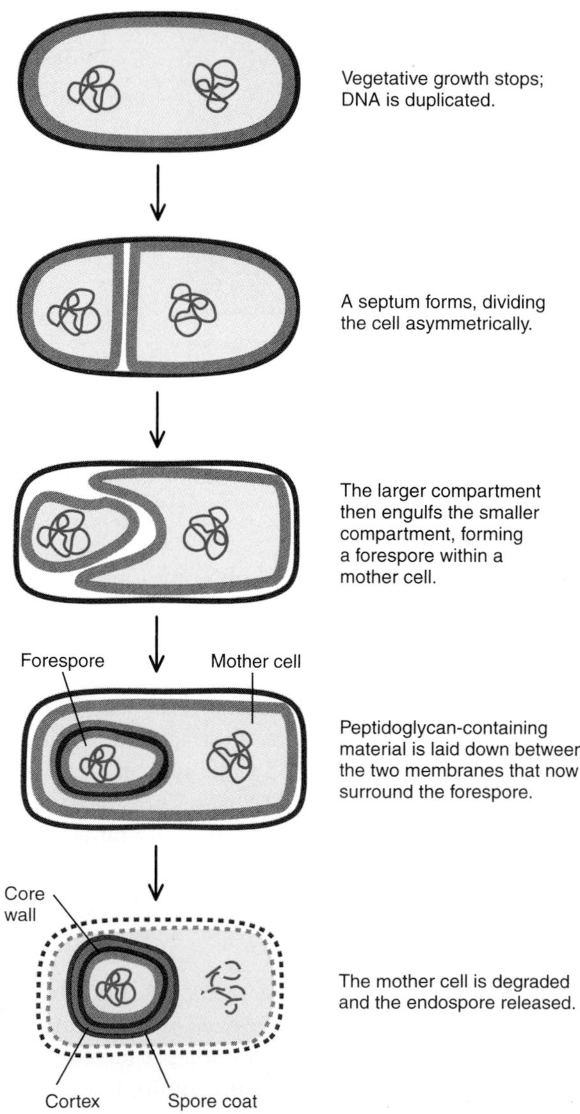
Vegetative growth stops; DNA is duplicated.
A septum forms, dividing the cell asymmetrically.
The larger compartment then engulfs the smaller compartment, forming a forespore within a mother cell.
Forespore
Mother cell
Peptidoglycan-containing material is laid down between the two membranes that now surround the forespore.
Core wall
Cortex
Spore coat
The mother cell is degraded and the endospore released.

Figure 3.48 The Process of Sporulation

After vegetative growth stops, DNA is duplicated and then a septum forms between the two chromosomes, dividing the cell asymmetrically. The larger compartment then engulfs the smaller compartment, forming a **forespore** within a **mother cell**. These two portions take on different roles in synthesizing the components that will make up the endospore. The forespore, which is enclosed by two membranes, will ultimately become the **core** of the endospore. Peptidoglycan-containing material is laid down between these two membranes, forming the **core wall** and the **cortex**. Meanwhile, the mother cell makes proteins that will form the **spore coat**. Ultimately, the mother cell is degraded and the endospore released.

The layers of the completed endospore function together, protecting the structure from damage. The spore coat is thought to function as a sieve, excluding molecules such as lysozyme. The cortex helps maintain the core in a dehydrated state, which protects it from the effects of heat.

Germination

Endospores can be activated to **germinate**, or leave their dormant state. This process can be triggered by a brief exposure to heat or certain chemicals. Following such exposure, the endospore takes on water and swells. The spore coat and cortex then crack open, and a vegetative cell grows out. Since one vegetative cell gives rise to one endospore, sporulation is not a means of cell reproduction.

MICROCHECK 3.9

The prokaryotic chromosome is usually a circular, double-stranded DNA molecule that contains all of the genetic information required by a cell. Plasmids encode information that is advantageous to a cell in certain conditions. Ribosomes are the structures that facilitate the joining of amino acids to form a protein. Storage granules are polymers synthesized from a nutrient that a cell has in relative excess. Gas vesicles provide buoyancy to a cell. An endospore is a highly resistant dormant stage produced by members of the genera *Bacillus* and *Clostridium*.

- Explain how glycogen granules benefit a cell.
- Explain why endospores are an important consideration for the canning industry.
- Why are the processes of sporulation and germination not considered a mechanism of multiplication?

THE EUKARYOTIC CELL

Eukaryotic cells are generally much larger than prokaryotic cells, and their internal structures are far more complex **(figure 3.49)**. One of their most distinguishing characteristics is the abundance of membrane-enclosed compartments or **organelles**. These can take up half the total cell volume and enable the cell to perform complex functions in spatially separated regions. For example, degradative enzymes are contained within an organelle, isolated from other cell contents, where they digest food and other material without posing a threat to the integrity of the cell itself.

Organelles are functionally beneficial, but they also create transit problems. Each organelle contains a variety of proteins and other molecules, many of which are synthesized at other locations. To deliver these to the **lumen**, or interior, of the appropriate organelle, an elaborate transportation system is

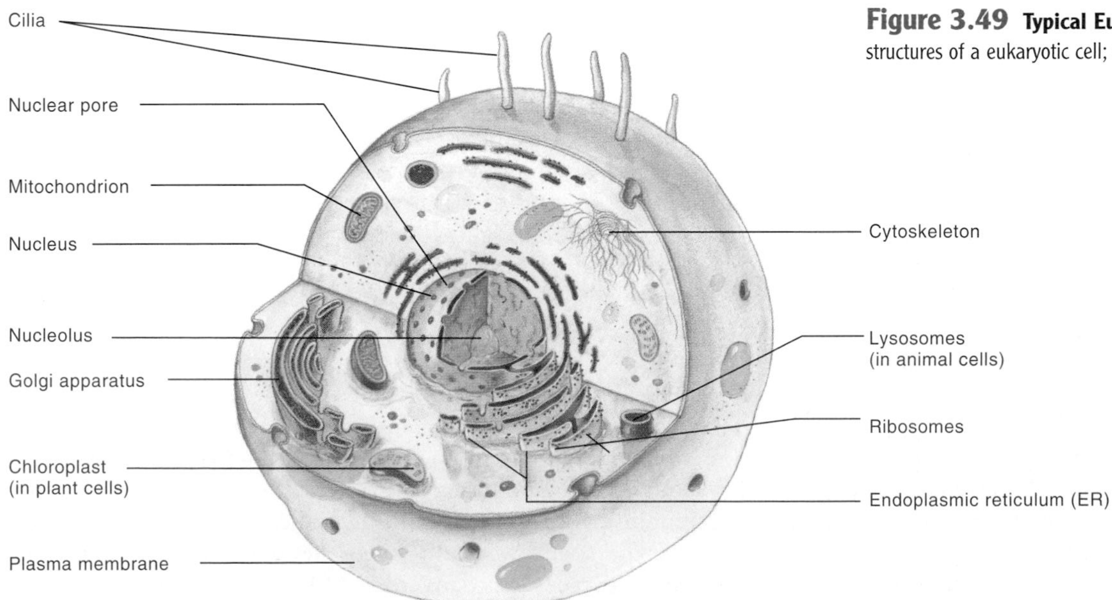

Figure 3.49 Typical Eukaryotic Cell A representation of typical structures of a eukaryotic cell; not all cells have all of the structures shown.

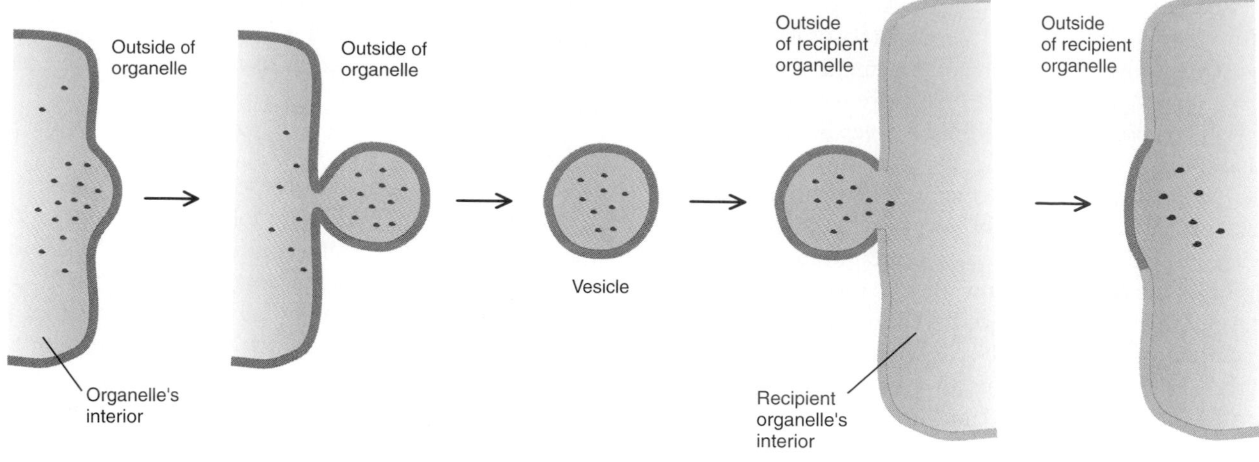

Figure 3.50 Vesicle Formation and Fusion A vesicle forms when a section of an organelle buds off. The mobile vesicle can then move to other parts of the cell, ultimately fusing with the membrane of another organelle.

required. To transfer material, a section of an organelle will bud or pinch off, forming a small membrane-enclosed **vesicle (figure 3.50)**. This mobile vesicle, containing a sampling of the contents of the organelle, can move to other parts of the cell. When that vesicle encounters the lipid membrane of the appropriate organelle, the two membranes will fuse to become one contiguous unit. By doing so, the vesicle introduces its contents to the lumen of that organelle. In this way, the contents of one membrane-enclosed compartment can be transferred to another. A similar process is used to export molecules synthesized within an organelle to the external environment.

As a group, eukaryotic cells are highly variable in many aspects. For example, protozoa, which are single-celled organisms, must function exclusively as a self-contained unit that seeks and ingests food. These cells must be mobile and flexible to take in food particles, including bacteria. Consequently, they lack a cell wall that would otherwise provide rigidity. Animal cells also lack a cell wall, because they too must be flexible to accommodate movement. Yeasts and most molds, on the other hand, are stationary and benefit from the protection provided by a rigid cell wall. Their cell wall is composed of chitin, a polymer of *N*-acetylglucosamine that is also found in crustaceans and insects. Plant cells, which are also stationary, have cell walls composed of cellulose, a polymer of glucose.

The individual cells of a multicellular organism may be distinctly different from one another. Mammals, for example, are composed of several hundred different types of cells, and it is obvious that a liver cell is quite different from a bone cell. Cells of plants and animals function in cooperative associations called **tissues**. The tissues in your body include muscle, connective, nerve, epithelial, blood, and lymphoid. Each of these provides a different function. Connective tissue, for example, which includes bone and cartilage, provides structure and support. Muscle provides movement. Combinations of various tissues function together to make up larger units, **organs**, including skin, heart, and liver. Organs and the tissues that constitute them will be covered in detail in the chapters on infectious diseases. At this point, however, it is helpful to recognize that a generic discussion of eukaryotic cells encompasses many functional varieties. ■ **types of epithelium, p. 459** ■ **lymphoid tissue, p. 373**

A comprehensive coverage of all aspects of eukaryotic cells is beyond the scope of this textbook. Instead, this section will focus on key characteristics, particularly those that directly affect the interaction of a microbe with a human host. These characteristics are summarized in **table 3.5**. A comparison of functional aspects of prokaryotic and eukaryotic cells is presented in **table 3.6**.

The Plasma Membrane

All eukaryotic cells have a cytoplasmic membrane, or **plasma membrane**, which is similar in chemical structure and function to that of prokaryotic cells. It is a typical phospholipid bilayer embedded with proteins. The lipid and protein composition of the leaflet that faces the cytoplasm, however, differs significantly from that facing the outside of the cell **(figure 3.51)**. The same is true for the membranes that surround the organelles. The leaflet that faces the lumen of the organelle is similar to its counterpart that faces the exterior of the cell. This lack of symmetry reflects the important role these membranes play in the complex processes that occur within the eukaryotic cell.

The proteins in the lipid bilayer perform a variety of functions. Some are involved in transport and others are attached to internal structures, helping to maintain cell integrity. Those in the outer leaflet often function as receptors. Typically, these receptors are **glycoproteins**, proteins that have various sugars attached. A given receptor binds a specific molecule, which is referred to as its **ligand**. These receptor-ligand interactions are extremely important in multicellular organisms because they allow cells to communicate with each other—a process called **signaling**. For example, in our bodies, some phagocytic cells secrete a specific protein when they encounter certain compounds perceived as dangerous. Other cells of the immune system have receptors for that protein on their surface. When the

Table 3.5 A Summary of Eukaryotic Cell Structures

	Characteristics
Plasma Membrane	Asymmetric lipid bilayer embedded with proteins. Selective permeability, conduit to external environment.
Internal Protein Structures	
Cilia	Appear to project out of a cell. Beat in synchrony to provide movement. Composed of microtubules in a 9 + 2 arrangement.
Cytoskeleton	Dynamic filamentous network that provides structure to the cell.
Flagella	Appear to project out of a cell. Propel or push the cell with a whiplike or thrashing motion. Composed of microtubules in a 9 + 2 arrangement.
Ribosomes	Two subunits, 60S and 40S, join to form the 80S ribosome.
Membrane-Bound Organelles	
Chloroplast	Harvests the energy of sunlight to generate ATP. Within the stroma are chlorophyll-containing, disclike thylakoids. The membranes of these contain the components of the electron transport chain and the proteins that use proton motive force to synthesize ATP.
Endoplasmic reticulum	Site of synthesis of macromolecules destined for other organelles or the external environment.
Rough	Attached ribosomes extrude the proteins they are synthesizing through pores that lead into the lumen of the organelle.
Smooth	Site of lipid synthesis and degradation, and Ca^{2+} storage.
Golgi apparatus	Modifies macromolecules that are synthesized in the endoplasmic reticulum and then sorts and packages them in vesicles for transport.
Lysosome	Digestion of foodstuffs.
Mitochondria	Use the energy released during the degradation of organic compounds to generate ATP. Within the highly folded inner membrane are the components of the electron transport chain and the proteins that use proton motive force to synthesize ATP.
Nucleus	Contains the DNA.
Peroxisome	Oxidation of lipids and toxic chemicals occurs.

Table 3.6 Comparison of Prokaryotic and Eukaryotic Cell Structures/Functions

Characteristic	Prokaryotic	Eukaryotic
Degradation of extracellular substances	Enzymes are secreted that degrade macromolecules outside of the cell. The resulting small molecules are transported into the cell.	Macromolecules are brought into the cell by pinocytosis or, in the case of protozoa and phagocytes, phagocytosis. Lysosomes carry digestive enzymes.
Chromosome	Single, circular DNA molecule is typical.	Multiple, linear DNA molecules. DNA is wrapped around histones.
Chromosome location	Located in a region of the cell called the nucleoid, which is not membrane-bound.	Contained within the membrane-bound nucleus.
Motility	Generally involves flagella, which are composed of protein subunits. Flagella rotate like a propeller, using proton motive force for energy.	Involves cilia and flagella, which are made up of a 9 + 2 arrangement of microtubules. Cilia move in synchrony; flagella propel a cell with a whiplike motion or thrash back and forth to pull a cell forward. Both use ATP for energy.
Protein secretion	A characteristic signal sequence marks proteins for secretion by the general secretory pathway. The precise mechanisms of translocation are still poorly understood.	Secreted proteins are translocated to the lumen of the rough endoplasmic reticulum as they are being synthesized. From there, they are transported to the Golgi apparatus for processing and packaging.
Protein synthesis	Uses 70S ribosomes, which are made up of 50S and 30S subunits.	Uses 80S ribosomes, which are made up of 60S and 40S subunits. Mitochondria and chloroplasts have 70S ribosomes.
Strength and rigidity	Peptidoglycan-containing cell wall.	Cytoskeleton composed of microtubules, intermediate filaments, and microfilaments. Some have a cell wall; some have sterols in the membrane.
Transport proteins	Primarily active transport; mechanisms include major facilitator superfamily and ABC transport. Group translocation.	Facilitated diffusion and active transport; mechanisms include major facilitator superfamily and ABC transport. Ion channels.

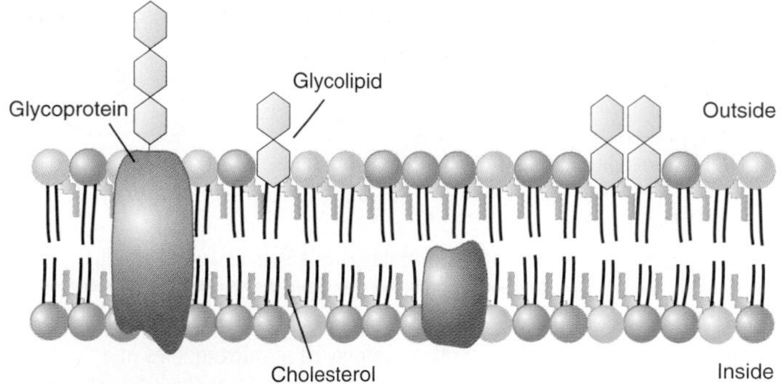

Figure 3.51 Plasma Membrane The eukaryotic membrane is asymmetric; the lipid and protein content of the outer leaflet differs significantly from that of the inner leaflet.

protein binds its receptor, those cells recognize the signal as a call for help and respond accordingly. This type of cell-to-cell communication enables the multicellular organism to function as a cohesive unit. ■ phagocytic cells, p. 372

Some of the lipids of the outer leaflet are lipids with various sugars attached, called **glycolipids**. Other short chains of sugars, **oligosaccharides**, are often attached to these. This creates a carbohydrate-rich layer, or a glycocalyx, surrounding the cell. This layer retains moisture and helps protect the cell from physical damage.

The membranes of many eukaryotic cells contain sterols, which provide strength to the otherwise fluid structure. Recall that *Mycoplasma*, a group of bacteria that lack a cell wall, also have sterols in their membranes. The sterol found in animal cell membranes is **cholesterol**, whereas fungal membranes contain **ergosterol**. This difference is exploited by antifungal medications that act by interfering with ergosterol synthesis or function. ■ antifungal medications, p. 514

The plasma membrane plays no direct role in energy generation; instead, that task is performed in an organelle. Although proton motive force is not generated across the membrane, an electrochemical gradient is maintained by energy-consuming mechanisms that expel either sodium ions or protons.

MICROCHECK 3.10

The plasma membrane is an asymmetric lipid bilayer embedded with proteins. Specific receptors on the outer leaflet mediate cell-to-cell signaling. Sterols provide strength to the fluid membrane.

- What is the medical significance of ergosterol in the fungal membranes?
- Describe why signaling is important in animal cells.
- How could one argue that the lumen of an organelle is "outside" of a cell?

Transfer of Molecules Across the Plasma Membrane

Foodstuffs, signaling molecules, and waste products pass through the plasma membrane. Some of these enter and exit the cell via transport proteins. Others are taken in through a process called **endocytosis**. **Exocytosis**, which is the reverse of endocytosis, can be used to expel material.

Transport Proteins

The transport proteins of eukaryotic cells may function as either a **channel** or a **carrier**. Channels are pores in the membrane. These are so small that only specific ions can diffuse through. They allow ions to move with the concentration gradient; they do not create such a gradient. To control ion passage, the channel has a gate, which can be either opened or closed, depending on the environmental conditions. Carriers are analogous to the proteins in prokaryotic cells that mediate facilitated diffusion and active transport.

Cells of multicellular organisms can often take up nutrients by facilitated diffusion, because the nutrient concentration of surrounding environments can be controlled. For example, glucose levels in the blood are maintained at a concentration that is higher than in most tissues. Consequently, animal cells generally do not need to expend energy transporting glucose.

The active transport mechanisms of eukaryotic cells are structurally analogous to the major facilitator superfamily and ABC transporter systems found in prokaryotic cells. Some of these are of medical interest because they can eject drugs from the cell. For example, some strains of *Plasmodium falciparum*, the eukaryotic parasite that causes a deadly form of malaria, use an ABC transporter system to eject the antimalarial drug chloroquine. These chloroquine resistant strains are a global health concern. Some human cancer cells use an ABC transporter system that ejects therapeutic drugs intended to kill those cells. ■ malaria, p. 731

Endocytosis and Exocytosis

Endocytosis is the process by which eukaryotic cells take material from the surrounding environment. They do this by enclosing those materials in a fluid-filled compartment **(figure 3.52)**. The

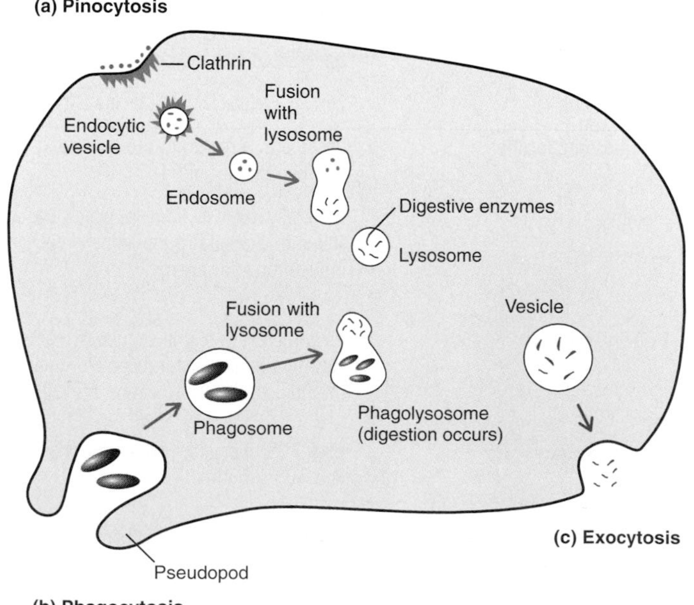

Figure 3.52 Endocytosis and Exocytosis Endocytosis includes **(a)** pinocytosis (receptor-mediated endocytosis) and **(b)** phagocytosis.

type of endocytosis common to most animal cells is **pinocytosis**. In this process a cell internalizes and pinches off small pieces of its own membrane, bringing along a small volume of liquid and any material attached to the membrane. This **endocytic vesicle** becomes a membrane-enclosed, low-pH compartment called an **endosome**. This then fuses with a digestive organelle called a **lysosome.** The characteristics of lysosomes will be discussed shortly.

In animal cells, pinocytosis is generally a type called **receptor-mediated endocytosis.** This process allows cells to internalize extracellular ligands that bind to the cell's receptors. Certain regions of the cell membrane are lined with a protein called **clathrin** and studded with receptors. These regions are internalized to form an endocytic vesicle, bringing with them the receptors along with their bound ligands. The low pH of the endosome frees the ligands from the receptors, which are often recycled. The endosome then fuses with a lysosome. Many viruses, including those that cause influenza and rabies, exploit receptor-mediated endocytosis to enter animal cells. By binding to a specific receptor, they too are taken up during pinocytosis. ■ influenza, p. 573 ■ rabies, p. 682

Protozoa and phagocytes, both of which ingest bacteria and large debris, illustrate a specific type of endocytosis called **phagocytosis**. Recall that phagocytes are important cells of our innate immune system. These cells send out armlike extensions, **pseudopods**, which surround and enclose extracellular material, including bacteria. This action envelops the material, bringing it into the cell in an enclosed compartment called a **phagosome**. These ultimately fuse with a lysosome to form a **phagolysosome**. Phagocytes have a greater abundance of lysosomes than do other animal cells, which reflects their specialized function. In addition, their lysosomes contain a wider array of powerful digestive enzymes. Thus, most microbes are readily dispatched within the phagolysosome. Those that resist the killing effects are able to cause disease. ■ professional phagocytes, p. 372 ■ survival within a phagocyte, p. 373

The process of **exocytosis** is the opposite of endocytosis. Membrane-bound vesicles inside the cell fuse with the plasma membrane and release their contents into the external medium. The processes of endocytosis and exocytosis result in the exchange of material between the inside and outside of the cell.

Secretion

Proteins destined for a non-cytoplasmic region, either outside of the cell or the lumen of an organelle, must be translocated across a membrane. Ribosomes that are synthesizing a secreted protein attach to the membrane of the **endoplasmic reticulum** (**ER**). The characteristics of this organelle will be described shortly. As the protein is being made, it is threaded through the membrane and into the lumen of the ER. The lumen of any organelle can be viewed as equivalent to an extracellular space. Once a protein or any substance is there, it can readily be transported by vesicles to the lumen of another organelle, or to the exterior of the cell.

MICROCHECK 3.11

Gated channels allow specific ions to pass across the membrane. Carriers facilitate the passage of molecules across the membrane and often use energy. Receptor-mediated endocytosis allows cells to internalize specific small molecules. Protozoa and phagocytes internalize bacteria and debris by phagocytosis. Exocytosis is used to expel material. Secreted proteins are translocated across the membrane of the endoplasmic reticulum as they are being made.

- How does a cell bring in ligands?
- How is the formation of an endocytic vesicle different from that of a phagosome?
- How might a bacterium resist the killing effects of a phagolysosome?

Protein Structures Within the Cytoplasm

Eukaryotic cells have unique protein structures that distinguish them from prokaryotic cells. These include the **cytoskeleton,** a dynamic filamentous network that provides structure and shape to the cell, and characteristic **flagella** and **cilia**, which allow the cell to move.

Another structure that characterizes the eukaryotic cell is the size and density of its ribosomes. The 80S eukaryotic ribosome is made up of a 60S and a 40S subunit.

Cytoskeleton

The threadlike proteins that make up the cytoskeleton continually reconstruct to adapt to the cell's constantly changing needs. The network is composed of three elements; **microtubules**, **microfilaments**, and **intermediate filaments (figure 3.53)**.

Microtubules, the thickest of the cytoskeleton structures, are long hollow cylinders composed of protein subunits called **tubulin**. Microtubules form the **mitotic spindles**, the machinery that partitions chromosomes between two cells in the process of cell division. Without mitotic spindles, cells could not reproduce. They also are the main structure that makes up the **cilia** and **flagella**, the mechanisms of locomotion in certain eukaryotic cells. In addition, microtubules also function as the framework along which organelles and vesicles move within a cell. The antifungal

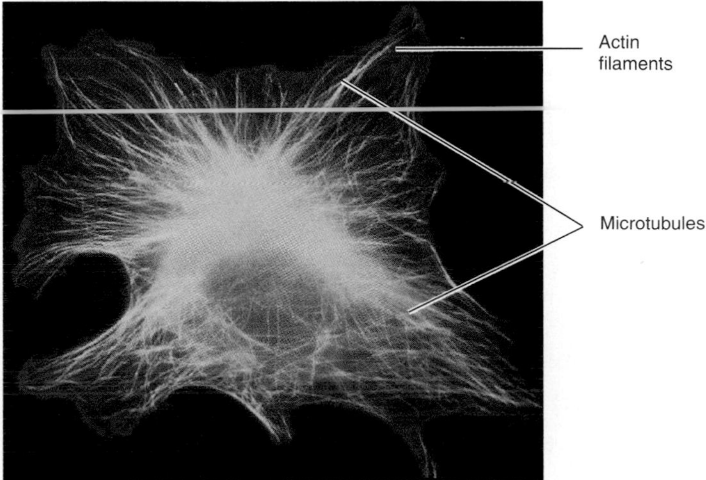

Figure 3.53 Cytoskeleton Photomicrograph of a eukaryotic cell cytoskeleton double-labeled with fluorescent antibodies to show actin filaments and microtubules.

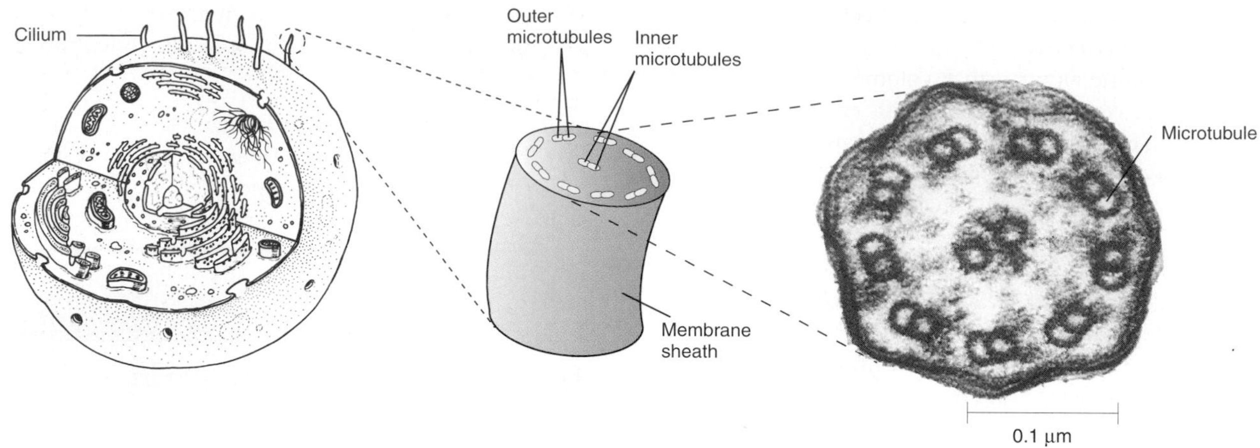

Figure 3.54 **Cilia** Diagram of a cilium; there is one pair of tubules in the center and nine pairs on the outside (9 + 2 arrangement). The electron photomicrograph shows a cross section of a flagellum from the protozoan *Trichonympha*.

drug griseofulvin is thought to interfere with the structural integrity of the microtubules of some fungi. ■ mitosis, p. 303

Microfilaments enable the cell cytoplasm to move. They are composed of a polymer of actin that can rapidly assemble and subsequently disassemble, which causes motion. For example, pseudopod formation relies on actin polymerization in one part of the cell and depolymerization in another. Some intracellular pathogens exploit this process and trigger a rapid polymerization of actin, creating an "actin tail," which moves them within that cell. Sometimes, this mechanism propels them with enough force to be ejected into an adjacent cell.

Intermediate filaments function like ropes, strengthening the cell mechanically. They enable cells to resist physical stresses.

Flagella and Cilia

Flagella and **cilia** are flexible structures that appear to project out of a cell yet are covered by an extension of the cytoplasmic membrane (**figure 3.54**). Both are composed of long microtubules grouped in what is called a 9 + 2 arrangement: nine pairs of microtubules surrounding two individual ones. Eukaryotic flagella function in motility, but are otherwise completely different from their prokaryotic counterparts. Using ATP as a source of energy, they either propel the cell with a whiplike motion or thrash back and forth to pull the cell forward. Recall that flagella in bacteria use proton motive force as a source of energy and rotate like a propeller to push the organism.

Cilia are shorter than flagella, often covering a cell and moving in synchrony. This movement can move a cell forward in an aqueous solution, or propel surrounding material along a stationary cell. For example, epithelial cells that line the respiratory tract have cilia that beat together in a directed fashion. This propels the mucus film that covers those cells, directing it upward toward the mouth, where it can be swallowed. This action removes microorganisms that have been inhaled before they can enter the lungs.

MICROCHECK 3.12

The 80S eukaryotic ribosome is composed of 60S and 40S subunits. The cytoskeleton is a dynamic filamentous network that provides structure to the cell; it is composed of microtubules, actin filaments, and intermediate filaments. Flagella function in motility. Cilia either propel a cell or move material along a stationary cell.

- Explain how actin microfilaments are related to phagocytosis.
- Explain what is meant by the 9 + 2 structure of cilia and flagella.

Membrane-Bound Organelles

The presence of membrane-bound organelles is an important feature that sets eukaryotic cells apart from their prokaryotic counterparts.

The Nucleus

The predominant distinguishing feature of the eukaryotic cell is the **nucleus**, which contains the DNA. Enclosing this structure are two concentric lipid bilayer membranes: the **inner membrane** and the **outer membrane**. These membranes make up the **nuclear envelope**. Spanning the membranes are complex protein structures that form **nuclear pores**, allowing large molecules such as ribosomal subunits and proteins to be transported into and out of the nucleus (**figure 3.55**). The **nucleolus** is a region within the nucleus where ribosomal RNAs are synthesized. These RNAs , along with ribosomal proteins that are synthesized in the cytoplasm and then transported into the nucleolus, are then assembled into ribosomal subunits. These then pass through the nuclear pores into the cytoplasm, where the subunits combine to form 80S ribosomes.

The nucleus contains multiple chromosomes, each one encoding different genetic information. Unlike the situation in most prokaryotic cells, the double-stranded chromosomal DNA is linear. To add structure and order to the long DNA molecule, it is packed by winding it around positively charged proteins called **histones**. These bind tightly to the negatively charged DNA molecule. One packing unit, called a **nucleosome**, consists of a complex of histones around which the linear DNA

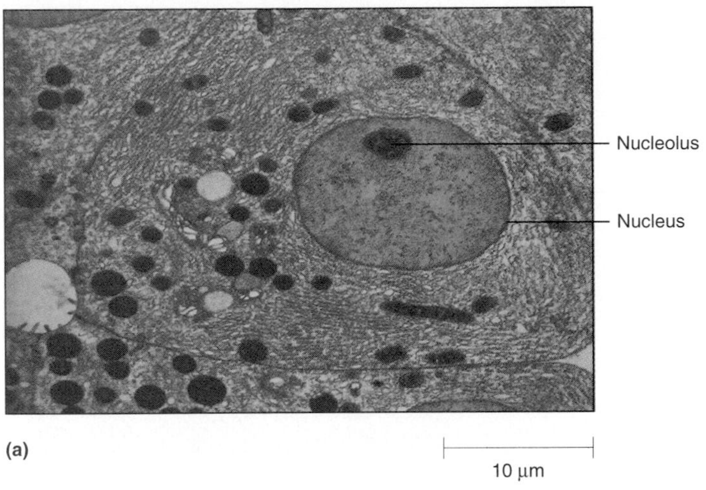

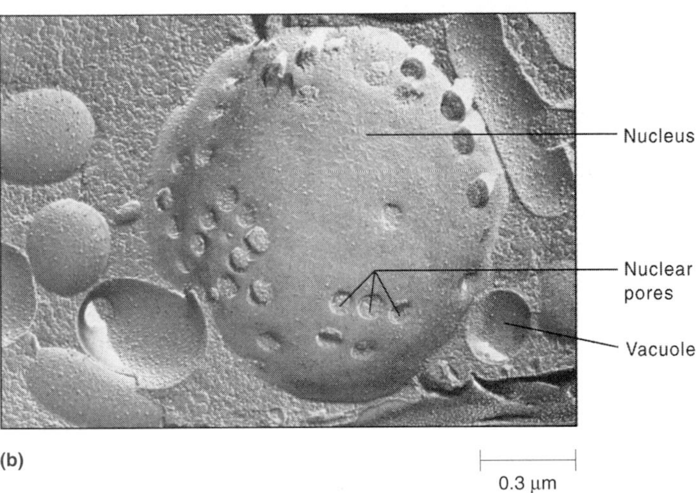

Figure 3.55 Nucleus **(a)** Electron micrograph of an animal cell (rat pancreas) showing the nucleus and nucleolus. **(b)** Electron micrograph of a yeast cell (*Geotrichum candidum*) by the freeze-fracture technique, showing a surface view of the nucleus and nuclear pores.

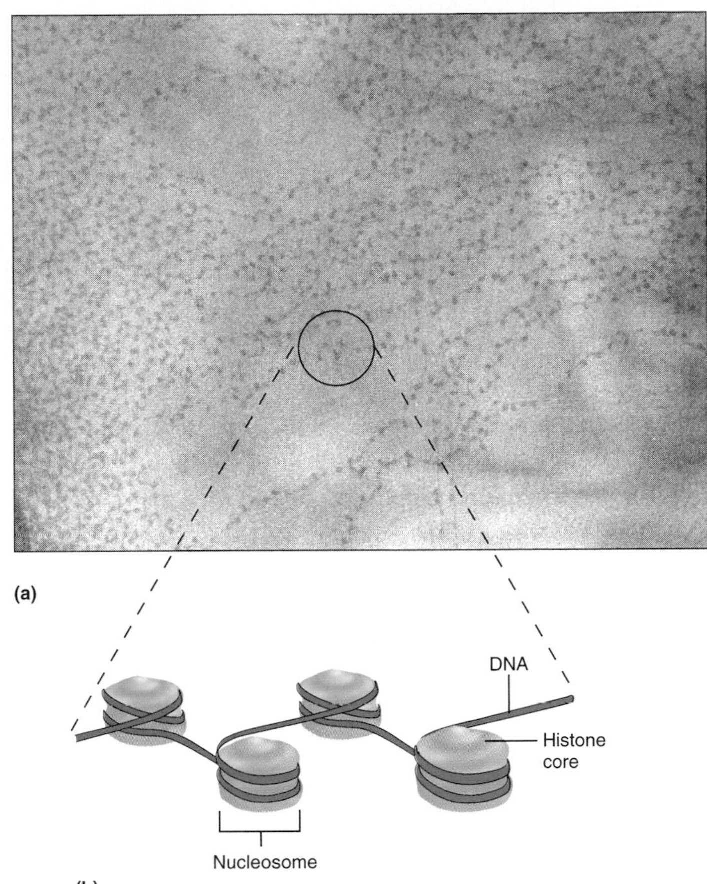

Figure 3.56 Chromatin **(a)** Chromatin fibers; the threads as well as the "beads" on the threads consist of DNA wrapped around histones. **(b)** Diagram of a nucleosome, the fundamental packing unit of eukaryotic DNA.

wraps twice (**figure 3.56**). The complex of DNA and proteins that together form the chromosomes is called **chromatin**.

The events that take place in the nucleus during cell division distinguish eukaryotes from prokaryotes. In eukaryotic cells, after the DNA is replicated the chromosomes go through a nuclear division process called **mitosis**, which ensures the daughter cells receive the same number of chromosomes as the original parent. Through mitosis, a cell that is **diploid**, or has two copies of each chromosome, will generate two diploid daughter cells. A different process, **meiosis**, generates haploid daughter cells.

Mitochondria and Chloroplasts

Mitochondria and **chloroplasts** are the organelles that function as the powerhouses that generate ATP, the universal form of energy used by all cells. Mitochondria are found in nearly all eukaryotic cells, whereas chloroplasts are found exclusively in plants and algae. These harvest the energy of sunlight to synthesize organic compounds—a process called **photosynthesis**.

■ **photosynthesis, p. 158**

Mitochondria and chloroplasts generate ATP using mechanisms analogous to those occurring in the cytoplasmic membrane of prokaryotes. Contained within their extensive internal membranes are the components of the electron transport chain. Recall that these transfer electrons and, in the process, eject protons to generate proton motive force. Proteins embedded within the membranes then use proton motive force to synthesize ATP. What sets mitochondria apart from chloroplasts is the source of the electrons transferred by the electron transport chain.

Mitochondria and chloroplasts do have several other similarities. They have two membranes: an outer and an inner membrane. Within the region enclosed by the inner membrane, called the **matrix** in mitochondria and the **stroma** in chloroplasts, they have DNA, ribosomes, and other molecules necessary for protein synthesis. Notably, their ribosomes are 70S rather than the 80S ribosome found in the cytoplasm of eukaryotic cells.

Mitochondria

Mitochondria use the energy released during the degradation of organic compounds to generate ATP. The electrons extracted during the degradation are transferred to the electron transport chain embedded within the inner membrane. This membrane is highly folded within the organelle, forming invaginations called **cristae** (**figure 3.57**). These increase the surface area of the inner membrane tremendously, expanding its ATP-generating capabilities.

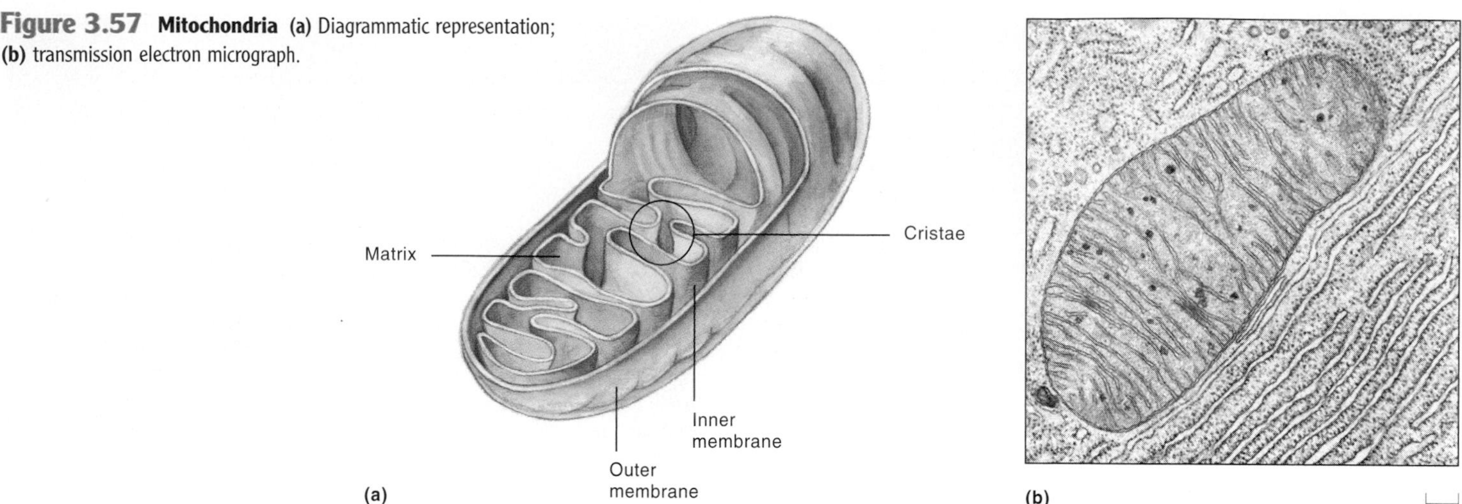

Figure 3.57 Mitochondria (a) Diagrammatic representation; (b) transmission electron micrograph.

Chloroplasts

Chloroplasts harvest the energy of sunlight to generate ATP. Within the stroma of the chloroplasts are membrane-bound disclike structures, called **thylakoids (figure 3.58)**. These structures contain chlorophyll, a pigment that emits electrons when stimulated by the energy of sunlight. These electrons are then transferred to the electron transport chain embedded in the membrane of the thylakoids. Meanwhile, chlorophyll returns to its original state when water is oxidized to produce oxygen.

The ATP generated in the chloroplast is used to convert CO_2 to organic compounds such as sugar and starch. These are the very compounds that many nonphotosynthetic organisms degrade as a source of energy.

Endoplasmic Reticulum

The **endoplasmic reticulum (ER)** is a complex three-dimensional internal membrane system of flattened sheets, sacs, and tubes **(figure 3.59)**. The **rough endoplasmic reticulum** has a characteristic bumpy appearance due to the multitude of ribosomes that coat it. It is the site where proteins that are not destined for the cytoplasm are synthesized. These include proteins destined for the lumen of an organelle or for secretion outside the cell. Membrane proteins such as receptors are also synthesized on the rough ER. The ribosomes making these proteins attach to the rough ER. During the process of synthesis, the protein is extruded through a special gated pore in the membrane of the ER. Within the lumen, they fold to assume their three-dimensional shape. Vesicles that bud off from the ER transfer the newly synthesized molecules to the Golgi apparatus for further modification and sorting.

Some regions of the ER are smooth. This **smooth endoplasmic reticulum** provides a variety of functions including lipid synthesis and degradation, and calcium ion storage. The quantity of the smooth ER varies according to cell type; it is particularly prevalent in cells that specialize in the synthesis and secretion of steroid hormones. As with material made in the rough ER, vesicles transfer compounds from the smooth ER to the Golgi apparatus.

The Golgi Apparatus

The **Golgi apparatus** consists of a series of membrane-bound flattened sacs **(figure 3.60)**. One of its roles is to modify the macromolecules that are synthesized in the endoplasmic reticu-

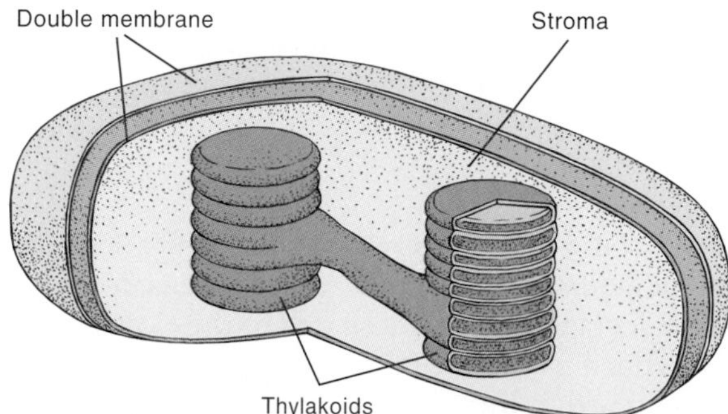

Figure 3.58 Chloroplast Diagrammatic representation of a chloroplast, which harvests the energy of light to produce ATP.

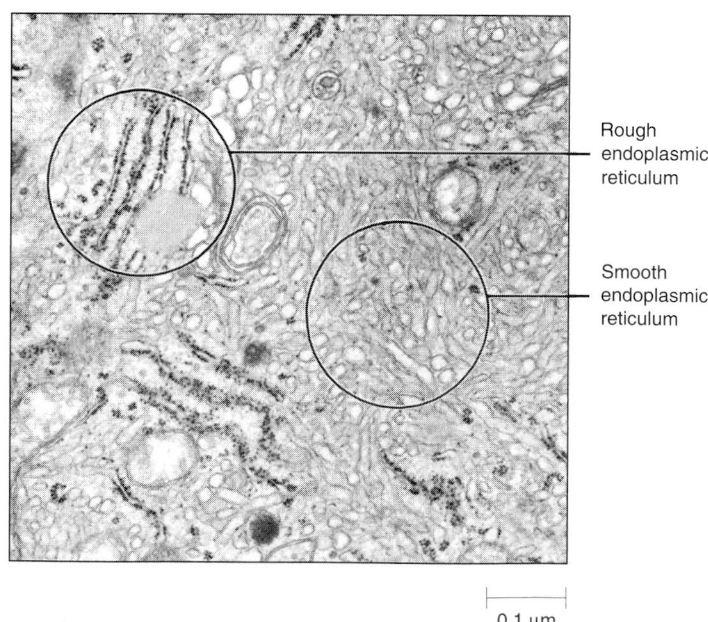

Figure 3.59 Endoplasmic Reticulum Electron micrograph showing the rough and smooth endoplasmic reticulum.

Perspective 3.1 The Origins of Mitochondria and Chloroplasts

Mitochondria and chloroplasts bear such a striking similarity to prokaryotic cells that it is no wonder scientists have speculated for many decades that these organelles evolved from bacteria. The **endosymbiont theory** states that the ancestors of mitochondria and chloroplasts were bacteria that had been residing within other cells in a mutually beneficial partnership. The intracellular bacterium in such a partnership is called an **endosymbiont**. As time went on, each partner became indispensable to the other, and the endosymbiont eventually lost key features such as a cell wall and the ability to replicate independently.

Several early observations have supported the endosymbiont theory. Mitochondria and chloroplasts, unlike other eukaryotic organelles, both carry some of the genetic information necessary for their function. These include genes for some of the ribosomal proteins and ribosomal RNAs that make up their 70S ribosomes. These ribosomes contrast with the typical 80S ribosomes that characterize eukaryotic cells and, in fact, are equivalent to the prokaryotic 70S ribosomes. Interestingly, cellular DNA encodes some of the components that make up these ribosomes. Another characteristic that supports the model that mitochondria and chloroplasts were once intracellular bacteria is the double membrane that surrounds these organelles. Present-day endosymbionts retain their cytoplasmic membranes and live within membrane-bound compartments in their eukaryotic host cell.

Evidence in favor of the endosymbiont theory continues to accumulate. Recent technology enables scientists to readily determine the precise order, or **sequence**, of nucleotides that make up DNA. This allows comparison of the nucleotide sequences of organelle DNA with genomes of different bacteria. It has become apparent that some mitochondrial DNA sequences bear a striking resemblance to DNA sequences of members of a group of obligate intracellular parasites, the rickettsias. These are quite likely relatives of modern-day mitochondria.

A tremendous effort is now under way to determine the nucleotide sequence of mitochondria from a wide variety of eukaryotic cells, including plants, animals, and protists. While the size of mitochondrial DNA varies a great deal among these different eukaryotic cells, common sequence themes are emerging. Today, researchers are no longer discussing "if" but "when" these organelles evolved from intracellular prokaryotes.

lum before they are transported to other destinations. These modifications, such as the addition of carbohydrate and phosphate groups, take place in a sequential order in different Golgi sacs. Much like an assembly line, the molecules are transferred in vesicles from one Golgi sac to another. These various molecules are then sorted and delivered in vesicles destined either for specific cellular compartments or to the outside of the cell.

Lysosomes and Peroxisomes

Lysosomes are organelles that contain a number of powerful degradative enzymes. These include various proteases and nucleases that could destroy the cell if not contained within the organelle. Recall that material brought into the cell through endocytosis is digested when the endosome or phagosome fuses with the lysosome. In a similar manner, exhausted organelles can fuse with a lysosome so that their contents are digested.

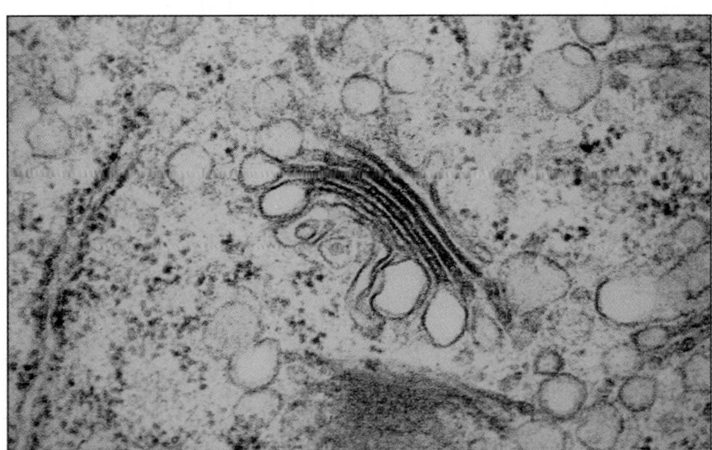

Figure 3.60 Golgi Apparatus Molecules are transferred in vesicles from one Golgi sac to another.

Peroxisomes are the organelles in which oxygen is used to oxidize substances, breaking down lipids and detoxifying certain chemicals. As a consequence, their enzymes generate highly reactive molecules such as hydrogen peroxide and superoxide. The peroxisome contains these molecules and ultimately degrades them, protecting the cell from their toxic effects. ■ hydrogen peroxide, p. 98 ■ superoxide, p. 98

MICROCHECK 3.13

The nucleus, which contains DNA, is the predominant distinguishing feature of eukaryotes. Mitochondria and chloroplasts generate energy in the form of ATP; mitochondria oxidize organic compounds to obtain energy, whereas chloroplasts harvest the energy of sunlight. The rough endoplasmic reticulum is the site where proteins not destined for the cytoplasm are synthesized. The smooth endoplasmic reticulum functions in lipid synthesis and degradation, and calcium ion storage. The Golgi apparatus modifies and sorts molecules synthesized in the rough ER. Lysosomes are the structures within which digestion takes place; peroxisomes are the organelles in which oxygen is used to oxidize substances.

- What is the function of a histone?
- How does the function of the rough endoplasmic reticulum differ from that of the smooth endoplasmic reticulum?
- If enzymes contained in a peroxisome are to act on a substrate, what must first occur?

FUTURE CHALLENGES

Understanding and Exploiting Transport Systems

Unraveling the complex mechanisms that prokaryotic and eukaryotic cells use to transport materials across their membranes can potentially aid in the development of new antimicrobial medications. Armed with a precise model of the structure and function of bacterial trans-

porter proteins, scientists might be able to design new drugs that exploit these systems. One strategy would be to design compounds that irreversibly bind to transporter molecules and jam the mechanism. If the microbes can be prevented from bringing in nutrients and removing wastes, their growth would cease. Another strategy would be to enhance the uptake or decrease the efflux of a specific compound that interferes with intracellular processes. This is already being done to some extent as new derivatives of current antibiotics are being produced, but more precise understanding of the processes by which bacteria take up or remove compounds could expedite drug development.

A more thorough understanding of eukaryotic uptake systems could be used to develop better antiviral drugs. Recall that viruses exploit the process of receptor-mediated endocytosis to gain entry into the cell. Once they are enclosed within the endosome, their protective protein coat is removed, releasing their genetic material. New drugs can potentially be developed that block these steps, preventing the entry or uncoating of infectious viral particles.

SUMMARY

MICROSCOPY AND CELL MORPHOLOGY

Microscopic Techniques: The Instruments (Table 3.1)

Principles of Light Microscopy: The Bright-Field Microscope

1. In light microscopy, visible light passes through the specimen. The most common type of microscope is the **bright-field microscope.** (Figure 3.1)
2. The **objective lens** and the **ocular lens** in combination magnify an object by a factor equal to the product of the magnification of each of the individual lenses.
3. The usefulness of a microscope depends on its **resolving power**. (Figure 3.2)

Light Microscopes that Increase Contrast

1. The **phase contrast** microscope amplifies differences in refraction. (Figure 3.4)
2. The **interference microscope** combines two light beams that pass through the specimen separately, causing the specimen to appear as a three-dimensional image. (Figure 3.5)
3. The **dark-field microscope** directs light toward a specimen at an angle. (Figure 3.6)
4. The **fluorescence microscope** is used to observe cells that have been stained with fluorescent dyes. It projects ultraviolet light through the objective lens and onto the specimen. (Figure 3.7)
5. The **confocal scanning laser microscope** is used to construct a three-dimensional image of a thick structure and to provide detailed sectional views of the interior of an intact cell. (Figure 3.8)

Electron Microscopes

1. Electron microscopes use electromagnetic lenses, electrons, and fluorescent screens to produce a magnified image. (Figure 3.9)
2. **Transmission electron microscopes (TEMs)** transmit electrons through a specimen that has been prepared by **thin sectioning, freeze fracturing**, or **freeze etching**. (Figure 3.10)
3. **Scanning electron microscopes** scan a beam of electrons back and forth over the surface of a specimen, producing a three-dimensional effect. (Figure 3.11)

Scanning Probe Microscopes

1. **Scanning probe microscopes** map the bumps and valleys of a surface on an atomic scale. (Figure 3.12)

Microscopic Techniques: Dyes and Staining (Table 3.2)

Differential stains

1. The **Gram stain** is the most widely used procedure for staining bacteria; Gram-positive bacteria stain purple and Gram-negative bacteria stain pink. (Figure 3.14)
2. The **acid-fast stain** is used to stain organisms such as *Mycobacteria*, which do not take up stains readily; **acid-fast** organisms stain pink and all other organisms stain blue. (Figure 3.15)

Special Stains to Observe Cell Structures

1. The **capsule stain** is an example of a **negative stain**; it colors the background, allowing the **capsule** to stand out as a halo around an organism. (Figure 3.16)
2. The **spore stain** uses heat to facilitate the staining of **endospores**. (Figure 3.17)
3. The **flagella stain** employs a **mordant** that enables the stain to adhere to and coat the otherwise thin **flagella**. (Figure 3.18)

Fluorescent Dyes and Tags

1. Some fluorescent dyes bind compounds that characterize all cells; others bind to compounds specific to only certain cell types. (Figure 3.19)
2. **Immunofluorescence** is used to tag a specific protein of interest with a fluorescent compound.

Morphology of Prokaryotic Cells

Shapes

1. Most common prokaryotes are either **cocci** or **rods**; other shapes include **coccobacilli**, **vibrios**, **spirilla,** and **spirochetes**. **Pleomorphic** bacteria have variable shapes. (Figure 3.20)

Groupings

1. Cells adhering to one another following division form a characteristic arrangement that depends on the plane in which the bacteria divide. (Figure 3.22)

Multicellular Associations

1. Some types of bacteria, such as myxobacteria, typically live in associations containing multiple cells.

2. Cells within **biofilms** often alter their activities when a critical number of cells are present.

THE STRUCTURE OF THE PROKARYOTIC CELL (Figure 3.23) (Table 3.3)

The Cytoplasmic Membrane

Structure and Chemistry of the Cytoplasmic Membrane (Figure 3.24)

1. The cytoplasmic membrane is a **phospholipid bilayer** embedded with a variety of different proteins. It serves as a barrier between the cell and the surrounding environment, allowing relatively few molecules to pass through freely.
2. Some membrane proteins function in transport; others provide a mechanism by which cells can sense and adjust to their surroundings.
3. The membrane lipids of the Archaea are distinctly different from those of the Bacteria.

Permeability of the Cytoplasmic Membrane

1. The cytoplasmic membrane is selectively permeable; water, gases, and small hydrophobic molecules are among the few compounds that can pass through by **simple diffusion.**
2. The inflow of water into the cell exerts more **osmotic pressure** on the cytoplasmic membrane than it can generally withstand; however, the rigid cell wall can withstand the pressure. (Figure 3.25)

The Role of the Cytoplasmic Membrane in Energy Transformation

1. The **electron transport chain** within the membrane expels protons, generating an **electrochemical gradient**, which contains a form of energy called **proton motive force.** (Figure 3.26)

Directed Movement of Molecules Across the Cytoplasmic Membrane

Transport Systems (Table 3.4)

1. **Facilitated diffusion**, or **passive transport**, moves impermeable compounds from one side of the membrane to the other by exploiting the concentration gradient. (Figure 3.27)
2. **Active transport** mechanisms use energy to accumulate compounds against a concentration gradient.
3. Members of the **major facilitator superfamily**, which includes **symporters**, **antiporters**, and **uniporters**, use proton motive force for energy. (Figure 3.28)
4. **ABC transport systems** require ATP for energy. (Figure 3.29)
5. **Group translocation** chemically modifies a molecule during its passage through the cytoplasmic membrane. (Figure 3.30)

Secretion

1. The **general secretory pathway** is the primary mechanism used to secrete proteins.
2. The presence of a characteristic **signal sequence** targets proteins for secretion.

Cell Wall

Peptidoglycan (Figure 3.32)

1. **Peptidoglycan** is a macromolecule found only in the Bacteria and provides rigidity to the cell wall.
2. Peptidoglycan is composed of **glycan strands**, which are alternating subunits of ***N*-acetylmuramic acid** (**NAM**) and ***N*-acetylglucosamine** (**NAG**), interconnected via the tetrapeptide chains on NAM.

The Gram-Positive Cell Wall (Figure 3.33)

1. The Gram-positive cell wall contains a relatively thick layer of peptidoglycan.
2. **Teichoic acids** and **lipoteichoic acids** stick out of the peptidoglycan layer.

The Gram-Negative Cell Wall (Figure 3.34)

1. The Gram-negative cell wall has a relatively thin layer of peptidoglycan sandwiched between the cytoplasmic membrane and an **outer membrane**.
2. **Periplasm** contains a variety of proteins, including those involved in nutrient degradation and transport.
3. The outer membrane contains **lipopolysaccharides**. The **Lipid A** portion of the lipopolysaccharide molecule is toxic, which is why **LPS** is called **endotoxin**. (Figure 3.35)
4. **Porins** form small channels that permit small molecules to pass through the outer membrane.

Antibacterial Compounds that Target Peptidoglycan

1. Penicillin binds to proteins involved in cell wall synthesis and, subsequently, prevents the cross-linking of adjacent glycan chains.
2. **Lysozyme** breaks the bond that links alternating NAG and NAM molecules, destroying the structural integrity of the glycan chain.

Differences in Cell Wall Composition and the Gram Stain

1. The Gram-positive, but not the Gram-negative, cell wall retains the crystal violet–iodine dye complex even when subjected to the trauma of acetone-alcohol treatment.

Characteristics of Bacteria that Lack a Cell Wall

1. Because *Mycoplasma* species do not have a cell wall, they are extremely variable in shape and are not affected by lysozyme or penicillin. (Figure 3.36)

Cell Walls of the Domain Archaea

1. Archaea have a greater variety of cell wall types than do the Bacteria.

Surface Layers External to the Cell Wall

Glycocalyx

1. A **capsule** is a distinct and gelatinous layer; a **slime layer** is diffuse and irregular. Both are usually made of polysaccharide. (Figure 3.37)
2. Capsules and slime layers enable bacteria to adhere to surfaces. Some capsules allow disease-causing microorganisms to thwart the innate defense system.

Sheaths

1. A **sheath** is a tube that holds a linear chain of cells; it is thought to protect the enclosed organisms from disruption. (Figure 3.38)

Filamentous Protein Appendages

Flagella (Figure 3.39)

1. The **flagellum** is a long protein structure, composed of a **filament**, a **hook**, and a **basal body**, that is responsible for most types of bacterial motility. (Figure 3.40)
2. **Chemotaxis** is the directed movement toward an **attractant** or away from a **repellent**; bacteria do this by adjusting the frequencies of their **runs** and **tumbles**. (Figure 3.41)
3. **Phototaxis, aerotaxis**, and **magnetotaxis** are directed movements toward light, oxygen, and a magnetic field, respectively.

Pili (Figure 3.43)

1. Many types of **pili (fimbriae)** enable attachment of cells to specific surfaces.
2. Some pili play a role in specific types of motility—**twitching** and **gliding**—both of which occur on solid surfaces and require cell-to-cell contact.
3. **Sex pili** are involved in **conjugation**, which enables DNA to be transferred from one cell to another.

Internal Structures

The Chromosome (Figure 3.44)

1. The **chromosome** of prokaryotes resides in the **nucleoid** rather than within a membrane nucleus.
2. The typical chromosome is a single, circular, double-stranded DNA molecule that contains all the genetic information required by a cell.

Plasmids

1. **Plasmids** are circular, double-stranded DNA molecules that typically encode genetic information that may be advantageous, but not required by the cell.
2. Populations of cells can gain and lose plasmids, depending on the relative advantages.

Ribosomes (Figure 3.45)

1. **Ribosomes** facilitate the joining of amino acids. The 70S bacterial ribosome is composed of a 50S and a 30S subunit.

Storage Granules (Figure 3.46)

1. **Storage granules** are dense accumulations of high molecular weight polymers, which are synthesized from a nutrient that a cell has in relative excess.

Gas Vesicles (Figure 3.47)

1. **Gas vesicles** are gas-permeable, water-impermeable rigid structures that provide buoyancy to aquatic cells, enabling the cell to float or sink to an ideal position in the water column.

Endospores

1. **Endospores** are a dormant stage produced by members of *Bacillus* and *Clostridium*; they can **germinate** to become a **vegetative cell**. (Figure 3.47)
2. Endospores are extraordinarily resistant to conditions such as heat, desiccation, toxic chemicals, and ultraviolet irradiation.
3. **Sporulation** is an 8-hour process initiated when cells are grown in nutrient-limiting conditions. (Figure 3.48)
4. **Germination** is the process by which an endospore leaves its dormant state.

THE EUKARYOTIC CELL

(Figure 3.49) (Table 3.5)

The Plasma Membrane

1. The **plasma membrane** is a phospholipid bilayer embedded with proteins. (Figure 3.51)
2. Proteins in the membrane are involved in transport, structural integrity, and **signaling**.

Transfer of Molecules Across the Plasma Membrane

Transport Proteins

1. **Channels** are pores in the membrane that are so small that only specific ions can pass through. These channels are gated.
2. Cells of multicellular organisms often take up nutrients by facilitated diffusion, because the nutrient concentration of their surrounding environment can be controlled.
3. **Carriers** involved in active transport include members of the major facilitator superfamily and ABC transporters.

Endocytosis and Exocytosis (Figure 3.52)

1. **Receptor-mediated endocytosis** is the most common form of endocytosis in animal cells. Receptors and their **ligands** are taken up this way. The **endocytic vesicle** fuses with an **endosome**, which then fuses with a **lysosome**.
2. Protozoa and phagocytes take up bacteria and debris through the process of **phagocytosis**. The **phagosome** fuses with the **lysosome**, where the material is digested.
3. **Exocytosis** expels products and is the reverse of endocytosis.

Secretion

1. Proteins that are destined for a non-cytoplasmic region are made by ribosomes bound to the **endoplasmic reticulum**. As the protein is made, it is threaded through the membrane and into the **lumen** of the endoplasmic reticulum.

Protein Structures Within the Cytoplasm

1. The **80S** eukaryotic ribosome is composed of **60S** and **40S subunits**.

Cytoskeleton (Figure 3.53)

1. **Microtubules** are the thickest of the cytoskeleton structures and are long hollow cylinders.
2. **Microfilaments** allow the cytoplasm to move and are composed of **actin**.
3. **Intermediate filaments** strengthen the cell mechanically.

Flagella and Cilia (Figure 3.54)

1. **Flagella** and **cilia** are composed of microtubules in a 9 + 2 arrangement.
2. Flagella propel a cell or pull the cell forward.

3. Cilia often cover the surface of a cell and move in synchrony to either propel a cell or move material along a stationary cell.

Membrane-Bound Organelles

The Nucleus (Figure 3.55)

1. The **nucleus**, which contains DNA, is the predominant distinguishing feature of eukaryotes.
2. Two membranes make up the **nuclear envelope**, which encloses the nucleus.
3. **Nuclear pores** allow large molecules to be transported into and out of the nucleus.
4. The **nucleolus** is where ribosomal RNAs are synthesized and, along with ribosomal proteins, are assembled into ribosomal subunits.

Mitochondria and Chloroplasts

1. Contained within the inner membrane of mitochondria and chloroplasts are the proteins of the electron transport chain and proteins that use proton motive force to generate ATP.
2. Mitochondria use the energy released during the degradation of organic compounds to generate ATP. (Figure 3.57)
3. Chloroplasts contain chlorophyll, which captures the energy of sunlight; this is then used to synthesize ATP. (Figure 3.58)

Endoplasmic Reticulum (Figure 3.59)

1. The **rough endoplasmic reticulum** is lined with ribosomes and serves as the site where proteins not located in the cytoplasm are synthesized.
2. Within the **smooth endoplasmic reticulum**, lipids are synthesized and degraded, and calcium is stored.

Golgi Apparatus (Figure 3.60)

1. The **Golgi apparatus** modifies and sorts molecules synthesized in the endoplasmic reticulum.
2. From the Golgi apparatus, molecules are delivered in **vesicles** to specific cellular compartments or to outside the cell.

Lysosomes and Peroxisomes

1. **Lysosomes** carry digestive enzymes.
2. **Peroxisomes** are the organelles in which oxygen is used to oxidize certain substances.

REVIEW QUESTIONS

Short Answer

1. Explain why resolving power is important in microscopy.
2. Explain why basic dyes are used more frequently than acidic dyes in staining.
3. Describe what happens at each step in the Gram stain.
4. Compare and contrast symporters, antiporters, and uniporters.
5. Give two reasons that the outer membrane of Gram-negative bacteria is significant medically.
6. Compare and contrast penicillin and lysozyme.
7. Describe how a plasmid can help and hinder a cell.
8. How is an organ different from tissue?
9. How is receptor-mediated endocytosis different from phagocytosis?
10. Explain how the Golgi apparatus cooperatively functions with the endoplasmic reticulum.

Multiple Choice

1. Which of the following is most likely to be used in a typical microbiology laboratory?
   A. Bright-field microscope
   B. Confocal scanning microscope
   C. Phase-contrast microscope
   D. Scanning electron microscope
   E. Transmission electron microscope
2. Which of the following stains is used to detect *Mycobacterium* species?
   A. Acid-fast stain
   B. Capsule stain
   C. Endospore stain
   D. Gram stain
   E. Simple stain
3. Penicillin
   1. is most effective against Gram-positive bacteria.
   2. is most effective against Gram-negative bacteria.
   3. functions in the cytoplasm of the cell.
   4. is effective against mycoplasma.
   5. kills only growing cells.

   A. 1,2 B. 2,3 C. 3,4 D. 4,5 E. 1,5
4. Endotoxin is associated with
   1. Gram-positive bacteria.
   2. Gram-negative bacteria.
   3. the cell wall.
   4. the endospore.
   5. the cytoplasmic membrane.

   A. 1,2 B. 2,3 C. 3,4 D. 4,5 E. 1,5
5. In prokaryotes, lipid bilayers are associated with the
   1. Gram-positive cell wall.
   2. Gram-negative cell wall.
   3. cytoplasmic membrane.
   4. capsule.
   5. nuclear membrane.

   A. 1,2 B.2,3 C. 3,4 D. 4,5 E. 1,5

6. In bacteria, the cytoplasmic membrane functions in
   1. protein synthesis.
   2. ribosome synthesis.
   3. generation of ATP.
   4. transport of molecules.
   5. attachment.

   A. 1,2 B. 2,3 C. 3,4 D. 4,5 E. 1,5
7. Attachment is mediated by the
   1. capsule.
   2. cell wall.
   3. cytoplasmic membrane.
   4. periplasm.
   5. pilus.

   A.1,2 B. 2,3 C. 3,4 D. 4,5 E. 1,5
8. Endocytosis is associated with
   1. mitochondria.
   2. prokaryotic cells.
   3. eukaryotic cells.
   4. clathrin.
   5. ribosomes.

   A. 1,2 B. 2,3 C. 3,4 D. 4,5 E. 1,5
9. Protein synthesis is associated with
   1. lysosomes.
   2. the cytoplasmic membrane.
   3. the Golgi apparatus.
   4. rough endoplasmic reticulum.
   5. ribosomes.

   A. 1,2 B. 2,3 C. 3,4 D. 4,5 E. 1,5
10. All of the following are composed of tubulin, *except*:

    A. actin

    B. cilia

    C. flagella

    D. microtubules

    E. more than one of the above

Applications

1. You are working in a pharmaceutical laboratory producing new antibiotics for human and veterinary use. One compound with potential value functions by inhibiting the action of prokaryotic ribosomes. The compound, however, was shown to inhibit the growth of animal cells in culture. What is one possible explanation for its effect on animal cells?
2. A research laboratory is investigating environmental factors that would inhibit the growth of Archaea. One question they have is if adding the antibiotic penicillin would be effective in controlling their growth. Explain the probable results of an experiment in which penicillin is added to a culture of Archaea.

Critical Thinking

1. The graph below shows facilitated diffusion of a compound across a cytoplasmic membrane and into a cell. As the external concentration of the compound is increased, the rate of uptake increases until it reaches a point where it slows and then begins to plateau. This is not the case with passive diffusion, where the

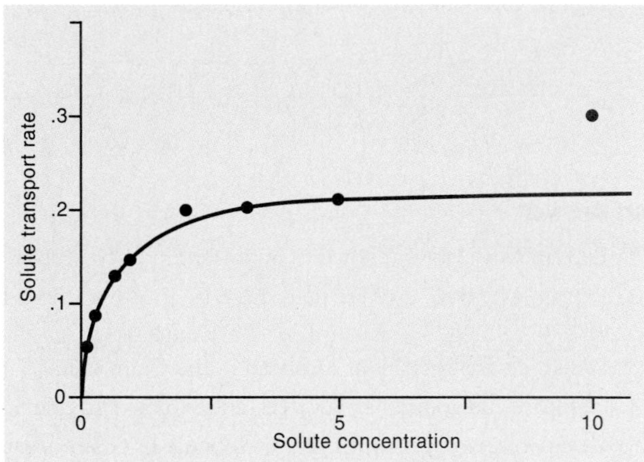

   rate of uptake continually increases. Why does the rate of uptake slow and then eventually plateau with facilitated diffusion?
2. Explain how the cell wall protects bacteria against a decrease, but not an increase, in the concentration of external solute.

Dynamics of Prokaryotic Growth

The greatest contributor to methods of culturing bacteria was Robert Koch (1843–1910), a German physician who combined a medical practice with a productive research career for which he received a Nobel Prize in 1905. Koch was primarily interested in identifying disease-causing bacteria. To do this, however, he soon realized it was necessary to have simple methods to isolate and grow these particular species. He recognized that a single bacterial cell could multiply on a solid medium in a limited area and form a distinct visible mass of descendants.

Koch initially experimented with growing bacteria on the cut surfaces of potatoes, but he found that a lack of nutrients in the potatoes prevented growth of some bacteria. To overcome this difficulty, Koch realized it would be advantageous to be able to solidify any liquid nutrient medium. Gelatin was used initially, but there were two major drawbacks—it melts at the temperature preferred by many medically important organisms and some bacteria can digest it. In 1882, Koch and others experimented with using agar. This solidifying agent was used to harden jelly at the time and proved to be the perfect answer.

Today, we take pure culture techniques for granted because of their relative ease and simplicity. Their development in the late 1800s, however, had a major impact on microbiology. Within 20 years, the agents causing most of the major bacterial diseases of humans were isolated and characterized. The concept that bacteria divide to form cells of similar shape and size to the original was established once and for all. The recognition that shapes, sizes, and metabolic properties characterize different bacterial species had a tremendous impact on the growth of the young science of bacteriology.

—A Glimpse of History

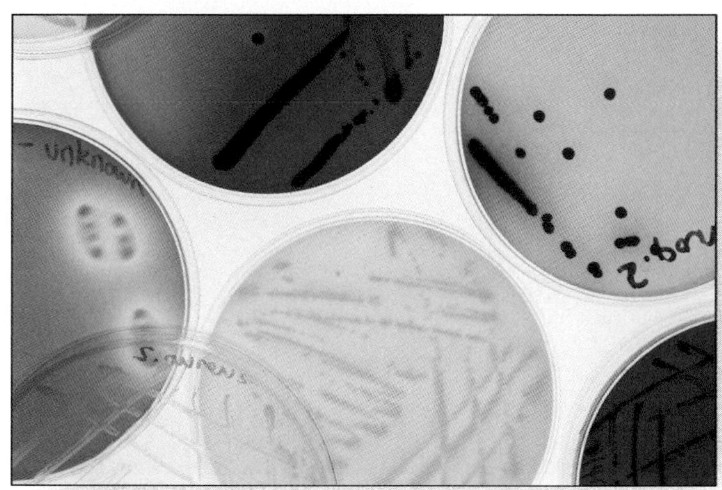

PROKARYOTES CAN BE FOUND GROWING EVEN IN the harshest climates and the most severe conditions. Environments that no unprotected human could survive, such as the ocean depths, volcanic vents, or the polar regions, have thriving species of prokaryotes. Indeed many scientists believe that if life exists on other planets, it may resemble these microorganisms. Each individual species, however, actually has a limited set of environmental conditions in which it can grow; even then, it will grow only if specific nutrients are available. Some prokaryotes can grow at temperatures above the boiling point of water but not at room temperature. Many species can only grow within an animal host, and then only in specific areas of that host.

Because of the medical significance of some bacteria, as well as the nutritional and industrial use of microbial by-products, microbiologists must be able to identify, isolate, and cultivate many species. To do this, one needs to understand the basic principles involved in prokaryotic growth while recognizing that there is a vast sea of information yet to be discovered.

Obtaining a Pure Culture

In nature, many different organisms, including bacteria, live together as a mixed population, jointly contributing to numerous activities and processes in their surroundings (**figure 4.1**). In the laboratory, bacteria are isolated and grown as a **pure culture** in order to study the functions of a particular species. A pure culture is defined as a population of organisms that descended from a single cell and is therefore separated from all other species. Only an estimated 0.1% of all prokaryotes, however, can currently be cultured successfully. This makes it exceedingly difficult to study the vast majority of environmental microorganisms. Fortunately for humanity, most known medically significant bacteria can be grown in pure culture.

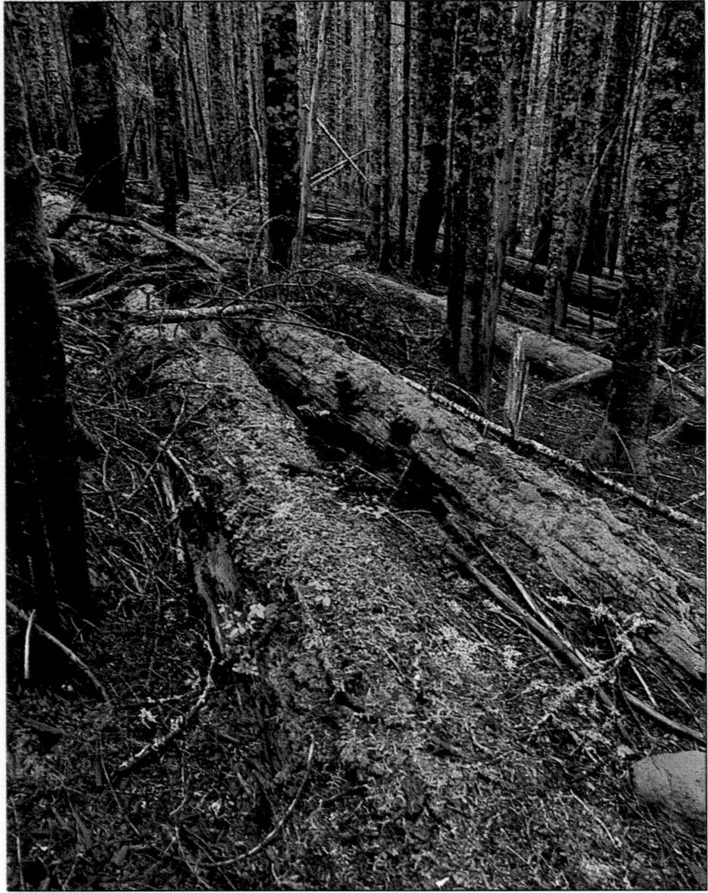

Figure 4.1 Bacteria Contribute to Numerous Activities in the Environment

Pure cultures are obtained using a variety of special techniques to isolate a single species from its natural environment. All glassware, media, and instruments must be **sterile**, or free of microbes, prior to use. These are then handled using **aseptic techniques**, which are procedures that minimize the chance of other organisms being accidentally introduced. The medium that the cells are grown in, or on, is a mixture of nutrients dissolved in water and may be in a liquid broth or a solidified gel-like form. ■ aseptic techniques, p. 115 ■ sterilization, p. 113

This section will describe the general techniques used to isolate and grow bacteria. A later section will explain the more specific requirements of various groups of prokaryotes.

Cultivating Bacteria on a Solid Medium

The basic requirements for obtaining a pure culture are a solid medium, a media container that can be maintained in an aseptic condition, and a method to separate individual cells. A single bacterium, supplied with the right nutrients, will multiply on the solid medium in a limited area to form a **colony**, which is a mass of cells all descended from the original one (**figure 4.2**). About 1 million cells are required for a colony to be easily visible to the naked eye.

Agar, a polysaccharide extracted from marine algae, is employed to solidify a specific nutrient solution to create the ideal medium on which to grow cells. Unlike other gelling agents such as gelatin, very few bacteria can degrade agar. It is not destroyed at high temperatures and can therefore be sterilized by heating, a process that also liquefies it. Melted agar will stay liquid until it is cooled to a temperature below 45°C. Therefore, nutrients that might be destroyed at high temperatures can be added at lower temperatures before the agar hardens. Once solidified, an agar medium will remain so until it is heated above 95°C. Thus, unlike gelatin, which is liquid at 37°C, agar remains solid over the entire temperature range at which the majority of bacteria grow. Agar is also translucent, enabling colonies imbedded in the solid medium to be seen more easily. ■ polysaccharide, p. 34

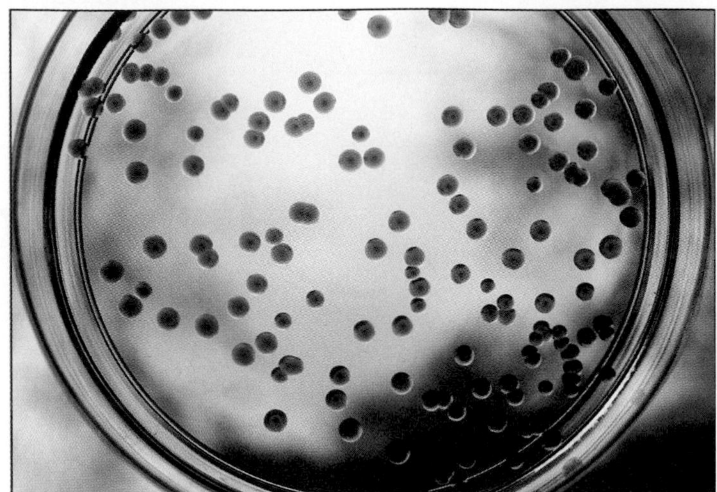

Figure 4.2 Colonies Growing on an Agar Medium

The culture medium is contained in a **Petri dish**. A Petri dish is a two-part, glass or plastic, covered container. While not airtight by design, the Petri dish does exclude airborne microbial contaminants. Once a Petri dish contains a medium, it is most commonly referred to as a plate of that medium type—for example, an **agar plate**, or a nutrient agar plate.

The Streak-Plate Method

The **streak-plate** method is the simplest and most commonly used technique for isolating bacteria (**figure 4.3**). A sterilized inoculating loop is dipped into a solution containing the organism of interest and is then lightly drawn several times across an agar plate, creating a set of parallel streaks covering approximately one-third of the plate. The loop is then sterilized and a new series of parallel streaks are made across and at an angle to the previous ones, covering another one-third of the plate. This action drags some of those cells streaked over the first portion of the plate to an uninoculated portion, creating a region containing a more dilute inoculum. The loop is sterilized again, and another set of parallel streaks are made, dragging into a third area some of the organisms that had been moved into the second section. The object of this is to reduce the number of cells being spread with each successive series of streaks, effectively diluting the concentration of cells. By the third set of streaks, cells should be separated enough so that distinct well-isolated colonies will form.

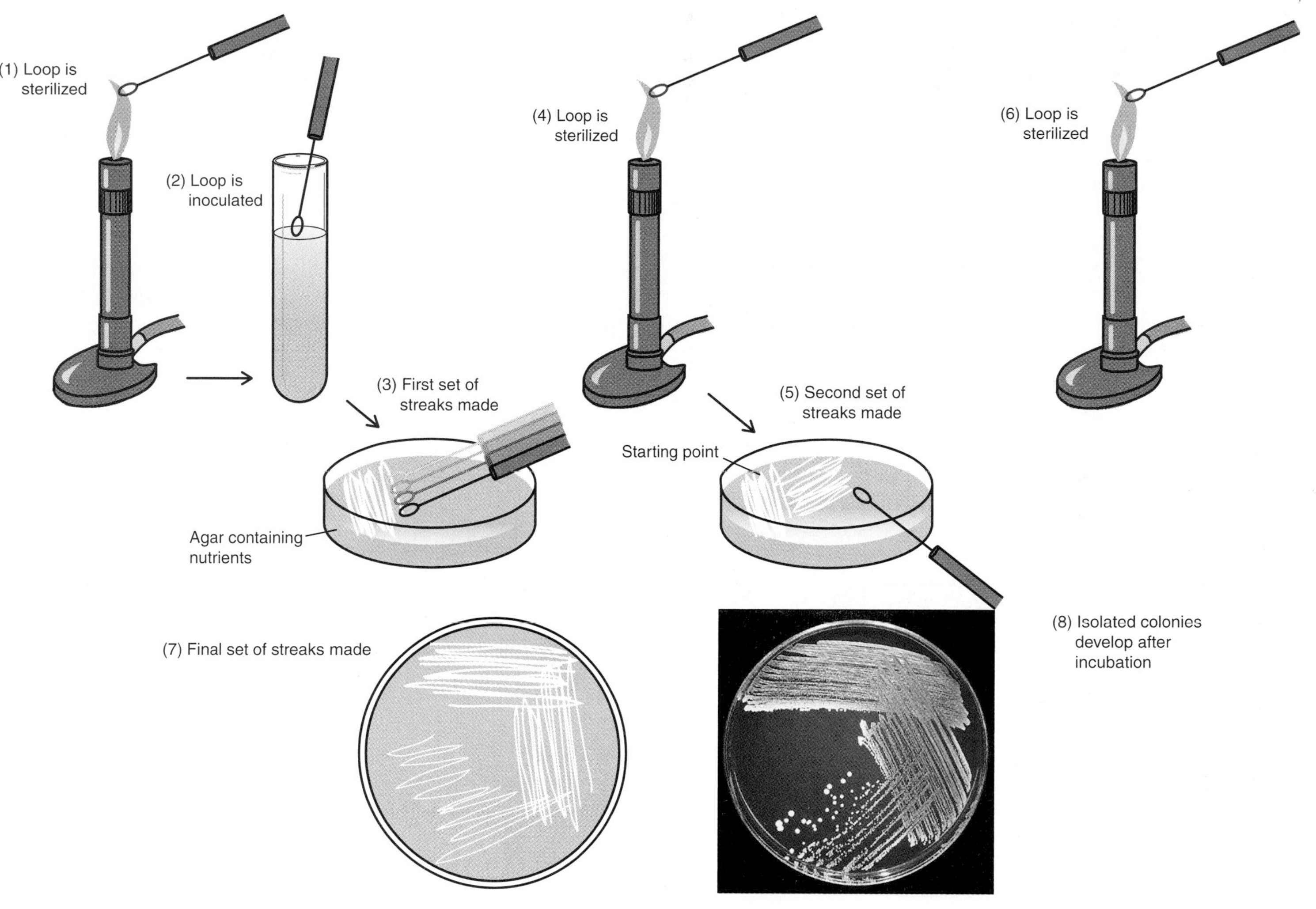
(1) Loop is sterilized
(2) Loop is inoculated
(3) First set of streaks made
Agar containing nutrients
(4) Loop is sterilized
(5) Second set of streaks made
Starting point
(6) Loop is sterilized
(7) Final set of streaks made
(8) Isolated colonies develop after incubation

Figure 4.3 The Streak-Plate Method A sterilized inoculating loop **(1)** is dipped into a culture **(2)** and is then lightly drawn several times across an agar plate **(3)**. The loop is sterilized again **(4)**, and a new series of streaks is made at an angle to the first set **(5)**. The loop is sterilized a final time **(6)**, and another set of parallel streaks is made **(7)**. The successive streaks dilute the concentration of cells. By the third set of streaks, cells should be separated enough so that isolated colonies develop after incubation **(8)**.

Maintaining Stock Cultures

Once a pure culture has been obtained, it can be maintained as a **stock culture**, a culture stored for use as an inoculum in later procedures. Often, stock cultures are stored in the refrigerator as growth on an **agar slant**. This is agar medium stored in a tube that was held at a shallow angle as the medium solidified, creating a larger surface area. For long-term storage, stock cultures can be frozen at −70°C in a glycerol solution. The glycerol prevents ice crystals from damaging cells. Alternatively, cells can be **lyophilized**, or freeze-dried. ■ **lyophilization, p. 128**

MICROCHECK 4.1

Only an estimated 0.1% of prokaryotes can be cultured in the laboratory. Agar is used to solidify nutrient-containing broth. The streak-plate method is used to obtain a pure culture.

- What properties of agar make it ideal for use in bacteriological media?
- How does the streak-plate method separate individual cells?
- What might be a reason that medically significant bacteria can be grown in pure culture more often than environmental organisms?

Principles of Bacterial Growth

Bacteria generally multiply by the process of **binary fission**. After a bacterial cell has increased in size and doubled the amounts of each of its parts, it divides (**figure 4.4**). One cell divides to become two, those two divide to become four, those four become eight, and so on. In other words, the increase in cell numbers is exponential. Because it is neither practical, nor particularly relevant, to determine the relative size of the cells in a given population, **microbial growth** is defined as an increase in the number of cells in a population.

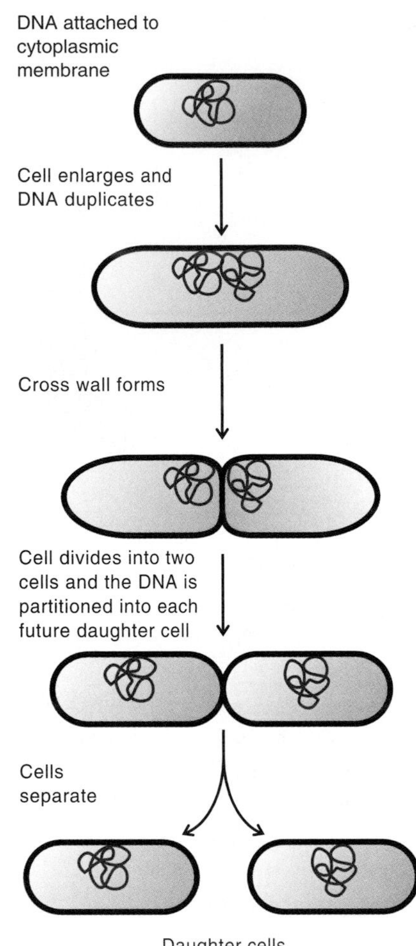
DNA attached to cytoplasmic membrane
Cell enlarges and DNA duplicates
Cross wall forms
Cell divides into two cells and the DNA is partitioned into each future daughter cell
Cells separate
Daughter cells

Figure 4.4 Binary Fission The chromosomal DNA is attached to the cytoplasmic membrane. As the membrane increases in length, the DNA is replicated and then partitioned into each of the two daughter cells.

The time it takes for a population to double in number is the **generation** or **doubling time**. This varies greatly depending on the on the species of the organism and the conditions in which it is grown. Some common organisms, such as *Escherichia coli*, can double in approximately 20 minutes; others, such as the causative agent of tuberculosis, *Mycobacterium tuberculosis*, require at least 12 to 24 hours to double even under the most favorable conditions. The environmental and nutritional factors that affect the rate of growth will be discussed shortly.

The exponential multiplication of bacteria has important health consequences. For example, a mere 10 cells of a foodborne pathogen in a potato salad, sitting for 4 hours in the warm sun at a picnic, may multiply to more than 40,000 cells. A simple equation expresses the relationship between the number of cells in a population at a given time (N_t), the original number of cells in the population (N_0), and the number of divisions those cells have undergone during that time (n). If any two values are known, the third can be easily calculated from the equation:

$$N_t = N_0 \times 2^n$$

In the above example, let us assume that we know that 10 cells of a disease-causing organism were initially added to the potato salad and we also know that the organism has a generation time of 20 minutes. The first step is to determine the number of cell divisions that will occur in a given time. Because the organism divides every 20 minutes, 3 times every hour, we know that in 4 hours it will divide 12 times. Now that we know the original number of cells and the number of divisions, we can solve for N_t:

$$10 \times 2^{12} = N_t = 40{,}960$$

Thus, after 4 hours the potato salad in the example will have 40,960 cells of our pathogen. Keep this in mind, and your potato salad in a cooler, the next time you go to a picnic!

MICROCHECK 4.2

Most bacteria multiply by binary fission. Microbial growth is an increase in the number of cells in a population. The time required for a population to double in number is the generation time.

- Explain why microbial growth refers to a population rather than a cell size.
- If a bacterium has a generation time of 30 minutes, and you start with 100 cells at time 0, how many cells will you have in 30, 60, 90, and 120 minutes?
- Why would placing potato salad in a cooler affect cell growth?

Methods to Detect and Measure Bacterial Growth

A variety of techniques are available to monitor bacterial growth, either by determining the number of cells in the population or their total mass, or by detecting their products. The choice depends on various characteristics of the sample and the goals of the measurements. The characteristics of the common methods for measuring bacterial growth are summarized in **table 4.1**.

Direct Cell Counts

Direct cell counts are particularly useful for determining the numbers of those bacteria that cannot be grown in culture. Unfortunately, they generally do not distinguish between living and dead cells. The simplest of these, the **direct microscopic count**, requires a relatively high concentration of bacteria in the sample being examined. More powerful methods use sophisticated equipment that count cells and other particles suspended in liquid.

Direct Microscopic Count

One of the most rapid methods of determining the number of cells in a suspension is the direct microscopic count. The number of bacteria in a measured volume of liquid is counted using special glass slides, **counting chambers**, that hold a known volume of liquid (**figure 4.5**). These can be viewed under the light microscope, and the number of bacteria contained in the liquid can be counted precisely. At least 10 million bacteria (10^7) per milliliter, however, are required to gain an accurate estimate. Otherwise, few, if any, cells will be seen in the microscope field.

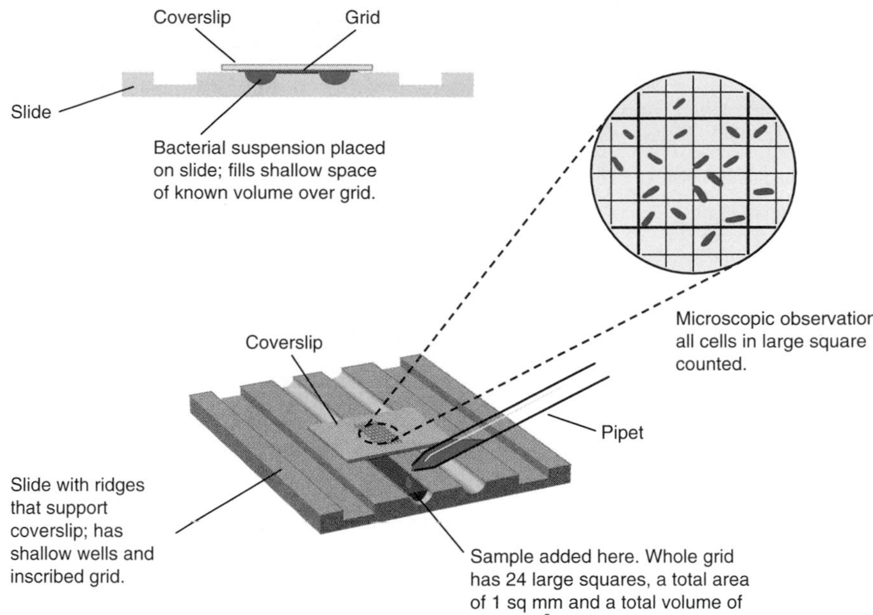

Figure 4.5 A Counting Chamber This special glass slide holds a known volume of liquid. The number of bacteria in that volume can be counted precisely.

Cell-Counting Instruments

A **Coulter counter** is an electronic instrument that counts cells in a suspension as they pass single file through a minute aperture **(figure 4.6)**. The suspending liquid must be saline or another conducting fluid, because the machine actually detects and subsequently counts brief changes in resistance that occur when nonconducting particles such as bacteria pass by.

A **flow cytometer** is similar in principle to a Coulter counter except that it measures the scattering of light by cells as they pass by a laser. The instrument can be used to count either total cells or, by using special techniques, a specific population of cells. This is done by first staining the cells with a fluorescent dye or tag that binds only to the cells of interest; the flow cytometer can count only those cells that carry the fluorescent marker. ■ **fluorescent dyes and tags, p. 52**

Viable Cell Counts

Viable cell counts are used to quantify the number of cells capable of multiplying. These methods require knowledge of the appropriate growth conditions for a particular microorganism as well as the time to allow growth to occur. They are invaluable for monitoring bacterial growth in samples, such as food and water, that often contain numbers too low to be seen using a direct microscopic count.

Plate counts and **membrane filtration** both measure the concentration of cells by determining the number of colonies that arise from a sample added to an agar plate. The **most probable number method** gives a statistical estimate of the number of cells based on a series of dilutions. With all of these methods, the numbers of particular species of bacteria can sometimes be determined by using selective and differential media. Characteristics of these types of media will be discussed later in the chapter.

Plate Counts

Plate counts measure the number of viable cells in a sample by exploiting the fact that an isolated cell on a nutrient agar plate will give rise to one colony. A simple count of the colonies determines how many cells were in the initial sample **(figure 4.7)**. The **pour-plate** and **spread-plate methods** differ in how the suspension of bacteria is applied to the agar plate. As the ideal number of colonies to count is between 30 and 300, and samples frequently contain many more bacteria than this, it is usually necessary to dilute the samples before plating out the

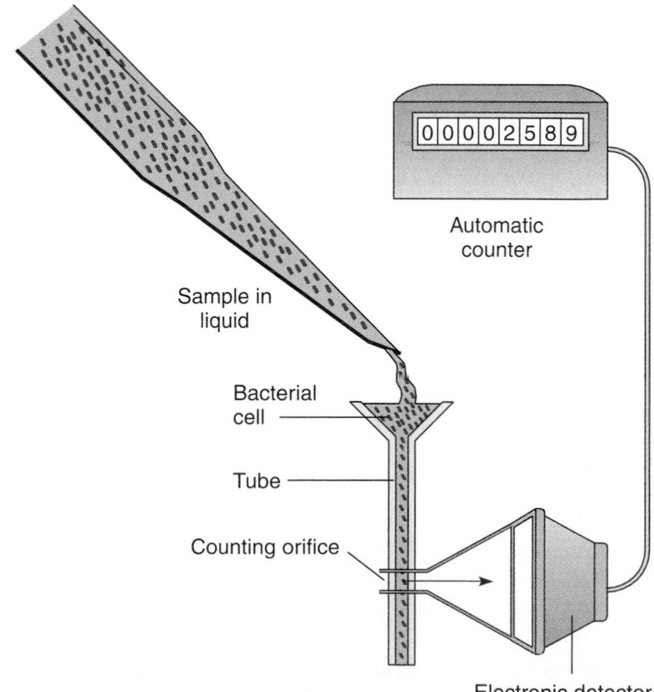

Figure 4.6 A Coulter Counter This instrument counts cells as they pass through a minute aperture. The bacteria, which are suspended in a conducting liquid, cause a brief change in resistance as they pass by the counter.

TABLE 4.1 Methods Used to Measure Bacterial Growth

Method	Characteristics and Limitations
Direct Cell Counts	Used to determine total number of cells; can be used for those bacteria that cannot be cultured.
Direct microscopic count	Rapid, but at least 10^7 cells/ml must be present to be effectively counted. Counts include living and dead cells.
Cell-counting instruments	Coulter counters and flow cytometers count total cells in dilute solutions. Flow cytometers can also be used to count organisms to which fluorescent dyes or tags have been attached.
Viable Cell Counts	Used to determine the number of viable bacteria in a sample, but only includes those that can grow in given conditions. Requires an incubation period of approximately 24 hours or longer. Selective and differential media can be used to enumerate specific species of bacteria.
Plate count	Time-consuming but technically simple method that does not require sophisticated equipment. Generally used only if the sample has at least 10^2 cells/ml.
Membrane filtration	Concentrates bacteria by filtration before they are plated; thus can be used to count cells in dilute environments.
Most probable number	Statistical estimation of likely cell number; it is not a precise measurement. Can be used to estimate numbers of bacteria in relatively dilute solutions.
Measuring Biomass	Biomass can be correlated to cell number.
Turbidity	Very rapid method; used routinely. A one-time correlation with plate counts is required in order to use turbidity for determining cell number.
Total weight	Tedious and time-consuming; however, it is one of the best methods for measuring the growth of filamentous microorganisms.
Chemical constituents	Uses chemical means to determine the amount of a given element, usually nitrogen. Not routinely used.
Measuring Cell Products	Methods are rapid but must be correlated to cell number. Frequently used to detect growth, but not routinely used for quantitation.
Acid	Titration can be used to quantify acid production. A pH indicator is often used to detect growth.
Gases	Carbon dioxide can be detected by using a molecule that fluoresces when the medium becomes slightly more acidic. Gases can be trapped in an inverted Durham tube in a tube of broth.
Luminescence	Firefly luciferase catalyzes light-emitting reaction when ATP is present.

cells. The sample is normally diluted in 10-fold increments, making the resulting math relatively simple. The **diluent**, or sterile solution used to make the dilutions, is generally physiological saline (0.85% NaCl in water). Distilled water can be used, but some bacteria may lyse in this hypotonic environment.

In the pour-plate method, 0.1–1.0 ml of the final dilution is transferred into a sterile Petri dish and then overlaid with melted nutrient agar that has been cooled to 50°C. Recall that at this temperature, agar is still liquid. The dish is then gently swirled to mix the bacteria with the liquid agar. When the agar hardens, the individual cells are fixed in place and, after incubation, form distinguishable colonies.

In the spread-plate method, 0.1–0.2 ml of the final dilution is transferred directly onto a plate already containing a solidified nutrient agar medium. This solution is then spread over the surface of the agar with a sterilized bent glass rod, which resembles a miniature hockey stick.

In both methods the plates are then incubated for a specific time to allow the colonies to form, which can then be counted. By knowing how much the sample was diluted prior to being plated, along with the amount of the dilution used in plating, the concentration of viable cells per milliliter in the original sample can then be calculated. Cells attached to one another form a single colony and are counted as a single cell or **colony-forming unit**.

Pour plates and spread plates are generally only used if a sample contains more than 100 organisms/ml. Otherwise, few if any cells will be transferred to the plates. In these situations, alternative methods give more reliable results.

Membrane Filtration

Membrane filtration is used when the numbers of organisms in a sample are relatively low, as might occur in dilute environments such as natural waters. This method concentrates the bacteria by filtration before they are plated. A known volume of liquid is passed through a sterile membrane filter, which has a pore size that causes retention of bacteria **(figure 4.8)**. The filter is subsequently placed on an appropriate agar medium and then incubated. The number of colonies that grow on the filter indicates the number of bacteria that were in the volume filtered.

Most Probable Number (MPN)

The **most probable number** (**MPN**) method is a statistical assay of cell numbers based on the theory of probability. The

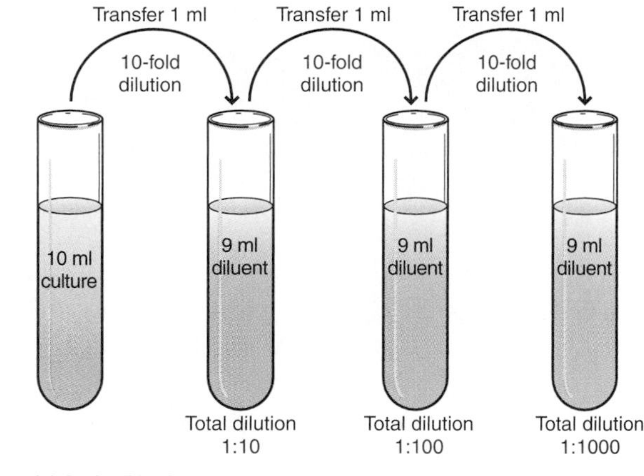

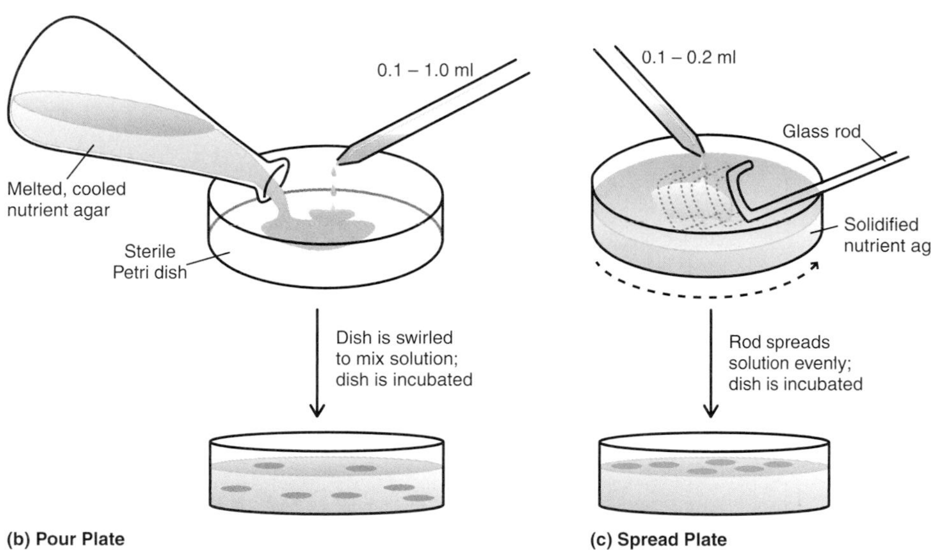

Figure 4.7 **Plate Counts** **(a)** A sample is first diluted in 10-fold increments. **(b)** In the pour-plate method, 0.1–1.0 ml of a dilution is transferred to a sterile Petri dish and overlaid with melted, cooled nutrient agar. When the agar hardens, the plate is incubated and distinguishable colonies form on the surface and within the agar. **(c)** In the spread-plate method, 0.1–0.2 ml of a dilution is spread on a hardened agar dish with a sterile glass rod. After incubation, distinguishable colonies form only on the surface of the agar.

goal is to successively dilute a sample and determine the point at which subsequent dilutions receive no cells.

To determine the MPN, three sets of three or five tubes containing the same growth medium are prepared **(figure 4.9)**. Each set receives a measured amount of a sample such as water, soil, or food. The amount added is determined, in part, by the expected bacterial concentration in that sample. What is important is that the second set receives 10-fold less than the first, and the third set 100-fold less. In other words, each set is inoculated with an amount 10-fold less than the previous set. After incubation, the presence or absence of growth in each tube in each set is noted; in some cases, growth along with a characteristic visible change such as gas production is noted. The results are then compared against an MPN table, which gives a statistical estimate of the cell concentration. The MPN method is most commonly used to determine the approximate number of **coliforms** in a water sample. Coliforms are lactose-fermenting, Gram-negative rods that typically reside in the intestine and thus serve as a bacterial indicator of fecal contamination. ■ coliforms, p. 791

Measuring Biomass

Instead of measuring the number of cells, the cell mass can be determined. This can be done by measuring the turbidity, the total weight, or the precise amount of chemical constituents such as nitrogen. These all relate to the number of cells present.

Turbidity

Cloudiness or **turbidity** of a bacterial suspension such as a broth culture is due to the scattering of light passing through the liquid by cells. The amount scattered is proportional to the number of cells. To measure turbidity, a **spectrophotometer** is used. This instrument transmits light through a specimen and measures the percentage that reaches a light detector **(figure 4.10)**. That number is inversely proportional to the optical density. To use turbidity to estimate cell numbers, a one-time correlation between optical density and cell concentration for the specific organism under study must be made. Once this correlation has been determined, the turbidity measurement becomes a rapid and relatively accurate assay.

One limitation of assaying turbidity is that a medium must contain relatively high numbers of bacteria in order to be cloudy. One milliliter of a solution containing 1 million bacteria (10^6) is still perfectly clear, and if it contains 10 million cells (10^7), it is barely turbid. Thus, although a turbid culture indicates that bacteria are present, a clear solution does not guarantee their absence. Not recognizing these facts can have serious consequences in the laboratory as well as outside. Experienced hikers, for example, know that the clarity of mountain streams does not necessarily mean that the water is free of *Giardia* or other harmful organisms. ■ giardiasis, p. 613

Total Weight

Determining the total weight of a culture is a tedious and time-consuming method that is not used routinely. It can be invaluable, however, for measuring the growth of filamentous organisms. These do not readily break up into individual cells, which are necessary for a valid plate count. To measure the **wet weight**, cells growing in liquid culture are centrifuged down and the liquid removed. The weight of the resulting packed cell mass is proportional to the number of cells in the culture. The **dry weight** of the cell mass can be determined by drying the centrifuged cells at

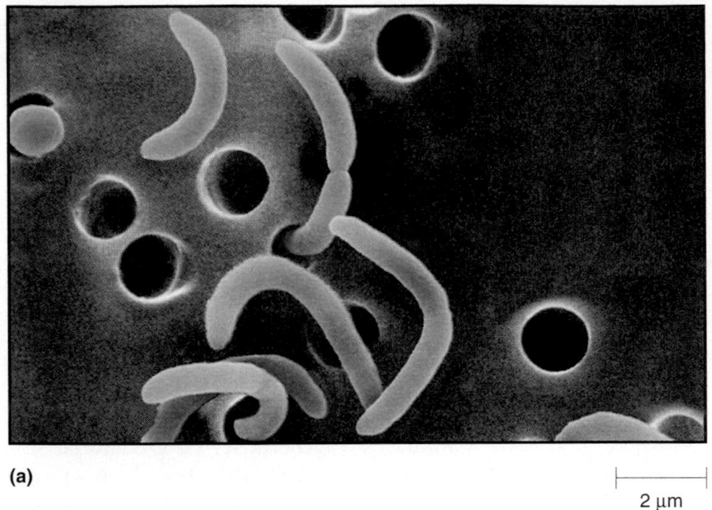

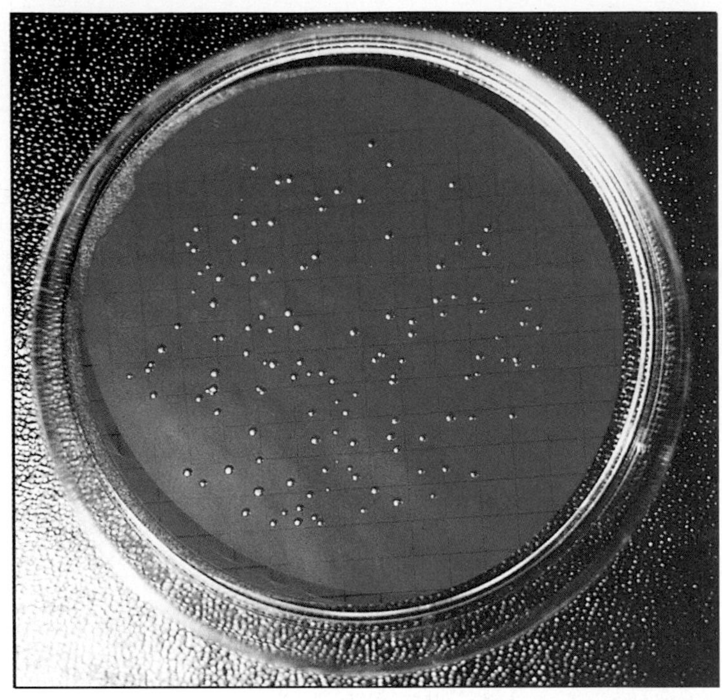

Figure 4.8 Membrane Filtration This technique concentrates bacteria before they are plated. **(a)** A known volume of liquid is passed through a sterile membrane filter, which has a pore size that retains bacteria. **(b)** The filter is then placed on an appropriate agar medium and incubated. The number of colonies that grow on the filter indicates the number of bacteria that were in the volume filtered.

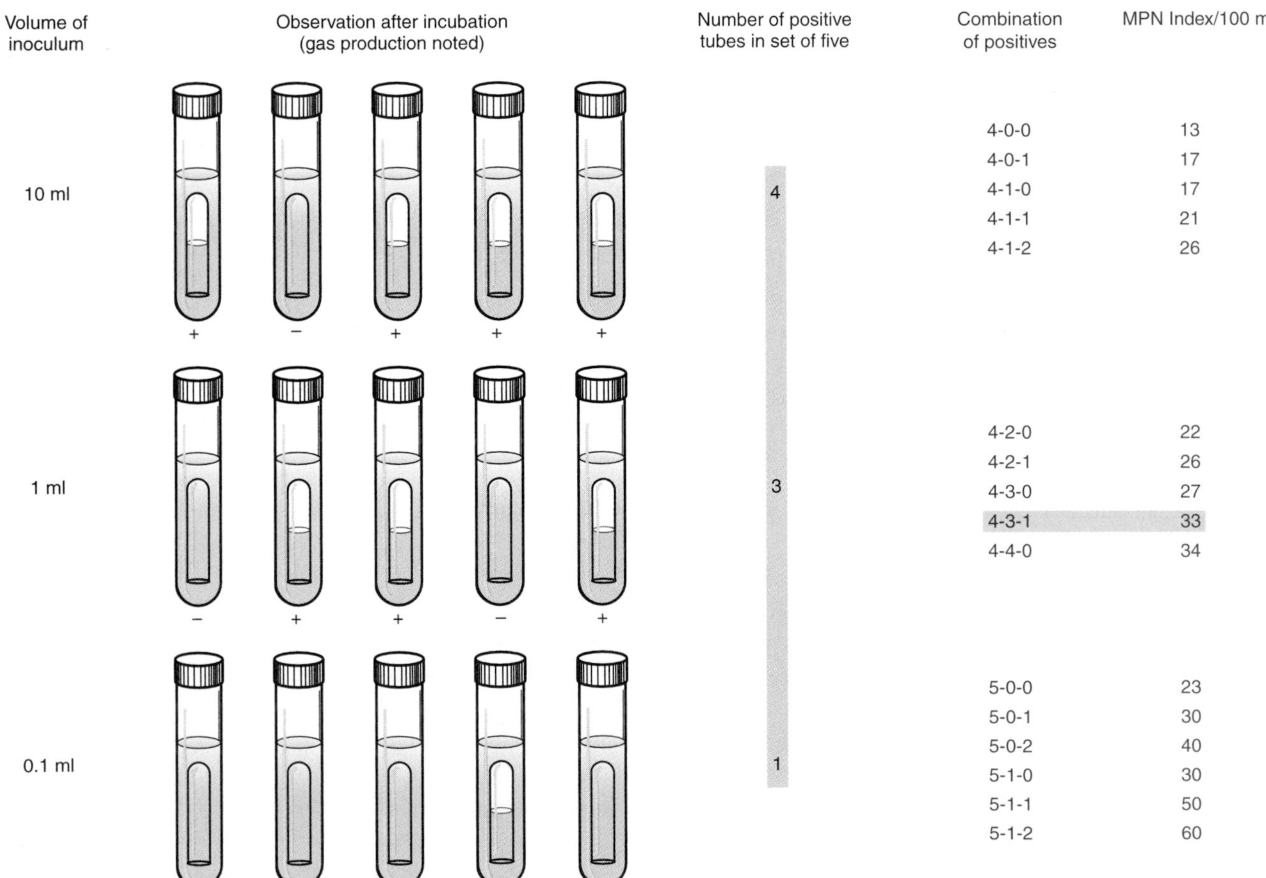

Figure 4.9 The Most Probable Number (MPN) Method Three sets of three or five tubes containing the same growth medium are prepared. Each set receives a measured amount of a sample; each set receives an amount 10-fold less than the first. After incubation the presence or absence of growth in each tube is noted. The results are then compared to an MPN table, which gives a statistical estimate of cell concentration.

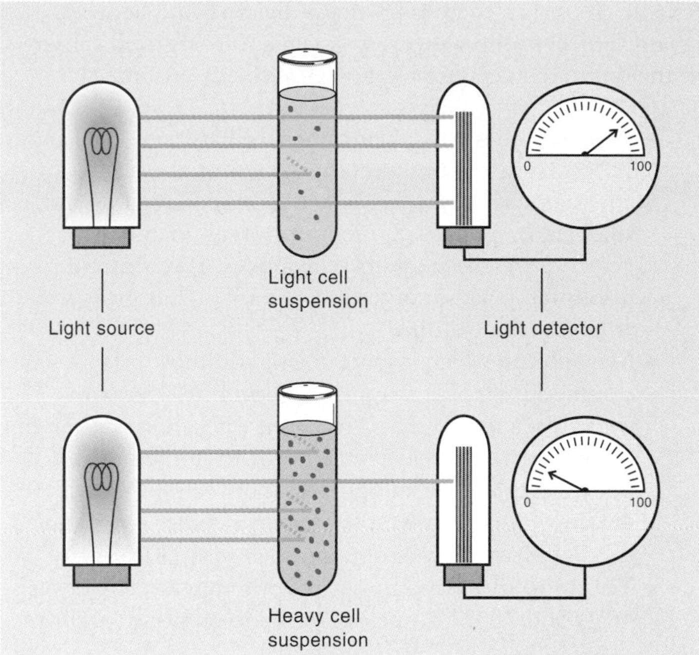

Figure 4.10 Measuring Turbidity with a Spectrophotometer The percentage of light that reaches the detector is inversely proportional to the optical density. To use turbidity to estimate cell number, a one-time experiment must be done to determine the correlation between cell concentration and optical density of a culture.

approximately 100°C for 8 to 12 hours before weighing them. About 70% of the weight of a cell is water.

Chemical Constituents

The quantity of a chemical constituent of a cell, typically nitrogen, can be determined and then used to calculate the biomass. For example, cells can be treated with sulfuric acid, which converts cellular nitrogen to ammonia. The amount of ammonia can then be easily assayed. Because cells are typically composed of 14% nitrogen, the biomass can be mathematically derived from the amount of ammonia released.

Detecting Cell Products

The products of microbial growth can be used to estimate the number of microorganisms or, more commonly, to confirm their presence. These products include acids, gases such as CO_2, and ATP.

Acid Production

As a consequence of the breakdown of sugars, which are used as an energy source, microorganisms produce a variety of acids. The precise amount of acid can be measured using chemical means such as titration. Most commonly, however, acid production is used to detect growth by incorporating a **pH indicator** into a medium. A pH indicator changes from one color to another as the pH of a medium changes. Several pH indicators are available, and they differ in the value at which their color changes.

Gases

The production of gases can be monitored in several ways. A method used in clinical labs employs a fluorescent molecule to detect bacteria growing in blood taken from patients who are suspected of having a bloodstream infection. The slight decrease in pH that accompanies the production of CO_2 increases the fluorescence. A method routinely used in the laboratory is to include an inverted small tube, called a **Durham tube**, in a broth of sugar-containing media. If bacteria produce gas as a result of degradation of the sugar, bubbles will be trapped in the tube.

ATP

The presence of ATP, the universal form of energy, can be detected by adding the firefly enzyme luciferase. The enzyme catalyzes a chemical reaction that uses ATP as an energy source to produce light. This method is sometimes used to assess the effectiveness of chemical agents formulated to kill bacteria. Light is produced only if viable organisms remain.

MICROCHECK 4.3

Direct microscopic counts and cell-counting instruments generally do not distinguish between living and dead cells. Plate counts determine the number of cells capable of multiplying; membrane filtration can be used to concentrate the sample. The most probable number is a statistical assay based on the theory of probability. Turbidity of a culture is a rapid measurement that can be correlated to cell number. The total weight of a culture and the amount of certain cell constituents can be correlated to the number of cells present. Microbial growth can be detected by the presence of cell products such as acid, gas, and ATP.

- Why is an MPN an estimate rather that an accurate number?
- Why would a direct microscopic count yield a higher number than a pour plate if a sample of seawater was examined by both methods?
- The nitrogen in microorganisms will typically be present in what molecules?

Environmental Factors that Influence Microbial Growth

Prokaryotes, as a group, inhabit nearly every conceivable environment on earth. Many live in habitats that humans consider quite comfortable. These are the organisms we associate with disease and food spoilage. Some prokaryotes preferentially live and multiply in harsh environments that would quickly kill most other organisms. Most of these, called **extremophiles**, are members of the Domain Archaea.

Recognizing the environmental factors that influence microbial growth, such as temperature, amount of oxygen, pH, and water availability, enables scientists to study microorganisms in the laboratory and helps them gain an understanding of the role they play in the complex ecology of the planet. The major environmental conditions that influence the growth of microorganisms are summarized in **table 4.2**.

Temperature Requirements

Each species of prokaryote has a well-defined upper and lower temperature limit within which they grow and outside of which growth stops. The temperature span between these limits is usually about 25°C. Within this range lies their **optimum growth temperature**, the temperature at which the organism multiplies most rapidly. As a general rule, this optimum temperature is close to the upper limit of the organism's range. This is because the speed of enzymatic reactions in the cell approximately doubles for each 10°C rise in temperature. At a critical point, however, the temperature becomes too high and enzymes required for growth are denatured and can no longer function. As a result, the cells die.

Prokaryotes are commonly divided into four groups based on their optimum growth temperatures (**figure 4.11**). Note, however, that this merely represents a convenient organization scheme. In reality, there is no sharp dividing line between each group. Furthermore, not every organism in a group can grow in the entire temperature range typical for its group.

- **Psychrophiles** have their optimum between –5°C and 15°C. These organisms are usually found in such environments as the Arctic and Antarctic regions and in lakes fed by glaciers. Some, but not all, members of the genus *Pseudomonas* are psychrophiles. Psychrotrophs have a temperature optimum of >15°C, but grow well at lower temperatures.
- **Mesophiles**, which include *E. coli* and most other common bacteria, have their optimum temperature within the range of 25°C to about 45°C. Disease-causing bacteria, which are adapted to growth in the human body, typically have an optimum between 35°C and 40°C. Mesophiles that inhabit soil, a colder environment, generally have a lower optimum, close to 30°C.
- **Thermophiles** have an optimum temperature between 45°C and 70°C. These organisms commonly occur in

Table 4.2 Environmental Factors that Influence Microbial Growth

Environmental Factor/ Descriptive Terms	Characteristics
Temperature	Thermostability appears to be due to protein structure.
Psychrophile	Optimum temperature between –5°C and 15°C.
Mesophile	Optimum temperature between 25°C and 45°C.
Thermophile	Optimum temperature between 45°C and 70°C.
Hyperthermophile	Optimum temperature between 70°C and 110°C.
Oxygen (O_2) Availability	Oxygen (O_2) requirement/tolerance reflects the organism's energy-converting mechanisms (aerobic respiration, anaerobic respiration, and fermentation) and its ability to detoxify O_2 derivatives.
Obligate aerobe	Requires O_2.
Obligate anaerobe	Cannot multiply in the presence of O_2.
Facultative anaerobe	Grows best if O_2 is present, but can also grow without it.
Microaerophile	Requires small amounts of O_2, but higher concentrations are inhibitory.
Aerotolerant anaerobe (obligate fermenter)	Indifferent to O_2.
pH	Prokaryotes that live in pH extremes appear to maintain a near neutral internal pH by transporting protons across the membrane.
Neutrophile	Multiplies in the range of pH 5 to 8.
Acidophile	Grows optimally at a pH below 5.5.
Alkaliphile	Grows optimally at a pH above 8.5.
Water Availability	Prokaryotes that can grow in high solute solutions increase their internal solute concentration by either pumping ions into the cell or synthesizing certain small organic compounds.
Osmotolerant	Can grow in relatively high salt solutions, up to approximately 10% NaCl.
Halophile	Requires sodium chloride; some require concentrations of 20% or greater.

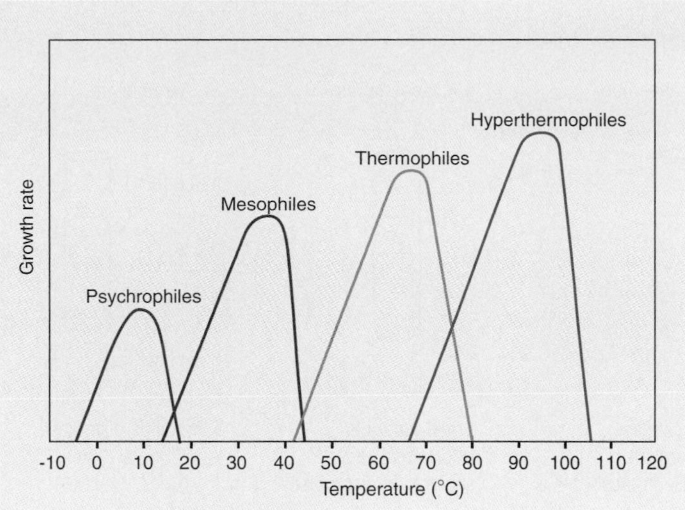

Figure 4.11 Temperature Requirements for Growth Prokaryotes are commonly divided into four groups based on their optimum growth temperatures. This graph depicts the typical temperature ranges of these groups. Note that the optimum temperature, the point at which the growth rate is highest, is near the upper limit of the range.

hot springs and compost heaps. They also are found in artificially created thermal environments such as water heaters and nuclear power plant cooling towers. *Lactobacillus delbrueckii* subspecies *bulgaricus* is a thermophile used in yogurt production. ■ **composting, p. 799** ■ **yogurt production, p. 808**

- **Hyperthermophiles** have an optimum growth temperature between 70°C and 110°C. These are usually members of the Archaea. *Pyrolobus fumarimii*, which was isolated from the wall of a hydrothermal vent at an ocean depth of 3650 meters, has a maximum growth temperature of 113°C, the highest yet recorded.

Why can some prokaryotes withstand such high temperatures but most cannot? As a general rule, proteins from thermophiles are not denatured at high temperatures. This thermostability is due to the sequence of the amino acids in the protein. These control the number and position of the bonds that form within the protein, which in turn determines its three-dimensional structure. For example, the formation of many covalent bonds, as well as many hydrogen and other weak bonds, prevents denaturation of proteins. Heat-stable proteins, which include enzymes that degrade fats and other proteins, are being used in high-temperature detergents. ■ **protein denaturation, p. 32**

Temperature and Food Preservation

Storage of fruits, vegetables, and cheeses at refrigeration temperatures (approximately 4°C) retards food spoilage because it limits the growth of otherwise fast-growing mesophiles. Psychrophiles and psychrotrophs, however, can still multiply at these temperatures, and consequently spoilage will still occur, albeit more slowly. Because of this, foods and other perishable products that can withstand below-freezing temperatures should be frozen for long-term storage. Microorganisms, which require liquid water to grow, cannot multiply under these conditions. It is important to recognize, however, that freezing is not an effective means of destroying microbes. Recall that freezing is routinely used to preserve stock cultures. ■ **low-temperature storage, p. 128**

Temperature and Disease

Wide variations exist in the temperature of various parts of the human body. Although the heart, brain, and gastrointestinal tract are near 37°C, the temperature of the extremities may be much less. For these reasons, some microorganisms can cause disease in certain body parts but not in others. For example, Hansen's disease (leprosy) typically involves the coolest regions of the body (ears, hands, feet, and fingers) because the causative organism, *Mycobacterium leprae*, grows best at these lower temperatures. The same situation applies to syphilis, in which lesions appear on the genitalia and then on the lips, tongue, and throat. Indeed, for more than 30 years the major treatment of syphilis was to induce fever by deliberately introducing the agent that causes malaria, which induces very high fevers. Hot-bath spas were commonly recommended as a less drastic treatment. ■ **Hansen's disease, p. 672** ■ **syphilis, p. 647**

Oxygen (O_2) Requirements

The oxygen (O_2) level in different environments varies greatly, providing many different niches with respect to its availability. Gaseous oxygen accounts for about 20% of the earth's atmosphere. Beneath the surface of soil and in swamps, however, very limited amounts, if any, may be available. The human body alone provides many different niches. While the surface of the skin is exposed to the atmosphere, the stomach and intestines are **anaerobic**, or contain no O_2.

Like humans, some bacteria have an absolute requirement for O_2. Others thrive in anaerobic environments, and many of these are even killed if O_2 is present. The O_2 requirements of some organisms can be determined by growing them in a **shake tube**. To prepare a shake tube, a tube of nutrient agar is boiled, which both melts the agar and drives off the O_2. The agar is then allowed to cool to 50°C. The test organism is then added and dispersed by gentle shaking or swirling. The agar is allowed to harden and the tube is incubated at an appropriate temperature. Because the solidified agar impedes the diffusion of O_2, the level of O_2 in the tube is stratified—O_2 levels are high at the top, whereas the bottom portion is anaerobic. The bacteria grow in the region that has the level of O_2 that suits their requirements **(table 4.3)**.

Based on their O_2 requirements, prokaryotes can be separated into the following groups:

- **Obligate aerobes** have an absolute or obligate requirement for oxygen (O_2). They use it to generate energy in the process of aerobic respiration. This and other energy-generating pathways will be discussed in detail in chapter 6. Obligate aerobes include members of the genus *Pseudomonas*, a diverse group of Gram-negative rods that are common in the environment. ■ **aerobic respiration, p. 140**

Table 4.3 Oxygen (O_2) Requirements of Prokaryotes

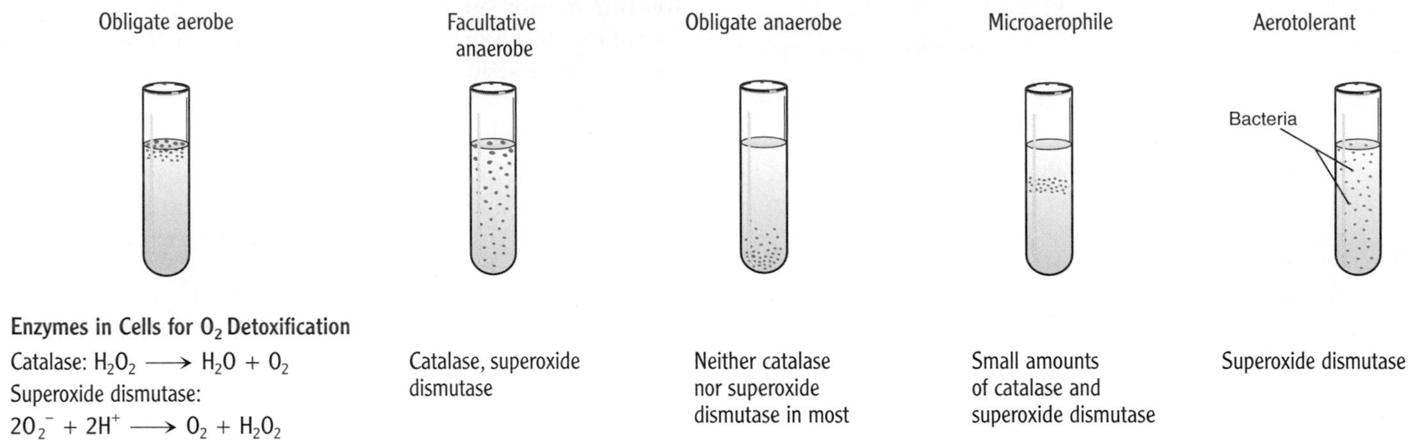

	Obligate aerobe	Facultative anaerobe	Obligate anaerobe	Microaerophile	Aerotolerant
Enzymes in Cells for O_2 Detoxification	Catalase: $H_2O_2 \longrightarrow H_2O + O_2$ Superoxide dismutase: $2O_2^- + 2H^+ \longrightarrow O_2 + H_2O_2$	Catalase, superoxide dismutase	Neither catalase nor superoxide dismutase in most	Small amounts of catalase and superoxide dismutase	Superoxide dismutase

- **Obligate anaerobes** cannot multiply if any O_2 is present; in fact, they are often killed by traces of O_2 because of its toxic derivatives, which will be discussed shortly. Obligate anaerobes may transform energy by fermentation or by anaerobic respiration; the details of these processes will be discussed in chapter 6. Obligate anaerobes include members of the genus *Bacteroides*, which are the major inhabitants of the large intestine. Another obligate anaerobe is *Clostridium botulinum*, the causative agent of botulism. It is estimated that one-half of all the cytoplasm on earth is in anaerobic bacteria! ■ **fermentation, p. 140** ■ **anaerobic respiration, p. 140**
- **Facultative anaerobes** grow better if O_2 is present, but can also grow without it. The term facultative means that the organism is flexible, in this case in its requirements for O_2. Facultative anaerobes use aerobic respiration if oxygen is available, but use fermentation or anaerobic respiration in its absence. Growth is more rapid when oxygen is present because aerobic respiration yields the most ATP of all these processes. Examples of facultative anaerobes include *E. coli*, a common inhabitant of the large intestine, and the yeast *Saccharomyces* (a eukaryote), which is used to make bread and alcoholic beverages.
- **Microaerophiles** require small amounts of O_2 (2% to 10%) for aerobic respiration; higher concentrations are inhibitory. Examples include *Spirillum volutans*, which is common in aquatic habitats, and *Helicobacter pylori*, which causes gastrointestinal ulcers.
- **Aerotolerant anaerobes** are indifferent to O_2. They can grow in its presence, but they do not use it to transform energy. Because they do not use aerobic or anaerobic respiration, they are also called **obligate fermenters**. They include *Lactobacillus bulgaricus*, which is used in cheese-making, and *Streptococcus pyogenes*, which causes strep throat.

Toxic Derivatives of Oxygen (O_2)

Although not toxic itself, O_2 can be converted into a number of compounds that are highly toxic. Some of these, such as **superoxide** (O_2^-), are produced both as a part of normal metabolic processes and as chemical reactions involving oxygen in light. Others, such as hydrogen peroxide (H_2O_2), result from metabolic processes involving oxygen. To survive in an environment containing O_2, cells must have enzymes that can convert these toxic compounds to nontoxic forms. The enzyme **superoxide dismutase** degrades superoxide to produce hydrogen peroxide. **Catalase** breaks down hydrogen peroxide to H_2O and O_2. Together, these two enzymes detoxify these reactive products of O_2.

$$\underset{\text{Superoxide}}{2O_2^-} + 2H^+ \xrightarrow{\text{Superoxide dismutase}} O_2 + \underset{\text{Hydrogen peroxide}}{H_2O_2} \qquad \underset{\text{Hydrogen peroxide}}{H_2O_2} \xrightarrow{\text{Catalase}} H_2O + O_2$$

Although most strict anaerobes do not have superoxide dismutase, some do, while a few aerobes lack it. Therefore, other factors must also be playing a role in protecting organisms from the toxic forms of oxygen.

pH

Each bacterial species can survive within a range of pH values; within this range, it has a pH optimum. Despite the pH of the external environment, cells maintain a constant internal pH, typically near neutral. ■ **pH, p. 25**

Most bacteria can live and multiply within the range of pH 5 (acidic) to pH 8 (basic) and have a pH optimum near neutral (pH 7). These bacteria are called **neutrophiles**. Preservation methods that acidify foods, such as pickling, are intended to inhibit these organisms. Surprisingly, some neutrophiles have adapted special mechanisms that enable them to grow at a very low pH. For example, *Helicobacter pylori* grows in the stomach, where it can cause ulcers. To maintain the pH close to neutral in its immediate surroundings, *H. pylori* produces the enzyme

urease, which splits urea in the stomach into carbon dioxide and ammonia. The ammonia neutralizes the stomach acid in the bacterium's immediate surroundings. ■ pickling, p. 808

Acidophiles grow optimally at a pH below 5.5. For example, *Thiobacillus ferroxidans* grows best at a pH of approximately 2.0. This bacterium obtains its energy by oxidizing sulfur compounds, producing sulfuric acid in the process. It maintains its internal pH near neutral by pumping out protons (H^+) as quickly as they enter the cell. *Picrophilus oshimae*, which is a member of the Archaea, has an optimum pH of less than 1! This prokaryote, which was isolated from the dry, acid soils of a gas-emitting volcanic fissure in Japan, appears to cope with the low pH by virtue of its unusual cytoplasmic membrane, which is unstable at a pH above 4.0.

Alkaliphiles grow optimally at a pH above 8.5. For example, *Bacillus alcalophilus* grows best at pH 10.5. It appears alkaliphiles maintain a relatively neutral internal pH by using an antiporter that exchanges internal sodium ions for external protons. Alkaliphiles often live in alkaline lakes and soils. ■ antiporter, p. 60

Water Availability

All microorganisms require water for growth. Even if water is present, however, it may not be available in certain environments. For example, dissolved substances such as salt (NaCl) and sugars interact with the water molecules and make that water unavailable to the cell. In any environment, particularly in certain natural habitats such as salt marshes, prokaryotes are faced with this situation. If the solute concentration is higher in the medium than in the cell, water diffuses out of the cell. This can cause the cytoplasm to dehydrate and shrink from the cell wall, a phenomenon called **plasmolysis (figure 4.12)**. ■ solute, p. 58

Prokaryotes can maintain the availability of water in a high salt environment by increasing the solute concentration inside the cell. This can be done in two ways. The organism can pump ions, most commonly potassium (K^+), from the outside to the inside of the cell, or it can synthesize certain small organic compounds such as the amino acid proline and sugar alcohols, which have no effect on normal cellular activity. Bacteria that can tolerate relatively high salt concentrations, up to approximately 10% NaCl, are called **osmotolerant**. *Staphylococcus* species, which reside on the dry salty environment of the skin, are osmotolerant. Certain members of the Archaea can live in very high salt solutions. These organisms, called **halophiles** (*halo* means "salt" and *phile* means "loving"), are found in such environments as the salt flats in Utah and the Dead Sea. ■ proline, p. 29

This growth-inhibiting effect of high concentrations of salt and sugars is used in food preservation. High levels of salt are added to preserve such foods as bacon, salt pork, and anchovies. High concentrations of sugars can also inhibit the growth of bacteria. Many foods with high sugar content, such as jams, jellies, honey, preserves, and sweetened condensed milk, are naturally preserved. ■ drying of foods, p. 128

MICROCHECK 4.4

A prokaryotic species can be grouped as a psychrophile, mesophile, thermophile, or hyperthermophile, according to its optimum temperature. A species can also be grouped according to its oxygen requirements as an obligate aerobe, obligate anaerobe, facultative anaerobe, microaerophile, or aerotolerant anaerobe. Most species are neutrophiles and grow best near neutral pH; acidophiles prefer acidic conditions, and alkaliphiles grow best in alkaline conditions. All organisms require water for growth, but halophiles grow best in high-salt conditions.

- Describe four environmental factors that influence the growth of bacteria.
- List the categories into which bacteria can be classified according to their requirements for oxygen.
- Why would small organic compounds affect the water content of cells?

Nutritional Factors that Influence Microbial Growth

The growth of any bacterium depends not only on a suitable physical environment, but also an available source of chemicals to use as nutrients. From these, the cell must synthesize all of the cell components discussed in the previous chapter, including lipid membranes, cell walls, proteins, and nucleic acids. These components are made from building blocks such as fatty acids, sugars, amino acids, and nucleotides. In turn, each of these building blocks is composed of a variety of elements, including carbon and nitrogen. What sets the prokaryotic world apart from all other forms of life is their remarkable ability to utilize diverse sources of these elements. For example, prokaryotes are the only organisms able to utilize atmospheric nitrogen (N_2) as a nitrogen source.

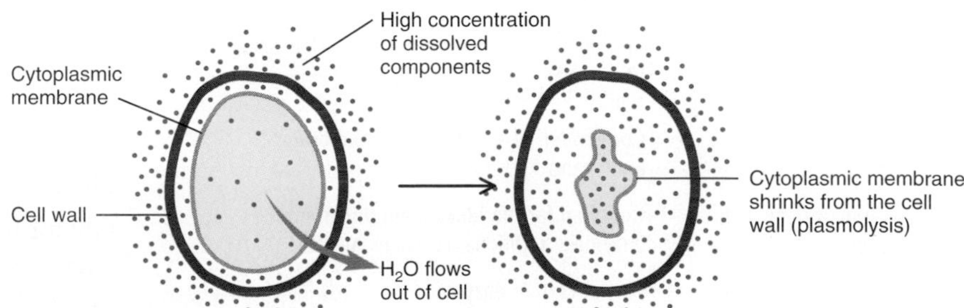

Figure 4.12 Effects of Solute Concentration on Cells The cytoplasmic membrane allows water molecules to pass through freely. If the solute concentration is higher outside of the cell, water moves out. The dehydrated cytoplasm shrinks from the cell wall, a process called plasmolysis.

Required Elements

The elements that make up cell constituents are called **major elements**. These include carbon, oxygen, hydrogen, nitrogen, sulfur, phosphorus, potassium, magnesium, calcium, and iron. They are the essential components of proteins, carbohydrates, lipids, and nucleic acids **(table 4.4)**.

The source of carbon, the most abundant of the major elements, distinguishes different groups of prokaryotes. Those that use organic carbon are called **heterotrophs** (*hetero* means "different"). Medically important bacteria typically use an organic source of carbon such as glucose. **Autotrophs** (*auto* means "self") utilize inorganic carbon in the form of carbon dioxide as their carbon source. They play a critical role in the cycling of carbon in the environment because they can convert inorganic carbon (CO_2) to an organic form, the process of **carbon fixation**. Without carbon fixation, the earth would quickly run out of organic carbon, which is essential to humans and other animals. ■ carbon cycling. p. 775

In addition to carbon, organisms require that the other major elements be supplied in a form they can utilize. Most microorganisms can use inorganic salts as a source of each of these elements. For example, ammonium sulfate supplies both nitrogen and sulfur. Some prokaryotes can convert inorganic nitrogen gas (N_2) to ammonia, the process of **nitrogen fixation**. Like carbon fixation, this process is essential to life on this planet. ■ nitrogen cycling, p. 777

Some elements, termed **trace elements**, are required in very minute amounts by all cells. They include cobalt, zinc, copper, molybdenum, and manganese. These elements form parts of enzymes or may be required for enzyme function. Very small amounts of these trace elements are found in most natural environments, including water.

Growth Factors

Some bacteria cannot synthesize some of their cell constituents, such as amino acids, vitamins, purines, and pyrimidines, from the major elements. Consequently, these organisms can only grow in environments where such compounds are available. Those low molecular weight compounds that must be provided to a particular bacterium are called **growth factors**. ■ purines, p. 35 ■ pyrimidines, p. 35

Microorganisms display a wide spectrum in their growth factor requirements, reflecting differences in their biosynthetic capabilities. The fewer enzymes an organism has for the biosynthesis of small molecules such as amino acids, the more growth factors must be provided. For example, *E. coli* is quite versatile and does not require any growth factors. It grows in a medium containing only glucose and six different inorganic salts. In contrast, species of *Neisseria* require at least 40 additional ingredients, including 7 vitamins and all of the 20 amino acids. Bacteria such as *Neisseria* that require a range of growth factors are called **fastidious**.

Fastidious bacteria are exploited to determine the quantity of specific vitamins in food products. To do this, a well-characterized species of *Lactobacillus* is grown in a medium that lacks a specific vitamin, but has been supplemented with a measured amount of the food product. The amount of growth of the bacterium, which can be measured by turbidity or acid production, is related to the amount of the test vitamin in the product.

Table 4.4 Representative Functions of the Major Elements

Chemical	Function
Carbon, oxygen, and hydrogen	Component of cellular constituents including amino acids, lipids, nucleic acids, and sugars.
Nitrogen	Component of amino acids and nucleic acids.
Sulfur	Component of some amino acids.
Phosphorus	Component of nucleic acids, membrane lipids, and ATP.
Potassium, magnesium, and calcium	Required for the functioning of certain enzymes; additional functions as well.
Iron	Part of certain enzymes.

Energy Sources

Organisms derive energy either from sunlight or from the oxidation of chemical compounds. These processes will be discussed in chapter 6. Organisms that harvest the energy of sunlight are called **phototrophs** (*photo* means "light" and *troph* means "nourishment"). These include plants, algae, and photosynthetic bacteria. Organisms that obtain energy by oxidizing chemical compounds are called **chemotrophs** (*chemo* means "chemical"). Mammalian cells, fungi, and many types of bacteria oxidize organic compounds such as sugars, amino acids, and fatty acids. Some prokaryotes can extract energy from seemingly unlikely sources such as hydrogen sulfide, hydrogen gas, and other inorganic compounds, an ability that distinguishes them from eukaryotes. Microbiologists often group prokaryotes according to the energy and carbon sources they utilize **(table 4.5)**.

Nutritional Diversity

Prokaryotes can thrive in virtually every conceivable environmental niche, because they are able to use diverse sources of carbon and energy.

Photoautotrophs

Photoautotrophs use the energy of sunlight and the carbon in the atmosphere to make the organic compounds required by many other organisms, including humans. Because of this, they are called **primary producers**. Cyanobacteria are important photoautotrophs that inhabit both freshwater and saltwater habitats. Many can fix nitrogen, providing another indispensable role in the biosphere.

Chemolithoautotrophs

Chemolithoautotrophs, commonly referred to simply as **chemoautotrophs**, use inorganic compounds for energy and derive their carbon from CO_2. These prokaryotes live in seemingly inhospitable environments such as sulfur hot springs,

Perspective 4.1 Can Prokaryotes Live on Only Rocks and Water?

Prokaryotes have been isolated from diverse environments that previously were thought to be incapable of sustaining life. For example, members of the Archaea have been isolated from environments 10 times more acidic than that of lemon juice. Other Archaea have been isolated from oil wells a mile below the surface of the earth at temperatures of 70°C and pressures of 160 atmospheres (at sea level, the pressure is 1 atmosphere). The isolation of these organisms suggests that thermophiles may be widespread in the earth's crust.

Perhaps the most unusual environment from which prokaryotes have been isolated are the volcanic rocks 1 mile below the earth's surface near the Columbia River in Washington state. What do these organisms use for food? They apparently get their energy from hydrogen gas that is produced chemically in a reaction between the iron-rich minerals in the rock and the groundwater. The groundwater also contains dissolved CO_2, which the bacteria can use as a source of carbon. Thus, these bacteria apparently exist on nothing more than rocks and water.

which are rich in reduced inorganic compounds such as hydrogen sulfide. In some regions of the ocean depths, hydrothermal vents have been discovered. Here, chemoautotrophs serve as the primary producers, supporting rich communities of life in these habitats utterly devoid of sunlight (**figure 4.13**). ■ hydrothermal vents, p. 773

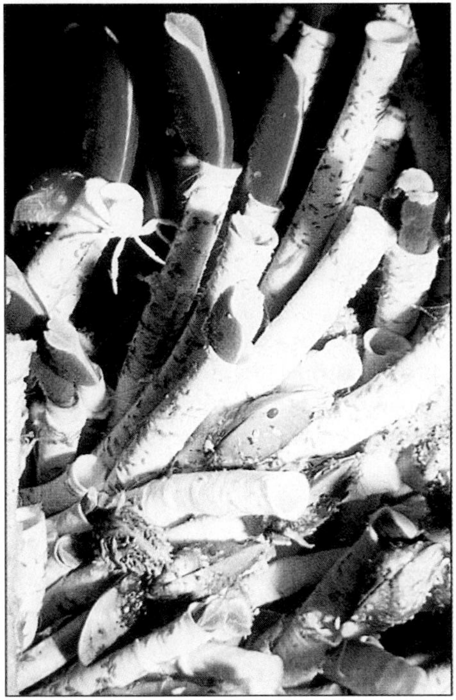

Figure 4.13 Hydrothermal Vent Community This diverse community of life is supported by the metabolic activities of chemoautotrophs.

Photoheterotrophs

Photoheterotrophs use the energy of sunlight and derive their carbon from organic compounds. Some are facultative in their nutritional capabilities. For example, some members of a group of bacteria called the purple nonsulfur bacteria can grow anaerobically using light as an energy source and organic compounds as a carbon source (photoheterotrophs). They can also grow aerobically in the dark using organic sources of carbon and energy (chemoheterotrophs). ■ purple nonsulfur bacteria, p. 277

Chemoorganoheterotrophs

Chemoorganoheterotrophs, commonly referred to as **chemoheterotrophs**, use organic compounds for energy and as a carbon source. They are by far the most common group associated with humans and other animals. Some play beneficial roles such as providing a source of vitamin K in the gut, whereas others cause disease.

As a group, chemoheterotrophs can degrade a wide range of organic chemicals. In fact, most natural organic molecules, regardless of their complexity, can be degraded by at least one species of microorganism. A large number of human-made compounds, however, such as certain herbicides and plastics, are degraded very slowly, and some may not be degraded at all.

Individual species of chemoheterotrophs differ in the number of organic compounds they can use. For example, certain members of the genus *Pseudomonas* can derive carbon and/or energy from more than 80 different organic compounds, including such unusual compounds as naphthalene (the ingredient associated with the smell of mothballs). At the other extreme, some organisms can degrade only a few compounds. For example, *Bacillus fastidiosus* can use only urea and certain of its derivatives as a source of both carbon and energy.

Table 4.5 Energy and Carbon Sources Used by Different Groups of Prokaryotes

Type	Energy Source	Carbon Source
Photoautotroph	Sunlight	CO_2
Photoheterotroph	Sunlight	Organic compounds
Chemolithoautotroph	Inorganic chemicals (H_2S, NH_3, NO_2^-, Fe^{2+}, H_2S)	CO_2
Chemoorganoheterotroph	Organic compounds (sugars, amino acids, etc.)	Organic compounds

MICROCHECK 4.5

Organisms require a source of major and trace elements. Heterotrophs use an organic carbon source, and autotrophs use CO_2. Bacteria that lack the ability to synthesize certain small molecules require these for growth. Phototrophs harvest the energy of sunlight, and chemotrophs obtain energy by degrading chemicals.

- List the major elements other than carbon required for growth of bacteria.
- What is the carbon source in a photoautotroph? In a chemoautotroph?
- Why would human-made materials (such as plastics) be degraded only slowly or not at all?

Cultivating Prokaryotes in the Laboratory

By knowing the environmental and nutritional factors that influence the growth of specific prokaryotes, it is often possible to provide the appropriate conditions for their cultivation. These include a medium on which to grow the organisms and a suitable atmosphere.

General Categories of Culture Media

Considering the diversity of bacteria, it is not surprising that a wide variety of media is used to culture them. For routine purposes, one of the many types of **complex media** is used. In contrast, **chemically defined media** are generally used for specific research experiments when nutrients must be precisely controlled. **Table 4.6** summarizes the characteristics of various types of media.

Complex Media

A **complex medium** contains a variety of ingredients such as meat juices and digested proteins, making what might be viewed as a tasty soup for microbes. Although a specific amount of each ingredient is in the medium, the exact chemical composition of these ingredients can be highly variable. One common ingredient is **peptone**. This is protein taken from any of a variety of sources that has been hydrolyzed to amino acids and short peptides by treatment with enzymes, acids, or alkali. **Extracts**, which are the water-soluble components of a substance, are also used. For example, beef extract is a water extract of lean meat and provides vitamins, minerals, and other nutrients. A commonly used complex medium, **nutrient broth,** consists of only 5 grams of peptone and 3 grams of beef extract per liter of distilled water. If agar is added, then **nutrient agar** results.

Many medically important bacteria are fastidious, requiring a medium that is even richer than nutrient agar. One rich medium commonly used in clinical laboratories is **blood agar**.

Table 4.6 Characteristics of Media Used to Cultivate Bacteria

Medium	Characteristic
Categories	
Complex	Composed of ingredients such as peptones and extracts, which may vary in their chemical composition.
Chemically defined	Composed of precise mixtures of pure chemicals such as ammonium sulfate.
Selective	Medium to which additional ingredients have been added that inhibit the growth of many organisms other than the one being sought.
Differential	Medium that contains an ingredient that can be changed by certain bacteria in a recognizable way.
Representative Types of Agar Media	
Blood agar	Complex medium used routinely in clinical labs. Not selective. Differential because colonies of hemolytic organisms are surrounded by a zone of clearing of the red blood cells.
Chocolate agar	Complex medium used to culture fastidious bacteria, particularly those found in clinical specimens. Not selective or differential.
Glucose-salts	Chemically defined medium. Used in laboratory experiments to study nutritional requirements of bacteria. Not selective or differential.
MacConkey agar	Complex medium used to isolate Gram-negative rods that typically reside in the intestine. Selective because bile salts and dyes inhibit Gram-positive organisms and Gram-negative cocci. Differential because the pH indicator turns red when the sugar in the medium, lactose, is fermented.
Nutrient agar	Complex medium used for routine laboratory work. Supports the growth of a variety of nonfastidious bacteria.
Thayer-Martin	Complex medium used to isolate *Neisseria* species, which are fastidious. Selective—contains antibiotics that inhibit most organisms except *Neisseria* species.

This contains red blood cells, which supply a variety of nutrients including hemin, in addition to other ingredients. A medium used to culture even more fastidious bacteria is **chocolate agar**, named for its appearance rather than the ingredients. Chocolate agar contains lysed red blood cells and additional nutrients.

Additional ingredients are often incorporated into complex media to counteract compounds that may be toxic to some exquisitely sensitive bacteria. For example, cornstarch is included in some types of media used to culture *Neisseria* species because it binds fatty acids, which may otherwise be toxic to these organisms.

Several biological supply companies manufacture hundreds of different types of media, each one specially formulated to permit the plentiful growth of one or several groups of organisms. Even with the availability of all of these different media, however, many organisms, including *Treponema pallidum*, the spirochete that causes syphilis, have yet to be successfully grown on culture media. ■ syphilis, p. 647

Chemically Defined

Chemically defined media are composed of mixtures of pure chemicals. They are generally not practical for use in most routine laboratory work, but they are invaluable when studying nutritional requirements of bacteria. **Glucose-salts**, which supports the growth of *E. coli*, contains only those chemicals listed in **table 4.7**. More elaborate recipes containing as many as 46 different ingredients can be used to make chemically defined media that support the growth of fastidious bacteria such as *Neisseria gonorrhoeae*, the bacterium that causes gonorrhea. ■ gonorrhea, p. 644

To maintain the pH near neutrality, **buffers** are often added to the medium. A common buffer is a mixture of two salts of phosphoric acid—the sodium phosphates, Na_2HPO_4 and NaH_2PO_4. These salts limit pH changes, because they can combine chemically with the H^+ ions of strong acids and the OH^- ions of strong bases to produce neutral compounds. They are often included in a defined medium because some bacteria can produce enough acid as a by-product of their metabolism to inhibit their own growth. This typically is not as much of a problem in complex media because the amino acids and other natural components provide at least some buffering function.

Table 4.7 Ingredients in Two Representative Types of Media that Support the Growth of *E. coli*

Nutrient Broth (complex medium)	Glucose-Salts (defined medium)
Peptone	Glucose
Meat extract	Dipotassium phosphate
Water	Monopotassium phosphate
	Magnesium sulfate
	Ammonium sulfate
	Calcium chloride
	Iron sulfate
	Water

Special Types of Culture Media

To detect or isolate an organism that is part of a mixed bacterial population, it is often necessary to make it more prevalent or more obvious. For these purposes **selective** and **differential media** are used. These media can be either complex or chemically defined, depending on the needs of the microbiologist.

Selective Media

Selective media inhibit the growth of organisms other than the one being sought. For example, **Thayer Martin agar** is used to isolate *Neisseria gonorrhoeae* from clinical specimens. This is a variation of chocolate agar to which three or more antimicrobial drugs have been added. The antimicrobials inhibit fungi, Gram-positive bacteria, and Gram-negative rods. Because these drugs do not inhibit most strains of *N. gonorrhoeae*, they allow growth with little competition from these other organisms.

MacConkey agar is used to isolate Gram-negative rods that typically reside in the intestine from various clinical specimens such as urine. This complex medium contains, in addition to peptones and other nutrients, two inhibitory compounds: crystal violet, a dye, inhibits Gram-positive bacteria, and bile salts inhibit most non-intestinal bacteria.

Differential Media

Differential media contain a substance that certain bacteria change in a recognizable way. For example, blood agar, in addition to being nutritious, is differential; it is used to detect bacteria that produce a **hemolysin**, a substance that lyses red blood cells (**figure 4.14**). The lysis appears as a zone of clearing around the colony growing on the blood agar plate. The type of hemolysis is used as an identifying characteristic. For example, species of *Streptococcus* that reside harmlessly in the throat often cause a type of hemolysis called **alpha hemolysis**, which is characterized by a zone of greenish clearing around the colonies. In contrast, *Streptococcus pyogenes*, which causes strep throat, causes **beta hemolysis**, which is characterized by a clearer zone of hemolysis. Still other bacteria have no effect on red blood cells. ■ *Streptococcus pyogenes*, p. 553

MacConkey agar, which is selective, is also differential (**figure 4.15**). In addition to peptones and other nutrients, it contains the sugar lactose and a pH indicator. Bacteria that ferment the sugar produce acid, which turns the pH indicator pink. Thus, those lactose-fermenting bacteria that can grow on MacConkey agar, such as *E. coli*, form pink colonies. Lactose-negative bacteria form tan or colorless colonies.

Providing Appropriate Atmospheric Conditions

To cultivate bacteria in the laboratory, appropriate atmospheric conditions must be provided. For example, broth cultures of obligate aerobes grow best when the tubes or flasks containing the media are shaken, providing maximum aeration. Special methods create atmospheric environments such as increased CO_2, microaerophilic, and anaerobic conditions.

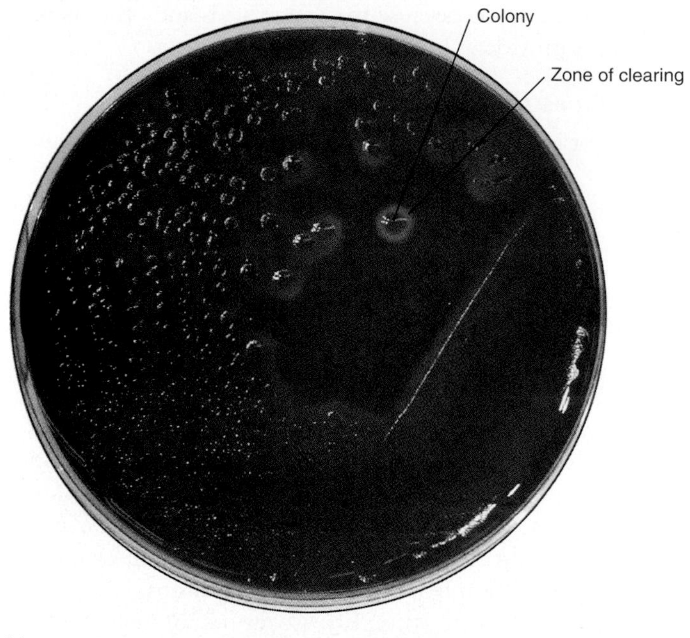

(a)

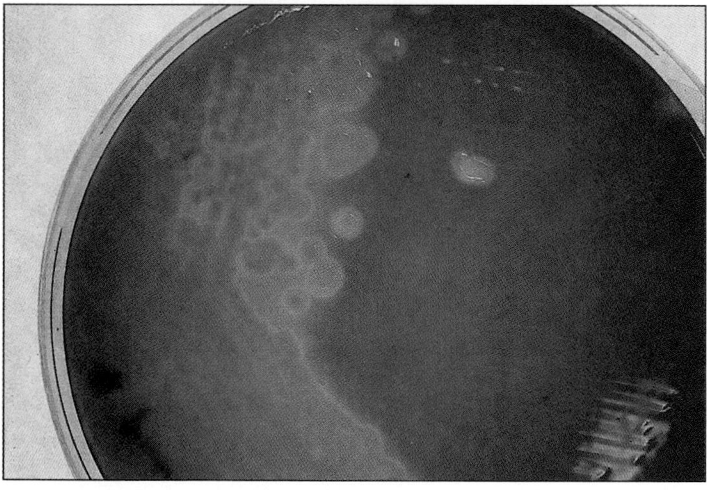

(b)

Figure 4.14 Blood Agar This complex medium is differential for hemolysis. **(a)** A zone of complete clearing around a colony growing on blood agar is called beta hemolysis. **(b)** A zone of greenish clearing is called alpha hemolysis.

Increased CO_2

Providing an environment that has increased levels of CO_2 enhances the growth of many medically important bacteria, including species of *Neisseria* and *Haemophilus.* Organisms that require increased CO_2, along with approximately 15% oxygen, are called **capnophiles**. One of the simplest ways to provide this atmosphere is to incubate them in a closed **candle jar**. A lit candle in the jar converts some of the O_2 in the air to CO_2; it soon extinguishes because of insufficient oxygen. Although a candle jar atmosphere contains about 3.5% CO_2, enough O_2 remains to support the growth of obligate aerobes and prevents the growth of obligate anaerobes. Special incubators are also available that maintain CO_2 at prescribed levels.

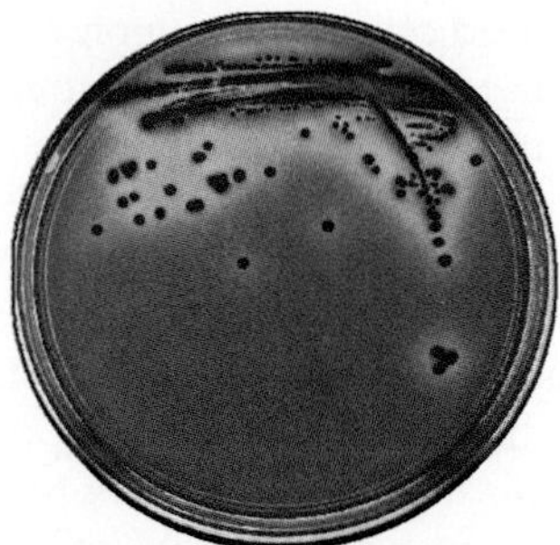

Figure 4.15 MacConkey Agar This complex medium is differential for lactose fermentation and selective for Gram-negative rods that typically reside in the intestine. Bacteria that ferment the sugar produce acid, which turns the pH indicator pink, resulting in pink colonies. Lactose-negative colonies are tan or colorless. The bile salts and dyes in the media inhibit all but certain Gram-negative rods.

Microaerophilic

Microaerophilic bacteria typically require O_2 concentrations that are less than what is achieved in a candle jar. Therefore, these bacteria are often incubated in a gastight jar with a special disposable packet containing chemicals that react to generate hydrogen and carbon dioxide. A catalyst in the jar speeds up the reaction of hydrogen with atmospheric oxygen to form water. The amount of hydrogen generated, however, is not enough to combine with all of the O_2, so that conditions do not become anaerobic.

Anaerobic

The cultivation of obligate anaerobes presents a great challenge to the microbiologist, because the cells may be killed if they are exposed to O_2 for even a short time. Obviously, special techniques that exclude O_2 are required for their cultivation. As techniques for cultivating anaerobes have improved in recent years, many more of these organisms are being isolated from various habitats, including wound and blood infections in which they had previously appeared to be absent.

Anaerobes that can tolerate a brief exposure to O_2 are cultivated in an **anaerobe jar (figure 4.16)**. This is the same type of jar used to incubate microaerophiles, but the chemical composition of the disposable packet converts all of the atmospheric oxygen to water.

Another method to cultivate anaerobes incorporates **reducing agents** into the culture medium. These chemicals react with O_2 and thus eliminate dissolved O_2; they include sodium thioglycollate, cysteine, and ascorbic acid. In some cases, immediately before the bacteria are inoculated, the medium is boiled to drive out dissolved O_2. Media that employ reducing agents frequently contain an O_2-indicating dye such as methylene blue.

A more elaborate method for working with anaerobes is an enclosed chamber that can be maintained in an anaerobic environment (**figure 4.17**). A special port, which can be filled with an inert gas, is used to add or remove items. Airtight gloves enable researchers to handle items within the chamber.

Enrichment Cultures

An **enrichment culture** provides conditions in a broth that preferentially enhance the growth of one particular organism in a

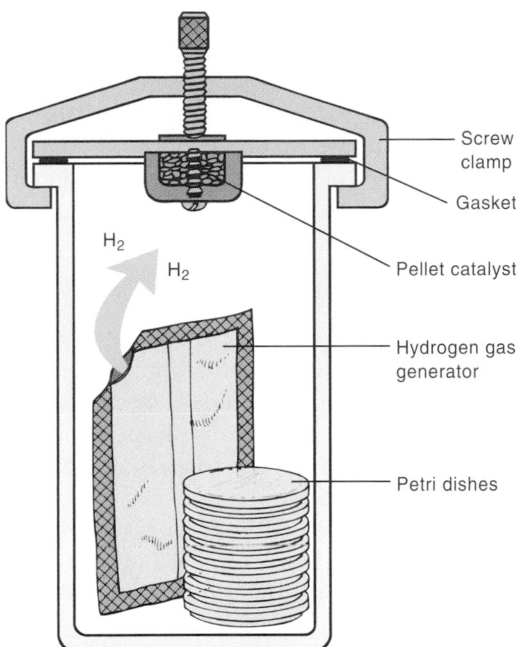

Figure 4.16 Anaerobe Jar The hydrogen released from the generator combines with any O_2 to form water, thereby producing an anaerobic environment.

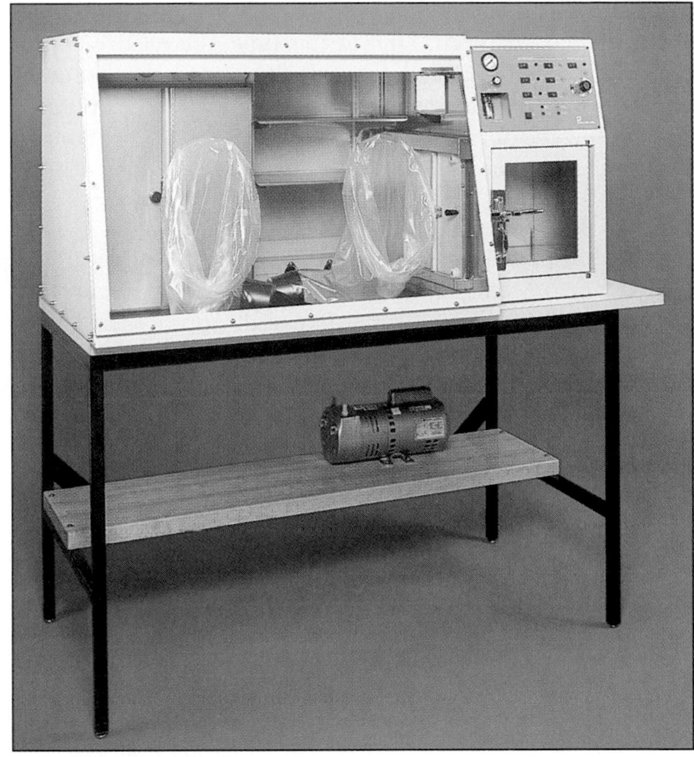

Figure 4.17 Anaerobic Chamber The enclosed chamber can be maintained in anaerobic conditions. A special port (visible on the right side of this device), which can be filled with inert gas, is used to add or remove items. The airtight gloves enable the researcher to handle items within the chamber.

mixed population (**figure 4.18**). This method is helpful in isolating an organism from natural sources when the bacterium is present in relatively small numbers. For example, if an organism is present at a concentration of only 1 cell/ml and it is outnumbered 10,000-fold by other organisms, isolating it using the streak-plate method would be difficult, even if a selective medium were used.

To enrich for an organism, a sample such as pond water is placed into a liquid medium that favors the growth of the desired organism over others. For example, if the target organism can grow using atmospheric nitrogen as a source of nitrogen, then nitrogen is left out of the medium. If it can use an unusual carbon source such as phenol, then that is added as the only available carbon source. In some cases selective agents such as bile are added; the procedure is then referred to as a **selective enrichment**. The culture is then incubated under temperature and

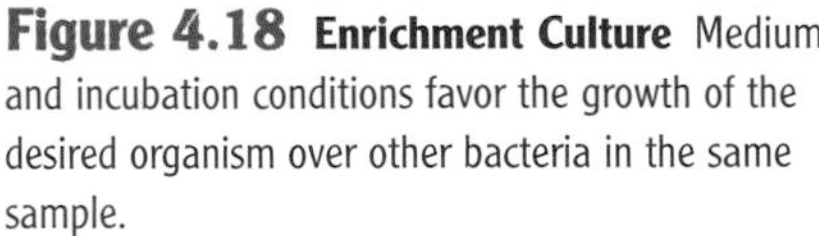

Figure 4.18 Enrichment Culture Medium and incubation conditions favor the growth of the desired organism over other bacteria in the same sample.

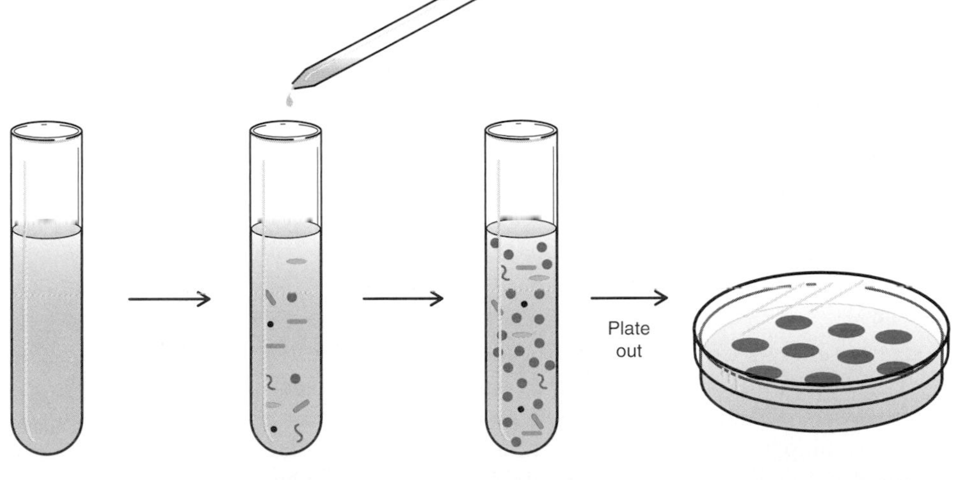

Plate
out
Medium contains select nutrient sources chosen because few bacteria, other than the organism of interest, can use them.
Sample that contains a wide variety of organisms, including the organism of interest, is added to the medium.
Organism of interest can multiply, whereas most others cannot.
Enriched sample is plated onto appropriate agar medium. A pure culture is obtained by selecting a single colony of the organism of interest.

atmospheric conditions that preferentially promote the growth of the desired organism. During this time, the relative concentration of a microorganism that initially made up only a minor fraction of the population can increase dramatically. A pure culture can then be obtained by inoculating the enrichment onto an appropriate agar medium and selecting a single colony.

MICROCHECK 4.6

Culture media may be either complex or chemically defined. Media may contain additional ingredients that make them selective or differential. Appropriate atmospheric conditions must be provided to isolate microaerophiles and anaerobes. An enrichment culture increases the relative concentration of an organism growing in a broth.

- Distinguish between a selective medium and a differential medium.
- Describe two methods used to create anaerobic conditions.
- Would bacteria that cannot utilize lactose be able to grow on MacConkey agar?

Bacterial Growth in Laboratory Conditions

In the laboratory, bacteria are typically grown in broth contained in a tube or flask, or on an agar plate. These are considered **closed** or **batch systems** because nutrients are not renewed, nor are waste products removed. Under these conditions, the cell population increases in number in a predictable fashion and then eventually declines. As the population in a closed system grows, it follows a pattern of stages, called a **growth curve**. This growth pattern is most distinct in a shaken broth culture, because all cells are exposed to the same environment. In a colony, cells on the outer edge of a colony experience very different conditions from those at the center.

To maintain cells in a state of continuous growth, nutrients must be continuously added and waste products removed. This is called an **open system**, or **continuous culture**.

The Growth Curve

This growth curve is characterized by four distinct stages—the **lag phase**, the **exponential** or **log phase**, the **stationary phase**, and the **death phase** (**figure 4.19**).

Lag Phase

When a culture of bacteria is diluted and then transferred into a different medium, the number of viable cells does not immediately increase. They go through a "tooling up" or lag phase prior to active multiplication. During this time they synthesize the macromolecules required for multiplication, including enzymes, ribosomes, and nucleic acids, and they generate energy in the form of ATP.

The length of the lag phase depends on the conditions in the original culture and the medium into which they are transferred. If cells are transferred from a rich medium to a chemically defined one, the lag time tends to be longer. This is because the cells must begin making enzymes to synthesize the components missing in the new medium. A similar situation occurs when a stock culture that has been stored in the refrigerator for several weeks is inoculated into fresh medium. In contrast, if young cells are transferred to a medium that is similar in composition, the lag time is quite short.

Exponential Phase (Log Phase)

During the exponential or log phase, cells divide at a constant rate and their numbers increase by the same percentage during each time interval. The generation time is measured during this period of active multiplication. Because bacteria are most susceptible to antibiotics and other chemicals during this time, the log phase is important medically.

During the initial phase of exponential growth, all the cells' activities are directed toward increasing cell mass. Cells produce compounds such as amino acids and nucleotides, the respective building blocks of proteins and nucleic acids. Cells are remarkably precise in their ability to regulate the synthesis of these compounds, ensuring that each is made in the appropriate relative amount for efficient assembly into macromolecules. Compounds synthesized during this period of active multiplication are called **primary metabolites.** A metabolite is any product of a chemical reaction in a cell and includes compounds required for growth, as well as waste materials. Some primary metabolites are commercially valuable as flavoring agents and food supplements. Understandably, industries that harvest these compounds are working to develop methods to manipulate bacteria to overproduce certain primary metabolites. ■ **regulation of gene expression, p. 179**

Cells' activities shift as they enter a stage called **late log phase**, which marks the transition to stationary phase. This change occurs in response to multiple factors that are inevitable in a closed system, such as a gradual increase in population density, depletion of nutrients, and buildup of waste products. As their surrounding environment changes, cells begin synthesizing different enzymes and other proteins, which collectively give rise to a new group of metabolites, called **secondary metabolites** (**figure 4.20**). Secondary metabolites, in contrast to primary metabolites, are not required for growth. Instead, they appear to make the cell more resistant to certain environmental conditions. For example, *Shigella flexneri* begins synthesizing several new proteins, some of which appear to be involved with its ability to survive acidic conditions. Commercially, the most valuable secondary metabolites are antibiotics. These are produced by many members of the genus *Streptomyces* and are medically important because they kill or inhibit other bacteria. It is during the late log phase when the endospore-formers *Bacillus* and *Clostridium* initiate the process of sporulation, and myxobacteria begin to aggregate and form fruiting bodies. ■ **antibiotics, p. 496** ■ **endospores, p. 72** ■ **fruiting bodies, p. 284**

Stationary Phase

Bacteria stop growing and enter the stationary phase after they have depleted a required nutrient, when O_2 is in short supply, or when toxic metabolites accumulate to high levels. The total

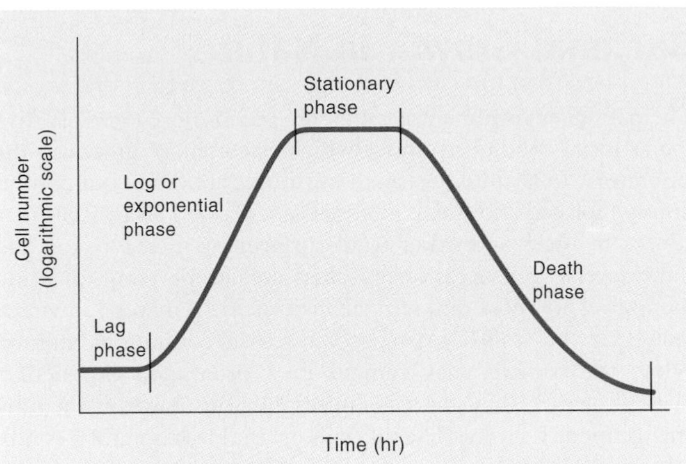

Figure 4.19 Growth Curve The growth curve is characterized by four distinct stages: lag phase, exponential or log phase, stationary phase, and death phase.

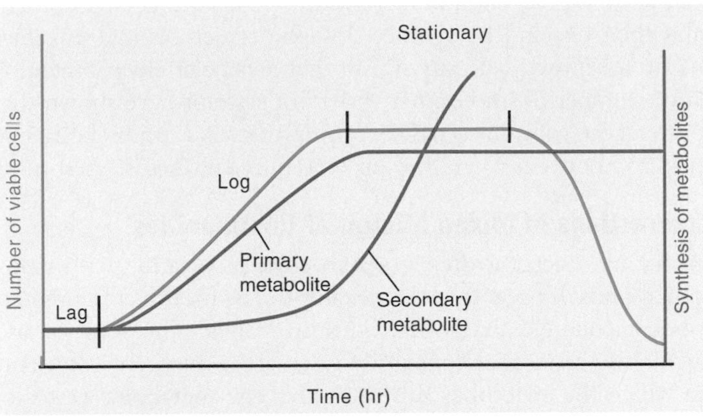

Figure 4.20 Primary and Secondary Metabolite Production Primary metabolites are synthesized during the period of active multiplication. Late in the log phase, cells begin synthesizing secondary metabolites. These compounds, which continue to be synthesized in stationary phase, appear to make the cells more resistant to environmental conditions.

number of viable cells in the population usually remains relatively constant. Some cells may die, but others divide to replace them.

Many of the metabolic activities started in the late log phase continue into the stationary phase and improve the cells' chance for survival. The cells become far more resistant to antibiotics, radiation, and harmful chemicals than they were in the log phase. Secondary metabolites continue to be synthesized. The length of time cells remain in the stationary phase varies, depending on the species and on environmental conditions. Some organisms remain in the stationary phase for only a few hours, whereas others remain for days. Generally, the more toxic chemicals that have accumulated as by-products of metabolism, the more rapidly the next phase is entered.

Death Phase

Death phase is the period when the total number of viable cells in the population decreases as cells die off at a constant rate. Like bacterial growth, death is exponential. A population of cells, however, generally dies much more slowly than it multiplies. The generation time of the organism bears no relationship to how fast the population will die off.

Colony Growth

The growth of a bacterial colony on a solid medium involves many of the same features as bacteria growing in liquid, but it is marked by some important differences. After a lag phase, the cells multiply exponentially and eventually compete with one another for available nutrients and become very crowded. Unlike a liquid culture, the position of a single cell within a colony markedly determines its environment. The cells multiplying on the edge of the colony face relatively little competition and can use O_2 in the air and obtain nutrients from the agar medium. In contrast, in the center of the colony the high density of cells rapidly depletes available O_2 and nutrients. Toxic metabolic wastes such as acids accumulate. As a consequence, cells at the edge of the colony may be growing exponentially, whereas those in the center may be in the death phase. Cells in locations between these two extremes may be in stationary phase.

The appearance of colonies of different species of bacteria can differ considerably (**figure 4.21**). For example, *Bacillus subtilis* grows in chains, forming colonies that spread outward from the initial inoculum. Depending on the medium, these can form crystalline snowflake patterns or dramatic swirls. Colonies of *Proteus* species typically spread over the entire plate, radiating outward in concentric rings. In contrast, *Staphylococcus aureus* and many other organisms form discrete round colonies with a smooth edge. *Streptococcus pneumoniae* colonies are typically small and flat, due in part to the enzyme produced by the cells, which causes their lysis. Those that produce a capsule, however, characteristically form larger colonies that appear gooey, or **mucoid**.

Figure 4.21 Colony Morphology The appearance of colonies of different species of bacteria can differ considerably. Variation can also occur when the same species is grown on different media.

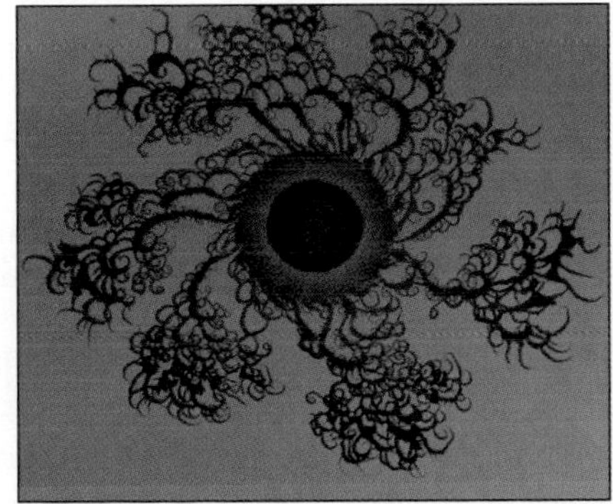

These diverse morphologies of colonies in size, shape, and texture help microbiologists identify the species. ■ capsule, p. 66

Continuous Culture

Bacteria can be maintained in a state of continuous exponential growth by using a **chemostat (figure 4.22)**. This device continually drips fresh medium into a liquid culture contained in a growth chamber. With each drop that enters, an equivalent volume, containing cells, wastes, and spent medium, leaves through an outlet. By manipulating the concentration of nutrients in the medium and the rate at which it enters the growth chamber, a constant cell density and generation time of log phase cells can be maintained. This makes it possible to study a uniform population of cells over a long period of time. The effect of adding various supplements to the medium or altering the cellular environment on long-term cell growth can be determined.

MICROCHECK 4.7

When grown in a closed system, a population of bacteria goes through four distinct phases: lag, log, stationary, and death. Cells within a colony may be in any one of the growth phases, depending on their relative location.

- Explain the difference between the two stages of the exponential phase.
- Describe how a chemostat keeps a culture in a continuous stage of growth.
- Why would bacteria be more susceptible to antibiotics during the log phase?

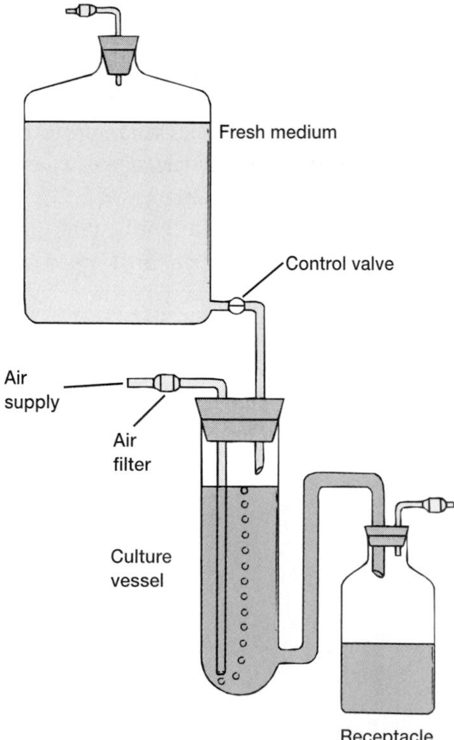

Figure 4.22 Chemostat This device can maintain a culture in a steady state of exponential growth by allowing fresh medium to drip in while an equivalent volume of spent medium, cells, and wastes leaves.

Bacterial Growth in Nature

The natural environment is generally much more dynamic than the artificial conditions under which bacteria are grown in the laboratory. In a running stream, nutrients are dilute but continuously replaced and waste materials are washed away. Cells may adhere to rocks and other solid surfaces by means of pili and slime layers, but occasionally they are swept away and find themselves in yet a different environment. A natural environment may be similar in many ways to a continuous culture. While the bacteria may remain in a prolonged exponential phase, however, they generally multiply more slowly in a natural environment than in the artificially favorable laboratory conditions. ■ pili, p. 69 ■ slime layer, p. 66

When in their natural environment, bacteria frequently synthesize structures, such as slime layers, that they may not produce when growing in the laboratory. This occurs because cells can sense various compounds found in the natural environment and then respond by synthesizing the structures and enzymes useful for growth and survival in that particular environment. A large number of studies in a variety of systems have shown that bacteria can grow in complex communities and exhibit behaviors that do not occur when they grow as pure cultures in a test tube.

Interactions of Mixed Microbial Communities

In nature, bacteria often grow in close association with many other kinds of organisms. For example, the mouth contains aerobes, facultative anaerobes, and anaerobes. The aerobes and facultative anaerobes consume O_2, creating microenvironments in which the anaerobes can multiply. The metabolic wastes of one species may serve as a nutrient or energy source of another. For example, members of the genus *Syntrophomonas* generate propionate as a product of the breakdown of fatty acids. *Syntrophobacter* uses this compound as an energy source. To complicate matters, both of these organisms are obligate anaerobes that generate energy through an unusual process, creating hydrogen gas as a byproduct. This process, however, will only continue if another bacterium is present to consume that gas, making the chemical reactions involved energetically favorable. Thus, *Syntrophomonas* and *Syntrophobacter* typically reside in close association with a methanogen, a chemoautotroph that uses hydrogen gas as an energy source and generates methane gas. Understandably, the conditions in these close associations are exceedingly difficult to reproduce in the laboratory, which is one reason that so few environmental organisms have been isolated in pure culture. ■ fatty acid, p. 37 ■ methanogen, p. 269

Biofilms

Bacteria may live suspended in an aqueous environment, but many attach to surfaces and live in a polysaccharide-encased community called a **biofilm (figure 4.23)**. Biofilms cause the slipperiness of rocks in a stream bed, the slimy "gunk" that coats kitchen drains, and the scum that gradually accumulates in toilet bowls. Biofilm formation begins when a bacterium adheres to a surface, where it multiplies and synthesizes a loose glycocalyx to which unrelated cells may attach and grow. Cells may move

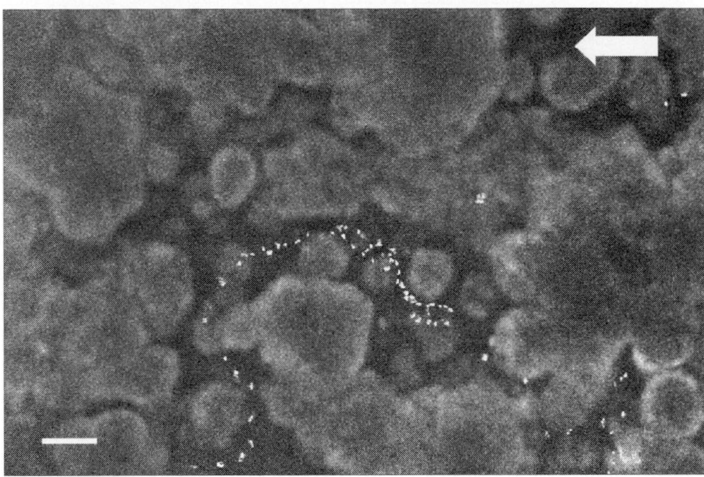

Figure 4.23 **Biofilm** Superimposed time sequence image shows a single latex bead moving through a biofilm water channel. The large light gray shapes are clusters of bacteria.

within the growing biofilm by twitching motility. ■ glycocalyx, p. 66 ■ pili, p. 69 ■ twitching motility, p. 69

Surprisingly, biofilms are not generally a haphazard mixture of microbes in a layer of slime; rather they have characteristic architectures with open channels through which nutrients and waste materials can pass. Cell-to-cell communication through the mechanism of quorum sensing appears to be important in establishing structure. ■ quorum sensing, p. 184

Biofilms are more than just an unsightly annoyance. The plaque on teeth that leads to tooth decay and gum disease is due to bacteria encased in biofilms. Troublesome persistent ear infections that resist antibiotic treatment and the complications of cystic fibrosis are thought to be due to bacteria that grow as a biofilm. In fact, it is estimated that 65% of human bacterial infections involve biofilms. Biofilms are also important in industry, where their growth in pipes, drains, and cooling water towers can interfere with various processes and damage equipment. Biofilms are particularly troublesome because they protect the organisms against harmful chemicals such as disinfectants. The bacteria encased in a biofilm may be hundreds of times more resistant to these compounds. ■ disinfectants, p. 114

While biofilms can be damaging, they can also be beneficial. Many **bioremediation** efforts, which use bacteria to degrade harmful chemicals, are enhanced by biofilms. Thus, as some industries are exploring ways to destroy biofilms, others, such as sewage treatment facilities, are looking for ways to foster their development. ■ bioremediation, p. 781

MICROCHECK 4.8

The metabolic wastes of one species may serve as a nutrient or energy source of another. Biofilms have characteristic architectures, with open channels though which nutrients and waste products can pass.

- Describe a situation in which the activities of one species of bacteria benefit another.
- Give three examples of biofilms.
- Why would bacteria in a biofilm be more resistant to harmful chemicals?

FUTURE CHALLENGES

Seeing How the Other 99.9% Lives

One of the biggest challenges for the future is the development of methodologies to cultivate and study a wider array of environmental bacteria. Without these bacteria, humans and other animals would not be able to exist. Yet considering their importance, we still know very little about most species, including the relative contributions of each to such fundamental processes as O_2 generation and nitrogen fixation.

Studying environmental microorganisms can be difficult. Much of our understanding of bacterial processes comes from work with pure cultures. Yet over 99% of bacteria have never been successfully grown in the laboratory. At the same time, when organisms are removed from their natural habitat, and especially when they are separated from other organisms, their environment changes drastically. Consequently, the study of pure cultures may not be the ideal for studying natural situations, even though it has historically been the method of choice.

Technological advances such as flow cytometry and fluorescent labeling, along with the recombinant DNA techniques discussed in chapter 9, may make it easier to study environmental bacteria. This may well lead to a better understanding of the diversity and the roles of microorganisms in our ecosystem. Scientists have learned a great deal about microorganisms since the days of Pasteur, but most of the microbial world is still a mystery.

SUMMARY

Obtaining a Pure Culture

1. Only an estimated 0.1% of bacteria can be cultured in the laboratory.

Cultivating Bacteria on a Solid Medium

1. A single bacterial cell will multiply to form a visible **colony**. (Figure 4.2)
2. **Agar** is used to solidify nutrient-containing broth; it melts at approximately 95°C and solidifies at approximately 45°C.

The Streak-Plate Method (Figure 4.3)

1. The **streak-plate** method is used to isolate bacteria in order to obtain a **pure culture**.

Maintaining Stock Cultures

1. A **stock culture** can be used as an inoculum in later experiments.
2. Stock cultures can be stored on an **agar slant** in the refrigerator, frozen in a glycerol solution, or **lyophilized**.

Principles of Bacterial Growth

1. Most bacteria multiply by **binary fission**; after a cell has increased in size and doubled the amounts of each of its parts, it divides to become two cells. (Figure 4.4)
2. **Microbial growth** is an increase in the number of cells in a population.
3. The time required for a population to double in number is the **generation time**.

Methods to Detect and Measure Bacterial Growth (Table 4.1)

Direct Cell Counts

1. Direct cell counts generally do not distinguish between living and dead cells.
2. One of the most rapid methods of determining the number of cells in a suspension is the **direct microscopic count.** (Figure 4.5)
3. Both a **Coulter counter** and a **flow cytometer** count cells as they pass through a minute aperture. (Figure 4.6)

Viable Cell Counts

1. **Plate counts** measure the number of viable cells in a sample by exploiting the fact that an isolated cell will form a single colony; samples are usually diluted before they are plated. (Figure 4.7)
2. **Membrane filtration** concentrates bacteria by filtration; the filter is then incubated on an agar plate. (Figure 4.8)
3. The most probable number (MPN) method is a statistical assay based on the theory of probability and is used to estimate cell numbers. (Figure 4.9)

Measuring Biomass

1. **Turbidity** of a culture is a rapid measurement that can be correlated to the number of cells; a **spectrophotometer** is used to measure turbidity. (Figure 4.10)
2. **Wet weight** and **dry weight** are proportional to the number of cells in a culture.
3. The quantity of a cell constituent such as nitrogen can be used to calculate biomass.

Measuring Cell Products

1. pH indicators can be used to monitor acid production.
2. Gas production can be detected in blood cultures by using a molecule that fluoresces more brightly when the pH decreases slightly or by using an inverted tube in culture media to trap gas.
3. ATP is detected by employing luciferase, which uses ATP to produce light.

Environmental Factors that Influence Microbial Growth (Table 4.2)

Temperature Requirements (Figure 4.11)

1. **Psychrophiles** have an optimum between –5°C and 15°C.
2. **Mesophiles** have an optimum between 25°C and 45°C.
3. **Thermophiles** have an optimum between 45°C and 70°C.
4. **Hyperthermophiles** have an optimum between 70°C and 110°C.
5. Storage of foods at refrigeration temperatures retards spoilage because it limits the growth of mesophiles.
6. Some microorganisms can inhabit certain parts of the body but not others because of temperature differences.

Oxygen (O_2) Requirements (Table 4.3)

1. **Obligate anaerobes** cannot multiply if O_2 is present.
2. **Facultative anaerobes** grow best if O_2 is present but can also grow without it.
3. **Microaerophiles** require small amounts of O_2, but higher concentrations are inhibitory.
4. **Aerotolerant anaerobes** are indifferent to O_2.
5. Although O_2 itself is not toxic, it can be converted to **superoxide** and **hydrogen peroxide**, both of which are toxic. **Superoxide dismutase** and **catalase** can break these down.

pH

1. Most bacteria are **neutrophiles**, which live within the pH range of 5 to 8.
2. **Acidophiles** grow optimally at a pH below 5.5.
3. **Alkaliphiles** grow optimally at a pH above 8.5.

Water Availability

1. All microorganisms require water for growth.
2. If the solute concentration is higher in the medium than in the cell, water diffuses out of the cell, causing **plasmolysis**. (Figure 4.12)
3. **Halophiles** have adapted to live in high salt environments.

Nutritional Factors that Influence Microbial Growth

Required Elements (Table 4.4)

1. The **major elements** make up cell constituents and include carbon, nitrogen, sulfur, and phosphorus.
2. **Heterotrophs** use organic carbon; **autotrophs** fix CO_2.
3. **Trace elements** are required in very minute amounts.

Growth Factors

1. Bacteria that cannot synthesize cell constituents such as amino acids and vitamins require these as **growth factors**.

Nutritional Diversity (Table 4.5)

1. Prokaryotes can thrive in every conceivable environmental niche because they are able to use diverse sources of carbon and energy.
2. **Photoautotrophs** use the energy of sunlight and the carbon in the atmosphere to make organic compounds.
3. **Chemolithoautotrophs** use inorganic compounds for energy and derive their carbon from CO_2.
4. **Photoheterotrophs** use the energy of sunlight and derive their carbon from organic compounds.
5. **Chemoorganoheterotrophs** use organic compounds for energy and as a carbon source.

Cultivating Prokaryotes in the Laboratory

General Categories of Culture Media (Table 4.6)

1. A **complex medium** contains a variety of ingredients such as peptones and extracts; examples include **nutrient agar**, **blood agar**, and **chocolate agar**.
2. A **chemically defined** medium is composed of precise mixtures of pure chemicals; an example is **glucose-salts** medium.

Special Types of Culture Media

1. A **selective medium** inhibits organisms other than the one being sought; examples include **Thayer-Martin agar** and **MacConkey agar**.
2. A **differential medium** contains a substance that certain bacteria change in a recognizable way; examples include **blood agar** and **MacConkey agar**. (Figures 4.14 and 4.15)

Providing Appropriate Atmospheric Conditions

1. A **candle jar** provides increased CO_2, which enhances the growth of many medically important bacteria.
2. Microaerophilic bacteria are incubated in a gastight jar along with a packet that generates hydrogen. This in turn combines with atmospheric oxygen, forming water.
3. Anaerobes may be cultivated in either an **anaerobe jar** or a medium that incorporates a reducing agent. (Figure 4.16)

Enrichment Cultures (Figure 4.18)

1. An **enrichment culture** provides conditions in a broth that enhance the growth of one particular organism in a mixed population.
2. Selective agents can be used to make a **selective enrichment**.

Growth Characteristics of Bacteria in the Laboratory Conditions

The Growth Curve (Figure 4.19)

1. When grown in a **closed system**, a population of bacteria goes through four phases: **lag**, **log**, **stationary**, and **death**.
2. The number of cells does not increase during the **lag phase**; during this period, bacteria synthesize the macromolecules required for multiplication.
3. During the **exponential** or **log phase**, the cells divide at a constant rate. Initially these bacteria synthesize **primary metabolites**, but they begin synthesizing **secondary metabolites** in the **late log phase**. (Figure 4.20)
4. Bacteria stop growing and enter stationary phase after they have used up a required nutrient, when oxygen is in short supply, or when toxic metabolites accumulate to high levels. Cells are more resistant to antibiotics, radiation, and harmful chemicals during this phase.
5. The total number of viable cells in the population decreases at a constant rate during the **death phase**.

Colony Growth

1. The position of a single cell within a colony markedly determines its environment; cells on the edge may be in log phase, whereas those in the center may be in death phase.
2. The appearance of colonies of different species of bacteria can differ considerably. (Figure 4.21)

Continuous Culture

1. Bacteria can be maintained in a state of continuous exponential growth by using a **chemostat**. (Figure 4.22)

Growth Characteristics of Bacteria in Nature

Interactions of Mixed Microbial Communities

1. Bacteria often grow in close associations with other kinds of organisms; the metabolic activities of one organism may facilitate the growth of another organism.

Biofilms (Figure 4.23)

1. Bacteria may live suspended in an aqueous environment, but many attach to surfaces and live as a **biofilm**, a polysaccharide-encased community.
2. From the human perspective, biofilms can be unsightly, damaging, or beneficial, depending on the situation.
3. Cell-to-cell communication through quorum sensing appears to be important in establishing the structure of a biofilm.

REVIEW QUESTIONS

Short Answer

1. Define a pure culture.
2. If the number of bacteria in lake water were determined using both a direct microscopic count and a plate count, which method would most likely give a higher number? Why?
3. List the four categories of optimum temperature, and describe the corresponding environment in which a representative might thrive.
4. Explain why obligate anaerobes are significant to the canning industry.
5. Explain why O_2-containing atmospheres kill some bacteria.
6. Explain why photoautotrophs are the primary producers.
7. Distinguish between a selective medium and a differential medium.
8. Explain what occurs during each of the four phases of growth.
9. Explain how the environment of a colony differs from that of a liquid broth.
10. Describe a detrimental and a beneficial effect of biofilms.

Multiple Choice

1. *E. coli* is present in a liquid sample at a concentration of between 10^4 and 10^6 bacteria per ml. To determine the precise number of living bacteria in the sample, it would be best to…
   A. use a counting chamber.
   B. plate out an appropriate dilution of the sample on nutrient agar.
   C. determine cell number by using a spectrophotometer.
   D. Any of the above three methods would be satisfactory.
   E. None of the above three methods would be satisfactory.
2. *E. coli,* a facultative anaerobe, is grown on the same solid medium, but under two different conditions: one aerobic, the other anaerobic. The size of the colonies that grow would be…
   A. the same under both conditions.
   B. larger when grown under aerobic conditions.
   C. larger when grown under anaerobic conditions.
3. A soil sample is placed in liquid and the number of bacteria in the sample determined in two ways: (1) by colony count and

(2) by counting the cells in a counting chamber (slide). How would the results compare?

A. Methods 1 and 2 would give approximately the same number of bacteria.

B. Many more bacteria would be estimated by method 1.

C. Many more bacteria would be estimated by method 2.

D. Depending on the soil sample, sometimes method 1 would be higher and sometimes method 2 would be higher.

4. Nutrient broth is an example of a…

A. synthetic medium.

B. complex medium.

C. selective medium.

D. indicator medium.

E. defined medium.

5. *E. coli* does not require vitamin E in the medium in which it grows. This is because *E. coli*…

A. does not require vitamin E for growth.

B. gets vitamin E from its host.

C. can substitute another vitamin for vitamin E in its metabolism.

D. can synthesize vitamin E from the simple compounds provided in the medium.

E. is a chemoheterotroph.

Below is a typical growth curve of bacteria with the various stages numbered 1–4. Answer questions 6 and 7 with regard to this growth curve.

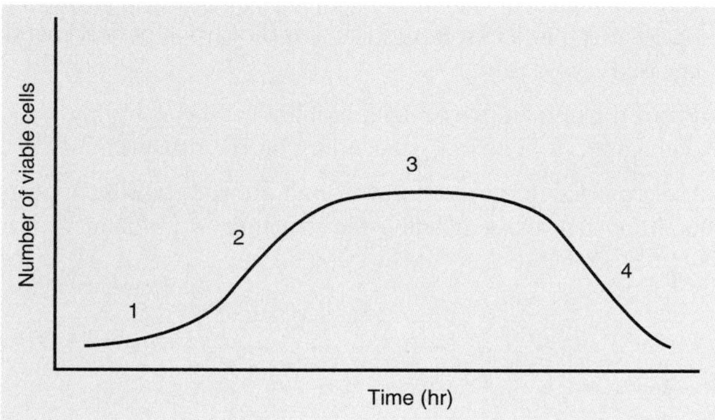

6. The cells are most sensitive to penicillin during…

A. stage 1.

B. stage 2.

C. stage 3.

D. stage 4.

E. more than one stage.

7. *Streptomyces* cells would most likely synthesize antibiotics during…

A. stage 1.

B. stage 2.

C. stage 3.

D. stage 4.

E. more than one stage.

8. In general, bacteria in nature, compared with their growth in the laboratory,…

A. grow more slowly.

B. grow faster.

C. usually grow at the same rate.

9. If there are 10^3 cells per ml at the middle of log phase, and the generation time of the cells is 30 minutes, how many cells will there be 2 hours later?

A. 2×10^3

B. 4×10^3

C. 8×10^3

D. 1.6×10^4

E. 1×10^7

10. The major effect of a temperature of 60°C on a mesophile is to…

A. destroy the cell wall.

B. denature proteins.

C. destroy nucleic acids.

D. destroy the cytoplasmic membrane.

E. cause the formation of endospores.

Applications

1. You are a microbiologist working for a pharmaceutical company and discover a new metabolite that can serve as a human medication. Your company asked you to oversee the production of the metabolite. What are some of the factors you must consider if you need to grow 5,000-liter cultures of the bacteria?

2. High-performance boat manufacturers know that bacteria can collect on a boat, ruining the boat's hydrodynamic properties. Periodic cleaning of the boat's surface and repainting eventually ruin that surface and do not solve the problem. A boat-manufacturing facility recently hired you to help with this problem because of your microbiology background. What strategies can you use to come up with a long-term remedy for the problem?

Critical Thinking Questions

1. The figure below shows a growth curve plotted on a non-logarithmic, or linear, scale. Compare this with figure 4.19. In both figures, the number of cells increases dramatically during the log or exponential phase. In this phase, the cell number increases more and more rapidly (this effect is more apparent in the figure below). Why should the increase be speeding up?

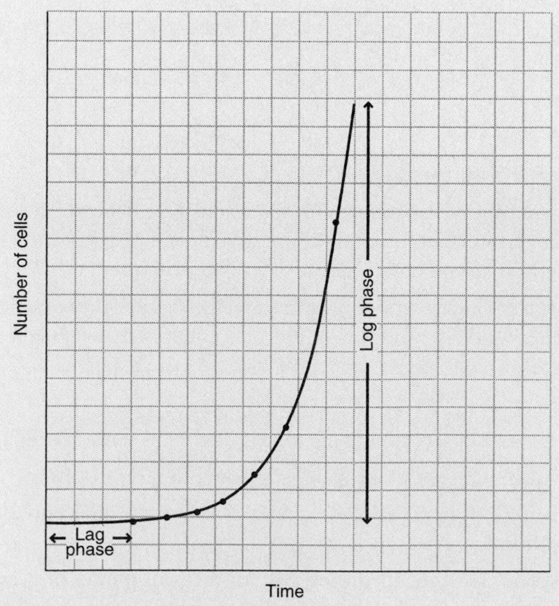

2. In the question above, how would the curve appear if the availability of nutrients were increased?

CHAPTER 5

Control of Microbial Growth

The British Medical Journal *stated that the British physician Joseph Lister (1827–1912) "saved more lives by the introduction of his system than all the wars of the 19th century together had sacrificed." He revolutionized surgery by developing effective methods that prevent surgical wounds from becoming infected. Impressed with Pasteur's work on fermentation (said to be caused by "minute organisms suspended in the air"), Lister wondered if "minute organisms" might be responsible also for the pus that forms in surgical wounds. He then experimented with a phenolic compound, carbolic acid, introducing it at full strength into wounds by means of a saturated rag. Lister was particularly proud of the fact that, after carbolic acid wound dressings became standard for his patients, they no longer developed gangrene. His work provided impressive evidence for the germ theory of disease, even though microorganisms specific for various diseases were not identified for another decade.*

Later, Lister improved his methods by introducing surgical procedures that excluded bacteria from wounds by maintaining a clean environment in the operating room and by sterilizing instruments. These procedures were preferable to killing the bacteria after they had entered wounds because they avoided the toxic effects of the disinfectant on the wound.

Lister was knighted in 1883 and subsequently became a baron and a member of the House of Lords.

—A Glimpse of History

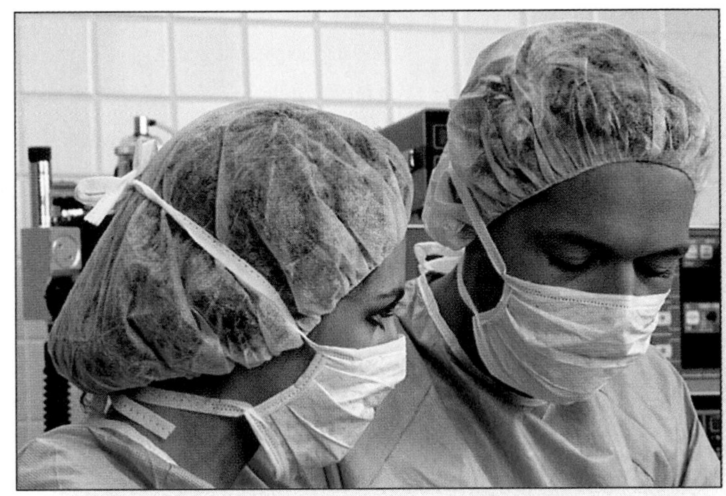

UNTIL THE LATE NINETEENTH CENTURY, PATIENTS undergoing even minor surgery were at great risk of developing fatal infections due to unsanitary medical practices and hospital conditions. Physicians did not know that their hands could pass diseases from one patient to the next. Nor did they understand that airborne microscopic organisms could cause infection in open wounds. Fortunately, today's modern hospitals use rigorous procedures to avoid microbial contamination, allowing operations, including major invasive surgery, to be performed with relative safety.

The growth of microorganisms affects more than our health. Producers of a wide variety of goods recognize that, unless microbial growth is controlled, the quality of their products can be compromised. This ranges from undesirable changes in the safety, appearance, taste, or odor of food products to the decay of untreated lumber.

This chapter will cover the methods that have been developed to destroy, remove, and inhibit the growth of microorganisms.

Approaches to Control

The processes used to control microorganisms are either physical or chemical, though a combination of both may be employed. **Physical methods** include heat treatment, irradiation, filtration, and mechanical removal (washing). **Chemical methods** use any one of a variety of antimicrobial chemicals. The method chosen depends on the circumstances and the resulting degree of control required.

Definitions related to microbial control are summarized in **table 5.1**.

Principles of Control

The process of removing or destroying all microorganisms and viruses on or in a product is called **sterilization**. These procedures include removing microbes by filtration, or destroying them using heat, certain chemicals, or irradiation. Destruction of microorganisms means they cannot be "revived" even when transferred from the sterilized product to an ideal growth medium. A **sterile** item

TABLE 5.1 Terms Commonly Used to Describe Antimicrobial Agents and Processes

Term	Definition
Antiseptic	A disinfectant that is nontoxic enough to be used on skin.
Aseptic technique	Use of specific methods to exclude contaminating microorganisms from an environment.
Bactericidal	Kills bacteria.
Bacteriostatic	Prevents the growth of, but does not kill, bacteria.
Decontamination	Treatment used to reduce the number of disease-causing microbes to a level that is considered safe to handle.
Disinfectant	A chemical used to destroy many microorganisms and viruses.
Disinfection	A process that eliminates most or all disease-causing microorganisms and viruses on or in a product.
Fungicide	Kills fungi.
Germicide	Kills microorganisms and inactivates viruses.
Pasteurization	A brief heat treatment used to reduce the number of spoilage organisms and to kill disease-causing microbes.
Preservation	The process of inhibiting the growth of microorganisms in products to delay spoilage.
Sanitize	To reduce the number of microorganisms to a level that meets public health standards; implies cleanliness as well.
Sterilant	A chemical used to destroy all microorganisms and viruses in a product, rendering it sterile.
Sterile	Completely free of all microorganisms and viruses; an absolute term.
Sterilization	The process of destroying all microorganisms and viruses, through physical or chemical means.
Viricide	Inactivates viruses.

is one that is absolutely free of microbes, including endospores and viruses. It is important to note, however, that the term **sterile** does not encompass **prions**. These infectious particles, which are thought to be proteins, are not destroyed by standard sterilization procedures. In fact, the minimum treatment required to destroy prions is still unknown. ■ endospores, p. 72 ■ prions, p. 364

Disinfection is the process that eliminates most or all disease-causing microorganisms and viruses on or in a material. Unlike sterilization, disinfection suggests that some viable microbes may persist. In practice, the term disinfection generally implies the use of antimicrobial chemicals. Those used for disinfecting inanimate objects are called **disinfectants**. Disinfectants are **biocides** (*bio* means "life," and *cida* means "to kill"). Although they are at least somewhat toxic to many forms of life, they are typically used in quantities and concentrations that target microscopic organisms, including bacteria and their endospores, fungi, and viruses. Thus, they are often called **germicides**. When disinfectants are formulated for use on skin they are called **antiseptics**. Antiseptics are routinely used to decrease the number of bacteria on skin to prepare for invasive procedures such as surgery.

Pasteurization utilizes a brief heat treatment to reduce the number of spoilage organisms and kill disease-causing organisms. Foods and inanimate objects can be pasteurized.

The term **decontaminated** is used to describe an item that has been treated to reduce the number of disease-causing organisms to a level that it is considered safe to handle. The treatment may be as simple as thorough washing, or it may involve the use of heat or disinfectants.

Sanitized generally implies a substantially reduced microbial population that meets accepted health standards. Most people also expect a sanitized object to be clean in appearance. Note that this term does not denote any specific level of control.

Preservation is the process of delaying spoilage of foods or other perishable products by adding growth-inhibiting ingredients or adjusting storage conditions to impede growth of microorganisms.

Situational Considerations

The methods used to control microbial growth vary greatly depending on the situation and the degree of control required **(figure 5.1)**. Control measures that are adequate for the routine circumstances of daily life might not be sufficient for situations such as hospitals, microbiology laboratories, foods and food production facilities, and other industries.

Daily Life

Washing and scrubbing with soaps and detergents achieves routine control of undesirable microorganisms and viruses. In fact, simple handwashing with plain soap and water is considered the single most important step in preventing the spread of many infectious diseases. Plain soap itself generally does not destroy organisms; it simply aids in the mechanical removal of transient microbes, including most pathogens, as well as dirt, organic material, and cells of the outermost layer of skin. Regular handwashing and bathing does not adversely impact the beneficial normal skin flora, which reside more

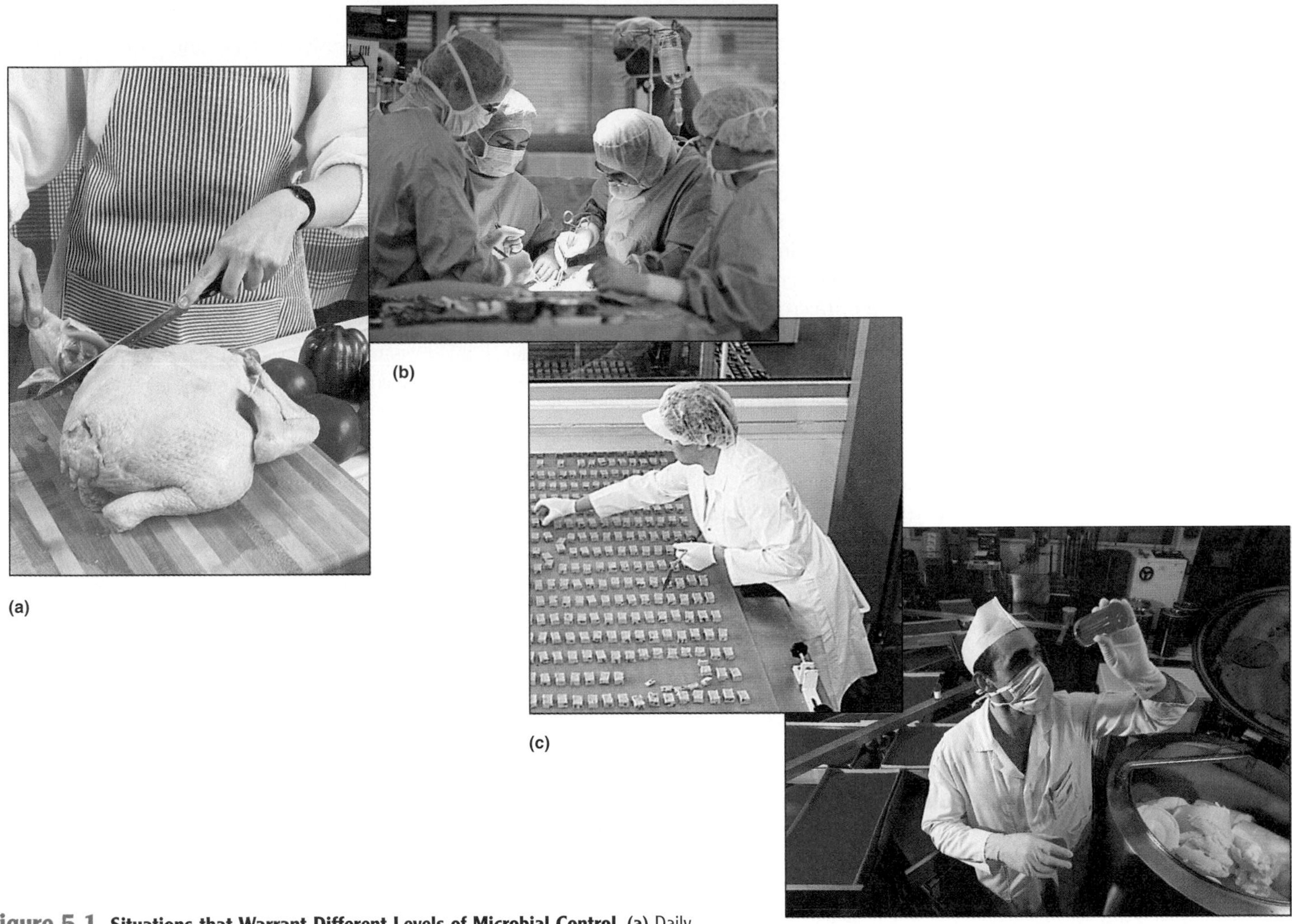

(a) (b) (c) (d)

Figure 5.1 Situations that Warrant Different Levels of Microbial Control **(a)** Daily home life; **(b)** hospitals; **(c)** foods and food production facilities; **(d)** other industries.

deeply on underlying layers of skin cells and in hair follicles. ■ **normal flora of the skin, p. 523**

Other methods used to control microorganisms in daily life include cooking of foods, cleaning of surfaces, and refrigeration of foods.

Hospitals

Minimizing the numbers of microorganisms in a hospital is particularly important because of the danger of hospital-acquired, or **nosocomial**, infections. Hospitalized patients are often more susceptible to infectious agents because of their weakened condition. In addition, patients may be subject to invasive procedures such as surgery, which breaches the intact skin that would otherwise help prevent infection. Finally, disease-causing organisms are more likely to be found in hospitals because of the high concentration of patients with infectious disease. These patients may shed disease-causing organisms in their feces, urine, respiratory droplets, or other body secretions. Thus, hospitals must be scrupulous in their control of microorganisms. Nowhere is this more important than in the operating rooms, where instruments used in invasive procedures must be sterile to avoid introducing even normally benign microbes into deep body tissue where they could easily establish infection. ■ **nosocomial infections, p. 488**

Prions are a relatively new concern for hospitals. Fortunately, disease caused by prions is thought to be exceedingly rare, less than 1 case per 1 million persons per year. Hospitals, however, must take special precautions when handling tissue that may be contaminated with prions, because these infectious particles are more difficult to destroy.

Microbiology Laboratories

Microbiology laboratories routinely work with microbial cultures and consequently must utilize rigorous methods of controlling microorganisms. To work with pure cultures, all media and instruments that contact the culture must first be rendered sterile to avoid contaminating the culture with environmental bacteria. All materials that have been used to culture microorganisms must again be treated before disposal in order to avoid contamination of workers and the environment. The use of specific methods to exclude contaminating microorganisms from an environment is called **aseptic technique**. Although all microbiology laboratories must use these prudent measures, those that work with known disease-causing microbes must be even more diligent.

Foods and Food Production Facilities

Foods and other perishable products retain their quality longer when the growth of contaminating microorganisms is prevented. This can be accomplished by physically removing or destroying microorganisms or by adding chemicals that impede their growth. Heat treatment is the most common and reliable method employed, but heating can alter the flavor and appearance of food. Irradiation can be used to kill microbes without causing perceptible changes in food, but the Food and Drug Administration (FDA) has approved this technology to treat only certain foods. Chemicals can be used to prevent the growth of microorganisms, but the risk of toxicity must always be a concern. Because of this, the FDA regulates chemical additives used in food and must deem them safe for consumption.

Food-processing facilities need to keep surfaces relatively free of microorganisms to avoid contamination. Machinery used to grind meat for example, if not cleaned properly, can create an environment in which bacteria multiply, eventually contaminating large quantities of product.

Other Industries

Many diverse industries have specialized concerns regarding microbial growth. Manufacturers of cosmetics, deodorants, or any other product that will be applied to the skin must avoid microbial contamination that could affect the product's quality or safety. Even finished products that are impervious to microbial actions may have steps or processes during their manufacture that could be affected by the growth of microorganisms.

MICROCHECK 5.1

Physical methods of control include heat treatment, irradiation, filtration, and mechanical removal. Chemical methods use any of a variety of antimicrobial chemicals. The methods used to control microbial growth depend on the situation and the degree of control required.

- How is sterilization different from disinfection?
- What is an antiseptic?
- Why would the term sterilization not encompass prions?

Selection of an Antimicrobial Procedure

The selection of an effective antimicrobial procedure is complicated by the fact that every procedure has parameters and drawbacks that limit its use. The ideal, multipurpose, nontoxic method simply does not exist. The ultimate choice depends on many factors including the type of microbe, the extent of the contamination, environmental conditions, and the potential risk of infection associated with use of the item.

Type of Microorganism

One of the most critical considerations in selecting a method of destroying microorganisms and viruses is the type of microbial population present on or in the product. Products contaminated with microorganisms more resistant to killing require a more rigorous heat or chemical treatment. Some of the more resistant microbes include:

- Bacterial endospores. The endospores of *Bacillus* and *Clostridium* are by far the most resistant forms of life. Only extreme heat or chemical treatment ensures their complete destruction. Chemical treatments that kill vegetative bacteria in 30 minutes may require 10 hours to destroy their endospores. ■ **endospores, p. 72**
- *Mycobacterium* species. The waxy cell walls of mycobacteria make them resistant to many chemical treatments. Thus, stronger, more toxic disinfectants must be used to disinfect environments that may contain *Mycobacterium tuberculosis*, the causative agent of tuberculosis. ■ **tuberculosis, p. 569**
- *Pseudomonas* species. These common environmental organisms are not only resistant to some chemical disinfectants, but in some cases can actually grow in them. *Pseudomonas* species are of particular importance in a hospital setting, where they are a common cause of infection. ■ ***Pseudomonas* infections, p. 697**
- Naked viruses. Viruses such as poliovirus that lack a lipid envelope are more resistant to disinfectants. Conversely, enveloped viruses, such as HIV, tend to be very sensitive to heat and chemical disinfectants. ■ **naked viruses, p. 324** ■ **enveloped viruses, p. 324**

Numbers of Microorganisms Initially Present

The time it takes for heat or chemicals to kill a population of microorganisms is dictated in part by the number of organisms initially present. It takes more time to kill a large population of bacteria than it does to kill a small population, because only a fraction of organisms die during a given time interval. For example, if 90% of a bacterial population is killed during the first 3 minutes, then approximately 90% of those remaining will be killed during the next three minutes, and so on. Recall that during the death phase of a cell population, cells die at a constant rate. Exposure to heat or disinfectants increases that rate. ■ **death phase, p. 107**

In the commercial canning industry, the **decimal reduction time** or **D value**, is the time required for killing 90% of a population of bacteria under specific conditions **(figure 5.2)**. The temperature of the process may be indicated by a subscript, for example D_{121}. A one D process reduces the number of organisms by one exponent. Thus, if the D value for an organism is 2 minutes, then it would take 4 minutes (2 D values) to reduce a population of 100 (10^2) organisms to only one (10^0) survivor. It would take 20 minutes (10 D values) to reduce a population of 10^{10} cells to only one survivor. Removing organisms by washing or scrubbing can minimize the time necessary to sterilize or disinfect a product.

Environmental Conditions

Factors such as pH, temperature, and presence of fats and other organic materials strongly influence microbial death rates. For example, a solution of sodium hypochlorite (household bleach) can kill a suspension of *M. tuberculosis* in 150 seconds at a

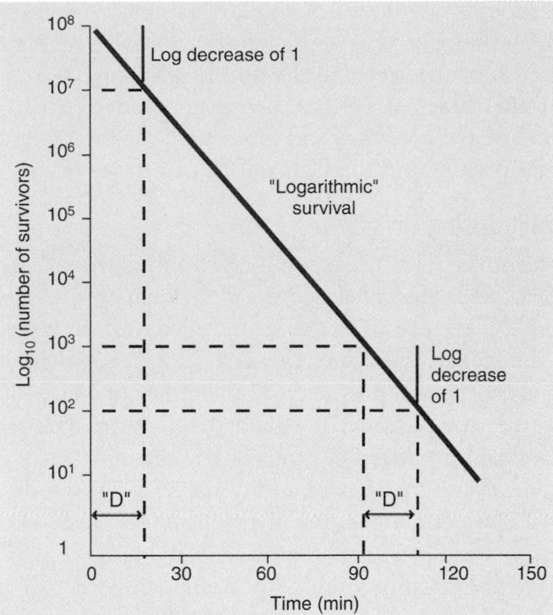

Figure 5.2 The Relationship Between the Numbers of Initial Microorganisms and the Time It Takes to Kill Them The D value is the time it takes to reduce the population by 90%.

temperature of 50°C; whereas it takes only 60 seconds to kill the same suspension with chlorine if the temperature is increased to 55°C. That hypochlorite solution would be even more effective at a low pH.

The presence of dirt, grease, and organic compounds such as blood and other body fluids can interfere with heat penetration and the action of chemical disinfectants. For this reason, it is important to thoroughly clean items before disinfection or sterilization. Meticulously cleaning a product also substantially decreases the number of microorganisms, which, as mentioned previously, shortens the time required to sterilize or disinfect the product.

Potential Risk of Infection

To guide medical biosafety personnel in their selection of germicidal procedures, items such as surgical instruments, endoscopes, and blood pressure cuffs are categorized according to their potential risk of transmitting infectious agents. Those that pose the greatest threat of transmitting disease must be subject to more rigorous sterilization procedures.

- **Critical items** come into direct contact with body tissues. These items, including needles, scalpels, and biopsy forceps, must be sterilized to avoid transmission of all infectious agents.
- **Semicritical items** come into contact with mucous membranes, but do not penetrate body tissue. These items, including gastrointestinal endoscopes, endotracheal tubes, and vaginal specula, must be free of all viruses and vegetative bacteria including mycobacteria. Low numbers of endospores that may remain on semicritical items pose little risk of infection because mucous membranes are effective barriers against their entry into deeper tissue.
- **Noncritical instruments** and surfaces pose little risk of infection because they only come into contact with unbroken skin. Countertops, stethoscopes, and blood pressure cuffs are examples of noncritical items.

MICROCHECK 5.2

The types and numbers of microorganisms initially present, environmental conditions, and the potential risks associated with use of the item must all be considered when determining which sterilization or disinfection procedure to employ.

- What form of microbial life is the most difficult to destroy?
- Define the term D value.
- Would it be safe to say that if all bacterial endospores had been killed, then all other microorganisms had also been killed?

Using Heat to Destroy Microorganisms and Viruses

Even before the discovery of the microbial world, heat was used to control microorganisms. For example, Aristotle (384–322 B.C.) advised Alexander the Great to have his armies boil their drinking water to prevent disease. Today heat treatment is still one of the most useful methods of microbial control, because it is fast, reliable, and inexpensive relative to most other methods and it does not introduce potentially toxic substances into the material being treated.

Some types of heat treatment, such as autoclaving, are used to sterilize products, whereas other treatments, such as pasteurization, are used to decrease the numbers of microorganisms. **Table 5.2** summarizes the methods, limitations, and uses of heat to control microorganisms and viruses.

Moist Heat

Moist heat, such as boiling water and pressurized steam, destroys microorganisms by causing the irreversible coagulation of their proteins. Boiling (100°C) for 10 minutes readily destroys most microorganisms and viruses. The notable exception is endospores, including those of the food-poisoning bacteria *Clostridium perfringens* and *C. botulinum*, which can survive many hours of boiling. Pressurized steam that reaches a temperature of 121°C, sustained for 15 minutes, is the only heat treatment that reliably kills endospores.

Pasteurization

Louis Pasteur developed the brief heat treatment we now call pasteurization as a way of avoiding spoilage of wine. The production of fine wines relies on yeast converting grape sugars into alcohol. If bacteria or molds grow in the wine, they metabolize the alcohol and produce undesirable flavors. Pasteur found that by moderately heating the wine at just the right conditions of time and temperature, spoilage microorganisms were eliminated without significantly changing the taste of the wine.

Table 5.2 Using Heat to Control Microorganisms

Method	Characteristics and Typical Uses
Boiling	Fast, reliable, and inexpensive method to destroy most bacteria and viruses. Unreliable for killing endospores. Can be used to treat drinking water.
Pasteurization	Significantly decreases the numbers of heat-sensitive microorganisms, including spoilage microorganisms and pathogens (except sporeformers). Milk and juices are routinely pasteurized.
Pressurized steam	Achieves temperatures above 100°C, which is necessary to destroy endospores. Used to sterilize surgical instruments and other items that can be penetrated by steam. Also used to produce canned foods.
Dry heat	High temperatures for long lengths of time can be used to sterilize items. Flame is used to sterilize bacteriological loops. Ovens are used to sterilize glassware and items that are not readily penetrated by steam.

Pasteurization does not sterilize substances but should significantly reduce the numbers of heat-sensitive organisms, including pathogens. Today, pasteurization is still used to destroy spoilage organisms in wine, vinegar, and a few other foods, but it is most widely used for killing disease-causing microorganisms in milk and juices. Pasteurization increases the shelf life of foods and protects consumers by killing organisms that cause diseases such as tuberculosis, brucellosis, salmonellosis, and typhoid fever, without significantly altering the quality of the food.

Original pasteurization methods utilized a relatively low temperature (62°C) for long time periods (30 minutes). Today, most pasteurization protocols employ the **high-temperature-short-time (HTST) method**. Using this method, milk is heated to 72°C and held for only 15 seconds. The parameters must be adjusted to the individual food product. For example, ice cream, which is richer than milk in fats, requires a pasteurization process of 82°C for about 20 seconds.

The single-serving containers of cream served in restaurants are processed using the **ultra-high-temperature (UHT) method**. Because this process is designed to render the product free of all microorganisms that can grow under normal storage conditions, it is technically not a type of pasteurization. The milk is rapidly heated to a temperature of 140°C to 150°C, held for several seconds, then rapidly cooled. The product is then aseptically packaged in containers that have been treated with the chemical germicide hydrogen peroxide. Boxed juices are processed and packaged in a similar manner.

Although one does not think of pasteurizing things such as cloth and rubber, this can easily be done by regulating the temperature of the water in a washing machine. For example, hospital anesthesia masks can be pasteurized at 80°C for 15 minutes. The temperatures and times used vary according to the organisms present and the heat stability of the material.

Sterilization Using Pressurized Steam

Pressure cookers and their commercial counterpart, the **autoclave**, heat water in an enclosed vessel that achieves temperatures above 100°C **(figure 5.3)**. As heated water in the vessel forms steam, the steam causes the pressure in the vessel to increase beyond atmospheric pressure. The higher pressure, in turn, increases the temperature at which steam forms **(figure 5.4)**. Whereas steam produced at atmospheric pressure never exceeds 100°C, steam produced at an additional 15 psi (pounds/square inch) is 121°C, a temperature that kills endospores. Note that the pressure itself plays no direct role in the killing.

Typical conditions used for sterilization are 15 psi and 121°C for 15 minutes. Longer time periods are necessary when sterilizing large volumes because it takes longer for heat to completely penetrate the liquid. For example, it takes longer to sterilize 4 liters of liquid in a flask than it would if the same volume were distributed into small tubes. When rapid sterilization is important, such as in operating rooms when sterile instruments must always be available, **flash autoclaving** at higher temperature can be used. By increasing the temperature to 135°C, sterilization is achieved in only 3 minutes. This same temperature, applied for 1 hour, is thought to destroy prions.

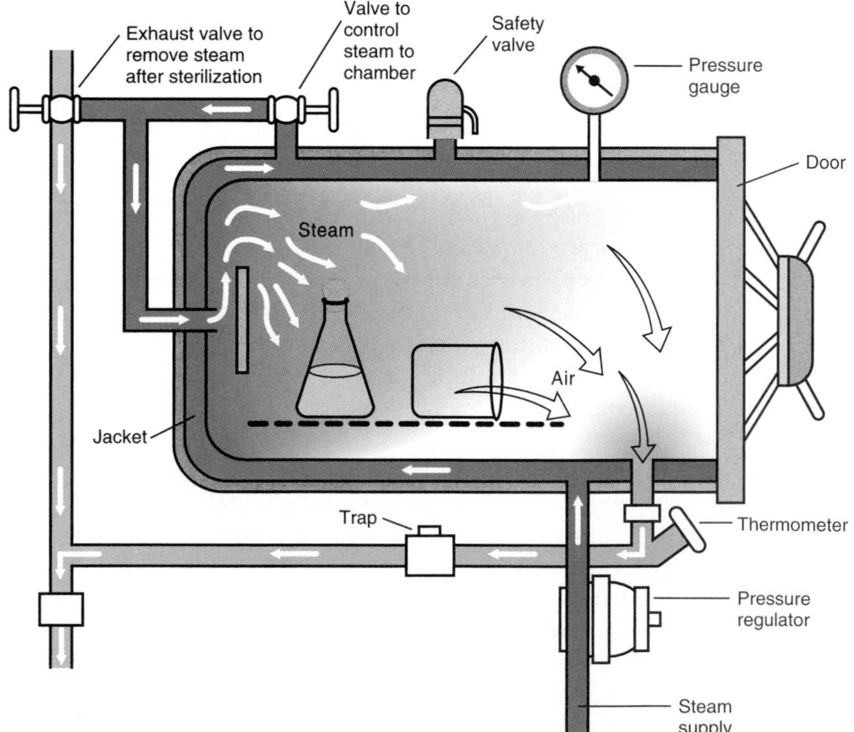

Figure 5.3 Steam-Jacketed Autoclave Steam first travels in an enclosed layer, or jacket, surrounding the chamber. It then enters the autoclave, displacing the air downward and out through a port in the bottom of the chamber.

Autoclaving is a consistently effective means of sterilizing most objects, provided the process is done correctly. The temperature and pressure gauge should both be monitored to ensure proper operating conditions. It is also critical that steam can enter items and displace the air. Long, thin containers should be placed on their sides. Likewise, containers and bags should never be closed tightly.

Attaching tape that contains a heat-sensitive indicator to objects when they are autoclaved can provide a visual signal that items have been heated. The indicator turns black during autoclaving **(figure 5.5a)**. A changed indicator, however, does not always mean that the object is sterile, because heating may not have been uniform.

Biological indicators are used to ensure that the autoclave is working properly (figure 5.5b). A tube containing the heat-resistant endospores of *Bacillus stearothermophilus* is placed near the center of an item or package being autoclaved. After autoclaving, the endospores are mixed with a growth medium by crushing a container within the tube. Following incubation, a change of color of the medium indicates growth of the organisms and thus faulty autoclaving.

The Commercial Canning Process

The commercial canning process uses pressurized steam in an industrial-sized autoclave called a **retort**. The conditions of the process are designed to ensure that endospores of *Clostridium botulinum* are destroyed. This is critical because surviving spores can germinate and the resulting vegetative cells can grow in the anaerobic conditions of low-acid canned foods, such as vegetables and meats, and produce botulinum toxin, one of the most potent toxins known. Because ingestion of even minute amounts of botulinum toxin can be lethal, the canning process of low-acid foods is designed to kill all endospores of *C. botulinum*. In doing so, the process also kills all other organisms capable of growing under normal storage conditions. Endospores of some thermophilic bacteria may survive the canning process, but these are usually of no concern because they only can grow at temperatures well above those of normal storage. Because of this, canned foods are called **commercially sterile** to reflect the fact that the endospores of some thermophiles may survive. **Figure 5.6** shows the steps involved in the commercial canning of foods.

Several factors dictate the time and temperature of the canning process. First, as discussed earlier, the higher the temperature, the shorter the time needed to kill all organisms.

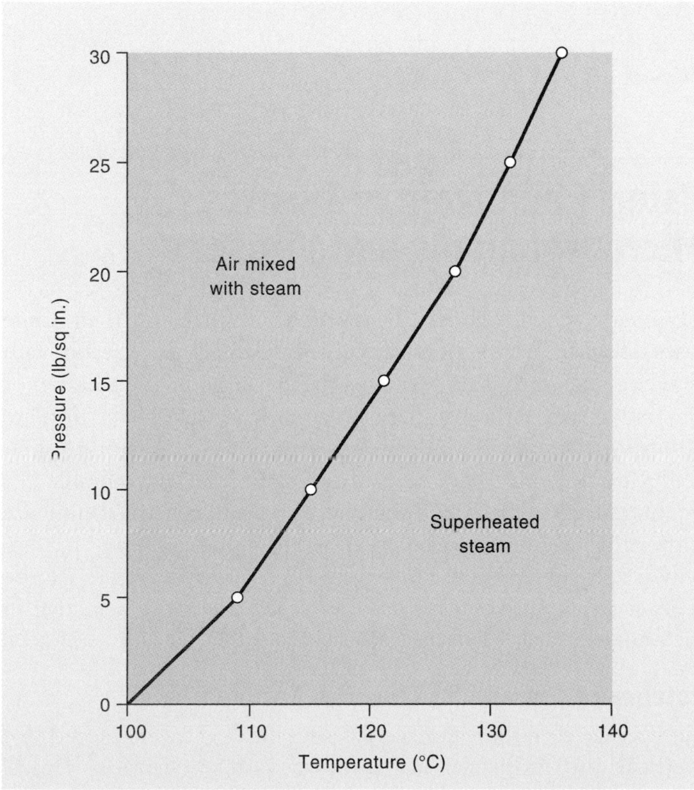

Figure 5.4 Relationship Between Pressure and the Temperature at Which Steam Is Formed Autoclaving conditions that generate air mixed with steam or superheated steam do not reliably sterilize substances.

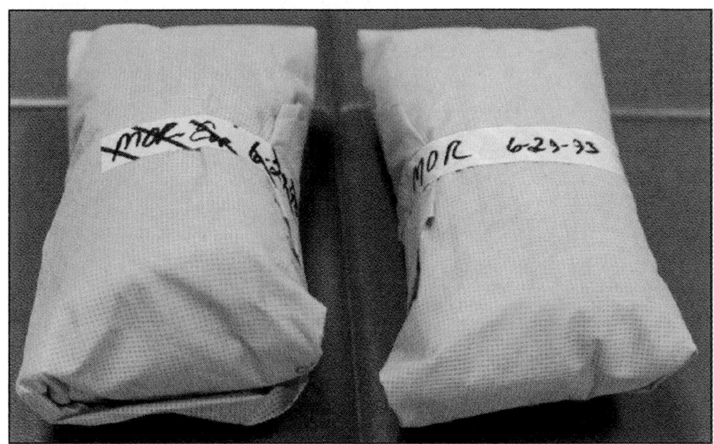

(a)

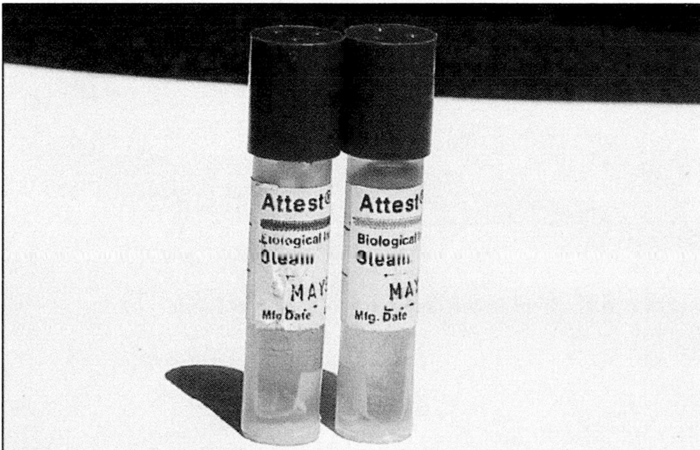

(b)

Figure 5.5 Indicators Used in Autoclaving **(a)** Chemical indicators. The pack on the left has been autoclaved. Diagonal marks on the tape have turned black with heat, indicating that the object was exposed to heat. **(b)** Biological indicators. Following incubation, a change of color to yellow indicates growth of endospore-forming organisms. Why would a biological indicator be better than other indicators to detect if sterilization had been completely effective?

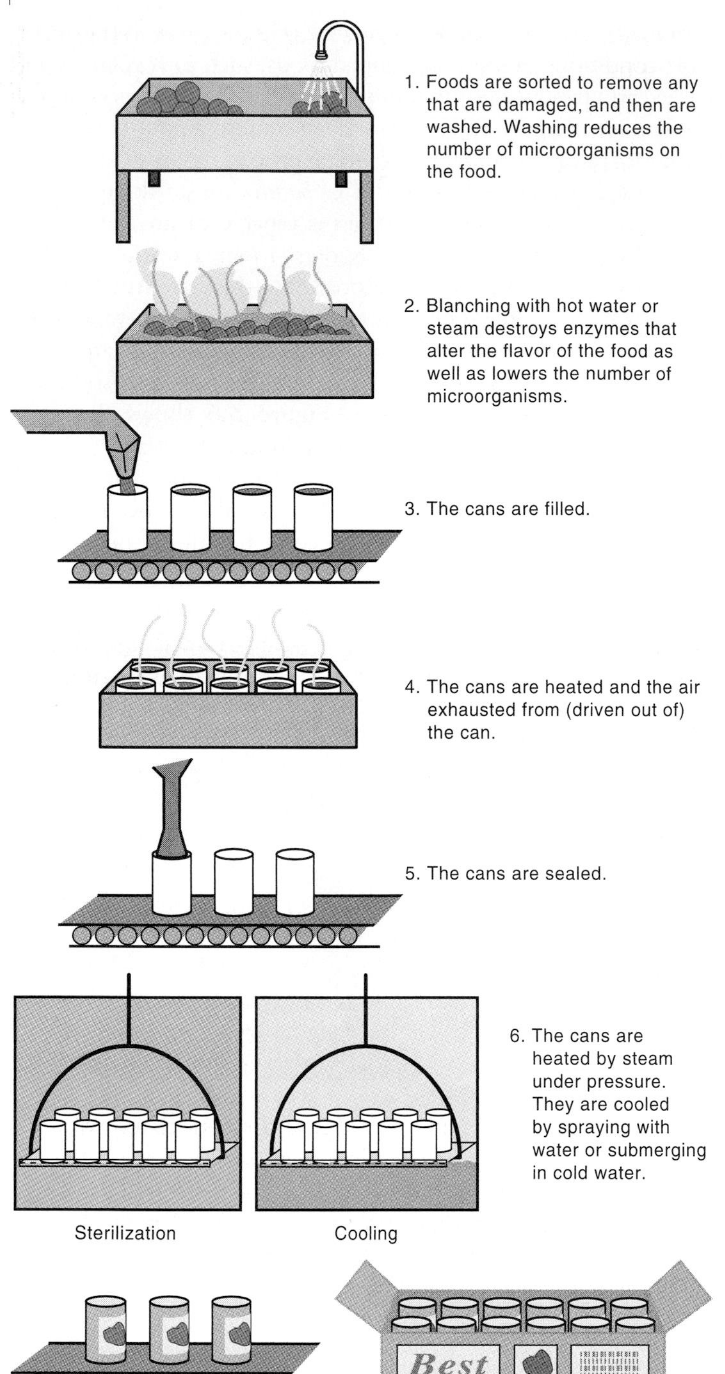

1. Foods are sorted to remove any that are damaged, and then are washed. Washing reduces the number of microorganisms on the food.
2. Blanching with hot water or steam destroys enzymes that alter the flavor of the food as well as lowers the number of microorganisms.
3. The cans are filled.
4. The cans are heated and the air exhausted from (driven out of) the can.
5. The cans are sealed.
6. The cans are heated by steam under pressure. They are cooled by spraying with water or submerging in cold water.
Sterilization
Cooling
Best
7. Cans are labeled, packaged, and shipped.

Figure 5.6 Steps in the Commercial Canning of Foods

Second, the higher the concentration of bacteria, the longer the heat treatment required to kill all organisms. To provide a wide margin of safety, the commercial canning process is designed to reduce a population of 10^{12} *C. botulinum* endospores to only one. In other words, it is a 12 D process. It is virtually impossible for a food to have this high a level of initial concentration of endospores, and so the process has a wide safety margin.

Dry Heat

Heating in the absence of moisture is frequently used to destroy microorganisms. For example, in microbiology laboratories, the wire loops that are continually reused to transfer bacterial cultures can be sterilized by heating them in a flame until they are red hot. Glass Petri dishes and glass pipets are sterilized in ovens with noncirculating air at temperatures of 160°C to 170°C for 2 to 3 hours. The process essentially burns the cell constituents, oxidizing them to ashes or irreversibly denaturing proteins. Dry heat has an advantage over moist heat in that it can be used for powders and other anhydrous materials that moist heat does not effectively penetrate. It also does not dull sharp instruments or corrode metals.

Dry heat takes much longer than wet heat to kill microorganisms. For example, 200°C for $1\frac{1}{2}$ hours of dry heat is the killing equivalent of 121°C for 15 minutes of moist heat. Ovens with a fan that circulates the hot air can sterilize in half the time because of the more efficient transfer of heat.

MICROCHECK 5.3

Moist heat such as boiling water destroys vegetative bacteria and viruses. Pasteurization significantly reduces the numbers of heat-sensitive organisms. Autoclaves generate pressurized steam that kills microbes, including endospores. The commercial canning process is designed to destroy the endospores of *Clostridium botulinum*. Dry heat takes longer than moist heat to kill microbes.

- Why is it important that the commercial canning process destroys the endospores of *Clostridium botulinum*?
- What are two purposes of pasteurization?
- What pressure would be necessary to produce steam at 135°C?

Using Chemicals to Destroy Microorganisms and Viruses

Germicidal chemicals can be used to disinfect and, in some cases, sterilize. Most chemical germicides react irreversibly with vital enzymes and other proteins, the cytoplasmic membrane, or viral envelopes, although their precise mechanisms of action are often not completely understood (**figure 5.7**). Although generally less reliable than heat, these chemicals are suitable for treating large surfaces and many heat-sensitive items. Some are sufficiently nontoxic to be used as antiseptics. Those that are only weakly germicidal, but have a bacteriostatic action, can be used as preservatives. Antibiotics, which are more selective in their toxicity, will be discussed in chapter 21.

Potency of Germicidal Chemical Formulations

Numerous different germicidal chemicals are marketed for medical and industrial use under a variety of trade names. Frequently, they contain more than one antimicrobial chemical as well as other chemicals such as buffers that can influence their antimicrobial activity. In the United States, the Food and

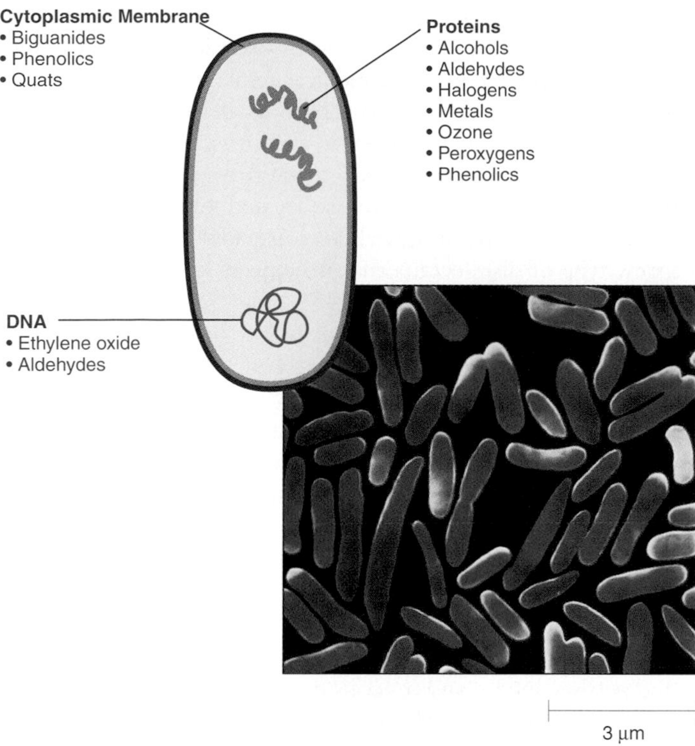

Figure 5.7 Sites of Action of Germicidal Chemical

Drug Administration (FDA) has the responsibility for regulating chemicals that can be used to process medical devices in order to ensure they perform as claimed. Most chemical disinfectants are considered pesticides and, as such, are regulated by the Environmental Protection Agency (EPA). To be registered with either the FDA or EPA, manufacturers of germicidal chemicals must document the potency of their products using testing procedures originally defined by the EPA. Germicides are grouped according to their potency (**table 5.3**):

- **Sterilants** can destroy all microorganisms, including their endospores, and viruses. Destruction of endospores usually requires a 6- to 10-hour treatment. These can be used to treat heat-sensitive critical items such as scalpels.
- **High-level disinfectants** destroy all viruses and vegetative microorganisms, but they do not reliably kill endospores. Most high-level disinfectants are simply sterilants used for time periods as short as 30 minutes, a time period not long enough to ensure endospore destruction. These can be used to treat semicritical items such as gastrointestinal endoscopes.
- **Intermediate-level disinfectants** destroy all vegetative bacteria including mycobacteria, fungi, and most, but not all, viruses. They do not kill endospores even with prolonged exposure. They can be used to disinfect noncritical instruments such as stethoscopes.
- **Low-level disinfectants** destroy fungi, vegetative bacteria except mycobacteria, and enveloped viruses. They do not kill endospores, nor do they reliably destroy naked viruses. Intermediate-level and low-level disinfectants are also called **general-purpose disinfectants**.

To perform properly, germicides must be used strictly according to the manufacturer's directions, especially as they relate to dilution, temperature, and the amount of time they must be in contact with the object being treated. It is extremely important that the object be thoroughly cleaned and free of organic material before the germicidal procedure is begun.

Selecting the Appropriate Germicidal Chemical

Selecting the appropriate germicide is a complex decision. Some of the points that must be considered include:

- Toxicity. By nature, germicides are at least somewhat toxic to humans and the environment. Therefore, the benefit of disinfecting or sterilizing an item or surface must be weighed against the risks associated with the use of germicidal procedure. For example, the risk of being exposed to a pathogenic microorganism in a hospital environment warrants the use of the most effective chemical germicides, even considering the potential risks of their use. The microbiological risks associated with typical household and office situations, however, may not justify the use of many of those same germicides.
- Activity in the presence of organic matter. Many germicidal chemicals, such as hypochlorite, are readily inactivated by organic matter and would not be appropriate to use in situations where organic material is present. Chemicals such as phenolics, however, tolerate the presence of some organic matter.

Table 5.3 Usual Pattern of Antimicrobial Activity of Germicides

	Bacteria			Fungi	Viruses	
Category of Germicide	**Endospores**	**Mycobacteria**	**Others**	**All**	**Nonlipid**	**Lipid**
Sterilant	yes	yes	yes	yes	yes	yes
High Level	some	yes	yes	yes	yes	yes
Intermediate Level	no	varies	yes	yes	varies	yes
Low Level	no	no	varies	varies	no	varies

- Compatibility with the material being treated. Items such as electrical equipment often cannot tolerate liquid chemical germicides, and so gaseous alternatives must be employed. Likewise, corrosive germicides such as hypochlorite often damage some metals and rubber.
- Residue. Many chemical germicides leave a residue that is toxic or corrosive. If a germicide that leaves a residue is used to sterilize or disinfect an item, the item must be thoroughly rinsed to entirely remove the residue. Obviously, sterile items must be rinsed with sterile water.
- Cost and availability. Some germicides are less expensive and more readily available than others. For example, hypochlorite can easily be purchased in the form of household bleach. On the other hand, ethylene oxide gas is not only more expensive, but it must be used in a special chamber, which influences the cost and practicality of the procedure.
- Storage and stability. Germicides such as phenolics and iodophores are available in concentrated stock solutions, decreasing the required storage space. The stock solutions are simply diluted according to the manufacturer's instructions before use. Germicides such as glutaraldehyde and chlorine dioxide come in two-component systems that, once mixed, have a limited shelf life.
- Environmental risk. Germicides that retain their antimicrobial activity after use can interfere with sewage treatment systems that utilize bacteria to degrade sewage. The activity of those germicides must be neutralized before disposal.

Classes of Germicidal Chemicals

Germicides are represented in a number of chemical families. Each type has characteristics that make it more or less appropriate for specific uses (**table 5.4**).

Alcohols

Aqueous solutions of 60% to 80% ethyl or isopropyl alcohol rapidly kill vegetative bacteria and fungi. They do not, however, reliably destroy bacterial endospores and some naked viruses. Alcohol is thought to act by coagulating enzymes and other essential proteins and by damaging lipid membranes. Proteins are more soluble and denature more easily in alcohol mixed with water, which is why aqueous solutions are more effective than pure alcohol.

Solutions of alcohol are commonly used as antiseptics to prepare for procedures such as injections that break intact skin and as disinfectants for treating instruments and surfaces. They are also used to enhance the antimicrobial activity of some other chemicals. They may damage some materials, however, such as rubber and some plastics. They are relatively nontoxic and inexpensive, and do not leave a residue, but they evaporate quickly, which limits their effective contact time and, consequently, their germicidal effectiveness.

Aldehydes

The aldehydes **glutaraldehyde**, **formaldehyde**, and **ortho-phthalaldehyde** (OPA) destroy microorganisms and viruses by inactivating proteins and nucleic acids. A 2% solution of alkaline glutaraldehyde is one of the most widely used liquid chemical sterilants for treating heat-sensitive medical items. Immersion in a glutaraldehyde solution for 10 to 12 hours destroys all forms of microbial life, including endospores, and viruses. Soaking times as short as 10 minutes can be used to destroy vegetative bacteria. Glutaraldehyde is toxic, however, so treated items must be thoroughly rinsed with sterile water before use. Ortho-phthalaldehyde is a new type of disinfectant that is being studied as an alternative to glutaraldehyde.

Formaldehyde is available as a gas or an aqueous 37% solution called **formalin**. It is an extremely effective germicide that kills most forms of microbial life within minutes. Formalin is used to kill bacteria and to inactivate viruses for use as vaccines. It has also been used to preserve biological specimens. Formaldehyde's irritating vapors and suspected carcinogenicity, however, now limit its use.

Biguanides

Chlorhexidine, the most effective of a group of chemicals called **biguanides**, is extensively used in antiseptic products. It adheres to and persists on skin and mucous membranes, is of relatively low toxicity, and destroys a wide range of microbes, including vegetative bacteria, fungi, and some enveloped viruses. Since the mid-1970s, chlorhexidine has been incorporated into a multitude of products including antiseptic skin creams, disinfectants, and mouthwashes. In the 1990s the FDA approved the use of chlorhexidine-impregnated catheters and implanted surgical mesh. Even tiny chips have been developed that can be inserted into periodontal pockets, where they slowly release chlorhexidine to treat periodontal gum disease. Adverse side effects of chlorhexidine are rare, but severe allergic reactions have been reported.

Ethylene Oxide

Ethylene oxide is an extremely useful gaseous sterilizing agent that destroys all microorganisms, including endospores and viruses, by reacting with proteins. As a gas, it penetrates well into fabrics, equipment, and implantable devices such as pacemakers and artificial hips. It is particularly useful for sterilizing heat- or moisture-sensitive items such as electrical equipment, pillows, and mattresses. Many disposable laboratory items, including plastic Petri dishes and pipets, are also sterilized with ethylene oxide.

A special chamber that resembles an autoclave is used to sterilize items with ethylene oxide. This allows the careful control of factors such as temperature, relative humidity, and ethylene oxide concentration, all of which influence the effectiveness of the gas. Because ethylene oxide is explosive, it is generally mixed with a nonflammable gas such as carbon dioxide. Under these carefully controlled conditions, objects can be sterilized in 3 to 12 hours. The toxic ethylene oxide must then be eliminated from the treated material using heated forced air for 8 to 12 hours. Absorbed ethylene oxide must be allowed to dissipate because of its irritating effects on tissues and its persistent antimicrobial effect, which, in the case of Petri dishes and other items used for culturing bacteria, is undesirable.

Ethylene oxide is mutagenic and therefore potentially carcinogenic. Indeed, studies have shown a slightly increased risk

Table 5.4 Chemicals Used in Sterilization and Disinfection and Preservation of Nonfood Substances

Chemical (examples)	Characteristics
Alcohols (ethanol and isopropanol)	Aqueous solutions are used as antiseptics and for disinfecting some instruments and surfaces. Alcohols leave no residue, are easy to obtain, and are inexpensive. They evaporate quickly, which limits their contact time.
Aldehydes (formaldehyde and glutaraldehyde)	Formalin is used in vaccine preparation. Glutaraldehyde is more effective than formaldehyde against endospores and is commonly used as a sterilant. Aldehydes inactivate proteins and nucleic acids. They are irritating to the respiratory system, skin, and eyes.
Biguanides (chlorhexidine)	Chlorhexidine is widely used as an antiseptic in soaps and lotions, and more recently, impregnated into catheters and surgical mesh.
Ethylene Oxide Gas	Commonly used as a sterilant. It easily penetrates hard-to-reach places and fabrics and does not damage moisture-sensitive material. It is explosive, toxic, and potentially carcinogenic.
Halogens (chlorine and iodine)	Chlorine is used to disinfect inanimate objects, surfaces, drinking water, and wastewater. Organic compounds and other impurities neutralize its activity. Tincture of iodine and iodophores can be used as disinfectants or antiseptics. Chlorine and iodine are both oxidizing agents.
Metals (silver)	Silver sulfadiazine is used in topical dressings to prevent infection of burns. Silver nitrate drops can be used to prevent gonococcal eye infections of newborns. Most other metal compounds are too toxic to be used medically.
Ozone	Used to disinfect drinking water and wastewater. This unstable molecular form of oxygen readily breaks down.
Peroxygens (peracetic acid and peroxide)	Strong oxidizing agents that are increasingly being used as sterilants.
Phenolics (hexachlorophene and triclosan)	Phenolic disinfectants are used to treat floors and other surfaces. They remain active in the presence of organic contaminants. Triclosan is used in toothpastes, lotions, and deodorant soaps. Hexachlorophene is highly effective against *Staphylococcus aureus,* but its use as an antiseptic is limited because it can cause neurological damage.
Quaternary Ammonium Compounds (benzalkonium chloride and cetylpyridinium chloride)	Widely used to disinfect clean inanimate objects and to preserve nonfood substances. Nontoxic enough to be used on food preparation surfaces. Easily inactivated by anionic soaps and detergents.

of malignancies in long-term users of the gas. Less toxic gaseous alternatives are currently being explored.

Halogens

Chlorine and iodine are common disinfectants that are thought to act by oxidizing proteins and other essential cell components.

Chlorine Chlorine destroys all types of microorganisms and viruses but is too irritating to skin and mucous membranes to be used an antiseptic. Chlorine-releasing compounds such as sodium hypochlorite can be used to disinfect waste liquids, swimming pool water, instruments, and surfaces, and at much lower concentrations, to disinfect drinking water.

Chlorine solutions are inexpensive, readily available disinfectants. An effective disinfection solution can easily be made by diluting liquid household bleach (5.25% sodium hypochlorite) 1:100 in water, resulting in a solution of 500 ppm (parts per million) chlorine. This concentration is several hundred times the amount required to kill most pathogenic microorganisms and viruses, but it is usually necessary for fast, reliable killing. In situations when excessive organic material is present, a 1:10 dilution of bleach may be required. This is because chlorine also readily reacts with organic compounds and other impurities in water. These compounds consume free chlorine, reducing the germicidal activity of chlorine-releasing compounds. The use of high concentrations, however, should be avoided when possible, because chlorine is both corrosive and toxic. Diluted solutions of liquid bleach deteriorate over time; thus, fresh solutions need to be prepared regularly. More stable forms of chlorine, including sodium dichloroisocyanurate and chloramines, which are available in powders and tablets, are often used in hospitals.

The chlorination of drinking water has saved hundreds of thousands of lives by preventing transmission of waterborne illnesses such as cholera. Properly chlorinated drinking water contains approximately 0.5 ppm chlorine, much less than that used for disinfectant solutions. The exact amount of chlorine that must be added depends on the amount of organic material in the water. The presence of organic compounds is also problematic because chlorine may react with some organic compounds to form trihalomethanes, which are potential carcinogens. ■ **cholera, p. 599**

Chlorine dioxide (ClO_2) is a strong oxidizing agent that is increasingly being used as a disinfectant and sterilant. It has an advantage over chlorine-releasing compounds in that it does not react with organic compounds to form trihalomethanes or other toxic chlorinated products. Compressed chlorine dioxide gas, however, is explosive and liquid solutions decompose readily, so that it must be generated on-site. It is used to treat drinking water, wastewater, and swimming pools.

Iodine Iodine, unlike chlorine, does not reliably kill endospores, but it can be used as a disinfectant. It is used as a **tincture**, in which the iodine is dissolved in alcohol, or more commonly as an **iodophore**, in which the iodine is linked to a carrier molecule that releases free (unbound) iodine slowly. Iodophores are not as irritating to the skin as tincture of iodine nor are they as likely to stain. Iodophores used as disinfectants contain more free iodine (30 to 50 ppm) than do those used as antiseptics (1–2 ppm). Stock solutions must be strictly diluted according to the manufacturer's instructions because dilution affects the amount of free iodine available.

Surprisingly, some *Pseudomonas* species survive in the concentrated stock solutions of iodophores. The reasons are unclear, but it may be due to inadequate levels of free iodine in concentrated solutions, the iodine being released from the carrier only with dilution. *Pseudomonas* species also can form biofilms, which are less permeable to chemicals. Nosocomial infections can result if a *Pseudomonas*-contaminated iodophore is unknowingly used to disinfect instruments. ■ **biofilm, p. 108**

Metal Compounds

Metal compounds kill microorganisms by combining with sulfhydryl groups of enzymes and other proteins, thereby interfering with their function. Unfortunately, most metals at high concentrations are too toxic to human tissue to be used medically. Silver is one of the few metals still used as a disinfectant. Dressings containing silver sulfadiazine are used to prevent infection of burns. For many years, doctors were required by law to instill drops of another silver compound, 1% silver nitrate, into the eyes of newborns to prevent **opthalmia neonatorum**, an eye infection caused by *Neisseria gonorrhoeae*, which is acquired from infected mothers during the birth process. Drops of antibiotics have now largely replaced use of silver nitrate because they are less irritating to the eye and more effective against another genitally acquired pathogen, *Chlamydia trachomatis.* ■ ***Neisseria gonorrhoeae*, p. 644** ■ ***Chlamydia trachomatis*, p. 646**

Compounds of mercury, tin, arsenic, copper, and other metals were once widely used as preservatives in industrial compounds and to prevent microbial growth in recirculating cooling water. Their extensive use resulted in serious pollution of natural waters, which has prompted strict controls. The FDA is currently examining the list of antiseptics and other products that contain mercury compounds, such as mercurochrome, thimerosal, and phenylmercuric acetate, in order to evaluate their safety.

Ozone

Ozone (O_3) is an unstable form of oxygen that is a powerful oxidizing agent. It decomposes quickly, however, so that it must be generated on-site, usually by passing air or oxygen between two electrodes. Many countries currently use ozone as an alternative to chlorine for disinfecting drinking water and wastewater. Advances in ozone-generating technology may improve the economic feasibility of its use, making it more widespread.

Peroxygens

Hydrogen peroxide and peracetic acid are powerful oxidizing agents that under controlled conditions can be used as sterilants. They are readily biodegradable and, in normal concentrations of use, appear to be less toxic than the traditional alternatives, ethylene oxide and glutaraldehyde.

Hydrogen Peroxide The effectiveness of hydrogen peroxide (H_2O_2) as a germicide depends in part on whether it is used on living tissue, such as a wound, or on an inanimate object. This is because all cells that use aerobic metabolism, including our body's cells, produce the enzyme **catalase**, which inactivates hydrogen peroxide by breaking it down to water and oxygen gas. Thus, when a solution of 3% hydrogen peroxide is applied to a wound, our cellular enzymes quickly break it down. When the same solution is used on an inanimate surface, however, it overwhelms the relatively low concentration of catalase produced by microscopic organisms. ■ **catalase, p. 98**

Hydrogen peroxide is particularly useful as a disinfectant because it leaves no residue and does not damage stainless steel, rubber, plastic, or glass. Hot solutions are commonly used in the food industry to yield commercially sterile containers for aseptically packaged juices and milk. Vapor-phase hydrogen peroxide is more effective than liquid solutions and can be used as a sterilant.

Peracetic Acid Peracetic acid is an even more potent germicide than hydrogen peroxide. A 0.2% solution of peracetic acid, or a combination of peracetic acid and hydrogen peroxide, can be used to sterilize items in less than 1 hour. It is effective in the presence of organic compounds, leaves no residue, and can be used on a wide range of materials. It has a sharp, pungent odor, however, and like other oxidizing agents, it is irritating to the skin and eyes.

Phenolics

Phenol (carbolic acid) is important historically because it was one of the earliest disinfectants, but its use is now limited because it has an unpleasant odor and is irritating to the skin. A group of structurally related compounds, however, called **phenolics**, have increased germicidal activity, which enables effective use of more dilute and therefore less irritating solutions. Phenolic compounds are the active ingredients in Lysol™.

Phenolics destroy cytoplasmic membranes of microorganisms and denature proteins. They kill most vegetative bacteria and, in high concentrations (from 5% to 19%), many can kill *Mycobacterium tuberculosis.* They do not, however, reliably inactivate all groups of viruses. The major advantages of phenolic compounds include their wide range of activity, reasonable cost, and ability to remain effective in the presence of detergents and organic contaminants. They also leave an active antimicrobial residue, which in some cases is desirable.

Some phenolics, such as **hexachlorophene** and **triclosan**, are sufficiently nontoxic to be used in soaps and lotions. Hexachlorophene has substantial activity against *Staphylococcus aureus*, the leading cause of wound infections, but high levels have been associated with symptoms of neurotoxicity. Although once widely used in hospitals, antiseptic skin cleansers containing hexachlorophene are now available only with a prescription. Triclosan is used as an ingredient in personal care products such as deodorant soaps, lotions, and toothpaste.

Perspective 5.1 Contamination of an Operating Room by a Bacterial Pathogen

A patient with burns infected with *Pseudomonas aeruginosa* was taken to the operating room for cleaning of the wounds and removal of dead tissue. After the procedure was completed, samples of various surfaces in the room were cultured to determine the extent of contamination. *P. aeruginosa* was recovered from all parts of the room. **Figure 1** shows how readily and extensively an operating room can become contaminated by an infected patient. Operating rooms and other patient care rooms must be thoroughly cleaned after use, in a process known as terminal cleaning.

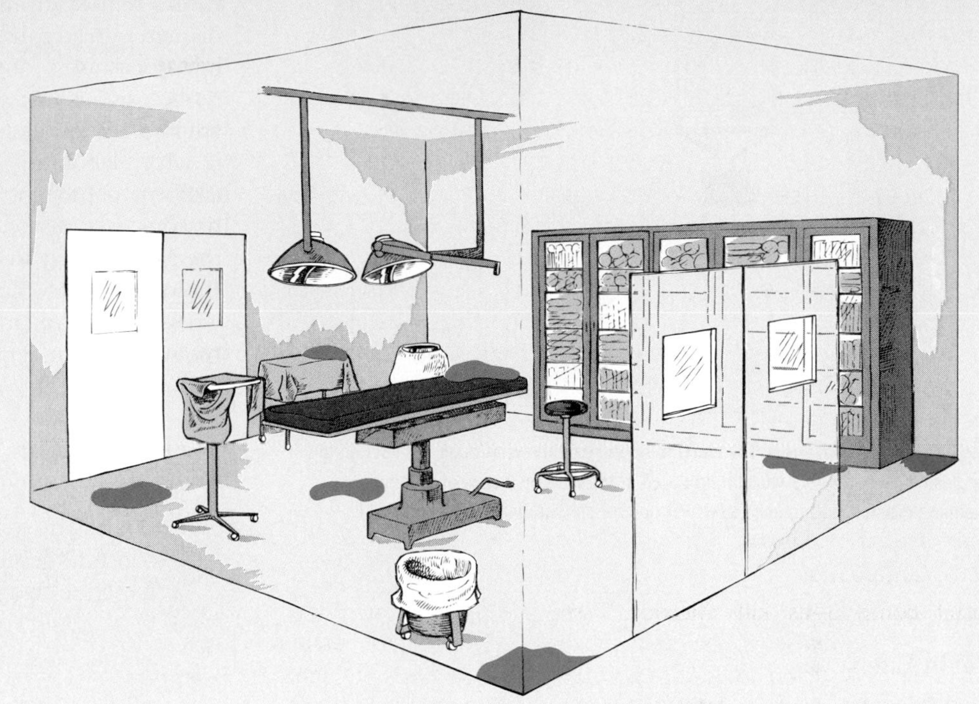

Figure 1 Diagram of an operating room in which dead tissue infected with *Pseudomonas aeruginosa* was removed from a patient with burns. Reddish areas indicate places where *P. aeruginosa* was recovered following the surgical procedure.

Quaternary Ammonium Compounds (Quats)

Quaternary ammonium compounds, also commonly called **quats**, are cationic (positively charged) detergents that are nontoxic enough to be used to disinfect food preparation surfaces. Like all detergents, quats have both a charged hydrophilic region and an uncharged hydrophobic region. This enables them to reduce the surface tension of liquids and help wash away dirt and organic material, facilitating the mechanical removal of microorganisms from surfaces. Unlike most common household soaps and detergents, however, which are anionic (negatively charged) and repelled by the negatively charged microbial cell surface, quats are attracted to the cell surface. They react with membranes, destroying many vegetative bacteria and enveloped viruses. They are not effective, however, against endospores, mycobacteria, or naked viruses.

Quaternary ammonium compounds are economical and effective agents that are widely used to disinfect clean inanimate objects and to preserve nonfood substances. The ingredients of many personal care products include quats such as benzalkonium chloride or cetylpyridinium chloride. They also enhance the effectiveness of some other disinfectants. Cationic soaps and organic material such as gauze, however, can neutralize their effect. In addition, *Pseudomonas*, a troublesome cause of nosocomial infections, resists the effects of quats and can even grow in solutions preserved with them.

MICROCHECK 5.4

Germicidal chemicals can be used to disinfect and, in some cases, sterilize, but they are less reliable than heat. They are especially useful for destroying microorganisms and viruses on heat-sensitive items and large surfaces.

- Describe four factors that must be considered when selecting a germicidal chemical.
- Explain why it is essential to dilute iodophores properly.
- Why would a heavy metal be a more serious pollutant than most organic compounds?

Removal of Microorganisms by Filtration

Organisms in heat-sensitive fluids can be removed by filtration (**figure 5.8**). This procedure has many uses such as in space technology, the production of unpasteurized beer, and the sterilization of sugar solutions. Filtration of the air entering specialized hospital rooms removes organisms that can otherwise cause disease in patients who have been extensively burned or have undergone chemotherapy.

Filtration of Fluids

Filters capable of removing bacteria from fluids were developed during the last decade of the Nineteenth century. They were made from materials such as porcelain, glass particles, diatomaceous earth, and asbestos. Today, two types of filters are in general use, **depth filters** similar to those used over the last century, and **membrane filters**.

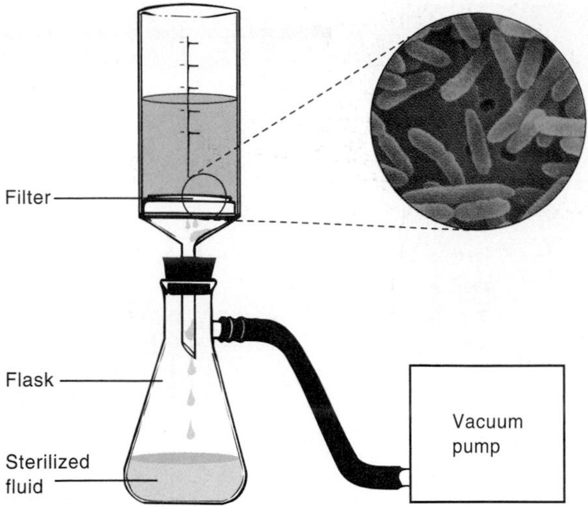

Figure 5.8 Filtration The liquid to be sterilized flows through the filter on top of the flask in response to a vacuum produced in the flask by means of a pump. Scanning electron micrograph (5,000×) shows a polycarbonate nucleopore filter retaining cells of *Pseudomonas*.

Depth Filters

Depth filters have complex, tortuous passages that retain microorganisms while letting the suspending fluid pass through the small holes. The diameter of the passages is often considerably larger than that of the microorganisms they retain, and trapping of microbes is partly by electrical charges on the walls of the filter passages.

Although the filters developed in the Nineteenth century remove most species of bacteria from their suspending media, they have several drawbacks. For example, they cannot reliably remove viruses. They also cannot be used to separate bacteria from desired products such as enzymes because they may retain proteins in addition to bacteria. The presence of proteins can also interfere with the filtering action by neutralizing electrostatic charges on the filter. Too much pressure applied to speed filtration can also overcome filtering action, allowing bacteria to pass through. Membrane filters overcome some of those drawbacks.

Membrane Filters

Membrane filters composed of compounds such as cellulose acetate, cellulose nitrate, polycarbonate, and polyvinylidene fluoride have been developed more recently and have largely replaced the use of depth filters. The filters are relatively inert chemically and absorb very little of the suspending fluid or its biologically important constituents such as enzymes. Paper-thin membrane filters are produced with graded pore sizes extending below the dimensions of the smallest known viruses. The most commonly used pore sizes for bacterial filters are 0.4 and 0.2 micrometers (μm). Membrane filtration of beer and wine results in a clear liquid free of spoilage microorganisms. Since pasteurization is unnecessary, filtered beer can be marketed as "draft" beer in bottles and cans. Some heat-sensitive medications are also sterilized by membrane filtration.

Filtration of Air

Special filters called **high-efficiency particulate air (HEPA) filters** remove from air nearly all microorganisms that have a diameter greater than 0.3 μm. These filters are used for keeping microorganisms out of specialized hospital rooms in which reside patients who are exquisitely susceptible to infection. They are also used in biological safety cabinets, **laminar flow hoods**, in which laboratory personnel work with potentially dangerous airborne pathogens such as *Mycobacterium tuberculosis*. A continuous flow of incoming and outgoing air is filtered through the HEPA filters to contain microorganisms within the cabinet. Biological safety cabinets are used to not only protect the worker from contamination by the sample, but also to protect the sample from environmental contamination.

MICROCHECK 5.5

Filters can be used to remove microorganisms and viruses from liquids and air.

- What is the difference between the mechanism of a depth filter and that of a membrane filter?
- Describe two uses of HEPA filters.
- How could too much pressure overcome the filtering action of a depth filter?

Using Radiation to Destroy Microorganisms and Viruses

Electromagnetic radiation can be thought of as waves having energy but no mass. Examples include X rays, gamma rays, and ultraviolet and visible light rays (see figure 8.8). The energy that electromagnetic rays possess is proportional to the frequency of the radiation (waves per second). Electromagnetic radiation of short wavelength, such as gamma rays, has much more killing power than that of long wavelength, such as visible light. Irradiation with gamma rays and ultraviolet light are valuable tools for microbial control. Microwaves do not kill microorganisms directly, but by the heat they generate.

Gamma Irradiation

Gamma rays are an example of ionizing radiation that causes biological damage by producing reactive molecules such as superoxide (O_2^-) and hydroxyl free radicals (OH•) when the rays transfer their energy to a microorganism. Gamma radiation from the radioisotope cobalt-60 is effective for controlling most microorganisms. Bacterial endospores are among the most radiation-resistant microbial forms, whereas the Gram-negative bacteria such as *Salmonella* and *Pseudomonas* are among the most susceptible. Radiation is used extensively to sterilize heat-sensitive materials including medical equipment, disposable items, and drugs such as penicillin. It provides an alternative for ethylene oxide sterilization and can even be carried out after packaging. ■ **superoxide, p. 98**

Foods can be either sterilized or pasteurized with radiation, depending on the doses employed. Treatments designed to sterilize food can cause undesirable flavor changes, however, which limits their usefulness. More commonly, food is irradiated as a method of pasteurization, eliminating pathogens and decreasing the numbers of spoilage organisms. For example, it can be used to kill pathogens such as *Salmonella* species in poultry with little or no change in taste of the product.

In the United States, irradiation has been used for many years to control microorganisms on spices and herbs. The Food and Drug Administration (FDA) has also approved irradiation of fruits, vegetables, and grains to control insects, pork to control the trichina parasite, and most recently, meats including poultry, beef, lamb, and pork to control pathogens such as *Salmonella* species and *E. coli* O157:H7.

Many consumers have been reluctant to accept irradiated products, even though the FDA and officials of the World Health and the United Nations Food and Agriculture Organizations have endorsed the technique. Some people erroneously believe that irradiated products are radioactive, which they are not. Others have lingering doubts about the possibility of irradiation-induced toxins or carcinogens being present in food, even though available scientific evidence indicates that consumption of irradiated food is safe. Another argument raised against irradiation is that it will cause a relaxation of other prudent food handling practices. Irradiation, however, is intended to complement, not replace, proper food handling procedures by producers, processors, and consumers.

Ultraviolet Radiation

Ultraviolet (UV) light in wavelengths of 220 to 300 nm kills microorganisms by damaging their DNA. The absorption of these wavelengths causes covalent bonds to form between adjacent thymine molecules in the DNA, creating thymine dimers. Actively multiplying organisms are the most easily killed; bacterial endospores are the most resistant.

Ultraviolet rays penetrate very poorly. A thin film of grease on the UV bulb or extraneous materials covering microorganisms may markedly reduce effective microbial killing. Most types of glass and plastic also screen out ultraviolet radiation, and so UV light is most effective when used at close range against exposed microorganisms in air or on clean surfaces. Caution should be used, however, because UV rays can also damage the skin and eyes and promote the development of skin cancers.

Microwaves

Microwaves do not affect microorganisms directly, but they can kill microorganisms by the heat they generate in an item. Organisms often survive microwave cooking, however, because the food heats unevenly.

MICROCHECK 5.6

Gamma irradiation can be used to sterilize products and to decrease the number of microorganisms in foods. Ultraviolet light can be used to disinfect surfaces. Microwaves do not kill microorganisms directly, but by the heat they generate.

- What is the purpose of irradiating fresh meat?
- How does ultraviolet light kill microorganisms?
- Why could sterilization by gamma irradiation be carried out even after packaging?

Preservation of Perishable Products

Preventing or slowing the growth of microorganisms extends the shelf life of products such as food, soaps, medicines, deodorants, cosmetics, and contact lens solutions. Their ingredients often include preservative chemicals (**figure 5.9**). The purpose of these agents is to prevent or slow the growth of microbes that are inevitably introduced from the environment. Other common methods of decreasing the growth rate of microbes are using low-temperature storage such as refrigeration or freezing, and reducing available water. These methods are particularly important in the preservation of foods.

Chemical Preservatives

Some of the germicidal chemicals previously described can be used to preserve nonfood items. For example, shampoo may contain formaldehyde, mouthwash may contain a quaternary ammonium compound, contact lens solutions may contain thimerosal, and leather belts may be treated with one or more phenol derivatives. Food preservatives, however, must be nontoxic for repeated safe ingestion.

Benzoic, sorbic, and propionic acids are organic acids that are sometimes added to foods such as bread, cheese, and juice to prevent microbial growth. In the undissociated form that predominates at a low pH, these weak acids alter cell membrane functions and interfere with energy generation. The low pH at which they are most effective is itself sufficient to prevent the growth of most bacteria, so that these preservatives are primarily added to acidic foods to prevent the growth of fungi. These organic acids also occur naturally in some foods such as cranberries and Swiss cheese.

Another preservative, nitrate, and its reduced form, nitrite, serve a dual purpose in processed meats. From a microbiological

Figure 5.9 Typical Products that Contain Preservatives

viewpoint, their most important function is inhibition of germination of endospores and subsequent growth of *Clostridium botulinum*. Without the addition of low levels of nitrate or nitrite to cured meats such as bologna, ham, bacon, and smoked fish, there is a risk that *C. botulinum* will grow and produce deadly botulinum toxin. At higher concentrations than are required for preservation, nitrate and nitrite react with myoglobin in the meat to form a stable pigment that gives a desirable pink color associated with fresh meat. Nitrates and nitrites also pose a potential hazard, however, because they can be converted to nitrosamines during the frying of meats in hot oil or by the metabolic activities of intestinal bacteria. Nitrosoamines have been shown to be potent carcinogens, which has caused concern regarding the use of nitrate and nitrite as preservatives.

Low-Temperature Storage

The growth of microorganisms is temperature-dependent. At low temperatures above freezing, many enzymatic reactions are very slow or nonexistent. Thus, low temperature storage is extremely useful in preservation. Some psychrophilic organisms, however, can grow at normal refrigeration temperatures. ■ **psychrophiles, p. 96**

Commercially, some fruits and vegetables such as apples and potatoes are held in cold storage for many months. Products are stored at temperatures near 10°C, in the dark, and at appropriate humidity and oxygen concentration. Prior to refrigeration, foods are sometimes irradiated in UV light to reduce the number of spoilage organisms.

Freezing is also an important means of preserving foods and other products. Freezing essentially stops all microbial growth. The formation of ice crystals can cause irreversible damage to microbial cells, killing up to 50% of microorganisms. The remaining organisms, however, can grow and spoil foods once they are thawed.

Reducing the Available Water

For many years, salting and drying have been used to preserve food. Both processes decrease the water activity (a_w) of a food below the limits required for growth of most microorganisms. The high-solute environment causes plasmolysis, which damages cells. ■ **a_w, p. 805** ■ **plasmolysis, p. 99**

Adding Salt or Sugar

Sugar and salt draw water out of cells and essentially dry them, thereby decreasing the available water in the food, which prevents the growth of microorganisms. High concentrations of sugars or salts are added to many foods as preservatives. For example, fruit is made into jams and jellies by adding sugar, and fish and meats are cured by soaking them in salty water, or **brine**. Some caution should be exercised when using salt as a preservative, however, because the food-poisoning bacterium *Staphylococcus aureus* can grow under quite high salt conditions.

Drying Food

Drying of foods by natural means (sun-drying) or by artificial means is often supplemented by other methods, such as salting or adding high concentrations of sugar or small amounts of chemical preservatives. For example, meat jerkies usually have added salt and sometimes sugar.

Lyophilization (freeze-drying) is widely used for preserving foods such as coffee, milk, meats, and vegetables. In the process of freeze-drying, the food is first frozen and then dried in a vacuum. When water is added to the lyophilized material, it reconstitutes. The quality of the reconstituted product is often much better than that of products treated with ordinary freezing or drying methods. The light weight and stability without refrigeration of freeze-dried foods make them popular with hikers.

Although drying stops microbial growth, it does not reliably kill bacteria and fungi in or on foods. For example, numerous cases of salmonellosis have been traced to dried eggs. Eggshells and even egg yolks may be heavily contaminated with *Salmonella* species from the gastrointestinal tract of the hen. To prevent the transmission of such pathogens, some states have laws requiring dried eggs to be pasteurized before they are sold.

MICROCHECK 5.7

Preservation techniques slow or halt the growth of microorganisms to delay spoilage.

- What organism that causes food poisoning is able to grow under high-salt conditions?
- What is the risk of consuming nitrate- or nitrite-free cured meats?
- Preservation by freezing is sometimes compared to drying. Why would this be so?

FUTURE CHALLENGES

Finding the Fine Line Between Use and Overuse

In our complex world, the solution to one challenge may inadvertently lead to the creation of another. Scientists have long been pursuing less toxic alternatives to many traditional biocidal chemicals. For example, glutaraldehyde has now largely replaced the more toxic formaldehyde, chlorhexidine is generally used in place of hexachlorophene, and gaseous alternatives to ethylene oxide are now being sought. Meanwhile, ozone and hydrogen peroxide, which are both readily biodegradable, may eventually replace glutaraldehyde. While these less toxic alternatives are better for human health and the environment, their widespread acceptance and use may be unwittingly creating an additional problem—the overuse and misuse of germicidal chemicals. Many products, including soaps, toothbrushes, and even clothing and toys are marketed with the claim of containing antimicrobial ingredients. Already there are reports of bacterial resistance to some of the chemicals included in these products.

The issues surrounding the excessive use of antimicrobial chemicals are complicated. On the one hand, there is no question that some microorganisms cause disease. Even those that are not harmful to human health can be troublesome because they produce metabolic end products that ruin the quality of perishable products. Based on that information, it seems prudent to destroy or inhibit the growth of microorganisms whenever possible. The role of microorganisms in our life, however, is not that simple. Our bodies actually harbor more microbial cells than human cells, and this normal flora plays an important role in the maintenance of our health. Excessive use of antiseptics or other antimicrobials may actually predispose a person to infection by damaging the normal flora.

An even more worrisome concern is that overuse of disinfectants and other germicidal chemicals will select for microorganisms that are more resistant to those chemicals, a situation analogous to our current problems with antibiotic resistance. By using antimicrobial chemicals indiscriminately, we may eventually make these useful tools obsolete. Excessive use of disinfectants may even be contributing to the problems of antibiotic resistance. Early indications suggest that disinfectant-resistant bacteria overproduce efflux pumps that expel otherwise damaging chemicals, including antibiotics from the cell. Thus, by overusing disinfectants, we may be inadvertently increasing antibiotic resistance.

Another concern is over the misguided belief that "nontoxic" or "biodegradable" chemicals cause no harm, and the common notion that "if a little is good, more is even better." For example, concentrated solutions of hydrogen peroxide, though biodegradable, can cause serious damage, even death, when used improperly. Other chemicals, such as chlorhexidine, can elicit severe allergic reactions in some people.

As less toxic germicidal chemicals are developed, the challenge will be to educate people on the appropriate use of these alternatives.

SUMMARY

Approaches to Control

1. The methods used to destroy or remove microorganisms and viruses can be **physical,** such as heat treatment, irradiation, and filtration, or **chemical**.

Principles of Control (Table 5.1)

2. **Sterilization** removes or destroys all microorganisms and viruses on or in a product.
3. **Disinfection** eliminates most or all disease-causing microorganisms or viruses on or in a material.
4. Chemicals used for disinfecting inanimate objects are called **disinfectants**; those formulated for use on skin are called **antiseptics**.
5. **Pasteurization** utilizes a brief heat treatment to reduce the number of spoilage organisms or kill disease-causing microbes.
6. A **decontaminated** item has been treated to reduce the number of disease-causing microbes to a level that is safe to handle.
7. A **sanitized** item has a substantially reduced microbial population that meets accepted health standards.

Situational Considerations (Figure 5.1)

1. In daily life, washing and scrubbing with soaps and detergents achieve routine control of undesirable microorganisms and viruses.
2. Hospitals must be scrupulous in controlling microorganisms because of the danger of **nosocomial infections**.
3. Microbiology laboratories must use **aseptic technique** to avoid contaminating cultures with extraneous microbes and to protect workers and the environment from contamination.
4. Foods and other perishable products retain their quality and safety when the growth of contaminating microorganisms is prevented.

Selection of an Antimicrobial Procedure

Type of Microorganism

1. One of the most critical considerations in selecting a method of destroying microorganisms and viruses is the type of microbial population present on or in the product.
2. The endospores of *Bacillus* and *Clostridium* are by far the most resistant forms of life.
3. The waxy cell wall of mycobacteria makes them resistant to many chemical treatments.
4. *Pseudomonas* species are common environmental organisms that not only are resistant to some chemical disinfectants, but can actually grow in them.
5. Viruses that lack a lipid envelope are more resistant to disinfectants than are enveloped viruses.
6. Chemical disinfectants are categorized according to their germicidal activity against resistant microbes.

Numbers of Microorganisms Initially Present

1. The amount of time it takes for heat or chemicals to kill a population of microorganisms is dictated in part by the number of organisms initially present.
2. Microbial death generally occurs at a constant rate; thus, only a fraction of the organisms die during a given time interval.
3. In the commercial canning industry, the **D value**, or **decimal reduction time**, is defined as the time it takes to kill 90% of a population of bacteria under specific conditions. (Figure 5.2)

Environmental Conditions

1. Factors such as pH or presence of fats and other organic materials strongly influence microbial death rates.
2. The presence of dirt, grease, and organic compounds such as blood and other body fluids can interfere with heat penetration and the action of chemical disinfectants.

Potential Risk of Infection

1. To guide medical biosafety personnel in their selection of germicidal procedures, items are categorized according to their potential risk of transmitting infectious agents.
2. **Critical items** come into direct contact with body tissues.
3. **Semicritical items** come into contact with mucous membranes, but do not penetrate body tissue.
4. **Noncritical instruments** and surfaces only come into contact with unbroken skin.

Using Heat to Destroy Microorganisms and Viruses (Table 5.2)

1. Heat can be used to destroy vegetative microorganisms and viruses, but temperatures above boiling are required to kill endospores.

Moist Heat

1. Moist heat, such as boiling water and pressurized steam, destroys microorganisms by causing irreversible coagulation of their proteins.
2. **Pasteurization** utilizes a brief heat treatment to destroy spoilage and disease-causing organisms, increasing the shelf life of products and protecting consumers.
3. Pressure cookers and **autoclaves** heat water in an enclosed vessel that causes the pressure in the vessel to increase beyond atmospheric pressure, increasing the temperature of steam, which kills endospores. (Figure 5.3)
4. The most important aspect of the commercial canning process is to ensure that endospores of *Clostridium botulinum* are destroyed. (Figure 5.6)

Dry Heat

1. Direct flame and ovens generate dry heat, which destroys microorganisms by oxidizing cells to ashes or irreversibly denaturing their proteins.
2. Dry heat takes much longer than wet heat to kill microorganisms.

Using Chemicals to Destroy Microorganisms and Viruses

1. Germicidal chemicals can be used to disinfect and, in some cases, sterilize, but they are less reliable than heat. They are especially useful for destroying microbes on heat-sensitive items and large surfaces.
2. Most chemical germicides react irreversibly with vital enzymes and other proteins, the cytoplasmic membrane, or viral envelopes. (Figure 5.7)

Potency of Germicidal Chemical Formulations

1. Germicides registered with either the FDA or EPA are grouped according to their potency as **sterilants**, **high-level disinfectants**, **intermediate-level disinfectants**, or **low-level disinfectants.** (Table 5.3)

Selecting the Appropriate Germicidal Chemical

1. Factors that must be included in the selection of an appropriate germicidal chemical include toxicity, residue, activity in the presence of organic matter, compatibility with the material being treated, cost and availability, storage and stability, and ease of disposal.

Classes of Germicidal Chemicals (Table 5.4)

1. Solutions of 60% to 80% ethyl or isopropyl alcohol in water rapidly kill vegetative bacteria and fungi by coagulating enzymes and other essential proteins, and by damaging lipid membranes.
2. **Glutaraldehyde** and **formaldehyde** destroy microorganisms and viruses by inactivating proteins and nucleic acids. A 2% solution of alkaline glutaraldehyde is one of the most widely used chemical sterilants.
3. **Chlorhexidine** is a **biguanide** extensively used in antiseptic products.
4. **Ethylene oxide** is a gaseous sterilizing agent that penetrates well and destroys microorganisms and viruses by reacting with proteins.
5. Sodium hypochlorite (liquid bleach) is one of the least expensive and most readily available forms of chlorine. **Chlorine dioxide** is used as a sterilant and disinfectant. **Iodophores** are iodine-releasing compounds used as antiseptics.
6. Metals interfere with protein function. Silver-containing compounds are used to prevent wound infections.
7. Ozone is used as an alternative to chlorine in the disinfection of drinking water and wastewater.
8. Peroxide and peracetic acid are both strong oxidizing agents that can be used alone or in combination as sterilants.
9. **Phenolics** destroy cytoplasmic membranes and denature proteins. **Triclosan** is used in lotions and deodorant soaps. **Hexachlorophene** has been associated with neurotoxicity, and antiseptic lotions containing it are only available with a prescription.
10. **Quaternary ammonium compounds** are cationic detergents; they are nontoxic enough to be used to disinfect food preparation surfaces.

Removal of Microorganisms by Filtration

Filtration of Fluids (Figure 5.8)

1. **Depth filters** have complex, tortuous passages that retain microorganisms while letting the suspending fluid pass through the small holes.
2. **Membrane filters** are produced with graded pore sizes extending below the dimensions of the smallest known viruses.

Filtration of Air

1. **High-efficiency particulate air (HEPA)** filters remove nearly all microorganisms.
2. HEPA filters are used in specialized hospital rooms to protect patients who are exquisitely susceptible to infection.
3. HEPA filters are used in biological safety cabinets, **laminar flow hoods**, which protect laboratory personnel who work with airborne pathogens.

Using Radiation to Destroy Microorganisms and Viruses

Gamma Irradiation

1. Gamma rays cause biological damage by producing superoxide and hydroxyl free radicals.
2. Irradiation can be used to sterilize heat-sensitive materials and to decrease the numbers of microorganisms in foods.
3. Irradiation has been approved by the FDA to control insects in fruits, vegetables, and grains, to destroy the trichina parasite in pork, and to control *Salmonella* species and *E. coli* O157:H7 in meats.

Ultraviolet Radiation

1. Ultraviolet light damages the structure and function of nucleic acids by causing the formation of covalent bonds between adjacent thymine molecules in DNA, creating thymine dimers.
2. UV light is used to disinfect surfaces.

Microwaves

1. Microwaves do not affect microorganisms directly, but they can kill microorganisms by the heat they generate in a product.

Preservation of Perishable Products

1. Preservation techniques slow or halt the growth of microorganisms to delay spoilage.

Chemical Preservatives

1. Benzoic, sorbic, and propionic acids are organic acids sometimes added to foods to prevent microbial growth.
2. Nitrate and nitrite are added to some foods to inhibit the germination and subsequent growth of *Clostridium botulinum* endospores. They also react with myoglobin to form a stable pigment that gives a pink color associated with fresh meat.

Low-Temperature Storage

1. Low temperatures above freezing inhibit microbial growth because many enzymatic reactions are rendered slow or nonexistent.
2. Freezing essentially stops all microbial growth.

Reducing the Available Water

1. Sugar and salt draws water out of cells, preventing the growth of microorganisms.
2. **Lyophilization** is used for preserving food. The food is first frozen and then dried in a vacuum.

REVIEW QUESTIONS

Short Answer

1. What is the primary reason that milk is pasteurized?
2. What is the primary reason that wine is pasteurized?
3. What is the most chemically resistant non-spore-forming pathogen?
4. Why are low acid foods processed at higher temperatures than high acid foods?
5. Explain why it takes longer to kill a population of 10^9 cells than it does to kill a population of 10^3 cells.
6. How is an iodophore different from a tincture of iodine?
7. How does microwaving a food product kill bacteria?
8. How is preservation different from pasteurization?
9. How are heat-sensitive liquids sterilized?
10. Name two products that are commonly sterilized using ethylene oxide gas.

Multiple Choice

1. Unlike a disinfectant, an antiseptic
   A. sanitizes objects rather than sterilizes them.
   B. destroys all microorganisms.
   C. is nontoxic enough to be used on human tissue.
   D. requires heat to be effective.
   E. can be used in food products.
2. The D value is defined as the time it takes to kill
   A. all bacteria in a population.
   B. all pathogens in a population.
   C. 99.9% of bacteria in a population.
   D. 90% of bacteria in a population.
   E. 10% of bacteria in a population.
3. Which of the following is the most resistant to destruction by chemicals and heat?
   A. Bacterial endospores
   B. Fungal spores
   C. *Mycobacterium tuberculosis*
   D. *E. coli*
   E. HIV
4. Ultraviolet light kills bacteria by
   A. generating heat.
   B. damaging nucleic acids.
   C. inhibiting protein synthesis.
   D. damaging cell walls.
   E. damaging cytoplasmic membranes.
5. Which concentration of ethyl alcohol is the most effective germicide?
   A. 100%
   B. 75%
   C. 50%
   D. 25%
   E. 5%
6. Which of the following chemical agents can most reliably be used to sterilize objects?
   A. Alcohol
   B. Phenolic compounds
   C. Ethylene oxide gas
   D. Formaldehyde
7. All of the following are routinely used to preserve foods, except
   A. high concentrations of sugar.
   B. high concentrations of salt.
   C. benzoic acid.
   D. freezing.
   E. ethylene oxide.
8. Aseptically boxed juices and cream containers are processed using which of the following heating methods?
   A. Canning
   B. High-temperature-short-time (HIST) method
   C. Low-temperature-long-time (LTLT) method
   D. Ultra-high-temperature (UHT) method

9. Commercial canning processes are designed to ensure destruction of which of the following?
   A. All vegetative bacteria
   B. All vegetative bacteria and their endospores
   C. Endospores of *Clostridium botulinum*
   D. *E. coli*
   E. *Mycobacterium tuberculosis*
10. Which of the following is false?
    A. A chemical that is a high-level disinfectant cannot be used as a sterilant.
    B. Heat-sensitive critical items must be treated with a sterilant.
    C. Low numbers of endospores may remain on semicritical items.
    D. Standard sterilization procedures do not destroy prions.
    E. Quaternary ammonium compounds can be used to disinfect food preparation surfaces.

Applications

1. An agriculture extension agent is preparing pamphlets on preventing the spread of disease in the household. In the pamphlet, he must explain the appropriate situations for using disinfectants around the house. What situations should the agent discuss?
2. As a microbiologist representing a food corporation, you have been asked to serve on a health food panel to debate the need for chemical preservatives in foods. Your role is to prepare a statement that compares the benefits of chemical preservatives and the risks. What points must you bring up that indicate the benefits of chemical preservatives?

Critical Thinking

1. The graph below shows the time it takes to kill populations of the same microorganism under different conditions. What conditions would explain the differences in lines a, b, and c?

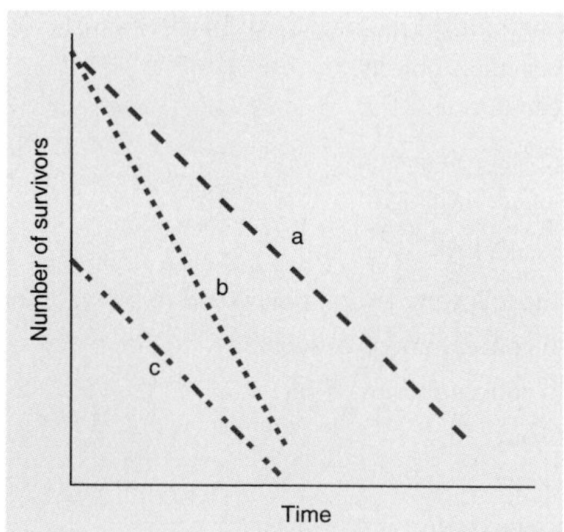

2. The diagram below shows the filter paper method used to evaluate the inhibitory effect of chemical agents, heavy metals, and antibiotics on bacterial growth. A culture of test bacteria is spread uniformly over the surface of an agar plate. Small filter paper discs containing the material to be tested are then placed on the surface of the medium. A disc that has been soaked in sterile distilled water is sometimes added as a control. After incubation, a film of growth will cover the plate, but a clear zone will surround those discs that contain an inhibitory compound. The size of the zone reflects several factors, one of which is the effectiveness of the inhibitory agent. What are two other factors that might affect the size of the zone of inhibition? What is the purpose of the control disc? If a clear area were apparent around the control disc, how would you interpret the observation?

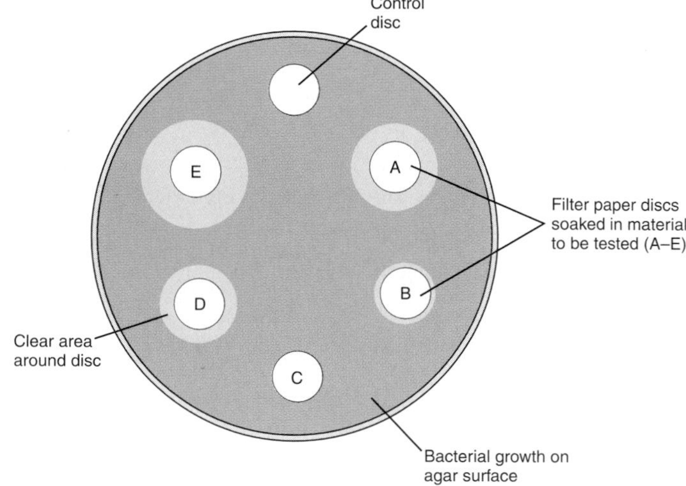

CHAPTER 6

Metabolism: Fueling Cell Growth

In the 1850s, Louis Pasteur, a chemist, enthusiastically accepted the challenge of studying how alcohol arises from grape juice. Biologists had already observed that when grape juice is held in large vats, alcohol and carbon dioxide are produced and the number of yeast cells increases. They argued that the multiplying yeast cells convert the sugar in the juice to alcohol and carbon dioxide. Pasteur agreed, but he could not convince two very powerful and influential German chemists, Justus von Liebig and Friedrich Wöhler, who refused to believe that the activities of microorganisms caused the breakdown of sugar. Both men lampooned the hypothesis and tried to discredit it by publishing pictures of yeast cells looking like miniature animals taking in grape juice through one orifice and eliminating carbon dioxide and alcohol through the other.

Pasteur studied the relationship between yeast and alcohol production using a strategy commonly employed by scientists today—that is, simplifying the experimental system so that relationships can be more easily identified. First, he prepared a clear solution of sugar, ammonia, mineral salts, and trace elements. He then added a few yeast cells. As the yeast grew, the sugar level decreased and the alcohol level increased, indicating that the sugar was being converted to alcohol as the cells grew. This strongly suggested that living cells caused the chemical transformation. Liebig, however, still would not believe the process was actually occurring inside microorganisms. To convince him, Pasteur tried to extract something from inside the yeast cells that would convert sugar into alcohol and carbon dioxide. He failed, like many others before him.

In 1897, Eduard Buchner, a German chemist, showed that crushed yeast cells could convert sugar to ethanol and CO_2. We now know that the active ingredients of the crushed cells that carried out this transformation were enzymes. For these pioneering studies, Buchner was awarded the Nobel Prize in 1907. He was the first of many investigators who received Nobel Prizes for studies on the processes by which cells degrade sugars.

—A Glimpse of History

TO GROW, ALL CELLS MUST ACCOMPLISH TWO FUNdamental tasks. They must continually synthesize new components including cell walls, membranes, ribosomes, nucleic acids, and surface structures such as flagella. While some of these components may be used to replace worn or damaged cell structures, sufficient quantities allow the cell to enlarge and eventually divide. In addition, cells need to harvest energy and convert it to a form that is usable to power biosynthetic reactions, transport nutrients and other molecules, and in some cases, move. The sum total of chemical reactions used for biosynthetic and energy-harvesting processes is called **metabolism**.

Bacterial metabolism is important to humans for a number of reasons. Many bacterial products are commercially or medically important. For example, the metabolic waste products of *Clostridium acetobutylicum* are the solvents, acetone and butanol. Cheese-makers intentionally add *Lactococcus* and *Lactobacillus* species to milk because the metabolic wastes of these bacteria contribute to the flavor and texture of various cheeses. Yet those same products are detrimental when related bacteria are growing on teeth, because they contribute to tooth decay. Microbial metabolism is also important in the laboratory, because products that are characteristic of a specific group of microorganisms can be used as identifying markers. The metabolic end products of *Escherichia coli* distinguish it from related Gram-negative rods such as *Klebsiella* and *Enterobacter* species. In addition, the metabolic pathways of organisms such as *E. coli* have served as an invaluable model for studying analogous processes in eukaryotes, including humans. Metabolic processes unique to prokaryotes are potential targets for antimicrobial drugs.

Principles of Metabolism

Metabolism can be viewed as having two components—**catabolism** and **anabolism (figure 6.1)**. Catabolism encompasses those processes that harvest the energy released during the disassembly or breakdown of compounds such as glucose and use that energy to synthesize **ATP**, the energy currency of all cells. In contrast, anabolism, or **biosynthesis**, includes the processes that utilize the energy stored in ATP to synthesize and assemble the subunits of macromolecules that make up the cell. These subunits include amino acids, nucleotides, and lipids.

Although catabolism and anabolism are often discussed separately, they are intimately linked. As mentioned, the ATP generated during catabolism is utilized in anabolism. In addition, some of the molecules produced in steps of the catabolic processes can be diverted by the cell and used as precursors of the subunits employed in anabolic processses. ■ ATP, p. 137

Harvesting Energy

Energy is defined as the capacity to do work. It can exist as **potential energy**, which is stored energy, and **kinetic energy**, which is energy of motion **(figure 6.2)**. Potential energy can be stored in a variety of forms including chemical bonds, a rock on a hill, or water behind a dam. The fundamental principles of energy relationships are embodied in the laws of thermodynamics. The **first law of thermodynamics** recognizes that the energy in the universe can never be created or destroyed; however, it can be changed from one form to another. In other words, while energy cannot be created, potential energy can be converted to kinetic energy and vice versa, and one form of potential energy can be converted to another. For example, hydroelectric dams unleash the potential energy of water stored behind a dam, creating the kinetic energy of moving water; this can then be used to generate an electrical current, which can then be used to charge a battery.

During every energy conversion, some energy is always "lost" as heat. This is recognized in the **second law of thermodynamics**, which states that everywhere in the universe, disorder, or **entropy**, always increases. Potential energy, which can be thought of as an orderly but unstable form of energy, readily converts to the disorder of random molecular motion, or heat. While heat is a form of energy, it is the most random form. It quickly dissipates and becomes less useful as an energy source. Because the natural tendency is to create randomness, an input of additional energy is always required to maintain order. This explains why cells require energy to assemble and organize their components.

Consistent with the laws of thermodynamics, cells do not create energy but instead convert available energy in the universe into a form that can be used by the cell. They can do this in a variety of different ways. **Photosynthetic** organisms, or **phototrophs**, harvest the energy of sunlight, using it to power the synthesis of organic compounds such as glucose **(figure 6.3)**. In other words, they convert the kinetic energy of photons to the potential energy of chemical bonds. In contrast, **chemoorganotrophs** obtain energy by degrading organic compounds such as glucose, releasing the potential energy of their chemical bonds. Thus, most chemoorganotrophs ultimately depend on the solar energy harvested by phototrophs, because this is what is used to power the synthesis of glucose. ■ phototroph, p. 100 ■ chemoorganotroph, p. 101

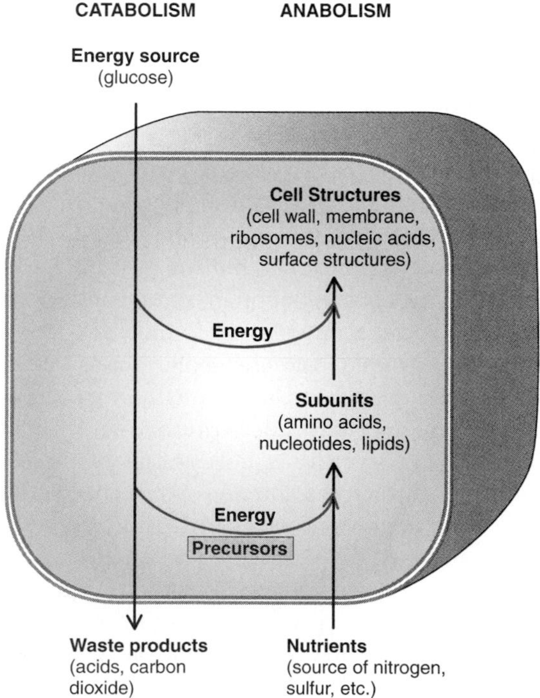

Figure 6.1 The Relationship Between Catabolism and Anabolism Catabolism encompasses those processes that harvest energy during the disassembly of compounds, using it to synthesize ATP; it also provides the precursor metabolites used in biosynthesis. Anabolism, or biosynthesis, includes the processes that utilize ATP and precursor metabolites to synthesize and assemble the subunits of the cell.

Figure 6.2 Forms of Energy Potential energy is stored energy, such as water held behind a dam. Kinetic energy is the energy of motion, such as the movement of water from behind the dam.

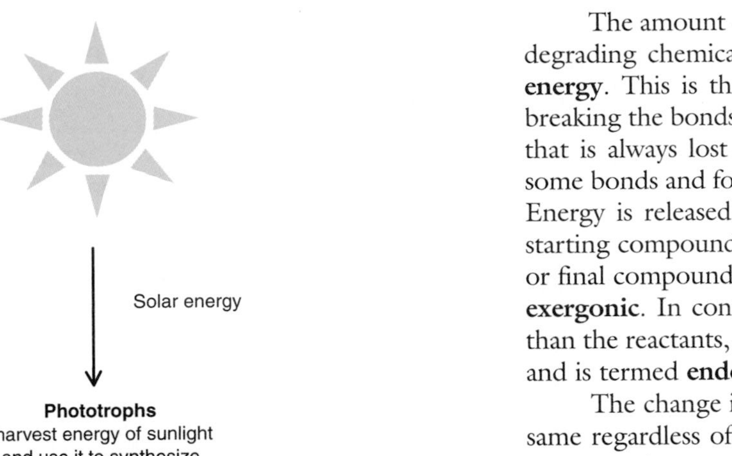

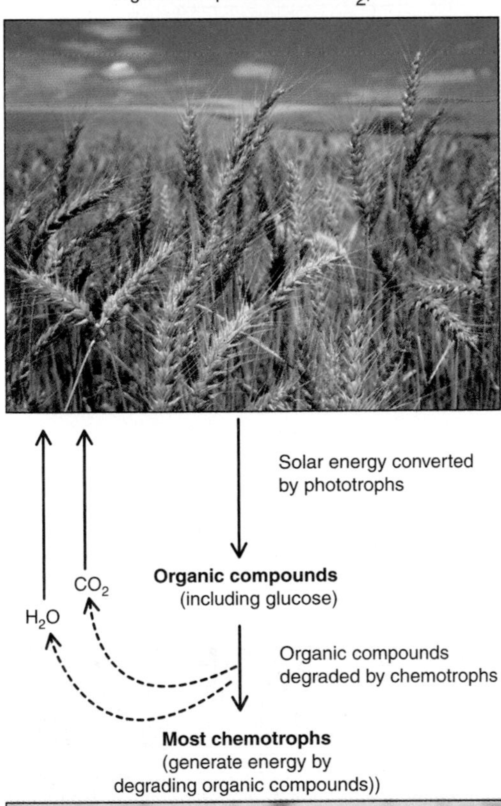

Figure 6.3 Most Chemoorganotrophs Depend on the Solar Energy Harvested by Phototrophs Phototrophs use the energy of sunlight to power the synthesis of organic compounds; chemoorganotrophs can then use those organic compounds as an energy source.

The amount of energy that can be released and utilized by degrading chemicals can be explained by the concept of **free energy**. This is the amount of energy that can be gained by breaking the bonds of a chemical; it does not include the energy that is always lost as heat. A chemical reaction, which breaks some bonds and forms others, results in a change in free energy. Energy is released in a chemical reaction if the **reactants**, or starting compounds, have more free energy than the **products**, or final compounds **(figure 6.4a)**. Such a reaction is said to be **exergonic**. In contrast, if the products have more free energy than the reactants, then the reaction requires an input of energy and is termed **endergonic** (figure 6.4b).

The change in free energy for a given reaction will be the same regardless of the number of steps involved (figure 6.4c). For example, converting glucose to carbon dioxide and water in a single step by combustion releases the same amount of energy as degrading it in a series of steps. Cells exploit this fact to slowly release free energy from compounds, harvesting the energy that is released at each step. A specific energy-releasing reaction is used to power an energy utilizing reaction. This coupling of exergonic and endergonic reactions conserves chemical energy that would otherwise be released as heat.

Components of Metabolic Pathways

Metabolic processes often occur as a series of sequential chemical reactions, which constitute a **metabolic pathway (figure 6.5)**. A series of **intermediates** are produced as the starting compound is gradually converted into the final product, or **end product**. A metabolic pathway may be linear, branched, or cyclical, and like the flow of a river that is controlled by dams, its activity may be modulated at certain points. In this way, a cell can regulate certain processes, ensuring that specific molecules are produced in precise quantities when needed. If a metabolic step is blocked, all products "downstream" of that blockage will be affected.

The intermediates and end products of metabolic pathways are sometimes organic acids, which are weak acids. Depending on the pH, these may exist primarily as either as the undissociated form or the dissociated (ionized) form. Biologists often use the names of the two forms interchangeably, for example pyruvic acid and pyruvate. Note, however, that at the near-neutral pH inside the cell, the ionized form predominates, whereas outside of the cell, the acid may predominate. ■ pH, p. 25

To recognize what metabolic pathways accomplish, it is helpful to first understand the critical components—enzymes, ATP, the energy source, electron carriers, and precursor metabolites **(table 6.1)**.

The Role of Enzymes

A specific **enzyme** facilitates each step of a metabolic pathway. Enzymes function as biological catalysts, accelerating the conversion of one substance, the **substrate**, into another, the **product**. Without enzymes, energy-yielding reactions would still occur, but at a rate so slow it would be imperceptible.

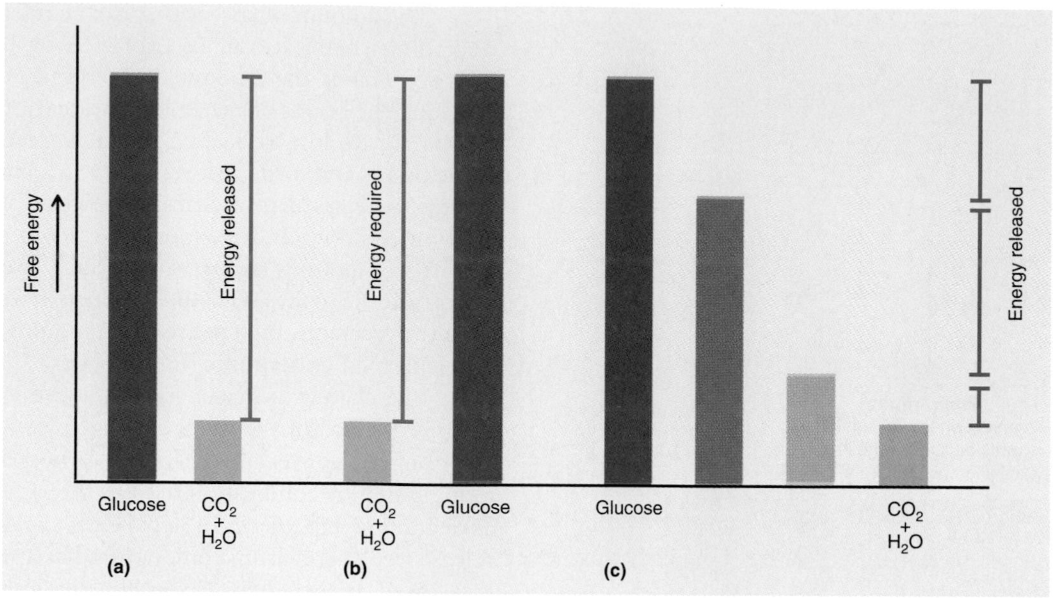

Figure 6.4 Energetics of Chemical Reactions **(a)** Exergonic reactions release energy. **(b)** Endergonic reactions consume energy. **(c)** The change in free energy for a given reaction will be the same, regardless of the number of steps involved.

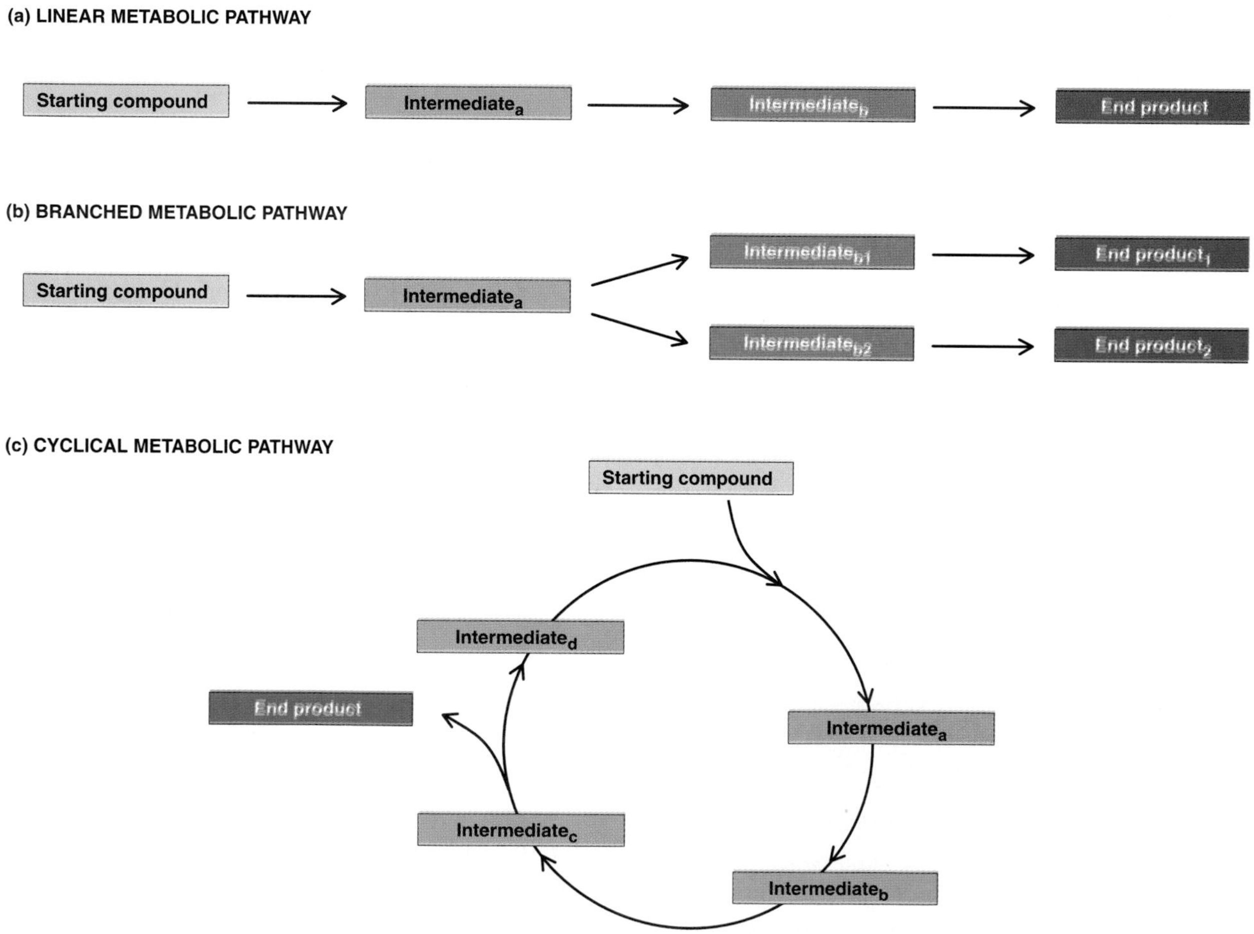

(a) LINEAR METABOLIC PATHWAY

Starting compound → Intermediate_a → Intermediate_b → End product

(b) BRANCHED METABOLIC PATHWAY

Starting compound → Intermediate_a → Intermediate_{b1} → End product_1

Intermediate_{b2} → End product_2

(c) CYCLICAL METABOLIC PATHWAY

Starting compound

Intermediate_a

Intermediate_b

Intermediate_c

Intermediate_d

End product

Figure 6.5 Metabolic Pathways Metabolic processes often occur as a series of sequential chemical reactions that convert starting compounds into intermediates and then, ultimately, into end products. A metabolic pathway can be **(a)** linear, **(b)** branched, or **(c)** cyclical.

TABLE 6.1 Components of Metabolism

Component	Function
Enzymes	Biological catalysts. A specific enzyme facilitates each step of a metabolic pathway by lowering the activation energy of a reaction that converts a substrate into a product.
Adenosine triphosphate (ATP)	Serves as the energy currency of the cell. Hydrolysis of its high-energy phosphate bonds can be used to power endergonic reactions.
Substrate-level phosphorylation	Synthesis of ATP from ADP and inorganic phosphate (P_i) using the energy released in an exergonic reaction.
Oxidative phosphorylation	Synthesis of ATP from ADP and inorganic phosphate (P_i) using the energy of the proton motive force. Proton motive force is generated as electrons are passed along the electron transport chain.
Energy source	The compound that is oxidized to release energy; also called an electron donor. The cell harvests that energy to synthesize ATP.
Electron carriers	Molecules including NAD^+, $NADP^+$, and FAD; the reduced forms, NADH, NADPH, and $FADH_2$, carry the electrons that are removed during the oxidation of the energy source. The reduced forms are called reducing power. Depending on the carrier, the reducing power can be used either to generate proton motive force or in biosynthesis.
Precursor metabolites	Metabolic intermediates that link anabolic and catabolic pathways. They are compounds that can either be used to make the subunits of macromolecules, or be oxidized to generate ATP.

An enzyme catalyzes a chemical reaction by lowering the **activation energy** of that reaction **(figure 6.6)**. This is the initial energy it takes to break chemical bonds; even exergonic chemical reactions have an activation energy. By lowering the activation energy barrier, enzymes allow chemicals to undergo rearrangements.

The Role of ATP

Adenosine triphosphate (**ATP**) is the energy currency of a cell, serving as the ready and immediate donor of free energy; it is composed of the sugar ribose, the nitrogenous base adenine, and three phosphate groups **(figure 6.7)**. Its counterpart, **adenosine diphosphate** (**ADP**), can be viewed as an acceptor of free energy. An input of energy is required to add an inorganic phosphate group (P_i) to ADP, forming ATP; energy is released when that group is removed from ATP, yielding ADP (figure 6.7b).

The phosphate groups of ATP are arranged in tandem. Because of their negative charges, these groups repel each other, making the bonds that join them unstable. The bonds are readily hydrolyzed, releasing the phosphate group and a sufficient amount of energy to power an endergonic reaction. Because of the relatively high amount of free energy released when the bonds between the phosphate groups are hydrolyzed, they are called **high-energy phosphate bonds**, denoted by the symbol ~. ■ **hydrolysis, p. 26**

Cells constantly turn over ATP, powering biosynthetic reactions by hydrolyzing the high-energy phosphate bond, and

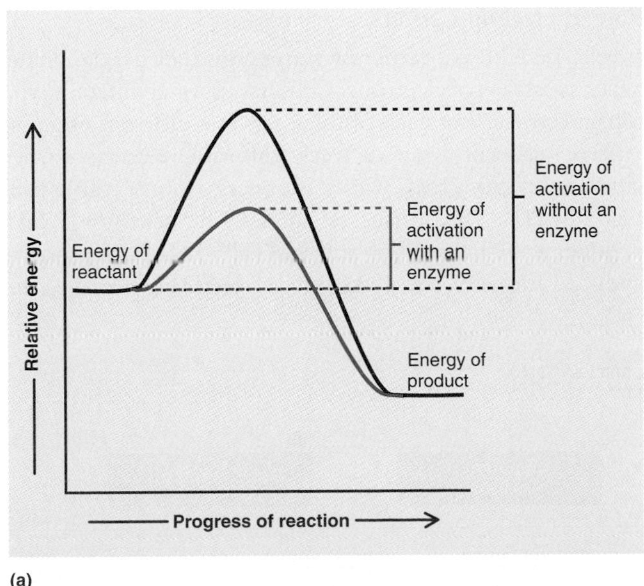

(a)

Starting compound --Enzyme a--> Intermediate$_a$ --Enzyme b--> Intermediate$_b$ --Enzyme c--> End product

(b)

Figure 6.6 The Role of Enzymes Enzymes function as biological catalysts. **(a)** An enzyme catalyzes a chemical reaction by lowering the activation energy of the reaction. **(b)** A specific enzyme facilitates each step of a metabolic pathway.

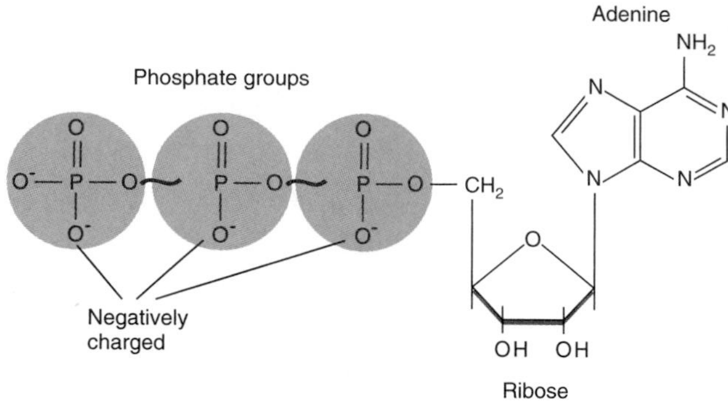

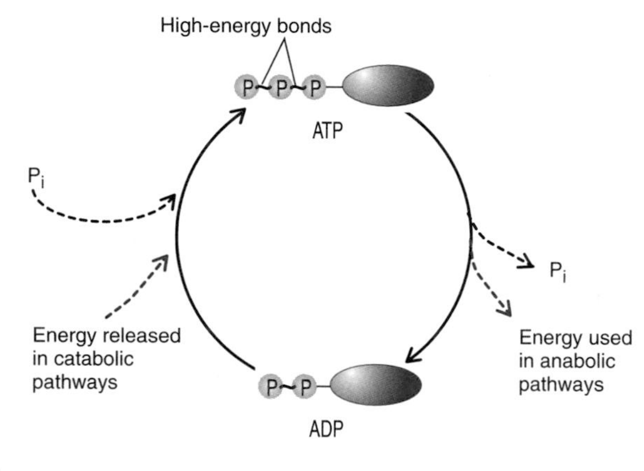

Figure 6.7 The Role of ATP Adenosine triphosphate (ATP) serves as the energy currency of a cell. **(a)** ATP is inherently unstable because of the tandem arrangement of its negatively charged phosphate groups; the bonds that join these groups are called high-energy phosphate bonds. **(b)** Hydrolysis of a high-energy phosphate bond releases a phosphate group and a sufficient amount of energy to power an endergonic reaction. An input of energy is required to re-form the high-energy phosphate bond, converting ADP back to ATP.

then exploiting energy-releasing reactions to form it again. Two different processes are used by chemoorganotrophs to provide the energy necessary to re-form the high-energy phosphate bond. **Substrate-level phosphorylation** uses the chemical energy released in an exergonic reaction to add P_i to ADP; **oxidative phosphorylation** harvests the energy of the proton motive force to do the same thing. Recall from chapter 3 that **proton motive force** is the form of energy that results from the electrochemical gradient established as protons are expelled from the cell. It powers the rotation of flagella, transport of nutrients, and synthesis of ATP. The electron transport chain that generates this type of energy will be discussed later in this chapter. Phototrophs can generate ATP using the process of **photophosphorylation**, utilizing radiant energy of the sun to drive the formation of a proton motive force. This is then harvested to generate ATP. ■ **proton motive force, p. 59**

The Role of the Energy Source

The compound broken down by a cell to release energy is called the **energy source**. As a group, prokaryotes show remarkable diversity in the variety of energy sources they can use. Many use organic compounds including, but not limited to glucose. Others use inorganic compounds including hydrogen sulfide and ammonia. Harvesting energy from a compound involves a series of coupled oxidation-reduction reactions.

Oxidation-Reduction Reactions **Oxidation-reduction reactions**, or **redox reactions**, transfer one or more electrons from one substance to another (**figure 6.8**). The compound that loses electrons becomes **oxidized**; the compound that gains those electrons becomes **reduced**.

When electrons are removed from a compound, protons often follow. In other words, an electron-proton pair, or hydrogen atom, is often removed. Thus, the removal of a hydrogen atom is an oxidation; correspondingly, the addition of a hydrogen atom is a reduction. An oxidation reaction in which an electron and an accompanying proton are removed can be called a **dehydrogenation**. A reduction reaction in which an electron and an accompanying proton are added can be called a **hydrogenation**.

When electrons are removed from the energy source, or **electron donor**, during catabolism, they are temporarily transferred to a specific molecule that serves as an **electron carrier**. That carrier can also be viewed as a **hydrogen carrier** if a proton accompanies the electron. Protons (H^+), however, unlike electrons, do not require carriers when in an aqueous solution. Because of this characteristic, the whereabouts of protons in biological reactions are often ignored.

The Role of Electron Carriers

Just as cells use ATP as a carrier of free energy, they use designated molecules as carriers of electrons. Cells have several different types of electron carriers, and each of these serves a different function.

Three different types of freely diffusible electron carriers participate in reactions that oxidize the energy source (**table 6.2**). They are **NAD⁺** (nicotinamide adenine dinucleotide), **FAD** (flavin adenine dinucleotide), and **NADP⁺** (NAD phosphate). The reduced forms of these carriers are **NADH**, **$FADH_2$**, and

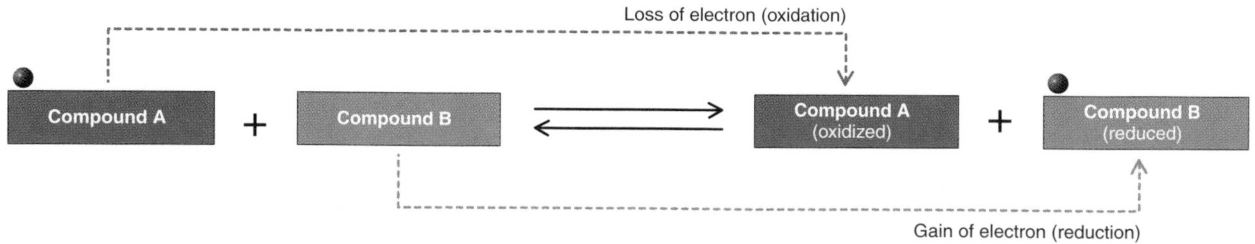

Figure 6.8 Oxidation-Reduction Reactions The compound that loses one or more electrons becomes oxidized; the compound that gains those electrons becomes reduced.

Table 6.2 Diffusible Electron Carriers

Carrier	Oxidized Form		Reduced Form	Typical Fate of Electrons Carried
			electron / electron carrier	
Nicotinamide adenine dinucleotide (carries 2 electrons and 1 proton)	$NAD^+ + 2e^- + 2H^+$	⇌	$NADH + H^+$	Used to generate a proton motive force that can drive ATP synthesis
Flavin adenine dinucleotide (carries 2 hydrogen atoms, i.e., 2 electron-proton pairs)	$FAD + 2e^- + 2H^+$	⇌	$FADH_2$	Used to generate a proton motive force that can drive ATP synthesis
Nicotinamide adenine dinucleotide phosphate (carries 2 electrons and 1 proton)	$NADP^+ + 2e^- + 2H^+$	⇌	$NADPH + H^+$	Biosynthesis

NADPH, respectively. The diffusible electron carriers can also be considered as hydrogen carriers because along with electrons, they carry protons. NAD^+ and $NADP^+$ can each carry a **hydride ion**, which consists of two electrons and one proton; $FADH_2$ carries two electrons and two protons.

Reduced electron carriers represent **reducing power** because their bonds contain a form of usable energy. The reducing power of NADH and $FADH_2$ is used to generate the proton motive force, which drives the synthesis of ATP in the process of oxidative phosphorylation. Ultimately the electrons are transferred to a compound such as O_2 that functions as a **terminal electron acceptor**. The reducing power of NADPH has an entirely different fate; it is used in biosynthetic reactions when a reduction is required.

Precursor Metabolites

Precursor metabolites are metabolic intermediates that link anabolic and catabolic pathways **(table 6.3)**. They are produced at specific steps in catabolic pathways and can be further oxidized to release their energy; alternatively, they can be siphoned off and used in anabolic pathways, serving as building blocks used to make the subunits of macromolecules (see figure 6.1). For example, the important precursor metabolite **pyruvate** can be further oxidized or it can be converted to the amino acid alanine.

Some prokaryotes, including *Escherichia coli*, can make all of their cell components, including proteins, lipids, carbohydrates, and nucleic acids using only 12 precursor metabolites **(table 6.3)**. Recall from chapter 4 that *E. coli* can grow in glucose

Table 6.3 Precursor Metabolites, Their Source, and Their Use in Biosynthesis in *E. coli*

Precursor Metabolite	Pathway Generated	Biosynthetic Role
Glucose 6-phosphate	Glycolysis	Lipopolysaccharide
Fructose 6-phosphate	Glycolysis	Peptidoglycan
Glyceraldehyde 3-phosphate	Glycolysis	Lipids (glycerol component)
3-phosphoglycerate	Glycolysis	Protein (the amino acids cysteine, glycine, and serine)
Phosphoenolpyruvate	Glycolysis	Protein (the amino acids phenylalanine, tryptophan, and tyrosine)
Pyruvate	Glycolysis	Proteins (the amino acids alanine, leucine, and valine)
Ribose 5-phosphate	Pentose phosphate cycle	Nucleic acids and proteins (the amino acid histidine)
Erythrose 4-phosphate	Pentose phosphate cycle	Protein (the amino acids phenylalanine, tryptophan, and tyrosine)
Acetyl-CoA	TCA cycle	Lipids (fatty acids)
α-ketoglutarate	TCA cycle	Protein (the amino acids arginine, glutamate, glutamine, and proline)
Succinyl-CoA	TCA cycle	Heme
Oxaloacetate	TCA cycle	Protein (the amino acids aspartate, asparagine, isoleucine, lysine, methionine, and threonine)

salts medium, which contains only glucose and a few inorganic salts. Eukaryotes and fastidious prokaryotes, however, are not as versatile as *E. coli* with respect to their biosynthetic capabilities. Any essential compounds that a cell cannot synthesize from the appropriate precursor metabolite must be provided from an external source. ■ **glucose salts medium, p. 103**

Scheme of Metabolism

Three key metabolic pathways—**glycolysis**, the **pentose phosphate pathway**, and the **tricarboxylic acid cycle** (**TCA cycle**)—are sometimes called the **central metabolic pathways**.Glycolysis is also called the **Embden-Meyerhoff** pathway to honor the scientists who described it, or the **glycolytic pathway**. The TCA cycle is sometimes called the **Krebs cycle** to recognize the scientist who initially characterized most of it, or the **citric acid cycle**. The central metabolic pathways are used together to gradually oxidize glucose, the preferred energy source of many cells, completely to carbon dioxide (**figure 6.9**). Because the free energy of glucose is much higher than that of CO_2, the pathways release energy that can be harvested to generate ATP and accumulate reducing power. In addition, during the course of the stepwise breakdown of glucose, precursor metabolites are formed. The central metabolic pathways are catabolic, but the precursor metabolites and reducing power they generate can also be diverted for use in biosynthesis. To reflect the dual role of these pathways, they are sometimes called **amphibolic pathways** (*amphi* means "both kinds").

The most common pathway that initiates the breakdown of sugars is glycolysis (*glycos* means "sugar," *lysis* means "dissolution"). This multistep pathway gradually oxidizes the 6-carbon sugar glucose to form two molecules of pyruvate, a 3-carbon compound. At about midpoint in the pathway, a 6-carbon derivative of glucose is split into two 3-carbon molecules. Both of these latter molecules then undergo the same series of transformations to produce the pyruvate molecules. Glycolysis provides the cell with a small amount of energy in the form of ATP, some reducing power in the form of NADH, and a number of precursor metabolites. Some bacteria have a different pathway called the **Entner-Doudoroff pathway** instead of or in addition to the glycolytic pathway; some archaea have a slightly modified version of this pathway. Like glycolysis, the Entner-Doudoroff pathway generates pyruvate, but it uses different enzymes, generates reducing power in the form of NADPH, and yields less ATP.

The pentose phosphate pathway also converts glucose to pyruvate, but its primary role in metabolism is the production of compounds used in biosynthesis, including reducing power in the form of NADPH and precursor metabolites. It operates as a secondary pathway in conjunction with the other glucose-degrading pathways (glycolysis and the Entner-Doudoroff pathway).

The pyruvate generated in any of the preceeding pathways must be converted into a specific two-carbon fragment to enter the Krebs cycle. This is accomplished in a complex reaction called the **transition step**, which removes CO_2, generates reducing power, and joins the resulting acetyl group to a compound called coenzyme A, forming acetyl-CoA. Note that the transition step is repeated twice for each molecule of glucose that is broken down.

The 2-carbon acetyl group of acetyl-CoA then enters the TCA cycle, initiating a series of oxidation steps that result in the release of two molecules of CO_2. For every acetyl-CoA that enters the TCA cycle, the cyclic pathway "turns" once. Therefore, it must "turn" twice to complete the oxidation of one molecule of glucose. The TCA cycle generates precursor metabolites, a great deal of reducing power, and a compound that is the functional equivalent of ATP.

Respiration uses the reducing power accumulated in glycolysis, the transition step, and the TCA cycle to generate ATP by oxidative phosphorylation (**table 6.4**). The diffusible electron carriers NADH and $FADH_2$ transfer their electrons to the electron transport chain, which exports protons to generate a proton motive force. This transfer of electrons also serves to recycle the diffusible carriers so they can once again accept electrons from catabolic reactions. In **aerobic respiration**, the electrons are ultimately passed to molecular oxygen (O_2), the terminal electron acceptor, producing water. Some organisms use **anaerobic respiration**. This is analogous to aerobic respiration, generating ATP by oxidative phosphorylation, but it employs an inorganic molecule other than O_2, such as nitrate (NO_3^-), as a terminal electron acceptor. Organisms that use respiration, either aerobic or anaerobic, are said to **respire**. Note that respiration employs an inorganic molecule as the terminal electron acceptor.

Cells that cannot respire are limited by their relative inability to recycle reduced electron carriers. A cell only has a limited number of carrier molecules; if electrons are not removed from the reduced carriers, none will be available to accept electrons. As a consequence, subsequent catabolic processes cannot occur. **Fermentation** provides a solution to this problem, but it results in only the partial oxidation of glucose. Thus, compared with respiration, fermentation produces relatively little ATP. It is used by facultative anaerobes when a suitable inorganic terminal electron acceptor is not available and by organisms that lack an electron transport chain. These cells must stop short of oxidizing glucose completely to avoid generating even more reducing power. Otherwise, they would run out of the oxidized form of the electron carriers very quickly. Instead of oxidizing pyruvate in the TCA cycle, they use pyruvate or a derivative of it as a terminal electron acceptor. By transferring the electrons carried by NADH to pyruvate or a derivative, NAD^+ is regenerated so it can once again accept electrons in the steps of glycolysis. Note that fermentation always utilizes an organic molecule as the terminal electron acceptor. Although fermentation does not use the TCA cycle, organisms that ferment still employ certain key steps of the cycle in order to generate the necessary precursor molecules. ■ **facultative anaerobes, p. 98**

MICROCHECK 6.1

Catabolic pathways gradually oxidize an energy source in a controlled manner so that the energy released can be harvested. A specific enzyme catalyzes each step. Substrate-level phosphorylation uses chemical energy to synthesize ATP; oxidative phosphorylation employs the proton motive force to synthesize ATP. Reducing power

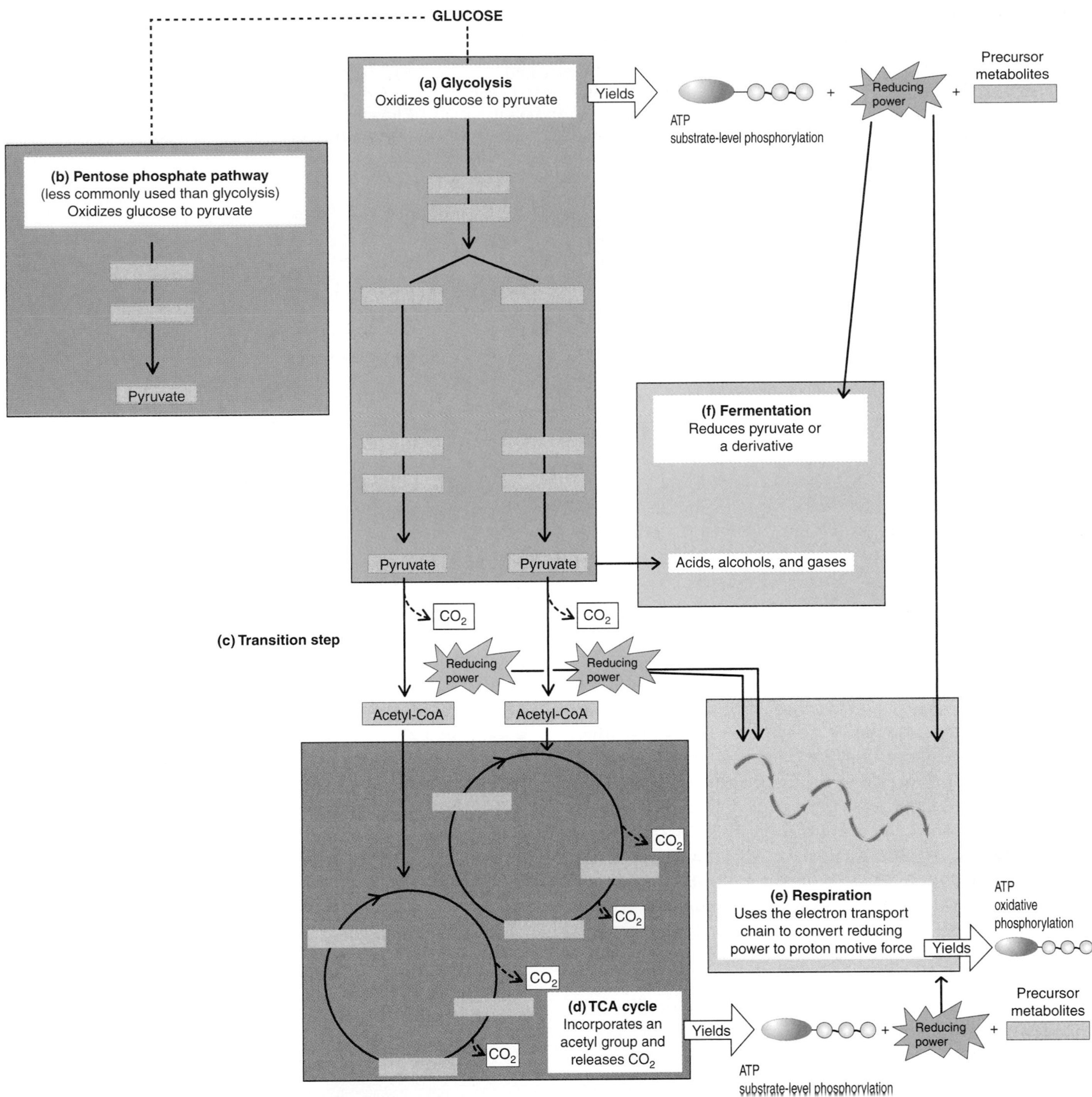

Figure 6.9 Scheme of Metabolism **(a)** Glycolysis, **(b)** the pentose phosphate pathway, **(c)** the transition step, and **(d)** the tricarboxylic acid cycle (TCA cycle) are used to gradually oxidize glucose completely to CO_2. Together, these pathways produce ATP, reducing power, and intermediates that function as precursor metabolites (depicted as gray bars). **(e)** Respiration uses the reducing power to generate ATP by oxidative phosphorylation, employing an inorganic molecule as a terminal electron acceptor. **(f)** Fermentation stops short of oxidizing glucose completely, and instead uses pyruvate or a derivative as an electron acceptor.

in the form of NADH and $FADH_2$ are used to generate the proton motive force; the reducing power of NADPH is utilized in biosynthesis. Precursor metabolites can either be further oxidized or used in biosynthesis. The central metabolic pathways generate ATP, reducing power, and precursor metabolites.

- How does the fate of electrons carried by NADPH differ from those carried by NADH?
- Why are the central metabolic pathways called amphibolic pathways?
- Why does fermentation release less energy than respiration?

Table 6.4 Metabolic Processes of Prokaryotic Chemoorganoheterotrophs

Metabolic Process	Pathways Used	Terminal Electron Acceptor	ATP Generated by Substrate-Level Phosphorylation (Theoretical Maximum)	ATP Generated by Oxidative Phosphorylation (Theoretical Maximum)	Total ATP Generated (Theoretical Maximum)
Aerobic respiration	Glycolysis, TCA cycle	O_2	2 in glycolysis (net) 2 in the TCA cycle 4 total	34	38
Anaerobic respiration	Glycolysis, TCA cycle	Inorganic molecule other than O_2 such as nitrate (NO_3^-), nitrite (NO_2), sulfate (SO_4^{2-})	2 in glycolysis (net) 2 in the TCA cycle 4 total	32 (actual number varies; however, the ATP yield of anaerobic respiration is less than that of aerobic respiration)	36 or less
Fermentation	Glycolysis	Organic molecule (pyruvate or a derivative)	2 in glycolysis (net) 2 total	0	2

Enzymes

Enzymes are proteins that act as biological catalysts, facilitating the conversion of a substrate into a product. They do this with extraordinary specificity and speed, usually acting on only one, or a very limited number of, substrates. They are neither consumed nor permanently changed during a reaction, allowing a single enzyme molecule to be rapidly used over and over again. In only one second, the fastest enzymes can transform more than 10^4 substrate molecules to products. More than a thousand different enzymes exist in a cell; most are given a common name that reflects their function and ends with the suffix -ase. For example, those that degrade proteins are collectively called proteases.

Mechanisms and Consequences of Enzyme Action

An enzyme has on its surface an **active site** or **catalytic site**, typically a relatively small crevice **(figure 6.10)**. This is the critical site to which a substrate binds by weak forces. The binding of the substrate to the active site causes the shape of the flexible enzyme to change slightly. This mutual interaction, or **induced fit**, results in a temporary intermediate called the **enzyme-substrate complex**. The substrate is held within this complex in a specific orientation so that the activation energy for a given reaction is lowered, allowing the products to be formed. The products are then released, leaving the enzyme unchanged and free to combine with new substrate molecules. Note that enzymes may also catalyze reactions in which two substrates are joined to create one product. In addition, theoretically, all enyzme-catalyzed reactions are reversible. The free energy change of certain reactions, however, makes them effectively nonreversible.

The interaction of the enzyme with its substrate is very specific. The substrate fitting into the active site may be likened to a hand fitting into a glove. Not only must it fit spatially, but the appropriate chemical interactions such as hydrogen and ionic bonding need to occur to induce the fit. This requirement for a precision fit and interaction explains why, with minor exceptions, it takes a different enzyme to catalyze every reaction in a cell. Very few molecules of any particular enzyme are needed, however, as each is swiftly reused again and again. ■ hydrogen bonds, p. 23 ■ ionic bonds, p. 22

In some cases, two different enzymes can recognize the same substrate. This occurs at the branch step in some pathways of biosynthesis where a particular precursor metabolite can be used to make several different products. Although different enzymes may recognize the same substrate, they hold

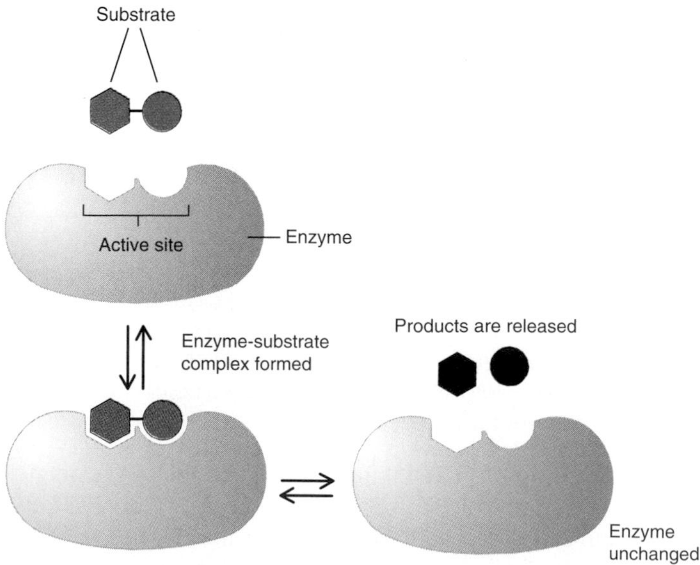

Figure 6.10 Mechanism of Enzyme Action The substrate binds to the active site, forming an enzyme-substrate complex. The products are then released, leaving the enzyme unchanged and free to combine with new substrate molecules.

the substrate in different ways, lowering the activation energy for only their designated reaction. Consequently, the enzyme that binds the substrate determines which of several possible bond-breaking or bond-forming reactions will occur.

Allosteric Regulation

Cells can rapidly fine-tune or regulate the activity of certain key enzymes using other molecules that bind to and distort them (**figure 6.11**). This has the effect of regulating the activity of metabolic pathways. These enzymes can be controlled because they are **allosteric enzymes** (*allo* means "other"), which have a binding site called an **allosteric site** that is separate from their active site. When a regulatory molecule, or **effector**, binds to the allosteric site, the shape of the enzyme changes. This distortion alters the relative **affinity**, or chemical attraction, of the enzyme for its substrate.

Allosteric enzymes generally catalyze the step that either initiates or commits to a given pathway. Because their activity can be controlled, they provide the cell with a means to modulate the pace of metabolic processes, turning off some pathways and activating others. Cells can also control the amount of enzyme they synthesize; this control mechanism, which will be discussed in the next chapter, also involves allosteric proteins. ■ regulation, p. 179

The end product of a given biosynthetic pathway generally acts as an allosteric inhibitor of the first enzyme of that pathway—a mechanism called **feedback inhibition**. This mechanism allows the product of the pathway to effectively modulate its own synthesis. For example, the first enzyme of the multistep pathway used to synthesize the amino acid tryptophan is an allosteric enzyme that is inhibited by the binding of tryptophan. The amino acid must be present at a relatively high concentration, however, to bind and inhibit the enzyme. Thus, the pathway will only be shut down when a cell has accumulated sufficient tryptophan to fill its immediate protein synthesis needs. Because the binding of the effector is reversible, the enzyme can again become active when tryptophan levels decrease.

Compounds that reflect a cell's relative energy stores often regulate allosteric enzymes of catabolic pathways, enabling cells to modulate the flow of those pathways in response to changing energy needs. High levels of ATP inhibit certain enzymes and, as a consequence, slow catabolic processes. In contrast, high levels of ADP warn that a cell's energy stores are low, and they function to stimulate the activity of some enzymes.

Some proteins other than enzymes are also subject to allosteric inhibition and activation. An example includes allosteric proteins that are involved in regulating gene expression, which will be discussed in the next chapter.

Cofactors and Coenzymes

Enzymes are proteins, but they sometimes act with the assistance of a non-protein component called a **cofactor (figure 6.12)**. **Coenzymes** are organic cofactors that act as loosely bound carriers of molecules or electrons (**table 6.5**). They include the electron carriers FAD, NAD^+, $NADP^+$ and **coenzyme A (CoA)**. Recall that the transition step between glycolysis and the TCA cycle generates acetyl-CoA; this compound is actually coenzyme A carrying an acetyl group. Other cofactors attach tightly to enzymes. For example, magnesium, zinc, copper, and other trace elements required for growth often function as cofactors. ■ trace elements, p. 100

All coenzymes transfer substances from one compound to another, but they function in different ways. Some remain bound to the enzyme during the transfer process, whereas others separate from the enzyme, carrying the substance being transferred along with them. The same coenzyme can assist different enzymes. Because of this, far fewer different coenzymes are required than enzymes. Like enzymes, coenzymes are recycled as they function and, consequently, are needed only in small quantities.

Most coenzymes are synthesized from vitamins (see table 6.5). Some bacteria, such as *E. coli,* can synthesize all their required vitamins and convert them to the necessary coenzymes. In contrast, humans and other animals must be provided with vitamins from external sources. Most often they must be supplied in the diet,

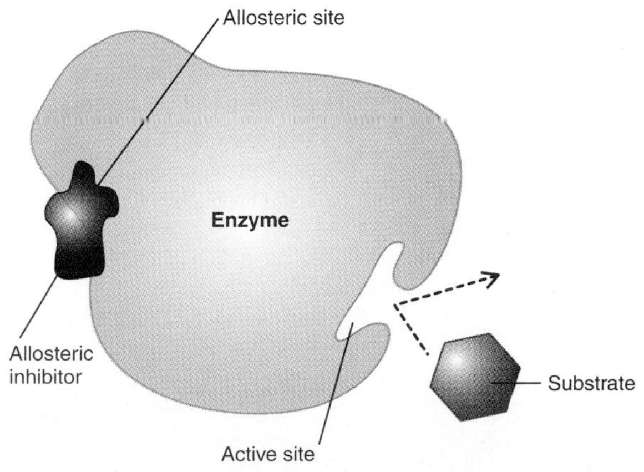

Figure 6.11 Regulation of Allosteric Enzymes Allosteric enzymes have, in addition to the active site, an allosteric site. The binding of an effector to the allosteric site causes the shape of the enzyme to change, altering the relative affinity of the enzyme for its substrate.

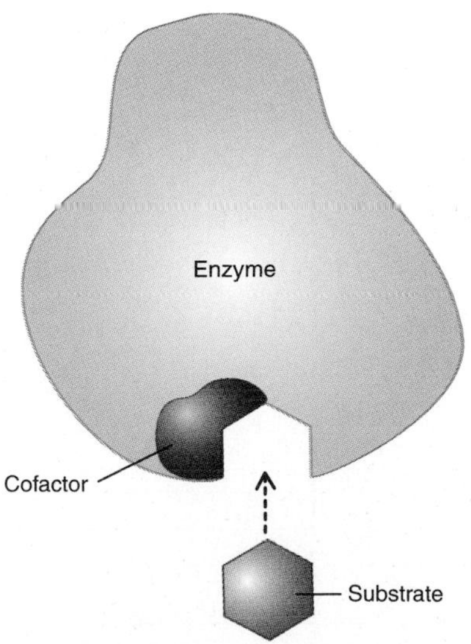

Figure 6.12 Enzymes May Act in Conjunction with a Cofactor Cofactors are non-protein components and may be either coenzymes or trace elements.

Table 6.5 Some Coenzymes and Their Function

Coenzyme	Vitamin from Which It Is Derived	Substance Transferred	Example of Use
Nicotinamide adenine dinucleotide (NAD^+)	Niacin	Hydride ions (2 electrons and 1 proton)	Carrier of reducing power
Flavin adenine dinucleotide (FAD)	Riboflavin	Hydrogen atoms (2 electrons and 2 protons)	Carrier of reducing power
Coenzyme A	Pantothenic acid	Acyl groups	Carries the acetyl group that enters the TCA cycle
Thiamin pyrophosphate	Thiamine	Aldehydes	Facilitates the removal of CO_2 from pyruvate in the transition step
Pyridoxal phosphate	Pyridoxine	Amino groups	Transfers amino groups in amino acid synthesis
Tetrahydrofolate	Folic acid	1-carbon molecules	Used in nucleotide synthesis

but in some cases vitamins synthesized by bacteria residing in the intestine can be absorbed. If an organism lacks a vitamin, the functions of all the different enzymes whose activity requires the corresponding coenzyme are impaired. Thus, a single vitamin deficiency has serious consequences.

Environmental Factors that Influence Enzyme Activity

The growth of any organism depends on the proper functioning of its enzymes. Several features of the environment influence how well enzymes function and in this way determine how rapidly bacteria multiply **(figure 6.13)**. Each enzyme has a narrow range of environmental factors—including temperature, pH, and salt concentration—at which it operates optimally. A 10°C rise in temperature approximately doubles the speed of enzymatic reactions, until optimal activity is reached; this explains why bacteria tend to grow more rapidly at higher temperatures. If the temperature gets too high, however, proteins will become denatured and no longer function. Most enzymes function best at low salt concentrations and at pH values slightly above 7. Not surprisingly then, most bacteria that have been studied grow fastest under these same conditions. Some prokaryotes, however, particularly certain members of the Archaea, are found in environments where conditions are extreme. They may require high salt concentrations, grow under very acidic conditions, or be found where temperatures are near boiling. ■ **pH, p. 98** ■ **temperature and growth, p. 96**

Enzyme Inhibition

Enzymes can be inhibited by a variety of compounds other than the effectors normally used by the cell for regulation. The site on the enzyme to which these molecules bind determines whether they function as competitive or noncompetitive inhibitors.

Competitive Inhibition

In **competitive inhibition**, the inhibitor competes with the normal substrate for binding to the active site **(figure 6.14)**. Generally this occurs because the inhibitor has a chemical structure similar to the normal substrate.

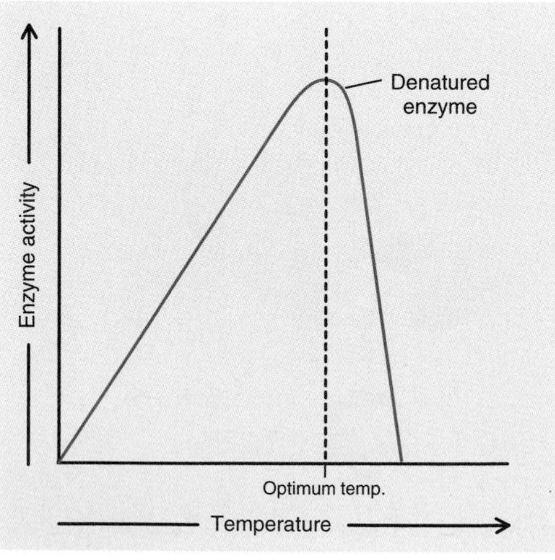

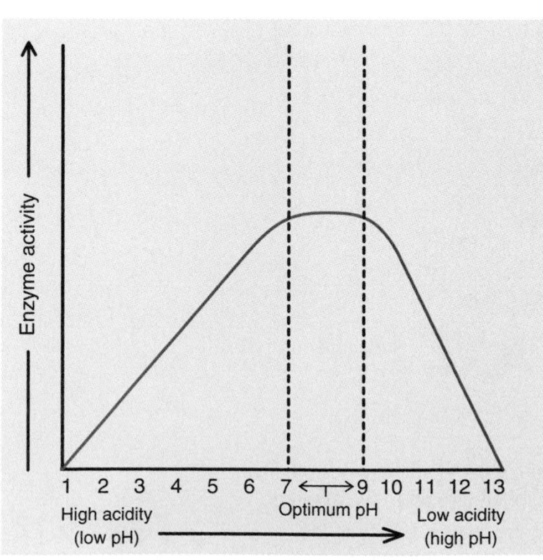

Figure 6.13 The Effects of Temperature and pH on Enzyme Activity Every enzyme has a narrow range of environmental factors at which it operates optimally.

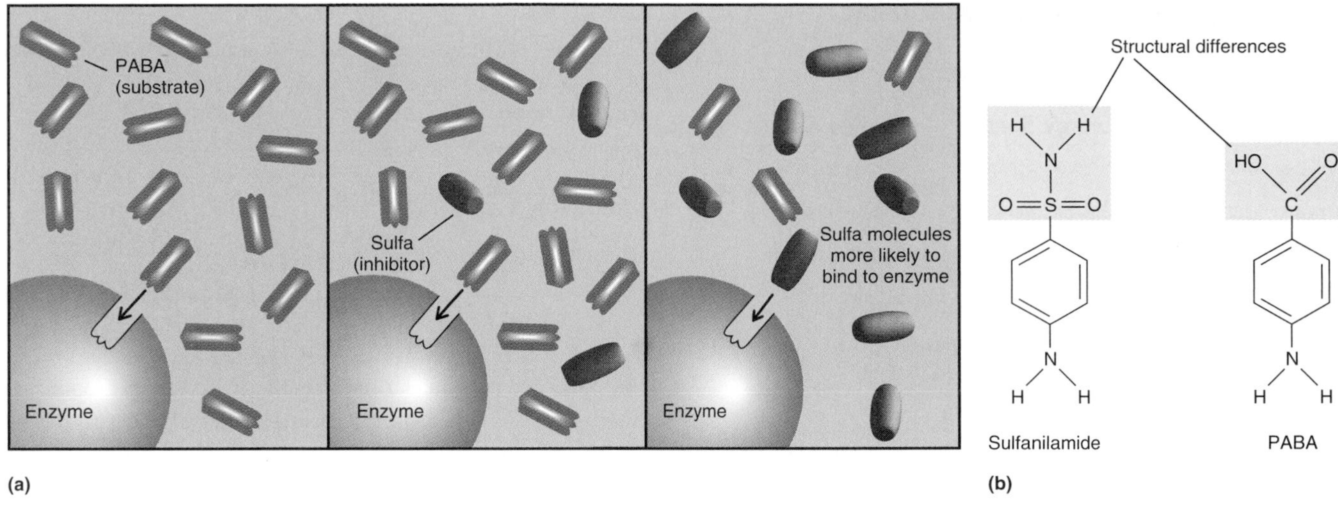

Figure 6.14 Competitive Inhibition of Enzymes (a) The inhibitor competes with the normal substrate for binding to the active site. **(b)** A competitive inhibitor generally has a chemical structure similar to the normal substrate.

A good example of competitive inhibition is the action of sulfanilamide, one of the sulfa drugs used as an antimicrobial medication. Sulfa drugs inhibit an enzyme in the pathway that bacteria use to synthesize the vitamin folic acid by binding to the active site of the enzyme. The drug does not affect human metabolism because humans cannot synthesize folic acid; it must be provided in the diet. Sulfa drugs have a structure similar to **para-aminobenzoic acid (PABA)**, an intermediate in the bacterial pathway for folic acid synthesis. Because of this, they fit into the active site of the enzyme that normally uses PABA as a substrate, preventing the attachment of PABA. The greater the proportion of sulfa molecules relative to PABA molecules, the more likely the active site of the enzyme will be occupied by a sulfa molecule. Once the sulfa is removed, the enzyme functions normally with PABA as the substrate. ■ **sulfa drugs, p. 505**

Noncompetitive Inhibition

Noncompetitive inhibition occurs when the inhibitor and the substrate act at different sites on the enzyme. Allosteric inhibition, discussed previously, can be considered an example of noncompetitive reversible inhibition that is exploited by the cell to modulate its processes (see figure 6.11). Noncompetitive, nonreversible inhibitors cause damaging effects by permanently altering the enzyme so that it can no longer function. For example, mercury in the antibacterial compound mercurochrome inhibits growth because it oxidizes the S—H groups of the amino acid cysteine in proteins. This converts cysteine to cystine, which cannot form the important covalent disulfide bond (S—S). As a result, the protein cannot achieve its proper shape.

MICROCHECK 6.2

Enzymes facilitate the conversion of a substrate into a product with extraordinary speed and specificity; furthermore, they are neither consumed nor permanently changed in the reaction. The activity of allosteric enzymes can be regulated. Some enzymes act with the assistance of a cofactor. Environmental factors influence enzyme activity and, by doing so, determine how rapidly bacteria multiply. A variety of different compounds adversely affect enzyme activity.

- Explain why sulfa drugs preferentially inhibit the growth of bacteria and not the host.
- Explain the function of a coenzyme.
- Why is it important for a cell that allosteric inhibition be reversible?

Pathways and Processes that Fuel Aerobic Growth of Chemoorganotrophs

Most bacteria and all eukaryotic organisms, with the exception of plants and algae, are chemoorganotrophs, transforming energy by oxidizing organic compounds **(table 6.6)**. Because they obtain carbon from these same compounds, they are also called **chemoheterotrophs**. They can be called **chemoorganoheterotrophs**, but for the sake of simplicity, they are usually referred to as chemoorganotrophs or chemoheterotrophs, depending on whether the discussion relates to energy generation or carbon utilization.

Aerobic growth of chemoorganotrophs generally requires the integrated use of the three central metabolic pathways—glycolysis, the pentose phosphate pathway, and the tricarboxylic acid cycle. These pathways modify organic molecules in a stepwise fashion to form:

- Intermediates with high-energy bonds that can be used to synthesize ATP by substrate-level phosphorylation
- Intermediates that can be oxidized to generate reducing power
- Intermediates and end products that function as precursor metabolites

Table 6.6 Categorizing Organisms According to Their Energy and Carbon Source

Category	Energy Source	Carbon Source
Chemolithoautotroph	Inorganic chemicals	Carbon Dioxide
Chemoorganoheterotroph	Organic chemicals	Organic chemicals
Photoautotroph	Sunlight	Carbon dioxide
Photoheterotroph	Sunlight	Organic chemicals

Note that the precursor metabolites can be siphoned off from these pathways for use in biosynthesis. The rate at which they are removed will affect the overall energy gain of catabolism. This is generally overlooked in descriptions of the ATP-generating functions of these pathways for the sake of simplicity. Recognize, however, that because these pathways serve more than one function, the energy yields are only theoretical.

The pathways of central metabolism are compared in **table 6.7**. The entire pathways with chemical formulas and enzyme names are illustrated in Appendix 2.

Glycolysis

Glycolysis is the primary pathway used by nearly all organisms to convert glucose to pyruvate **(figure 6.15)**. In the 9-step pathway, one molecule of glucose is converted into two molecules of pyruvate. This generates a net gain of two molecules of ATP and two molecules of NADH. The overall process can be summarized as follows:

$$\text{glucose (6 C)} + 2\ NAD^+ + 2\ ADP + 2\ P_i \longrightarrow 2\ \text{pyruvate (3 C)} + 2\ NADH + 2\ H^+ + 2\ ATP$$

In addition to generating ATP and reducing power (NADH), the pathway produces 6 of the 12 precursor molecules needed by *E. coli* (see table 6.3). Because none of the reactions in the pathway require molecular oxygen, it can occur under either aerobic or anaerobic conditions.

- **Step 1:** A high-energy bond is expended to initiate the pathway when a phosphate group from ATP is transferred to glucose, forming glucose 6-phosphate. In bacteria, the high-energy phosphate bond is expended as glucose is being transported into the cell.
- **Step 2:** A chemical rearrangement occurs to generate fructose 6-phosphate.

Table 6.7 Comparison of the Central Metabolic Pathways

Pathway	Characteristics
Glycolysis	Generally used in aerobic respiration, anaerobic respiration, and fermentation to obtain energy and precursor metabolites. Reducing power is also generated. In respiration, this reducing power can be used to drive the synthesis of ATP; in fermentation, it is consumed to recycle the electron carrier without an additional gain in energy. Glycolysis generates: • 2 ATP (net) by substrate-level phosphorylation • 2 NADH + 2 H^+ • 6 different precursor metabolites
Pentose phosphate cycle	May be used in aerobic respiration, anaerobic respiration, and fermentation, primarily to obtain precursor metabolites and reducing power in the form of NADPH. The pentose phosphate cycle generates: • ATP by substrate-level phosphorylation (amount varies) • NADPH + H^+ (amount varies) • 2 different precursor metabolites
Transition step	Used in aerobic respiration and anaerobic respiration. The reducing power produced in this step can be used to drive the synthesis of ATP by oxidative phosphorylation. The transition step, repeated twice to oxidize two molecules of pyruvate to acetyl-CoA, generates: • 2 NADH + 2 H^+ • 1 precursor metabolite
TCA cycle	Used in aerobic respiration and anaerobic respiration. The reducing power produced in this cycle can be used to drive the synthesis of ATP by oxidative phosphorylation. The TCA cycle, repeated twice to incorporate two acetyl groups, theoretically generates: • 2 ATP by substrate-level phosphorylation (from conversion of GTP) • 6 NADH + 6 H^+ • 2 $FADH_2$ • 3 different precursor metabolites

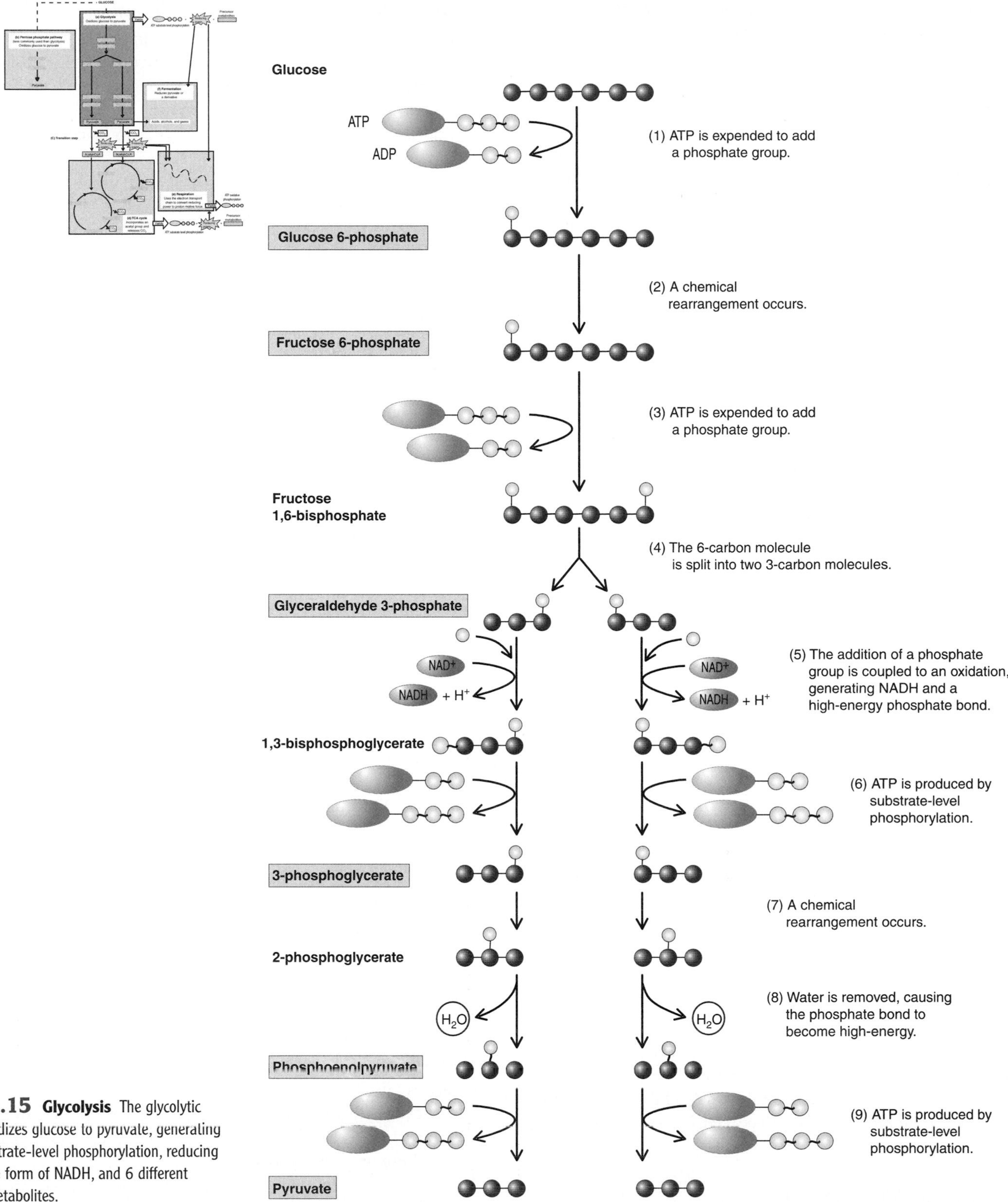

Figure 6.15 Glycolysis The glycolytic pathway oxidizes glucose to pyruvate, generating ATP by substrate-level phosphorylation, reducing power in the form of NADH, and 6 different precursor metabolites.

- **Step 3:** A second high-energy phosphate bond from ATP is expended to form fructose 1, 6-bisphosphate, priming the molecule so it can be more readily split into two molecules. The step is non-reversible and serves as an important control point for the pathway. High levels of ATP inhibit the allosteric enzyme that catalyzes the step.

- **Step 4:** The 6-carbon compound formed in the previous step is split to form two 3-carbon molecules—glyceraldehyde 3-phosphate (G3P) and dihydroxyacetone phosphate The latter molecule is readily converted into G3P and, therefore, we will consider it equivalent to G3P.

- **Step 5:** This is a coupled reaction; the energy released by the oxidation of G3P is used to add an inorganic phosphate group, creating 1,3-bisphosphoglycerate (BPG), which has a high-energy phosphate bond. During the oxidation, the electron carrier NAD^+ is reduced to form NADH + H^+. Note that this and subsequent steps of glycolysis each occur twice, once for each of the two G3P formed as a result of the previous step.
- **Step 6:** Substrate-level phosphorylation then occurs, as the energy in the high-energy phosphate bond of BPG is used to generate ATP. Removal of the phosphate group from BPG forms 3-phosphoglycerate (3PG). At this point in the glycolytic pathway, the net energy gain is 0, because this step simply replenishes the ATP that was expended in steps 1 and 2.
- **Step 7:** This chemically rearranges 3PG to form 2-phosphoglycerate (2PG).
- **Step 8:** A water molecule is removed from 2PG, creating **phosphoenolpyruvate** (PEP), a molecule that contains an unstable (and therefore) high-energy phosphate bond.
- **Step 9**: Substrate-level phosphorylation occurs once more as the energy of PEP is used to generate ATP, giving rise to a net gain of ATP. Removal of the phosphate group from PEP forms pyruvate, the end product of glycolysis.

Yield of Glycolysis

For every glucose molecule degraded, the steps of glycolysis produce:

- **ATP**—The maximum possible energy gain as ATP in glycolysis is:

Energy expended	2 ATP molecules (steps 1 and 3)
Energy harvested	4 ATP molecules (steps 6 and 9)
Net gain	2 ATP molecules

- **Reducing power**—An oxidation takes place at step 5, which occurs twice in glycolysis, converting 2 NAD^+ to 2 NADH + $2H^+$.
- **Precursor metabolites**—Five intermediates of glycolysis as well as the end product, pyruvate, are precursor metabolites used by *E. coli*. The use of any of these for biosynthesis reduces the amount of ATP and reducing power generated during glycolysis.

Pentose Phosphate Pathway

The other pathway used by prokaryotes and eukaryotes to convert glucose to pyruvate is the pentose phosphate pathway. This can provide energy, but its greatest significance is in its contribution to biosynthesis; the reducing power it generates is in the form of NADPH, and two of its intermediates, ribose 5-phosphate and erythrose 4-phosphate, are important precursor metabolites. The pentose phosphate pathway, like glycolysis, can operate in the presence or absence of molecular oxygen.

Yield of the Pentose Phosphate Pathway

The yield of the pentose phosphate pathway varies, depending on which of several possible alternatives are taken. Because it is used primarily to support biosynthesis, and its yields can vary, its contributions to ATP generation are not included in the calculation of ATP yields. It can, however, produce:

- **ATP**—A variable amount of ATP can be produced by substrate-level phosphorylation.
- **Reducing power**—A variable amount of reducing power in the form of NADPH is produced.
- **Precursor metabolites**—Two intermediates of the pentose phosphate pathway are precursor metabolites.

Transition Step

The transition step links glycolysis to the TCA cycle (see figure 6.16). In prokaryotic cells, the entire oxidation process takes place in the cytoplasm. In eukaryotic cells, however, the 3-carbon pyruvate must first enter the mitochondria. This is because the enzymes that facilitate the glycolytic pathway are located in the cytoplasm, whereas those that catalyze the steps of the TCA cycle are found only within the matrix of the mitochondria. ■ mitochondria, p. 79

The transition step involves several integrated reactions catalyzed by a group of enzymes that form a large multi-enzyme complex. In the concerted series of reactions, carbon dioxide is first removed from the pyruvate, a process called **decarboxylation**. Then, an oxidation occurs, reducing NAD^+ to form NADH + H^+. Finally, the remaining 2-carbon acetyl group is joined to the coenzyme A to form **acetyl-CoA**, a precursor metabolite used in fatty acid synthesis.

Yield of the Transition Step

- **Reducing power**—The transition step, which occurs twice for every molecule of glucose that enters glycolysis, is an oxidation. This converts 2 NAD^+ to 2 NADH + 2 H^+.
- **Precursor metabolites**—The end product of the transition step, acetyl-CoA, is an important precursor metabolite.

Tricarboxylic Acid Cycle

The eight steps of the tricarboxylic acid cycle complete the oxidation of glucose (**figure 6.16**). The pathway itself does not directly use O_2. A large amount of reducing power is generated, however, and those electrons are ultimately passed to molecular oxygen. Allosteric enzymes regulate the cycle at several steps, but these will not be discussed.

The TCA cycle incorporates the acetyl groups that result from the transition step, releasing CO_2 in the following net reaction:

$$\text{2 acetyl groups (2 C)} + 6\ NAD^+ + 2\ FAD + 2\ ADP + 2\ P_i \longrightarrow 4\ CO_2 + 6\ NADH + 6\ H^+ + 2\ FADH_2 + 2\ ATP$$

In addition to generating ATP and reducing power, the steps of this conversion form 4 more of the 12 precursor metabolites used by *E. coli* (see table 6.3).

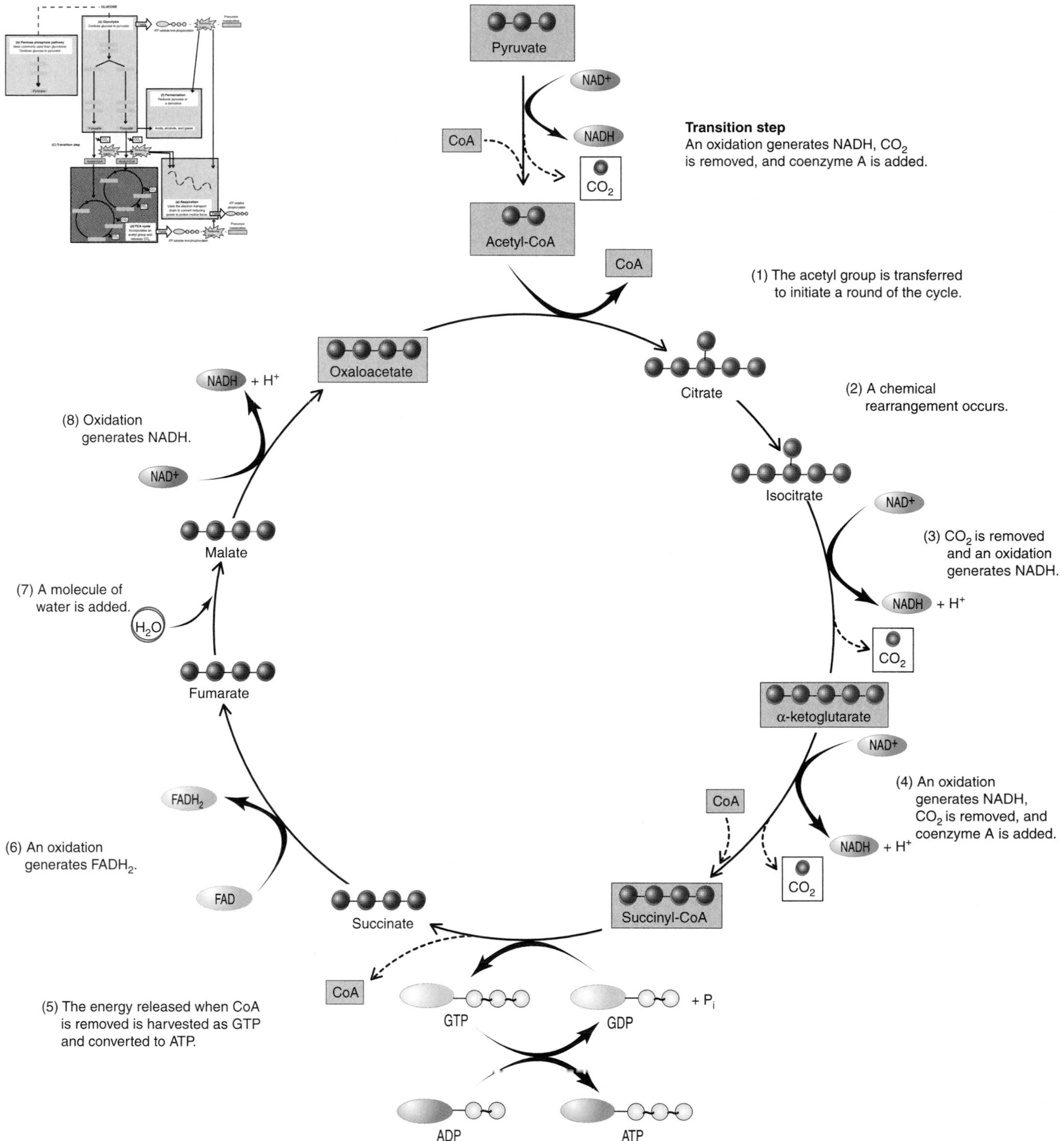

Pyruvate
NAD+
NADH
CoA
CO_2
Transition step
An oxidation generates NADH, CO_2 is removed, and coenzyme A is added.
Acetyl-CoA
CoA
(1) The acetyl group is transferred to initiate a round of the cycle.
Oxaloacetate
Citrate
(2) A chemical rearrangement occurs.
NADH + H^+
(8) Oxidation generates NADH.
NAD+
Isocitrate
NAD+
(3) CO_2 is removed and an oxidation generates NADH.
Malate
(7) A molecule of water is added.
H_2O
NADH + H^+
CO_2
Fumarate
α-ketoglutarate
NAD+
(4) An oxidation generates NADH, CO_2 is removed, and coenzyme A is added.
$FADH_2$
CoA
(6) An oxidation generates $FADH_2$.
NADH + H^+
CO_2
FAD
Succinate
Succinyl-CoA
CoA
(5) The energy released when CoA is removed is harvested as GTP and converted to ATP.
GTP
GDP
+ P_i
ADP
ATP

Figure 6.16 The Transition Step and the Tricarboxylic Acid Cycle The transition step links glycolysis and the TCA cycle, converting pyruvate to acetyl-CoA; it generates reducing power and 1 precursor metabolite. The TCA cycle incorporates the acetyl group of acetyl-CoA and using a series of steps, releases CO_2; it generates GTP, which can be converted to ATP, reducing power in the form of both NADH and $FADH_2$, and 3 different precursor metabolites.

- **Step 1**: The cycle begins when CoA transfers its acetyl group to the 4-carbon compound oxaloacetate, thereby forming the 6-carbon compound citrate.
- **Step 2:** Citrate is chemically rearranged to form an isomer, isocitrate. ■ **isomer, p. 33**
- **Step 3**: Isocitrate is oxidized and a molecule of CO_2 is removed leaving the 5-carbon compound α-ketoglutarate. During the oxidation step, NAD^+ is reduced to form NADH + H^+.

- **Step 4**: Like the transition step that converts pyruvate to acetyl-CoA, this reaction involves a group of reactions catalyzed by a complex of enzymes. In this step, α-ketoglutarate is oxidized, CO_2 is removed, and CoA is added, producing the 4-carbon compound succinyl-CoA. During the oxidation step, NAD^+ is reduced to form NADH + H^+.
- **Step 5**: This removes CoA from succinyl-CoA, harvesting the energy to generate guanosine triphosphate (GTP), which can be used to make ATP. The reaction forms succinate.
- **Step 6**: Succinate is oxidized to form fumarate. During the oxidation, FAD is reduced to form $FADH_2$.
- **Step 7**: A molecule of water is added to fumarate to form malate.
- **Step 8**: Malate is oxidized to form oxaloacetate; note that oxaloacetate is the starting compound to which acetyl-CoA is added to start the cycle. During the oxidation, NAD^+ is reduced to form NADH + H^+.

Yield of the TCA Cycle

The tricarboxylic acid cycle "turns" once for each acetyl-CoA that enters. Because two molecules of acetyl-CoA are generated for each glucose molecule that enters glycolysis, the breakdown of one molecule of glucose causes the cycle to "turn" twice. Assuming no precursors leave the cycle, these two "turns" generate:

- **ATP**—2 ATP; the GTP formed in step 5 can be converted to ATP.
- **Reducing power**—6 NADH + 6 H^+ and 2 $FADH_2$. NAD^+ is reduced to NADH + H^+ during the oxidations in the TCA cycle at steps 3, 4, and 8. FAD is reduced to $FADH_2$ during the oxidation at step 6.
- **Precursor metabolites**—Three precursor metabolites used by *E. coli* are formed as a result of steps 3, 4, and 8.

Oxidative Phosphorylation

Oxidative phosphorylation uses the reducing power generated in glycolysis and the TCA cycle to synthesize ATP. It does this by passing the electrons carried by NADH to the electron transport chain, which uses them to generate the proton motive force that drives the synthesis of ATP. In aerobic respiration, those electrons are ultimately transferred to O_2, which functions as the terminal electron acceptor. In 1961, the British scientist Peter Mitchell originally proposed the **chemiosmotic theory** that describes the remarkable mechanism by which ATP synthesis is linked to electron transport, but his hypothesis was widely dismissed. Only through many years of self-funded research was he finally able to convince others of its validity, being awarded the Nobel Prize in 1978.

The Electron Transport Chain

The membrane-embedded electron transport chain is a series of electron carriers that sequentially pass electrons from one to another (**figure 6.17**). In prokaryotes, this is located in the cytoplasmic membrane, whereas in eukaryotic cells, it is in the inner membrane of the mitochondria. The carriers are arranged in three large proton-pumping complexes, with

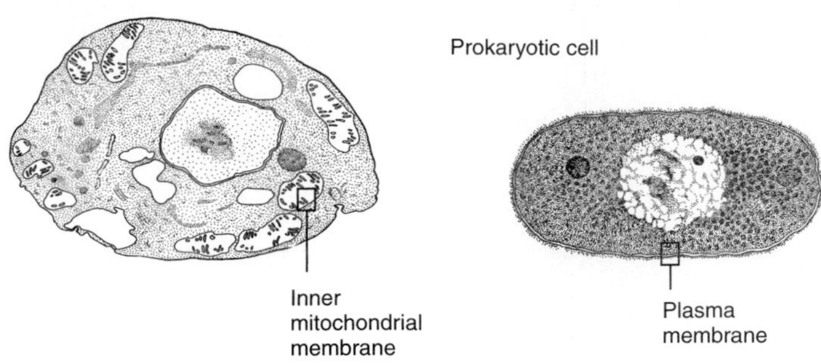

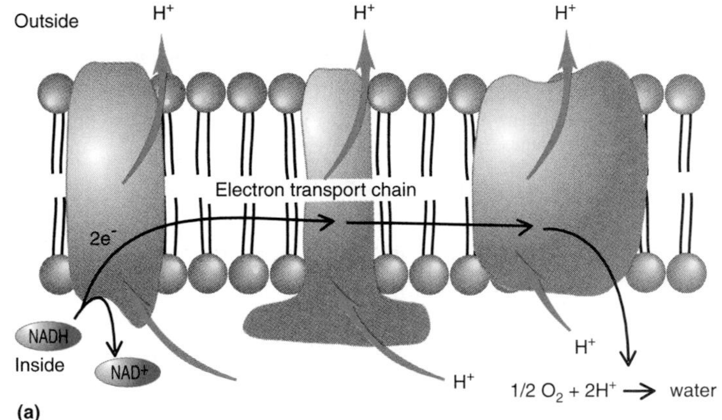

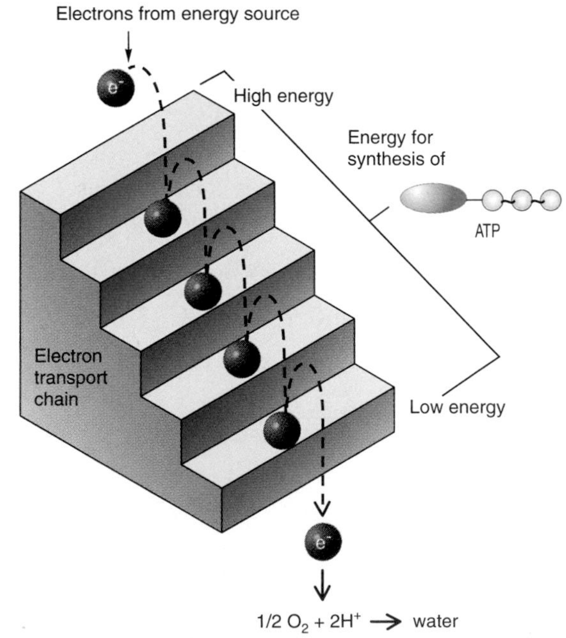

Figure 6.17 The Electron Transport Chain **(a)** The carriers are arranged in three proton-pumping complexes; other carriers transfer electrons between the complexes. The spatial arrangement of the carriers causes protons to be translocated across the membrane. **(b)** Energy is released as electrons are passed along the carriers of the electron transport chain.

other smaller carriers to shuttle electrons between the complexes. ■ mitochondria, p. 79

Electrons enter the electron transport chain when a diffusible electron carrier such as NADH or $FADH_2$ transfers its electrons to a membrane-embedded carrier. The electrons are then sequentially transferred along the electron transport chain. An important characteristic of the carriers is that some accept only hydrogen atoms, or electron-proton pairs, whereas others accept only electrons. The spatial arrangement of these two types of carriers in the membrane causes protons to be shuttled from the inside of the membrane to the outside. This occurs because a hydrogen carrier that receives electrons from an electron carrier must pick up protons; because of the hydrogen carrier's relative location in the membrane, those protons come from the inside. When that hydrogen carrier passes the electrons to a carrier that accepts electrons but not protons, the free protons are released to the outside. The net effect of this process is that a concentration gradient of protons across the membrane is established. This **chemiosmotic gradient**, called the **proton motive force**, can be used to not only fuel the synthesis of ATP but also drive certain transport processes and power the rotation of flagella.

Several other details of the electron transport chain are worth noting. First, the electrons carried by $FADH_2$ and NADH enter the electron transport chain at different sites. The place where the electrons carried by $FADH_2$ enter is "downstream" of where those carried by NADH enter. Because of this, electrons carried by NADH result in more protons being expelled from the cell than a pair carried by $FADH_2$. An additional detail is that prokaryotes are diverse in their types and arrangement of membrane-embedded carriers. Different groups of organisms often have distinct carriers, and a single species may vary its carriers according to growth conditions. The presence of one type of carrier, **cytochrome c**, is used to distinguish among certain types of bacteria. It can be detected using a rapid biochemical test called the **oxidase test**, which is commonly used in the identification scheme of *Pseudomonas* and *Campylobacter* species.

ATP Synthesis

The complex membrane structure **ATP synthase** harvests the energy of proton motive force to synthesize ATP **(figure 6.18)**. It allows protons to pass back into the cell and uses the energy released to fuel the addition of a phosphate group to ADP. The precise mechanism of how this occurs is not well understood.

Theoretical ATP Yield of Oxidative Phosphorylation

The complexity of oxidative phosphorylation makes it exceedingly difficult to determine the actual maximum yield of ATP. Unlike the yield of substrate-level phosphorylation, which can be calculated based on the stoichiometry of relatively simple chemical reactions, oxidative phosphorylation involves processes that have many variables. This is particularly true for prokaryotic cells because they use proton motive force to drive processes other than ATP synthesis, including flagella rotation and membrane transport. In addition, as a group, they employ different carriers in their electron transport chain, and these may vary in the number of protons that are ejected per pair of electrons passed. Another complicating factor in energy yield

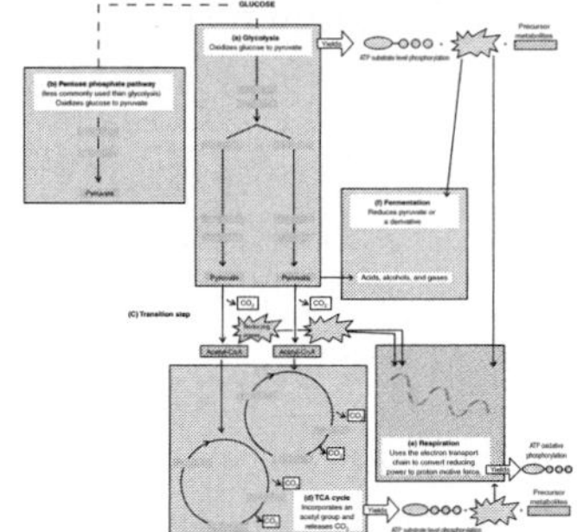

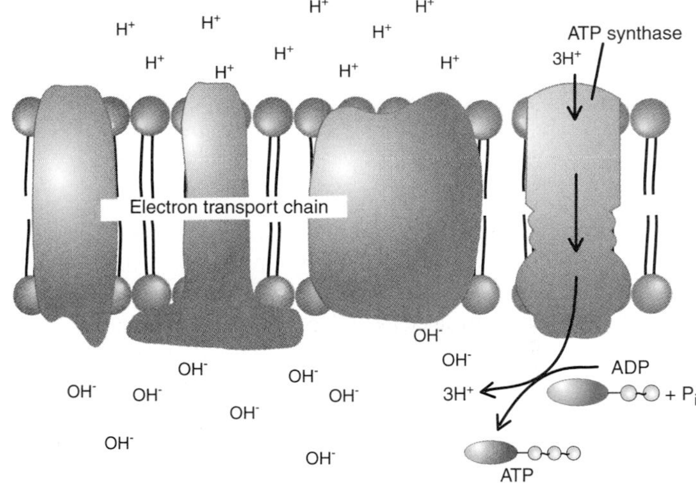

Figure 6.18 ATP Synthesis The membrane structure ATP synthase allows protons to move back across the membrane, driving the synthesis of ATP.

calculations is that the number of ATP generated per reduced electron carrier is not necessarily a whole number.

For each pair of electrons transferred to the electron transport chain by NADH, between 2 and 3 ATP may be generated; for each pair transferred by $FADH_2$, the maximum yield is between 1 and 2 ATP. Although experimental studies using rat mitochondria indicate that the yield is approximately 2.5 ATP/NADH and 1.5 ATP/$FADH_2$, for simplicity we will use whole numbers (3 ATP/NADH and 2 ATP/$FADH_2$) to calculate the maximum ATP gain of oxidative phosphorylation in a prokaryotic cell. Note, however, that these numbers are only theoretical and serve primarily as a means of comparing the relative energy gains of respiration and fermentation.

The ATP gain as a result of oxidative phosphorylation will be at least slightly different in eukaryotic cells than in prokaryotic cells because of the fate of the reducing power (NADH) generated during glycolysis. Recall that in eukaryotic cells, glycolysis takes place in the cytoplasm, whereas the electron transport chain is located in the mitochondria. Consequently, the electrons carried by cytoplasmic NADH must be translocated across the mitochondrial membrane before they can enter the electron transport chain. As a result, between 1 and 2 ATP are produced from each NADH generated during glycolysis.

The maximum theoretical energy yield for aerobic oxidative phosphorylation in prokaryotic cells, is summarized next. This yield does not include the gains of the pentose phosphate cycle; recall that the reducing power of this pathway is primarily used in biosynthesis.

- **From glycolysis**:
  2 NADH ⟶ 6 ATP (assuming 3 for each NADH)
- **From the transition step**:
  2 NADH ⟶ 6 ATP (assuming 3 for each NADH)
- **From the TCA cycle**:
  6 NADH ⟶ 18 ATP (assuming 3 for each NADH)
  2 $FADH_2$ ⟶ 4 ATP (assuming 2 for each $FADH_2$)

ATP Yield of Aerobic Respiration in Prokaryotes

Now that the ATP-yielding components of the central metabolic pathways have been considered, we can calculate the theoretical maximum ATP yield of aerobic respiration in prokaryotes. This yield is illustrated in **figure 6.19**.

- **Substrate-level phosphorylation**:
  2 ATP (from glycolysis; net gain)
  2 ATP (from the TCA cycle)
  4 ATP (total; substrate-level phosphorylation)
- **Oxidative phosphorylation**:
  6 ATP (from the reducing power gained in glycolysis)
  6 ATP (from the reducing power gained in the transition step)
  22 ATP (from the reducing power gained in the TCA cycle)
  34 (total; oxidative phosphorylation)
- **Total ATP gain (theoretical maximum) = 38**

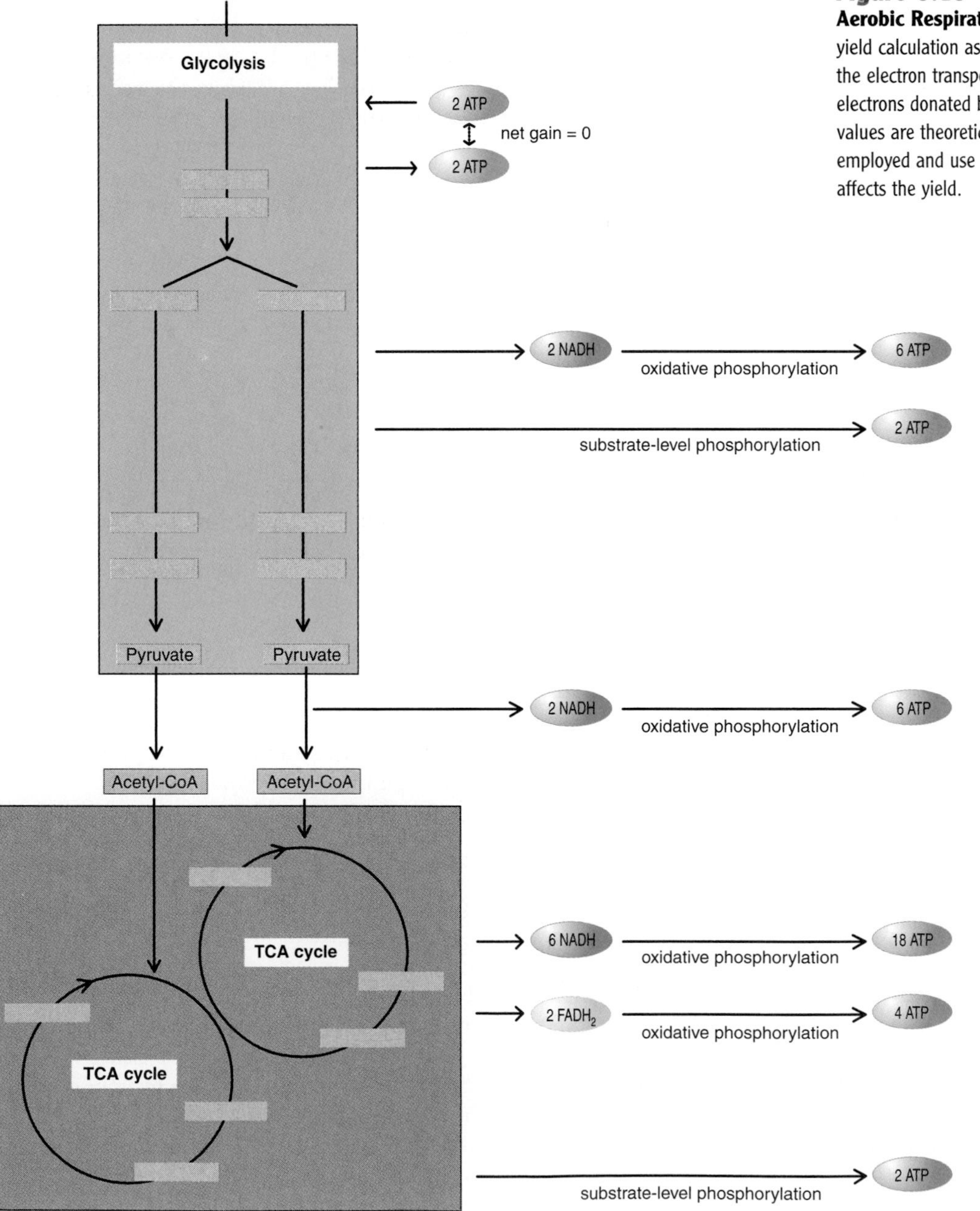

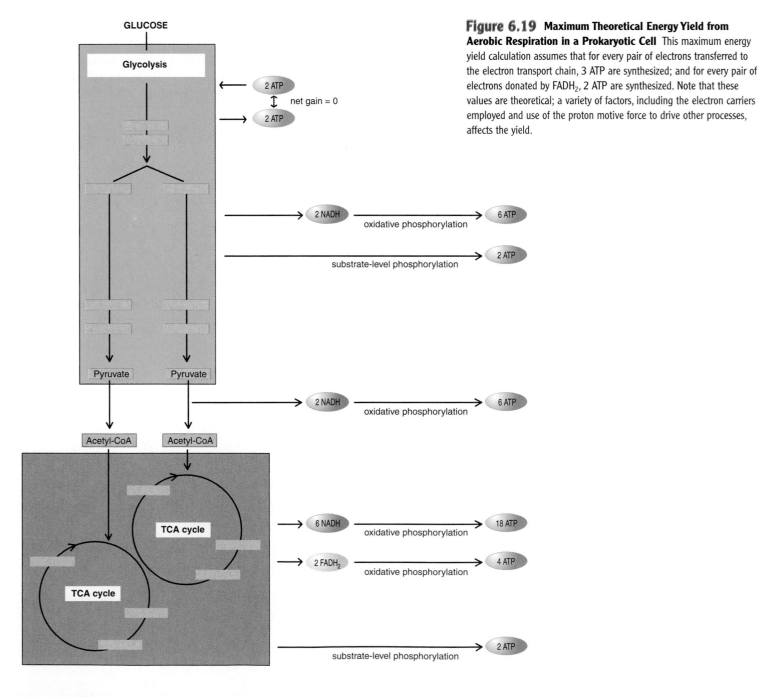

Figure 6.19 Maximum Theoretical Energy Yield from Aerobic Respiration in a Prokaryotic Cell This maximum energy yield calculation assumes that for every pair of electrons transferred to the electron transport chain, 3 ATP are synthesized; and for every pair of electrons donated by $FADH_2$, 2 ATP are synthesized. Note that these values are theoretical; a variety of factors, including the electron carriers employed and use of the proton motive force to drive other processes, affects the yield.

MICROCHECK 6.3

Glycolysis oxidizes glucose to pyruvate, yielding some ATP and NADH and 6 precursor metabolites. The pentose phosphate pathway oxidizes glucose to pyruvate; its greatest significance is its contribution of 2 precursor metabolites and NADPH for biosynthesis. The transition step and the TCA cycle, repeated twice, complete the oxidation of glucose, yielding some ATP, a great deal of reducing power, and 4 precursor metabolites. Oxidative phosphorylation uses the electron transport chain to convert reducing power into proton motive force, and then harvests this to synthesize ATP. In aerobic respiration, O_2 serves as the terminal electron acceptor.

- Explain why the TCA cycle ultimately results in a greater ATP gain than glycolysis.
- How does the electron transport chain generate proton motive force?
- Which compound contains more free energy—pyruvate or oxaloacetate? Why?

Pathways and Processes that Fuel Anaerobic Growth of Chemoorganotrophs

Facultative anaerobes, aerotolerant anaerobes, and obligate anaerobes use pathways that do not require oxygen (O_2). Recall from chapter 4 that facultative anaerobes grow best when using molecular oxygen as a terminal electron acceptor, but also can grow in its absence. In contrast, aerotolerant anaerobes are indifferent to molecular oxygen; they can grow in its presence, but lack the ability to use it as a terminal electron acceptor. Obligate anaerobes are actually killed in the presence of O_2 because of toxic derivatives such as superoxide and hydrogen peroxide. ■ **oxygen requirements, p. 97**

Some of the same pathways used to generate energy during aerobic growth are also used to fuel anaerobic growth. Just as glycolysis and the pentose phosphate pathway are used by nearly all organisms growing in the presence of oxygen (O_2), they are also used in its absence. Recall that neither of these pathways uses oxygen. What distinguishes the two types of anaerobic metabolism—anaerobic respiration and fermentation—is the fate of the pyruvate. Anaerobic respiration oxidizes pyruvate to CO_2, extracting even more energy, whereas fermentation uses pyruvate or a derivative as a terminal electron acceptor.

Anaerobic Respiration

Unlike eukaryotes, some prokaryotes can respire using an inorganic molecule other than oxygen (O_2) as a terminal electron acceptor **(figure 6.20)**. Some facultative anaerobes such as *E. coli* can transfer the electrons to nitrate (NO_3^-), reducing it to form nitrite (NO_2^-). Other facultative organisms can reduce nitrate even further, forming compounds such as nitrous oxide (N_2O), and nitrogen gas (N_2). A group of obligate anaerobes called the **sulfate-reducers** use sulfate (SO_4^{2-}) as a terminal electron acceptor, producing hydrogen sulfide as an end product. The diversity and ecology of many of these organisms will be discussed in chapter 11. ■ **sulfate-reducers, p. 272**

Anaerobic respiration is a less efficient form of energy transformation than is aerobic respiration. This is partly due to the lesser amount of energy released in reactions that involve the reduction of inorganic compounds other than molecular oxygen. Because of this, alternative electron carriers are used in the electron transport chain.

Fermentation

Fermentation is used by organisms that cannot respire, either because a suitable inorganic terminal electron acceptor is not available or because they lack an electron transport chain. *Escherichia coli* is a facultative anaerobe that has the ability to use any of three ATP-generating options—aerobic respiration, anaerobic respiration, and fermentation; the choice depends in part on the availability of terminal electron acceptors. In contrast, members of a group of aerotolerant anaerobes called the **lactic acid bacteria** lack the ability to respire; they only ferment, regardless of the presence of oxygen (O_2). Because they can grow in the presence of oxygen but never use it as a terminal electron acceptor, they are sometimes called **obligate fermenters**. The situation is entirely different for obligate anaerobes that use fermentation pathways; they cannot even grow in the presence of O_2, and many are rapidly killed in its presence.

In general, the only ATP-yielding reactions of fermentation are those of glycolysis, and involve substrate-level phosphorylation. The other steps function primarily to consume reducing power, providing a mechanism for recycling NADH **(figure 6.21)**. If this reduced carrier were not recycled, no NAD^+ would be available to accept electrons in subsequent rounds of glycolysis, blocking that ATP-generating pathway. To consume reducing power, fermentation pathways use an organic molecule as a terminal electron acceptor (figure 6.21). In some cases, pyruvate

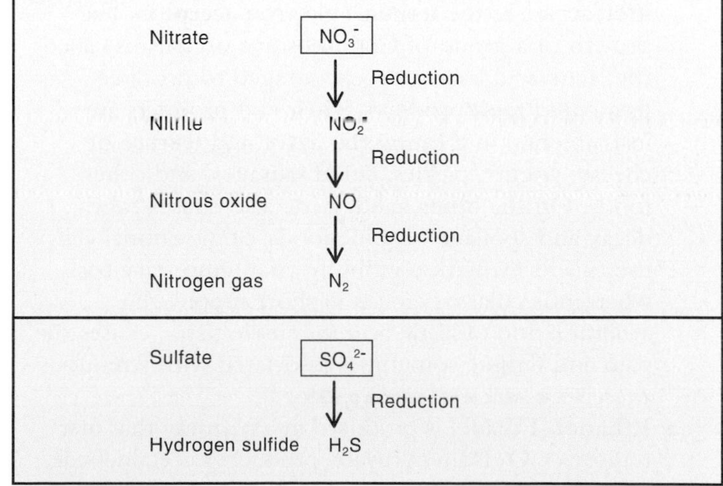

Figure 6.20 Anaerobic Respiration Anaerobic respiration employs an inorganic molecule other than oxygen as a terminal electron acceptor.

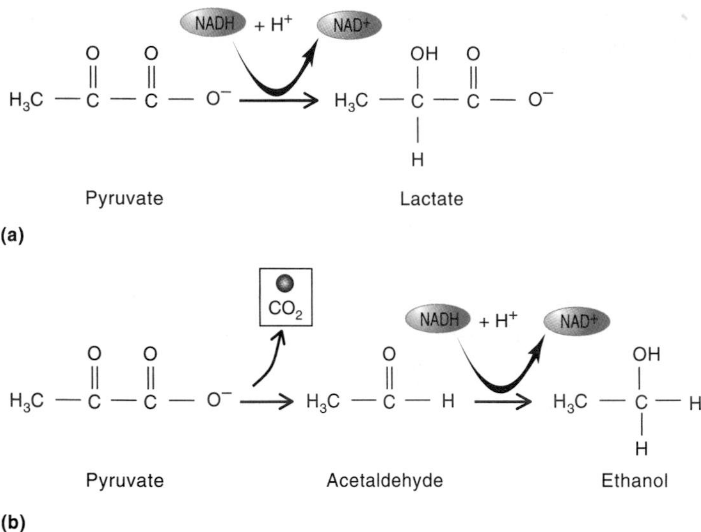

Figure 6.21 Fermentation Pathways Use Pyruvate or a Derivative As a Terminal Electron Acceptor **(a)** In lactic acid fermentation, pyruvate serves directly as a terminal electron acceptor, producing lactate. **(b)** In ethanol fermentation, pyruvate is first converted to acetaldehyde, which then serves as the terminal electron acceptor, producing ethanol.

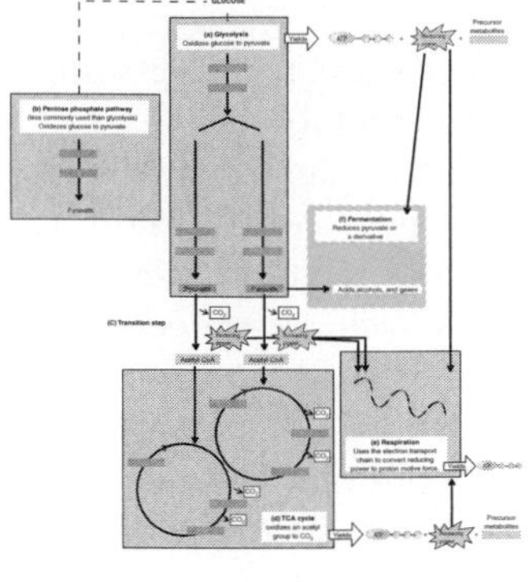

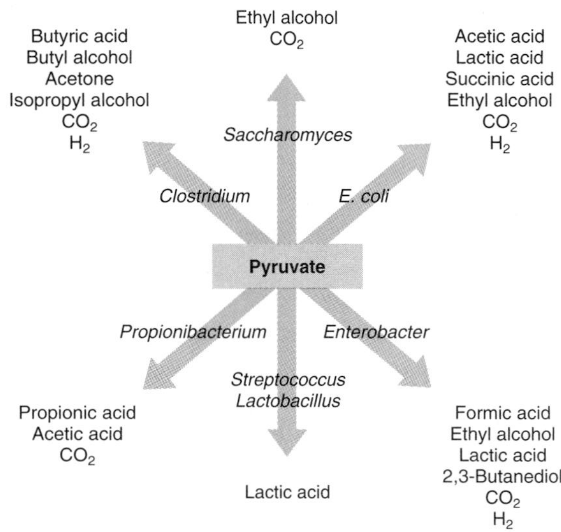

Figure 6.22 End Products of Fermentation Pathways Because a given type of organism uses a characteristic fermentation pathway, the end products can be used as an identifying marker. Some end products are commercially valuable.

serves directly as the electron acceptor; in others, it is first converted to another compound before being reduced.

The end products of fermentation are significant for a number of reasons. Because a given type of organism uses a characteristic fermentation pathway, the end products can sometimes be used as a marker to aid in identification. Some end products are commercially valuable. In fact, much of chapter 32 is devoted to the fermentations that produce beverages and food products. Note that the organic acids produced during fermentation are traditionally referred to by the name of their undissociated form. For example, lactate, which is produced by reducing pyruvate, is referred to as lactic acid. Important end products of fermentation pathways include **(figure 6.22)**:

- **Lactic acid**. Lactic acid is produced when pyruvate itself serves as the terminal electron acceptor. The growth of a group of Gram-positive organisms called the lactic acid bacteria is encouraged to produce fermented food products. Their end products are instrumental in creating the flavor and texture of cheese, yogurt, pickles, cured sausages, and other foods. On the other hand, lactic acid causes tooth decay and spoilage of some foods. Some animal cells use this fermentation pathway on a temporary basis when molecular oxygen is in short supply; the accumulation of lactic acid in muscle tissue causes the pain and fatigue sometimes associated with strenuous exercise. ■ **lactic acid bacteria, p. 275**
- **Ethanol**. Ethanol is produced in a pathway that first removes CO_2 from pyruvate, producing acetaldehyde, which then serves as the terminal electron acceptor. The end products of these sequential reactions are ethanol and CO_2, which are used to make wine, beer, spirits, and bread. Members of *Saccharomyces* (yeast) and *Zymomonas* (bacteria) use this pathway.
- **Butyric acid**. Butyric acid (the ionized form is butyrate) and a variety of other end products are produced in a complex multistep pathway used by species of *Clostridium*, an obligate anaerobe. Under certain conditions, some species use a variation of this pathway to produce the organic solvents butanol and acetone.
- **Propionic acid**. Propionic acid (the ionized form is propionate) is generated in a multistep pathway that first removes CO_2 from pyruvate, generating a compound that then serves as a terminal electron acceptor. After NADH reduces this, it is further modified to form propionate. Members of the genus *Propionibacterium* use this pathway; their growth is encouraged in the production of Swiss cheese. The CO_2 they form makes the holes, and the propionic acid gives the cheese its characteristic flavor.

- **2, 3-Butanediol**. This is produced in a multistep pathway that uses two molecules of pyruvate to generate two molecules of CO_2 and acetoin. The latter then serves as the terminal electron acceptor. The primary significance of this pathway is that it differentiates certain members of the family *Enterobacteriaceae*. The **Voges-Proskauer test** is used to detect acetoin, distinguishing bacteria that use this pathway, such as *Klebsiella* and *Enterobacter*, from those that do not, such as *E. coli*. ■ Voges-Proskauer test, p. 253
- **Mixed acids**. These are produced when several branches are used, generating a variety of different fermentation end products including lactic acid, succinic acid (the ionized form is succinate), ethanol, acetic acid (the ionized form is acetate), CO_2, and gases. The primary significance of this pathway is that it differentiates certain members of the family *Enterobacteriaceae*. The **methyl-red test** is used to detect the end products of the pathway, distinguishing bacteria that use this pathway, such as *E. coli*, from those that do not, such as *Klebsiella* and *Enterobacter*. ■ methyl-red test, p. 253

MICROCHECK 6.4

Glycolysis and the pentose phosphate pathway do not require molecular oxygen and are used to fuel anaerobic growth. Some prokaryotes can respire using an inorganic molecule other than molecular oxygen as a terminal electron acceptor. Fermentation stops short of the TCA cycle, using pyruvate or a derivative of it as a terminal electron acceptor. Many end products of fermentation are commercially valuable.

- What is the end product when sulfate is used as a terminal electron acceptor?
- Compare and contrast the fermentation pathways that generate lactic acid and propionic acid.
- Fermentation is used as a means of preserving foods. Why would it slow spoilage?

Catabolism of Organic Compounds Other than Glucose

Cells can use a variety of organic compounds other than glucose as energy sources, including macromolecules such as polysaccharides, lipids, and proteins. To break these down into their respective sugar, amino acid, and lipid subunits, cells synthesize **hydrolytic enzymes**, which break bonds by adding water. To use a macromolecule in the surrounding medium, a cell must secrete the appropriate hydrolytic enzyme and then transport the resulting subunits into the cell. Inside the cell, the subunits are further degraded to form the appropriate precursor metabolites (**figure 6.23**). Recall that precursor metabolites can be either oxidized in one of the central metabolic pathways or used in biosynthesis. ■ hydrolysis, p. 26

Polysaccharides and Disaccharides

Starch and cellulose are both polymers of glucose, but different types of chemical bonds join their subunits. The nature of this difference profoundly affects the mechanisms by which they are degraded. Enzymes called **amylases** are produced by a wide variety of organisms to digest starches. In contrast, cellulose is digested by enzymes called **cellulases**, which are produced by relatively few organisms. Among the organisms that can degrade cellulose are bacteria that reside in the rumen of animals, and many types of fungi. Considering that cellulose is the most abundant organic compound on earth, it is not surprising that fungi are important decomposers in terrestrial habitats. The glucose subunits released when polysaccharides are hydrolyzed can then enter glycolysis to be oxidized to pyruvate. ■ polysaccharides, p. 34 ■ cellulose, p. 34

Disaccharides including lactose, maltose, and sucrose are hydrolyzed by specific **disaccharidases**. For example, the enzyme β-galactosidase breaks down lactose, forming glucose and galactose. Glucose can enter glycolysis directly, but the other monosaccharides must first be modified. ■ disaccharides, p. 34

Lipids

The most common simple lipids are fats, which are a combination of fatty acids joined to glycerol. Fats are hydrolyzed by enzymes called **lipases**. The glycerol component is then converted to the precursor metabolite glyceraldehyde 3-phosphate; this enters the glycolytic pathway. The fatty acids are degraded using a series of reactions collectively called **β-oxidation**. Each sequential reaction transfers a 2-carbon unit from the end of the fatty acid to coenzyme A, forming acetyl-CoA; this can enter the TCA cycle. Each reaction is an oxidation, generating one $NADH + H^+$ and one $FADH_2$. ■ simple lipids, p. 37

Proteins

Proteins are hydrolyzed by enzymes called **proteases**, which break the peptide bonds that join the amino acid subunits. The amino group of the resulting amino acids is removed by a reaction called a **deamination**. The remaining carbon skeletons are then converted into the appropriate precursor molecules. ■ protein, p. 28

MICROCHECK 6.5

In order for polysaccharides, lipids, and proteins to be used as energy sources, they are first hydrolyzed to release their respective subunits. These are then converted to the appropriate precursor metabolites so they can enter a central metabolic pathway.

- Why do cells secrete hydrolytic enzymes?
- Explain the process used to degrade fatty acids.
- How would cellulose-degrading bacteria in the rumen of a cow benefit the animal?

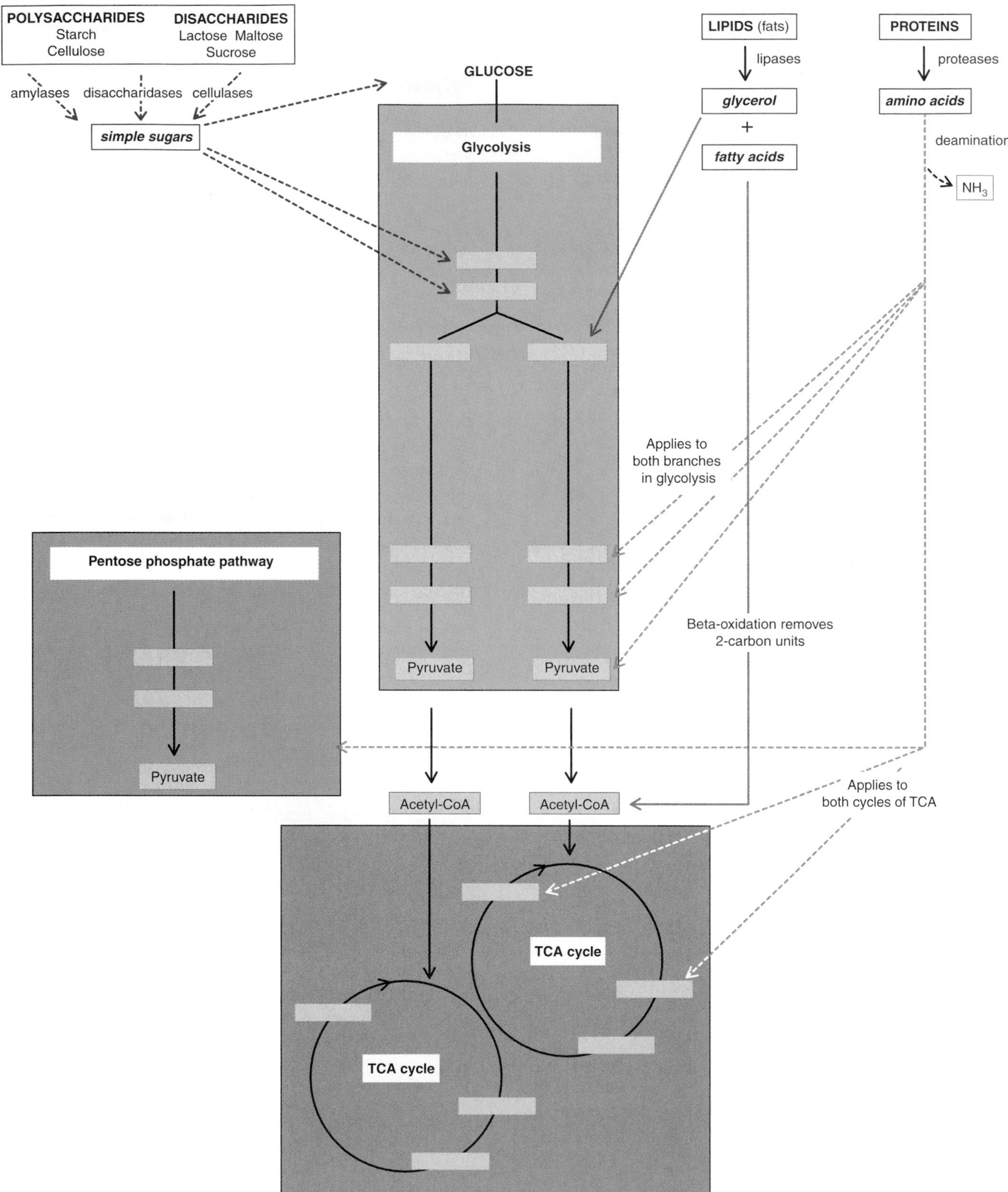

Figure 6.23 Catabolism of Organic Compounds Other than Glucose The subunits of macromolecules are degraded to form the appropriate precursor metabolites. These metabolites can then either be oxidized in one of the central metabolic pathways or be used in anabolism.

Perspective 6.1 Mining with Microbes

Microorganisms have been used for thousands of years in the production of bread and wine. It is only in the past several decades, however, that they are being used with increasing frequency in another area of biotechnology, the mining industry. Although mining is a centuries-old industry, modern technologies have not yet been employed in mining as they have in medicine and agriculture. The mining process consists of digging crude ores from the earth, crushing them, and then extracting the desired minerals from the contaminants. The extraction process of such minerals as copper and gold frequently involves harsh conditions, such as smelting, and burning off the contaminants before extracting the metal, such as gold, with cyanide. Such activities are expensive and deleterious to the environment. With the development of biomining, some of these problems are being solved.

In the process of biomining copper, the low grade ore is dumped outside the mine and then treated with sulfuric acid. The acid conditions encourage the growth of the acidophilic bacterium *Thiobacillus ferrooxidans*, which occurs naturally in the ore. This organism uses CO_2 as a source of carbon and gains its energy by oxidizing sulfides of iron first to sulfur and then to sulfuric acid. The sulfuric acid dissolves the insoluble copper and gold from the ore. Currently, about 25% of all copper produced in the world comes from the process of biomining. Similar processes are being applied to gold mining.

The current process of biomining employs only the bacteria that are indigenous to the ore. Many improvements should be possible. For example, the oxidation of the minerals generates heat to the point that the bacteria may be killed. The use of thermophiles should overcome this problem. Further, many ores contain heavy metals, such as mercury, cadmium, and arsenic, which are toxic to the bacteria. It should be possible to isolate bacteria that are resistant to these metals. Biomining is still in its infancy.

Chemolithotrophs

Prokaryotes as a group are unique in their ability to use reduced inorganic compounds such as hydrogen sulfide (H_2S) and ammonia (NH_3) as a source of energy. Note that these are the very compounds produced as a result of anaerobic respiration, when inorganic molecules such as sulfate and nitrate serve as terminal electron acceptors. This is one important example of how nutrients are cycled; the waste products of one organism serve as an energy source for another.

Unlike organisms that use organic molecules to fill both their energy and carbon needs, chemolithotrophs do not require an external source of organic carbon. Instead, they incorporate inorganic carbon, CO_2, into an organic form. This process, called **carbon fixation**, will be described later. Organisms that can fix CO_2 are called **autotrophs**. Because chemolithotrophs can fix carbon, they are most accurately described as **chemolithoautotrophs**. This term is rarely used, however, and they are generally called either chemolithotrophs or **chemoautotrophs**.

Chemolithotrophs generally thrive in very specific environments where reduced inorganic compounds are found. For example, *Thiobacillus ferrooxidans* is found in certain acidic environments that are rich in sulfides. Because these organisms oxidize metal sulfides, they can be used to enhance the recovery of metals (see **Perspective 6.1**). Thermophilic chemolithotrophs thrive near the hydrothermal vents of the deep ocean, harvesting the energy of reduced inorganic compounds that spew from the vents. The characteristics of some chemolithotrophs are summarized in **table 6.8**; their diversity and ecology will be discussed in chapter 11.

MICROCHECK 6.6

Chemolithotrophs use reduced inorganic compounds as an energy source. They use carbon dioxide as a carbon source.

- Describe the roles of hydrogen sulfide and carbon dioxide in chemolithoautotrophic metabolism.
- Why can the term chemolithotroph be used interchangeably with chemoautotroph when they mean very different things?

Table 6.8 Metabolism of Chemolithoautotrophs

Common Name of Organism	Source of Energy	Oxidation Reaction (Energy Yielding)	Important Features of Group	Common Genera in Group
Hydrogen bacteria	H_2 gas	$H_2 + \frac{1}{2} O_2 \rightarrow H_2O$	Can also use simple organic compounds for energy	*Hydrogenomonas*
Sulfur bacteria (nonphotosynthetic)	H_2S	$H_2S + \frac{1}{2} O_2 \rightarrow H_2O + S$ $S + 1\frac{1}{2} O_2 + H_2O \rightarrow H_2SO_4$	Some organisms of this group can live at a pH of less than 1	*Thiobacillus* *Beggiatoa* *Thiothrix*
Iron bacteria	Reduced iron (Fe^{2+})	$2\ Fe^{2+} + \frac{1}{2} O_2 + H_2O \rightarrow 2\ Fe^{3+} + 2\ OH^-$	Iron oxide present in the sheaths of these bacteria	*Sphaerotilus* *Gallionella*
Nitrifying bacteria	NH_3	$NH_3 + 1\frac{1}{2} O_2 \rightarrow HNO_2 + H_2O$	Important in nitrogen cycle	*Nitrosomonas*
	HNO_2	$HNO_2 + 1\frac{1}{2} O_2 \rightarrow HNO_3$	Important in nitrogen cycle	*Nitrobacter*

Phototrophs

Plants, algae, and several distinct groups of bacteria are **phototrophs**; they harvest the energy of sunlight and use it to power the synthesis of ATP. The conversion of solar energy to chemical energy is called **photosynthesis**. The general reaction for photosynthesis can be summarized as follows (X depends on the source of electrons for reducing power and can be a number of different compounds):

$$6\,CO_2 + 12\,H_2X \xrightarrow{\text{Light energy}} C_6H_{12}O_6 + 12\,X + 6\,H_2O$$

Photosynthetic processes are generally considered in two distinct stages. The **light-dependent reactions**, often simply called the **light reactions**, are those used to derive energy from sunlight. The **light-independent reactions**, also termed the **dark reactions**, then use that energy to fix CO_2. The mechanism used to fix CO_2 will be discussed later.

The Role of Photosynthetic Pigments

Photosynthetic organisms are highly visible in their natural habitats because they possess **chlorophylls** and **carotenoids**, colored pigments used to capture light energy. These pigments vary in color because they absorb different wavelengths of light.

The primary light-absorbing pigments are the chlorophylls. Most photosynthetic organisms, including an important group of bacteria called cyanobacteria, use **chlorophyll *a***. A different group of photosynthetic bacteria, the purple and green bacteria, use **bacteriochlorophyll**. This absorbs wavelengths not absorbed by chlorophyll *a*, enabling the purple and green bacteria to grow in habitats where other photosynthetic organisms cannot. The carotenoids are accessory pigments that increase the efficiency of light utilization. They do this by absorbing wavelengths of light not absorbed by the chlorophylls and then transferring that energy to the chlorophylls. ■ **cyanobacteria, p. 278** ■ **purple bacteria, p. 276** ■ **green bacteria, p. 277**

Chlorophyll and other light-gathering pigments are organized in protein complexes called **photosystems** located in special photosynthetic membranes. In cyanobacteria, these are found in stacked membranes called **thylakoids**. Plants and algae also have thylakoids, in the stroma of their chloroplasts (see figure 3.58). The photosynthetic pigments in purple and green bacteria are either contained within the cytoplasmic membrane or within structures called **chlorosomes** attached to the membrane.

Photophosphorylation

Within the photosystem are two components that work together to harvest energy **(figure 6.24)**. The **antenna complex** is composed of hundreds of light-gathering pigments. This complex acts as a funnel, capturing the energy of light and then transferring it to a **reaction-center chlorophyll**. When this chlorophyll absorbs that energy, one of its electrons becomes excited, or raised to a higher energy level. In a manner similar to oxidative phosphorylation, the high-energy electron is sequentially passed along a series of carriers in a membrane-embedded electron transport chain. This results in the translocation of protons across the membrane, generating the proton motive force. ATP synthase permits the flow of protons back across the membrane, and uses that energy to generate ATP.

The high-energy electrons from reaction-center chlorophyll that are passed along the electron transport chain may or may not return to their original source. In **cyclic photophosphorylation**, electrons are returned directly to the chlorophyll, being used only to generate the proton motive force. In **noncyclic photophosphorylation**, proton motive force is produced but, in addition, high-energy electrons are drawn off to generate reducing power in the form of NADPH. Electrons must still be returned to chlorophyll, but they must come from a different source, such as water or hydrogen sulfide.

Electron Source

Photosynthetic organisms require a source of electrons to generate reducing power in the form of NADPH. The compound used to obtain those electrons dictates whether or not oxygen (O_2) is evolved during photosynthesis.

Oxygenic Photosynthesis

When plants, algae, and cyanobacteria use noncyclic photophosphorylation to make reducing power, they use water as the source of the electrons that are returned to chlorophyll. When electrons are extracted from water, protons and oxygen (O_2) are released. It is becaue of oxygenic photosynthesis that obligate aerobes, including humans and other animals, are able to inhabit the earth.

Anoxygenic Photosynthesis

Anoxygenic phototrophs, such as the green and purple bacteria, lack the sophisticated photosystem that the oxygenic photosynthetic organisms use to generate NADPH. Instead, they must use alternative mechanisms such as reversing the flow of their electron transport chain. The electrons come from compounds such as hydrogen sulfide, hydrogen gas, and reduced organic molecules. Because water is not used as an electron source, anoxygenic phototrophs do not evolve oxygen (O_2).

MICROCHECK 6.7

Phototrophs harvest the energy of sunlight and use it to drive the synthesis of ATP. Chlorophylls and carotenoids harvest solar energy. When reaction-center chlorophyll absorbs the energy of light, one of its electrons becomes excited. This is then passed along an electron transport chain to generate proton motive force, which is harvested to synthesize ATP. Oxygenic phototrophs use water as a source of electrons when they generate reducing power. Anoxygenic phototrophs obtain electrons from a reduced compound other than water.

- How does bacteriochlorophyll differ from chlorophyll *a*?
- What is the role of the antenna complex?
- It requires energy to reverse the flow of the electron transport chain. Why would this be so?

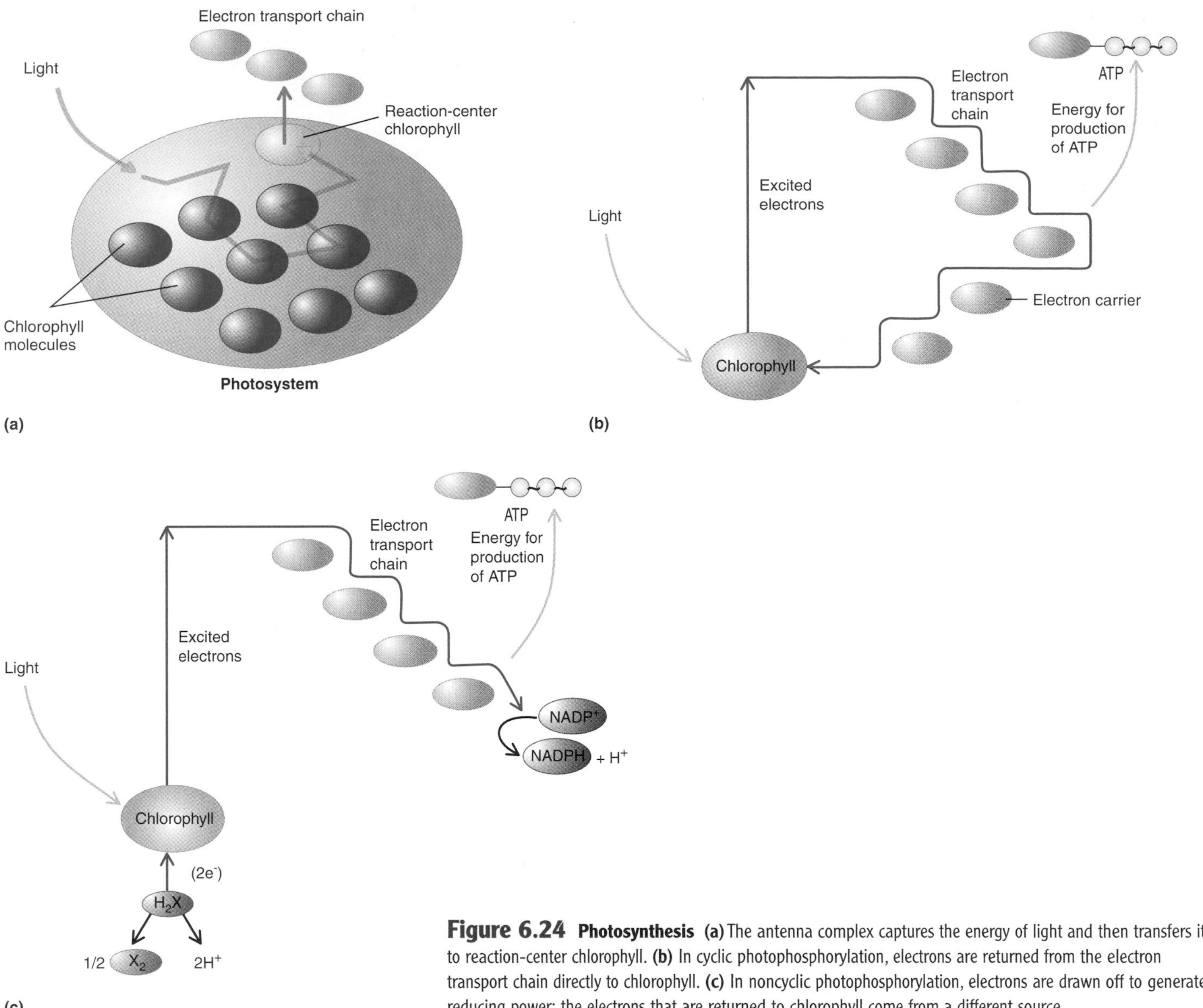

Figure 6.24 Photosynthesis **(a)** The antenna complex captures the energy of light and then transfers it to reaction-center chlorophyll. **(b)** In cyclic photophosphorylation, electrons are returned from the electron transport chain directly to chlorophyll. **(c)** In noncyclic photophosphorylation, electrons are drawn off to generate reducing power; the electrons that are returned to chlorophyll come from a different source.

Carbon Fixation

Chemolithoautotrophs and photoautotrophs convert carbon dioxide into an organic form, the process of **CO_2 fixation** or **carbon fixation**. In photosynthetic organisms, the process occurs in the light-independent reactions. CO_2 fixation consumes a great deal of ATP and reducing power, which should not be surprising considering that the reverse process—oxidizing those same compounds to CO_2—liberates a great deal of energy. The Calvin cycle is by far the most common pathway used to fix carbon, but some prokaryotes incorporate CO_2 using other mechanisms. For example, the green sulfur bacteria and some members of the Archaea use a pathway that effectively reverses the steps of the TCA cycle.

Calvin Cycle

The **Calvin cycle**, or Calvin-Benson cycle, named in honor of the scientists who described much of it, is a complex cycle that can be viewed as having three essential stages—fixation of CO_2 into an organic form, reduction of the resulting molecule, and regeneration of the starting compound **(figure 6.25)**. Because of the complexities of the cycle, it is easiest to consider the process as consisting of six turns of the cycle. Together, these six "turns" generate a net gain of two molecules of glyceraldehyde 3-phosphate, which can be converted into one molecule of fructose 6-phosphate. The Calvin cycle consists of three stages:

- **Stage 1**: Carbon dioxide enters the cycle when the enzyme ribulose bisphosphate carboxylase, commonly called **rubisco** joins it to a 5-carbon compound,

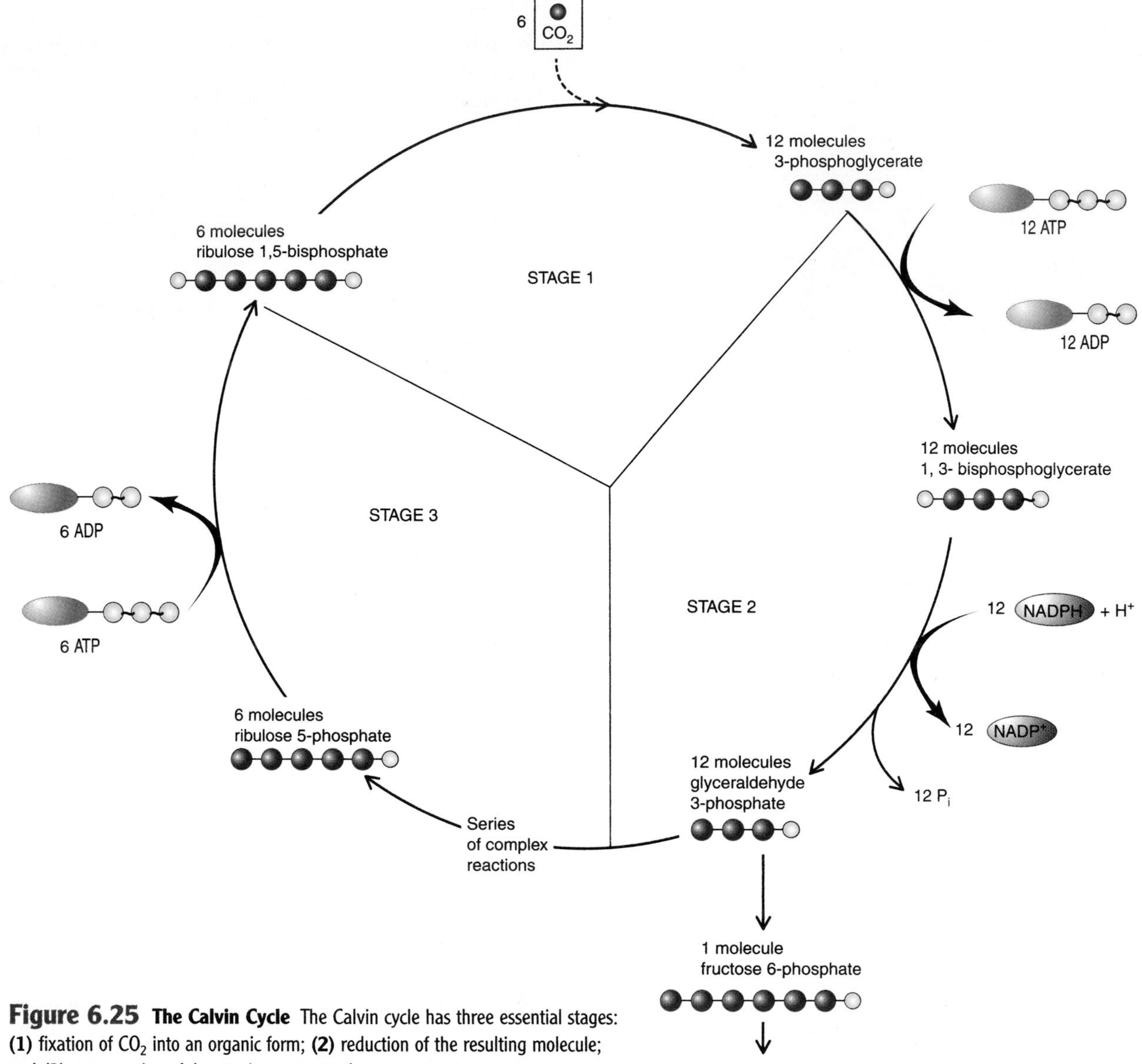

Figure 6.25 **The Calvin Cycle** The Calvin cycle has three essential stages: **(1)** fixation of CO_2 into an organic form; **(2)** reduction of the resulting molecule; and **(3)** regeneration of the starting compound.

ribulose 1, 5-bisphosphate. The resulting compound spontaneously hydrolyzes to produce two molecules of a 3-carbon compound, 3-phosphoglycerate (3PG). Interestingly, although rubisco is unique to autotrophs, it is thought to be the most abundant enzyme on earth.

- **Stage 2**: A sequential input of energy (ATP) and reducing power (NADPH) is used in steps that, together, convert 3PG to glyceraldehyde 3-phosphate (G3P). This compound is identical to the precursor metabolite formed as an intermediate in glycolysis. It can be converted to a number of different compounds used in biosynthesis, oxidized to make other precursor compounds, or converted to a 6-carbon sugar. A critical aspect of the pathway of CO_2 fixation, however, stems from the fact that it operates as a cycle—ribulose 1, 5-bisphosphate must be regenerated from G3P for the process to continue. Consequently, in six cycles, a maximum of 2 G3P can be converted to a 6-carbon sugar; the rest is used to regenerate ribulose 1, 5-bisphosphate.
- **Stage 3**: Many of the steps that are used to regenerate ribulose 1, 5-bisphosphate involve reactions of the pentose phosphate cycle.

Yield of the Calvin Cycle

One molecule of the 6-carbon sugar fructose can be generated for every six "turns" of the cycle. These six "turns" consume 18 ATP and 12 NADPH + H^+.

MICROCHECK 6.8

The process of carbon dioxide fixation consumes a great deal of ATP and reducing power. The Calvin cycle is the most common pathway used to incorporate inorganic carbon into an organic form.

- What is the role of rubisco?
- What would happen if ribulose 1, 5-bisphosphate were depleted in a cell?

Anabolic Pathways—Synthesizing Subunits from Precursor Molecules

While prokaryotes as a group are highly diverse with respect to the compounds they use for energy, they are remarkably similar when comparing their biosynthetic processes. Using precursor metabolites, reducing power in the form of NADPH, and ATP, cells synthesize the necessary subunits using specific anabolic pathways. Organisms that lack one or more enzymes in a given pathway must have the end product of that pathway provided from an external source. This is why fastidious bacteria such as the lactic acid bacteria require many different growth factors. Once the subunits are either synthesized or supplied, they can be assembled to make macromolecules. Various different macromolecules can then be joined to form the structures that make up the cell. ■ fastidious, p. 100

Lipid Synthesis

Synthesis of most lipids in microorganisms can be viewed as having two essential components—fatty acid synthesis and glycerol synthesis. Synthesis starts with the transfer of the acetyl group of acetyl-CoA to a carrier protein called **acyl carrier protein (ACP)**. This carrier serves to hold the fatty acid chain as it is elongated by progressively adding 2-carbon units. When the newly synthesized fatty acid reaches its required length, usually 14, 16, or 18 carbons long, it is released from ACP. The glycerol component of the fat is synthesized from glyceraldehyde 3-phosphate.

Amino Acid Synthesis

Proteins are composed of various combinations of 20 different amino acids. Amino acids can be grouped into structurally related families that share common pathways of biosynthesis. Some are synthesized from precursor metabolites formed during glycolysis, while others are derived from compounds of the TCA cycle (see table 6.3).

Branched pathways are used to synthesize some families of amino acids. These are controlled at key points by allosteric enzymes, regulating the flow of certain branches as well as the common initial steps of the pathway. The amino acids that are the end products of the various branches serve as feedback inhibitors.

Glutamate

Although all amino acids are necessary for protein synthesis, glutamate is especially important because it is used to form many other amino acids. In addition, its synthesis provides a mechanism for bacteria to incorporate nitrogen into an organic material. Recall from chapter 4 that many bacteria utilize ammonium (NH_4^+) salts provided in the medium as their source of nitrogen; it is primarily through the synthesis of glutamate that they do this.

Bacteria that synthesize glutamate use a single-step reaction that incorporates ammonia into the precursor metabolite α-ketoglutarate, produced in the TCA cycle **(figure 6.26a)**. Once glutamate has been produced, its amino group can be transferred to other carbon compounds to produce amino acids such as aspartate (figure 6.26b). This transfer of the amino group, a **transamination**, regenerates α-ketoglutarate from glutamate. The α-ketoglutarate can then be used again to incorporate more ammonia. ■ amino group, p. 27

Aromatic Amino Acids

The synthesis of aromatic amino acids such as tyrosine, phenylalanine, and tryptophan requires a multistep branching pathway **(figure 6.27)**. This serves as an excellent illustration of many important features of the regulation of amino acid synthesis.

The pathway begins with the formation of a 7-carbon compound, resulting from the joining of two precursor metabolites, erythrose 4-phosphate (4-carbon) and phosphoenolpyruvate (3-carbon). These precursors originate in the pentose phosphate pathway and glycolysis, respectively. The 7-carbon compound is converted through a series of steps until a branch point is reached. At this juncture, two options are possible. If synthesis proceeds in one direction, tryptophan is produced. In the other direction, another branch point is reached; from there, either tyrosine or phenylalanine can be made.

When a given amino acid is provided to a cell, it would be a waste of carbon, energy, and reducing power for that cell to continue synthesizing it. But when only one product of a branched pathway is present, how does the cell control synthesis? In the pathway for aromatic acid biosynthesis, this partly occurs by regulating the enzymes at the branch points. Tryptophan acts as a feedback inhibitor of the enzyme that directs the

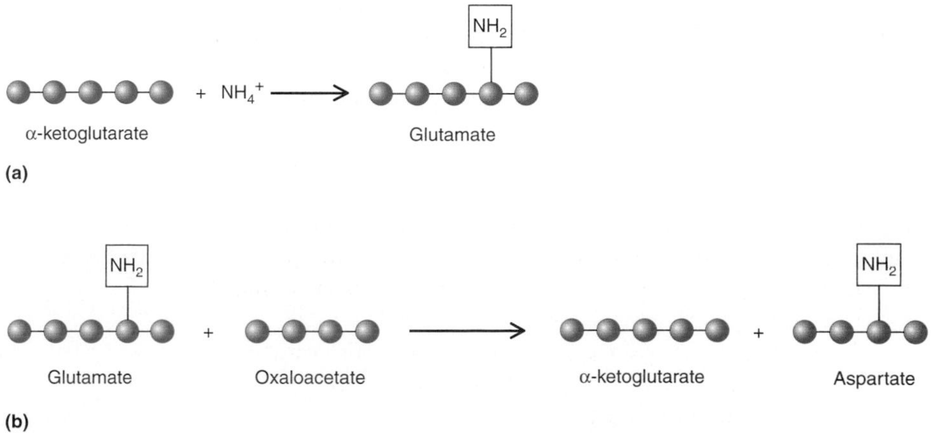

Figure 6.26 Glutamate (a) Glutamate is synthesized in a single-step reaction that incorporates ammonia into the precursor metabolite α-ketoglutarate. **(b)** The amino group of glutamate can be transferred to other carbon compounds in order to produce other amino acids. For example, transferring it to oxaloacetate produces aspartate.

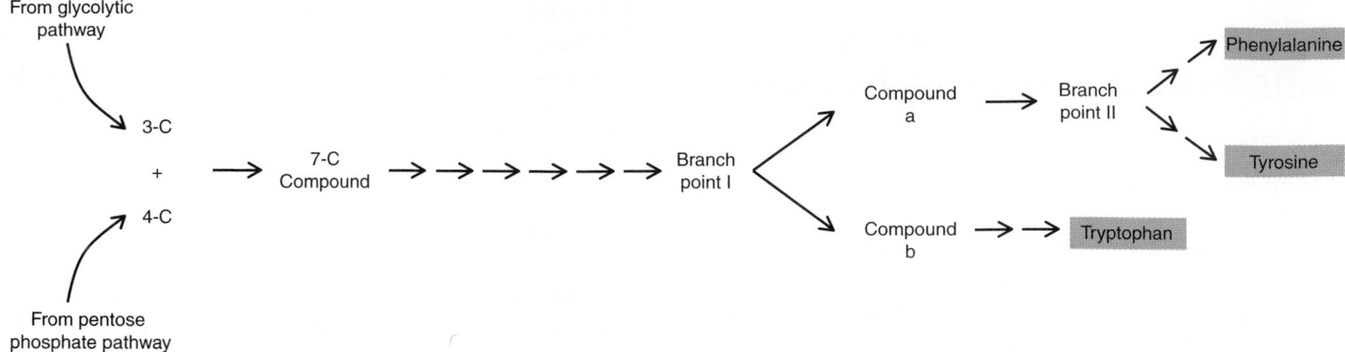

Figure 6.27 **Synthesis of Aromatic Amino Acids** A multistep branching pathway is used to synthesize aromatic amino acids. The end product of a branch inhibits the first enzyme of that branch; in addition, the end product inhibits one of the three enzymes that catalyze the first step of the pathway.

branch to its synthesis; this sends the pathway to the steps leading to the synthesis of the other amino acids, tyrosine and phenylalanine. Likewise, these two amino acids each inhibit the first enzyme of the branch leading to their synthesis.

In addition, the three amino acids each control the first step of the full pathway, the formation of the 7-carbon compound. Three different enzymes can catalyze this step; each has the same active site, but they have different allosteric sites. Each aromatic amino acid acts as a feedback inhibitor for one of the enzymes. If all three amino acids are present in the environment, then very little of the 7-carbon compound will be synthesized. If only one or two of those amino acids are present, then proportionally more of the compound will be synthesized.

Nucleotide Synthesis

The nucleotide subunits of DNA and RNA are composed of three units: a 5-carbon sugar, a phosphate group, and a nitrogenous base, either a purine or a pyrimidine (see figure 2.30). They are synthesized as ribonucleotides, but these can then be converted to deoxyribonucleotides by replacing the hydroxyl group on the 2′carbon of the sugar with a hydrogen atom. ■ **purine, p. 35** ■ **pyrimidine, p. 35**

The purine nucleotides are synthesized in a distinctly different manner from the pyrimidine nucleotides. Purine nucleotides are synthesized on the sugar phosphate component in a very complex process. In fact, nearly every carbon and nitrogen of the purine ring comes from a different source (**figure 6.28**). The starting compound is ribose 5-phosphate, a precursor metabolite generated in the pentose phosphate pathway. Then, in a highly ordered sequence, atoms from the other sources are added. Once this purine is formed, it is converted to adenylic and guanylic acid, which are components of the nucleic acids. To synthesize pyrimidine nucletotides, the pyrimidine ring is made first, and then attached to ribose 5-phosphate. After one pyrimidine nucleotide is formed, the base component can be converted into one of the other pyrimidines. ■ **adenylic acid, p. 35** ■ **guanylic acid, p. 35**

MICROCHECK 6.9

Biosynthetic processes of different organisms are remarkably similar, using precursor metabolites, NADPH, and ATP to form subunits. Synthesis of the amino acid glutamate provides a mechanism for bacteria to incorporate nitrogen in the form of ammonia into organic material. Allosteric enzymes are used to regulate certain biosynthetic pathways. The purine nucleotides are synthesized in a very different manner from the pyrimidine nucleotides.

- Explain why the synthesis of glutamate is particularly important for a cell.
- What are three general requirements that the cell must fulfill in order to carry out biosynthesis?
- With a branched biochemical pathway, why would it be important for a cell to shut down the first step as well as branching steps?

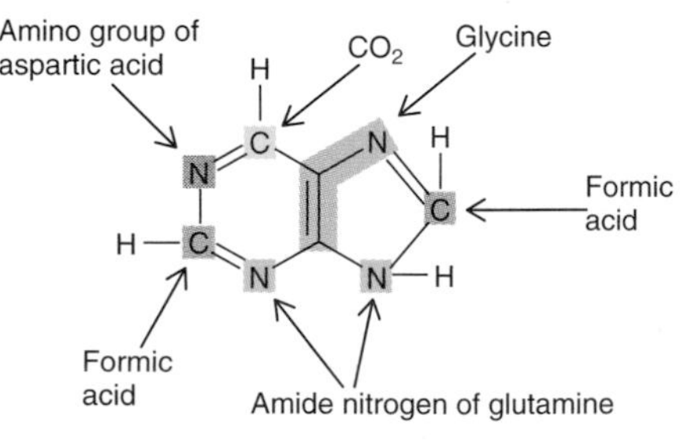

Figure 6.28 **Source of the Carbons and Nitrogen Atoms in Purine Rings** Nearly every carbon and nitrogen of the purine ring comes from a different source.

FUTURE CHALLENGES

Discovering and Exploiting Novel Enzymes

The remarkable speed and precision of enzyme activity is already exploited in a number of different processes. For example, the enzyme glucose isomerase is used to modify corn syrup, converting some of the glucose into fructose, which is much sweeter than glucose. The resulting high-fructose corn syrup is used in the commercial production of a variety of beverages and food products. Other enzymes, including proteases, amylases, and lipases, are used in certain laundry detergents to facilitate stain removal. These enzymes break down proteins, starches, and fats, respectively, which otherwise adhere strongly to fabrics. Similar enzymes are being added to some dishwashing detergents, decreasing the reliance on chlorine bleaching agents and phosphates that can otherwise pollute the environment. Enzymes are also used by the pulp and paper industry to facilitate the bleaching process.

Even with the current successes of enzyme technology, however, only a small fraction of enzymes in nature have been characterized. Recognizing that the field is still in its infancy, some companies are actively searching diverse environments for microorganisms that produce novel enzymes, hoping that some may be commercially valuable. Among the most promising enzymes are those produced by the extremophiles, members of the Archaea that preferentially live in conditions inhospitable to other forms of life. Because these organisms live in severe environments, it is expected that their enzymes can withstand the harsh conditions that characterize certain processes. For example, the enzymes of the extreme thermophiles will likely withstand temperatures that would quickly inactivate enzymes of mesophiles.

With the aid of enzymes, many of which are still to be discovered, scientists may eventually be able to precisely control a greater variety of commercially important chemical processes. This, it is hoped, will result in fewer unwanted by-products and a decreased reliance on harsh chemicals that damage the environment.

SUMMARY

Principles of Metabolism

1. **Catabolism** encompasses those processes that capture and store energy by breaking down complex molecules. (Figure 6.1)
2. **Anabolism** includes the processes that utilize energy to synthesize and assemble the building blocks of a cell.

Harvesting Energy

1. **Energy** is the ability to do work.
2. The **first law of thermodynamics** states the energy in a system can never be created or destroyed; the **second law of thermodynamics** states that **entropy** always increases.
3. **Phototrophs** harvest the energy of sunlight, using it to power the synthesis of organic compounds. **Chemoorganotrophs** transform and use energy contained in organic compounds. (Figure 6.3)
4. **Free energy** is the amount of energy that can be gained by breaking the bonds of a chemical. **Exergonic** reactions release energy; **endergonic** reactions utilize energy. (Figure 6.4)

Components of Metabolic Pathways (Table 6.1)

1. A **metabolic pathway** can be linear, branched, or cyclical, generating a sequential series of **intermediates** and, ultimately, an **end product**. (Figure 6.5)
2. A specific **enzyme** facilitates each step of a metabolic pathway by lowering the activation energy of the reaction that converts a **substrate** into a **product**. (Figure 6.6)
3. **ATP** is the energy currency of the cell. **Substrate-level phosphorylation** uses the chemical energy released in an exergonic reaction to add P_i to ADP. **Oxidative phosphorylation** harvests the energy of proton motive force to do the same thing. (Figure 6.7)
4. The **energy source** is **oxidized** to release its energy; this **oxidation-reduction** reaction **reduces** an electron carrier. (Figure 6.8)
5. **NAD^+**, **$NADP^+$**, and **FAD** are electron carriers. Their reduced forms function as **reducing power**. **NADH** and **$FADH_2$** are used to provide electrons for the generation of proton motive force. **NADPH** is used in biosynthesis. (Table 6.2)
6. **Precursor metabolites** are building blocks that can be used to make the subunits of macromolecules, and they can also be oxidized to generate energy. (Table 6.3)

Scheme of Metabolism (Figure 6.9)

1. The **central metabolic pathways** are **glycolysis** (also called the **glycolytic pathway** and the **Embden-Meyerhoff pathway**), the **pentose phosphate pathway**, and the **tricarboxylic acid cycle** (**TCA cycle**) (also called the **Krebs cycle** and the **citric acid cycle**). Because they play a role in both catabolism and anabolism, they are sometimes called **amphibolic pathways**.
2. **Glycolysis** oxidizes glucose to pyruvate, producing ATP, reducing power, and precursor metabolites.
3. The **pentose phosphate pathway** oxidizes glucose to pyruvate. Its primary role is the production of precursor metabolites and reducing power essential for biosynthesis.
4. The **transition step** forms acetyl-CoA, which then enters the **tricarboxylic acid** (**TCA**) **cycle**.
5. For every molecule of glucose degraded, the TCA cycle "turns" twice, generating precursor metabolites, some ATP, and a great deal of reducing power.

6. **Respiration** uses the reducing power accumulated in the central metabolic pathways to generate ATP by oxidative phosphorylation. **Aerobic respiration** uses O_2 as a terminal electron acceptor; **anaerobic respiration** uses an inorganic molecule other than O_2 as a terminal electron acceptor. (Table 6.4)
7. **Fermentation** uses pyruvate or a derivative as a terminal electron acceptor rather than oxidizing it further in the TCA cycle; this recycles the reduced electron carrier NADH.

Enzymes

1. **Enzymes** function as biological catalysts; they are neither consumed nor permanently changed during a reaction.

Mechanisms and Consequences of Enzyme Action (Figure 6.10)

1. The substrate binds to the **active site** or **catalytic site** to form a temporary intermediate called an **enzyme-substrate complex**.

Allosteric Regulation (Figure 6.11)

1. Cells can fine-tune the activity of an **allosteric enzyme** by using an **effector** that binds to the **allosteric site** of the enzyme. This binding alters the relative **affinity** of the enzyme for its substrate.

Cofactors and Coenzymes (Figure 6.12, Table 6.5)

1. Enzymes sometimes act in conjunction with **cofactors** such as coenzymes and trace elements.

Environmental Factors that Influence Enzyme Activity (Figure 6.13)

1. The factors most important in influencing enzyme activities are temperature, pH, and salt concentration.

Enzyme Inhibition

1. **Competitive inhibition** occurs when the inhibitor competes with the normal substrate for the active binding site. (Figure 6.14)
2. **Noncompetitive inhibition** occurs when the inhibitor and the substrate act at different sites on the enzyme.

Pathways and Processes that Fuel Aerobic Growth of Chemooraganotrophs (Table 6.6)

1. The central metabolic pathways chemically alter organic molecules to form high-energy bonds that can be used to synthesize ATP, intermediates that can be oxidized to generate reducing power, and precursor metabolites. (Table 6.7)

Glycolysis (Figure 6.15)

1. Glycolysis is a nine-step pathway that converts one molecule of glucose into two molecules of pyruvate; the theoretical net yield is 2 ATP, 2 NADH + H^+, and 6 different precursor metabolites.

Pentose Phosphate Pathway

1. The pentose phosphate pathway can generate some ATP, but its greatest significance is that it forms NADPH and 2 different precursor metabolites.

Transition Step (Figure 6.16)

1. The transition step results in the decarboxylation and oxidation of pyruvate, and joins the resulting acetyl group to coenzyme A, forming acetyl-CoA. This produces NADH + H^+ and 1 precursor metabolite.

Tricarboxylic Acid Cycle (Figure 6.16)

1. The eight steps of the TCA cycle complete the oxidation of glucose; the theoretical yield is 6 NADH + 6 H^+, 2 $FADH_2$, 2 ATP, and 3 different precursor metabolites.

Oxidative Phosphorylation

1. The reducing power accumulated in glycolysis, the transition step and the TCA cycle is used to drive the synthesis of ATP.
2. The **electron transport chain** sequentially passes electrons, and, as a result, ejects protons; this mechanism generates the **chemiosmotic gradient** called the **proton motive force**. (Figure 6.17)
3. **ATP synthase** harvests the energy released by the electron transport chain as it allows protons to move back across the membrane, driving the synthesis of ATP. (Figure 6.18)

ATP Yield of Aerobic Respiration in Prokaryotes (Figure 6.19)

1. The theoretical maximum yield of ATP of aerobic respiration is 38 ATP.

Pathways and Processes that Fuel Anaerobic Growth of Chemoorganotrophs

1. Glycolysis and the pentose phosphate pathway are used anaerobically to oxidize glucose to pyruvate.

Anaerobic Respiration

1. Unlike eukaryotes, some prokaryotes can respire using an inorganic molecule other than molecular oxygen as a terminal electron acceptor. (Figure 6.20)
2. Anaerobic respiration generates less energy than aerobic respiration, and alternative electron carriers are used in the electron transport chain.

Fermentation

1. Fermentation is used by organisms that cannot respire, either because a suitable inorganic terminal electron acceptor is not available or because they lack an electron transport chain.
2. In general the only ATP-yielding reactions of fermentations are those of the glycolytic pathway; the other steps provide a mechanism for recycling NADH. (Figure 6.21)
3. Some end products of fermentation are commercially valuable. Lactic acid is important in the production of foods such as cheese and yogurt. Ethanol is used to make alcoholic beverages and breads. (Figure 6.22)
4. Because a given type of organism uses a specific fermentation pathway, the end products can be used as markers that aid in identification.

Catabolism of Organic Compounds Other than Glucose (Figure 6.23)

1. **Hydrolytic enzymes** break down macromolecules into their respective subunits.

Polysaccharides and Disaccharides

1. **Amylases** digest starch, releasing glucose subunits, and are produced by many organisms. **Cellulases** degrade

cellulose; relatively few groups of organisms produce these enzymes.

2. The sugar subunits released when polysaccharides are broken down can then enter glycolysis to be oxidized to pyruvate.

Lipids

1. Fats are hydrolyzed by **lipase**, releasing glycerol and fatty acids.
2. Glycerol is converted to the precursor metabolite glyceraldehyde 3-phosphate; fatty acids are degraded by a **β-oxidation**, generating reducing power and the precursor metabolite acetyl-CoA.

Proteins

1. Proteins are hydrolyzed by **proteases**.
2. **Deamination** removes the amino group; the remaining carbon skeleton is then converted into the appropriate precursor molecule.

Chemolithotrophs

1. Prokaryotes, as a group, are unique in their ability to use reduced inorganic compounds such as hydrogen sulfide (H_2S) and ammonia (NH_3) as a source of energy.
2. Chemolithotrophs are autotrophs; they do not require an external source of organic carbon because they can fix carbon dioxide.

Phototrophs

1. Phototrophs harvest the energy of sunlight and use it to drive the synthesis of ATP.

The Role of Photosynthetic Pigments

1. **Chlorophylls** are the primary pigments used to harvest solar energy. Most photosynthetic organisms, including cyanobacteria, use **chlorophyll *a***; photosynthetic purple and green bacteria use **bacteriochlorophyll**.
2. **Carotenoids** are accessory pigments that absorb wavelengths of light not absorbed by the chlorophylls and then transfer that energy to the chlorophylls.
3. The light-gathering pigments are in **photosystems** located in special membranes.

Photophosphorylation (Figure 6.24)

1. The **antenna complex** acts as a funnel, capturing the energy of light and transferring it to **reaction-center chlorophyll**. An electron in reaction-center chlorophyll becomes excited, enabling it to be passed along an electron transport chain, which generates proton motive force.
2. ATP synthase permits the flow of protons back across the membrane, and uses that energy to power the synthesis of ATP.

Electron Source

1. Plants, algae and cyanobacteria, which extract electrons from water, are **oxygenic phototrophs**; the green and purple bacteria, which extract electrons from reduced compounds other than water, are **anoxygenic phototrophs**.

Carbon Fixation

Calvin Cycle (Figure 6.25)

1. The most common pathway used to incorporate CO_2 into an organic form is the **Calvin cycle**.
2. Many steps of the Calvin cycle are used to regenerate the 5-carbon starting compound; only one molecule of a 6-carbon sugar is made for every six "turns" of the cycle.

Anabolic Pathways—Synthesizing Subunits from Precursor Molecules

Lipid Synthesis

1. The fatty acid components of fat are synthesized by progressively adding 2-carbon units to an acetyl group. The glycerol component is synthesized from glyceraldehyde 3-phosphate.

Amino Acid Synthesis

1. Structurally related families of amino acids share common pathways of biosynthesis.
2. The synthesis of glutamate from α-ketoglutarate and ammonia provides a mechanism for cells to incorporate nitrogen into organic molecules. (Figure 6.26)
3. The synthesis of aromatic amino acids requires a multistep branching pathway. Allosteric enzymes regulate key steps of the pathway. (Figure 6.27)

Nucleotide Synthesis

1. Purine nucleotides are synthesized on the sugar-phosphate component; the pyrimidine ring is made first and then attached to the sugar-phosphate. (Figure 6.28)

REVIEW QUESTIONS

Short Answer

1. Explain the difference between catabolism and anabolism.
2. How does ATP serve as a carrier of free energy?
3. How do enzymes catalyze chemical reactions?
4. Explain how precursor molecules serve as junctions between catabolic and anabolic pathways.
5. How do cells use effectors to regulate enzyme activity?
6. Why do the electrons carried by $FADH_2$ result in less ATP production than those carried by NADH?
7. Name three food products produced with the aid of microorganisms.
8. In photosynthesis, what is encompassed by the term light independent reactions?
9. Unlike the oxygenic phototrophs, the anoxygenic phototrophs do not evolve oxygen (O_2). Why not?
10. What is the role of transamination in amino acid biosynthesis?

Multiple Choice

1. Which of these environmental factors does not affect general enzyme activity?

   A. Temperature

   B. Inhibitors

C. Coenzymes
D. Humidity
E. pH

2. Which of the following statements is false? Enzymes…
A. bind to substrates.
B. lower the energy of activation.
C. convert coenzymes to products.
D. speed up biochemical reactions.
E. can be named after the kinds of reaction they catalyze.

3. Which of these is not a coenzyme?
A. FAD
B. Coenzyme A
C. NAD^+
D. ATP
E. $NADP^+$

4. What is the end product of glycolysis?
A. Glucose
B. Citrate
C. Oxaloacetate
D. α-ketoglutarate
E. Pyruvate

5. The major pathway(s) of central metabolism are…
A. glycolysis and the TCA cycle only.
B. glycolysis, the TCA cycle, and the pentose phosphate pathway.
C. glycolysis only.
D. glycolysis and the pentose phosphate pathway only.
E. the TCA cycle only.

6. Which of these pathways has the potential to produce the most ATP?
A. TCA cycle
B. Pentose phosphate pathway
C. Lactic acid fermentation
D. Alcohol fermentation
E. Glycolysis

7. In fermentation, the terminal electron acceptor is…
A. oxygen (O_2).
B. hydrogen (H_2).
C. carbon dioxide (CO_2).
D. an organic compound.

8. In the process of oxidative phosphorylation, the energy of proton motive force is used to generate…
A. NADH.
B. ADP.
C. ethanol.
D. ATP.
E. glucose.

9. In the TCA cycle, the carbon atoms contained in acetate are converted into…
A. lactic acid.
B. glucose.
C. glycerol.
D. CO_2.
E. all of the above.

10. Degradation of fats as an energy source involves all of the following, except…
A. β-oxidation.
B. acetyl-CoA.
C. glycerol.
D. lipase.
E. transamination.

Applications

1. A worker in a cheese-making facility argues that whey, a nutrient-rich by-product of cheese, should be dumped in a nearby pond where it could serve as fish food. Explain why this proposed action could actually kill the fish by depleting the oxygen in the pond.
2. Scientists working with DNA *in vitro* often store it in solutions that contain EDTA, a chelating agent that binds magnesium (Mg^{2+}). This is done to prevent enzymes called DNases from degrading the DNA. Explain why EDTA would interfere with enzyme activity.

Critical Thinking

1. A student argued that aerobic and anaerobic respiration should produce the same amount of ATP. He reasoned that they both use basically the same process; only the terminal electron acceptor is different. What is the primary error in this student's argument?
2. Chemolithotrophs near hydrothermal vents support a variety of other life forms there. Explain how their role there is analogous to that of photosynthetic organisms in terrestrial environments.

CHAPTER 7

The Blueprint of Life, from DNA to Protein

In 1866, the Czech monk Gregor Mendel showed that traits are inherited by means of physical units, which we now call genes. It was not until 1941, however, that the precise function of genes was revealed when George Beadle, a geneticist, and Edward Tatum, a chemist, published a scientific paper reporting that genes determine the structure of enzymes. Biochemists had already shown that enzymes catalyze the conversion of one compound into another in a biochemical pathway.

Beadle and Tatum studied Neurospora crassa, *a common bread mold that grows on a very simple medium containing sugar and simple inorganic salts. Beadle and Tatum created* N. crassa *strains with altered properties,* ***mutants****, by treating cells with X rays, which were known to alter genes. Some of these mutants could no longer grow on the glucose-salts medium unless growth factors such as vitamins were added to the medium. To isolate these Beadle and Tatum had to laboriously screen thousands of progeny to find the relatively few that required the growth factors. Each mutant presumably contained a defective gene.*

The next task for Beadle and Tatum was to identify the specific biochemical defect of each mutant. To do this, they added different growth factors, one at a time, to each mutant culture. The one that allowed a particular mutant to grow had presumably bypassed the function of a defective enzyme. In this manner, they were able to pinpoint in each mutant the specific step in the biochemical pathway that was defective. Then, using these same mutants, Beadle and Tatum showed that the requirement for each growth factor was inherited as a single gene, ultimately leading to their conclusion that a single gene determines the production of one enzyme. Their conclusion has been modified somewhat, because we now know that some enzymes are made up of more than one protein. A single gene determines the production of one protein. In 1958, Beadle and Tatum shared the Nobel Prize in Medicine, largely for these pioneering studies that ushered in the era of modern biology.

As so often occurs in science, the answer to one question raised many more questions. How do genes specify the synthesis of enzymes? What are genes made of? How do genes replicate? Numerous other investigators won more Nobel Prizes for answering these questions, many of which are covered in this chapter.

—A Glimpse of History

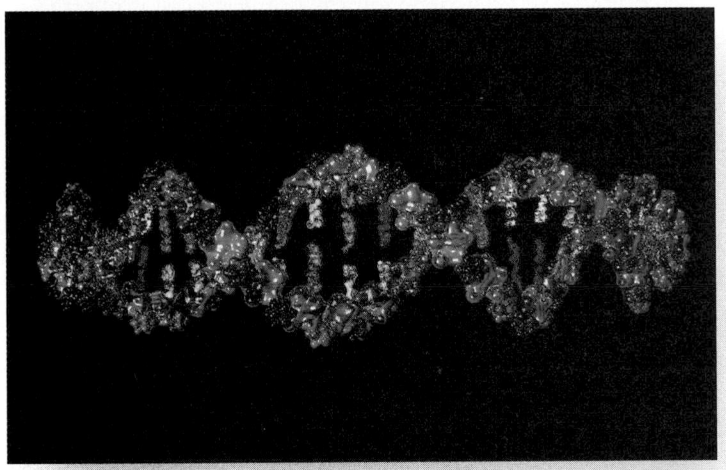

CONSIDER FOR A MOMENT THE VAST DIVERSITY OF cellular life forms that exist. Our world contains a remarkable variety of microorganisms and specialized cells that make up plants and animals, including your own body. Every characteristic of each of these cells, from its shape to its function, is dictated by information contained in its deoxyribonucleic acid (DNA). DNA encodes the master plan, the blueprint, for all cell structures and processes. Yet for all the complexity this would seem to require, DNA is a string composed of only four different nucleotide bases: adenine (A), thymine (T), cytosine (C), and guanine (G). ■ nucleotide bases, p. 35

While it might seem improbable that the vast array of life forms can be encoded by a molecule consisting of only four different units, think about how much information can be transmitted by binary code, the language of all computers, which has a base of only two. A simple series of ones and zeros can code for each letter of the alphabet. String enough of these series together in the right sequence and the letters become words, and the words can become complete sentences, chapters, books, or even whole libraries.

The four nucleotides of a DNA molecule create information in a similar fashion. A set of three nucleotides encodes a specific amino acid. In turn, a string of amino acids makes up a protein, the function of which is dictated by the order of the amino acid subunits. Some proteins serve as structural components of a cell. Others, such as enzymes, mediate cellular activities including biosynthesis and energy conversion. Together, the proteins synthesized by a cell are responsible for every aspect of that cell. Thus, the sequential order of nucleotide bases in a cell's DNA ultimately dictates the characteristics of that cell. ■ amino acids, p. 28 ■ protein structure, p. 31 ■ flagella, p. 67 ■ enzymes, p. 135

This chapter will focus on the cellular process of converting the information encoded within DNA into proteins, concentrating primarily on the mechanisms used by prokaryotic cells. The eukaryotic processes have many similarities, but they are considerably more complicated and will only be discussed briefly.

Overview

The complete set of genetic information for a cell is referred to as its **genome**. Technically, this includes plasmids as well as the chromosome; however, the term genome is often used interchangeably with chromosome. The genome of all cells is composed of DNA, but some viruses have an RNA genome. The functional unit of the genome is a **gene**. A gene encodes a product, the **gene product**, most commonly a protein. The study of the function and transfer of genes is called **genetics**, whereas the study and analysis of the nucleotide sequence of DNA is called **genomics**. ■ chromosome, p. 70 ■ plasmid, p. 70

All living cells must accomplish two general tasks in order to multiply. The double-stranded DNA must be duplicated prior to cell division so that its encoded information can be passed on to future generations. This is the process of **DNA replication**. In addition, the information encoded by the DNA must be deciphered and utilized, or **expressed**, so that the cell can synthesize the necessary gene products at the appropriate time. Gene expression involves two interrelated molecular processes, transcription and translation. **Transcription** copies the encoded information into a slightly different molecule, RNA. The RNA serves as a transitional, temporary form of the genetic information and is the one that is actually deciphered. **Translation** interprets the information carried by the RNA to synthesize the encoded protein. The chemistry and structure of DNA and RNA ensure that each of these processes can occur seamlessly and with great accuracy.

The flow of information from DNA to RNA to protein is often referred to as the **central dogma of molecular biology** (**figure 7.1**). It was once believed that information flow proceeded only in this direction. Although this direction is by far the most common, certain viruses, such as the one that causes AIDS, have an RNA genome but copy that information into the form of DNA.

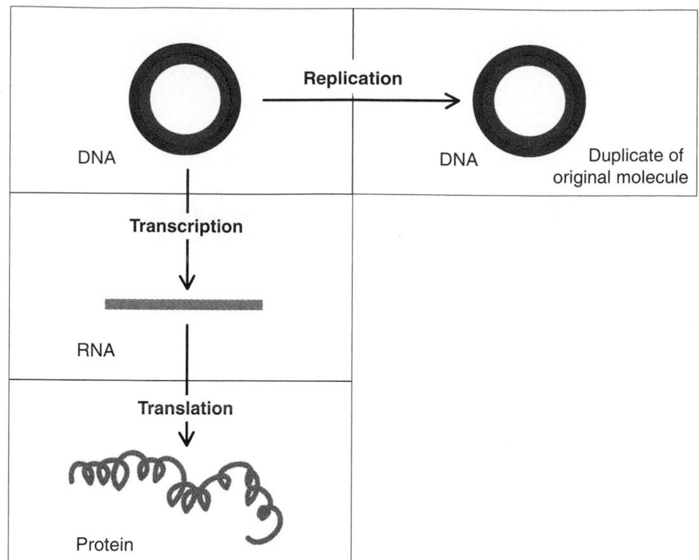

Figure 7.1 Overview of Replication, Transcription, and Translation DNA replication is the process that duplicates DNA so that its encoded information can be passed on to future generations. Transcription is the process that copies the genetic information into a transitional form, RNA. Translation is the process that deciphers the encoded information to synthesize a specific protein.

Characteristics of DNA

A single strand of DNA is composed of a series of deoxyribonucleotide subunits, more commonly called nucleotides. These are joined in a chain by a covalent bond between the $5'PO_4$ (5 prime phosphate) group of one nucleotide and the 3′OH (3 prime hydroxyl) group of the next. Note that the designations 5′ and 3′ refer to the numbered carbon atoms of the pentose sugar of the nucleotide. The joining of the nucleotides in this manner creates a series of alternating sugar and phosphate moieties, called the **sugar-phosphate backbone**. Connected to each sugar is one of the nitrogenous bases, an adenine (A), thymine (T), guanine (G), or cytosine (C). Because of the chemical structure of the nucleotides and how they are joined, a single strand of DNA will always have $5'PO_4$ group at one end and a 3′OH group at the other. These ends are often referred to as the **5′ end** and the **3′ end** and have important implications in DNA and RNA synthesis that will be discussed later. ■ deoxyribonucleic acid (DNA), p. 35 ■ nucleotides, p. 35

The DNA in a cell usually occurs as a double-stranded, helical structure (**figure 7.2**). The two strands are held together by weak hydrogen bonds between the nitrogenous bases of the opposing strands. While individual hydrogen bonds are readily broken, the duplex structure of double-stranded DNA is quite stable because of the sheer number of bonds that occurs along its length. Because short fragments of DNA have correspondingly fewer hydrogen bonds, they are readily separated into single-stranded pieces. The separating of the two strands is called **denaturing**. ■ hydrogen bonds, p. 23

The two strands of double-stranded DNA are complementary. Wherever there is an adenine in one strand, there is a thymine in the other; these two opposing nucleotides are held together by two hydrogen bonds between them. Similarly, wherever there is a cytosine in one strand, there is a guanine in the other. These are held together by the formation of three hydrogen bonds, a slightly stronger attraction than that of an A:T pair. The characteristic bonding of A to T and G to C is called **base-pairing** and is fundamental to the remarkable functionality of DNA. Because of the rules of base-pairing, one strand can always be used as a **template** for the synthesis of the complementary opposing strand. ■ complementary, p. 36

While the two strands of DNA in the double helix are complementary, they are also **antiparallel**. That is, they are oriented in opposite directions. One strand is oriented in the 5′ to 3′ direction and its complement is oriented in the 3′ to 5′ direction (**figure 7.3**). Again, this has important implications in the function and synthesis of nucleic acids.

A highly coiled line is used to depict genomic DNA.

Red and blue lines placed in a helical arrangement depict the two complementary strands and highlight the three-dimensional structure of DNA.

A circular arrangement of the red and blue lines is used as the simplified form of prokaryotic DNA.

Red and blue lines separated by a thin black line are used as a simple representation of the double-stranded DNA molecule.

Two parallel lines are used to emphasize the base-pairing interactions and nucleotide sequence characteristics of the two complementary strands. The "tracks" between the lines are not intended to depict a specific number of base pairs, only the general interaction between complementary strands.

Either a red or a blue line can be depicted as the "top" strand, since DNA is a three-dimensional structure.

Denatured DNA is depicted as separate red and blue lines to emphasize its single-stranded nature.

Figure 7.2 Diagrammatic Representations of the Structure of DNA Although DNA is a double-stranded helical structure, explanatory diagrams may depict it in a number of different ways. In this chapter we will use these color-coded representations.

Characteristics of RNA

RNA is in many ways comparable to DNA, but with some important exceptions. One difference is that RNA is made up of ribonucleotides rather than deoxynucleotides, although in both cases these are usually referred to simply as nucleotides. Another distinction is that RNA contains the nitrogenous base uracil in place of the thymine found in DNA. Like DNA, RNA consists of a sequence of nucleotides, but RNA usually exists as a single-stranded linear molecule and is much shorter than DNA. ■ **ribonucleic acid (RNA), p. 36** ■ **nucleotide, p. 35**

A fragment of RNA, a **transcript**, is synthesized using a region of one of the two strands of DNA as a template. In making the RNA transcript, the same base-pairing rules of DNA apply except uracil, rather than thymine, base-pairs with adenine. This base-pairing is only transient, however, and the molecule quickly leaves the DNA template. Numerous different RNA transcripts can be generated from a single chromosome using specific regions as templates. In theory, either strand may serve as the template. In a region the size of a single gene, however, only one of the two strands is generally transcribed. Because of this important characteristic, two complementary strands of RNA are not normally generated.

There are three different functional groups of RNA molecules, each one transcribed from different genes. Most genes encode proteins and are transcribed into **messenger RNA (mRNA)**. These are the molecules that are translated during protein synthesis. The encrypted information on mRNA is deciphered according to the **genetic code**, which correlates each set of three nucleotides, called a **codon**, to a particular amino acid. Some genes are never translated into proteins; instead the RNAs themselves are the ultimate products. These genes encode either

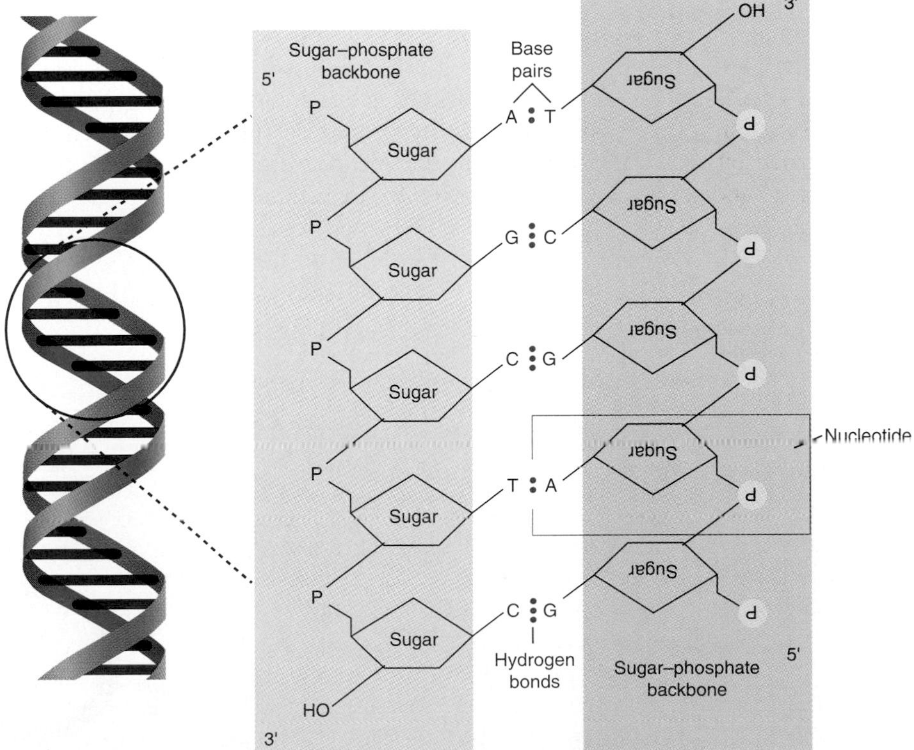

Figure 7.3 The Double Helix of DNA The two strands of DNA are antiparallel; one strand is oriented in the 5′ to 3′ direction, and its complement is oriented in the 3′ to 5′ direction. Hydrogen bonding occurs between the complementary base pairs; three bonds form between a G-C base pair, and two bonds form between an A-T base pair.

ribosomal RNA (rRNA) or **transfer RNA (tRNA)**, each of which plays a different but critical role in protein synthesis. Unlike genes that encode proteins, the rRNA and tRNA genes are often redundant, or in more than one copy.

Regulating the Expression of Genes

While the basic structure of DNA and RNA is relatively simple, it contains extensive and complex information. The nucleotide sequence contains genes that encode the amino acid sequence of proteins, and it also must encode mechanisms to regulate the expression of those genes. Not all proteins are required in the same quantity and at all times; therefore, mechanisms to determine the extent and duration of their synthesis are needed.

One of the key mechanisms a cell uses to control protein synthesis is to regulate the synthesis of mRNA molecules. Unless a gene is transcribed into mRNA, the encoded protein cannot be synthesized. The number of mRNA copies of the gene also influences the level of expression. Meanwhile, if transcription of a gene ceases, the level of gene expression rapidly declines. This is because mRNA is short-lived, often only minutes, due to the activity of enzymes called **RNases** that readily degrade it.

MICROCHECK 7.1

Replication is the process of duplicating double-stranded DNA. Transcription is the process of copying the information encoded in DNA into RNA. Translation is the process of interpreting the information carried by messenger RNA in order to synthesize the encoded protein.

- How does the 5′ end of DNA differ from the 3′ end?
- If the nucleotide sequence of one strand of DNA is 5′ ACGTTGCA 3′, what is the sequence of the complementary strand?
- Why is a short-lived RNA important in cell control mechanisms?

DNA Replication

The purpose of DNA replication is to create a second DNA molecule, identical to the original, so that each of the two cells generated during binary fission receives a complete copy. DNA molecules such as the circular chromosome and certain plasmids are replicated once during each cell division, whereas multiple-copy plasmids must be replicated numerous times. ■ binary fission, p. 89

DNA replication is generally **bidirectional**. From a distinct starting point in circular DNA, replication proceeds in both directions, creating an ever-expanding "bubble" of two identical replicated portions of the chromosome **(figure 7.4)**. Bidirectional

Replication
DNA
DNA
Duplicate of original molecule
Transcription
RNA
Translation
Protein

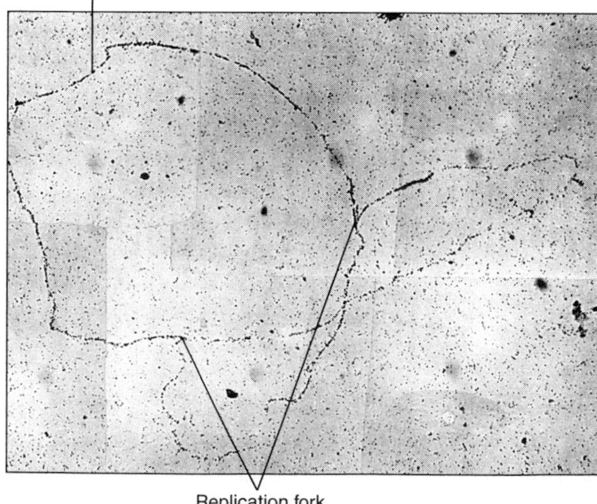

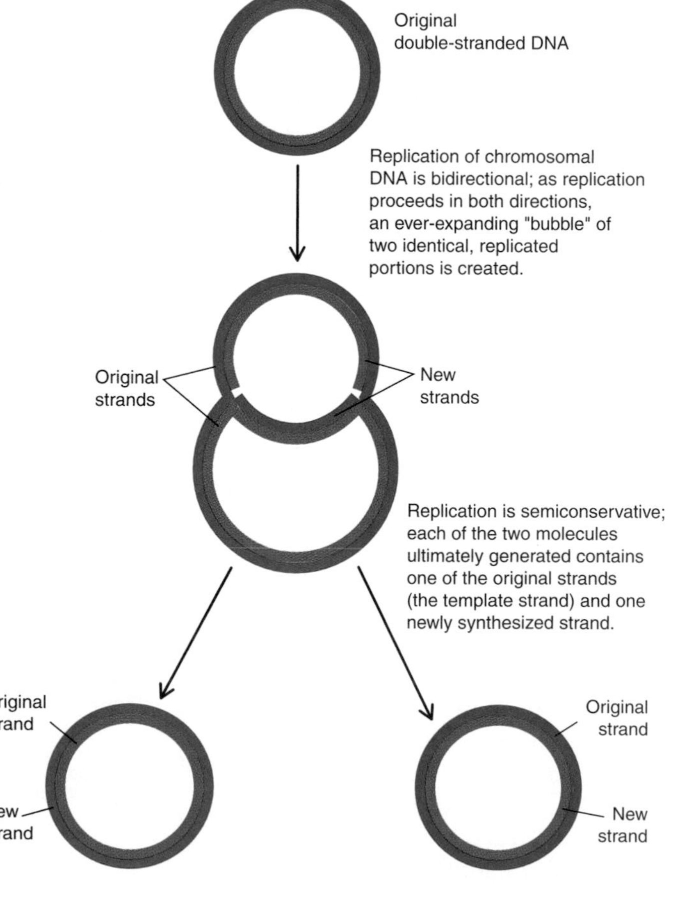

Figure 7.4 Replication of Chromosomal DNA of Prokaryotes

replication allows an entire chromosome to be replicated in half the time it would take if replication were unidirectional.

Replication of double-stranded DNA is **semiconservative**. Each of the two molecules generated contains one of the original strands (the template strand) and one newly synthesized strand. Thus, the two cells produced as a result of division each have one of the original strands of DNA paired with a new complementary strand.

The process of DNA replication requires the coordinated action of many different enzymes and other proteins, the most critical of which exist together as a complex called a **replisome**. The replisome appears to act as a fixed DNA-synthesizing factory, reeling in the DNA to be replicated. **DNA polymerases** are enzymes in the replisome that synthesize DNA, using one strand as a template to generate the complementary strand. These enzymes can only add nucleotides onto a preexisting fragment of nucleic acid, either DNA or RNA. Thus, the fragment serves as a **primer** from which synthesis can continue.

DNA is synthesized one nucleotide at a time as the deoxynucleotides (dATP, dGTP, dCTP, and dTTP) are covalently joined to the nucleotide at the 3′ end of the growing strand. DNA polymerase always elongates the chain in the 5′ to 3′ direction; but because the two DNA strands are antiparallel, the enzyme must "read" the template strand in the 3′ to 5′ direction (**figure 7.5**). The base-pairing rules determine the specific nucleotides that are added.

The process of replication is very accurate, resulting in only one mistake approximately every billion nucleotides. Part of the reason for this remarkable precision is the proofreading ability of some DNA polymerases. If an incorrect nucleotide is incorporated into the growing chain, the enzyme can edit the mistake by replacing that nucleotide before moving on.

It takes approximately 40 minutes for the chromosome of *E. coli* to be replicated, regardless of the environmental conditions. How, then, can *E. coli* sometimes multiply with a generation time of only 20 minutes? Under favorable growing conditions, a cell will sometimes initiate replication before the preceding round of replication is completed. In that way, the two progeny that result from cell division each will get one complete chromosome that has already started another round of replication.

■ generation time, p. 90

Initiation of DNA Replication

To begin the process of DNA replication, specific proteins must recognize and bind to a distinct region on the DNA, called an **origin of replication**. All molecules of DNA, including chromosomes and plasmids, must have this region of approximately 250 nucleotides in length in order for replication to be initiated. The binding of the proteins causes localized denaturation, or **melting**, of a specific region within the origin. Using the exposed single strands as templates, small fragments of RNA are synthesized to serve as primers for DNA synthesis. The enzymes that synthesize RNA do not require a primer.

The Replication Fork

The bidirectional progression of replication around a circular DNA molecule creates two advancing Y-shaped regions where active replication is occurring. Each of these is called a **replication fork**. The template strands continue to "unzip" at each fork due to the activity of enzymes called **helicases**. Synthesis of one new strand proceeds continuously in the 5′ to 3′ direction, as fresh single-stranded template DNA is exposed (**figure 7.6**). This strand is called the **leading strand**. Synthesis of the opposing strand, the **lagging strand**, is considerably more complicated because the DNA polymerase cannot add nucleotides to the 5′ end of DNA. Instead, synthesis must be reinitiated periodically as advancement of the replication fork exposes more of the template DNA. Each initiation event must be preceded by the synthesis of an RNA primer by the enzyme **primase**. The result is the synthesis of a series of fragments, called **Okazaki fragments**, each of which begins with a short stretch of RNA. As DNA polymerase adds nucleotides to the 3′ end of an Okazaki fragment, it eventually reaches the initiating point of the previous fragment. A different type of DNA polymerase then removes those RNA primer nucleotides and simultaneously replaces them with deoxynucleotides. The enzyme **DNA ligase** seals the gaps between adjacent fragments by catalyzing the formation of a covalent bond between them.

Several other proteins are also involved in DNA replication. Among them is **DNA gyrase**, an enzyme that helps relieve the tension caused by the unwinding of the two strands of the DNA helix. This enzyme is one of the targets

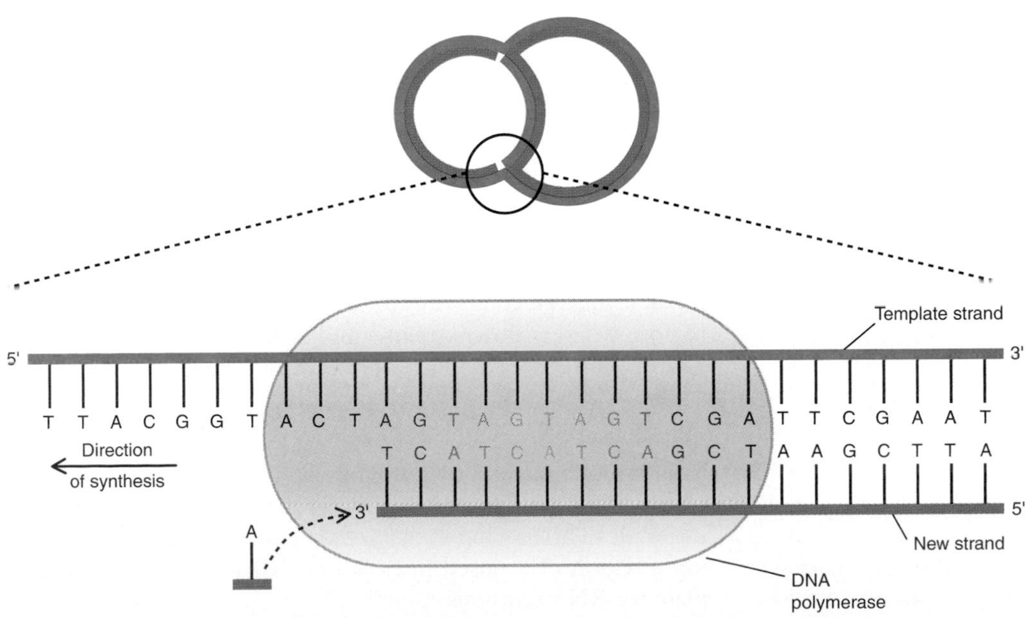

Figure 7.5 The Process of DNA Synthesis DNA polymerase synthesizes a new strand by adding one nucleotide at a time to the 3′ end of the elongating strand. Because DNA is synthesized in the 5′ to 3′ direction, the enzyme must "read" the template strand in the 3′ to 5′ direction. The base-pairing rules determine the specific nucleotides that are added.

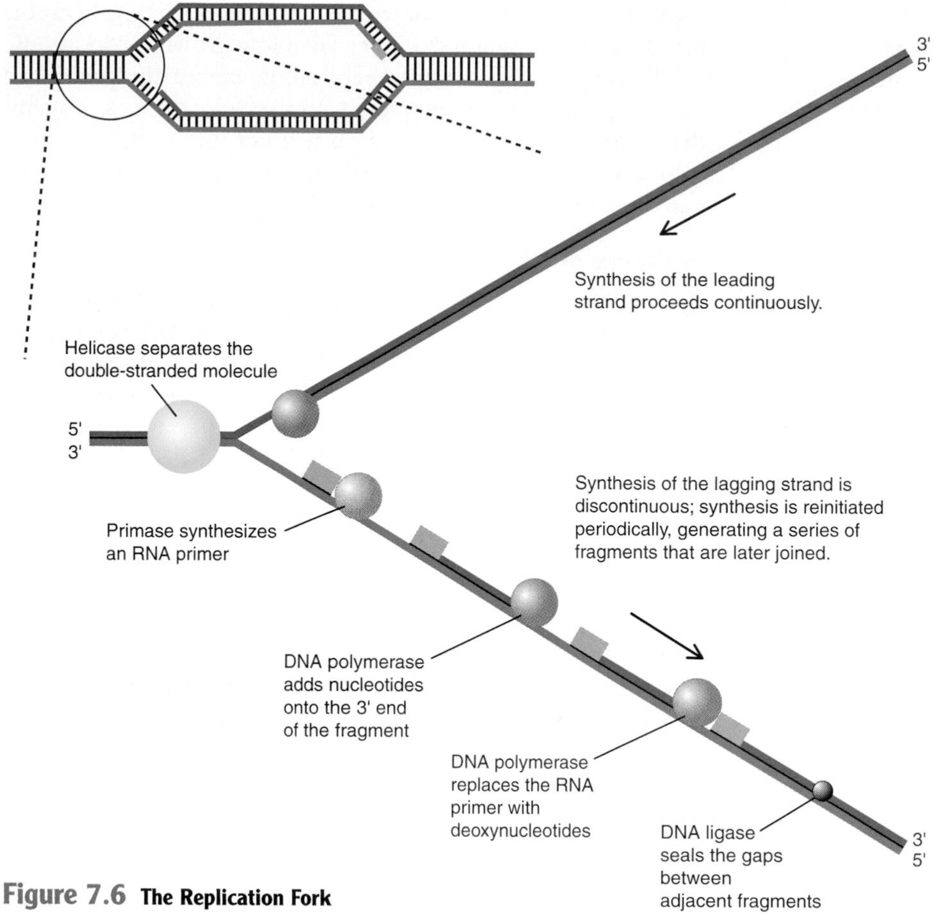
3'
5'
Synthesis of the leading strand proceeds continuously.
Helicase separates the double-stranded molecule
5'
3'
Synthesis of the lagging strand is discontinuous; synthesis is reinitiated periodically, generating a series of fragments that are later joined.
Primase synthesizes an RNA primer
DNA polymerase adds nucleotides onto the 3' end of the fragment
DNA polymerase replaces the RNA primer with deoxynucleotides
DNA ligase seals the gaps between adjacent fragments
3'
5'

Figure 7.6 The Replication Fork

of a class of antibacterial drugs called fluoroquinolones. By inhibiting the function of gyrase, the fluoroquinolones interfere with bacterial DNA replication and prevent the growth of bacteria. ■ **fluoroquinolones, p. 505**

MICROCHECK 7.2

DNA polymerases synthesize DNA in the 5′ to 3′ direction, using one strand as a template to generate the complementary strand. Replication of DNA begins at a specific sequence, called the origin of replication, and then proceeds bidirectionally, creating two replication forks.

- Why is a primer required for DNA synthesis?
- How does synthesis of the lagging strand differ from that of the leading strand?
- If DNA replication were shown to be "conservative," what would this mean?

Gene Expression

Gene expression involves two separate but interrelated processes, transcription and translation. Transcription is the process of synthesizing RNA from a DNA template. During translation, the information encoded on an mRNA transcript is deciphered to synthesize a protein.

Transcription

The enzyme RNA polymerase catalyzes the process of transcription, producing a single-stranded RNA molecule that is complementary and antiparallel to the DNA template (**figure 7.7**). To describe the two strands of DNA in a region that is transcribed into RNA, the terms **minus (−) strand** and **plus (+) strand** are sometimes used. The strand that serves as the template for RNA synthesis is called the minus (−) strand, whereas its complement is called the plus (+) strand. Recall that the base-pairing rules of DNA and RNA are the same, except that RNA contains uracil in place of thymine. Therefore, because the RNA is complementary to the (−) strand, its nucleotide sequence is analogous to that of the (+) strand. Likewise, the RNA transcript has the same 5′ to 3′ direction, or **polarity**, as the (+) strand.

In prokaryotes, a mRNA molecule can carry the information for one or multiple genes. A transcript that carries one gene is called **monocistronic** (a cistron is synonymous with a gene). Those that carry multiple genes are called **polycistronic**. Generally, the proteins encoded on a polycistronic message are all involved in a single biochemical pathway. This enables the cell to express these related genes in a coordinated manner.

Transcription of RNA from the DNA template begins when RNA polymerase recognizes a sequence of nucleotides on the DNA called a **promoter**. Thus, the promoter identifies the region of the DNA molecule that will be transcribed into RNA. In addition, the promoter orients the RNA polymerase in one of the two possible directions. This dictates which of the two DNA strands is used as a template (**figure 7.8**). Like DNA polymerase, RNA polymerase can only synthesize nucleic acid in the 5′ to 3′ direction and must "read" the template in the 3′ to 5′ direction. Unlike DNA polymerase, however, RNA polymerase can begin synthesis of a new chain without a primer.

The transcribed RNA molecule can be used as a reference point to describe direction on the analogous DNA. **Upstream** implies the direction toward the 5′ end of the transcribed region, whereas **downstream** implies the direction toward the 3′ end. Thus, a promoter is upstream of a gene.

Initiation of RNA Synthesis

Transcription of RNA begins when RNA polymerase recognizes and binds to a promoter region on the double-stranded DNA molecule. The binding melts a short stretch of the DNA, creating a region of exposed nucleotides that then serves as a template for RNA synthesis.

A particular subunit of RNA polymerase recognizes the promoter region prior to the initiation of transcription. This subunit, **sigma (σ) factor**, is only a loose component of the enzyme. Soon

Perspective 7.1 Making Sense of Antisense RNA

The knowledge gained through basic science research is fundamental to commercially valuable applications. For example, by understanding how genes are transcribed and translated, scientists can develop methods to suppress the expression of certain genes. We know that only one strand of DNA is transcribed into a single strand of mRNA. This mRNA, the **sense strand** or **plus (+) strand,** is translated into a sequence of amino acids. An RNA molecule that is the complement of the sense strand is called an **antisense strand** or a **minus (−) strand**. Antisense RNA, which is not typically made by a cell, can base-pair with the sense strand to form a double-stranded RNA molecule, which cannot be translated.

Short fragments of antisense RNA can be chemically synthesized and used to interfere with gene expression. Recently, the first therapeutic drug based on antisense technology, fomivirsen, was approved for treating eye infections by cytomegalovirus (CMV) in AIDS patients. Fomivirsen is antisense RNA that is complementary to the mRNA of CMV; it prevents the expression of two proteins required for viral replication. ■ **cytomegalovirus, p. 354**

Cells can also be genetically engineered to produce antisense RNA by introducing a copy of the gene with the promoter upstream from the antisense strand rather than from the sense strand **(figure 1)**. Exploiting this principle, a plant biotechnology company genetically engineered tomato plants to synthesize antisense RNA of the gene that codes for the enzyme polygalacturonase. This plant enzyme breaks down plant cell walls and is responsible for the mushiness of ripe tomatoes a few days after they are picked. As a result of the genetic engineering, the tomatoes with antisense RNA to polygalacturonase do not get mushy for several weeks after they are picked, since the antisense RNA prevents polygalacturonase from being synthesized. Such technological achievement, however, does not guarantee economic success; commercially, the bioengineered tomatoes were a failure. ■ **genetic engineering, p. 222**

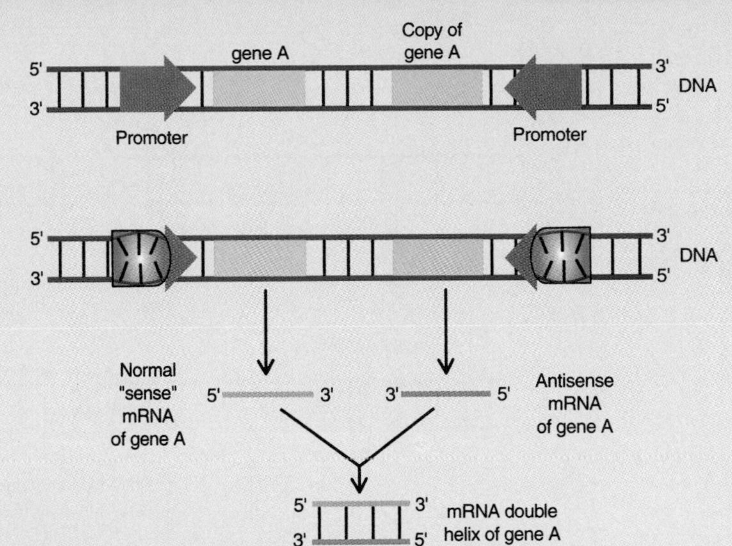

Figure 1 Formation of a Double-Stranded RNA Molecule The two copies of Gene A are oriented in opposite directions with respect to their promoters. When these genes are both transcribed, two complementary copies of mRNA (sense and antisense) are generated. These two molecules base-pair, forming a double-stranded molecule that cannot be translated into protein.

after transcription is initiated, the σ factor dissociates from the enzyme, leaving the remaining portion of RNA polymerase, called the **core enzyme**, to complete transcription. A cell may have different types of σ factors; these recognize different promoters.

Elongation

In the elongation phase, the RNA polymerase moves along the template strand of the DNA, synthesizing the complementary single-stranded RNA molecule. The RNA molecule is synthesized in the 5′ to 3′ direction as the enzyme adds nucleotides to the 3′OH group at the end of the growing chain. The RNA polymerase advances along the DNA, melting a new stretch and allowing the previous stretch to close **(figure 7.9)**. This exposes a new region of the template, permitting the elongation process to continue. The rate of polymerization is about 30 nucleotides per second.

Once elongation has proceeded far enough for RNA polymerase to clear the promoter, another molecule of RNA polymerase can bind to that promoter, initiating a new round of transcription.

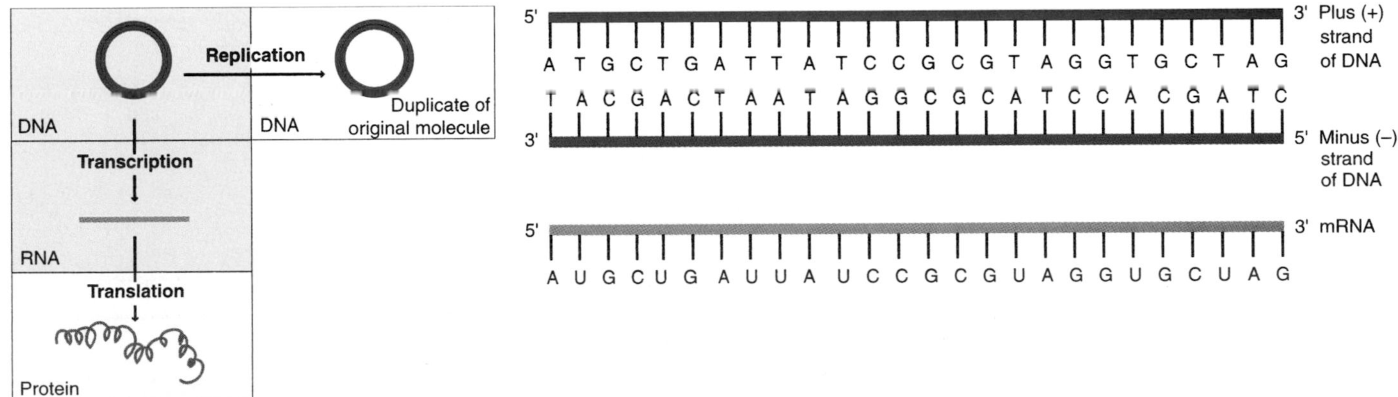

Figure 7.7 RNA Is Transcribed from a DNA Template The DNA strand that serves as a template for RNA synthesis is called the (–) strand of DNA. The nucleotide sequence of the transcript is analogous to that of the (+) strand, with uracil (U) occurring in place of thymine (T) in the mRNA.

Replication
DNA
DNA
Duplicate of original molecule
Transcription
RNA
Translation
Protein

A sequence of nucleotides, called a promoter, identifies the region of DNA that will be transcribed into RNA.
Gene 1
Gene 2
5'
3'
3'
5'
DNA
Promoter 1
Promoter 2
The promoter orients RNA polymerase, determining the direction of transcription.
Template strand
RNA polymerase
DNA
Template strand
The direction of transcription dictates which strand of DNA is used as the template.
RNA 3'
5'
5'
3' RNA

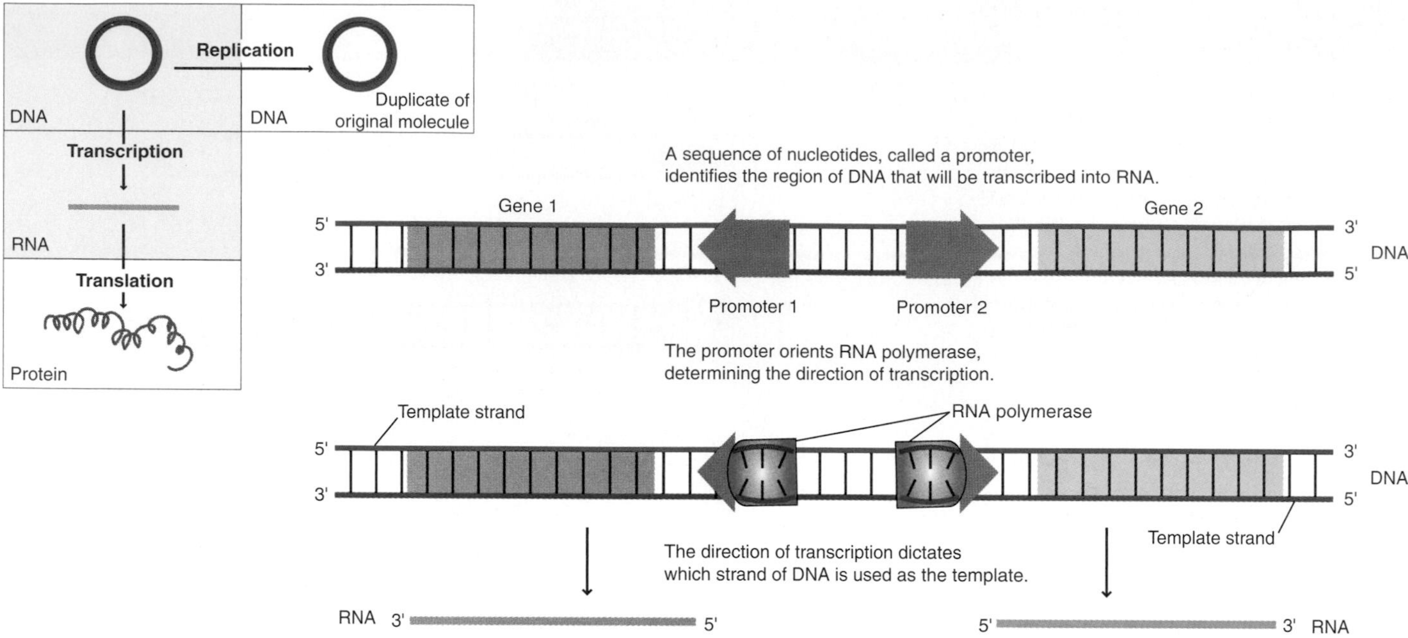

Figure 7.8 Promoters Direct Transcription A promoter not only identifies the region of DNA that will be transcribed into RNA, its orientation determines which strand will be used as the template.

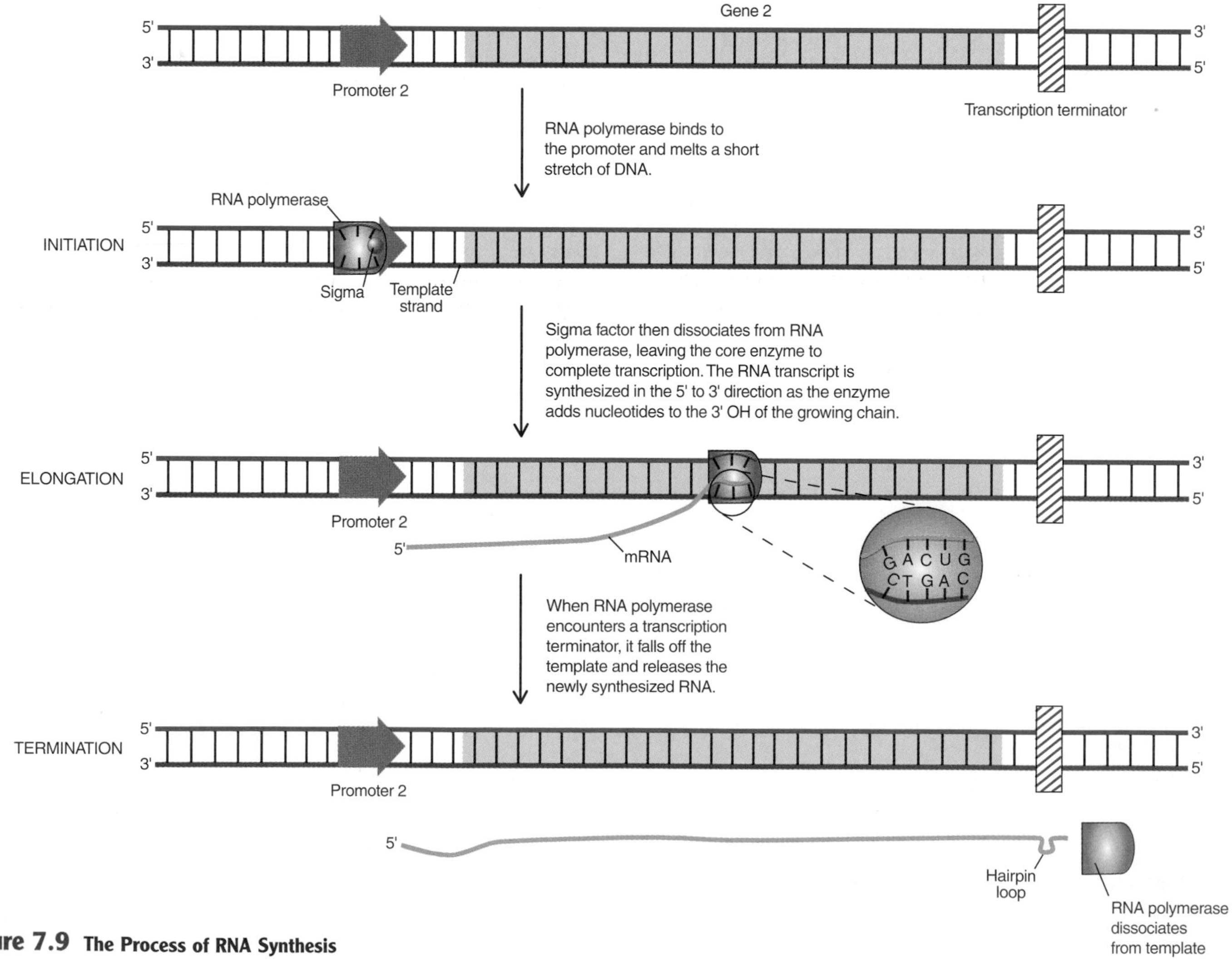

Gene 2
5'
3'
3'
5'
Promoter 2
Transcription terminator
RNA polymerase binds to the promoter and melts a short stretch of DNA.
RNA polymerase
INITIATION
Sigma
Template strand
Sigma factor then dissociates from RNA polymerase, leaving the core enzyme to complete transcription. The RNA transcript is synthesized in the 5' to 3' direction as the enzyme adds nucleotides to the 3' OH of the growing chain.
ELONGATION
Promoter 2
5'
mRNA
GACUG
CTGAC
When RNA polymerase encounters a transcription terminator, it falls off the template and releases the newly synthesized RNA.
TERMINATION
Promoter 2
5'
Hairpin loop
RNA polymerase dissociates from template

Figure 7.9 The Process of RNA Synthesis

Thus, a single gene can be transcribed multiple times in a very short time interval.

Termination

Just as the initiation of transcription occurs at a distinct site on the DNA, so does termination. When RNA polymerase encounters a **transcription terminator**, it falls off the DNA template and releases the newly synthesized RNA. The transcription terminator is a sequence of nucleotides in the DNA that, when transcribed, permits two complementary regions of the resulting RNA to base-pair, forming a hairpin loop structure. For reasons that are not yet understood, this causes the RNA polymerase to stall, resulting in its dissociation from the DNA template and release of the RNA.

Translation

Translation is the process of decoding the information carried on the mRNA to synthesize the specified protein. The proteins are synthesized by adding amino acid subunits sequentially to the carboxyl group at the end of an elongating polypeptide chain. Each amino acid added is specified by one codon on the mRNA, as directed by the genetic code. The process of translation requires three primary components: mRNA, ribosomes, and tRNAs. ■ carboxyl group, p. 27

First Letter	Middle Letter: U (Reading frame 5′ 3′)	Middle Letter: C (Reading frame 5′ 3′)	Middle Letter: A (Reading frame 5′ 3′)	Middle Letter: G (Reading frame 5′ 3′)	Last Letter
U	UUU Phenylalanine	UCU Serine	UAU Tyrosine	UGU Cysteine	U
	UUC Phenylalanine	UCC Serine	UAC Tyrosine	UGC Cysteine	C
	UUA Leucine	UCA Serine	UAA (Stop)	UGA (Stop)	A
	UUG Leucine	UCG Serine	UAG (Stop)	UGG Tryptophan	G
C	CUU Leucine	CCU Proline	CAU Histidine	CGU Arginine	U
	CUC Leucine	CCC Proline	CAC Histidine	CGC Arginine	C
	CUA Leucine	CCA Proline	CAA Glutamine	CGA Arginine	A
	CUG Leucine	CCG Proline	CAG Glutamine	CGG Arginine	G
A	AUU Isoleucine	ACU Threonine	AAU Asparagine	AGU Serine	U
	AUC Isoleucine	ACC Threonine	AAC Asparagine	AGC Serine	C
	AUA Isoleucine	ACA Threonine	AAA Lysine	AGA Arginine	A
	AUG Methionine (Start)	ACG Threonine	AAG Lysine	AGG Arginine	G
G	GUU Valine	GCU Alanine	GAU Aspartate	GGU Glycine	U
	GUC Valine	GCC Alanine	GAC Aspartate	GGC Glycine	C
	GUA Valine	GCA Alanine	GAA Glutamine	GGA Glycine	A
	GUG Valine	GCG Alanine	GAG Glutamine	GGG Glycine	G

Figure 7.10 The Genetic Code The genetic code correlates each series of three nucleotides, a codon, with one amino acid. Three of the codons do not code for an amino acid and instead serve as a stop codon, terminating translation. AUG functions as a start codon.

The Role of mRNA

The mRNA is a temporary copy of genetic information; it carries the encoded instructions for the synthesis of a specific polypeptide, or in the case of a polycistronic message, a specific group of polypeptides. That information is deciphered using the **genetic code (figure 7.10)**. This is a universal code, used by all living things, and correlates each series of three nucleotides, a codon, with one amino acid. Because a codon is a sequence of any combination of the four nucleotides, there are 64 different codons (4^3). Three of these are stop codons; these will be discussed later. The remaining 61 translate to the 20 different amino acids. This means that more than one codon can encode a specific amino acid. For example, both ACA and ACG encode the amino acid threonine. Because of this redundancy, the genetic code is said to be **degenerate**.

An equally important aspect of mRNA is that it carries the information that indicates where the coding region actually begins. This is critical because the genetic code is read as groups of three nucleotides. Thus, any given sequence has three possible **reading frames**, or ways in which triplets can be grouped **(figure 7.11)**. If translation occurs in the wrong reading frame, a very different, and generally nonfunctional, polypeptide would be synthesized.

The Role of Ribosomes

Ribosomes serve as the site of translation, and their structure functions to facilitate the joining of one amino acid to another. Ribosomes do this by bringing each amino acid into a favorable position so that a specific enzyme can catalyze the formation of a peptide bond between them. The ribosome also helps to identify key punctuation sequences on the mRNA molecule, such as the point at which protein synthesis should be initiated. The ribosome moves along the mRNA in the 5′ to 3′ direction, "presenting" each codon in a sequential order for deciphering, helping to maintain the correct reading frame.

Prokaryotic ribosomes are composed of a 30S subunit and a 50S subunit, each of which is made up of protein and rRNA; the "S" stands for Svedberg unit, which is a unit of size **(figure 7.12)**. Some of the ribosomal components are important

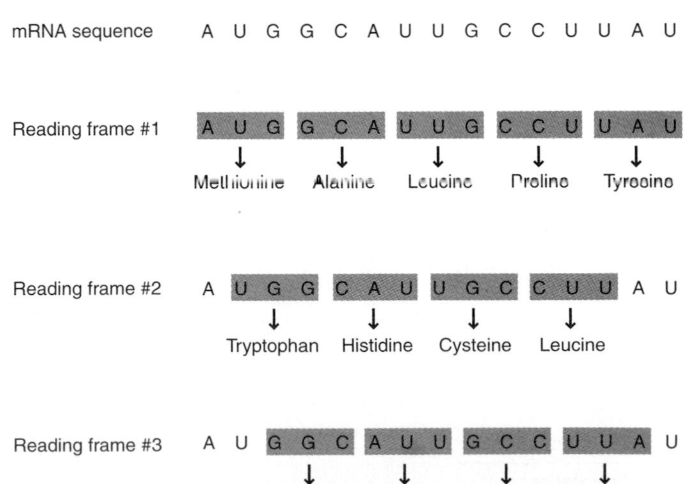

Figure 7.11 Reading Frames A nucleotide sequence has three potential reading frames. Because each reading frame encodes a very different order of amino acids, translation of the correct reading frame is important.

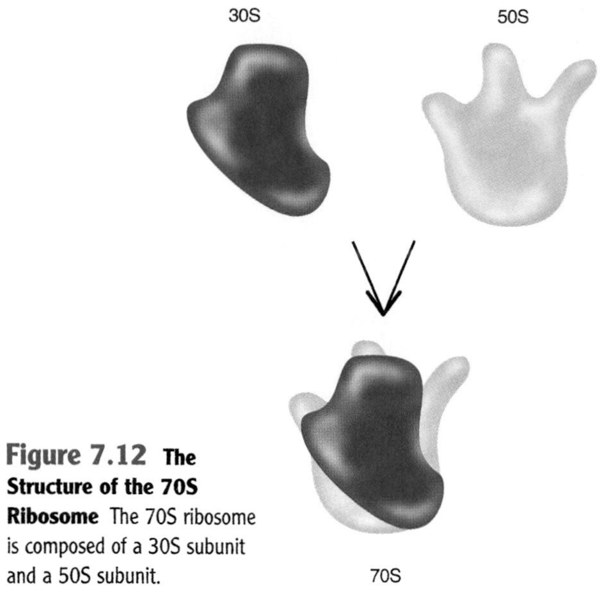

Figure 7.12 The Structure of the 70S Ribosome The 70S ribosome is composed of a 30S subunit and a 50S subunit.

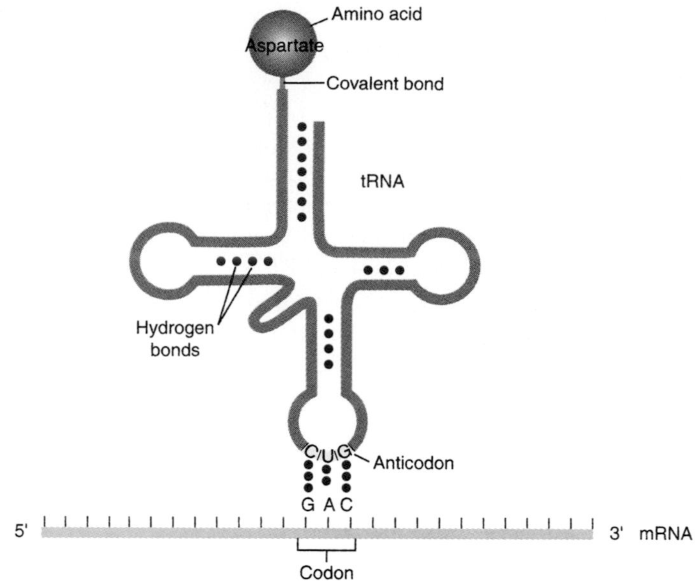

Figure 7.13 The Structure of Transfer RNA (tRNA) The anticodon of the tRNA base-pairs with a specific codon on the mRNA; by doing so, the appropriate amino acid is delivered to the site. The amino acid that the tRNA carries is dictated by the genetic code. The tRNA that recognizes the codon GAC carries the amino acid aspartic acid.

in other aspects of microbiology as well. For example, comparison of the nucleotide sequence of rRNA molecules is playing an increasingly prominent role in the establishment of the genetic relatedness of various organisms. Medically, ribosomal proteins and rRNA are significant because they are the targets of several groups of antimicrobial drugs. ■ **ribosomal subunits, p. 70** ■ **rRNA sequencing, p. 255** ■ **antimicrobial drugs, p. 497**

The Role of Transfer RNA

The tRNAs are segments of RNA that can carry specific amino acids, thus acting as keys that interpret the genetic code. They each recognize and base-pair with a specific codon and in the process deliver the appropriate amino acid to that site. This recognition is made possible because each tRNA has an **anticodon**, three nucleotides complementary to a particular codon in the mRNA. The amino acid each tRNA carries is dictated by its anticodon and the genetic code (**figure 7.13**).

Although there are 64 different codons, there are not that many different tRNA molecules. For example, **stop codons**, which signal the end of the protein, have no corresponding tRNA molecules that recognize them. Also, the anticodon of some tRNA molecules can recognize more than one codon. It appears that a certain amount of "wobbling" is tolerated in the base-pairing so that the recognition of the third nucleotide of the codon is not always precise. Due to the degeneracy of the genetic code, however, the correct amino acid is still incorporated.

Initiation of Translation

Initiation of translation begins even as the mRNA is still being synthesized (**figure 7.14**). The small 30S subunit of the ribosome binds to a sequence on mRNA called the **ribosome-binding site**. The first time the codon for methionine (AUG) appears after that site, translation generally starts. That first AUG is typically 7 nucleotides downstream of the ribosome-binding site. Note that AUG functions as a **start codon** only when preceded by a ribosome-binding site; at other sites, it simply encodes methionine. The position of the first AUG is critical, as it determines the reading frame used for translation of the remainder of that protein.

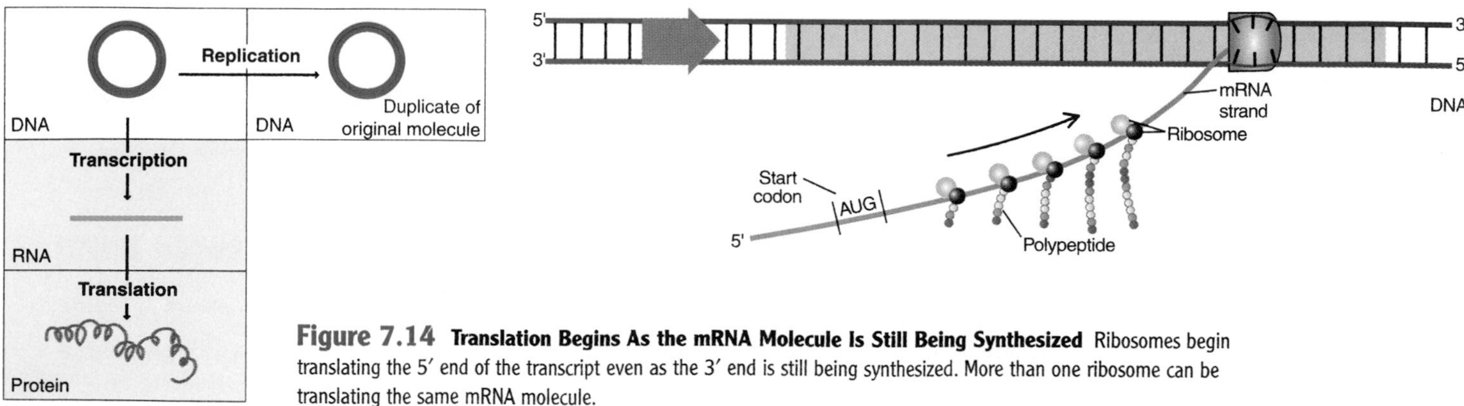

Figure 7.14 Translation Begins As the mRNA Molecule Is Still Being Synthesized Ribosomes begin translating the 5′ end of the transcript even as the 3′ end is still being synthesized. More than one ribosome can be translating the same mRNA molecule.

At that first AUG, the ribosome begins to assemble. An **initiation complex**, consisting of the 30S ribosomal subunit, a tRNA that carries a chemically altered form of the amino acid methionine, *N*-formylmethionine or **f-Met**, and proteins called **initiation factors,** comes together. Shortly thereafter, the 50S subunit of the ribosome joins that complex and the initiation factors leave, forming the 70S ribosome. The elongation phase then begins.

Elongation

The 70S ribosome has two sites to which tRNA-carrying amino acids can bind **(figure 7.15)**. One is called the **P-site** (peptidyl site), and the other is called the **A-site** (aminoacyl site, commonly referred to as the acceptor site). The initiating tRNA, carrying the f-Met, binds to the P-site. A tRNA that recognizes the next codon on the mRNA then fills the unoccupied A-site. The f-Met carried by the tRNA in the P-site is then covalently joined to the amino acid carried by the tRNA that just entered the A-site. This transfers the amino acid from the initiating tRNA to the amino acid carried by the incoming tRNA. **Translocation** now occurs. This results in the ribosome advancing a distance of one codon, and the tRNA that carried the f-Met is released through a recently discovered adjacent site called the **E-site** (exit site). Translocation requires several different proteins, called **elongation factors**. As a result of translocation the remaining tRNA, which now carries the two-amino-acid chain, occupies the P-site; the A-site is transiently vacant. A tRNA that recognizes the next codon then quickly fills the empty A-site, and the process repeats.

Once translation has progressed far enough for the ribosome to clear the ribosome-binding site and the first AUG, another ribosome can bind to initiate another round of synthesis of the encoded polypeptide. Thus, at any one time, multiple ribosomes can be translating a single mRNA molecule. This allows the maximal expression of protein from a single mRNA template. The assembly of multiple ribosomes attached to a single mRNA molecule is called a **polyribosome** or a **polysome**.

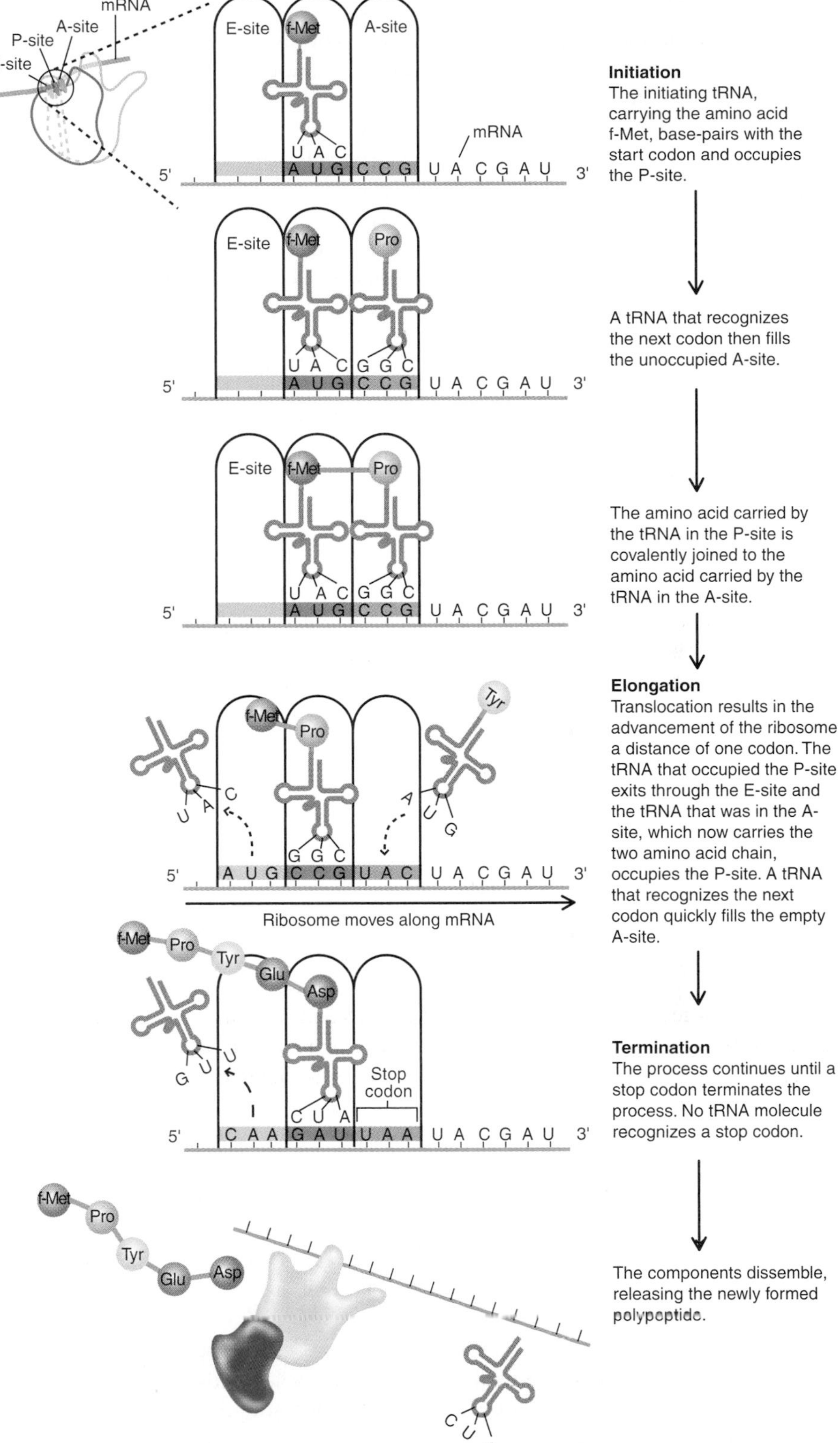

mRNA
A-site
P-site
E-site
E-site
f-Met
A-site
U A C
A U G C C G U A C G A U
5'
3'
mRNA
Initiation
The initiating tRNA, carrying the amino acid f-Met, base-pairs with the start codon and occupies the P-site.
Pro
U A C G G C
A tRNA that recognizes the next codon then fills the unoccupied A-site.
The amino acid carried by the tRNA in the P-site is covalently joined to the amino acid carried by the tRNA in the A-site.
Tyr
A U G C C G U A C U A C G A U
Ribosome moves along mRNA
Elongation
Translocation results in the advancement of the ribosome a distance of one codon. The tRNA that occupied the P-site exits through the E-site and the tRNA that was in the A-site, which now carries the two amino acid chain, occupies the P-site. A tRNA that recognizes the next codon quickly fills the empty A-site.
Glu
Asp
Stop codon
C U A
C A A G A U U A A U A C G A U
Termination
The process continues until a stop codon terminates the process. No tRNA molecule recognizes a stop codon.
The components dissemble, releasing the newly formed polypeptide.

Figure 7.15 The Process of Translation

Termination

Elongation of the polypeptide is terminated when the ribosome reaches a **stop codon**, a codon that does not code for an amino acid and is not recognized by a tRNA. At this point, proteins called **release factors** free the newly synthesized polypeptide by breaking the covalent bond that joins it to the tRNA. The ribosome falls off the mRNA and dissociates into its two component subunits, 30S and 50S. These can then be reused to initiate translation at other sites.

Post-Translational Modification

Proteins must often be modified after they are synthesized. For example, some proteins require the assistance of another protein, a **chaperone**, to fold into the final functional shape. Those proteins destined for transport outside of the cell membrane also must be modified. Such proteins have a characteristic series of hydrophobic amino acids, a **signal sequence,** at their amino terminal end, which "tags" them for transport through the membrane. The signal sequence is removed when the protein leaves the cytoplasm. ■ **chaperones, p. 32** ■ **hydrophobic amino acids, p. 28**

MICROCHECK 7.3

RNA polymerase recognizes sequences on DNA called promoters and at those sites initiates synthesis of RNA using one strand of DNA as a template. The RNA is synthesized in the 5′ to 3′ direction and is terminated when RNA polymerase encounters a transcription terminator. Translation occurs as ribosomes move along mRNA in the 5′ to 3′ direction, with the ribosomes serving as the structure that facilitates the joining of one amino acid to another. tRNAs carry specific amino acids, thus acting as keys that interpret the genetic code.

- How does the orientation of the promoter dictate which strand is used as a template for RNA synthesis?
- Explain why it is important for the translation machinery to recognize the correct reading frame.
- Could two mRNAs have different nucleotide sequences and yet code for the same protein?

Differences Between Eukaryotic and Prokaryotic Gene Expression

Eukaryotes differ significantly from prokaryotes in several aspects of transcription and translation (**table 7.1**). For example, in eukaryotic cells, most mRNA molecules are extensively modified, or **processed**, in the nucleus during and after transcription. Shortly after transcription begins, the 5′ end of the transcript is modified, or **capped**, by the addition of a methylated guanine derivative, creating what is called a **cap**. The cap is thought to serve several functions including stabilizing the transcript and enhancing translation. The 3′ end of the molecule is also modified, even before transcription has been terminated. This process, called **polyadenylation**, involves cleaving the transcript at a specific sequence of nucleotides and then adding a series of approximately 200 adenine derivatives to the newly exposed 3′ end. This creates what is called a **poly A tail**, which is thought to stabilize the transcript as well as enhance translation. Another important modification is **splicing**, a process that removes specific segments of the transcript (**figure 7.16**). Splicing is necessary because eukaryotic genes are not always contiguous; they are often interrupted by noncoding sequences. These intervening sequences, or **introns**, are transcribed along with the expressed regions, or **exons**, generating what is called **precursor mRNA**. The introns must be removed from precursor mRNA to form the mature mRNA that can be successfully translated.

Table 7.1 Major Differences Between Prokaryotic and Eukaryotic Transcription and Translation

Prokaryotes	Eukaryotes
mRNA is not processed.	A cap is added to the 5′ end of mRNA, and a poly A tail is added to the 3′ end.
mRNA does not contain introns.	mRNA contains introns, which are removed by splicing.
Translation of mRNA begins as it is being transcribed.	The mRNA transcript is transported out of the nucleus so that it can be translated in the cytoplasm.
mRNA is often polycistronic; translation begins at the first AUG that follows a ribosome-binding site.	mRNA is monocistronic; translation begins at the first AUG.

The mRNA in eukaryotic cells must be transported out of the membrane-bound nucleus before it can be translated in the cytoplasm. Thus, the same mRNA molecule cannot be transcribed and translated at the same time or even in the same cellular location. Unlike in prokaryotes, the mRNA of eukaryotes is generally monocistronic. Translation of the message begins at the first occurrence of AUG in the molecule.

As mentioned in chapter 3, the ribosomes of eukaryotes are different from those of prokaryotes. Whereas the prokaryotic ribosome is 70S, made up of 30S and 50S subunits, the eukaryotic ribosome is 80S, made up of 40S and 60S subunits. The differences in ribosome structure account for the ability of certain types of antibiotics to kill bacteria without causing significant harm to mammalian cells.

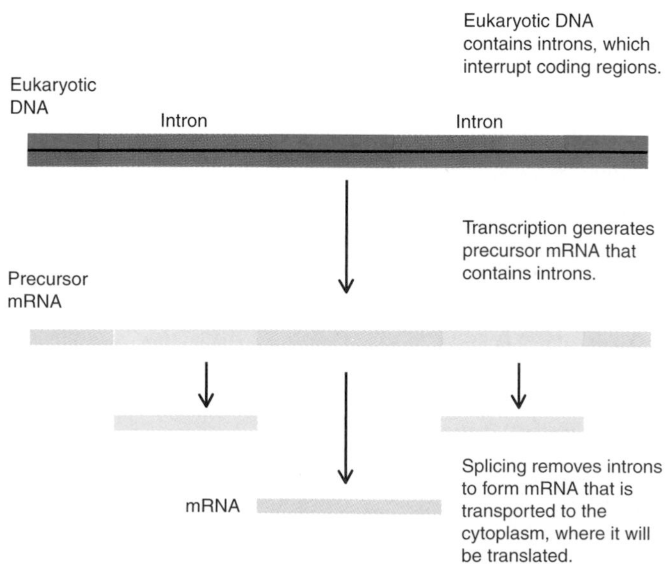

Figure 7.16 **Splicing of Eukaryotic RNA**

Perspective 7.2 RNA: The First Macromolecule?

The 1989 Nobel Prize in Chemistry was awarded to two Americans, Sidney Altman of Yale University and Thomas Cech of the University of Colorado, who independently made the surprising and completely unexpected observation that RNA molecules can act as enzymes. Before their studies, it was believed that only proteins had enzymatic activity. The key observation was made by Cech in 1982 when he was trying to understand how introns were removed from mRNA that coded for ribosomal RNA in the eukaryotic protozoan *Tetrahymena*. Since he was convinced that proteins were responsible for cutting out these introns, he added all of the protein in the cells' nuclei to the mRNA that still contained the introns.

As expected, the introns were cut out. As a control, Cech looked at the ribosomal RNA to which no nuclear proteins had been added, fully expecting that nothing would happen. Much to his surprise, the introns were also removed. It did not make any difference whether the protein was present—the introns were removed regardless. Thus, Cech could only conclude that the RNA acted on itself to cut out pieces of RNA.

The question remained of how widespread this phenomenon was. Did RNA have catalytic properties other than that of cutting out introns from rRNA? The studies of Altman and his colleagues, carried out simultaneously to and independently of Cech's, provided answers to these further questions. Altman's group found that RNA could convert a tRNA molecule from a precursor form to its final functional state. Additional studies have shown that enzymatic reactions in which catalytic RNAs, termed **ribozymes**, play a role are very widespread. Ribozymes have been shown to occur in the mitochondria of eukaryotic cells and to catalyze other reactions that resemble the polymerization of RNA. Whether catalytic RNA cuts out introns from mRNA in the nucleus is not known.

These observations have profound implications for evolution: Which came first, proteins or nucleic acids? The answer seems to be that nucleic acids came first, specifically RNA, which acted both as a carrier of genetic information as well as an enzyme. Billions of years ago, before the present universe in which DNA, RNA, and protein are found, probably the only macromolecule that existed was RNA. Once tRNA became available, these adapters could carry amino acids present in the environment to specific nucleotide sequences on a strand of RNA. In this scenario, the RNA functions as the genes as well as the mRNA.

Some of the proteins that play essential roles in translation differ between eukaryotic and prokaryotic cells. Diphtheria toxin, which selectively kills eukaryotic but not prokaryotic cells, illustrates this difference. This toxin is produced by *Corynebacterium diphtheriae*; it binds to and inactivates one of the elongation factors of eukaryotes. Since this protein is required for translocation of the ribosome, translation ceases and the eukaryotic cell dies, producing the typical symptoms of diphtheria. ■ diphtheria toxin, p. 556

MICROCHECK 7.4

Eukaryotic mRNA must be processed, which involves capping, polyadenylation, and splicing. In eukaryotic cells, the mRNA must be transported out of the nucleus before it can be translated in the cytoplasm. Eukaryotic mRNA is monocistronic.

- What is an intron?
- Explain the mechanism of action of diphtheria toxin.
- Would a deletion of two base pairs have a greater consequence if it occurred in an intron or in an exon?

Principles of Regulating Gene Expression

To cope with changing conditions in their environment, microorganisms have evolved elaborate control mechanisms to synthesize the maximum amount of cell material from a limited supply of energy. This is critical, because generally a microorganism must reproduce more rapidly than its competitors in order to be successful.

Consider the situation of *Escherichia coli*. For over 100 million years, it has successfully inhabited the gut of mammals, where it reaches concentrations of 10^6 cells per milliliter. In this niche, it must cope with alternating periods of feast and famine. For a limited time after a mammal eats, *E. coli* in the large intestine prosper, wallowing in the milieu of amino acids, vitamins, and other nutrients. The cells actively take up these compounds they would otherwise synthesize, expending minimal energy. Simultaneously, these cells shut down their biosynthetic pathways, channeling the conserved energy into the rapid synthesis of macromolecules, including DNA, RNA, and protein. Under these conditions, the cells divide at their most rapid rate. Famine, however, follows the feast. Between meals, which may be many days in the case of some mammals, the rich source of nutrients is depleted. Now the cells' biosynthetic pathways must be activated, utilizing energy and markedly slowing cell division. Cells that divide several times an hour in a nutrient-rich environment may divide only once every 24 hours in a famished mammalian gut.

A cell controls its metabolic pathways by two general mechanisms. The most immediate of these is the allosteric inhibition of enzymes. The most energy-efficient strategy, however, is to control the actual synthesis of the enzymes, making only what is required. To do this, cells have the ability to control the expression of certain genes. ■ allosteric inhibition, p. 143

Not all genes are subjected to the same mechanisms of regulation. Many are routinely expressed, whereas some are either turned on or off by certain conditions. Enzymes that are continuously synthesized at the same levels are **constitutive**; the genes encoding them are always active. These enzymes usually play indispensable roles in the pathways of central metabolism. For example, the enzymes of glycolysis are constitutive. Some enzymes are not produced routinely but can be **induced**; their synthesis is turned on only by certain conditions. Inducible enzymes are often involved in the utilization of specific energy sources. For example, the sole function of the enzyme **β-galactosidase** is to break down disaccharide lactose into its two component monosaccharides, glucose and galactose. It would be wasting precious energy for a cell to synthesize β-galactosidase when lactose is not present. Not surprisingly, β-galactosidase is an inducible enzyme. Other enzymes are ordinarily synthesized but can be **repressed**; their synthesis is turned off by certain conditions. This is typically a characteristic of enzymes of biosynthetic pathways. A cell requires the product of these enzymes; thus, the product must be either synthesized or available as a component of the medium. For example, cells must have a source of the amino acid tryptophan. If tryptophan is not available in the medium, a cell must produce the enzymes for its synthesis. When tryptophan is available in sufficient quantity, however, it would be a waste of energy for a cell to produce more. Thus, cells have evolved mechanisms by which tryptophan not only inhibits the activity of enzymes for its synthesis, but also prevents their continued production.

MICROCHECK 7.5

Genes encoding constitutive enzymes are always active. Genes encoding enzymes that can be induced are turned on only by certain conditions. Genes encoding enzymes that can be repressed are turned off by certain conditions.

- Explain the difference between an enzyme that is constitutive and one that is inducible.
- What is the function of β-galactosidase?
- Why would it be advantageous for a cell to control the activity of an enzyme as well as its synthesis?

Mechanisms to Control Transcription

Cells have mechanisms to control the frequency with which a particular gene is transcribed. One feature that allows a gene to routinely be expressed at a level that is different from other genes is the nucleotide sequence of its promoter. Not all promoter sequences are identical; rather, they have a common identifying theme, called a **consensus sequence**. In general, the more closely the actual nucleotide sequence resembles the consensus sequence, the more effective the promoter and the more frequently that particular RNA will be transcribed.

Cells also have a variety of mechanisms to modulate the expression of certain genes, generally by either preventing or facilitating transcription. Because these mechanisms are reversible, they can effectively control the relative number of transcripts made. Some control mechanisms involve proteins that bind to specific DNA sequences, whereas others exploit novel sigma factors that direct RNA polymerase to recognize different promoters. Proteins such as DNA-binding proteins and sigma factors that facilitate transcription are called **transcription factors**.

The mechanisms that control transcription can, in some cases, affect only a limited number of genes. In other cases, a wide array of genes is coordinately controlled. The simultaneous regulation of numerous unrelated genes is called **global control** or **global regulation**. For example, in *E. coli* the expression of more than 300 different genes is affected by the availability of glucose as an energy source.

DNA-Binding Proteins

Many genes have a **regulatory region** near their promoter to which a specific protein can bind, acting as a sophisticated on/off switch that controls transcription. When a **regulatory protein** binds to DNA, it can either block or enhance the function of RNA polymerase, depending on the location of its binding site relative to the promoter.

These regulatory proteins can control the transcription of one or more genes. The term **operon** refers to a set of genes that are coordinately controlled by a regulatory protein and transcribed as a single polycistronic message. In contrast, a **regulon** is a set of related genes that are transcribed as separate units but are controlled by the same regulatory protein.

Repressors

A **repressor** is a regulatory protein that blocks transcription. It does this by binding to DNA at a region, called the **operator**, located immediately downstream of a promoter. This effectively prevents RNA polymerase from progressing past that region. Regulation involving a repressor is called **negative control**.

Specific molecules bind to the repressor and, by doing so, alter the ability of the repressor to bind DNA. This can occur because repressors are allosteric proteins. They each have a distinct site to which another molecule can bind. When that molecule binds, the shape of the repressor is altered. In turn, this affects the ability of the repressor to bind DNA. As shown in **figure 7.17**, there are two general mechanisms by which different repressors can function:

1. The repressor is synthesized as a form that alone cannot bind to the operator. The binding to it of a molecule, termed a **corepressor**, however, alters its shape so that it can bind to the operator, blocking transcription.
2. The repressor is synthesized as a form that effectively binds to the operator, blocking transcription. The binding to it of a molecule that functions as an **inducer**, however, alters the shape of the repressor so that it no longer binds the operator.

Activators

An **activator** is a regulatory protein that facilitates transcription. Genes that are controlled by an activator have an ineffective promoter that is preceded by an **activator-binding site**. The binding of the activator to the DNA enhances the ability of RNA polymerase to initiate transcription at that promoter. Regulation involving an activator is sometimes called **positive control**.

Like repressors, activators are allosteric proteins whose function can be modulated by the binding of other molecules. A molecule that binds to an activator and alters its shape so it can effectively bind to the activator-binding site functions as an inducer (**figure 7.18**). Thus, the term inducer applies to a molecule that turns on transcription, either by stimulating the function of an activator or interfering with the function of a repressor.

Alternative Sigma Factors

Simultaneous activation of multiple genes sometimes involves specific promoter sequences that are only recognized by **alternative sigma (σ) factors**. Recall that the σ subunit of RNA polymerase is the portion that enables the enzyme to recognize a certain promoter sequence. While the cell's primary σ factor initiates transcription of most genes, alternative forms can direct the polymerase to transcribe more specialized sets of genes that are only required under specific environmental conditions such as nutrient deprivation.

Alternative σ factors are a common mechanism of global regulation. For example, many genes involved in the formation of endospores have promoters recognized by alternative σ factors. These alternative σ factors are only expressed during a certain stage of the 8-hour process of sporulation. As a consequence, a wide variety of genes whose promoters are only recognized by the alternative σ factor are activated during sporulation. ■ endospores, p. 72

MICROCHECK 7.6

A repressor blocks transcription when it binds to an operator. An activator enhances transcription when it binds to an activator-binding site. The functioning of specific activators and repressors may require or be blocked by other molecules. Alternative sigma factors direct RNA polymerase to recognize specific promoter sequences.

- Describe the differences between positive and negative control.
- Explain the specific role of an inducer in positive and negative control.
- Promoters are sometimes referred to by their strength (i.e., "strong" or "weak"). How would the strength of a promoter relate to its consensus sequence?

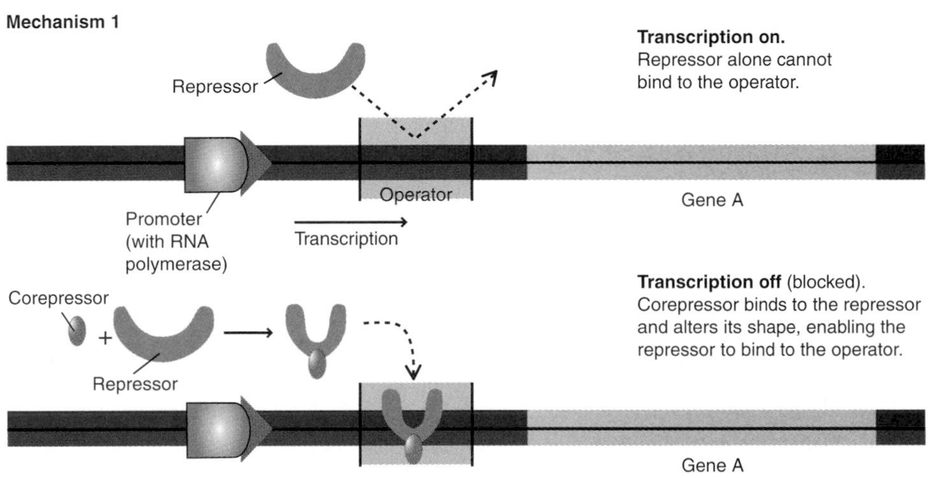
Mechanism 1
Repressor
Transcription on.
Repressor alone cannot bind to the operator.
Operator
Gene A
Promoter (with RNA polymerase)
Transcription
Corepressor
Repressor
Transcription off (blocked).
Corepressor binds to the repressor and alters its shape, enabling the repressor to bind to the operator.
Gene A

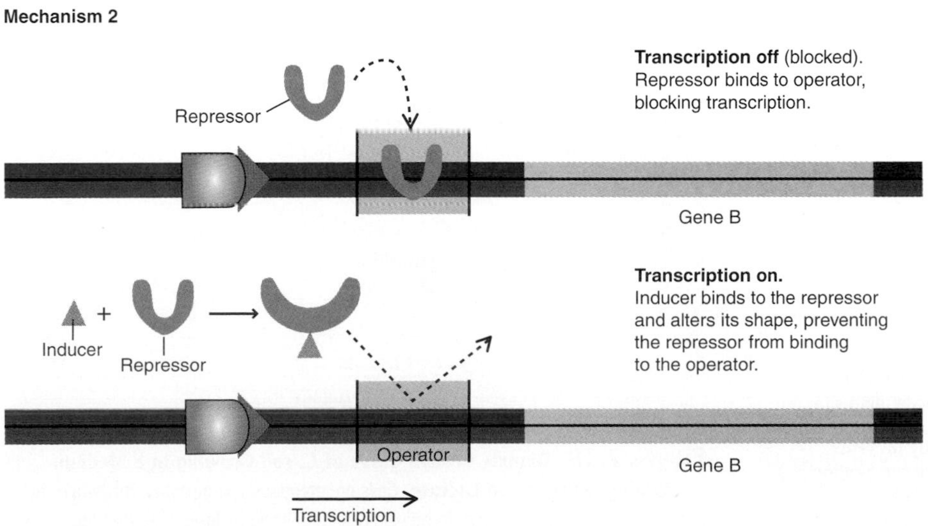
Mechanism 2
Repressor
Transcription off (blocked).
Repressor binds to operator, blocking transcription.
Gene B
Inducer
Repressor
Transcription on.
Inducer binds to the repressor and alters its shape, preventing the repressor from binding to the operator.
Operator
Gene B
Transcription

Figure 7.17 **Transcriptional Regulation by Repressors**

Modulating Expression in Response to Environmental Conditions

Microorganisms respond to changing conditions, such as the presence or absence of specific nutrients, by altering the level of expression of certain genes **(table 7.2)**. This ability requires a means of sensing the conditions in both their internal and external environments. Any change must then be communicated to the regulatory proteins so that the cell can turn on or off the appropriate genes accordingly. There are various strategies that cells use to sense what compounds are available.

Repression of Biosynthetic Pathways

The product of a biosynthetic pathway generally plays a pivotal role in regulating the genes that encode the enzymes required for its synthesis. It may do this by serving as a corepressor. One of the best characterized examples is the ***trp* operon**, which encodes five enzymes that catalyze the sequential steps of the pathway for tryptophan biosynthesis. When present at a high concentration, tryptophan serves as a corepressor by binding to the repressor, which is expressed continuously. This binding allows the repressor to bind to the operator and, as a consequence, block transcription of the operon. Other regulatory mechanisms help to further fine-tune the expression of the *trp* operon.

Induction of Degradative Pathways

Transcription of genes that encode enzymes of degradative pathways is turned on in response to an inducer. The inducer is generally the substrate of the first enzyme of that particular pathway. The mechanism by which the inducer promotes expression depends on whether an operon is under negative or positive control—that is whether the regulatory protein is a repressor that binds to an operator, or an activator that binds to an activator-binding site.

The lactose operon, or ***lac* operon,** encodes proteins required for the degradation of lactose and has served as one of the most important models for studying gene regulation. The operon is under negative control by a repressor. Unlike the repressor of the *trp* operon, however, this repressor can bind to the operator without the aid of another molecule. Because the repressor

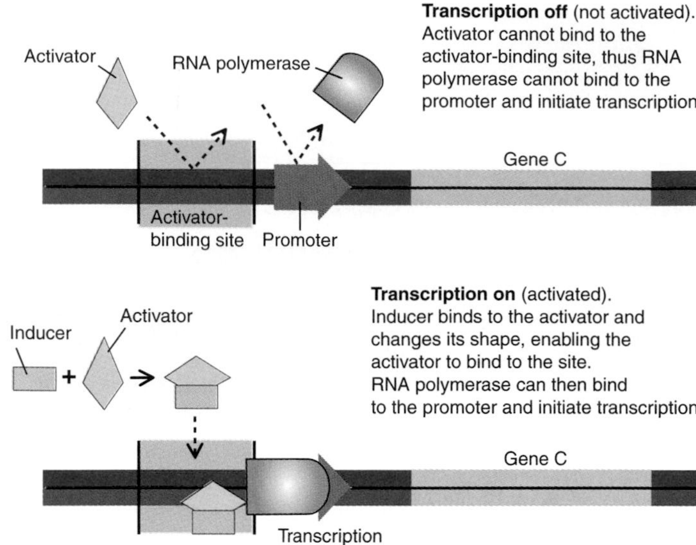
Activator
RNA polymerase
Transcription off (not activated). Activator cannot bind to the activator-binding site, thus RNA polymerase cannot bind to the promoter and initiate transcription.
Gene C
Activator-binding site
Promoter
Inducer
Activator
Transcription on (activated). Inducer binds to the activator and changes its shape, enabling the activator to bind to the site. RNA polymerase can then bind to the promoter and initiate transcription.
Gene C
Transcription

Figure 7.18 Transcriptional Regulation by Activators

alone can bind and is always expressed, transcription of the *lac* operon is normally blocked. Lactose can bind to the repressor, however, preventing the repressor from binding to the operator. Without the repressor bound to the operator, RNA polymerase can then transcribe the genes for lactose degradation.

The maltose regulon, which encodes the enzymes for the degradation of the sugar maltose, is under positive control by an activator. The genes are only expressed when the activator binds to the activator-binding site, facilitating transcription. In a situation that is in some ways similar to repression of the *trp* operon, the activator alone cannot bind to DNA. Only when the activator is bound by maltose does it assume the proper shape to attach to the activator-binding site. Thus, maltose plays a fundamental role in facilitating the expression of the genes required for its utilization.

Table 7.2 Examples of Positive and Negative Control of Gene Expression

Type of Control/Example		Mechanism
Negative Control		
Tryptophan operon	Repression	Tryptophan enables repressor to bind to operator.
Lactose operon	Induction	Lactose prevents repressor from binding to operator.
Positive Control		
Maltose regulon	Induction	Maltose enables activator to bind to activator-binding site.
Catabolite repression	Repression	Glucose indirectly prevents the facilitation of the binding of the activator to the activator-binding site.

While the preceding examples illustrate the general phenomenon of induction, the control of degradative pathways often includes an additional layer of complexity. This is because cells must respond not only to whether or not a particular energy source is present, but also to whether a more suitable alternative is available. For example, the enzymes for glucose degradation are constitutive; it would be wasteful for a cell to synthesize additional enzymes required to use other sugars when glucose is present.

To cope with the need for different responses, cells often use separate but interconnected strategies to control the expression of genes that encode degradative enzymes. While induction responds to the presence of the substrate, the strategy discussed next indirectly detects the availability of glucose.

Catabolite Repression

E. coli preferentially uses glucose over other sugars. This can readily be demonstrated by observing growth and sugar utilization of *E. coli* in a medium containing glucose and lactose. Cells actively grow, metabolizing only glucose until its supply is exhausted **(figure 7.19)**. Growth then ceases for a short period until the cells begin utilizing lactose. At this point, the cells again start actively multiplying. This two-step growth response, **diauxic growth**, represents the ability of glucose to repress the enzymes of lactose degradation—a phenomenon called **catabolite repression.**

The regulatory mechanism of catabolite repression does not directly sense glucose in a cell **(figure 7.20)**. Instead, it recognizes the concentration of a nucleotide derivative, **cyclic AMP** (**cAMP**), which is low when glucose is present. When

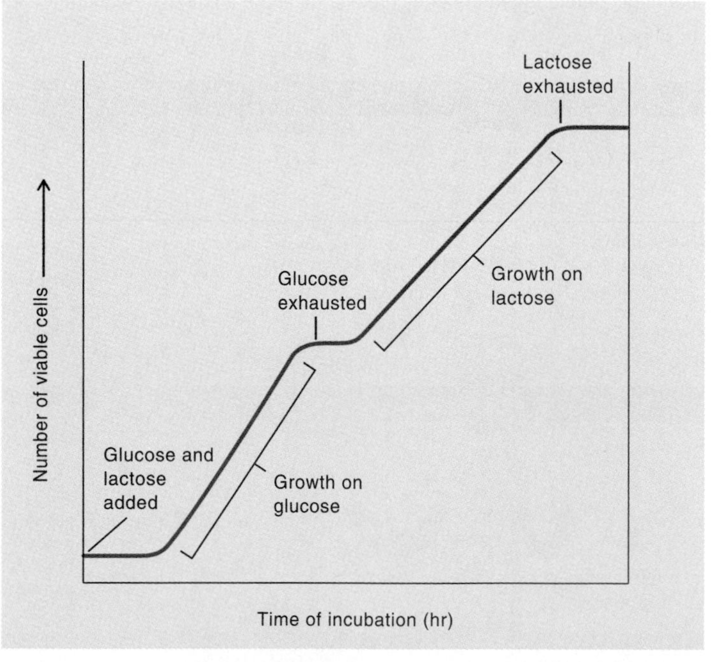
Lactose exhausted
Glucose exhausted
Growth on lactose
Number of viable cells
Glucose and lactose added
Growth on glucose
Time of incubation (hr)

Figure 7.19 Diauxic Growth Curve of *E. coli* Growing in a Medium Containing Glucose and Lactose Cells preferentially use glucose. Only when the supply of glucose is exhausted do cells start metabolizing lactose. Note that the growth on lactose is slower than it is on glucose.

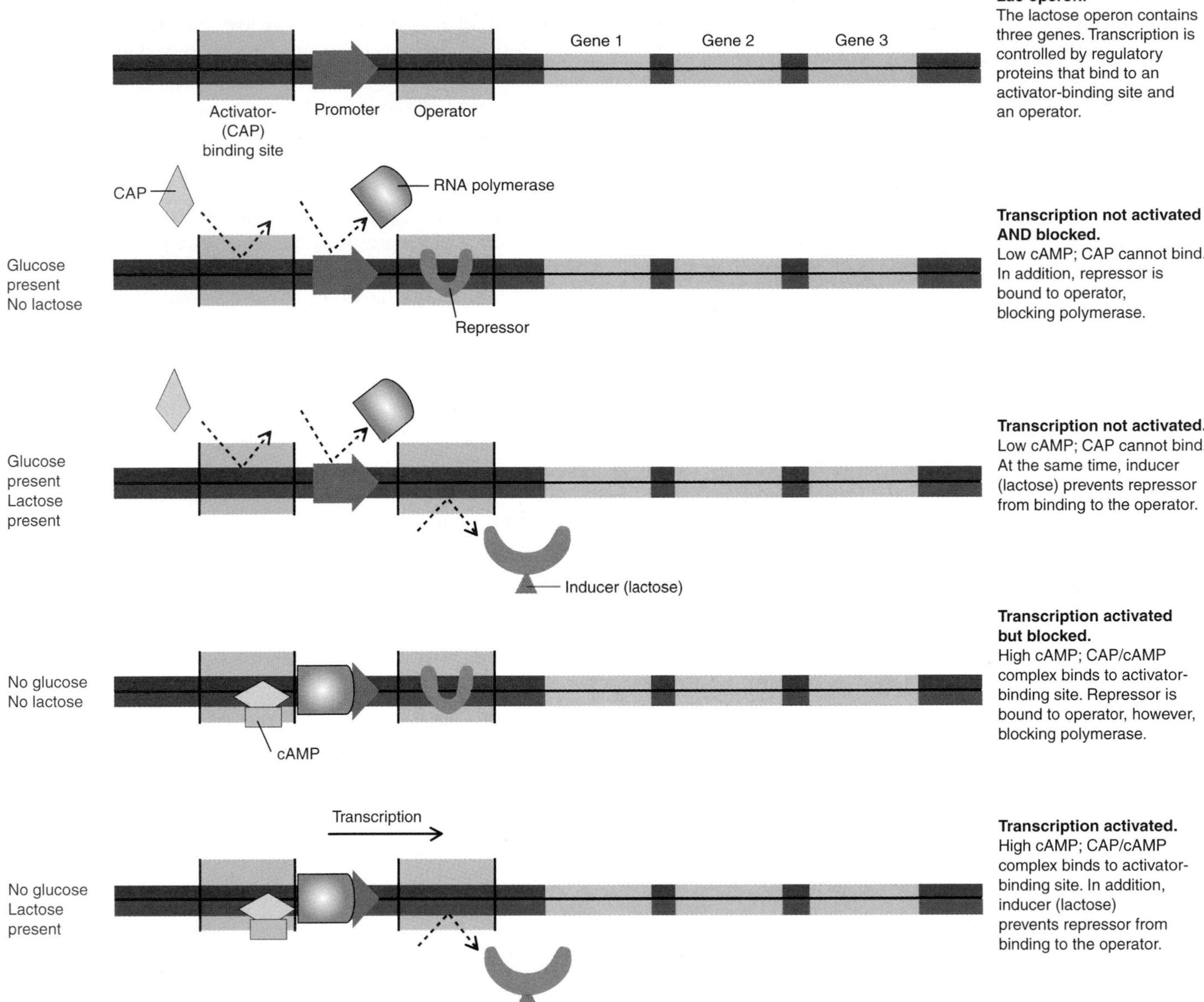
Gene 1
Gene 2
Gene 3
Activator-
(CAP)
binding site
Promoter
Operator
Lac operon.
The lactose operon contains three genes. Transcription is controlled by regulatory proteins that bind to an activator-binding site and an operator.
CAP
RNA polymerase
Glucose present
No lactose
Repressor
Transcription not activated AND blocked.
Low cAMP; CAP cannot bind. In addition, repressor is bound to operator, blocking polymerase.
Glucose present
Lactose present
Inducer (lactose)
Transcription not activated.
Low cAMP; CAP cannot bind. At the same time, inducer (lactose) prevents repressor from binding to the operator.
No glucose
No lactose
cAMP
Transcription activated but blocked.
High cAMP; CAP/cAMP complex binds to activator-binding site. Repressor is bound to operator, however, blocking polymerase.
Transcription
No glucose
Lactose present
Transcription activated.
High cAMP; CAP/cAMP complex binds to activator-binding site. In addition, inducer (lactose) prevents repressor from binding to the operator.

Figure 7.20 **Regulation of the *lac* Operon**

glucose levels are high, cAMP levels are low, and enzymes for the degradation of other sugars are repressed. In contrast, when glucose levels are low, and therefore cAMP levels are high, these enzymes can be synthesized, provided that their substrate is also available.

Expression of enzymes that are subject to catabolite repression is under the control of an activator called **CAP** (catabolite activator protein). The genes are not efficiently transcribed unless CAP binds near the promoter region, enhancing the ability of RNA polymerase to initiate transcription; this is an example of positive control. CAP alone, however, does not have the proper shape to bind to the promoter region. It is the binding of cAMP to CAP that changes the shape of that activator and allows it to bind. Thus, these genes can only be transcribed when cAMP levels are high, which occurs when glucose levels are low. Because CAP controls a wide variety of genes, catabolite repression is an example of global regulation.

Note that genes controlled by CAP may also be subjected to other control mechanisms. For example, the lactose operon is also affected by the presence of its inducer. The *lac* operon is only transcribed when cAMP is bound to CAP and lactose is bound to the repressor.

Signal Transduction

Signal transduction is a process that transmits information from outside a cell to the inside, allowing that cell to respond to changing environmental conditions. For example, cells turn on or off certain genes in response to variations in such factors as osmotic pressure, cell concentration, and nitrogen availability.

Two-Component Regulatory Systems

One mechanism that cells use to relay information about the external environment to the relevant genes is a **two-component regulatory system**. This system relies on the coordinated activities of two different proteins, a sensor and a response regulator. The **sensor** spans the cytoplasmic membrane so that the part that recognizes changes in the environment is outside the cell. In response to specific changes in the external environment, the sensor chemically modifies a region on its internal portion, usually by phosphorylating a specific amino acid. The phosphoryl group is transferred to a **response regulator (figure 7.21)**. The modified response regulator can then act as either an activator or a repressor, turning on or off genes, depending on the system.

Bacteria use different two-component regulatory systems to detect and respond to a wide variety of environmental cues. *E. coli*, for example, uses such systems to control the expression of genes for its alternative types of metabolism. When nitrate is present in anaerobic conditions, cells activate genes required to use nitrate as a terminal electron acceptor. Some pathogens use two-component regulatory systems to sense environmental magnesium concentrations, and then activate specific genes in response. Because the magnesium concentration within certain tissue cells is generally lower than that of extracellular sites, these pathogens are able to recognize whether or not they are within a cell. In turn, they can activate appropriate genes that facilitate their survival against the onslaught of host defenses intended to protect that relative site.

Quorum Sensing

Some organisms can "sense" the density of cells within their own population—a phenomenon called **quorum sensing**. This enables them to activate genes that are only beneficial when expressed by a critical mass of cells. The most studied example of quorum sensing is bacterial bioluminescence. The marine bacterium *Vibrio fischeri* can emit light, but only when growing in a critical mass. Light production by this organism is an intensely energy-consuming process and is presumably only beneficial when enough cells participate.

Bacteria that utilize quorum sensing synthesize one or more varieties of a **homoserine lactone** (**HSL**) (or AHL for **acylated homoserine lactone**). These small molecules can move freely in and out of a cell. When few cells are present, the concentration of a given HSL is very low. As the cells multiply in a confined area, however, the concentration of that HSL increases proportionally. Only when it reaches a critical level does it induce the expression of specific genes.

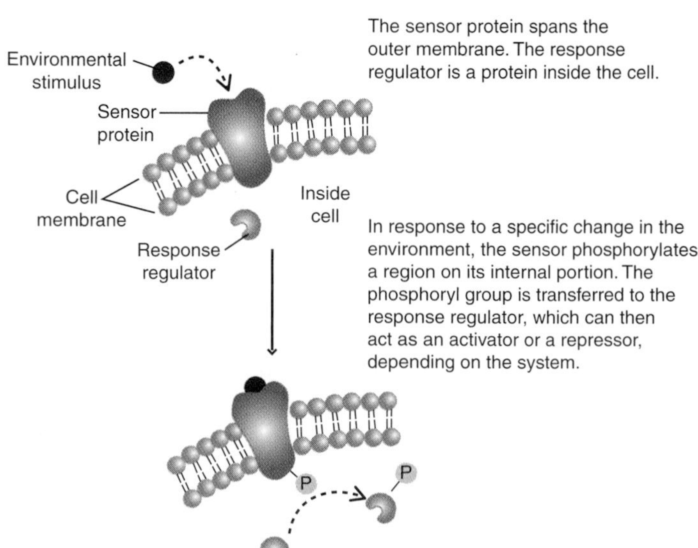
Environmental stimulus
Sensor protein
Cell membrane
Inside cell
Response regulator
The sensor protein spans the outer membrane. The response regulator is a protein inside the cell.
In response to a specific change in the environment, the sensor phosphorylates a region on its internal portion. The phosphoryl group is transferred to the response regulator, which can then act as an activator or a repressor, depending on the system.
P
P

Figure 7.21 Two-Component Regulatory System

Natural Selection

Natural selection can also play a role in the control of gene expression. The expression of some genes changes randomly, presumably enhancing the chances of survival of at least a subset of a population under certain environmental conditions. This is most readily apparent in bacteria that undergo **antigenic variation**, a routine alteration in the characteristics of certain surface proteins such as flagella, pili, and outer membrane proteins. Disease-causing organisms that are able to change these proteins can stay one step ahead of the body's defenses by altering the very molecules our immune systems must learn to recognize. One of the best characterized examples of bacteria that undergo antigenic variation is *Neisseria gonorrhoeae*. This organism is highly successful at disguising itself from the immune system because it can adeptly change several of its surface proteins. One of these is **pilin**, the protein subunit that makes up pili. *N. gonorrhoeae* appears to have many different genes for this protein, yet most are not expressed. The only one that is expressed resides in a particular chromosomal location called an **expression locus**. *N. gonorrhoeae* has a mechanism to shuffle the pilin genes, randomly moving different ones in and out of the expression locus. In a population of 10^4 cells, at least one cell is expressing a different type of pilin. It appears that expression of different pilin genes is not regulated in any controlled manner but instead occurs randomly. Only some of the changes, however, are advantageous to a cell's survival. When the body's immune system eventually begins to respond to a specific pilin type, those cells that have already "switched" to produce a different type will survive and then multiply. Eventually, the immune system will learn to recognize those, but by that time, another subpopulation will have "switched" its pilin type. Thus, natural selection serves to indirectly regulate the changes. ■ **pili, p. 69** ■ ***Neisseria gonorrhoeae*, p. 644**

Another mechanism of randomly altering gene expression is **phase variation**, the routine switching on and off of certain genes. Presumably, phase variation helps an organism adapt to selective pressures. By routinely altering the expression of certain critical genes, at least a part of the population is apt to be poised for change and thus able to survive and multiply. For example, phase variation of genes that encode fimbriae may allow some members of a population to attach to a surface, while permitting others to detach and colonize surfaces elsewhere.

MICROCHECK 7.7

End products of biosynthetic pathways typically repress expression of the associated enzymes, whereas substrates of degradative pathways typically induce the expression of the associated enzymes. Signal transduction allows a cell to respond to changing conditions outside of that cell. The expression of some genes changes randomly, presumably enhancing the chances of survival of at least a subset of a population of cells under varying environmental conditions.

- Explain the mechanism by which glucose represses the lactose operon.
- Explain the mechanism by which certain bacteria can "sense" the density of cells.
- Why would it be advantageous for a bacterium to be able to synthesize more than one type of homoserine lactone?

Genomics

The increasingly rapid methods of determining the nucleotide sequence of DNA have led to exciting advancements in genomics. Fueled by the commitment to sequence the entire human genome, scientists honed the methodologies by first sequencing the genomes of select microorganisms. In 1995, the sequence of the chromosome of *Haemophilus influenzae* was published, marking the first complete genomic sequence ever determined. By January 2000, the genome of more than 30 other organisms had been completely sequenced **(table 7.3)**.

Although sequencing methodologies are becoming more rapid, analyzing the resulting data and extracting the pertinent information is far more complex than it might initially seem. One of the most difficult steps is to locate and characterize the potential protein-encoding regions. Imagine trying to determine the amino acid sequence of a protein encoded by a 1,000-base-pair (bp) stretch of DNA without knowing anything about the orientation of the promoter or the reading frame of the transcribed mRNA. Since either strand of the double-stranded DNA molecule could potentially be the template strand, two entirely different mRNA molecules could potentially code for the protein. In turn, each of those two molecules has three reading frames, for a total of six reading frames. Yet only one of these actually codes for the protein. Understandably, computers are an invaluable aid and are used extensively in deciphering the meaning of the raw sequencing data. In turn, this has resulted in the emergence of a new field, **bioinformatics**, to create the computer technology to store, retrieve, and analyze nucleotide sequence data.

Analyzing a DNA Sequence

When analyzing a DNA sequence, the nucleotide sequence of the (+) strand is used to infer information contained in the corresponding RNA transcript. Because of this, terms like start codon, which actually refers to a sequence on mRNA, are used to describe sequences on DNA. For example, to locate the start codon AUG, which would be found on mRNA, one would look for the analogous sequence, ATG, on the (+) strand of the DNA molecule. In most cases it is not initially known which of the two strands is actually used as a template for RNA synthesis, so that both strands are potentially a (+) strand. Only after a promoter is located is it known which strand in a given region is actually the (+) strand.

Table 7.3 Representative Microorganisms Whose Complete Genomes Have Been Sequenced

Name of Organism	Genome Size (10^6 base pairs)	Important Characteristics
Bacillus subtilis	4.20	Endospore-former; has served as a model for studies of Gram-positive bacteria.
Borrelia burgdorferi	1.44	Important human pathogen; causative agent of Lyme disease.
Deinococcus radiodurans	3.28	Radiation-resistant bacterium. Genome consists of two chromosomes, a large plasmid and a small plasmid.
Escherichia coli	4.6	Has served as model for studies of Gram-negative bacteria. Common inhabitant of the intestinal tract.
Haemophilus influenzae	1.83	First bacterial genome sequenced; important human pathogen; causes ear and respiratory infections and meningitis, mostly in children.
Helicobacter pylori	1.66	Important human pathogen; causes gastric diseases, including stomach ulcers.
Lactococcus (Streptococcus) lactis	2.35	Important bacterium to the dairy industry; used to make cheeses and other fermented milk products.
Methanococcus jannaschii	1.75	First archaeal genome sequenced; also the first autotrophic organism sequenced; hyperthermophile isolated from a hydrothermal vent; strict anaerobe; methane producer.
Mycobacterium tuberculosis	4.40	Important human pathogen; causative agent of tuberculosis.
Mycoplasma genitalium	0.58	Smallest known bacterial genome; represents what might be the minimal genome; human pathogen.
Saccharomyces cerevisiae	13	Yeast; first eukaryotic genome completed.
Synechocystis species	3.57	Cyanobacterium; has served as a model for studies of photosynthesis.
Treponema pallidum	1.14	Important human pathogen; causes syphilis; has not been cultured *in vitro*.

To locate protein-encoding regions, computers are used to search for **open reading frames (ORFs)**, stretches of DNA, generally longer than 300 bp, that begin with a start codon and end with a stop codon. An ORF potentially encodes a protein. Other characteristics, such as the presence of an upstream sequence that can serve as a ribosome-binding site, also indicate that an ORF encodes a protein. Another helpful indicator is **codon usage**. Recall that more than one codon can code for the same amino acid. Most organisms have a bias toward the use of certain codons, and that tendency differs among groups of organisms. Thus, an ORF that is consistent with the typical codon usage of that particular organism is more likely to encode a protein.

The nucleotide sequence of the ORF or the deduced amino acid sequence of the encoded protein can be compared with other known sequences by searching computerized databases of published sequences. Not surprisingly, as the genomes of more organisms are being sequenced, the information contained in these databases is growing at a remarkable rate. If the encoded protein shows certain amino acid similarities, or **homology**, to characterized proteins, a putative function can sometimes be assigned. For example, DNA-binding proteins share amino acid sequences in certain regions. Likewise, regulatory regions on DNA such as promoters and operators can sometimes be identified based on the nucleotide homologies to known sequences.

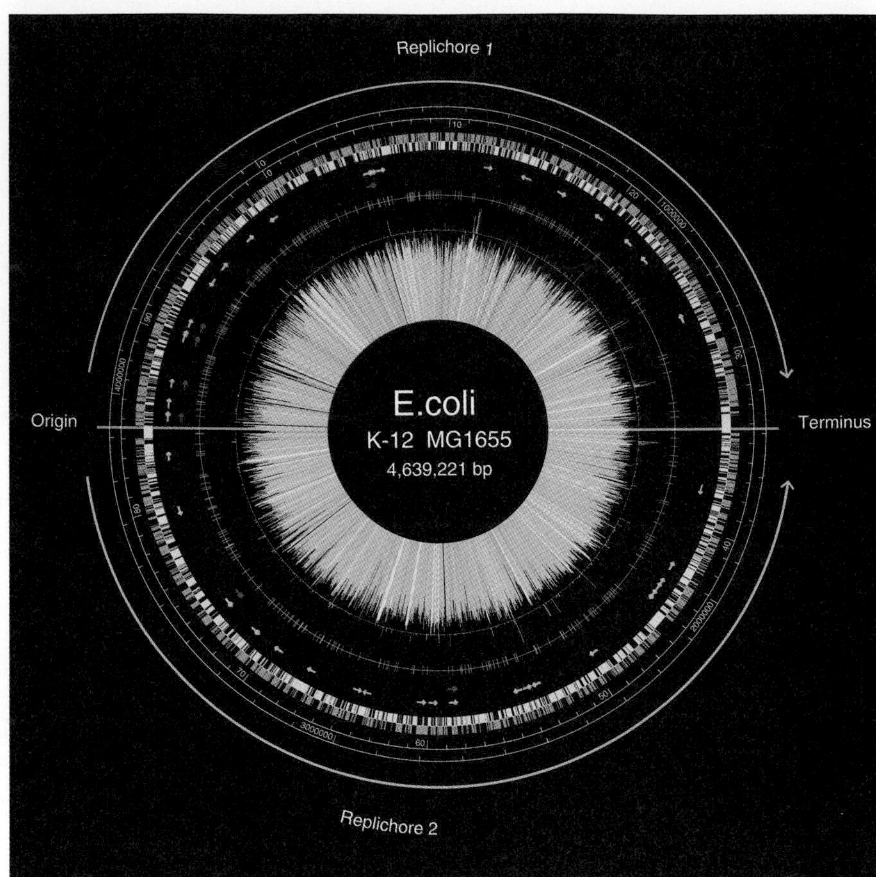

Figure 7.22 The *E. coli* Genome

The *E. coli* Genome

One of the earliest candidates for complete genomic sequencing was *E. coli*. Because this bacterium has been used as a model in so many scientific studies, a great deal is already known about it; sequencing data can be used to further enhance our understanding.

In 1997, the 4,639,221-base-pair genome of *E. coli* was published. According to the analysis of that data, protein-coding genes account for 88% of the genome, whereas less than 1% codes for rRNA and tRNA **(figure 7.22)**. Of the 4,288 proteins presumably encoded, only 1,853 had previously been characterized. Of the uncharacterized genes, less than one-third of the encoded proteins could be assigned a putative function based on the predicted amino acid sequence. Consistent with previous findings, the largest functional groups of proteins are involved with (1) transport and binding and (2) energy metabolism.

MICROCHECK 7.8

The first genomic sequence of a microorganism was completed in 1995. The sequencing methodologies are quickly becoming more rapid, but analyzing the data and extracting the pertinent information is difficult.

- What is an open reading frame?
- Describe two things that can be learned by searching a computerized database for sequences that have homologies to a newly sequenced gene.
- There are some characteristic differences in the nucleotide sequences of the leading and lagging strands. Why might this be so?

FUTURE CHALLENGES

Capitalizing on Genomics

From a medical standpoint, one of the most exciting challenges will be to capitalize on the rapidly accruing genomic information and use that knowledge to develop new drugs and therapies. The potential gains are tremendous, particularly in the face of increasing resistance to current antimicrobial drugs. For example, by studying the genomes of pathogenic microorganisms, scientists can learn more about specific genes that enable an organism to cause disease. Already we know that many of these are encoded in large segments called ***pathogenicity islands*** *and that the genes in these segments are often coordinately regulated. By learning more about the signals and mechanisms that turn these genes on and off, scientists may be able to one day design a drug that prevents the synthesis of critical bacterial proteins. Such a drug could interfere with that pathogen's ability to survive within our body and thereby render it harmless.* ■ **resistance to antimicrobial drugs, p. 509**

Learning more about the human genome provides another means of developing drug therapies. Already, companies are searching genomic databases, a process called ***genome mining,*** *to locate ORFs that may encode proteins of medical value. What they generally look for are previously uncharacterized proteins that have certain sequence similarities to proteins of proven therapeutic value. Some of their discoveries are now in clinical trials to test their efficacy. For example, a protein involved in bone-building is being tested as a treatment for osteoporosis. Likewise, another protein discovered through genome mining may facilitate the healing of wounds. Genes encoding many other medically useful proteins are probably still hidden, waiting to be discovered.*

SUMMARY

Overview (Figure 7.1)

1. **Replication** is the process of duplicating double-stranded DNA.
2. **Transcription** is the process of copying the information encoded on DNA into RNA.
3. **Translation** is the process of interpreting the information carried by **messenger RNA** to synthesize the encoded protein.

Characteristics of DNA (Figure 7.3)

1. A single strand of DNA has a 5′ end and a 3′ end.
2. The two strands of DNA in the double helix are **antiparallel**; they are oriented in opposite directions.
3. The separating of double-stranded DNA is called **denaturing** or **melting.**

Characteristics of RNA

1. A single-stranded RNA fragment is transcribed from one of the two strands of DNA.
2. There are three different functional groups of RNA molecules: **messenger RNA** (**mRNA**), **ribosomal RNA** (**rRNA**), and **transfer RNA** (**tRNA**).

Regulating the Expression of Genes

1. Protein synthesis is generally controlled by regulating the synthesis of mRNA molecules.
2. mRNA is short-lived because **RNases** degrade it within minutes.

DNA Replication

1. DNA replication is generally **bidirectional** from a distinct starting point in circular DNA. (Figure 7.4)
2. Replication of double-stranded DNA is **semiconservative**; each of the two molecules generated contains one of the original strands (the template strand) and one newly synthesized strand.
3. The DNA chain always elongates in the 5′ to 3′ direction; the base-pairing rules determine the specific nucleotides that are added. (Figure 7.5)

Initiation of DNA Replication

1. DNA replication begins at the **origin of replication**. **DNA polymerase** synthesizes DNA in the 5′ to 3′ direction, using one strand as a **template** to generate the complementary strand.

The Replication Fork (Figure 7.6)

1. The bidirectional progression of replication around a circular DNA molecule creates two replication forks.
2. Numerous enzymes and other proteins are involved in DNA replication, including DNA polymerase, **primase**, **DNA ligase**, and **DNA gyrase**.
3. **DNA polymerase** adds nucleotides to the 3′ OH group at the end of the growing chain.

Gene Expression

Transcription

1. The enzyme **RNA polymerase** catalyzes the process of transcription, producing a single-stranded RNA molecule that is complementary and antiparallel to the DNA template. (Figure 7.7)
2. In prokaryotes, a mRNA molecule can be **monocistronic** or **polycistronic**.
3. **Transcription** begins when RNA polymerase recognizes and binds to a **promoter**; it is the **sigma factor** of the enzyme that recognizes the promoter sequence. (Figure 7.8)
4. RNA is synthesized in the 5′ to 3′ direction as polymerase adds nucleotides to the 3′ OH group at the end of the growing chain. (Figure 7.9)
5. When **RNA polymerase** encounters a **transcription terminator**, it falls off the DNA template and releases the newly synthesized RNA.

Translation

1. The information encoded on mRNA is deciphered using the genetic code. (Figure 7.10)
2. A nucleotide sequence has three potential **reading frames**. (Figure 7.11)
3. **Ribosomes** function as the site of translation occurs. (Figure 7.12)
4. **tRNAs** carry specific amino acids and act as keys that interpret the genetic code. (Figure 7.13)
5. Initiation of translation begins when the ribosome binds to the **ribosome-binding site** on the mRNA molecule; this occurs even while the mRNA is still being synthesized. Translation starts at the first AUG downstream of that site. (Figure 7.14)
6. The ribosome moves along mRNA in the 5′ to 3′ direction so that one codon is translated at a time. Translation terminates when the ribosome reaches a **stop codon.** (Figure 7.15)
7. Proteins are often modified after they are synthesized; those that contain a signal sequence are transported to the outside of the cell.

Differences Between Eukaryotic and Prokaryotic Gene Expression (Table 7.1)

1. Eukaryotic mRNA is **processed**; a cap and a **poly A tail** are added.

2. Eukaryotic genes often contain **introns** which are removed from **precursor mRNA** by a process called **splicing**. (Figure 7.16)
3. In eukaryotic cells, the mRNA must be transported out of the nucleus before it can be translated in the cytoplasm. Eukaryotic mRNA is typically **monocistronic**.

Principles of Regulating Gene Expression

1. Genes encoding **constitutive** enzymes are always active.
2. Genes encoding enzymes that can be **induced** are turned on only by certain conditions; those that can be **repressed** are turned off by certain conditions.

Mechanisms to Control Transcription

DNA-Binding Proteins (Table 7.2)

1. Many genes have a **regulatory region** near their promoter to which a specific protein can bind, thereby controlling transcription.
2. A **repressor** is a regulatory protein that blocks transcription. (Figure 7.17)
3. An **activator** is a regulatory protein that enhances transcription. (Figure 7.18)

Alternative Sigma Factors

1. **Alternative sigma factors** direct RNA polymerase to recognize specific promoter sequences; they are a common mechanism of **global regulation**.

Modulating Expression in Response to Environmental Conditions

Repression of Biosynthetic Pathways

1. The product of a biosynthetic pathway, for example, tryptophan, regulates the genes that encode the enzymes required for its synthesis by functioning as a **corepressor**.

Induction of Degradative Pathways

1. The substrate of a degradative pathway typically regulates the genes that encode the enzymes required for its degradation by serving as an **inducer**.
2. Lactose activates the ***lac* operon** by binding to a repressor. This prevents the repressor from blocking transcription.
3. Maltose activates the genes required for its degradation by binding to an activator. This allows the activator to bind to the **activator-binding site**.

Catabolite Repression

1. **Catabolite repression** turns off certain genes when more readily degradable energy sources, such as glucose, are available. (Figure 7.20)

Signal Transduction

1. **Two-component regulatory systems** utilize a **sensor** that recognizes changes outside the cell and then transmits that information to a **response regulator.** (Figure 7.21)
2. Bacteria that utilize **quorum sensing** synthesize a soluble compound, a **homoserine lactone**, which can move freely in and out of a cell. Only when that compound reaches a critical concentration does it activate specific genes.

Natural Selection

1. The expression of some genes changes randomly, presumably enhancing the chances of survival of at least a subset of a population under varying environmental conditions.
2. **Antigenic variation** is a routine change in the expression of surface proteins such as flagella, pili, and outer membrane proteins.
3. **Phase variation** is the routine switching on and off of certain genes.

Genomics

Analyzing a DNA Sequence

1. When analyzing a DNA sequence, the nucleotide sequence of the (+) strand is used to infer information carried by the corresponding RNA transcript.
2. Computers are used to search for **open reading frames** (**ORFs**).
3. The nucleotide sequence of the ORF or the amino acid sequence of the encoded protein can be compared with other known sequences by searching a computerized database that contains all published sequences.

The *E. coli* Genome (Figure 7.22)

1. In 1997, the 4,639,221-base-pair genome of *E. coli*, which has served as a model bacterium in many scientific studies, was published.

REVIEW QUESTIONS

Short Answer

1. Explain what the term *semiconservative* means with respect to DNA replication.
2. How can *E. coli* have a generation time of only 20 minutes when it takes 40 minutes to replicate its chromosome?
3. What is the function of primase in DNA replication? Why is this enzyme necessary?
4. What is polycistronic mRNA?
5. Explain why knowing the orientation of a promoter is critical when determining the amino acid sequence of an encoded protein.
6. What is characteristic about the nucleotide sequence of a transcription terminator?
7. What happens to a polypeptide that has a signal sequence?

8. Compare and contrast the regulation of the genes that encode enzymes for the degradation of lactose and maltose.
9. Explain how bacteria sense the density of cells in their own population.
10. Explain why it is sometimes difficult to locate genomic regions that encode a protein.

Multiple Choice Questions

1. All of the following are involved in transcription, *except*
   A. polymerase.
   B. primer.
   C. promoter.
   D. sigma factor.
   E. uracil.
2. All of the following are involved in DNA replication, *except*
   A. elongation factors.
   B. gyrase.
   C. polymerase.
   D. primase.
   E. primer.
3. All of the following are directly involved in translation, *except*
   A. promoter.
   B. ribosome.
   C. start codon.
   D. stop codon.
   E. tRNA.
4. Using the DNA strand depicted below as a template, what will be the sequence of the RNA transcript?

   5′ GCGTTAACGTAGGC 3′

   ⟶ promoter

   3′ CGCAATTGCATCCG 5′

   A. 5′ GCGUUAACGUAGGC 3′
   B. 5′ CGGAUGCAAUUGCG 3′
   C. 5′ CGCAAUUGCAUCCG 3′
   D. 5′ GCCUACGUUAACGC 3′
5. A ribosome binds to the following mRNA at the site indicated by the dark box. What are the first three amino acids that will be incorporated into the resulting polypeptide?

   5′ ■ GCCGGAAUGCUGCUGGC

A. alanine	aspartic acid	alanine
B. methionine	threonine	cysteine
C. methionine	leucine	leucine
D. alanine	glycine	methionine

6. Maltose induces the *mal* regulon by binding to a(n)...
   A. operator.
   B. repressor.
   C. activator.
   D. CAP protein.
7. Under which of the following conditions will transcription of the *lac* operon occur?
   A. Lactose present/glucose present
   B. Lactose present/glucose absent
   C. Lactose absent/glucose present
   D. Lactose absent/glucose absent
   E. A and B
8. Which of the following statements about gene expression is *false*?
   A. More than one RNA polymerase can be transcribing a specific gene at a given time.
   B. More than one ribosome can be translating a specific transcript at a given time.
   C. Translation begins at a site called a promoter.
   D. Transcription stops at a site called a terminator.
   E. Some amino acids are coded for by more than one codon.
9. Which of the following is *not* characteristic of eukaryotic gene expression?
   A. 5′ cap is added to the mRNA.
   B. A poly A tail is added to the 3′ end of mRNA.
   C. Introns must be removed to create the mRNA that is translated.
   D. The mRNA is often polycistronic.
   E. Translation begins at the first AUG.
10. Which of the following statements is *false*?
   A. Tryptophan serves as a corepressor of the *trp* operon.
   B. Signal transduction provides a mechanism for a cell to sense the conditions of its external environment.
   C. The function of homoserine lactone is to enable a cell to sense the density of like cells.
   D. An example of a two-component regulatory system is the maltose operon, which is controlled by a repressor and an activator.
   E. An ORF is a stretch of DNA that may encode a protein.

Applications

1. A graduate student is trying to isolate the gene coding for an enzyme found in a species of *Pseudomonas* that degrades trinitrotoluene (TNT). The student is frustrated to find that the organism does not produce the enzyme when grown in nutrient broth, making it is difficult to collect the mRNA needed to help identify the gene. What could the student do to potentially increase the amount of the desired enzyme?
2. A student wants to remove the introns from a segment of DNA coding for protein X. Devise a strategy for how this could be accomplished.

Critical Thinking

1. The study of protein synthesis often uses a cell-free system where cells are ground with an abrasive to release the cell contents and then filtered to remove the abrasive. The following materials are added to the system, generating the indicated results:

Materials Added	Results
Radioactive amino acids	Radioactive protein produced
Radioactive amino acids *and* RNase (an RNA-digesting enzyme)	No radioactive protein produced

What is the best interpretation of these observations?

2. In a variation of the experiment in the previous question, the following materials were added to three separate cell-free systems, generating the indicated results:

Materials Added	Results
Radioactive amino acids	Radioactive protein produced
Radioactive amino acids *and* DNase (a DNA-digesting enzyme)	Radioactive protein produced
Several hours after grinding: Radioactive amino acids *and* DNase	No radioactive protein produced

What is the best interpretation of these observations?

CHAPTER 8

Bacterial Genetics

Barbara McClintock (1902–1992) was a remarkable scientist who made several crucial discoveries in genetics dealing with chromosome structure. Her studies were carried out before the age of large interdisciplinary research teams and before the sophisticated tools of molecular genetics were available. Her tools consisted of a clear mind that could make sense of confusing and revolutionary observations, and a consuming curiosity that led her to work 12-hour days, 6 days a week in a small laboratory at Cold Spring Harbor on Long Island, New York.

In 1983, at age 81, McClintock received the Nobel Prize in Medicine or Physiology largely for her discovery 40 years earlier of transposable elements, or transposons, popularly called "jumping genes." Her experimental system consisted of ears of corn. She observed various colored kernels that were produced by different enzymes (see figure 8.5). If the gene coding for an enzyme responsible for color formation was inactivated, the kernel was not pigmented. If the enzyme was only partially inactivated, then the kernel was partially pigmented. Thus, by looking at kernel colors, McClintock could detect changes in genes. She concluded that fragments of DNA must be capable of moving from one site on the chromosome to another, because the kernels had different colors. A transposable element moving into a gene would inactivate it. When the element left a gene, and the gene was restored, it would function properly again.

At the time McClintock published her results, most scientists believed that chromosomes were very stable and unchanging. Consequently, most geneticists were very skeptical of McClintock's heretical ideas. As a result, she stopped publishing many of her observations. It was not until the late 1970s that her ideas were accepted. By that time, transposable elements had been discovered in many organisms, including bacteria. Although first discovered in plants, once transposons were found in bacteria, the field moved ahead very quickly. The techniques of molecular biology, biochemistry, and genetics made it possible to isolate, characterize, and understand the movement of these remarkable pieces of DNA.

—A Glimpse of History

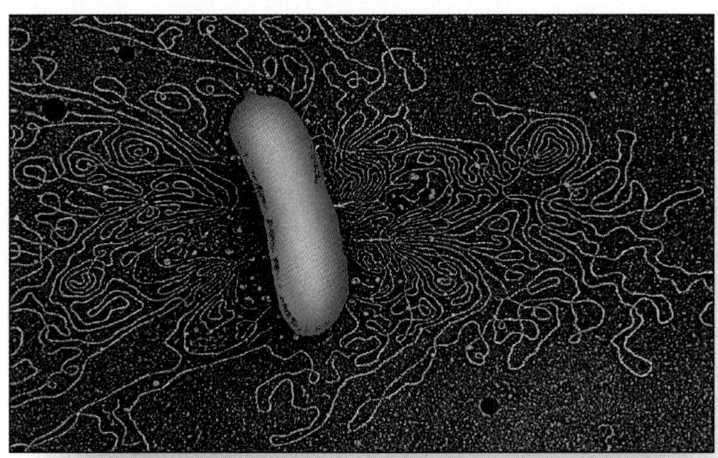

AT THE END OF WORLD WAR II, BACTERIAL DYSENtery (caused by members of the genus *Shigella*) was a common disease in Japan. With the introduction of the sulfa drugs, the number of cases of dysentery decreased, as this antimicrobial medication proved highly successful. After 1949, however, the incidence of dysentery began to rise again, and many of the *Shigella* strains causing dysentery were resistant to sulfa. Fortunately, these organisms remained susceptible to such antibiotics as streptomycin, chloramphenicol, and tetracycline. A new problem arose, however. In 1955, a Japanese woman returning from Hong Kong developed *Shigella* dysentery that did not respond to any antimicrobial treatment.

In the next several years, a number of dysentery epidemics developed in Japan. In some patients, the causative organisms were sensitive to antimicrobial medications; in others, they were resistant. Curiously, bacteria isolated from certain patients treated with a single antimicrobial medication abruptly became resistant not only to that medication but also to other antimicrobials as well. In addition, cells of *Escherichia coli* residing in the large intestine of many of the patients who carried antimicrobial-resistant organisms were also resistant to the same antimicrobial medications. How can these observations be explained? A reasonable conclusion is that the genes responsible for antimicrobial resistance spread from one bacterium to another. The phenomenon of genes being able to jump from one site on DNA to another, discovered by McClintock, plays an important role in the movement of antimicrobial-resistance genes between bacteria.

Antimicrobial medications have reduced the incidence of many microbial infections. The development and spread of resistant strains of bacteria, however, have created unique problems for the treatment of infectious diseases. Understanding the mechanisms by which organisms become resistant to antimicrobials and how this resistance is transferred to sensitive cells may enable the medical community to keep one step ahead of disease-causing bacteria and maintain the usefulness of antimicrobials.

Bacterial genetics encompasses the study of heredity—how genes function, how they can change, and how they are transferred to other cells in the population. The functioning of genes was discussed in chapter 7. In this chapter we will focus on how the chemical structure of DNA can change or mutate, thereby conferring new properties on the cell. These changes allow bacteria to adapt to changing environmental conditions. We will also discuss how genes are transferred from one bacterium to another, thereby changing the properties of cells into which the genes are transferred.

Sources of Diversity in Microorganisms

The tremendous diversity seen in different microorganisms in the biological world stems from two major factors. First, nucleotide sequences of DNA can undergo changes in their chemical structure and arrangement, resulting in organisms with different properties. These alterations are called **mutations**, and the organisms are **mutants**. Second, the expression of genetic information that organisms possess is regulated so that environmental conditions turn some genes on and others off, depending on the particular environment. The environmental influences on gene expression have been discussed in chapter 7. ■ DNA structure, p. 35

The properties of a cell determined by its DNA composition are its **genotype** and, if the DNA is altered, the genotype changes. For example, a cell with a mutation in a gene required for the synthesis of this amino acid must be supplied with histidine in order to grow. The cell is **auxotrophic** for the growth factor histidine; a cell that requires a growth factor is an **auxotroph** (*auxo-* means "increase," as an increase in requirements). The genotype is designated by a three-letter abbreviation, in this case, *his*. The first letter is not in caps, and the three letters are italicized. A superscript minus is used when the cell requires the compound to grow. Thus the genotype of a cell requiring histidine for growth because of a lack of a functional gene for histidine synthesis is *his*$^-$. If a cell has all of the genes functional for histidine synthesis, its genotype is *his*$^+$. The cell is said to be **prototrophic** for histidine (*proto-* means "lowest," as in the least number of requirements). The term **allele** is used to indicate that there are two different forms of the gene, in this case *his*$^+$ and *his*$^-$. To indicate all the genes that a cell possesses is cumbersome, and so the genotype of a cell is designated by what is lacking rather than what is present. A cell is assumed to be prototrophic for all growth requirements unless otherwise indicated. ■ growth factor, p. 100

The observable characteristics displayed by an organism in any given environment make up its **phenotype**. Thus, the phenotype results from the genotype plus the properties conferred on the cell by the environment. The phenotype of a cell that requires histidine is abbreviated His$^-$ where the first letter is capitalized and the word is not italicized. To summarize, the genotype refers to the DNA; the phenotype describes the actual properties of the organism. Changes in either the DNA or the expression of its genes will alter the properties of the cell.

MICROCHECK 8.1

The properties of bacteria can change either through mutations, a change in genotype, or by alterations in gene expression, a change in phenotype.

- Contrast genotype and phenotype.
- Which has a longer-lived effect on a cell, a change in the genotype, or a change in the phenotype? Explain.

GENE MUTATION

The organism in which a mutation occurs is the **parent strain**, which is commonly called the **wild-type** strain. The change in a nucleotide or a nucleotide sequence may lead to an altered phenotype such as the requirement for histidine, or it may not lead to any observable change in which case it is called a **silent mutation**. Any change in phenotype results from altering the ability of the protein coded by the gene to function. The substitution of even one amino acid for another in a critical location in the protein may cause the protein to be nonfunctional, thereby changing the properties of the cell. For example, if any gene of the tryptophan operon is altered so that the enzyme for which it codes no longer functions, the cells will grow only if this amino acid is in its environment. ■ enzymes, p. 135 ■ protein structure, p. 30 ■ operon, p. 180

Spontaneous Mutations

Spontaneous mutations occur without the addition of agents, called **mutagens**, that are known to cause mutations. Genes mutate infrequently and randomly, but each gene mutates spontaneously at a characteristic frequency. The **rate of mutation** is defined as the probability that a mutation will occur in a given gene each time a cell divides; this rate is generally expressed in an exponential form. The mutation rate of different genes usually varies between 10^{-4} and 10^{-12} per cell division. In other words, the chances that any single gene will undergo a mutation when one cell divides into two are between one in 10,000 (10^{-4}) and one in a trillion (10^{-12}). ■ exponents, Appendix I

Mutations are stable, so that the progeny of streptomycin-sensitive cells that mutate to streptomycin resistance will remain streptomycin-resistant. On rare occasions, however, the nucleotide will change back to its original state, resulting in the streptomycin-resistant cells becoming streptomycin-sensitive. These cells would be killed if streptomycin were present in the medium. The change in a cell's genotype and phenotype to its original state through a change in the mutated gene is termed **reversion.** It occurs spontaneously at low frequencies that can be increased by mutagens.

Base Substitution

The most common type of mutation results from a mistake during DNA synthesis, when an incorrect base is incorporated into the DNA, an event called **base substitution (figure 8.1)**. If only one base pair is changed, the mutation is called a **point mutation**. Base substitution errors occur because purines and pyrimidines exist in two structural forms. The most common form results in proper base-pairing between adenine and thymine and between guanine and cytosine. The H atoms, however, can move to form a rare form of the base that alters its hydrogen-bonding properties **(figure 8.2)**. This is termed a **tautomeric shift**. Thus, a shift in the hydrogen atom of adenine allows it to bond to cytosine, and a shift in the H atom in thymine allows this base to hydrogen bond to guanine **(see figure 8.2)**. This mistake in incorporation then is passed on to the cell's progeny (offspring), resulting in an incorrect amino acid being incorporated into the protein coded by the gene (see figure 8.1). Frequently, the incorporation of one incorrect amino acid results in the synthesis of a protein that still functions partially. For example, a mutation in a gene of tryptophan biosynthesis may result in a strain that grows slowly in the absence of tryptophan. Such a mutation is termed **leaky.** If the mutation totally inactivates the gene resulting in a strain that is unable to grow at all unless tryptophan is added, it is termed a **null** or **knockout** mutation. ■ hydrogen bonding, p. 23 ■ purines and pyrimidines, p. 35

Removal or Addition of Nucleotides

The deletion (removal) or addition of one or several purine or pyrimidine nucleotides also changes the nucleotide sequence. Because translation of a gene begins at a specific codon and proceeds one codon at a time, the deletion or addition of a nucleotide shifts the codons of the DNA when it is transcribed into mRNA **(figure 8.3)**. This type of mutation is termed a **frameshift mutation**. If three nucleotides are added or removed, however, then a single amino acid usually will be added to or subtracted from the protein encoded by the DNA. ■ codon, p. 169 ■ translation, p. 175

A frameshift affects all amino acids incorporated beyond the original site at which the addition or deletion occurred. Frequently, one of the many new codons generated by the frameshift will be a **stop codon,** so that the protein synthesized will be incomplete and nonfunctional. Further, not only will the gene in which the frameshift occurred be affected, but all genes in an operon downstream of the mutation site, may also be inactivated. Frameshift mutations change many more codons than are changed by the substitution of a single base, and they most likely result in a knockout mutation. ■ stop codon, p. 177 ■ operon, p. 180, downstream genes, p. 172

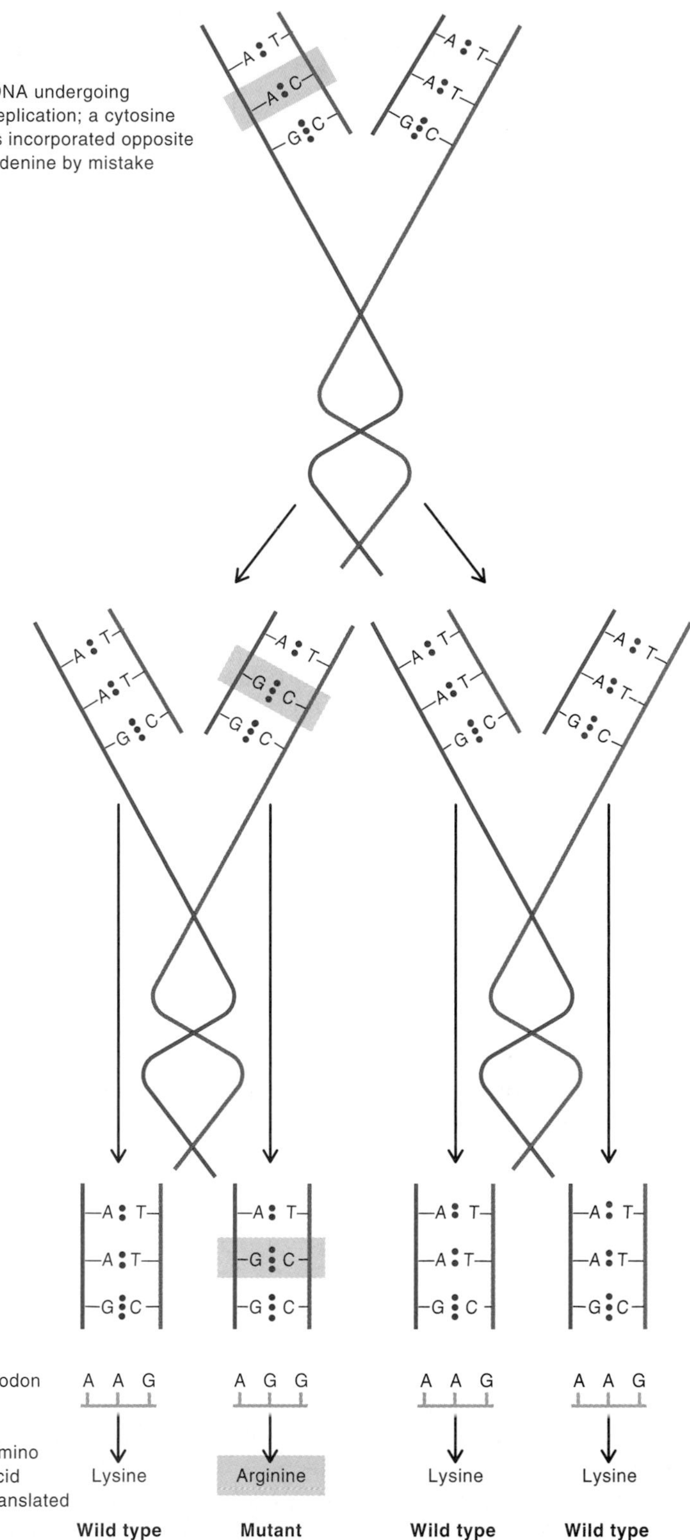

Figure 8.1 Base Substitution Shown here is the generation of a mutant organism as a result of the incorporation of a pyrimidine base (cytosine) in place of thymine in DNA replication.

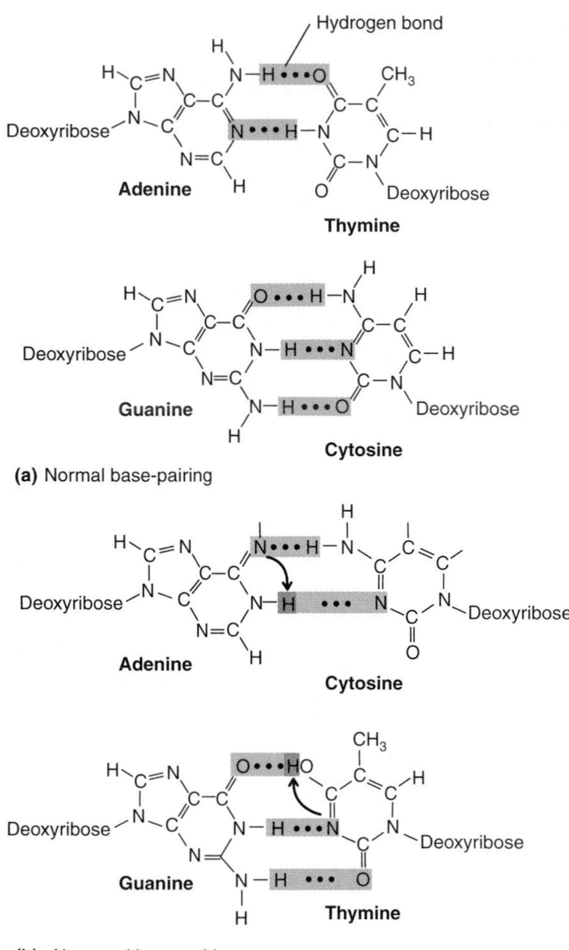

Figure 8.2 Base-Pairing by Hydrogen Bonding **(a)** The usual base-pairing between adenine and thymine and between guanine and cytosine. **(b)** Tautomeric shift. On rare occasions, the H in adenine moves from its common position to a new position, so that hydrogen bonding is possible between adenine and cytosine. A similar movement of H in thymine allows hydrogen bonding between thymine and guanine.

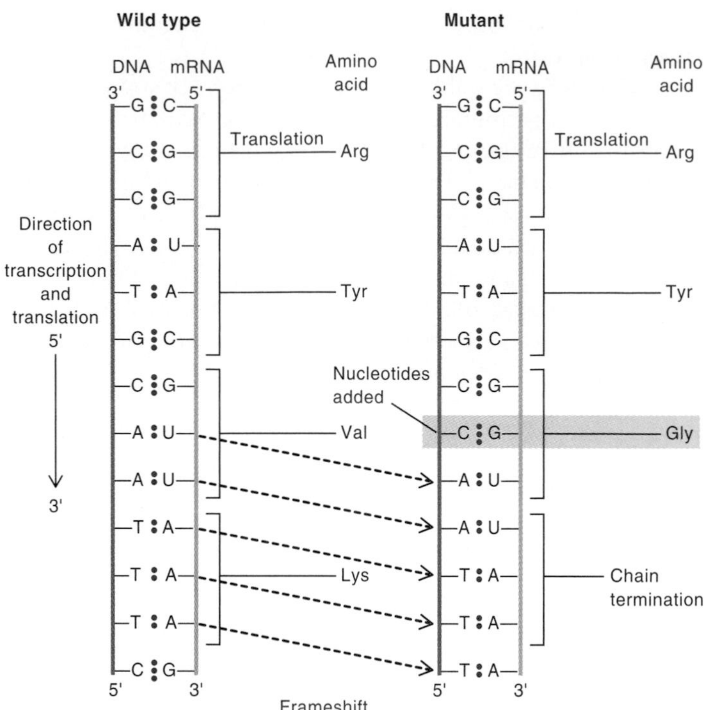

Figure 8.3 Production of Mutation As a Result of Base Addition The addition of a nucleotide to the DNA results in a frameshift in the transcription of the DNA and a new triplet code word, which is translated as a new amino acid. The deletion of a nucleotide would have essentially the same effect. The protein chain terminates when a stop codon appears in the DNA.

Transposable Elements (Jumping Genes)

Transposable elements, also called **transposons** or jumping genes, are special segments of DNA that can move spontaneously from one site to another in the same or different DNA molecules, a process called **transposition (figure 8.4)**. Any gene into which a transposable element inserts itself is disrupted, so that it no longer codes for a functional protein. The gene usually suffers a knockout mutation in transposition. The structure and biology of transposons will be considered later in this chapter.

The classic studies of transposition were carried out by Dr. Barbara McClintock (see A Glimpse of History). She observed variation in the colors of kernels of corn as a result of transposons moving into and out of genes concerned with pigment synthesis **(figure 8.5)**.

MICROCHECK 8.2

Mutations, changes in the nucleotide sequence of DNA, may result in proteins that are nonfunctional, thereby altering the properties of the cell. A leaky mutation results in a partially functional protein; a knockout mutation results in a nonfunctional protein. Mutations most commonly occur spontaneously as a result of mistakes in DNA replication.

- What is the genotype of a silent mutation in a gene involved in histidine biosynthesis? What is the phenotype?
- Would the addition of three bases to DNA always lead to a frameshift mutation? Explain your answer.

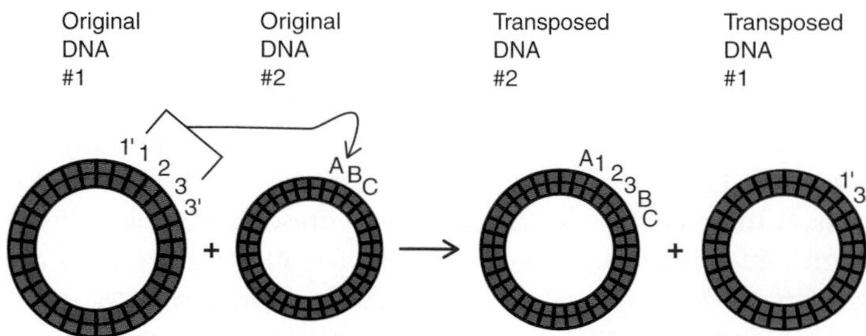

Figure 8.4 Transposition The transposon has the ability to "jump" from one piece of DNA to another piece of DNA in the same cell, where it becomes integrated in its new location. The first piece of DNA loses the genes, and the second piece of DNA gains them. The transposon may "jump" from the chromosome to a plasmid or to different sites on the chromosome. The above circular DNA molecule can represent plasmids or chromosomes.

Figure 8.5 **Transposition Detected by Color Changes** Variegation in color observed in the kernels of corn is caused by the insertion of transposable elements into genes involved in the synthesis of different pigments, thereby altering the synthesis of the pigments.

Induced Mutations

Geneticists use mutations to identify and study the functions of genes. Creating mutations allows one to observe how they affect the phenotype of the cell. Thus, geneticists spend a great deal of time creating and isolating mutants. One major advantage of working with bacteria in genetic studies is that mutants are easy to isolate. Bacteria grow rapidly and can reach enormous numbers in a small volume. Because the frequency of spontaneous mutations is so low, investigators trying to isolate certain mutants generally resort to using mutagens, that can increase the frequency of mutations at least 1,000-fold. Such mutations are said to be **induced**. The inducing agents can be certain chemicals or radiation.

Chemical Mutagens

Any chemical treatment that alters the hydrogen-bonding properties of a purine or pyrimidine base already in the DNA will increase the frequency of mutations as the DNA replicates.

Chemical Modification of Purines and Pyrimidines

One powerful mutagen that modifies purines and pyrimidines in DNA is nitrous acid (HNO_2). This chemical primarily converts amino ($—NH_2$) to keto (C=O) groups—for example, converting cytosine to uracil. Uracil then pairs with adenine rather than guanine when the DNA is replicated. ■ **purines and pyrimidines, p. 35**

The largest group of chemical mutagens consists of the **alkylating agents**, highly reactive chemicals that add **alkyl groups**, short chains of carbon atoms, onto purines and pyrimidines, thereby altering their hydrogen-bonding properties. A common alkylating agent used in research laboratories is **nitrosoguanidine**. Many compounds formerly used in cancer therapy, such as nitrogen mustard, are in this group. These compounds kill rapidly dividing cancer cells, but they also damage DNA in normal cells. As a result, these agents have caused cancers to arise more than 10 years after they were used to treat the original cancer.

Base Analogs

Base analogs are compounds that resemble the purine or pyrimidine bases closely enough that they are incorporated into DNA in place of the natural bases during DNA replication **(figure 8.6)**. Base analogs such as 5-bromouracil and 2-amino purine, however, do not have the same hydrogen-bonding properties as the natural bases, thymine and adenine respectively. This difference increases the probability that, once incorporated into DNA, the base analog will pair with the wrong base as the complementary strand is being synthesized.

Certain base analogs are also useful in treating diseases because of their ability to inhibit DNA replication. For example, azidothymidine (AZT), an analog of thymidine, inhibits replication of the human immunodeficiency virus (HIV) and is used in treating HIV-infected people. The analog does not have an —OH on the third carbon of deoxyribose and therefore the growth of the DNA chain cannot proceed. Note that the sugar deoxyribose and not the base thymine is modified in this base analog. ■ **thymidine, p. 35** ■ **AZT, p. 752**

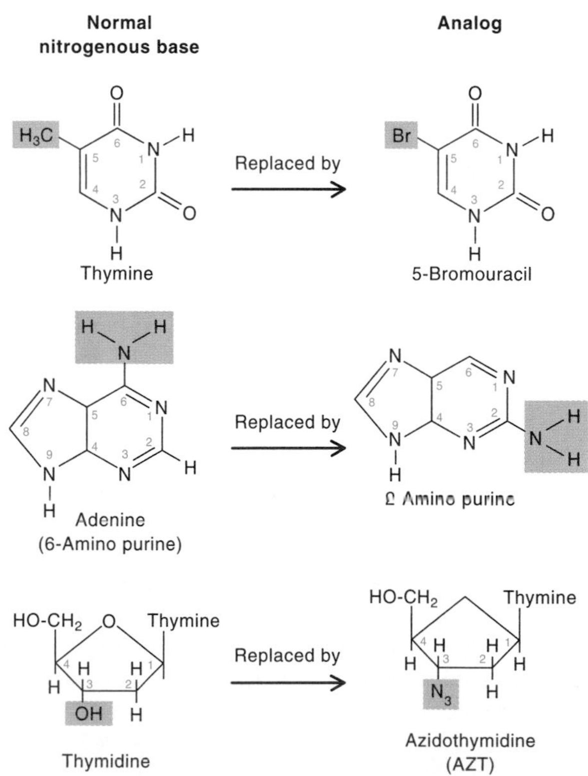

Figure 8.6 **Common Base Analogs and the Normal Bases They Replace in DNA** The important differences between the normal bases and the analogs are in boxes. The numbers refer to the carbon atoms in the ring, which are not shown. Thymidine consists of thymine and deoxyribose.

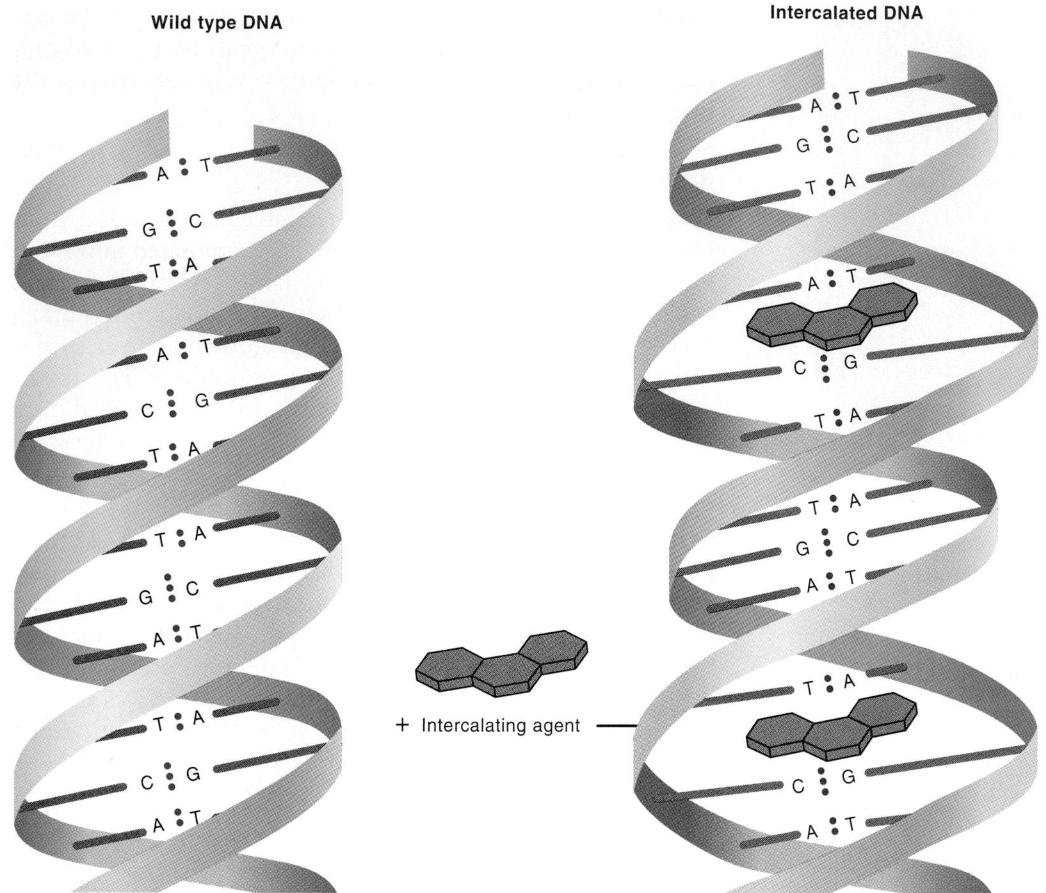

Figure 8.7 Intercalation The insertion of the intercalating agent moves the adjacent base pairs farther from one another, so that base additions occur. This results in frameshift mutation.

Intercalating Agents

A number of chemical mutagens, termed **intercalating agents**, are three-ringed planar molecules of about the same size as a pair of nucleotides in DNA. These molecules do not alter hydrogen-bonding properties of the bases; rather, they insert or intercalate between adjacent base pairs in the replication fork in DNA synthesis (**figure 8.7**). This pushes the nucleotides apart, producing enough space between bases that an extra base often is added in the strand being synthesized. The result is a **frameshift mutation**. Less frequently, the intercalation results in a newly synthesized strand that lacks a base. As in spontaneous frameshift mutants, the addition or subtraction of a nucleotide often results in a stop codon being generated prematurely, and a shortened protein being synthesized. An intercalating agent commonly used in the laboratory to isolate plasmids is **ethidium bromide**. This intercalating agent changes the density of the two structurally dissimilar DNA molecules to different degrees, and so the DNA molecules can be separated by centrifugation. After the ethidium bromide intercalates, the DNA will fluoresce when ultraviolet light shines on it. The manufacturer now warns users that ethidium bromide should be used with great care because it is a **carcinogen**, a cancer-causing agent. ■ replication fork, p. 171

Transposition

A common procedure to generate mutants in research laboratories is to introduce a transposon into a cell where the transposon must integrate into the cell's genome in order to replicate. The gene into which the transposon has inserted will usually suffer a knockout mutation, termed an **insertion mutation**.

Radiation

Two kinds of radiation are mutagens: ultraviolet (UV) light and X rays. Their wavelengths are shown in **figure 8.8**.

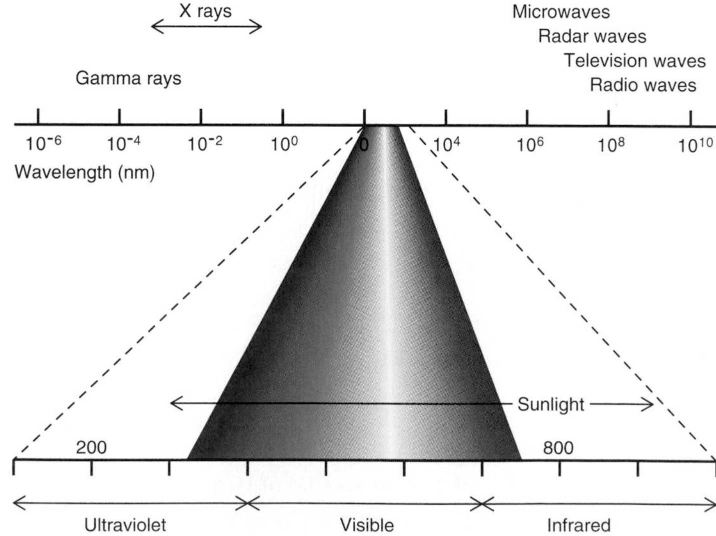

Figure 8.8 Wavelengths of Radiation The visible wavelengths include the colors of the rainbow.

Ultraviolet Irradiation

Irradiation of cells with ultraviolet light causes covalent bond formation between adjacent thymine molecules on the same strand of DNA (intrastrand bonding), resulting in the formation of **thymine dimers (figure 8.9)**. The covalent bonding distorts the DNA strand so much that the dimer cannot fit properly into the double helix, and the DNA is damaged. DNA cannot be replicated nor can genes be transcribed beyond this site of damage. The major mutagenic action of UV light, however, does not result from the damaged DNA directly, but rather from the cell's attempt to repair the damage by a mechanism termed **SOS repair** (see the next section).
■ gene transcription, p. 172

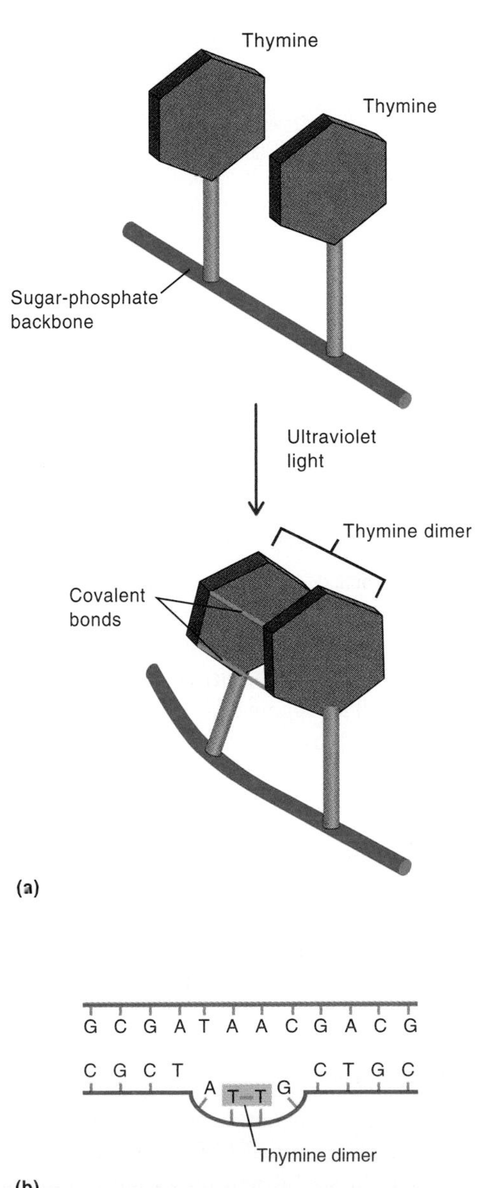

Figure 8.9 Thymine Dimer Formation (a) Covalent bonds form between adjacent thymine molecules on the same strand of DNA. This distorts the shape of the DNA and prevents replication past the dimer. **(b)** Distortion of the DNA helix.

X Rays

X rays cause several types of damage: single- and double-strand breaks in DNA, and alterations to the bases. Double-strand breaks often result in deletions that are lethal.

Table 8.1 summarizes information on the common mutagens.

MICROCHECK 8.3

The frequency of mutations can be increased significantly by treating cells with chemicals and radiation. These treatments induce mutations. Mutations most frequently result from altering the hydrogen-bonding properties of the nitrogenous bases.

- How does UV light affect cells?
- If you wished to isolate a point mutant, what mutagen would you use? To isolate a knockout mutant?

Repair of Damaged DNA

No molecule is more important to the cell than DNA. The amount of spontaneous and mutagen-induced damage to DNA that occurs in cells is enormous. Every 24 hours, the DNA in every cell in the human body is spontaneously damaged more than 10,000 times. This damage, if not repaired, can lead to cell death and, in animals, cancer. In humans, two breast cancer susceptibility genes code for enzymes that repair damaged DNA. Mutations in either one result in a high (80%) probability of breast cancer. A major reason that mutations are so rare is that they are repaired before they can alter the properties of cells. It is not surprising that, in the course of many millions of years of evolution, all cells, both prokaryotic and eukaryotic, have developed several different mechanisms for repairing any damage that their DNA might suffer. Except for the light repair of UV dimers, the mechanisms that we will discuss operate on all types of damaged DNA.

Repair of Errors in Base Incorporation

A major cause of spontaneous mutation occurs during DNA replication—the incorporation of the wrong base by DNA polymerase. This complex enzyme selects the proper base to hydrogen bond to the template strand and then adds (polymerizes) the nucleotide to the 3′ OH group on the strand of DNA being synthesized. On rare occasions, however, the wrong base is selected, so that proper hydrogen bonding to the base in the template strand does not occur. This leads to a distortion in the DNA helix and, if allowed to remain, results in a mutation. Cells have developed two ways of dealing with these errors. One major process is carried out by the DNA polymerase itself and is called **proofreading**. A second mechanism, called **mismatch repair,** involves an enzyme that cuts out these errors in base incorporation. ■ DNA replication, p.170 ■ DNA polymerase, p. 171

Proofreading by DNA Polymerase

DNA polymerase is a complex enzyme that not only is involved in the synthesis of DNA but also has a proofreading function.

Table 8.1 Common Mutagens

Agent	Action	Result
Chemical Agent		
Base analog Example: 5-bromouracil	Incorporates in place of thymine in DNA	Base substitution
Intercalating agents Example: ethidium bromide	Inserts between base pairs	Addition or subtraction of base pairs
Chemical modification of bases Examples: nitrous acid	Converts amino group to keto group in adenine and cytosine	Base substitution
alkylating agents	Adds alkyl groups (CH_3 and others) to nitrogenous bases such as guanine	Base substitution
Transposons	Random insertion into any gene	Knockout mutations
Radiation		
Ultraviolet (UV)	Intrastrand thymine dimer formation	Base substitution
X rays	Single- and double-strand breaks in DNA	Deletion of bases Knockout mutations

One part of this enzyme can back up and then excise the base if it is not correctly hydrogen bonded to the base in the template strand. DNA polymerase recognizes improper base-pairing of the newly introduced base to the template strand. Following excision, the DNA polymerase then selects the proper base and incorporates it into the growing DNA strand.

Mismatch Repair

The proofreading function of DNA polymerase is very efficient. Cells, however, have a backup repair system that recognizes incorrect bases missed by the proofreading of DNA polymerase. This repair mechanism is called mismatch repair. In this process, the mismatch repair enzyme, an endonuclease, recognizes the pair of improperly hydrogen-bonded bases and excises a short stretch of nucleotides from the newly synthesized strand **(figure 8.10)**. The gap is then filled in with the proper nucleotides by DNA polymerase. **DNA ligase** joins the end of the repaired strand to the newly synthesized one. ■ DNA ligase, p. 171

The cell is faced with a potential problem in this repair process. If the original strand of DNA were excised, the misincorporated base would not be removed and the mistake would not be repaired. How does the mismatch repair enzyme recognize which strand should be excised? The repair enzyme can recognize the original and newly synthesized strands because they differ in one respect—**methylation**, the presence of methyl ($—CH_3$) groups on the DNA. In bacteria, methylation of DNA serves as a code for different kinds of identification. In the present case, it distinguishes newly synthesized DNA from the template strand. An enzyme methylates certain adenine nucleotides in the DNA long after the DNA strand is synthesized. The strand containing the misincorporated base, being newly synthesized, is not yet methylated. The mismatch repair endonuclease recognizes the difference in methylation and excises the DNA from the unmethylated strand. Mismatch repair also occurs in humans. Defects in this repair system have been shown to lead to an increased incidence of colorectal cancers.

Repair of Thymine Dimers

Since UV light is a part of sunlight, cells are frequently exposed to this mutagenic agent in their natural environment (see figure 8.8). Many bacteria and other organisms, including people, have developed several mechanisms to combat the harmful effects of these rays. The repair of UV dimers is the best understood of all repair mechanisms.

Light Repair

An enzyme, called **photolyase,** can break the covalent bond of thymine dimers, but only in the presence of visible light. The light is absorbed by the enzyme, which then uses the light energy to cleave the covalent bond. This mechanism is called **light repair** or **photoreactivation** and restores the DNA to its original state **(figure 8.11)**.

Excision (Dark) Repair

Some bacteria also have an enzyme that excises the damaged segment from a single strand of DNA and another enzyme that repairs the resulting break by synthesizing a strand complementary to the undamaged strand. The repaired strand is then joined to the one end of the undamaged region by DNA ligase **(figure 8.12)**. This process, termed **excision repair**, can also cut out pieces of DNA that contain mismatched bases caused by

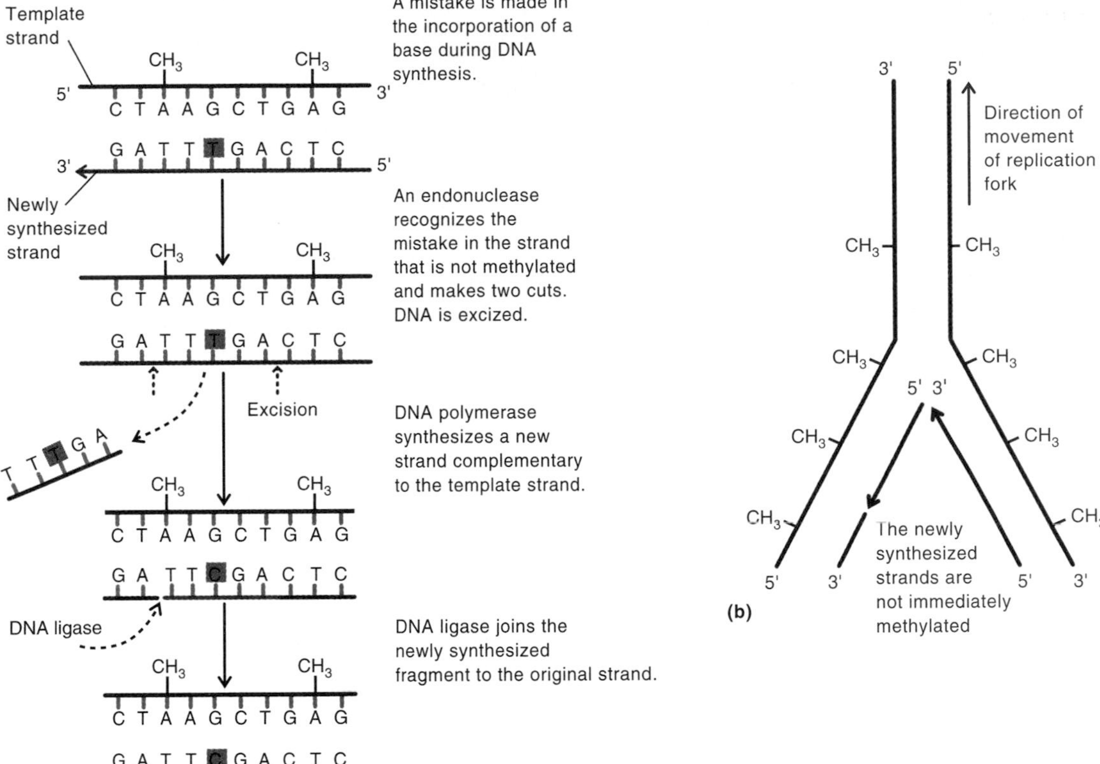

Figure 8.10 Mismatch Repair (a) The endonuclease excises a piece of DNA containing the misincorporated nucleotide. A new complementary strand is then synthesized and joined to the original strand by DNA ligase. (b) The endonuclease recognizes the strand of DNA that is not yet methylated as the strand to excise.

chemical mutagens or spontaneous mutations. Because visible light is not required for the action of these enzymes, such repair is also called **dark repair**.

This same repair mechanism is also important to the well-being of humans. People exposed to the sun, with its UV light, for long periods have a higher incidence of skin cancer than people who are not exposed. Also, people who have a defective dark repair enzyme have an increased incidence of skin cancer. This illustrates both the damage UV light can cause to DNA and the importance of repair mechanisms in overcoming the damage.

SOS Repair

If DNA is very heavily damaged by UV light such that it contains many thymine dimers, then light and excision repair may not be able to correct all of the dimers. Therefore, bacteria have a mechanism, termed **SOS repair**, that is a last ditch mechanism to bypass the damaged DNA and allow replication to continue. The damaged DNA induces the SOS system, which comprises about 20 genes. One of the activated gene products binds to DNA polymerase and enables the enzyme to synthesize DNA across the damaged DNA. The altered enzyme, however, also loses its proofreading ability. As a result, the DNA that is synthesized contains many misincorporated bases. This is why UV light is such an effective mutagen. The SOS system also is induced when DNA is damaged by chemicals and is a major mechanism by which chemical mutagens induce mutations. Recall that the other mechanisms that repair damaged DNA copy the template strand of DNA with great accuracy, so that mutations are not induced by these repair mechansims.

Table 8.2 summarizes the key features of the major DNA repair systems in bacteria. Note that the cell cannot repair all types of mutations, such as insertion mutations caused by transposition.

MICROCHECK 8.4

Bacteria can repair damaged DNA that contains errors resulting from the incorporation of wrong nucleotides by a variety of mechanisms. These include proofreading

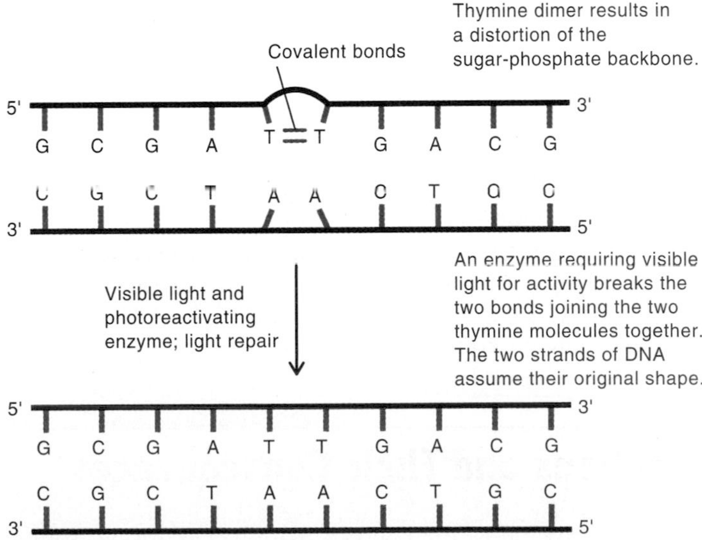

Figure 8.11 Repair of a Thymine Dimer by a Light-Requiring Enzyme (photoreactivation)

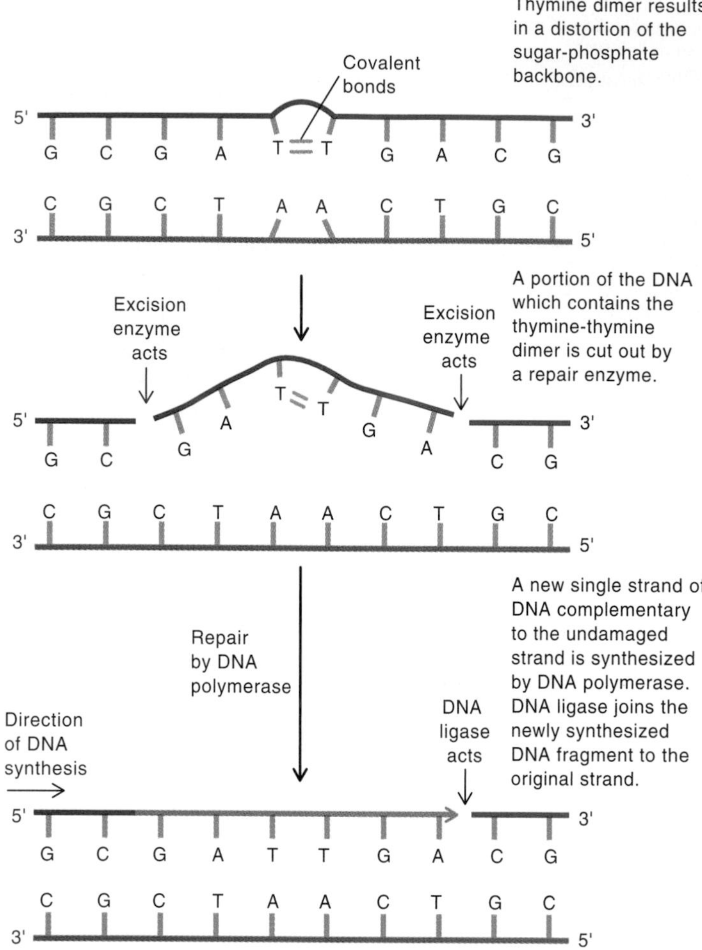

Thymine dimer results in a distortion of the sugar-phosphate backbone.
Covalent bonds
5'
3'
G C G A T T G A C G
C G C T A A C T G C
3'
5'
A portion of the DNA which contains the thymine-thymine dimer is cut out by a repair enzyme.
Excision enzyme acts
Excision enzyme acts
Repair by DNA polymerase
A new single strand of DNA complementary to the undamaged strand is synthesized by DNA polymerase.
DNA ligase acts
DNA ligase joins the newly synthesized DNA fragment to the original strand.
Direction of DNA synthesis

Figure 8.12 Dark, or Excision, Repair of a Thymine Dimer The single strand of DNA containing the thymine dimer is removed and destroyed. The newly synthesized strand is joined to the end of the original strand by the enzyme DNA ligase.

by DNA polymerase, and excising the nucleotide errors by mismatch repair. Thymine dimers can be repaired through light and dark repair mechanisms and also by bypassing the damaged DNA by the SOS system.

- How does UV light cause mutations?
- Distinguish between light and dark repair of thymine dimers.
- If you wish to maximize the number of mutations following UV irradiation, should you incubate the irradiated cells in the light or in the dark, or does it make any difference? Explain your answer.

Mutations and Their Consequences

Because of mutations, the concept that all cells arising from a single cell are identical is not strictly true, since every large population contains mutants. Even the cells in a single colony that contains about 1 million cells are not completely identical because of random mutations. These mutations provide a mechanism by which organisms, with their altered characteristics, can respond to a changing environment. This is the process of **natural selection**. The environment, however, does not cause the mutation. The environment selects those cells that can grow under its conditions. Thus, a spontaneous mutation to antimicrobial resistance, though rare, will result in the mutants becoming the dominant organisms in a hospital environment where the antimicrobial medication is present, because only the resistant cells can survive. The antimicrobial kills the sensitive cells and thereby allows the resistant cells to take over the population.

Since genes mutate independently of one another, the chance that two given mutations will occur within the same cell is very low. Indeed, the actual occurrence is the product of the individual rates of mutation of the two genes (given by taking the sum of the exponents). For example, if the mutation rate to streptomycin resistance is 10^{-6} per cell division and the mutation rate to penicillin resistance is 10^{-8} per cell division, the probability that both mutations will occur within the same cell is 10^{-6} X 10^{-8}, or 10^{-14}. For this reason, two or more drugs may be administered simultaneously in the treatment of some diseases such as tuberculosis and AIDS. This is called **combination therapy.** Any mutant cell or virus resistant to one antimicrobial medication is likely to be sensitive to the other and therefore will be killed by the combination of the two antimicrobials. ■ combination therapy, p. 571

Expression of Mutations

One advantage of using bacteria in genetic studies is that results can be quickly observed. One reason is that mutations are expressed rapidly. Except for the short time just before a cell divides, most bacteria contain only a single chromosome, with each gene present in only one copy. They are **haploid.** Therefore, any mutation will be observed soon after it occurs. For example, if a mutation knocks out a gene coding for an enzyme of histidine biosynthesis, the mutant bacterium will soon stop growing unless histidine is in its environment.

MICROCHECK 8.5

Mutations occur randomly—they are not caused by agents that allow mutants to grow. In bacteria, all mutations are expressed immediately because they are haploid.

- If the rate of mutation to streptomycin resistance is 10^{-6} and to penicillin resistance is 10^{-4}, what is the rate of mutation to simultaneous resistance to both antibiotics?
- When do bacteria have several copies of all their genes?

Mutant Selection

Even when mutagens are used, mutations that appear in the population are rare. This presents a major challenge to the investigator who wants to isolate a desired mutant. Several clever techniques have simplified this process. As discussed in

Table 8.2 Repair of Damaged DNA

	Type of Defect	Repair Mechanism	Biochemical Mechanism	Result
Spontaneous				
	Wrong base incorporated during DNA replication	Proofreading by DNA polymerase	Removal of mispaired base by DNA polymerase	Potential mutation eliminated
		Mismatch repair	Excision of short stretch of unmethylated single-stranded DNA and synthesis of new strand by DNA polymerase	Potential mutation eliminated
		Excision (dark) repair	Excision of short stretch of DNA and synthesis of new strand by DNA polymerase	Potential mutation eliminated
Mutagens				
Chemical	Wrong base incorporated during DNA replication	Same as for spontaneous mutations	Same as for spontaneous mutations	Same as for spontaneous mutations
UV light	Thymine dimer formation	Photoreactivation (light repair)	Breaking of covalent bond forming thymine molecules	Original DNA molecule restored
		Excision repair (dark repair)	Excision of a short stretch of single-stranded DNA containing thymine dimer and synthesis of a new strand by DNA polymerase	Mutation eliminated
		SOS repair	DNA synthesis by a modified DNA polymerase bypasses site of damaged DNA	Cell survives but mutations common

chapter 4, bacteria can multiply on simple media and produce several billion cells per milliliter of medium in less than 24 hours. In such a large population, every gene should be mutant in at least one cell in the population. The major problem becomes how to find and identify the bacteria containing the desired mutation. Depending on the type of mutant being sought, one of two simple techniques can be used, direct or indirect selection.

Direct Selection

Direct selection involves inoculating cells onto a medium on which the mutant, but not the parent, can grow. For example, mutants resistant to the antibiotic streptomycin can be easily selected directly by inoculating cells onto a medium containing streptomycin. Only the rare resistant cells in the population will form a colony **(figure 8.13)**. Antimicrobial-resistant mutants are usually very easy to isolate by direct selection.

Indirect Selection

Indirect selection is required to isolate an auxotrophic mutant, one that requires a growth factor, such as histidine, which the parent strain does not. On a medium that contains histidine, both the parent and histidine auxotroph grow. On a medium lacking histidine, the histidine-requiring mutant will not grow, but the parent will. There is no medium on which the mutant will grow and the parent will not.

Replica Plating

An ingenious technique for indirect selection of auxotrophic mutants, **replica plating**, was devised by the husband-and-wife team of Joshua and Esther Lederberg in the early 1950s

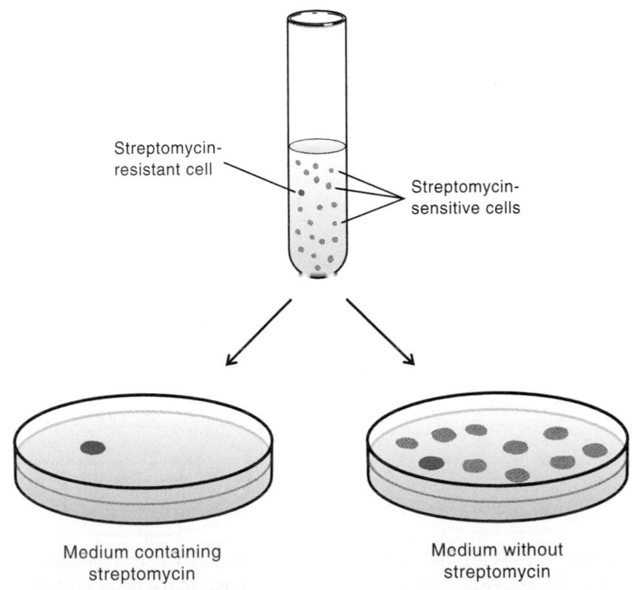

Figure 8.13 Direct Selection of Mutants Only the streptomycin-resistant cells will grow on the streptomycin-containing medium. All cells will grow on media without streptomycin and have the same appearance.

(figure 8.14). In this technique, a master plate containing isolated colonies of all cells growing on an enriched medium is pressed onto sterile velvet, a fabric with tiny threads that stand on end like tiny bristles. This operation transfers some cells of every bacterial colony onto the velvet. Next, two sterile plates, one containing a glucose-salts (minimal) medium and the second an enriched, complex medium, are pressed in succession onto the same velvet. This procedure transfers cells imprinted on the velvet from the master plate to both the glucose-salts medium and the enriched medium. All prototrophs will form colonies on both the enriched and the glucose-salts medium, but auxotrophs will only form colonies on the enriched medium. ■ **glucose-salts medium, p. 103**

By keeping the orientation of the two plates the same as they touch the velvet, any colony on the master plate that can grow on the enriched medium but not on the glucose-salts medium can be identified by the pattern the colonies form on the agar. The cells in such a colony must require a growth factor present in the complex medium. The particular growth factor required can then be determined by adding the various factors individually to the glucose-salts medium and determining which one promotes cell growth. Before the invention of replica plating, each colony had to be individually transferred from the complex to the glucose-salts medium in order to identify the auxotrophs and determine their growth requirements, a very laborious job indeed. ■ **growth factor, p. 100**

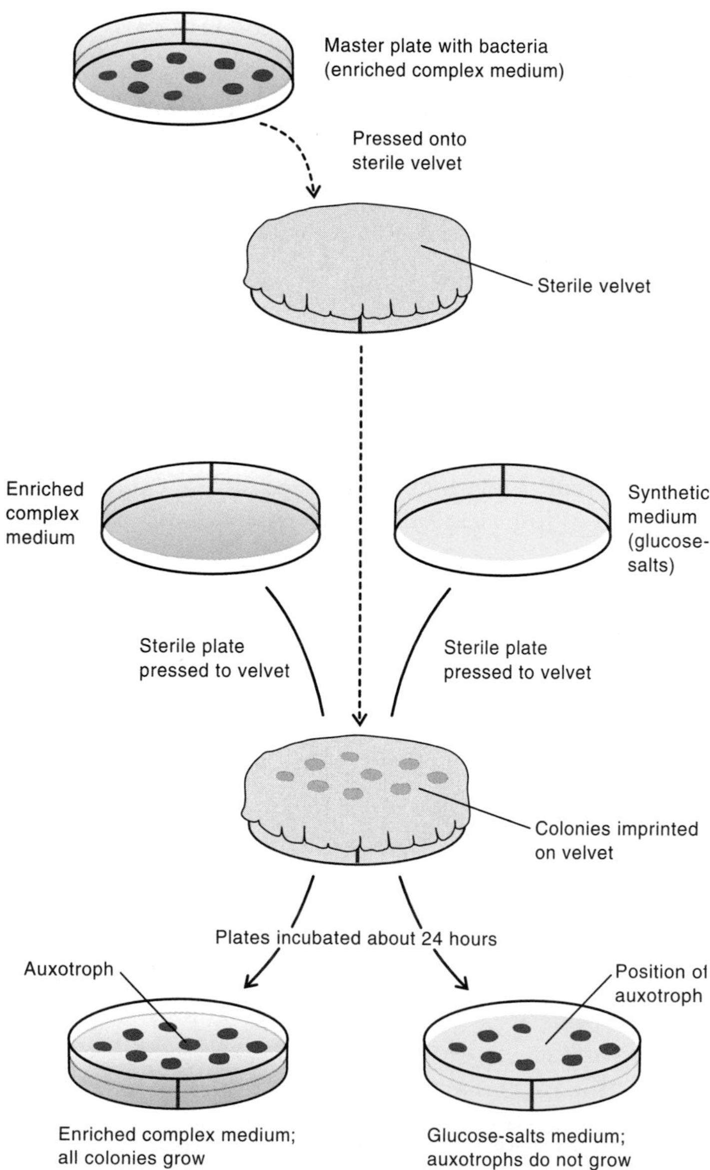

Figure 8.14 Indirect Selection of Mutants by Replica Plating

Penicillin Enrichment

Even using mutagenic agents, the frequency of mutation in a particular gene is low, ranging perhaps from less than one in 1,000 to one in 100 million cells. In cases where the parent cell is sensitive to penicillin, the proportion of auxotrophic mutants in the population can be increased by a technique called **penicillin enrichment.** Following treatment with a mutagen, the cells are grown in a glucose-salts medium containing penicillin. Since penicillin kills only growing cells, most of the growing prototrophs will be killed, while the non-multiplying auxotrophs will survive **(figure 8.15)**. The enzyme **penicillinase** is then added to destroy the penicillin, and the cells are plated on an enriched medium. This plate can then be replica plated onto a glucose-salts medium supplemented with the growth factor for which a specific mutation is being sought and also to an unenriched glucose-salts medium. The mutant being sought grows only on the supplemented medium. ■ **action of penicillin, p. 65**

Conditional Lethal Mutants

The mutants discussed so far will grow if a required growth factor they cannot synthesize is added to the medium. Mutants, however, also may be defective in the enzymes required for

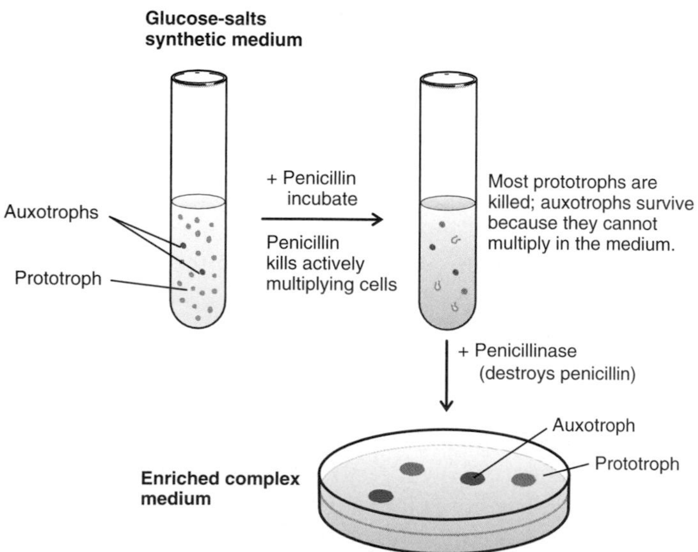

Figure 8.15 Penicillin Enrichment of Mutants

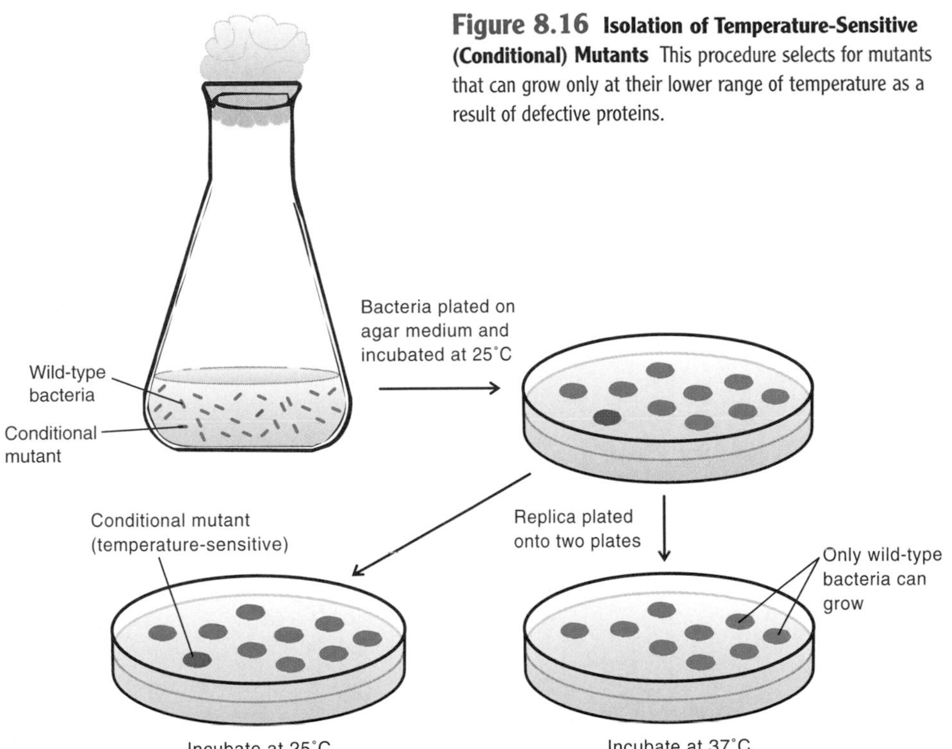

Figure 8.16 Isolation of Temperature-Sensitive (Conditional) Mutants This procedure selects for mutants that can grow only at their lower range of temperature as a result of defective proteins.

macromolecule synthesis, such as factors involved in the translation of mRNA in protein synthesis or the enzymes of DNA replication. These mutations are ordinarily lethal to the cell, because their defects cannot be overcome by adding growth factors to the medium. Therefore, they cannot be isolated by the techniques previously described. Mutants in essential genes, however, given the name **conditional lethal mutants**, can be isolated since mutant proteins often function at a low temperature such as 25°C but not at a higher temperature such as 37°C. They are called **temperature-sensitive mutants**, one class of conditional lethal mutants.

Temperature-sensitive mutants can be isolated by incubating a master plate at a low temperature (25°C) at which all cells, both normal and temperature-sensitive, will grow **(figure 8.16)**. The master plate is then replica plated onto two plates containing an enriched medium. One is incubated at 37°C, the other at 25°C. Colonies growing at the low temperature but not the high temperature may be mutant in genes required for macromolecule synthesis or other growth functions that cannot be overcome by adding growth factors to the medium.

Mutant Screens

The mutants that we have discussed thus far can be selected, either directly or indirectly, because of differences in their ability to grow on certain media. Another class of mutants are those that are not selected but **screened**. These mutants remain prototrophs but have defects in genes that alter their phenotype in a way that can be seen when colonies are grown on a particular medium. For example, many bacteria synthesize enzymes that degrade polysaccharides, proteins, and nucleic acids. These bacteria show zones of clearing around the colonies when the medium contains the appropriate macromolecule **(figure 8.17)**. A mutation in a gene that codes for one of these enzymes results in cells that cannot degrade the macromolecule and therefore a colony does not show the zone of clearing.

In screening for mutants, one must plate out the bacteria on a medium on which all of the cells will grow but on which the desired mutant will grow in an identifiable way. Such mutants are easier to isolate than a specific auxotrophic mutant because replica plating is not required in a screen. Large numbers of bacterial colonies may have to be screened, however, before the mutant being sought is identified.

Testing of Chemicals for Their Cancer-Causing Ability

Strong evidence exists that a substantial proportion of all cancers are caused by chemicals in the environment called carcinogens. How can the thousands of chemicals released into the environment, such as pesticides, herbicides, and the byproducts of manufacturing processes, be tested for their carcinogenic activity? Testing in animals takes 2 to 3 years and may cost $100,000 or more for the testing of a single compound. Today a number of much less expensive, more rapid, and simpler tests have been devised. All are based on assaying the effect of the potential carcinogen on DNA in a microbiological system. The first one was devised by Bruce Ames and his colleagues in the 1960s and illustrates the concept of such tests. This test takes only a few days and is based on three facts: (1) the reversion of a mutant gene in a biosynthetic pathway, such as histidine biosynthesis, can be readily measured; (2) the frequency of reversions is increased by mutagens; and (3) most carcinogens are mutagens. Specifically, the Ames test measures the rate of

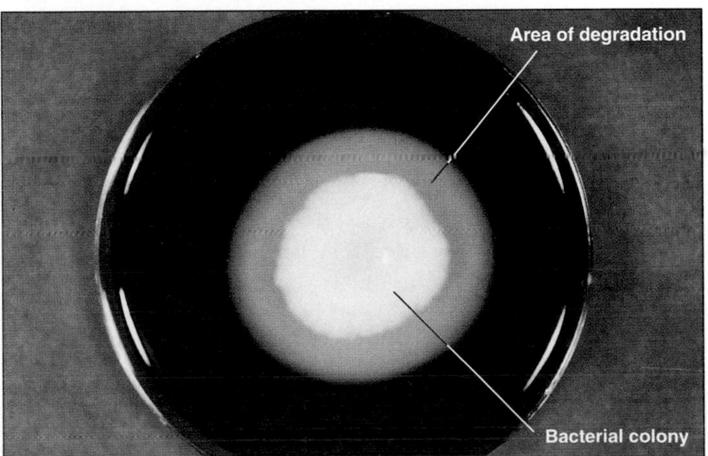

Figure 8.17 Breakdown of a polysaccharide (starch) incorporated into the medium by a degradative enzyme secreted by the colony.

reversion of a histidine auxotroph of *Salmonella* to prototrophy in both the presence and absence of the chemical being tested **(figure 8.18)**. If the chemical is mutagenic, and by implication carcinogenic, it will increase the reversion rate. The test also gives some idea about how powerful the mutagen is, and therefore how potentially hazardous the chemical is by the number of revertants that arise.

The Ames test as just discussed fails to detect many carcinogens, because some substances are not carcinogenic themselves but can be converted to active carcinogens by a metabolic reaction that occurs in animals but not in bacteria. Therefore, an extract of ground-up rat liver, which has the enzymes to carry out these conversions, is added to the Petri plates containing the mutagen (carcinogen) being analyzed by the Ames test.

Additional testing must be done on any mutagenic agent identified in the Ames test to confirm that it is actually carcinogenic in animals. Although data are not available on the percentage of mutagens that are carcinogens, it is clear that the Ames test is useful as a rapid screening test to identify those compounds that have a high probability of being carcinogenic. Thus far, no compound with a negative Ames test has been shown to be carcinogenic in animals.

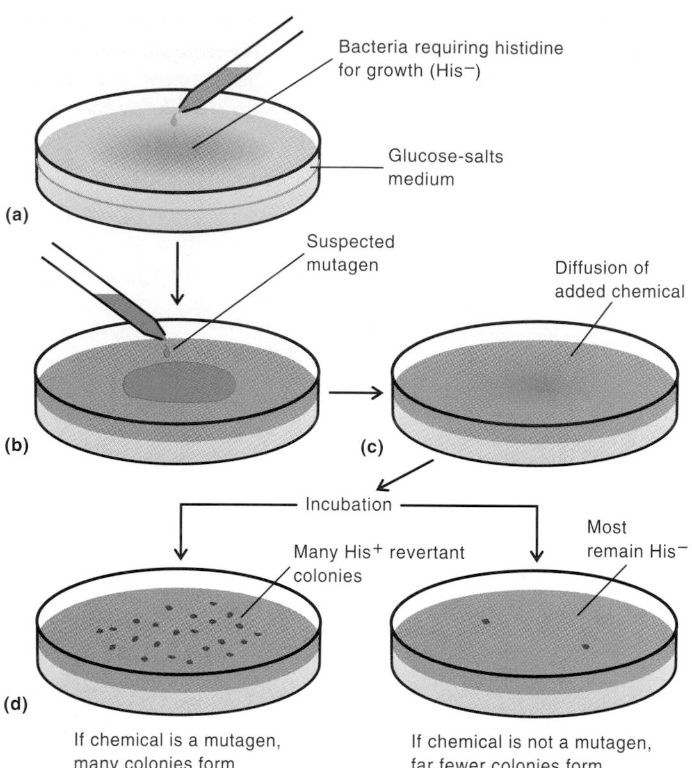

Figure 8.18 Ames Test to Screen for Mutagens The chemical will increase the frequency of reversion of His^- to His^+ cells if it is a mutagen and, therefore, most likely a carcinogen.

MICROCHECK 8.6

Mutants can be selected using either direct techniques or indirect techniques such as replica plating. Penicillin enrichment can often help in the isolation of auxotrophic mutants by killing multiplying cells. Mutants that do not involve the biosynthesis of growth factors can be isolated using other techniques such as mutant screens and incubating cells at low temperatures.

- Distinguish between direct and indirect selection.
- When does penicillin enrichment not work?
- If you have the option of isolating a particular mutant either through direct selection or screening, which would you choose? Why?
- The Ames test measures the reversion of a mutant gene. Would it be just as good to test for the generation of a mutant gene rather than its reversion? Explain.

MECHANISMS OF GENE TRANSFER

In addition to mutation, the genetic information in a cell can be altered if the cell gains genes from other cells. The movement of DNA from one cell to another accounts for the rapid spread of resistance to antimicrobial medications and heavy metals in bacterial populations, as described for *Shigella* earlier in this chapter.

Mutants make it possible to determine whether **genetic recombination**, the combining of DNA or genes from the same or from two different cells, has occurred in bacteria. Recombination in the DNA from two different bacteria can be readily recognized because the resulting cells, termed **recombinants**, have certain properties of each of the bacteria. For example, when cells that are streptomycin resistant (Str^R) and require histidine (His^-) and tryptophan (Trp^-) to grow are mixed with cells that are killed by streptomycin (Str^S) and require leucine (Leu^-) and threonine (Thr^-), rare recombinants appear that are resistant to streptomycin and grow on a glucose-salts medium—that is, the recombinants are His^+, Trp^+, Leu^+, Thr^+ and Str^R **(figure 8.19)**. These recombinant cells contain a mixture of genes from the two original types of cells. Since the bacteria are usually haploid, such recombinants can arise only if DNA has been transferred from one cell, the **donor**, to another cell, the **recipient**. Reversion of the auxotrophs cannot account for the result because at least two genes would have had to revert in the same cell, an extremely rare event.

Genes are naturally transferred between bacteria by three different mechanisms:

1. **DNA-mediated transformation**, in which DNA is transferred as "naked" DNA
2. **Transduction**, in which bacterial DNA is transferred by a bacterial virus

3. **Conjugation**, in which DNA is transferred between bacteria that are in contact with one another

Although these processes differ from one another in how the DNA is delivered, they share certain features.

- Only a part of the chromosome is usually transferred. This contrasts with the situation in eukaryotic cells, in which cells, such as the egg and the sperm, fuse so that two sets of all chromosomes are present.
- Although DNA may be transferred as either a single- or double-stranded molecule, single-stranded DNA replaces the **homologous,** or identical genes in the recipient cell in the integration process. This is the process of **homologous recombination**.
- DNA is generally transferred to no more than a small fraction of the recipient cells.

To detect gene transfer, one can select directly for recombinant cells, by inoculating cells on a medium on which only the recombinants will grow. Non-recombinant bacteria grow very slowly or die, so that only recombinants are found in the final population. Since several billion bacteria can be inoculated onto the agar contained in a single Petri dish, rare recombinants can be detected readily.

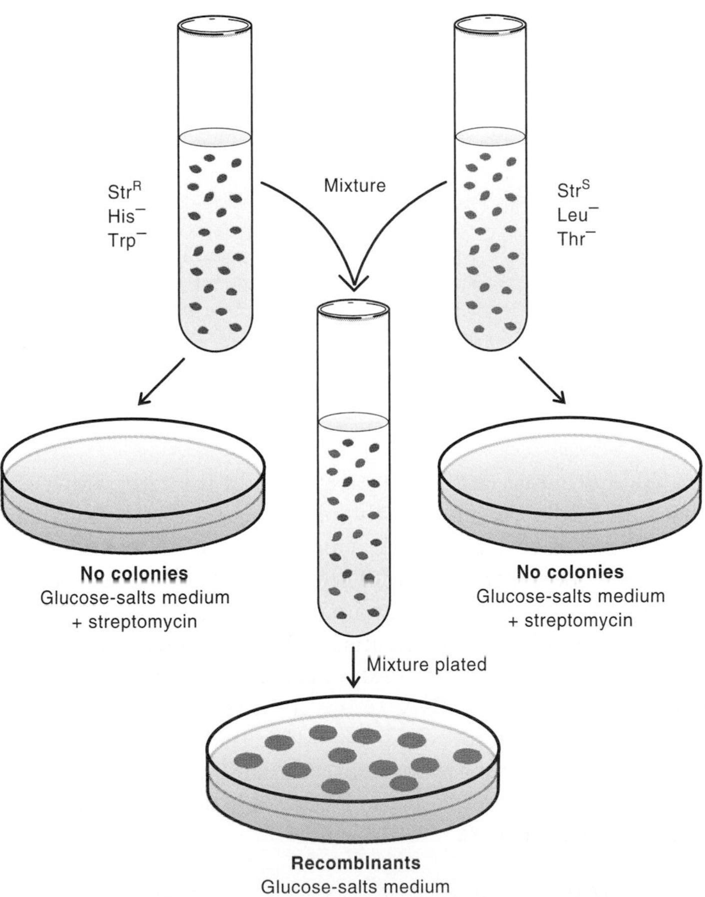

Figure 8.19 General Experimental Approach for Detecting Gene Transfer in Bacteria

DNA-Mediated Transformation

DNA-mediated transformation, referred to commonly as DNA transformation, involves the transfer of "naked" DNA in the environment to recipient cells (**figure 8.20**). If the cell walls of bacteria rupture, as frequently occurs in the stationary and death phase of bacterial growth, the long, circular molecules of chromosomal DNA that are tightly jammed into the bacteria break up into several hundred pieces as they explode through the broken cell walls. The pieces of DNA average about 20 genes each. Some of these DNA macromolecules pass through the cell walls and cytoplasmic membranes of the recipient cells

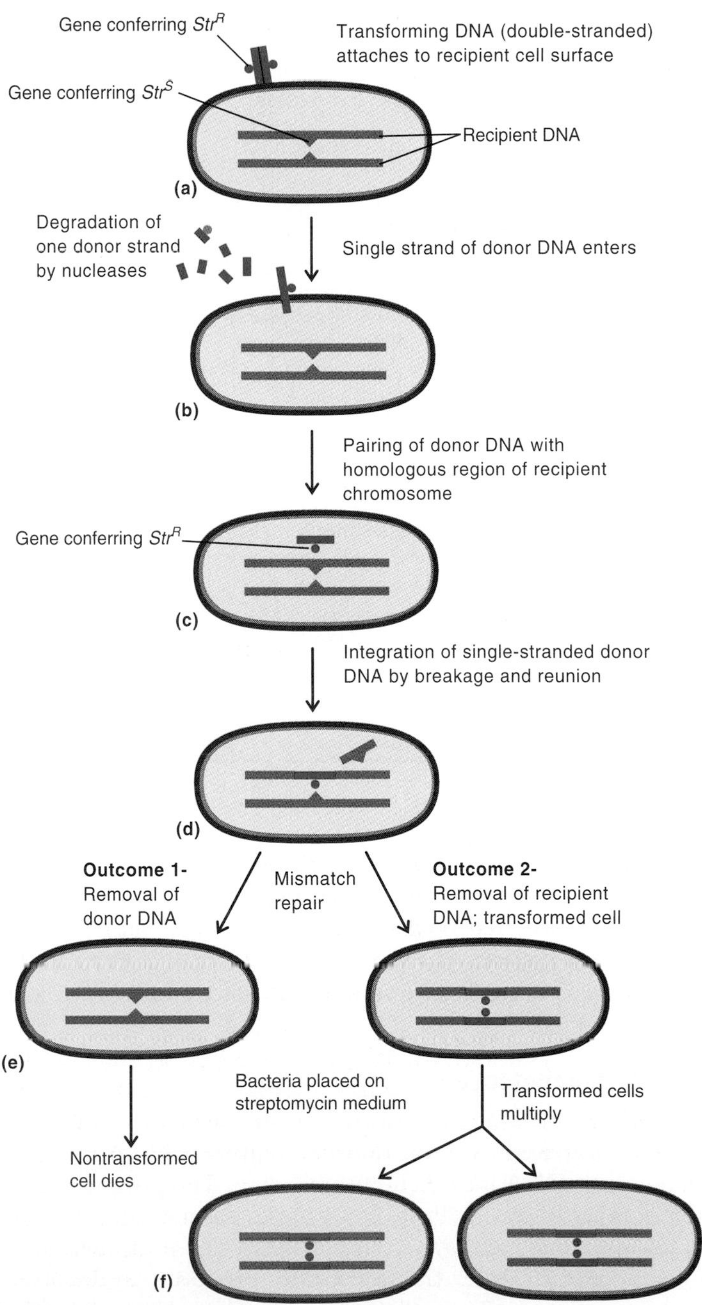

Figure 8.20 DNA-Mediated Transformation The donor DNA comes from a cell that is streptomycin resistant (Str^R). The recipient cell is streptomycin sensitive (Str^S).

PERSPECTIVE 8.1 The Biological Function of DNA: A Discovery Ahead of Its Time

In the 1930s, it was well known that DNA occurred in all cells, including bacteria. Its function, however, was a mystery. Since DNA consisted of only four repeating subunits, most scientists believed that it could not be a very important molecule. Its crucial biological role in the cell was discovered through a series of experiments conducted during a 20-year period by scientists in England and the United States.

In the 1920s, Frederick Griffith, an English bacteriologist, was studying pneumococci, the bacteria that cause pneumonia. It was known that pneumococci could cause this disease only if they made a polysaccharide capsule. In trying to understand the role of this capsule in the disease, Griffith killed encapsulated pneumococci and mixed them with living mutant pneumococci that could not synthesize a polysaccharide capsule. When he inoculated this mixture of organisms into mice, much to his surprise, they developed pneumonia and died (**figure 1**). Griffith isolated living encapsulated pneumococci from the dead mice. When he injected the killed encapsulated organisms and living nonencapsulated organisms into separate mice, they did not develop pneumonia.

Two years after Griffith reported these findings, another investigator, M. H. Dawson, lysed heat-killed encapsulated pneumococci and passed the suspension of ruptured cells through a very fine filter, through which only the cytoplasmic contents of the bacteria could pass. When he mixed the filtrate (the material passing through the filter) with living bacteria that were unable to make a capsule, some of these bacteria gained the ability to synthesize a capsule. Moreover, these bacteria passed on this ability to all of their offspring. Something in the filtrate was "transforming" the harmless, unencapsulated bacteria into bacteria with the ability to make a capsule.

What was this transforming principle? In 1944, after years of painstaking chemical analyses of lysates capable of transforming pneumococci, three investigators from the Rockefeller Institute, Oswald T. Avery, Colin MacLeod, and Maclyn McCarty, submitted one of the most important papers ever published in biology. In it, they reported that the molecule that could change (transform) a cell's properties was DNA.

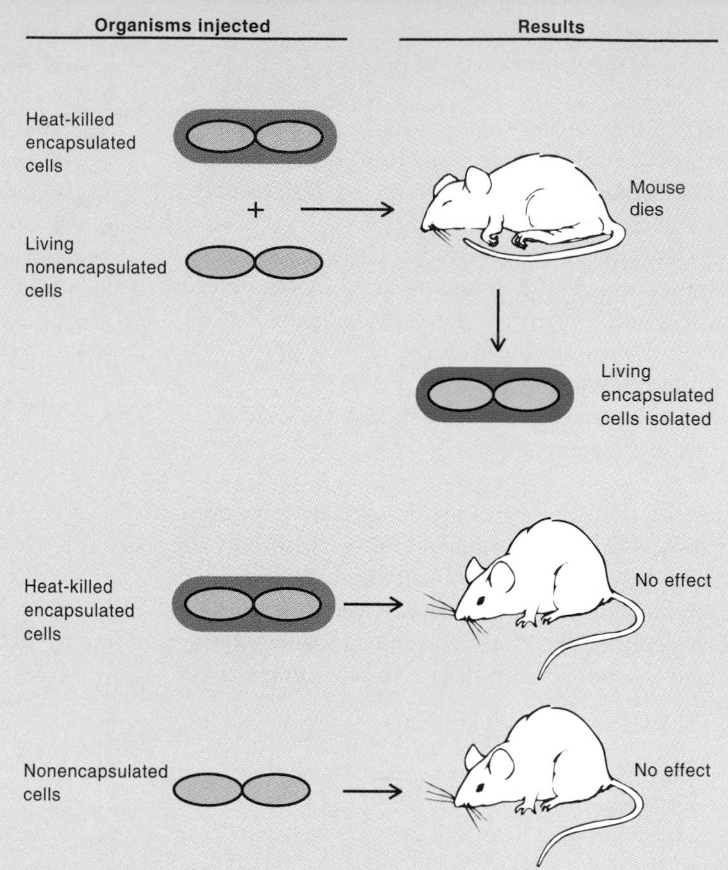

Figure 1 Demonstration of the Transforming Principle

The significance of their discovery was not appreciated at the time. Perhaps the discovery was premature, and scientists were slow to recognize its significance and importance. None of the three investigators received a Nobel Prize, although many scientists believe that they deserved it. Their studies pointed out that DNA is a key molecule in the scheme of life and led to James Watson and Francis Crick's determination of its structure, which they published in 1953. The understanding of the structure and function of DNA revolutionized the study of biology and ushered in the era of molecular biology. Microbial genetics serves as molecular biology's foundation.

and are then integrated into their chromosomes, replacing the homologous genes. ■ **bacterial growth curve, p. 106** ■ **death phase, p. 107**

DNA-mediated transformation can occur naturally in a wide variety of Gram-positive and Gram-negative bacteria. In addition, the yeast *Saccharomyces cerevisiae*, a eukaryote, can be transformed.

Natural Competence

The process of DNA transformation involves recipient cells, termed **competent** with the unusual ability to take up and integrate donor DNA into their chromosome. The process occurs in a number of stages. First, DNA passes through the cell wall and cytoplasmic membrane. This is one of the most interesting aspects of the DNA transformation process. As discussed already in chapter 3, usually only molecules whose molecular weight is no greater than a few hundred can pass through the cell envelope. It is still a mystery how DNA molecules, which are 100,000 times larger, can pass through. It is clear, however, that only under certain growth conditions can bacteria take up such large molecules of DNA. This usually occurs near the end of the log phase of growth. The mechanism by which a cell becomes competent is poorly understood, but apparently the cell undergoes a series of changes. These include modification of the cell wall and the synthesis of protein receptors on the cell surface that can bind DNA. Other competence proteins inside the cell are also synthesized, including those involved in the integration of donor DNA. Cells also can be made competent through artificial means in the laboratory, as will be discussed shortly. ■ **log phase, p. 106**

Entry of DNA

Although double-stranded DNA molecules bind to competent cells, single-stranded DNA enters. As one strand enters, nucleases at the cell surface degrade the other (figure 8.20a and 8.20b).

Integration of Donor DNA

Once inside the recipient cell, the single-stranded donor DNA becomes positioned, by hydrogen bonding, next to the complementary region of the recipient DNA (see figure 8.20c). The two regions are said to be homologous. The donor and recipient strands are held together by hydrogen bonds. Then, a nuclease cleaves the recipient cell's single-stranded DNA on either side of the donor DNA. This fragment of DNA is released into the cytoplasm, where it is degraded by nucleases. The donor DNA then replaces the recipient DNA precisely, by the process of homologous recombination (see figure 8.20d). This mechanism of recombination in which the homologous region of the recipient DNA is replaced by donor DNA is termed **breakage and reunion.** The same mechanism also accounts for the integration of donor DNA into recipient cells in both conjugation and transduction.

Mismatch Repair

Once the DNA is integrated, the site of the mutation in the single strand of recipient DNA will not hydrogen bond to the corresponding base on the donor DNA, thereby causing a mismatch in base-pairing. Therefore, the mismatch repair system comes into play. This repair system will remove either the donor or the recipient single-stranded DNA, which differ from each other in at least one nucleotide. (see figure 8.20e). Which strand is removed depends on chance. In the laboratory, however, the cells are plated on selective media, on which only transformants will grow; hence, if the donor DNA is removed, a colony will not form (see figure 8.20f). ■ **mismatch repair, p. 198**

Cell Multiplication

The transformed cells multiply under selective conditions in which the nontransformed cells cannot grow (see figure 8.20f). For example, if the donor cells are *str*^{R} and the recipient cells are *str*^{S}, then only cells transformed to *str*^{R} will grow on medium that contains streptomycin. Although many other donor genes besides *str*^{R} will be transferred and integrated into the chromosome of the recipient cells, these transformants will go undetected since the donor and recipient cells are identical in these other genes.

In addition to transformants, any recipient cells that mutate from *str*^{S} to *str*^{R} will also grow on medium containing streptomycin. Therefore, proper controls must be carried out before any cells growing on streptomycin can be considered transformants. An appropriate control is the addition of deoxyribonuclease to the mixture of donor and recipient cells. This enzyme will degrade the DNA released from the donor cells before it can be taken up by recipient cells and thereby inhibit all transformation. Any streptomycin-resistant colonies that arise must be revertants, or must have arisen by a mechanism of gene transfer other than transformation.

Artificial Competence

Although not all bacteria become naturally competent, DNA can be introduced into most cells including bacteria, animals, and plants through a special treatment of the recipient cells. In one technique called **electroporation**, bacteria and DNA are mixed together and the mixture is subjected to an electric current (**figure 8.21**). The current apparently makes holes in the bacterial cell wall and cytoplasmic membrane through which the DNA enters. Once inside the cytoplasm, the DNA becomes integrated into the recipient chromosome by homologous recombination, presumably by the same mechanism as was described for DNA transformation.

Transduction

Bacterial viruses, called **bacteriophage** or simply **phage**, will also transfer bacterial genes from one cell to another. Phage have a protein coat that surrounds the genetic material of the virus. The phage infect bacteria, multiply inside the cells, and then are released by lysing the cells. Some of the released phage may carry bacterial genes in place of phage genes inside their protein coats. The phage containing bacterial genes then infect other bacteria and can transfer bacterial genes from the first infected bacterium to the second. This is the mechanism termed transduction. (**figure 8.22**). There are two types of transduction; **generalized** and **specialized.** In generalized transduction, any gene of the donor cell can be transferred. In specialized transduction, only a few specific genes can be transferred. We will briefly cover generalized transduction now and both will be discussed in more detail when the replication cycles of bacteriophage are considered in chapter 13.

In generalized transduction, after the phage infects the bacterium, the phage DNA codes for the enzyme deoxyribonuclease. This enzyme cleaves the replicated phage DNA into the proper size fragments to be enclosed in the coat protein of the phage. This deoxyribonuclease also cleaves the bacterial chromosome into fragments, however, which can by chance become

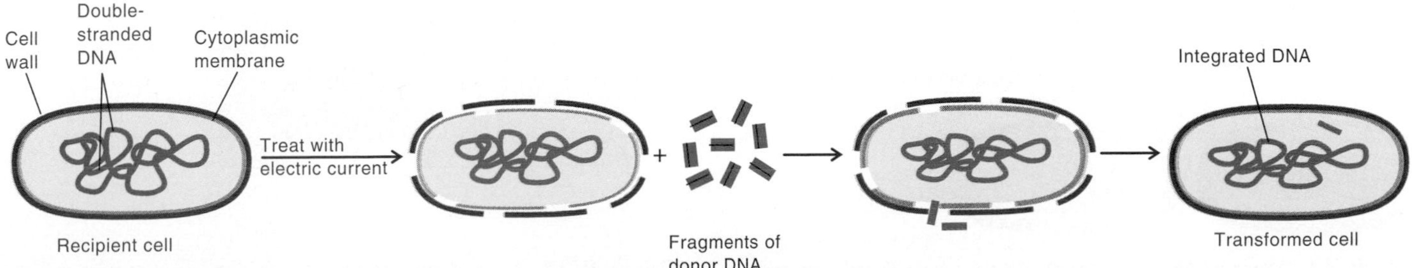

Figure 8.21 Electroporation The electric current makes holes in both the cell wall and the cytoplasmic membrane through which the DNA can pass. These holes are then repaired by the cell, and the DNA becomes incorporated into the chromosome of the cell.

surrounded by the phage coat, taking the place of phage DNA. Once released from the infected cell, the phage containing the bacterial DNA may infect other nearby cells and thereby transfer bacterial genes. Inside the recipient cell, the bacterial DNA becomes integrated by homologous recombination, replacing genes in the recipient cell. Transduction is a common mechanism of gene transfer and occurs in a wide variety of Gram-positive and Gram-negative bacteria.

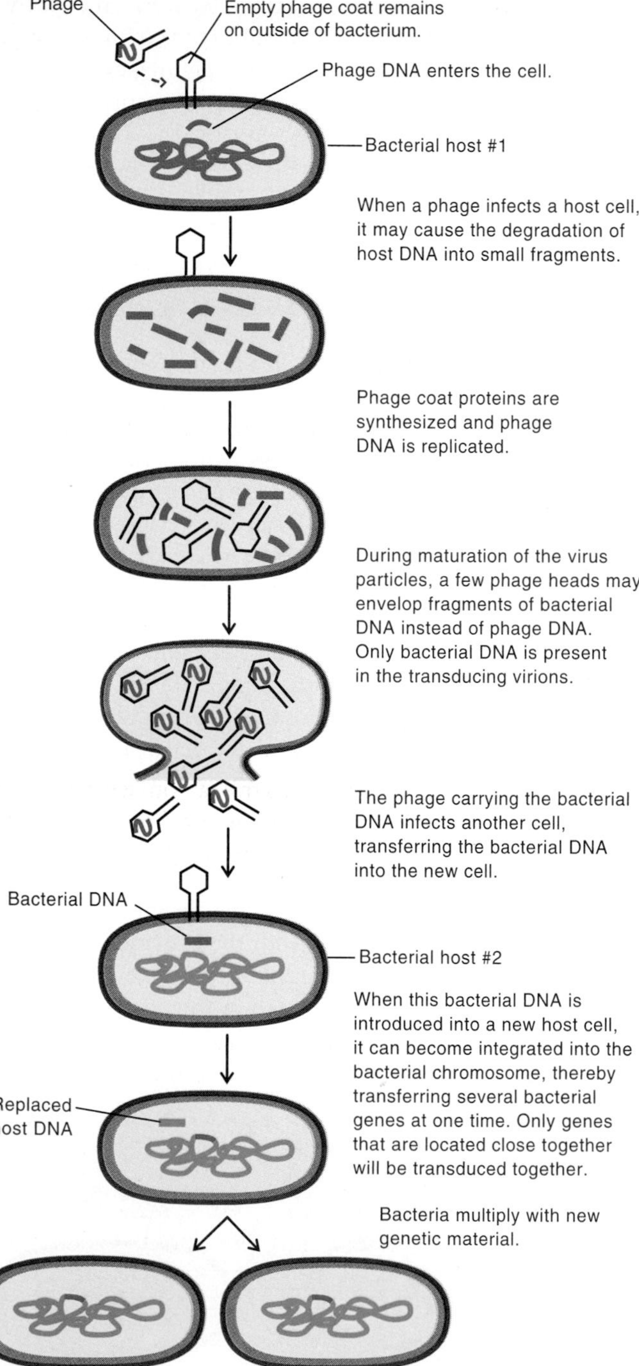

Figure 8.22 Transduction (Generalized) Any piece of the chromosomal DNA of the donor cell can be transferred in this process. All of the DNA molecules, of the bacterial virus and of the bacteria, are double-stranded.

Conjugation

Another important and common mechanism of gene transfer is conjugation. Some *E. coli* can transfer a plasmid, and other *E. coli*, chromosomal DNA. Conjugation, in contrast to transformation and transduction, requires contact between donor and recipient cells. This requirement can be shown through the following experiment. If two different auxotrophic mutants are placed on either side of a filter through which fluids, but not bacteria, can pass, recombination does not occur. If the filter is removed, however, allowing cell-to-cell contact, recombination takes place. Conjugation is a complex process and many aspects are not understood even though it was first observed in *E. coli* more than 50 years ago. ■ **plasmid, p. 70**

Plasmid Transfer

Populations of *E. coli* can be divided into two types of cells. One, the **donor** cell, contains an **F** or **fertility plasmid** and is designated **F⁺**. The other, recipient cells, do not contain this plasmid and are called **F⁻**. DNA is transferred only in one direction, from F^+ to F^-, i.e., in a polar fashion. Consequently, F^+ cells are often referred to as males and the F^- cells as females. The F plasmid codes for the synthesis of a structure, the **sex or F pilus**, the protein appendage that attaches the donor to the recipient cell (**figure 8.23**). The plasmid also carries information required for its own transfer. Donor cells can transfer their F plasmid but not their chromosome into recipient cells. ■ **F pilus, p. 69**

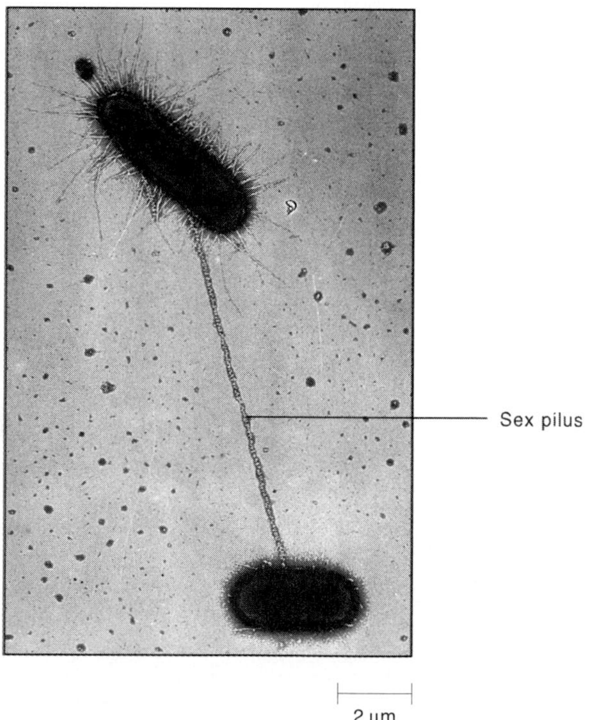

Figure 8.23 Sex or F Pilus Holding Together Donor and Recipient Cells of *E. coli* During DNA Transfer The tiny knobs on the pilus are bacterial viruses that have adsorbed to the pilus.

Plasmid transfer can be divided into four steps (**figure 8.24**).

Step 1: Contact between donor and recipient cells. The sex pili of the donor cells recognize and bind to specific **receptor sites** on the cell wall of the recipient cell. When donor and recipient cells are mixed together, the sex pili act as grappling hooks, pulling the two cells together.

Step 2: Mobilization or activation of DNA transfer. The plasmid becomes mobilized for transfer when a plasmid-encoded enzyme cleaves the bottom strand of the plasmid at a specific nucleotide sequence, termed the **origin of transfer.**

Step 3: Plasmid transfer. Within minutes of the F^+ cell contacting the F^- cell, a single strand of the F plasmid, beginning at the origin of transfer, enters the F^- cell. This transfer takes about 2 minutes. Note that a single strand of the F plasmid remains in the donor cell.

Step 4: Synthesis of a functional plasmid inside the recipient and donor cells. Once inside the recipient cell, a complementary strand to the single-stranded transferred DNA is synthesized. Likewise, a strand complementary to the single-stranded plasmid DNA remaining in the donor is synthesized. Thus, both the donor and recipient cells contain a copy of the F plasmid and are therefore F^+. Both cells can act as donors of the F plasmid. This explains why a single F^+ cell of *E. coli* mixed with a population of F^- cells can convert the entire population to F^+ after overnight growth of the culture. Cells can spontaneously lose their plasmids, which explains why not all cells are F^+. Cells that have lost their plasmid are said to be **cured**.

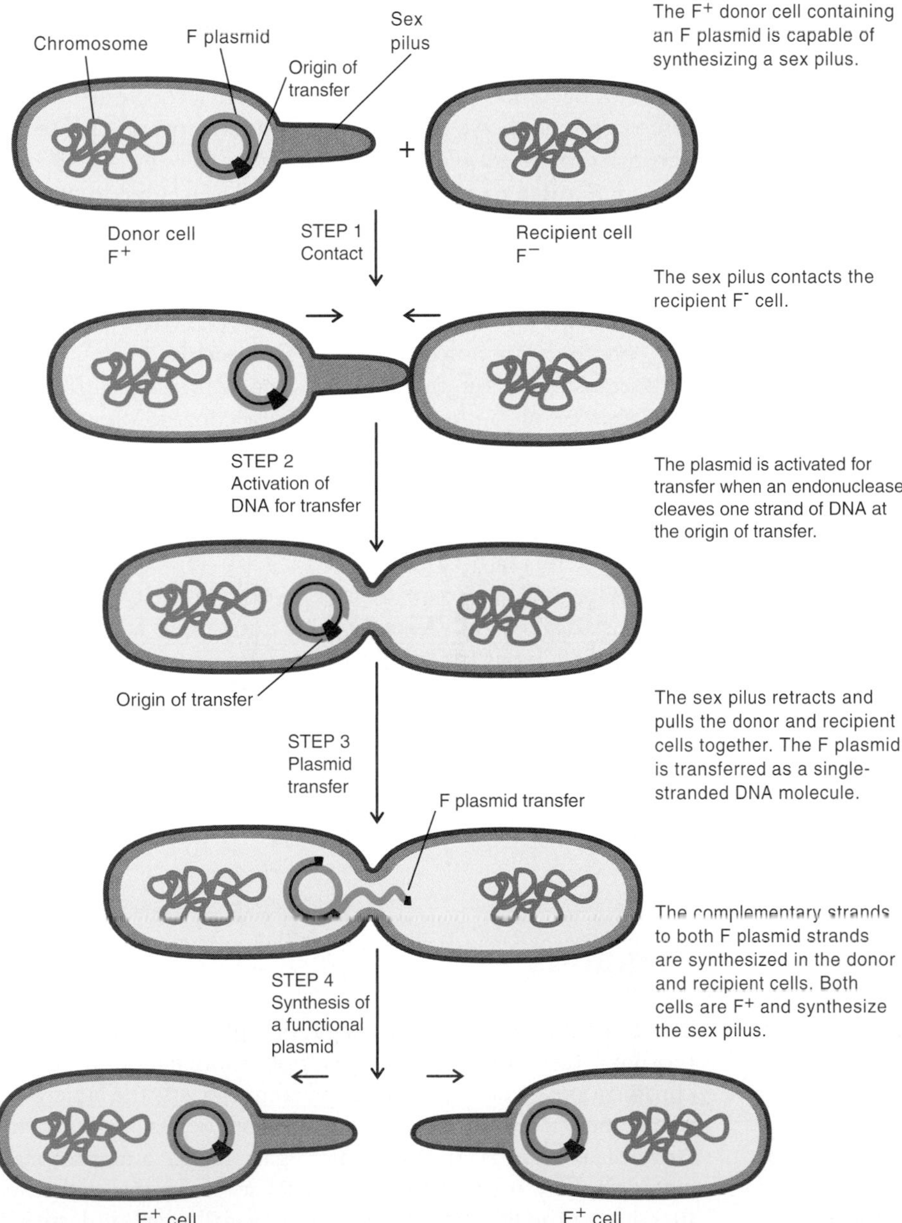

Figure 8.24 Conjugation—Transfer of the F Plasmid The exact process by which the donor DNA passes to the recipient cells is not known.

Transfer of Plasmids Other than F

Plasmids other than F are also transferred by conjugation. The F plasmid is termed a **self-transmissible plasmid** because it carries all the genetic information necessary for its own transfer. The plasmid codes for two functions; the synthesis of the sex pilus, and the mobilization of its DNA for transfer. Other plasmids are not self-transmissible—they do not code for both functions. A self-transmissible plasmid, however, can often provide the missing function to a non-self-transmissible plasmid if both are in the same cell. Thus, a non-self-transmissible plasmid in the same cell as the F plasmid will often be transferred.

Chromosome Transfer

Thus far, we have discussed only the transfer of plasmids by conjugation. However, the bacterial chromosome is sometimes transferred to F^- cells by a few cells in the F^+ population. These donor cells are not in fact F^+ but arise from F^+ cells as a result of the F plasmid becoming integrated on rare occasions, into the donor cell chromosome at specific sites (**figure 8.25**). This integration occurs by recombination at sites on the chromosome that are homologous to sites on the plasmid. These sites are frequently insertion sequences. The cells able to transfer the chromosome are called **Hfr** for **high frequency of recombination**. In the Hfr cell, the F plasmid replicates as part of the chromosome. ■ **insertion sequence, p. 214**

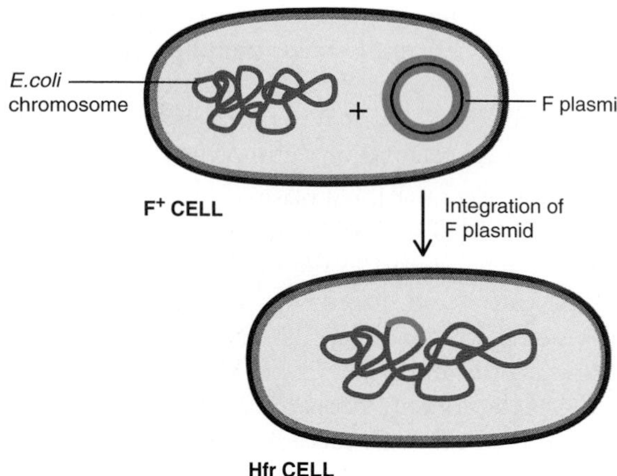

Figure 8.25 Hfr Formation Integration of the F plasmid into the bacterial chromosome to form Hfr. There are homologous sites on the F plasmid and the chromosome that allow the integration to occur. There is no replacement of DNA in the chromosome; the F plasmid increases the size of the chromosome.

Many features of chromosome transfer are similar to those involved in plasmid transfer. (**figure 8.26**). These similarities include the following:

- The Hfr cell contacts the F$^-$ cell by means of the sex pilus.
- DNA is transferred as a single strand and is made double stranded in the recipient cell.
- Donor cells remain Hfr, just as F$^+$ donor cells remain F$^+$.
- Transfer is very efficient, and all of the donor cells can transfer DNA.

There are, however, two major differences in the transfer of the chromosome compared to plasmid transfer.

First, most recipient cells that receive chromosomal DNA remain F$^-$, in contrast to recipient cells conjugating with F$^+$ cells. Consequently, a recipient cell receiving chromosomal DNA from an Hfr cell cannot transfer the chromosomal DNA to other cells in the population. The transfer of the chromosome starts at the origin of transfer in such a way that a piece of the F plasmid integrated into the donor chromosome is transferred first (see figure 8.26). The rest of the F plasmid is at the other end of the chromosome, however, so that the entire chromosome must be transferred if the recipient cell is to become Hfr. It takes 100 minutes for the entire chromosome to be transferred, and the rate of transfer is constant. Thus, it takes 25 minutes to transfer one-quarter of the chromosome and 50 minutes to transfer one-half. This can be readily demonstrated by placing the conjugating cells in a blender at various times. The stirring action breaks the pilus, stopping DNA transfer immediately. One then plates the cells on media supplemented with certain growth factors to determine which genes have been transferred. The two cells must remain in close proximity for transfer to occur, and they rarely remain in contact for the length of time required for the transfer of the entire donor chromosome. Therefore, only genes near the origin of transfer are transferred at a high frequency, and the recipient cells usually remain F$^-$.

Second, once inside the recipient cell, the donor DNA integrates into the recipient chromosome by replacing homologous genes in the recipient cell through recombination (see figure 8.26). This contrasts with the transfer of plasmids, which replicate independently of the chromosome and therefore do not integrate into the recipient cell chromosome.

F′ Donors

The F plasmid in the Hfr strain can be excised from the chromosome; thus, the process of F plasmid incorporation into the chromosome is reversible. In some instances in the process of excision, however, a small piece of the bacterial chromosome remains attached to the F plasmid (**figure 8.27**). This F plasmid with its piece of attached chromosome is called **F′** (F prime), and like the F plasmid it is rapidly and efficiently transferred to all F$^-$ cells in the population. Consequently, any chromosomal genes attached to the F plasmid are also transferred to all cells. The F′ plasmid usually remains **extrachromosomal**—that is, it does not become a part of the recipient cell's chromosome. On rare occasions, however, it can become incorporated into the chromosome of recipient cells, which then become Hfr.

The three mechanisms of DNA transfer are compared in **Table 8.3.**

MICROCHECK 8.7

In nature, DNA is transferred between bacteria by three distinct mechanisms: DNA-mediated transformation, transduction, and conjugation. In all cases, only a portion of the donor chromosome is transferred. In transformation and conjugation, DNA is transferred as a single strand.

- By what mechanism of DNA transfer can the largest piece of DNA be transferred?
- What mechanism of transfer involves polarity of transfer? Explain why DNA is transferred in a polar fashion.
- If a recipient cell lacks a mismatch repair system, how would this affect the frequency of DNA transformation?
- What happens to chromosomal DNA that is transferred in conjugation if it does not integrate into the recipient cell's chromosome? Explain.

Plasmids

The F plasmid is only one example of a **plasmid**, an extrachromosomal piece of DNA that multiplies independently of the chromosome. Plasmids are very common in the microbial world, being found in most members of the Bacteria and the Archaea as well as in eukaryotic algae, fungi, and protozoa. They play key roles in the lives of these organisms with their function in the life of bacteria being especially well understood. One of the most interesting plasmids is the tumor-inducing (Ti) plasmid of *Agrobacterium*. It allows *Agrobacterium* to conjugate with plants (see **Perspective 8.2**). ■ plasmid, p. 70

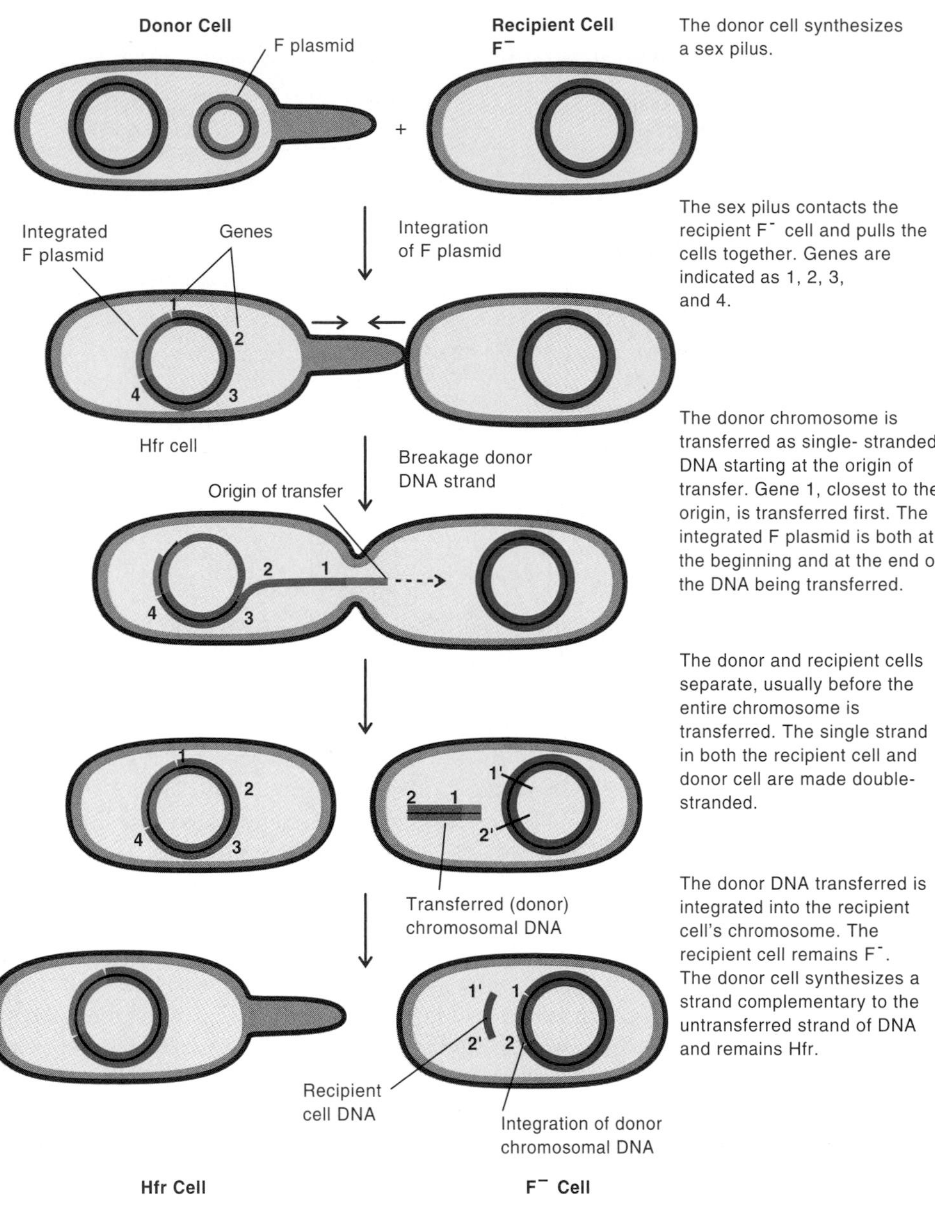

Donor Cell
F plasmid
Recipient Cell
F⁻
The donor cell synthesizes a sex pilus.
Integrated F plasmid
Genes
Integration of F plasmid
The sex pilus contacts the recipient F⁻ cell and pulls the cells together. Genes are indicated as 1, 2, 3, and 4.
Hfr cell
Breakage donor DNA strand
Origin of transfer
The donor chromosome is transferred as single- stranded DNA starting at the origin of transfer. Gene 1, closest to the origin, is transferred first. The integrated F plasmid is both at the beginning and at the end of the DNA being transferred.
The donor and recipient cells separate, usually before the entire chromosome is transferred. The single strand in both the recipient cell and donor cell are made double-stranded.
Transferred (donor) chromosomal DNA
The donor DNA transferred is integrated into the recipient cell's chromosome. The recipient cell remains F⁻. The donor cell synthesizes a strand complementary to the untransferred strand of DNA and remains Hfr.
Recipient cell DNA
Integration of donor chromosomal DNA
Hfr Cell
F⁻ Cell

Figure 8.26 Conjugation—Transfer of Chromosomal DNA The DNA is transferred as a single-stranded DNA molecule. The recipient genes are designated with a prime (′). The corresponding forms in the donor lack the prime (′).

A study of a wide variety of plasmids in many different bacteria has led to the following generalizations.

- Most prokaryotes contain one or more different plasmids; their functions, however, are often unknown.
- The traits coded by plasmids generally provide the bacterial cell with useful but not indispensable capabilities. For example, the enzymes of the glycolytic pathway are never encoded on plasmids, since they are indispensable to the cell. Many members of the genus *Pseudomonas*, however, can grow not only on such common sugars as glucose, but also on unusual compounds, such as camphor, if they have a certain plasmid. Thus, the plasmid extends the cell's biochemical capabilities for degrading certain compounds. Plasmids code for a wide variety of other traits, some of which are listed in **table 8.4**.
- Plasmids vary in size from a few genes to several hundred. The only function that all plasmids must carry out is replication and so all of them are **replicons**, pieces of DNA that have the genetic information required to replicate. The smallest plasmids contain only the information necessary for replication. The larger ones code for more functions, such as antibiotic resistance.
- The number of copies per cell of each kind of plasmid varies, depending on the plasmid. For example, the F plasmid is present in only one to two copies per cell and is called a **low-copy-number plasmid**. Other plasmids, called **high-copy-number plasmids,** may be present in more than 500 copies per cell.
- Although plasmids can be transferred to other bacteria by the three methods we have discussed, conjugation is by far the most important means of transfer for many Gram-negative species of bacteria.
- Most plasmids can multiply only in one species of bacteria. They are termed **narrow host range plasmids.** The F plasmid is an example of such a plasmid. A few, however, termed **wide host range plasmids**, can multiply in many different species of bacteria.

R Plasmids

Among the best studied and most important groups of plasmids are the **resistance** or **R plasmids**. They are so named because they confer resistance to many different antimicrobial medications and

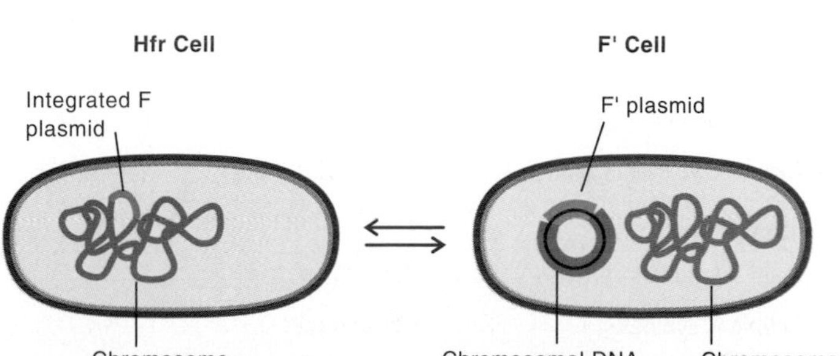

Figure 8.27 Formation of F′ Plasmid This F′ plasmid has properties of the F plasmid but carries chromosomal DNA. This process is reversible.

Table 8.3 Comparison of Mechanisms of DNA Transfer

Mechanism	Main Features	Size of DNA Transferred	Polarity?	Sensitivity to DNase addition
Transformation	Naked DNA transferred	About 20 genes	No	Yes
Transduction	DNA enclosed in a bacteriophage coat	Usually part of the chromosome	No	No
Conjugation				
Chromosomal DNA	Cell-to-cell contact required	Small fraction of chromosome	Yes	No
F plasmid	Cell-to-cell contact required	Entire F plasmid	Yes	No

heavy metals, such as mercury and arsenic. Like antimicrobial medications, heavy metals are found in the hospital environment, and it is in this environment that R plasmids are commonly found. Many of these plasmids are composed of two parts: the **resistance genes** or **R genes**, which code for the resistance traits, and the **resistance transfer factor** or **RTF**, which codes for the transfer of the plasmid to other bacteria by conjugation **(figure 8.28)**. In a self-transmissible plasmid, the RTF portion contains genes that code for pilus synthesis, the origin of transfer, and mobilization genes, all of which are required for plasmid transfer.

The R genes code for resistance to widely used antimicrobial medications, which can include sulfanilamide, streptomycin, chloramphenicol, and tetracycline, and a number of different heavy metals. Certain R plasmids confer resistance to some, but not all, of these antimicrobial substances; yet others confer resistance to more than these particular ones.

These plasmids may have arisen by recombination between two plasmids in the same cell, one carrying the R genes and the other encoding the functions required for plasmid transfer. If each plasmid has certain sequences identical to those in the other plasmid, then homologous recombination can occur and a single plasmid is formed. This is similar to the formation of an Hfr cell by the integration of the F plasmid into the chromosome.

Perhaps the most important feature of many R plasmids is that they can be transferred to antimicrobial– and heavy metal–sensitive bacteria, thereby conferring simultaneous resistance to several antimicrobials and heavy metals encoded by the R genes. Furthermore, many R plasmids have wide host ranges and can multiply in a wide variety of different Gram-negative genera. These include *Shigella*, *Salmonella*, *Escherichia*, *Yersinia*, *Klebsiella*, *Vibrio*, and *Pseudomonas*. If R plasmids are present in one species in a mixed population of these organisms, other cells will receive the R plasmid and thus become resistant to a variety of antimicrobial agents. This transfer of R plasmids helps explain why so many different organisms in a hospital environment are resistant to many different antimicrobials. ■ nosocomial infections, p. 488

Not surprisingly, antimicrobial-resistant organisms are most common in locations where antimicrobials are in greatest use. The antimicrobials kill the sensitive cells and allow the few antimicrobial-resistant cells in the population to grow. Soon, most of the cells in this environment are antimicrobial resistant. This process is why hospitals harbor many antimicrobial-resistant bacteria. It is not surprising that antimicrobial-resistant organisms are more common in certain parts of the world, particularly in developing countries. This probably results from the overuse of antimicrobials in these countries, where antimicrobials can be sold without a doctor's prescription. Also, in areas that lack proper hygiene, bacteria are more readily transferred to people through contaminated food and water. Non-disease-causing bacteria such as *E. coli* that may be carried by healthy

Table 8.4 Some Plasmid-Coded Traits

Trait	Organisms in Which Trait Is Found
Antibiotic resistance	*Escherichia coli*, *Salmonella* sp., *Neisseria* sp., *Staphylococcus* sp., *Shigella* sp., and many other organisms
Pilus synthesis	*E. coli*, *Pseudomonas* sp.
Tumor formation in plants	*Agrobacterium tumefaciens*
Nitrogen fixation (in plants)	*Rhizobium* sp.
Oil degradation	*Pseudomonas* sp.
Gas vacuole production	*Halobacterium* sp.
Insect toxin synthesis	*Bacillus thuringiensis*
Plant hormone synthesis	*Pseudomonas* sp.
Antibiotic synthesis	*Streptomyces* sp.
Increased virulence	*Yersinia enterocolitica*

PERSPECTIVE 8.2 Bacteria Can Conjugate with Plants: A Natural Case of Genetic Engineering

For more than 50 years, scientists have known that DNA can be transferred between bacteria. Twenty-five years ago, it was shown that a bacterium can even transfer its genes into plant cells, such as tobacco, carrots, and cedar trees, through a process analogous to conjugation. What led to this discovery started almost 100 years ago in the laboratory of the plant pathologist, Dr. Erwin Smith. He showed that the causative agent of a common plant disease, termed **crown gall** is a bacterium, *Agrobacterium tumefaciens*. This disease is characterized by large galls or swellings that occur on the plant at the site of infection, usually near the soil line, the crown of the plant. When other investigators cultured the diseased plant tissue on agar plates containing nutrients necessary for the growth of plant tissue, it displayed properties that differed from normal plant tissue. Whereas normal tissue requires several plant hormones for growth, crown gall tissue grows in the absence of these added hormones. In addition, crown gall tissue synthesizes large amounts of a compound termed an **opine,** which neither normal plant tissue nor *Agrobacterium* synthesizes. The most surprising observation was that the plant cells maintained their altered nutritional requirements and the ability to synthesize opine even after the bacteria were killed by penicillin. Investigators concluded that the crown gall plant cells are permanently transformed. Although *Agrobacterium* is required to start the infection, they are not necessary to maintain the altered nutritional requirements and biosynthetic capabilities of the plant cells.

The explanation of the process by which *Agrobacterium* causes crown gall tumors and transforms plant cells was established in 1977 following a report from a team of Belgian investigators that all strains of *Agrobacterium* capable of causing crown gall tumors contained a large plasmid termed the **tumor inducing** or **Ti plasmid**. A group of microbiologists at the University of Washington showed that a specific piece of the Ti plasmid, termed the **transferred DNA**, or **T-DNA**, is transferred from the bacterial cell to the plant cell, where it becomes incorporated into the plant chromsome **(figure 1)**. Like conjugation between bacteria, a pilus is required for DNA transfer.

The studies of investigators from around the world yielded many surprising observations. This bacterial DNA acts like plant DNA because the promoter regions of the transferred genes resemble those of plants rather than those of bacteria. Therefore, the genetic information in the T-DNA is expressed in the plants but not in *Agrobacterium*. This DNA encodes enzymes for the synthesis of the plant hormones as well as for the opine. The expression of these genes supplies the plant cells with the plant hormones, explaining why the transformed plant cells can grow in the absence of added hormones and are able to synthesize opine. Thus once incorporated into the plant chromosome, the DNA provides the transformed cell with additional genetic information that confers new properties on the plant cell.

Why does *Agrobacterium* transform plants? This bacterium has the ability to use the opine as a source of carbon and energy, whereas most other bacteria in the soil, as well as plants, cannot. Therefore, *Agrobacterium* subverts the metabolism of the plant to produce food that only *Agrobacterium* can use. Thus, *Agrobacterium* is a natural genetic engineer of plants.

The *Agrobacterium*–crown gall system is of great interest for several reasons. First, it shows that DNA can be transferred from prokaryotes to eukaryotes. Many people believed that such transfer would be impossible in nature and could only occur in the laboratory. Second, this system has spawned an industry of plant biotechnology dedicated to improving the quality of higher plants. Thus, it is possible to replace the genes of hormone synthesis in the Ti plasmid with any other genes, which will then be transferred and incorporated into the plant. With this technology, genes conferring resistance to bacteria, viruses, insects, and different herbicides have been incorporated into a wide variety of plants. Rice has been transformed to synthesize high levels of β-carotene, the precursor of vitamin A. Edible vaccines are being synthesized in bananas following transformation by *Agrobacterium*.

Genetic engineering of plants became a reality once scientists learned how a common soil bacterium caused a well-recognized and serious plant disease. This system serves as a beautiful example of how solving a riddle in basic science can lead to major industrial applications.

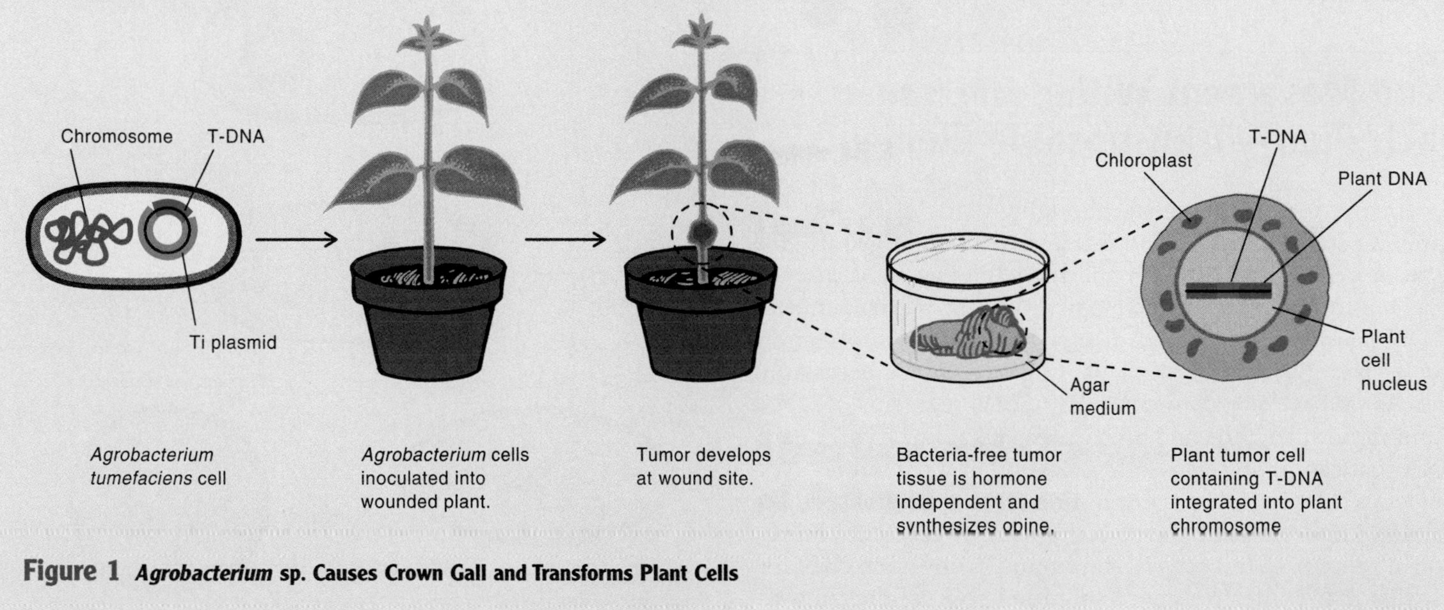

Figure 1 ***Agrobacterium* sp. Causes Crown Gall and Transforms Plant Cells**

people can serve as a reservoir for R plasmids, which then can be transferred to disease-causing organisms.

MICROCHECK 8.8

All members of the microbial world contain plasmids, which in most cases code for unknown functions. Plasmids vary in size, copy number, host range, their genetic composition, and their ability to be transferred to other cells. One of the most important plasmids is the R plasmid, which codes for resistance to various antimicrobial medications and heavy metals.

- What functions must a plasmid code for in order to be self-transmissible?
- What does the phrase, "reservoir for R plasmids" mean when referring to plasmids carried by non-disease-causing bacteria?

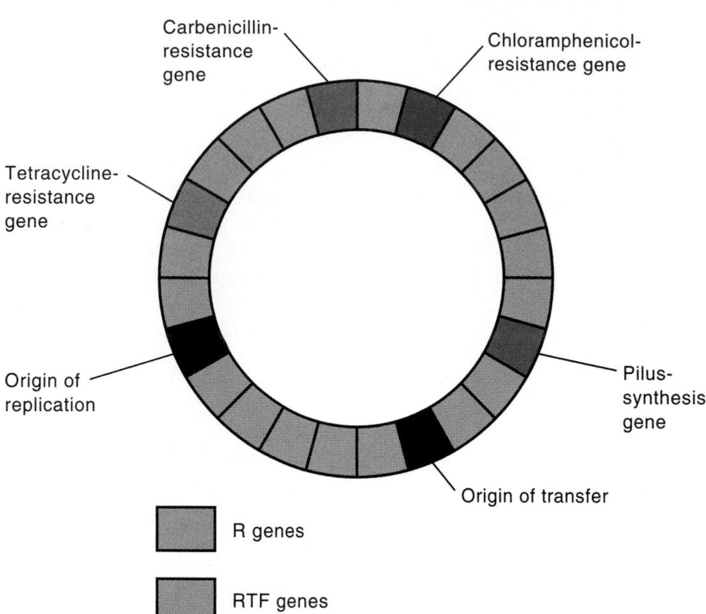

Figure 8.28 Two Regions of an R Plasmid The R (resistance) genes code for resistance to various antimicrobials; the RTF (resistance transfer factor) region codes for plasmid replication and the transfer of the plasmid to other bacteria.

Gene Movement Within the Same Bacterium—Transposable Elements

For many years, biologists thought that DNA was an extremely stable molecule and that genes did not move around inside a cell. Since DNA is the most important carrier of genetic information, it seemed logical that any rearrangements of DNA inside a cell would disrupt this genetic information and be harmful to the cell. This, however, is not necessarily true. As we have already mentioned, special segments of DNA found in both bacteria and archaea as well as eukaryotes have DNA that can jump to any other location in the genome. Not only does this process of transposition create mutations, it also plays an important role in the movement of genes of antimicrobial resistance. In bacteria, since transposons often code for proteins that confer resistance to antimicrobials, their movement can be very important to humans and animals. Indeed, the formation of R factor plasmids with their multiple antimicrobial-resistance genes probably results from the coming together of transposable elements coding for resistance to different antimicrobial medications.

Transposons and Transfer of Genes to Unrelated Bacteria

Transposons that code for antimicrobial resistance can move from sites on the chromosome of a cell into plasmids in the same cell. Plasmids are then readily transferred by conjugation to other cells. If the plasmid happens to be a self-transmissible, wide host range plasmid, it can transfer into a number of bacteria unrelated to the donor **(figure 8.29)**. The combination of inserting transposable elements into broad host range plasmids with their subsequent transfer into unrelated bacteria explains how the same antimicrobial-resistance genes can be found in unrelated bacteria.

Structure of Transposons

What is the structure of transposons, and what gives them the ability to move? Several types of transposons exist, varying in the complexity of their structure. The simplest kind, termed an **insertion sequence (IS) element**, consists of a gene that codes for the enzyme, **transposase.** This enzyme is required for transposition and is bound on either side by about 15 to 25 base pairs, which have the structure shown in **figure 8.30a**. The key feature of this sequence of nucleotides is that the sequence in one strand of DNA is the same as in the other strand, only going in the opposite direction (see figure 8.30a). Such a sequence is termed an **inverted repeat**. The *E. coli* chromosome contains more than a dozen different IS elements, labeled IS1, IS2, and so on. Several copies of each are present in the chromosome.

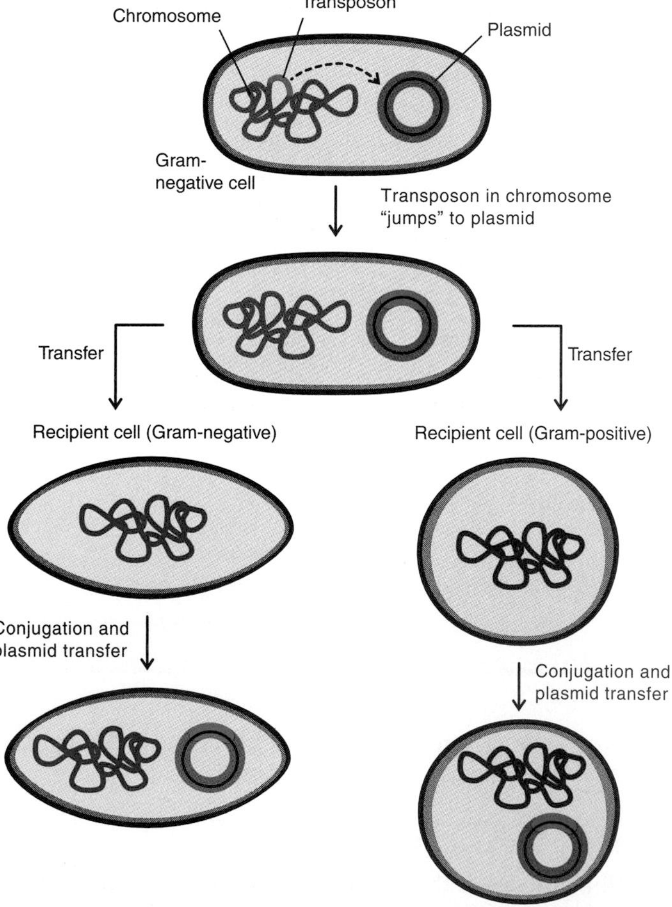

Figure 8.29 The Movement of a Transposon Throughout a Bacterial Population Note that transfer can be between bacteria belonging to different genera and between Gram-negative and Gram-positive bacteria.

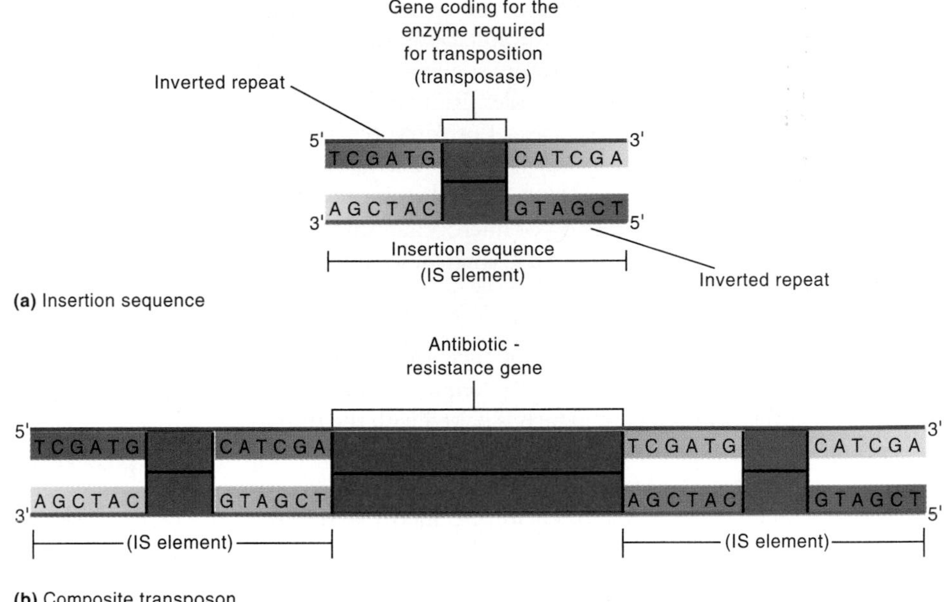

Figure 8.30 Transposable Elements (a) Insertion sequence. Note that the sequence of bases at the end of the IS element in one strand is the same as the sequence in the opposite strand reading in the same direction (5′ to 3′). Only 6 of the 15 to 25 base pairs are shown. **(b)** Composite transposon. Antibiotic-resistance locus flanked by two insertion elements.

They are frequently on plasmids also. They often are responsible for the recombination that occurs between two replicons within the same cell (see figure 8.25).

A more complex type of transposable element, called a **composite transposon**, consists of a gene whose product is easily recognized, such as a gene coding for antimicrobial resistance, flanked by IS elements (figure 8.30b). The importance of this structure is that the antimicrobial-resistance gene can move from the chromosome to the plasmid, which can be readily transferred to other cells. In theory, any gene or group of genes can move to another site if they are bounded by IS elements.

MICROCHECK 8.9

Genes can transpose from one location in the genome to another, thereby causing mutations. Transposition requires that the gene transposing have inverted repeat sequences on its ends and transpose as part of the insertion sequence.

- What are two features common to all insertion sequences?
- How does transposition promote gene transfer between bacteria?
- What would you think is the substrate for transposase?

Barriers to Gene Transfer

As a general rule, a recipient bacterium will take up DNA by transformation, conjugation, or transduction only if the donor DNA is from the same species of bacteria. If other DNA, termed **foreign DNA,** enters the cell, a deoxyribonuclease in the recipient cell will degrade it. This mechanism allows cells to maintain stability of their genome against invasion by foreign DNA.

Restriction of DNA

The enzymes that degrade entering DNA are called **restriction enzymes** because they "restrict" the entry of foreign DNA into cells. The enzymes recognize specific short sequences of 4 to 8 nucleotides in foreign DNA and then cleave one strand of DNA at these sites. One cleavage site is on one strand, and the other is on the opposite strand **(figure 8.31)**. Enzymes in different

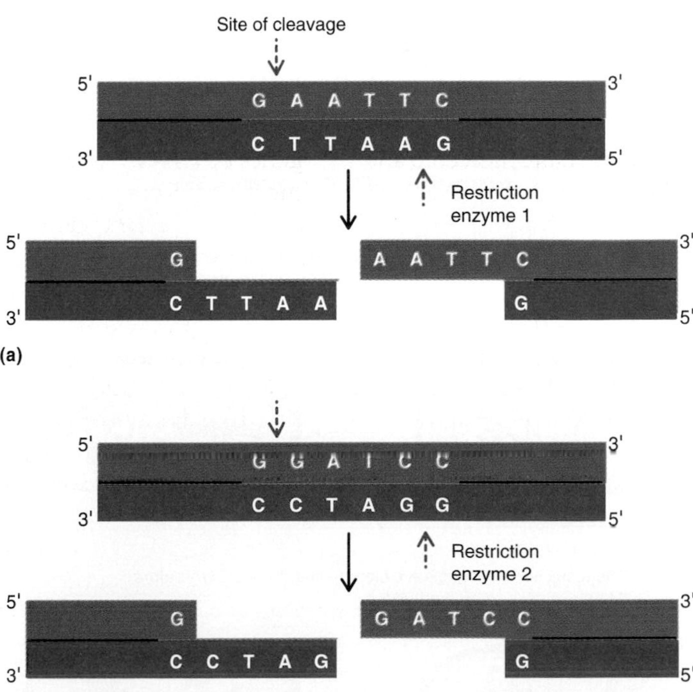

Figure 8.31 Action of Restriction Endonucleases Cleaving of double-stranded DNA at specific sites indicated by the red arrows. The few hydrogen bonds between the two cleavage sites cannot hold the strands together. Consequently, the DNA splits into two molecules frequently with overlapping ends. **(a)** Site of cleavage by a restriction enzyme, isolated from *E. coli*. **(b)** Site of cleavage by a restriction enzyme isolated from *Bacillus amyloliquefaciens*.

organisms recognize and cleave different nucleotide sequences (see figure 8.31a, b). Other deoxyribonucleases that degrade DNA without regard to its nucleotide sequence then break down the DNA into individual nucleotides. Restriction enzymes have been found in virtually all species of prokaryotes, but not in any eukaryotes. Thus, eukaryotes lack this barrier to gene entry. Restriction enzymes, which have proved to be extremely important in the genetic engineering of microorganisms, will be considered further in chapters 9 and 13. ■ restriction enzyme, p. 225

Modification Enzymes

How do bacteria prevent destruction of their own DNA? The partial answer to this question is that bacteria contain another enzyme, a **modification enzyme**, which is paired with the restriction enzyme. The modification enzyme adds methyl ($—CH_3$) groups to adenine and cytosine in the section of DNA recognized by the restriction enzyme. These bases are also methylated in the recipient strain. Once the DNA is methylated, the restriction enzymes do not recognize the specific base sequence as a substrate and so do not cleave the DNA. For example, if cytosine has a methyl group, then it is not a substrate for a restriction enzyme. Since cytosine and methylcytosine have the same hydrogen-bonding characteristics, this alteration does not induce mutations. Note that not all of the adenine and cytosine molecules in the DNA are methylated, only those recognized by the restriction enzyme of the cell that is paired with the modification enzyme. The pattern of methylation of DNA defines the DNA as **self** or foreign and whether or not the restriction enzyme will degrade it.

When foreign DNA enters a cell, a race begins between methylation and restriction. If the DNA is methylated before it is degraded by the restriction enzyme, it becomes self and will not be degraded. In most cases, however, it is degraded before it has a chance to be methylated **(figure 8.32)**. The two enzymes, restriction enzyme and methylase are called the **restriction-modification system.** It can be considered a crude immune system. ■ restriction-modification system, p. 335

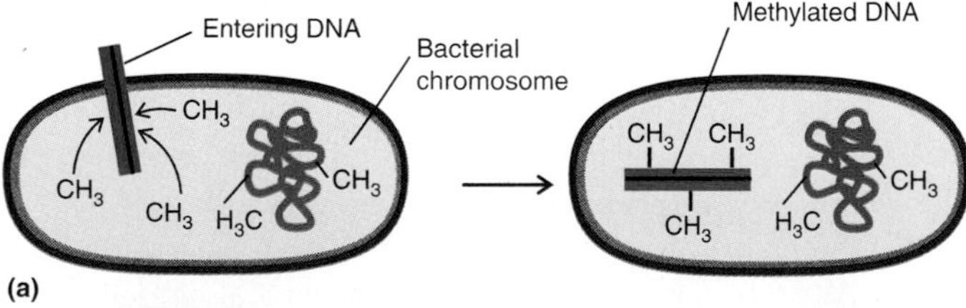
The entering DNA is methylated by a modification enzyme and becomes resistant to degradation by the restriction endonuclease.
Entering DNA
Bacterial chromosome
Methylated DNA
CH3
CH3
CH3
H3C
CH3
CH3
CH3
CH3
H3C
CH3
(a)

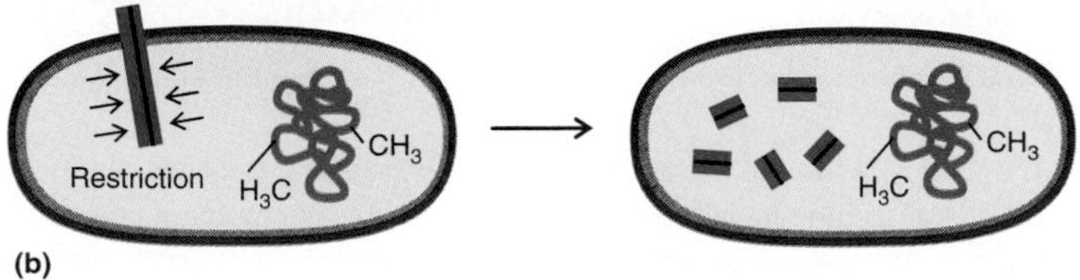
The restriction endonuclease cleaves the foreign DNA before it has a chance to be methylated.
Restriction
H3C
CH3
H3C
CH3
(b)

Figure 8.32 Restriction and Modification of Entering DNA

MICROCHECK 8.10

Transposons, often coding for antimicrobial resistance, can move from one site on the genome to any other site. Genes on broad host range plasmids can be transferred to unrelated bacteria. Bacteria can inhibit the entry of foreign DNA into cells by restriction enzymes.

- How does a restriction enzyme tell the difference between its own and foreign DNA?
- What effect does a lack of a restriction-modification system have on the entry of foreign DNA?

Importance of Gene Transfer to Bacteria

Acquiring genes through gene transfer provides new genetic information to microorganisms, which may allow them to survive changing environments. The major source of variation within a bacterial species is mutation. In mutations, usually only a single gene changes at any one time to provide new genetic information to the cell. In contrast, gene transfer results in many genes being transferred simultaneously, giving the recipient cell much more additional genetic information. For example, in a hospital environment in which a large variety of antimicrobials are constantly used, a mutation usually confers a selective advantage for resistance to only a single antimicrobial medication. If other antimicrobials are in the environment, the mutant will still die. In contrast, transfer of a resistance plasmid confers resistance simultaneously to many different antimicrobials: thus, the organism will achieve resistance to many more antimicrobial agents and is more likely to survive.

Transposons greatly expand the opportunity for gene movement. Any gene or genes flanked by insertion sequences can move to different sites in the genome. It has been shown that a chromosomal gene flanked by insertion sequences can move to a broad host range plasmid in the same cell, which can then be transferred to a large number of different genera of bacteria, both Gram-positive and Gram-negative. This process explains why the same antimicrobial-resistance genes occur in *E. coli* (Gram-negative) and *Clostridium* (Gram-positive).

It is not surprising that many if not all bacteria have evolved mechanisms for gene transfer and also for spontaneous gene mutation. Both mechanisms play important roles in altering the genotype of bacteria, which allows them to adapt to changing environmental conditions through the process of natural selection.

MICROCHECK 8.11

Gene transfer into a cell can alter the cell in a number of different ways simultaneously whereas a mutation usually results in the change of a single gene.

- Give two ways in which transposable elements can alter the phenotype of bacteria.
- What is the most likely mechanism for the transfer of antibiotic resistance from *E. coli* to *Clostridium*? Why?

FUTURE CHALLENGES

Targeting the Genome

Because of the increasing resistance of microorganisms to current antimicrobial medications, the demand for new antimicrobials that will kill these resistant organisms is rapidly increasing. It is surprising and sobering to realize that only one new antimicrobial has been introduced in the past 35 years. The challenge is great to develop new antimicrobial agents that strike at targets different from the few used by current antimicrobials. Such antimicrobials should not be susceptible to existing resistance mechanisms that microorganisms have developed.

With the revolution that has been occurring in biology over the past 10 years, the development of new antimicrobial agents may become a reality. What are these new strategies that hold such great promise? They are based on knowing the entire genomic sequence of microbial pathogens. More than 25 microbial genomes have now been sequenced, mostly model organisms such as E. coli *or human pathogens. This DNA sequencing is the science of* ***genomics.*** *The next step is to identify genes necessary for the survival of the microorganism or required to cause disease. To gain some understanding of the function of any gene, one must compare its DNA sequence with the sequence of all other genes that have been put into a database, called a* ***genebank.*** *If the gene that has been sequenced is similar in sequence to any other gene, then it is assumed that the two genes have similar functions. Thus, if the function of a gene that has been sequenced in any organism is known, then the function of all genes with a similar sequence can be estimated. This is the science of* ***bioinformatics,*** *or one aspect of* ***functional genomics,*** *the science of understanding the function of a gene through its nucleotide sequence.*

The genes of a pathogen that are predicted to be essential to cause disease can then be mutated and the mutants tested for survival or ability to cause disease. This will confirm or disprove the bioinformatics analysis. In a pathogen, the ideal targets for new antimicrobials are gene products that are:

- *Necessary for survival of the pathogen or required to produce disease*
- *Present in a wide range of pathogens*
- *Absent or very different in humans*

Another approach to identifying potential targets for antimicrobials involves protein analysis. Once the function of an essential protein encoded by a gene is known, several approaches can be tried to inhibit its function. One approach is to design a molecule that interferes with its function. A well-known example is sulfa, which competes with the normal substrate of an essential enzyme, thereby inhibiting its activity. Recall that the development of sulfa was fortuitous; the methods we are discussing did not exist when it was first used. ■ **sulfa, p. 145**

To develop successful enzyme inhibitors requires considerable information about the enzyme, how it folds, what its three-dimensional structure is, and whether or not it interacts with other proteins. Protein analysis of this type is called ***proteomics.***

Over the past 50 years, the search for antimicrobial medications has relied on random screening. Scientists looked for growth inhibition of a pathogen by unknown organisms isolated from soil samples collected from various environments around the world. New technologies based on microbial genomics should identify new targets and provide a rational approach to developing new antimicrobials. There certainly will not be a hiatus of 35 years between the introduction of a new antimicrobial medication this year and the next introduction.

SUMMARY

Sources of Diversity in Microorganisms

1. The properties of bacteria can change through **mutations**, changes in the chemical structure of DNA, or by changes in gene expression.

GENE MUTATION

Spontaneous Mutations

1. **Spontaneous mutations** are changes in the nucleotide sequences in DNA that occur without the addition of agents known to cause mutations.
2. These mutations are rare and occur at a characteristic frequency for each gene.
3. They are stable but on rare occasion can undergo a change back to the nonmutant form—a **reversion**.

Base Substitution

1. **Base substitutions** usually occur during DNA replication; **point mutations** occur when only one base pair changes. **(Figure 8.1)**
2. **Tautomeric shift** occurs when hydrogen atoms move to alter the bonding of the base pairs. **(Figure 8.2)**

Removal or Addition of Nucleotides **(Figure 8.3)**

1. **Frame shift mutations** involve the addition or deletion of nucleotides, rendering all genes downstream of a **stop codon** and in the same operon nonfunctional.

Transposable Elements (Jumping Genes) **(Figure 8.4)**

1. Certain genes, called **transposons**, have the ability to move to any other location in the genome.
2. Introduction of a transposon into another gene inactivates that gene. **(Figure 8.5)**

Induced Mutations **(Table 8.1)**

Chemical Mutagens

1. Chemical **mutagens** alter hydrogen-bonding properties of purines and pyrimidines, increasing the frequency of mutations.
2. **Base analogs** with different hydrogen-bonding properties can be incorporated into DNA in place of the usual purines and pyrimidines. **(Figure 8.6)**

3. **Intercalating agents** are planar molecules that insert into the double helix and push nucleotides apart, resulting in a frameshift mutation. (Figure 8.7)

Transposition

1. An **insertion mutation** results when a transposon integrates into a recipient cell's genome.

Radiation (Figure 8.8)

1. Ultraviolet irradiation results in **thymine dimers** due to the formation of covalent bonds between adjacent thymine molecules on the same strand of DNA. (Figure 8.9)
2. X rays causes single-strand breaks, double-strand breaks, and alterations to the DNA bases.

Repair of Damaged DNA (Table 8.2)

Repair of Errors in Base Incorporation

1. DNA polymerase has a **proofreading** function.
2. In **mismatch repair**, an endonuclease cuts out the damaged fragment and a new DNA strand is synthesized. (Figure 8.10)

Repair of Thymine Dimers

1. In **light repair**, a photoreactivating enzyme breaks the bonds of the thymine dimer, thereby restoring the original molecule. (Figure 8.11)
2. In **excision** or **dark repair**, the damaged segment is excised by an endonuclease. A new strand is synthesized by DNA polymerase. (Figure 8.12)

SOS Repair

1. **SOS repair** is a last ditch repair mechanism in which about 20 enzymes are induced by damaged DNA, resulting in DNA polymerase bypassing the damaged part of the strand, but without proofreading the DNA product. Consequently, the synthesized DNA contains many mutations.
2. SOS repair accounts for the mutagenic activity of UV irradiation.

Mutations and Their Consequences

1. Genes mutate independently of one another, and the chance that two mutations will occur within the same cell is the product of the individual mutation rates.
2. Mutations provide a mechanism for altering the population of an organism to adapt to a changing environment—a process called **natural selection**.

Expression of Mutations

1. Mutations are expressed rapidly in bacteria because bacteria are **haploid**.

Mutant Selection

Direct Selection

1. **Direct selection** involves inoculating cells onto a medium on which the mutant but not the parent can grow; these are the easiest kinds of mutants to isolate. (Figure 8.13)

Indirect Selection

1. **Indirect selection** is required when the desired mutant does not grow on a medium on which the parent grows.
2. **Replica plating** involves the simultaneous transfer of all the colonies on one plate to another and the comparison of the growth of individual colonies on both plates. (Figure 8.14)
3. **Penicillin enrichment** increases the proportion of mutants in a population by killing growing bacteria in a medium on which only nonmutants will grow. (Figure 8.15)

Conditional Lethal Mutants

1. **Conditional lethal mutations** cannot be overcome by adding growth factors to the medium. (Figure 8.16)

Mutant Screens

1. Mutant **screens** identify mutations that result in observable changes to media on which colonies grow. (Figure 8.17)

Testing of Chemicals for Their Cancer-Causing Ability

1. The Ames test measures whether a suspected **carcinogen** increases the frequency of reversion; a positive test indicates the subject chemical is a mutagen and therefore a likely carcinogen. (Figure 8.18)

MECHANISMS OF GENE TRANSFER

DNA-Mediated Transformation (Table 8.3)

1. **DNA-mediated transformation** involves the transfer of "naked" DNA. (Figure 8.20)

Natural Competence

1. Natural **competence** is the ability of a cell to take up DNA.
2. DNA enters the cell as a single-stranded molecule and is integrated by replacing recipient cell genes via **homologous recombination**.

Artificial Competence

1. Cells can be made artificially competent by **electroporation;** in this process, the cells are treated with an electric current, which makes holes in the cell envelope through which DNA can pass. (Figure 8.21)

Transduction

1. **Transduction** involves the transfer of bacterial DNA by a **bacteriophage**. (Figure 8.22)

Conjugation

1. **Conjugation** requires cell-to-cell contact. (Figures 8.23, 8.24)

Plasmid Transfer

1. The **donor cells** synthesize a **sex pilus** encoded on a **F plasmid**, which recipient cells do not have; the F plasmid is transferred from an $\mathbf{F^+}$ to an $\mathbf{F^-}$ cell.

Transfer of Plasmids Other Than F

1. The F plasmid can often mobilize and transfer other plasmids to the recipient cell.

Chromosome Transfer

1. Chromosome transfer occurs when the F plasmid integrates into the chromosome and the resulting cell can transfer a portion of the chromosome into a recipient cell. (Figure 8.26)

F′ Donors

1. In an F′ donor, the F plasmid excises a fragment from the chromosome and transfers it as the F plasmid is transferred. (Figure 8.27)

Plasmids (Table 8.4)

1. **Plasmids** are extrachromosomal replicons that code for nonessential information.

R plasmids

1. **R plasmids** code for antibiotic resistance; many are **self-transmissible** to other bacteria. (Figure 8.28)

Gene Movement Within the Same Bacterium—Transposable Elements

Transposons and Transfer of Genes to Unrelated Bacteria (Figure 8.29)

1. Transposable elements often transfer antimicrobial-resistance genes from a chromosome to a plasmid. Plasmids can be readily transferred by conjugation to unrelated bacteria.

Structure of Transposons

1. There are different types of transposons, which differ in complexity; all have inverted repeat sequences at their ends.
2. An **insertion sequence (IS) element** contains a gene that codes for the enzyme **transposase**. (Figure 8.30a)
3. A **composite transposon** consists of one or more genes flanked by insertion sequences. (Figure 8.30b)

Barriers to Gene Transfer

Restriction of DNA

1. DNA entering unrelated bacteria is recognized as foreign and is degraded by a class of deoxyribonucleases, called **restriction enzymes**, that recognize specific nucleotide sequences. (Figure 8.31)

Modification Enzymes

1. The DNA inside a cell is not cleaved, because it is methylated at the potential cleavage sites by a **modification enzyme** that is paired with the restriction enzyme to form a **restriction-modification system**. (Figure 8.32)

Importance of Gene Transfer to Bacteria

1. Gene transfer allows bacteria to survive changing environments by providing cells with a set of new genes.
2. In contrast, mutations only result in a single gene being modified.

REVIEW QUESTIONS

Short Answer

1. What is the one activity that all plasmids must be able to carry out?
2. What is the term that describes a plasmid that can transfer itself from *E. coli* into *Pseudomonas*, where it can replicate?
3. What is the enzyme that is responsible for protecting cells against invasion of foreign DNA?
4. What are the two necessary features of an insertion sequence?
5. What enzyme has proofreading ability? How does it function in proofreading?
6. Give an example of gene transfer that involves homologous recombination. Give one that involves nonhomologous recombination.
7. Single-strand integration of DNA has been shown to be a feature of which mechanism of DNA transfer?
8. Methylation of DNA plays a role in two biological processes discussed in this chapter. What are they?
9. What environmental condition is useful in isolating lethal conditional mutants? Explain.
10. Name three ways in which plasmids differ from bacterial chromosomes.

Multiple Choice

1. A culture of *E. coli* is irradiated with ultraviolet (UV) light. Answer questions 1 and 2 based on this statement. The effect of the UV light is to specifically…
   A. join the two strands of DNA together through covalent bonds.
   B. join the two strands of DNA together through hydrogen bonds.
   C. form covalent bonds between thymine molecules on the same strand of DNA.
   D. form covalent bonds between guanine and cytosine.
   E. prevent transcription
2. The highest frequency of mutations would be obtained if, after irradiation, the cells were immediately…
   A. placed in the dark.
   B. exposed to visible light.
   C. shaken vigorously.
   D. incubated at a temperature below their optimum for growth.
   E. The frequency would be the same no matter what the environmental conditions are after irradiation.
3. Penicillin enrichment of mutants works on the principle that…
   A. only Gram-positive cells are killed.
   B. cells are most sensitive to antimicrobial medications during the lag phase of growth.

C. most Gram-negative cells are resistant to penicillin.

D. penicillin only kills growing cells.

E. penicillin inhibits formation of the lipopolysaccharide layer.

4. The two repair mechanisms that take place in the course of DNA synthesis are…
   1. mismatch repair.
   2. proofreading by DNA polymerase.
   3. light repair.
   4. SOS repair.
   5. excision repair.

   A. 1,2 B. 2,3 C. 3,4 D. 4,5 E. 1,5

5. You are trying to isolate a mutant of wild-type *E. coli* that requires histidine for growth. This can best be done using…
   1. direct selection.
   2. replica plating.
   3. penicillin enrichment.
   4. a procedure for isolating conditional mutants.
   5. reversion.

   A. 1,2 B. 2,3 C. 3,4 D. 4,5 E. 1,5

6. The properties that all plasmids share are that they…
   1. all carry genes for antimicrobial resistance.
   2. are self-transposable to other bacteria.
   3. always occur in multiple copies in the cells.
   4. code for nonessential functions.
   5. replicate in the cells in which they are found.

   A. 1,2 B. 2,3 C. 3,4 D. 4,5 E. 1,5

7. The addition of deoxyribonuclease to a mixture of donor and recipient cells will prevent gene transfer via…

   A. DNA transformation.

   B. chromosome transfer by conjugation.

   C. plasmid transfer by conjugation.

   D. generalized transduction.

8. A sex pilus is essential for …
   1. DNA transformation.
   2. chromosome transfer by conjugation.
   3. plasmid transfer by conjugation.
   4. generalized transduction.
   5. cell movement.

   A. 1,2 B. 2,3 C. 3,4 D. 4,5 E. 1,5

9. A plasmid that can replicate in *E. coli* and *Pseudomonas* is most likely a…

   A. broad host range plasmid.

   B. self-transmissible plasmid.

   C. high-copy-number plasmid.

   D. essential plasmid.

   E. low-copy-number plasmid.

10. The frequency of transfer of an F′ DNA molecule by conjugation is closest to the frequency of transfer of…

    A. chromosomal genes by conjugation.

    B. an F plasmid by conjugation.

    C. an F plasmid by transformation.

    D. an F plasmid by transduction.

    E. an R plasmid by DNA transformation.

Applications

1. Some bacteria are more resistant to UV light than other bacteria. Discuss two reasons why this might be the case. What experiments could you do to determine whether each of the two possibilities could be correct?
2. A pharmaceutical researcher is disturbed to discover that the major ingredient of a new drug formulation causes frameshift mutations in bacteria. What other information would the researcher want to obtain before looking for a substitute chemical?

Critical Thinking

1. You have the choice of different kinds of mutants for use in the Ames test to determine the frequency of reversion by suspected carcinogens. You can choose a deletion, point mutation, or a frameshift mutation. Would it make any difference which one you chose? Explain.
2. You have isolated a strain of *E. coli* that is resistant to penicillin, streptomycin, chloramphenicol, and tetracycline. You also observe that when you mix this strain with cells of *E. coli* that are sensitive to the four antibiotics, they become resistant to streptomycin, penicillin, and chloramphenicol but remain sensitive to tetracycline. Explain what is going on.

CHAPTER 9

Biotechnology and Recombinant DNA

In September of 1976, Argentinean newspapers reported a violent shootout that had occurred between soldiers and the occupants of a house in suburban Buenos Aires, leaving the five "extremists" inside dead. Conspicuously absent from those reports was the identity of the "extremists"—a young couple and their three children, ages six years, five years, and six months. Over the next seven years, similar scenarios recurred as the military junta that ruled Argentina eliminated thousands of its citizens who were perceived as threats. This "Dirty War," as it came to be known, finally ended in 1983 with the collapse of the military junta and the election of a democratic government. The new leaders opened previously sealed records, which confirmed what many had already suspected—that more than 200 children had survived the carnage, and had in fact been kidnapped and placed with families in favor with the junta.

Dr. Mary-Claire King was at the University of California at Berkeley when her help was enlisted in the effort to return the kidnapped children to the surviving members of their biological families. Dr. King and others recognized that the rapidly developing methods of DNA technology could be put to use for this important humanitarian cause. DNA is composed of a string of nucleotides, the sequence of which can serve as a molecular "fingerprint" of an individual. By analyzing certain DNA sequences, blood and tissue samples from one individual can be distinguished from those of another. These same principles can also be used to show that a particular child is the progeny of a given set of parents. Because a person has two copies of each chromosome—one inherited from each parent—one half of a child's DNA will represent maternal sequences and the other half will represent paternal traits. The case of the Argentinean children presented an even greater challenge, however, as most of the parents were dead or missing. Often, the only surviving relatives were aunts and grandmothers. It is sometimes possible to use chromosomal DNA to show genetic relatedness between a child and such relatives, but it is more difficult and the results may not be conclusive. Dr. King responded to that challenge by investigating mitochondrial DNA (mtDNA). This organelle DNA, unlike chromosomal DNA, appears to be inherited only from the mother. A child will have the same nucleotide sequence of mitochondrial DNA as his or her siblings, the mother and her siblings, as well as the maternal grandmother.

To determine if mtDNA sequences could be used as an identifying marker, Dr. King studied the nucleotide sequence of an approximately 1,200-base-pair region adjacent to the origin of replication of mtDNA. Portions of this region had previously been shown to be among the most variable of the human genome. By comparing the nucleotide sequences of this region in different individuals, Dr. King was able to locate key positions that varied extensively among unrelated people, but were the same or markedly similar in maternal relatives. Dr. King's technique, borne out of a desire to help reunite families victimized by war, has now found many uses. Today her lab, which is now at the University of Washington, remains very active using molecular biology techniques for humanitarian efforts, identifying the remains of victims of atrocities around the world.

—*A Glimpse of History*

A REVOLUTION HAS OCCURRED IN MOLECULAR biology over the past several decades—the science has been transformed from a descriptive study of what cells are, to an intricate study of how cells function. A driving force in that revolution was the development of simple methods to extract and manipulate DNA, the blueprint of life.

Biotechnology is the use of microbiological and biochemical techniques to solve practical problems and produce more useful products. In the past, this usually meant laboriously searching for naturally occurring mutants that produced maximal product or expressed other desirable characteristics. Today, the rapid developments in **recombinant DNA techniques,** the methods we use to study and manipulate DNA, have made it possible to genetically alter organisms to give them more useful traits. Biotechnology is now nearly synonymous with **genetic engineering**, the process of deliberately altering an organism's genetic information. Researchers can isolate genes from one organism, manipulate the purified DNA *in vitro*, and then transfer the genes into another organism, a process called **gene cloning**. The more that scientists learn about genes and their functions, the more the field of biotechnology advances. In this chapter, we will explore some of the applications of biotechnology and describe several of the most powerful techniques.

Applications of Recombinant DNA Technology

Gene cloning and other recombinant DNA techniques provide powerful tools for genetically manipulating organisms for medical, industrial, or research uses. Organisms can be engineered to produce proteins more efficiently, acquire commercially valuable traits, or to produce increased amounts of specific sequences of DNA for a variety of purposes, including research **(figure 9.1)**.

Protein Production

Microorganisms can be altered to produce commercially important proteins more efficiently. A gene coding for a valuable protein can be inserted into a **high-copy-number plasmid** and then introduced into the same or a different organism; high-copy-number plasmids are those that are present in numerous copies per cell. The organism will then make increased amounts of the protein, because each gene copy can be transcribed and translated **(figure 9.2)**. Thus, bacteria can be engineered to produce medically and commercially valuable proteins that are normally produced only by animal cells.

Pharmaceutical Proteins

A variety of proteins can be used to treat people suffering from certain diseases **(table 9.1)**. In the past, these proteins were extracted from live animal or cadaver tissues, which made them expensive and limited the supply.

Human insulin, used in treating diabetics, was one of the first important pharmaceutical proteins to be produced through genetic engineering. The original commercial product, extracted from pancreatic glands of cattle and pigs, caused allergic reactions in approximately 1 in 20 patients. Once the gene for human insulin was cloned, microorganisms became the major source of insulin sold in the United States. A 2,000-liter culture of *E. coli* yields 100 grams of purified insulin, an amount that would require 1,600 pounds of pancreatic glands. The microbial production of insulin is safer and more economical than extracting it from animal tissues.

Vaccines

Genetically engineered microorganisms are being used in the production of a number of different vaccines. Vaccines protect against disease by harmlessly exposing a person's immune system to a killed or weakened form of the disease-causing agent, or to a part of the agent. Although a vaccine is generally composed of whole bacterial cells or viral particles, only specific proteins, or parts of the proteins, are actually necessary to induce protection, or immunize, against the disease. The genes coding for these proteins can be cloned in yeast or bacteria so that a large amount of the pure immunizing proteins can be produced. This type of vaccine is currently used to prevent human hepatitis B and foot-and-mouth disease of domestic animals. ■ vaccines, p. 425

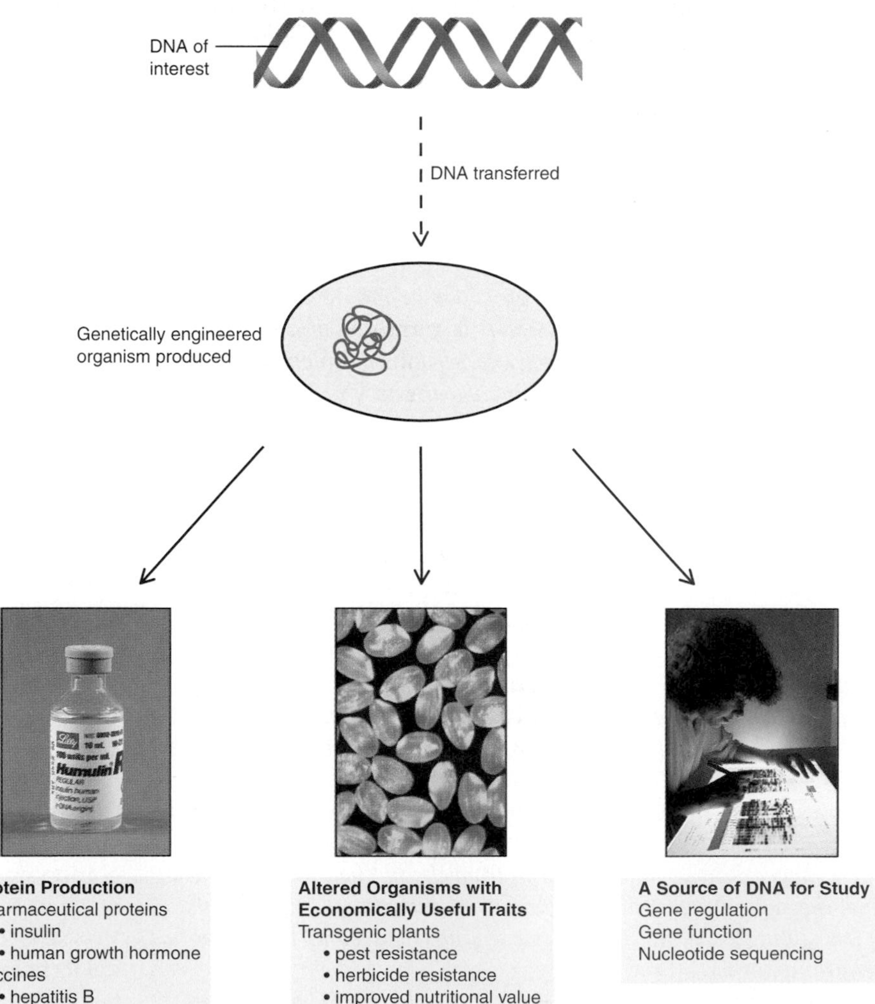

Figure 9.1 Applications of Recombinant DNA Technology

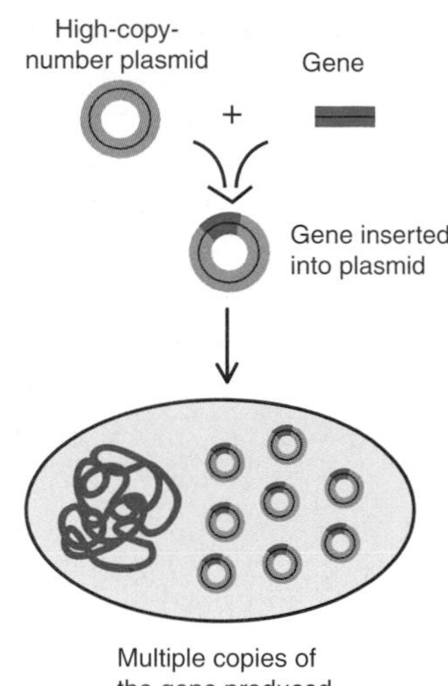

Figure 9.2 Cloning into a High-Copy-Number Plasmid When a gene is inserted into a high-copy-number plasmid, multiple copies of that gene will be present in a single cell, resulting in the synthesis of many more molecules of the encoded protein.

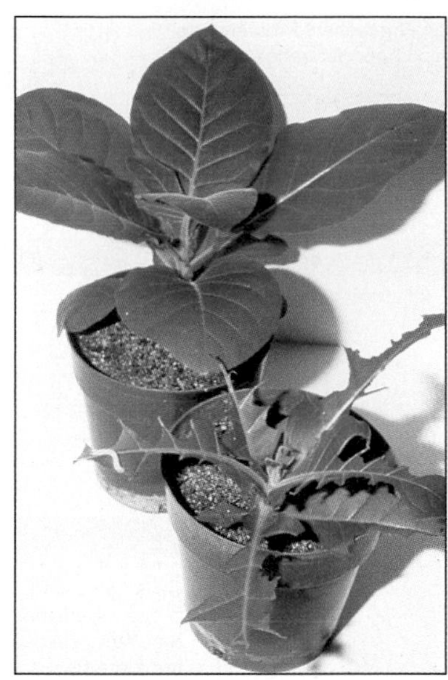

Figure 9.3 Pest-Resistant Transgenic Plant The plant in back has been genetically engineered to produce Bt-toxin. The control plant in front has been ravaged by pests.

Altering Organisms to Give Them Economically Useful Traits

Genetic engineering can be used to alter organisms, giving them new traits. The most commercially promising applications so far involve plants, which have been successfully engineered to synthesize their own insecticide, to resist a common herbicide, and to improve their nutritional value. Plants and animals to which new DNA has been introduced are called **transgenic**.

Pest Resistant Plants

Plants have been engineered to resist natural pests **(figure 9.3)**. Some plants, including corn, cotton, and potatoes, have been engineered to produce a biological insecticide. The insecticide, a protein crystal called **Bt-toxin**, is naturally produced by the bacterium *Bacillus thuringiensis* as it forms endospores. Bt-toxin is actually a **protoxin**, an inactive protein that must be processed in the insect gut to form the active toxin. There, the toxin binds to specific receptors, causing the gut to dissolve. Unlike many chemically synthesized insecticides, Bt-toxin is toxic only to insects, including their larvae. Crystal/spore preparations have been applied to a wide variety of plants around the world for over 30 years without any apparent harm to the environment. ■ endospore, p. 72

Herbicide Resistant Plants

Plants such as soybeans, cotton, and corn are being engineered to resist the effects of the herbicide glyphosate (Roundup™). This enables growers to apply this herbicide, which kills all other plants but is readily biodegradable, in place of more persistent alternatives. Also, because the herbicide can be applied throughout the growing season to suppress weed growth, the soil can be tilled less frequently, preventing erosion. In 1999, over 50% of the soybeans planted were seed varieties that had been genetically modified to resist this herbicide.

Plants with Improved Nutritional Value

Researchers around the world are attempting to use genetic engineering to improve the nutritional value of plants that are used as food. One of the most recent examples is the introduction of the genes that code for the synthesis of β-carotene, a precursor of vitamin A, into a variety of rice (see figure 9.1, middle panel). The rice was also engineered to provide more dietary iron. Considering that the diet of a significant proportion of the world's population is deficient in these essential nutrients, advances such as this could have a profound impact on world health.

Providing a Source of DNA for Study

A great deal of the information described in this textbook has been revealed using recombinant DNA technology. By cloning a segment of DNA into a bacterium, a ready source of that sequence is available for study or further manipulation. For example, regulation of gene expression can be studied by creating a fusion between the gene being studied and a **reporter gene (figure 9.4)**. The reporter gene encodes a readily observable phenotype such as fluorescence, making it possible to directly observe the expression

TABLE 9.1 Medically Important Proteins Produced by Genetically Engineered Microorganisms

Protein	Medical Use
Alpha interferon	Treating cancer
Beta interferon	Treating multiple sclerosis
Deoxyribonuclease	Treating cystic fibrosis
Factor VIII	Treating hemophilia
Gamma interferon	Treating cancer
Growth hormone	Treating dwarfism
Insulin	Treating diabetes
Interferon α-2b	Treating hepatitis B infection
Streptokinase	Dissolving blood clots
Tissue plasminogen activator	Dissolving blood clots

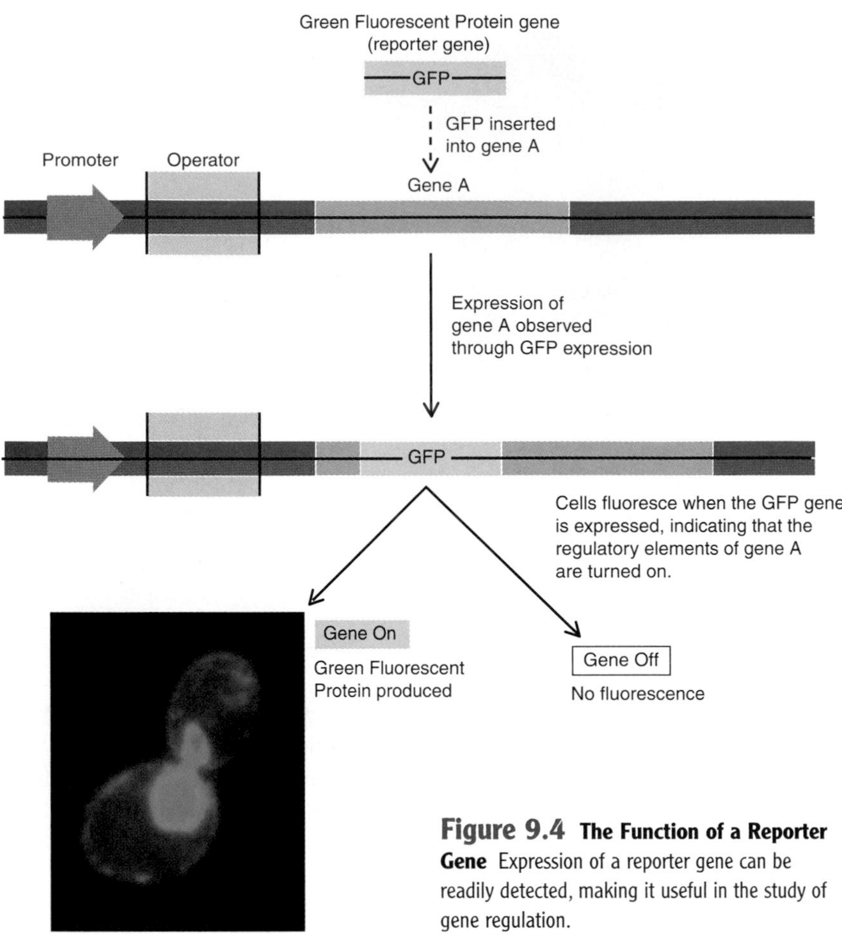

Figure 9.4 The Function of a Reporter Gene Expression of a reporter gene can be readily detected, making it useful in the study of gene regulation.

of the gene. This, in turn, makes it possible to determine the conditions that affect gene activity. One widely used reporter gene encodes a product called **green fluorescent protein** (**GFP**).

MICROCHECK 9.1

Organisms can be engineered to produce proteins more efficiently, to acquire commercially valuable traits, or to produce increased amounts of specific sequences of DNA for research purposes.

- Name two diseases that are treated using proteins produced by genetically engineered microorganisms.
- Describe the attributes of two different types of transgenic plants.
- Eukaryotic proteins are often processed in the Golgi apparatus after synthesis. What problems might this pose when engineering a bacterium to produce a mammalian protein?

Cloning DNA

Cloning a gene involves removing a fragment of DNA from one organism and introducing it into another one, most commonly *E. coli*. As the new host grows, the cloned fragment replicates and is passed to daughter cells, generating a population of cells each of which contains a DNA fragment identical to the original. If the original gene was in a human cell and is cloned in *E. coli*, then the quantity of that gene can be increased dramatically because it is easier to grow bacteria than human cells. The number of copies can be increased 10- to 100-fold or more, if it is inserted into a high-copy-number plasmid.

The cloning process is a complicated challenge for several reasons. As described in the previous chapter, DNA must have an origin of replication in order to replicate in a cell and be passed on to daughter cells (**figure 9.5**). Therefore, the fragment being cloned must be joined to an independently replicating DNA molecule such as a plasmid. To enable the cloned DNA to replicate in a cell, it is attached to a piece of DNA called a **vector**, to form a chimeric or **recombinant molecule** that is part vector and part cloned DNA. The cloned DNA is called an **insert**.

An even more formidable challenge is the fact that a bacterial genome of over 1 million nucleotides long can encode more than 1,000 different genes. The human genome is much larger, 3 billion nucleotides. Thus, a specific gene represents only a tiny fraction of the total genome, a needle in a haystack. An approach frequently used to clone a specific bacterial gene is to first clone into a population of *E. coli* cells a set of DNA fragments that together make up the entire genome of the organism being studied. While each cell in the resulting population contains only one fragment of the genome, the entire genome is represented in the population as a whole. Because each cell can be viewed as containing one "book" of the total genetic information, the population is called a **DNA library** (**figure 9.6**). For example, to clone the gene for bacterial luciferase, the enzyme that causes bioluminescence in *Vibrio fischeri*, a researcher first clones numerous DNA fragments that together contain all genes of that organism and then identifies the cells that contain the bioluminescence gene. Determining which cells in a DNA library contain the gene of interest is considerably easier when the gene encodes a readily observable phenotype such as bioluminescence. ■ **origin of replication, p. 171**

The general steps in any cloning experiment are summarized in **figure 9.7**. Because the DNA cloning procedures involving bacterial cells and their DNA are less complicated than those involving eukaryotes, they will be described first.

Isolating DNA

The first step of a cloning experiment is to obtain the DNA that will be cloned. To isolate bacterial DNA, cells in a broth culture are lysed by adding a detergent. As the cells burst open, the relatively fragile DNA is inevitably sheared into many pieces of varying lengths (**figure 9.8**). These long DNA fragments generate a characteristically viscous solution upon their release. The solution can be treated with specific enzymes such as protease

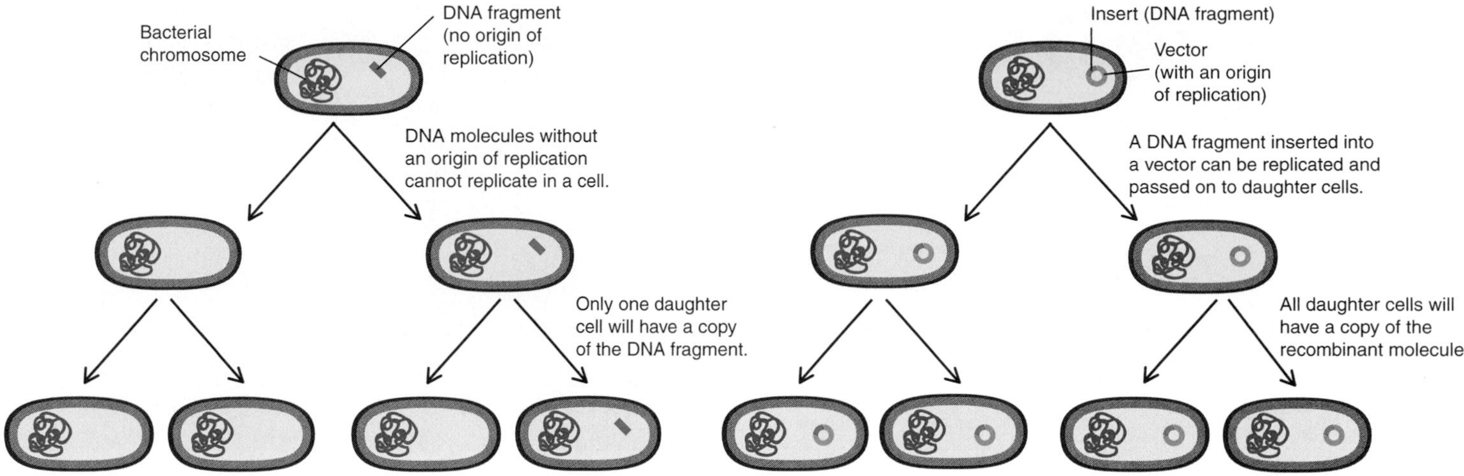

Figure 9.5 **DNA Must Replicate in a Cell in Order to be Maintained in a Population of Cells**

and RNase that degrade cell protein and RNA, respectively, to purify the DNA. Because the original broth culture contained a billion or more cells per ml, the resulting solution of DNA contains a like number of copies of the chromosome. This is important to recognize when trying to envision how a cloning experiment works. While descriptions and diagrams only show one or a few DNA molecules, in reality there are billions. The same holds true for the molecules of DNA depicted in most of the techniques covered in this chapter.

Using Restriction Enzymes to Generate Fragments of DNA

The purified DNA is cut into smaller fragments by treating it with a **restriction enzyme**. Restriction enzymes act as molecular scissors that recognize and cleave specific sequences of DNA. Molecular biologists have found that these naturally occurring enzymes are remarkable tools because they cut DNA in a predictable and controllable manner. Cutting, or **digesting**, DNA with these enzymes generates pieces called **restriction fragments**. Without restriction enzymes, many recombinant DNA techniques would be exceedingly difficult or impossible.

Hundreds of different restriction enzymes have been discovered. Each enzyme has been given a seemingly peculiar name, but the name simply represents the bacterium from which the enzyme was initially isolated. The first letter represents the first letter of the genus name, the next two letters are derived from the species name; any other numbers or letters designate the strain and order of discovery. For example, a restriction enzyme from *E. coli* strain RY13 is called *Eco*RI and an enzyme from *Staphylococcus aureus* strain 3A is called *Sau*3A.

Most restriction enzymes recognize a sequence that is a palindrome. In other words, the sequence of

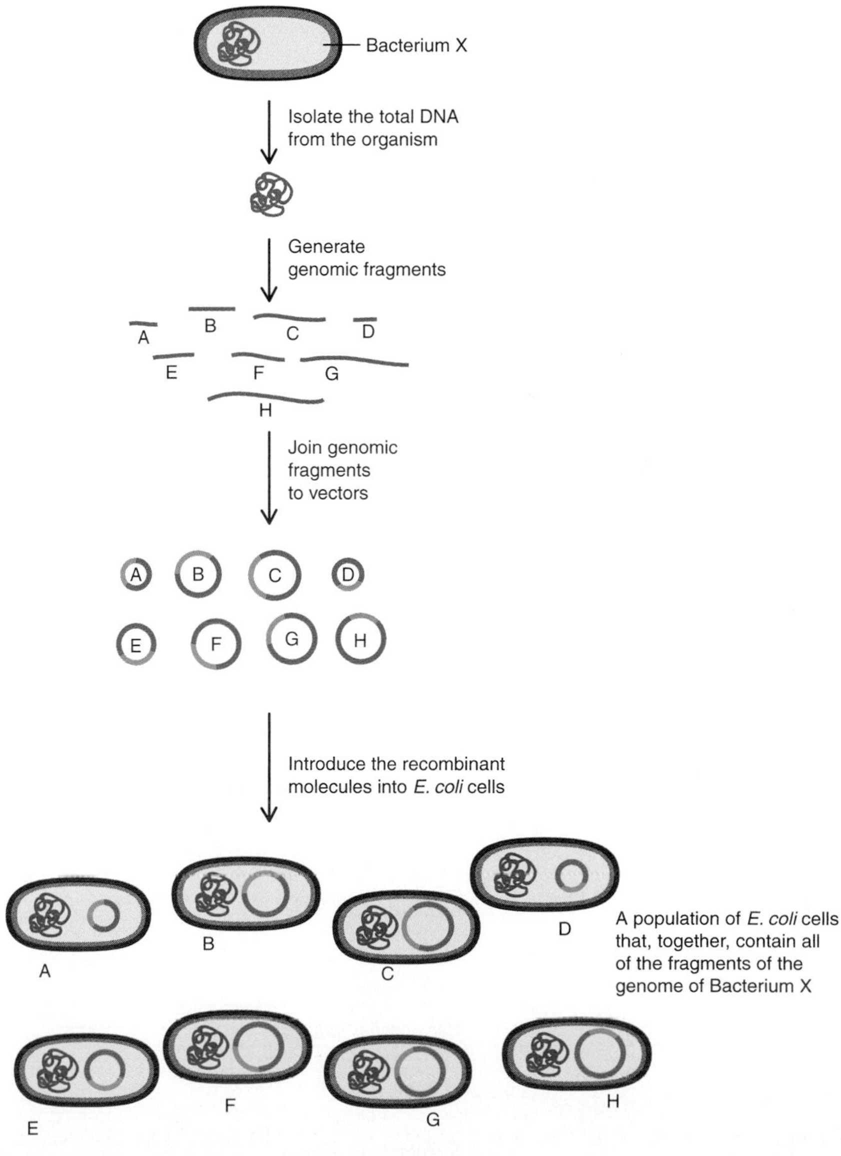

Figure 9.6 **A DNA Library** Each cell in a DNA library contains one fragment of a given organism's genome.

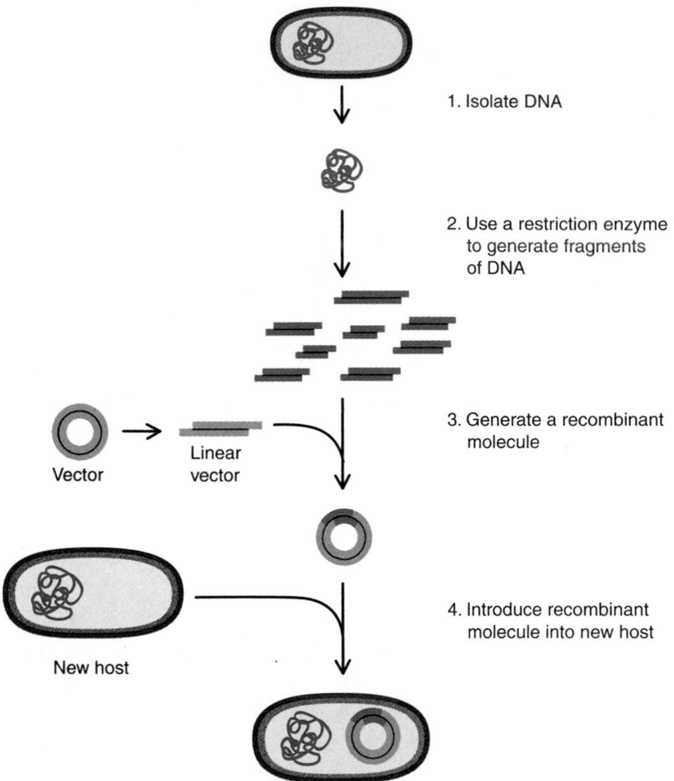

Figure 9.7 The Steps of a Cloning Experiment

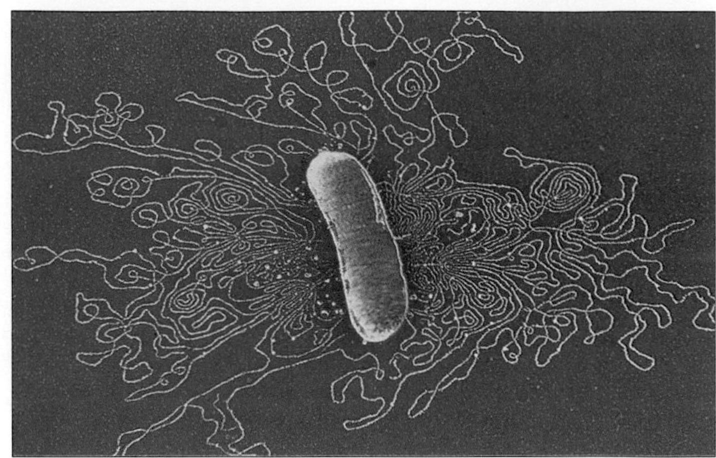

Figure 9.8 Cells Release Their DNA when They Burst The relatively fragile DNA is inevitably sheared into many pieces of varying lengths.

one strand is identical to the sequence of the complementary strand when each is read in the 5′ to 3′ direction. Recall that the two DNA strands are antiparallel; thus, they are read in opposite directions. The restriction enzyme recognizes and cuts the identical sequence on each strand of the double-stranded molecule. The sequence that a particular enzyme recognizes is called a **recognition sequence (table 9.2)**. ■ **antiparallel, p. 168**

Digestion of DNA with a restriction enzyme generates either a **sticky** or a **blunt end**, depending on where the enzyme cuts in the sequence. An enzyme such as *Alu* I that cuts in the middle of the recognition sequence generates blunt ends, whereas an enzyme such as *Bam* HI that produces a staggered cut generates ends with a short overhang of usually four bases. The overhangs are called sticky ends or **cohesive ends** because they will form base pairs, or **anneal**, with one another. Because any two complementary cohesive ends can anneal, regardless of the source of DNA, these ends can be used to facilitate the joining of DNA fragments from different sources (**figure 9.9**).

Table 9.2 Examples of Common Restriction Enzymes

Enzyme	Microbial Source	Recognition Sequence (arrows indicate cleavage sites)
Alu I	*Arthrobacter luteus*	5′ A G ↓ C T 3′ 3′ T C ↑ G A 5′
Bam HI	*Bacillus amyloliquefaciens* H	5′ G ↓ G A T C C 3′ 3′ C C T A G ↑ G 5′
Eco RI	*Escherichia coli* RY13	5′ G ↓ A A T T C 3′ 3′ C T T A A ↑ G 5′
Hind III	*Haemophilus influenzae*	5′ A ↓ A G C T T 3′ 3′ T T C G A ↑ A 5′
Sal I	*Streptomyces albus*	5′ G ↓ T C G A C 3′ 3′ C A G C T ↑ G 5′

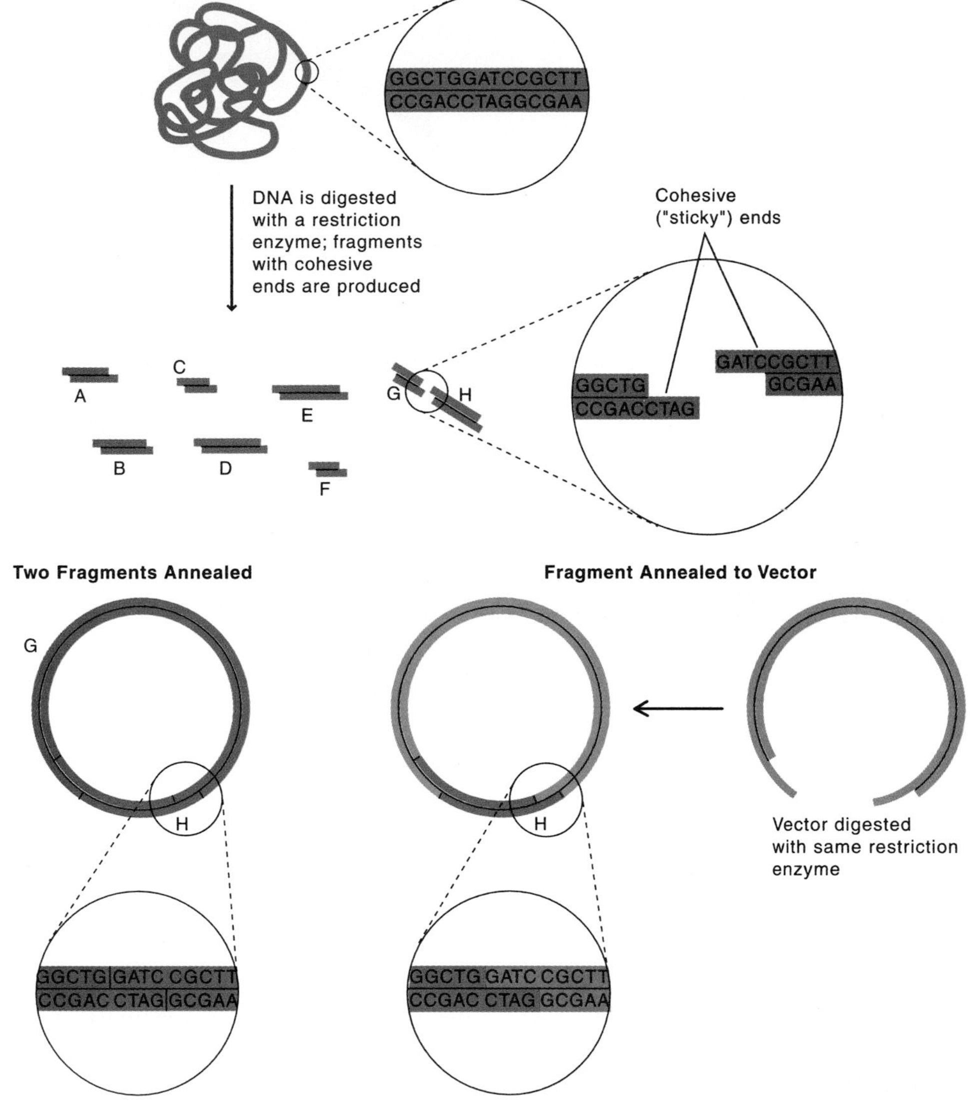

Figure 9.9 Cohesive ends Restriction fragments that were cut with the same restriction enzyme can anneal, regardless of their original source.

Generating a Recombinant DNA Molecule

The vector, usually a modified plasmid or bacteriophage, has an origin of replication and functions as a carrier of the cloned DNA. Because the recombinant molecule can replicate in a cell, the creation of the vector-insert hybrid allows the cloned DNA to be passed on to daughter cells. ■ plasmid, p.210, bacteriophage, p. 323

Properties of an Ideal Vector

In addition to having the ability to replicate inside a cell, a vector usually has other characteristics that make it useful. Some of these traits are natural features of many plasmids, but most vectors used today have been extensively modified to give them even more helpful properties.

Most vectors encode some type of **selectable marker**, generally resistance to an antibiotic such as ampicillin. The selectable marker makes it possible to select for cells that contain the vector sequence, either as part of a recombinant plasmid or as an intact circular molecule. This attribute is important because when DNA is added to a host, most cells do not take up that DNA. The selectable marker is used to eliminate cells that have not taken up vector sequences. If the selectable marker is antibiotic resistance, then only those cells that have taken up the vector or a recombinant molecule can grow on antibiotic-containing medium. ■ antibiotic, p. 496 ■ ampicillin, p. 502

Vectors also must have at least one restriction enzyme recognition site so that the circular vector can be cut to form a linear molecule to which the insert can be joined. Many vectors have been engineered to contain a short sequence called a **multiple-cloning site** that has the recognition sites of several different restriction enzymes. The value of a multiple-cloning site is its versatility; a fragment obtained by digesting with any of a number of different restriction enzymes can be inserted into the site.

Most vectors have a second genetic marker, in addition to the selectable marker, that is used to differentiate cells containing recombinant plasmids from those that contain intact vector. This is important because when the vector and insert DNA are both cut with the same restriction enzyme and the fragments are mixed together, not all molecules will join to form the desired recombinant plasmids. If a vector anneals to itself, the intact vector is regenerated and will replicate when introduced into a cell. This second marker is a gene that spans the cloning site (**figure 9.10**). Insertion of a DNA fragment into the vector disrupts the gene, resulting in a nonfunctional gene product and an altered phenotype. One common vector, pUC18, has a multiple cloning site in the middle of the ***lacZ′* gene**. The product of the *lacZ′* gene enables cells to cleave a colorless chemical, x-gal, to form a blue compound. The creation of a vector-insert hybrid disrupts the *lacZ′* gene, resulting in a nonfunctional gene product. Colonies of cells that harbor intact vector have a functional *lacZ′* gene and are blue, whereas those that contain a vector-insert chimera are white (**figure 9.11**). ■ phenotype, p. 192

Using DNA Ligase to Join the Vector and Insert

In creating the recombinant molecule, the enzyme **DNA ligase** is used. If restriction enzymes are viewed as scissors that cut DNA into fragments, then the enzyme DNA ligase is the glue that pastes the fragments together. The complementary cohesive ends created by restriction enzymes can anneal, but the relatively weak hydrogen bonds that hold the strands together are only temporary. DNA ligase can form covalent bonds between the sugar-phosphate residues of adjacent nucleotides, thus joining the two fragments. The combined actions of restriction enzymes and DNA ligase enable a researcher to take fragments of DNA from diverse sources and then join them to generate a recombinant DNA molecule. ■ **ligase, p. 171**

Introducing the Recombinant DNA into a New Host

Once a group of recombinant plasmids is generated, they must be transferred into a suitable host where the molecules can replicate.

Selecting a Suitable Host

For routine cloning experiments, one of the many well-characterized laboratory strains of *E. coli* is generally used as a host. *E. coli* is a desirable host because it is easy to grow and much is known about its genetics and biochemistry. An important feature of these strains is that they cannot grow in normal environmental conditions found outside of the laboratory. This minimizes the chances of a genetically engineered strain "escaping" from the laboratory environment. These strains also have known phenotypic characteristics such as sensitivity to specific antibiotics. ■ **phenotype, p. 192** ■ **antibiotics, p. 496**

If protein production is the goal of the experiment, a host must be selected that can express the cloned gene product. As a general rule, the more closely two organisms are related, the more likely the genes of one will be expressed in the cells of the other. For reasons that are not clear, *E. coli* seems to be especially able to express foreign genes from other bacteria. Like all Gram-negative organisms, however, *E. coli* has an outer membrane that contains endotoxin, which is toxic when injected into humans and other animals. Therefore, great pains must be taken to purify any cloned gene product used intravenously, such as insulin, lest any outer membrane product contaminate the preparation. ■ **endotoxin, p. 64**

Introducing DNA into Cells

A common method of introducing DNA into a bacterial host is DNA-mediated transformation. In this process, cells in a physiological state called **competence** take up naked DNA from their surrounding environment. While some bacteria are naturally competent, *E. coli* must be specially treated in a dilute calcium chloride solution to induce them to take up DNA. Even though this method is used routinely by a multitude of laboratories, the exact mechanism by which it enables cells to become competent is not clear. An alternative technique is to introduce the DNA by **electroporation**, a procedure that involves treating the cells with an electric current. This procedure works with *E. coli* as well as most other bacteria. ■ **DNA-mediated transformation, p. 205** ■ **electroporation, p. 207**

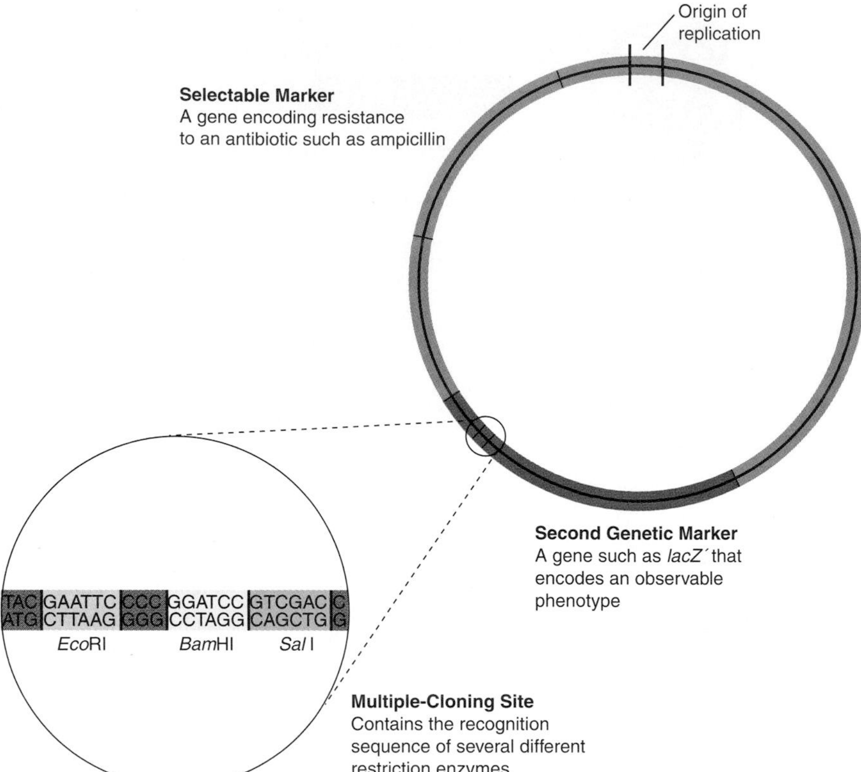

Figure 9.10 Typical Properties of an Ideal Vector Most vectors have an origin of replication, a selectable marker, and a multiple-cloning site. A second genetic marker, used to differentiate cells containing recombinant plasmids from those that contain intact vector, spans the multiple-cloning site.

Selecting for Transformants

After the DNA is introduced into the new host, the transformed bacteria are cultivated on a medium that both selects for cells containing vector sequences and differentiates those that carry recombinant plasmids. The medium exploits the selective marker encoded on the vector to permit the growth of only those cells that have taken up plasmids containing vector sequences. If, for example, the vector encodes resistance to the antibiotic ampicillin, then the transformed cells are grown on ampicillin-containing medium. Cells that have not been transformed with a vector sequence are killed by the ampicillin. To differentiate cells that carry a recombinant plasmid, the medium exploits the second marker. In the case of vectors that use the *lacZ′* gene as a second marker, the chemical x-gal is added to the medium. Colonies of cells containing intact vector turn blue on the medium, whereas those containing recombinant plasmid are white. The white colonies are then further characterized to determine if they carry the gene of interest. Most often, this is done using a **colony blot**, a technique that will be discussed later in the chapter. ■ **ampicillin, p. 502**

MICROCHECK 9.2

Restriction enzymes recognize and cleave specific sequences of DNA, cutting DNA into fragments. To clone a fragment, it must be inserted into a vector,

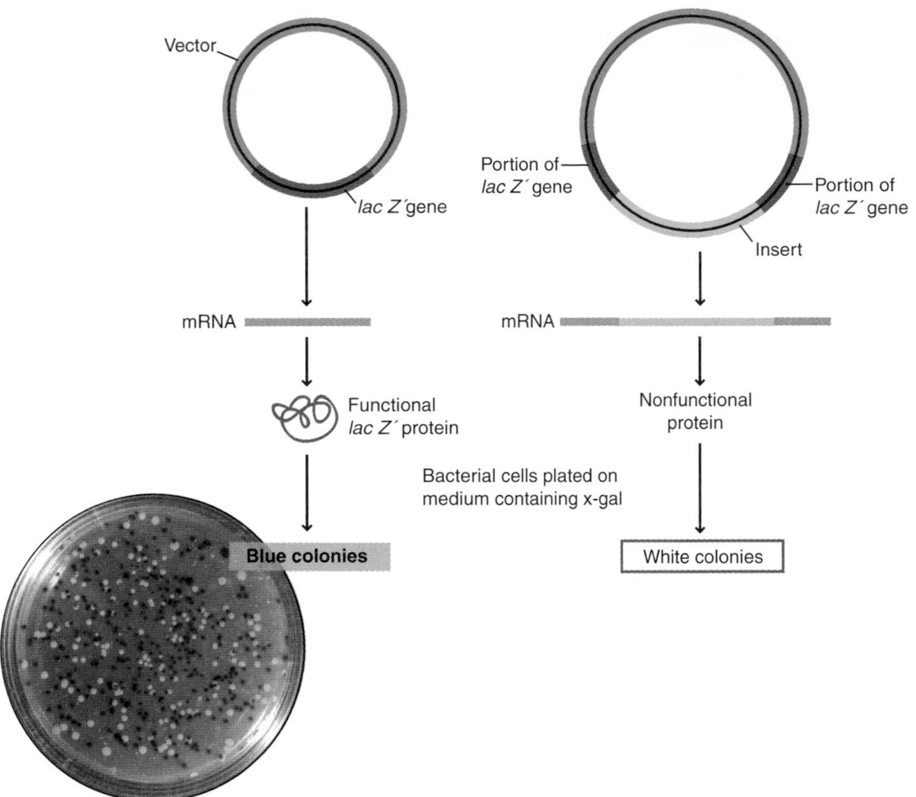

Figure 9.11 The Function of the *lacZ′* Gene in a Vector The *lacZ′* gene is used to differentiate cells that contain recombinant plasmid from those that contain vector alone. A functional *lacZ′* gene results in blue colonies when bacteria harboring intact vector are plated on a medium containing x-gal. Because disruption of the gene by an insert generates a nonfunctional product, cells harboring recombinant plasmids form white colonies.

creating a recombinant molecule that can replicate in a specific host cell.

- Name three properties of an ideal vector.
- Explain how the *lacZ′* gene is used to distinguish colonies that harbor a recombinant plasmid.
- What characteristics other than antibiotic resistance could be used as a selectable marker?

Overcoming the Obstacles to Cloning DNA from Eukaryotic Cells

The complexity of eukaryotic cells and their genomes presents special technical obstacles that must be overcome in the cloning process, which is why it is considerably more difficult to clone DNA from eukaryotes.

Removing Introns from Eukaryotic DNA

Most eukaryotic genes have intervening sequences, or **introns**, that are removed from precursor messenger RNA by the eukaryotic cell machinery before the mRNA is translated into protein. This presents a problem when genes of eukaryotic cells are cloned in bacteria, because bacteria have no means to remove introns. To overcome this obstacle, the eukaryotic genes are not inserted directly into the vector. Instead, mRNA from which the eukaryotic cell has already removed the introns is first isolated from the appropriate eukaryotic tissue. Then, a strand of DNA complementary to the mRNA is synthesized *in vitro* using **reverse transcriptase**, an enzyme encoded by retroviruses that can copy the sequence of mRNA into a complementary sequence of DNA. That single strand of DNA is then used as a template for synthesis of its complement, creating double-stranded DNA. The resulting copy of DNA, or **cDNA** encodes the same protein as the original DNA, but it lacks the introns **(figure 9.12)**. ■ **introns, p. 178** ■ **retrovirus, p. 355**

Vectors Used to Clone Eukaryotic Genes into Prokaryotic Cells

The type of vector used to clone eukaryotic DNA into a bacterial cell depends largely on the ultimate purpose of the procedure. If the goal of cloning is to produce the protein that is encoded in the DNA, then a vector designed to optimize transcription and translation of the insert DNA is used. To create a DNA library of a human or other eukaryotic genome, however, a vector that can carry a large insert is generally used.

Expression Vectors

With the human insulin gene or any other gene cloned for the purpose of protein production, cloning is only successful if the gene is expressed in the new host. This can be a difficult problem, depending on the source of the cloned gene and the relatedness of its new host. For transcription to occur, the promoter of the cloned gene must be recognized by the host's RNA polymerase. Meanwhile, sequences recognized by the host's ribosomes are required for translation to occur. If the cloned gene comes from an organism unrelated to the new host, such as a human gene cloned in *E. coli*, then a promoter and ribosome-binding site recognized by the expression machinery of the new host must be provided. Special vectors called **expression vectors** have been developed to facilitate transcription and translation of insert DNA. These vectors have been engineered to contain an appropriate promoter and ribosome-binding site adjacent to the multiple-cloning site. Thus, the insert DNA can be transcribed and translated when the recombinant plasmid is introduced into that host, regardless of the original source. The vector may also contain regulatory regions such as an operator, which enables a researcher to control the level of gene expression. ■ **promoter, p. 172** ■ **transcription, p. 172** ■ **translation, p. 175** ■ **ribosome-binding site, p. 176** ■ **operator, p. 180**

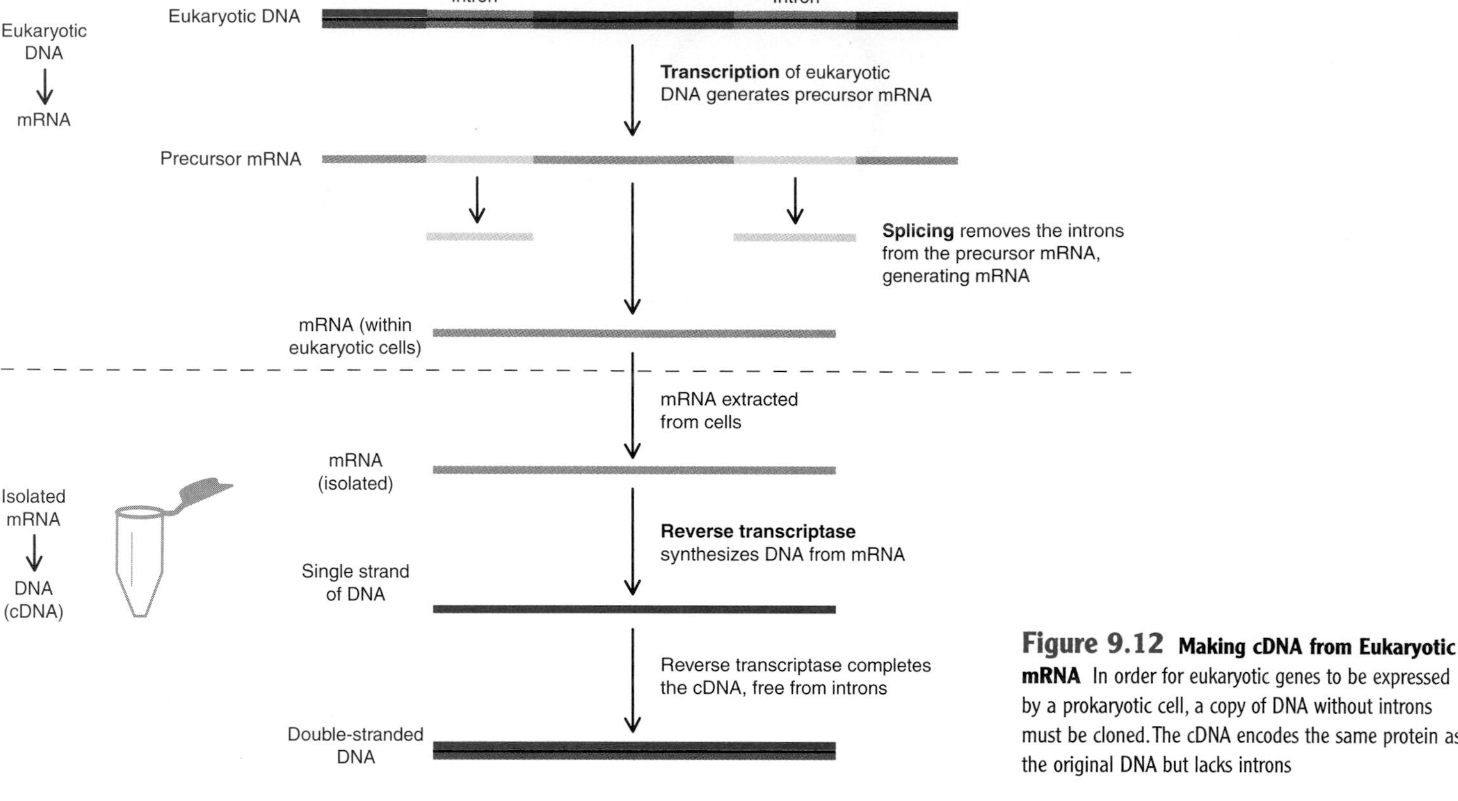

Figure 9.12 Making cDNA from Eukaryotic mRNA In order for eukaryotic genes to be expressed by a prokaryotic cell, a copy of DNA without introns must be cloned. The cDNA encodes the same protein as the original DNA but lacks introns

Bacterial Artificial Chromosomes

To create a DNA library of a eukaryotic organism using *E. coli* as a host, a vector that can accept extremely large inserts must be used. To serve this purpose, special vectors called **bacterial artificial chromosomes** (**BACs**) have been developed. These vectors are derivatives of the F plasmid and can contain inserts as large as 300,000 nucleotides in length. The cloned DNA can then be used for a variety of purposes, including the sequencing reactions that will be discussed later in the chapter. ■ F plasmid, p. 208

MICROCHECK 9.3

Introns must first be removed if eukaryotic DNA is to be expressed in a prokaryotic cell. The type of vector used to clone eukaryotic DNA depends largely on the ultimate purpose of the procedure.

- Explain the role of reverse transcriptase in the procedure to clone eukaryotic DNA.
- What are two critical components of an expression vector?
- What would be the result if introns were not removed before cloning eukaryotic DNA?

Transferring Genes into Eukaryotic Cells

Transferring genes into eukaryotic cells is considerably more difficult than transferring them into prokaryotic cells.

Vectors Used to Transfer DNA into Eukaryotic Cells

To transfer a fragment of DNA into a eukaryotic cell, a vector capable of delivering the DNA and enabling its replication in that cell must be used. Like the vectors used for cloning DNA in bacterial cells, these vectors generally have a selectable marker and a unique restriction enzyme site. Frequently, they also have a characteristic that allows the researcher to determine whether or not the vector has acquired an insert. Vectors are generally specific for the type of eukaryotic cell used as a host. For example, the vectors used to genetically engineer plants are different from those used to introduce DNA into yeast cells.

The Ti Plasmid of *Agrobacterium*

The workhorse for genetically engineering plants is the plant pathogen *Agrobacterium tumefaciens*. This bacterium, which causes the plant disease called crown gall, naturally has a large plasmid, the **tumor-inducing plasmid** or **Ti plasmid**. This plasmid is unique because when *A. tumefaciens* comes into contact with any of a wide variety of plant cells, a portion of the Ti plasmid is transferred to that cell and is integrated into its chromosome. The transferred DNA, **T-DNA**, encodes proteins that cause the crown gall tumors in the plant. Scientists recognized that they could exploit the natural processes of the Ti plasmid by removing the tumor-producing genes and inserting other genes in their place. These new genes can then be transferred and integrated into the plant DNA. The manipulation of the Ti plasmid has made it possible to genetically engineer plants (**figure 9.13**). ■ *Agrobacterium*, p. 285

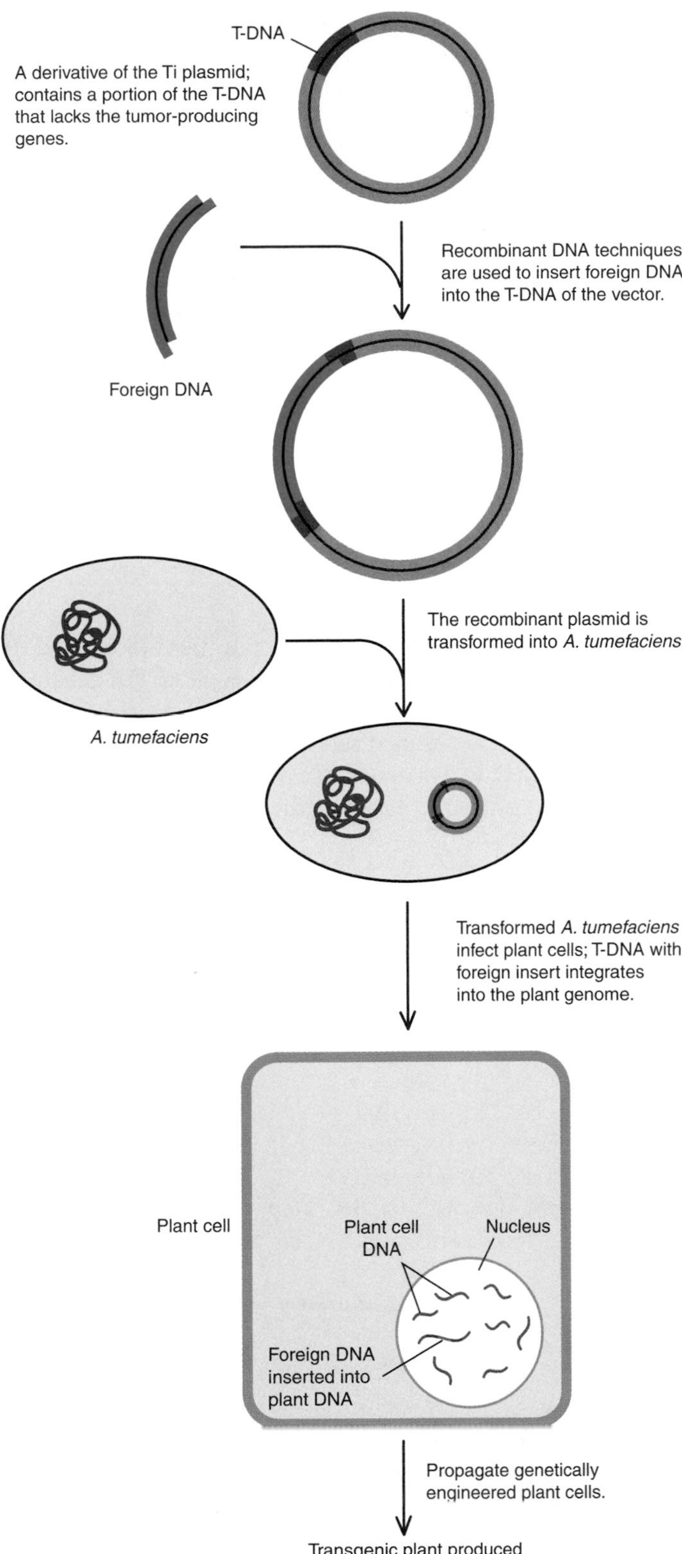
T-DNA
A derivative of the Ti plasmid; contains a portion of the T-DNA that lacks the tumor-producing genes.
Recombinant DNA techniques are used to insert foreign DNA into the T-DNA of the vector.
Foreign DNA
The recombinant plasmid is transformed into A. tumefaciens.
A. tumefaciens
Transformed A. tumefaciens infect plant cells; T-DNA with foreign insert integrates into the plant genome.
Plant cell
Plant cell DNA
Nucleus
Foreign DNA inserted into plant DNA
Propagate genetically engineered plant cells.
Transgenic plant produced

Figure 9.13 A Derivative of the Ti Plasmid Is Used to Introduce Foreign DNA into Plants Cells

Artificial Chromosomes

Artificial chromosomes are segments of DNA that have the key elements of a chromosome required for replication in certain cells and are stripped of other genetic information. This creates sites into which large pieces of insert DNA can be cloned. **Yeast artificial chromosomes** can be used to clone segments of DNA as large as 1,000,000 nucleotides in length. These vectors are used to introduce foreign DNA into the yeast *Saccharomyces cerevisiae*. Most recently, **mammalian artificial chromosomes** have been developed that can be used to clone DNA into animal cells.

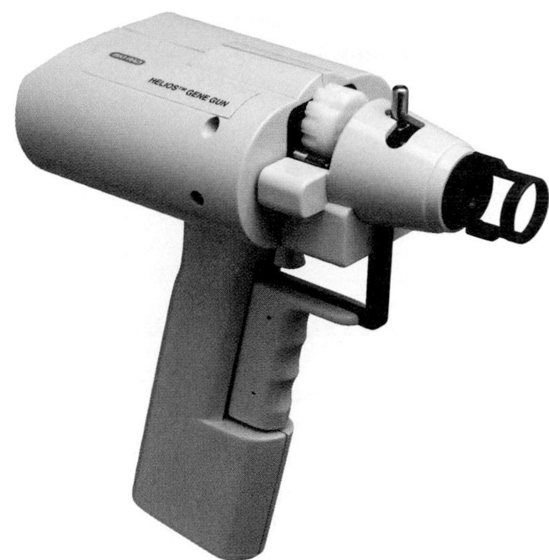
HELIOS™ GENE GUN

Figure 9.14 A Gene Gun This handheld device uses a gas pulse to propel DNA-coated gold particles directly into a cell.

Viruses

Certain viruses have been engineered for use as vectors to carry DNA into eukaryotic cells. Some of these can replicate independently of the chromosome for at least short periods of time, but they are not maintained for extended periods. Those that integrate into the DNA of the host cell remain as a permanent part of the cell. Because these vectors insert randomly into the chromosome, however, they have the undesirable characteristic of potentially disrupting the function or regulation of existing genes. ■ viruses, p. 13

Introducing DNA into Eukaryotic Cells

A vector is only useful if the DNA it carries can be delivered into a cell. An enormous advantage of using the Ti plasmid to engineer plant cells is that the T-DNA, as well as any genes inserted into it, is naturally transferred from *Agrobacterium* into the new host cell. Likewise, the natural ability of a virus to infect a cell can be exploited to deliver virus-derived vectors into a cell. Another method of introducing DNA into eukaryotic cells is electroporation, a method also used to introduce DNA into prokaryotic cells. The newest and perhaps most versatile method uses a **gene gun**, a handheld device that uses a gas pulse or other mechanism to propel DNA-coated gold particles directly into a cell (**figure 9.14**).

MICROCHECK 9.4

Vectors used to clone DNA into eukaryotic cells are often specific for that cell type. The Ti plasmid is used as a vector to introduce DNA into plant cells. Artificial chromosomes and certain viruses are used to carry DNA into mammalian cells.

- Explain the significance of the Ti plasmid.
- Describe one drawback of using a virus as a vector.
- How could insertion of a vector randomly into the chromosome disrupt the regulation of existing genes?

Nucleic Acid Hybridization

The high specificity of base-pairing interactions between two strands of DNA can be used to locate a specific nucleotide sequence in a sample. Double-stranded DNA, when exposed to high temperature or a high pH solution, will **denature**, or separate into two single strands. When the temperature is lowered and the pH is neutral, the two strands will anneal because of the base-pairing interactions of the complementary strands. Two complementary strands from different sources will also anneal, but the process may then be called **hybridization** to reflect the fact that each strand originated from a different source to create a hybrid molecule. Regions of double-stranded DNA in which the respective single strands hybridize to one another are **homologous**, or have similar or identical nucleotide sequences, and probably encode similar gene products. Thus, nucleic acid hybridization can be used to locate homologous sequences (**figure 9.15**). The two general procedures that use nucleic acid hybridization to locate a sequence are:

- **Colony blot**—detects a given sequence in a crude preparation of genomic DNA obtained by replica plating colonies onto a membrane filter. This technique is most often used to identify the cells in a DNA library that contain the desired cloned DNA. ■ **replica plating. p. 201**
- **Southern blot**—detects a given sequence in restriction fragments that have been separated according to size and transferred to a membrane filter. This technique is used to locate and compare genes in other organisms that are homologous to one being studied. It is also used to observe subtle genomic variations that characterize different strains and is a fundamental part of **DNA fingerprinting** (see **Perspective 9.1**).

DNA Probes

The heart of both a colony blot and a Southern blot is the use of a **probe** to locate a specific nucleotide sequence in a denatured DNA sample affixed to a solid surface such as a nylon membrane. The probe is a single-stranded piece of nucleic acid that has been tagged, or **labeled**, with a detectable marker such as the radioisotope ^{32}P and is complementary to either of the two strands being

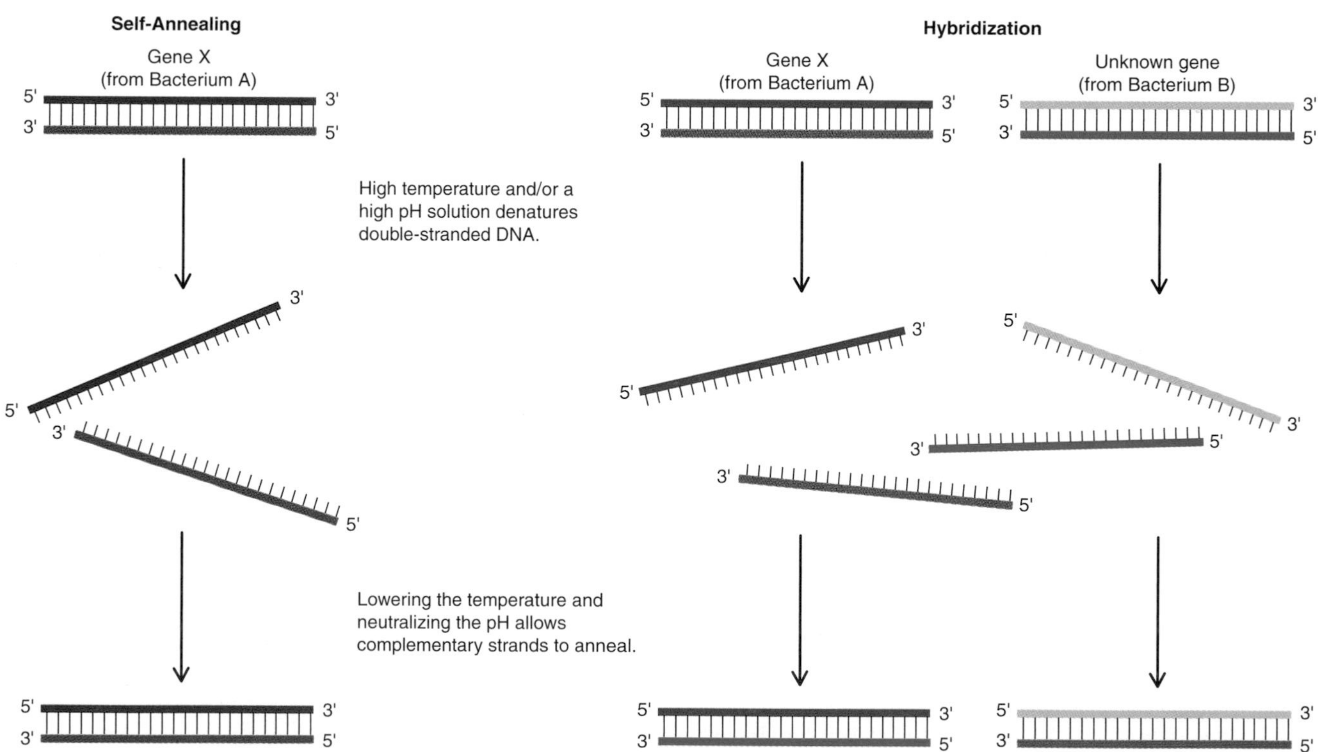

Figure 9.15 Nucleic Acid Hybridization Can Be Used to Locate Homologous Sequences

detected. Once the probe has been given enough time to hybridize, its location is determined by placing the membrane on a piece of X-ray film overnight. The radioactivity of the probe exposes the film, which upon development generates a black spot. Using film to detect a radioactive molecule is called **autoradiography**. ■ autoradiography, p. 25 ■ isotope, p. 25

Obtaining a DNA Probe

A DNA probe can be obtained in a number of ways. In some cases, a similar gene from another organism has already been cloned, and so can be used as a probe. For example, the Shiga toxin gene of *Shigella* can be used to probe for similar toxin genes, such as the one encoded by *E. coli* 0157:H7. The double-stranded DNA is labeled and then denatured to create the single-stranded probe. Because the probe originated from a double-stranded DNA molecule, it contains each of the two strands of complementary DNA. Although these two probe strands can anneal to each other, the concentration of probe molecules ensures that sufficient numbers will remain single-stranded until they can hybridize to the DNA of interest. ■ Shiga toxin, p. 602 ■ *E. coli* 0157:H7, p. 603

Another way to obtain a probe is to synthesize a short sequence of nucleotides, an **oligonucleotide**. Rapid and efficient DNA synthesis machines now make this a relatively simple task. Synthesizing the appropriate probe, however, requires at least some advance knowledge about the sequence of nucleotides that are to be located. To gain this, the protein encoded by the gene is first isolated and then a portion of its amino acid sequence determined. Knowing the amino acid sequence makes it possible to deduce the potential nucleotide sequences using the genetic code. For example, if the protein has the amino acid methionine followed by a tryptophan, then one strand of the corresponding nucleotide sequence must be ATGTGG. In practice, several different oligonucleotides are usually synthesized, because the redundancy of the genetic code means that several slightly different sequences can encode the same short stretch of amino acids. ■ genetic code, p. 175

Colony Blot

To do a colony blot, a plate of colonies is replica-plated onto a nylon membrane, creating a pattern of colonies identical to that of the original plate (**figure 9.16**). The membrane is then soaked in an alkaline solution, which simultaneously lyses the cells and denatures their DNA. A liquid solution containing the labeled single-stranded probe is then added to the membrane and incubated under conditions that allow the probe to hybridize to complementary sequences on the filter. The unbound probe is then washed off. Autoradiography is used to determine the location of the hybridized probe. The dark spots that appear on the film can then be correlated with a specific colony on the original plate. ■ replica plating, p. 201

Southern Blot

The **Southern blot**, developed by and named for Dr. E. M. Southern, not only verifies the presence or absence of the sequence but also determines the size of any restriction fragments that contain the sequence. The principles of the Southern blot spawned two analogous techniques using a probe to detect a molecule that has been separated according to size and then transferred to a filter. To reflect their resemblance to the Southern blot, each of these was named after other compass points. A **Northern blot** uses a nucleic acid probe to detect sequences of RNA; a **Western blot** uses a labeled antibody to detect specific proteins. ■ Western blot, p. 422

Digestion and Gel Electrophoresis of DNA

The first step of the Southern blot technique is the isolation and restriction enzyme digestion of the DNA to be studied (**figure 9.17**). The DNA fragments are then separated according to size using a technique called **gel electrophoresis**. This technique uses a slab of gel that has the consistency of very firm gelatin and is made of **agarose**, a highly purified form of agar. The digested DNA samples are put into slots or wells in the gel. As a means to eventually determine the size of the various fragments in the restriction digest, a **size standard** is routinely put

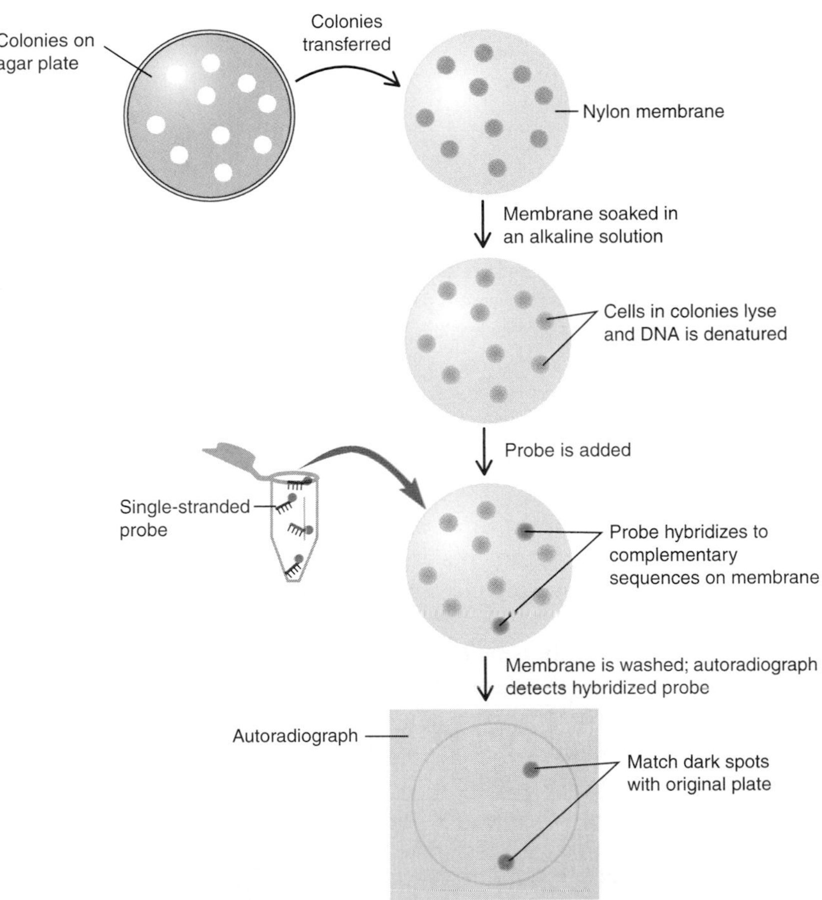

Figure 9.16 The Steps of a Colony Blot This technique is used to determine which colonies on a plate contain a given DNA sequence.

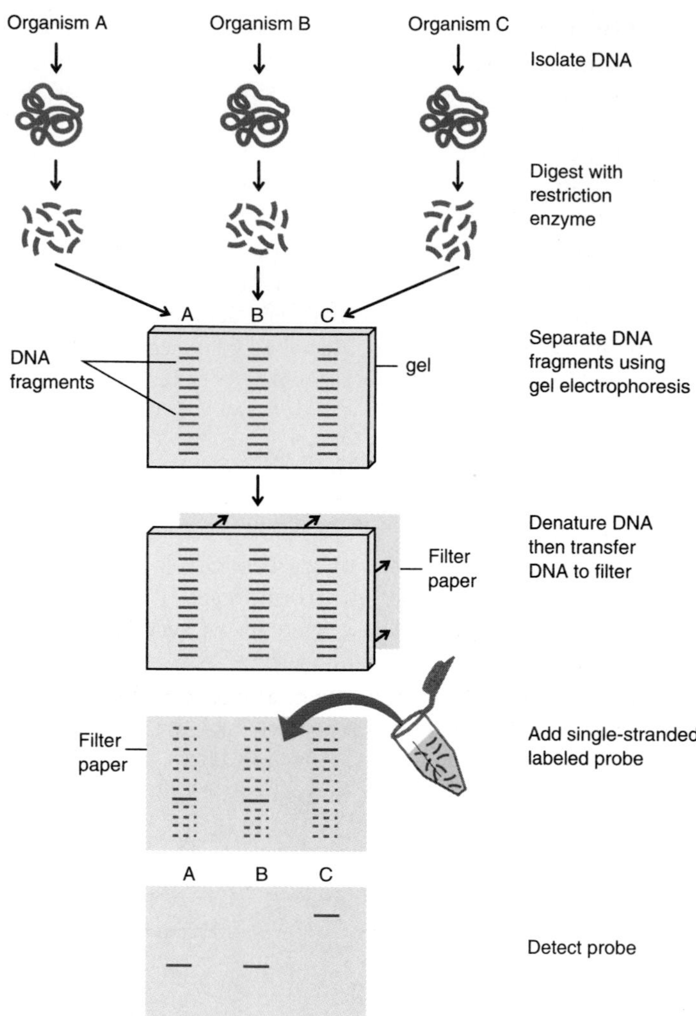

Figure 9.17 The Steps of a Southern Blot This technique is used to verify the presence or absence of a given sequence and determine the size of any restriction enzyme fragments that contain the sequence. If the DNA fragments contain the same genes, why are the fragments different in length?

into a well in the same gel. The size standard is simply a series of DNA fragments of known sizes that can be used as a basis for later comparison.

The gel is then placed in an electric field. DNA is negatively charged, and the fragments are attracted to and migrate toward the positively charged electrode. As the DNA moves in the electric field, the gel acts as a sieve, impeding the large fragments while allowing the smaller ones to pass through more quickly. Because of the sievelike effect of the gel, the restriction fragments can be separated according to their size.

The DNA is not directly visible on the gel; to view it the gel must first be immersed in a solution containing a dye, **ethidium bromide,** that binds to nucleic acid. Ethidium bromide–stained DNA is fluorescent when viewed with UV light. Each fluorescent band represents a specific-sized fragment of DNA **(figure 9.18)**. Staining the gel, however, only confirms that different-sized bands are present; it gives no information to suggest which band encodes the nucleotide sequence of interest. ■ **ethidium bromide, p. 196**

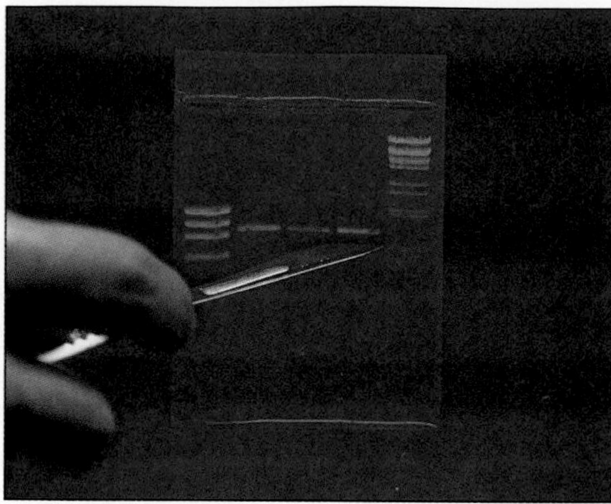

Figure 9.18 Bands of DNA on an Ethidium Bromide-Stained Agarose Gel The stained DNA is visible because it fluoresces when viewed with UV light. Each band represents a specific-sized fragment of DNA.

Transferring the DNA to a Solid Membrane Support

Because the gel is too fragile to withstand much handling, the DNA fragments are transferred in-place to a nylon membrane. To do this, the gel is first soaked in an alkaline solution, which denatures the DNA. Then, by placing the membrane directly on the gel in a special apparatus, the DNA migrates to the membrane in the same relative position as it occupied on the gel.

Detecting Specific DNA Sequences

A liquid solution containing the radioactively labeled probe is added to the membrane and incubated under conditions that allow the probe to hybridize to any complementary sequences. The unbound probe is then washed off, and the location of the bound probe is detected using autoradiography. The dark bands that appear on the film correspond to specific bands on the original gel.

Applications of a Southern Blot

The most obvious use of a Southern blot is to locate sequences that are homologous to the one being studied, which is how many functionally related genes have been discovered and characterized. This information can also be used to simplify a cloning experiment. For example, if the gene of interest is shown to be encoded on a *Bam* HI fragment that is 6,000 base pairs in length, then fragments of that size can be separated from the rest and cloned. To separate the fragments, gel electrophoresis is used. The band of DNA on the gel corresponding to 6,000 base pairs can then be cut out and the DNA extracted.

A less apparent but equally important use of the Southern blot is to distinguish different strains of a given species by detecting subtle variations in their nucleotide sequences. Certain mutations will create, or destroy, restriction enzyme recognition sequences at particular sites in the genome. Thus, when genomic DNA of different strains is digested with the same restriction enzyme, each will give rise to a slightly different

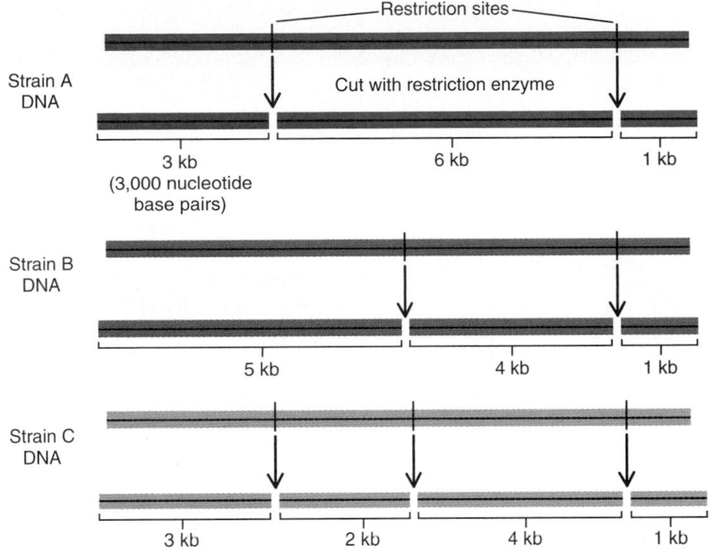

Figure 9.19 Restriction Fragment Length Polymorphism (RFLP)
Different strains of species may have subtle variations in nucleotide sequences that may give rise to a slightly different assortment of restriction fragment sizes. What if different restriction enzymes were used; what would be the effect?

assortment of restriction fragment sizes (**figure 9.19**). The difference is called a **restriction fragment length polymorphism** (**RFLP**) and is the basis for DNA fingerprinting. While these RFLPs are useful for characterizing strains, they are generally not noticeable in the pattern of bands that appear on an ethidium bromide–stained agarose gel. This is because thousands of fragments of different sizes are generated when genomic DNA is digested with most common restriction enzymes, and thus a difference in one or two bands is not obvious. Therefore, to detect RFLPs, a probe is generally used to selectively visualize only certain fragments. The probe that is employed is one that has already been shown by trial and error to hybridize to fragments that are likely to vary in size, demonstrating maximal differences between strains. One probe that has been successfully used to detect RFLP differences in a variety of organisms hybridizes to the genes that code for ribosomal DNA. Categorizing strains using this probe is called **ribotyping**. ■ **ribotyping, p. 257**

MICROCHECK 9.5

Nucleic acid probes detect a specific nucleotide sequence in a sample. Colony blots are used to identify colonies that contain a given sequence of DNA. Southern blots are used to determine the size of the restriction fragments that contain a sequence of interest and to detect subtle variations in nucleotide sequences that occur in different strains of a given species.

- Describe two ways to obtain a DNA probe.
- What is the purpose of gel electrophoresis?
- In what situation would it be important to distinguish between different bacterial strains?

DNA Sequencing

The initiation of the Human Genome Project has resulted in DNA sequencing techniques that are increasingly automated and more efficient. In turn, scientists are now more readily able to determine the chromosomal sequence of other organisms, including both prokaryotes and eukaryotes.

Applications of DNA Sequence Analysis

The sequence of nucleotides of a segment of DNA can give very useful information. Because the genetic code is universal, the nucleotide sequence of a gene can be used to determine the amino acid sequence of the protein it encodes. Based on this information, the folding or shape of the protein can be at least partially deduced. In addition, some idea of its function can be determined by comparing its sequence wtih that of other proteins whose functions are known. Those with similar amino acid sequences likely have analogous functions. Known gene sequences are now stored in computerized databases that are available to all scientists. ■ **protein shapes, p. 31**

Knowing the DNA sequence of a functional human gene helps scientists identify genetic alterations that result in certain diseases. For example, sickle cell anemia is due to a single base-pair change in the gene that encodes a part of the hemoglobin protein. This alteration results in a distorted protein that does not function properly. Likewise, the genetic disease cystic fibrosis is most often caused by a three-base-pair deletion in a protein involved with chloride transport in and out of cells.

DNA sequence analysis is also used to study evolutionary relatedness of organisms. Those that have extensive sequence homology are likely to be closely related. Conversely, those that have vastly different sequences are probably only distantly related. ■ **determining evolutionary relatedness, p. 261**

Dideoxy Chain Termination Method

One widely used technique for determining the nucleotide sequence of DNA is the **dideoxy chain termination method**. The fundamental part of the technique is an *in vitro* DNA synthesis reaction. Like any DNA synthesis reaction, a sequencing reaction requires:

- A single-stranded piece of DNA, the **template**, from which a complementary copy is synthesized. ■ **template, p. 168**
- A short single-stranded piece of DNA, a **primer,** that anneals to its complementary sequence on the single-stranded template and from which DNA synthesis can begin. Oftentimes, the DNA being sequenced has been cloned into a vector. Because the nucleotide sequences of vectors are known, a primer that anneals to a portion of the vector adjacent to the DNA to be sequenced can be readily obtained. ■ **primer, p. 171**
- DNA polymerase, the enzyme that catalyzes DNA synthesis. ■ **DNA polymerase, p. 171**
- Each of the four deoxynucleotides that are used in DNA synthesis—dATP, dGTP, dCTP, and dTTP; to facilitate

Perspective 9.1 Science Takes the Witness Stand

Fingerprints have been used to identify criminals and solve crimes for more than 100 years. More recently, DNA fingerprinting is gaining prominence for its role in criminal investigations. If a criminal leaves a blood drop or semen at the scene of a crime, the DNA can be extracted and compared with the DNA from the blood of a suspect.

It is not feasible to compare the entire nucleotide sequence of two samples; instead, restriction fragment length polymorphisms (RFLPs) are compared. The probes that are used hybridize to regions of the chromosome that vary greatly among individuals. These regions are located between genes and consist of a repetitive core sequence, a **tandem repeat.** The number of times that this sequence is repeated in a various chromosomal locations differs from one individual to the next. For example, in a given chromosomal location, one individual may have a series of 8 tandem repeats whereas another individual may have 10. The relative number of repeats is reflected in the size and number of DNA fragments generated by digestion with a restriction enzyme, giving rise to the RFLPs. These variations can be detected by Southern blot hybridization using the core sequence as a probe **(figure 1)**.

DNA fingerprinting compares the hybridization pattern of DNA in samples from different sources. By comparing the DNA taken from a sample found at the scene of a crime with that of DNA from a suspect, it is possible to determine the likelihood that the samples of DNA came from the same person.

It is essential that the results of DNA testing be interpreted with caution. For example, it is difficult to state with certainty that two bands are identical in size. Further, during electrophoresis DNA fragments of the same size can migrate at different rates because of variations in the preparation of the gel, concentration of the DNA, or other unknown variables. Even with these technical problems, however, there seems little doubt that molecular biology will play an increasingly important role in criminal investigations as science takes the witness stand.

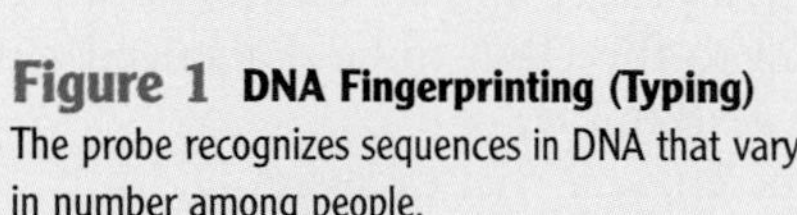

Evidence #1 (victim's blood)
Evidence #2 (unidentified blood)
Suspect's blood
Evidence is gathered. Work begins on matching evidence with the victim and the suspect.
DNA is extracted from victim's blood, the unidentified blood, and the suspect's blood.
Treat with restriction enzyme
A restriction enzyme is used to cleave the DNA into specific-sized fragments.
1.
2.
3.
Electric field
Electrophoresis process. The DNA fragments are separated according to their length, and end up as bands.
The band patterns are transferred to a nylon membrane.
A radioactive probe binds to the target fragments.
X-ray film
Probe treated DNA is then x-rayed. The result is an array of patterns that resemble a retail bar code.
1. Evidence #1 (victim's blood)
2. Evidence #2 (unidentified blood)
3. Suspect's blood
A comparison can then be made between resulting patterns from the suspect and from the unidentified blood found at the scene of the crime. In this example, they match, indicating that the suspect was at the scene of the crime.
DNA fingerprinting (typing). The probe recognizes sequences in DNA that vary in number among people.

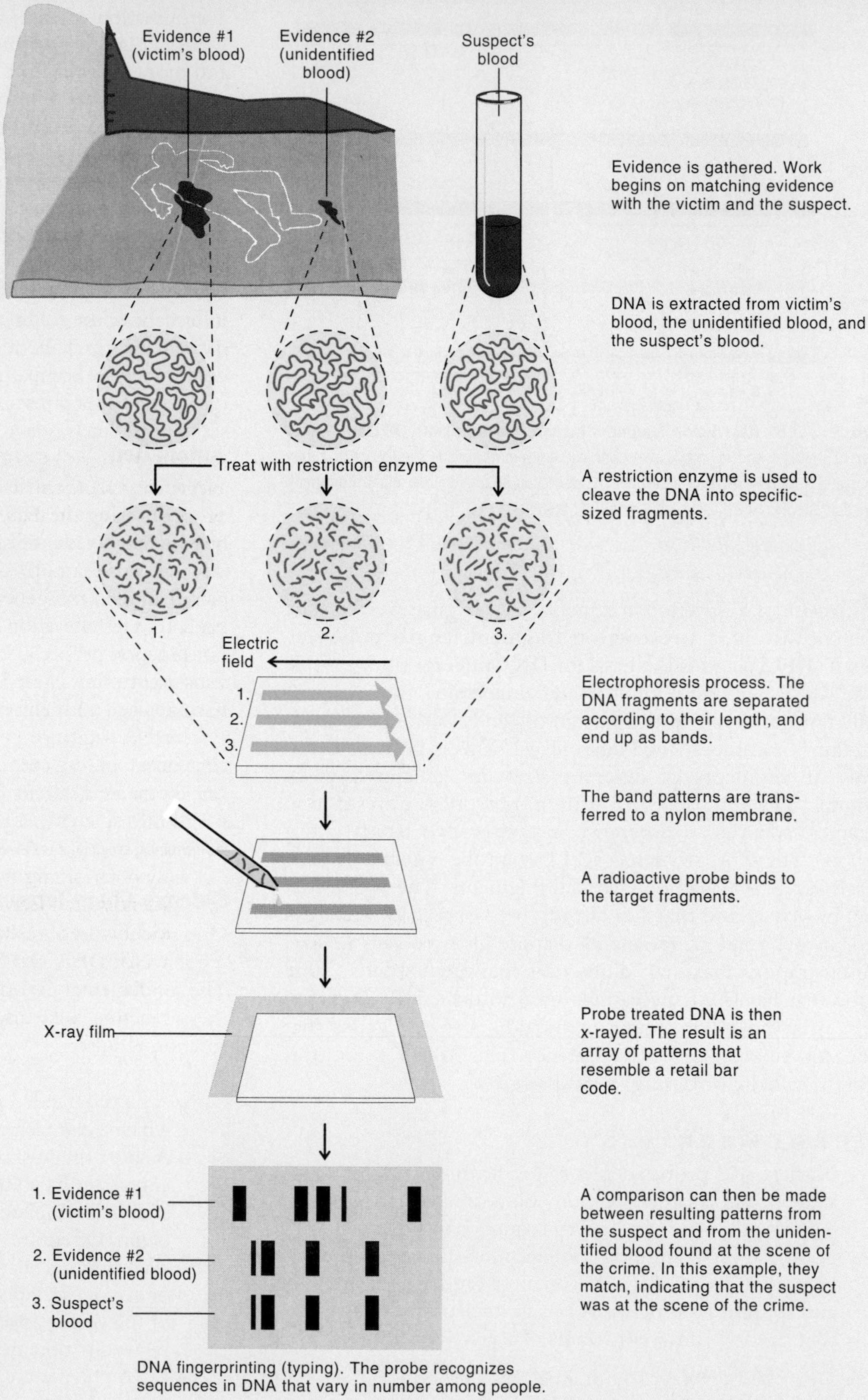

Figure 1 DNA Fingerprinting (Typing)
The probe recognizes sequences in DNA that vary in number among people.

detection of the newly synthesized DNA, one of these is labeled with a detectable marker such as a radioisotope.

■ nucleotide, p. 35

If these were the only ingredients in the reaction, then a full-length molecule complementary to the template DNA would be synthesized. A key additional ingredient is added, however, which can terminate DNA synthesis before a full-length complement is made. This key ingredient is a **dideoxynucleotide (ddNTP)**. Dideoxynucleotides are identical to their deoxynucleotide (dNTP) counterparts except they lack the 3′OH group, the portion of a nucleotide required for the addition of subsequent nucleotides during DNA synthesis (**figure 9.20**). Because they lack the 3′OH group, dideoxynucleotides are **chain terminators.** When one of these nucleotides is incorporated into a growing strand of DNA, no additional nucleotides can be added, and elongation of that strand ceases.

To perform a sequencing reaction, DNA polymerase, template DNA, primer, and each of the four deoxynucleotides are mixed together and then apportioned equally into four separate tubes. Each tube then receives a very small amount of a different dideoxynucleotide (ddATP, ddGTP, ddCTP, or ddTTP). In other words, ddATP is added to one tube, ddGTP to the next, and so on. When the reaction is incubated at an appropriate temperature, the primer will anneal to the template and DNA synthesis will begin. Chain elongation proceeds until a dideoxynucleotide is incorporated, terminating synthesis at that point. Termination, however, is an infrequent event because of the numerous template molecules and the small amount of the ddNTP relative to its dNTP counterpart. The result is that a tube will contain a set of newly synthesized DNA strands of various lengths. Each of these strands will have been terminated at the point that the specific ddNTP was incorporated into the chain. For example, in the tube that contained ddATP, numerous fragments of different lengths will be obtained, but each fragment will have been terminated at a position that called for the incorporation of the nucleotide base adenine. The sizes of the fragments in a tube indicate the positions of a specific nucleotide base in the synthesized DNA strand (**figure 9.21**).

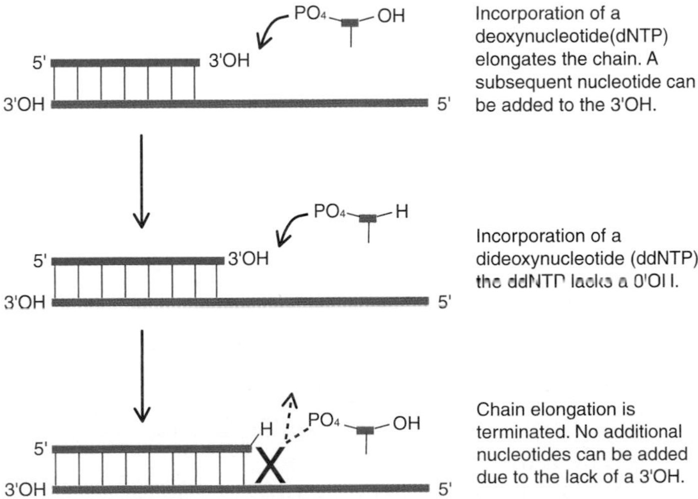
PO4
OH
5'
3'OH
3'OH
5'
Incorporation of a deoxynucleotide(dNTP) elongates the chain. A subsequent nucleotide can be added to the 3'OH.
PO4
H
5'
3'OH
3'OH
5'
Incorporation of a dideoxynucleotide (ddNTP); the ddNTP lacks a 3'OH.
H
PO4
OH
X
5'
3'OH
5'
Chain elongation is terminated. No additional nucleotides can be added due to the lack of a 3'OH.

Figure 9.20 Chain Termination by a Dideoxynucleotide Once a dideoxynucleotide is incorporated into a growing strand, subsequent nucleotides cannot be added due to the lack of a 3′OH.

A special type of gel electrophoresis is used to separate the DNA fragments and determine their relative size. The gel has four separate wells arranged side-by-side. Each well receives the contents of a specific tube from the sequencing reaction. The gel conditions (pH, temperature, and gel concentration) denature the DNA and enable separation of fragments that differ by only one nucleotide in length. Autoradiography is then used to detect the relative positions of the newly synthesized fragments. The order that the terminating nucleotides were incorporated can then be determined by "reading" the gel from the bottom up (**figure 9.22**). ■ autoradiography, p. 25

Automated DNA Sequencing

Newer developments in DNA sequencing techniques make the process more automated and consequently more rapid. Most automated techniques do not use a radioactive label to detect the newly synthesized DNA, but instead use a fluorescent dye attached directly to the ddNTP. Different dye colors are used for each of the four ddNTPs. By having each different ddNTP labeled with a distinct color, the reaction can be done in a single tube and run in one lane of the gel. Again, gel electrophoresis is used to separate the bands. A laser is used to detect the colors of fluorescent bands as they run off the gel, recording their intensity as a peak. The order of the colored peaks reflects the nucleotide sequence of the DNA (**figure 9.23**).

MICROCHECK 9.6

By determining the nucleotide sequence of a gene, scientists can learn about the protein it encodes and identify genetic alterations that result in certain diseases. DNA sequence analysis can also be used to study evolutionary relatedness. The dideoxy chain termination method is a widely used sequencing technique.

- Describe the chemical difference between a ddNTP and a dNTP.
- How does a ddNTP terminate DNA synthesis?
- What would happen in the sequencing reaction if the relative concentration of a dideoxynucleotide were increased?

Polymerase Chain Reaction

The **polymerase chain reaction (PCR)** has revolutionized research by making it possible to create millions of copies of a given region of DNA in only hours. Using this relatively

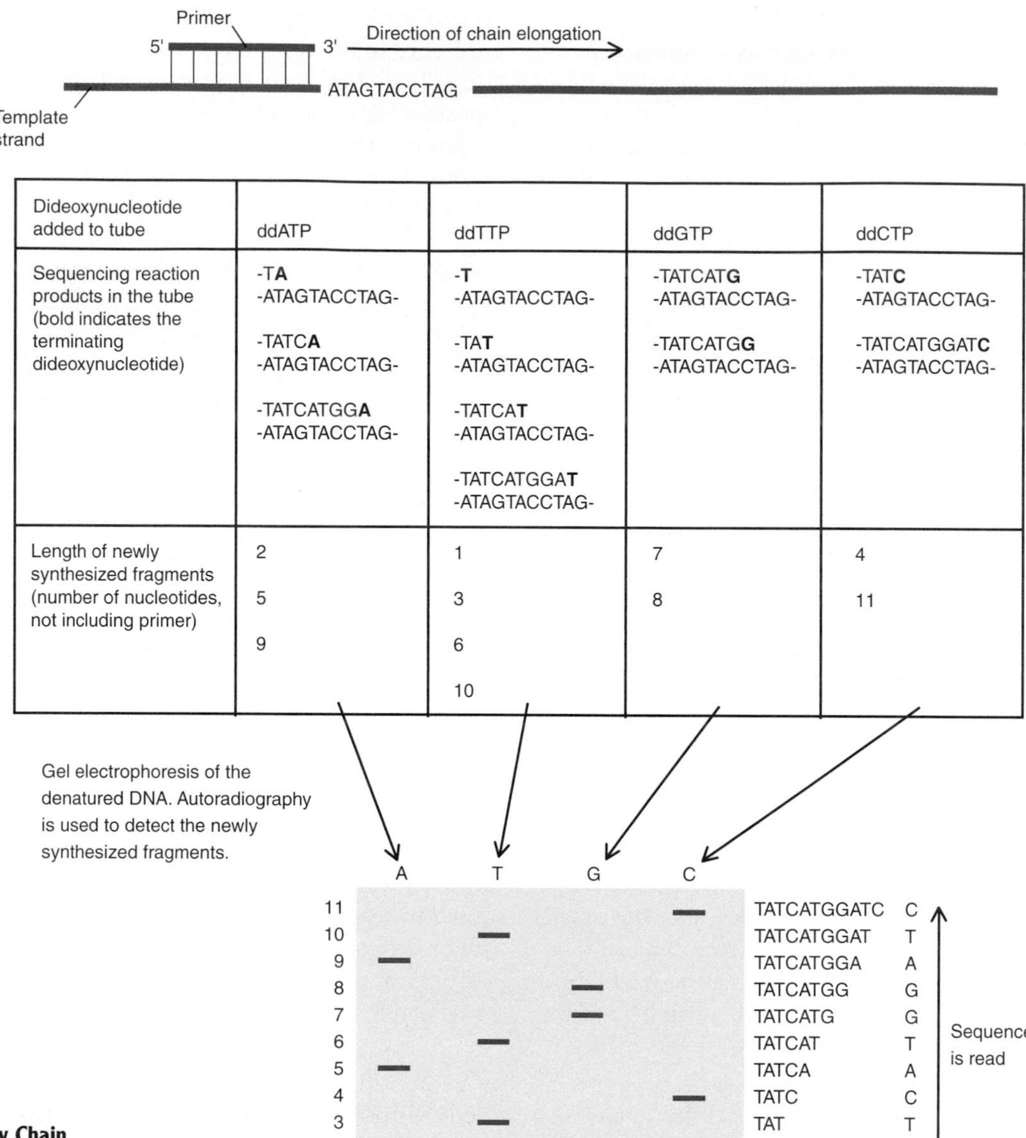

Dideoxynucleotide added to tube	ddATP	ddTTP	ddGTP	ddCTP
Sequencing reaction products in the tube (bold indicates the terminating dideoxynucleotide)	-T**A** -ATAGTACCTAG- -TATC**A** -ATAGTACCTAG- -TATCATGG**A** -ATAGTACCTAG-	-**T** -ATAGTACCTAG- -TA**T** -ATAGTACCTAG- -TATCA**T** -ATAGTACCTAG- -TATCATGGA**T** -ATAGTACCTAG-	-TATCAT**G** -ATAGTACCTAG- -TATCATG**G** -ATAGTACCTAG-	-TAT**C** -ATAGTACCTAG- -TATCATGGAT**C** -ATAGTACCTAG-
Length of newly synthesized fragments (number of nucleotides, not including primer)	2 5 9	1 3 6 10	7 8	4 11

Figure 9.21 Dideoxy Chain Termination Method for Determining the Nucleotide Sequence of DNA

simple technique, even bacteria that have never been grown in culture can now be identified. ■ **identification of non-culturable organisms, p. 255**

The critical aspect of PCR is the ability to amplify a specifically selected sequence, even in an impure sample such as soil or saliva, so that the resulting fragment can be visualized on an ethidium bromide–stained gel (**figure 9.24**). By amplifying a chosen sequence that is unique to a given organism, it is possible to detect the presence of that organism in a specimen. For example, if a sequence found only in *Neisseria gonorrhoeae* can be amplified from a vaginal secretion, then *N. gonorrhoeae* must be present in that specimen. Likewise, PCR can be used to detect HIV in a sample of blood cells. The exquisite sensitivity of PCR, however, is also its greatest drawback. Great care must be taken not to contaminate a sample with an external source of the DNA sequence to be amplified. In the example of HIV detection, inadvertently contaminating a specimen with even a minute amount of HIV proviral DNA from another source would give rise to a false positive test. ■ **provirus, p. 355**

The PCR reaction starts with a double-stranded DNA molecule, from which millions of identical copies of a select region can be produced. The process involves a sequential series of DNA synthesis reactions and the following key ingredients:

- Double-stranded DNA containing the region to be amplified, the **target DNA.**
- A heat-stable DNA polymerase, ***Taq*** **polymerase**, which is from *Thermus aquaticus.*
- Each of the four nucleotides (dATP, dGTP, dCTP, dTTP).
- Short single-stranded segments of DNA, generally about 20 nucleotides in length, to serve as primers**.** The selection of these primers will be discussed shortly.

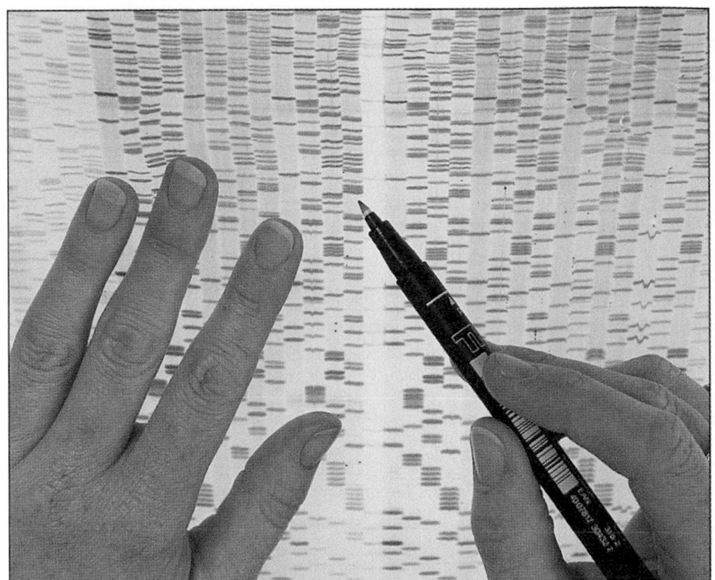

Figure 9.22 An Autoradiograph of a Sequencing Gel Each dark band represents a specific-sized fragment. The sequence of the newly synthesized strand can be determined by "reading" the gel from the bottom up.

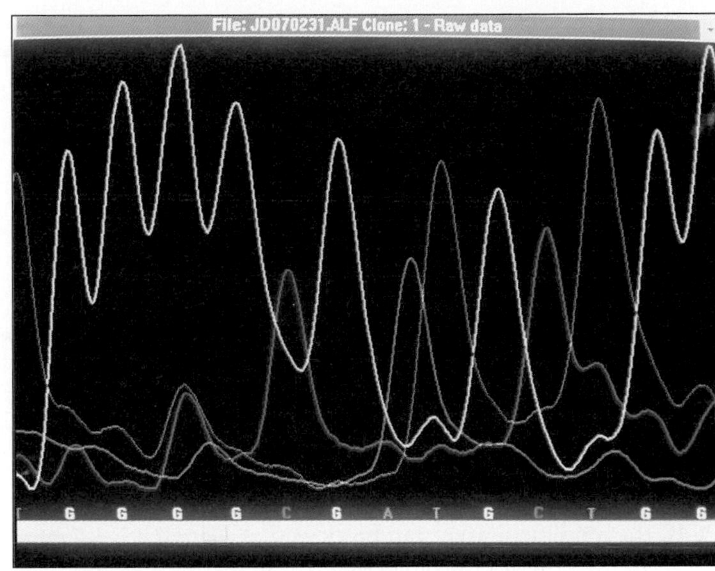

Figure 9.23 Results of Automated DNA Sequencing The order of the colored peaks reflects the nucleotide sequence of the DNA.

The Three-Step Amplification Cycle

PCR requires a repeating cycle consisting of three steps (**figure 9.25**). In the first step, the double-stranded DNA is denatured by heating the sample to approximately 95°C. In the second step, the temperature is lowered to approximately 50°C; within seconds, the primers anneal to their complementary sequences on the denatured target DNA. In the third step, DNA synthesis occurs when the temperature is raised to the optimal temperature for *Taq* DNA polymerase, approximately 70°C. The DNA polymerase adds nucleotides to the 3′ end of the DNA primer using the opposing strand as a template. The net result is the synthesis of two new strands of DNA, each of which is complementary to the other. In other words, the three-step cycle results in the duplication of the original target DNA. Since each of the newly synthesized strands can then serve as a template strand for the next cycle, the DNA is amplified exponentially. After a single cycle of the three-step reaction, there will be two double-stranded DNA molecules for every original double-stranded target; after the next cycle, there will be four, after the next cycle there will be eight, and so on.

A critical factor in PCR is the heat-stable DNA polymerase of a thermophile, *Thermus aquaticus. Taq* polymerase, unlike the DNA polymerase of *E. coli*, is not destroyed at the high temperature used to denature the DNA in the first step of each

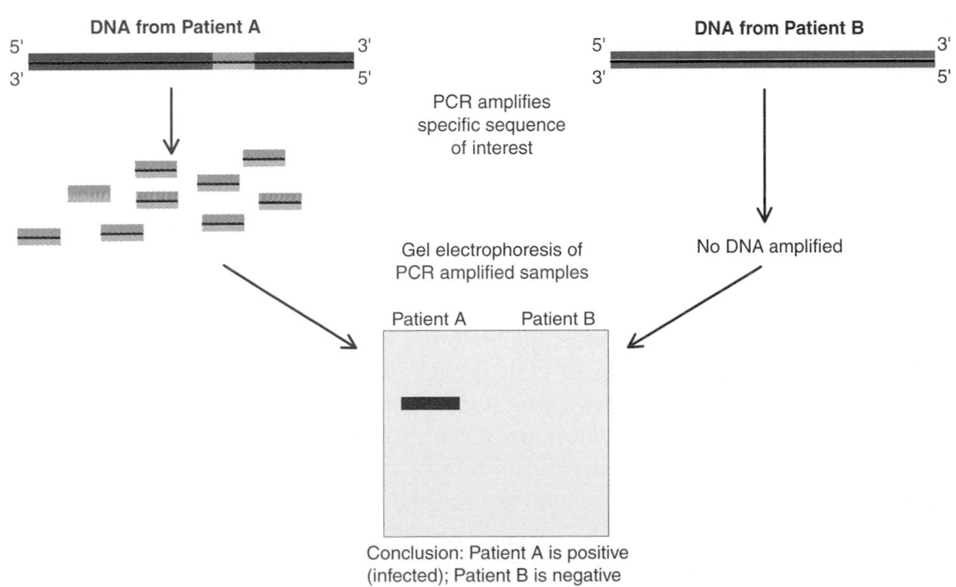

Figure 9.24 PCR Amplifies Selected Sequences By amplifying a chosen sequence that is unique to a given organism, it is possible to detect the presence of minute amounts of that organism in a sample.

amplification cycle. If a heat-stable polymerase were not used, fresh polymerase would need to be added for every cycle of the reaction. Thus, the discovery and characterization of *T. aquaticus* through basic research was instrumental in developing this widely used and commercially valuable method. ■ **thermophile, p. 96**

Generating a Discrete-Sized Fragment

While the proceeding description explains how PCR amplifies the target DNA exponentially, it does not clarify how a discrete-length fragment becomes the ultimate product. The generation of fragments of a particular size is important, because it enables a researcher to use PCR to readily detect the presence of target DNA in a sample. After PCR, the amplified target can be viewed as a single band on the ethidium bromide–stained gel.

To understand how discrete-sized fragments of target DNA are generated, you must consider the exact sites to which the primers anneal and visualize at least three cycles of replication (**figure 9.26**). In the first cycle, two new fragments are generated. Note, however, that these fragments are shorter than the original full-length template molecules but longer than the target DNA. Their 5′ end is primer DNA. These mid-length products will be generated whenever the original full-length molecule is used as a template, which is one time each replication cycle.

In the next cycle, the full-length molecules will again be used as templates, repeating the process described above. More importantly, the mid-length fragments created during the first cycle will be used as templates for DNA synthesis. As before, the primers will anneal to these fragments and then nucleotides will be added to the 3′ end. Elongation, however, will stop at the 5′ end of the template molecule, because DNA synthesis requires a template. Recall that the 5′ end of the template is primer DNA. Thus, whenever the mid-length fragments are used as templates, a short fragment is generated. The 5′ and 3′ ends of this fragment are determined by the sites to which the primers initially annealed.

In the third round of replication, the full-length and the mid-length fragments again will be used as templates, repeating the processes described above. The short fragments generated in the preceding round will also be used as templates, however, generating short double-stranded molecules. Continuing to follow the events in further rounds of replication will reveal that it is this fragment that is exponentially amplified. Ultimately, enough of these short fragments are generated to be detectable using gel electrophoresis followed by ethidium bromide staining.

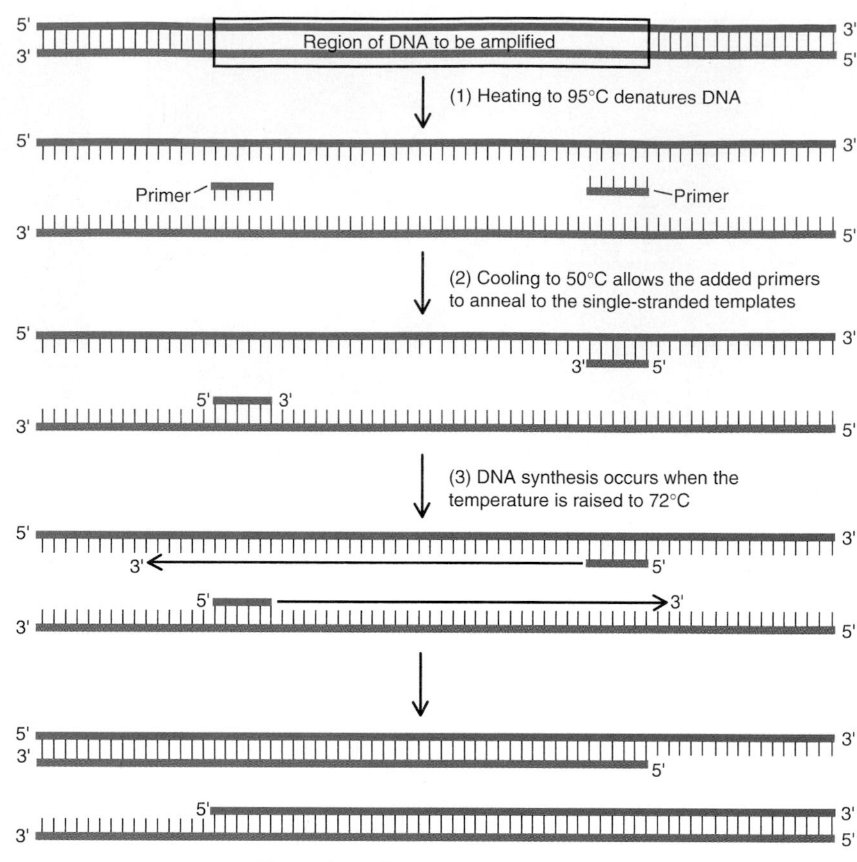

Figure 9.25 Steps of a Single Cycle of PCR

Selecting Primer Pairs

The nucleotide sequences of the two primers are critical because it is the primers that dictate which portion of the DNA is amplified. Each must be complementary to a sequence on the appropriate strand, flanking the region to be amplified. The synthesis reaction will then add nucleotides onto the primers, elongating the DNA chain so that the target DNA between those primers is copied. Thus, if a researcher wants to amplify a DNA sequence that encodes a specific protein, then the researcher must first determine the nucleotide sequences flanking the gene that encodes the protein and then synthesize the appropriate pair of oligonucleotides to serve as primers.

MICROCHECK 9.7

The polymerase chain reaction (PCR) is used to rapidly increase the amount of either a specific segment or total DNA in a sample.

- What happens during each of the three temperature steps of PCR (95°C, 50°C, and 70°C)?
- Explain why it is important to use a polymerase from a thermophile in the PCR reaction.
- Sequencing reactions can be done using PCR. In this case would two primers be necessary?

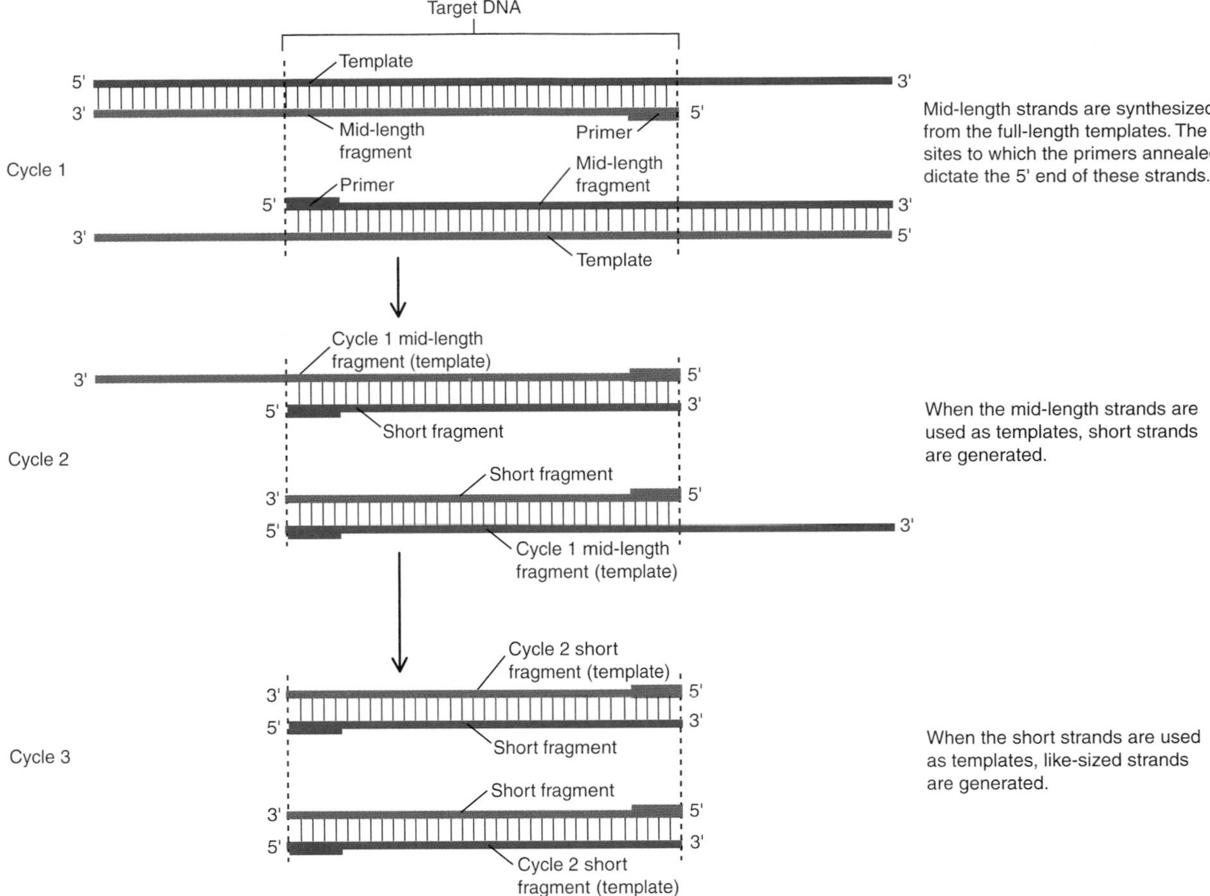

Figure 9.26 The Final PCR Product Is a Fragment of Discrete Size The positions to which the primers anneal to the original template molecule dictate the size and sequence of the fragment that is amplified exponentially.

FUTURE CHALLENGES

Fishing the Chips for Nucleotide Sequences

A key challenge for the future will be to fully exploit one of the newest additions to the arsenal of biotechnology tools, ***nucleotide array technology.*** *A nucleotide array is solid support to which a two-dimensional arrangement of numerous different single-stranded DNA fragments of known sequences has been attached. One of the most sophisticated types of arrays is a* ***DNA chip,*** *a silicon chip of less than an inch in diameter that can carry an array of tens or hundreds of thousands of oligonucleotides. The power of these arrays is that each nucleic acid fragment functions in a manner analogous to a probe, enabling a researcher to screen a single sample for a vast range of different sequences simultaneously. Unlike typical probes, however, the arrays do not carry a detectable label. Instead, the label must be attached to the nucleic acid of interest—for example, genomic DNA from a bacterium. The DNA of interest is digested into small fragments, denatured, and then added to the array, where it will hybridize to complementary sequences. After washing the array to remove unhybridized DNA, the locations of the labeled DNA are detected. Because the sequence of each of the fragments in the array is known, the location of the label can be used to determine the presence of specific sequences in the DNA of interest.*

Microbiologists are already using nucleotide array technology to study gene expression in those bacteria where the genome has been sequenced. First, a series of oligonucleotides, each one specific for a given open reading frame, is made. These oligonucleotides are then attached to a support to make an array that contains a probe specific for every gene in the genome. Then, mRNA is isolated from a culture grown under a specific set of conditions. By labeling the mRNA and hybridizing it to the array, a researcher can determine which genes are being transcribed under those culture conditions. By repeating the experiment using a culture grown under a different set of conditions, perhaps an environment that mimics what a pathogen would encounter in the body, differences in gene expression can be ascertained.

SUMMARY

Applications of Recombinant DNA Technology (Figure 9.1)

Protein Production

1. The ability to clone the genes encoding medically important proteins and have them expressed in microorganisms ensures production in large quantities and with reliable quality. (Table 9.1)

Altering Organisms to Give Them Economically Useful Traits

1. Plants have been genetically engineered to synthesize their own insecticide, resist a common herbicide, and improve their nutritional value. (Figure 9.3)

Providing a Source of DNA for Study

1. By cloning a segment of DNA into a bacterium, a ready source of that nucleotide sequence is available for study or further manipulation.

Cloning DNA (Figure 9.7)

1. An approach frequently used to clone a specific bacterial gene is to first clone into a population of *E. coli* cells a set of DNA fragments that together make up the entire genome of the organism being studied. The resulting population is called a **DNA library**. (Figure 9.6)

Isolating DNA

1. To isolate bacterial DNA, cells in a broth culture are lysed by adding a detergent.

Using Restriction Enzymes to Generate Fragments of DNA

1. Purified DNA is cut into smaller fragments by digesting it with a **restriction enzyme**. These enzymes recognize and cleave specific sequences of DNA. (Figure 9.9)

Generating a Recombinant DNA Molecule

1. To enable cloned DNA to replicate in a cell, it is attached to a piece of DNA called a **vector**, forming a **recombinant molecule** that is part vector and part cloned DNA.
2. A vector typically has an **origin of replication**, a **selectable marker**, and a second genetic marker that contains within it a **multiple-cloning site**. (Figures 9.10 and 9.11)
3. The enzyme **DNA ligase** is used to join the vector and the insert.

Introducing the Recombinant DNA into a New Host

1. For routine cloning experiments, a strain of *E. coli* is used as a host.
2. The recombinant molecule is introduced into the new host using **transformation** or **electroporation**.
3. The transformed cells are cultivated on medium that both selects for cells containing vector sequences and differentiates those that carry recombinant plasmids.

Overcoming the Obstacles to Cloning DNA from Eukaryotic Cells

Removing Introns from Eukaryotic DNA (Figure 9.12)

1. Introns must first be removed if eukaryotic DNA is to be expressed in a prokaryotic cell.
2. Instead of cloning eukaryotic DNA directly into the vector, **cDNA** is produced by using **reverse transcriptase** to make a copy of DNA from an mRNA template.

Vectors Used to Clone Eukaryotic Genes into Prokaryotic Cells

1. The type of vector used to clone eukaryotic DNA depends largely on the ultimate purpose of the procedure.
2. Special vectors called **expression vectors** have been developed that have the appropriate promoter and ribosome-binding site to enable *E. coli* to express cloned genes.
3. Bacterial artificial chromosomes are used to clone very large inserts.

Transferring Genes into Eukaryotic Cells

1. Vectors used to clone DNA into eukaryotic cells are often specific for that cell type.

Vectors Used to Transfer DNA into Eukaryotic Cells

1. The **Ti plasmid** of *Agrobacterium* is used to genetically engineer plant cells. (Figure 9.13)
2. **Artificial chromosomes** can contain very large fragments of DNA.
3. Certain viruses have been engineered for use as vectors to carry DNA into eukaryotic cells.

Introducing DNA into Eukaryotic Cells

1. Ti plasmids and viruses naturally carry DNA into eukaryotic cells.
2. **Electroporation** and a **gene gun** can be used to move DNA into eukaryotic cells. (Figure 9.14)

Nucleic Acid Hybridization

DNA Probes

1. A **probe**, which is a labeled single-stranded piece of nucleic acid, is used to locate a specific nucleotide sequence in a DNA sample affixed to a nylon membrane.

Colony Blot (Figure 9.16)

1. **Colony blots** are used to identify colonies that contain a given sequence of DNA.
2. Colonies are replica plated onto a nylon membrane; a DNA probe is then used to identify colonies that contain the sequence of interest.

Southern Blot (Figure 9.17)

1. **Southern blots** are used to determine the size of the restriction fragments that contain a sequence of interest and to detect subtle variations in nucleotide sequences that occur in different strains of a given species.
2. **Gel electrophoresis** is used to separate DNA fragments according to size. (Figure 9.18)
3. The separated DNA is transferred in-place to a nylon membrane.
4. A DNA probe is added to the membrane to locate specific nucleotide sequences.
5. Southern blots are used to locate functionally related genes and to detect **restriction fragment length polymorphisms** (**RFLPs**), which are used to characterize strains. (Figure 9.19)

DNA Sequencing

Applications of DNA Sequence Analysis

1. The nucleotide sequence of a gene can be used to determine the amino acid sequence of the protein for which it codes.
2. Knowing the DNA sequence of a functional human gene helps scientists identify genetic alterations that result in certain diseases.
3. DNA sequence analysis can be used to study evolutionary relatedness.

Dideoxy Chain Termination Method (Figure 9.22)

1. A key ingredient in a sequencing reaction is a **dideoxynucleotide**, a nucleotide that lacks the 3′OH and therefore functions as a **chain terminator**. (Figure 9.21)
2. Each tube of a sequencing reaction contains a different ddNTP.
3. The sizes of fragments synthesized in a tube indicate the positions of that nucleotide base in the synthesized DNA strand. (Figure 9.22)

Automated DNA Sequencing (Figure 9.23)

1. Each different ddNTP is labeled with a different color of fluorescent dye.
2. The reactions can be done in one tube and run on one lane of a gel; a laser detects the color of the band as it runs off the gel.

Poylmerase Chain Reaction

1. The **polymerase chain reaction** (**PCR**) is used to rapidly increase the amount of a specific DNA segment in a sample. (Figure 9.24)

The Three-Step Amplification Cycle (Figure 9.25)

1. Double-stranded DNA is denatured.
2. **Primers** anneal to their complementary sequences.
3. DNA is synthesized, thus amplifying the target sequence.

Generating a Discrete-Sized Fragment (Figure 9.26)

1. A discrete-sized fragment that is amplified exponentially is obtained after three cycles of replication.
2. The size of the amplified fragment is dictated by the positions to which the primers annealed.

Selecting Primer Pairs

1. The two primers that are selected dictate which portion of the DNA is amplified.

REVIEW QUESTIONS

Short Answer

1. What should the first restriction enzyme that is isolated from *Serratia marcescens* be called?
2. Name the two general types of enzymes used to create a recombinant DNA molecule.
3. Define the term gene library.
4. Explain how a PCR eventually generates a discrete-sized fragment from a much longer piece of DNA.
5. What is cDNA? Why is it used when cloning eukaryotic genes?
6. What is the benefit of using a bacterial artificial chromosome when making a DNA library of a eukaryotic genome?
7. What information can you determine from a Southern blot that cannot be determined from a colony blot?
8. Explain how gel electrophoresis separates DNA fragments.
9. How many different temperatures are used in each cycle of the polymerase chain reaction?
10. Explain how nucleotide array technology can be used to study gene expression.

Multiple Choice

1. Which of the following could be a restriction enzyme recognition site?
   A. AAAGGG
   B. ATCCTA
   C. ATGCAT
   D. AUCCUA
   E. ATCATC
2. An ideal vector has all of the following,…
   A. an origin of replication.
   B. a gene encoding a restriction enzyme.
   C. a gene encoding resistance to an antibiotic.
   D. a multiple-cloning site.
   E. the *lacZ′* gene.
3. Which of the following describes the function of the *lacZ′* gene in a cloning vector?
   A. means of selecting for cells that contain vector sequences
   B. means of distinguishing recombinant molecules
   C. site required for the vector to replicate
   D. mechanism by which cells take up the DNA
   E. gene for a critical nutrient required by transformed cells
4. The Ti plasmid of *Agrobacterium tumefaciens* is used as a vector to genetically engineer which of the following cell types?
   A. Animals
   B. Bacteria
   C. Plants
   D. Yeast
   E. All of the above
5. Southern blots, but not colony blots, require which of the following?
   A. Autoradiography
   B. DNA hybridization
   C. Gel electrophoresis
   D. Labeled probe
   E. Membrane filter

6. Ribotyping utilizes which technique to distinguish strains?
   A. Cloning
   B. Colony blot
   C. Polymerase chain reaction
   D. Sequencing
   E. Southern blot
7. Which of the following does a dideoxynucleotide lack?
   A. 5′PO_4
   B. 3′OH
   C. 5′OH
   D. 3′PO_4
   E. C and D
8. In a sequencing reaction, the dATP was left out of the tube to which ddATP was added. What would be the result of this error?
   A. No synthesis would occur.
   B. Synthesis would always stop at the position that the first A was incorporated.
   C. Synthesis would not stop until the end of the template.
   D. Synthesis would terminate randomly, regardless of the nucleotide incorporated.
   E. The error would have no effect.
9. The polymerase chain reaction uses *Taq* polymerase rather than a DNA polymerase from *E. coli*, because *Taq* polymerase…
   A. introduces fewer errors during DNA synthesis.
   B. is heat-stable.
   C. can initiate DNA synthesis at a wider variety of sequences.
   D. can denature a double-stranded DNA template.
   E. is easier to obtain.
10. The polymerase chain reaction generates a fragment of a distinct size even when an intact chromosome is used as a template. What determines the boundaries of the amplified fragment?
    A. The concentration of one particular deoxynucleotide in the reaction.
    B. The duration of the elongation step in each cycle.
    C. The position of a termination sequence, which causes the *Taq* polymerase to fall off the template.
    D. The sites to which the primers anneal.
    E. The temperature of the elongation step in each cycle.

Applications

1. Two students in a microbiology class are arguing about the origins of biotechnology. One student argued that biotechnology started with the advent of genetic engineering. The other student disagreed, saying that biotechnology was as old as ancient civilization. What was the rationale for the argument by the second student?
2. A student wants to clone Gene X. On both sides of the DNA encoding Gene X are the recognition sequences for *Alu* I and *Bam* HI. Which enzyme would be easier to use for the cloning experiment and why?

Critical Thinking

1. Viruses are frequently used to deliver new genetic material into a cell. What special abilities of viruses make them well suited to carry out this function?
2. An effective DNA probe can sometimes be developed by knowing the amino acid sequence of the protein encoded by the gene. A student argued that this process is too time-consuming since it is necessary to determine the complete amino acid sequence in order to create the probe. Does the student have a valid argument? Why or why not?

Classification and Identification of Prokaryotes

In the early 1870s, the German botanist Ferdinand Cohn published several papers on bacterial classification. Cohn grouped microorganisms according to shape: spherical, short rods, elongated rods, and spirals. This classification scheme implied that bacteria are not plastic and flexible but maintain a constant shape. Thus, spherical organisms give rise to spherical organisms; they do not become rods following binary fission. At the same time, Cohn recognized that classification based solely on shapes was not adequate for categorizing all of the different bacteria. There were too many kinds and too few shapes.

The second major attempt at bacterial classification was initiated by Sigurd Orla-Jensen. Orla-Jensen's early training in Copenhagen was in the field of chemical engineering, but he soon became interested in microbiology and fermentations. In 1908, he proposed that bacteria be classified according to their physiological properties rather than their morphological properties. He considered organisms that gained their energy from inorganic sources as the most primitive.

A quarter of a century later, two Dutch microbiologists, Albert Kluyver and C. B. van Niel, proposed two classification systems that were based on presumed evolutionary relationships. They recognized a very serious problem in their classification scheme, however: There was no way to distinguish between "resemblance" and "relatedness." The fact that two prokaryotes look alike does not mean they are genetically related.

In 1970, Roger Stanier, a microbiologist at the University of California, Berkeley, pointed out that relationships could be determined by comparing either gene products, such as proteins and cell walls, or nucleotide sequences. At that time, leading microbiologists, including Stanier, assumed that all prokaryotes are basically similar. When the chemical compositions of a wide variety of prokaryotes were examined in detail, however, it was found that many had features that differed from those present in the intensively studied Escherichia coli, *considered a "typical" bacterium. These "unusual" features pertained to the chemical nature of the cell wall, cytoplasmic membrane, and ribosomal RNA.*

In the late 1970s, Carl Woese and his colleagues at the University of Illinois took advantage of nucleic acid technologies, which allowed investigators to determine the sequence of bases in ribosomal RNA. When they compared such sequences from a wide variety of organisms, it became clear that prokaryotes could be divided into two major groups that differ from one another as much as they differ from the eukaryotic cell. Thus, prokaryotes, once believed to be a part of the plant kingdom, have now been separated into two domains, the Archaea and the Bacteria. Each of these is on the same level as the Eukarya, which includes the animals, plants, and fungi; these are all eukaryotes.

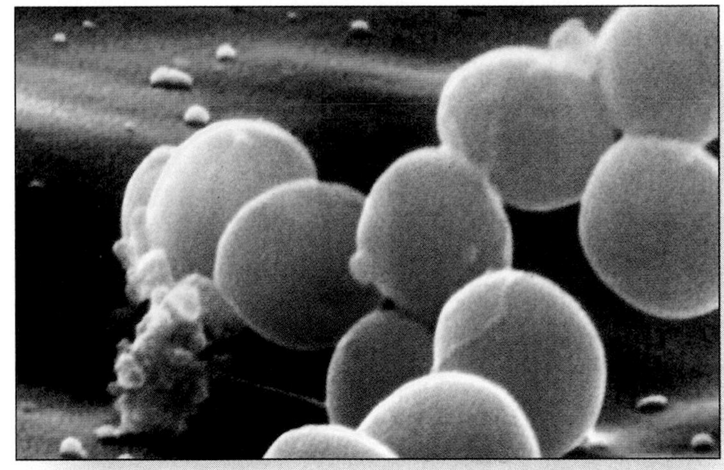

—*A Glimpse of History*

INFORMATION THAT IS LOGICALLY ORGANIZED IS easier to both retrieve and understand. Newspapers, for instance, do not scatter various subjects throughout the paper; rather they are divided into sections such as local news, sports, and entertainment. A large library would be extremely difficult to use if the multitude of books were not split into sections by subject matter. Likewise, scientists have divided living organisms into different groups, the better to understand the relationships among the species.

Take a moment and think about how you would group bacteria if you were to arrange a classification system. Would you group them according to shape? Or would it make more sense to group them according to their motility? Perhaps you would group them according to their medical significance. But then, how would you classify two similar bacteria that differ in their disease-causing potential? And how would you classify a newly identified organism of unknown medical significance?

Principles of Taxonomy

Taxonomy is the science that studies organisms in order to arrange them into groups; those organisms with similar properties are grouped together and separated from those that are different. Taxonomy can be viewed as three separate but interrelated areas:

- **Identification**—the process of characterizing organisms
- **Classification**—the process of arranging organisms into similar or related groups, primarily to provide easy identification and study
- **Nomenclature**—the system of assigning of names to organisms

Strategies Used to Identify Prokaryotes

In practical terms, identifying the genus and species of a prokaryote may be more important than understanding its genetic relationship to other microbes. For example, a food manufacturer is most interested in detecting the presence of microbial contaminants that can spoil a food product. In a clinical laboratory, it is critical to quickly identify the microbes that are isolated from patients so the best possible treatment can be given.

To characterize and identify microorganisms, a wide assortment of technologies is used including microscopic examination, culture characteristics, biochemical tests, and nucleic acid analysis. While each of these methods is best suited for detecting certain kinds of organisms in specific types of specimens, combinations of the tests may provide the most accurate identification. The number and type of tests used depends on the microbe to be detected or identified and the specimen being examined.

In a clinical laboratory, the patient's disease symptoms play an important role in identifying the infectious agent. For example, pneumonia in an otherwise healthy adult is typically caused by *Streptococcus pneumoniae*, an organism that is easily differentiated from others using a few specific tests. In contrast, diagnosing the cause of a wound infection is often more difficult, because many different microorganisms could be involved. Often, however, it is only necessary to rule out the presence of organisms known to cause a particular disease, rather than to conclusively identify each and every organism in the specimen. For example, a fecal specimen from a patient complaining of a diarrhea and fever would generally only be tested for the presence of specific organisms that cause those symptoms. ■ ***Streptococcus pneumoniae*, p. 564**

The various methods used to identify prokaryotes will be discussed in detail later in the chapter.

Strategies Used to Classify Prokaryotes

Understanding the evolutionary relatedness, or **phylogeny**, of prokaryotes is important in constructing a classification scheme that reflects the actual evolution and biology of these organisms. Such a scheme is more useful than one that simply groups organisms by arbitrary characteristics, because it is less prone to the bias of human perceptions. It also makes it easier to classify newly recognized organisms and allows scientists to make predictions, such as which genes are likely to be transferred between organisms.

Unfortunately, determining genetic relatedness among prokaryotes is more difficult than it is for plants and animals. Not only do prokaryotes have few differences in size and shape, they do not undergo sexual reproduction. In higher organisms such as plants and animals, the basic taxonomic unit, a **species**, is generally considered to be a group of morphologically similar organisms that are capable of interbreeding to produce fertile offspring. Obviously, it is not possible to apply these same criteria to prokaryotes, thus making classification problematic.

Historically, taxonomists have relied heavily on phenotypic attributes to classify prokaryotes; however, the development and application of molecular techniques such as nucleotide sequencing is finally making it possible to determine the genetic relatedness of microorganisms. These techniques used to classify prokaryotes based on phenotype and genotype will be discussed in detail later in the chapter. Regardless of the methods employed to assess relatedness, the objective of classification is to arrange organisms into structured categories that reflect the similarities of individuals within the groups. **■ genotype, p. 192 ■ phenotype, p. 192**

Taxonomic Hierarchies

Taxonomic classification categories are arranged in a hierarchical order, with the species being the basic unit. The species designation gives a formal taxonomic status to a group of related isolates or **strains**, which, in turn, permits their identification. Without classification, scientists and others would not be able to communicate about organisms with any degree of accuracy. Taxonomic categories include:

- **Species**—a group of related isolates or strains. Note that members of a species are not all identical; individual strains may vary in minor properties. The difficulty for the taxonomist is to decide how different two isolates must be in order to be classified as separate species rather than strains of the same species.
- **Genus**—a collection of related species.
- **Family**—a collection of similar genera. In prokaryotic nomenclature, the name of the family ends in the suffix *-aceae.*
- **Order**—a collection of similar families. In prokaryotic nomenclature, the name of the family ends in the suffix *-ales.*
- **Class**—a collection of similar orders. In prokaryotic nomenclature, the name of the family ends in the suffix *-ia.*
- **Phylum** or **Division**—a collection of similar classes.
- **Kingdom**—a collection of similar phyla or divisions. The number of different kingdoms varies according to the classification system used.
- **Domain**—a collection of similar kingdoms. The domain is a relatively new taxonomic category that reflects the characteristics of the cells that make up the organism.

Note, however, that microbiologists often group prokaryotes into informal categories rather than utilizing the higher taxonomic ranks such as order, class, and phylum. Examples of such informal groupings include the lactic acid bacteria, the anoxygenic phototrophs, the endospore-formers and the sulfate reducers. Organisms within these informal groupings share sim-

Table 10.1 Taxonomic Ranks of the Bacterium *Escherichia coli*

Formal Rank	Example
Domain	Bacteria
Phylum	Proteobacteria
Class	Gammaproteobacteria
Order	*Enterobacteriales*
Family	*Enterobacteriaceae*
Genus	*Escherichia*
Species	*coli*

ilar phenotypic and physiological characteristics, but may not be genetically related. ■ lactic acid bacteria, p. 275 ■ anoxygenic phototrophs, p. 276 ■ sulfate reducers, p. 272

An example of how a particular bacterial species is classified is shown in **table 10.1**. Note that the table intentionally omits the taxonomic category of kingdom. This is because the use of kingdoms within the Bacteria is still in a state of flux.

Classification Systems

Taxonomy is still an evolving discipline, with systems of classification that change over the years as new information is discovered. There is no such thing as an "official" classification system, and, as new ones are introduced, others fall into disfavor. The classification scheme currently favored by most microbiologists is the three-domain system. This designates all organisms as belonging to one of the three domains—Bacteria, Archaea, and Eukarya (**figure 10.1**). The system is based on the work of Carl Woese and colleagues who compared the sequences of nucleotide bases in ribosomal RNA from a wide variety of organisms. They showed that prokaryotes could be divided into two major groups that differ from one another as much as they do from the eukaryotic cell. The ribosomal RNA data are consistent with other observed differences between the Archaea and Bacteria, including the chemical compositions of their cell wall and cytoplasmic membrane (**table 10.2**). ■ Bacteria, p. 9 ■ Archaea, p. 9 ■ Eukarya, p. 10

Before the three-domain classification system was introduced, the most widely acccepted system was the five-kingdom system, proposed by R. H. Whittaker in 1969. The five kingdoms in this system are the Plantae, Animalia, Fungi, Protista (mostly single-celled eukaryotes) and Prokaryotae (**figure 10.2**). While the five-kingdom system recognizes the obvious morphological differences between plants and animals, it does not reflect the recent genetic insights of the ribosomal RNA data, which indicates that plants and animals are more closely related to each other than Archaea are to Bacteria.

The genetic insights provided by the ribosomal RNA techniques are causing significant upheaval in classification of microorganisms, as evidenced by the creation of the new taxonomic rank, domain. This new information is resulting in the need to shuffle some organisms within taxonomic categories, causing both confusion and controversy. In addition, the transition from the five-kingdom system to the three-domain system of classification causes significant change in itself. This is because the new system simultaneously elevates and splits what was once a single kingdom into two separate domains, permitting the designation of multiple new kingdoms within each domain.

Bergey's Manual of Systematic Bacteriology

While there is no "official" classification of prokaryotes, microbiologists generally rely on the reference text *Bergey's Manual of Systematic Bacteriology* as a guide. All known species are described here, including those that have not yet been cultivated. If the properties of a newly isolated organism do not agree with any description in *Bergey's Manual*, then presumably

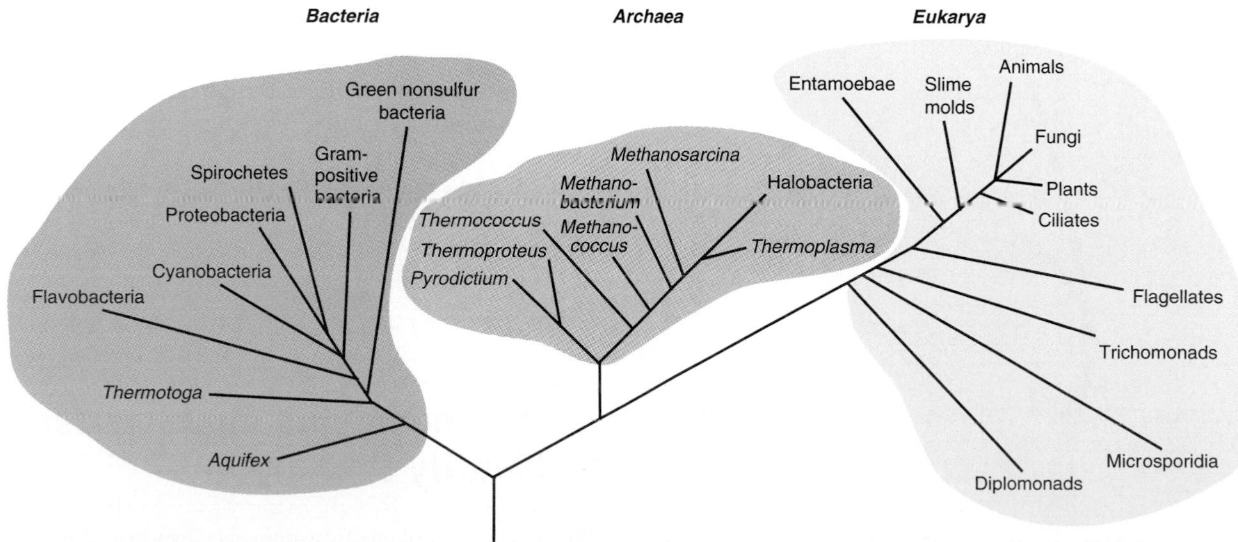

Figure 10.1 **The Three-Domain System of Classification** This classification system separates prokaryotic organisms into two domains—Bacteria and Archaea. The third domain, Eukarya, contains all organisms composed of eukaryotic cells. This system of classification is based on ribosomal RNA sequence data.

Table 10.2 A Comparison of Some Properties of the Three Domains—Archaea, Bacteria, and Eukarya

Cell Feature	Archaea	Bacteria	Eukarya
Peptidoglycan Cell Wall	No	Yes	No
Cytoplasmic Membrane Lipids	Hydrocarbons (not fatty acids) linked to glycerol by ether linkage	Fatty acids linked to glycerol by ester linkage	Fatty acids linked to glycerol by ester linkage
Ribosomes	70S	70S	80S
Presence of Introns	Sometimes	No	Yes
Membrane-Bound Nucleus	No	No	Yes

a new organism has been isolated. The newest edition of this comprehensive manual is being published in five volumes and classifies prokaryotes according to the most recent information on their genetic relatedness. In some cases, this classification differs substantially from that of the previous edition, which grouped organisms according to their phenotypic characteristics. The outline of the newest edition (the 2nd edition) is presented in Appendix IV.

In addition to containing descriptions of organisms, all five volumes contain information on the ecology, methods of enrichment, culture, and isolation of the organisms as well as methods for their maintenance and preservation. However, the heart of the work is a description of all of the prokaryotes and their groupings.

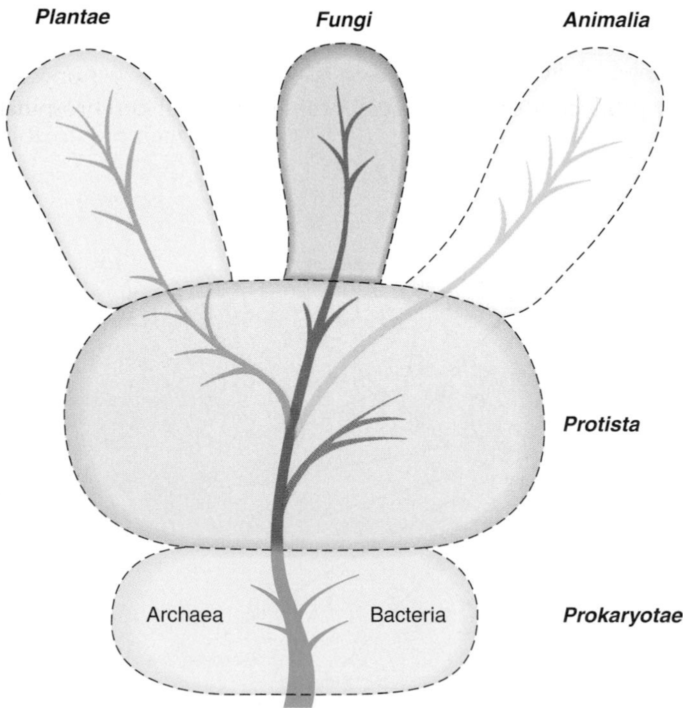

Figure 10.2 The Five-Kingdom System of Classification This classification system splits the eukaryotes into four different kingdoms—Plantae, Animalia, Fungi, and Protista (mostly single-celled eukaryotes); it groups the Archaea and Bacteria into a fifth kingdom, Prokaryotae.

Nomenclature

Bacteria are given names according to an official set of internationally recognized rules, the *International Code for the Nomenclature of Bacteria.* Bacterial names may originate from any language, but they must be given a Latin suffix. In some cases the name reflects a characteristic of an organism such as its habitat, but often bacteria are named in honor of a prominent researcher.

Just as classification is always in a state of flux, so is the assignment of names to taxonomic groups of bacteria. While revision of names is desirable from a scientific perspective, practically it is often a great source of confusion, particularly when the names of medically important bacteria are changed. To ease the transition of nomenclature changes, the former name is sometimes included in parentheses. For example, *Lactococcus lactis*, a bacterium that until recently was included in the genus *Streptococcus*, is often indicated as *Lactococcus* (*Streptococcus*) *lactis.*

MICROCHECK 10.1

Taxonomy consists of three interrelated areas: identification, classification, and nomenclature. In clinical laboratories, identifying the genus and species of an organism is more important than understanding its evolutionary relationship to other organisms.

- Why is it generally easier to determine the cause of pneumonia than it is to determine the cause of a wound infection?
- Why do microbiologists prefer the three-domain system of classification?
- Some biologists have been reluctant to accept the three-domain system. Why might this be?

Using Phenotypic Characteristics to Identify Prokaryotes

Phenotypic characteristics such as cell morphology, colony morphology, biochemical traits, and the presence of specific proteins can all be used in the process of identifying microorganisms. Most of these methods do not require sophisticated equipment and can easily be done anywhere in the world. Methods used to identify prokaryotes are summarized in **table 10.3**.

Microscopic Morphology

An important initial step in identifying a microorganism is to view it using a microscope to determine its size, shape, and staining characteristics. Microscopic examination gives information very quickly and is sometimes enough to make a presumptive identification.

Size and Shape

The size and shape of a microorganism can readily be determined by microscopically examining a wet mount. Based only on the size and shape, one can readily decide whether the organism in question is a prokaryote, fungus, or protozoan. In a clinical lab, this can sometimes provide all the information needed for diagnosis of certain infections. For example, a wet mount of vaginal secretions is routinely used to diagnose infections caused by yeast or by the protozoan *Trichomonas* (**figure 10.3**). A wet mount of stool is examined for the eggs of parasites when certain roundworms are suspected. The size, shape, and special features of the eggs are often sufficient to allow identification of the intestinal parasite. ■ **wet mount, p. 52** ■ **vaginal infections, p. 639** ■ **roundworms, p. 618**

Gram Stain

The Gram stain is a differential stain that distinguishes between Gram-positive and Gram-negative bacteria. This relatively rapid test narrows the possible identities of an organism by excluding numerous others and provides suggestive information that can be helpful in the identification process. ■ **Gram stain, p. 49**

In a clinical lab, the Gram stain of a specimen by itself is generally not sensitive or specific enough to diagnose the cause of most infections, but it is still an extremely useful tool. The clinician can see the Gram reaction, the shape and arrangement of the bacteria, and whether the organisms appear to be growing as a pure culture or with other bacteria and/or cells of the host. However, most medically important bacteria do not have distinctive shapes or staining characteristics and usually cannot be identified by Gram stain alone. For example, *Streptococcus pyogenes*, which causes strep throat, cannot be distinguished microscopically from the other streptococci that are part of the normal flora of the throat. A Gram stain of a stool specimen cannot distinguish *Salmonella* species from *E. coli*. These organisms may be isolated in pure culture and tested for their biochemical attributes to provide precise identification. ■ **strep throat, p. 553**

In certain cases, the Gram stain gives enough information to start appropriate antimicrobial therapy while awaiting more accurate identification. For example, if a urine sample from an otherwise healthy person reveals more than one Gram-negative rod per oil immersion field, the clinician will suspect a urinary tract infection caused by *E. coli*, the most common cause of such infections. Likewise, a Gram stain of sputum showing numerous white blood cells and Gram-positive encapsulated diplococci is highly suggestive of *Streptococcus pneumoniae*, an organism that causes pneumonia (**figure 10.4a**). In certain other cases, the result of a Gram stain is enough for accurate diagnosis. For example, the presence of Gram-negative diplococci clustered in white blood cells in a sample of a urethral secretion from a male is considered diagnostic for gonorrhea, the sexually transmitted disease caused by *Neisseria gonorrhoeae* (figure 10.4b). This diagnosis can be made because *N. gonorrhoeae* is the only Gram-negative diplococcus found inhabiting the normally sterile urethra of a male. ■ **pneumonia, p. 564** ■ **gonorrhea, p. 644**

Special Stains

Certain microorganisms have unique characteristics that can be detected with special staining procedures. For example, *Filobasidiella* (*Cryptococcus*) *neoformans* is one of the few types of yeast that produce a capsule. Thus, a capsule stain of cerebrospinal

Table 10.3 Methods Used to Identify Prokaryotes

Method	Comments
Phenotypic Characteristics	Most of these methods do not require sophisticated equipment and can easily be done anywhere in the world.
Microscopic morphology	Size, shape, and staining characteristics such as Gram stain can give suggestive information as to the identity of the organism. Further testing, however, is needed to confirm the identification.
Metabolic differences	Culture characteristics can give suggestive information. A battery of biochemical tests can be used to confirm the identification.
Serology	Proteins and polysaccharides that make up a bacterium are sometimes characteristic enough to be considered identifying markers. These can be detected using specific antibodies.
Fatty acid analysis	Cellular fatty acid composition can be used as an identifying marker and is analyzed by gas chromatography.
Genotypic Characteristics	These methods are increasingly being used to identify microorganisms.
DNA hybridization	Probes can be used to identify bacteria grown in culture. In some cases, the method is sensitive enough to detect the organism directly in a specimen.
Amplifying DNA using PCR	Even an organism that occurs in very low numbers in a mixed culture can be identified.
Sequencing rRNA genes	This method requires amplifying, cloning, and then sequencing rRNA genes, but it can be used to identify unculturable organisms.

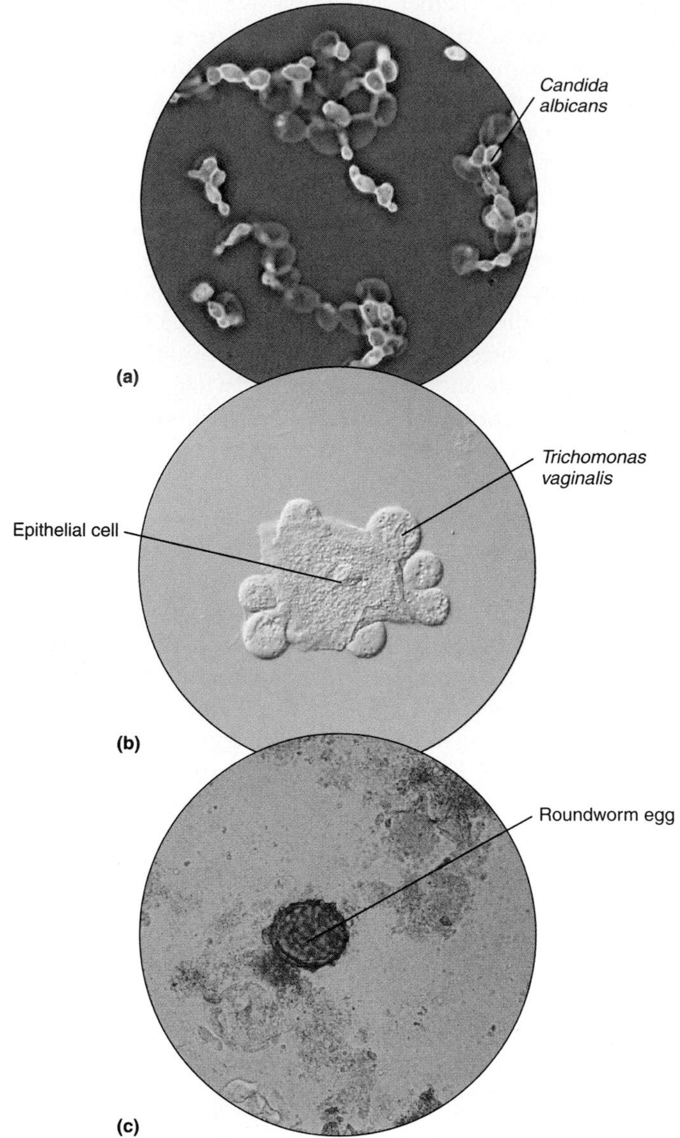

Figure 10.3 Wet Mounts of Clinical Specimens **(a)** Vaginal secretions containing yeast (*Candida albicans*, 410×); **(b)** *Trichomonas vaginalis* attached to squamous epithelium, as viewed through a Nomarski microscope (400×); and **(c)** roundworm (*Ascaris*) eggs in a stool (400×).

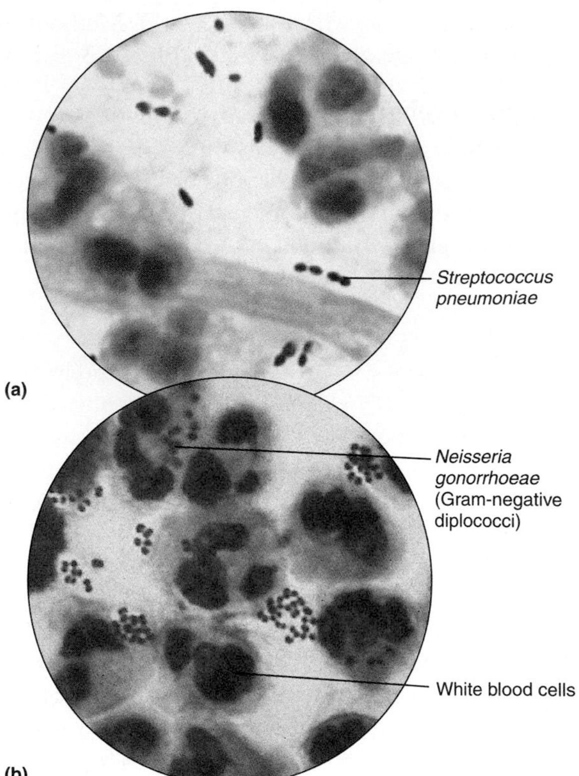

Figure 10.4 Gram Stains of Clinical Specimens **(a)** Sputum showing Gram-positive *Streptococcus pneumoniae* and **(b)** male urethra secretions showing Gram-negative *Neisseria gonorrhoeae* inside white blood cells.

fluid that shows the presence of encapsulated yeast is diagnostic for cryptococcal meningitis. Members of the genus *Mycobacterium* are some of the few microorganisms that are acid-fast. If a patient has symptoms of tuberculosis, then an acid-fast stain will be done on a sample of their sputum to determine whether *Mycobacterium tuberculosis* can be detected. ■ **capsule stain p. 52** ■ **acid-fast stain, p. 51**

Metabolic Differences

The identification of most bacteria relies on analyzing their metabolic capabilities such as the types of sugars utilized or the end products produced. In some cases these characteristics are revealed by the growth and colony morphology on cultivation media, but most often they are demonstrated using biochemical tests. ■ **growth of colonies, p. 107**

Culture Characteristics

Microorganisms that can be grown in pure culture are the easiest to identify, because it is possible to obtain high numbers of a single type of microorganism. Even the colony morphology can give initial clues to the identity of the organism. For example, colonies of streptococci are generally fairly small relative to many other bacteria such as staphylococci. Colonies of *Serratia marcescens* are red when incubated at 22°C due to the production of a pigment. *Pseudomonas aeruginosa* often produces a soluble greenish pigment, which discolors the growth medium. In addition, cultures of *P. aeruginosa* have a distinct fruity odor. ■ ***Pseudomonas aeruginosa* pigment production, p. 281**

The use of selective and differential media in the isolation process can provide additional information that helps identify an organism. For example, if a soil sample is plated onto medium that lacks a nitrogen source and is then incubated aerobically, any resulting colonies are likely members of the genus *Azotobacter*. The ability to fix nitrogen under aerobic conditions is an identifying characteristic of these bacteria. ■ ***Azotobacter*, p. 284** ■ **nitrogen fixation, p. 100** ■ **selective media, p. 103** ■ **differential media, p. 103**

In a clinical lab, where rapid but accurate diagnosis is essential, specimens are plated onto media specially designed to provide important clues as to the identity of the disease-causing organism. For example, a specimen taken by swabbing the throat of a patient complaining of a sore throat is inoculated onto blood agar, a nutritionally rich medium containing red blood cells. This differential medium enables the detection of the characteristic β-hemolytic colonies that are typical of *Streptococcus pyogenes*. Urine collected from a patient suspected of having a urinary tract infection is plated onto MacConkey agar, which is both selective and differential. MacConkey agar has bile salts, which inhibit the growth of most non-intestinal organisms, and lactose along with a pH indicator, which differentiates lactose-fermenting organisms. *E. coli*, the most common cause of urinary tract infections, forms characteristic red colonies on MacConkey agar due to its ability to ferment lactose. Other bacteria can also grow and ferment lactose on this medium, however, so colony appearance alone is not enough to conclusively identify *E. coli*. ■ **blood agar, p. 102** ■ **urinary tract infection, p. 635** ■ **MacConkey agar, p. 103**

Biochemical Tests

Growth characteristics on culture media can narrow down the number of possible identities of an organism, but biochemical tests are generally necessary for a more conclusive identification. One of the simplest of these is to assay for the enzyme catalase (**figure 10.5a**). Nearly all bacteria that grow in the presence of oxygen are catalase positive. Important exceptions are the lactic acid bacteria, which include members of the genus *Streptococcus*. Thus, if a throat culture yields β-hemolytic colonies but further testing reveals they are all catalase positive, then *Streptococcus pyogenes* has been ruled out. ■ **catalase, p. 98** ■ ***Streptococcus pyogenes*, p. 553**

Most biochemical tests rely on a pH indicator or chemical reaction that results in a color change when a compound is degraded. For example, to test for the ability of an organism to ferment a given sugar, a broth medium containing that sugar and a pH indicator is employed. Fermentation of the sugar results in acid production, which lowers the pH, resulting in a color change from pink to yellow; an inverted tube traps any gas that is simultaneously produced (figure 10.5b). A medium designed to detect the enzyme **urease**, an enzyme that degrades urea to produce carbon dioxide and ammonia, utilizes a different pH indicator that turns bright pink in alkaline conditions (figure 10.5c). The characteristics of these and other important biochemical tests are summarized in **table 10.4**.

The basic strategy for identifying bacteria based on biochemical tests relies on the use of a **dichotomous key**, which is essentially a flow chart of tests that give either a positive or negative result (**figure 10.6**). Because each test often requires an incubation period, however, it would be too time-consuming to proceed one step at a time. In addition, relying on a single biochemical test at each step could lead to misidentification. For example, if a strain that normally gives a positive result for a certain test lost the ability to produce a key enzyme, it would instead produce a negative result. Therefore, simultaneously inoculating a battery of different tests identifies the organism faster and more conclusively.

In certain cases, biochemical testing can be done without culturing the organism. *Helicobacter pylori*, the cause of most stomach ulcers, can be detected using the **breath test**, which assays for the presence of urease. The patient drinks a solution containing urea that has been labeled with an isotope of carbon. If *H. pylori* is present, its urease breaks down the urea, releasing labeled carbon dioxide, which escapes through the airway. Sev-

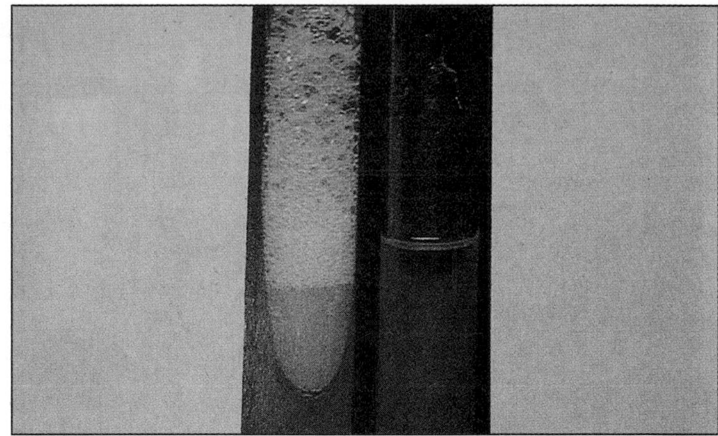

(a)

(b)

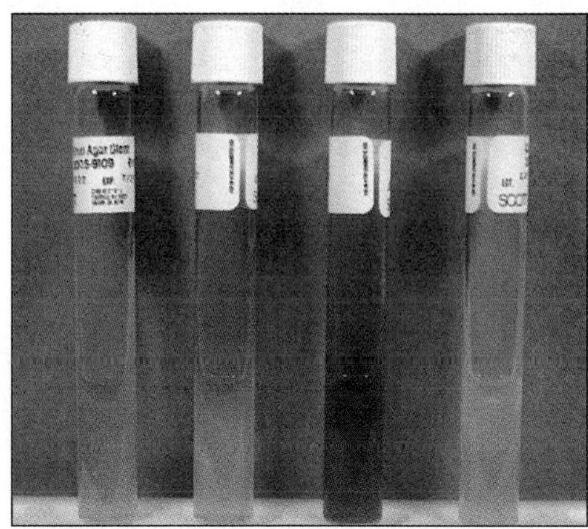

(c)

Figure 10.5 Biochemical Tests **(a)** Catalase production. Bacteria that produce catalase break down hydrogen peroxide (H_2O_2) to release oxygen gas (O_2) which causes the bubbling. ($2\ H_2O_2 \longrightarrow 2\ H_2O + O_2$). A negative catalase test is shown in the right tube. **(b)** Sugar fermentation. The tubes on the left and right show acid (yellow color) and gas. The center tube shows no color change, indicating that the sugar was not utilized. **(c)** Urease production. Breakdown of urea releases ammonia, which turns the pH indicator pink.

Table 10.4 Characteristics of Some Important Biochemical Tests

Biochemical Test	Principle of the Test	Positive Reaction
Catalase	Detects the activity of the enzyme catalase, which causes the breakdown of hydrogen peroxide to produce oxygen and water.	Bubbles.
Citrate	Determines whether or not citrate can be used as a sole carbon source.	Growth, which is usually accompanied by the color change of a pH indicator.
Gelatinase	Detects enzymatic breakdown of gelatin to polypeptides.	The solid gelatin is converted to liquid.
Hydrogen Sulfide Production	Detects the enzymatic conversion of the sulfur-containing amino acids to H_2S.	A black precipitate forms due to the reaction of H_2S with iron salts in the medium.
Indole	Detects the enzymatic removal of the amino group removed from tryptophan.	The product, indole, reacts with a chemical reagent that is added, turning the reagent a deep red color.
Lysine Decarboxylase	Detects the enzymatic removal of the carboxyl group from lysine.	The medium becomes more alkaline, causing a pH indicator to change color.
Methyl Red	Detects mixed acids, the characteristic end products of a particular fermentation pathway. ■ **mixed acids, p. 155**	The medium becomes acidic (pH <4.5); a red color develops upon the addition of a pH indicator.
Oxidase	Detects the activity of cytochrome c, a component of the electron transport chain of specific organisms. ■ **cytochrome c, p. 151**	A dark color develops upon the addition of a specific reagent.
Phenylalanine Deaminase	Detects the enzyme that removes amino group from phenylalanine.	The product of the reaction, phenylpyruvic acid, reacts with ferric chloride to give the medium a green color.
Sugar Fermentation	Detects the acidity resulting from fermentation of the sugar incorporated into the medium. Also detects gas production.	The color of a pH indicator incorporated into the medium changes if acid if produced. An inverted tube traps any gas that is made.
Urease	Detects the enzyme that degrades urea to produce carbon dioxide and ammonia.	The medium becomes alkaline, causing a pH indicator to change color.
Voges-Proskauer	Detects acetoin, an intermediate of the fermentation pathway that leads to the production of a 2, 3- butandiol. ■ **2, 3- butanediol, p. 155**	A red color develops upon addition of chemicals that detect acetoin.

eral hours after drinking the solution, the patient exhales into a balloon. The expired air is then tested for labeled carbon dioxide. This test is less invasive and, consequently, much cheaper and faster than the stomach biopsy that would otherwise need to be performed to culture the organism. ■ ***Helicobacter pylori*, p. 592** ■ **isotope, p. 25**

Commercial Modifications of Traditional Biochemical Tests

Several less labor-intensive commercial modifications of traditional biochemical tests are available (**figure 10.7**). For example, the API™ system utilizes a strip holding a series of cupules that contain dehydrated media. A liquid suspension of the isolated test bacterium is inoculated into each of the compartments, thus rehydrating the media. Because the formulations of the media are similar in composition to those used in traditional tests, the positive results give rise to similar color changes. After a 16-hour incubation of the inoculated test strip, the results are read manually. The pattern of results is converted to a numerical score, which can then be entered into a computer to identify the organism. A similar system is the Enterotube™, a tube that has small compartments each containing a different type of medium. One end of a metal rod that runs through the tube is used to touch a bacterial colony. When the rod is withdrawn it inoculates each of the compartments.

Highly automated systems also are available. The Vitek™ system uses a miniature card that contains multiple wells with different formulations of dehydrated media. A computer then reads the growth pattern in the wells after a relatively short incubation period.

Serology

The proteins and polysaccharides that make up a bacterium are sometimes characteristic enough to be considered identifying markers. The most useful of these are the molecules that make up surface structures including the cell wall, glycocalyx, flagella, and pili. For example, some species of *Streptococcus* contain a unique carbohydrate molecule as part of their cell wall that can be used to distinguish them from other species. These carbohydrates, as well as any other distinct protein or polysaccharide, can be detected using techniques that rely on the specificity of interaction between antibodies and antigens. Methods that exploit these interactions are called **serology** and will be discussed in more detail in later chapters. Some serological tests, such as those used to confirm the identity of *S. pyogenes*, are quite simple and rapid. ■ **serology, p. 414** ■ **antibodies, p. 390** ■ **antigens, p. 390**

Fatty Acid Analysis

Bacteria differ in the type and relative quantity of fatty acids that make up their membranes; thus, the cellular fatty acid composition can be used as an identifying marker. In the case of Gram-negative bacteria, the fatty acids are contained in both the cytoplasmic and outer membranes. Gram-positive bacteria, however, lack an outer membrane; the cytoplasmic membrane is the source of their fatty acids. To analyze their fatty acid composition, bacterial cells are grown under standardized conditions. The cells are then chemically treated with sodium hydroxide and methanol to release the fatty acids and to convert those acids to their more volatile methyl ester form. The resulting fatty acid methylated esters can then be separated and analyzed using gas chromatography. By comparing the pattern of peaks, or **chromatogram** to those of known species, an isolate can be identified (**figure 10.8**).

MICROCHECK 10.2

The size, shape, and staining characteristics of a microorganism, all of which can be viewed using a microscope, yield important clues to its identity. Conclusive identification generally requires a battery of biochemical tests that assay for compounds indicating the presence of specific biochemical pathways. The proteins and polysaccharides that make up a bacterium are sometimes unique

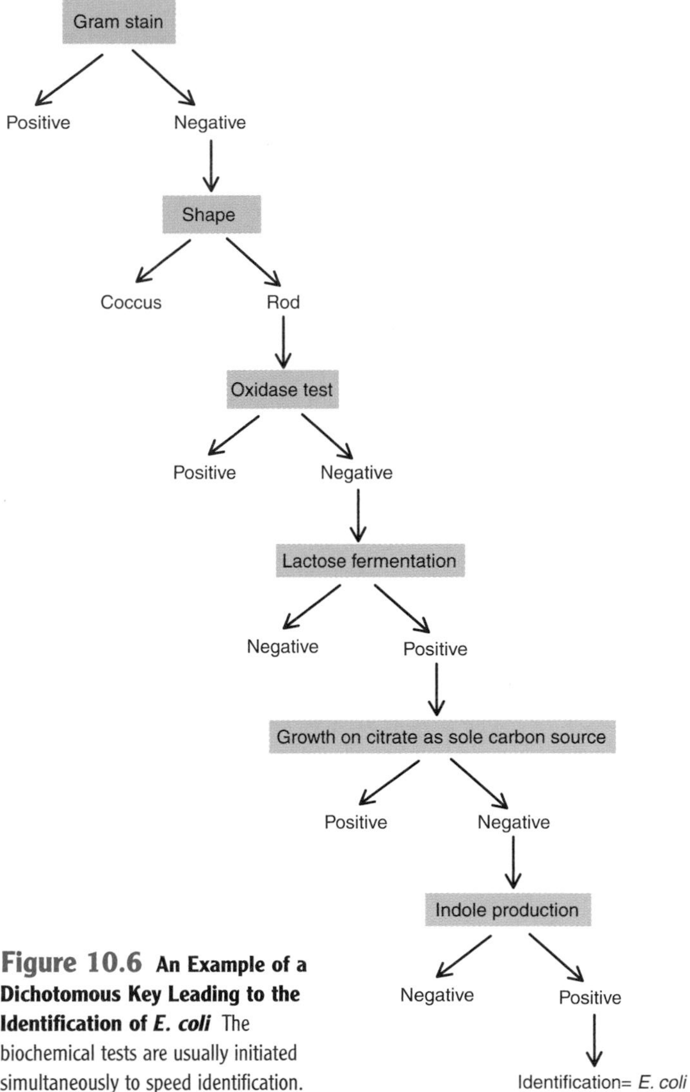

Figure 10.6 An Example of a Dichotomous Key Leading to the Identification of *E. coli* The biochemical tests are usually initiated simultaneously to speed identification.

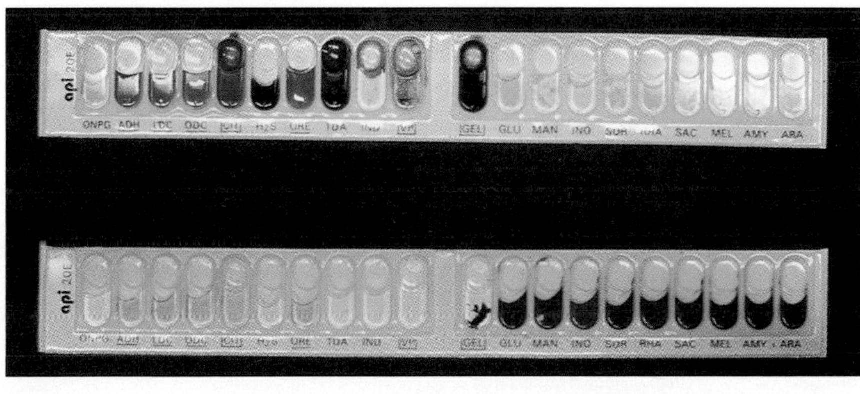

(a)

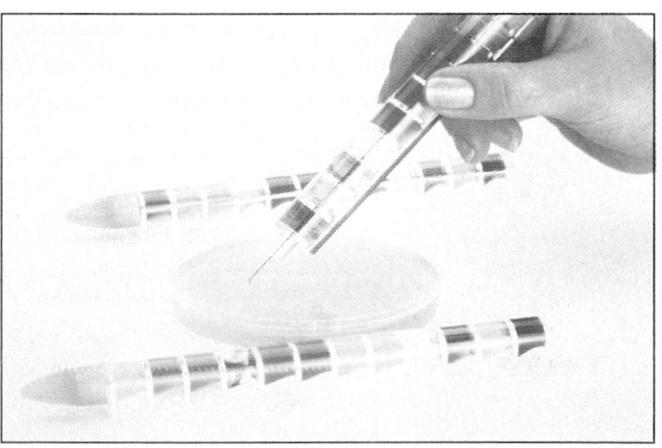

(b)

Figure 10.7 Commercial Modifications of Traditional Biochemical Tests These methods are less labor-intensive than traditional tests. **(a)** An API™ test strip. Each of the cupules contains dehydrated medium similar in formulation to the traditional tests. A liquid suspension of the isolated test bacterium is added to each compartment; after incubation, the results are read manually. The upper panel shows positive results, and the lower panel shows negative results. **(b)** An Enterotube.™ Each compartment contains a different type of medium. One end of a metal rod that runs through the tube is used to touch a bacterial colony. When the rod is withdrawn it inoculates each of the compartments. After incubation, the results are read manually.

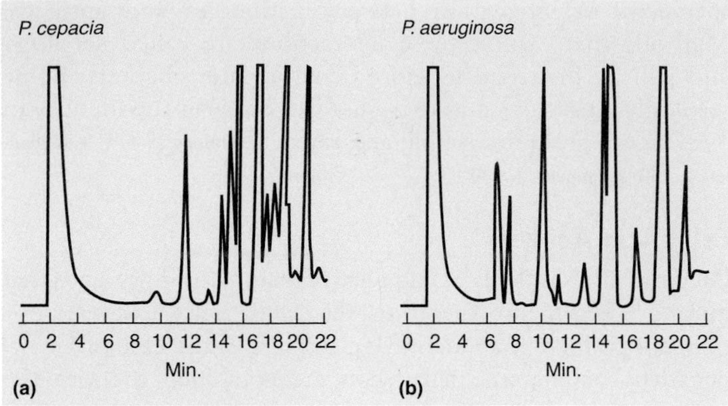

P. cepacia
P. aeruginosa
0 2 4 6 8 10 12 14 16 18 20 22
Min.
(a)
0 2 4 6 8 10 12 14 16 18 20 22
Min.
(b)

Figure 10.8 Chromatogram of the Fatty Acid Profiles Each peak represents a different fatty acid; the size of the peak correlates with the relative amount of that particular molecule. **(a)** A chromatogram of the fatty acids of *Pseudomonas cepacia* and **(b)** a chromatogram of the fatty acids of a closely related species, *Pseudomonas aeruginosa*.

enough to be considered identifying markers. Cellular fatty acid composition can be used as an identifying characteristic.

- How does MacConkey agar help to identify the cause of a urinary tract infection?
- Describe two methods that can be used to test for the enzyme urease.
- Why must a sample contain many microorganisms for any to be viewed by microscopic examination?

Using Genotypic Characteristics to Identify Prokaryotes

Increasingly, genotypic characteristics are being used to identify microorganisms, particularly those that are difficult to cultivate. DNA probes and polymerase chain reaction (PCR) can both be used to detect nucleotide sequences that are unique to a given organism. DNA sequencing has made identification possible of organisms that cannot be grown in culture. ■ **DNA probe, p. 232** ■ **polymerase chain reaction, p. 237** ■ **sequencing, p. 235**

Nucleic Acid Probes to Detect Specific DNA Sequences

A nucleic acid probe can be used to locate a unique nucleotide sequence that identifies a particular known species of microorganism (**figure 10.9**). The probe is a single-stranded piece of nucleic acid, usually DNA, that has been labeled with a detectable tag, such as a radioisotope. It is complementary to the sequence of interest. In most cases, probe technology is not sensitive enough to detect the relatively few bacterial cells of a given species that might be present in a clinical specimen such as a throat swab. Instead, that specimen must first be cultured on an agar plate so that each cell can multiply to form a colony containing well over a million cells. Each probe is very specific and, therefore, will only confirm or rule out the presence of one particular organism.

In a few cases, hybridization assays have been developed that are sensitive enough to detect organisms directly in the clinical specimen. For example, cervical swabs can be screened for the presence of the sexually transmitted pathogens *Neisseria gonorrhoeae* and *Chlamydia trachomatis.* Throat swabs can be tested for the presence of *Streptococcus pyogenes.* These types of hybridization tests usually rely on a preliminary step that amplifies the DNA in the sample *in vitro*, thus increasing its concentration. **Table 10.5** lists some of the medically important bacteria that can now be identified using probe technology.

Amplifying Specific DNA Sequences Using the Polymerase Chain Reaction

The polymerase chain reaction (PCR) can be used to amplify a specific nucleotide sequence present in nearly any environment (see figure 9.24). This includes DNA in samples such as body fluids, soil, food, and water. The technique can be used to detect organisms that are present in extremely small numbers as well as those that cannot be grown in culture.

To use PCR to detect a microbe of interest, a sample is first treated to release and denature the DNA. Specific primers,

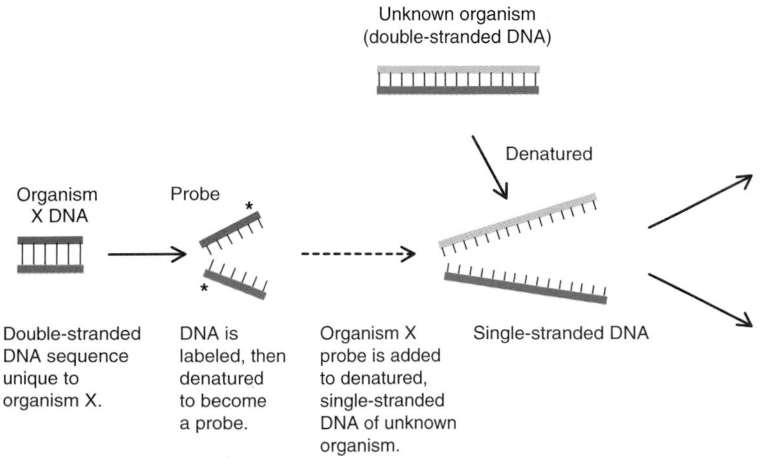

Unknown organism
(double-stranded DNA)
Denatured
Organism
X DNA
Probe
Double-stranded
DNA sequence
unique to
organism X.
DNA is
labeled, then
denatured
to become
a probe.
Organism X
probe is added
to denatured,
single-stranded
DNA of unknown
organism.
Single-stranded DNA

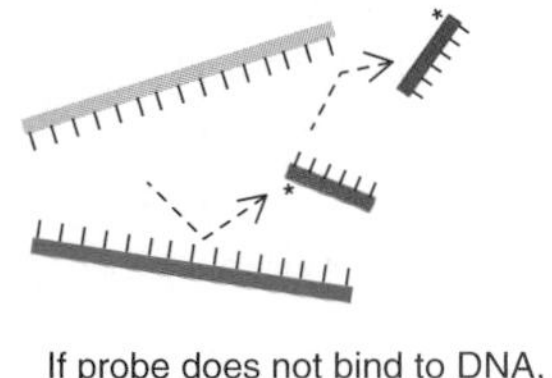

If probe does not bind to DNA,
then unknown organism
is not organism X.

If probe binds to DNA,
then unknown organism
is organism X.

Figure 10.9 Nucleic Acid Probes to Detect Specific DNA Sequences The probe, which is a single-stranded piece of nucleic acid that has been labeled with a detectable marker, is used to locate a unique nucleotide sequence that identifies a particular known species of microorganism.

Table 10.5 Representative Bacteria that Can Be Identified Using Probe Technology

Bacteria that Can Be Identified in Culture Using a Probe	
Campylobacter species	■ p. 606
Enterococcus species	■ p. 636
Haemophilus influenza	■ p. 558
Listeria monocytogenes	■ p. 671
Neisseria gonorrhoeae	■ p. 644
Staphylococcus aureus	■ p. 694
Streptococcus agalactiae (Group B strep)	■ p. 668
Streptococcus pneumoniae	■ p. 564
Streptococcus pyogenes (Group A strep)	■ p. 553
Bacteria that Can Be Identified Directly from a Specimen Using a Probe	
Chlamydia trachomatis	■ p. 646
Mycobacterium tuberculosis	■ p. 569
Neisseria gonorrhoeae	■ p. 644
Streptococcus pyogenes (Group A strep)	■ p. 553

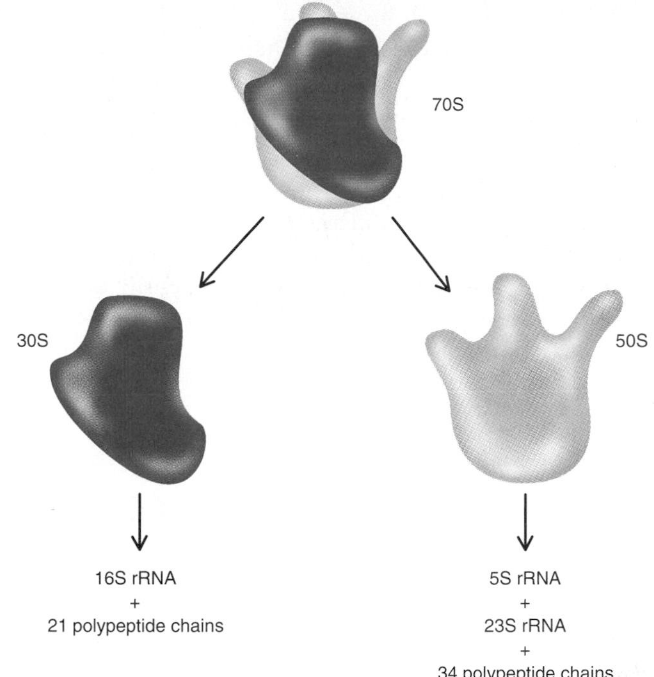

Figure 10.10 Ribosomal RNA The 70S ribosome of prokaryotes has three types of rRNA: 5S, 16S, and 23S.

Taq polymerase, and nucleotides are then added to the denatured DNA forming the components of the PCR reaction (see figure 9.25). Some information about the nucleotide sequence of the organism must be known in order to select the appropriate primers. After approximately 30 cycles of PCR, the DNA region flanked by the primers will have been amplified approximately a billion-fold (see figure 9.26). In most cases, this results in a sufficient quantity for the amplified fragment to be readily visible as a discrete band on an ethidium bromide–stained agarose gel. Alternatively, a DNA probe can be used to detect the amplified DNA. ■ ethidium bromide, p. 234 ■ gel electrophoresis, p. 233 ■ *Taq* polymerase, p. 238

Sequencing Ribosomal RNA Genes

The nucleotide sequence of ribosomal RNA (rRNA) may be used to identify prokaryotes, particularly those that are difficult or currently impossible to grow in culture. The prokaryotic 70S ribosome, which plays an indispensable role in protein synthesis, is composed of protein and three different rRNAs (5S, 16S, and 23S) **(figure 10.10)**. Because of their highly constrained and essential function, the nucleotide sequence changes that can occur in the rRNAs, yet still allow the ribosome to operate, are limited. This is why they have proved so useful in microbial classification and, more recently, identification. While earlier methods relied on determining the sequence of the rRNA molecule itself, newer techniques sequence **rDNA**, the DNA that encodes rRNA.

Of the different rRNAs, the 16S molecule has proved most useful in taxonomy because of its moderate size (approximately 1,500 nucleotides). The 5S molecule lacks the critical amount of information because of its small size (120 nucleotides), whereas the larger size of the 23S molecule (approximately 3,000 nucleotides) has made it more difficult to sequence in the past.

Some regions of the 16S rRNA molecule are virtually the same in all prokaryotes, whereas others are quite variable. It is the variable regions that are used to identify an organism. Once the nucleotide sequence of that region has been determined, it can be compared with the 16S rDNA sequences of known organisms by searching extensive computerized databases.

Using rDNA to Identify Unculturable Organisms

In any environment, including soil, water, and the human body, a multitude of organisms exist that cannot yet be grown in culture. Some scientists estimate that every gram of fertile soil contains 4,000 to 5,000 different species of prokaryotes, the vast majority of which have not been identified. With current technologies that enable amplification of specific portions of DNA, followed by the cloning and then sequencing of those fragments, it is possible not only to detect such organisms, but also to obtain information about their identity.

One of the first examples of using molecular biology to identify an unculturable organism was the characterization of the causative agent of a rare illness called Whipple's disease. This was done by using PCR to amplify bacterial 16S rDNA from intestinal tissue of patients who had symptoms of Whipple's disease. That DNA was then cloned and sequenced. The nucleotide sequence suggested that the causative agent was an actinomycete unrelated to any of those previously identified. It was given the name *Tropheryma whippelii.* Even though it had never been grown in culture, a specific probe was then developed that can detect it in intestinal tissue.

MICROCHECK 10.3

A microorganism can be identified by using a probe or PCR to detect a nucleotide sequence that is unique to that particular organism. Ribosomal RNA genes can be sequenced to identify an organism that cannot be grown in culture.

- When using a probe to identify an organism, why is it necessary to have some idea as to the organism's identity?
- How can ribosomal DNA be used to identify bacteria that cannot be grown in culture?
- Why are molecular methods especially useful when bacteria are difficult to cultivate?

Characterizing Strain Differences

In some situations it is helpful to distinguish among different strains of a given species. This is particularly useful when only certain strains cause disease. For example, of the multitude of *E. coli* strains, relatively few have been shown to cause intestinal disease. This is because only a few strains have the virulence factors, such as toxin production, necessary to cause disease. Thus, finding *E. coli* in feces is of no significance, unless the strain is identified as one that causes disease. Detecting strain differences is also helpful in tracing the spread of a given organism. For example, *Listeria monocytogenes* is a common environmental organism that can cause foodborne illness. The fact that it is possible to distinguish among strains of this organism made it possible to trace the source of a recent outbreak of listeriosis. When contaminated hot dogs were implicated, an immediate recall of the product took place, halting the spread of the disease. ■ ***Listeria monocytogenes*, p. 671**

The methods used to characterize different strains are summarized in **table 10.6**.

Biochemical Typing

Biochemical tests are used to identify various species of bacteria, but they can also be used to distinguish strains. If the biochemical variation is uncommon, it can be used for tracing the source of certain disease outbreaks. A strain that has a characteristic biochemical pattern is called a **biovar** or a **biotype**. For example, a biochemical variant of *Vibrio cholerae* called Eltor is increasingly being implicated in outbreaks of cholera around the world. Because this biovar can be readily distinguished, its spread can be traced. ■ ***Vibrio cholerae*, p. 599**

Serological Typing

Proteins and carbohydrates that vary among strains can be used to differentiate strains. For example, *E. coli* and other Gram-negative bacteria vary in the antigenic structure of certain parts of the lipopolysaccharide portion of the cell wall, the **O antigen** (see figure 11.16). The composition of the flagella, the **H antigen**, can also vary. Thus, the strain designation of *E. coli* O157:H7 refers to the structure of its lipopolysaccharide and its flagella. A strain that differs serologically from other strains is sometimes called a **serovar** or a **serotype**. ■ **lipopolysaccharide, p. 63** ■ ***E. coli* O157:H7, p. 603**

Genomic Typing

Molecular methods can be used to detect genomic variations that characterize certain strains. In some cases, these differences include genes that encode toxins or other proteins related to disease. For example, *E. coli* O157:H7 owes its virulence, in part, to the production of a toxin that is not encoded by most other strains of *E. coli*, which are normal intestinal inhabitants. The toxin gene can be detected using a probe that consists of a specific nucleotide sequence unique to that gene. Probes can also be used to detect genes that encode resistance to the antibiotic penicillin. Infections caused by organisms that have this gene must be treated with an alternative antimicrobial medication.

Table 10.6 Summary of Methods Used to Characterize Different Strains

Method	Characteristics
Biochemical Typing	Biochemical tests are most commonly used to identify various species of bacteria, but in some cases they can be used to distinguish strains. A strain that has a characteristic biochemical pattern is called a biovar or a biotype.
Serological Typing	Proteins and carbohydrates that vary among strains can be used to differentiate strains. A strain that has a characteristic serological type is called a serovar or a serotype.
Genomic Typing	Molecular methods can be used to detect restriction fragment length polymorphisms (RFLPs).
Pulsed-field gel electrophoresis	The genomic DNA is digested with an enzyme that cuts the chromosome into 10 to 20 large fragments. These are then separated using a special modification of gel electrophoresis. The pattern of sizes can be determined directly by looking at the stained gel.
Ribotyping	The genomic DNA is digested into many small fragments. These are then separated by gel electrophoresis and transferred to a membrane, which is then probed with labeled rDNA, resulting in a distinct pattern of bands.
Phage Typing	Strains of a given species sometimes differ in their susceptibility to various types of bacteriophage.
Antibiogram	Antibiotic susceptibility patterns can be used to characterize strains.

Perspective 10.1 Tracing the Source of an Outbreak of Foodborne Disease

In October 1996, physicians at the Children's Hospital and Medical Center in Seattle, Washington, noticed a slight increase in cases of children with severe bloody diarrhea. They quickly recognized that the children were infected with *Escherichia coli* O157:H7, a strain that can cause a fatal disease. Within days, the source of the outbreak was traced to contaminated unpasteurized apple juice, prompting the manufacturer to immediately recall that and related products from all stores in their seven-state distribution area. The prompt action by physicians, health officials and the juice manufacturer ended the outbreak and averted a potential widespread tragedy. ■ ***E. coli* O157:H7, p. 603**

While the sequence of events that led to the recognition and cessation of the *E. coli* O157:H7 outbreak may sound quite simple, they are actually very complex. For one thing, most strains of *E. coli* are normal inhabitants of the intestine and can be found in nearly any stool sample. How was this particular strain separated and distinguished from the hundreds of *E. coli* strains that do not cause diarrheal disease? It is also true that sporadic cases of *E. coli* O157:H7 infections occur regularly, originating from unrelated sources. How was it recognized that these cases were connected?

To identify *E. coli* O157:H7 in a stool specimen, the sample is plated onto a special agar medium designed to distinguish it from those that typically inhabit the large intestine. One such medium is sorbitol-MacConkey, a modified version of MacConkey agar in which the lactose is replaced with the carbohydrate sorbitol. On this medium, most *E. coli* O157:H7 isolates are colorless because they do not ferment sorbitol. In contrast, common strains of *E. coli* ferment the carbohydrate, giving rise to pink colonies. Serology is then used to determine if the colorless colonies are serotype O157; those that test positive are then generally tested to confirm they are serotype H7.

The next task is to determine whether or not two isolates of *E. coli* O157:H7 originated from the same source. DNA is extracted and purified from each isolate and is then digested with restriction enzymes. Pulsed-field gel electrophoresis is generally used to compare the resulting restriction fragment length polymorphism (RFLP) patterns of the isolates. Those that have identical patterns are presumed to have originated from the same source. The patients from whom those isolates originated can then be questioned to determine their likely point of contact with the disease-causing organism. Culture methods are then used to try to isolate the organism from the suspected source. If that attempt is successful, the RFLP pattern of that isolate is then compared with those of the related cases.

Subtle differences in DNA sequences can be used to distinguish among strains that are phenotypically identical. This is particularly important in tracing epidemics of foodborne illness in which a single source of contaminated food is shipped to several states. The ability to link geographically distant cases can enable public health officials to trace the source of the epidemic, leading to the recall of the implicated product and preventing further cases of disease. One method of doing this is to compare the pattern of fragment sizes produced when the same restriction enzyme is used to digest DNA from each organism. When the lengths of the restriction fragments vary among organisms, the fragments are said to be polymorphic. The different patterns of fragment sizes obtained by digesting DNA with restriction enzymes are called **restriction fragment length polymorphisms (RFLPs)** (see figure 9.19). Two isolates of the same species that have different RFLPs are considered different strains. Two isolates that have an identical RFLP may be the same strain. It cannot be concluded with absolute certainty, however, that they are indeed the same strain. ■ **restriction enzymes, p. 225** ■ **restriction fragment length polymorphisms, p. 235**

Common methods used to look for restriction fragment length polymorphisms include:

- **Pulsed-field gel electrophoresis** This method uses a restriction enzyme that cuts infrequently along with a special type of gel electrophoresis that can separate very large fragments. For example, bacterial genomic DNA can be cut with the enzyme *Not*I, which recognizes a sequence of 8 bases, to yield approximately 10 to 20 large fragments. Pulsed-field gel electrophoresis is used to separate these fragments, which are generally over 100,000 base pairs in length. The resulting bands can be visualized by staining the gel with ethidium bromide **(figure 10.11)**.
- **Ribotyping** This method uses a restriction enzyme that cuts genomic DNA into many small fragments. Southern blot hybridization is then done using a probe that hybridizes to only those fragments that have sequences encoding ribosomal RNA (see figure 9.17). Because bacteria generally have several rRNA genes, the probe hybridizes to several different restriction fragments, the pattern of which varies among strains **(figure 10.12)**. ■ **Southern blot hybridization, p. 233**

To facilitate the tracking of foodborne disease outbreaks, the Centers for Disease Control recently established the **National Molecular Subtyping Network for Foodborne Disease Surveillance (PulseNet)**, which catalogues the RFLPs of certain pathogenic organisms. Laboratories from around the country can submit RFLP patterns to a computer database and quickly receive

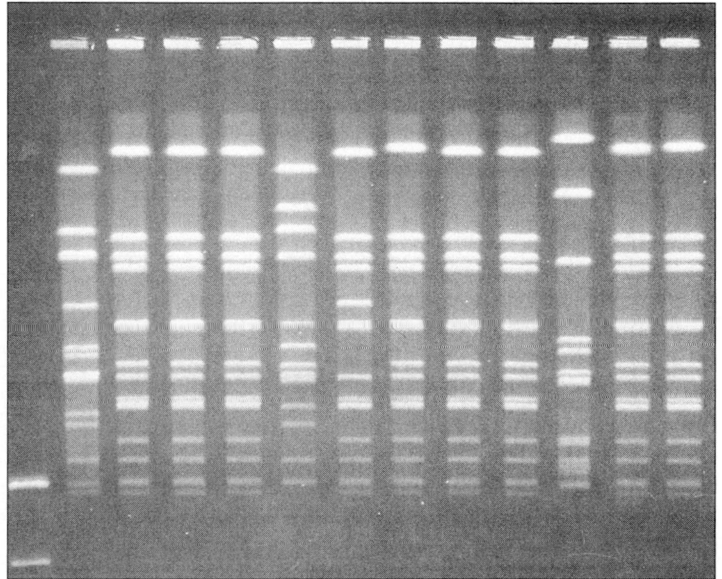

Figure 10.11 Detecting Restriction Fragment Length Polymorphisms (RFLPs) Using Pulsed-Field Gel Electrophoresis Genomic DNA is digested with a restriction enzyme that cuts infrequently; the resulting fragments are then separated by pulsed-field gel electrophoresis.

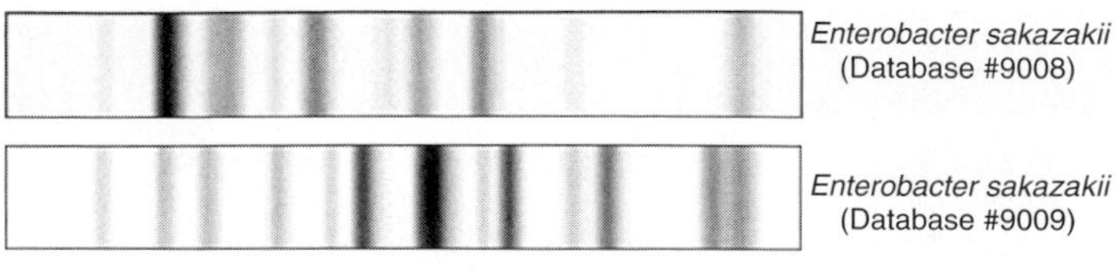

Figure 10.12 Ribotyping Genomic DNA is digested with a restriction enzyme that cuts frequently. Southern blot hybridization is then done using a probe that hybridizes only to those fragments that have sequences encoding ribosomal RNA. Because bacteria generally have several rRNA genes, the probe hybridizes to several different restriction fragments, the pattern of which varies among strains.

information about other isolates showing the same patterns. Using this database, multistate foodborne disease outbreaks can more readily be recognized and traced.

Phage Typing

Strains of a given species sometimes differ in their susceptibility to various types of **bacteriophages**. A bacteriophage, or **phage**, is a virus that infects and multiplies within bacteria and usually lyses the infected cells; each type of phage has a limited host range. Lysis of the infected cell releases more phages, which in turn infect neighboring cells. The susceptibility of an organism to a particular type of phage can be readily demonstrated in the laboratory. First, a culture of the test organism is inoculated into melted, cooled nutrient agar and poured onto the surface of an agar plate, thus creating a uniform layer of cells. Then drops of different types of bacteriophages are carefully placed on the surface of the agar. During incubation, the bacteria multiply, forming a visible haze of cells. A clear area will form at each spot where bacteriophage was added, however, if the organism is susceptible to the type of phage. The patterns of clearing indicate the susceptibility of the test organism to different phages and it is these patterns that are compared to determine strain differences (**figure 10.13**). Bacteriophage typing has now largely been replaced by molecular methods that detect genomic differences, but it is still a useful tool for laboratories that lack sophisticated equipment. ■ **bacteriophage, p. 323** ■ **host range, p. 335**

Antibiograms

Antibiotic susceptibility patterns, or **antibiograms**, can also be used to distinguish among different strains. Again, this method has now largely been replaced by molecular techniques. To determine the antibiogram, a culture is uniformly inoculated onto the surface of a nutritional agar medium. Paper discs, each of which has been impregnated with a given antibiotic, are then placed on the surface of the agar. During incubation, the organism will multiply to form a visible film of cells. A clear area, indicating lack of growth, will form around each antibiotic disc that inhibits the organism. Different strains will have different patterns of clearing (**figure 10.14**). ■ **antibiotic, p. 496**

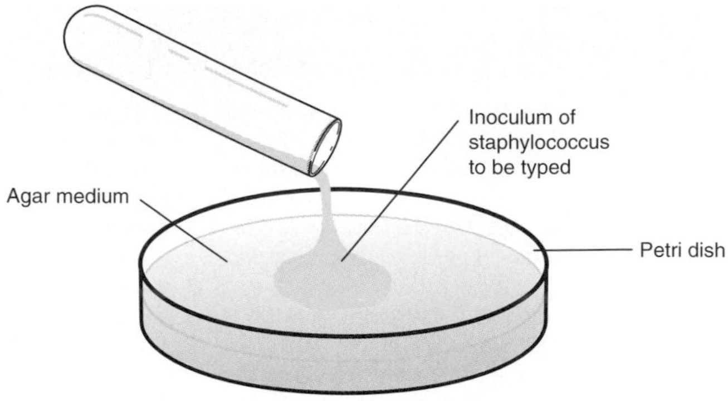

(a) An inoculum of *S. aureus* is spread over the surface of agar medium.

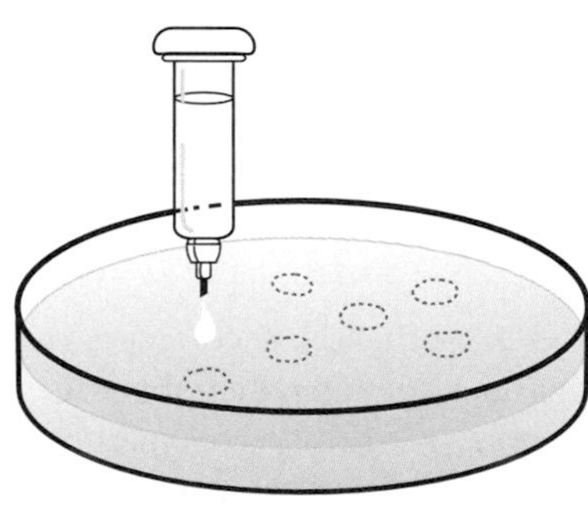

(b) 31 different bacteriophage suspensions are deposited in a fixed pattern.

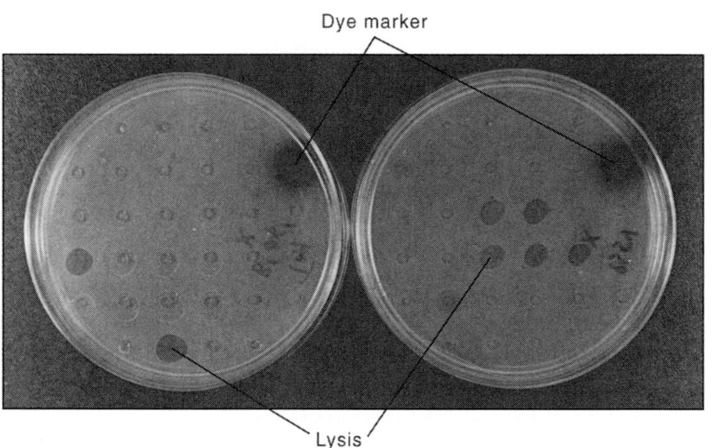

(c) After incubation, different patterns of lysis are seen with different strains of *S. aureus*.

Figure 10.13 Phage typing

MICROCHECK 10.4

Strains of a given species may differ in phenotypic attributes such as biochemical capabilities, protein and polysaccharide components, susceptibility to bacteriophages, and sensitivity to antimicrobial drugs. Molecular techniques can be used to detect subtle genomic differences between strains that are phenotypically identical.

- Explain the difference between a biotype and serotype.

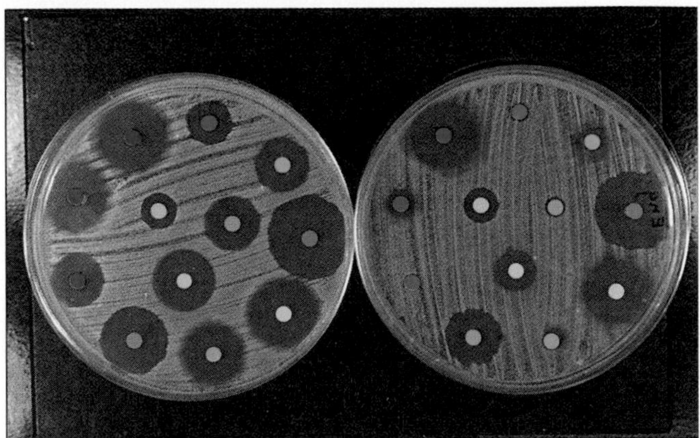

Figure 10.14 An Antibiogram In this example, 12 different antimicrobial drugs incorporated in paper discs have been placed on two plates containing different cultures of *Staphylococcus aureus*. Clear areas represent zones of inhibited growth. The different patterns of clearing indicate that these are two different strains of *S. aureus*.

- Describe the technique of ribotyping.
- What observations would you expect when *E. coli* O157:H7 is cultured on a medium containing sorbitol and a pH indicator?

Difficulties in Classifying Prokaryotes

Fossilized **stromatolites**, coral-like mats of filamentous microorganisms, suggest that prokaryotes have existed on the earth for at least 3.5 billion years **(figure 10.15)**. More recent data based on isotopic analysis of rocks in Greenland suggest that prokaryotes existed on this planet 3.85 billion years ago. Because of the relatively few sizes and shapes of bacteria, however, such fossilized remains do little to help identify or understand these ancient organisms. It thus remains difficult to place the diverse types of prokaryotes into their proper place in the evolution of living beings.

Historically, prokaryotes have been grouped according to phenotypic attributes such as size and shape, staining characteristics, and metabolic capabilities. Using this system, a species can loosely be defined as group of organisms that share many properties and differ significantly from other groups. While the use of phenotypic characteristics is a convenient approach to prokaryotic taxonomy, there are several drawbacks. For example, observable differences may be due to only a few gene products, and a single mutation resulting in a nonfunctional enzyme can dramatically alter that phenotypic property. In addition, organisms that are phenotypically similar may, in fact, be only distantly related. Conversely, those that appear dissimilar may be closely related.

Newer molecular approaches circumvent some of the problems associated with phenotypic classification schemes while also giving greater insights into evolutionary relatedness of microorganisms. The more similar the nucleotide sequences, the more closely related are two organisms. Differences in DNA sequences can also be used to determine the point in time at which two organisms diverged from a common ancestor. This is because random mutations cause sequences to change over time. Thus, the more time that has elapsed since two organisms diverged, the greater the difference in the sequences of their DNA. Sequence differences can therefore be viewed as a form of a chronological measurement, an evolutionary chronometer. The fact prokaryotes can transfer DNA to other species, called **horizontal** or **lateral gene transfer**, can complicate such insights. ■ **phenotype, p. 192**

Some of the methods used to classify prokaryotes by determining their relatedness are summarized in **table 10.7**.

(a)

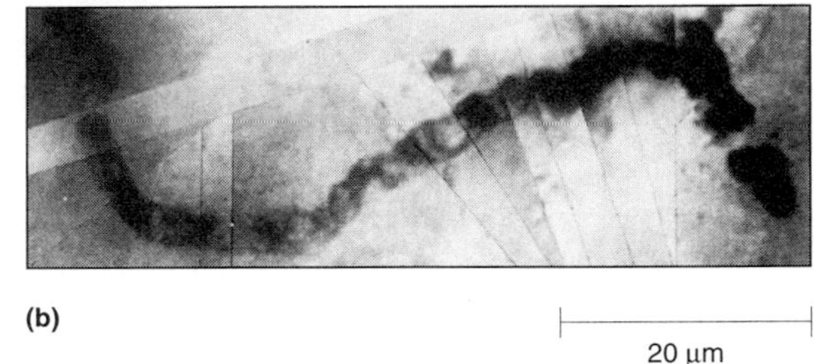

(b)

Figure 10.15 Stromatolites and Fossilized Bacteria (a) Stromatolites in Australia. **(b)** Fossils of bacteria in sections of rock 3.5 billion years old.

MICROCHECK 10.5

Historically prokaryotes have been grouped according to phenotypic attributes. Molecular approaches give greater insights into the relatedness of microorganisms.

- Describe one drawback of using phenotypic characteristics to classify bacteria.
- Some scientists that work on bacterial classification are called "lumpers" and others are called "splitters." What do these terms likely indicate?

Table 10.7 Methods Used to Determine the Relatedness of Different Prokaryotes for Purposes of Classification

Methods	Comments
Phenotypic Characteristics	Traditionally, relatedness of different bacteria has been decided by comparing properties such as ability to degrade lactose and the presence of flagella. These characteristics, however, do not necessarily reflect the evolutionary relatedness of organisms.
Numerical taxonomy	A large number of phenotypic traits are examined, resulting in the generation of a similarity coefficient.
Molecular Characteristics	Differences in DNA sequences can be used to determine the point in time at which two organisms diverged from a common ancestor.
DNA base composition	Determining the G + C content offers a crude comparison of genomes. Organisms with identical G + C contents can be entirely unrelated, however.
DNA hybridization	The extent of nucleotide sequence similarity can be determined by measuring how completely single strands of their DNA will hybridize to one another.
Comparing the sequences of 16S rDNA	This technique has revolutionized classification. Certain regions of the 16S rDNA can be used to determine distant relatedness of diverse organisms; other regions can be used to determine more recent divergence.

Using Phenotypic Characteristics to Classify Prokaryotes

Traditionally, relatedness of different prokaryotes has been decided by comparing their phenotypic characteristics. **Numerical taxonomy** determines relatedness based on the percentage of characteristics that two groups have in common. In this approach, the investigator conducts a large number of tests to determine whether certain features are present or absent in an organism. These tests include such characteristics as the ability to degrade lactose, the ability to form endospores, and the presence of flagella. Some characteristics, such as motility and spore-forming ability, depend on many genes and therefore are more fundamental and important properties of an organism than are a number of biochemical traits. For example, only a few genes are required to degrade lactose into its monosaccharide components. If a large enough number of different traits are examined, however, then there is no need to consider some features as more important. Thus, one of the advantages of numerical taxonomy is that the results are unbiased by subjective judgments of the investigator as to which characteristics are more important. Different investigators should arrive at the same conclusion.

The final result of classification by numerical taxonomy is expressed in terms of a **similarity coefficient.** This is calculated by dividing the number of common positive results from two stains by the number of characteristics tested that were not negative for both; it is expressed as a percentage (**figure 10.16**). The greater the number of positive characteristics that two organisms have in common, the more closely they are related. After numerous characteristics are examined in a variety of strains, a computer program is used to calculate a similarity coefficient for each pair of strains and to construct a similarity chart. On the basis of this chart, the strains can be ordered in such a way that those that have

Strain Number	1	2	3	4	5	6	7
1	100						
2	5	100					
3	10	95	100				
4	0	90	95	100			
5	80	15	35	15	100		
6	70	25	40	10	80	100	
7	95	10	20	10	90	75	100

(a)

		A				B		
	Strain Number	1	7	5	6	3	2	4
A	1	100						
A	7	95	100					
A	5	80	90	100				
A	6	70	75	80	100			
B	3	10	20	35	40	100		
B	2	5	10	15	25	95	100	
B	4	0	10	15	10	95	90	100

(b)

Figure 10.16 Numerical Taxonomy Seven strains of bacteria were tested for 100 different traits, each test producing a positive or negative result. Each strain was then compared with the other strains by determining the similarity coefficient, the percentage of the total positive characteristics tested held in common in each of the strains. Note that each strain holds all characteristics (100%) in common with itself. In **(a)**, the strains are arranged randomly; in **(b)**, they are arranged so that the organisms with similar properties are grouped together. Note that the seven organisms fall into two unrelated groups, designated here as A and B, each of which is probably composed of a single species.

more than a 70 to 80% similarity coefficient are classified as a single species, while other more distinct groups are classified as separate species and perhaps even different genera.

MICROCHECK 10.6

Numerical taxonomy uses a battery of phenotypic characteristics to classify bacteria.

- Two bacteria that form endospores are more likely to be closely related to each other than are two that ferment lactose. Why would this be so?
- How would you classify two isolates that have a similarity coefficient of 90%?

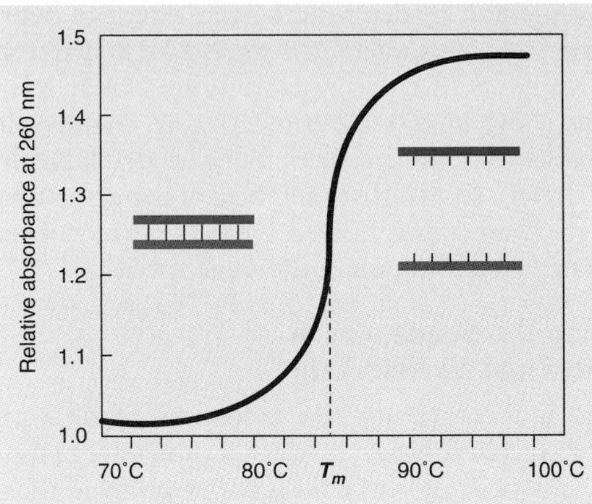

Figure 10.17 A DNA Melting Curve The absorbance (relative absorbance at 260 nm) rapidly increases as double-stranded DNA denatures, or melts. The T_m is the temperature at which 50% of the DNA has melted; it reflects the GC content.

Using Genotypic Characteristics to Classify Prokaryotes

The most recent methods of classification are based on the comparison of the nucleotide sequences of the DNA of different organisms. Because the complete genomes of relatively few microorganisms have been determined, other, less precise methods are used to rapidly compare and contrast general similarities. Some of these methods crudely examine total DNA and are used to broadly sort organisms as to their evolutionary relatedness; others more closely inspect specific nucleotide segments as a mechanism of fine-tuning the evolutionary scale.

DNA Base Ratio (G + C Content)

One way to roughly compare the genomes of different bacteria is to determine their base ratio, which is the relative portion of adenine (A), thymine (T), guanine (G), and cytosine (C). Because of nucleotide base-pairing rules, the number of molecules of G in double-stranded DNA always equals the number of molecules of C. Likewise, the number of molecules of T equals that of A. The base ratio of an organism is usually expressed as the percent of guanine plus cytosine, termed the **G + C content**, or more simply, the **GC content**. If an organism has a GC content of 60%, then the remaining 40% must be the AT content.

The GC content is often measured by determining the temperature at which the double-stranded DNA denatures, or **melts**. Because three hydrogen bonds occur between G-C base pairs, and only two hydrogen bonds between A-T, DNA that has a high GC content melts at a higher temperature. The temperature at which double-stranded DNA melts can readily be determined by monitoring the absorbance of a solution of DNA as it is heated. The absorbance rapidly increases as double-stranded DNA denatures (**figure 10.17**).

The GC content varies widely among different kinds of bacteria, with numbers ranging from about 22% to 78% (**table 10.8**). Organisms that are related by other criteria have DNA base compositions that are similar or identical. Thus, if the GC content of two organisms differs by more than a small percent, they cannot be closely related. For example, although *Proteus*, *Escherichia*, and *Pseudomonas* are all Gram-negative, rod-shaped organisms, they cannot be closely related because their GC contents are so different. A similarity of base composition, however, does not necessarily mean that the organisms are related, since many arrangements of the bases are possible. For example, although the bacterium *Bacillus subtilis* and humans both have 40% GC in their DNA, they are obviously vastly different organisms. The genome size and actual nucleotide sequence differ greatly. The wide range of GC content in some groups, such as *Bacillus* and *Pseudomonas*, suggests that many organisms in these groups are unrelated to one another and that their members likely will be classified into other genera as more is learned about them.

DNA Hybridization

The extent of nucleotide sequence similarity can be determined by measuring how completely single strands of their DNA will hybridize to one another. Just as two complementary strands of DNA from one organism will base-pair, or **anneal**, so will the similar DNA of a different organism. The extent of hybridization reflects the degree of similarity. DNA from organisms that share many of the same sequences will hybridize more completely than DNA from those that do not.
■ **DNA hybridization, p. 232**

Table 10.8 Range of G + C Content of Species in Selected Genera

Bacteria	% G + C
Bacillus	32–62
Chlamydia	41–44
Proteus	39–42
Escherichia	50–53
Pseudomonas	58–70

One method of determining the extent of hybridization utilizes the steps illustrated in **figure 10.18**. Extensive DNA similarity suggests that two organisms are closely related. Two strains that show at least 70% similarity are often considered to be members of the same species. Surprisingly, DNA hybridization studies have shown that members of the genus *Shigella* and *E. coli*, which are quite different based on biochemical tests, should actually be grouped in the same species.

Comparing the Sequences of 16S Ribosomal Nucleic Acid

Analyzing and comparing the sequences of ribosomal RNA (rRNA) has revolutionized the classification of organisms. In the 1970s, Carl Woese and his colleagues determined the sequence of 16S ribosomal RNA of more than 400 organisms, discovering characteristic sequences, called **signature sequences**, that are almost always found in particular groups. Based on these sequences as well as other properties, they proposed the three-domain system of classification, which is now believed to reflect the primary lines of evolutionary descent. Since that time, the genes encoding the 16S rRNA of thousands of prokaryotes have been sequenced. To study the phylogeny of eukaryotes, 18S rRNA is used, which is analogous in function to the 16S prokaryotic structure. In either case, it is generally the rDNA, which encodes the rRNA, that is sequenced.

Long after organisms have diverged so much that they appear to be unrelated by nucleic acid hybridization, portions of their rDNAs are still similar. Changes in these highly conserved regions occur very slowly over time and are thus useful for determining even distant relationships of diverse organisms. At the same time, certain regions of rDNAs are relatively variable. Comparing these sequences can be used to determine more

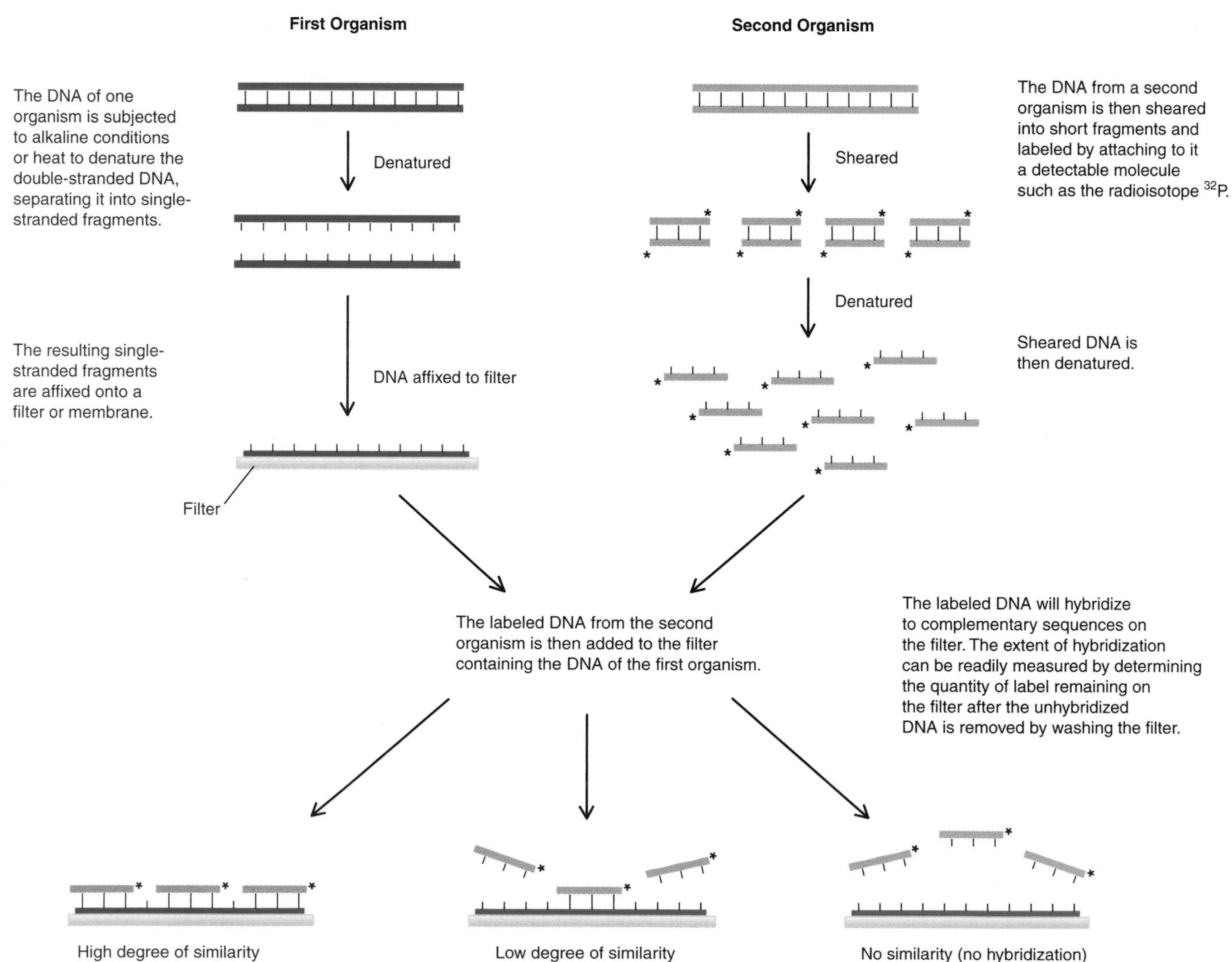

Figure 10.18 Using DNA Hybridization to Assess Relatedness

recent divergence. Thus, depending on the region compared, the sequence of rDNA can be used to assess distant as well as close relationships between organisms.

Determining Evolutionary Relatedness

Determining the base sequence of 16S rDNA is the most accurate and reliable procedure yet available for identifying evolutionary relationships of organisms. It is better than DNA hybridization, because it is far more quantitative. Organisms in which DNA hybridization can barely be detected are related, but how closely is not certain. Comparing the sequence of their 16S ribosomal RNA genes can give an accurate estimate. Some organisms that appear to be quite different based on their phenotypic characteristics are actually related. For example, the mycoplasmas, cell wall-deficient bacteria, are related to members of the genus *Clostridium*, anaerobic spore-forming rods. As another example, some photosynthetic bacteria are included in a group with many nonphotosynthetic species.

A genomic analysis such as 16S ribosomal RNA or DNA sequencing enables one to more accurately construct a **phylogenetic tree**. These trees are somewhat like a family tree, tracing the evolutionary heritage of organisms. Each line, or **branch,** of the tree represents the evolutionary distance between two species (**figure 10.19**). Individual species are represented as **nodes**. An **external node**, one that includes a species name, represents an organism that still exists. In contrast, an **internal node,** a branch point, represents an ancestor to today's organisms. Ancient prokaryotes, those that branch at an early point in evolution, are sometimes called **deeply branching** to reflect their position in the phylogenetic tree.

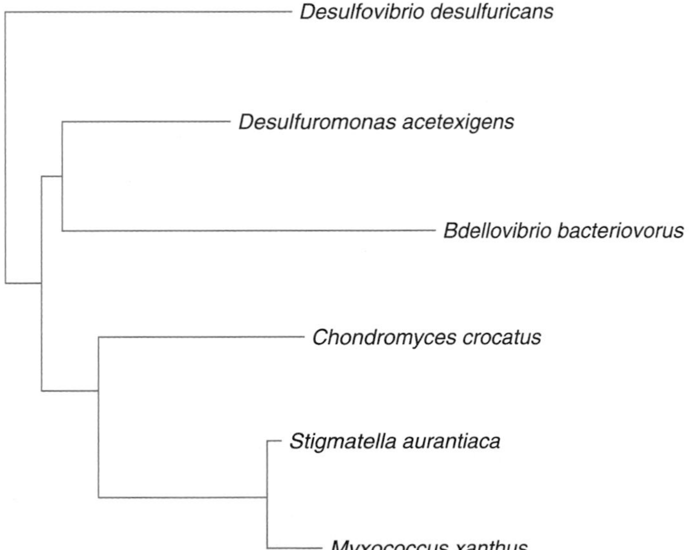

Figure 10.19 A Phylogenetic Tree Each branch represents the evolutionary distance between two species. An external node, marked with a species name, represents an organism that still exists. An internal node, a branch point, represents an ancestor to today's organisms.

MICROCHECK 10.7

The most recent methods of prokaryotic classification are based on the comparison of the nucleotide sequence of the DNA of different organisms. Differences in DNA sequences can to used to determine the point in time at which two organisms diverged from a common ancestor. The base-pair ratio is a crude comparison of genomic differences. DNA hybridization measures the extent of nucleotide similarity. Analyzing and comparing the sequences of ribosomal RNA have revolutionized the classification of organisms; these have been used to construct a phylogenetic tree.

- Explain why two organisms that have identical GC contents might be completely unrelated.
- Explain why 16S rDNA has proved so useful in determining evolutionary relatedness.
- Why would it be easier to sequence rDNA than rRNA?

FUTURE CHALLENGES

Tangled Branches in the Phylogenetic Tree

While 16S rDNA sequencing has given remarkable new insight into the evolutionary relatedness of organisms, total genome sequencing promises to reveal myriad additional information. In some cases this will serve to clarify the evolutionary picture, providing solid evidence that two organisms are, in fact, closely related. Genomic sequencing, however, can also give information that conflicts with that of rDNA data, creating an uncomfortable confusion. Already, examples exist that challenge the current rDNA phylogenetic tree. For example, sequences of some genes for highly conserved proteins show a closer relationship between some members of the Archaea and Bacteria than their rDNA sequences suggest. This conflicting data might be accounted for by evidence suggesting that horizontal DNA transfer has occurred even between distantly related organisms. Due to gene transfer, an organism from one domain may contain genes acquired from an organism in a different domain. As a result, early hopes that rDNA data could help construct an accurate evolutionary tree may, in some cases, be tempered by conflicting genomic data.

The vast amount of genetic knowledge being obtained is affecting bacterial nomenclature. As more genomes are sequenced, the evolutionary relatedness among organisms will become apparent. Nomenclature is bound to change because present-day classification is based on shared properties and not evolutionary relationships. Already, some species of what were formerly included in the genus Streptococcus *have been removed to create two relatively new genera,* Enterococcus *and* Lactococcus. *The genera* Pseudomonas *and* Bacillus *will likely be similarly subdivided. Meanwhile, other bacteria will be moved from one existing genus to another. All of these changes will create a potential source of confusion for the scientist and layperson alike.*

SUMMARY

Principles of Taxonomy

1. **Taxonomy** consists of three interrelated areas: **identification**, **classification**, and **nomenclature**.

Strategies Used to Identify Prokaryotes

1. To characterize and identify microorganisms, a wide assortment of technologies is used, including microscopic examination, cultural characteristics, biochemical tests, and nucleic acid analysis.
2. In a clinical laboratory, the patient's disease symptoms play an important role in identifying the infectious agent.

Strategies Used to Classify Prokaryotes

1. Taxonomic classification categories are arranged in a hierarchical order, with the **species** being the basic unit.
2. Taxonomic categories include **species**, **genus**, **order**, **class**, **phylum** (or division), **kingdom**, and **domain**. Individual **strains** within a species vary in minor properties.
3. Microbiologists favor emphasizing the domains rather than using kingdoms. (Figures 10.1, 10.2 and Table 10.2)
4. Detailed descriptions of taxonomic groups of prokaryotes are contained in the reference text *Bergey's Manual of Systematic Bacteriology.*

Nomenclature

1. Prokaryotes are given names according to an official set of rules.

Using Phenotypic Characteristics to Identify Prokaryotes (Table 10.4)

Microscopic Morphology

1. The size, shape, and staining characteristics of a microorganism, all of which can be viewed using a microscope, yield important clues as to its identity.
2. The size and shape of a microorganism can readily be determined by examining a **wet mount**. (Figure 10.3)
3. The Gram stain is a differential stain that distinguishes between Gram-positive and Gram-negative bacteria. (Figure 10.4)
4. Certain microorganisms have unique identifying characteristics that can be detected with special staining procedures.

Metabolic Differences

1. The identification of most bacteria relies on analyzing their metabolic capabilities.
2. The use of selective and differential media in the isolation process can provide information that helps identify an organism.
3. Most biochemical tests rely on a pH indicator or chemical reaction that shows a color change when a compound is degraded. (Figure 10.5 and Table 10.3)
4. The basic strategy for identification using biochemical tests relies on the use of a **dichotomous key**. (Figure 10.6)
5. Several less labor-intensive commercial modifications of traditional biochemical tests are available. (Figure 10.7)

Serology

1. The proteins and polysaccharides that make up a prokaryote's surface are sometimes characteristic enough to be considered identifying markers.

Fatty Acid Analysis

1. Prokaryotes differ in the type and relative quantity of fatty acids that make up their membranes; thus, the cellular fatty acid composition can be used as an identifying marker. (Figure 10.8)

Using Genotypic Characteristics to Identify Prokaryotes

Nucleic Acid Probes to Detect Specific DNA Sequences

1. By selecting a probe that is complementary to a sequence unique to a given microbe, researchers can use DNA hybridization to detect the presence of specific organisms. (Figure 10.9)

Amplifying Specific DNA Sequences Using the Polymerase Chain Reaction

1. By selecting primers that can be used to amplify a nucleotide sequence unique to a microbe of interest, researchers can apply PCR to determine if a particular agent is present in environments such as body fluids, soil, or water.

Sequencing Ribosomal RNA Genes

1. The nucleotide sequence of ribosomal RNA (rRNA) can be used to identify prokaryotes. (Figure 10.10)
2. Early methods relied on determining the sequence of the rRNA molecule itself, but newer techniques simply sequence rDNA, the DNA that encodes rRNA.
3. Organisms that cannot be cultured in the laboratory can be identified by amplifying, cloning, and then sequencing specific regions of prokaryotic rDNA.

Characterizing Strain Differences (Table 10.6)

Biochemical Typing

1. A strain that has a characteristic biochemical variation is called a **biovar** or a **biotype**.

Serological Typing

1. A strain that differs serologically from other strains is called a **serovar** or a **serotype**.

Genomic Typing

1. Probes can be used to detect genomic differences such as toxin production.
2. Two isolates of the same species that have different **restriction fragment length polymorphisms (RFLPs)** are considered different strains. **Pulsed-field gel electrophoresis** and **ribotyping** are two methods of detecting RFLPs. (Figures 10.11 and 10.12)

Phage Typing

1. The patterns of susceptibility to various types of bacteriophages can be used to demonstrate strain differences. (Figure 10.13)

Antibiograms

1. Antibiotic susceptibility patterns can be used to distinguish strains. (Figure 10.14)

Difficulties in Classifying Prokaryotes

1. Historically, prokaryotes have been grouped according to phenotypic attributes. Molecular approaches give greater insight into the relatedness of microorganisms.

Using Phenotypic Characteristics to Classify Prokaryotes

1. Numerical taxonomy relies on a battery of phenotypic characteristics and classifies bacteria based on their similarity coefficients. (Figure 10.16)

Using Genotypic Characteristics to Classify Prokaryotes (Table 10.7)

1. The most recent methods of bacterial classification are based on comparison of the nucleotide sequences of the DNA of different organisms.
2. Differences in DNA sequences can be used to determine the point in time at which two organisms diverged from a common ancestor.

DNA Base Ratio (G + C Content)

1. The base ratio of an organism is usually expressed as the **G + C content**; the G + C content can be measured by determining the temperature at which double-stranded DNA melts. (Figure 10.17)
2. If the G + C content of two organisms differs by more than a few percent, then they are not closely related.

DNA Hybridization

1. The extent of nucleotide sequence similarity can be determined by measuring how completely single strands of their DNA will anneal to one another. (Figure 10.18)
2. Two strains that show at least 70% homology are often considered to be members of the same species.

Comparing the Sequences of 16S Ribosomal Nucleic Acid

1. Analyzing and comparing the sequences of ribosomal RNA has revolutionized the classification of organisms.
2. Depending on the region compared, the sequence of 16S rDNA can be used to assess distant as well as close relationships between organisms.
3. A genomic analysis enables a more accurate construction of a **phylogenetic tree**. (Figure 10.19)

REVIEW QUESTIONS

Short Answer

1. Name and describe each of the areas of taxonomy.
2. Compare and contrast the five-kingdom and three-domain systems of classification.
3. Describe how a dichotomous key is used in the identification of bacteria.
4. Describe the difference between using a probe and using PCR to detect a specific sequence.
5. Explain how signature sequences are useful when amplifying rDNA.
6. Describe ribotyping.
7. Describe how the GC content of DNA can be measured.
8. Describe a method used to measure the extent of DNA hybridization between two different bacterial isolates.
9. What is the difference between an external and internal node of a phylogenetic tree?
10. Why is it preferable for a classification scheme to reflect the phylogeny of organisms?

Multiple Choice

1. Which of the following is the newest taxonomic unit?
   A. Strain
   B. Family
   C. Order
   D. Species
   E. Domain
2. An acid-fast stain can be used to detect which of the following organisms?
   A. *Filobasidiella* (*Cryptococcus*) *neoformans*
   B. *Mycobacterium tuberculosis*
   C. *Neisseria gonorrhoeae*
   D. *Streptococcus pneumoniae*
   E. *Streptococcus pyogenes*
3. The "breath test" for *Helicobacter pylori* infection assays for the presence of which of the following?
   A. Antigens
   B. Catalase
   C. Hemolysis
   D. Lactose fermentation
   E. Urease
4. The "O" of *E. coli* O157:H7 refers to the…
   A. Biotype
   B. Serotype
   C. Phage type
   D. Ribotype
   E. Antibiogram

5. PulseNet catalogs which of the following in an attempt to trace spread of foodborne illness?
   A. Biotype
   B. Serotype
   C. Phage type
   D. RFLP
   E. Antibiogram
6. Numerical taxonomy relies on all of the following characteristics, except...
   A. ability to form endospores.
   B. 16S ribosomal RNA sequence.
   C. sugar degradation.
   D. motility.
7. If the GC content of two organisms is 70%, which of the following is true?
   A. The organisms are definitely related.
   B. The organisms are definitely not related.
   C. The AT content is 30%.
   D. The organisms likely have extensive DNA homology.
   E. The organisms likely have many characteristics in common.
8. Which of the molecular methods of assessing similarity gives the crudest approximation of relatedness?
   A. DNA hybridization
   B. PCR
   C. 16S rDNA sequencing
   D. DNA base composition
9. The sequence of which ribosomal genes are most commonly used for establishing phylogenetic relatedness?
   A. 5S
   B. 16S
   C. 23S
   D. All of the above are commonly used.
10. Which of the following statements is false?
   A. *Tropheryma whippelii* can be identified even though it has never been grown in culture.
   B. The GC content of DNA can be measured by determining the temperature at which double-stranded DNA melts.
   C. Sequence differences between organisms can be viewed as an evolutionary chronometer to assess their relatedness.
   D. Based on DNA homology studies, members of the genus *Shigella* should be in the same species as *Escherichia coli.*
   E. Pulsed-field gel electrophoresis is used to determine the ribotype of an organism.

Applications

1. Microbiologists debate the use of biochemical similarities and cell features as a way of determining the taxonomic relationships among prokaryotes. Explain why some microbiologists believe these similarities and differences are a powerful taxonomic indicator while others think it is not very useful for that purpose.
2. The mitochondria of eukaryotic cells are theorized to have bacterial origins. A researcher interested in investigating the relationship of mitochondria to bacteria must now decide on the best method to make this determination. What advice would you give the researcher?

Critical Thinking

1. In figure 10.17, how would the curve appear if the GC content of the DNA sample were increased? How would the curve appear if the AT content were increased?
2. When DNA probes are used to detect specific sequence similarities in bacterial DNA, the probe is heated and the two strands of DNA are separated. Why must the probe DNA be treated in this way before detection is possible?

The Diversity of Prokaryotic Organisms

In his native country, the Netherlands, Cornelis B. van Niel (1897–1985) earned a degree in chemical engineering from the Technological University at Delft. At the Delft School, as it is often called, an outstanding general and applied microbiology program within the Department of Chemical Technology was chaired in succession by two prominent microbiologists—Martinus Beijerinck and Albert Kluyver.

After earning his degree in 1923, van Niel accepted a position as assistant to Kluyver, caring for an extensive culture collection and helping prepare demonstrations for lecture courses. Kluyver was relatively new to the school, but he had a vast knowledge of microbiology and biochemistry. Although little was known at the time about metabolic pathways, Kluyver believed that biochemical processes were fundamentally the same in all cells and that microorganisms, which can be grown in pure culture, could be an important research tool, serving as a model to study biochemical processes. Thirty years later, Kluyver and van Niel would present lectures that would be published in a book entitled The Microbe's Contribution to Biology. *Under Kluyver's direction, van Niel began studying the photosynthetic activities of vividly colored purple bacteria such as* Chromatium *species, a subject for which he developed a lifelong interest.*

Shortly after earning his Ph.D. in 1928, van Niel moved to the United States, bringing with him the intense appreciation for general microbiology that had been fostered at the Delft School. Settling at the Hopkins Marine Station in California, he continued his work on purple photosynthetic bacteria. Using systematic methods, van Niel conclusively showed that the growth of these organisms is light dependent, yet they do not evolve O_2. Furthermore, his experiments showed that in order to incorporate CO_2 into cellular material, these anoxygenic phototrophs oxidize hydrogen sulfide. He noted that the reaction stoichiometry of this process was remarkably similar to that of the photosynthesis of green plants and algae, except hydrogen sulfide was used in place of water, and oxidized sulfur compounds were produced instead of O_2. This finding raised the possibility that O_2 generated by plants did not come from carbon dioxide, as was believed at the time, but rather from water.

In addition to his scientific contributions, van Niel was recognized as an outstanding teacher. During the summers at Hopkins Marine Station, he taught a bacteriology course, inspiring many microbiologists with his enthusiasm for the diversity of microorganisms and their importance in nature. His keen memory and knowledge of the literature, along with his appreciation for the remarkable abilities of microorganisms, enabled him to successfully impart the awe and wonder of the microbial world to his students.

—A Glimpse of History

SCIENTISTS ARE ONLY BEGINNING TO UNDERSTAND the vast diversity of microbial life. Although a million species of prokaryotes are thought to exist, only approximately 6,000 of these, grouped into 850 genera, have been actually described and classified. Traditional culture and isolation techniques have not supported the growth, and subsequent study, of the vast majority. Not surprisingly, most effort has been put into the study of microbes intimately associated with the human population, especially those causing disease, and these have been most extensively described. This situation is changing as new molecular techniques aid in the discovery and characterization of previously unrecognized species. The sheer volume of the rapidly accruing information made possible by this modern technology, however, can be daunting for scientists and students alike.

The phylogenetic relationships being elucidated by the ribosomal RNA studies discussed in the previous chapter are causing significant upheaval in prokaryotic classification schemes. Some organisms, once grouped together based on their phenotypic similarities, have now been split into different taxonomic units based on their ribosomal RNA differences.

The genera *Staphylococcus* and *Micrococcus*, for example, were once included in the same family, but recent phylogenetic classification schemes now separate them into different classes. Some organisms that for many years were thought to be entirely unrelated are now grouped more closely together because of their shared ribosomal RNA sequences. As an example, *Rhizobium*, a genus of nitrogen-fixing bacteria that form a symbiotic relationship with certain plants, and *Rickettsia*, a genus of tick-borne human pathogens that are obligate intracellular parasites, have been moved into the same class. Ribosomal RNA analysis even makes it possible to assess the genetic relatedness of organisms that cannot be grown in culture, which has made it possible to study the relationships of a wider variety of archaea. The current phylogenetic classification of prokaryotes used in the *Bergey's Manual of Systematic Bacteriology*, second edition, is presented in Appendix XX. ■ **classification schemes, p. 246**

This chapter covers a wide spectrum of prokaryotes, focusing particularly on their extraordinary diversity rather than concentrating solely on their phylogenetic relationships, which are still being elucidated. To highlight the remarkable abilities of prokaryotes and convey a sense of appreciation for the ecological niches they inhabit and the essential role that they play in our biosphere, groups of microorganisms are described according to their metabolic characteristics and other physiological traits. **Table 11.1** summarizes the characteristics of prokaryotes that are covered in this chapter and serves as an outline for the more complete description within the text. **Table 11.2** indicates the medical importance of representative bacteria and serves as a directory for information both about the organisms covered in this chapter and the respective diseases covered elsewhere in the text. Note, however, that no single chapter could describe all known prokaryotes and, consequently, only a relatively small selection is presented.

METABOLIC DIVERSITY

As a group, prokaryotes use an impressive array of compounds to obtain energy in the form of ATP. This remarkable versatility enables them to occupy a wide variety of environmental habitats, including those where eukaryotes cannot exist (**figure 11.1**). Recall from chapter 6 that nonphotosynthetic cells obtain energy by removing electrons from a reduced compound, the **energy source,** and transferring them to another compound, the **terminal electron acceptor**. These organisms are called **chemotrophs**, to distinguish them from **phototrophs**, which harvest energy from sunlight. Unlike chemotrophic eukaryotes, prokaryotes are not confined to using only organic compounds as an energy source. Those that do obtain energy by oxidizing organic chemicals are termed **chemoorganotrophs**; those that obtain energy by oxidizing inorganic chemicals are **chemolithotrophs**. These terms and others used to describe organisms according to their metabolic capabilities are defined in **table 11.3**. ■ **chemotroph, p. 100** ■ **phototroph, p. 100** ■ **terminal electron acceptor, p. 139**

Figure 11.1 An Extreme Environment Inhabited by Certain Prokaryotes
A hot spring in Yellowstone National Park.

Prokaryotes are also unique in the wide array of electron acceptors they may employ. Aerobic respiration, using molecular oxygen (O_2) as a terminal electron acceptor, generates the most ATP. Prokaryotes may also use anaerobic respiration, however, transferring electrons to an inorganic molecule such as sulfur, sulfate, nitrate, or nitrite rather than O_2. Many others are fermentative, using an internally generated organic compound such as pyruvate as a terminal electron acceptor. ■ **anaerobic respiration, p. 153** ■ **fermentation, p. 153**

Understanding the impact of metabolic diversity involves recognizing the relative energy gains of different types of metabolism, which depend on both the energy source and the electron acceptor. As illustrated in **figure 11.2**, the oxidation of an energy source that has a high tendency to lose electrons, coupled with the reduction of an electron acceptor that has a high tendency to gain electrons, releases the most energy. The released energy can be captured in the form of ATP. Recognizing the differences in energy gain helps explain why some types of metabolism are prevalent in some environments but not others. ■ **oxidation-reduction reactions, p. 138**

Anaerobic Chemotrophs

For approximately the first 1.5 billion years that prokaryotes inhabited earth, the atmosphere was **anoxic**, or devoid of O_2. In the anaerobic environment, some early chemotrophs probably used a pathway of anaerobic respiration, employing electron acceptors such as carbon dioxide or elemental sulfur, which were plentiful in the environment. Others may have used fermentation, passing the electrons to an organic molecule such as pyruvate.

Today, anaerobic habitats still abound. Mud and tightly packed soil limit the diffusion of gases, and any O_2 that does penetrate is rapidly consumed by aerobically respiring chemotrophs. This creates anaerobic conditions just below the surface. Aquatic environments may also become anaerobic if they contain high levels of nutrients, permitting the rapid growth of O_2-consuming microbes. This is evident in polluted lakes, where fish may die because of a lack of dissolved O_2. The bodies of humans and other animals also provide numerous anaerobic environments. It is estimated that 99% of the bacteria that inhabit the intestinal tract are obligate anaerobes. Even the skin and the oral cavity, which are routinely exposed to O_2, have anaerobic microenvironments. These are created via the localized depletion of O_2 by aerobes. ■ **obligate anaerobes, p. 98**

Anaerobic Chemolithotrophs

Chemolithotrophs oxidize reduced inorganic chemicals such as hydrogen gas (H_2) to transform energy. Those that grow anaerobically obviously cannot use O_2 as a terminal electron acceptor and instead must use an alternative such as carbon dioxide or sulfur. Relatively few anaerobic chemolithotrophs have been discovered, and most are members of the domain Archaea. Some that inhabit aquatic environments will be discussed later.

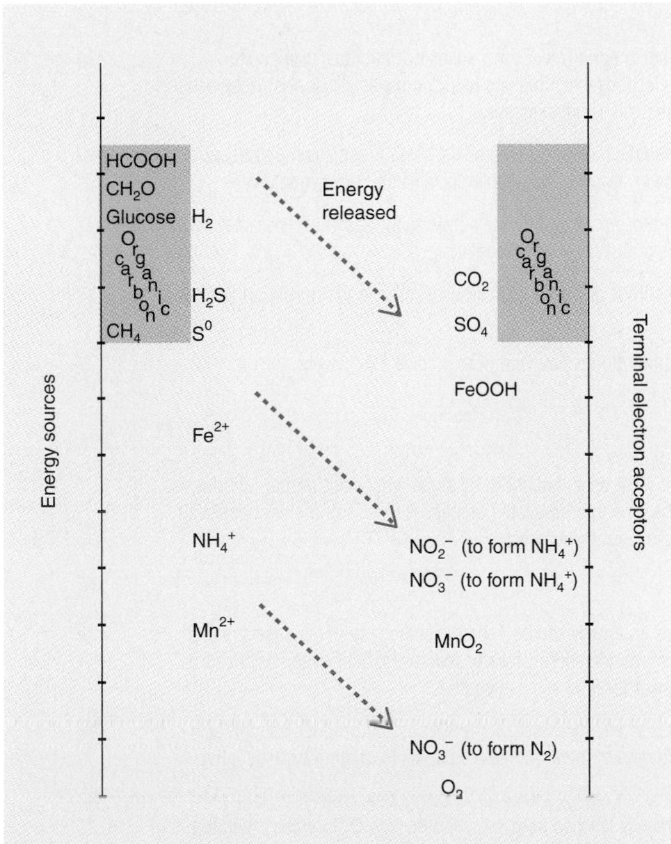

Figure 11.2 Relative Energy Gain of Different Types of Metabolism The left axis shows potential energy sources, ordered according to their relative tendency to give up electrons; those at the top lose electrons most easily. The right axis shows potential terminal electron acceptors, ordered according to their relative tendency to gain electrons; those at the bottom accept electrons most readily. Energy is released only when electrons are transferred from an energy source to a terminal electron acceptor that is lower on the chart; the greater the downward slope, the more energy that can be harvested to make ATP.

The Methanogens

The **methanogens** are a group of archaea that generate ATP by oxidizing hydrogen gas, using CO_2 as a terminal electron acceptor. This process generates methane (CH_4), a colorless, odorless, flammable gas:

$$4\ H_2 + CO_2 \longrightarrow CH_4 + H_2O$$

Many methanogens can also use alternative energy sources such as formate; some can use methanol or acetate as well. A wide variety of morphological varieties of methanogens have been described, including rods, cocci, and spirals (**figure 11.3**). Representative genera of methanogens include *Methanococcus* and *Methanospirillum.*

Methanogens are found in anaerobic environments where hydrogen gas and carbon dioxide are available. Because these gases are generated by chemoorganotrophs during the fermentation of organic material, methanogens often grow in association with these microorganisms. Methanogens, however, are generally not found in environments that contain high levels of sulfate, nitrate, or other inorganic electron acceptors. This is because microorganisms that oxidize hydrogen gas using these electron acceptors have a competitive advantage; the use of CO_2 as an electron acceptor releases comparatively little energy (see figure 11.2). Environments from which methanogens are commonly isolated include sewage, swamps, marine sediments, rice paddies, and the digestive tracts of humans and other animals. The methane produced can present itself as bubbles rising in

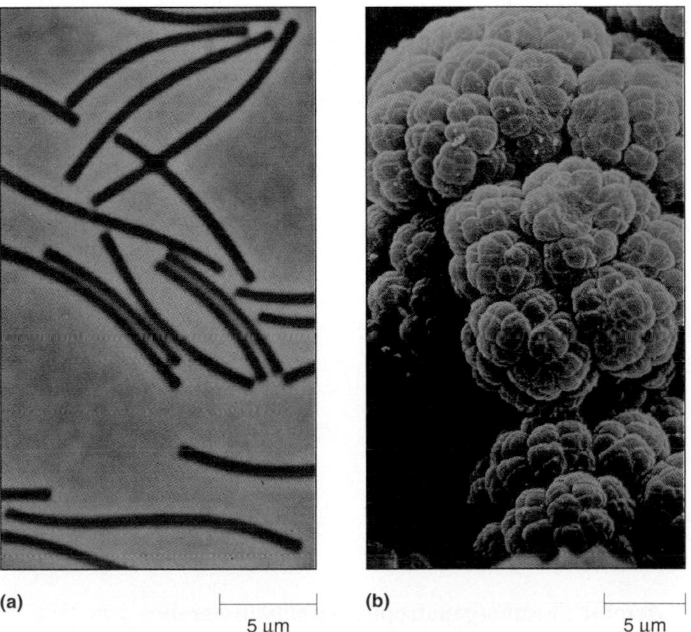

Figure 11.3 Methanogens (a) Phase-contrast micrograph of a *Methanospirillum* species. **(b)** Scanning electron micrograph of a *Methanosarcina* species.

Table 11.1 Prokaryotes Notable for Their Environmental Significance

Group/Genera	Characteristics	Page Number
METABOLIC DIVERSITY		268
Anaerobic Chemolithotrophs		269
Methanogens *Methanococcus, Methanospirillum*	Members of the Archaea that oxidize hydrogen gas, using CO_2 as a terminal electron acceptor to generate methane.	269
Anaerobic Chemoorganotrophs—Anaerobic Respiration		272
Sulfur- and sulfate-reducing bacteria *Desulfovibrio*	Use sulfate as a terminal electron acceptor, generating hydrogen sulfide. Found in anaerobic muds that are rich in organic material. Gram-negative.	272
Anaerobic Chemoorganotrophs—Fermentation		272
Clostridium	Endospore-forming obligate anaerobes. Common inhabitants of soil. Some species are medically important. Gram-positive.	272
Lactic acid bacteria *Streptococcus, Enterococcus, Lactococcus, Lactobacillus, Leuconostoc*	Produce lactic acid as the major end product of their fermentative metabolism. Although most can grow in the presence of O_2, they can only ferment. Several genera are exploited by the food industry; some species are medically important. Gram-positive.	275
Propionibacterium	Obligate anaerobes that produce propionic acid as their primary fermentation end product. Used in the production of Swiss cheese. Gram-positive.	276
Anoxygenic Phototrophs		276
Purple sulfur bacteria *Chromatium, Thiospirillum, Thiodictyon*	Grow in colored masses in sulfur springs and other sulfur-rich habitats, using sulfur compounds as a source of electrons when making reducing power. Most accumulate sulfur granules contained within the cell. Gram-negative.	276
Purple nonsulfur bacteria *Rhodobacter, Rhodopseudomonas*	Grow in a wide variety of aquatic habitats, preferentially using organic compounds as a source of electrons for reducing power. Many are metabolically versatile. Gram-negative.	277
Green sulfur bacteria *Chlorobium, Pelodictyon*	Found in habitats similar to those preferred by the purple sulfur bacteria. Those that accumulate sulfur form granules outside of the cell. Gram-negative.	277
Green nonsulfur bacteria *Chloroflexus*	Characterized by their filamentous growth. Metabolically similar to the purple nonsulfur bacteria. Gram-negative.	277
Others *Heliobacterium*	Other types of anoxygenic phototrophs have not been studied extensively.	277
Oxygenic Phototrophs—Cyanobacteria		278
Anabaena, Synechococcus, Trichodesmium	Photosynthetic bacteria that were once thought to be algae. Important primary producers. Those that fix N_2 support the growth of unrelated organisms in environments that would otherwise be nitrogen-deficient. Gram-negative.	
Aerobic Chemolithotrophs		279
Filamentous sulfur oxidizers *Beggiatoa, Thiothrix*	Oxidize sulfur compounds as an energy source. Found in sulfur springs, sewage-polluted waters, and on the surface of marine and freshwater sediments. Their overgrowth causes bulking in sewage at treatment facilities. Gram-negative.	279
Unicellular sulfur oxidizers *Thiobacillus*	Oxidize sulfur compounds as an energy source. Some species can produce enough acid to lower the pH to 1.0. Oxidation of metal sulfides causes bioleaching. Gram-negative.	280
Nitrifiers *Nitrosomonas, Nitrosococcus, Nitrobacter, Nitrococcus*	Oxidize ammonia or nitrate as an energy source. In so doing, they convert certain fertilizers to a form that is readily leached from soils, and deplete O_2 in waters polluted with ammonia-containing wastes. Genera that oxidize nitrite prevent the toxic buildup of this compound in soils. Gram-negative.	280
Hydrogen-oxidizing bacteria *Aquifex, Hydrogenobacter*	Thermophilic bacteria that oxidize hydrogen gas as an energy source. According to 16S rRNA studies, they were one of the earliest bacterial forms to exist on earth.	281
Aerobic Chemoorganotrophs—Obligate Aerobes		281
Micrococcus	Widely distributed; common contaminants on bacteriological media. Gram-positive.	281
Mycobacterium	Waxy cell wall resists staining; acid-fast. Some species are medically important.	281

Table 11.1 *(Continued)*

Group/Genera	Characteristics	Page Number
Pseudomonas	Common environmental bacteria that, as a group, can degrade a wide variety of compounds. Some species are medically important. Gram-negative.	281
Thermus	*Thermus aquaticus* is the source of *Taq* polymerase, the heat-resistant polymerase used in PCR. Unusual cell wall stains Gram-negative.	282
Deinococcus	Extraordinarily resistant to the damaging effects of gamma radiation. Unusual cell wall stains Gram-positive, but has multiple layers.	282
Aerobic Chemoorganotrophs—Facultative Anaerobes		282
Corynebacterium	Widespread in nature; form metachromatic granules. Some species are medically important. Gram-positive.	282
The *Enterobacteriaceae* *Escherichia coli, Enterobacter, Klebsiella, Proteus, Salmonella, Shigella, Yersinia*	Most reside in the intestinal tract. Those that ferment lactose are coliforms; their presence in water serves as an indicator of fecal pollution. Some species are medically important. Gram-negative.	282
ECOPHYSIOLOGY		283
Thriving in Terrestrial Environments		283
Endospore-formers *Bacillus, Clostridium*	Endospores are the most resistant life form known. *Bacillus* species include both obligate aerobes and facultative anaerobes; *Clostridium* species are obligate anaerobes. Some species are medically important. Gram-positive.	283
Azotobacter	Form a resting stage called a cyst. Notable for their ability to fix nitrogen in aerobic conditions. Gram-negative.	284
Myxobacteria *Chondromyces, Myxococcus, Stigmatella*	Congregate to form a fruiting body; cells within this differentiate to form dormant microcysts. Gram-negative.	284
Streptomyces	Resemble fungi in their pattern of growth, forming dormant conidia. Naturally produce a wide array of medically useful antibiotics. Gram-positive.	284
Agrobacterium	Cause plant tumors. Scientists use their plasmid to introduce desired genes into plant cells. Gram-negative.	285
Rhizobium	Fix nitrogen; form a symbiotic relationship with legumes. Gram-negative.	285
Thriving in Aquatic Environments		286
Sheathed bacteria *Sphaerotilus, Leptothrix*	Form chains of cells enclosed within a protective sheath. Swarmer cells move to new locations. Gram-negative.	286
Prosthecate bacteria *Caulobacter, Hyphomicrobium*	Appendages increase their surface area. *Caulobacter* species serve as a model for cellular differentiation. *Hyphomicrobium* species have a distinctive method of reproduction. Gram-negative.	286
Bdellovibrio	Predator of *E. coli* and other bacteria, multiplying within the periplasm of the prey. Gram-negative.	287
Bioluminescent bacteria *Photobacterium, Vibrio fischeri*	Some bioluminescent species form a symbiotic relationship with specific species of squid and fish.	287
Legionella	Often reside within protozoa. Some species are medically important. Gram-negative.	289
Free-living spirochetes *Spirochaeta, Leptospira* (some species)	Long spiral-shaped bacteria that move by means of an axial filament. Some species are medically important. Gram-negative.	289
Magnetospirillum	Contain a string of magnetic crystals that enable them to move up or down in water and sediments. Gram-negative.	290
Spirillum	Spiral-shaped, microaerophilic bacteria; some species form metachromatic granules. Gram-negative.	290
Sulfur-oxidizing, nitrate-reducing marine bacteria *Thioploca, Thiomargarita*	Use novel mechanisms to compensate for the fact that their energy source (reduced sulfur compounds) and terminal electron acceptor (nitrate) do not coexist.	290
Animals as Habitats—see table 11.2		

(Continued)

Table 11.1 *(Continued)*

Group/Genera	Characteristics	Page Number
ARCHAEA		293
Methanogens *Methanococcus, Methanospirillum*	Generate methane when they oxidize hydrogen gas as an energy source, using CO_2 as a terminal electron acceptor.	293
Extreme halophiles *Halobacterium, Halorubrum, Natronobacterium, Natronococcus*	Found in salt lakes, soda lakes, and brines. They produce pigments and can be seen as pink blooms in concentrated salt water ponds.	293
Methane-generating thermophiles *Methanothermus*	Found near hydrothermal vents; can grow at temperatures near 100°C.	293
Sulfur- and sulfate-reducing hyperthermophiles *Thermococcus, Archaeoglobus, Thermoproteus, Pyrodictium, Pyrolobus*	Obligate anaerobes that use sulfur or, in one case, sulfate as an a terminal electron acceptor, generating hydrogen sulfide. *Thermococcus* and *Archaeoglobus* (Euryarchaeota) oxidize organic compounds as an energy source; *Thermoproteus, Pyrodictium,* and *Pyrolobus* (Crenarchaeota) oxidize H_2 as an energy source.	293
Sulfur oxidizers *Sulfolobus*	Oxidize sulfur as a source of energy, using O_2 as a terminal electron acceptor to generate sulfuric acid. They grow only at a temperature above 50°C and at a pH between 1 and 6.	293
Thermophilic extreme acidophiles *Thermophilus, Picrophilus*	Grow only in extremely hot, acidic environments.	294

swamp waters or the 10 cubic feet of gas discharged from a cow's digestive system each day. As a by-product of sewage treatment plants, methane gas can be collected and used for heating, cooking, and even the generation of electricity (see Perspective 30.1).

The study of methanogens is quite challenging because they are exquisitely sensitive to O_2, as are many of their enzymes. Special techniques, including anaerobe jars and anaerobic chambers, are used for their cultivation. ■ **culturing anaerobes, p. 104**

Anaerobic Chemoorganotrophs—Anaerobic Respiration

Chemoorganotrophs oxidize organic compounds such as glucose to obtain energy. Like the chemolithotrophs, chemoorganotrophs that grow anaerobically must employ a terminal electron acceptor other than O_2. Sulfur and sulfate are common inorganic compounds used as a terminal electron acceptor by these organisms.

Sulfur- and Sulfate-Reducing Bacteria

When sulfur compounds are used as a terminal electron acceptor, they become reduced to form hydrogen sulfide, the compound responsible for the rotten-egg smell of many anaerobic environments. In addition to the unpleasant odor, the H_2S is a problem to industry because it reacts with iron to form iron sulfide, which corrodes pipes and other metals. Ecologically, however, prokaryotes that reduce sulfur compounds are an indispensable component of the sulfur cycle. ■ **sulfur cycle, p. 779**

Sulfate- and sulfur-reducing bacteria generally live in mud that is rich in organic material and oxidized sulfur compounds. The H_2S they produce causes mud and water to turn black when it reacts with iron molecules. At least a dozen genera are recognized in this group, the most extensively studied of which are species of the Gram-negative curved rod *Desulfovibrio.*

Some representatives of the Archaea also use sulfur compounds as a terminal electron acceptor, but they generally do not inhabit the same ecological niche as their bacterial counterparts. While most of the sulfur-reducing bacteria are mesophiles or thermophiles, these sulfur-reducing archaea are hyperthermophiles, inhabiting such extreme environments as hydrothermal vents. They will be discussed later in the chapter. ■ **thermophiles, p. 96** ■ **hyperthermophiles, p. 97**

Anaerobic Chemoorganotrophs: Fermentation

Numerous types of anaerobic bacteria transform energy using the process of fermentation, producing ATP only by substrate-level phosphorylation. The end products of fermentation include a variety of acids and gases that are generally characteristic for a given species. Consequently, they can often be used as identifying markers.

The Genus *Clostridium*

Members of the genus *Clostridium* are Gram-positive rods that can form endospores. They are common inhabitants of soil, where the vegetative cells live in the anaerobic microenvironments created when other, aerobic organisms consume available O_2. Their endospores, a dormant form, are indifferent to O_2 and can survive for long periods by withstanding measures of heat, desiccation, chemicals, and irradiation that would kill all vegetative bacteria. When the appropriate conditions are renewed, these endospores germinate, and the resulting vegetative bacteria may once again multiply (see figure 3.48). Vegetative cells that arise from soil-borne endospores are responsible for a variety of diseases, including tetanus (caused by *C. tetani*), gas gangrene (caused by *C. perfringens*), and botulism (caused by *C. botulinum*). Some species of *Clostridium* are normal inhabitants of the intestinal tract of humans and other animals. ■ **endospores, p. 72** ■ **tetanus, p. 698** ■ **gas gangrene, p. 701** ■ **botulism, p. 675**

Table 11.2 Medically Important Chemoorganotrophs

Organism	Medical Significance	Page Number	Disease Description Page Number
Gram-Negative Rods			
Bacteroides	Obligate anaerobes that commonly inhabit the mouth, intestinal tract, and genital tract. Causes abscesses and bloodstream infections.	290	589
Enterobacteriaceae			
Enterobacter species	Normal flora of the intestinal tract.	282	
Escherichia coli	Normal flora of the intestinal tract. Some strains cause urinary tract infections; some strains cause specific types of intestinal disease. Causes meningitis in newborns.	282	603, 636
Klebsiella pneumoniae	Normal flora of the intestinal tract. Causes pneumonia.	282	565
Proteus species	Normal flora of the intestinal tract. Causes urinary tract infections.	282	636
Salmonella Enteritidis	Causes gastroenteritis. Grows in the intestinal tract of infected animals; acquired by consuming contaminated foods.	282	604
Salmonella Typhi	Causes typhoid fever. Grows in the intestinal tract of infected humans; transmitted in feces.	282	604
Shigella species	Causes dysentery. Grows in the intestinal tract of infected humans; transmitted in feces.	282	602
Yersinia pestis	Causes bubonic plague, which is transmitted by fleas, and pneumonic plague, which is transmitted in respiratory droplets of infected individuals.	282	724
Haemophilus influenzae	Causes ear infections, respiratory infections, and meningitis in children.	291	558, 668
Haemophilus ducreyi	Causes chancroid, a sexually transmitted disease	291	651
Legionella pneumophila	Causes Legionnaires' disease, a lung infection. Grows within protozoa; acquired by inhaling contaminated water droplets.	289	571
Pseudomonas aeruginosa	Causes burns, urinary tract, and bloodstream infections. Ubiquitous in the environment. Grows in nutrient-poor aqueous solutions and is resistant to many disinfectants and antimicrobial medications.	281	697
Gram-Negative Rods—Obligate Intracellular Parasites			
Chlamydia pneumoniae	Causes atypical pneumonia, or "walking pneumonia." Acquired from an infected person.	292	656
Chlamydia psittaci	Causes psittacosis, a form of pneumonia. Transmitted by birds.	292	
Chlamydia trachomatis	Causes a sexually transmitted disease that mimics the symptoms of gonorrhea. Also causes trachoma, a serious eye infection, and conjunctivitis in newborns.	292	646
Coxiella burnetii	Causes Q fever. Acquired by inhaling organisms shed by infected animals.	292	
Ehrlichia chaffeenis	Causes human ehrlichiosis. Transmitted by ticks.	292	
Orientia tsutsugamushi	Causes scrub typhus. Transmitted by mites.	292	
Rickettsia prowazekii	Causes epidemic typhus. Transmitted by lice.	292	521
Rickettsia rickettsii	Causes Rocky Mountain spotted fever. Transmitted by ticks.	292	529
Gram-Negative Curved Rods			
Campylobacter jejuni	Causes gastroenteritis. Grows in the intestinal tract of infected animals; acquired by consuming contaminated food.	291	606
Helicobacter pylori	Causes stomach and duodenal ulcers. Neutralizes stomach acid by producing urease, resulting in the breakdown of urea to form ammonia.	291	593
Vibrio cholerae	Causes cholera, a severe diarrheal disease. Grows in the intestinal tract of infected humans; acquired by drinking contaminated water.	289	599
Vibrio parahaemolyticus	Causes gastroenteritis. Acquired by consuming contaminated seafood.	289	

(Continued)

Table 11.2 *(Continued)*

Organism	Medical Significance	Page Number	Disease Description Page Number
Gram-Negative Cocci			
Neisseria meningitidis	Causes meningitis.	291	669
Neisseria gonorrhoeae	Causes gonorrhea, a sexually transmitted disease.	291	644
Gram-Positive Rods			
Bacillus anthracis	Causes anthrax. Acquired by inhaling endospores in soil, animal hides, and wool. Possible biological warfare agent.	283	491
Bifidobacterium species	Predominant member of the intestinal tract in breast-fed infants. Thought to play a protective role in the intestinal tract of infants by excluding pathogens.	291	
Clostridium botulinum	Causes botulism. Acquired by ingesting contaminated foods, typically canned foods that have been improperly processed.	272	657
Clostridium perfringens	Causes gas gangrene. Acquired when soil-borne endospores contaminate a wound.	272	701
Clostridium tetani	Causes tetanus. Acquired when soil-borne endospores are inoculated into deep tissue.	272	698
Corynebacterium diphtheriae	Toxin-producing strains cause diphtheria, a frequently fatal throat infection.	282	557
Gram-Positive Cocci			
Enterococcus species	Normal intestinal flora. Causes urinary tract infections.	275	636
Micrococcus species	Found on skin as well as in a variety of other environments; often contaminate bacteriological media.	281	
Staphylococcus aureus	Leading cause of wound infections. Causes food poisoning and toxic shock syndrome.	290	525, 694
Staphylococcus epidermidis	Normal flora of the skin.	290	524
Staphylococcus saprophyticus	Causes urinary tract infections.	290	636
Streptococcus pneumoniae	Causes pneumonia and meningitis.	275	564
Streptococcus pyogenes	Causes pharyngitis (strep throat), rheumatic fever, wound infections, glomerulonephritis, and streptococcal toxic shock.	275	527, 553, 696
Acid-Fast Rods			
Mycobacterium tuberculosis	Causes tuberculosis, typically a chronic respiratory infection.	281	569
Mycobacterium leprae	Causes Hansen's disease (leprosy); peripheral nerve invasion is characteristic.	281	672
Spirochetes			
Treponema pallidum	Causes syphilis, a sexually transmitted disease that can spread throughout the body. The organism has never been grown in culture.	291	647
Borrelia burgdorferi	Causes Lyme disease, a tick-borne disease that initially causes a rash and then spreads throughout the body.	291	532
Borrelia recurrentis and *B. hermsii*	Causes relapsing fever. Transmitted by arthropods.	291	
Leptospira interrogans	Causes leptospirosis, a waterborne disease that can spread throughout the body. Excreted in urine of infected animals.	289	637
Cell Wall-less			
Mycoplasma pneumoniae	Causes atypical pneumonia ("walking pneumonia"). Not susceptible to penicillin because it lacks a cell wall.	291	566

Table 11.3 Terms Used to Describe Microorganisms According to Their Metabolic Capabilities

Term	Definition	Comments
Chemotroph	An organism that obtains energy by oxidizing chemicals; the same process provides reducing power for biosynthesis.	Aerobic respiration uses O_2 as a terminal electron acceptor, anaerobic respiration uses an inorganic compound other than O_2 as a terminal electron acceptor, and fermentation uses an organic compound such as pyruvate as a terminal electron acceptor.
Chemolithotroph	Inorganic compounds such as H_2S are used as an energy source (*litho-* means "rock").	Generally, a chemolithotroph obtains carbon from CO_2 and is therefore a chemoautotroph; because of this, the terms chemolithotroph and chemoautotroph are often used interchangeably. Alternatively, and more correctly, the term chemolithoautotroph is used.
Chemoorganotroph	Organic compounds such as glucose are used as an energy source.	A chemoorganotroph obtains carbon from organic compounds and is therefore a chemoheterotroph; because of this, the terms chemoorganotroph and chemoheterotroph are often used interchangeably. The term chemoorganoheterotroph is rarely used, although it is technically more correct.
Phototrophs	Organisms that harvest energy from sunlight; cells need a source of electrons to make reducing power for biosynthesis.	Anoxygenic phototrophs use reduced compounds such as H_2S as a source of electrons for reducing power. Oxygenic phototrophs use H_2O as a source of electrons for reducing power, generating O_2.
Photoautotroph	Energy is harvested from sunlight; carbon is obtained from CO_2.	
Photoheterotroph	Energy is harvested from sunlight; carbon is obtained from organic compounds.	

As a group, *Clostridium* species ferment a wide variety of compounds, including sugars, cellulose, and ethanol. Some of the end products are commercially valuable; for example, *C. acetobutylicum* produces acetone and butanol. Some species can ferment amino acids by an unusual process that oxidizes one amino acid, using another as a terminal electron acceptor. This generates a variety of foul-smelling end products associated with putrefaction.

The Lactic Acid Bacteria

Gram-positive bacteria that generate lactic acid as a major end product of their fermentative metabolism make up a group called the **lactic acid bacteria**. They include the genera *Streptococcus*, *Enterococcus*, *Lactococcus*, *Lactobacillus* and *Leuconostoc*. Most can grow in aerobic environments, but all are obligate fermenters and thus derive no benefit from O_2. They can be readily distinguished from other bacteria that grow in the presence of O_2 because they lack the enzyme catalase, which converts hydrogen peroxide, a toxic derivative of oxygen, into water and O_2. ■ **obligate fermenters, p. 98** ■ **catalase, p. 98**

Streptococcus species are cocci that typically grow in chains of varying lengths (**figure 11.4**). They inhabit the oral cavity, generally as normal flora. Some, however, are notable for their adverse effect on human health. One of the most important is *S. pyogenes* (Group A strep), which causes pharyngitis (strep throat) and other diseases. Unlike the streptococci that typically inhabit the throat, *S. pyogenes* is β-hemolytic, an important characteristic used in its identification (see figure 4.14). ■ ***Streptococcus pyogenes*, p. 553** ■ **hemolysis, p. 103**

Species of *Lactococcus* and *Enterococcus* were at one time included in the genus *Streptococcus*. The genus *Lactococcus* now includes those species used by the dairy industry to produce fermented milk products such as cheese and yogurt. Species of the

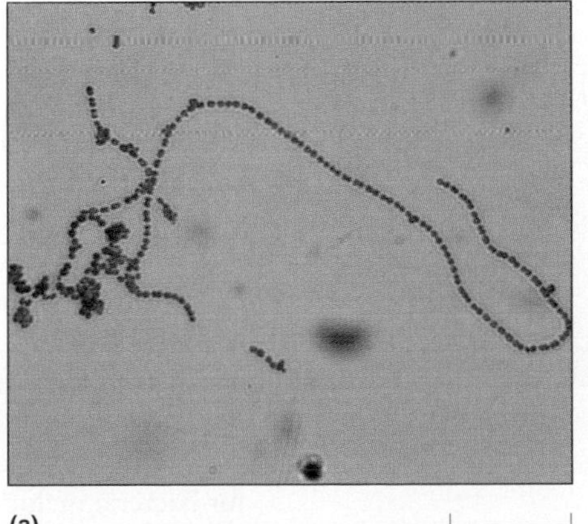

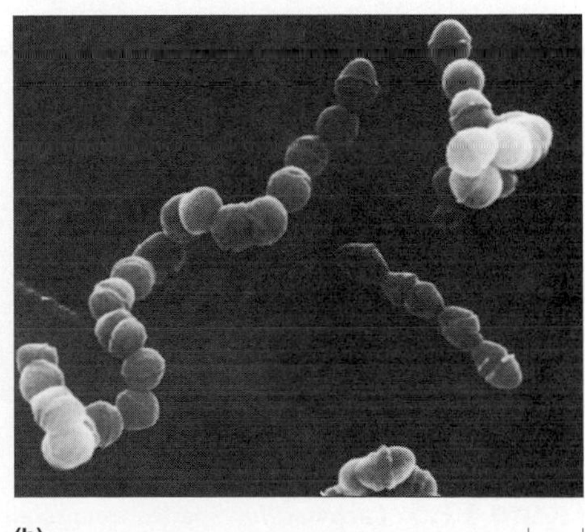

Figure 11.4 ***Streptococcus*** **(a)** Gram stain. **(b)** Scanning electron micrograph.

genus *Enterococcus* typically inhabit the intestinal tract of humans and other animals. ■ fermentation of dairy products, p. 806

Members of the genus *Lactobacillus* are rod-shaped organisms that grow as single cells or loosely associated chains. They are common among the microflora of the mouth and of the healthy human vagina during child-bearing years. In the vagina, they metabolize glycogen, which has been deposited in the vaginal lining in response to the female sex hormone, estrogen. The low pH that results helps the vagina resist infection. Lactobacilli are also often present in decomposing plant material, milk, and other dairy products. Like the lactococci, they are important in the production of fermented foods (**figure 11.5**).

The Genus *Propionibacterium*

Propionibacterium species are Gram-positive pleomorphic (irregular-shaped) rods that produce propionic acid as their primary fermentation end product. Significantly, they can also ferment lactic acid. Thus, *Propionibacterium* species can extract residual energy from a waste product of another group of bacteria.

Propionibacterium species are important to the dairy industry, because their fermentation end products play an indispensable role in the production of Swiss cheese. The propionic acid is responsible for the typical nutty flavor of Swiss cheese. The CO_2, also a product of the fermentation, creates the signature holes in the cheese. *Propionibacterium* species are also found growing in the intestinal tract and in anaerobic microenvironments on the skin. ■ Swiss cheese production, p. 807

MICROCHECK 11.1

The methanogens oxidize hydrogen gas, using CO_2 as a terminal electron acceptor, to generate methane. The sulfur- and sulfate-reducing bacteria oxidize organic compounds, with sulfur compounds serving as terminal electron acceptors, to generate hydrogen sulfide. *Clostridium* species, the lactic acid bacteria, and *Propionibacterium* species oxidize organic compounds, with an organic compound serving as a terminal electron acceptor.

- What metabolic process creates the rotten-egg smell characteristic of many anaerobic environments?
- Describe two beneficial contributions of the lactic acid bacteria.
- Relatively little is known about many obligate anaerobes. Why would this be so?

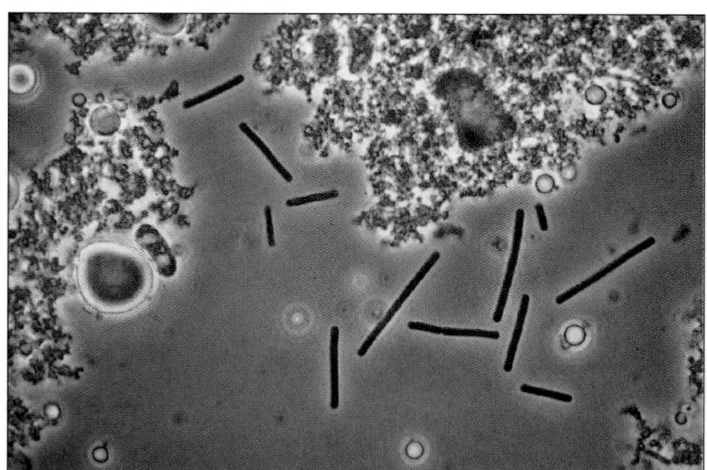

Figure 11.5 ***Lactobacillus*** **Species from Yogurt**

Anoxygenic Phototrophs

The earliest photosynthesizing organisms were likely **anoxygenic phototrophs**. These use hydrogen sulfide or organic compounds rather than water as a source of electrons when making reducing power in the form of NADPH, and therefore they do not generate O_2. Modern anoxygenic phototrophs are a phylogenetically diverse group of bacteria that inhabit a restricted ecological niche that provides adequate light penetration yet little or no O_2; most often, they are found in aquatic habitats such as bogs, lakes, and the upper layer of muds. ■ reducing power, p. 139

The photosynthetic systems of the anoxygenic phototrophs are fundamentally different from those of plants, algae, and cyanobacteria. They have a unique type of chlorophyll called **bacteriochlorophyll**. This pigment absorbs wavelengths of light that penetrate to greater depths and are not used by other photosynthetic organisms. ■ photosynthetic pigments, p. 158

The Purple Bacteria

The **purple bacteria** are Gram-negative organisms that appear red, orange, or purple due to their light-harvesting pigments. Unlike other anoxygenic phototrophs, the components of their photosynthetic apparatus are all contained within the cytoplasmic membrane. Invaginations in this membrane effectively increase the surface area available for the photosynthetic processes.

Purple Sulfur Bacteria

Purple sulfur bacteria can sometimes be seen growing as colored masses in sulfur-rich habitats such as sulfur springs (**figure 11.6a**). The cells are relatively large, sometimes in excess of 5 μm in diameter, and some are motile by flagella. They may also have gas vesicles, which enable them to move up or down to their preferred level in the water column. Most accumulate sulfur in granules that are readily visible microscopically and appear to be contained within the cell (figure 11.6b).

The purple sulfur bacteria preferentially use hydrogen sulfide to generate reducing power, although some species can use other inorganic molecules, such as H_2, or organic compounds, such as pyruvate. Many species of purple sulfur bacteria are strict anaerobes and phototrophs, but some can grow in absence of light aerobically, oxidizing reduced inorganic or organic compounds as a source of energy. Representative genera of purple sulfur bacteria include *Chromatium*, *Thiospirillum*, and *Thiodictyon*.

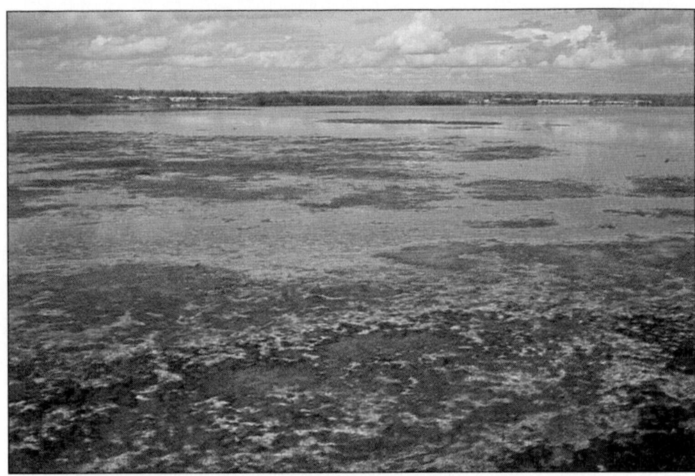

(a)

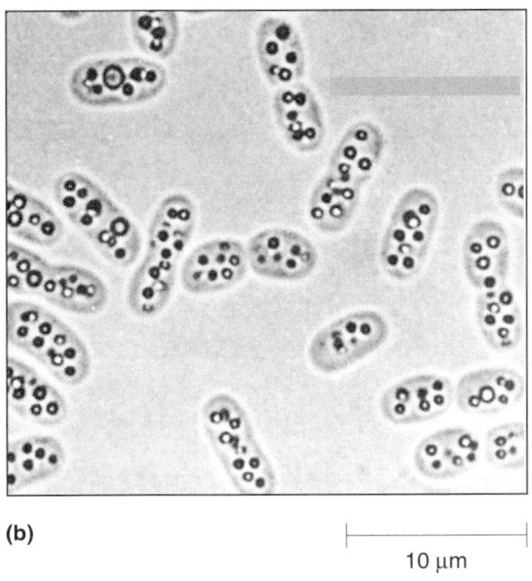

(b) 10 μm

Figure 11.6 Purple Sulfur Bacteria (a) Photograph of bacteria growing in a bog. **(b)** Photomicrograph showing intracellular sulfur granules.

Purple Nonsulfur Bacteria

The purple nonsulfur bacteria are found in a wide variety of aquatic habitats, including moist soils, bogs, and paddy fields. One important characteristic that distinguishes them from the purple sulfur bacteria is that they preferentially use a variety of organic molecules rather than hydrogen sulfide as a source of electrons for reducing power. In addition, they lack gas vesicles. If sulfur does accumulate, the granules form outside of the cell.

Purple nonsulfur bacteria are remarkably versatile metabolically. Not only do they grow as phototrophs using organic molecules to generate reducing power, but many can use a metabolism similar to the purple sulfur bacteria, employing hydrogen gas or hydrogen sulfide as an energy source. In addition, most can grow aerobically in the absence of light using chemotrophic metabolism. Representative genera of purple sulfur bacteria include *Rhodobacter* and *Rhodopseudomonas*.

The Green Bacteria

The **green bacteria** are Gram-negative organisms that are typically green or brownish in color. Unlike the purple bacteria, their light-harvesting pigments are located in structures called **chlorosomes** and their cytoplasmic membranes do not have extensive invaginations.

Green Sulfur Bacteria

The green sulfur bacteria are found in habitats similar to those preferred by the purple sulfur bacteria. Like the purple sulfur bacteria, they use hydrogen sulfide as a source of electrons for reducing power and they form sulfur granules. Those granules, however, form outside of the cell (**figure 11.7**). The green sulfur bacteria lack flagella, but many have gas vesicles. All are strict anaerobes, and none can use a chemotrophic metabolism. Representative genera include *Chlorobium* and *Pelodictyon*.

Green Nonsulfur Bacteria

The green nonsulfur bacteria are characterized by their filamentous growth. Metabolically, they resemble the purple nonsulfur bacteria, preferentially using organic compounds to generate reducing power. As an alternative, they can use hydrogen gas or hydrogen sulfide. In addition, they can grow in the dark aerobically using chemotrophic metabolism. *Chloroflexus* is the only genus in this group that has been grown in pure culture.

Other Anoxygenic Phototrophs

While the green and purple bacteria have been studied most extensively, other types of anoxygenic phototrophs have recently been discovered. Among these is the genus *Heliobacterium*, a Gram-positive endospore-forming rod related to members of the genus *Clostridium*.

MICROCHECK 11.2

The anoxygenic phototrophs harvest the energy of sunlight, but do not generate O_2 because they do not use water as a source of electrons when making

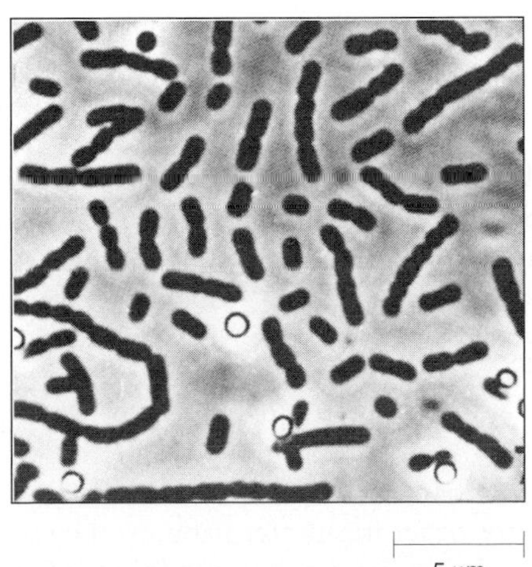

5 μm

Figure 11.7 Green Sulfur Bacteria Note that the sulfur granules are extracellular.

reducing power in the form of NADPH. The purple sulfur bacteria and the green sulfur bacteria use hydrogen sulfide as a source of electrons; the purple nonsulfur and green nonsulfur bacteria preferentially use organic compounds.

- Describe a structural characteristic that distinguishes the purple sulfur bacteria from the green sulfur bacteria.
- What is the function of gas vesicles?
- Why is it beneficial to the anoxygenic phototrophs to have light-harvesting pigments that absorb wavelengths of light that penetrate to greater depths?

Oxygenic Phototrophs

Nearly 3 billion years ago, the earth's atmosphere began changing as O_2 was gradually introduced to the previously anoxic environment. This change was probably due to the evolution of the cyanobacteria, thought to be the earliest **oxygenic phototrophs**. These are photosynthetic organisms that use water as a source of electrons for reducing power, liberating O_2.

$$6\ CO_2 + 6\ H_2O \longrightarrow C_6H_{12}O_6 + 6\ O_2$$

Today, cyanobacteria still play an essential role in the biosphere. As **primary producers**, they harvest the energy of sunlight, using it to convert CO_2 into organic compounds. They were initially thought to be a form of algae and were called blue-green algae until electron microscopy revealed their prokaryotic structure. Unfortunately, this early misunderstanding generated long-lasting confusion in their classification and nomenclature. ■ **primary producers, p. 100**

The Cyanobacteria

The cyanobacteria are a diverse group of more than 60 genera of Gram-negative bacteria. They inhabit a wide range of environments, including freshwater and marine habitats, soils, and the surfaces of rocks. In addition to being photosynthetic, many are able to convert nitrogen gas (N_2) to ammonia, which can then be incorporated into cell material. This process, called **nitrogen fixation**, is an exclusive ability of prokaryotes. ■ **nitrogen fixation, p. 100**

General Characteristics of Cyanobacteria

Cyanobacteria are morphologically diverse (**figure 11.8**). Some genera are unicellular, with typical prokaryotic shapes such as cocci, rods, and spirals. Others form filamentous multicellular associations called **trichomes** that may or may not be enclosed within a sheath. Motile trichomes glide as a unit. Cyanobaceria that inhabit aquatic environments often have gas vesicles, enabling them to move vertically within the water column. When large numbers of buoyant cyanobacteria accumulate in stagnant lakes or other freshwater habitats, they may form mats on the surface. In the bright, hot conditions of summer, these cells lyse and decay, creating an odiferous scum called a **nuisance bloom** (**figure 11.9**).

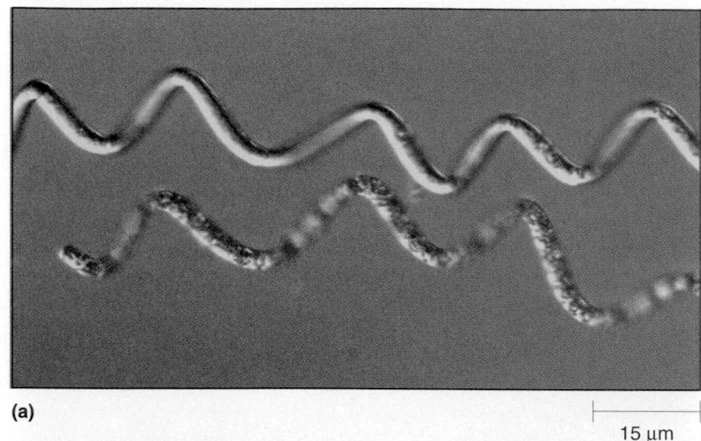

Figure 11.8 Cyanobacteria **(a)** The spiral shapes of *Spirulina* species. **(b)** Differential interference contrast photomicrograph of a species of *Oscillatoria*. Note the multicellular association called a trichome.

The photosynthetic systems of the cyanobacteria are like those contained within the chloroplasts of algae and plants. This is not surprising in light of the genetic evidence indicating chloroplasts evolved from a species of cyanobacteria that once resided as an endosymbiont within a eukaryotic cell (see Perspective 3.1). In addition to light-harvesting chlorophyll pigments, cyanobacteria have **phycobiliproteins**. These pigments absorb energy from wavelengths of light that are not well absorbed by chlorophyll. They contribute to the blue-green, or sometimes reddish, color of the cyanobacteria.

Nitrogen-Fixing Cyanobacteria

Nitrogen-fixing cyanobacteria are critically important ecologically. Because they can incorporate both N_2 and CO_2 into organic material, they generate a form of these nutrients that can then be used by other organisms. Thus, their activities can ultimately support the growth of a wide range of organisms in environments that would otherwise be devoid of usable nitrogen and carbon. As an example, nitrogen-fixing cyanobacteria that inhabit the oceans are essential primary producers that support other sea life. Also, like all cyanobacteria, they help control atmospheric carbon dioxide buildup by utilizing the gas as a carbon source.

Figure 11.9 Nuisance Bloom Excessive growth of cyanobacteria causes buoyant masses to rise to the surface. The eventual death and decay of the cells creates objectionable odors.

Nitrogenase, the enzyme that mediates the process of nitrogen fixation, is destroyed by O_2; therefore, nitrogen-fixing cyanobacteria must protect the enzyme from the O_2 they generate. Species of *Anabaena*, which are filamentous, isolate nitrogenase by confining the process of nitrogen fixation to a specialized thick-walled cell called a **heterocyst (figure 11.10)**. Heterocysts are not photosynthetic and, consequently, do not generate O_2. The heterocysts of some species form at very regular intervals within the filament, reflecting the ability of cells within a trichome to communicate. One species of *Anabaena*, *A. azollae*, forms an intimate relationship with the water fern *Azolla*. The bacterium grows and fixes nitrogen within the protected environment of a special sac in the fern, providing *Azolla* with a source of available nitrogen. *Synechococcus* species, which are unicellular, fix nitrogen only in the dark. Consequently, nitrogen fixation and photosynthesis are temporally separated. Remarkably, members of the genus *Trichodesmium* fix nitrogen even while they are photosynthesizing. How they protect their nitrogenase enzyme from O_2 is still not understood.

Other Notable Characteristics of Cyanobacteria

The cyanobacteria have various other notable characteristics—some are beneficial, others damaging. Filamentous cyanobacteria appear to be responsible for maintaining the structure and productivity of soils in cold desert areas such as the Colorado Plateau. Their sheaths persist in soil, creating a sticky fibrous network that prevents erosion. In addition, these bacteria provide an important source of nitrogen and organic carbon in otherwise nutrient-poor soils. Cyanobacteria growing in freshwater lakes and reservoirs used as a source of drinking water can impart an undesirable taste to the water. This is due to their production of a compound called **geosmin**, which has a distinctive "earthy" odor. Finally, some aquatic species such as *Microcystis aeruginosa* can produce toxins. These can be deadly to animals that consume them.

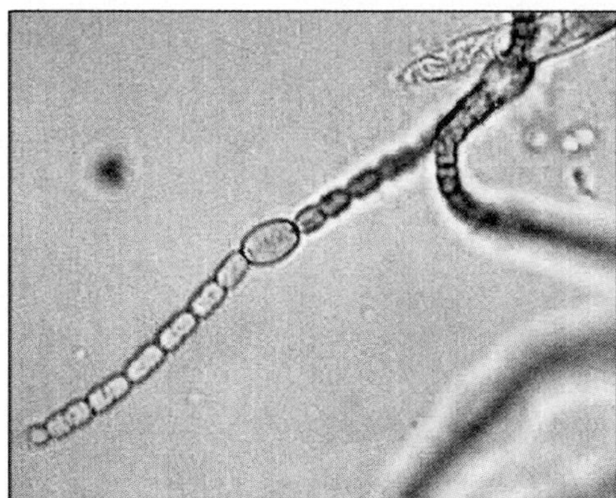

Figure 11.10 Heterocyst of an *Anabaena* Species Nitrogen fixation occurs within these specialized cells.

MICROCHECK 11.3

The photosynthetic systems of cyanobacteria generate O_2 and are similar to those of algae and plants. Many cyanobacteria species can fix nitrogen.

- What is the function of a heterocyst?
- Why might cyanobacteria prevent erosion in cold desert regions?
- How could heavily fertilized lawns foster the development of nuisance blooms?

Aerobic Chemolithotrophs

Aerobic chemolithotrophs obtain energy by oxidizing reduced inorganic compounds. These compounds most commonly include sulfur, ammonia, nitrite, and hydrogen gas. Some chemolithotrophs can also oxidize iron and manganese.

The Sulfur-Oxidizing Bacteria

The sulfur-oxidizing bacteria are Gram-negative rods or spirals, which sometimes grow in filaments. They obtain energy by oxidizing one or a variety of sulfur compounds, including hydrogen sulfide, elemental sulfur, and thiosulfate. Molecular oxygen serves as a terminal electron acceptor, generating sulfuric acid. These bacteria play an important role in the sulfur cycle. ■ sulfur cycle, p. 779

Filamentous Sulfur Oxidizers

Species of the filamentous sulfur oxidizers *Beggiatoa* and *Thiothrix* live in sulfur springs, in sewage-polluted waters, and on the surface of marine and freshwater sediments. They accumulate sulfur, depositing it as intracellular granules. Members of the genera *Beggiatoa* and *Thiothrix* differ in the nature of their filamentous growth **(figure 11.11)**. The filaments of *Beggiatoa* species move by gliding motility, the mechanism of which is poorly understood. The filaments may flex or twist to form a tuft. In contrast, the filaments of *Thiothrix* species are immobile; they fasten at one end to rocks or other solid surfaces. Often

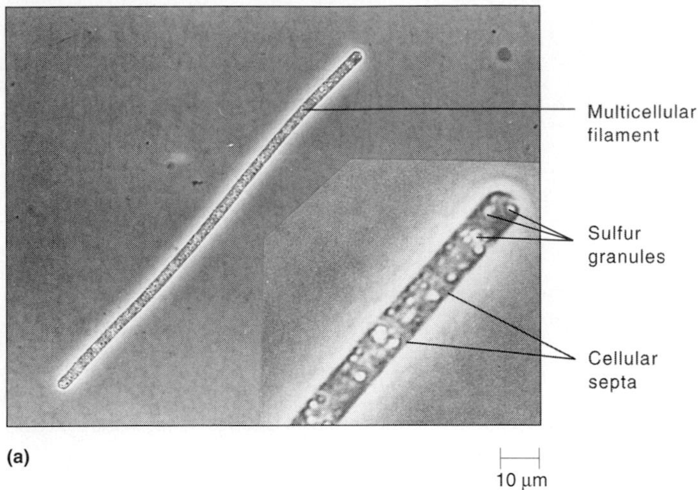

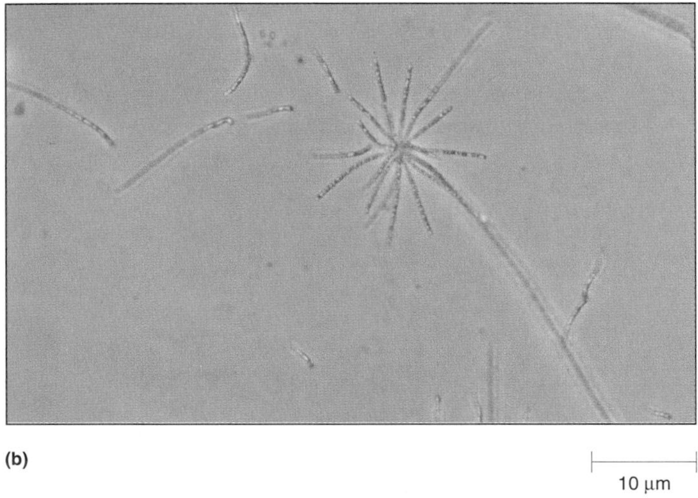

Figure 11.11 Filamentous Sulfur Bacteria Phase-contrast photomicrographs. **(a)** Multicellular filament of a *Beggiatoa* species. **(b)** Multicellular filaments of a *Thiothrix* species, forming a rosette arrangement.

they attach to other cells, causing the filaments to form a characteristic rosette arrangement. Progeny cells detach from the ends of these filaments and use gliding motility to disperse to new locations, where they form new filaments. Overgrowth of these filamentous organisms in sewage at treatment facilities causes a problem called **bulking**. Because the masses of filamentous organisms do not settle easily, bulking interferes with the separation of the solid sludge from the liquid effluent. ■ **sewage treatment, p. 793**

Unicellular Sulfur Oxidizers

Thiobacillus species are found in both terrestrial and aquatic habitats, where their ability to oxidize metal sulfides is responsible for a process called **bioleaching.** In this process, insoluble metal sulfides are oxidized, producing sulfuric acid while converting the metal to a soluble form; some species can produce enough acid to lower the pH to 1.0. Bioleaching can cause severe environmental problems. For example, the strip mining of coal exposes metal sulfides, which can then be oxidized by *Thiobacillus* species to produce sulfuric acid. The resulting runoff can acidify nearby streams, killing trees, fish, and other wildlife **(figure 11.12)**. The runoff may also contain toxic metals made soluble by the bacteria. Under controlled conditions, however, bioleaching can enhance the recovery of metals. For example, gold can be extracted from deposits of gold sulfide. The metabolic activities of *Thiobacillus* species can also be used to prevent acid rain, which results from the burning of sulfur-containing coals and oils. After allowing them to oxidize the sulfur to sulfate, the latter can be extracted from the fuel. One species of *Thiobacillus*, *T. ferrooxidans*, can oxidize iron as well as sulfur.

The Nitrifiers

The **nitrifiers** are a diverse group of Gram-negative bacteria that obtain energy by oxidizing inorganic nitrogen compounds such as ammonia or nitrite. These bacteria are of particular interest to farmers who fertilize their crops with ammonium nitrogen, a form of nitrogen that is retained by soils because its positive charge enables it to adhere to negatively charged soil particles. The potency and longevity of the fertilizer are affected by nitrifying bacteria converting the ammonia to nitrate. While plants use this form of nitrogen more readily, it is rapidly leached from soils. Nitrifying bacteria are also an important consideration in disposal of sewage or other wastes with a high ammonia concentration. As nitrifying bacteria oxidize nitrogen compounds, they consume O_2. Because of this, waters polluted with nitrogen-containing wastes can quickly become anaerobic.

The nitrifiers encompass two metabolically distinct groups of bacteria that typically grow in close association. Together, they can oxidize ammonia to form nitrate. The **ammonia oxidizers**, which include the genera *Nitrosomonas* and *Nitrosococcus*, convert ammonia to nitrite. The **nitrite oxidizers**, which include the genera *Nitrobacter* and *Nitrococcus*, then convert nitrite to nitrate. The latter group is particularly important in preventing the buildup of nitrite in soils, which is toxic and can leach into groundwater. The oxidation of ammonia to nitrate is called **nitrification** and is an important part of the nitrogen cycle. ■ **nitrogen cycle, p. 777**

Figure 11.12 Acid Drainage From a Mine Sulfur-oxidizing bacteria oxidize exposed metal sulfides, generating sulfuric acid. The yellow-red color is due to insoluble iron oxides.

The Hydrogen-Oxidizing Bacteria

Members of the Gram-negative genera *Aquifex* and *Hydrogenobacter* are among the few hydrogen-oxidizing bacteria that are obligate chemolithotrophs. These related organisms are thermophilic and typically inhabit hot springs. Some *Aquifex* species have a maximum growth temperature of 95°C, the highest of any members of the Domain Bacteria. The hydrogen-oxidizing bacteria are deeply branching in the phylogenetic tree, meaning that according to 16S rRNA studies, they were one of the earliest bacterial forms to exist on earth. The fact that they require O_2 seems contradictory to the presumed evolutionary position, but in fact, the low amount they require might have been available early on in certain niches due to photochemical processes that split water.

A wide range of aerobic chemoorganotrophs can also oxidize hydrogen gas. These organisms switch between energy sources as conditions dictate.

MICROCHECK 11.4

The sulfur oxidizers use sulfur compounds as an energy source, generating sulfuric acid. The nitrifiers oxidize nitrogen compounds such as ammonium or nitrite.

- How does the growth of *Thiobacillus* species cause bioleaching?
- Why would farmers be concerned about nitrifying bacteria?
- Why would sulfur-oxidizing bacteria accumulate sulfur?

Aerobic Chemoorganotrophs

Aerobic chemoorganotrophs oxidize organic compounds to obtain energy and use O_2 as a terminal electron acceptor. They include a tremendous variety of bacteria, ranging from those that inhabit very specific environments to those that are ubiquitous. This section will profile only representative genera that are found in a variety of different environments. Later sections will describe other chemotrophs, emphasizing those adapted to live in specific habitats.

Obligate Aerobes

Obligate aerobes obtain energy using respiration exclusively; none can use fermentation as an alternative.

The Genus *Micrococcus*

Members of the genus *Micrococcus* are Gram-positive cocci found in soil and on dust particles, inanimate objects, and skin. Because they are often airborne, they easily contaminate bacteriological media. There, they typically form pigmented colonies, a characteristic that aids in their recognition. The colonies of *M. luteus*, for example, are generally yellow (**figure 11.13**). Like members of the genus *Staphylococcus*, which will be discussed later, they tolerate arid conditions and can grow in 7.5% NaCl.

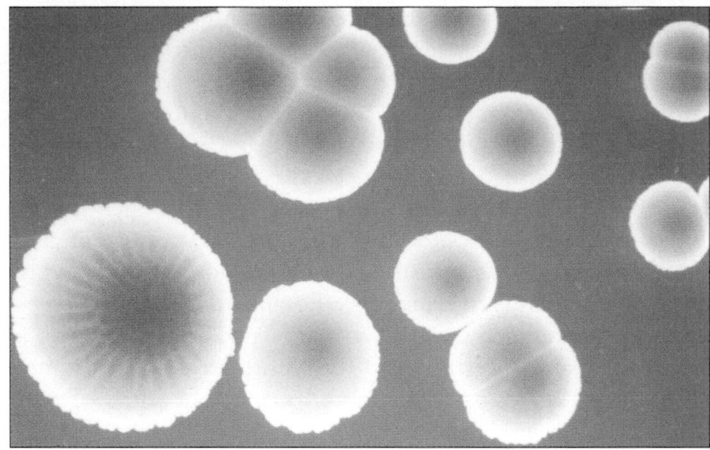

Figure 11.13 ***Micrococcus luteus*** **Colonies**

The Genus *Mycobacterium*

Mycobacterium species are widespread in nature and include harmless **saprophytes**, which live on dead and decaying matter, as well as organisms that produce disease in humans and domestic animals. Although they have a Gram-positive type of cell wall, these bacteria stain poorly because of a waxy lipid in their cell wall. Special procedures can be used to enhance the penetration of the stain. Once they are stained, they resist destaining, even when acidic decolorizing solutions are used. Because of this, *Mycobacterium* species are called **acid-fast**, and the acid-fast staining procedure is an important step in their identification (see figure 3.15). *Nocardia* species, a related group of bacteria that commonly reside in the soil, are also acid-fast. ■ acid-fast, p. 51

Mycobacterium species are generally pleomorphic rods; they often occur in chains that sometimes branch, or bunch together to form cordlike groups. Two species are notable for their effect on human health—*M. tuberculosis*, as the name implies, causes tuberculosis, and *M. leprae* causes Hansen's disease (leprosy). *Mycobacterium* species are more resistant to disinfectants than most other vegetative bacteria. In addition, they differ from other bacteria in their susceptibility to antimicrobial drugs.

The Genus *Pseudomonas*

Pseudomonas species are Gram-negative rods that are motile by polar flagella and that often produce pigments (**figure 11.14**).

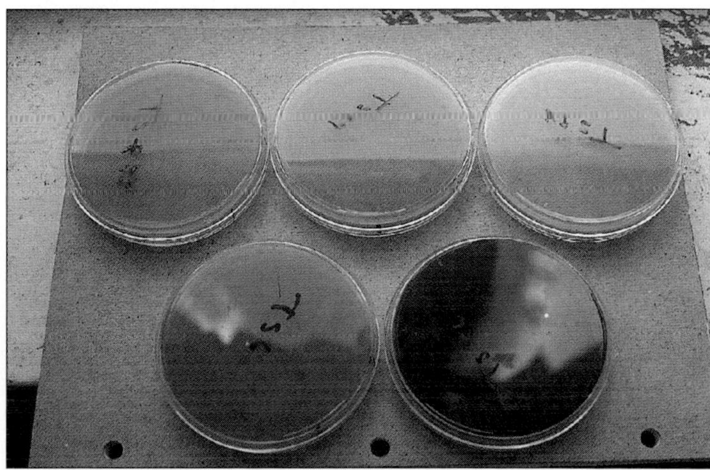

Figure 11.14 Pigments of *Pseudomonas* Species Cultures of different strains of *Pseudomonas aeruginosa*. Note the different colors of the water-soluble pigments.

Although most are strict aerobes, some can grow anaerobically if nitrate is available as a terminal electron acceptor. The fact that they do not ferment and are oxidase positive serve, in the laboratory, as important characteristics to distinguish them from members of the family *Enterobacteriaceae,* which will be discussed in a later section. ■ oxidase, p. 253

As a group, *Pseudomonas* species have extremely diverse biochemical capabilities. Some can use more than 80 different substrates, including unusual sugars, amino acids, and compounds containing aromatic rings. Because of this, *Pseudomonas* species play an important role in the degradation of many synthetic and natural compounds that resist breakdown by most other microorganisms. The ability to carry out some of these degradations is encoded by plasmids.

Pseudomonas species are widespread, typically inhabiting soil and water. While most are harmless, some cause disease in plants and animals. Medically, the most significant species is *P. aeruginosa*. It is a common **opportunistic pathogen**, meaning that it primarily infects people who have underlying medical conditions. Unfortunately, it can grow in nutrient-poor environments, such as water used in respirators, and it is resistant to many disinfectants and antimicrobial medications. Because of this, hospitals must be diligent to prevent its spread among patients. Many species that were formerly classified as *Pseudomonas* have recently been removed and placed in one of several new genera. This revision was prompted in part by ribosomal RNA studies. ■ *Pseudomonas aeruginosa*, p. 697

The Genera *Thermus* and *Deinococcus*

Thermus and *Deinococcus* are related genera that have scientifically and commercially noteworthy characteristics. *Thermus* species are thermophilic, as their name implies. This trait has proven to be extremely valuable because of their heat-stable enzymes, including DNA polymerase. An integral part of the polymerase chain reaction (PCR) is ***Taq* polymerase**, the DNA polymerase of *T. aquaticus.* Although they have an unusual cell wall, they stain Gram-negative. *Deinococcus* species are unique in their extraordinary resistance to the damaging effects of gamma radiation. For example, *D. radiodurans* can survive exposure to a dose several thousand times that lethal to a human being. The dose literally shatters the organism's genome into many fragments, yet enzymes are able to repair the extensive damage. Scientists anticipate that through genetic engineering, *Deinococcus* species may help clean up the soil and water contaminated by the 10 million cubic yards of radioactive waste that have accumulated in the United States. Although their unusual cell wall has multiple layers, they stain Gram-positive. ■ PCR, p. 237

Facultative Anaerobes

Facultative anaerobes preferentially use aerobic respiration if O_2 is available. As an alternative, however, they can use fermentative metabolism.

The Genus *Corynebacterium*

Members of the genus *Corynebacterium* commonly inhabit soil, water, and the surface of plants. They are Gram-positive pleomorphic rods that are often club-shaped and arranged to form V shapes or palisades (*koryne* is Greek for "club") **(figure 11.15)**. Bacteria that exhibit this characteristic microscopic morphology are sometimes referred to **coryneforms** or **diphtheroids.** *Corynebacterium* species are generally facultative anaerobes, although some are strict aerobes. They form volutin granules, which are storage forms of phosphate. These are sometimes called **metachromatic granules** to reflect their characteristic staining with the dye methylene blue. Many species of *Corynebacterium* reside harmlessly in the throat, but toxin-producing strains of *C. diphtheriae* can cause the disease diphtheria. ■ volutin, p. 71 ■ metachromatic granules, p. 71 ■ diphtheria, p. 556

The Family *Enterobacteriaceae*

Members of the family *Enterobacteriaceae*, frequently referred to as **enterics** or **enterobacteria**, are Gram-negative rods. Their name reflects the fact that most reside in the intestinal tract of humans and other animals (Greek *enteron* means "intestine"), although some thrive in rich soil. Enterics that are part of the normal flora of the intestine include *Enterobacter*, *Klebsiella*, and *Proteus* species as well as most strains of *E. coli*. Those that cause diarrheal disease include *Shigella* species, *Salmonella* Enteritidis, and some strains of *E. coli.* Life-threatening systemic diseases include typhoid fever, caused by *Salmonella* Typhi, and both the bubonic and pneumonic forms of plague, caused by *Yersinia pestis.* ■ diarrheal disease, p. 598 ■ typhoid fever, p. 606 ■ plague, p. 480

Members of the family *Enterobacteriaceae* are facultative anaerobes that ferment glucose and, if motile, have peritrichous flagella. The family includes about 40 recognized genera that can be distinguished using biochemical tests. A battery of such tests is often used to simultaneously determine the genus and species. Within a given species, many different strains have been described. These are often distinguished using serological tests that detect differences in their cell walls, flagella, and capsules **(figure 11.16)**. ■ peritrichous flagella, p. 68

Enterics that ferment lactose are included in a group called **coliforms**. This is an informal grouping of certain common

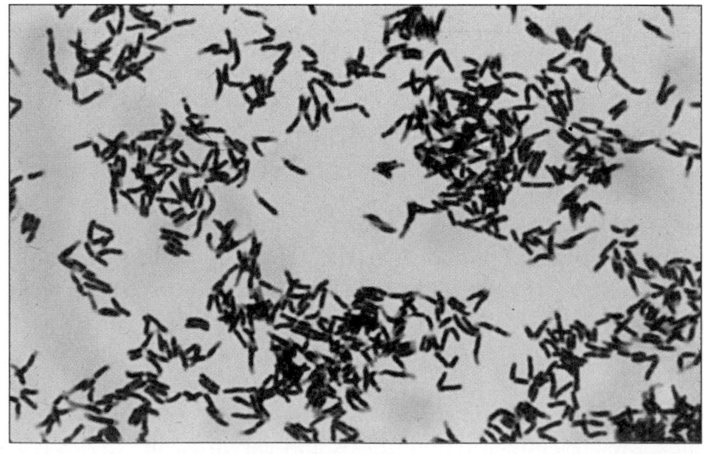

Figure 11.15 ***Corynebacterium*** The Gram-positive pleomorphic rods are often arranged to form V shapes or palisades.

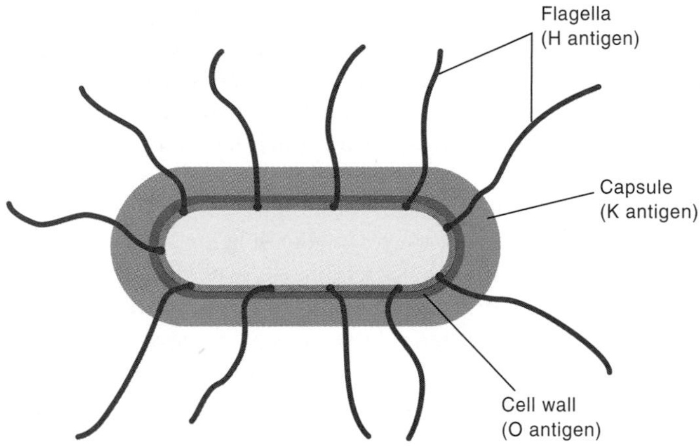

Figure 11.16 Schematic Drawing of a Member of the Family *Enterobacteriaceae* The cell structures used to distinguish different strains are shown.

intestinal inhabitants such as *E. coli* that are easy to detect in food and water; they are used by regulatory agencies as an indicator of fecal pollution. Their presence indicates a potential health risk because fecal-borne pathogens might also be present. Typically, coliforms are enumerated using membrane filtration or the most probable number (MPN) method. ■ **coliform, p. 791** ■ **membrane filtration, p. 91** ■ **MPN, p. 92**

MICROCHECK 11.5

Members of the genera *Micrococcus*, *Mycobacterium*, *Pseudomonas*, *Thermus*, and *Deinococcus* are obligate aerobes that generate ATP by degrading organic compounds, using O_2 as a terminal electron acceptor. Most *Corynebacterium* species and all members of the family *Enterobacteriaceae* are facultative anaerobes.

- What is the significance of coliforms being present in drinking water?
- What unique characteristic makes members of the genus *Deinococcus* noteworthy?
- Why would it be an advantage for a *Pseudomonas* species to encode enzymes for degrading certain compounds on a plasmid rather than the chromosome?

E C O P H Y S I O L O G Y

As a group, prokaryotes show remarkable diversity in their physiological adaptations to a range of habitats. From the hydrothermal vents of the deep ocean to the frozen expanses of Antarctica, prokaryotes have evolved to thrive in virtually all environments, including many that most plants and animals would find inhospitable.

Thriving in Terrestrial Environments

Microorganisms that inhabit soil must endure a variety of conditions. Daily and seasonally, soil is an environment that can routinely alternate between wet and dry as well as warm and cold. The availability of nutrients can also cycle from abundant to sparse. To thrive in this ever-changing environment, microbes have evolved mechanisms to cope with adverse conditions and to exploit unique sources of nutrients.

Bacteria that Form a Resting Stage

Several genera that inhabit the soil can form a resting stage that enables them to survive the dry periods that are typical in many soils. Of these various types of dormant cells, endospores are by far the most resistant to environmental extremes.

Endospore-formers

Bacillus and *Clostridium* species are common Gram-positive rod-shaped bacteria that form endospores; the position of the spore in the cell can be use as an aid in identification (**figure 11.17**). *Clostridium* species, which are obligate anaerobes, were

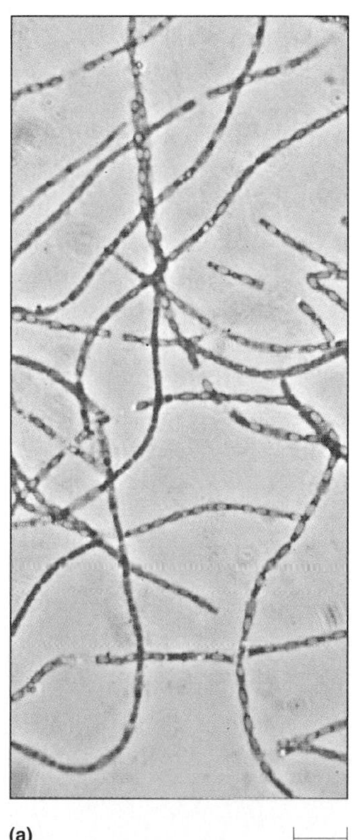

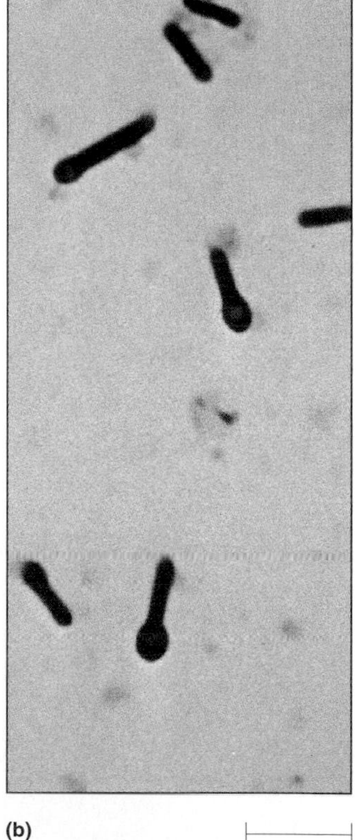

Figure 11.17 Endospore-Formers (a) Endospores forming in the midportion of the cells of *Bacillus anthracis*. **(b)** Endospores forming at the ends of the cells in *Clostridium tetani*. Both of these species can cause fatal disease, but many other species of endospore-formers are harmless.

discussed earlier. *Bacillus* species may be either obligate aerobes or facultative anaerobes, and many have notable characteristics. A suspension of endospores of *B. stearothermophilus*, a thermophilic species, is used as a biological indicator in quality control procedures to ensure that an autoclave is working properly. *Bacillus anthracis* causes the disease anthrax, which can be acquired from contacting its endospores in soil or in animal hides or wool. If conditions allow, the endospores will germinate into vegetative, disease-causing bacteria. ■ autoclave, p. 118 ■ anthrax, p. 491

The Genus *Azotobacter*

Azotobacter species are Gram-negative pleomorphic, rod-shaped bacteria that live in alkaline soil. They can form a type of resting cell called a **cyst (figure 11.18)**. These have negligible metabolic activity and can withstand drying and ultraviolet radiation but are not highly resistant to heat. Their formation involves changes in the cell wall of a vegetative bacterium, causing the cell to shorten and the wall to thicken; the enzyme content of the cyst is similar to that of the cell from which it arises. Cysts differ from endospores in both their manner of formation and their lesser degree of resistance to harmful agents.

Azotobacter species are also notable for their ability to fix nitrogen in aerobic conditions; recall that the enzyme nitrogenase is inactivated by O_2. Apparently, the exceedingly high respiratory rate of *Azotobacter* species consumes O_2 so rapidly that an anaerobic environment is produced inside the cell.

Myxobacteria

The myxobacteria are a group of aerobic Gram-negative rods that have a unique developmental cycle as well as a resting stage. When conditions are favorable, cells secrete a slime layer that other cells then follow, creating a **swarm** of cells. But then, when nutrients are exhausted, the behavior of the group changes. The cells begin to move toward each other and congregate. These then pile up to form a complex structure called a **fruiting body**, which is often brightly colored **(figure 11.19)**. In some species the fruiting body is quite elaborate, consisting of a mass of cells elevated and supported by a stalk made of a hardened slime. The cells within the fruiting body differentiate to become a spherical, dormant form called a **microcyst**. These are considerably more resistant to heat, drying, and radiation than are the vegetative cells of myxobacteria, but they are much less resistant than bacterial endospores.

Myxobacteria are important in nature as degraders of complex organic substances; they can digest living or dead bacteria and certain algae and fungi. Scientifically, even though the mechanisms the cells use for group communication are still poorly understood, these bacteria serve as an important model for developmental biology. Included in the myxobacteria are members of the genera *Chondromyces*, *Myxococcus*, and *Stigmatella*.

The Genus *Streptomyces*

The genus *Streptomyces* encompasses more than 500 species of Gram-positive bacteria that resemble fungi in their pattern of growth. Like the fungi, they form a **mycelium**, which is a visible mass of branching filaments. The filaments are called **hyphae**. At the tips of the hyphae, chains form of characteristic spores called

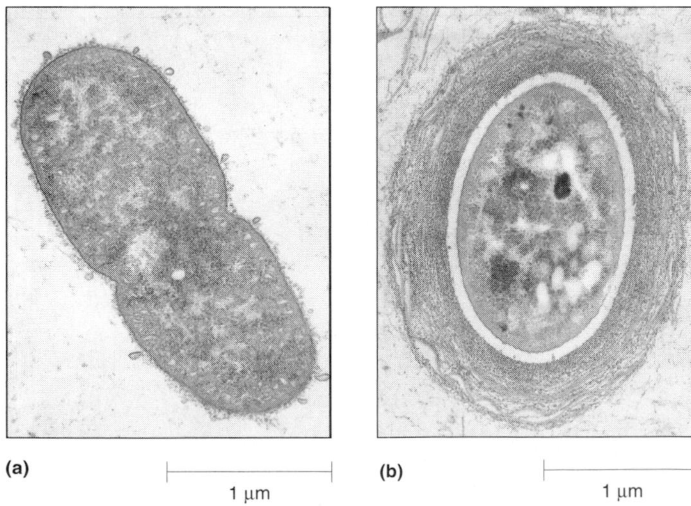

Figure 11.18 ***Azotobacter*** **(a)** Vegetative cells; **(b)** cyst.

(a)

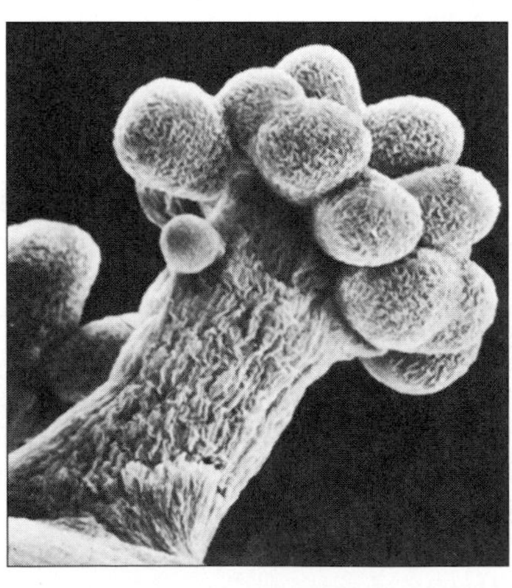

(b)

Figure 11.19 Fruiting Bodies of Myxobacteria These are the elaborate fruiting bodies of a species of *Chondromyces*: **(a)** photograph; **(b)** scanning electron micrograph.

conidia (figure 11.20). These dormant spores are desiccation resistant and are readily dispersed in air currents to potentially more favorable locations. Note that while this pattern of growth resembles fungi, which are eukaryotes, *Streptomyces* species are much smaller and are prokaryotes.

Streptomyces species are obligate aerobes that produce a variety of extracellular enzymes, which enables them to degrade a variety of organic compounds. They are also responsible for the characteristic "earthy" odor of soil; like the cyanobacteria, they produce geosmin. One species of *Streptomyces, S. somaliensis*, can cause an infection of subcutaneous tissue called an actinomycetoma.

Streptomyces species naturally produce a wide array of medically useful antibiotics, including streptomycin, tetracycline, and erythromycin. The role that these antimicrobial compounds play in the life cycle of *Streptomyces* has not been proven, but it is quite possible that they provide the organism a competitive advantage.

Bacteria that Associate with Plants

Members of two related genera use very different means to obtain the nutrients needed for growth from plants. *Agrobacterium* species are plant pathogens that cause tumorlike growths, whereas *Rhizobium* species form a mutually beneficial relationship with certain types of plants.

The Genus *Agrobacterium*

Agrobacterium species have an unusual mechanism of gaining a competitive advantage in soil. They cause plant tumors, which are a manifestation of their ability to genetically alter a plant for their own benefit **(figure 11.21)**. These Gram-negative rod-shaped bacteria enter the plant via a wound, and they then transfer to the plant a portion of a plasmid; in *Agrobacterium tumefaciens* that plasmid is called the **Ti plasmid** (for "tumor-inducing"). The transferred DNA encodes the ability to synthesize a specific plant growth hormone, causing uncontrolled growth of the plant tissue and resulting in a tumor. The transferred DNA also encodes enzymes that direct the synthesis of an **opine,** an unusual amino acid derivative; *Agrobacterium* can then use this compound as a nutrient source (see Perspective 8.2).

Scientists have now modified the Ti plasmid, turning it into a commercially valuable tool. By removing those genes that cause tumor formation, the plasmid can be used as a vector to introduce DNA into plant cells. ■ vector, p. 224

The Genus *Rhizobium*

Rhizobium species can fix nitrogen, a trait that is exploited in their intimate relationship with **legumes**, a group of plants that bear seeds in pods. *Rhizobium* species live within cells in nodules formed on the root **(figure 11.22)**. The plant synthesizes the protein **leghemoglobin**, which binds and controls the levels of O_2. Within the resulting microaerobic confines of the nodule, the bacteria are able to fix nitrogen. A *Rhizobium* strain within a plant cell is an example of an **endosymbiont**, providing a benefit to the cell it resides within. *Rhizobium* species are Gram-negative. ■ symbiotic nitrogen fixers, p. 780

MICROCHECK 11.6

Bacillus and *Clostridium* species produce endospores, the most resistant life form known. *Azotobacter* species, myxobacteria, and *Streptomyces* species all produce dormant forms that tolerate some adverse conditions but are less resistant than endospores. Members of the genera *Agrobacterium* and *Rhizobium* derive nutrients from plants, although the former are plant pathogens and the latter benefit the plant.

- How does an endospore differ from a cyst of *Azotobacter*?
- How does *Agrobacterium* derive a benefit from inducing a plant tumor?
- If you wanted to determine the number of endospores in a sample of soil, what could you do before plating it?

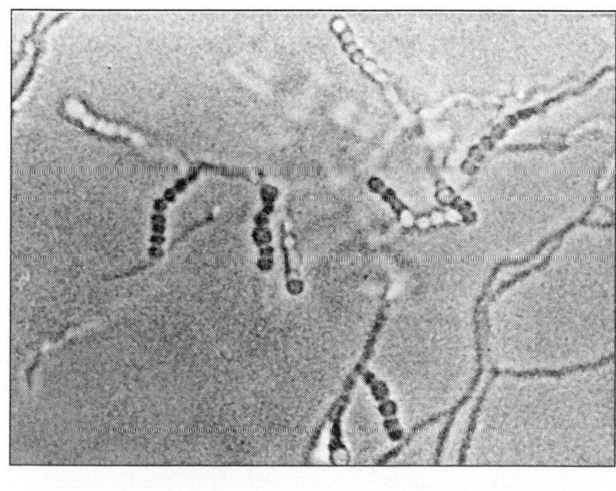

Figure 11.20 ***Streptomyces*** A photomicrograph showing the spherical conidia at the ends of the filamentous hyphae.

Figure 11.21 Plant Tumor Caused by *Agrobacterium tumefaciens*

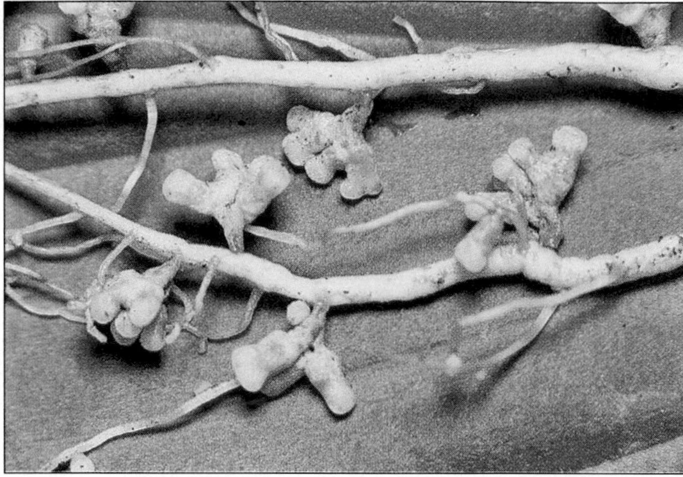
(a)

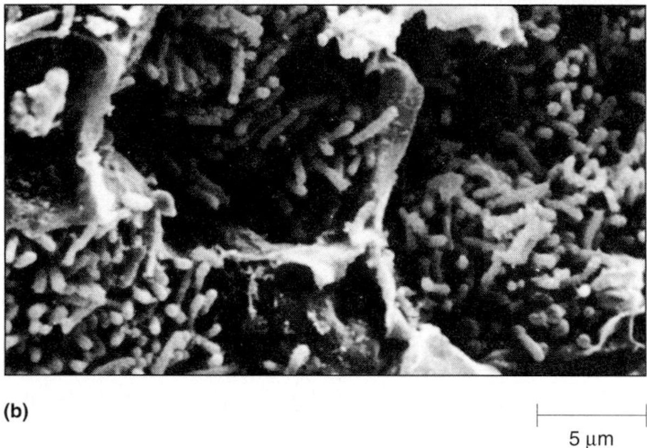
(b)

Figure 11.22 Symbiotic Relationship Between *Rhizobium* Species and Certain Plants **(a)** Root nodules that contain a *Rhizobium* strain. **(b)** Scanning electron micrograph of *Rhizobium* cells within a nodule.

Thriving in Aquatic Environments

Most aquatic environments lack a steady supply of nutrients. To thrive in these habitats, bacteria have evolved various mechanisms to maximize nutrient acquisition and retention. This can mean clustering within a sheath that can attach to solid objects in favorable locations, producing extensions called prosthecae that maximize the absorptive surface area, exploiting nutrients provided by other organisms, efficiently moving to more favorable locations, or forming storage granules. ■ **storage granules, p. 71**

Sheathed Bacteria

Sheathed bacteria form chains of cells encased within a tube, or **sheath (figure 11.23)**. This is thought to provide a protective function, helping the bacteria attach to solid objects located in favorable habitats while sheltering them from attack by predators. Masses of these filamentous sheaths can often be seen streaming from rocks or wood in flowing water polluted by nutrient-rich effluents. They often interfere with sewage treatment and other industrial processes by clogging pipes. Sheathed

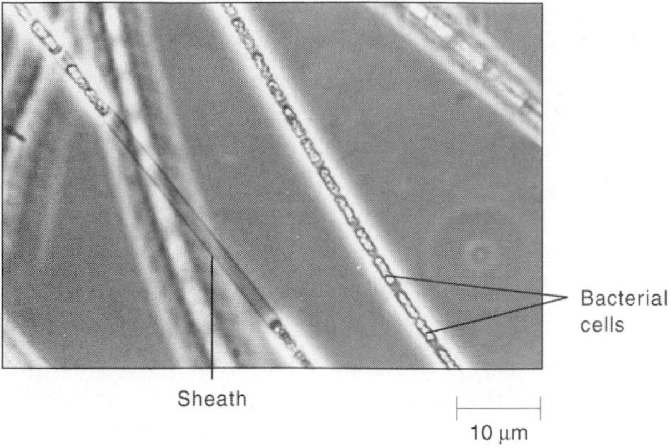

Figure 11.23 Sheathed Bacteria Phase-contrast photomicrograph of a *Sphaerotilus* species.

bacteria include species of *Sphaerotilus* and *Leptothrix*, which are Gram-negative rods.

Sheathed bacteria disperse themselves by forming **swarmer cells**, which have polar flagella and exit through the unattached end of the sheath. These motile cells move to a new solid surface, where they attach. If enough nutrients are present, they can multiply and form a new sheath, which elongates as the chain of cells grows.

Prosthecate Bacteria

A diverse group of Gram-negative bacteria possess projections called **prosthecae**, which are extensions of the cytoplasm and cell wall. These extensions are thought to provide increased surface area to facilitate absorption of nutrients. Some enable the organisms to attach to a solid surface.

The Genus *Caulobacter*

Because of their remarkable life cycle, *Caulobacter* species have served as a model for the research of cellular differentiation. Entirely different events occur in an orderly fashion at opposite ends of the cell. *Caulobacter* cells have a single polar prostheca, commonly called a **stalk (figure 11.24)**. At the tip of the stalk is an adhesive **holdfast**, which provides a mechanism for attachment. To multiply, the cell elongates and divides by binary fission at the end opposite its stalk, producing a motile swarmer cell. This swarmer cell has a single polar flagellum, located at the pole opposite the site of division. This motile cell detaches and moves to a new location, where it adheres via a holdfast near the base of its flagellum. It then loses its flagellum, replacing it with a stalk. Only then can the daughter cell replicate its DNA and repeat the process. In favorable conditions, a single cell can divide and produce daughter cells many times. With each division a ring remains at the site of division, enabling a researcher to count the number of progeny.

The Genus *Hyphomicrobium*

Hyphomicrobium species are in many ways similar to *Caulobacter* species, except they have a distinct method of reproduction. The single polar prostheca of the parent cell enlarges at the tip to form a bud **(figure 11.25)**. This continues enlarging and

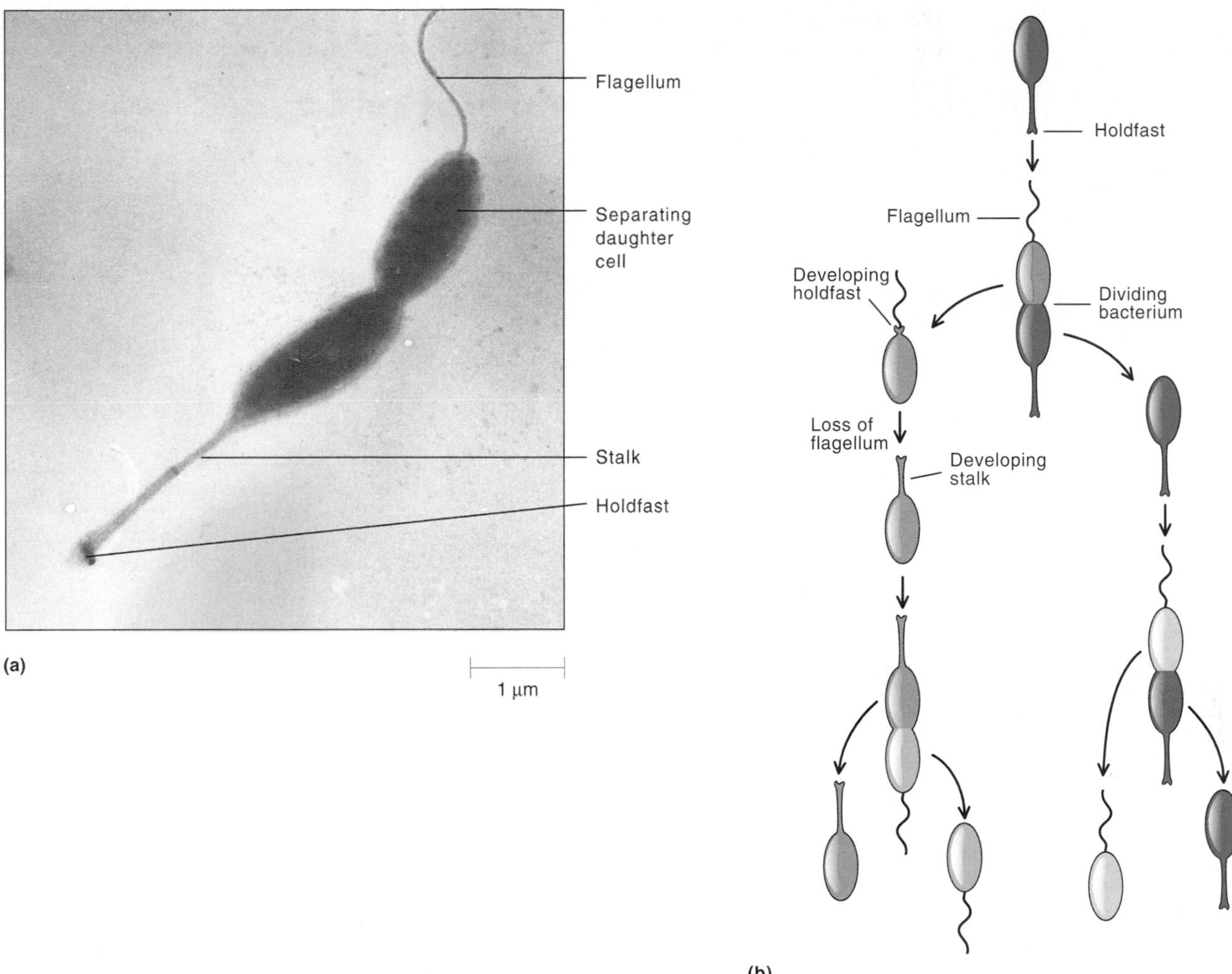

Figure 11.24 ***Caulobacter*** **(a)** Photomicrograph; **(b)** life cycle.

develops a flagellum, eventually giving rise to a motile daughter cell. The daughter cell then detaches and moves to a new location, eventually losing its flagella and forming a polar prostheca at the opposite end to repeat the cycle. As with *Caulobacter* species, a single cell can sequentially produce multiple daughter cells.

Bacteria that Derive Nutrients from Other Organisms

Some bacteria obtain nutrients directly from other organisms. *Bdellovibrio* species do this by preying on other bacteria, attacking them and digesting their contents, eventually killing them. Bioluminescent bacteria use another tactic; they have established relationships with fish and squid in which the animal provides nutrients and protection while the bacterium provides a source of luminescence. *Legionella* species can live intracellularly within the protected confines of protozoa.

The Genus *Bdellovibrio*

Bdellovibrio species (*bdello*, from the Greek word for "leech") are highly motile Gram-negative bacteria that prey on *E. coli* and other Gram-negative bacteria **(figure 11.26)**. They are small, curved rods, approximately 0.25 μm wide and 1 μm long, with a polar flagellum. When a *Bdellovibrio* cell attacks, it strikes its prey with such a high velocity that it actually propels the prey a short distance. The parasite then attaches to its host and rotates with a spinning motion. At the same time, it synthesizes digestive enzymes that break down lipids and peptidoglycan. Within 10 minutes, a hole in the cell wall of the prey is formed. This allows the parasitic bacterium to penetrate, lodging in the periplasm between the cytoplasmic membrane and the peptidoglycan layer. There, over a period of several hours, *Bdellovibrio* degrades and utilizes the prey's cellular contents. It derives energy by aerobically oxidizing amino acids and acetate. The parasite increases in length as it resides in the periplasm, ultimately dividing to form several motile daughter cells. When the host cell lyses, the *Bdellovibrio* progeny are released to find new hosts, repeating the growth and reproduction cycle.

Bioluminescent Bacteria

Some species of *Photobacterium* and *Vibrio* can emit light **(figure 11.27)**. This phenomenon, called **bioluminescence**, plays an important role in the symbiotic relationship between some of these bacteria and specific types of fish and squid. For example, certain types of squid have a specialized organ within the ink sac that is colonized by *Vibrio fischeri*. The light produced in the organ is thought to serve as a type of camouflage, obscuring the squid's contrast against the light from above and

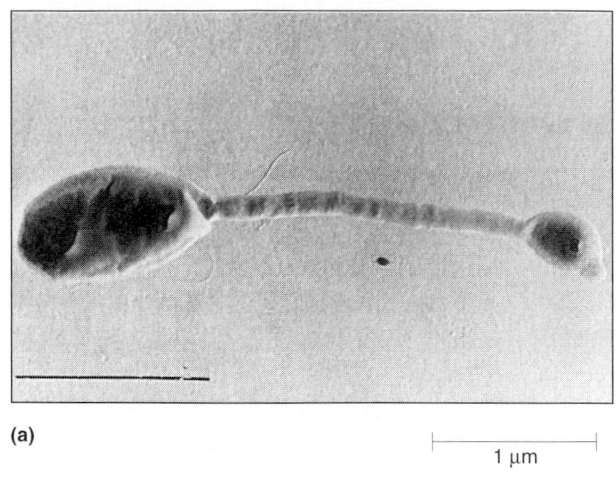

(a) 1 μm

Hypha forming

New nucleoid moving into hypha

Young bud

Swarmer cell with subpolar to lateral flagellum (one to three)

Hypha lengthens more and produces another bud

(b)

Figure 11.25 ***Hyphomicrobium*** **(a)** Photomicrograph. Note the bud forming at the tip of the polar prostheca. **(b)** Life cycle.

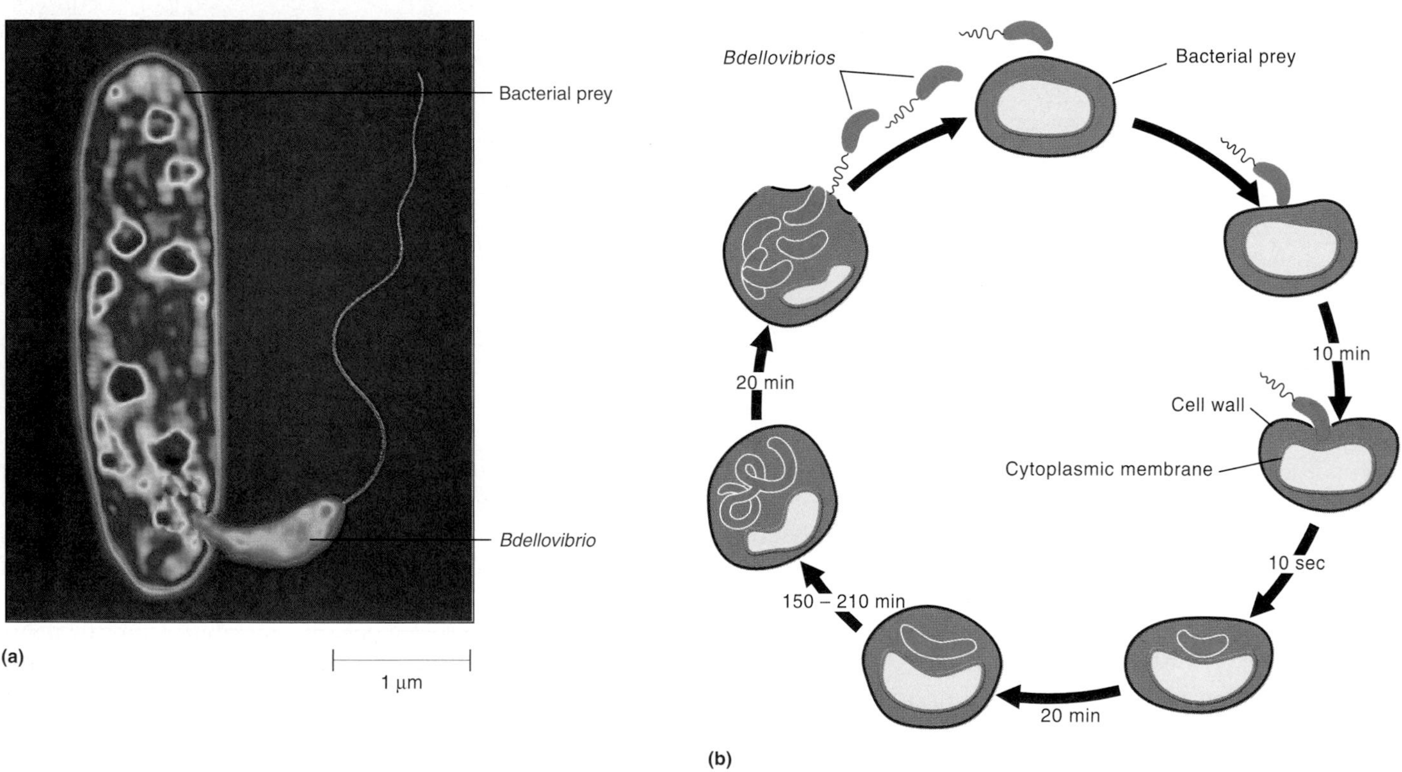

Figure 11.26 ***Bdellovibrio*** **(a)** Color-enhanced transmission micrograph of *Bdellovibrio bacteriovorus* attacking its prey. **(b)** Life cycle of *B. bacteriovorus*. Note that the diagram exaggerates the size of the space in which *B. bacteriovorus* multiplies.

any shadow it might otherwise cast. It appears that the squid provides nutrients to the symbiotic bacteria, facilitating their growth. Another example is the flashlight fish, which has a light organ in a specialized pouch, below its eye that harbors bioluminescent bacteria. By opening and closing a lid that covers the pouch the fish can control the amount of light released, which is believed to confuse predators and prey.

Luminescence is catalyzed by the enzyme **luciferase**. Studies of the regulation of its synthesis revealed that the genes encoding it are only expressed when the density of the bacterial population reaches a critical point. This phenomenon of **quorum sensing** is now recognized as an important mechanism used by a variety of different bacteria to regulate the expression of certain genes. ■ **quorum sensing, p. 184**

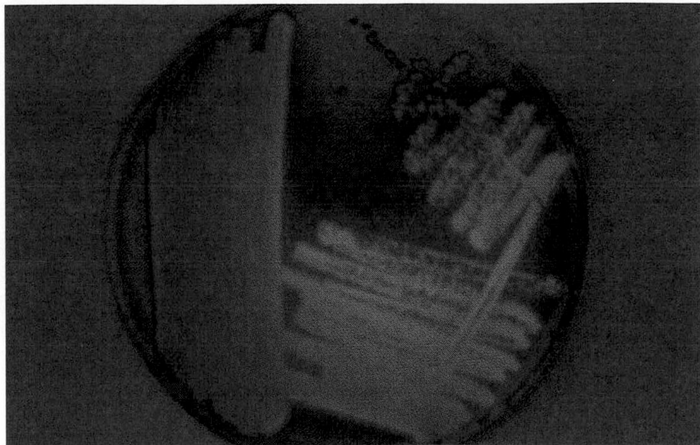

(a)

(b)

Figure 11.27 **Luminescent Bacteria** **(a)** Plate cultures of luminescent bacteria. **(b)** Photograph of a flashlight fish; under the eye is a light organ colonized with luminescent bacteria.

Members of the genera *Photobacterium* and *Vibrio* are Gram-negative rods (the rods of *Vibrio* species are curved) with polar flagella. They are facultative anaerobes and typically inhabit aqueous environments. Those species that require sodium for growth are usually found in marine environments. Not all are luminescent, and some species of *Vibrio* cause human disease. Medically important species include *V. cholerae*, which causes cholera, and *V. parahaemolyticus*, which causes gastrointestinal disease.

The Genus *Legionella*

Legionella species are commonly found in aquatic environments, where they often reside within protozoa. They have even been isolated from water in air conditioners and produce misters. They are Gram-negative obligate aerobes that utilize amino acids but not carbohydrates as a source of carbon and energy. *Legionella pneumophila* can cause respiratory disease when it is inhaled in aerosolized droplets.

Bacteria that Move by Unusual Mechanisms

Some bacteria have unique mechanisms of motility, which enable them to easily move to desirable locations. These organisms include the spirochetes and the magnetotactic bacteria.

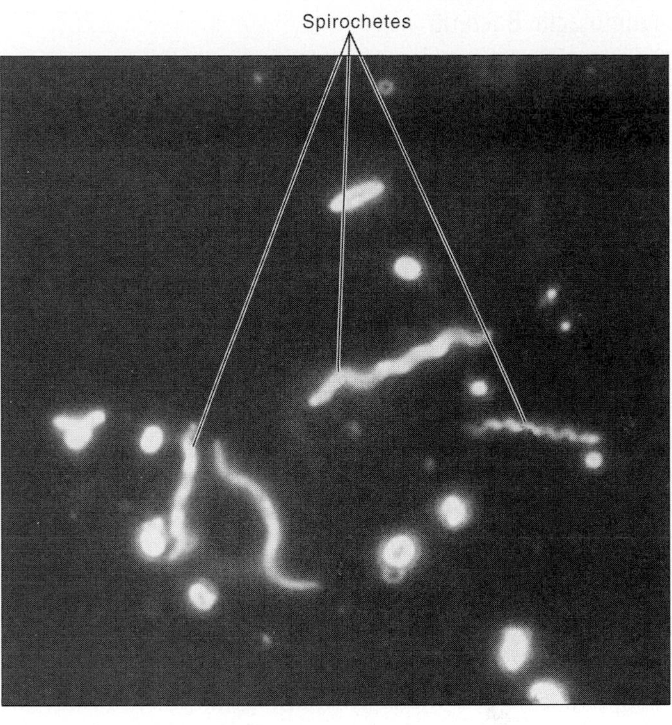

Figure 11.28 **Spirochetes** Dark-field photomicrograph of spirochetes.

Spirochetes

The **spirochetes** (Greek *spira* for "coil" and *chaete* for "hair") are a group of Gram-negative bacteria with a unique motility mechanism that enables them to move through thick, viscous environments such as mud. Distinguishing characteristics include their spiral shape, flexible cell wall, and motility by means of an **axial filament**. The axial filament is composed of sets of flagella that originate from both poles of the cell; unlike typical flagella, these are contained within the periplasm. The opposing sets of flagella extend toward each other, overlapping in the mid-region of the cell. Rotation of the flagella within the confines of the outer membrane causes the cell to move like a corkscrew, sometimes deviating into flexing motions. Many spirochetes are very slender and can only be seen using special methods such as dark-field microscopy (**figure 11.28**). Many are also difficult or impossible to cultivate, and their classification is based largely on their morphology and ability to cause disease.

Spirochetes include free-living bacteria that inhabit aquatic environments, as well as species that reside on or in animals. Members of the genus *Spirochaeta* are anaerobes or facultative anaerobes that thrive in muds and anaerobic waters. *Leptospira* species are aerobic; some species are free-living in aquatic environments, whereas others thrive within animals. *Leptospira interrogans* causes the disease leptospirosis, which can be transmitted in the urine of infected animals. Spirochetes that are adapted to reside in body fluids of humans and other animals will be discussed later. ■ **leptospirosis, p. 636**

Magnetotactic Bacteria

Magnetotactic bacteria such as *Magnetospirillum* (*Aquaspirillum*) *magnetotacticum* contain a string of magnetic crystals that align them with the earth's magnetism (see figure 3.42). This enables them to efficiently move up or down in the water or sediments. It is thought that this unique type of movement enables them to locate the microaerophilic niches they require. *Magnetospirillum* species are Gram-negative spiral-shaped organisms. ■ **magnetotaxis, p. 69**

Bacteria that Form Storage Granules

A number of aquatic bacteria form granules that serve to store nutrients. Recall that anoxygenic phototrophs often store sulfur granules, which can later be used as a source of electrons for reducing power. Some bacteria store phosphate, and others store compounds that can be used to generate ATP.

The Genus *Spirillum*

Members of the genus *Spirillum* are Gram-negative spiral-shaped, microaerophilic bacteria. *Spirillum volutans* forms volutin granules, which are storage forms of phosphate. These are sometimes called **metachromatic granules** to reflect their characteristic staining with the dye methylene blue. The cells of *S. volutans* are typically large, over 20 μm in length. In wet mounts, *Spirillum* species may be seen moving to a narrow zone near the edge of the coverslip, where O_2 is available in the optimum amount. ■ **volutin, p. 71**

Sulfur-Oxidizing, Nitrate-Reducing Marine Bacteria

Some marine bacteria store both sulfur, which can be oxidized as an energy source, and nitrate, which can serve as an electron acceptor. This provides an advantage to the bacteria, because anaerobic marine sediments are often abundant in reduced sulfur compounds but deficient in suitable terminal electron acceptors. In contrast, waters above those sediments lack reduced sulfur compounds but provide a source of nitrate. In other words, the energy source and terminal electron acceptor do not coexist. *Thioploca* species respond by forming long sheaths within which filamentous cells shuttle between the sulfur-rich sediments and nitrate-rich waters, storing reserves of sulfur and nitrate in order to obtain energy.

Scientists recently discovered a huge bacterium in the ocean sediments off the coast of the African country of Namibia that they named *Thiomargarita namibiensis*, "sulfur pearl of Namibia" (see Perspective 1.1). The cells are a pearly white color due to globules of sulfur in their cytoplasm. They each contain a large nitrate storage vacuole. It is thought that these cells, which may reach a diameter of ¾ mm, can store a 3-month supply of sulfur and nitrate. The cells are not motile and instead appear to rely on occasional disturbances such as storms to bring them into contact with the nitrate-rich water.

MICROCHECK 11.7

Prokaryotes have evolved a variety of mechanisms to help them thrive in aquatic habitats. Sheathed bacteria cluster within a tube that can attach to solid objects in favorable locations. Prosthecate bacteria produce extensions that maximize the absorptive surface area. *Bdellovibrio* species, bioluminescent bacteria, and *Legionella* species exploit nutrients provided by other organisms. Spirochetes and magnetotactic bacteria move by unusual mechanisms to more favorable locations. Some organisms form storage granules that contain compounds such as phosphate and sulfur.

- What characteristic of *Caulobacter* species makes them an important model for research?
- How do squid benefit from having a light organ colonized by luminescent bacteria?
- The genomes of free-living spirochetes are larger than those of ones that live within an animal host. Why would this be so?

Animals As Habitats

The bodies of animals, including humans, provide a wide variety of ecological niches in which prokaryotes reside—from arid, O_2 rich surfaces to moist, anaerobic recesses.

Bacteria that Inhabit the Skin

The skin is typically dry and salty, providing an environment inhospitable to many microorganisms. Members of the genus *Staphylococcus*, however, thrive under these conditions. The propionic acid bacteria, which were discussed earlier, inhabit anaerobic microenvironments of the skin.

The Genus *Staphylococcus*

Staphylococcus species are Gram-positive cocci that are facultative anaerobes. Most, such as *S. epidermidis*, reside harmlessly as a component of the normal flora of the skin. Like most bacteria that aerobically respire, *Staphylococcus* species are catalase positive. This distinguishes them from *Streptococcus*, *Enterococcus*, and *Lactococcus* species, which are also Gram-positive cocci but are obligate fermenters that lack the enzyme catalase. Several species of *Staphylococcus* are notable for their medical significance. *Staphylococcus aureus* causes a variety of diseases, including skin and wound infections, as well as food poisoning. *Staphylococcus saprophyticus* causes urinary tract infections. Even *Staphylococcus* species that are typically part of the normal flora can cause infections in people who have underlying medical problems.

Bacteria that Inhabit Mucous Membranes

Mucous membranes of the respiratory tract, genitourinary tract, and respiratory tract provide a habitat for numerous kinds of bacteria, many of which have already been discussed. For example, *Streptococcus* and *Corynebacterium* species reside in the respiratory tract, *Lactobacillus* species inhabit the vagina, and *Clostridium* species and members of the family *Enterobacteriaceae* thrive in the intestinal tract. Some of the other genera are discussed next.

The Genus *Bacteroides*

Members of the genus *Bacteroides* are small, strictly anaerobic Gram-negative rods and coccobacilli. They inhabit the mouth, intestinal tract, and genital tract of humans and other animals. *Bacteroides fragilis* and related species constitute about 30% of the bacteria in human feces and are often responsible for abscesses and bloodstream infections that follow appendicitis and abdominal surgery. Since many are killed even by brief exposure to O_2, they are difficult to study.

The Genus *Bifidobacterium*

Bifidobacterium species are irregular Gram-positive rod-shaped anaerobes that reside primarily in the intestinal tract of humans and other animals. They are the predominant members of the intestinal flora of breast-fed infants and are thought to provide a protective function by excluding disease-causing bacteria. Formula-fed infants are also colonized with members of this genus, but generally the concentrations are lower. *Bifidobacterium* species ferment sugars to produce lactic acid and acetic acid.

The Genera *Campylobacter* and *Helicobacter*

Members of the genera *Campylobacter* and *Helicobacter* are genetically related curved Gram-negative rods that include species of major medical importance. As microaerophiles, they require specific atmospheric conditions to be successfully cultivated. *Campylobacter jejuni* causes diarrheal disease in humans. It typically resides in the intestinal tract of domestic animals, particularly poultry. *Helicobacter pylori* inhabits the stomach, where it can cause stomach and duodenal ulcers. It has also been implicated in the development of stomach cancer. An important factor in its ability to survive in the stomach is its production of the enzyme **urease**. This enzyme breaks down urea to produce ammonia, which neutralizes the acid in the immediate surroundings. **■ stomach ulcers, p. 592**

The Genus *Haemophilus*

Members of the genus *Haemophilus* are tiny Gram-negative coccobacilli that, as their name reflects, are "blood loving." They require one or more compounds found in blood, such as hematin and NAD. Many species are common flora of the respiratory tract, but *H. influenzae* can also cause ear infections, respiratory infections, and meningitis, primarily in children. *Haemophilus ducreyi* causes the sexually transmitted disease chancroid. **■ ear infections, p. 558 ■ meningitis, p. 668 ■ chancroid, p. 651**

The Genus *Neisseria*

Neisseria species are Gram-negative kidney bean shaped cocci that typically occur in pairs. They are common inhabitants of animals including humans, growing in the oral cavity and other mucous membranes. *Neisseria* species are obligate aerobes, and many are nutritionally fastidious. Those noted for their medical significance include *N. gonorrhoeae*, which causes the sexually transmitted disease gonorrhea, and *N. meningitidis*, which causes meningitis. **■ fastidious, p. 100 ■ gonorrhea, p. 644 ■ meningitis, p. 669**

The Genus *Mycoplasma*

Members of the genus *Mycoplasma* are part of the taxonomic class characterized by the absence of a cell wall. Because these bacteria lack cell walls, they are pliable and can pass through the pores of filters that retain other bacteria. Most have sterols in their membrane, which provides added strength and rigidity, protecting the cells from osmotic lysis. They are among the smallest forms of life, and their genomes are thought to be the minimum size for encoding the essential functions for a free-living organism; the genome of *Mycoplasma genitalium* is only 5.8×10^5 base pairs, which is approximately one-eighth the size of the *E. coli* genome.

Medically, the most significant member of this group is *M. pneumoniae*, which, as its name implies, causes a form of pneumonia. This type of pneumonia cannot be treated with penicillin and other antimicrobial drugs that interfere with peptidoglycan synthesis, because these organisms lack a cell wall. Colonies of *Mycoplasma* species growing on solid media produce a characteristic "fried egg" appearance **(figure 11.29)**.

The Genera *Treponema* and *Borrelia*

Members of the genera *Treponema* and *Borrelia* are spirochetes that typically inhabit body fluids and mucous membranes of humans and other animals. Recall that spirochetes are characterized by their corkscrew shape and axial filaments. Although they have a Gram-negative cell wall, they are often too thin to be viewed by conventional microscopy. *Treponema* species are obligate anaerobes or microaerophiles that often inhabit the mouth and genital tract. Study of the species that causes syphilis, *T. pallidum*, is difficult because it has never been grown in culture. Its genome has been sequenced, however, providing evidence that it is a microaerophile with a metabolism that is highly dependent on its host. It lacks critical enzymes of the TCA cycle and a variety of other pathways.

Three species of *Borrelia* are medically significant microaerophiles that are transmitted by arthropods such as ticks and lice. *B. recurrentis* and *B. hermsii* both cause relapsing fever; *Borrelia burgdorferi* causes Lyme disease. A striking feature of *Borrelia* species is their genome, which is composed of a linear chromosome and many linear and circular plasmids.

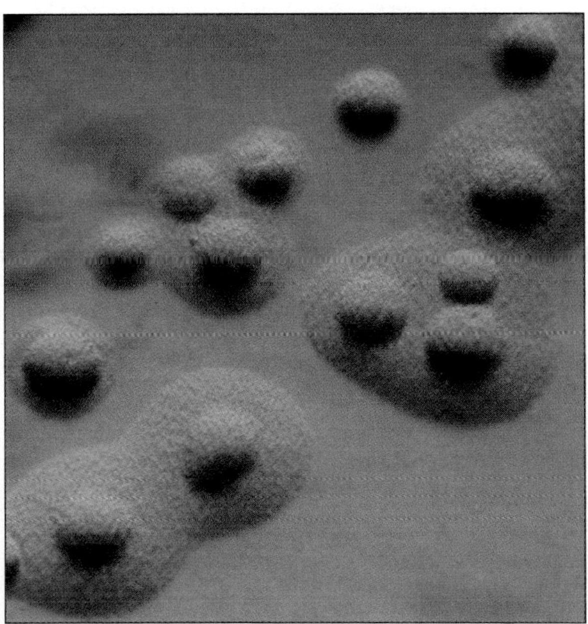

Figure 11.29 ***Mycoplasma pneumonia* Colonies** Note the dense central portion of the colony, giving it the typical "fried egg" appearance.

Obligate Intracellular Parasites

Obligate intracellular parasites are unable to reproduce outside a host cell. In general, the term only refers to those that infect eukaryotic cells. By living within eukaryotic cells, these bacteria are supplied with a readily available source of compounds they would otherwise need to synthesize for themselves. As a result, most species of these genera have lost the ability to synthesize substances needed for extracellular growth. Bacterial examples include members of the genera *Rickettsia*, *Orientia*, *Ehrlichia*, *Coxiella*, and *Chlamydia*, which are all tiny Gram-negative rods or coccobacilli.

The Genera *Rickettsia*, *Orientia*, and *Ehrlichia*

Species of *Rickettsia*, *Orientia*, and *Ehrlichia* are responsible for several serious human diseases that are spread when a blood-sucking arthropod such as a tick or louse that has fed on an infected animal transfers bacteria to a human from whom it takes another blood meal. *Rickettsia rickettsii* causes Rocky Mountain spotted fever, *R. prowazekii* causes epidemic typhus, *O. tsutsugamushi* causes scrub typhus, and *E. chaffeenis* causes human ehrlichiosis.

The Genus *Coxiella*

The only species of *Coxiella, C. burnetii,* is an obligate intracellular bacterium that survives well outside the host cell and is transmissible from animal to animal without necessarily involving a blood-sucking parasite. During its intracellular growth, *C. burnetii* forms sporelike structures that later allow it to survive in the environment, although the structures lack the extreme resistance to heat and disinfectants characteristic of most endospores (**figure 11.30**). *Coxiella burnetii* causes Q fever of humans, a disease most often acquired by inhaling bacteria shed from infected animals.

The Genus *Chlamydia*

Chlamydia species are quite different from the other obligate intracellular parasites. They are transmitted directly from person to person rather than through the bite of a blood-sucking arthropod, and they have a unique growth cycle (**figure 11.31**). Inside the host cell, they initially exist as a fragile non-infectious form called a **reticulate body** that reproduces by binary fission. Later in the infection, the bacteria differentiate into a smaller, dense-appearing infectious form called an **elementary body** that is released upon death and rupture of the host cell. The cell wall of *Chlamydia* species is highly unusual among the Bacteria in that it lacks peptidoglycan, although it has the general appearance of a Gram-negative type of cell wall. *Chlamydia trachomatis* causes eye infections and a sexually transmitted disease that mimics gonorrhea, *C. pneumoniae* causes atypical pneumonia, and *C. psittaci* causes psittacosis, a form of pneumonia. ■ *Chlamydia trachomatis*, p. 646

MICROCHECK 11.8

Staphylococcus species are able to thrive in the dry, salty conditions of the skin. *Bacteroides* and *Bifidobacterium* species reside in the gastrointestinal tract; *Campylobacter* and *Helicobacter* species may cause disease when they reside there. *Neisseria* species, mycoplasma, and spirochetes inhabit other mucous membranes. Obligate intracellular parasites including *Rickettsia*, *Orientia*, *Ehrlichia*, *Coxiella*, and *Chlamydia* species are unable to reproduce outside of a host cell.

- What characteristic of *Mycoplasma* species separates them from other bacteria?
- How is *Helicobacter pylori* able to withstand the acidity of the stomach?
- Why would breast feeding affect the composition of a baby's intestinal flora?

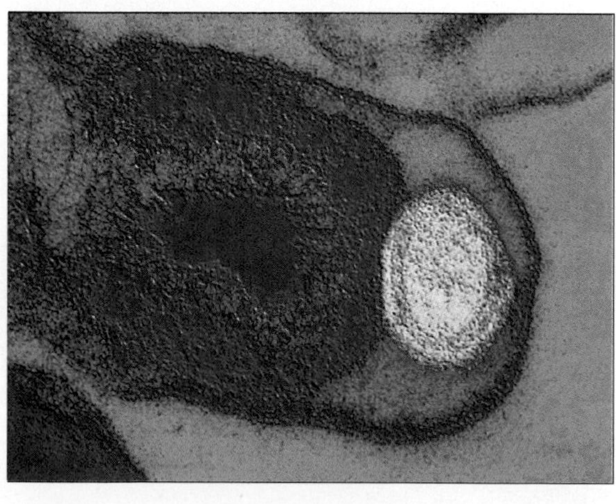

Figure 11.30 ***Coxiella*** Color-enhanced transmission electromicrographs of *C. burnetti*. The oval copper-colored object is the sporelike structure.

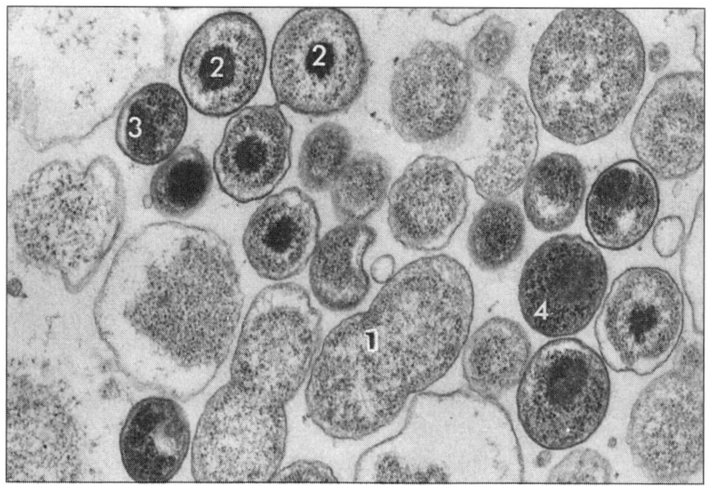

Figure 11.31 ***Chlamydia* Growing in Tissue Cell Culture** The numbers indicate the development from the dividing reticulate body to an infectious elementary body.

The Archaea—Thriving in Extreme Conditions

Members of the Archaea that have been characterized typically thrive in extreme environments that are otherwise devoid of life. These include conditions of high heat, acidity, alkalinity, and salinity. An exception to this attribute is the methanogens, which inhabit anaerobic niches shared with members of the Bacteria. Because of their intimate association with bacteria, the methanogens were discussed earlier in the chapter.

The Archaea fall into two broad phylogenetic groups—the **Euryarchaeota** and the **Crenarchaeota**. The Euryarchaeota include all of the known methanogens and extreme halophiles as well as some of the extreme thermophiles, or hyperthermophiles. The Crenarchaeota include other hyperthermophiles and all of the known thermophilic extreme acidophiles. In addition to the characterized genera, many other archaea have been detected in a variety of environments using DNA probes that bind to rRNA genes. ■ **halophiles, p. 99** ■ **hyperthermophiles, p. 97**

Extreme Halophiles

The extreme halophiles are found in very high numbers in high-salt environments such as salt lakes, soda lakes, and brines used for curing fish. Most can grow well in a saturated salt solution (32% NaCl), and they require a minimum of about 9% NaCl. Because they produce pigments, their growth can be seen as red patches on salted fish and pink blooms in concentrated salt water ponds (**figure 11.32**).

Extreme halophiles are aerobic or facultatively anaerobic chemoheterotrophs, but some also can obtain some energy from light. These organisms have the light-sensitive pigment **bacteriorhodopsin**, which absorbs energy from sunlight and uses it to expel protons from the cell. This creates a proton gradient that can be used to drive flagella or synthesize ATP.

Extreme halophiles come in a variety of shapes, including rods, cocci, discs, and triangles. They include genera such as *Halobacterium*, *Halorubrum*, *Natronobacterium*, and *Natronococcus*; members of these latter two genera are extremely alkaliphilic as well as halophilic.

Figure 11.32 **Typical Habitat of Extreme Halophiles Solar Evaporation Ponds.** The red color is due to the pigments of halophilic organisms such as *Halobacterium* species.

Extreme Thermophiles

The extreme thermophiles include members of the Euryarchaeota and the Crenarchaeota. Many are found in regions of volcanic vents and fissures that exude sulfurous gases and other hot vapors. Because these regions are thought to closely mimic the environment of early earth, scientists are particularly interested in studying the prokaryotes that thrive there. Others are found in hydrothermal vents in the deep sea and hot springs.

Methane-Generating Hyperthermophiles

Recall that methanogens are archaea, but may live in anaerobic environments inhabited by a diverse assortment of bacteria. Some methanogens, however, are extreme thermophiles. Like the mesophilic methanogens, these organisms oxidize H_2, using CO_2 as a terminal electron acceptor, to yield methane. For example, *Methanothermus* species, which can grow in temperatures as high as 97°C, grow optimally at approximately 84°C.

Sulfur- and Sulfate-Reducing Hyperthermophiles

The sulfur- and sulfate-reducing hyperthermophiles are split between the Euryarchaeota and Crenarchaeota. They are obligate anaerobes that use sulfur or, in one case, sulfate as an electron acceptor, generating H_2S.

Members of the genera *Thermococcus* and *Archaeoglobus* are Euryarchaeota. *Thermococcus* species use organic compounds as an energy source and thrive in hot springs at temperatures of 90°C. *Archaeoglobus* species have an optimum growth temperature of approximately 80°C and have been isolated from near hydrothermal vents. They can use hydrogen gas and organic compounds as energy sources.

The sulfur-reducing Crenarchaeota include the genera *Thermoproteus*, *Pyrodictium*, and *Pyrolobus*. All are obligate anaerobes that oxidize H_2 as an energy source. *Thermoproteus* species are found in the anaerobic environments of acidic sulfur-containing hot springs and water holes. Their optimum temperature is approximately 85°C. In addition to oxidizing hydrogen gas, they can utilize organic compounds as an energy source. *Pyrodictium* species are found in hydrothermal vents of the deep sea. Fibers they produce form a network that connects the disc- or dish-shaped cells. The optimum temperature of one species is approximately 105°C, and it cannot grow below 82°C. *Pyrodictium* species preferentially grow in acidic conditions, with an optimum pH of approximately 5.5. *Pyrolobus fumarii* was found growing in a "black smoker" 3,650 m (about 12,000 feet) deep in the Atlantic Ocean. It grows between 90°C and 113°C.

Sulfur Oxidizers

Sulfolobus species are found at the surface of acidic sulfur-containing hot springs such as many of those found in Yellowstone National Park (**figure 11.33**). They are obligate aerobes that oxidize sulfur compounds, using O_2 as a terminal electron acceptor to generate sulfuric acid. They are thermoacidophilic, only growing above 50°C and at a pH between 1 and 6.

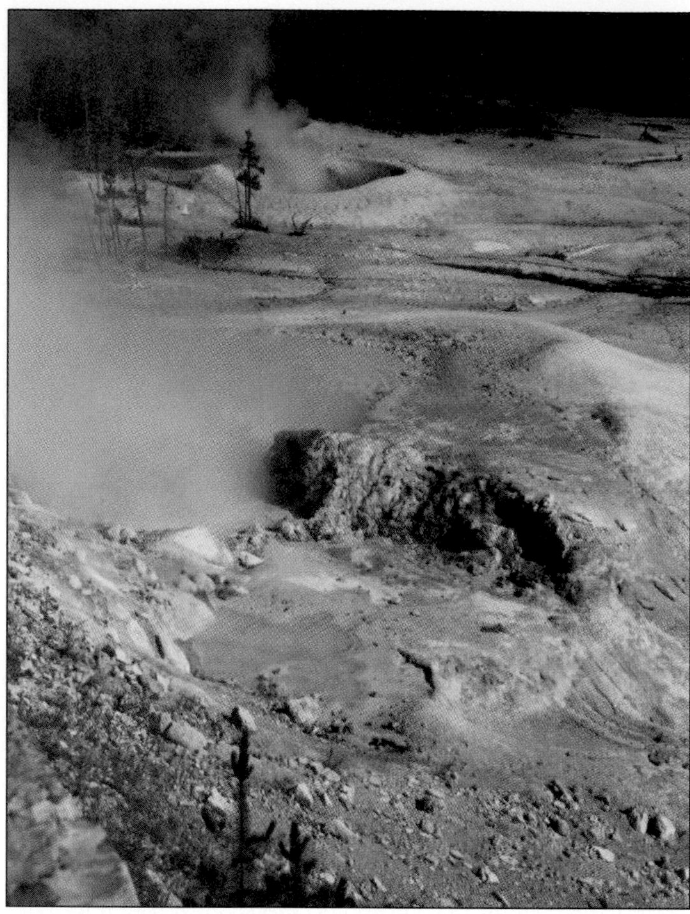

Figure 11.33 Typical Habitat of *Sulfolobus* Sulfur hot spring in Yellowstone National Park.

Thermophilic Extreme Acidophiles

Members of two genera, *Thermoplasma* and *Picrophilus*, are notable for their preference of growing in extremely acidic, hot environments. *Thermoplasma* species grow optimally at pH 2; in fact, *T. acidophilum* lyses at neutral pH. It was originally isolated from coal refuse piles. *Picrophilus* species tolerate conditions that are even more acidic, growing optimally at a pH below 1. Two species have been isolated in Japan from acidic areas in regions that exude sulfurous gases.

MICROCHECK 11.9

Archaea typically inhabit extreme environments that are otherwise devoid of life. These include conditions of high salinity, heat, acidity, and alkalinity.

- Why do seawater ponds sometimes turn pink as the water evaporates?
- At which relative depth in a sulfur hot spring would a sulfur reducer likely be found? How about a sulfur oxidizer?
- What characteristic of the methanogens makes it logical to discuss them with the Bacteria rather than the other Archaea?

FUTURE CHALLENGES

Astrobiology: The Search for Life on Other Planets

If life as we know it exists on other planets, one form it will take will likely be microbial. The task, then, is to figure out how to find and detect such extraterrestrial microorganisms. Considering that we still know relatively little about the microbial life on our own planet, coupled with the extreme difficulty of obtaining or testing extraterrestrial samples, this is a daunting challenge with many as yet unanswered questions. For example, what is the most likely source of life on other planets? What is the best way to preserve specimens for study on earth? What will be the culture requirements to grow such organisms? ***Astrobiology****, the study of life in the universe, is a new field that is bringing together scientists from a wide range of disciplines, including microbiology, geology, astronomy, biology, and chemistry, to begin answering some of these questions. The goal is to determine the origin, evolution, distribution, and destiny of life in the universe. These Astrobiologists are also given the task of developing lightweight, dependable, and meaningful testing devices to be used in future space missions.*

Astrobiologists believe that within our solar system, life would most likely be found either on Europa, a moon of Jupiter, or on Mars. This is because Europa and Mars appear to have, or have had, water, which is crucial for all known forms of life. Europa has an icy crust, beneath which may be liquid water or even a liquid ocean. Mars is the planet that is closest to Earth, and it has the most similar environment. Photographs suggest that flowing water once existed there. Besides missions to both these bodies, NASA also has future plans to return material from both a comet and an asteroid.

To prepare for researching life on other planets, microbiologists have turned to some of the most extreme environments here on Earth. These include glaciers and ice shelves, hot springs, deserts, volcanoes, deep ocean hydrothermal vents, and subterranean features such as caves. Because select microorganisms can survive in these environs, which are analogous to conditions expected on other planets, they are good testing grounds for the technology to be used on future missions.

SUMMARY

METABOLIC DIVERSITY (Table 11.3, Figures 11.1, 11.2)

Anaerobic Chemotrophs

1. For approximately the first 1.5 billion years that prokaryotes inhabited the earth, the atmosphere was **anoxic**.
2. Modern anaerobic environments can be found in mud and tightly packed soils, nutrient-rich waters that foster the growth of O_2-consuming microbes, and certain body parts of animals.

Anaerobic Chemolithotrophs

1. The **methanogens** are a group of archaea that generate energy by oxidizing hydrogen gas (H_2), using CO_2 as a terminal electron acceptor. (Figure 11.3)

Anaerobic Chemoorganotrophs—Anaerobic Respiration

1. Chemoorganotrophs oxidize organic compounds such as glucose to obtain energy; those that grow anaerobically must employ a terminal electron acceptor other than O_2.
2. *Desulfovibrio* reduces sulfur compounds to form hydrogen sulfide.

Anaerobic Chemoorganotrophs—Fermentation

1. The end products of fermentation include a variety of acids and gases.
2. *Clostridium* species are Gram-positive rods that produce endospores.
3. The **lactic acid bacteria** are a group of Gram-positive organisms that produce lactic acid as their primary fermentation end product. (Figures 11.4, 11.5)
4. *Propionibacterium* species are Gram-positive pleomorphic rods that produce propionic acid as their primary fermentation end product.

Anoxygenic Phototrophs

1. The earliest photosynthesizers were likely the **anoxygenic phototrophs**.
2. Anoxygenic phototrophs are a phylogenetically diverse group of bacteria that harvest the energy of sunlight, using photosynthesis to synthesize organic materials; they have **bacteriochlorophyll**.

The Purple Bacteria

1. The **purple bacteria** are Gram-negative organisms that appear red, orange, or purple; the components of their photosynthetic apparatus are all contained within the cytoplasmic membrane, which has extensive invaginations.
2. The purple sulfur bacteria preferentially use sulfur as a source of electrons for reducing power; most accumulate sulfur in intracellular granules. (Figure 11.6)
3. The purple nonsulfur bacteria preferentially use organic molecules as a source of electrons for reducing power.

The Green Bacteria

1. The **green bacteria** are Gram-negative organisms that are typically green or brownish in color. Their light-harvesting pigments are located in structures called **chlorosomes**, and their cytoplasmic membranes do not have extensive invaginations.
2. The green sulfur bacteria use hydrogen sulfide as a source of electrons for reducing power. (Figure 11.7)
3. The green nonsulfur bacteria are characterized by their filamentous growth; metabolically, they resemble the purple nonsulfur bacteria.

Other Anoxygenic Phototrophs

1. Other types of anoxygenic phototrophs have recently been discovered, including a Gram-positive rod that forms endospores.

Oxygenic Phototrophs

1. An estimated 3 billion years ago, O_2 was introduced into the atmosphere, probably due to the evolution of the cyanobacteria, thought to be the first **oxygenic photosynthesizers**.
2. Today, cyanobacteria are still essential **primary producers**; unlike eukaryotic photosynthesizers, some can fix nitrogen.

The Cyanobacteria (Figures 11.8, 11.9)

1. The cyanobacteria are a diverse group of Gram-negative bacteria.
2. Genetic evidence indicates that chloroplasts of plants and algae evolved from a species of cyanobacteria.
3. Nitrogen-fixing cyanobacteria are critically important ecologically, because they provide an available source of both carbon and nitrogen. (Figure 11.10)
4. Filamentous cyanobacteria may maintain the structure and productivity of some soils.
5. Some species of cyanobacteria produce toxins that can be deadly to animals that ingest heavily contaminated water.

Aerobic Chemolithotrophs

1. Aerobic chemolithotrophs generate energy by oxidizing reduced inorganic compounds, using O_2 as a terminal electron acceptor.

The Sulfur-Oxidizing Bacteria (Figure 11.12)

1. Sulfur-oxidizing bacteria are Gram-negative rods or spirals, sometimes growing in filaments.
2. The filamentous sulfur oxidizers *Beggiatoa* and *Thiothrix* live in sulfur springs, in sewage-polluted waters, and on the surface of marine and freshwater sediments. (Figure 11.11)
3. *Thiobacillus* are unicellular prokaryotes found in both terrestrial and aquatic habitats.

The Nitrifiers

1. Ammonia oxidizers convert ammonia to nitrite and include *Nitrosomonas* and *Nitrosococcus*; nitrite oxidizers oxidize nitrite to nitrate and include *Nitrobacter* and *Nitrococcus*.

The Hydrogen-Oxidizing Bacteria

1. The genera *Aquifex* and *Hydrogenobacter* are Gram-negative thermophilic bacteria that are thought to be among the earliest bacterial forms to exist on earth.

Aerobic Chemoorganotrophs

1. Aerobic chemoorganotrophs oxidize organic compounds for energy, using O_2 as a terminal electron acceptor.

Obligate Aerobes

1. Obligate aerobes generate energy exclusively by respiration.
2. *Micrococcus* species are Gram-positive cocci found in soil and on dust particles, inanimate objects, and skin. (Figure 11.13)
3. *Mycobacterium* species are widespread in nature. Although they have a Gram-positive type of cell wall, they stain poorly; they are **acid-fast**. *Mycobacterium* species cause tuberculosis and Hansen's disease (leprosy).
4. *Pseudomonas* species are Gram-negative rod-shaped bacteria that are widespread in nature and have extremely diverse metabolic capabilities. (Figure 11.14)
5. *Thermus aquaticus* is the source of *Taq* polymerase, which is an essential component in the polymerase chain reaction.
6. *Deinococcus radiodurans* can survive exposure to a dose of gamma radiation several thousand times that lethal to a human being.

Facultative Anaerobes

1. *Corynebacterium* species are Gram-positive pleomorphic, rod-shaped organisms that commonly inhabit soil, water, and the surface of plants. (Figure 11.15)
2. Members of the family *Enterobacteriaceae* are Gram-negative rods that typically inhabit the intestinal tract of animals, although some reside in rich soil. Enterics that ferment lactose are included in the group **coliforms** and are used as indicators of fecal pollution. (Figure 11.16)

ECOPHYSIOLOGY

Thriving in Terrestrial Environments

Bacteria that Form a Resting Stage

1. Of the various types of dormant cells produced by soil organisms, endospores are by far the most resistant to environmental extremes.
2. Endospore-forming genera include *Bacillus* and *Clostridium*. (Figure 11.17)
3. *Azotobacter* species are Gram-negative pleomorphic rods that form a resting cell called a **cyst** and they are notable for their ability to fix nitrogen in aerobic conditions. (Figure 11.18)
4. The **myxobacteria** aggregate to form a **fruiting body** when nutrients are exhausted; within the fruiting body, cells differentiate to form a dormant **microcyst**. (Figure 11.19)
5. *Streptomyces* species are Gram-positive bacteria that resemble fungi in their pattern of growth; they form chains of **conidia** at the end of **hyphae**. Many species naturally produce antibiotics. (Figure 11.20)

Bacteria that Associate with Plants

1. *Agrobacterium* species cause plant tumors. They transfer a portion of a specific plasmid to plant cells, genetically engineering the plant cells to produce opines and plant growth hormones.
2. *Rhizobium* species reside as **endosymbionts** within cells in nodules formed on the roots of **legumes**. In these protected confines, they fix nitrogen. (Figure 11.22)

Thriving in Aquatic Environments

Sheathed Bacteria

1. Sheathed bacteria such as *Sphaerotilus* and *Leptothrix* form chains of cells encased in **sheath**, which enables cells to attach to solid objects in favorable habitats while sheltering them from attack by predators. (Figure 11.23)

Prosthecate Bacteria

1. *Caulobacter* species have a single polar prostheca called a **stalk**; at the tip of the stalk is a **holdfast**. The cells divide by binary fission. (Figure 11.24)
2. *Hyphomicrobium* species divide by forming a bud at the tip of their single polar prostheca. (Figure 11.25)

Bacteria that Derive Nutrients from Other Organisms

1. *Bdellovibrio* species are highly motile Gram-negative rods that prey on other bacteria. (Figure 11.26)
2. Certain species of bioluminescent bacteria of the Gram-negative genera *Photobacterium* and *Vibrio* establish a symbiotic relationship with specific types of squid and fish. (Figure 11.27)
3. *Legionella* species often reside within protozoa and can cause respiratory disease when inhaled in aerosolized droplets.

Bacteria that Move by Unusual Mechanisms

1. **Spirochetes** are a group of Gram-negative spiral-shaped bacteria that move by means of an axial filament. (Figure 11.28)
2. Magnetotactic bacteria such as *Magnetospirillum magnetotacticum* contain a string of magnetic crystals that enable them to move up or down in water or sediments to the microaerophilic niches they require.

Bacteria that Form Storage Granules

1. *Spirillum volutans* stores polyphosphate granules.
2. *Thiomargarita namibiensis* is the largest bacterium known; it stores granules of sulfur and has a nitrate-containing vacuole.

Animals as Habitats (Table 11.2)

Bacteria that Inhabit the Skin

1. *Staphylococcus* species are Gram-positive cocci that are facultative anaerobes.

Bacteria that Inhabit Mucous Membranes

1. *Bacteroides* species are strictly anaerobic Gram-negative rods that inhabit the mouth, intestinal tract, and genital tract of humans and other animals.
2. *Bifidobacterium* species are irregular Gram-positive rod-shaped organisms that reside primarily in the intestinal tract of animals, including humans, particularly breast-fed infants.
3. *Campylobacter* and *Helicobacter* species are microaerophilic, curved Gram-negative rods.

4. *Haemophilus* species are Gram-negative coccobacilli that require compounds found in blood for growth.
5. *Neisseria* species are nutritionally fastidious, obligate aerobes that grow in the oral cavity and genital tract.
6. *Mycoplasma* species lack a cell wall; they often have sterols in their membrane that provide strength and rigidity. **(Figure 11.29)**
7. *Treponema* and *Borrelia* species are spirochetes that typically inhabit mucous membranes and body fluids of humans and other animals.

Obligate Intracellular Parasites

1. Obligate intracellular parasites are unable to reproduce outside a host cell; most have lost the ability to synthesize substances needed for extracellular growth.
2. Species of *Rickettsia*, *Orientia*, and *Ehrlichia* are tiny Gram-negative rods that are spread when a blood-sucking arthropod transfers bacteria during a blood meal.
3. *Coxiella burnetii* is a Gram-negative rod that survives well outside the host due to the production of sporelike structures. **(Figure 11.30)**
4. *Chlamydia* species are transmitted directly from person to person. **(Figure 11.31)**

The Archaea—Thriving in Extreme Conditions

1. The Archaea fall into two broad phylogenetic groups—the **Euryarchaeota** and the **Crenarchaeota**.

Extreme Halophiles

1. Extreme halophiles are found in salt lakes, soda lakes, and brines used for curing fish; they can grow well in saturated salt solutions. They include the genera *Halobacterium*, *Halorubrum*, *Natronobacterium*, and *Natronococcus*. **(Figure 11.32)**
2. Some extreme halophiles can obtain limited amounts of energy from light.

Extreme Thermophiles

1. *Methanothermus* species are hyperthermophiles that generate methane.
2. Sulfur- and sulfate-reducing hyperthermophiles are obligate anaerobes that use sulfur (or, in one case, sulfate) as a terminal electron acceptor. They include the genera *Thermococcus*, *Archaeoglobus*, *Thermoproteus*, *Pyrodictium*, and *Pyrolobus*.
3. Sulfur-oxidizing hyperthermophiles oxidize sulfur compounds, using O_2 as a terminal electron acceptor, to generate sulfuric acid. The are exemplified by *Sulfolobus* species, which are obligate aerobes. **(Figure 11.33)**

Thermophilic Extreme Acidophiles

1. *Thermoplasma* and *Picrophilus* species have an optimum pH of 2 or below.

REVIEW QUESTIONS

Short Answer

1. What kind of bacteria might compose the subsurface scum of polluted ponds?
2. What kind of bacterium might be responsible for plugging the pipes in a sewage treatment facility?
3. Give three examples of energy sources used by chemolithotrophs.
4. Name two genera of endospore-forming bacteria. How do they differ?
5. Serological tests involving which structures help distinguish the different enterobacteria from each other?
6. What unique motility structure characterizes the spirochetes?
7. In what way does the metabolism of *Streptococcus* species differ from that of *Staphylococcus* species?
8. How have species of *Streptomyces* contributed to the treatment of infectious diseases?
9. What characteristic of *Azotobacter* species protects their nitrogenase enzyme from inactivation by O_2?
10. Compare and contrast the relationships of *Agrobacterium* and *Rhizobium* species with plants.

Multiple Choice

1. A catalase-negative colony growing on a plate that was incubated aerobically could be which of the following genera?
   A. *Bacillus*
   B. *Escherichia*
   C. *Micrococcus*
   D. *Staphylococcus*
   E. *Streptococcus*
2. All of the following genera are spirochetes, except…
   A. *Borrelia*
   B. *Caulobacter*
   C. *Leptospira*
   D. *Spirochaeta*
   E. *Treponema*
3. Which of the following genera would you most likely find growing in acidic runoff from a coal mine?
   A. *Clostridium*
   B. *Escherichia*
   C. Lactic acid bacteria
   D. *Thermus*
   E. *Thiobacillus*
4. The dormant forms of which of the following genera are the most resistant to environmental extremes?
   1. *Azotobacter*
   2. *Bacillus*
   3. *Clostridium*
   4. *Myxobacteria*
   5. *Streptomyces*

   A. 1, 2 B. 2, 3 C. 3, 4 D. 4, 5 E. 1, 5

5. Which of the following genera are coliforms?
   A. *Bacteroides*
   B. *Bifidobacterium*
   C. *Clostridium*
   D. *Escherichia*
   E. *Streptococcus*
6. Which of the following genera preys on other bacteria?
   A. *Bdellovibrio*
   B. *Caulobacter*
   C. *Hyphomicrobium*
   D. *Photobacterium*
   E. *Sphaerotilus*
7. All of the following genera are obligate intracellular parasites, except…
   A. *Chlamydia*
   B. *Coxiella*
   C. *Ehrlichia*
   D. *Mycoplasma*
   E. *Rickettsia*
8. Which of the following genera fix nitrogen?
   1. *Anabaena*
   2. *Azotobacter*
   3. *Deinococcus*
   4. *Mycoplasma*
   5. *Rhizobium*

   A. 1, 3, 4 B. 1, 2, 5 C. 2, 3, 5
   D. 2, 4, 5 E. 3, 4, 5
9. Which of the following archaea are most likely to be found coexisting with bacteria?
   A. *Archaeoglobus*
   B. *Halobacterium*
   C. *Methanococcus*
   D. *Picrophilus*
   E. *Sulfolobus*
10. Unlike other genera of archaea, *Thermoplasma* and *Picrophilus* grow best in which of the following extreme conditions?
    A. High pH
    B. High salt
    C. High temperature
    D. Low pH
    E. Low temperature

Applications

1. A student argues that it makes no sense to be concerned about coliforms in drinking water because coliforms are harmless members of our normal intestinal flora. Explain why regulatory agencies are concerned about coliforms.
2. A friend who has lakefront property and cherishes her lush green lawn complains of the green odiferous scum on the lake each summer. Explain how her lawn might be contributing to the problem.

Critical Thinking

1. Soil often goes through periods of extreme dryness and extreme wetness. What characteristics of *Clostridium* species make them well suited for these conditions?
2. Some organisms use sulfur as an electron donor (a source of energy), whereas others use sulfur as an electron acceptor. How can this be if there must be a difference between the electron affinity of electron donors and acceptors for an organism to obtain energy?

C H A P T E R 12

The Eukaryotic Members of the Microbial World

In Salem Village, Massachusetts, in 1692, a group of teenage girls was afflicted with symptoms of "bewitchment." They suffered convulsions; had sensations of being pinched, pricked, or bitten; and had feelings of being torn apart. Some became temporarily blind, deaf, or speechless and were sick to their stomachs. During the inquest into their bewitchment, the girls accused several women of being instruments of the devil. These women were tried and then executed as witches.

Recently, some historians have reexamined this event and, with the help of the medical and biological community, have concluded that the Salem witch hunts may have been more a microbiological phenomenon, than a social phenomenon. A member of the microbial world may have been the culprit. The fungus Claviceps purpurea, the rye smut, produces the poison ***ergot****, which when ingested causes symptoms similar to those experienced by the girls in Salem. Apparently, several conditions existed in New England at this time that would have predisposed the population to ergot poisoning. First, the years 1690 to 1692 were particularly wet and cool. Second, during those years, rye grass had replaced wheat in New England as the principal grain because the wheat was seriously affected by another fungus, the wheat rust. In addition, most of the victims were children and teenagers who would have been more affected by ergot poisoning than adults because they would have ingested more of the poison per body weight than adults.*

This is only one highly publicized example of the way in which microbes have played a large role in the history of the world.

—A Glimpse of History

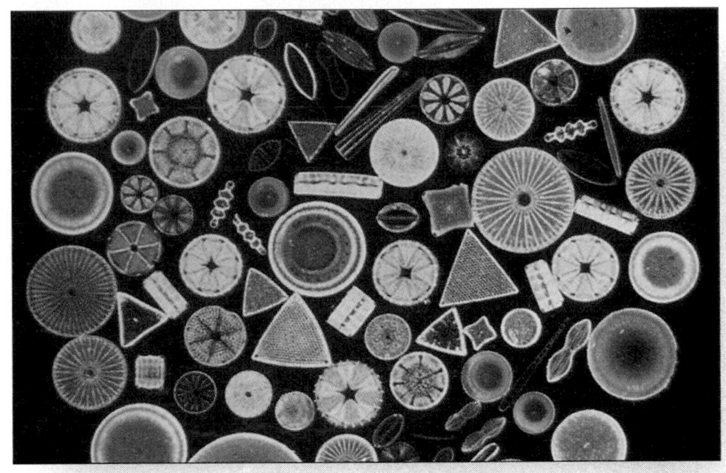

AS NOTED AT THE BEGINNING OF THIS BOOK, WHEN the ribosomal RNA (rRNA) sequences of organisms are compared, the living world can be divided into three divisions: Bacteria, Archaea, and Eukarya. In this chapter we will consider the **Eukarya**. These organisms, which include the algae, fungi, protozoa, multicellular parasites, and insect vectors, have one feature in common: they are all eukaryotic organisms. Recall from chapter 3 that the basic cell structure of the Eukarya is distinctly different from that of the Bacteria and Archaea. They are included in a textbook of microbiology because many members of these groups are microscopic and are studied with techniques that are similar to those used to study bacteria and archaea. In addition, many of these organisms cause disease in humans as well as plants and animals. ■ **Eukarya, p. 10**

Classification using gross anatomical characteristics of algae, fungi, protozoa, and even some multicellular parasites has always been problematic. Now, however, with modern techniques that examine these organisms at the molecular and ultrastructural levels, it has been discovered that some of the organisms that were traditionally grouped together were more dissimilar than similar. Instead, they arose at various times along a continuum of evolution. Therefore, in classification schemes that describe an accurate evolutionary history of organisms and are based on molecular and ultrastructure examination, the words algae, fungi, and protozoa are no longer really accurate. For the purposes of this book, however, we will use the term algae to describe the photosynthetic members and fungi and protozoa to describe the nonphotosynthetic members that are discussed in this chapter. In addition, a discussion of arthropods and helminths is included because these eukaryotes are also implicated in human disease. **Figure 12.1** shows a phylogeny based on the ribosomal RNA sequences of the eukaryotic organisms. We will refer to this phylogeny throughout this chapter, highlighting where the organisms fit on this evolutionary scale.

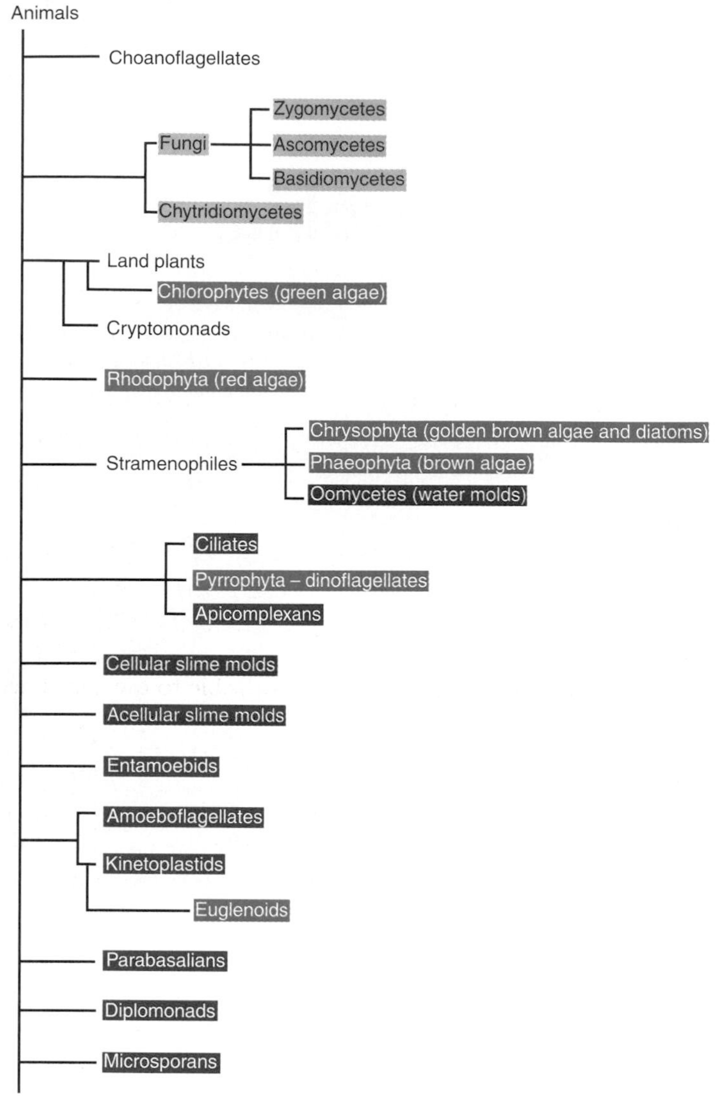

Figure 12.1 A Phylogeny of the Eukaryotes Based on Ribosomal RNA Sequence Comparison

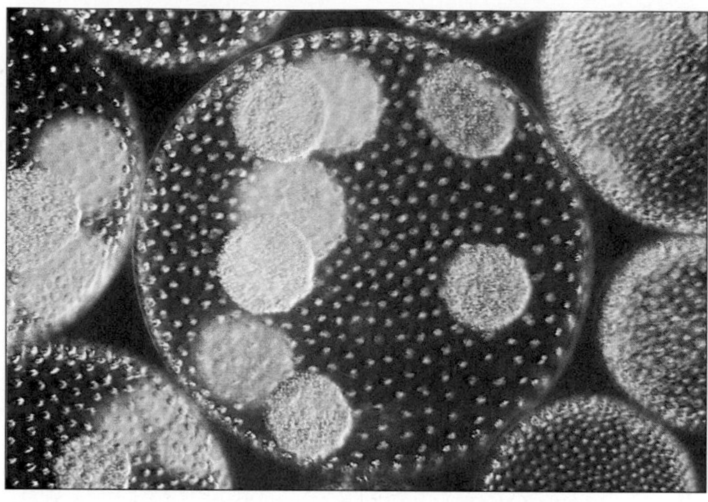

(a)

(b)

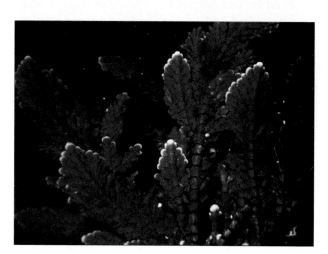

Figure 12.2 Algae (a) *Volvox* sp. a colony of cells formed into a hollow sphere (125×). The yellow-green circles are reproductive cells that will eventually become new colonies. **(b)** *Corallina gracilis,* red coral algae.

Algae

The algae are a diverse group of eukaryotic organisms that share some fundamental characteristics but are not related on the phylogenetic tree. These organisms are studied by **algologists** in a field known as **algology**. Algae are organisms that use light energy to convert CO_2 and H_2O to carbohydrates and other cellular products with the release of oxygen. Algae contain chlorophyll *a*, which is necessary for photosynthesis. In addition, many algae contain other pigments that extend the range of light waves that can be used by these organisms for photosynthesis. Algae include both microscopic unicellular members and macroscopic multicellular organisms. **(figure 12.2)**. Algae, however, differ from other eukaryotic photosynthetic organisms such as land plants in their lack of an organized vascular system and their relatively simple reproductive structures. ■ **photosynthesis, p. 158**

Algae do not directly infect humans, but some produce toxins that cause paralytic shellfish poisoning. Some of these toxins do not cause illness in the shellfish that feed on the algae but accumulate in their tissues and when eaten by humans cause nerve damage. ■ **paralytic shellfish poisoning, p. 303**

As one of the primary producers of carbohydrates and other cellular products, the algae are essential in the food chains of the world. In addition, they produce a large proportion of the oxygen in the atmosphere.

Classification of Algae

As noted above, *algae* is not a strict classification term; nevertheless, organisms considered under the general heading of algae are grouped for identification by a number of properties. These include the principal photosynthetic pigments of each group, cell wall structure, type of storage products, mechanisms of motility, and mode of reproduction. The names of the different algal groups are derived from the major color displayed by most of the algae in that group. Note that the different algal groups lie in different places along the evolutionary tree **(figure 12.3)**.

Some of the general characteristics of the algal groups are summarized in **table 12.1**.

Algal Habitats

Algae are found in both fresh and salt water, as well as in soil. Since the oceans cover more than 70% of the earth's surface, aquatic algae are major producers of oxygen as well as important users of carbon dioxide. Unicellular algae make up a significant part of the **phytoplankton**, (*phyto* means "plant" and *plankton* means "drifting"), the free-floating, photosynthetic organisms that are found in marine environments. More oxygen is produced by the phytoplankton than by all forests combined. Phytoplankton is a major food source for many animals, both large and small. Microscopic animals in the **zooplankton** (*zoo* means "animal") graze on this phytoplankton, and then both the zooplankton and phytoplankton become food for the benthic whales, some of the world's largest mammals. The unicellular algae of the phytoplankton are well adapted to this aquatic environment. As single cells, they have large, adsorptive surfaces relative to their volume and can move freely about, thus effectively using the dilute nutrients available.

Because one or more algal species can grow in almost any environment, algae often grow where other forms of life cannot thrive. Frequently, algae are among the first organisms to become established in barren environments, where they synthesize the organic materials necessary for the subsequent invasion and survival of other plants and animals. They are often found on rocks, preparing the surface for the growth of more complex members of the biological community.

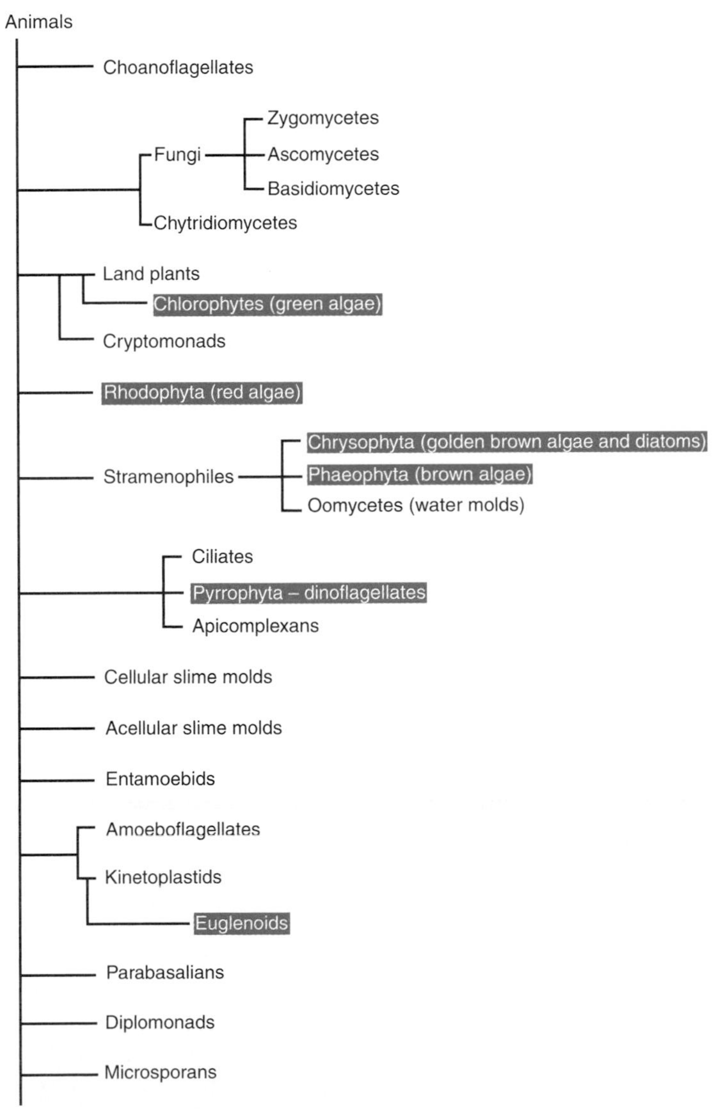

Figure 12.3 Phylogeny of Algae The algal groups are highlighted in green. Note that the algae are not directly related to one another but are found all along the evolutionary continuum.

Structure of Algae

Algae can be both microscopic and macroscopic. Microscopic algae can be single-celled organisms floating free or propelled by flagella, or they can grow in long chains or filaments. Some microscopic algae such as Volvox form colonies of 500 to 60,000 biflagellated cells, which can be visible to the naked eye (see figure 12.2a).

Macroscopic algae are multicellular organisms with a variety of specialized structures that serve specific functions **(figure 12.4)**. Some possess a structure called a **holdfast**, which looks like a root system but primarily serves to anchor the organism to a rock or some other firm substrate. Unlike a root, it is not used to obtain water and nutrients for the organism. Nutrients

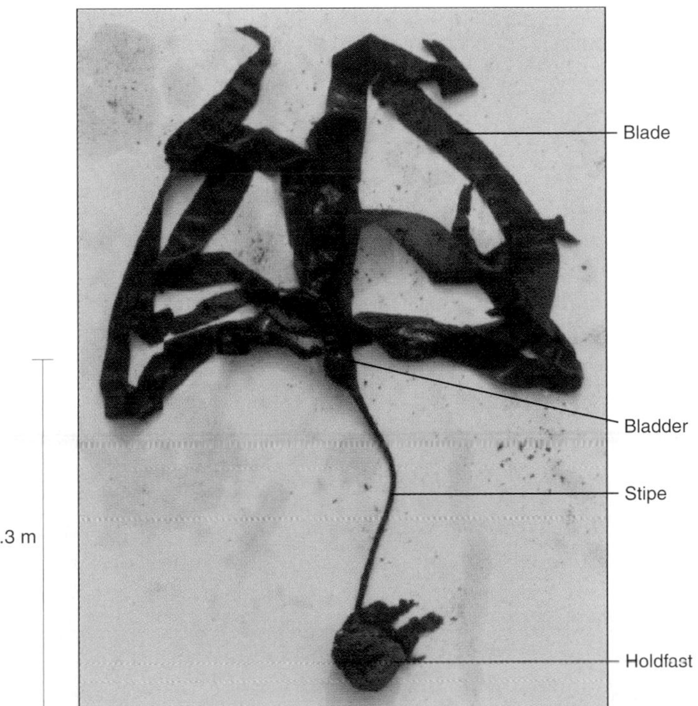

Figure 12.4 A Young *Nereocystis luetkeana*, the Bladder Kelp The alga has a large bladder filled with gas. This bladder keeps the blades floating on the surface of the water to maximize exposure to sunlight. The blades are the most active sites for photosynthesis. The holdfast anchors the kelp to rocks or other surfaces. In a single season, kelp can grow to lengths of 5 to 15 m.

Table 12.1 Characteristics of Major Groups of Algae

Group and Representative Member(s)	Usual Habitat	Principal Pigments (in addition to chlorophyll *a*)	Storage Products	Cell Walls	Mode of Motility (if present)	Mode of Reproduction
Chlorophyta						
Green algae	Fresh water; salt water; soil; tree bark; lichens	Chlorophyll *b*; carotenes; xanthophylls	Starch (–1,4-glucan)	Cellulose and pectin	Mostly nonmotile except one order, but some reproductive elements are flagellated	Asexual by multiple fission; spores or sexual
Phaeophyta						
Brown algae	Salt water	Xanthophylls, especially fucoxanthin	Starchlike carbohydrates; annitol; fats	Cellulose and pectin; alginic acid	Two unequal, lateral flagella	Asexual, motile zoospores; sexual,motile gametes
Rhodophyta						
Red algae corallines	Mostly salt water; several genera in fresh water	Phycobilins including phycoerythrin and phycocyanin; carotenes; xanthophylls	Starchlike carbohydrates	Cellulose and pectin; agar; carrageenan	Nonmotile	Asexual spores; sexual gametes
Chrysophyta						
Diatoms, golden brown algae	Fresh water; salt water; soil; higher plants	Carotenes	Starchlike carbohydrates (–1, 3-glucan)	Pectin, often impregnated with silica or calcium	Unique diatom motility; one, two, or more unequal flagella	Asexual or sexual
Pyrrophyta						
Dinoflagellates	Mostly salt water but common in fresh water	Carotenes; xanthophylls	Starch; oils	Cellulose and pectin	Two unequal, lateral flagella in different planes	Asexual; rarely sexual
Euglenophyta						
Euglena	Fresh water	Chlorophyll *b*; carotenes; xanthophylls	Fats; starchlike carbohydrates	Lacking, but elastic pellicle present	One to three anterior flagella	Asexual only by binary fission

and water surround the organism and do not need to be drawn up from the soil. The stalk of an alga, known as the **stipe**, usually has leaflike structures or **blades** attached to it. The blades are the principal photosynthetic portion of an alga, and some also bear the reproductive structures. Many large algae have gas-containing **bladders** or floats that help them maintain their blades in a position suitable for obtaining maximum sunlight.

Cell Walls

Algal cell walls are rigid and for the most part composed of cellulose, often associated with pectin. Some multicellular species of algae such as some red algae contain large amounts of other compounds in their cell walls. Compounds such as carrageenan and agar are harvested commercially and commonly used in foods as stabilizing compounds. As described earlier, agar is also used to solidify growth media in the laboratory. It is useful because although it melts at 100°C, it stays in a liquid state at relatively low temperatures (45°C–50°C) so that nutrients can be added, and yet is solid at room temperatures to act as a growing surface in a Petri dish. ■ **agar, p. 88** ■ **Petri dish, p. 88**

Diatoms are algae that have silicon dioxide incorporated into their cell walls. When these organisms die, their shells sink to the bottom of the ocean, and the silicon-containing material does not decompose. Deposits of diatoms that formed millions of years ago are mined for a substance known as diatomaceous earth, used for filtering systems, abrasives in polishes, insulation, and many other purposes.

Eukaryotic Cell Structures

As is true in all eukaryotes, algae have a membrane-bound **nucleus**. The genetic information is contained in a number of chromosomes that are tight packages of DNA with their associated basic protein. These chromosomes are enclosed in a nuclear membrane. ■ **nucleus, p. 78**

In addition, algae have other organelles in their cytoplasm such as **chloroplasts** and **mitochondria**. Chloroplasts contain chlorophyll as well as other light-trapping pigments such as carotenoids and phycocyanin. Photosynthesis occurs in the chloroplast. Respiration and oxidative phosphorylation occur in the mitochondria. ■ **chloroplasts, p. 80** ■ **mitochondria, p. 79**

Algal Reproduction

Most single-celled algae reproduce asexually by binary fission, as do most bacteria **(figure 12.5)**. The major difference between prokaryotic and eukaryotic fission involves events that take place with the genetic material within the cell. Recall from chapter 3 that in prokaryotic fission, the circular DNA replicates and each daughter cell receives half the original double strand of DNA plus a newly replicated strand. In eukaryotic organisms with multiple chromosomes, after the DNA is replicated, the chromosomes go through a nuclear division process called **mitosis**. This process ensures that the daughter cells receive the same number of chromosomes as the original parent.

Some algae, especially multicellular filamentous species, reproduce asexually by fragmentation. In this type of reproduction, portions of the parent organisms break off to form new organisms (see figure 12.5), and the parent organism survives.

Sexual reproduction also regularly occurs in most algae. During the process known as **meiosis, haploid** cells with half the chromosome content are formed. These cells are called **gametes** and when they fuse together they form a **diploid** cell with a full complement of chromosomes known as a **zygote**. Gametes are often flagellated and highly motile. Many algae alternate between a haploid generation and a diploid generation. Sometimes, as is the case with Ulva (sea lettuce), the generations look physically similar and can only be told apart by microscopic examination. In other cases, the two forms look quite different.

Paralytic Shellfish Poisoning

Although algae do not directly cause disease in humans, they do so indirectly. A number of algae produce toxins that are poisonous to humans and other animals. Several dinoflagellates of the group Pyrrophyta cause red tide, or algal blooms in the ocean. Red tide was reported in the Bible along the Nile and today seems to be spreading worldwide. In the warm waters of Florida and Mexico, the red tides result from an abundance of *Gymnodinium breve* **(figure 12.6)**. This dinoflagellate discolors the water about 17 to 75 km from shore and produces **brevetoxin**, which kills the fish that feed on the phytoplankton. It is unclear why algae suddenly grow in such large numbers, but it is thought that sudden changes in conditions of the water are responsible. The runoff of fertilizers along the waterways and coastlines may also be the cause of red tides. In addition, an upwelling of the water often brings more nutrients as well as the cysts that are resting, resistant stages of the *G. breve* from the ocean bottom to the surface. When these cysts encounter warmer waters and additional nutrients, they are released from their resting state and begin to multiply rapidly. Persons eating fish that have ingested *G. breve* and thus contain brevetoxin may suffer a tingling sensation in their mouths and fingers, a reversal of hot and cold sensations, reduced pulse rate, and diarrhea. The symptoms may be unpleasant but are rarely deadly, and people recover in 2 to 3 days. ■ **toxins, p. 460**

Red tides caused by the dinoflagellate of the genus *Gonyaulax* are much more serious. *Gonyaulax* species produce **neurotoxins** such as **saxitoxin** and **gonyautoxins**, some of the most potent non-protein poisons known. Shellfish such as clams, mussels, scallops, and oysters feed on these dinoflagellates without apparent harm and, in the process, accumulate the neurotoxin in their tissues. Then when humans eat the shellfish, they suffer symptoms of paralytic shellfish poisoning including general numbness, dizziness, general muscle weakness, and impaired respiration. Death can result from respiratory failure. Gonyaulax species are found in both the North Atlantic and the North Pacific. They have seriously affected the shellfish industry on both coasts over the years. At least 200 manatees died along the coast of Florida in the spring of 1996 as a result of red tide poisoning.

Another dinoflagellate, *Pfiesteria piscida*, usually is found as a nontoxic cyst or ameba in marine sediments. This organism changes when a school of fish approaches. The fish secrete a chemical cue that alerts the *Pfiesteria* to transform into a flagellated zoospore. It then releases two toxins. One stuns the fish and the other causes its skin to slough away. The *Pfiesteria* then enjoys a meal on the fish's red blood cells and sexually reproduces.

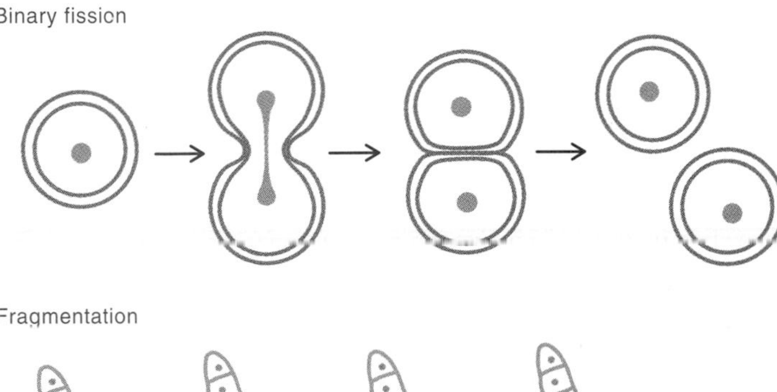

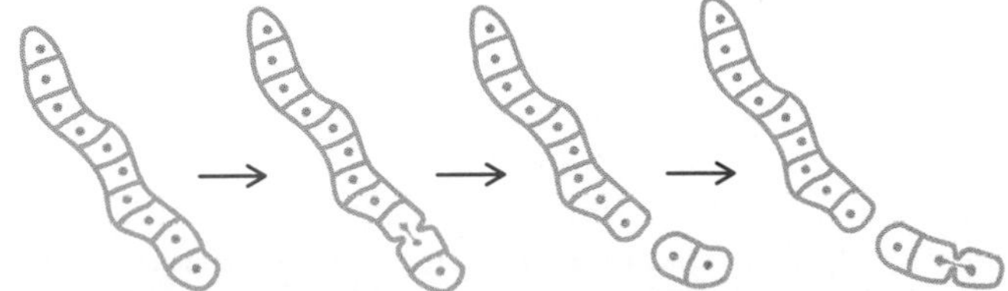

Figure 12.5 Binary Fission Is an Asexual Reproduction Process in Which a Single Cell Divides into Two Independent Daughter Cells
Fragmentation is a form of asexual reproduction in which a filament composed of a string of cells breaks into pieces to form new organisms.

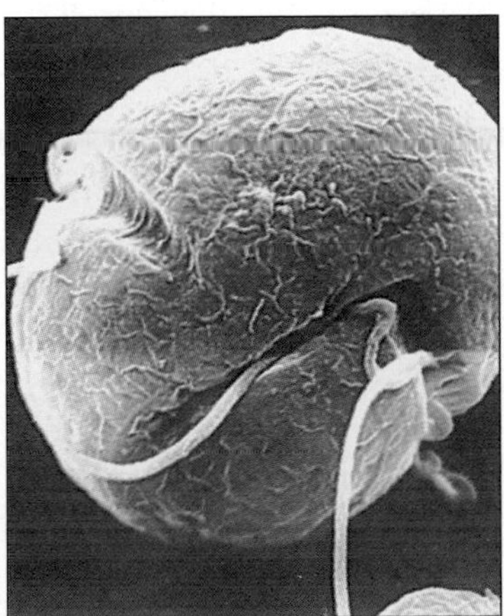

Figure 12.6 The Dinoflagellate *Gymnodinium* Scanning Electron Micrograph (4000×)

The toxins are so potent that researchers working with them in a laboratory were seriously affected. This organism now must be studied using the same precautions that are used to study the virus that causes AIDS.

Another algal toxin found in some species of diatoms has more recently been recognized to cause paralytic shellfish poisoning. This poison is domoic acid and is most often associated with mussels and other shellfish and crabs that accumulate the poison. Persons eating the shellfish suffer nausea, vomiting, diarrhea, and abdominal cramps as well as some neurological symptoms such as loss of memory.

State agencies constantly monitor for algal toxins, and it is wise to check with the local health department before harvesting shellfish for human consumption. Algal toxins may be present even when the water is not obviously discolored. Cooking the shellfish does not destroy these toxins.

MICROCHECK 12.1

Algae are a diverse group of eukaryotic organisms that all contain chlorophyll *a* and carry out photosynthesis. Algae are found in both fresh and salt water and are a significant part of the phytoplankton. Algae do not cause disease directly, but they can produce toxins that are harmful when ingested by humans.

- What are the primary characteristics used to distinguish algae from other organisms?
- What harmful effect can algae have on humans?
- Why should single cells have a large absorptive surface relative to their surface area?
- Would organisms that reproduce by binary fission necessarily be genetically identical? Why or why not?

Protozoa

Along with the algae, the protozoa constitute another group of eukaryotic organisms that traditionally have been considered part of the microbial world. The protozoa are microscopic, unicellular organisms that lack photosynthetic capability, usually are motile at least at some stage in their life cycle, and reproduce most often by asexual fission.

Classification of Protozoa

As with algae, classification of protozoa according to rRNA and ultrastructure shows that they are not a unified group, but appear along the evolutionary continuum. (**figure 12.7**). The primary reason that they are lumped together in a field known as **protozoology** is because they are all single-celled eukaryotic organisms that lack chlorophyll. Some members of this group cause disease. We will concentrate on these organisms.

Protozoa have traditionally been divided into groups primarily based on their mode of locomotion. **Table 12.2** and **figure 12.7** show where each group fits in these schemes.

The phylum **Sarcomastigophora** includes two subphyla in which most of the human disease-causing protozoa are found. The subphylum **Mastigophora** includes the flagellated protozoa.

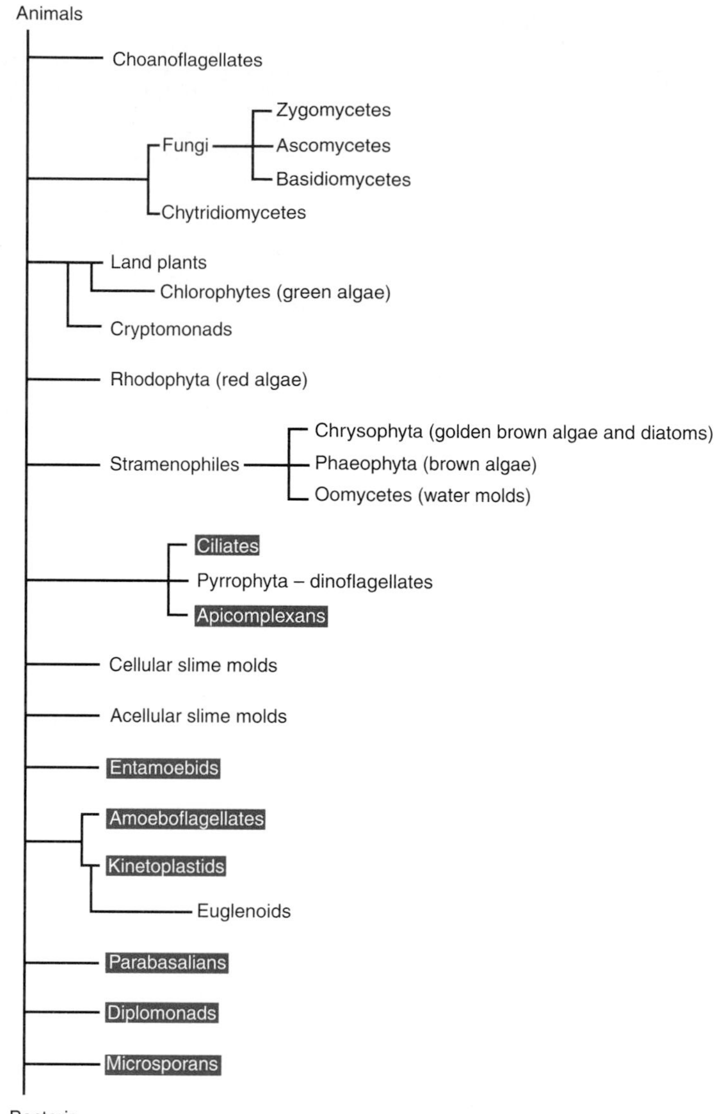

Figure 12.7 A Phylogenetic Scheme of Eukaryotes Based on rRNA Sequence Comparisons The protozoan groups are highlighted in red.

They are mostly unicellular and have one or more flagella at some time in their life cycle. These flagella are used for locomotion and food gathering as well as sensory receptors. Some of the organisms included within the Mastigophora are among the ancestors of some of the oldest organisms on earth. The most important disease-causing Mastigophora are *Giardia lamblia, Leshmania* species, *Trichomonas vaginalis, Trypanosoma brucei rhodesiense* and *Trypanosoma brucei gambiense*. (see table 12.2). Each of these diseases will be discussed later in this book. ■ **flagella, p. 78**

Members of the subphylum **Sarcodina** move by means of pseudopodia. The Sarcodina change shape as they move. *Entamoeba histolytica* infects humans, causing diarrhea ranging from mild asymptomatic disease to severe dysentery. ■ **diarrhea, p. 598**

The phylum **Ciliophora** or the **ciliates**, includes organisms that have cilia. The cilia are similar in construction to the flagella and usually completely cover the surface of an organism. Most often, they are arranged in distinct rows and are connected to one another by fibrils known as kinetodesma. Cilia beat in a

Table 12.2 Protozoa of Medical Importance

Traditional Classification	18s rRNA Classification	Genus of Disease Causing Protozoa	Disease Caused by Protozoa	Mode of Motility	Mode of Reproduction	Page for Additional Information
Phylum: Sarcomastigophora						
Subphylum:						
Mastigophora	Kinetoplastid	*Trypanosoma*	African sleeping sickness	Flagella	Longitudinal fission	p. 686
	Diplomonad	*Giardia*	Giardiasis			p. 613
	Parabasalian	*Trichomonas*	Trichomoniasis			p. 657
	Kinetoplastic	*Leshmania*	Leshmaniasis			
Sarcodina	Entamoebids	*Entamoeba*	Amebiasis (diarrhea)	Pseudopodia	Binary fission	p. 616
Phylum: Ciliophora	Ciliates	*Balantidium*	Dysentery	Cilia	Transverse fission	
Phylum: Apicomplexa	Apicomplexans	*Plasmodium*	Malaria	Flagella	Multiple fission	p. 731
		Toxoplasma	Toxoplasmosis			p. 757
		Cryptosporidium	Cryptosporidiosis			p. 614
Phylum: Microspora	Microsporans	*Microsporidium*	Immunosupressed disease	Polar filament	?	

coordinated fashion in waves across the body of the protozoan. A beat of one cilium affects the cilia immediately around it, but there is no evidence that the connecting fibrils aid in this coordination. The cilia found near the oral cavity propel food into the opening. Paramecia are members of the Ciliophora. *Balantidium coli* is the only known ciliate to cause human disease. It produces ulcers in the large intestines, and pigs are its major reservoir.

Organisms in the phylum **Apicomplexa,** also referred to as **sporozoa**, cause some of the most serious protozoan diseases of humans. Malaria is caused by one of four *Plasmodium* species. It is transmitted by the female *Anopheles* mosquito. Cats are the primary host for *Toxoplasma gondii*, with humans serving as secondary hosts. Another Apicomplexa is *Cryptosporidium parvum*, which causes the diarrheal disease cryptosporidiosis. ■ malaria, p. 731 ■ toxoplasmosis, p. 757 ■ cryptosporidiosis, p. 614

The phylum **Microspora** includes the intracellular protozoa that infect immunocompromised humans, especially persons with AIDS. There are other protozoan phyla such as Labyrinthomorpha, Ascetospora, and Myxozoa, but they are not implicated in human disease and so we will not consider them here. They are most often found in marine habitats and are parasitic on fish and other sea life.

Protozoan Habitats

A majority of protozoa are free-living and found in marine, freshwater, or terrestrial environments. They are essential as decomposers in many ecosystems. Some species, however, are parasitic, living on or in other host organisms. The hosts for protozoan parasites range from simple organisms, such as algae, to complex vertebrates, including humans. All protozoa require large amounts of moisture, no matter what their habitat.

In marine environments, protozoa make up part of the zooplankton, where they feed on the algae of the phytoplankton and are an important part of the aquatic food chains. On land, protozoa are abundant in soil as well as in or on plants and animals. Specialized protozoan habitats include the guts of termites, roaches, and ruminants such as cattle.

Protozoa are an important part of the food chain. They eat bacteria and algae and, in turn, serve as food for larger species. The protozoa help maintain an ecological balance in the soil by devouring vast numbers of bacteria and algae. For example, a single paramecium can ingest as many as 5 million bacteria in one day. Protozoa are important in sewage disposal because most of the nutrients they consume are metabolized to carbon dioxide and water, resulting in a large decrease in total sewage solids. ■ sewage treatment, p. 793

Structure of Protozoa

Cell Wall

Protozoa lack the rigid cellulose cell wall found in algae. Most protozoa do, however, have a specific shape determined by the rigidity or flexibility of the material lying just beneath the plasma membrane. **Foraminifera** have distinct hard shells composed of silicon or calcium compounds (**figure 12.8**). The foraminifers, which secrete a calcium shell, have through the course of millions of years formed limestone deposits such as the white cliffs of Dover on England's southern coast.

Eukaryotic Cell Structures

Protozoa are eukaryotic organisms and as such have a membrane-bound nucleus as well as other membrane-bound organelles such as mitochondria. Protozoa are not photosynthetic and thus lack chloroplasts. ■ eukaryotic organelles, p. 78

Protozoa have specialized structures for movement such as **cilia**, **flagella**, or **pseudopodia**. As described in chapter 3,

Figure 12.8 A Group of Protozoa with Hard Silicon Shells

eukaryotic flagella and cilia are distinctly different in construction from prokaryotic flagella (see figure 3.54). Protozoa are grouped by their mode of locomotion. For example, the Mastigophora have flagella, and Ciliophora have cilia during at least some part of their life cycle, and Sarcodina use pseudopodia for movement (see table 12.2).

Feeding in Protozoa

Since protozoa live in an aquatic environment, water, oxygen, and other small molecules readily diffuse through the cell membrane. In addition, as described in chapter 3, protozoa use either pinocytosis or phagocytosis to obtain food and water (see figure 3.52).

Protozoan Reproduction

The life cycles of protozoa are sometimes complex, involving more than one habitat or host. Morphologically distinct forms of a single protozoan species can be found at different stages of the life cycle. Such organisms are said to be **polymorphic (figure 12.9)**. This polymorphism is comparable in some respects to the differentiation of various cell types that form plant and animal tissues.

The ability to exist in either a **trophozoite** (vegetative or feeding form) or **cyst** (resting form) is characteristic of many protozoa. Certain environmental conditions, such as the lack of nutrients, moisture, oxygen, low temperature, or the presence of toxic chemicals may trigger the development of a protective cyst wall within which the cytoplasm becomes dormant. Cysts provide a means for the dispersal and survival of protozoa under adverse conditions and can be compared to the bacterial endospore. Protozoan cysts, however, are not as resistant to heat and other adverse conditions as are bacterial endospores. When the cyst encounters a favorable environment, the trophozoite emerges. Thus, a number of parasitic protozoa are disseminated to new hosts during their cyst stage. ■ **endospores, p. 72**

Both asexual and sexual reproduction are common in protozoa and may alternate during the complicated life cycle of some organisms. Binary fission takes place in many groups of protozoa (**figure 12.10**). In the flagellates, it usually occurs longitudinally, and in the ciliates, it occurs transversely. Since some protozoa possess both cilia and flagella, their method of asexual reproduction determines in which group they are classified.

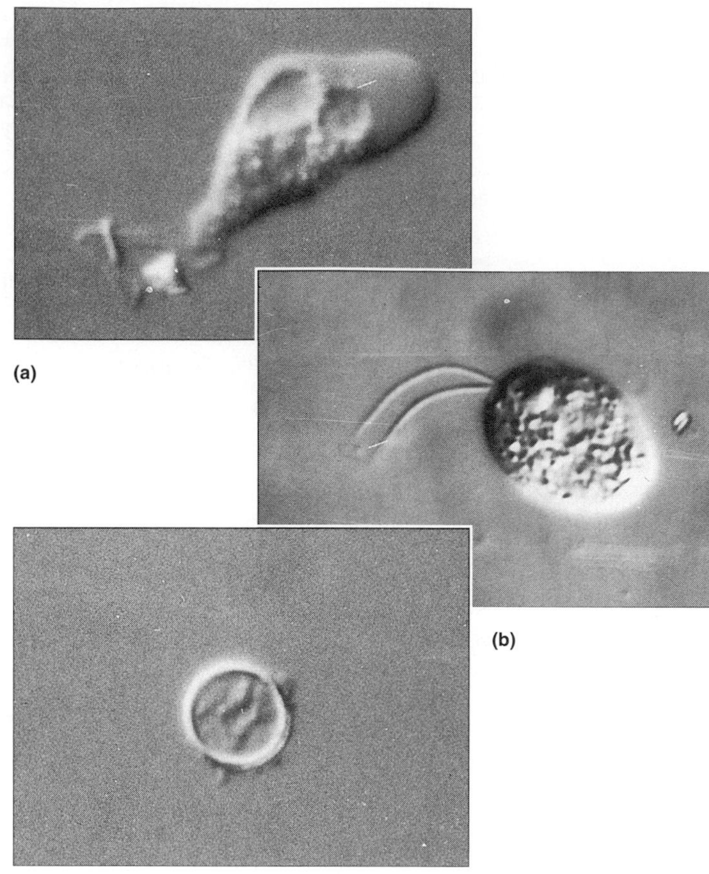

Figure 12.9 Polymorphism in a Protozoan The species of *Naegleria* may infect humans. **(a)** In human tissues, the organism exists in the form of an ameba (10–11 μm at its widest diameter). **(b)** After a few minutes in water, the flagellate form appears. **(c)** Under adverse conditions, a cyst is formed.

Some protozoa divide by multiple fissions, or **schizogony**, in which the nucleus divides a number of times and then the cell produces many small single-celled organisms, each one capable of infection. Multiple fission of the asexual forms in the human host results in large numbers of parasites released into the host's circulation at regular intervals, producing the characteristic cyclic symptoms of malaria. A more detailed account of this process is described in chapter 28.

Protozoa and Human Disease

The major threat posed by protozoa results from their ability to parasitize and often kill a wide variety of animal hosts. Human infections with the protozoans *Toxoplasma gondii*, which causes toxoplasmosis, *Plasmodium* species, which are responsible for malaria, *Trypanosoma* species, which cause sleeping sickness, and *Trichomonas vaginalis*, which causes vaginitis, are common in many parts of the world. Malaria has been one of the greatest killers of humans through the ages. At least 150 million people in the world contract malaria each year, and 1.5 million die of it. Protozoan infections of animals are so common that it is difficult to estimate their extent or overestimate their economic

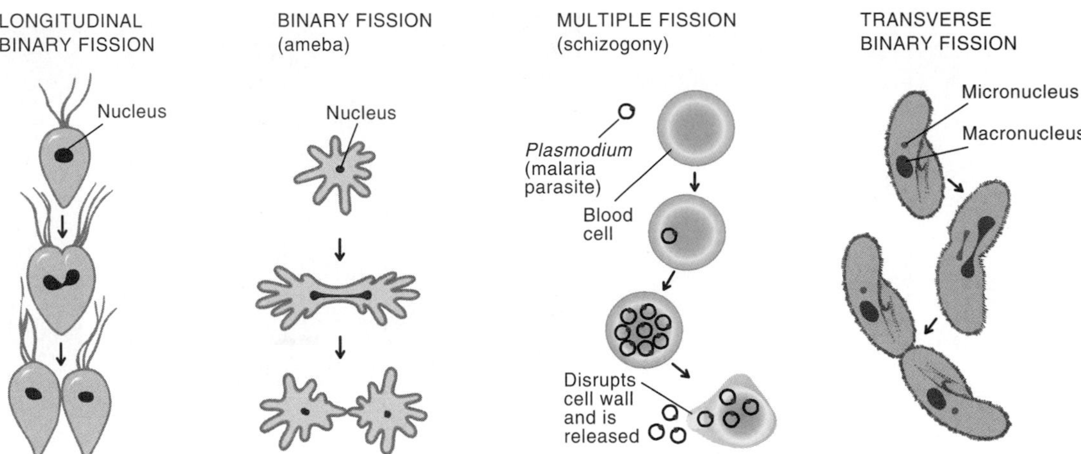

Figure 12.10 Various Forms of Asexual Reproduction in Protozoa

importance. Some parts of tropical Africa are uninhabitable, due in large part to the presence of the tsetse fly, the carrier of the trypanosomes that cause African sleeping sickness. Humans have no natural defense mechanisms against this infection. Some animals, including cows and horses, are reservoirs for these organisms.

Some diseases caused by protozoa are given in table 12.2 and discussed more extensively in the chapters dealing with the organ systems that are affected.

MICROCHECK 12.2

Protozoa are microscopic, single-celled, nonphotosynthetic, motile organisms. Most are free-living, but some can cause serious human disease. Protozoa are a very important part of the food chain.

- What are the primary characteristics that distinguish protozoa from other eukaryotic organisms?
- What are some important diseases caused by protozoa?
- Why would all protozoa be expected to require large amounts of water?

Fungi

The term fungi describes a taxonomic classification of organisms but no longer includes organisms such as slime molds and water molds that had traditionally been considered to be **Fungi** **(figure 12.11)**. The slime molds and water molds once thought to be related to the fungi now appear to have occurred much earlier and will be considered separately in this chapter.

Fungi require organic compounds for energy and as a carbon source, often from dead organisms. Most fungi are aerobic or facultatively anaerobic. Only a few fungi are anaerobic. ■ **aerobic, p. 97** ■ **facultatively anaerobic, p. 98** ■ **anaerobic, p. 97**

A large number of fungi cause disease in plants. Fortunately, only a few species cause disease in animals and humans. As modern medicine has advanced to treat once-fatal diseases, however, it has left many individuals with impaired immune systems. It is these immunocompromised individuals who are most vulnerable to the fungal diseases.

The study of fungi is known as **mycology**, and a person who studies fungi is known as a **mycologist**. Some species of fungi are useful sources of food and industrial products, whereas

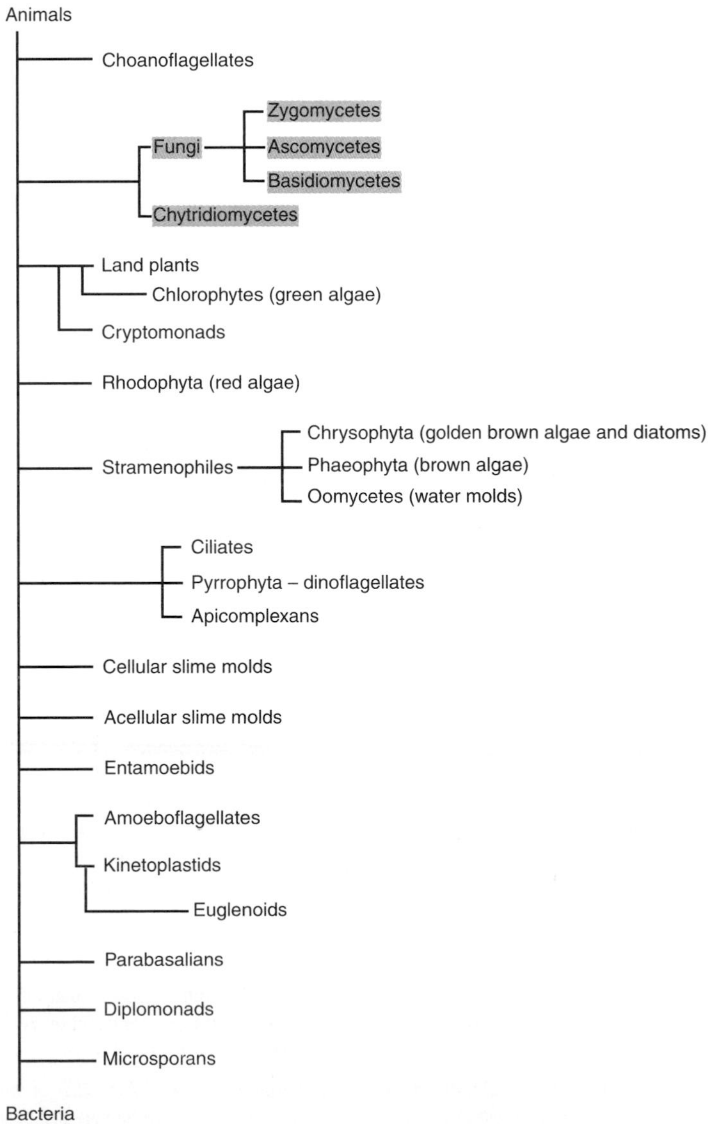

Figure 12.11 A Phylogenetic Scheme of Eukaryotes Based on rRNA Sequence Comparisons The fungal groups are highlighted in yellow.

(a) (b) (c)

Figure 12.12 Fungi Range in Size from Microscopic to Macroscopic Forms (a) Microscopic *Candida albicans* , showing the chlamydospores (large, round circles) that are highly resistant to adverse conditions. **(b)** *Polyporus sulphureus* (chicken of the woods), a shelflike fungus growing on a tree. **(c)** *Amanita muscaria,* a highly poisonous mushroom, growing in a cranberry bog on the Oregon coast.

others can spoil almost any organic material. Along with bacteria, fungi are the principal decomposers of carbon compounds on earth. This decomposition releases carbon dioxide into the atmosphere and nitrogen compounds into the soil, which are then taken up by plants and converted into organic compounds. Without this breakdown of organic material, the world would quickly be overrun with organic waste.

Classification of Fungi

Many fungi are microscopic and can be examined using basic microbiological techniques; others are macroscopic **(figure 12.12)**. All fungi have chitin in their cell walls and no flagellated cells at any time during their life cycle. There are four groups of true fungi, the **Zygomycetes, Basidiomycetes, Ascomycetes,** and **Deuteromycetes** or **Fungi Imperfecti**. The classification of the first three groups is based on their method of sexual reproduction. For the fourth group, the Deuteromycetes, sexual reproduction has not been observed, and so these fungi have traditionally been lumped together. With additional rRNA analysis, however, most Deuteromycetes can now be placed in one of the other three fungal groups. Most are either Ascomycetes or Basidiomycetes who have lost the sexual part of their life cycle **(table 12.3)**.

The Zygomycetes include the common bread mold (Rhizopus) and other food spoilage organisms. The Ascomycetes include

Table 12.3 Characteristics of Major Groups of Fungi

Group and Representative Member	Usual Habitat	Some Distinguishing Characteristics	Asexual Reproduction	Sexual Reproduction
Zygomycetes *Rhizopus stolonifer* (black bread mold)	Terrestrial	Multicellular, coenocytic mycelia (with many haploid nuclei)	Asexual spores develop in sporangia on the tips of aerial hyphae	Sexual spores known as zygospores can remain dormant in adverse environment
Basidiomycetes *Agaricus campestris* (meadow mushroom) *Cryptococcus neoformans*	Terrestrial	Multicellular, uninucleated mycelia. Group includes mushrooms, smuts, rusts that affect the food supply	Commonly absent	Produce basidiophores that are borne on club-shaped structures at the tips of the hyphae
Ascomycetes				
Neurospora, Saccharomyces cerevisiae (baker's yeast)	Terrestrial, on fruit and other organic materials	Unicellular and multicellular with septated mycelia	Is common by budding; conidiospores	Involves the formation of an ascus (sac) on specialized hyphae
Deuteromycetes (Fungi Imperfecti) *Penicillium, Aspergillus*	Terrestrial	A number of these are human pathogens	Budding	Absent or unknown

fungi that cause Dutch elm disease and rye smut. Smuts got this common name because the black spores that are produced give the appearance of soot. The Basidiomycetes include the common mushrooms and puffballs. Many medically and economically important species of fungus, including the one that produces penicillin, belong to the Deuteromycetes. In addition to the above four groups of fungi, the **Chytridiomycetes** are a close relative on the evolutionary scale. They have flagellated sexual spores and more variable life cycles, however, than the fungi. Most of these organisms live in water or soil, and a few are parasitic. Black wart disease of potatoes is caused by a chytridiomycete.

Common Groupings of Fungal Forms

When people talk about fungi, they frequently use terms such as yeast, mold, and mushroom. These terms have nothing to do with the classification of fungi but instead indicate their morphological forms.

Yeasts are single-celled fungi (**figure 12.13**). Yeasts can be spherical, oval, or cylindrical and are usually 3 to 5 µm in diameter. Some yeasts reproduce by binary fission, whereas others reproduce by budding, in which a small outgrowth on the cell produces a new cell (**figure 12.14**).

Molds are filamentous fungi. A single filament is known as a **hypha** (plural, **hyphae**), and a collection of hyphae growing in one place is known as a **mycelium**. Hyphae develop from fungal spores. A fungal reproductive spore is typically a single cell about 3 to 30 µm in diameter, depending on the species. When a fungal spore lands on a suitable substrate, it germinates and sends out a projection called a **germ tube** (**figure 12.15**). This tube grows at the tip and develops into a hypha. The cells divide and form new cells. In some fungi, the cell wall does not completely close off one cell from another. In that case the cells become multinucleated. The white mass seen inside the potato or on moldy bread (**figure 12.16**) is an example of a mycelium. Only a small portion of the mycelium is actually visible on the surface of the bread; the rest is buried deep within. Some mycelia appear above the surface of the substrate as a mushroom, or puffball (**figure 12.17**). These macroscopic structures produce reproductive spores. Some large mushrooms are edible.

Hyphae are well adapted to absorb food. They are narrow and threadlike. With their high surface-to-volume ratio, they can absorb large amounts of nutrients. Hyphae release enzymes that break down the material into readily absorbed smaller organic compounds. In addition, these enzymes act to repel the growth of other hyphae near it. As a result, hyphae spread throughout the food source, ensuring that each hypha will have access to adequate nutrients.

Parasitic fungi have specialized hyphae called **haustoria**, which can penetrate animal or plant cell walls to gain nutrients. Saprophytic fungi sometimes have specialized hyphae called **rhizoids**, which anchor them to the substrate.

Dimorphic Fungi

Dimorphic fungi are capable of growing either as yeastlike cells or as mycelia, depending on the environmental conditions. Many of the fungi that cause disease in humans are dimorphic. Certain fungi such as *Coccidioides immitis* grow in the soil as molds. When their spores, which are readily carried

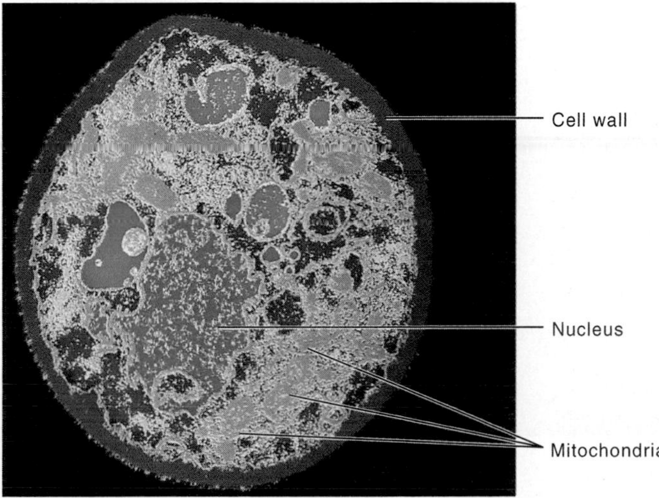

Figure 12.13 Morphology of a Yeast Cell As Seen with an Electron Microscope

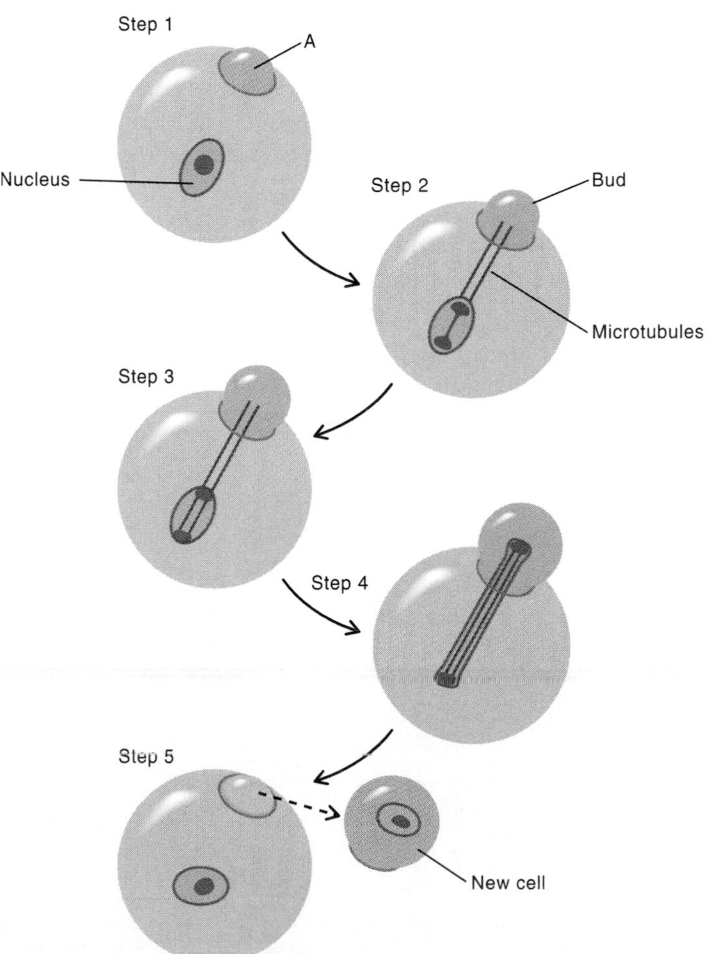

Figure 12.14 Budding in Yeast **(1)** Cell wall softens at point A, allowing the cytoplasm to bulge out. **(2, 3)** Nucleus divides by mitosis, and **(4)** one of the nuclei migrates into the bud. **(5)** Cell wall grows together and the bud breaks off, forming a new cell.

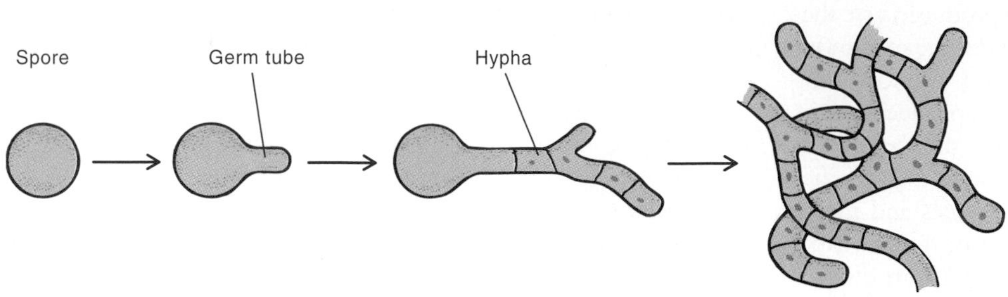

Figure 12.15 Formation of Hyphae and Mycelium Spores of fungi germinate to form a projection from the side of the cell called a germ tube, which elongates to form hyphae. As the hyphae continue to grow, they form a tangled mass called a mycelium.

in the air, are inhaled into the warm, moist environment of the lungs, they develop into the yeast form of the organism and cause disease.

Fungal Habitats

Fungi are found in virtually every habitat on the earth where organic materials exist. Whereas algae and protozoa grow primarily in aquatic environments, the fungi are mainly terrestrial organisms. Some species occur only on a particular strain of one genus of plants, whereas others are extremely versatile in what they can attack and use as a source of carbon and energy. Materials such as leather, cork, hair, wax, ink, jet fuel, and even some synthetic plastics like the polyvinyls can be attacked by fungi. Some species can grow in concentrations of salts, sugars, or acids strong enough to kill most bacteria. Thus, fungi are often responsible for spoiling pickles, fruit preserves, and other foods. Some fungi are resistant to pasteurization and others can grow at temperatures below the freezing point of water, rotting bulbs and destroying grass in frozen ground. Fungi are found in the thermal pools at Yellowstone National Park, in volcanic craters, and in lakes with very high salt content, such as the Great Salt Lake and the Dead Sea. ■ **food spoilage, p. 813**

Fungal reproductive cells, or spores, are found throughout the earth. They also occur in tremendous numbers in the air near the earth's surface as well as at altitudes of more than 7 miles. Although not as resistant as bacterial endospores, fungal spores are generally resistant to the ultraviolet rays of sunlight. Sunlight will, however, sometimes kill fungal vegetative cells. Fungal spores are a major cause of asthma. ■ **asthma, p. 436**

Growth Requirements of Fungi

Most fungi prefer a slightly moist environment with a relative humidity of 70% or more, and various species can grow at temperatures ranging from –6°C to 50°C. The optimal temperature for the majority of fungi is in the range of 20°C to 35°C.

The pH at which different fungi can grow varies widely, ranging from as low as 2.2 to as high as 9.6, but fungi usually grow well at an acid pH of 5.0 or lower. This explains why fungi grow well on fruits and many vegetables that tend to be acidic.

As heterotrophs, fungi secrete a wide variety of enzymes that degrade organic materials, especially complex carbohydrates, into small molecules that can be readily absorbed. Most fungi are aerobic, but some of the yeasts are facultative anaerobes and carry out alcoholic fermentation. Facultatively anaerobic

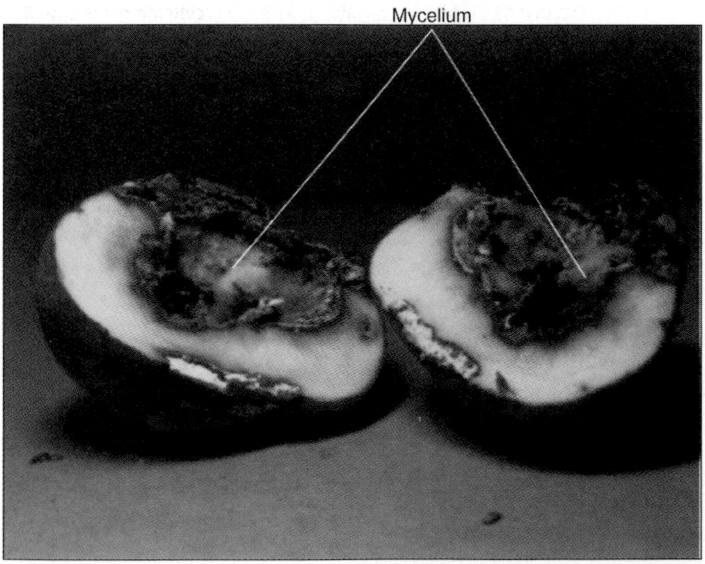

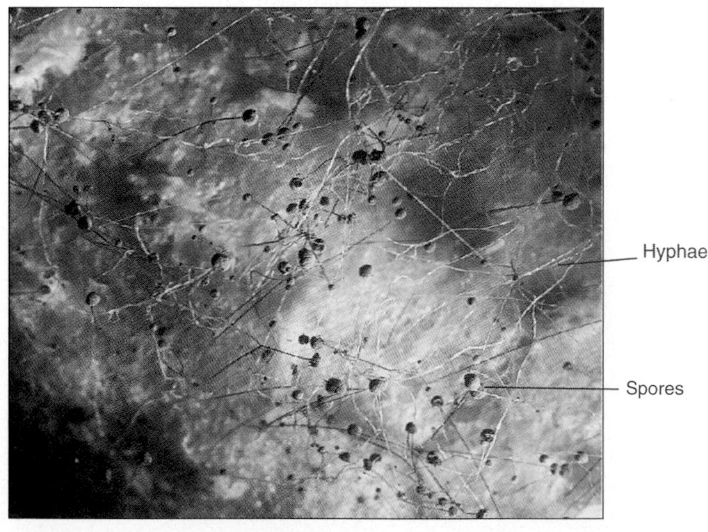

Figure 12.16 Examples of a Mycelium on Various Foods **(a)** The cottony white mass inside the potato is an example of a mycelium. **(b)** Magnified hyphae of *Rhizopus stolonifer*, black bread mold, showing the hyphae and the spores.

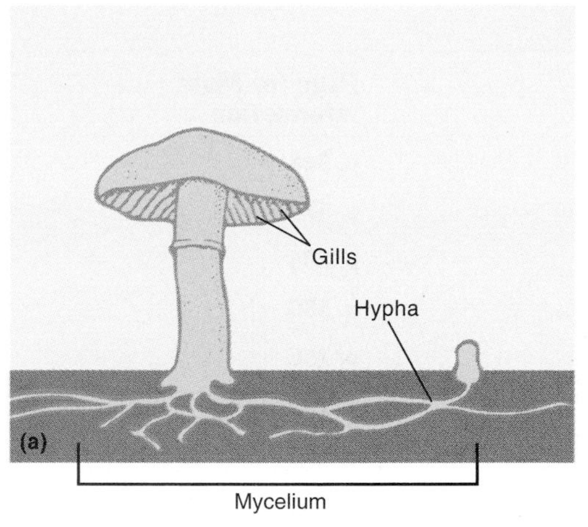

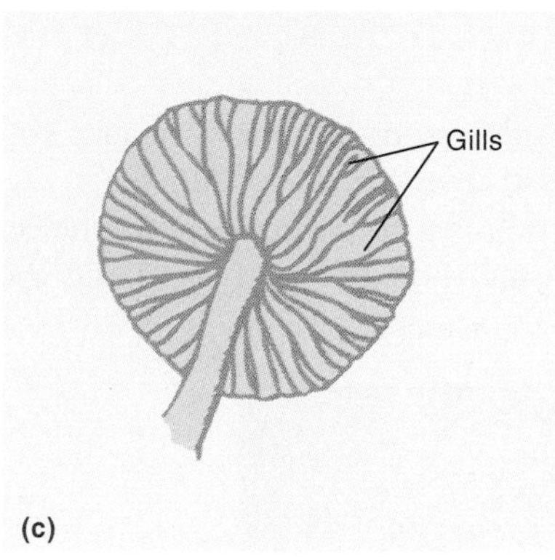

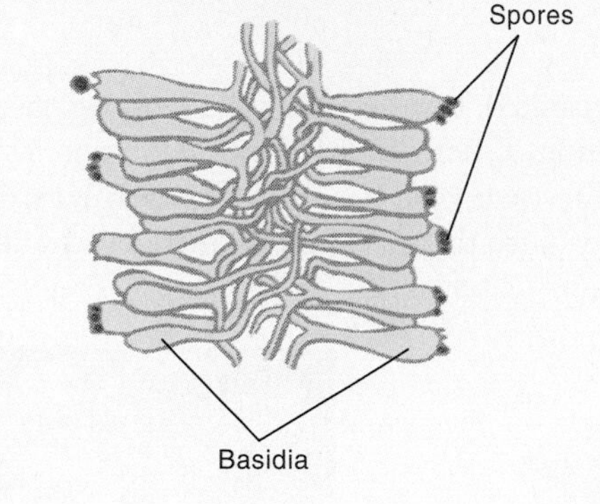

Figure 12.17 A Meadow Mushroom, *Agaricus campestri* **(a)** A drawing shows the extensive underground mycelium with a fruiting body emerging as a small button. **(b)** A photograph of *Lepiota rachodes,* a similar mushroom, showing the fruiting bodies and gills. **(c)** The underside of the cap is composed of radiating gills. **(d)** Magnified view of the surface of the gill showing a mass of basidia, bearing spores.

fungi live in the intestines of certain species of fish and help degrade algae. Some fungi living in the rumen of cows and sheep are known to be obligately anaerobic. They are important in the digestion of the plant material that these animals ingest.

Fungal Disease in Humans

Fungi cause disease in humans in one of four ways. First, a person may develop an allergic reaction to fungal spores or vegetative cells. Second, a person may react to the toxins produced by fungi. Third, the fungi may actually grow on or in the human body, causing disease or **mycoses**. Fourth, fungi can destroy the human food supply, causing starvation and death.

Allergic Reactions in Humans

Medical mycologists study fungi that affect humans, including fungi that cause allergic reactions. Allergic diseases such as hay fever and asthma can result from inhaling fungi or their spores if exposed humans have become sensitized. Sometimes, severe, long-term allergic lung disease results from these allergic reactions. ■ **hay fever, p. 435** ■ **asthma, p. 436**

Effect of Fungal Toxins

For their hallucinogenic properties, certain mushrooms have long been used as part of religious ceremonies in some cultures. The lethal effects of many mushrooms have also been known for centuries. The poisonous effects of a rye smut called **ergot** were known during the Middle Ages, but only recently has the active chemical been purified from this fungus to yield the drug ergot, which is now used to control uterine bleeding, relieve migraine headaches, and assist in childbirth.

Some fungi produce toxins that are carcinogenic. The most thoroughly studied of these carcinogenic toxins, produced by species of *Aspergillus*, are called **aflatoxins**. Ingestion of aflatoxins in moldy foods, such as grains and peanuts, has been implicated in the development of liver cancer (hepatoma) and thyroid cancer in hatchery fish. Governmental agencies monitor levels of aflatoxins in foods such as peanuts, and if a certain level is exceeded, the food cannot be sold. ■ **aflatoxin, p. 814**

Mycoses

Fungal diseases are called **mycoses**. The names of the individual diseases often begin with the names of the causative fungi. Thus, **histoplasmosis**, a disease seen worldwide, is a mycosis caused by the fungus *Histoplasma capsulatum*. Similarly, **coccidioidomycosis** is a mycosis caused by *Coccidioides immitis*, a fungus unique to certain arid regions of the Western Hemisphere. Diseases caused by the yeast *Candida albicans* are called **candidiasis** and are among the most common mycoses. ■ **histoplasmosis, p. 579** ■ **coccidioidomycosis, p. 580** ■ **candidiasis, p. 640**

Infections can also be referred to by the parts of the body that they affect. **Superficial mycoses** affect only the hair, skin, or nails. **Intermediate mycoses** are limited to the respiratory tract or the skin and subcutaneous tissues. **Systemic mycoses** affect tissues deep within the body. Some diseases caused by fungi are given in **table 12.4** and are discussed in the chapters dealing with the organ systems that are affected.

Symbiotic Relationships Between Fungi and Other Organisms

Fungi form several types of symbiotic relationships with other organisms. For example, lichens result from the association of a fungus with a photosynthetic organism such as an alga or a cyanobacterium (**figure 12.18**). These associations are extremely close and, in some cases, the fungal hyphae actually penetrate the cell wall of the photosynthetic partner. The fungus provides the protection and growing platform for the pair. In addition, the fungus absorbs water and minerals for the association. The photosynthetic member supplies the fungus with organic nutrients. It is possible to grow each partner of the lichen association separately.

Table 12.4 Some Medically Important Fungal Diseases

Disease	Causative Agent	Page for More Information
Candidial skin infection	*Candida albicans*	p. 544
Coccidioidomycosis	*Coccidioides immitis*	p. 579
Cryptococcal meningoencephalitis	*Filbasidiella neoformans*	p. 684
Histoplasmosis	*Histoplasma capsulatum*	p. 580
Pneumocytosis	*Pneumocystis carinii*	p. 756
Sporotrichosis	*Sporothrix schenckii*	p. 709
Vulvovaginal candidiasis	*Candida albicans*	p. 640

Usually the algal partner is able to grow well when separated, but the fungal partner does not. Because of this association, lichens can grow in extreme ecosystems where neither could survive on its own. Lichens are often a good indicator of air quality since they are very sensitive to sulfur dioxide, ozone, and toxic metals. You will not find very many lichens in cities with air pollution.

Mycorrhizas, fungal symbioses of particular importance, are formed by the intimate association between fungi and the

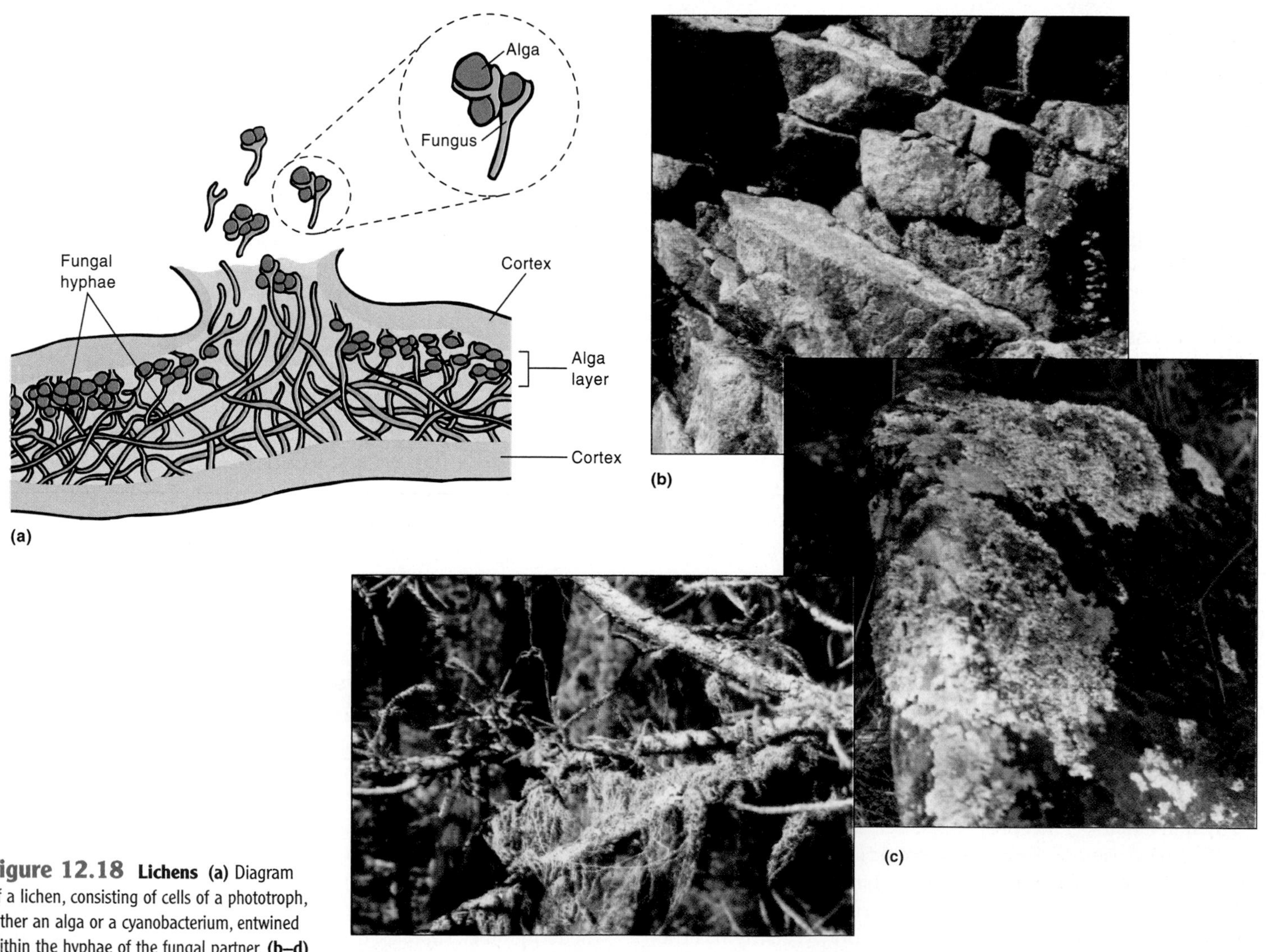

Figure 12.18 Lichens (a) Diagram of a lichen, consisting of cells of a phototroph, either an alga or a cyanobacterium, entwined within the hyphae of the fungal partner. **(b–d)** Photographs of lichens.

roots of certain plants such as the Douglas fir. It is estimated that 80% of the vascular plants have some type of mycorrhizal association with their roots. There are more than 5,000 species of fungi that form mycorrhizal associations. By increasing the absorptive power of the root, they often allow their plant partners to grow in soils where these plants could not otherwise survive. With the world facing major shortages of food and forest products, a better understanding of these symbiotic relationships is urgently needed. Many trees, including the conifers, are able to live in sandy soil because of their extensive mycelial mycorrhizas. In Puerto Rico, for example, the pine tree industry almost perished before the proper fungi were introduced to form mycorrhizal relationships with the trees. Now the industry is flourishing. Similarly, orchids cannot grow without mycorrhizal association of a fungus that helps provide nutrients to the young plant.

Certain insects also depend on symbiotic relationships with fungi. For example, leaf cutter ants are estimated to bring about 15% of the tropical vegetation into their nests to use for food. The leaves are used as food by a fungus that the ants cultivate in their nests. The fungus removes the poison from the leaves of the plant, and then the ants use the fungus as their food source.

Economic Importance of Fungi

Many fungi are important commercially. The yeast *Saccharomyces* has long been used in the production of wine, beer, and bread. Other fungal species are useful in making the large variety of cheeses that are found throughout the world. Penicillin, griseofulvin, and other antimicrobial medicines are synthesized by fungi.

Ironically, fungi are also among the greatest spoilers of food products, and large amounts of food are thrown away each year because they have been made inedible by species of *Penicillium, Rhizopus*, and others.

Fungi also cause many diseases of plants. Dutch elm disease caused by *Ceratocystis ulmi* is transmitted by beetles. It has destroyed the American elm trees that once shaded the streets in many cities. The wheat rust (*Puccinia graminis*) destroys tons of wheat yearly. Rust-resistant varieties have been bred to reduce the losses, but mutations in the rusts have made these advantages short-lived.

Fungi have been very useful tools for genetic and biochemical studies. *Neurospora crassa*, a common mold, has been widely used for modern genetic studies as well as for investigating biochemical reactions. More recently, yeasts have been genetically engineered to produce human insulin and the human growth hormone somatostatin, as well as a vaccine against hepatitis B. ■ **biotechnology, p. 221**

MICROCHECK 12.3

True fungi are organisms that have chitin in their cell walls and reproduce by both asexual and sexual methods. They are saprophytes and do not have motile cells during any stage in their life cycle. They are a source of food as well as a source of disease in plants and animals including humans.

- What are the primary characteristics that distinguish true fungi from other eukaryotic organisms?
- What are some ways that fungi cause disease in humans?
- What kind of symbiotic relationships do fungi form with other organisms? How are these relationships beneficial to each partner?
- How would the narrow, threadlike structure of hyphae indicate that they have a high surface-to-volume ratio?

Slime Molds and Water Molds

The slime molds and water molds used to be considered types of fungus. They are, however, completely unrelated to the true fungi and are good examples of **convergent evolution.** Convergent evolution occurs when two organisms develop similar characteristics because of adaptations to similar environments and yet are not related on a molecular level **(figure 12.19)**.

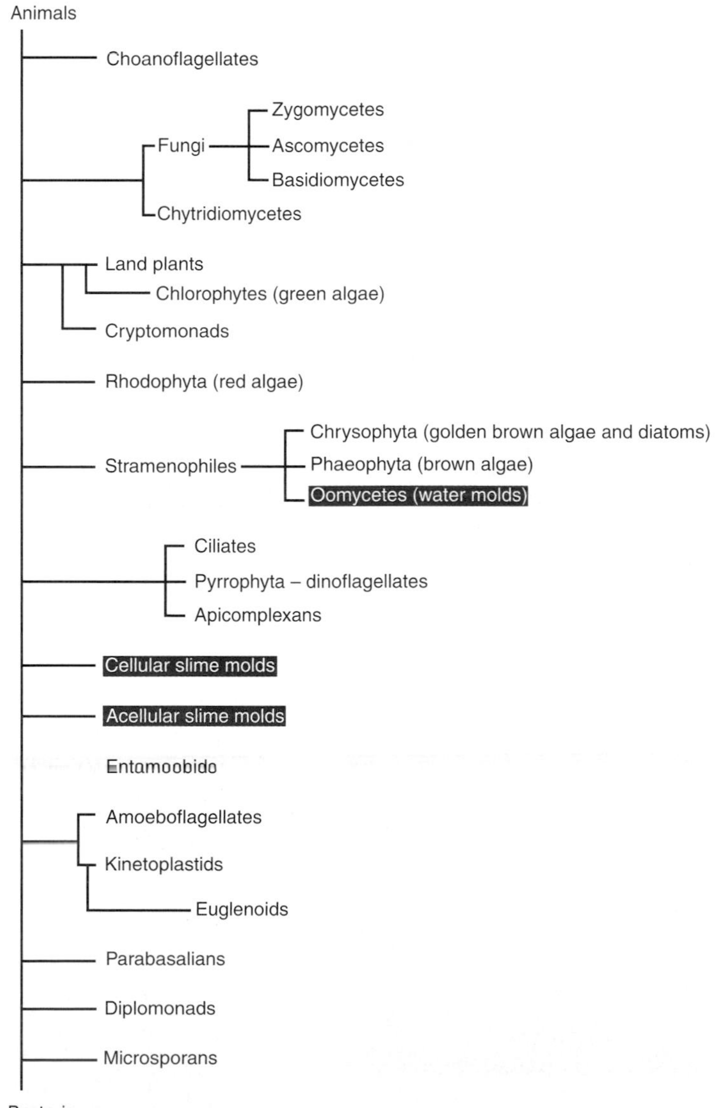

Figure 12.19 A Phylogenetic Scheme of Eukaryotes Based on rRNA Sequence Comparisons The slime molds and water molds are highlighted in blue.

Acellular and Cellular Slime Molds

There are two groups of slime molds, the **acellular slime molds** and the **cellular slime molds.** The slime molds are terrestrial organisms living on soil, leaf litter, or the surfaces of decaying leaves or wood. They are nonmotile, and reproduction depends on the formation of spores that can dispersed. Acellular slime molds are widespread and readily visible in their natural environment. Following sporulation and germination of the spores, ameboid cells fuse to form a myxameba. The nucleus of this ameba divides repeatedly, forming a multinucleated stage called a **plasmodium.** The plasmodium oozes like slime over the surface of decaying wood and leaves. As it moves, it ingests microorganisms, spores of other fungi, and any other organic material with which it comes in contact. Eventually the plasmodium is stimulated to form a spore-bearing fruiting body, and the process begins again (**figure 12.20a**).

The cellular slime mold has a vegetative form composed of single, ameba-like cells. When cellular slime molds run out of food, the single cells congregate into a mass of cells called a **slug** that then forms a fruiting body and spores. These fruiting bodies and spores look very much like fungal fruiting bodies and spores (figure 12.20b).

Slime molds are important links in the food chain in the soil. They ingest bacteria, algae, and other organisms and, in turn, serve as food for larger predators. Slime molds have been valuable as unique models for studying cellular differentiation during their aggregation and formation of their fruiting bodies.

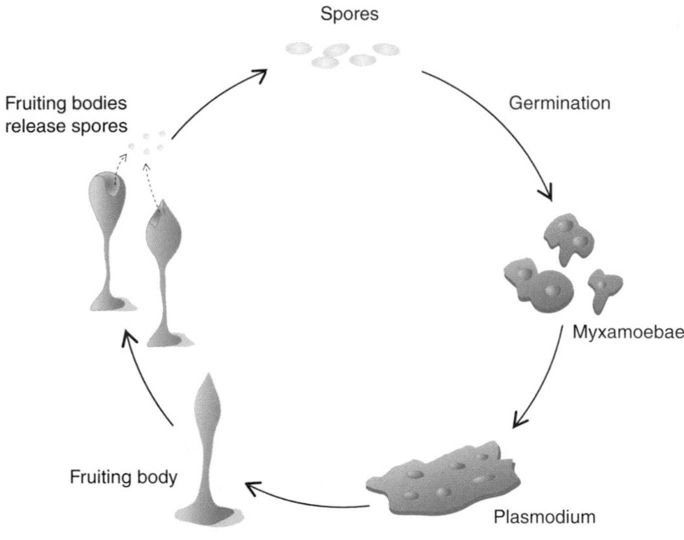

Figure 12.20 Slime Molds
(a) Life cycle of an acellular slime mold. **(b)** Fruiting body of a cellular slime mold.

Oomycetes (Water Molds)

The **oomycetes** or **water molds** are members of a group of organisms known as **heterokonts**. The oomycetes do not have chlorophyll, whereas other heterokonts, which include the grass green algae, diatoms, and brown algae, have chlorophyll and other photosynthetic pigments. Oomycetes were once considered fungi because they look like fungi. They form masses of white threads on decaying material. They have flagellated reproductive cells known as **zoospores**. Oomycetes cause some serious diseases of food crops. The late blight of potato (*Phytophthora infestans*) and downy mildew of grapes are included in this group. The late blight of potato was the cause of the potato famine in Ireland in the 1840s that sent waves of immigrants to the United States.

MICROCHECK 12.4

There are three other types of fungilike organisms, the acellular slime molds, the cellular slime molds, and the water molds. They are examples of convergent evolution.

- Why were the slime molds and water molds once considered to be fungi?
- When cellular slime molds run out of food, they form a fruiting body. What would be the advantage of this reproductive strategy?

Multicellular Parasites: Arthropods and Helminths

A number of disease-causing multicellular organisms are also studied using the same microscopic and immunological techniques that are used to study microorganisms and viruses. As a result, they are included here. Most of the medically important multicellular parasites fall into one of two groups: **arthropods** and **helminths**. The arthropods are more highly advanced on the evolutionary scale and include the insects, ticks, lice, and mites. Their main medical importance is that they serve as **vectors** that may transmit microorganisms and viruses to humans. The helminths, which include the **nematodes** (roundworms), **cestodes** (tapeworms), and the **trematodes** (flukes), are more primitive animals. In only a few instances do they transmit microbial infections to their host animal. Instead, they cause disease by invading the host's tissues or robbing it of nutrients.

Most multicellular parasites have been well controlled in the industrialized nations, but they still cause death and misery to many millions in the economically underdeveloped areas of the world. Our need to know about these problems has come about because more people are traveling farther, more people are moving from one place to another, and more goods are being exchanged worldwide. A clear example of this occurred in New York City in the summer of 1999 when West Nile fever was contracted by a number of people. At least 61 persons suffered serious disease and seven people died. A significant number of crows died at the same time and were found to be carrying the disease. In addition to birds and people, horses, cats, and dogs were also found to carry the virus. It is not clear how the virus arrived in New York City, but it perhaps could have been carried by a traveler from Africa, West Asia, or the Middle East, where it is commonly found. It could possibly have been brought by an imported bird from the same areas. Worldwide travel makes us more vulnerable to diseases from other parts of the world.

In addition, world wide climatic conditions are changing and bringing increases in certain insect populations to areas that were previously free of them. As a result, more cases of multicellular parasitic infections are being seen by physicians in the United States than previously.

Arthropods

The arthropods include insects, ticks, fleas, and mites. Arthropods act as vectors for transmitting diseases. In some instances, an arthropod such as a fly simply picks up a pathogen on its feet from some contaminated material such as feces and then lands on food that is then eaten by humans, thus transmitting the pathogen. In this case, the fly acts as a **mechanical vector**. In other cases, such as with *Plasmodium* sp., the cause of malaria, the vector, a mosquito, is a host for the organism before it transfers that organism to a human through a bite on the skin. In this case, the vector acts as an essential part of the life cycle of the organism and is known as a **biological vector**. The pathogen actually multiplies in number within the vector. ■ **mechanical vector, p. 479** ■ **biological vector, p. 479**

Examples of some important arthropods, the agents they transmit, and the resulting diseases are shown in **table 12.5**.

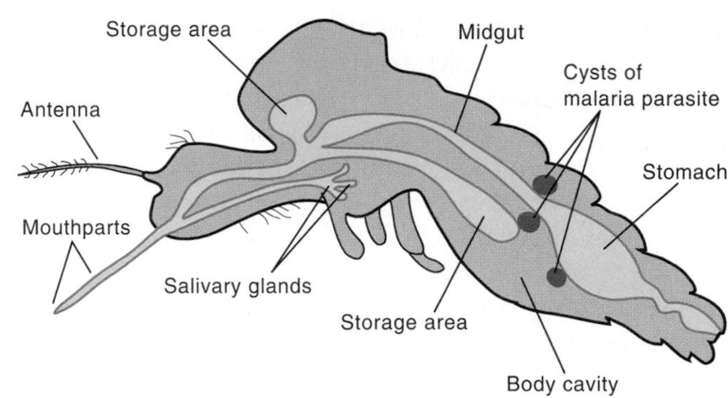

Figure 12.21 **Internal Anatomy of a Mosquito** Note the storage areas that allow ingestion of large amounts of blood and the salivary glands that discharge pathogens into the host.

Mosquitoes

The female mosquito needs the blood of a warm-blooded animal for the proper development of her eggs. To get this, she needs to bite such an animal. The mosquito can take in as much as twice its body weight in blood, thus giving it a relatively good chance of picking up infectious agents such as malarial parasites circulating within the host's capillaries. The anatomy of a mosquito is particularly adapted to transmit disease (**figure 12.21**). The mouthparts of the female mosquito consist of sharp stylets that are forced through the host's skin to the subcutaneous capillaries. One of these needlelike stylets is hollow, and the mosquito's saliva is pumped through it. The saliva increases blood flow and prevents clotting as the victim's blood is sucked into a tube formed by the other mouthparts of the insect. The saliva can also cause allergic reactions (the itch of a mosquito bite). After the mosquito has taken more than one blood meal, she can transmit disease from one animal to the next. Viruses found in the blood of the first animal are then transmitted to the next, and so on.

Table 12.5 Some Arthropods That Transmit Infectious Agents

Arthropod	Infectious Agent	Disease and Characteristic Features	Page to Find More Information
Insects			
Mosquito (*Anopheles* species)	*Plasmodium* species	Malaria—chills, bouts of recurring fever	**p. 731–733**
Mosquito (*Culex* species)	Togavirus	Equine encephalitis—fever, nausea, convulsions, coma	**p. 678–679**
Mosquito (*Aedes aegypti*)	Flavivirus	Yellow fever—fever, vomiting, jaundice, bleeding	**p. 729–730**
Flea (*Xenopsylla cheopis*)	*Yersinia pestis*	Plague—fever, headache, confusion, enlarged lymph nodes, skin hemorrhage	**p. 724–725**
Louse (*Pediculus humanus*)	*Rickettsia prowazekii*	Typhus—fever, hemorrhage, rash, confusion	**p. 521**
Arachnids			
Tick (*Dermacentor* species)	*Rickettsia rickettsii*	Rocky Mountain spotted fever—fever, hemorrhagic rash, confusion	**p. 521, 530**
Tick (*Ixodes* species)	*Borrelia burgdorferi*	Lyme disease—fever, rash, joint pain, nervous system impairment	**p. 532**
Mite (*Sarcoptes scabei*)	*Sacroptes scabei*	Scabies—itchy rash	**p. 659**

Mosquitoes in an area of arthropod-borne disease can be trapped and identified microscopically, and the blood they have ingested can be tested to see on which kinds of animals the different species are feeding. Precise identification of species and subspecies of these genera is important because different species of mosquitoes differ greatly in their breeding areas, time of feeding, and choice of host. Identification depends largely on microscopic examination of antennae, wings, claws, mating apparatus, and other features. Correct identification is often essential in designing specific control measures. ■ **epidemiology, p. 474**

Fleas

Fleas are wingless insects that depend on powerful hind legs to jump from place to place. Points of importance in identifying fleas include the spines (combs) about the head and thorax, the muscular pharynx, the long esophagus, and the spiny valve composed of rows of teethlike cells. Fleas are generally more of a nuisance than a health hazard, but they can transmit the bacterium *Yersinia pestis,* which causes plague, and a rickettsial disease, murine typhus, to humans. Larval fleas have a chewing type of mouth for feeding on organic matter. They ingest eggs of the common dog and cat tapeworm, *Dipylidium caninum*, serving as its intermediate host. Children acquire this tapeworm when they accidentally swallow fleas. Fleas can live in vacant buildings in a dormant stage for many months. When the building becomes inhabited, the fleas quickly mature and hungrily greet the new hosts. ■ **plague, p. 723** ■ **tapeworm, p. 624**

Lice

Like fleas, lice are small, wingless insects that prey on warm-blooded animals by piercing their skin and sucking blood. The legs and claws of lice, however, are adapted for holding onto body surfaces and clothing rather than for jumping. Human lice generally survive only a few days away from their hosts.

Pediculus humanus, the most notorious of the lice, is 1 to 4 mm long, with a characteristically small head and thorax, and a large abdomen **(figure 12.22)**. This louse has a membrane-like lip with tiny teeth that anchor it firmly to the skin of the host. Within the floor of the mouth is a piercing apparatus somewhat similar to that of fleas and mosquitoes. *Pediculus humanus* has only one host—humans—but easily spreads from one person to another by direct contact or by contact with personal items, especially in areas of crowding and poor sanitation.

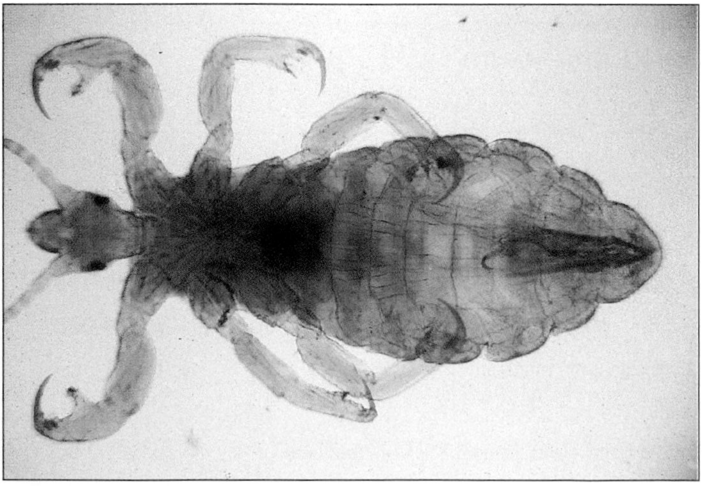

Figure 12.22 ***Pediculus humanus*** A body louse, which is the vector for *Rickettsia prowazekii*, the cause of typhus.

There are two subspecies, popularly termed head lice and body lice. Body lice can transmit trench fever, which is caused by the bacterium, *Bartonella quintana*; epidemic typhus, which is caused by the bacterium, *Rickettsia prowazekii*; and relapsing fever, caused by the bacterium *Borrelia recurrentis.* Trench fever occurs episodically among severe alcoholics and the homeless of large American and European cities.

The crab louse, *Phthirus pubis,* is commonly transmitted among young adults during sexual intercourse. It is not a vector of infectious disease, but it can cause an unpleasant itch.

Ticks

Ticks are arachnids. Arachnids differ from insects in their lack of wings and antennae, and their thorax and abdomen are fused together. Although like insects the immature ticks have three pairs of legs, the adults have four pairs. *Dermacentor andersoni*, the wood tick, is the vector for Rocky Mountain spotted fever caused by the bacterium *Rickettsia rickettsii.* Another tick, *Ixodes scapularis,* transmits with its saliva *Borrelia burgdorferi*, the spirochete that causes Lyme disease. In addition, the saliva of several genera of ticks can produce a profound paralysis, especially in children on whom the tick feeds for several days. Paralyzed humans and animals usually recover rapidly following removal of the tick. ■ **Rocky Mountain spotted fever, p. 530** ■ **Lyme disease, p. 531**

Mites

Mites, like ticks, are arachnids. They are generally tiny, fast moving, and live on the outer surfaces of animals and plants. *Demodex folliculorum* and *D. brevis* are elongated microscopic mites that live in the hair follicles or oil-producing glands usually of the face, typically without producing symptoms. Other species of mites cause human disease.

The disease scabies, caused by a mite, *Sarcoptes scabiei,* is characterized by an itchy rash most prominent between the fingers, under the breasts, and in the genital area. Scabies is easily transmitted by personal contact, and the disease is commonly acquired during sexual intercourse. The female mites burrow into the outer layers of epidermis **(figure 12.23)**, feeding and laying eggs over a lifetime of about 1 month. Allergy to the mites is largely responsible for the itchy rash. The diagnosis can only be made by demonstrating the mites, since scabies mimics other skin diseases. Treatment of scabies is easily accomplished with medication applied to the skin. *Sarcoptes scabiei* is not known to transmit infectious agents.

Mites of domestic animals and birds can cause an itchy rash in humans, as can mites sometimes present in hay, grain, cheese, or dried fruits. The dust mites that often live in large numbers in bedrooms can sometimes cause asthma when the mites and their excreta are inhaled.

The mites of rodents can transmit rickettsial diseases to humans. Rickettsial pox, caused by *Rickettsia akari* transmitted

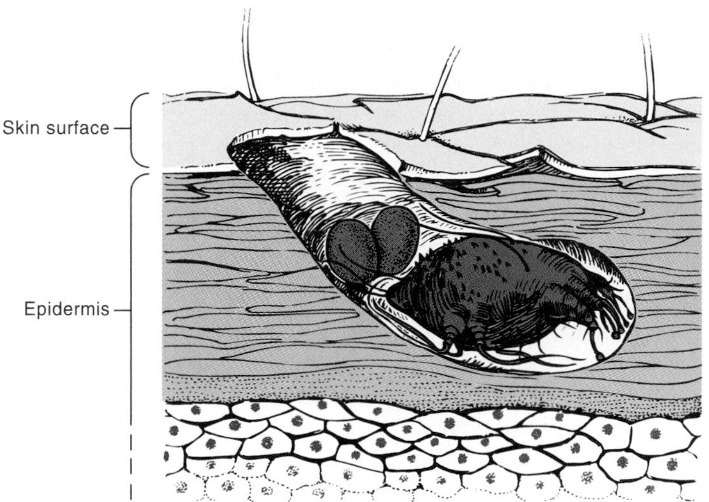

Figure 12.23 ***Sarcoptes scabiei*** **(Scabies Mite)** The female burrows into outer skin layers to lay her eggs, causing an intensely itchy rash.

by mouse mites, is a mild disease characterized by fever and rash. Epidemics occur periodically in cities of the eastern United States. Serious rickettsial diseases of other parts of the world such as scrub typhus are transmitted by rodent mites.

Helminths

In addition to the arthropods that can lead to disease in humans, the other group of multicellular animals that causes human disease are the **helminths**. In humans, the helminths that cause disease generally belong to one of three classes: the **nematodes,** or roundworms; the **cestodes**, or the tapeworms; and the **trematodes**, or the flukes (see figure 1.11).

Nematodes or Roundworms

The nematodes or roundworms have a cylindrical tapered body with a tubular digestive tract that extends from the mouth to the anus. There are both male and female nematodes. Nematodes include a large number of species. Many nematodes are free-living in soil and water. Others are parasites of human and other animals and plants and produce serious disease.

The nematodes that cause disease can be divided into two groups—the ones that inhabit the gastrointestinal tract of the host, and the ones that are found in the blood and other tissues of the host. Generally, diagnosis of worm infestation depends on microscopic identification of the worms or their ova (eggs), or on blood tests for antibody to the worms. **Table 12.6** summarizes features of these parasites and lists where a more detailed discussion of the diseases caused by nematodes can be found.

Cestodes or Tapeworms

Cestodes or tapeworms have flat, ribbon-shaped bodies that are segmented. The head of the tapeworm has suckers for attachment and sometimes has hooks **(figure 12.24)**. Directly behind the head is a region that produces the reproductive segments. Each segment has both male and female sex organs. The tapeworm does not have a digestive system but rather absorbs nutrients

Table 12.6 Nematodes, Cestodes, and Trematodes

Infectious Agents	Disease	Disease Characteristics	Page for More Information
Nematodes (roundworms)			
Pinworms (*Enterobius vermicularis*)	Enterobiasis	Anal itching, restlessness, irritability, nervousness, poor sleep	**p. 618**
Whipworm (*Trichuris trichiura*)	Trichuriasis	Abdominal pain, bloody stools, weight loss	**p. 619**
Hookworm (*Necator americanus*) and (*Ancylostoma duodenale*)	Hookworm disease	Anemia, weakness, fatigue, physical and mental retardation in children	**p. 621**
Threadworm (*Strongyloides stercoralis*)	Strongyloidiasis	Skin rash at site of penetration, cough, abdominal pains, weight loss	**p. 622**
Ascaria (*Ascaris lumbricoides*)	Ascariasis	Abdominal pain, live worms vomited or passed in stools	**p. 620**
Trichinella (*Trichinella spiralis*)	Trichinosis	Fever, swelling of upper eyelids, muscle soreness	**p. 623**
Filaria (*Wushereria bancrofti*) (*Brugia malayi*)	Filariasis	Fever, swelling of lymph glands, genitals, and extremities	**p. 734**
Cestodes (tapeworms)			
Fish tapeworm (*Diphyllobothrium latum*)	Tapeworm disease	Few or no symptoms, sometimes anemia	**p. 625**
Beef tapeworm (*Taenia saginata*)			
Pork tapeworm (*Taenia solium*)			
Trematodes (flukes)			
Cercaria (*Schistosoma mansoni*)	Schistosomiasis	Liver damage, malnutrition, weakness, and accumulation of fluid in the abdominal cavity	**p. 735**
Cercaria of birds and other animals	Swimmer's itch	Inflammation of the skin, itching	**p. 736**

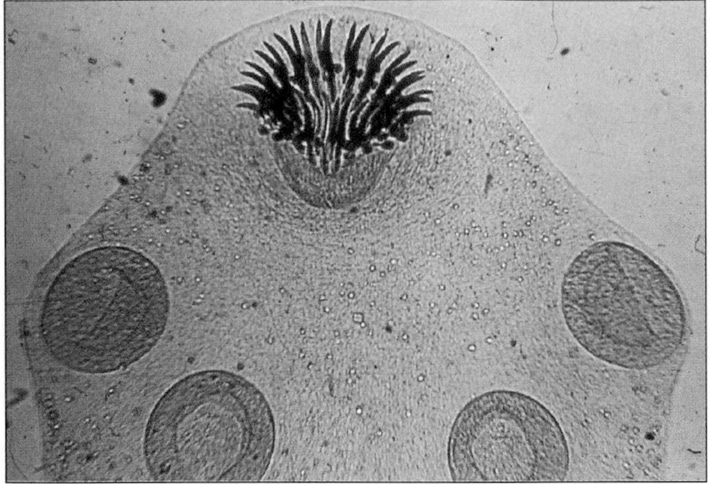

Figure 12.24 ***Taenia solium* Scolex** This tapeworm (25×) is acquired from eating inadequately cooked pork. Notice the hooks and suckers on the scolex (head).

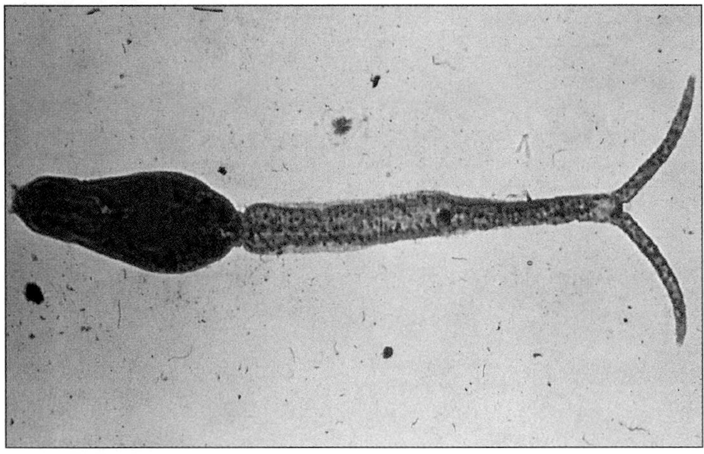

Figure 12.25 ***Schistosoma mansoni* Cercaria** The form that penetrates the skin and initiates the infection. Water bird schistosomes may penetrate the skin of swimmers and then die, causing swimmers itch.

directly. Tapeworms are often associated with beef, lamb, pork, and fish. Transmission of these organisms to humans often occurs when the flesh of these animals is eaten either uncooked or undercooked. Some tapeworms are transmitted to humans from ingesting fleas infected with dog or cat tapeworms. Table 12.6 gives information about specific tapeworm diseases and the locations of further discussion in this book.

Trematodes or Flukes

Trematodes or flukes are bilaterally symmetrical, flat, and leaf-shaped. They have suckers that hold the organism in place as well as suck fluids from the host. Most species are hermaphroditic (have both sex organs in the same worm). Most trematodes have a complicated life cycle, which may include one or more intermediary hosts. Usually, the worms begin with a larval form developing within the egg. These larvae escape into the environment, where they are taken up by one or more intermediate hosts such as a snail. Eventually, the last stage is a tail-bearing larva known as a **cercaria**, which is released from the snail and is ready to attach to the susceptible host (**figure 12.25**). For example, if a human is wading in water and the cercaria of *Schistosoma mansoni* have been discharged into that water, the cercaria can penetrate the skin and work its way through the circulation to the liver and intestine, where it matures and lays eggs. ■ **schistosomes, p. 734**

MICROCHECK 12.5

Arthropods such as mosquitoes, fleas, lice, and ticks act primarily as vectors for the spread of disease. On the other hand, the helminths, which include the roundworms, tapeworms, and flukes, cause serious disease in humans.

- How are arthropods able to spread disease in humans?
- What are the major differences among nematodes, cestodes, and trematodes?
- How would increased travel lead to increased spread of multicellular parasites?

FUTURE CHALLENGES

Algae: An Alternative to Gasoline

The burning of fossil fuels appears to be a major contributor to global warming and a major source of pollutants in the environment. Accordingly, scientists have been searching for alternative sources of energy. Hydrogen is one such energy source, since when it burns water is its only by-product. One promising source of hydrogen comes from a green alga, Chlamydomonas reinhardtii, found around the world as the green pond scum. This alga potentially could produce plenty of hydrogen, because it can directly split water into hydrogen and oxygen using an enzyme, hydrogenase. It appears that this alga evolved to take advantage of radically different environments. When this alga lives in ordinary air and sunlight, it converts sunlight, water, and carbon dioxide into oxygen and the other products the alga needs for growth. If, however, this alga is deprived of an essential nutrient such as sulfur or forced to live in an anaerobic environment, it switches to another mechanism of metabolism and produces hydrogen instead. The problem of getting this alga to produce large amounts of hydrogen arises when there is oxygen in the environment. The alga then shuts off its hydrogen-producing enzyme and produces only trace amounts. Since this alga is photosynthetic, it normally produces oxygen as a by-product of photosynthesis.

Scientists have found that when they deprive the alga of essential sulfate salts, the alga no longer maintains the protein complex needed for producing oxygen photosynthetically and instead goes into its hydrogen-producing metabolic pathway. Two major problems need to be solved, however, before large quantities of hydrogen could be produced by this alga. The first problem is that the alga cannot grow in this sulfur-deprived mode for very long before the organism needs to revert to the oxygen-producing mode. The alga must replenish the proteins that are burned up in the course of metabolism in order to survive. One laboratory reported that they could grow the alga for 4 days before they needed to add sulfate so that the organism could return to its normal metabolic processes. During the period of sulfate deprivation, the alga produced 3 ml of hydrogen for each liter of growing culture. Although this amount of hydrogen is not large, it is an improvement over the amount recovered in the past.

A second problem is that when oxygen is available, the alga produces very small amounts of hydrogen. To solve this problem, workers in another laboratory were able to put nitrogen gas above the bottled algae so that the oxygen leaving the water did not stop the reaction. When the oxygen built up in the air above the water, they could flush it out with more nitrogen. This method of producing hydrogen was sustained for 58 days. Scientists hope that they will be able to develop oxygen-tolerant algae and thereby increase hydrogen production. These research efforts offer a future challenge to scientists as they explore alternative sources of energy for the world.

SUMMARY

The Eukaryotic Members of the Microbial World (Figure 12.1)

1. Cell structure in the **Eukarya** is different from that seen in the Bacteria or the Archaea.
2. *Algae*, *fungi*, and *protozoa* are not accurate classification terms when the rRNA sequences of these organisms are considered. (Figure12.2)

Algae

1. Algae are a diverse group of photosynthetic organisms that contain chlorophyll.

Classification (Table 12.1 and Figure 12.3)

1. Classification of algae is based on their major photosynthetic pigments.
2. Organisms are placed on the phylogenetic tree according to rRNA sequences.

Algal Habitats

1. Algae are found in fresh and salt water as well as in soil.
2. Unicellular algae make up a significant part of the **phytoplankton**.

Structure of Algae

1. Algae may be microscopic or macroscopic.
2. Their cell walls are made of cellulose and other commercially important materials such as agar and carrageenan.
3. They have membrane-bound organelles including a **nucleus**, **chloroplasts**, and **mitochondria**.

Algal Reproduction

1. Algae produce asexually as well as sexually. (Figure 12.5)

Paralytic Shellfish Poisoning

1. Algae do not directly cause disease, but produce toxins that are ingested by fish and shellfish. (Figure 12.6)
2. When these fish and shellfish are eaten by humans, dizziness, muscle weakness, and even death may result; cooking does not destroy the toxins.

Protozoa

1. Protozoa are microscopic, unicellular organisms that lack chlorophyll, are motile during at least one stage in their development, and reproduce most often by binary fission.

Classification of Protozoa (Figure 12.7)

1. In classification schemes based on rRNA, protozoa are not a single group of organisms.
2. Protozoa have traditionally been put into groups based on their mode of locomotion. (Table 12.2)
3. **Sarcomastigophora** include the **Mastigophora,** the flagellated protozoa, and **Sarcodina**, which move by means of pseudopodia.
4. **Ciliophora** move by means of cilia.
5. **Apicomplexa,** also referred to as the sporozoa, include *Plasmodium* sp., the cause of malaria.
6. **Microsporidia**, an intracellular protozoa, causes disease in immunocompromised individuals.

Protozoan Habitats

1. Most protozoa are free-living and are found in marine and fresh water as well as terrestrial environments.
2. They are important decomposers in many ecosystems and are a key part of the food chain.

Structure of Protozoa

1. Protozoa lack a cell wall, but most maintain a definite shape using the underlying ectoplasm.
2. Life cycles are often complex and include more than one habitat. (Figure 12.9)
3. Protozoa feed by either phagocytosis or pinocytosis.

Protozoan Reproduction (Figure 12.10)

1. Reproduction is often by binary fission; some reproduce by multiple fissions or **schizogony.**

Protozoa and Human Disease (Table 12.2)

1. Protozoa cause some serious diseases such as malaria, African sleeping sickness, toxoplasmosis, and vaginitis.

Fungi

1. Fungi can cause serious disease, primarily in plants.
2. Fungi produce useful food products.

Classification of Fungi (Table 12.3 and Figure 12.11)

1. **Zygomycetes**, **Ascomycetes**, **Basidiomycetes**, and **Deuteromycetes** or **Fungi Imperfecti** are the four groups of true fungi.
2. **Chytridiomycetes** are a close relative.
3. **Yeast, mold**, and **mushroom** are common terms that indicate morphological forms of fungi. (Figures 12.12, 12.13, and 12.17)
4. Fungal filaments are called **hyphae** and a group of hyphae is called a **mycelium**. (Figures 12.15, 12.16)
5. **Dimorphic fungi** can grow either as a single cell (yeast) or as mycelia.

Fungal Habitats

1. Fungi inhabit just about every ecological niche and can spoil a large variety of food materials because they can grow in high concentrations of sugar, salt, and acid.
2. Fungi can be found in moist environments at temperatures from –6°C to 50°C and pH from 2.2 to 9.6.
3. Fungi are heterotrophs with enzymes that can degrade most organic materials.

Fungal Disease in Humans (Table 12.4)

1. Fungi may produce an allergic reaction.
2. They may produce toxins that make humans ill. These include **ergot**, those in poisonous mushrooms, and **aflatoxin.**
3. They cause **mycoses** such as **histoplasmosis**, **coccidioidomycosis**, and **candidiasis**.

Symbiotic Relationships Between Fungi and Other Organisms

1. **Lichens** result from an association of a fungus with a photosynthetic organism such as an alga or a cyanobacterium. (Figure 12.18)
2. **Mycorrhizas** are the result of an intimate association of a fungus and the roots of a plant.

Economic Importance of Fungi

1. The yeast *Saccharomyces* is used in the production of beer, wine, and bread.
2. *Penicillium* and other fungi synthesize antibiotics.
3. Fungi spoil many food products.
4. Fungi cause diseases of plants such as Dutch elm disease and wheat rust.
5. Fungi have been useful tools in genetic and biochemical studies.

Slime molds and Water Molds (Figure 12.19)

1. **Acellular** and **cellular slime molds** are important links in the terrestrial food chain. (Figure 12.20)
2. **Oomycetes**, also known as **water molds**, cause some serious diseases of plants.

Multicellular Parasites: Arthropods and Helminths

Arthropods

1. **Arthropods** act as **vectors** for disease. (Table 12.5)
2. Mosquitoes spread disease by picking up disease-causing organisms when the mosquito bites, and later injecting these organisms into subsequent animals that it bites. (Figure 12.21)
3. Fleas transmit disease such as plague; lice can transmit trench fever, epidemic typhus, and relapsing fever. (Figure 12.22)
4. Ticks are implicated in Rocky Mountain spotted fever and Lyme disease.
5. Mites cause scabies, and dust mites are responsible for allergies and asthma. (Figure 12.23)

Helminths (Table 12.6)

1. Most **nematodes** or roundworms are free-living, but they may cause serious disease such as pinworm disease, whipworm disease, hookworm disease, and ascariasis.
2. **Cestodes** are tapeworms with segmented bodies and hooks to attach to the wall of the intestine. (Figure 12.24)
3. Most tapeworm infections occur in persons who eat uncooked or undercooked meats; some tapeworms are acquired by ingesting fleas infected with dog or cat tapeworms.
4. **Trematodes**, or flukes, often have complicated life cycles that necessarily involve more than one host.
5. *Schistosoma mansoni* **cercaria** can penetrate the skin of persons wading in infected waters and cause serious disease. (Figure 12.25)

REVIEW QUESTIONS

Short Answer

1. What are the major differences between the Eukarya, the Bacteria, and the Archaea?
2. What distinguishes algae from all the other eukaryotic organisms?
3. Why are algae economically important?
4. Contrast the various modes of locomotion in protozoa.
5. Why are protozoa economically important?
6. What is the difference between a yeast, a mold, and a mushroom?
7. What are fungal diseases called?
8. Why are fungi economically important?
9. Discuss the differences and similarities in the ways algae, protozoa, fungi, helminths, and arthropods cause disease in humans.
10. What is a vector? Give two examples.

Multiple Choice

Choose one or more of these organisms that best answers the question.

1. Members of this group have chitinous cell walls.
   A. Algae
   B. Protozoa
   C. Fungi
   D. Helminths
   E. Arthropods
2. Members of this group are photosynthetic.
   A. Algae
   B. Protozoa
   C. Fungi
   D. Helminths
   E. Arthropods

3. All members of this group are single-celled.
   A. Algae
   B. Protozoa
   C. Fungi
   D. Helminths
   E. Arthropods
4. This group can have both single-celled and multicellular members.
   A. Algae
   B. Protozoa
   C. Fungi
   D. Helminths
   E. Arthropods
5. This group can only cause disease in humans through toxins.
   A. Algae
   B. Protozoa
   C. Fungi
   D. Helminths
   E. Arthropods
6. This group lacks a cell wall.
   A. Algae
   B. Protozoa
   C. Fungi
   D. Helminths
   E. Arthropods
7. These groups are found in plankton.
   A. Algae
   B. Protozoa
   C. Fungi
   D. Helminths
   E. Arthropods
8. Red tides are caused by this group.
   A. Algae
   B. Protozoa
   C. Fungi
   D. Helminths
   E. Arthropod
9. This group helps produce many of the foods that we eat.
   A. Algae
   B. Protozoa
   C. Fungi
   D. Helminths
   E. Arthropod
10. Without these groups the food chains of the world would not exist.
   A. Algae
   B. Protozoa
   C. Fungi
   D. Helminths
   E. Arthropod

Applications

1. A molecular biologist working for a government-run fishery in Vietnam is interested in controlling *Pfisteria* in fish farms. He needs to come up with a treatment that kills *Pfisteria* without harming the fish and other protista and beneficial algae that serve as food for the young fish. What strategy should the biologist consider for developing a selective treatment?
2. Paper recycling companies refuse to collect paper products that are contaminated with food or have been sitting wet for a day. A college sorority member who is running a recycling program on campus wishes to know the reason for this. What reason did the chemist who works for the recycling company probably give her for this policy?

Critical Thinking

1. Explain why it may be more difficult to treat diseases in humans caused by members of the Eukarya than diseases caused by the Bacteria.
2. Fungi are known for growing and reproducing a wide range of environmental extremes in temperature, pH, and osmotic pressure. What does this tolerance for extremes indicate about fungal enzymes?

CHAPTER 13

Viruses of Bacteria

During the late Nineteenth century, many bacteria, fungi, and protozoa were identified as infectious organisms. Most of these organisms could be readily seen with a microscope, and they generally could be grown in the laboratory. In the 1890s, D. M. Iwanowsky and Martinus Beijerinck found that a disease of tobacco plants, called ***mosaic disease****, was caused by an agent different from anything that was known. About 10 years later, F. W. Twort in England and F. d'Herelle in France showed that infectious agents existed that could destroy bacteria. Both groups of agents had the same unusual properties. They were so small that they could not be seen with the light microscope, and they passed through filters that retained almost all known bacteria. These same agents could only be grown in media if it contained living cells. The agents were called* ***filterable viruses*** *by Beijerinck. Virus means "poison," a term that once had been applied to all infectious agents. With time, the adjective* filterable *was dropped and only the word* virus *was retained.*

In many ways, viruses had features that were more characteristic of complex chemicals than of cells. For example, tobacco mosaic virus (TMV) could be precipitated from a suspension with ethyl alcohol and would still remain infective. A similar treatment destroyed the infectivity of bacteria. Further, in 1935, Wendell Stanley of the University of California, Berkeley, crystallized tobacco mosaic virus. Its physical and chemical properties obviously differed from those of cells, which cannot form crystals. Surprisingly, the crystallized tobacco mosaic virus could still cause the disease described by Iwanowsky and Beijerinck 40 years before.

—A Glimpse of History

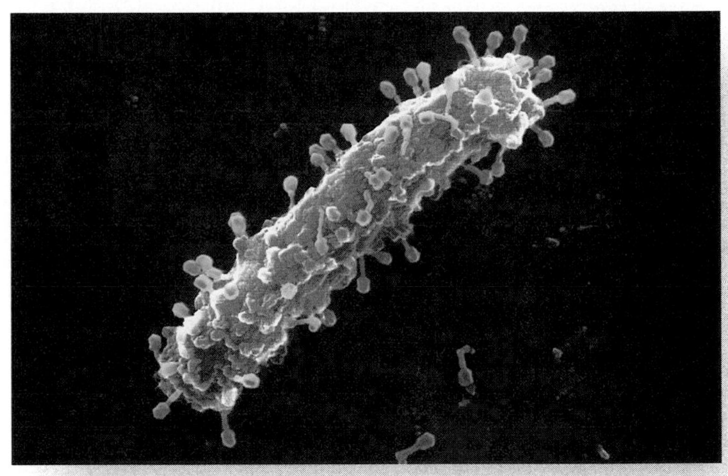

VIRUSES POSED A MYSTERY TO SCIENTISTS AS recently as 50 years ago. They were clearly smaller than known bacteria and possessed properties different from those of cells. The nature of these curious agents, some of which infected animals and others plants, became clearer through the study of viruses that infect bacteria. Because bacterial cells can be grown easily and multiply rapidly, viruses that infect them, called **bacteriophage** or **phage** (phago means "to eat"), can be cultivated much more readily than the viruses that infect other organisms. Bacteriophages have been studied extensively, and the knowledge gained from these studies has contributed enormously to an understanding both of the viruses and of the molecular biology of all organisms.

General Characteristics of Viruses

Viruses are nonliving agents that can infect all forms of life, including all members of the Bacteria, Archaea, and Eukarya. They commonly are referred to by the organisms they infect. The word *bacteriophage* or *phage* that infect bacteria is both singular and plural when referring to one type of virus. The word *phages* is used when different types of phages are being referenced. This is the same usage as *fish* and *fishes.* Animal viruses infect animals and plant viruses infect plants. These organisms are the **hosts** for the viruses.

Virus Architecture

Each virus particle, often called a **virion** when it is on the outside of its host cell, consists of nucleic acid (DNA or RNA) surrounded by a protective protein coat, termed a **capsid.** Different viruses have different shapes **(figure 13.1)**. Some are **isometric,** having **icosahedral** symmetry made up of flat surfaces, forming equilateral triangles. They appear spherical when viewed with the electron microscope; others are **helical,** which gives the virion a filamentous or rodlike appearance. Most phages are more **complex** in shape, having an isometric head with a long helical component, the **sheath** or **tail**. It is estimated that more than 10^{23} phages of this shape exist in the world, making it the most prevalent biological agent.

The shape of a virus is determined by the shape of the protein capsid, either helical or spherical, that encloses the viral nucleic acid genome. The viral capsid together with the nucleic acid that is tightly packed within the protein coat is called the **nucleocapsid (figure 13.2)**. Each capsid is composed of many identical protein subunits, called **capsomeres.** All bacterial and animal viruses, but not plant viruses, must be able to attach (adsorb) to specific receptor sites on host cells. In tailless isometric viruses, **attachment proteins** or **spikes** project from the capsid and are involved in attaching the virus to the host cell (see figure 13.2). In isometric viruses with tails (sheaths), tail fibers serve to attach the virus to the host cell (see figure 13.1).

There are two basic types of virions. The outer coat of most phages consists of the protein capsid. This type of virion is called **naked**. Virtually all phages are naked. Some virions that infect humans and other animals, however, have an additional covering over the capsid protein. This consists of a double layer of lipid similar in structure to the cell membrane of the eukaryotic cell. These are termed **enveloped viruses**. Just inside the lipid envelope is a protein, the **matrix protein**. The attachment spikes project from the envelope (see figure 13.2).

Viruses are notable for their small size **(figure 13.3)**. They are approximately 100- to 1,000-fold smaller than the cells they infect. The smallest viruses are approximately 10 nm in diameter, while the largest animal viruses are about 500 nm, the size of the smallest bacterial cells. The smallest viruses contain very little nucleic acid, perhaps as few as 10 genes.

The Viral Genome

The structure of the viral genome is unusual. Viruses contain either RNA or DNA—but never both. Depending on the type of nucleic acid they contain, viruses are frequently referred to

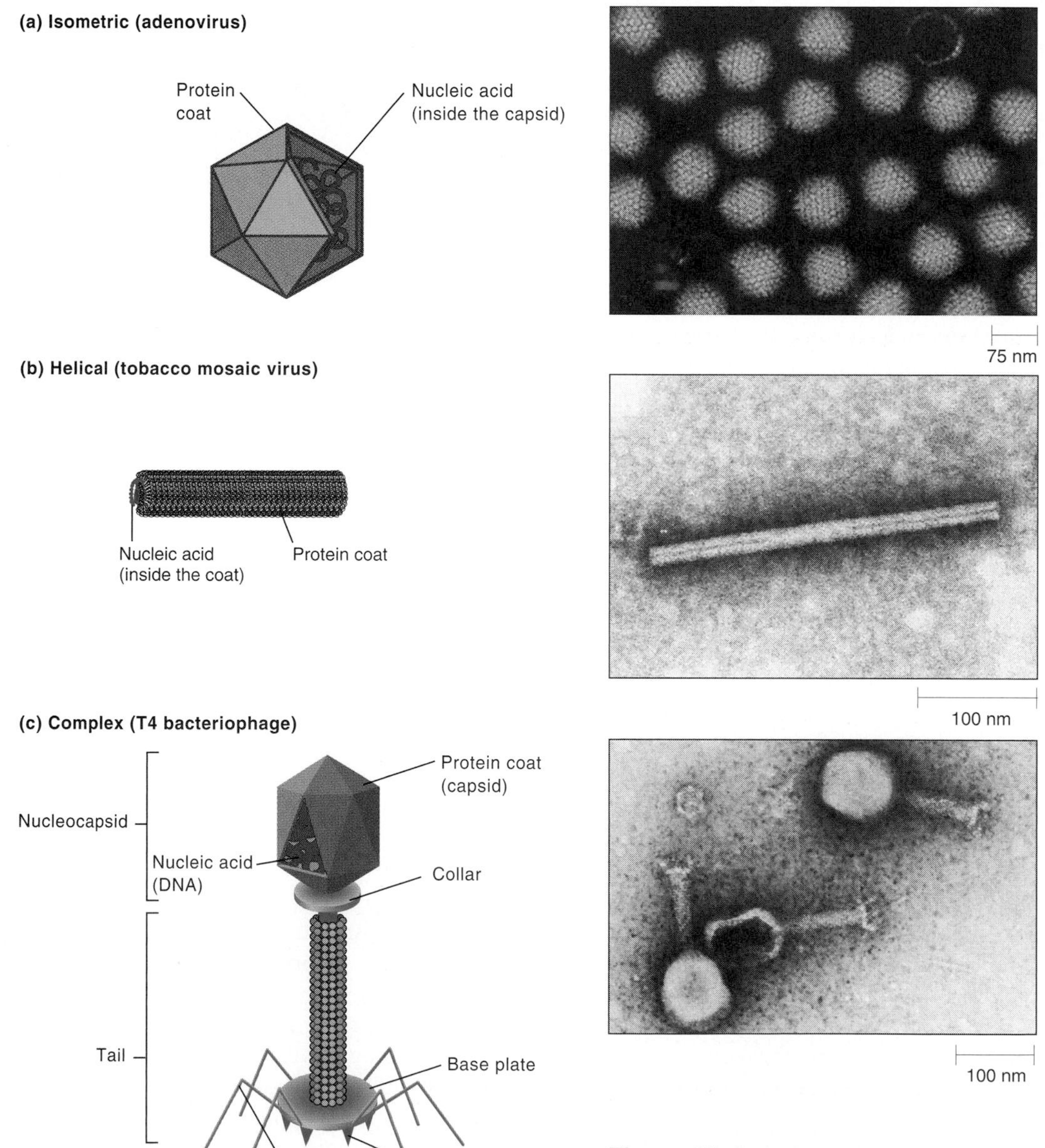

Figure 13.1 Common Shapes of Viruses (a) Isometric adenovirus virions; spikes protruding from the capsid are not visible. **(b)** Helical, tobacco mosaic virus. **(c)** Complex, T4 bacteriophage.

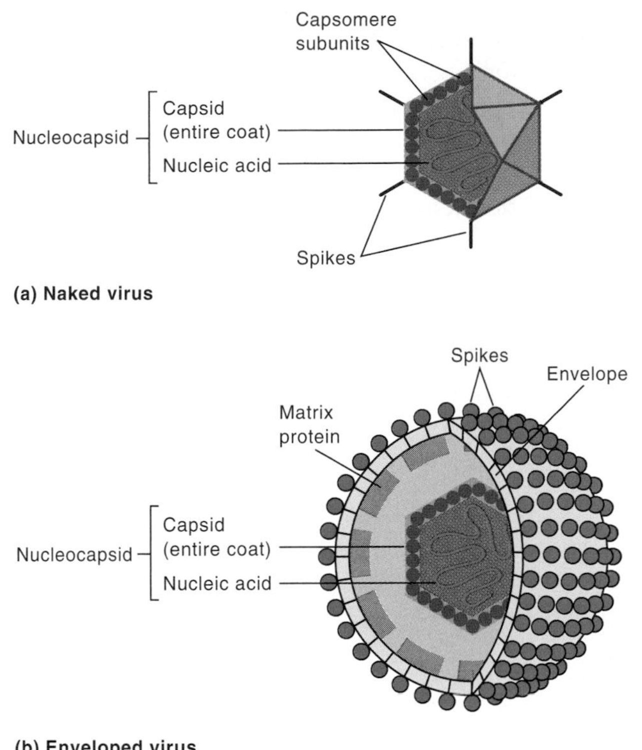

Figure 13.2 Two Different Types of Virions **(a)** Naked, not containing an envelope around the capsid and **(b)** enveloped, containing an envelope around the capsid.

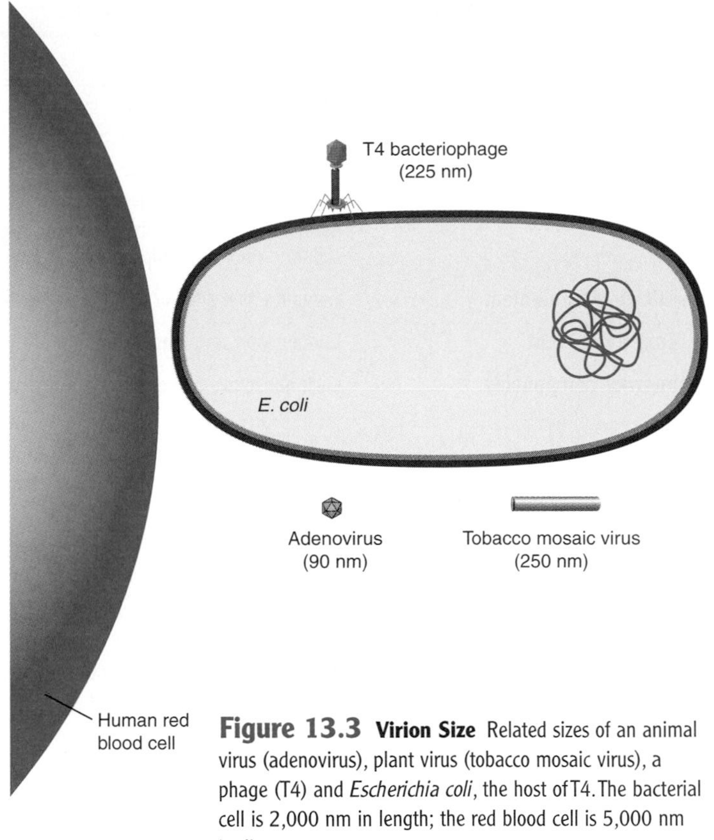

Figure 13.3 Virion Size Related sizes of an animal virus (adenovirus), plant virus (tobacco mosaic virus), a phage (T4) and *Escherichia coli*, the host of T4. The bacterial cell is 2,000 nm in length; the red blood cell is 5,000 nm in diameter.

as RNA or DNA viruses. The nucleic acids can occur in one of several different forms characteristic of the virus. DNA may be linear or circular, either double-stranded or single-stranded. RNA is usually single-stranded but may rarely form a double-stranded molecule.

Replication Cycle

Viruses can only multiply within living cells that are actively metabolizing; viruses use structures and enzymes of such cells to support their own reproduction. Viruses lack the cellular components necessary to generate energy and synthesize protein. They have no mitochondria or ribosomes, or even the enzymes necessary to perform these functions. In all viruses, the nucleic acid separates from its coat before replication begins. Because viruses contain so little nucleic acid, their few genes can code for only a very limited number of proteins, including enzymes. Every virus, however, must contain a minimum amount of genetic information. Their genes must be able to encode the proteins required to (1) make the viral protein coat, (2) assure replication of viral nucleic acid, and (3) move the virus into and out of the host cell. ■ **mitochondria, p. 79** ■ **ribosomes, p. 70.**

Viruses exist in two distinct phases. Outside of living cells, they persist in the form called virions, which are metabolically inert. Essentially, they are only macromolecules. Inside susceptible cells, viruses are in a replication form using the metabolic machinery and pathways of their host cells along with their own genetic information to produce new virions. Viruses and cells are compared in **table 13.1**.

MICROCHECK 13.1

Viruses are nonliving agents that consist of nucleic acid surrounded by a protein coat. Viruses only multiply within living cells, because they need certain functions of the host for their own replication.

- List four features of viruses that distinguish them from free-living cells.
- List three functions that all viruses must perform.
- An antibiotic is added to a growing culture of *E. coli*, resulting in death of the cells. Bacteriophage is then added. Would the phage replicate in the *E. coli* cells? Explain your answer.

Virus Interactions with Host Cells

The virion and host cell can interact in several different ways. Some types of viruses take over the metabolism of the host cell completely with resulting lysis of the cell. An example is the phage T4, which infects *E. coli*. Others do not take over the host cells' metabolism; the host cell continues to multiply and the virions leak out of the cell. Examples are the filamentous phages. Other viruses live in harmony with the host cell and multiply as the host cell multiplies. An example is the phage lambda (λ), which also infects *E. coli*. This phage confers new properties on the host bacterium.

Although viruses infect all kinds of cells, the relationships between even the most highly studied animal viruses and the animal host cells they invade are poorly understood in comparison

Table 13.1 Comparison Between Viruses and Cells

	Viruses	Cells
Size	10–500 nm	Bacteria: 1,000 nm Animal: 200,000 nm Plant: 400,000 nm
Multiplication	Multiply only within cells; outside the cell they are metabolically inert; protein coat and nucleic acid separate prior to multiplication	Most are free-living and multiply in the absence of other cells; entire cell remains intact during multiplication
Nucleic Acid Content	Contain either DNA or RNA, but never both	Always contain both DNA and RNA
Enzyme Content	Contain very few, if any, enzymes	Contain many enzymes
Internal Components	Lack ribosomes and enzymes for generating energy	Contain ribosomes and enzymes for generating energy

with phage-bacteria systems. This is largely because the eukaryotic host cells of animal viruses are far more complex and grow much more slowly than the prokaryotic host cells of the phage.

We will first focus on bacteriophages, which serve as excellent models for all other viruses. What you learn about them will be of immense help in understanding similar relationships between viruses and the animal cells they infect, described in the next chapter.

How phages affect the cells they infect depends primarily on the phage. Some phages multiply inside the cells they invade and escape by lysing the host cell. Since more virus is produced, this interaction is termed a **productive infection** and the phages that lyse the cell are termed **lytic**. These viruses take over the metabolism of the cell and direct it to phage replication. Another type of productive infection is carried out by phage that multiply inside the cells they invade, and then leak out or **extrude** without killing the host cells. These phages do not take over all the metabolic processes of the cell. An example is the filamentous phage, M13. Still other phages, termed **temperate,** integrate their DNA into the genome of the bacteria they infect or the DNA replicates as a plasmid. The phage replicates as the bacterial DNA replicates. The infection is termed **latent** because there may be no sign that the cells are infected. The bacteria carrying the phage DNA is a **lysogen** and the cells are in the **lysogenic state.** The phage DNA often codes for proteins that modify the properties of the host, the phenomenon of **lysogenic conversion**. An example of a temperate phage that infects *E.coli* is lambda (λ). These relationships are shown in **figure 13.4**.

Some of the phages that undergo these three kinds of relationship with their host bacteria are listed in **table 13.2**.

■ plasmid, p. 210

Lytic Phage Replication by Double-Stranded DNA Phages

In all lytic phage infections, the phage nucleic acid enters the host cell while its protein coat remains on the outside. The nucleic acid is replicated along with phage proteins, resulting in the formation of many virions. At the end of the replication cycle, phage exit by lysing the bacterium. Phage that go through this productive life cycle are termed **virulent**. One of the most intensively studied virulent phages is the double-stranded DNA phage, T4 (see figure 13.1). Its replication cycle illustrates how a phage can take over the life of a cell and reprogram its activities solely to the synthesis of phage **(figure 13.5)**.

Step 1: Attachment

When a suspension of T4 phage is mixed with a susceptible strain of *E. coli*, the phages collide by chance with the bacteria. Since viruses do not have flagella, they cannot swim toward their hosts. Protein fibers at the end of the phage tail attach to specific receptors on the bacterial cell wall (Step 1). These receptors, either protein or carbohydrate in structure, generally perform other functions for the cell, such as transport functions. The phage use these structures for their own purposes.

Step 2: Penetration

A few minutes after attachment, an enzyme (lysozyme) located in the tip of the phage tail degrades a small portion of the bacterial

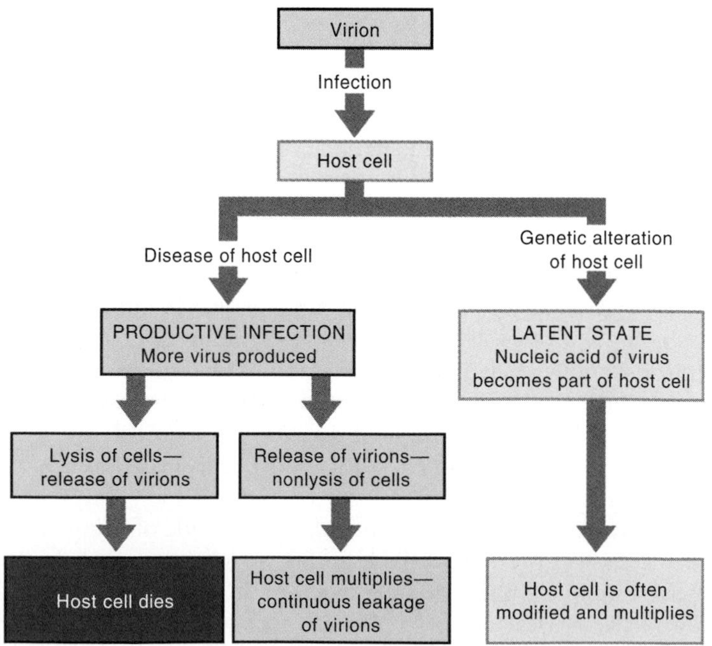

Figure 13.4 **Major Types of Relationships Between Viruses and the Host Cells They Infect**

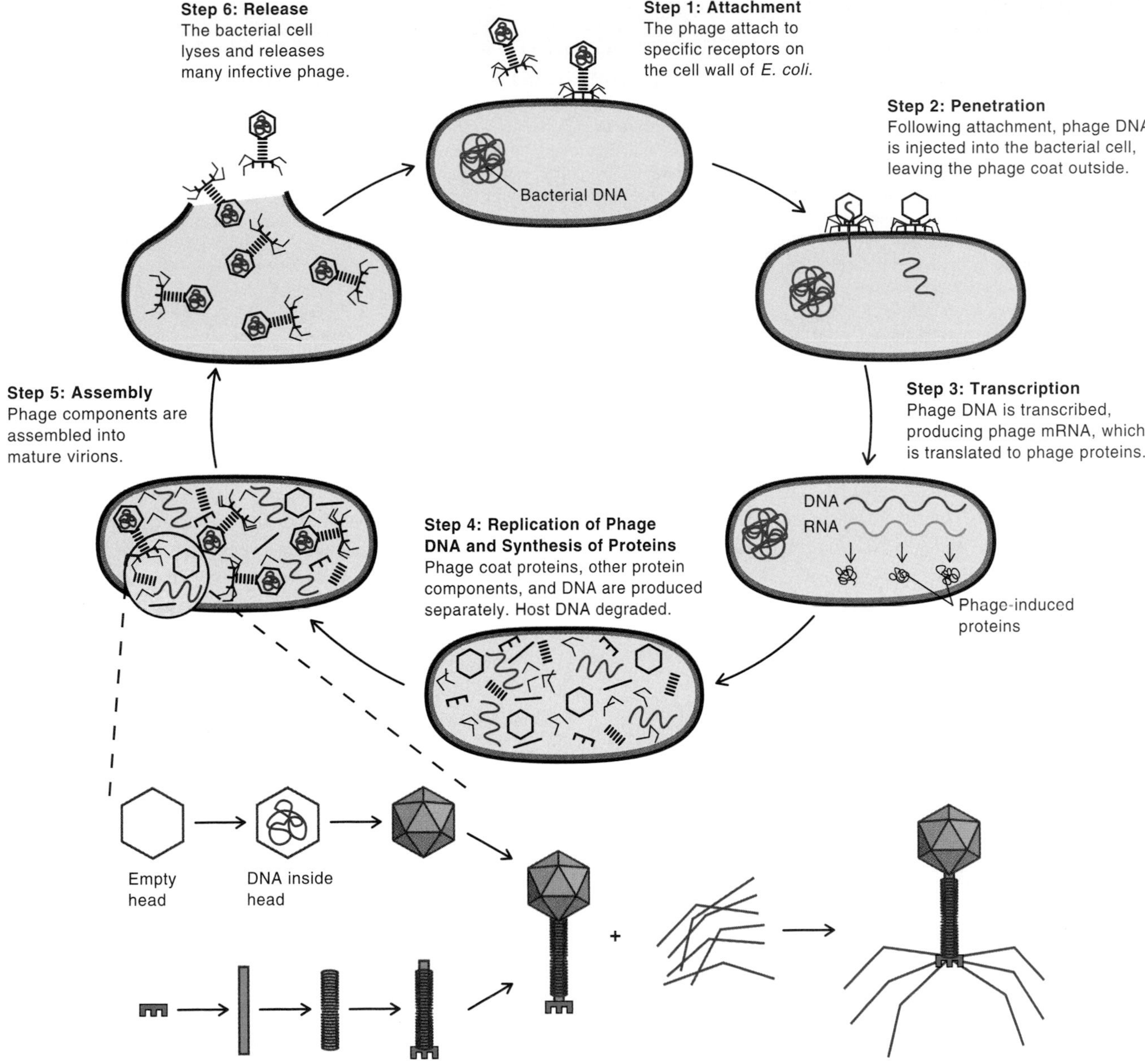

Step 6: Release
The bacterial cell lyses and releases many infective phage.
Step 1: Attachment
The phage attach to specific receptors on the cell wall of E. coli.
Bacterial DNA
Step 2: Penetration
Following attachment, phage DNA is injected into the bacterial cell, leaving the phage coat outside.
Step 5: Assembly
Phage components are assembled into mature virions.
Step 3: Transcription
Phage DNA is transcribed, producing phage mRNA, which is translated to phage proteins.
DNA
RNA
Step 4: Replication of Phage DNA and Synthesis of Proteins
Phage coat proteins, other protein components, and DNA are produced separately. Host DNA degraded.
Phage-induced proteins
Empty head
DNA inside head
+

Figure 13.5 **Steps in the Replication of T4 Phage in *E. coli***

cell wall. The tail contracts, and the tip of the tail opens. The double-stranded linear viral DNA in the head passes through the open channel of the phage tail (Step 2). The DNA is literally injected through the cell wall and, by some unknown mechanism, passes through the cytoplasmic membrane and enters the interior of the cell. The protein coat of the phage remains on the outside of the cell. Thus, the protein and nucleic acid separate, a feature of all virus replication.

Step 3: Transcription

Within minutes of the entry of phage DNA into a host cell, a portion is transcribed into mRNA, which is then translated into proteins that are specific for the infecting phage (Step 3). The first proteins produced, termed **early proteins,** are enzymes that are not normally present in the uninfected host cell. They are also known as **phage-induced proteins**, because they are coded by phage DNA. Some are essential for the synthesis of phage DNA, which contains an unusual pyrimidine, and others are involved in the synthesis of phage coat protein. One phage-induced early protein is a nuclease that degrades the DNA of the host cell. As a result, soon after infection, host cell DNA cannot be transcribed and all mRNA synthesized is transcribed from the phage DNA. In this way, the phage takes over the metabolism of the bacterial cell for its own purpose, namely the synthesis of more phages. Preexisting bacterial enzymes, however, continue to function. They supply the energy necessary for phage replication through the breakdown of energy sources in the environment. The host enzymes also synthesize amino acids and nucleotides for the production of phage proteins and nucleic acids, the former being carried out on the bacterial ribosomes.

Not all phage-induced enzymes are synthesized simultaneously. Rather, they are made in a sequential manner during

Perspective 13.1 Viral Soup

For many years, viruses have been recognized to cause many very deadly diseases of humans, animals, and plants. Within the past few years, scientists have discovered that they may play another important role in nature—that of being a major component of **bacterioplankton**, the bacteria found in the ocean. In the 1970s and 1980s, bacterial cells were found to be much more abundant than had been believed previously. Earlier estimates of the number of bacteria in aquatic environments were based on the numbers that could be cultured in the laboratory. When direct microscopic counts were made, however, the number of heterotrophic bacteria was estimated to be 10^5 to 10^7 bacteria per ml, much higher than previously estimated. Indeed, most carbon fixation in the ocean is carried out by bacteria. In addition, the concentration of cyanobacteria and single-celled eukaryotic algae is much greater than previously believed. What was most surprising, though, was the number of bacteriophages in aquatic environments. To get an accurate estimate of their numbers, scientists centrifuged large volumes of natural unpolluted waters from various locations. They then counted the number of particles in the pellet. Much to their surprise, up to 2.5×10^8 phages per ml were counted, about 1,000 to 10 million times as high as previous estimates. Why are these new numbers important? First, they may answer the question of why bacteria have not saturated the ocean. The bacteria are most likely held in check by the phages. From the estimates of the numbers of bacteria and phages, it is calculated that one-third of the bacterial population may experience a phage attack each day. Second, the interaction of phages with bacteria in natural water implies that the phages may be actively transferring DNA from one bacterium to another, by the process of transduction.

Thus, the smallest agents in natural waters may have important consequences for the ecology of the aquatic environment. Perhaps viruses have a similar function for eukaryotic organisms.

the course of infection. Those required early are synthesized near the beginning of infection; others that are not required until later in infection are made later on. Examples of early-induced enzymes are the nuclease that degrades the host chromosome and the enzymes of phage DNA synthesis. Late-induced enzymes include those concerned with the assembly of capsids and the phage lysozyme that lyses the bacteria to release the newly formed phage.

Step 4: Replication of Phage DNA and Synthesis of Proteins

Phage protein and nucleic acid replicate independently of one another. The DNA of the entering phage serves two independent functions: the template for replication of more phage DNA and the template for transcription of mRNA, which is then translated into phage-induced enzymes and the proteins that form the phage. The DNA and proteins are synthesized by the mechanisms described for bacterial DNA and protein synthesis (Step 4). This mechanism of phage reproduction is termed **vegetative reproduction.** ■ **DNA synthesis, p. 170** ■ **protein synthesis, p. 172, 175**

Step 5: Assembly

The **assembly** or **maturation**, process involves the assembly of phage protein with phage DNA to form intact or mature phage (Step 5). This is a complex, multistep process in the T4 phage. All of the protein structures of the phage, such as the heads, tails (sheaths), tail spikes, and tail fibers, are synthesized independently of one another. Once the phage head is formed, it is packed with DNA; the tail is then attached, followed by the addition of the tail spikes. The synthesis of some of these various components involve a **self-assembly** process, in which the protein components come together spontaneously without any enzyme catalyst to form a specific structure. In other steps, certain phage proteins serve as scaffolds on which various protein components associate. The scaffolds themselves do not become a part of the final structure, much as scaffolding required to build a house does not become part of the house.

Step 6: Release

During the latter stages of the infection period, the phage-induced enzyme **lysozyme** is synthesized. This enzyme digests the host cell wall from within, resulting in cell lysis and the release of phage (Step 6). This enzyme must not be synthesized early in the infection process, because it would lyse the host cell before any mature phage could be formed. This is a good example of why phage genes are expressed only when the enzymes for which they code are needed.

In the case of the T4 phage, the **burst size**, the number of phage released per cell, is about 200. The time required for the entire cycle from adsorption to release is about 30 minutes. These phage then infect any susceptible cells in the environment, and the process of phage replication is repeated.

Table 13.2 Representative Important Bacteriophages

Bacteriophage	Host	Shape	Genome Structure*	Relationship to Host Cell
T4, T1–T7	*Escherichia coli*	Complex	ds DNA	Lytic
M13, fd	*Escherichia coli*	Filamentous	ss DNA	Exit by extrusion
ϕ174	*Escherichia coli*	Isometric	ss DNA	Lytic
Lambda (λ)	*Escherichia coli*	Complex	ds DNA	Latent
MS2, Qβ	*Escherichia coli*	Isometric	ss RNA	Lytic
Beta (β)	*Corynebacterium diphtheriae*	Isometric	ds DNA	Latent; codes for diphtheria toxin

*ds, double-stranded; ss, single-stranded.

Lytic Phage Replication by Single-Stranded RNA Phages

Another group of lytic phages that undergo a productive infection are the single-stranded RNA phages. The ones most intensively studied are those that infect *E. coli*. These include MS2 and Qβ, which show many similarities to one another and share several unusual properties (see table 13.2). First, they infect only F$^+$ strains of *E. coli* because they attach to the sides of the sex pilus. Second, they replicate rapidly and have a burst size of about 10,000. They also all replicate by a similar mechanism. ■ **sex pilus, p. 208**

RNA Phage Replication

The RNA phages contain only a single **positive (+) sense strand** of RNA (**figure 13.6**). A positive-sense strand is one that acts directly as mRNA; a **negative (−) sense strand** does not. To replicate more positive-sense strands, the positive-sense RNA strands first serve as a template for the synthesis of negative-sense strands by an unusual RNA polymerase that is phage-induced. RNA synthesis using RNA as the template is not an enzymatic reaction that occurs inside an uninfected cell, and so the entering positive-sense strands must code for the synthesis of this enzyme. The negative-sense strands in turn serve as the templates for positive-sense strand synthesis, which are then incorporated into the phage during assembly. A phage-induced lysozyme lyses the cell, releasing a large number of phage. ■ **RNA polymerase, p. 172**

No RNA phage with a single negative-sense strand of RNA has been discovered, and double-stranded RNA phages are known but are extremely rare.

Phage Replication in a Latent State—Phage Lambda

Some phages can go through a productive lytic infection like T4 or live in harmony with their host bacteria. These latter are known as temperate phages. Ninety percent of all phages are temperate. In the harmonious relationship, the entering phage DNA most commonly becomes integrated into the chromosome of the host cell, where it replicates as the host chromosome replicates. When integrated into the host cell DNA, the phage DNA is called a **prophage** and the bacterial cell carrying a prophage is a lysogenic cell or lysogen. Some temperate phages multiply as plasmids inside the host cell, another kind of latency. ■ **plasmid, p. 210**

The most thoroughly studied temperate phage whose DNA is integrated into the *E. coli* chromosome is lambda (λ). Its size and shape are similar to T4 phage. Once the phage DNA enters the host cell, one of two events occur. Some of the bacteria will be lysogenized, while others will be lysed (**figure 13.7**). Whether the lambda phage lyses the majority of the cells or lysogenizes them depends largely on chance events that can be tipped in favor of one or the other by modifying the external environment. If the bacteria are growing slowly, for example, because of nutrient deprivation, then the likelihood of the

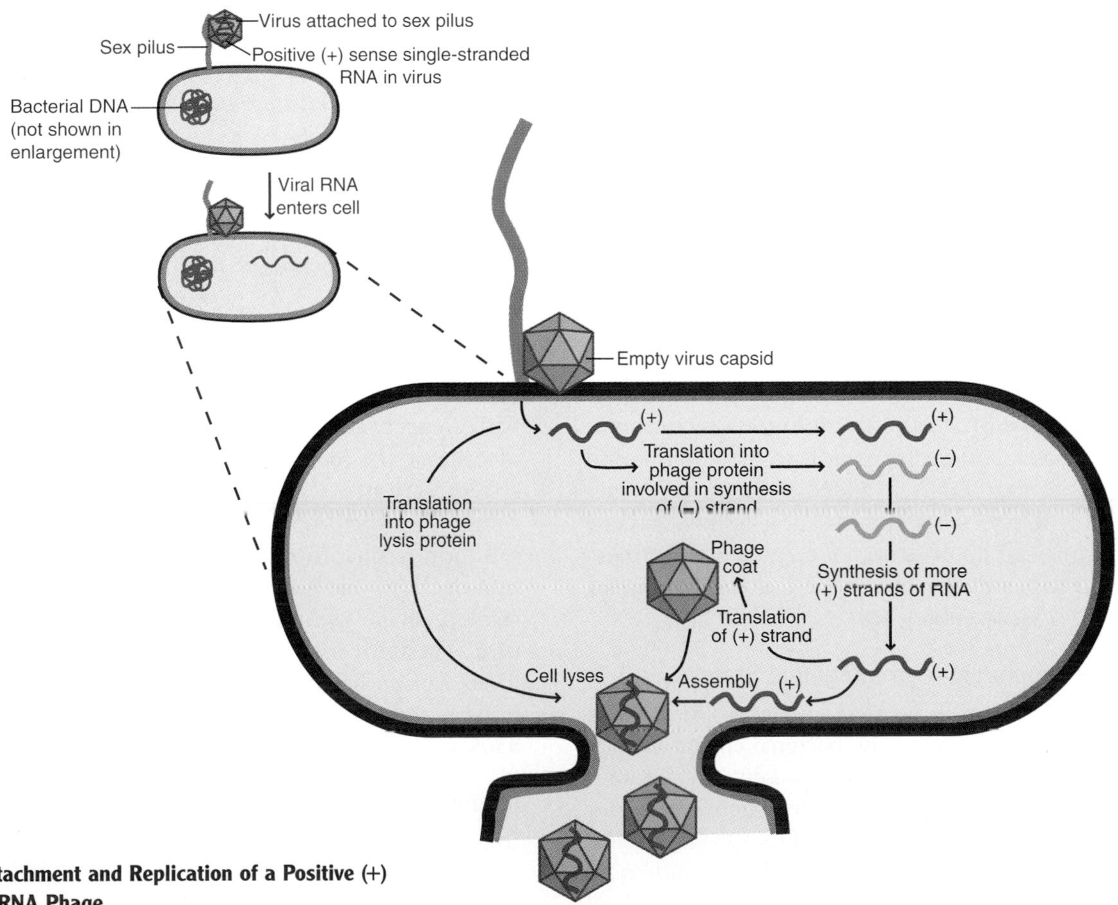

Figure 13.6 **Attachment and Replication of a Positive (+) Sense Strand Lytic RNA Phage**

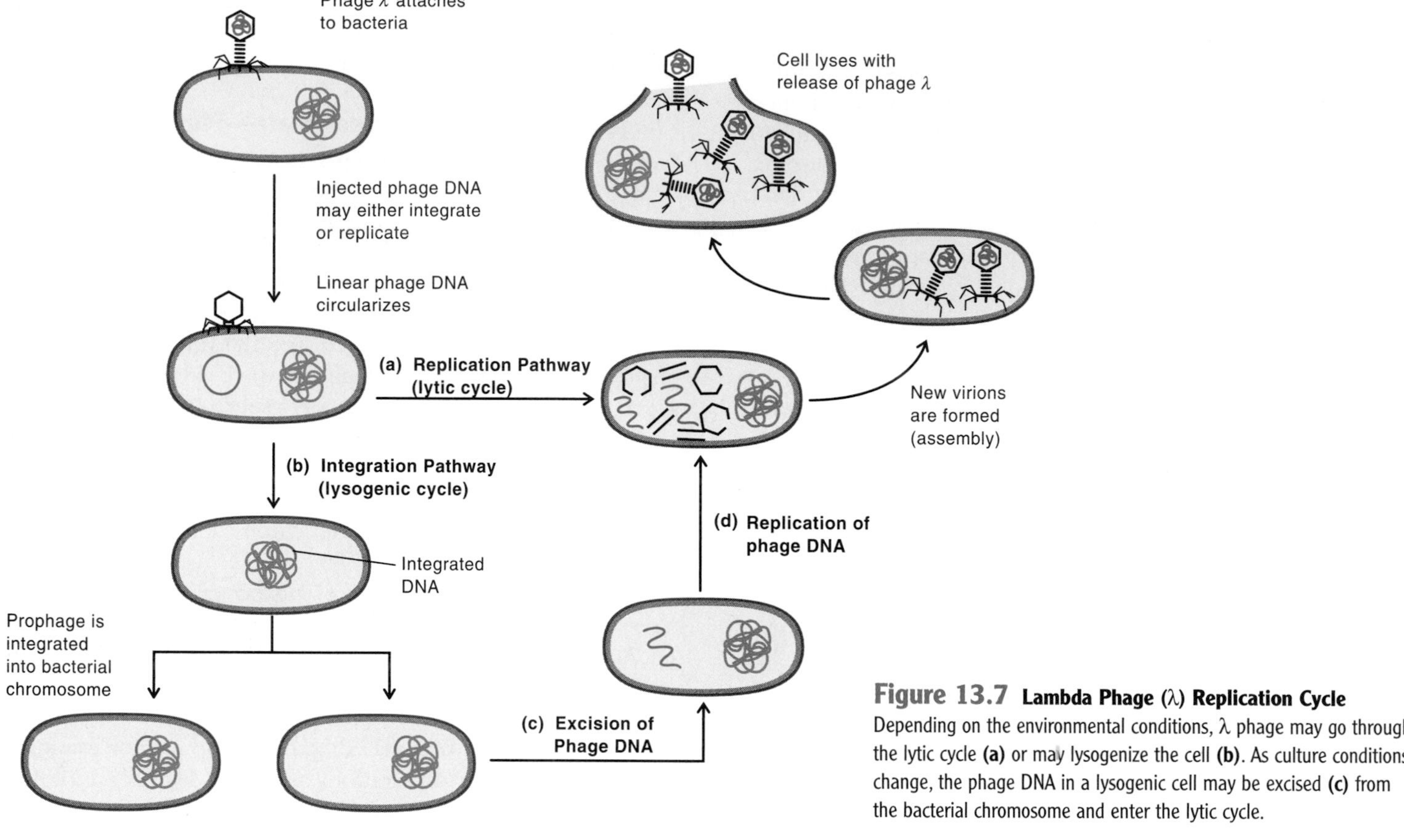
Phage λ attaches to bacteria
Injected phage DNA may either integrate or replicate
Linear phage DNA circularizes
(a) Replication Pathway (lytic cycle)
(b) Integration Pathway (lysogenic cycle)
Integrated DNA
Prophage is integrated into bacterial chromosome
(c) Excision of Phage DNA
(d) Replication of phage DNA
New virions are formed (assembly)
Cell lyses with release of phage λ

Figure 13.7 Lambda Phage (λ) Replication Cycle Depending on the environmental conditions, λ phage may go through the lytic cycle **(a)** or may lysogenize the cell **(b)**. As culture conditions change, the phage DNA in a lysogenic cell may be excised **(c)** from the bacterial chromosome and enter the lytic cycle.

phage DNA becoming integrated is increased because phage replicate vegetatively only in actively metabolizing cells.

The Integration of Phage DNA into the Bacterial Chromosome

Phage λ DNA becomes part of the genome of *E. coli* at a specific site through a process called **site specific recombination.** Short nucleotide sequences in the DNA of phage λ termed the *att* site, P, P′ and the *E. coli* host, the B, B′ site, are identical (homologous), allowing the phage and the bacterial DNA to **synapse (figure 13.8)**. The B, B′ region is located between the genes coding for galactose metabolism and biotin synthesis. Following synapsis, the phage DNA becomes integrated into the bacterial chromosome through the process of **homologous recombination,** between the genes coding for galactose metabolism and biotin synthesis. Note that DNA is added without any genes being replaced. This differs from the situation following gene transfer by transformation, transduction or conjugation. In these cases, genes in the recipient cells are replaced. ■ **DNA transformation, p. 205** ■ **transduction, p. 207** ■ **conjugation, p. 208**

Maintenance of the Prophage in an Integrated State

The phage DNA can remain integrated in the prophage state indefinitely and replicate along with the bacterial chromosomes. For this to occur, however, a set of genes present on the integrated phage DNA must be repressed. These genes code for enzymes that **excise** (remove) the integrated DNA from the host chromosome. Once excised, the viral DNA replicates and codes for the synthesis of phage proteins, and the vegetative life cycle of a productive phage infection ensues. How are these genes repressed to maintain the lysogenic state? One gene in the integrated viral DNA codes for a repressor that binds to a viral operator, thereby preventing transcription of the integrated genes required for excision. As long as this repressor is produced, the integrated phage DNA is not excised. If the repressor is no longer synthesized or is inactivated, however, viral genes are transcribed and an enzyme, an **excisase** which excises the viral DNA from the bacterial chromosome, is synthesized. A productive infection ensues (see figure 13.7d). ■ **repressor, p. 180** ■ **operator, p. 180**

Under ordinary conditions of growth, the phage DNA is excised from the chromsome only about once in 10,000 divisions of the lysogen. If, however, a lysogenic culture is treated with an agent that damages the bacterial DNA (such as ultraviolet light), the repair system, called SOS, comes into play. This system activates a protease that destroys the repressor. As a result, all of the prophage enter the lytic cycle, and a productive infection results. In this way the phage escape from a host that is in deep trouble because of damage to its DNA. This process, termed **phage induction**, results in complete lysis of the culture. The term *induction* as it is used here should not be confused with induced enzyme synthesis as discussed in chapter 7 or with phage-induced enzymes discussed earlier in this chapter. ■ **SOS system, p. 199** ■ **ultraviolet light, p. 196**

Immunity of Lysogens

In addition to maintaining the prophage in the integrated state, the repressor protein produced by a prophage prevents the infection of a lysogenic cell by phage of the same type as the phage DNA already carried by the lysogenic cell. Infections are blocked

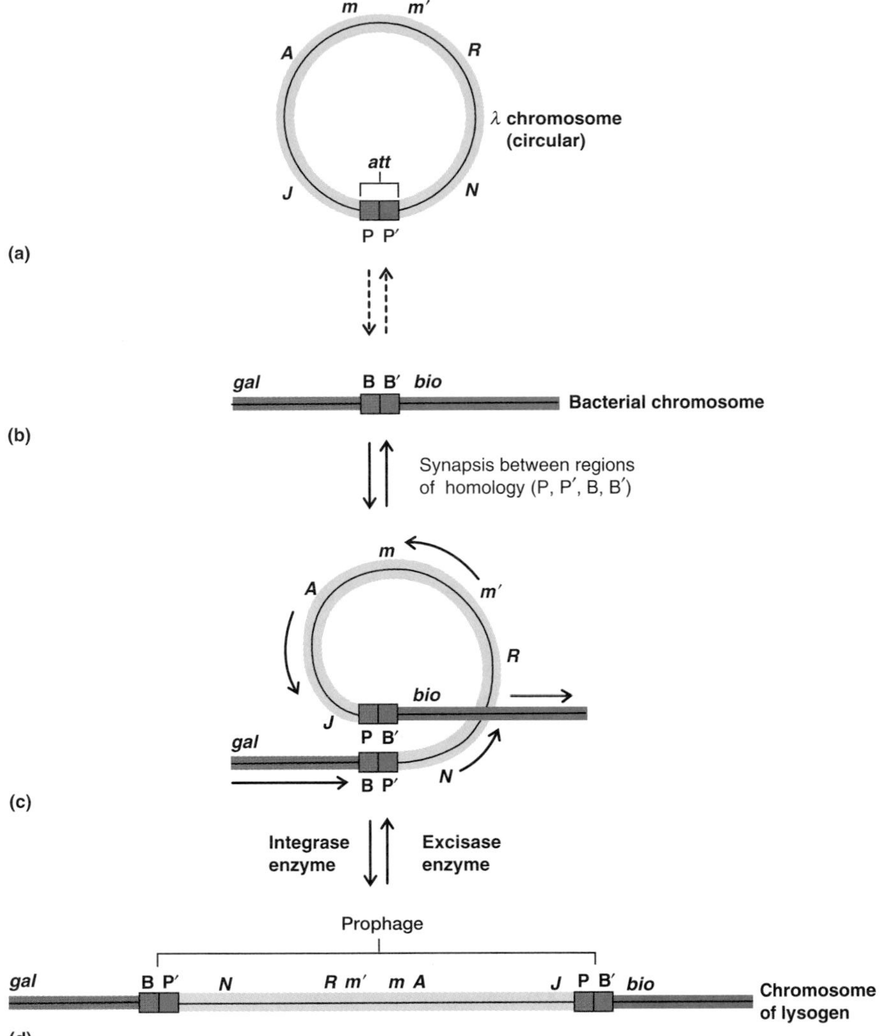

Figure 13.8 Reversible Insertion and Excision of Lambda (λ) Phage After the λ DNA circularizes **(a)**, the phage *att* site P, P′ synapses with an identical bacterial sequence B, B′ located between *gal* and *bio* **(b)** and is integrated between the *gal* and *bio* operons **(c)** to form the prophage, **(d)**. If the process is reversed, the lambda chromosome is excised from the bacterial chromosome and can then replicate and code for the synthesis of phage proteins.

because the repressor binds to the operator in the phage DNA as it enters the cell and inhibits its replication. Consequently, the cell is immune to infection by the same phage but not to infection by other phages, to whose DNA the repressor cannot bind. In this way the phage is protecting its turf from closely related phages.

Lysogenic Conversion

Lysogenic cells may also differ from their nonlysogenic counterparts in other important ways. The prophage can confer new properties on the cell, the phenomenon of lysogenic conversion. For example, strains of *Corynebacterium diphtheriae* that are lysogenic for a certain phage (β phage) synthesize the toxin that causes diphtheria. Similarly, lysogenic strains of *Streptococcus pyogenes* and *Clostridium botulinum* manufacture toxins that are responsible for scarlet fever and botulism, respectively. In all of these cases, if the prophage is eliminated from the bacterium, the cells cannot synthesize the toxin. The genes that code for these toxins are phage genes, which are expressed only when the phage DNA is integrated into the bacterial chromosome. Some properties of bacteria conferred by prophage are given in **table 13.3**.

Extrusion Following Phage Replication—Filamentous Phages

In addition to the phages that develop productive (lytic) infections and latency, some phage can develop other relationships with their host cells. A few closely related bacterial viruses that appear as long thin fibers are known as the **filamentous phages.** These include M13 and fd **(figure 13.9)**. They are single-stranded DNA phages that do not take over the metabolism of the host cell. They do not lyse the cell; rather, these phages are released by extrusion. This process does not destroy the host cell which continues to multiply. Indeed, these few phages are the only ones that do not lyse their hosts, although they make the cells unhealthy.

Replication of Filamentous Phages

In the replication cycle of filamentous phages that are extruded, the phage first adsorbs to the tip of the F⁺ pilus of *E. coli* and therefore only infects male cells **(figure 13.10)**. It is clear that the single-stranded DNA that enters the cell does not completely take over their host's metabolism, because the bacteria continue to multiply. The phage DNA replicates as discussed below and also codes for the synthesis of the phage coats. Interestingly, neither mature filamentous phage nor their coats can be detected in the cytoplasm of the host cells. It seems likely that after the phage coats are synthesized, they

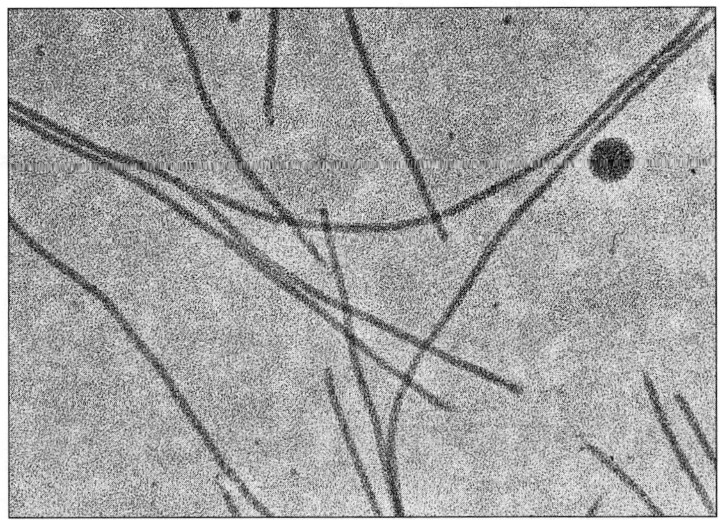

Figure 13.9 Electronphotomicrograph of fd phage This phage only infects cells that have a sex pilus.

Table 13.3 Some Properties Conferred by Prophage

Microorganism	Medical Importance	Property Coded by Phage
Corynebacterium diphtheriae	Causes diphtheria	Synthesis of diphtheria toxin
Clostridium botulinum	Causes botulism	Synthesis of botulinum toxin
Streptococcus (*β-hemolytic*)	Causes scarlet fever, strep throat, nephritis	Erythrogenic toxin responsible for scarlet fever
Salmonella	Causes food poisoning	Modification of lipopolysaccharide of cell wall
Vibrio cholerae	Causes cholera	Synthesis of cholera toxin

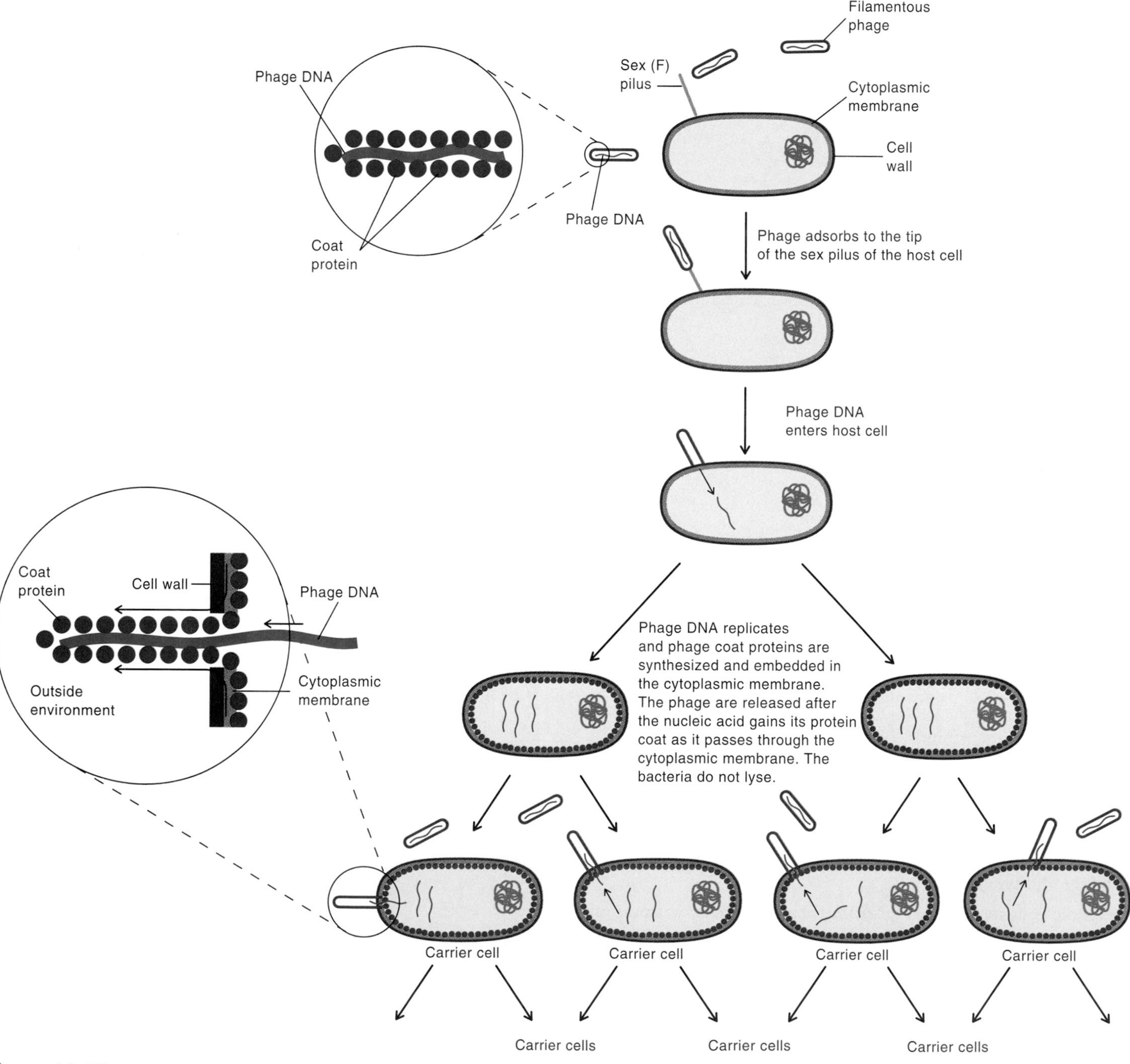

Figure 13.10 Replication of a filamentous phage The phage coat protein is embedded in the cytoplasmic membrane of the bacterium. Maturation of the phage occurs as the DNA is extruded through the cytoplasmic membrane. Carrier cells continue to extrude phage as they multiply.

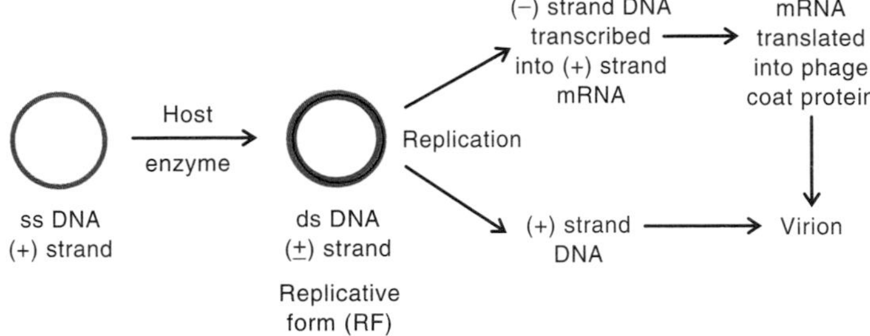

Figure 13.11 **Macromolecule Synthesis in Filamentous Phage Replication**

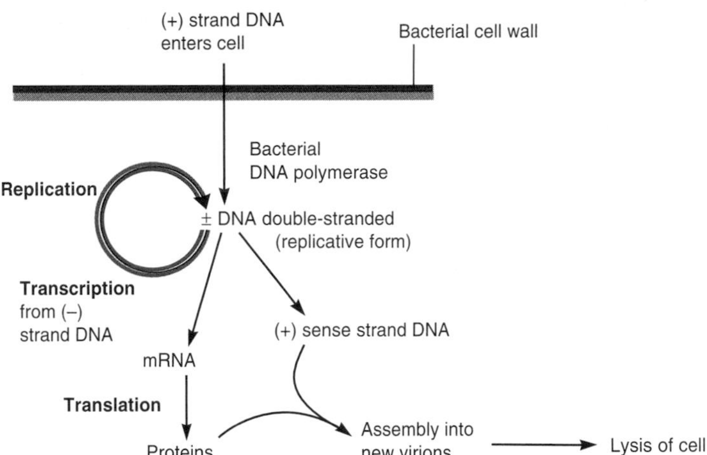

Figure 13.12 **Reproduction of ϕX174, an Isometric Positive (+) Sense Single-Stranded DNA Phage that Lyses Its Host**

are stored in the cytoplasmic membrane of the bacterium. The phage are assembled as they are extruded from the cell. The extrusion occurs continuously, and the phage are not released in a burst. Infected cells are termed **carrier cells.** They can be subcultured and stored in the same way as non-infected cells.

Replication of Single-Stranded DNA of Filamentous Phage

The replication of the single-stranded DNA of a filamentous phage has some features that are similar but others that are different from double-stranded DNA replication. The DNA that enters the cell is a positive (+) sense molecule that is then converted to a double-stranded form by enzymes of the host cell **(figure 13.11)**. The double-stranded, **replicative form** consists of a positive (+) and a negative (−) sense strand. The double-stranded replicative form replicates in much the same way as double-stranded DNA of bacteria. This replicative form gives rise to the single-stranded positive (+) sense strand and a negative (−) sense strand. The negative-sense DNA strand is the strand that is transcribed into mRNA, which then gives rise to phage proteins. The single-stranded DNA that is incorporated into the coat protein as the virus is extruded from the cell is a positive (+) sense strand, and it is derived from the double-stranded replicative form. ■ **DNA replication, p. 170**

Lytic Infection by Single-Stranded DNA Phage

Not all phages with single-stranded DNA are extruded. The isometric phage, ϕX174 also contains a positive (+) sense strand but differs in shape from the filamentous phages and is not male-specific. This phage goes through a lytic cycle **(figure 13.12)**. In the early stages of infection, the DNA of ϕX174 also replicates as a double-stranded DNA molecule by the same semiconservative mechanism by which DNA is synthesized in uninfected cells. In the later stages of infection, DNA is synthesized by a process termed **rolling circle replication,** a mechanism of DNA replication by which a single strand of DNA is synthesized. The positive (+) sense strands are assembled into the phage heads. The cells are then lysed by the phage by a mechanism that is not understood.

Table 13.4 summarizes the salient features of the various interactions of the phages with their host cells.

MICROCHECK 13.2

Some phages lyse their host cells and some are extruded without killing the host. Most live in harmony with them. Phage that live in harmony with their host cells often code for gene products that confer new properties on the host, which is termed lysogenic conversion.

- How does T4 phage prevent the infected bacterium from synthesizing bacterial proteins?
- What are two ways that phage can replicate in harmony with their host?

TABLE 13.4 Features of Interactions of Phage with Host Cells

Phage	Type of Interaction	Type of Nucleic Acid*	Result of Interaction
T4	Productive	ds DNA	Cell lyses
MS2, Qβ	Productive	ss RNA	Cell lyses
Lambda (λ)	Latent	ds DNA	Phage and host live in harmony
M13, fd	Productive	ss DNA	Phage extrudes from cell, which continues to multiply
ϕX174	Productive	ss DNA	Cell lyses

*ds, double-stranded; ss, single-stranded.

- Have phage that are extruded from the bacterial cell surface undergone a productive infection? Are they virulent? Explain your answers.
- In what way is phage induction an advantage for the phage?

Transduction

Phages play an important role in the transfer of bacterial genes from one bacterium to another. As briefly discussed in chapter 8, DNA can be transferred from one bacterial cell, the donor, to another, the recipient, by phage in the process called **transduction.** There are two types of transduction. In one type, any bacterial gene can be transferred through the process called **generalized transduction.** In this process, the phages are termed **generalized transducing phages.** The second type is termed **specialized transduction** since only a few specific genes can be transferred by the phages, which are **specialized transducing phages.** ■ transduction, p. 207

Generalized Transduction

Virulent as well as temperate phage can serve as generalized transducing phage. Recall that some phage in their life cycle degrade the bacterial chromosome into many fragments at the beginning of a productive infection. These short DNA fragments can be incorporated into the phage head in place of phage DNA in the process of phage maturation. Following lysis and release of the phage carrying the bacterial DNA, the phage binds to another bacterial cell and injects the bacterial DNA. Once inside the new host, the DNA from the donor cell can integrate into the recipient cell DNA by homologous recombination. These recipient cells do not lyse but are transduced. Since the genetic information transferred can be any gene of the donor cell, this gene transfer mechanism is called generalized transduction.

Why do the transducing virulent phage not lyse the cells they invade? The answer is that because the bacterial DNA replaces the phage DNA inside the phage's head, the genetic information necessary for the synthesis of many phage-induced proteins is lacking. The phage is termed **defective** because it lacks all the DNA necessary to form complete phage and lyse the recipient cell.

Specialized Transduction

Specialized transduction involves the transfer of only a few specific genes. The most actively studied specialized transducing phage is λ, which integrates only at specific sites in the chromosomes of *E. coli* (**figure 13.13**).

Usually, only the phage λ DNA is excised following induction. On rare occasions, however, a piece of bacterial DNA remains attached to the piece of phage DNA that is excised and a piece of phage DNA is left behind in the bacterial chromosome. This creates a defective phage. The bacterial genes attached to the phage DNA replicate as the phage DNA replicates. This DNA, mostly of bacterial but also some phage genes, then becomes incorporated into mature phage in the maturation process. The defective phage is released from the lysed cells. When the phage infects another bacterial cell, both phage and bacterial DNA enter the new host and can again become integrated into the chromosome. Thus, the resulting lysogen contains bacterial

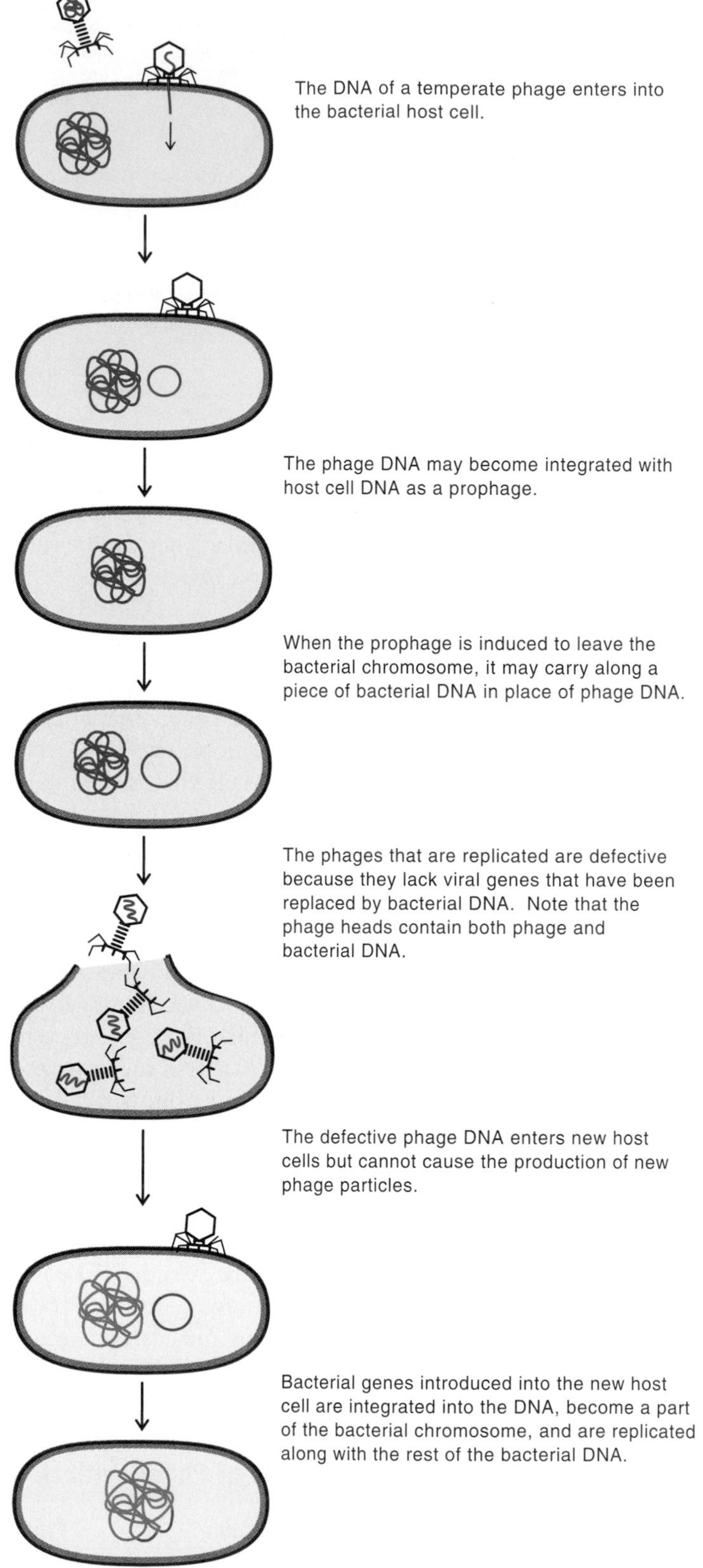

The DNA of a temperate phage enters into the bacterial host cell.
The phage DNA may become integrated with host cell DNA as a prophage.
When the prophage is induced to leave the bacterial chromosome, it may carry along a piece of bacterial DNA in place of phage DNA.
The phages that are replicated are defective because they lack viral genes that have been replaced by bacterial DNA. Note that the phage heads contain both phage and bacterial DNA.
The defective phage DNA enters new host cells but cannot cause the production of new phage particles.
Bacterial genes introduced into the new host cell are integrated into the DNA, become a part of the bacterial chromosome, and are replicated along with the rest of the bacterial DNA.

Figure 13.13 Specialized Transduction by Temperate Phage Only bacterial genes near the site where the prophage is integrated can be transduced.

genes from the previously lysogenized cell. Only bacterial genes located near the site of integration of the phage DNA can be transduced, a property of specialized transduction restricted to temperate phages.

MICROCHECK 13.3

Bacterial genes from a donor can be transferred to recipient bacteria following the incorporation of bacterial genes in place of phage genes in the virion. After the phage lyse the donor bacteria and infect the recipient cells, the bacterial genes are integrated into the chromosome of the recipient cell, the process of transduction. There are two types of transduction: generalized, in which any gene of the host can be transferred; and specialized, in which only the gene near the site at which the phage DNA integrates into the host chromosome can be transferred. The latter process is carried out only by temperate phages.

- What is meant by a defective phage?
- Most temperate phages integrate into the host chromosome, whereas some replicate as plasmids. Which kind of relationship would you think would be more likely to maintain the phage in the host cell? Why?

Host Range of Phages

The number of different bacteria that a particular phage can infect defines its **host range.** Several thousand different phages have been isolated, and the host range of any particular phage is usually limited to a single bacterial species and often to only a few strains of that species. Several factors limit the host range of phage. The two most important are the requirements that (1) the phage must attach to specific receptors on the host cell surface to start infection and (2) the restriction-modification system of the host cell.

These strict requirements that phage must pass in order to infect host cells means that phages seldom transfer DNA between unrelated bacteria. Thus, transduction is an unlikely mechanism in horizontal gene transfer. ■ **horizontal gene transfer, p. 510**

Receptors on the Bacterial Surface

Receptor sites vary in chemical structure and location. Receptors are usually on the bacterial cell wall, although a few phages attach to pili and a few others attach to sites on flagella (**figure 13.14**). Receptor sites can be altered by two distinct mechanisms, thereby creating a resistant cell. As we have already discussed, some but not all temperate phages that have lysogenized a bacterial cell can alter the cell surface, an example of lysogenic conversion. As a result another virulent phage that binds to the same receptor site as the original phage cannot adsorb. Thus, the prophage protects its host and, in turn, is able to continue replicating inside it.

The receptor sites can also be modified by mutation. In any large population of susceptible cells, some will have a modified (mutant) receptor site that confers resistance to any given phage.

Restriction-Modification System

Virtually all prokaryotes have two genes located next to each other in the cell that are involved in the **restriction-modification system**. The restriction gene codes for a **restriction enzyme**, an

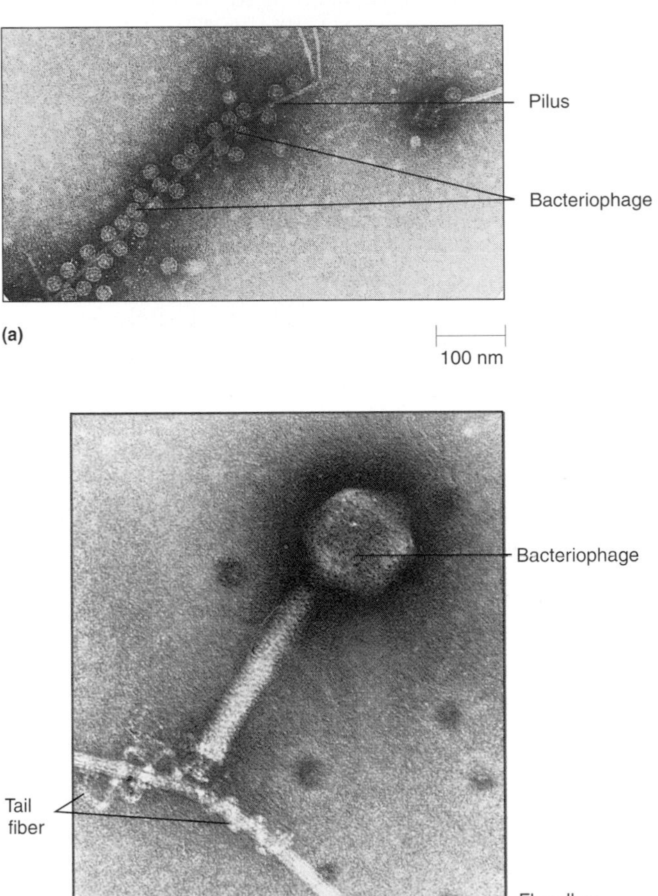

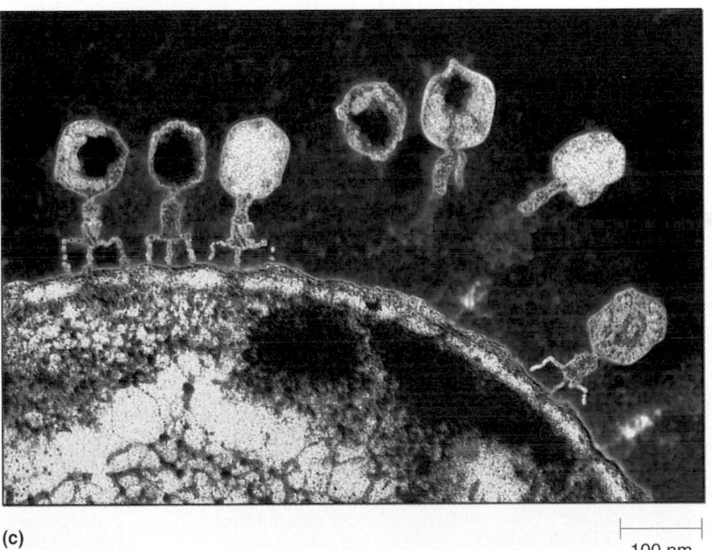

Figure 13.14 Adsorption of Bacteriophages on Various Cell Structures **(a)** Bacteriophage on pilus of *E. coli*. **(b)** Bacteriophage tail fiber entwined around flagellum. **(c)** Bacteriophage T4 attached to cell wall of *E. coli*.

endonuclease that recognizes short base sequences in DNA and cleaves the DNA at these sequences. The second gene codes for a **modification enzyme**, which attaches methyl groups to the bases of the sequence recognized by the restriction enzyme. These methylated bases are not recognized by the restriction enzyme, and so it does not cleave the DNA. This is the mechanism that the cell uses to protect its own DNA from being degraded by the restriction enzyme present in the cell. Its own DNA is methylated and therefore not degraded, but foreign DNA entering the cell will be degraded unless it is methylated first. ■ **restriction enzyme, p. 215, 226**

The restriction-modification system explains why a recipient bacterium will take up DNA by transformation, conjugation, or transduction only if the donor DNA and recipient cells are of the same species or even subspecies. Recipient cells restrict the entry of any DNA if the methylation pattern of the entering donor DNA differs from its own. Foreign DNA from different species, however, does enter recipient cells through the gene transfer systems of DNA transformation, conjugation, and transduction. How does this occur in light of the cells having restriction enzymes? How the restriction-modification system functions in keeping most, but not all, donor DNA from entering the cell is illustrated by the following story. ■ **restriction-modification, p. 216**

Phage that infect, multiply, and lyse *E. coli* strain K-12 can infect the same strain again with a very high efficiency (**figure 13.15**). These same phage infect the B strain of *E. coli* with a very low efficiency. If the phage that are released from the few infected B cells are added again to the B strain, however, infection now occurs at a high frequency. These observations can be explained by the presence of different restriction enzymes and different modification enzymes in the *E. coli* B and K-12 strains. The sequence of bases that is recognized and cleaved by the restriction enzyme of K-12 is different from the sequence of bases cleaved by the restriction enzyme of B. Likewise, the methylation of the bases in DNA by the modification enzyme of K-12, which protects the DNA from degradation in K-12, does not protect it from degradation in B. This explains why a strain always has a combination of the two genes that code for a restriction enzyme and the corresponding methylating enzyme.

The reason that phage released from only a few B cells can then infect B cells at a high efficiency can also be explained by the restriction-modification system. In most cases, foreign DNA entering a cell is degraded by the restriction enzyme before it has a chance to go through its replication cycle. In rare cases, however, the entering foreign DNA is methylated by the modification enzyme of the strain before the restriction enzyme has the chance to degrade it. Once the specific bases are methylated, the foreign DNA is converted to nonforeign DNA. This DNA will replicate, and phage will be released that can now infect and lyse the same host strain at a high efficiency.

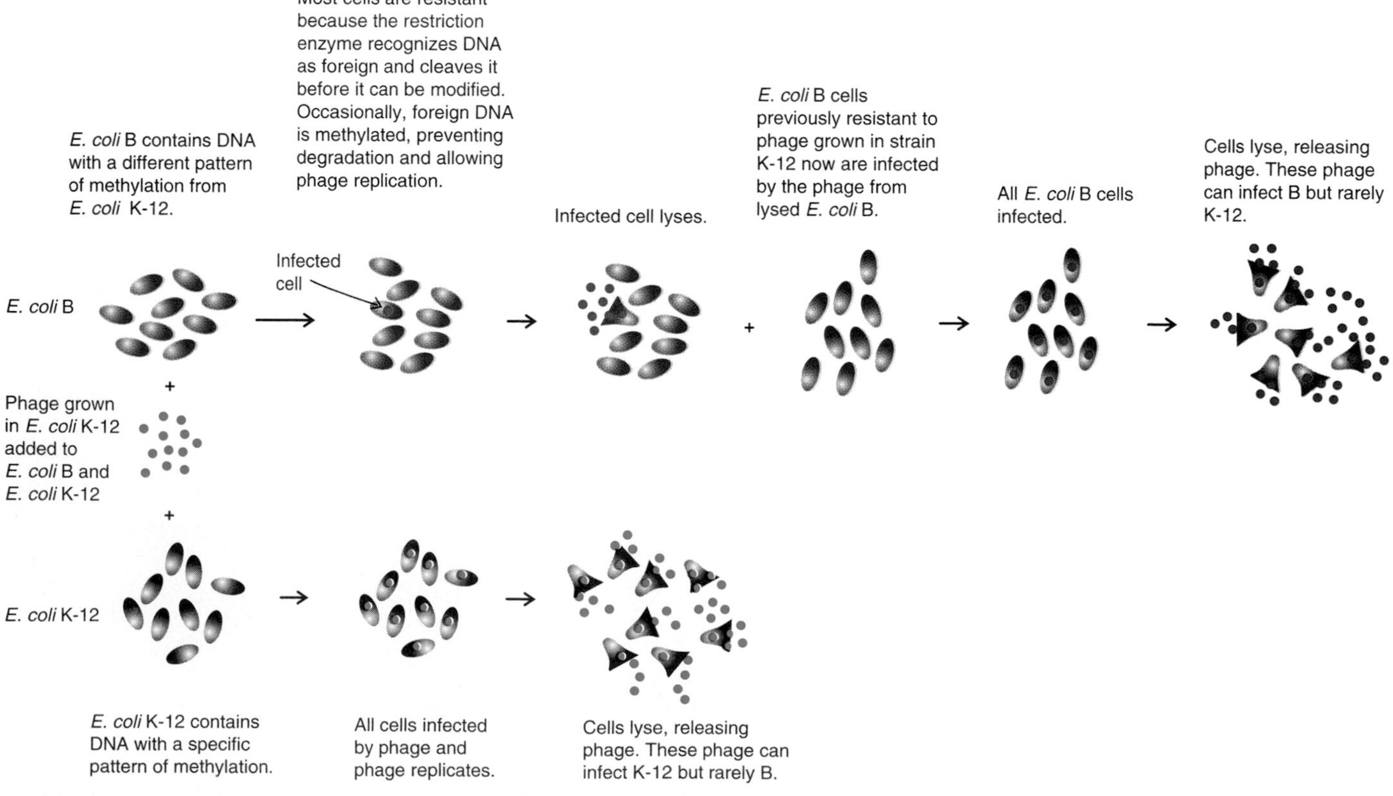

Figure 13.15 Restriction-Modification System The ability of phage to infect and replicate in a strain of *E. coli* other than the one it last infected is limited by the restriction-modification system.

MICROCHECK 13.4

The host range of bacteriophages is determined by the attachment proteins on the virus recognizing receptors on the bacterial host and by the ability of the entering phage DNA to escape degradation by host restriction-enzymes. Most receptor sites for phage are on the cell wall of the bacterium, although some are on pili and flagella.

- What kind of enzymatic activity does the modification system have?
- You add an unknown phage to a mixture of F^+ and F^- cells of *E. coli* and plate out the bacteria. The bacterial colonies that grow are all F^-. How can you explain this phenomenon?
- A mutation in *E. coli* results in the loss of both restriction and modification enzymes. Would you expect any difference in the frequency of gene transfer by transduction from *Salmonella*? Explain.

FUTURE CHALLENGES

"Take Two Phage and Call Me in the Morning"

More and more bacteria are becoming resistant to an ever greater number of antibiotics, and so the effectiveness of these medications has been greatly reduced. One novel approach to treating infectious diseases is to use phage to kill the disease-causing bacteria. This is not a new idea. Felix d'Herelle, the co-discoverer of phage, believed that they were mainly responsible for natural immunity. He also claimed that phages would be a universal prophylaxis and therapy. So convinced was he that while he was a Professor of Bacteriology at Yale, he also owned a commercial laboratory in Paris that produced preparations of phage that he sold to the pharmaceutical industry. After leaving Yale, he went to the Bacteriophage Research Institute in the USSR state of Georgia. Today, much of the work on the possible use of phage as therapeutic agents is being carried out in this same Institute.

Many problems are associated with this technology. First, a specific phage will only infect a single species or only a specific strain of a given species. Since it would take too long to isolate the causative infectious bacterium and isolate a phage that could attack it, "cocktails" containing many different phages must be used. Thus, to treat intestinal diseases, a "cocktail" of 17 different phages that attack intestinal pathogens has been developed. To treat burns that may be subject to infection by Pseudomonas, a kind of bandage saturated with a cocktail of 5 to 9 different phages has been employed. Whether phages are present that will attack all of the bacteria responsible for the disease, however, is questionable.

Another concern is that great care must be taken in purifying the phage before the preparation is administered. When phages lyse bacteria, a great deal of debris from the lysed cells is mixed in with the phage. This material could be very dangerous if it got into the bloodstream.

A third problem that must be overcome is that phages are recognized as foreign by the body's immune system, and therefore they are quickly eliminated. Some success has been achieved in increasing the time the phages stick around by altering the phage coat.

The fourth hurdle that must be overcome is the need for in-depth animal testing and human trials. No clinical trials involving human patients have been carried out and not even trials on animals have been conducted. Enough statements of success have been made in the use of phage in Russia and countries in the former Soviet Union, however, that phages deserve to be seriously considered as therapeutic agents in the Western countries. It is the only therapy that increases in amount as it successfully carries out its job.

SUMMARY

General Characteristics of Viruses (Table 13.1)

1. **Viruses** are nonliving agents associated with all forms of life. Each **virion** consists of nucleic acid surrounded by a protein coat. They are approximately 100- to 1,000-fold smaller than the cells they infect. (Figures 13.2, 13.3)

Virus Architecture

1. Different viruses have different shapes. Some are **isometric**, others are **helical** and still others are **complex**. (Figure 13.1)
2. The shape is determined by the **protein coat** (or **capsid**) that surrounds the nucleic acid. These make up the **nucleocapsid**. Each capsid is composed of **capsomeres**; **attachment proteins** project from the capsid. (Figures 13.2, 13.1)
3. Some animal viruses have a lipid bilayer surrounding the coat. These viruses are **enveloped**. Viruses without this envelope are **naked**. Virtually all viruses that infect bacteria, **bacteriophages**, are complex and naked. (Figure 13.2)

The Viral Genome

1. Viruses contain either RNA or DNA, but never both. The nucleic acid may be single-stranded or double-stranded.

Replication Cycle

2. Viruses only multiply within living cells and use the machinery of the cells to support their own multiplication.
3. Some viruses take over the metabolism of the host cell and kill it; others live in harmony with their hosts.
4. Viruses exist in two states. Outside the cell they are metabolically inert. Inside infected cells, they replicate.

Virus Interactions with Host Cells (Table 13.4, Figure 13.4)

1. Bacteriophages (phages) have the same relationships to their host as do animal viruses. Phages are much easier to study and serve as excellent model systems.
2. Some phages multiply inside bacteria and lyse the cells; this is a **productive infection**; the phages are virulent and lytic.
3. Other phages multiply but are **extruded** from the cell and do not kill it.
4. **Temperate** phages incorporate their DNA into the host cell, where it multiplies either as a plasmid or more commonly is integrated into the chromosome of the host.
5. A **latent** infection may show no sign that cells are infected.

Lytic Phage Replication by Double-Stranded DNA Phages (Figure 13.5)

1. This type of productive infection is one in which the host cell metabolism is taken over by a **virulent** phage and lyses. The infection proceeds through a number of defined steps.
2. **Attachment—protein tail fibers** on the tail of the phage adsorb to specific receptors on the cell wall.
3. **Penetration**—the DNA passes through the open channel of the **tail** and is injected into the cell. The phage coat remains on the outside.
4. **Transcription**—the phage DNA is **transcribed**. Some is transcribed early in infection and other regions are transcribed later. The mRNA is then **translated** into **phage proteins**.
5. **Replication** of phage DNA and proteins—the phage DNA and proteins replicate independently of one another. A phage enzyme degrades the bacterial chromosome so that only phage proteins are synthesized.
6. **Assembly** (**maturation**)—this is a highly complex and ordered series of processes, some catalyzed by enzymes, others not. The net result is the assembly of the phage components into a complete virion.
7. **Release**—a **phage induced** lysozyme lyses the cells resulting in the release of many virions per infected cell. This is the **burst size.**

Lytic Phage Replication by Single-Stranded RNA Phages (Figure 13.6)

1. These phages attach to the **sex pilus**.
2. Replication is unusual in that a phage-induced enzyme uses viral RNA that enters the cell as a template and synthesizes a complementary strand. This latter strand serves as a template for the synthesis of single-stranded RNA that enters the phage head.

Phage Replication in a Latent State—Phage Lambda (Figure 13.7)

1. The temperate phage λ can either go through a lytic cycle similar to T4 or integrate its DNA into a specific site in the bacterial chromosome.
2. **Integration** of phage DNA into the bacterial chromosome as a prophage occurs by means of site-specific recombination. (Figure 13.8)
3. The **prophage** is maintained in an integrated state because a repressor prevents expression of genes coding for an enzyme (excisase) that **excises** the prophage from the chromosome.
4. **Lysogens,** the bacteria carrying prophage, are immune to the same phage whose DNA they carry
5. Prophage often code for proteins that confer unique properties on the bacteria, a process called **lysogenic conversion**. (Table 13.3)

Extrusion Following Phage Replication—Filamentous Phages (Figures 13.10, 13.9)

1. The **filamentous phage** attach to the sex pilus of *E. coli*, and the single-stranded DNA enters the cell.
2. The entering positive-sense single strand of DNA is converted to a double-stranded **replicative form**. This DNA gives rise to the two single-stranded forms, one **positive sense**, the other **negative sense**.
3. The negative sense strand gives rise to its complement, which serves as mRNA for the synthesis of phage proteins.
4. Filamentous phages do not take over the metabolism but multiply productively as the host multiplies.
5. Phage are released by **extrusion** through the cell wall, a process that does not kill the bacteria.
6. As the positive sense strand of DNA is extruded, the DNA becomes surrounded by a protein coat.

Lytic Infection by Single-Stranded DNA Phages (Figure 13.12)

1. In **rolling circle replication**, a single strand of DNA is synthesized and incorporated into phages that are released by lysis.

Transduction

1. There are two types of transduction: **generalized** and **specialized**.

Generalized Transduction (Figure 8.22)

1. Generalized transduction involves the transfer of any piece of the bacterial chromosome from one cell to another cell of the same species.
2. Generalized transduction can be carried out by virulent and temperate phages.

Specialized Transduction (Figure 13.13)

1. Specialized transduction involves the transfer of specific genes which is carried out only by termperate phages. Lambda is a well-studied **specialized transducing phage** that transfers only genes of galactose and biotin metabolism.
2. Only genes located near the site at which the temperate phage integrates its DNA are transduced. Bacterial genes attached to the phage DNA may result when the phage DNA excises from the bacterial chromosome.
3. Phage genes along with bacterial DNA replicate and code for the phage coat protein. The coat surrounds the bacterial and phage DNA, and, following lysis, the bacterial DNA is transferred to other bacteria in the environment.

Host Range of Phages

1. Several factors determine the **host range** of phage. These include the requirement that phage attachment proteins must bind to specific receptors on the bacteria, and the need to circumvent the **restriction-modification** system found in all prokaryotes that degrades **foreign DNA**.

Receptors on the Bacterial Surface (Figure 13.14)

1. Most receptors are found primarily on the bacterial cell wall, but some phage attach to pili and flagella.

Restriction-Modification System (Figure 13.15)

1. All prokaryotes have **restriction enzymes** that recognize short sequences of bases in DNA and cleave, and thereby degrade the DNA at those sites. All prokaryotes have **modification enzymes** that add methyl groups to the bases in the short sequence so that they are not recognized by the endonuclease and are not cleaved. This is the mechanism that cells use to protect their own DNA from degradation.
2. When DNA from one strain of *E. coli* enters another strain, a race is on between the restriction enzyme and the modification enzyme. If the modification enzyme can methylate the sequences that the restriction endonuclease recognizes and cleaves, then the DNA will be spared from degradation and will replicate. Its pattern of methylation will be identical to the DNA of the cell in which it is replicating, but different from the cell that it came from.

REVIEW QUESTIONS

Short Answer

1. What is the name given to the phage whose DNA can be integrated into the host chromosome? What is the name of the host cell containing the phage DNA?
2. Name the process by which prophages confer new properties on their host cells.
3. Name the two types of transduction. Explain how they differ from each other in the DNA that is transferred.
4. Do all productive infections result in the complete takeover of host metabolism by the phage? Explain your answer.
5. Name three structures of bacteria that contain receptors for phage.
6. List the two mechanisms that allow bacteria to resist phage infection.
7. What is the most common shape of phages?
8. Compare double-stranded DNA phages with single-stranded DNA and single-stranded RNA phages in terms of cell lysis and number of phages produced.
9. How does UV light induce the synthesis of virions from lysogenic bacteria?
10. In virulent phage infection, why is it important that not all phage-induced enzymes be synthesized simultaneously? What general classes of enzymes are synthesized first? What class is synthesized last?

Multiple Choice Questions

1. Capsids are composed of
   A. DNA.
   B. RNA.
   C. protein.
   D. lipids.
   E. polysaccharides.
2. Temperate phages often
   1. lyse their host cells.
   2. change properties of their hosts.
   3. integrate their DNA into the host DNA.
   4. kill their host cells on contact.
   5. are rare in nature.

   A. 1,2 B. 2,3 C. 3,4 D. 4,5 E. 1,5
3. All phages must have the ability to
   1. have their nucleic acid enter the host cell.
   2. kill the host cell.
   3. Multiply in the absence of living bacteria
   4. lyse the host cell.
   5. have their nucleic acid replicate in the host cell.

   A. 1,2 B. 2,3 C. 3,4 D. 4,5 E. 1,5
4. The tail fibers on phages are associated with
   A. attachment.
   B. penetration.
   C. transcription of phage DNA.
   D. assembly of virus.
   E. lysis of host.
5. The RNA phages Qβ and MS2
   1. contain double-stranded RNA.
   2. infect all strains of *E. coli*.
   3. contain an RNA strand that is translated into protein.
   4. have very large burst sizes.
   5. contain single-stranded DNA.

   A. 1,2 B. 2,3 C. 3,4 D. 4,5 E. 1,5
6. T4 is
   1. a phage that contains double-stranded DNA.
   2. a phage that contains single-stranded DNA.
   3. a phage that can lysogenize cells.
   4. a phage that can carry out specialized transduction.
   5. a virulent phage.

   A. 1,2 B. 2,3 C. 3,4 D. 4,5 E. 1,5
7. The induction of a temperate phage by ultraviolet light results from
   A. damage to the phage.
   B. formation of thymine dimers.
   C. destruction of excision enzymes.
   D. destruction of a repressor.
   E. killing of the host cell.
8. Filamentous phages
   A. attach to bacterial receptors in the cell wall.
   B. take over metabolism of the host cell.
   C. are extruded from the host cell.
   D. undergo assembly in the cytoplasm of the host cell.
   E. degrade the host cells' DNA.
9. Phages have
   1. only one kind of nucleic acid.
   2. a protective protein coat.
   3. many enzymes.
   4. a single shape.
   5. a wide host range.

   A. 1,2 B. 2,3 C. 3,4 D. 4,5 E. 1,5
10. *E. coli* most likely becomes resistant to T4 phage through mutations in
    A. the cell wall.
    B. a restriction enzyme.
    C. a modification enzyme.
    D. the structure of pili.
    E. the cytoplasmic membrane.

Applications

1. A researcher discovered a mutation in *E. coli* that prevented phage T4 but not lambda from lysing the cells. What would you surmise is the nature of this mutation?
2. A public health physician isolated large numbers of phage from rivers used as a source of drinking water in western Africa. The physician is very concerned about humans becoming ill from drinking this water although she knows that phages specifically attack bacteria. Why is she concerned?

Critical Thinking

1. Would you expect transduction or conjugation to be the most likely mechanism of gene transfer from a Gram-negative to a Gram-positive organism? Explain.
2. A filter capable of preventing passage of bacteria is placed at the bottom of a U tube to separate the two sides. Streptomycin-resistant cells of a bacterium are placed on one side of the filter and streptomycin-sensitive cells are placed on the other side. After incubation for 24 hours, the side of the tube that originally contained only streptomycin-sensitive cells now contains some streptomycin-resistant cells. Give three possible reasons for this observation. What further experiments would you do to determine the correct explanation?
3. A suspension of phage when added to a culture of bacteria lyses the cells. When a suspension of the same phage is added to bacteria that have been previously agitated in a blender, no lysis occurs. Explain.

CHAPTER 14

Viruses of Animals and Plants

Although scientific reports as early as the Eighteenth century suggested that invisible agents might cause tumors, not until the early Twentieth century did this idea gain strong experimental support. At that time, Dr. Peyton Rous of the Rockefeller Institute caused tumors in healthy chickens by injecting them with a filtered suspension of ground-up cells from tumors of other chickens. These studies were not taken very seriously, because it was not generally accepted that viruses might cause tumors. As years passed, however, the idea gained favor as other investigators made similar observations. For example, in the early 1930s, Dr. Richard Shope observed skin tumors (papillomas) in wild rabbits in Iowa and Kansas. He showed that these tumors were caused by an agent that was not trapped in filters that removed bacteria. When he injected the filtrate into domestic rather than wild rabbits, the agent did not cause tumors. This phenomenon of a virus causing a disease in one animal but not in another has since been observed more recently with many other viruses.

In 1957, almost 50 years after Peyton Rous's study, Dr. Sarah Stewart and Dr. Bernice Eddy at the National Institutes of Health in Washington, D.C., studied two tumor-causing viruses, polyoma, a DNA virus, and SV40, an RNA virus. They noted that when the polyoma virus, grown in monkeys or mouse embryo tissues, was inoculated into a wide variety of animals such as rats, hamsters, and rabbits, it caused many different kinds of tumors. Eddy also pioneered the study of SV40, a virus that could be isolated from certain monkeys in which it did not cause any tumors. When injected into other animals, such as baby hamsters, however, the virus did cause tumors. Thus, SV40, like the papilloma virus studied by Shope, appears to be "silent" in its natural host but causes tumors when inoculated into certain foreign hosts.

We now know that the ability of a virus to form tumors depends on whether or not the virus undergoes a lytic cycle of reproduction. If it does, then it will lyse the cells it infects and no tumor will result. If the viral genome is only partially expressed, however, then a non-productive infection with tumor formation can result.

—A Glimpse of History

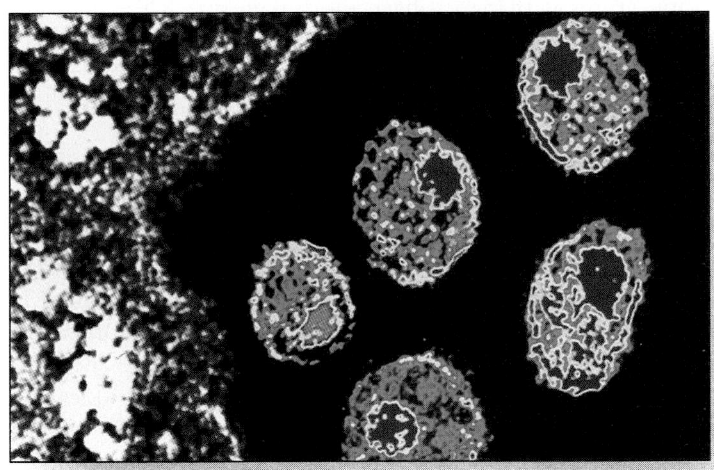

MUCH OF THE BASIC BIOLOGY OF BACTERIOPHAGES also applies to animal and plant viruses; however, each viral group has certain unique properties. This chapter presents a general approach to the classification of animal viruses, followed by a discussion of their modes of replication and effects on host cells, including their role in causing certain tumors. A brief discussion of plant viruses as well as viroids and prions is also included.

Structure and Classification of Animal Viruses

The structure of viruses was covered in chapter 13 (see figure 13.2); we give a brief review here. The structures of phage, and animal and plant viruses are similar, namely nucleic acid, either DNA or RNA, surrounded by a protein coat, the **capsid**. The capsid and nucleic acid together are called the **nucleocapsid**. The capsid is composed of a defined number of units called **capsomeres**, which are held together by noncovalent bonds. If there is no additional covering, the virus is termed **naked**. Many viruses that infect humans and other animals have a lipid membrane or **envelope** that surrounds the protein coat. Such a virus, called an **enveloped** virus, is rarely found among phage. The viral envelope is usually acquired from the cytoplasmic membrane of the infected cell during the release of the virus from the cell. Thus, the structure of the viral envelope is similar to the membrane of the cell, a lipid bilayer containing various proteins. Just inside the lipid envelope is a protein called the **matrix protein**. The **attachment proteins**, or **spikes**, that bind the virus to the cell project from the envelope or the capsid. Plant viruses, like tobacco mosaic virus, do not bind to specific sites on the plant cell wall; rather, they enter through wounds and have no protruding attachment proteins.

Another distinguishing feature of some RNA animal and plant viruses is that they have more than one RNA molecule enclosed in their capsids. The RNA molecules may not be identical. For example, the influenza virus has eight RNA molecules, each carrying different genetic information. Such viruses are termed **segmented viruses.**

The virion can have a number of shapes **(figure 14.1)**. **Isometric** in which the protein subunits are arranged in groups of equilateral triangles, the most common arrangement being **icosahedral symmetry**. These viruses appear spherical when viewed with the electron microscope. A less common shape in animal viruses is the helical- or rod-shaped structure. Other virions are **pleomorphic**—they have an irregular shape. The most common type of phage, the complex tailed form, does not occur in animal and plant viruses. ■ complex phages, p. 323

Classification of Animal Viruses

The taxonomy of animal viruses changes as more is learned about their properties. The taxonomy likely will continue to evolve, and so only general principles are considered here. The most widely employed taxonomic criteria for animal viruses are based on a number of characteristics:

1. Genome structure—DNA or RNA, single-stranded or double-stranded, segmented or a single molecule
2. Virus particle structure—isometric (icosahedral), helical (rod-shaped), or pleomorphic (irregular in shape)
3. Presence or absence of a viral envelope

Based on these major criteria, animal viruses are divided into a number of families, whose names end in *-viridae*. Fourteen families of RNA-containing viruses and seven families of DNA-containing viruses infect vertebrates (**tables 14.1** and **14.2**, respectively). The members of each family are derived from a common ancestor, as shown by nucleic acid hybridization. Evolutionary relationships between families however, cannot be inferred from this taxonomic scheme. The names of the families come from a variety of sources (see tables 14.1 and 14.2). In some cases, the name comes from the appearance of the virion, for example, *Coronaviridae* coming from corona, which means crown. In other cases, the name comes from the geographic area the virion was first isolated. *Bunyaviridae* is derived from Bunyamwera, a locality in Uganda, Africa. Each family contains numerous genera whose names end in *-virus*, making it a single word, for example, *Enterovirus*. The species name is the name of the disease the virus causes (for example, polio or poliovirus). The species name is one or two words. In contrast to bacterial nomenclature in which an organism is referred to by its genus and species name, viruses are commonly referred to only by their species name, the name of the disease they cause. In the classification of viruses, the family and genus are capitalized and written in italics. The species names are neither capitalized nor italicized.

The classification of the family *Picornaviridae*, which infects humans, is shown in **table 14.3**. This family contains three genera. The genus *Enterovirus* contains four species. Each species contains numerous "types." Whether these "types" should really be different species is in dispute. *Rhinovirus* contains a single species, rhinovirus, which in turn contains 100 "types." ■ nucleic acid hybridization, p. 232

In general, viruses with a similar genome structure replicate in a similar way. For example, animal viruses generally follow the same replication strategies as phages with similar genomes.

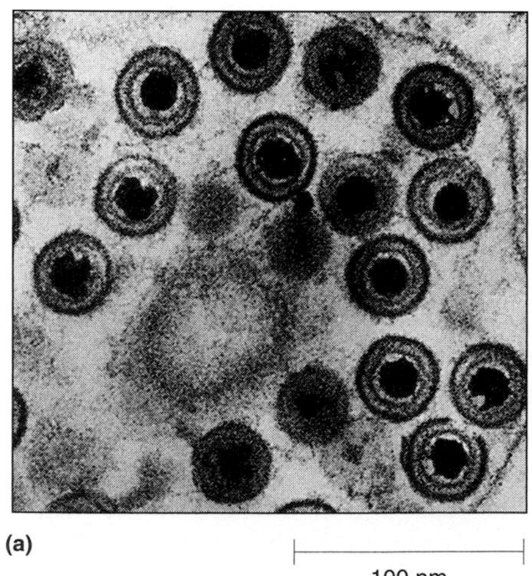

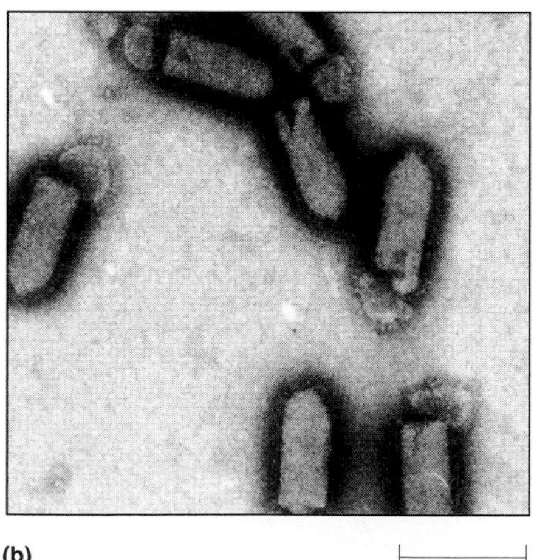

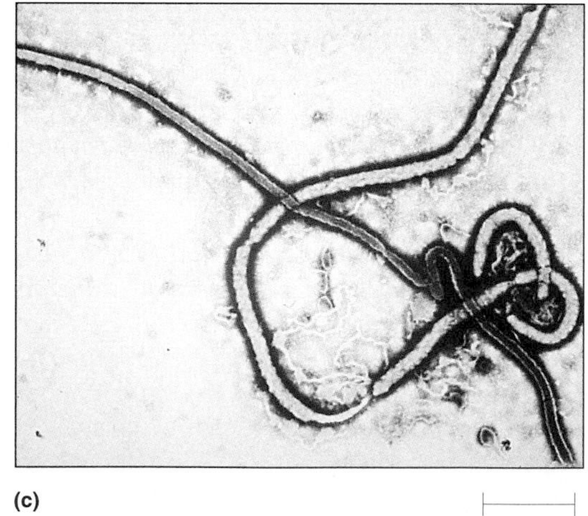

Figure 14.1 Shapes of Viruses **(a)** EM of human papillomavirus, an isometric virus whose capsomeres can be clearly seen. **(b)** EM of rhabdovirus particles, with their characteristic bullet shape. **(c)** Ebola virus, a filamentous virus that occurs in a number of shapes.

TABLE 14.1 Classification of RNA Viruses Infecting Vertebrates

Family	Drawing of Virion	Virion Structure	Genome Structure*	Characteristics and Representative Pathogenic Members
Picornaviridae (*pico*, micro; *rna*, ribonucleic acid)		Naked isometric	1 molecule, ss RNA	Poliovirus; rhinovirus
Caliciviridae (*calix*, cup)		Naked isometric	1 molecule, ss RNA	Norwalk virus; many members cause gastroenteritis
Togaviridae (*toga*, cloak)		Lipid-containing envelope	1 molecule, ss RNA	Many multiply in arthropods and vertebrates; encephalitis in humans
Flaviviridae (*flavus*, yellow)		Lipid-containing envelope	1 molecule, ss RNA	Yellow fever virus; dengue virus
Coronaviridae (*corona*, crown)		Lipid-containing envelope	1 molecule, ss RNA	Colds and respiratory tract infections
Rhabdoviridae (*rhabdos*, rod)		Bullet-shaped; lipid-containing envelope	1 molecule, ss RNA	Rabies virus
Filoviridae, (*filo*, threadlike)		Long filamentous; sometimes circular; lipid-containing envelope	1 molecule, ss RNA	Marburg virus; ebola virus
Paramyxoviridae, (*para*, by the side of; *myxa*, mucus)		Pleomorphic; lipid-containing envelope	1 molecule, ss RNA	Mumps virus; parainfluenza virus
Orthomyxoviridae (*orthos*, straight; *myxa*, mucus)		Pleomorphic; lipid-containing envelope	7–8 segments of linear ss RNA	Influenza virus
Bunyaviridae (*Bunya*mwera, a locality in Uganda)		Lipid-containing envelope	3 molecules of ss RNA	Hantaan virus
Arenaviridae (*arena*, sand)		Pleomorphic; lipid-containing envelope	2 molecules of ss RNA with hydrogen-bonded ends	Lassa virus
Reoviridae (respiratory enteric orphan virus)		Naked virion	Linear ds RNA divided into 10, 11, or 12 segments	Diarrhea in animals
Birnaviridae (*bi*, two)		Naked virion; icosahedral	2 segments of linear ds RNA	No human pathogens; diseases in chickens and fish
Retroviridae (*retro*, backwards)		Icosahedral; lipid-containing envelope	2 identical molecules ss RNA	HIV

* ss, single-stranded; ds, double-stranded.

Groupings Based on Routes of Transmission

Viruses that cause disease are often grouped according to their routes of transmission from one individual to another. These are not taxonomic groupings, and members of more than one family may be included in the same group. The following groupings, summarized in **table 14.4**, provide examples of such a scheme.

The **enteric viruses** are usually ingested on material contaminated by feces, the **fecal-oral route**. They replicate primarily in the intestinal tract, where they usually remain localized. They often cause **gastroenteritis**, an inflammation of the stomach and intestine. Some, however, such as the poliovirus, replicate first in the intestines but do not cause gastroenteritis. Rather, they cause a **systemic** disease.

TABLE 14.2 Classification of DNA Viruses Infecting Vertebrates

Family	Drawing of Virion	Virion Structure	Genome Structure*	Characteristics and Representative Pathogenic Members
Hepadnaviridae (*hepa*, liver; *dna*, deoxyribonucleic acid)		Lipid-containing envelope	1 molecule, mainly ds DNA but with a single-stranded gap	Hepatitis B virus
Parvoviridae (*parvus*, small)		Naked	1 molecule, ss DNA	Outbreaks of gastroenteritis following eating of shellfish
Papovaviridae (papilloma, polyoma, vacuolating agent)		Naked	1 molecule, circular ds DNA	Human papillomaviruses associated with genital and oral carcinomas
Adenoviridae (*adenos*, gland)		Naked	1 molecule, ds DNA	Some cause tumors in animals
Herpesviridae (*herpes*, creeping)		Enveloped with surface projections	1 molecule, ds DNA	Herpes simplex virus; cytomegalovirus
Poxviridae (*poc*, pustule)		Enveloped; large brick-shaped	ds DNA; covalently closed ends	Smallpox virus; vaccinia virus
Iridoviridae (*irid*, rainbow)		Enveloped	1 molecule, ds DNA	No known human pathogens; only animal pathogens

* ss, single-stranded; ds, double-stranded.

Respiratory viruses usually enter the body in inhaled droplets and replicate in the respiratory tract. The respiratory viruses include only those viruses that remain localized in the respiratory tract. Viruses that infect via the respiratory tract but then cause systemic diseases are not considered respiratory viruses. These latter include the viruses that cause mumps and measles.

Viral **zoonoses** are diseases that are transmitted from an animal to a human or another animal. Many viruses, such as rabies, are transmitted directly from animals to humans. Others, such as canine distemper, can be transmitted from dogs to African lions. One group of viruses, the arboviruses, are so named because they infect arthropods such as mosquitoes, ticks, and sandflies, where they replicate. Arthropods then bite vertebrates and transmit the virus. Thus, the viruses are arthropod-borne. In many cases, viruses can invade and replicate in widely different species. The same arthropod may bite birds, reptiles, and mammals and transfer viruses among these widely different groups. More than 500 arboviruses are known, and about 80 are known to infect humans. Twenty cause significant diseases such as West Nile fever, yellow fever, Western equine encephalitis, and dengue fever.

Sexually transmitted viruses cause lesions in the genital tract. These include herpesviruses and papillomaviruses. Other viruses that cause systemic infections are often transmitted during sexual activity. They include human immunodeficiency virus (HIV) and hepatitis viruses.

TABLE 14.3 Classification of Human Picornaviruses

Family	Genus	Species
Picornaviridae	*Enterovirus*	Polioviruses 1–3 Coxsackieviruses A1–A24 (no A23), B1–B6 Echoviruses 1–34 (no 10 or 28)Enteroviruses 68–71
	Rhinovirus	Rhinoviruses 1–100
	Hepatovirus	Hepatitis A virus

MICROCHECK 14.1

The genomic structure and shapes of animal viruses are similar to bacteriophages, except the complex shape is not represented. Many animal viruses have an envelope derived from the host cell membrane surrounding their capsid. Animal viruses are classified based on whether they contain DNA or RNA, whether they are enveloped, and their shape. Viruses that cause disease are often grouped by their routes of transmission.

- Give two differences in the structure of some animal viruses and bacteriophages.
- List four ways in which viruses can be transmitted from one organism to another.
- Why do animal viruses have envelopes and phages do not?

TABLE 14.4 Grouping of Human Viruses Based on Route of Transmission

Virus Group	Mechanism of Transmission	Common Viruses Transmitted
Enteric	Fecal-oral route	Enteroviruses (polio, coxsackie B); rotaviruses (diarrhea)
Respiratory	Respiratory or salivary route	Influenza; measles; rhinoviruses (colds)
Zoonoses	Vector (such as arthropods) Animal to human directly	Sandfly fever; dengue Rabies; cowpox
Sexually transmitted	Sexual contact	Herpes simplex virus-2 (genital herpes); HIV

Methods Used to Study Viruses

A variety of techniques are available to recognize the presence of viruses, identify them, and grow them in large quantities. The focus here is on methods for studying animal viruses, which are far more expensive and time-consuming than those used in studying phage.

Cultivation of host cells

Since viruses can multiply only inside living cells, such cells must be available to study virus growth. The study of bacterial viruses has advanced much more rapidly than investigations on animal and plant viruses, in large part because bacteria are much easier to grow in large quantities in short time periods. The primary difficulty in studying animal viruses is not so much in purifying the virions as it is in obtaining enough cells to infect. Some viruses can only be cultivated in living animals. Others may be grown in **embryonated chicken eggs,** those that contain developing chicks. Some animal viruses can be grown in cells taken from a human or another animal.

When animal viruses can be grown in isolated animal cells, the host cells are cultivated in the laboratory by a technique called **cell culture** or **tissue culture**. To prepare cells for growth outside the body of the animal (*in vitro*), a tissue is removed from an animal and minced into small pieces (**figure 14.2**). The cells are separated by treating them with a protease enzyme, such as trypsin, that breaks down protein. The suspension of cells is then placed into a screw-capped flask in a medium containing a mixture of amino acids, minerals, vitamins, and sugars as a source of energy. The growth of animal cells also requires a number of additional growth factors that have yet to be identified but are present in blood serum. Tissue cultures prepared directly from the tissues of an animal are termed **primary cultures**.

The cells bathed in the proper nutrients attach to the bottom of the flask and divide every several days, eventually covering the surface of the dish with a single layer of cells, a **monolayer**. When the cells become crowded, they stop dividing and enter a resting state. One can continue to propagate the cells by treating them with trypsin, removing them from the primary culture, diluting them, and putting the diluted suspension

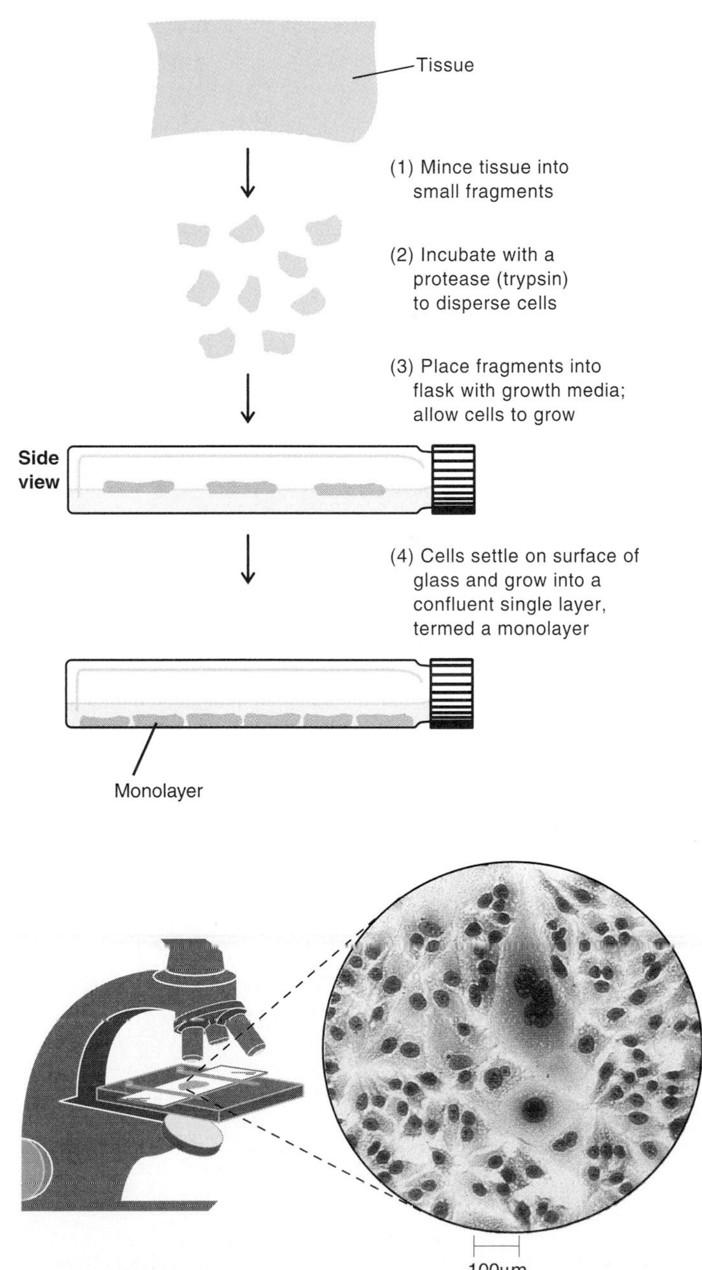
Tissue
(1) Mince tissue into small fragments
(2) Incubate with a protease (trypsin) to disperse cells
(3) Place fragments into flask with growth media; allow cells to grow
Side view
(4) Cells settle on surface of glass and grow into a confluent single layer, termed a monolayer
Monolayer
100μm

Figure 14.2 Preparation of Primary Cell Culture

into another flask containing the required nutrients. This results in an **established** cell line.

Cells taken from normal vertebrate tissue die after a certain number of divisions in culture. For example, human skin cells divide 50 to 100 times and then die, even when they are diluted into fresh media. Accordingly, cells must once again be taken from the animal and a new primary culture started.

Cells taken from a tumor, however, can be cultivated *in vitro* indefinitely. They are much easier to use for growing viruses than normal tissue, and several tissue lines have been established from tumors.

Tissue culture is important for growing viruses in the laboratory. The virus is mixed with susceptible cells, and the mixture is incubated until the infected cells are lysed. Following lysis, the unlysed cells and the cell debris are removed by centrifugation. The cells and debris go to the bottom of the centrifuge tube while the light, small virions remain in the liquid, the **supernatant**. This liquid containing the virions is also termed a **lysate**.

Tissue culture cells can also be used to detect the presence of a virus. When a virus is propagated in tissue culture cells, it often changes the cells' appearance. Often, these changes are characteristic for a particular virus and are referred to as the **cytopathic effect** of the virus (**figure 14.3**). Thus, the presence of a virus in an unknown sample and some idea of its identity can often be gained by culturing the specimen on cells in tissue culture.

Quantitation

Tissue culture is also used in virology to study the number of virions in a sample. The most commonly used method for detecting and quantifying the amount of virus present in any sample is the **plaque** assay. A number of other methods can be used for quantitating the number of virions in a sample. These include the **counting** of virions using an electron microscope, **quantal assays**, and in the case of some animal viruses, **hemagglutination**.

Plaque Assay

The plaque assay involves determining the number of viruses in solution by adding a known volume of the solution to actively metabolizing cells in a Petri dish. The infection, lysis, and subsequent infection of surrounding cells leads to a clear zone or plaque surrounded by the uninfected cells (**figure 14.4**). Each plaque represents one virion, initially infecting one cell, and so

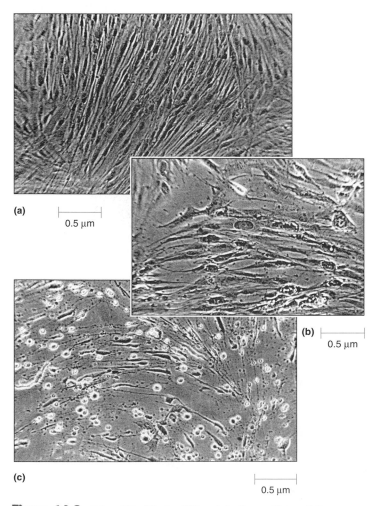

(a) 0.5 μm

(b) 0.5 μm

(c) 0.5 μm

Figure 14.3 Cytopathic Effects of Virus Infection on Tissue Culture **(a)** Fetal tonsil diploid fibroblasts growing as a monolayer, uninfected. **(b)** Same cells infected with adenovirus. **(c)** Same cells infected with herpes simplex virus. Note that the monolayer is totally destroyed.

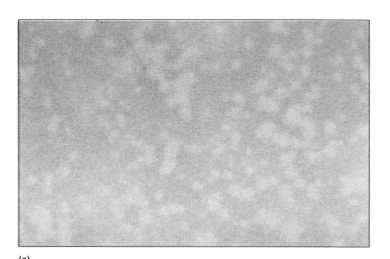

(a)

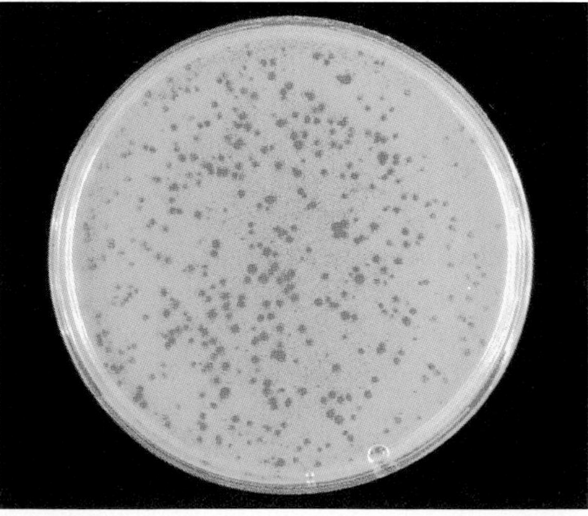

(b)

Figure 14.4 Viral Plaques **(a)** Plaques formed by the polio virus infecting a monolayer of cells that have been stained. **(b)** Plaques formed by bacteriophage.

the number of virions in the original solution can be readily determined. Plaques are only formed by infective virues and can be used with any viruses that lyse their host cells, including bacteriophage (see figure 14.4b).

It is possible to prepare large numbers of viruses for future studies by adding some liquid to the plate, scraping the surface of the plate with a glass rod, and harvesting the virions.

Counting of Virions with the Electron Microscope

If reasonably pure preparations of virions are obtainable, their concentration may be readily determined by counting the number of virions in a specimen prepared for the electron microscope **(figure 14.5)**. Unfortunately, this method cannot distinguish between infective and non-infective virions. From the shape and size, however, it does provide clues as to the identity of the virus.

Quantal Assays

Quantal assays can often provide an approximate virus concentration. In this assay, several dilutions of the virus preparation are administered to a number of animals, cells, or chick embryos, depending on the host specificity of the virus. The **titer** of the virus, or the **endpoint**, is the dilution at which 50% of the inoculated hosts are infected or killed. This titer can be reported as either the $\mathbf{ID_{50}}$, **infective dose**, or the $\mathbf{LD_{50}}$, **lethal dose**.

Hemagglutination

Some animal viruses are able to clump or **agglutinate** red blood cells because they interact with the surfaces of the cells. This phenomenon is called hemagglutination. In this process, a virion attaches to two red blood cells simultaneously and causes them to clump **(figure 14.6a)**. Sufficiently high concentrations of virus cause large readily visible aggregates of red blood cells to form (figure 14.6b). Hemagglutination can be measured by mixing serial dilutions of the viral suspension with a standard amount of red blood cells. The highest dilution showing maximum agglutination is the titer of the virus. One group of animal viruses that can agglutinate red blood cells is the myxoviruses, of which the influenza virus is a member. ■ **hemagglutination, p. 419**

MICROCHECK 14.2

Various hosts are required to grow different viruses. These include whole animals, embryonated chicken eggs, and cells taken from vertebrates. Cells taken from tumors divide indefinitely and so are convenient to use. Many animal viruses lyse their host cells, and so their number can be determined by counting plaques.

- Which of the methods used in quantitating viruses requires an electron microscope?
- Why is it necessary to continually make new primary cultures of normal cells?
- Would you expect the number of virions to be the same if you measured them by the plaque assay or by counting using the electron microscope? Explain your answer.

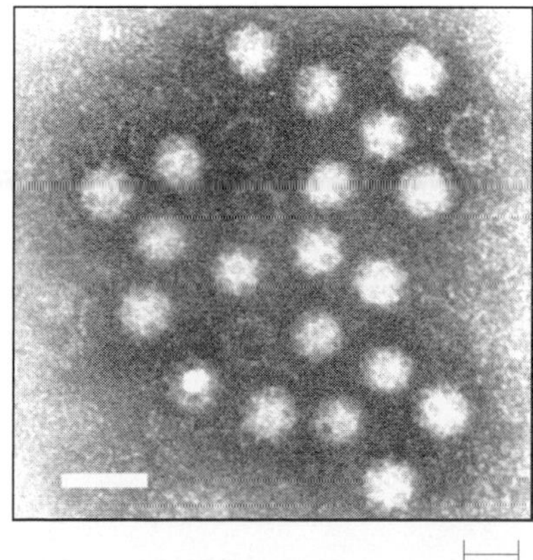

Figure 14.5 Electron Micrograph of *Calicivirus* The number of virions can be counted, and the number of empty capsids also can be readily determined.

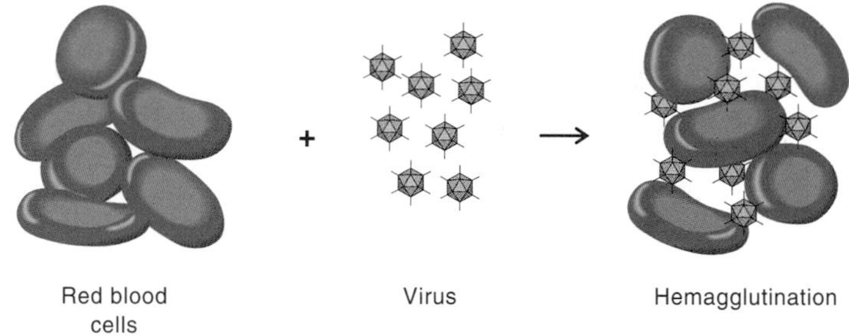

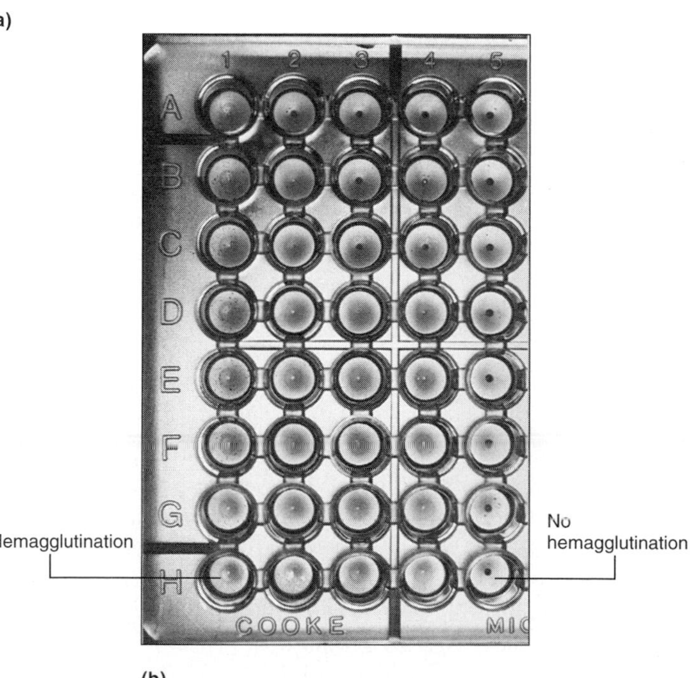

Figure 14.6 Hemagglutination (a) Diagram showing virions combining with red blood cells resulting in hemagglutination. **(b)** Assay of viral titer by hemagglutination. Each horizontal series of cups in horizontal lanes (A–H) represents serial twofold dilutions of different preparations of influenza virus mixed with a suspension of red blood cells.

Interactions of Animal Viruses with Their Hosts

For bacterial viruses, the host organism is a single cell, and so other kinds of cells do not affect the course of the infection. In the case of animals, however, the outcome of viral infection depends on many factors that are independent of the infected cell. Of special importance are the defense mechanisms of the host, such as the presence of protective antibodies that can confer immunity against a virus ordinarily lethal to an individual without such immunity. Devastating epidemics of measles and smallpox, which decimated the indigenous native population following the arrival of Europeans to the Americas, are good examples of the consequences of the lack of immunity to particular viruses.

Sudden epidemics causing widespread deaths are the most dramatic events of virus interactions with humans. The death of the host, however, also means that the virus can no longer multiply. Obviously, this is not in the best interests of the virus. Just as with the vast majority of bacterial viruses, animal viruses may develop a relationship with their normal hosts in which they cause no obvious harm or disease. Thus, the virus infects and persists within the host, in a state of **balanced pathogenicity**, in which neither the virus nor the host is in serious danger. Indeed, most healthy animals, including humans, carry a number of viruses as well as antibodies against these viruses without suffering any ill effects. If, however, a virus is transmitted to an animal that has no immunity against it, disease may result.

Many viruses that are carried by one group of organisms without causing disease may cause serious disease when transferred to another group. For example, Lassa fever virus does not cause disease in rodents, but when the virus is transferred to humans, a large percentage of the infected population dies.

The relationship between disease-causing viruses and their hosts can be divided into two major categories based on the disease and the state of the virion in the host. These two categories of infection are **acute** and **persistent**. Acute infections are usually self-limited diseases in which the virus often remains localized **(figure 14.7)**. In persistent infection, the virus establishes infections that persist for years or even life, often without any disease symptoms.

Acute Infections

Acute infections are usually of relatively short duration, and the host organism may develop long-lasting immunity. Viruses that cause acute infections result in productive infections. The infected cells die and may or may not lyse with the release of virions. Viruses that cause lysis of host cells are usually naked. Those that do not cause lysis are frequently enveloped. Although infected cells die, however, this does not mean that the host dies. Disease symptoms result from localized or widespread tissue damage. With recovery, the defense mechanisms of the host gradually eliminate the virus over a period of days to months. Examples of acute infections are mumps, measles, influenza, and poliomyelitis. ■ **mumps, p. 596** ■ **measles, p. 537** ■ **influenza, p. 573** ■ **polio, p. 679**

The life cycle of an animal virus that results in an acute infection with cell lysis can be compared with the productive infection of a bacterium with a virulent phage. The steps are the same. The two systems, however, vary in minor details. ■ **phage life cycle, p. 326**

Step 1: Attachment

The process of attachment (adsorption) is basically the same in all virus-cell interactions, except that the process is more complex in animal viruses than in phages. Animal viruses usually do not contain a single specific attachment appendage, a tail with fibers, as does phage T4, for instance. Rather, surface projections containing attachment proteins or spikes protrude all over the surface of a virion (see figure 13.2). Frequently, there are several different attachment proteins. The receptors to which the viral attachment proteins bind are usually **glycoproteins** located on the **plasma membrane**, and often more than one receptor is required for effective attachment. For example, the HIV virus must bind to two key molecules on the cell surface before it can enter the cell. The receptors number in the tens to hundreds of thousands per host cell. As with bacterial phage receptors, the normal function of these glycoproteins is completely unrelated to their role in virus attachment. For example, many receptors are immunoglobulins; other viruses use hormone receptors and permeases. ■ **immunoglobulins, p. 391** ■ **glycoproteins, p. 32**

Different viruses may use the same receptor, and related viruses may use different receptors. Certain viruses can bind to more than one type of receptor and thus be able to invade different kinds of cells. The binding of the attachment proteins to their receptors often changes the shape of viral proteins concerned with entry of the virion and facilitates their entry. Because a virion must bind to specific receptors, frequently a particular virus can infect only a limited number of cell types, and most viruses can infect only a single species and some only a single cell type within a host species. This may account for the resistance that some animals have to certain diseases. For example, dogs do not contract measles from humans, and humans do not contract distemper from cats. Some viruses however—for example, those that cause zoonoses—can infect unrelated animals such as horses and humans with serious consequences.

Step 2: Penetration

The mechanism of entry of animal viruses into host cells depends on whether the virion is enveloped or naked. In the case of enveloped viruses, two mechanisms exist. In one mechanism, after attachment to a host cell receptor, the envelope of the virion fuses with the plasma membrane of the host **(figure 14.8a)**. This

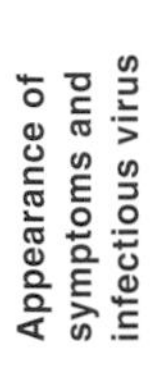

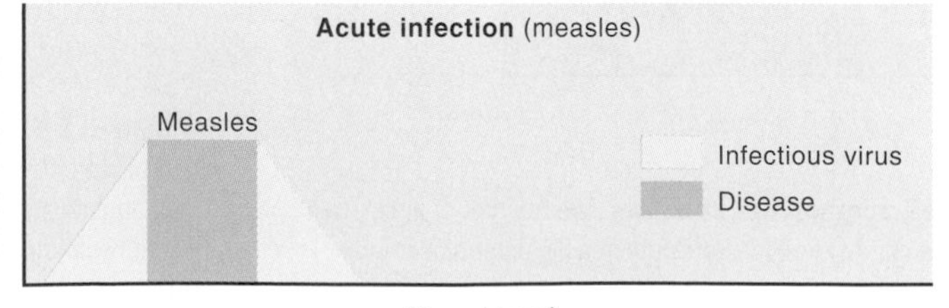

Figure 14.7 Time Course of Appearance of Symptoms of Measles and the Measles Virions

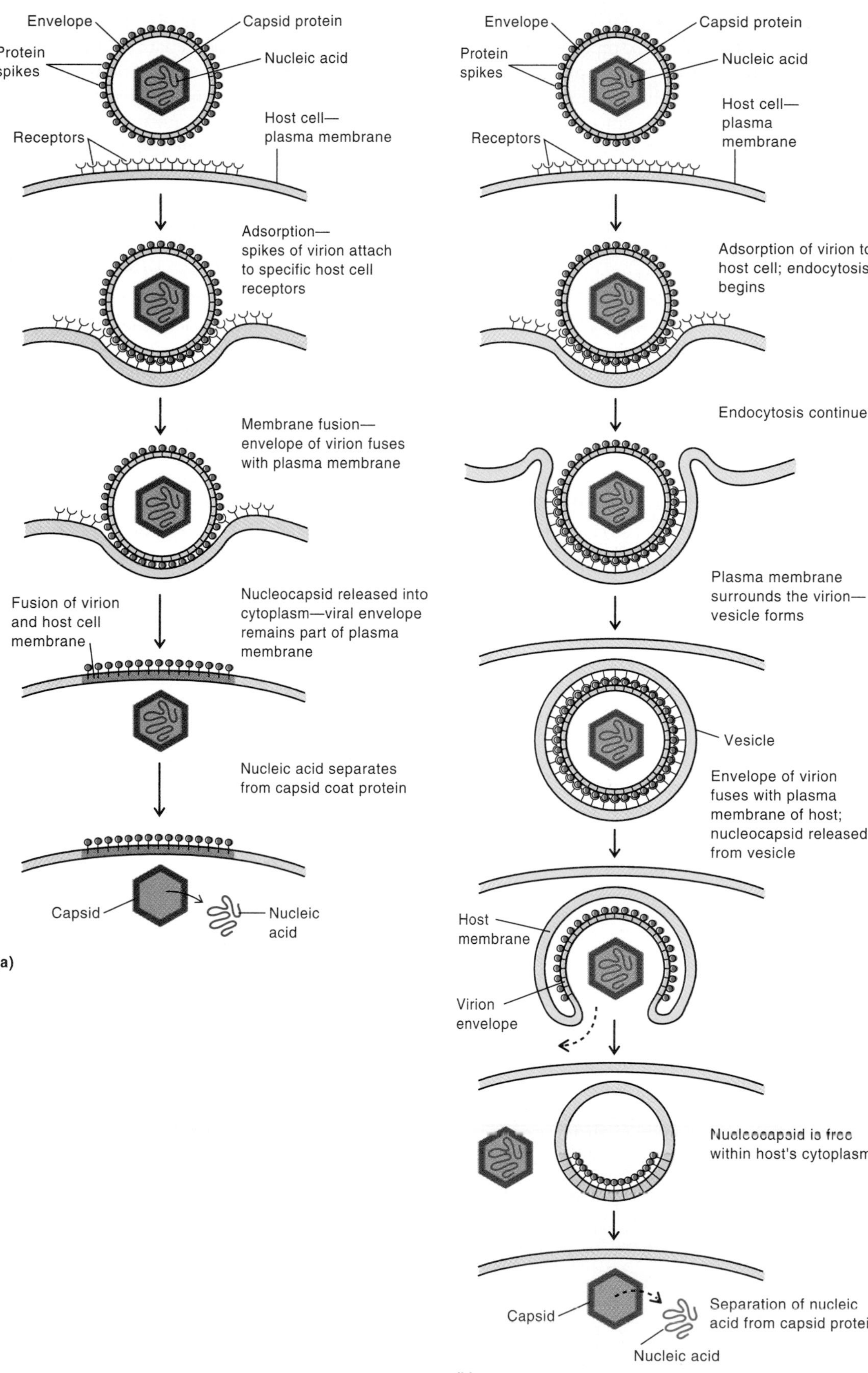

Figure 14.8 Entry of Enveloped Animal Viruses into Host Cells
(**a**) Entry following membrane fusion and (**b**) entry by endocytosis.

fusion is promoted by a **specific fusion protein** on the surface of the virion. In some viruses, such as measles, mumps, influenza, and HIV, the protein, which recognizes a target protein on the cell, changes its shape when it contacts the host cell. Following fusion, the nucleocapsid is released directly into the cytoplasm, where the nucleic acid separates from the protein coat.

In another mechanism, enveloped viruses adsorb to the host cell with their protein spikes, and the virions are taken into the cell in a process termed **endocytosis** (figure 14.8b). In this process, the host cell plasma membrane surrounds the whole virion and forms a vesicle. Then, the envelope of the virion fuses with the plasma membrane of the vesicle. The nucleocapsid is then released into the host's cytoplasm. ■ endocytosis, p. 76

In the case of naked virions, the virion also enters by endocytosis. Since the virus has no envelope, however, it cannot fuse with the plasma membrane. Rather, after being engulfed, the virus dissolves the vesicle, resulting in release of the nucleocapsid into the cytoplasm, where the nucleic acid separates from its protein coat prior to replication.

Penetration by animal viruses differs from phage penetration in two ways. First, the envelope of the virion and the plasma membrane of the host may fuse. Such fusion is not possible when the outside covering of the host has a rigid cell wall. Second, the entire virion is taken into the cell, whereas in the case of phages, the protein coat remains on the outside of the bacterium. In all viruses, the nucleic acid separates from its protein coat prior to the start of replication.

Step 3: Transcription

Diverse strategies are followed by viruses of different families for the synthesis of mRNA. In large part, the transcription strategy depends on whether the virus is RNA or DNA and whether the nucleic acid is single- or double-stranded **(figure 14.9)**. For example, in the case of single-stranded RNA viruses of a positive sense, the RNA itself functions as a messenger, whereas in the case of single-stranded RNA viruses of a negative sense, the RNA must be transcribed into a positive-sense strand. In this case, the RNA-dependent RNA polymerase required for transcription enters the cell as part of the virion.

In all cases, the patterns of transcription of the viral genome follow the same patterns as phages having the same type of genome follow. If a viral enzyme is required for transcription, it may be carried into the cell in the virion or the viral genome may encode the enzyme early in replication. In the case of phages, enzymes do not enter the cell; thus, if host cell enzymes are not available, the enzymes are encoded by the entering viral genome.

Step 4: Replication of Virus DNA and Proteins

In all virus systems, phage, animal, and plant, the nucleic acids and proteins replicate independently of one another. The replication of viral nucleic acid depends to varying degrees on enzymes present in the host cell prior to infection and virus-induced enzymes. Whether enzymes of the host or phage-induced enzymes are used varies with the virus. In some cases, viral enzymes concerned with virus replication enter the host cells along with the virion. As a general rule, the larger the viral genome, the fewer host cell enzymes are involved in replication. This is not surprising since enough DNA is present in large viral genomes to encode most enzymes of nucleic acid synthesis. For example, the largest of the DNA animal viruses, the poxviruses, like the T4 phages, are totally independent of host cell enzymes for the replication of their nucleic acid. On the other hand, the very small parvoviruses depend so completely on the biosynthetic machinery of the host cell that they require that the host cell actually be synthesizing its own DNA at the time of infection so that viral DNA can also be synthesized. Most animal viruses are between these two extremes. For example, small DNA viruses such as the papovaviruses code for a protein that is involved in initiating DNA synthesis, but once started, the rest of the process is carried out by host cell enzymes. This replication is similar to that of bacteriophage ϕX174, a single-stranded DNA virus that codes for proteins necessary for the initiation of phage DNA replication. Once replication has started, however, the viral DNA is synthesized by the same enzymes, such as DNA

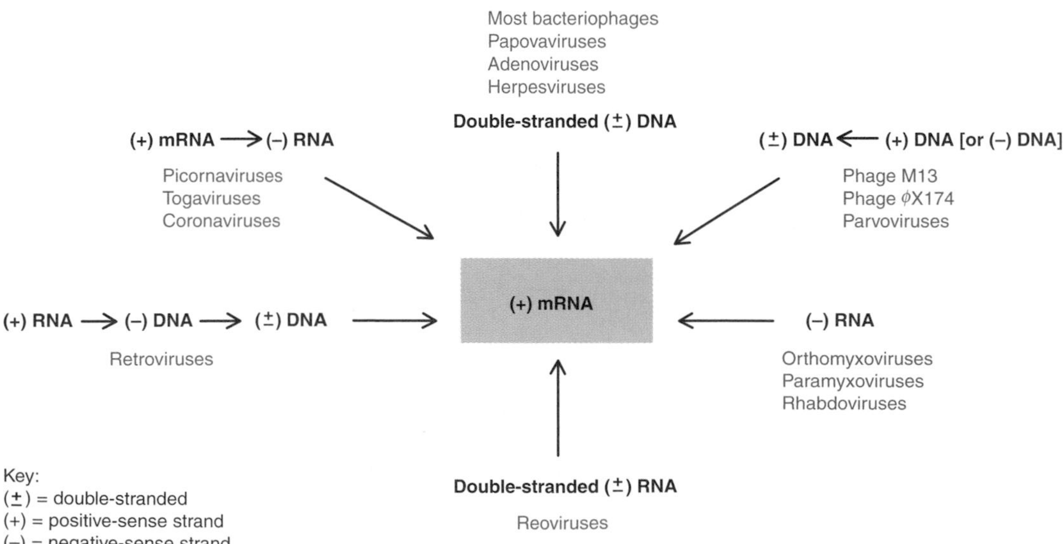

Figure 14.9 Strategies of Transcription Employed by Different Viruses Viruses that have the same genome structure follow the same strategy for the synthesis of (+)mRNA which is then translated into protein.

polymerase, that replicate the host cell *E. coli* DNA. The replication of the genome of many RNA viruses requires enzymes that are not found in the uninfected cell. Obviously, these must be encoded by the virus. The varying degrees of dependency of different animal DNA viruses on host cell enzymes for DNA replication are summarized in **table 14.5**. ■ ϕX174, p. 333

Step 5: Assembly

The final assembly of the nucleic acid with its coat protein, the process of maturation, is preceded by formation of the protein capsid structure that surrounds the viral genome. The maturation process and multistep formation of the viral coat involve the same general principles in all kinds of viruses. This process has already been discussed for bacteriophage T4. ■ assembly, p. 328

The maturation of a tobacco mosaic virus (TMV), a cylindrically shaped plant virus, has been studied extensively and serves as a model for both animal and plant viruses (**figure 14.10**). For TMV many identical protein structural subunits, the capsomeres, are first formed and then are added one by one to the growing coat structure that surrounds the viral RNA. The coat elongates in both directions, starting from a specific site on the single-stranded viral RNA. The RNA interacts with each protein disc as it is added, and when the end of the long RNA molecule is reached, the discs are no longer added. Enzymes are not required for the process, since it is a self-assembly process. Recall that the maturation of bacteriophage T4 is also, in part, a self-assembly process. It is a far more complicated process in T4 since this virion has many more different parts than the coat of TMV.

Step 6: Release

Most nonenveloped viruses accumulate within the cytoplasm or nucleus following their assembly. Unlike virulent phages, animal viral nucleic acid does not code for enzymes that lyse the host cells. Infected cells often die because viral DNA and proteins rather than host cell material are synthesized. Thus, functions required for cell survival are not carried out, and cells die. Cell degradation and lysis may also result from the release of degradative enzymes normally contained in cellular lysosomes. This degradation and release of the virions give rise to the cytopathic effects. The dead cells lyse, releasing the virions, which may then invade any healthy cells in the vicinity. ■ lysosomes, p. 81

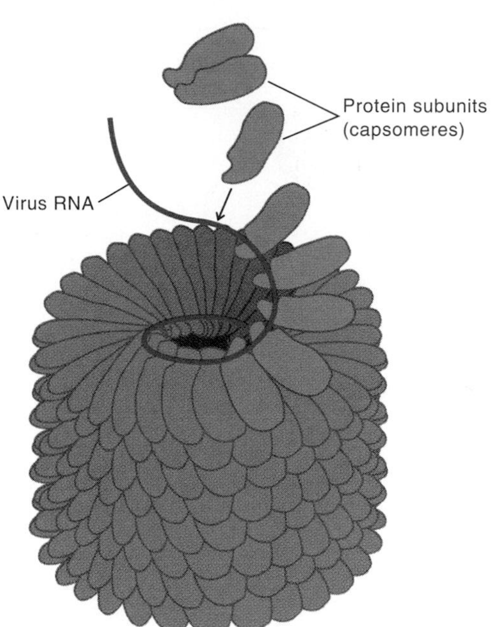

Figure 14.10 Tobacco Mosaic Virus Assembly The capsomeres are added, one by one, to the coat structure to enclose the viral nucleic acid (RNA).

Another mechanism for release is by **budding** from the plasma membrane (**figure 14.11**). This process is frequently associated with persistent infections, but the process often kills infected cells. An example of the latter is the killing by HIV as it buds from cells of the immune system. This process involves a number of steps. First, the region of the host cell plasma membrane where budding is going to take place acquires the protein spikes coded by the viruses, which eventually are attached to the outside of the virion. Then, the inside of the plasma membrane becomes coated with the matrix protein of the virus. In the next step, the nucleocapsid becomes completely enclosed by the region of the plasma membrane into which the spikes and matrix protein are embedded. Most enveloped viruses obtain their envelopes as they exit the cell through the plasma membrane. Some viruses, however, bud through the Golgi apparatus or rough endoplasmic reticulum. Vesicles containing the virus

TABLE 14.5 Relationship of Virus Size to Dependency on Host Cell Enzymes for DNA Replication

Virus*	Increasing Size of Virus	Dependence on Host Cell DNA Synthetic Enzymes
Parvovirus ss DNA	↓	Depends totally on host cell enzymes.
Polyomavirus ds DNA		Codes for a protein that is involved in the start of DNA synthesis; rest depends on host enzymes.
Adenovirus ds DNA		Codes for proteins involved in initiation of DNA synthesis and also DNA polymerase.
Poxvirus ds DNA		Is totally independent of host cell enzymes.

* ss, single-stranded; ds, double-stranded.

then migrate to the plasma membrane, with which they fuse. The virions are released by **exocytosis**. Thousands of virions can be released over hours or days, often without significant cell damage. For all enveloped viruses, budding is part of the maturation process. The process of budding may not lead to cell death, because the plasma membrane can be repaired following budding. As discussed in chapter 13, filamentous phages also are released from bacterial cells by budding or extrusion, without killing the bacterial cells. ■ **filamentous phage, p. 331** ■ **Golgi apparatus, p. 80** ■ **rough endoplasmic reticulum, p. 80**

Differences in the various steps in the replication cycle of virulent animal viruses and phages are listed in **table 14.6**.

Persistent Infections

In persistent infections, the viruses are continually present in the body and are released from infected cells by budding. Persistent infections can be divided conveniently into four categories. These are (a) **late complications following an acute infection**, (b) **latent infections**, (c) **chronic infections**, and (d) **slow infections**. The categories are distinguished from one another largely by whether a virus can be detected in the body during the long period of persistence **(figure 14.12)**.

Persistent infection may or may not cause disease, but since the infected person carries the virus, he or she is a potential source of infection to others. In all cases, the person is a **carrier** and therefore able to spread disease. Some persistent infections have features of more than one of these categories. These depend on such circumstances as the time after infection and the cell type in which the virus is located. For example, infection by HIV has features of latent, chronic, and slow infections.

Late Complications that Follow an Acute Infection

An example of a late complication that follows an acute infection is **subacute sclerosing panencephalitis (SSPE)**, which follows an acute measles infection (see figure 14.12a). This invariably fatal brain disorder occurs in about one in 300,000 people up to 10 years after a person has had measles.

Latent Infections

Latent infections are those in which infectious virus particles cannot be detected until reactivation of the disease occurs (see figure 14.12b). The viruses causing latent infections can be either DNA or RNA viruses. The best known examples are members of the herpesvirus family (*Herpesviridae*), which is divided into two herpes simplex types: **HSV-1** and **HSV-2**. The latter, frequently called **genital herpes**, is an important and common sexually transmitted disease. ■ **HSV-1, p. 594** ■ **HSV-2, p. 594**

Initial infection of young children with herpes simplex type 1 (HSV-1) may not lead to any symptoms, but cold sores

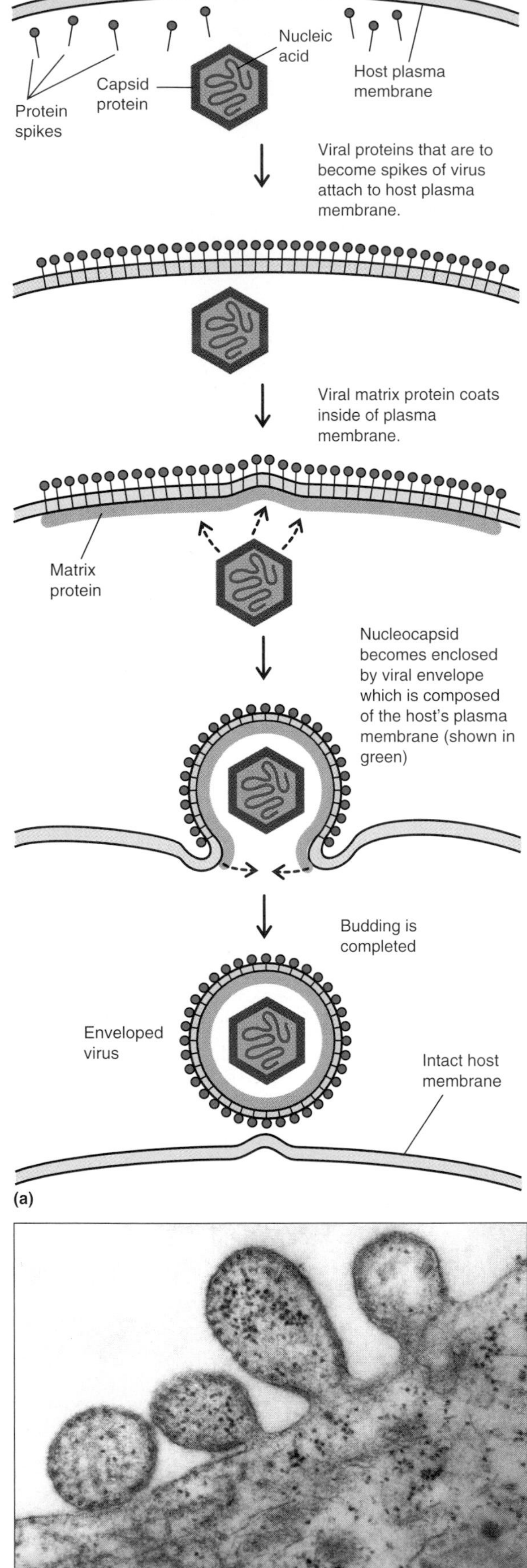
Nucleic acid
Host plasma membrane
Capsid protein
Protein spikes
Viral proteins that are to become spikes of virus attach to host plasma membrane.
Viral matrix protein coats inside of plasma membrane.
Matrix protein
Nucleocapsid becomes enclosed by viral envelope which is composed of the host's plasma membrane (shown in green)
Budding is completed
Enveloped virus
Intact host membrane
(a)
(b)
100 nm

Figure 14.11 Mechanism for Releasing Enveloped Virions **(a)** Process of budding. **(b)** Electron micrograph of virus particles budding from the surface of a human cell. The virion on the left has completed the process. The other three are in various degrees of completion. It is clear from the micrograph how the virions gain the plasma membrane of the host cell. Note that the membrane of the host remains intact after budding has been completed.

(Courtesy of J. Griffith)

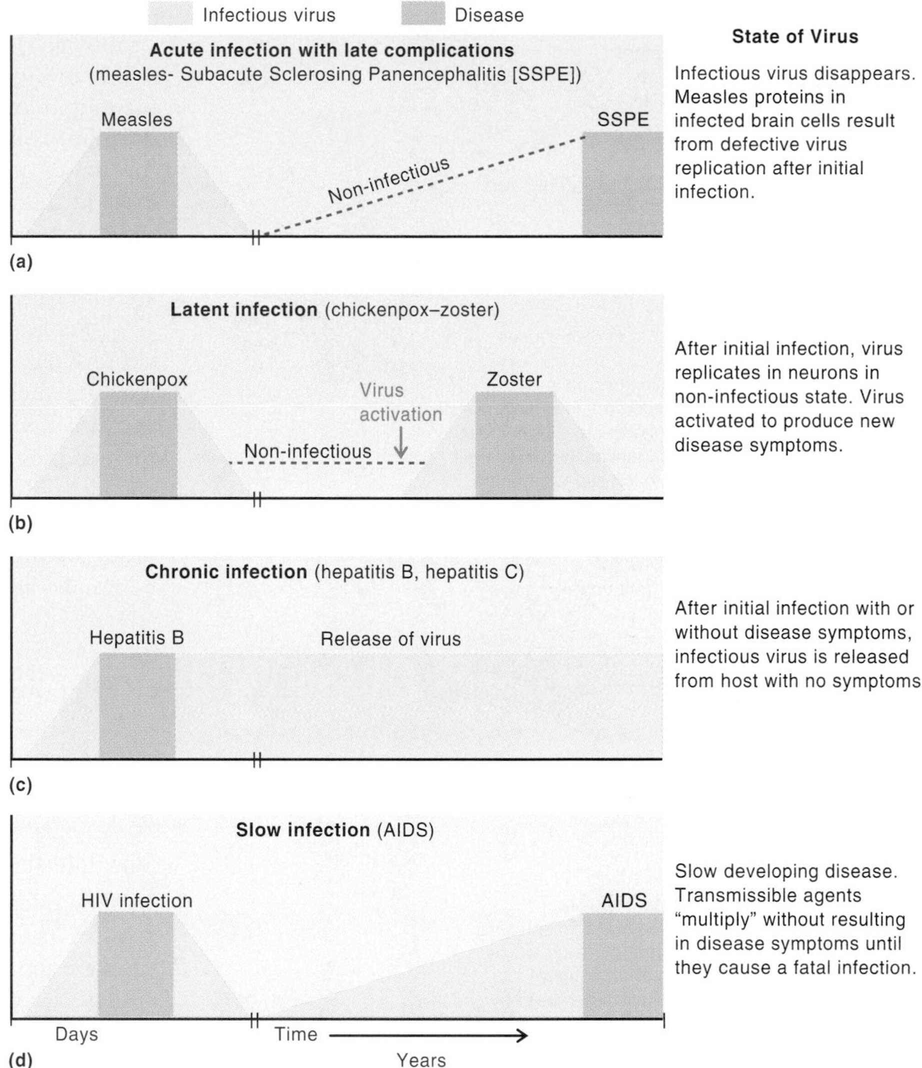
Infectious virus
Disease
Acute infection with late complications
(measles- Subacute Sclerosing Panencephalitis [SSPE])
Measles
SSPE
Non-infectious
(a)
State of Virus
Infectious virus disappears. Measles proteins in infected brain cells result from defective virus replication after initial infection.
Latent infection (chickenpox–zoster)
Chickenpox
Virus activation
Zoster
Non-infectious
(b)
After initial infection, virus replicates in neurons in non-infectious state. Virus activated to produce new disease symptoms.
Chronic infection (hepatitis B, hepatitis C)
Hepatitis B
Release of virus
(c)
After initial infection with or without disease symptoms, infectious virus is released from host with no symptoms.
Slow infection (AIDS)
HIV infection
AIDS
Days
Time
Years
(d)
Slow developing disease. Transmissible agents "multiply" without resulting in disease symptoms until they cause a fatal infection.

Figure 14.12 Time Course of Appearance of Disease Symptoms and Infectious Virions in Various Kinds of Viral Infections

and fever blisters often result. After this initial acute infection, the HSV-1 infects the sensory nerve cells, where it remains in a non-infectious form without causing symptoms of disease (**figure 14.13**). Replication of this virus in the nerve cells is repressed by some unknown mechanism but can be activated by such conditions as menstruation, fever, or sunburn. Following the initiation of replication, mature infectious virions are produced and are carried to the skin or mucous membranes by the nerve cells, once again resulting in cold sores. After these sores have healed, the virus and host cells once again exist in harmony and, as with other latent infections, no virions are synthesized until the disease recurs. ■ **herpes simplex type 1, p. 594**

Another example of a latent infection is provided by another member of the herpesvirus family, varicella-zoster virus, the cause of **chickenpox (varicella)**. Initial infection of normal children results in a rash termed chickenpox. This virus can remain latent for years without producing any disease symptoms. It can then be reactivated and produce the disease called **shingles**, or **herpes zoster**. Most infections, however, never reactivate. Thus, chicken pox and shingles are different diseases caused by the same virus. ■ **chickenpox, p. 534** ■ **shingles, p. 536**

Most herpesviruses, including HSV-2, tend to become latent under various conditions. It appears that part or all of the viral DNA becomes integrated into the genome of the host, or copies of the nucleic acid of some herpesviruses may replicate as plasmids in the host cell. Note the similarity to infection by temperate phages. **Table 14.7** gives some examples of latent infections. ■ **temperate phages, p. 329** ■ **plasmids, p. 210**

Chronic Infections

In chronic infections, the infectious virus can be demonstrated at all times. (see figure 14.12c). Disease may be present or absent during an extended period of time or may develop late. The best known chronic human infection is caused by the **hepatitis B virus**, formerly called **serum hepatitis virus**. This disease is transmitted sexually or from the blood of a chronic carrier who shows no symptoms. Some people who contract the virus develop an acute illness marked by nausea, fever, and jaundice.

TABLE 14.6 Comparison of Replication Cycle of Bacteriophages and Animal Viruses in Virulent Infections

Stage	Bacteriophages	Animal Viruses
Attachment	Fusion of capsid with host membrane does not occur.	Fusion of viral envelope and host membrane common.
Entry	Only nucleic acid enters cell—no enzymes.	Entire virion enters cell, including enzymes of replication.
Replication cycle	Depends on whether nucleic acid is DNA or RNA, double- or single-stranded.	Same pattern of replication as phage with the same genome.
Exit	In lytic infection, phage codes for lytic enzyme, which lyses the cell. Budding rare—cells not killed.	Cell dies and lyses with release of virus. Budding common—cells may or may not be killed.

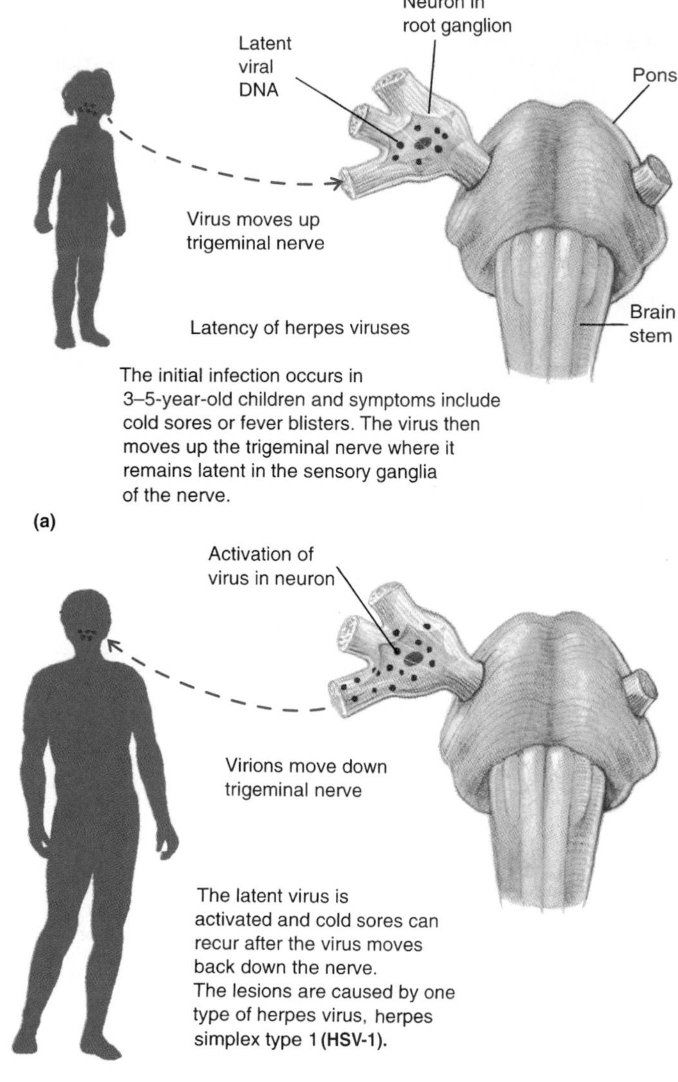
Neuron in root ganglion
Latent viral DNA
Pons
Virus moves up trigeminal nerve
Latency of herpes viruses
Brain stem
The initial infection occurs in 3–5-year-old children and symptoms include cold sores or fever blisters. The virus then moves up the trigeminal nerve where it remains latent in the sensory ganglia of the nerve.
(a)
Activation of virus in neuron
Virions move down trigeminal nerve
The latent virus is activated and cold sores can recur after the virus moves back down the nerve.
The lesions are caused by one type of herpes virus, herpes simplex type 1 (HSV-1).
(b)

Figure 14.13 Infection Cycle of Herpes Simplex Virus, HSV-1

About 300 million people are carriers of the virus, and a significant number develop cirrhosis or cancer of the liver; more than 1 million people die each year from hepatitis B. ■ hepatitis B, p. 609

In the **carrier state**, infectious virions of hepatitis B are continually produced and can be detected in the bloodstream, saliva, and semen. The viral DNA genome may also occur as a plasmid in liver cells (hepatocytes), where it replicates and produces many infectious virions. Following replication in this plasmid state, the genome can also integrate into the cells of the liver. In this state, only portions of the protein components of the virion are synthesized and infectious virions are not produced. Some examples of viruses that cause chronic infections are summarized in **table 14.8**.

Replication of DNA is very unusual in the hepatitis B virus. Unlike all other double-stranded DNA viruses, which replicate their DNA using the viral DNA as a template, the hepatitis B virus synthesizes RNA from the DNA. The RNA is then used as a template for the synthesis of DNA, using a viral-encoded DNA polymerase. This enzyme acts as a reverse transcriptase, an unusual enzyme found only in the retroviruses, a group of viruses that will be discussed shortly. Replication can be simply diagrammed as follows:

$$\text{DNA} \longrightarrow \text{RNA} \xrightarrow{\text{Reverse transcriptase}} \text{DNA}$$

A more detailed discussion of the replication of hepatitis B virus is presented in chapter 26. ■ hepatitis B replication, p. 610

Slow Infections

In slow infections, the infectious agent gradually increases in amount over a very long time during which no significant symptoms are apparent. Eventually, a slowly progressive lethal disease ensues (see figure 14.12d). AIDS has features of slow virus infections. ■ AIDS, p. 655, 741

Two groups of unusual agents that cause slow infections have been identified. One group is the *Lentivirus* (*lenti* means "slow"), which are in the family *Retroviridae* (retroviruses; *retro* means backwards). Members of this family cause tumors in animals. The second is the protein infectious agents called **prions**. Both groups cause diseases that have long preclinical phases and result in progressive, invariably fatal diseases. In both groups, the infectious agent can be recovered from infected animals during

TABLE 14.7 Examples of Latent Infections

Virus	Primary Disease	Recurrent Disease	Cells Involved in Latent State
Herpes Simplex Virus			
HSV-1	Primary oral herpes	Recurrent herpes simplex	Neurons of sensory ganglia
HSV-2	Genital herpes	Genitalis	Neurons of sensory ganglia
Varicella-zoster virus (herpesvirus family)	Chicken pox	Herpes zoster (shingles)	Satellite cells of sensory ganglia
Cytomegalovirus (CMV; herpesvirus family)	Usually subclinical except in fetus or immunocompromised host	CMV pneumonia, eye infections, mononucleosis-like symptoms	Salivary glands, kidney epithelium, leukocytes
Epstein-Barr virus (herpesvirus family)	Mononucleosis	Burkitt's lymphoma	B cells, which are involved in antibody production

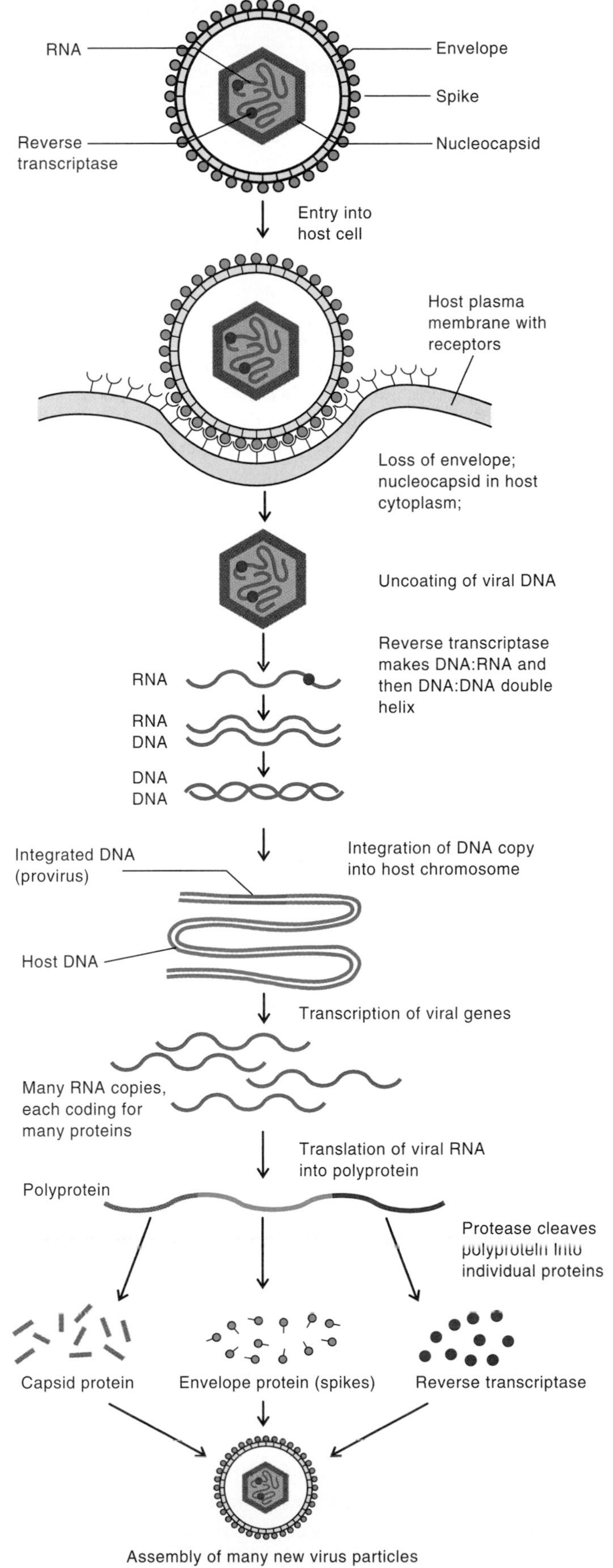

Figure 14.14 Replication Cycle of a Retrovirus A retrovirus is the cause of AIDS and many tumors in animals.

both the preclinical years when no symptoms are evident and the time that clinical symptoms are present. ■ **prions, p. 364**

Complex Infections

Some viral infections cannot be neatly pigeonholed into a single type of infection; they have features associated with more than one type. We call such infections **complex**. One example is the infection caused by a member of the *Retroviridae*: the HIV virus, which causes AIDS. AIDS results from the invasion and subsequent destruction of T lymphocytes and macrophages, important components in the immune system of the body. Without a healthy level of T lymphocytes and macrophages, the body becomes susceptible to a wide variety of infectious diseases. AIDS and HIV infection will be covered in detail in chapter 29. ■ **macrophages, p. 372, T lymphocytes, p. 374**

The retroviruses are single-stranded enveloped RNA viruses, many of which infect humans. Many members cause tumors in animals, one causes a leukemia in humans, but the most prominent member is HIV. The HIV genome is unusual in that it consists of two copies of RNA.

The replication of retroviruses is unusual and has no counterpart in phages. The major feature is that the genetic information in its RNA is converted into DNA, which is then integrated into the genome of the host cell. Its replication is shown in **figure 14.14**. A more complete diagram of all of the steps of the replication cycle is presented in chapter 29. ■ **HIV replication cycle, p. 748**

Like all retroviruses, the replication of HIV requires that its single-stranded RNA genome be converted into a double-stranded DNA copy. Two Americans, Howard Temin and David Baltimore, independently demonstrated in 1970 that retroviruses contain in their capsid an unusual enzyme, **reverse transcriptase**, that enters the host cell as part of the entering nucleocapsid at the time of infection. Reverse transcriptase is not found in uninfected cells. This enzyme copies the single-stranded viral RNA into a complementary strand of DNA. A second strand of DNA complementary to the first DNA strand is then synthesized. The double-stranded DNA is integrated permanently into a chromosome of the host cell as a **provirus**. This provirus is superficially analogous to the phage lambda (λ) when it is present as a prophage in *E. coli*. Recall, however, that lambda integrates at specific sites in the chromosome of *E. coli* whereas HIV DNA integrates randomly into host cell chromosomes. ■ **prophage λ, p. 329**

The details of how the activity of the provirus is regulated are not nearly so clear as they are in the case of lambda. It is known that some of the infected cells continuously synthesize new virions that bud from the cell. In this situation, the RNA is transcribed by the host cell RNA polymerase to produce one long mRNA molecule that contains all of the viral information, a **polygenic** or **polycistronic** mRNA. This mRNA molecule is translated into a long **polyprotein**, which is then cleaved by a viral-encoded protease to yield the individual proteins that make up the virion. If the action of this protease is inhibited, then the

TABLE 14.8 Examples of Chronic Infections

Virus	Site of Infection	Location of Infectious Virions in Carrier State	Disease
Hepatitis B	Liver	Plasma, saliva, genital secretions	Hepatitis, cirrhosis, carcinoma
Hepatitis C	Liver	Plasma, saliva, genital secretions	Hepatitis, cirrhosis, carcinoma
Rubella virus	Many organs	Urine, saliva	Congenital rubella syndrome

virus cannot be assembled. Consequently, inhibitors of this viral protease, termed **protease inhibitors**, created in the laboratory are a major weapon against HIV. ■ polycistronic mRNA, p. 172

Many cells carry the provirus in the latent state, and no virions are produced. A variety of agents can activate the provirus, however, so that it results in a productive infection in which the virions are released by budding. What these agents are in nature is not known. Small amounts of the virus are present continuously or intermittently in the blood and genital secretions, and carriers can transmit the infection through sexual contact.

DNA polymerase makes very few mistakes in the replication of DNA because of its proofreading ability. Reverse transcriptase, however, has no proofreading activity and makes many mistakes when copying RNA into DNA. As a result, the DNA that codes for a variety of different capsid proteins codes for altered proteins. Many of these proteins are no longer recognized by antibodies that recognized the protein capsid of the original virus. These errors in copying help explain why the virus becomes resistant very quickly to antiviral drugs such as AZT and probably protease inhibitors. AIDS is covered in great detail in chapter 29. ■ proofreading, DNA polymerase, p. 197 ■ AZT, p.195, 752

MICROCHECK 14.3

Most animal viruses live in harmony with their natural hosts and do not cause serious illness. If the virus infects an unnatural host, a serious disease may result. The various kinds of relationships of animal viruses with their hosts are in general similar to those seen with bacterial viruses and bacteria. Replication of viral nucleic acid depends to varying degrees on the enzymes of nucleic acid replication of the host cells.

A retrovirus, HIV, is responsible for the disease AIDS, which has features of a latent, chronic, and slow infection. The replication of HIV involves an enzyme, reverse transcriptase, which copies the single-stranded RNA of the virion into DNA which then becomes integrated into certain cells of the immune system.

- Name the two major kinds of infections that viruses cause.
- What is a major difference in the entry of animal viruses and bacterial viruses into their host cells?
- How do animal viruses cause lysis of the host cell?
- Explain why HIV becomes resistant to drugs so quickly.

Tumors—General Aspects

The word **tumor**, or **neoplasm**, indicates a swelling that results from the abnormal growth of cells. If the growth remains within a defined region and is not carried to other areas by the circulatory system, it is a **benign tumor**. If the abnormal growth spreads or **metastasizes** to other parts of the body, it is a **malignant tumor** or **cancer**. It is estimated that less than one in 10,000 tumor cells that escape the primary tumor survives to colonize another tissue. Because this discussion is concerned only with primary tumors, the term *tumor* exclusively will be synonymous with abnormal growth. The term *cancer* denotes a life-threatening tumor.

Tumors result when the controls for cell growth and differentiation do not function properly. To grow and differentiate properly, a cell must synthesize key proteins at the proper times and in the correct amounts. These proteins include those that are involved in cell-to-cell communication as well as those that regulate key activities within cells. Such synthesis requires the appropriate activation of genes at specific times in the cell cycle (up-regulation) and their silencing at other times (down-regulation). If these key genes do not function properly, then abnormal growth resulting in tumor formation may result.

Two classes of host cell regulatory genes are commonly involved in tumor formation, **proto-oncogenes** and **tumor-suppressor genes**. Proto-oncogenes code for proteins that activate gene transcription. (Tumor viruses have similar genes, termed **oncogenes,** which will be discussed shortly.) Proto-oncogenes were identified by two microbiologists, Dr. Michael Bishop and Dr. Harold Varmus, who in 1989 were jointly awarded the Nobel Prize for their studies. Mutations in proto-oncogenes result in altered proteins that cause the cells to grow abnormally. Fusing a normal cell with a tumor cell containing an abnormal proto-oncogene results in a cell that grows abnormally; the mutation is dominant.

The second class of regulatory genes associated with tumor formation are the tumor-suppressor genes. These genes are responsible for putting the brakes on abnormal or excessive growth and maintaining normal growth. Some tumor-suppressor genes code for enzymes that repair damaged DNA. Mutations in these genes are recessive, since fusing a normal cell with a tumor cell having a defective tumor-suppressor gene results in a normal cell. The normal cell supplies the braking system required for normal cell growth and repairs damaged DNA.

Proto-oncogenes and tumor-suppressor genes are most commonly altered in function through mutation. If the proto-oncogenes or the tumor-suppressor genes fail to function properly, abnormal growth and tumor formation may result. Mutations can arise if the repair systems that operate in DNA replication or mismatch repair are defective. How these repair systems function has already been discussed for bacteria. When mutagens in the environment mutate the genes, the frequency of tumor formation increases. Recall that Bruce Ames pointed out that all carcinogens are mutagens because they affect the DNA of the organism. If the altered genes are essential for growth, such as the genes of macromolecule synthesis, then the cell will die. If the genes are one of the proto-oncogenes or tumor-suppressor genes, then a tumor may well develop. It is not surprising that mutations in genes concerned with the repair of damaged DNA lead to an increased rate of tumor formation. It is estimated that 50% of all human tumors result from mutations in tumor-suppressor genes. ■ **mismatch repair, p. 198** ■ **Ames test, p. 203** ■ **mutations, p. 192**

Many proto-oncogenes and tumor-suppressor genes exist. Some human cancers, such as those of the colon, require that several of the proto-oncogenes and tumor-suppressor genes be mutant simultaneously. A mutation in only one of these genes leads to slightly abnormal growth; a second mutation results in somewhat more abnormal growth, and so on. This explains why certain cancers develop in stages.

Abnormal growth can also be caused by viruses. Tumor viruses, like mutations, alter the activity of the proto-oncogenes and the tumor-suppressor genes. Viruses are recognized as a frequent cause of tumors in animals, but they are not a frequent cause of tumors in humans. Only 15% of all human tumors are estimated to be caused by viruses. This number, however, will likely be revised upward because of AIDS. At least 30% of people with AIDS also develop cancer. A popular view is that the majority of these cancers are caused by viruses.

MICROCHECK 14.4

Two classes of host regulatory genes, proto-oncogenes and tumor-suppressor genes, are commonly involved in tumor formation. Viruses can alter the activity of the genes.

- Differentiate between the terms *tumor* and *cancer*.
- Are all mutagens carcinogens and all carcinogens mutagens? Discuss.

Viruses and Animal Tumors

As pointed out in the Glimpse of History, viruses were implicated in causing tumors in chickens in the early Twentieth century. Retroviruses are the most significant tumor viruses in animals, whereas in humans, DNA viruses are the most important.

Tumor Viruses and Cell Transformation

An understanding of the mechanism by which viruses cause tumors in animals was given a big boost when it was observed that tumor viruses could rapidly change the properties of human cells growing in cell culture. These changes are inheritable and are readily observed. Such changed cells are referred to as **transformed cells**. Note that this phenomenon is quite different from what was discussed already for the transformation of bacteria by naked DNA. The properties of normal and transformed cells in culture are compared in **table 14.9**.

The properties of the transformed cells in culture are similar to those of tumor cells in the body. Transformed cells do not respond to signals that limit their growth. Thus, whereas normal cells grow as a monolayer on a glass surface, transformed cells grow in multiple layers. Also, normal cells go through only a limited number of cell divisions and then die. In contrast, transformed cells growing in tissue culture multiply indefinitely. They are immortal. Further, normal cells stick tightly to the surface of the glass culture dish, whereas transformed cells readily detach from the surface, a phenomenon analogous to metastasizing in the body. Also, when transformed cells in culture are injected into animals, most, but not all, cause tumors. ■ **bacterial transformation, p. 205**

Once it was discovered that most tumor viruses such as the DNA viruses SV40 and polyoma as well as tumor retroviruses could transform normal cells in culture, experimentation on tumor viruses became much simpler, cheaper, and faster. To be able to assay the tumor-inducing ability of viruses quickly and cheaply simply by mixing the virions with tissue culture cells greatly expedited research. Within a very short time, the viral genes responsible for cell transformation, the oncogenes, were identified.

How Retroviruses Transform Animal Cells

Retroviruses, the most common cause of tumors in animals, transform cells by inserting transforming genes of the virus into the genome of the host cell (see figure 14.14). The transformation requires that the genetic information in the entering single-stranded RNA molecule be converted into a double-stranded DNA molecule by the viral enzyme reverse

TABLE 14.9 Some Properties of Normal Cells and Tumor Cells in Culture

Normal Cells

1. Cells grow as a monolayer (single layer).
2. Cells grow attached to one another and to glass surfaces.
3. Cells multiply for a limited number of generations and then die, even if diluted into fresh medium.
4. Cells do not form tumors in susceptible animals.

Tumor (Transformed) Cells

1. Cells grow in an unorganized pattern and in multiple layers.
2. Cells attach to surfaces less firmly.
3. Cells continue to multiply indefinitely.
4. Cells may form tumors in susceptible animals.

transcriptase. This DNA is then integrated into the host cell genome, where it is expressed.

What is the nature of the transforming genes? These genes are termed oncogenes, (from the word *onkos*, which means "mass" or "lump"). They are mutant forms of the normal cell's proto-oncogenes. The mutation modifies the biochemical properties of the protein coded by the oncogene so that an oncogene integrated into the genome of a cell interferes with the normal intracellular control functions of the proto-oncogene. This disruption results in tumor formation. More than 60 oncogenes have been discovered in retroviruses thus far; a few are listed in **table 14.10**. Different tumor viruses contain different oncogenes that function somewhat differently. They are, however, all involved in regulating normal cell growth.

The best available evidence suggests that oncogenes originated from proto-oncogenes that were "captured" by the retrovirus in the course of being excised from the host genome. Once inside the virus, the proto-oncogene underwent mutations that converted it into an oncogene. The temperate phage system in bacteria provides an excellent analogy for this "capture" hypothesis. Recall that temperate phages integrated into the bacterial host chromosome can incorporate a piece of bacterial DNA into their chromosome when the phage DNA is excised from the host chromosome. The virus can then transfer the bacterial DNA to other bacteria by transduction. By analogy, retroviruses, which also can integrate into the chromosome of the host, could very well carry along a proto-oncogene from the host in the process of excision from the host chromosome. ■ **temperate phage, p. 326, 329** ■ **transduction, p. 207, 334**

In summary, tumors develop from alterations in the function of genes that play critical roles in regulating normal cell growth. These alterations can result from mutations in proto-oncogenes or the tumor-suppressor genes of the cell, or by the introduction of oncogenes of a transforming virus. In all cases, normal cell growth is disrupted.

TABLE 14.10 Some Oncogenes Associated with Tumors

Oncogene	Virus	Function of Oncogene
src	Rous sarcoma	Tyrosine protein kinase*
fFps	Fujinami sarcoma virus (chickens)	Tyrosine protein kinase
H-ras	Mouse sarcoma virus	Mutant guanosine triphosphate–binding protein
raf	Mouse sarcoma virus	Serine protein kinase
jun	Avian sarcoma virus	Mutant sequence-specific gene activator
skv	Sloan-Kettering virus	Mutant sequence-specific DNA-binding protein

Oncogenes identified in different retroviruses. Each of these oncogenes corresponds to a proto-oncogene in the host.
*An enzyme that adds a phosphate molecule to tyrosine in a regulatory protein modifying the activity of the protein.

MICROCHECK 14.5

Animal tumor viruses can transform animal cells growing in culture. Retroviruses cause tumors in animals by inserting an oncogene, which alters normal cell growth.

- Distinguish between oncogenes and proto-oncogenes.
- List four properties of transformed cells in culture.
- Why is reverse transcriptase not found in normal cells?

Viruses and Human Tumors

Most human tumors are not caused by viruses, despite intensive efforts to prove otherwise. These numbers are increasing, however, because of the common occurrence of a viral-induced tumor, Kaposi's sarcoma, in AIDS patients. The majority of human tumors appear to be caused by mutations in either proto-oncogenes or tumor-suppressor genes. About 30% are estimated to be due to activation of a particular mutant proto-oncogene, and 50% of all human tumors result from mutations in tumor-suppressor genes. A particular tumor-suppressor gene ***p53***, (*p* stands for the protein for which the gene codes, and *53* is its molecular weight in thousands), seems especially important in preventing tumors, since mutations in this gene are frequently found in tumor cells. The *p53* protein is a transcription factor activated by damaged DNA that blocks the division of cells that contain damaged DNA. It also activates genes that initiate **apoptosis**, programmed cell death. Selectively removing cells that have damaged DNA protects an organism from tumor development, and mutation in p53 prevents cell death. Tumor viruses also can inhibit the function of tumor-suppressor genes, and this contributes to certain human cancers such as colorectal cancers. ■ **transcription factor, p. 180** ■ **apoptosis, p. 747**

Considering all viruses, double-stranded DNA viruses are the main cause of tumors in humans (**table 14.11**). DNA tumor viruses interact with their host cells in one of two ways. They can go through a productive infection in which they lyse the cells, or they can transform the cells without killing them. The cancers caused by the DNA viruses result from the integration of all or part of the virion genome into the host chromosome. Following integration, the transforming genes are expressed, resulting in uncontrolled growth of the host cells. Thus, these cases of abnormal growth are analogous to lysogenic conversion observed in certain temperate phage infections of bacteria. In both cases, the expression of viral genes integrated into the host's chromosome confers new properties on the host cells. ■ **lysogenic conversion, p. 331**

In the case of some DNA viruses, such as papillomaviruses and herpesviruses, the viral DNA is not integrated but apparently replicates as a plasmid. In the case of the papillomaviruses, on rare occasions, the plasmid may integrate into the host chromosome and this may cause tumors. Certain types of human papillomavirus are linked with most cases of cervical cancer, as well as with vulval, penile, and anal cancers. Although the viral oncogenes are necessary for development of cancer, their expression

TABLE 14.11 Viruses Associated with Cancers in Humans

Virus	Type of Nucleic Acid	Kind of Tumor
Human papillomaviruses (HPV)	DNA	Different kinds of tumors, including squamous cell and genital carcinomas, caused by different HPV types
Hepatitis B virus	DNA	Hepatocellular carcinoma
Epstein-Barr virus	DNA	Burkitt's lymphoma; nasopharyngeal carcinoma; B-cell lymphoma
Hepatitis C virus	DNA	Hepatocellular carcinoma
Human herpes, virus 8	DNA	Kaposi's sarcoma
HTLV-1	RNA (retrovirus)	Adult T-cell leukemia (rare)

is not sufficient. It appears that products of the oncogenes must interact with the tumor-suppressor protein, p53. Apparently this interaction results in the degradation of p53, and tumor formation is promoted. ■ **papillomavirus, p. 654** ■ **herpesvirus, p. 352**

Kaposi's sarcoma, a cancer of the skin and internal organs common in AIDS patients, is caused by a herpesvirus. How this particular virus causes normal cells to become tumorous is not known. Note that in all cases of virus-induced tumors, the virus does not kill the host cell but instead changes its properties. ■ **Kaposi's sarcoma, p. 754**

The various interactions that viruses display with their hosts are illustrated in **figure 14.15**.

Retroviruses and Human Tumors

Although retroviruses are the main class of viruses causing tumors in animals, it was not until 1980 that a human leukemia was also shown to be caused by a retrovirus. It was named human T-cell lymphotrophic virus type 1 (HTLV-1), and this tumor is restricted to certain geographic areas. The virus causes tumors by a somewhat different mechanism than discussed thus

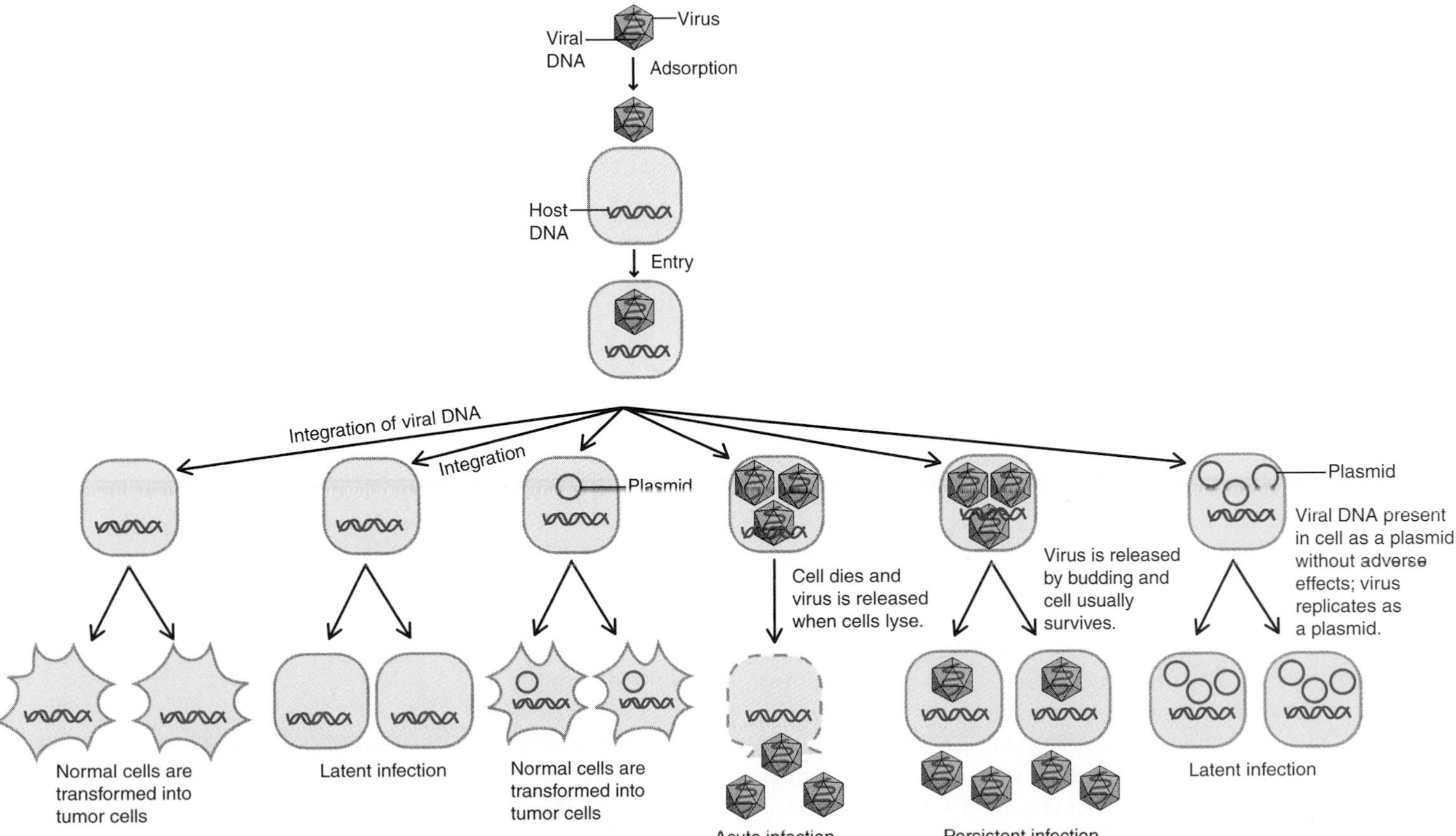

Figure 14.15 Various Effects of Animal Viruses on the Cells They Infect

far. The virus carries an oncogene that, when integrated into the host cell genome, codes for an activator protein that activates regulatory genes of the cell.

MICROCHECK 14.6

Most human tumors are caused by mutations in tumor-suppressor genes such as *p53*. The most common viral cause of tumors in humans is DNA tumor viruses. One retrovirus is known to cause a rare human tumor.

- Name 3 viruses that cause tumors in humans.
- Name a tumor common in AIDS patients and the virus that causes it.
- Why is it not surprising that AIDS patients frequently suffer a viral-induced tumor?

Viral Host Range

Most viruses can infect a single species or even only certain cells within an organism. Some viruses, however, those that cause zoonoses, can multiply in widely divergent species. For example, the West Nile virus that killed seven people in the New York City area in the summer of 1999 was also responsible previously for the deaths of numerous crows and in the winter of 2000 a red-tailed hawk. The virus also multiplies in mosquitoes, which then transmit the virus when they bite.

Viruses can have their host range modified when two virions that differ in their genome infect the same cell. One modification results from an exchange of protein coats called **phenotypic mixing**. The other modification stems from an exchange of genetic information, **genetic reassortment.**

Exchange of Protein Coats—Phenotypic Mixing

Animal cells sometimes can be infected simultaneously by more than one virus, even when the viruses are from different genera. As the different viruses synthesize their different protein coats and replicate their nucleic acids in such multiply-infected cells, an exchange of protein coats can occur in a phenomenon termed phenotypic mixing or **transcapsidation (figure 14.16)**. Since the host range of any virus is determined in large part by its coat protein, a virus that was once unable to infect a certain cell type may gain the ability to do so with its new coat. Thus, its host range can be expanded by this mechanism. Once inside the cell, however, the nucleic acid codes for its original coat protein and not the one that it wore when it infected the cell. Therefore, the virions that exit the cell cannot infect the same cell type that was just infected. They return to their original host range. Phenotypic mixing can occur among viruses of the same or different genera. The rhabdo-, herpes-, retro-, and paramyxoviruses can undergo phenotypic mixing among themselves.

How this phenomenon works can be shown using the following example of two retroviruses, one of which normally infects mice and the other, chickens **(figure 14.17)**. Both retroviruses can infect duck cells. When both retroviruses infect the same cells, some of the progeny of the viruses exchange their outside coats. These new viruses have the host range of the virus

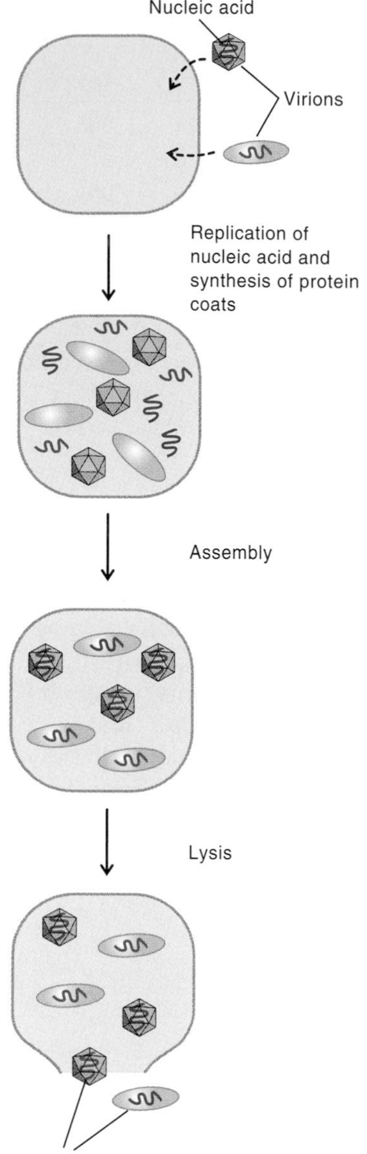

Figure 14.16 Phenotypic Mixing (Transcapsidation) The viruses that are released from the lysed cells cannot infect the same type of host cells from which they are released. Rather, they can only infect their original host cells.

from which they derived their coat. Now the virus that normally could only infect chickens can transfer its RNA to mice but not chickens. Conversely, the virus that could only infect mice cells can now infect a variety of birds that previously were resistant. The virus that has gained the ability to infect mice, however, loses this ability after it has multiplied in mice. Now it can only infect chickens. The other virus also loses its ability to infect the alternative host.

Genome Exchange in Segmented Viruses

Segmented viruses may alter their properties by genetic reassortment **(figure 14.18)**. This phenomenon likely explains how the virulence of the human influenza virus changes so dramatically every 10 to 30 years. These changes result in deadly worldwide epidemics, called **pandemics**, because the global

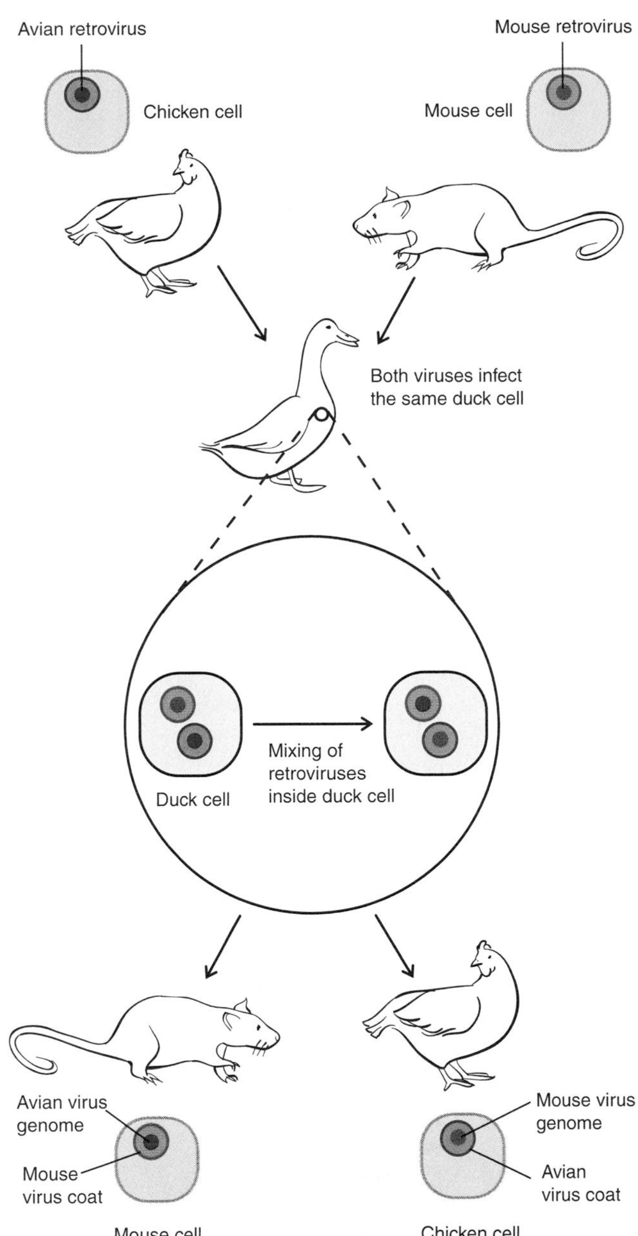

Figure 14.17 Phenotypic Mixing of Two Retroviruses When a cell is infected by two different retroviruses, one from a chicken and one from a mouse, exchange of the virus envelopes can occur. The avian virus genome carries the mouse virus coat, and the mouse virus genome is enclosed in the coat of the chicken virus.

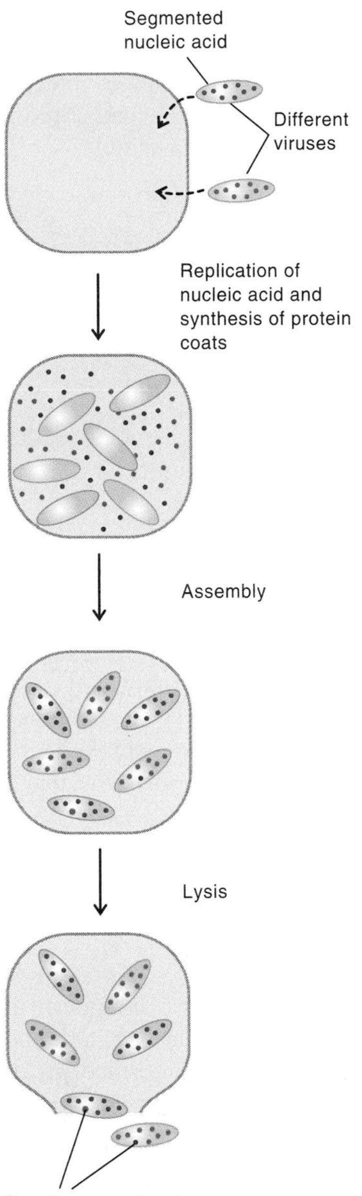

Figure 14.18 Genetic Reassortment The virion that undergoes genetic reassortment continues to give rise to progeny with the same characteristics.

population does not have antibodies that protect against the new strain of the virus.

There are a number of different strains of the influenza virus, which can be distinguished by the fact that they are specific for infecting different species. The genome of the influenza virus is divided into eight segments of RNA, each containing different genetic information. One of the segments codes for **hemagglutinin**, a key protein in causing influenza. A human with antibodies against hemagglutinin from the human strain is protected against the disease. If, however, the structure of the hemagglutinin gene changes, then the antibodies will not recognize the protein and will not protect against it. Experimental data suggest that avian and human influenza viruses can simultaneously infect the same cells in a pig, which then serves as a mixing vessel for the 16 RNA segments of the two virions. On occasion, the RNA segment coding for the avian hemagglutinin is incorporated into the protein coat along with the seven RNA segments from the human strain. Such a strain is still able to infect humans but its avian hemagglutinin makes it a new strain now able to evade the host's antibody defense. This is called an **antigenic shift**. In addition to the hemagglutinin gene experiencing this major genetic change, the hemagglutinin gene, along with other viral genes, can undergo point mutations that result in relatively small changes in the protein. These changes are termed **antigenic drift**. Both of these processes will be discussed later in terms of the epidemiology of influenza. ■ **antigenic drift and shift, p. 574** ■ **antibody, p. 391** ■ **antigen, p. 390**

MICROCHECK 14.7

The host range of animal viruses can be altered if two viruses with different host ranges infect the same cell. Genes of one virus can be incorporated into the protein coat of the other virus in assembly.

- Why does the exchange of protein coats alter the host range of a virus?
- Differentiate between antigenic shift and antigenic drift in the influenza virus.
- Can phenotypic mixing alter other properties of a virus other than host range? Explain.

Plant Viruses

A great number of plant diseases are caused by viruses. These can be of major economic importance, particularly when they occur in crop plants such as corn, wheat, and rice. Virus infections are especially prevalent among perennial plants (those that live for many seasons), such as tulips and potatoes, and those propagated vegetatively (not by seeds), such as potatoes. Other crops in which viruses cause considerable damage are soybeans and sugar beets. A serious virus infection may reduce yields of these crops by more than 50%.

Infection of plants by viruses may be recognized through various outward signs (**figure 14.19**). Localized abnormalities may result in a loss of green pigment, and entire leaves may turn color. In many cases, rings or irregular lines appear on the leaves and fruits of the plant. Individual cells or specialized organs of the plant may die, and tumors may appear. Usually, infected plants become stunted in their growth, although in a few cases growth is stimulated, leading to deformed structures. In the vast majority of cases, plants do not recover from viral infections, for unlike animals, plants are not capable of developing specific immunity to rid themselves of invading viruses. On occasion, however, infected plants produce new growth in which visible signs of infection are absent, even though the infecting virus is still present. The reasons for this are not understood.

In severely infected plants, virions may accumulate in enormous quantities. For example, as much as 10% of the dry weight of a TMV-infected tobacco plant may consist of virus.

In a few instances, plants have been purposely maintained in a virus-infected state. The best known example of this involves tulips, in which a virus transmitted through the bulbs can cause a desirable color variegation of the flowers (**figure 14.20**). The infecting virus was transmitted through bulbs for a long time before the cause of the variegation was even suspected. The multiplication of viruses in plants is analogous to that of bacterial and animal viruses in most respects.

Spread of Plant Viruses

In contrast to phages and animal viruses, plant viruses do not attach to specific receptors on host plants. Instead, they enter through wound sites in the cell wall, which is very tough and rigid. Once started, infection in the plant can spread from cell to cell through openings, the **plasmadesmata,** that interconnect cells.

(a)

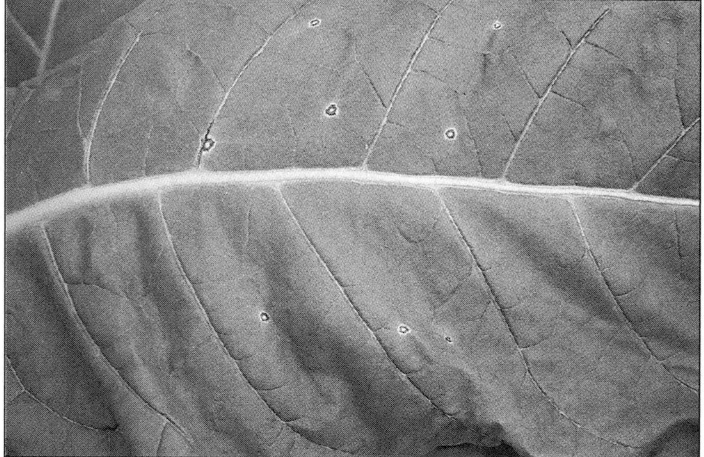

(b)

(c)

Figure 14.19 Symptoms of Viral Diseases of Plants **(a)** A healthy wheat leaf can be seen in the center. The yellowed leaves on either side are infected with wheat mosaic virus. **(b)** Typical ring lesions on a tobacco plant leaf resulting from infection by tobacco mosaic virus. **(c)** Stunted growth (right) in a wheat plant caused by wheat mosaic virus.

Perspective 14.1 Viral Insecticides

Insects can cause enormous losses to agricultural crops and forests. For example, the East African army worm has been known to denude 300 square miles of grassland in a few weeks. The tussock moth is responsible for major defoliation of Douglas-fir trees. To counter these insects, certain insect viruses are being used with some success as biological control agents. These viruses belong to the **baculovirus** group, large double-stranded DNA viruses that are infective to invertebrates, primarily insects. They have been reported to infect more than 600 insect species, with great specificity, such that a particular baculovirus will only infect one or very few species. Further, they are harmless to non-hosts including humans.

Some serious problems with this control system, however, have limited its use. First, it is difficult to achieve a high titer of the virus economically. Second, baculoviruses take a relatively long time to kill, and so infected insects can cause considerable damage before they die. To solve this problem, insect-specific toxins, such as from *Bacillus thuringiensis* (Bt), are being genetically engineered into the virus to speed up its killing action.

One possible role for baculoviruses is in **integrated pest management**. The baculovirus might be used to achieve long-term population control of an insect population. For short-term effects in which rapid insect killing is necessary, biological insecticides such as Bt or chemical insecticides could be used.

Baculoviruses are very persistent in the soil in the absence of their insect hosts. They have been reported to remain infective for periods of up to 5 years, apparently because the viruses are protected by a viral protein called **polyhedrin** in which the virions are embedded (**figure 1**). These **occlusion bodies** are eaten by the larvae of insects and dissolve in their gut, releasing the still infective virions. Baculoviruses represent one more weapon in the armamentarium of people attempting to protect plant life without resorting to nonspecific chemical insecticides.

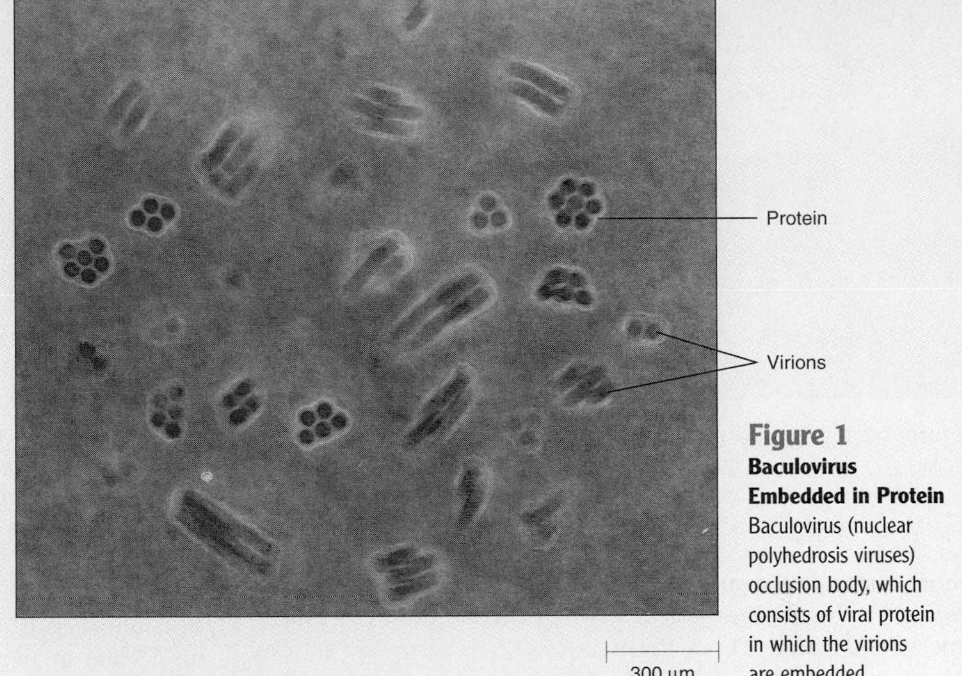

Figure 1 Baculovirus Embedded in Protein Baculovirus (nuclear polyhedrosis viruses) occlusion body, which consists of viral protein in which the virions are embedded.

Many plant viruses are extraordinarily resistant to inactivation. Tobacco mosaic virus apparently retains its infectivity for up to 50 years, which explains why it is usually difficult to eradicate the virions from a contaminated area. Smokers who garden can transmit TMV to susceptible plants from their cigarette tobacco. The virions are very stable, which is important in maintaining the virus because the usual processes of infection are generally inefficient.

Some viruses are transmitted through soil contaminated by prior growth of infected plants. Some 10% of the known plant viruses transmit disease through contaminated seeds or tubers, or by pollination of flowers on healthy plants with contaminated pollen from diseased plants. Virus infections may also spread through grafting of healthy plant tissue onto diseased plants. Another more exotic transmission mechanism is employed via the parasitic vine, **dodder (figure 14.21)**. This vine can establish simultaneous connections with the vascular tissues of two host plants, which serve as conduits for transfer of viruses from one host plant to the other.

Other important infection mechanisms involve vectors of various types. These include insects, worms, fungi, and humans. For example, TMV, which causes a serious disease of tobacco, has no known insect vector. Humans are the major vectors of this disease. Viruses are transmitted to healthy seedlings on the hands of workers who have been in contact with the virus from infected plants or by people who smoke. The most important plant virus vectors are probably insects; thus, insect control is a potent tool for controlling the spread of plant viruses.

Insect Transmission of Plant Viruses

Plant viruses can be transmitted by insects in several ways. First, in **external** or **temporary transmission,** a virus is associated with the external mouthparts of the vector. In this case, the ability to transmit the virus lasts only a few days. Second, in **circulative**

Figure 14.20 Tulips The variegated colors of tulips result from viral infection. The virus is transmitted directly from plant to plant.

Figure 14.21 Dodder Orange-brown twining stems of dodder wrap around two different hosts so that virions can pass from one host to the other through dodder.

transmission, the virus circulates but does not multiply in the body of the insect; the virus may be transmitted during the lifetime of the insect. Third, the transmission may involve actual multiplication of the virus within the insect. In this case, the virus is infectious for both the insect and a plant cell.

In many instances of insect infection with plant viruses, the viruses are passed from generation to generation of the insect and may be transmitted to plants at any time. The existence of insect-transmitted plant viruses raises several interesting questions about virus evolution. In particular, plant and animal viruses may not be as different as they first appeared.

MICROCHECK 14.8

Plant viruses cause many plant diseases and are of major economic importance. They invade through wound sites in the cell wall, and humans are major vectors for their transmission. Some viruses can multiply in both plants and insects.

- Why is it especially important for plant viruses to be stable outside the plant?
- How does a plant virus penetrate the tough outer coat of the plant cell?
- Why is it surprising that some viruses have the ability to replicate in both plants and animals?

Viruslike Agents

Although viruses are composed of only one type of nucleic acid surrounded by a protective protein coat, other agents that cause serious diseases are even simpler in structure. These are the prions and the viroids.

Prions

In addition to viruses, another group of agents that cause slow diseases are **prions**, proteinaceous infectious agents that apparently contain only protein and no nucleic acid. (see figure 1.14). These agents have been linked to a number of fatal human diseases as well as to diseases of animals. In all of these afflictions, brain function degenerates as neurons die, and brain tissue develops spongelike holes. Thus, the general term **transmissible spongiform encephalopathies** has been given to all of these diseases. Some of the slow but always fatal infections that have been attributed to prions are listed in **table 14.12**. Prions will be covered in greater detail in chapter 28. ■ prions, p. 13, 687

Viroids

The term **viroid** defines a group of pathogens that are much smaller and are distinctly different from viruses (see figure 1.13). Viroids that have been characterized consist solely of a small, single-stranded RNA molecule that varies in size from 246 to 375 nucleotides. This is about one-tenth the size of the smallest infectious viral RNA known. They have no protein coat.

Their other properties include the following:

- Viroids replicate autonomously within susceptible cells. No other virions or viroids are required.
- A single viroid RNA molecule is capable of infecting a cell.
- The viroid RNA is circular and is resistant to digestion by nucleases.

All viroids that have been identified infect plants, where they cause serious disease. These diseases include potato spindle tuber, chrysanthemum stunt, citrus exocortis, cucumber pale fruit, hopstunt, and cadang-cadang.

A great deal is known about the structure of viroid RNA, but many questions remain. How do viroids replicate? How do they cause disease? How did they originate? Do they have counterparts in animals, or are they restricted to plants? The answers

TABLE 14.12 Slow Infections Caused by Prions

Agent	Host	Site of Infection	Disease
Scrapie agent	Sheep	Central nervous system	Scrapie spongiform encephalopathy
Kuru agent	Humans	Central nervous system	Kuru spongiform encephalopathy
Creutzfeldt-Jakob agent	Humans	Central nervous system	Creutzfeldt-Jakob spongiform encephalopathy
Mad cow agent	Cows and humans	Central nervous system	Mad cow spongiform encephalopathy

to these questions will provide insights into new and fascinating features of unusual members of the microbial world.

MICROCHECK 14.9

Two infectious agents that are structurally simpler than viruses are prions and viroids. Prions contain protein and no nucleic acid; viroids contain only single-stranded RNA and no protein.

- Distinguish between a prion and a viroid.
- What are the hosts of prions? Of viroids?
- Why are viroids resistant to nucleases?

FUTURE CHALLENGES

Great Promise, Greater Challenge

Gene therapy, the treatment of disease by introducing new genetic information into the body, is a procedure with tremendous promise, but with disappointing results thus far. In large part, this lack of success results from a number of basic biological and technological problems that have yet to be solved. Advances are being made, however, that suggest gene therapy will ultimately be a powerful approach to treating many inherited metabolic diseases such as hemophilia, and cystic fibrosis as well as certain tumors.

Gene therapy is rooted in the advances that have been made in microbial genetics, molecular biology, and virology. The idea that it might be possible to introduce genes into mammalian cells and correct genetic defects was first suggested by a number of scientists who worked with microbes and studied gene transfer in bacteria. These included Joshua Lederberg and Edward Tatum, codiscoverers of conjugation in E. coli. *Lederberg also discovered transduction by the phage lambda. It was a small step for these investigators to suggest the possibility of gene transfer into mammalian cells by an animal virus.*

The major challenge of gene therapy is to design a viral vector that can deliver and express genes in mammalian cells with great efficiency and absolute safety. Considerable progress is being made to develop suitable vectors that have these properties.

One current approach is to use a vector that can deliver, integrate, and express genes in nondividing cells, since many cells in the body are not multiplying. One potential vector that can do this is a lentivirus. Lentiviruses include the retrovirus HIV that causes AIDS. The genetic information in its single-stranded RNA is converted into DNA, which in turn is integrated into the nucleus of the host cell where it is maintained as part of the host's chromosome. It is ironic that one of the deadliest viruses known is a potential agent to prevent killing.

Certain problems are associated with this virion. First, lentiviruses only bind to certain cells in the body concerned with the immune response, $CD4^+$ T cells and macrophages. To expand its host range to other cells, the viral envelope has been modified by incorporating the gene of another enveloped virus, allowing the virus to attach to a wider range of cells.

It is also necessary to protect the virus-invaded cell from being destroyed. The body has protective mechanisms that recognize and kill cells infected with virus. To trick the body's defenses, a lentivirus has been constructed that synthesizes no viral proteins. The infected cell, however, must synthesize the protein that overcomes the genetic defect. In normal people this protein would not be foreign to the host and would not set off an allergic reaction. In the person who has not synthesized this protein before, however, it is a foreign protein and, therefore, may set off an allergic reaction. Great progress is being made in developing regulatable promoter systems that would function only in certain tissues, only at certain times, and under the control of chemicals that could be readily delivered into the body. Many of the cells in the body can shut off the viral promoters that have been used to express the introduced genes.

The safety issue is especially important because in the fall of 1999 an 18-year-old volunteer died soon after an adenovirus vector was administered in a gene therapy trial. A number of approaches are being considered to produce safe vectors. One approach is to use lentiviruses that do not infect humans. Another is to remove the genes of the virus that are required for virus replication.

The problems that must still be solved before clinical trials are attempted again are formidable. However, the advances that have been made recently in basic studies of virology, immunology and gene expression in the laboratory should go a long way in transforming a great promise into reality.

SUMMARY

Structure and Classification of Animal Viruses

1. Phages and animal and plant viruses are nucleic acid surrounded by a protein coat, the **capsid**. Many animal viruses have an additional covering, an **envelope**, a lipid bilayer similar to the plasma membrane of the host. **(Figure 13.2)**
2. Animal and plant viruses containing more than one RNA molecule are **segmented viruses.**

Classification of Animal Viruses

1. The criteria for classifying animal viruses are **genome structure, particle structure**, and the presence or absence of a **viral envelope**. **(Tables 14.1, 14.2)**

Groupings Based on Routes of Transmission

1. Viruses that cause disease can be grouped according to their **routes of transmission**—the **enteric** viruses, **respiratory** viruses, **zoonoses**, (transferred from animals to other animals), and **sexually transmitted diseases**. **(Table 14.4)**

Methods Used to Study Viruses

Cultivation of Host Cells

1. Some viruses can only be cultivated in living animals; others can be grown in **tissue culture**. **(Figures 14.2, 14.3)**

Quantitation

1. The **plaque assay** is commonly used to determine the number of infective virions in a sample. (Figure 14.4)
2. Virions can be counted with an **electron microscope**. (Figure 14.5)
3. **Quantal assays** determine the **infective** or **lethal** dose in a viral preparation.
4. Some viruses can clump red blood cells and their concentration can be measured by determining the dilution of virus able to clump the cells, **hemagglutination**. (Figure 14.6)

Interactions of Animal Viruses with Their Hosts

1. Most viruses infect and persist within their hosts in a state of **balanced pathogenicity** in which the host is not killed.
2. Viruses cause diseases that can be classified as **acute** or **persistent**.

Acute Infections

1. Acute infections are self-limited; the virus remains localized, and diseases are of short duration and lead to lasting immunity. (Figure 14.7)
2. The replication cycle of virulent animal virus that causes an acute infection is similar to the lytic cycle shown by phage T4. (Table 14.6, Figure 13.5)
3. The steps in the infection process include **attachment** to specific receptors; **penetration** of the entire virion (Figure 14.8); **transcription** of the viral genome into mRNA (Figure 14.9); **replication** of virus DNA and proteins; **assembly** of the virion (Figure 14.10); **release** of the virion so that it infects other cells in the vicinity.

Persistent Infections

1. In persistent infections, the virions are continually present in the body and are released from cells by budding. (Figure 14.11)
2. Persistent infections can be (1) **late complications following an acute infection,** (2) **latent infections** (Table 14.7) (3) **chronic infections** (Table 14.8) (4) **slow infections.** These categories are distinguished from one another largely by whether a virus can be detected in the body during the period of persistence. (Figure 14.12)
3. Infections with characteristics of more than one category are **complex infections.** An example is HIV, infection which has characteristics of a chronic, latent, and slow viral infection.
4. In HIV infection, the viral RNA is converted to double-stranded DNA by the viral enzyme **reverse transcriptase**. The DNA is then integrated into chromosomes of the host cell concerned with the immune response, where it can remain latent or give rise to intact virions.
5. Unlike DNA polymerase, the enzyme reverse transcriptase has no proofreading ability, and so the virion undergoes numerous uncorrected mutations.

Tumors—General Aspects

1. Two classes of host cell regulatory genes commonly involved in tumor formation are **proto-oncogenes** and **tumor-suppressor genes**.
2. Mutations in either of these gene classes predisposes the host to tumor formation. Mutations in proto-oncogenes are dominant; those in tumor-suppressor genes are recessive.
3. Proto-oncogenes are **transcriptional activators**. Tumor-suppressor genes code for a number of functions, such as **DNA repair** and **apoptosis**.
4. Viruses can alter the activity of proto-oncogenes and tumor-suppressor genes.

Viruses and Animal Tumors

1. Retroviruses are the most important tumor viruses in animals.

Tumor Viruses and Cell Transformation

1. **Oncogenes** of tumor viruses can modify the properties of cells growing in tissue culture. (Table 14.9) These genes are mutants of the host cell's proto-oncogenes. (Table 14.10)

How Retroviruses Transform Animal Cells

1. Retroviruses transform cells by inserting oncogenes into the genome of the host cell, where they interfere with the normal intracellular control functions of the cell's proto-oncogenes. (Figure 14.15)

Viruses and Human Tumors

1. The relatively few human tumors caused by viruses are primarily caused by double-stranded DNA tumor viruses. (Table 14.11)
2. Tumor formation is promoted by the interaction of products of the oncogenes with tumor-suppressor proteins.

Retroviruses and Human Tumors

1. A rare tumor, a leukemia, is caused by a retrovirus.

Viral Host Range

1. Most viruses can infect only a single species and only certain cells within an organism.
2. Some viruses, those causing **zoonoses**, can multiply in widely divergent species such as birds, mosquitoes, and humans.
3. Viruses can modify their host range if two viruses with different host ranges can infect the same cell.

Exchange of Protein Coats—Phenotypic Mixing (Figures 14.16, 14.17)

1. In phenotypic mixing, the genome of one virus may be surrounded by the coat of the other virus and the virus achieves the host range of the virus whose coat it has.
2. The genome of the resulting virus will code for its original coat, and therefore it reverts to its original host range after it infects the new cells.

Genome Exchange in Segmented Viruses (Figure 14.18)

1. **Segmented viruses** can expand their host range through an exchange of genomes following infection of the same cell by two viruses with different host ranges, the process of **genetic reassortment**.
2. This type of exchange likely accounts for the pandemics caused by the influenza virus every 30 years.

Plant Viruses

1. Many plant diseases are caused by viruses. (Figure 14.19)
2. Virions do not bind to receptor sites on plant cells, but enter through wound sites.

Spread of Plant Viruses

1. Many plant viruses are very resistant to inactivation.
2. Viruses are spread in large part by humans, by planting seeds in contaminated soils, through transfer from infected plants by grafting, and through the parasitic plant dodder, which can establish connections between an infected and uninfected plant. (Figure 14.21)
3. Various vectors, like insects and worms, can also spread virions.

Insect Transmission of Plant Viruses

1. Viruses may be associated with the external mouthparts of an insect or they may multiply within the insect.

Viruslike Agents

Prions (Figure 1.14, Table 14.2)

1. Prions consist of **protein** and no **nucleic acid**, and they have been linked to a number of fatal neurodegenerative diseases called **transmissible spongiform encephalopathies**.

Viroids (Figure 1.13)

1. Viroids are plant pathogens that consist of circular, single-stranded RNA molecules; they are about one-tenth the size of the smallest infectious viral RNA known.
2. Many unanswered questions remain regarding their origin, how they multiply, and how they cause disease.

REVIEW QUESTIONS

Short Answer

1. Name three differences between normal and transformed cells.
2. What are the substrate and product of reverse transcriptase?
3. What is the difference between an oncogene and a proto-oncogene?
4. What is the major mechanism for the spread of plant viruses from one plant to another?
5. Compare the mechanisms by which phages, and animal and plant viruses adsorb to their host cells.
6. What is the major difference between a latent and chronic infection in terms of the virion?
7. Name two DNA viruses that cause tumors in humans. Name one retrovirus.
8. Distinguish between a prion and a viroid. What hosts does each infect?
9. What family of animal viruses is most closely related to temperate phages in their interaction with host cells?
10. Define genetic reassortment. Explain how it can alter the infectivity of a virus.

Multiple Choice

Questions 1 to 7 concern the differences and similarities between animal, plant, and bacterial viruses. Answer each question based on the following possibilities by circling the correct letter:

A. Bacteriophages only
B. Animal viruses only
C. Plant viruses only
D. Bacteriophages and animal viruses
E. Animal and plant viruses

1. Bind to specific receptors on the host cell to initiate infection.
   A. B. C. D. E.
2. Only the nucleic acid enters the host cell.
   A. B. C. D. E.
3. Lipid envelopes are common.
   A. B. C. D. E.
4. Reverse transcriptase is involved in the replication cycle.
   A. B. C. D. E.
5. Integration of viral nucleic acid leads to changes in properties of the host.
   A. B. C. D. E.
6. Tumors are caused by these viruses.
   A. B. C. D. E.
7. Restriction endonucleases play a role in the host range of these viruses.
   A. B. C. D. E.
8. Prions
   A. contain only nucleic acid without a protein coat.
   B. replicate like HIV.
   C. integrate their nucleic acid into the host genome.
   D. cause diseases of humans.
   E. cause diseases of plants.
9. Viroids...
   1. contain only single-stranded RNA and no protein coat.
   2. use reverse transcriptase in their replication.
   3. are similar in structure to bacteriophages.
   4. cause diseases in animals.
   5. cause diseases in plants.
   A. 1,2 B. 2,3 C. 3,4 D. 4,5 E. 1,5
10. Acute infections in animals
    1. are a result of productive infection.
    2. generally lead to long-lasting immunity.
    3. result from integration of viral nucleic acid into the host.
    4. are usually followed by chronic infections.
    5. often lead to tumor formation.
    A. 1,2 B. 2,3 C. 3,4 D.4,5 E. 1,5

Applications

1. You are a scientist at a pharmaceutical company in charge of developing drugs against HIV. Discuss four possible targets for drugs that might be effective against this virus.
2. Researchers debate the evolutionary value of a virus's ability to cause disease. Many argue that viruses accidentally cause disease and only in animals that are not the natural host. They state that

this strategy may eventually prove fatal to the virus's future in that host. It is reasoned that the animals will eventually develop immune mechanisms to combat the virus and prevent its spread. Another group of researchers supports the view that disease is a way to enhance the survival of the virus. What rationale would this group use to support its view?

Critical Thinking

1. Would ID50 and LD50 necessarily be the same for a given virus? Why or why not?
2. The observation that viruses can agglutinate red blood cells suggests to some people that both viruses and red blood cells must have multiple binding sites. Is this a good argument? Why or why not?
3. An agricultural scientist is investigating ways to prevent viral infection of plants. Is preventing the specific attachment of the virus to its host cells a possible way to prevent infection? Why or why not?

CHAPTER 15

Nonspecific Immunity

During the latter part of the Nineteenth century, the studies of Louis Pasteur, Robert Koch, and others created much interest in microorganisms and the diseases that they caused. Once microorganisms were shown to cause disease, scientists worked to explain how the body defended itself against invasion by microorganisms. Elie Metchnikoff, a Russian-born scientist, theorized that there were specialized cells within the body that could destroy invading organisms. His ideas arose from observations that he made while studying the transparent immature larval form of starfish in Sicily in 1882. As he looked at the larvae in the microscope, he could see amebalike cells within their bodies. He described his observations as follows:

> *...I was observing the activity of the motile cells of a transparent larva, when a new thought suddenly dawned on me. It occurred to me that similar cells must function to protect the organism against harmful intruders.... I thought that if my guess was correct a splinter introduced into the larva of a starfish should soon be surrounded by motile cells much as can be observed in a man with a splinter in his finger. No sooner said than done. In the small garden of our home.... I took several rose thorns that I immediately introduced under the skin of some beautiful starfish larvae which were as transparent as water. Very nervous, I did not sleep during the night, as I was waiting for the results of my experiment. The next morning, very early, I found with joy that it had been successful.*

Metchnikoff reasoned that certain cells present in animals were responsible for ingesting and destroying foreign material. He called these cells ***phagocytes****, meaning "cells that eat," and he proposed that these cells were primarily responsible for the body's ability to destroy invading microorganisms.*

When Metchnikoff returned to Russia, he looked for a way to study the ingestion of materials by phagocytes, called ***phagocytosis****. The water flea* Daphnia *spp., which could be infected with a yeast, provided a vehicle for such studies. He observed phagocytes ingesting and destroying invading yeast cells within the experimentally infected, transparent water fleas. In 1884,*

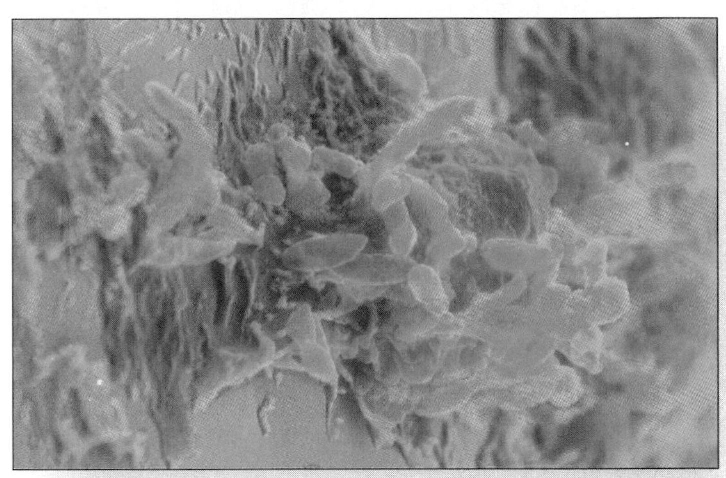

Metchnikoff published a paper that strongly supported his contention that phagocytic cells were primarily responsible for destroying disease-causing organisms. He spent the rest of his life studying phagocytosis and other biological phenomena; in 1908, he was awarded the Nobel Prize in Physiology or Medicine for these early studies of immunity.

—A Glimpse of History

FROM BEFORE ELIE METCHNIKOFF'S DISCOVERIES in the late Nineteenth century to the present day, scientists have been studying ways in which animals, especially humans and other vertebrates, protect themselves against invasion by foreign organisms or substances. The field of **immunology** is devoted to studying these many mechanisms of defense.

Although the science of immunology arose from studies of **immunity**, or protection, against infectious agents, it encompasses many other areas. Immune mechanisms operate not only against infectious agents, but also against cancers. They are also responsible for rejection of transplanted cells and organs. Under some conditions, immune responses may be directed against the cells of one's own body, causing **autoimmune** responses. Sometimes they cause damage as well as provide protection, thereby resulting in **allergic** or **hypersensitivity reactions**. When a host lacks effective immunological defense mechanisms, even the most innocuous microorganisms may become deadly **opportunists**, taking advantage of an opportunity to cause disease, as seen in the current AIDS epidemic.

Immune reactions can be used in the laboratory to identify substances with a high degree of specificity. For example, immunological techniques are used to type blood and tissue cells, diagnose diseases, classify bacteria, and identify suspects in the investigation of criminal cases.

The Host Defends Itself: An Overview

Invertebrates have many nonspecific defenses but cannot make a specific immune response; however, vertebrate hosts, including humans, have two lines of defense against foreign invaders. In addition to nonspecific defenses, vertebrates can make highly specific immune responses to materials their body recognizes as foreign.

Nonspecific Defense Mechanisms

When infection occurs, the first host defenses are those of innate, nonspecific immunity. **Nonspecific immunity** can be defined as immune responses that are not directed specifically to the infectious agent or other material involved, and these responses are not affected by prior exposures. Nonspecific immunity is usually innate or inborn, rather than occurring in response to external stimuli. It operates constantly to prevent the establishment of any infection. Nonspecific protective mechanisms include physical barriers such as the skin, mucous membranes, and secreted mucus that prevent organisms from entering internal tissues. The flushing action of the urinary tract washes out potential infecting microbes. Similarly, saliva flow removes many organisms from the oral cavity. In addition, enzymes and chemical antimicrobial factors are active in blood, saliva, mucus, milk, phagocytes, and other body sites.

A very important defense system is the nonspecific tissue response to injury called **inflammation**. This includes the release of chemical factors that attract phagocytes and other cells to the area of injury, where the phagocytes can ingest and destroy foreign material by the process of phagocytosis. Inflammation is nonspecific because it is directed against any foreign material or organism that tries to invade the host. As we shall see, many of the mechanisms of inflammation involve recognition of materials generally not found in the host, but they are not specific with regard to a particular invader. These nonspecific defense mechanisms are often found across the animal world. Recall that Metchnikoff observed his phagocytic cells in the invertebrate starfish and water fleas.

The body also has physiological defense mechanisms, such as the increase in internal body temperature called **fever**, which acts in several ways to discourage infections. The host's ability to restrict availability of iron also may limit microbial growth. Metabolic changes such as increases in protein and carbohydrate metabolism are essential in providing nutrients and energy for the body's response to infections.

Specific Defense Mechanisms

In addition to many nonspecific defense mechanisms, vertebrates have evolved a second line of defense, the acquired **specific immune response**, which depends on specialized cells called **lymphocytes**. This response is not innate, it is acquired in response to substances the body can recognize as foreign. The specific immune response is most highly developed in birds and mammals, which have efficient circulation through which cells and antimicrobial fluids can travel to all tissues of the body. Lymphocytes and glycoprotein molecules called **antibodies** or **immunoglobulins** respond specifically to organisms or other foreign materials called **antigens** or **immunogens**. When these same antigens are encountered again, there is an enhanced and antigen-specific immune response called the **memory response**. The specific immune response takes a week or more to develop and only becomes effective after the initial nonspecific reaction to the infection or injury. The lymphocytes of the specific response must multiply and mature extensively to be effective.

Cells and Tissues Involved in Host Defense

The nonspecific and specific arms of the immune response involve many of the same cells and molecules that communicate between these cells. The bloodstream and the lymphatic vessels are like an extensive highway system through which many cells and materials move from one part of the body to another. Cells important in body defense are always found in normal blood, but their numbers usually increase during acute bacterial infections. Fluids and some blood cells leave the bloodstream and enter the tissues through the vessels of the lymphatic system (see figure 15.4). In the lymphatic system, proteins and other materials from the tissues are removed from the fluid called lymph. The lymph is filtered through **lymphoid tissues and organs** (collections of lymphocytes and related cells) along the way, such as the many lymph nodes and the spleen, and it empties back into the blood circulatory system at a large vein usually behind the left collarbone. Many of the materials filtered from lymph are antigens that can interact with the immune system in lymph nodes and other lymphoid tissues.
■ blood and lymphatic systems, p. 716

Some of the cells important in body defense are mature as they circulate in the bloodstream, but others **differentiate**, that is, gain functional properties, after they leave the blood and enter the tissues. All blood cells originate from the same type of cell, the **hematopoietic stem cell**, found in the bone marrow. The word hematopoietic comes from Greek words that mean "bloodmaking." Under the influence of **colony-stimulating factors (CSFs)**, specific types of blood cells are produced. Red blood cells called erythrocytes carry oxygen in the blood. **Platelets** (also called thrombocytes; fragments arising from large cells called megakaryocytes) are important for blood clotting. White blood cells called **leukocytes** are important in immunity **(figure 15.1)**.

Leukocytes are the cells primarily responsible for the defense of the body against microorganisms, viruses, multicellular parasites, and even tumor cells. There are several subsets of leukocytes, each with special functions. They are the *granulocytes*, including eosinophils, basophils, and neutrophils; the *mononuclear phagocytes*, including monocytes, macrophages, and dendritic cells; and the *lymphocytes* **(table 15.1)**.

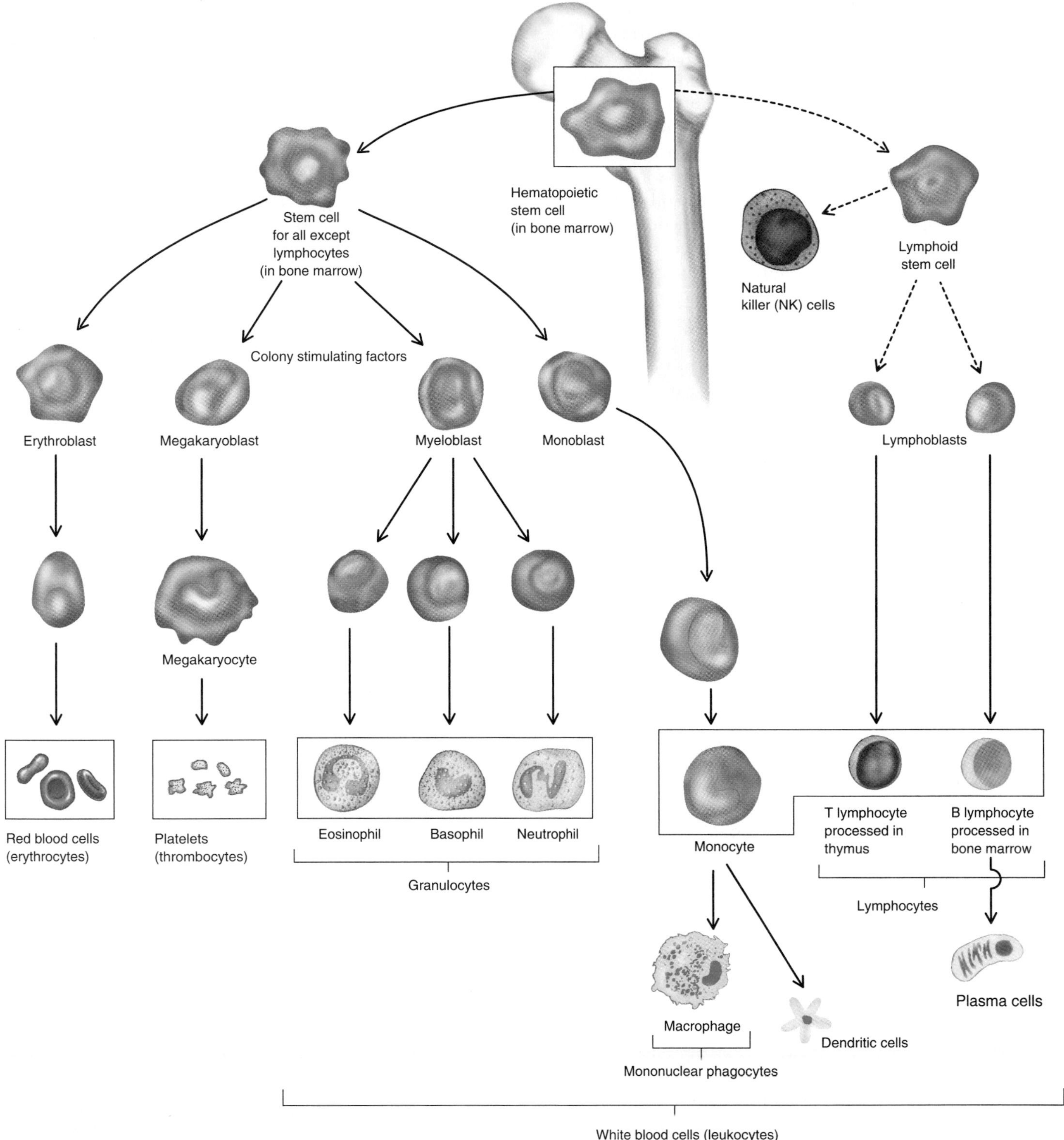

Figure 15.1 Blood and Lymphoid Cells All these types of cells are derived from precursor stem cells found in the bone marrow. Some of the steps not yet clearly defined are indicated by dotted arrows. Multiple steps occur between the stem cell and the final cells produced. The role of these cells in the immune response will be explained in the next two chapters.

Granulocytes

The **granulocytes** are classified as eosinophils, basophils, or neutrophils based on the staining properties of their prominent cytoplasmic granules. The granules contain biologically active chemicals that are important in the function of each cell type. **Eosinophil** granules are stained red by the acidic dye eosin. They contain antimicrobial substances and also histaminase, an enzyme that breaks down histamine. The number of eosinophils in the blood may increase during allergic reactions, and eosinophils are important in rejecting parasitic worms. The granules in **basophils** are stained dark purplish blue by the basic dye methylene blue. They contain **histamine** and other chemicals that increase blood flow during inflammation. **Mast cells** are similar in appearance and function to basophils of the blood, but they are found in virtually all tissues, rather than in blood, and they do not come from the same precursor cells as basophils. Mast cells are important in the inflammatory response and are responsible for many allergic reactions.

Neutrophils are actively phagocytic cells and are extremely important in the first encounter with infective agents. Their granules, which stain poorly, contain many antimicrobial substances and degradative enzymes that are essential for the destruction of engulfed materials. They are sometimes called **polymorphonuclear neutrophils, polys** or **PMNs** because they have a single nucleus with lobes. Neutrophils normally account for over 50% of circulating leukocytes, and their number increases during most acute bacterial infections. It is estimated that for every PMN in the circulation, 100 are in reserve in the bone marrow, ready to be mobilized when needed. Mature neutrophils are end cells that do not reproduce. They usually exist for about 10 hours in the circulation, and only 1 to 2 days in tissues after leaving the bloodstream. They contain a predetermined amount and variety of granules. Once the granules have been used, the cell dies.

Mononuclear Phagocytes

Although many types of cells in the body can phagocytize to a limited extent, polymorphonuclear neutrophils and mononuclear phagocytes, which include **monocytes** and **macrophages**, are the major phagocytic cells of the body. These are the so-called **professional phagocytes**, and they are highly efficient at phagocytizing and destroying foreign materials. Mononuclear phagocytes are found in virtually all parts of the body and constitute the **mononuclear phagocyte system (MPS)**. Monocytes are found in normal blood, where they represent about 3% to 8% of circulating leukocytes; they also migrate into tissues, where they develop into tissue macrophages. Tissue macrophages may live several weeks to several years. Within the tissues, macrophages develop or differentiate to perform various functions and, hence, take on slightly different appearances according to the tissue in which they are found. Although macrophages are found in virtually all tissues, they are present in especially large numbers in the liver, the spleen and lymph nodes, the lungs, and the peritoneal (abdominal) cavity **(figure 15.2)**.

TABLE 15.1 Human Leukocytes

Cell Type (% of Blood Leukocytes)	Morphology	Location in body	Functions
Granulocytes			
Neutrophils (polymorphonuclear neutrophils or PMNs, often called polys; 55%–65%)	Lobed nucleus; granules in cytoplasm; ameboid appearance	Account for most of the circulating leukocytes; few in tissues except during inflammation and in reserve locations	Phagocytize and digest engulfed materials
Eosinophils (2%–4%)	Large eosinophilic granules; nonsegmented or bilobed nucleus	Few in tissues except in certain types of inflammation and Ig E-mediated allergies	Participate in inflammatory reaction and immunity to some parasites
Basophils (0%–1%), mast cells	Lobed nucleus; large basophilic granules	Basophils in circulation; mast cells present in most tissues	Release histamine and other inflammation-causing chemicals from the granules
Mononuclear phagocytes			
Monocytes (3%–8%), macrophages, dendritic cells	Single nucleus; abundant cytoplasm	Macrophages present in all tissues and in lining of vessels; monocytes less-mature circulating forms	Phagocytize and digest engulfed materials; can participate in killing foreign cells that are engulfed; play vital role in development of specific immunity
Lymphocytes			
Several types (25%–35%)	Single nucleus; little cytoplasm before differentiation	In lymphoid tissues (such as lymph nodes, spleen, thymus, appendix, tonsils); also in circulation	Participate in specific immunological responses

Macrophages engulf and degrade foreign materials, but unlike neutrophils, macrophages can become activated and induced to make specific enzymes needed to break down engulfed particles and macromolecules **(figure 15.3)**. For example, when large amounts of fat are engulfed, lipase enzymes that degrade fat are made. If polysaccharides containing galactosides are ingested, galactosidase is produced to break down the large molecules. If these mechanisms of intracellular destruction fail and chronic infection ensues, large numbers of macrophages can fuse together to form giant cells that gather in collections called **granulomas** to retain and localize particularly resistant engulfed organisms, such as the bacteria causing tuberculosis or some of the fungi that cause persistent lung disease.

In addition to functioning in nonspecific immunity, *macrophages play key roles in specific immunity.* They can communicate with other cells, especially lymphocytes, via chemicals called cytokines. **Dendritic cells** are branched cells, some of which are related to macrophages. Both dendritic cells and macrophages are important antigen-presenting cells in specific immune responses.

Lymphocytes and Lymphoid Tissues

Lymphocytes are the cells primarily responsible for specific immune responses. In addition, some large, granular lymphocytes called **natural killer (NK) cells** act nonspecifically against abnormal cells. Although they are nonspecific with respect to antigen, NK cells can detect and interact with virus-infected cells and some other abnormal cells. They attach to target cells, release some of their granule contents, such as **perforin** (a hole-forming protein), close to the target cell membrane, and thereby kill the target.

Most lymphocytes found in normal blood are small, nondividing cells with only a little cytoplasm surrounding the nucleus. Lymphocytes circulate in the blood and lymphatic vessels to reach lymphoid tissues, where they are the principal cells of the **lymphoid system (figure 15.4)**.

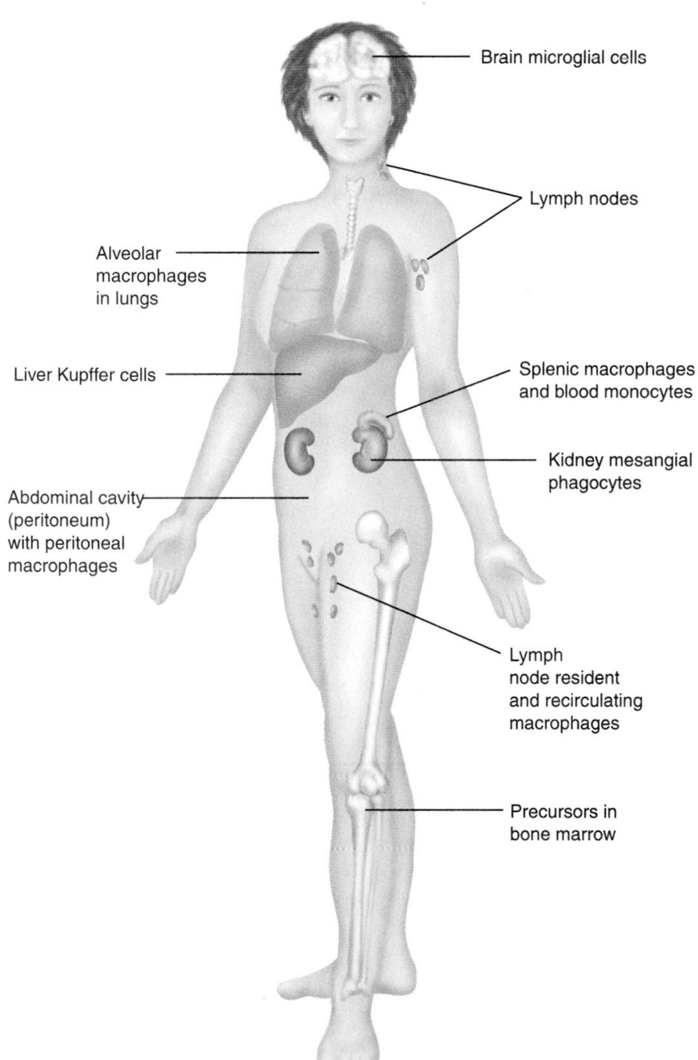

Figure 15.2 Monocytes and Macrophages Make Up the Mononuclear Phagocyte System, Formerly Known as the Reticuloendothelial System. Many of these cells have special names to denote their location—for example, Kupffer cells in the liver, and alveolar macrophages in the lung.

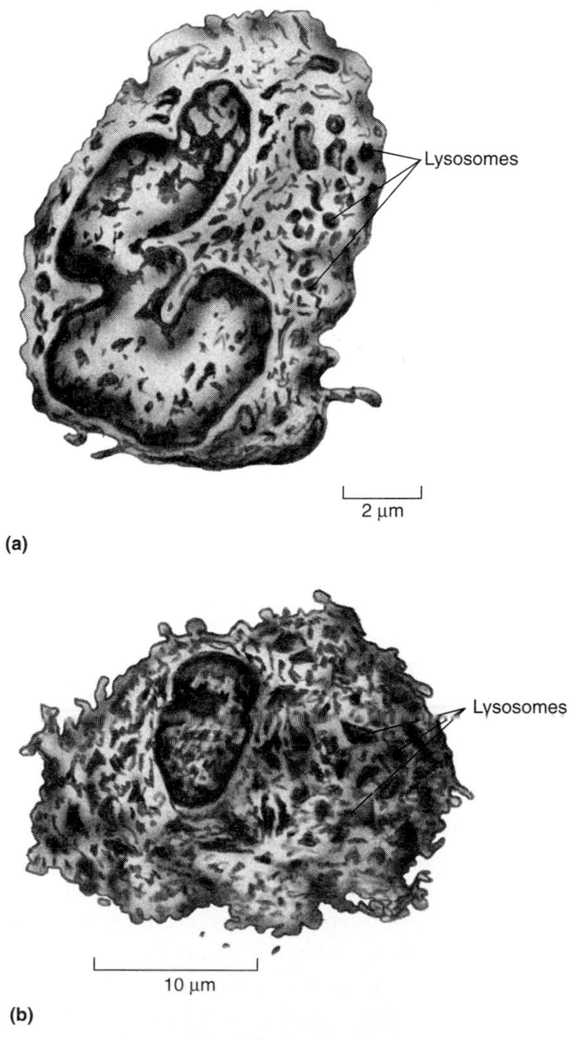

Figure 15.3 Normal and Activated Macrophages In a normal lung macrophage **(a)**, few phagocytic vesicles are present. In an activated lung macrophage **(b)**, there are abundant phagocytic vesicles and lysosomes (dark areas) in the cytoplasm.

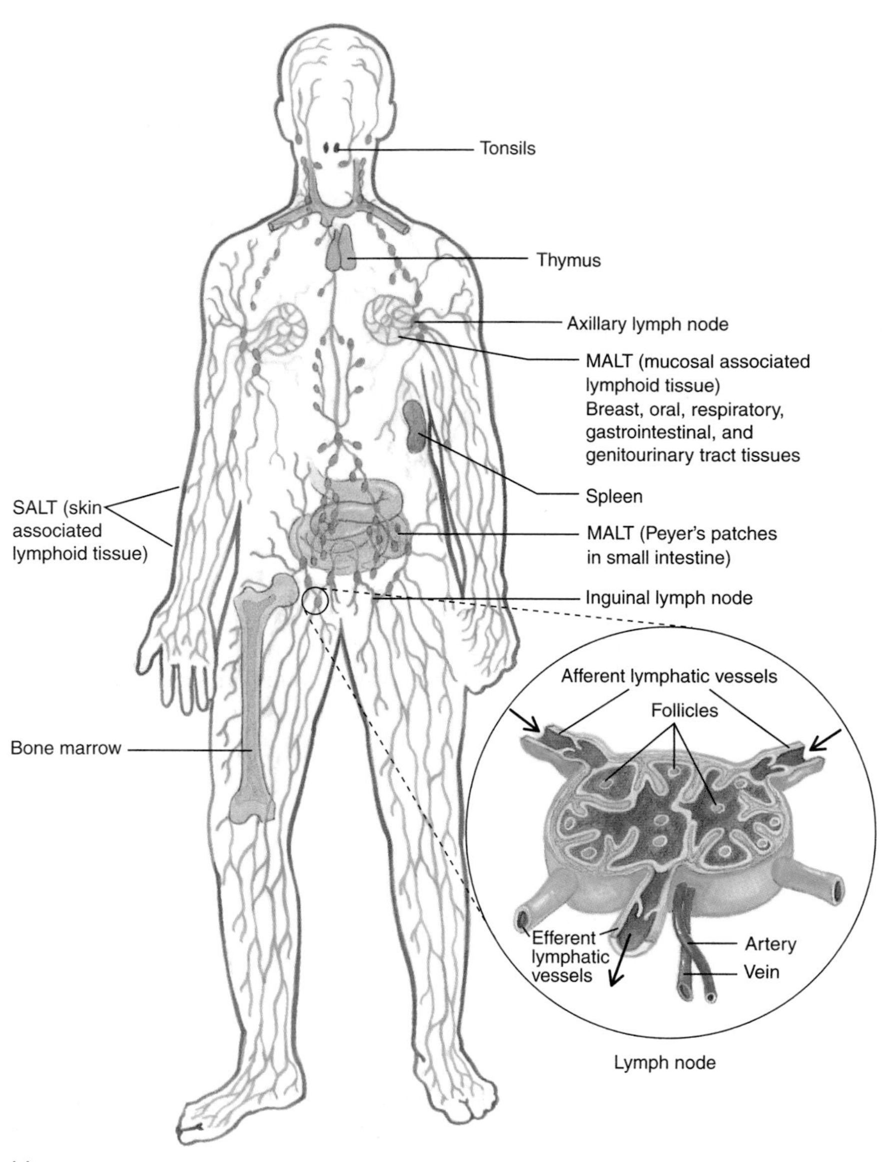

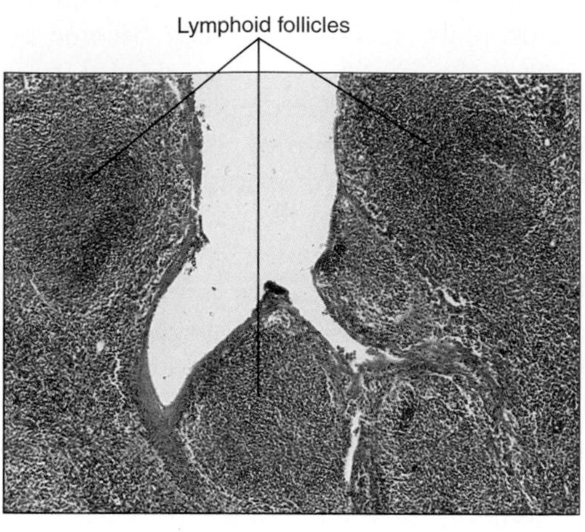

Figure 15.4 Distribution of the Lymphoid Tissues (a) Lymph is distributed through a system of lymphatic vessels, passing through many lymph nodes and lymphoid tissues. For example, lymph enters a lymph node (inset) through the afferent lymph vessels, percolates through and around the follicles in the node, and leaves through the efferent lymphatic vessels. The lymphoid follicles are the site of cellular interactions and extensive immunologic activity. **(b)** Light micrograph of a section through a human tonsil (12×). The tonsils are lymphoid tissues found in two small clumps at the back of the mouth. Note the circular lymphoid follicles.

The tissues of the lymphoid system are distributed at strategic locations throughout the body. The bone marrow and thymus are **primary lymphoid organs** in which lymphoid stem cells mature. Lymphocytes that mature in the **bone marrow** are called **B cells** and are the cells responsible for producing antibodies. Lymphocytes that mature in the **thymus** are called **T cells** and are responsible for specific cellular immune responses.

Secondary lymphoid organs include the adenoids, tonsils, spleen, appendix, and lymph nodes. Lymph nodes occur at frequent intervals along lymphatic vessels, filtering the draining lymph. The spleen filters materials from the blood. Other secondary lymphoid organs are the collections of lymphocytes that are plentiful under the skin, where they are called the **skin-associated lymphoid tissue (SALT)**, and those abundant beneath mucous membranes in the respiratory, genitourinary, and gastrointestinal tracts, where they are called the **mucosal-associated lymphoid tissue (MALT)**. Included in the MALT are Peyer's patches and other lymphoid tissue in the gut (called gut-associated lymphoid tissue, or GALT), lymphoid tissue in the bronchi (called bronchial-associated lymphoid tissue, or BALT), and lymphoid tissue in the breast.

Lymphoid tissues are situated so that defensive immune responses can be initiated at almost any location in the body. Secondary lymphoid tissues are areas where lymphocytes make contact with specific antigen and respond, both by proliferating to form clones of cells specific for that antigen and by producing memory cells. In the secondary lymphoid tissues, lymphocytes may become metabolically active and divide, and then they have larger nuclei and more abundant cytoplasm. The familiar lymph node ("gland") swelling that occurs in nodes draining a site of infection, as in the neck area during an episode of sore throat, represents the dividing lymphoid cells actively engaged in a specific immune response against the agent causing the infection. Lymphocytes will be discussed in more detail in the next chapter which examines specific immunity.

MICROCHECK 15.1

Nonspecific immunity is the body's first line of defense. Physical and chemical barriers, inflammation, and physiologic mechanisms such as fever contribute to early defense. Later, lymphocytes and antibodies provide a second line of defense, which is acquired and is highly specific and effective.

Hematopoietic stem cells in the bone marrow give rise to leukocytes active in both nonspecific and specific immune responses. The granulocytes are eosinophils, basophils, and neutrophils (PMNs). Monocytes and macrophages, along with the neutrophils, are the professional phagocytes. Lymphocytes are primarily responsible for specific immune responses; macrophages and dendritic cells also play essential roles. Natural killer cells are lymphocytes that kill target cells nonspecifically with respect to antigen.

- What is the mononuclear phagocyte system? Where is it found?
- How do nonspecific and specific immune responses differ? Which would you expect in mammals and which in insects?
- With respect to nonspecific immunity, what would the phrase "not affected by prior exposure" mean?

Physical Barriers and Antimicrobial Factors in Nonspecific Immunity

The importance of skin and mucous membranes as physical barriers to infection cannot be overemphasized. All exposed surfaces of the body are covered with epithelial cells (**figure 15.5**). These cells are packed tightly together and rest on a thin layer of fibrous material, the basement membrane. The parts of the body that are exposed to the outside world include not only the *skin*, which is in direct contact, but also the *mucous membranes* of the alimentary tract, the respiratory tract, and the genitourinary tract. Although they are inside the body, they are nevertheless exposed to the external environment through intake of food, water, and air.

Not only do these tissues form barriers, but both skin and mucous membranes are endowed with **antimicrobial secretions**. Sweat and other secretions of skin are quite acidic and inhibit the growth of most disease-producing bacteria (the pH of the skin varies between 3.5 and 5.8). Saliva, tears, and mucus produced by mucous membranes are rich in a variety of antimicrobial substances (**figure 15.6**), and their movements eliminate many organisms from mucous membranes. Hydrochloric acid creates a very low pH in the stomach, which kills most organisms, and the rapid pH change that occurs when stomach contents reach the upper intestine is also inhibitory to many microbes. The flow of urine through the urinary tract is important in maintaining sterility there, as witnessed by the frequent occurrence of infection when urinary flow is impeded by anesthesia or other means. To counteract these mechanisms for preventing infection, microorganisms must be able to attach firmly to the tissues in order to colonize successfully.

A large variety of nonspecific antimicrobial factors are found in body fluids and cells. A few of the important ones are listed in **table 15.2**. **Lysozyme** is found in serum, tears, and mucus, and in high concentrations in neutrophils and macrophages. It is released during inflammation—for example, when bacteria enter the body. Lysozyme degrades the peptidoglycan layer of the bacterial cell wall. Therefore, it is particularly effective against gram-positive bacteria, whose peptidoglycan is more likely to be exposed and accessible to the enzymes than is that of gram-negative bacteria.

Peroxidase enzymes, together with hydrogen peroxide and halide ions such as chloride and iodide, make up a very effective antimicrobial system. The peroxidase enzymes are found in neutrophil granules, saliva, and milk, and the peroxide is formed during oxygen metabolism, either by the host cell or by the invading organism. The interaction of hydrogen peroxide with peroxidase and chloride ions results in the formation of chlorine and hypochlorite, the active ingredient in bleach. It would seem that organisms would have no chance of surviving a system that produces these powerful chemicals. Bacteria that produce the enzyme catalase, however, may escape this mechanism of killing

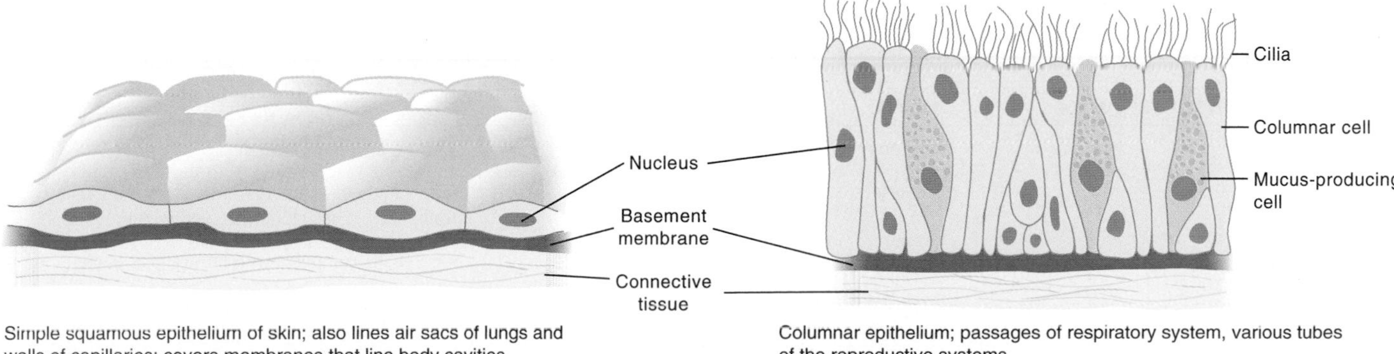

Figure 15.5 Epithelial Barriers Common features of epithelium are the tight junctions between cells that make entry of microorganisms virtually impossible under normal conditions and the presence of basement membranes that also discourage entry of materials through the epithelium. Note the cilia on some epithelial cells that function to sweep materials to areas where they are expelled or destroyed.

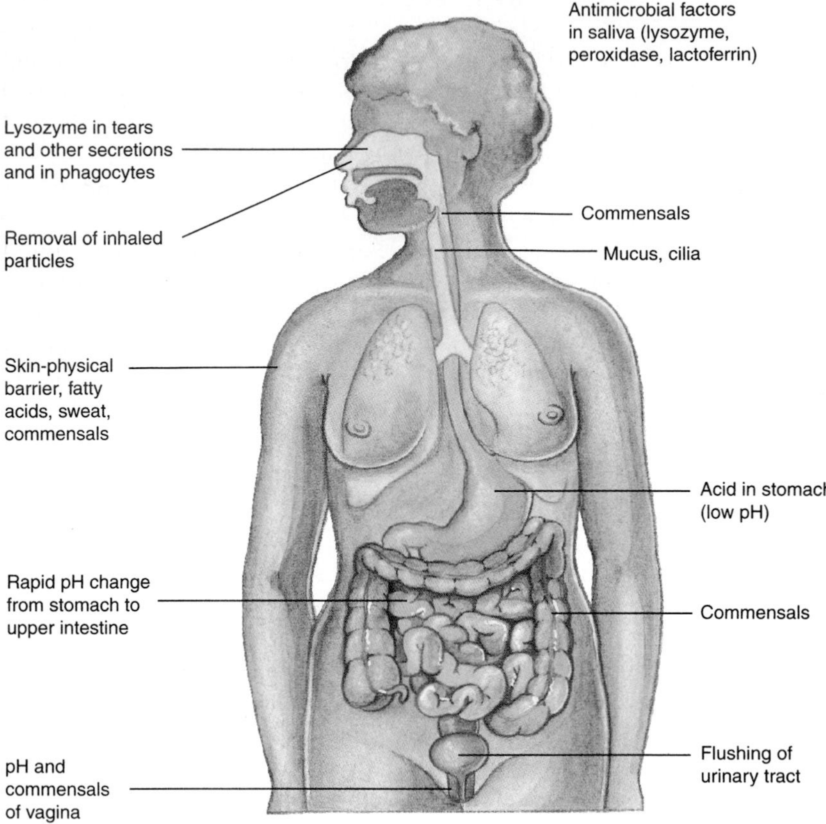

Figure 15.6 Nonspecific Defense Mechanisms in Humans Physical barriers, such as skin and mucous membranes, antimicrobial secretions, and commensals (normal flora organisms that are neither harmful nor beneficial to the host) work together to prevent entry of microorganisms into the host's tissues.

because the catalase breaks down the peroxide in a different manner than peroxidase does. Catalase-negative organisms are more sensitive to peroxidase killing. **Lactoferrin**, a protein found in milk, neutrophils, saliva, and mucus, binds iron firmly and prevents microorganisms from getting enough iron for growth.

The **normal flora** of the human body also protects against infection by competing with potential **pathogens**, or disease-producing organisms. Normal flora organisms are often **commensals**, which are neither harmful nor especially beneficial to the human host. For example, lactobacilli are harmless commensals that predominate in the normal flora of the vagina during the childbearing years, where they are responsible for the acid pH that discourages growth of many pathogens. ■ commensal organisms, p. 452

Nonspecific defenses work together to protect the body against infection. For example, in the respiratory tract the nose removes up to 90% of inhaled microscopic particles, such as bacteria, fungal spores, and dusts. Mucus in the nose is rich in lysozyme and other antimicrobial substances. The motion of ciliated cells in the upper respiratory tract, called the mucociliary escalator, propels the blanket of mucus away from the lungs and prevents the entry of most extraneous material into the lungs. Normal flora in the upper respiratory tract inhibits the growth of pathogens. These and other defense mechanisms are discussed more fully in the chapters dealing with each body system. In spite of all these protective defenses of the host, microorganisms are extremely clever in their abilities to evade protective mechanisms, as described in chapter 19.

MICROCHECK 15.2

Skin and mucous membranes present strong barriers to infection, aided in their effectiveness by the flushing action of secretions. Important antimicrobial factors in secretions, body fluids, and cells include lysozyme, the peroxidase system, and the iron-binding protein lactoferrin.

Competition by normal flora prevents infection with many potential pathogens.

- Name several defense mechanisms that help to protect the respiratory system from infection.
- How does the peroxidase system operate?
- How would normal flora inhibit growth of pathogens?
- What acid would the lactobacilli produce that would result in a low pH?

TABLE 15.2 Some Important Nonspecific Antimicrobial Factors

Antimicrobial or Antiviral Factor	Source	Effects
Lysozyme	Most body fluids; also within phagocytes	Destroys bacterial cell walls
Peroxidase enzymes	Leukocytes, saliva, milk	Kill a variety of microorganisms; important killing mechanism in saliva, in milk, and within neutrophils
Lactoferrin	Neutrophils, saliva, mucus, milk	Successfully competes with intracellular bacteria for iron
Complement system	Produced by macrophages and other host cells	Proteins acting in special sequence to produce effects such as chemotaxis, increased phagocytosis, and cell lysis
Interferons	Produced by leukocytes and tissue cells	Interfere with the multiplication of viruses by causing the formation of antiviral proteins

The Complement System

Complement is a reactant in the body's defenses, nonspecific because it can be activated by many different foreign invaders. It destroys invading bacteria and foreign cells by disrupting their cytoplasmic membranes. Complement is also involved in the inflammatory response by contributing to vascular permeability, stimulating chemotaxis, and enhancing phagocytosis.

Complement consists of a group of at least 26 different blood proteins. Some of these proteins are inactive enzymes, or proenzymes. Once activated, these act in sequence one after another in a cascade reaction. For example, one complement component is converted to an active enzyme that acts on the next component and converts it from a proenzyme to an active enzyme, often splitting off a biologically active small molecule that contributes to the inflammatory response. Each active enzyme molecule produced acts on many molecules of the next complement component, and so forth. The reaction is greatly amplified at each step. Other complement proteins act to regulate and control complement activity. The major components of complement are given a number along with the letter C for complement. Major components are C1 through C9. These were numbered in the order in which they were discovered and not the order in which they react.

Complement can be activated in two major ways, either by the classical pathway, which requires a specific immune reaction of antigen and antibody for activation, or by pathways that do not require a specific immune reaction. Nonspecific activation can occur by the alternative pathway or by the lectin pathway. The C3 molecule is the key component in the sequence of events occurring in any of these pathways.

Classical and Alternative Pathways of Complement Activation

In the **classical pathway**, antibody that is combined with antigen interacts with the inactive enzyme C1, making it an active enzyme (**figure 15.7**). This active enzyme then splits, or cleaves, C4 and C2, resulting in an enzyme called C3 convertase, which cleaves C3 into C3a and C3b. The antigen-antibody activation of a single C1 molecule can ultimately cause the cleavage of millions of C3 molecules, making this a very efficient mechanism ([1] in the figure).

In the nonspecific **alternative pathway**, endotoxin or polysaccharides, substances from the capsules and cell walls of microorganisms, react directly with a tiny amount of C3b that is constantly present in the blood (see [2] in figure 15.7). It is available because native C3 contains an unstable bond that for microseconds allows C3b to be split and active, before regulatory mechanisms inactivate it. This provides a constant trickle of a few C3b molecules to start the alternative pathway. Properdin and Factors B and D lead to production of C3 convertase, which cleaves C3 to C3a and C3b. It is estimated that one C3b molecule deposited on a microorganism can become 4 million in four minutes by the activity of the alternative pathway.

In the other nonspecific pathway, the **lectin pathway**, mannose in the pathogen's surface binds to a lectin, a protein that binds carbohydrates. This substance, called mannan-binding lectin (MBL), is found normally in blood but in increased amounts early in the inflammatory response. Mannose is present in many bacterial cell walls, but usually it is not present on mammalian cell membranes; thus, this pathway acts against many bacteria but spares host cells. The structure of MBL is similar to part of the C1 component of complement. The combination of MBL and mannose of the pathogen activates serum enzymes that in turn initiate complement activity.

The classical pathway remains the most efficient, because many more C3 molecules are involved in the reactions; however, the alternative and lectin pathways have the advantage of operating immediately without the need for a specific immune response. It is likely that the alternative and lectin pathways of complement activation evolved early to provide nonspecific protection against infections. Later, the antibody response developed and offered a specific, highly efficient means of complement activation. ■ **endotoxin, p. 461**

All pathways of complement activation follow the same sequence after cleavage of C3 (see [3] in the figure). The result may be (1) production of molecules that contribute to the inflammatory response, (2) opsonization, and/or (3) lysis of cell membranes.

When C3 is cleaved, C3b combines with other complement components to become an active enzyme that cleaves C5 into C5a and C5b. The small molecules *C3a* and *C5a*, which split off from C3 and C5 respectively, are very important in inflammation. They trigger the release from mast cells of a number of biologically active chemicals. These include factors that contribute to the vascular permeability of inflammation, and chemotactic factors that cause phagocytes to migrate into areas of inflammation.

C3a and C5a also directly stimulate the metabolic activity of phagocytes and cause the phagocytes to produce more receptors for C3b on their surfaces. The C3b formed when C3 is cleaved quickly attaches to the surface of the microorganism or other material that caused the cleavage. The C3b-coated material then adheres to the C3b receptors on phagocytes, which enhances phagocytosis of the material. An enhancement of phagocytosis by coating the surface of material to be ingested is called **opsonization**, and it can be caused in other ways besides C3b activity.

Following the cleavage of C5 to release C5a, C5b molecules on the surface of bacteria or foreign cells bind together loosely and form a complex with C6 and C7. This complex of C5b, C6, and C7 causes C7 to insert into the cell membrane; C8 joins the complex, creating changes in C9 such that it polymerizes and forms a **membrane attack complex (MAC)** that inserts through the membrane of eukaryotic cells, resulting in cell lysis (**figure 15.8**). The multilayers of peptidoglycan in Gram-positive bacteria may limit complement access to the cell membrane and thus interfere with lysis, but C3b still increases phagocytosis and intracellular destruction of these bacteria. Gram-negative bacteria are more susceptible to complement lysis than are Gram-positive organisms. Specific antibodies can

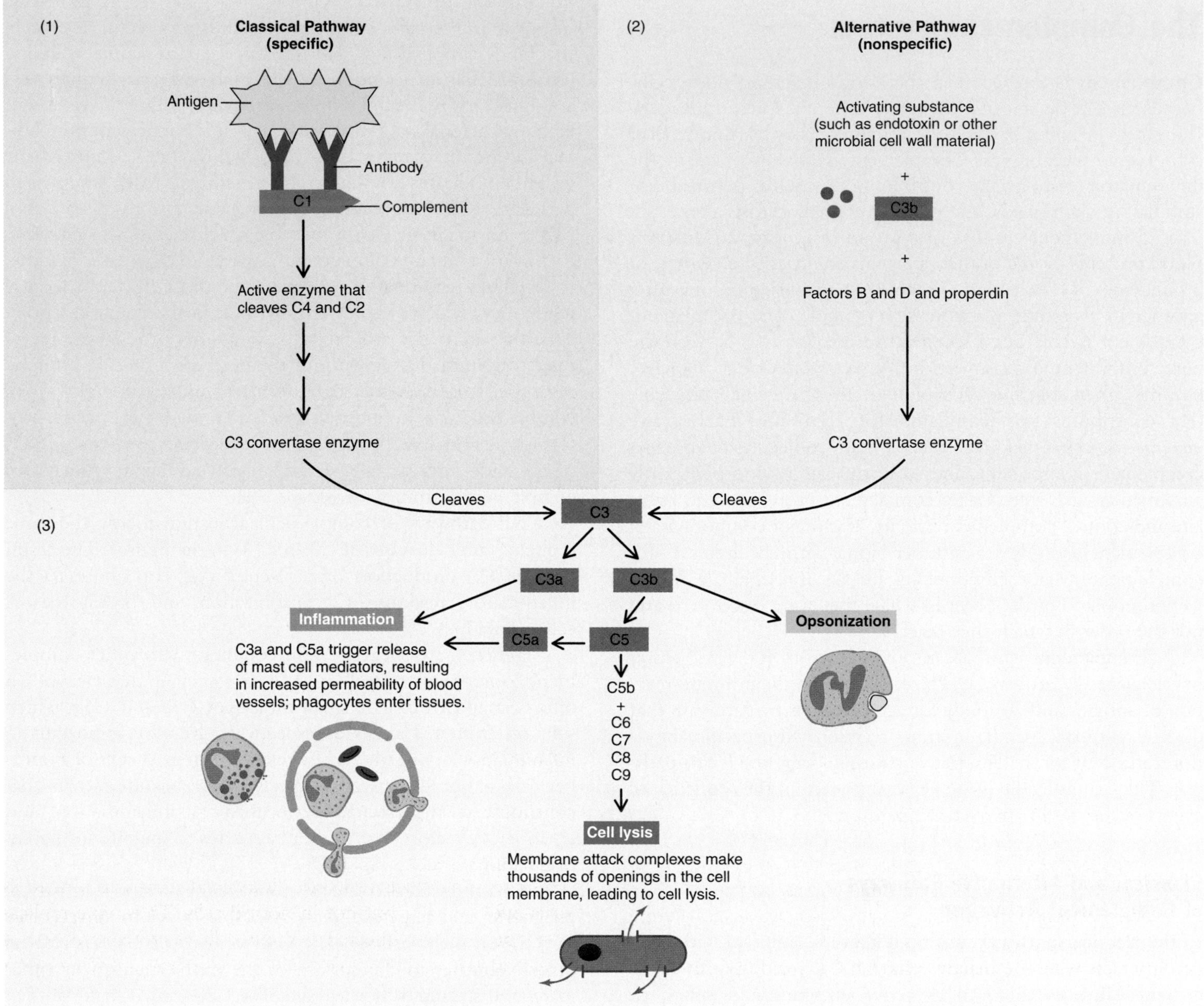

Figure 15.7 **Simplified Scheme of the Activation of Complement**

combine with components of the outer membrane of Gram-negative bacteria and thus activate complement, but lysozyme from phagocytes is required to completely lyse these cells.

As would be expected, stringent mechanisms operate to control the complement system at various points under normal circumstances. Should these controls not function, disease results. ■ **complement abnormalities, p. 444**

MICROCHECK 15.3

Complement is a group of blood proteins that act in sequence. Some complement components are proenzymes that become enzymes acting on the next component, and so on, in a cascading fashion, leading to an amplified effect. The net result may be (1) production of molecules that contribute to the inflammatory response, (2) opsonization, and/or (3) lysis of cell membranes. Complement is activated by a specific antigen-antibody reaction in the classical pathway, or nonspecifically by bacterial products in the alternative pathway or the lectin pathway.

- How could complement cause lysis of foreign red blood cells?
- Could the complement system be activated by injection of a material? How?

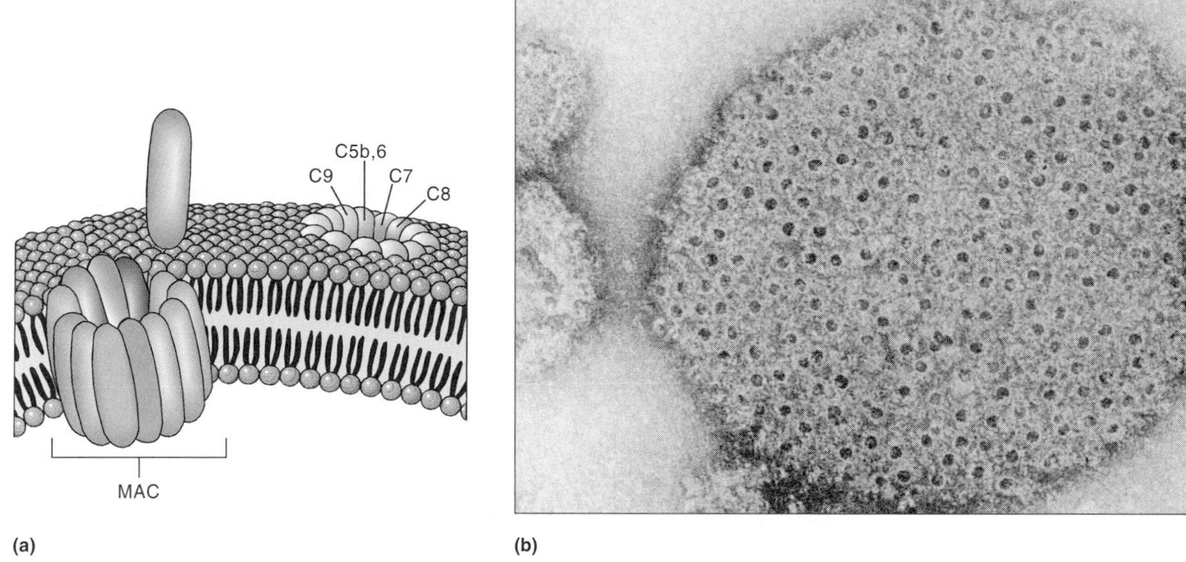

Figure 15.8 Membrane Attack Complex of Complement (MAC) **(a)** The MAC is formed after C5b, C6, and C7 combine into a complex on the cell surface of bacteria or other foreign cells. This complex, together with C8, causes changes in C9, allowing it to polymerize with the complex and form a MAC. The MAC inserts through the membrane, resulting in lysis of the cell. **(b)** An electron micrograph of MAC

Cytokines

Cytokines are low molecular weight regulatory proteins made by cells that affect the behavior of other cells **(table 15.3)**. They are essential for communication between cells. They bind to cytokine receptors to induce new cellular activities, such as growth, differentiation, or death. Cytokines include interferons, interleukins, and colony-stimulating factors, among others. Cytokines are very powerful, acting at extremely low concentrations and usually at short range. They often act together or in sequence, in complex fashion, and their actions are short-lived.

Interferons

Interferons are antiviral glycoproteins. Three types are known. One type, interferon alpha, is a family of closely related proteins produced by various white blood cells, and aside from its antiviral activity it also contributes to fever production. A second type, interferon beta, is made by cells of the fibrous supporting tissue called fibroblasts. Both alpha and beta interferons are made by many cell types upon their infection with viruses. A third type, interferon gamma, is made by lymphocytes. In addition to being antiviral, interferon gamma is very important in activating macrophages and functions in the development and regulation of the specific immune response.

The antiviral effects of all types of interferon are of value very early during a viral infection, before antibodies or immune lymphocytes are produced. Interferon is made soon after the start of a viral infection, reaches a peak amount at about 3 days, and then decreases to almost nothing by a week after infection. It is nonspecific with respect to the virus; thus, it protects against most viral infections.

Virus infection of a cell results in the accumulation of double-stranded RNA, which signals the cell to produce interferon **(figure 15.9)**. The molecules of interferon act on nearby cells to prevent replication of the virus within those cells. First, the interferon molecules attach to receptors on the neighboring cell surface. Second, the cell responds by activating genes responsible for producing enzymes that degrade messenger RNA and inhibit protein synthesis. The action of these enzymes requires the presence of double-stranded RNA that is formed in virus-infected cells, but not in normal cells. Thus, viral replication is prevented in infected cells, but cellular protein synthesis is also inhibited. Other mechanisms of interferon activity against viruses are under study. ■ viral replication, p. 350

Clearly, it is of value to be able to use interferon in the treatment or prevention of viral infections, but for a time this was not possible because interferons, although nonspecific against viruses, are quite species-specific with regard to the host. Many other biologically active substances can be made in animal species and still be active in human cells, but only human interferons are effective in human cells. This problem has been overcome by using genetic engineering techniques. Human genes for interferons have been inserted into bacteria and yeast cells, which are then grown in huge quantities to produce large amounts of human interferons. These interferons can then be purified and used therapeutically, not just against virus infections but also against cancers and in various other ways. Their use is limited because of unpleasant and sometimes dangerous side effects. Nevertheless, interferon treatment is useful for controlling life-threatening serious viral diseases where the benefits outweigh the deleterious side effects. Interferon has been approved in the United States for treatment of Kaposi's sarcoma in AIDS patients, chronic hepatitis B, hepatitis

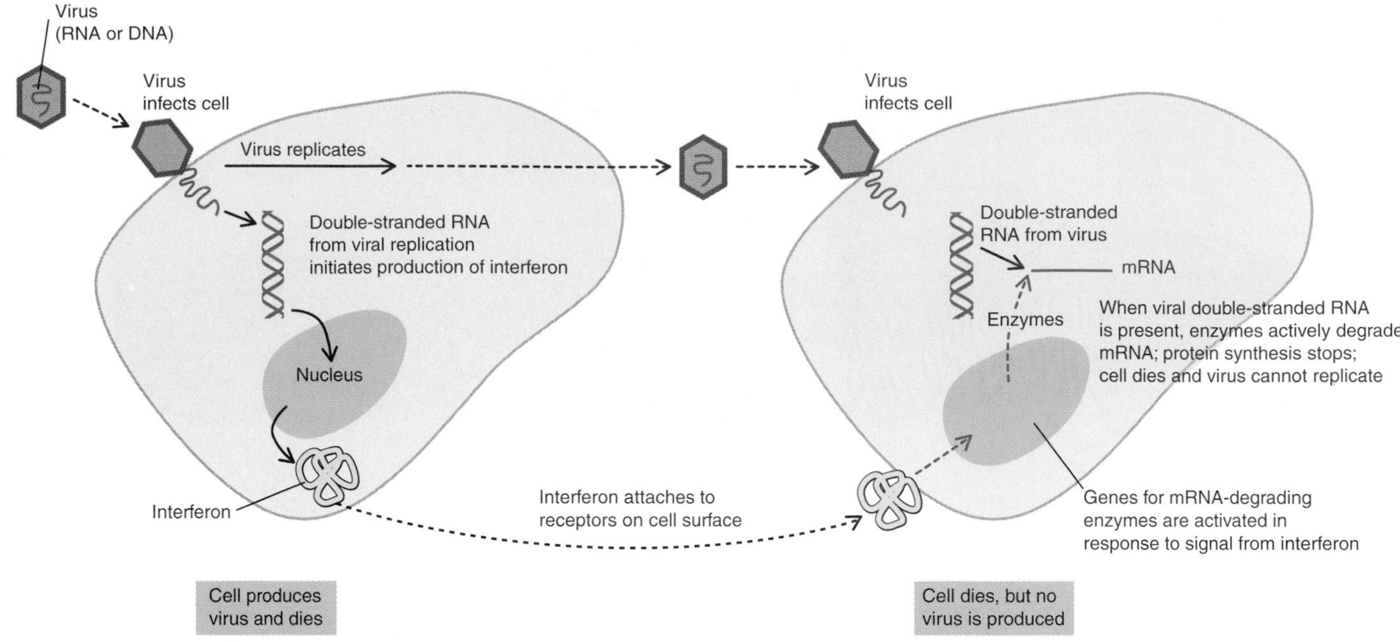

Figure 15.9 Mechanism of the Antiviral Activity of Interferons Interferons are small glycoproteins made by virus-infected cells that act on nearby cells, causing them to produce antiviral proteins. These proteins are enzymes that inhibit virus replication in various ways.

C infections, and several other diseases. Although some success has been achieved in treating a few kinds of cancer with interferon, recurrences of the tumors are common.

Interleukins

Interleukins (IL) are cytokines produced by leukocytes. They contribute to the inflammatory response and to the specific immune response, and they function in many other ways. At least 18 interleukins have been studied. A few examples are mentioned in table 15.3.

Interleukin-1 induces fever, and has many effects on the local area of inflammation. It signals the release of neutrophils from the bone marrow in large numbers; the cells then bind to the walls of blood vessels in the area of injury and migrate into local tissues. IL-1 causes macrophages to produce other cytokines, and it induces the production of acute-phase proteins. IL-2 and IL-3 function more in specific immune responses. Activities of the interleukins often overlap.

Colony-Stimulating Factors

Colony-stimulating factors (CSFs) are important in the multiplication and differentiation of leukocytes. During the immune response when more leukocytes are needed, a variety of colony-stimulating factors direct immature cells into the appropriate maturation pathways.

Tumor Necrosis Factors

Tumor necrosis factor (TNF) alpha, aside from killing some tumor cells, acts in a variety of ways to influence immune responses. For example, TNF-alpha is stored in mast cell granules and released during inflammation. It recruits increased numbers of neutrophils into sites of inflammation, and it can be antiviral. It is also important in inducing fever. TNF-beta is made by lymphocytes and functions in killing target cells.

Functional Groups of Cytokines

Cytokines can be grouped according to their major functions. Some are particularly important in inflammation, some in promoting antibody responses, and others in cell-mediated immune responses. The cytokines contributing to inflammation include TNF-alpha, IL-1, and IL-6, among others. These cause the liver to make an **acute-phase response**, releasing proteins important in early inflammation, including the lectin (MBL) that can nonspecifically initiate complement activation. C-reactive protein is another acute-phase substance; it can opsonize some bacteria and fungi, and also interact with complement. Cytokines in this group also lead to the release of neutrophils from the bone marrow, they cause metabolic changes and fever, and they stimulate cells that will participate in specific immune responses. Cytokines especially involved in promoting antibody responses include IL-4, IL-5, IL-10, and IL-14. Those that promote specific cell-mediated immune responses include IL-2, and interferon gamma, among others.

We will return to the cytokines later as we discuss various facets of the immune response.

MICROCHECK 15.4

Cytokines are small regulatory proteins essential for communication between cells. They include the interferons that are antiviral and have other important effects, the interleukins, colony-stimulating factors, and tumor necrosis factors. Cytokines have many

TABLE 15.3 Some Important Cytokines

Cytokine	Source	Effects
Interferons		
Interferon alpha	Leukocytes	Antiviral; induces fever; contributes to inflammation
Interferon beta	Fibroblasts	Antiviral
Interferon gamma	T lymphocytes	Antiviral; macrophage activation; development and regulation of specific immune responses
Interleukins (ILs)		
IL–1	Macrophages, epithelial cells	Proliferation of lymphocytes; macrophage production of cytokines, induce receptors for PMNs on blood vessel cells; induce fever
IL–2 (T-cell growth factor)	T lymphocytes	Changes in growth of lymphocytes; activation of natural killer cells; promote specific cell-mediated immune responses
IL–3	T lymphocytes, mast cells	Changes in growth of precursors of blood cells and also of mast cells
IL–4, IL–5, IL–10, IL–14	T lymphocytes, mast cells, other cells	Promote antibody responses
IL–6	T lymphocytes, macrophages	T- and B-cell growth; production of acute-phase proteins; fever
Colony-Stimulating Factors (CSFs)	Fibroblasts, endothelium, other cells	Stimulation of growth and differentiation of different kinds of leukocytes
Tumor Necrosis Factors (TNFs)		
Alpha	Macrophages, T lymphocytes, other cell types, mast cell granules	Cytotoxicity for some tumor cells; regulation of certain immune functions; induce fever; chemotactic for granulocytes
Beta	T lymphocytes	Killing of target cells by T cytotoxic cells and natural killer (NK) cells

actions in the inflammatory response and they contribute to fever production and to development of specific immune responses.

- List at least three actions of interferons.
- Name three interleukins that can contribute to fever production.

Phagocytosis

Phagocytosis involves a series of complex steps necessary for phagocytes to engulf and kill invading microorganisms (**figure 15.10**). First, the phagocyte must be attracted to the microorganism or virus. Certain chemical products of microorganisms, components of the complement system (C5a), and phospholipids released by injured mammalian cell membranes act as chemoattractants to the phagocytes, which then move toward the microorganism by a process called **chemotaxis**.

A second step is **attachment** (see chapter opening photo). Once the phagocytic cell comes into contact with the microorganism, it must attach to it in order to engulf it efficiently. The C3b component of complement coats bacteria or other particles, allowing them to be attached to phagocytes via C3b receptors on the phagocytic cells. This process of coating to enhance phagocytosis is called opsonization. After antibodies have been produced, they also cause opsonization. Capsules can greatly inhibit attachment of organisms to phagocytes; the encapsulated organisms are not phagocytized until they are opsonized by antibody or complement and can attach.

Third, the microorganism is engulfed by the phagocyte into a vacuole known as a **phagosome**. Other membrane-bound bodies within the phagocyte (called granules in neutrophils, and lysosomes in mononuclear phagocytes) contain a variety of digestive enzymes including lysozyme and proteases. Fourth, the granules or lysosomes come into contact with and then fuse with the phagosome and release their enzymes into the vacuole resulting from this fusion, known as a **phagolysosome**.

Fifth, within the phagolysosome, organisms are killed, and sixth, they are digested. Oxygen consumption increases enormously as sugars are oxidized via the TCA cycle, with the production of highly toxic oxygen products such as superoxide anions, hydrogen peroxide, singlet oxygen, and hydroxyl radicals. In mouse leukocytes, toxic products of nitrogen metabolism, such as nitric oxide, are also produced and are highly antimicrobial; these products probably function in human leukocytes, as well. As the oxygen level declines, the metabolic pathway switches

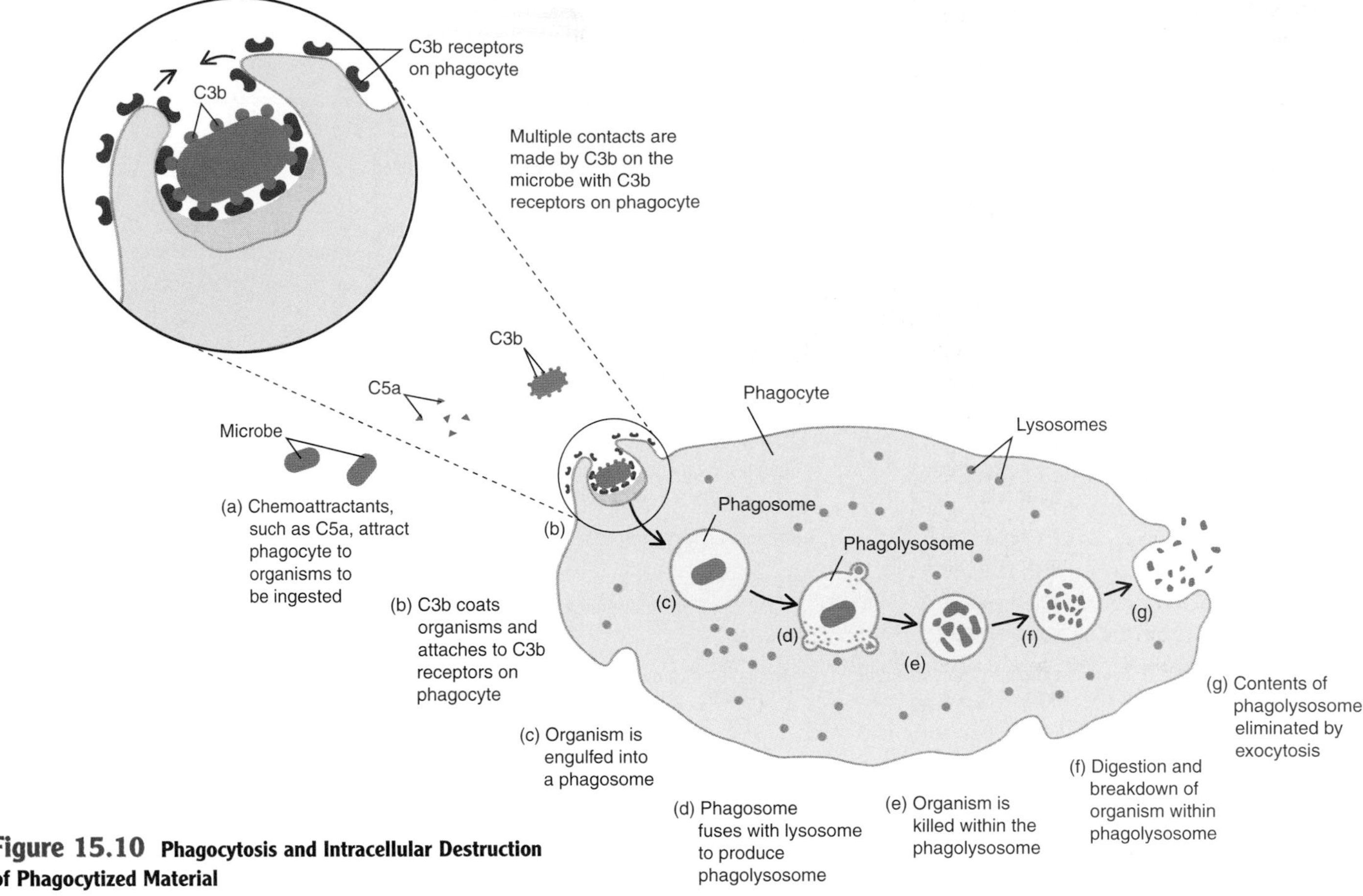

Figure 15.10 Phagocytosis and Intracellular Destruction of Phagocytized Material

to fermentation with the production of lactic acid, which lowers the pH. Within the phagolysosome, various enzymes degrade the peptidoglycan of the bacterial cell walls and other macromolecular components of the cell. ■ **TCA cycle, p. 148**

Seventh, and finally, following the digestion of the microorganisms, the contents of the phagolysosome are eliminated by **exocytosis**. As mentioned previously, once neutrophils have used their granules, they do not make more and the cells die, but mononuclear phagocytes can regenerate their lysosomes in order to meet the challenge of another bacterial attack. Surprisingly, some microorganisms are not only able to resist all these defenses, but they can actually multiply within mononuclear phagocytes. This ability will be discussed in more detail later. ■ **intracellular microorganisms, p. 454**

MICROCHECK 15.5

Phagocytosis occurs in several steps: (1) chemotaxis, (2) attachment, (3) ingestion, (4) fusion to produce phagolysosomes, (5) killing within phagolysosomes, (6) breakdown of dead materials, and (7) exocytosis of the resulting debris. Some organisms resist killing and continue to live within mononuclear phagocytes.

- What are the steps in phagocytosis that lead to destruction of bacteria?
- What are some differences between phagocytosis by neutrophils and that by macrophages?
- During the digestion of ingested materials within the phagolysosome, the metabolic pathways switch to fermentation. Why should this be?

Inflammation

Once an infectious agent has penetrated external barriers such as the skin or mucous membranes and has entered the tissues, the first host response is a nonspecific reaction to injury called the inflammatory response, or inflammation. Everyone is familiar with the signs of inflammation; in fact, the four cardinal signs were described by the Roman physician Celsus in the first century A.D. They are swelling, redness, heat, and pain. A fifth sign, loss of function, is sometimes present.

The same sequence of events occurs in response to any injury, whether caused by invading bacteria, burns, trauma, or other stimuli. **Figure 15.11** and **table 15.4** give the general sequence of events that take place during inflammation. Very early, cytokines, C3a and C5a components of complement, and other substances cause the release of chemical mediators from tissue mast cell granules **(table 15.5)**. These chemicals, in turn, dilate

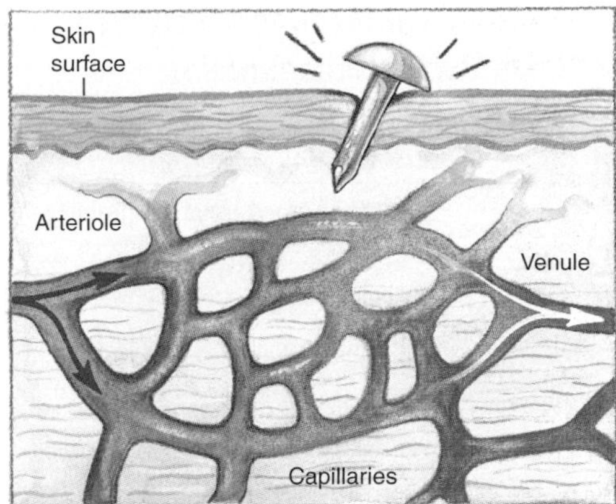

(a) Normal blood flow in the tissues as injury occurs.

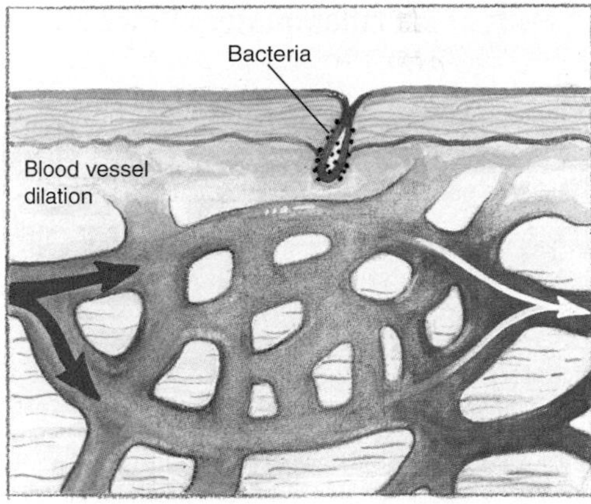

(b) Substances released from injured mast cells and as a result of the inflammatory reaction cause dilation of small blood vessels and increased blood flow in the immediate area.

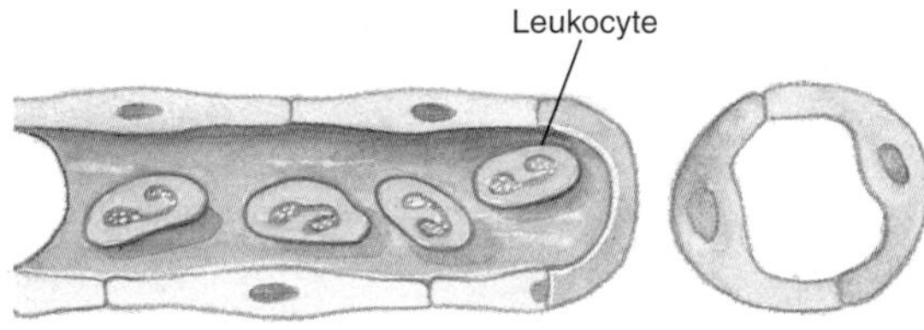

(c) Small blood vessels normally have tight junctions between the lining cells and blood stays within the vessels.

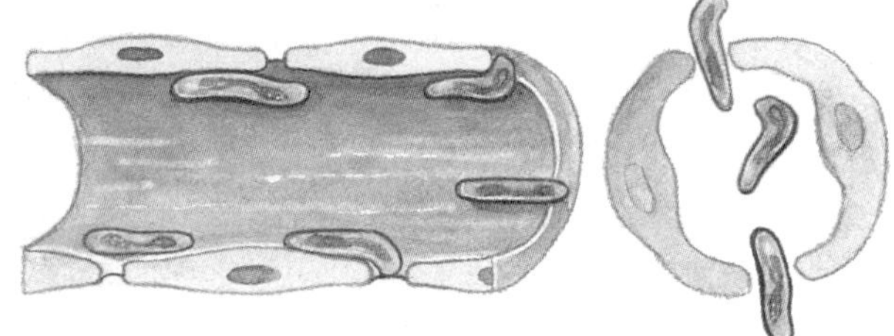

(d) During inflammation, the dilation of vessels makes openings between cells lining the vessels. Leukocytes attach to the altered vessels and migrate through the openings between the cells into surrounding tissues. Some substances released during inflammation are chemotactic and attract certain leukocytes to the area of inflammation.

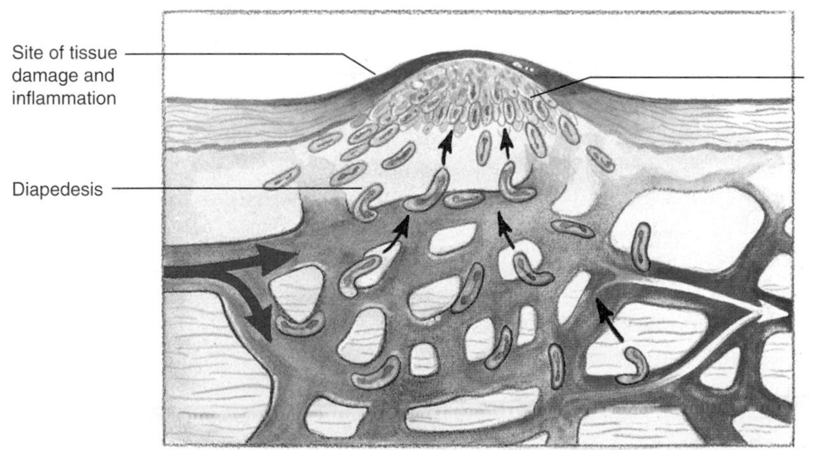

(e) The attraction of leukocytes causes them to move to the site of damage and inflammation. Collections of dead leukocytes and tissue debris make up the pus often found at sites of an active inflammatory response.

Figure 15.11 The Inflammatory Response

small blood vessels, making them permeable. Some of the chemicals cause the production of receptors for leukocytes on blood vessel walls in the area of the inflammation. Circulating leukocytes adhere to these receptors on the inner walls of the altered blood vessels and then migrate through the dilated permeable vessel walls by the process called **diapedesis** into the tissues in response to chemical attractants. This response is termed chemotaxis. Plasma leaks into the tissues, accounting for the **swelling** (1) noted in inflammation, but also delivering antimicrobial substances and possibly diluting any toxic materials at the site. The **heat** (2) and **redness** (3) in the region result from increased blood flow through the dilated vessels. **Pain** (4) is caused by the increased fluid in the tissues and by direct effects of chemicals on sensory nerve endings. ■ **chemotaxis, p. 68**

TABLE 15.4 Principal Events of the Inflammatory Process

Event	Effects
Tissue injury	Release of chemical mediators that increase the permeability of adjacent small blood vessels
Blood vessel dilation and increased permeability to plasma, which may clot	**1.** Swelling of the tissues resulting from the leakage of plasma; **2.** elevated temperature of the region as a result of increased blood flow through the dilated vessels; **3.** redness for the same reason; **4.** pain from increased fluid in the tissues and from direct effect of chemicals on sensory nerve endings
Circulating white blood cells adhere to the walls of the altered blood vessels	White blood cell chemotactic migration through the vessel walls and to the area of injury to induce phagocytosis of foreign material and tissue debris and to initiate antibody production

TABLE 15.5 Chemical Mediators Released from Mast Cells During Inflammation

Chemical	Effects
Histamine	Dilation and increase of permeability of small blood vessels; constriction of the bronchi
Chemotactic factors	Eosinophil and neutrophil (PMN) chemotaxis
Interleukins 3, 4, 5, and 6	Many interactions
Tumor necrosis factor alpha	Recruitment of granulocytes to area of inflammation; inducement of fever
Leukotrienes	Dilation of small blood vessels; constriction of bronchi; chemotaxis of leukocytes
Prostaglandins	Increase in vascular permeability; regulation of immune responses

The first kind of leukocyte to be lured from the circulation is the neutrophil (PMN). Soon after injury, neutrophils predominate in the area of inflammation. After the influx of neutrophils, monocytes and some lymphocytes accumulate in the area. Both the neutrophils and the monocytes (which mature into macrophages at the site of infection) actively phagocytize foreign materials. Clotting factors are in the fluid that leaks into the tissues, and some clotting occurs. This helps to prevent further bleeding and to halt dissemination of invading organisms that get caught in the clot. As the inflammatory process continues, large quantities of dead neutrophils accumulate. Along with tissue debris, these dead cells make up pus. A large amount of pus constitutes a **boil** or **abcess**.

As inflammation subsides, healing occurs. During healing, new capillary blood vessels grow into the area and destroyed tissues are replaced, producing a reddish material called **granulation tissue**. Eventually, scar tissue is formed. The extent of inflammation varies, depending on the nature of the injury, but the response is local, begins immediately upon injury, and increases in intensity usually over a short period of time. When acute inflammation cannot limit the infection, chronic (long-lasting) inflammation occurs and can last for years. An attempt is made to wall off the invaders with macrophages and giant cells in collections called **granulomas**. The granulomatous response is important in many chronic diseases, such as tuberculosis, syphilis, and some fungal infections. ■ **granulomatous response in respiratory tract infections, p. 570**

MICROCHECK 15.6

Inflammation is a nonspecific tissue response to injury of any type, characterized by swelling, redness, heat, and pain. Upon injury, cytokines are released that lead to local dilation of local small blood vessels, chemotaxis of leukocytes, and leakage of fluids into the tissues. Acute inflammation is local, begins immediately upon injury, and increases in intensity over a short time. Chronic inflammation involves the formation of granulomas and may last for years.

- What are the four cardinal signs of inflammation and what causes each one?
- What role do mast cells play in inflammation?

Physiological Changes and Immune Responses

The host's metabolism is profoundly altered in response to infection, and most of the changes are geared to meet the needs of host defenses. The most significant changes include an alteration of temperature regulation leading to fever, redistribution of iron in the body, and changes in protein and carbohydrate metabolism.

Fever

Fever is one of the strongest indications of infectious disease, especially bacterial, although significant infections can occur without it. There is abundant evidence that fever is an important host defense mechanism in a number of vertebrates, including humans. Within the human body, the temperature is normally kept within a narrow range around 37° C by a temperature-regulating center in the hypothalamus of the brain. The hypothalamus controls temperature by regulating the blood flow to the skin, the amount of sweating, and respiration. In this way, the heat produced by metabolism is conserved or lost in order to maintain a fairly constant temperature. In infection, the regulating center continues to function, but the body's thermostat is "set" at higher levels. An oral temperature above 37.8° C is regarded as fever.

A higher setting occurs because certain components of invading microorganisms attach to receptors on phagocytic cells, thereby stimulating these cells to release interleukin-1 (IL-1), tumor necrosis factor alpha, and interferon alpha, cytokines that can induce fever. The cytokines are carried by the bloodstream to the hypothalamus, where they act as messages to the hypothalamus that microorganisms have invaded the body. These cytokines and other fever-inducing substances are called **pyrogens** from the Greek roots pyr ("fire") and gennan ("to produce" or "engender"). Fever-inducing cytokines are **endogenous pyrogens**, made by the body, whereas microbial products, such as bacterial endotoxins, are **exogenous pyrogens**, introduced by external sources. The temperature-regulating center responds to pyrogens by raising body temperature. The resulting fever inhibits the growth of many pathogens by at least two mechanisms: (1) elevating the temperature above the optimum growth temperature of the pathogen, and (2) activating and speeding up a number of other body defenses.

The adverse effect of fever on pathogens correlates in part with their ideal growth temperature. Bacteria that grow best at 37°C are less likely to cause disease in people with fever. The growth rate of bacteria often declines sharply as the temperature rises above their optimum growth temperature. A slower growth rate allows more time for other host defenses to destroy the invaders. ■ effects of temperature on bacterial metabolism, p. 96

A moderate increase in temperature increases the rate of enzymatic reactions. It is thus not surprising that fever has been shown to enhance the inflammatory response, phagocytic killing by leukocytes, the multiplication of lymphocytes, the release of substances that attract neutrophils, and the production of interferons and antibodies. Release of leukocytes into the blood from the bone marrow is also enhanced. Further, fever decreases the host's absorption of iron from food. For all these reasons, one generally should avoid using drugs to reduce the fever of infectious diseases except when absolutely necessary, as when the fever is very high and sustained.

Changes in Iron Availability

The ability to limit the amount of iron available to invading bacteria is a major nonspecific defense mechanism. Bacteria as well as humans require iron for the functioning of some important enzymes. Iron is found in humans attached to specific proteins such as transferrin and lactoferrin. Bacteria must compete with these proteins for iron in order to obtain enough to grow. Some bacteria can successfully compete for the iron. For example, *Neisseria gonorrhoeae*, the cause of gonorrhea, can get iron directly from the carrier molecules, transferrin and lactoferrin. Unusually high concentrations of iron make it easier for the bacteria to obtain the iron they need.

Intracellular bacteria also must have iron. Iron bound to transferrin gets into cells of the host by means of transferrin receptors on the cell surface. These receptors bind the transferrin-iron complex and allow it into the cell. When mononuclear phagocytes ingest bacteria, they decrease the number of transferrin receptors on their surfaces, thereby decreasing the amount of iron available within the cell for use of the microorganisms. Also, lactoferrin in the phagolysosomes can compete successfully with most bacteria for intracellular iron. Iron administered in the form of medication can increase the susceptibility of people to infection, because it increases the amount of iron available to bacteria, as well as to the host.

Changes in Protein and Carbohydrate Metabolism

Both nonspecific and specific immune responses require increased carbohydrate and protein metabolism by the host. It is essential that the host rapidly synthesize many different defensive proteins, including interferons and other cytokines, structural cell proteins and enzymes for proliferation of more leukocytes, enzymes involved in the lysis of microorganisms, and antibodies. These syntheses and the demands of phagocytosis require that cells of the host have abundant supplies of energy and raw materials for biosynthetic reactions. Energy is derived from carbohydrates and other constituents by both glycolysis and the TCA cycle as well as from metabolism of fatty acids released during breakdown of fat tissue. Amino acids used in the synthesis of various proteins are generated primarily from hydrolysis of the proteins of muscle and other body tissue, as witnessed by the weight loss and weakness commonly seen in individuals with prolonged infectious diseases. The utilization of foodstuffs for energy increases the need for oxygen. Because vitamins are converted to coenzymes for the various enzymes of metabolism, their levels decline in the blood. Severe infections can cause vitamin deficiency diseases, especially in people who are already malnourished.

MICROCHECK 15.7

Fever is caused by pyrogens that act on the hypothalamus to increase body temperature. Fever decreases the growth rate of many bacteria, and it increases the rate of production of host defense cells and substances. Changes in iron availability and in protein and carbohydrate metabolism also contribute to body defense. Iron is made less available to microorganisms, while host protein and carbohydrate metabolism is increased to meet the demands of the active immune response.

- Differentiate between endogenous and exogenous pyrogens.
- How does fever affect pathogens in the tissues?

FUTURE CHALLENGES

Polishing the Magic Bullet

Very early in the Twentieth century, Paul Ehrlich looked for a "magic bullet," a drug that could affect only invading microorganisms and not host cells, one that could be delivered directly to the area of infection or inflammation. In spite of extensive investigations, Ehrlich did not find a magic bullet, but others carried on the search. Through the years many attempts have been made to attach drugs to various molecules, such as specific antibodies, that could home to the area where the drugs were needed.

Currently, a promising approach to finding the elusive magic bullet is based on ongoing investigations of the mechanisms of inflammation. Early during the inflammatory response, a family of cell surface receptors called ***selectins*** *is expressed on cells of small blood vessels in the area of injury. Selectins bind to carbohydrate molecules on circulating leukocytes. As the leukocytes roll rapidly* *through the blood vessels, selectin molecules bind them, stopping them in their tracks. The leukocytes are held captive in the region of inflammation. Other molecules in that area attract the captured white blood cells across the vessel wall and into the injured tissue. Selectin activity is a marker for tissue damage and, therefore, a logical target for delivery of drugs into areas of inflammation resulting from infections. The drugs could be antimicrobials to act against invading microorganisms, or anti-inflammatory agents to lessen the symptoms caused by inflammation.*

In some diseases, such as arthritis and other autoimmune diseases, the tissue damage caused by inflammation is the major factor in the disease process. In addition to delivering anti-inflammatory drugs to the area, blocking the actions of selectins by a selectin inhibitor might prevent the development of inflammatory damage.

Scientists working in this area are studying the exact structure of the various selectin molecules and their characteristics. One selectin is stored in cells of the blood vessel and is expressed on the cell surface within minutes of tissue injury. Another is synthesized after injury and expressed on cell surfaces after about 4 hours. The structure of leukocyte surface molecules that will bind to selectins has also been determined. It should be possible to attach antimicrobial or anti-inflammatory drugs to the small portions of binding molecules that actually bind to the selectin. This combination would then be selectively removed from the circulation in the areas of inflammation, the only areas where selectin is produced. It is estimated that preparations of selectin-binding antimicrobials or anti-inflammatory drugs could be available within several years, making Ehrlich's long-ago dream a reality.

SUMMARY

The Host Defends Itself: An Overview

1. **Nonspecific immunity** is the body's first line of defense.
2. Physical and chemical barriers, **inflammation**, and physiological mechanisms such as **fever** contribute to early defense.
3. **Lymphocytes** and **antibodies** provide a second line of defense, which is acquired and is highly specific and effective.
4. **Hematopoietic stem cells** in the bone marrow give rise to leukocytes active in both nonspecific and specific immune responses.

Nonspecific Defense Mechanisms

1. Physical barriers include the skin and mucous membranes.
2. Secretions, such as mucus, saliva, and urine, as well as antimicrobial substances in body fluids and phagocytes assist in nonspecific defense.
3. **Inflammation** is directed against any foreign material or organism.
4. Physiological defenses, such as **fever**, discourage infections.

Specific Defense Mechanisms

1. The specific immune response depends on **lymphocytes**.
2. **Antibodies** and lymphocytes respond to specific antigens.
3. Memory responses occur upon second and subsequent exposures to an antigen.

Cells and Tissues Involved in Host Defense (Figure 15.1, Table 15.1)

1. The **granulocytes** are **eosinophils**, **basophils**, and **neutrophils** (PMNs).
2. The mononuclear phagocytes, **monocytes** and **macrophages**, along with the neutrophils, are the professional phagocytes. (Figure 15.2)
3. Macrophages and **dendritic cells** also play essential roles in specific immune responses.
4. **Natural killer cells** are lymphocytes that kill abnormal cells in a nonantigen-specific manner.
5. **Lymphoid tissues** are distributed all over the body and are sites where lymphocytes interact with foreign materials. (Figure 15.4)
6. **T cells** are responsible for specific cellular immune responses; **B cells** are responsible for producing antibodies.

Physical Barriers and Antimicrobial Factors in Nonspecific Immunity (Figure 15.6, Table 15.2)

1. Skin and mucous membranes present strong barriers to infection, aided in their effectiveness by the flushing action of secretions.
2. Important antimicrobial factors in secretions, body fluids, and cells include **lysozyme**, the **peroxidase** system, and the iron-binding protein **lactoferrin**.
3. Competition by **normal flora** prevents infection with many potential **pathogens**.

The Complement System

1. **Complement** is a group of blood proteins that act in sequence.
2. Some complement components are proenzymes that become enzymes acting on the next component, and so on, in a cascading fashion, leading to an amplified effect.

Classical and Alternative Pathways of Complement Activation (Figure 15.7)

1. Complement is activated by a specific antigen-antibody reaction in the **classical pathway**, or nonspecifically by bacterial products in the **alternative pathway** and the **lectin pathway**.
2. Ultimately, complement activation may result in (1) production of molecules that contribute to the inflammatory response, (2) phagocytosis, and/or (3) lysis of cell membranes.

Cytokines (Table 15.3)

1. **Cytokines** are small regulatory proteins essential for communication between cells.
2. Cytokines have many actions in the inflammatory response, and they contribute to fever production and to development of specific immune responses.

Interferons (Figure 15.9)

1. **Interferons** are antiviral glycoproteins.
2. Interferon gamma is very important in activating macrophages and also functions in development and regulation of specific immune responses.

Interleukins

1. **Interleukins** are cytokines produced by leukocytes.
2. There are at least 18 interleukins; they function in many ways, including inducing fever, signaling release of PMNs from bone marrow, attracting leukocytes into areas of inflammation, and inducing proliferation of lymphocytes.

Colony-Stimulating Factors

1. Colony-stimulating factors direct immature cells into appropriate maturation pathways.

Tumor Necrosis Factors

1. Tumor necrosis factor alpha is stored in mast cell granules. Upon release, this antiviral protein can induce fever and recruit neutrophils into areas of inflammation.
2. TNF-beta can kill target cells.

Functional Groups of Cytokines

1. Some cytokines are particularly important in inflammation, some in promoting antibody responses, and others in cell-mediated specific immune responses.

Phagocytosis (Figure 15.10)

1. **Phagocytosis** occurs in several steps: (1) **chemotaxis**, (2) attachment, (3) ingestion, (4) fusion to produce **phagolysosomes**, (5) killing within phagolysosomes, (6) breakdown of dead materials, and (7) **exocytosis** of the resulting debris.
2. Some organisms resist killing and continue to live within mononuclear phagocytes.

Inflammation (Figure 15.11, Table 15.4)

1. Inflammation is a nonspecific tissue response to injury of any type, characterized by **swelling**, **heat**, **redness**, and **pain**.
2. Upon injury, cytokines are released causing mast cell degranulation and dilation of local small blood vessels, chemotaxis of leukocytes, and leakage of fluids into the tissues. (Table 15.5)
3. Acute inflammation is local, begins immediately upon injury, and increases in intensity over a short time.
4. Chronic inflammation involves the formation of **granulomas** and may last for years.

Physiological Changes and Immune Responses

1. Physiological changes contribute to body defense.

Fever

1. Fever is caused by pyrogens that act on the hypothalamus to increase body temperature.
2. Fever decreases the growth rate of many bacteria, and increases the rate of production of host defense cells and substances.

Changes in Iron Availability

1. Iron is made less available to microorganisms.

Changes in Protein and Carbohydrate Metabolism

1. Host protein and carbohydrate metabolism is increased to meet the demands of the active immune response.

REVIEW QUESTIONS

Short Answer

1. Why is iron metabolism important in body defense?
2. How do white blood cells get into the tissues during an inflammatory response?
3. Describe ways in which the skin protects against infection.
4. What are the benefits of saliva in protection against infection? What factors found in saliva aid in protection?
5. Is lysozyme more effective against Gram-positive bacteria, Gram-negative bacteria, or viruses? Why?
6. Contrast the classical and alternative pathways of complement activation.
7. How can the activation of a few molecules of C1 result in the cleavage of millions of molecules of C3?
8. How does complement cause cell lysis?
9. Name at least two important cytokines and give their effects.
10. What is meant by *professional phagocytes*? Which cells are they?
11. Describe mast cells and their functions.
12. Describe the various kinds of granulocytes and their functions.

Multiple Choice

1. All of the following are characteristic of granulocytes, except they…
   A. have prominent cytoplasmic granules.
   B. include the PMN phagocytes.
   C. are most important in specific immune responses.
   D. are important in the inflammatory response.
   E. are derived from hematopoietic stem cells in the bone marrow.
2. Under normal circumstances, the respiratory tract is protected by all of the following except…
   A. mucus.
   B. motion of ciliated cells.
   C. presence of normal flora.
   D. lysozyme in secretions.
   E. large numbers of neutrophils in secretions.
3. Eosinophils…
   A. lack granules.
   B. stain blue with commonly used blood stains.

C. are mononuclear phagocytes.

D. act against some parasitic worms.

E. are most active against fungi.

4. All of the following are part of the mucosal-associated lymphoid tissues, except…

A. Peyer's patches.

B. the thymus.

C. lymphoid tissue in the bronchi.

D. lymphoid tissue in the gastrointestinal tract.

E. lymphoid tissue in the genital tract.

5. Interleukins…

A. are cytokines produced by leukocytes.

B. are enzymes.

C. are polysaccharides.

D. act only at high concentrations.

E. usually act on distant tissues.

6. Within the macrophage phagolysosome…

A. oxygen consumption decreases as sugars are oxidized.

B. the pH increases as metabolism progresses.

C. highly toxic oxygen products are made.

D. most bacteria multiply.

E. lactic acid increases the pH.

7. All of the following are characteristic of acute inflammation, except…

A. chemicals released from mast cell granules cause dilation of small blood vessels.

B. receptors for leukocytes are produced on the dilated vessels.

C. leukocytes adhere to receptors and migrate through the vessels.

D. monocytes are the first cells to migrate into tissues.

E. cytokines contribute to the reaction.

8. Physiological changes that are important in defense include…

A. an increase in iron absorption by mononuclear phagocytes.

B. release of reserves of neutrophils from bone marrow into blood.

C. decrease in body temperature.

D. decrease in energy metabolism by the host.

E. decrease in host protein synthesis.

9. All of the following are true of fever-inducing cytokines, except…

A. they are exogenous pyrogens.

B. they act on the hypothalamus.

C. TNF-alpha is a pyrogen.

D. interferon alpha is a pyrogen.

E. IL-1 can induce fever.

8. Primary lymphoid tissues include the…

A. spleen.

B. lymph nodes.

C. bone marrow.

D. MALT.

E. appendix.

Applications

1. Physicians regularly have to treat recurrent urinary tract infections in paralyzed paraplegic patients. What explanation would the physician provide to a patient who asked why the condition keeps coming back in spite of repeated treatment?

2. A cattle farmer sees a sore on the leg of one of his cows. The farmer feels the sore and notices that the area just around the sore is warm to the touch. A veterinarian examines the wound and explains that the warmth may be due to inflammation. The farmer wants an explanation of the difference between the localized warmth and fever. What would be the vet's explanation to the farmer?

3. A lawsuit brought against a factory owner claims that an increase in infectious diseases noticed among schoolchildren in the area is caused by water pollution from the factory. A biologist discovers that factory waste entering the water inhibits the development of hematopoietic cells in rats. How can the biologist argue that the polluting waste could be the cause of increased infectious diseases in the children?

Critical Thinking

1. A student argues that phagocytosis is a wasteful process because after engulfed organisms are digested and destroyed, the remaining material is excreted from the cell (see figure 15.10). A more efficient process would be to release the digested material *inside* the cell. This way, the material and enzymes could be reused by the cell. Does the student have a valid argument? Why or why not?

2. According to figure 15.9 *any* cell infected by viruses may die due to the action of interferons. This strategy, however, seems counterproductive. The same result would occur without interferon—any cell infected by a virus might die directly from the virus. Is there any apparent benefit from the interferon action?

3. Is fever deleterious or beneficial? Why?

Specific Immunity

Near the end of the Nineteenth century, diphtheria was a terrifying disease that killed many infants and small children. The first symptom of the disease was a sore throat, often followed by the development of a gray membrane in the throat that made breathing impossible. Death occurred rapidly, even in the absence of a membrane. Frederick Loeffler, working in Robert Koch's laboratory in Berlin, found club-shaped bacteria growing in the throat of people with the disease but not elsewhere in their bodies. He guessed that the organisms were making a poison that spread through the bloodstream. In Paris, at the Pasteur Institute, Emile Roux and Alexandre Yersin followed up by growing the bacteria in quantity and extracting the poison, or toxin, from strained culture fluids. When injected into guinea pigs, the toxin killed the animals.

Back in Berlin, Emil von Behring injected the diphtheria toxin into guinea pigs that had been previously inoculated with the bacteria and had recovered from diphtheria. These guinea pigs did not become ill from the toxin, suggesting to von Behring that something in their blood, which he called antitoxin, protected against the toxin. To test this theory, he mixed toxin with serum from a guinea pig that had recovered from diphtheria and injected this mixture into an animal that had not had the disease. The guinea pig remained well. In further experiments, he cured animals with diphtheria by giving them antitoxin.

The results of these experiments in animals were put to the test in people in late 1891, when an epidemic of diphtheria occurred in Berlin. On Christmas night of that year, antitoxin was first given to an infected child, who then recovered from the dreaded diphtheria. The substances in blood with antitoxin properties soon were given the more general name of antibodies, and materials that could generate antibodies were called antigens.

Emil von Behring received the first Nobel Prize in Medicine in 1901 for this work on antibody therapy. It took many more decades of investigation before the biochemical nature of antibodies was elucidated. In 1972, Rodney Porter and Gerald Edelman were awarded the Nobel Prize for their part in determining the chemical structure of antibodies.

—***A Glimpse of History***

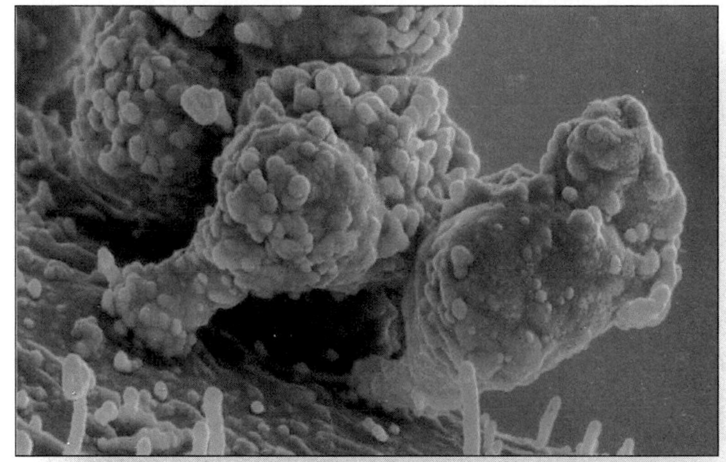

IN THE PREVIOUS CHAPTER, MECHANISMS OF **NON-specific immunity** were discussed. Nonspecific immunity is innate, is present in all normal people, and is ready to act whenever the body is infected or injured. Mechanisms of innate immunity are nonspecific with respect to any offending agent. They are equally effective on the first encounter or in any subsequent encounter with an invading organism. They provide an immediate response as a first line of defense, ever ready to protect. While this early defense is at work, another, even more effective response is being brought into play. This is the **specific acquired immune response**.

Fundamental Features of Specific Acquired Immunity

The specific immune response is directed only against the offending agent. In addition to being **specific**, this response has **memory**, so that it is more rapid and effective against a second or subsequent exposure to the same agent than it is during the first encounter with the agent. The memory response is also called an **anamnestic** or **secondary response**. For example, a first exposure to the measles virus causes a specific immune response that takes about a week to develop fully and that usually halts the measles virus infection, but only after the virus has caused the disease known as measles. Subsequent exposure to the same virus, however, leads to a much faster and more effective memory response, so that the virus is destroyed before it can cause disease; a long-lasting or permanent state of immunity against measles exists. Similarly, infection with one type of *Streptococcus pyogenes* induces long-lasting immunity to the same type but not to other types of *S. pyogenes.*

An **antigen** is a molecule that reacts specifically with an antibody or immune lymphocyte. An **immunogen** is an antigen that can induce an immune response. **Antibodies** are glycoprotein molecules that react specifically with chemical structures in the antigen that induced them; they bind those chemical structures with a high degree of specificity. **Activated (immune) T lymphocytes** also react specifically with the antigen. The specificity of antigens, antibodies, and T lymphocytes can only be defined in a functional relationship with each other.

Specific immune responses occur along two pathways, the antibody response and the specific cellular response. The antibody response is often called the **humoral response**, because the antibodies are found cell-free in blood and other body fluids that used to be called humors. The antibody response is made by **B cells**. Specifically reactive **T cells** are responsible for the **cellular immune response**. Macrophages and dendritic cells play essential roles in both antibody and cellular immune responses. They are **antigen-presenting cells** that provide antigen and stimuli required for activation of T cells. B cells can also be antigen-presenting cells.

MICROCHECK 16.1

Acquired immune responses have the following fundamental features: (1) They are highly antigen-specific. (2) The responses have memory. Once the host has responded to a specific antigen, a subsequent exposure to the same antigen is more rapid and effective, thereby quickly eliminating that antigen. (3) The responses involve two sets of lymphocytes, acting together with macrophages and dendritic cells. B cells produce an antibody response, and T cells produce a cellular immune response.

- What are the major differences between humoral and cell-mediated immunity ?
- What is meant by the anamnestic response?

The Nature of Antigens and Haptens

Antigens react specifically with antibodies or immune cells, and generally they are also immunogens, able to induce the specific response. Antigens are usually foreign to the host and are recognized as such by the immune system.

Most antigens are macromolecules of high molecular weight, such as proteins, polysaccharides, and occasionally, lipids or nucleic acids. Molecules differ in their effectiveness in stimulating an immune response. Proteins and polysaccharides generally induce a strong response, whereas lipids and nucleic acids are rarely antigenic alone. When complexed, as in a lipoprotein, lipids become better antigens. Substances with a molecular weight of less than 10,000 daltons do not make good antigens.

Although antigens are large molecules, the immune response is not directed toward the entire molecule, but rather to particular chemical groups on the molecule known as **antigenic determinants** or **epitopes (figure 16.1)**. On large protein molecules, sequences of 10 to 25 amino acids act as epitopes, which are recognized by the antibodies. Complex structures, such as most bacterial cell walls, may have 100 or more epitopes. A cell will usually have many macromolecules on its surface, each with a number of different epitopes, and it may have multiple copies of any of these epitopes.

An antigen may be virtually any substance recognized by an individual's body, such as bacterial cells, fungi, protozoa, plant pollens, chemicals, drugs, or foods. For example, red blood cells from other people may be recognized as having foreign epitopes and be destroyed by the recipient's immune system in a transfusion reaction. Notice that the red blood cells are not antigenic in the donor, but they may be antigenic to others of different blood types. Similarly, one person's lymphocytes will have surface epitopes that are antigenic in another person or an animal. Antibodies are protein molecules with many different epitopes on their surfaces, some of which are recognized in other people or animals when injected or otherwise introduced into the body. This explains why antibodies can themselves act as antigens.

While low molecular weight molecules by themselves are not immunogenic, some small molecules called **haptens** can combine with large "carrier" molecules such as protein or polysaccharide to become immunogens able to induce an immune response (**figure 16.2**). The hapten alone can react with the antibodies or immune cells produced in response to the hapten-carrier complex. Penicillin, for example, has a molecular weight of about 350 daltons and is a hapten that by itself is incapable of causing antibody production. In the body, however, the penicillin is changed by enzymes so that it can combine with large protein carrier molecules. This immunogenic hapten-carrier complex induces the formation of antibodies specific for the hapten. The reaction of these antibodies with the hapten penicillin can result in allergic reactions in a low percentage of the population, precluding the further use of penicillin in these reactive individuals.

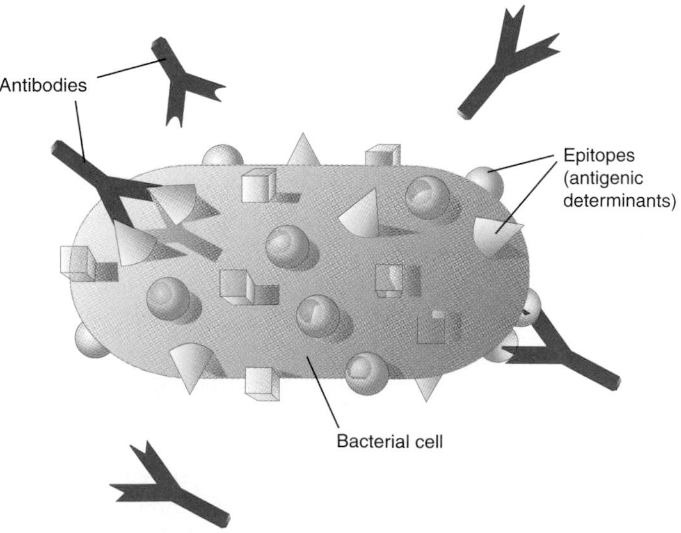

Figure 16.1 Antibodies and Antigen Epitopes on a Bacterial Cell

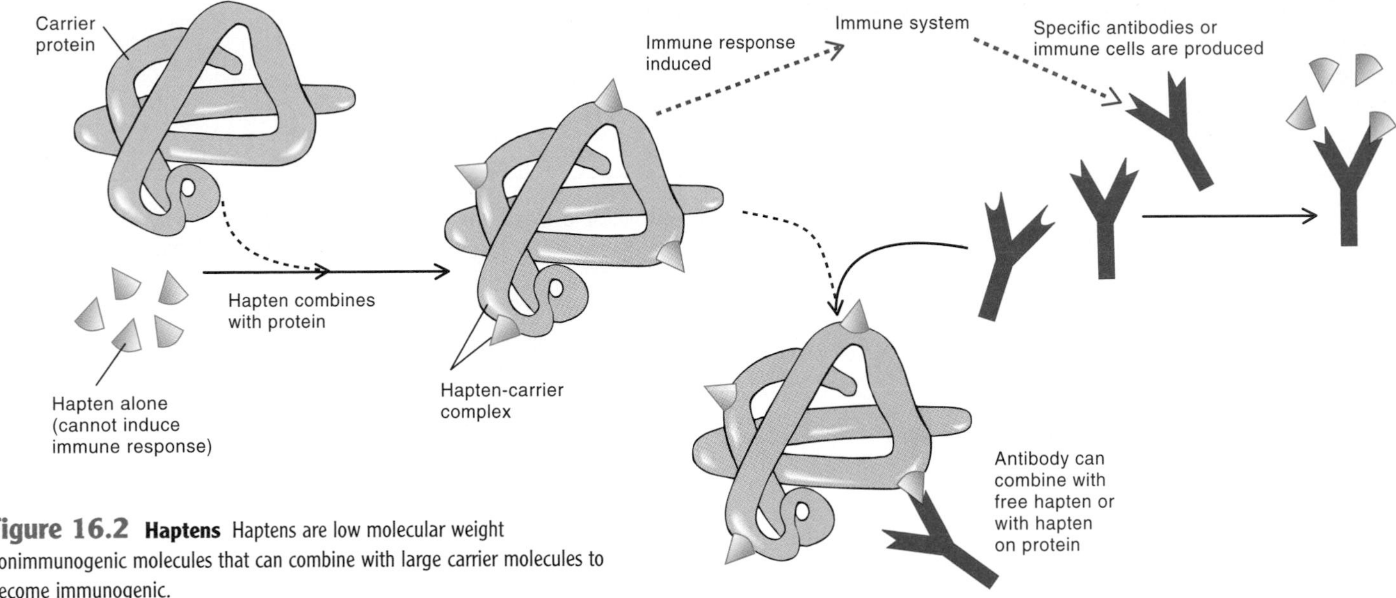
Carrier protein
Immune response induced
Immune system
Specific antibodies or immune cells are produced
Hapten combines with protein
Hapten alone (cannot induce immune response)
Hapten-carrier complex
Antibody can combine with free hapten or with hapten on protein

Figure 16.2 Haptens Haptens are low molecular weight nonimmunogenic molecules that can combine with large carrier molecules to become immunogenic.

MICROCHECK 16.2

Antigens are large molecules that (1) contain many surface epitopes, (2) are usually foreign to the host, and (3) react specifically with antibodies and immune cells. Most antigens are immunogens that can induce the production of specific antibodies or immune cells. Haptens are small molecules that cannot induce an immune response unless complexed with large carrier molecules, but then the hapten alone can react specifically with the antibodies or immune cells produced in response to the hapten-carrier complex.

- What is an epitope?
- What is a hapten?
- Would nucleic acids make good antigens? Why or why not?

The Nature of Antibodies

Antibodies are glycoprotein molecules called **immunoglobulins** that react specifically with the antigen that induced their production. They protect by (1) recognizing and binding to foreign substances and facilitating their removal, (2) increasing phagocytosis, (3) neutralizing toxins or viruses, and (4) activating complement, among other ways.

Properties of Antibodies

Immunoglobulins all have a similar molecular structure, consisting of two heavy (high molecular weight) polypeptide chains (H chains) and two light (lower molecular weight) polypeptide chains (L chains) joined by disulfide bonds to form a single molecule, the immunoglobulin (antibody) monomer. There are five classes of human immunoglobulins (Ig): IgG, IgM, IgA, IgD, and IgE. All of these have the monomeric four-chain structure. IgM usually forms a pentamer, composed of five monomeric subunits, and IgA characteristically forms a dimer of two subunits. Each class of antibody differs in some characteristics of its heavy chains and in its corresponding biological properties. **Table 16.1** gives some properties of the different classes of immunoglobulins.

IgG has been the most extensively studied, and so we will first examine the structure of IgG molecules. They are Y-shaped molecules consisting of two identical halves. Each half contains an H polypeptide chain and an L polypeptide chain. These two chains are held together by disulfide bonds, as are the two halves of the molecule (**figure 16.3**).

The Y shape of the IgG antibody molecule, as diagrammed in figure 16.3, can actually be seen in electron micrographs. At very high magnification, a flexible or "hinge" region of the molecule is revealed at the fork of the Y. Antibody molecules combined with antigen molecules form different angles around the hinge, and the arms of the Y move. The arms can move out far enough to give the molecule a T shape. In fact, recent evidence indicates that antibody molecules are not only flexible in the hinge region, but they may be pliable in other places as well. For example, some antibodies have been shown to bind to proteins inside crevices on the surface of rhinovirus virions, cause of the common cold.

Each polypeptide chain in IgG molecules consists of a variable region and a constant region. The constant regions of H chains for all molecules of the same immunoglobulin class have virtually the same amino acid sequence and are responsible for the characteristic functions and properties of that class. Part of this area of the immunoglobulin molecule was given the designation Fragment crystallizable, or **Fc region**, because these fragments of IgG were observed to crystallize, a property only of proteins that have identical amino acid sequences (see figure 16.3). The different classes of immunoglobulins differ in their constant regions.

The variable regions on the outer end of each arm of the Y on both the H and L chains contain the section of the molecule that interacts with the epitopes of antigens, and together they are called Fragment antigen binding, or **Fab region**. Since each epitope differs from other epitopes, the variable regions of

TABLE 16.1 Characteristics of the Various Classes of Human Immunoglobins

Class and Molecular Weight (daltons)	Structure	Percent of Total Serum Immunoglobulin (Half-life in serum)	Properties	Functions
IgG 146,000	Monomer	80%-85% (21 days)	Specific attachment to phagocytes; complement fixation; ability to cross placenta	Agglutination; precipitation; opsonization; ADCC; complement fixation
IgM 970,000	Pentamer	5%-10% (10 days)	Complement fixation; major class produced to some polysaccharides; first antibody produced during immune response	Agglutination; precipitation; complement fixation leads to opsonization and inflammation
IgA monomer 160,000; secretory IgA 390,000	Dimer in secretions	10% (6 days)	Secreted into saliva, milk, mucus, and other secretions; secretory form resists enzymatic degradation	Protection of mucous membranes by preventing attachment of organisms
IgD 184,000	Monomer	<1% (3 days)	Antigen receptor on B lymphocytes	Facilitates development and maturation of the antibody response
IgE 188,000	Monomer	<0.01% (2 days)	Attaches to mast cells and basophils; antigen reacts with cell-bound IgE to release granule contents	Involved in many allergic reactions; functions in ADCC; helps to expel parasites

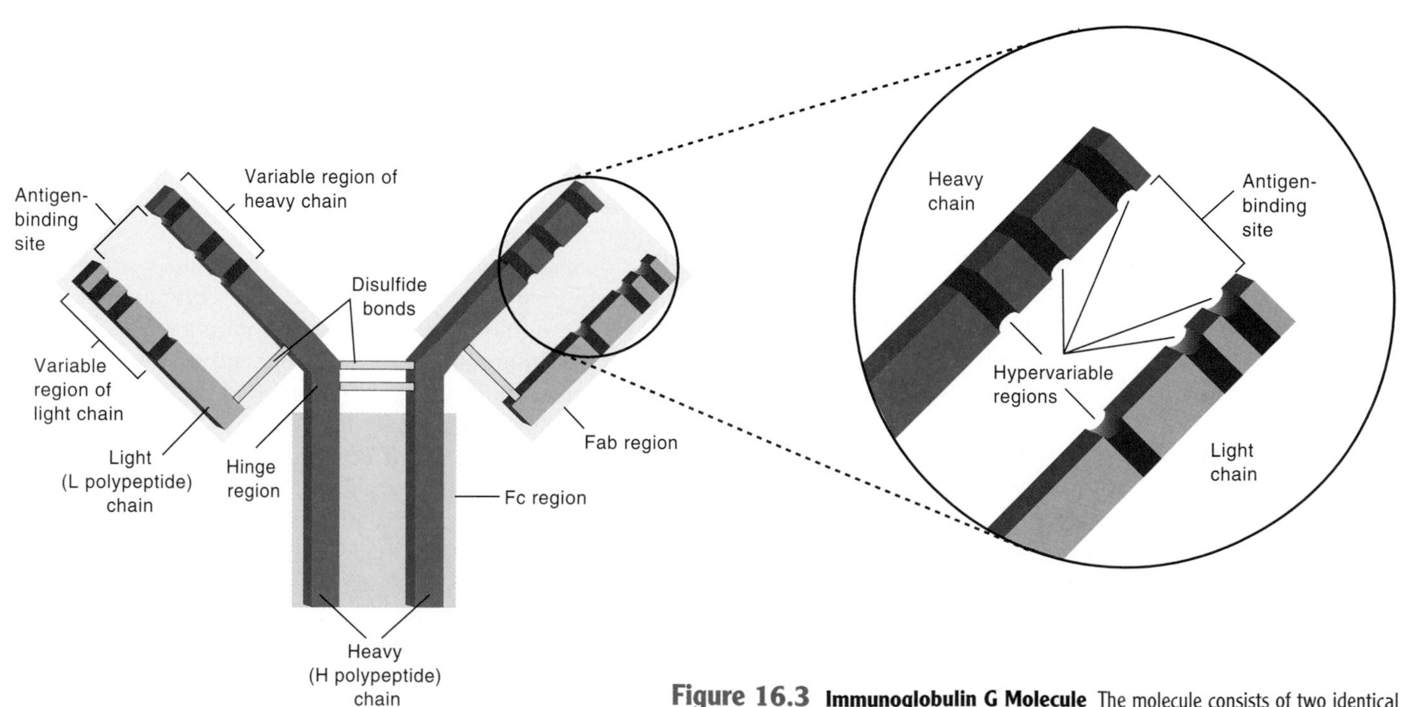

Figure 16.3 Immunoglobulin G Molecule The molecule consists of two identical light chains and two identical heavy chains held together by disulfide bonds.

the antibody also differ in their amino acid sequences. This difference accounts for the specificity of antibodies.

An IgG molecule has two identical **antigen-binding sites**, one at the end of each arm of the Y. The differences in the variable regions of different antibodies are primarily found in three small **hypervariable regions** in the H chain and three in the L chain (see figure 16.3). Because of the folding of the polypeptide chains, these areas lie near the end of the arms, where they can easily interact with antigen.

The globular shapes of immunoglobulins are shown in **figure 16.4**. The model is consistent with molecules of the class IgG, IgD, or IgA, with two globular regions in the L chain and four in the H chain. Both IgE and IgM have light chains with two globular regions in the L chains, but the H chains each contain five globular regions, making them larger. The hinge is lacking in IgE and IgM; however, these molecules are still flexible to some extent.

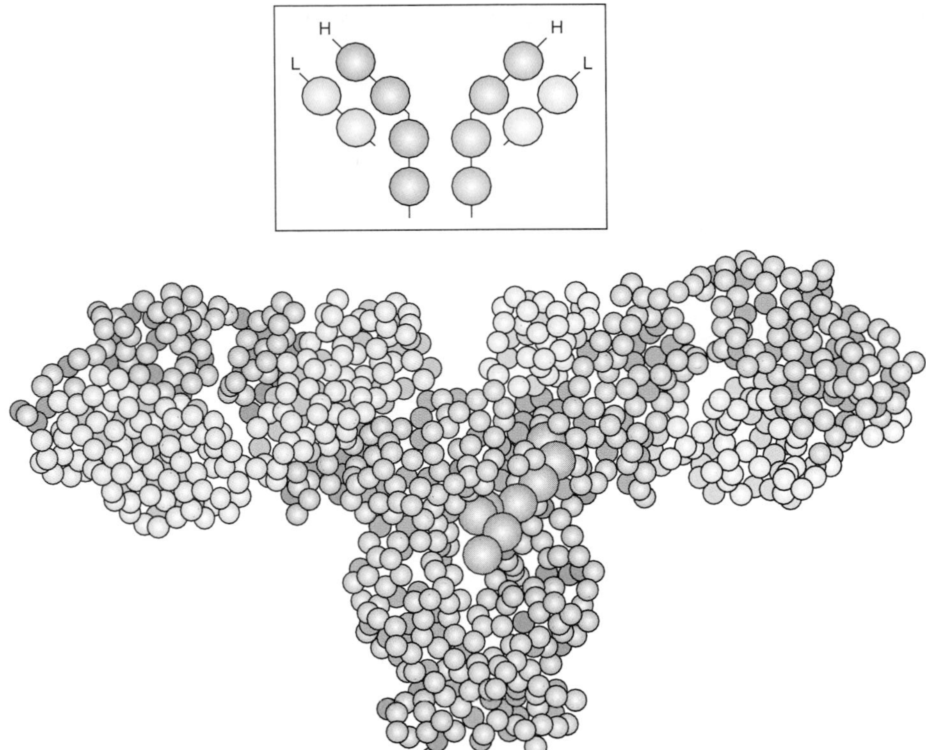

Figure 16.4 Model of an IgG Molecule The inset shows that each L chain consists of two globular regions, and each H chain of four globular regions. In the model, one H chain is red and the other is blue; one L chain is green and the other is yellow. A complex sugar attached to the protein is gray.

The interaction of antigens and antibodies depends on a close complementarity between the antigen-binding sites and the specific antigenic epitopes **(figure 16.5)**. The fit must be close and complementary because the bonds that hold the antibody and the antigen together are noncovalent, weak, and of short range, such as hydrogen bonds, Van der Waals forces, electrostatic charges, and hydrophobic interactions. Although the forces holding the antigen and antibody together are weak, many such bonds are formed, usually keeping the two together very effectively. Nevertheless, the antigen-antibody reaction is reversible. Upon reversal, both antigen and antibody are unchanged.

Antibodies are referred to by their class, such as IgM or IgG, but they may also be known by their function. Thus, an

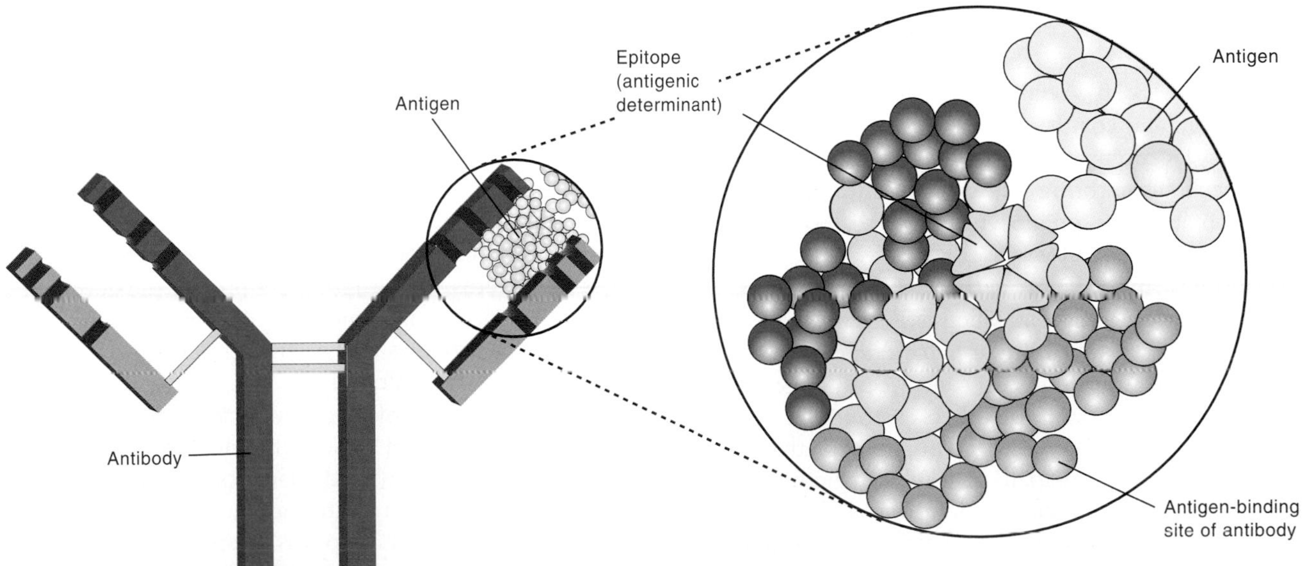

Figure 16.5 Antigen Epitope with a Complementary Antigen-Binding Site of an Antibody The schematic drawing shows how a protruding part of the antigen molecule, the epitope, fits into the antigen-binding site of the antibody molecule to bring the atoms of each into close proximity. With such closeness, weak chemical bonds such as hydrogen bonds can form and hold the antigen and antibody molecules together. The closer the fit, the more bonds that are formed and the greater the affinity (strength of binding) of the reaction.

IgG antibody that causes the antigen to precipitate is a precipitin; one that causes it to agglutinate is an agglutinin; one that coats the antigen so that it is more readily phagocytized is an opsonin; one that leads to complement lysis is a complement-fixing antibody; one that neutralizes the effects of a toxin is an antitoxin; and one that neutralizes the ability of a virus to infect cells is a neutralizing antibody. Some of the important functions carried out by the different classes of antibodies within the human body are listed in table 16.1.

Immunoglobulin Classes

As noted, all five classes of immunoglobulin molecules have the same basic structure: two identical light polypeptide chains connected by disulfide bonds to two identical heavy polypeptide chains. The amino acid sequences and carbohydrate content of the heavy chains vary with each class. Polymers of the basic subunit can form in IgM and IgA.

Immunoglobulin G (IgG)

IgG accounts for about 80% to 85% of the total serum immunoglobulin in normal people over the age of two years. Most of the IgG produced is in the circulation, where it can persist for several weeks before being degraded. Antibodies of this class are effective in *agglutination*, *precipitation*, *neutralization*, *opsonization*, *antibody-dependent-cellular cytotoxicity* (*ADCC*), and *complement fixation*.

IgG antibodies cause agglutination and precipitation of antigens into clumps, which are removed by phagocytes more efficiently than single particles. IgG molecules also can neutralize toxins, and they can prevent infection by blocking adherence of viruses and microorganisms to host cells.

IgG opsonizes, increasing phagocytosis in two ways; both of these activities occur through specific sites in the constant (Fc) region of the IgG heavy chain. First, after antigen reacts with specific IgG antibodies, the Fc portion of the IgG readily interacts with complement, producing complement components that coat the antigen and then attach to receptors on phagocytes (see figure 15.10). Second, the Fab portions of IgG react with the antigen and coat it; and the Fc region of the IgG attaches directly to Fc receptors on phagocytes **(figure 16.6)**. Both of these mechanisms bring the antigen-antibody molecules in close contact with phagocyte surfaces, resulting in highly effective phagocytosis. ■ **opsonization, p. 381**

IgG antibodies may act together with neutrophils, eosinophils, macrophages, or natural killer (NK) cells to kill target cells in **antibody-dependent cellular cytotoxicity (ADCC)**. Target cells may be virus-infected cells, cancer cells, large parasites that cannot be phagocytized, or others. Antibodies attach to the specific target cells, leaving the Fc portions of the antibodies exposed. Cells that participate in ADCC, whether neutrophils, eosinophils, macrophages, or NK cells, have Fc receptors, so that they attach via the Fc portion of the antibodies and thus come in close contact with the target cells. Contents of granules and lysosomes within the attacking cells are released in close proximity to the target cell membranes, resulting in target cell killing **(figure 16.7)**. ■ **neutrophils, eosinophils, NK cells, p. 372**

IgG is the only class of immunoglobulin that can *cross the placenta from the mother to the fetus*. It does this by means of an amino acid sequence in the Fc region of the heavy chain that is recognized by receptors in the placenta, permitting the transport across to the fetus. These antibodies protect the developing fetus and the newborn against infections, but this protection wanes as the maternal antibodies decline over the first months of life and is no longer effective after about 3 to 6 months **(figure 16.8)**. During this time, the infant produces its own antibodies, beginning with IgM and later IgG. By the time the maternal antibodies have undergone normal degradation at three to six months, the infant has its own protective antibodies; however, IgG levels are low from the age of 3 months to one year.

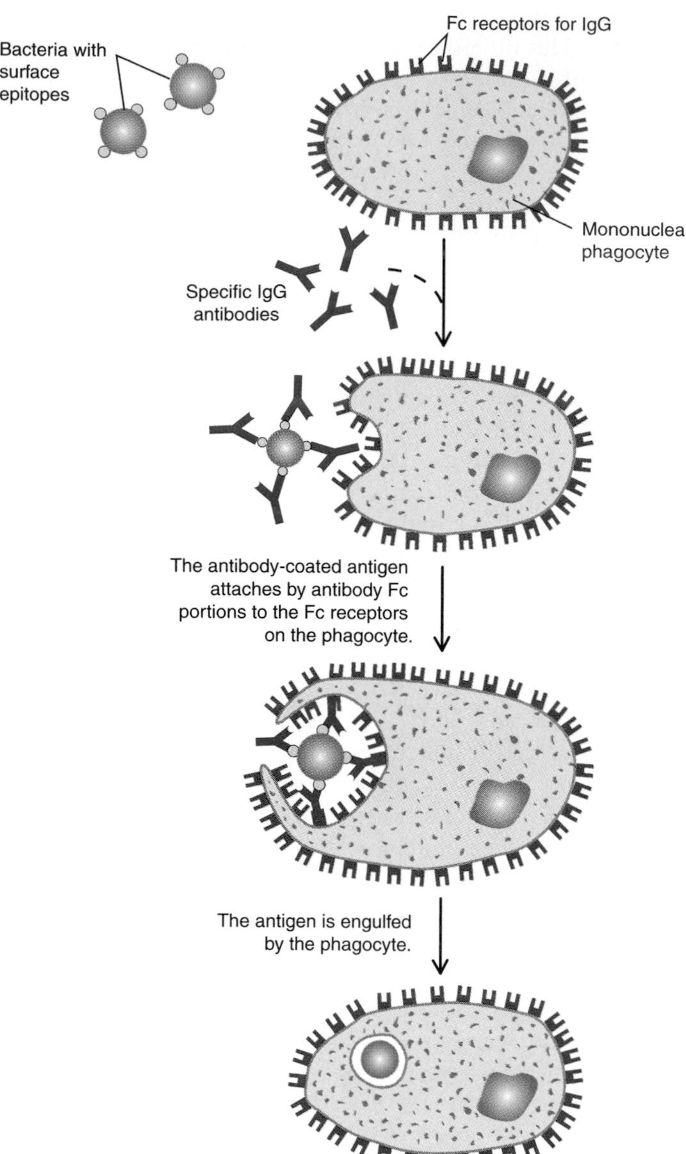

Figure 16.6 Opsonization by IgG Antibodies Facilitates Uptake of Antigens by Phagocytes

Immunoglobulin M (IgM)

IgM is the first immunoglobulin class produced after exposure to an antigen. In its monomeric form, it is an important antigen receptor on the surface of potential antibody-producing cells that permits the cells to recognize antigen and begin to make

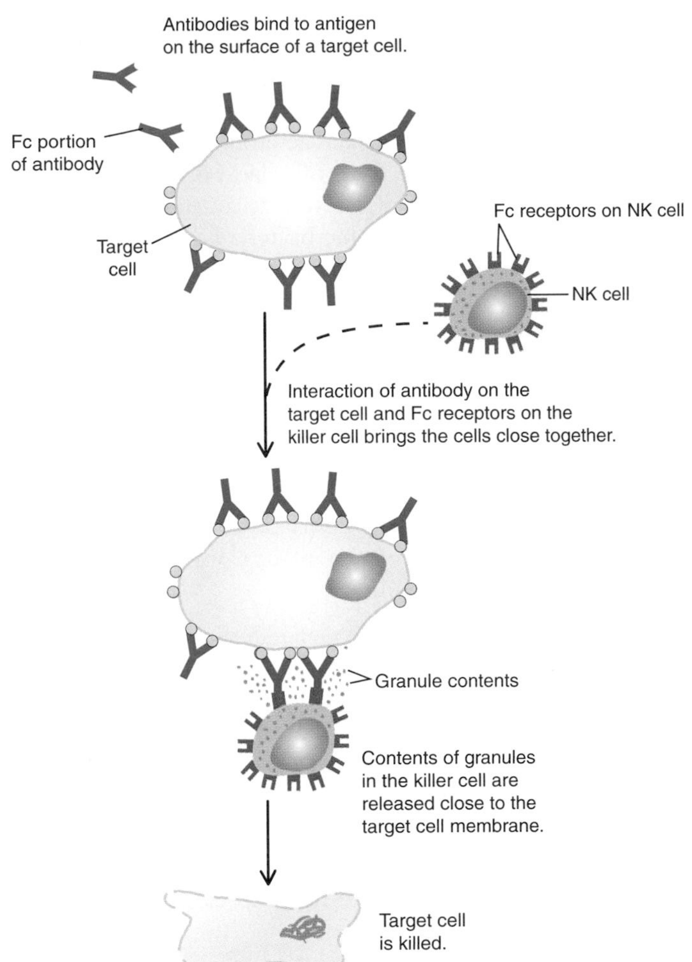

Figure 16.7 Antibody-Dependent Cellular Cytotoxicity (ADCC) Effector cells include NK cells, macrophages, neutrophils, and eosinophils, all of which bear receptors for the Fc portion of immunoglobulins. The antibodies bring effector cells and target cells close together. Target cells are killed by perforin or other granule contents released from effector cells in close proximity to target cells, or by apoptosis.

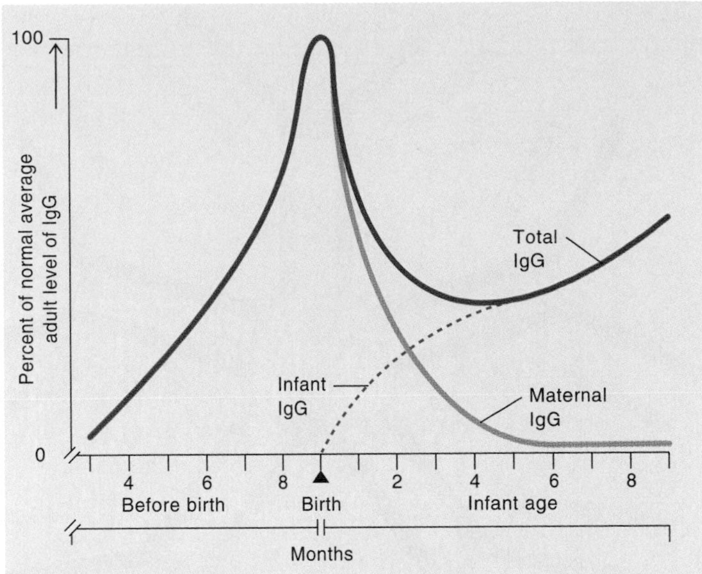

Figure 16.8 Immunoglobulin G Levels in the Fetus and Infant During gestation, maternal IgG crosses the placenta to the fetus. Normally, the fetus does not make appreciable amounts of immunoglobulin; it depends on the maternal antibodies that are transferred passively. After birth, the infant begins to produce immunoglobulins to replace the maternal antibodies, which gradually disappear over a period of about 6 months. Usually by 3 to 6 months, most of the antibodies present are those produced by the infant.

specific antibodies. IgM accounts for about 5% to 10% of the circulating immunoglobulins. The IgM found circulating in the blood is a pentamer of five subunits and has a potential of 10 antigen-binding sites **(figure 16.9a)**. IgM is actually more than five times larger than the IgG molecule because (1) IgM contains, in addition to the five monomers, a polypeptide chain called the J chain that initiates the joining of monomers to form the pentamer; (2) it has a higher carbohydrate content, and (3) its H chains are longer than those of IgG.

The multiple antigen-binding sites of IgM make it very efficient in agglutination and precipitation; the clumps of antigen and antibody that form are rapidly removed by phagocytes. IgM is also an opsonizing antibody that increases phagocytosis by activating complement. It is the most efficient antibody in interacting with complement, which aids in phagocytosis of foreign materials and also functions in a number of ways in inflammation. IgM protects against many extracellular organisms and soluble antigens, such as bacterial toxins. It is the major or only antibody class produced in response to some antigens, such as capsular polysaccharides of bacteria. The presence of IgM in the circulation early during infection causes removal of many microorganisms that enter the bloodstream. IgM thereby helps to prevent the development of septicemia, acute illness caused by infectious agents or their products circulating in the bloodstream.

Because IgM is the first antibody made during the antibody response, it is important in newborn infants. As the maternal antibody decreases, much of the antibody produced during the first months of life is of the class IgM. In the case of infections of the fetus in the uterus, IgM antibodies can be made by the fetus, but normal non-infected babies begin to make antibodies about the time of birth.

Immunoglobulin A (IgA)

IgA accounts for about 10% to 13% of antibodies in the blood of normal individuals, mostly in the form of a monomer. Most IgA antibodies, however, are associated with mucous membranes, or in saliva and milk, where they occur as dimers. In this form they are called secretory IgA (sIgA). Large quantities of secretory IgA are exported onto mucosal surfaces of the gastrointestinal, genitourinary, and respiratory tracts. Secretory IgA molecules are composed of two basic subunits similar to the IgG monomer and they have a J chain that initiates joining of the monomers **(see figure 16.9b)**. This part of the molecule is made by cells of the B-cell lineage, chiefly in the mucosal-associated lymphoid tissues (MALT). A polypeptide called the secretory component, made by epithelial cells of the mucosa, is added; it permits transport of the IgA molecule across and onto the surface of the mucous membrane. Once on mucous membranes, the secretory component protects the antibody from being destroyed by most proteolytic enzymes, including those formed by a majority of pathogenic bacteria. Because IgA is deposited along mucosal surfaces and is present in secretions, it

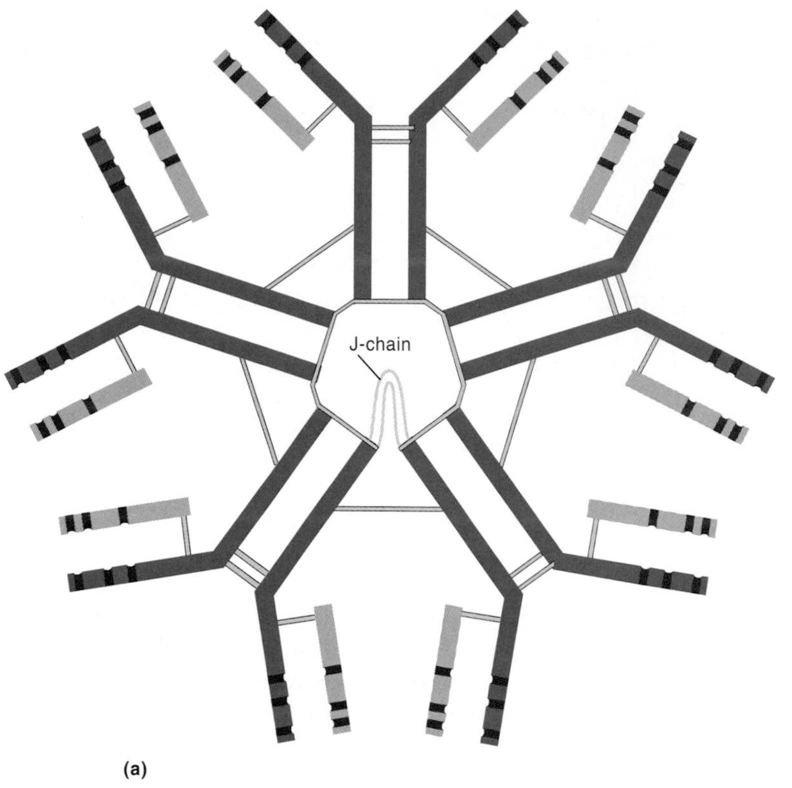

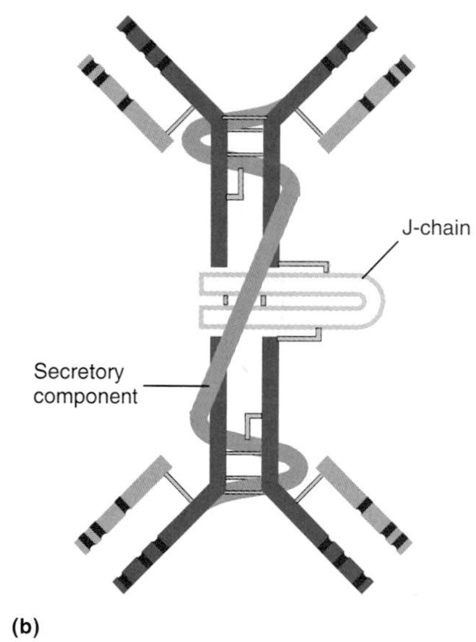

Figure 16.9 Immunoglobulin Structures **(a)** Pentameric structure of human IgM. **(b)** Secretory IgA occurs as dimers. Note the secretory piece that aids the immunoglobulin transport across the mucous membrane.

is actually the most abundant immunoglobulin made by the body. ■ **MALT, p. 374**

Antibodies of the class IgA bind to microorganisms and inhibit their adherence to host cells. For example, specific secretory IgA antibodies inhibit attachment of *Vibrio cholerae*, the bacterium that causes the severe diarrheal disease cholera, to cells in the gastrointestinal tract. Also, the sexually transmitted disease gonorrhea depends on the adherence of the causative organism, *Neisseria gonorrhoeae*, to cells in the genitourinary tract. In both of these cases, secretory IgA antibodies protect by preventing attachment. In the mouth, specific secretory IgA antibodies can prevent the adherence of a large variety of bacteria, some of which cause tooth decay and other oral diseases. A few pathogenic bacteria have found a way to fight back; they produce an enzyme that degrades the IgA, making it ineffective, so the bacteria can attach and invade. IgA does not activate complement by the classical pathway; however, aggregated IgA may activate the alternative pathway.

Young infants that are breast-fed are protected against intestinal pathogens by secretory IgA in breast milk. This IgA is not absorbed into the bloodstream of the child, but it is important in protecting the gastrointestinal tract of the infant from infection and diseases, such as diarrhea.

Immunoglobulin D (IgD)

IgD accounts for less than 1% of all serum immunoglobulins. It is a monomer similar in structure to IgG. IgD functions during the development and maturation of the antibody response. Most of the IgD is found on the surface of mature B lymphocytes, where it serves as a membrane receptor for the specific epitope it recognizes. Its functions have not been clearly defined as yet.

Immunoglobulin E (IgE)

IgE is a monomer, slightly larger than the IgG molecule. IgE binds strongly to basophils and mast cells, so most of the IgE is not free in the circulation. In fact, it is barely detectable in normal blood, being present in a very small fraction of the amounts of other classes of immunoglobulins. IgE is especially important in many allergic, or hypersensitivity, reactions. These include diseases such as hayfever, hives, and asthma. When a specific antigen combines with IgE bound to cells, it causes cross-linking of the antibodies and release of potent chemicals from the granules of the basophils or mast cells, leading to immediate reactions, such as coughing, sneezing, and muscular contractions that are protective when they help to expel parasites, infectious agents or other foreign materials from the lungs or gut but that cause allergic symptoms in response to normally harmless materials, such as dusts, weeds, and pollens. ■ **basophils and mast cells, p. 372** ■ **allergies, p. 433**

Both IgE and IgG can function in ADCC against parasites. This has been studied in some detail in the parasitic disease schistosomiasis. IgE and IgG attach to the parasite by their antigen-binding sites and also attach by their Fc portions to Fc receptors on the eosinophil, bringing the eosinophil and the parasite close together. The target parasite is killed by a combination of the peroxidase–halide-ion antimicrobial system and other antimicrobial proteins released from the eosinophil granules.

MICROCHECK 16.3

Immunoglobulin antibodies react specifically with the antigen that induced their formation. The five classes of immunoglobulins are IgA, IgD, IgE, IgG, and IgM. All of them are proteins with two heavy and two light polypeptide chains that bind antigen in a close complementary fit. IgD, IgE, and IgG are monomers, IgA a dimer, and IgM a pentamer. The Fc part of the molecule accounts for many of the biological functions of the antibody, unique to each class.

- How does IgM differ from IgG?
- In opsonization by IgG antibodies, why is it necessary that antigen *first* reacts with specific IgG?
- How would IgA antibodies prevent attachment of pathogens to cells?

The Role of Lymphocytes in Specific Immunity

Lymphocytes are the cells responsible for specific immune responses. Most lymphocytes are either T cells or B cells.

Lymphocytic stem cells in the bone marrow divide and give rise to cells that multiply and mature in primary lymphoid tissues in one of two locations: in the thymus gland or in the bone marrow. Lymphocytes that mature in the thymus become **T lymphocytes (T cells)** and play an important role in cell-mediated immunity, the direct destruction of body cells that have been invaded by parasites or that undergo degeneration. T cells also play a significant regulatory role in the development and activation of all types of immune responses. Lymphocytes that mature in the bone marrow become **B lymphocytes (B cells)** and are the cells responsible for the antibody response and **antibody-mediated immunity**. B cells can also be antigen-presenting cells, under certain conditions. B lymphocytes can leave the circulation, go into lymphoid tissues, and develop into **plasma cells**, which are the terminally differentiated, active antibody-producing cells.

T and B lymphocytes cannot be distinguished by appearance. **Figure 16.10a** shows a lymphocyte as seen by light microscopy and figure 16.10b a lymphocyte as seen by scanning electron microscopy. Plasma cells are larger and have a different appearance, with extensive rough endoplasmic reticulum, as seen in the transmission electron micrograph figure 16.10c. Both B and T lymphocytes recognize antigens and participate in specific immune responses. B and T cells recognize antigen in different ways, however, which will be discussed in more detail later. Both the cell-mediated immune response and the antibody-mediated immune response are specific acquired immune responses.

All lymphocytes are identified by proteins on the surfaces of the cells (see Perspective 16.1). For example, B cells are identified by the presence of immunoglobulin on their surfaces. Each B cell expresses only the single specificity of antibody molecule it will eventually secrete. Similarly, T cells express unique T-cell receptors for antigen on their surfaces. During maturation, a variety of **cluster of differentiation**, or **CD**, protein molecules (or markers) are expressed on the cell membranes of subpopulations of cells, enabling them to be identified.

CD molecules are especially useful in identifying lymphocytes. All T cells carry CD3 molecules, which are associated with the T-cell receptors for antigen. There are different subgroups of T cells, however, that can be further identified by other CD molecules. Most of the mature T cells are either CD4 cells ($CD4^+$), bearing the CD4 protein, or CD8 cells ($CD8^+$), with the CD8 protein. There are two functional subsets of CD4 lymphocytes. **Th1 cells** produce cytokines that drive the development of CD8 cytotoxic cells and activate macrophages; they are sometimes called **inflammatory T cells**. **Th2 cells** produce cytokines that stimulate B cells to produce antibodies. The latter

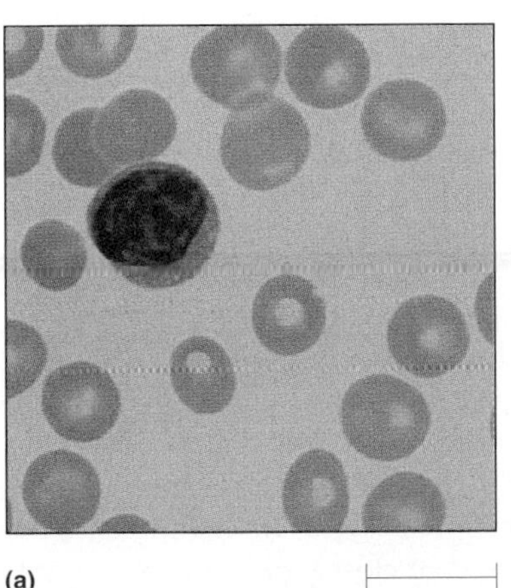

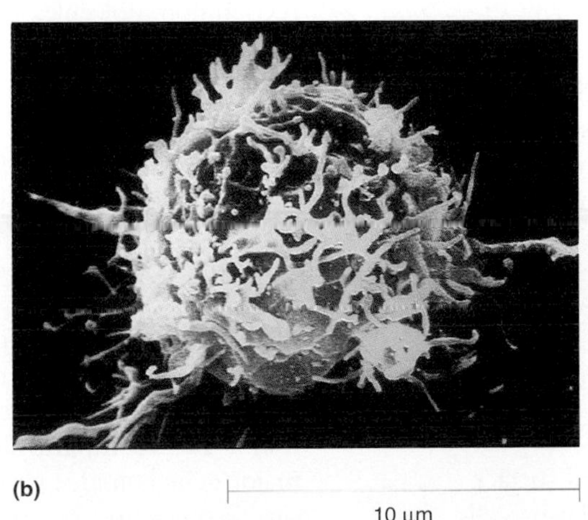

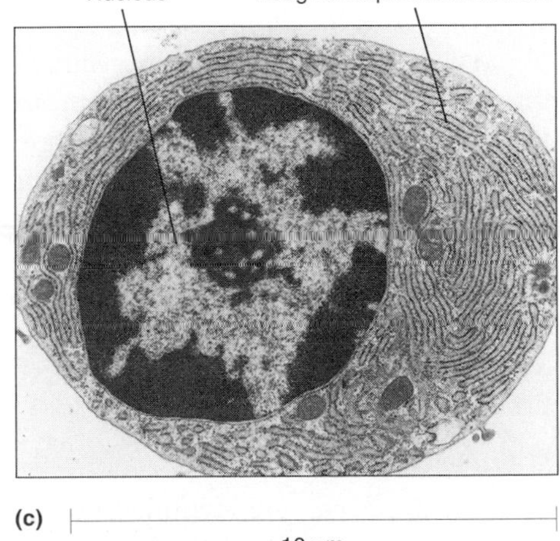

Figure 16.10 Lymphocytes and Plasma Cells (a) Light micrograph of a T lymphocyte. The morphology is the same as that of a B lymphocyte. **(b)** Scanning electron micrograph of a T lymphocyte. T and B lymphocytes cannot be distinguished by microscopy. **(c)** Plasma cell, a form of B cell that is highly differentiated to produce large amounts of antibody. Note the extensive rough endoplasmic reticulum, the site of protein synthesis. All of the antibodies produced by a single plasma cell have the same specificity.

Perspective 16.1 Don't I Know You?

How do we recognize our friends and acquaintances? A relatively few physical characteristics generally will permit us to distinguish between them. Tall or short; blonde, carrot-top, or dark hair; curly hair or straight hair; blue or brown eyes; dark skin, fair skin; or other external features are a big help. Sometimes fingerprints, blood types, or similar characteristics must be used to identify people.

It is the same with lymphoid cells. Easily identifiable characteristics, mostly cell surface proteins, are found on lymphocytes that enable them to be classified as B cells, plasma cells, or various kinds of T cells (Table 1). Plasma cells have extensive rough endoplasmic reticulum (ER), which is easily seen in electron micrographs and with stains that react with the ER.

TABLE 1 Some Characteristics Used to Distinguish Lymphoid Cells

Cell Type	Surface Ig	Extensive Rough ER	CD3	CD4	CD8	MHC
B cells	+	–	–	–	–	Class I and II
Plasma cells	+	+	–	–	–	Class I and II
Th cells	–	–	+	+	–	Class I
Tc cells	–	–	+	–	+	Class I

CD4 lymphocytes function as helper T (Th) cells in specific antibody responses. Many CD8 lymphocytes act as cytotoxic T (Tc) cells in the cellular immune response. Some other lymphocytes act as **suppressor (Ts) cells**, or **T-regulatory cells**, to suppress immune responses, probably by means of secreted factors. Although some Ts lymphocytes appear to be CD8 cells, Ts cells are not a well-defined subpopulation of T lymphocytes. Similarly, other T cells are active in delayed hypersensitivity reactions, to be discussed later. Although these have been called Td cells, they are not a well-defined population and are probably part of the CD4 group.

Within the lymph nodes and tonsils (see figure 15.4) and other secondary lymphoid tissues, the lymphoid follicles serve as areas of T- and B-cell interaction. The follicles are like busy lymphoid coffee shops where many cellular meetings take place and information is exchanged. The anatomy of the node permits various types of cells, including lymphocytes, dendritic cells, macrophages, and other cells important in the immune response, to come into contact with each other and with antigen, or to communicate by means of cytokine molecules. Both antibody and cell-mediated immune responses develop in the lymphoid follicles of any secondary lymphoid tissue such as lymph nodes, spleen, tonsils, and Peyer's patches in the intestine, among others. ■ **lymphoid tissues, figure 15.4, p. 374**

MICROCHECK 16.4

Lymphocytes are identified by various cell surface molecules. B lymphocytes mature into plasma cells, the principal antibody producers. Many CD4 T cells are helper (Th) cells in antibody and cellular immune responses, while many CD8 T cells are cytotoxic (Tc) cells in the cellular response. Follicles in lymph nodes and other lymphoid tissues are centers of activity in which immune responses develop.

- From which cells do plasma cells arise? What is their function?
- Give the principal functions of CD4 T cells and CD8 T cells.

Major Histocompatibility Complex

In addition to CD markers and immunoglobulins, many other recognition molecules are found on cell surfaces. One interesting group is composed of **histocompatibility antigens**. The discovery of histocompatibility antigens came over half a century ago, soon after World War II. During that war, bombing raids resulted in many patients with serious burns and consequent problems in treatment, and this stimulated research into the transplantation of skin to replace burned tissue. This research quickly spread to the study of transplantation of a variety of other organs besides skin, including bone marrow, which contains the stem cell precursors of all blood cells in the body. It was soon recognized that cell surface molecules of primary importance in the rejection of transplants were coded for by a cluster of genes now known as the **major histocompatibility complex (MHC)**. In humans, this major gene complex is composed of closely linked genes located on chromosome 6, and the cell surface molecules are called **human leukocyte antigens (HLA)**. ■ **transplantation, p. 443**

Individuals who are genetically the same, such as identical twins, have the same MHC molecules on their cells. When a transplant is made between identical twins, the immune system fails to recognize the MHC molecules on the transplanted cells as foreign, and the transplant is accepted. When cells are transplanted between nonidentical people, the immune system recognizes the histocompatibility molecules on the transplanted cells and makes a vigorous immune response, leading to rejection of the tissue. These observations seemed strange at the time of discovery. Since transplanting tissues is very unnatural, why should the body have such a complex system for rejecting the transplant? A great deal of research in many areas of immunology has uncovered some reasons. The MHC gene products are important in many ways that do not involve tissue transplantation. They participate in cell-to-cell interactions, essential for the functioning of the immune system. Two classes of MHC molecules, class I and class II, are particularly important (**figure 16.11**). Both of them are surface proteins that span the cell membrane. Both class I and class II MHC molecules have a peptide-binding groove or cleft where peptide antigen fragments can be inserted. As indicated in Perspective 16.1, MHC markers are among those that can be used to distinguish between human T and B lymphocytes.

Class I MHC

Class I MHC molecules (see figure 16.11a) are found on all human cells except mature red blood cells, which lack a nucleus. They are found in large quantities on white blood cells. The CD8 protein previously discussed binds to MHC class I molecules.

The groove on class I molecules accommodates a peptide fragment of 8 to 10 amino acids. Antigens activate T cytotoxic cells only when they are presented as peptides complexed with MHC class I molecules in this manner, and in conjunction with a signal given by antigen-presenting cells. If the peptides are nonself, such as those produced by intracellular pathogens, Tc cells can identify and destroy the infected cell. In a cell infected by a virus, small peptides from the virus protein attach to the MHC I molecule and are then transported to the infected cell membrane. These complexes serve as a flag to the Tc lymphocyte that the cell is infected and must be sacrificed for the sake of the neighboring cells. Therefore, any cell in the body that becomes infected or altered can be detected readily by the immune system because of expression of peptides from the pathogen or altered genes on class I molecules.

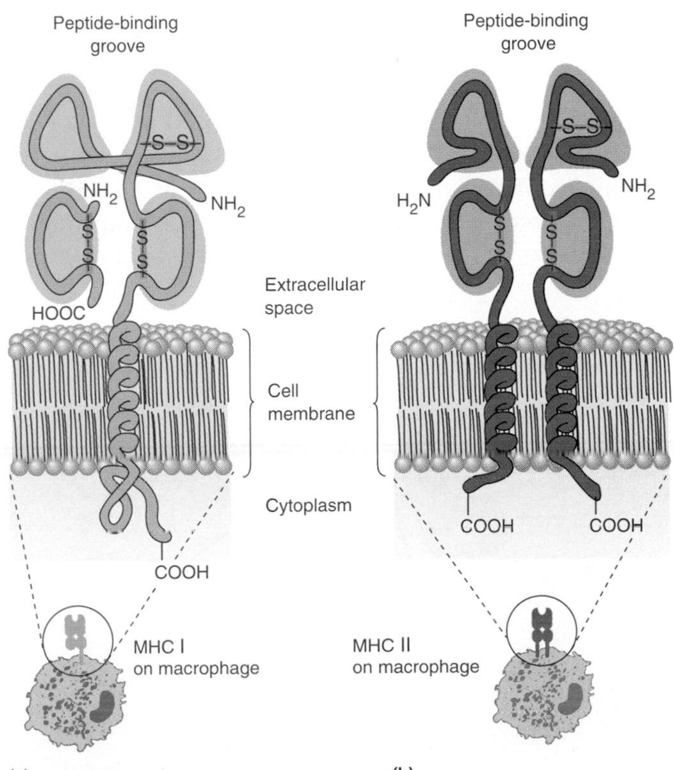

Figure 16.11 MHC Molecules Class I and class II MHC molecules. **(a)** MHC Class I **(b)** MHC Class II.

Class II MHC

Class II molecules (see figure 16.11b) are found principally on macrophages and dendritic cells, the "professional" antigen-presenting cells, and on B cells, which can also present antigen. CD4 protein binds to MHC class II molecules. Class II grooves can hold peptides from 13 up to 25 amino acids ln length. MHC II molecules display antigenic peptides from ingested proteins, often part of phagocytized pathogens. In host macrophages and dendritic cells, the protein antigens are broken down into peptides within vesicles. The nonphagocytic B cells can internalize protein attached to their surface immunoglobulin, degrade it intracellularly, and present it as peptides complexed with class II MHC on the cell surface. Th cells recognize antigen fragments only when they are complexed with class II molecules. This recognition by Th cells will recruit and activate other Th, B, and Tc cells to multiply and quickly respond to antigen.

MICROCHECK 16.5

Class I and class II major histocompatibility proteins are located on cell surfaces. MHC class I molecules are found on almost all human cells, whereas MHC class II molecules are expressed only on antigen-presenting cells such as macrophages, dendritic cells, and B cells. Cells expressing peptides using MHC class I molecules can be recognized by antigen-specific CD8-bearing T cells. Cells expressing peptides using MHC class II molecules can be recognized by antigen-specific CD4-bearing T cells.

- Which human cells bear MHC class I molecules? Which ones have class II molecules?
- What is the function of the groove found in both class I and class II MHC molecules?
- If tissue were transplanted between *nonidentical* twins, what would be observed?

Development of the Antibody Response

We will first examine the events that occur during the development of an antibody response. Later we will discuss the second type of specific immunity, cell-mediated immunity, and consider the similarities and differences between the antibody and cell-mediated responses.

B Lymphocytes and the Antibody Response

During an antibody response, B lymphocytes recognize an antigen epitope by means of specific antibody molecules on the surface of the B cell, the only antibody that particular B cell is committed to make. There are generally only a few B lymphocytes with surface antibodies that recognize one of the foreign epitopes of an antigen, another small set of B lymphocytes that recognize another epitope, and so on, for the millions of epitopes each person can recognize. Each B lymphocyte will interact only with the one epitope recognized by its unique surface-bound antibody.

T-Cell-Dependent Antigens

Protein antigens are T-cell-dependent; that is, they require the cooperation of CD4 T-helper (Th2) cells in order for the B cells to respond to the antigen and eventually make antibodies **(figure 16.12)**. An antibody response may be induced by soluble molecules or by surface epitopes on cells. A soluble antigen could be a protein toxin secreted by bacteria, whereas an epitope on the surface of a bacterium is an example of antigen that is part of a cell. In either case, the antigen, being a large molecule, will probably have a number of different epitopes on its surface.

1. To induce a response, extracellular antigens, such as proteins secreted by invading bacteria, must be taken up and processed by dendritic cells or macrophages, collectively referred to as antigen-presenting cells. The large antigen molecules are broken down in the cell into smaller fragments, attached to the groove of MHC class II molecules, and presented on the cell surface.
2. When the Th cell recognizes and binds to the fragment presented in the class II molecule groove, the Th cell secretes cytokines such as IL-4, IL-5, and IL-10 that activate the B cell to multiply and begin to secrete specific antibodies. B-cell recognition of an antigen alone does not result in antibody formation. A variety of other molecules, such as CD4, aid in binding the B cell and antigen, and cytokine activities are essential in causing cellular proliferation and differentiation.
3. With T-cell-dependent antigens, signals from Th cells are necessary for the B cells to proliferate and differentiate to a state where they secrete a significant quantity of antibody. Memory cells are produced, as well as antibody-synthesizing plasma cells. Antibodies that are produced either enter the circulation or become fixed to cells; IgE, for example, is usually fixed to mast cells, whereas IgM generally is found in the circulation. In either instance, the antibodies are then available to protect against the specific antigen they recognize. Some of the ways they protect were outlined in table 16.1. ■ cytokines, p. 379

T-Cell-Independent Antigens

T-cell-independent antigens can stimulate an IgM antibody response without the aid of T-helper cells. People and animals that are deficient in T cells can still make antibodies against a number of bacterial lipopolysaccharides and polysaccharides. B cells can produce antibodies to two kinds of T-cell-independent antigens **(figure 16. 13)**. Bacterial lipopolysaccharide endotoxins are examples of type 1 T-cell-independent antigens. These large molecules have O antigens that are recognized by B cells. The lipid A part of the lipopolysaccharide molecule stimulates the B cells to divide, thus substituting for the role of Th cells in stimulating B-cell proliferation. Type 2 T-cell-independent antigens are large molecules with the same antigenic determinant repeated many times, such as the polysaccharide capsules of pneumococci. At least 12 to 16 evenly spaced repeating epitopes on the antigen are needed to combine with antibodies on the B-cell surface to induce a good response. The cross-linking of surface antibodies activates enzymes that lead to cell division, substituting for the Th stimulus given in responding to T-cell-dependent antigens. ■ lipopolysaccharide endotoxins, p. 461

Responses to T-cell-independent antigens produce only IgM antibodies and do not lead to a memory response. Both the memory response and the switch from IgM to other classes of immunoglobulins depend on the activities of Th cells. Many pathogenic bacteria have antiphagocytic lipopolysaccharides and polysaccharides in their cell walls, and the ability of the host to

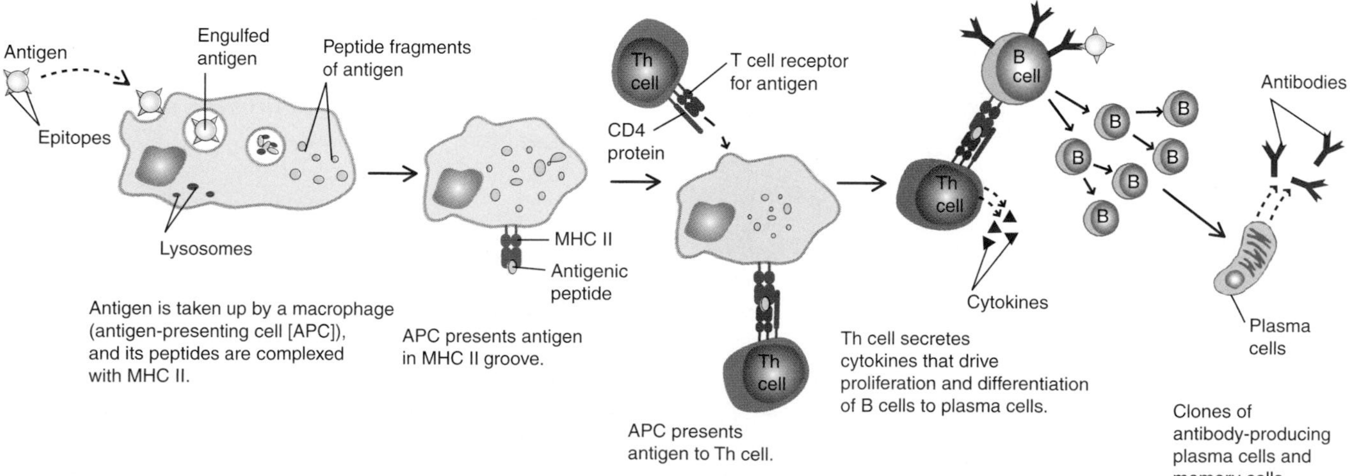

Antigen
Epitopes
Engulfed antigen
Peptide fragments of antigen
Lysosomes
Antigen is taken up by a macrophage (antigen-presenting cell [APC]), and its peptides are complexed with MHC II.
MHC II
Antigenic peptide
APC presents antigen in MHC II groove.
Th cell
T cell receptor for antigen
CD4 protein
Th cell
APC presents antigen to Th cell.
B cell
Th cell
Cytokines
Th cell secretes cytokines that drive proliferation and differentiation of B cells to plasma cells.
B
Antibodies
Plasma cells
Clones of antibody-producing plasma cells and memory cells.

Figure 16.12 Cellular Interactions in the Antibody Response to T-Cell-Dependent Antigens

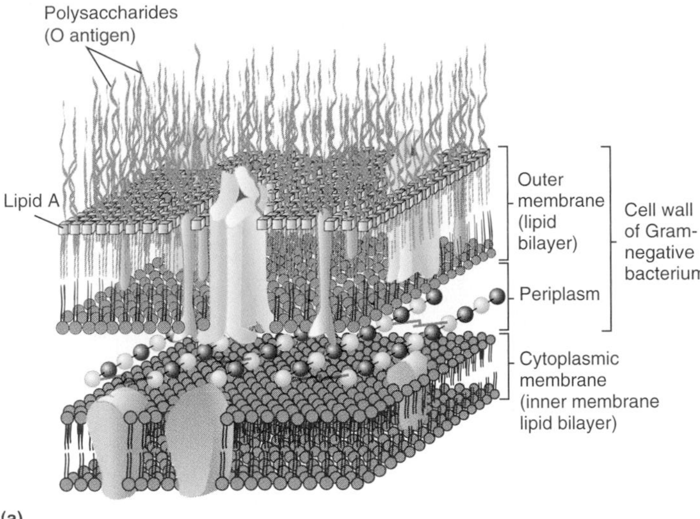

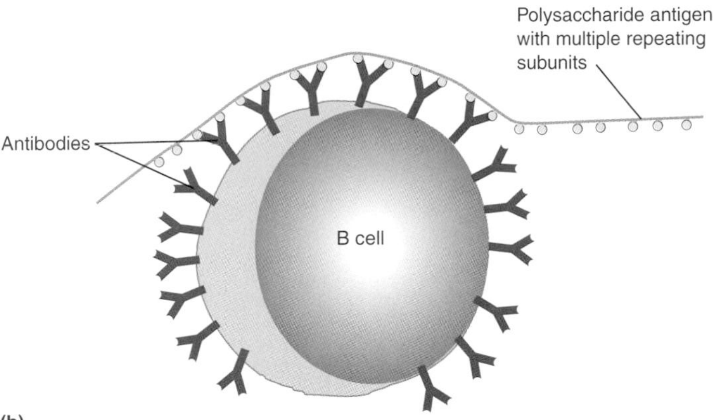

Figure 16.13 T-Cell-Independent Antigens **(a)** Type 1. Gram-negative bacteria contain endotoxins that are type 1 T-cell-independent antigens. The polysaccharide O antigen is recognized by B cells, and the lipid A portion of the endotoxin stimulates the B cells to proliferate and produce antibodies to the O antigen. **(b)** Type 2 antigens, such as some polysaccharides, have multiple repeating epitopes that cause extensive cross-linking of antibody molecules in the B-cell surface, signaling the B cell to proliferate and produce antibodies to the polysaccharide antigen.

make quick antibody responses to these molecules, even before Th cells develop, is an important early specific defense.

Clonal Selection and Expansion

The **clonal selection theory** states that acquired specific immunity depends on individual antigen-specific lymphocytes that proliferate in response to antigen, a reaction termed **clonal selection**, and differentiate into antigen-specific **effector cells** and **memory cells**. Proliferation leads to formation of expanded **clones** of cells, a process called **clonal expansion**. The effector cells eliminate the antigen, while the memory cells maintain immunity against later exposure to that antigen. During the primary antibody response, B lymphocytes recognize the antigen epitope by means of specific antibody molecules on their surfaces **(figure 16.14)**. This recognition alone does not initiate antibody production; rather, it primes the B cells to respond to co-stimulatory signals given by the Th cells, inducing the B cells to proliferate. Each cell divides to give two cells with the same antibody specificity, then four, then eight, and so on to form large clones of B lymphocytes producing the same antibody. After a period of division, some B cells differentiate, resulting in the production of plasma cells. These are highly specialized antibody-producing cells, capable of producing thousands of molecules of antibody per second. Plasma cells are terminally differentiated cells; after they produce quantities of antibody, they die. Further, some cells of the developing clone respond to cytokine signals to differentiate into memory cells with the same specificity as the B cells from which they arose. These memory cells persist in the body for years and are present in numbers sufficient to give a prompt and effective protective anamnestic, or memory, secondary response when the same antigen is encountered again at a later time. The activities of memory cells are responsible for (1) long-lasting immunity to a particular disease once a person has recovered from it and for (2) the effectiveness of vaccines.

The Primary and Secondary (Memory) Responses

The first time a particular antigen is introduced into the body, a primary response occurs; about 10 days to 2 weeks is required before a substantial amount of antibody is detected in the blood. The amount of antibody, or titer, rises with time until the antigen is removed and the response wanes **(figure 16.15)**. A slow rise in titer is typical of antibody production during the **primary antibody response**. If, however, the same antigen is introduced at a later date, the titer of antibodies rises much faster and is much higher, due to rapid activation of the clones of memory cells present. This is typical of the **secondary (memory) antibody response**.

IgM is the first class of antibody produced during an immune response, and it is the only class produced in response to T-cell-independent antigens. With T-cell-dependent antigens, though, there is a switch to other classes of immunoglobulin as the response develops. Cytokines produced by Th cells play a major role in causing the switch. In the secondary antibody response, essentially the same sort of IgM response occurs as in the primary response, but the IgG response is greatly accelerated and enhanced. In other words, memory develops for IgG production but not for IgM. The production of antibodies of all classes contributes to the elimination of pathogens, because each class can confer beneficial clearance mechanisms.

Memory Th and Tc cells are produced during the primary response. The development of memory cells is the basis for immunization. Both cell numbers and the potential to respond are increased in a memory response, so that protection is prompt and effective. ■ immunization, p. 425

The Diversity of Immune Specificity

Normal humans can respond immunologically to many millions of different antigen epitopes. Each epitope is recognized by specific antibodies on the surface of B lymphocytes, or specific antigen receptors on T-lymphocyte surfaces. Any lymphocyte will recognize only one specific antigen. During development, as lymphocytes differentiate from bone marrow stem cells into either B cells or T cells, each cell will commit to making a surface receptor for one given antigenic epitope. Once the precursor cell commits to its specificity, all clonal daughters of that cell will continue

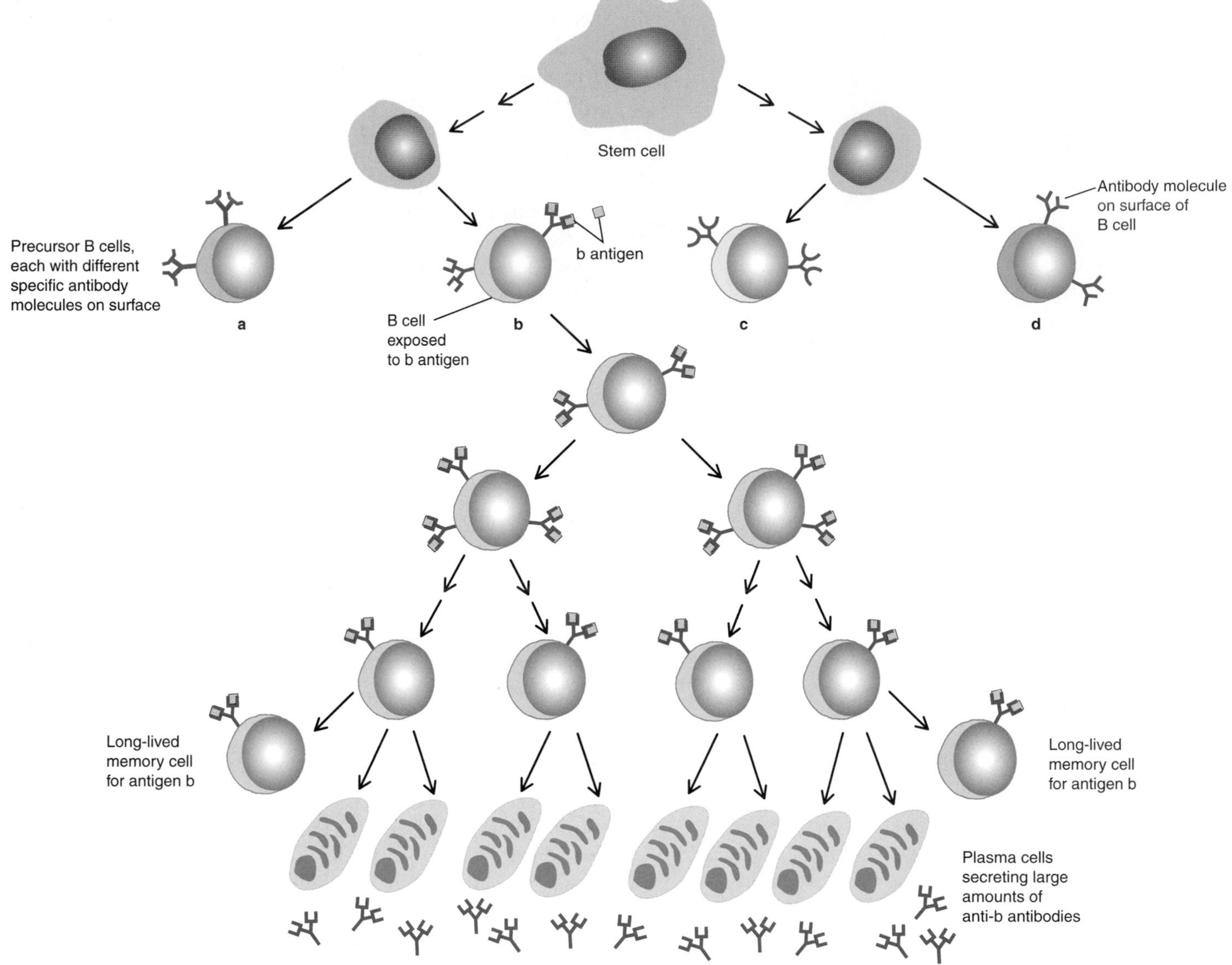

Figure 16.14 **Clonal Selection and Expansion During the Antibody Response**

to produce the same receptor specificity. Thus, out of millions of cells only a small number will make a receptor for that one epitope. A small number of lymphocytes will produce surface receptors for epitope d, another few lymphocytes will have receptors for epitope e, and so on, for millions of different epitopes.

For years, one of the mysteries of antibody formation was how B cells and plasma cells, using the limited genetic information available to them, produce antibody molecules specific for the millions of antigenic molecules known to exist. Human cells do not have enough DNA to have separate genes for each antibody molecule. Dr. Susumu Tonegawa won the Nobel Prize for Medicine in 1988 for solving the mystery of this antibody diversity. In fact, anyone aspiring to win the Nobel Prize would do well to enter the fields of microbiology or immunology. Many of the prizes in Medicine and Physiology have been awarded for work in these areas. Since 1960 about half of these prizes have been given in recognition of the astounding progress in these areas of investigation (see inside cover).

Tonegawa showed that the DNA contains a wardrobe of different genes that can be mixed and matched, much as shirts, slacks, and jackets can be mixed to produce different outfits. As the lymphocyte divides during maturation, its genes are rearranged. DNA sequences for each light chain are assembled by the random selection of a V (variable) gene segment and a J (joining) gene segment. DNA sequences for each heavy chain are assembled by the random selection of a V gene segment, a J gene segment, and a D (diversity) gene segment. The constant region of the molecule is encoded by a single C (constant) gene segment **(figure 16.16)**. The genes for H chains are given the Greek letters for their names, gamma for IgG, mu for IgM, alpha for IgA, delta for IgD, and epsilon for IgE. In this example, the H chain will form an IgG molecule because the gamma gene was selected. Only one of the constant region genes for H chains is selected.

Because of the large number of ways in which these genes can be rearranged, the various B lymphocytes together have the

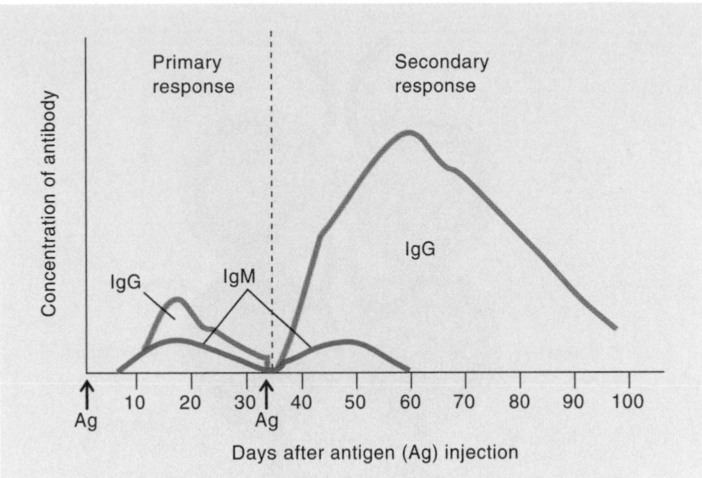

Figure 16.15 The Memory Response The first exposure to antigen causes a primary response, during which small amounts of antibody are produced for a short time. Subsequent doses of the same antigen stimulate a secondary (memory) response, during which much more antibody is produced over a longer period. During the secondary response, IgG production is accelerated and enhanced, while the IgM response is essentially the same as in a primary response.

capacity to produce a very large number of antibody proteins. Also, mutation can occur in immunoglobulin V genes, and several other means exist to increase variability. In mice, for example, in which extensive studies have been made, it is known that different arrangements of these DNA sequences can code for more than 10^{11} (100 billion) different antibody molecules. This number of different antibody molecules is ample enough to account for all antigens that might be encountered in the environment. A similar mechanism of gene rearrangement for generation of diversity occurs in T-cell receptors, leading to production of T-cell populations that can recognize all antigens that might be encountered.

MICROCHECK 16.6

Specific antibodies on B-cell surfaces recognize a single epitope to induce an antibody response. T-cell-dependent antigens require the cooperation of CD4 T-helper cells to elicit an antibody response. T-cell-independent antigens can induce antibody formation without help from T cells.

The clonal selection theory states that an antigen selects the lymphocytes specific for that antigen, resulting

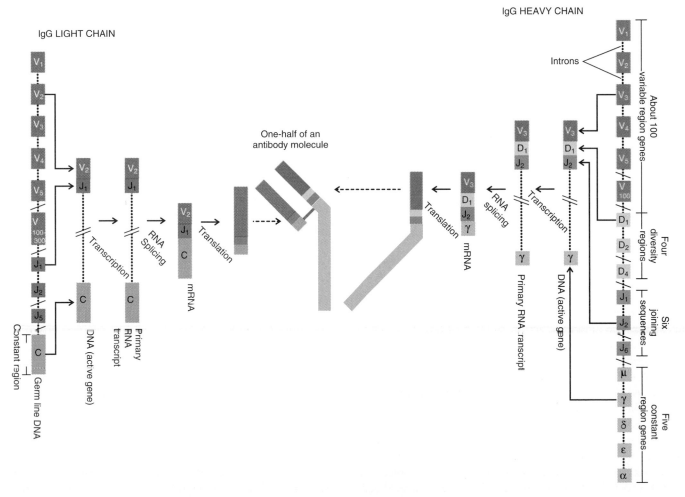

Figure 16.16 Antibody Diversity Immunoglobulin (IgG) gene arrangement in an immature lymphocyte and the mechanism of active gene formation. Shown are a heavy chain, a light chain, and formulation of one-half an antibody molecule.

in proliferation of expanded clones of antigen-specific effector and memory cells. The IgM response is the same in the primary and the secondary responses, but the IgG response is greater in secondary responses due to the quick response of memory cells.

Both B and T lymphocytes generate diversity in their antigen receptors by gene rearrangement. B lymphocytes also generate diversity by mutation in their variable genes.

- Why is clonal selection and expansion of B cells necessary?
- What are some major differences between the primary and secondary immune response?
- With type 2 T-cell-independent antigens, why are 12 to 16 epitopes necessary to elicit a good response? Would a smaller number work just as well?

Development of the Cell-Mediated Immune Response

Cell-mediated immunity is directed toward eliminating certain dysfunctional host cells, such as those infected by a pathogen or cancerous cells. Cell-mediated immunity is initiated by interaction of the antigen with a few T lymphocytes capable of recognizing that specific antigen, and it operates in two ways. Either a CD8 **T cytotoxic cell (Tc)** population develops that can kill antigen-bearing target cells, or macrophages are activated by cytokines produced by CD4 cells of the Th1 subpopulation. Such **activated macrophages** can kill or inhibit many intracellular microorganisms.

T Lymphocytes and the Cell-Mediated Immune Response

Recall that, in antibody formation, the antigen interacts with specific receptors, antibodies, on the surface of B cells. In cell-mediated immunity, antigen is recognized by specific molecules called **T-cell receptors (TCRs)** on the T-cell surface. The T-cell receptors for antigen are not immunoglobulins; rather, they are proteins with certain similarities to antibodies **(figure 16.17)**. The TCR consists of two polypeptide chains, each with variable and constant regions. As in the antibody molecule, the variable regions supply the antigen-reaction sites. The genes that code for T-cell receptors are rearranged genes, with diversity similar to that of antibodies.

The Tc-cell response is directed only to antigens that are on whole cells, such as virus-infected cells or tumor cells. This is because Tc cells are CD8 lymphocytes that only recognize antigen plus class I MHC antigen in a complex on the host cell membrane. The Th1 cells that activate macrophages to control intracellular microorganisms are CD4 lymphocytes that only recognize antigen complexed with MHC class II molecules. Both CD4 and CD8 lymphocytes develop from naive T cells that have specific TCRs but have not yet encountered antigen **(figure 16.18)**. The cytokine IL-2 drives the proliferation of naive T cells into clones, which differentiate under the influence of other cytokines into effector cells and memory cells.

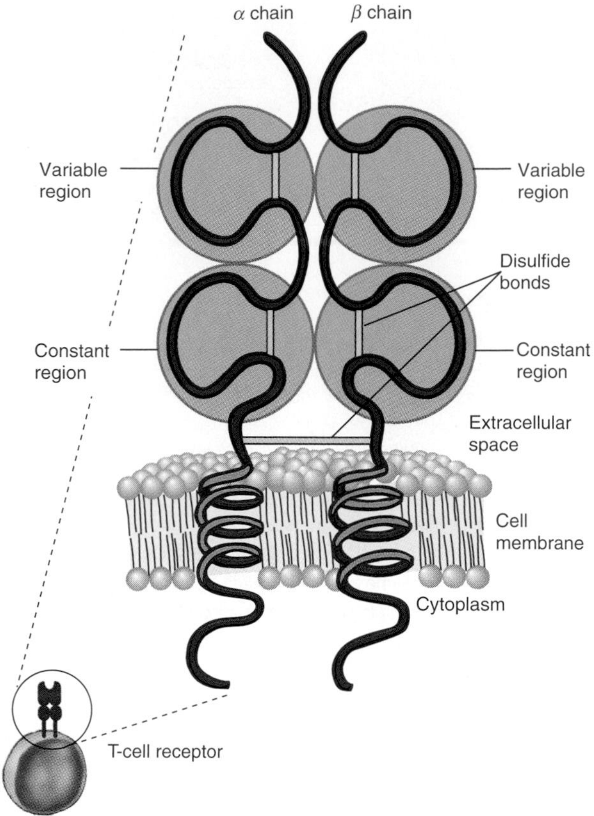

Figure 16.17 T-Cell Receptor Composed of an Alpha and a Beta Polypeptide Chain Each chain has one variable (V) and one constant (C) region. Unlike antibodies, which have two binding sites for antigen, T-cell receptors have only one. The icon for a T-cell receptor reflects the structure shown here.

Cytotoxic T Cells Against Cells Recognized by the Immune System

During cell-mediated immunity, a cancer cell or a virus-infected cell becomes a target to be destroyed. All nucleated cells routinely sample a small portion of the protein synthesized within the cell. The protein is degraded in the cellular recycling center (proteosome) and the resulting peptide fragments inserted into the grooves of newly synthesized class I MHC molecules for display on the cell surface. Nonself peptides presented with a co-stimulatory signal lead to the activation of T cells and induction of their cytotoxic capacity. Subsequent recognition of the same peptides on body cells leads to their destruction by these activated T cells or their daughters. For example, in the virus-infected cell, the virus causes the host cell to synthesize proteins of the viral coat. Just as with normal cell proteins, some of these are enzymatically degraded into short peptide sequences. Then, within the cell, they complex with MHC class I molecules being formed there. The complex of MHC class I with antigen peptide is then moved to the cell surface, where it can be presented to specific responding Tc cells **(figure 16.19)**. Only protein antigens can induce cell-mediated immunity, and the epitopes recognized are short peptides.

What is the advantage of having antigenic peptides presented on the cell surface together with MHC class I molecules? It could be disastrous if Tc cells destroyed normal cells that were not really infected but only had a molecule of foreign antigen

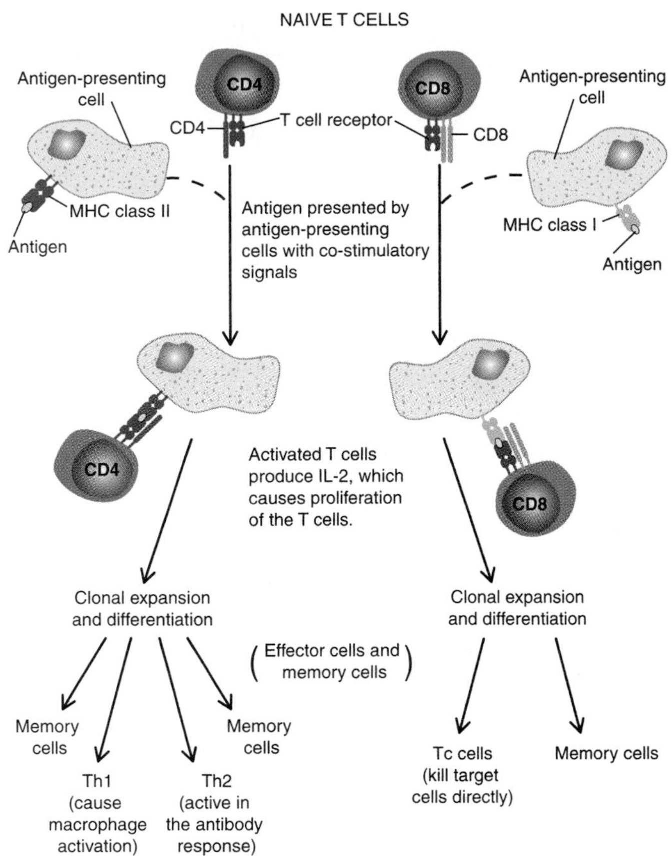

Figure 16.18 The Development of Activated T Cells from Naive T Cells that Have Specific Antigen Receptors but Have Not Yet Responded to Antigen

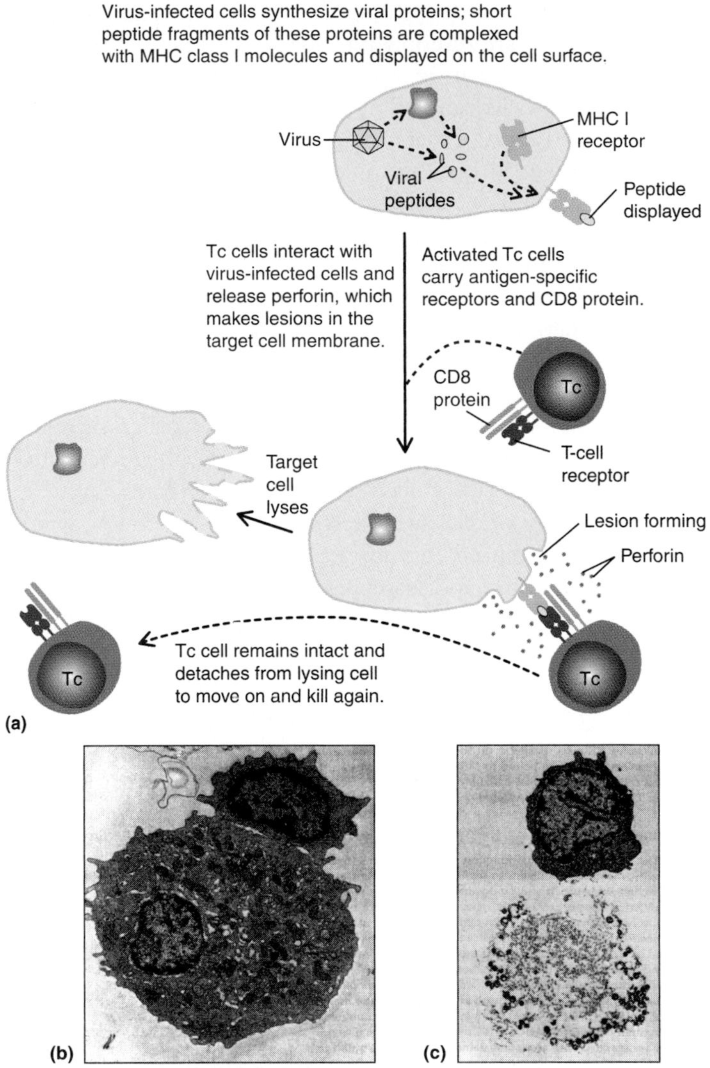

Virus-infected cells synthesize viral proteins; short peptide fragments of these proteins are complexed with MHC class I molecules and displayed on the cell surface.

Virus
Viral peptides
MHC I receptor
Peptide displayed
Tc cells interact with virus-infected cells and release perforin, which makes lesions in the target cell membrane.
Activated Tc cells carry antigen-specific receptors and CD8 protein.
CD8 protein
Tc
T-cell receptor
Target cell lyses
Lesion forming
Perforin
Tc cell remains intact and detaches from lysing cell to move on and kill again.
Tc
Tc
(a)
(b)
(c)

Figure 16.19 Tc-Cell Activity Against Target Cells **(a)** The same mechanisms are effective against virus-infected cells, tumor cells, and transplanted cells. **(b and c)** Electron micrograph shows Tc cell killing a target cell in culture.

attached to their surfaces. The fact that the antigen is recognized by Tc cells only when complexed with MHC I prevents uninfected healthy cells from being perceived as foreign and destroyed.

The responding Tc cells not only have antigen receptors, they also have CD8 receptors for MHC class I molecules. Specific reaction of the epitope and the T-cell antigen receptor and interaction between the MHC class I molecules and the CD8 molecules that serve as receptors for class I help to hold the Tc and the target cells together (see figure 16.19); in addition, a variety of antigen-nonspecific adhesion molecules are required to couple the effector and target cells together. Contents of granules in the Tc cell, such as perforin, are released in close contact with the target cell membrane. Perforin, a molecule produced by Tc cells (and the immunologically nonspecific natural killer [NK] cells), kills target cells by forming a pore similar to that formed by the membrane attack complex of complement. The Tc cell survives and can go on to kill other targets (see figure 16.19). NK cells are not specific with respect to antigen, but they do recognize abnormal cells that lack MHC I (see chapter 19). ■ **membrane attack complex, p. 377**

Since tumor (cancer) cells arise from host tissue, it might be expected that they would not incite an immune response. It has clearly been shown, however, that they produce antigens that differ from normal host tissue and that lead to immune responses. When tumors are induced in animals by certain chemicals, each tumor has its own unique antigens not shared by other tumors produced in the same way. When tumors are induced by a virus, however, they have shared surface antigens. This lends hope that it may be possible eventually to immunize against a number of human tumors, such as those induced by viruses. In fact, the hepatitis B vaccine is expected to give protection against some liver cancer in humans. In animals, effective immunization is available for Marek's disease, a virus-induced tumor of chickens. Tumor cells are destroyed largely by Tc-cell activity, although other mechanisms also act against these immunologically foreign cells.

In addition to virus-infected cells and tumor cells, transplanted cells and organs, such as a kidney transplant from one person to another, are also recognized as foreign and are destroyed largely by cell-mediated mechanisms (see chapter 18).

Control of Microbial Growth in Macrophages

Different mechanisms of cell-mediated immunity operate against some pathogens that multiply in host macrophages, where they are shielded from antibodies and many other host defenses. If the immune system were to kill the macrophages in which organisms are multiplying, this would result in the infection of many more cells by the released organisms. Instead, the subset of CD4 cells called Th1 produce a variety of cytokines that cause activation of macrophages. The CD4 (Th1) cells recognize microbial antigens complexed with class II MHC molecules and respond by making a variety of cytokines, such as interferon gamma, and tumor necrosis factor, that activate macrophages, producing dramatic changes. The cells enlarge, their membranes become ruffled and irregular, and they become metabolically very active, with greatly increased numbers of lysosomes containing antimicrobial substances (see figure 15.3). The activated macrophages can then kill the intruders within phagolysosomes (**figure 16.20**). The other subset of CD4 cells, the Th2 cells, do not cause macrophage activation because they do not produce the activating cytokines.

This form of cell-mediated immunity operates against many diseases caused by intracellular infections, including tuberculosis, Hansen's disease (leprosy), brucellosis, tularemia, histoplasmosis, and others. For example, in tuberculosis, *Mycobacterium tuberculosis* bacilli usually are inhaled and infect the lung. Like all other extraneous inhaled material, the organisms are engulfed by the many alveolar macrophages in the lung, but these bacteria are difficult for the macrophages to destroy. The chemical composition of their cell walls, with a very high concentration of lipids complexed with peptides and polysaccharides, makes the organisms resistant to destruction within normal macrophages. In fact, once inside the macrophages, the bacilli multiply for a time. Then, as cell-mediated immunity to the bacilli develops, the macrophages become activated. They increase in size and membrane activity. Their lysosomes increase in number and contain greatly increased amounts of degradative enzymes. New enzymes are also induced. All of this activity is directed toward the destruction of the intracellular bacteria. If the immune response does not keep up with the growth of the bacteria, disease results. This kind of cell-mediated immunity, however, is often enough to control a primary infection and prevent progression to a disease state. A few organisms usually remain in cells, enough to maintain a state of responsiveness to the antigens, known as delayed hypersensitivity (see chapter 18).

If the macrophage response described is not sufficient to control the infection, other mechanisms come into play. Activated macrophages that cannot destroy intracellular microorganisms are able to fuse together. As many as 50 to 100 macrophages may fuse their membranes to form a **giant cell**. Macrophages and giant cells form foci, called **granulomas**, that wall off and retain organisms that cannot be destroyed by the cells, thereby preventing the microbes from escaping to infect other cells. Granulomas are part of the disease process in tuberculosis and similar diseases, sometimes called granulomatous diseases.

When the immune system is compromised, as in AIDS or other immunosuppressed states, the few tubercle bacilli that remain in the body from a primary infection with *Mycobacterium tuberculosis* can multiply, causing active tuberculosis. This can also happen in older people who were infected at an early age and were able to contain the infection without developing disease. Later in life, when immune function declines, as it does with aging, the few bacilli resume multiplication and cause disease. Reactivation tuberculosis, due to immunosuppression, is a major problem in some parts of the world, such as Africa, where inactive infection with tubercle bacilli is common (in some areas estimated to occur in more than 80% of the population), and immunosuppression caused by AIDS, malaria, and other diseases is rampant.

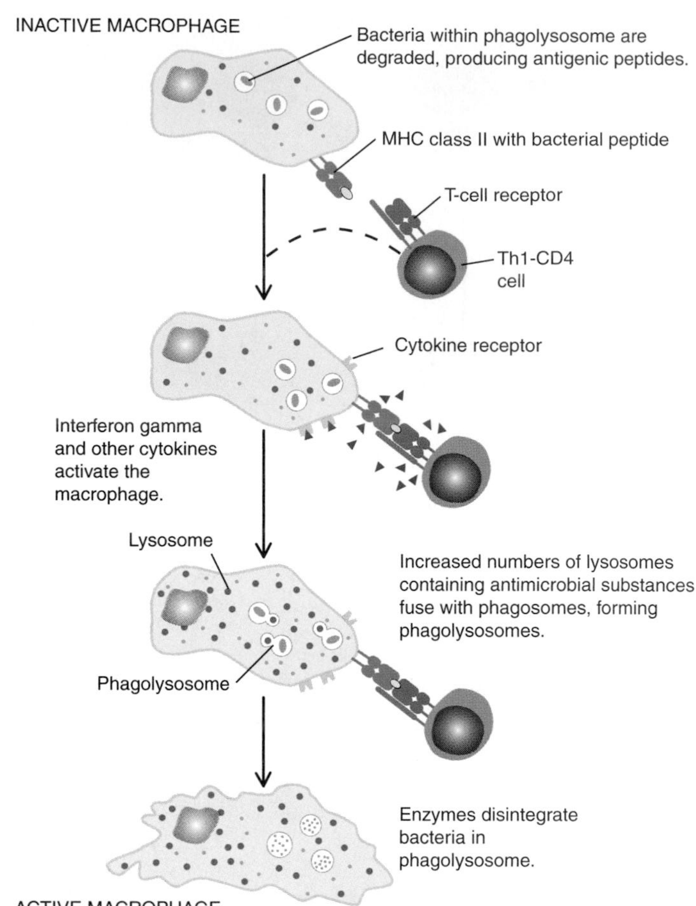

Figure 16.20 **Control of Microbial Growth in Macrophages**

The cell-mediated immune response differs in many ways from the antibody response. **Table 16.2** compares the two responses. ■ **tuberculosis, p. 569** ■ **AIDS, p. 655**

MICROCHECK 16.7

T cytotoxic (Tc) cells respond to protein antigens on intact cells, complexed with MHC I antigen. T-cell receptors recognize the antigen. Tc cells attack and destroy virus-infected cells and cancer cells by releasing contents of Tc-cell granules at close range. Cytokines produced by CD4 (Th1) cells activate macrophages, controlling growth of pathogens within them.

- How do T cytotoxic cells kill target cells, such as virus-infected or cancer cells?
- Describe how cell-mediated immunity controls growth of pathogens within macrophages.
- What would be the consequences if dysfunctional host cells were not eliminated by cell-mediated immunity?
- In T-cell killing, why is close contact with the target cell necessary? What if the Tc cell was farther away?

TABLE 16.2 Comparison of Antibody-Mediated and Cell-Mediated Immune Responses

	Antibody-Mediated		Cell-Mediated	
	T Dependent	**T Independent**	**T Cytotoxic**	**Activated Macrophages**
Lymphocytes Responsible	B cells, plasma cells	B cells, plasma cells	CD8T cells	CD4T cells (Th1)
Lymphocyte Receptor for Antigen	Immunoglobulin	Immunoglobulin	T-cell receptor	T-cell receptor
T-Cell Help	CD4Th2	None	CD4Th1	CD4Th1
MHC Requirement	Class II + antigen peptide	None	Class I + antigen peptide	Class II + antigen peptide
Cytokine Activities	IL-4, IL-5, IL-10	None	Interferon gamma, tumor necrosis factors	Interferon gamma, tumor necrosis factors
Effector Functions	Production of antibodies of all classes	Production of IgM antibodies only	Killing of target cells by apoptosis or by perforin action	Intracellular organisms in macrophages inhibited or killed

Immunological Tolerance

A state of *specific* unresponsiveness is called **immunological tolerance**. Failure to respond to self-antigens is an example of immunological tolerance. When tolerance to self is lost, immune responses can damage the body, a condition known as autoimmunity. Autoimmune diseases are common, including rheumatoid arthritis, lupus erythematosus, some forms of diabetes and thyroiditis, and many others. It is possible under some circumstances to develop immunological tolerance to antigens other than self. This is desirable in the case of transplantation, when the transplanted tissues must be spared immunological rejection, but inducing tolerance is not readily accomplished. It is important to distinguish between tolerance, which is specific, and **immunosuppression**, which is a *nonspecific* general suppression of acquired immune responses. ■ autoimmune diseases, p. 446 ■ transplantation, p. 443

One mechanism leading to immunological tolerance is **clonal deletion**, the destruction of cells that recognize and respond to a particular antigen. During development, immature T cells in the thymus first express neither CD4 nor CD8 proteins. As they develop they express both CD4 and CD8, before maturing into either CD4 or CD8 cells. At the immature stage of $CD4^+$ $CD8^+$ expression, deletion of cells may occur. The immature T cells interact with the specific self-antigen they recognize. When many T-cell receptors strongly bind to the self-antigen:MHC complex shown by antigen-presenting cells, a signal is sent that results in deletion of that clone of cells by the programmed cell death known as apoptosis. Self-tolerance results from such clonal deletion. Some immature B cells are also deleted in the bone marrow. ■ apoptosis, p. 358

Another means of inducing immunological tolerance, called **clonal anergy**, occurs when clones of potentially responsive cells cannot make an active response. Anergy occurs when a **naive** lymphocyte—that is, a lymphocyte (either CD4 or CD8) that has not yet met and responded to its specific antigen—encounters that antigen:MHC complex in the absence of co-stimulatory signals. The co-stimulation is normally provided by certain CD molecules on antigen-presenting cells interacting with other molecules on the responding lymphocytes. In clonal anergy, the T cells do not make IL-2, the cytokine that drives clonal proliferation and differentiation to active effector cells. Lack of response can occur also when Th-cell function is lacking, so that B cells do not get the help they need to produce antibodies. Also, active cells can be suppressed by various means, including by T-suppressor cells.

MICROCHECK 16.8

Immunological tolerance is a state of specific unresponsiveness. It may be induced by clonal deletion, or by clonal anergy. T-suppressor cells may also prevent responsiveness.

- What is the difference between immunosuppression, as in malnourished people, and immunological tolerance?
- What is clonal deletion?

Control of the Immune Response

Clearly, a process as complex as the immune response must have equally complex regulatory mechanisms to prevent overproduction of antibodies or immune T lymphocytes. One of the more obvious agents of control is the antigen itself. As the immune response increases in intensity, antigen is effectively removed. When antigen is no longer available, B and T cells are no longer stimulated and the response wanes. Also, as antibodies and immune T cells are formed, they can inhibit the specific response, not only by removing antigen, but also by preventing the activation of naive B and T cells. For example, anti-Rh antibody is used clinically to prevent Rh-negative mothers from responding to the Rh-positive red cells of their babies. The soluble specific antibodies bind to antigen and then can cross-link receptors on the surface of the naive B cells, preventing activation. ■ Rh disease (hemolytic disease of the newborn), p. 438

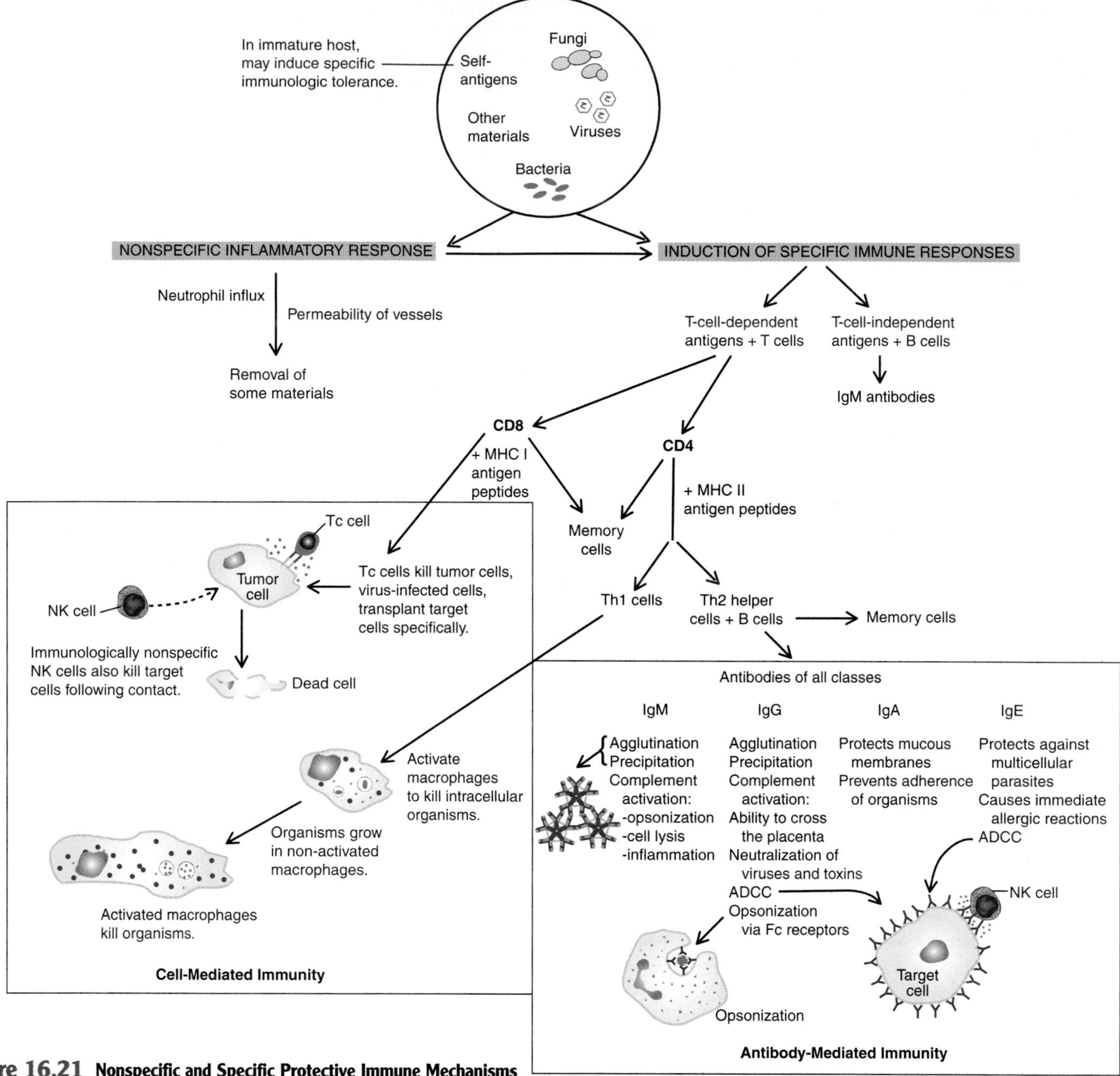

Figure 16.21 **Nonspecific and Specific Protective Immune Mechanisms**

Immune responses are also regulated by T-cell activities. Complex interactions of cytokines are critically important in regulating and controlling immune responses. Some of the cytokines produced by T cells stimulate and some inhibit other immunologically active cells. As mentioned earlier, T-suppressor cells may block antibody production and Tc-cell development, probably by producing cytokines and other soluble factors that suppress the other cells.

The genetic makeup of an individual has an influence over immune responses. Some people lack the genetic capability to respond or they respond poorly to certain antigens, whereas others have the genetic makeup to respond well to those antigens. It is possible to breed animals to produce offspring that are either high- or low-antibody producers to various antigens. Genes for MHC that influence the presentation of antigens also help to regulate the response.

It is well established that general body fitness, good nutrition, well-balanced endocrine function, and lack of stress are important in maintaining optimal immune function. This is seen, for example, in the failure of immunity in severely malnourished persons.

A summary of immune processes is shown in **figure 16.21**.

MICROCHECK 16.9

Control of immune responses is complex. It includes removal of antigen, inhibition by specific antibodies, cytokine activities, genetic influences, and general health and body function.

- How could antigen control the immune response?
- How could immune responses be regulated by T cells?

FUTURE CHALLENGES

Cell Signaling and Gene Regulation in Immune Responses

Questions constantly arise in the study of immunology. How does this cell perform its function? Why does the antigen cause one sort of response and not another? How do cells send their signals, and do they always send them in the same manner? Many of these puzzles have been solved by research done over the past decade, but many more challenging questions remain unanswered.

Some investigators are examining the exact ways in which lymphocytes develop from immature cells into mature B or T lymphocytes. Others are looking for details of the structural and molecular mechanisms involved in the interaction of MHC molecules and antigen leading to antigen recognition. Another challenge is offered in working out the details of T-cell activation. What signals are sent to the cells, and how do these signals operate to cause the cell to become active? Although some of the signals are known, much remains to be learned about how they work and about how the many different signals interact.

Cytokines present another fertile field for research. Some cytokines send different signals to different types of cells, and most of them have complex interactions with cells. It is a challenge to learn more about the exact structure of the cytokine molecules and the receptors where cytokines interact with cells. How do these molecules function in regulation of immune responses, and in the effector activities of immune cells? What are the genes that regulate such diverse immunological phenomena as cytokines, the programmed cell death known as apoptosis, and the switch from IgM to other classes of antibodies, among others?

One key challenge is presented in finding ways to use basic scientific knowledge in the treatment of patients. For example, an IL-2 receptor has been found on leukemia cells of some patients that binds IL-2 more firmly than the receptor on normal cells. This offers an opportunity to attack the leukemia cells by attaching radioactive substances to antibodies specific for the patient's IL-2 receptors. The antibodies then could deliver radioactivity directly and selectively to the leukemia cells. Another example is found in attempts to induce Tc lymphocytes specific for molecules common in human cancers; the Tc cells could then be used in cancer treatment. Studies are under way to find how genes are regulated in normal and cancerous lymphocytes, and how this knowledge can be used to control cancer cells.

Obviously, the challenges are great, but the tools to study and find answers to many of these questions are at hand. We can expect that many more Nobel Prizes will be awarded in immunology in the coming years.

SUMMARY

Fundamental Features of Specific Acquired Immunity

1. Acquired immune responses are highly antigen-specific.
2. Once the host has responded to a specific **antigen**, the **memory** response to subsequent exposure to the same antigen is more rapid and effective, thereby quickly eliminating that antigen.
3. The responses involve two sets of lymphocytes, acting with macrophages and dendritic cells. **B cells** produce an **antibody response**, and **T cells** produce a **cellular immune response**.

The Nature of Antigens and Haptens

1. Antigens are large molecules that contain many surface **epitopes**, are usually foreign to the host, and react specifically with antibodies and immune cells.
2. Most antigens are **immunogens** that can induce the production of specific antibodies or immune cells.
3. **Haptens** are small molecules that cannot induce antibody formation unless complexed with large carrier molecules, but the hapten alone can then react specifically with the antibodies or immune cells produced in response to the hapten-carrier complex. (Figure 16.2)
4. The immune response is directed toward epitopes on the antigen, sequences of 10 to 25 amino acids in the case of proteins. (Figure 16.1)
5. Antigens may be virtually any substance recognized by the host's immune system.

The Nature of Antibodies

1. **Immunoglobulin** antibodies react specifically with the antigen that induced their formation.
2. Immunoglobulins are proteins with two heavy and two light polypeptide chains that bind antigen in a close complementary fit. (Figure 16.3)
3. The Fc part of the molecule accounts for many of the biological functions of the antibody, unique to each class.

Properties of Antibodies

1. Antibody monomers have a Y shape with an **antigen-binding (Fab) site** at the end of each arm of the Y. The tail of the Y is the **Fc region**.
2. Noncovalent, short-range bonds hold antigen and antibody together. The bonding is reversible.

Immunoglobulin Classes (Table 16.1)

1. IgG is a monomer that can cause opsonization, agglutination, precipitation, complement fixation, **ADCC**, and neutralization of toxins and viruses. It is the only class of immunoglobulins that can cross the placenta. (Figures 16.4, 16.6, 16.7, and 16.8)
2. IgM, usually a pentamer, is the first class of immunoglobulins produced during an immune response. It is very efficient in agglutination, precipitation, opsonization, and complement activation. (Figure 16.9)
3. IgA is abundant as a dimer in secretions. It prevents infection by inhibiting adherence of organisms to host cells, and it is important in protecting mucous membrane surfaces. (Figure 16.9)
4. IgD is a monomer found on B-cell surfaces, where it is a membrane receptor for the specific antigen it recognizes.
5. IgE is a monomer that binds strongly to mast cells and basophils. Reaction of cell-bound IgE helps protect against some

multicellular parasites and also contributes significantly to many allergic reactions.

The Role of Lymphocytes in Specific Immunity

1. B lymphocytes mature into plasma cells, the principal antibody producers.
2. Many CD4 T cells are **T-helper (Th) cells** in antibody and cellular immune responses, whereas many CD8 T cells are **cytotoxic (Tc) cells** in the cellular response.
3. T lymphocytes multiply and mature in the thymus, B lymphocytes in the bone marrow.
4. T and B lymphocytes cannot be distinguished by their appearance. A variety of cell surface molecules are used to identify the various cells. These include immunoglobulins, T-cell receptors, and CD molecules.
5. Lymphoid follicles in secondary lymphoid tissues, such as lymph nodes, provide the environment for immune cells to interact and for the immune response to develop.

Major Histocompatibility Complex (Figure 16.11)

1. Class I and class II **major histocompatibility complex** proteins are located on cell surfaces.
2. MHC class I molecules are found on almost all human cells, whereas MHC class II molecules are expressed only on **antigen-presenting cells** such as macrophages, dendritic cells, and B cells.

Class I MHC

1. Cells expressing peptides using MHC class I molecules can be recognized by antigen-specific CD8-bearing T cells.

Class II MHC

1. Cells expressing peptides using MHC class II molecules can be recognized by antigen-specific CD4-bearing T cells.

Development of the Antibody Response

1. A single epitope selects B cells by binding to specific antibodies on the B cell surface to induce an antibody response.
2. T-cell-dependent antigens require the cooperation of CD4 T-helper cells to elicit an antibody response. T-cell-independent antigens can induce antibody formation without help from T cells.
3. Both T and B lymphocytes generate diversity in their antigen receptors by gene rearrangement.

B Lymphocytes and the Antibody Response

1. Specific antibodies on B-cell surfaces recognize a single antigen. (Figure 16.12)

T-Cell-Dependent Antigens

1. T-cell-dependent antigens require the cooperation of CD4 Th cells to elicit an antibody response.
2. Extracellular antigens must be broken down into fragments, complexed with class II MHC, and presented to Th cells by macrophages or other antigen-presenting cells. These cells secrete cytokines that activate B cells to proliferate into clones, and to synthesize and secrete antibodies.

T-Cell-Independent Antigens (Figure 16.13)

1. T-cell-independent antigens can induce IgM antibody formation without help from T cells.

Clonal Selection and Expansion (Figure 16.14)

1. The **clonal selection theory** states that antigen selects the lymphocytes specific for that antigen, resulting in proliferation of expanded **clones** of antigen-specific **effector** and **memory cells**.
2. When B lymphocytes respond to antigen, differentiation leads to production of plasma cells that make antibody, and to the production of memory cells.

The Primary and Secondary (Memory) Responses

1. A slow rise in first IgM and later IgG antibodies occurs in the **primary response**. In the **secondary response**, IgM production is the same, but there is a rapid rise in IgG antibodies, due to memory cell activity. (Figure 16.15)

The Diversity of Immune Specificity

1. The few genes required for producing immunoglobulin molecules can be rearranged in many millions of combinations to provide antibodies of millions of specificities. (Figure 16.16)
2. Mutation in genes for the variable region also contributes to antibody diversity.

Development of the Cell-Mediated Immune Response (Table 16.2)

T Lymphocytes and the Cell-Mediated Immune Response

1. Surface receptors on T cells contain variable regions that recognize antigen. (Figure 16.17)
2. Tc-cell receptors recognize antigen complexed with class I MHC, and Th-cell receptors recognize antigen complexed with class II MHC.

Cytotoxic T Cells Against Cells Recognized by the Immune System

1. Tc cells attack and destroy virus-infected cells by releasing contents of granules at close range.
2. Cancer cells and transplanted cells are also attacked and destroyed by Tc cells.

Control of Microbial Growth in Macrophages

1. Cytokines produced by CD4 (Th1) cells activate macrophages, resulting in killing of intracellular organisms (Figure 16.20)

Immunological Tolerance

1. **Immunological tolerance** is a state of specific unresponsiveness. It may be induced by **clonal deletion**, or by **clonal anergy**. **T-suppressor cells** may also prevent responsiveness.

Control of the Immune Response

1. Control of immune responses is complex. It includes removal of antigen, inhibition by specific antibodies, cytokine activities, genetic influence, and general health and body function.

REVIEW QUESTIONS

Short Answer

1. Which immunoglobulin classes can activate complement? How does this ability relate to their function?
2. Why is IgE so scarce in the circulation?
3. How can T lymphocytes be distinguished from macrophages?
4. How do natural killer cells differ from T cytotoxic cells?
5. What is the major difference between responses to T-cell-dependent and T-cell-independent antigens?
6. What is the major histocompatibility complex (MHC)? What is its importance?
7. Describe clonal selection and expansion in the immune response.
8. Is there reason to hope for a vaccine to protect against all cancers or any cancers? Explain.
9. How does the antigen receptor on T cells differ from the antigen receptor on B cells?
10. What are the "professional" antigen-presenting cells? How do they present antigen?
11. Diagram an IgG molecule and label (a) the Fc area and (b) the areas that combine with antigen.

Multiple Choice

1. Newborn infants…
   A. lack any immunological protection.
   B. benefit from the mother's IgM that crossed the placenta.
   C. gain protection from IgA in breast milk and IgG that crossed the placenta.
   D. are immunologically mature
   E. have adult levels of IgE
2. The interaction of antigens and antibodies…
   A. is nonspecific.
   B. depends on the formation of covalent bonds.
   C. cannot be reversed.
   D. only occurs in the body.
   E. can lead to many biological results, such as phagocytosis.
3. B lymphocytes resemble T lymphocytes in…
   A. that they both have CD8 molecules on their surfaces.
   B. that they both have immunoglobulin on their surfaces.
   C. their ability to mature into plasma cells.
   D. their cytotoxic capacities.
   E. their appearance in micrographs.
4. T-cell-independent antigens act in at least two ways to…
   A. stimulate B cells to multiply and produce antibodies.
   B. induce cytotoxic T cells to function.
   C. restrict antibody formation.
   D. activate macrophages.
   E. activate Th cells.
5. Immune responses may be regulated by all of the following mechanisms, *except*…
   A. availability of antigen.
   B. presence of specific antibodies.
   C. activity of T-suppressor cells.
   D. genetic influences.
   E. production of natural killer cells.
6. Acquired immune responses have all of the following features, *except*…
   A. they are highly specific.
   B. they have memory.
   C. they involve B cells.
   D. they involve T cells.
   E. anamnestic responses are not antigen-specific.
7. Antigens reacted to by a host are usually…
   A. self-proteins.
   B. small molecules.
   C. molecules with a single epitope.
   D. lipids.
   E. proteins or polysaccharides.
8. Secretory IgA…
   A. is a monomer.
   B. is a trimer with a J chain.
   C. lacks a J chain.
   D. is found largely in the circulation and lymph nodes.
   E. gains a secretory component at the mucous membrane.
9. The major histocompatibility complex has all the following characteristics, *except*…
   A. is a gene complex.
   B. codes for class I molecules found on most body cells.
   C. codes for class II molecules found on T lymphocytes.
   D. codes for cell surface proteins.
   E. in humans is referred to as HLA.
10. All antibody molecules…
    A. consist of H and L polypeptide chains.
    B. can fix complement.
    C. can cross the placenta to the fetus.
    D. contain J polypeptide chains.
    E. may be secreted onto mucous membranes.

Applications

1. Currently there is much debate about keeping active smallpox virus stored, since the disease has been fully eradicated. What would be an argument for keeping the virus? What should be done to protect against possible use of smallpox virus in biological warfare?
2. A researcher discovers a drug that prevents allergic reactions in people sensitive to chocolate. It works without suppressing the

rest of the immune system and does not affect antibody actions. How might such a drug work?

3. What kinds of diseases would be expected to occur as a result of lack of T or B lymphocytes?

Critical Thinking

1. The development of primary and secondary immune responses to an antigen differ significantly in their timing. The primary response may take a week or more to develop fully and establish memory. The secondary response is rapid and relies on the activation of clones of memory cells. Wouldn't the body be better served if clones of reactive cells were maintained regardless of prior exposure? In this way, the body could always respond rapidly to *any* antigen exposure. Would there be any disadvantages to this approach? Why?
2. Contrary to the human immune system, the immune system in sharks relies heavily on V, J, and D gene segments that produce "pre-joined" antibodies. In other words, the shark system lacks much of the immunologic flexibility found in humans (although sharks do have some genes that can recombine). What hypothesis could explain why the antibody production system of sharks depends so strongly on genes with fixed antigen specificity?
3. Early investigators proposed two hypotheses to explain the specificity of antibodies. The clonal selection hypothesis states that each lymphocyte can produce only one specificity of antibody. When an antigen appears that binds to that antibody, the lymphocyte is selected to give rise to a clone of plasma cells producing the antibody. The template hypothesis states that any antigen can interact with any lymphocyte and act as a template, causing newly forming antibody to be specific for that antigen. In one experiment to test these hypotheses, an animal was immunized with two different antigens. After several days, lymphocytes were removed from the animal and individual cells placed in separate small containers. Then, the original two antigens were placed in the containers with each cell. What result would support the clonal selection hypothesis? The template hypothesis?

Applications of Immune Responses

Vaccination, the practice of deliberately stimulating the immune system in order to protect individuals against a disease, has become routine all over the world. As a result, many of the diseases that once caused widespread death and disability, such as smallpox, measles, whooping cough, mumps, and poliomyelitis, are now nonexistent or much less common.

Even before people knew that microorganisms caused disease, it was recognized that individuals who recovered from a disease such as smallpox rarely contracted it a second time. Old Chinese writings dating from the Sung dynasty (A.D. 960–1280) describe a procedure known as variolation, in which small amounts of the powdered crusts of smallpox pustules were inhaled or placed into a scratch made in the skin. Usually the resulting disease was mild, and a permanent immunity to smallpox resulted. Occasionally, however, severe disease developed, often resulting in death. Nevertheless, the risk of serious disease using this procedure was rare enough that people were willing to take the risk.

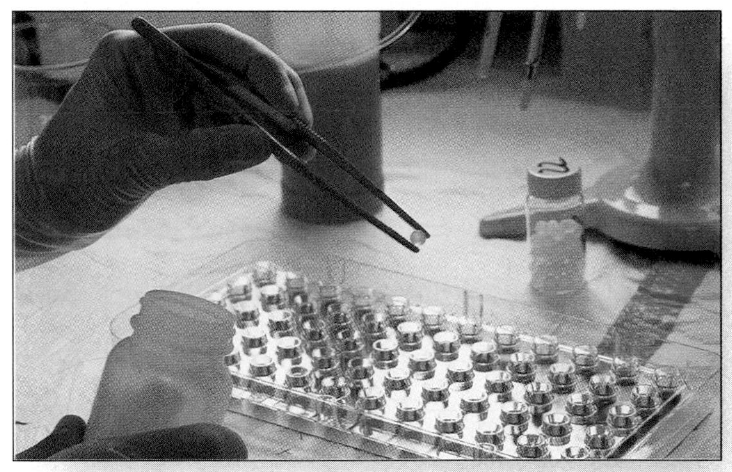

Although variolation was practiced in China and the Mideast a thousand years ago, it was not widely used in Europe until after 1719. At that time, Lady Mary Wortley Montagu, wife of the British ambassador to Turkey, had their children immunized against smallpox in this way. Since smallpox was disfiguring, producing deep scars ("pocks"), and was often fatal, variolation subsequently became popular in Europe.

Although a person exposed to smallpox through variolation would usually completely recover and so was not at great risk, he or she would become contagious. Because of the danger of contagion and because the procedure was reasonably expensive, large segments of the population in Europe remained unprotected.

As an apprentice physician, Edward Jenner noted that milkmaids who had suffered cowpox infections rarely got smallpox. Cowpox was a disease of cows that caused few or no symptoms in humans. In 1796, long before bacteria and viruses had been discovered, Jenner conducted a classic experiment in which he deliberately transferred material from a cowpox lesion on the hand of a milkmaid, Sarah Nelmes, to a scratch on the arm of a young boy named James Phipps. Six weeks later, when exposed to pus from a smallpox victim, Phipps did not develop the disease. The boy had been made immune to smallpox when he was inoculated with pus from the cowpox lesion. Using the less dangerous cowpox material in place of the pustules from smallpox cases, Jenner and others worked to spread the practice of variolation. Later, Pasteur used the word vaccination *(from the Latin* vacca *for "cow") to describe any type of protective inoculation. By the Twentieth century, most of the industrialized world was generally free of smallpox as the result of routine vaccination of large populations.*

In 1967, the World Health Organization (WHO) initiated a program of intensive smallpox vaccination. Since there were no animal hosts and no nonimmune humans to which it could be spread, the disease died out. The last case of naturally contracted smallpox occurred in Somalia, Africa, in 1977, and in 1979 in a ceremony in Nairobi, Kenya, WHO declared the world free of smallpox. Nevertheless, a few laboratories around the world still have the virus. In this age of potential bioterrorism, some see smallpox as a major threat should the deadly virus ever be released into the largely unprotected populations of the world. Only small stores of the vaccine are currently available.

It is now known that the smallpox and cowpox viruses are closely related. The vaccinia virus used in recent years for vaccination against smallpox is neither the cowpox nor the smallpox virus, and it is not clear where this virus originated. Some scientists believe that the vaccinia virus is a hybrid of the cowpox and smallpox viruses. Quite possibly it developed during Jenner's time, when calves injected with cowpox virus to produce vaccine material accidentally received the smallpox virus at the same time. In addition to its importance in defeating smallpox, the vaccinia virus has been genetically engineered in recent years to make experimental vaccines against other diseases such as AIDS and viral hepatitis. Thus, Edward Jenner's legacy extends beyond the eradication of smallpox.

—A Glimpse of History

IN THE LAST TWO CHAPTERS, WE DISCUSSED BASIC defense mechanisms, both nonspecific and specific, and became acquainted with antibodies and immune cells. This chapter will explore some useful applications of immunological reactions in diagnostic tests. We will also consider how immunization can be used to enhance the immune response and how immunization techniques have advanced remarkably in recent years to become safer and more effective. In fact, immunization has had probably the greatest impact on human health of any medical procedure, and we shall see that even better means of immunization are likely in the near future.

Principles of Immunological Testing

The specificity of immunological reactions is useful in many diagnostic tests. During most infectious diseases, antibodies and immune cells specific for the infecting agent are produced. In a number of diseases other than infections, antibodies or immune cells are also formed. These can be detected and measured to aid in diagnosis and management of the diseases. Specific antibodies are found in many locations; they are plentiful in the blood and are easily obtained from serum, the fluid from clotted blood. **Serology** refers to the use of serum antibodies to detect and measure antigens, or conversely, the use of antigens to detect serum antibodies. **Immunoassay** is another term used to describe assays, or tests, using immunological reagents such as antigens and antibodies. While in many serological tests known antigens are used to detect serum antibodies, it is important to remember that known antibodies can also be used to identify antigens. Monoclonal antibodies are often used in immunoassays (see **Perspective 17.1**).

Quantifying Antigen-Antibody Reactions

Individuals exposed to infectious agents for the first time usually do not have detectable specific antibodies in the blood serum until about a week or 10 days after infection. The change from negative serum without specific antibodies to serum positive for specific antibodies is called **seroconversion.** As the infection progresses, more and more antibodies are formed and the amount of antibody in the blood, the **titer**, increases. Small amounts of antibody that do not increase during the disease could result from infection at a previous time or for other reasons, but a **rise in titer** of antibodies is characteristic of an active infection.

Specimens that can be tested for the presence of antigens or antibodies include blood serum (the fluid portion remaining after blood clots) and plasma (the fluid portion of blood, treated with an anticoagulant to prevent clotting). In addition, urine, cerebrospinal fluid, sputum, or other body constituents may contain antigens or antibodies. Also, antigens and antibodies can be detected in solid tissues in thin sections examined with tagged reagents, such as fluorescent-labeled antibodies. Blood serum is probably the specimen most often used. A single sample of serum or other specimen can be used for many different tests, due to the availability of micromethods and automated techniques.

The amount of antibodies or the titer of a serum sample is usually determined in **serial dilutions.** Serum is diluted in either 2-fold or 10-fold dilutions, antigen is added to each dilution, and the titer is assigned to the last dilution that gives a detectable antigen-antibody reaction. For example, undiluted serum is diluted with a saline solution half and half, to give a 1:2 dilution; this dilution is again mixed with an equal volume of saline to give a 1:4 dilution; the 1:4 dilution is again diluted by half to give a 1:8 dilution; and so on, to produce dilutions of 1:2, 1:4, 1:8, 1:16, 1:32, 1:64, 1:128, 1:256, 1:512, 1:1,024, and so forth. To each of these dilutions of antibody-containing serum is added an equal volume of antigen, further diluting the amount of serum present. For example, if the volume of serum dilution is 0.1 ml, then 0.1 ml of antigen is added. The final serum dilutions would then be 1:4, 1:8, and so on through 1:2,048. After waiting enough time for the antigen-antibody interaction to occur, a lab worker examines each dilution for evidence of a reaction. The reciprocal of the last dilution showing a positive reaction is taken as the titer. Thus, if a positive reaction is observed in the dilution 1:256 but not in 1:512, the titer is 256. In any such test, controls must always be included to be sure all the reagents are reacting properly. A positive control contains known antigen and known antibody. A negative control contains the antigen with no antibodies, to check for false positive reactions.

Serology tests can be set up in test tubes, but many tubes and large quantities of reagents would be required. Therefore, the tests are usually done using plastic microtiter plates (**figure 17.1**). Proteins, either antigen or antibody, adsorb to the tiny plastic wells. These proteins can then be treated with reagents and washed without being removed from the plastic surface. The volumes used in each well are a mere fraction of the volumes needed for even a small test tube, so that tests can be done on very small samples. A standard microtiter plate has 96 wells, eight rows of 12 wells each; a single row is used for each sample tested, and 8 samples can be tested on a single plate. The 12 wells in a row allow two controls and 10 dilutions of sample. Thus, dilutions of 1:4 through 1:2,048 can be examined in a single test. Special equipment allows rapid dilution and mixing of reagents, and accurate reading of results.

MICROCHECK 17.1

Serology uses antibodies, usually in serum or other body fluids, to detect and identify antigens. Conversely, known antigens are used to detect and identify antibodies. The change from negative to positive for specific antibodies during an infection is seroconversion; a rise in titer is characteristic of an active infection. Serial dilution of specimens permits quantification of antibodies in the sample.

- Why is blood serum used in many immunological tests?
- What is the significance of a rise in titer of specific antibodies in serum samples taken early and late during an infectious disease?
- Why is the sample usually diluted when measuring the titer of antibodies in a serum sample? What would be observed if no antibodies were present in the serum?
- In a serological test, what is meant by a false positive reaction?

Perspective 17.1 Monoclonal Antibodies

In 1975, an exciting breakthrough occurred in immunology. Georges Köhler and Cesar Milstein developed techniques that fused normal antibody-producing B lymphocytes with malignant plasma cells (myeloma tumor cells), resulting in clones of cells they termed **hybridomas**. Since these hybridomas are clones, they produce antibodies with a single specificity that, therefore, are known as **monoclonal antibodies (figure 1)**.

Plasma cells produce large amounts of antibody, and when they become malignant myeloma cells grow profusely and indefinitely. Special myeloma cells are used to make hybridomas; they have lost the ability to make their own specificity of antibody but have retained the ability to produce large amounts of immunoglobulin. The normal B cell in the hybridoma supplies the genes for the specific antibody to be produced; the myeloma cell supplies the cellular machinery, the rough endoplasmic reticulum, for producing the antibodies.

Usually, when an animal is injected with an immunizing agent, it responds by making a variety of antibodies directed against different epitopes on the antigen. Therefore, even though there is a single antigen, the result is a mixture of different antibodies. When these antisera are used in immunological tests, standardizing the results is difficult, since there are differences each time the antiserum is made. Monoclonal antibodies, however, will be of the same immunoglobulin class and have the same variable regions and, thus, the same specificity and other characteristics. With such specificity, tests can be standardized much more easily and with greater reliability.

It was hoped that monoclonal antibodies would also become a useful therapeutic tool. In theory, it should be possible to make monoclonal antibodies that are specific for, say, a cancer cell. Radioactive materials that destroy cancer cells could be attached to the monoclonals making a sort of "magic bullet." These antibodies would search and find the cancer cells and attach specifically to them, allowing the radioactive material to destroy the cells. In reality, many problems have arisen when the monoclonal antibodies have been used to treat humans. One problem was that mouse cells were used to produce the hybridomas, and humans reacted to the mouse antigens on the hybridoma antibodies, causing them to be rapidly removed from the body. This limited the effectiveness of the monoclonal antibodies to one or a few doses. This and many other unforeseen difficulties are being addressed now, with some encouraging results. For example, genetic engineering is being used to construct antibody genes from human DNA that can be put into hybridomas to produce monoclonal antibodies useful in treating humans.

In the laboratory, monoclonal antibodies are the basis of a number of diagnostic tests. For example, monoclonal antibodies against a hormone can detect pregnancy only 10 days after conception. Specific monoclonal antibodies are used for rapid diagnosis of hepatitis, influenza, herpes simplex, and chlamydia infections. Köhler and Milstein won the Nobel Prize in 1984 for their work.

Hyperimmunize mouse with antigen a to produce many lymphocytes making anti-a antibodies.
Myeloma cell culture
Remove spleen and make a suspension of lymphocytes.
Myeloma cells
Antibody-producing lymphocytes (B cells) from spleen.
Mix lymphocytes with special cultured myeloma cells that have lost the ability to produce their specific antibody and that lack a particular enzyme.
Add a chemical that induces fusion of the two cell types to form hybridomas.
Grow hybridomas in a special medium. Spleen lymphocytes only grow a few days in culture. Myeloma cells will not survive in the special medium because of the lacking enzyme.
Select single hybridoma cells producing anti-a antibody and clone the hybridomas.
Hybridomas grow indefinitely because missing enzyme was supplied by the fused lymphocyte.
Grow hybridomas in cultures and purify large amounts of antibody from the culture medium.

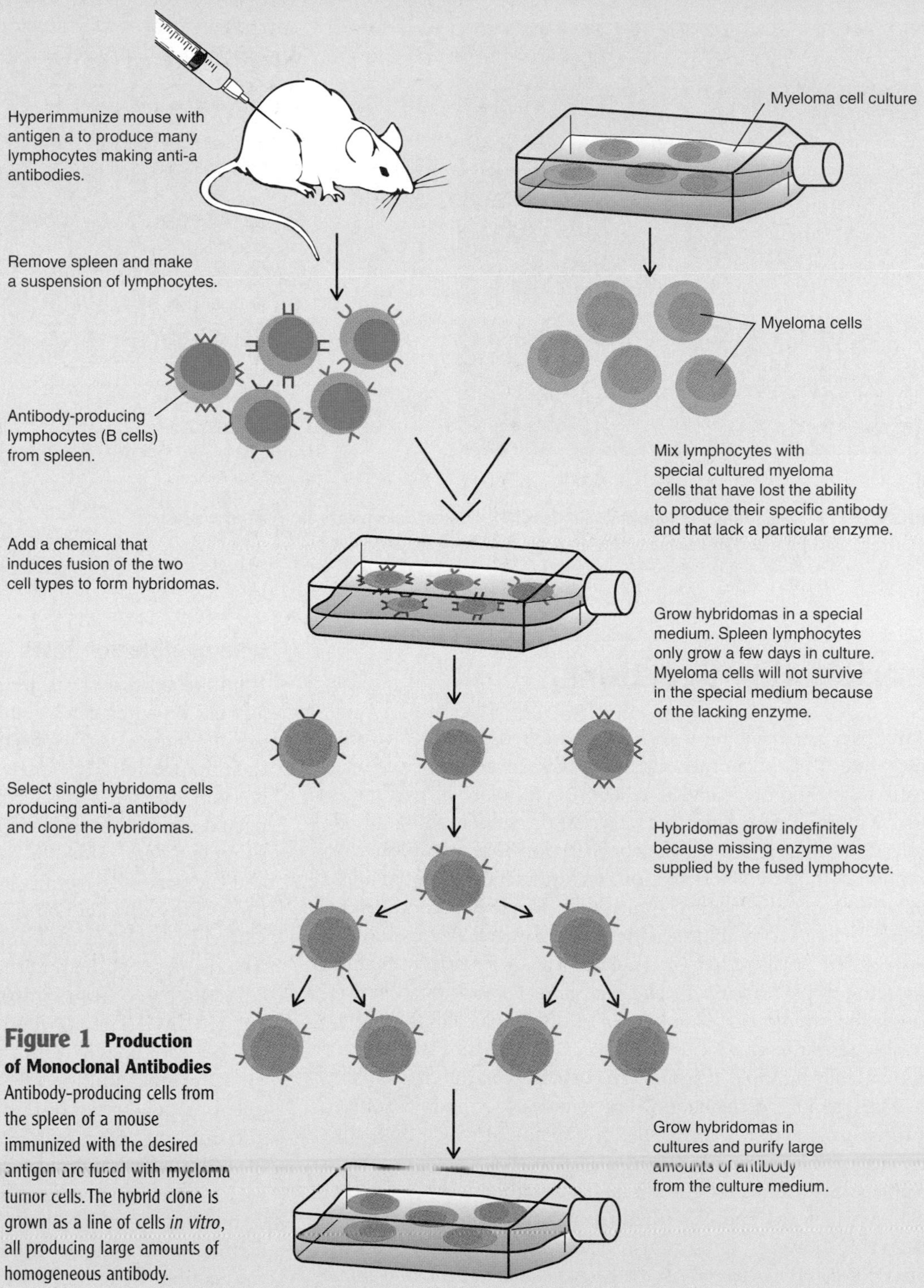

Figure 1 Production of Monoclonal Antibodies Antibody-producing cells from the spleen of a mouse immunized with the desired antigen are fused with myeloma tumor cells. The hybrid clone is grown as a line of cells *in vitro*, all producing large amounts of homogeneous antibody.

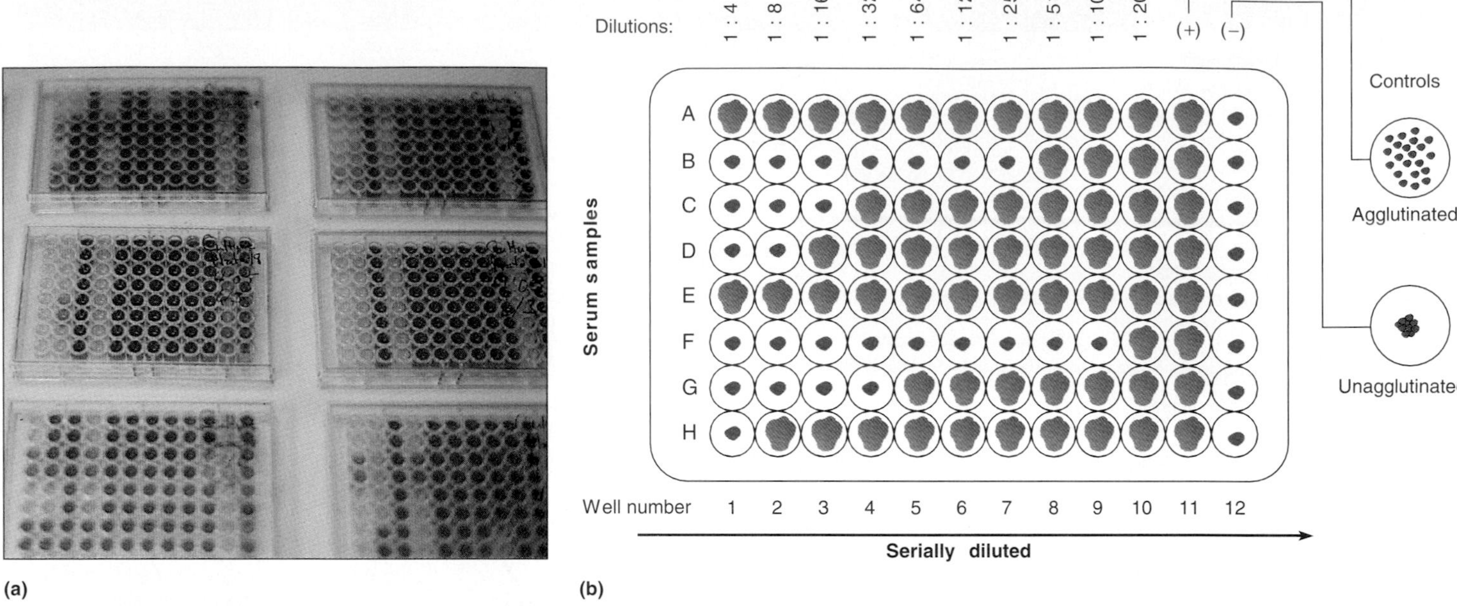

Figure 17.1 **Quantitation of Immunologic Tests** **(a)** Microtiter plates used to quantitate immunoassays. **(b)** Hemagglutination tests done in a microtiter plate. Eight tests (A–H) can be done in a single plate.

Precipitation Reactions

Antibodies can combine with soluble antigens to form a visible precipitate. This phenomenon is the basis for a number of diagnostic tests. The precipitation reaction takes place in two stages. First, within seconds after they are mixed, antigen and antibody molecules collide and form small primary **immune complexes**. An immune complex is, by definition, a combination of antigen and antibody, sometimes with complement included. Second, over a period lasting from minutes to hours, latticelike cross-linking between the small primary complexes causes large precipitating immune complexes to form. The mechanism of precipitation reactions is the extensive cross-linking of soluble molecules to produce a visible insoluble product. Of course, only antigens with two or more, usually multiple, epitopes can participate in the cross-linking necessary to give precipitation. Precipitation occurs in a zone of **optimal proportion**—that is, the proportion at which both antigen and antibody are fully incorporated in the precipitate. When soluble antigen and antibody are mixed together in tubes, the precipitate will form and fall to the bottom of the tube (**figure 17.2a**). If, however, antiserum is placed in a small tube and the antigen is carefully layered over it to avoid mixing, the precipitate will form in a line where antigen and antibody meet in optimal proportions (figure 17.2b). This is called the **ring test**, and it illustrates an important principle of antigen-antibody precipitation reactions: if there is a great excess of either antigen or antibody, a cross-linking lattice cannot form; consequently, no precipitate is formed. When these same interactions occur in the body, rather than in a laboratory test, large complexes are removed from the circulation by phagocytes, but small primary immune complexes with a slight antigen excess tend to remain in the circulation and can cause disease. ■ **immune complex diseases, p. 440**

Immunodiffusion Tests

Immunodiffusion tests are precipitation reactions carried out in agarose or other gels. A simple application of the immunodiffusion technique is the **radial immunodiffusion test**, which is a quantitative test. It can be used, for example, to measure the amount of IgG in the serum if an immune abnormality is suspected. Antibodies specific for human IgG molecules are added to melted agar. After the agar hardens, wells are cut in the agar and the person's serum is added to a well. In this case the IgG in the serum is the antigen being tested for. A precipitin ring forms where the antigen and antibody meet in optimal proportion. The diameter of the ring that forms is compared with a standard curve prepared from samples with known concentrations of the antigen IgG. The concentration of IgG in the patient's serum can be read directly from the standard curve (**figure 17. 3**).

Another immunodiffusion test, the **double diffusion in gel**, is a qualitative test. It involves diffusion of both antigen and antibody (**figure 17.4**). Antigen and antibody are placed in separate wells cut in the gel and are allowed to diffuse toward each other. When a specific antigen and its antibody meet at optimal proportions between the wells, a line of precipitate forms. Since there is often more than one antigen present in the sample being tested and many antibodies in the serum, more than one line can form, each in its area of optimal proportions. This method is useful for identifying unknown substances and can be used to identify more than one antigen or antibody in a mixture.

Immunoelectrophoresis

Immunoelectrophoresis is a variation of the precipitation in gel technique, combining precipitation with electrophoresis. Electrophoresis is used to separate mixtures of protein antigens on the basis of their movement through an electrical field (**figure 17.5**).

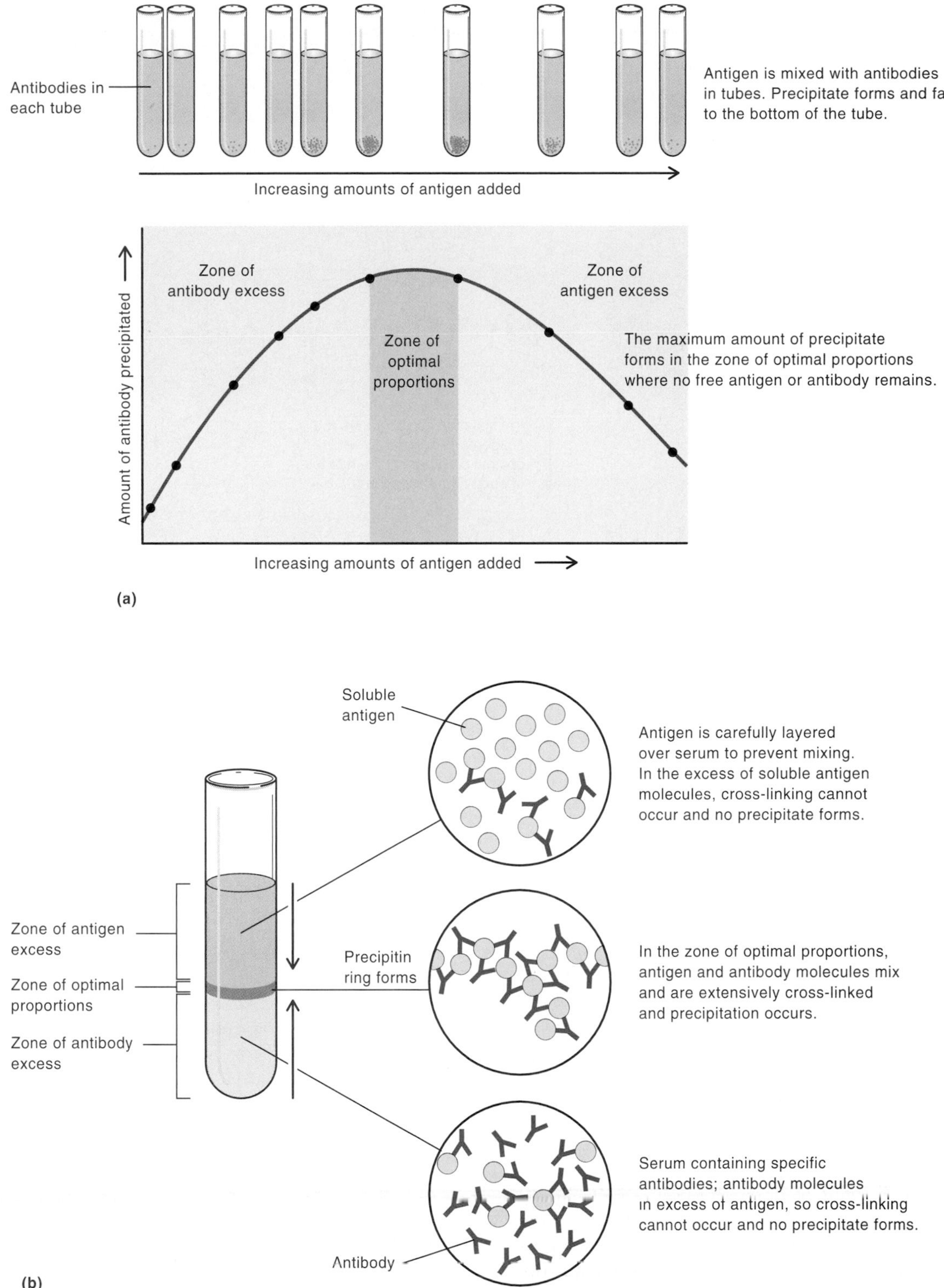

Figure 17.2 Antigen-Antibody Precipitation Reactions **(a)** Precipitin curve in the tube precipitation test. **(b)** The ring precipitation test. Why does the amount of precipitate first increase and then decrease as more antigen is added?

Once the proteins have been separated, the antibodies are placed in a trough and allowed to diffuse toward the separated protein antigens. A line of precipitate forms at the location of each antigen that is recognized by antibodies.

One of the many applications of immunoelectrophoresis is to determine whether people have normal immunoglobulins. The person's serum being tested for immmunoglobulins, the antigens in this case, is placed in the well; then, electrophoresis is used to separate the proteins in the sample. Anti-immunoglobulin antibodies are placed in the trough and allowed to diffuse toward the antigen. Where each antibody meets an antigen in the area of optimal proportions, a line of

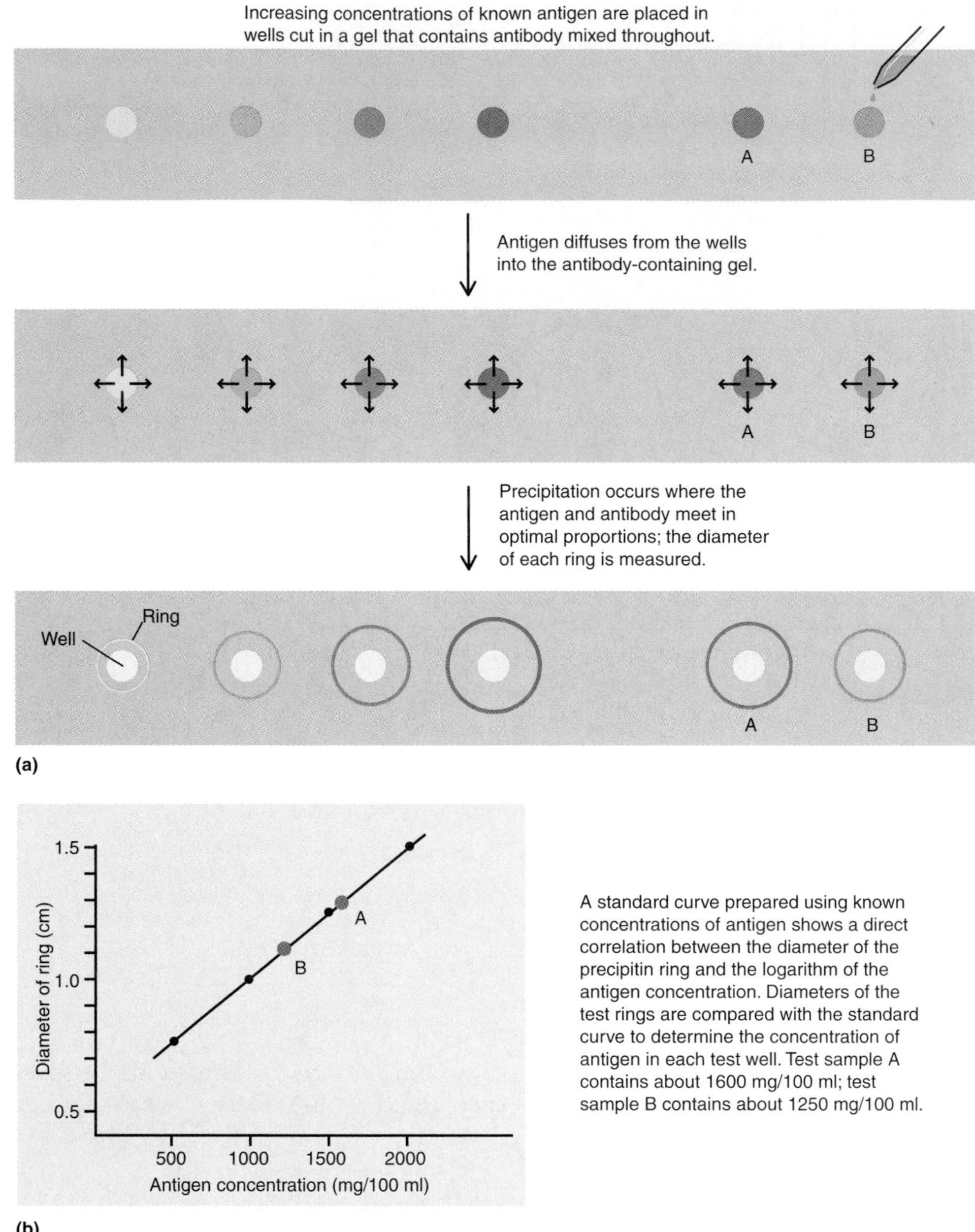

Increasing concentrations of known antigen are placed in wells cut in a gel that contains antibody mixed throughout.
Unknown samples A and B are placed in wells.
A
B
Antigen diffuses from the wells into the antibody-containing gel.
A
B
Precipitation occurs where the antigen and antibody meet in optimal proportions; the diameter of each ring is measured.
Ring
Well
A
B
(a)
1.5
1.0
0.5
Diameter of ring (cm)
A
B
500
1000
1500
2000
Antigen concentration (mg/100 ml)
A standard curve prepared using known concentrations of antigen shows a direct correlation between the diameter of the precipitin ring and the logarithm of the antigen concentration. Diameters of the test rings are compared with the standard curve to determine the concentration of antigen in each test well. Test sample A contains about 1600 mg/100 ml; test sample B contains about 1250 mg/100 ml.
(b)

Figure 17.3 Radial Immunodiffusion is Used to Measure the Concentration of Antigen in a Sample
(a) Antigen diffusing from a well into antibody-containing gel forms a ring of precipitation in the zone of optimal proportions. **(b)** The amount of antigen in a sample is determined by comparing the diameter of the precipitin ring with a standard curve prepared using known concentrations of antigen.

precipitation is formed. In normal serum, lines will form for IgA, IgG, and IgM. IgD and IgE are normally present in such small quantities that they do not yield visible precipitates. Some patients lack immunoglobulins, and their sera will give no lines of precipitation with anti-human antibodies. Other patients have myeloma tumors, in which a single plasma cell has given rise to a tumor. All the cells of the tumor produce the same immunoglobulin that the original cell produced. In this case, the serum from the myeloma patient will give a heavy, thick line for the class of immunoglobulin being produced. Thus, abnormally high or low production of immunoglobulins is readily detected by immunoelectrophoresis. ■ **electrophoresis, p. 233**

MICROCHECK 17.2

Precipitation occurs when soluble antigens interact with antibodies in optimal proportions to cause cross-linking into a large insoluble lattice. Precipitation tests are often done in gels where the precipitate can be readily seen. In immunoelectrophoresis, mixtures are separated by electrophoresis before adding antibodies to identify the separated antigens.

- What is the advantage of immunoelectrophoresis over radial immunodiffusion in gel, and vice versa?
- Why can't a cross-linking lattice form when there is an excess of antibody?

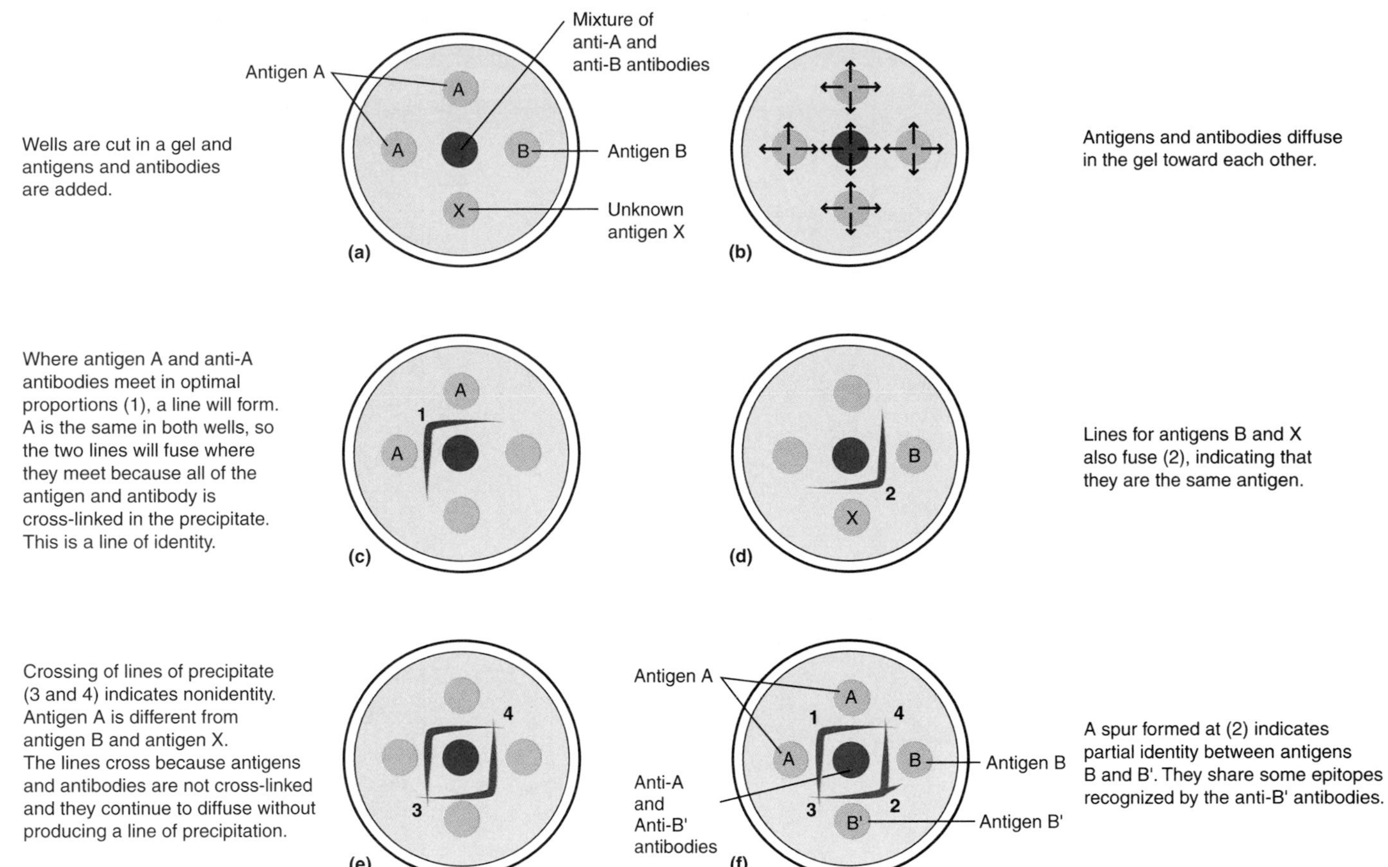

Figure 17.4 Double Immunodiffusion Double immunodiffusion in gel can detect antigens that are the same, or partially the same, as recognized by antibodies in the serum.

Agglutination Reactions

Agglutination and precipitation reactions are similar in principle; both depend on cross-linking and lattice formation. In agglutination reactions, however, the antigen consists of relatively large particles rather than molecules, so that larger aggregates of antigen and antibody are formed, which are much easier to see **(figure 17.6)**.

Direct Agglutination Tests

In direct agglutination tests, antibodies are measured against particulate antigens, such as red blood cells, bacteria, or fungi. The serum is serially diluted and tested against a known amount of antigen to give an estimation of the amount of antibody present. The procedure is essentially the same as for precipitation, except that the antigen is composed of particles instead of molecules.

Indirect Agglutination Tests

It is possible for soluble antigen to be used in an agglutination test if the antigen is first attached to red blood cells (indirect hemagglutination), latex beads, or other particles. Then, when antibody is added in the test, agglutination occurs and the reaction is much more easily seen than a precipitate of soluble molecules would be. Similarly, specific antibodies can be coated onto latex beads and used to detect antigens. Antibody-coated latex bead products are commercially available to test for a variety of bacteria, fungi, viruses, and parasites, as well as hormones, drugs, and other substances. The beads are mixed with a drop of serum, urine, spinal fluid, or other specimen. If the specific antigen is present, easily visible clumps will form.

Hemagglutination Inhibition

Hemagglutination is the clumping of red blood cells (see figure 17.6). Some viruses, such as those causing infectious mononucleosis, influenza, measles, and mumps, agglutinate red blood cells by an interaction between surface components of the virus and the red cell. Since specific antiviral antibodies inhibit the agglutination, antibodies to the virus can be measured by **hemagglutination inhibition**. Serum is serially diluted and mixed with the virus and red blood cells. Dilutions of serum containing specific antibodies inhibit the usual agglutination of virus and red cells. The titer of antibodies is determined by the highest dilution of serum in which agglutination does not occur.

MICROCHECK 17.3

Agglutination tests depend on cross-linking of particulate antigen by antibody molecules to form readily visible clumps. Soluble antigens can be coated onto particles to give an indirect agglutination.

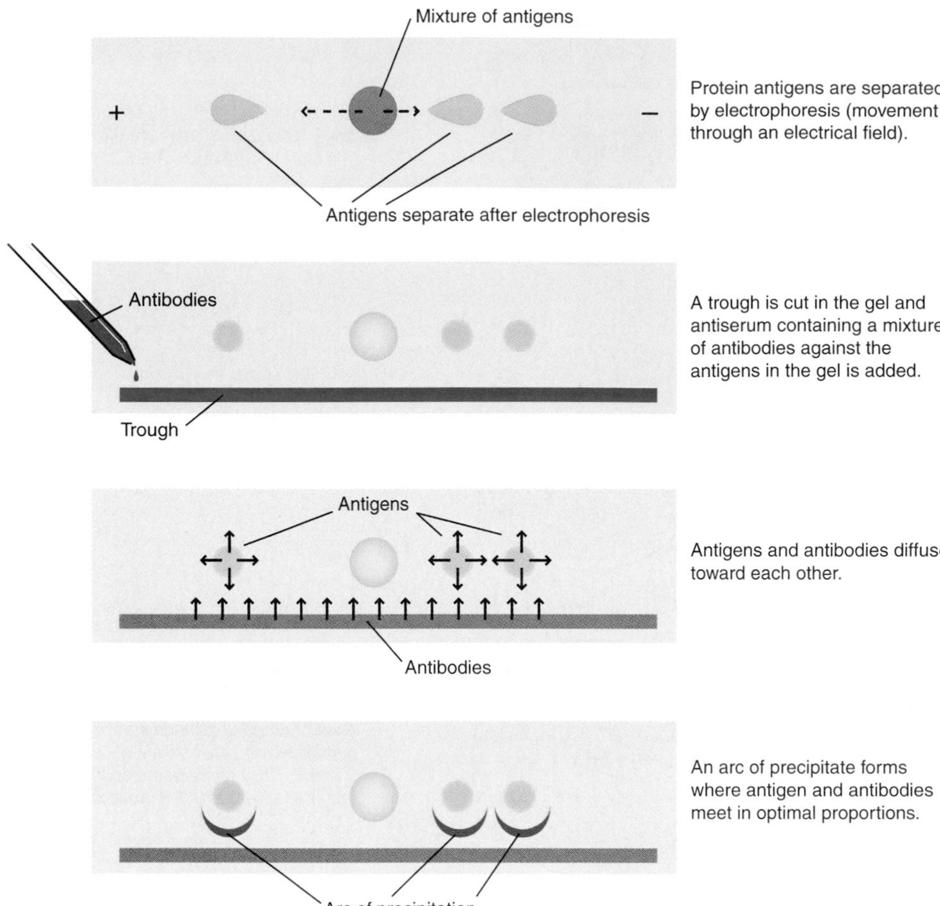

Mixture of antigens
+
−
Protein antigens are separated by electrophoresis (movement through an electrical field).
Antigens separate after electrophoresis
Antibodies
A trough is cut in the gel and antiserum containing a mixture of antibodies against the antigens in the gel is added.
Trough
Antigens
Antigens and antibodies diffuse toward each other.
Antibodies
An arc of precipitate forms where antigen and antibodies meet in optimal proportions.
Arc of precipitation

Figure 17.5 Immunoelectrophoresis Permits Identification of Antigens in a Mixture Electrophoresis is used to separate the antigens; then, immunodiffusion is used to identify them.

Antibodies interfere with viral agglutination of red cells in the hemagglutination inhibition test.

- Compare the advantages and disadvantages of precipitation and agglutination tests.
- In the hemagglutination inhibition test for antibodies against measles: (1) the measles virus when mixed with red cells causes hemagglutination; but (2) when anti-measles antibody is added to the virus before mixing with red cells, no hemagglutination occurs. Are the reactions in (1) and (2) antigen-specific? Why is hemagglutination inhibited?

Immunofluorescence Tests

Fluorescent dyes such as fluorescein or rhodamine can be attached to known specific antibodies, which are then used to detect the presence of an antigen in serum or microorganisms in a sample. There are two basic versions of these quick and sensitive tests.

Direct Fluorescent Antibody Test

In this test, the microorganism to be tested is fixed to a slide. Known antibodies labeled with fluorescein are added, the mixture is incubated, and the slide is washed. Antibody that binds to the microorganism will not wash away. The slide is then examined under a fluorescence microscope with a special light source to permit light of the desired wavelength to excite the dye. The dye then emits light of a different wavelength, which is viewed by using special filters in the microscope. If the microorganisms have bound the antibody, they will glow a yellow-green color **(figure 17.7a)**. If rhodamine is used instead of fluorescein, the microorganisms will glow a red color. By using various fluorescent dyes, it is possible to locate different antigens in the same cell or preparation.

Indirect Fluorescent Antibody Test

This test is used to find a specific antibody in human serum following an infection. A known organism is fixed to the slide, a sample of the test serum is added, and the mixture on the slide is incubated. If specific antibodies are present in the serum, they will complex with the antigens of the organism; but the complex cannot be seen until it is labeled or tagged in some way, as with a fluorescent dye. In order for the antigen-antibody complexes formed to be seen, any unfixed material is washed off and an anti–human-gamma-globulin (anti-HGG) antiserum that has been labeled with fluorescein is added to the slide. Anti-HGG is prepared by immunizing animals with the gamma globulin portion of human serum, the portion that contains the immunoglobulin antibodies. Such anti-human antibodies are useful in many immunological tests. The anti-HGG attaches to any antibodies remaining on the slide from the unknown serum. Then the slide is examined in the fluorescence microscope. Again, the organisms appear as yellow-green (figure 17.7b). A positive indirect fluorescent antibody test for *Treponema pallidum* is shown in figure 17.7c.

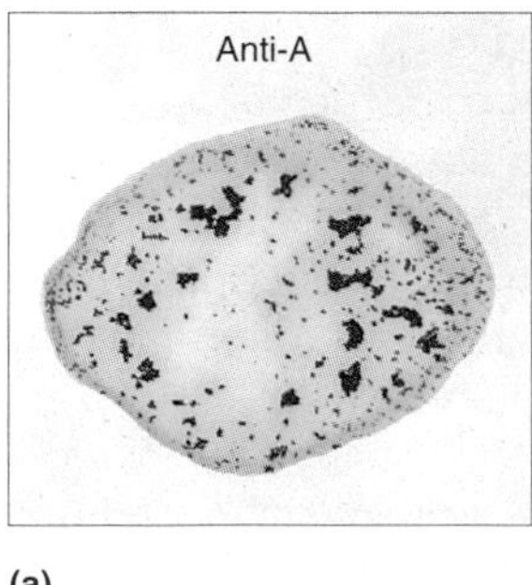

(a)

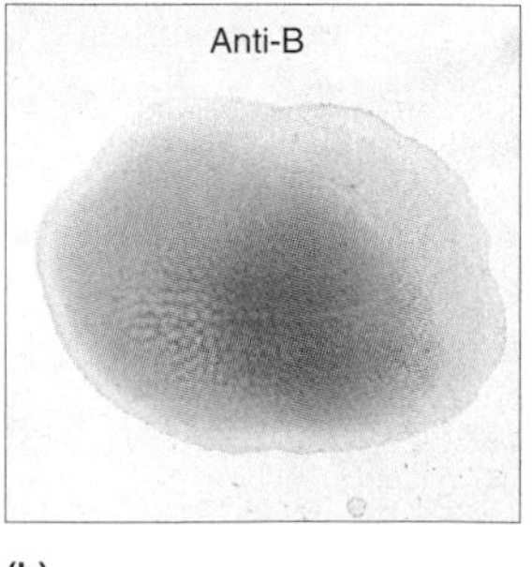

(b)

Figure 17.6 Agglutination of Erythrocytes in an ABO Blood Typing **(a)** The red cells are agglutinated when mixed with anti-A antibodies. **(b)** Anti-B antibodies do not agglutinate the red cells. The blood group is Type A.

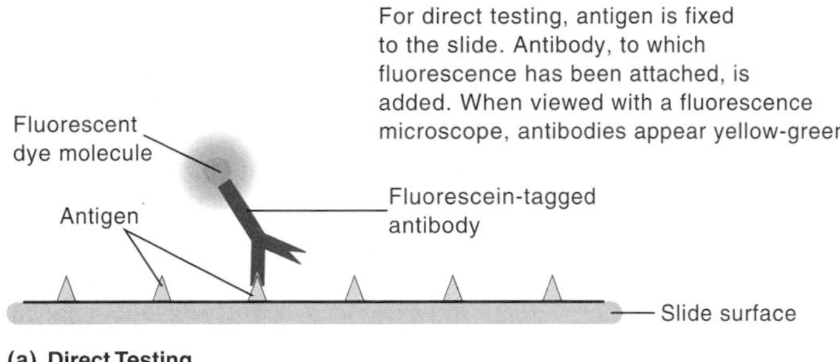
For direct testing, antigen is fixed to the slide. Antibody, to which fluorescence has been attached, is added. When viewed with a fluorescence microscope, antibodies appear yellow-green.
Fluorescent dye molecule
Antigen
Fluorescein-tagged antibody
Slide surface

(a) Direct Testing

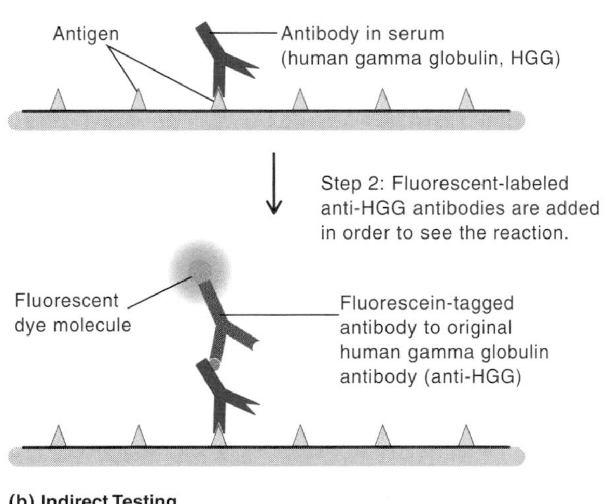
Step 1: Antigen fixed to the slide reacts with antibodies (human gamma globulin, HGG) in the test sample.
Antigen
Antibody in serum (human gamma globulin, HGG)
Step 2: Fluorescent-labeled anti-HGG antibodies are added in order to see the reaction.
Fluorescent dye molecule
Fluorescein-tagged antibody to original human gamma globulin antibody (anti-HGG)

(b) Indirect Testing

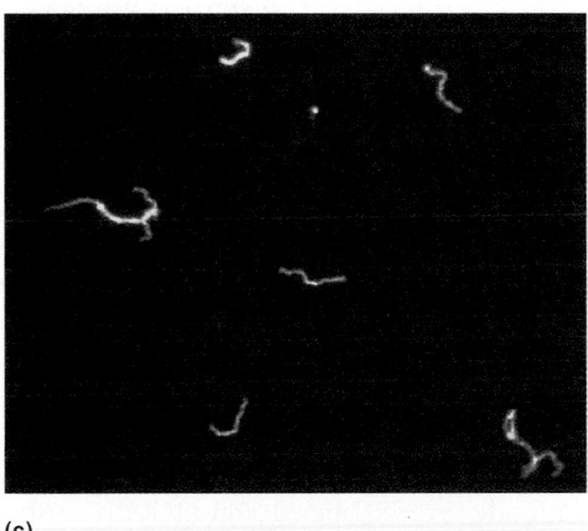

(c)

Figure 17.7 Fluorescent Antibody Tests (a) Direct testing. **(b)** Indirect testing. **(c)** *Treponema pallidum* spirochetes stained with fluorescein-tagged antibodies in an indirect test.

MICROCHECK 17.4

Fluorescent dyes are used to visualize antibodies under the fluorescence microscope, either directly or indirectly.

- How do the direct and indirect immunofluorescent antibody tests differ?
- Why is anti-human-gamma-globulin antisera labeled with detectable substances used in many immunological tests, such as indirect fluorescent antibody tests?
- How could different antigens in tissues be located?

The Complement Fixation Test and Neutralization Tests

Bacteria, red blood cells, or other cells may sometimes lyse as the result of complement activity. Recall that complement is a complex system of proteins; they interact with antibody bound to antigenic components of cells (a process known as complement fixation) and cause the cells to lyse. This phenomenon is the basis for the **complement fixation test**, which is used to test for specific antibodies in a patient's serum. The test involves several steps **(figure 17.8)**. For example, to test for antibodies against a specific virus:

1. The patient's serum sample is heated to inactivate any complement present.
2. The heat-inactivated serum is serially diluted, known virus is added, a known amount of animal complement is added, and the mixture is incubated, allowing the virus antigen and the serum antibody to interact and bind, or fix, the added complement. If no specific antibodies are present, the complement will remain free in the mixture.
3. To visualize the reaction, an indicator system is added to detect any free complement in the mixture. The indicator system usually consists of sheep red blood cells and anti-sheep-red-blood-cell antibodies.

If a specific virus antigen-antibody reaction has fixed or bound the complement (in step 2), the red cells in the indicator system will not be lysed, but will clump and settle; this is a positive test for the antiviral antibody. If complement has not been fixed, the red cells in the indicator system will be lysed, releasing their hemoglobin. This is a negative reaction for antiviral antibodies.

Once widely used, the complement fixation test is complicated and lengthy. Consequently, it has been often superseded by newer, easier techniques, such as ELISA.

Neutralization tests are another example of useful tests that are often superseded by other, easier methods. In the virus neutralization test, serum is mixed with a known viral suspension. If antibodies to that particular virus are present, they will bind to it and neutralize the virus, preventing its attachment to and subsequent infection of cells. When the virus is then added to an appropriate cell culture, it is unable to replicate and cause cell damage. Neutralizing antibodies can be demonstrated in herpesvirus infections and some other viral infections.

Neutralization tests can also be used to test for toxin. Specific antibodies against the toxin (antitoxin) will neutralize the effects of the toxin. For example, in a suspected case of botulism, a lethal form of food poisoning, leftover food thought to be the source of the toxin can be tested by injecting some of the food into mice. The antigenic type of toxin is determined by mixing part of the food sample with specific antitoxin before injection.

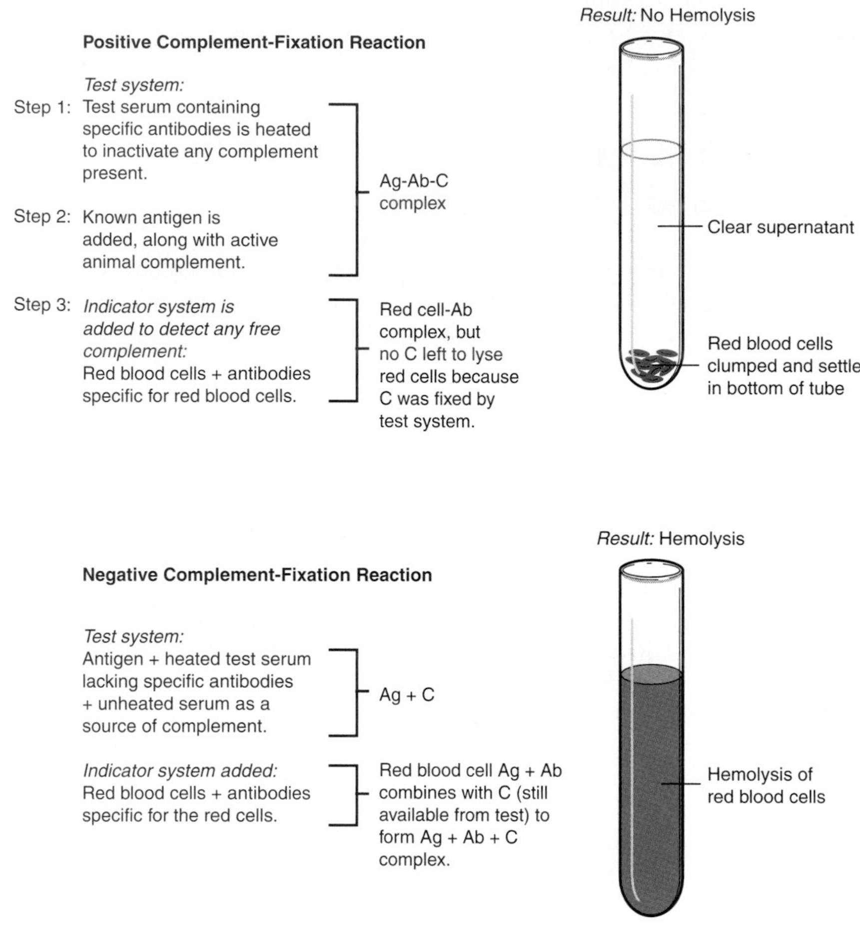

Figure 17.8 Complement Fixation Test Procedure

Untreated food containing botulinum toxin kills the mice, but they are protected if the food is first mixed with specific antitoxin that neutralizes the toxin. Toxin may also be detected by indirect hemagglutination or radioimmunoassay. ■ botulism, p. 675, 815

MICROCHECK 17.5

The complement fixation test measures the binding of complement by an antigen-antibody interaction, using an indicator system. Antibodies neutralize viruses or toxins in neutralization tests.

- How can neutralization tests detect antiviral antibodies?
- How can the antigenic type of a toxin be determined?
- In the virus neutralization test, why would the virus be unable to replicate?

Radioimmunoassay (RIA), Enzyme-Linked Immunosorbent Assay (ELISA), and Western Blot

RIA and **ELISA** are widely used, because these extremely sensitive tests can be used to analyze a large number of samples quickly using very little sample and reagents. For example, the ELISA technique is now routinely used to test donated blood for antibodies against the human immunodeficiency virus, or HIV (the virus that causes AIDS), before the blood is used for transfusion. The presence of specific antibodies implies that the virus is present. The Western blot combines electrophoresis with ELISA to give a colorimetric reaction, or with a radioactive label for detection of the reaction.

Radioimmunoassay (RIA)

RIA is a competitive inhibition assay used to measure antigens or antibodies that can be radioactively labeled. For example, to test for antigen in a sample, a fixed amount of unlabeled antibody is adsorbed onto plastic wells, and then the ability of a test sample to inhibit the binding of a labeled specific antigen is measured. Antigen binding is measured by the amount of radioactivity retained in the well, determined by using a gamma counter. The test is based on competition for specific antibody between known amounts of radioactively labeled antigen and unknown amounts of unlabeled antigen in the sample being tested. The concentration of the unknown is determined by comparison with a standard curve prepared using several concentrations of a known standard unlabeled antigen. This highly sensitive method is used to measure extremely small amounts of hormones or drugs in a clinical sample. A special RIA called **radioallergosorbent test (RAST)** is used to measure the very minute amounts of IgE antibody that react with a specific antigen. This method can detect as little as 1 to 50 ng/ml of serum antibody.

Enzyme-Linked Immunosorbent Assay (ELISA)

In the indirect ELISA, the known antigen is put into a plastic test tube or on a plastic microtiter plate. Some of the antigen adsorbs to the plastic surface, and the excess is washed away. The serum to be tested is added to the tube or plate and incubated for a short time. If specific antibodies are present, they bind to the antigen. To detect the antigen-antibody reaction, the ELISA uses anti-HGG chemically coupled with an enzyme label such as peroxidase. This binds to the test antibody, excess labeled anti-HGG is washed away, and chromogen is added. Chromogen is a colorless substrate that produces a colored end product when acted on by an enzyme such as peroxidase. This color change can be measured and is directly related to the amount of antibody bound (**figure 17.9**).

Direct ELISA tests are also useful in studying infectious diseases. Modifications of ELISA tests are available commercially to test for group A streptococci, antibiotics, cytokines, and pregnancy, among many other uses (**figure 17.10**). ELISA tests are more widely used than RIA because they are cheaper and there is no radioactive material to be discarded.

Western Blot

Western blot, or immunoblot, is a technique that combines electrophoresis with ELISA to separate and identify protein antigens

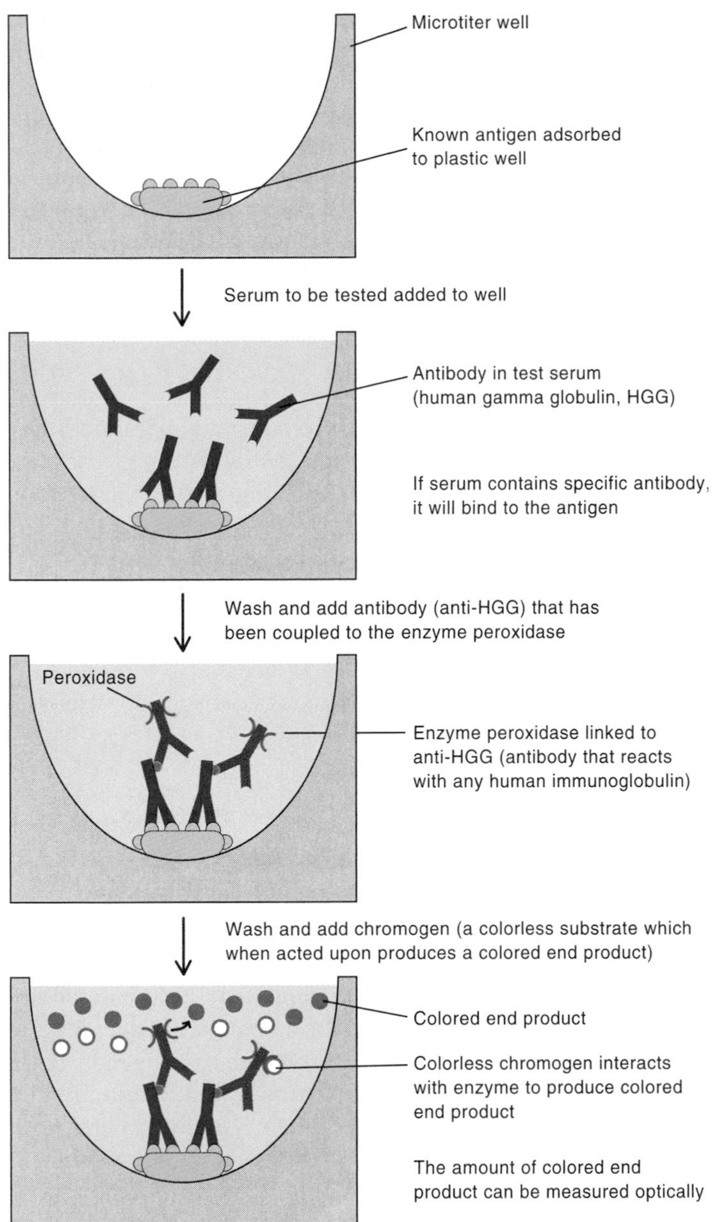
Microtiter well
Known antigen adsorbed to plastic well
Serum to be tested added to well
Antibody in test serum (human gamma globulin, HGG)
If serum contains specific antibody, it will bind to the antigen
Wash and add antibody (anti-HGG) that has been coupled to the enzyme peroxidase
Peroxidase
Enzyme peroxidase linked to anti-HGG (antibody that reacts with any human immunoglobulin)
Wash and add chromogen (a colorless substrate which when acted upon produces a colored end product)
Colored end product
Colorless chromogen interacts with enzyme to produce colored end product
The amount of colored end product can be measured optically

Figure 17.9 **Enzyme-Linked Immunosorbent Assay (ELISA)**

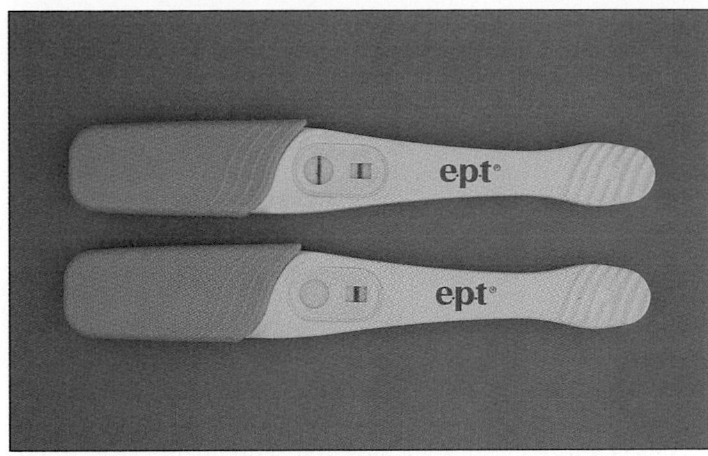
ept
ept

Figure 17.10 **ELISA Test for Pregnancy** The test detects human chorionic gonadotropin (HCG), an antigen present only in pregnant women. A urine sample is applied on the left. Two pink lines indicate reaction of HCG with antibodies, a positive test. A single line indicates absence of HCG in the urine, a negative test.

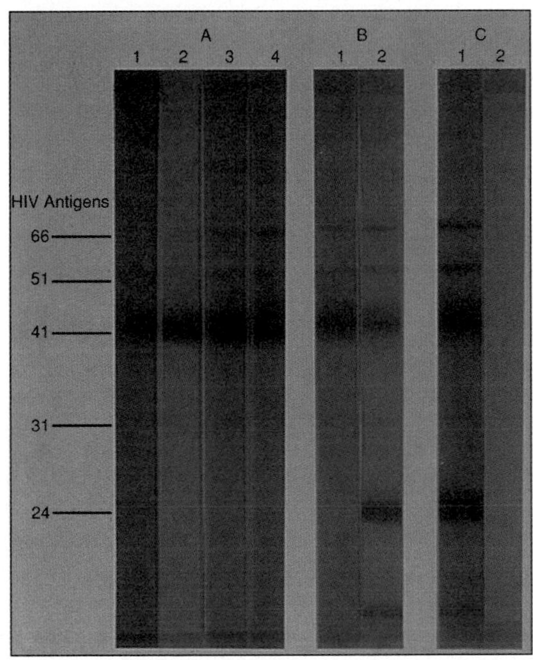
A
B
C
1
2
3
4
1
2
1
2
HIV Antigens
66
51
41
31
24

Figure 17.11 **The Western Blot Test**

in a sample. It has many research applications, but its main clinical use is the identification of antibodies against HIV. Western blot is used to confirm positive ELISA screening tests of blood before transfusion, to reduce the risk of transmitting HIV. The ELISA tests are simpler and cheaper, but they give a small percentage of false positive results; therefore, positive results need to be confirmed by a different method. In the Western blot technique, a mixture of antigens is separated by electrophoresis in a polyacrylamide gel. Smaller proteins migrate faster than larger ones. The resulting bands of separated antigens can react with specific antibodies, but antibodies do not diffuse well into these gels. Therefore, it is necessary to transfer the antigens present in the bands from the gel by blotting them onto a filter (**figure 17.11**). To test for antibodies against HIV, commercial kits are available in which antigens of the HIV virus have been electrophoresed and blotted onto the filter, which is then cut into strips for each test. Each sample to be tested for antibodies to HIV is applied to a strip and incubated to allow specific antibodies to react with the viral antigens. To detect the antigen-antibody combinations formed, a label is used, usually an enzyme-labeled anti-HGG, as in ELISA, or sometimes a radioactive label. The Western blot is a more laborious and expensive technique than the ELISA; however, it offers greater specificity, since antigens are identified by two criteria: their size and their reactivity with antibodies.

MICROCHECK 17.6

Radioimmunoassay uses a radioactive label to detect antigens or antibodies, whereas enzymes that give a color reaction are used as labels in the enzyme-linked immunosorbent assay, ELISA. The Western blot technique combines electrophoresis with ELISA to separate and identify protein antigens in a mixture.

- Give examples of uses for RIA, ELISA, and Western blot tests.
- Why is the RIA highly sensitive?

Tests Used in Cellular Immunology

In general, serological tests are cheaper and easier than tests using lymphocytes or other cells, and they are more widely used in clinical settings. There are, however, a number of cellular tests that are useful clinically and many more that are essential for research purposes. A few of these are described here.

Identification of Subsets of Lymphocytes

Subsets of lymphocytes, such as CD4 Th cells and CD8 Tc cells, are identified by molecules on the cell surface. Fluorescent-labeled monoclonal antibodies will distinguish these subsets. A fluorescence microscope can be used, but a **fluorescence-activated cell sorter (FACS)** is more efficient, as it counts and separates the cells quickly. Cells in a mixture are reacted with specific labeled antibody and then forced into a stream so fine that single cells move in the stream and through a laser beam. The laser causes the fluorescent label to emit light, which is measured with sensitive equipment to give information about each cell. By using various antibodies labeled with different dyes, several types of cells in a mixture can be distinguished and separated.

Lymphocyte Response to Mitogens

A mitogen is a substance known to stimulate cellular proliferation. One test used clinically to determine if a person is immunodeficient or is immunologically competent is mitogen stimulation of lymphocytes from the blood. The cells are cultured with any of several different substances known to stimulate proliferation of normal lymphocytes. For example, a substance from kidney beans called phytohemagglutinin causes normal T lymphocytes to proliferate, as measured by the incorporation of radioactive thymidine into their DNA. Failure of lymphocytes to proliferate in response to phytohemagglutinin indicates a deficiency in T-cell responses. Other mitogens are used to test for B-cell and T-cell function.

Cytotoxic T Cell Function

The ability of Tc cells to kill target cells is measured by labeling target cells with radioactive chromate and mixing these labeled cells with lymphocytes. If the lymphocytes are cytotoxic for the target cells, they will kill the targets, resulting in release of the chromate. Counting the radioactivity in the released chromate gives a measure of the cytotoxicity.

Cell-Mediated Immunity to Infectious Agents

Cell-mediated immunity to infectious agents can be assayed by measuring the uptake of radioactive thymidine into DNA in lymphocyte cultures stimulated by a specific antigen. Whereas mitogens stimulate large proportions of the cells, an antigen stimulates only the small percentage of cells responding to that specific antigen. Skin testing is also helpful in assessing cell-mediated immune responsiveness. For example, people with cell-mediated immunity to tubercle bacilli will give a positive delayed skin reaction to tuberculin antigens of the bacilli (discussed in chapter 18). Similarly, people immune to various fungi and to the leprosy bacilli will react to these specific antigens. Failure to respond to any of a set of antigens that commonly cause delayed skin reactions indicates defective cell-mediated immunity.

MICROCHECK 17.7

Among the many tests used to assess cellular immunity are fluorescent techniques to identify and/or separate subsets of lymphocytes, tests for mitogen stimulation of lymphocytes, chromate-release tests to measure lymphocyte cytotoxicity, and the stimulation of lymphocyte proliferation by specific antigens.

- How does mitogen stimulation of lymphocytes differ from antigen stimulation?
- Describe a test for Tc function against specific virus-infected cells.

Principles of Immunization

Immunity can be either natural or artificial. Natural immunity occurs through natural exposure to infectious agents or other immunizing agents that induce an immune response; immunity following an attack of measles is an example. Artificial immunity results from deliberate exposure of the host to antigen in order to protect, as with routine immunization of children. Immunity can also be either active or passive. Active immunity is produced by the individual in response to an antigenic stimulus, whereas passive immunity results from the transfer of immune serum or cells produced by other individuals or animals **(figure 17.12)**. This section will focus on artificial immunity deliberately induced, both passively and actively.

Passive Immunity

Passive immunity occurs naturally during pregnancy; the mother's IgG antibodies cross the placenta and protect the fetus. These antibodies remain active in the newborn infant during the first few months of life, when the neonate's own immune responses are weak. Consequently, a number of infectious diseases normally do not occur until a baby is three to six months of age, by which time the maternal antibodies have been degraded. Obviously, there is no memory for passive immunity. Once the transferred antibodies are degraded, the protection is lost.

Passive immunity can also be artificial, involving the transfer of antibodies produced by other people or animals. For example, a number of diseases, such as tetanus, are caused by microbial toxins. Antibodies specific against toxin are protective, and they can be obtained from humans who have been immunized against the toxin.

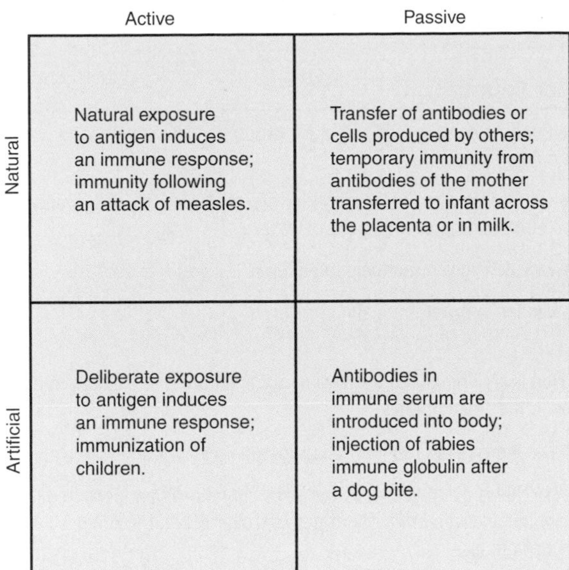

Active
Passive
Natural
Natural exposure to antigen induces an immune response; immunity following an attack of measles.
Transfer of antibodies or cells produced by others; temporary immunity from antibodies of the mother transferred to infant across the placenta or in milk.
Artificial
Deliberate exposure to antigen induces an immune response; immunization of children.
Antibodies in immune serum are introduced into body; injection of rabies immune globulin after a dog bite.

Figure 17.12 Acquired Immunity Acquired immunity may be natural or artificial, active or passive.

Passive immunization preparations currently in use include **immune serum globulin**, or gamma globulin, which is the immunoglobulin G (IgG) fraction of pooled blood plasma from many donors. This contains a variety of antibodies that the various donors have made because of infections and immunization procedures. This may be given to travelers who visit areas where sanitation is substandard to offer some protection against hepatitis A and other common diseases associated with poor hygiene. It is also useful for immunosuppressed people who lack antibodies. In addition, **hyperimmune globulin** is prepared from the sera of donors with high titers of antibodies to certain diseases, and these are used to prevent or treat specific diseases. Examples are human tetanus immune globulin, rabies immune globulin, and hepatitis A and hepatitis B immune globulin. These preparations given during the incubation period—after exposure, but before disease develops—can often prevent severe diseases from developing.

Active Immunity

Active immunity is produced by an individual in response to antigen. Active immunity involves the activation and proliferation of antigen-specific B and T lymphocytes and thus confers lasting protection or memory to the host. Active immunity can develop from either an actual infection or administration of a vaccine.

MICROCHECK 17.8

Immunity is either natural or artificial, passive or active. Active immunity occurs naturally in response to infections or other natural exposure to antigens, and artificially in response to vaccine administration.

- Distinguish between natural, artificial, passive, and active immunity.
- What would be a primary advantage of passive immunity with diseases such as tetanus?

Vaccines and Immunization Procedures

A **vaccine** is a preparation of living or inactivated microorganisms or viruses or their components used to induce active specific immunity. Sometimes vaccines contain substances that help to induce a better response. These substances are called **adjuvants**.

Effective vaccines should be safe, with few side effects, while giving lasting protection against the specific illness. They should induce specific antibodies or immune cells, or both, as appropriate. For example, polio vaccine should induce antibodies that neutralize the virus, thus preventing it from reaching and attaching to nerve cells to cause the paralysis of severe poliomyelitis. On the other hand, an effective vaccine against tuberculosis would induce cellular immunity that can limit growth of the intracellular bacteria. Of course, vaccines ideally should be low in cost, stable with a long shelf life, and easy to administer. **Table 17.1** lists a number of human diseases for which immunizing agents are available. As the table indicates, some immunizing agents are routinely used, whereas others are used only in special circumstances.

Attenuated Immunizing Agents

Attenuated agents (either organisms or viruses) cause an infection with undetectable or mild disease and usually a solid and long-lasting immunity. Attenuated means that the agent, though able to replicate, has been modified to be incapable of causing disease under ordinary circumstances. Modification of growth conditions can cause attenuation. For example, Pasteur first produced successful vaccines of attenuated anthrax and chicken cholera by growing the organisms at higher than normal temperatures and in other unusual conditions. Viruses of humans may be attenuated by growing them in cells of an animal species; mutations occur so that the virus then grows poorly in human cells. Genetic recombination is now being used to produce strains of pathogens with low virulence. Specific genes are mutated and used to replace wild-type genes. The inserted mutant genes are engineered so they cannot revert to the wild type. Also, genes can be deleted from vaccine virus strains, making them safer and giving the added advantage of being able to trace the virus and distinguish it from wild strains.

A single dose of an attenuated agent can be sufficient to induce immunity. If it is given orally, mucosal immunity is induced. Since the agent multiplies in the body, the antigen is present for a longer period and in greater amounts than with inactivated agents. Such a preparation has the added potential of being spread from an individual being immunized to other nonimmune people, who can become immunized by the inadvertent spread of the attenuated organisms. This helps to develop **herd immunity**, the inability of an infectious disease to spread in a population because of the lack of a critical concentration of susceptible nonimmune hosts. Herd immunity is responsible for dramatic declines in childhood diseases, both in the United States and in developing countries.

The disadvantage of using attenuated agents to immunize is that they have the potential to cause disease in immunosuppressed

TABLE 17.1 Some Important Immunizing Agents for Humans

Disease	Type of Vaccine	Other Information
Anthrax	Inactivated bacteria	Given only to persons who might be exposed, such as veterinary workers and military personnel.
Cholera	Inactivated bacteria	Given to people who live or travel in affected areas; vaccine is only 50% effective in preventing the disease.
Diphtheria	Toxoid	This vaccine is given routinely to children.
Haemophilus influenzae type b	Polysaccharide toxoid conjugate	This vaccine is given routinely to children.
Hepatitis A	Inactivated virus	Vaccine is recommended for routine administration to children in high-risk areas in the United States.
Hepatitis B	Purified antigen from recombinant yeast	This vaccine is given routinely to children.
Influenza	Inactivated virus, usually given by injection in the United States, but by nosedrops in parts of Europe.	This vaccine is given yearly, as the virus changes its antigens frequently. Vaccination is given primarily to persons over 65 and to other high-risk groups
Mumps	Live attenuated virus	This vaccine is given routinely to children.
Pertussis (whooping cough)	Acellular vaccine given together with diphtheria and tetanus toxoids (DTaP)	This vaccine is given routinely to children.
Plague	Inactivated bacteria	This vaccine is given to agricultural workers in affected areas, as well as to laboratory workers and others at high risk.
Pneumococcal infection	Purified polysaccharide of many serotypes	This vaccine is given once to people over 65 and to other high-risk groups.
Rabies	Inactivated virus grown in human or rhesus monkey cells	Used primarily for people at high-risk such as veterinarians and other animal handlers.
Rubella (German measles)	Live attenuated virus	This vaccine is given routinely to children.
Rubeola (measles)	Live attenuated virus	This vaccine is given routinely to children.
Tetanus	Toxoid	This vaccine is given routinely to children; adults receive a booster every 10 years.
Tuberculosis	Live attenuated BCG strain of tuberculosis bacteria	Widely used in other countries but only used in special circumstances in the United States.
Typhoid fever	Inactivated bacteria, injected; or live attenuated bacteria, orally	Short-term partial protection; given to high-risk workers and to people in affected areas.
Varicella-zoster (chicken pox)	Live attenuated virus	Given routinely to susceptible children; may be given to susceptible adults.
Yellow fever	Live attenuated virus	Given to travelers to affected areas; gives protection for 10 years.

people, and rarely they can revert or mutate to strains that cause serious disease. Care must be taken to avoid giving attenuated vaccines to pregnant women, because the vaccines may cross the placenta and cause damage to the developing fetus. Another disadvantage of attenuated vaccines, especially in developing countries where they are desperately needed, is that they usually require refrigeration to keep them active.

Attenuated vaccines currently in widespread use include those against measles, mumps, rubella, and yellow fever. The Sabin poliovirus vaccine that has given protection against polio to millions of people over many years is no longer recommended for use in the United States because of a small number of cases of active polio resulting from its use.

Inactivated Immunizing Agents

Inactivated immunizing agents include **whole agent vaccines,** made of inactivated bacteria, viruses, or bacterial toxins, and **subunit vaccines**, made of products or portions of an agent.

Inactivated whole agent vaccines include those against cholera, plague, influenza, and rabies and the Salk vaccine against polio. The agents are inactivated by treatment, often with formalin, a chemical that does not significantly change the surface epitopes. Such treatments leave the agent antigenic even though it cannot reproduce. These vaccines are advantageous in that they cannot cause infections or revert to dangerous forms. They also have several disadvantages. Since they do not replicate, it is usually necessary to give several booster doses of the vaccine to induce solid immunity. Parts of the agents that are not concerned with the immune response are included in the vaccines and can sometimes cause unwanted reactions. For example, the whooping cough (pertussis) killed vaccine that was previously used routinely for immunizing babies and young children often caused reactions such as pain, tenderness at the site of the injection, fever, and occasionally, convulsions. Currently, this killed whole cell vaccine is replaced by a subunit (acellular) vaccine, made of fragments of the bacteria, that does not cause these side effects.

In the case of polio, there are three types of poliovirus, any of which cause the disease poliomyelitis. The Salk vaccine, developed in the mid-1950s, consists of inactivated viruses of all three types. It was a huge success in lowering the rate of poliomyelitis, but it had the disadvantage of requiring a series of injections over a period of time for maximum protection. In 1960 the Sabin vaccine became available, with the advantage of cheaper oral administration. Even though this attenuated poliovirus vaccine replicates in the intestine, however, it still has to be given in a series of three doses rather than one because of interactions among the three types of virus included in the vaccine. Both attenuated and inactivated polio vaccines induce antibodies and protect against viral invasion of the central nervous system and consequent paralytic poliomyelitis. If only the inactivated vaccine is given, wild-type virus can still replicate in the gastrointestinal tract and be transmitted to others who are susceptible. The attenuated virus also induces mucosal immunity, however, so that polioviruses cannot replicate in the intestines. Because of the very rare incidence of paralytic polio occurring in people given the attenuated vaccine (one case/2.4 million doses of vaccine administered), only the inactivated vaccine is used routinely in the United States. Worldwide, there is a big push to eradicate poliomyelitis completely. This effort has been interrupted by war in some countries, such as Afghanistan, but it has been possible at times to arrange a cease-fire for National Immunization Days, to permit this vital public health program to continue. ■ **poliomyelitis, p. 679**

Vaccines against diphtheria and tetanus are **toxoids**, whole agent vaccines. These are prepared from the bacterial toxins, treated to destroy the toxic part of the molecules while retaining the antigens. These vaccines contain alum adjuvant along with the toxoid.

Subunit vaccines are made by isolating the antigens or antigenic fragments of the agent and administering only a subunit of the total agent. Some subunit vaccines in use are those directed against meningococci, pneumococci, pertussis, and *Haemophilus influenzae* type b. The success of the latter is seen in the statistics. The incidence of invasive *H. influenzae* infections has dropped more than 99% since this vaccine was introduced into general use. ■ ***Haemophilus influenzae* type b, p. 668**

Genetic engineering is used to produce another kind of subunit vaccine, **recombinant vaccine**. An example is the vaccine against the hepatitis B virus. Genes for a part of the viral protein coat are inserted into yeast cells, which then produce the proteins. These make a safe vaccine but one that requires several doses to be effective. Genetically engineered bacteria have been used to produce vaccines, and genetically altered vaccinia virus is also being used for vaccines still in the experimental or clinical trial stages.

The Use of Adjuvants

Currently, alum (aluminum hydroxide) is the only adjuvant approved for use in vaccines for humans, although several others are being tested in clinical trials. Antigens alone tend to diffuse rapidly from the area of injection, but antigens given with the adjuvant adhere to the alum, causing a persistent deposition of antigen, with a slow but relatively constant release of antigen to the tissues and surrounding blood vessels. Macrophages accumulate at the site and are activated by the alum. The activated macrophages produce increased amounts of interleukin-1 and other cytokines, leading to B-cell proliferation and increased numbers of lymphocytes in the area. Alum also stimulates T-helper cells. The total result is a much stronger immune response with adjuvant than without.

The Importance of Routine Immunizations for Children

Although there is some risk associated with almost any medical procedure, with routine immunizations the benefits greatly outweigh the very slight risks. **Table 17.2** shows the effectiveness of immunization in the United States over the Twentieth century. There is ample evidence that the benefits of immunization outweigh the risks. For example, data show that a child with measles has a 1:2,000 chance of developing serious encephalitic involvement of the nervous system, compared with a 1:1,000,000 chance from measles vaccine. Even so, parents sometimes worried about the rare chance that immunization procedures might be harmful, and refused to have their children vaccinated. As a result, between 1989 and 1991 measles immunization rates dropped 10% and an outbreak of 55,000 cases occurred, with 120 deaths. Now that immunization is again widespread, measles outbreaks are rarely seen. The suggestion that the measles, mumps, and rubella (MMR) vaccine is associated with autism in young children,

TABLE 17.2 The Effectiveness of Universal Immunization of Children in the United States During the Twentieth Century

Disease	Cases per Year Before Immunization	Decrease After Immunization
Smallpox	48,164 (1900–1904)	100%
Diphtheria	175,885 (1920–1922)	100%
Pertussis (whooping cough)	147,271 (1922–1925)	95.7%
Tetanus	1,314 (1922–1926)	100%
Paralytic poliomyelitis	16,316 (1951–1954)	100%
Measles	503,282 (1958–1962)	100%
Mumps	152,209 (1968)	99.6%
Rubella (congenital syndrome)	823 (estimated)	99.4%
Haemophilus influenzae type b infections	20,000 (estimated)	99.7%

however, is again threatening the acceptance of immunization. Studies so far have not shown evidence of this association, but more work is under way to be sure. Routine immunization against pertussis (whooping cough) caused a marked decrease in its incidence in the United States and saved many lives. Because of some adverse reactions to the killed whole cell vaccine being used, however, many parents refused to allow their babies to get this vaccine. By 1990, this refusal of vaccination resulted in the highest incidence of pertussis cases in 20 years and the deaths of some children, mostly under one year of age. Currently an acellular subunit pertussis vaccine is used, usually in combination with diphtheria and tetanus toxoids (DTaP). Several large-scale studies have shown the acellular pertussis vaccine to be more effective and have fewer side effects than the whole cell vaccine.

Consider that before vaccination was available for common childhood diseases, thousands of children died or were left with permanent disability from these diseases. Even now, nearly 20% of American children under age two are not fully immunized and many people in the United States become ill or even die every year from diseases like measles and hepatitis B that are readily prevented by vaccines. The recommendations of the U.S. Center for Disease Prevention and Control for childhood and adolescent immunizations are shown in **table 17.3**. Since children need a minimum of 15 separate injections to complete the 2000 recommended childhood immunization schedule from birth to six years, it is desirable that several vaccines be combined into a single preparation. More of these combination vaccines are being licensed for use in the United States each year.

MICROCHECK 17.9

Vaccines may be attenuated or inactivated agents. Inactivated whole agent vaccines are made of whole bacteria, viruses, or bacterial toxins. Inactivated subunit vaccines are made of products or portions of an agent. Recombinant vaccines are made by genetic engineering, inserting genes for antigenic proteins into yeasts or other cells that produce the antigen. Adjuvants cause a stronger immune response to the antigen in a vaccine. Routine childhood immunizations have prevented millions of cases of disease and many deaths during the past decades. Universal immunization is essential to eradicate some diseases and to preserve herd immunity against others.

- Compare and contrast attenuated and inactivated vaccines.
- Since many childhood diseases such as measles and mumps are rare now, why is it important for children to be immunized against them?
- What would be a primary advantage of using an attenuated agent rather than just an antigen from that agent?
- Why might several doses of a genetically engineered recombinant vaccine be required?

Current Progress in Immunization

Much remains to be done in developing vaccines against some serious and widespread infectious diseases, such as malaria and AIDS. **Table 17.4** shows the pressing need for effective vaccines against several common diseases. While vaccination is widely used to protect against infectious diseases, it is also under study in other diseases. Attempts are being made to develop vaccines to control fertility and hormone activity, and to prevent diabetes and cancer, among other conditions. Vaccines are also being used experimentally in the treatment of cancer.

Many new methods for vaccine development are being intensively studied. **Table 17.5** lists some of these experimental approaches. Among the new immunizing agents are peptides. The peptides from pathogenic organisms are either isolated or

TABLE 17.3 Recommended Childhood Immunization Schedule in the United States (2000)

Vaccine	Birth	1 mo	2 mo	4 mo	6 mo	12 mo	15 mo	18 mo	4–6 yrs	11–12 yrs	14–16 yrs
Hepatitis B											
Diphtheria, tetanus (Td) acellular pertussis (DTaP)			DTaP	DTaP	DTaP		DTaP			Td	
Haemophilus infuenzae type b (Hib)											
Poliovirus (IPV-inactivated polio vaccine)											
Measles-mumps-rubella (MMR)											
Varicella (chickenpox-Var)											

Range of acceptable ages for vaccination indicated by colors:
First dose · Second dose · Third dose · Subsequent doses · To be assessed and given if necessary

TABLE 17.4 Some Diseases for Which New or Improved Vaccines Are Sought

Disease	Estimated Incidence	Number of Cases or Deaths
HIV/AIDS	Infects 16,000 daily, worldwide	30.6 million cases, worldwide
Malaria	Up to 500 million cases/yr, worldwide	Up to 3 million deaths/yr
Influenza	30 to 50 million cases/yr	20,000 deaths/yr in United States
Strep throat	20 million cases/yr in United States	
Genital herpes	Infects 500,000/yr in United States	60 million cases in United States
Hepatitis C		170 million carry the virus, worldwide
Cancer	1 in 3 in United States may get cancer	560,000 deaths/yr in United States

TABLE 17.5 Some Experimental Vaccines Under Study

Vaccine	Characteristics
Peptide	Peptides isolated or synthesized; stable, but expensive and weakly immunogenic
DNA	Selected portions of genes for antigens injected directly into humans, who then synthesize protein coded for by the DNA
Edible vaccines	Genes for antigens introduced into foods, leading to immune responses
Cholera B subunit with killed cholera vibrios	Oral vaccine effective in trials; may be able to use other antigens along with cholera B subunit as oral vaccine
Human fertility control	Vaccines aimed at immunizing against sperm, egg, or reproductive hormones
Modulation of hormone action	Enhancement or depression of activity of hormones, such as growth hormones

synthesized in the laboratory and made into vaccines. Such vaccines are stable to heat and other environmental influences, do not contain extraneous materials to cause unwanted reactions or side effects, and have several other desirable features. Peptide vaccines, however, are weakly immunogenic and are very expensive to prepare. Although under active investigation, no peptide vaccines are yet approved for human use.

Another new and very promising approach is the use of DNA alone as an immunizing agent. Segments of naked DNA from infectious organisms can be introduced directly into muscle tissue. There, the DNA replicates and codes for production of microbial antigens that then induce an immune response. Thus the host actually produces the particular antigen to which it will respond. It is not yet known how this occurs, but it has been shown that it is effective. This procedure eliminates the possibility of infection with the immunizing agent, as only a small portion of the microbe's DNA is used.

MICROCHECK 17.10

There is a pressing need for vaccines to protect against a variety of diseases. Many experimental vaccines are under study or in clinical trials.

- What are some advantages and disadvantages of peptide vaccines?
- How can DNA be used as an immunizing agent?

FUTURE CHALLENGES

Global Immunization

In addition to developing new safe and effective vaccines, a major challenge for the future is delivering available vaccines to worldwide populations. When this was done in the case of smallpox, the disease was eliminated from the world; and poliomyelitis is almost eradicated now. The World Health Organization and the governments that support it deserve much of the credit for these achievements. Much remains to be done, however, and a recent gift of a billion dollars from Bill Gates, one of the founders of Microsoft, and his wife will go a long way to implementing further progress in this area.

To immunize universally, it is necessary to have vaccines that are easily administered, inexpensive, stable under a variety of environmental conditions, and preferably, painless. Instead of expensive needles and syringes and painful injections, vaccines may be delivered in a number of easy ways. For example, naked DNA is coated onto microscopic gold pellets, which are shot from a gunlike apparatus directly through the skin into the muscle, painlessly. Skin patches deliver antigens slowly through the skin. Vaccines against mucous membrane pathogens can be delivered by a nasal spray, as in some new influenza vaccines, or by mouth. Time-release pills introduce antigen steadily to give a sustained immune response.

In addition to sprays and pills that deliver antigens to mucous membranes in order to induce mucosal immunity, techniques are being developed to get antigens directly to M cells in the gastrointestinal mucosa. These are the cells that can endocytose antigens and deliver them across the membrane to the lymphoid tissue of the Peyer's patches, where immune responses occur. Antigens are incorporated into substances known to bind to M cells, thereby facilitating entry into the M cells and the Peyer's patches. ■ M cells, p. 458

Another promising means of immunization, especially for developing nations, is the use of edible vaccines produced in plants. Preliminary studies have shown that genes from various pathogenic organisms can be introduced into plants. For example, potato plants were genetically engineered with a gene for part of an Escherichia coli *exotoxin that affects the gastrointestinal tract. The potatoes expressing the harmless fragment of the exotoxin were then fed to mice, inducing the production of serum and secretory antibodies against the toxin. It is hoped that appropriate genes can be introduced into common plants, such as tomatoes and bananas, resulting in very low cost immunization for whole populations. These investigations, however, are in their early stages.*

One of the major remaining challenges is development of an effective vaccine against HIV that can be administered universally. Although many anti-HIV vaccines are under investigation and in clinical trials, there is no indication that a truly effective immunization program for HIV disease is imminent. It is also probable that when HIV disease is brought under control, a new and currently unforeseen challenge will arise during the Twenty-first century.

SUMMARY

Principles of Immunological Testing

1. **Serology** uses antibodies, usually in serum or other body fluids, to detect and identify antigens, or conversely, uses known antigens to detect antibodies.

Quantifying Antigen-Antibody Reactions

1. The change from negative to positive for specific antibodies during an infection is **seroconversion**; a **rise in titer** is characteristic of an active infection.
2. **Serial dilution** of specimens permits quantification of antibodies in the sample.

Precipitation Reactions

1. Precipitation occurs when soluble antigens interact with antibodies in **optimal proportions** to cause cross-linking into a large insoluble lattice. **(Figure 17.2)**

Immunodiffusion Tests

1. Precipitation tests are done in gels where the precipitate can be readily seen. **(Figure 17.4)**

Immunoelectrophoresis **(Figure 17.5)**

1. In immunoelectrophoresis, mixtures are separated by electrophoresis before the addition of antibodies to identify the separated antigens.

Agglutination Reactions

1. Agglutination tests depend on cross-linking of particulate antigen by antibody molecules to form readily visible clumps. **(Figure 17.6)**

Direct Agglutination Tests

1. Particulate antigen reacts directly with antibodies.

Indirect Agglutination Tests

1. Soluble antigen is coated onto particles to give indirect agglutination.

Hemagglutination Inhibition

1. Antibodies interfere with viral agglutination of red blood cells.

Immunofluorescence Tests

1. Fluorescent dyes are used to visualize antibodies under the fluorescence microscope. **(Figure 17.7)**

Direct Fluorescent Antibody Test

1. Antibodies tagged with fluorescent dyes react directly with antigen.

Indirect Fluorescent Antibody Test

1. Antigen and antibody interact, and the resulting complex is detected with fluorescent-labeled antibodies against the immunoglobulin in the complex.

The Complement Fixation Test and Neutralization Tests

1. The **complement fixation test** measures the binding of complement by an antigen-antibody interaction, using an indicator system. **(Figure 17.8)**
2. Antibodies neutralize viruses or toxins in **neutralization tests**.

Radioimmunoassay (RIA), Enzyme-Linked Immunosorbent Assay (ELISA), and Western Blot

1. These tests are extremely sensitive and efficient.

Radioimmunoassay (RIA)

1. **RIA** is based on competition for specific antibody in a test sample between known amounts of radioactively labeled antigen and unknown amounts of unlabeled antigen.
2. Concentration of the unknown is determined by comparison with a standard curve.

Enzyme-Linked Immunosorbent Assay (ELISA)

1. Enzymes that give a color reaction are used as labels in the **ELISA** test. **(Figure 17.9)**

Western Blot

1. The Western blot technique combines electrophoresis with ELISA to separate and identify protein antigens in a mixture. **(Figure 17.11)**

Tests Used in Cellular Immunology

Identification of Subsets of Lymphocytes

1. Fluorescent-labeled monoclonal antibodies will distinguish subsets, either by microscopy or by separation in a cell sorter.

Lymphocyte Response to Mitogens

1. A variety of substances that stimulate cellular proliferation can distinguish subsets of lymphocytes.

Cytotoxic T Cell Function

1. Release of radioactive chromate from target cells measures killing by Tc cells.

Cell-Mediated Immunity to Infectious Agents

1. Lymphocyte proliferation in response to specific antigens is measured by incorporation of radioactive thymidine into DNA.

Principles of Immunization

1. Immunity is either natural or artificial, passive or active. (Figure 17.12)

Passive Immunity

1. Passive immunity occurs naturally from mother to fetus, and artificially by transfer of preformed antibodies, as in **hyperimmune globulin**.

Active Immunity

1. Active immunity occurs naturally in response to infections or other natural exposure to antigens, and artificially in response to vaccine administration.

Vaccines and Immunization Procedures

1. A **vaccine** is a preparation of living or inactivated microorganisms or viruses or their components used to induce active immunity. (Table 17.1)

Attenuated Immunizing Agents

1. Attenuated agents are antigenic and can replicate, but they are modified to be incapable of causing disease under normal circumstances.

Inactivated Immunizing Agents

1. Inactivated agent vaccines may contain inactivated **whole agents** or **subunits** of the agent. **Recombinant** vaccines are genetically engineered.

The Use of Adjuvants

1. **Adjuvants** increase the intensity of the immune response to the antigen in a vaccine.

The Importance of Routine Immunizations for Children (Table 17.2)

1. Routine childhood immunizations have prevented millions of cases of disease and many deaths during the past decades.
2. Universal immunization is essential to eradicate some diseases and to preserve **herd immunity** against others. (Table 17.3)

Current Progress in Immunization

1. There is a pressing need for vaccines to protect against a variety of diseases. (Table 17.4)
2. Many experimental vaccines are under study or in clinical trials. (Table 17.5)

REVIEW QUESTIONS

Short Answer

1. What are the advantages of the ELISA test?
2. Describe how both active and passive immunization can be used to combat tetanus.
3. Which would be expected to be more cost-efficient, the complement fixation test or the direct fluorescent antibody test? (Costs include reagents and labor.)
4. Why are acellular subunit vaccines replacing some whole cell vaccines? Discuss an example.
5. Describe five ways in which vaccines can be delivered painlessly, without the need for injections.
6. In a precipitation reaction, what is meant by optimal proportions?
7. To determine a person's blood type, antibodies against red blood cells bearing the A and/or B antigens are mixed with the person's red cells, which are examined for agglutination. Is this a direct or an indirect agglutination test?
8. During cryptococcal meningitis, capsular material is shed from the fungi into the spinal fluid. Known antibodies against antigenic molecules of *Cryptococcus neoformans* are coated onto latex particles, which are then mixed with the cerebrospinal fluid of a patient suspected of having cryptococcal meningitis. When agglutination occurs, is this a direct or an indirect agglutination test?
9. In the example in question 8, what is being detected when agglutination occurs?
10. In the example in question 8, what would you deduce about the patient's diagnosis?
11. An ELISA test is used to screen blood for HIV before transfusing it. A positive ELISA test is confirmed by a Western blot test. Why not the other way around, with the ELISA second?
12. What test is sensitive enough to detect the extremely small quantities of specific IgE antibodies in blood?
13. What is meant by an acellular vaccine?
14. What are some ways in which the number of injections needed for childhood immunization could be lessened?
15. Can DNA vaccines cause the disease they are meant to protect against?
16. Which would be expected to be more effective against common childhood diseases, active or passive immunity? Why? Answer in terms of protection and cost-effectiveness.

Multiple Choice

1. All of the following are attenuated vaccines, *except* that against
   A. measles.
   B. mumps.
   C. rubella.
   D. Salk polio.
   E. yellow fever.
2. Alum adjuvant has all of the following properties, *except*...
   A. antigens adhere to the alum.
   B. antigen is slowly released from the alum.
   C. lymphocytes accumulate around the alum.
   D. macrophages are activated by the alum.
   E. alum stimulates Th cells.
3. Disease may be caused in immunosuppressed individuals by administration of
   A. inactivated whole agent vaccines.
   B. toxoids.

C. subunit vaccines.

D. genetically engineered vaccine against hepatitis B.

E. attenuated vaccines.

4. Vaccines ideally should be all of the following, *except*

A. effective in protecting against the disease.

B. inexpensive.

C. stable.

D. living.

E. easily administered.

5. Antibodies against viruses can be detected by all of the following tests, *except*

A. hemagglutination inhibition.

B. complement fixation.

C. neutralization of virus replication in cell culture.

D. toxin neutralization.

E. fluorescent antibody.

6. Examples of active immunization include

A. giving antibodies against diphtheria.

B. gamma globulin injections to prevent hepatitis.

C. Sabin polio immunization.

D. rabies immune globulin.

E. tetanus immune globulin.

7. Precipitation tests include all of the following, *except*

A. radial immunodiffusion.

B. ring test.

C. complement fixation.

D. immunoelectrophoresis.

E. double diffusion in gel.

8. Characteristics of both radioimmunoassay and enzyme-linked immunosorbent assay are

A. they are sensitive and efficient tests.

B. a radioactive label is used.

C. they are competitive inhibition assays.

D. they require large amounts of serum to be tested.

E. peroxidase and chromogen are added.

9. An important subunit vaccine that is widely used is the

A. pertussis vaccine.

B. Sabin vaccine.

C. Salk vaccine.

D. measles vaccine.

E. mumps vaccine.

10. In quantifying antibodies in a patient's serum

A. total protein in the serum is measured.

B. the antibody is usually measured in grams per ml.

C. the serum is serially diluted.

D. both antigen and antibody are diluted.

E. the titer refers to the amount of antigen added.

Applications

1. A chemist working for the U.S. Department of Agriculture is interested in using simple, fast tests to gather information about the quality of prepared foods sold for human consumption. Explain how immunodiffusion can be used in the food industry to test for food quality.
2. Many dairy operations keep cow's milk for sale and use formula and feed to raise any calves. One farmer noticed that calves raised on the formula and feed needed to be treated for diarrhea more frequently than calves left with their mothers to nurse. He had some tests run on the diets and discovered no differences in the calories or nutritional content. The farmer called a veterinarian and asked him to explain the observations. What was the vet's response?
3. A medical researcher is interested in developing a vaccine against a Gram-negative bacterium noted for causing a common form of diarrhea. Explain how the vaccine could be designed and administered to effectively combat the organism.

Critical Thinking

1. In figure 17.2, how would the curve change if the concentration of antibody in the original sample were increased? (Would the shape of the curve change? Would the curve be shifted left, right, up, or down?) Briefly explain your answer.
2. In double diffusion in gel (figure 17.4), how can a reaction indicate identity or nonidentity of the antigens being detected?
3. A cell biologist suggested measuring cell-mediated immunity by measuring the incorporation of adenine (instead of thymine) into DNA in lymphocyte cultures stimulated by antigen. Would this give different results compared with measuring thymidine incorporation? Why or why not?

CHAPTER 18

Immunologic Disorders

Pasteur is widely quoted as saying, "Chance favors the prepared mind." This was certainly the case with Charles Richet in his discovery of ***hypersensitivities****, commonly called allergies, near the end of the Nineteenth century. Richet was a well-known physiologist who had discovered, among other things, that the acid in the stomach is hydrochloric acid. Richet also performed early experiments with antisera. He and his colleague, Paul Portier, while cruising in the South Seas on Prince Albert of Monaco's yacht, decided that the Portuguese man-of-war jellyfish must have a toxin responsible for its ugly stings. They made an extract of the tentacles of one such jellyfish and showed that it was very toxic to rabbits and ducks.*

Upon returning to France, they were no longer able to study the toxins from the Portuguese man-of-war jellyfish, so they began studying the effects of toxins from the tentacles of sea anemones. Richet and Portier noted that some dogs in their experiments survived the potent toxin, but when these animals were tested with the toxin again, they died a few minutes after receiving a small dose. One very healthy dog survived the first challenge with the toxin. When given a second dose 22 days later, the dog became very ill within a few seconds. It could not breathe, lay on its side, had violent diarrhea, vomited blood, and died within 25 minutes.

Because of the insights they gained from their early work with antisera, Richet and Portier had prepared minds. They recognized that the dogs' reactions were probably immunologically mediated. The reactions, however, represented the opposite effect of prevention of disease, ***prophylaxis****, conferred by antisera. Therefore, they named this development of hypersensitivity to relatively harmless substances* ***anaphylaxis****, the extreme opposite of prophylaxis. For this and other outstanding contributions to medicine, Richet received the Nobel Prize in 1913.*

—A Glimpse of History

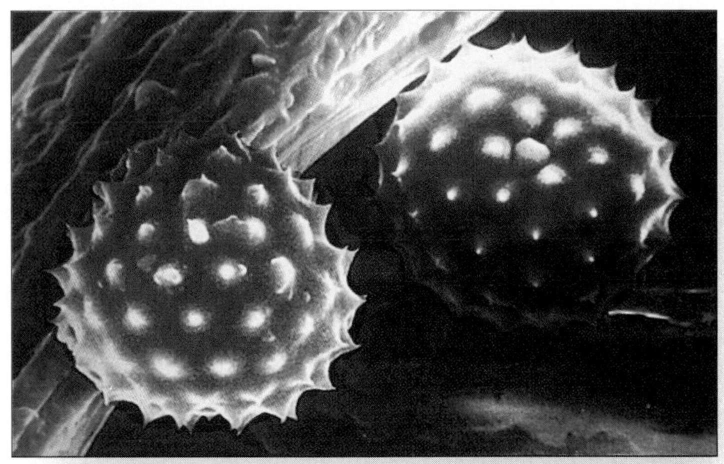

FOR THE MOST PART, THE IMMUNE SYSTEM DOES A superb job protecting the body from invasion by various microorganisms and viruses; however, as Richet showed, the *same mechanisms* that are so effective in protecting us can, under some circumstances, be detrimental. The protective responses are called immunity; immune responses that cause tissue damage are referred to as hypersensitivities. In addition to hypersensitivities, a second type of immunologic disorder occurs when the immune system responds too little, resulting in **immunodeficiency**. A third type of immunologic disorder is **autoimmune disease**, resulting from responses against self-antigens.

Exactly the same mechanisms occur in immunological responses, whether the reactions are protective or damaging. The immune response can be likened to fire. Fire is essential for warmth and cooking and in many ways it is beneficial, but exactly the same fire is destructive if it starts in the wrong place or becomes uncontrolled. Similarly, the immune response is essential for protection, but it can be destructive if it is out of control.

Hypersensitivity reactions to usually harmless substances are often called **allergies** or allergic reactions. Antigens that cause allergic reactions are **allergens**. Hypersensitivities are categorized according to which parts of the immune response are involved and how quickly the response occurs. Most allergic or hypersensitivity reactions fall into one of four major types:

- **Immediate IgE-mediated**
- **Cytotoxic**
- **Immune complex–mediated**
- **Delayed cell-mediated**

TABLE 18.1 Some Characteristics of the Major Types of Hypersensitivities

Characteristic	Type I hypersensitivity Immediate; IgE-mediated	Type II hypersensitivity Cytotoxic	Type III hypersensitivity Immune complex–mediated	Type IV hypersensitivity Delayed cell-mediated
Cell type responsible	B cells	B cells	B cells	T cells
Type of antigen	Soluble	Cell-bound	Soluble	Soluble or cell-bound
Type of antibody	IgE	IgG, IgM	IgG	None
Other cells involved	Basophils, mast cells	Red blood cells, white blood cells, platelets	Various host cells	Various host cells
Mediators	Histamine, serotonin, leukotrienes	Complement, ADCC	Complement, neutrophil proteases	Cytokines
Transfer of hypersensitivity	By serum	By serum	By serum	By T cells
Time of reaction after challenge with antigen	Immediate, up to 30 minutes	Hours to days	Hours to days	Peaks at 48 to 72 hours
Skin reaction	Wheal and flare	None	Arthus	Induration, necrosis
Examples	Anaphylactic shock, hay fever, hives	Transfusion reaction, hemolytic disease of newborns	Serum sickness, farmer's lung, malarial kidney damage	Tuberculin reaction, contact dermatitis, tissue transplant rejection

The main characteristics of the various types of hypersensitivities are shown in **table 18.1**. Allergic reactions occur only in **sensitized** individuals—that is, those who have been immunized or sensitized by prior exposure to that specific antigen.

Immediate IgE-Mediated Hypersensitivities (Type I)

Antibodies of the class IgE are important in protective immunity. For example, during a worm infestation, soluble worm antigens react with specific IgE antibodies, causing mast cells to degranulate, releasing mediators that cause an increase in vascular permeability and attract inflammatory cells. Also, IgE antibodies and eosinophils together can kill some parasites by antibody-dependent cellular cytotoxicity (ADCC). Thus, IgE has a protective role, especially against parasites. IgE, however, also causes immediate (type I) hypersensitivity reactions, characterized by a reaction in a sensitized individual within minutes of exposure to antigen. Some people develop such a reaction when exposed to substances such as dust, pollens, animal dander, and molds, which do not evoke a response in nonsensitized people. ■ ADCC, p. 394

Sensitization occurs when the antigen makes contact with some part of the body (pollen grains contact the mucosa of the nose, for example) and induces a response **(figure 18.1a)**. The antigen is taken up, processed by antigen-presenting cells, and presented to T cells. The cytokines they produce activate B cells to become antibody-producing cells. Although B cells making any of the classes of immunoglobulin are probably induced to make antibodies, the tissues under the mucous membranes are rich in B cells committed to IgA and IgE production, and IgE-producing cells are more abundant in allergic than in nonallergic individuals. These cells have been switched from making other classes of immunoglobulin to making IgE by the action of cytokines. As IgE is produced in this area, which is rich in mast cells, the IgE molecules attach via their Fc portion to receptors on the mast cells and also on circulating basophils. Mast cells and basophils contain many large cytoplasmic granules packed with histamine, enzymes, and other potent chemical mediators of the type I hypersensitivity reaction. Once attached, the IgE molecules can survive for many weeks with their reaction sites waiting to interact with antigen. The individual is sensitized to that antigen.

People with a type I hypersensitivity will have many IgE antibodies fixed to mast cells throughout their bodies. When such a person is not exposed to antigen, these antibodies are harmless. Upon exposure, however, the antigen readily combines with the cell-fixed IgE antibodies. At least two cell-bound IgE molecules must react with specific antigen, cross-linking the IgE molecules, in order for a reaction to occur (figure 18.1b). Within seconds, the IgE-antigen attachment and cross-linking of IgE in the cell membrane cause the mast cell to **degranulate**, that is to release histamine and other preformed mediators from its cytoplasmic granules. The cells can rapidly synthesize *leukotrienes*, *prostaglandin*, and *cytokines* such as IL-4, also stored within the granules. The chemical mediators released from the granules are the direct causes of urticaria (hives), allergic rhinitis (hay fever), asthma, anaphylactic shock, and other allergic manifestations. ■ mast cell mediators, p. 384

The tendency to have type I allergic reactions is inherited. The reactions occur in at least 20% to 30% of the population of the United States. Susceptible people often have higher than normal levels of IgE in the circulation, and increased numbers of eosinophils.

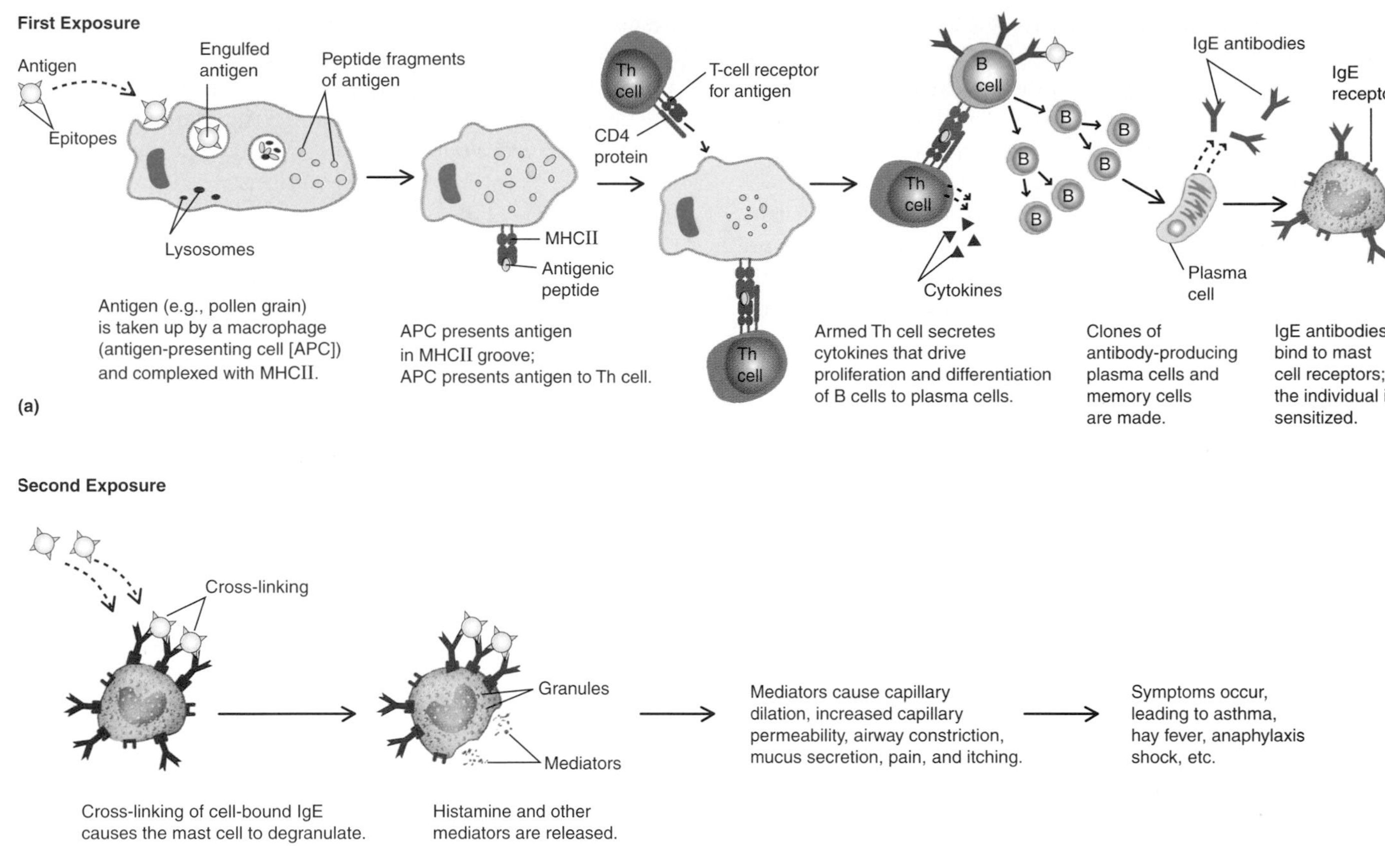
First Exposure
Antigen
Epitopes
Engulfed antigen
Peptide fragments of antigen
Lysosomes
Antigen (e.g., pollen grain) is taken up by a macrophage (antigen-presenting cell [APC]) and complexed with MHCII.
(a)
MHCII
Antigenic peptide
APC presents antigen in MHCII groove; APC presents antigen to Th cell.
Th cell
T-cell receptor for antigen
CD4 protein
B cell
Cytokines
Armed Th cell secretes cytokines that drive proliferation and differentiation of B cells to plasma cells.
Clones of antibody-producing plasma cells and memory cells are made.
Plasma cell
IgE antibodies
IgE receptor
IgE antibodies bind to mast cell receptors; the individual is sensitized.
Second Exposure
Cross-linking
Cross-linking of cell-bound IgE causes the mast cell to degranulate.
Granules
Mediators
Histamine and other mediators are released.
Mediators cause capillary dilation, increased capillary permeability, airway constriction, mucus secretion, pain, and itching.
Symptoms occur, leading to asthma, hay fever, anaphylaxis shock, etc.
(b)

Figure 18.1 Mechanisms of Immediate IgE-Mediated (Type I) Hypersensitivity **(a)** First exposure to antigen induces an IgE antibody response leading to sensitization. **(b)** A reaction is triggered when antigen combines with and cross-links cell-bound IgE on mast cells in a sensitized individual.

Localized Anaphylaxis

Anaphylaxis is the name given to allergic reactions caused by IgE-mediated release of mast cell granules. Although anaphylaxis may be generalized or local, by far the most usual allergic reactions are examples of localized anaphylaxis. **Urticaria (hives)** is an allergic skin condition characterized by the formation of a wheal and flare; the wheal is a fluid-filled itchy swelling generally resembling a mosquito bite, surrounded by the red flare. The wheal and flare reaction is seen also in positive skin tests for allergens (**figure 18.2**). Hives may occur, for example, when a person allergic to lobster eats some of the seafood. Lobster antigen absorbed from the intestinal tract enters the bloodstream and is carried to tissues such as skin, where it reacts with mast cells that have anti-lobster IgE antibody attached to them. Reaction between antibody and antigen on the mast cell surface releases histamine, which in turn causes dilation of tiny blood vessels and the leaking of plasma into skin tissues. Life-threatening respiratory obstruction is possible during a reaction if there is extensive tissue swelling in the throat and larynx. Because histamine is a major mediator in this situation, the reaction is blocked by antihistamines.

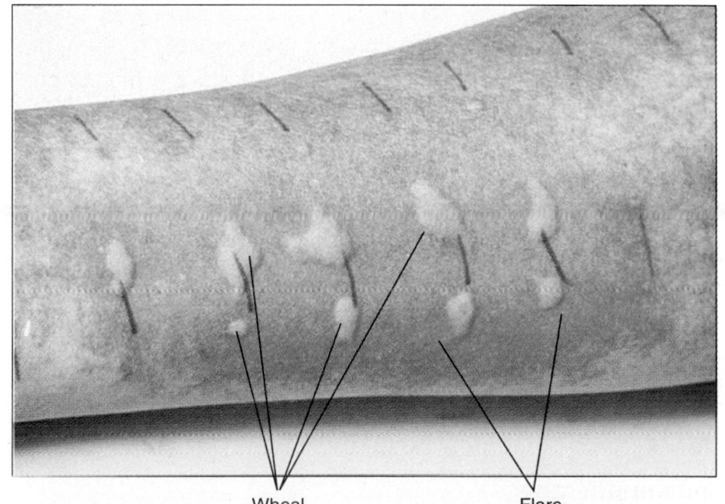
Wheal
Flare

Figure 18.2 The Wheal and Flare Skin Reaction of IgE-Mediated Allergies A variety of antigens are injected or placed in small cuts in the skin to test for sensitivity. Immediate wheal and flare reactions occur with antigens to which the person is sensitive.

Allergic rhinitis (hay fever), marked by itching, teary eyes, sneezing, and runny nose, occurs when allergic persons inhale an antigen such as ragweed pollen to which they are sensitive. The mechanism is similar to that of hives and is also blocked by antihistamines.

Asthma is another type of immediate respiratory allergy. Antigen-induced release of chemical mediators from IgE-sensitized mast cell granules and from eosinophils attracted into the area of inflammation causes increased mucus secretion and spasms of the bronchi, markedly interfering with breathing. Mediators other than histamine, mainly lipids such as leukotrienes and prostaglandins and protein cytokines, are responsible for the bronchospasm and increased mucus production. Antihistamines are therefore not effective in treating asthma, but several other drugs are available to block the reaction, such as bronchodilating drugs that relax constricted muscles and relieve the bronchospasm. An example is albuterol, a bronchodilator usually delivered by an inhaler. Steroids are often used to decrease the inflammatory reaction. An anti-IgE therapy, described later, is currently in clinical trials.

Generalized Anaphylaxis

Generalized systemic anaphylaxis is a rare but serious form of IgE-mediated allergy. This is the form of anaphylaxis described by Richet. Antigen enters the bloodstream and becomes widespread. Instead of being localized in areas of the skin or respiratory tract, the reaction affects almost the entire body. Loss of fluid from the blood vessels into tissues causes not only swelling, but also a condition called **shock**. Shock is a state in which there is insufficient blood pressure to supply the blood flow required to meet the oxygen needs of vital body tissues. In the case of generalized anaphylaxis, large amounts of released mediators cause extensive blood vessel dilation and loss of fluid from the blood, so that blood pressure falls dramatically and there is insufficient blood flow to vital organs such as the brain. Anaphylactic reactions may be fatal within minutes. Bee stings, peanuts, and penicillin injections probably account for most cases of generalized anaphylaxis. As mentioned in chapter 16, penicillin molecules are changed in the body to a haptenic form that can react with body proteins. In a small percentage of people, this hapten-protein complex causes the formation of IgE antibodies that react with penicillin. Fortunately, only a tiny fraction of people who make the antibodies are prone to anaphylactic shock. Peanuts have been the cause of many of the reactions reported recently. Peanuts and their products, such as peanut oil, are found in so many foods that it is often hard to predict where they will be encountered. The problem is widespread enough that peanuts have been barred by some airlines as a snack during flights. Generalized anaphylaxis can usually be controlled by epinephrine injected immediately. Anyone who has had a local allergic reaction to bee sting, penicillin, peanuts, or other substance is at risk of anaphylactic shock if exposed to the same substances and should wear a medical-alert bracelet and carry emergency medications.

Immunotherapy

In view of the danger of severe generalized anaphylaxis upon exposure to even tiny amounts of the offending antigen, it might seem paradoxical that one way of preventing immediate hypersensitivity reactions is to inject the person with extremely dilute solutions of the antigen. This form of therapy is called **desensitization** or **hyposensitization**. The concentration of antigen in the injected solution is very gradually increased over a period of months in a series of injections. The patient gradually becomes less and less sensitive to the antigen and may even lose the hypersensitivity entirely. During treatment, there is an increase in the levels of IgG and a decrease in the IgE response to the antigen. The IgG antibodies produced are thought to protect the patient by binding to the offending antigen, thus preventing its attachment to cell-bound IgE **(figure 18.3)**. Also, with immunotherapy there is an increase in T-suppressor (Ts) cells. The activities of the Ts cells may suppress further production of IgE antibodies. ■ **T-suppressor cells, p. 398**

It might seem desirable to use anti-IgE antibodies to counteract and remove IgE that is causing allergies, but most IgE molecules are attached to receptors on mast cells and basophils. Injected anti-IgE antibodies would bind to IgE attached to cell receptors, causing cross-linking of receptors and massive disastrous mediator release. Advances in understanding the structure of IgE and receptor molecules, however, along with new technologies, have now made anti-IgE therapy feasible. In order to use this form of immunotherapy, a safe and effective anti-IgE molecule had to be *engineered.* A first step was finding the amino acid sequences in the Fc portion of human IgE that bind to the IgE receptor on human cells. Then a monoclonal antibody (Mab) was produced in mouse cells that bound circulating human IgE specifically at that receptor-binding site. This Mab could not react with cell-bound IgE to cause mediator release because the binding site was already blocked by its attachment to cell receptors. Although the Mab blocked the IgE attachment to cells, mouse monoclonal antibodies were not effective in humans for very long because they are foreign molecules and are removed quickly from the body. The next step was to "humanize" the Mab by making a recombinant antibody with the antigen-binding site (anti-IgE) from the mouse Mab and the rest of the molecule from a human antibody. This hybrid recombinant molecule is called **rhuMab** (recombinant human Monoclonal antibody). Trials have shown that rhuMab injections are well tolerated by asthmatic people and are remarkably effective in treating their asthma. Amounts of free IgE in the circulation of treated people were reduced more than 90%, and their mast cells and basophils were much less sensitive to IgE binding. The anti-IgE also bound to IgE-producing B cells, causing apoptosis of these cells and thereby preventing more IgE synthesis. Advanced clinical trials of this promising therapy are continuing.

MICROCHECK 18.1

Hypersensitivity reactions are immunological reactions that cause tissue damage. Type I hypersensitivity reactions mediated by cell-bound IgE antibodies occur immediately after exposure to antigen. The reactions are caused by the release of mediators from mast cell granules. Localized anaphylactic reactions include hives, hay fever, and asthma; generalized reactions lead to anaphylactic shock. Immunotherapy is directed toward inducing IgG rather than IgE. Also, anti-IgE is a promising new therapy.

- How do localized and generalized anaphylactic reactions differ?

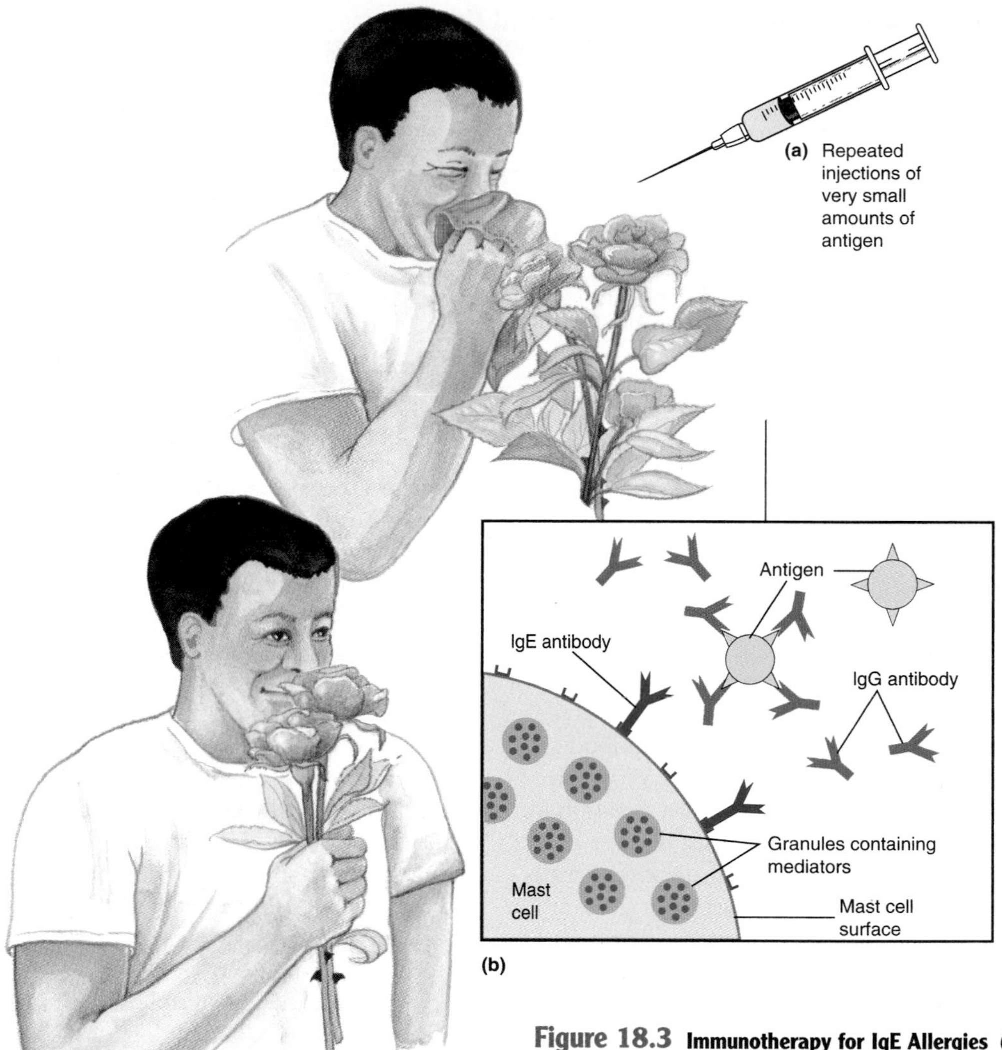

Figure 18.3 Immunotherapy for IgE Allergies **(a)** Repeated injections of very small amounts of antigen are given over several months. **(b)** This regimen leads to the formation of specific IgG antibodies. The IgG reacts with antigen before it can bind to IgE, and therefore it blocks the IgE reaction.

- Define allergen and give five common examples of substances that act as allergens.
- Why would leaking of plasma from blood vessels during an allergic reaction cause a wheal?

Cytotoxic Hypersensitivities (Type II)

In type II hypersensitivity reactions, complement-fixing antibody reacting with cell surface antigens may cause injury or death of the cell. This phenomenon is especially striking in the case of red blood cells, because the reaction results in rupture of the cells and a visible release of hemoglobin. The reaction can occur in response to foreign antigenic or haptenic material such as a drug that attaches to erythrocytes or to platelets, but more common examples are transfusion reactions and hemolytic disease of the newborn. In these cases, normal red cells are destroyed as a result of antibodies reacting with erythrocyte epitopes. Cells can be destroyed in type II reactions not only by complement lysis, but also by **antibody-dependent cellular cytotoxicity (ADCC)**. ■ **epitope, p. 390** ■ **complement, p. 377**

Transfusion Reactions

Normal erythrocytes have many different antigens on their surfaces, and these antigens differ from one individual to another. When a person receives transfused red blood cells that are antigenically different from his or her own, immune lysis (by antibody and complement) of the red blood cells often results. This lytic reaction is especially likely when the antigens involved are those of the ABO blood group system. People are designated as having blood type A, B, AB, or O, depending on which, if any, ABO polysaccharide is present on the red blood cells. The possible ABO blood types are shown in **table 18.2**.

A locus on human chromosome 9 determines which of the ABO antigens will be present on an individual's red blood cells. Since one gene of the chromosome pair is originally derived from the mother and the other is from the father, the genes of the pair may differ. In the case of the ABO system, there are three possible types of genes for the chromosomal locus: A, B, and O. A and

TABLE 18.2 Antigens and Antibodies in Human ABO Blood Groups

Blood Type	Antigen Present on Erythrocyte Membranes	Antibody in Plasma	Incidence of Type in United States		
			Among Whites	Among Asians	Among Blacks
A	A	Anti-B	41%	28%	27%
B	B	Anti-A	10%	27%	20%
AB	A and B	Neither anti-A nor anti-B	4%	5%	7%
O	Neither	Anti-A and anti-B	45%	40%	46%

B genes each code for different glycosyltransferases responsible for attaching either A or B polysaccharide antigen to the red cell surface. The O gene does not code for either, indicating a lack of A or B glycosyltransferases. When both an A and a B gene are present, they are codominant, meaning that both antigens are expressed on the red blood cell surface.

One important and unexpected feature of the ABO system is that people who lack A or B erythrocyte polysaccharides have antibodies against the antigen that they lack. Thus, people with blood type O have both anti-A and anti-B antibodies; those of blood type A have anti-B; and those of type B have anti-A. Type AB people have neither anti-A nor anti-B antibodies. These antibodies are called natural antibodies, because they are present without any obvious or deliberate stimulus (the anti-A and anti-B antibodies are not present at birth but generally appear before the age of six months). These natural antibodies are mostly of the class IgM and are capable of fixing complement and lysing red cells. Since they are IgM, they cannot cross the placenta. They most probably arise because of multiple exposures to small amounts of substances similar to the blood type antigens, substances known to be found in many environmental materials such as bacteria, dust, and foods.

A **transfusion reaction** can occur if a patient receives erythrocytes differing antigenically from his or her own during a blood transfusion. Cross-matching the bloods and other techniques are used to ensure compatibility of donor and recipient. In the case of ABO incompatibility, IgM antibodies cause a type II hypersensitivity reaction. The symptoms of a typical transfusion reaction include fever, low blood pressure, pain, nausea, and vomiting. These symptoms occur because the foreign erythrocytes are agglutinated by the recipient's antibody, complement is activated, and red blood cells are lysed.

Hemolytic Disease of the Newborn

There are a number of other human red cell antigen systems; an important one is the Rhesus, or Rh system (first described in rhesus monkeys). It is the practice to test donor and recipient bloods for Rh and other possible incompatibilities in addition to ABO.

The Rh (Rhesus) blood group system is complex and involves various antigens. If the Rh antigen is present on a person's erythrocytes, he or she is Rh-positive; if it is lacking, the person is Rh-negative. Rh-positive people have the Rh antigen and do not make antibodies to it. An Rh-negative adult, however, may have a transfusion reaction as a result of being immunized against the Rh antigen, through a blood transfusion, organ graft, or pregnancy that might lead to introduction of the antigen and consequent development of anti-Rh antibodies. The Rh-positive cells are destroyed by macrophages in antibody-dependent cellular cytotoxicity, not by complement lysis.

Anti-Rh antibodies formed by a pregnant woman may cross the placenta and damage her baby. The resulting disease is called **hemolytic disease of the newborn** or simply Rh disease (**figure 18.4**). Blood incompatibilities other than Rh may be responsible for hemolytic disease of the newborn, but such cases are generally less severe.

While an Rh-negative mother is carrying an Rh-positive fetus, a few fetal red blood cells enter the mother's circulation via the placenta, but usually not enough to cause a primary antibody response. At the time of birth, however, enough of the Rh-positive baby's erythrocytes may enter the mother's circulation to cause an immune response. Induced or spontaneous abortions may also be responsible for immunizing an Rh-negative mother to the Rh antigen. The anti-Rh antibodies formed by the mother cause her no harm, because her red blood cells lack the Rh antigen. With the second and each subsequent Rh-positive fetus, however, even a few Rh-positive cells that might enter the mother's circulation from the fetus are enough to provoke a secondary response. The result is that the mother produces large quantities of anti-Rh antibodies of the class IgG. Like other IgG antibodies, anti-Rh antibodies readily cross the placenta, enter the circulation of the fetus, and cause extensive fetal red cell damage. Although miscarriage with loss of the fetus can result, the disease often is not apparent until shortly after birth.

The fetus survives the anti-Rh antibody attack while still in utero because harmful products of red cell destruction are eliminated from its system by enzymes of the mother. These enzymes are present in only scant amounts in the newborn. As early as 36 hours after birth, jaundice may appear and result in permanent brain damage or even death. Not only does the baby become seriously ill from the toxic products of red cell destruction, but it also develops a severe anemia. In this critical situation, it may be necessary to withdraw a portion of the baby's blood and replace it with Rh-negative blood. Also, light treatment of the baby detoxifies the erythrocyte breakdown products. Irradiation with light at 420 to 480 nm makes the red blood cell breakdown products more readily excretable.

Most cases of hemolytic disease caused by Rh incompatibility are preventable by injecting the Rh-negative mother with

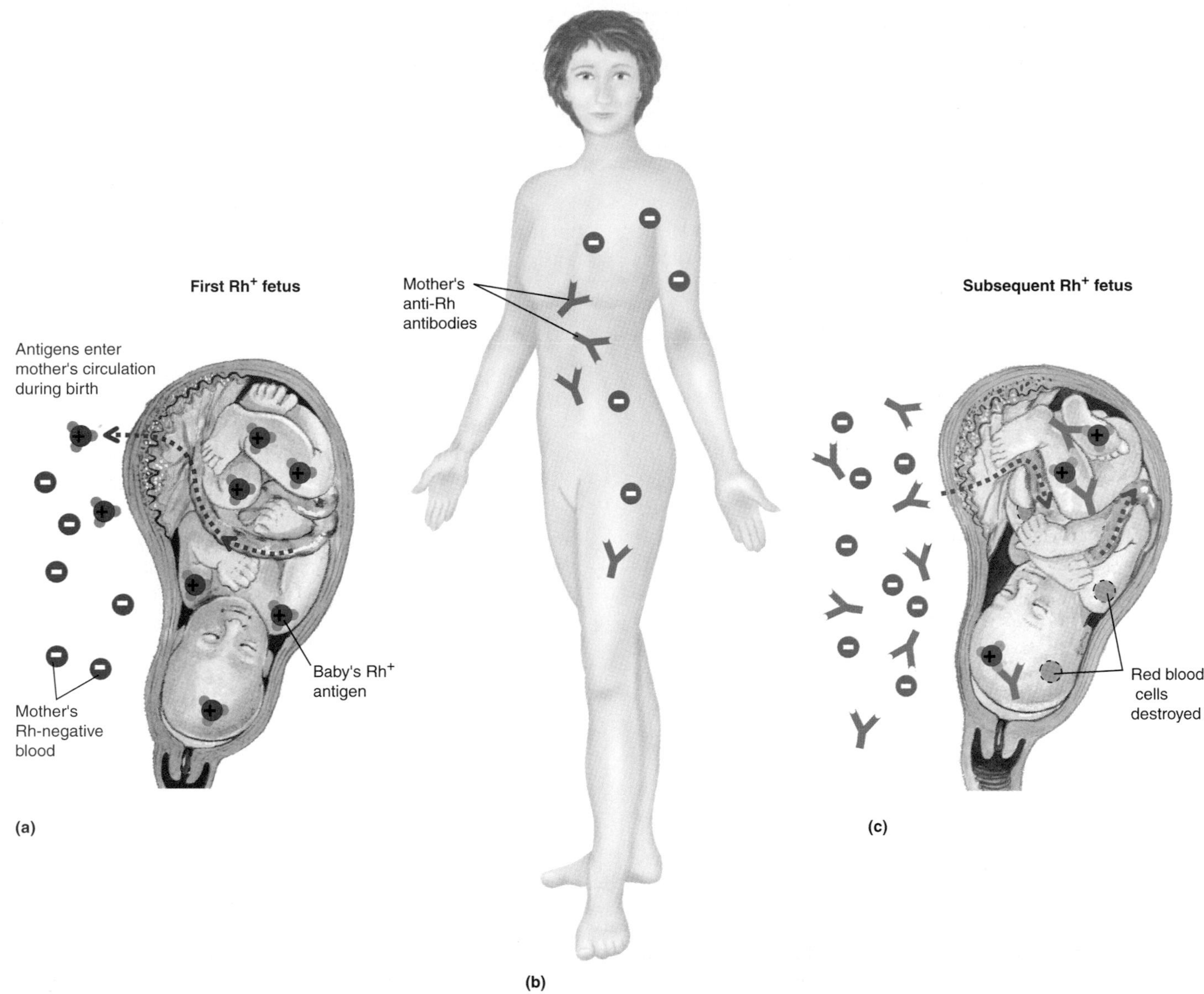

Figure 18.4 Hemolytic Disease of the Newborn (a) The fetus of an Rh-negative (Rh^-) mother may inherit paternal genes for Rh antigens and have Rh-positive (Rh^+) blood. During delivery of a first Rh^+ baby, enough Rh^+ red cells can enter the Rh^- mother to induce an anti-Rh response. **(b)** The mother makes a primary response to Rh antigens and develops memory cells for anti-Rh antibodies. **(c)** During subsequent pregnancies with an Rh^+ fetus, the very few fetal red cells that cross the placenta can cause a vigorous secondary response. The IgG anti-Rh antibodies cross the placenta and destroy the baby's red blood cells.

anti-Rh antibodies within the first 24 to 48 hours (and preferably within the first few hours) after abortion or delivery. These administered antibodies bind to the baby's Rh-positive red cells that have entered the mother's circulation and prevent the development of a primary response. This treatment does not inhibit the activation of B memory cells or a secondary immune response, and therefore is not helpful if the mother is already sensitized to the Rh antigens. ■ **control of immune responses, p. 407**

MICROCHECK 18.2

Type II cytotoxic hypersensitivity reactions are mediated by antibodies, either by complement lysis of cells or by antibody-dependent cellular cytotoxicity. Blood transfusion reactions and hemolytic disease of the newborn are examples.

- Describe the mechanism of cell damage in a blood transfusion where ABO antigens are mismatched.
- Why do Rh-negative but not Rh-positive mothers sometimes have babies with hemolytic disease of the newborn?
- Why is the finding surprising that people lacking the A or B antigen are found to have antibodies to the corresponding antigen (anti-A or anti-B)?

Immune Complex–Mediated Hypersensitivities (Type III)

An **immune complex** consists of antigen and antibody bound together, often with some complement components. Immune complexes usually adhere to Fc receptors on cells of the mononuclear phagocyte system and are engulfed and destroyed intracellularly, especially if they are large complexes. Under ordinary circumstances, they are rapidly removed from circulation. In conditions where a moderate excess of antigen over antibody exists, however, the small complexes formed are not quickly removed and destroyed but persist in circulation or at their sites of formation in tissue. Immune complexes possess considerable biological activity. They initiate the blood clotting mechanism, and they activate components of complement that attract neutrophils into the area and contribute to inflammation (**figure 18.5**). Proteases released from the neutrophils are especially active in inducing tissue damage. Circulating immune complexes are commonly deposited in skin, in joints, and in the kidneys, where they cause glomerulonephritis. Immune complexes are responsible for the rashes, joint pains, and other symptoms seen in a number of diseases, such as farmer's lung, bacterial endocarditis, early rubella infection, and malaria. Immune complexes can also precipitate a devastating condition, **disseminated intravascular coagulation**, in which clots form in small blood vessels, leading to failure of vital organs. ■ **disseminated intravascular coagulation, p. 720** ■ **glomerulonephritis, p. 528, 529**

Immune complex disease may arise during a variety of bacterial, viral, and protozoan infections, as well as from inhaled dusts or bacteria and injected medications such as penicillin. The steps in the formation of immune complex pathology are summarized in **table 18.3**.

Immune complex formation is also responsible for the localized injury or death of tissue, known as the **Arthus** reaction, that occurs if antigen is injected into the tissue of a previously immunized animal or person with high levels of circulating specific antibody. The immune complexes formed locally activate complement, producing complement components that attract neutrophils. The release of neutrophil contents and inflammation results in a local reaction that peaks at about 6 to 12 hours.

Serum sickness is an immune complex disease caused by passive immunization where an antibody-containing serum from a horse or other animal is injected into humans to prevent or treat a disease such as diphtheria or tetanus. The recipient of the animal serum may make an immune response to antigens in the foreign serum, and after 7 to 10 days enough immune complexes form to cause signs of disease, which include fever, inflammation of blood vessels, arthritis, and kidney damage. The disease generally resolves as the antigens of the animal serum are cleared. Of course, horses are no longer used to produce antibodies against diphtheria and tetanus; instead, hyperimmune human sera are used. The serum sickness form of hypersensitivity is rarely seen now, but it can occur following

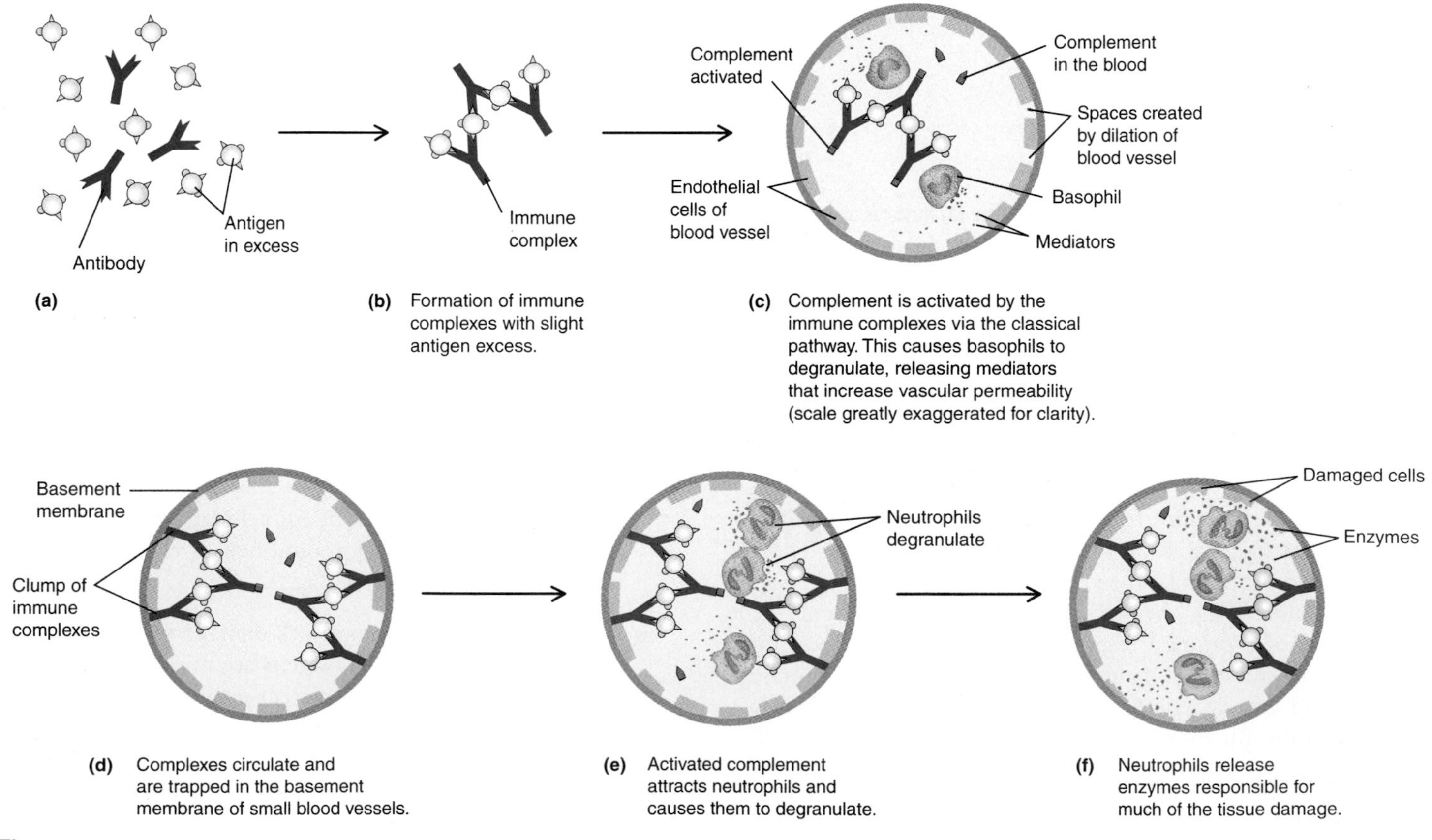

(a)

(b) Formation of immune complexes with slight antigen excess.

(c) Complement is activated by the immune complexes via the classical pathway. This causes basophils to degranulate, releasing mediators that increase vascular permeability (scale greatly exaggerated for clarity).

(d) Complexes circulate and are trapped in the basement membrane of small blood vessels.

(e) Activated complement attracts neutrophils and causes them to degranulate.

(f) Neutrophils release enzymes responsible for much of the tissue damage.

Figure 18.5 Immune Complex (Type III) Hypersensitivity

TABLE 18.3 Pathogenesis of Immune Complex Disease

1. Antibody combines with excess soluble antigen.
2. The antibody-antigen combination reacts with complement.
3. Complexes are deposited in sites such as skin, kidney, and joints.
4. Fragments of complement cause release of histamine and other mediator substances from mast cells or basophils and also attract neutrophils.
5. Release of the mediators causes increased permeability of blood vessel walls.
6. Immune complexes penetrate or form in blood vessel walls.
7. Neutrophils enter the vessel walls chemotactically.
8. Neutrophils release lysosomal enzymes, especially proteases, that induce tissue injury.

treatment of heart attack patients with the bacterial enzyme streptokinase to dissolve clots, after the use of serum from horses immunized with snake venom to treat snakebites in people, or in a few other rare instances. ■ passive immunization, p. 424

MICROCHECK 18.3

Type III immune complex–mediated hypersensitivities are caused by small complexes of antigen-antibody and complement that persist in tissues or blood and that cause inflammation.

- Why do immune complexes remain in the circulation or in the tissues in immune complex diseases?
- How can immune complexes cause tissue damage?
- Why would the kidneys be particularly prone to immune complex damage?

Delayed Cell-Mediated Hypersensitivities (Type IV)

Harmful effects produced by the mechanisms of cell-mediated immunity are referred to as **delayed hypersensitivity**. The name reflects the slowly developing response to antigen; reactions peak at 2 to 3 days rather than in minutes as in immediate hypersensitivity. As would be expected with cell mediated responses, T cells are responsible and antibodies are not involved. Delayed hypersensitivity reactions can occur almost anywhere in the body. They are wholly or partly responsible for **contact dermatitis** (such as from poison ivy and poison oak), tissue damage in a variety of infectious diseases, rejection of tissue grafts, and some autoimmune diseases directed against antigens of self.

Tuberculin Skin Test

A familiar example of a delayed hypersensitivity reaction is the positive reaction to a tuberculin skin test that occurs in most people who have been infected with *Mycobacterium tuberculosis.* This test involves the introduction of very small quantities of protein antigens of the tubercle bacillus into the skin. In those with delayed hypersensitivity to *M. tuberculosis*, the site of injection reddens and gradually becomes indurated (thickened) within 6 to 24 hours. The reaction reaches its peak at 2 to 3 days (**figure 18.6**). There is no wheal formation, as would be seen with IgE-mediated reactions. The redness and induration of delayed skin hypersensitivity reactions are the result mainly of the reaction of sensitized T cells with specific antigen, followed by the release of cytokines and the influx of macrophages to the injection site.

Contact Hypersensitivities

Contact hypersensitivity is mediated by T cells. The T cells that have become sensitized to a particular antigen release cytokines when they come into contact again with the same antigen. These cytokines cause inflammatory reactions that attract macrophages to the site. The macrophages then release mediators that add to the inflammatory response, resulting in allergic dermatitis.

Familiar examples of contact hypersensitivity (contact allergy or contact dermatitis) are poison ivy and poison oak (**figure 18.7**) and allergic reactions to the nickel of metal jewelry, the chromium salts in certain leather products, or components of some cosmetics. In the case of poison oak or ivy, the antigen is an oily product of the plant. In the case of a metal, a soluble salt of the metal acts as a hapten.

In recent years latex products have become a frequent cause of contact hypersensitivity reactions, although they can also cause IgE-mediated reactions. Latex, a product of the rubber tree, contains a plant protein that readily induces sensitization. Many products contain latex, such as fabrics, elastics, toys, and contraceptive condoms, but latex gloves probably account for the greatest increase in occurrence of latex sensitization. Gloves are used extensively by health care and laboratory workers, food preparers, and many others. Typically, a person will

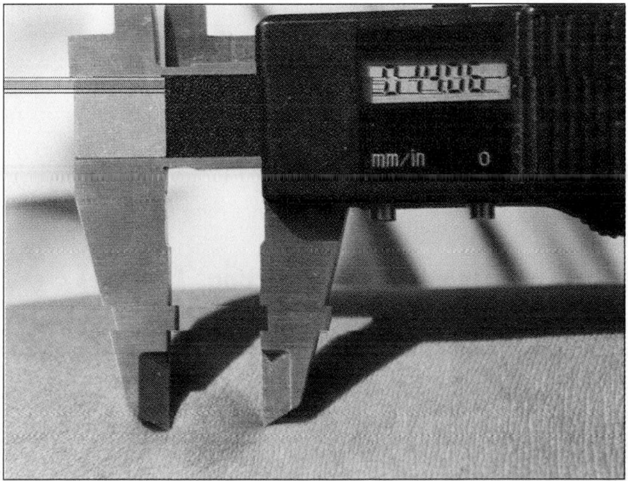

Figure 18.6 Positive Delayed (Type IV) Hypersensitivity Skin Test
Injection of tuberculin protein into the skin of a person sensitized to the tubercle bacilli causes a hardened red lump to form by 48 to 72 hours. A reaction greater than 10 mm in diameter is considered a positive reaction.

Hapten from the poison oak

Carrier protein

MHCII

Hapten peptide

APC

Th 1

T-cell receptor

CD4 protein

In a nonsensitized person no visible reaction

(b) Antigen-presenting cells (APC) present the hapten peptide complexed with MHCII to Th1 (inflammatory) CD4 T cells.

(c) A primary response ensues, resulting in the presence of primed Th1 T cells.

(a) Exposure to poison oak

In a sensitized person visible delayed reaction

Cytokines

Cytokine receptor

Macrophage

APC

Mediators

(d) Antigen-presenting cells present the hapten-peptide complex to sensitized Th1 cells, which secrete cytokines and attract macrophages; the macrophages are activated and secrete mediators of inflammation that cause skin lesions.

(e) Characteristic skin lesions appear after 24 hours, reaching their peak at 48–72 hours after exposure to the plant.

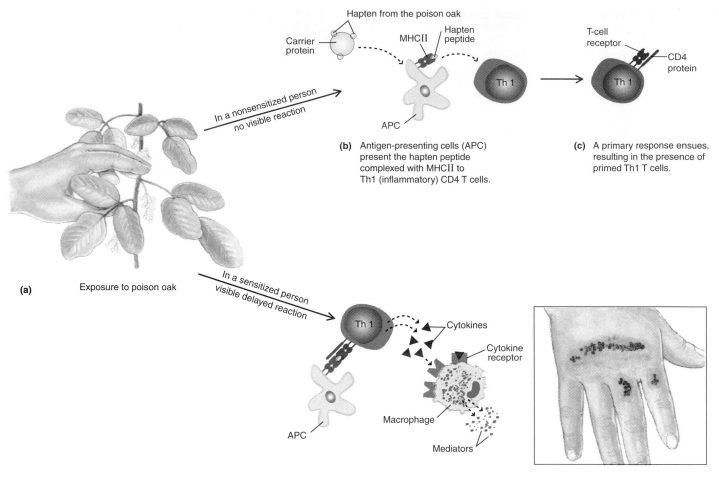

Figure 18.7 Poison Oak Dermatitis Is an Example of Delayed Cell-Mediated (Type IV) Hypersensitivity

notice redness, itching, and a rash on the hands after wearing gloves. To control the reaction, latex gloves should be replaced by vinyl or other synthetic gloves, and cortisone cream helps heal the skin.

Contact hypersensitivity is commonly detected by patch tests, in which the suspect substance is applied to the skin under an adhesive bandage. Positive reactions reach their maximum in about 3 days and consist of redness, itching, and blisters of the skin. **Figure 18.8** shows a severe contact hypersensitivity skin rash.

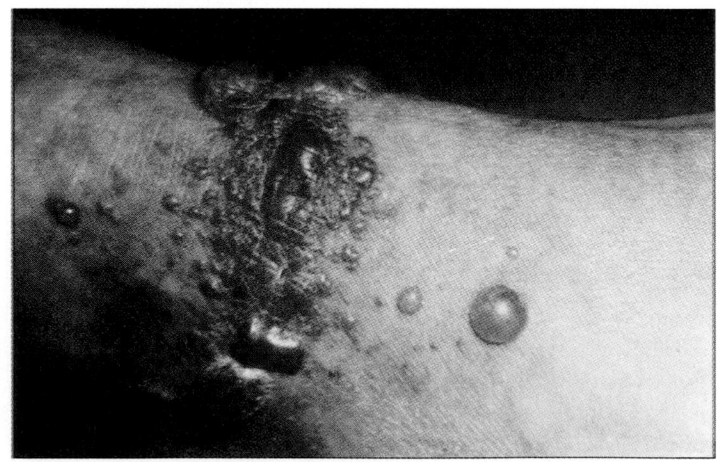

Figure 18.8 Severe Contact Hypersensitivity Skin Rash, Showing Redness, Blisters, and Scaling of the Skin

Delayed Hypersensitivity in Infectious Diseases

The role of cell-mediated immunity in combating intracellular infections through the cell-destroying activity of activated macrophages and T lymphocytes was discussed previously. Although these functions are protective, tissue damage or hypersensitivity also results. These infections may be caused by viruses, mycobacteria and certain other bacteria, protozoa, and fungi. They include leprosy, tuberculosis, leishmaniasis, and herpes simplex, among many others. In particularly slowly progressing infections, delayed hypersensitivity causes extensive cell destruction and progressive impairment of tissue function, such as the damaged sensory nerves in leprosy. The immune response is a two-edged sword, protecting on the one side, but causing damage on the other. ■ **immunity to intracellular infections, p. 405**

Perspective 18.1 The Fetus As an Allograft

Grafts between nonidentical members of the same species are **allografts**, and they are normally rejected by immunological mechanisms. The rejection time for grafts that differ in their major histocompatibility antigens is measured in days, about 10 to 14 days in most systems. The rejection process is complex, but it depends principally on host T-cell destruction of grafted cells. For example, when skin allografts are transplanted, the grafted skin appears normal for about a week. Gradually it begins to look bruised and unhealthy, until 10 days to 2 weeks after transplantation. By that time the grafted skin becomes dried and is sloughed off. Microscopic examination of the graft shows that by the end of the first week T cells invade the tissues and within the next few days the lymphocytes have killed the grafted cells. The same events occur in allografts of kidney or other organs, unless effective immunosuppressive therapy is given. There is one kind of allograft that does not follow this sequence and that is not rejected, however—the human or other mammalian fetus.

The fetus is an allograft, with half of its antigens of paternal origin and likely to differ from the other half contributed by the mother. In spite of these major immunological differences, the fetus lives and thrives in the uterus for 9 months and is not rejected. In fact, over a period of years a mother often has several (or even occasionally as many as 20!) children by the same father without showing any signs of immunological rejection of the fetus. The mechanisms for survival of the fetal allograft have been the subject of research for many years, but they are not yet fully understood.

It cannot be that paternal antigens from the fetus do not reach the mother's immune system to cause a response. Mothers are known to make antibodies to paternal antigens, such as the Rhesus red blood cell antigens. Also, the antibodies used for typing major histocompatibility antigens have long been obtained from women who have borne several children by the same father and have made antibodies to his MHC antigens. Furthermore, various techniques show the presence of small numbers of fetal cells in the maternal circulation during pregnancy. Clearly, paternal antigens can reach the mother's immune system and cause a response. The placenta, however, does prevent most fetal cells from entering the mother and most maternal T cells from reaching the fetus.

The outer layer of the placenta, the trophoblast, forms sort of a buffer zone between the fetus and the mother. The trophoblast does not express MHC class I or II antigens and is not subject to T-cell attack; it also has a mechanism for avoiding destruction by natural killer cells. Thus, the fetus is protected by being in an immunologically privileged site. A few other areas in the body—the brain, the eyes, and the testes—are also immunologically privileged sites. Antigens leaving these sites do not drain through lymphatic vessels and reach lymphoid tissues where antigen-presenting cells are abundant. Also, antigen leaves these privileged sites accompanied by cytokines that are immunosuppressive and direct the immune response toward tolerance, rather than toward active harmful responses. It is well recognized that maternal immune responses are suppressed to some extent during pregnancy, though the reasons for this are not clear.

Thus, one major factor responsible for preventing the rejection of the fetal allograft is the location of the fetus in the pregnant uterus, protected by the placental barrier. A second factor is the ability of the pregnancy to cause an immunosuppressive response in the mother.

MICROCHECK 18.4

Type IV delayed hypersensitivity depends on the action of sensitized T cells. The reaction peaks 2 to 3 days after exposure to antigen. Examples are contact dermatitis, damage in a variety of infectious diseases, rejection of tissue grafts, and some autoimmune diseases.

- Explain the events that occur in the skin during a positive delayed hypersensitivity skin test.
- Describe a patch test for contact hypersensitivity.
- In the tuberculin skin test, why would there be no reaction if there is no infection?

Transplantation Immunity

Transplantation of some organs and tissues between genetically nonidentical humans is a well-established clinical procedure. Such grafts are called **allografts**. The major drawback to allograft transplantation is possible immunological rejection of the transplant. Body cells vary antigenically from individual to individual, and differences between tissues of the transplant donor and recipient lead to rejection of the graft. Transplantation rejection in this situation is predominantly a type IV cellular immunological reaction showing both specificity and memory. Killing of the graft cells occurs through a complex combination of mechanisms, including direct contact with sensitized T cytotoxic lymphocytes and natural killer (NK) cells. The ability to overcome immunologic rejection by treatment with immunosuppressive agents allows the grafts to survive. Perspective 18.1 discusses a very special transplantation situation, the fetus as an allograft. ■ **T cytotoxic cells, p. 404** ■ **NK cells, p. 373**

There are many different tissue antigens, but those of the major histocompatibility complex (MHC) system are the ones most commonly involved in transplantation graft rejections. Although these MHC antigens are found on many human cells, they are abundant on leukocytes and so have been called human leukocyte antigens or HLAs. MHC tissue typing is done in an effort to ensure that no major tissue incompatibility exists between a prospective tissue donor and the recipient patient. In addition to carefully matching donor and recipient tissue antigens, it is necessary to use immunosuppressant drugs indefinitely to prevent graft rejection. These drugs are needed because many minor antigens exist and it is impossible to find a donor compatible in all of these tissue antigens. Only genetically identical siblings, such as identical twins, have tissues that match in almost every way. ■ **MHC antigens, p. 398**

Radiation and various cytotoxic immunosuppressive drugs interfere with the rejection process, but at the same time they make the patient highly susceptible to opportunistic infections and also more likely to develop cancer. Cyclosporin A (produced by a fungus) and tacrolimus or FK506 (produced from a species of *Streptomyces*) are more effective immunosuppressants. They interfere with cellular signaling and thereby inhibit clonal expansion of activated T lymphocytes. These drugs specifically suppress T-cell proliferation, and thus they have fewer side effects than other immunosuppressants that affect many cell types.

Combinations of agents are commonly used to prevent allograft rejection. For example, cyclosporin A and steroids may be given, along with a monoclonal antibody preparation called basiliximab. This preparation binds to interleukin-2 receptors and blocks IL-2 binding.

MICROCHECK 18.5

Successful organ and tissue transplantation depend on matching major histocompatibility antigens and using immunosuppressive agents such as cyclosporin and tacrolimus to minimize the immune response to other antigens of the graft. Rejection of transplants is complex, but type IV cellular immune responses are the major mechanism of rejection for allografts.

- What are the antigens primarily responsible for allograft rejection?
- How are allografts rejected?
- Why is matching of transplant donors and recipients important?
- What would happen if administration of the antirejection drugs were discontinued?

Immunodeficiency Disorders

In immunodeficiencies, the body is incapable of making or sustaining an adequate immune response. There are two basic types of immunodeficiency diseases: primary, or congenital; and secondary, or acquired. Primary immunodeficiency can be inborn as the result of a genetic defect or can result from developmental abnormalities. Secondary immunodeficiency can be acquired as the result of infection or other stresses on the immune system such as malnutrition. People with either type of immunodeficiency are subject to repeated infections. The types of these infections will often depend on which part of the immune system is absent or malfunctioning. Some of the more important immunodeficiency diseases are listed in **table 18.4.**

Primary Immunodeficiencies

The genetic or developmental abnormalities that cause primary immunodeficiencies affect B cells, T cells, or both. Some primary immunodeficiencies affect natural killer (NK) cells, phagocytes, or complement components. Primary immunodeficiencies are generally rare. Agammaglobulinemia, a disease in which few or no antibodies are produced, occurs in one in about 50,000 people, and severe combined immunodeficiency (SCID), where neither T nor B lymphocytes are functional, occurs in only about one of 500,000 live births.

Selective IgA deficiency, in which very little or no IgA is produced, is an exception in that it is not rare. It is the most common primary immunodeficiency known. It has been reported in different studies to occur as often as one per 333 to 700 people. Although people with this disorder may appear healthy, many have repeated bacterial infections of the respiratory, gastrointestinal, and genitourinary tracts, where secretory IgA is normally protective.

Primary deficiencies may occur in various components of the complement system. For example, the few individuals who lack C3 are prone to develop severe, life-threatening infections with encapsulated and pyogenic bacteria. Patients with deficiencies in the early components of complement such as C1 and C2, may develop immune complex diseases, because these components normally help to clear immune complexes from the circulation. Patients who lack late components of the classical pathway of C activation (C5, C6, C7, C8) have recurrent *Neisseria* infections. Immunity to these bacteria is associated with destruction of the organisms by complement-dependent bactericidal antibodies. The effects of defects in other parts of the complement system are reflected by various symptoms, depending on the deficient component, but in general there is an increase in infections. People who lack one of the important control proteins of the sequence, C1-inhibitor, experience uncontrolled complement activation. This causes fluid accumulation and potentially fatal tissue swelling, a condition called **hereditary angioneurotic edema.** ■ complement cascade, p. 377

In children with DiGeorge syndrome, the thymus fails to develop in the embryo. As a result, T cells do not differentiate and are absent. Affected individuals have other developmental defects as well, such as heart and blood vessel abnormalities, and

TABLE 18.4 Immunodeficiency Diseases

	Disease	Part of the Immune System Involved
Primary Immunodeficiencies		
	DiGeorge syndrome	T cells (deficiency)
	Congenital agammaglobulinemia	B cells (deficiency)
	Infantile X-linked agammaglobulinemia	Early B cells (deficiency)
	Selective IgA deficiency	B cells making IgA (deficiency)
	Severe combined immunodeficiency (SCID)	Bone marrow stem cells (defect)
	Chediak-Higashi disease	Phagocytes (defect)
	Chronic granulomatous disease	Phagocytes (defect)
Secondary Immunodeficiencies		
	Acquired immunodeficiency syndrome (AIDS)	T cells (destroyed by virus)
	Monoclonal gammopathy	B cells (multiply out of control)

a characteristic appearance of the upper lip and ears. As expected from a lack of T cells, affected people are very susceptible to infections by eukaryotic pathogens, such as fungi, as well as viruses, obligate intracellular bacteria, and *Pneumocystis carinii*. ■ ***Pneumocystis carinii*, p. 756**

Severe combined immunodeficiency (SCID) results when neither T nor B lymphocytes are produced from bone marrow stem cells. Children with SCID die of infectious diseases at an early age unless they are successfully treated by receiving a bone marrow transplant to reconstitute the bone marrow with healthy cells. There are a variety of gene defects that can cause SCID. One defect is in an enzyme necessary for V, D, and J chain recombination to form B- and T-cell receptors for antigen. Without these receptors, there are no functioning B and T cells. SCID has also been shown to result from mutation in a gene for the interleukin-2 receptor on lymphocytes, such that the cells could not receive the signal to proliferate. Other individuals with SCID lack adenosine deaminase, an enzyme important in the proliferation of B and T cells. A number of these people have responded well to repeated replacement of the adenosine deaminase enzyme. It has been possible to correct this condition temporarily in a few children by collecting their own defective T cells, inserting the adenosine deaminase gene linked to a retrovirus vector, and returning the cells to them. Unfortunately, the genetically altered cells do not live long, and the treatment must be repeated. Still, these results are promising, and there is much excitement about the possibility of treating other severe disorders with gene therapy. Many of the gene defects that cause primary immunological disorders are known and work is under way to correct them. **Table 18.5** lists some of the primary immunodeficiencies for which gene defects have been identified. As the table indicates, deficiencies and defects can occur at any point in the complex steps that lead to an effective immune response. ■ **interleukin-2, p. 380** ■ **antigen receptors on B and T cells, p. 404** ■ **gene therapy, p. 365**

Chronic granulomatous disease involves the phagocytes, which fail to produce hydrogen peroxide and certain other active products of oxygen metabolism, due to a defect in an oxidase system normally activated by phagocytosis. Hence, the phagocytes are defective in their ability to kill some organisms, especially the catalase-positive *Staphylococcus aureus*. In Chediak-Higashi disease, the phagocyte lysosomes are deficient in certain enzymes and thus cannot destroy phagocytized bacteria. People with this condition suffer from recurring pyogenic bacterial infections. ■ **pyogenic bacterial infections, p. 693**

Secondary Immunodeficiencies

Secondary, or acquired, immunodeficiency diseases result from environmental rather than genetic factors. Malignancies, advanced age, certain infections (especially viral infections), immunosuppressive drugs, or malnutrition may all lead to secondary immunodeficiencies. Often, an infection will cause a depletion of certain cells of the immune system. The measles virus, for example, replicates in lymphoid cells, killing many of them and leaving the body temporarily open to other infections.

TABLE 18.5 Some Primary Immunodeficiency Diseases for Which Genetic Defects Are Known

Severe combined immunodeficiency (SCID)	X-linked hyper-IgM syndrome
X-linked SCID	Wiscott-Aldrich syndrome
MHC class II deficiency	Ataxia telangiectasia
CD3 deficiency	Chronic granulomatous disease
CD8 deficiency	Leukocyte adhesion deficiency
X-linked agammaglobulinemia	Many complement deficiencies

Syphilis, leprosy, and malaria affect the T-cell population and also macrophage function, causing defects in cell-mediated immunity. Malnutrition also causes decreased immune responses, especially the cell-mediated response.

Malignancies involving the lymphoid system often decrease effective antibody-mediated immunity. For example, multiple myeloma is a malignancy arising from a single plasma cell that proliferates out of control and produces large quantities of its specific immunoglobulin. This overproduction of a single kind of molecule results in the body using its resources to produce a single specificity of immunoglobulin at the expense of those needed to fight infection. The result is an overall immunodeficiency. Other lymphoid disorders include macroglobulinemia (overproduction of IgM) and some forms of leukemia.

One of the most serious and widespread secondary immunodeficiencies is AIDS (acquired immunodeficiency syndrome), caused by human immunodeficiency virus (HIV). This RNA virus of the retrovirus group infects and destroys helper T cells, leaving the affected person highly susceptible to infections, especially with opportunistic agents. AIDS and opportunistic infections are covered in chapter 29. ■ **retrovirus, p. 743**

MICROCHECK 18.6

Immunodeficiencies may be primary, either genetic or developmental defects in any components of the immune response, or they may be secondary, acquired as a result of infection or environmental influences.

- What is the defect in severe combined immunodeficiency? What could cause it?
- Multiple myeloma is a plasma cell tumor in which a clone of malignant plasma cells produces large amounts of immunoglobulin. With all this excess immunoglobulin, how can a person with multiple myeloma be immunodeficient?
- Why is infection with opportunistic agents a particular problem in individuals with AIDS?

Autoimmune Diseases

Usually, the body's immune system recognizes its self-antigens and deletes clones of cells that would respond and attack its own tissues. A growing number of diseases are suspected of being caused by an autoimmune process, however, meaning that the immune system of the body is responding to the tissues of the body as if they are foreign. Some of these diseases are listed in **table 18.6**. Susceptibility to many of them is influenced by the major histocompatibility makeup of the patient, and so, not surprisingly, they often occur in the same family. The table notes some of the diseases in which an association with MHC genes (such as DR genes) has been found. ■ clonal deletion, p. 407

Autoimmune diseases may result from reaction to antigens that are similar though not identical to antigens of self. Some bacterial and viral agents try to evade destruction by the immune system by developing amino acid sequences that are similar to self-antigens. As a result, the immune system is unable to discriminate between the agent and self. The immune system may then destroy the substances of self as well as those of the bacteria or virus. It has been found that the likeness of amino acid sequences need not be exact for this self-destruction; even a 50% likeness may lead to an autoimmune response. Autoimmune responses may also occur after tissue injury in which self-antigens are released from the injured organ, as in the case of a heart attack. The autoantibodies formed react with heart tissue and cause further damage.

The Spectrum of Autoimmune Reactions

Autoimmune reactions occur over a spectrum ranging from organ-specific to widespread responses not limited to any one tissue. Examples of organ-specific autoimmune reactions are several kinds of thyroid disease, in which only the thyroid is affected. Widespread responses include lupus erythematosus and rheumatoid arthritis. Lupus is a disease in which autoantibodies are made against nuclear constituents of all body cells. In rheumatoid arthritis, an immune response is made against the collagen protein of supporting connective tissues. In these widespread diseases, many organs are affected. In both organ-specific and widespread autoimmune diseases, the damage may be caused either by antibodies, immune cells, or both.

Myasthenia gravis, characterized by muscle weakness, is an example of an **autoantibody-mediated disease**. The disease is caused by the production of antibodies to the acetylcholine receptor proteins that are present on muscle membranes where the nerve contacts the muscle. Normally, transmission of the impulses from the nerve to the muscle takes place when acetylcholine is released from the end of the nerve and crosses the gap to the muscle fiber, causing muscle contraction. Immunofluorescent tests have shown autoantibodies that bind to the acetylcholine receptors, blocking access of acetylcholine to the receptors. These antibodies along with complement cause many of the receptors to be taken into the muscle cells and degraded, so that fewer receptors are present on the muscle membranes. Further evidence that IgG antibodies are involved comes from babies born to mothers with myasthenia gravis. The babies also experience muscle weakness, since IgG antibodies cross the placenta. Fortunately, the effect is not permanent, as these IgG antibodies decay within a few months and the babies are no longer affected. Treatment of myasthenia gravis includes the administration of drugs that inhibit the enzyme cholinesterase, allowing acetylcholine to accumulate so some contact with receptors can occur. Immunosuppressive medications and thymectomy are helpful in many cases. The role of the thymus in this disease is not understood.

Insulin-dependent diabetes mellitus is a common autoimmune disease caused by *cellular mechanisms.* In diabetes there is a lack of insulin, resulting in an increase of sugar in the blood. This leads to marked thirst and increase in urine production. The peak onset of this form of diabetes is about 12 years of age; thus, it is also called juvenile-onset diabetes. In the United States more than 100,000 children with this disease require daily insulin replacement. After many years of diabetes, blindness, kidney failure, or other severe or fatal complications often occur. In insulin-dependent diabetes mellitus, although autoantibodies are present, the major damage is destruction of the insulin-producing cells of the pancreas by infiltrating cytotoxic T cells.

TABLE 18.6 Characteristics of Some Autoimmune Diseases

Disease (Known MHC Relationship)	Organ Specificity	Major Mechanism of Tissue Damage
Graves' disease (DR3)	Thyroid	Autoantibodies bind thyroid-stimulating hormone receptor, causing overstimulation of thyroid
Myasthenia gravis (DR3)	Muscle	Autoantibodies bind to acetylcholine receptor on muscle, preventing muscle contraction
Insulin-dependent diabetes mellitus (DR3/DR4)	Pancreas	T-cell destruction of pancreatic cells
Autoimmune hemolytic anemia	Red blood cells	Antibody, complement, and phagocyte destruction of red cells
Rheumatoid arthritis (DR4)	Widespread, especially joints	Lymphocyte destruction of joint tissues; immune complexes of IgG and anti-IgG
Systemic lupus erythematosus (DR3)	Widespread (glomerulonephritis, vasculitis, arthritis)	Autoantibodies to DNA and other nuclear components form immune complexes in small blood vessels

A combination of antibody and cellular mechanisms is active in some autoimmune diseases, such as **rheumatoid arthritis (figure 18.9)**. This crippling inflammatory condition is one of the most common autoimmune diseases, occurring in both sexes and in adults and children all over the world. About 1% of males and 3% of females in the United States are affected. Rheumatoid arthritis is most common in women aged 30 to 50. There is an infiltration of Th1 cells, the inflammatory CD4 T cells, into the joints. When stimulated by specific antigens there, the T cells release cytokines that cause inflammation. Autoantibodies are also formed and contribute to widespread tissue damage by forming immune complexes.

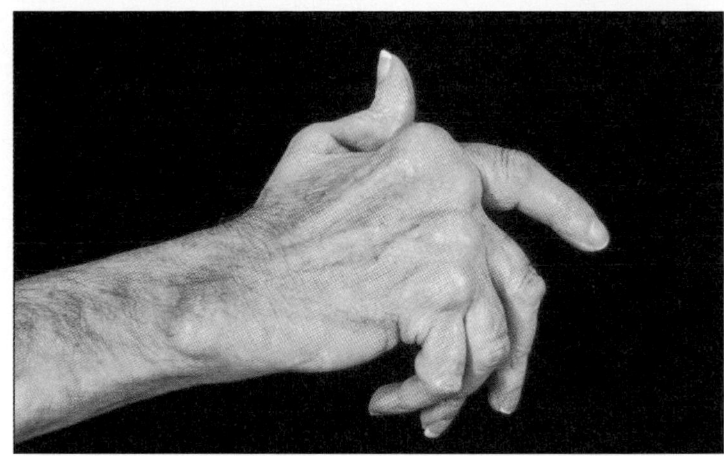

Figure 18.9 The Autoimmune Reactions of Rheumatoid Arthritis Cause Chronic Inflammation and Destruction of the Joints

Treatment of Autoimmune Diseases

Autoimmune diseases are usually treated with immunosuppressants that kill dividing cells and thus control the response, or drugs that interfere with T-cell signaling, such as cyclosporin. Also, steroids and other anti-inflammatory drugs are often used. In some autoimmune diseases, especially organ-specific ones, replacement therapy is necessary; for example, insulin in diabetes or thyroid hormone in some of the autoimmune thyroid diseases.

In insulin-dependent diabetes mellitus, attempts are being made to cure the disease by replacing the tissues destroyed by immune cells. Transplantation of the pancreas or insulin-producing cells of the pancreas offers hope for a cure. In fact, many people have had successful pancreas transplants, but the method is not widely used because dangerous immunosuppressive agents must be given to prevent rejection. Therefore at present, only people with advanced diabetes who require a kidney transplant, and must have the immunosuppressive drugs anyway, are given pancreas transplants. Research efforts are directed toward developing better methods of transplantation, such as enclosing the pancreatic cells within semipermeable membranes to protect them from attack by immune cells.

Ideally, it would be better to induce tolerance to the specific antigen causing the autoimmune response. Interesting experimental approaches are being tested in some autoimmune diseases. In a mouse model of diabetes, mice fed with insulin were protected from diabetes. Rheumatoid arthritis patients often have an active immune response to collagen, a protein prominent in the joints and surrounding tissues. An experimental treatment involves feeding solutions of animal collagen daily to the patients. This experimental approach should not be confused with the popular fad of taking pills containing collagen to treat the degenerative osteoarthritis of aging, which results from the wearing-down of cartilage and other collagen-containing compounds. The pills are taken with the hope of rebuilding cartilage and repairing the damage caused by aging. Osteoarthritis is not an immunologically mediated disease. The rationale of the experiments in the immunologically mediated rheumatoid arthritis depends on a well-known phenomenon called **feeding tolerance**. Antigen introduced by the oral route can cause a local intestinal immune response and suppress a response when the same antigen is later introduced by other routes. The mechanisms of feeding tolerance are not yet clear. It will be interesting to see if this approach will prove effective in the treatment of autoimmune diseases. ■ **immunological tolerance, p. 407**

MICROCHECK 18.7

Autoimmune disease can result when the immune system reacts to substances of self. Autoimmune diseases cover a spectrum from organ-specific to generalized. They are treated with drugs that suppress immune or inflammatory responses. Attempts are being made to induce specific tolerance to the causative substances.

- How could viruses or bacteria be implicated in causing autoimmune diseases?
- Explain what is meant by the spectrum of autoimmune diseases, from organ-specific to generalized.
- How would semipermeable membranes protect transplanted pancreas cells?

FUTURE CHALLENGES

New Approaches to Correcting Immunologic Disorders

In recent years many of the genes responsible for immunodeficiency diseases have been identified. It has been possible to correct some of these gene defects in cells in the laboratory and, rarely, in patients. In the near future, research will be directed toward developing the existing technology for gene transfer to make it more effective in correcting these gene defects in human patients. It is important also to continue the search for other defective genes; it is likely that with increasing knowledge of the human genome more will be found soon. A continuing challenge is finding ways to overcome graft rejection to make bone marrow and other transplants more acceptable. The challenge of treating cancer and of preventing rejection of essential transplants may be met, at least in part, by the development of gene transfer technology and by better understanding of the mechanisms of cellular immunological mechanisms.

The identification of immune system molecules, such as cytokines, and progress in understanding their functions will offer new means of treating autoimmune diseases. Research into understanding the regulation of immune responses will also help in controlling autoimmune and immunodeficiency disease, and it will permit development of improved vaccines.

A promising and challenging area is the development of human stem cell research. Stem cells have an almost unlimited capacity to divide, and some of them can differentiate into most of the tissues in the body. They could be used to generate cells for transplantation and to replace defective or injured tissues such as nerve tissue. They might also be used to test the effects of drugs on human cells, without the danger of testing on human beings. A major stumbling block is the fact that the stem cells come from fetal material, either from early stage embryos obtained from fertility treatments or from nonliving fetuses from terminated pregnancies. Currently, controversy rages about legal and ethical guidelines for the use of these cells.

SUMMARY

Immediate IgE-Mediated Hypersensitivities (Type I)

1. IgE attached to mast cells or basophils reacts with specific antigen, resulting in the release of powerful mediators of the allergic reaction. (Figure 18.1)

Localized Anaphylaxis

1. Localized anaphylactic (type I) reactions include **urticaria (hives), allergic rhinitis (hay fever)**, and **asthma**. (Figure 18.2)

Generalized Anaphylaxis

1. Generalized or systemic anaphylaxis is a rare but serious reaction that can lead to **shock** and death.

Immunotherapy

1. **Desensitization** or immunotherapy is often effective in decreasing the type I hypersensitivity state. (Figure 18.3)
2. A new treatment, using an engineered anti-IgE, is proving effective in treating asthma.

Cytotoxic Hypersensitivities (Type II)

1. Type II hypersensitivity reactions, or cytotoxic reactions, are caused by antibodies that can destroy normal cells by complement lysis or by **antibody-dependent cellular cytotoxicity (ADCC)**.

Transfusion Reactions (Table 18.2)

1. The ABO blood groups have been the major cause of **transfusion reactions**.

Hemolytic Disease of the Newborn (Figure 18.4)

2. The Rhesus blood groups are usually responsible for this disease.

Immune Complex–Mediated Hypersensitivities (Type III)

1. Type III hypersensitivity reactions are mediated by small antigen-antibody complexes that activate complement and other inflammatory systems, attract neutrophils, and contribute to inflammation.
2. The small **immune complexes** are often deposited in small blood vessels in organs, where they cause inflammatory disease—for example, glomerulonephritis in the kidney or arthritis in the joints. (Figure 18.5, Table 18.3)

Delayed Cell-Mediated Hypersensitivities (Type IV)

1. **Delayed hypersensitivity** reactions depend on the actions of sensitized T lymphocytes.

Tuberculin Skin Test (Figure 18.6)

1. A positive reaction to protein antigens of the tubercle bacillus introduced under the skin peaks 2 to 3 days after exposure to antigen.

Contact Hypersensitivities

1. Contact allergy, or **contact dermatitis**, occurs frequently in response to substances such as poison ivy, nickel in jewelry, and chromium salts in leather products. (Figures 18.7, 18.8)

Delayed Hypersensitivity in Infectious Diseases

1. Delayed hypersensitivity is important in responses to many chronic, long-lasting infectious diseases.

Transplantation Immunity

1. Transplantation rejection of **allografts** is caused largely by type IV cellular reactions.

Immunodeficiency Disorders

1. **Immunodeficiencies** may be primary genetic or developmental defects in any components of the immune response, or they may be secondary and acquired. (Table 18.4)

Primary Immunodeficiencies

1. B-cell immunodeficiencies result in diseases involving a lack of antibody production, such as agammaglobulinemias and selective IgA deficiency.
2. T-cell deficiencies result in diseases such as DiGeorge syndrome.
3. Lack of both T- and B-cell functions results in combined immunodeficiencies, which may be severe.
4. Defective phagocytes are found in chronic granulomatous disease and Chediak-Higashi disease.

Secondary Immunodeficiencies

1. Acquired immunodeficiencies can result from malnutrition, immunosuppressive agents, infections (such as AIDS), and malignancies.

Autoimmune Diseases

1. Responses against substances of self can lead to **autoimmune diseases.** (Table 18.6)

The Spectrum of Autoimmune Reactions

1. Autoimmune diseases may be organ-specific or widespread.
2. Some autoimmune diseases are caused by antibodies produced to body components, and others result from cell-mediated reactions. (Table 18.5)

Treatment of Autoimmune Diseases

1. Autoimmune diseases are usually treated with drugs that suppress the immune and/or inflammatory responses.

REVIEW QUESTIONS

Short Answer

1. Why are antihistamines useful for treating many IgE-mediated allergic reactions but not effective in treating asthma?
2. Penicillin is a very small molecule, yet it can cause any of the types of hypersensitivity reactions, especially type I. How can this occur?
3. What are some major differences between an IgE-mediated skin reaction, such as hives, and a delayed hypersensitivity reaction, such as a positive tuberculin skin test?
4. Why might malnutrition lead to immunodeficiencies?
5. What is the most common primary immunodeficiency disorder?
6. How can genetic abnormalities leading to immunodeficiency disorders be corrected? Give an example.
7. What causes insulin-dependent diabetes mellitus?
8. Give at least two lines of evidence showing that myasthenia gravis results from antibody activities.
9. Give an example of an organ-specific autoimmune disease and one that is widespread, involving a variety of tissues and organs.
10. Describe an Arthus reaction.

Multiple Choice

1. An IgE-mediated allergic reaction…
   A. reaches a peak within minutes after exposure to antigen.
   B. occurs only to polysaccharide antigens.
   C. requires complement activation.
   D. requires considerable macrophage participation.
   E. is characterized by induration.
2. Which of the following is true of the ABO blood group system in humans?
   A. A antigen is present on type O red cells.
   B. B antigen is the most common antigen in the population of the United States.
   C. Natural anti-A and anti-B antibodies are of the class IgG.
   D. People with blood group O do not have natural antibodies against A and B antigens.
   E. In blood transfusions, incompatibilities cause complement lysis of red blood cells.
3. All of the following are true of immune complexes, *except*…
   A. the most common complexes consist of antigen-IgE-complement.
   B. an immune complex consists of antigen attached to antibody.
   C. usually complement components are included in antigen-antibody complexes.
   D. immune complexes activate strong inflammatory reactions.
   E. immune complexes deposit in kidneys, joints, and skin.
4. Delayed hypersensitivity reactions in the skin…
   A. are characterized by a wheal and flare reaction.
   B. peak at 4 to 6 hours after exposure to antigen.
   C. require complement activation.
   D. show induration because of the influx of sensitized T cells and macrophages.
   E. depend on activities of the Fc portion of antibodies.
5. Organ transplants, such as of kidneys…
   A. are experimental at present.
   B. can be successful only if there are exact matches between donor and recipient.
   C. survive best if radiation is used for immunosuppression.
   D. survive best if B cells are suppressed.
   E. are rejected by a complex process in which cellular mechanisms predominate.
6. Patients with primary immunodeficiencies in the complement system…
   A. who lack late-acting components (C5, C6, C7, C8) show increased susceptibility to *Neisseria* infections.
   B. who lack C3 are prone to develop tuberculosis.
   C. generally have no symptoms.
   D. only show defects in the major components, C1 through C9.
   E. usually handle infections normally.
7. One of the most serious of the secondary immunodeficiencies is…
   A. acquired immunodeficiency syndrome, caused by the human immunodeficiency virus.
   B. severe combined immunodeficiency.

C. DiGeorge syndrome.
D. chronic granulomatous disease.
E. Chediak-Higashi disease.

8. All of the following are true of autoimmune diseases, *except...*
   A. many of them show association with particular major histocompatibility types.
   B. some of them often occur in members of the same family.
   C. during heart attacks heart antigens are released, but no response occurs to them.
   D. disease may result from reaction to viral antigens that are similar to antigens of self.
   E. some are organ-specific and some are widespread in the body.
9. Autoantibody-induced autoimmune diseases...
   A. can sometimes be passively transferred from mother to fetus.
   B. include diabetes mellitus.
   C. are always organ-specific.
   D. are never organ-specific.
   E. cannot be treated.
10. All of the following are approaches being used to treat autoimmune diseases, *except...*
   A. cytotoxic drugs to prevent lymphoid cell proliferation.
   B. cyclosporin.
   C. antibiotics.
   D. steroid with anti-inflammatory action.
   E. replacement therapy, as with insulin in diabetes.

Applications

1. Patients with advanced leprosy do not give positive type IV reactions to any of a variety of antigens that normal people are exposed to and usually respond to with a positive delayed hypersensitivity skin reaction. Why might this be? What is another disease where lack of the ability to give a positive delayed hypersensitivity reaction could be a useful diagnostic criterion?
2. The BCG vaccine against tuberculosis is not very effective, and great efforts are being made to produce a better vaccine. Nevertheless, BCG vaccine does give some protection and it is routinely used in many countries of the world, but not the United States. Since tuberculosis is far from eradicated in the United States, why would the BCG vaccine not be used?

Critical Thinking

1. Hypersensitivity reactions, by definition, lead to tissue damage. Can they also be beneficial? Explain.
2. What hypothesis could explain why primary immunodeficiencies are generally rare?
3. Why does blood need to be cross-matched before transfusion? Why not just type the blood and use compatible ABO types?

CHAPTER 19

Host-Microbe Interactions

The ancients thought that epidemics and diseases were divine punishment of the people for their sins. By the time of Moses, however, the Egyptians and Hebrews had come to believe that leprosy could be transmitted by contact with lepers. In Europe, around 430 B.C. Thucydides had concluded that some plagues were contagious. By the Middle Ages, many believed this, and they fled cities to escape the diseases. Fracastorius, in 1546, first proposed a theory that communicable diseases were caused by living agents, passed from one person or animal to another. He had no way to test this theory, however.

With the discovery of microorganisms in the late Seventeenth century by Leeuwenhoek, people began to suspect that the microorganisms might cause disease, but the techniques of the times could not prove this. It was not until 1876 that Robert Koch offered convincing proof of the "germ theory" of disease when he showed that Bacillus anthracis *is the cause of anthrax, an often fatal disease of humans, sheep, and other animals. With his microscope, he observed* B. anthracis *bacteria in the blood and spleen of dead sheep. He then inoculated mice with the infected sheep blood and was able to recover* B. anthracis *from the blood of the mice. In addition, he grew the bacteria in pure culture and showed that they caused anthrax when injected into healthy mice. From these experiments and later work with* Mycobacterium tuberculosis, *Koch formalized a group of criteria for establishing the cause of an infectious disease, known as* Koch's Postulates.

—A Glimpse of History

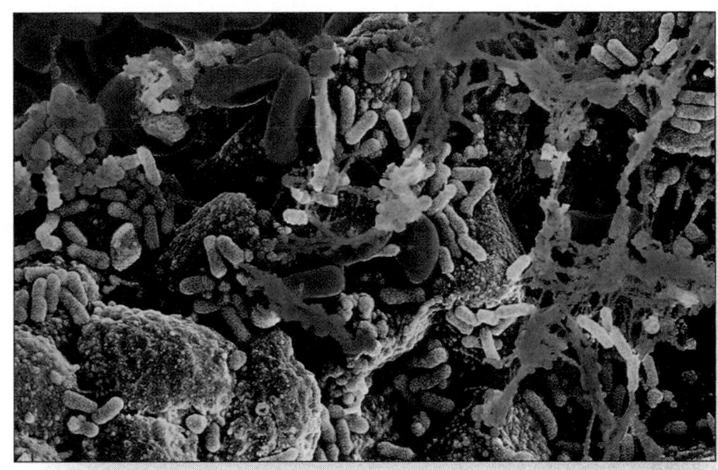

AFTER LEARNING SOME OF THE MANY WAYS IN which human beings can protect themselves against organisms and other foreign materials, it may seem surprising that microorganisms are actually able to infect people and cause disease. Microbes and viruses, however, have a vast multitude of ploys for gaining entry into people, living in or on the human body, and countering body defenses. In this chapter, we will explore some of the ways in which microorganisms and viruses colonize the human host and either live in harmony with the host in the normal flora or cause disease. We will also see how the host responds to some of the disease-producing strategies of microbes and how microbes sometimes evade host responses. The study of host-microbe interactions has become very exciting, as techniques developed for genetics, molecular biology, and immunology have been used to determine how human and animal hosts interact with microbes and how microorganisms cause disease.

Normal Flora of the Human Body

In nature, all organisms, including humans and microbes, interact with one another as part of a biological community called an **ecosystem**. Because microorganisms live in a wide range of natural habitats, it is not surprising that some of them grow abundantly on various surfaces of the human body, such as the skin and parts of the genitourinary, respiratory, and gastrointestinal tracts. The microorganisms that grow on the external and internal surfaces of the body without producing obvious harmful effects to these tissues are known as the **normal microbial flora**, or **normal flora**. Microorganisms that only occasionally inhabit the body are termed **transient microbial flora**.

Symbiotic Relationships Between Microorganisms and Hosts

Organisms that live within biological communities interact both positively and negatively, or in a neutral manner, with each other. The term **symbiosis** (for "living together") describes interactions that occur between different organisms that live close together on more or less a permanent basis, called **symbionts**. Microorganisms of the normal flora have a variety of symbiotic relationships with the human body. These relationships can take on different characteristics depending on the closeness of the association, the relative advantage to each partner in the association, and the dependence of partners on each other. Symbiotic associations are either commensal, mutualistic, or parasitic.

Commensalism

Commensalism refers to an association in which one partner benefits but the other remains unaffected. Most microorganisms on the surfaces of the human body are commensals. For example, the *Corynebacterium* sp. found living on the surface of skin are generally neither harmful nor beneficial to the human host, but the bacteria benefit by obtaining food and other necessities from the host.

Mutualism

Mutualism refers to an association in which both partners benefit. For example, in the large intestine of humans, some bacteria such as *Escherichia coli* synthesize vitamin K and certain B vitamins. These nutrients are absorbed by the large intestine, which benefits the human. In turn, the large intestine provides nutrients and shelter to the bacteria. Most normal flora organisms are commensals or mutualists. Either they are mutually beneficial or they do not live at the expense of the host.

Parasitism

Parasitism refers to an association in which one organism, the **parasite**, benefits at the expense of the other organism, the **host**. All living organisms can act as hosts for parasites. Most often, the parasites are members of the microbial world: the bacteria, fungi, protozoa, and viruses. Some invertebrates, such as roundworms and flatworms, also can be parasitic.

Some parasites do very little damage to the host, but others do great damage, often killing the host. Effective parasites will not kill their hosts until they have a reliable means to enter a new host. Indeed, the most successful parasites will live in harmony with their hosts. Most disease-causing organisms are parasites, but their disease-producing potential depends not only on their properties but also on the interactions that occur between the organisms and host defenses.

Establishment of Normal Flora

Normally, a human fetus has no resident microbial population. During passage through the vagina at birth, the baby is exposed to a variety of microorganisms and viruses that take up residence on its skin and in its gastrointestinal tract. Further, various microorganisms in food or on other humans and in the environment become established as permanent residents on the newborn. These microorganisms then become part of the newborn's normal flora and are said to have colonized the host. **Colonization** implies that microbial growth has become established on or within the host.

Importance of Normal Flora

The normal flora is important to the overall health of the human host. In humans lacking a well-balanced diet, the intestinal flora may be a significant source of essential nutrients. One of the most beneficial effects of normal flora is inhibiting other potentially more harmful organisms. Frequently, the growth of one organism is prevented or slowed down by the growth of another organism living close by. Such mutual inhibition or antagonism often occurs among members of the normal flora found on the human body as well as between organisms of the established normal flora and potentially harmful organisms that appear. Normal flora inhibit other organisms in several ways:

1. They may cover potential attachment sites for invading organisms and thus *prevent attachment* of the invader.
2. One microorganism may successfully *compete for essential nutrients*, such as vitamins, amino acids, and iron, thereby inhibiting the growth of other organisms. Microorganisms may consume large amounts of oxygen, reducing the available supply, inhibiting the growth of aerobic microorganisms.
3. Normal flora organisms may *produce antimicrobial substances*. For example, *Corynebacterium diphtheriae*, the bacterium causing diphtheria, is inhibited by hydrogen peroxide produced by some groups of streptococci, organisms that form part of the normal flora of the nasopharynx and oral cavities. Microorganisms may produce toxic metabolic by-products such as acids. Additionally, *Staphylococcus epidermidis* and *Propionibacterium* sp. that grow all over the skin break down lipids to fatty acids that inhibit most other bacteria. Some bacteria produce antibiotics or bacteriocins, antimicrobial substances that inhibit the growth of other bacteria. When organisms of the normal flora are removed or their growth suppressed, as can happen during treatment of the host with antibiotics, potentially harmful antibiotic-resistant organisms can colonize and possibly cause disease. For example, oral administration of antibiotics can suppress normal intestinal flora and allow disease-producing bacteria to become established and cause diarrhea.
4. Organisms of the normal flora can *stimulate the immune system*. This effect is shown dramatically in mice reared in a microbe-free environment; the animals lack a normal flora and they have greatly underdeveloped mucosal-associated lymphoid tissues.
5. Normal flora organisms frequently cause the *production of antibodies that cross-react* with potentially pathogenic organisms and prevent the harmful organisms from colonizing or invading. ■ **antibiotics, p. 496** ■ **bacteriocins, p. 496**

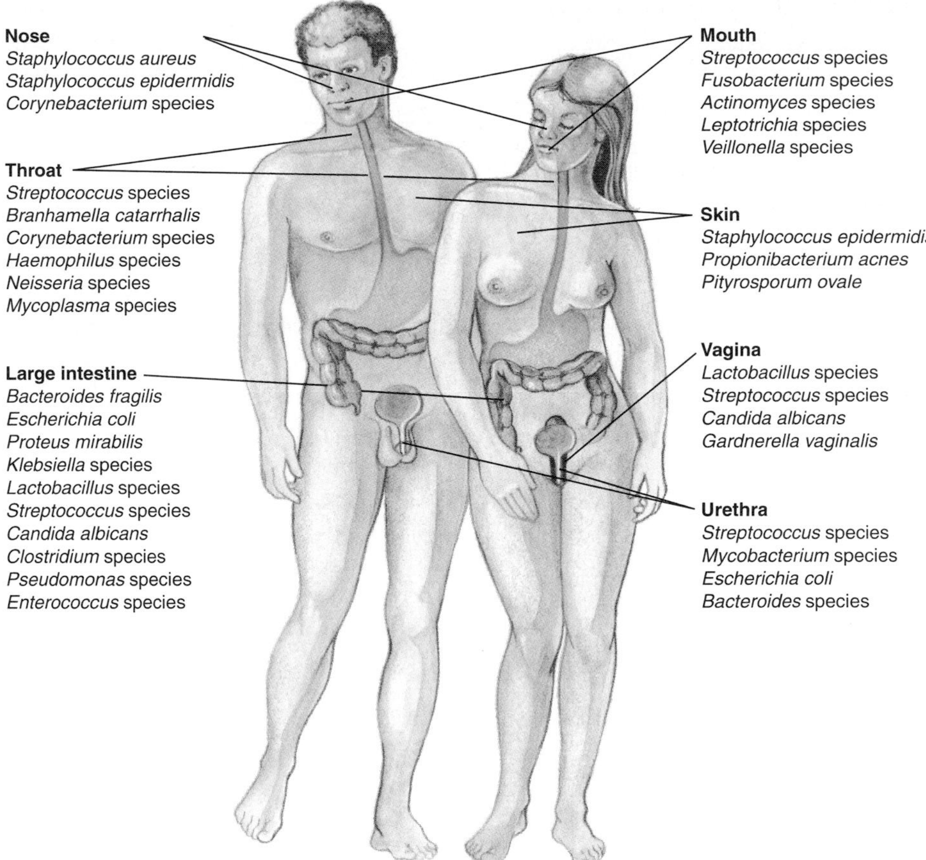

Figure 19.1 Normal Flora Location of many of the organisms that are part of the normal flora in and on the male and the female human body.

Changes in the Normal Flora

Many different species of microorganisms make up the normal microbial flora, and they occur in large numbers. In fact, there are more bacteria in just one person's mouth than there are people in the world. **Figure 19.1** shows some of the organisms that are part of the normal flora and their locations on the human body. The figure lists only a few representative species. At any one time, the composition of this complex ecosystem represents a dynamic balance of many forces that may alter the bacterial population quantitatively as well as qualitatively. The microbial composition is constantly changing, both in response to external influences and also as a direct result of the activities of the human host. Further, each member of this ecosystem is influenced by the other members and, in turn, exerts its influence on others in the system. For example, the lactobacilli that predominate in the female vagina normally suppress growth of the yeast *Candida albicans* found there in small numbers; during intensive treatment with antibiotics such as tetracycline, however, normal bacterial flora are inhibited, allowing the fungi to overgrow and cause disease.

Intestinal microbial populations change with variations in diet, acidity of the stomach, ingestion of antibiotics, and peristalsis, the wavelike contractions that move intestinal contents, including microorganisms, in the proper direction. For example, people who eat a substantial amount of meat will have more *Bacteroides* and other Gram-negative rods in their large intestine than will people on vegetarian diets. Also, many antibiotics will kill or inhibit not only pathogens, but also some of the normal flora, allowing unwanted organisms to proliferate. In hospitalized patients, the suppressed normal flora may be replaced by antibiotic-resistant organisms such as *Pseudomonas* or *Staphylococcus*, which can cause serious infections.

MICROCHECK 19.1

The microorganisms that grow in or on the body without producing obvious harmful effects make up the normal flora. They live in symbiotic relationships with the host. The normal flora is acquired at birth. It inhibits potentially harmful organisms by preventing attachment, competing for essential nutrients, producing antimicrobial substances, stimulating the immune system, and inducing the production of antibodies that cross-react with the potental pathogens. The normal flora changes in response to changing conditions.

- When and how is the normal flora of a human established?
- List three ways in which the normal flora helps to protect the body against invasion by microorganisms.
- What is the advantage to the normal flora in producing antimicrobial substances?

Microorganisms and the Human Host

The interaction of many host and microbial factors influences the effects that microorganisms have on the human body, regardless of whether these microorganisms are established members of the normal flora or new arrivals from the environment or from other hosts. The terms used to describe the most common interactions or results of these interactions are given in **table 19.1**.

Infection

Infection occurs when a parasitic organism grows and multiplies on or in the body of the host. The term **disease** describes a noticeable impairment of body function. A disease caused by a

TABLE 19.1 Terms Used in the Study of Infectious Diseases

Term	Definition
Parasite	An organism that benefits at the expense of another organism, the host
Colonization	The establishment of microbial growth on a body surface
Disease	Noticeable impairment of body function
Infectious disease	Disease caused by an infecting microorganism or virus
Primary infection	Infection in a previously healthy person
Secondary infection	Infection that occurs as a result of a primary infection and that occurs during or immediately following the primary infection
Pathogen	Any disease-causing microorganism or virus
Pathogenic	Disease-causing
Opportunistic pathogens	Organisms that cause disease only when introduced into an unusual location or into an immunologically compromised host
Immunologically compromised host	Host with lowered resistance to infection
Virulence	Attributes of a microorganism or virus that promote pathogenicity
Avirulent	Lacking the ability to cause disease
Systemic infection	Widespread infection through blood or lymph
Viremia	Viruses circulating in the bloodstream
Bacteremia	Bacteria circulating in the bloodstream
Septicemia	Acute illness caused by infectious agents or their products circulating in the bloodstream
Toxemia	Toxin circulating in the bloodstream
Inapparent infection	Infection with mild or no symptoms that goes unnoticed
Latent infection	Microorganisms or viruses living in host tissues for long terms without causing symptoms

microorganism or virus is an **infectious disease**. It is important to remember that infection does not always lead to disease; disease only occurs when body function is significantly impaired. Infection in a previously healthy person is a **primary infection**, as measles in a previously healthy child. The primary infection may cause a situation that allows another infection, a **secondary infection**, to occur along with or immediately following the primary infection. An example is streptococcal infections that follow measles because of damage done by the measles virus.

Pathogenicity and Pathogens

Pathogenicity refers to the ability to cause disease. A **pathogen** is any disease-causing microorganism or virus; **pathogenic** means disease-causing. This term is sometimes restricted to those microorganisms that most commonly cause disease in normal people. **Opportunistic pathogens**, or **opportunists**, are organisms that cause disease only when introduced into an unusual location or in **immunologically compromised** individuals. Immunologically compromised hosts are those with lowered resistance to infection, which may be due to many causes including malnutrition, cancer, AIDS or other diseases, surgery, wounds, genetic defects, alcohol or drug dependencies, or effects of immunosuppressive therapy. The opportunists are often members of the normal flora or occur in the environment. For example, *E. coli*, normally harmless in the large intestine, may cause disease if it gets into the urinary bladder. ■ **opportunistic infections, p. 522, 589**

Pathogens can be described as **extracellular**, **facultatively intracellular**, or **obligately intracellular**, depending on their interaction with host cells. An extracellular pathogen reproduces in the spaces and fluids surrounding the cells and tissues in their hosts. In cholera, the most extreme example, the toxin-producing organism never penetrates the outer layer of the intestinal epithelium and causes no visible damage to it. The symptoms are produced entirely by the toxin the bacterium produces. For the most part, extracellular pathogens invade and cause damage to host tissues ranging from the superficial to the profound but grow and reproduce outside of cells. For example, staphylococci enter tissues and cause formation of a boil, all the while multiplying extracellularly.

In sharp contrast, obligate intracellular pathogens, such as rickettsia, chlamydia, viruses, and certain protozoa, can only reproduce inside cells. Therefore, they cannot be grown in the laboratory except in living cells. To varying degrees, these infectious agents lack the ability to grow and replicate independently. They depend on host cells for the missing components or capabilities. Members of a third category, facultative intracellular pathogens, have the ability to multiply both intracellularly and extracellularly in the host. In the laboratory they can be grown on culture media in the absence of living cells. *Mycobacterium tuberculosis* and the other mycobacteria are examples of facultatively intracellular pathogens.

Virulence

Infections do not necessarily result in disease. Even when disease occurs, it may vary in severity. The term **virulence** describes attributes of a microorganism or virus that promote pathogenicity (**table 19.2**). Different strains within a species may vary in virulence. For example, pneumococci that have a capsule are virulent and often cause disease; those lacking a capsule are **avirulent** and do not cause disease. The word *virulent* is also used in a quantitative sense. A highly virulent organism has more disease-promoting attributes than do other less virulent strains of the same species. Such an organism is more likely to cause disease, particularly severe disease, than is a less virulent strain.

MICROCHECK 19.2

Infection occurs when parasitic organisms grow in or on the host's body. If the infection causes damage, infectious disease is present. Pathogens are disease-producing organisms. They may be extracellular, living outside of cells; obligately intracellular, living only inside of cells; or facultatively intracellular, able to live

TABLE 19.2 Some Important Virulence Factors of Bacteria

Adhesins of pili and other surface molecules that facilitate attachment of bacteria to host cells
Ability to trigger actin arrangement in nonphagocytic cells, leading to endocytosis of the bacteria
Capsules, protein A, and enzymes that interfere with phagocytosis by professional phagocytes
Endotoxins
Exotoxins
Cytolytic toxins and toxic enzymes
Superantigens
Proteases that break down secretory IgA
Ability to evade host responses in various ways

in either location. Virulence factors are responsible for the pathogenicity of microorganisms.

- What is an opportunistic pathogen?
- What is a virulence factor?
- Could a virus ever be an extracellular pathogen? Why or why not?

Infections and Infectious Diseases

Infectious diseases that are spread from an infected animal or person to another animal or person are called **communicable** or **contagious** diseases. Contagious diseases, such as colds and measles, are highly communicable and are spread from one host to another very readily. The interval between entrance of an organism into the susceptible host and the onset of illness is the **incubation period (figure 19.2)**. The incubation period may vary considerably, from only a few days for the common cold, to several weeks for hepatitis A, to many months for rabies, and even years for leprosy. The length of the incubation period depends on the infecting dose, the condition of the host, and many other factors.

A period of **illness** follows the incubation period. During this period symptoms and signs of disease occur. **Symptoms** are effects of the disease experienced by the patient, such as headache, pain, or nausea. **Signs** are effects of the disease that can be observed by examining the patient, such as rashes, pus formation, vomiting, or swelling.

After the illness subsides there is a period of **convalescence**, which is the stage of recuperation and recovery from the illness. Even though there is no indication of disease during the incubation and convalescence periods, many infectious agents may be spread during those times. Some individuals, called **carriers**, may harbor infectious agents for months or years and continue to spread disease-producing organisms, even though they themselves show no further signs or symptoms of the disease. Gastrointestinal diseases often spread by carriers include dysentery and typhoid fever. Other people may carry infectious agents in their blood and can transmit diseases via blood for long periods of time; examples of such infectious agents are the viruses that cause hepatitis B, hepatitis C, and AIDS. ■ **infectious disease carriers, p. 477**

Infections may be **acute**, in which the symptoms have a rapid onset and are usually severe, often with fever, but lasting only a short time. Other infections are **chronic**. They develop more slowly and last longer, for months or years. Strep throat is an acute disease; tuberculosis is a chronic disease. ■ **streptococcal infections, p. 528, 589** ■ **tuberculosis, p. 569**

Infections may be **localized**, as is the case with a boil caused by staphylococci, or they may be **systemic** with the infectious agent spreading throughout the body, as in measles. A **viremia** results when viruses circulate in the bloodstream. Similarly, a **bacteremia** results when bacteria enter the circulation. **Septicemia** is an acute illness caused by infectious agents or their products circulating in the bloodstream. Some organisms remain localized but produce toxins that spread through the circulation, a **toxemia**, as in tetanus. ■ **measles, p. 537** ■ **tetanus, p. 698**

Infections may be **inapparent** or **subclinical**, in which case symptoms do not occur or are mild enough to go unnoticed, as in many instances of infection with polioviruses and the

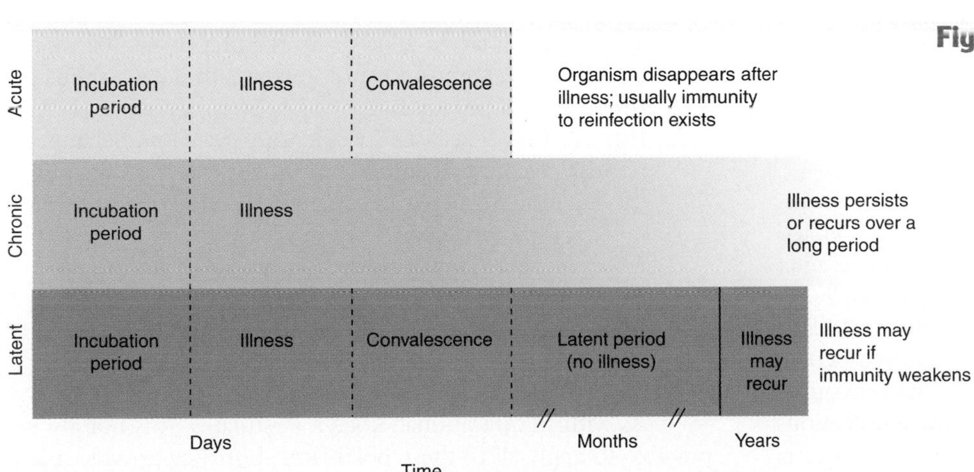

Figure 19.2 The Course of Infectious Diseases

hepatitis A virus. In these instances specific antibodies can be demonstrated and protective immunity often develops. ■ polioviruses, p. 679 ■ hepatitis A virus, p. 608

Long-term **latent infections** occur when viruses or microorganisms live in host tissues, often within host cells, for years without causing any symptoms. If there is a decrease in immune response, the latent infection may become reactivated and symptomatic. For example, the varicella-zoster virus infects and produces the characteristic rash and illness of chicken pox. The disease is halted by an effective immune response, leaving the patient immune against reinfection. The virus, however, is not eliminated, but takes refuge in sensory nerves, where its DNA remains, although virions are held in check by cellular immunity. Later in life, if there is a decrease in cellular immunity, the viral DNA can again be expressed as complete virions, causing the skin disease called shingles (herpes zoster). Cold sores, genital herpes, typhus, and tuberculosis are other diseases in which the causative agents often become latent. For example, in tuberculosis, the mycobacteria are often confined within a small area of tissue by host defense mechanisms, but they may begin growing and destroying tissue if the host becomes immunocompromised. ■ varicella virus, p. 536

Following **recovery from infection**, or after immunization, the host normally will have developed specific antibodies or immune cells that prevent reinfection with the same organism or virus. The host is no longer susceptible to infection with that agent.

MICROCHECK 19.3

Infectious diseases that are communicable or contagious are spread between people or animals. Stages of infectious diseases are the incubation period, the period of illness, and convalescence. Infections may be acute or chronic, localized or systemic, inapparent, or latent. Following recovery from an infectious disease, the host is usually immune to that disease.

- Give examples of a localized and a systemic infection.
- How does a latent infection differ from an inapparent infection?
- What factors might contribute to a long incubation period?

Establishing the Cause of an Infectious Disease

The fact that microorganisms (or "germs") cause infectious disease is now commonly accepted. The evidence that linked microorganisms to disease was discovered not much more than a century ago, however, when Robert Koch showed that *Bacillus anthracis* causes anthrax (see A Glimpse of History). This was an important step forward, because it often allowed people to avoid or control infection and it led to the development of antimicrobial drugs and other therapies to combat infectious diseases. From his experiments Koch formalized a set of criteria for determining the cause of infectious diseases, known as **Koch's Postulates**. These are now supplemented and often supplanted by the **Molecular Postulates**, which use precise genetic techniques to determine the pathogenicity, or disease-producing capabilities, of microorganisms. In this approach, genes become more important than the species in identifying potential pathogens.

Koch's Postulates

Koch proposed that in order to determine that a microorganism causes a particular disease, the following postulates must be fulfilled:

1. The microorganism must be present in every case of the disease.
2. The organism must be grown in pure culture from the diseased host.
3. The same disease must be produced when a pure culture of the organism is introduced into a susceptible host.
4. The organism must then be recovered from the experimentally infected host.

Koch used these postulates to prove that *B. anthracis* causes anthrax in sheep. He grew *B. anthracis* from all cases of anthrax tested; he introduced pure cultures of the organisms grown in the laboratory into healthy susceptible mice, again causing the disease anthrax; and finally, he recovered the causative bacilli from the experimentally infected mice. There are many situations in which these postulates cannot be carried out, however, either because the organism cannot be cultured in the laboratory, suitable animal hosts are not available to test the postulates, or for other reasons. In addition, some diseases such as pneumonia and urinary tract infection can be caused by a variety of microorganisms, and some microorganisms such as streptococci can cause a number of different diseases.

Molecular Postulates

A group of criteria for determining the cause of an infectious disease by using molecular techniques has been incorporated into a new set of postulates called the Molecular Postulates. These newer techniques deal directly with the virulence factors of microorganisms, their properties that enhance disease production. The Molecular Postulates state that:

1. The virulence factor gene or its product should be found in pathogenic strains but not in nonpathogenic strains of the suspected organism.
2. Introduction of a cloned virulence gene should change a nonpathogenic to a pathogenic strain, whereas disrupting the function of the virulence gene should reduce the virulence of the organism.
3. The genes for virulence must be expressed during the disease process.
4. Antibodies or immune cells specific for the virulence gene products should be protective.

As with the traditional Koch's Postulates, it is not always possible to apply all of these postulates, but they provide many new approaches to determining the causative agents of infectious diseases. The genes in different strains of organisms are more important than the species in determining the pathogenic potential of that organism.

MICROCHECK 19.4

Koch's Postulates are used with organisms that can be cultured in the laboratory to prove the cause of an infectious disease. The Molecular Postulates use genes to identify virulence factors in pathogenic strains of organisms.

- How were Koch's Postulates used to prove the cause of anthrax?
- List the Molecular Postulates.
- In two of Koch's Postulates (#2 and #3), it is indicated that a pure culture of the organism is required. Why is this a requirement? What would happen if the culture were not pure?

The Sources of Infectious Agents and Their Modes of Transmission

The sources of particular infectious agents and their transmission to humans will be discussed in more detail in following chapters. In brief, infectious agents can originate from soil, water, numerous animals including insects, and human beings, including a person's own normal flora **(table 19.3)**.

Infectious disease agents are transmitted or communicated from one individual to another by **direct contact**, such as touching, handshaking, or sexual intercourse, or **indirectly** through the air, by droplet transmission as a result of coughing, sneezing, and talking, through food and water, or by insect bites. Organisms are also transmitted indirectly by touching inanimate objects called **fomites**, such as contaminated clothing or dishes.

Transmission is successful if enough organisms are transferred to colonize and infect. The number of organisms necessary to ensure infection, termed the **infectious dose**, varies with the pathogen. For example, as many as 10^6 *Salmonella* sp. must be ingested to establish infection in humans, but only 10 to 100 *Shigella* sp., in part because shigellae are more resistant to stomach acids. The establishment of infection does not necessarily mean that disease develops, but it is an essential early step in the disease process.

The **portal of entry** or entry point is important in determining whether an organism will cause disease. Entry points may be the skin, respiratory tract, the eye, gastrointestinal tract, or genitourinary tract. For example, *Enterococcus* will cause disease if it enters the normally sterile urinary bladder, but it is a harmless member of the normal flora when it enters the gastrointestinal tract.

The **portal of exit** is critical for transmission of pathogens to new susceptible hosts. Organisms may exit from the respiratory tract, the alimentary tract, the genitourinary tract, the conjunctiva of the eye, the skin, and various discharges. ■ **portals of exit, p. 477**

TABLE 19.3 Sources and Transmission of Infectious Agents

Source	Examples and Mode of Transmission
Soil	*Clostridium* sp., enter dirty wounds; *Coccidioides immitis,* airborne fungal spores are inhaled
Water	*Legionella pneumophila,* inhaled in contaminated aerosols; *Vibrio cholerae*, ingested in contaminated drinking water
Animals	*Brucella abortus,* from cattle and milk; *Salmonella enteritidis,* ingested from poultry and other animals; *Plasmodium vivax,* injected by mosquito bites; Fungi that cause ringworm, by contact with infected dogs, cats, or other animals
Human beings	*Staphylococcus aureus,* from normal flora or from infected persons by contact; Common cold viruses, by inhalation of sneezed particles or by contact with infected people; HIV virus of AIDS, through direct sexual contact or injection of contaminated blood products

MICROCHECK 19.5

Infectious agents are transmitted from soil, water, animals, and normal flora, either indirectly or by direct contact. The portal of entry is important in determining whether an organism will cause disease, and the portal of exit is critical for transmission of organisms to a new host.

- What is the infectious dose?
- How can organisms be transmitted indirectly?
- Why is the portal of entry important in the disease process?

Mechanisms of Pathogenesis and Damage to the Host

Successful pathogens have the ability to carry out a number of steps essential for their virulence and their survival. These include

1. transmission of the causative agent to a susceptible host,
2. adherence of the agent to a target tissue,
3. colonization and usually invasion of tissues or cells,
4. damage to the host by the activities of toxins or other mechanisms,
5. exit from the host, and
6. survival outside the host long enough for step 1 to occur.

The transmission of microbes, their portals of entry and exit from the body, and their survival between hosts are topics discussed later. The abilities of microorganisms to become established and to cause disease in the host constitute their mechanisms of pathogenicity. The microbial properties responsible for pathogenicity are virulence factors.

Adherence in the Establishment of Infection

Humans as well as plants and animals have very effective mechanisms to keep most microorganisms and viruses from entering the interior organs of the body and causing damage. For example, skin cells are shed regularly, removing any organisms that have adhered to them. Ciliated cells of the respiratory tract transfer mucus-entrapped microorganisms out of the lungs and into the throat, where they are swallowed and then destroyed by stomach acid. The peristaltic movement in the intestine moves potential pathogens out of the gastrointestinal tract, and the flushing action of the urinary tract is its main defense against potential pathogens. ■ **barriers to infection, p. 375**

Because these mechanisms are very effective in removing organisms, a pathogen must adhere or attach to host cells as a necessary first step in the establishment of infection. As a general rule, both pathogens and host cells are negatively charged and therefore tend to repel each other. For attachment to occur, these repulsive forces must be circumvented. In many instances, the pili on the surface of the microorganism with their adhesins bind specifically to surface receptors on the host cells (**figure 19.3**). Viruses have spikes on their surfaces that act as adherence factors and carry out the same function. ■ **adhesins, p. 69** ■ **viral spikes, p. 324**

The adhesin and receptor molecules are held together by short-range, weak bonds that include hydrogen bonds. In most cases studied, the adhesins are proteins and the receptors are glycoproteins (glyco = sugar = carbohydrate complexed with protein). For example, the adhesins on the pili of *E. coli* and *Vibrio cholerae* are proteins, and their receptors are glycoproteins containing mannose and fucose, respectively. Similarly, an adherence factor on the surface of *Mycoplasma pneumoniae* (a microbe lacking a cell wall) interacts with receptors containing sialic acid. More than one kind of adherence factor may occur on the same pathogen, and the cells of different tissues of the same host may have different receptors. Some pathogens produce substances that allow the bacteria to attach to surfaces nonspecifically to form a biofilm. For example, the streptococci responsible for tooth decay produce polysaccharides that form a sticky network of glucan fibrils called the glycocalyx, which allows the bacteria to attach and form a biofilm on tooth surfaces. ■ **glycocalyx, p. 66, 694,** ■ **biofilm, p. 108** ■ **pili, p. 69**

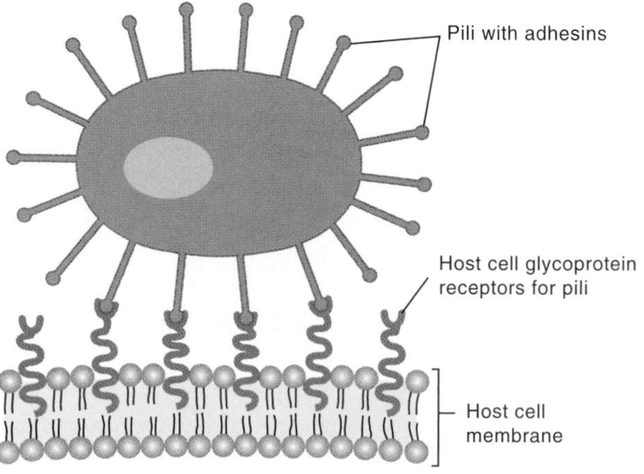

Figure 19.3 Pili Attachment to Host Cell Protein adhesins of microorganisms attach to glycoprotein receptors on host cells. (See figure 3.43.)

Different strains of the same species of pathogen may have different adherence factors. This helps account for the variety of diseases caused by different strains of the same organism. For example, certain strains of *Haemophilus influenzae* selectively infect the eye and cause conjunctivitis ("pink eye"), whereas most *H. influenzae* strains infect the upper respiratory tract. The adherence factors on the strains causing conjunctivitis bind specifically to receptors found on the eye. Similarly, certain *Streptococcus pyogenes* strains attach to throat epithelial cells causing "strep throat," whereas others attach to the skin and cause impetigo and other diseases.

Just because a microorganism attaches to cells does not mean it will cause disease. Many members of the normal flora of the mouth and intestine attach strongly to cells yet do not cause disease. Indeed, most bacteria live attached to surfaces, rather than as free-living independent cells. Other factors such as toxin production and invasive ability of the microorganisms must come into play before disease results.

Colonization and Invasion of Host Tissues and Cells

An organism must multiply in order to establish itself on or within the host; that is, it must colonize the host. Colonization is a prerequisite for the production of an infectious disease. To colonize, bacteria must not only overcome many obstacles but must also find conditions that allow them to multiply. They must compete successfully with the normal flora inhabiting the site for space and nutrients and overcome toxic products that the normal flora may produce such as fatty acids or bacteriocins. Nonspecific and specific components of the host's immune response have to be overcome. The organisms also require a supply of nutrients and a physical environment that includes the proper moisture, gaseous content, and acidity in order to multiply.

Most pathogens must invade tissues to cause disease, but entry into the intact host is exceedingly difficult. When a bite, abrasion, wound, or burn damages the epithelial layer, microorganisms are then able to gain direct entry into the tissues. In addition, some microorganisms produce enzymes or other substances that destroy the basement membrane that underlies epithelial cells. Still others disrupt the cell membranes or degrade the carbohydrate-protein complexes between cells, thereby allowing the epithelial cells to separate and create a passageway through the barrier. To penetrate mucous membranes, organisms must either go between or through the cells of the membranes. Some organisms enter cells from the outside and emerge on the other side of the membrane, extracellular but within the tissues; others enter cells and remain as intracellular parasites.

Other invaders cross the mucous membrane by way of M cells. Within the lymphoid tissue in the gastrointestinal tract (part of the mucosal-associated lymphoid tissue, or MALT) there are specialized cells called M cells, capable of phagocytosis, that are part of the mucous membrane (**figure 19.4**). Although they can phagocytize, M cells are not true phagocytes as are the macrophages and neutrophils. The normal function of M cells is to take up antigens in the intestines and convey them

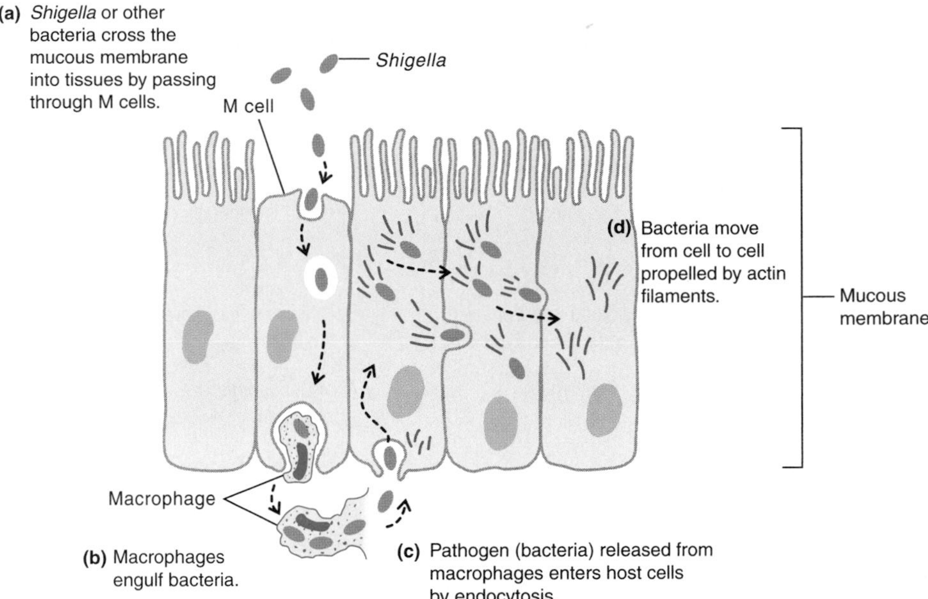
Intestinal Space
(a) Shigella or other bacteria cross the mucous membrane into tissues by passing through M cells.
M cell
Shigella
(d) Bacteria move from cell to cell propelled by actin filaments.
Mucous membrane
Macrophage
(b) Macrophages engulf bacteria.
(c) Pathogen (bacteria) released from macrophages enters host cells by endocytosis.
Tissue Inside Mucous Membrane

Figure 19.4 Mucosal-Associated Lymphoid Tissue (MALT) in the Gastrointestinal Tract Is the Route of Entry for Some Bacteria

across the mucosa to the antibody-producing lymphoid tissues of the Peyer's patches. Peyer's patches are collections of lymphoid follicles, each follicle containing M cells, macrophages, B cells, and T cells. Some bacteria, such as *Shigella* sp., can cross the mucosal barrier by being phagocytized by M cells and then released on the other side of the mucous membrane, where they are taken up by macrophages. The infected macrophages release IL-1, causing an intense inflammatory reaction. Shigella organisms released from the macrophages can attach to epithelial cells in that area and enter by endocytosis **(figure 19.5)**.

Following attachment, other microorganisms and viruses also enter host cells by endocytosis. **Table 19.4** gives some examples of pathogens that can enter cells by this process. Eukaryotic cells normally polymerize soluble actin molecules in the cytoplasm into insoluble filaments that give the cell its shape. When macrophages and neutrophils engulf materials, the actin filaments rearrange to make the pseudopods that surround materials and bring them into the cell. In endocytosis, the pathogen attaches to the normally nonphagocytic cell and is able to induce the cell to rearrange its actin and ingest the attached pathogen. If the pathogen cannot induce the cell to take it up, it cannot cause disease. This actin rearrangement is important also in helping organisms to evade host responses, as we shall see later. ■ endocytosis, p. 76, 350

Microorganisms in high concentration are more likely to penetrate and invade surrounding tissues than those in low concentration. In part this is because nonspecific host defenses can eliminate small numbers of organisms. Also important is the fact that some bacteria can sense their cell density by cell-to-cell communication. Only when the cell density is high are their virulence genes expressed. Human cells communicate by means of cytokine molecules; bacteria also communicate between themselves by means of particular small molecules. When these molecules reach a sufficient concentration, the bacteria sense that there is a large enough "quorum" present and transcription of the virulence genes is activated. This is called **quorum sensing**. *Pseudomonas aeruginosa* uses quorum sensing to regulate the expression of many virulence genes. Analogy has been made to invasion by an army. The pathogen masses the troops but does not reveal its weapons (toxins, other virulence factors) until there are enough to destroy the opposition. By not producing virulence factors at low cell densities but waiting until large amounts of the factors can be released, host immune responses can be more readily overcome. In the human host, small numbers of *Pseudomonas* organisms frequently come in contact with the eye

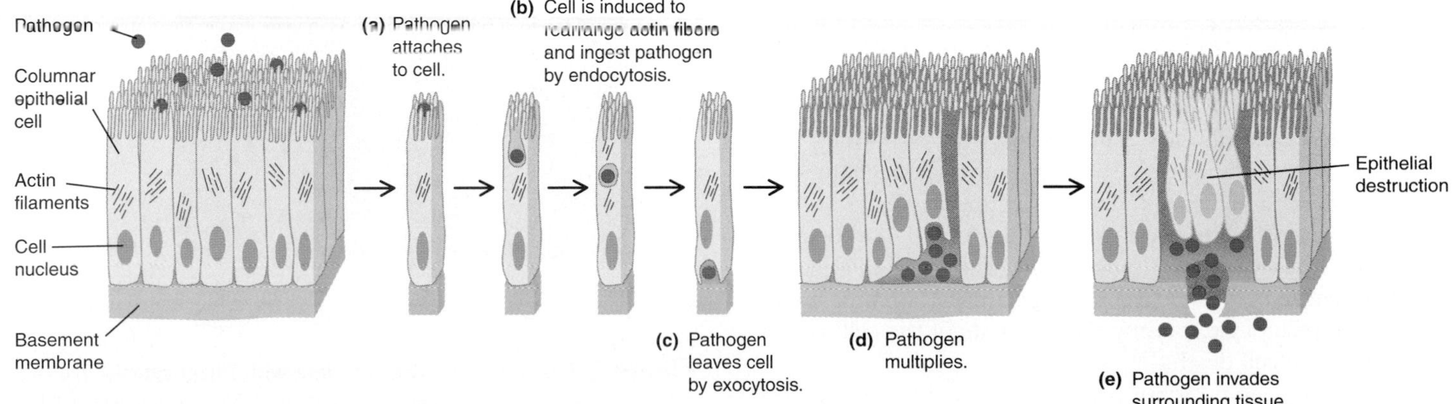
Pathogen
Columnar epithelial cell
Actin filaments
Cell nucleus
Basement membrane
(a) Pathogen attaches to cell.
(b) Cell is induced to rearrange actin fibers and ingest pathogen by endocytosis.
(c) Pathogen leaves cell by exocytosis.
(d) Pathogen multiplies.
Epithelial destruction
(e) Pathogen invades surrounding tissue.

Figure 19.5 Infection of Epithelium Involving Endocytosis and Exocytosis

TABLE 19.4 Some Pathogens that Enter Epithelial Cells by Endocytosis

Pathogen	Disease	Characteristics
Rhinovirus	Common cold	Infected cells of the respiratory tract lose contact with the basement membrane and are carried away.
Chlamydia sp.	Conjunctivitis, urethritis, pneumonia	Obligate intracellular bacteria infect the eye, urethra, or other surface, depending on the strain.
Escherichia coli (some strains), *Shigella* sp., *Campylobacter jejuni*	Dysentery	Infection is generally confined to the intestinal mucosa and supporting tissues.
Salmonella sp. (some strains)	Blood infections with or without diarrhea	Bacteria may pass through the intestinal epithelial cells but cause damage mostly to other parts of the body.
Neisseria gonorrhoeae	Gonorrhea	As with *Salmonella* sp., the bacteria are discharged from the host cell by exocytosis and may then spread to other tissues via the bloodstream.

with no evidence of any infection. If, however, contaminated eye medications in which these bacteria have grown to large numbers are introduced into the eyes, the organisms can invade and destroy the eyes with their destructive enzymes such as proteases.

Thus, organisms enter tissues by (1) taking advantage of trauma, such as a wound, that opens tissues to the outside, (2) by going between cells in tissues, using enzymes or other ways to separate the cells, or (3) by entering cells directly and either traversing through those cells and exiting into the tissues or remaining within cells. High concentrations of some bacteria are more likely to be successful in invading than small numbers, because only at high density are their virulence genes expressed. ■ quorum sensing, p. 184

Interference with Phagocytosis and Complement Activity

Once a pathogen has entered the body, to establish an infection it must cope with phagocytes. Pathogens interfere with the activities of phagocytes in many ways. For example, *Streptococcus pneumoniae* has a large capsule that inhibits the attachment of phagocytes and thus inhibits killing **(figure 19.6)**. These capsules, which are important virulence factors, are antigenic and induce an antibody response. When specific antibodies are present, they coat the capsule and permit the organism to be ingested.

Both *Streptococcus pyogenes* and *Staphylococcus aureus* produce surface proteins that are antiphagocytic. The strep make protein G and the staph protein A, molecules that combine with the Fc portion of antibodies. This prevents the antibodies that have complexed with antigen from binding to Fc receptors on host phagocytes and being ingested **(figure 19.7)**. *S. aureus* and other bacteria also produce proteins called **leukocidins** (*leuko*, meaning "white cell," plus *cidin*, "kill") that kill phagocytes. One particular leukocidin damages the cytoplasmic membrane of the phagocyte, causing the contents of the cell to leak out, resulting in cell death.

Recall that an important part of the inflammatory process is the movement of phagocytes toward the C5a component of complement. *S. pyogenes* produces an enzyme called C5a peptidase that breaks down C5a and interferes with its activity. As a result, phagocytes do not migrate effectively into the area and phagocytosis is impeded. Also, some capsules prevent complement activation, again interfering with phagocytosis. Further, some Gram-negative bacteria can modify their lipopolysaccharide surface molecules in a way that interferes with complement lysis of these cells. All of these ploys interfere with phagocytosis and complement action. ■ complement, p. 377

Activities of Toxins

A number of bacterial pathogens produce harmful substances called **toxins**. These toxins can be subdivided on the basis of their chemical nature into lipopolysaccharide **endotoxins** and protein **exotoxins**. Nearly all Gram-negative bacteria, whether pathogens or not, possess endotoxins. By contrast, only a relatively small number of species of Gram-positive and Gram-negative bacteria produce exotoxins. Except in certain cases of septic shock in which large numbers of bacteria enter the bloodstream,

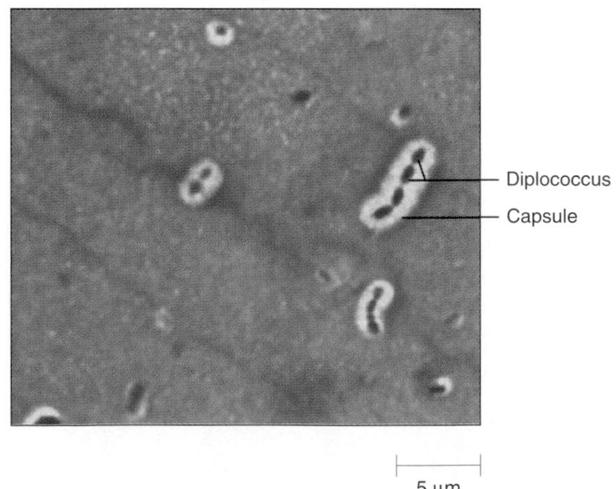

Figure 19.6 Capsules Can Interfere with Phagocytosis The large polysaccharide capsule of the diplococcus *Streptococcus pneumoniae* effectively prevents phagocytosis. The inhibition can be overcome by specific antibodies that combine with the capsule and allow phagocytosis to occur.

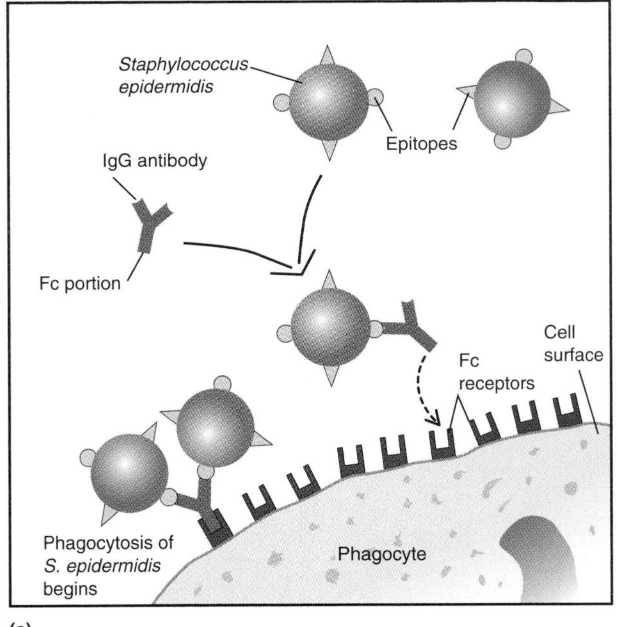

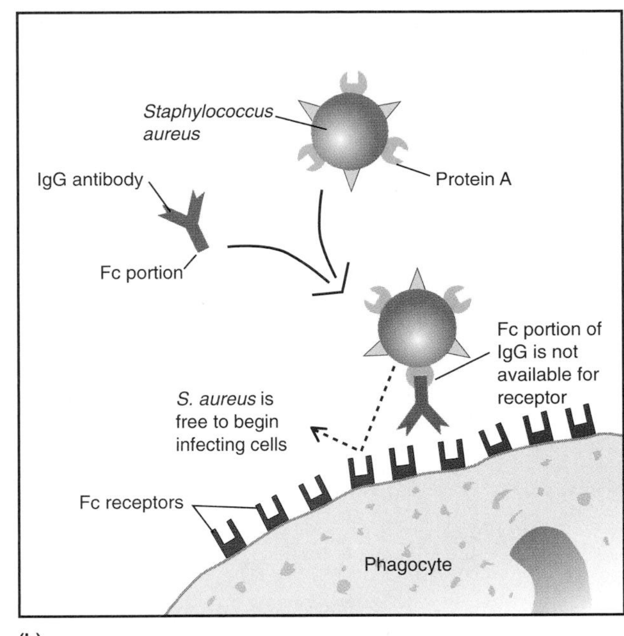

Figure 19.7 Protein A of *Staphylococcus aureus* Inhibits Opsonization **(a)** *S. epidermidis* that does not produce protein A reacts with specific antibodies. The Fc portion of the IgG antibody then reacts with Fc receptors on phagocytes to facilitate phagocytosis. **(b)** *S. aureus* produces protein A, which coats the staphylococci and binds to the Fc region of IgG, preventing interaction of Fc with Fc receptors.

endotoxins generally play a contributing rather than a primary role in pathogenesis. Exotoxins, on the other hand, are often the major cause of damage by organisms that produce them. Antibodies against exotoxins are called **antitoxins** and usually protect against the harmful effects of the toxin. Exotoxins treated to make them nontoxic but still immunogenic are **toxoids**, and they are used to immunize against the toxin. **Table 19.5** summarizes the properties of bacterial toxins.

Endotoxins of Gram-Negative Cell Walls

All endotoxins are lipopolysaccharides, part of the outer membrane of cell walls of Gram-negative bacteria (**figure 19.8**). Endotoxins consist of a two-part molecule, a polysaccharide and a lipid. The **lipid A** component is embedded in the outer portion of the bacterial cell wall, where it constitutes about half of the outer membrane. This lipid is responsible for the toxic properties of the endotoxin and has essentially the same toxic effects no matter which bacterium produced it. Like most lipids, lipid A alone is not immunogenic and does not induce a specific immune response. The polysaccharide portion of the molecule, or the **O antigen**, consists of chains that stick out into the environment like little hairs. The O antigens are different for each bacterial strain. The specific immune response is made to the O antigen. Some Gram-negative bacteria interfere with complement lysis by adding on to the O antigen chains, making them long enough that complement membrane attack complexes cannot penetrate the bacterial cell to cause lysis.

Endotoxins are released only when the bacterial cell is damaged or divides. After the endotoxin is freed from the bacterial cell, it is a powerful inducer of cytokine synthesis. It binds to macrophages and causes them to release proteins, including the cytokine interleukin-1. These proteins from macrophages are

TABLE 19.5 Important Properties of Bacterial Toxins

Property	Exotoxins	Endotoxins
Bacterial Source	Gram-positive and Gram-negative species	Gram-negative species only
Location in bacterium	Synthesized in cytoplasm and released from cell	Component of cell wall
Chemical nature	Protein	Lipopolysaccharide containing lipid A
Ability to form a toxoid	Present	Absent
Stability	Generally heat-labile, 60°C–100°C for 30 minutes	Heat-stable
Action	A distinctive effect from each	Same effect: fever, circulatory collapse, and characteristic damage to tissues

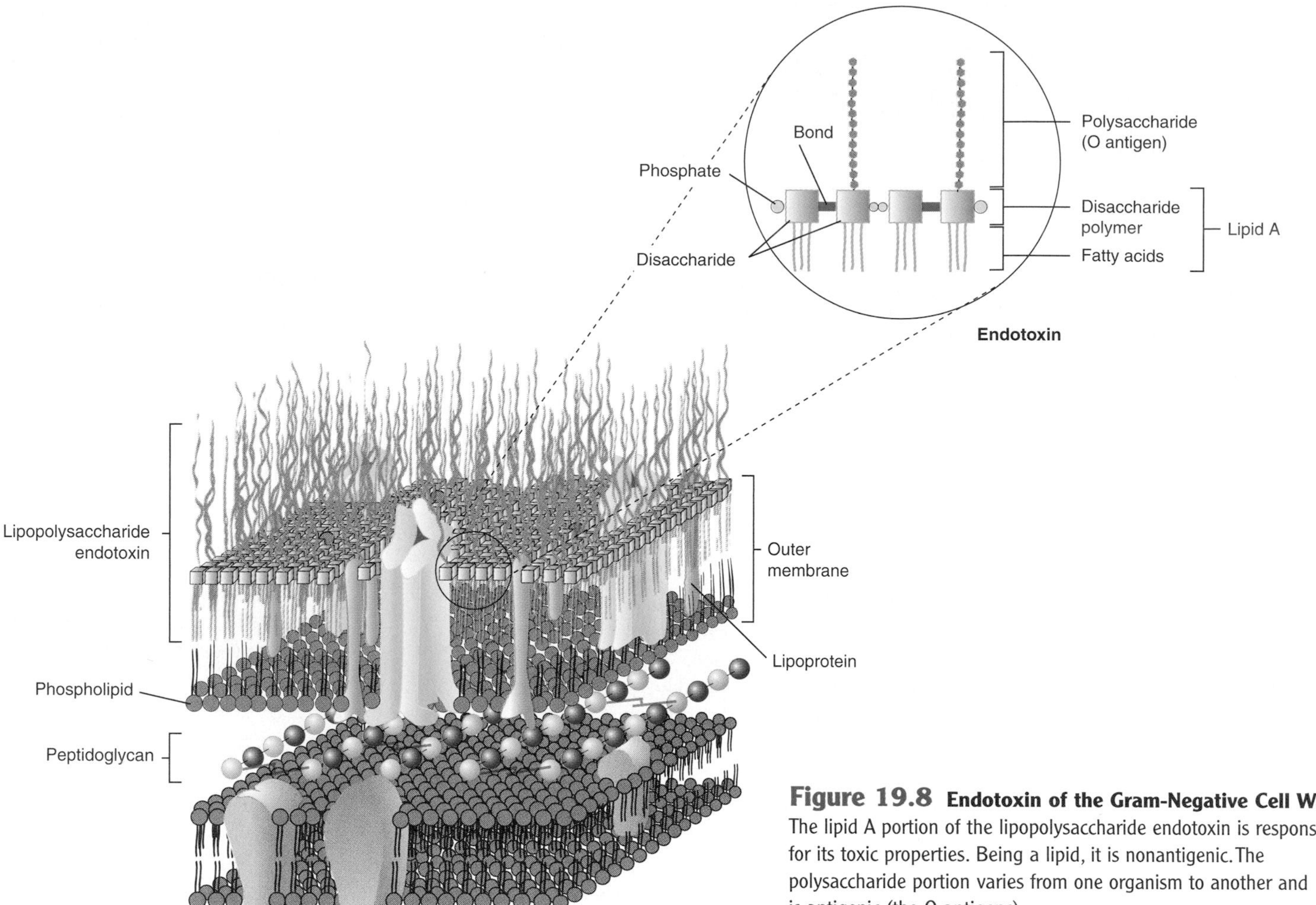

Figure 19.8 Endotoxin of the Gram-Negative Cell Wall
The lipid A portion of the lipopolysaccharide endotoxin is responsible for its toxic properties. Being a lipid, it is nonantigenic. The polysaccharide portion varies from one organism to another and is antigenic (the O antigens).

responsible for the symptoms of infections with Gram-negative bacteria; these include **fever**, weakness, generalized aching, and in the case of an overwhelming infection, **shock** due to the release of large amounts of endotoxin. ■ **fever and shock, p. 384, 720**

Endotoxins are resistant to heat, cannot be converted to effective toxoids for immunization, and have the same mode of action regardless of the species of Gram-negative bacterium from which they were derived. Since endotoxins are heat-stable, they are not affected by autoclaving. Consequently, solutions for intravenous administration must not only be sterile, but free of endotoxin as well. Disastrous results including death have resulted from giving intravenous fluids contaminated with minute amounts of endotoxin. A very sensitive test known as the *Limulus* assay is used to verify that fluids are not contaminated. This test uses material from the horseshoe crab *Limulus polyphemus* that precipitates in the presence of endotoxin. The precipitate is examined with a spectrophotometer, and as little as 10 to 20 picograms (1 picogram = 10^{-12} grams) of endotoxin per milliliter can be detected.

Exotoxins

Exotoxins have diverse effects, and some are among the most powerful poisons known. They are active in extremely small quantities. As a group, exotoxins can affect a wide variety of different cellular structures and functions, although generally each has a specific action. Some species of both Gram-positive and Gram-negative bacteria produce exotoxins. Exotoxins are either actively secreted by the organism or leak into the surrounding fluid following lysis of the bacterial cell wall. Most exotoxins are inactivated by heating at 60°C to 100°C for 30 minutes, a notable exception being the staphylococcal toxins that cause food poisoning. Generally, exotoxins are carried by the bloodstream from the point of infection to distant parts of the body, where they can cause dramatic effects. In the case of botulism poisoning, bacterial colonization and infection are not necessary prerequisites for the bacteria to cause disease. Ingestion of minute amounts of the toxin in food is sufficient to cause the paralysis that characterizes the disease botulism (see **Perspective 19.1**). Most exotoxins of Gram-positive bacteria can be converted to toxoids, which are effective immunizing agents. The modes of action of various exotoxins and the effects of these in diseases such as diphtheria, botulism, and tetanus are discussed in more detail in later chapters. ■ **toxoids in immunization, p. 427**

A-B Toxins

Some exotoxins are called **A-B toxins** because they consist of two parts, A and B. The A portion is an enzyme and constitutes the toxic part, and the B portion binds to host cell receptors,

Perspective 19.1 Botox for Beauty and Pain Relief

The exotoxin produced by *Clostridium botulinum* is one of the most powerful poisons known. These anaerobic soil bacteria are sporeformers found naturally on many foods. They survive usual cooking methods and inadequate canning procedures. Should conditions become anaerobic in the food, toxin can be produced, and if ingested it causes botulism. A few milligrams of this exotoxin is sufficient to kill the entire population of a large city. The botulinum toxin blocks transmission of acetylcholine nerve signals to the muscles, resulting in paralysis and often in death. This is a very dangerous toxin. Yet surprisingly, in recent years the toxin from type A organisms, known as botox, has become useful in treating various conditions.

The most unusual use of botox is cosmetic treatment to remove facial lines, such as frown lines. Extremely dilute botox is injected directly into the area, inactivating the muscles that are causing the frown or other lines. While the lines are erased, the effect is temporary and the treatment must be repeated after several months.

Easier to understand is the use of botox to relieve a number of very painful and disabling conditions involving muscle contractions, such as dystonia (severe muscle cramping). For example, cervical dystonia is a painful disease in which muscles in the neck and shoulders contract involuntarily, causing jerky movements, muscle pain, and tremors. Injections of botox directly into the affected areas give relief for 3 to 4 months, after which the treatment may be repeated. With time, some people become resistant to the effects of botox prepared from type A organisms, and now type B toxin is being used with good effect in those cases. Of course, very dilute toxin is injected, yet it is sufficient to paralyze the muscles for weeks or months. Another example is Parkinson's disease, a condition in which certain nerve cells are lost, resulting in tremor, impaired movement, and in some cases, dystonia. Botox injected into the affected muscles can give dramatic, although temporary, relief. So in spite of its powerful and dangerous properties, botulinum toxin when used with great care can be a useful therapeutic agent.

allowing the toxic enzyme to enter the cell (**figure 19.9**). A-B toxins cause diphtheria, cholera, tetanus, and *Shigella* dysentery. The A portion of many of these toxins are enzymes that remove the ADP-ribosyl group from the coenzyme NAD and attach it to some host cell protein, inhibiting the function of that protein. For example, the A portion of cholera toxin activates a G protein that in turn activates the enzyme adenylate cyclase, which converts ATP to cyclic AMP. This constant activation leads to increased cAMP levels in the cell, causing uncontrolled water and ion loss into the bowel and resulting in a severe watery diarrhea that can be fatal. The A portion of diphtheria toxin acts on a protein that is essential in host cell protein synthesis, thereby stopping protein synthesis and killing the host cell. Other A-B toxins function differently. The action of various exotoxins, including A-B toxins, is summarized in **table 19.6**.

■ NAD, p. 138 ■ cholera, p. 599

Cytolytic Toxins and Toxic Enzymes

Cytolytic toxins disrupt cell membranes and are produced by pathogens such as *S. aureus* and *Clostridium perfringens*. The cell-lysing alpha toxin of *S. aureus* is a protein that inserts into host cell membranes, forming pores that allow water to enter, thereby disrupting the cell. Leukocidins are proteins that damage the cell membranes of phagocytes, causing lysis of the cells. They are produced by various bacteria, including *S. aureus*. The *C. perfringens* toxin is phospholipase, an enzyme that disrupts

TABLE 19.6 Examples of Bacterial Exotoxins

Bacterium	Gram Stain	Disease	Exotoxin Effect
Corynebacterium diphtheriae	+	Diphtheria	Enzyme; blocks protein synthesis; cells most severely affected are those of nerves, heart, and kidney
Clostridium botulinum	+	Botulism	Inhibits acetylcholine release from motor nerve endings, causing paralysis; resists gastrointestinal proteases
Clostridium tetani	+	Tetanus	Blocks function of certain nerves that normally prevent muscle spasm; spastic paralysis results
Bacillus anthracis	+	Anthrax	Causes swelling and hemorrhage of tissue and circulatory failure
Staphylococcus aureus	+	Scalded skin syndrome; food poisoning; toxic shock syndrome	Causes skin layers to separate, causing blisters and skin sloughing; toxin produced by *S. aureus* growing in food; heat resistant; diarrhea results from small intestinal hypersecretion; stimulates vomiting center of brain by acting on visceral receptors
Streptococcus pyogenes	+	Scarlet fever	Toxin absorbed from strep throat acts on skin
Bordetella pertussis	–	Pertussis (whooping cough)	Damages cilia of respiratory mucosal cells
Escherichia coli	–	Diarrhea	Two kinds of toxin, one similar to cholera toxin, the other heat resistant; production controlled by plasmid genes
Pseudomonas aeruginosa	–	Septic shock	Enzyme; acts like diphtheria toxin but affects different tissues
Vibrio cholerae	–	Cholera	Binds to cell receptors of the small intestinal epithelium and causes abnormally high cAMP production with consequent copious diarrhea

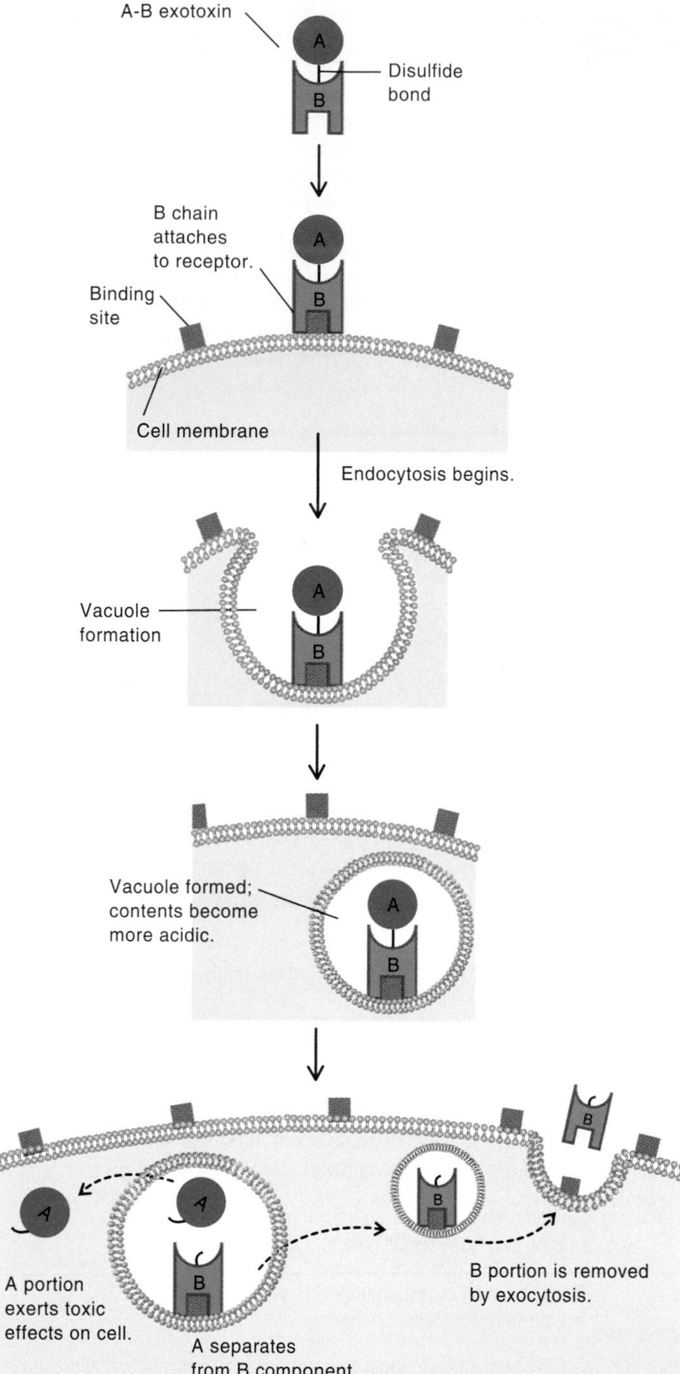

Figure 19.9 The Action of A-B Exotoxins The A portion is toxic and the B portion gets the toxin into host cells by binding to receptors on host cells and thus determining which cells will be affected by the toxin. After binding, the A-B molecule is taken up by endocytosis. After endocytosis, the contents of the vacuole become more acidic, causing the A and B portions to separate. The B portion remains in the vacuole and is removed by exocytosis, while the A portion enters the cytoplasm of the cell and exerts its toxic effect.

cell membranes. Other toxic enzymes break down tissue components such as collagen, hyaluronic acid, fat, DNA, and fibrin clots. Some of these membrane-disrupting toxins and toxic enzymes are described in **table 19.7**.

TABLE 19.7 Some Cytolytic Toxins and Toxic Enzymes that Contribute to Virulence

Product	Description	Function
Alpha toxin	Protein	Forms pores in membranes
Collagenase	Enzyme	Breaks down collagen, tissue fiber
Deoxyribonuclease	Enzyme	Breaks down DNA
Hemolysin	Oxygen-sensitive protein	Destroys red blood cells
Hyaluronidase	Enzyme	Breaks down hyaluronic acid, a tissue component
Leukocidin	Protein	Destroys cell membranes of phagocytes
Lipase	Enzyme	Breaks down fat
Phospholipase	Enzyme	Breaks down lecithin, a lipid component of mammalian cell membranes
Streptokinase	Enzyme	Breaks down fibrin and dissolves clots

Superantigens

Superantigens make up a third type of bacterial exotoxin. Bacterial superantigens include the staphylococcal enterotoxin that causes food poisoning and the staphylococcal toxin that causes toxic shock syndrome. Superantigens are not processed intracellularly as other antigens are and presented to specific responding cells; instead, they bind directly to the outer portion of major histocompatibility class II antigen on helper T cells and also to a region of the T-cell receptor distinct from the usual antigen-binding site **(figure 19.10)**. Rather than binding as antigens normally do to only the cells specific for that antigen, which represent about one in 10,000 T cells, the superantigen binds without antigen specificity to as many as 2% to 20 % of the T cells. The result is the release and activity of excessive amounts of interleukin-2 (IL-2) and other cytokines that enter the bloodstream instead of only acting locally as they normally do. This leads to the symptoms of fever, nausea, vomiting, diarrhea, and sometimes shock, with failure of many organ systems, circulatory failure, and often death. In addition, the immune response is suppressed because many T cells die following the excessive stimulation. Superantigens are also suspected of contributing to autoimmune diseases, by stimulating the few cells that might respond to self, and activating them to form large clones of cells reacting against self. Superantigens act as toxins for large populations of lymphocytes, and not as antigens that only stimulate the lymphocytes specifically recognizing them.
■ MHC, p. 398 ■ cytokines, p. 379

MICROCHECK 19.6

Important steps in the pathogenesis of infectious diseases are adherence to body cells or tissues,

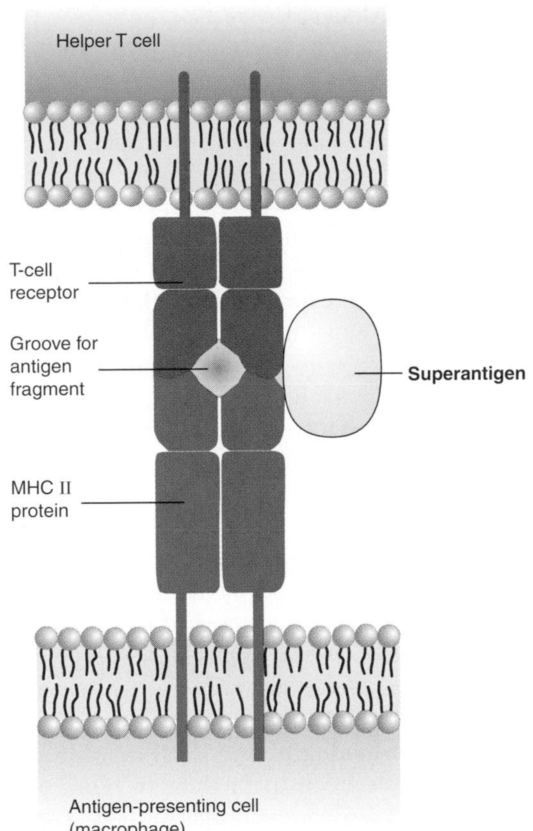

Figure 19.10 Superantigens These exotoxins are not processed and put into grooves in Class II MHC proteins to be presented to T cells. Instead, they attach to the outside of MHC II proteins and T-cell receptors. Rather than stimulating only one T cell in 10,000 or more, superantigens stimulate 2% to 20% of T cells, resulting in a huge overproduction of cytokines, with toxic effects.

Source: Adapted from *Arousing the Fury of the Immune System* ©1998 Howard Hughes Medical Institute.

colonization and invasion either between or through cells, and interference with phagocytosis and complement activities. Damage may be caused by toxins.

- How do microorganisms adhere to host cells and tissues?
- How do organisms enter hosts by endocytosis?
- How would different adherence factors lead to different diseases?
- Why would endotoxins be resistant to heat while exotoxins are not?

Genetic Transfer of Virulence Factors

The ability to produce toxins and other virulence factors is frequently transferred genetically from one strain of bacteria to another by transduction, transformation, or conjugation. In transduction, DNA is transferred from one bacterium to another by phages (bacterial viruses). The genes for the virulence factor, such as a toxin, are part of the phage. In lysogenic conversion, the phage becomes integrated into the DNA of the bacterial cell, the cell becomes lysogenic, and the bacteria gain the ability to produce the toxin. For example, the organisms that cause diphtheria are lysogenic for the beta phage that carries the gene for diphtheria exotoxin. Phages in lysogenized strains of other bacteria carry the genes for exotoxins that cause scarlet fever and botulism. Transformation is the transfer of genes by uptake of naked DNA from the donor cell into the recipient cell, bestowing new properties on the recipient. The prominent capsule of the pneumococcus *Streptococcus pneumoniae* protects it from phagocytosis and destruction. The ability to produce the capsule can be transferred from one strain of pneumococcus to another by transformation. Nonencapsulated avirulent strains of pneumococci become encapsulated pathogenic strains following uptake of DNA coding for capsules, the critical virulence factor.

Conjugation requires cell-to-cell contact, with DNA passed from a donor bacterium to a recipient bacterium. Both chromosomal and plasmid DNA may be transferred. Thus, genes for virulence factors carried on plasmids, as well as chromosomal genes, can be transferred by conjugation. For example, a gene of *E. coli* that codes for a particular adherence factor called colonization factor is carried on plasmids and can be transferred from one strain to another. While usually a benign member of the normal flora of the large intestine, *E. coli* can acquire a plasmid containing genes that code for the colonization factor and for the production of toxins that affect the gastrointestinal tract. The colonization factor is a pilus protein that allows the *E. coli* to attach to and colonize the human small intestine. Once the cells have attached they produce the toxins, resulting in severe diarrhea, sometimes called "traveler's diarrhea." Many other plasmid-transferable virulence factors account for the pathogenicity of various *E. coli* strains. ■ **plasmids, p. 70, 210** ■ **transduction, transformation, conjugation, p. 204, 205**

Genes for toxins and other virulence factors are found in many different locations on the bacterial chromosome. In Gram-negative bacteria, however, many of the genes required for pathogenicity are located close together on the chromosome in large regions called **pathogenicity islands**. For example, some strains of *E. coli* are known to cause urinary tract infections, whereas other strains cause gastrointestinal disease. The pathogenicity island in the strains causing urinary tract infections codes for a cytolysin that can destroy cells in the urinary tract; the pathogenicity island in the strains that cause gastrointestinal disease codes for secreted proteins that lead to intestinal cell damage. Usually, a pathogenicity island contains genes for a number of different virulence factors. A single bacterial strain may contain several different pathogenicity islands. The entire set of genes in a pathogenicity island is transferred as a unit between bacteria by conjugation, transformation, or transduction. It is clear that the virulence of microbial strains can be changed rapidly and dramatically by gene transfer.

MICROCHECK 19.7

Genes for virulence factors are often transferred between bacteria by the usual means of transduction, transformation, and conjugation. Plasmids may carry one or more virulence genes. In addition, groups of virulence genes clustered on the chromosome form pathogenicity islands that are transferred together.

- What role do bacteriophages play in the pathogenicity of some bacteria?
- How could dead bacteria of one strain confer virulence to another strain?
- Why should a pathogenicity island contain several virulence factors and not just one?

Pathogenic Effects of Viruses and Other Nonbacterial Agents

Viruses, algae, fungi, protozoa, and larger parasites such as the helminths all have developed many ways to cause disease in human hosts. The algae are not usually invaders, but they do produce some toxins that are ingested by people, with dire results. Algae are important inhabitants of the sea, where they sometimes produce toxins that are taken up by shellfish. Beaches are often closed to clammers, crabbers, and collectors of mussels because of algal neurotoxins from *Gonyaulax*, the red tide, or domoic acid made by other algae. ■ **algal toxins, p. 303**

Fungi, such as yeasts and molds, are widespread in the environment but are not pathogenic as frequently as bacteria are. One group of molds can cause superficial infections of human hair, skin, and nails but do not invade deeper tissues. These fungi have keratinase enzymes that break down the keratin in superficial tissues for use by the fungi, resulting in diseases such as ringworm and athlete's foot. Often, these are accidental infections in humans, who contract the fungi from their usual animal hosts. Fungi in the normal flora, especially the yeast *Candida albicans*, can cause disease in compromised hosts. This happens when the normal flora is out of balance because of antibiotic treatment or hormonal influences, as a result of diabetes, or from a wide range of other factors. Usually, the result is an infection of the mucous membrane, as vaginitis, but in severe situations the yeast can invade systemically. Other fungi can exist as either molds or in a tissue form. These infect but usually do not cause serious disease unless there is an overwhelming infection or an immunologic deficiency. Some of them are molds with airborne spores that are inhaled (*Histoplasma* or *Coccidioides*). Within the lung, the fungi assume a tissue form and take up residence in macrophages. Normal hosts overcome the infection with little or no disease. In immunologically compromised hosts or with an overwhelming dose, however, the disease progresses and causes granulomatous damage to the lung and sometimes to other organs.

Protozoa are responsible for millions of infections and deaths worldwide each year. They have many pathogenic tricks. Some protozoa actively penetrate cells using their own enzymes and motility; for example, the protozoan *Toxoplasma gondii* makes a small hole in the host cell and then enters. Malarial parasites use mosquitoes to transmit them from one human to another. They cause extensive destruction of red blood cells, with disastrous results. The trypanosomes that cause sleeping sickness depend on the biting tsetse fly for their transmission. Trypanosomes can produce new surface antigens repeatedly by activating different genes. As a result the host makes antibodies that soon become ineffective, but immune complexes form that contribute to the disease process.

In protozoan infection with the virulent ameba *Entamoeba histolytica*, a whole armory of offensive weapons combine to account for its pathogenicity. The amebas attach to special receptors on target cells of the mucous membrane. They then release proteins that form a pore in the target cell. Virulent strains of the amebas can express three different forms of pore-forming proteins that not only kill host cells but also disrupt bacterial cells. These proteins resemble some antibacterial peptides found in granules of human neutrophils, and some cell lysins found in NK and cytotoxic T cells. It appears that the primary function of these pore-forming proteins is to kill bacteria the amebas have phagocytosed for nutrients, but they incidentally damage human cells as well. Enzymes produced by the amebas are also virulence factors that destroy host cells. Clearly, if simple amebas have these and many other virulence factors, more complex pathogens are expected to have even more. Discussions of multicellular parasites in later chapters will show this to be true. ■ ***Toxoplasma gondii*, p. 758**

A virus, on the other hand, is not an organism, but just nucleic acid and surrounding capsid, with or without an envelope. Yet viruses are frequent pathogens of humans, and they have innumerable ways to enter and harm their hosts. Even so, many of them can live in perfect peace and harmony with their hosts in latent states. Viruses have several ways of entering host cells. Many enter by endocytosis, whereas the enveloped viruses usually enter cells by a process involving fusion of the viral envelope with the cell membrane. Many viruses cause tissue damage and disease by direct cytopathic effects on the cells they inhabit. The viruses take over cellular metabolism to produce more viruses, until eventually the cell dies, releasing viral progeny to infect more cells. Viruses also induce the production of immune cells that kill virus-infected cells, limiting the infection while at the same time causing disease. ■ **viral entry into cells, p. 348, 746**

Viruses have developed many ingenious ways to evade host immune responses. Normally when a virus replicates within a host cell, fragments of the virus are complexed with class I major histocompatibility complex (MHC) proteins and moved to the surface of the cell **(figure 19.11)**. Constantly patrolling killer T cells recognize the viral fragments as foreign and kill the infected cell. The cytomegalovirus, a herpesvirus that causes severe problems in immunocompromised people, has developed several ways to evade this protective mechanism. First, the virus suppresses production of class I protein, so that the killer T cells do not recognize the cell as being infected. The host counters with another defense, the natural killer (NK) cells. These respond to cells that are different, such as tumor cells and cells that lack class I proteins, killing such cells. Not to be outdone, the virus then induces the cell to make a fake class I protein that prevents killing by NK cells but does not display the foreign viral fragment that would lead to T-cell killing. Similarly, adenovirus has genes that suppress expression of class I proteins and thus interfere with killing by T cells. But adenovirus also has a way to get its host cell to divide at the direction of the virus, which

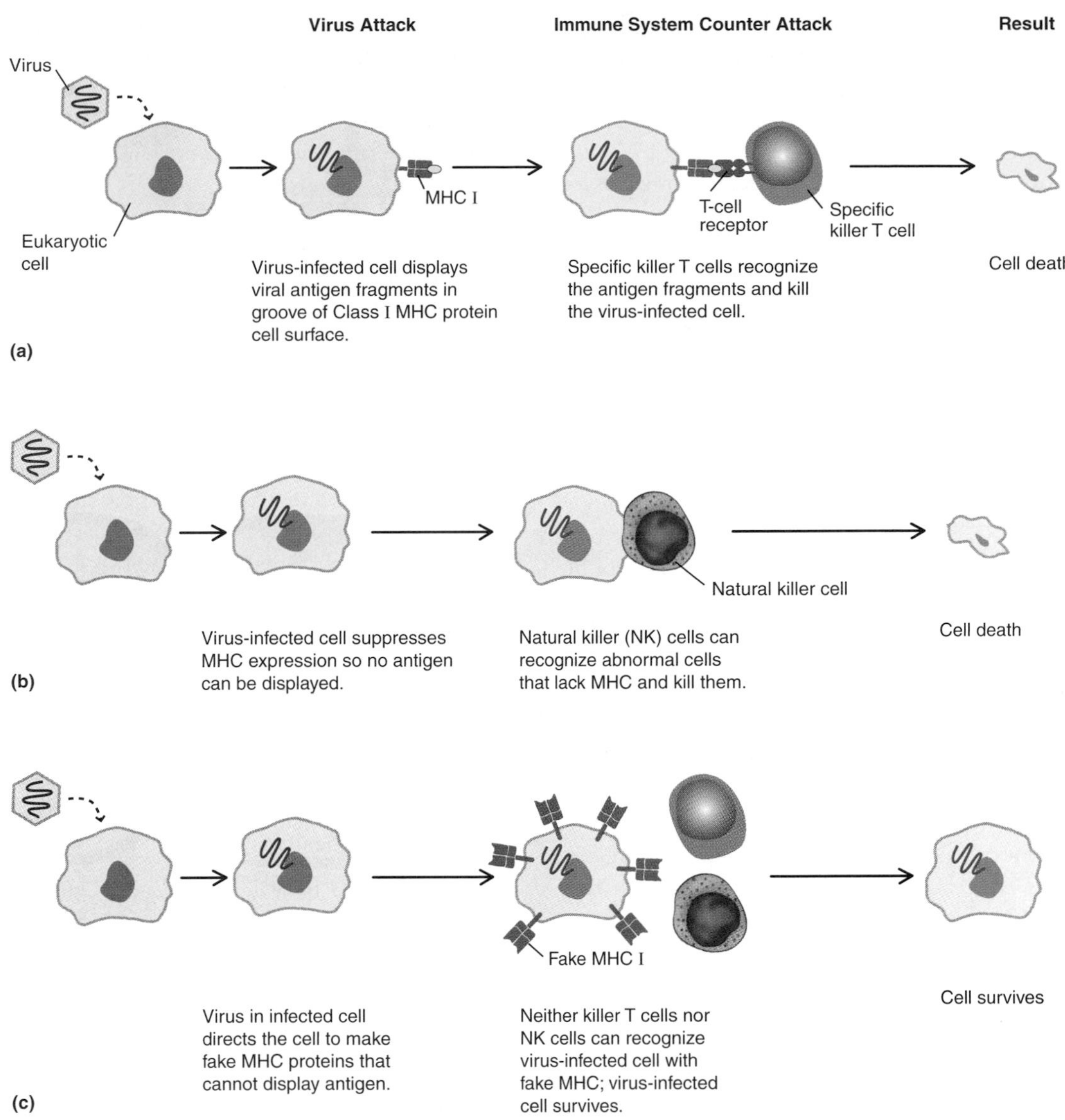
Virus Attack
Immune System Counter Attack
Result
Virus
Eukaryotic cell
MHC I
T-cell receptor
Specific killer T cell
Virus-infected cell displays viral antigen fragments in groove of Class I MHC protein cell surface.
Specific killer T cells recognize the antigen fragments and kill the virus-infected cell.
Cell death
(a)
Natural killer cell
Virus-infected cell suppresses MHC expression so no antigen can be displayed.
Natural killer (NK) cells can recognize abnormal cells that lack MHC and kill them.
Cell death
(b)
Fake MHC I
Virus in infected cell directs the cell to make fake MHC proteins that cannot display antigen.
Neither killer T cells nor NK cells can recognize virus-infected cell with fake MHC; virus-infected cell survives.
Cell survives
(c)

Figure 19.11 **Cytomegalovirus–Immune Cell Interactions** **(a)** Normal killing of virus-infected cell by killer T cell. **(b)** Some viruses can evade killer cell destruction by suppressing MHC expression, but NK cells of host can kill cells lacking MHC protein. **(c)** Virus causes cell to produce fake MHC that neither killer T cells nor NK cells can recognize, and so the virus-infected cell survives.

could be deadly for the host. The human cells counter with the desperate measure of cell suicide, a ploy that destroys virus, but also the host cell. Again, the adenovirus fights back; it has a gene that interferes with the cell suicide. Other viruses have equally ingenious ways to counter host immune responses.

MICROCHECK 19.8

Algae, fungi, protozoa, multicellular parasites, and viruses each have multiple ways of causing disease.

- How do some molds cause disease of skin, hair, and nails?
- What are some virulence factors of pathogenic amebas?
- Describe several ways that viruses can cause disease and also evade host responses.

Host Responses that Contribute to Pathogenesis

The outcome of any infection depends on the pathogenic capabilities of the infecting organism, and also on the defensive capabilities of the host. Protective immune responses were discussed previously. In addition, host responses can contribute to the pathogenesis of infectious diseases.

Hypersensitivity and Autoimmune Responses

Tissue damage and disease can result from immune responses and can contribute substantially to the progression of infectious diseases. For example, during the host response immune complexes may form, causing kidney and joint disease, as in infections caused by *Streptococcus pyogenes*, hepatitis B, and malaria, among others. Some infections lead to the development of autoimmune antibodies. *Mycoplasma pneumoniae* infections can induce antibodies against red blood cells of the host, causing hemolytic anemia. *S. pyogenes* infections can cause rheumatic fever, probably by inducing antibodies that cross-react with heart tissue. ■ *Streptococcus pyogenes*, p. 528 ■ *Mycoplasma* infections, p. 568

In many infections the disease results from hypersensitivity reactions to the infecting organisms. In tuberculosis, for example, the tissue response is a delayed (type IV) hypersensitivity reaction, leading to granuloma formation. The cellular immune response attempts to limit the infection, but at the same time it causes tissue damage and disease. ■ tuberculosis, p. 569

Effects of Impaired Body Defenses

Both nonspecific and specific immune responses are vitally important in maintaining the health of the host. Therefore, impairment in any one of these mechanisms makes the host more susceptible to invasion by microbes or viruses. For example, some people are born unable to produce functional phagocytes or antibodies. Others lose these capabilities later in life. Still others, such as those receiving heart or kidney transplants from other people, are given drugs that impair their inflammatory response, immune response, or both, in order to prevent rejection of the transplant. Certain conditions, including cancer, diabetes, AIDS, and malnutrition, may cause defects in protective mechanisms of the host. In all of these instances, microorganisms of the normal flora can invade, and latent infections are reactivated more frequently than they would be in hosts having normal defenses.

Host Genetic Factors

Different strains of laboratory animals have marked differences in susceptibility to certain infections. For example, some strains of inbred mice are much more susceptible to *Salmonella* infections than are other strains. Humans also show genetic differences in susceptibility. For example, people with genes for the sickle cell trait are more resistant to malaria but more susceptible to *Salmonella* infections than are people in the general population.

Host cell receptors for adherence factors often vary in number and kind among individuals and among species. This variety accounts in part for differences in susceptibility to infection by various microorganisms and viruses. For example, most humans are not susceptible to diseases of cats and dogs, such as the viral disease distemper. Human cell receptors are not compatible with the adherence factors on the surface of the virus and, therefore, the virus cannot attach.

Effects of Age and Stress

Infants and the aged have less effective immune systems than the population as a whole; thus, both the very young and the very old are more susceptible to infections. Receptors for adherence factors vary with the host's age. Immune capabilities vary with age, as do many physiological conditions affecting host-parasite interactions. It has long been recognized that stress can adversely affect the outcome of infection. In response to stress, high levels of corticosteroid hormones are produced, and these are well known to be immunosuppressive. Also, lymphoid tissue has been shown to have receptors for catecholamines, compounds that act on nerve cells, and catecholamines can suppress T-cell activity in lab animals. These and other findings suggest that messages go, not only between immune cells, but also between the immune system and the nervous system, affecting the overall immune response.

MICROCHECK 19.9

The outcome of infection depends on both the virulence of the pathogen and the response of the host. The genetic makeup and the immune status of the host are important.

- How can host genetic factors influence infectious diseases?
- Describe two ways that host hypersensitivity reactions may contribute to pathogenesis of infectious diseases.

Evasion of Host Responses

Organisms must evade protective host responses if they are to persist in the body and cause serious disease. Microorganisms evade host responses in many ways (**table 19.8**). One way is by inducing genetic variations from one substance to another; for example, they may vary from one type of pilus to another. Thus,

TABLE 19.8 Some Mechanisms Microorganisms Use to Evade Host Responses

Evasion of Complement Activity	Evasion of Phagocytic Activity	Evasion of Antibody Activity
Some capsules prevent complement activation	C5a peptidase production destroys C5a and inhibits the attraction of phagocytes to the area of infection.	Production of proteases destroys secretory IgA.
Some Gram-negative rods can lengthen O chains in their lipopolysaccharide to prevent complement membrane attack.	Leukocidins produced are toxic for phagocytes.	Staphylococcal protein A and similar proteins interfere with antibody-mediated opsonization.
	Some microorganisms can escape from phagosomes before they fuse with lysosomes.	Variation in surface antigens of bacteria, including pili, makes antibodies useless.
	Some microorganisms resist antimicrobial lysosomal contents.	Some microorganisms produce capsules that are not antigenic.

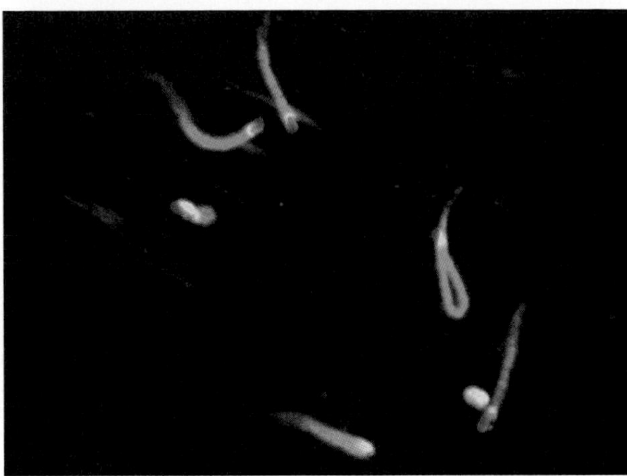

Figure 19.12 ***Listeria monocytogenes*** **(red) in Infected Cell, Showing Polymerized Actin "Comet Tails" (green) that Propel the Bacteria**

TABLE 19.9 Some Ways Organisms Persist Intracellularly

Mechanisms	Examples
Dissolve the phagosome membrane and live in the cytoplasm of host cells	*Shigella, Listeria,* some rickettsia
Survive bactericidal agents in lysosomes and live in phagolysosomes	*Coxiella*
Prevent acidification of the vacuole containing the bacteria	*Mycobacterium, Legionella*
Formation of large "spacious" phagosomes that appear to dilute host antimicrobial factors	*Salmonella*
Prevent fusion of the phagosome with lysosomes	*Salmonella*
Produce enzymes that neutralize host antimicrobial substances	Many different bacteria

as antibodies are produced to the old pili, the new pili still can cause adherence and colonization.

Other microbes make substances that mimic host materials and are not recognized as foreign. *Streptococcus pyogenes* produces capsules of hyaluronic acid, and some strains of meningococci have sialic acid capsules. Both of these substances are part of normal human bodies, are not recognized as foreign, and do not induce immune responses. Schistosoma organisms carry this defense one step beyond when they coat themselves with actual host proteins to avoid having their antigens detected by immune cells. Disguised in this manner, they can live for many years in the host's circulation. ■ *Schistosoma* sp., p. 734

Several pathogens that successfully colonize mucous membranes make a protease that specifically breaks down secretory IgA, permitting the bacteria to attach to cells. Microorganisms have evolved many such ways to evade the immune response. A number of these are summarized in table 19.8.

Some bacteria, such as the nonmotile *Shigella* sp., have a gene that allows them to take over the cell's actin and be endocytosed. Once within the host cell, the bacteria can direct the actin to form what has been described as a "comet's tail" that propels the shigella organisms from one cell to the next. The bacteria can move about without venturing outside the cells into an environment where antibodies or immune cells can attack (see figure 19.5). When this gene was introduced into a nonpathogenic strain of *E. coli*, these bacteria also grew cometlike actin tails and moved from one cell to another in a manner described as "like earthworms through dirt." Another organism with this capability to induce actin comet tails is *Listeria monocytogenes* (**figure 19.12**). Without inducing actin tails, the shigella and listeria infections cannot progress. With the actin tails, they are able to remain within cells, and to move between cells, protected from immune forces.

Bacteria that are phagocytized or enter other cells may become intracellular parasites if they can escape the normal intracellular destruction that occurs, especially in phagocytes. Some organisms are able to prevent fusion of the phagosome with lysosomes and so remain protected within their own vacuole. Others can prevent their vacuoles from becoming acidified, thereby blocking the activity of many degradative enzymes. **Table 19.9** gives examples of some ways organisms persist intracellularly. Often intracellular pathogens are released when the host cell dies, and they are then phagocytized or invade other cells. Microorganisms living within macrophages remain viable and may be carried to other parts of the body. By remaining in the phagocyte, they are protected from some antimicrobial drugs and the immune system of the host.

MICROCHECK 19.10

Organisms evade host responses by genetic variation of virulence factors, mimicking host substances to escape immune responses, degrading host antibodies, taking over the host cell actin and thereby remaining undetected in host cells, using various means to escape destruction by phagocytes, and many other ways.

- What is the effect of a bacterial protease that breaks down secretory IgA?
- How can genetic variation in bacteria help them evade host responses?
- Could *Schistosoma* attach to specific cell sites with adhesins?

FUTURE CHALLENGES

Genetics of Pathogenicity

Study of the pathogenesis of infectious diseases has come a long way since the first bacterial toxins were discovered little more than a century ago. Many questions, however, remain unanswered. Advances in genetic techniques have opened many areas of investigation. The availability of genetic sequence data has shown similarities in the virulence factors of diverse organisms. The discovery of pathogenicity islands bearing genes for a number of virulence factors in a cluster that is readily transferred opens the door to new approaches to understanding virulence.

One challenge is to understand how these virulence genes are regulated. It is known that bacteria express only the virulence genes appropriate for each host cell environment. Control is complex, and its study offers a fertile field of research.

Other challenges arise in trying to better understand interactions between microorganisms and host cells and tissues. Tools such as "knock-out mice" will be useful. These are mice in which specific genes have been inactivated or "knocked-out". It may be possible to use the host's imune system to modify the infectious process.

Of course, microorganisms are constantly changing and evolving. New pathogens will continue to emerge as a result of transfers of genetic information, such as pathogenicity islands. Use of the Molecular Postulates will permit identification of new pathogens. New techniques will produce better diagnostic methods and new aproaches to control and treatment of infectious diseases.

SUMMARY

Normal Flora of the Human Body

1. The microorganisms that grow in or on the body without producing obvious harmful effects make up the **normal flora**. (Figure 19.1)

Symbiotic Relationships Between Microorganisms and Hosts

1. **Symbiosis** describes the living together of two dissimilar organisms. Symbiotic relationships include commensalism, mutualism, and parasitism.
2. In **commensalism**, one partner benefits while the other is unaffected.
3. Both partners benefit in **mutualism**.
4. The **parasite** benefits at the expense of the **host**.

Establishment of Normal Flora

1. Acquisition of the normal flora begins at birth and the flora changes in response to changing conditions.

Importance of Normal Flora

1. The normal flora inhibits potentially harmful organisms by preventing attachment, competing for essential nutrients, producing antimicrobial substances, stimulating the immune system, and inducing the production of antibodies that cross-react with potential pathogens.

Changes in the Normal Flora

1. The normal flora is dynamic, and it changes in response to variations in the environment, such as changes in diet, acidity, or antibiotic intake.

Microorganisms and the Human Host

1. The interactions of many host and microbial factors influence effects the microbes have on the human body. (Table 19.1)

Infection

1. **Infection** occurs when parasitic organisms grow in or on the host.
2. If the infection causes damage, then infectious **disease** is present.

Pathogenicity and Pathogens

1. **Pathogens** are disease-producing organisms. They may be extracellular, living outside of host cells; obligately intracellular, living only inside cells; or facultatively intracellular, able to live in either location. Opportunistic pathogens cause disease only in immunologically compromised hosts or when introduced into an unusual location.

Virulence

1. **Virulence** describes the properties of an infectious agent that promote its pathogenicity.
2. Virulence factors are responsible for the pathogenicity of microorganisms. (Table 19.2)

Infections and Infectious Diseases (Figure 19.2)

1. Infectious diseases are **communicable** or **contagious** and spread between people or animals.
2. Stages of infectious diseases are the **incubation period**, the period of **illness**, and **convalescence**.
3. Infections may be **acute** or **chronic**, **localized** or **systemic**, **inapparent**, or **latent**.
4. Following recovery from an infectious disease, the host is usually immune to that disease.

Establishing the Cause of an Infectious Disease

1. It is not always possible to establish the cause of an infectious disease, but two major approaches are helpful.

Koch's Postulates

1. These postulates are used with organisms that can be cultured in the laboratory to prove the cause of an infectious disease.

Molecular Postulates

1. These postulates use genes to identify virulence factors in pathogenic strains of organisms.

The Sources of Infectious Agents and Their Modes of Transmission

1. Infectious agents are transmitted from soil, water, animals, and normal flora, either indirectly or by direct contact. (Table 19.3)
2. The **portal of entry** is important in determining whether an organism will cause disease, and the **portal of exit** is critical for transmission of organisms to a new host.

Mechanisms of Pathogenesis and Damage to the Host

Adherence in the Establishment of Infection

1. Microorganisms adhere to host cell receptors by means of adhesins, such as some protein molecules of pili.

2. The adherence factors determine to which host cells the microbes will attach, and consequently what sort of disease may follow. (Figure 19.3)

Colonization and Invasion of Host Tissues and Cells

1. Organisms must multiply and colonize the host to become established.
2. To invade tissues, a pathogen may enter through a break in the skin or mucous membranes, destroy the basement membrane underlying epithelial cells, or go between or through the cells of the membrane. Some may be endocytosed or phagocytized. (Figures 19.4, 19.5)
3. Some virulence genes are expressed only at high bacterial cell concentrations.

Interference with Phagocytosis and Complement Activity

1. Pathogens make a variety of substances that impede phagocytosis, and enzymes that break down complement components, thereby inhibiting phagocytosis and complement activities. (Figures 19.6, 19.7)

Activities of Toxins (Tables 19.5, 19.6)

1. **Endotoxins** of Gram-negative cell walls are lipopolysaccharides composed of a toxic **lipid A** portion and an antigenic polysaccharide **O antigen**.
2. Endotoxins have many effects, causing fever and sometimes shock.
3. **Exotoxins** are powerful toxins with diverse effects.
4. The A portion of **A-B exotoxins** is a toxic enzyme and the B portion binds to cell receptors, allowing the toxin to enter. (Figure 19.9)
5. **Cytolytic toxins** and toxic enzymes break down cells or tissue components.
6. **Superantigens** bind directly to many T cells at sites distinct from the regular antigen receptor sites. This causes the release of toxic amounts of cytokines. (Figure 19.10)

Genetic Transfer of Virulence Factors

1. Virulence genes are frequently transferred between microorganisms by transduction, transformation, or conjugation.
2. Large regions of genes coding for a number of virulence factors may be linked in **pathogenicity islands** and transferred as a unit.

Pathogenic Effects of Viruses and Other Nonbacterial Agents

1. Algae, fungi, protozoa, multicellular parasites, and viruses each have multiple ways of causing disease.

Host Responses that Contribute to Pathogenesis

1. The outcome of any infection depends on virulence factors of the organisms and on host responses.
2. Although immune hosts can usually withstand infection, some host responses actually contribute to pathogenesis.

Hypersensitivity and Autoimmune Responses

1. These responses cause tissue damage and contribute to disease.

Effects of Impaired Body Defenses

1. Impaired body defenses allow even organisms of low virulence to cause serious disease.

Host Genetic Factors

1. Susceptibility to some infections is genetically determined.

Effects of Age and Stress

1. Age and stress influence the body's ability to contain infections.

Evasion of Host Responses

1. Organisms evade host responses by genetic variation of virulence factors, mimicking host substances to escape immune responses, degrading host antibodies, taking over the host cell actin and thereby remaining undetected in host cells, using various means to escape destruction by phagocytes, and using many other ways. (Tables 19.8, 19.9, Figures 19.11, 19.12)

REVIEW QUESTIONS

Short Answer

1. Describe three symbiotic relationships between microorganisms and human hosts.
2. What can cause changes in the normal flora?
3. Distinguish among extracellular, obligately intracellular, and facultatively intracellular pathogens.
4. What are the characteristics of a successful pathogen?
5. Why are capsules virulence factors?
6. Decribe endotoxins and their activities.
7. How do exotoxins differ from endotoxins?
8. List ways in which organisms can evade immune responses of the host.
9. What effects might antibiotic treatment have on normal flora?
10. Is the incubation period for an infectious disease always the same? Why or why not?
11. Why are Koch's Postulates not sufficient to establish the cause of all infectious diseases?
12. List several routes organisms may use to exit the host
13. What is the normal function of M cells in Peyer's patches? How do bacteria take advantge of M cells to invade the body?
14. Describe two ways in which bacteria interfere with complement activity.
15. List at least five toxins that either disrupt cell membranes or harm cells by enzyme action.
16. Name three major mechanisms for genetic transfer of virulence factors between bacteria.
17. Describe one way that viruses can evade host responses.
18. How can amebas cause disease in humans?
19. Why are hyaluronic acid capsules of advantage to the streptococci that produce them?
20. How can some bacteria persist within macrophages?

Multiple Choice

1. Adhesins of *Escherichia coli*…
   A. are proteins on the pili of the bacteria.
   B. contain mannose.
   C. react with sialic acid in receptors.
   D. are the same as the ones on vibrios.
   E. are polysaccharides.
2. Bacteria commonly invade humans in all of the following ways, *except*…
   A. through intact skin.
   B. by separating epithelial cells and passing between them.
   C. by destroying the basement membrane.
   D. by passing through M cells in mucous membranes.
   E. by using enzymes to penetrate host cells.
3. The C5a peptidase enzyme of *Streptococcus pyogenes* breaks down C5a, resulting in…
   A. lysis of the strep.
   B. lack of opsonization of the strep.
   C. killing of phagocytes.
   D. inhibition of chemotaxis and phagocytosis.
   E. inhibition of membrane attack complexes.
4. Endotoxins affect the host by…
   A. their enzyme activity.
   B. binding to lymphocytes.
   C. interfering with antibody production.
   D. disrupting cell membranes.
   E. causing macrophages to release interleukin-1 and other proteins.
5. All of the following are true of A-B toxins, *except*…
   A. the A portion is an enzyme.
   B. the B portion binds to host cells and allows the toxin to enter the cells.
   C. they are exotoxins.
   D. the B portion is responsible for the toxic properties.
   E. some of them cause cholera, diphtheria, and tetanus.
6. Pathogenicity islands…
   A. contain genes for one or two virulence factors at most.
   B. are always found in only one spot on the bacterial chromosome.
   C. occur as a single island per bacterial strain.
   D. are sets of genes for virulence transferred as a unit.
   E. do not affect the pathogenicity of an organism.
7. *Listeria* and *Shigella* bacteria evade host immune responses by…
   A. moving directly from cell to cell with the help of the host cell actin.
   B. producing protein A, which inhibits phagocytosis.
   C. producing protein G, which inhibits phagocytosis.
   D. coating themselves with host cell proteins.
   E. preventing fusion of phagosomes and lysosomes.
8. Superantigens…
   A. are exceptionally large antigen molecules.
   B. cause a very large antibody response.
   C. attach to a large number of T cells, causing cytokine release.
   D. attach specifically to B-cell antigens.
   E. assist in a protective immune response.
9. All of the following are true of viruses, *except* they…
   A. may enter host cells by endocytosis.
   B. may enter host cells by fusion of viral envelope with cell membrane.
   C. may kill host cells directly by commandeering host cell machinery.
   D. cannot evade host immune responses.
   E. may suppress class I MHC on host cells.
10. Opportunistic pathogens are likely to affect all of the following, *except*…
   A. AIDS patients.
   B. cancer patients.
   C. healthy college students.
   D. drug addicts.
   E. transplant recipients.

Applications

1. A group of smokers suffering from the outcome of severe *Staphylococcus aureus* infections are suing the cigarette companies for extent of their injuries. They claim that the disease was aggravated by cigarette smoking. The group is citing studies indicating that phagocytes are inhibited in their action by compounds in cigarette smoke. A statement prepared by their lawyers states that the *S. aureus* would not have caused such a severe disease if the phagocytes and immune system overall were functioning properly. During the proceedings, a microbiologist was callled in as a professional witness for the court. What were her conclusions about the validity of the claim?
2. A microbiologist put forth a grant proposal to study the molecules bacteria use to communicate. His principal rationale was that the damaging effects of many pathogenic microorganisms could be prevented by inactivating the molecules these bacteria use to communicate. Is this a reasonable proposal? Why or why not?

Critical Thinking

1. A student argued that no distinction should be made between commensalism and parasitism. Even in commensalism, the microorganisms are gaining some benefit (such as nutrients) from the host and this represents a loss to the host. In this sense the host is being damaged. Does the student have a valid argument? Why or why not?
2. A microbiologist argued that there is no such thing as "normal" flora in the human body, since the population is dynamic and is constantly changing depending on diet and external environment. What would be an argument against this microbiologist's view?
3. Adhesins on the pili of microorganisms are typically proteins, while the receptors are usually glycoproteins. Why should this be the case? Would it be just as logical for adhesins to be glycoproteins and receptors to be proteins?

CHAPTER 20

Epidemiology

For centuries, it was accepted that many mothers developed fever and died following childbirth. Puerperal fever, an illness that we now recognize is the result of a bacterial infection of the uterus following childbirth, had occurred at least since the time of Hippocrates (460–377 B.C.). In the Eighteenth century, when it became popular for women to deliver their babies in hospitals, the incidence of the disease rose to epidemic proportions. In Prussia between 1816 and 1875, more than 363,600 women died of puerperal fever. In the middle of the Nineteenth century in the hospitals of Vienna, the major medical center of the world at that time, about one of every eight women died of puerperal fever following childbirth.

In 1841, Ignaz Semmelweis, a Hungarian, traveled to Vienna to study medicine. After finishing medical school, he became the first assistant to Professor Johann Klein at the Lying-In Hospital. There were two sections of the hospital. The first section was under the management of Professor Klein and the medical students. Midwives and midwifery students served the second section. Being an astute observer, Semmelweis soon noticed that the incidence of puerperal fever in the first section often rose as high as 18%, four times that in the second section. When he investigated the conditions in the two sections, they appeared to be the same except in terms of their management. Semmelweis was dismissed from his post when he implicated Professor Klein and the students in the spread of the disease, but he was reinstated a few months later when others intervened on his behalf.

Semmelweis recognized that disease symptoms that led to the death of a friend who had incurred a scalpel wound while doing an autopsy were similar to puerperal fever. He reasoned that the "poison" that killed his friend probably also contaminated the hands of the medical students who did autopsies. These students were transferring the "poison" from the cadavers to the women in childbirth. Midwives did not perform autopsies. Since this was before Pasteur and Koch established the germ theory of disease, Semmelweis had no way of knowing that the "poison" being transferred was probably Streptococcus pyogenes, *a common cause of many infections, including puerperal fever. Those being autopsied had likely died of streptococcal disease as well. To test his hypothesis, Semmelweis instituted the practice of having physicians and students wash their hands with a solution of chloride of lime, a strong disinfectant, before attending their patients. The result was a drop in the incidence of puerperal fever to one-third its previous level. Instead of accepting these findings and the new techniques that Semmelweis employed, his colleagues in Vienna refused to accept responsibility for the deaths of so many patients. His work was so fiercely attacked that he was forced to leave Vienna and return to his native Hungary. There he was again able to use disinfection techniques to achieve a remarkable reduction in the number of deaths from puerperal fever.*

Semmelweis became increasingly outspoken and bitter, finally becoming so deranged that he was confined to a mental institution. Ironically, he died one month later of a generalized infection similar to the kind that had killed his friend and the many women who had contracted puerperal fever following childbirth. The infection originated from a finger wound received before his confinement. Some said he deliberately infected himself from a cadaver while performing an autopsy.

The accomplishments of Ignaz Semmelweis would be the envy of today's hospital epidemiologists. Without knowledge of microbiology or the germ theory of disease, he rigorously defined the affected population, located the source of the causative agent and its mode of transmission, and showed how to stop the spread of the disease.

—A Glimpse of History

EPIDEMIOLOGY COMBINES DIVERSE DISCIPLINES including ecology, microbiology, sociology, statistics, and psychology to study the cause and distribution of health states, both positive and negative, in populations. **Epidemiologists** are the "health detectives" who do most of this work. They collect and compile data to describe disease outbreaks, much as a criminal detective describes the scene of a crime. Many habits in the daily routine of life, from handwashing to waste disposal, reflect a current understanding of epidemiology.

Society as a whole depends on epidemiology for its collective well-being. Governments direct public health agencies and epidemiologists to investigate the status of disease at all social levels. Through the cooperation of physicians, clinics, hospitals, and schools, data are accumulated in towns, counties, states, and countries. The monitoring, study, control, and containment of infectious diseases are major global concerns, made even more so as population densities rise, increasing the risks of spread of disease-causing agents.

Principles of Epidemiology

Diseases that can be transmitted from one person to another, such as measles, colds, and influenza, are **communicable diseases**. In order for a communicable disease to spread, a specific chain of events must occur. First, the pathogen must have a suitable environment in which to live. That natural habitat, the **reservoir**, may be on or in an animal, including humans, or in an environment such as soil or water (**figure 20.1a**). A pathogen must then leave its reservoir in order to be transmitted to the susceptible host. If the reservoir is an animal, the body orifice or surface from which a microbe is shed is called the **portal of exit**. Disease-causing organisms must then be transmitted to the next host, usually through direct contact or via contaminated food, water, or air (figure 20.1b). They enter the next host through a body surface or orifice called the **portal of entry** (figure 20.1c).

Diseases that cannot spread from person to person, such as botulism (which results from the ingestion of a foodborne toxin), are called **non-communicable diseases**. Microorganisms that cause these diseases most often arise from an individual's normal flora or from an environmental reservoir. Prevention of non-communicable illness is often disease-specific. Botulism, for example, can be prevented through the use of proper canning methods.

Rate of Disease in a Population

Epidemiologists are most concerned with the **rate** of a disease, the percentage of a given population infected, rather than absolute number of cases. For example, 100 people in a large city developing disease X in a given period may not be abnormal, whereas 100 people in a small rural community developing the same disease would indicate a much higher rate and thus be of greater concern to the epidemiologist. A related concept is the **attack rate**, which is the number of cases developing in a group of people who were exposed to the infectious agent. For example, if 100 people at a party ate chicken that was contaminated with *Salmonella*, and 10 people came down with symptoms of salmonellosis, then the attack rate was 10%. The attack rate reflects many factors, including the infectious dose of the organism and the immunity of the population.

Other rates are also often used to express the effect of a disease on a population. **Morbidity rate** is calculated as the number of cases of illness in a given time period divided by the population at risk. Contagious diseases such as influenza often have a high morbidity rate because each infected individual may transmit the infection to several others. **Mortality rate** reflects the percentage of a population that dies from the disease. Diseases such as plague and Ebola hemorrhagic fever are feared because of their very high mortality rate. The **incidence** of a disease reflects the number of new cases in a specific time period in a given population at risk, whereas the **prevalence** reflects the number of total existing cases both old and new in a given population at risk. Usually these rates are expressed as the number of cases per 100,000 people. The prevalence is useful to assess the overall impact of the disease on society because it takes the duration of the disease into account, whereas the incidence rate provides a means of measuring the risk of an individual contracting the disease.

Diseases that are constantly present in a given population are called **endemic**. For example, both the common cold and influenza are endemic in the United States. An unusually large number of cases in a population constitutes an **epidemic**. Epidemics may be caused by diseases that are not normally present in a population, such as cholera being reintroduced to the Western Hemisphere, or by diseases that are normally endemic, such as influenza and pneumonia (**figure 20.2**). A related term is **outbreak**, which generally implies a cluster of cases occurring during a brief time interval and affecting a specific population; an outbreak may herald the onset of an epidemic. When an epidemic spreads worldwide, such as we have seen with AIDS, it is called a **pandemic**. Terms commonly used to describe the epidemiology of disease are summarized in **table 20.1**.

MICROCHECK 20.1

Epidemiologists study the frequency and distribution of disease in order to identify the cause, source, and route of transmission. Epidemiologists are most concerned with the rate of disease rather than absolute numbers.

- Define the term reservoir as it relates to epidemiology.
- Distinguish between a disease that is endemic and one that is epidemic.
- If a drug were developed to block the transmission of HIV, which would dramatically drop, the incidence of HIV infection or the prevalence?

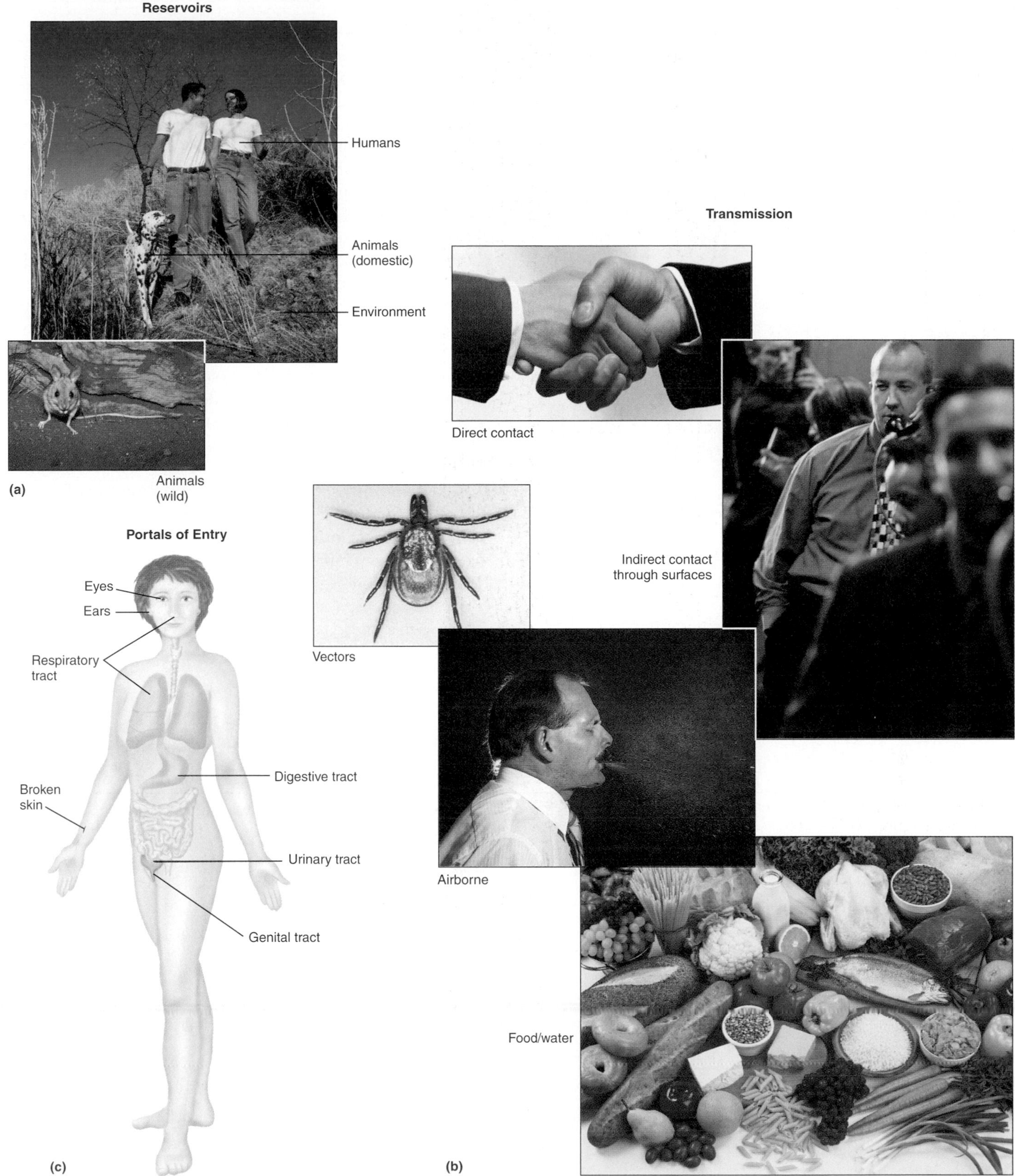

Figure 20.1 Spread of Pathogens **(a)** Reservoirs, **(b)** transmission, and **(c)** portals of entry are necessary for the spread of communicable diseases.

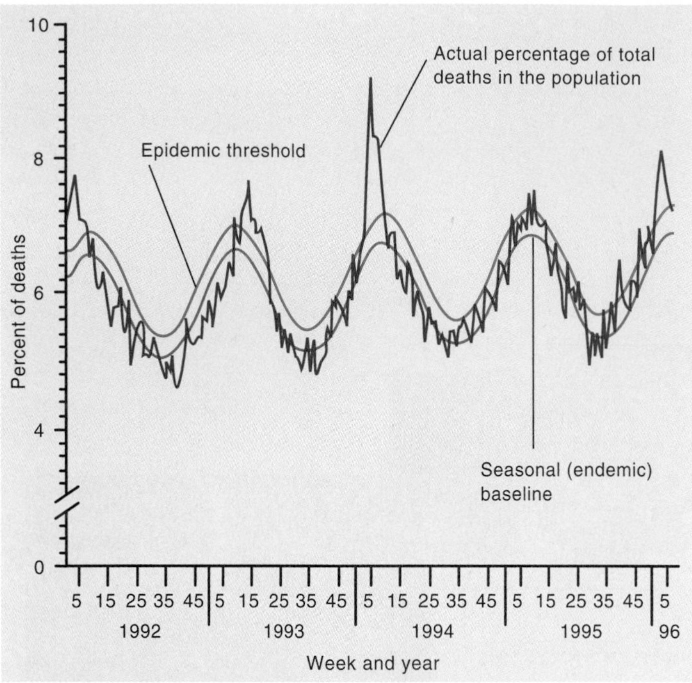

Figure 20.2 Incidence of Influenza, an Endemic Disease that Can Be Epidemic Weekly pneumonia and influenza mortality as a percentage of all deaths for 121 cities of the United States, January 1, 1992, to February 3, 1996.

Reservoirs of Infection

The reservoir of a pathogen is important because it affects the extent and distribution of a disease. For example, coccidioidomycosis is endemic only in the hot, dry, dusty areas of the Western Hemisphere because that is the only habitat of the causative agent, the fungus *Coccidioides immitis.* Determining the reservoir can help protect a population from disease, because measures can then be instituted to prevent the people from coming into contact with that source. The fact that the United States does not have epidemics of plague, the disease that killed over one-quarter of the European population in the Fourteenth century, is in part because we recognize the importance of controlling its rodent reservoir. Wild rats, mice, and prairie dogs are the natural reservoir of *Yersinia pestis,* the bacterium that causes plague. By avoiding a buildup of garbage in cities and homes, we prevent rodent infestation of our living quarters. ■ *Coccidioides immitis*, p. 579 ■ *Yersinia pestis*, p. 724

Human Reservoirs

Infected humans are the most significant reservoir of the majority of communicable diseases. In some cases, humans are the only reservoir. In other instances, the pathogenic organism can exist not only in humans but in other animals and, occasionally, the environment as well. If infected humans are the only reservoir, then theoretically the disease is easier to control. This is because it is more feasible to institute prevention and control

TABLE 20.1 Common Terms in Epidemiology

Term	Definition
Attack rate	The proportional number of cases developing in the population that was exposed to the infectious agent
Communicable disease	An infectious disease that can be transmitted from one person to another
Endemic	A disease or other occurrence that is constantly present in a population
Epidemic	A disease or other occurrence whose incidence is higher than expected
Herd immunity	A phenomenon that occurs when a critical concentration of immune hosts prevents the spread of an infectious agent
Incidence	The number of new cases of a disease in a population at risk during a specified period of time
Index case	The first identified case of a disease in an outbreak or epidemic
Morbidity	Illness. Most often expressed as the rate of illness in a given population at risk
Mortality	Death. Most often expressed as a rate of death in a given population at risk
Non-communicable disease	A disease that cannot be transmitted from one person to another
Outbreak	A cluster of cases occurring during a brief time interval and affecting a specific population; an outbreak may herald the onset of an epidemic
Pandemic	A worldwide epidemic
Portal of entry	Surface or orifice from which a disease-causing agent enters the body
Portal of exit	Surface or orifice through which a disease-causing agent exits and disseminates
Prevalence	The total number of cases in a given population at risk at a point in time
Reservoir	The natural habitat of a disease-causing organism

programs in humans than it is in wild animals. The eradication of smallpox is an excellent example. The combined effects of widespread vaccination programs, which resulted in fewer susceptible people, and the successful isolation of those who did become infected, eliminated the smallpox virus from nature. The virus no longer had a reservoir in which to multiply.

Symptomatic Infections

People who have symptomatic illnesses are an obvious source of infection, and ideally they understand the importance of taking precautions to avoid transmitting their illness to others. Staying home and resting while ill both helps the body recover and protects others from exposure to the disease-causing agent. Even conscientious people, however, can unintentionally be a source of infection to others. For example, people who are in the incubation period of mumps still shed virus, as do those who have recovered symptomatically from poliomyelitis. ■ **mumps, p. 596** ■ **incubation period, p. 455** ■ **poliomyelitis, p. 679**

Asymptomatic Carriers

A more problematic source of infection occurs when a person can harbor a pathogen with no ill effects whatsoever, acting as a **carrier** of the disease agent. These people may shed the organism intermittently or constantly for months, years, or even a lifetime.

Some carriers have an **asymptomatic infection**; their immune system is actively responding to the invading microorganism, but they have no obvious clinical symptoms. Because these people often have no reason to consider that they are a reservoir of infection, they move freely about, spreading the pathogen. People with asymptomatic infections are a significant complicating factor in the control of sexually transmitted diseases such as gonorrhea. Up to 50% of women infected with *Neisseria gonorrhoeae* have no obvious symptoms, which means they often unknowingly transmit the disease to their sexual partners. In contrast, most infected men are symptomatic and therefore seek medical treatment. Gonorrhea infections can be treated with antibacterial drugs, but tracking down sexual contacts of infected people is difficult and costly. As another example, people who recover from typhoid fever without benefit of antibiotic treatment may harbor the causative strain of *Salmonella* in their gallbladder, shedding it in their feces. Even after typhoid fever has been controlled in a population, the presence of chronic carriers poses a threat of disease recurrence for many years. ■ ***Neisseria gonorrhoeae*, p. 644** ■ **typhoid fever, p. 604**

Some pathogens can colonize the skin or mucosal surfaces, establishing itself as part of the person's microbial flora. For example, at least 60% of the population carry *Staphylococcus aureus* as a part of their nasal or skin flora during some period of their life. At least 20% of these are chronic carriers who harbor the organism for a year or more. *Staphylococcus aureus* carriers may never have any illness or disease as a result of the organism, but they remain a potential source of infection to themselves and others. Likewise, some people are carriers of the pathogen *Streptococcus pyogenes.* Health care workers who carry this bacterium have been linked to outbreaks in hospitals. Ridding a colonized carrier of the infectious organism is often difficult, even with the use of antimicrobial drugs. ■ ***Staphylococcus aureus*, p. 525** ■ ***Streptococcus pyogenes*, p. 553**

Nonhuman Animal Reservoirs

Nonhuman animal reservoirs are the source of some pathogens. For example, poultry are a reservoir of gastrointestinal pathogens such as species of *Campylobacter* and *Salmonella.* In the United States, raccoons, skunks, and bats are the reservoir of the rabies virus. Recall that rodents are the reservoir for *Yersinia pestis*, the causative agent of plague. Occasional transmission of plague to humans is still reported in the southwestern states where *Y. pestis* is endemic in prairie dogs and other rodents, but rodent control has helped prevent epidemics of the disease. Rodents, particularly the deer mouse, are also the reservoir for hantavirus. ■ ***Campylobacter*, p. 606** ■ ***Salmonella*, p. 604**

Diseases such as plague and rabies that can be transmitted to humans but primarily exist in other animals are called **zoonotic** diseases or **zoonoses**. Zoonotic diseases are often more severe in humans than in the normal animal host because the infection in humans is accidental; there has been no evolution toward the balance that normally exists between a host and parasite. ■ **parasitism, p. 452**

Environmental Reservoirs

Some pathogens have environmental reservoirs. For example, *Clostridium botulinum*, which causes botulism, and *Clostridium tetani*, which causes tetanus, are both widespread in soils. *Legionella pneumophila* is found in water in association with amebas. Unfortunately, pathogens that have environmental reservoirs are probably impossible to eliminate. ■ ***Clostridium botulinum*, p. 675** ■ ***Clostridium tetani*, p. 699** ■ ***Legionella pneumophila*, p. 571**

MICROCHECK 20.2

Preventing susceptible people from coming in contact with the reservoir can prevent infectious disease. The reservoir for a disease can be infected people, other animals, or the environment.

- Explain why smallpox was successfully eradicated but rabies probably never will be.
- Why are colonized carriers of *Streptococcus pyogenes* important in a health care setting?
- From a pathogen's perspective, explain why it might be disadvantageous to kill a host quickly.

Portals of Exit

Microorganisms must leave one host in order to be transmitted to another. Those that inhabit the intestinal tract are routinely shed in the feces. Pathogens such as *Vibrio cholerae* that cause massive volumes of watery diarrhea may have an evolutionary advantage because the large volumes discharged may enhance their dispersal. Respiratory organisms are expelled in droplets of saliva when people talk, laugh, sing, sneeze, or cough (see figure 20.1b). Pathogens such as *Mycobacterium tuberculosis* and various respiratory viruses exit the body via this route. Meanwhile, organisms that inhabit the skin are constantly shed on skin cells. Even as you read this text you are shedding skin cells, some of which may have *Staphylococcus aureus* on their surface. Genital pathogens such as *Neisseria gonorrhoeae* can be carried in semen and vaginal secretions. Some pathogens, such as the eggs of the helminth *Schistosoma hematobium*, can exit in urine. Hantavirus is found in the saliva, urine, and droppings of the deer mouse. ■ *Vibrio cholerae*, p. 599 ■ *Neisseria gonorrhoeae*, p. 644 ■ *Mycobacterium tuberculosis*, p. 569 ■ *Staphylococcus aureus*, p. 525

MICROCHECK 20.3

In order to spread, infectious microorganisms must exit a host.

- Identify three portals of exit and indicate why they are significant sources of microorganisms.
- Considering that circulating blood is not normally released from the body, describe how blood-borne microorganisms might exit.

Transmission

A successful pathogen must somehow be passed from its reservoir to the next susceptible host. Transmission of a disease-causing organism from one person to another through contact, ingestion of food or water, or via a living agent such as an insect is called **horizontal transmission**. This contrasts with **vertical transmission**, which is the transfer of a pathogen from a pregnant woman to the fetus, or from a mother to her infant during childbirth. Certain microbes can cross the placenta and damage a developing fetus, causing diseases such as congenital syphilis. Others, such as group B streptococci, can infect the newborn as it passes through the birth canal. ■ syphilis, p. 647 ■ group B streptococci, p. 668

Contact

Transmission of a pathogen from one person to another often involves some form of contact. This can be through direct touch, inhalation of droplets of respiratory secretions or saliva, or indirect contact by way of a nonliving object.

Direct Contact

Direct contact occurs when one person physically touches another. It can be through an act as simple as a handshake, or a more intimate contact such as sexual intercourse. In some cases, direct contact is the primary way in which an organism is transmitted. This is particularly true if the transfer of even very low numbers of an organism can initiate an infection. For example, the infectious dose of the intestinal pathogen *Shigella* is approximately 10 to 100 organisms, a number easily passed when shaking hands. Once on the hands, the organisms can inadvertently be ingested. This is just one example of how **fecal-oral transmission**, the inadvertent consumption of organisms that originate from the intestine, can occur. Handwashing, a fairly simple routine that physically removes organisms, is important in preventing this type of spread of disease. Even washing in plain water reduces the numbers of potential pathogens on the hands, which in turn decreases the possibility of transferring or ingesting sufficient numbers of an organism to establish an infection. In fact, routine handwashing is considered to be the single most important measure for preventing the spread of infectious disease. ■ *Shigella* sp., p. 602

Pathogens that cannot survive for extended periods in the environment must generally, because of their fragile nature, be transmitted through direct contact. For example, *Treponema pallidum*, which causes syphilis, and *Neisseria gonorrhoeae*, which causes gonorrhea, both die quickly when exposed to a relatively cold dry environment and thus require intimate sexual contact for their transmission. ■ *Treponema pallidum*, p. 648 ■ *Neisseria gonorrhoeae*, p. 644

Indirect Contact

Indirect transmission involves transfer of pathogens via inanimate objects, or **fomites**, such as clothing, tabletops, doorknobs, and drinking glasses. For example, carriers of *Staphylococcus aureus* may inoculate their hands with the organism when touching a skin lesion or colonized nostril. Organisms on the hands can then easily be transferred to a fomite. Another person can readily acquire the microbes when handling that object. Again, handwashing is an important control measure.

Droplet Transmission

Large microbe-laden respiratory droplets generally fall to the ground no farther than a meter (approximately 3 feet) from release. People in close proximity can inhale those droplets, however, resulting in the spread of respiratory disease via **droplet transmission**. Although physical contact is not necessary, droplet transmission is considered direct transmission because of the close range involved. Droplet transmission is particularly important as a source of contamination in densely populated buildings such as schools and military barracks. Desks or beds in such locations ideally should be spaced more than 4 feet and preferably 8 to 10 feet apart to minimize the transfer of infectious agents. Another way to minimize the spread of respiratory diseases is to educate people about the importance of covering their mouths with a tissue when they cough or sneeze.

Food and Water

Pathogens, particularly those that infect the gastrointestinal tract, can be transmitted through contaminated food or water. Foods can become contaminated in a number of different ways. Animal products such as meat and eggs may harbor pathogens that originated from the animal itself. This is the case with poultry that is contaminated with species of *Salmonella* or *Campylobacter* and hamburger that is contaminated with *E. coli* O157:H7. Pathogens can also be inadvertently added during food preparations. For example, typhoid carriers who do not wash their hands thoroughly after defecating and prior to preparing food can easily contaminate the food. This is yet another example of how fecal-oral transmission can occur. **Cross-contamination** results when pathogens from one food are transferred to another. A cutting board used first to carve raw chicken and then to cut cooked potatoes can serve as a fomite, transferring *Salmonella* sp. from the chicken onto the potatoes. Because many foods are a rich nutrient source, microorganisms can multiply to high numbers if the contaminated food is improperly stored. Sound food-handling methods, including sanitary preparation as well as thorough cooking and proper storage, can prevent foodborne diseases. ■ food storage, p. 97

Waterborne disease outbreaks can involve large numbers of people because municipal water systems distribute water to large areas. For example, the 1993 waterborne outbreak of *Cryptosporidium parvum*, an intestinal parasite, in Milwaukee, Wisconsin, was estimated to have involved approximately 400,000 people. Prevention of waterborne diseases requires chlorination and filtration of drinking water and proper disposal and treatment of sewage. ■ *Cryptosporidium parvum*, p. 615 ■ drinking water treatment, p. 793 ■ sewage treatment, p. 795

Air

Respiratory diseases can also be transmitted through the air. When particles larger than 10 μm are inhaled, they are usually trapped in the mucus lining of the nose and throat and eventually swallowed. Smaller particles, however, can enter the lungs, where any pathogens they carry can potentially cause disease.

As mentioned earlier, when people talk, laugh, or sneeze they continually discharge microorganisms in liquid droplets. While large droplets quickly fall to the ground, the smaller droplets dry, leaving one or two organisms attached to a thin coat of the dried material, creating **droplet nuclei**. The droplet nuclei can remain suspended indefinitely in the presence of even slight air currents. Other airborne particles, including dead skin cells, household dust, and soil disturbed by the wind, may also carry respiratory pathogens. An air conditioning system may be a source of infectious agents, because it can distribute air contaminated by people or with organisms growing within the system.

The number of viable organisms in air can be estimated by using a machine that pumps a measured volume of air, including any suspended dust and particles, against the surface of a nutrient-rich medium in a Petri dish. This technique has shown that the number of bacterial colonies in the air sampled rises in proportion to the number of people in a room (**figure 20.3**). The survival of organisms in air varies greatly with the type of organism and with air conditions such as humidity, temperature, and degree of light. In general, Gram-positive organisms survive longer in air than do Gram-negative organisms, and survival in dim light is greater than that in bright light.

Understandably, airborne transmission of pathogens is very difficult to control. To prevent the buildup of airborne pathogens, modern public buildings have ventilation systems that constantly change the air. Hospital microbiology laboratories can be kept under a slight vacuum so that air flows in from the corridors, preventing microorganisms and viruses from being swept out of the lab to other parts of the building. Air in some laboratories, specialized hospital rooms, and jetliners is circulated through high-efficiency particulate (HEPA) filters to remove airborne organisms that may be present. ■ HEPA filters, p. 126

(a)

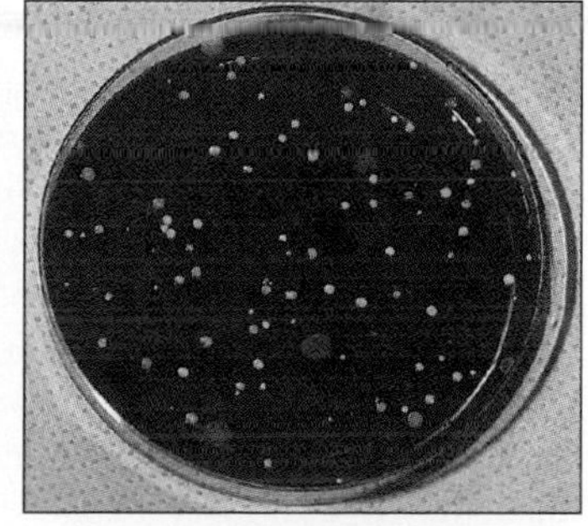
(b)

Figure 20.3 Air Sample Cultures **(a)** Air from a clean, empty hospital room. **(b)** Air from a small room containing 12 people. In both situations, 5 cubic feet of air was sampled.

Vectors

A **vector** such as a mosquito or flea can transmit some diseases. The term vector applies to any living organism that can carry a disease-causing microbe, but most commonly these are arthropods such as mosquitoes and ticks. A vector may carry a pathogen externally or internally.

Flies that land on feces can pick up intestinal pathogens such as *Escherichia coli* O157:H7 and *Shigella* sp. on their legs. If the fly then moves to a food, it transfers the microorganisms to a source that could be consumed. In this case, the fly serves as a **mechanical vector**, carrying the microbe on its body from one place to another.

Diseases such as such as malaria, plague, and Lyme disease are transmitted through arthropods that harbor the pathogen internally. The vector either injects the infectious agent while taking a blood meal or defecates, depositing the pathogen onto a person's skin where it can then be inadvertently inoculated when the individual scratches the bite. For example, fleas inject *Yersinia pestis* while attempting to take a blood meal. In the case of malaria, caused by species of the eukaryotic pathogen *Plasmodium*, the mosquito is not only the transmitter of the parasite but also serves as an essential part of its reproductive life cycle. A vector that is required as a part of a parasite's life cycle is called a **biological vector**. An important significance of a biological vector is that the pathogen can multiply to high numbers within the vector.

Prevention of vector-borne disease relies on control of mosquitoes, ticks, and other arthropods. The success of such controls was demonstrated when malaria, once endemic in the continental United States, was successfully eliminated from the nation. This was accomplished through a combination of mosquito elimination and prompt treatment of infected patients. Unfortunately, worldwide eradication efforts that initially showed great promise ultimately failed, in part due to the decreased vigilance that accompanied the dramatic but short-lived decline of the disease.

MICROCHECK 20.4

Handwashing can prevent diseases that are spread through direct or indirect contact, as well as those that are spread via contaminated food. Airborne transmission of pathogens is difficult to control. Prevention of vector-borne diseases relies largely on control of the arthropod vectors.

- Explain the difference between horizontal and vertical transmission.
- What is a fomite?
- Why might populations become less vigilant in disease elimination efforts as the incidence of the disease wanes?

Portal of Entry

To cause disease, not only must a pathogen be transmitted from its reservoir to a new host, it must also enter or colonize a surface of that new host. Colonization is generally a prerequisite for causing disease. For example, *Shigella* may be transferred via a handshake, but it will only cause disease if the person then transfers it to his or her mouth or to food and inadvertently ingests it. This allows the pathogen the opportunity to establish itself in the intestinal tract. Respiratory pathogens that are released into the air during a cough generally cause disease only when someone inhales them. Many organisms that cause disease if they enter one body site are harmless if they enter another. For example, *Enterococcus faecalis* may cause a bladder infection if it enters the normally sterile urinary tract, but it is harmless in the intestine where it frequently resides as a member of the normal flora.

The importance of the route of entry in disease development is illustrated in the case of plague transmission. When a flea that normally resides on the rodent reservoir becomes infected and then bites a person, *Yersinia pestis* is injected and the form of plague called **bubonic plague** develops. The bacteria multiply rapidly inside lymph nodes, resulting in a disease that is not contagious but has a mortality rate of 50% to 75% if not treated promptly. In a small percentage of people with bubonic plague, however, the organism spreads to the lungs, resulting in a different manifestation of the disease, **pneumonic plague**. Pneumonic plague presents a much more severe situation, because it is nearly always fatal and is readily transmitted from person to person through respiratory droplets. ■ *Yersinia pestis*, p. 724

MICROCHECK 20.5

The portal of entry of a pathogen can affect the outcome of disease.

- Through which portal must *Shigella* enter to cause disease?
- Contrast the transmission of bubonic plague with that of pneumonic plague.
- Plague epidemics are usually due to the pneumonic form of the disease. Why would this be so?

Factors that Influence the Epidemiology of Disease

An infectious agent that is successfully transmitted from a reservoir to a new host can potentially cause disease. The outcome of such transmission events, however, is affected by many different factors including the dose, the incubation period, and characteristics of the host population.

The Dose

The probability of infection and disease is generally lower when an individual is exposed to small numbers of a pathogen. This is because there must be a certain minimum number of pathogens in the body to produce enough damage to cause disease symptoms. For example, if only 30 *Salmonella typhi* are ingested in contaminated drinking water yet 1,000,000 are required to produce typhoid fever symptoms, then it will take some time for the bacterial population to increase to that number. Because host defenses are being mobilized at the same time and are racing to eliminate the bacteria, small doses often result in a higher percentage of asymptomatic infections. The immune system sometimes eliminates the organism before symptoms appear. On the other hand, there are few if any infections for which immunity is absolute. An unusually large exposure to a pathogen, such as can occur in a laboratory accident, may produce serious disease in a person who has immunity to ordinary doses of the pathogen. Therefore, even immunized persons should take precautions to minimize exposure to infectious agents. This principle is especially important for medical workers who attend patients with infectious diseases.

The Incubation Period

The extent of the spread of an infectious agent is influenced by the incubation period. Diseases with typically long incubation periods such as AIDS can spread extensively before the first cases appear. An excellent example of the importance of this factor was the spread of typhoid fever from a ski resort in Switzerland in 1963. As many as 10,000 people had been exposed to drinking water containing small numbers of *Salmonella* Typhi, the causative agent of the disease. The long incubation period of the disease, 10 to 14 days, allowed widespread dissemination of the organisms by the skiers, since they flew home to various parts of the world before they became ill. As a result, there were more than 430 cases of typhoid fever in at least six countries. ■ incubation period, p. 455

Population Characteristics

Certain population groups are more likely to be affected by a given disease-causing agent. In some situations, the reasons are obvious. For example, vaccination decreases the likelihood of disease. In other cases, the reasons are not as clear and probably involve multiple factors.

Immunity to the Pathogen

Previous exposure or immunization of the population to a disease agent or antigenically related agents influences the number of people who become ill from a disease. A disease is unlikely to spread very widely in a population in which 90% of the people are immune to the disease agent. If humans are the only reservoir, then a continuous source of susceptible people is required or the disease will disappear from the population. Recall that the requirement for a susceptible host allowed smallpox to be eradicated. When an infectious agent cannot spread in a population because it lacks a critical concentration of nonimmune hosts, a phenomenon called **herd immunity** results. The nonimmune individuals are essentially protected by the lack of a reservoir of infection. Unfortunately, some infectious agents are able to undergo antigenic variation so that they can continue to propagate even in a previously exposed population. ■ antigenic variation, p. 184

General Health

Malnutrition, overcrowding, and fatigue increase the susceptibility of people to infectious diseases and enhance the diseases' spread. Infectious diseases have generally been more of a problem in poor areas of the world where individuals are crowded together without proper food or sanitation. Factors that promote good general health result in increased resistance to diseases such as tuberculosis. When infection does occur in a healthy individual, it is more likely to be asymptomatic or to result in mild disease.

Age

The very young and the elderly are generally more susceptible to infectious agents. The immune system of young children is not fully developed, and consequently, they are predisposed to certain diseases. For example, young children are particularly susceptible to meningitis caused by *Haemophilus influenzae*. The elderly are more prone to disease because the immune system wanes over time. Influenza outbreaks in nursing homes can have fatal consequences. **■ meningitis, p. 666 ■ influenza, p. 573**

Gender

In some cases, gender influences disease distribution. For example, women are more likely to develop urinary tract infections because their urethra, the tube that connects the bladder to the external environment, is relatively short. Microbes can ascend the urethra into the bladder. Women who are pregnant are more susceptible to listeriosis, caused by *Listeria monocytogenes*. **■ *Listeria monocytogenes*, p. 671**

Religious and Cultural Practices

The distribution of disease is also influenced by religious and cultural practices. For example, infants who are breast-fed are less likely to have diarrhea caused by infectious agents, presumably because of the protective effects of antibodies in the mother's milk. Groups who eat traditional dishes made from raw freshwater fish are more likely to acquire tapeworm, a parasite normally killed by cooking.

Genetic Background

Natural immunity can vary with genetic background, but it is usually difficult to determine the relative importance of genetic, cultural, and environmental factors. In a few instances, however, the genetic basis for resistance to infectious disease is known. For example, many people of black African ancestry are not susceptible to malaria caused by *Plasmodium vivax* because they lack a specific red blood cell receptor used by the organism. Some populations of Northern European ancestry are less susceptible to HIV infection because they lack a certain receptor on their white blood cells. Studies of the incidence of tuberculosis in identical twins suggest that genetic factors play a role, and there is also evidence of a genetic influence in the development of paralytic poliomyelitis.

MICROCHECK 20.6

Diseases with long incubation periods can spread extensively before the first cases appear. The probability of infection and disease is generally lower if an individual is exposed to small numbers of pathogens. Population characteristics including prior vaccination or exposure to the disease, general health, age, gender, religious and cultural practices, and genetic background can influence susceptibility to disease.

- Why would a small dose of an infecting agent result in an asymptomatic infection?
- Explain the phenomenon of herd immunity.
- The magnitude of the dose can affect the incubation period and, consequently, the spread of disease. Explain why this might be so.

Epidemiological Studies

Epidemiologists investigate a disease outbreak to determine both the causative agent and also its reservoir and route of transmission, so as to recommend ways to minimize the spread. They are concerned with emerging diseases such as that caused by *Escherichia coli* O157:H7, and also with diseases that have been recognized for centuries, such as tuberculosis and cholera. The studies that epidemiologists conduct are in many ways similar to criminal investigations. After a disease outbreak, investigators conduct a **descriptive study** to determine the characteristics of the persons involved and the place and the time of the outbreak. This information gives clues as to the possible cause, reservoir, and transmission of the illness. Once the occurrence of the outbreak has been fully described, an **analytical study** is done to identify specific conditions, or **risk factors**, associated with high frequencies of disease. Finally, **experimental studies** are sometimes done to assess the effectiveness of measures to prevent or treat disease.

The British physician John Snow illustrated the power of a well-designed epidemiological study over a century ago. Years before the relationship between microbes and disease was accepted, he documented that the cholera epidemics plaguing England from 1849 to 1854 were due to contaminated water supplies. He did this by carefully comparing the conditions of households that were affected by cholera and those that were not, eventually determining that the primary difference was their water supply. At one point, he ordered the removal of the handle of a public water pump in the neighborhood of an outbreak; this simple act helped halt an epidemic that in 10 days had killed more than 500 people. **■ cholera, p. 599**

Descriptive Studies

After a disease outbreak, epidemiologists race to define characteristics such as the person, the place, and the time. These data are then used to compile a list of possible risk factors involved in the spread of disease. The significance of these hypothesized risk factors can be tested in an analytical study.

The Person

Determining the profile of those who become ill is critical to defining the population at risk. Variables such as age, sex, race/ethnicity, occupation, personal habits, previous illnesses, socioeconomic class, and marital status may all yield clues about risk factors for developing the disease. For example, the fact that adults are more likely than children to develop tetanus may indicate that they are not receiving adequate immunizations, which are recommended once every 10 years **(figure 20.4)**.

The Place

The geographic location of disease acquisition identifies the general site of contact between the person and the infectious agent. This helps pinpoint the exact source. The location may also give clues about potential disease reservoirs, vectors, or geographical boundaries that may affect disease transmission. For example, malaria can only be transmitted in regions that have the appropriate mosquito vector.

The Time

The timing of the outbreak may also yield helpful clues. A rapid rise in the numbers of people who became ill suggests that they were all exposed to a single common source of the infectious agent, such as contaminated chicken at a picnic. This type of outbreak is called a **common-source epidemic**. If the numbers of ill people rise gradually, the disease is likely contagious, with one person transmitting it to several others, who each then transmit it to several more, and so on. This type of pattern is called a **propagated epidemic (figure 20.5)**. The first case in an outbreak is called the **index case**. In a propagated epidemic, if a direct chain of contacts can be established, the time between the onset of symptoms in one case and the next reflects the incubation period of the disease.

The season of the year in which the epidemic occurs may also be significant. Respiratory diseases including influenza, respiratory syncytial virus infections, and the common cold are more easily transmitted in crowded indoor conditions during the winter **(figure 20.6)**. Conversely, vector- and foodborne diseases are more often transmitted in warm weather when people are more likely to be exposed to mosquitoes and ticks, or eating picnic food that has not been stored properly **(figure 20.7)**.

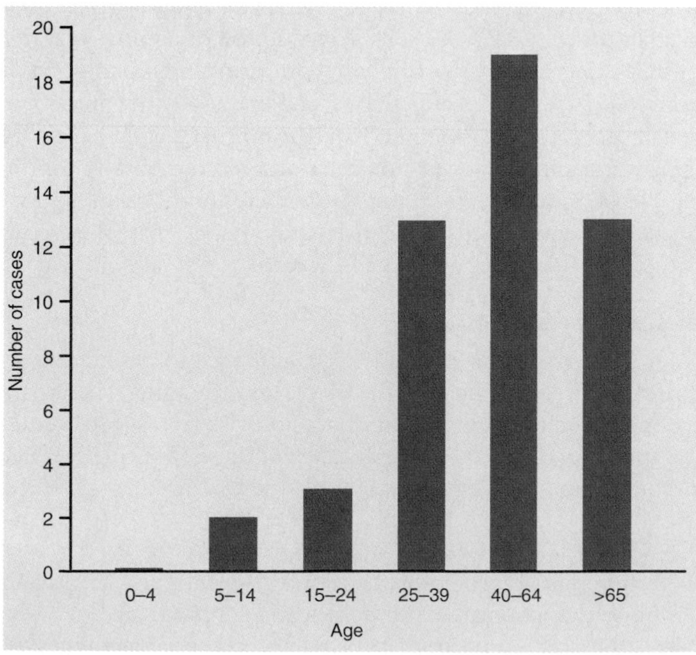

Figure 20.4 Incidence of Tetanus by Age Group

Analytical Studies

Analytical studies are designed to determine which of the potential risk factors identified by the descriptive studies are actually relevant in the spread of the disease.

Cross-Sectional Studies

A **cross-sectional study** surveys a range of people to determine the prevalence of any of a number of characteristics including disease, risk factors associated with disease, or previous exposure to a disease-causing agent. This survey provides a rapid assess-

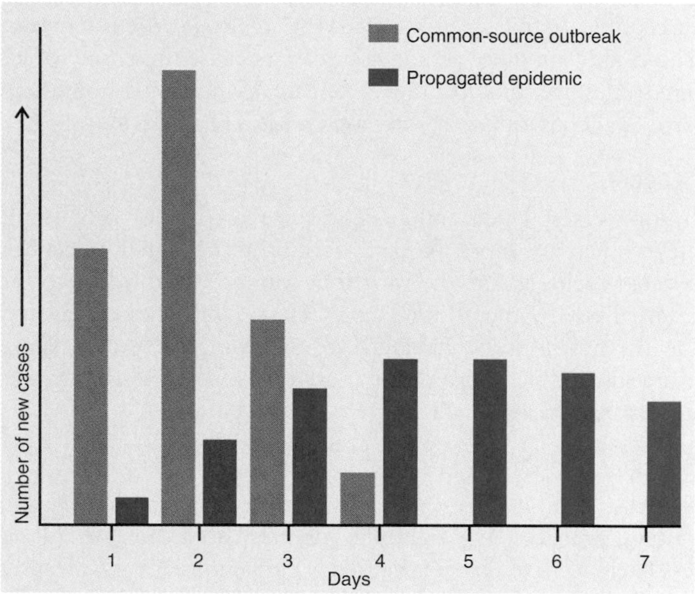

Figure 20.5 Comparison of Propagated Versus Common Source Epidemics The graph depicts the number of new cases that develop over a period of days.

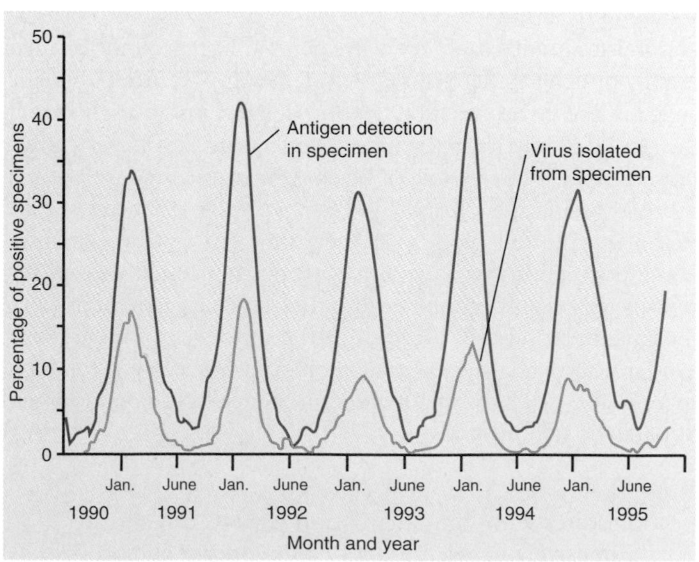

Figure 20.6 Seasonal Occurrence of Respiratory Infections Caused by Respiratory Syncytial Virus

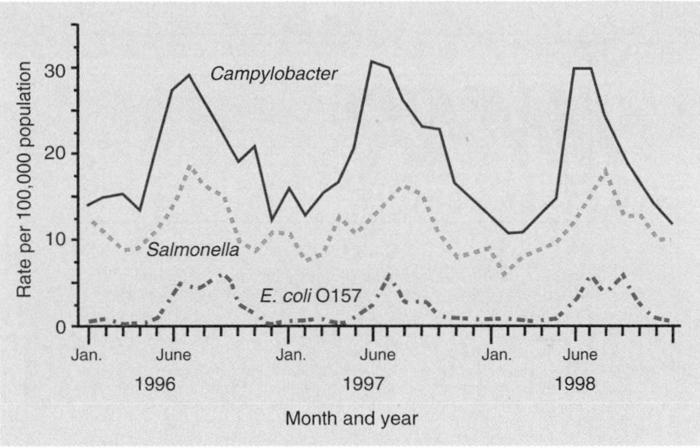

Figure 20.7 **Seasonal Occurrence of Gastrointestinal Diseases**

ment of the features of a population at a given point in time and may suggest associations between risk factors and disease. These associations can then be explored using other types of analytical studies. The cross-sectional survey does not attempt to follow a certain group, nor does it establish cause of disease.

Retrospective Studies

A **retrospective study** is done following a disease outbreak. This type of study compares the actions and events surrounding clinical cases (individuals who developed the disease) against appropriate controls (those who remained healthy). Thus, a case-control study starts by looking at the effect, which is the disease, and attempts to identify the causative chain of events. The activity or event that was common among the cases but not the controls is likely to have been a factor in the development of the disease. It is important to select controls that match the cases with respect to variables not thought to be associated with disease. Matching these variables, which might include factors such as age, sex, and socioeconomic status, ensures that all controls had equal probability of coming in contact with the disease agent. Otherwise, the bias can skew the results.

Prospective Studies

A **prospective study** is one that looks ahead to see if the risk factors identified by the retrospective study predict a tendency to develop the disease. **Cohort groups**, which are study groups that have a known exposure to the risk factor, are selected and then followed in time. The incidence of disease in those who were exposed to the risk factor and those who were not are then compared. By following cohort groups, the study focuses on the hypothesized cause and attempts to determine if it does indeed correlate with the expected effect. This type of study is less prone to bias than the retrospective study because the groups are selected before disease occurs. It is generally more time-consuming and expensive, however, particularly when examining a disease that has a long incubation period. Also, an error in the initial identification of the risk factor renders the entire study useless.

Experimental Studies

An **experimental study** is used to judge the cause and effect relationship of the risk factors or, more commonly, the preventative factors and the development of disease. Experimental studies are done most frequently to assess the value of a particular intervention or treatment, such as antimicrobial drug therapy. The effectiveness of the treatment is compared with one of known value or with a **placebo**. A placebo is a mock drug; it looks and tastes like the experimental drug but has no medicinal value. To assess the value of the experimental drug, a group of patients is divided into two subgroups, one of which will be given the treatment and the other an alternative or a placebo. To avoid bias, an experimental study should ideally be **double-blind**, where neither the physicians nor the patients know who is receiving the actual treatment. Ethical issues sometimes necessitate the use of experimental animals rather than patients in experimental studies.

MICROCHECK 20.7

Descriptive epidemiological studies attempt to identify the potential risk factors that lead to disease. Analytical studies try to determine which factors are actually relevant to disease development. Experimental studies are generally used to evaluate the effectiveness of a treatment or intervention in preventing disease.

- On what three factors does a descriptive study focus?
- What is the value of a double-blind experimental study?
- Why is it important to include a placebo in a scientific study to assess the effectiveness of a drug?

Identifying the Source of an Epidemic

Determining the source of an epidemic is sometimes complicated by the facts that different disease agents may produce the same disease symptoms and a single agent may produce a wide variety of different disease symptoms. For example, more than 100 different virus strains can cause the common cold, whereas poliovirus can produce an illness with symptoms ranging from those of a cold to fatal paralysis. Thus, to determine if potentially related illnesses are indeed connected, it is often necessary to identify the causative agents of diseases. ■ **strains, p. 255**

The sources of epidemics that are due to common normal flora or environmental organisms are particularly difficult to trace because of the widespread prevalence, or **ubiquity**, of the organisms. In these cases, it is sometimes necessary to identify strains of the organisms in order to identify the actual origins of the outbreak. The most sensitive methods for distinguishing different strains utilize newer molecular biology techniques such as analyzing restriction fragment length polymorphisms (RFLPs). Traditional methods, however, such as determining the antibiotic susceptibility pattern, or antibiogram, and bacteriophage typing, in which various strains are tested for their sensitivity to a series of different types of bacteriophage, can also be used. While less reliable, these traditional methods can be faster and require simpler technology. ■ **RFLP, p. 256** ■ **antibiogram, p. 258** ■ **bacteriophage typing, p. 257**

The ability to characterize different strains made it possible to link more than 40 cases of listeriosis that occurred in 10 different states over a 5-month period in 1998. Listeriosis is a foodborne disease that occurs sporadically and can cause a fatal infection, primarily in the immunocompromised and the elderly. Traditional biochemical tests could identify the causative agent, *Listeria monocytogenes*, but could not differentiate between the multitude of different strains. Using molecular methods, scientists determined that the 40-odd cases were caused by the same strain, prompting health officials to conduct a multistate case-control study, which ultimately linked the strain to consumption of certain hot dogs. The strain was then isolated from a package of these hot dogs, leading to the immediate recall of the product.

MICROCHECK 20.8

In some cases it is necessary to distinguish different strains within a species in order to determine the original source of an epidemic.

- Why is it difficult to trace the source of an epidemic caused by a common bacterium?
- Describe two techniques that can be used to differentiate bacterial strains.

Infectious Disease Surveillance

Infectious disease surveillance both nationally and worldwide is one of the most important aspects of disease prevention.

National Disease Surveillance Network

Infectious disease control nationwide depends heavily on a network of people and agencies across the country that monitors disease development. It is partly because of the success of this network that infectious diseases do not claim more lives in the United States.

National Centers for Disease Control and Prevention

The National Centers for Disease Control and Prevention (CDC) is part of the U.S. Department of Health and Human Services and is located in Atlanta, Georgia. It provides support for infectious disease laboratories in the United States and abroad and collects data on diseases of public health importance. Each week, the CDC publishes a booklet, the *Morbidity and Mortality Weekly Report* (*MMWR*), which summarizes the status of a number of diseases **(figure 20.8)**. The *MMWR* is now available online (http://www2.cdc.gov/mmwr/), making it readily accessible to anyone in the world.

The number of new cases of 52 **notifiable diseases** is reported to the CDC by individual states **(table 20.2)**. The diseases that are considered notifiable are determined through collaborative efforts of the CDC and state health departments. Typically they are diseases of relatively high incidence or otherwise a potential danger to public health. Diseases caused by pathogens such as *E. coli* O157:H7 have recently been added to the list, whereas others have been dropped. The data collected by the CDC are published in the *MMWR* along with historical numbers to reflect any trends. Potentially significant case reports, such as the 1981 report of a cluster of opportunistic infections in young gay men that heralded the AIDS epidemic, and the 1995 report on the Ebola outbreak in Zaire, are also

CDC — February 4, 2000 / Vol. 49 / No. 4

MORBIDITY AND MORTALITY WEEKLY REPORT

73 Outbreaks of *Salmonella* Serotype Enteritidis Infection Associated with Eating Raw or Undercooked Shell Eggs — United States, 1996–1998
79 Prevalence of Selected Risk Factors for Chronic Disease and Injury Among American Indians and Alaska Natives — United States, 1995–1998
91 National Child Passenger Safety Week — February 13–19, 2000

Outbreaks of *Salmonella* Serotype Enteritidis Infection Associated with Eating Raw or Undercooked Shell Eggs — United States, 1996–1998

During the 1980s and 1990s, *Salmonella* serotype Enteritidis (SE) emerged as an important cause of human illness in the United States. The rate of SE isolates reported to CDC increased from 0.6 per 100,000 population in 1976 to 3.6 per 100,000 in 1996 (Figure 1). Case-control studies of sporadic infections and outbreak investigations found that this increase was associated with eating raw or undercooked shell eggs (*1*). From 1996 to 1998, the rate of culture-confirmed SE cases reported to CDC declined to 2.2 per 100,000; however, outbreaks of illness caused by SE continue to occur. This report describes four SE outbreaks during 1996–1998 associated with eating raw or undercooked shell eggs and discusses measures that may be contributing to the decline in culture-confirmed SE cases.

Los Angeles County, California

In August 1997, the Los Angeles County Department of Health Services (LACDHS) received reports of gastrointestinal illness in members of a Girl Scout troop and some of their parents. The ill persons had eaten food prepared in a private residence by the scouts. Stool cultures taken from 12 ill persons yielded SE; selected isolates tested were phage type 4.

An investigation by LACDHS found that of 17 persons at the dinner, 13 had gastrointestinal illness consistent with salmonellosis. Cheesecake served at the dinner was associated with illness; all 13 ill persons and two well persons ate the cheesecake (attack rate=87%; relative risk [RR]=undefined; p=0.04). The cheesecake contained raw egg whites and egg yolks that were cooked in a double boiler until slightly thickened. California Department of Health Services and Department of Food and Agriculture investigated the farm that supplied the eggs and found SE contamination. Of 476 environmental cultures taken from manure, feed, and water, 21 (4.4%) yielded SE; all positive cultures were from manure. Nineteen isolates were phage type 4, and two were phage type 7. SE also was isolated from one (0.5%) of 200 pooled egg samples obtained at the farm. On the basis of these findings, the layer flock was depopulated to prevent further SE cases.

U.S. DEPARTMENT OF HEALTH & HUMAN SERVICES

Figure 20.8 ***Morbidity and Mortality Weekly Report*** A publication of the Centers for Disease Control and Prevention.

included in the *MMWR*. This publication is an invaluable aid to physicians, public health agencies, teachers, students, and anyone else studying infectious disease or public health. In fact, many of the epidemiological charts and stories in this textbook are extracted from the *MMWR*.

The CDC also conducts research relating to infectious diseases and can dispatch teams worldwide to assist with identifying and controlling epidemics. In addition, the CDC provides refresher courses that update the knowledge of laboratory and infection control personnel.

Public Health Departments

Each state has a public health laboratory that is involved in infection surveillance and control as well as other health-related activities. Individual states have the authority to mandate which diseases must be reported by physicians to the state laboratory. The prompt response of health authorities in Washington State, which led to the cessation of the 1993 outbreak of *Escherichia coli* O157:H7 caused by contaminated hamburger patties, was partly because Washington then was one of the few states with active surveillance and reporting measures for the organism. The epidemic had actually started in other states but had gone unrecognized.

State laboratories also examine specimens and cultures submitted by physicians, local health departments, hospital laboratories, and others. They also deal with environmental health matters, testing water supplies for potential pathogens, giving

TABLE 20.2 Notifiable Infectious Diseases

Individual states and territories require physicians to report cases of these notifiable diseases. In turn, the number of cases is reported to the CDC, where they are collated and published in the *MMWR*.

1. AIDS
2. Anthrax
3. Botulism
4. Brucellosis
5. Chancroid
6. *Chlamydia trachomatis,* genital infections
7. Cholera
8. Coccidioidomycosis
9. Cryptosporidiosis
10. Diphtheria
11. Encephalitis, California
12. Encephalitis, eastern equine
13. Encephalitis, St. Louis
14. Encephalitis, western equine
15. *Escherichia coli* O157:H7
16. Gonorrhea
17. *Haemophilus influenzae,* invasive disease
18. Hansen's disease (Leprosy)
19. Hantavirus pulmonary syndrome
20. Hemolytic uremic syndrome, post-diarrheal
21. Hepatitis A
22. Hepatitis B
23. Hepatitis, C/non-A, non-B
24. HIV infection, pediatric
25. Legionellosis (Legionnaires' disease)
26. Lyme disease
27. Malaria
28. Measles
29. Meningococcal disease
30. Mumps
31. Pertussis
32. Plague
33. Poliomyelitis
34. Psittacosis
35. Rabies, animal
36. Rabies, human
37. Rocky Mountain spotted fever
38. Rubella
39. Rubella, congenital syndrome
40. Salmonellosis
41. Shigellosis
42. Streptococcal disease, invasive, group A
43. *Streptococcus pneumoniae,* drug-resistant
44. Streptococcal toxic shock syndrome
45. Syphilis, congenital
46. Syphilis
47. Tetanus
48. Toxic shock syndrome
49. Trichinosis
50. Tuberculosis
51. Typhoid fever
52. Yellow fever

advice on laboratory safety and design, and assisting in handling outbreaks of infectious disease. Additional programs enhance the reliability of laboratories in the state and ensure that they meet standards for certification. These laboratories are often tested for their reliability by seeing if they can correctly identify known pathogens.

Other Components of the Public Health Network

The public health network also includes public schools, which report absentee rates, and hospital laboratories, which report on the isolation of pathogens that have epidemiological significance for the community. In conjunction with these local activities, the news media alert the general public of the presence of infectious disease.

Worldwide Disease Surveillance

The World Health Organization (WHO) is an international agency devoted to achieving the highest possible level of health for all peoples. An agency of the United Nations, the WHO has 191 member countries. It has four main functions: (1) to provide worldwide guidance in the field of health; (2) to set global standards for health; (3) to cooperatively strengthen national health programs; (4) to develop and transfer appropriate health technology. To accomplish its goals, the WHO provides education and technical assistance to member countries. One of their most recent successes has been the establishment of truces between various factions in war-torn countries in order to allow vaccination efforts to proceed.

The WHO disseminates information through a series of periodicals and books. For example, the *Weekly Epidemiological Record* reports timely information about epidemics of public health importance, particularly those of global concern.

MICROCHECK 20.9

Across the country, a network of people and agencies including the Centers for Disease Control and Prevention and state and local public health departments monitors disease development. It is partly because of the success of this network that infectious diseases do not claim more lives in the United States. The WHO is devoted to achieving the highest possible level of health for all peoples.

- What is the *MMWR*?
- Give one reason that public health authorities in Washington State were so prompt in their recognition of the 1993 outbreak of *E. coli* O157:H7.
- Explain why we have relatively accurate data on the number of cases of measles that occur in the United States but not on the number of cases of the common cold.

Trends in Disease

The rapid scientific advances made over the past several decades led some people to speculate that the war against infectious disease, particularly those caused by bacteria, had been won. Microorganisms have occupied this planet far longer than have humans, however, evolving to occupy every niche having the potential for life, including the human body. Perhaps it should be no surprise, then, that new and previously unrecognized pathogens are emerging, and that some of those we had previously controlled are now making a comeback.

Reduction and Eradication of Disease

Humans have been enormously successful in developing the means to eliminate or reduce the occurrence of certain diseases through such efforts as improved sanitation, reservoir and vector control, vaccination, and antibiotic treatment. One disease, smallpox, has been globally eradicated, eliminating the fear of a disease that had a 25% mortality rate, and left many of those who survived permanently disfigured (**figure 20.9**). The World Health Organization hopes to eradicate polio by the end of the year 2000 (**figure 20.10**). Dracunculiasis and measles are also targeted for elimination. Whether these efforts will be successful remains to be seen. Political and social upheaval, complacency, and lack of financial support can result in a resurgence of disease unless the pathogens are completely eliminated. ■ polio, p. 679 ■ measles, p. 537

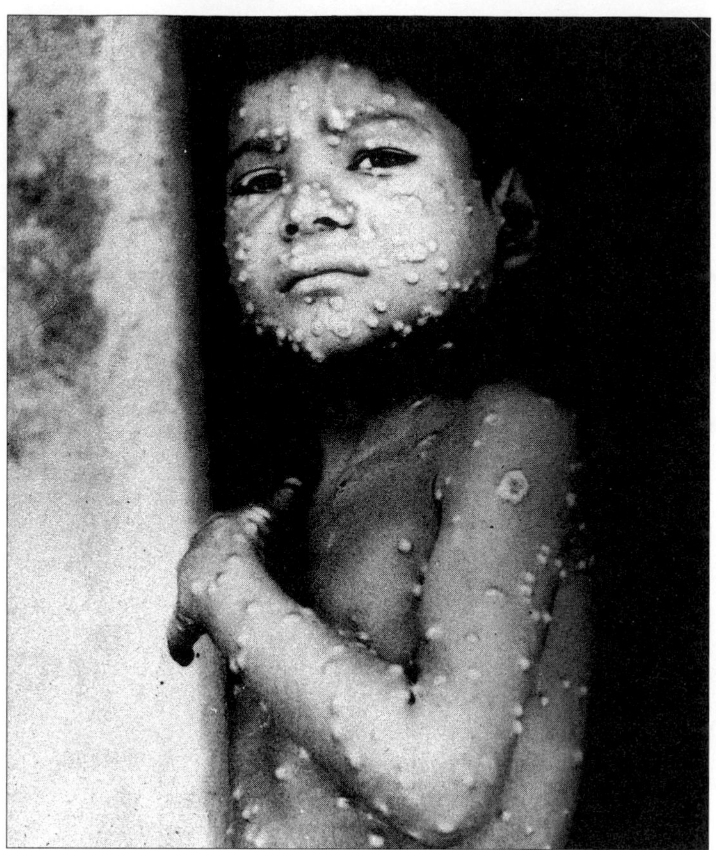

Figure 20.9 A Case of Smallpox, a Disease that has Now Been Eradicated

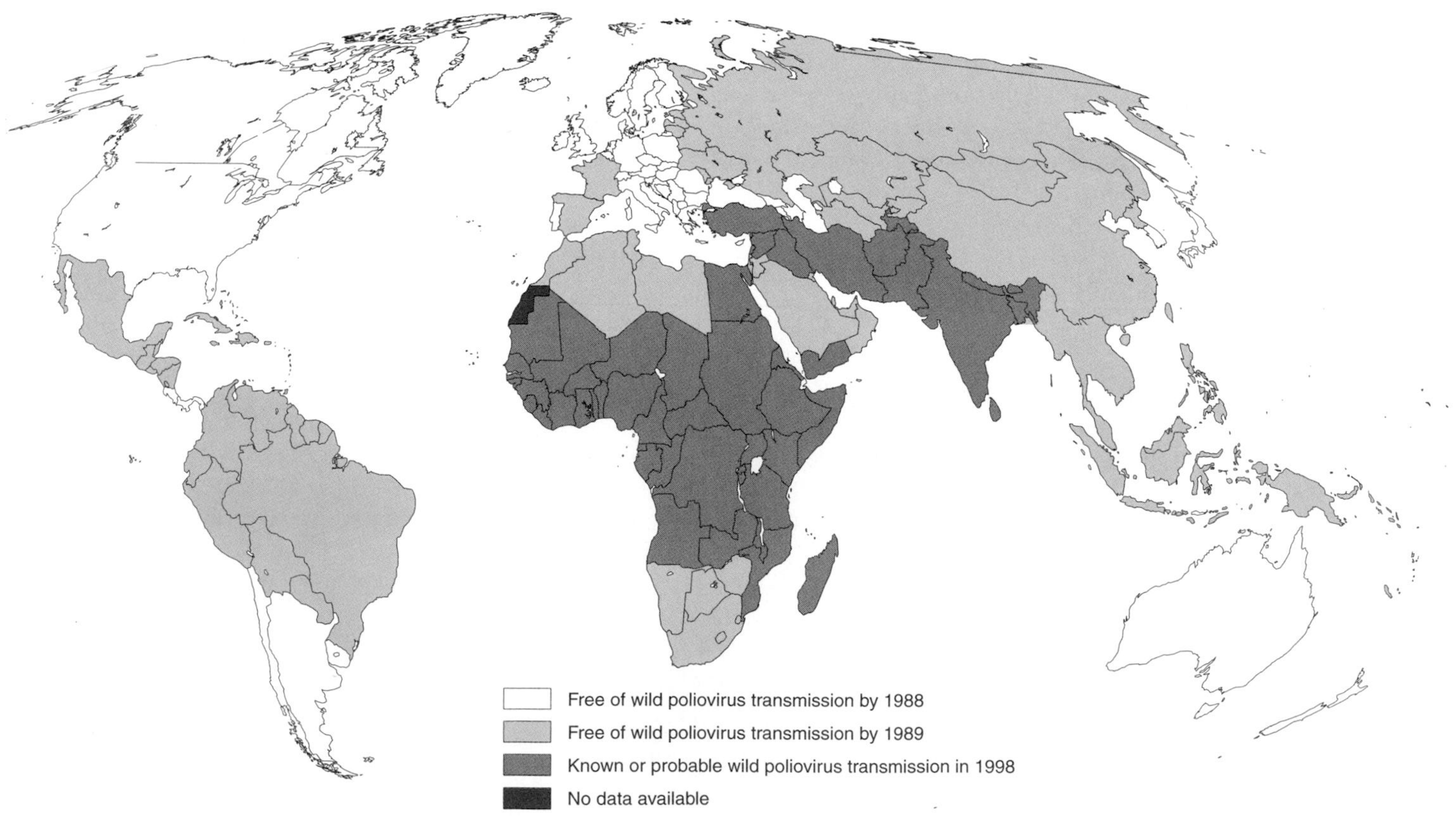

Figure 20.10 Reductions in Wild Poliovirus Transmissions Between 1988 and 1998

In the United States, many diseases that were once common and claimed many lives are now relatively rare. Successful vaccination programs have led to dramatic decreases in the number of deaths caused by *Haemophilus influenzae*, *Corynebacterium diphtheriae*, *Clostridium tetani*, and *Bordetella pertussis.* Meanwhile, recognizing and controlling the source of diseases such as malaria, plague, and cholera have been effective in limiting their spread.

Emerging Diseases

Just as humans have been successful in reducing and eliminating certain diseases, microorganisms are equally adept at taking advantage of new opportunities in which to thrive and multiply. As human lifestyles change due to advancing technologies, increasing populations, and shifting social behaviors, new diseases emerge while those that have been controlled in the past sometimes make a comeback.

Diseases that have increased in incidence in the past two decades are referred to as **emerging diseases**. These include new or newly recognized diseases such as Ebola as well as familiar ones such as malaria that are reemerging after years of decline. **Figure 20.11** shows the emergence of diseases around the globe. Some of the factors that contribute to the emergence and reemergence of diseases include the following:

- **Microbial evolution.** The emergence of some diseases is due to the natural evolution of microorganisms. For example, a new serotype of *Vibrio cholerae*, designated O139, appears to be nearly identical to the strain that most commonly causes cholera epidemics, *V. cholerae* O1, except that it has gained the ability to produce a capsule. The consequence of the new serotype is that even people who have immunity against the earlier strain are susceptible to the new strain. Resistance to the effects of antimicrobial drugs is contributing to the reemergence of many diseases, including malaria. ■ *V. cholerae*, p. 599 ■ malaria, p. 731
- **Complacency and the breakdown of public health infrastructure.** As infectious diseases are controlled and therefore of lessening concern, complacency can develop, paving the way for the resurgence of a disease. The preliminary success of the plan to eliminate tuberculosis in the United States by the year 2000 resulted in less public attention being paid to the disease. News reports, education, and research money were all diverted to more common diseases. Simultaneously, the AIDS crisis developed and funding for some social welfare programs was curtailed, resulting in an increase in the number of people at risk of developing active tuberculosis due to poor health and living conditions. Consequently, tuberculosis reemerged as an increasing threat. Fortunately, increased public health measures, including direct observation of drug therapy compliance, brought the disease back under at least temporary control. As another example, a decreased vigilance in ensuring childhood immunization resulted in the resurgence of measles in the 1980s. ■ tuberculosis, p. 569
- **Changes in human behavior.** Changes in society's norms and behavior can inadvertently create opportunities for microorganisms to spread and flourish. For example, day care centers, where diapered infants mingle, oblivious to sanitation and hygiene, are a relatively new component of American society. For obvious reasons, the centers can be hotbeds of contagious diseases. Many young children have not yet acquired immunity to common communicable diseases. As a consequence, illnesses such as colds and diarrhea are readily transmitted among this susceptible population. This is particularly true with intestinal pathogens such as *Giardia* and *Shigella* that have a low infectious dose, because infants often explore through taste and touch and are thus likely to ingest fecal organisms. ■ *Giardia*, p. 613 ■ *Shigella*, p. 602

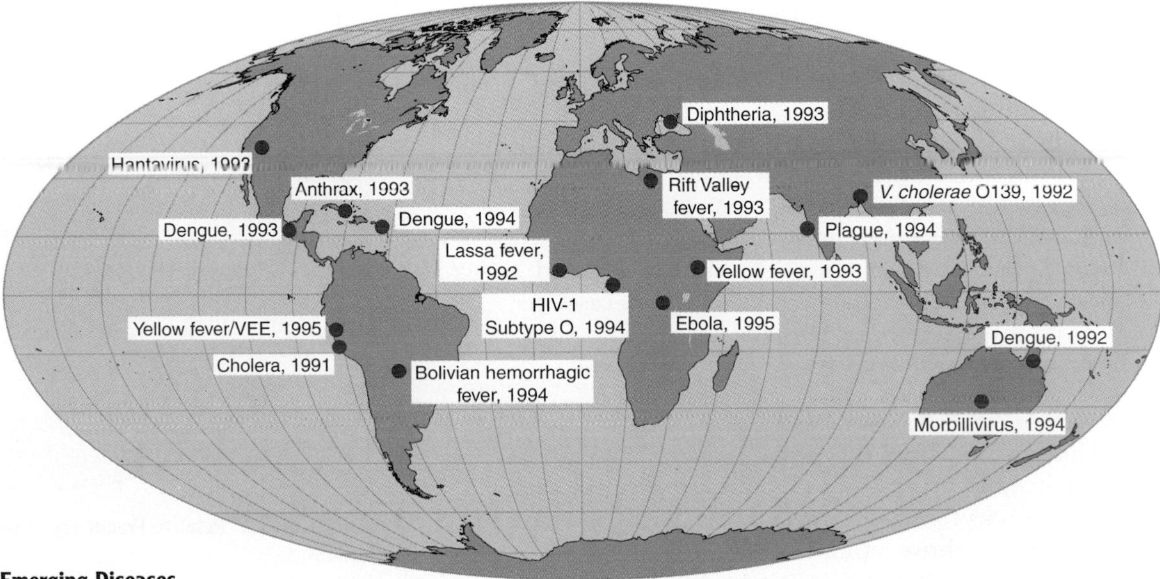

Figure 20.11 World Map of Emerging Diseases

- **Advances in technology.** Technology can make life easier but can inadvertently create new habitats for microorganisms. For example, the advent of contact lenses to correct vision gave microorganisms the opportunity to grow in a new niche, the lenses and storage solutions of users who did not employ proper disinfection techniques. In turn, this resulted in new types of eye infections.
- **Population expansion.** The increase in world population, and the subsequently denser habitats, create situations in which diseases can be more readily transmitted. In areas where the population has expanded outward, people are coming in contact with previously unknown reservoirs of disease such as that of the Ebola virus. The reservoir of Ebola still has not been determined.
- **Development.** Dams, which provide important sources of power necessary for economic development, have inadvertently extended the range of certain diseases. For example, transmission of the disease schistosomiasis relies on the presence of an aquatic snail that serves as a host for the *Schistosoma* parasite. Construction of dams such as the Aswan dam on the Nile River has increased the habitat for the snail, thus extending the distribution of the disease. ■ schistosomiasis, p. 734
- **Mass distribution and importation of food.** Foodborne illness has always existed, but the ease with which we can now transport items worldwide can create new problems. Widespread distribution of foods contaminated with pathogens can result in a similarly broad outbreak of disease. For example, contaminated raspberries grown in Guatemala were linked to a 1996 multistate outbreak of diarrheal disease in the United States, affecting more than 900 people, caused by the intestinal parasite *Cyclospora*. ■ *Cyclospora*, p. 616
- **War and civil unrest.** Wars and civil unrest can disrupt the infrastructure on which disease prevention relies. Refugee camps that crowd people into substandard living quarters lacking toilet facilities and safe drinking water are hotbeds of infectious disease. Epidemics of cholera, dysentery, and other infectious diseases are common in these situations. Unfortunately, war also disrupts disease eradication efforts. ■ cholera, p. 599, ■ dysentery, p. 598
- **Climate changes.** Changes in temperature and rainfall may affect the incidence of certain diseases. For example, warm temperatures favor the reproduction and survival of some arthropods, which in turn can serve as vectors for diseases such as malaria. The heavy rainfall and flooding that resulted in a surge of cholera cases in Africa may have been due to the effects of El Niño. ■ malaria, p. 731 ■ cholera, p. 599

MICROCHECK 20.10

Humans have been successful in reducing and eliminating certain diseases, but microorganisms are equally adept at taking advantage of new opportunities in which to thrive and multiply.

- Explain why diarrheal diseases spread so easily in day care centers.
- Describe two factors that can contribute to the reemergence of an infectious disease.
- What political and societal factors might lead to a decrease in childhood immunizations?

Nosocomial Infections

A hospital can be seen as a high-density population made up of unusually susceptible people where the most antimicrobial-resistant and virulent pathogens can potentially circulate. Considering this, it is not surprising that hospital-acquired infections, or **nosocomial infections**, have been a problem since hospitals began (nosocomial is derived from the Greek word for hospital). Modern medical practices, however, including the extensive use of antimicrobial drugs and invasive therapeutic procedures, have changed the nature of the problem. In the United States alone it is estimated that 5% to 6% of patients admitted to the hospital develop a nosocomial infection, adding over $4.5 billion to the price of health care. Nosocomial infections are so common that they make up at least half of all cases of infectious disease treated in the very hospitals where they were contracted. Many of these are from the patient's own normal flora, but approximately one-third of these infections are potentially preventable. **Figure 20.12** shows the relative frequency of different types of nosocomial infections.

Nosocomial infections may range from very mild to fatal. Sometimes, because of a long incubation period, the

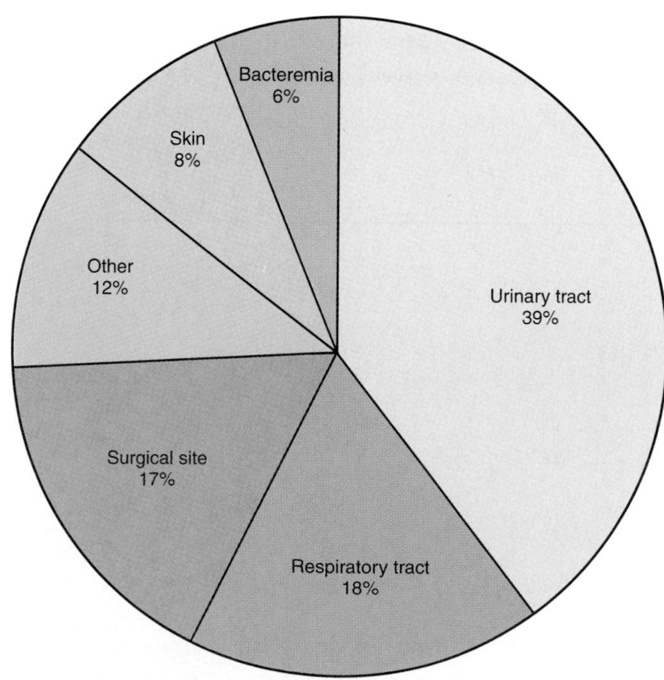

Figure 20.12 Relative Frequency of Different Types of Nosocomial Infections

infection may not be discovered until after the patient has been discharged. Many factors determine which microorganisms or viruses are responsible for these infections. These include the length of time the person is exposed, the manner in which a patient is exposed, the virulence and number of organisms, and the state of the patient's host defenses. The bacteria most commonly implicated in nosocomial infections include the following:

- *Enterococcus* sp. Enterococci are part of the normal intestinal flora. Some strains are resistant to all conventional antimicrobial drugs. Enterococci are a common cause of nosocomial urinary tract infections as well as wound and blood infections. ■ Enterococci, p. 636
- *Escherichia coli* and other members of the *Enterobacteriaceae*. *Escherichia coli* is a part of the normal intestinal flora. It is the most common cause of nosocomial urinary tract infections. ■ *E. coli*, p. 636 ■ *Enterobacteriaceae*, p. 282
- *Pseudomonas* species. These bacteria grow readily in many moist, nutrient-poor environments such as the water in the humidifier of a mechanical ventilator. *Pseudomonas* species are resistant to many disinfectants and antimicrobial drugs. They are a common cause of hospital-acquired pneumonia, and infections of the urinary tract and burn wounds. ■ *Pseudomonas*, p. 697
- *Staphylococcus aureus*. Many people including health care personnel are carriers of this organism. Because it survives for prolonged periods in the environment, it is readily transmissable on fomites. It is a common cause of nosocomial pneumonia and surgical site infections. Hospital strains are often resistant to a variety of antimicrobial drugs. ■ *Staphylococcus aureus*, p. 694
- *Staphylococcus* sp. other than *S. aureus*. These normal skin flora can colonize the tips of intravenous catheters, small plastic tubes inserted into the veins. The resulting biofilms continuously seed organisms into the bloodstream and increase the likelihood of a systemic infection. ■ biofilm, p. 108

Reservoirs of Infectious Agents in Hospitals

The organisms that cause nosocomial infections may originate from other patients, the hospital environment, medical personnel, or from the patient's own normal flora. Because of the widespread use of antimicrobial drugs in hospitals, many organisms that cause nosocomial infections are resistant to these medications.

Other Patients

Because of the very nature of hospitals, infectious agents are always present. Patients are often hospitalized because they have a severe infectious disease. The pathogens that these patients harbor can be discharged into the environment via skin cells, respiratory droplets, and other body secretions and excretions. Scrupulous cleaning and the use of disinfectants minimize the spread of these pathogens.

Hospital Environment

Some Gram-negative rods, particularly the common opportunistic pathogen *Pseudomonas aeruginosa*, can thrive in certain hospital environments such as sinks, respirators, and toilets. Not only is *P. aeruginosa* resistant to the effects of many disinfectants and antimicrobial drugs, it requires few nutrients, enabling it to multiply in environments containing little other than water. Many nosocomial infections have been traced to soaps, disinfectants, and other aqueous solutions that have become contaminated with the organism.

Health Care Workers

Outbreaks of nosocomial infections are sometimes traced to infected health care personnel. Clearly, those who report to work with even a mild case of influenza can expose patients to an infectious agent that has serious or fatal consequences to those with impaired health. A more troublesome source of infection is a health care worker who is a carrier of a pathogen such as *Staphylococcus aureus* or *Streptococcus pyogenes*. These personnel often do not recognize they pose a risk to patients until they have been implicated in an outbreak. Carriers who are members of a surgical team pose a particular threat to patients, because inoculation of a pathogen directly into a surgical site can result in a systemic infection.

Patient's Own Normal Flora

Many hospital-acquired infections originate from the patient's own flora. Nearly any invasive procedure can transmit microorganisms that are part of the normal flora to otherwise sterile body sites. When intravenous fluids are administered, for example, *Staphylococcus epidermidis*, a common member of the normal skin flora, can potentially gain access to the bloodstream. While the immune system can usually readily eliminate these normally benign organisms, the underlying illness of many hospitalized patients compromises their immunity and they can develop a bloodstream infection. Patients who undergo intestinal surgery are prone to surgical site infection by their normal bowel flora. Similarly, patients who are on certain medications or have impaired cough reflexes can inadvertently inhale their normal oral flora, resulting in hospital-acquired pneumonia.

Severely immunocompromised patients, such as people who have undergone cancer chemotherapy or are on immunosuppressive drugs, are prone to activation of latent infections that their immune system was previously able to control. For example, latent infections of *Toxoplasma gondii*, a protozoan parasite commonly acquired during childhood, can become activated and cause a life-threatening disease.

Transmission of Infectious Agents in Hospitals

Diagnostic and therapeutic procedures during hospitalization can potentially transmit infectious agents to patients. This is particularly true in intensive care units (ICUs), where patients generally have indwelling catheters used to deliver intravenous fluids or monitor the patient's condition (**figure 20.13**).

Medical Devices

Nosocomial infections most often result from medical devices that breach the first-line barriers of the normal host defense. For

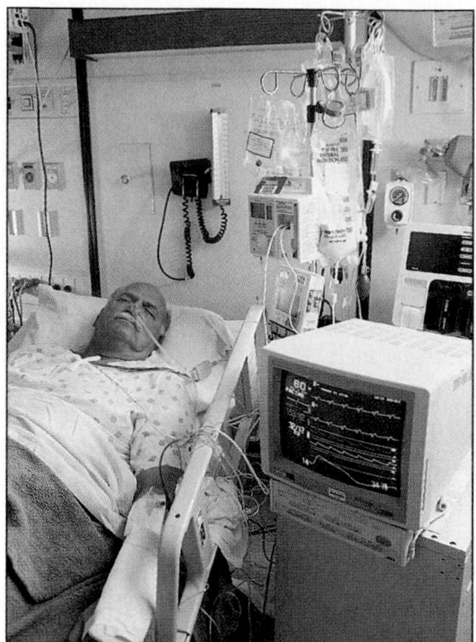

Figure 20.13 Patient in an Intensive Care Unit

example, catheterization of the urinary tract can readily introduce microorganisms into the normally sterile bladder. Because urine is an excellent growth medium, the urinary tract often becomes infected. Urinary tract infections are the most common type of nosocomial infection (see figure 20.12).

Just as urinary catheters can introduce bacteria into the bladder, intravenous (IV) catheters can introduce microorganisms into the bloodstream. This can happen when normal skin flora colonize the tip of an indwelling catheter or when environmental organisms contaminate IV fluids or the lines that deliver them. Even normally benign skin flora can cause life-threatening bacteremia when they gain access to the bloodstream.

Mechanical respirators that assist a patient's breathing by pumping air directly into the trachea can potentially deliver microorganisms to the lungs. This is particularly a problem if a nutritionally versatile organism such as *Pseudomonas* can gain access to water droplets in the machine, allowing it to multiply.

Inadequately sterilized instruments that are used in invasive procedures such as surgery or biopsy can also transmit infectious agents. Endoscopes and other heat-sensitive instruments are often treated with chemical sterilants to render them microbe-free. Improper use of these chemicals, however, can result in the survival of some organisms. ■ **sterilants, p. 121**

Health Care Personnel

Health care personnel must be extremely vigilant to avoid transmitting infectious disease agents, particularly from patient to patient. What Ignaz Semmelweis found to be true in the 1800s is equally true today—handwashing between contact with individual patients helps prevent the spread of disease. Unfortunately, this relatively simple procedure is too often overlooked.

Health care personnel should routinely wash their hands after touching one patient before going on to the next. A more thorough hand scrubbing, requiring 10 minutes and using a strong disinfectant, should be performed by nurses, physicians, and other personnel before participating in an operation, or when working in a nursery, or intensive care or isolation unit. Gloves are also worn whenever there is contact with blood, mucous membranes, broken skin, or body fluids.

Airborne

Most hospitals are designed to minimize the airborne spread of microorganisms. For example, the airflow is usually regulated so that it is supplied to the operating room under slight pressure, thereby preventing contaminated air in the corridors from flowing into the room. Floors are washed with a damp mop or floor washer rather than swept in order to avoid resuspending microbes into the air.

To exclude airborne microorganisms and viruses from rooms in which exquisitely susceptible patients reside, such as those who have recently undergone a bone marrow transplant, high-efficiency particulate filters (HEPA) are employed. These filter out most airborne particles, including microorganisms.

Preventing Nosocomial Infections

The most important steps in preventing nosocomial infections are to first recognize their occurrence and then establish policies to prevent their development. To do this, nearly every hospital has an Infection Control Committee, composed of representatives of the various professionals in the hospital, such as nurses, physicians, dietitians, housekeeping staff, and microbiology laboratory personnel. On this committee, and sometimes chairing it, is often a **hospital epidemiologist**, a professional specially trained in hospital infection control. Hospitals may also employ an **infection control practitioner (ICP)**, whose role is to perform active surveillance of the types and numbers of infections that arise in the hospital. The Infection Control Committee, in conjunction with the ICP, drafts and implements preventative policies following the guidelines suggested by the Standard Precautions and the Transmission-Based Precautions (see **Perspective 20.1**).

The CDC also takes an active role in preventing nosocomial infections and in the early 1990s established the Hospital Infection Control Program Advisory Committee (HICPAC). The role of this national committee is to provide advice to hospitals and recommend guidelines for surveillance, prevention, and control of nosocomial infections.

MICROCHECK 20.11

A hospital is a high-density population of unusually susceptible people, into which the most virulent and antibiotic-resistant microbial pathogens are continually introduced. Nosocomial infections may originate from other patients, the hospital environment, medical personnel, or the patient's own normal flora. Diagnostic and therapeutic procedures can potentially transmit infectious agents. The most important steps in preventing nosocomial infections are to first recognize their occurrence and then establish policies to prevent their development.

- Explain why an intravenous catheter can pose a risk to a patient.
- Describe two ways in which infectious agents can be transmitted to a patient.
- The rate of nosocomial infections is often relatively high in emergency room settings. Explain why this might be so.

Perspective 20.1 Protecting Patients and Health Care Personnel

One of the biggest challenges for a hospital has always been the prevention of spread of disease within that confined setting. A century ago, patients with infectious diseases were segregated in separate hospitals; those with similar diseases were sometimes housed in clusters on the same floor. In 1910, a cubicle system of isolation was introduced in which patients were placed in multiple-bed wards. Aseptic nursing procedures were aimed at preventing the transmission of infectious agents to other patients and personnel. These scrupulous measures were so successful that general hospitals were able to incorporate infectious disease patients, ultimately resulting in the closure of many infectious disease hospitals beginning in the 1950s. To assist general hospitals with isolation procedures, in 1970 the CDC began publishing a manual that recommended a category system of seven isolation procedures for patients based on their diagnosis. These procedures included Strict Isolation, Respiratory Precautions, Protective Isolation, Enteric Precautions, Wound and Skin Precautions, Discharge Precautions, and Blood Precautions. Over the years, some of these categories were changed or deleted.

In the early 1980s it became increasingly apparent that health care workers were acquiring hepatitis B and, later, HIV from contact with the blood or other body fluids of patients, including those who were not suspected of having blood-borne disease. Existing guidelines were primarily aimed at preventing patient-to-patient transmission of disease, rather than patient-to-personnel. In response, the CDC recommended an additional set of guidelines, the **Universal (Blood and Body Fluid) Precautions**, to be followed when working with any patient, regardless of the diagnosis. These defined the situations in which gloves, gowns, masks, and eye protection were required to prevent contact with blood. Many hospitals then broadened this concept, requiring the use of gloves to isolate all moist and potentially infectious body substances. This approach, **Body Substance Isolation**, made the traditional diagnosis-dependent isolation procedures largely unnecessary.

The advent of Universal Precautions and Body Substance Isolation, in addition to the previous diagnosis-dependent isolation procedures, resulted in a such a mix of recommendations that it generated a great deal of confusion. No existing single set of guidelines was sufficient, and it was not clear which one should be used when. In response, the CDC and the Hospital Infection Control Program Advisory Committee established a new set of guidelines in 1996 that incorporates the strength of each of the alternatives. These new guidelines have two tiers of isolation procedures. The fundamental measures are the **Standard Precautions**, designed for the care of all patients in all hospitals. The **Transmission-Based Precautions** are supplementary measures to be used in addition to the Standard Precautions if a patient is, or might be, infected with highly transmissible or epidemiologically important pathogen. These are separated into three sets—Airborne Precautions, Droplet Precautions, and Contact Precautions—which are used singly or in combination as appropriate.

The Standard Precautions can be summarized as follows:

Handwashing. Wash hands with soap and water after touching blood, body fluids, secretions, excretions, and contaminated items, whether or not gloves are worn.

Gloves. Clean disposable gloves are worn whenever there is possible contact with blood, body fluids, secretions, excretions, mucous membranes, skin wounds, and contaminated items. Change gloves between procedures on the same patient after contact with material that may contain high concentrations of microbes. Remove gloves immediately after use and wash hands thoroughly. Gloves should always be removed in a reverse manner so as not to touch the contaminated surface.

Mask, eye protection, face shield, gown. When doing a procedure that is likely to generate splashes or sprays of blood, body fluids, secretions, and excretions, wear a mask and eye protection or a face shield to protect the mucous membranes of the eyes, nose, and mouth. Wear a clean gown. Remove the soiled gown as soon as possible and wash hands.

Patient-care equipment. Handle used patient-care equipment soiled with blood, body fluids, secretions, and excretions in a manner that prevents skin and mucous membrane exposure, and contamination and transfer to other patients and environments. Reusable equipment must be cleaned and processed appropriately. Single-use items must be disposed of properly.

Environmental control. Use adequate procedures for the routine care, cleaning, and disinfection of environmental surfaces, bed rails, and other frequently touched surfaces.

Linen. Soiled, reusable items are placed in protective bags to prevent leaking and contamination.

Occupational health and blood-borne pathogens. All needles and sharp objects are discarded in a rigid, puncture-proof container without touching them or replacing needle caps.

Patient placement. Private rooms are used for any patient who contaminates the environment or might soil other people and their surroundings.

From: http://www.cdc.gov/ncidod/hip/isolat.htm
The Public Health Services, U.S. Department of Health and Human Services, Centers for Disease Control and Prevention. Garner, J. Hospital Infection Control Practices and Advisory Committee. Guidelines for Isolation Precautions in Hospitals. *Infection Control and Hospital Epidemiology* 1996; 17: 53–80, and *American Journal of Infection Control* 1996; 24: 24–52.

FUTURE CHALLENGES

Preparing for the Threat of Bioterrorism

Today, an unfortunate challenge in epidemiology is to prepare for the potential of biological warfare. Even as we work to control, and seek to eradicate, some diseases, we must be aware that these same microbes can be used as a potent biological weapon. Experts in the field of bioterrorism list four diseases—anthrax, botulism, plague, and smallpox—as the greatest threats.

While it is hoped that a biological weapons attack will never actually occur, it is crucial to be prepared for the possibility. Prompt recognition of such an event, followed by rapid and appropriate isolation and treatment procedures, can help to minimize the consequences. Already the CDC, in cooperation with the Association for Professionals in Infection Control and Epidemiology (APIC), has prepared a bioterrorism readiness plan to be used as a template by health care facilities. Many of the recommendations are based on the Standard Precautions already employed by hospitals to prevent the spread of infectious agents (see Perspective 20.1). Some of the most relevant characteristics of the diseases that are the most likely candidates for biological weapons include the following:

- ***Anthrax.*** *The most severe outcome results from the inhalation of the airborne endospores of* Bacillus anthracis, *which can rapidly result in a fatal pneumonia. Anthrax can be prevented by vaccination, but that option is not widely available. Post-exposure prophylaxis with antimicrobial medications is possible, but this requires prompt recognition of exposure. Fortunately, person-to-person transmission of anthrax is not likely.*
- ***Botulism.*** *Botulism is caused by the ingestion of botulinum toxin, produced by* Clostridium botulinum. *Aerosolized toxin could also be used as a weapon. Botulism can be prevented by vaccination, but that option is not widely available. An antitoxin is also available. Botulism is not contagious.*
- ***Plague.*** *Pneumonic plague, caused by inhalation of* Yersinia pestis, *is the most likely form of plague to result from a biological weapon. Although no effective vaccine is available, post-exposure prophylaxis with antimicrobial medications is possible. Special isolation precautions must be used for patients who have pneumonic plague because the disease is easily transmitted by respiratory droplets.*

- ***Smallpox.*** *Although a vaccine is available, routine immunization was stopped over 20 years ago because the natural disease has been eradicated. As is the case with nearly all infections caused by viruses, effective drug therapy is not available. Special isolation precautions must be used for smallpox patients because the virus may be acquired through droplet, airborne, or contact transmission.*

SUMMARY

Principles of Epidemiology

1. Epidemiologists study the frequency and distribution of disease in order to identify its cause, source, and route of transmission.

Rate of Disease in a Population

1. Epidemiologists are concerned not simply with the number of people affected by a disease, but more importantly, with the **rate** of disease.
2. Diseases that are constantly present in a population are **endemic**; an unusually large number of cases in a population constitutes an **epidemic.** (Figure 20.2)

Reservoirs of Infection

1. Preventing susceptible people from coming in contact with a **reservoir** can prevent infectious disease.

Human Reservoirs

1. Infected humans are the most significant reservoir of the majority of human diseases.
2. People who have asymptomatic infections or are colonized with a pathogen are **carriers** of the infectious agent.

Nonhuman Animal Reservoirs

1. **Zoonotic** diseases are those such as plague and rabies that can be transmitted to humans but exist primarily in other animals.

Environmental Reservoirs

1. Pathogens that have environmental reservoirs are probably impossible to eliminate.

Portals of Exit

1. To spread, infectious microorganisms must exit one host to be transmitted to another.
2. Pathogens may be shed in feces, in respiratory droplets, on skin cells, in genital secretions, and in urine.

Transmission

1. Handwashing is a key control measure in preventing diseases that are spread through direct or indirect contact, as well as those that spread via contaminated food.

Contact

1. **Direct contact** occurs when one person physically touches another. Diseases caused by pathogens that have a low infectious dose can be transmitted through direct contact. Pathogens that cannot survive for extended periods in the environment must generally be transmitted through direct contact.
2. **Indirect contact** involves transfer of pathogens via **fomites**.
3. **Droplet transmission** of respiratory pathogens is considered direct contact because of the close proximity involved.

Food and Water

1. Foodborne pathogens can originate from the animal reservoir or from contamination during food preparation.
2. Waterborne pathogens often originate from sewage contamination.

Air

1. **Droplet nuclei**, dead skin cells, household dust, and soil may carry airborne respiratory pathogens.
2. Airborne transmission of pathogens is the most difficult to control.

Vectors

1. **Mechanical vectors** carry the microbe from one place to another.
2. **Biological vectors** are a required part of the life cycle of a parasite.
3. Prevention of vector-borne disease relies on mosquito, tick, and insect control.

Portal of Entry

1. The portal of entry of a pathogen can affect the outcome of disease.

Factors that Influence the Epidemiology of Disease

The Dose

1. The probability of infection and disease is generally lower if an individual is exposed to small numbers of pathogens.

The Incubation Period

1. Diseases with a long incubation period can spread extensively before the first cases appear.

Population Characteristics

1. A disease is unlikely to spread very widely in a population in which 90% of the people are immune to the disease agent.
2. Malnutrition, overcrowding, and fatigue increase the susceptibility of people to infectious diseases.
3. The very young and the elderly are generally more susceptible to infectious agents.
4. Natural immunity can vary with genetic background, but it is difficult to determine the relative importance of genetic, cultural, and environmental factors.

Epidemiological Studies

Descriptive Studies

1. **Descriptive studies** attempt to identify potential risk factors that correlate with the development of disease by creating a profile of the persons who became ill.
2. Determining the geographical location may give clues about potential reservoirs, vectors, and geographical boundaries that may affect disease transmission.
3. Determining the time that the illness occurred helps distinguish a **common-source epidemic** from a **propagated epidemic.** (Figure 20.5)

Analytical Studies

1. **Analytical studies** try to determine which risk factors are actually relevant to disease development.
2. A **retrospective study** compares the activities of **cases** with **controls** to determine the cause of the epidemic.
3. A **prospective study** looks ahead, comparing **cohort groups**, to determine if the identified risk factors predict a tendency to develop disease.

Experimental Studies

1. **Experimental studies** are generally used to evaluate the effectiveness of a treatment or intervention in preventing disease; they examine the relationship between risk factors or preventative factors and the development of disease.

Indentifying the Source of an Epidemic

1. In some cases, it is necessary to distinguish different strains of an organism in order to determine the original source of an epidemic.

Infectious Disease Surveillance

National Disease Surveillance Network

1. The Centers for Disease Control and Prevention collects data on diseases of public health importance and summarizes their status in the ***Morbidity and Mortality Weekly Report*** (*MMWR*); other activities of the CDC include research, assistance in controlling epidemics, and support for infectious disease laboratories. (Figure 20.8)
2. State public health departments are involved in infection surveillance and control; they examine specimens submitted by physicians, local health departments, hospital laboratories, and others.

Worldwide Disease Surveillance

1. The World Health Organization (WHO) is an international agency devoted to achieving the highest possible level of health for all peoples.

Trends in Disease

Reduction and Eradication of Disease

1. Smallpox has been eradicated.
2. The World Health Organization hopes to soon eliminate other diseases including dracunculiasis, polio, and measles. (Figure 20.10)

Emerging Diseases

1. Emerging diseases include those that are new or newly recognized and familiar ones that are reemerging after years of decline. (Figure 20.11)
2. Factors that contribute to the emergence and reemergence of diseases include microbial evolution, the breakdown of public health infrastructure, changes in human behavior, advances in technology, population expansion, economic development, mass distribution and importation of food, war, and climate changes.

Nosocomial Infections

1. A hospital can be seen as a high-density population made up of unusually susceptible people, into which the most virulent and antibiotic-resistant microbial pathogens are continually introduced.

Reservoirs of Infectious Agents in Hospitals

1. The organisms that cause nosocomial infections may originate from other patients, the hospital environment, medical personnel, or the patient's own flora.

Transmission of Infectious Agents in Hospitals

1. Diagnostic and therapeutic procedures during hospitalization can potentially transmit infectious agents to patients.
2. Nosocomial infections most often result from medical devices that breach the first-line barriers of the normal host defense.
3. Health care personnel should routinely wash their hands after touching one patient before going on to the next in order to prevent transmission of disease-causing organisms.

Preventing Nosocomial Infections

1. The most important steps in preventing nosocomial infections are to first recognize their occurrence and then establish policies to prevent both their development and spread.

REVIEW QUESTIONS

Short Answer

1. Explain the difference between incidence and prevalence.
2. What is the epidemiological significance of people who have an asymptomatic infection?
3. Explain why zoonotic diseases are often severe when transmitted to humans.
4. Name the most important control measure for preventing person-to-person transmission of a disease.
5. Explain why *Shigella* and *Giardia* species are readily transmitted in day care centers.
6. Explain how smallpox was eradicated.
7. Describe three factors that contribute to the emergence of disease.
8. Draw a representative graph (time versus number of people ill) depicting a propagated epidemic and a common-source epidemic.

9. What are three important factors that a descriptive epidemiological study attempts to determine?
10. Describe the difference between a retrospective (case-control) study and a prospective (cohort) study.

Multiple Choice

1. Which of the following is an example of a fomite?
   A. Table
   B. Flea
   C. *Staphylococcus aureus* carrier
   D. Water
   E. Air
2. Which of the following would be the easiest to eradicate?
   A. A pathogen that is common in wild animals but sometimes infects humans
   B. A disease that occurs exclusively in humans, always resulting in obvious symptoms
   C. A mild disease of humans that often results in no obvious symptoms
   D. A pathogen that is found in marine sediments
   E. A pathogen that readily infects both wild animals and humans
3. Which of the following methods of disease transmission is the most difficult to control?
   A. Airborne
   B. Foodborne
   C. Waterborne
   D. Vector-borne
   E. Direct person to person
4. Which of the following statements is *false*?
   A. A botulism epidemic that results from improperly canned green beans is an example of a common-source outbreak.
   B. Bubonic plague has a higher mortality rate than pneumonic plague.
   C. Congenital syphilis is an example of a disease acquired through vertical transmission.
   D. Plague is endemic in the prairie dog population in parts of the United States.
   E. The first case in an outbreak is called the index case.
5. Which of the following statements is *false*?
   A. A disease with a long incubation period might spread extensively before an epidemic is recognized.
   B. A person who is exposed to a low dose of a pathogen might not develop disease.
   C. The young and the aged are more likely to develop certain diseases.
   D. Malnourished populations are more likely to develop certain diseases.
   E. Herd immunity occurs when a population does not engage in a given behavior, such as eating raw fish, that would otherwise increase their risk of disease.
6. The purpose of an analytical study is to…
   A. identify the person, place, and time of an outbreak.
   B. identify risk factors that result in high frequencies of disease.
   C. assess the effectiveness of preventative measures.
   D. determine the effectiveness of a placebo.
   E. None of the above
7. Which of the following causes of emerging diseases is thought to be a new pathogen?
   A. *Giardia*
   B. *Vibrio cholerae* O139
   C. *Mycobacterium tuberculosis*
   D. *Shigella dysenteriae*
   E. *Schistosoma*
8. All of the following are thought to contribute to the emergence of disease, *except*…
   A. advances in technology.
   B. breakdown of public health infrastructure.
   C. construction of dams.
   D. mass distribution and importation of food.
   E. widespread vaccination programs.
9. Which of the following common causes of nosocomial infections is an environmental organism that grows readily in nutrient-poor solutions?
   A. *Enterococcus*
   B. *Escherichia coli*
   C. *Pseudomonas aeruginosa*
   D. *Staphylococcus aureus*
10. What is the most common type of nosocomial infection?
    A. Bloodstream infection
    B. Gastrointestinal infection
    C. Pneumonia
    D. Surgical wound infection
    E. Urinary tract infection

Applications

1. The television news station reported a story about a potentially fatal epidemic disease occurring in a small Laotian village. An epidemiologist from the CDC was interviewed to discuss the disease and was very distressed that it was not being contained. Why did the epidemiologist feel the disease was a concern for people in North America?
2. An international research team was gathered to discuss how funding should be spent to eliminate human infectious disease. There is only enough for this project to eliminate one disease. How would the scientists go about choosing the next disease to be eliminated from the planet?

Critical Thinking

1. *Yersinia pestis* and hantavirus are both found in wild rodents in the southwestern United States. What is the risk of trying to stop a hantavirus epidemic by destroying rodents in that region?
2. A student disagreed with the presentation of the examples in figure 20.5. She claimed that the number of cases from a common-source outbreak could remain high over a much longer period of time in some cases and not decrease to zero. Is the student's claim reasonable? Why or why not?

CHAPTER 21

Antimicrobial Medications

Paul Ehrlich (1854–1915), a German physician and bacteriologist, was born into a wealthy family. As the only son after many daughters, family and servants indulged his interests, even though his large collection of frogs and snakes occasionally entered the laundry room. As an adult, he was rarely without a good cigar and habitually scribbled notes on his shirt cuffs. After receiving a degree in medicine in 1878, he became intrigued with the way various types of body cells differ in their ability to take up dyes and other substances. When he observed that certain dyes stain bacterial cells but not animal cells, indicating that the two cell types are somehow fundamentally different, it occurred to him that it might be possible to find a dye or chemical that selectively harms bacteria without affecting human cells.

Ehrlich began a systematic search attempting to find a "magic bullet," a term he used to describe a drug that would kill a microbial pathogen without harming the human host. He began looking for a chemical that would cure the sexually transmitted disease syphilis, which is caused by the bacterium Treponema pallidum. *Much of the mental illness during this time resulted from tertiary syphilis, a late stage of the disease. Ehrlich knew that an arsenic compound had shown some success in treating a protozoan disease of animals, and so he and his colleagues began tediously synthesizing hundreds of different arsenic compounds in search for a cure for syphilis. In 1910, the 606th compound tested, arsphenamine, proved to be highly effective in treating the disease in laboratory animals. Although the drug itself was potentially lethal for patients, it did cure infections that were previously considered hopeless. The drug was called Salvarsan, a name derived from the words salvation and arsenic. The use of Salvarsan to cure syphilis proved that chemicals could selectively kill pathogens without permanently harming the human host.*

—A Glimpse of History

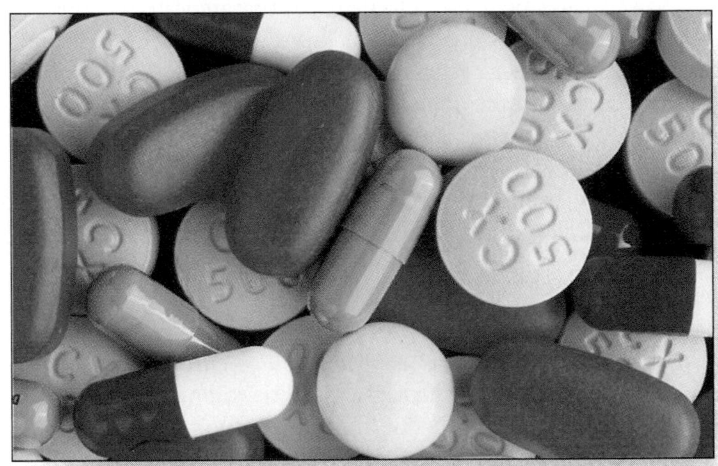

THINK BACK TO THE LAST TIME THAT YOU WERE prescribed an antimicrobial medication. Could you have recovered from the infection without the drug? The prognosis for people with common diseases such as bacterial pneumonia and severe staphylococcal infection was grim before the discovery and widespread availability of penicillin in the 1940s. Physicians were able to identify the cause of the disease, but they were generally unable to recommend treatments other than bed rest. Today, however, antimicrobials are routinely prescribed, and the simple cure they provide for so many infectious diseases is often taken for granted. Unfortunately, the misuse of these life-saving medications, coupled with bacteria's amazing ability to adapt, has led to an increase in the number of drug-resistant organisms. Some people even speculate that we are in danger of seeing an end to the era of antimicrobial medications. In response, scientists are scrambling to develop new drugs.

History and Development of Antimicrobial Drugs

To appreciate the unique antibiotic era in which we now live, it is helpful to understand the history and development of these life-saving remedies.

Discovery of Antimicrobial Drugs

The development of Salvarsan by Paul Ehrlich was the first documented example of a chemical used successfully as an antimicrobial medication. The next breakthrough in the developing science of antimicrobial chemotherapy came almost 25 years later. In 1932, the German chemist Gerhard Domagk, using the same dogged persistence demonstrated by Erlich, discovered that a red dye called Prontosil was dramatically effective in treating streptococcal infections in animals. Surprisingly, Prontosil had no effect on streptococci growing in test tubes. It was later discovered that enzymes in the blood of the animal split the Prontosil molecule, producing a smaller molecule called **sulfanilamide**; this breakdown product acted against the infecting streptococci. Thus, the discovery of sulfanilamide, the first **sulfa drug**, was based on luck as well as scientific effort. If Prontosil had only been screened against bacteria in test tubes and not given to infected animals, its effectiveness might never have been discovered.

Salvarsan and Prontosil were the first documented examples of chemicals used successfully as antimicrobial medications. Any chemical that is used to treat a disease is called a **chemotherapeutic agent**. One used to treat microbial infections can also be called an **antimicrobial drug** or, more simply, an **antimicrobial**. Antimicrobials can also be referred to by terms that indicate the range of microorganisms they are used to kill or inhibit. For example, one that is used to treat bacterial infections is called an **antibacterial drug**. These and other definitions are summarized in **table 21.1**.

Discovery of Antibiotics

Alexander Fleming, a British scientist, was working with cultures of *Staphylococcus* when he noticed that colonies growing near a contaminating mold looked odd, as if they were dissolving. Recognizing that the mold might be secreting a substance that killed the bacteria, he proceeded to study it more carefully. He identified the mold as a species of *Penicillium* and showed that it was indeed producing a bacteria-killing substance; he called this **penicillin**. Even though Fleming was unable to purify penicillin, he showed that it was remarkably effective in killing many different kinds of bacteria and could be injected into rabbits and mice without adverse effects. Although Fleming recognized the potential medical significance of his discovery, he became discouraged with his inability to purify the compound and eventually abandoned his study of it.

Ten years after Fleming's discovery, two other scientists in Britain, Ernst Chain and Howard Florey, were successful in their attempts to purify penicillin. In 1941, the drug was tested for the first time on a police officer with a life-threatening *Staphylococcus aureus* infection. He improved so dramatically that within 24 hours his illness seemed under control. Unfortunately, the supply of purified penicillin ran out, and the man eventually died of the infection. Later, with greater supplies of the drug, the experiment was repeated and two deathly ill patients were successfully cured. World War II spurred a cooperation of British and American scientists to determine the chemical structure of penicillin and to develop the means for its large-scale production so that it could be used to treat infected soldiers and workers. Several different penicillins were found in the mold cultures, and were designated alphabetically. **Penicillin G** (or benzyl penicillin) was found to be the most suitable for treating infections. This was the first of what we now call **antibiotics**, antimicrobial drugs naturally produced by microorganisms.

Soon after the discovery of penicillin, Selman Waksman isolated a bacterium from soil, *Streptomyces griseus*, that produced an antibiotic he called **streptomycin**. The realization that bacteria as well as molds could produce medically useful antimicrobial drugs prompted researchers to begin laboriously screening hundreds of thousands of different strains of microorganisms for antibiotic production. Even today, pharmaceutical companies examine soil samples from around the world for organisms that produce novel antibiotics.

TABLE 21.1 Terms that Relate to Antimicrobial Medications

Term	Definition
Acquired resistance	Resistance that develops through mutation or acquisition of new genes.
Antibacterial drug	An antimicrobial drug that is used to treat a disease caused by bacteria.
Antibiotic	A compound that is naturally produced by certain molds and bacteria that inhibits the growth of or kills other microorganisms.
Antimicrobial drug (or antimicrobial)	A chemical that inhibits the growth of or kills microorganisms. This term encompasses antibiotics and chemically synthesized drugs.
Antiviral drug	A drug that interferes with the replication of viruses. All antiviral drugs are chemically synthesized; none are antibiotics.
Bactericidal drug	An antimicrobial drug that kills bacteria.
Bacteriostatic drug	An antimicrobial drug that inhibits the growth of bacteria.
Broad-spectrum antimicrobial	An antimicrobial that is effective against a wide range of microorganisms, often including both Gram-positive and Gram-negative bacteria.
Chemotherapeutic agent	A chemical used to treat a disease.
Intrinsic resistance	Resistance due to an inherent characteristic of the microorganism.
Narrow-spectrum antimicrobial	An antimicrobial that is effective against a limited range of microorganisms.
R plasmid	A plasmid that encodes resistance to one or more antimicrobial drugs.
Therapeutic index	A measure expressing the relative toxicity of a chemotherapeutic drug. It is the lowest dose toxic to the patient divided by the dose used for therapy.

Development of New Generations of Drugs

In the 1960s scientists discovered that they could alter the chemical structure of drugs such as penicillin G, and give them new properties. For example, penicillin G, which is mostly active against Gram-positive bacteria, can be altered to produce ampicillin, a drug that kills additional Gram-negative species as well. Other changes to penicillin created the drug methicillin, which is less susceptible to enzymes used by some bacteria to inactivate penicillin. Thus, methicillin can be used to treat infections caused by certain penicillin-resistant organisms. Today a variety of penicillin-like medications exist, making up what is referred to as the family of penicillins (**figure 21.1**). All of these derivatives retain the core portion of penicillin G, called **6-aminopenicillanic acid (6-APA)**. Other unrelated antimicrobial drugs have also been altered to give them new characteristics.

MICROCHECK 21.1

Antimicrobials are chemotherapeutic agents that are effective in treating microbial infections. Antibiotics are antimicrobial chemicals naturally produced by particular microorganisms.

- How is the microbe that makes penicillin different from the one that makes streptomycin?
- Define and contrast the terms *chemotherapeutic agent*, *antimicrobial*, and *antibiotic*.
- How might *Streptomyces griseus* cells protect themselves from the effects of streptomycin?

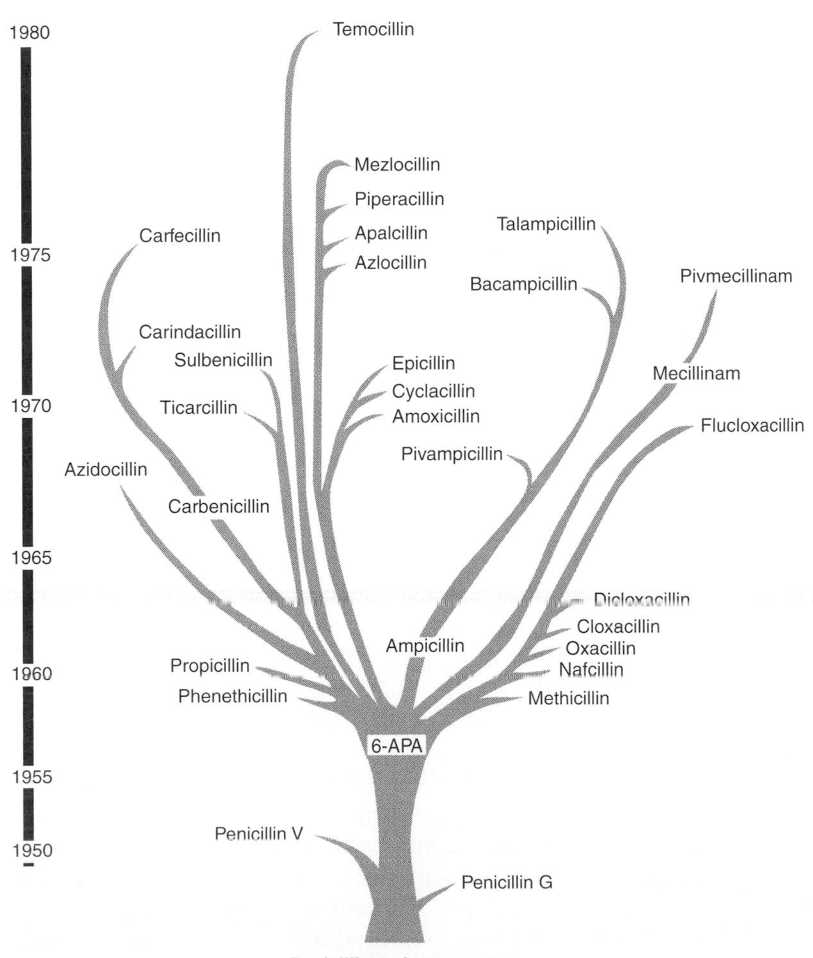

Figure 21.1 **Family Tree of Penicillins**

Features of Antimicrobial Drugs

Most modern antibiotics come from microorganisms that normally reside in the soil; these include species of the bacteria *Streptomyces* and *Bacillus*, and the eukaryotic fungi *Penicillium* and *Cephalosporium*. To commercially produce an antibiotic, a carefully selected strain of the appropriate species is inoculated into a broth medium and incubated in a huge vat. As soon as the maximum antibiotic concentration is reached, the drug is extracted from the medium and extensively purified. In many cases, the antibiotic is chemically altered after purification to impart new characteristics such as increased stability. These chemically modified compounds are called **semisynthetic**. In some cases, the entire drug can be synthesized in the laboratory. By convention, these partially or totally synthetic chemicals are still called antibiotics because microorganisms naturally produce them. A multitude of antimicrobial drugs is now available, each with characteristics that make it more or less suitable for a given clinical situation. Hundreds of tons and many millions of dollars' worth of antibiotics are now produced each year.

Selective Toxicity

Medically useful antimicrobial drugs exhibit **selective toxicity**, causing greater harm to microorganisms than to the human host. They do this by interfering with essential biological structures or biochemical processes that are common in microorganisms but not in human cells. Although a wide variety of antibiotics have been discovered, not all are selectively toxic, which is why only a small percentage can be safely used for treating diseases of humans or other animals.

While the ideal antimicrobial drug is nontoxic to humans, most can be harmful at high concentrations. In other words, selective toxicity is a relative term. The toxicity of a given drug is expressed as the **therapeutic index**, which is the lowest dose toxic to the patient divided by the dose typically used for therapy. Antimicrobials that have a high therapeutic index are less toxic to the patient, often because the drug acts against a vital biochemical process of bacteria that does not exist in human cells. For example, penicillin G, which interferes with bacterial cell wall synthesis, has a very high therapeutic index. When an antimicrobial that has a low therapeutic index is administered, the concentration in the patient's blood must be carefully monitored to ensure it does not reach a toxic level. Drugs that are too toxic for systemic use can sometimes be used for topical applications, such as first-aid antibiotic skin ointments.

Antimicrobial Action

Antimicrobial drugs may either kill microorganisms or inhibit their growth. Those that inhibit growth are called **bacteriostatic**. These drugs depend on the normal host defenses to kill or eliminate the pathogen after its growth has been inhibited. For example, sulfa drugs, which are frequently prescribed for urinary tract infections, inhibit the growth of bacteria in the bladder until they are eliminated during the normal process of urination. Drugs that kill bacteria are **bactericidal**. These drugs are particularly useful in situations in which the normal host defenses cannot be relied on to remove or destroy pathogens. A given drug can be bactericidal in one situation yet bacteriostatic in another, depending on the concentration of the drug and the growth stage of the microorganism.

Spectrum of Activity

Antimicrobial drugs vary with respect to the range of microorganisms they kill or inhibit. Some kill or inhibit a narrow range of microorganisms, such as only Gram-positive bacteria, whereas others affect a wide range, generally including both Gram-positive and Gram-negative organisms. Antimicrobials that affect a wide range of bacteria are called **broad-spectrum** antimicrobials. These are very important in the treatment of acute life-threatening diseases when immediate antimicrobial therapy is essential and there is no time to culture and identify the disease-causing agent. The disadvantage of broad-spectrum antimicrobials is that, by affecting a wide range of organisms, they disrupt the normal flora that play an important role in excluding pathogens. This in turn can leave the patient predisposed to other infections. Antimicrobials that affect a limited range of bacteria are **narrow-spectrum** antimicrobials. Their use requires identification of the pathogen, but they cause less disruption to the normal flora. ■ normal flora, p. 452

Tissue Distribution, Metabolism, and Excretion of the Drug

Antimicrobials differ not only in their action and activity, but also in how they are distributed, metabolized, and excreted by the body. For example, only some drugs are able to cross from the blood into the cerebrospinal fluid, an important factor for a physician to consider when prescribing a drug to treat meningitis. Drugs that are unstable in acid are destroyed by stomach acid when taken orally, and so these drugs must instead be administered through intravenous or intramuscular injection. ■ meningitis, p. 666

Another important characteristic of an antimicrobial is its rate of elimination, which is expressed as the **half-life**. The half-life of a drug is the time it takes for the body to eliminate one-half of the original dosage in the serum. The half-life of a drug dictates the frequency of doses required to maintain an effective level in the body. For example, penicillin V, which has a very short half-life, needs to be taken four times a day, whereas azithromycin, with a half-life of over 24 hours, is taken only once a day or less. Patients who have kidney or liver dysfunction often excrete or metabolize drugs more slowly, and so their drug dosages must be adjusted accordingly.

Effects of Combinations of Antimicrobial Drugs

Combinations of antimicrobials are sometimes used to treat infections, but care must be taken when selecting the combinations because some drugs will counteract the effects of others. When the action of one drug enhances the activity of another, the combination is called **synergistic**. In contrast, combinations in which the activity of one interferes with the other are called **antagonistic**. Combinations that are neither synergistic nor antagonistic are called **additive**.

Adverse Effects

As with any medication, several concerns and dangers are associated with antimicrobial drugs. It is important to remember, however, that antimicrobials are extremely valuable drugs that have saved countless lives when properly prescribed and used.

Allergic Reactions

Some people develop hypersensitivities or allergies to certain antimicrobials. An allergic reaction to penicillin or other related drugs usually results in a fever or rash but can abruptly cause life-threatening anaphylactic shock. For this reason, people who have allergic reactions to antimicrobials must alert their physicians and pharmacists so that alternative drugs can be prescribed. A bracelet or necklace that records that information should also be worn in case of emergency. ■ anaphylactic shock, p. 436

Toxic Effects

Several antimicrobials are toxic at high concentrations or occasionally cause adverse reactions. For example, aminoglycosides can damage kidneys, impair the sense of balance, and even cause irreversible deafness. Patients taking these drugs must be closely monitored because of the very low therapeutic index. Some antimicrobials have such severe potential side effects that they are reserved for only life-threatening conditions. For example, in rare cases, chloramphenicol causes the potentially lethal condition **aplastic anemia**, in which the body is unable to make white and red blood cells. For this reason, chloramphenicol is usually used only when no other alternatives are available.

Suppression of the Normal Flora

The normal flora plays an important role in host defense by excluding pathogens. When the composition of the normal flora is altered, which happens when a person takes an antimicrobial, pathogens normally unable to compete may multiply to high numbers. Patients who take broad-spectrum antibiotics orally sometimes develop the life-threatening disease called **antibiotic-associated colitis**, caused by the growth of toxin-producing strains of *Clostridium difficile*. This organism is not usually able to establish itself in the intestine due to competition from other bacteria. When the normal intestinal flora are inhibited or killed, however, *C. difficile* can sometimes flourish and cause disease. ■ normal flora, p. 452 ■ *Clostridium difficile*, p. 589

Resistance to Antimicrobials

Just as humans are assembling a vast array of antimicrobial drugs, microorganisms have their own genetic toolbox of mechanisms to avoid their effects. In some cases, certain types of bacteria are inherently resistant to the effects of a particular drug; this is called **innate** or **intrinsic resistance**. For example, members of the genus *Mycoplasma* lack a cell wall, so, not surprisingly, they are resistant to any drug such as penicillin that exerts its action by interfering with cell wall synthesis. Many

Gram-negative organisms are intrinsically resistant to certain drugs because the lipid bilayer of their outer membrane excludes entry of the drug. In other instances, previously sensitive organisms develop resistance through spontaneous mutation or the acquisition of new genetic information; this is called **acquired resistance**. The mechanisms and acquisition of resistance will be discussed later.

Cost

While the price of a drug is not a concern from a purely scientific standpoint, it can significantly affect the overall expense of health care. In general, newly introduced drugs are far more expensive than their traditional counterparts. For example, one relatively new addition to the family of penicillins is 20 times as costly as the original version. This higher price partially reflects the inherent expense of the research and development of new drugs. Unfortunately, the rapidly increasing resistance of microorganisms to traditional antimicrobials is affecting the cost of health care.

MICROCHECK 21.2

A multitude of antimicrobial drugs is available, each with characteristics that affect its suitability for a given clinical situation. When choosing an antimicrobial to prescribe, a physician must consider a variety of factors including the therapeutic index, antimicrobial action, spectrum of activity, effects of combinations of antimicrobials, tissue distribution, half-life, adverse effects, resistance of the microbe, and cost of the drug.

- Which would you rather take: an antimicrobial that has a low therapeutic index or one that has a high therapeutic index? Why?
- In what clinical situation is it most appropriate to use a broad-spectrum antimicrobial?
- Why would antimicrobials that have toxic side effects be used at all?

Mechanisms of Action of Antibacterial Drugs

A number of bacterial processes utilize enzymes or structures that are either different, absent, or not commonly found in eukaryotic cells. Several microbial processes, including the synthesis of bacterial cell walls, proteins, and nucleic acids, metabolic pathways, and the integrity of the cytoplasmic membrane, are the targets of most antimicrobial drugs (**figure 21.2**).

This section will discuss the bacterial processes commonly targeted by antimicrobial medications. To illustrate how these targets are affected, the mechanism of action of some of the most widely used antimicrobials will be described (**table 21.2**). A group of antibiotics called β-lactam drugs will be covered in the greatest detail, because they serve as excellent examples of some of the important features of antimicrobials.

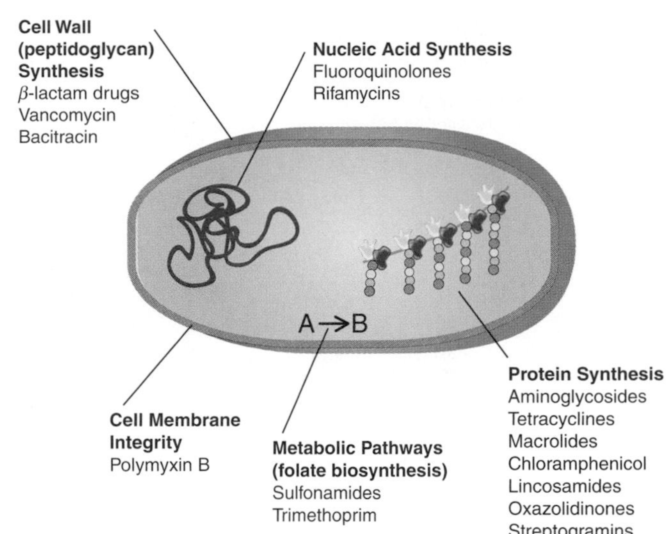

Figure 21.2 Targets of Antibacterial Medications

Antibacterial Medications that Inhibit Cell Wall Synthesis

Bacterial cell walls are unique in that they contain peptidoglycan, a three-dimensional structure composed of strands of alternating subunits of *N*-acetylglucosamine (NAG) and *N*-acetylmuramic acid (NAM) (see figure 3.32). These strands, called **glycan chains**, are interconnected through peptide bridges between the amino acid side chains of NAM. The intact structure provides the cell the rigidity to maintain integrity. Because peptidoglycan is specific to bacteria, drugs that interfere solely with cell wall synthesis do not affect eukaryotic cells, and they usually have a very high therapeutic index. ■ peptidoglycan, p. 62

A number of medically useful antimicrobial drugs interfere with cell wall biosynthesis (**figure 21.3**). Among these, the penicillins and the cephalosporins are the most widely used. They are members of an even larger group, collectively referred to as **β-lactam drugs**, which also includes the monobactams and the carbapenems; these drugs all have a shared chemical structure called a **β-lactam ring** (**figure 21.4**). Other drugs that target peptidoglycan synthesis include vancomycin and bacitracin.

The β-Lactam Drugs

The β-lactam drugs irreversibly inhibit enzymes involved in the final steps of cell wall synthesis. They have been extensively studied because they are such important antimicrobial agents, having very few side effects other than allergic reactions. Their properties such as acid stability, spectrum of activity, and half-life can be modified by chemically altering their structure.

The enzymes inhibited by β-lactam drugs mediate the formation of the peptide bridges between adjacent strands of peptidoglycan. They are called **penicillin-binding proteins** (**PBPs**), reflecting the fact that they bind penicillin and were initially discovered during experiments to study the effects of penicillin. Unfortunately, the name causes some confusion because it seems to imply that the natural biological function of the enzymes is to bind penicillin, when in fact their function is associated with peptidoglycan biosynthesis. In addition, the PBPs may bind any of a number of β-lactam drugs, not just the penicillins.

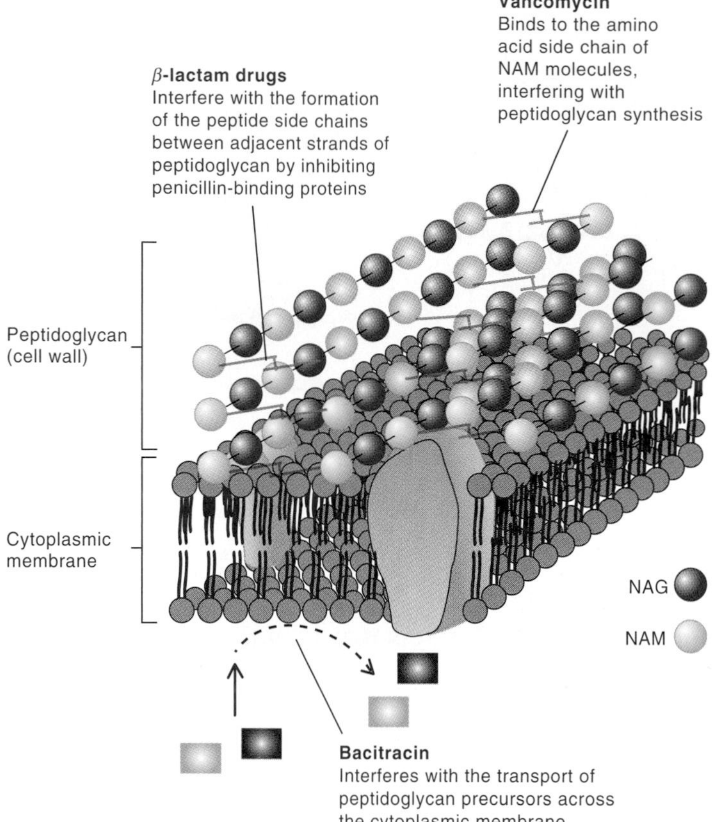
Vancomycin
Binds to the amino acid side chain of NAM molecules, interfering with peptidoglycan synthesis
β-lactam drugs
Interfere with the formation of the peptide side chains between adjacent strands of peptidoglycan by inhibiting penicillin-binding proteins
Peptidoglycan (cell wall)
Cytoplasmic membrane
NAG
NAM
Bacitracin
Interferes with the transport of peptidoglycan precursors across the cytoplasmic membrane

Figure 21.3 Antibacterial Medications that Interfere with Cell Wall Biosynthesis

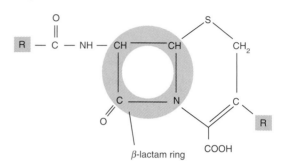

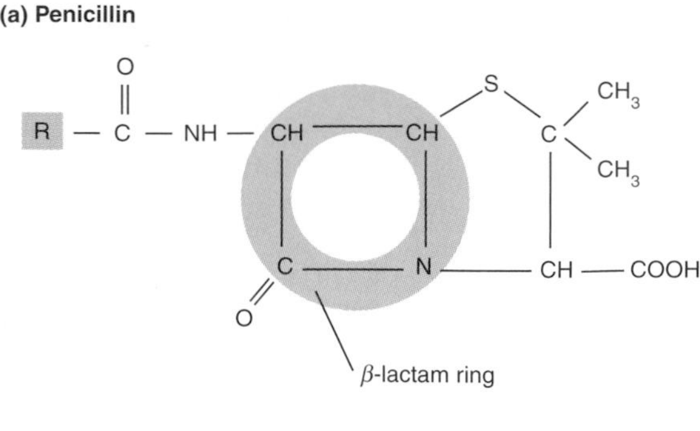
(a) Penicillin
β-lactam ring
(b) Cephalosporin
β-lactam ring

Figure 21.4 The β-Lactam Ring of Penicillins and Cephalosporins The core chemical structure of **(a)** a penicillin **(b)** a cephalosporin. The β-lactam rings are marked by an orange circle. The R groups vary among different penicillins and cephalosporins.

The β-lactam ring of the penicillins and other similar drugs bears structural similarity to the normal substrate of the PBPs. By mimicking that substrate, the β-lactam drugs are bound by PBPs and thus competitively inhibit their enzymatic activity. This causes a disruption in cell wall biosynthesis, leading to a series of events that ultimately causes the cells to lyse (**figure 21.5**). Because cell walls are only synthesized in actively multiplying cells, β-lactam drugs are only effective against growing bacteria. ■ **competitive enzyme inhibition, p. 144**

The different β-lactam drugs vary in their spectrum of activity. Some are more active against Gram-positive bacteria, whereas others are more active against Gram-negative organisms. One reason for this difference arises from the architecture of the cell wall. The peptidoglycan layer of Gram-positive organisms directly contacts the outside environment, making it readily accessible to drugs. In contrast, the outer membrane of Gram-negative bacteria excludes many antimicrobials, making many of these organisms innately resistant to many medications, including certain β-lactam drugs. Another difference is the affinity of an organism's penicillin-binding proteins for a particular β-lactam drug. The PBPs of Gram-positive bacteria differ somewhat from those of Gram-negative bacteria, and the PBPs of obligate anaerobes differ from those of aerobes. Differences in affinity can even exist among related organisms such as Gram-positive cocci.

Some bacteria can resist the effects of certain β-lactam drugs by synthesizing an enzyme called a **β-lactamase** that breaks the critical β-lactam ring, destroying the activity of the antibiotic. Just as there are many β-lactam drugs, there are various β-lactamases; these differ in the range of drugs they destroy. A β-lactamase that was originally detected in some strains of staphylococci only inactivates members of the penicillin family. To reflect this fact, it is often called a **penicillinase**. In contrast, some of the β-lactamases produced by Gram-negative organisms inactivate a wide variety of β-lactam drugs. The **extended-spectrum β-lactamases** inactivate both the penicillins and the cephalosporins. As a whole, Gram-negative bacteria can produce a much more extensive array of β-lactamases than can Gram-positive organisms.

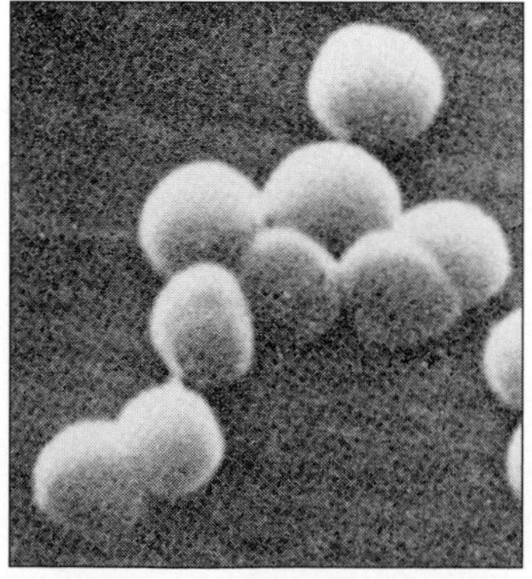

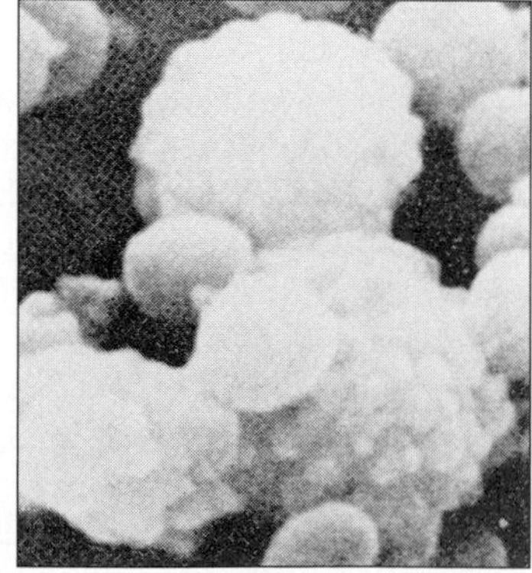

Figure 21.5 Effect of a β-Lactam Drug on a Cell The drug disrupts cell wall biosynthesis, leading to a series of steps that ultimately causes the cells to lyse.

TABLE 21.2 Characteristics of Antibacterial Drugs

Target/Drug	Comments/Characteristics
Cell Wall Synthesis	
β-lactam drugs	Bactericidal against a variety of bacteria; inhibits penicillin-binding proteins. Resistance is due to synthesis of β-lactamases, decreased affinity of penicillin-binding proteins, or decreased uptake.
Penicillins	A family of antimicrobial medications; different groups vary in their spectrum of activity and their susceptibility to β-lactamases. Some must be injected, but others can be taken orally.
Natural penicillins: penicillin G, penicillin V	Active against Gram-positive organisms and Gram-negative cocci. Penicillin G is destroyed by stomach acid, and so it usually must be administered by injection. Penicillin V can be taken orally.
Penicillinase-resistant: methicillin, dicloxicillin	Similar to natural penicillins, but resistant to inactivation by the penicillinase of staphylococci.
Broad-spectrum: ampicillin, amoxicillin	Similar to the natural penicillins, but more active against Gram-negative organisms.
Extended-spectrum: ticarcillin, piperacillin	Increased activity against Gram-negative rods, including *Pseudomonas.*
Cephalosporins Cephalexin, cephradine, cefaclor, cefprozil, cefixime, cefibuten, cefepime	Some cephalosporins can be taken orally; others must be injected because of their instability in stomach acid. The later generations are generally more effective against Gram-negative bacteria and less susceptible to destruction by β-lactamases.
Carbapenems Imipenem, meropenem	Resistant to inactivation by β-lactamases. Imipenem must be given in combination with a drug that inhibits certain kidney enzymes in order to avoid its inactivation.
Monobactams Aztreonam	Resistant to β-lactamases and can be given to patients who are allergic to penicillin. Primarily active against members of the family *Enterobacteriaceae*.
Vancomycin	Bactericidal against Gram-positive bacteria; binds to the peptide side chain of *N*-acetylmuramic acid. Used to treat serious systemic infections and antibiotic-associated colitis. In enterococci, resistance is due to a plasmid-encoded altered target.
Bacitracin	Bactericidal against Gram-positive bacteria; interferes with the transport of peptidoglycan precursors. Common ingredient of non-prescription antibiotic ointments.
Protein Synthesis	
Aminoglycosides Streptomycin, gentamicin, tobramycin, amikacin, neomycin	Bactericidal against aerobic and facultative bacteria; binds to the 30S ribosomal subunit, blocking the initiation of translation and causing the misreading of mRNA. Toxicity limits their use. Resistance is due to a plasmid-encoded inactivating enzyme, alteration of the target molecule, or decreased uptake by a cell. Neomycin is commonly used in non-prescription topical antibiotic ointments
Tetracyclines Tetracycline, doxycycline	Bacteriostatic against some Gram-positive and Gram-negative bacteria; binds to the 30S ribosomal subunit, blocking the attachment of tRNA. Resistance is generally due to decreased accumulation, either through decreased uptake or increased efflux.
Macrolides Erythromycin, clarithromycin, azithromycin	Bacteriostatic against many Gram-positive organisms as well as the most common causes of atypical pneumonia; binds to the 50S ribosomal subunit, preventing the continuation of protein synthesis. Used for treating patients who are allergic to β-lactam drugs. Resistance is due to an inactivating enzyme, alteration of the target molecule, or decreased uptake by a cell.
Chloramphenicol	Bacteriostatic and broad-spectrum; binds to the 50S ribosomal subunit, preventing peptide bonds from being formed. Generally used only as a last resort for life-threatening infections. Resistance is often due to a plasmid-encoded inactivating enzyme.
Lincosamides Lincomycin, clindamycin	Bacteriostatic against a variety of Gram-positive and Gram-negative bacteria, including the anaerobe *Bacteroides fragilis;* binds to the 50S ribosomal subunit, preventing the continuation of protein synthesis. Associated with an even greater risk of developing antibiotic-associated colitis.

(continued)

TABLE 21.2 Characteristics of Antibacterial Drugs *(continued)*

Target/Drug	Comments/Characteristics
Oxazolidinones Linezolid	A new class of broad-spectrum drugs; binds to the 50S ribosomal subunit and is thought to interfere with the initiation of protein synthesis. Effective against a variety of Gram-positive bacteria.
Streptogramins Quinupristin, dalfopristin	A class of drugs recently introduced as a synergistic combination; these bind to two different sites on the 50S ribosomal subunit, inhibiting distinct steps of protein synthesis. Effective against a variety of Gram-positive bacteria.
Nucleic Acid Synthesis	
Fluoroquinolones Ciprofloxacin, ofloxacin	Bactericidal against a wide variety of Gram-positive and Gram-negative bacteria; inhibits topoisomerases. Resistance is most often due to structural alterations in the topoisomerase target.
Rifamycins Rifampin	Bactericidal against Gram-positive and some Gram-negative bacteria. Binds RNA polymerase, blocking the initiation of RNA synthesis. Primarily used to treat infections caused by *Mycobacterium tuberculosis* and as prophylaxis for patients who have been exposed to *Neisseria meningitidis*.
Folate Biosynthesis	
Sulfonamides	Bacteriostatic against a variety of Gram-positive and Gram-negative organisms. Structurally similar to para-aminobenzoic acid (PABA) and therefore inhibits the enzyme for which PABA is a substrate. Resistance is most commonly due to a plasmid-encoded alternative enzyme.
Trimethoprim	Often used in combination with a sulfa drug for a synergistic effect; inhibits the enzyme that catalyzes a step following the one inhibited by the sulfonamides. Resistance is commonly due to a plasmid-encoded alternative enzyme; the genes that encode resistance to sulfa drugs are often carried on the same plasmid.
Cell Membrane Integrity	
Polymyxin B	Bactericidal against Gram-negative cells by damaging cell membranes. Its toxicity limits its use primarily to topical applications, but it is a common ingredient in non-prescription antibiotic ointments.
Mycobacterium tuberculosis	
Ethambutol	Inhibits the synthesis of a component of the mycobacterial cell wall.
Isoniazid	Inhibits synthesis of mycolic acid, a major component of the mycobacterial cell wall.
Pyrazinamide	Mechanism unknown.

The Penicillins Each member of the family of penicillins shares a common portion, 6-aminopenicillanic acid (6-APA). Only the side chain has been modified in the laboratory to create penicillin derivatives, each with unique characteristics **(figure 21.6)**. Currently the family of penicillins can be loosely grouped into several categories, each of which consists of several different drugs:

- **Natural penicillins.** These are the original penicillins produced naturally by the mold *Penicillium chrysogenum*. Natural penicillins are narrow-spectrum antibiotics, effective against Gram-positive bacteria and some Gram-negative cocci. Strains of bacteria that produce penicillinase are resistant to the natural penicillins. Natural penicillins include penicillin G and penicillin V. Penicillin V is more stable in acid and, therefore, better absorbed than penicillin G when taken orally.
- **Penicillinase-resistant penicillins.** These drugs were developed in the laboratory as a response to the problem of penicillinase-producing staphylococci. Their side chains prevent penicillinase from inactivating them. Unfortunately, some strains of penicillinase-producing *Staphylococcus aureus* have responded by synthesizing altered PBPs to which β-lactam drugs, including the penicillins, no longer bind. Penicillinase-resistant penicillins include methicillin and dicloxicillin.
- **Broad-spectrum penicillins.** The modified side chains of these drugs give them a broad spectrum of activity. They retain their activity against penicillin-sensitive Gram-positive bacteria, yet they are also active against Gram-negative organisms. They can be inactivated, however, by many β-lactamases. Broad-spectrum penicillins include ampicillin and amoxicillin.

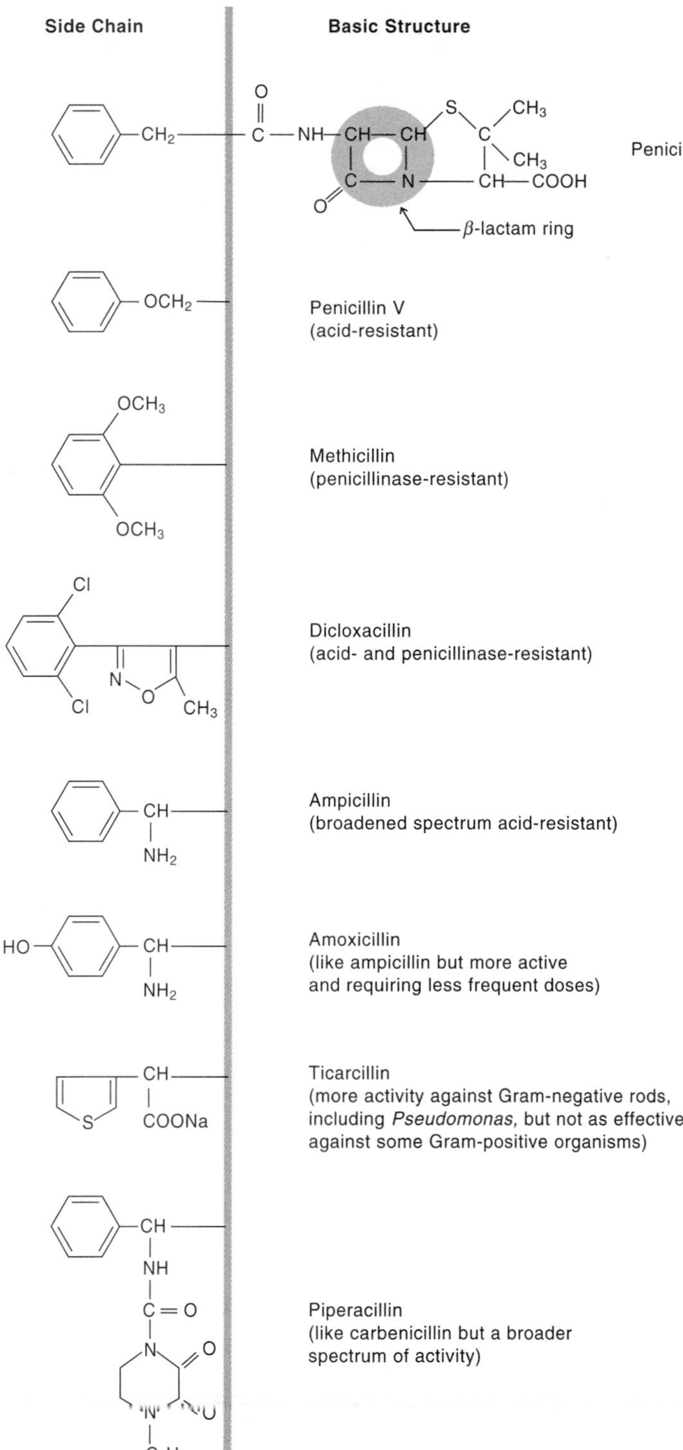

Figure 21.6 Chemical Structures and Properties of Representative Members of the Penicillin Family The entire structure of penicillin G and the side chain of each of the other penicillins are shown.

- **Extended-spectrum penicillins.** These have greater activity against *Pseudomonas* species, Gram-negative bacteria that are unaffected by many conventional antimicrobial drugs. The extended-spectrum penicillins, however, have less activity against Gram-positive organisms. Like the other broad-spectrum penicillins, they are destroyed by many β-lactamase-producing organisms. Extended-spectrum penicillins include ticarcillin and piperacillin.
- **Penicillins + β-lactamase inhibitor.** Rather than a new drug, this is a novel combination of therapeutic agents. β-lactamase inhibitors are chemicals that interfere with the activity of some types of β-lactamases. When a β-lactamase inhibitor is administered with one of the penicillins, the medication is protected against enzymatic destruction.

The Cephalosporins The **cephalosporins** are derived from an antibiotic produced by the fungus *Acremonium cephalosporium* (formerly called *Cephalosporium acremonium*). Generally included in this family of drugs are a closely related group of antibiotics that are made by a genus of filamentous bacteria related to *Streptomyces*. The chemical structure of the cephalosporins makes them resistant to inactivation by certain β-lactamases, but some have a low affinity for penicillin-binding proteins of Gram-positive bacteria, thus limiting their effectiveness against these organisms. Like the penicillins, the cephalosporins have been chemically modified to produce a family of various related antibiotics. They are grouped as the first-, second-, third-, and fourth-generation cephalosporins. These include cephalexin and cephradine (first generation), cefaclor and cefprozil (second generation), cefixime and cefibuten (third generation), and cefepime (fourth generation). The later generations are generally more effective against Gram-negative bacteria and are less susceptible to destruction by β-lactamases.

Other β-Lactam Antibiotics Two other groups of β-lactam drugs, **carbapenems** and **monobactams**, are very resistant to β-lactamases. The carbapenems are effective against a wide range of Gram-negative and Gram-positive bacteria. Two types are available, imipenem and meropenem. Imipenem is rapidly destroyed by a kidney enzyme and is therefore administered in combination with a drug that inhibits that enzyme. The only monobactam used therapeutically, aztreonam, is primarily effective against members of the family *Enterobacteriaceae*, which are Gram-negative rods. Structurally, it is slightly different from other β-lactam drugs; this characteristic is important because aztreonam can be given to patients who have developed an allergy to penicillin. ■ ***Enterobacteriaceae*, p. 282**

Vancomycin

Vancomycin binds to the terminal amino acids of the peptide side chain of NAM molecules that are being assembled to form glycan chains. By doing so, it blocks synthesis of peptidoglycan, resulting in weakening of the cell wall and, ultimately, cell lysis. Vancomycin does not cross the outer membrane of Gram-negative bacteria; consequently, these organisms are innately resistant. It is, however, a very important medication for treating infections caused by Gram-positive bacteria that are resistant to β-lactam drugs. In addition, it is sometimes the preferred drug for treating severe cases of antibiotic-associated colitis. Because vancomycin is poorly absorbed from the intestinal tract, it must be administered intravenously except when used to treat intestinal infections. Acquired resistance to vancomycin is most often due to an alteration in the peptide side chain of the NAM molecule that prevents vancomycin from binding.

Bacitracin

Bacitracin inhibits cell wall biosynthesis by interfering with the transport of peptidoglycan precursors across the cytoplasmic membrane. Its toxicity limits its use to topical applications; however, it is a common ingredient in non-prescription first-aid ointments.

Antibacterial Medications that Inhibit Protein Synthesis

Several types of antibacterial drugs inhibit prokaryotic protein synthesis (**figure 21.7**). While all cells synthesize proteins, the structure of the prokaryotic 70S ribosome, which is composed of a 30S and a 50S subunit, is different enough from the eukaryotic 80S ribosome to make it a suitable target for selective toxicity. The mitochondria of eukaryotic cells also have 70S ribosomes, however, which may partially account for the toxicity of some of these drugs. ■ ribosome structure, p. 175 ■ protein synthesis, p. 175

The major classes of antibiotics that inhibit protein synthesis are the aminoglycosides, the tetracyclines, and the macrolides. Others include the lincosamides and chloramphenicol. Of these, only the aminoglycosides are bactericidal; the others are all bacteriostatic. Two classes of drugs that have recently been approved for use in the United States are the oxazolidinones and the streptogramins.

The Aminoglycosides

The **aminoglycosides** irreversibly bind to the 30S ribosomal subunit, causing it to distort and malfunction. This blocks the initiation of translation and causes misreading of mRNA by ribosomes that have already passed the initiation step. Aminoglycosides are actively transported into bacterial cells by a process that requires respiratory metabolism. Consequently, they are generally not effective against anaerobes, enterococci, and streptococci. To extend their spectrum of activity, the aminoglycosides are sometimes used in a synergistic combination with a β-lactam drug. The β-lactam drug interferes with cell wall synthesis, which, in turn, allows the aminoglycoside to more easily enter cells that would otherwise be resistant. Examples of aminoglycosides include streptomycin, gentamicin, tobramycin, and amikacin. Unfortunately, these all can cause severe side effects including hearing loss and kidney damage; consequently, they are generally used only when other alternatives are not available. Recently, a form of tobramycin that can be administered through inhalation rather than intravenously was developed, making treatment of lung infections in cystic fibrosis patients caused by *Pseudomonas aeruginosa* safer and more effective. Another aminoglycoside, neomycin, is too toxic for systemic use; however, it is a common ingredient in non-prescription topical ointments. ■ active transport, p. 60

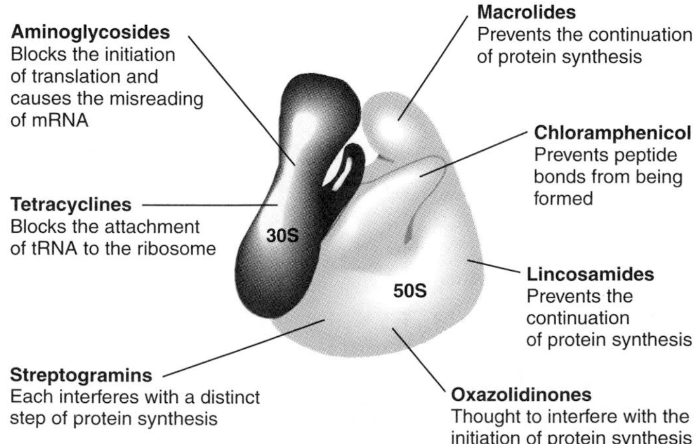
Aminoglycosides
Blocks the initiation of translation and causes the misreading of mRNA
Macrolides
Prevents the continuation of protein synthesis
Chloramphenicol
Prevents peptide bonds from being formed
Tetracyclines
Blocks the attachment of tRNA to the ribosome
30S
50S
Lincosamides
Prevents the continuation of protein synthesis
Streptogramins
Each interferes with a distinct step of protein synthesis
Oxazolidinones
Thought to interfere with the initiation of protein synthesis

Figure 21.7 Antibacterial Medications that Inhibit Prokaryotic Protein Synthesis These medications bind to the 70S ribosome.

The Tetracyclines

The **tetracyclines** reversibly bind to the 30S ribosomal subunit, blocking the attachment of tRNA to the ribosome and preventing the continuation of protein synthesis. These drugs are actively transported into prokaryotic but not animal cells, which effectively concentrates them inside bacteria. This, in part, accounts for their selective toxicity. The tetracyclines are effective against certain Gram-positive and Gram-negative bacteria. The newer tetracyclines such as doxycycline have a longer half-life, allowing less frequent doses. Resistance to the tetracyclines is primarily due to a decrease in their accumulation by the bacterial cell, either by decreased uptake or increased excretion. Tetracyclines can cause discoloration in teeth when used by young children.

The Macrolides

The **macrolides** reversibly bind to the 50S ribosomal subunit and prevent the continuation of protein synthesis. Macrolides as a group are effective against a variety of bacteria, including many Gram-positive organisms as well as the most common causes of atypical pneumonia ("walking pneumonia"). They often serve as the drug of choice for patients who are allergic to penicillin. Macrolides are not effective against members of the family *Enterobacteriaceae*, however, because the outer membrane of these organisms excludes the drug. Examples of macrolides include erythromycin, clarithromycin, and azithromycin. Both clarithromycin and azithromycin have a longer half-life than erythromycin, so that they can be taken less frequently. Resistance to all the macrolides can occur through modification of the ribosomal RNA target. Other mechanisms of resistance include the production of an enzyme that chemically modifies the drug and alterations that result in decreased uptake of the drug. ■ walking pneumonia, p. 504

Chloramphenicol

Chloramphenicol binds to the 50S ribosomal subunit, preventing peptide bonds from being formed and, consequently, blocking protein synthesis. Although it is active against a wide range of bacteria, it is generally only used as a last resort for life-threatening infections in order to avoid a rare but lethal side effect. This complication, aplastic anemia, is not related to dose and is characterized by the inability of the body to form white and red blood cells.

The Lincosamides

The **lincosamides** bind to the 50S ribosomal subunit and prevent the continuation of protein synthesis, inhibiting a variety of Gram-negative and Gram-positive organisms. They are particularly useful for treating infections resulting from intestinal perforation because they inhibit *Bacterioides fragilis*, a member of normal intestinal flora that is frequently resistant to other antimicrobials. Unfortunately, the risk for people taking lincosamides developing antibiotic-associated colitis is greater than for some other antimicrobials because *Clostridium difficile* is generally resistant to these drugs. The most commonly used lincosamide is clindamycin.

The Oxazolidinones

The **oxazolidinones** are a promising new class of antimicrobial drugs. They bind to the 50S ribosomal subunit and are thought to interfere with the initiation of protein synthesis. They are effective against a variety of Gram-positive bacteria and should prove useful in treating infections caused by bacteria that are resistant to β-lactam drugs and vancomycin. Because the oxazolidinones have only recently been developed, it is anticipated that resistance will be rare, at least initially. Linezolid is the first drug of this class to be approved for use.

The Streptogramins

Two **streptogramins**, quinupristin and dalfopristin, are administered together in a recently approved medication called Synercid®. These are a synergistic combination, binding to two different sites on the 50S ribosomal subunit and inhibiting distinct steps of protein synthesis. Synercid® is effective against a variety of Gram-positive bacteria, including some of those that are resistant to β-lactam drugs and vancomycin.

Antibacterial Medications that Inhibit Nucleic Acid Synthesis

Enzymes that are required for nucleic acid synthesis are the targets of some groups of antimicrobial drugs. These include the fluoroquinolones and the rifamycins.

The Fluoroquinolones

The synthetic drugs called the **fluoroquinolones** inhibit one or more of a group of enzymes called **topoisomerases**, which maintain the supercoiling of closed circular DNA within the bacterial cell. One type of topoisomerase, called **DNA gyrase** or **topoisomerase II**, breaks and rejoins strands to relieve the strain caused by the localized unwinding of DNA during replication and transcription. Consequently, inhibition of this enzyme prevents these essential cell processes. The fluoroquinolones are bactericidal against a wide variety of bacteria, including both Gram-positive and Gram-negative organisms. Examples of fluoroquinolones include ciprofloxacin and ofloxacin. Acquired resistance is most commonly due to an alteration in the DNA gyrase target. ■ **supercoiled DNA, p. 70**

The Rifamycins

The **rifamycins** block prokaryotic RNA polymerase from initiating transcription. Rifampin, which is the most widely used rifamycin, exhibits bactericidal activity against many Gram-positive and Gram-negative bacteria as well as members of the genus *Mycobacterium*. It is primarily used to treat tuberculosis and Hansen's disease (leprosy) and to prevent meningitis in people who have been exposed to *Neisseria meningitidis*. In some patients, a reddish-orange pigment appears in urine and tears. Resistance to rifampin develops rapidly and is due to a mutation in the gene that encodes RNA polymerase.

Antibacterial Medications that Inhibit Metabolic Pathways

Relatively few antibacterial medications interfere with metabolic pathways. Among the most useful are the folate inhibitors—sulfonamide and trimethoprim. These each inhibit different steps in the pathway that leads initially to the synthesis of folic acid and ultimately to the synthesis of a coenzyme required for nucleotide biosynthesis (**figure 21.8**). Animal cells lack the enzymes in the folic acid synthesis portion of the pathway, which is why folic acid is a dietary requirement. ■ **coenzyme, p. 143**

The Sulfonamides

Sulfonamide and related compounds, collectively referred to as **sulfa drugs**, inhibit the growth of many Gram-positive and Gram-negative bacteria. They are structurally similar to para-aminobenzoic acid (PABA), a substrate in the pathway for folic acid biosynthesis. Because of this similarity, the enzyme that normally binds PABA preferentially binds sulfa drugs, resulting in its competitive inhibition. Human cells lack this enzyme, providing the basis for the selective toxicity of the sulfonamides. Resistance of bacteria to the sulfonamides is most often due to the acquisition of a plasmid-encoded enzyme that has a lower affinity for the drug. ■ **competitive inhibition, p. 144**

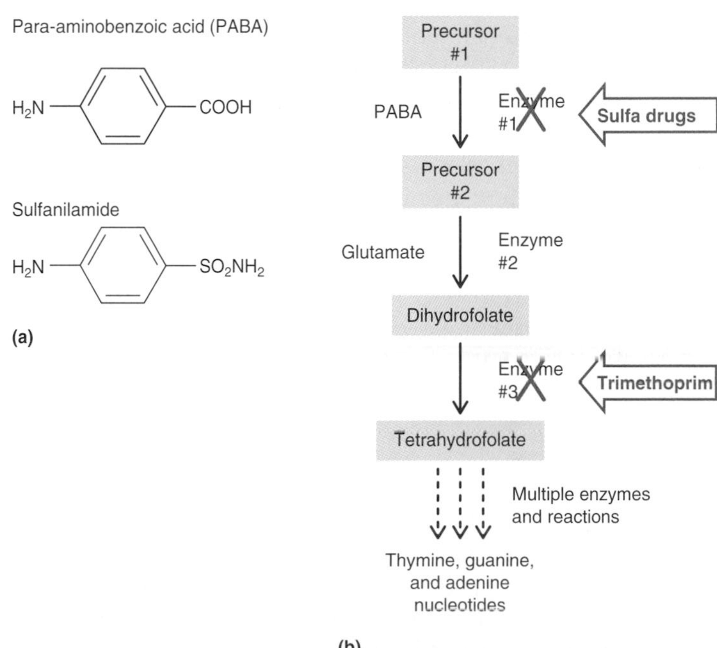

Figure 21.8 Inhibitors of the Folate Pathway **(a)** The chemical structure of PABA and a sulfa drug (sulfanilamide). **(b)** The sulfonamides and trimethoprim interfere with different steps of the pathway that leads initially to the synthesis of folic acid and ultimately to the synthesis of a coenzyme required for nucleotide biosynthesis.

Trimethoprim

Trimethoprim inhibits the bacterial enzyme that catalyzes a metabolic step following the one inhibited by sulfonamides. Fortunately, the drug has little effect on the analogous enzyme in human cells. The combination of trimethoprim and a sulfonamide has a synergistic effect, and they are often used to treat urinary tract infections. The most common mechanism of resistance is a plasmid-encoded alternative enzyme that has a lower affinity for the drug. Unfortunately, the genes encoding resistance to trimethoprim and sulfonamide are often carried on the same plasmid.

Antibacterial Medications that Interfere with Cell Membrane Integrity

A few antimicrobial drugs damage the bacterial membranes. For example, **polymyxin B**, a common ingredient in first-aid skin ointments, binds to the membrane of Gram-negative cells and alters their permeability, leading to leakage of cellular contents and eventual death of the cells. Unfortunately, these drugs also bind to eukaryotic cells, though to a lesser extent, which generally limits their use to topical applications.

Antibacterial Medications that Interfere with Processes Essential to *Mycobacterium tuberculosis*

Only a limited range of antimicrobials can be used to treat infections caused by *Mycobacterium tuberculosis.* This is due to several factors, including the chronic nature of the disease caused by these organisms (tuberculosis), their slow growth, and their waxy cell wall, which is impervious to many drugs. A group of five medications, called the **first-line** drugs, are preferred because they are the most effective as well as the least toxic. These are generally given in combination to prevent the development of resistant mutants; a small fraction of the infecting population might spontaneously develop resistance to one drug, but the other drug will eliminate those. The **second-line** medications can be used if the first-line drugs are not an option; however, they are either less effective or more toxic.
■ *Mycobacterium tuberculosis*, p. 569

Of the first-line medications, some specifically target the unique cell wall that characterizes the mycobacteria. **Isoniazid** inhibits the synthesis of mycolic acids, a primary component of the cell wall. **Ethambutol** inhibits enzymes that are required for synthesis of other mycobacterial cell wall components. The mechanism of **pyrazinamide** is unknown. Other first-line drugs include rifampin and streptomycin, which have already been discussed.

MICROCHECK 21.3

Bacterial processes that utilize enzymes or structures that are either different, absent, or not commonly found in eukaryotic cells are the targets of most medically useful antimicrobial drugs. The targets of antimicrobial drugs include biosynthetic pathways for peptidoglycan, protein, nucleic acid, and folic acid and the integrity of membranes. Drugs used to treat tuberculosis often interfere with processes unique to *Mycobacterium tuberculosis.*

- Explain the normal biological role of penicillin-binding proteins.
- What is the target of the macrolides?
- Why would co-administration of a bacteriostatic drug interfere with the effects of penicillin?

Determining the Sensitivity of Bacteria to an Antimicrobial Drug

In many cases, susceptibility of a pathogenic organism to a specific antimicrobial drug is unpredictable. Unfortunately, it has often been the practice to try one drug after another until a favorable response is observed or, if the infection is very serious, to give several together. Both approaches are undesirable. With each unnecessary drug given, needless risks of toxic or allergic effects arise and the normal flora may be altered, permitting the overgrowth of pathogens that are resistant to the drug. A better approach is to determine the susceptibility of the specific pathogen to various antimicrobial drugs and then choose the drug that acts against the offending pathogen but against as few other bacteria as possible.

Quantitative tests determine the lowest concentration of a specific drug that prevents the growth of an organism *in vitro.* This concentration is called the **minimum inhibitory concentration** (**MIC**). Interpreting the significance of a given MIC requires knowledge of the level of the drug that can be achieved in the patient. Other simpler tests can be used to determine whether an organism is sufficiently susceptible to a drug for a person to be successfully treated.

Determining the Minimum Inhibitory and Bactericidal Concentrations

The minimum inhibitory concentration is determined by assaying the ability of bacteria to grow in broth cultures containing different concentrations of the antimicrobial. Serial dilutions generating decreasing concentrations of the drug are first prepared in tubes containing a suitable growth medium (**figure 21.9**). Then, a known concentration of the organism is added to each tube. The tubes are incubated for at least 16 hours and then are examined for visible growth or turbidity. The lowest concentration of the drug that prevents growth of the microorganism is the minimum inhibitory concentration. The fact that an organism is inhibited by a given concentration of drug, however, does not necessarily mean that an infection can be successfully treated with the drug. For example, an organism with a MIC of 20 μg/ml of a drug is susceptible to that concentration *in vitro.* Nevertheless, the organism would be considered resistant for the purposes of treatment if the level that can be achieved in blood were only 5 μg/ml. Microorganisms requiring inhibitory concentrations on the borderline between sensitive (treatable) and resistant (untreatable) are called intermediate.

The **minimum bactericidal concentration** (**MBC**) is the lowest concentration of a specific antimicrobial drug that kills 99.9% of a given strain of bacteria. The MBC is determined by assaying for live organisms in those tubes from the MIC test that

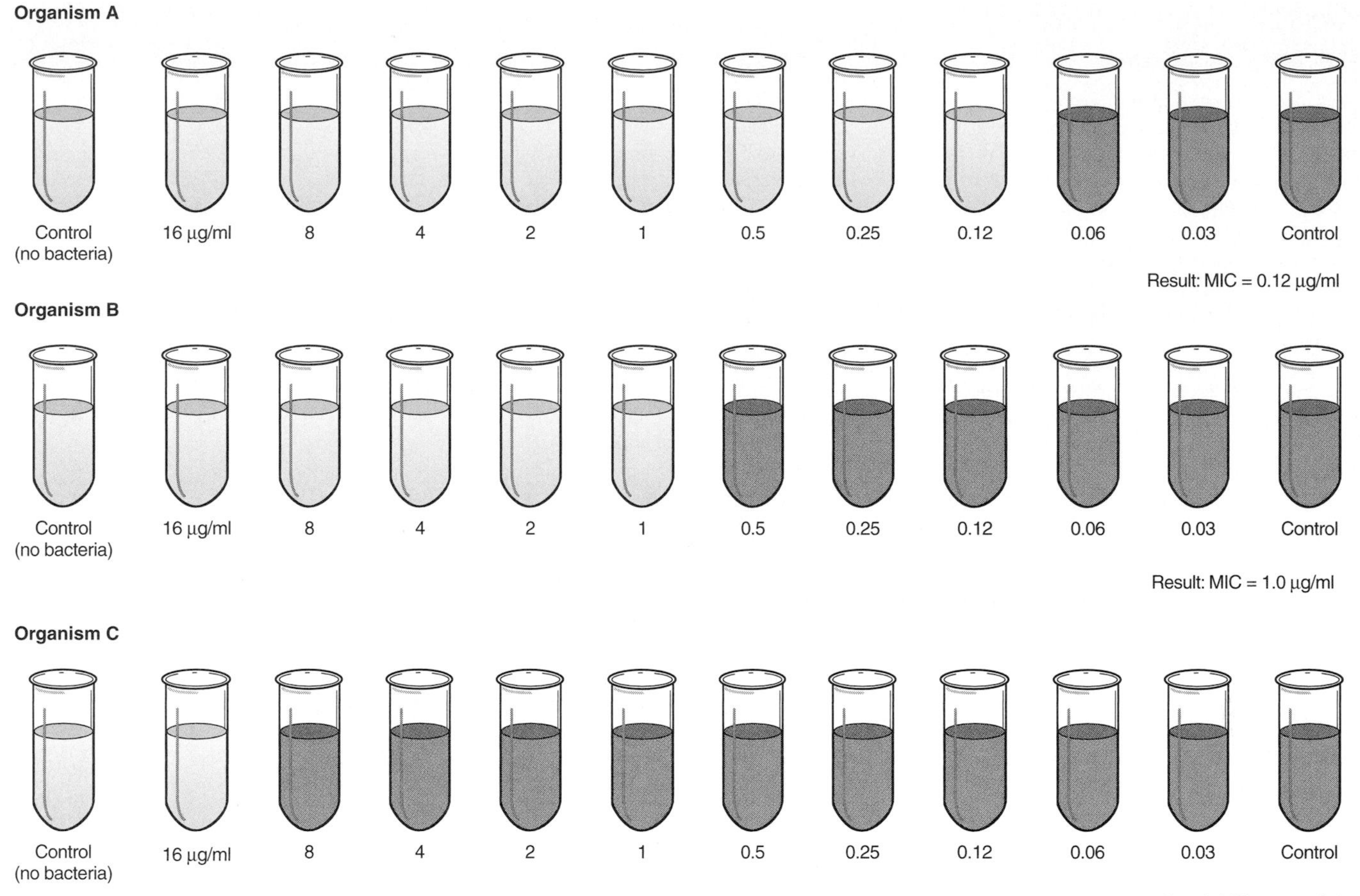

Figure 21.9 Determining the Minimum Inhibitory Concentration (MIC) of an Antimicrobial Drug The lowest concentration of drug that prevents growth of the culture is the MIC.

showed no growth. A small sample from each of those tubes is transferred to fresh, antibiotic-free medium. If growth occurs, then living organisms remained in the original tube. Conversely, if no growth occurs, then no living organisms remained, indicating that the antibiotic was bactericidal at that concentration.

Determining the MIC and MBC using these conventional methods gives precise information regarding an organism's susceptibility. The techniques, however, are labor-intensive and consequently expensive. In addition, individual sets of tubes must be inoculated to determine susceptibility to different antimicrobials.

Conventional Disc Diffusion Method

The **Kirby-Bauer disc diffusion test** is routinely used to qualitatively determine the susceptibility of a given organism to a battery of antimicrobial drugs. A given concentration of a bacterial strain is first uniformly spread on the surface of an agar plate. Then 12 or so discs, each impregnated with a specified concentration of a selected antimicrobial drug, are placed on the surface of the medium (**figure 21.10**). During incubation, the various drugs diffuse outward from the discs at a rate inversely proportional to their size, forming a concentration gradient of drug around each disc. A clear **zone of inhibition** around an antimicrobial disc reflects, in part, the degree of susceptibility of the organism to the drug. The zone size is also influenced by characteristics of the drug including its molecular weight, stability, and concentration in the disc.

Special charts have been prepared correlating the size of the zone of inhibition to susceptibility of the drug. Based on the size of the zone, organisms can be described as sensitive, intermediate, or resistant to the drug. These seemingly simple categories reflect the extensive work of researchers who tested the antimicrobial sensitivity of a variety of bacteria that had a wide range of drug susceptibilities. For each organism, they determined the minimum inhibitory concentration of a particular antimicrobial drug and the size of the zone of inhibition around that antimicrobial disc. The correlation of MIC and zone size, along with the determination of the achievable blood levels of each of the various drugs, enabled the Kirby-Bauer disc diffusion test to become a standard method to guide physicians in selecting an appropriate antimicrobial drug.

Commercial Modifications of Antimicrobial Susceptibility Testing

Commercial modifications of the conventional methods offer certain advantages. They are less labor-intensive, and the results can be read in as little as 4 hours. One system utilizes a small card with miniature wells containing specific antimicrobial concentrations. The highly automated system inoculates and incubates the

Perspective 21.1 Measuring the Concentration of an Antimicrobial Drug in Blood or Other Body Fluids

There are many situations in which it is necessary to determine the concentration of an antimicrobial drug in a patient's blood or other body fluid. For example, patients who are being administered an aminoglycoside must often be carefully monitored to ensure that the concentration of drug in their blood does not reach an unsafe level, particularly if they have kidney or liver dysfunction that interferes with normal elimination. Likewise, new drugs must be tested to determine achievable levels in the blood, urine, or other body fluids.

A technique called the **diffusion assay** is used to measure the concentration of an antimicrobial in a fluid specimen. The test relies on the same principle as the Kirby-Bauer test, except in this case it is the concentration of drug, not the sensitivity of organism, being assayed. A culture of a stock organism that is highly susceptible to the drug is added to melted cooled agar, and the mixture is poured into an agar plate and allowed to solidify. This results in a solid medium that is uniformly inoculated throughout with the sensitive organism. Cylindrical holes are then punched out of the agar, creating wells. **Standards**, or fluids containing known concentrations of the drug, are then added to some of the wells, while others are filled with the body fluid being tested. Following overnight incubation, zones of inhibition form around the agar wells, the sizes of which correspond to the concentrations of the drug (figure 1)The higher the concentration of antimicrobial, the larger is the zone of inhibition. The zone sizes around the standards are measured, and from this a **standard curve** is constructed by plotting the zone sizes against the corresponding drug concentration. A line relating zone size to concentration is obtained, from which the concentration of the antimicrobial drug in the body fluid can be read.

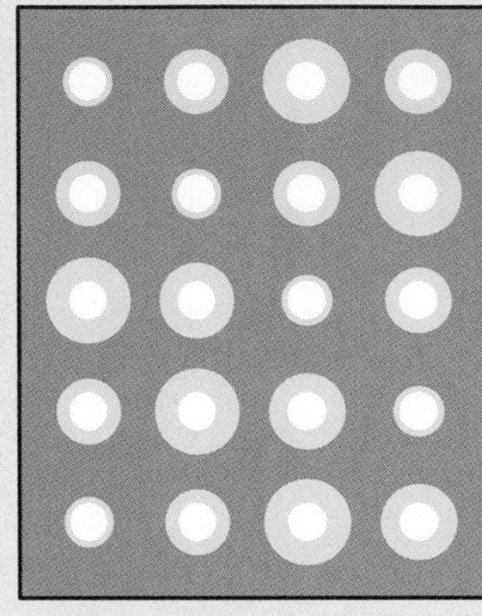

(a) Standards and patient's serum are added to agar that has been seeded with susceptible strain of bacteria.

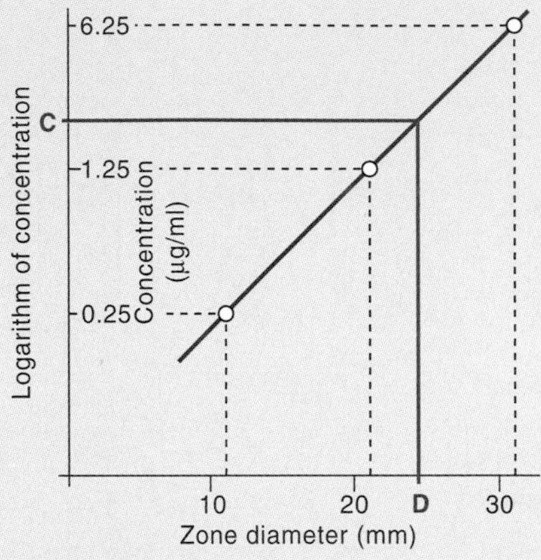

(b) A standard curve that correlates the size of the zone with the concentration of antibiotic is constructed. The concentration of the antimicrobial drug in the body fluid can be read from the line relating zone size to concentration.

Figure 1 Biological Assay to Determine the Concentration of an Antimicrobial Drug in a Patient's Body Fluid

cards, determines the growth rate by reading the turbidity, and uses mathematical formulas to interpret the results and derive the MICs in 6 to 15 hours (**figure 21.11**). Another system uses a microdilution tray containing various concentrations of antimicrobials that is manually inoculated.

The **E test**, a modification of the disc diffusion test, utilizes a strip impregnated with a gradient of concentrations of an antimicrobial drug. Multiple strips, each containing a different drug, are placed on the surface of an agar medium that has been uniformly inoculated with the test organism (**figure 21.12**). During incubation the test organism will form a zone of inhibition around the strip, but because of the gradient of concentrations, the zone of inhibition is shaped somewhat like a teardrop that will intersect the strip at some point. The MIC is determined by reading a number off the numerical scale printed on the strip at the point where the bacterial growth intersects it.

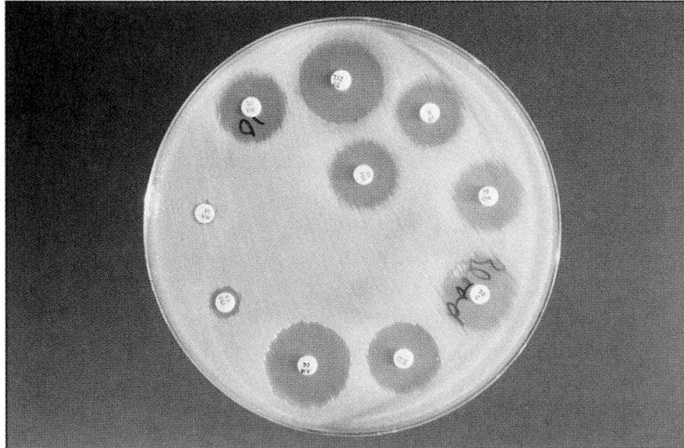

Figure 21.10 Kirby-Bauer Method for Determining Drug Susceptibility The size of the zone of inhibition surrounding the disc reflects, in part, the sensitivity of the microorganism to the drug. Because zone size is influenced by characteristics of the drug such as molecular weight, however, a chart must be consulted that correlates zone size to susceptibility.

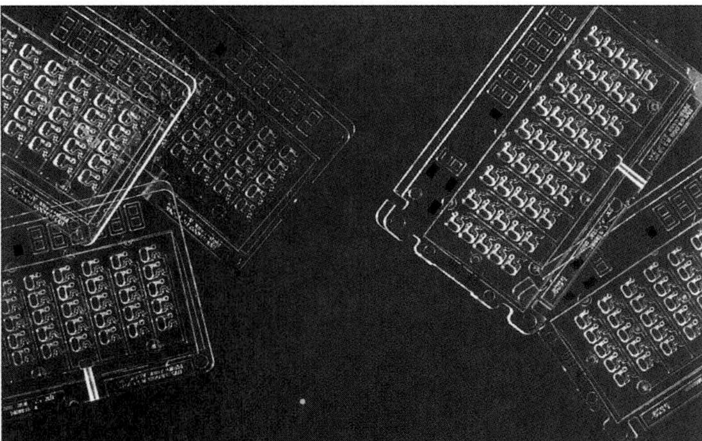

Figure 21.11 Automated Tests Used to Determine Antimicrobial Susceptibility The miniature wells in the card contain specific concentrations of an antimicrobial drug. An automated system inoculates and incubates the cards, determines the growth rate by reading turbidity, and uses mathematical formulas to interpret the results and derive the MICs.

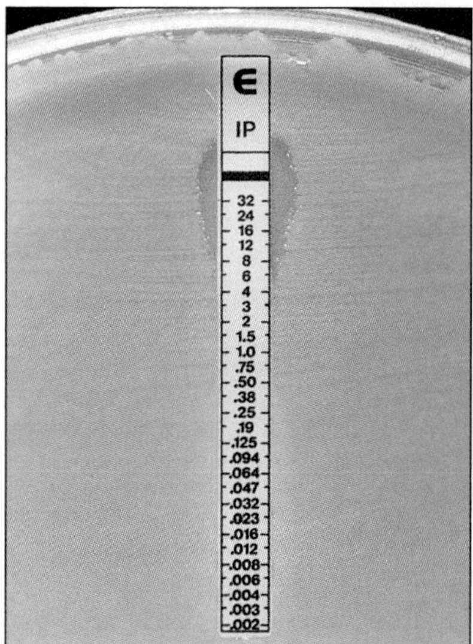

Figure 21.12 The E Test The strip is impregnated with a gradient of concentrations of a given antimicrobial drug. The MIC is determined by reading the number on the strip at the point at which growth intersects the strip.

MICROCHECK 21.4

The minimum inhibitory concentration and minimum bactericidal concentration are quantitative measures of an organism's susceptibility to an antimicrobial drug. Disc diffusion tests can determine whether an organism is sufficiently susceptible to a specific drug for it to successfully be used in treatment. Commercial tests for determining antimicrobial sensitivity are less labor-intensive and often more rapid.

- Explain the difference between the MIC and the MBC.
- List two factors other than an organism's sensitivity to a drug that can influence the size of the zone of inhibition around an antimicrobial disc.
- Why would it be important for the Kirby-Bauer disc diffusion test to use a standard concentration of the bacterial strain being tested?

Resistance to Antimicrobial Drugs

After the introduction of sulfa drugs and penicillin, there was great hope that such drugs would soon eliminate most bacterial diseases. It is now recognized, however, that drug resistance limits the usefulness of all known antimicrobials. As these drugs are increasingly used and misused, the bacterial strains that are resistant to their effects have a selective advantage over their sensitive counterparts (**figure 21.13**). For example, when penicillin G was first introduced, less than 3% of *Staphylococcus aureus* strains were resistant to its effects. Heavy use of the drug, measured in hundreds of tons per year, progressively eliminated sensitive strains, so that 90% or more are now resistant. This development is understandably of great concern to health professionals because of the impact on the cost, complications, and outcomes of treatment. Understanding the mechanisms and the spread of antimicrobial resistance is an important step in curtailing the problem, and it may also allow pharmaceutical companies to develop new drugs that foil common resistance mechanisms.

Mechanisms of Resistance

Bacteria can resist the effects of antimicrobials through a variety of mechanisms. In some cases this resistance is innate, but in many others it is acquired. **Figure 21.14** depicts the most common mechanisms of acquired antimicrobial resistance.

Drug-Inactivating Enzymes

Some organisms produce enzymes that chemically modify a specific drug in such a way as to render it ineffective. Recall that bacteria that synthesize the enzyme penicillinase are resistant to the bactericidal effects of penicillin. As another example, the enzyme chloramphenicol acetyltransferase chemically alters the antibiotic chloramphenicol.

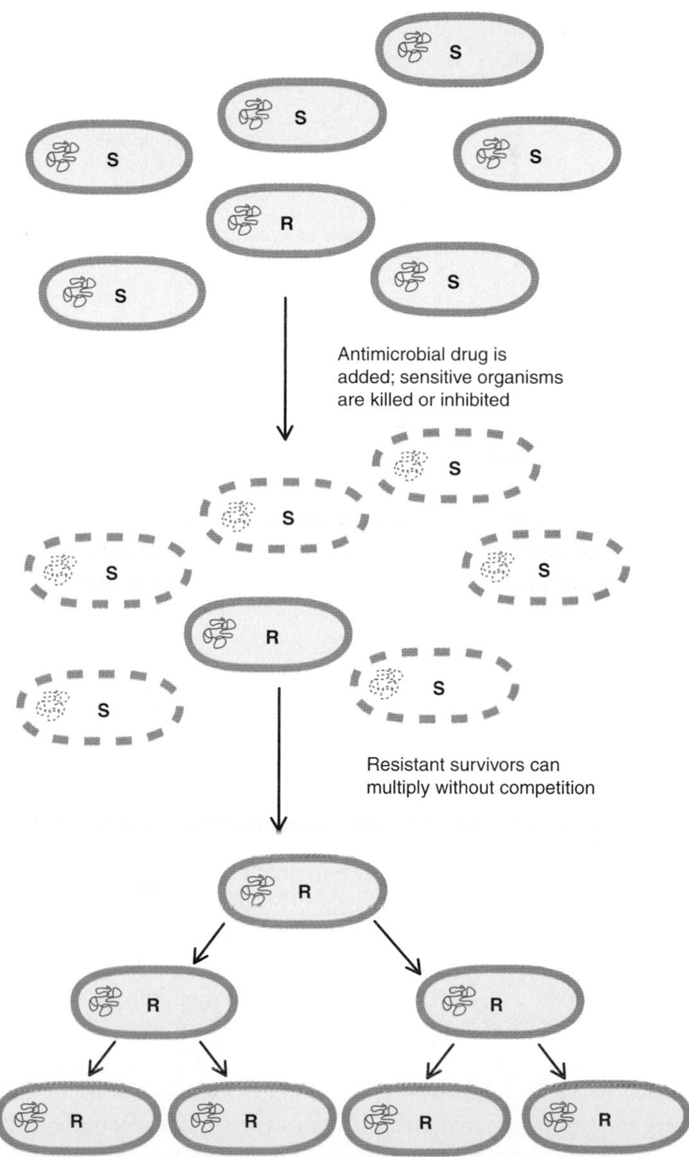

Figure 21.13 The Selective Advantage of Drug Resistance When antimicrobial drugs are used, bacterial strains that are resistant (R) to their effects have a selective advantage over their sensitive (S) counterparts.

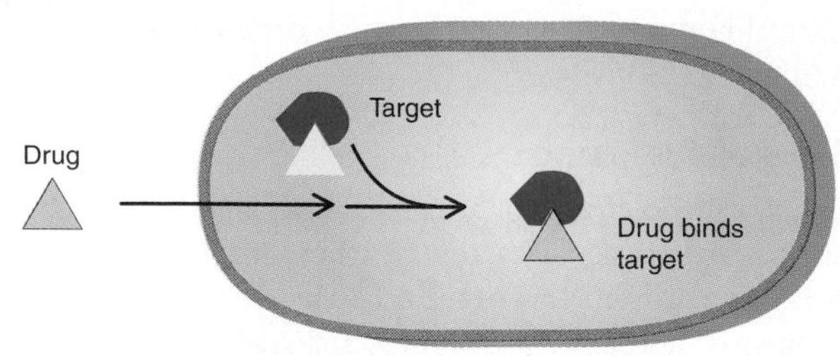

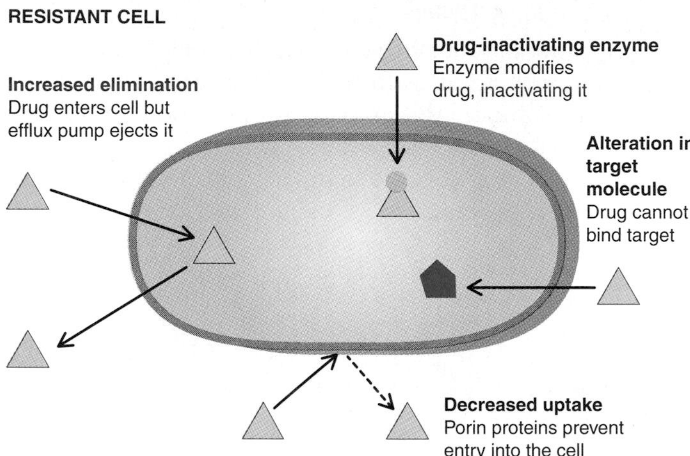

Figure 21.14 **Common Mechanisms of Antimicrobial Drug Resistance**

Alteration in the Target Molecule

An antimicrobial drug generally recognizes and binds to a specific target molecule in a bacterium, interfering with its function. Minor structural changes in the target, which result from mutation, can prevent the drug from binding. For example, alterations in the penicillin-binding proteins prevent β-lactam drugs from binding to them. Similarly, a change in the ribosomal RNA, the target for the macrolides, prevents those drugs from interfering with ribosome function.

Decreased Uptake of the Drug

The porin proteins in the outer membrane of Gram-negative bacteria selectively permit small hydrophobic molecules to enter a cell. Alterations in these proteins can therefore alter permeability and prevent certain drugs from entering the cell. By excluding entry of a drug, an organism avoids its effects. ■ **porins, p. 63**

Increased Elimination of the Drug

The systems that bacteria use to transport detrimental compounds out of a cell are called **efflux pumps**. Alterations that result in the increased expression of these pumps can increase the overall capacity of an organism to eliminate a drug, thus enabling the organism to resist higher concentrations of that drug. In addition, structural changes might influence the array of drugs that can be actively pumped out. Resistance that develops by this mechanism is particularly worrisome because it potentially enables an organism to become resistant to several drugs simultaneously. ■ **efflux pumps, p. 59**

Acquisition of Resistance

Antimicrobial resistance can be due to either spontaneous mutation, which alters existing genes, or acquisition of new genes (**figure 21.15**). Acquisition of resistance through spontaneous mutation is called **vertical evolution**, because it affects only the progeny of the altered cell. In contrast, acquisition of resistance through gene transfer is called **horizontal evolution**; even entirely unrelated organisms can gain new traits this way.

Spontaneous Mutation

As cells replicate, spontaneous mutations occur at a relatively low rate. Even at a low rate, however, such mutations can ultimately have a profound effect on the resistance of a bacterial population to an antimicrobial drug. For example, resistance to the aminoglycoside streptomycin results from a single base-pair change in the gene encoding the ribosomal protein to which streptomycin binds; that point mutation alters the target sufficiently to render the drug ineffective. When a streptomycin-sensitive bacterium is grown in streptomycin-free medium to a population of 10^9 cells, it is probable that at least one cell in the population has that particular mutation in its genome. Through spontaneous mutation, that particular cell has acquired resistance to streptomycin. If streptomycin is then added to the medium, that cell but no others will be able to replicate, eventually generating a population of streptomycin-resistant clones.

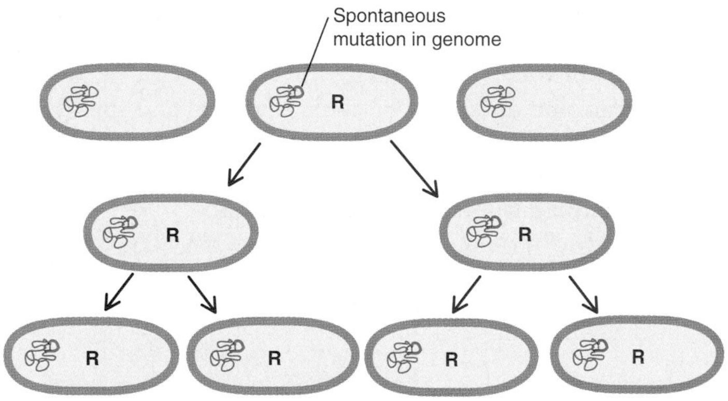

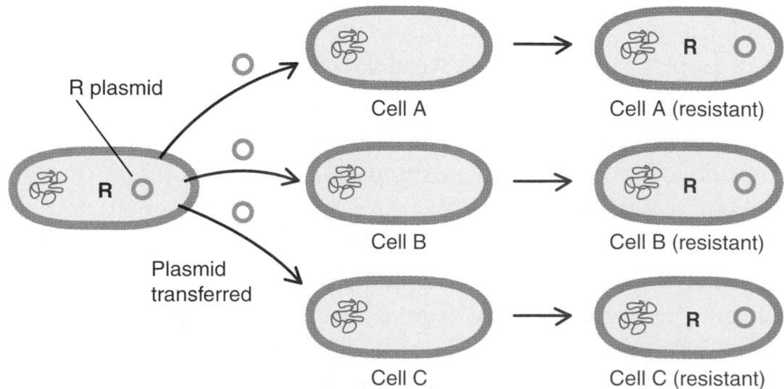

Figure 21.15 **The Acquisition of Antimicrobial Resistance**
(a) Spontaneous mutation, leading to vertical evolution. Only the progeny of the mutant cell are resistant (R). **(b)** Gene transfer, leading to horizontal evolution. Even unrelated bacteria can acquire resistance.

When an antimicrobial drug has several different potential targets or has multiple binding sites on a single target, it is more difficult for an organism to develop resistance through spontaneous mutation. This is because several different mutations are required to prevent binding of the drug. Unlike streptomycin, the newer aminoglycosides bind to several sites on the ribosome, making resistance due to spontaneous mutation less likely.

Drugs such as streptomycin to which single point mutations can confer resistance are sometimes used in combination with another drug to prevent survival of resistant mutants. The chance of an organism simultaneously developing mutational resistance to each drug is extremely low. If any organism spontaneously develops resistance to one drug, the other drug will still kill it.

Gene Transfer

Given the mobility of DNA, genes encoding resistance to an antimicrobial drug can spread to different strains, species, and even genera. The most common mechanism of transfer of resistance is through the conjugative transfer of **R plasmids**. R plasmids frequently carry several different resistance genes, each one mediating resistance to a specific antimicrobial drug. Thus, when an organism acquires an R plasmid, it acquires resistance to several different medications simultaneously. Many of the resistance genes on R plasmids are carried on transposons that can move from a plasmid to the chromosome, from one plasmid to another, or from the chromosome to a plasmid. Thus, if one organism has two different plasmids, an antibiotic-resistance gene can move from one to the other. In this way a gene could move from a **narrow host range plasmid** to a **wide host range plasmid**. Wide host range plasmids, in contrast to narrow host range plasmids, can replicate even if they are transferred to unrelated bacteria. ■ **R plasmid, p. 211** ■ **transposon, p. 194**

In some cases, the resistance gene transferred from one organism to another originated through spontaneous mutation of a common bacterial gene, such as one encoding the target of the drug. In other cases, the gene may have originated from the soil microbe that naturally produces that antibiotic. For example, a gene coding for an enzyme that chemically modifies an aminoglycoside likely originated from the *Streptomyces* species that produces that aminoglycoside.

Examples of Emerging Antimicrobial Resistance

Some of the problems associated with the increasing resistance of bacteria to antimicrobial drugs are highlighted by the following examples.

Enterococci

One of the most dramatic examples of antimicrobial resistance is the enterococci, a group of bacteria that are part of the normal intestinal flora and a common cause of nosocomial infections. Enterococci are intrinsically less susceptible to many common antimicrobials. For example, their penicillin-binding proteins have low affinity for certain β-lactam antibiotics. In addition, many enterococci have plasmid-encoded resistance genes. Some strains, called **vancomycin-resistant enterococci (VRE)**, are now resistant to vancomycin. This drug is usually reserved as a last resort for treating life-threatening infections caused by Gram-positive organisms that are resistant to all β-lactam drugs. Because vancomycin resistance in these strains is encoded on a plasmid, the resistance is potentially transferable to other organisms. ■ **nosocomial infections, p. 488**

Staphylococcus aureus

Staphylococcus aureus, another common cause of nosocomial infections, is becoming increasingly resistant to antimicrobials. Over the past 50 years, most strains have acquired resistance to penicillin due to their acquisition of a gene encoding the enzyme penicillinase. Up until recently, infections caused by these strains could be treated with methicillin or other penicillinase-resistant penicillins. New strains have emerged, however, that not only produce penicillinase, but also have penicillin-binding proteins with low affinity for all β-lactam drugs. These strains, called **methicillin-resistant *Staphylococcus aureus*** or **MRSA**, are resistant to methicillin as well as all other β-lactam drugs. Infections caused by these strains are generally treated with vancomycin. A few hospitals, however, have reported isolates that are no longer susceptible to normal levels of vancomycin. So far, strict hospital guidelines designed to immediately halt the spread of these **vancomycin-intermediate *S. aureus***, or **VISA**, have been successful. ■ ***Staphylococcus aureus*, p. 694**

Streptococcus pneumoniae

Until recently, *Streptococcus pneumoniae*, the leading cause of pneumonia in adults, has remained exquisitely sensitive to penicillin. Some isolates, however, are now resistant to the drug. This acquired resistance is due not to the production of a β-lactamase, but rather to modifications in the chromosomal genes coding for four different penicillin-binding proteins, decreasing their affinities for the drug. The nucleotide changes do not appear to have arisen through a series of point mutations as one might expect; instead, they are due to the acquisition of chromosomal DNA from other species of *Streptococcus*. As you may recall from earlier reading, *S. pneumoniae* undergoes DNA-mediated transformation. ■ ***Streptococcus pneumoniae*, p. 564** ■ **DNA-mediated transformation, p. 205**

Mycobacterium tuberculosis

Treatment of *Mycobacterium tuberculosis* active infections has always been a long and complicated process, requiring a combination of two or three different drugs taken for a period of 6 months or more. Unfortunately, the first-line drugs are often those to which spontaneous mutations readily occur. Because large numbers of organisms are found in an active infection, the probability that one has developed spontaneous resistance to a single drug is likely, which is why multiple drug therapy is required. The length of treatment is due to the tediously slow growth of *M. tuberculosis*. ■ ***Mycobacterium tuberculosis*, p. 569**

Many tuberculosis patients do not comply with the complex treatment regime, skipping doses or stopping treatment too soon, and as a result strains of *M. tuberculosis* are developing resistance to the first-line drugs. This results in even lengthier, more costly treatments that are also less effective. Strains that are resistant to two of the favored drugs for tuberculosis treatment, isoniazid and rifampin, are called **multiple drug resistant *M. tuberculosis*** or **MDR-TB**. To prevent the emergence of these strains, some cities are utilizing **directly observed therapy** in which health care workers routinely visit patients in the community and watch them take their drugs to ensure they comply with their prescribed antimicrobial treatment.

Slowing the Emergence and Spread of Antimicrobial Resistance

To reverse the alarming trend of increasing antimicrobial resistance, everyone must cooperate. On an individual level, physicians as well as the general public must take more responsibility for the appropriate use of these life-saving drugs. On a global scale, countries around the world need to make important policy decisions about what is, and what is not, an appropriate use of these medications.

The Responsibilities of Physicians and Other Health Care Workers

Physicians and other health care workers need to increase their efforts to identify the causative agent of infectious diseases and, only if appropriate, prescribe suitable antimicrobials. They must also educate their patients about the proper use of prescribed drugs in order to increase patient compliance. While these efforts may be more expensive in the short term, they will ultimately save both lives and money.

The Responsibilities of Patients

Patients need to carefully follow the instructions that accompany their prescriptions, even if those instructions seem inconvenient. It is essential to maintain the concentration of the antimicrobial in the blood at the required level for a specific time period. When a patient skips a scheduled dose of a drug, the blood level of the drug may not remain high enough to inhibit the growth of the least sensitive members of the population. If these less sensitive organisms then have a chance to grow, they will give rise to a population that is not as sensitive as the original. Likewise, failure to complete the prescribed course of treatment may not kill the least sensitive organisms, allowing their subsequent multiplication. Misusing antimicrobials by skipping doses or failing to complete the prescribed duration of treatment promotes the gradual emergence of resistant organisms. In essence, the patient is selecting for the stepwise development of resistant mutants.

The Importance of an Educated Public

A greater effort must also be made to educate the public about the appropriateness and limitations of antibiotics in order to ensure that they are utilized wisely. First and foremost, people need to understand that antibiotics are not effective against viruses. Taking antibiotics will not cure the common cold or any other viral illness. A few antiviral drugs are available, but they are effective against only a limited group of viruses such as HIV and herpesviruses. Unfortunately, surveys indicate that far too many people erroneously believe that antibiotics are effective against viruses and often seek prescriptions to "cure" viral infections. This misuse only selects for antibiotic-resistant bacteria in the normal flora. Even though these organisms are not pathogenic themselves, they can serve as a reservoir for R plasmids, eventually transferring their resistance genes to an infecting pathogen.

Global Impacts of the Use of Antimicrobial Drugs

Worldwide, there is growing concern about the overuse of antimicrobial drugs. Countries may vary in their laws and social norms, but antimicrobial resistance recognizes no political boundaries. An organism that develops resistance in one country can quickly be transported globally. In many parts of the world, particularly in developing countries, antimicrobial drugs are available on a non-prescription basis. Because of the consequences of inappropriate use, there is a growing opinion that over-the-counter availability of these drugs should be curtailed or eliminated. Another worldwide concern is the use of antimicrobial drugs in animal feeds. Low levels of these drugs in feeds result in larger, more economically productive animals, and therefore, less expensive meat, a seemingly attractive option. This use, like any other, however, selects for drug-resistant organisms, which has caused some scientists to question its ultimate wisdom. In fact, infections caused by drug-resistant *Salmonella* strains have been linked to animals whose feed was supplemented with those drugs. In response to these concerns, there is growing pressure worldwide to ban the use of antimicrobial drugs in animal feeds.

MICROCHECK 21.5

Mutations and transfer of genetic information have enabled microorganisms to develop resistance to each new antimicrobial drug that has been developed. Drug resistance affects the cost, complications, and outcomes of medical treatment. Slowing the emergence and spread of bacteria involves the cooperation of health care personnel, educators, and the general public.

- Explain how using a combination of two antimicrobial drugs helps prevent the development of spontaneously resistant mutants.
- Explain the significance of a member of the normal flora that harbors an R plasmid.
- A student argued that "spontaneous mutation" meant that a drug could cause mutations. Is the student correct? Why or why not?

Mechanisms of Action of Antiviral Drugs

Viruses rely almost exclusively on the host cell's metabolic machinery for their replication, making them extremely difficult targets for selective toxicity. Viruses are simply nucleic acid, either DNA or RNA, that is surrounded by a protein coat. They have no cell wall, ribosomes, or any other structure targeted by commonly used antibiotics. Thus, viruses are completely unaffected by antibiotics. Some viruses encode their own polymerases, however, and

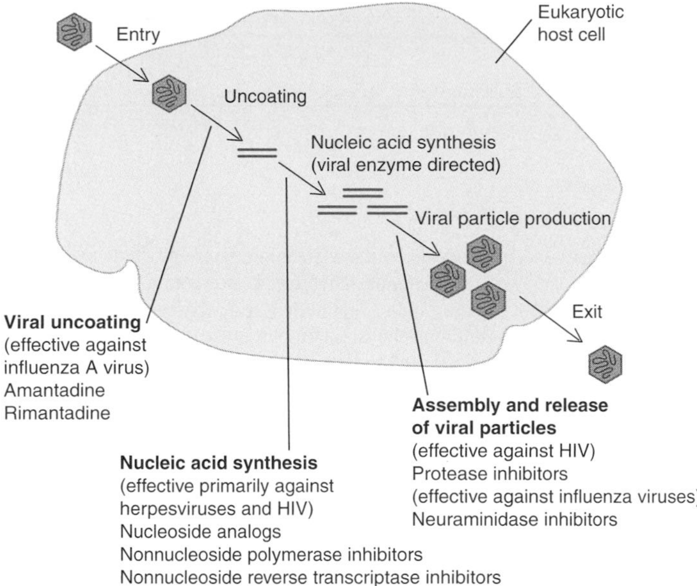

Figure 21.16 Targets of Antiviral Drugs

these are potential targets of antiviral drugs (**figure 21.16**). Relatively few other targets have been discovered.

Many researchers and pharmaceutical companies are currently trying to develop more effective antiviral drugs. The relatively few drugs available are generally effective against only a specific type of virus; none can eliminate latent viruses. **Table 21.3** summarizes the characteristics of the most common antiviral drugs. ■ latent viruses, p. 352

Viral Uncoating

After a virus enters a host cell, the protein coat must dissociate from the nucleic acid in order for replication to occur. Drugs that interfere with the uncoating step thus prevent viral replication. The only effective drugs that target this step, however, are those that block influenza A viruses. ■ influenza A virus, p. 573

Amantadine and Rimantadine

Two drugs, **amantadine** and **rimantadine**, which are similar in both their chemical structure and mechanism of action, block the uncoating of influenza A virus after it enters a cell. Consequently, they prevent or reduce the severity and duration of the disease. The drugs are only used until active immunization can be achieved, since immunization is important for long term protection. Unfortunately, resistance develops frequently and may eventually limit the usefulness of the drugs.

Nucleic Acid Synthesis

Many of the most effective antiviral drugs exploit the error-prone virally encoded enzymes used to replicate viral nucleic acid. With few exceptions, however, the use of these drugs is generally limited to treating infections caused by the herpesviruses and HIV.

Nucleoside Analogs

A growing number of antiviral drugs are **nucleoside analogs**, compounds similar in structure to a nucleoside. These analogs can be phosphorylated *in vivo* by a virally encoded or normal cellular enzyme to form a **nucleotide analog**, a chemical structurally similar to the nucleotides of DNA and RNA. In some cases, incorporation of the nucleotide analog results in termination of the growing nucleotide chain, in a manner analogous to the *in vitro* technique used to determine the sequence of DNA. In other cases, incorporation of the analog results in a defective strand with altered base-pairing properties. ■ nucleotide, p. 35 ■ DNA sequencing, p. 235

The basis for the selective toxicity of most nucleoside analogs is the fact that virally encoded enzymes are much more prone to incorporating the corresponding nucleotide analogs than are host polymerases. Thus, more damage is done to the rapidly replicating viral genome than to host cells. Nucleoside analogs, however, are only effective against replicating viruses. Because viruses such as herpesvirus and HIV can remain latent in cells, the drugs do not cure these infections; they simply limit the duration of the active infection. The virus can still undergo reactivation, causing a reoccurrence of symptoms.

Most nucleoside analogs are reserved for severe infections because of their significant side effects. An important exception to this principle is **acyclovir**, a drug used to limit herpesvirus infections that causes little harm to infected cells. This is because the conversion of acyclovir to a nucleotide analog does not utilize a normal cellular enzyme but instead requires a virally encoded enzyme. The enzyme is only present in cells infected by herpesviruses such as herpes simplex virus (HSV) and varicella-zoster virus (VZV). Thus, acyclovir does not function as a nucleotide analog in uninfected cells. Other nucleoside analogs include ganciclovir, which is used to treat life-threatening or sight-threatening cytomegalovirus (CMV) infections in immunocompromised patients, and ribavirin, which is used to treat respiratory syncytial virus infections (RSV) in newborns. ■ herpes simplex virus, p. 594 ■ varicella zoster, p. 534

A number of different nucleoside analogs are used to treat HIV infection by interfering with the activity of reverse transcriptase. Unfortunately, the virus rapidly develops mutational resistance against these drugs, which is why they are often used in combination with other anti-HIV drugs. Nucleoside analogs that inhibit reverse transcriptase include zidovudine (AZT), didanosine (ddI), and lamivudine (3TC). Two of these are often used in combination for HIV therapy.

Nonnucleoside Polymerase Inhibitors

Nonnucleoside polymerase inhibitors are compounds that inhibit the activity of viral polymerases by binding to a site other than the nucleotide-binding site. One example, foscarnet, is used to treat infections caused by ganciclovir-resistant CMV and acyclovir-resistant HSV.

Nonnucleoside Reverse Transcriptase Inhibitors

Nonnucleoside reverse transcriptase inhibitors inhibit the activity of reverse transcriptase by binding to a site other than the

TABLE 21.3 Characteristics of Antiviral Drugs

Target/Drug Examples	Comments/Characteristics
Viral Uncoating	
Amantadine and rimantadine	Used to treat influenza A infections.
Nucleic Acid Synthesis	
Nucleoside analogs Acyclovir, ganciclovir, ribavirin, zidovudine (AZT), didanosine (ddI), lamivudine (3TC)	Primarily used to treat infections caused by herpesviruses and HIV; they do not cure latent infections. The drugs are converted within eukaryotic cells to a nucleotide analog; virally encoded enzymes are prone to incorporate these, resulting in premature termination of synthesis or improper base-pairing of the viral nucleic acid. Acyclovir is used to treat active herpes simplex virus (HSV) and varicella-zoster virus (VZV) infections. Ganciclovir is used to treat cytomegalovirus infections in immunocompromised patients. Ribavirin is used to treat respiratory syncytial virus (RSV) infections in newborns. Combinations of nucleoside analogs such as zidovudine (AZT), didanosine (ddI), and lamivudine (3TC) are used to treat HIV infections.
Nonnucleoside polymerase inhibitors Foscarnet	Primarily used to treat infections caused by herpesviruses. They inhibit the activity of viral polymerases by binding to a site other than the nucleotide-binding site. Foscarnet is used to treat ganciclovir-resistant cytomegalovirus (CMV) and acyclovir-resistant herpes simplex virus (HSV).
Nonnucleoside reverse transcriptase inhibitors Nevirapine, delavirdine, efavirenz	Used to treat HIV infections. They inhibit the activity of reverse transcriptase by binding to a site other than the nucleotide-binding site and are often used in combination with nucleoside analogs.
Assembly and Release of Viral Particles	
Protease inhibitors Indinavir, ritonavir, saquinavir, nelfinavir	Used to treat HIV infections. They inhibit protease, an essential enzyme of HIV, by binding to its active site.
Neuraminidase inhibitors Zanamivir, oseltamivir	Used to treat influenza virus infections.

nucleotide-binding site. They are often used in combination with nucleoside analogs to treat HIV infections. The medications include nevirapine, delavirdine, and efavirenz.

Assembly and Release of Viral Particles

Virally encoded enzymes that are necessary for the production and release of viral particles are the targets of medications used to treat certain viral infections.

Protease Inhibitors

Protease inhibitors are used to treat HIV infections. These medications inhibit the HIV-encoded enzyme protease, which plays an essential role in the production of viral particles. When HIV replicates, several of its proteins are translated as a single amino acid chain, or polyprotein; protease then cleaves the polyprotein into individual proteins. The various protease inhibitors, including indinavir, ritonavir, saquinavir, and nelfinavir, differ in several aspects including dosage and side effects. ■ **HIV protease, p. 745**

Neuraminidase Inhibitors

Neuraminidase inhibitors inhibit **neuraminidase**, an enzyme encoded by influenza viruses that is essential for the release of infectious viral particles from infected cells. Two neuraminidase inhibitors are currently available—zanamivir, which is administered by inhalation, and oseltamivir, which is administered orally. Both limit the duration of influenza infections when taken within two days of the onset of symptoms.

MICROCHECK 21.6

Viral replication generally uses host cell machinery; thus, there are few targets for selectively toxic antiviral drugs. Most antiviral drugs are virus-specific and target nucleic acid synthesis.

- Explain why acyclovir has fewer side effects than do other nucleoside analogs.
- Why are nucleoside analogs active only against replicating viruses?

Mechanisms of Action of Antifungal Drugs

Eukaryotic pathogens such as fungi more closely resemble human cells than do bacteria. It is not surprising, therefore, that relatively few drugs are available for systemic use against fungal pathogens. Most of these interfere with the synthesis or function of a fungal membrane component. The few drugs that have been available to treat internal infections are quite toxic. Fortunately, however, a

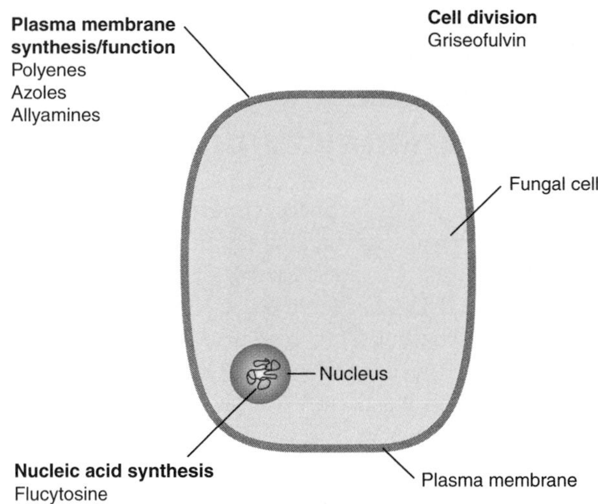

Figure 21.17 **Targets of Antifungal Drugs**

number of less toxic alternatives have recently been developed. The targets of antifungal drugs are illustrated in **figure 21.17**.

In general, there are no routine methods to determine the sensitivity of fungi to various drugs; instead, the drug that previous experience indicates will most likely eliminate the pathogen is generally prescribed. **Table 21.4** summarizes the characteristics of the most common antifungal drugs.

Plasma Membrane Synthesis and Function

The target of most antifungal drugs is the chemical compound **ergosterol**, which is found in the plasma membrane of fungal but not human cells. Unfortunately, these drugs sometimes have a low therapeutic index because they interfere with similar compounds found in eukaryotic cells, which limits their internal use to the treatment of life-threatening infections. Some, however, can be applied topically to treat skin, hair, and nail infections.

Polyenes

The polyenes, a group of antibiotics produced by *Streptomyces*, bind to ergosterol. This disrupts the fungal membrane, causing leakage of the cytoplasmic contents and leading to cell death. Unfortunately, the polyenes are quite toxic to humans, which limits their systemic use to life-threatening infections. Amphotericin B causes severe side effects, but it is the most effective drug for treating many systemic infections. Newer lipid-based emulsions are less toxic but are more expensive. Nystatin is too toxic to be given systemically, but it is used topically.

Azoles

The **azoles** are a large family of chemically synthesized drugs, some of which have antifungal activity. They include two classes, the imidazoles and the newer triazoles; the latter are generally less toxic. Both classes inhibit the synthesis of ergosterol, resulting in defective fungal membranes that leak cytoplasmic contents. Fluconazole and itraconazole, which are triazoles, are increasingly being used to treat systemic fungal infections. Ketoconazole, an imidazole, is also used systemically, but it is associated with more severe side effects. Other imidazoles, including miconazole and clotrimazole, are commonly

TABLE 21.4 Characteristics of Antifungal Drugs

Drug Target/Drug	Comments
Plasma Membrane	
Polyenes	Binds to ergosterol, disrupting the plasma membrane and causing leakage of the cytoplasm. Amphotericin B is very toxic but the most effective drug for treating life-threatening infections; newer lipid-based emulsions are less toxic but very expensive. Nystatin is too toxic for systemic use, but it can be used topically.
Amphotericin B, nystatin	
Azoles	Interferes with ergosterol synthesis, leading to defective cell membranes; active against a wide variety of fungi. Used to treat a variety of systemic and localized fungal infections. Triazoles are less toxic than imidazoles.
Imidazoles: ketoconazole, miconazole, clotrimazole	
Triazoles: fluconazole, itraconazole	
Allylamines	Inhibits an enzyme in the pathway of ergosterol synthesis. Administered topically to treat dermatophyte infections. Terbinafine can be taken orally.
Naftifine, terbinafine	
Cell Division	
Griseofulvin	Used to treat skin and nail infections. Taken orally for months; concentrates in the dead keratinized layers of the skin; taken up by fungi invading those cells and inhibits their division. Active only against fungi that invade keratinized cells.
Nucleic Acid Synthesis	
Flucytosine	Used to treat systemic yeast infections; enzymes within yeast cells convert the drug to 5-fluorouracil, which inhibits an enzyme required for nucleic acid synthesis; not effective against most molds; resistant mutants are common.

used in nonprescription creams, ointments, and suppositories to treat vaginal yeast infections. They are also used topically to treat dermatophyte infections. ■ **dermatophyte, p. 545**

Allylamines

The **allylamines** inhibit an enzyme in the pathway of ergosterol synthesis. Naftifine and terbinafine can be administered topically to treat dermatophyte infections. Terbinafine can also be taken orally.

Cell Division

The target of one antifungal drug, griseofulvin, is cell division.

Griseofulvin

The exact mechanism of **griseofulvin** is unknown, but it appears to interfere with the action of tubulin, a necessary factor in nuclear division. Because tubulin is a part of all eukaryotic cells, the selective toxicity of this drug may be due to its greater uptake by fungal cells. When the drug is taken orally for months, it is absorbed and eventually concentrated in the dead keratinized layers of the skin. The fungi that then invade keratin-containing structures such as skin and nails take up the drug, which prevents their multiplication. It is only active against fungi that invade keratinized cells and is used to treat skin and nail infections. ■ **tubulin, p. 77** ■ **keratin, p. 522**

Nucleic Acid Synthesis

Nucleic acid synthesis is a common feature of all eukaryotic cells, which generally makes it a poor target for antifungal drugs. The drug flucytosine, however, is taken up by yeast cells and then converted by yeast enzymes to an active, inhibitory form.

Flucytosine

Flucytosine is a synthetic derivative of cytosine, one of the pyrimidines found in nucleic acids. Enzymes within infecting yeast cells convert flucytosine to 5-fluorouracil, which inhibits an enzyme required for nucleic acid synthesis. Unfortunately, resistant mutants are common, and therefore, flucytosine is used mostly in combination with amphotericin B or as an alternative drug for patients with systemic yeast infections who are unable to tolerate amphotericin B. Flucytosine is not effective against molds.

MICROCHECK 21.7

Because fungi are eukaryotic cells, there are relatively few targets for selectively toxic antifungal drugs. Most antifungal drugs interfere with the function or synthesis of ergosterol, which is found in the membrane of fungal but not human cells. Some antifungal drugs target cell division and nucleic acid synthesis.

- Why is amphotericin B, a polyene, used only for treating life-threatening infections?
- Why is flucytosine generally used only in combination with other drugs?
- If griseofulvin were not concentrated in keratin-containing structures, would it still be toxic to fungi that invade these structures? Why or why not?

Mechanisms of Action of Antiprotozoan and Antihelminthic Drugs

Most antiparasitic drugs probably interfere with biosynthetic pathways of protozoan parasites or the neuromuscular function of worms. Unfortunately, compared with antibacterial, antifungal, and antiviral drugs, little research and development goes into these drugs, because most parasitic diseases are concentrated in the poorer areas of the world where people simply cannot afford to spend money on expensive medications.

Some of the most important antiparastic drugs and their characteristics are summarized in **table 21.5**.

FUTURE CHALLENGES

War with the Superbugs

With respect to antimicrobial drugs, the future challenge is already upon us. The challenge is to maintain the effectiveness of antimicrobials by (1) preventing the continued spread of resistance, and (2) developing new drugs that have even more desirable properties. New ways must be developed to fight infections caused by the ever greater numbers of bacterial strains that are resistant to the effects of conventional antimicrobial drugs. Several strategies are being used to develop potential weapons against these new "superbugs."

One way to combat resistance is to continue developing modifications of existing antimicrobial drugs. Researchers constantly work to modify drugs chemically, trying to keep at least one step ahead of bacterial resistance. Another method to foil drug resistance is to interfere with the resistance mechanism itself, as is done currently in using β-lactamase inhibitors in combination with β-lactam drugs to protect the antimicrobial drug from enzymatic destruction. Likewise, it may be possible to thwart resistance using other mechanisms—for example, developing chemicals that can be used to inactivate or interfere with bacterial efflux systems.

Other researchers are focusing on developing new drugs that are entirely unrelated to conventional antimicrobials. One example is a class of compounds called ***defensins****, which are short peptides, approximately 29 to 35 amino acids in length, produced naturally by a variety of eukaryotic cells to fight infections. Various defensins and related compounds are being intensively studied as promising antimicrobials.*

Accumulating knowledge of the genetic sequences of pathogens and identification of genes associated with pathogenicity may allow development of antimicrobials that interfere directly with those processes. The "master switch" that controls expression of virulence determinants is one such potential target. Nucleotide sequence information and determination of the three-dimensional conformation of proteins may uncover new targets for antimicrobial drug therapy.

TABLE 21.5 Characteristics of Antiprotozoan and Antihelminthic Drugs

Causative Agent/Drug	Comments
Intestinal protozoa	
Iodoquinol	Mechanism unknown. Poorly absorbed but taken orally to eliminate amebic cysts in the intestine.
Nitroimidazoles Metronidazole	Activated by the metabolism of anaerobic organisms. Interferes with electron transfer and alters DNA. Does not reliably eliminate the cyst stage. Metranidazole is also used to treat infections caused by anaerobic bacteria.
Quinacrine	Mechanism of action is unknown, but it may be due to interference with nucleic acid synthesis.
Plasmodium* (Malaria) and *Toxoplasma	
Folate antagonists Pyrimethamine, sulfonamide	Interferes with folate metabolism. Used to treat toxoplasmosis and malaria.
Quinolones Chloroquine, mefloquine, primaquine, quinine	The mechanism of action is not completely clear. Chloroquine is concentrated in infected red blood cells and is the drug of choice for preventing or treating the red blood cell stage of the malarial parasite. Its effects may be due to inhibition of an enzyme that protects the parasite from the toxic by-products of hemoglobin degradation. Primaquine destroys the liver stage of the parasite and must be used to treat relapsing forms of malaria. Mefloquine or quinine is used to treat infections caused by chloroquine-resistant strains of the malarial parasite.
Trypanosomes and *Leishmania*	
Eflornithine	Used to treat infections caused by some types of *Trypanosoma*. It inhibits the enzyme ornithine decarboxylase.
Heavy metals Melarsoprol, sodium stibogluconate, meglumine antimonate	These inactivate sulfhydryl groups of parasitic enzymes, but they are very toxic to host cells as well. Melarsoprol is used to treat trypanosomiasis, but the treatment itself is often lethal. Sodium stibogluconate and meglumine antimonate are used to treat leishmaniasis.
Nitrofurtimox	Widely used to treat acute Chagas' disease; it forms reactive oxygen radicals that are toxic to the parasite as well as the host.
Intestinal and Tissue Helminths	
Avermectins Ivermectin	Ivermectin causes neuromuscular paralysis in parasites. It is used to treat infections caused by *Strongyloides* and tissue nematodes.
Benzimidazoles Mebendazole, thiabendazole, albendazole	Mebendazole binds to tubulin of helminths, blocking microtubule assembly and inhibiting glucose uptake. It is poorly absorbed in the intestine, making it effective for treating intestinal, but not tissue, helminths. Thiabendazole may have a similar mechanism, but it is well absorbed and has many toxic side effects. Albendazole is used to treat tissue infections caused by *Echinococcus* and *Taenia solium*.
Phenols Niclosamide	Absorbed by cestodes in the intestinal tract, but not by the human host.
Piperazines Piperazine, diethylcarbamazine	Piperazine causes a flaccid paralysis in worms and can be used to treat infections caused by *Ascaris*. Diethylcarbamazine immobilizes filarial worms and alters their surface, which enhances killing by the immune system. The resulting inflammatory response, however, causes tissue damage.
Pyrazinoisoquinolines Praziquantel	A single dose of praziquantel is effective in eliminating a wide variety of trematodes and cestodes. It is taken up but not metabolized by the worm, ultimately causing tetanic contractions of the worm.
Tetrahydropyrimidines Pyrantel pamoate, oxantel	Pyrantel pamoate interferes with neuromuscular activity of worms, causing a type of paralysis. It is not readily absorbed from the gastrointestinal tract and is active against intestinal worms including pinworm, hookworm, and *Ascaris*. Oxantel can be used to treat *Trichuris* infections.

SUMMARY

History and Development of Antimicrobial Drugs

Discovery of Antimicrobial Drugs (Table 21.1)

1. Salvarsan, developed by Paul Ehrlich, was the first documented example of an antimicrobial medication.
2. Chemotherapeutic agents are chemicals used as therapeutic drugs.
3. **Antimicrobial drugs**, or **antimicrobials**, are chemotherapeutic agents that are effective against microbial infections.

Discovery of Antibiotics

1. Alexander Fleming discovered that a species of the fungus *Penicillium* produces **penicillin**, which kills some bacteria.
2. **Antibiotics** are antimicrobial chemicals that are naturally produced by microorganisms.

Development of New Generations of Drugs

1. Antimicrobial drugs can be chemically modified to give them new properties.
2. Penicillin has been altered to create a family of drugs with a variety of new characteristics; the derivatives retain the core portion, called **6-aminopenicillanic acid** (**6-APA**). (Figure 21.1)

Features of Antimicrobial Drugs

1. Most modern antibiotics come from the bacteria *Streptomyces* and *Bacillus*, and the eukaryotic fungi *Penicillium* and *Cephalosporium*.

Selective Toxicity

1. Medically useful antimicrobials are **selectively toxic**, meaning they are more toxic to the pathogen then they are to humans.
2. The relative toxicity of a drug is expressed as the **therapeutic index**, which is the lowest dose toxic to the patient divided by the dose typically used for therapy.

Antimicrobial Action

1. **Bacteriostatic** drugs inhibit the growth of bacteria; drugs that kill bacteria are **bactericidal**.

Spectrum of Activity

1. **Broad-spectrum** antimicrobials affect a wide range of bacteria; those that affect a narrow range are called **narrow-spectrum**.

Tissue Distribution, Metabolism, and Excretion of the Drug

1. Some antimicrobials cross the blood-brain barrier into the cerebrospinal fluid; these can be used to treat meningitis.
2. Drugs that are unstable in acid cannot be taken orally and therefore must be administered through injection.
3. Drugs that have a long **half-life** need to be administered less frequently.

Effects of Combinations of Antimicrobial Drugs

1. **Synergistic** combinations of drugs result in enhanced antimicrobial activity.
2. Some antimicrobials interfere with the activity of others. These combinations are called **antagonistic**.
3. Combinations that are neither synergistic nor antagonistic are called **additive**.

Adverse Effects

1. Some people develop allergies to certain antimicrobials.
2. Some antimicrobials can have potentially damaging side effects such as kidney damage.
3. When the composition of the normal flora is altered, which happens when a person takes antimicrobials, pathogens normally unable to compete may grow to high numbers.

Resistance to Antimicrobials

1. The resistance of certain types of bacteria to a particular drug is **intrinsic** or **innate**.
2. Previously sensitive microorganisms can develop resistance through spontaneous mutation or the acquisition of new genetic information.

Cost

1. Newly introduced drugs are generally far more expensive than their traditional counterparts.

Mechanisms of Actions of Antibacterial Drugs (Table 21.2)

1. Bacterial processes utilizing enzymes or structures that are either different, absent, or not commonly found in eukaryotic cells are the targets of most medically useful antimicrobial drugs. (Figure 21.2)

Antibacterial Medications that Inhibit Cell Wall Synthesis (Figure 21.3)

1. Bacterial cell walls are unique in that they contain peptidoglycan.
2. The β-lactam drugs, which include the penicillins, **cephalosporins**, **carbapenems**, and **monobactams**, irreversibly inhibit **penicillin-binding proteins (PBPs)**, ultimately leading to cell lysis. (Figure 21.4)
3. The different β-lactam drugs differ in their spectrum of activity; some bacteria destroy the activity of β-lactam antibiotics by synthesizing a **β-lactamase**.
4. **Vancomycin** binds to the terminal amino acids of the peptide side chain of NAM, blocking peptidoglycan synthesis; it is important for treating infections caused by bacteria that are resistant to β-lactam drugs.
5. **Bacitracin** interferes with the transport of peptidoglycan precursors across the cytoplasmic membrane; it is a common ingredient in non-prescription ointments.

Antibacterial Medications that Inhibit Protein Synthesis (Figure 21.7)

1. The prokaryotic 70S ribosome, composed of 30S and 50S subunits, is sufficiently different from the 80S eukaryotic ribosome to serve as a target for selective toxicity.
2. Antibiotics that inhibit protein synthesis by binding to the 70S ribosome include the **aminoglycosides**, the **tetracyclines**, the **macrolides**, **chloramphenicol**, the **lincosamides**, the **oxazolidinones**, and the **streptogramins**.

Antibacterial Medications that Inhibit Nucleic Acid Synthesis

1. The **fluoroquinolones** interfere with DNA replication and transcription by inhibiting one or more topoisomerases.
2. The **rifamycins** block prokaryotic RNA polymerase from initiating transcription.

Antibacterial Medications that Inhibit Metabolic Pathways (Figure 21.8)

1. **Sulfa drugs** competitively inhibit an enzyme in the metabolic pathway leading to folic acid synthesis because they are structurally similar to a substrate in the pathway.

2. **Trimethoprim** inhibits the enzyme that catalyzes a metabolic step following the one inhibited by sulfonamides.

Antibacterial Medications that Interfere with Cell Membrane Integrity

1. **Polymyxin B** damages bacterial membranes.

Antibacterial Medications that Interfere with Processes Essential to *Mycobacterium tuberculosis*

1. **First-line** medications that specifically target species of *Mycobacterium* include **isoniazid**, **ethambutol**, and **pyrazinamide**.

Determining the Sensitivity of Bacteria to an Antimicrobial Drug

Determining the Minimum Inhibitory and Bactericidal Concentrations (Figure 21.9)

1. The **minimum inhibitory concentration** (**MIC**) is determined by assaying the ability of the bacterium to grow in broth cultures containing different concentrations of the antimicrobial.
2. The **minimum bactericidal concentration** (**MBC**) is a measurement of the lowest concentration of a specific antimicrobial drug that kills 99.9% of a given strain of bacteria.

Conventional Disc Diffusion Method (Figure 21.10)

1. The **Kirby-Bauer disc diffusion test** is routinely used to qualitatively determine the susceptibility of a given organism to a battery of antimicrobial drugs.

Commercial Modifications of Antimicrobial Susceptibility Testing (Figure 21.11)

1. Automated methods can determine antimicrobial susceptibility in as little as 4 hours.
2. The **E test** utilizes a strip impregnated with a gradient of concentrations of an antimicrobial drug to determine the MIC. (Figure 21.12)

Resistance to Antibacterial Drugs

1. As antimicrobials are increasingly used and misused, the bacterial strains that are resistant to their effects have a selective advantage over their sensitive counterparts. (Figure 21.13)

Mechanisms of Resistance (Figure 21.14)

1. Some organisms produce enzymes that chemically modify a drug, rendering it ineffective.
2. Minor structural changes in the target can prevent the drug from binding.
3. Altered porin proteins can prevent drugs from entering the cell.
4. Efflux pumps can actively pump antimicrobial drugs out of the cell.

Acquisition of Resistance (Figure 21.15)

1. **Vertical evolution** is the acquisition of resistance through spontaneous mutation; **horizontal evolution** is the acquisition of resistance through gene transfer.
2. The most common mechanism of transfer of antibiotic resistance genes is through the conjugative transfer of **R plasmids**.

Examples of Emerging Antimicrobial Resistance

1. **Vancomycin-resistant enterococci** (**VRE**) are resistant to the drug usually reserved as the last resort for treating infections caused by organisms resistant to all β-lactam drugs.
2. **Methicillin-resistant *Staphylococcus aureus*** (**MRSA**) are resistant to all β-lactam drugs.

Slowing the Emergence and Spread of Antimicrobial Resistance

1. Physicians should prescribe antimicrobial medications only when appropriate.
2. The public must be educated about the appropriateness and limitations of antimicrobial therapy. Patients need to carefully follow prescribed instructions when taking antimicrobials.

Mechanisms of Action of Antiviral Drugs (Figure 21.16)

1. Viruses rely almost exclusively on host cell machinery, making them extremely difficult targets for selectively toxicity.
2. The relatively few antiviral drugs available are generally effective against only a specific type of virus; none can eliminate latent viruses. (Table 21.3)

Viral Uncoating

1. **Amantadine** and **rimantadine** block the uncoating of influenza A virus after it enters a cell.

Nucleic Acid Synthesis

1. Most antiviral drugs take advantage of the error-prone virally encoded enzymes used to replicate viral nucleic acid.
2. **Nucleoside analogs** are phosphorylated *in vivo* to form **nucleotide analogs**; when these are incorporated into viral DNA they interfere with replication.

Assembly and Release of Viral Particles

1. **Protease inhibitors** bind to and inhibit protease, the enzyme required for the production of infectious HIV particles.
2. **Neuraminidase inhibitors** interfere with the release of influenza virus particles from a cell.

Mechanisms of Action of Antifungal Drugs (Figure 21.17, Table 21.4)

1. Because fungi are eukaryotic cells, there are relatively few targets for selectively toxic antifungal drugs.

Plasma Membrane Synthesis and Function

1. The **polyenes** disrupt fungal cell membranes by binding to ergosterol.
2. The **azoles** inhibit the synthesis of ergosterol.
3. The **allylamines** inhibit an enzyme in the pathway of ergosterol synthesis.

Cell Division

1. **Griseofulvin** is concentrated in keratinized skin cells, where it inhibits fungal cell division.

Nucleic Acid Synthesis

1. **Flucytosine** is taken up by yeast cells and converted by yeast enzymes to an active form.

Mechanisms of Action of Antiprotozoan and Antihelminthic Drugs (Table 21.5)

1. Most antiparasitic drugs are thought to interfere with biosynthetic pathways of protozoan parasites or the neuromuscular function of worms.

REVIEW QUESTIONS

Short Answer

1. Describe the difference between what is implied by the terms antibiotic and antimicrobial.
2. Define therapeutic index and explain its importance.
3. Explain the role of penicillin-binding proteins in drug susceptibility.
4. Name three of the first-line drugs used to treat tuberculosis.
5. Name three antimicrobial medications that target ribosomes.
6. Compare and contrast the method for determining the minimum inhibitory concentration of an antimicrobial drug by serial dilution with the Kirby-Bauer disc diffusion test.
7. Name three targets that can be altered sufficiently via spontaneous mutation to result in resistance to an antimicrobial drug.
8. What is MRSA? Why is it significant?
9. Why is it difficult to develop antiviral drugs?
10. Explain the difference between the mechanism of action of an azole and that of a polyene.

Multiple Choice

1. Which of the following targets would you expect to be the most selective with respect to toxicity?
   A. Cytoplasmic membrane function
   B. DNA synthesis
   C. Glycolysis
   D. Peptidoglycan synthesis
   E. 70S ribosome
2. Penicillin has been modified to make derivatives that differ in all of the following, *except*...
   A. spectrum of activity.
   B. resistance to β-lactamases.
   C. potential for allergic reactions.
   D. A and C.
3. Which of the following is the target of β-lactam antibiotics?
   A. Peptidoglycan synthesis
   B. DNA synthesis
   C. RNA synthesis
   D. Protein synthesis
   E. Folic acid synthesis
4. Which of the following is *false*?
   A. A bacteriostatic drug stops the growth of a microorganism.
   B. The lower the therapeutic index, the less toxic the drug.
   C. Broad-spectrum antibiotics are associated with the development of antibiotic-associated colitis.
   D. Azithromycin has a longer half-life than does penicillin V.
   E. Chloramphenicol can cause a life-threatening type of anemia.
5. All of the following interfere with the function of the ribosome, *except*...
   A. fluoroquinolones.
   B. lincosamides.
   C. macrolides.
   D. streptogramins.
   E. tetracyclines.
6. The target of the sulfonamides is...
   A. cytoplasmic membrane proteins.
   B. folic acid synthesis.
   C. gyrase.
   D. peptidoglycan biosynthesis.
   E. RNA polymerase.
7. Routine antimicrobial therapy to treat tuberculosis involves taking...
   A. one drug for 10 days.
   B. three drugs for 10 days.
   C. one drug for at least 6 months.
   D. three drugs for at least 6 months.
   E. five drugs for 2 years.
8. Methicillin-resistant *Staphylococcus aureus* are sensitive to...
   A. Methicillin.
   B. Penicillin.
   C. Cephalosporin.
   D. Vancomycin.
   E. None of the above.
9. Acyclovir is a...
   A. nucleoside analog.
   B. nonnucleoside polymerase inhibitor.
   C. protease inhibitor.
   D. None of the above.
10. The antifungal drug griseofulvin is used to treat...
   A. vaginal infections.
   B. systemic infections.
   C. nail infections.
   D. hair infections.

Applications

1. A physician was treating one young female and one elderly patient for urinary tract infections caused by the same type of bacterium. Although the patients had similar body dimensions and weight, the physician gave a smaller dose of drug to the older patient. What was the physician's rationale for this decision?
2. An advocacy group in Washington, D.C., is petitioning the United States Department of Agriculture (USDA) to stop the use of low-dosage antimicrobial agents used to enhance the growth of cattle and chickens. Why is the group against this practice? Why does the USDA permit it?

Critical Thinking

1. Figure 21.12 shows the E-test procedure for determining an MIC value. How would the zone of inhibition appear if the drug concentrations in the strip were decreased slightly?
2. Why is acyclovir converted to a nucleotide analog only in cells infected with herpes simplex virus?

CHAPTER 22

Skin Infections

Howard T. Ricketts was born in Ohio in 1871. He studied medicine in Chicago, and then specialized in pathology, the study of the nature of disease and its causes. In 1902, he was appointed to the faculty of the University of Chicago, where his research interests turned to an often fatal and little understood disease characterized by a dramatic rash, Rocky Mountain spotted fever. The disease could be transmitted to laboratory animals by injecting them with blood from an infected person, and Ricketts noticed that people and laboratory animals with the disease had tiny bacilli in their blood. Ricketts was sure that these tiny bacteria were the cause of the disease, but he was never able to cultivate them on laboratory media. Based on observations of victims with the disease, Ricketts and others suspected that Rocky Mountain spotted fever was contracted from tick bites, and Ricketts went on to prove that certain species of ticks could transmit the disease from one animal to another. The infected ticks remained healthy but capable of transmitting the disease for long periods of time, and oftentimes the offspring of infected ticks were also infected. Ricketts was able to explain this by showing that the eggs of infected ticks often contained large numbers of the tiny bacilli, an example of transovarial, meaning via the eggs, passage of an infectious agent. Frustrated by his inability to cultivate the bacilli for further studies, Ricketts declined to give them a scientific name and went off to Mexico to study a very similar disease, louse-borne typhus (Rocky Mountain spotted fever is also known as tick-borne typhus). Unfortunately, Ricketts contracted the disease and died at the age of 39. Five years later a European scientist, Stanislaus Prowazek, studying the same disease in Serbia and Turkey met the same fate at almost the same age. The martyrdom of the two young scientists struggling to understand infectious diseases is memorialized in the name of the louse-borne agent, Rickettsia prowazekii. *Both the genus and species names of the Rocky Mountain spotted fever agent,* Rickettsia rickettsii, *recognize Howard Ricketts. We now know that these bacteria are obligate intracellular parasites, which explains why they could not be cultivated on ordinary laboratory media. Antibiotics, which could have saved these men, had not yet been discovered.*

—A Glimpse of History

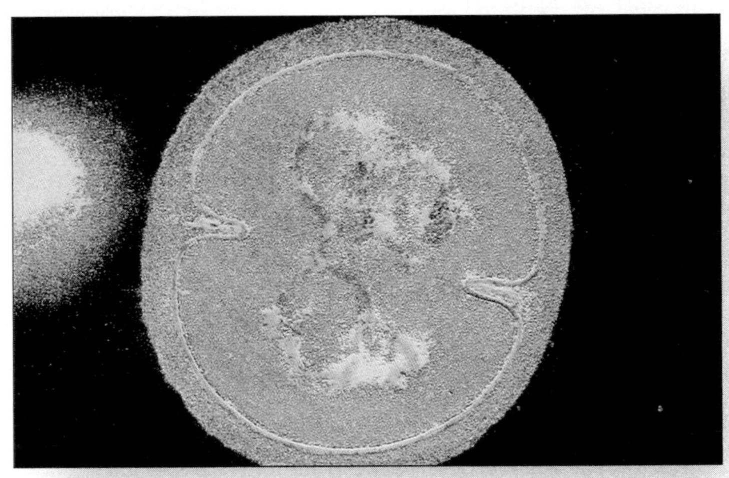

A MAJOR PART OF THE BODY'S CONTACT WITH THE outside world occurs at the surface of the skin. As long as skin is intact, this tough, flexible outer covering is remarkably resistant to infection. Because of its exposed state, however, it is frequently subject to cuts, punctures, burns, chemical injury, hypersensitivity reactions, and insect or tick bites. These skin injuries provide a way for pathogens to enter and infect the skin and underlying tissues. Skin infections also occur when microorganisms or viruses are carried to the skin by the bloodstream after entering the body from another site, such as the respiratory or gastrointestinal systems.

Anatomy and Physiology

The skin is far more than an inert wrapping for the body. Control of body temperature and prevention of loss of fluid from body tissues are vitally important functions of the skin. It also plays an important role in the synthesis of vitamin D, which is needed for normal teeth and bones. Numerous sensory receptors of various types occur in the skin, providing the central nervous system with information about the environment. The skin also produces cytokines that aid the development and function of cell-mediated immunity, and collections of lymphocytes are closely associated with the skin. Because of its exposed location, the temperature of the skin is generally lower than that of the rest of the body. ■ cytokines, p. 379 ■ cell-mediated immunity, p. 404 ■ skin-associated lymphoid tissue, p. 374

The **epidermis**, the surface layer of the skin **(figure 22.1)**, ranges from 0.007 to 0.12 mm thick. The outer portion is composed of scaly material made up of flat cells containing **keratin**, a durable protein also found in hair and nails. The cells on the skin surface are dead and, along with any resident organisms, continually peel off, replaced by cells from deeper in the epidermis. These cells, in turn, become flattened and die as keratin is formed within them. This process results in a complete regeneration of the skin about once a month. Dandruff represents excessive shedding of skin cells, but mostly the shedding process is unnoticed and represents one of the skin's defenses against infection.

The epidermis is supported by the **dermis**, a second, deeper layer of skin cells through which many tiny nerves, blood vessels, and lymphatic vessels penetrate. The dermis adheres in a very irregular fashion to the fat and other cells that make up the subcutaneous tissue (see figure 22.1).

Fine tubules of sweat glands and hair follicles traverse the dermis and epidermis (see figure 22.1). Since sweat is a salty solution, high concentrations of salt occur as it evaporates. Evaporation of sweat and regulation of the amount of blood flow through the skin's blood vessels are critically important in controlling body temperature. A hair follicle is like a tiny tunnel lined with squamous epithelium from which a hair, which can be coarse or so fine as to be almost invisible, arises **(figure 22.2)**. Sebaceous glands produce an oily secretion called **sebum** that feeds into the sides of the hair follicles. This secretion flows up through the follicles and spreads out over the skin surface, keeping the hair and skin soft, pliable, and water-repellent. The hair follicles provide passage for certain salt-tolerant bacteria to penetrate the skin and reach deeper tissues.

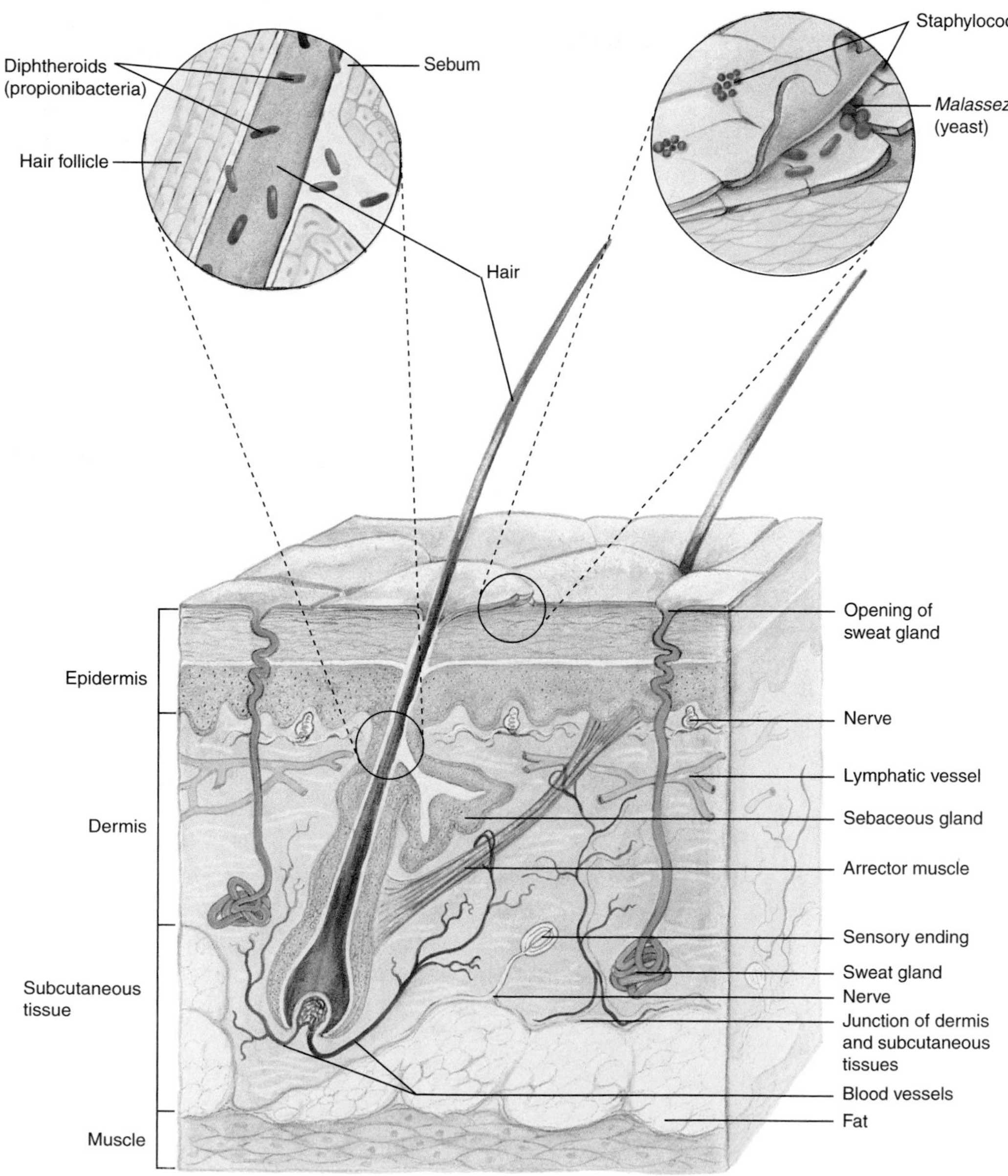

Figure 22.1 Microscopic Anatomy of the Skin Notice that the sebaceous unit, composed of the hair follicle and the attached sebaceous gland, almost reaches the subcutaneous tissue.

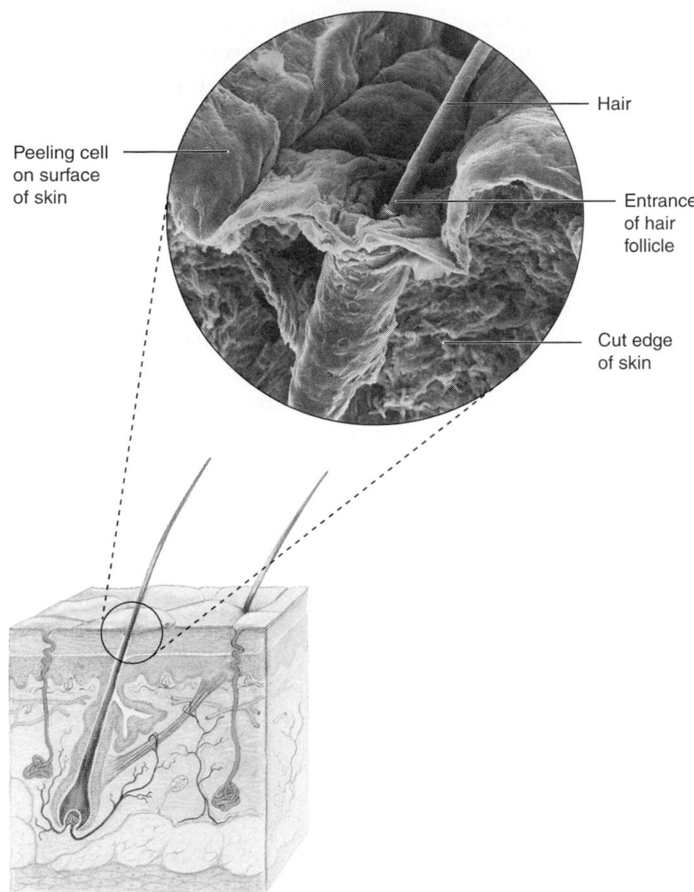

Figure 22.2 Scanning Electron Micrograph of the Surface and Cut Edge of Normal Human Skin Notice peeling of some of the surface cells, and the entrance of the hair follicle.

The secretions of the sweat and sebaceous glands are essential to the normal microbial population of the skin because they supply water, amino acids, and lipids, which serve as nutrients for microbial growth. The normal pH of skin ranges from 4.0 to 6.8. Breakdown of the lipids by the microbial residents of normal skin results in fatty-acid by-products that inhibit the growth of many potential disease-producers. In fact, the normal skin surface is an unfriendly habitat for most potential pathogens, being too dry, salty, unstable because of shedding, acidic, and toxic for their survival.

MICROCHECK 22.1

The skin is a large, complex organ that covers the external surface of the body. Properties of the skin cause it to resist colonization by most microbial pathogens, pose a physical barrier to infection, provide sensory input from the environment, and assist the body's regulation of temperature and fluid balance.

- Give three routes by which microorganisms invade the skin.
- Give four characteristics of skin that help it resist infection.
- Would a person living in the tropics or in the desert have larger numbers of bacteria living on the surface of their skin?

Normal Flora of the Skin

The skin represents a distinct ecological niche, analogous to a cool desert, compared to the warm, moist tropical conditions that exist in other body systems. Large numbers of microorganisms live on and in the various components of the normal skin. For example, depending on the body location and amount of skin moisture, the number of bacteria on the skin surface may range from only about 1,000 organisms per square centimeter on the back to more than 10 million in the groin and armpit, where moisture is more plentiful. The numbers actually increase after a hot shower because of increased flow from the skin glands where many reside. Most of the microbial skin inhabitants can be categorized in three groups, diphtheroids, staphylococci, and yeasts (**table 22.1**, and see figure 22.1). Although generally harmless, skin organisms are opportunistic pathogens, meaning that they can only cause disease in people with impaired body defenses. AIDS patients and others with impaired immunity are especially vulnerable. ■ normal flora, p. 451

Diphtheroids

Diphtheroids are a group of bacteria named for their resemblance to the diphtheria bacillus, *Corynebacterium diphtheriae*. Their distinctive characteristics are Gram-positive staining, variation in shape, and low virulence. Unlike *C. diphtheriae*, they do not produce exotoxin. Diphtheroids are responsible for body odor. A diphtheroid found on the skin in large numbers is *Propionibacterium acnes*, which is present on virtually all humans. Surprisingly, most strains of *P. acnes* are anaerobic, although some strains are aerotolerant. This bacterium grows primarily within the hair follicles, where conditions are anaerobic. Growth of *P. acnes* is enhanced by the oily secretion of the sebaceous glands, and the organisms are usually present in large numbers only in areas of the skin where these glands are especially well developed—on the face, upper chest, and back. These are also the areas of the skin where acne most commonly develops, and the frequent association of *P. acnes* with acne inspired its name, even though most people who carry the organisms do not have acne. ■ aerotolerance, p. 98

Acne in its most common form begins at puberty in association with a rise in sex hormones, enlargement of the sebaceous glands, and enhanced secretion of sebum. The hair follicle epithelium thickens and sloughs off in cohesive clumps, causing increasing obstruction to the flow of sebum to the skin surface. The follicle becomes distended with sebum, which causes the

TABLE 22.1 Principal Members of the Normal Skin Flora

Name	Characteristics
Diphtheroids	Variably shaped nonmotile, Gram-positive rods of the *Corynebacterium* and *Propionibacterium* genera
Staphylococci	Gram-positive cocci arranged in packets or clusters; coagulase negative; facultatively anaerobic
Malassezia sp.	Small yeasts that require oily substances for growth

epidermis to bulge outward, producing a whitish lesion called a whitehead. Continued sebum production by the gland can force a plug of material to the surface, where it is visible as a blackhead. The *P. acnes* that normally reside in the gland multiply to enormous numbers in the trapped sebum. Lipases of the bacteria degrade the sebum, releasing fatty acids and glycerol, a growth requirement of the organisms. The metabolic products of the bacteria cause an inflammatory response, attracting leukocytes (white blood cells) whose enzymes damage the wall of the distended follicle. The inflammatory process can cause the follicle to rupture, releasing the follicle contents into the surrounding tissue. The result is an **abscess**, a collection of white blood cells, bacteria, and cellular debris, which eventually heals and leaves a scar. Squeezing acne lesions is ill-advised, because it promotes rupture of the inflamed follicles and therefore more acne scars. Usually acne can be controlled until it goes away by itself, by using medications such as antibiotics and benzoyl peroxide that inhibit the growth of *P. acnes*, or by those such as azelaic acid (Azelex) and isotretinoin (Accutane) that act primarily to reverse the hair follicle abnormalities. The latter medication is reserved for the most serious cases of acne because it has potentially serious side effects. ■ **inflammation, p. 382** ■ **leukocytes, p. 370**

Staphylococci

The second group of microorganisms universally present on the normal skin is composed of members of the genus *Staphylococcus*. They are salt-tolerant organisms and grow well on the salty skin surface. As with the diphtheroids, most of these bacteria have little virulence, although they certainly can cause serious disease if host defenses are impaired. Generally, staphylococci are the most common of the skin bacteria able to grow aerobically. The principal species is *Staphylococcus epidermidis*. ■ **staphylococci, p. 290**

Important functions of the skin's staphylococci are to prevent colonization by pathogens and to maintain a balance among the microbial inhabitants of the skin ecosystem. These Gram-positive cocci compete for nutrients with other potential skin colonizers, and they also produce antimicrobial substances highly active against *P. acnes* and other Gram-positive bacteria.

Fungi

Tiny lipophilic, meaning oil-requiring, yeasts almost universally inhabit the normal human skin from late childhood onward. Their shape varies with different strains, being round, oval, or sometimes short, rods. These yeasts can be cultivated on laboratory media containing fatty substances such as olive oil. They belong to the genus *Malassezia*, formerly *Pityrosporum*, and are generally harmless. In some people, however, they cause skin conditions such as a scaly face rash, dandruff, or **tinea versicolor (figure 22.3)**. The latter is a common skin disease that causes a patchy scaliness and increased pigment in light-skinned persons, or a decrease in pigment in dark-skinned people. Scrapings of the affected skin show large numbers of *Malassezia furfur* both in its yeast form and as short filaments called pseudohyphae. Unknown factors, probably relating to the host, are important in these diseases since most people carry *Malassezia* sp. on their skin without any disease. AIDS patients often have a severe rash with pus-filled pimples caused by *Malassezia* yeasts, and the organisms may even infect internal organs in patients receiving fat-containing intravenous feedings. ■ **yeasts, p. 309**

MICROCHECK 22.2

The normal skin flora are important because they help protect against colonization by pathogens. Occasionally, they cause disease when body defenses are

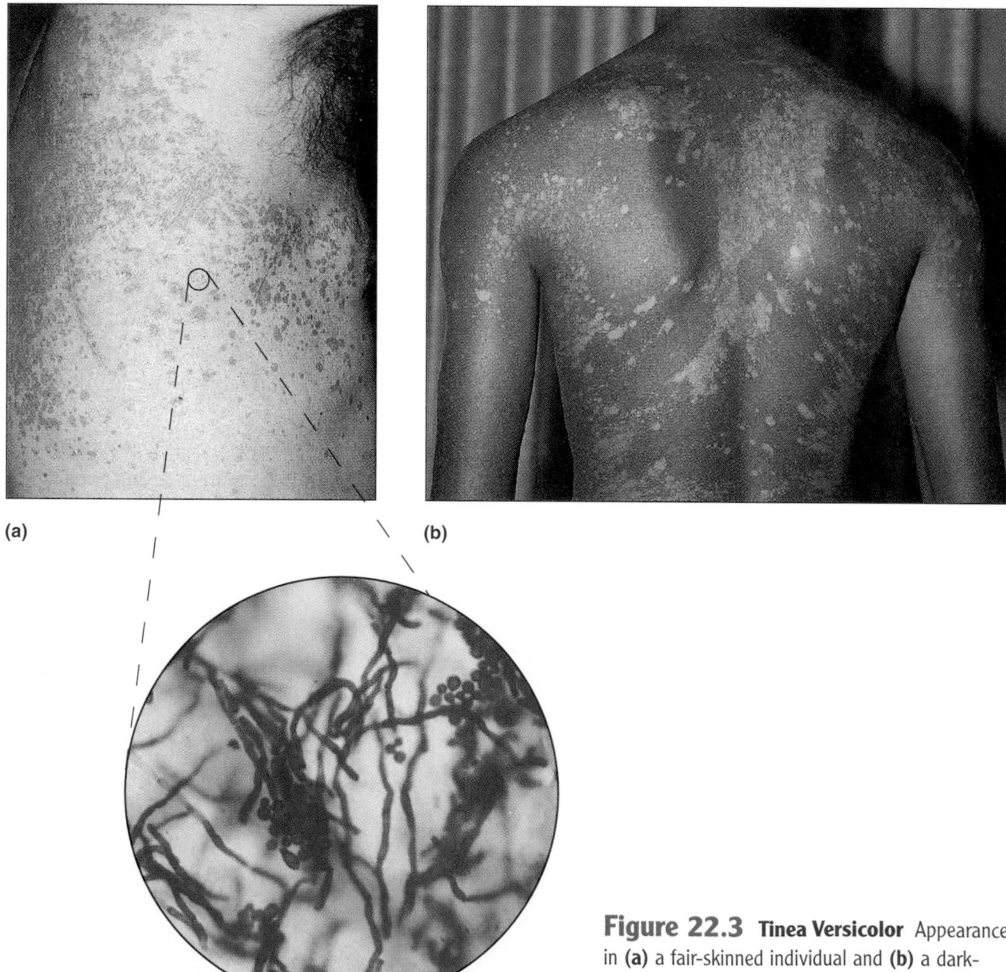

Figure 22.3 Tinea Versicolor Appearance in **(a)** a fair-skinned individual and **(b)** a dark-skinned individual. **(c)** Microscopic appearance of skin scraping showing *Malassezia furfur* yeast and filamentous forms.

impaired. They are responsible for body odor, and probably contribute to acne.

- Name and describe the three groups of organisms generally present on normal skin.
- Under what circumstances is *Malassezia furfur* most likely to be pathogenic?
- Would frequent showering tend to increase or decrease the numbers of *Staphylococcus* on the surface of the skin? Why?

Bacterial Skin Diseases

Only a few species of bacteria commonly invade the intact skin directly, which is not surprising in view of the anatomical and physiological features discussed earlier. Hair follicle infections exemplify direct invasion.

Hair Follicle Infections

Infections originating in hair follicles commonly clear up without treatment. In some instances, however, they progress into severe or even life-threatening disease.

Symptoms

Folliculitis, furuncles, and carbuncles represent different outcomes of hair follicle infections. In **folliculitis**, a small red bump, or pimple, develops at the site of the involved hair follicle. Often, the hair can be pulled from its follicle, accompanied by a small amount of pus, and then the infection goes away without further treatment. If, however, the infection extends from the follicle to adjacent tissues, causing localized redness, swelling, severe tenderness, and pain, the lesion is called a **furuncle** or boil. Pus may drain from the boil along with a plug of inflammatory cells and dead tissue. A **carbuncle** is a large area of redness, swelling, and pain punctuated by several sites of draining pus. Carbuncles usually develop in areas of the body where the skin is thick, such as the back of the neck. Fever is often present, along with other signs of a serious infection.

Causative Agent

Most furuncles and carbuncles, as well as many cases of folliculitis, are caused by *Staphylococcus aureus*, a staphylococcus that produces coagulase and is therefore called "coagulase-positive." It is much more virulent than the staphylococci normally found on the skin. The name derives from *staphyle*, "a bunch of grapes," referring to the arrangement of the bacteria as seen on stained smears, and *aureus*, "golden," referring to the color of the *S. aureus* colonies. This bacterium is an extremely important pathogen and is mentioned frequently throughout this text as the cause of a number of medical conditions. ■ **coagulase, p. 694** ■ ***Staphylococcus aureus*, p. 641, 694**

Pathogenesis

Infection begins when virulent staphylococci attach to the cells of a hair follicle, multiply, and spread downward to involve the follicle and sebaceous glands. The infection induces an inflammatory response with swelling and redness, followed by attraction and accumulation of polymorphonuclear leukocytes. If the infection continues, the follicle becomes a plug of inflammatory cells and necrotic tissue overlying a small abscess (**figure 22.4**). The infectious process spreads deeper, reaching the subcutaneous tissue where a large abscess forms. Without effective treatment, pressure within the abscess increases, causing it to expand to other hair follicles, causing a carbuncle. If organisms enter the bloodstream, the infection can spread to other parts of the body, such as the heart, bones, or brain.

The properties of *S. aureus* that contribute to its virulence are shown in **table 22.2**. Virtually all strains possess an unusual cell wall component called **protein A**. This protein, some of which is released from the cell, binds to the Fc region of antibody molecules, thereby preventing the antibody from attaching to Fc receptors on phagocytes. Thus, a major effect of protein A is to interfere with phagocytosis. Many strains of *S. aureus* growing in body tissues synthesize a polysaccharide capsule that also inhibits phagocytosis. The *S. aureus* genes responsible for capsule formation are activated following invasion of tissue; the same strains generally do not synthesize capsules when grown on enriched laboratory media. Besides producing these cellular components, *S. aureus* produces numerous extracellular products that might contribute to virulence. These products include **leukocidins**, which kill white blood cells; **hyaluronidase**, which degrades hyaluronic acid, a component of host tissue that helps hold the cells together; **proteases**, which degrade various host proteins including collagen, the white fibrous protein found in skin, tendons, and connective tissue; and **lipases**, which degrade lipids. Lipases may assist colonization of the oily hair follicles by strains of *S. aureus* that cause follicle infections. ■ **protein A, p. 460** ■ **Fc receptors, p. 394**

Epidemiology

Staphylococcus aureus inhabits the nostrils of virtually everyone at one time or another, each nostril containing as many as 10^8

TABLE 22.2 Virulence Factors of *Staphylococcus aureus*

Product	Effect
Capsule	Inhibits phagocytosis
Coagulase	May impede progress of leukocytes into infected area by producing clots in the surrounding capillaries
Exfoliatin	Separates layers of epidermis, causing scalded skin syndrome
Hyaluronidase	Breaks down hyaluronic acid component of tissue, thereby promoting extension of infection
Leukocidin	Kills white blood cells by producing holes in their cytoplasmic membrane
Lipase	Breaks down fats by hydrolyzing the bond between glycerol and fatty acids
Proteases	Degrade collagen and other tissue proteins
Protein A	Binds to Fc portion of antibody, inhibiting phagocytosis (blocks attachment to Fc receptors on white blood cells)
Toxic shock syndrome toxin	Causes rash, diarrhea, and shock

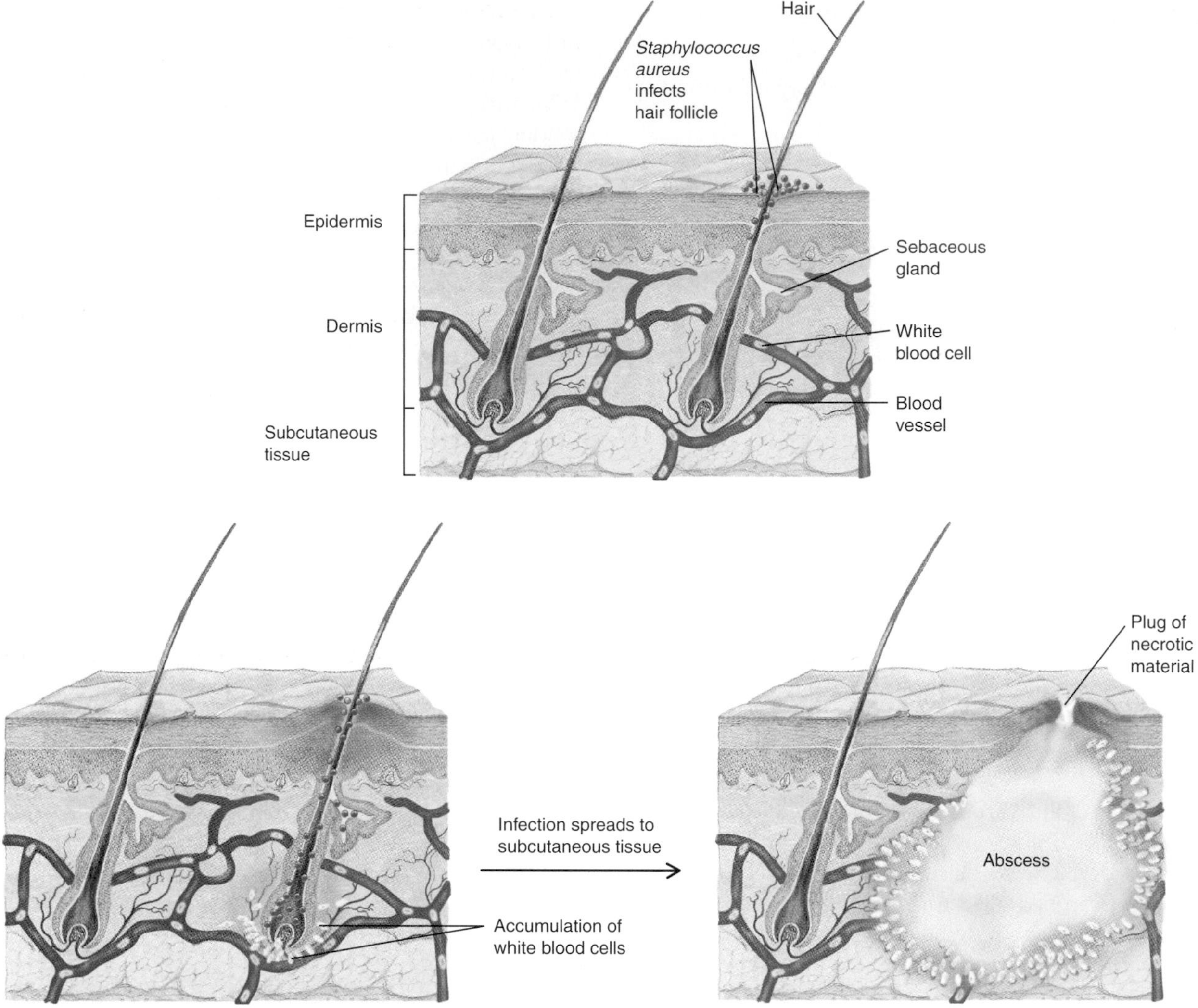

Figure 22.4 Pathogenesis of a Boil *Staphylococcus aureus* infects a hair follicle through its opening on the skin surface. The infection produces a plug of necrotic material, a small abscess in the dermis, and a larger abscess in the subcutaneous tissue.

bacteria. About 20% of healthy adults have continually positive nasal cultures for a year or more, while over 60% will be colonized at some time during a given year. The organisms are mainly disseminated to other parts of the body and to the environment by the hands. Although the nasal chambers seem to be the preferred habitat of *S. aureus*, moist areas of skin are also frequently colonized. People with boils and other staphylococcal infections shed large numbers of *S. aureus* and should not work with food, or near patients with surgical wounds or chronic illnesses. Staphylococci survive well in the environment, which favors their transmission from one host to another. Various domestic animals carry *S. aureus*, but animal strains do not commonly cause human infections. Since *S. aureus* is so commonplace and there are many different strains in the population, epidemics of staphylococcal disease can generally be traced to their sources only by precise identification of the epidemic strain. Techniques for characterizing strains of *S. aureus* include the pattern of susceptibility to multiple antibiotics, bacteriophage typing, and plasmid identification. All of these techniques, however, have their limitations. A more reliable method is to compare the electrophoretic patterns of the DNA fragments produced by treatment with a restriction enzyme. ■ **bacteriophage typing, p. 257** ■ **restriction enzymes, p. 215, 225, 336**

Prevention and Treatment

Prevention of staphylococcal skin disease is very difficult. Attempts are made to eliminate the carrier state by applying an antistaphylococcal cream to the nostrils, and using soaps containing an antistaphylococcal agent such as hexachlorophene to bathe the skin. Effective treatment of boils and carbuncles often requires that the pus be surgically drained from the lesion and an antistaphylococcal medicine be given. Antibiotic treatment is complicated by the fact that about 90% of *S. aureus* strains produce the penicillin-destroying enzyme penicillinase, a β-lactamase, and so penicillin cannot routinely be used in treatment. Some strains are resistant to multiple antibiotics including β-lactamase-resistant penicillins, cephalosporins, and vancomycin. ■ **cephalosporins, p. 503**

Scalded Skin Syndrome

Staphylococcal scalded skin syndrome, SSSS, is a potentially fatal toxin-mediated disease that occurs mainly in infants but can also occur in children and adults.

Symptoms

As the name suggests, the skin appears to be scalded (**figure 22.5**). SSSS begins as a generalized redness of the skin affecting 20% to 100% of the body. Other symptoms, such as **malaise**—a vague feeling of discomfort and uneasiness—irritability, and fever are also present. The nose, mouth, and genitalia may be painful for one or more days before the typical features of the disease become apparent. Within 48 hours after the redness appears, the skin becomes wrinkled, and large blisters filled with clear fluid develop. The skin is tender to the touch and looks like sandpaper.

Causative Agent

Staphylococcal scalded skin syndrome is caused by toxins called **exfoliatins** produced by certain strains of *Staphylococcus aureus.* These toxins destroy material that binds together the layers of skin. At least two kinds of exfoliatins exist: one is coded by a plasmid gene, and the other is chromosomal. Some strains of *S. aureus* produce both kinds of toxins. ■ plasmids, p. 70, 208, 210

Pathogenesis

Exfoliatin, released by *S. aureus* at the site of infection, is absorbed and carried by the bloodstream to large areas of the skin. In the skin, it causes a split in the cellular layer of the epidermis just below the dead keratinized outer layer. The toxins have esterase activity, which is thought to break ester bonds that hold these cells together. *Staphylococcus aureus* is usually not present in the blister fluid but may be cultivated from the primary site of infection or from the blood, skin, or nose. Because the outer layers of skin are lost as in a severe burn, there is marked loss of body fluid and danger of secondary infection with Gram-negative bacteria such as *Pseudomonas* sp., or with fungi such as *Candida albicans.* **Secondary infection** means invasion by a new organism of tissues damaged by an earlier infection. Mortality can range up to 40%, depending on how promptly the disease is diagnosed and treated, and the patient's age and general health. ■ ester bonds, p. 37 ■ *Candida albicans*, p. 308, 311

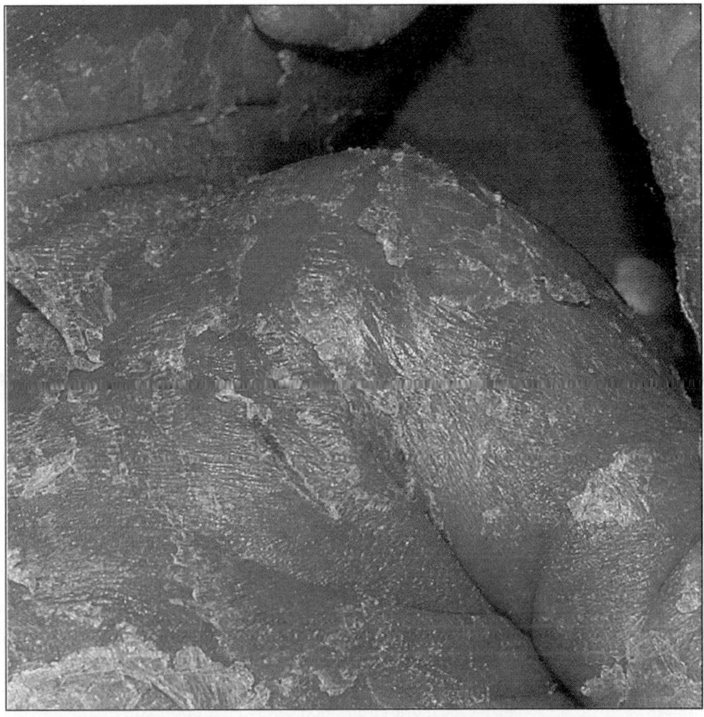

Figure 22.5 Staphylococcal Scalded Skin Syndrome (SSSS) A toxin called exfoliatin, produced by certain strains of *Staphylococcus aureus*, causes the outer layer of skin to separate.

Epidemiology

About 5% of *S. aureus* strains produce exfoliatins. The disease can appear in any age group but occurs most frequently in newborn infants, the elderly, and immunocompromised adults. Staphylococcal scalded skin syndrome usually appears in isolated cases, although small epidemics in nurseries sometimes occur.

Prevention and Treatment

There are no preventive measures except to place patients suspected of having SSSS in protective isolation. This means putting the patient in a private room attended by staff who use gloves, breathing masks, and protective gowns. These measures help to limit spread of the pathogen to others and help prevent secondary infection of the isolated patient. Initial therapy includes a bactericidal antistaphylococcal antibiotic such as methicillin, a penicillinase-resistant derivative of penicillin. All dead skin and other tissue are removed to help prevent secondary infection.

Table 22.3 describes the main features of this disease.

Streptococcal Impetigo

A skin infection characterized by pus production is called **pyoderma**. Pyodermas can result from infection of an insect bite, burn, scrape, or other wound. Sometimes, the injury is so slight

TABLE 22.3 Staphylococcal Scalded Skin Syndrome

Symptoms	Tender red rash with sandpaper texture, malaise, irritability, fever, large blisters, peeling of skin
Incubation period	Variable, usually days
Causative agent	Strains of *Staphylococcus aureus* that produce exfoliatin toxin
Pathogenesis	Exfoliatin toxin is produced by staphylococci at an infection site, usually of the skin, and carried by the bloodstream to the epidermis, where it causes a split in a cellular layer; loss of body fluid and secondary infections contribute to mortality
Epidemiology	Person-to-person transmission; seen mainly in newborns, but can occur at any age
Prevention and treatment	Isolation of the victim to protect from environmental potential pathogens; penicillinase-resistant penicillin; removal of dead tissue

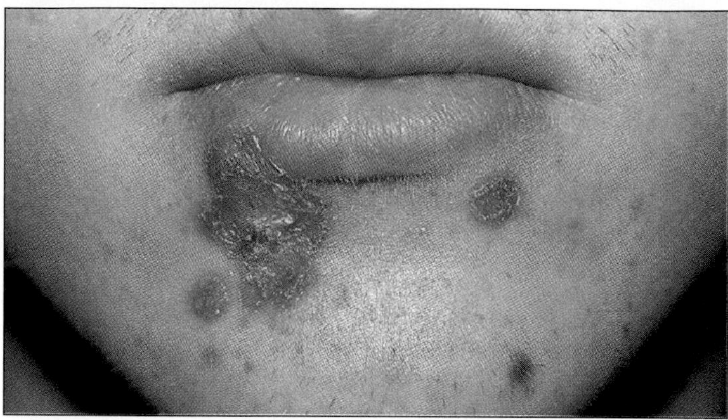

Figure 22.6 Impetigo This type of pyoderma is often caused by *Streptococcus pyogenes* and may result in glomerulonephritis.

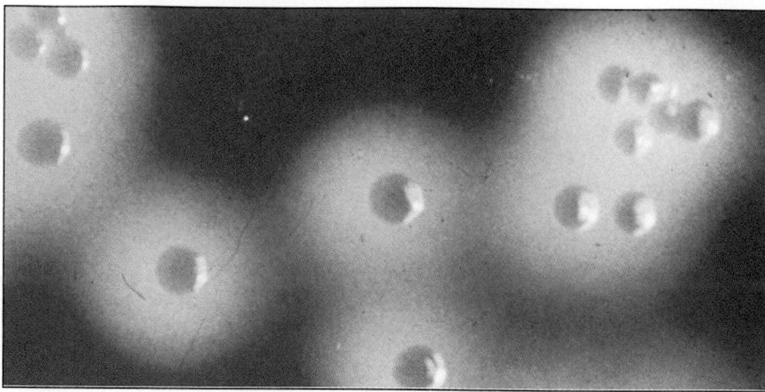

Figure 22.7 *Streptococcus pyogenes* Growing on Blood Agar The colonies are small and surrounded by a wide zone of β-hemolysis.

that it is not apparent. **Impetigo** is the most common type of pyoderma **(figure 22.6)**.

Symptoms

Impetigo is a superficial skin infection, involving patches of epidermis just beneath the dead, scaly outer layer. Thin-walled blisters first develop, then break, and are replaced by yellowish crusts that form from the drying of plasma that weeps through the skin. Usually, little fever or pain develop, but lymph nodes near the involved areas often enlarge, indicating that bacterial products have entered the lymphatic system and an immune response is occurring.

Causative Agent

Although *Staphylococcus aureus* often causes impetigo, many cases, even epidemics, are due to *Streptococcus pyogenes*. These Gram-positive, chain-forming cocci are β-hemolytic **(figure 22.7)** and are frequently referred to as β-hemolytic group A streptococci because their cell walls contain a polysaccharide called group A carbohydrate. A more detailed description of *S. pyogenes* is found in chapters 23 and 27. ■ **hemolytic streptococci, p. 103** ■ ***Streptococcus pyogenes*, p. 553, 696**

Table 22.4 compares *Staphylococcus aureus* and *Streptococcus pyogenes*.

Pathogenesis

Many different strains of *Streptococcus pyogenes* exist, some of which can colonize the skin. Infection is probably established by scratches or other minor injuries that introduce the bacteria into the deeper layer of epidermis. In impetigo, even though the infection is limited to the epidermis, streptococcal products are absorbed into the circulation.

As with *Staphylococcus aureus*, a number of extracellular products may contribute to the virulence of *Streptococcus pyogenes*. These products include enzymes such as proteases that degrade protein, nucleases that degrade nucleic acids, and hyaluronidase, which degrades the hyaluronic acid component of host tissues. As with staphylococci, it is probable that such enzymes contribute to streptococcal pathogenicity. None of them appear to be essential, however, since antibody against them fails to protect experimental animals. On the other hand, the surface components of *S. pyogenes*, notably a hyaluronic acid capsule and a cell wall component known as the M protein, are very important in enabling this organism to cause disease because they interfere with phagocytosis. ■ **M protein, p. 553**

Acute glomerulonephritis is a serious complication of *S. pyogenes* pyoderma. This condition appears abruptly during convalescence from untreated *S. pyogenes* infections, with fever, fluid retention, high blood pressure, and blood and protein in the urine. Acute glomerulonephritis is caused by inflammation of structures within the kidneys, the glomeruli (singular: glomerulus), small tufts of tiny blood vessels, and the nephrons, responsible for the forma-

TABLE 22.4 *Streptococcus pyogenes* vs. *Staphylococcus aureus*

	Streptococcus pyogenes	***Staphylococcus aureus***
Characteristics	Gram-positive cocci in chains; β-hemolytic colonies; cell wall contains group A polysaccharide and M protein	Gram-positive cocci in clusters; cream-colored colonies; cell wall contains protein A
Extracellular Products	Hemolysins: streptolysins O and S; streptokinase, DNase, hyaluronidase, and others	Hemolysins, leukocidin, hyaluronidase, nuclease, protease, penicillinase, and others
Disease Potential	Causes impetigo, strep throat, wound infections, scarlet fever, puerperal fever, toxic shock, and flesh-eating fasciitis. Complications: glomerulonephritis, rheumatic fever, and chorea	Causes boils, scalded skin syndrome, wound infections, abscesses, impetigo, food poisoning, and toxic shock syndrome

Figure 22.8 Pathogenesis of Post-Streptococcal Acute Glomerulonephritis Immune complexes are deposited in the kidney glomeruli, inciting an inflammatory response.

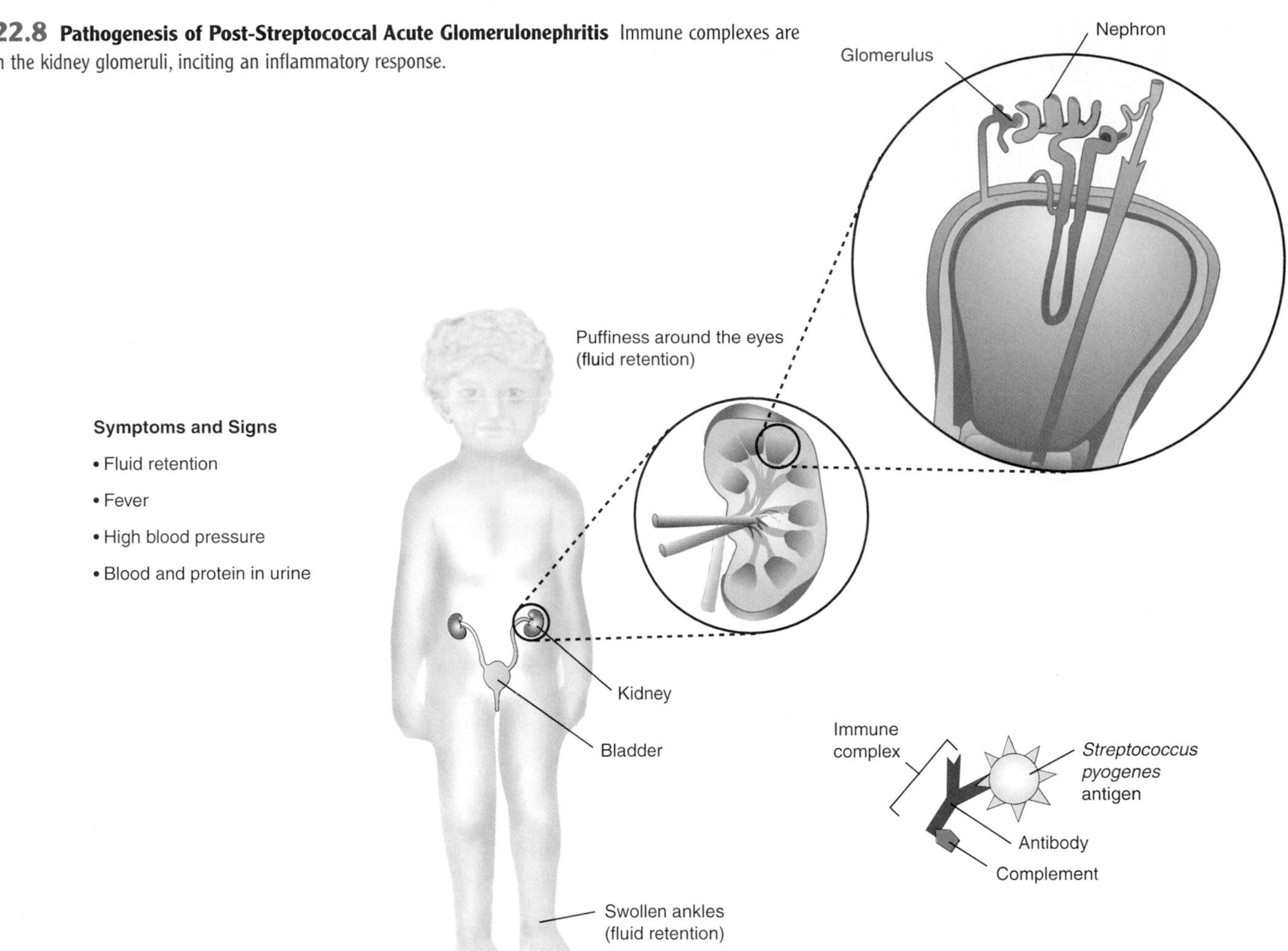

tion and composition of urine (**figure 22.8**). Only a few of the many *S. pyogenes* strains cause the condition. Streptococci are absent from the urine and diseased kidney tissues. Indeed, the bacteria have generally been eliminated from the infection site in the skin by the immune response by the time symptoms of glomerulonephritis appear. Damage to the kidney is caused by immune complexes that are deposited in the glomeruli and provoke an inflammatory reaction. Both streptococcal skin and throat infections can sometimes cause acute glomerulonephritis, but rheumatic fever, a serious complication of strep throat, is not generally a complication of streptococcal pyoderma. ■ **immune complexes, p. 440** ■ **rheumatic fever, pp. 554–555**

Epidemiology

Impetigo is most prevalent among poor children of the tropics or elsewhere during the hot, humid season. Children two to six years are mainly afflicted. Person-to-person contact spreads the disease, as do flies and other insects, and fomites—inanimate objects such as toys or towels. Impetigo patients often become throat and nasal carriers of *S. pyogenes*.

Prevention and Treatment

General cleanliness and avoiding people with impetigo help prevent the disease. Prompt cleansing of wounds and application of antiseptic probably also decrease the chance of infection. So far, *S. pyogenes* strains remain susceptible to penicillin. In patients allergic to penicillin, erythromycin can be substituted.

Table 22.5 summarizes the main features of impetigo.

TABLE 22.5 Impetigo

Symptoms	Blisters that break and "weep" plasma and pus; formation of golden-colored crusts; lymph node enlargement
Incubation period	2 to 5 days
Causative organisms	*Streptococcus pyogenes, Staphylococcus aureus*
Pathogenesis	Initiated by organisms entering the skin through minor breaks; certain strains of *S. pyogenes are* prone to cause impetigo; some *S. aureus* strains that make exfoliatin produce large blisters called bullae
Epidemiology	Spread by direct contact with carriers or patients with impetigo
Prevention and treatment	Cleanliness; care of skin injuries. Oral penicillin if cause is known to be *S. pyogenes;* otherwise, an antistaphylococcal antibiotic orally or topically

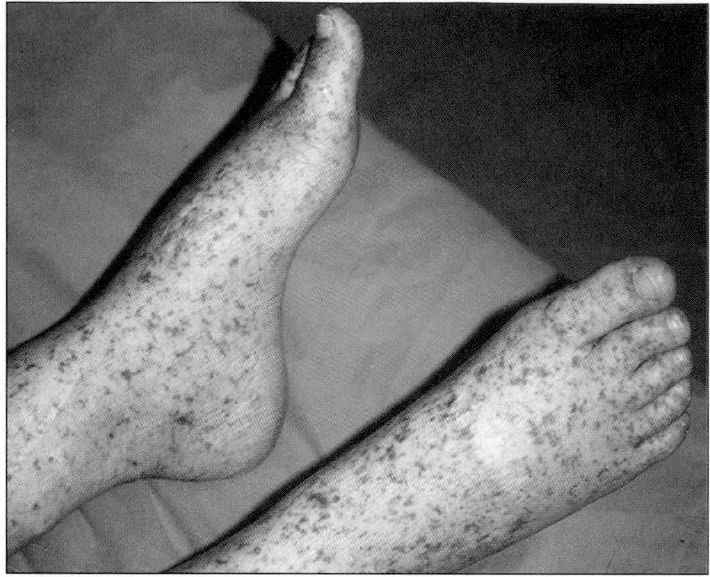

Figure 22.9 Rash Caused by Rocky Mountain Spotted Fever
Characteristically, the rash begins on the arms and legs, spreads centrally, and as shown in this photo, becomes hemorrhagic.

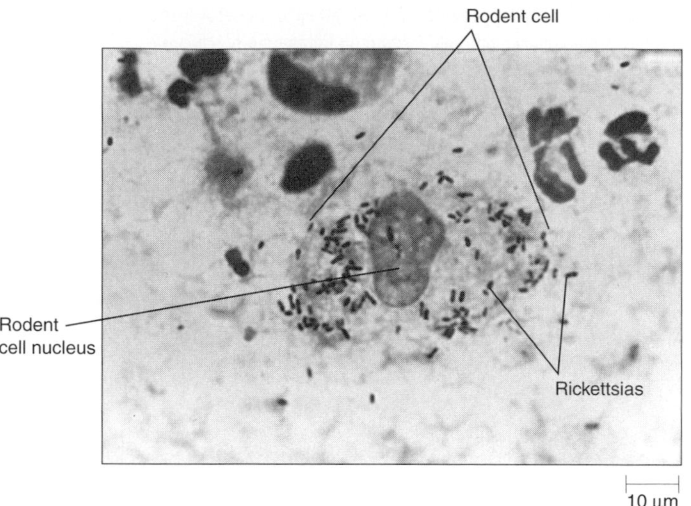

Figure 22.10 *Rickettsia rickettsii* Growing Within a Rodent Cell

Rocky Mountain Spotted Fever

Rocky Mountain spotted fever was first recognized in the Rocky Mountain area of the United States—thus its name. The disease is representative of a group of serious rickettsial diseases that occur worldwide and are transmitted by certain species of ticks, mites, or lice.

Symptoms

Rocky Mountain spotted fever generally begins suddenly with a headache, pains in the muscles and joints, and fever. Within a few days, a rash consisting of faint pink spots appears on the palms, wrists, ankles, and soles. This rash spreads up the arms and legs to the rest of the body and becomes raised and hemorrhagic (**figure 22.9**), meaning that it is due to blood leaking from damaged blood vessels. Bleeding may occur at various other sites, such as the mouth and nose. Involvement of the heart, kidneys, and other body tissues can result in shock and death unless treatment is given promptly.

Causative Agent

Rocky Mountain spotted fever is caused by *Rickettsia rickettsii* (**figure 22.10**), an obligate intracellular bacterium. The organisms are tiny, Gram-negative, nonmotile coccobacilli. Their cell walls contain a lipopolysaccharide that cross-reacts immunologically with various other bacteria. *Rickettsia rickettsii* is difficult to see well in Gram-stained smears but can be seen using special stains such as Giemsa. The organism can be cultivated in embryonated hens' eggs and in cell cultures. *Rickettsia rickettsii* can sometimes be identified early in an infection by demonstrating the organisms in **biopsies**—bits of tissue removed surgically—of skin lesions. Also, their DNA can be magnified by the polymerase chain reaction (PCR) and identified with a probe. ■ **cell, or tissue cultures, p. 345** ■ **PCR, p. 237**

Pathogenesis

Rocky Mountain spotted fever is acquired from the bite of a tick infected with *R. rickettsii*. The bite is usually painless and unnoticed; the tick remains attached for hours while it feeds on capillary blood. Rickettsias are not immediately released into tick saliva from the tick's salivary glands. Therefore, the infection is not usually transmitted until the tick has fed for 4 to 10 hours. When the organisms are released into capillary blood with the tick saliva, they are taken up preferentially by the cells lining the small blood vessels. Following attachment to host cells, *R. rickettsii* is taken into the cells by endocytosis. Inside the cell, the bacteria leave their phagosome and multiply in both the cytoplasm and nucleus without being enclosed in vacuoles. Some exit the cell early in the infection by entering, and then lysing, fingerlike host cell cytoplasmic projections. Eventually, the integrity of the cell membrane suffers from this process, the cell takes in water, lyses, and releases the remaining rickettsias. These seed the bloodstream, infecting even more cells. Infection can also extend into the walls of the small blood vessels, causing an inflammatory reaction, clotting of the blood vessels, and small areas of necrosis, or death of tissue. This process is readily apparent in the skin as a hemorrhagic rash but, more ominously, occurs throughout the body, resulting in damage to vital organs such as the kidneys and heart. Potentially even more serious is the release of endotoxin into the bloodstream from the rickettsial cell walls, causing shock and generalized bleeding because of **disseminated intravascular coagulation**. ■ **endotoxin, p. 461** ■ **disseminated intravascular coagulation, p. 440, 720**

Epidemiology

Rocky Mountain spotted fever is an example of a **zoonosis**, a disease that exists primarily in animals other than humans. It occurs in a spotty distribution across the contiguous United States and extends into Canada, Mexico, and a few countries of South America. The involved areas change over time, but despite the name of the disease, in the United States the highest incidence has generally been in the south Atlantic and south-central states

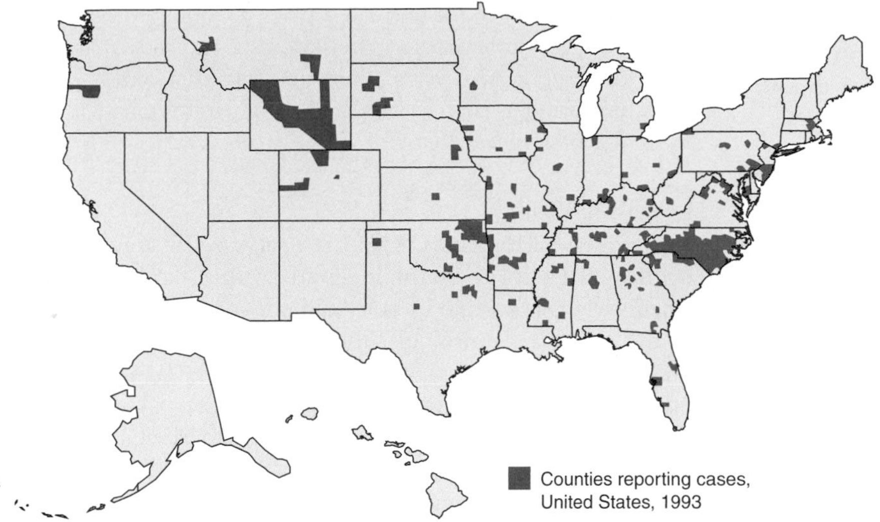

Figure 22.11 **Counties Reporting Cases of Rocky Mountain Spotted Fever, United States, 1993**

(**figure 22.11**). Rocky Mountain spotted fever is maintained in nature in various species of ticks and mammals. Generally, little or no illness develops in these natural hosts, but humans, being an accidental host, often develop severe disease. Several species of ticks transmit the disease to humans. The main vector in the western United States is the wood tick, *Dermacentor andersoni* (**figure 22.12**), while in the East it is the dog tick, *Dermacentor variabilis*. Once infected, ticks remain infected for life, transmitting *R. rickettsii* from one generation to the next through their eggs. Ticks are most active from April to September, and it is during this time period that most cases of Rocky Mountain spotted fever occur.

Prevention and Treatment

No vaccine against Rocky Mountain spotted fever is currently available to the public, although promising genetically engineered vaccines are under development. The disease can be prevented if people take the following precautions: (1) avoid tick-infested areas when possible; (2) use protective clothing; (3) use tick repellents such as dimethyltoluamide; (4) carefully inspect their bodies, especially the scalp, armpits, and groin, for ticks several times daily; and (5) remove attached ticks carefully to avoid crushing them and thereby contaminating the bite wound with their infected tissue fluids. Gentle traction with tweezers applied near the mouth parts is the safest method of removal. Touching the tick with gasoline or whiskey may help to loosen its attachment. After removal of the tick, the site of the bite should be treated with an antiseptic.

The antibiotics tetracycline and chloramphenicol are highly effective in treating Rocky Mountain spotted fever if given early in the disease, before irreversible damage to vital organs has occurred. Without treatment, the overall mortality from the disease is about 20%, but it can be considerably higher in elderly patients. With early diagnosis and treatment, the mortality rate is less than 5%.

The main features of Rocky Mountain spotted fever are summarized in **table 22.6**.

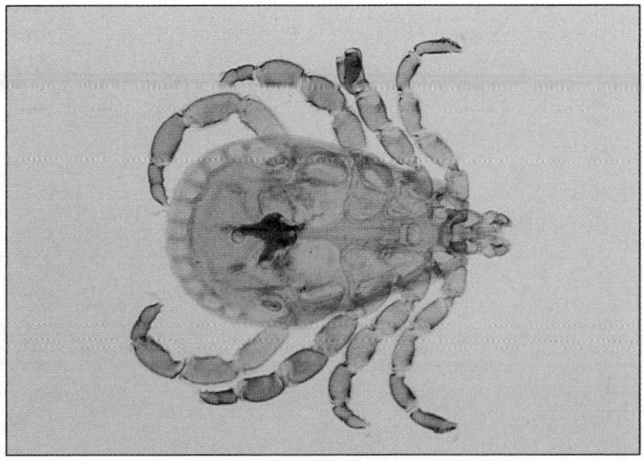

Figure 22.12 ***Dermacentor andersoni*, the Wood Tick** The wood tick is the principal vector of Rocky Mountain spotted fever in the western United States.

Lyme Disease

In the mid-1970s, studies of a group of cases in Lyme, Connecticut, led to the recognition of Lyme disease as a distinct entity. It was not until 1982 that the cause was first identified, in ticks from New York state, by Dr. Willy Burgdorfer at the Rocky Mountain Laboratories, National Institute of Allergy and Infectious Diseases, in Hamilton, Montana. We now know that Lyme disease was present in many areas of the world long before its identification at Lyme. The ecology of the disease is complex and still incompletely understood, but we are beginning to get some answers as to why the disease has increased and extended its range.

TABLE 22.6 Rocky Mountain Spotted Fever

Symptoms	Headache, pains in muscles and joints, and fever, followed by a hemorrhagic rash that begins on the extremities
Incubation period	4 to 8 days
Causative organism	*Rickettsia rickettsii*, an obligate intracellular bacterium
Pathogenesis	Organisms multiply at site of bite; the bloodstream is invaded and endothelial cells of blood vessels are infected; vascular lesions and endotoxin account for pathologic changes
Epidemiology	Transmitted by bite of infected tick, usually *Dermacentor* sp.
Prevention and treatment	Avoidance of tick-infested areas, use of tick repellent, removal of ticks within 4 hours of exposure. Treatment: tetracycline or chloramphenicol

Symptoms

Symptoms of Lyme disease can be divided roughly into three stages, although individual patients may lack symptoms in one or more of the three.

- The first stage typically begins a few days to several weeks after a bite by an infected tick. It is characterized by a skin rash called **erythema migrans (figure 22.13)**. The rash begins as a red spot or bump at the site of the tick bite and slowly enlarges to a median diameter of 15 cm (about 6 inches). The advancing edge is bright red, while the redness of the central portion fades as the lesion enlarges. About half of these cases develop smaller satellite lesions that behave similarly. The characteristic rash is the hallmark of Lyme disease but is present in only two-thirds of the cases. Most of the other symptoms that occur during this stage are influenza-like—fatigue, chills, fever, headache, stiff neck, joint and muscle pains, and backache. Without treatment, these symptoms slowly subside, commonly overlapping the next stage.
- Symptoms of the second stage generally begin 2 to 8 weeks after the appearance of erythema migrans and involve the heart and the nervous system. Electrical conduction within the heart is impaired, leading to dizzy spells or fainting, and a temporary pacemaker is sometimes required to maintain a normal heartbeat. Involvement of the nervous system can cause one or more of the following symptoms: paralysis of the face, severe headache, pain when moving the eyes, difficulty concentrating, emotional instability, fatigue, and impairment of the nerves of the legs or arms. These symptoms usually subside slowly without any treatment and overlap symptoms of the third stage.
- The symptoms of the third stage are characterized by arthritis, manifest as joint pain, swelling, and tenderness, usually of a large joint such as the knee. These symptoms develop in 60% of untreated cases, beginning on the average 6 months after the skin rash, and slowly disappear over subsequent years. Chronic nervous system impairments such as localized pain, paralysis, and depression can occur.

Causative Agent

Lyme disease is caused by *Borrelia burgdorferi*, a large spirochete **(figure 22.14)**,11 to 25 μm in length, with a number of axial filaments wrapped around its body and enclosed in the outer sheath of the cell wall. Surprisingly, the *Borrelia* genome is linear and present in multiple copies, completely unlike *E. coli* and most other prokaryotes, which have a single copy of a circular chromosome. The organism also contains plasmidlike elements, both circular and linear, peculiar in that they contain genes usually found on bacterial chromosomes. These findings may lead to an understanding of how *B. burgdorferi* can infect such widely differing species as mice, lizards, and ticks. The bacterium can be cultivated in the laboratory on a special medium under microaerophilic conditions. ■ spirochetes, p 289 ■ microaerophilic conditions, p. 104

Pathogenesis

The spirochetes are introduced into the skin by an infected tick, multiply, and migrate outward in a radial fashion. The cell walls of the organisms cause an inflammatory reaction in the skin, which produces the expanding rash. The host's immune response is initially suppressed, allowing continued multiplication of the spirochete. The organisms then enter the bloodstream and become disseminated to all parts of the body but generally do not cross the placenta of pregnant women. Wide dissemination of the organisms accounts for the influenza-like symptoms of the first stage. After the first few weeks, an intense immune response occurs, and thereafter, it becomes very difficult to recover *B. burgdorferi* from blood or body tissues. The immune response against the bacterial antigens is probably responsible for the symptoms of the

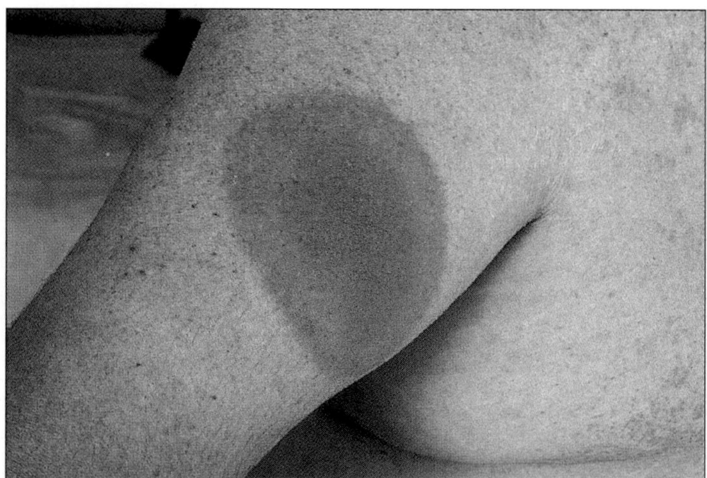

Figure 22.13 Erythema Migrans, the Characteristic Rash of Lyme Disease The rash usually has a targetlike or bull's-eye appearance. It generally causes little or no discomfort. While highly suggestive of Lyme disease, many victims of the disease fail to develop the rash.

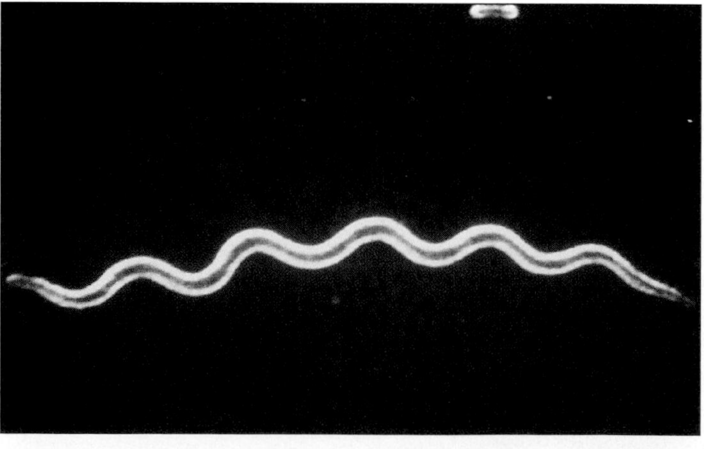

Figure 22.14 Scanning Electron Micrograph of *Borrelia burgdorferi*, the Cause of Lyme Disease

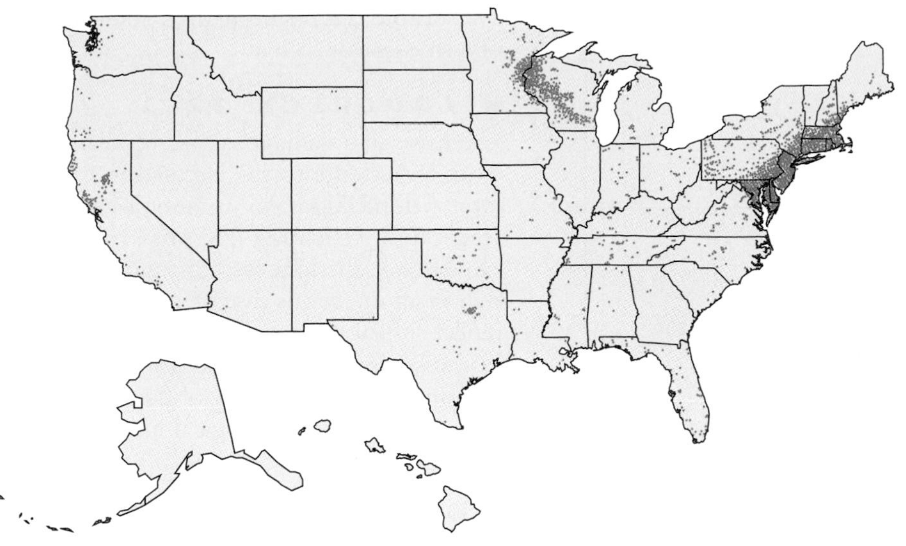

Figure 22.15 **Distribution of the Reported Cases of Lyme Disease, United States, 1997**

second stage. The third stage of Lyme disease is characterized by arthritis, and the affected joints have high concentrations of highly reactive immune cells and immune complexes. The joint and chronic nervous system symptoms of the third stage probably result from immune responses against persisting bacterial antigens, but evidence suggests a role for autoimmunity in some cases.

Epidemiology

Like Rocky Mountain spotted fever, Lyme disease is a zoonosis, and humans are accidental hosts. The disease is widespread in the United States **(figure 22.15),** and its incidence depends on complex factors that place humans and infected ticks in close proximity. Several species of ticks have been implicated as vectors, but the most important in the eastern United States is the black-legged tick, *Ixodes scapularis* **(figure 22.16)**. In some areas of the East Coast, 80% of these ticks are infected with *Borrelia burgdorferi.* Because of their small size (1–2 mm before feeding; 3–5 mm when fully engorged with blood), these ticks often feed and drop off their host without being detected, so that two-thirds of Lyme disease patients are unable to recall a tick bite. The ticks mature during a 2-year cycle **(figure 22.17)**. A six-legged larval form emerges from the egg. After growing, it molts, shedding its outer covering to become an eight-legged form called a nymph. After another molt as the tick grows in size, the nymph becomes the sexually mature adult form. The nymph avidly seeks blood meals and is therefore mainly responsible for transmitting Lyme disease. The preferred host of *I. scapularis* is the white-footed mouse, which acquires *Borrelia burgdorferi* from an infected tick and develops a sustained bacteremia (meaning the bacteria circulate in the mouse's bloodstream for long periods of time). The mouse thus becomes a source of infection for other ticks. Passage of the spirochete from adult tick to its offspring via its eggs rarely occurs. Infected ticks and mice constitute the main reservoir of *B. burgdorferi*, but deer, while not a significant reservoir, are important because they are the preferred host of the adult ticks and the site where mating occurs. Moreover, deer can quickly spread the disease over a wide area. Tick nymphs are the most active from May to September, corresponding to the peak occurrence of Lyme disease cases. Adult ticks sometimes bite humans late in the season and transmit the disease. Infectious ticks can be present in well-mowed lawns as well as in wooded areas. Expanding human populations continually intrude into the zoonotic life cycle. ■ **reservoirs, p. 476**

Prevention and Treatment

General preventive measures for Lyme disease are the same as those for Rocky Mountain spotted fever. A vaccine was approved in December 1998 for use in persons 15 to 70 years of age, but it fails to prevent the disease in one out of five vaccinees. This means that even though vaccinated, persons are advised to use every other means to avoid infection. A vaccine is also available to prevent Lyme disease in dogs. A biological control program for *Ixodes* ticks using a tiny parasitic wasp has not consistently been successful. Several antibiotics are effective in patients with early disease. In late disease, the response to treatment is less satisfactory, presumably because the spirochetes are not actively multiplying and antibacterial medications are usually ineffective against nongrowing bacteria. Nevertheless, prolonged treatment with intravenous ampicillin or ceftriaxone has been curative in many cases.

Figure 22.16 **The Black-Legged Tick, *Ixodes scapularis*, Adult and Nymph** This tick is the most important vector of Lyme disease in the eastern and north-central United States.

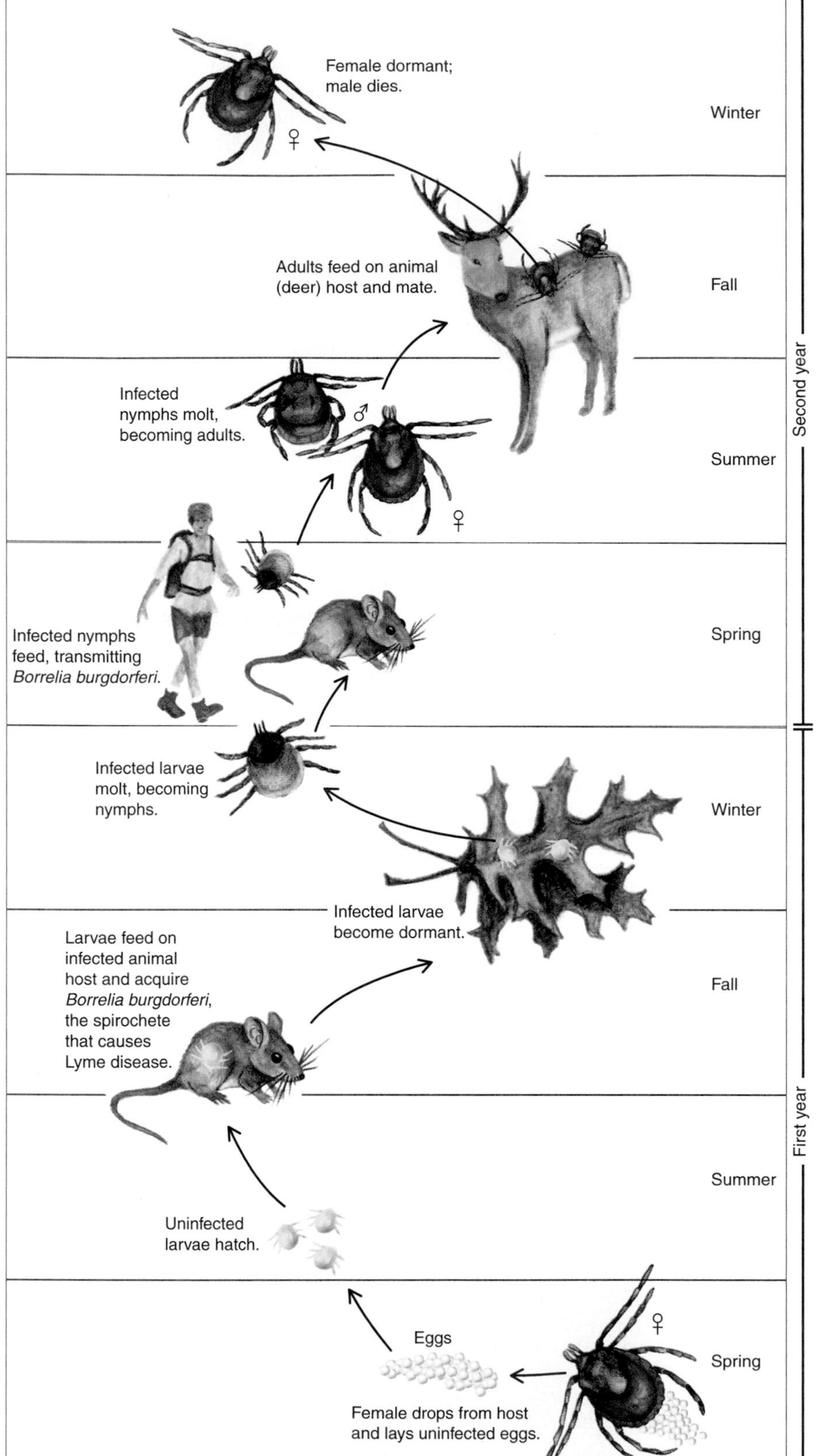
Female dormant; male dies.
Winter
Adults feed on animal (deer) host and mate.
Fall
Infected nymphs molt, becoming adults.
Summer
Infected nymphs feed, transmitting Borrelia burgdorferi.
Spring
Second year
Infected larvae molt, becoming nymphs.
Winter
Infected larvae become dormant.
Larvae feed on infected animal host and acquire Borrelia burgdorferi, the spirochete that causes Lyme disease.
Fall
Uninfected larvae hatch.
Summer
First year
Eggs
Female drops from host and lays uninfected eggs.
Spring

Figure 22.17 Life Cycle of the Black-Legged Tick, *Ixodes scapularis*, the Principal Vector of *Borrelia burgdorferi*, Cause of Lyme Disease Note that the life cycle covers 2 years, during which the tick obtains three blood meals. The males die soon after mating, the females after depositing their eggs in the following spring. Variations in the life cycle occur, probably dependent on climate and food availability.

Table 22.7 summarizes some features of Lyme disease.

MICROCHECK 22.3

Extensive skin damage can result from a toxin absorbed into the circulation from a localized infection. An immunological reaction to circulating microbial products can damage the kidneys. Changes in the skin in an infectious disease commonly reflect similar changes in other body tissues. Zoonoses involving ticks and small mammals pose a widespread danger to humans. Complex ecological factors can govern the incidence of infectious diseases.

- List four extracellular products of *Staphylococcus aureus* that contribute to its virulence.
- Describe the characteristic rash of Lyme disease.
- The existence of extensive scalded skin syndrome does not indicate that *Staphylococcus* is growing in all the affected areas. Why?

Skin Diseases Caused by Viruses

Several childhood diseases are characterized by distinctive skin rashes, **exanthems**, caused by viruses carried to the skin by the blood from sites of infection in the upper respiratory tract. This group of diseases is usually diagnosed by inspection of the rash and other clinical findings. When the disease is not typical, however, tests can be performed to identify specific antibody against the virus, and the virus can often be cultivated from skin lesions, upper respiratory secretions, or other material. Viruses, like obligate intracellular bacteria, can only reproduce in living cells, most conveniently in cultures of cells originally derived from human or other tissues and maintained in the laboratory.

Chickenpox (Varicella)

Chickenpox is the common name for **varicella**, the most common of the viral rashes of childhood. The causative virus is a member of the herpesvirus family and, like others in that group, produces a latent infection that can reactivate long after recovery from the initial illness. ■ **latent infections, p. 352**

TABLE 22.7 Lyme Disease

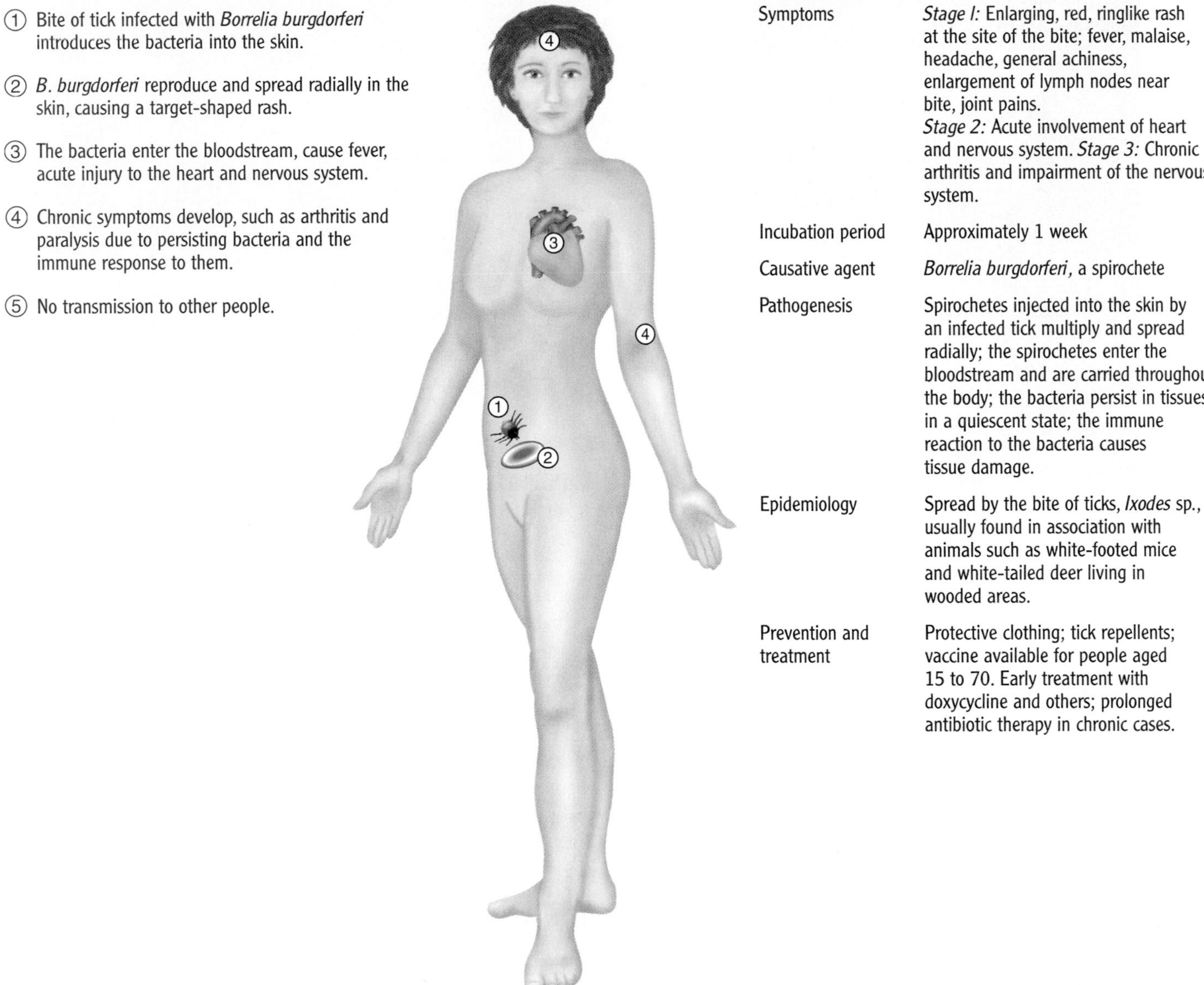

1. Bite of tick infected with *Borrelia burgdorferi* introduces the bacteria into the skin.
2. *B. burgdorferi* reproduce and spread radially in the skin, causing a target-shaped rash.
3. The bacteria enter the bloodstream, cause fever, acute injury to the heart and nervous system.
4. Chronic symptoms develop, such as arthritis and paralysis due to persisting bacteria and the immune response to them.
5. No transmission to other people.

Symptoms	*Stage I:* Enlarging, red, ringlike rash at the site of the bite; fever, malaise, headache, general achiness, enlargement of lymph nodes near bite, joint pains. *Stage 2:* Acute involvement of heart and nervous system. *Stage 3:* Chronic arthritis and impairment of the nervous system.
Incubation period	Approximately 1 week
Causative agent	*Borrelia burgdorferi*, a spirochete
Pathogenesis	Spirochetes injected into the skin by an infected tick multiply and spread radially; the spirochetes enter the bloodstream and are carried throughout the body; the bacteria persist in tissues in a quiescent state; the immune reaction to the bacteria causes tissue damage.
Epidemiology	Spread by the bite of ticks, *Ixodes* sp., usually found in association with animals such as white-footed mice and white-tailed deer living in wooded areas.
Prevention and treatment	Protective clothing; tick repellents; vaccine available for people aged 15 to 70. Early treatment with doxycycline and others; prolonged antibiotic therapy in chronic cases.

Symptoms

Most cases of chickenpox (varicella) are mild, sometimes unnoticed, and recovery is usually uncomplicated. The typical case has a rash that is diagnostic. It begins as small, red spots called macules, little bumps called papules, and small blisters called vesicles, surrounded by a narrow zone of redness. The lesions can erupt anywhere on the body, although usually they first appear on the back of the head, then the face, mouth, main body, and arms and legs, ranging from only a few lesions to many hundreds. The lesions appear at different times, and within a day or so they go through a characteristic evolution from macule to papule to vesicle to pustule, a pus-filled blister. After the pustules break, leaking virus-laden fluid, a crust forms, and then healing takes place. At any time during the rash, lesions are at various stages of evolution **(figure 22.18)**. The lesions are pruritic, meaning itchy, and scratching may lead to serious, even fatal, secondary infection by *Streptococcus pyogenes* or *Staphylococcus aureus*.

Symptoms of varicella tend to be more severe in older children and adults. In about 20% of adults, pneumonia develops, causing rapid breathing, cough, shortness of breath, and a dusky skin color. The pneumonia subsides with the rash, but respiratory symptoms often persist for weeks. Varicella is also a major threat to newborn babies if the mother develops the disease within 5 days before delivery to 2 days afterward. Mortality in these babies has been as high as 30%. Also, **congenital varicella syndrome** develops in a fraction of a percent of babies whose mothers contract varicella earlier in pregnancy. These babies are born with such defects as underdeveloped head and limbs, and cataracts. In

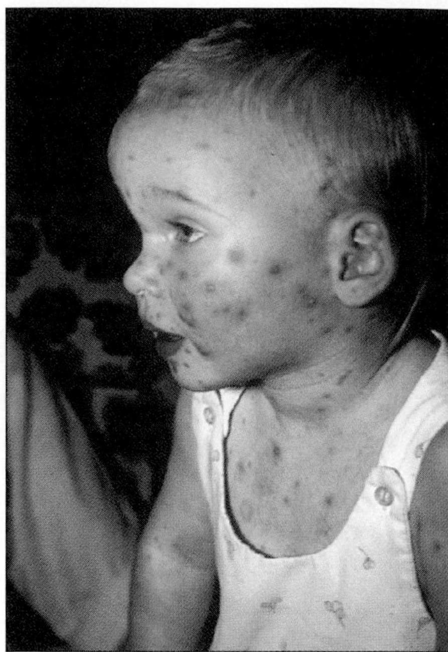

Figure 22.18 A Child with Chickenpox (Varicella) Characteristically, lesions in various stages of evolution—macules, papules, vesicles, and pustules—are present.

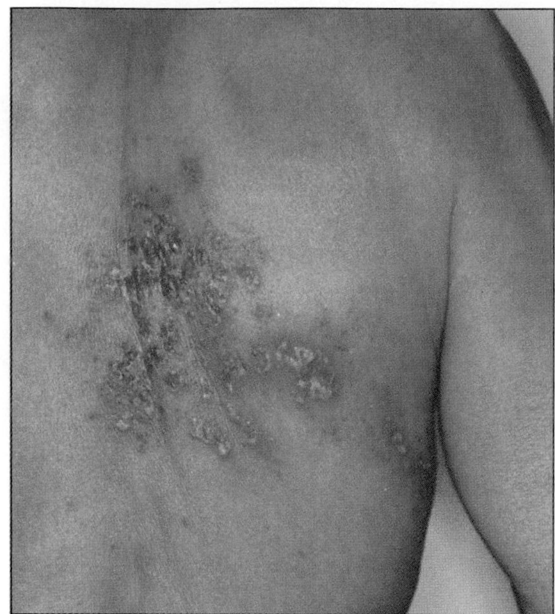

Figure 22.19 Shingles (Herpes Zoster) The rash mimics that of chickenpox, except that it is limited to a sensory nerve distribution on one side of the body.

addition, the disease is a threat to immunocompromised patients of any age. The virus can damage the lungs, heart, liver, kidneys, and brain, resulting in death in about 20% of the cases.

Reactivation of chickenpox is called **shingles**, or **herpes zoster**. It can occur at any age but becomes increasingly common with advancing age. It begins with pain in the area supplied by a nerve of sensation, often on the chest or abdomen but sometimes on the face or an arm or leg. After a few days to 2 weeks, a rash characteristic of chickenpox appears, but unlike chickenpox the rash is usually restricted to an area supplied by the branches of the involved sensory nerve (**figure 22.19**). The rash generally subsides within a week, but pain may persist for weeks, months, or longer. In people with AIDS or other serious immunodeficiency, instead of being confined to one area the rash often spreads to involve the entire body, as in a severe case of chickenpox.

A curious affliction known as **Reye's syndrome** occasionally occurs in association with chickenpox, usually within 2 to 12 days of the onset of the infection. The patients begin vomiting and slip into a coma. The syndrome occurs predominantly in children between 5 and 15 years old and is characterized by liver and brain damage. It occurs uncommonly, with a general trend to declining incidence, but the death rate has been around 30%. Reye's syndrome is also seen in association with a number of other viral infections including influenza A and B. Epidemiologic evidence suggesting that aspirin therapy increases the risk of Reye's syndrome has led physicians to use this drug sparingly in children with fever.

Causative Agent

Chickenpox is caused by the **varicella-zoster virus**, a member of the herpesvirus family. It is an enveloped, medium-sized (150–200 nm), double-stranded DNA virus, indistinguishable from other herpesviruses in appearance (**figure 22.20**). It can be cultivated in cell cultures and identified using specific antiserum.

Pathogenesis

The virus enters the body by the respiratory route, establishes an infection, replicates, and disseminates to the skin via the bloodstream. After the living layers of skin cells are infected, the virus spreads directly to adjacent cells, and the characteristic skin lesions appear. Stained preparations of infected cells show **intranuclear inclusion bodies**, visible as pink staining bodies at the place in the nucleus where the virus reproduces. Some infected cells fuse together, forming multinucleated giant cells. The infected cells swell and ultimately lyse. The virus enters the sensory nerves, presumably when an area of skin infection advances to involve a sensory nerve ending. Conditions inside the nerve cell do not permit full expression of the viral genome;

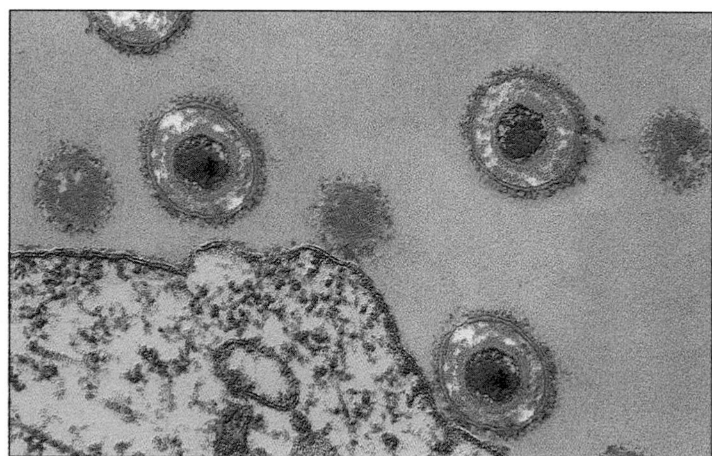

Figure 22.20 Electron Micrograph of Varicella-Zoster Virus, Cause of Chickenpox and Shingles

however, viral DNA is present in the ganglia (singular: ganglion) of the nerves and is fully capable of coding for mature infectious virus. Ganglia are small bulges in sensory nerves located near the spine; they contain the nuclei and cell bodies of the nerves. The mechanism of suppression of viral replication within the nerve cell is not known but is probably under the control of immune cells.

The occurrence of shingles correlates with a decline in cell-mediated immunity. With the decline, infectious varicella-zoster virus is presumably produced in the nucleus of the nerve cells and is carried to the skin by the normal circulation of cytoplasm within the nerve cell. With the appearance of the skin lesions, a prompt, intense anamnestic boost of both cellular and humoral immunity ensues. A marked inflammatory reaction occurs in the ganglion with an accumulation of immune cells, and shingles quickly disappears, although sometimes leaving scars and chronic pain. ■ anamnestic response, p. 401

Epidemiology

The annual incidence of chickenpox (varicella) in the United States is estimated at 3.7 million. Reporting the disease is not required, so that most cases go unreported, and many are so mild that they go unnoticed. As with many diseases transmitted by the respiratory route, most cases occur in the winter and spring months. Humans are the only reservoir, and because the disease is highly contagious, about 90% of people are infected by the age of 15. The incubation period of the disease averages about 2 weeks, with a range of 10 to 21 days. Cases are infective from 1 to 2 days before the rash appears until all the lesions have crusted (usually 4 days after the onset).

The mechanism by which the varicella-zoster virus persists in the body allows it to survive indefinitely in small isolated populations. By contrast, when a virus such as measles is introduced into an isolated community, it spreads quickly and infects most of the susceptible individuals, who either become immune or die. If susceptible victims are unavailable, the measles virus will disappear from the community. On the other hand, varicella-zoster virus will reappear from cases of shingles whenever sufficient numbers of susceptible children have been born. Shingles occurs in about 1% of elderly people.

Prevention and Treatment

In March 1995, a live, attenuated chickenpox vaccine was licensed in the United States. The virus was originally isolated from a varicella case in Japan in 1970 and was attenuated by subculturing it many times in various cell cultures. It has proven safe with use in over 2 million people in various countries, since about 1984. The vaccine is recommended for all healthy persons age 12 months or older who do not have a history of chickenpox or who lack laboratory evidence of immunity to the disease. Immunization should be done sometime before one's 13th birthday because of the increased likelihood of serious complications from chickenpox in older children and adults. It is not given during pregnancy, and pregnancy should be avoided for 3 months after vaccination because of fear the vaccine might rarely result in congenital varicella syndrome. In general, the vaccine should not be administered to people with malignant or immunodeficiency diseases. Healthy, nonimmune contacts of such people, however, should be vaccinated. By preventing chickenpox, the vaccine markedly decreases the chance of developing shingles. ■ attenuated vaccines, p. 425

Increasing numbers of individuals with impaired immunity are at risk of severe disseminated varicella-zoster virus infections. These include persons with cancer, AIDS, and organ transplants and newborn babies whose mothers contracted chickenpox near the time of delivery. They can be partially protected from severe disease if they are passively immunized by injecting them with zoster immune globulin (ZIG) derived from the blood of recovered herpes zoster patients. The antiviral medications acyclovir and famciclovir, among others, are helpful in preventing and treating varicella-zoster infections. ■ passive immunization, p. 424

The main features of chickenpox are summarized in **table 22.8**.

Measles (Rubeola)

Measles, "hard measles," and "red measles" are common names for **rubeola**. One of the great success stories of the last half of the Twentieth century has been the dramatic reduction in measles cases by immunizing children with a live vaccine against the disease. Now there is reason to hope that the disease can be entirely eliminated from the world.

Symptoms

Measles begins with fever, runny nose, cough, and swollen, red, weepy eyes. Within a few days, a fine red rash appears on the forehead and spreads outward over the rest of the body (**figure 22.21**). Unless complications occur, symptoms generally disappear in about 1 week.

Unfortunately, many cases are complicated by secondary infections caused by bacterial pathogens, mainly *Staphylococcus aureus*, *Streptococcus pneumoniae*, *Streptococcus pyogenes*, and *Haemophilus influenzae*. These pathogens readily invade the body because measles damages the normal body defenses. Secondary

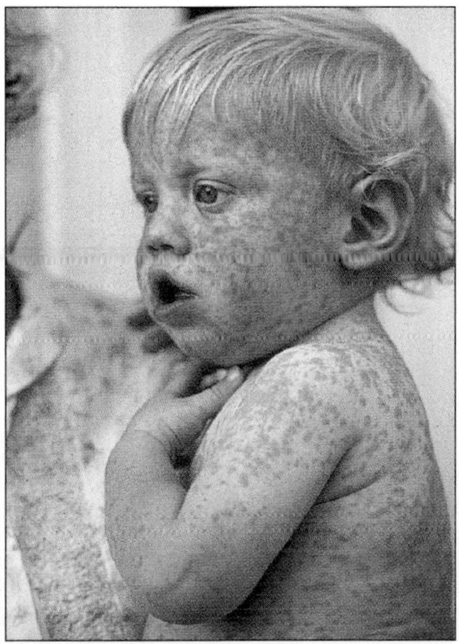

Figure 22.21 A Child with Measles (Rubeola) The rash is usually accompanied by fever, runny nose, and a bad cough.

TABLE 22.8 Chickenpox (Varicella)

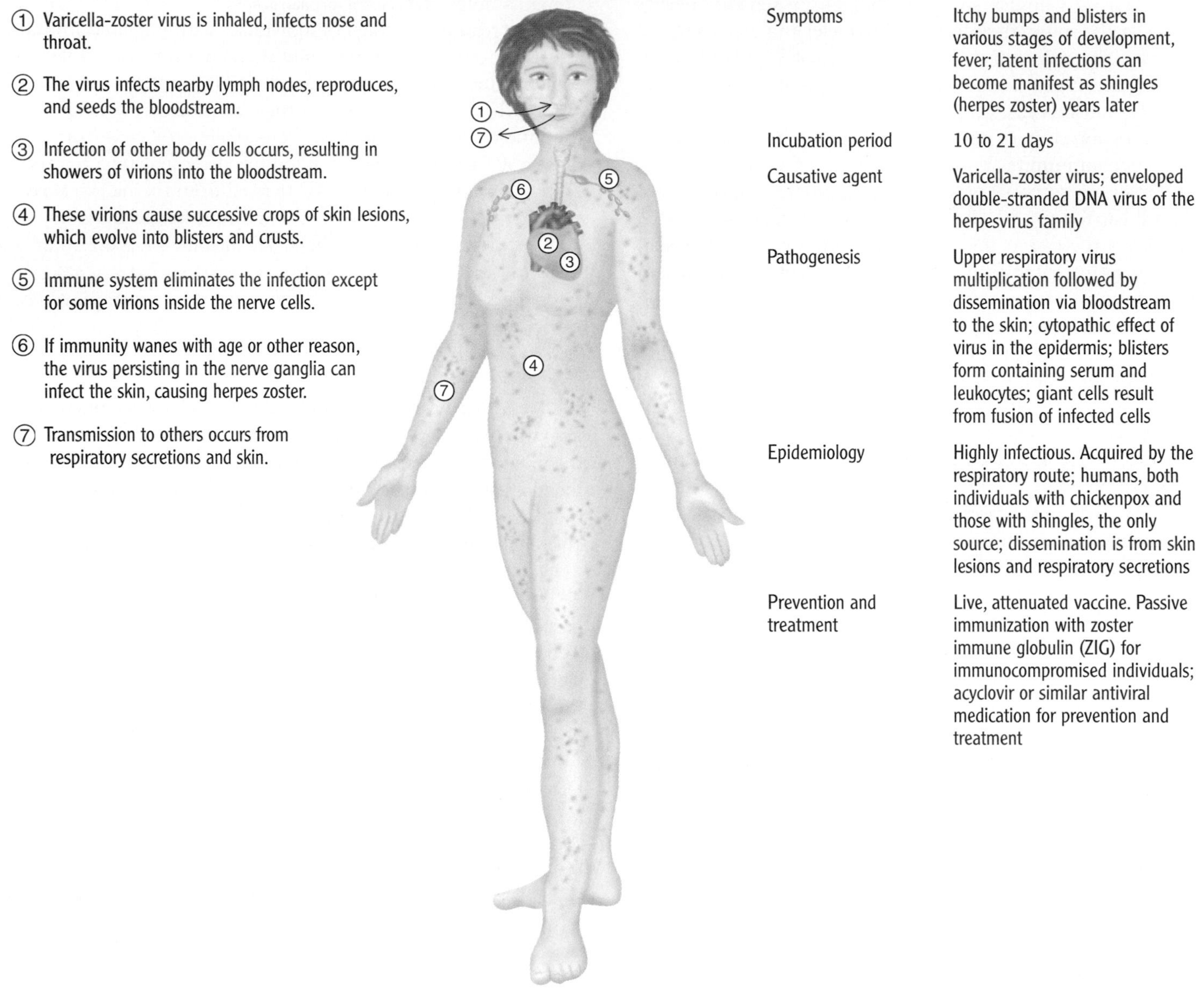

1. Varicella-zoster virus is inhaled, infects nose and throat.
2. The virus infects nearby lymph nodes, reproduces, and seeds the bloodstream.
3. Infection of other body cells occurs, resulting in showers of virions into the bloodstream.
4. These virions cause successive crops of skin lesions, which evolve into blisters and crusts.
5. Immune system eliminates the infection except for some virions inside the nerve cells.
6. If immunity wanes with age or other reason, the virus persisting in the nerve ganglia can infect the skin, causing herpes zoster.
7. Transmission to others occurs from respiratory secretions and skin.

Symptoms	Itchy bumps and blisters in various stages of development, fever; latent infections can become manifest as shingles (herpes zoster) years later
Incubation period	10 to 21 days
Causative agent	Varicella-zoster virus; enveloped double-stranded DNA virus of the herpesvirus family
Pathogenesis	Upper respiratory virus multiplication followed by dissemination via bloodstream to the skin; cytopathic effect of virus in the epidermis; blisters form containing serum and leukocytes; giant cells result from fusion of infected cells
Epidemiology	Highly infectious. Acquired by the respiratory route; humans, both individuals with chickenpox and those with shingles, the only source; dissemination is from skin lesions and respiratory secretions
Prevention and treatment	Live, attenuated vaccine. Passive immunization with zoster immune globulin (ZIG) for immunocompromised individuals; acyclovir or similar antiviral medication for prevention and treatment

infections most commonly involve the middle ear and the lungs, causing earaches and pneumonia, respectively.

In about 5% of cases, the rubeola virus itself causes pneumonia, with rapid breathing, shortness of breath, and dusky skin color from lack of adequate oxygen exchange in the lungs. Encephalitis, inflammatory disease of the brain, is another serious complication, marked by fever, headache, confusion, and seizures. This complication occurs in about one out of every 1,000 cases of measles. Permanent brain damage, with mental retardation, deafness, and epilepsy, commonly results from measles encephalitis.

Very rarely, rubeola is followed 2 to 10 years later by a disease called **subacute sclerosing panencephalitis (SSPE)**, which is marked by slowly progressive degeneration of the brain, generally resulting in death within 2 years. A defective measles virus can be detected in the brains of these patients, and high levels of measles antibody are present in their blood. This is an example of a "slow virus" disease. It has all but disappeared from the United States with widespread vaccination against measles. ■ **slow virus, p. 364**

Measles that occurs during pregnancy results in an increased risk of miscarriage, premature labor, and low birth weight. Birth defects, however, are generally not seen.

Causative Agent

Measles is caused by rubeola virus, a pleomorphic, medium-sized (120–200 nm diameter), single-stranded, negative-sense RNA virus of the paramyxovirus family. The virus can be cultivated from the patient's nasal secretions, blood, and urine until the second or third day of the rash. The viral envelope has two biologically active projections. One, H, is responsible for viral attachment to host

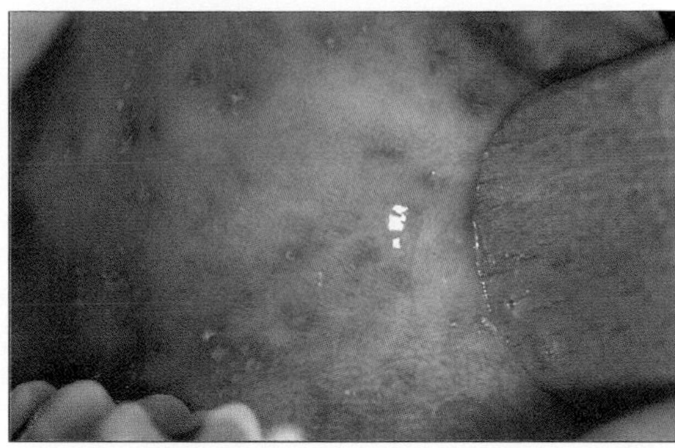

Figure 22.22 Koplik's Spots, Characteristic of Measles (Rubeola), Are Usually Transitory They resemble grains of salt on a red base.

cells, and the other, M, is responsible for fusion of the viral outer membrane with the host cell. The M antigen also causes adjacent infected host cells to fuse together, producing multinucleated giant cells. ■ positive- and negative-sense viruses, p. 329

Pathogenesis

Rubeola virus is acquired by the respiratory route. It presumably replicates in the upper respiratory epithelium, spreads to lymphoid tissue, and following further replication, eventually spreads to all parts of the body. Mucous membrane involvement is responsible for an important diagnostic sign, **Koplik's spots (figure 22.22)**, which are usually best seen opposite the molars, the teeth that grind food, located in the back part of the mouth. Koplik's spots look like grains of salt lying on an oral mucosa that is red and rough, resembling red sandpaper. Damage to the respiratory mucous membranes partly explains the markedly increased susceptibility of measles patients to secondary bacterial infections, especially infection of the middle ear and lung. Involvement of the intestinal epithelium may explain the diarrhea that sometimes occurs in measles and contributes to high measles death rates in impoverished countries. In the United States, deaths from measles occur in about one to two of every 1,000 cases, mainly from pneumonia and encephalitis.

The skin rash of measles results from the effect of rubeola virus replication in skin cells and the cellular immune response against the viral antigen in the skin. It is not known why the rash characteristically spreads outward, often clearing on the face before it reaches the lower extremities. The measles virus temporarily suppresses cellular immunity, causing cold sores to appear and latent tuberculosis to activate. ■ cold sores, p. 594 ■ tuberculosis, p. 569

Epidemiology

Humans are the only natural host of rubeola virus. Before vaccination became widespread in the 1960s, probably less than 1% of the population escaped infection with this highly contagious virus. Continued use of measles vaccine resulted in a progressive decline in cases, and there was hope in the 1970s that measles would be quickly eradicated in the United States. Small outbreaks of the disease, however, continue to occur. For example, 33 cases were reported in Alaska in 1998 following introduction of a case in a Japanese visitor.

The continued occurance of measles in the United States is due to the presence of nonimmune populations and the influx of immigrants and travelers from other countries where the disease is still uncontrolled. Nonimmune individuals include (1) children too young to be vaccinated; (2) preschool children never vaccinated; (3) children and adults inadequately vaccinated; and (4) persons not vaccinated for religious or medical reasons. Worldwide, measles ranks among the leading causes of death and disability among the impoverished, where the mortality rate may reach 15% and secondary infections may reach 85%.

Prevention and Treatment

Measles can be prevented by injecting a live, attenuated rubeola virus vaccine. At the time of the introduction of the vaccine in the early 1960s, there were about 400,000 cases of the disease reported each year in the United States, probably representing one-tenth of the actual number of victims. At present, less than 100 are generally reported, many due to importations from other countries. Globally, vaccination programs have lowered the number of cases from about 100 million with 5.8 million deaths in 1980 to around 30 million cases and 888,000 deaths currently. The decline in incidence has not been uniform among the different regions of the world (**figure 22.23**), largely due to wars and poverty, but substantial

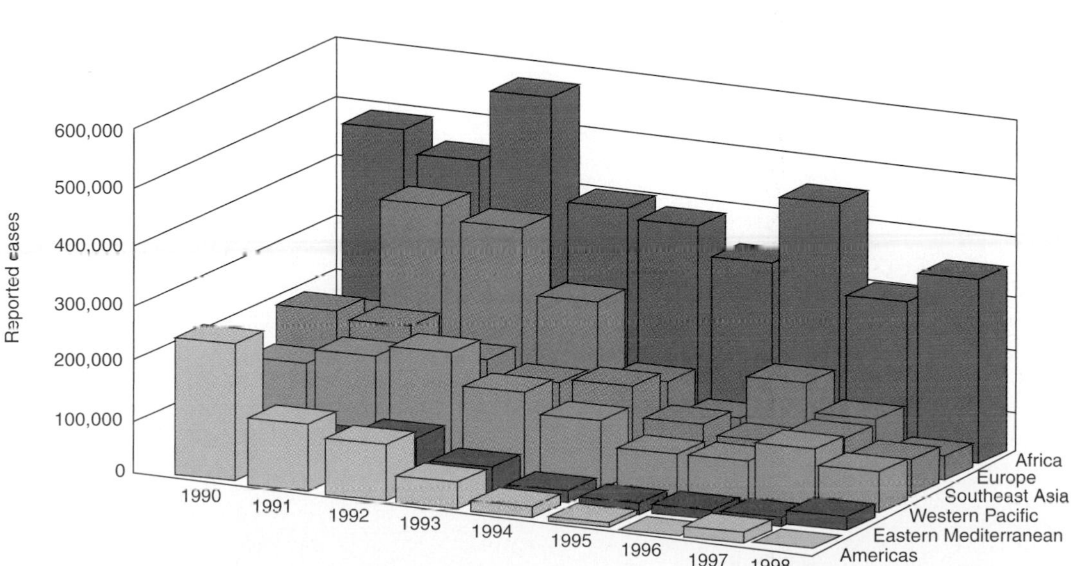

Figure 22.23 Reported Incidence of Measles in Different Regions of the World, 1990 to 1998 Reported cases represent only a small fraction of the total. The lack of progress in measles control in some regions reflects difficulties in delivering measles vaccine.

progress has been made, and complete elimination of measles is now a realistic goal.

The measles vaccine is usually given together with mumps and rubella vaccines, MMR. The first injection of vaccine is given near an infant's first birthday. Since 1989, a second injection of vaccine is given at entry into elementary school. The two-dose regimen has resulted in at least 99% of the recipients becoming immune. In an epidemic, vaccine is given to babies as young as 6 months, who are then reimmunized before their second birthday. Students entering high school or college are advised to get a second dose of vaccine if they have not received one earlier. Those at special risk of acquiring rubeola, such as medical personnel, should be immunized regardless of age, unless they definitely have had measles or have laboratory proof of immunity.

No antiviral treatment exists for rubeola at present. Some features of rubeola are summarized in **table 22.9**.

German Measles (Rubella)

German measles and three-day measles are common names for **rubella**. The term *German measles* arose because the disease was first described in Germany. In contrast to varicella and rubeola, rubella is typically a mild, often unrecognized disease that is difficult to diagnose. Nevertheless, infection of pregnant women can have tragic consequences.

Symptoms

Characteristic symptoms of German measles are slight fever, mild cold symptoms, and enlarged lymph nodes behind the ears and on the back of the neck. After about a day, a faint rash consisting

TABLE 22.9 Measles (Rubeola)

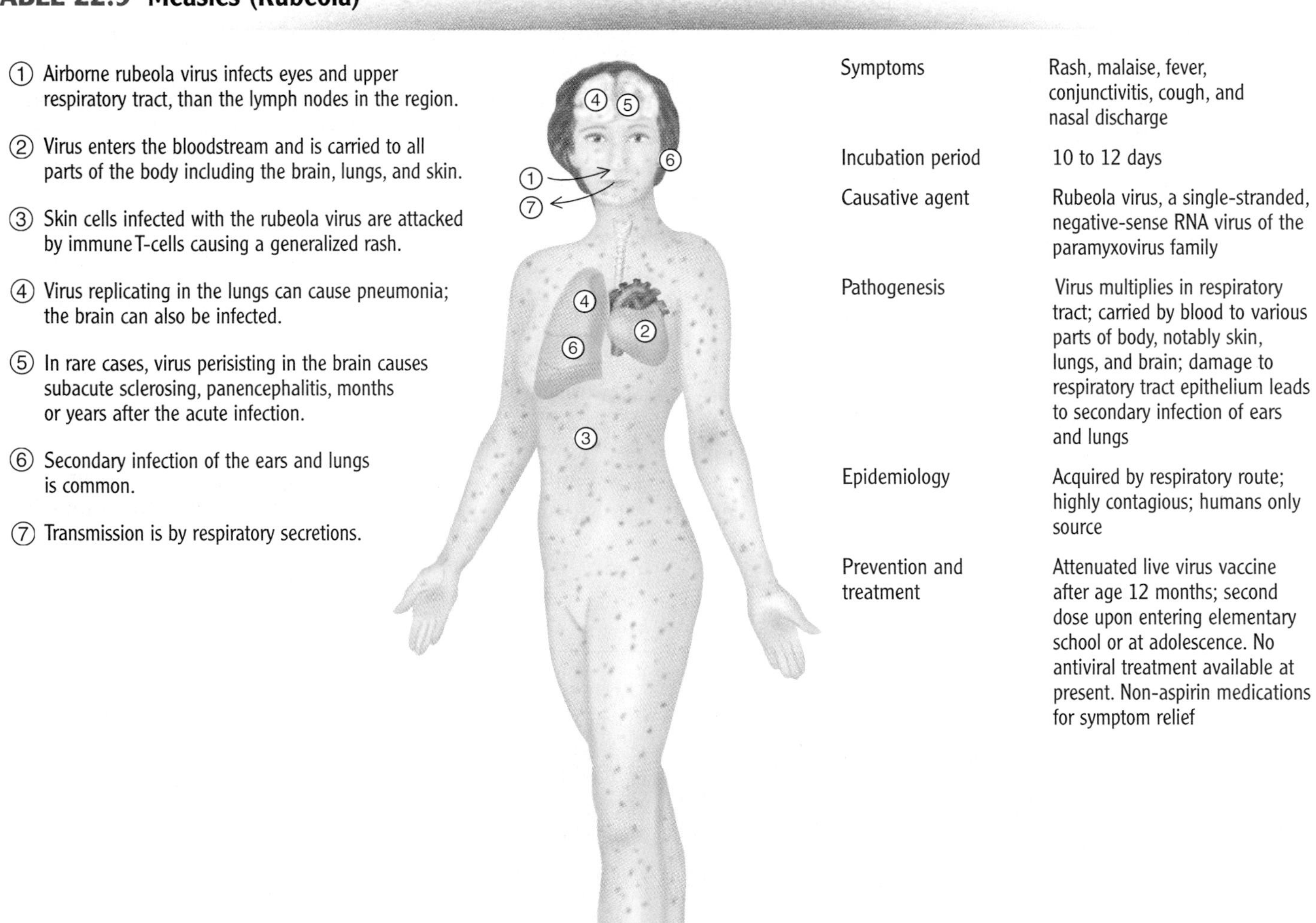
① Airborne rubeola virus infects eyes and upper respiratory tract, than the lymph nodes in the region.
② Virus enters the bloodstream and is carried to all parts of the body including the brain, lungs, and skin.
③ Skin cells infected with the rubeola virus are attacked by immuneT-cells causing a generalized rash.
④ Virus replicating in the lungs can cause pneumonia; the brain can also be infected.
⑤ In rare cases, virus perisisting in the brain causes subacute sclerosing, panencephalitis, months or years after the acute infection.
⑥ Secondary infection of the ears and lungs is common.
⑦ Transmission is by respiratory secretions.
④ ⑤
⑥
①
⑦
④
②
⑥
③

Symptoms	Rash, malaise, fever, conjunctivitis, cough, and nasal discharge
Incubation period	10 to 12 days
Causative agent	Rubeola virus, a single-stranded, negative-sense RNA virus of the paramyxovirus family
Pathogenesis	Virus multiplies in respiratory tract; carried by blood to various parts of body, notably skin, lungs, and brain; damage to respiratory tract epithelium leads to secondary infection of ears and lungs
Epidemiology	Acquired by respiratory route; highly contagious; humans only source
Prevention and treatment	Attenuated live virus vaccine after age 12 months; second dose upon entering elementary school or at adolescence. No antiviral treatment available at present. Non-aspirin medications for symptom relief

CASE PRESENTATION

The patient was a twenty-year-old asymptomatic man who was immunized against measles as a requirement for starting college. He had received his first dose of measles vaccine at approximately 1 year of age.

Past medical history revealed that he was a hemophiliac and had contracted the human immunodeficiency virus from clotting factor (a blood product given to control bleeding) contaminated with the virus.

Laboratory tests showed that he had a very low CD4+ lymphocyte count, indicating a severely damaged immune system.

About a month after his precollege immunization, he developed pneumocystosis, a lung infection characteristic of AIDS, was hospitalized, had a good response to treatment, and was discharged. Ten months later, he was again hospitalized for symptoms of a severe lung infection. He had no rash. Multiple laboratory tests to determine the cause of his infection were negative. Finally, a lung biopsy was performed and revealed "giant cells," very large cells with multiple nuclei. Cytoplasmic and intranuclear inclusion bodies were also present. This picture was highly suggestive of measles pneumonia, and measles virus subsequently was recovered from cell cultures of the biopsy material. Subsequent studies showed it to be the measles vaccine virus. The patient received intravenous gamma globulin and an experimental antiviral medication, ribavirin, and improved. Subsequently, however, his condition deteriorated, and he died of presumed complications of AIDS.

1. Is measles immunization a good idea for people with immunodeficiency?
2. Is it surprising that the vaccine virus was still present in this patient 11 months after vaccination? Explain.
3. Despite the severe infection, there was no rash. Why?

Discussion

1. Measles is often disastrous for persons with AIDS or other immunodeficiencies. They should be immunized as soon as possible in their illness, before the immune system becomes so weakened it cannot respond effectively to the vaccine. Also, as this and other cases have shown, the vaccine virus can itself be pathogenic when immunodeficiency is severe. With the worldwide effort to eliminate measles, the risk of exposure to the wild-type measles virus, as opposed to the laboratory-derived vaccine virus, is declining, but outbreaks in colleges and other institutions still occur. A severely immunodeficient individual can be passively immunized against measles with gamma globulin if exposure to the wild-type virus occurs.
2. Measles is often given as an example of a persistent viral infection, meaning that following infection the virus can persist in the body for months or years in a slowly replicating form. It has been suggested but not proven that this explains the lifelong immunity conferred by measles infection in normal people. In rare presumably normal individuals and more commonly in malnourished or immunodeficient individuals, persisent infection leads to damage to the brain, lung, liver, and possibly, the intestine. Subacute sclerosing panencephalitis can follow measles vaccination, but at a much lower rate than after wild virus infection.
3. Following acute infection, the measles virus floods the bloodstream and is carried to various tissues of the body, including the skin. The rash of measles is caused by T lymphocytes attacking measles virus antigen lodged in the skin capillaries. In the absence of functional T lymphocytes, the rash does not occur.

of innumerable pink spots appears over the face, chest, and abdomen (**figure 22.24**). Unlike rubeola, there are no diagnostic mouth lesions. Adults commonly develop painful joints, with pain generally lasting 3 weeks or less. Other symptoms generally last only a few days. The significance of rubella, however, lies not with these symptoms but rather with rubella's threat to the fetuses of pregnant women.

Causative Agent

German measles is caused by the rubella virus, a member of the togavirus family. It is a small, about 60 nm in diameter, enveloped, single-stranded, positive-sense RNA virus that can readily be cultivated in cell cultures. Surface glycoproteins give the virus *in vitro* hemagglutinating ability, which is inhibited by specific antibody, allowing serological identification of the virus.

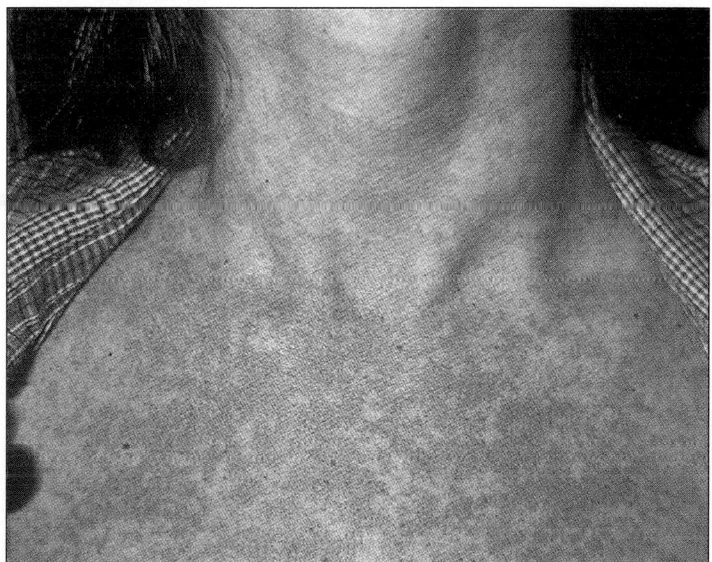

Figure 22.24 Adult with German Measles (Rubella) Symptoms are often very mild, but the effects on a fetus can be devastating.

Pathogenesis

The rubella virus enters the body via the respiratory route. It multiplies in the nasopharynx and enters the bloodstream, causing a sustained viremia (meaning viruses circulating in the bloodstream). The blood transports the virus to various body tissues, including the skin and joints. Humoral and cell-mediated immunity develop against the virus, and the resulting antibody-antigen reactions probably account for the rash and joint symptoms and then for the rapid disappearance of the disease.

In pregnant women, the placenta becomes infected during the period of viremia. Early in pregnancy, the virus readily crosses the placenta and infects the fetus, but as the pregnancy progresses, fetal infection becomes increasingly less likely. Virtually all types of fetal cells are susceptible to infection; some cells are killed, while others develop a persistent infection in which cell division is impaired and chromosomes are damaged. The result is a characteristic pattern of fetal abnormalities that is referred to as the **congenital rubella syndrome**. The abnormalities include cataracts and other abnormalities of the eyes, brain damage, deafness, heart defects, and low birth weight despite normal gestation. Babies may be stillborn. Those that live continue to excrete rubella virus in throat secretions and urine for many months. The viral persistence occurs despite the presence of neutralizing antibody. The babies produce antiviral IgM and IgG antibodies, as well as a cellular immune response that, in time, eliminates the persistently infected cells. The likelihood of the syndrome varies according to the age of the fetus

Perspective 22.1 Epidemic on a Cruise Ship

The Centers for Disease Control and Prevention was notified by a Florida cruise line of an outbreak of illness among the crew of one of its ships. The ship sailed regularly between Florida and the Bahamas, carrying 8,000 passengers per week, with a crew of 385. The sick individuals complained of a rash and fever.

Examination suggested that the illness was German measles (rubella), and blood tests for IgM antibody against rubella virus were carried out on the crew members. Sixteen individuals were confirmed to have the acute disease, half of whom were asymptomatic. Eighty-five percent of the crew members were foreign born, coming from 50 different countries, and three-fourths of them had no history of German measles or vaccination against the disease. A survey of about 3,600 passengers was carried out to estimate the risk of their contracting the disease, especially pregnant women. This survey indicated that 0.8% of all passengers were pregnant women, half of whom were in the first trimester of pregnancy.

Assuming that all the 8,000 passengers in a week were exposed to German measles, that 10% of women of childbearing age lack immunity to German measles, and that 85% of pregnant women infected during the first trimester will have a baby with congenital rubella syndrome, about how many cases of the syndrome would you expect as a result? What simple measure could you advise the cruise line to undertake to minimize the risk of crew members transmitting German measles to its passengers in the future?

when infection occurs. Infections occurring during the first 6 weeks of pregnancy result in almost 100% of the fetuses having a detectable injury, most commonly minor deafness. Even infants who are apparently normal, however, excrete rubella virus for extended periods and thus can infect others. ■ **neutralizing antibody, p. 421**

Epidemiology

Humans are the only natural host for rubella virus. The disease is highly contagious although less so than rubeola; it is estimated that in the prevaccine era, 10% to 15% of people reached adulthood without being infected. Complicating the epidemiology of rubella is the fact that over 40% of infected individuals fail to develop symptoms. People with few or no symptoms continue their usual activities and can spread the virus. People who develop typical rubella can be infectious for as much as 7 days before the rash appears until 7 days afterward. Before widespread use of the vaccine began in 1969, periodic major epidemics arose. One epidemic in 1964 resulted in about 30,000 cases of congenital rubella syndrome.

Prevention and Treatment

Prevention of German measles depends on subcutaneous injection of a live attenuated rubella virus administered to babies at 12 to 16 months of age with a second dose at age 4 to 6 years. The vaccine produces long-lasting immunity in about 95% of recipients. To reduce the risk of rubella transmission, most states require documentation of vaccination before entry into school, but many preschool children are underimmunized. Among women of child-bearing age, 6% to 25% of different groups lack rubella immunity. The vaccine is not given to pregnant women for fear that it might result in congenital defects. As an added precaution, women are advised not to become pregnant for 3 months after receiving the vaccine.

Use of the vaccine has markedly reduced the incidence of rubella in the United States to generally less than 250 cases per year (**figure 22.25**). Nevertheless, outbreaks of the disease continue to occur because of the presence of underimmunized populations. Gamma globulin, given as passive immunization after a person has been exposed to rubella, does not prevent the disease. No specific antiviral therapy is available.

Some features of German measles are summarized in **table 22.10**.

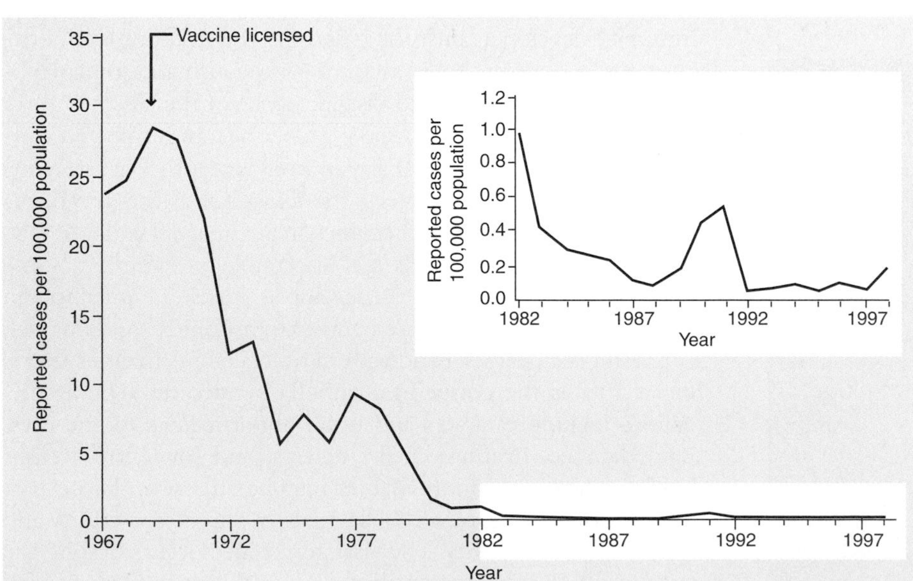

Figure 22.25 Reported Cases of German Measles (Rubella), United States, 1967 to 1997
Most rubella virus infections now occur in people more than 20 years old. Can you explain why?

Other Viral Rashes of Childhood

The kinds of viruses that can cause childhood rashes probably number in the hundreds. One group alone, the enteroviruses, has about 50 members that have been associated with skin lesions. In the early 1900s the causes of the common childhood rashes were largely unknown, and it was common practice to number them 1 to 6 as follows: (1) rubeola, (2) scarlet fever, (3) rubella, (4) Duke's disease—a mild disease with fever and bright red generalized rash, now thought to have been due to an enterovirus, (5) erythema infectiosum, and (6) exanthem subitum. The causes of two common childhood rashes, erythema infectiosum and exanthem subitum, have only been established in recent years.

Fifth disease (**erythema infectiosum**) occurs in both children and young

TABLE 22.10 German Measles (Rubella)

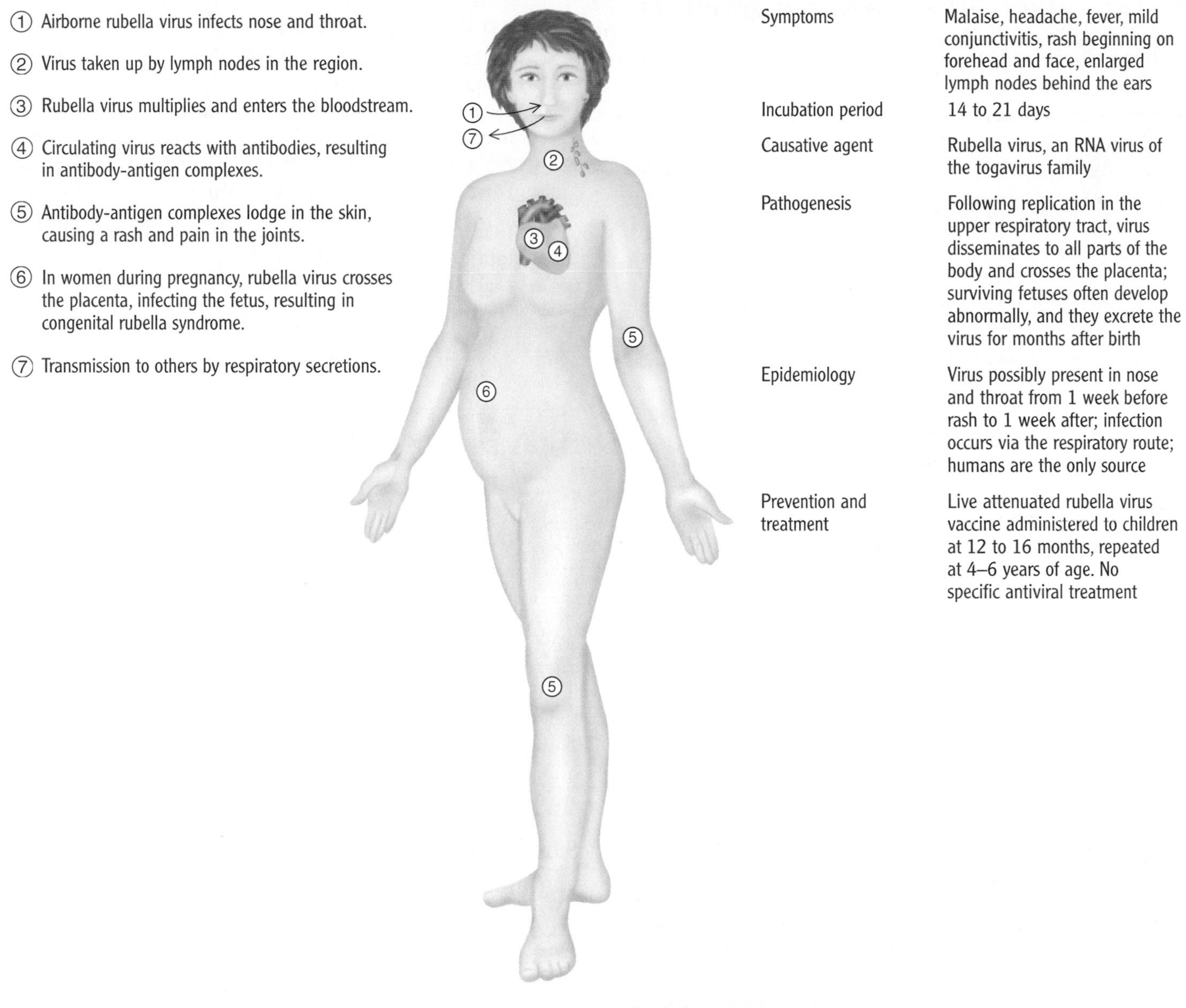
① Airborne rubella virus infects nose and throat.
② Virus taken up by lymph nodes in the region.
③ Rubella virus multiplies and enters the bloodstream.
④ Circulating virus reacts with antibodies, resulting in antibody-antigen complexes.
⑤ Antibody-antigen complexes lodge in the skin, causing a rash and pain in the joints.
⑥ In women during pregnancy, rubella virus crosses the placenta, infecting the fetus, resulting in congenital rubella syndrome.
⑦ Transmission to others by respiratory secretions.
①
⑦
②
③
④
⑤
⑥
⑤

Symptoms	Malaise, headache, fever, mild conjunctivitis, rash beginning on forehead and face, enlarged lymph nodes behind the ears
Incubation period	14 to 21 days
Causative agent	Rubella virus, an RNA virus of the togavirus family
Pathogenesis	Following replication in the upper respiratory tract, virus disseminates to all parts of the body and crosses the placenta; surviving fetuses often develop abnormally, and they excrete the virus for months after birth
Epidemiology	Virus possibly present in nose and throat from 1 week before rash to 1 week after; infection occurs via the respiratory route; humans are the only source
Prevention and treatment	Live attenuated rubella virus vaccine administered to children at 12 to 16 months, repeated at 4–6 years of age. No specific antiviral treatment

adults. The illness begins with fever, malaise, and head and muscle aches. A diffuse redness appears on the cheeks, giving the appearance of the face as if it were slapped. The rash commonly spreads in a lacy pattern to involve other parts of the body, especially the extremities. The rash may come and go for 2 weeks or more before recovery. Joint pains are a prominent feature of some adult infections. The disease is caused by parvovirus B-19, a small (18–28 nm), nonenveloped, single-stranded DNA virus of the parvovirus family. The virus preferentially infects certain bone marrow cells and is a major threat to persons with sickle cell and other anemias because the infected marrow sometimes stops producing blood cells, a condition known as **aplastic crisis**. Also, about 10% of women infected with the virus during pregnancy suffer spontaneous abortion.

Roseola (**exanthem subitum, roseola infantum**), is a common disease in infants six months to three years old. It causes a great deal of parental anxiety because it begins abruptly with fever that may reach 105°F and cause convulsions. The children generally do not appear ill, however. After several days, the fever vanishes and a transitory red rash appears, mainly on the chest and abdomen. The patient has no symptoms at this point, and the rash vanishes in a few hours to 2 days. This disease is caused by herpesvirus, type 6. There is no vaccine against the disease, and no treatment except to reduce the risk of seizures by sponging with lukewarm water and using medication to keep the temperature below 102°F.

Warts

Papillomaviruses, cause of warts, can infect the skin through minor abrasions. **Warts** are small tumors called papillomas that consist of multiple nipplelike protrusions of tissue covered by skin or mucous membrane. Warts rarely become cancers, although some of the many papillomaviruses are strongly associated with cervical cancer. About 50% of the time, warts on the skin disappear within 2 years without any treatment. Papillomaviruses (**figure 22.26**) belong to the papovavirus family. They are small (about 50 nm diameter), nonenveloped, double-stranded DNA viruses. More than 50 different papillomaviruses are known to infect humans. Warts of other animals are generally not infectious for humans.

Papillomaviruses have been very difficult to study because they fail to grow in cell cultures or experimental animals. Wart viruses can survive on inanimate objects such as wrestling mats, towels, and shower floors, and infection can be acquired from such contaminated objects. The virus infects the deeper cells of the epidermis and reproduces in the nuclei. Some of the infected cells grow abnormally and produce the wart. The incubation period ranges from 2 to 18 months. Infectious virus is present in the wart and can contaminate fingers or objects that pick or rub the lesions. Like other tumors, warts can only be treated effectively by killing or removing all of the abnormal cells. This can usually be accomplished by freezing the wart with liquid nitrogen, by cauterization, meaning burning the tissue usually with an electrically heated needle, or by surgical removal. Virus generally remains in the adjacent normal-appearing skin, however, and may cause additional warts. Warts that grow on the soles of the feet are called **plantar warts** (often mistakenly called planter's warts; *plantar* is a word meaning "referring to the sole of the foot"). These warts are very difficult to get rid of because the pressure of standing on them causes them to grow wide and deep. ■ cervical cancer, p. 653, 755

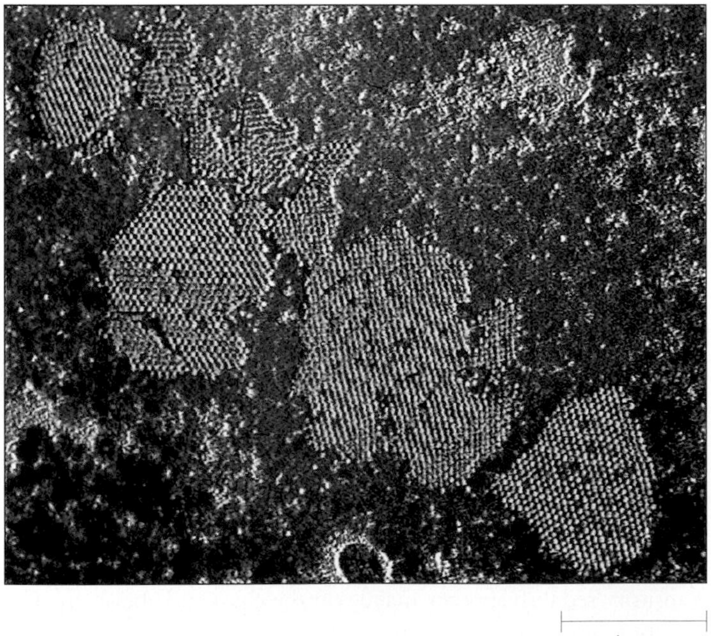

1 µm

Figure 22.26 Wart Virus The virions appear yellow in this color-enhanced transmission electron micrograph.

MICROCHECK 22.4

Chickenpox, measles, and rubella can be controlled by vaccines. Viral diseases that may only inconvenience a pregnant woman can be disastrous to her fetus. A viral infection acquired in childhood can remain latent for years only to reactivate in a different form. One group of viruses causes benign skin tumors.

- What important diagnosis sign is often present in the mouth of measles (rubeola) victims?
- What is the epidemiological significance of shingles?
- Why is it a good idea to immunize little boys against rubella?

Skin Diseases Caused by Fungi

Diseases caused by fungi are called **mycoses**. Earlier in this chapter, we mentioned the role of normal flora yeast of the genus *Malassezia* in causing mild skin diseases, such as tinea versicolor. Other fungi are responsible for more serious infections of the skin, although even in these cases the condition of the host's defenses against infection is often crucial. The yeast *Candida albicans* (**figure 22.27**) may live harmlessly among the normal flora of the skin, but in some people it invades the deep layers of the skin and subcutaneous tissues. In many people with candidal skin infections, no precise cause for the invasion can be determined. Certain molds also cause cutaneous mycoses, but they are not as likely as *C. albicans* to invade the deep skin layers.

Superficial Cutaneous Mycoses

Certain species of molds can invade hair, nails, and the keratinized portion of the skin. The resulting mycoses have colorful names such as jock itch, athlete's foot, and ringworm, and more traditional latinized names that describe their location: tinea capitis—scalp, tinea barbae—beard, tinea axillaris—armpit, tinea corporis—body, tinea cruris—groin, and tinea pedis—feet, to list a few. Tinea just means "worm," which probably reflects early ideas about the cause.

Symptoms

Most people colonized by these molds have no symptoms at all. Others complain of itching, a bad odor, or a rash. In ringworm, a rash occurs at the site of the infection and consists of a scaly area surrounded by redness at the outer margin, producing irregular rings or a lacy pattern on the skin. On the scalp, patchy areas of hair loss can occur, with a fine stubble of short hair left behind. Involved nails become thickened and brittle and may separate from the nailbed. Sometimes, a rash consisting of fine papules and vesicles develops distant from the infected area. This rash is referred to as a dermatophytid, or "id" reaction, a reflection of allergy to products of the infecting fungus.

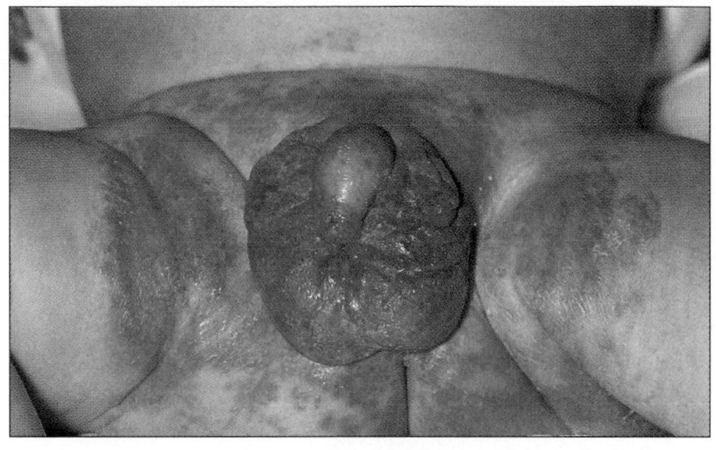

(a)

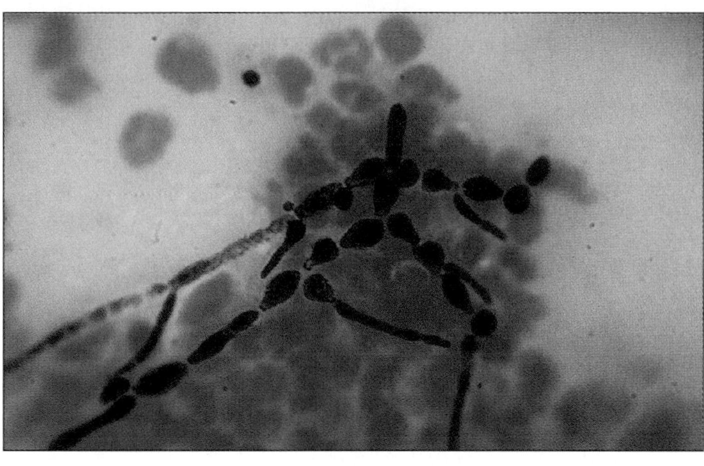

(b) 10 μm

Figure 22.27 ***Candida albicans*** **(a)** Causing a diaper rash; **(b)** Gram stain of pus showing *C. albicans* yeast forms and filamentous forms called pseudohyphae.

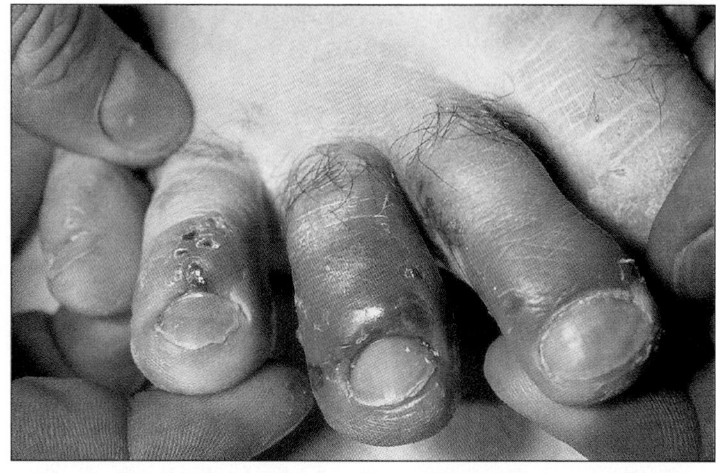

(a)

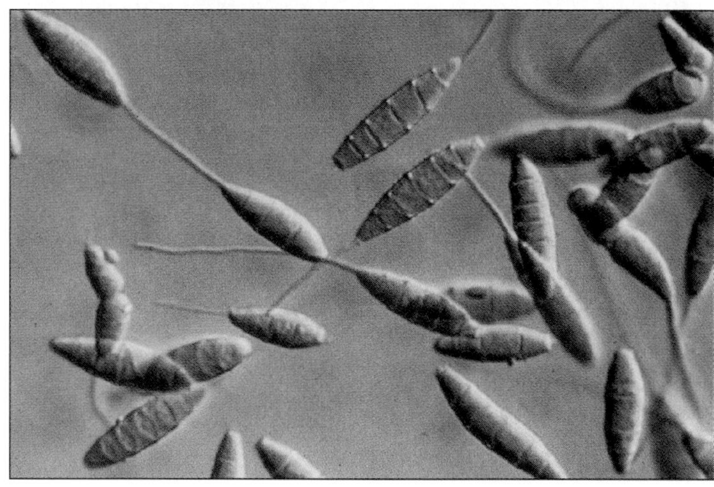

(b) 20 μm

Figure 22.28 **Dermatophytosis** **(a)** Tinea pedis, usually caused by species of *Trichophyton*. **(b)** Large boat-shaped spores of *Microsporum gypseum*, a cause of scalp ringworm in children.

Causative Agents

The skin-invading molds belong mainly to the genera *Epidermophyton*, *Microsporum*, and *Trichophyton* and are collectively termed **dermatophytes (figure 22.28)**. They can be cultivated on media especially designed for molds and are identified by their colonial and microscopic appearance, their nutritional requirements, and biochemical tests.

Pathogenesis

The normal skin is generally resistant to invasion by dermatophytes. Some species, however are relatively virulent and can even cause epidemic disease, especially in children. In conditions of excessive moisture, dermatophytes can invade keratinized structures, including the epidermis down to the level of the keratin-producing cells. A keratinase enables them to dissolve keratin and use it as a nutrient. Hair is invaded at the level of the hair follicle because the follicle is relatively moist. Fungal products diffuse into the dermis and provoke an immune reaction, which probably explains why adults tend to be more resistant to infection than children. It also explains why some people develop the allergic "id" reactions.

Epidemiology

As mentioned, age, virulence of the infecting strain of mold, and excessive moisture are important factors in causing infections. Common causes of excessive moisture are obesity causing folds of skin to lie together, tight clothing, and plastic or rubber footware. Potentially pathogenic molds may be present in soil and on pets such as young cats and dogs.

Prevention and Treatment

Attention to cleanliness and maintenance of normal dryness of the skin and nails effectively prevent most dermatophyte infections. Powders, open shoes, changing of socks, and application of alcohol after bathing may help prevent toenail infections. Numerous prescription and over-the-counter medications are promoted for treating dermatophytoses, and most are effective for treating superficial skin infections. Nail infections are often very difficult to cure, requiring taking medication by mouth for months, and sometimes surgical removal of the nail.

MICROCHECK 22.5

Fungal causes of skin mycoses commonly colonize skin without causing symptoms. The best protection against fungal skin infections is to maintain normal skin dryness.

- What is a mycosis?
- What kinds of structures are invaded by dermatophytes?

FUTURE CHALLENGES

The Ecology of Lyme Disease

Lyme disease is often referred to as one of the emerging diseases. Unrecognized in the United States before 1975, it is now the most commonly reported vector-borne disease. Because of the seeming explosion in the numbers of Lyme disease cases, and its apparent extension to new geographical areas, the ecology of Lyme disease is under intense study. In the northeastern United States, large increases in white-footed mouse populations occur in oak forests during years in which there is a heavy acorn crop, with a corresponding increase in Ixodes scapularis *ticks. Both deer and mice feed on the acorns and subsequently spread the disease to adjacent areas. Variations in weather conditions, and their effect on food supply for these animals, might therefore be an important ecological factor, although it is not clear that weather cycles completely explain the emerging nature of the disease. The presence of animals other than white-footed mice for the ticks to feed on is another factor. Aternative tick hosts usually do not have a sustained* Borrelia burgdorferi *bacteremia following infection from a tick, and the blood of a common lizard host along the West Coast even kills the spirochetes. The role of snakes, foxes, and birds of prey that control mouse populations and that of birds, spiders, and wasps that feed on ticks are also under study. The challenge is to define more completely the ecology of Lyme and other tick-borne diseases in order to predict their emergence and find new ways for their prevention.*

SUMMARY

Anatomy and Physiology (Figures 22.1, 22.2)

1. The skin is a large complex organ with many functions, including temperature regulation, vitamin D synthesis, and aiding cell-mediated immunity.
2. The skin repels potential pathogens by shedding and being dry, acidic, and toxic.

Normal Flora of the Skin

1. Skin is inhabited by large numbers of bacteria of little virulence that help prevent colonization by more dangerous species. (Table 22.1)

Diphtheroids

1. **Diphtheroids** are Gram-positive, pleomorphic, rod-shaped bacteria that play a role in acne and body odor. Fatty acids, produced from their metabolism of the oily secretion of sebaceous glands, keeps the skin acidic.

Staphylococci

1. Staphylococci are Gram-positive cocci arranged in clusters. Universally present, they help prevent colonization by potential pathogens and maintain the balance among flora of the skin. The principal species, *Staphylococcus epidermidis*, can sometimes be pathogenic.

Fungi

1. *Malassezia* sp. are single-celled yeasts found universally on the skin. Usually harmless, they can cause tinea versicolor, probably some cases of dandruff, and serious skin disease in AIDS patients. (Figure 22.3)

Bacterial Skin Diseases

Hair Follicle Infections (Figure 22.4)

1. Boils and **carbuncles** are caused by *Staphylococcus aureus* (Table 22.2), which is coagulase-positive and often resists penicillin and other antibiotics. A carbuncle is more serious because the infection is more likely to be carried to the heart, brain, or bones.

Scalded Skin Syndrome (Figure 22.5, Table 22.3)

1. Staphylococcal scalded skin syndrome results from exotoxins produced by certain strains of *Staphylococcus aureus.*

Streptococcal Impetigo (Figure 22.6, Table 22.5)

1. Impetigo is a superficial skin infection caused by *Streptococcus pyogenes* and *Staphylococcus aureus.*
2. **Acute glomerulonephritis**, a kidney disease, caused by an antibody-antigen reaction, is an uncommon complication of *S. pyogenes* infections. (Figure 22.8)

Rocky Mountain Spotted Fever (Figure 22.9, Table 22.6)

1. Rocky Mountain spotted fever, caused by the obligate intracellular bacterium *Rickettsia rickettsii*, is an often fatal disease transmitted to humans by the bite of an infected tick. (Figures 22.10, 22.11, 22.12)

Lyme Disease (Figure 22.13, Table 22.7)

1. Lyme disease can imitate many other diseases. It is caused by a spirochete, *Borrelia burgdorferi*, transmitted to humans by

certain ticks. A target-shaped rash, present in most victims, is the hallmark of the disease. (Figures 22.13, 22.14, 22.15, 22.16, 22.17)

Skin Diseases Caused by Viruses

Chickenpox (Varicella) (Figure 22.18, Table 22.8)

1. Chickenpox is a common disease of childhood caused by the varicella-zoster virus, a member of the herpesvirus family (Figure 22.20). **Shingles**, or **herpes zoster**, can occur months or years after chickenpox, a reactivation of the varicella-zoster virus infection in the distribution of a sensory nerve. Shingles cases can be sources of chickenpox epidemics. (Figure 22.19)

Measles (Rubeola) (Figures 22.21, 22.22, Table 22.9)

1. Measles (**rubeola**) is a potentially dangerous viral disease that can can lead to serious secondary bacterial infections, and fatal lung or brain damage.
2. Measles can be controlled by immunizing young children and susceptible adults with a live attenuated vaccine. (Figure 22.23)

German Measles (Rubella) (Figure 22.24, Table 22.10)

1. German measles (**rubella**), if contracted by a woman in the first 8 weeks of pregnancy, results in a 90% chance of birth defects making up the **congenital rubella syndrome**.
2. Immunization with a live attenuated virus protects against this disease. (Figure 22.25)

Other Viral Rashes of Childhood

1. Numerous other viruses can cause rashes. **Fifth disease** (**erythema infectiosum**) is characterized by a "slapped cheek" rash. It is caused by parvovirus B-19. It can be fatal to people with certain anemias.
2. **Roseola** (**exanthem subitum**) is marked by several days of high fever and a transitory rash that appears as the temperature returns to normal. It occurs mainly in infants six months to three years old. The disease is caused by human herpesvirus, type 6.

Warts

1. Warts are skin tumors caused by a number of papillomaviruses. They are generally benign, but some sexually transmitted papillomaviruses are associated with cancer of the uterine cervix. (Figure 22.26)

Skin Diseases Caused by Fungi

Superficial Cutaneous Mycoses

1. Invasive skin infections, such as diaper rashes, are sometimes caused by the yeast *Candida albicans*. (Figure 22.27)
2. Certain species of molds that feed on keratin cause athlete's foot, ringworm, and invasions of the hair and nails. (Figure 22.28)

REVIEW QUESTIONS

Short Answer

1. What is the difference between a furuncle and carbuncle?
2. In what age group is staphylococcal scalded skin syndrome most likely to occur?
3. Why is the blister fluid of staphylococcal scalded skin syndrome free of staphylococci?
4. Give three ways in which *Streptococcus pyogenes* resembles *Staphylococcus aureus*, and three ways in which it differs.
5. What is the epidemiological importance of passage through the eggs of *Rickettsia rickettsii*?
6. Describe the causative agent of Lyme disease.
7. Outline the life cycle of the main vector of Lyme disease in the eastern and north-central United States.
8. What is the relationship between chickenpox (varicella) and shingles (herpes zoster)?
9. Why do so many people suffer permanent damage or die from measles?
10. What is characteristic about the rash of varicella?
11. What viral disease might be associated with an aplastic crisis? Describe its characteristic rash.
12. What is the significance of rubella viremia during pregnancy?
13. How does a person contract warts?
14. What is the allergic rash called that appears in response to ringworm and distant to it?

Multiple Choice

1. Which of the following conditions are important in the ecology of the skin?
   A. Temperature
   B. Salt concentration
   C. Lipids
   D. pH
   E. All of the above
2. *Staphylococcus aureus* can be responsible for which of the folllowing conditions?
   A. Impetigo
   B. Food poisoning
   C. Toxic shock syndrome
   D. Scalded skin syndrome
   E. All of the above
3. The main effect of staphylococcal protein A is to...
   A. interfere with phagocytosis.
   B. enhance the attachment of the Fc portion of antibody to phagocytes.
   C. coagulate plasma.
   D. kill white blood cells.
   E. degrade collagen.

4. Which of the following is essential for the virulence of *Streptococcus pyogenes*?
   A. Protease
   B. Hyaluronidase
   C. DNase
   D. All of the above
   E. None of the above.
5. Which of the following is true of streptococcal acute glomerulonephritis?
   A. It is a streptococcal infection of the kidneys.
   B. It is caused by immune complexes containing streptococcal antigen.
   C. It is caused by most strains of β-hemolytic group A streptococci.
   D. It is the result of a streptococcal toxin directed against the kidneys.
   E. All of the above.
6. All of the following are true of Rocky Mountain spotted fever, *except...*
   A. the disease is most prevalent in the western United States.
   B. it is caused by an obligate intracellular bacterium.
   C. it is a zoonosis transmitted to human beings by ticks.
   D. those with the disease characteristically develop a hemorrhagic rash.
   E. antibiotic therapy is usually curative if given early in the disease.
7. All of the following are true of Lyme disease, *except*
   A. it is caused by a spirochete.
   B. it is transmitted by certain species of ticks.
   C. it occurs only in the region around Lyme, Connecticut.
   D. most cases get a rash that looks like a target.
   E. it can cause heart and nervous system damage.
8. Which of the following is more likely to be true of measles (rubeola) than German measles (rubella)?
   A. Koplik's spots are present.
   B. It causes birth defects.
   C. It causes only a mild illness.
   D. Human beings are the only natural host.
   E. Live, attenuated virus vaccine is available for prevention.
9. All of the following must be cultivated in cell cultures instead of cell-free media, *except...*
   A. *Rickettsia rickettsii.*
   B. Rubella virus.
   C. Varicella-zoster virus.
   D. *Borrelia burgdorferi.*
   E. Rubeola virus.
10. All of the following might contribute to development of ringworm or other superficial cutaneous mycoses, *except*
   A. Obesity.
   B. playing with kittens.
   C. rubber boots.
   D. using skin powder.
   E. dermatophyte virulence.

Applications

1. A school administrator in a small Iowa community prohibited a child with chickenpox from attending school. He claimed that this was the first case of chickenpox seen in the school in 6 years and that he did not want to have an outbreak at the school. Several parents argued to the school board that an outbreak would benefit the school in the long term. Discuss the pros and cons of allowing this child to attend school.
2. A public health official was asked to speak about immunization during a civic group luncheon. One parent asked if rubella was still a problem. In answering the question, the official cautioned women planning to have another child to have their present children immunized against rubella. Why did the official make this statement to the group?

Critical Thinking

1. A microbiology instructor stated that the presence of large numbers of *Propionibacterium acnes* in the same areas where acne develops illustrates that occurrence of a bacterium and a disease together does not necessarily imply cause and effect. Why would the instructor make this statement?
2. When Lyme disease was first being investigated, the observation that frequently only one person in a household was infected was a clue leading to the discovery that the disease was spread by arthropod bites. Why was this so?
3. Why might it be more difficult to eliminate a disease like Lyme disease or Rocky Mountain spotted fever from the earth than rubeola or rubella?

CHAPTER 23

Respiratory System Infections

Rebecca Lancefield's mother was a direct descendent of Lady Mary Wortley Montagu, who promoted smallpox vaccination in England more than 75 years before Jenner. Lancefield was educated in various schools as her parents moved from one army base to another. In 1912, she entered Wellesley College and became interested in biology. She was awarded a scholarship to Columbia University and studied under the famous microbiologist Hans Zinsser, a pioneer in the science of immunology. She received a Master's degree, but her studies were interrupted when her husband was drafted into the armed services in World War I. Fortunately he was assigned to the Rockefeller Institute, where Rebecca got a job as a laboratory technician for the distinguished microbiologists O. T. Avery and A. R. Dochez.

Well known for their studies of pneumococci, Avery and Dochez had been commissioned to study streptococcal cultures from personnel at army camps. At that time, classification of the numerous kinds of streptococci was based largely on whether their colonies on blood agar produced β-hemolysis or a green discoloration. Studies of the army camp cultures at the Rockefeller Institute were among the first to use immunological methods to classify streptococci. By using different streptococcal strains to immunize mice and produce antisera, Avery and Dochez showed that their cultures could be divided into groups based on their antigens. Later, the Lancefields returned to Columbia, where Rebecca received her Ph.D. in 1925 for her studies of α-hemolytic streptococci. At that time, α-hemolytic streptococci were considered a possible cause of rheumatic fever, a common cause of serious heart disease, but she showed that these streptococci occurred just as often in healthy people as in those with rheumatic fever. Thereafter, Rebecca returned to the Rockefeller Institute where she spent the remainder of her scientific career studying streptococci. Building on the discoveries of others, she showed that almost all the strains of β-hemolytic streptococci from human infections had the same cell wall carbohydrate "A," which could easily be extracted from the cells by hot acid. Streptococci from other sources had different carbohydrates: "B" from cattle infections; "C" from cattle, horses, and guinea pigs; "D" from cheese and human normal flora; and so forth. The grouping of streptococci by their cell wall carbohydrates, now referred to as "Lancefield grouping," proved to be a much better predictor of pathogenic potential than hemolysis on blood agar. Lancefield received numerous honors for her work and, in 1960, became the first woman president of the American Association of Immunologists. In 1970, Lancefield was elected to the prestigious National Academy of Sciences. She died in 1981 at the age of 86. ■ **Mary Wortley Montagu, p. 413**

—A Glimpse of History

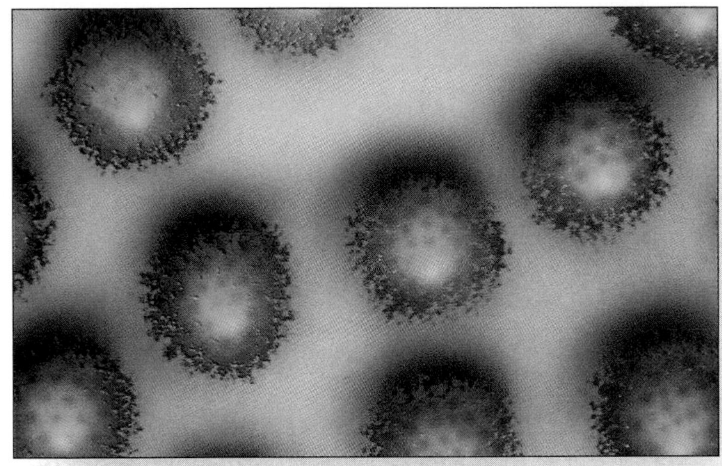

RESPIRATORY INFECTIONS ENCOMPASS AN ENORmous variety of illnesses ranging from the trivial to the fatal. For convenience, they can be divided into infections of the upper part of the respiratory system, primarily the head and neck, and infections of the lower respiratory system, in the chest. Most upper respiratory infections are uncomfortable, but not life threatening and go away without treatment in about a week. However they are so common that they far outweigh other infections in terms of the cumulative misery they cause. Some illnesses such as the childhood rashes discussed in the last chapter, have a minor upper respiratory component but injure the skin, lung, nervous system, or other parts of the body. Still others cause major symptoms involving the eye, nose, throat, middle ear, sinuses, and other body systems. The lower respiratory system is usually sterile, well protected from colonization by microorganisms. However pathogens sometimes evade the body's defenses and cause serious diseases, such as pneumonia, tuberculosis, or whooping cough.

Anatomy and Physiology

Microorganisms that colonize the respiratory system enter an environment quite different than the one faced by skin organisms. The dark caverns and tortuous passages of the respiratory system are relatively hot and humid, and the atmosphere contains less oxygen and more carbon dioxide than occurs in air. The eyes and the nose are the two gateways, or "portals of entry," to the respiratory system. They are more than entrances; respiratory infections often first establish themselves there and then spread to other parts of the system.

Structures commonly involved in infections of the upper respiratory system **(figure 23.1)** are:

- the conjunctiva, composing the moist surfaces of the eyes and eyelids. Infection of the conjunctiva is called conjunctivitis;
- the nasolacrimal, or tear ducts, from the eyes to the nasal chamber. Infection of the tear ducts is called dacryocystitis;
- the middle ear. Infection of the middle ear is called otitis media. The external ear canal is not part of the respiratory system. It is a blind tunnel lined with skin and is part of the skin ecosystem; infection there is called external otitis;
- the air-filled chambers of the skull, the sinuses and mastoid air cells. Infections of these chambers are called sinusitis and mastoiditis, respectively;
- the nose. Infection of the nose is called rhinitis;
- the throat, or pharynx. Infections of the throat are called pharyngitis;
- the epiglottis is a little muscular flap that covers the opening to the lower respiratory system during swallowing, thereby preventing material from entering. Infection of the epiglottis is called epiglottitis;

The tonsils are composed of lymphoid tissue, strategically located to come into contact with and respond immunologically to incoming microorganisms. The tonsils are important producers of antibodies to infectious agents, but paradoxically they can also be the sites of infection, resulting in tonsillitis. Enlargement of the pharyngeal tonsil, also called "adenoids," can contribute to ear infections by interfering with normal drainage from the eustachian tubes. These tubes, which extend from the middle ear to the nasopharynx, are 3 to 4 cm long. They equalize the pressure in the middle ear and drain normal mucous secretions. ■ lymphoid tissue, p. 374

A person normally breathes about 16 times per minute, inhaling about 0.5 liters of air with each breath, or more than 11,500 liters of air per day, with any accompanying microorganisms or other pollutants. The air enters the respiratory system at the nostrils, flows into the nasal cavity, and is deflected downward through the throat. The nasal cavity is incompletely divided into right and left halves by a vertical wall extending almost back to the throat. Spongy masses of tissue bulge into each half of this chamber from its outside walls. These tissues are similar to other erectile tissues of the body in that they expand and contract with alterations in blood flow controlled by nervous reflexes. If they are enlarged, these spongy tissues contribute to nasal congestion or obstruction of the airways. Many stimuli can cause these changes, including variations in temperature and humidity, emotional factors, and the irritating effects of smoke and infections.

One of the main functions of the upper respiratory tract is to regulate the temperature and water content of inspired air. When cold air enters the upper respiratory tract, nervous reflexes immediately increase the blood flow to the spongy tissues in the nose, thereby transferring heat to the air. This mechanism usually adjusts the temperature to within 2 to 3 degrees of body temperature by the time it reaches the lungs. Inspired air also becomes saturated with water vapor, the nasal tissues giving up as much as a quart of water per day to keep the air humidified. The warmth and moisture provide optimum conditions for the body's defenses against infection.

The air then enters the lower respiratory tract below the epiglottis. Structures commonly involved in lower respiratory system infections include:

- the voice box, or larynx (see figure 23.1). Inflammation of the larynx is called laryngitis, and is manifest as hoarseness.
- The larynx is continuous with the windpipe, or trachea, which branches into two bronchi. Inflammation of the bronchi is called bronchitis, commonly the result of infection or smoking tobacco.
- The bronchi branch repeatedly, becoming bronchioles, site of an important viral infection called bronchiolitis.
- The smallest branches of the bronchioles end in the tiny, thin-walled air sacs called alveoli that comprise the bulk of lung tissue. Inflammation of the lungs is called pneumonitis, often the result of viral infections. Pneumonitis that results in the filling of alveoli with pus and fluid is called pneumonia. Macrophages are numerous in the lung tissues and readily move into the alveoli and airways to engulf infectious agents, thus helping to prevent pneumonia from developing.
- The lungs are surrounded by two membranes: One adheres to the lung and the other to the chest wall and diaphragm. These membranes, termed pleura, normally slide against each other as the lung expands and contracts. Inflammation of the pleura is called pleurisy, characterized by severe chest pain aggravated by breathing or coughing.

The Mucociliary Escalator

The moist membranes of the respiratory system are coated with a slimy glycoprotein material called **mucus**, which is produced by specialized unicellular glands called goblet cells. **Goblet cells**, so called because of their shape, narrow at the base and wide at surface, are sprinkled among the other cells composing the membrane. Most of the respiratory system mucous membranes are composed of ciliated epithelium. Ciliated cells have tiny hairlike projections called cilia along their exposed free border. The cilia beat synchronously at a rate of about 1,000 times a minute, continually propelling the mucous film out of the mastoids, middle ear, nasolacrimal duct, sinuses, and the lungs.

Figure 23.1 Anatomy and Infections of the Respiratory System **(a)** Lateral view, upper respiratory system, showing the inside of the nasal chamber and its connections with the eyes, middle ears, and sinuses, and details of the respiratory epithelium. **(b)** Frontal view of the upper and lower respiratory system, including details of the alveoli.

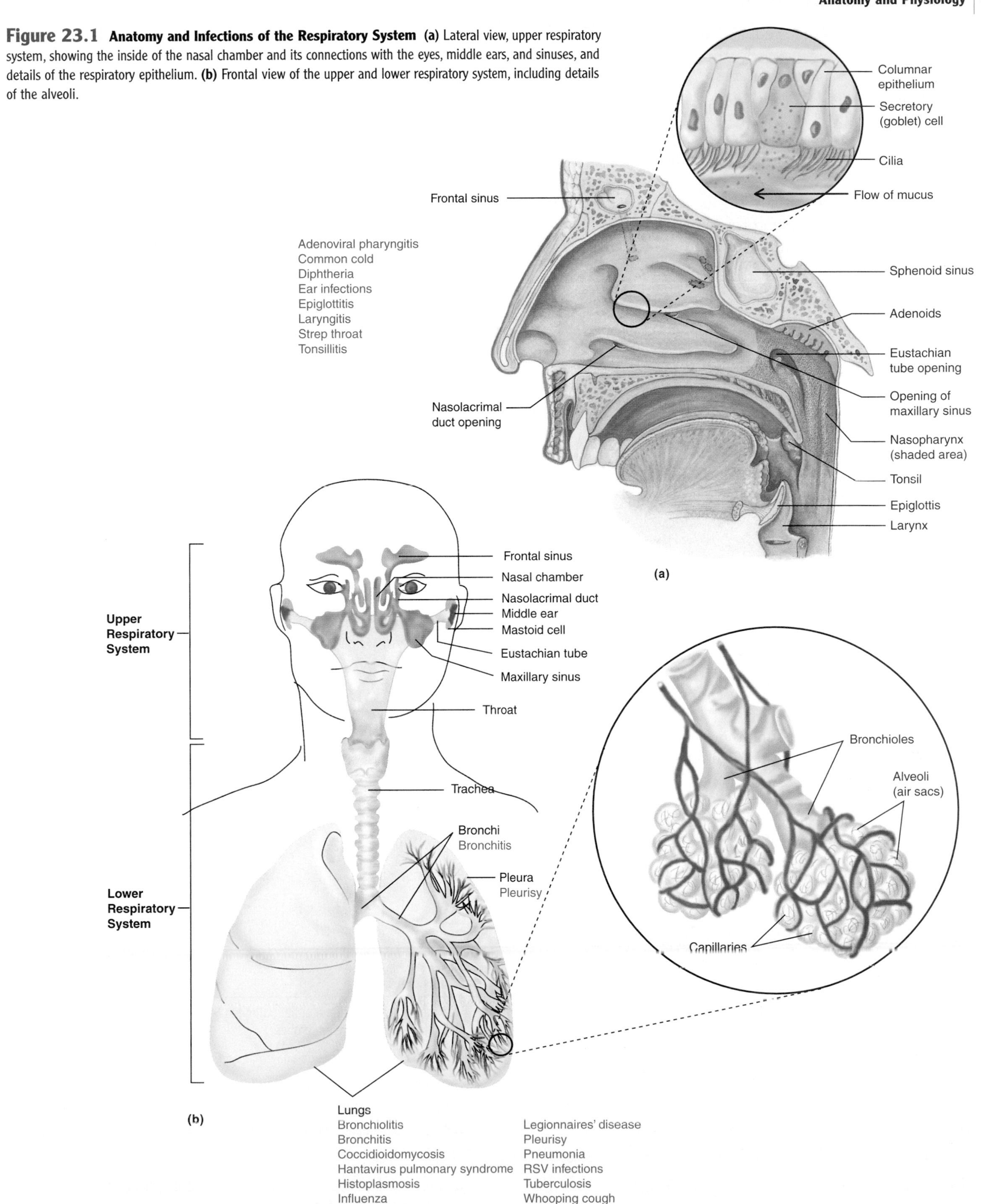

The mucus is then swallowed, and any entrapped microorganisms and viruses are exposed to the killing action of stomach acid and enzymes. This mechanism, called the **mucociliary escalator**, normally keeps the middle ears, sinuses, mastoids, and lungs completely free of microorganisms. Viral infection, tobacco smoke, alcohol, and narcotics, however, impair ciliary movement and increase the chance of infection.

MICROCHECK 23.1

The respiratory system provides a habitat for microorganisms that is warm and moist, with an atmosphere containing less oxygen and more carbon dioxide than the external environment. It is guarded by tonsils and adenoids representing the immune system, and by the mucociliary escalator.

- What is the normal function of the tonsils and adenoids?
- Why would the respiratory passages contain less oxygen than air?

Normal Flora

The mastoids, middle ear, sinuses, trachea, bronchi, bronchioles, and alveoli are normally sterile. The nasal cavity, nasopharynx, and pharynx are colonized by numerous bacterial species. Aerobes, facultative anaerobes, aerotolerant organisms, and anaerobes are all represented. Although generally harmless, members of the normal bacterial flora are opportunists and can cause disease when host defenses are impaired. ■ **oxygen requirements, p. 98** ■ **opportunists, p. 454**

Surprisingly, even though the eyes are constantly exposed to a multitude of microorganisms, normal healthy people commonly have no bacteria on their conjunctivae, the moist membrane covering the eye and inner surfaces of the eyelids. Presumably, this results from the frequent automatic washing of the eye with lysozyme-rich tears and from the eyelid's blinking reflex, which cleans the eye surface like a windshield wiper sweeps a windshield. Unless they are able to attach to the epithelium, the viruses and microorganisms that impinge on the conjunctiva are swept into the nasolacrimal duct and into the nasopharynx. Organisms recovered from the normal conjunctiva are usually few in number and originate from the skin flora. They are adapted to live in a different environment and are generally unable to colonize the respiratory system. ■ **lysozyme, p. 65**

Some of the bacterial genera that inhabit the upper respiratory system are shown in **table 23.1**. The secretions of the nasal entrance usually contain diphtheroids and staphylococci. About 20% of healthy people constantly carry *Staphylococcus aureus* in their noses, and an even higher percentage of hospital personnel are likely to be carriers of these important opportunistic pathogens. Farther inside the nasal passages, the microbial population increasingly resembles that of the nasopharynx. The nasopharynx contains large numbers of microorganisms, mostly α-hemolytic streptococci of the viridans group, nonhemolytic streptococci, *Moraxella catarrhalis* representing Gram-negative diplococci, and diphtheroids. Anaerobic Gram-negative bacteria, including species of *Bacteroides*, are also present in large numbers in the nasopharynx. In addition, commonly pathogenic bacteria such as *Streptococcus pneumoniae*, *Haemophilus influenzae*, and the Gram-negative diplococcus *Neisseria meningitidis* are often found, especially during the cooler seasons of the year.

MICROCHECK 23.2

Except in parts of the upper respiratory tract, the respiratory system is free of a normal microbial flora. The upper respiratory tract flora is highly diverse, including aerobes, anaerobes, facultative anaerobes, and aerotolerant bacteria. Although most of them are of low virulence, these organisms can sometimes cause disease.

- What are some possible advantages to the body of providing a niche for normal flora in the upper respiratory tract?
- How can strict anaerobes exist in the upper respiratory tract?

TABLE 23.1 Normal Flora of the Respiratory System

Genus	Characteristics	Comments
Staphylococcus	Gram-positive cocci in clusters	Commonly includes the potential pathogen *Staphylococcus aureus*, inhabiting the nostrils. Facultative anaerobes.
Corynebacterium	Pleomorphic, Gram-positive rods; nonmotile; non-spore-forming genera	Aerobic. Diphtheroids include anaerobic and aerotolerant organisms.
Moraxella	Gram-negative diplococci and diplobacilli	Aerobic. Some resemble pathogenic *Neisseria* species such as *N. meningitidis*.
Haemophilus	Small, Gram-negative rods	Facultative anaerobes. Commonly include the potential pathogen *H. influenzae*.
Bacteroides	Small, pleomorphic, Gram-negative rods	Strict anaerobes.
Streptococcus	Gram-positive cocci in chains	α (especially viridans), β, and γ types; the potential pathogen, *S. pneumoniae* is often present. Aerotolerant.

INFECTIONS OF THE UPPER RESPIRATORY SYSTEM

Bacterial Infections of the Upper Respiratory System

A number of different species of bacteria can infect the upper respiratory system. Some, such as *Haemophilus influenzae* and β-hemolytic streptococci of Lancefield group C, can cause sore throats but generally do not require treatment because the bacteria are quickly eliminated by the immune system. Other infections require treatment because they are not so easily eliminated and can cause serious complications.

Strep Throat (Streptococcal Pharyngitis)

Sore throat is one of the most common reasons that people in the United States seek medical care, resulting in about 27 million doctor visits per year. Many of these visits are due to a justifiable fear of streptococcal pharyngitis, commonly known as strep throat.

Symptoms

Streptococcal pharyngitis typically is characterized by redness of the throat, with patches of adhering pus and scattered tiny hemorrhages, and fever. The lymph nodes in the neck are enlarged and tender. Abdominal pain or headache may be prominent in older children and young adults. Not usually present are red, weepy eyes, cough, or runny nose. Most patients with streptococcal sore throat recover spontaneously after about a week. In fact, many infected people have only mild symptoms or no symptoms at all.

Causative Agent

Streptococcus pyogenes, the cause of strep throat, is a Gram-positive coccus that grows in chains of varying lengths (**figure 23.2**). It can be differentiated from other steptococci that normally inhabit the throat by its characteristic colonial morphology when grown on blood agar. *Streptococcus pyogenes* produces hemolysins, enzymes that lyse red blood cells, which result in the colonies being surrounded by a zone of β-hemolysis. Because of their characteristic hemolysis, *S. pyogenes* and other streptococci that show a similar phenotype are called β-hemolytic streptococci. In contrast, species of *Streptococcus* that are typically part of the normal throat flora are either nonhemolytic, or they produce α-hemolysis, characterized by a zone of incomplete, often greenish clearing around colonies grown on blood agar. ■ **streptococcal hemolysis, p. 103**

Streptococcus pyogenes is commonly referred to as the group A streptococcus. The group A carbohydrate in the cell wall of *S. pyogenes* differs antigenically from that of most other streptococci and serves as a convenient basis for identification. Lancefield grouping uses antibodies to differentiate the various species of streptococci based on their cell wall carbohydrate. These antibodies are also the basis for many of the rapid diagnostic tests done on throat specimens in a physician's office. Other rapid tests utilize a DNA probe.

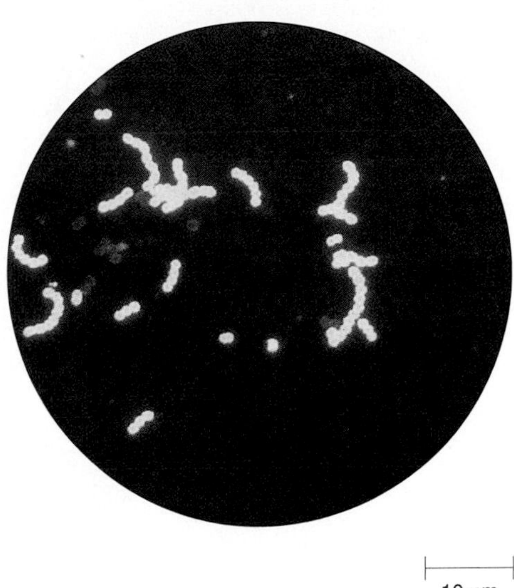

Figure 23.2 ***Streptococcus pyogenes*** Chain formation in fluid culture is revealed by fluorescence microscopy.

Pathogenesis

Streptococcus pyogenes causes a wide variety of illnesses, which is not surprising considering its vast arsenal of virulence factors. Some of these disease-causing mechanisms are structural components of the outer cell wall that enable the bacterium to foil host defenses (**figure 23.3**), while others are destructive enzymes and toxins released by the bacterial cell. Some of the virulence factors identified to date in *S. pyogenes* are listed below:

- Protein F of the cell wall mediates attachment of *S. pyogenes* to the throat by adhering to a protein found on the surface of epithelial cells.
- M protein interferes with complement component C3b, an opsonin that would otherwise promote phagocytosis of the bacteria. Antibody to M protein neutralizes its ability to bind a C3b-destroying protein and allows the bacterium to be killed by phagocytes. M protein is of central importance in the virulence of *S. pyogenes*, because antibody to it prevents infection from occurring. Unfortunately, more than 80 different kinds of M protein exist among the many strains of *S. pyogenes*, and antibody to one type of M protein does not prevent infection by a *S. pyogenes* strain that has another kind of M protein.
- A capsule composed of hyaluronic acid, present on only some strains of *S. pyogenes*, also inhibits phagocytosis. Hyaluronic acid is a normal component of human tissue, and its presence as a *S. pyogenes* capsule may impair recognition of the bacterium by the body's immune system.
- Protein G of the *S. pyogenes* cell wall has a function like that of the protein A of *Staphylococcus aureus* in binding the Fc segment of immunoglobulin G. The effect is to

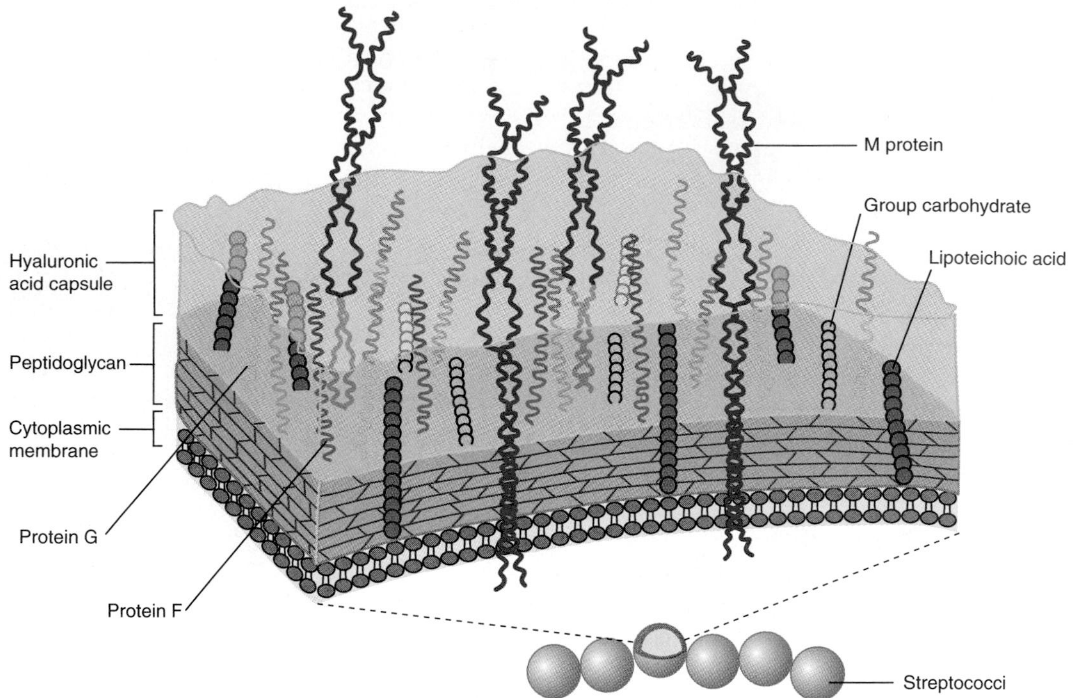

Figure 23.3 Components of the Cell Envelope of *Streptococcus pyogenes* The M protein is important in virulence, while the group carbohydrate helps identify the species. Protein F binds the bacterium to epithelial cells, while protein G helps interfere with phagocytosis by binding to the Fc portion of antibodies.

prevent phagocytosis mediated by specific antibody against the bacterium.

- C5a peptidase is an enzyme released by *S. pyogenes*. This enzyme destroys the complement component C5a, a substance responsible for attracting phagocytes to the site of a bacterial infection.
- A number of other enzymes produced by *S. pyogenes*, including DNase, protease, streptokinase, and hyaluronidase, degrade tissue and probably enhance the spread of the bacterium. The tissue destruction these enzymes cause may trigger the inflammatory response, which in turn leads to the symptoms of strep throat.
- Streptolysins O and S are the proteins responsible for β-hemolysis. These hemolysins lyse erythrocytes and leukocytes by making holes in their cell membranes.

Streptococcal pyrogenic exotoxins, (**SPEs**) are a family of genetically similar exotoxins produced by strains of *S. pyogenes* that cause severe streptococcal diseases frequently unrelated to strep throat. These diseases include scarlet fever, streptococcal toxic shock syndrome, and "flesh eating" necrotizing fasciitis. Only scarlet fever is usually preceded by strep throat symptoms. Some of these exotoxins are encoded by temperate bacteriophages, which can probably be transferred to other streptococcal strains and species. Several of these exotoxins are superantigens, causing massive activation of T cells. The resulting uncontrolled release of cytokines is probably responsible for the seriousness of these infections. ■ **complement, p. 377** ■ **staphylococcal protein A, p. 460** ■ **temperate bacteriophages, p. 329** ■ **superantigens, p. 464** ■ **cytokines, p. 379** ■ **opsonins, p. 377, 394**

Complications of streptococcal pharyngitis can occur during the acute illness. One example, **scarlet fever**, is due to erythrogenic toxin, an SPE released from the infection site, that enters the bloodstream and circulates throughout the body, causing a redness of the skin and a whitish coating of the tongue. Another example is **quinsy**, a painful abscess that develops around one of the tonsils.

Other complications develop days or weeks after the acute illness, often times when *S. pyogenes* has already been eliminated from the body. One of these late complications, acute glomerulonephritis, was discussed in chapter 22. Another late complication is acute rheumatic fever. ■ **acute glomerulonephritis, p. 528**

The symptoms of **acute rheumatic fever**, including fever, joint pain, chest pain, nodules under the skin, rash, and uncontrollable jerky movements, reflect an inflammatory process involving various tissues, especially the joints, heart, skin, and brain. Heart failure and death can occur during the acute illness, but usually symptoms subside with rest and anti-inflammatory medicines such as aspirin. **Chorea**, the condition manifest as jerky movements, reflects brain involvement and generally begins after the other symptoms subside. Rheumatic fever often results in damage to the heart valves, causing one or more to leak and lead to heart failure later in life. Significant damage occurs primarily on the left side of the heart (**figure 23.4**). Normally, oxygenated blood returning from the lungs enters the left ventricle through the mitral, or bicuspid, valve. This valve closes as the heart begins to contract, while the aortic, or semilunar, valve is pushed open as the contracting heart muscle of the left ventricle forces the blood into the circulation. When the heart finishes contracting and begins to relax, the aortic valve closes and the mitral valve opens. If either of the valves fails to open and close properly, the heart muscle has to work hard to deliver enough blood to the circulation. Heart failure results when it can no longer do so. The damaged valves are also subject to infection, usually by bacteria from the normal skin or mouth flora, causing **subacute bacterial endocarditis**. ■ **inflammation, p. 382** ■ **subacute bacterial endocarditis, p. 718**

Acute rheumatic fever is a complication of throat, not skin, *S. pyogenes* infections. Despite many years of study, its pathogenesis is not understood. Symptoms usually begin about 3 weeks after the onset of untreated streptococcal pharyngitis. Cultures of the blood, heart, and joint tissues are negative for *S. pyogenes*, and antibodies against the bacterium are present in the blood of the patient. One theory is that these antibodies could attack a tissue antigen that is similar to a *S. pyogenes* antigen, starting an autoimmune process. Another theory is that some streptococcal product, present in only a small percentage of strains, damages the tissue directly or makes it susceptible to attack by the immune system. Genetic factors probably play a role in the development of rheumatic fever, because it occurs more commonly in individuals with certain MHC types. Overall, the risk of developing acute rheumatic fever after severe, untreated streptococcal pharyngitis is 3% or less. Those with untreated mild pharyngitis have a lower risk. Mild infections result in many cases of rheumatic fever, however, because they are common and much more likely to go untreated. In the United States, rheumatic fever has generally declined in incidence over many years, probably as a result of timely treatment of streptococcal pharyngitis with antibiotics, and a decline in the streptococcal strains associated with the disease. In 1994, the last year in which it was a nationally reportable disease, there were only 112 cases reported, down from about 10,000 in 1961. ■ **major histocompatibility complex (MHC), p. 398**

Epidemiology

Streptococcal infections spread readily by respiratory droplets generated by yelling, coughing, and sneezing, especially in the range of about 2 to 5 feet from an infected individual. If strep throat is untreated, the person may be an asymptomatic carrier for weeks. People who carry the organism in their nose spread the streptococci more effectively than do pharyngeal carriers. Anal carriers are not common but can be a dangerous source of nosocomial infections. Epidemics of strep throat can originate from food contaminated by *S. pyogenes* carriers. Some people become long-term carriers of *S. pyogenes*. In these cases, the infecting strain usually becomes deficient in M protein and is not a threat to the carrier or to others. The peak incidence of strep throat occurs in winter or spring and is highest in grade school children. Among students visiting a clinic because of sore throat at a large West Coast university, less than 5% had strep throat. With some groups of military recruits, however, the incidence has been above 20%.

Prevention and Treatment

Adequate ventilation and avoidance of crowding help to control the spread of streptococcal infections. Because so many strains exist, preparing a vaccine has been impractical. A basis for a vaccine is being sought, however, by identifying conserved epitopes among the different M proteins. Persons with fever and sore

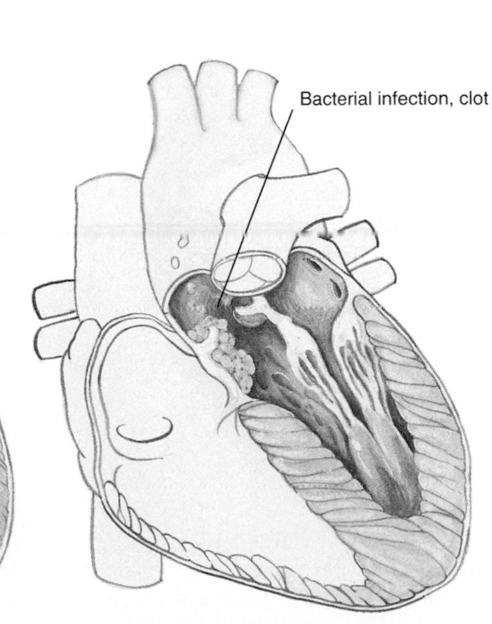

Figure 23.4 Rheumatic Heart Disease Damage to the heart valves results from an immune reaction, the nature of which is still controversial. The damaged area sometimes becomes infected by mouth or skin organisms years after acute rheumatic fever, resulting in subacute bacterial endocarditis.

throat should have a throat culture for *S. pyogenes* so that antibiotic treatment can be given promptly and complications prevented. Individuals with a history of having had acute rheumatic fever with cardiac damage take penicillin daily for years to prevent recurrence of the disease. Confirmed streptococcal pharyngitis is treated with a full 10 days of therapy with penicillin or erythromycin, which eliminates the organism in about 90% of the cases. Treatment given as late as 9 days after the onset prevents rheumatic fever from developing. ■ **epitopes, p. 390**

Table 23.2 summarizes some important facts about streptococcal pharyngitis.

Diphtheria

Diphtheria, a deadly toxin-mediated disease, is now rare in the United States because of childhood immunization; only a few cases are reported each year. Events in other parts of the world, however, are a reminder of what can happen when public health is neglected. In 1990, a diphtheria epidemic began in the Russian Federation. Over the next 5 years, it spread to all the newly independent states of the former Soviet Union. By the end of 1995, 125,000 cases and 4,000 deaths had been reported. The social and economic disruption following the breakup of the Soviet Union had allowed diphtheria to reemerge after being well controlled over the previous quarter of a century.

Symptoms

Diphtheria usually begins with a mild sore throat and slight fever, accompanied by a great deal of fatigue and malaise. Swelling of the neck is often dramatic. A whitish gray membrane forms on the tonsils and the throat or in the nasal cavity. Heart and kidney failure and paralysis may follow these symptoms.

TABLE 23.2 Strep Throat (Streptococcal Pharyngitis)

① *Streptococcus pyogenes* enters by inhalation (nose), or by ingestion (mouth).

② Pharyngitis, fever, enlarged lymph nodes; sometimes tonsillitis, abcess; scarlet fever with strains that produce erythrogenic toxin

③ Symptoms go away.

④ *S. pyogenes* exits by nose and mouth.

Late complications appear:

⑤ glomerulonephritis

⑥ rheumatic fever

⑦ chorea

Complications subside.

⑧ Damaged heart valves leak, heart failure develops.

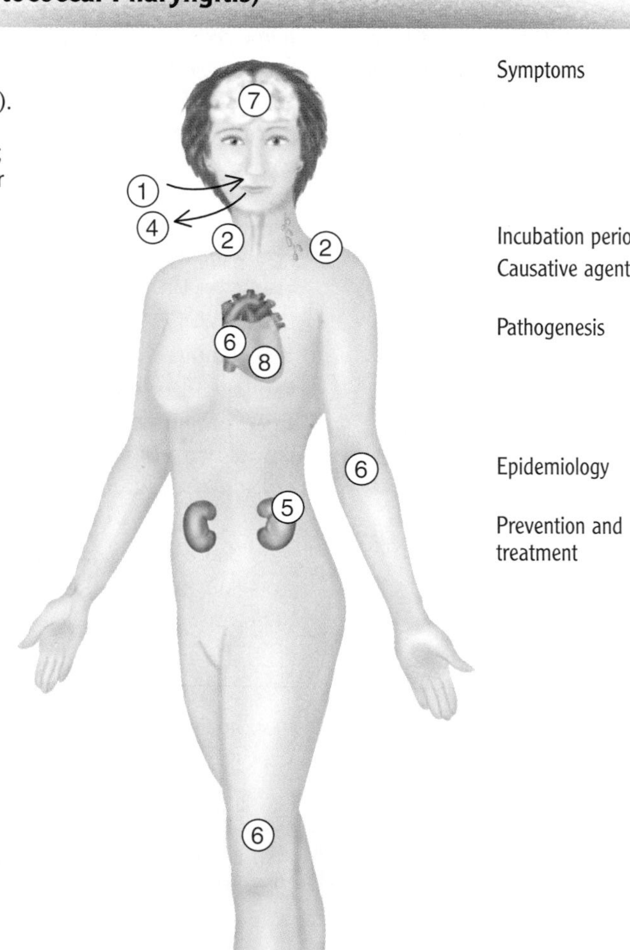

Symptoms	Red throat, often with pus and tiny hemorrhages, enlargement and tenderness of lymph nodes in the neck; less frequently, abscess formation involving tonsils; occasionally, rheumatic fever and glomerulonephritis as sequels
Incubation period	2 to 5 days
Causative agent	*Streptococcus pyogenes,* Lancefield group A β-hemolytic streptococci
Pathogenesis	Virulence associated with hyaluronic acid capsule and M protein, both of which inhibit phagocytosis; protein G binds Fc segment of IgG; protein F for mucosal attachment; multiple enzymes
Epidemiology	Direct contact and droplet infection; ingestion of contaminated food
Prevention and treatment	Avoidance of crowding; adequate ventilation; daily penicillin to prevent recurrent infection in those with a history of rheumatic heart disease. Treatment: 10 days of penicillin or erythromycin

Causative Agent

Diphtheria is caused by *Corynebacterium diphtheriae*, a variably shaped, nonmotile, non-spore-forming, Gram-positive rod that often stains irregularly. Slide preparations of *C. diphtheriae* commonly show the organisms arranged in "Chinese letter" patterns **(figure 23.5)** or side by side in "palisades." *Corynebacterium diphtheriae* grows aerobically on many media. Generally the bacterium grows rapidly, although some strains do not develop visible colonies for several days. Different colony types can be identified, a property that is useful in tracing epidemic spread of the organisms. These bacteria require special media and microbiological techniques to aid their recovery from the mixture of bacteria present in throat material. It is unlikely that they would be recovered by the usual methods for culturing *S. pyogenes* from sore throats. This organism is identified by its appearance after staining with methylene blue dye, its colonial appearance, its biochemical activity, and by demonstrating diphtheria toxin.

Most but not all strains of *C. diphtheriae* release diphtheria toxin, a powerful exotoxin that diffuses from the organism and is responsible for the seriousness of diphtheria. Production of this toxin requires that the bacterium be lysogenized by one of a certain group of bacteriophages, an example of lysogenic conversion. Strains that are nontoxigenic become toxin producers when lysogenized with the bacteriophage. Toxin production can be demonstrated *in vitro*, or *in vivo* by its effect on guinea pigs. The toxin is produced only when the medium on which *C. diphtheriae* is growing has too little iron for optimal growth. The gene for toxin production is under the control of a repressor that is only active when bound to iron. In a medium containing adequate iron, the repressor shuts down toxin production; when the concentration of iron is very low, the iron molecule leaves the repressor and the toxin gene becomes active. ■ **lysogenic conversion, p. 331** ■ **repressors, p. 180**

Pathogenesis

Corynebacterium diphtheriae has little invasive ability, rarely entering the blood or tissues, but the powerful exotoxin it releases is absorbed by the bloodstream. The gray-white membrane that forms in the throat of people with diphtheria is made up of clotted blood along with epithelial cells of the host mucous membrane and the leukocytes that congregate during inflammation, killed by exotoxin released from *C. diphtheriae*. This membrane may come loose and obstruct the airways, causing the patient to suffocate. Entry of the toxin into the bloodstream results in damage to the heart, nerves, and kidneys.

The diphtheria toxin is a large protein released from the bacterium in an inactive form. It is cleaved extracellularly into two chains, A and B, which remain joined by a disulfide bond. The B chain attaches to specific receptors on a host cell membrane, and the entire molecule is taken into the cell by endocytosis **(figure 23.6)**. Cells that lack the appropriate receptors do not take up the toxin and are unaffected by it. This receptor specificity explains why some tissues of the body are not affected in diphtheria, while others such as the heart, kidneys, and nerves are severely damaged.

Once the toxin molecule is inside the cell, the A chain separates from the B chain and becomes an active enzyme. This enzyme catalyses a chemical reaction that inactivates elongation factor 2 (EF-2), a substance required for movement of the ribosome on mRNA (see figure 23.6). This halts protein synthesis, and the cell dies. Since the toxin A chain is an enzyme, it is not consumed in the process, so one or two molecules of toxin can inactivate essentially all the cell's elongation factor 2. This explains the extreme potency of diphtheria toxin. *In vitro*, however, it is easily inactivated by heat and certain chemicals. ■ **A-B toxins, p. 462** ■ **elongation factor, p. 177**

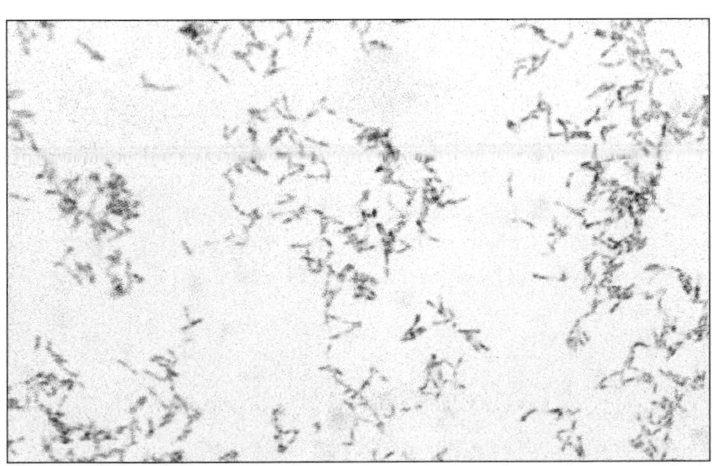

Figure 23.5 ***Corynebacterium diphtheriae*** Note the "Chinese letter" pattern.

Epidemiology

Humans are the primary reservoir for *C. diphtheriae*. Sources of infection include carriers who have recovered from an infection, new cases not exhibiting symptoms, and people with active disease. *Corynebacterium diphtheriae* can colonize people who have been immunized, making them asymptomatic carriers. The bacterium is carried in chronic skin ulcers in some indigent populations. Typically, the organisms are spread by air and acquired by inhalation.

Prevention and Treatment

Toxoid, prepared by formalin treatment of diphtheria toxin, is used for immunization against diphtheria. Toxoid administration causes the body to produce antibodies that specifically neutralize the diphtheria toxin. Since the disease results primarily from toxin absorption rather than microbial invasion, its control can be accomplished most effectively by immunization with toxoid. The well-known childhood vaccination DPT consists of diphtheria and tetanus toxoids and pertussis vaccine, all three generally given together. Unfortunately, these immunizations have often been neglected, particularly among socioeconomically disadvantaged groups, and serious epidemics of the diseases have occurred periodically. Since the 1980s, there has been an active campaign in most of the United States to ensure that children who are entering school are immunized against diphtheria. As a result, today only a few cases of diphtheria are reported annually in the United States, as compared with 30,000 cases reported in 1936. Of concern is the finding that widespread childhood

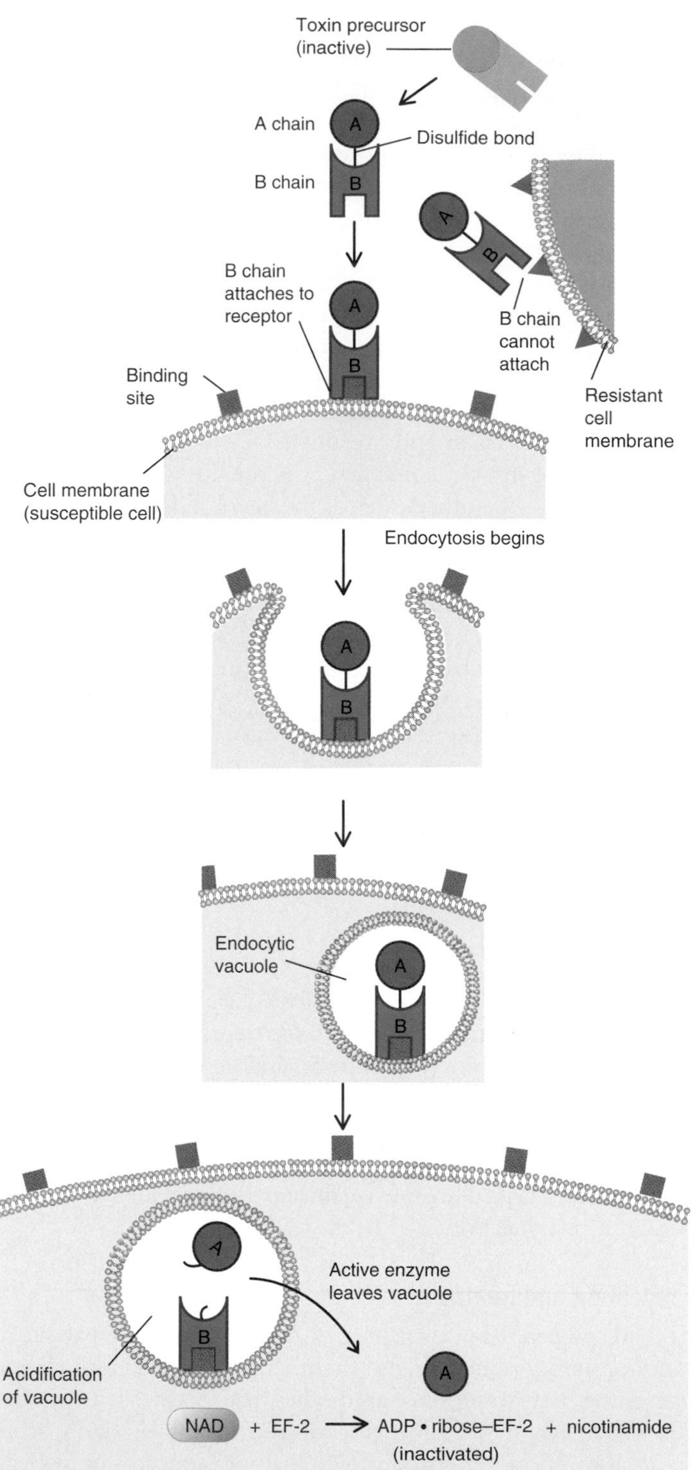

Figure 23.6 Mode of Action of Diphtheria Toxin The toxin precursor released from the bacterium is cleaved extracellularly into A and B chains joined only by a disulfide bond; the B chain attaches to a specific receptor on the cell membrane of a susceptible cell. The toxin enters the cell by endocytosis. With acidification of the endocytic vacuole, the A chain separates from the B chain and then enters the cytoplasm as an active enzyme that inactivates EF-2 (elongation factor 2) by ADP ribosylation. Since EF-2 is required for moving the ribosome on mRNA, protein synthesis ceases and the cell dies.

immunization against diphtheria has shifted susceptibility to the disease to adolescents and adults. Immunity wanes after childhood because of lack of exposure to now scarce *C. diphtheriae* to boost immunity. To prevent buildup of a large population of susceptible older individuals, booster injections should be given every 10 years following childhood immunization.

Effective treatment of diphtheria depends on injecting antiserum against diphtheria toxin into the patient as soon as possible. If the disease is suspected, antiserum must be administered without waiting several days for culture results, because the delay could be fatal. The bacteria are sensitive to antibiotics such as erythromycin and penicillin, but such treatment only stops transmission of the disease; it has no effect on toxin that has already been absorbed. Even with treatment, about 1 of 10 diphtheria patients dies (range 3.5% to 22% depending on age, severity of disease, promptness of treatment, etc.).

Table 23.3 summarizes some important facts about diphtheria.

Pinkeye, Earache, and Sinus Infections

Bacterial infection of the eye's surface, generally known as conjunctivitis or "pinkeye," and infections of the middle ear (otitis media) and sinuses are very common, often occur together, and frequently have the same causative agent. This is not surprising in view of their intimate association with the nasal chamber and nasopharynx (see figure 23.1). Otitis media is a formidable problem, especially in children, responsible for 30 million doctor visits per year in the United States, at an estimated cost of $1 billion. Pinkeye, because of it high communicability, is a frequent source of panic when it appears in a day care facility or classroom. Sinusitis, inflammation of the sinuses, is common in both adults and children.

Symptoms

The symptoms of acute bacterial conjunctivitis include increased tears, redness of the conjunctiva, swelling of the eyelids, sensitivity to bright light, and large amounts of pus. Acute bacterial conjunctivitis must be distinguished from conjunctivitis caused by a large number of viruses. In the latter cases, eyelid swelling and pus are usually minimal. In a typical case of otitis media, a young child wakes from sleep screaming with earache pain. Fever is generally mild or absent. Vomiting often occurs at the height of the pain. Sometimes the pain ends abruptly because the eardrum has ruptured, and drainage of fluid appears in the external ear canal. In sinusitis, pain and a pressure sensation characteristically occur in the region of the involved sinus. Tenderness is also present over the sinus in many instances. Sinus infections are commonly associated with a headache and severe malaise.

Causative Agents

In all three conditions the most common bacterial pathogens are *Haemophilus influenzae*, a tiny Gram-negative rod, and *Streptococcus pneumoniae*, the Gram-positive encapsulated diplococcus known as the pneumococcus. Strains that infect the conjunctiva have ligands that allow firm attachment to the epithelium. ■ **ligands, p. 74** ■ ***Streptococcus pneumoniae*, p. 564** ■ ***Haemophilus influenzae*, p. 668**

TABLE 23.3 Diphtheria

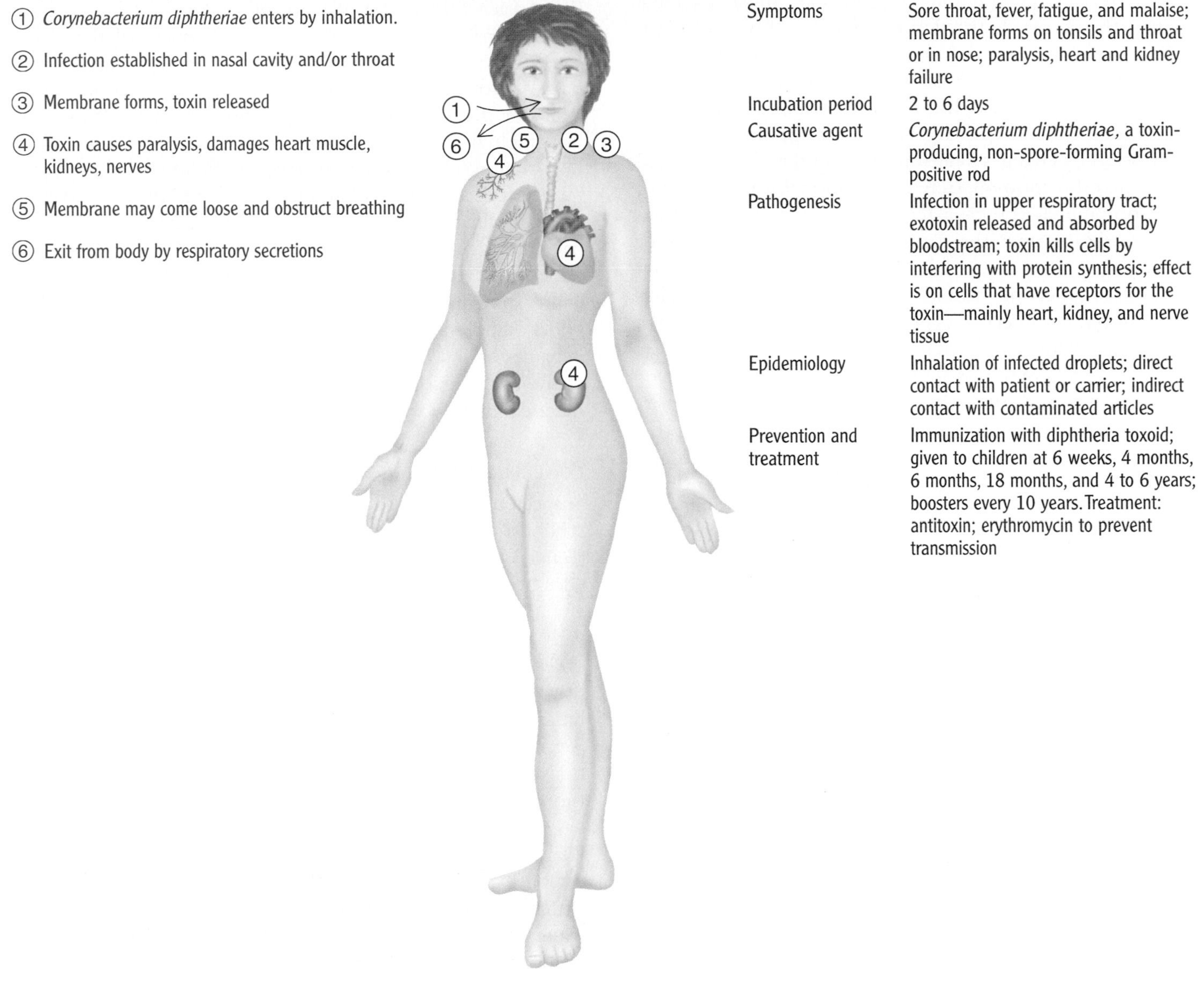
① *Corynebacterium diphtheriae* enters by inhalation.
② Infection established in nasal cavity and/or throat
③ Membrane forms, toxin released
④ Toxin causes paralysis, damages heart muscle, kidneys, nerves
⑤ Membrane may come loose and obstruct breathing
⑥ Exit from body by respiratory secretions
1
6
5
2
3
4
4
4

Symptoms	Sore throat, fever, fatigue, and malaise; membrane forms on tonsils and throat or in nose; paralysis, heart and kidney failure
Incubation period	2 to 6 days
Causative agent	*Corynebacterium diphtheriae,* a toxin-producing, non-spore-forming Gram-positive rod
Pathogenesis	Infection in upper respiratory tract; exotoxin released and absorbed by bloodstream; toxin kills cells by interfering with protein synthesis; effect is on cells that have receptors for the toxin—mainly heart, kidney, and nerve tissue
Epidemiology	Inhalation of infected droplets; direct contact with patient or carrier; indirect contact with contaminated articles
Prevention and treatment	Immunization with diphtheria toxoid; given to children at 6 weeks, 4 months, 6 months, 18 months, and 4 to 6 years; boosters every 10 years. Treatment: antitoxin; erythromycin to prevent transmission

Less commonly, *Moraxella lacunata,* enterobacteria, *Neisseria gonorrhoeae*, and many other microorganisms and viruses infect the conjunctiva. Among these are agents that can contaminate eye medications and contact lens solutions:

Bacillus sp., free-living aerobic Gram-positive spore-forming rods.
Pseudomonas sp., free-living Gram-negative aerobic rods.
Acanthamoeba sp., free-living protozoa that live in soil, and fresh and salt water. They have a cyst form that is resistant to drying and is carried by dust.

Infections caused by these organisms are very uncommon, but when they occur they are usually serious and can lead to destruction of the eye. Contact lens wearers need to be very careful to follow exactly the instructions on use of cleansing solutions, and cloudy or outdated solutions and eye medications should be discarded.

Besides *Streptococcus pneumoniae* and *Haemophilus influenzae*, *Mycoplasma pneumoniae*, *Streptococcus pyogenes*, and *Staphylococcus aureus* can cause otitis media. About one-third of the cases are caused by respiratory viruses. This helps explain why some infections fail to respond to antimicrobial medications, which have no effect on viruses. The same infecting agents are often involved in sinusitis.

Pathogenesis

Few details are known about the pathogenesis of acute bacterial conjunctivitis. Probably, the organisms are inoculated directly onto the conjunctiva from airborne respiratory droplets or rubbed in from contaminated hands. The bacterial ligands attach them to the conjunctiva, and like most bacterial pathogens, they resist destruction by lysozyme. Otitis media is sometimes developing at the time conjunctivitis is diagnosed. Generally, otitis media and sinusitis are preceded by infection of the nasal chamber and nasopharynx. Probably, infection spreads upward through the eustachian tube to the middle ear **(figure 23.7)**, assisted by changes in airway pressure that force infected secretions in a direction opposite to the normal flow of mucus. The infection damages the ciliated cells, resulting in inflammation and swelling. Since the damaged eustachian tube cannot move secretions from the middle ear, pressure builds up from fluid and pus collecting behind the eardrum. The throbbing ache of a middle ear infection is produced by pressure on nerves supplying the middle ear. Bacterial infections of the middle ear can cause the eardrum to perforate, giving discharge of blood or pus from the ear and immediate relief from pain. With treatment, the eardrum perforations usually heal promptly. The pressure in the middle ear can force infected material into the mastoids, resulting in mastoiditis. The fluid behind the eardrum impairs movement of the drum, thereby decreasing ability to hear. Because of the fluid, some young children cannot hear clearly and show a delay in normal speech development. Spread of infection from the nasopharynx to the sinuses probably occurs by a mechanism similar to that in otitis media. Both middle ear and sinus infections sometimes spread to the coverings of the brain, causing meningitis.

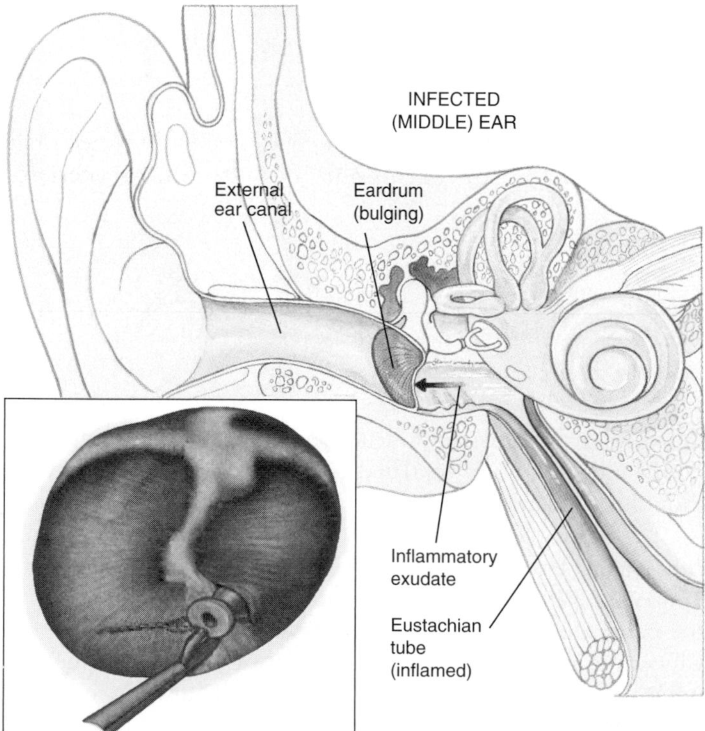

Figure 23.7 Otitis Media A stylized view of the infected middle ear showing accumulation of fluid, swelling of the eustachian tube, and outward bulging of the eardrum. The photograph shows a ventilation tube placed in the eardrum to equalize middle ear pressure in individuals with chronically malfunctioning eustachian tubes.

Epidemiology

The carrier rates for *H. influenzae* and *S. pneumoniae* can sometimes reach 80% in the absence of disease, and the ecological factors involved in the appearance and spread of the diseases caused by these organisms are largely unknown. Epidemics of bacterial conjunctivitis commonly occur among schoolchildren. Probably the virulence of the bacterium, crowding of susceptible children, and the presence of respiratory viruses all play important roles. A preceding or concomitant viral illness is common in otitis media and sinusitis; probably the virus damages the mucociliary mechanism that would normally protect against bacterial infection. Otitis media is rare in the first month of life. It becomes very common in early childhood, especially during winter and spring when viral respiratory diseases are frequent. Older children develop immunity to *H. influenzae*, and it becomes an increasingly uncommon cause of otitis media after about five years of age. Sinusitis tends to involve adults and older children in whom the sinuses are more fully developed.

Prevention and Treatment

Suspected cases of bacterial conjunctivitis are removed from school or day care settings for diagnosis and start of treatment. General preventive measures include handwashing, and avoidance of rubbing or touching the eyes and sharing towels.

Use of pacifiers beyond the age of two years is associated with a substantial increase in otitis media. Viral infections and other conditions that cause inflammation of the nasal mucosa play a role in some cases. Nasal allergies and exposure to air pollution and cigarette smoke are examples of nasal irritants. Administration of influenza vaccine to infants in day care facilities substantially decreases the incidence of otitis media during the "flu" season. Pneumococcal vaccine as presently constituted has proved disappointing in preventing otitis media because infants under two years of age fail to get an adequate immune response. New, presently experimental, vaccines will probably be better. Ampicillin or sulfasoxazole given continuously over the winter and spring are useful preventives in people who have three or more bouts of otitis media within a six-month period. Surgical removal of enlarged adenoids improves drainage from the eustachian tubes and can be helpful in preventing recurrences in certain patients. In those with chronically malfunctioning eustachian tubes, plastic ventilation tubes are sometimes installed in the eardrums so that pressure can equalize on both sides of the drum (see figure 23.7). In the United States, this is the most common operation performed on children, but its value in preventing acute otitis media and hearing loss has been questioned. There are no proven preventive measures for sinusitis.

Bacterial conjunctivitis is effectively treated with eyedrops or ointments containing an antibacterial medicine to which the infecting strain is sensitive. Antibacterial therapy with ampicillin is generally effective against otitis media. Alternative medications are available for communities where antibiotic-resistant strains of *H. influenzae* and *S. pneumoniae* are common. Occasionally, pus must be drained surgically by making an opening in the eardrum in otitis or by inserting a large needle through the bone in the case of sinusitis. In general, antibacterial treatment has been highly successful in preventing serious complications such as mastoiditis and meningitis. Decongestants and antihistamines generally are ineffective and can be harmful.

MICROCHECK 23.3

Untreated *Streptococcus pyogenes* infections can sometimes cause injury to other parts of the body that appear long after the infections have healed. In diphtheria, toxin is absorbed into the bloodstream and circulates throughout the body, selectively damaging certain tissues such as heart, kidneys, and nerves that have appropriate receptors. Viral infections often pave the way for bacterial upper respiratory infections.

- What is scarlet fever?
- What is implied by the observation that symptoms of some diseases are observed only after the organisms are eliminated from the body?
- What purpose would adequate ventilation serve in preventing the spread of streptococcal infections?

Viral Infections of the Upper Respiratory System

The average person in the United States suffers two to five episodes of viral upper respiratory infections each year. They generally subside without any treatment and rarely cause permanent damage. Most of the causative agents are highly successful parasites, using us to replicate themselves to astronomical numbers, then moving on without killing us or even leaving long-standing immunity. Although hundreds of kinds of viruses are involved, the range of symptoms they produce is similar. Their major importance to health is that they damage respiratory tract defenses and thus pave the way for more serious bacterial diseases. Respiratory viruses, like all other viruses, are obligate intracellular parasites, meaning they can only reproduce inside living cells. They are most conveniently studied by inoculating them into laboratory cultures of cells originally obtained from tissues of humans or other animals. Usually, viral infection of the cultured cells is detectable by a cytopathic effect, meaning observable changes in the cells caused by a viral infection. ■ **cell or tissue cultures, p. 345**

The Common Cold

Medical practitioners have been taught for generations that the symptoms of the common cold, "if treated vigorously, will go away in seven days, whereas if left alone, they will disappear over the course of a week." So far, little has occurred to change this assessment. On average, a person in the United States has two to three colds per year. Colds are the leading cause of absences from school, and result in loss of about 150 million work days per year. Immunity generally lasts only a few years or less.

Symptoms

Colds generally begin with malaise, followed by a scratchy or mildly sore throat, runny nose, cough, and hoarseness. The nasal secretions, initially profuse and watery, thicken in a day or two, then become cloudy and greenish. Unless secondary bacterial infection occurs, there is no fever; symptoms are mostly gone within a week, but a mild cough sometimes continues for longer periods of time.

Causative Agents

Between 30% and 50% of colds are caused by the 100 or more types of rhinoviruses (*rhino* means "nose," as in rhinoceros, or "horny nose"), members of the picornavirus family (*pico* is Italian for "small" and *rna* is ribonucleic acid [RNA]; thus, "small RNA viruses") **(figure 23.8)**. The viruses are nonenveloped and their RNA is single-stranded, positive sense. The rhinoviruses were initially difficult to cultivate in the laboratory, since most failed to infect laboratory animals or cell cultures. In the 1950s, researchers in England accidentally discovered that if cell cultures were incubated at 33°C instead of at body temperature (37°C) and at a slightly acid pH instead of at the alkaline pH of body tissues, many rhinovirus strains would infect them. These conditions of lower temperature and pH normally exist in the upper respiratory tract. Rhinoviruses, however, are killed if the pH drops below 5.3, and therefore, they are usually destroyed in the human stomach. Many other viruses and some bacterial species can also produce the symptoms of a cold.

Pathogenesis

Rhinoviruses attach to specific receptors on respiratory epithelial cells and infect them. After replicating, large numbers of virions are released and infect other cells. Infected cells cease ciliary motion and may slough off. The injury causes the release of inflammatory mediators and stimulates nervous reflexes. The result is an increase in nasal secretions, sneezing, and swelling of the mucosa and nasal erectile tissue. This swelling partially or completely obstructs the airways. Later, in the inflammatory response, dilation of blood vessels, oozing of plasma, and congregation of leukocytes in the infected area occur. Secretions from the area may then contain pus and blood. The infection is eventually halted by the inflammatory response, interferon

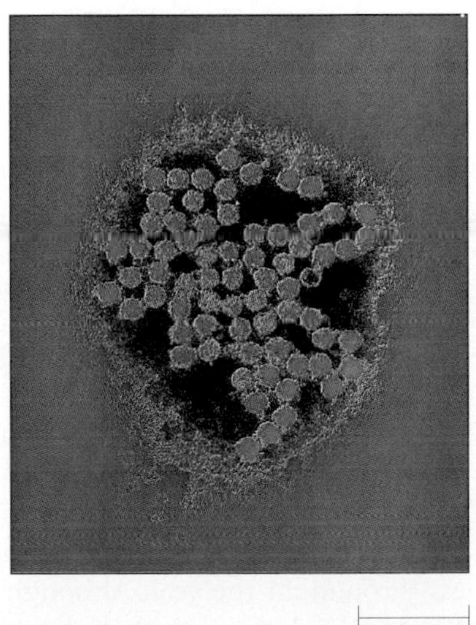

Figure 23.8 Rhinovirus Transmission electron micrograph. The red spheres are the rhinovirus virions, and the blue object is the cell in which they are replicating.

release, and cellular and humoral immunity, but it can extend into the ears, sinuses, or even the lower respiratory tract before it is stopped. Rhinoviruses can even cause life-threatening pneumonia in individuals with AIDS. ■ interferon, p. 379

Epidemiology

Humans are the only source of cold viruses, and close contact with an infected person is generally necessary for viruses to be transmitted. A person with severe symptoms early in the course of a cold is much more likely to transmit a cold than is someone whose symptoms are mild or who is in the late stage of a cold. Very high concentrations of virus are found in the nasal secretions and often on the hands of infected people during the first 2 or 3 days of a cold. By the fourth or fifth day, virus levels are often undetectable, but low levels can be present for 2 weeks. A few virions are sufficient to infect the nasal mucosa, but the mouth is quite resistant to infection. In adults the disease is usually contracted when airborne droplets containing virus particles are inhaled. Transmission can also occur when virus-containing secretions are unwittingly rubbed into the eyes or nose by contaminated hands. Virus introduced into the eye is promptly transmitted to the nasal passage via the nasolacrimal duct. With reasonable caution, however, colds are not highly contagious. In a study of nonimmune adults exposed in a family or dormitory setting, less than half contracted colds. Young children, however, transmit colds and other respiratory viruses very effectively because they are often careless with their respiratory secretions. Experimental and epidemiological studies show that there is no relationship between exposure to cold temperature and development of colds, contrary to popular belief. Emotional stress, however, can almost double the risk of catching a cold.

Prevention and Treatment

Since such a large number of immunologically different viruses cause colds, vaccines are impractical. Rhinoviruses, like all other viruses, are not affected by antibiotics and other antibacterial medications. Prevention of the spread of rhinoviruses includes handwashing, even in plain water, which readily removes rhinoviruses, and keeping hands away from the face. In addition, one should avoid crowds and crowded places such as commercial airplanes when respiratory diseases are prevalent, and especially avoid people with colds during the first couple of days of their symptoms. Except for the estimated 10% of colds caused by bacteria, there is no effective medicine for treating colds. In experimental rhinovirus infections, aspirin and acetaminophen, commonly present in cold remedies, both somewhat prolonged symptoms, antibody production, and duration of virus excretion.

Table 23.4 summarizes some facts about the common cold.

TABLE 23.4 The Common Cold

Symptoms	Scratchy throat, nasal discharge, malaise, headache, cough
Incubation period	1 to 2 days
Causative agent	Mainly rhinoviruses—more than 100 types; many other viruses, some bacteria
Pathogenesis	Viruses attach to respiratory epithelium, starting infection that spreads to adjacent cells; ciliary action ceases and cells slough; mucus secretion increases, and inflammatory reaction occurs; infection stopped by interferon release and antibody production
Epidemiology	Inhalation of infected droplets; transfer of infectious mucus to nose or eye by contaminated fingers; children initiate many outbreaks in families because of lack of care with nasal secretions
Prevention and treatment	Handwashing; avoiding people with colds and touching face. No treatment except for control of symptoms

Adenoviral Pharyngitis

Adenoviruses are widespread and can cause epidemics of upper respiratory illness throughout the year. Although they typically produce sore throat and enlarged lymph nodes with or without conjunctivitis, they can sometimes also cause pneumonia or diarrhea. They are representative of the many viruses that cause upper respiratory symptoms with fever.

Symptoms

Adenoviral infections often cause a runny nose and sore throat, but unlike the common cold, fever is commonly present. Typically, the throat is sore, and gray-white pus present on the pharynx and tonsils can cause confusion with strep throat. The lymph nodes of the neck enlarge and become tender. In some epidemics, conjunctivitis is a prominent symptom, sometimes showing multiple conjunctival nodules composed of infiltrating lymphocytes, and long-lasting opacities in the cornea, the clear central portion of the eye. Other strains of the virus cause epidemics of hemorrhagic conjunctivitis. A mild cough is common; sometimes a severe cough develops, indicating possible pneumonia or whooping cough, or chest pain indicating pleurisy. With or without treatment, recovery usually occurs in 1 to 3 weeks.

Causative Agent

More than 45 antigenic types of adenoviruses infect humans. The viruses are nonenveloped, 70 to 90 nm in diameter, with double-stranded DNA. They can be cultivated in a variety of cell cultures, where they produce a characteristic cytopathic effect. They can remain infectious in the environment for long periods of time and can be transferred readily from one patient to another on medical instruments. They are easily inactivated, however, by heat at 56°C, chlorine, and various other disinfectants.

Pathogenesis

Little is known about the pathogenesis of adenoviral disease. Humans are the only source of infection; however, related viruses infect animals. Once inside the cells of the host, the virus multiplies in the nuclei. In severe infections, extensive cell destruction and inflammation occur. Different types of adenoviruses vary in the tissues they affect. For example, adenovirus types 4, 7, and 21 typically cause illness characterized by sore throat and enlarged lymph nodes, whereas type 8 is likely to cause extensive eye infection.

Perspective 23.1 Outbreak of Acute Adenoviral Disease

In the spring of 1997, an outbreak of acute adenoviral respiratory disease occurred involving 146 of the 240 students in a job training school in South Dakota. The range of symptoms and signs and the percentage of sick individuals involved was as follows:

Coldlike symptoms	96%
Sore throat	95%
Headache	95%
Fever	52%
Enlarged lymph nodes	23%
Conjunctivitis	21%
Abnormal chest X ray	19%
Pleurisy	18%
Vomiting	15%
Urinary discomfort or blood	2%

Outbreaks of adenoviral disease have been recognized mainly in children and in military recruits. This outbreak was unusual because it occurred in adult civilians, and it was caused by adenovirus type 11, which usually causes hemorrhagic conjunctivitis or urinary bladder involvement. The differing symptoms among individuals is common in epidemic diseases and probably reflects variations in host susceptibility, infectious dose, and route of entry. Crowding was probably an important factor in this outbreak since all the students lived in dormitories, 6 to 10 persons to a room, and ate together in the school cafeteria.

Source: Centers for Disease Control and Prevention. 1998. *Morbidity and Mortality Weekly Report* 47:567.

Epidemiology

Adenoviral illness is prevalent among schoolchildren, usually in sporadic cases, but occasionally occurring as outbreaks of respiratory sickness in winter and spring. Summertime epidemics occur as a result of transmission of the viruses in inadequately chlorinated swimming pools. These cases typically have fever, conjunctivitis, pharyngitis, enlargement of the lymph nodes of the neck, a rash, and diarrhea. Adenoviral disease is likely to occur in groups of young people living together in crowded conditions and has been a big problem in military recruits. Epidemic spread is fostered by a high percentage of asymptomatic infections. The viruses are shed from the upper respiratory tract during the acute illness and continue to be eliminated in the feces for months thereafter.

Prevention and Treatment

Formerly, a live attenuated vaccine administered orally was helpful for preventing acute respiratory disease in military recruits, but the vaccine is no longer manufactured. As with colds, there is no treatment, and most patients get well on their own. Secondary bacterial infections, however, may occur and require antibacterial medication.

Table 23.5 summarizes some important facts about adenoviral pharyngitis.

MICROCHECK 23.4

Many different kinds of infectious agents can produce the same symptoms and signs of respiratory disease. Emotional stress significantly increases the risk of contracting a cold, but exposure to cold temperatures probably does not. Persons suffering a cold are most likely to transmit it if symptoms are severe, and during the first few days of illness. Adenovirus infections can mimic colds, pinkeye, strep throat, and whooping cough.

- How is an adenovirus infection treated?
- People who staff polar ice stations where they are isolated from other human contact for long periods often do not develop colds. Is this an expected observation? Why or why not?

TABLE 23.5 Adenoviral Pharyngitis

Symptoms	Fever, very sore throat, severe cough, swollen lymph nodes of neck, pus on tonsils and throat, and conjunctivitis; less frequently, pneumonia
Incubation period	5 to 10 days
Causative agent	Adenoviruses—more than 45 types
Pathogenesis	Virus multiplies in epithelial cells; cell destruction and inflammation occur; latent infections occur with some strains
Epidemiology	Inhalation of infected droplets; possible spread from gastrointestinal tract
Prevention and treatment	Live virus vaccine formerly used by the military is no longer produced. No treatment except for relief of symptoms

INFECTIONS OF THE LOWER RESPIRATORY SYSTEM

Bacterial Infections of the Lower Respiratory System

Bacterial infections of the lower respiratory system are less common than those of the upper system, largely because they are stopped by body defenses at the portal of entry. Lower tract infections, however, are generally much more serious. An earache or sore throat is unlikely to be life threatening, but their causative organisms can endanger life when they infect the lung. Distinctive patterns of signs and symptoms are produced by the different kinds of organisms that infect the lower respiratory system. The pneumonias are inflammatory diseases of the lung in which fluid fills the alveoli. They top the list of infectious killers in the general population of the United States, and they are important as nosocomial, meaning hospital-acquired, infections. Whooping cough, tuberculosis, and Legionnaires' disease are other distinctive types of infection. ■ **body defenses, nonspecific immunity, p. 370**

Pneumococcal Pneumonia

Pneumococci are an important cause of pneumonia acquired in the community, accounting for about 60% of the adult pneumonia victims requiring hospitalization.

Symptoms

The typical symptoms of pneumococcal pneumonia are cough, fever, chest pain, and sputum production. These symptoms are usually preceded by a day or two of runny nose and upper respiratory congestion, ending with an abrupt rise in temperature and a single body-shaking chill. The sputum quickly becomes pinkish or rust colored, and severe chest pain develops, aggravated by each breath or cough. As a result of the pain, breathing becomes shallow and rapid. The patient becomes short of breath and develops a dusky color because of poor oxygenation. Without treatment, after 7 to 10 days people who survive show profuse sweating and a rapid fall in temperature to normal.

Causative Agent

Streptococcus pneumoniae, the pneumococcus, is a Gram-positive diplococcus (**figure 23.9**). The most striking characteristic of *S. pneumoniae* is its thick polysaccharide capsule, which is responsible for the organism's virulence. There are more than 80 different types of *S. pneumoniae*, as determined by differing capsular antigens. ■ **capsules, p. 66**

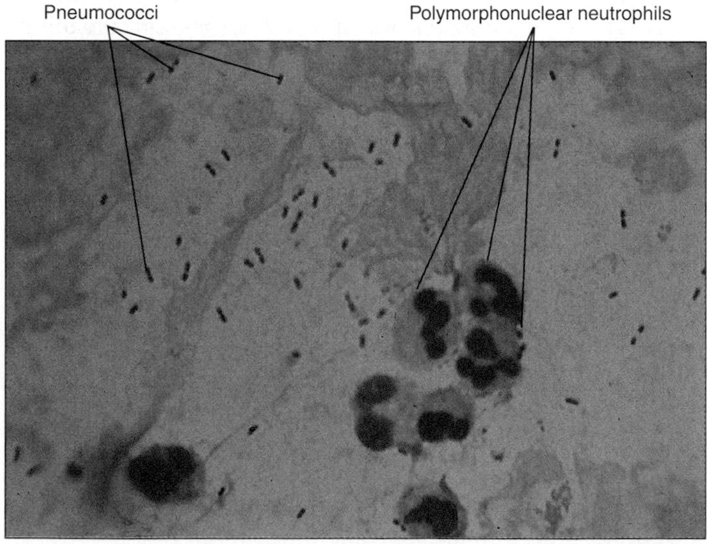

Figure 23.9 ***Streptococcus pneumoniae*** Gram stain of sputum from a person with pneumococcal pneumonia showing Gram-positive diplococci and polymorphonuclear neutrophils (PMNs).

Pathogenesis

Pneumococcal pneumonia develops when encapsulated pneumococci, the virulent form, are inhaled into the alveoli of a susceptible host, multiply rapidly, and cause an inflammatory response. The bacteria are resistant to phagocytosis because their capsules interfere with the action of C3b, the fraction of complement responsible for opsonization. Serum and phagocytic cells pour into the air sacs of the lung, causing difficulty breathing. This increase in fluid produces abnormal shadows on chest X-ray films of patients with pneumonia (**figure 23.10**). Material coughed from the lungs, called **sputum**, increases in amount and contains pus, blood, and many pneumococci. ■ **opsonization by C3b, p. 377**

The inflammatory response to the infection often involves nerve endings, causing pain; the condition in which pain arises from an inflamed pleura is called **pleurisy**. Pneumococci that enter the bloodstream from the inflamed lung are responsible for three often fatal complications: **septicemia**, a symptomatic infection of the bloodstream; **endocarditis**, an infection of the heart valves; and **meningitis**, an infection of the membranes covering the brain and spinal cord. Individuals who do not develop such complications usually develop sufficient specific anti-capsular antibodies within about a week to permit phagocytosis and destruction of the infecting organisms. Complete recovery usually results. Most pneumococcal strains do not destroy lung tissue.

Epidemiology

Up to 30% of healthy people carry encapsulated pneumococci in their throat. Because of the effectivness of the mucociliary escalator, these bacteria rarely reach the lung. The risk of pneumococcal pneumonia rises dramatically when this defense mechanism is impaired, however, as it is with alcohol and narcotic use, and with respiratory viral infections such as influenza. There is also an increased risk of the disease with underlying heart or lung disease, diabetes, and cancer, and with age over 50.

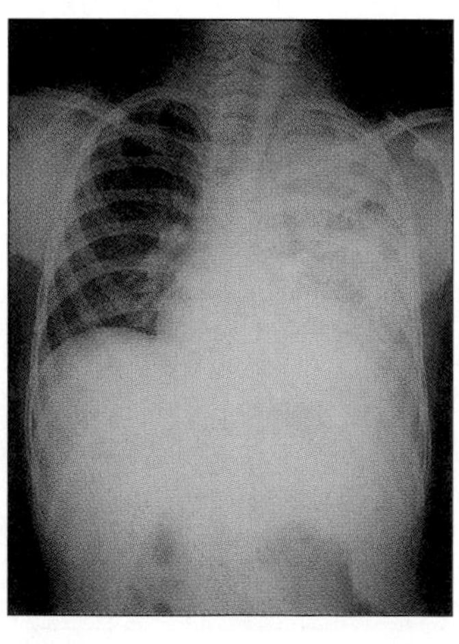
(a)

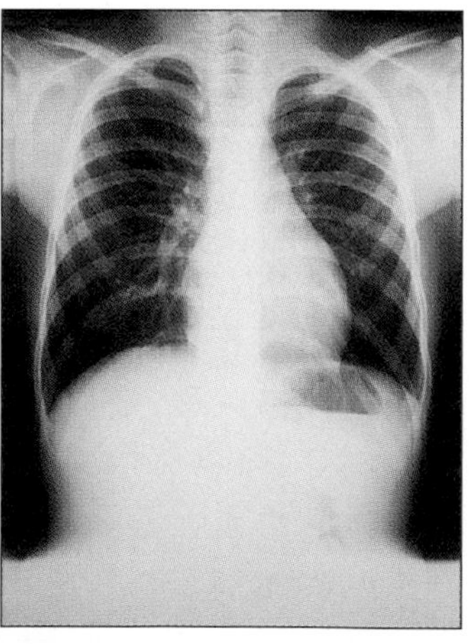
(b)

Figure 23.10 Chest X-Ray Appearance in Pneumococcal Pneumonia
(a) Pneumonia. The left lung (right side of figure) appears white because fluid-filled alveoli stop the X-ray beam from reaching the X-ray film and turning it black.
(b) Normal X-ray film after recovery. The X-ray beam passes through air-filled alveoli and turns the film black.

Prevention and Treatment

A vaccine is available that stimulates production of anticapsular antibodies and gives immunity to 23 strains that account for over 90% of serious pneumococcal disease. Immunization is especially important for certain high-risk individuals, such as those suffering from chronic heart or lung disease, or from alcoholism. Unfortunately, some of the people in these high-risk groups fail to develop protective antibody responses to the current vaccine, probably because their immune systems are impaired. Most pneumococcal infections can be cured if penicillin or erythromycin is given early in the illness. Strains of pneumococci resistant to one or more antibiotics, however, are increasingly being encountered.

See table 23.6 for a description of some features of pneumococcal pneumonia.

Klebsiella Pneumonia

Enterobacteria such as *Klebsiella* sp., and other Gram-negative rods, can cause pneumonia, especially if host defenses are impaired. These pneumonias attack the very old, the very young, alcoholics, nursing home patients, those debilitated by other diseases, and immunocompromised persons. Pneumonias caused by Gram-negative rods cause most of the deaths from nosocomial infections. *Klebsiella* pneumonia is an example of pneumonias caused by Gram-negative rods. ■ **enterobacteria, p. 282** ■ **nosocomial infections, p. 488**

Symptoms

In general, the symptoms of *Klebsiella* pneumonia—cough, fever, and chest pain—are indistinguishable from those of pneumococcal pneumonia. *Klebsiella* pneumonia patients typically have repeated chills, however, and their sputum is red and gelatinous, resembling currant jelly. Their mortality rate without treatment is 50% to 80%, and they tend to die sooner than other pneumonia patients.

Causative Agent

Several species of *Klebsiella* cause pneumonia, but *Klebsiella pneumoniae* is the best known. It is a Gram-negative rod with a large capsule and big, strikingly mucoid colonies (**figure 23.11**). Identification depends on biochemical testing and demonstrating that it is nonmotile at 37°C.

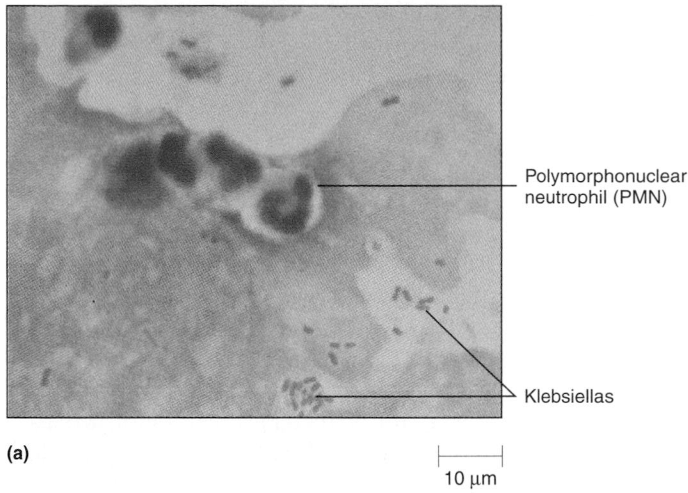

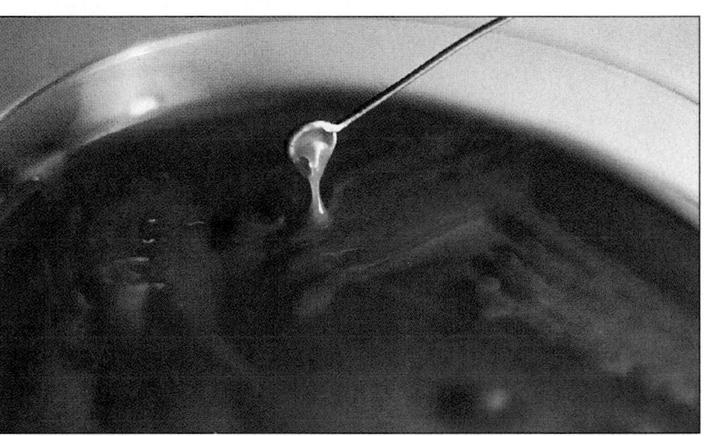

Figure 23.11 ***Klebsiella pneumoniae*** **(a)** In sputum from a pneumonia victim. **(b)** Colonies. When the colony is touched with a microbiological loop, it "strings out."

Pathogenesis

The organisms first colonize the mouth and throat and then are carried to the lung by inspired air or a ball of aspirated mucus. Evidence indicates that specific adhesins aid colonization, but their exact nature is not yet known. The capsule is an essential virulence factor, probably functioning like the pneumococcal capsule in limiting the availability of C3b, a critically important opsonin before the appearance of antibody. Unlike *Streptococcus pneumoniae*, *K. pneumoniae* causes death of tissue and rapid formation of lung abscesses. The infection often enters the bloodstream, causing abscesses in other tissues, release of endotoxin, and shock. Therefore, even with effective antibacterial medication, the lung can be permanently damaged and death may occur. ■ **endotoxin, p. 461**

Epidemiology

Klebsiellas are part of the normal flora of the intestine in a small percentage of individuals. Colonization of the mouth and throat is common in debilitated individuals, especially in an institutional setting. In hospitals and nursing homes, the organisms are often resistant to antimicrobial medications, and they readily colonize patients taking the medications. *Klebsiella* resistance to antimicrobial medications can be chromosomal or plasmid-mediated. The organisms easily acquire and are a source of R factors, all of which contain transposons. This allows resistance genes to spread readily to other species and genera when antibacterial medications are used. ■ **R plasmids and transposons, p. 194, 211, 214, 511**

Prevention and Treatment

There are no specific preventive measures. Disinfection of the environment, use of sterile respiratory equipment, and use of antimicrobial medications only when necessary help control the organisms in hospitals.

See table 23.6 for a description of the main features of this disease.

Mycoplasmal Pneumonia

Mycoplasmal pneumonia is the leading kind of pneumonia in college students and is also common among military recruits. The disease is generally mild, and difficult to distinguish from pneumonias caused by *Chlamydia pneumoniae* and various viruses. These pneumonias are often referred to by their popular name, "walking pneumonia," and usually do not require hospitalization.

Symptoms

The onset of mycoplasmal pneumonia is typically gradual. The first symptoms are fever, headache, muscle pain, and fatigue. After several days, a dry cough begins, but later mucoid sputum may be produced. About 15% of cases have middle ear infections.

Causative Agent

The causative agent of mycoplasmal pneumonia is *Mycoplasma pneumoniae*, a small (0.2 μm diameter), easily deformed bacterium lacking a cell wall. Distinguishing characteristics are slow growth and aerobic metabolism. As with other mycoplasmas, the central portion of *M. pneumoniae* colonies grows down into the medium, producing a fried egg appearance (**figure 23.12**).

■ **mycoplasmas, p. 291**

Pathogenesis

Only a few inhaled organisms are necessary to start an infection with *M. pneumoniae*. The organisms attach to specific receptors on the respiratory epithelium (**figure 23.13**), interfere with ciliary action, and cause the ciliated cells to slough off. An inflammatory response characterized by infiltration of lymphocytes and macrophages causes the walls of the bronchial tubes and alveoli to thicken.

Epidemiology

The mycoplasmas are spread by aerosolized droplets of respiratory secretions. Transmission from person to person is aided by the long time period in which *M. pneumoniae* is present in respiratory secretions, ranging from about 1 week before symptoms begin to many weeks afterward. Mycoplasmal pneumonia accounts for about one-fifth of bacterial pneumonias, and it has a peak incidence in young people. Immunity after recovery is not permanent, and repeat attacks have occurred within 5 years.

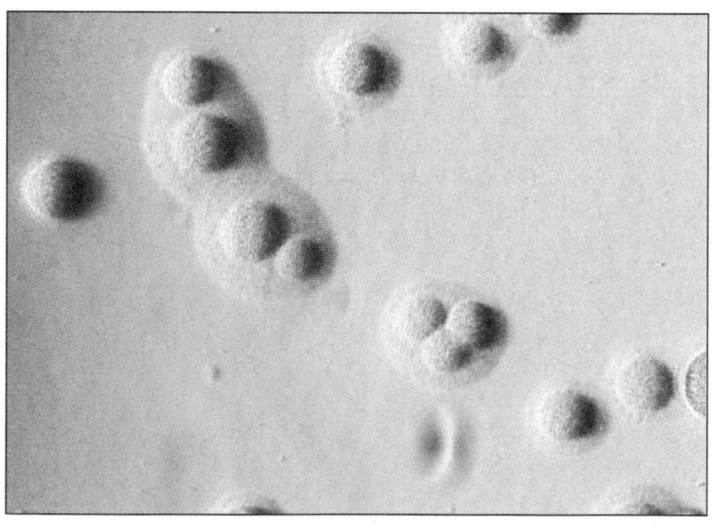

Figure 23.12 ***Mycoplasma pneumoniae*** Typical "fried egg" appearance of the colonies.

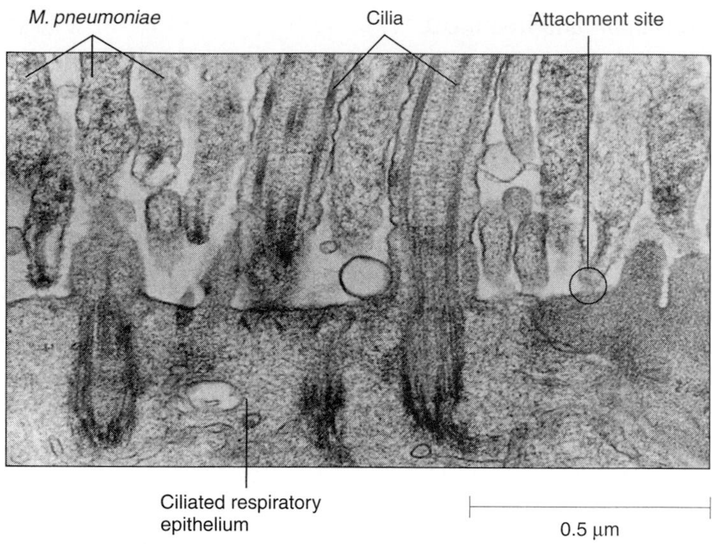

Figure 23.13 ***Mycoplasma pneumoniae* Infecting Respiratory Epithelium** Transmission electron micrograph. Notice the distinctive appearance of the tips of the mycoplasmas adjacent to the host epithelium. The tips probably represent a site on the microorganism that is specialized for attachment.

Prevention and Treatment

No practical preventive measures exist for mycoplasmal pneumonia, except avoiding crowding in schools and military facilities. Antibiotics that act against bacterial cell walls, such as the penicillins and cephalosporins, are of course not effective. Tetracycline and erythromycin shorten the illness if given early, but they are bacteriostatic, meaning they only inhibit the growth of the bacteria without killing them. They do not readily eliminate *M. pneumoniae* from the respiratory secretions.

Some features of pneumococcal, *Klebsiella*, and mycoplasmal pneumonias are compared in **table 23.6**.

Whooping Cough (Pertussis)

Whooping cough is the common name for **pertussis**, a disease that is now uncommon in the United States because of childhood immunization. The causative bacterium is widespread, however, and remains a threat to those people who lack immunity. Worldwide, the disease causes 300,000 to 500,000 deaths yearly.

Symptoms

Pertussis typically begins with a number of days of runny nose followed by sudden bouts of violent, uncontrollable coughing. This symptom, termed **paroxysmal coughing**, is severe enough to rupture small blood vessels in the eyes. The coughing spasm is followed by forceful inspiration, the "whoop." Vomiting and seizures may occur during this phase of the illness.

Causative Agent

The disease is caused by *Bordetella pertussis* (**figure 23.14**), a tiny, encapsulated, strictly aerobic, Gram-negative rod. These

TABLE 23.6 Pneumococcal, *Klebsiella*, and Mycoplasmal Pneumonias Compared

	Pneumococcal Pneumonia	*Klebsiella* Pneumonia	Mycoplasmal Pneumonia
Symptoms	Cough, fever, single shaking chill, rust-colored sputum from degraded blood, shortness of breath, chest pain	Chills, fever, cough, chest pain, and grossly bloody, mucoid sputum	Cough, fever, sputum production, headache, fatigue, and muscle aches
Incubation period	1 to 3 days	1 to 3 days	2 to 3 weeks
Causative agent	The pneumococcus, *Streptococcus pneumoniae,* encapsulated strains	*Klebsiella pneumoniae,* an entero-bacterium	*Mycoplasma pneumoniae*
Pathogenesis	Inhalation of infected droplets of mucus; colonization of the alveoli incites inflammatory response; plasma, blood, and inflammatory cells fill the alveoli; pain results from involvement of nerve endings	Aspiration of colonized mucus droplets from the throat. Colonization of the alveoli incites inflammatory response; plasma, blood, and inflammatory cells fill the alveoli; destruction of lung tissue and abscess formation common; infection spreads via blood to other body tissues	Cells attach to specific receptors on the respiratory epithelium; inhibition of ciliary motion and destruction of cells follow
Epidemiology	High carrier rates for *S. pneumoniae.* Risk of pneumonia increased with conditions such as alcoholism, narcotic use, chronic lung disease, and viral infections that impair the mucociliary escalator. Other predisposing factors are chronic heart disease, diabetes, and cancer	*Klebsiella* sp. and other Gram-negative rods are common causes of fatal nosocomial pneumonias	Inhalation of infected droplets; mild infections common
Prevention and treatment	Vaccine available contains 23 capsular antigens. Treatment: penicillin, erythromycin, and others	No vaccine available. A cephalosporin with an aminoglycoside	No vaccine available; avoidance of crowding in schools and military facilities advisable; tetracycline or erythromycin for treatment

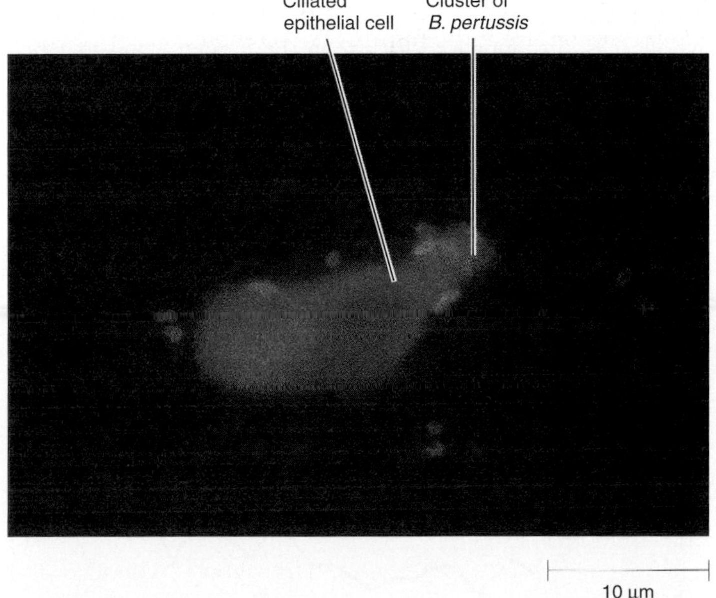

Figure 23.14 ***Bordetella pertussis*** Fluorescent antibody stain of respiratory secretions from an individual with whooping cough. The bacteria stain a greenish-yellow color. The large orange object is a ciliated epithelial cell. Notice the cluster of *B. pertussis* at one end of the cell, the location of the cell's cilia.

organisms do not tolerate drying or sunlight and die quickly outside the host.

Pathogenesis

Bordetella pertussis enters the respiratory tract with inspired air and attaches specifically to ciliated cells of the respiratory epithelium. Among the proteins present on the bacterial surface, pertussis toxin (Ptx) and filamentous hemagglutinin (Fha) are probably the main ones involved in attachment.

The areas colonized by *B. pertussis* include the nasopharynx, trachea, bronchi, and bronchioles. The violent symptoms arise because of involvement of the passageways of the lower respiratory tract. The organisms grow in dense masses on the epithelial surface, but they do not invade. Mucus secretion increases markedly, while ciliary action declines precipitously, and patches of ciliated cells slough off. Only the cough reflex remains for clearing the secretions. Some of the bronchioles become completely obstructed, resulting in small areas of collapsed lung. Others, because of spasm or mucus plugging, let air enter but not escape, causing hyperinflation. Paroxysmal coughing causes hemorrhages in the brain, and seizures can occur. Pneumonia due to *B. pertussis* or, more commonly, secondary bacterial infection is the chief cause of death.

A number of toxic products of *B. pertussis* probably play a role in its pathogenesis. Pertussis toxin (Ptx), mentioned earlier for its role in colonization, is an A-B toxin. **Figure 23.15** shows the steps involved in the toxin's action. First, the B portion attaches specifically to receptors on the host cell surface. Second, the A portion is transferred through the cytoplasmic membrane of the host cell, becoming an active ribosylating enzyme in the process. The activated A portion of the Ptx toxin inactivates the G protein normally responsible for initiating a decrease in cyclic adenosine monophosphate (cAMP) synthesis, thereby allowing maximum production of cAMP. Unregulated cAMP production causes marked increase in mucus output, decreased killing ability of phagocytes, massive release of lymphocytes into the bloodstream, ineffectiveness of natural killer cells, and low blood sugar. Invasive adenylate cyclase, another *B. pertussis* toxin, also causes an increase in cAMP production. ■ **A-B toxins, p. 462** ■ **cAMP, p. 182**

Tracheal cytotoxin produced by *B. pertussis* acts with endotoxin to cause release of nitric oxide (NO) from goblet cells. The NO causes ciliated epithelial cells to die and slough off. Tracheal cytotoxin also causes the release of interleukin-1 (IL-1), a fever-causing cytokine. ■ **interleukin-1, p. 380**

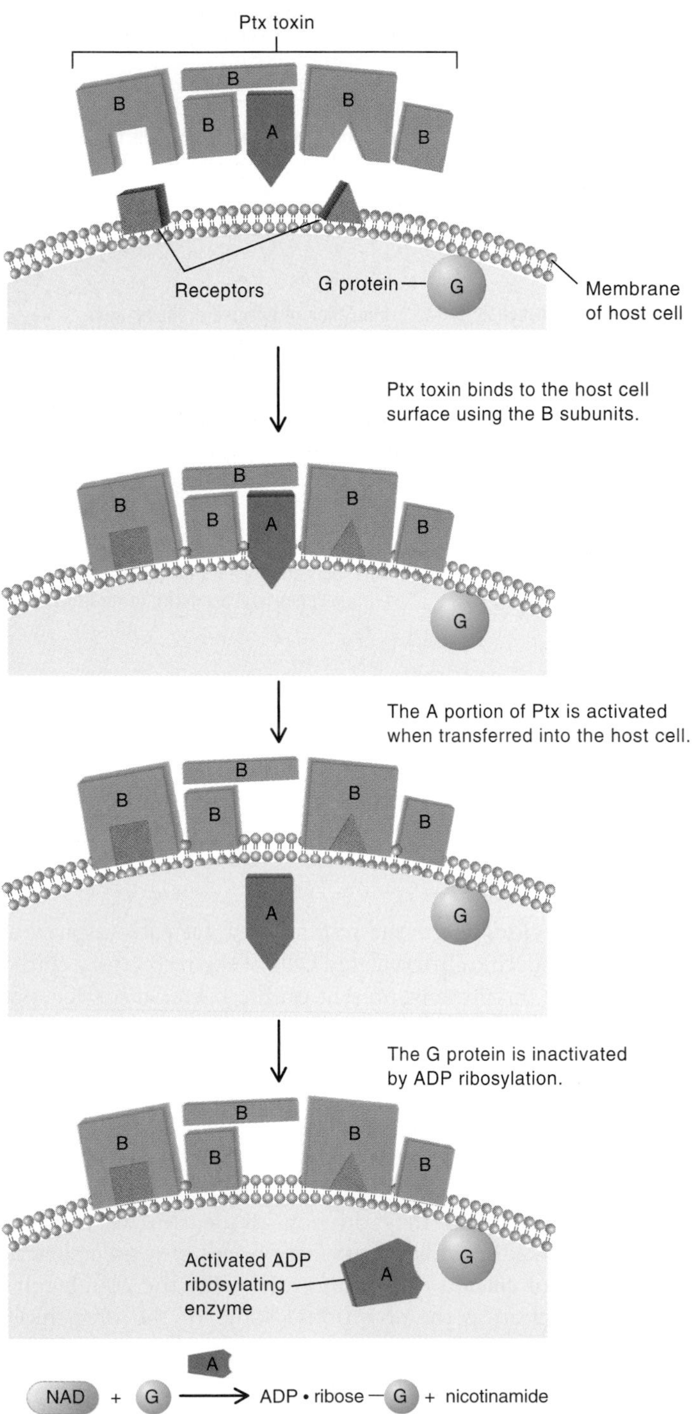

Figure 23.15 Mode of Action of Pertussis Toxin The B portion of this A-B toxin attaches to receptors on the cell membrane. The B portion stays behind as the A portion enters the cell, becoming an enzyme activated by intracellular conditions. Activated A enzyme inactivates a G protein by ADP ribosylation, thereby maximizing cAMP production.

Epidemiology

Pertussis spreads via respiratory secretions suspended in air. Patients are most infectious during the runny nose period, the numbers of expelled organisms decreasing substantially with the onset of violent coughing. Pertussis is classically a disease of infants. It can occur in a mild form in older children and adults, however, overlooked as a cold, thus fostering transmission.

Prevention and Treatment

Intensive vaccination of infants with killed *B. pertussis* cells can prevent the disease in about 70% of individuals. Its widespread administration to young children is responsible for the drop from 235,239 reported cases of pertussis in the United States in 1934 to only about 1,000 in 1981. Since then, reported cases have generally ranged from 2,000 to 6,000 per year **(figure 23.16)**.

Pertussis vaccine is given along with diphtheria and tetanus toxoids by injection at 6 weeks, and 4, 6, and 18 months of age, with a booster dose between 4 and 6 years. Common reactions to the vaccine include pain and tenderness at the injection site and fever, occasionally high enough to cause convulsions. Acellular vaccines that use only part of the bacterium instead of the whole cells, first approved for use in the United States in 1991, are now advised for all five doses and give as good or better immunity with fewer side effects. Erythromycin given for treatment reduces the duration of symptoms if given early in

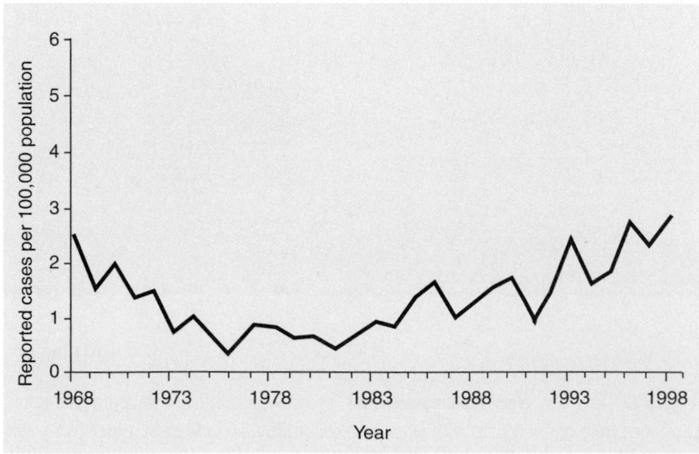

Figure 23.16 Incidence of Whooping Cough (Pertussis), United States

TABLE 23.7 Pertussis

Symptoms	Nasal congestion and mild cough followed by spasms of violent coughing; vomiting and possible convulsions
Incubation period	7 to 21 days
Causative agent	*Bordetella pertussis,* a tiny Gram-negative rod.
Pathogenesis	Colonization of the surfaces of the upper respiratory tract and tracheobronchial system; ciliary action slowed; toxins released by *B. pertussis* cause death of epithelial cells and increased cAMP; fever, excessive mucus output, and a rise in the number of lymphocytes in the bloodstream result
Epidemiology	Inhalation of infected droplets; older children and adults have mild symptoms
Prevention and treatment	Acellular vaccines, for immunization of infants and children; erythromycin, somewhat effective if given before coughing spasms start, eliminates *B. pertussis*

the disease, and the antibiotic usually eliminates *B. pertussis* from the respiratory secretions. ■ **acellular vaccines, p. 425**

The main features of this disease are shown in **table 23.7**.

Tuberculosis

Over the last hundred or more years, tuberculosis gradually declined in industrialized countries in association with improved living standards. In 1985, however, the rate of decline slowed, and then the incidence of tuberculosis began to rise (**figure 23.17**). The rise continued for the next 6 years in association with the expanding AIDS epidemic and increases in treatment-resistant cases. Fortunately, starting in 1993, with better funding of control measures and more aggressive treatment, tuberculosis resumed its slow retreat. In 1989, a Strategic Plan for the Elimination of Tuberculosis in the United States was published by the Centers for Disease Control and Prevention. The plan depends largely on increased efforts in identifying and treating people with the disease among the high-risk groups, particularly poor people, people with AIDS, prisoners, and immigrants from countries with high rates of tuberculosis. The high rates of tuberculosis and increasing percentage of treatment-resistant cases worldwide have put the strategic plan in jeopardy.

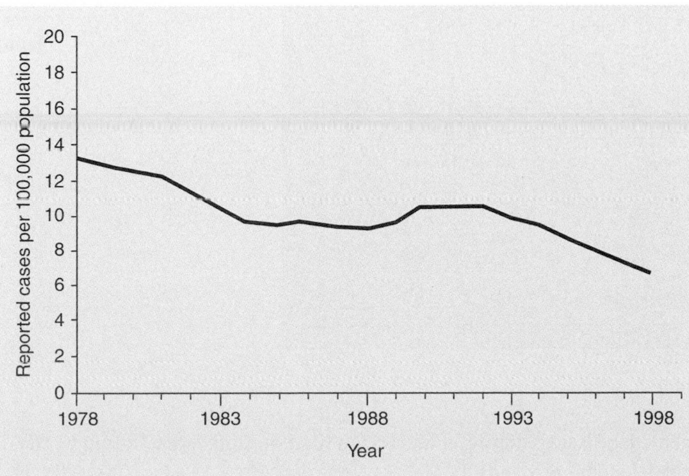

Figure 23.17 Incidence of Tuberculosis, United States Resurgence 1988 to 1994 responded to better funding of control measures.

Symptoms

Tuberculosis is a chronic illness characterized by slight fever, progressive weight loss, sweating at night, and chronic cough, often productive of blood-streaked sputum.

Causative Agent

Today, most cases of tuberculosis are caused by *Mycobacterium tuberculosis*, although in the years before widespread milk pasteurization, the cattle-infecting species *Mycobacterium bovis* was a common cause. *Mycobacterium tuberculosis*, commonly called the tubercle bacillus, is a slender acid-fast, rod-shaped bacterium (**figure 23.18**). The organism is a strict aerobe that grows very slowly, with a generation time of 12 hours or more. This slow growth makes it difficult to diagnose tuberculosis quickly. Although easily killed by pasteurization, the tubercle bacillus is unusually resistant to drying, disinfectants, and strong acids and alkali because of its unique cell wall. The bacterium has a typical Gram-positive cell wall, but bound covalently to the cell wall peptidoglycan is a thick layer composed of complex glycolipids, and other lipids compose much of the bacterial surface. Up to 60% of the dry weight of the *M. tuberculosis* cell wall consists of lipids, a much higher percentage than that in most other bacteria. ■ **pasteurization, p. 117** ■ **lipids, p. 37**

Pathogenesis

Tuberculosis is usually contracted by inhaling airborne organisms from a person with tuberculosis. In the lungs, the bacteria are taken up by pulmonary macrophages, where they resist

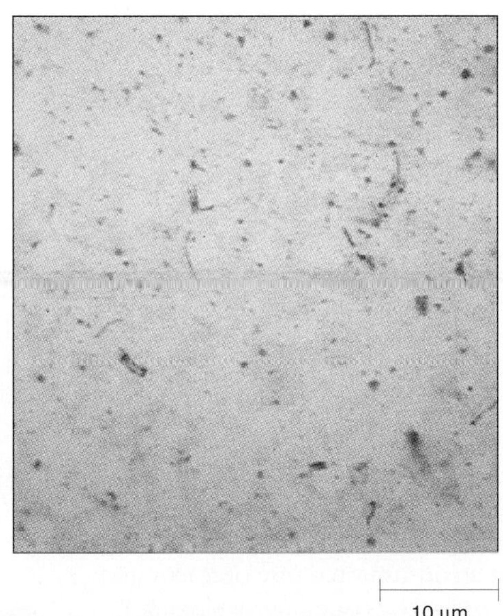

Figure 23.18 *Mycobacterium tuberculosis* as Seen in Sputum from an Individual with Tuberculosis The sputum has been "digested," meaning that it has been treated with strong alkali to kill the other bacteria that are invariably present.

destruction and multiply. Within the macrophages, they are carried to nearby lymph nodes. At this stage, there is little inflammatory response to the infection. The organisms continue to multiply, lyse the macrophages, and spread throughout the body. After about 2 weeks, delayed hypersensitivity to the tubercle bacilli develops. An intense reaction then occurs at sites where the bacilli have lodged. Macrophages, now activated, collect around the bacteria, and some macrophages fuse together to form large multinucleated giant cells. Lymphocytes and macrophages then collect around these multinucleated cells and wall off the infected area from the surrounding tissue. The localized collection of inflammatory cells is called a **granuloma** (**figure 23.19**), which is the characteristic response of the body to microorganisms and other foreign substances that resist digestion and removal. The granulomas of tuberculosis are called tubercles. Although they may not kill the mycobacteria, they generally halt growth of the organisms, and usually no significant illness develops. Commonly, *M. tuberculosis* remains alive in the tubercles for many years. ■ delayed hypersensitivity, p. 441

Sometimes the mycobacteria are not contained by the inflammatory response. Lysis of activated macrophages attacking *M. tuberculosis* releases their enzymes into the infected tissue. The result is death of tissue, with the formation of a cheesy material by a process called **caseous necrosis**. If this process involves a bronchus, the dead material may discharge into the airways, causing a large lung defect called a **cavity** and spread of the bacteria to other parts of the lung. Lung cavities characteristically persist, slowly enlarging for months or years and shedding tubercle bacilli into the bronchi. Coughing and spitting transmit the organisms to other people. In areas of caseous necrosis, *M. tuberculosis* often remains alive, although multiplication ceases. Wherever living organisms persist, they can resume growing if the person's immunity becomes impaired by stress, advanced age, or AIDS. Disease resulting from renewed growth of the organisms is called reactivation tuberculosis.

Epidemiology

An estimated 10 million Americans are infected by *M. tuberculosis.* Infection rates are highest among nonwhites and elderly poor people. Transmission of tuberculosis occurs almost entirely by the respiratory route; 10 or even fewer inhaled organisms are enough to cause infection. Factors important in transmission include the frequency of coughing, the adequacy of ventilation (transmission is unlikely to occur outdoors), and the degree of crowding. Immunodeficiency can result in activation of latent tuberculosis, and over 5% of AIDS patients have developed active tuberculosis. Outbreaks are commonly seen in families, schools, nursing homes, or hospitals, usually starting with an unsuspected case of reactivated tuberculosis spread by coughing or singing.

The tuberculin test, also known as the Mantoux (pronounced man-too) test, is an extremely important tool for studying the epidemiology of the disease and in detecting those who are infected with *M. tuberculosis.* The test is carried out by injecting into the skin a very small amount of a sterile fluid called purified protein derivative, or PPD, derived from cultures of *M. tuberculosis.* People who are infected with the bacterium develop redness and a firm swelling at the injection site, reaching a peak intensity after 48 to 72 hours (**figure 23.20**). This reaction is due to the congregation of macrophages and T lymphocytes at the injection site, a manifestation of delayed hypersensitivity to the tubercle bacillus. A strongly positive reaction to the test generally indicates that living *M. tuberculosis* bacilli are present somewhere in the body of the person tested. A positive tuberculin test does not necessarily mean that the person has tuberculosis; only that he or she has been infected by *M. tuberculosis* at some time in the past.

Prevention and Treatment

Vaccination against tuberculosis has been widely used in many parts of the world with varying success. The vaccinating agent, a living attenuated mycobacterium known as Bacille Calmette-Guérin, or BCG, is derived from *M. bovis.* Repeated subculture in the laboratory over many years resulted in selection of this

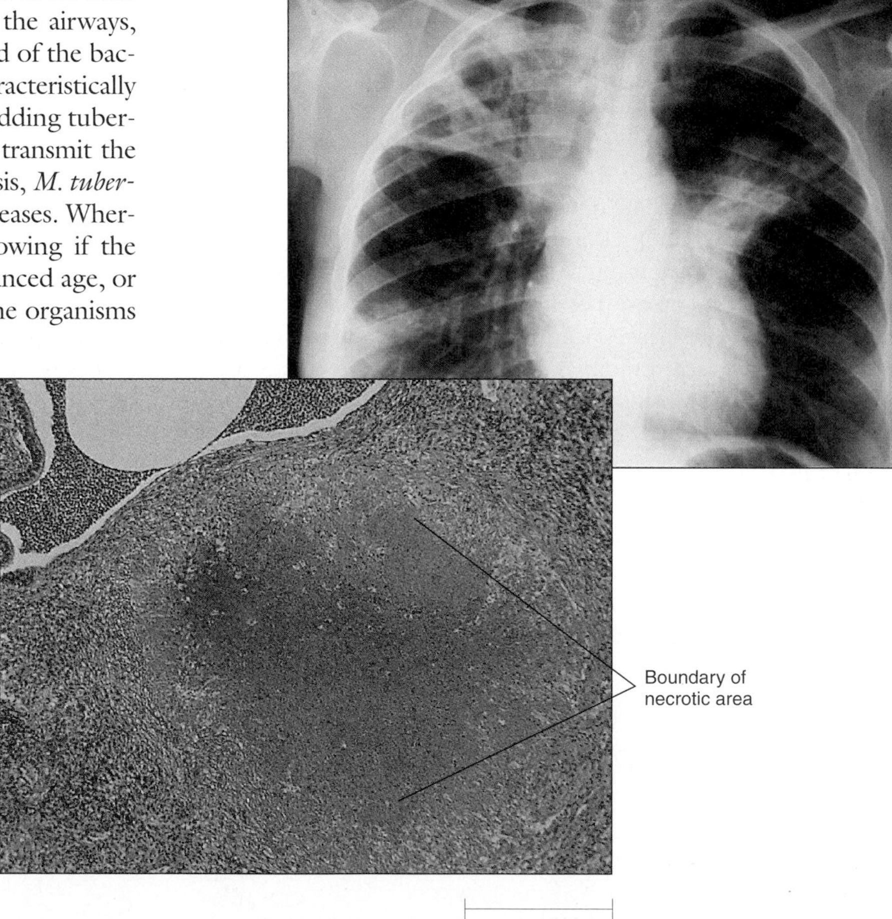

Figure 23.19 Stained Lung Tissue Showing a Tubercle, a Kind of Granuloma Caused by the Body's Reaction to *Mycobacterium tuberculosis* The innumerable dark dots around the outer portion of the picture are nuclei of lung tissue and inflammatory cells. Centrally and extending to the right of the photograph, most of the nuclei have disappeared because the cells are dead and the tissue has begun to liquefy. The photograph depicts a chest X-ray film of an individual with tuberculosis.

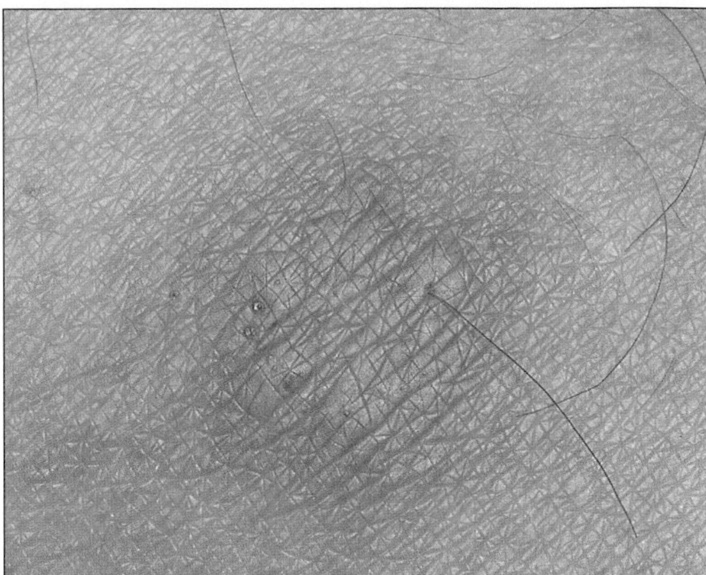

Figure 23.20 Tuberculin Test A positive test is the result of delayed hypersensitivity to *Mycobacterium tuberculosis* antigens injected into the skin, and it gives evidence of past or current infection with the bacterium.

strain of *M. bovis*, which has little virulence in humans but produces some immunity to tuberculosis. Use of the vaccine is discouraged in the United States, because people who receive the vaccine usually develop a positive tuberculin test. By causing a positive test, BCG vaccination eliminates an important way of diagnosing tuberculosis early in the disease when it can most easily be treated. BCG is not safe to use in severely immunocompromised patients because the vaccine bacillus can spread throughout the body and cause disease. Safer and more effective genetically engineered vaccines are under development. Control of tuberculosis is aided by identifying unsuspected cases using skin tests and lung X rays. Individuals with active disease are then treated, thus interrupting the spread of *M. tuberculosis*. People whose Mantoux tests have changed from negative to positive are also treated, even when no evidence of active disease exists. Treatment reduces the risk (estimated to be about 12%) that these people will develop active disease later in life.

Mutants resistant to antibacterial medications frequently occur among sensitive *M. tuberculosis* strains. Since mutants simultaneously resistant to more than one antimicrobial medication occur with a very low frequency, two or more of the medications are always given together in treating tuberculosis. The combination of rifampin and isoniazid (INH) is favored because both drugs are bactericidal against actively growing organisms in cavities as well as metabolically inactive intracellular organisms. Because of the long generation times of *M. tuberculosis* and its resistance to destruction by body defenses, drug treatment of tuberculosis must generally be continued for a minimum of 6 months to cure the disease. During the prolonged treatment symptoms usually disappear, and unfortunately many individuals become careless about taking their medications, or they discontinue them prematurely. Such negligence allows the rare resistant mutants to multiply and can lead to high rates of relapse. Up to two-thirds of the *M. tuberculosis* strains obtained from inadequately treated patients are resistant to one or more antitubercular medications. These resistant strains can infect others in the community and are increasingly responsible for new infections. The problem of drug-resistant *M. tuberculosis* strains has reached alarming proportions in some areas and has motivated the search for new vaccines against the disease. ■ **antibiotic resistance, p. 509**

The main features of tuberculosis are shown in **table 23.8**.

Legionnaires' Disease

Legionnaires' disease was unknown until 1976, when a number of people attending an American Legion Convention in Philadelphia developed a mysterious pneumonia that was fatal in many cases. Months of scientific investigation eventually paid off when the cause was discovered to be a previously unknown bacterium commonly present in the natural environment.

Symptoms

Legionnaires' disease typically begins with headache, malaise, rapid rise in temperature, confusion, and shaking chills. A dry cough develops that later produces small amounts of sputum, sometimes streaked with blood. Pleurisy, manifest as chest pain brought on by coughing or deep breathing, can also occur. About one-fourth of the cases also have some alimentary tract symptoms such as diarrhea, abdominal pain, and vomiting. Shortness of breath is common, and oxygen therapy is often needed. Recovery is slow, and weakness and fatigue last for weeks.

Causative Agent

The causative agent of Legionnaires' disease is *Legionella pneumophila* (**figure 23.21**), a member of the γ-proteobacteria. The organism is rod-shaped, Gram-negative, and requires a special medium for laboratory culture, which partly explains why it escaped detection for so long. Also, in tissue, *L. pneumophila* stains poorly with many of the usual microbiological stains, probably the result of some unusual fatty acids and other chemical characteristics of the cell wall. There are a number of different strains of *L. pneumophila*, some of which cause symptoms differing from Legionnaires' disease. ■ **γ-proteobacteria, p. A–10**

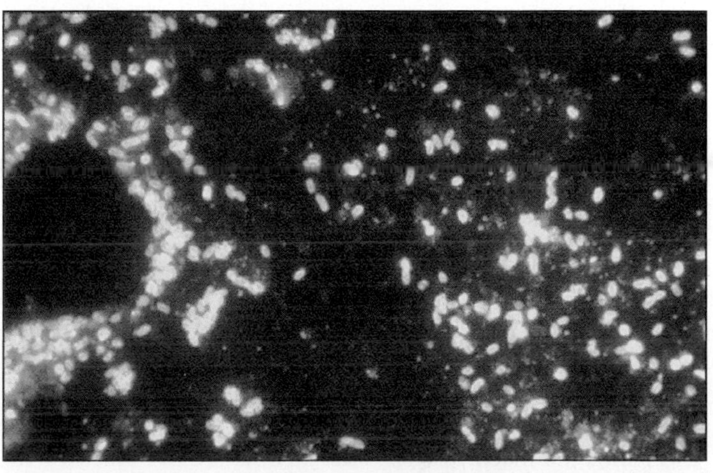

Figure 23.21 ***Legionella pneumophila*** **Stained with Fluorescent Antibody** When present in tissue or sputum, the bacterium fails to stain with most of the usual microbiological stains.

TABLE 23.8 Tuberculosis

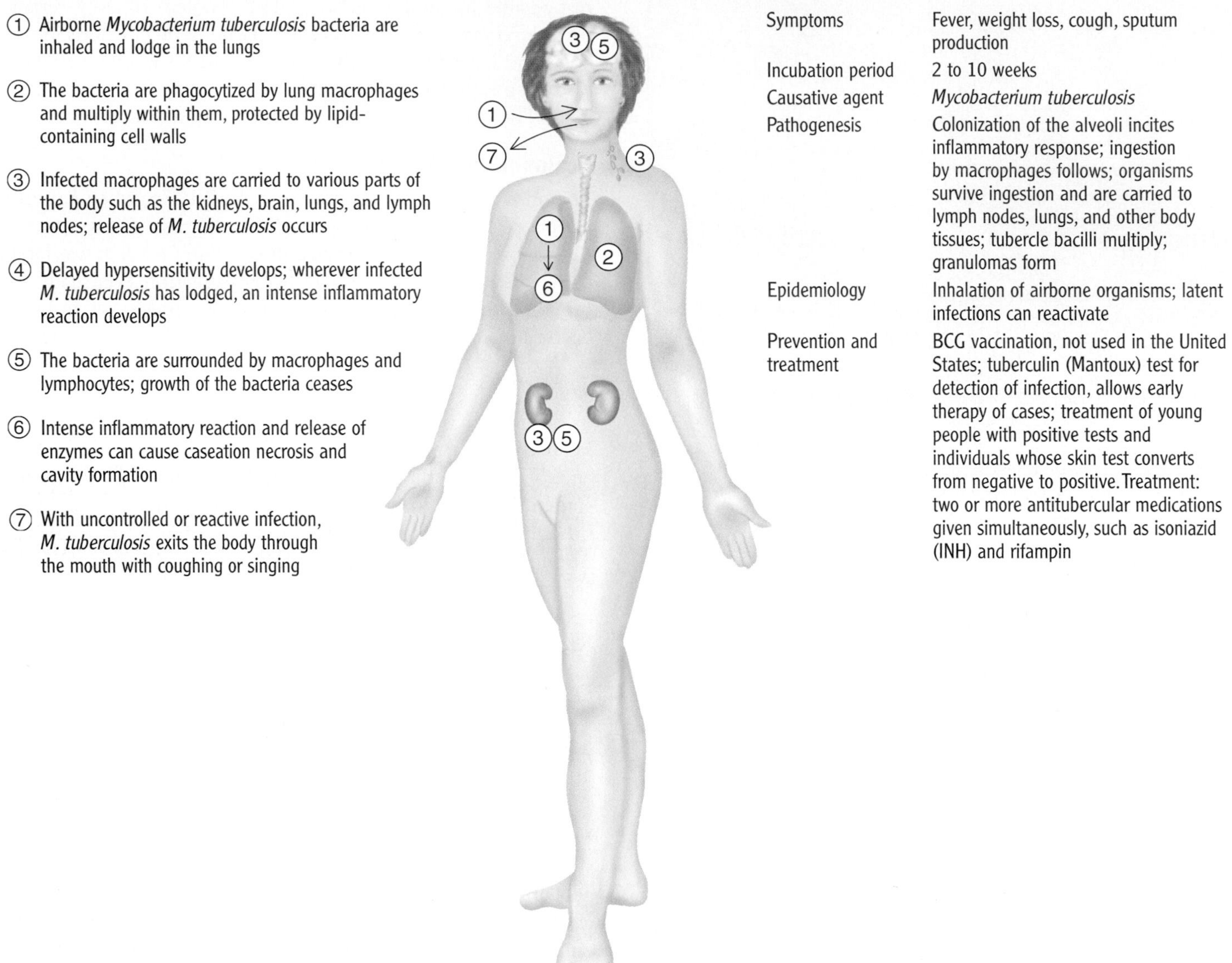

① Airborne *Mycobacterium tuberculosis* bacteria are inhaled and lodge in the lungs
② The bacteria are phagocytized by lung macrophages and multiply within them, protected by lipid-containing cell walls
③ Infected macrophages are carried to various parts of the body such as the kidneys, brain, lungs, and lymph nodes; release of *M. tuberculosis* occurs
④ Delayed hypersensitivity develops; wherever infected *M. tuberculosis* has lodged, an intense inflammatory reaction develops
⑤ The bacteria are surrounded by macrophages and lymphocytes; growth of the bacteria ceases
⑥ Intense inflammatory reaction and release of enzymes can cause caseation necrosis and cavity formation
⑦ With uncontrolled or reactive infection, *M. tuberculosis* exits the body through the mouth with coughing or singing

Symptoms	Fever, weight loss, cough, sputum production
Incubation period	2 to 10 weeks
Causative agent	*Mycobacterium tuberculosis*
Pathogenesis	Colonization of the alveoli incites inflammatory response; ingestion by macrophages follows; organisms survive ingestion and are carried to lymph nodes, lungs, and other body tissues; tubercle bacilli multiply; granulomas form
Epidemiology	Inhalation of airborne organisms; latent infections can reactivate
Prevention and treatment	BCG vaccination, not used in the United States; tuberculin (Mantoux) test for detection of infection, allows early therapy of cases; treatment of young people with positive tests and individuals whose skin test converts from negative to positive. Treatment: two or more antitubercular medications given simultaneously, such as isoniazid (INH) and rifampin

Pathogenesis

Legionella pneumophila infection is acquired by breathing aerosolized water contaminated with the organism. Healthy people are quite resistant to infection, but smokers and those with impaired host defenses from chronic diseases such as cancer and heart or kidney disease are susceptible. The organisms lodge in and near the alveoli of the lung, and their porin proteins bind complement component C3b, which aids phagocytosis of *L. pneumophila* by macrophages. A surface protein of *L. pneumophila*, macrophage invasion potentiator (Mip), aids entry into the macrophages. The bacteria are not killed by the phagocytes, but multiply within them and are released upon death of the macrophages to infect other tissues. Necrosis (tissue death) of alveolar cells and an inflammatory response result, causing multiple small abscesses, pneumonia, and pleurisy. Bacteremia is often present. Fatal respiratory failure, meaning that the lungs can no longer adequately oxygenate the blood or expel carbon dioxide, occurs in about 15% of hospitalized cases. Curiously, *L. pneumophila* infections remain confined to the lung in most cases. ■ **porin proteins, p. 63**

Epidemiology

The organism is widespread in warm natural waters where other microorganisms are present, and where it is taken up and multiplies within amebas in the same way it grows in human macrophages. *Legionella pneumophila* is considerably more resistant to chlorine than the other Gram-negative bacterial pathogens and survives well in the water systems of buildings, particularly in hot water systems, where chlorine levels are generally low. Legionnaires' disease cases have originated from

TABLE 23.9 Legionnaires' Disease

Symptoms	Muscle aches, headache, fever, cough, shortness of breath, chest and abdominal pain, diarrhea
Incubation period	2 to 10 days
Causative agent	*Legionella pneumophila*, a Gram-negative bacterium that stains poorly in clinical specimens; member of the γ-proteobacteria
Pathogenesis	Organism multiplies within phagocytes; death of these and other cells followed by sloughing of cells lining the alveoli, tissue necrosis, and formation of microabscesses
Epidemiology	Originates mainly from warm water contaminated with other microorganisms, such as found in air conditioning systems
Prevention and treatment	Avoidance of contaminated water aerosols; regular cleaning and disinfection of humidifying devices. Treatment: erythromycin and rifampin

contaminated aerosols from air conditioner cooling towers, nebulizers, and even from showers and water faucets. Direct person-to-person spread, however, does not occur. Monoclonal antibodies against different *L. pneumophila* strains help trace the source of epidemics. ■ **monoclonal antibodies, p. 415**

Prevention and Treatment

Most efforts at control have focused on designing equipment to minimize the risk of infectious aerosols and on disinfecting procedures once the source of an epidemic has been identified. Environmental surveillance is not practical because of the lack of a simple method for detecting virulent strains. Legionnaires' disease is treated with high doses of erythromycin, sometimes concurrently with rifampin. Like many other pathogens, *L. pneumophila* produces the enzyme β-lactamase, which makes it resistant to many penicillins and cephalosporins. ■ **β-lactamase, p. 500**

The main features of Legionnaires' disease are given in **table 23.9**.

MICROCHECK 23.5

Pneumococcal pneumonia is typically acquired in the community and leads the list of pneumonias in adults requiring hospitalization. Pneumonia due to *Klebsiella* sp. and other Gram-negative rods is mainly nosocomial and leads the causes of death from nosocomial infections. Mycoplasmal pneumonia usually does not require hospitalization. Whooping cough (pertussis) is mainly a threat to infants; childhood immunization against the disease protects them, but immunity often does not persist to adulthood. Tuberculosis is a chronic disease spread from one person to another by coughing and singing. Most *Mycobacterium tuberculosis* infections become latent, posing the risk of reactivation throughout life. *Legionella pneumophila*, the bacterium that causes Legionnaires' disease, originates from waters containing other microorganisms, where it can grow within protozoa. In the human lung, the bacterium readily multiplies within macrophages.

- What feature of the *S. pneumoniae* cell is responsible for its virulence?
- Why isn't pertussis toxin eliminated by the mucociliary escalator?
- Why should people infected with *M. tuberculosis* react positively to a skin injection of PPD? Wouldn't a negative reaction seem more likely if the person is already infected?

Viral Infections of the Lower Respiratory Tract

DNA viruses such as varicella-zoster virus and the adenoviruses sometimes cause serious pneumonias, and some adenovirus infections can mimic pertussis. RNA viruses are of greater overall importance, however, because of the large number of people they infect and their potential for serious outcomes. The following examples come from the orthomyxovirus, paramyxovirus, and bunyavirus families of RNA viruses. ■ **RNA viruses, p. 324**

Influenza

Influenza is a good example of the constantly changing interaction between people and the agents that infect them. Antigenic changes in the influenza viruses are responsible for serious epidemics of the disease that recur in human populations. Epidemics of influenza also occur in animals such as birds, seals, and pigs, and it is likely that antigenic variability of influenza viruses is partly due to movement of viral genes from one animal species to another. There are three major influenza types, designated A, B, and C, based on differences in their nucleoprotein antigens. Type A, considered here, causes the most severe disease and its epidemics are the most widespread. Outbreaks due to type B strains occur each year, but they are less extensive and the severity of the disease is less. Type C strains are of relatively little importance.

Symptoms

After a short incubation period, averaging 2 days, influenza typically begins with headache, fever, and muscle pain, which reach a peak in 6 to 12 hours. A dry cough worsens over a few days. Usually these acute symptoms go away within a week, leaving the patient with a lingering hacking cough, fatigue, and generalized weakness for additional days or weeks. Infection by influenza viruses is associated with Reye's syndrome on rare occasions. ■ **Reye's syndrome, p. 536**

Causative Agents

Influenza A virus belongs to the orthomyxovirus family; its genome consists of eight segments of single-stranded, negative-sense RNA. The virion is surrounded with a lipid-containing envelope derived from the host cell membrane **(figure 23.22)**. Projecting from the envelope are two kinds of glycoprotein

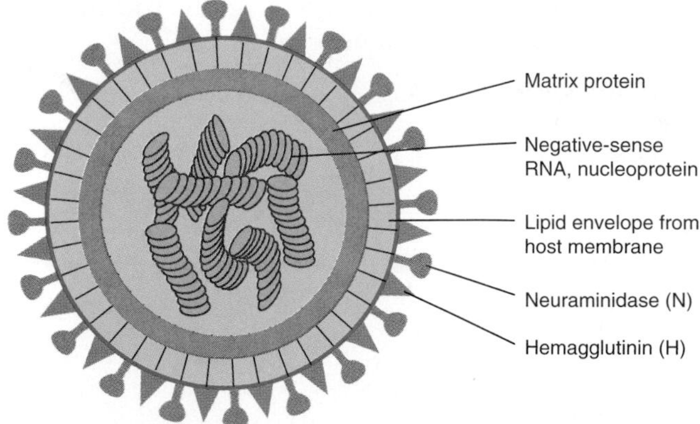

Figure 23.22 Diagrammatic Representation of Influenza Virus Note the surface structures, hemagglutinin and neuraminidase, the eight-segment genome, and the nucleoprotein, which distinguishes the three viral types, A, B, and C.

spikes, hemagglutinin (H) and neuraminidase (N). Hemagglutinin attaches the virus to specific receptors on ciliated epithelial cells of the host and thus initiates infection. Neuraminadase, on the other hand, is an enzyme that destroys the cell receptor to which the hemagglutinin attaches. This aids in the release of newly formed virions from the infected host cell and fosters the spread of the virus to uninfected host cells.

Pathogenesis

Influenza is acquired by inhaling aerosolized respiratory secretions from a person with the disease. The virions attach by their hemagglutinin to specific receptors on ciliated epithelial cells, their envelope fuses with the cell membrane, and the virus enters the cell. Host cell protein and nucleic acid synthesis cease, and rapid synthesis of viral nucleoproteins begins. Regions of the cell membrane become embedded with viral glycoproteins, specifically hemagglutinin and neuraminidase. Within 6 hours, mature virions bud from the host cell, receiving an envelope of cell membrane containing viral hemagglutinin and neuraminidase as they are released. The virus spreads rapidly to nearby cells, including mucus-secreting cells and cells of the alveoli. Infected cells ultimately die and slough off, thus destroying the mucociliary escalator and severely impairing one of the body's major defenses against infection. The immune response of the body quickly controls the infection in the vast majority of cases, although complete recovery of the respiratory epithelium may take 2 months or more. Usually, only a small percentage of people with influenza die, but even so, epidemics are so widespread that the total number of deaths is high. Influenza virus infection alone can kill apparently normal, healthy people. More often, however, death occurs because of bacterial secondary infections, usually by *Staphylococcus aureus*, *Streptococcus pyogenes*, or *Haemophilus influenzae*. Influenza takes a heavy toll on people whose hearts or lungs are weak.

Epidemiology

Outbreaks of influenza occur every year in the United States and are associated with an estimated 10,000 to 40,000 deaths. Figure 20.2 on p. 476 shows the variations in the percentage of total deaths attributed to pneumonia and influenza. Pandemics occur periodically over the years, marked by rapid spread of influenza viruses around the globe and higher than normal morbidity. ■ **pandemics, p. 474**

Although other factors are involved in the spread of influenza viruses, major attention has focused on their antigenic changeability. Two types of variation are seen: antigenic drift and antigenic shift **(figure 23.23)**. Antigenic drift is seen in interepidemic years and consists of minor mutations in the hemagglutinin (H) antigen that make immunity developed during prior years less effective and ensure that enough susceptible people are available for the virus to survive. This type of change is exemplified by A/Texas/77 (H3N2) and A/Bangkok/79 (H3N2), wherein there have been mutations that have altered the H antigen slightly. (The geographic names are the places where the virus was first isolated, the next two numbers represent the year.) The antibody produced by people who have recovered from A/Texas/77 (H3N2) is only partially effective against the mutant H3 antigen of A/Bangkok/79 (H3N2). Thus, the newer Bangkok strain might be able to spread and cause a minor epidemic in a population previously exposed to the Texas strain.

Antigenic shifts are represented by more dramatic changes; virus strains appear that are markedly different antigenically from the strains previously seen. Most likely they arise as a result of rare events in which two different viruses infect a cell at the same time. As can be demonstrated in the laboratory, dual infections result in an exchange of large segments of the viral genomes, so that totally different viruses result. Animal strains of influenza virus can occasionally infect humans and cause dual infections with human strains. If genetic mixing results in a new virus that is infectious, virulent, and possesses a hemagglutinin for which a population had no immunity, it could cause widespread disease before all the susceptible people had been infected. Ecological studies show that all the known influenza A virus types exist in aquatic birds, generally causing asymptomatic chronic intestinal infections and fecal elimination of the viruses. These bird influenza viruses readily infect domestic fowl, and from them can infect other domestic animals and humans. A well-studied episode in Hong Kong in 1997 involved an H5N1 virus from chickens that caused fatal infections in humans, but fortunately it did not spread easily from person to person. These ecological considerations indicate that operations in which high densities of domestic fowl or other animals are raised should be strictly isolated from contact with wild fowl. Influenza epidemics have devastated human populations in the past, and the emergence of virulent new strains of influenza virus continues to be an extremely serious threat to humankind.

Prevention and Treatment

Killed vaccines can be 80% to 90% effective in preventing disease when the vaccine is produced from the epidemic strain. Split vaccines consisting mostly of hemagglutinin are now widely used instead of whole virus vaccines. Because of antigenic drift, vaccines produced from earlier strains having the same H type may be decreasingly effective, so that new vaccines must be produced each year. It takes 6 to 9 months from the appearance of a new influenza strain before adequate amounts of vaccine can be manufactured.

A serious reaction was observed during a nationwide immunization program with the swine influenza vaccine produced

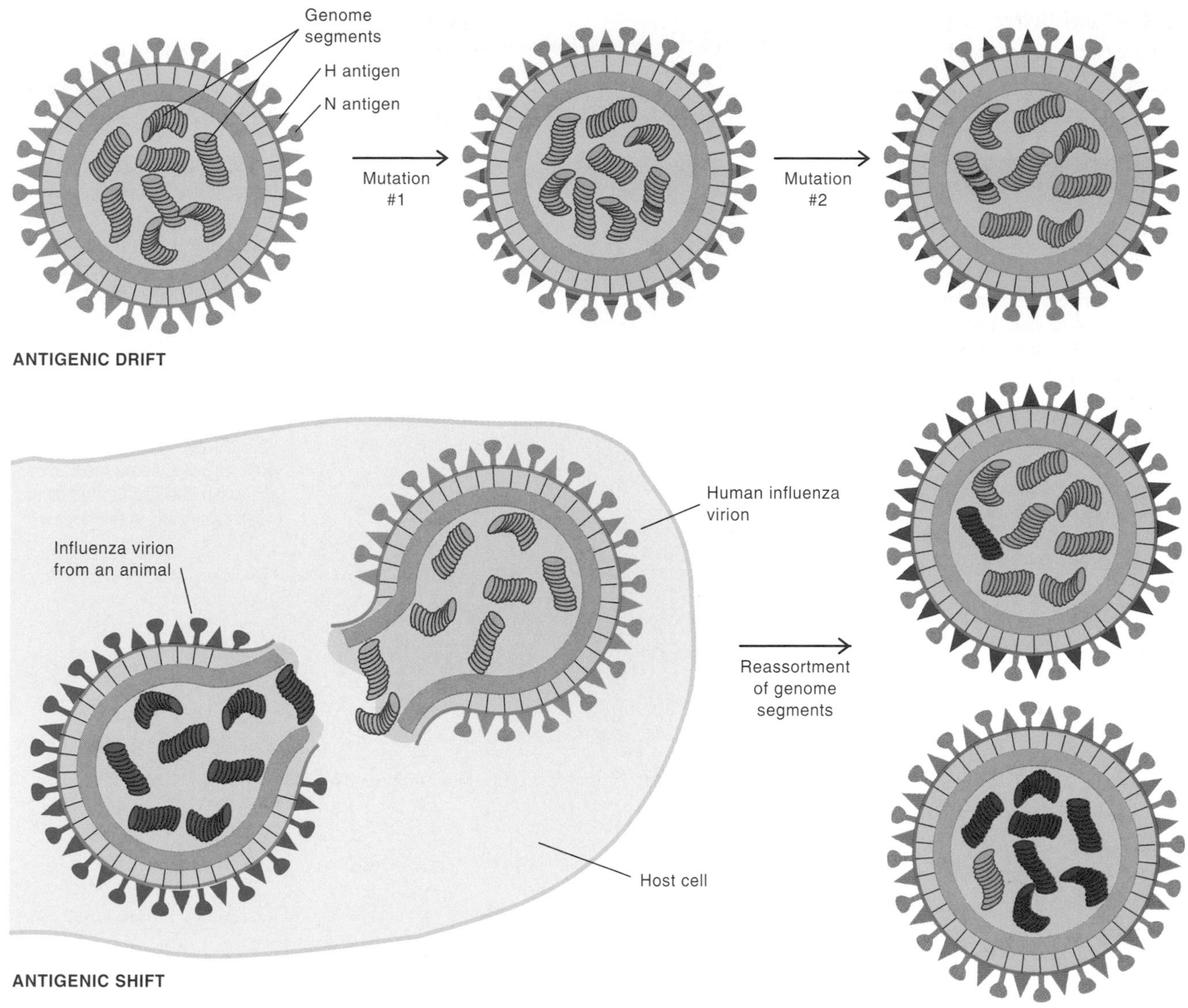

Figure 23.23 Influenza Virus: Antigenic Drift and Antigenic Shift With drift, repeated mutations cause a gradual change in the antigens composing the hemagglutinin, so that antibody against the original virus becomes progressively less effective. With shift, there is an abrupt, major change in the hemagglutinin antigens because the virus acquires a new genome segment, which in this case codes for hemagglutinin. Changes in neuraminidase could occur by the same mechanism.

against the 1976 Fort Dix H1N1 influenza A strain. This reaction, called the **Guillain-Barré syndrome**, occurred in about 1 of every 100,000 persons vaccinated. It is characterized by severe paralysis, and although most people recover completely, about 5% die of the paralysis. The Guillain-Barré syndrome occurs following a variety of illnesses, and 1 to 2 cases occur per million vaccinations with the current influenza vaccines. The reason for its apparent association with the swine influenza vaccine is not known.

Medications such as amantidine and rimantidine employed for short-term prevention of influenza A disease when vaccine is not available are 70% to 90% effective. These medicines are best used in conjunction with vaccination to protect exposed individuals until they can develop immunity. They also reduce the severity and duration of illness if given within 48 hours of the onset of influenza A infections. Promising new medications that inhibit neuraminidase are effective against both type A and B viruses, and live vaccines used nasally are under evaluation.

The main features of influenza are summarized in **table 23.10**.

Respiratory Syncytial Virus Infections

Respiratory syncytial virus (**RSV**) infection is first among serious lower respiratory tract infections of infants and young children, causing an estimated 90,000 hospitalizations and 4,500 deaths in the United States each year. It is also responsible for serious disease in elderly people, and for nosocomial epidemics.

Symptoms

Symptoms begin after an incubation period of 1 to 4 days with runny nose followed by cough, wheezing, and difficulty breathing.

TABLE 23.10 Influenza

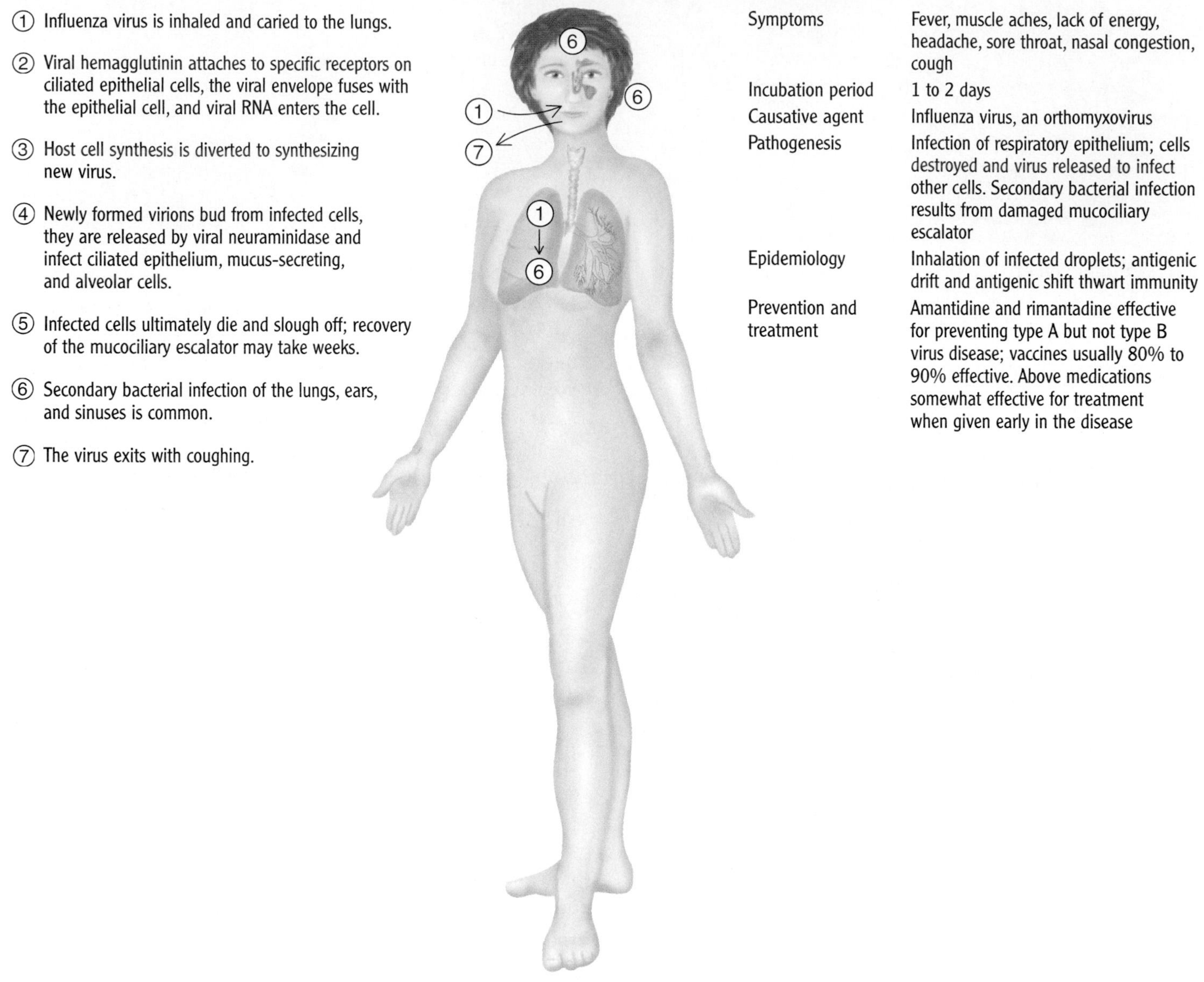

① Influenza virus is inhaled and caried to the lungs.

② Viral hemagglutinin attaches to specific receptors on ciliated epithelial cells, the viral envelope fuses with the epithelial cell, and viral RNA enters the cell.

③ Host cell synthesis is diverted to synthesizing new virus.

④ Newly formed virions bud from infected cells, they are released by viral neuraminidase and infect ciliated epithelium, mucus-secreting, and alveolar cells.

⑤ Infected cells ultimately die and slough off; recovery of the mucociliary escalator may take weeks.

⑥ Secondary bacterial infection of the lungs, ears, and sinuses is common.

⑦ The virus exits with coughing.

Symptoms	Fever, muscle aches, lack of energy, headache, sore throat, nasal congestion, cough
Incubation period	1 to 2 days
Causative agent	Influenza virus, an orthomyxovirus
Pathogenesis	Infection of respiratory epithelium; cells destroyed and virus released to infect other cells. Secondary bacterial infection results from damaged mucociliary escalator
Epidemiology	Inhalation of infected droplets; antigenic drift and antigenic shift thwart immunity
Prevention and treatment	Amantidine and rimantadine effective for preventing type A but not type B virus disease; vaccines usually 80% to 90% effective. Above medications somewhat effective for treatment when given early in the disease

Fever may or may not be present. Victims often develop a dusky color, indicating that they are not getting enough oxygen. Death occurs in 0.5% to 1% of hospitalized infants but is much higher in those with underlying diseases such as heart and lung disease, cancer, and immunodeficiency. RSV is one of the causes of **croup**, seen commonly in older infants and manifest as a loud high-pitched cough and noisy inspiration due to airway obstruction at the larynx. Infections of healthy older children and adults generally cause symptoms of a bad cold.

Causative Agent

Respiratory syncytial virus is a member of the paramyxovirus family and contains single-stranded, negative-sense RNA. Its genome, unlike influenza virus, is nonsegmented. Its virion is enveloped, but lacks hemagglutinin and neuraminidase. It causes cells in cell cultures to fuse together. The clumps of fused cells are known as **syncytia**, thus the name of the virus. ■ paramyxovirus family, p. 343

Pathogenesis

The virus enters the body by inhalation and infects the respiratory tract epithelium, causing death of the cells and sloughing from the basement membrane. Bronchiolitis is a common feature of the disease; the inflamed bronchioles become partially plugged by sloughed cells, mucus, and clotted plasma that has oozed from the walls of the bronchi. The initial obstruction causes wheezing when air rushes through the narrowed passageways, sometimes causing the condition to be confused with asthma. The obstruction often acts like a one-way valve, allowing air to enter the lungs, but not leave them. In many cases the inflammatory process extends into the alveoli, causing pneumonia. There is a high risk of secondary infection because of the damaged mucociliary escalator. ■ mucociliary escalator, p. 550

Epidemiology

RSV outbreaks are common from late fall to late spring, peaking in mid-winter. Recovery from infection produces only weak and short-lived immunity, so that infections can recur throughout life. Healthy children usually have mild illness and readily spread the virus to others.

Prevention and Treatment

No vaccines are available. Preventing nosocomial RSV illness requires strict isolation technique. Subjects with underlying illnesses can be protected from the disease by injecting them with a monoclonal antibody called palivizumad. ■ monoclonal antibody, p. 415

The main features of RSV infections are summarized in **table 23.11**.

Hantavirus Pulmonary Syndrome

In the spring of 1993 a previously unrecognized disease made a dramatic appearance in the American Southwest near the place where the corners of four states, Arizona, Colorado, New Mexico, and Utah, come together. The initial outbreak of the disease was alarming. It involved less than a dozen people, mostly Navajo, and most were vigorous young adults who started with influenza-like symptoms, but were dead in a few days. Scientists from the Centers for Disease Control and Prevention dropped what they were doing and rushed to join local epidemiologists and health officials investigating the outbreak. Their studies quickly established that the disease was associated with exposure to mice. By allowing the serum from victims to react with many different known viruses, they learned within a month that antibodies in the blood of the victims reacted with a hantavirus that had plagued American troops during the 1950s Korean War. Using the polymerase chain reaction (PCR), the scientists recovered a viral genome from the lungs of patients who had died, and they showed it was identical to one from mice captured in the area. This virus proved to be a close relative of the Korean virus. By the fall of 1993, they had grown the new hantavirus in cell cultures. ■ polymerase chain reaction, p. 237

Symptoms

Hantavirus pulmonary syndrome usually begins with fever, muscle aches (especially in the lower back), nausea, vomiting, and diarrhea. Unproductive cough and increasingly severe shortness of breath appear within a few days, followed by shock and death.

Causative Agents

The causative agents are the Sin Nombre, Spanish for "no name," virus and various related hantaviruses that exist at different locations in the Western Hemisphere. The hantaviruses are enveloped viruses of the bunyavirus family. Their genome consists of three segments of single-stranded, negative-sense RNA. In nature these viruses cause lifetime infections, primarily of rodents, without any apparent harm to the animals. ■ bunyavirus family, p. 343

Pathogenesis

The virus enters the body by inhalation of air containing dust contaminated with the urine, feces, or saliva of infected rodents. By an unknown mechanism, the virus enters the circulation and is carried throughout the body, infecting the cells that line tissue capillaries. Massive amounts of the viral antigen appear in lung capillaries, but the antigen is also demonstrable in capillaries of the heart and other organs. The inflammatory response to the viral antigen causes the capillaries to leak large amounts of plasma into the lungs, suffocating the patient and causing the blood pressure to fall. Shock and death occur in more than 40% of the cases.

Epidemiology

Hantavirus pulmonary syndrome is called an emerging disease because of its recent discovery and apparent increase in frequency. Nevertheless, this zoonosis has undoubtedly existed for centuries, occasionally claiming human victims when mice and men get too close together. Since the description of the syndrome, cases have been identified from Canada to Argentina and Chile, including more than 200 cases from the United States. Most of the cases have occurred west of the Mississippi River and were due to Sin Nombre virus carried by deer mice (**figure 23.24**). Others have been caused by related viruses carried by other rodents. Outbreaks of the disease correlate with marked increases in mouse populations adjacent to impoverished communities with substandard housing. The virus spreads more easily when the mouse population density is high. Thirty percent or more of the mice can become carriers of the disease. These mice eagerly and easily invade the houses of the poor. Complex ecological factors control the size of mouse populations. The numbers of foxes, owls, snakes, and other predators play a role. Moreover, two outbreaks of hantavirus pulmonary syndrome occurred in association with El Niño weather patterns, which brought increased rainfall to the area, yielding increased plant growth and seed production for mice to feed on.

TABLE 23.11 Respiratory Syncytial Virus (RSV) Infections

Symptoms	Range from life-threatening bronchiolitis and pneumonia to croup and common cold
Incubation period	1 to 4 days
Causative agent	RSV, a paramyxovirus that produces syncytia
Pathogenesis	Sloughing of respiratory epithelium and inflammatory response plug bronchioles, cause bronchiolitis; pneumonia results from bronchiolar and alveolar inflammation, or secondary infection
Epidemiology	Yearly epidemics during the cool months; readily spread by healthy older children and adults who often have mild symptoms; no lasting immunity
Prevention and treatment	No vaccine. Preventable by injections of a monoclonal antibody; no satisfactory antiviral treatment

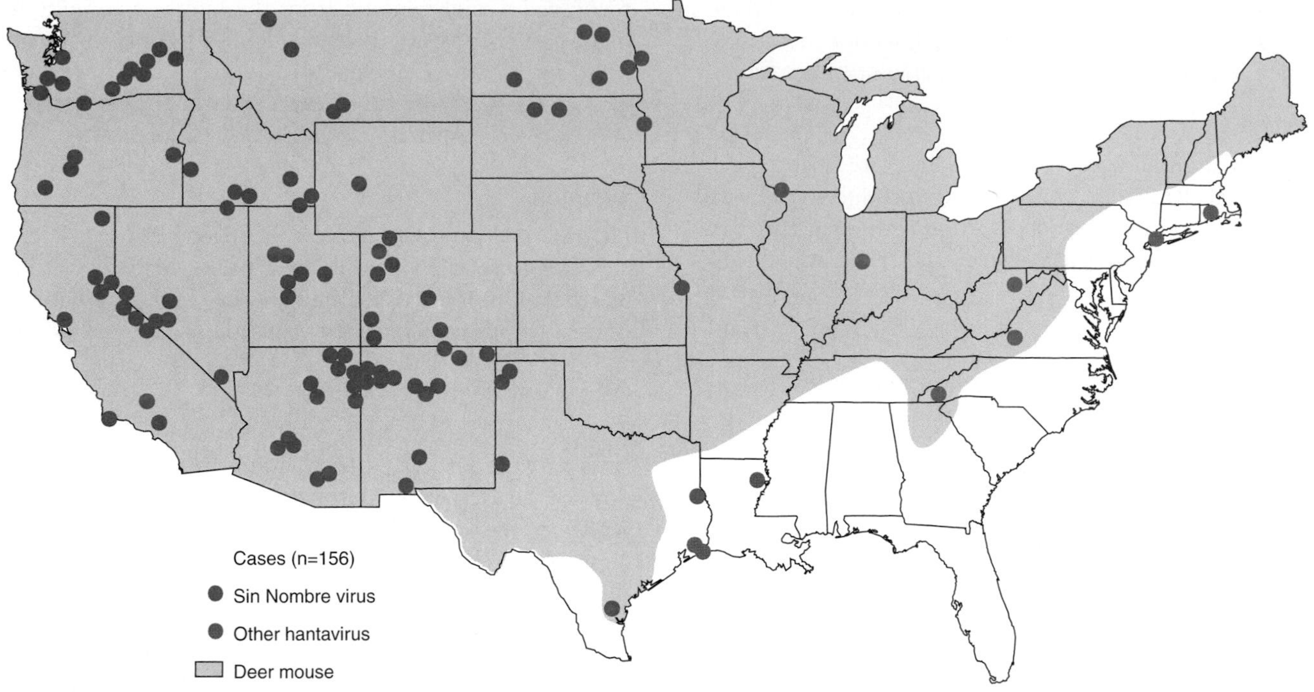

Figure 23.24 Distribution of the Deer Mouse and the 156 Cases of Hantavirus Pulmonary Syndrome Reported as of 1997

The emergence of hantavirus pulmonary syndrome is a convincing example of how environmental change can result in infectious human disease. Luckily, despite the large amount of viral antigen in the lung capillaries, few mature infectious virions enter the air passages of the lung; thus, person-to-person transmission occurs rarely, if ever. ■ emerging diseases, p. 6, 487

Prevention and Treatment

Prevention of hantavirus pulmonary syndrome is based on minimizing exposure to rodents and dusts contaminated by them. Food for humans and pets should be kept in containers. Buildings should be made mouse-proof if possible, and maximal ventilation is advisable when cleaning a rodent-infested area. Mopping with a disinfectant solution is preferable to use of brooms and vacuum cleaners, because the latter two stir up dust. Lethal traps and poisons may be necessary to decrease the rodent population in the area. When camping, a tent with a floor should be used. There is no proven antiviral treatment for this highly fatal disease.

The main features of hantavirus pulmonary syndrome are summarized in **table 23.12**.

MICROCHECK 23.6

Because of the speed with which influenza travels around the world, and the potential for development of virulent influenza virus strains, the disease poses an extremely serious threat to humankind. Most deaths from influenza are caused by secondary bacterial infections. Respiratory syncytial virus is the leading cause of serious respiratory disease in infants and young children. Hantavirus pulmonary syndrome, first recognized in 1993, is often fatal. It is contracted from inhalation of dust contaminated by mice infected with certain hantaviruses.

- Why are there so many deaths from influenza when it is generally a mild disease?
- What is the source of the virus that cases hantavirus pulmonary sundrome?

TABLE 23.12 Hantavirus Pulmonary Syndrome

Symptoms	Fever, muscle aches, vomiting, diarrhea, cough, shortness of breath, shock
Incubation period	3 days to 6 weeks
Causative agent	Sin Nombre and related hantaviruses of the bunyavirus family
Pathogenesis	Viral antigen localizes in capillary walls in the lungs; inflammation
Epidemiology	Zoonosis likely to involve humans in proximity to booming mouse populations; generally no person-to-person spread
Prevention and treatment	Avoid contact with rodents; seal access to houses and food supplies; good ventilation, avoid dust, use disinfectants in cleaning rodent-contaminated areas. No proven antiviral treatment

Fungal Infections of the Lung

Serious lung diseases caused by fungi are quite unusual in healthy, immunocompetent individuals. Symptomatic and asymptomatic infections that subside without treatment, however, are common. One organism, probably a fungus, infects most of us in childhood without causing symptoms. This organism, *Pneumocystis carinii*, will be discussed in chapter 29 on AIDS. Coccidioidomycosis and histoplasmosis are two other examples of widespread mycoses.

Valley Fever (Coccidioidomycosis)

In the United States, coccidioidomycosis occurs mainly in California, Arizona, Nevada, New Mexico, Utah, and West Texas. People who are exposed to dust and soil, such as farm workers, are most likely to become infected, but only 40% develop symptoms.

Symptoms

"Flu"-like symptoms such as fever, cough, chest pain, and loss of appetite and weight are common symptoms of coccidioidomycosis. About 10% of victims experience symptoms caused by hypersensitivity to the fungal antigens, manifested by tender nodules often localized to the shins and pain in the joints. The majority of people afflicted with coccidioidomycosis recover spontaneously within a month. A small percentage of victims develop chronic disease.

Causative Agent

The causative agent of coccidioidomycosis, *Coccidioides immitis*, is a dimorphic fungus. The mold form of the organism grows in soil. The mold's hyphae develop numerous barrel-shaped, highly infectious arthrospores (**figure 23.25a**), which separate easily and become airborne. Infected tissues contain *C. immitis* in the form of thick-walled spheres that may contain several hundred small cells called endospores (**figure 23.25b**). ■ **dimorphic fungi, p. 309**

Pathogenesis

Arthrospores enter the lung with inhaled air and develop into thick-walled spheres that mature and rupture, spilling numerous endospores. The endospores mature and repeat the process, each time provoking an inflammatory response and subsequent immune response. Symptoms and tissue injury are caused mainly by the host's immune response to coccidioidal antigens. Usually, the organisms are eliminated by body defenses, but in a small percentage of individuals caseation necrosis can occur and result in a lung cavity. Rarely, organisms are carried throughout the body by the bloodstream and infect the skin, mucous membranes, brain, and other organs. This disseminated form of the disease occurs more often in people with AIDS or other immunodeficiency and is fatal without treatment. ■ **caseation necrosis, p. 570**

Epidemiology

Coccidioides immitis grows only in semi-arid desert areas of the Western Hemisphere (**figure 23.26**). In these areas, infections occur only during the hot, dry, dusty seasons when airborne spores are easily dispersed from the soil. Dust stirred up by earthquakes can result in epidemics. Rainfall promotes growth of the fungus, which then produces increased numbers of spores when dry conditions return. People can contract coccidioidomycosis by simply traveling through the endemic area. Infectious spores have unknowingly been transported to other areas, but the organism apparently is unable to establish itself in moist climates.

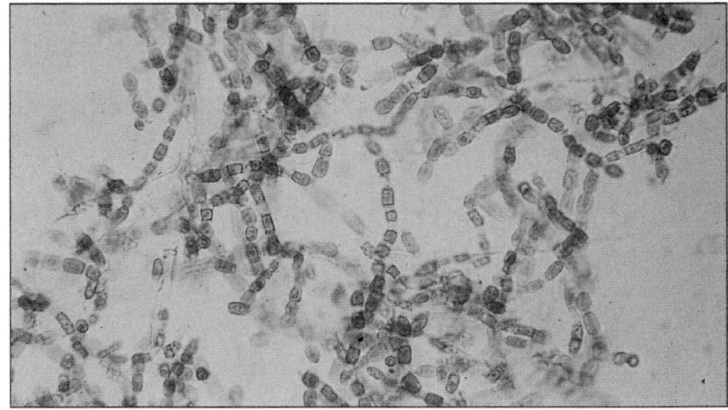

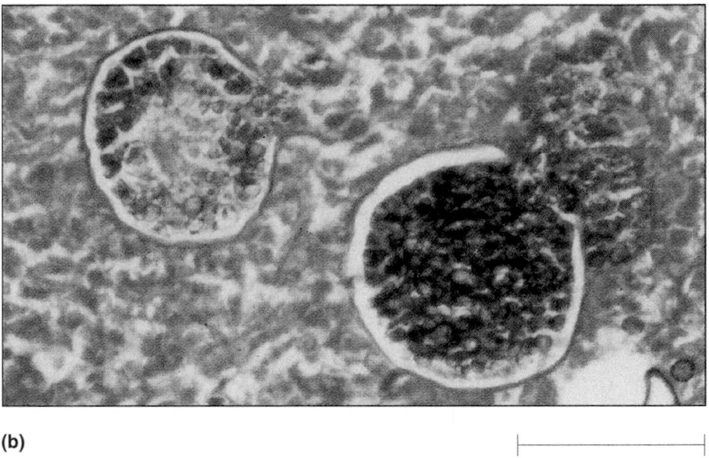

Figure 23.25 ***Coccidioides immitis*** **(a)** Mold-phase hyphae fragmenting into barrel-shaped arthrospores, the infectious form of the fungus. **(b)** Spherical spores containing endospores, the form of the fungus found in tissues.

Prevention and Treatment

Preventive measures include avoiding dust in the endemic areas. Watering and planting vegetation aid in dust control. Medications approved for treatment of the serious cases include amphotericin B and fluconazole. They markedly improve the prognosis but must be given for long periods of time, and cause troublesome side effects. Even with treatment, disseminated disease can reactivate months or years later.

Table 23.13 describes the main features of coccidioidomycosis.

Spelunkers' Disease (Histoplasmosis)

Histoplasmosis, like coccidioidomycosis, is usually benign but occasionally mimics tuberculosis. Rare, serious forms of the disease

Figure 23.26 Area of Distribution of *Coccidioides immitis*

suggest that AIDS or other immunodeficiency may also be present. The distribution is more widespread than that of coccidioidomycosis and is associated with different soil and climate.

Symptoms

Most infections are asymptomatic. Fever, cough, and chest pain are the most common symptoms, sometimes with shortness of breath. Mouth sores may develop, especially in children.

TABLE 23.13 Coccidioidomycosis

Symptoms	Fever, cough, chest pain, loss of appetite and weight; less frequently, painful nodules on extremities, pain in joints; skin, mucous membranes, brain, and internal organs sometimes involved
Incubation period	2 days to 3 weeks
Causative agent	*Coccidioides immitis,* a dimorphic fungus
Pathogenesis	After lodging in lung, arthrospores develop into spheres that mature and discharge endospores, each of which then develops into another sphere; inflammatory response damages tissue; hypersensitivity to fungal antigens causes painful nodules and joint pain
Epidemiology	Inhalation of airborne *C. immitis* spores with dust from soil growing the organism. Occurs only in certain semi-arid regions of the Western Hemisphere
Prevention and treatment	Dust control methods such as grass planting and watering. Treatment: amphotericin B and fluconazole

Causative Agent

Histoplasmosis is caused by the dimorphic fungus *Histoplasma capsulatum.* This organism prefers to grow in soils contaminated by bat or bird droppings, but it is not pathogenic for these animals. In pus or tissue from people with active disease, *H. capsulatum* is a tiny oval yeast that grows within host macrophages **(figure 23.27a)**. Contrary to its name, the fungus is not encapsulated. The mold form of the organism characteristically produces two kinds of spores: large conidia, which often have numerous projecting knobs (figure 23.27b) and tiny pear-shaped or spherical conidia.

Pathogenesis

Infectious conidia of *H. capsulatum* are inhaled with dust from contaminated soils and develop into the yeast form. The organ-

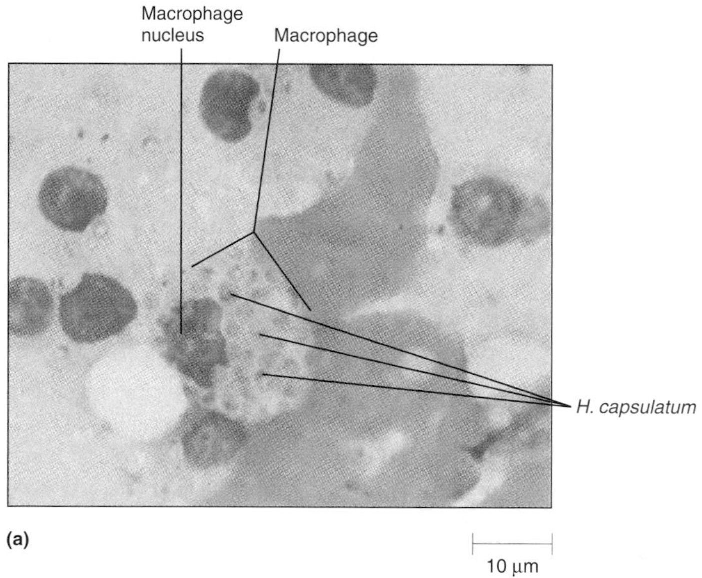

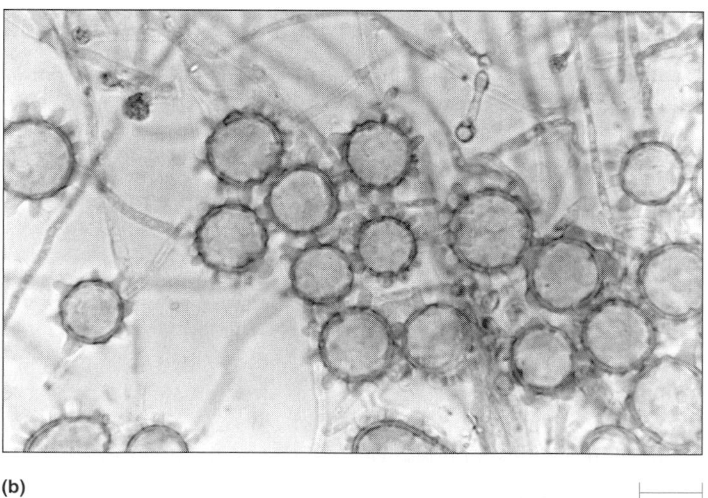

Figure 23.27 *Histoplasma capsulatum* **(a)** Yeast-phase organisms packing the cytoplasm of a macrophage. **(b)** Mold phase, showing large conidia with projecting knobs.

CASE PRESENTATION

A 25-year-old previously healthy man was admitted to a Colorado hospital with a 12-day history of fever, weight loss, fatigue, joint pain, and productive cough. His private physician had treated him with two antibiotics over the prior 8 days, but there was no response to the treatment.

Seven and 14 days before his symptoms began, the man was employed in a prairie dog relocation project, digging up and destroying the abandoned burrows. The work required using a gasoline-powered device that created clouds of dust, and also using hand trowels, which brought his face close to the soil. Above average rainfall had occurred in the 2 weeks before the work was done.

A few days after the man was admitted, a coworker entered the hospital with similar symptoms.

Initial laboratory studies including examination of the man's sputum were not helpful in making the diagnosis. A CT scan (an X-ray procedure) was then performed and revealed multiple small dense shadows throughout both lungs.

1. What are the diagnostic considerations in this individual?
2. How is the diagnosis to be made?
3. What is the natural history of the causative organism?
4. Could the disease have been prevented? Is treatment indicated?

Discussion

1. The CT scan, showing multiple shadows in both lungs, could not distinguish between various possible causes, including certain cancers, allergic lung diseases, tuberculosis, and fungal diseases. The acute nature of the illness, and the fact that his coworker contracted a similar illness at almost the same time, suggested an infectious cause. Exposure to dust arising from the soil 14 and 7 days before symptoms began pointed to a soil organism, and the lack of response to antibiotic therapy suggested that it might be fungal. Colorado, however, is not in the geographical areas where histoplasmosis, coccidioidomycosis, or other fungal lung diseases generally occur.
2. A lung biopsy was performed. The man was anaesthetized, an incision made in the chest to expose the lung, an area of abnormality identified, and a small portion excised for microscopic studies and culture. Microscopic examination revealed cells of a large spherical fungus, reproducing by a single bud, typical of *Blastomyces dermatitidis*, confirmed 10 days later when colonies of the organism first appeared in culture and were identified by a nucleic acid probe. These findings established the diagnosis of blastomycosis.
3. *Blastomyces dermatitidis* is a dimorphic soil fungus occurring mainly in the Mississippi and Ohio River valleys, and in the southeastern states, thus largely overlapping the distribution of *Histoplasma capsulatum*. Even in these areas it uncommonly causes human disease, and it rarely does so in Colorado. In the present case, recent heavy rainfall and soil heavily enriched with feces and other material from the animal burrows provided conditions likely to enhance growth of the fungus. The conditions under which the people worked allowed for possible inhalation of large numbers of *B. dermatitidis* spores.
4. Blastomycosis is probably preventable by using face masks that reliably exclude dust, and other dust control measures when working with potentially contaminated soil. Individuals with less extensive lung infection than this man may heal their disease without treatment, but the disease is prone to spread via the bloodstream to the skin, where it causes large disfiguring chronic ulcers, and less commonly to bones and other body tissues. In this instance, both men were treated for 10 days with intravenous amphotericin B, followed by 6 weeks of itraconazole.

Source: Centers for Disease Control and Prevention. 1999. *Morbidity and Mortality Weekly Report* 48: 98.

isms promptly enter macrophages and grow intracellularly. Granulomas develop in infected areas, closely resembling those seen in tuberculosis, sometimes even showing caseation necrosis. Eventually, the lesions are replaced with scar tissue, and many calcify, meaning that the body deposits calcium compounds in them that show up on X rays. In rare cases, the disease is not controlled and spreads throughout the body. This suggests AIDS or other immunodeficiency.

Epidemiology

The distribution of histoplasmosis is quite different from that of coccidioidomycosis. **Figure 23.28** shows the distribution of histoplasmosis in the United States, but the disease also occurs in tropical and temperate zones scattered around the world. Cave explorers, spelunkers, are at risk for contracting the disease because many caves contain soil enriched with bat droppings. Most cases of histoplasmosis in the United States have occurred in the Mississippi and Ohio River drainage area and in South Atlantic states. Skin tests reveal that millions of people living in these areas have been infected. As with *C. immitis*, only a small fraction of people infected with *H. capsulatum* develop serious illness.

Prevention and Treatment

No proven preventive measures are known other than to avoid areas where soil is heavily enriched with bat, chicken, and bird droppings, especially if they have been left undisturbed for a long period. Some researchers have recommended that several inches of clay soil be placed over soils containing large quantities of old droppings. Treatment of histoplasmosis is similar to that of coccidioidomycosis. Amphotericin B and itraconazole are used for treating severe disease, but both medications have potentially serious side effects.

The main features of histoplasmosis are described in **table 23.14**.

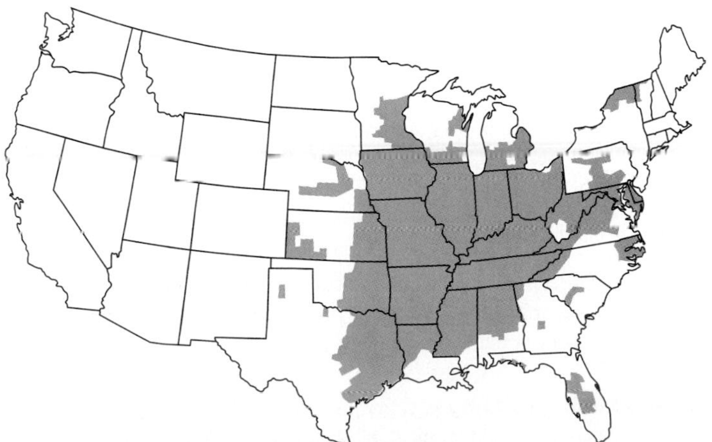

Figure 23.28 Geographic Distribution *of Histoplasma capsulatum* in the United States as Revealed by Positive Skin Tests Human histoplasmosis has been reported in more than 40 countries besides the United States, including Argentina, Italy, South Africa, and Thailand.

TABLE 23.14 Histoplasmosis

Symptoms	Mild respiratory symptoms; less frequently, general malaise, fever, chest pain, cough, chronic sores
Incubation period	5 to 8 days
Causative agent	*Histoplasma capsulatum*, a dimorphic fungus
Pathogenesis	Spores inhaled, change to tissue phase, multiply in macrophages; granulomas form; disease spreads in individuals with AIDS or other immunodeficiencies
Epidemiology	The fungus prefers to grow in soil contaminated by bird or bat droppings, especially in Ohio and Mississippi River valleys, and in the U.S. Southeast. Spotty distribution in many other countries around the world. Spelunkers are at risk of infection
Prevention and treatment	Avoidance of soils contaminated with chicken, bird, or bat droppings. Treatment: amphotericin B and itraconazole for serious infections

MICROCHECK 23.7

Coccidioidomycosis and histoplasmosis are two diseases caused by fungi that live in the soil. The body responds to these infections by forming granulomas, mimicking tuberculosis. Each fungus has its own ecological niche, *Coccidioides immitis* in semi-arid regions of the Western Hemisphere, and *Histoplasma capsulatum* in moist soils enriched with bird or bat droppings around the world.

- Why should an immunodeficient person avoid traveling through hot, dry, dusty areas of the Southwest?
- Why might cave exploration increase the risk of histoplasmosis?

FUTURE CHALLENGES

TB or Not TB?

Almost one-third of the world's population is infected with Mycobacterium tuberculosis, *with about 8 million developing active disease and 2 million dying of the disease each year. Contributing to the problem is a raging AIDS pandemic that causes activation of latent tuberculosis. Moreover, the worldwide incidence of* M. tuberculosis *strains resistant to medications is rising steadily. Countries of Asia, Africa, and Latin America have tuberculosis rates 5 to 30 times as high as the United States; currently, more than 40% of tuberculosis cases in the United States occur in foreign-born individuals, and this percentage is rising. There is little prospect of new cheap, effective antitubercular medications, and BCG, the most widely used of all vaccines, has not controlled the disease. The short-term challenge is to apply tuberculosis control methods shown to be successful in some Western countries to the rest of the world. The long-term challenge is to apply the methods of molecular biology to study the pathogenesis of tuberculosis in order to develop effective vaccines and new antitubercular medications.*

SUMMARY

Anatomy and Physiology (Figure 23.1)

1. The moist lining of the eyes (conjunctiva), nasolacrimal duct, middle ears, sinuses, mastoid air cells, nose, and throat make up the main structures of the upper respiratory system.
2. The functions of the upper respiratory tract include temperature and humidity regulation of inspired air and removal of microorganisms.
3. The lower respiratory system includes the trachea, bronchi, bronchioles, and alveoli. Pleural membranes cover the lungs and line the chest cavity.

The Mucociliary Escalator

1. Ciliated cells line much of the respiratory tract and remove microorganisms by constantly propelling mucus out of the respiratory system.

Normal Flora (Table 23.1)

1. Secretions of the nasal entrance are often colonized by diphtheroids and *Staphylococcus aureus*, a coagulase-positive staphylococcus.
2. Viruses and microorganisms are normally absent from the lower respiratory system.

INFECTIONS OF THE UPPER RESPIRATORY SYSTEM

Bacterial Infections of the Upper Respiratory System

Strep Throat (Streptococcal Pharyngitis) (Table 23.2)

1. *Streptococcus pyogenes* causes strep throat, a significant bacterial infection that may lead to **scarlet fever**, rheumatic fever, toxic shock, or glomerulonephritis. (Figure 23.2)

Diphtheria (Table 23.3)

1. Diphtheria, caused by *Corynebacterium diphtheriae*, is a toxin-mediated disease that can be prevented by immunization. (Figures 23.5, 23.6)

Pinkeye, Earache, and Sinus Infections

1. Conjunctivitis (pinkeye) is usually caused by *Haemophilus influenzae* or *Streptococcus pneumoniae*, the pneumococcus. Viral causes, including adenoviruses and rhinoviruses, usually result in a milder illness.
2. Otitis media and sinusitis develop when infection extends from the nasopharynx. (Figure 23.7)

Viral Infections of the Upper Respiratory System

The Common Cold (Table 23.4)

1. The common cold can be caused by many different viruses, rhinoviruses being the most common. (Figure 23.8)

Adenoviral Pharyngitis (Table 23.5)

1. Adenoviruses cause illnesses varying from mild to severe, which can resemble a common cold or strep throat.

INFECTIONS OF THE LOWER RESPIRATORY SYSTEM

Bacterial Infections of the Lower Respiratory System

Pneumococcal Pneumonia (Table 23.6, Figures 23.9, 23.10)

1. *Streptococcus pneumoniae*, one of the most common causes of pneumonia, is virulent because of its capsule.

Klebsiella Pneumonia (Table 23.6)

1. *Klebsiella* pneumonia, a Gram-negative bacterial pneumonia, is representative of many nosocomial pneumonias that cause permanent damage to the lung. (Figure 23.11)
2. Serious complications such as lung abscesses and bloodstream infection are more common than with many other bacterial pneumonias.
3. Treatment is more difficult, partly because klebsiellas often contain R factor plasmids.

Mycoplasmal Pneumonia (Table 23.6, Figures 23.12, 23.13)

1. Mycoplasmal pneumonia is often called walking pneumonia; serious complications are rare.
2. Penicillins and cephalosporins are not useful in treatment because the cause, *M. pneumoniae*, lacks a cell wall.

Whooping Cough (Pertussis) (Table 23.7, Figures 23.14, 23.15)

1. Whooping cough is characterized by violent spasms of coughing and gasping.
2. Childhood immunization against the Gram-negative rod, *Bordetella pertussis*, prevents the disease. (Figure 23.16)

Tuberculosis (Table 23.8)

1. Tuberculosis, caused by the acid-fast rod *Mycobacterium tuberculosis*, is generally slowly progressive or heals and remains latent, presenting the risk of later reactivation.

Legionnaires' Disease (Table 23.9, Figure 23.21)

1. Legionnaires' disease occurs when there is a high infecting dose of the causative microorganisms or an underlying lung disease. The cause, *Legionella pneumophila*, is a rod-shaped bacterium common in the environment.

Viral Infections of the Lower Respiratory Tract

Influenza (Table 23.10, Figure 23.22)

1. Widespread epidemics are characteristic of influenza A viruses. Antigenic shifts and drifts are responsible. (Figure 23.23)
2. Deaths are usually but not always caused by secondary infection.
3. Reye's syndrome may rarely occur during recovery from influenza and other viral infections but is probably not caused by the virus itself.

Respiratory Syncytial Virus Infections (Table 23.11)

1. RSV is the leading cause of serious respiratory disease in infants and young children.

Hantavirus Pulmonary Syndrome (Table 23.12, Figure 23.24)

1. Hantavirus pulmonary syndrome is contracted from inhalation of dust contaminated by mice infected with the hantavirus and is often fatal.

Fungal Infections of the Lung

Valley Fever (Coccidioidomycosis) (Table 23.13, Figure 23.26)

1. Coccidioidomycosis occurs in hot, dry areas of the Western Hemisphere and is initiated by airborne spores of the dimorphic soil fungus *Coccidioides immitis*.

Spelunker's Disease (Histoplasmosis) (Table 23.14)

1. Histoplamosis is similar to coccidioidomycosis but occurs in tropical and temperate zones around the world. (Figure 23.28)
2. The causative fungus, *Histoplasma capsulatum*, is dimorphic and found in soils contaminated by bat or bird droppings. (Figure 23.27)

REVIEW QUESTIONS

Short Answer

1. How does contamination of the eye lead to upper respiratory infection?
2. What potentially pathogenic bacterium commonly colonizes the nostrils?
3. After you recover from strep throat, can you get it again? Explain why or why not.
4. Where is the gene for diphtheria toxin production located?
5. Describe the process by which diphtheria toxin enters host cells and kills them.
6. Describe two ways to decrease the chance of contracting a cold.
7. What kinds of diseases are caused by adenoviruses?
8. How do alcoholism and cigarette smoking predispose a person to pneumonia?
9. Give a mechanism by which *Klebsiella* sp. become antibiotic-resistant.
10. Which group of bacteria accounts for most cases of nosocomial pneumonia?
11. Why does the incidence of whooping cough rise promptly when pertussis immunizations are stopped?
12. Why start immunization against *B. pertussis* at two months of age or less?
13. Why are two or more antitubercular medications used together to treat tuberculosis?
14. Why did it take so long to discover the cause of Legionnaires' disease?

15. What are the differences between antigenic drift and shift of influenza virus?
16. Name the infectious form of *C. immitis*, and of *H. capsulatum*.

Multiple Choice

1. The following are all complications of streptococcal pharyngitis, *except...*
   A. quinsy.
   B. scarlet fever.
   C. chorea.
   D. acute rheumatic fever.
   E. Reye's syndrome.
2. All of the following are true of diphtheria, *except...*
   A. a membrane that forms in the throat can cause suffocation.
   B. a toxin is produced that acts by ADP ribosylation.
   C. the causative organism typically invades the bloodstream.
   D. immunization with a toxoid prevents the disease.
   E. nerve injury with paralysis is common.
3. Adenoviral infections generally differ from the common cold in all the following ways, except adenoviral infections are...
   A. not caused by picornaviruses.
   B. often associated with fever.
   C. likely extensively to involve the cornea and conjunctiva.
   D. much more likely to cause pneumonia.
   E. associated with negative cultures for *Streptococcus pyogenes*.
4. All are true of mycoplasmal pneumonia, *except...*
   A. it is a mycosis.
   B. it usually does not require hospitalization.
   C. penicillin is ineffective for treatment.
   D. it is the leading cause of bacterial pneumonia in college students.
   E. the infectious dose of the causative organism is low.
5. All of the following are true of Legionnaires' disease, *except...*
   A. the causative organism can grow inside amebas.
   B. it spreads readily from person to person.
   C. it is more likely to occur in long-term cigarette smokers than in nonsmokers.
   D. it is often associated with diarrhea or other intestinal symptoms.
   E. it can be contracted from household water supplies.
6. Which of the following infectious agents is most likely to cause a pandemic?
   A. Influenza A virus
   B. *Streptoccus pyogenes*
   C. *Histoplasma capsulatum*
   D. Sin Nombre virus
   E. *Coccidioides immitis*
7. Respiratory syncytial virus
   A. is a leading cause of bronchiolitis in infants.
   B. is an enveloped DNA virus of the adenovirus family.
   C. attaches to host cell membranes by means of neuraminidase.
   D. poses no threat to elderly people.
   E. mainly causes disease in the summer months.
8. In the United States, hantaviruses...
   A. are limited to southwestern states.
   B. are carried only by deer mice.
   C. infect human beings with a fatality rate above 40%.
   D. were first identified in the early 1970s.
   E. are contracted mainly in bat caves.
9. All of the following are true of coccidioidomycosis, *except...*
   A. it is contracted by inhaling arthrospores.
   B. it is caused by a dimorphic fungus.
   C. endospores are produced within a spore.
   D. it is more common in Maryland than in California.
   E. it is often associated with painful nodules on the legs.
10. The disease histoplasmosis...
    A. is caused by an encapsulated bacterium.
    B. is contracted by inhaling arthrospores.
    C. occurs mostly in hot, dry, and dusty areas of the American Southwest.
    D. is a threat to AIDS patients living in areas bordering the Mississippi River.
    E. is commonly fatal for pigeons and bats.

Applications

1. A physician is consulting with family on the condition of a diphtheria patient. After being told about the toxin produced by the bacterium, the family wants to know why the toxin affects some tissues and not others. What would be the physician's explanation of this observation?
2. A mother is questioning a physician about the future health of her daughter. The child just recovered from a severe bout of pneumococcal pneumonia. Her mother wants to know whether she can get the disease again. What should the physician say to the mother?
3. An epidemiologist for the city of Houston was asked to explain to the mayor why tuberculosis has been on the rise in the city. What did the epidemiologist explain in his report to the mayor? What knowledge would he need to make suggestions about controlling further spread of the disease?

Critical Thinking

1. What would happen if the A and B chains of the diphtheria toxin became separated before rather than after entering a cell?
2. If all transmission of *Mycobacterium tuberculosis* from one person to another were stopped, how long would it take for the world to be rid of the disease?
3. New medications show promise for preventing and treating influenza by binding to neuraminidase on the viral surface. These new medications act against all the kinds of influenza viruses that infect humans. What does this imply about the nature of the interaction between the medications and the neuraminidase molecules?

Alimentary System Infections

"The face was sunken as if wasted by lingering consumption, perfectly angular, and rendered peculiarly ghastly by the complete removal of all the soft solids, in their places supplied by dark lead-colored lines. The hands and feet were bluish white, wrinkled as when long macerated in cold water; the eyes had fallen to the bottom of their orbes, and envinced a glaring vitality, but without mobility, and the surface of the body was cold."

This vivid description of cholera was written by Army surgeon S. B. Smith in 1832 when the disease first appeared in the United States. "Lingering consumption" refers to the marked wasting of the body seen with chronic tuberculosis.

Cholera is a very old disease and is thought to have originated in the Far East thousands of years ago. Sanskrit writings indicate that it existed endemically in India many centuries before Christianity. With the increased shipping of goods, and mobility of people during the Nineteenth century, cholera spread from Asia to Europe and then to North America. Cholera was a major epidemic disease of the Nineteenth century, appearing in almost every part of the world.

In 1854, John Snow, a London physician, demonstrated that cholera was transmitted by contaminated water. He observed that almost all people who contracted cholera got their water from a well on Broad Street. When the handle of the Broad Street pump was removed, people were forced to obtain their water elsewhere and the cholera epidemic in that area subsided. Snow's explanation was not generally accepted by other doctors, mostly because disease-causing "germs" had yet to be discovered. It was not until 1883 that Robert Koch isolated Vibrio cholerae, *the bacterium that causes cholera.*

In the United States, cholera epidemics occurred in 1832, 1849, and 1866. The disease seemed to infect poorer people, those unfortunate enough to be crowded together in cities where cleanliness was impossible to maintain. Many doctors were of the opinion that not only were the personal habits of these people "rash and excessive," but also that they insisted on taking the "wrong" medicines. By 1866, however, it was evident that where cholera appeared, the lack of sanitation was at fault. Public health agencies then played a major role in the elimination of epidemic cholera.

—A Glimpse of History

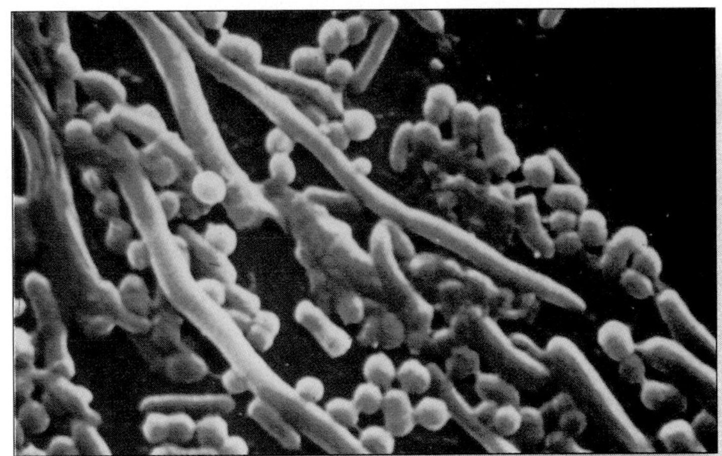

IN THE SPRING OF 1997, AN EPIDEMIC OF SEVERE diarrhea broke out among 90,000 sick and malnourished refugees in the Democratic Republic of Congo. More than 1,500 deaths occurred over a 3-week period. Volunteers from the medical relief organization Medecins Sans Frontieres rushed to set up medical facilities, while people from the World Health Organization, Red Cross, and local health agencies worked to provide a filtered and chlorinated water source, construct latrines, and educate the people about sanitary measures. Treatments were started using oral and intravenous rehydration fluids. Unfortunately, at that point, the refugees were all scattered and driven away by unidentified soldiers.

Fear generated by a diarrhea epidemic, compounded by ignorance, can defeat efforts based on reason and understanding of disease.

Anatomy and Physiology

The alimentary tract, sometimes referred to as the gastrointestinal or GI tract, is the passageway running from the mouth to the anus. Like the skin, it is one of the body's boundaries with the environment, and it is one of the major routes into the body for invading microbial pathogens. The alimentary tract and its appendages together compose the alimentary system, the main purpose of which is to provide nourishment for the body. The upper part is composed of the mouth, salivary glands, esophagus, and stomach, and the lower part includes the intestines, pancreas, and liver. **Figure 24.1** illustrates the relationships among the various parts of the alimentary system. The main functions of the system are:

- To grind food in the mouth into small particles that can readily react with digestive juices that break down large, complex molecules into absorbable components
- To move the food through the esophagus to the stomach, where it undergoes preliminary treatment with acid and enzymes
- To discharge stomach contents at a controlled rate into the small intestine, where they undergo digestion under alkaline conditions and absorption of nutrients into the bloodstream

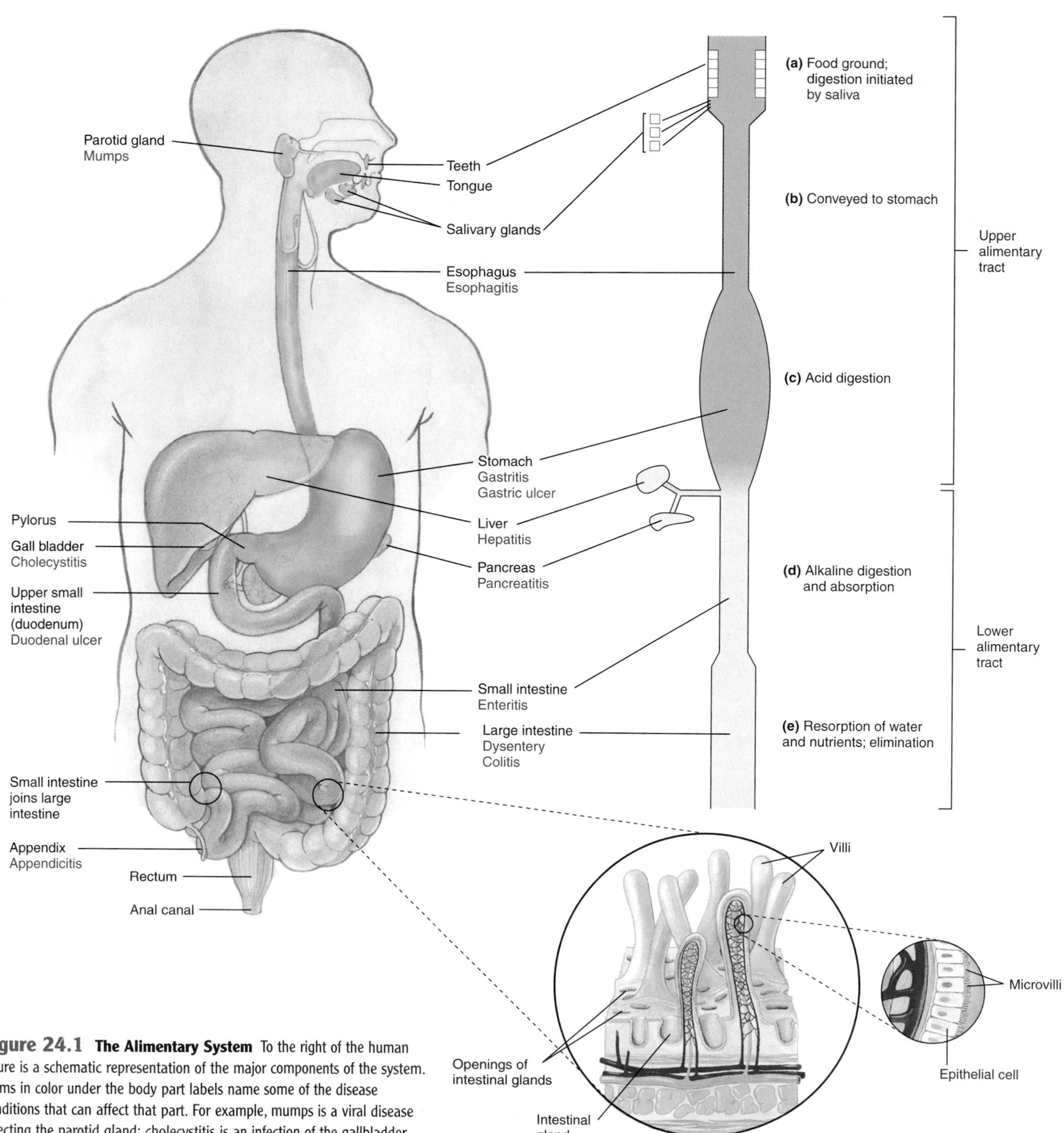

Figure 24.1 The Alimentary System To the right of the human figure is a schematic representation of the major components of the system. Terms in color under the body part labels name some of the disease conditions that can affect that part. For example, mumps is a viral disease affecting the parotid gland; cholecystitis is an infection of the gallbladder.

- To move undigested material into the large intestine, where water is absorbed from the intestinal contents along with any remaining vitamins, minerals, or nutrients
- To discharge the waste as feces.

The Mouth

The mouth is mainly a grinding apparatus, the tongue and the cheeks moving bites of food to be ground up by the teeth. The teeth (**figure 24.2**) are made up largely of crystals of hydroxyapatite, a calcium phosphate compound. The outer portion, called the **enamel**, is comprised of a hard, protective layer of densely packed crystals. Pits and crevices normally present on the enamel surfaces of the teeth collect food particles and offer protected sites for microbial colonization. Beneath the enamel lies the **dentin**, composed of softer, more easily penetrated hydroxyapatite. The **gingival crevice**, the space between the tooth and gum, is important because inflammation at this site can lead to loss of the tooth.

Salivary Glands

Salivary glands are located under the tongue and laterally in the floor of the mouth. There are two additional saliva-producing glands, called the **parotids**, one on either side of the face below the ears. About 1,500 ml of saliva is secreted from the salivary glands each day. It moistens and lubricates the mouth, contains the enzyme **amylase**, which begins the breakdown of starches present in food, and is extremely important in protecting the teeth from decay because it is saturated with calcium and contains buffers that help neutralize acids. It also contains antibacterial substances including **lysozyme**, the enzyme that attacks the peptidoglycan of bacterial cell walls, **lactoferrin**, a substance that binds iron ions critically required by bacteria, and specific IgA antibodies that inhibit bacterial attachment. ■ lactoferrin, p. 376 ■ IgA, p. 395

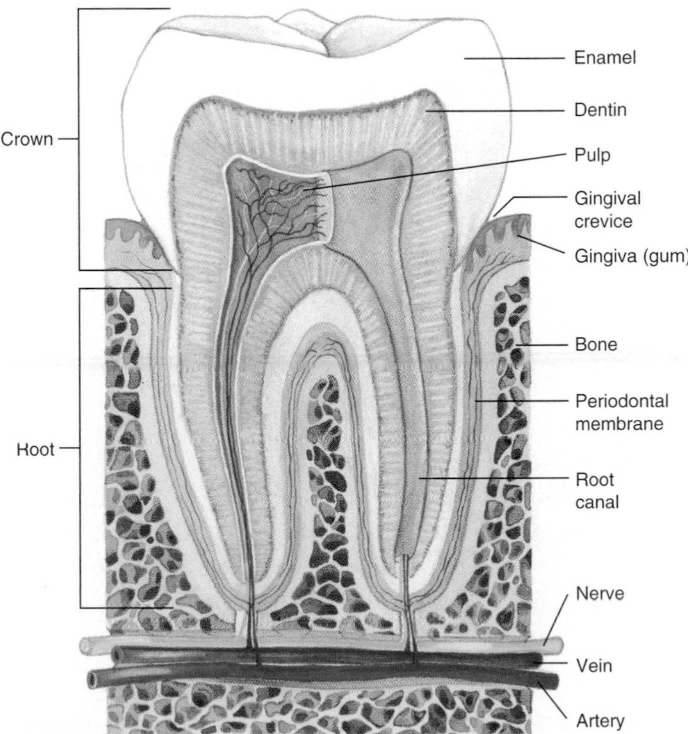

Figure 24.2 Structure of a Tooth and Its Surrounding Tissues

The Esophagus

The tongue and throat muscles work together in swallowing to propel food from the mouth and throat into the esophagus, a collapsible tube about 10 inches long, located behind the windpipe. The esophagus, like the intestines, has a muscular wall that contracts rhythmically, in a process called **peristalsis**, to move food and liquid. The lower portion of the esophagus normally closes to prevent regurgitation of stomach contents. The esophagus rarely becomes infected except in individuals with AIDS or other immunodeficiencies.

The Stomach

The lower portion of the esophagus leads into the stomach, an elastic saclike structure with a muscular wall. Some of the cells that line it produce hydrochloric acid, while others produce pepsinogen, which becomes the protein-splitting enzyme **pepsin** upon contact with acid. The stomach itself is protected from the acid and enzymes by a thick layer of **mucus** secreted by the stomach lining. Stomach emptying is controlled by a complex system of nerve impulses and hormones acting on a muscular valve, the **pylorus**, which determines the rate at which the stomach contents enter the intestine.

The Small Intestine

Although it is only 8 or 9 feet long, the small intestine has an enormous surface area, about 30 square feet, which facilitates nutrient absorption. The inside surface of the intestine is covered with many small, fingerlike projections called **villi**, each 0.5 to 1 mm long. Each of these villi are, in turn, covered with cells that have cytoplasmic projections called **microvilli** (see figure 24.1). Attached to the microvilli are many intestinal digestive enzymes. This epithelial surface has both secretory and absorptive functions. Tiny intestinal glands continuously secrete large amounts of fluid into the **lumen**, the space enclosed by the walls of the intestine. This fluid is then reabsorbed along with the products of food digestion.

Small intestine function depends on active transport of nutrients and electrolytes across the plasma membranes of the epithelial cells. For example, sodium ions (Na^+) are taken up by the cell when glucose or amino acids are absorbed, while hydrogen ions (H+) and bicarbonate ions (HCO_3^-) are secreted to adjust the pH of the intestinal contents. Movement of these substances is accompanied by water molecules. Digestion takes place as the cells secrete juices rich in enzymes such as the ones that break peptides into amino acids, and disaccharidases that convert complex sugars into simpler ones. The major part of nutrient absorption, including absorption of monosaccharides, amino acids, fatty acids, and vitamins, takes place in the small intestine. Some minerals, such as iron, can only be absorbed there. In addition, the small intestine reabsorbs approximately 9 liters of fluid per day. The lining cells of the small intestine are continuously shed and replaced by new cells, so that cells poisoned by a microbial toxin are soon replaced. Intestinal epithelial cells are completely replaced every 9 days, one of the fastest turnover rates in the body.

The Pancreas

Some cells of the pancreas produce hormones, and others produce digestive enzymes. The hormones are released directly into the bloodstream and help regulate various body functions, as for example, **insulin** helps regulate the amount of glucose in the blood. On the other hand, about 2 liters pancreatic digestive juices discharge directly into the upper portion of the small intestine each day. These fluids, as well as digestive juices of the small intestine itself, are alkaline and neutralize the stomach acid as it passes into the intestine.

The Liver

The liver aids digestion, neutralizes poisons, degrades medications, and removes a normal breakdown product of hemoglobin from the bloodstream. This yellowish product, called bile pigment, is discharged with the bile, the fluid produced by the liver. Each day, about 500 ml of bile flows through a system of tubes into the upper small intestine. Severe liver disease and obstruction of the bile ducts can produce jaundice, a yellow color of the skin and eyes caused by buildup of bile pigment in the blood. The gallbladder is a saclike structure in which bile is concentrated and stored. Substances in the bile called bile salts help the intestine absorb oils and fats and the fat-soluble vitamins A, D, E, and K. The normal brown color of feces results from the action of intestinal bacteria on bile pigment.

The liver also inactivates poisonous substances that enter the bloodstream. For example, ammonia produced by intestinal bacteria could poison the body if it were not detoxified by the liver. The liver also chemically alters and excretes many medications; if the liver is damaged, as it might be by a viral infection, lower medication doses might be needed to avoid the buildup of toxic levels.

The Large Intestine

The main function of the large intestine is to recycle water and to absorb nutrients. Because the small intestine absorbs so much water, only 300 to 1,000 ml of fluid normally reach the large intestine per day. From this volume of fluid, the large intestine absorbs water, electrolytes, vitamins, and amino acids. The semisolid feces, composed of indigestible material and bacteria, remain. Infection of the large intestine can interfere with absorption, and stimulate the painful peristaltic contractions known as "stomach cramps."

MICROCHECK 24.1

The alimentary tract is a major route for pathogenic microorganisms to enter the body. People with poor saliva production risk severe tooth decay. Infections of the esophagus are so unusual in normal people that their occurrence suggests immunodeficiency. The stomach is responsible for acid digestion of food, the small intestine for alkaline digestion and absorption of nutrients. Undigested material becomes feces in the large intestine.

- What makes the surface area of the small intestine so large, when it is only 8 to 9 feet long?
- What is the main function of the large intestine?

Normal Flora

The normal flora is important in protecting the body against invasion by pathogens. This section focuses mainly on flora of the oral cavity and intestine. The esophagus has a relatively sparse population, consisting mostly of bacteria from the mouth and upper respiratory tract. When empty of food, the normal stomach is devoid of microorganisms because they are killed by the action of acid and pepsin. ■ normal flora, p. 451

The Mouth

A variety of bacterial species live on the mouth's mucous membranes and teeth, and in saliva. The ecology of this community of microorganisms is complex. Of all the species of bacteria introduced into the mouth from the time of birth onward, relatively few can colonize the oral cavity. Streptococcal species, Gram-positive, chain-forming cocci that produce lactic acid as a by-product of carbohydrate metabolism, are the most numerous. One species of *Streptococcus* preferentially colonizes the upper part of the tongue, another colonizes the teeth, while still another colonizes the mucosa of the cheek. Streptococci and other bacteria have pili that attach specifically to receptors on host tissues, allowing the microorganisms to resist the scrubbing action of food and the tongue and the flushing action of salivary flow. The fact that the cells lining different parts of the mouth have differing receptors accounts for the distribution of the various species of streptococci. The host limits the numbers of bacteria on its mucous membranes by constantly shedding the superficial layers of cells and replacing them with new ones; the rate of this shedding correlates with the number of microorganisms present on the surface.

Because teeth are a non-shedding surface, large collections of bacteria can build up on them in a biofilm. These masses of bacteria, called **dental plaque (figure 24.3)**, form because the bacteria attach to specific receptors on each other or on the tooth, and they may be bound together by extracellular polysaccharides.

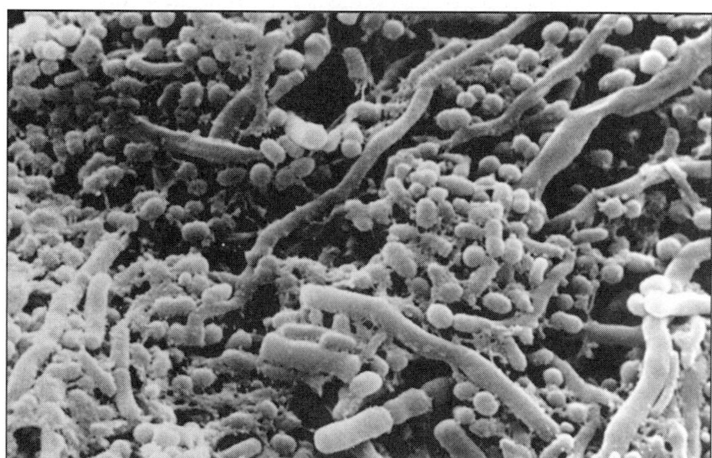

Figure 24.3 Scanning Electron Micrograph of Dental Plaque The many kinds of bacteria composing the plaque exhibit specific attachments to the tooth and to each other and may be bound together by extracellular polysaccharides.

There can be up to 100 billion bacteria per gram of plaque. Metabolic by-products of one species are utilized by another. Plaque organisms consume oxygen, thereby creating conditions that permit the growth of strict anaerobes. In fact, colonization of the mouth by strictly anaerobic flora requires the presence of teeth, because only teeth provide sufficiently anaerobic habitats for growth of these organisms. Dental plaque, gingival crevices, and fissures in the teeth are such habitats.

Intestines

Only small numbers of bacteria live in the upper small intestine because they are continually flushed away by the rapid passage of digestive juices. The predominant organisms are usually aerobic and facultatively anaerobic Gram-negative rods and some streptococci. Lactobacilli and yeasts such as *Candida albicans* are found in small numbers. The bacterial population increases as the intestinal contents move toward the large intestine.

In contrast to the relatively scanty numbers of organisms in the small intestine, the large intestine contains very high numbers of microorganisms, approximately 10^{11} bacteria per gram of feces. These large numbers occur because of the abundance of nutrients in undigested and indigestible food material. Bacteria make up about one-third of the fecal weight. The numbers of anaerobic bacteria, notably including members of the genus *Bacteroides*, generally exceed other organisms by about 100-fold. Of the fecal microorganisms able to grow in the presence of air, facultatively anaerobic, Gram-negative rods, particularly *Escherichia coli* and other enterobacteria, predominate. Fecal organisms are an important source of opportunistic pathogens, especially for the urinary tract. The enormous population of bacteria comprises species that can work together enzymatically to change numerous materials. The currently recommended high-fiber diets contain substances that are indigestible by the gastric and intestinal juices but are readily degraded by intestinal organisms, which often produce large amounts of carbon dioxide, hydrogen, and methane gas. Abdominal discomfort and discharge of intestinal gas, flatus, from the anus is an unfortunate result. Bacterial enzymes can also convert various substances in food to carcinogens and therefore may be involved, along with diet, in the production of intestinal cancer. ■ **opportunistic pathogens, p. 454**

Intestinal bacteria synthesize a number of useful vitamins, including niacin, thiamine, riboflavin, pyridoxine, vitamin B_{12}, folic acid, pantothenic acid, biotin, and vitamin K. These vitamins are important when a person's diet is inadequate.

The normal flora helps prevent colonization of the large intestine by pathogens. A sometimes life-threatening disease called **antibiotic-associated** or **pseudomembranous colitis** can follow antibiotic therapy. The disease is caused by the toxin-producing anaerobic bacterium, *Clostridium difficile*, which readily colonizes the intestine of people whose normal intestinal flora has been reduced by antimicrobial chemotherapy. The toxins of *C. difficile* are lethal to intestinal epithelium and cause small patches called pseudomembranes, composed of dead epithelium, inflammatory cells, and clotted blood, to form on the intestine. Fever, abdominal pain, and profuse diarrhea result. Suppression of intestinal flora with antibacterial medications can also increase susceptibility to other pathogens such as *Salmonella enterica*, to be discussed later.

MICROCHECK 24.2

The distribution of bacterial flora in the mouth is governed by differing receptors on the epithelium. The quantity of flora on mouth epithelium is limited by shedding. The teeth, being a non-shedding surface, allow enormous populations of bacteria, manifest as dental plaque. A properly functioning stomach destroys most microorganisms before they reach the intestine. The normal intestinal flora helps protect the body from infection; disruption of the flora caused by antibiotics can lead to infection. Metabolic activity of intestinal flora produces vitamins, but it causes gas and can contribute to cancer.

- Why must teeth be present in order for anaerobic bacteria to colonize the mouth?
- What is the advantage to the body of having pepsin becoming active only on exposure to acid?
- Why is it that the tongue and cheek epithelium doesn't provide sufficiently anaerobic conditions for plaque anaerobes to grow?

UPPER ALIMENTARY SYSTEM INFECTIONS

Bacterial Diseases of the Upper Alimentary System

It may be surprising that the most common bacterial disease of human beings occurs in the mouth, and that bacterial infections of the stomach are not only common but can lead to ulcers and cancer. These infections often go unnoticed for years but have consequences both locally and for the rest of the body. Oral flora, for example, can enter the bloodstream during dental procedures and cause subacute bacterial endocarditis, and some studies suggest that chronic gum infections play a role in hardening of the arteries and arthritis. The most important bacterial diseases of the upper alimentary system concern the teeth, the gums, and the stomach. ■ **subacute bacterial endocarditis, p. 718**

Tooth Decay (Dental Caries)

Dental caries, generally known as tooth decay, is the most common infectious disease of human beings. The cost to restore and replace teeth damaged by this disease is estimated to be about $20 billion per year in the United States. Dental caries is the main reason for tooth loss. In the different states of the United States, from 14% to 47% of people aged 65 or older have lost all their teeth.

Symptoms

Dental caries is usually far advanced before any symptoms develop. The severe throbbing pain of a toothache is often the first symptom. Sometimes there is a detectable roughness or defect in a tooth, and a tooth can break during chewing.

Causative Agent

Dental caries is caused principally by *Streptococcus mutans* and closely related species. The bacteria live only on the teeth and cannot colonize the mouth in the absence of teeth. They produce lactic acid as a by-product of their metabolism of sugars, and unlike many other bacteria, they thrive under acidic conditions below pH 5. Another important feature is that they produce insoluble extracellular **glucans** from sucrose, but not from other sugars. Glucans, polysaccharides composed of repeating subunits of glucose, are essential for the production of dental caries on smooth tooth surfaces.

Pathogenesis

The first step in the formation of a **cariogenic plaque**, meaning a plaque that causes tooth decay, is the adherence of oral streptococci to specific receptors on the tooth pellicle. The pellicle is a thin film of proteinaceous material adsorbed on the tooth from the saliva. Other species of bacteria attach specifically to earlier arrivals. Two species that might not attach to each other may attach to a third. If dietary sucrose is present, *S. mutans* attaches to the bacterial mass and produces glucans from sucrose through the action of extracellular enzymes. Sucrose is split by the enzymes to the monosaccharides glucose and fructose. The glucose is polymerized, yielding glucan, and the fructose is metabolized, producing lactic acid. The glucans bind the organisms together and to the tooth, and make the plaque impenetrable to saliva.

When sugar enters the mouth, the pH of cariogenic plaques drops from its normal value of about 7 to below 5 within minutes (**figure 24.4**); with more than a 100-fold increase in the acidity of the plaque, the calcium phosphate of teeth dissolves. The duration of this acidic state depends on how long the teeth are exposed to sugars and on the concentration of the sugars. After food leaves the mouth, the pH of the plaque rises slowly to neutrality. The delay in return of the pH to neutrality is due to the ability of *S. mutans* to store a portion of its food as an intracellular, starchlike polysaccharide that is later metabolized with the production of acid. Cariogenic plaque thus acts as a tiny, acid-soaked sponge closely applied to the tooth. Both *S. mutans* and a suitable sucrose-rich diet are required to produce dental caries on smooth surfaces of the teeth. In deep fissures or pits in the teeth, plaque can accumulate in the absence of *S. mutans* but still be cariogenic as long as lactic acid–producing bacteria and fermentable substances are present.

Epidemiology

Dental caries is worldwide in distribution, but the incidence varies markedly depending mainly on dietary sucrose and access

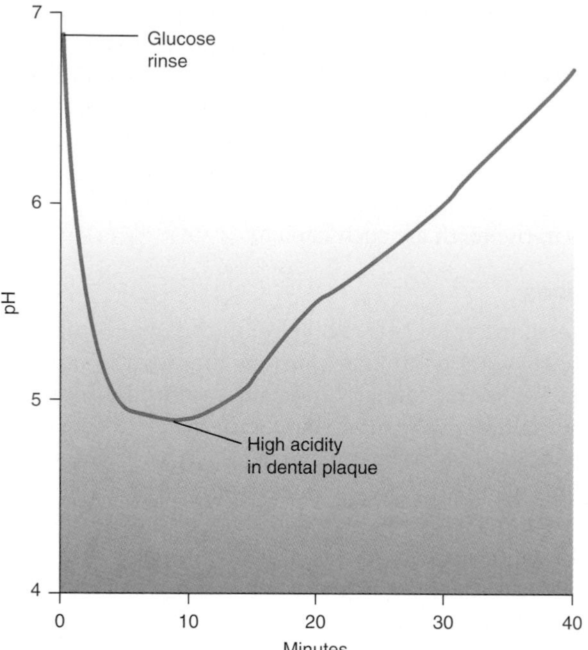

Figure 24.4 Increase in Acidity in Cariogenic Dental Plaque After Rinsing the Mouth with a Glucose Solution Tooth enamel begins to dissolve at about pH 5.5, continuing as pH decreases.

to preventive dental care. Heredity also plays an important role, some individuals inheriting resistance to the disease. Young people are generally much more susceptible than older people probably because the pits and fissures that are sites for dental caries wear down with time.

Prevention and Treatment

The most important method for controlling dental caries is restricting sucrose and other refined dietary carbohydrates, thereby reducing *Streptococcus mutans* colonization of teeth and acid production by cariogenic plaques. Dental caries can be reduced by 90% if sucrose-containing sweets are eliminated from the diet. It is not, however, simply the quantity of sugar in the diet that is important. The frequency of eating and the length of time food stays on the teeth is more critical than the actual quantity of sucrose ingested. Interestingly, chewing paraffin or sorbitol-sweetened gum reduces dental caries, probably because it increases the flow of saliva.

Trace amounts of fluoride are required for teeth to resist the acid of cariogenic plaques. Fluoride makes tooth enamel harder and more resistant to dissolving in acid. In the United States, more than 100 million people are currently supplied by fluoridated public drinking water, which has resulted in a 60% reduction in dental caries. In areas where fluoridated drinking water is not available, fluoride tablets or solutions can be used. To have optimum effect, children should begin receiving fluoride before the permanent teeth erupt. Fluoride applied to tooth surfaces in the form of mouthwashes, gels, or toothpaste is generally less effective.

Mechanical removal of plaque by toothbrushing and use of dental floss is another important preventive measure, reducing the incidence of dental caries by about 50%. Toothbrush bristles cannot remove plaque from the pits and fissures normally present in children's teeth, however, because they are too deep and narrow. Caries commonly develops in these pits and fissures and can be prevented by using a **sealant**, a kind of epoxy glue that seals the fissures, kills the plaque, and prevents bacterial recolonization. Older people have less fissure-related caries; however, receding gums can expose root surfaces that become sites for dental caries. Treatment requires drilling out the carious lesion, filling the defect with **amalgam** or other material, and restoring the contour of the tooth.

Periodontal Disease

Periodontal disease is a chronic inflammatory process involving the gums and tissues around the roots of the teeth. It usually develops slowly over many years, and it is an important cause of tooth loss from middle age onward.

Symptoms

The majority of individuals with periodontal disease are asymptomatic. Common symptoms are bleeding gums, sensitivity of the gums, bad breath, and loosening of the teeth. Discoloration, ranging from yellowish to black, occurs at the base of the teeth. The gums generally recede and expose the roots of the teeth to dental caries.

Causative Agent

Periodontal disease is caused by dental plaque that forms at the point where the gum joins the tooth. The plaque may or may not be cariogenic. Hundreds of kinds of bacteria have been identified in plaques associated with periodontal disease.

Pathogenesis

Plaque forms on teeth at the gum margin, especially in hard to clean areas between the teeth. Calcium salts deposited in the plaque result in hard to remove **dental calculus**, often referred to as tartar. Plaque gradually extends into the gingival crevice and bacterial products incite an inflammatory and immune response manifested by swelling and redness of the gingiva (**figure 24.5**). This gum inflammation is known as **gingivitis**. If the plaque remains small, polymorphonuclear leukocytes along with cellular and humoral immunity limit the process at this stage. Large populations of microorganisms, however, release the enzymes collagenase and hyaluronidase, which weaken the gingival tissue and cause the gingival crevice to widen and deepen. As the plaque enlarges, the proportion of anaerobic Gram-negative bacteria such as *Porphyromonas gingivalis* increases. These organisms release endotoxin and a variety of exotoxins that attack leukocytes and host tissue, but they generally do not invade the tissues of the host. The membrane that attaches the root of the tooth to the bone weakens, and the bone surrounding the tooth gradually softens. The tooth becomes loose and may be lost.

Epidemiology

Periodontal disease is mainly a disease of those over 35 years of age, and after age 65 almost 90% of individuals have some degree of periodontal disease. Persons with AIDS and other immunodeficiencies, and those with defective polymorphonuclear leukocytes, often have severe periodontal disease.

Prevention and Treatment

Careful flossing and toothbrushing can prevent periodontal disease, especially if combined with twice yearly polishing and removal of calculus at a dental office. Periodontal disease can be treated in its earlier stages by cleaning out the inflamed gingival crevice and removing plaque and calculus. In advanced cases, surgery is usually required to expose and clean the roots of the teeth.

Trench Mouth

Trench mouth (**figure 24.6**), also known as Vincent's disease, or acute necrotizing ulcerative gingivitis (ANUG), is a severe, acute condition distinct from other forms of periodontitis . The disease was rampant among soldiers living in trenches during World War I because they were unable to attend to mouth care—thus its name.

Symptoms

Trench mouth is characterized by an abrupt onset, fever, bleeding and painful gums, and a foul odor.

Causative Agent

The suspected causative agent is an oral unculturable spirochete of the *Treponema* genus, antigenically related to *Treponema pallidum*, the cause of syphilis. These trench mouth spirochetes probably act synergistically with other anaerobic species, with which they are always associated. ■ **syphilis, p. 647**

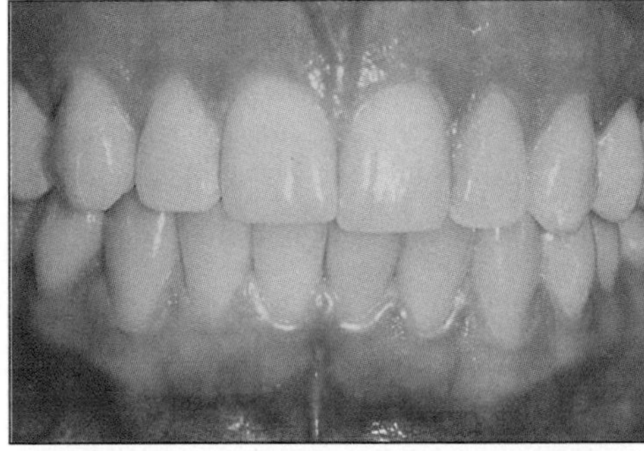

(a)

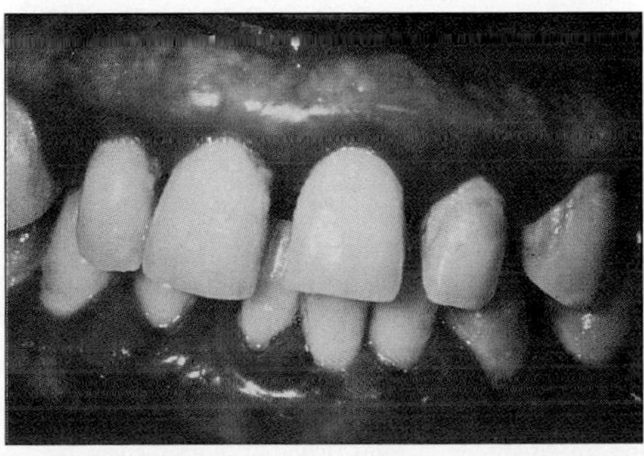

(b)

Figure 24.5 Periodontal Disease **(a)** Normal gingiva. **(b)** Periodontal disease, with plaque, inflammatory changes, bleeding, and shortening of the gingiva between the teeth.

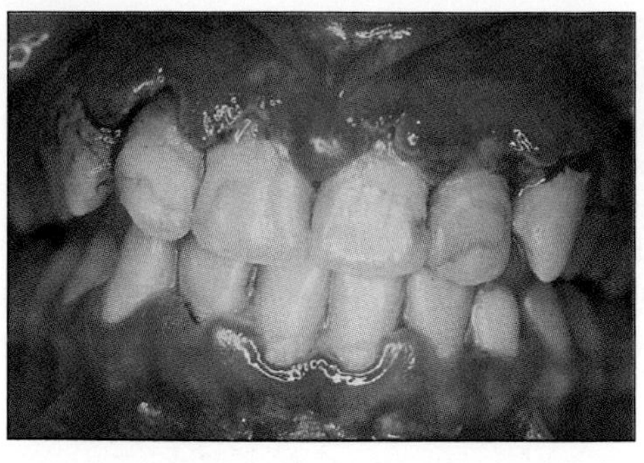

(a)

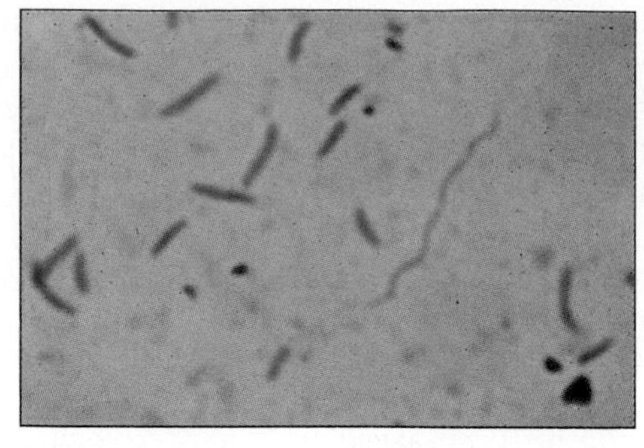

(b)

Figure 24.6 Trench Mouth (Acute Necrotizing Ulcerative Gingivitis, or ANUG) **(a)** Red, swollen gingiva with loss of tissue, especially between the teeth. **(b)** Gram stain of exudate showing a spirochete and rod-shaped bacteria.

Pathogenesis

The spirochetes and the other anaerobes are presumed to act together to destroy tissue, but the precise mechanisms are unknown. Plaque is always present, but its bacterial composition shows much larger numbers of spirochetes and other anaerobes than in chronic periodontal disease. The spirochetes invade the tissue causing necrosis and ulceration, mainly of the gums between the teeth.

Epidemiology

The disease can occur at any age in association with poor mouth care, especially with stress, malnutrition, or immunodeficiency. It is not contagious.

Prevention and Treatment

Control of plaque by daily brushing and flossing, and twice yearly professional cleaning are preventive. Antibacterial treatment directed against the spirochetes and anaerobic rods rapidly relieves the acute symptoms, but this must be followed by extensive removal of plaque and calculus.

The main features of teeth and gum infections are presented in **table 24.1**.

Helicobacter pylori Gastritis

Gastritis, meaning inflammation of the stomach, is commonly present in otherwise healthy asymptomatic people. It was not until the early 1980s that a bacterial cause of this condition was

TABLE 24.1 Important Infections of the Teeth and Gums

	Dental Caries	Periodontal Disease	Trench Mouth
Symptoms	None until advanced disease. Late: discoloration, roughness, broken tooth, throbbing pain	Most cases asymptomatic until advanced disease. Bleeding, sensitive gums, bad breath, loosening of the teeth. Receding gums with exposed discolored tooth roots	Abrupt onset of fever, painful bleeding gums, and a foul mouth odor
Incubation period	1 to 24 months before cavity is detectable	Months or years	Undetermined
Causative agent	Dental plaque populated with *Streptococcus mutans* or similar species	Dental plaque, cariogenic or not	Probably a spirochete of the genus *Treponema* acting with *Fusobacterium, Prevotella,* or other anaerobes
Pathogenesis	Plaque produces acid from dietary sugars; slowly dissolves the calcium phosphate crystals composing the tooth	Plaque forms at the gum margins and gradually extends into the gingival crevices. Bacterial products incite an inflammatory response. The crevices widen and deepen, and the proportion of anaerobes increases. Toxins and enzymes weaken the tissues holding the teeth and cause them to become loose	The spirochetes and certain other anaerobes act synergistically to cause death of tissue, ulceration, and tissue invasion by spirochetes
Epidemiology	Worldwide distribution, incidence depending on dietary sucrose, natural or supplemental fluoride. The young are more susceptible than the old	Primarily a disease of those older than 35 years. Immunodeficient individuals are at increased risk of severe disease	All ages are susceptible in association with poor mouth care, malnutrition, or immunodeficiency. It is not contagious
Prevention and treatment	Restriction of dietary sucrose, supplemental fluoride, mechanical removal of plaque, sealing pits and fissures in childhood teeth	Avoid buildup of plaque. Surgical treatment in severe cases to expose tooth roots and remove plaque and calculus	Daily attention to removal of plaque and professional cleaning to remove calculus. Antibiotic treatment acutely, followed by treatment of periodontal disease

identified and its association with ulcers and gastric malignancy became apparent. These ulcers occur in the stomach and uppermost part of the duodenum, and they are localized, roughly circular erosions of the epithelium that can extend deeply into the underlying tissue.

Symptoms

The initial infection sometimes causes symptoms ranging from belching and mild abdominal distress to vomiting. Most people, however, have no symptoms associated with the infection except when it is complicated by ulcers or cancer. Localized abdominal pain, tenderness, and bleeding are manifestations of these complications.

Causative Agent

Helicobacter pylori is a short, spiral Gram-negative bacterium with multiple unusual-appearing sheathed polar flagella **(figure 24.7)**. Cultivation requires a special medium, one reason it took so long to identify the organism.

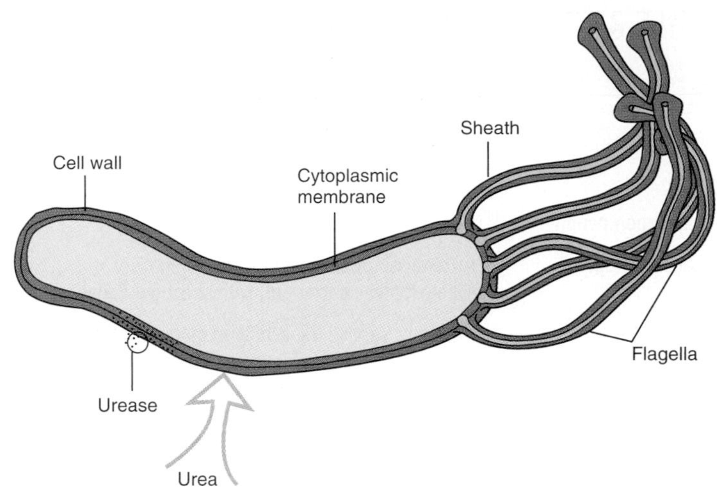

Figure 24.7 ***Helicobacter pylori*** This bacterium is unusual in having sheathed flagella with terminal knobs. It produces a powerful urease that allows it to survive stomach acid by converting urea to ammonia, which neutralizes acidity.

Pathogenesis

These remarkable organisms survive the extreme acidity of the stomach because of their powerful urease. This enzyme creates an alkaline microenvironment by hydrolyzing urea to ammonia. Urea is a waste product of protein catabolism by the body's cells and is normally present in the gastric juices.

Once the bacteria reach the mucus that coats the stomach or intestinal lining, they use their flagella to corkscrew through the mucus to the epithelial cells. In this location the pH of the mucus is nearly neutral, and the bacteria attach to the mucus-secreting epithelium or multiply adjacent to it. Bacterial products including at least one exotoxin incite an inflammatory response in the wall of the stomach, and mucus production decreases. Once infection occurs it persists for years, often for life. From 10% to 20% of infected persons develop ulcers; 65% to 80% of patients with gastric ulcers and 95% of those with duodenal ulcers are infected with *H. pylori*. The thinning of the protective mucus layer at the site of infection **(figure 24.8)**

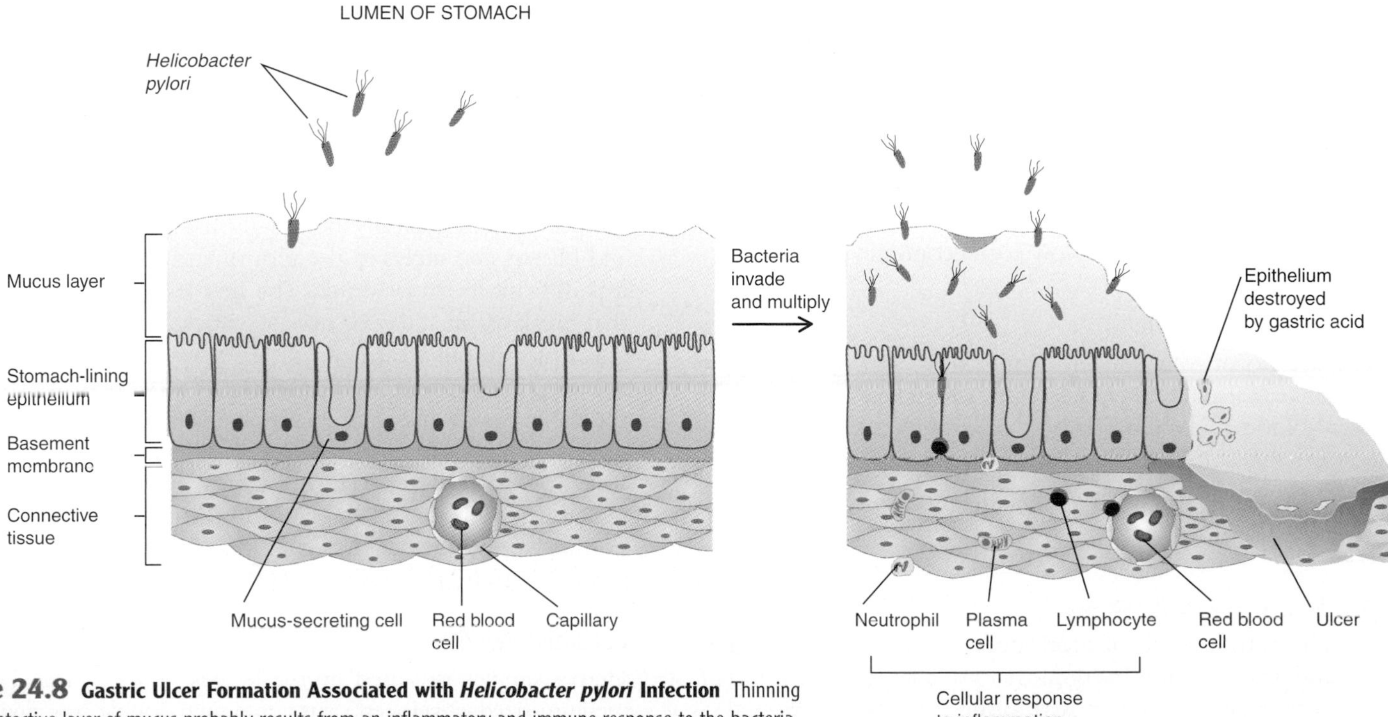

Figure 24.8 Gastric Ulcer Formation Associated with *Helicobacter pylori* Infection Thinning of the protective layer of mucus probably results from an inflammatory and immune response to the bacteria growing on or near the epithelium. The acid and pepsin of the stomach juices can then attack the epithelium. Neutrophils are phagocytic early responders to inflammation; lymphocytes and plasma cells represent the immune response.

TABLE 24.2 *Helicobacter pylori* Gastritis

Symptoms	Initial infection: belching and mild abdominal distress to vomiting. Localized abdominal pain and tenderness, bleeding when complicated by ulcer or cancer
Incubation period	Usually indeterminate
Causative agent	*Helicobacter pylori*, a spiral Gram-negative microaerophilic bacterium, with sheathed flagella
Pathogenesis	Organisms survive the acidity of stomach juices by producing a powerful urease. Upon reaching the layer of mucus, they penetrate to the epithelial surface, where bacterial products incite an inflammatory response. Thinning of the mucus layer occurs, and 10 to 20% of infected individuals develop ulcerations. Only a small percentage develop cancer, but more than 90% of individuals with stomach cancers are infected with *H. pylori*
Epidemiology	Probably fecal-oral transmission. Progressive increase with age, reaching almost 80% of those over 75
Prevention and treatment	No proven preventive. Most infections are cured using two antibiotics together, plus a medication to suppress stomach acid

probably accounts for the development of peptic ulcers of the stomach and duodenum. A very small percentage of individuals infected with *H. pylori* develop cancer of the stomach, but more than 90% of those with stomach cancer are infected by the bacterium.

Epidemiology

Infections tend to cluster in families. Transmission of *H. pylori* probably occurs by the fecal-oral route, and the bacteria have been found in well water. Also, flies are capable of transmitting the organisms. Overall, about 20% of the adult U.S. population is infected with the bacterium, but the incidence progressively increases with age, reaching almost 80% for those over age 75. Infection rates are highest in low socioeconomic groups.

Prevention and Treatment

There are no proven preventive measures. *Helicobacter pylori* infections can usually be eradicated by combined treatment with two antibiotics and a medication that inhibits stomach acid production, with complete clearing of the gastritis and healing of any ulcers.

The main features of *H. pylori* gastritis are shown in **table 24.2**.

MICROCHECK 24.3

The pathogenesis of tooth decay depends on both diet and acid-forming bacteria that are adapted to colonize hard, smooth surfaces within the mouth. Periodontal disease is a chronic condition, caused by plaque at the gum margin, that can lead to loosening of teeth. Trench mouth is an acute illness with destruction of gingival tissue. Chronic infection by *Helicobacter pylori* is a key factor in the development of gastric and duodenal ulcers.

- What enzyme produced by *Heliobacter pylori* assists its survival in stomach acid?
- A sucrose-rich diet is important in tooth decay, but figure 24.4 shows the effect of glucose. Are sucrose and glucose equivalent? Why or why not?
- Discuss the pros and cons of the statement "*Helicobacter pylori* causes stomach cancer."

Viral Diseases of the Upper Alimentary System

Some viral diseases involve the upper alimentary system but produce more dramatic symptoms elsewhere in the body. For example, measles produces Koplik's spots in the mouth and a dramatic skin rash and respiratory symptoms; chickenpox causes oral blisters and ulcers, but a striking skin rash; infectious mononucleosis can cause multiple oral ulcers and bleeding gums, but impressively enlarged lymph nodes and spleen. In this section, we focus on herpes simplex, with its characteristically painful oral ulcers, and mumps, with its enlarged, painful parotid glands. ■ **measles, p. 537** ■ **chickenpox, p. 534** ■ **infectious mononucleosis, p. 726**

Herpes Simplex

Herpes simplex is an extremely widespread disease with many manifestations. In its most common form, it begins in the mouth and throat. Involvement of the esophagus is suggestive of AIDS or other immunodeficiency. The infection persists for life, its causative virus transmissible with saliva. Usually insignificant, the disease can have tragic consequences.

Symptoms

Herpes simplex typically begins during childhood with fever, and blisters and ulcers in the mouth and throat so painful that it is difficult to eat or drink. The first lesions are small blisters that break within a day or two, leaving superficial painful ulcers that heal without treatment within about 10 days. Thereafter, the infection becomes latent and the affected person may suffer the recurrent disease, herpes simplex labialis (the word *labialis* indicating its location on the lips), otherwise known as "cold sores" or "fever blisters" (**figure 24.9**). The symptoms of recurrences usually begin on the lips and include a tingling, itching, burning, or painful sensation. Blisters then appear, followed by painful ulcerations. Healing occurs within 7 to 10 days.

Causative Agent

Herpes simplex is caused by the herpes simplex virus (HSV), a medium-sized, enveloped virus containing double-stranded, linear DNA. There are two types of the virus, HSV-1 and HSV-2. Most oral infections are due to HSV-1; HSV-2 usually causes genital infections. ■ **herpes simplex virus, p. 352, 354** ■ **genital herpes simplex, p. 652**

CASE PRESENTATION

The patient was a 35-year-old man who consulted his physician because of upper abdominal pain. The pain was described as a steady burning or gnawing sensation, like a severe hunger pain. Usually it came on 1 1/2 to 3 hours after eating, and sometimes it woke him from sleep. Generally, it was relieved in a few minutes by food or antacid medicines.

On examination, the patient appeared well, without evidence of weight loss. The only positive finding was tenderness slightly to the right of the midline in the upper part of the abdomen.

A test of the patient's feces was positive for blood. The remaining laboratory tests were normal.

Endoscopy, a procedure which employs a long flexible fiber-optic device passed through the mouth, showed a patchy redness of parts of the stomach lining. A biopsy was taken. The endoscopy tube was passed through the pylorus and into the duodenum. About 2 cm into the duodenum, there was a lesion 8 mm in diameter that lacked a mucous membrane and appeared to be "punched out." The base of the lesion was red and showed adherent blood clot. After the endoscopy, a biopsy portion was placed on urea-containing medium. Within a few minutes, the medium began to turn color, indicating a developing alkaline pH.

1. What is the patient's diagnosis?
2. What would you expect microscopic examination and culture of the gastric mucosa biopsy to show?
3. Outline the pathogenesis of this patient's disease.
4. Why did it take so long for doctors to accept that this condition had an infectious etiology?

Discussion

1. This patient had a duodenal ulcer. The ulcer had penetrated deeply beyond the mucosa, involving small blood vessels and causing bleeding. This was apparent from the clot that was visualized at endoscopy and the positive test for blood in the stool.
2. Microscopic examination of the biopsy showed curved bacteria, confirmed by culture to be *Helicobacter pylori*.
3. *Helicobacter pylori* enters the gastrointestinal tract by the fecal-oral route. In the stomach, they escape the lethal effect of gastric acid because they produce urease. Highly motile, they enter the gastric mucus and follow a gradient of acidity ranging from pH 2 in the gastric juices to pH 7.4 at the epithelial surface. Mutant strains that lack the ability to produce urease are only infectious if they are introduced directly into the mucus layer. Multiplication occurs just above the epithelial surface, but some of the bacteria attach to the epithelial cells and cause a loss of microvilli and thickening at the site of attachment. An inflammatory reaction develops beneath the affected mucosa. Two genes, *vacA* and *cagA*, correlate with virulence. The gene product VacA is a toxin similar to the adenylate cyclase of *Bordetella pertussis*. CagA is not itself toxic but provokes a strong immune response. Once established, *H. pylori* infections persist for years and often for a lifetime. It is not known why some people develop gastric or duodenal ulcers and others do not. Both host and bacterial factors are almost certainly involved. For example, strains of *H. pylori* isolated from peptic ulcer patients tend to be more virulent than those from patients who just have gastritis; patients with blood group O have more receptors for the bacterium and a higher incidence of peptic ulcers than do other people. Stomach acid and peptic enzymes probably play a role in ulcer formation by acting on damaged epithelium unprotected by normal mucus.
4. Claude Bernard, a scientist of Pasteur's time, put it this way: "It is that which we do know which is the greatest hindrance to our learning that which we do not know." In 1983, when Dr. Barry J. Marshall proclaimed before an international gathering of infectious disease experts that a bacterium caused stomach and duodenal ulcers, everyone "knew" it could not be true because no organism was known to exist that could survive stomach acidity and enzymes. Indeed, almost everyone already "knew" the cause of ulcers to be psychosomatic. There is much still to be learned about the cause of ulcers, however, and Bernard's statement remains true.

Pathogenesis

The virus multiplies in the epithelium of the mouth or throat, and it destroys the cells. The blisters that form contain large numbers of infectious virions. Some epithelial cells fuse together, producing large, multinucleated giant cells. Nuclei of infected cells characteristically contain a deeply staining area called an intranuclear inclusion body (see figure 24.9), which is the site of earlier viral replication. Some of the virus particles are carried by lymph vessels to the nearby lymph nodes; an immune response develops and quickly limits the infection. Some of the virions enter the sensory nerves in the area. Viral DNA persists in these nerve cells in a non-infectious non-replicating form. This latent virus can from time to time become infectious, be carried by the nerves to skin or mucous membranes, and produce recurrent disease. Stresses that can precipitate recurrences include menstruation, sunburn, and any illness associated with fever. Small numbers of infectious virions are frequently released into the saliva without any symptoms.

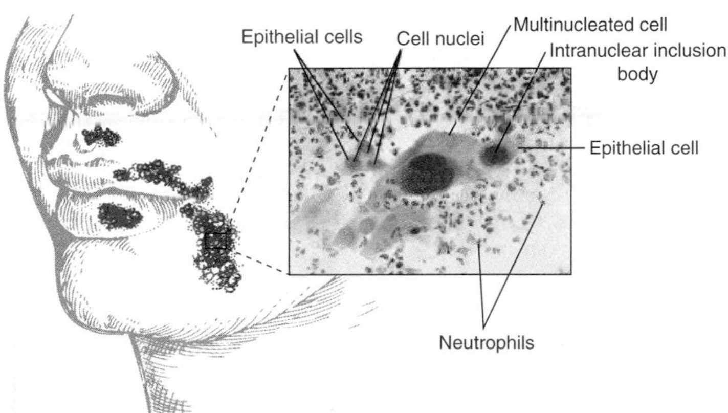

Figure 24.9 Herpes Simplex Labialis, Also Known As Cold Sores or Fever Blisters The photomicrograph is of stained material from a herpes simplex lesion. It shows a multinucleated giant cell and intranuclear inclusion bodies. The pink areas within the epithelial cell nuclei are inclusion bodies, the sites of viral replication.

Epidemiology

HSV is extremely widespread and infects up to 90% of some U.S. inner-city populations, usually resulting in mild, if any, symptoms. An estimated 20% to 40% of Americans suffer recurrent herpes simplex. The virus is transmitted primarily by close physical contact, although it can survive for several hours on plastic and cloth. The greatest risk of infection is from contact with lesions or saliva from patients within a few days of disease onset, because at this time large numbers of virions are present. The saliva of people with no symptoms can be infectious, however, posing a risk to therapists who have contact with saliva, such as nurses and dental workers. HSV can infect almost any body tissue. For example, **herpetic whitlow**, a painful finger infection, is not uncommon among nurses, and wrestlers can develop infections at almost any skin site because saliva containing HSV can contaminate wrestling mats and get rubbed into abrasions. The virus rubbed into the eye can cause blindness, and HSV is the most frequently identified cause of sporadic, viral encephalitis, a serious brain disease.

Prevention and Treatment

Acyclovir and similar medications inhibit HSV DNA polymerase and halt the progress of herpes simplex, but they do not affect the latent virus and thus cannot rid the body of HSV infection. These medications are useful for treating severe cases and for preventing disabling recurrences. Since the ultraviolet portion of sunlight can trigger recurrent disease, sunscreens are sometimes a helpful preventive. ■ **DNA polymerase, p. 171, 197**

Some of the main features of herpes simplex are shown in **table 24.3**.

Mumps

Mumps is an acute viral illness that preferentially attacks glands such as the parotids. Formerly common in the United States, the disease is now rare because of routine childhood immunization.

Symptoms

The onset of mumps is marked by fever, loss of appetite, and headache. Typically these symptoms are followed by painful swelling of one or both parotid glands **(figure 24.10)**. Spasm of the underlying muscle makes it difficult to chew or talk. The word *mumps* probably derives from the verb *mump*, meaning "to mumble" or "whisper." Symptoms of mumps usually disappear in about a week.

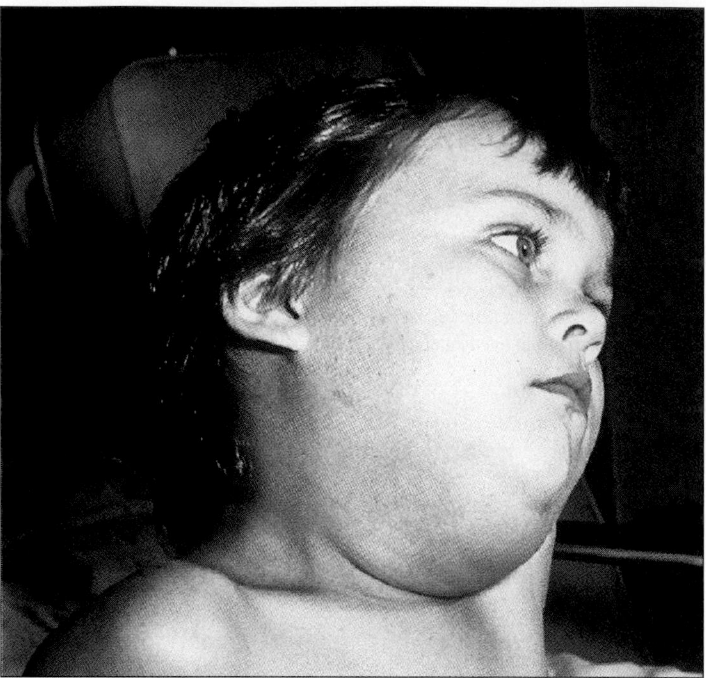

Figure 24.10 A Child with Mumps The swelling directly below the earlobe is due to enlargement of the parotid gland.

Although painful parotid swelling is characteristic of mumps, up to half of the cases of mumps virus infection show no obvious parotid involvement. Symptoms can arise elsewhere in the body with or without parotid swelling. For example, headache and stiff neck indicate that the virus is causing **meningitis**, infection of the coverings of the brain, a common manifestation of mumps virus infection. Generally, mumps symptoms are much more severe in individuals past the onset of puberty. For example, about one-quarter of cases of mumps in postpubertal boys and men are complicated by rapid intensely painful swelling of one or both testicles to three to four times their normal size. Atrophy, or shrinkage, of the involved testicles commonly develops after recovery from the illness, and in rare cases, sterility is a sequel. In women and postpubertal girls, ovarian involvement occurs in about one of 20 cases, manifested by pelvic pain. Pregnant women with mumps commonly miscarry, but birth defects do not result from mumps as they do from rubella. Serious consequences of mumps are rare and are most likely to occur in older people. These consequences include deafness and death from **encephalitis**, or brain infection. ■ **rubella, p. 540**

Causative Agent

Mumps virus is similar in size to herpes simplex virus (about 200 nm diameter) and also has a lipid-containing envelope. Otherwise, the viruses are quite different. Mumps virus is classed in the paramyxovirus family, a group of single-stranded RNA viruses that includes the rubeola and respiratory syncytial viruses. Only one antigenic type of the mumps virus is known.

TABLE 24.3 Herpes Simplex

Symptoms	Initial infection: fever, severe throat pain, ulcerations of the mouth and throat. Recurrences: itching, tingling or pain usually localized to the lip, followed by blisters that break leaving a painful sore, which usually heals in 7 to 10 days
Incubation period	2 to 20 days
Causative agent	Herpes simplex virus (HSV), usually type 1
Pathogenesis	The virus multiplies in the epithelium, producing cell destruction and blisters containing large numbers of infectious virions. An immune response quickly limits the infection, but non-infectious HSV DNA persists in sensory nerves. Associated with a variety of stresses, this DNA becomes the source of infectious virions that are carried to the skin or mucous membranes, usually of the lip, causing recurrent sores
Epidemiology	Widespread virus, transmitted by close physical contact. The saliva of asymptomatic individuals is commonly infectious
Prevention and treatment	Acyclovir, penciclovir, and similar medications that inhibit HSV DNA polymerase can shorten the illness or prevent recurrences. Sunscreens are helpful in preventing recurrences due to ultaviolet exposure

Pathogenesis

Infection occurs when virus-laden droplets of saliva are inhaled by a person who lacks immunity to mumps. The incubation period is long, generally 15 to 21 days, because the virus reproduces first in the respiratory tract, then is spread throughout the body by the bloodstream, and produces symptoms only after infecting tissues such as the parotid glands, meninges, pancreas, ovaries, or testicles. In the salivary glands, the virus multiplies in the epithelium of ducts that convey saliva to the mouth. This destroys the epithelium, which releases enormous quantities of virus into the saliva. The body's inflammatory response to the infection is responsible for the severe swelling and pain. A similar sequence of events occurs in the testicles, where the virus infects the system of tubules that convey the sperm. The marked swelling and pressure often impair the blood supply, leading to hemorrhages and death of testicular tissue. Not much is known about infection of other body structures, but kidney tubules are infected, and the virus can be cultivated from the urine for 10 or more days following the onset of illness. The immune system of the host eliminates the infection, and only rarely does significant permanent damage result.

Epidemiology

Humans are the only natural host of mumps virus, and natural infection confers lifelong immunity. Individuals sometimes claim to have had mumps more than once, probably because other infectious and noninfectious diseases can cause parotid swelling. The virus is spread by the high percentage, about 30%, of individuals who have asymptomatic infections and continue to mingle with other people while infectious. In symptomatic patients, the virus can be present in saliva from almost a week before symptoms appear to 2 weeks afterward, although peak infectivity is from 1 to 2 days before parotid swelling until the gland begins to return to normal size.

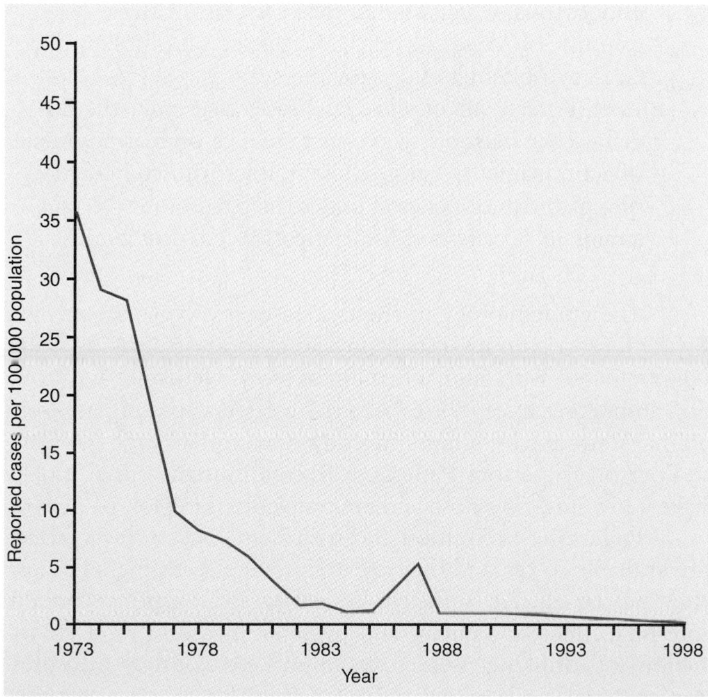

Figure 24.11 Reported Cases of Mumps, United States, 1973 to 1998
Mumps vaccine was licensed in 1967.

Prevention and Treatment

An effective live attenuated mumps vaccine has been available in the United States since 1967. **Figure 24.11** shows how the incidence of mumps has generally declined, although it increased in the 1980s because of a decline in funding for vaccinations. Since its host range is limited to humans, there is only a single viral serotype, and latent recurrent infections do not occur, mumps should be a candidate for eradication just like smallpox, which was eliminated in 1977.

The main features of mumps are presented in **table 24.4**.

MICROCHECK 24.4

Herpes simplex is characterized by acute infection followed by lifelong latency and the possibility of recurrent disease. Infectious virus is often present in saliva in the absence of symptoms. Mumps virus infections characteristically cause enlargement of the parotid glands, but they can involve the brain, testicles, and ovaries, and cause miscarriages. Mumps is a good candidate for eradication.

- What infections are caused by the two types of herpes simplex viruses?
- Why would you expect acyclovir to be ineffective against latent herpes simplex viral infections?
- Why is mumps a good candidate for eradication from the world?

TABLE 24.4 Mumps

Symptoms	Fever, headache, loss of appetite, typically followed by painful swelling of one or both parotid glands. Painful enlargement of the testicles, pelvic pain in women, and symptoms arising from brain involvement are likely to occur in individuals past the age of puberty
Incubation period	Generally 15 to 21 days
Causative agent	Mumps virus, a single-stranded RNA virus of the paramyxovirus family
Pathogenesis	The virus initially replicates in the upper respiratory system, then disseminates to the parotids and other organs of the body via the bloodstream. The inflammatory response to cell destruction is responsible for the marked swelling and pain
Epidemiology	Humans are the only source of the virus. Transmission is favored by a high percentage of asymptomatic infections
Prevention and treatment	An effective live attenuated vaccine has been used since 1967. No antiviral therapy is available

LOWER ALIMENTARY SYSTEM INFECTIONS

Bacterial Diseases of the Lower Alimentary System

Each year, in developing countries, one out of five children dies of diarrhea before the age of 5 years; most of the 5 million individuals who die of diarrhea each year are infants, but no age group is spared. Fatal cases are less common in the United States than in the developing world, but there are millions of diarrhea cases annually. Diarrheal illness is a common result of bacterial infection of the intestine, but bacteria can also use the intestine as an entryway to the rest of the body, thereby causing other types of illness. Because of the wide variety of bacterial intestinal diseases, we will first consider some generalities.

The all too familiar symptoms of lower alimentary system disease include diarrhea, loss of appetite, nausea and vomiting, and sometimes fever—or any combination of these symptoms can occur. Members of the medical profession often loosely ascribe these symptoms to **gastroenteritis** (*gastro-*, "stomach," *entero-*, "intestine," *-itis*, "inflammation"), while others prefer the term "stomach flu." Diarrhea can be copious and watery with infections of the small intestine, or in smaller amounts containing mucus, pus, and sometimes blood when the large intestine is involved. The name **dysentery** is given to diarrheal illnesses when pus and blood are present in the feces.

A minority of bacterial pathogens that invade the body from intestinal infections can cause severe headache, high temperature, abscesses throughout the body, intestinal rupture, shock, and death. This kind of illness is called **enteric fever** and is generally seen after infection by certain strains of *Salmonella*. Typhoid fever is an example.

Various causative agents can be responsible for these symptoms, including both poisonings by microbial toxins in food, foodborne intoxication, and infections. Chapter 32 gives examples of microbial poisonings, while this chapter will focus on infectious causes. Bacterial causes will be described here, and viral, protozoal, and multicellular parasitic causes will be discussed later. ■ **foodborne intoxication, p. 814**

Only three kinds of bacteria account for almost all bacterial intestinal infections:

- *Vibrio* species. These are salt-tolerant, curved or straight, Gram-negative rods associated with the sea.
- *Campylobacter jejuni*. This is a small Gram-negative comma- or S-shaped microaerophilic organism, commonly contracted from fowl. Not surprisingly, it has a high optimal growth temperature corresponding to the high body temperatures of chickens and other fowl.
- Enterobacteria. Three very closely related genera of enterobacteria, *Salmonella*, *Shigella*, and *Escherichia*, contain most of the pathogens in this group. ■ **enterobacteria, p. 282**

The pathogenesis of bacterial intestinal disease involves a number of mechanisms. Different strains within the same species can differ in the way they cause disease, and the same strain can employ more than one mechanism. Moreover, the responsible genes can be transferred by conjugation or by bacteriophages. Microbiologists are discovering exciting details at the molecular level of how bacteria cause disease, and finding unexpected complexity in these seemingly simple creatures. Many of these pathogens, upon contact with a host cell, activate a type III **secretion system** that employs a structure resembling a short flagellum to join the bacterial cytoplasm with that of the host cell (**figure 24.12**). Gene products from pathogenicity islands then pass directly into the host cell and activate endocytosis or other processes to the benefit of the bacterium. ■ **pathogenicity islands, p. 6, 186, 465**

The main pathogenic mechanisms can be summarized as follows:

- Attachment. Infecting bacteria must have a means of attaching to intestinal cells to avoid being swept away by the flow of intestinal contents. Initial attachment is often by means of pili, but other adhesins are usually required for intimate attachment. The secretion system sometimes transfers substances into the host cell that become receptors for bacterial attachment.
- Toxin production. Toxins involved in intestinal infections fall into two groups: (1) Toxins that increase secretion of water and electrolytes; (2) Toxins that cause cell death by halting protein synthesis, as seen classically with *Shigella dysenteriae*.
- Cell invasion. The bacterium activates the process of endocytosis, which takes it into the cell. In the classic example of *Shigella* species, the bacteria multiply in the cell's cytoplasm and destroy the cell.
- Loss of microvilli or villi. Substances that enter the host cell via the bacterial secretion system cause rearrangments of actin filaments, causing loss of microvilli and creation of a platform or pedestal under the bacterium. Certain strains of *E. coli* cause these effects (see figure 24.12).

The epidemiology of these diseases involves transmission by the fecal-oral route, usually ingesting food or drinking water contaminated with animal or human feces. Generally, with certain important exceptions, bacterial intestinal pathogens from animal sources have a high infecting dose and are not transmitted person to person. Pathogens from a human source usually have a low infecting dose and are transmitted person to person.

Regarding prevention and treatment, only a few vaccines are available so far, and they are of limited effectiveness because they fail to elicit a sufficient secretory IgA response on the intestinal mucosa. Sanitary and hygienic measures, and chlorination of drinking water are important control measures. Antibacterial medications are not helpful for most cases originating from animal sources, and they often prolong the illness because they depress the normal flora. On the other hand, antibacterial medications can be life-saving in cases where the

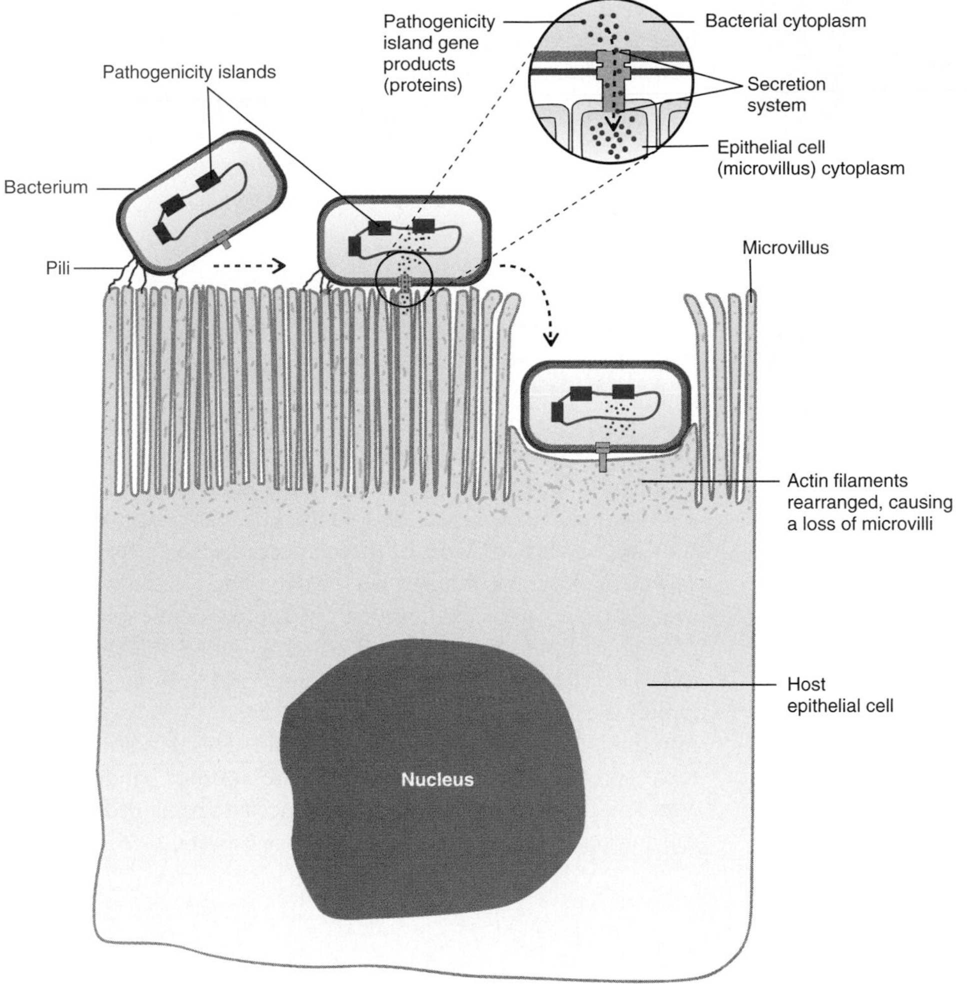

Figure 24.12 Type III Secretion System, an Important Mechanism of Pathogenicity in Gram-Negative Bacteria Following contact of the pili of the bacterium with the host cell, a structure resembling a short flagellum or a microscopic syringe penetrates the cell's cytoplasmic membrane, thereby establishing communication between the bacterial and host cell cytoplasms. Proteins coded by pathogenicity islands then pass directly from the bacterium into the host cell and alter the cell to the benefit of the bacterium. In this case, the transferred proteins act on the actin filaments that normally link together to support the microvilli. Rearrangement of the filaments results in loss of the microvilli.

bacterium invades beyond the intestine. Cases of severe watery diarrhea can cause rapid loss of water and electrolytes, so that the blood becomes reduced in volume and there is insufficient blood flow to keep vital organs, such as the kidneys, working properly. Severe watery diarrhea can be rapidly fatal unless the lost fluid can be replaced promptly. Fortunately, despite the inability of the small intestine to control fluid secretion, it is still able to absorb fluid in most diarrheal diseases. An oral rehydration solution (ORS) that consists of water, glucose, and electrolytes is a highly effective life saver. ■ **IgA antibodies, p. 395**

Cholera

There have been seven cholera pandemics since the early 1800s. Between 1832 and 1836, more than 200,000 Americans died when the second and fourth pandemics of cholera swept across North America. The seventh pandemic began in 1961 in Indonesia, spreading to South Asia, the Middle East, and parts of Europe and Africa. South America had remained cholera-free for a hundred years, until January 1991, when the disease abruptly appeared in Peru. Introduction probably occurred when a freighter discharged bilgewater into a Lima harbor. The municipal water supply was not chlorinated and quickly became contaminated. The disease then spread rapidly, so that in 2 years more than 700,000 cases and 6,323 deaths had been reported from South and Central America. Cases also appeared in 14 states in the United States. ■ **pandemic disease, p. 474**

Symptoms

Cholera is the classic example of severe watery diarrhea. The diarrheal fluid can amount to 20 liters a day and because of its appearance, has been described as "rice water stool." Vomiting also occurs in most people at the onset of the disease, and many people suffer muscle cramps caused by loss of fluid and electrolytes.

Causative Agent

The causative agent is *Vibrio cholerae*, a curved Gram-negative rod able to tolerate strong alkaline conditions and high salt concentrations.

Pathogenesis

Vibrio cholerae is killed by acid, and so large numbers must be ingested before enough survive stomach passage to establish infection. The organisms adhere to the small intestinal epithelium by means of pili and other surface proteins. The organisms multiply on the epithelial cells **(figure 24.13)** but do no visible damage to them. The bacteria, however, produce the potent exotoxin, **cholera toxin**, which is responsible for the symptoms of cholera. As with a number of other bacterial exotoxins, cholera toxin is heat-labile, and its protein molecule is composed of two parts, A and B, a classic A-B toxin **(figure 24.14)**. As in all such toxins, the B fragment has no toxic activity but serves to bind the toxin, irreversibly in this case, to specific receptors on the microvilli of the epithelial cells. The A fragment, responsible for toxicity, causes the activation of the enzyme adenylate cyclase, which converts ATP to cyclic adenosine monophosphate (cAMP). This is another example of ADP ribosylation, in which a toxin causes the ADP ribose part of nicotinamide adenine dinucleotide (NAD) to react with a cellular target. In this case, the target of the toxin is a regulatory substance called a **G protein** that under normal conditions switches back and forth from active to inactive forms to regulate adenylate cyclase activity and, therefore, fluid secretion. ADP ribosylation of this G protein makes it permanently active, thus causing maximal cAMP production by the cell. Accumulation of cAMP in the intestinal cells converts them into little pumps that continuously

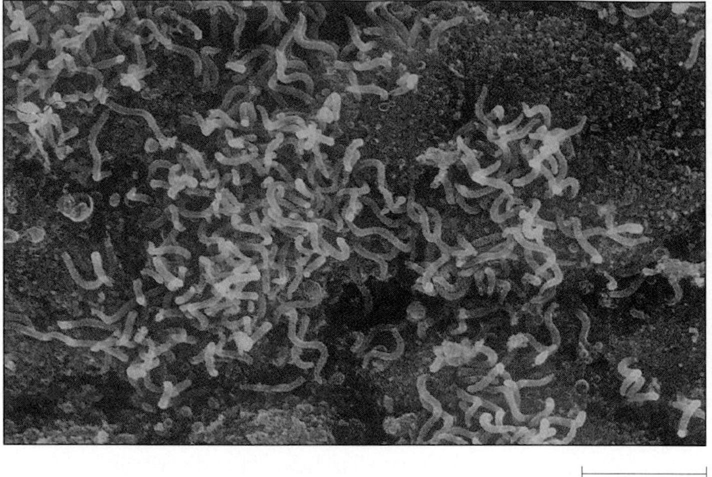
10 μm

Figure 24.13 Scanning Electron Micrograph of *Vibrio cholerae* Attached to Small Intestinal Mucosa Attachment occurs by means of pili.

pour water and electrolytes from the blood into the intestine. Although the large intestine is not affected by the toxin, it cannot absorb the huge volume of fluid that rushes through it, and diarrhea results. Over time, the normal shedding of intestinal cells gets rid of the toxin. ■ A-B toxins, ADP ribosylation, p. 462, 463 ■ NAD, p. 138

In mid-1996, scientists at Harvard Medical School showed that cholera toxin was encoded by a filamentous bacteriophage that infects *V. cholerae* via the same pili by which the bacteria attach themselves to the intestinal cells. Synthesis of both cholera toxin and the pili is regulated by the same bacterial gene, so that both toxin and pili, the two factors required for disease production, are synthesized simultaneously. Synthesis of cholera toxin by lysogenic *V. cholerae* is another example of lysogenic conversion. ■ lysogenic conversion, p. 331

Epidemiology

Fecally contaminated water is the most common source of cholera infection, although foods such as crab and vegetables fertilized with human feces have also been implicated in outbreaks. A person with cholera may discharge a million or more *V. cholerae* organisms in each milliliter of feces. Although cholera is relatively common worldwide, few cases due to the pandemic strain have been acquired in the United States since the early 1900s. Since 1973, however, sporadic cases have occurred along the Gulf of Mexico. Most of these infections have been traced to coastal marsh crabs that had been eaten. These cases were due to a strain of *V. cholerae* different from the pandemic strain. Ominously, in September 1992, a new strain of *V. cholerae*, belonging to a different O group (O139), appeared in India and spread rapidly across South Asia, attacking even those people with immunity to the seventh pandemic strain. This new strain and its rapid spread suggested it could initiate another pandemic, but after about a year it declined in incidence and remained endemic, causing resurgent localized epidemics. ■ O antigen, p. 256

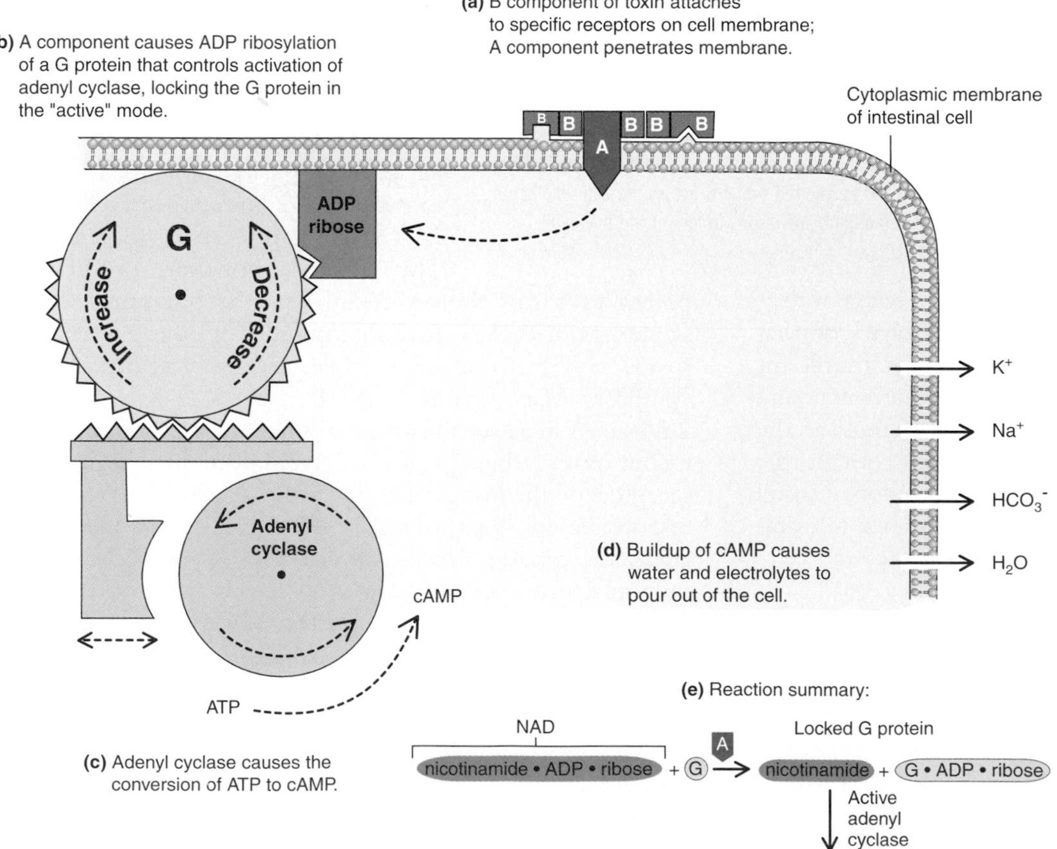
(a) B component of toxin attaches to specific receptors on cell membrane; A component penetrates membrane.
(b) A component causes ADP ribosylation of a G protein that controls activation of adenyl cyclase, locking the G protein in the "active" mode.
Cytoplasmic membrane of intestinal cell
B
B
A
B
B
B
ADP ribose
G
Increase
Decrease
K+
Na+
HCO3-
H2O
Adenyl cyclase
cAMP
ATP
(c) Adenyl cyclase causes the conversion of ATP to cAMP.
(d) Buildup of cAMP causes water and electrolytes to pour out of the cell.
(e) Reaction summary:
NAD
Locked G protein
nicotinamide • ADP • ribose + G → nicotinamide + G • ADP • ribose
A
Active adenyl cyclase
ATP → cAMP

Figure 24.14 Mode of Action of Cholera Toxin As with other A-B toxins, the B portion attaches the toxin to the host cell, and the A portion penetrates the cell and causes toxicity. In this case, the target of the A portion is a G protein responsible for regulating production of cAMP. By splitting off the ADP-ribose portion of NAD and attaching it to G, the toxin makes it impossible for G to down-regulate cAMP production.

Prevention and Treatment

Control of cholera depends largely on adequate sanitation and the availability of safe, clean water supplies. Travelers to areas where cholera is occurring are advised to cook food immediately before eating it. Crabs should be cooked for no less than 10 minutes. No fruit should be eaten unless peeled personally by the traveler, and ice should be avoided unless it is known to be made from boiled water. Orally administered vaccines are commercially available in a number of countries outside the United States. One consists of a living, genetically altered strain of *V. cholerae*, and another, killed *V. cholerae* in combination with the purified recombinant B subunit of cholera toxoid. Treatment of cholera depends on the rapid replacement of electrolytes and water before irreversible damage to vital organs can occur. The prompt administration of intravenous or oral rehydration fluid decreases the mortality of cholera to less than 1%.

The main features of cholera are summarized in **table 24.5**.

Shigellosis

Shigellosis is distributed worldwide wherever sanitary practices are lacking. Reported cases in the United States have averaged about 21,500 per year, but the true prevalence is much higher.

Symptoms

The classic symptom of shigellosis is dysentery. Other symptoms are headache, stiff neck, convulsions, and joint pain. Shigellosis is commonly fatal for infants in developing countries.

TABLE 24.5 Cholera

1. *Vibrio cholerae,* the causative bacterium, enters the mouth with fecally contaminated food or drink
2. The bacteria attach to epithelial cells of the small intestine
3. *V. cholerae* toxin enters the cells and prevents them from down-regulating secretion of water and electrolytes
4. The epithelial cells pump water and electrolytes from the blood into the intestinal lumen, causing watery diarrhea
5. Shock and death occur because of fluid loss from the circulatory system, unless the fluid can be replaced
6. The bacteria exit the body with feces

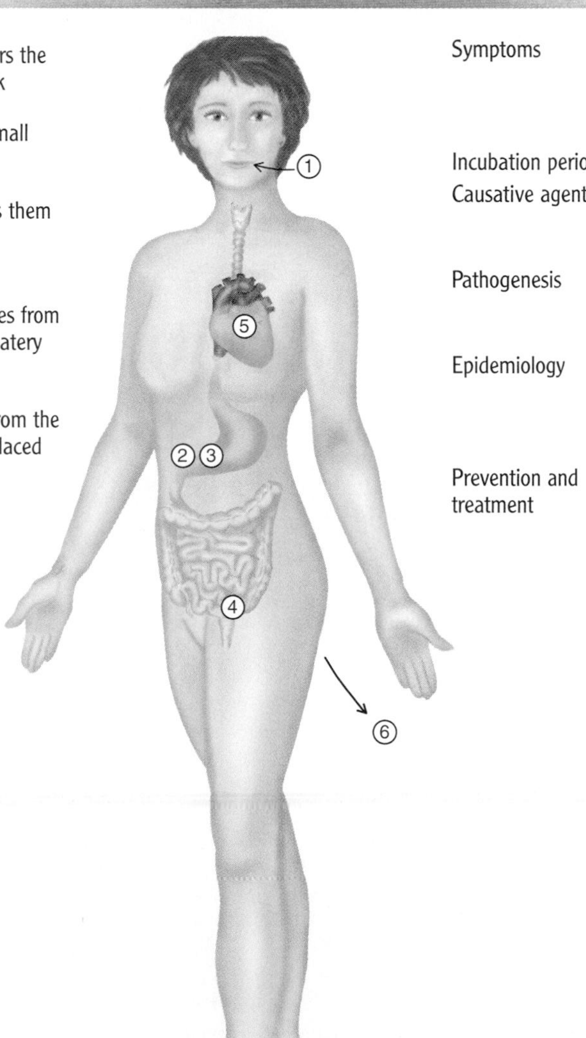

Symptoms	Abrupt onset of massive diarrhea, generally followed by vomiting without nausea, muscle cramps, and when severe, dehydration and shock
Incubation period	Short, generally 12 to 48 hours
Causative agent	*Vibrio cholerae,* a lysogenic, curved, Gram-negative rod bacterium with a polar flagellum
Pathogenesis	Heat-labile exotoxin causes excessive secretion of water and electrolytes by the intestinal epithelium
Epidemiology	Ingestion of fecally contaminated food or water; sometimes natural sources associated with zooplankton and marine crustaceans; few asymptomatic carriers
Prevention and treatment	Purification of water, careful handwashing; vaccination. Treatment: Rehydration with a solution of electrolytes and glucose, given intravenously in severe cases; or similar electrolyte solution containing a glucose source given by mouth in milder cases; an orally administered tetracycline antibiotic can shorten the illness

Perspective 24.1 Ecology of Cholera

The ecology of cholera is fascinating but still incompletely understood. The O139 *Vibrio cholerae* strain arose in India in the fall of 1992 from a serotype O1 strain, the same serotype as the pandemic and American Gulf Coast strains. Interestingly, the change in O antigen was accompanied by acquisition of a capsule, these changes probably resulting from genes acquired by transduction or conjugation. Various strains of *V. cholerae*, most of which are not pathogenic for humans, live in the coastal seas around the world largely in association the zooplankton. These zooplankton can sometimes also harbor the *V. cholerae* strains responsible for pandemic cholera. Moreover, the zooplankton feed on phytoplankton and therefore increase markedly in number when warm seas and abundant nutrients cause explosive growth of phytoplankton. These findings could explain (1) how new pandemic strains could arise through genetic interchange, (2) an association between cholera and climatic changes such as El Niño, and (3) the onset and rapid spread of epidemic cholera by ocean currents along coastal areas. ■ **zooplankton, p. 301, 305**

Causative Agents

There are four species of *Shigella*, *S. flexneri*, *S. boydii*, *S. sonnei*, and *S. dysenteriae*. Virulent strains contain a large plasmid that is essential for the attachment and entry of *Shigella* into host cells.

Pathogenesis

The first step in the infection is phagocytosis of the bacteria by M cells associated with gastrointestinal associated lymphoid tissue (GALT) in the large intestine. These cells transport the bacteria beneath the epithelium, where they are discharged by exocytosis and taken up by macrophages. Instead of destroying the bacteria, the macrophages die and release the organisms. The shigellas attach to specific receptors located at or near the bases of the epithelial cells. Attachment induces the process of endocytosis, which takes the bacteria into the epithelial cells enclosed in a phagosome. Intracellularly, they break out of the phagosome, multiply at a high rate in the cytoplasm, and kill the cells. Even though *Shigella* sp. are nonmotile, they are moved efficiently from cell to cell by a process first described in 1995. A chromosomal gene of the bacterium codes for a protein that causes the host cell's actin filaments to form a progressively lengthening tail attached to one end of the bacterium (**figure 24.15**). Actin tails push the bacteria from one cell to another. The final result of the infectious process is sloughed areas of epithelium surrounding lymphoid tissues. The denuded areas are intensely inflamed, pus covered, and bleeding, the source of blood and pus in the diarrheal stool. ■ **actin, p. 77, 459** ■ **GALT, p. 374**

Some strains of *Shigella dysenteriae*, a species rarely encountered in the United States, produce a potent toxin known as the **Shiga toxin**. It is a chromosomally coded A-B toxin structurally similar to cholera toxin but otherwise unrelated. The toxin does not cause ADP ribosylation. Instead, the A subunit reacts with the host cell ribosome, thereby halting protein synthesis. The importance of this toxin is that it has a strong association with the **hemolytic uremic syndrome**, an often fatal condition that can follow *Shigella dysenteriae* dysentery. In this syndrome, red blood cells break up in the tiny blood vessels of the body, resulting in anemia and kidney failure, sometimes accompanied by paralysis or other signs of nervous system injury. Shiga toxin is also produced by strains of *Escherichia coli* that cause the hemolytic uremic syndrome.

Epidemiology

Shigellosis generally has a human source. The bacteria are not easily killed by stomach acid, so that the infectious dose is small and as few as 10 organisms can initiate infection. Transmission occurs most readily in overcrowded populations with poor sanitation. It is a common problem in day care centers and among homosexual men. Fecally contaminated food and water have also caused outbreaks. Fortunately, the most virulent species, *Shigella dysenteriae*, is also the least prevalent in the United States.

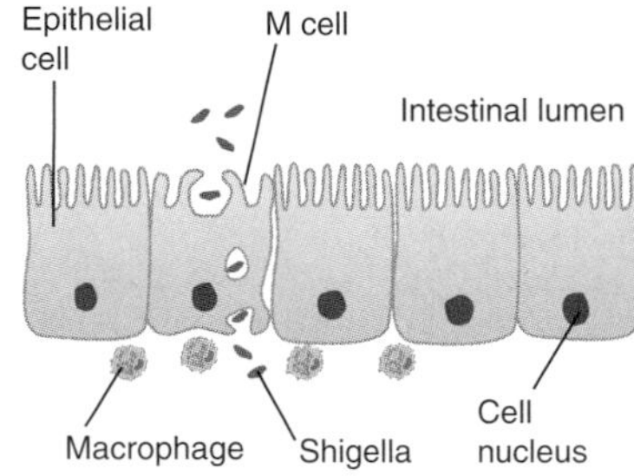

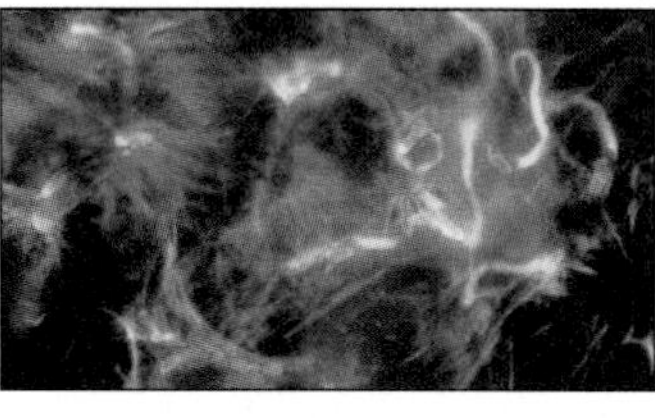

(1) Shigellas are taken up by M cells and transported beneath the epithelium.

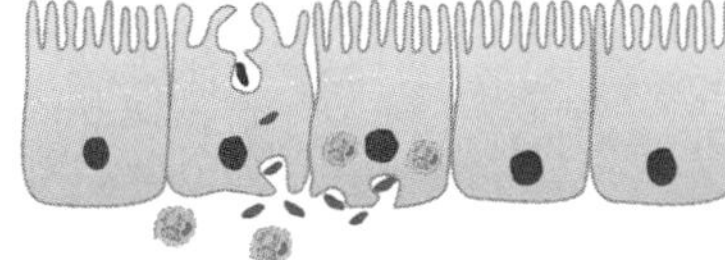

(2) The bacteria enter the inferior and lateral aspects of the epithelial cells by inducing endocytosis. The endosomes are quickly lysed, leaving the shigellas free in the cytoplasm.

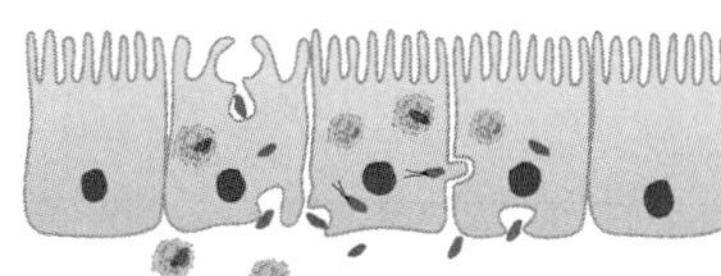

(3) Actin filaments quickly form a tail, pushing the shigellas into the next cell. Macrophages that take up shigellas are killed.

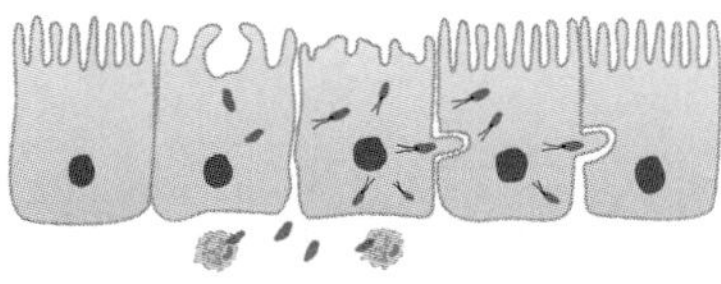

(4) Shigellas multiply in the cytoplasm, and the infection extends to the next cell.

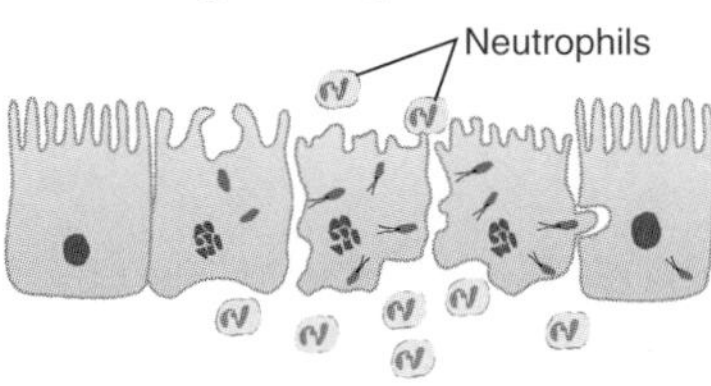

(5) Infected cells die and slough off. Intense response of acute inflammatory cells (neutrophils), bleeding and abscess formation.

Figure 24.15 Pathogenesis of Shigellosis The bacteria do not invade the intestinal epithelium directly, but are transported to the area underneath the epithelial cells by M cells. They then invade the epithelial cells from their inferior and lateral borders. The photomicrograph shows the actin tails (green) that form on intracellular shigellas (orange) and rapidly push these nonmotile bacteria from cell to cell.

Prevention and Treatment

The spread of *Shigella* sp. is controlled by sanitary measures and surveillance of food handlers and water supplies. There is no vaccine. Antimicrobial medications such as ampicillin and cotrimoxazole are useful against susceptible strains because they shorten the duration of symptoms and the time during which shigellas are discharged in the feces. Almost 20% of *Shigella* strains, however, are resistant to these two commonly employed medications. R factor plasmid–mediated antibiotic resistance is common, requiring that sensitivity testing be done to choose the best antimicrobial medication. Two specimens of feces, collected at least 48 hours after stopping antimicrobial medicines, must be negative for *Shigella* sp. before a person is allowed to return to a day care center or food-handling job. ■ **R plasmids, p. 511**

Table 24.6 describes the main features of shigellosis.

Escherichia coli Gastroenteritis

Escherichia coli generally ferments lactose, in contrast to most shigellas and salmonellas. It is an almost universal member of the normal intestinal flora of humans and a number of other animals. Long ignored as a possible cause of gastrointestinal disease, certain strains were shown in 1945 to cause life-threatening epidemic gastroenteritis in hospitalized infants. Later, *E. coli* strains were shown to cause gastroenteritis in adults, notably as agents responsible for traveler's diarrhea (also called "Delhi belly," "Montezuma's revenge," "Turkey trots," etc.). Still later, *E. coli* strains were identified as causative agents in dysentery, cholera-like illnesses, and diarrheas associated with the hemolytic uremic syndrome.

Symptoms

Symptoms depend largely on the virulence of the infecting *E. coli* strain. They range from a few loose bowel movements, to profuse watery diarrhea, to severe cramps and bloody diarrhea. Fever is not usually prominent and recovery usually occurs within 10 days. Hemolytic uremic syndrome, however, develops after some bloody diarrhea cases.

TABLE 24.6 Shigellosis

Symptoms	Fever, diarrhea, vomiting, pus and blood in feces; less frequently, headache, stiff neck, convulsions, and painful joints
Incubation period	3 to 4 days
Causative agent	Species of *Shigella,* Gram-negative, nonmotile enterobacteria
Pathogenesis	Invasion of and multiplication within intestinal epithelial cells; death of cells, intense inflammation and ulcerations of intestinal lining
Epidemiology	Transmission via fecal-oral route; sometimes by fecally contaminated food or water; asymptomatic carriers occur; humans generally the only source
Prevention and treatment	Sanitary precautions including careful handwashing Treatment: Antibacterial medications such as ampicillin and cotrimoxazole (trimethoprim plus sulfamethoxazole) shorten duration of symptoms and time shigellas are excreted; multiple R factor resistances may be present

Causative Agent

Most of the diarrhea-causing *E. coli* fall into four groups: enterotoxigenic (ETEC), enteroinvasive (EIEC), enteropathogenic (EPEC), and enterohemorrhagic *E. coli* (EHEC). In contrast to other strains of *E. coli*, strains in these groups possess virulence factors that allow them to be intestinal pathogens.

Enterotoxigenic *E. coli* (ETEC) is a common cause of traveler's diarrhea and diarrhea in infants. Some members of this group are responsible for significant mortality in young livestock due to diarrhea. ETEC strains usually possess adhesins that allow them to adhere to and colonize the intestinal epithelium, where they secrete one or more toxins. One such toxin is nearly identical to cholera toxin in action and antigenicity.

Enteroinvasive *E. coli* (EIEC) infections result in a disease closely resembling that caused by *Shigella* sp.

Enteropathogenic *E. coli* (EPEC) strains cause diarrheal outbreaks in hospital nurseries and in bottle-fed infants in developing countries and also can cause chronic diarrhea in infants. They possess plasmid-dependent adhesins and cause loss of microvilli and a thickening of the cell surface at the site where the organisms attach. Enteropathogenic *E. coli* strains are easily distinguished from other *E. coli* strains by using fluorescent antibodies to identify their somatic and capsular antigens.

The fourth group of diarrhea-producing strains, enterohemorrhagic *E. coli* (EHEC), was discovered in 1982. Almost all of these organisms belong to a single serological type, O157:H7. They often produce severe illness including bloody diarrhea. This strain has been shown to produce a potent group of toxins that cause the death of intestinal epithelium by interfering with protein synthesis. The toxins are closely related to the Shiga toxins of *Shigella dysenteriae*, mentioned earlier. Unlike shigellas, however, enterohemorrhagic *E. coli* generally do not penetrate intestinal epithelial cells. Their toxin production depends on lysogenic conversion by a distinct bacteriophage. Many infected individuals develop hemolytic uremic syndrome, marked by lysis of red blood cells and kidney failure. Fatalities are common in infants and the elderly. Outbreaks of O157 *E. coli* disease have occurred commonly in Canada and Great Britain, as well as in a number of states in the United States. A large 1993 epidemic in Washington State was traced to inadequately cooked hamburgers served by a fast-food restaurant chain. Mass production of hamburger has created conditions for large outbreaks, and several recalls of processed meat have been necessary. Food and drink contaminated with cattle feces have been the source of a number of epidemics. ■ **enterohemorrhagic *E. coli* epidemic, p. 257**

Characteristics of diarrhea-causing *E. coli* are summarized in **table 24.7**.

Pathogenesis

Of hundreds of different strains of *E. coli*, only those possessing certain virulence factors cause gastrointestinal disease. Some of these factors were discussed in the previous section; there is strong evidence that not all factors have been identified. Two

TABLE 24.7 Characteristics of Diarrhea-Causing *Escherichia coli*

Designation	Characteristic Features	Clinical Picture
Enterotoxigenic *E. coli* (ETEC)	Can release two kinds of toxin, one similar to cholera toxin; small intestinal location	Nausea, vomiting, abdominal cramps, massive watery diarrhea leading to dehydration
Enteroinvasive *E. coli* (EIEC)	Entry into and growth in intestinal epithelium, with cell destruction; large intestinal location	Fever, cramps, blood and pus in the feces
Enteropathogenic *E. coli* (EPEC)	Attachment of the bacterium is followed by loss of microvilli and formation of a platform or pedestal of actin fibrils under the bacterium; small intestinal location	Fever, vomiting, watery diarrhea containing mucus; associated with a limited number of serotypes
Enterohemorrhagic *E. coli* (EHEC)	Same as above, except large intestine location and release of Shigalike toxin	Fever, abdominal cramps, bloody diarrhea without pus; 2% to 7% develop hemolytic uremic syndrome; most cases due to serotype 0157:H7

important virulence factors, enterotoxin production and the ability to adhere to the small intestine, are coded by plasmids. These plasmids can be transferred to other *E. coli* organisms by conjugation, through which virulence is conferred on the recipient strain. Gastroenteritis-producing strains of *E. coli* often have more than one type of virulence plasmid. It is interesting that the heat-labile toxin of *E. coli* is antigenically closely related to the cholera toxin of *Vibrio cholerae*. This similarity suggests that the genes responsible for the two toxins have a common ancestry. ■ **conjugation, p. 208** ■ **plasmids, p. 70, 208, 210**

Epidemiology

Epidemics occur from person-to-person spread, from contamination of foods, unpasteurized milk and juices, and water sources contaminated with feces. Human beings, and domestic and wild animals can all be sources of pathogenic strains.

Prevention and Treatment

Treatment of *E. coli* gastroenteritis includes replacing the fluid lost from vomiting and diarrhea. In addition, infants may require antibiotics such as gentamicin or polymyxin for a few days. Traveler's diarrhea can be prevented with bismuth preparations (such as Pepto-Bismol) or an antibacterial medication such as fluoroquinolone if the *E. coli* strains in the geographic area visited are sensitive to the antibiotic. The widespread use of an antibiotic to prevent diarrhea, however, promotes the appearance of resistant strains by fostering the spread of R plasmids, and thus antibiotics should not be used routinely.

Some features of *E. coli* gastroenteritis are summarized in **table 24.8**.

Salmonellosis

Salmonellosis, a disease caused by bacteria of the *Salmonella* genus, can be contracted from many animal sources. On the average, roughly 47,500 cases are reported per year in the United States, but most cases go unreported, and the actual number is estimated at well over 2,000,000 per year.

Humans are the only source for a few *Salmonella* strains, which are now rare in the United States. These human strains generally cause severe symptoms, as in **typhoid fever**.

Symptoms

Salmonellosis is generally characterized by diarrhea, abdominal pain, nausea, vomiting, and fever. The symptoms vary depending on the virulence of the strain of *Salmonella* and the number of infecting organisms. The symptoms are usually short-lived and mild.

A few *Salmonella* strains cause progressively increasing fever over a number of days, severe headache, and abdominal pain, followed in some cases by intestinal rupture, internal bleeding, shock, and death.

Causative Agents

It is generally recognized that there are only two species of *Salmonella*, *S. enterica* and *S. bongori*, the latter species rarely isolated from humans. The *Salmonella* are subdivided into more than 2,400 serotypes based on differences in their somatic (O), flagellar (H), and capsular (K) antigens. They are referred to as serotypes because the *Salmonella* strains are identified by using serum from the blood of laboratory animals that have been injected with known strains of the bacterium and therefore have antibodies against the strains. Formerly, many of these serotypes were assigned species names; for example, some names, such as Dublin and Heidelberg, reflected the place where they were first isolated. Such names are still commonly used in place of the species name to indicate specific serotypes. (In this discussion we will not italicize the serotype name, to distinguish it from a species name, as in *Salmonella* Dublin and *Salmonella* Heidelberg.) *Salmonella* Typhimurium and *Salmonella* Enteritidis are the serotypes most commonly isolated in the United States. Typhoid fever, an example of enteric fever, is caused by *Salmonella* Typhi.

Pathogenesis

Like *Vibrio cholerae*, most salmonellas are killed by acid, and so a large number must generally be ingested to survive passage through the stomach. Upon reaching the lower small intestine, an adhesin on the bacterial surface attaches the bacteria to specific receptors on the surface of the epithelial cells. Contact with intestinal epithelium activates a type III secretion system. In only a matter of minutes, transfer of bacterial substances into the

TABLE 24.8 *Escherichia coli* Gastroenteritis

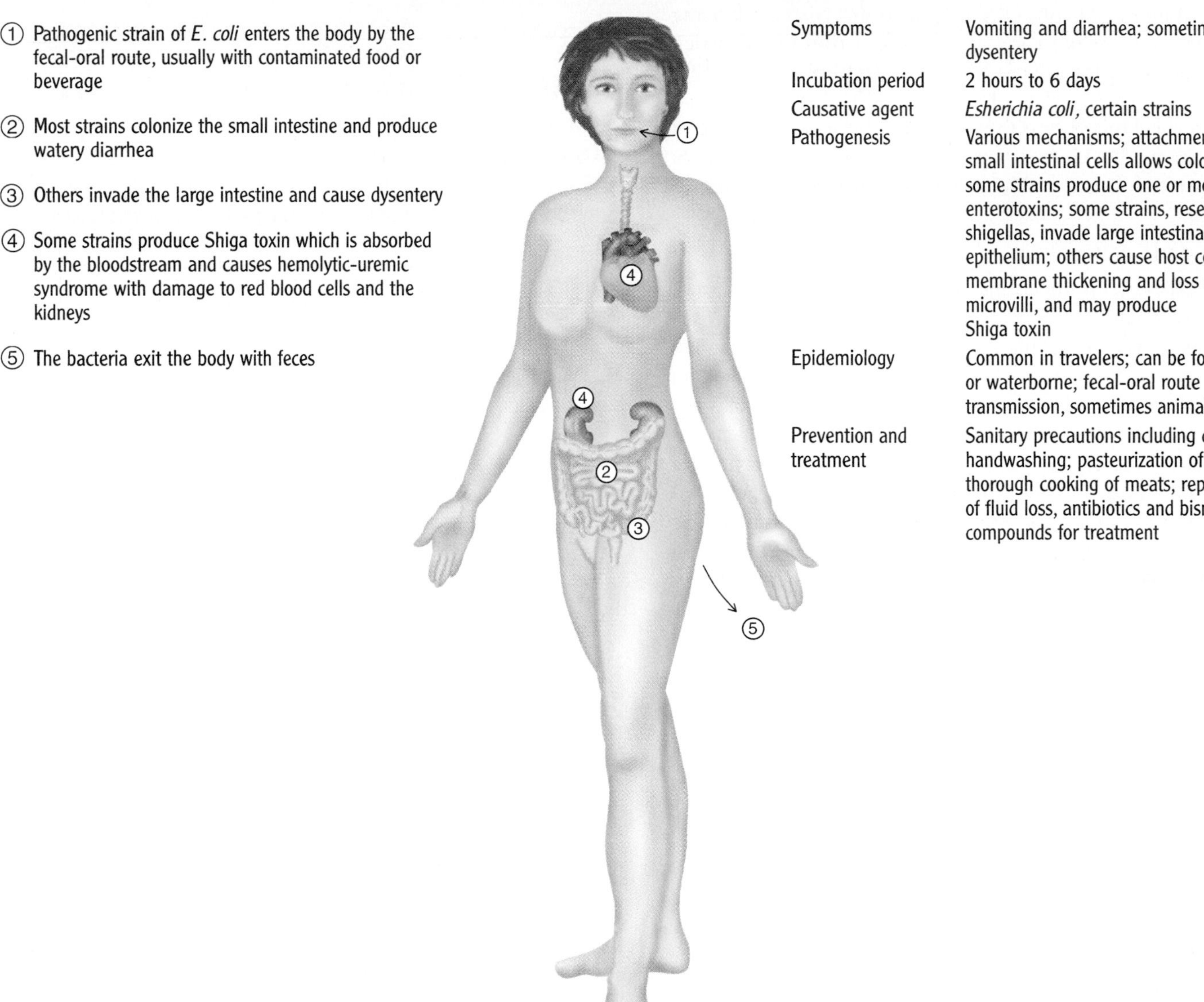

1. Pathogenic strain of *E. coli* enters the body by the fecal-oral route, usually with contaminated food or beverage
2. Most strains colonize the small intestine and produce watery diarrhea
3. Others invade the large intestine and cause dysentery
4. Some strains produce Shiga toxin which is absorbed by the bloodstream and causes hemolytic-uremic syndrome with damage to red blood cells and the kidneys
5. The bacteria exit the body with feces

Symptoms	Vomiting and diarrhea; sometimes dysentery
Incubation period	2 hours to 6 days
Causative agent	*Esherichia coli,* certain strains
Pathogenesis	Various mechanisms; attachment to small intestinal cells allows colonization; some strains produce one or more enterotoxins; some strains, resembling shigellas, invade large intestinal epithelium; others cause host cell membrane thickening and loss of microvilli, and may produce Shiga toxin
Epidemiology	Common in travelers; can be foodborne or waterborne; fecal-oral route transmission, sometimes animal source
Prevention and treatment	Sanitary precautions including careful handwashing; pasteurization of drinks, thorough cooking of meats; replacement of fluid loss, antibiotics and bismuth compounds for treatment

epithelial cell tricks it into taking in the bacterium by endocytosis. The bacteria multiply within a phagosome and are discharged from the base of the cell by exocytosis. In most instances the bacteria are quickly taken up and killed by macrophages, but the inflammatory response to the infection increases fluid secretion, causing diarrhea. ■ **endocytosis, exocytosis, p. 459**

Some strains, such as *Salmonella* Typhi, are not so easily eliminated by host defenses. They resist killing by macrophages, multiply within them, and are carried by the bloodstream throughout the body. Released by death of the macrophages, the bacteria can invade tissues, cause prolonged fever, abscesses, septicemia, and shock, often with little or no diarrhea. *Salmonella* Typhi can also cause destruction of Peyer's patches, leading to intestinal rupture and hemorrhage. ■ **septicemia, p. 716** ■ **Peyer's patches, p. 459**

Epidemiology

Salmonellas survive for long periods in the environment. Children are commonly infected by seemingly healthy pets that discharge salmonellas in their feces, such as turtles, iguanas, baby chickens, and ducks. Eggs and poultry are often contaminated with *Salmonella* strains. Other outbreaks have resulted from contaminated brewer's yeast, alfalfa sprouts, protein supplements, dry milk, and even a red dye used to diagnose intestinal disease. Most cases of *Salmonella* gastroenteritis have an animal source rather than a human source. Enteric fever strains such as *Salmonella* Typhi are generally the exception.

While fecal discharge of gastroenteritis-causing strains is usually of short duration, carriers of *Salmonella* Typhi can eliminate up to 10 billion of the bacteria per gram of their feces for years. The source of these organisms is usually the gallbladder, which *Salmonella* Typhi colonizes free of competition since much of the normal flora is killed or inhibited by concentrated bile. Mary Mallone, "Typhoid Mary," a young Irish cook living in New York state in the early 1900s, was a notorious carrier. She was responsible for at least 53 cases of typhoid fever over a 15-year period. In her day, about 350,000 cases of typhoid occurred in the United States each year. The low incidence of typhoid fever in the United States today can be attributed to improved sanitation and public health surveillance measures.

Prevention and Treatment

Control of *Salmonella* infections depends on reporting cases of salmonellosis, tracing sources, sanitary handling of animal carcasses, treating animal products with pasteurization and irradiation, and testing them for contamination. Adequate cooking effectively kills salmonellas and is very important, especially for frozen fowl. The heat penetration to the center of the carcass may be inadequate to kill salmonellas even when the outside appears "well done." Unfortunately, the number of reported cases of salmonellosis has tended to rise, due in part to mass distributions of food and water that become contaminated from time to time. Most people with salmonellosis recover without antimicrobial treatment; this is fortunate, because *S. enterica* has shown increasing plasmid-mediated resistance to antibacterial treatment. About 35% of *Salmonella* Typhimurium isolates are resistant to five or more antibacterial medications. This resistance is partly due to selection of resistant strains of *Salmonella* by the widespread, ill-advised addition of antibiotics to animal feeds. Antibacterial medications are not advised for most cases of salmonellosis, because they are ineffective and prolong the discharge of the causative strain.

A live attenuated oral vaccine for preventing typhoid fever is about 50% to 75% effective. Surgical removal of the gallbladder is often necessary to rid *Salmonella* Typhi carriers of their infection.

Table 24.9 summarizes some of the features of salmonellosis.

Campylobacteriosis

It was not until 1972 that *Campylobacter jejuni* was isolated from a diarrheal stool. It took another five years for a suitable culture medium to be developed and widely used, which led to recognition of campylobacteriosis as the leading bacterial diarrheal illness in the United States, with an estimated 2.1 to 2.4 million cases per year. There are generally less than 1,000 fatalities per year, mostly in the elderly and those with AIDS or other immunodeficiency. ■ ***Campylobacter*, p. 291**

Symptoms

Symptoms are similar to those with *Shigella* sp., with dysentery occurring in about half the cases.

Causative Agent

Campylobacter jejuni is a motile, curved Gram-negative rod (**figure 24.16**) that can be cultivated from feces on a selective medium under microaerophilic conditions.

TABLE 24.9 Salmonellosis

Symptoms	Diarrhea and vomiting; rarely, prolonged fever, headache, abdominal pain, abscesses, and shock
Incubation period	Usually 6 to 72 hours; can be 1 to 3 weeks in typhoid fever
Causative agent	*Salmonella enterica*, motile, Gram-negative, lactose nonfermenting, enterobacteria
Pathogenesis	Invasion of the lining cells of lower small and large intestine, with penetration to underlying tissues; body's inflammatory response causes increase in fluid secretion. Sometimes survival within macrophages and spread throughout the body, destruction of Peyer's patches
Epidemiology	Ingestion of food contaminated by animal feces, especially poultry. Human fecal source in typhoid fever–like illnesses
Prevention and treatment	Adequate cooking and handling of food; attenuated vaccine against typhoid fever. Usually no antimicrobial advised unless invasion of tissues or blood occurs

Pathogenesis

The number of organisms required to produce symptomatic infection is 500 or less. The bacteria penetrate the intestinal epithelial cells of the small and large intestine, multiply within and beneath the cells, and cause an inflammatory reaction. As with most strains of *Shigella* and *Salmonella*, penetration into the bloodstream is uncommon. A mysterious sequel to *C. jejuni* infections, **Guillain-Barré syndrome**, occurs in about 0.1% of cases. While *C. jejuni* infections are probably not a direct cause of this syndrome, almost 40% of all cases are preceded by campylobacteriosis, most of the rest by viral infections, and in one group of cases, immunization with swine influenza vaccine. The syndrome begins abruptly within about 10 days of the

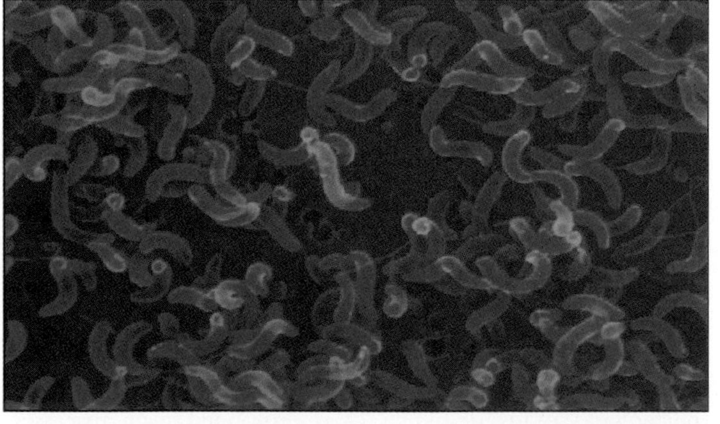

Figure 24.16 Scanning Electron Micrograph of *Campylobacter jejuni*, the Most Common Bacterial Cause of Diarrhea in the United States

TABLE 24.10 Campylobacteriosis

Symptoms	Diarrhea, fever, abdominal cramps, nausea, vomiting, bloody stools
Incubation period	Usually 3 days (range, 1 to 5 days)
Causative agent	*Campylobacter jejuni,* a curved Gram-negative, microaerophilic rod
Pathogenesis	Low infecting dose. The bacteria multiply within and beneath the epithelial cells and incite an inflammatory response. Bloodstream invasion is uncommon. Complicated by a generalized paralysis, Guillain-Barré syndrome, on rare occasions
Epidemiology	Large foodborne and waterborne outbreaks originating from chickens, cows and other animals. Person-to-person spread rarely occurs
Prevention and treatment	Water chlorination and pasteurization of beverages are effective control measures. Avoiding contamination of hands and food preparation areas with uncooked poultry, and cooking of poultry until it is no longer pink are also effective. Most victims recover within 10 days without antibacterial medications

onset of diarrhea, with tingling of the feet followed by progressive paralysis of the legs, arms, and rest of the body. Most victims require hospitalization, but over 80% recover completely. About 5% of patients die of the syndrome despite treatment.

Epidemiology

Numerous foodborne and waterborne outbreaks of *C. jejuni* have been reported, involving as many as 3,000 people. Most cases, however, are sporadic. Like the salmonellas, *C. jejuni* lives in the intestines of a variety of domestic animals. Poultry is a common source of infection, and as many as 89% of raw poultry products have harbored the organism. One drop of juice from raw chicken meat can easily contain an infectious dose. Cats have also been implicated as a source. Epidemics have resulted from ingesting unpasteurized cow and goat milk and from drinking nonchlorinated surface water. Despite its low infectious dose, person-to-person spread of *C. jejuni* rarely occurs.

Prevention and Treatment

Prevention of outbreaks depends mostly on pasteurization of milk and on chlorination of water. Proper cooking and handling of raw poultry to avoid contamination of hands and kitchen surfaces can prevent most of the other cases. Chicken should be cooked until it is no longer pink. Most cases of *C. jejuni* gastroenteritis subside without antimicrobial treatment in 10 days or less and leave the infected individual immune to further infection. The antibiotic erythromycin is recommended for severe cases.

The main features of campylobacteriosis are shown in **table 24.10**.

MICROCHECK 24.5

Vibrios, *Campylobacter jejuni*, and enterobacteria account for most intestinal bacterial infections. Pathogenic mechanisms include attachment, toxin production, cell invasion, and destruction of microvilli or villi. Gastroenteritis can be caused either by infectious microorganisms consumed with the food or by toxic products of microbial growth in food. Some microbial toxins alter the secretory function of cells in the small intestine without killing or visibly damaging them. Unsuspected human carriers can remain sources of enteric infections for many years. Most vaccines are of limited value because they induce little IgA production.

- Explain how *Vibrio cholerae* causes cholera without apparent damage to the intestinal epithelium.
- How does the epidemiology of *Salmonella* Typhi differ from that of most other *S. enterica*?
- What is the leading cause of bacterial diarrhea in the United States? What is its source?
- How would you devise a selective medium for *Vibrio cholerae*?

Viral Diseases of the Lower Alimentary System

Viral infections of the lower alimentary system represent a major illness burden on all age groups. Both the intestine and accessory organs such as the liver can be affected. At least five different groups of viruses can cause epidemic gastroenteritis, resulting in millions of cases each year in the United States. More than six different viruses can cause **hepatitis**, meaning inflammation of the liver. Hepatitis A virus (HAV), hepatitis B virus (HBV), and hepatitis C virus (HCV) account for most cases of viral hepatitis.

Rotaviral Gastroenteritis

Rotaviruses cause most cases of viral gastroenteritis in infants and children. In the United States, they cause an estimated 2.7 million cases each year in individuals less than five years old, resulting in 500,000 emergency department or clinic visits, and 49,000 hospital admissions. Worldwide, 600,000 deaths are attributed to rotaviruses each year.

Symptoms

Illness begins abruptly with vomiting and slight fever, followed in a short time by profuse watery diarrhea. Symptoms generally are gone in about a week. Unless adequate replacement fluid is given, death can occur from dehydration.

Causative Agents

Rotaviruses (**figure 24.17**) are about 70 nm in diameter, have a double-walled capsid, and a double-stranded, 11-segment RNA genome. The viruses represent a major subgroup of the reovirus family. Their name comes from *rota*, Latin for "wheel."

■ **reovirus family, p. 343**

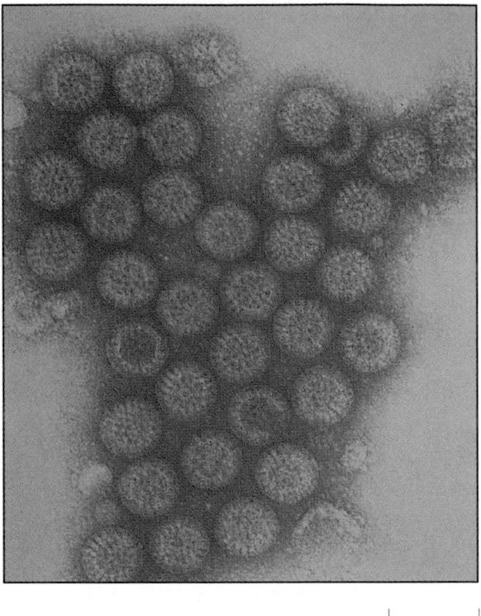

Figure 24.17 **Transmission Electron Micrograph of Rotavirus in Human Feces**

Pathogenesis

The viruses infect mainly the epithelial cells that line the upper part of the small intestine, causing cell death and decreased production of digestive enzymes. These changes gradually return to normal over a number of weeks.

Epidemiology

Rotaviruses spread by the fecal-oral route. Childhood epidemics generally occur in winter in temperate climates, perhaps because children are more apt to be confined indoors in groups where the viruses can spread easily. Generally, by the age of four years, children have acquired some immunity to these viruses. Rotaviruses also cause about 25% of traveler's diarrhea cases. Rotaviruses that infect a wide variety of young wild and domestic animals do not cause human disease. Experimentally, however, reassortment of genetic segments can occur with dual infections. It is not known whether new human pathogenic rotaviruses arise by this mechanism.

Prevention and Treatment

Handwashing, disinfectant use, and other sanitary measures help limit the spread of rotaviruses. An effective live attenuated vaccine was approved in 1998, but this was withdrawn the next year because of concerns that it increased the risk of a rare kind of bowel obstruction. Efforts are continuing to develop a safe and effective vaccine.

Norwalk Virus Gastroenteritis

Norwalk viruses are responsible for almost half the cases of viral gastroenteritis in the United States. Their name comes from Norwalk, Ohio, the place where they were first implicated in an epidemic of gastroenteritis.

Symptoms

The symptoms of nausea and vomiting of abrupt onset are identical to those with rotaviruses. Symptoms with Norwalk virus infections, however, generally last only 1 or 2 days.

Causative Agent

Norwalk viruses are small (about 30 nm diameter), nonenveloped, single-stranded RNA viruses. Their spherical surface is covered with small depressions **(figure 24.18)**. These viruses represent a group of gastroenteritis-producing viruses within the calcivirus family, none of which have been cultivated in the laboratory. ■ **calcivirus family, p. 343**

Pathogenesis

Norwalk viruses infect the upper small intestinal epithelium and produce changes similar to those with rotaviruses. The epithelium generally recovers fully within about 2 weeks.

Epidemiology

Transmission is by the fecal-oral route. Although some viral strains cause gastroenteritis in infants, Norwalk viruses typically infect children and adults. Infected individuals generally eliminate virus in their feces for only a few days. The illness is sometimes contracted from eating shellfish, and in some of these instances, the shellfish have apparently been free of exposure to human feces. Calciviruses are widespread among marine animals, raising the question whether the sea could be involved in the ecology of Norwalk viruses.

Prevention and Treatment

Handwashing, disinfectants, and other sanitary measures can minimize transmission of Norwalk viruses. There is no vaccine, and no proven antiviral medication.

Table 24.11 compares Norwalk and rotaviruses and the illnesses they produce.

Hepatitis A

Hepatitis A, formerly called **infectious hepatitis**, is endemic around the world. The distribution of reported cases in the United States is shown in **figure 24.19**. With its lack of an animal

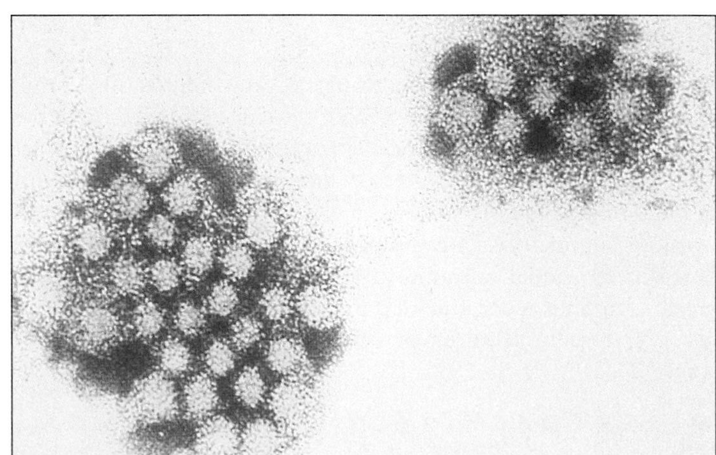

Figure 24.18 **Electron Micrograph of Norwalk Virus**

TABLE 24.11 Rotavirus and Norwalk Virus Compared

Causative Agent	Rotavirus	Norwalk Virus
Characteristics	70 nm diameter double-walled capsid; double-stranded RNA	30 nm diameter, single-stranded, positive-sense RNA
Incubation period	24 to 48 hrs	18 to 72 hrs
Typical symptoms	Vomiting, abdominal cramps, diarrhea, lasting 5 to 8 days	Vomiting, abdominal cramps, diarrhea, lasting 24 to 48 hours
Prevention	Attenuated vaccine	No vaccine

reservoir for the causative virus, its antigenic simplicity, and the availability of an effective vaccine, the disease could be considered for eradication. Hepatitis A is only one example of viral hepatitis; at least five other kinds of viral hepatitis, B, C, D, E, and G, are known.

Symptoms

Typical symptoms of hepatitis A are fatigue, fever, loss of appetite, nausea, right-sided abdominal pain, dark-colored urine, clay-colored feces, and jaundice. In children less than six years old, 70% have no symptoms and those with symptoms rarely have jaundice. Older children and adults usually develop symptomatic infections and most of them develop jaundice. Usually patients recover within 2 months, but 10% to 15% require up to 6 months for full recovery.

Causative Agent

Hepatitis A is caused by the hepatitis A virus (HAV), a small (27 nm) single-stranded, positive-sense RNA virus of the picornavirus family. HAV differs from other members of the family and has been given the name hepatovirus. There is only one serotype. ■ **picornaviruses, p. 343**

Pathogenesis

Following ingestion, the virus reaches the liver by an unknown route. The liver is the main site of replication and the only tissue known to be damaged by the infection. The virus is released into the bile and eliminated with the feces.

Epidemiology

Hepatitis A virus spreads by the fecal-oral route, principally through fecal contamination of hands, food, or water. Many outbreaks of the disease have originated from restaurants because food handlers who carried the virus failed to wash their hands. Eating raw shellfish is a frequent source of infection since these animals concentrate the hepatitis A virus from fecally polluted seawater. A high percentage of hepatitis A occurs in low socioeconomic groups of people because of crowding and inadequate sanitation. Other groups at high risk of hepatitis A include attendees of day care centers and nursing homes, and homosexual men. Infants and children with hepatitis A can eliminate the virus in their feces for several months after symptoms begin, but the amount of virus in feces usually drops markedly with the appearance of jaundice.

Prevention and Treatment

An effective inactivated virus vaccine against hepatitis A has been available since 1995. It is indicated for travelers to economically deprived regions, homosexual men, those who work on sewers, health care workers, individuals with chronic liver disease, and food handlers. Gamma globulin containing antibody to HAV can be given by injection for passive immunization of people exposed to the virus. It gives short-term protection against the disease if given within 2 weeks of exposure. No antiviral treatment is available.

Hepatitis B

Hepatitis B, formerly known as **serum hepatitis**, represents over 40% of the cases of viral hepatitis in the United States. Approximately 240,000 new cases occur each year, and as many as 1% to 6% of the infected adults will become carriers of the disease.

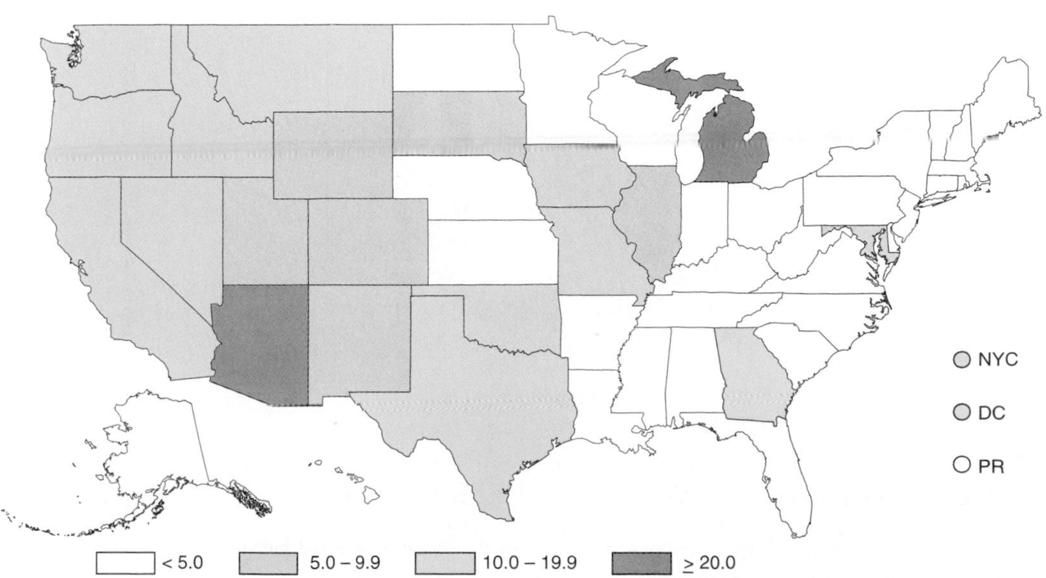

Figure 24.19 Distribution of Hepatitis A in the United States Reported cases per 100,000 population.

About 1.25 million people in the United States have chronic hepatitis B and therefore an increased risk of death from cirrhosis and liver cancer. Hepatitis B infects at least 5% of the world's population and is the ninth leading cause of death worldwide.

Symptoms

Symptoms of hepatitis A and B are similar, except that hepatitis B tends to be more severe than hepatitis A, causing death from liver failure in 1% to 10% of hospitalized cases.

Causative Agent

Hepatitis B is caused by hepatitis B virus (HBV) **(figure 24.20a)**, a member of the hepadnavirus family (*hepa-*, referring to the liver, and *-dna-* "DNA"). The virus contains double-stranded DNA and has a lipid-containing outer envelope. Three important HBV antigens are (1) surface antigen (HBsAg), (2) core antigen (HBcAg), a protein of the nucleocapsid, and (3) e antigen (HBeAg), a soluble component of the viral core. HBsAg is produced during viral replication in amounts far in excess of that needed for virus production. It occurs in the bloodstream on small spheres and filaments of viral envelope material empty of DNA (figure 24.20b) in quantities often 1,000 or more times as great as complete virions. HBsAg is responsible for the ability of the virus to attach to and infect host cells; antibody to surface antigen (anti-HBsAg) confers immunity.

Pathogenesis

Following entry into the body, HBV is carried to the liver by the bloodstream. The mechanism by which HBV causes liver injury is not known but most likely results from the body's immune system attacking the infected liver cells. After entering the liver cell, the virus replicates by a mechanism involving reverse transcriptase **(figure 24.21)**, unusual for a DNA virus. The double-stranded viral DNA genome is transported to the host cell nucleus, where messenger RNA and a RNA copy of the genome are synthesized. This RNA copy of the genome is subsequently transcribed into DNA by reverse transcriptase, and when a double strand is formed by DNA polymerase, this DNA becomes the genome for the newly forming virions. Double-strand formation ceases when the virions bud from the host cell, however, generally leaving part of the genome single-stranded. ■ **reverse transcriptase, p. 229**

HBsAg appears in the bloodstream days or weeks after infection, often long before signs of liver damage are evident. In 1% to 6% of cases, the infection smolders in the liver, unsuspected for years. About 40% of chronically infected people eventually die from cirrhosis, meaning scarring, of the liver, or from liver cell cancer. Evidence indicates that these cancers result from HBV transformation of liver cells. ■ **transformed cells, p. 357**

Epidemiology

From 1965 to 1985, a progressive rise in reported hepatitis B cases occurred. Since then, the incidence of the disease has appeared to plateau and decline **(figure 24.22)**. HBV is spread mainly by blood, blood products, and semen. Persisting viremia, meaning virus circulating in the bloodstream, can follow both symptomatic and asymptomatic cases, and the virus may continue to circulate in the blood for many years. Carriers are of major importance in the spread of hepatitis B because they are often unaware of their infection. If only a minute amount of blood from an infected person is injected into the bloodstream or rubbed into minor wounds, infection can result. Blood and other

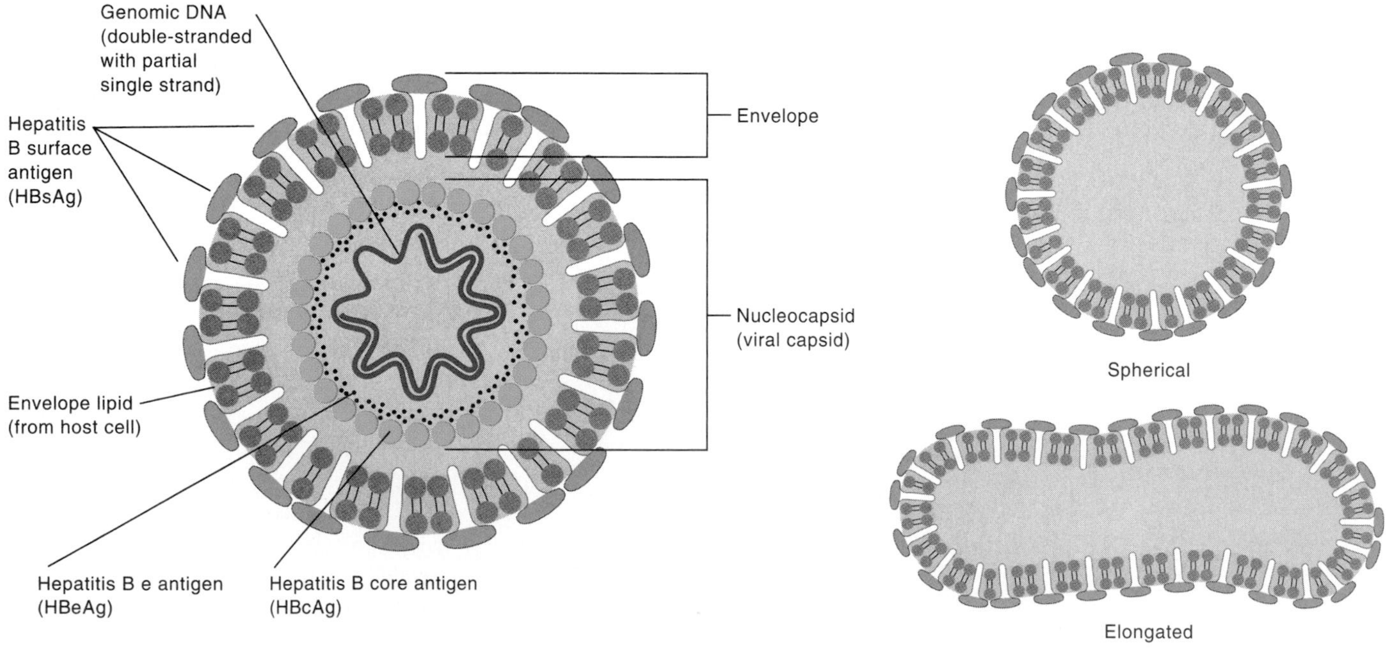

Figure 24.20 Hepatitis B Virus Components Found in the Blood of Infected Individuals **(a)** Complete infectious virion. **(b)** Smaller spherical and elongated envelope particles lacking DNA.

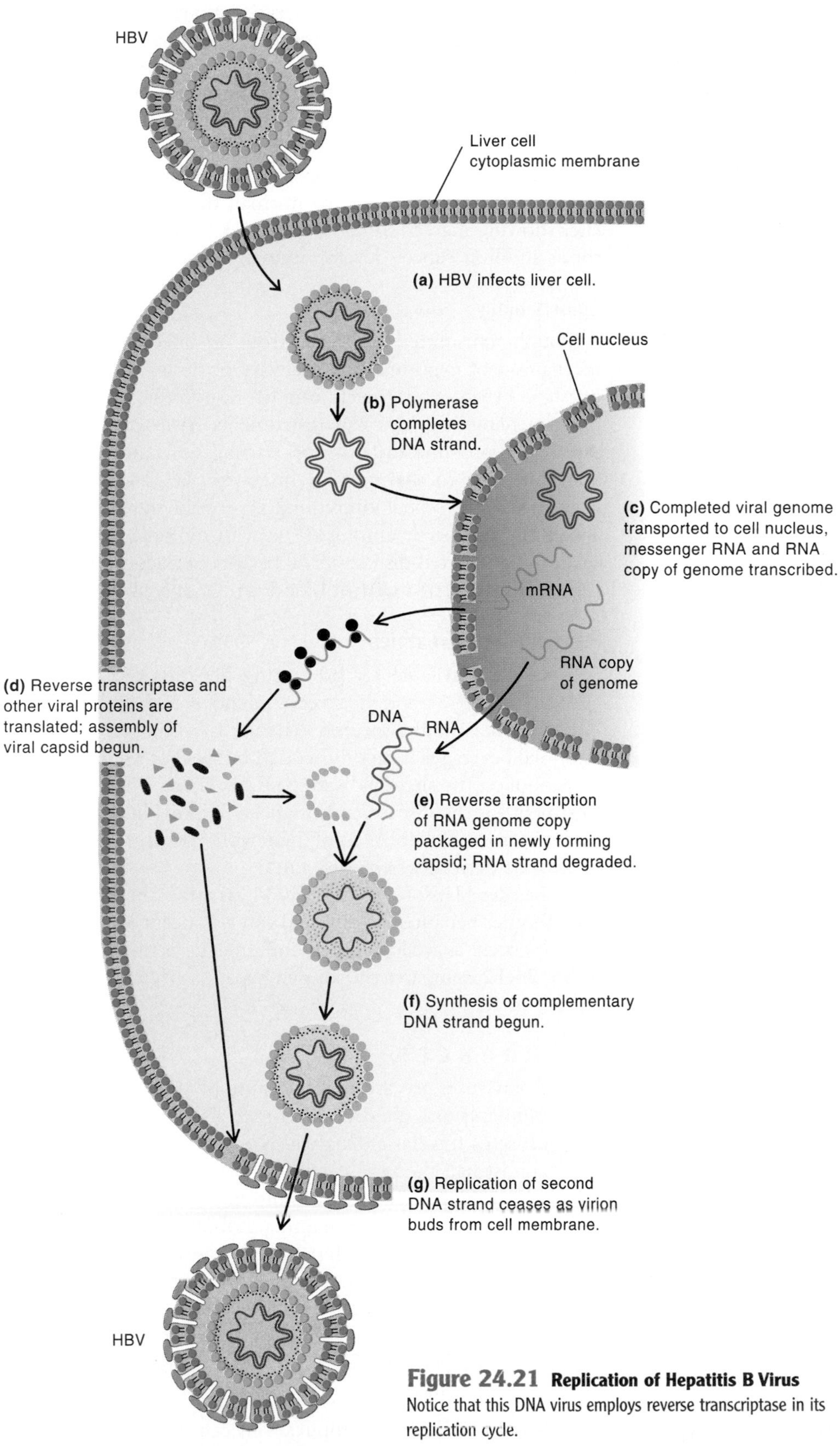

Figure 24.21 Replication of Hepatitis B Virus
Notice that this DNA virus employs reverse transcriptase in its replication cycle.

body fluids can be infectious by mouth, the virus probably infecting the recipient through small scratches or abrasions. Many hepatitis B virus infections result from sharing of needles by drug abusers. Unsterile tattooing and ear-piercing instruments and shared toothbrushes, razors, or towels can also transmit HBV infections.

Sexual intercourse is responsible for transmission in nearly half of hepatitis B cases in the United States. HBV antigen is often present in saliva and breast milk, but the quantity of infectious virus and risk of transmission is low. Five percent or more of pregnant women who are HBV carriers transmit the disease to their babies at delivery, and more than two-thirds of women who develop hepatitis late in pregnancy or soon after delivery do so. Most of these babies have asymptomatic infections and become long-term carriers, but some die of liver failure.

Prevention and Treatment

The first vaccine against hepatitis B, approved in the early 1980s, consisted of HBsAg obtained from the blood of chronic carriers; 1 ml of their blood often had enough antigen to immunize eight people! Since 1986, however, an effective genetically engineered vaccine produced in yeast has been available against hepatitis B. Infants born to an infected mother are injected at birth with hepatitis B immune globulin (HBIG) containing a high titer of antibody to HBV. Active immunization is also started, with injections of the vaccine at one week, one month, and six months of age. Passive immunization with HBIG offers immediate, partial protection against HBV infection until active immunity is achieved from the vaccine. Vaccination against hepatitis B prevents the disease, and it can also help prevent many of the 500,000 to 1 million new liver cell cancers that occur worldwide each year. Education of groups at high-risk of contracting hepatitis B can help prevent the disease. Those likely to be exposed to blood are taught to consider all blood to be infectious, to wash their hands contaminated with blood, to wear gloves and protective clothing, and to handle contaminated sharp objects such as needles and scalpel blades carefully. Teaching the importance and proper use of condoms also helps limit spread of the infection. There is no curative antiviral treatment yet for hepatitis B. Remarkable improvement can be achieved in many cases, however, by giving injections of genetically engineered interferon along with an antiviral medication such as lamivudine, a reverse transcriptase inhibitor. ■ **interferon, p. 379**

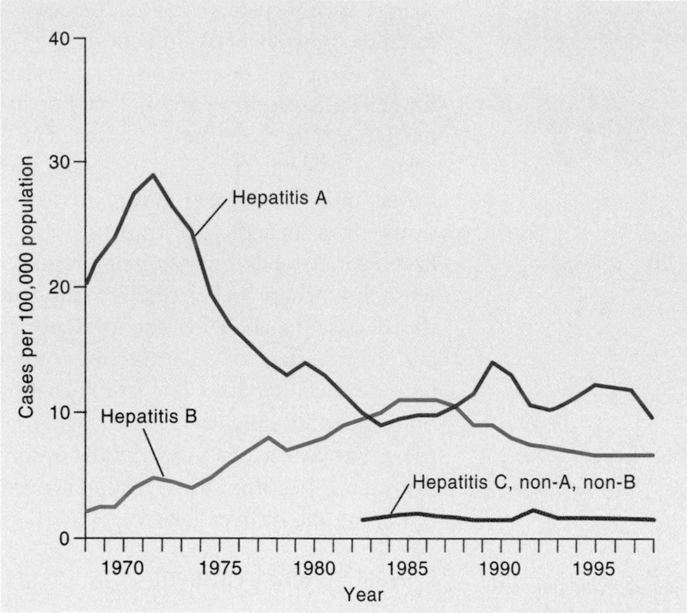

Figure 24.22 Incidence of Viral Hepatitis in the United States Most of the non-A, non-B hepatitis is caused by hepatitis C virus. The first hepatitis B vaccine was licensed in 1982, the first hepatitis A vaccine in 1995. A test for hepatitis C antibody became available in 1990.

Hepatitis C Virus (HCV) Disease

Even when blood that tested positive for HBV was excluded from transfusions, post-transfusion hepatitis continued to be common, indicating the presence of another bloodborne hepatitis virus. After a number of years of trying to isolate a non-A, non-B hepatitis virus, scientists in 1989 were able to clone parts of the genome of a transfusion-associated virus, now known as hepatitis C virus (HCV). One of the gene products of these clones proved to be an HCV antigen satisfactory for detecting antibody to HCV in the blood of prospective blood donors. By not using donated blood containing the HCV antibody, the incidence of post-transfusion hepatitis fell from about 8% to very low levels. The antibody test has revealed that more than 3 million Americans are infected with HCV and that over 30,000 new cases occur each year. It is now the most common chronic bloodborne infection in the United States.

Symptoms

The symptoms of hepatitis C are the same as hepatitis A and B except generally milder. About 65% of infected individuals have no symptoms relating to the acute infection, while only about 25% have jaundice. ■ **flavivirus family, p. 343**

Causative Agent

HCV is an enveloped, single-stranded, positive-sense RNA virus of the flavivirus family. Like HBV, it has not been cultivated *in vitro*. There is considerable genetic variability in the viruses from different people, leading to classification into types and subtypes having differing pathogenicity.

Pathogenesis

Few details are known about the pathogenesis of hepatitis C. Infection generally occurs from exposure to contaminated blood. The incubation period averages about 6 weeks (range, 2 weeks to 6 months). Despite the lack of symptoms in most people with acute infection, more than 80% develop chronic infections. The virus infects the liver and incites inflammatory and immune responses. Viral variants arise frequently, so that antibody developing in response to one variant fails to react with a newer variant. The disease process in the liver waxes and wanes, at times seeming to return to normal, then weeks or months later showing marked inflammation. After months or years, cirrhosis and liver cancer develop in many patients.

Epidemiology

Although transmitted by blood from an infected person, the mechanism of exposure is not always obvious. Sharing toothbrushes, razors, and towels can be responsible. Tattoos and body piercing with unclean instruments have transmitted the disease. Approximately 60% of transmissions in the United States are due to sharing of syringes by illegal drug abusers. Transmission by sexual intercourse is probably rare, although it apparently can occur among those with multiple partners and sexually transmitted diseases. The risk of contracting the disease from transfusion of a unit of blood is now only about 0.001%.

Prevention and Treatment

No vaccine is avilable for preventing hepatitis C. Vaccination against hepatitis A and B is recommended to help avoid dual infections that might severely damage the liver. Avoidance of alcoholic beverages is recommended because of the toxic effect of alcohol on the already damaged liver. There is no satisfactory treatment for the disease, although some individuals are helped by interferon injections combined with ribavirin, an antiviral nucleoside derivative. ■ **ribavirin, p. 513**

Besides HAV, HBV, and HCV, hepatitis can be caused by several other hepatitis viruses, and can also occur as part of viral diseases such as yellow fever and infectious mononucleosis. **Table 24.12** compares the known hepatitis viruses, A, B, C, D, E, and G.

MICROCHECK 24.6

Rotaviruses are the leading cause of viral gastroenteritis in infants and children, and they also are a common cause of traveler's diarrhea. Norwalk viruses cause almost half the viral gastroenteritis in children and adults. Hepatitis A is transmitted by the fecal-oral route and is preventable by gamma globulin and an inactivated vaccine. Hepatitis B, transmitted by exposure to blood and by sexual intercourse, is preventable by a vaccine produced in yeast. Hepatitis C is transmitted by blood and perhaps occasionally by sexual intercourse. Chronic hepatitis often results in cirrhosis and cancer.

- What two serious complications can occur late in the course of both chronic hepatitis B and C?
- Why might it be more difficult to prepare a vaccine against Norwalk viruses than rotaviruses?
- At what stage in the replication of hepatitis B would a reverse transcriptase inhibitor act?

TABLE 24.12 Viral Hepatitis

Disease	Hepatitis A	Hepatitis B	Hepatitis C
Causative agent	Non-enveloped, single-stranded RNA picornavirus, HAV	Enveloped, double-stranded DNA hepadnavirus, HBV	Enveloped, single-stranded RNA flavivirus, HCV
Mode of spread	Fecal-oral	Blood, semen	Blood, possibly semen
Incubation period	3 to 5 weeks (range, 2 to 7 weeks)	10 to 15 weeks (range, 6 to 23 weeks)	6 to 7 weeks (range, 2 to 24 weeks)
Prevention	Gamma globulin; inactivated vaccine	Recombinant vaccine	No vaccine
Comments	Usually mild symptoms, but often prolonged; full recovery; no long-term carriers	Symptoms often more severe than hepatitis A; progressive liver damage in 1% to 6% can lead to cirrhosis and cancer; chronic carriers; can cross the placenta	Usually few or no symptoms; progressive liver damage; chronic carriers; virus can cross the placenta

Disease	Hepatitis D	Hepatitis E	Hepatitis G
Causative agent	Defective single-stranded RNA virus, HDV	Non-enveloped, single-stranded RNA calcivirus, HEV	Single-stranded RNA flavivirus
Mode of spread	Blood, semen	Fecal-oral	Blood, possibly semen
Incubation period	2 to 12 weeks	2 to 6 weeks	Weeks
Prevention	No vaccine	No vaccine	No vaccine
Comments	Prior or concurrent HBV infection necessary; can cause worsening of hepatitis B; can cross the placenta	Similar to hepatitis A, except severe disease in pregnant women; same or related virus in rats	Usually mild symptoms; persistant viremia for months or years

Protozoan Diseases of the Lower Alimentary System

Protozoa are important causes of human intestinal disease. All the major protozoan motility groups are represented: Mastigophora, those with flagella; Ciliata, those with cilia; Sporozoa, usually immobile, and Sarcodina, those that move by pseudopods. All are transmitted by the fecal-oral route. ■ protozoa, p. 304

Giardiasis

Disease resulting from *Giardia lamblia* infection, called giardiasis, is the most commonly identified waterborne illness in the United States. It can be contracted from clear mountain streams or chlorinated city water and can be transmitted by person-to-person contact. Giardiasis occurs worldwide and is responsible for many cases of traveler's diarrhea. In Americans, *G. lamblia* is the most commonly identified intestinal parasitc, and in economically underdeveloped areas of the world, its incidence often exceeds 10%.

Symptoms

In the usual epidemic of giardiasis, about two-thirds of the exposed individuals develop symptoms. The incubation period is generally 6 to 20 days. Symptoms can range from mild (indigestion, "gas," and nausea) to severe (vomiting, explosive diarrhea, abdominal cramps, fatigue, and weight loss). The symptoms usually disappear without treatment in 1 to 4 weeks, but some cases become chronic. Both symptomatic and asymptomatic persons can become long-term carriers, unknowingly excreting infectious cysts with their feces.

Causative Agent

Giardia lamblia, a member of the Mastigophora, is a flagellated protozoan shaped like a pear cut lengthwise, with two side-by-side nuclei that resemble eyes. These features along with an adhesive disc on its undersurface give the organism an unmistakable appear ance (**figure 24.23**). *Giardia lamblia* can exist in two forms: as a vegetative **trophozoite** or as a resting form called a cyst. The trophozoite is the actively feeding form, and it colonizes the upper part of the small intestine. Generally, only trophozoites are present in the feces when an infected person is having diarrhea. When trophozoites are carried slowly by intestinal contents toward the large intestine, they develop into cysts. *Giardia* cysts have thick walls composed of chitin, a tough, flexible, nitrogen-containing polysaccharide. The cyst wall protects the organism from harsh environmental conditions. Curiously, *Giardia* lacks mitochondria, and its ribosomal RNA resembles that of prokaryotes. ■ chitin, p. 308

Figure 24.23 Scanning Electron Micrograph of *Giardia lamblia* Trophozoites in Human Intestine

Pathogenesis

The cyst form is responsible for infection because, unlike the trophozoite, it is resistant to stomach acid. Two trophozoites emerge from each cyst after it is ingested and reaches the small intestine. Some of these trophozoites attach to the epithelium by their adhesive disc, while others move freely in the intestinal mucus using their flagella. Some may even migrate up the bile duct to the gallbladder and cause crampy pain or jaundice. The protozoa do not destroy the host epithelium, but in severe infestations they may entirely cover the epithelial surface. They interfere with the ability of the intestine to absorb nutrients and secrete digestive enzymes. The result is malnutrition, bulky feces containing fat, and excessive intestinal gas from bacterial digestion of unabsorbed food material. Some of the intestinal impairment is probably a side effect of the host immune system attacking the parasites.

Epidemiology

Transmission of *G. lamblia* is usually by the fecal-oral route, especially via fecally contaminated water. Known or suspected sources of *G. lamblia* include beavers, raccoons, muskrats, dogs, cats, and humans. A single human stool can contain 300 million *G. lamblia* cysts; only 10 cysts are required to establish infection. The cysts can remain viable in cold water for more than 2 months. Usual levels of chlorination of municipal water supplies are ineffective against the cysts, and water filtration is necessary to remove them. Hikers who drink from streams, even in remote areas thought to contain safe water, are at risk of contracting giardiasis.

Although waterborne outbreaks are the most common, person-to-person contact can also transmit the disease. This mode of transmission is especially likely in day care centers where hands become contaminated in the process of diaper changing. People who promiscuously engage in anal intercourse and fellatio are also prone to contracting the disease. Transmission by fecally contaminated food has also been reported. Good personal hygiene, especially handwashing, decreases the chance of passing on the infection. ■ **day care centers, p. 487**

Prevention and Treatment

The best way to make drinking water safe from giardiasis is to boil it for 1 minute. Using a few drops per quart of water of household sodium hypochlorite bleach, tincture of iodine, or commercial water-purifying tablets, is also effective. As in all sterilization procedures, however, time and temperature are important. Only an hour may be necessary to treat warm water, but many hours are required to kill cysts in cold water. Individuals infested with *Giardia* can usually be identified by microscopically examining their feces for *Giardia* cysts. A more sensitive test employing antibody to *G. lamblia* is also available. Several medicines, including quinacrine (Atabrine) and metronidazole (Flagyl), effectively treat giardiasis.

Table 24.13 describes the main features of giardiasis.

Cryptosporidiosis

Cryptosporidiosis, disease caused by the *Cryptosporidium* genus of protozoa, was recognized in animals in the early 1900s, but not identified as a threat to human beings until the AIDS epidemic struck in the early 1980s. Since then, it has become clear that cryptosporidiosis is a major hazard not only to those with immunodeficiency, but to the public at large. One waterborne outbreak of the disease involved more than 403,000 people!

Symptoms

Cryptosporidiosis is characterized by fever, loss of appetite, nausea, crampy abdominal pain, and profuse watery diarrhea, beginning

TABLE 24.13 Giardiasis

Symptoms	Mild illness: indigestion, flatulence, nausea; severe: vomiting, diarrhea, abdominal cramps, weight loss
Incubation period	6 to 20 days
Causative agent	*Giardia lamblia,* a flagellated pear-shaped protozoan with two nuclei
Pathogenesis	Ingested cysts survive stomach passage; trophozoites emerge from the cysts in the small intestine, where some attach to epithelium and others move freely; mucosal damage occurs, which is reversible after parasite has been eradicated by medication
Epidemiology	Ingestion of fecally contaminated water; person-to-person, in day-care centers
Prevention and treatment	Boiling or disinfecting drinking water; filtration of community water supplies. Treatment: quinacrine hydrochloride (Atabrine) or metronidazole (Flagyl)

after an incubation period of 4 to 12 days. The symptoms generally last 10 to 14 days, but in people with immunodeficiency diseases, they can last for months and be life threatening.

Causative Agent

Cryptosporidium parvum, a coccidian member of the Apicomplexa, is the causative agent of cryptosporidiosis. *Cryptosporidium parvum* multiplies intracellularly in the small intestinal epithelium, its entire life cycle occurring in a single host. The oocyst stage is a tiny (4 to 6 μm) acid-fast sphere that contains four banana-shaped sporozoites (**figure 24.24**). ■ apicomplexa, p. 305

Pathogenesis

Few details are known about the pathogenesis of cryptosporidiosis. Following ingestion, the digestive juices of the small intestine release the sporozoites from the oocysts. The sporozoites invade the epithelial cells of the small intestine, causing deformity of the epithelium and intestinal villi and an inflammatory response beneath the cells. Secretion of water and electrolytes increases and absorption of nutrients decreases, but the mechanism is unknown. Cell-meditated immunity is important in controlling the infection.

Epidemiology

The oocysts of *C. parvum* are infectious when eliminated with the feces, and fewer than 30 of them are sufficient to cause an infection. Person-to-person spread readily occurs under conditions of poor sanitation. Infected individuals often have prolonged diarrhea and can continue to eliminate infectious cysts for 2 weeks or more after the diarrhea ceases. Moreover, the cysts can survive for long periods of time in food and water, are even more resistant to chlorine than *Giardia*, and are too small to be removed from drinking water by standard filtration methods. Equally troubling, *C. parvum* has a wide host range, infecting domestic animals such as dogs, pigs, and cattle. These animals, as well as humans, can contaminate food and drinking water. Epidemics have arisen from drinking water, swimming pools, a water slide, a zoo fountain, day care centers, unpasteurized apple juice, and other food and drink. The organism is responsible for many cases of traveler's diarrhea.

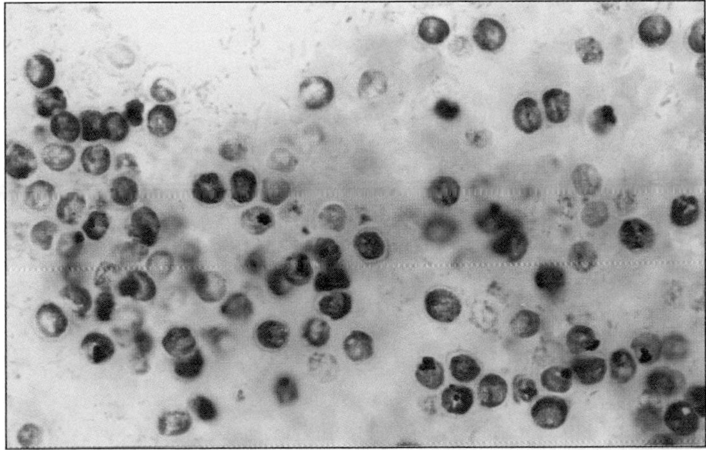

Figure 24.24 Oocysts of *Cryptosporidium parvum* in the Feces of an Infected Individual (Acid-fast Stain) Person-to-person and waterborne transmission are common.

Prevention and Treatment

Effective measures for preventing cryptosporidiosis are limited to careful monitoring of municipal water supplies, pasteurization of liquids for human consumption, and sanitary disposal of human and animal feces. Food handlers with diarrhea should not handle food until symptom-free, and all food handlers should adhere to handwashing and other sanitary measures. Immunodeficient individuals are advised to avoid contact with animals, to boil or filter drinking water using a one micrometer or smaller pore size filter, and to avoid recreational water activities. No satisfactory treatment exists for cryptosporidiosis, although a combination of two antibiotics, paromomycin and azithromycin, has helped control the disease in some AIDS patients.

The main features of cryptosporidiosis are shown in **table 24.14**.

Cyclosporiasis

Cyclosporiasis first came to medical attention in the late 1980s, with widely scattered epidemics of severe diarrhea. For several years, the causative organism was known only as a "cyanobacterium-like body." Later, the bodies were shown to be the oocysts of a coccidian protozoan tentatively named *Cyclospora cayetanensis*. ■ cyanobacteria, p. 278

TABLE 24.14 Cryptosporidiosis

Symptoms	Fever, loss of appetite, nausea, crampy abdominal pain, watery diarrhea, usually lasting 10 to 14 days.
Incubation period	Usually about 6 days (range, 4 to 12 days)
Causative agent	*Cryptosporidium parvum,* a coccidian member of the Apicomplexa. Its life cycle takes place entirely within the epithelial cells of the small intestine
Pathogenesis	Following ingestion of oocysts, the intestinal juices cause the release of four banana-shaped sporozoites. The sporozoites invade the epithelial cells, causing deformity of the cells and the intestinal villi and inciting an inflammatory response. Secretion of water and electrolytes increases, and absorption of nutrients decreases. The process is brought under control largely by cell-mediated immunity
Epidemiology	Oocysts of *C. parvum* are infectious when discharged with the feces. The infectious dose is small, and person-to-person spread readily occurs under unsanitary conditions. Infected individuals can discharge the oocysts for weeks after symptoms subside. The oocysts resist chlorination, and they pass through most municipal water filtration systems. The host range is wide, including cattle, pigs, and dogs.
Prevention and treatment	Pasteurization of beverages such as milk and apple juice, boiling of drinking water, or filtering it using a 1-μm pore size filter. Sanitary disposal of human and animal feces. No effective treatment exists

Symptoms

Cyclosporiasis begins after an average incubation period of about 1 week, with fatigue, loss of appetite, slight fever, vomiting, and watery diarrhea, followed by weight loss. The diarrhea usually subsides in 3 to 4 days, but relapses occur for up to 4 weeks.

Causative Agent

Cyclospora cayetanensis is also a coccidian member of the Apicomplexa, and its oocysts are similar to those of *Cryptosporidium parvum*, except *C. cayetanensis* oocysts are larger, about 8 to 10 μm in diameter, and when passed in feces do not yet contain sporozoites. After days under favorable conditions outside the body, two sporocysts each containing two sporozoites develop within each oocyst, which is then infectious. Structural features justify placing the protozoan with the *Cyclospora*, a genus described in 1881, but studies of its ribosomal RNA base sequence suggest that it belongs in the genus *Eimeria*. This is an example of the turmoil arising from molecular biological findings as the newer science challenges the old on the naming of microorganisms. ■ **molecular taxonomy, p. 260**

Pathogenesis

Little is known about the pathogenesis of cyclosporiasis, because laboratory animals are not susceptible to the disease. Scant information available from small intestinal biopsies of patients infected with *C. cayetanensis* confirms that both sexual and asexual forms of the protozoan are present in the intestinal epithelium.

Epidemiology

The epidemiology of *Cyclospora cayetanensis* differs in important ways from that of *Cryptosporidium parvum*. The oocysts of *C. cayetanensis* are immature when eliminated in the stool and are non-infectious, so that person-to-person spread does not occur. Most infections occur in spring and summer, probably because warm moist conditions favor maturation of the *C. cayetanensis* oocysts. Travelers to tropical areas are more likely than others to become infected. So far, no animal source has been identified. Fresh produce, especially imported raspberries, has been implicated in a number of epidemics, but the source of the contaminating organisms is unknown. In most instances, the produce has been imported from a tropical region. The protozoa have been identified in natural waters, but a human source could not be ruled out. Outbreaks have occurred in Canada, the United States, Nepal, Peru, Haiti, and various other countries of South and Central America, Southeast Asia, and Eastern Europe.

Prevention and Treatment

No preventive measures are available except boiling or filtering drinking water and and thoroughly washing produce such as berries and leafy vegetables during an epidemic. Cotrimoxazole (trimethoprim plus sulfamethoxazole) effectively treats cyclosporiasis.

The main features of cyclosporiasis are presented in **table 24.15**.

Amebiasis

Disease resulting from *Entamoeba histolytica* infection is called amebiasis. In the United States, cases occur mainly among male homosexuals who have many sexual partners, in poverty-stricken areas, and among migrant farm workers. Although usually a mild disease, worldwide it causes about 30,000 deaths per year, most of which occur in Mexico, parts of South America, Asia, and Africa. Life-threatening disease occurs in these areas because of the presence of virulent *E. histolytica* strains and crowded, unsanitary living conditions.

TABLE 24.15 Cyclosporiasis

Symptoms	Fatigue, loss of appetite, vomiting, watery diarrhea, and weight loss. Symptoms improve in 3 to 4 days, but relapses can occur for up to a month
Incubation period	Usually about 8 days (range, 1 to 12 days)
Causative agent	A coccidian tentatively named *Cyclospora cayetanensis*, formerly known as a "cyanobacterium-like body"
Pathogenesis	Biopsies of human intestine show sexual and asexual stages in the intestinal epithelium. Laboratory animals are not susceptible to the infection.
Epidemiology	The oocysts of *C. cayetanensis* are not infectious when discharged in the feces, and so person-to-person spread does not occur. Travelers to tropical countries are at risk of infection. Produce, especially raspberries, imported from tropical Central America, has been implicated in most North American outbreaks
Prevention and treatment	Use of boiled or filtered drinking water is advised in the tropics. Thorough washing of imported berries and leafy vegetables. Most victims are effectively treated with trimethoprim-sulfamethoxazole (trade names Bactrim, Septra)

Symptoms

Amebiasis is commonly asymptomatic, but symptoms ranging from chronic mild diarrhea lasting months or years, to acute dysentery and death, can occur.

Causative Agent

The causative organism, *Entamoeba histolytica* (see figure 24.25 inset), is a member of the Apicomplexa, ranging from about 20 to 40 μm in diameter. Like *Giardia*, it has a cyst form with a chitin-containing cyst wall. Immature cysts have a single nucleus. As the cyst matures, the nucleus divides twice to form a cyst with four nuclei, the infectious form for the next host. The life cycle of *E. histolytica* is shown in **figure 24.25**.

Pathogenesis

Ingested quadrinucleate cysts of *E. histolytica* survive passage through the stomach. The organisms are released from their cysts in the small intestine, whereupon the cytoplasm and nuclei divide, yielding eight trophozoites. Upon reaching the lower intestine, these trophozoites begin feeding on mucus and intestinal bacteria. Many, but not all, strains produce a cytotoxic enzyme that kills intestinal epithelium on contact, allowing the organisms to penetrate the lining cells and enter deeper tissues of the intestinal wall. Sometimes, they penetrate into blood vessels and are carried to

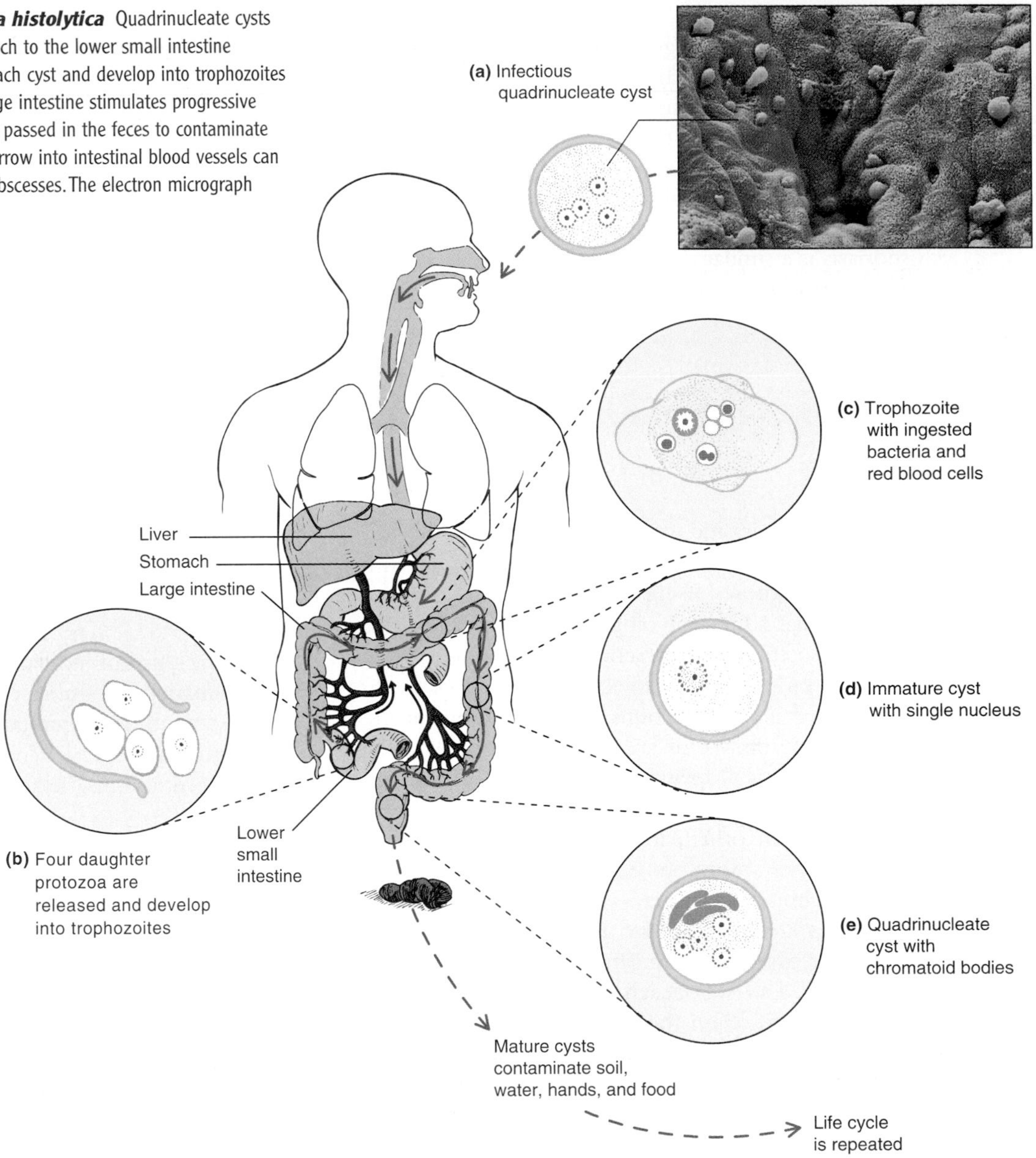

Figure 24.25 Life Cycle of *Entamoeba histolytica* Quadrinucleate cysts enter the mouth **(a)** and pass through the stomach to the lower small intestine **(b)**. Four daughter protozoa are released from each cyst and develop into trophozoites **(c)**, the feeding form **(d)**. Dehydration in the large intestine stimulates progressive stages of cyst development **(e)**. Mature cysts are passed in the feces to contaminate soil, water, hands, and food. Trophozoites that burrow into intestinal blood vessels can be carried to the liver or other organs, causing abscesses. The electron micrograph shows an infectious quadrinucleate cyst.

(a) Infectious quadrinucleate cyst
(b) Four daughter protozoa are released and develop into trophozoites
(c) Trophozoite with ingested bacteria and red blood cells
(d) Immature cyst with single nucleus
(e) Quadrinucleate cyst with chromatoid bodies
Liver
Stomach
Large intestine
Lower small intestine
Mature cysts contaminate soil, water, hands, and food
Life cycle is repeated

the liver or other body organs. Multiplication of the organisms and tissue destruction in the intestine and in other body tissues can result in amebic abscesses. The irritating effect of the amebas on the cells lining the intestine causes intestinal cramps and diarrhea. Due to intestinal ulceration, the diarrheal fluid is often bloody, and the condition is referred to as **amebic dysentery**.

Epidemiology

Amebiasis is distributed worldwide, but it is more common in tropical areas where sanitation is poor. Humans constitute the only important reservoir. Transmission is fecal-oral. The disease is mainly associated with poverty, male homosexuality, and migrant workers.

Prevention and Treatment

Prevention of amebiasis depends on sanitary measures and avoiding fecal contamination of drinking water. Metronidazole and other medications are available for treatment.

Table 24.16 gives the main features of amebiasis.

TABLE 24.16 Amebiasis

Symptoms	Diarrhea, abdominal pain, blood in feces
Incubation period	2 days to several months
Causative agent	*Entamoeba histolytica*
Pathogenesis	Ingested cysts liberate trophozoites in the small intestine; upon reaching the large intestine, trophozoites feed on mucus and cells lining the intestine; enzymes are produced that allow penetration of the intestinal epithelium, sometimes the intestinal wall and blood vessels, and thence to the liver and other organs, resulting in abscesses
Epidemiology	Ingestion of fecally contaminated food or water; disease associated with poverty, homosexual men, and migrant workers
Prevention and treatment	Good sanitation and personal hygiene. Treatment: metronidazole, paromomycin

MICROCHECK 24.7

Giardiasis, caused by the flagellated protozoan *Giardia lamblia*, is a common waterborne disease, but it can also be transmitted person-to-person. Cryptosporidiosis is another waterborne disease that is also transmissible person-to-person, but it is caused by a coccidian member of the Apicomplexa that multiplies within intestinal epithelial cells. There is no satisfactory treatment. Cyclosporiasis is a similar disease, but it is not transmissible person-to-person and it can be effectively treated. Amebiasis is caused by an ameba that ulcerates the large intestinal epithelium, resulting in dysentery.

- Explain why person-to-person spread does not occur in cyclosporiasis.
- Explain why individuals with severe giardiasis often have bulky feces containing fat.
- Would you expect an individual with giardiasis who has diarrhea to be more likely to transmit the disease than an individual with giardiasis who does not have diarrhea? Explain.

MULTICELLULAR PARASITES

A number of species of parasitic worms can inhabit the human intestinal tract and cause disease. While not microorganisms, they are studied and diagnosed using microscopic and immunologic techniques familiar to microbiologists. Some invade tissue and others do not, so that their presence in the body is often called an infestation rather than an infection. These worms, or **helminths**, are divided into two groups, the **roundworms** and the **flatworms**, and the flatworms are further divided into the **tapeworms** and the **flukes**. Tapeworms are ribbon-shaped and segmented, while flukes are generally shorter and leaf-shaped. Tapeworms differ from the other parasitic helminths in that they lack a digestive tract; they absorb nutrients through their skin. In most instances the helminths themselves are difficult to find and examine, so that they are identified by the microscopic appearance of their eggs, or ova, which are often present in the host's feces in large numbers. The ova of each helminth species are distinctive and can be identified on the basis of size, shape, thickness of the outer covering, embryonation, and other features. Some helminths have complex life cycles, involving one or more **intermediate hosts** where early stages of development occur, and a **definitive host** where the sexually mature forms occur. Generally, helminths have male and female forms, but most of the flatworms are **hermaphroditic**, meaning that both male and female sex organs reside in the same worm. ■ helminths, p. 317

Diseases Due to Intestinal Roundworms

Some notable examples of diseases due to roundworms (nematodes) are discussed in this section. Pinworm disease, whipworm disease, and ascariasis are acquired by ingesting ova. Hookworm disease and strongyloidiasis are contracted by skin penetration by worm larvae, and trichinosis by eating undercooked meat containing worm larvae. ■ nematodes, p. 317

Pinworm Disease (Enterobiasis)

Pinworm disease, enterobiasis, is caused by the pinworm, *Enterobius vermicularis*. It is the most common helminth disease seen by American physicians, and pinworms infest more than 200 million people worldwide.

Symptoms

About one-third of individuals infested with *E. vermicularis* are asymptomatic. The most common symptom in adults is anal itching, sometimes accompanied by vaginal irritation. Sleeplessness, nightmares, nervousness, irritability, anal pain, and weight loss are common in affected children.

Causative Agent.

Enterobius vermicularis is a tiny roundworm; the females average about 10 mm in length, and the males, 2 to 3 mm. The female has a sharply pointed tail end, which inspired its popular name pinworm. The worms can often be detected in the anal area by using a flashlight about 2 hours after retiring, or early in the morning. Ova can be collected from the anal area by pressing transparent tape against the anus. The pinworm ova are elongated and have a distinctive appearance, flat on one side and containing a larval worm (**figure 24.26a**).

Pathogenesis

The worms live mainly in the upper part of the large intestine. They may, however, migrate throughout the intestinal and female genital tracts during their 1- to 2-month life span, feeding on bacteria and other intestinal material. Tiny ulcerations and inflammation occur where the worms attach to the intestinal epithelium. The male worms die soon after mating. The mature females containing 5,000 to 15,000 ova migrate out of the anus, usually while their host is sleeping, deposit their eggs in a sticky matrix, and die. Figure 24.26b shows the enormous numbers of ova present in a single worm. The sensation caused by worms exiting the anus leads to scratching, bleeding, and bacterial infection. If the anal area is not kept clean, eggs deposited there hatch, and the larval worms crawl through the anus, enter the intestine, and perpetuate the infestation.

Epidemiology

Scratching deposits *E. vermicularis* ova on the fingers and under the fingernails. Because of the enormous numbers of ova, clothing,

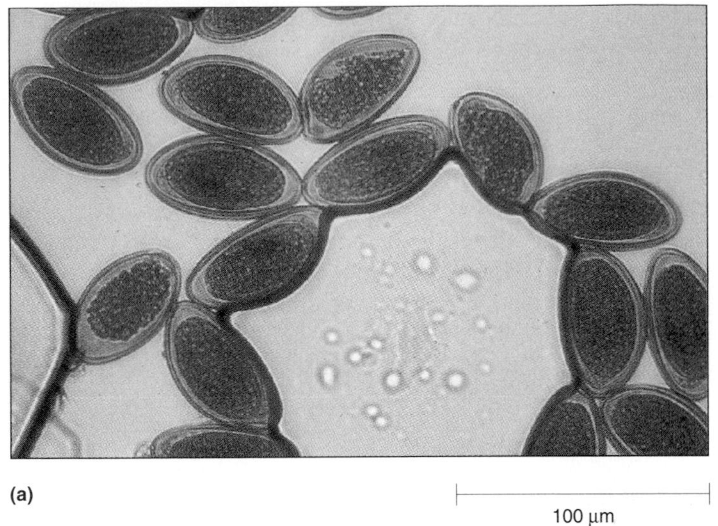

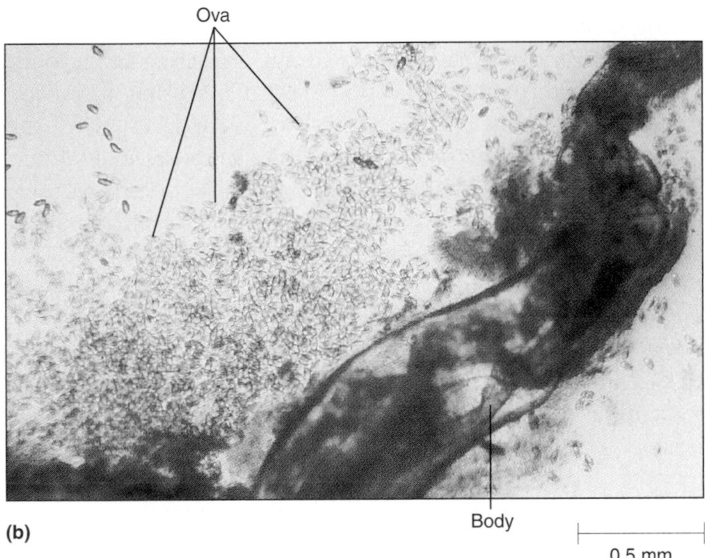

Figure 24.26 Pinworm, *Enterobius vermicularis* **(a)** Microscopic view of the ova, showing flattened side and developing worm larva. **(b)** Body of the adult worm surrounded by thousands of its ova.

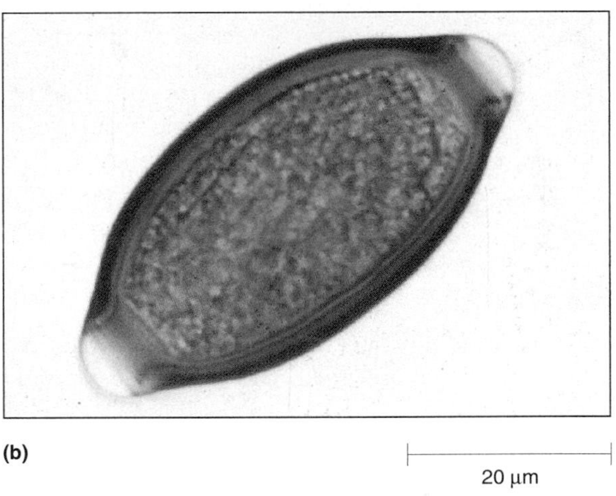

Figure 24.27 Whipworm, *Trichuris trichiura* **(a)** The slender anterior portion is embedded in the mucous membrane, while the thicker portion protrudes into the intestinal lumen. **(b)** Ovum. Notice the long oval shape, thick wall, and plugs at each end.

bedding, various surfaces, and even the air in the household become contaminated. Inhaled and ingested ova are swallowed and hatch in the small intestine, and the larval worms reach maturity in 2 to 6 weeks. Ova can survive for about a week in cool moist conditions, but they quickly die if conditions are hot and dry. Enterobiasis occurs worldwide and, unlike most other helminth diseases, is more common in temperate climates and does not spare upper socioeconomic groups. Humans are the only host for *E. vermicularis*.

Prevention and Treatment

Hygienic habits and good ventilation minimize the risk of acquiring *E. vermicularis*. Treatment is easy because effective medications are available, but preventing reinfestation is difficult. All members of a household must be treated simultaneously, even those without symptoms. Clothing and bedding should be washed in hot water, and household surfaces thoroughly cleaned. Ideally, heat in the household should be turned up and all members take a vacation for a week.

Whipworm Disease (Trichuriasis)

Whipworm disease, trichuriasis, affects an estimated 500 to 800 million people in warm and wet regions of the world where people defecate on the ground or use human feces for fertilizer.

Symptoms

Most infestations, even with as many as 100 worms, are asymptomatic. Very heavy infestations can cause anemia from blood loss, weight loss, abdominal pain, diarrhea, and protrusion of the bowel through the rectum from forcefully straining to expel the worms.

Causative Agent

The causative agent of trichuriasis is the whipworm, *Trichuris trichiura*. The worms are 3 to 5 cm long, with a slender anterior portion and a short, thicker posterior portion, hence the name whipworm (**figure 24.27a**). The unmistakable ova can be found by microscopic examination of the feces (figure 24.27b).

Pathogenesis

Trichuriasis is contracted by ingesting the mature *T. trichiura* ova. The ova hatch in the small intestine, and the emerging larvae burrow into pockets in the intestinal epithelium. Here they develop for about a week, then return to the bowel lumen and migrate to the large intestine. They become sexually mature and start producing eggs in 1 to 2 months. The slender anterior region of the worm is inserted into the intestinal mucosa, with the thicker portion protruding into the bowel lumen. The intestinal epithelium is damaged by the penetrating worm, causing blood loss and an inflammatory response. Sometimes people infested with *T. trichiuria* develop hives because of allergy to worm antigens.

Epidemiology

Infection is by the fecal-oral route. Each mature female worm produces about 5,000 ova per day. These ova are eliminated with the feces and can contaminate hands and foods, but they are not immediately infectious. The ova require about 3 weeks in a warm, moist, and shady place before they are capable of parasitizing a new host. The worms can live 3 to 8 years, so that repeated ingestions of eggs can cause a large buildup of the numbers of the worms in a person's intestine. Also, because of the long life of *T. trichiura*, the worms are frequently found in people who traveled to endemic areas years ago. Humans are the only reservoir for *T. trichiura.*

Prevention and Treatment

Trichuriasis is prevented by proper disposal of human feces and other sanitary measures. Several medications provide safe and effective treatment of the disease.

Ascariasis

Ascariasis is a helminthic disease caused by *Ascaris lumbricoides*, the largest and most prevalent human roundworm. Worldwide, an estimated 1 billion people are infested.

Symptoms

Most cases of ascariasis are asymptomatic. Many individuals infested with the worms, however, develop one or more bouts of fever, trouble breathing, coughing, and wheezing. Some infested people have the frightening experience of vomiting up a large worm, or passing one in their feces; others develop abdominal pain when the worms obstruct the intestine or gallbladder.

Causative Agent

Ascaris lumbricoides (see figure 24.28) is a roundworm prevalent worldwide in areas where human feces contaminate the soil. Female worms are enormous, ranging from 20 to 45 cm long and 3 to 6 mm wide, while males are somewhat smaller. The ova are nearly spherical, with a thick wall and irregular outer surface (see figure 24.28). The life cycle of *A. lumbricoides* is complex. The ova released by mature females are eliminated with the host's feces. The tiny worm larva inside the ova must develop in soil or other location outside the host's body for at least $1\frac{1}{2}$ weeks before they can survive and mature in a new host. After being swallowed by the new host, the ova hatch in the duodenum. At this point, the tiny larval worms do something totally unexpected. Instead of staying in the intestine, they burrow through the intestinal epithelium, enter the circulatory system, and are carried throughout the body. Those that reach the lungs push through the capillary walls into the alveoli. Here they undergo further development, reach 1 to 2 mm in length, and shed their covering twice to get ready for the rest of their journey. After about 10 days, they move out of the alveoli into the air passages of the lungs, and they are coughed up and swallowed (**figure 24.28**). The developmental changes that take place in the lungs make them resistant to stomach juices, and so they pass safely through the stomach into the small intestine. They mature to adulthood after about 2 months in the upper small intestine, feeding on intestinal contents. They live about 1 year.

Pathogenesis

Large masses of worms can obstruct the intestine, while wandering worms can obstruct or perforate various organs. Except for the larval worms that reach the lung, circulating larvae lodge in capillaries throughout the body and die, causing an inflammatory and immune response and consequent aches, pains, and fever. The larval penetration into the lungs causes microscopic bleeding, which if widespread, can predispose one to bacterial pneumonia; lesser amounts of damage provoke inflammatory and immune responses, coughing, wheezing, and pain.

Roundworms of some species of animals can penetrate human tissues but cannot complete their life cycle. Instead, their larvae wander through the tissues and eventually die, causing an intense inflammatory reaction. Preschool children are especially susceptible to this condition, called **visceral larva migrans**, because of their tendency to put rocks and dirt into their mouths. If the material is contaminated with certain roundworm ova, the eggs hatch, and the larvae migrate through body tissues. One example is *Toxocara canis*, an intestinal roundworm of dogs that resembles *A. lumbricoides.* This helminth can cause damage to various organs of a child, even blindness when its larvae lodge in the eyes.

Epidemiology

Humans are the only natural hosts of *A. lumbricoides.* A closely similar helmith of pigs can sometimes infect people. The helminth female produces more than 200,000 ova daily. These ova are remarkably resistant to heat, cold, drying, and disinfectants. Moreover, the ova remain infectious for up to 10 years, contaminating soil, dust, and food. In many areas of the world, human feces is used as fertilizer for food crops, and children defecate outdoors causing extensive soil contamination. Young children commonly place soiled objects into their mouth, and they are prone to eat soil if they are anemic or iron deficient. Children may have 50 or more worms, the record being well over 1,000.

Prevention and Treatment

Prevention of ascariasis depends on hygienic practices and sanitary disposal of human feces. Effective and safe medications are available for treating the disease.

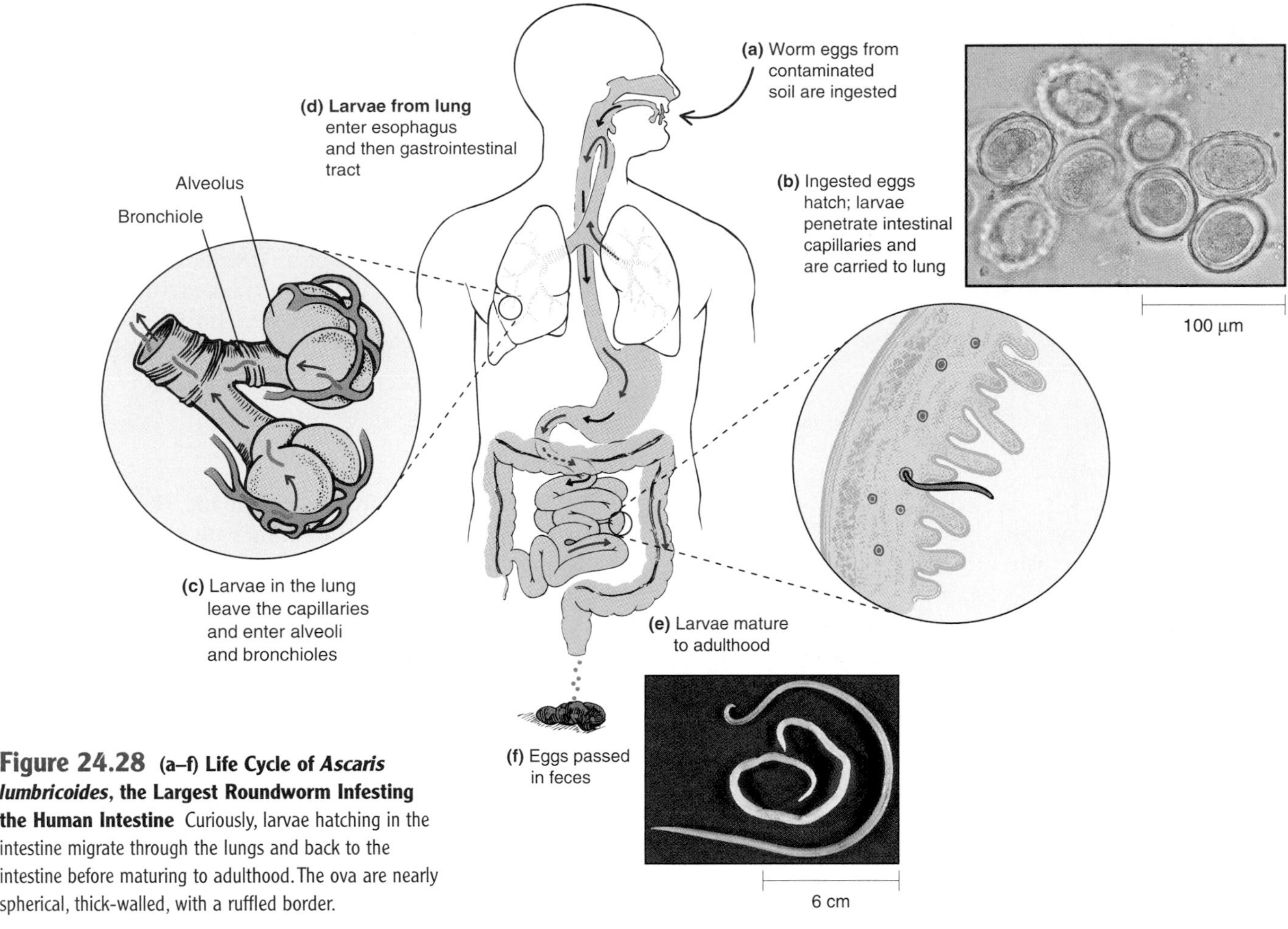
(a) Worm eggs from contaminated soil are ingested
(b) Ingested eggs hatch; larvae penetrate intestinal capillaries and are carried to lung
(c) Larvae in the lung leave the capillaries and enter alveoli and bronchioles
(d) Larvae from lung enter esophagus and then gastrointestinal tract
(e) Larvae mature to adulthood
(f) Eggs passed in feces
Alveolus
Bronchiole
100 µm
6 cm

Figure 24.28 (a–f) Life Cycle of *Ascaris lumbricoides*, the Largest Roundworm Infesting the Human Intestine Curiously, larvae hatching in the intestine migrate through the lungs and back to the intestine before maturing to adulthood. The ova are nearly spherical, thick-walled, with a ruffled border.

Hookworm Disease

Hookworms infest about 900 million people in warm, wet areas around the world where people defecate on the ground and do not wear shoes. Hookworm disease was once common in the southeastern United States. Asymptomatic individuals are still seen in most states, but they have contracted their hookworms in other countries.

Symptoms

The vast majority of hookworm infestations are asymptomatic. An itchy foot rash commonly occurs at the site of skin penetration by the worms. Cough, fever, and shortness of breath can occur, and later, diarrhea, nausea, and vomiting. The most important symptoms of hookworm disease are mental and physical retardation in children.

Causative Agents

Necator americanus and *Ancylostoma duodenale* account for most cases of human hookworm disease. The worms are about 10 mm long and live in the small intestine, attaching by means of small hooks or plates located about their mouth. The hookworm life cycle is shown in **figure 24.29**. The female worm releases her eggs, and they are discharged with the feces, contaminating soil in places that lack toilets. The eggs hatch in the soil, releasing tiny larvae that develop in the soil and become infectious. These larvae generally penetrate the skin of a person's feet, wiggle their way into the bloodstream, and are carried to the heart and lungs. In the lungs, as in ascariasis, they push through the capillary walls into the alveoli, migrate into the bronchial tubes, and then enter the gastrointestinal system when lower respiratory tract secretions are coughed up and swallowed. Once in the small intestine, the larvae attach and mature to complete their life cycle.

Pathogenesis

Hookworms feed by sucking the blood of the host. More than 1,000 worms may be found in a single person, and the constant loss of blood frequently produces anemia. Children who acquire hookworms may be undernourished already, and the anemia resulting from hookworm disease may cause weakness, fatigue, and physical and mental retardation. Initial attachment of the worms can cause nausea and vomiting. Migration of the larvae through the lungs causes microscopic bleeding and inflammatory and immune responses that result in coughing, wheezing, and shortness of breath. The itchy rash, called ground itch, that

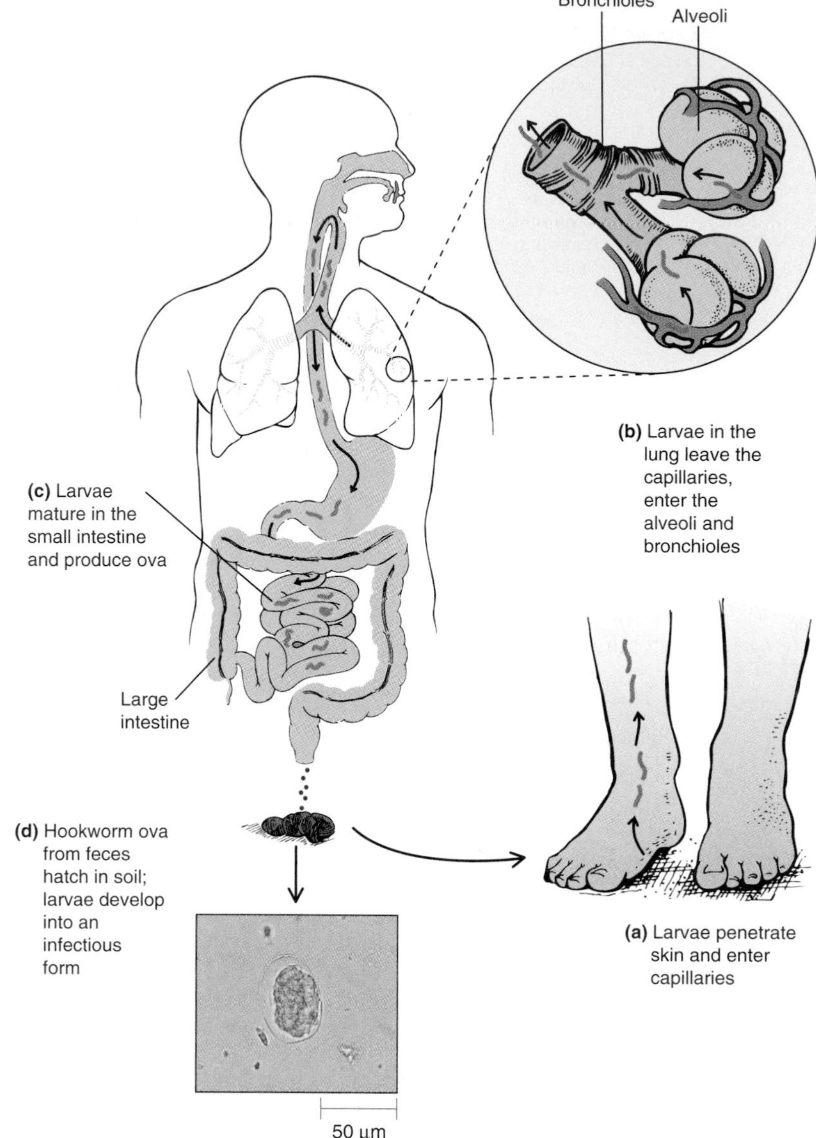
Bronchioles
Alveoli
(b) Larvae in the lung leave the capillaries, enter the alveoli and bronchioles
(c) Larvae mature in the small intestine and produce ova
Large intestine
(d) Hookworm ova from feces hatch in soil; larvae develop into an infectious form
(a) Larvae penetrate skin and enter capillaries
50 µm

Figure 24.29 (a–d) Life Cycle of a Hookworm Infection begins with penetration of the skin by the worm larvae. The micrograph shows the thin-walled ovum containing the developing worm embryo.

often develops between the toes at the site of larval penetration, is probably an allergic response to repeated exposures.

The larvae of dog and cat hookworms can contaminate soil and penetrate the skin, especially of people such as plumbers that sometimes have to crawl in dirt under buildings. The larvae wander just under the skin surface, causing itchy red tracts **(figure 24.30)** that advance as much as 1 to 2 cm per day for a month or more, before dying. This condition is called **cutaneous larva migrans**.

Epidemiology

Hookworm disease occurs around the world in temperate and tropical climates where the soil is moist, human feces is deposited on the ground, and people are barefooted. Each female worm can produce 25 million ova in her lifetime of 2 to 10 years.

Prevention and Treatment

Prevention of hookworm disease depends on sanitary disposal of human feces and wearing shoes. Highly effective medications are available to eliminate the infestation.

Strongyloidiasis

Strongyloidiasis, the disease caused by *Strongyloides stercoralis*, occurs in a spotty distribution in warm, wet areas of the world, with an estimated 400,000 cases in the southeastern United States and Puerto Rico. This disease differs from most other helminthic diseases in that it perpetuates itself indefinitely in the absence of reexposure to the causative agent.

Symptoms

Most infections are asymptomatic, but abdominal pain, diarrhea, recurrent rashes on the buttocks and lower back, and respiratory symptoms can occur with heavy infestations.

Causative Agent

The adult *Strongyloides stercoralis* **(figure 24.31)**, sometimes called a threadworm, is only about 2 mm long. Its life cycle is similar to that of hookworms with two important exceptions: (1) the worms can multiply sexually in the soil; and (2) ova produced by intestinal worms hatch before they are discharged in the feces. These intestinal larvae can mature to an infectious form, penetrate the intestine or the anal skin, and thereby perpetuate the infection indefinitely.

Pathogenesis

Most individuals with strongyloidiasis are asymptomatic because their immune system limits tissue invasion. Heavy infestations, however, can damage the lungs and intestine, sometimes leading to invasion by bacterial pathogens. Recurrent rashes result from hypersensitivity to the larvae at the sites of penetration. In people who are debilitated, alcoholic, or immunosuppressed, massive numbers of the worms arise, penetrating all parts of the body, including the brain. These massive infections are rapidly fatal unless treated.

Epidemiology

Strongyloidiasis occurs in much the same distribution as hookworm, but favoring tropical areas. Since the organisms can multiply in soil, contaminated areas may remain infectious even when there are no new deposits of human feces. Human infections can last for a lifetime, and they can be transmitted to other people through close physical contact. Many American soldiers contracted the disease during World War II while imprisoned in Southeast Asia.

Prevention and Treatment

Prevention of strongyloidiasis is the same as for hookworm disease. Effective medications are available but require a longer

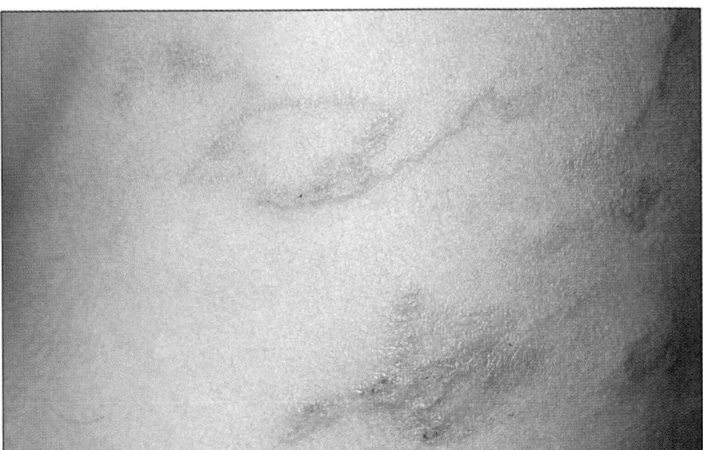

Figure 24.30 Cutaneous Larva Migrans, an Occupational Hazard of Plumbers and Others Who Crawl Under Buildings Where Dog or Cat Feces Have been Deposited Larval hookworms from the feces penetrate and wander under the skin, causing serpentine tracts of inflammation before dying.

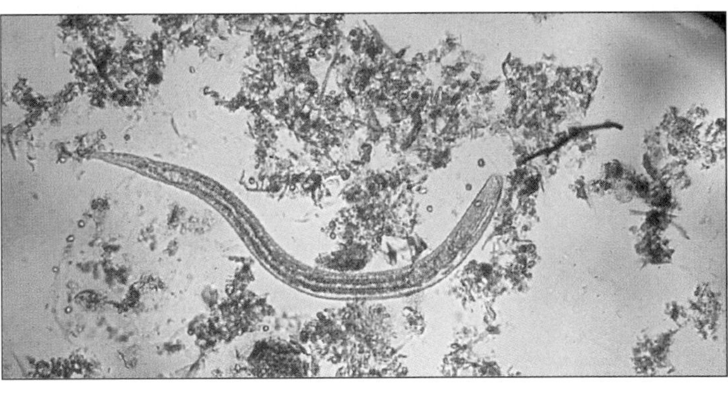

Figure 24.31 *Strongyloides stercoralis* Larva in Feces The adult worm is only about 2 mm long, and has a life cycle similar to hookworms. Unlike other intestinal roundworms, however, *S. stercoralis* can multiply in the soil and in the body of a single host.

period of treatment than for hookworm, and even so, relapses may occur.

Trichinosis (Trichinellosis)

Trichinosis, also known as trichinellosis, differs from all the roundworm diseases discussed previously in that it is contracted by eating inadequately cooked meat. Since pork is a leading source of infection, the incidence of the disease has fallen markedly in the United States over the years due largely to federal regulations on how pigs raised commercially are fed. Nevertheless, the potential for large outbreaks of the disease still exists because the meat of a number species of animals can be infectious.

Symptoms

Trichinosis is characterized by abdominal pain and diarrhea, followed in about a week by fever, muscle pain, swelling around the eyes, and sometimes a rash. Occasional cases are fatal because of damage to the heart or brain. The disease occurs worldwide except for Australia and some Pacific Islands.

Causative Agent

The cause of trichinosis is usually *Trichinella spiralis*, a 1 to 4 mm long roundworm that lives in the small intestine of meat-eating animals, especially rats, pigs, bears, dogs, and humans. Its life cycle is shown in **figure 24.32**. The female worm discharges her living young into the lymph and blood vessels of the host's intestine without an intervening egg stage, and those larvae are carried to all parts of the body. Most of the larvae are killed by body defenses, but some survive in the muscles of the host where they become encased with scar tissue. The worms then stay alive for months or years within the muscle. If the flesh of an infested pig or other carnivorous animal is eaten by humans or other animals, the digestive juices of the new host release the larvae, permitting them to burrow into the new host's intestinal lining. The larvae then mature, and the females begin producing larvae in the new host to complete the life cycle of the worm. Each female adult *Trichinella* may live 4 months or more and produce 1,500 young.

Pathogenesis

The penetration of larval worms into the host's tissues is responsible for the symptoms of trichinosis. Abdominal pain and diarrhea begin within the first week after eating meat containing *T. spiralis*, when the worms mature and begin discharging their larvae. Fever, muscle pain, rash, and facial swelling result from inflammatory and immune reactions to larvae lodged in various tissues. Only larvae that penetrate skeletal muscle survive.

Epidemiology

In trichinosis, the same animal generally serves as both the definitive and the intermediate host, human beings one hopes being the exception. Almost all warm-blooded carnivores can be hosts for *T. spiralis*. Many cases of the disease have been contracted from the meat of wild animals, including bear, wild boar, walrus, and cougar.

Prevention and Treatment

Prevention of trichinosis depends on thorough cooking of meat so that all parts reach at least 170°F (77°C). Pigs raised commercially must by law be fed only cooked garbage to ensure that they do not receive meat scraps containing viable *Trichinella* larvae. Government inspection of meat does not detect *Trichinella* infestation. Larvae in pork are generally killed at 5°F (−15°C) or lower for 3 weeks if the meat is less than 15 cm thick. Wild game, however, often harbors *Trichinella* species that are not killed by freezing. Medication is available that can probably prevent the disease in persons who have eaten meat shown to contain *Trichinella*, but there is no effective treatment.

MICROCHECK 24.8

Parasitic worms are divided into the roundworms (nematodes) and the flatworms. Flatworms are further divided into tapeworms (cestodes) and flukes

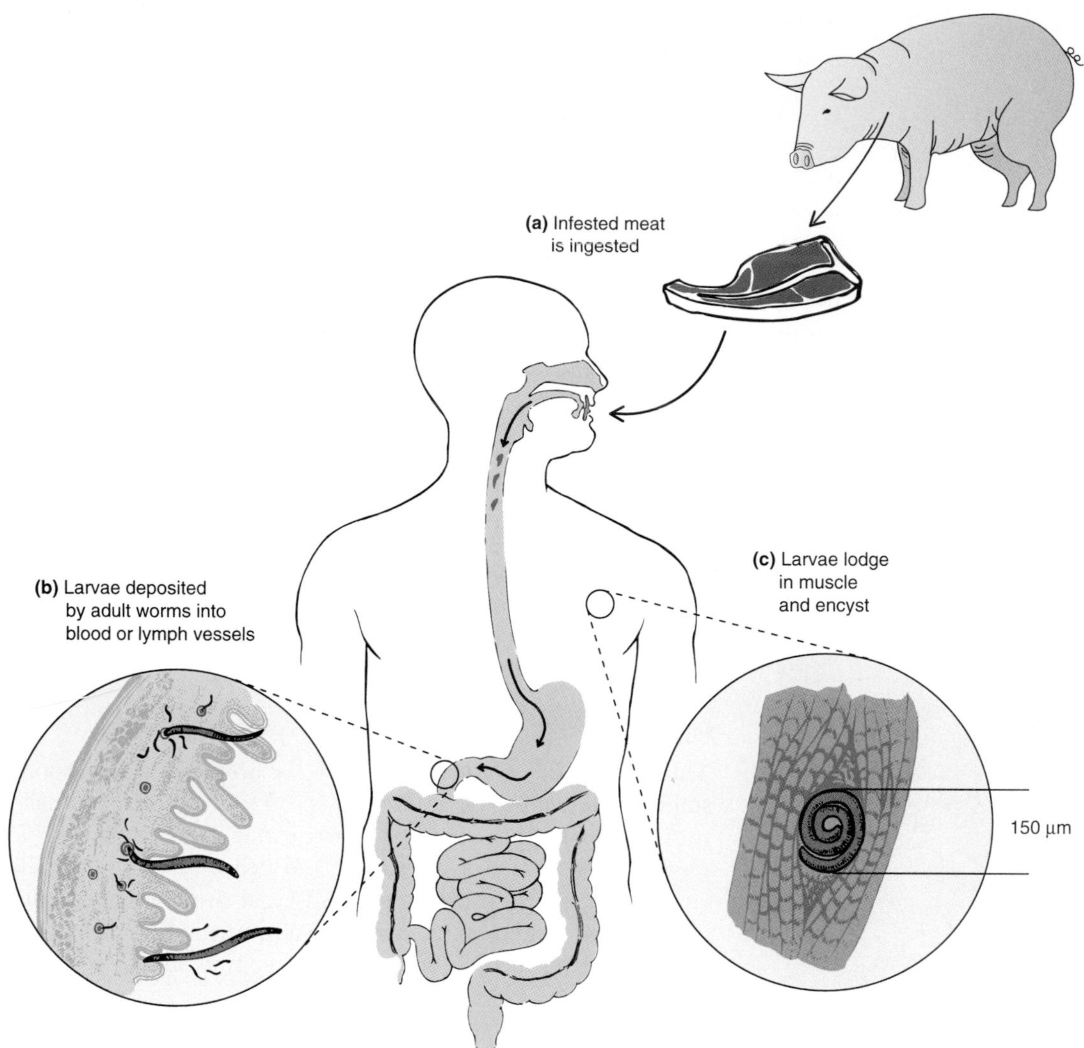
(a) Infested meat is ingested
(b) Larvae deposited by adult worms into blood or lymph vessels
(c) Larvae lodge in muscle and encyst
150 μm

Figure 24.32 **(a–c) Life cycle of *Trichinella spiralis*, the Cause of Trichinosis** Most cases of the disease result from eating inadequately cooked pork or bear meat.

(trematodes). Roundworm diseases include pinworm, whipworm, and ascariasis contracted by ingesting worm ova; hookworm disease and strongyloidiasis, contracted through skin penetration by the worm larvae; and trichinosis, acquired by ingesting inadequately cooked meat containing encysted worm larvae.

- Why might asymptomatic strongyloidiasis pose a serious threat to an individual who develops immunodeficiency?
- Would you be surprised to see new cases of ascariasis turning up 3 years after all the inhabitants of a Southeast Asian village began using outhouses to deposit their feces? Why or why not?

Diseases Due to Intestinal Flatworms

This section gives some examples of diseases due to tapeworms, cestodes, and an example of a disease caused by a very large intestinal fluke, or trematode. ■ cestodes, p. 317 ■ trematodes, p. 318

Tapeworm Disease

Tapeworms can exceed 10 m (more than 30 ft.) in length and have a ribbonlike appearance. Their structure is quite distinctive in that there is no digestive tract. The head end, called a **scolex**, is only 1 to 2 mm in diameter and is merely an attachment device to anchor the organism so that it is not eliminated by peristalsis. The neck region continuously produces segments called **proglottids** that become more mature the further they are from the scolex. The proglottids are little more than hermaphroditic egg factories; depending on the species, ova are discharged from an opening in the proglottid, or the proglottid separates from the rest of the worm and releases its ova by breaking apart. People become infested with tapeworms by eating inadequately cooked red meat or fish. Three important tapeworms are the beef tapeworm, the pork tapeworm, and the fish tapeworm. Beef, pork, and fish refer to the sources of human infection, not where the adult worms exist; human beings are the definitive host for all three.

Symptoms

Most tapeworm victims have few or no symptoms. An occasional individual will develop vomiting and abdominal pain.

Rarely, weakness, unsteadiness, mental abnormalities, tumorlike swellings, blindness, or epilepsy occur.

Causative Agents

The tapeworms *Taenia saginata* or beef tapeworm, *Taenia solium* or pork tapeworm, and *Diphyllobothrium latum* or fish tapeworm can be distinguished by the structure of their scolices, their proglottids, and their ova. *Taenia saginata* and *T. solium* each have a single intermediate host, cattle and pigs, respectively, while *D. latum* has two or more.

Cattle ingest *T. saginata* ova when they feed on pasture contaminated with feces from people who carry the adult worm. The ova hatch in the cattle intestines, releasing larvae that burrow through the intestinal wall and lodge in the flesh of the animals. Here each of the parasites develops into a **cysticercus**, a cystic structure containing an immature scolex. When the inadequately cooked beef is eaten by a person, the cysticercus is released from the meat by digestive enzymes, matures, attaches to the intestinal wall, and soon develops into an egg-producing adult tapeworm. The life cycle for the pork tapeworm, *Taenia solium*, is similar except pigs rather than cattle are the intermediate host. Cysticerci in inadequately cooked meat represent the only source of adult worms in people.

The intermediate hosts of *D. latum* are water fleas and fish, whereas the definitive host, the mature egg-producing form, occurs in humans; bears, dogs, and other animals can be definitive hosts for the same or similar tapeworms. When ova of *D. latum* passed with a person's feces enter fresh water, they soon release ciliated larvae, which are devoured by water fleas (**figure 24.33**). The larvae lose their cilia, claw their way into the water flea's body using tiny hooks, absorb nutrients from the flea's blood, and increase in size. These larvae can develop further only if the water flea is eaten by a fish. In a fish's muscle, the organism continues developing and becomes infectious for humans. Usually, however, it gets passed from one fish to another as bigger fishes eat smaller ones, before being eaten by the human host. In the human host, the flesh of the fish is digested away by stomach juices, releasing the wormlike larva. It attaches to the small intestine, rapidly matures, and begins producing ova.

Pathogenesis

Symptoms of tapeworm disease can arise from the presence of adult worms in the intestine. Nausea, vomiting, and abdominal pain sometimes arise because the worms partly obstruct the intestine. Also, large fish tapeworms can absorb almost 100% of vitamin B_{12} from intestinal juices, depriving the host of this essential vitamin and causing anemia, weakness, unsteadiness, and other manifestations of nervous system malfunction.

The intermediate developmental forms of tapeworms can also infect humans and cause symptoms. This occurs in the case of *D. latum* when infected water fleas are accidentally ingested, as might happen with drinking lake water. Infection can also result from the practice of applying raw meat or fish to open wounds in the mistaken belief that the meat will aid recovery. If larvae are present in the flesh, they penetrate the tissues via the wound.

Pork tapeworms are the most dangerous of the three types, because its eggs may hatch before being discharged in the persons's feces or ova contaminating fingers or foods can be ingested and hatch. Larvae emerging from the ova do the same as they would do in a pig. Although humans are a poor excuse

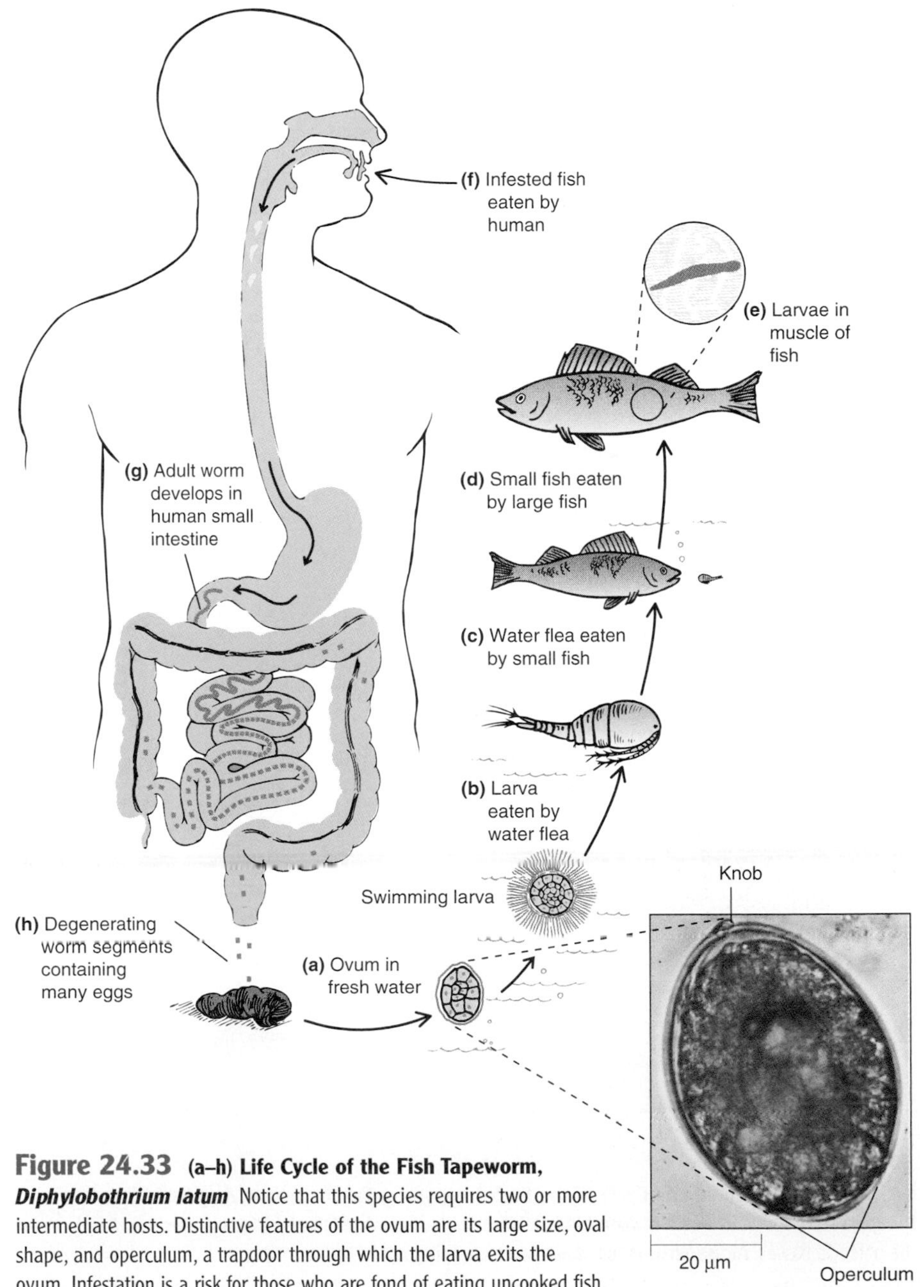

(f) Infested fish eaten by human
(e) Larvae in muscle of fish
(d) Small fish eaten by large fish
(c) Water flea eaten by small fish
(b) Larva eaten by water flea
(g) Adult worm develops in human small intestine
Swimming larva
Knob
(h) Degenerating worm segments containing many eggs
(a) Ovum in fresh water
20 µm
Operculum

Figure 24.33 (a–h) Life Cycle of the Fish Tapeworm, *Diphylobothrium latum* Notice that this species requires two or more intermediate hosts. Distinctive features of the ovum are its large size, oval shape, and operculum, a trapdoor through which the larva exits the ovum. Infestation is a risk for those who are fond of eating uncooked fish.

for an intermediate host, the larvae penetrate the intestinal wall by means of tiny hooklets. Then they enter the bloodstream, from which they invade tissues throughout the body, including the eyes, brain, heart, and muscles. In the tissues they develop into cysticerci, 0.5 to 5 or more cm in diameter. The condition resulting from cysticerci in the tissues is called **cysticercosis** and can mimic a brain tumor or cause epilepsy **(figure 24.34)**.

Epidemiology

Tapeworm disease occurs wherever people eat raw meat or fish. The beef tapeworm is the most common in the United States, reported from eight states in one survey. The fish tapeworm mainly occurs in the north central part of the United States, where formerly outhouses were constructed so as to discharge directly into the lakes. The pork tapeworm is rare in the United States, but it may occur in immigrants from parts of Mexico, India, the Philippines, and other parts of the world where eating raw pork is traditional.

Prevention and Treatment

Control of tapeworms depends on adequate cooking of meat until it is no longer pink in the center. Fish should also be thoroughly cooked, or if eaten raw should be deep frozen for a week before eating. Proper disposal of human feces and good hygienic habits are especially important for controlling pork tapeworm disease. Effective medicines are available both for ridding the body of adult worms and for treating cysticercosis.

Giant Intestinal Fluke Disease (Fasciolopsiasis)

Fasciolopsiasis, disease caused by *Fasciolopsis buski*, may be acquired in a number of countries of Southern Asia, including parts of China, India, Myanmar (Burma), Thailand, Laos, Cambodia, Vietnam, Malaysia, Indonesia, and the Philippines. An estimated 10 million people are infested worldwide.

Symptoms

Symptoms of fasciolopsiasis are diarrhea, abdominal pain, generalized weakness, and swelling of the face and other parts of the body.

Causative Agent

Fasciolopsis buski, also known as the giant intestinal fluke, can be as large as 7.5 cm long and 2.0 cm wide. This hermaphroditic helminth lives in the upper and middle small intestine; attachment

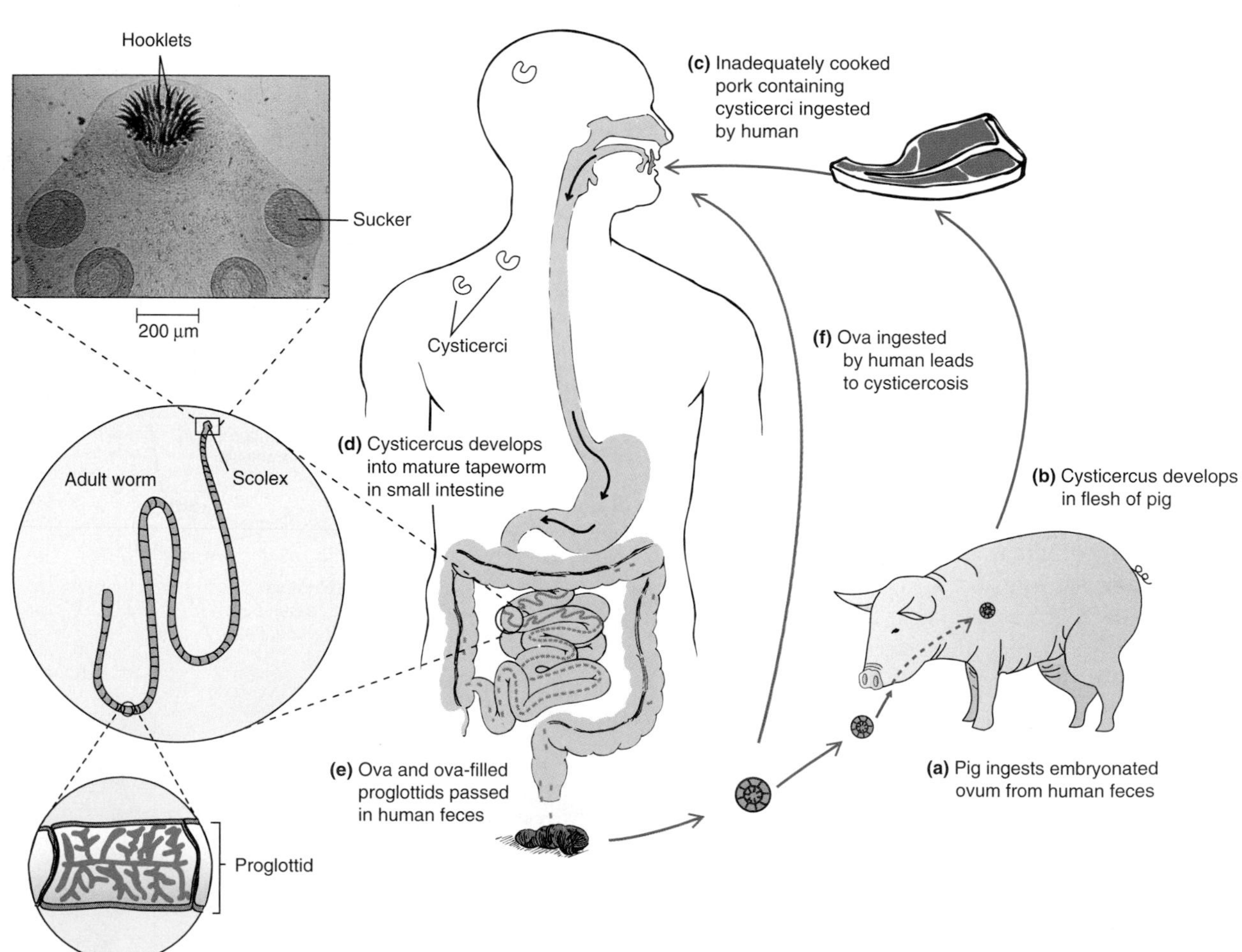

Figure 24.34 **(a–e) Life Cycle of the Pork Tapeworm, *Taenia solium*, Acquired by Eating Inadequately Cooked Pork** If worm ova hatch in the intestine, larval forms called cysticerci can develop throughout the person's tissues. A cysticercus in the brain can cause epilepsy.

and movement is aided by two suckers. Unlike tapeworms, flukes like *F. buski* have a digestive system. Ova are passed in the feces of the definitive hosts, humans and pigs, and the enclosed embryos mature in warm fresh water in about 6 weeks (**figure 24.35**). The ova then hatch, releasing ciliated larvae, which penetrate the flesh of certain snail species that serve as intermediate hosts. In the snail, the organisms develop and multiply asexually, eventually exiting the snail in the form of motile long-tailed larvae called **cercariae**. These cercariae attach to the outer covering of water plants such as water chestnuts, where they form cysts infectious for pigs and human beings. People generally become infected when they peel off the outer portion of water plants with their teeth in order to eat the edible inner portion of the plant. The worms exit from their cysts in the duodenum, attach, and after several months begin producing ova. ■ cercariae, p. 318

Pathogenesis

Inflammation occurs at the site of attachment of the helminths, followed by ulceration and bleeding. Heavy infestations can interfere with nutrition or even obstruct the intestinal tract. Severe symptoms also arise from toxicity and allergy to substances released by the worms and absorbed by the host's circulation.

Epidemiology

Fasciolopsiasis occurs in warm climates where people raise pigs and eat raw freshwater plants, and where a suitable snail intermediate host exists. Feces from infested humans or pigs must reach the ponds inhabited by the snails and edible water plants. Each adult worm produces more than 20,000 ova per day and lives about 1 year.

Prevention and Treatment

Fasciolopsiasis can be prevented by killing encysted *F. buski* on water plants by dipping them in boiling water for a few seconds before peeling and eating. Rinsing the edible portion thoroughly in clean water after peeling is probably also effective. The disease is effectively treated with praziquantel. **Table 24.17** compares some features of these parasitic helminths.

MICROCHECK 24.9

Tapeworms are ribbonlike creatures, consisting largely of a string of hermaphroditic egg-producing segments. The pork tapeworm, *Taenia solium*, is the most dangerous because its ova can hatch in the human intestine, the larvae entering brain and other tissues,

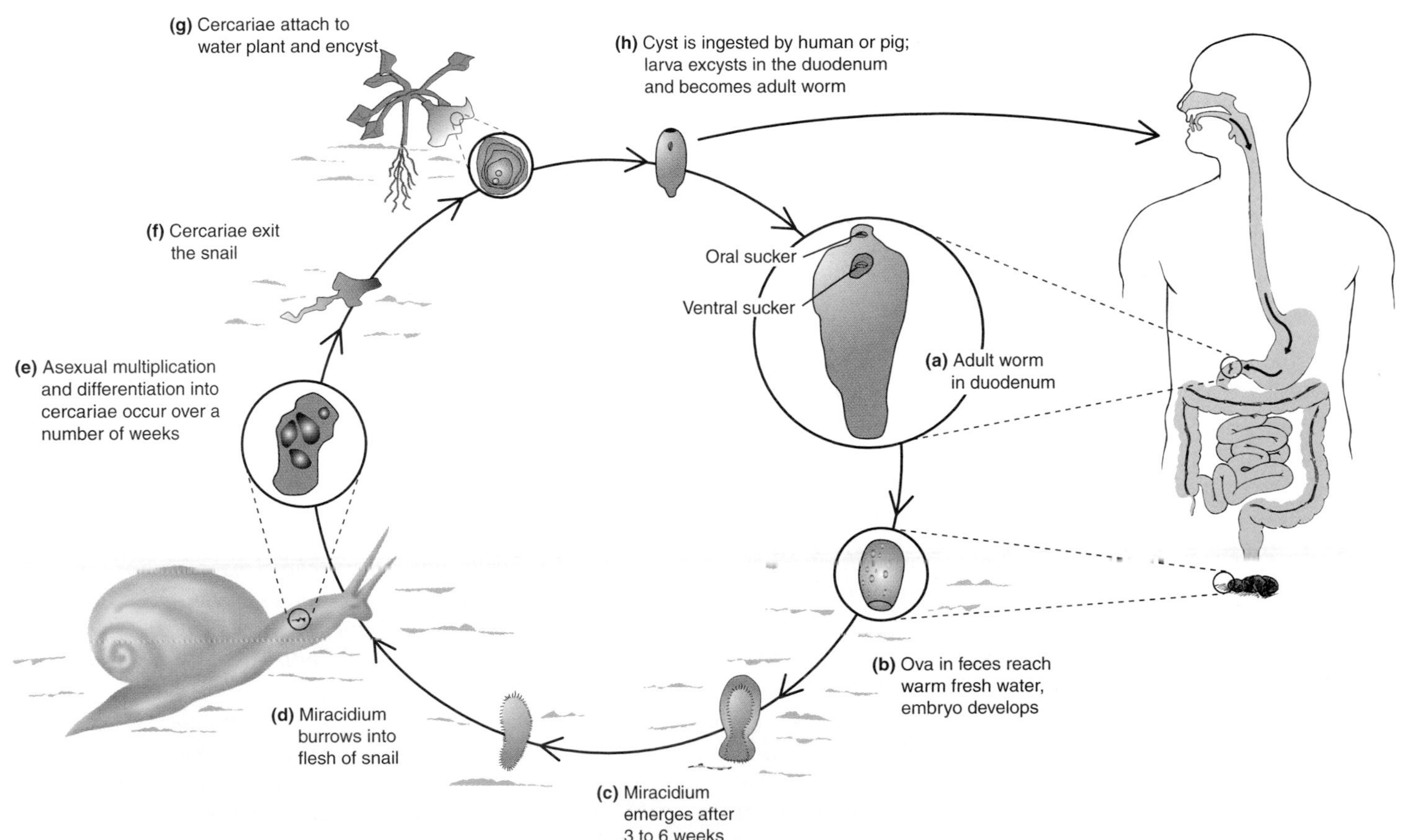
(g) Cercariae attach to water plant and encyst
(h) Cyst is ingested by human or pig; larva excysts in the duodenum and becomes adult worm
(f) Cercariae exit the snail
Oral sucker
Ventral sucker
(a) Adult worm in duodenum
(e) Asexual multiplication and differentiation into cercariae occur over a number of weeks
(b) Ova in feces reach warm fresh water, embryo develops
(d) Miracidium burrows into flesh of snail
(c) Miracidium emerges after 3 to 6 weeks

Figure 24.35 **(a–h) Life Cycle of the "Giant Intestinal Fluke," *Fasciolopsis buski*** The infectious larva of this worm encysts on certain edible water plants, including a kind of water chesnut. People become infested when they peel the edible portion with their teeth.

TABLE 24.17 Notable Helminthic Diseases of Human Beings

Disease and Causative Agent	How Acquired	Disease Characteristics	Prevention and Control
Enterobiasis (pinworm), *Enterobius vermicularis*	Hand to mouth; inhalation	Itching of anal region, restlessness, nervousness, irritability, poor sleep	Handwashing, daily change of underclothing and bed sheets
Trichuriasis, (whipworm) *Trichuris trichiura*	Ingestion of eggs of parasite along with contaminated food or water	Abdominal pain, bloody stools, diarrhea, and weight loss	Sanitary disposal of human feces; frequent handwashing
Ascariasis, *Ascaris lumbricoides*	Ingestion of eggs of parasite or water along with contaminated food	Abdominal pain, live worms vomited or passed in stool	Sanitary disposal of human feces
Hookworm disease, *Necator americanus* and *Ancylostoma duodenale*	Larvae penetrate bare feet	Anemia, weakness, fatigue, physical and mental retardation in children	Sanitary disposal of human feces; wearing shoes
Strongyloidiasis, *Strongyloides stercoralis*	Larvae penetrate bare feet	Skin rash at site of penetration, cough, abdominal pains, weight loss	Sanitary disposal of human feces; wearing shoes
Trichinosis, *Trichinella spiralis*	Eating raw or undercooked meat, usually pork	Fever, swelling of upper eyelids, muscle soreness	Adequate cooking of meat
Tapeworm disease, *Taenia solium*, *T. saginata*, *Diphyllobothrium latum*	Eating raw or undercooked meat or fish	Abdominal pain, anemia, cysticercosis	Thorough cooking or deep freezing of meat and fish
Fasciolopsiasis, *Fasciolopsis buski*	Eating raw water plants with encysted larvae	Diarrhea, abdominal pain, weakness, body swelling	Brief treatment of plants in boiling water

resulting in cysticercosis. Tapeworm disease is usually acquired by eating inadequately cooked beef, pork, or fish. Fasciolopsiasis is caused by the giant intestinal fluke, a relatively short, nonsegmented hermaphoditic flatworm usually acquired when a person uses their teeth to peel certain edible aquatic plants having encysted fluke larvae on their surface.

- Describe the anatomy of a tapeworm.
- How might you explain the development of epilepsy in a Mexican man fond of eating traditional dishes made from raw pork?

FUTURE CHALLENGES

Defeating Diarrhea

Development of better preventive and treatment techniques for alimentary tract diseases has an urgency arising from massive food and beverage production and distribution methods. For example, in 1994, an estimated 224,000 people became ill because a tanker truck used to transport ice cream mix had previously carried liquid eggs. One day's production from a ground beef factory can yield hundreds of thousands of pounds of hamburgers, which are soon sent to many parts of this or other countries. The challenge is to better educate the producers and transporters, to develop guidelines to help them avoid contamination, and to utilize newly approved methods such as meat irradiation.

Other challenges include the following:

- *Exploit the power of molecular biology techniques to produce effective vaccines against rotavirus, hepatitis C virus, and bacterial pathogens. There is special need for bacterial vaccines that can be administered by mouth and evoke long-lasting mucosal immunity.*
- *Find better ways of preventing and reversing resistance to antibacterial medications. In the early 1980s, 0.6% of* Salmonella *Typhimurium isolates were resistant to ampicillin, chloramphenicol, tetracycline, streptomycin, and sulfa drugs. Now, in some areas, over 30% of isolates of this common salmonellosis cause are resistant to all these medications.*
- *Explore further the influence of global warming on alimentary tract diseases. A study of Peruvian children by Johns Hopkins University scientists found an 8% increase in clinic visits for diarrhea with each 1°C increase in temperature from the normal.*
- *Explore new prevention and treatment options. Scientists at the University of Florida have developed a genetically engineered* Streptococcus mutans that *does not produce lactic acid but readily displaces wild strains of the dental decay causing bacterium. Researchers at the University of Alberta have custom-designed a molecule that binds circulating Shiga toxin, potentially preventing hemolytic uremic syndrome. Others work on finding an effective therapy for cryptosporidiosis.*

SUMMARY

Anatomy and Physiology (Figures 24.1, 24.2)

1. The alimentary tract is the passageway from the mouth to the anus. It is a major route for pathogens to enter the body.
2. The alimentary system is composed of the mouth and saliva producing glands, esophagus, stomach, small intestine, liver, pancreas, and large intestine.

Normal Flora

The Mouth

1. The species of bacteria that inhabit the mouth colonize different locations depending on their ability to attach to specific receptors.
2. **Dental plaque** consists of enormous quantities of bacteria of various species attached to teeth or each other. (Figure 24.3)
3. The presence of teeth allows for colonization by anaerobic bacteria.

Intestines

1. The normal fasting stomach is devoid of microorganisms.
2. Microorganisms make up about one-third of the weight of feces.
3. The biochemical activities of microorganisms in the large intestine include synthesis of vitamins, degradation of indigestible substances, competitive inhibition of pathogens, chemical alteration of medications, and production of carcinogens.

UPPER ALIMENTARY SYSTEM INFECTIONS

Bacterial Diseases of the Upper Alimentary System

Tooth Decay (Dental Caries) (Table 24.1)

1. Dental caries is caused mainly by *Streptococcus mutans* involved in formation of extracellular glucans from dietary sucrose.
2. Penetration of the calcium phosphate tooth structure depends on acid production by **cariogenic dental plaque**. *S. mutans* is not inhibited by acid and stores fermentable intracellular polysaccharide. (Figure 24.4)
3. Control of dental caries depends mainly on supplying fluoride and restricting dietary sucrose. Dental sealants fill the pits and crevices in children's teeth.

Periodontal Disease (Table 24.1)

1. Periodontal disease is caused by an inflammatory response to the plaque bacteria at the gum line; it is mainly responsible for tooth loss in older people. (Figure 24.5)

Trench Mouth (Table 24.1)

1. Trench mouth, or acute necrotizing ulcerative **gingivitis** (ANUG), can occur at any age in association with poor mouth care. (Figure 24.6)

Helicobacter pylori Gastritis (Table 24.2)

1. *H. pylori* predisposes the stomach and the uppermost part of the duodenum to peptic ulcers. (Figures 24.7, 24.8)
2. Treatment with antimicrobial medications can cure the infection and prevent peptic ulcer recurrence.

Viral Diseases of the Upper Alimentary System

Herpes Simplex (Table 24.3, Figure 24.9)

1. Herpes simplex is caused by an enveloped DNA virus. Infected cells show intranuclear inclusion bodies.
2. HSV persists as a latent infection inside sensory nerves; active disease occurs when the body is stressed.

Mumps (Table 24.4, Figure 24.10)

1. Mumps is caused by an enveloped RNA virus that infects not only the parotid glands, but the meninges, testicles, and other body tissues.
2. Mumps virus generally causes more severe disease in persons beyond the age of puberty; it can be prevented using a live attenuated vaccine. (Figure 24.11)

LOWER ALIMENTARY SYSTEM INFECTIONS

Bacterial Diseases of the Lower Alimentary System (Figure 24.12)

Cholera (Table 24.5)

1. Cholera is a severe form of diarrhea caused by a toxin of *Vibrio cholerae* that acts on the small intestinal epithelium. (Figures 24.13, 24.14)

Shigellosis (Table 24.6)

1. Shigellosis is caused by species of *Shigella*, common causes of **dysentery** because they invade the epithelium of the large intestine. (Figure 24.15)
2. One *Shigella* species produces **Shiga toxin**, which causes the **hemolytic uremic syndrome**.

Escherichia coli Gastroenteritis (Tables 24.7, 24.8)

1. Virulence factors often depend on plasmids, which can transfer virulence to other enteric bacteria.
2. Some *E. coli* strains, such as O157:H7, originating from wild and domestic animals, can cause the hemolytic uremic syndrome.

Salmonellosis (Table 24.9)

1. Strains of *Salmonella* that originate from animals usually cause **gastroenteritis**. The organisms are often foodborne, commonly with eggs and poultry.
2. Typhoid fever is caused by *Salmonella* Typhi, which only infects humans. The disease is characterized by high fever, headache, and abdominal pain. Untreated, it has a high mortality rate. An oral attenuated vaccine helps prevent the disease.

Campylobacteriosis (Table 24.10)

1. *Campylobacter jejuni* is the most common bacterial cause of diarrhea in the United States; it usually originates from domestic animals. (Figure 24.16)

Viral Infections of the Lower Alimentary System

Rotaviral Gastroenteritis (Table 24.11)

1. Rotaviral gastroenteritis is the main diarrheal illness of infants and young children, but also can involve adults, as in traveler's diarrhea.

2. Rotaviruses are segmented RNA viruses of the reovirus family. (Figure 24.17)

Norwalk Virus Gastroenteritis (Table 24.11)

1. Norwalk virus gastroenteritis accounts for almost half the cases of viral gastroenteritis in the United States.
2. Norwalk viruses are small RNA viruses of the calcivirus family. (Figure 24.18)

Hepatitis A (Table 24.12)

1. Hepatitis A is usually mild or asymptomatic in children; some cases are prolonged, with weakness, fatigue, and jaundice.
2. Hepatitis A virus (HAV) is a picornavirus spread by fecal contamination of hands, food, or water.
3. An injection of gamma globulin gives temporary protection from the disease; an inactivated vaccine is available actively to immunize against the disease.

Hepatitis B (Table 24.12)

1. Hepatitis B virus (HBV) is a hepadnavirus spread by blood, blood products, semen, and from mother to baby. (Figures 24.20, 24.21)
2. Asymptomatic carriers are common and can unknowingly transmit the disease.
3. Chronic infection is common and can lead to scarring of the liver (cirrhosis) and liver cancer.

Hepatitis C Virus (HCV) Disease (Table 24.12)

1. Hepatitis C virus (HCV) is a flavivirus transmitted mainly by blood; 60% of cases are acquired from needle sharing by injecting drug abusers.
2. Hepatitis C is asymptomatic in over 60% of acute infections; 80% of infections become chronic.

Protozoan Diseases of the Lower Alimentary System

Giardiasis (Table 24.13)

1. Transmission of *Giardia lamblia*, a mastigophoran, is usually by drinking water contaminated by feces. It is a common cause of traveler's diarrhea. (Figure 24.23)
2. Its cysts survive in chlorinated water and must be removed by filtration.

Cryptosporidiosis (Table 24.14)

1. The life cycle of *Cryptosporidium parvum*, a coccidium, a member of the Apicomplexa, takes place in the small intestinal epithelium. (Figure 24.24)
2. Its oocysts are infectious, resist chlorination, and are too small to be removed by most filters.
3. It is a cause of many water and foodborne epidemics, and traveler's diarrhea.

Cyclosporiasis (Table 24.15)

1. Transmission of *Cyclospora cayetanensis*, a coccidium, is fecal-oral, via water or produce such as berries; it causes traveler's diarrhea.
2. Oocysts are not infectious when passed in feces, and thus no person-to-person spread; no hosts other than humans are known.

Amebiasis (Table 24.16)

1. *Entamoeba histolytica*, a member of the Sarcodina group of protozoa, is an important cause of dysentery; often chronic; infection can spread to the liver and other organs. (Figure 24.25)
2. The organism exists in ameba and cyst forms; the quadrinucleate cyst is infectious.

MULTICELLULAR PARASITES

Diseases Due to Intestinal Roundworms

Pinworm Disease (Enterobiasis)

1. The **roundworm**, *Enterobius vermicularis*, also known as the pinworm, causes the most common **helminth** disease in the United States. (Figure 24.26)
2. Transmission is fecal-oral, sometimes airborne.
3. Main symptom is anal itching; also emotional distress; reinfection is common.

Whipworm Disease (Trichuriasis)

1. *Trichuris trichiura*, also known as the whipworm, is common in tropical climates where unsanitary conditions prevail. (Figure 24.27)
2. Ova mature in warm, moist soil before becoming infectious by the fecal-oral route.
3. Symptoms include abdominal pain, diarrhea, and protrusion of the bowel from the rectum with heavy infestations.

Ascariasis (Figure 24.28)

1. *Ascaris lumbricoides*, found worldwide, the largest and most prevalent of the roundworms.
2. Transmission by the fecal-oral route of infectious ova; ova can remain infectious in the soil for years.
3. Life cycle involves migration of larvae through the lungs, where they develop acid resistance before passage through the stomach and maturing in the small intestine.
4. Symptoms arise from trauma of seeing the worms, from migration of larvae through the lungs, and from adult worms obstructing or perforating various organs.

Hookworm Disease (Figure 24.29)

1. Hookworm disease is caused by two roundworms, *Necator americanus* and *Ancylostoma duodenale*, in moist, warm areas of the world where people defecate on the ground and do not wear shoes.
2. After penetrating the skin, hookworm larvae enter blood vessels and are carried to the lungs, and are coughed up and swallowed. In the small intestine, they attach and suck blood for nourishment.
3. Similar larvae of dog and cat hookworms can penetrate and wander under the skin of human beings, but not complete their life cycle, causing the condition called **cutaneous larva migrans**. (Figure 24.30)

Strongyloidiasis

1. The roundworm, *Strongyloides stercoralis*, is prevalent under the same conditions as hookworms; infectious larvae in the soil penetrate human skin. (Figure 24.31)

2. Infectious larvae penetrate blood vessels, are carried to the lungs, are coughed up and swallowed, and mature in the small intestine.

Trichinosis (Trichinellosis) (Figure 24.32)

1. *Trichinella spiralis* is transmitted by inadequately cooked meat; the adult worm lives in the small intestine of meat-eating warm-blooded animals such as rats, pigs, bears, dogs, and humans.
2. Adult *T. spiralis* deposit their larvae directly into intestinal blood vessels.
3. Larvae that arrive in muscle become encased in scar tissue and remain alive and infectious for years; the **definitive host** becomes an **intermediate host** when it is eaten by another animal.

Diseases Due to Intestinal Flatworms

Tapeworm Disease

1. Tapeworms, like species of *Taenia* and *Diphyllobothrium*, are ribbonlike **hermaphroditic** organisms that lack a digestive system, and can be well over 30 feet long. (Figures 24.33, 24.34)
2. Tapeworm disease is often unnoticed, but it can be serious if humans accidentally become intermediate hosts; this is most likely to occur with the pork tapeworm, *T. solium*.

Giant Intestinal Fluke Disease (Fasciolopsiasis)

1. *Fasciolopsis buski* is a large intestinal hermaphroditic fluke.
2. Humans and pigs are definitive hosts; certain species of snails are intermediate hosts. (Figure 24.35)
3. Infection is acquired by ingesting a larval form encysted on water plants.

REVIEW QUESTIONS

Short Answer

1. Name a disease caused by *Clostridium difficile*.
2. Name a heart disease caused by normal bacterial mouth flora.
3. What acid is produced when streptococci metabolize sugars?
4. Name three characteristics of *Streptococcus mutans* that contribute to its ability to cause dental caries.
5. Give the pathogenesis of periodontal disease.
6. Give three examples of stressful events that can reactivate herpes simplex.
7. At what stage of maturation of boys is mumps likely to be complicated by swelling of the testicles?
8. What is the hemolytic uremic syndrome? Name an organism that can cause it.
9. What is the usual source of *Campylobacter jejuni* infections?
10. What is ADP ribosylation?
11. How can shigellas move from one host cell to another even though they are nonmotile?
12. Name four different groups of *Escherichia coli* based on their pathogenic mechanisms.
13. Contrast the transmission of hepatitis A and hepatitis B.
14. Name two kinds of hepatitis that can be prevented by vaccines.
15. Contrast the cause and epidemiology of giardiasis and amebiasis.
16. Give two characteristics of *Cryptosporidium parvum* that allow it to contaminate municipal water supplies.
17. Which helminthic disease do American physicians see most commonly?
18. Which parasitic roundworm bears live offspring instead of depositing eggs?
19. Give the common name of a hermaphroditic flatworm that can cause vitamin B_{12} deficiency.

Multiple Choice

1. The following are all true of intestinal bacteria, *except…*
   A. they produce vitamins.
   B. they can produce carcinogens.
   C. they are mostly aerobes.
   D. they produce gas from indigestible substances in foods.
   E. they include potential pathogens.
2. All of the following attributes of *Streptococcus mutans* are important in tooth decay, *except…*
   A. it attaches specifically to tooth pellicle.
   B. it can grow at pH below 5.
   C. it produces lactic acid.
   D. it synthesizes glucan.
   E. it stores fermentable polysaccharide.
3. All of the following are true of *Helicobacter pylori*, *except…*
   A. it is a helical bacterium with sheathed flagella.
   B. it has not been cultivated *in vitro*.
   C. it produces a powerful urease.
   D. it causes stomach infections that last for years.
   E. it has an important role in the causation of stomach ulcers.
4. All of the following are probably important to the cholera-causing ability of *Vibrio cholerae*, *except…*
   A. it attaches firmly to small intestinal epithelium.
   B. it produces cholera toxin.
   C. lysogenic conversion.
   D. acid resistance.
   E. it survives in the sea in association with zooplankton.
5. Pick out which one of the following statements concerning *Salmonella* Typhi is *false*:
   A. It is commonly acquired from domestic animals.
   B. It can colonize the gallbladder for years.
   C. It is highly resistant to killing by bile.
   D. It can destroy Peyer's patches.
   E. It causes typhoid fever.
6. One of the subsequent five statements about rotaviral gastroenteritis is *false*. Which one?
   A. The name of the causative agent was suggested by its appearance.

B. Most of the 600,000 deaths occurring worldwide from this disease are due to dehydration.
C. Most cases of the disease occur in infants and children.
D. The causative agent infects mainly the stomach.
E. The disease is transmitted by the fecal-oral route.

7. Which of the following statements about hepatitis is *false*?
   A. Both RNA and DNA viruses can cause hepatitis.
   B. Some kinds of hepatitis can be prevented by vaccines.
   C. More than half of the new hepatitis C cases are the result of injected-drug abuse.
   D. Lifelong carriers of hepatitis A are common.
   E. At least six different viruses can cause hepatitis.
8. Choose the most accurate statement about cryptosporidiosis.
   A. Waterborne transmission is unlikely.
   B. The host range of the causative agent is narrow.
   C. It is prevented by chlorination of drinking water.
   D. Person-to-person spread does not occur.
   E. The life cycle of the causative agent occurs within small intestinal epithelial cells.
9. All of the following are true of *Ascaris lumbricoides*, *except*...
   A. infestation is detected by using transparent tape.
   B. its larvae migrate through the lungs.
   C. its ova must mature in the soil before becoming infectious.
   D. it is the largest roundworm that infests human beings.
   E. a single female worm can produce more than 100,000 ova per day.
10. Of the following statements about tapeworms, which one is most likely to be *false*?
    A. They have no digestive tract.
    B. Some can reach 30 or more feet in length.
    C. The male worm fertilizes the female in the small intestine.
    D. People can become infested by eating raw or inadequately cooked meat or fish.
    E. Some species have more than one intermediate host.

Applications

1. An international health commission has been asked to judge whether mumps virus or herpes simplex virus should be the next target for elimination from the world. What should they advise, and what facts should they take into account in making their decision?
2. One reason given by Peruvian officials for not chlorinating their water supply was that chlorine can react with substances in water or the intestine to produce carcinogens. How do you assess the relative risks of chlorinating or not chlorinating drinking water?
3. A medical scientist is designing a research program to determine the effectiveness of hepatitis B vaccine in preventing liver cell cancer. Since liver cell cancer probably has multiple causes, how would you measure the success of an anticancer vaccination program?
4. An official at the Centers for Disease Control noticed that the annual rate of new hepatitis C virus infections, which averaged 230,000 during the 1980s, had dropped to 36,000 by 1996. What did she determine were the probable reasons for this dramatic change?

Critical Thinking

1. What might the lack of a brown color of feces indicate?
2. Mutant strains of *Helicobacter pylori* that lack the ability to produce urease fail to cause infection when they are swallowed. Infection occurs, however, if a tube is used to introduce them directly into the layer of mucus that overlies the stomach epithelium. What does this imply about the role of urease in the bacterium's pathogenicity?
3. Mutant strains of *Vibrio cholerae* can be isolated that fail to produce cholera. A microbiologist postulated that a defect in either the A or the B portion of cholera toxin could explain the lack of virulence. Is the microbiologist's explanation reasonable? Why or why not?
4. Looking at the replication cycle of hepatitis B virus (see figure 24.21), a student wondered whether core protein might be necessary for the assembly of viral envelope. What observation supports or refutes this idea?

Genitourinary Infections

Although some historians argue that syphilis was transported to Europe from the New World by Columbus's crew, others find convincing evidence, including biblical references, that syphilis existed in the Old World for many years before Columbus returned. History and literature record that kings, queens, statesmen, and heroes degenerated into madness or mental incompetence or were otherwise seriously disabled as a result of the later stages of syphilis. Henry VIII (King of England, 1509–1547), Ivan the Terrible (Czar of Russia, 1547–1584), Catherine the Great (Empress of Russia, 1762–1796), Merriwether Lewis (Lewis and Clark Expedition, 1803–1808), and Benito Mussolini (Premier of Italy, 1883–1945) are a few of the famous people who probably suffered from syphilis.

Syphilis was first named the "French pox" or the "Neapolitan disease" because it was believed to have come from France or Italy. In 1530, Girolamo Fracastoro, an Italian physician who suggested the germ theory long before the discovery of microorganisms, wrote a poem about a shepherd named Syphilis who had ulcerating sores covering his body. The description matched the symptoms of syphilis, and from that time on, the disease was known by the shepherd's name. Initially, the method of transmission was unknown, but gradually it became generally recognized that the disease was sexually transmitted. The symptoms of syphilis were very severe in the early years of the epidemic, often causing death within a few months. Treatment, consisting of doses of mercury, and guaiacum, the resin from a tropical American tree of the genus Guaiacum, *had little beneficial effect. As the decades went by, mutations and natural selection resulted in more resistant hosts and a less virulent microbe, and the disease evolved into the chronic illness known today.*

By the late Nineteenth century, the disease was shown to be transmissible to certain laboratory animals, but although various bacteria could usually be cultivated from lesions on the infected animals, all proved to be contaminating organisms because when isolated in pure culture, they could not reproduce the disease. Thus, the causative agent of syphilis remained a mystery.

In 1905, Fritz Schaudinn, a German protozoologist, examined some fluid from a syphilitic sore and saw a faintly visible organism "twisting, drilling back and forward, hardly different from the dim nothingness in which it swam." When Schaudinn used dark-field illumination, he was able to see the organism much more clearly. It appeared very thin and pale, similar to a corkscrew without a handle. The spirochetes appeared in specimens from other cases of syphilis, and Schaudinn later succeeded in staining them. By this time, he felt certain he had discovered the cause of syphilis, but the organism could not be cultivated on laboratory media. Schaudinn named the organisms Spirochaeta pallida, *"the pale spirochete." This organism is now called* Treponema pallidum, *from treponema, "a turning thread."*

—A Glimpse of History

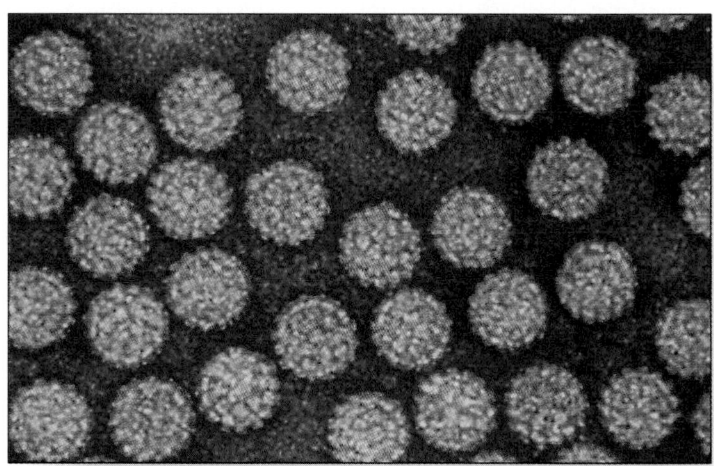

INFECTIONS OF THE REPRODUCTIVE AND URINARY tracts are very common, often uncomfortable or quietly destructive, and sometimes tragic in their consequences. Urinary infections lead the list of infections acquired in hospitals and are the chief source of fatal nosocomial bacterial bloodstream invasions. Nosocomial uterine infections, once a common cause of maternal deaths from childbirth, still require strict medical vigilance to be prevented. Much of this chapter will be devoted to sexually transmitted diseases (STDs).

The United States leads the industrialized nations in the incidence of STD even though other countries are more permissive in their attitudes toward sexual activity. About 1 million unintended pregnancies occur among the estimated 12 million college and university students each year. Only about 15% of the students have never engaged in sexual intercourse, and almost 35% of the remaining students have had six or more lifetime sexual partners. About 20% of women students say they have been forced to have sexual intercourse against their will. Approximately 70% of sexually active students report that they or their partner rarely or never use a condom. These data suggest a lack of knowledge or concern about the high risk of contracting and transmitting an STD.

Anatomy and Physiology

The reproductive and urinary systems are often considered together, and they are referred to as the genitourinary system because of their close proximity to one another. Also, both systems are commonly affected by the same pathogens. The genitourinary system is one of the portals of entry for pathogens, meaning the place where they get into the body.

The Urinary System

The urinary system consists of the kidneys, the ureters, the bladder, and the urethra (**figure 25.1**). The kidneys act as a specialized filtering system to clean the blood of many waste materials, selectively reabsorbing substances that can be reused. Waste materials are excreted in the urine, which is usually acidic because of excretion of excess hydrogen ions from foods and metabolism. Consistently alkaline urine suggests infection with a urease-producing bacterium that converts the urea in urine to ammonia. Antimicrobial medications are commonly excreted in the urine and reach concentrations higher than those in the bloodstream. Each kidney is drained by a ureter, which connects it with the urinary bladder. The bladder acts as a holding tank. Once filled, it empties through the urethra. Infections of the urinary tract occur far more frequently in women than in men, because the female urethra is short (about 4 cm or 1.5 inches, compared with 20 cm or 8 inches in the male) and is adjacent to openings of the genital and intestinal tracts. Special groups of muscles near the urethra keep the system closed most of the time and help prevent infection. The downward flow of urine also helps clean the system by flushing out microorganisms before they have a chance to multiply and cause infection.

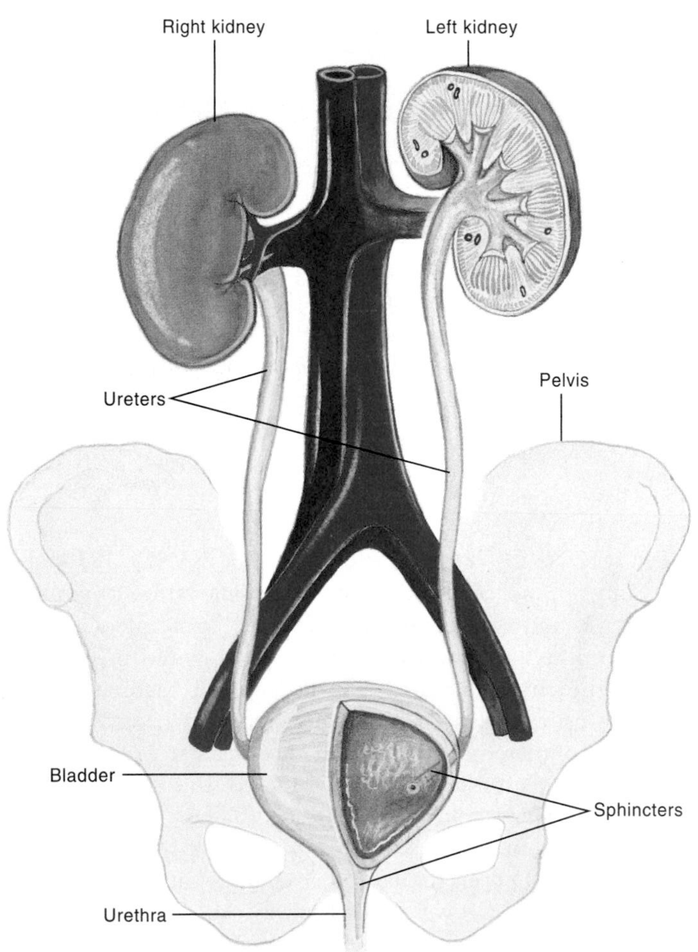

Figure 25.1 Anatomy of the Urinary System Urine flows from the kidneys, down the ureters, and into the bladder, which empties through the urethra. Sphincter muscles help to prevent microorganisms from ascending.

The urinary tract is protected from infection by a number of mechanisms besides its anatomy. Normal urine contains antimicrobial substances such as organic acids and small quantities of antibodies. During urinary tract infections, larger quantities of specific antibodies can be found in the urine. Antibody-forming lymphoid cells in the infected kidneys or bladder form protective antibodies locally at the site where they are needed. In addition, during infection an inflammatory response occurs in which phagocytes enter the bladder and are of the utmost importance in engulfing and destroying the invading microorganisms. ■ phagocytes, p. 372

The Genital System

The anatomy of the female and male genital systems is shown in **figure 25.2**. In women during the child-bearing years, an egg, or ovum, is expelled from one of the two ovaries each month and swept into the adjacent fallopian tube. Fertilization normally takes place in the fallopian tube; ciliated epithelium of the tube then moves the fertilized ovum to the uterus, where it implants itself in the epithelial lining. If fertilization does not occur, the epithelial lining of the uterus sloughs off, producing a menstrual period. Infection of the fallopian tubes can cause scarring and destruction of the ciliated epithelium, so that ova are not moved efficiently to the uterus. Note that the fallopian tubes are open on both ends, providing a pathway into the abdominal cavity where the liver and other structures can be infected. Also notice the uterine cervix, a common site of sexually transmitted infections, and a place where cancer can develop. Except during menstruation, the cervical opening is tiny and filled with mucus. The vagina is a portal of entry for a number of infections that can advance to the uterus and fallopian tubes at the time of menstruation. The vaginal opening is framed on each side by two labia, or lips. These plus the clitoris and entryway of the vagina constitute the female external genitalia, or **vulva**.

In men, the paired reproductive organs, the testes (testicles), exist outside the abdominal cavity in the scrotum. Sperm from each testis collect in a tightly coiled tubule called the epididymis and are conveyed by a long tube called the vas deferens that enters the abdomen in the groin to join the prostate gland.

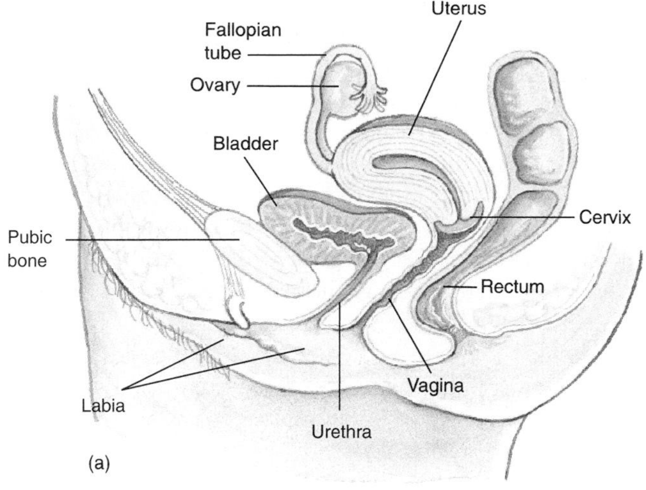

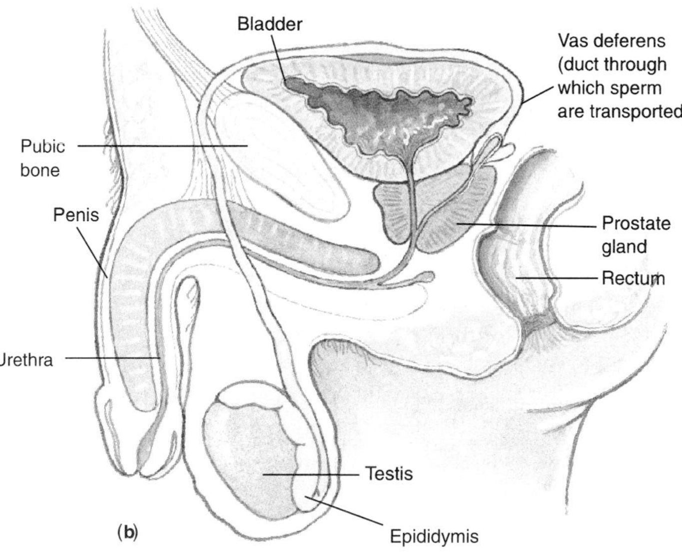

Figure 25.2 Anatomy of the Genital System (a) Female and **(b)** male.

The sperm and secretions of the prostate gland compose the **semen**. The urinary and reproductive systems join at the prostate gland. The prostate can be infected by urinary or sexually transmitted pathogens. In older men it often enlarges and hinders the flow of urine, thus fostering urinary infections.

MICROCHECK 25.1

The genitourinary system is one of the portals of entry for pathogens. Many antimicrobial medications are excreted in the urine in concentrations higher than in the blood. The flushing action of urination is a key defense mechanism against bladder infections. The fallopian tubes can provide a passageway for pathogens to enter the abdominal cavity. The uterine cervix is a common site of infection by sexually transmitted pathogens. Prostate enlargement predisposes men to urinary infection.

- Name two kinds of cells that help defend the bladder against infection.
- Would ammonia make the urine basic? How?

Normal Flora of the Genitourinary System

Normally, the urine and urinary tract above the entrance to the bladder are essentially free of microorganisms; the lower urethra, however, has a normal resident flora. Species of *Lactobacillus*, *Staphylococcus* (coagulase-negative), *Corynebacterium*, *Haemophilus*, *Streptococcus*, and *Bacteroides* are common inhabitants.

The normal flora of the genital tract of women is influenced by the action of estrogen hormones on the epithelial cells of the vaginal mucosa. When estrogens are present, glycogen is deposited in these cells. The glycogen is converted to lactic acid by lactobacilli, resulting in an acidic pH that inhibits the growth of many potential pathogens. Lactobacilli may also release hydrogen peroxide, a powerful inhibitor of some anaerobic bacteria, as a byproduct of metabolism. Thus, the normal flora and resistance to infection of the female genital tract vary considerably with the person's hormonal status. For example, prepubertal girls, having low estrogen levels, are much more susceptible to vaginal infections with *Streptococcus pyogenes* and *Neisseria gonorrhoeae* than women during the child-bearing years. ■ lactobacilli, p. 275

MICROCHECK 25.2

Normally the urine and urinary tract are free of microorganisms above the entrance to the bladder, but the lower urethra hosts a number of genera of bacteria. During the child-bearing years, a woman's hormones are important in the vagina's resistance to infection, because estrogen promotes growth of lactobacilli and acidic conditions.

- Which parts of the genitourinary system are normally sterile?
- Name two species of pathogenic bacteria to which the vagina is especially susceptible during childhood.

Urinary System Infections

Urinary tract infections account for about 7 million visits to the doctor's office each year in the United States. Urinary infections can involve the urethra, the bladder, or the kidneys, alone or in combination. Any situation in which the urine does not flow naturally increases the chance of infection. After anaesthesia and major surgery, for example, the reflex ability to void urine is often inhibited for a time, and urine accumulates and distends the elastic bladder. Even being too busy to empty the full bladder may predispose to infection. Catheterization of the bladder is another common cause of infection. Most cases, however, occur in otherwise healthy young women without impaired urinary flow.

The causative organisms of diseases involving areas of the body other than the urinary tract can also infect the urinary system. For example, in typhoid fever, *Salmonella* Typhi organisms disseminate throughout the body, infect the kidneys, and are excreted in the urine starting about 1 to 2 weeks after infection. ■ typhoid fever, p. 598, 604

Bacterial Cystitis

The most common type of urinary infection involves the bladder and is called bacterial **cystitis**, meaning inflammation of the bladder. Bacterial cystitis is common among otherwise healthy women, and it is a common nosocomial infection.

Symptoms

Typically the onset is abrupt, with a burning pain in urination, an urgent sensation to void, and frequent voiding in small amounts. The urine is cloudy due to the presence of leukocytes, often smells bad, and sometimes has a pale red color due to bleeding. Tenderness may be present in the area above the pubic bone due to the underlying inflamed bladder. Some cases, however, are asymptomatic, especially among children and the elderly. Others develop sudden elevation in temperature, chills, vomiting, back pain, and tenderness overlying the kidneys, indicating a potentially serious complication, a kidney infection called **pyelonephritis**.

Causative Agents

Bladder infections usually originate from the normal intestinal flora. Specific strains of *Escherichia coli* cause most cases of bacterial cystitis, accounting for 80% to 90% of cystitis cases in women during the reproductive years and about 70% of all bladder infections. Other enterobacteria such as *Klebsiella* and *Proteus* species account for 5% to 10%, and the Gram-positive species, *Staphylococcus saprophyticus*, accounts for another 5% to 10% of infections in young women. Hospitalized patients are more likely than outpatients to have bladder infections with bacteria such as the Gram-negative rods *Serratia marcescens* and *Pseudomonas aeruginosa* and the Gram-positive coccus *Enterococcus faecalis*, that are resistant to antibacterial medications. Those with long-standing bladder catheters are often chronically infected with multiple species of intestinal bacteria. ■ *Pseudomonas*, p. 281, 697

Pathogenesis

Generally, the causative agents reach the bladder by ascending from the urethra. The process is aided by motility of the organisms. Urine is a good growth medium for many species of bacteria. Species of *E. coli* that infect the urinary system possess pili that attach specifically to receptors on the cells that line the bladder. Experimental evidence indicates that attachment is followed by death and sloughing of the superficial layer of epithelium, followed by penetration of newly exposed cells by endocytosis. Bacteria ascend the ureters and cause pyelonephritis in many cases of cystitis. Repeated episodes of pyelonephritis cause scarring and shrinkage of the kidneys and are an important cause of kidney failure. ■ pili, p. 69, 208 ■ endocytosis, p. 459

Epidemiology

About 30% of women develop cystitis at some time during their life. Factors involved in urinary infections in women include:

- A relatively short urethra. The position of the urethra makes it subject to fecal contamination and colonization with potentially pathogenic intestinal bacteria that have only a few centimeters to traverse to the bladder.
- Sexual intercourse. About one-third of urinary tract infections in sexually active women are associated with sexual intercourse. Bladder infections in women often occur from the massaging effect of sexual intercourse on the urethra, which introduces bacteria from the urethra into the urinary bladder. Many women develop their first bladder infections following their first sexual intercourse, a condition referred to as "honeymoon cystitis."
- Use of a diaphram for contraception. The ring of the diaphragm compresses the urethra and impedes the flow of urine, increasing by two to three times the risk of urinary infection.

Urinary infections are unusual in men until about age 50, when enlargement of the prostate gland compresses the urethra and makes it difficult to completely empty the bladder.

Medical conditions may require insertion of a bladder catheter for periods ranging from several days to months. Bacteria can reach the bladder and establish urinary tract infections both via the catheter lumen and the mucus between the wall of the catheter and the urethra. Pathogens establish a biofilm on the catheter, making it difficult or impossible to kill them with antibacterial medications. The risk of infection increases about 5% each day the catheter remains in place. Paraplegics, individuals with paralysis of the lower half of the body, are almost always afflicted with urinary infections. Since they lack bladder control, paraplegics are unable to void normally and require a catheter indefinitely to carry urine to a container. In the United States, about 500,000 hospitalized patients develop bladder infections each year, mostly after catheterization.

Prevention and Treatment

General measures for preventing urinary infections include taking enough fluid to ensure voiding at least four or five times daily, voiding immediately after sexual intercourse, and wiping from front to back after defecation to minimize fecal contamination of the vagina and urethra. Preventing recurrent infections may require taking a dose of antibacterial medication immediately before or after sexual intercourse, or taking daily a small dose of an antibiotic that is concentrated in the urine. Treatment of cystitis is usually easily carried out with a few days of an antimicrobial medication to which the causative bacterium is susceptible. Pyelonephritis usually requires prolonged treatment and often hospitalization.

The main features of bacterial cystitis are presented in **table 25.1**.

Leptospirosis

Leptospirosis is a disease in which the urinary system is infected from the bloodstream rather than by bacteria ascending from the urethra. It is probably the most widespread of all the zoonoses, diseases that are transmissible to humans but exist mainly in other animals. Leptospirosis occurs around the world in all types of climates, although it is more common in the tropics. One or two fatal cases occur in the United States each year,

TABLE 25.1 Bacterial Cystitis

Symptoms	Abrupt onset, burning pain on urination, urgency, frequency, foul smell, red-colored urine; with pyelonephritis, fever, chills, back pain, and vomiting
Incubation period	Usually 1 to 3 days
Causative agents	Most due to *Escherichia coli;* other enterobacteria, *Staphylococcus saprophyticus* cause some cases; nosocomial infections with antibiotic-resistant strains of *Pseudomonas, Serratia,* and *Enterococcus* genera
Pathogenesis	Usually, bacteria ascend the urethra, enter the bladder, and attach by pili to receptors on urinary tract epithelium. Sloughing of cells and an inflammatory response ensue. Spread to the kidneys can occur via the ureters, causing pyelonephritis and potential kidney failure
Epidemiology	Bacterial cystitis is common in women, promoted by a relatively short urethra, use of a diaphragm, and sexual intercourse. Middle-aged men are prone to infection because enlargement of the prostate gland partially obstructs their urethra. Placement of a bladder catheter commonly results in infection
Prevention and treatment	Taking sufficient fluid to void urine four to five times daily, wiping from front to back. Single dose of antimicrobial medication with sexual intercourse may help prevent bacterial cystitis in women. Short-term antimicrobial therapy usually sufficient. Longer treatment for pyelonephritis

but most cases are mild, recover without treatment, and remain undiagnosed. It has not been a reportable disease since 1994.

Symptoms

Many infections are asymptomatic. A typical case begins after an incubation period averaging 10 days (range, 2 to 30 days), with abrupt onset of headache, spiking fever, chills, and severe muscle pain. The most characteristic feature of this phase of the disease is the development of redness of the eyes due to dilation of small blood vessels. In mild cases, which are the most common, symptoms usually subside within a week and all signs and symptoms of illness are gone in a month or so. Severe cases follow a similar course except after 1 to 3 days of feeling well, symptoms recur sometimes accompanied by bleeding from various sites, confusion, and evidence of severe heart, brain, liver, and kidney damage. This biphasic course of the illness is characteristic of cases that come to medical attention.

Causative Agent

Leptospirosis is caused by a slender spirochete with hooked ends, *Leptospira interrogans* (**figure 25.3**). Although some of the more than 200 antigenic types have been given different species names in the past, probably all belong to this single species. ■ **spirochetes, p. 289**

Pathogenesis

The organisms enter the body through mucous membranes and breaks in the skin, which may be trivial. No lesion develops at the site of entry, but the organisms multiply and spread throughout the body by way of the bloodstream, penetrating all tissues including the eyes and the brain. Severe pain is characteristic of this first (septicemic) phase, and it may lead to unnecessary surgery for suspected appendicitis or gallbladder infection. A lack of inflammatory changes or tissue damage is a striking feature. Within a week, type-specific antibody arises and with complement destroys the organisms present in most tissues, although they continue to multiply in the kidneys. The victim then characteristically enjoys 1 or more days of improvement before symptoms recur. The cause of the recurrent symptoms is presumed to be an immune response, and this second phase of the illness is often called the immune phase. This phase is characterized by injury to the cells that line the lumen of tiny blood vessels, causing clotting and impaired blood flow in tissues throughout the body. This accounts for most of the serious effects of the illness, including kidney failure, the main cause of fatalities.

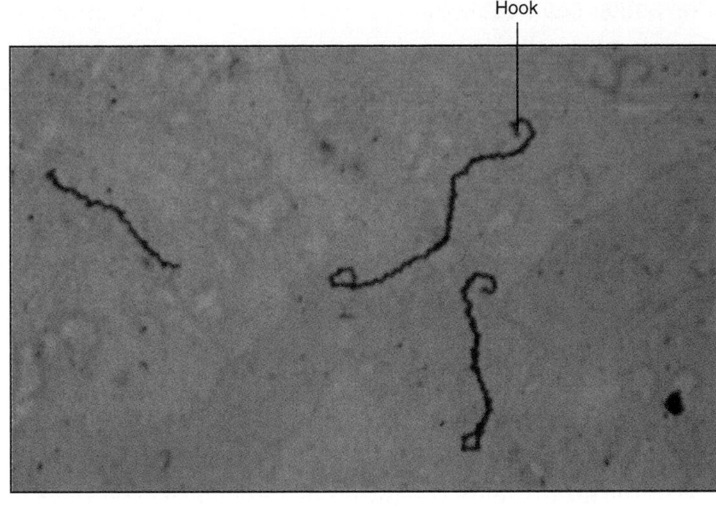

Figure 25.3 ***Leptospira interrogans*, the Cause of Leptospirosis** Note the hooked ends of this spirochete.

Epidemiology

Leptospira interrogans infects numerous species of wild and domestic animals, usually causing little or no apparent illness, but ranging to highly fatal epidemic disease. Characteristically, the organisms are excreted in their urine, which provides the principal mode of transmission to other hosts. Spots on the ground where urine has been deposited can remain infectious for as long as 2 weeks, while *Leptospira* in mud or water can survive for several weeks. Warm summer temperatures and neutral or slightly alkaline moist conditions promote survival of the bacteria. Swimming and getting splashed with urine-contaminated water account for many cases. Infected humans usually excrete the organisms for less than a few weeks, but sometimes excretion can continue for many months. Other animals, particularly rodents, often excrete *Leptospira* for a lifetime.

Prevention and Treatment

Because of the ubiquity of the causative organism, there are few effective preventive measures other than to avoid animal urine. Maintaining general sanitary conditions is helpful in the care of domestic animals raised for food. Multivalent vaccines, ones that contain a number of different serotypes of *L. interrogans*, are available for preventing the disease in domestic animals, but they do not consistently prevent the carrier state. In an epidemic situation, small doses of a tetracycline antibiotic can prevent the disease in exposed individuals. A number of antibiotics are effective for treating the disease, but only if started during the first 4 days of the illness. As in a number of other microbial diseases, the onset of treatment is often followed in 4 to 6 hours by a transitory worsening of symptoms called the **Jarisch-Herxheimer reaction**, probably due to massive release of antigens from organisms lysed by the antimicrobial medication.

The main features of leptospirosis are shown in **table 25.2**.

MICROCHECK 25.3

Situations that interfere with the normal flow of urine predispose to urinary system infections. Bladder infections are common, especially in women, caused by bowel bacteria ascending from the urethra. Pyelonephritis is a feared complication. Leptospirosis is a widespread zoonosis spread by urine, in which the kidneys are infected from the bloodstream. Usually mild, the disease can be a severe biphasic illness characterized by tissue invasion and pain in the first phase, tissue destruction in the second.

- What is the leading cause of bladder infections in otherwise healthy women? Where does this organism come from?
- How do people become infected with *Leptospira interrogans*?

TABLE 25.2 Leptospirosis

1. Water or animal urine contaminated with *Leptospira* sp. splashes onto mucous membrane or abraded skin
2. The bacteria infect the bloodstream and are carried throughout all the body tissue causing fever, intense pain
3. Symptoms subside and bacteria disappear from blood and tissues, except kidneys
4. Symptoms recur associated with severe damage to liver and kidneys
5. Complete recovery occurs if kidney failure can be effectively treated
6. Excretion of *Leptospira* continues in urine

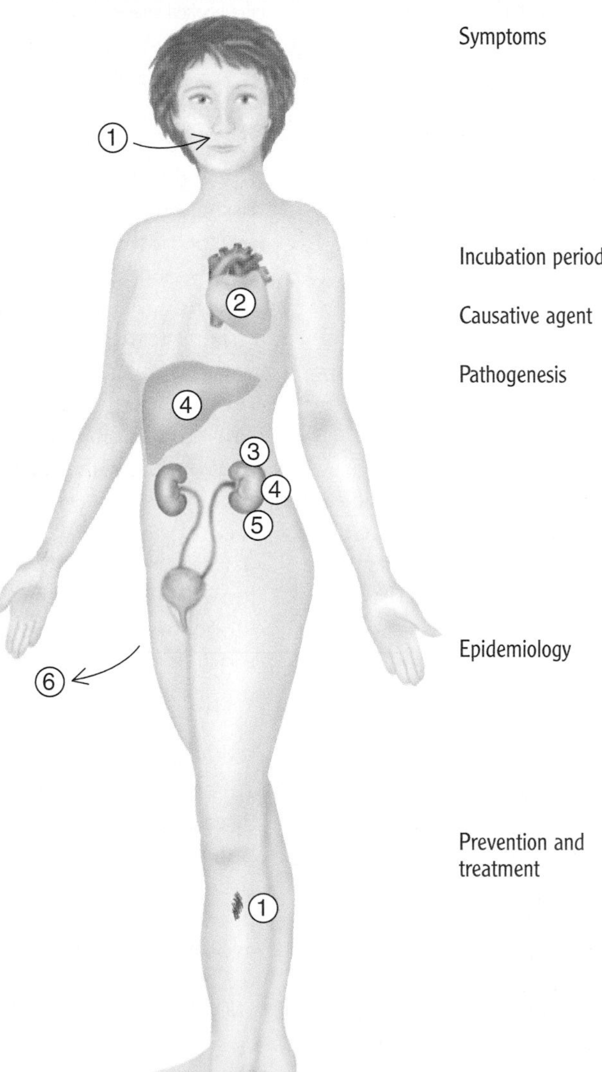

Symptoms	Many mild and asymptomatic cases. Others have a biphasic illness: spiking fever, headache, muscle pain, bloodshot eyes in the first (septicemic) phase, then 1 to 3 days of improvement; second (immune) phase has recurrence of severe symptoms with heart, brain, liver, and kidney damage
Incubation period	Usually about 10 days (range, 2 to 30 days)
Causative agent	*Leptospira interrogans,* a cultivable spirochete with many serotypes
Pathogenesis	The bacteria penetrate mucous membranes or breaks in the skin, multiply in the bloodstream, and are carried to all parts of the body. Severe pain with penetration of body tissues, but little or no tissue damage. Immune phase: damage to cells that line small blood vessels and clotting of blood. Causes severe damage to the liver, kidneys, heart, brain, and other organs
Epidemiology	Worldwide distribution. Wide range of animal hosts chronically excrete the bacteria in their urine, causing contamination of natural waters and soils. Organisms remain infectious for long periods of time under warm, moist, neutral or alkaline conditions.
Prevention and treatment	Avoid contact with animal urine. Vaccines prevent disease in domestic animals, may not prevent urinary carriage. Tetracycline antibiotics preventive in epidemics. Various antibacterial medications useful in treatment of leptospirosis, but only if given early in the disease. Symptoms often worsen transiently with effective treatment due to the massive release of leptospiral antigens

CASE PRESENTATION

The patient was a 32-year-old married woman complaining of 1 week of burning pain on urination, and frequent voiding of small amounts of bloody urine. About 8 days earlier, she completed 3 days of trimethoprim-sulfamethoxazole therapy for similar symptoms. Tests at that time showed that her urine was infected with *Escherichia coli* resistant only to amoxicillin. When the symptoms returned, she began drinking 12 ounces of cranberry juice three times daily but had only partial relief. She denied having chills, fever, back pain, nausea, or vomiting. She was approximately 12 weeks pregnant. ■ **trimethoprim, p. 506**

Her medical history revealed that she had suffered two or three similar episodes of urinary symptoms every year for a number of years. Sometimes the symptoms would go away just by forcing herself to take extra fluid, but at other times the symptoms would persist and she would obtain medical evaluation and treatment with an antibacterial medication. On one occasion several years before the present illness, chills, fever, back pain, nausea, and vomiting accompanied her symptoms and she was hospitalized for a "kidney infection." There was no history suggesting any underlying disease such as diabetes, cancer, or immunodeficiency.

On examination the patient appeared well, with no obvious distress. Her temperature was normal, as was the remainder of the physical examination.

The results of her laboratory tests included normal leukocyte count and kidney function tests. Microscopic examination of her urine showed numerous red and white blood cells. A Gram-stained smear of the uncentrifuged urine showed numerous rod-shaped Gram-negative bacteria and polymorphonuclear neutrophils. Culture of the urine revealed more than 100,000 colonies of *Escherichia coli* per ml. The bacterium was resistant to amoxicillin, but sensitive to the other antibacterials useful for treating urinary infections. ■ **polymorphonuclear neutrophils, p. 372**

1. What is the diagnosis?
2. What is the treatment?
3. What is the prognosis?
4. What future preventive measures might be undertaken?

Discussion

1. This woman's symptoms and signs clearly lead to a diagnosis of bacterial cystitis, but there are a number of clues in the presentation of this case that point to a significant complication. First, most patients with uncomplicated bacterial cystitis are cured by 3 days of an antibacterial medication to which the causative bacterium is susceptible. Her symptoms recurred only 1 day after completing her medication. Second, her symptoms had been present for a full week before she sought medical evaluation. Third, she was pregnant. Fourth, she gave a past history of being hospitalized for pyelonephritis.

   These clues make it highly likely that she has a condition called subclinical pyelonephritis, in which her bladder infection has spread to her kidneys but has not yet produced the symptoms of pyelonephritis. As many as 30% of patients with cystitis have subclinical pyelonephritis, depending on risk factors such as the clues mentioned in this case. Although contraindicated in this patient because of her pregnancy, the kidney infection can be demonstrated with a scintogram, an image of the kidneys produced following injection into the bloodstream of a tiny amount of radioactive material, which is removed from the blood by the kidneys and excreted.
2. The patient can be treated with trimethoprim-sulfamethoxazole, the same medication that was given before. Because infection in the kidneys takes much longer to cure than bladder infections, however, the treatment must be continued for 2 weeks or longer. This medication diffuses well into vaginal secretions and can eliminate *E. coli* colonization of the vagina, which may be present. Although safe to use early in pregnancy, it cannot be used late in pregnancy because it can worsen jaundice in the newborn. A urine culture was done 1 week after completion of treatment to be sure that the infection was truly gone.
3. The outlook is good for a full recovery without any permanent damage to the kidneys. The patient was advised, however, that repeated future infections of the kidneys could lead to kidney failure if not treated promptly.
4. For the future, all the usual methods for preventing urinary infections should be employed for this patient, including taking enough fluid intake to ensure voiding urine four or five times daily, avoiding delays in emptying the bladder, not using a diaphragm for contraception, voiding after intercourse, and taking a preventive antimicrobial medication. Two types of vaccine are under development for preventing urinary infections. One consists of pili antigens administered by injection, and the other, a vaginal suppository containing killed *E. coli*. Although encouraging experimental results have been reported, no vaccine has yet been approved for use in the United States.

Non-Venereal Genital System Diseases

The genital tract is the portal of entry for numerous infectious diseases, both venereal and non-venereal (*venereal*, from Venus, the goddess of love; *non-venereal* means not transmitted by making love). This section discusses some examples of non-venereal genital diseases. The term venereal disease (VD), formerly meaning disease transmitted almost exclusively by sexual intercourse, is now often used interchangeably with sexually transmitted disease (STD), which includes all diseases spread by sexual intercourse, even if commonly transmitted by other routes as well. STDs will be discussed later in this chapter.

Puerperal ("childbed") fever, which can still occur occasionally following childbirth, is an example of a non-venereal genital system infection. Other serious infections are associated with menstruation and spontaneous or induced abortions. Organisms from dirt and dust and from a woman's own fecal bacteria can attack the traumatized uterus. One of the most feared bacteria, *Clostridium perfringens*, causes uterine gas gangrene and has been responsible for many fatalities following abortions induced under unclean conditions. Also, the normal vagina can be colonized by various pathogens, producing symptoms that range from annoying to life threatening. ■ **puerperal fever, p. 473**

Bacterial Vaginosis

In the United States, bacterial vaginosis is the most common vaginal disease of women during the child-bearing years. It is termed *vaginosis* rather than *vaginitis* because inflammatory changes are absent. Pregnant women with this condition have a sevenfold increase in the risk of having a premature baby or other complications. Nevertheless, the risk is small (less than 10%).

Symptoms

Bacterial vaginosis is characterized by a thin, grayish-white, slightly bubbly vaginal discharge that has a characteristic pungent "fishy" odor. For many, the odor is more distressing than the vaginal discharge, which is often slight; half of those with the disease are asymptomatic.

Causative Agent

The cause or causes of bacterial vaginosis are unknown. A marked decrease in vaginal lactobacilli is a constant feature, suggesting that conditions supressing lactobacilli, or promoting other flora, play a causative role. *Gardnerella vaginalis*, a small, aerotolerant bacterium with a Gram-positive type of cell wall, is commonly present in large numbers, as are strictly anaerobic bacteria of the *Mobiluncus* and *Prevotella* genera, *Mycoplasma* sp., and anaerobic streptococci. Pure cultures of these species do not consistently produce bacterial vaginosis when healthy people are voluntarily inoculated, although the discharge from people with the condition can produce the disease. Each of these species of bacteria can occur in vaginal secretions of women without the disease, although in much smaller numbers.

Pathogenesis

The key changes in this disease are a decrease in the normal acidity of the vagina, a marked derangement of the normal vaginal flora, and a substantial increase in the numbers of clue cells (**figure 25.4**). Clue cells are epithelial cells that have sloughed off the vaginal wall and are covered with masses of bacteria. The pH of the vaginal secretions is elevated above the normal of 4.5. No inflammation occurs unless there is another, concurrent vaginal infection. The strong fishy odor is due to metabolic products of the anaerobes, including the amines putrescine and cadaverine.

Epidemiology

Since the cause of bacterial vaginosis is not known, its epidemiology is incomplete. The disease is most common among sexually active women and sometimes occurs in children who have been sexually abused. Sexual promiscuity and a new sex partner also increase risk of the disease. Virgins rarely get the disease. Proof is lacking, however, that bacterial vaginosis is a sexually transmitted infection.

Prevention and Treatment

There are no proven preventive measures. Studies on the use of yogurt by mouth or vaginally to restore vaginal lactobacilli have given conflicting results. Treatment of the male sex partners of patients with recurrent disease does not prevent recurrences. Most cases respond promptly to treatment with metronidazole and other antibacterial medications. The main features of bacterial vaginosis are summarized in **table 25.3**.

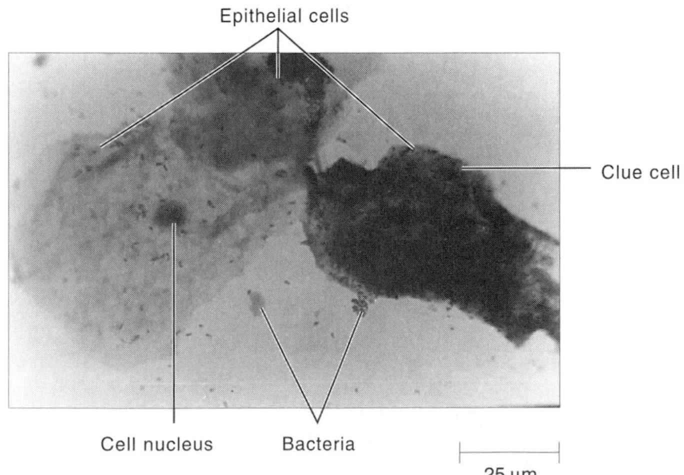

Figure 25.4 Clue Cell in an Individual with Bacterial Vaginosis The cells in the photograph are epithelial cells that have sloughed from the vaginal wall, one of which, the clue cell, is completely covered with adherent anaerobes.

Vulvovaginal Candidiasis

Vulvovaginal candidiasis is the second most common cause of vaginal symptoms after bacterial vaginosis. As with bacterial vaginosis, vulvovaginal candidiasis appears to follow a disruption of normal flora. As the name indicates, the infection often involves not only the vagina, but the woman's vulva, or external genitalia, as well.

Symptoms

The most common symptom of vulvovaginal candidiasis is itching, which can be intense and unremitting. Typically, the vaginal discharge is scanty and whitish, often occurring in curdlike clumps. The involved area is usually red and somewhat swollen.

Causative Agent

Vulvovaginal candidiasis is caused by *Candida albicans* (**figure 25.5**), a yeast that is part of the normal flora of the vagina in about 35% of women. Since *C. albicans* is a fungus, it has a eukaryotic cell structure. ■ yeast, p. 309

Pathogenesis

Normally, *C. albicans* causes no symptoms. The interaction between large numbers of the normal vaginal lactobacilli and small numbers of these fungi results in a balance between them probably based on competition for nutrients. When this balance is upset, however, as occurs during intensive antibacterial treatment, *C. albicans* multiplies without restraint, and the symptoms of vulvovaginitis occur. Other predisposing factors to *Candida* infection are late pregnancy, the use of oral contraceptives, and uncontrolled sugar diabetes when glucose is excreted in the urine. It is clear that other presently unknown factors play a role in pathogenesis.

TABLE 25.3 Bacterial Vaginosis

Symptoms	Gray-white vaginal discharge and unpleasant fishy odor
Incubation period	Unknown
Causative agent	Unknown
Pathogenesis	Uncertain. Marked distortion of the normal flora. Increased sloughing of vaginal epithelium in the absence of inflammation. Odor due to metabolic products of anaerobic bacteria. Association with complications of pregnancy, including premature births
Epidemiology	Associated with many sexual partners or a new partner, but can occur in the absence of sexual intercourse. Probably not a sexually transmitted disease
Prevention and treatment	No proven preventive measures. Treatment is with metronidazole or, in early pregnancy, clindamycin

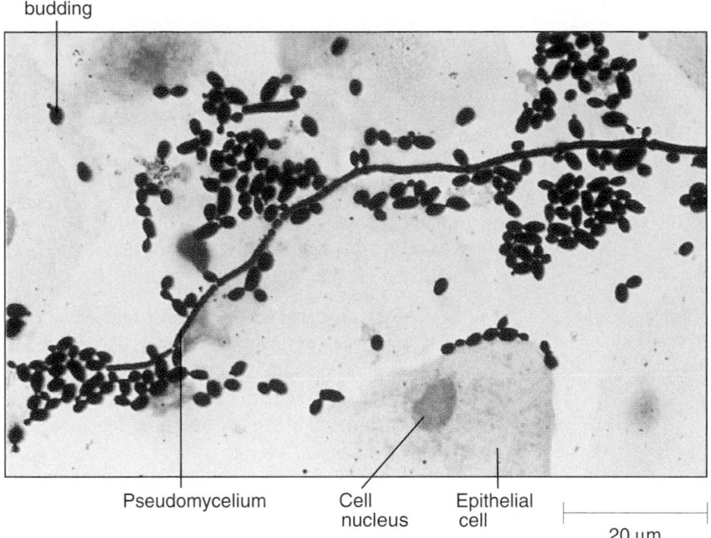

Figure 25.5 ***Candida albicans*** **in the Vaginal Discharge of a Woman with Vulvovaginal Candidiasis**

TABLE 25.4 Vulvovaginal Candidiasis

Symptoms	Itching, burning, thick white vaginal discharge, rash on vulva
Incubation period	Usually unknown. Generally 3 to 10 days when associated with antibacterial medications
Causative agent	*Candida albicans,* a yeast
Pathogenesis	Inflammatory response to overgrowth of the yeast normally present among the normal flora
Epidemiology	Not contagious. Usually not sexually transmitted. Associated with antibacterial therapy, use of oral contraceptives, pregnancy, and uncontrolled diabetes, but most cases have no identifiable predisposing factor
Prevention and treatment	No proven preventive measures. Intravaginal antifungal medications such as clotrimazole usually effective

Epidemiology

The disease does not spread person to person and is generally not sexually transmitted. Treatment with antibacterial medications poses an increased the risk of the disease, which increases with the duration of treatment.

Prevention and Treatment

Prevention depends on minimizing the use and duration of antibacterial medications, and on effective treatment of underlying conditions such as diabetes. Most patients, however, have no known predisposing factors. Vaginal treatment of *C. albicans* infections with antifungal medicines such as nystatin or clotrimazole creams are usually effective. Simultaneous treatment of any male sex partners is rarely helpful, except when a *Candida* infection of the penis is present.

The main features of vulvovaginal candidiasis are presented in **table 25.4**.

Staphylococcal Toxic Shock

Toxic shock syndrome was described in the late 1970s in several children with staphylococcal infections. In 1980, it became epidemic in young healthy menstruating women **(figure 25.6)** in association with use of a brand of high-absorbency tampon that has since been removed from the market. The term *toxic shock syndrome* was used to represent the symptoms and signs of their illness, which now that we know the cause, is appropriately called *staphylococcal toxic shock.* This was not a new disease, but emerged in a new form, and became much more common, as a result of a change in technology and human behavior.

Symptoms

Staphylococcal toxic shock is characterized by sudden onset of high temperature, headache, muscle aches, bloodshot eyes, vomiting, diarrhea, a sunburnlike rash, confusion, and without treatment, dropping blood pressure that can lead to kidney failure and death. Typically, the skin peels about a week after the onset.

Causative Agent

Staphylococcal toxic shock is caused by strains of *Staphylococcus aureus* that produce one or more of a particular kind of exotoxin.

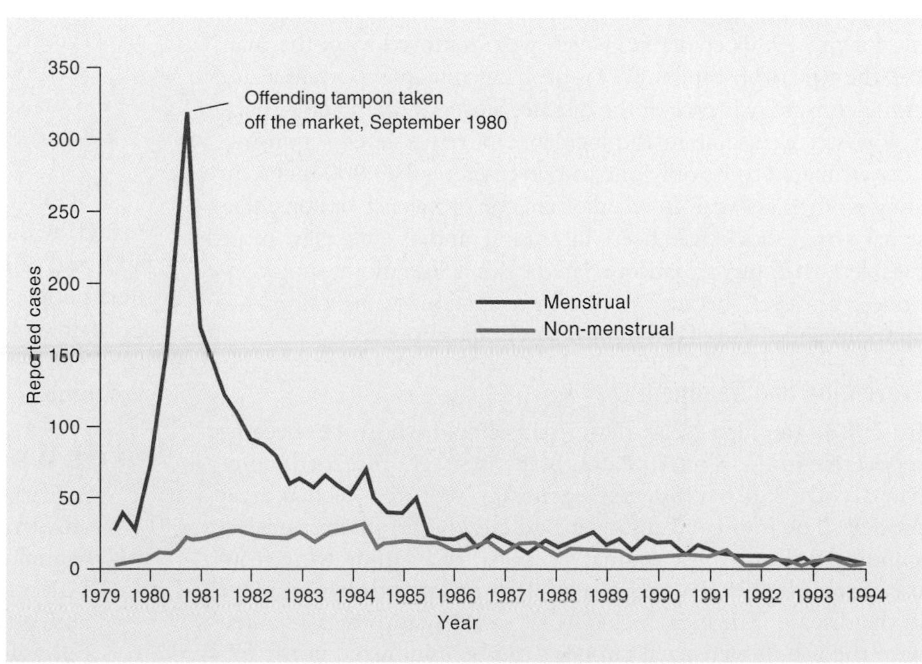

Figure 25.6 Staphylococcal Toxic Shock, United States, 1979 through 1993, Reported Every 3 Months A sharp drop in cases occurred when a brand of high-absorbency tampon was taken off the market. Since 1993, there has been further decine in cases, with non-menstrual cases (men, women, and children) sometimes exceeding the number of cases associated with menstruation (almost all of them tampon users).

The involved toxins are all superantigens. Toxic shock syndrome toxin-1 (TSST-1) accounts for about 75% of the cases; the remainder are due to staphylococcal exotoxins called enterotoxins because they cause gastrointestinal symptoms. ■ ***Staphylococcus aureus*, p. 525, 527, 694 ■ exotoxins, p. 462 ■ superantigens, p. 464 ■ staphylococcal enterotoxins, p. 462**

Pathogenesis

Tampon-associated toxic shock is not usually an infection, since the causative staphylococci grow in the menstrual fluid–soaked tampon and rarely spread throughout the body. Staphylococcal toxic shock results from absorption into the bloodstream of one or more of the toxins from the site where the organisms are growing, a situation analogous to staphylococcal scalded skin syndrome. Tampons that abrade the vaginal wall probably promote toxin absorption. Toxic shock syndrome toxin-1 and the other responsible toxins, being superantigens, cause a massive release of cytokines, which in turn cause a drop in blood pressure and kidney failure, the most dangerous aspect of the illness. **■ staphylococcal scalded skin syndrome, p. 527**

Epidemiology

Toxic shock syndrome can occur after any infection with *Staphylococcus aureus* strains that produce one of the responsible toxins. It does not spread person to person. The syndrome can occur after infected surgical wounds, infections associated with childbirth, and other types of staphylococcal infections. Using tampons increases the risk of staphylococcal toxic shock, and the higher absorbency tampons may pose a greater risk. Most of the menstruation-associated cases occur in women under 30, probably because younger women are less likely to have protective antibody against the toxin. Recovery from the disease does not consistently give rise to immunity. In fact, about 30% of those who recover will suffer a recurrence of the disease, although it is usually milder than the original illness. Except for the tampons responsible for the 1980s epidemic, which were removed from the market, the type of absorbent fiber is probably not an important factor in the current incidence of the disease. Since 1990, there has been a slow steady decline in the incidence of toxic shock syndrome, now estimated to be only one to two cases per 100,000 menstruating women per year. In recent years, the prevalence of non-menstrual toxic shock has been increasing and is generally more prevalent than menstruation-related cases. Close monitoring continues, however, because of the introduction of prolonged-use tampons and those of different fiber composition.

Prevention and Treatment

In 1983, the first year that toxic shock syndrome became reportable in the United States, 502 cases were recorded, compared with 150 to 200 cases generally reported each year at present. The improved figures result largely from better understanding of how to use vaginal tampons, and withdrawing from use certain highly absorbent types that promoted development of the disease. The incidence of toxic shock syndrome associated with the use of menstrual tampons can be minimized in the following ways:

- Hands should be washed thoroughly before and after inserting a tampon.
- Use tampons with the lowest absorbency that is practical. Since 1990 the law requires that absorbency be stated on the package, ranging from less than 6 to 15 ml.
- Tampons probably should be changed about every 6 hours and a pad instead of a tampon used while sleeping. Tampons for overnight use introduced in the last few years, however, appear so far to be safe.
- Trauma to the vagina should be avoided when inserting tampons.
- Tampon users should know and understand the symptoms of toxic shock syndrome, and remove any tampon immediately if they occur.
- Tampons should not be used by persons who have had toxic shock syndrome previously.

Staphylococcal toxic shock syndrome can be effectively treated with an antistaphylococcal medication, intravenous fluid, and other measures to prevent shock and kidney damage. Most people recover fully in 2 to 3 weeks. The mortality rate is approximately 3%.

Table 25.5 describes the main features of toxic shock syndrome.

TABLE 25.5 Staphylococcal Toxic Shock Syndrome

Symptoms	Fever, vomiting, diarrhea, muscle aches, low blood pressure, and a rash that peels
Incubation period	3 to 7 days
Causative agent	*Staphylococcus aureus,* certain toxin-producing strains
Pathogenesis	Toxin (TSST-1 and others) produced by certain strains of *S. aureus;* toxins are superantigens, causing cytokine release and drop in blood pressure
Epidemiology	Associated with certain high-absorbency tampons, leaving tampons in place for long periods of time, and abrasion of the vagina from tampon use. Also as a result of infection by certain toxin-producing *S. aureus* strains in other parts of the body, such as skin, bone, and lung, and surgical wounds
Prevention and treatment	Awareness of symptoms. Prompt treatment of *S. aureus* infections; frequent change of tampons by menstruating women. Antimicrobial medication based on laboratory tests of susceptibility of the causative *S. aureus* strain; intravenous fluids

MICROCHECK 25.4

A marked derangement of the normal vaginal flora characterizes bacterial vaginosis, the most prevalent vaginal disease of women in the child-bearing years. Vulvovaginal candidiasis often occurs as a result of antibacterial therapy suppressing normal vaginal flora, but many other cases arise for unknown reasons. Toxic shock syndrome is caused by certain strains of *Staphylococcus aureus* whose exotoxins are absorbed into the bloodstream, causing the massive release of cytokines responsible for shock.

- Why was the incidence of toxic shock syndrome higher in menstruating women than in other people during the early 1980s?
- Why is puerperal fever not regarded as a venereal disease? What spreads the disease?

Sexually Transmitted Diseases: Scope of the Problem

Despite the expenditure of approximately $8 billion each year for sexually transmitted disease (STD) control, 15 million Americans annually, including 3 million teenagers, become infected. An individual who acquires an STD is apt to have acquired others without knowing it, and he or she needs to be tested for that possibility. The risk of acquiring an STD rises steeply with the number of partners with whom an individual has unprotected sexual intercourse, and with the numbers of sexual partners of those partners (**figure 25.7**). The use of alcohol and other drugs that release inhibitions adds to the risk. **Table 25.6** lists some often overlooked symptoms that can indicate the possibility of a sexually transmitted disease and warrant clinical evaluation even if they go away without treatment, especially if they appear within a few weeks of sexual intercourse with a new partner. A number of STDs can cause few or no symptoms and yet produce serious effects and be transmissible to other people.

Simple measures, although generally not popular, are highly effective in preventing STDs. A small but increasing percentage of students are abstaining from sexual intercourse, or conducting a monogamous relationship with a non-infected person. For others, proper and consistent use of latex or polyurethane condoms, while not an absolute guarantee of protection, markedly reduces the risk of acquiring an STD.

TABLE 25.6 Symptoms that Suggest STD

1. Abnormal discharge from the vagina or penis
2. Pain or burning sensation with urination
3. Sore or blister, painful or not, on the genitals or nearby; swellings in the groin
4. Abnormal vaginal bleeding or unusually severe menstrual cramps
5. Itching in the vaginal or rectal area
6. Pain in the lower abdomen in women; pain during sexual intercourse
7. Skin rash or mouth lesions

The list of diseases that can be transmitted sexually is very long. Shigellosis, giardiasis, scabies, viral hepatitis, and many others that have nonsexual modes of transmission cannot be overlooked. The epidemiology can be obscure, as when hepatitis B is contracted from sharing a contaminated needle during drug abuse, smolders unrecognized for years, and then is transmitted to another person by semen during sexual intercourse.
■ shigellosis, p. 601 ■ giardiasis, p. 613 ■ hepatitis, pp. 608–613

Table 25. 7 lists some common sexually transmitted diseases discussed in subsequent sections of this chapter.

MICROCHECK 25.5

An individual who acquires an STD may well have acquired others without knowing it. The chance of a person acquiring an STD increases steeply with the

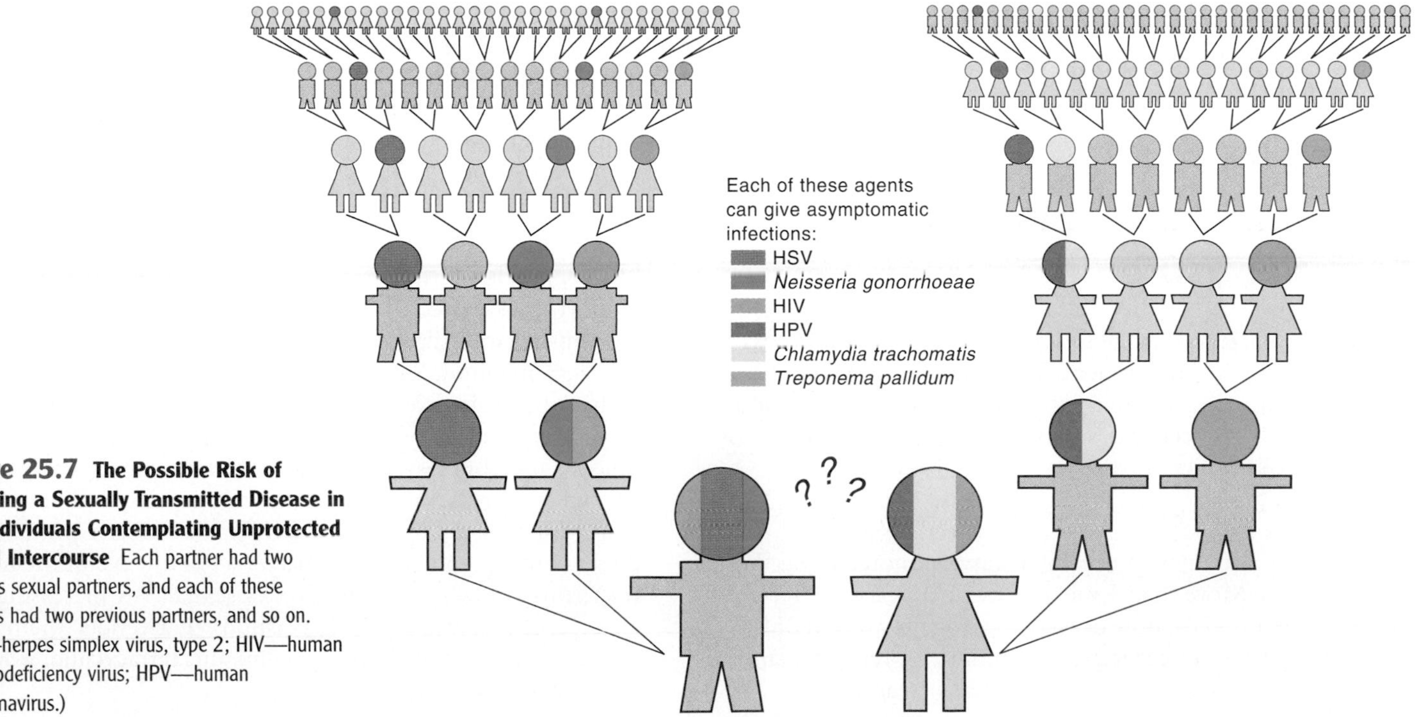

Figure 25.7 The Possible Risk of Acquiring a Sexually Transmitted Disease in Two Individuals Contemplating Unprotected Sexual Intercourse Each partner had two previous sexual partners, and each of these partners had two previous partners, and so on. (HSV—herpes simplex virus, type 2; HIV—human immunodeficiency virus; HPV—human papillomavirus.)

TABLE 25.7 Common Sexually Transmitted Diseases

Disease	Cause	Comment
Bacterial		
Gonorrhea ("clap")	*Neisseria gonorrhoeae*	Average reported cases per year 1995 to 1998 = 350,000. True incidence much higher.
Chlamydial infections	*Chlamydia trachomatis*	Average reported cases per year 1995 to 1998 = 527,000. True incidence much higher.
Syphilis	*Treponema pallidum*	Average reported cases (all stages of disease) per year, 1995 to 1998 = 51,600.
Chancroid	*Haemophilis ducreyi*	Average reported cases per year, 1995 to 1998 = 356. True incidence much higher.
Viral		
Genital herpes simplex	Herpes simplex virus (HSV)	Not reportable. Estimated 30 million Americans infected; about 85% HSV, type 2.
Papillomavirus infections	Human papillomavirus (HPV)	Not reportable. Estimated 40 million Americans infected.
AIDS	Human immunodeficiency virus (HIV)	Average reported cases per year, 1995 to 1998 = 60,860.
Protozoal		
Trichomoniasis ("trich")	*Trichomona vaginalis*	Not reportable. Estimated 2.5 million new infections per year.
Multicellular Parasites		
Pubic lice ("crabs")	*Phthirus pubis*	Common. No recent estimates.
Scabies ("seven-year itch")	*Sarcoptes scabiei*	2% to 4% of dermatology office visits. Often sexually transmitted in adults.

number of his or her sexual partners, and with the number of sexual partners of those partners. Symptoms of STD can sometimes easily be overlooked. Simple measures are highly effective for preventing STD.

- Name two diseases that have both sexual and nonsexual modes of transmission.
- Why might an individual with an STD need to be checked for other STDs even though there are no symptoms of any others?

Bacterial STD

Most of the bacteria that cause sexually transmitted diseases survive poorly in the environment. Transmission from one person to another usually requires intimate physical contact and is highly unlikely to occur via a handshake or contact with a contaminated toilet seat.

Gonorrhea

Gonorrhea is not a new problem; it was common among World War I recruits. More than 1 million cases per year were reported during the late 1970s, now down to an average around 350,000 per year. Today's concerns are its continued prevalence and its slowly increasing resistance to antibacterial treatment.

Symptoms

The incubation period of gonorrhea is generally only 2 to 5 days. Asymptomatic infections can occur in both sexes. In men, gonorrhea is characterized by urethritis, with pain during urination, and a thick, pus-containing discharge from the penis. These symptoms are dramatic and unpleasant, and they usually inspire prompt treatment. In women, the usual symptoms of painful urination and vaginal discharge tend to be milder than in men, and they may be overlooked. Therefore, women are more likely than men to become unknowing carriers of gonorrhea.

Causative Agent

Gonorrhea is caused by *Neisseria gonorrhoeae*, the gonococcus, a Gram-negative diplococcus that can be cultivated on chocolate agar medium. The organisms are typically found on and within the leukocytes in urethral pus (**figure 25.8**). Gonococci are parasites of humans only, preferring to live on the mucous membranes of their host. Most strains are susceptible to cold and drying and, hence, do not survive well outside the host. For this reason, gonorrhea is transmitted primarily by direct contact, and since the bacteria mainly live in the genital tract, this contact is almost always sexual. An increasing percentage of strains contain R plasmids that render them resistant to antibiotics such as penicillin and tetracycline. ■ **chocolate agar, p. 103** ■ **R plasmids, p. 511**

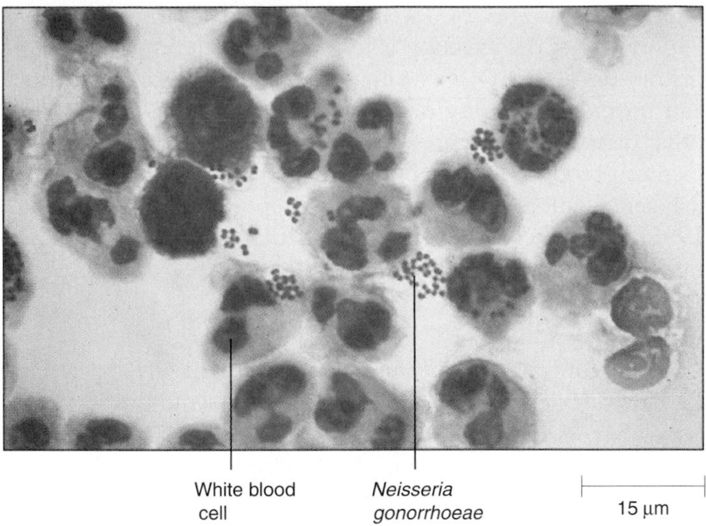

Figure 25.8 Appearance of *Neisseria gonorrhoeae* in Pus from the Urethra

Pathogenesis

Gonococci selectively attach to certain epithelial cells of the body, notably those of the urethra, uterine cervix, pharynx, and conjunctiva. Initial attachment is by pili (fimbriae) that project from the surface of the cocci **(figure 25.9)** and attach specifically to receptors on host cells. Besides playing a role in attachment, pili interfere with phagocytosis by macrophages. Pili, as well as certain other surface proteins involved in attachment, can either be expressed or not expressed, an example of phase variation. A single strain of gonococcus can express many different kinds of pili by chromosomal rearrangements within the pili genes. This variation in pili expression explains why cultures from different body sites in an infected individual and those from his or her sex partner may yield *N. gonorrhoeae* with different types of pili. The ability of the gonococcus to express different surface antigens allows it to attach to many different kinds of cell receptors and allows the organism to escape the effects of antibody formed against one kind of pilus. So far, the large variety of *N. gonorrhoeae* surface antigens has defeated efforts to develop a vaccine for prevention of gonorrhea. ■ phase variation, p. 184 ■ pili, p. 69, 208

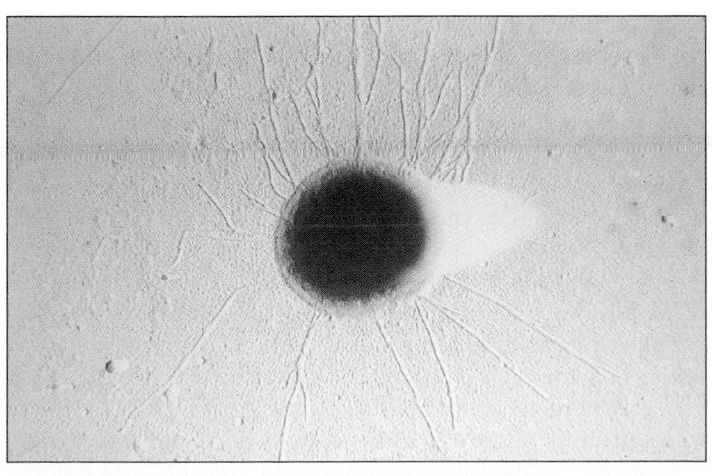

Figure 25.9 Electron Micrograph of *Neisseria gonorrhoeae* Showing Pili

The disease in men can sometimes be self-limiting and disappear on its own accord. Complications can occur, however, if the disease is not treated. Inflammatory reaction to the infection can lead to scar tissue formation that partially obstructs the urethra, predisposing the man to urinary tract infections. The infection may spread to the prostate gland and testes, producing hard-to-treat prostatic abscesses and **orchitis**. Sterility can result when scar tissue blocks the tubes that carry the sperm, or if testicular tissue is destroyed by the infection.

Gonorrhea follows a different course in women. Besides infecting the urethra, gonococci thrive in the cervix and fallopian tubes, as well as in glands in the vaginal wall and other areas of the genital tract. Some 15% to 30% of untreated infections progress upward through the uterus into the fallopian tubes, causing **pelvic inflammatory disease** (**PID**). Occasionally the infection exits the fallopian tube and passes into the abdominal cavity, where it attacks the surface of the liver or other abdominal organ. It is not clear how *N. gonorrhoeae* (which is nonmotile) can traverse the uterus to reach the fallopian tubes. One possibility is that the organisms hitch a ride on sperm, to which the bacteria are known to attach. Scar tissue formed as a result of the infection in the fallopian tubes can block normal passage of the ova through the tubes, causing sterility. Scarring of a fallopian tube can also lead to a dangerous complication—**ectopic pregnancy**—in which the ovum is fertilized and develops in the fallopian tube or even in the abdominal cavity outside the uterus. Ectopic pregnancy can lead to life-threatening internal hemorrhaging.

Ophthalmia neonatorum is a destructive *N. gonorrhoeae* infection of the eyes of newborn babies whose mothers have symptomatic or asymptomatic gonorrhea. The bacteria are transmitted during the infant's passage through the infected birth canal. This form of gonococcal infection is now unusual in the United States because laws require the use of a 1% silver nitrate solution or an antibiotic ointment such as 0.5% erythromycin placed directly into the eyes of all newborn infants. This treatment must be given within 1 hour of birth. Some mothers who feel certain that they do not have gonorrhea have challenged these laws because they do not want their babies to have the unpleasant experience of receiving potentially irritating eye medications. The long-standing and asymptomatic nature of gonorrhea in many women, however, makes omission of prophylactic treatment of a baby's eyes risky.

Infrequently, *Neisseria gonorrhoeae* can also cause **disseminated gonococcal infection** (**DGI**). Curiously, this complication is not usually preceded by urogenital symptoms. Disseminated gonococcal infections are characterized by one or more of the following: fever, rash, and arthritis caused by growth of gonococci within the joint spaces. Any joint may be affected, especially the larger ones. The heart valves can also be infected and destroyed in DGI.

Epidemiology

Gonorrhea is among the most prevalent of the sexually transmitted diseases. In the United States, its incidence is the highest of any reportable bacterial disease other than *Chlamydia* infection.

A steady rise in reported cases of gonorrhea from about 350,000 in 1966 to more than 1 million cases in 1976 occurred despite the availability of effective treatment. From 1976 through 1980, more than 1 million cases were reported each year, but since then the incidence has progressively declined, probably due largely to increased condom use from fear of AIDS.

Factors that influence the incidence of gonorrhea include:

- Birth control pills. About 35% of sexually active American college and university students say that they or their partner employs a contraceptive pill. Oral contraceptives without use of a condom offer no protection against venereal diseases and may increase susceptibility to them. Use of oral contraceptives leads to migration of gonorrhea-susceptible epithelial cells from the cervical lumen onto more exposed areas of the outer cervix. Oral contraceptives also tend to increase both the pH and the moisture content of the vagina, favoring infection with gonococci and other agents of sexually transmitted diseases. Besides being more vulnerable to gonorrhea, women taking oral contraceptives are also more likely to develop serious complications from the disease.
- Carriers. Carriers of gonococci, both male and female, can unknowingly transmit these bacteria over months or even years.
- Lack of immunity. There is little or no immunity following recovery from the disease. Individuals can contract gonorrhea repeatedly.

Prevention and Treatment

Prevention depends on abstinence, monogamous relationships, and consistent use of condoms. No vaccine is available against gonorrhea. The antigens are poorly immunogenic, and antibodies against them are not protective. Also, virtually all pathogenic *N. gonorrhoeae* produce an enzyme that destroys IgA antibody found on mucosal surfaces. A small percentage of gonococcal strains have developed resistance to most antimicrobial medications over the years, but aside from penicillin and tetracycline, widespread resistance has not yet occurred in the United States. Several fluoroquinolones and cephalosporins are effective against more than 95% of the *N. gonorrhoeae* strains encountered, and spectinomycin resistance is still rare. Resistance is much more common in parts of the world where antimicrobial usage is unrestricted, and the importation of resistant strains is inevitable. ■ **fluoroquinolones, p. 505** ■ **cephalosporins, p. 503**

Table 25.8 describes the main features of this disease.

Chlamydial Genital System Infections

Between 80% and 90% of college men with symptoms suggesting gonorrhea actually have other infections. From 25% to 40% of these students are infected with *Chlamydia trachomatis*, a common cause of sexually transmitted disease of men and women. Chlamydial infections mimic gonorrhea in several ways, including production of urethritis, and testicle and fallopian tube damage. Like *N. gonorrhoeae*, *C. trachomatis* attaches to sperm, and some have speculated that its rapid ascent to the fallopian tubes occurs by hitching a ride on sperm. The main importance of the infection is that it can produce pelvic inflammatory disease (PID) in women, damaging the fallopian tubes and promoting sterility or ectopic pregnancy. Most alarming, tubal damage can occur without symptoms. The infection in men can produce sterility by infecting the epididymis. Chlamydial genital infections have only been reportable nationally since 1995, and they substantially exceed the numbers of reported cases of gonorrhea. ■ **chlamydia, p. 292**

Symptoms

The symptoms of *C. trachomatis* infection generally appear 7 to 14 days after exposure. In men, the main symptom is a thin, gray-white discharge from the penis, sometimes with painful testes. Women most commonly develop an increased vaginal discharge, sometimes accompanied by painful urination, abnormal vaginal bleeding, and upper or lower abdominal pain. Many infections of men and women are asymptomatic.

Causative Agent

Chlamydia trachomatis is a spherical, obligate intracellular bacterium. Inclusion bodies containing a glycogen-like material form at the site of bacterial replication in the cytoplasm of host cells. These inclusions stain deeply with iodine and can provide a rapid way of identifying *C. trachomatis* infections. There are a number of different antigenic types of *C. trachomatis*, and different types may cause different diseases. Approximately eight types are responsible for most *C. trachomatis* sexually transmitted disease. Three other types cause **lymphogranuloma venereum**, a rare STD in the United States, in which lymph nodes in the groin swell up and drain pus; after years, gross swelling of the genitalia can occur. Four other types cause the chronic eye disease **trachoma**, an important preventable cause of blindness worldwide. ■ **glycogen, p. 34**

Pathogenesis

The infectious form of *C. trachomatis*, called an elementary body, attaches specifically to receptors on the surface of the host epithelial cell, thereby signaling the cell to take in the bacterium by endocytosis. In the endocytic vacuole, the bacterium enlarges and becomes a non-infectious form called a reticulate body. The reticulate body divides repeatedly by binary fission, resulting in numerous elementary bodies that are released from the host cell and infect nearby cells. The infected cells release cytokines that provoke an intense inflammatory reaction. Much of the tissue damage results from the cell-mediated immune response. The infection usually involves the urethra in both men and women. In men, it spreads to the tubules that collect sperm from the testicles, causing acute pain and swelling. In women, the infection commonly involves the cervix, making it bleed easily, often with sexual intercourse. The uterus is commonly involved, causing pain and bleeding. From the uterus, the infection spreads to the fallopian tubes (**figure 25.10**), causing PID, and like gonorrhea, it can exit the fallopian tubes and infect the surface of the liver. Subsequent scar tissue formation is responsible for many of the serious consequences of the disease such as sterility and ectopic pregnancy. ■ **endocytosis, p. 459**

TABLE 25.8 Gonorrhea

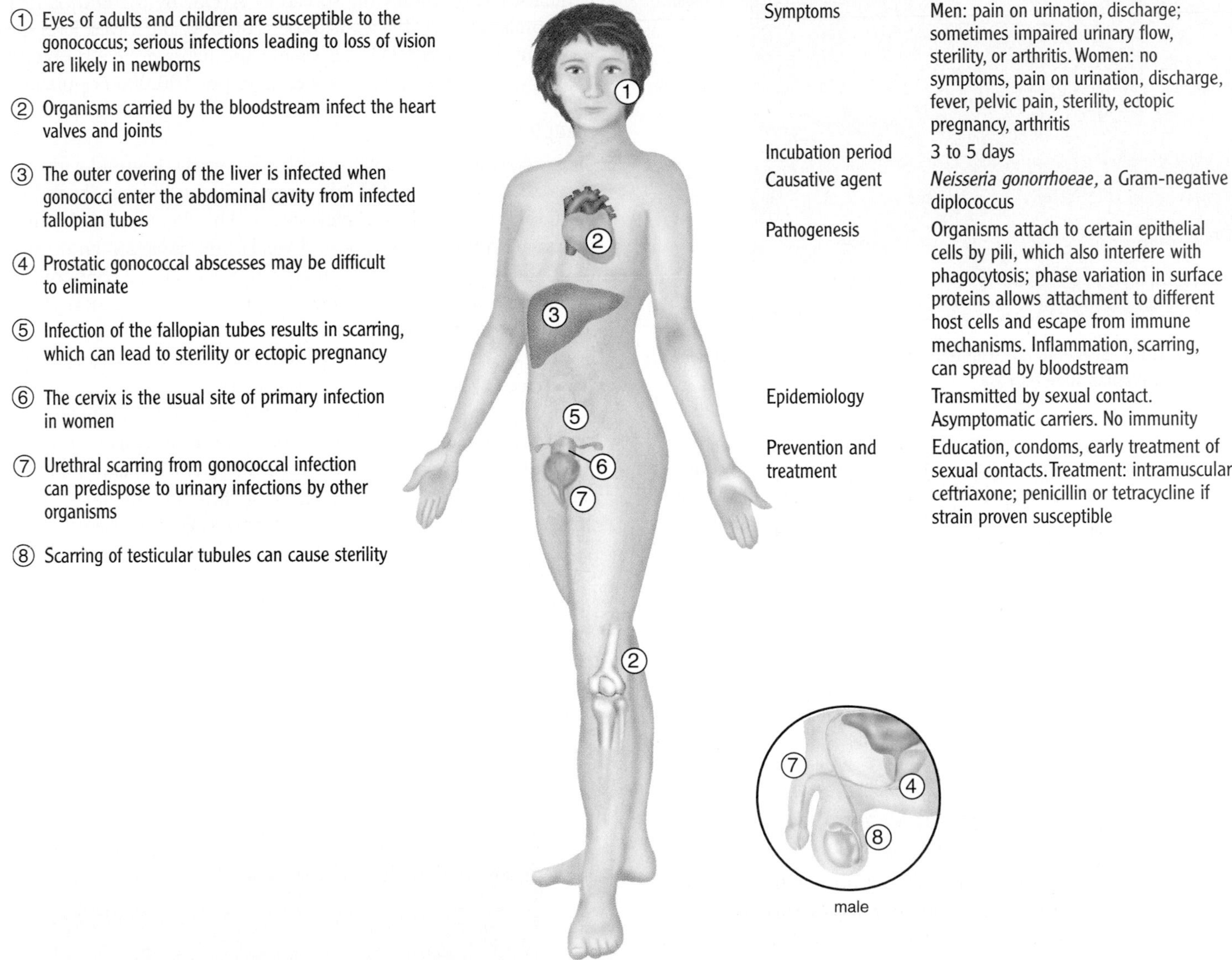
① Eyes of adults and children are susceptible to the gonococcus; serious infections leading to loss of vision are likely in newborns
② Organisms carried by the bloodstream infect the heart valves and joints
③ The outer covering of the liver is infected when gonococci enter the abdominal cavity from infected fallopian tubes
④ Prostatic gonococcal abscesses may be difficult to eliminate
⑤ Infection of the fallopian tubes results in scarring, which can lead to sterility or ectopic pregnancy
⑥ The cervix is the usual site of primary infection in women
⑦ Urethral scarring from gonococcal infection can predispose to urinary infections by other organisms
⑧ Scarring of testicular tubules can cause sterility
male

Symptoms	Men: pain on urination, discharge; sometimes impaired urinary flow, sterility, or arthritis. Women: no symptoms, pain on urination, discharge, fever, pelvic pain, sterility, ectopic pregnancy, arthritis
Incubation period	3 to 5 days
Causative agent	*Neisseria gonorrhoeae,* a Gram-negative diplococcus
Pathogenesis	Organisms attach to certain epithelial cells by pili, which also interfere with phagocytosis; phase variation in surface proteins allows attachment to different host cells and escape from immune mechanisms. Inflammation, scarring, can spread by bloodstream
Epidemiology	Transmitted by sexual contact. Asymptomatic carriers. No immunity
Prevention and treatment	Education, condoms, early treatment of sexual contacts. Treatment: intramuscular ceftriaxone; penicillin or tetracycline if strain proven susceptible

Epidemiology

Chlamydial genital infections lead all reportable bacterial infectious diseases. In contrast to gonorrhea, the number of reported chlamydial infections has tended to rise each year, probably a reflection of increased awareness of the disease and better diagnostic tests. Still only a fraction of the total cases, estimated at 4 million per year in the United States, is reported by physicians to public health departments. More than 14% of sexually active high school and college women were asymptomatic carriers of *C. trachomatis* in one study. Nonsexual transmission of this agent also occurs, for example, in nonchlorinated swimming pools. Newborn babies of infected mothers often develop chlamydial ophthalmia neonatorum, also called **inclusion conjunctivitis**, and pneumonia from *C. trachomatis* infection contracted during passage through the birth canal.

Prevention and Treatment

Condoms properly used can prevent transmission of the disease. All sexually active women are advised to get tested for *Chlamydia* each year, or twice yearly if they have multiple partners or if their partner has multiple partners. Several antibiotics offer effective treatment and prevent serious complications if the disease is diagnosed and treated promptly. Azithromycin can be given as a single dose, whereas tetracyclines and erythromycin are less expensive alternatives. The sexual partner is treated at the same time.

The main features of chlamydial genital infections are summarized in **table 25.9**.

Syphilis

During the first half of the Twentieth century, syphilis was a major cause of mental illness and blindness, and a significant contributor

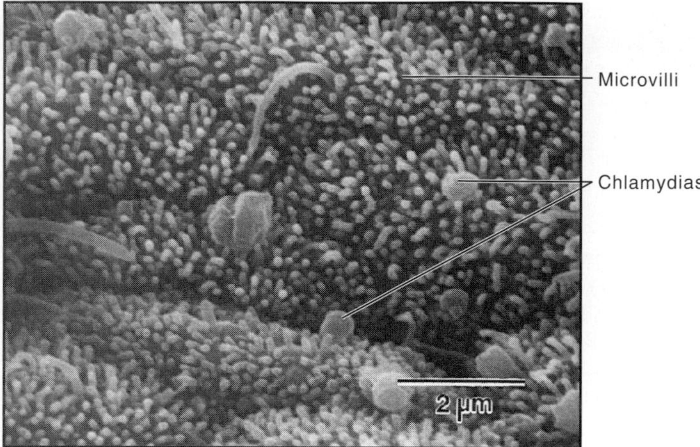

Figure 25.10 Scanning Electron Micrograph of *Chlamydia trachomatis* Attached to Fallopian Tube Mucosa

to the incidence of heart disease and stroke. Syphilis was common in the United States during World War II, present in about 5% of military recruits, but by the mid-1950s, it was almost eradicated. This was accomplished by aggressively locating syphilis cases and their sexual contacts and treating them with penicillin, which became generally available after the war. A number of factors, however, conspired to cause a resurgence of the disease. Inner-city poverty, prostitution, and drug use were linked to a high incidence of syphilis, which exceeded 100 new cases per 100,000 population in at least seven cities in 1990. Since then, renewed efforts in education, case finding, and treatment have caused a dramatic drop in new cases. The presence of the AIDS epidemic added urgency to syphilis control efforts, because syphilis, like other STDs that cause genital sores, promotes the spread of AIDS. By the end of 1998, the syphilis rate was down to 2.6 per 100,000, the lowest level since reporting began in 1941, surpassing the national health objective for the year 2000 of four or fewer cases per 100,000 population.

Symptoms

Syphilis occurs in so many forms that it is easily confused with other diseases and is often called "the great imitator." Generally, its manifestations occur in three clinical stages. The characteristic feature of **primary syphilis** occurs about 3 weeks after infection and consists of a painless, red ulcer with a hard rim called a **hard chancre** (pronounced "shanker") that appears at the site of infection (**figure 25.11**). Primary syphilis often goes unnoticed in women and homosexual men. After 2 to 10 weeks or longer, the manifestations of **secondary syphilis** usually appear, including runny nose and watery eyes, aches and pains, sore throat, a rash that includes the palms and soles, and whitish patches on the mucous membranes. After a latent period that can last for many years, manifestations of **tertiary syphilis** occur and typically include mental illness, blindness, stroke, and other nervous system disorders.

Causative Agent

Syphilis is caused by *Treponema pallidum*, an extremely slender, motile spirochete with tightly wound coils, which ranges up to 20 µm in length. Dark-field microscopy is used to see *T. pallidum* (**figure 25.12**) and to observe its characteristic slow rotational and flexing motions. The organism is difficult to study because it cannot be cultivated *in vitro* and must be grown in the testicles of

TABLE 25.9 Chlamydial Genital System Infections

Symptoms	Men: thin, gray-white penile discharge, painful testes. Women: vaginal discharge, vaginal bleeding, lower or upper abdominal pain
Incubation period	Usually 7 to 14 days
Causative agent	*Chlamydia trachomatis,* an obligate intracellular bacterium, certain serotypes
Pathogenesis	Elementary body attaches to specific receptors on the epithelial cell, causing endocytosis; transforms to reticulate body in the endocytic vacuole; repeated replication by binary fission and differentiation into elementary bodies; rupture of the vacuole and release of elementary bodies to infect adjacent cells; release of cytokines results in inflammatory response; cellular immune response against infection causes extensive damage; scar tissue formation responsible for ectopic pregnancy and sterility
Epidemiology	The leading reportable bacterial infection in the United States. Large numbers of asymptomatic men and women carriers
Prevention and treatment	Abstinence, monogamous relationship, condom use. Test sexually active men and women at least once yearly to rule out asymptomatic infection. Treatment: azithromycin, single dose; other antibacterial medications

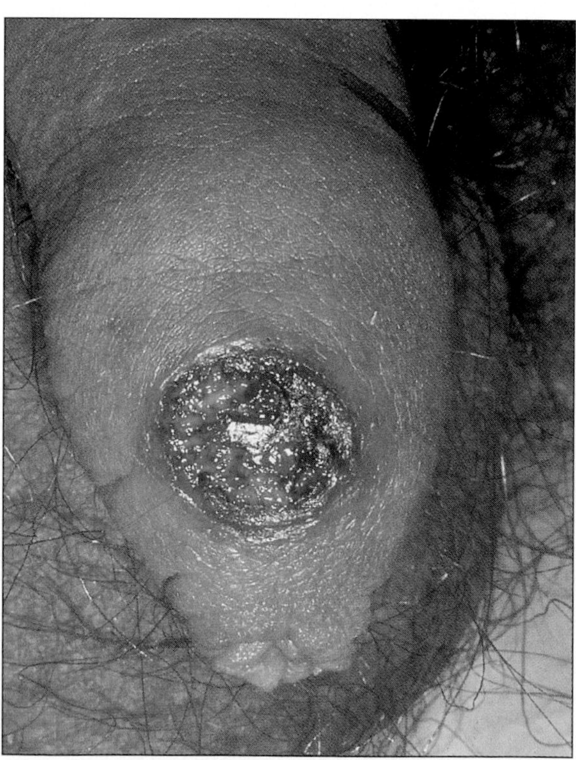

Figure 25.11 Syphilitic Chancre on the Foreskin of an Uncircumcised Man This is the site where *Treponema pallidum* entered the man's body.

Perspective 25.1 The Demise of Syphilis?

The arrival of the year 2000 saw the incidence of syphilis at an all-time low in the United States. The Centers for Disease Control and Prevention has announced the goal of lowering the number of cases to less than 1,000, and to have 90% of the 3,115 counties syphilis-free by the year 2005, with the aim of eliminating transmission of the disease entirely. Syphilis is a tempting target for elimination because it has no animal reservoir, infectious cases can generally be treated with a single injection of penicillin, and the incubation period is long enough so that sexual contacts of an infected person can be found and treated before they spread the disease any further. Eliminating syphilis has been tried before, but it has always returned in epidemic flare-ups. There are several reasons in addition to its historically low incidence, however, why it might be susceptible to a new attack. Public health experts have learned from the past that whenever control measures brought syphilis to its knees, public health personnel and funds were diverted to other problems, allowing resurgence of syphilis instead of killing it off. Moreover, instead of being widely endemic, most syphilis cases are now concentrated in a relatively few areas and certain populations. Also, new techniques are now available for rapid diagnosis, treatment, and epidemiological tracing of the disease. Finally, control of syphilis is closely linked to AIDS control efforts because the risk of HIV transmission is six times as great if either sex partner has syphilis.

One target population is persons detained in county and city jails, representing more than 500,000 individuals at any one time, with a known high incidence of STDs including syphilis. Screening these people for syphilis and getting them treated has proved frustrating because almost half of them are discharged within 48 hours, too soon to get the results of blood tests and arrange for medical evaluation and treatment. After release from jail, these individuals are notoriously hard to find and are often unwilling or unable to reveal the names and locations of their sexual contacts. Studies carried out in Illinois during 1996 showed that by using a screening blood test for syphilis that gave results in 15 minutes, and having a physician's assistant available for examination and treatment, almost 80% of jail inmates found to have syphilis got treatment the same day. Unfortunately, in the past, few jailers have had the interest or resources to focus on STD control to this extent.

If the goal of eliminating syphilis can be reached, it will mark the first time a disease has been conquered by finding cases and treating them. But the victory is bound to be short-lived unless the disease can also be eliminated in the many other countries where it is now endemic. To this end, scientists are hard at work attempting to develop a vaccine against *T. pallidum*, a quest that has been aided substantially by the complete sequencing of the bacterium's genome.

laboratory rabbits. Although the organism can be maintained *in vitro* and its metabolism studied, it undergoes little or no multiplication. Like most strains of gonococci, *T. pallidum* is killed by drying and chilling and is therefore transmitted almost exclusively by sexual or oral contact. Normally, *T. pallidum* is only a parasite of humans. We now know the sequence of bases in its entire genome, which it is hoped will give better understanding of its virulence, why it cannot be cultivated *in vitro*, where it originated, and how to make a vaccine against it. ■ **dark-field microscopy, p. 46**

Pathogenesis

Like the spirochetes that cause leptospirosis, *T. pallidum* readily penetrates mucous membranes and abraded skin. The infectious dose is very low, less than 100 organisms. In the first stage, called primary syphilis, *T. pallidum* grows and multiplies in a localized area of the genitalia, spreading from there to the lymph nodes and bloodstream. The hard chancre represents an intense inflammatory response to the bacterial invasion. Sometimes no chancre develops, only a pimple small enough to go unnoticed. Examination of a drop of fluid squeezed from the chancre reveals that it is teeming with infectious *T. pallidum*. Whether or not treatment is given, the chancre disappears within 4 to 6 weeks, and the victim may mistakenly believe that recovery from the disease has occurred. The organisms resist destruction by the body's defenses by an unknown mechanism, however, and progression of the disease can continue for years.

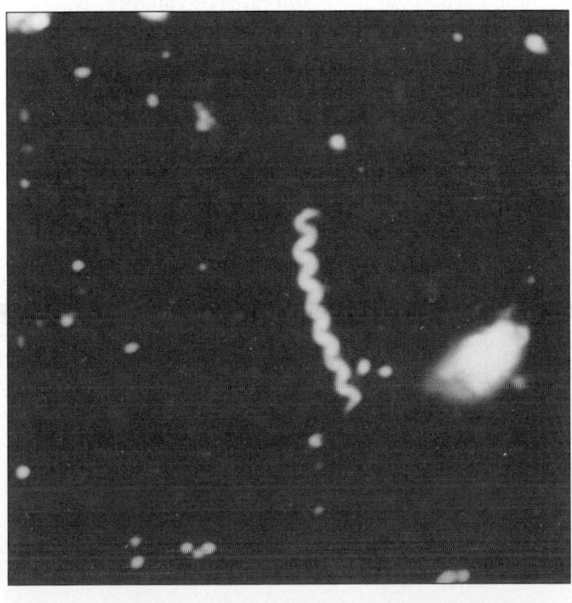

Figure 25.12 Appearance of *Treponema pallidum* with Dark-Field Illumination This technique readily detects the organism in skin and mucous membrane lesions of syphilis.

Many of the manifestations of secondary syphilis are due to the reaction of circulating *T. pallidum* with specific antibodies to form immune complexes. By this time, the spirochetes have spread throughout the body, and infectious lesions occur on the skin and mucous membranes in various locations, especially in the mouth (**figure 25.13**). Syphilis can be transmitted by kissing during this stage. The secondary stage lasts for weeks to months, sometimes as long as 1 year, and then gradually subsides. About 50% of untreated cases never progress past the secondary stage; however, after a latent period of from 5 to 20 years or even longer, some people with the disease develop tertiary syphilis. ■ **immune complexes, p. 440**

Tertiary syphilis, or the third stage of syphilis, represents a hypersensitivity reaction to small numbers of *T. pallidum* that grow and persist in the tissues. In this stage, the patient is no longer infectious. The remaining *T. pallidum* organisms may be present in almost any part of the body, and the symptoms of tertiary syphilis depend on where the hypersensitivity reactions occur. If they occur in the skin, bones, or other areas not vital to existence, the disease is not life threatening. If, however, they occur within the walls of a major blood vessel such as the aorta, the vessel may become weakened and even rupture, resulting in death. Hypersensitivity reactions to *T. pallidum* in the eyes cause blindness; central nervous system involvement most commonly manifests itself as a stroke. A granulomatous necrotizing mass called a **gumma (figure 25.14)**, analogous to the tubercle

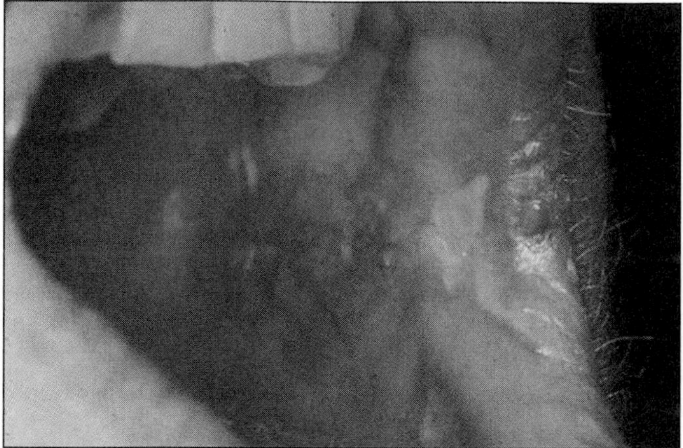

Figure 25.13 Appearance of the Mucous Patches of Secondary Syphilis These lesions are swarming with *Treponema pallidum* and are highly infectious.

of tuberculosis, can involve any part of the body. A characteristic pattern of symptoms and signs called **general paresis** develops an average of 20 years after infection. Typical findings include personality change, emotional instability, delusions, hallucinations, memory loss, impaired judgment, abnormalities of the pupils of the eye, and speech defects. ■ tubercle, p. 570

The main characteristics of the three stages of syphilis are summarized in **table 25.10**.

Congenital Syphilis During pregnancy, *T. pallidum* readily crosses the placenta and infects the fetus. This can occur at any stage of pregnancy, but damage to the fetus does not generally develop until the fourth month. Therefore, if the mother's syphilis is diagnosed and treated before the fourth month of pregnancy, the fetus will be treated too and will not develop the disease. Without treatment, the risk to the fetus depends partly on the stage of the mother's infection. Three-fourths or more of the fetuses become infected if the mother has primary or early secondary syphilis; risk decreases with the duration of the mother's infection but is still significant into the latent period. Fetal infections can occur in the absence of any symptoms of syphilis in the mother. About two out

TABLE 25.10 Stages of Syphilis

Stage of Disease	Main Characteristics	Infectious?
Primary	Firm painless ulcer (hard chancre) at site of infection; lymph node enlargement	Yes
Secondary	Rash, aches and pains; mucous membrane lesions	Yes
Tertiary	Gummas; damage to large blood vessels, eyes, nervous system; insanity	No

of every five infected fetuses are lost through miscarriage or stillbirth. The remainder are born with congenital syphilis. Although these infants frequently appear normal at the time of birth, some of them develop secondary syphilis, which is often fatal, within a few weeks. Others develop characteristic deformities of their face, teeth (**figure 25.15**), and other body parts later in childhood.

Epidemiology

There is no animal reservoir. Syphilis is usually transmitted by sexual intercourse, but infection can be contracted from kissing. Elimination of transmission in the United States is within reach, but depends on identifying and treating cases within high-incidence groups such as prostitutes, promiscuous homosexual men, and jail inmates, and guarding against reintroduction of the disease from countries where it is still rampant. Effective screening to detect infected individuals can easily be accomplished using a blood test.

Prevention and Treatment

No vaccine against syphilis is currently available. Monogamous relationships, condoms, and other safer sex practices decrease the risk of contracting the disease. Primary and secondary syphilis are effectively treated with an antibiotic such as penicillin. Treatment must be continued for a longer period for tertiary syphilis,

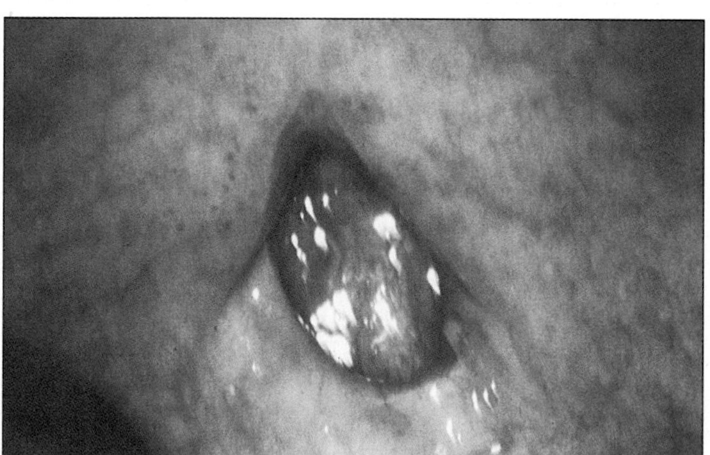

Figure 25.14 A Gumma of Tertiary Syphilis Gummas consist of an inflammatory mass which can perforate, as in this example in the roof of the mouth. Gummas can occur anywhere in the body.

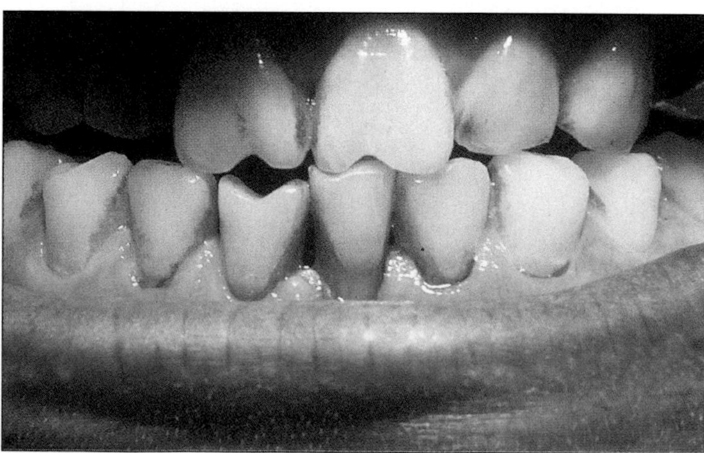

Figure 25.15 Hutchinson's Teeth Notice the notched, deformed incisors, a late manifestation of congenital syphilis.

TABLE 25.11 Syphilis

① *Treponema pallidum* enters the body through a microscopic abrasion or mucous membrane, usually genitalia, mouth, or rectum

② A chancre develops at site of entry

③ Organisms multiply locally and spread throughout the body by the bloodstream

④ Infectious mucous patches and skin rashes of secondary syphilis appear. A fetus will become infected, resulting in miscarriage or a live-born infant with congenital syphilis

⑤ An asymptomatic latent period occurs. *T. pallidum* disappears from blood, skin, and mucous membranes

⑥ After months or years, symptoms of tertiary syphilis appear:
- heart and great vessel defects
- gummas
- strokes
- eye abnormalities
- general paresis
- insanity

Symptoms	Chancre, fever, rash, stroke, nervous system deterioration; can imitate many other diseases
Incubation period	10 to 90 days
Causative agent	*Treponema pallidum,* a non-culturable spirochete
Pathogenesis	Primary lesion, or chancre, appears at site of inoculation, heals after 2 to 6 weeks; *T. pallidum* invades the blood vessel system and is carried throughout the body, causing fever, rash, mucous membrane lesions; damage to brain, arteries, and peripheral nerves appears years later
Epidemiology	Sexual contact with infected partner; kissing; transplacental passage
Prevention and treatment	Education, use of condoms, treatment of sexual contacts, reporting cases. Treatment: penicillin

however, probably because most of the organisms are not actively multiplying.

Table 25.11 summarizes the main features of syphilis.

Chancroid

Chancroid is another bacterial sexually transmitted disease that showed a marked increase in frequency, reaching 5,000 reported cases in 1988, followed by a progressive decline to the lowest frequency in decades. Although widespread in the United States, it is not commonly reported. It is another STD with genital sores that promote the spread of AIDS.

Symptoms

Chancroid is characterized by single or multiple genital sores called **soft chancres (figure 25.16a)**, which are painful, unlike the hard chancre of syphilis. The lymph nodes in the groin enlarge **(figure 25.16b)**; they become tender, and sometimes pus-filled.

Causative Agent

The causative organism of chancroid is *Haemophilus ducreyi*, a small, pleomorphic, Gram-negative rod that requires X-factor, a blood component called hematin, for growth. ■ ***Haemophilus*, p. 552**

Pathogenesis

Knowledge of the pathogenesis of chancroid is incomplete. Typically, a small pimple appears, presumably at the site of entry of the organisms. After a few days, this ulcerates and gradually enlarges, reaching an inch or more in diameter, often joining other lesions. Organisms that reach the lymph nodes incite an intense inflammatory response, with congregation of polymorphonuclear neutrophils, release of proteolytic enzymes, and liquefication of the

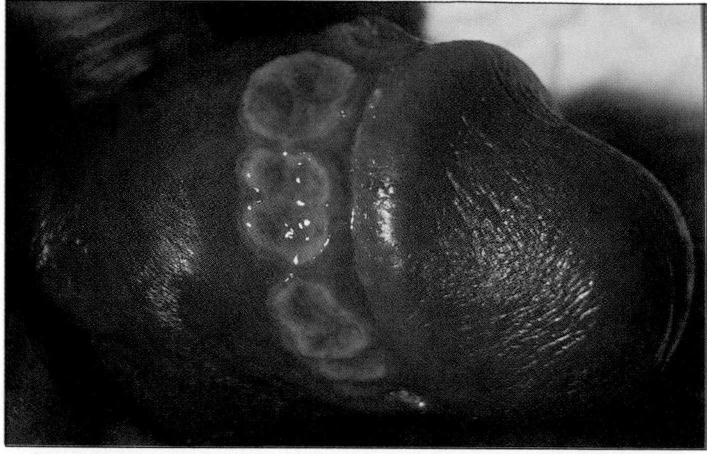

(a)

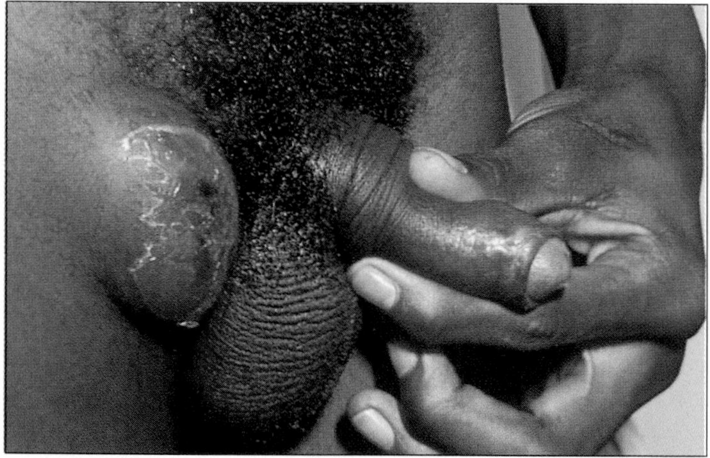

(b)

Figure 25.16 Chancroid **(a)** Lesions of the penis ("soft chancres"). These ulcerations are soft and painful, in contrast to the chancres of syphilis. **(b)** Swollen groin lymph nodes. The nodes are tender, often pus-filled, and may break open and drain.

lymphoid tissue. The pus-filled nodes enlarge, stretch the skin, and rupture through to the surface, discharging their contents.

Epidemiology

Epidemics in American cities are associated with prostitution; in some tropical countries, chancroid is second only to gonorrhea among the sexually transmitted diseases. Chancroid and other ulcerating diseases of the genitalia promote AIDS because they ease the entry of the human immunodeficiency virus (HIV).

Prevention and Treatment

Abstinence from sexual intercourse, monogamous relationships, and use of condoms are effective preventives. Chancroid usually responds well to treatment with the antibiotics erythromycin, azithromycin, or ceftriaxone. Some strains, however, possess R plasmids that code for resistance to multiple antibacterial medications. Effectiveness of therapy is also sharply reduced if the patient has AIDS. As in a number of bacterial infections, body defenses and antibacterial medications work together to defeat the disease. ■ **R plasmids, p. 211, 511**

TABLE 25.12 Chancroid

Symptoms	One or more painful, gradually enlarging, soft chancres on or near the genitalia; large, tender regional lymph nodes
Incubation period	3 to 10 days
Causative agent	*Haemophilus ducreyi*, a small, pleomorphic Gram-negative rod requiring X-factor for growth
Pathogenesis	A small pimple appears first, which ulcerates and gradually enlarges; multiple lesions may coalesce; lymph nodes enlarge, liquefy, and may discharge to the skin surface
Epidemiology	Sexual transmission. Common in prostitutes; fosters the spread of AIDS
Prevention and treatment	Abstinence from sex, monogamous relationships; safer sex practices, including avoidance of sexually promiscuous partners; proper use of condoms. Treatment: several antibacterial medications effective. Resistance can be a problem

Table 25.12 gives the main features of chancroid.

MICROCHECK 25.6

Transmission of bacterial STDs usually requires direct human-to-human contact. Unsuspected sexually transmitted infections of pregnant women pose a serious threat to fetuses and newborn babies. Even asymptomatic infections can cause genital tract damage and be transmissible to other people.

- Can a baby have congenital syphilis without its mother ever having had symptoms of syphilis? Explain.
- What, if any, is the relationship between chancroid and development of AIDS?
- Why should scarring of a fallopian tube raise the risks of an ectopic pregnancy?
- Why should scar tissue formation predispose males to urinary tract infections?

Viral STD

Viral sexually transmitted diseases are probably as common or more common than the bacterial diseases, but they are incurable. Their effects can be severe and long lasting. Genital herpes simplex can give recurrent symptoms for years, papillomavirus infections can lead to cancer, and HIV infections usually result in AIDS.

Genital Herpes Simplex

Genital herpes simplex is among the top three or four most common sexually transmitted diseases. An estimated 30 million Americans are infected with the causative virus and about 500,000 new cases occur each year.

Symptoms

Symptoms of genital herpes simplex begin 2 to 20 days (usually about a week) after exposure, with genital itching, burning, and in women, often severe pain. Infection of the urethra may imitate the symptoms of a bladder infection. Clusters of small red bumps appear on the genitalia, which then become blisters surrounded by redness **(figure 25.17)**. The blisters break in 3 to 5 days, leaving an ulcerated area. The ulcers slowly dry and become crusted and then heal without a scar. Symptoms are the worst in the first 1 to 2 weeks and disappear within 3 weeks. They recur in an irregular pattern; about 75% of patients will have at least one recurrence within a year, and the average is about four recurrences a year. The recurrent symptoms are usually not quite as severe as those of the first episode, and recurrences generally decrease in frequency with time. Some individuals may have recurrences for life.

Causative Agents

The causative agent of genital herpes is usually herpes simplex virus type 2 (HSV-2), a medium-sized, enveloped, double-stranded DNA virus. On electron micrographs it looks like herpes simplex virus type 1 (HSV-1), the cause of "cold sores" ("fever blisters," herpes simplex labialis), and the genomes of the two viruses are about 50% homologous. This disease, like cold sores, often recurs, because the virus becomes latent. Either virus can infect the mouth and the genitalia, but the type 2 virus causes more severe genital lesions whose frequency of recurrence is greater than that of the type 1 virus. During initial infection and for as long as 1 month after, the virus is found in genital secretions. During recurrence, the virus is usually present in large numbers for less than a week. ■ **HSV-1, p. 594**

Pathogenesis

Lesions begin with infection of a group of epithelial cells that are lysed following viral replication, creating a small fluid-filled blister or vesicle containing large numbers of infectious virions. Latency of the disease is incompletely understood. The viral DNA exists within nerve cells in a circular, non-infectious form during times when there are no symptoms. In this state, only a small portion of the HSV genome is transcribed. At times, however, the entire viral chromosome can be transcribed and complete infectious virions replicated. These reinfect the area supplied by the nerve and cause a recurrence. The mechanisms by which the latent infection is maintained or reactivated are not known in detail, but they probably depend on cellular immunity. ■ **latent infections, p. 352**

Genital herpes can pose a serious risk to newborn babies. If the mother has a primary infection near the time of delivery, the baby has about a one in three risk of acquiring the infection. The baby often dies from overwhelming infection or is permanently disabled by it. To prevent contact with the infected birth canal, these babies must be delivered by cesarean section. If the mother has recurrent disease, the risk to the baby is very low, presumably because transplacental anti-HSV antibody from the mother protects the baby, but physicians will often do a cesarean section to minimize the risk.

Genital herpes is associated with cancer of the uterine cervix, since women with genital herpes have more than five times the risk of precancerous changes of the cervix compared with women who do not have herpes. The association of herpes simplex type 2 infection with cervical cancer, however, does not necessarily mean that the virus causes the cancer.

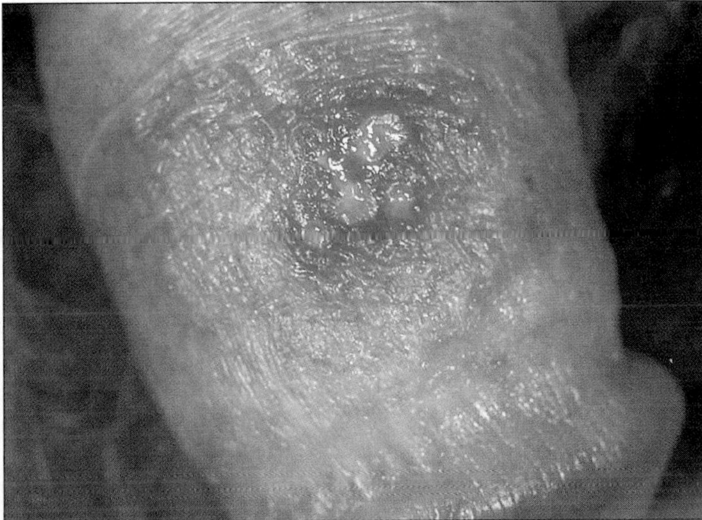

Figure 25.17 Genital Herpes Simplex on the Shaft of the Penis The lesions typically begin as a group of small vesicles surrounded by redness. Within 3 to 5 days, they break, leaving the small superficial ulcers as shown. The lesions can be very painful, especially in women.

Epidemiology

There are no animal reservoirs. The virus can survive for short periods on fomites or in bathwater, but nonsexual tranmission rarely occurs. Sexual transmission of HSV is most likely to occur when lesions are present, but it can happen in the absence of symptoms or signs of the disease. Once infected, there is a lifelong risk of transmitting the virus to another person. Herpes simplex, like other ulcerating genital diseases, promotes the spread of AIDS by increasing the risk of HIV infection.

Prevention and Treatment

Avoidance of sexual intercourse during active symptoms and the use of condoms with a spermicide reduce but do not eliminate the chance of transmission. Spermicidal jellies and creams can inactivate herpes simplex viruses. There is no cure for genital herpes, although medications such as acyclovir and famciclovir can decrease the severity of the first attack and the incidence of recurrences. Women with recurrent genital herpes should have a **Papanicolaou smear** every 12 months to detect cervical cancer in its early (curable) stage. The "Pap test" consists of removing cells from the cervix with a small brush or spatula, staining them on a microscope slide, and examining them for any abnormal cells **(figure 25.18)**.

The main features of genital herpes simplex are presented in **table 25.13**.

Papillomavirus STDs: Genital Warts and Cervical Cancer

Sexually transmitted human papillomaviruses (HPV) are probably the most common of the sexually transmitted disease agents, infecting an estimated 40 million Americans. Although they cannot be cultivated in the laboratory, their presence in human tissues can be studied by using nucleic acid probes. Some of them are responsible for **papillomas**, warty growths, of the

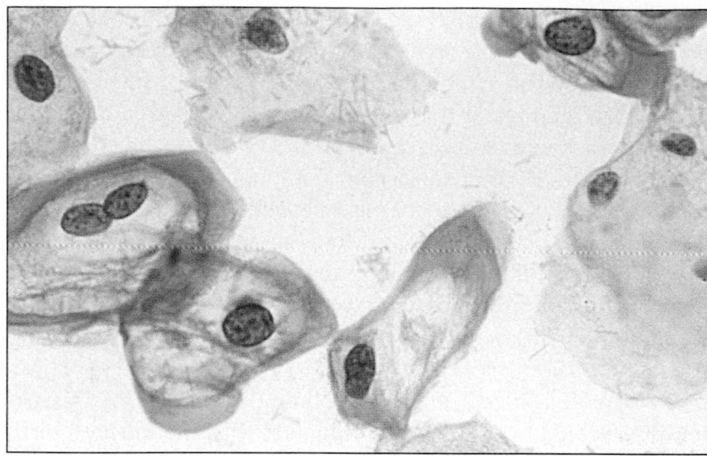

Figure 25.18 Abnormal Papanicolaou Smear The pink and blue objects are squamous epithelial cells; abnormalities include doubling of the nuclei and a clear area around them. Most abnormal smears in young women are due to human papillomavirus infection, when persistent an important factor in the development of cancer of the cervix.

external and internal gentalia; others cause nonwarty lesions of mucosal surfaces such as the uterine cervix and are a major factor in the development of cervical cancer. ■ nucleic acid probes, p. 232

Symptoms

Many infected individuals, including almost half of all infected women, have no symptoms. Warts are the most easily recognized manifestation of infection, with about 750,000 new cases in the United States each year. They often appear on the head or shaft of the penis (**figure 25.19**), at the vaginal opening, or around the anus. Occasionally, the warts become inflamed or bleed. They often have a large emotional impact because of an unjustified fear of cancer and the embarrassment of having such an obvious STD. Important precancerous lesions of the cervix are asymptomatic and can only be detected by vaginal examination and Papanicolaou test.

Causative Agents

Human papillomaviruses are small, nonenveloped, double-stranded DNA viruses of the papovavirus family. There are more than 80 types of HPV based on DNA homology, and about one-third of them infect the genitalia. Approximately 15 types are strongly associated with cancer of the cervix and other cancers. The different papillomavirus types infect different kinds of epithelium. For example, HPVs that cause the common warts of the hands and plantar warts of the feet generally do not infect the genitalia. The HPVs that cause genital warts may infect the cervix, but they do not generally predispose to cancer. ■ DNA homology, p. 186

Pathogenesis

HPVs are thought to enter and infect the deeper layers of epithelium through microscopic abrasions. A latent infection, without warts or microscopic lesions, is the most common result of infection. Latent infection can last for years, but studies in college age women reveal that about 90% of HPV infections are eliminated by body defenses within 2 years. Viral replication and release are intermittent and require that the deeper layer of epithelial cells enter a maturation process. Different kinds of lesions develop depending on the virus type and the location of the infection. Lesions can be flat, raised or unraised, cauliflower-like, or hidden within the epithelium.

The mechanism giving rise to warts is unknown. The virus exists in the cytoplasm of infected cells as closed DNA circles. Warts usually appear about 3 months after infection (range, 3 weeks to 8 months). Following removal or destruction of warts, HPV persists in surrounding normal-appearing epithelium and can give rise to additional warts. Warts can occasionally partly obstruct the urethra or, if very large, the birth canal. Newborn infants can become infected with HPV at birth and develop warts that obstruct their respiratory tract, a serious condition that occurs in less than one out of every 100,000 births.

Worldwide, cancer of the cervix is second in frequency only to breast cancer among the malignant tumors of women. In the United States, about 5,000 women die of cervical cancer each year. Most of these cancers are associated with certain HPV types. Cancer-associated HPV types differ from wart-causing HPV types in that the former can integrate into the chromosome of a host cell and code for a protein that permits excessive cell growth. These oncogenic types of HPV tend to cause infections that persist longer than other HPV types, and to cause precancerous

TABLE 23.13 Genital Herpes Simplex

Symptoms	Itching, burning pain at the site of infection, painful urination, tiny blisters with underlying redness. The blisters break, leaving a painful superficial ulcer, which heals without scarring. Recurrences common
Incubation period	Usually 1 week (range, 2 to 20 days)
Causative agent	Usually herpes simplex virus, type 2. The cold sore virus, herpes simplex type 1, can also be responsible. Herpesviruses are enveloped and contain double-stranded DNA
Pathogenesis	Lysis of infected epithelial cells results in fluid-filled blisters containing infectious virions. Rupture of these vesicles causes a painful ulceration. The acute infection is controlled by body defenses; genome persists within nerve cells in a non-infectious form beyond the reach of body defenses. Replication of infectious virions can occur and cause recurrent symptoms in the area supplied by the nerve
Epidemiology	No animal reservoirs. Transmission by sexual intercourse, oral-genital contact. Transmission risk greatest first few days of active disease. Transmission can occur in the absence of symptoms. Newborn infants can contract fatal generalized herpetic infection if their mother has a symptomatic infection at the time of delivery. Strong association between genital herpes simplex and cervical cancer. Herpes simplex increases the risk of contracting AIDS
Prevention and treatment	Abstinence, monogamy, and condoms help prevent transmission. Medications help prevent recurrences, shorten duration of symptoms. No cure. Women with genital herpes advised to get a Pap test at least once yearly

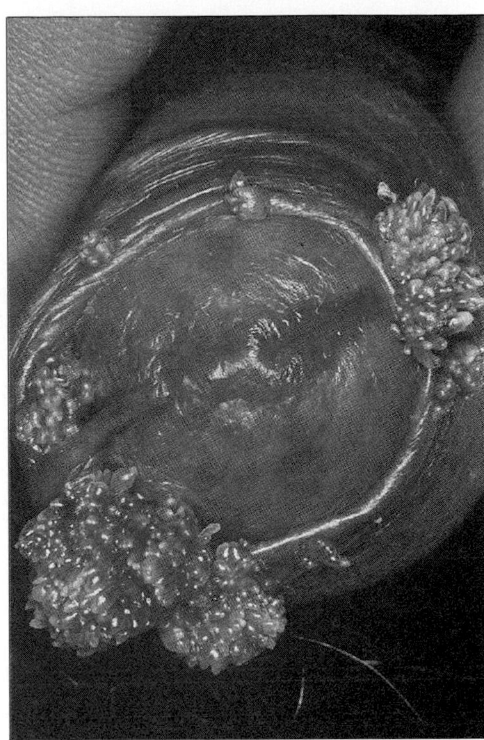

Figure 25.19 **Genital Warts on the Penis, a Manifestation of Human Papillomavirus Infection**

TABLE 25.14 Papillomavirus STDs

Symptoms	Usually no symptoms. Warts of the external and internal genitalia the most common symptom
Incubation period	Usually 3 months; (range, 3 weeks to 8 months)
Causative agents	Human papillomaviruses, many types, small, nonenveloped double-stranded DNA viruses of the papovavirus family. Different types infect different tissues and produce different lesions
Pathogenesis	Virus enters epithelium through abrasions, infects deep layer of epithelium; establishes latency; cycles of replication occur when host cell begins maturation; cancer-associated viral types can integrate into the host cell chromosome, cause precancerous lesions
Epidemiology	Asymptomatic individuals can transmit the disease; 60% transmission with a single sexual contact; multiple sex partners the greatest risk factor; warts can be transmitted to the mouth with oral sex, and to newborn babies
Prevention and treatment	Latex condoms advised to minimize transmission. Avoidance of sexual contact with those having multiple sex partners. Pap tests at least yearly for sexually active women. Wart removal by multiple techniques, does not cure the infection. Imiquimod useful in treating multiple warts about anus and external genitalia

lesions detectable by Papanicolaou tests. A cancer-associated HPV is present in almost all cervical cancers, but only a small percentage of infections by these viruses result in cancer. This indicates that other unknown factors must be present for cancer to develop. Cervical cancer–associated HPVs have also been found in vaginal cancers, cancers of the penis, and in anal cancers of men and women who engage in receptive anal intercourse.

Epidemiology

HPVs are readily spread by sexual intercourse; a single sexual exposure to an infected person transfers the infection 60% of the time. Asymptomatic individuals infected with HPV can transfer the virus to others. Forty-six percent of women university students seeking routine gynecological examinations were found to be infected with HPV in one study. HPV infection is the most common reason for an abnormal Papanicolaou test in teenage women. A history of having multiple sex partners is the most important risk factor for acquiring genital HPV infection. Warts can develop in the mouth from HPV infection transferred by oral sex.

Prevention and Treatment

HPV-infected individuals are advised to employ condoms to help decrease the chance of transmission. Cervical cancer is generally preceded by precancerous lesions, areas of abnormal cell growth that can be detected by the Papanicolaou test. The abnormal growth can be removed, thereby preventing development of a cancer. Warts can be removed by laser treatment or freezing with liquid nitrogen, but these measures do not cure the infection. It is not known whether treating genital warts decreases the risk of transmitting the virus to others.

Table 25.14 summarizes some important features of papillomavirus STDs.

AIDS

AIDS, acquired immunodeficiency syndrome, unquestionably the most important sexually transmitted disease of the Twentieth century, is covered briefly here and more fully in chapter 29. The AIDS epidemic was first recognized in the United States in 1981. Over the first decade and a half of the epidemic, 500,000 Americans developed AIDS and 300,000 died of the disease. Despite major advances in prevention and treatment of the disease, the epidemic continues its stealthy advance, and its full, devastating impact on human society will probably not be felt for another decade. During 1998 alone, an estimated 5.8 million people around the world first became infected with the causative agent, human immunodeficiency virus (HIV), and 2.5 million died of AIDS. Some important advances have occurred in the prevention and treatment of AIDS in Western countries, but these are largely unavailable to the vast majority of those around the world who suffer because of the disease **(figure 25.20)**. ■ AIDS, p. 741

Symptoms

AIDS is the end stage of what is more properly called **HIV disease**. Six days to 6 weeks after contracting the virus, some individuals develop a fever, head and muscle aches, enlarged lymph nodes, and a rash. These symptoms are often mild, attributed to the flu, and go away by themselves. Many individuals have no symptoms at all.

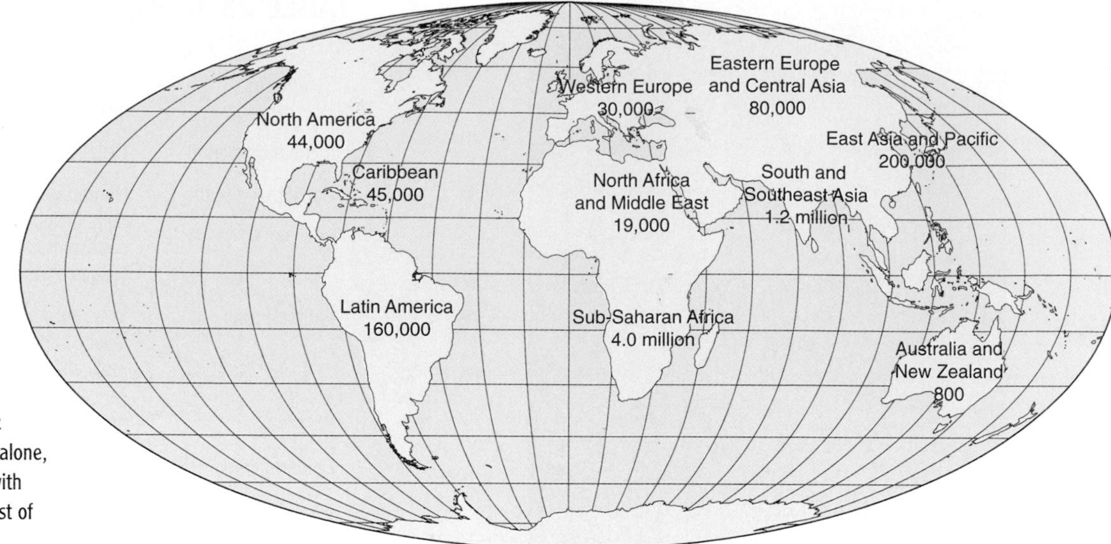

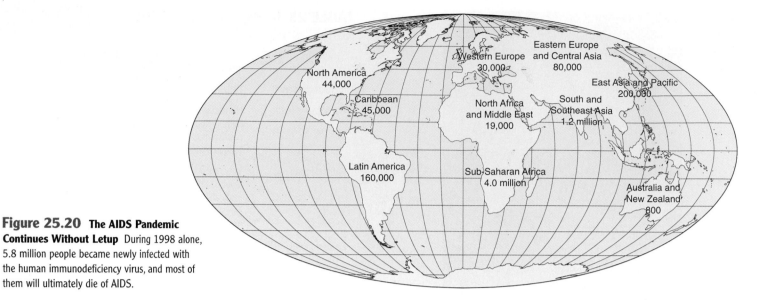

Figure 25.20 The AIDS Pandemic Continues Without Letup During 1998 alone, 5.8 million people became newly infected with the human immunodeficiency virus, and most of them will ultimately die of AIDS.

The typical case of HIV disease then smolders unnoticed for almost 10 years before immunodeficiency develops, as evidenced by certain malignancies and unusual microbial infections that often attack the lungs, intestines, eyes, or the central nervous system.

Causative Agents

Most AIDS cases in the United States are caused by human immunodeficiency virus type 1 (HIV-1), an enveloped single-stranded RNA virus of the retrovirus family. Each virion contains reverse transcriptase and two copies of the viral genome. In parts of western Africa, another retrovirus, HIV-2, is a common cause of AIDS. ■ retroviruses, pp. 354–355

Pathogenesis

HIV can attack a variety of types of cells in the human body, but most critically a subset of lymphocytes, the T-helper (Th) cells. These cells possess the CD4 surface antigen to which the virus attaches, and they are referred to as $CD4^+$ cells. In order for the virus to enter the cell, it must also attach to certain cytokine receptors on the cell surface. Following entry of the virus, a DNA copy is made using reverse transcriptase. This copy integrates itself into the cell's genome and hides there, inaccessible to antiviral chemotherapy. If the cell becomes activated, however, the virus leaves the cell genome, replicates, and destroys the cell, releasing virions to infect additional cells. Despite the body's ability to replace hundreds of billions of $CD4^+$ lymphocytes, the number of these lymphocytes slowly declines over many months or years. Infection of macrophages contributes to the decline in cellular immunity. Macrophages, also vitally important to the immune system, are also $CD4^+$ cells. Although they are generally not killed, their function is impaired both by the viral infection and the lack of interaction with Th lymphocytes. Eventually, the immune system becomes so impaired it can no longer respond to infections or cancers. ■ T-helper cells, p. 397 ■ macrophages, p. 372

Epidemiology

HIV disease spreads mainly by sexual intercourse, sharing hypodermic needles, or from mother to newborn. The virus is not highly contagious, and most transmissions could be halted by simple changes in human behavior.

Prevention and Treatment

One of the most important recent developments in the prevention of AIDS has been the finding that transmission from mother to newborn can be interrupted in about two-thirds of the cases by chemotherapy. Another effective preventive measure is to supply injected-drug abusers with sterile needles and syringes in exchange for used ones, proven to decrease transmissions of the virus without encouraging addiction. Specifically designed educational programs targeting certain groups of people have also been effective short term in decreasing risky sexual practices. Lastly, identification and treatment of other STDs decreases the risk of contracting HIV disease, as does consistent use of latex condoms.

Treatment is designed to block replication of HIV, but it does not affect viral nucleic acid integrated in the host genome. Two classes of medications are currently available, those that block reverse transcriptase activity, and those that act against viral protease. These two enzymes act at different stages of viral replication and are vital for its reproduction. Viral mutants resistant to a single medication, however, quickly arise. To help prevent the selection of resistant variants, a "cocktail" of several medications each acting in a different way are given simultaneously, usually a protease inhibitor along with two reverse transcriptase inhibitors that act at different sites on the reverse trancriptase molecule. In many cases this therapy can clear the patient's blood of viral nucleic acid, halt the progress of the disease, and even allow partial recovery of immune function. Therapy is very expensive and the medications often have serious side effects, two features that preclude use of the regimen in most of the world's HIV disease sufferers. While not curative, chemotherapy can markedly prolong and improve the quality of life for a lucky minority. ■ HIV replication, pp. 747–748

Each person has a role to play in controlling the AIDS pandemic by studying the disease and doing his or her part in preventing spread from one person to another. Those who, since the 1980s, have engaged in injected drug use or risky sexual behavior, and anyone having had unprotected sexual intercourse with them, could consider obtaining a blood test to rule out HIV infection. An individual who knows that his or her HIV test is positive can receive optimum treatment sooner and prevent transmission of the virus to others.

Table 25.15 presents the main features of HIV disease and AIDS.

MICROCHECK 25.7

Viral STDs are at least as common as bacterial STDs, but so far they are incurable. Genital herpes simplex is widespread and like most other STDs can be transmitted in the absence of symptoms. Human papillomaviruses cause genital warts; some play an important role in cancer of the cervix and probably cancers of the vagina, penis, and rectum. HIV disease generally ends in AIDS; despite advances in treatment, millions around the world die of the disease every year because they do not have access to therapy, and as yet there is no effective vaccine.

- What are the possible consequences of sexually transmitted papillomavirus infections for men, women, and babies?
- What kinds of sex partners present a high risk of transmitting human immunodeficiency virus?
- Why do you think AIDS is regarded as the most important sexually transmitted disease of the Twentieth century? (Other diseases are more widespread.)
- Why should a person be concerned about genital herpes simplex—is it not just a cold sore on the genitals? Explain.

TABLE 25.15 HIV Disease and AIDS

Symptoms	No symptoms, or "flu"-like symptoms early in the illness; an asymptomatic period typically lasting years; symptoms of lung, intestine, skin, brain, and other infections, and certain cancers
Incubation period	About 6 weeks for "flu"-like symptoms; many months or years for cancers and unusual infections
Causative agents	Usually human immunodeficiency virus, type 1 (HIV-1)
Pathogenesis	The virus infects $CD4^+$ lymphocytes and macrophages, thereby slowly destroying the ability of the immune system to fight infections and cancers
Epidemiology	HIV present in blood, semen, and vaginal secretions in symptomatic and asymptomatic infections; spread usually by sexual intercourse, sharing of needles by injected-drug abusers, and from mother to infant at childbirth. Other STDs foster transmission
Prevention and treatment	Abstinence from sexual intercourse and drug abuse; monogamy; consistent use of latex condoms; avoidance of sexual contact with injected-drug abusers, those with multiple partners or history of STDs. Anti-HIV medication for expectant mothers and their newborn infants. Treatment: reverse transcriptase and protease inhibitors in combination

Protozoan STD

A number of protozoan diseases are frequently sexually transmitted, even though they do not infect the genital system. For example, giardiasis and cryptosporidiosis are intestinal diseases transmitted by the fecal-oral route in individuals who engage in oral-genital and -anal contact as part of sexual activity. By contrast, trichomoniasis is a sexually transmitted protozoan disease that involves the genital system. ■ protozoa, p. 304 ■ cryptosporidiosis, p. 614

"Trich" (Trichomoniasis)

Trichomoniasis is caused by *Trichomonas vaginalis.* It ranks third after bacterial vaginosis and candidiasis among the diseases that commonly cause vaginal symptoms. Infected men usually are asymptomatic carriers of the organism. An estimated 2.5 to 3 million Americans contract this STD each year.

Symptoms

Most symptomatic *T. vaginalis* infections occur in women and are characterized by itching of the vulva and inner thighs, itching and burning of the vagina, a frothy, sometimes malodorous, yellowish-green vaginal discharge, at times accompanied by burning discomfort with urination. Although most infected men are free of symptoms, some have burning pain with urination, painful testes, or a tender prostate gland.

Causative Agent

Trichomonas vaginalis is a protozoan measuring about 10 by 30 μm, has four anterior flagella, and has a long posterior flagellum attached to an undulating membrane (**figure 25.21**). It also has a slender posteriorly protruding rigid rod called an axostyle. *Trichomonas vaginalis* has an unmistakable jerky motility on microscopic examination of the vaginal drainage. There is no cyst form. The organism is a eukaryote, but it lacks mitochondria. *Trichomonas vaginalis* has interesting cytoplasmic organelles called hydrogenosomes. Enzymes within these double membrane-bounded organelles remove the carboxyl group (COOH) from pyruvate and transfer electrons to hydrogen ions, thereby producing hydrogen gas. ■ mitochondria, p. 79

Pathogenesis

The pathogenesis of trichomoniasis has not yet been fully explained. The vulva and walls of the vagina are diffusely red and slightly swollen. Often there are scattered pinpoint hemorrhages. These changes have been attributed to mechanical trauma by the moving protozoa, but toxins or exoenzymes have not been ruled out. The frothy discharge is probable due to gas produced by the organisms.

Epidemiology

Trichomonas vaginalis is distributed worldwide as a human parasite and has no other reservoirs. Since it lacks a cyst form, it is easily killed by drying. Therefore, transmission is usually by sexual contact. The organism can survive for a time on moist objects such as towels and bathtubs, so that it can occasionally

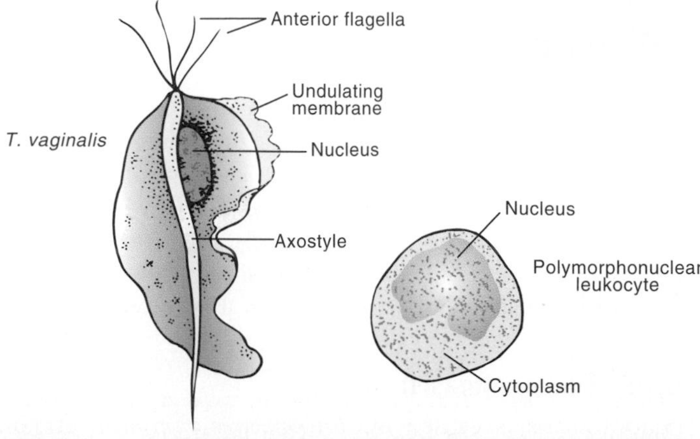

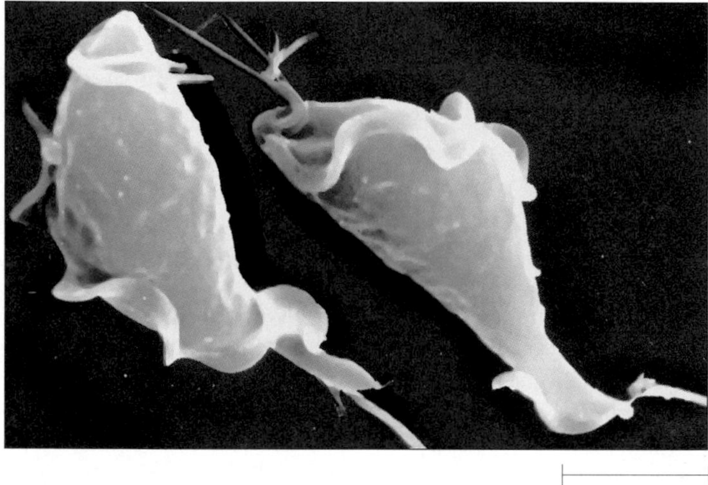

Figure 25.21 ***Trichomonas vaginalis*, a Common Sexually Transmitted Cause of Vaginitis** Diagram shows the relative sizes of the protozoan and a polymorphonuclear leukocyte. Although it is a flagellated protozoan, *T. vaginalis* often exhibits pseudopodia.

be transmitted nonsexually. Nevertheless, *T. vaginalis* infections in children should at least raise the question of sexual abuse and possible exposure to other sexually transmitted diseases. Newborn infants can contract the infection from infected mothers at birth. There is a high percentage of asymptomatic carriers, especially among men, and this fosters transmission of the disease. Infection rates are highest in men and women with multiple sex partners.

Prevention and Treatment

Transmission is prevented by the use of condoms. Most strains of *T. vaginalis* respond quickly to metronidazole treatment, but a few are resistant.

Table 25.16 gives the main features of trichomoniasis.

MICROCHECK 25.8

Trichomoniasis is a protozoan STD that involves the genital tract; other protozoan STDs infect the intestine. The causative agent is a peculiar flagellated protozoan that does not form cysts and lacks mitochondria. Organelles called hydrogenosomes produce hydrogen gas from pyruvate.

- What significance does the lack of a cyst form of *Trichomonas vaginalis* have on the epidemiology of trichomoniasis?
- Why is there a strong relationship between being "easily killed by drying" and "transmission usually by sexual contact"?

TABLE 25.16 Trichomoniasis

Symptoms	Women: itching, burning, swelling, vaginal redness, frothy, sometimes smelly, yellow-green discharge, and burning on urination. Men: discharge from penis, burning on urination, painful testes, tender prostate. Many women, most men asymptomatic
Incubation period	4 to 20 days
Causative agent	*Trichomonas vaginalis,* a protozoan with four anterior flagella, and a posterior flagellum attached to an undulating membrane; a rigid rodlike structure called an axostyle protrudes posteriorly; unmistakable jerky motility; no mitochondria; hydrogenosomes are present
Pathogenesis	Unexplained. Inflammatory changes and pinpoint hemorrhages suggest mechanical trauma from the motile organisms
Epidemiology	Worldwide distribution; asymptomatic carriers foster spread; easily killed by drying due to lack of cyst form, transmission by intimate contact; high rate of infection with multiple sex partners. Newborn infants of infected mothers can acquire the infection at birth
Prevention and treatment	Abstinence, monogamy, and consistent use of condoms prevent the disease. Treatment: metronidazole

STD Caused by Multicellular Parasites

Two common STDs are caused by lice or mites, multicellular skin parasites that are distributed worldwide and infect only human beings. Nonsexual transmission can sometimes occur in both instances. Besides the discomfort they cause, their chief importance is to raise the possibility that other, unrecognized, STDs might be present. Fortunately, neither of these parasites are vectors of microbial diseases.

"Crabs" (Pubic Lice, Pediculosis Pubis)

Pediculosis pubis results from infestation by a species of louse. Lice are wingless insects dependent for nutrition on sucking blood from their host.

Symptoms

The principal symptoms are itching in the pubic area, and psychological stress from seeing the tiny lice and their eggs, called nits, both visible to the unaided eyes. Sometimes, tiny spots of blood are seen on the underclothing. In heavy infestations, these lice may cling to hair of parts of the body other than the pubic area, sometimes even the eyebrows and eyelashes.

Causative Agent

Pediculosis pubis is caused by *Phthirus pubis*, a slow-moving insect resembling a microscopic crab (**figure 25.22**). The adult form is approximately 2 mm in diameter; smaller juvenile forms feed along with the adults. The females lay about six eggs daily, attached tightly to body hairs. Hatching occurs in about 1 week, and after several developmental stages, adulthood is reached in about 2 weeks. Their life span is approximately 1 month.

Pathogenesis

Crab lice have piercing mouthparts with which they penetrate the skin to obtain a blood meal. They usually feed for several hours twice daily, leaving a tiny puncture site that can ooze a little blood. Symptoms are due to an allergic reaction to the feeding lice, and they do not appear for about a week after the infestation begins. If the host becomes reinfested at a later date, symptoms begin almost immediately. The intense itching leads to scratching, which can in turn lead to secondary bacterial infection. ■ **immediate hypersensitivities, p. 434**

Epidemiology

Phthirus pubis is slow moving and prefers to cling to the host's hair using its clawlike hooks. Since it can live only about 24 hours away from a human host, transmission generally occurs during close physical contact, mainly but not exclusively during sexual intercourse. Fomites such as a towel or a theater seat can occasionally transmit the organisms.

Prevention and Treatment

Condoms do not prevent transmission. Several insecticides are approved for treating pediculosis pubis, but they must be used exactly as directed to avoid possible serious toxicity. Bedding and clothing is washed in hot water, dry cleaned, or kept from human contact for 3 days. Any sexual partners over the prior month are examined and treated if needed. All are reexamined after 1 week and retreated if necessary.

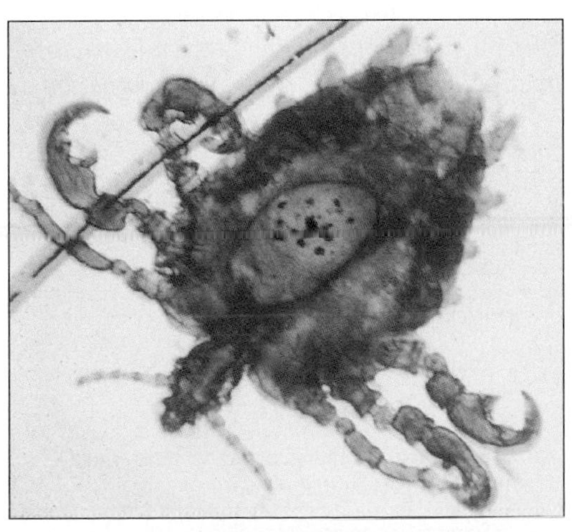

Figure 25.22 Crab Louse, *Phthirus pubis*, is Usually Sexually Transmitted It is not known to be a vector of infectious agents.

"Seven-Year Itch" (Scabies)

Scabies, also known as the "seven-year itch," results from infestation with a mite, a tiny member of the arachnids, the class of eight-legged creatures that includes spiders and scorpions. Other mites, such as the house mouse mite, can transmit microbial diseases, and the dust mite causes allergic asthma. The scabies mite is restricted to human beings, and it does not transmit microbial diseases. Scabies occurs worldwide predominantly associated with poverty and crowding, but anyone can contract the disease.

Symptoms

As with pediculosis pubis, the dominant symptom of scabies is intense itching. Besides the pubic area, the space between the fingers, the wrists, and the area under the breasts are often involved. Rarely, the entire body below the neck is infested, causing a scaly rash.

Causative Agent

The causative agent of scabies is *Sarcoptes scabiei*. The adult female (**figure 25.23a**) is only 330 to 450 μm long; the male is smaller. The organisms live on the surface of the human skin. The females make burrows into the outer layers of epidermis (figure 25.23b), forming tunnels where they lay several eggs daily over a life span of 1 to 2 months. The tunnels are visible to the unaided eyes as short, dark-colored wavy lines, sometimes with a vesicle overlying the end containing the mite. The mites can be scraped from the tunnels with a scalpel and identified microscopically, the only way accurately to diagnose scabies. Six-legged larvae hatch from the eggs and mature on the skin surface to the eight-legged form in about 2 weeks.

Pathogenesis

Usually it takes 2 to 6 weeks after contracting scabies for symptoms to appear. If following a cure scabies is contracted again, however, symptoms appear in 1 to 4 days. Thus as with pediculosis pubis, the host's allergic response to the mites and their feces is largely responsible for symptoms of the disease. In the usual individual with scabies, the number of mites is less than 100 and the disease goes away by itself in a matter of months. In individuals with AIDS and other immunodeficiencies, however, millions of the organisms can be present, spreading over much of the body and causing a severe rash with thickening and peeling of the skin. Opportunistic bacterial pathogens, introduced by scratching, are a serious threat to patients with scabies.

Epidemiology

Scabies is usually transmitted by close contact with a person who has the disease. In adults, transmission often occurs as a result of sexual contact; in children, this is rarely so. Adults with the disease should be checked for other possible STDs. Nonsexual transmission of scabies occurs readily among children, especially those less than five years old.

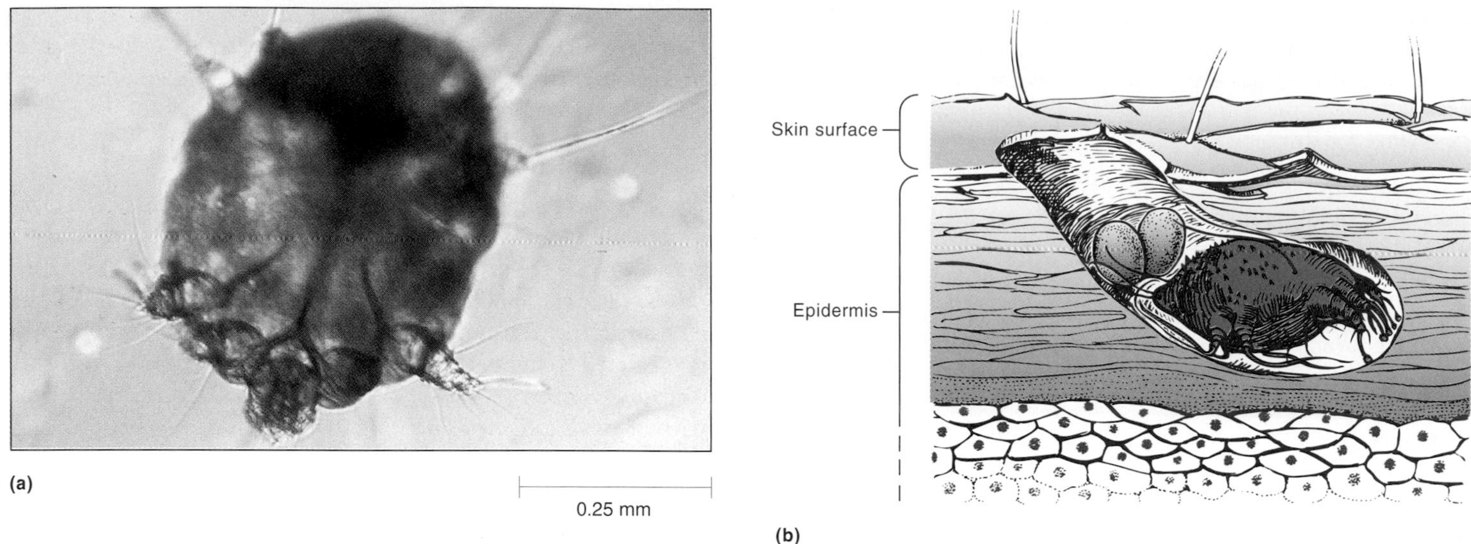

Figure 25.23 **Seven-Year Itch Mite, *Sarcoptes scabiei*, Commonly Sexually Transmitted**
(a) Microscopic appearance in scrapings of a scabies burrow. **(b)** The female mite burrows into the outer layer of skin to lay her eggs, causing an intensely itchy rash. *Sarcoptes scabiei* is not known to be a vector of infectious diseases.

Prevention and Treatment

Scabies is prevented by avoiding contact with those who have scabies, their bedding, and their clothing. Sexual and other close contacts within the month prior to diagnosis need to be examined and treated. Bedding and clothing is washed in hot water, dry cleaned, or removed from human contact for 3 days. Insecticides suitable for use on the skin are effective against *S. scabiei*, as is a sulfur-containing ointment, and ivermectin, an antiparasitic medication that paralyzes the organism. These medications must be used exactly as prescribed in order to minimize the risk of serious side effects.

The main characteristics of pediculosis pubis and scabies are presented in **table 25.17**.

MICROCHECK 25.9

The main importance of pubic lice and scabies is to raise the possibility of other sexually transmitted diseases. The possibility of secondary infection introduced by scratching the involved area is another important consideration.

- Explain why it takes 2 to 6 weeks for symptoms to appear after first contracting scabies, but only a few days after contracting the disease again.
- What further testing should be considered once pubic lice or scabies is diagnosed in an adult?
- What evidence can you present that the immune system plays a role in controlling scabies?
- Why will condoms not prevent the spread of crab lice?

FUTURE CHALLENGES

Getting Control of Sexually Transmitted Diseases

Few problems are as complicated as getting control of sexually transmitted diseases because of the psychological, cultural, religious, and economic factors that are involved, which vary from one population to another.

TABLE 25.17 Pediculosis Pubis and Scabies

	Pediculosis Pubis ("Crabs")	Scabies ("Seven-Year Itch")
Symptoms	Intense itching, visible lice and eggs	Intense itching
Incubation period	Usually about 1 week	Usually about 1 month (2 to 6 weeks)
Causative agent	*Phthirus pubis*, a louse	*Sarcoptes scabiei*, a mite
Pathogenesis	Skin penetration by a blood-sucking insect; allergic reaction to it	Burrowing into the epidermis by an arachnid; allergic reaction to it
Epidemiology	Transmitted by sexual intercourse and other close physical contact; sometimes by fomites	Same as for pediculosis pubis
Prevention and treatment	Avoidance of persons with the disease, their clothing and bedding; treatment of contacts; insecticide medications applied to the skin	Same as for pediculosis pubis. Additional treatment options include ivermectin and a sulfur ointment

Gaining control means focusing on diagnosis, interruption of transmission, education, and treatment. The challenge is to develop innovative approaches in each of these areas and to apply them worldwide on a sustained basis.

In the area of diagnosis, the challenge is to use molecular biological techniques to identify and mass produce specific antigens and antibodies that can be used to quickly identify causative agents and the extent of their spread. The development of nucleic acid probes to identify genes coding for resistance to therapeutic agents could assist in promptly choosing treatment.

Interruption of transmission by the consistent use of condoms can be highly effective, but in practice there are a number of problems. Infected college students, when asked why they did not use a condom, often state that "It takes away the romance," "It decreases pleasurable sensation," or "That's for sissies." Also many individuals are allergic to latex, but perhaps except for polyurethane, the alternatives are unreliable for preventing disease transmission. Worldwide, as many as one out of three condoms is defective, and the poorest countries cannot even afford them. The challenge is to develop cheaper, more reliable, and more acceptable barrier methods, perhaps using new space age polymers.

Effective vaccines could block the spread of STDs, but their development has been extremely problematic and remains a long-term goal. Sequencing the genomes of the causative agents is helping to identify appropriate antigens to include in vaccines, but using them to stimulate a protective immune response is a continuing challenge.

Educational efforts focusing on groups at high risk of STDs have had mixed success. Education can be a useful tool for gaining control of STDs, and the challenge is to better understand the reasons for its failures.

SUMMARY

Anatomy and Physiology

1. The genitourinary system is one of the portals of entry for pathogens to invade the body.

The Urinary System (Figure 25.1)

1. The kidneys, ureters, bladder, and urethra compose the urinary system.
2. Frequent complete emptying is an important bladder defense mechanism against infection.
3. Infections occur more frequently in women than in men because of the shortness of the female urethra and its closeness to the genital and intestinal tracts.

The Genital System (Figure 25.2)

1. The fallopian tube, open on both ends, provides a passageway for infection to enter the abdominal cavity.
2. The uterine cervix is a frequent site of infection and a place where cancer can develop. The vagina is a portal of entry for a number of infections. In men, the prostate can enlarge and partially obstruct the urinary flow.

Normal Flora of the Genitourinary System

1. The distal urethra is inhabited by various microorganisms, sometimes including potential pathogens.
2. The normal vaginal flora is influenced by estrogen hormones, which cause deposition of glycogen in the cells lining the vagina. Vaginal lactobacilli release lactic acid and hydrogen peroxide when they metabolize the glycogen, making the vagina more resistant to colonization by pathogens.

Urinary System Infections

1. Any condition that impairs normal bladder emptying increases the risk of infection.
2. Usually the urinary system becomes infected by organisms ascending from the urethra, but it can also be infected from the bloodstream.

Bacterial Cystitis (Table 25.1)

1. Urine is a nutritious medium for many bacteria.
2. Kidney infection, **pyelonephritis**, may complicate untreated bladder infection when pathogens ascend through the ureters and involve the kidneys.
3. Most urinary tract infections in healthy people are caused by *Escherichia coli* or other enterobacteria from the person's own normal intestinal flora.
4. Nosocomial urinary infections are common, caused by *Pseudomonas aeruginosa*, *Serratia marcescens*, and *Enterococcus faecalis*, organisms that are generally resistant to many antibiotics.

Leptospirosis (Table 25.2)

1. In leptospirosis, the urinary system is infected by *Leptospira interrogans* from the bloodstream. (Figure 25.3)
2. This biphasic illness causes fever, bloodshot eyes, and pain in the septicemic phase; then improvement; then recurrent symptoms with damage to multiple organs during the immune phase.
3. Many species of animals are chronically infected with *L. interrogans* and excrete the organism in their urine.

Non-Venereal Genital System Diseases

1. *Non-venereal* means that the infection is not transmitted by sexual intercourse.

Bacterial Vaginosis (Table 25.3, Figure 25.4)

1. Bacterial vaginosis is the most common cause of vaginal symptoms, including a gray-white discharge from the vagina and a pungent fishy odor; there is no inflammation.
2. The causative agent or agents are unknown.

Vulvovaginal Candidiasis (Table 25.4)

1. Vulvovaginal candidiasis is second among causes of vaginal disorders; symptoms include itching, burning, a vulvar rash, and a thick white discharge.
2. The causative agent is a yeast, *Candida albicans*, that is commonly part of the normal vaginal flora. (Figure 25.5)

3. Antibacterial treatment, uncontrolled diabetes, and oral contraceptives are predisposing factors, but in most cases no such factor can be identified.

Staphylococcal Toxic Shock (Table 25.5)

1. First described in children, it became widely known with a 1980 epidemic in menstruating women who used a certain kind of vaginal tampon, since removed from the market. (Figure 25.6)
2. Symptoms include sudden fever, headache, muscle aches, bloodshot eyes, vomiting, diarrhea, a sunburnlike rash that later peels, and confusion. The blood pressure drops, and without treatment kidney failure and death may occur.

Sexually Transmitted Diseases: Scope of the Problem (Tables 25.6, 25.7)

1. In the United States, 15 million new sexually transmitted infections occur each year, including 3 million in teenagers.
2. Among the 12 million college and university students, there are about 1 million unintended pregnancies each year, and about 70% of sexually active students or their partners rarely or never employ condoms.
3. Simple measures exist for controlling STDs: abstinence from sexual intercourse, a monogamous relationship with an uninfected person, and consistent use of latex or polyurethane condoms. (Figure 25.7)

Bacterial STD

1. Most of the bacteria that cause STDs survive poorly in the environment and require intimate contact for transmission.

Gonorrhea (Table 25.8)

1. Gonorrhea is caused by *Neisseria gonorrhoeae*, has been generally declining in incidence, but it is still one of the most commonly reported bacterial diseases. (Figures 25.8, 25.9)
2. Men usually develop painful urination and thick pus draining from the urethra; women may have similar symptoms, but they tend to be milder and are often overlooked.
3. Expression of different surface antigens allows attachment of *N. gonorrhoeae* to different types of cells, and frustrates development of a vaccine.
4. Inflammatory reaction to the infection causes scarring, which can partially obstruct the urethra or cause sterility in men and women.

Chlamydial Genital System Infections (Table 25.9, Figure 25.10)

1. Chlamydial genital infections, caused by *Chlamydia trachomatis*, are reported more often than any other bacterial disease.
2. Symptoms and complications of chlamydial infections are very similar to those of gonorrhea, but milder; asymptomatic infections are common, and readily transmitted.
3. Sexually active individuals should be tested at least once a year so that transmission can be halted and complications prevented.

Syphilis (Tables 25.10, 25.11)

1. Syphilis, caused by *Treponema pallidum*, is called the "great imitator" because its many manifestations can resemble other diseases. (Figure 25.12)
2. Primary syphilis is noted by a painless firm ulceration called a **hard chancre**; the organisms multiply and spread throughout the body. (Figure 25.11)
3. In secondary syphilis, skin and mucous membranes show lesions, which teem with the organisms; a latent period of months or years separates secondary from tertiary syphilis. (Figure 25.13)
4. Tertiary syphilis is not contagious. It is manifest mainly by damage to the eyes and the cardiovascular and central nervous systems. An inflammatory, necrotizing mass called a **gumma** can involve any part of the body. (Figure 25.14)
5. Syphilis in pregnant women can spread across the placenta to involve the fetus, resulting in congenital syphilis. (Figure 25.15)

Chancroid (Table 25.12)

1. Chancroid is another widespread bacterial STD, but it is not commonly reported because of difficulties making a bacterial diagnosis. Epidemics in the United States have been associated with prostitution.
2. Caused by *Haemophilus ducreyi*, chancroid is characterized by single or multiple soft, tender genital ulcers, and enlarged, painful groin lymph nodes. (Figure 25.16)
3. Some strains of *H. ducreyi* possess R factors, making treatment more difficult.

Viral STD

1. Viral STDs are at least as common as bacterial STDs, but they are not yet curable.

Genital Herpes Simplex (Table 25.13, Figure 25.17)

1. It is a very common disease, important because of the discomfort and emotional trauma it causes, its potential for causing death in newborn infants, its association with cancer of the cervix, and the increased risk it poses of transmitting HIV infection and AIDS.
2. Symptoms may include a group of vesicles with itching, burning, or painful sensations; these break, leaving a tender superficial ulcer. Local lymph nodes enlarge. Many infections have few or no symptoms; some have painful recurrences.
3. The virus establishes a latent infection in sensory nerves and cannot be cured; it can be transmitted in the absence of symptoms, but the risk is greatest when lesions are present.

Papillomavirus STDs: Genital Warts and Cervical Cancer (Table 25.14)

1. Papillomavirus STDs are probably more prevalent than any other kind of STD; HPVs are the main cause of abnormal Pap smears in young women. (Figure 25.18)
2. They are manifest in two ways, as warts on or near the genitalia (Figure 25.19), and as precancerous lesions. The latter are asymptomatic and can only be detected by medical examination.
3. The causative agents, human papillomaviruses, are small DNA viruses that have not been cultivated in the laboratory.

AIDS (Table 25.15)

1. AIDS is the end stage of disease caused by human immunodeficiency virus (HIV). Worldwide, more than 5 million new HIV infections and 2.5 million deaths from AIDS occur each year (Figure 25.20). Advances in treatment, available to HIV disease victims in some countries, are not available to most of the world's HIV victims.
2. HIV disease is usually first manifest as a "flu"-like illness that develops about 6 weeks after contracting the virus. An asymptomatic interval follows that typically lasts almost 10 years, during which the immune system is slowly and progressively

destroyed. Unusual cancers and infectious diseases then herald the onset of AIDS.

3. No vaccine or medical cure is yet available, but spread of infection could be significantly slowed by consistent use of condoms, and employment of sterile needles by injected-drug abusers. A marked reduction in mother to newborn transmission can be achieved with medication.

Protozoan STD

1. Intestinal protozoan diseases such as giardiasis and cryptosporidiosis are transmitted by the fecal-oral route in those individuals who engage in oral-genital and -anal contact as part of sexual activity.

"Trich" (Trichomoniasis) (Table 25.16)

1. Symptoms include itching, burning, swelling, and redness of the vagina, with frothy, sometimes smelly, yellow-green discharge, and burning on urination; men have discharge from the penis, burning on urination, sometimes accompanied by painful testes and a tender prostate gland. Many women and most men are asymptomatic.
2. The causative agent is *Trichomonas vaginalis*, a protozoan with four anterior flagella, and a long posterior flagellum attached to an undulating membrane. (Figure 25.21)

STD Caused by Multicellular Parasites

1. Two common sexually transmitted diseases are multicellular skin parasites that infest only humans. Nonsexual transmission of these diseases also occurs.

"Crabs" (Pubic Lice, Pediculosis Pubis) (Table 25.17)

1. Lice, *Phthirus pubis* (Figure 25.22), are wingless insects dependent for nutrition on sucking blood from their host. Eggs, called nits, may be visible.
2. Symptoms, including intensely itchy rash, are mainly due to an allergic reaction to these blood-sucking organisms and their feces.

"Seven-Year Itch" (Scabies) (Table 25.17)

1. This disease is caused by a tiny mite, *Sarcoptes scabiei*, is less than 0.5 mm in length. Females tunnel into the epidermis to lay their eggs, and allergy to the mites and their feces develops. (Figure 25.23)
2. Symptoms include an intensely itchy rash beginning about 1 month after exposure and often persist for several weeks after the organisms are killed by medications.

REVIEW QUESTIONS

Short Answer

1. Name two substances released by lactobacilli that help protect the vagina from potential pathogens.
2. List four things that predispose to the development of infection of the urinary bladder.
3. What is pyelonephritis?
4. Name two genera of bacteria that infect the kidneys from the bloodstream.
5. What possible danger lurks in a spot on the ground where an animal has urinated 1 week earlier?
6. What probably causes the Jarisch-Herxheimer reaction?
7. What causes the characteristic odor of bacterial vaginosis?
8. What is a clue cell?
9. How does gonorrhea threaten the normal functioning of the heart?
10. Name two bacterial causes of pelvic inflammatory disease (PID).
11. What is ophthalmia neonatorum?
12. List three diseases caused by different antigenic types of *Chlamydia trachomatis*.
13. Which bacterium leads all others as a cause of reportable infectious disease?
14. Why is dark-field microscopy used to visualize *Treponema pallidum*?
15. Give two ways in which the chancre of chancroid differs from the chancre of syphilis.
16. Give two ways in which genital herpes simplex due to HSV-1 differs from that of HSV-2.
17. What do the letters AIDS stand for? What is the relationship between AIDS and HIV disease?
18. What, if any, effect does laser removal of genital warts have on a genital HPV infection?
19. What is the reservoir of *Trichomonas vaginalis*?
20. Why is the vaginal discharge in trichomoniasis often frothy?

Multiple Choice

1. All of the following are true of bacterial cystitis, *except*...
   A. almost one-third of all women will have it at some time during their life.
   B. catheterization of the bladder markedly increases the risk of contracting the disease.
   C. individuals who have a bladder catheter in place indefinitely risk bladder infections with multiple species of intestinal bacteria at the same time.
   D. bladder infections occur as often in men as they do in women
   E. bladder infections can be asymptomatic.
2. Choose the one correct statement about leptospirosis.
   A. Humans are the only reservoir.
   B. Most infections produce severe symptoms.
   C. Transmission is by the fecal-oral route.
   D. It can lead to unnecessary abdominal surgery.
   E. Effective vaccine is generally available for preventing human disease.
3. Which one of the following statements about bacterial vaginosis is *false*?
   A. It is the most common vaginal disease in women of child-bearing age.

B. In pregnant women it is associated with a sevenfold increased risk of obstetrical complications.

C. Inflammation of the vagina is a constant feature of the disease.

D. The vaginal flora shows a marked decrease in lactobacilli and a marked increase in anaerobic bacteria.

E. The cause is unknown.

4. Pick the one *false* statement about vulvovaginal candidiasis.

A. It often involves the external genitalia.

B. It is readily transmitted by sexual intercourse.

C. It is caused by a yeast present among the normal vaginal flora in about one-third of healthy women.

D. It is associated with prolonged antibiotic use.

E. It involves increased risk late in pregnancy.

5. All of the following statements about staphylococcal toxic shock are true, *except...*

A. it can quickly lead to kidney failure.

B. the causative organism usually does not enter the bloodstream.

C. it occurs only in vaginal tampon users.

D. almost one-third of victims of the disease will suffer a recurrence sometime after recovery.

E. person-to-person spread does not occur.

6. Which of the following statements about gonorrhea is *false*?

A. The incubation period is only a few days.

B. Disseminated gonococcal infection (DGI) is almost invariably preceded by prominant urogenital symptoms.

C. DGI can result in arthritis of the knee.

D. Phase variation helps the causative organism evade the immune response.

E. Pelvic inflammatory disease (PID) is common in untreated women.

7. Which one of these statements about chlamydial genital infections is *false*?

A. The incubation period is usually shorter than in gonorrhea.

B. Infected cells develop inclusion bodies.

C. Pelvic inflammatory disease (PID) can be complicated by infection of the surface of the liver.

D. Tissue damage largely results from cell-mediated immunity.

E. Fallopian tube damage can occur in the absence of symptoms.

8. Which symptom is *least* likely to occur as a result of tertiary syphilis?

A. Gummas

B. White patches on mucous membranes

C. Emotional instability

D. Stroke

E. Blindness

9. Each year, worldwide, approximately how many people become infected with the AIDS-causing human immunodeficiency virus (HIV)?

A. 50,000

B. 100,000

C. 1 million

D. 5 million

E. 10 million

10. All of the following are true of "trich" (trichomoniasis), *except...*

A. it can cause burning pain on urination and painful testes in men.

B. it occurs worldwide.

C. asymptomatic carriers are rare.

D. transmission can be prevented by proper use of condoms.

E. individuals with multiple sex partners are at high risk of contracting the disease.

Applications

1. A medical student, working in the hospital late at night, noticed that the laboratory had reported staphylococci growing in a urine culture from one of his pregnant patients. Remembering that staphylococci produce toxins, he became alarmed and phoned the attending physician, waking her from sleep at 1:00 A.M. Is the medical student a hero, or should he be asked to repeat Microbiology?
2. Religious restrictions of a small North African community are preventing a World Health Organization project from reducing the incidence of gonorrhea. The community will not permit the testing of females for the disease. They can be treated, however, if they show outward evidence of the disease. Only males are allowed to participate fully in the project, with testing for the disease and treatment. The village elders argue that eradicating the disease from males would eventually remove it from the population. What would be the impact of these restrictions on the success of the project?
3. Former president Ronald Reagan once commented at a press conference that the best way to combat the spread of AIDS in the United States was to prohibit everyone from having sexual contact for 5 years. What would be the success of such a program if it were possible to carry it out?

Critical Thinking

1. The upper curve of figure 25.6 shows the occurrence of staphylococcal toxic shock in menstruating women from 1979 to 1994. What aspect of these data argue that high absorbency tampons were not the only cause of toxic shock syndrome associated with menstruation?
2. In early attempts to identify and isolate the cause of syphilis, various bacteria in the discharge from syphilitic lesions in experimental animals were isolated in pure culture. None of them, however, would cause the disease when used in attempts to infect healthy animals. Why was it considered a critical step to have the cultivated bacteria reproduce the disease in the healthy animals?
3. Vaginal yeast infections do not spread person to person, generally not even by sexual intercourse. What does this suggest about the cause of vaginal yeast infections?
4. Discuss the implications of sexually transmitted diseases for the unborn and newborn.

Nervous System Infections

Gerhard Henrik Armauer Hansen (1841–1912) was a burly Norwegian physician who lived in the southern part of that country in the bustling seaport of Bergen. He was a man of many interests, ranging from music, religion, and polar exploration, to marine biology. A vigorous trade of goods and ideas with other countries was going on at the time, especially with Germany and England, and like many of the city's inhabitants Dr. Hansen was fluent in several languages. He was well aware of the scientific advances of the day, including Darwin's studies, and Pasteur's proof that bacteria could cause fermentation. His Norwegian contemporaries included such luminaries as his friend, Edvard Grieg, the famous composer, Henrik Ibsen, a world-renowned playwright, and Bjornstijerne Bjornson, a Nobel Prize–winning writer and poet, a leader in the movement for an independent Norway.

When he was 32 years old, Hansen decided to go into medical research, and was named assistant to Dr. Daniel C. Danielson, a leading authority on leprosy. He married Danielson's daughter, Stephanie, but she died of tuberculosis some months later. In honor of his love, Hansen named a new marine organism that he had discovered Stephanostoma hansenii. *Danielson had developed the theory that leprosy was a disease of the blood, that once acquired, was hereditary. He had discarded the idea that the disease was contagious as a "peasant superstition," and, to "dispose of this foolish notion," injected himself with material from leprosy patients. Fortunately for him and others who had done the same thing, no disease developed, because effective treatment of leprosy did not become available for another three-quarters of a century. Luckily, Danielson and Hansen had great mutual respect, because Hansen in a meticulous series of 58 studies over a number of years, disproved Danielson's theory and showed that a unique bacillus was associated with the disease in every leprosy patient he studied. His findings, finally reported in 1873, almost a decade before Koch's proof of the cause of tuberculosis, linked for the first time a specific bacterium to a disease. His studies forever changed the way people looked at leprosy, overturning a viewpoint that had existed for many centuries.*

Today it is hard to appreciate the fear and loathing formerly attached to leprosy. In the Greek translation of the Bible from Hebrew, a variety of disfiguring skin diseases, including leprosy, are covered by the word lepros, *meaning "scaly," but having the connotation of being disgustingly filthy, outcast, condemned by God for sin. In the Bible, Moses calls lepers "unclean" and proclaims that they must live away from others. In the Middle Ages, lepers wore bells to warn others of their presence, and attended a symbolic burial of themselves before being sent away. Because the English word leprosy carries these innuendoes, many people prefer to use the term Hansen's disease after the discoverer of the causative organism. Even in the United States during the first half of the Twentieth century, persons diagnosed with leprosy risked having their houses burned to destroy contagion. Their names were changed to avoid embarrassing the family, and they were whisked off to a leprosarium such as the one at Carville, Louisiana, surrounded by a 12-foot fence topped with barbed wire. They were separated from spouses and children, and denied the right to marry or to vote. Those who attempted to escape were captured and brought back in handcuffs. The story of the Carville leprosarium is told by one who did escape, Betty Martin, in her wonderful book,* Miracle at Carville. *The leprosarium was finally closed and converted to a military-style academy for high school dropouts in 1999.*

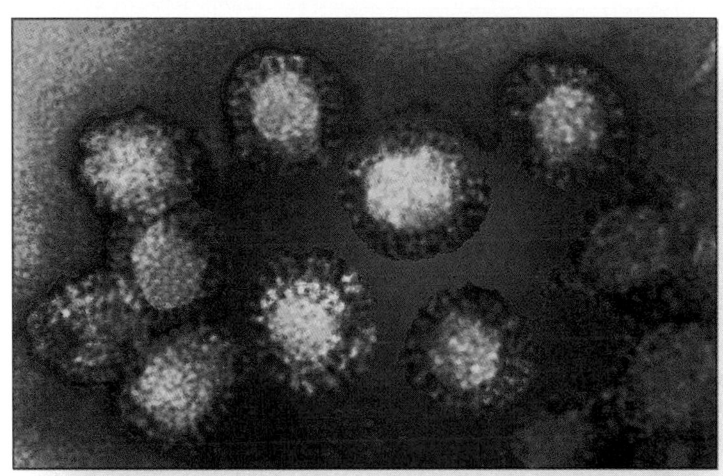

—A Glimpse of History

INFECTIONS OF THE NERVOUS SYSTEM ARE APT TO be serious because they threaten a person's ability to move, to feel, and to think normally. Just consider poliomyelitis, in which nerve cells are destroyed, leaving the victim with a paralyzed arm or leg, or unable to breathe without mechanical assistance. In Hansen's disease, nerve damage can result in complete loss of fingers or toes, or deformity of the face, while infections of the brain or its covering membranes can render a child deaf or mentally retarded. Fortunately, nervous system infections are uncommon. Examples given in this chapter, however, reveal a multiplicity of causes, including bacteria, viruses, fungi, and even some protozoa.

Anatomy and Physiology

The brain and spinal cord make up the central nervous system (CNS). Both are enclosed by bone—the brain by the skull and the spinal cord by the vertebral column **(figure 26.1)**. The network of nerves throughout the body, the peripheral nervous system, is connected with the CNS by bundles of nerve fibers that penetrate the protective bony covering. Nerves can be damaged if the bones are infected at sites of nerve penetration. **Motor nerves** carry messages from the CNS to different parts of the body and cause them to act; **sensory nerves** transmit sensations like heat, pain, light, and sound from the periphery to the central nervous system. All nerves are made up of cells with very long, thin extensions called **axons** that transmit electrical impulses. The sensory nerve cells are located in small bodies called **ganglia** (singular, ganglion) located near the vertebral column; motor nerve cells are located in the central nervous system. Bundles of axons make up a nerve. They can sometimes regenerate if severed or damaged, but they cannot be repaired if the nerve cells are killed, as occurs, for example, in poliomyelitis. Some viruses and toxins can move through the body within the cytoplasm of the nerve fibers, and some herpesviruses can remain latent in nerve cells for many years.

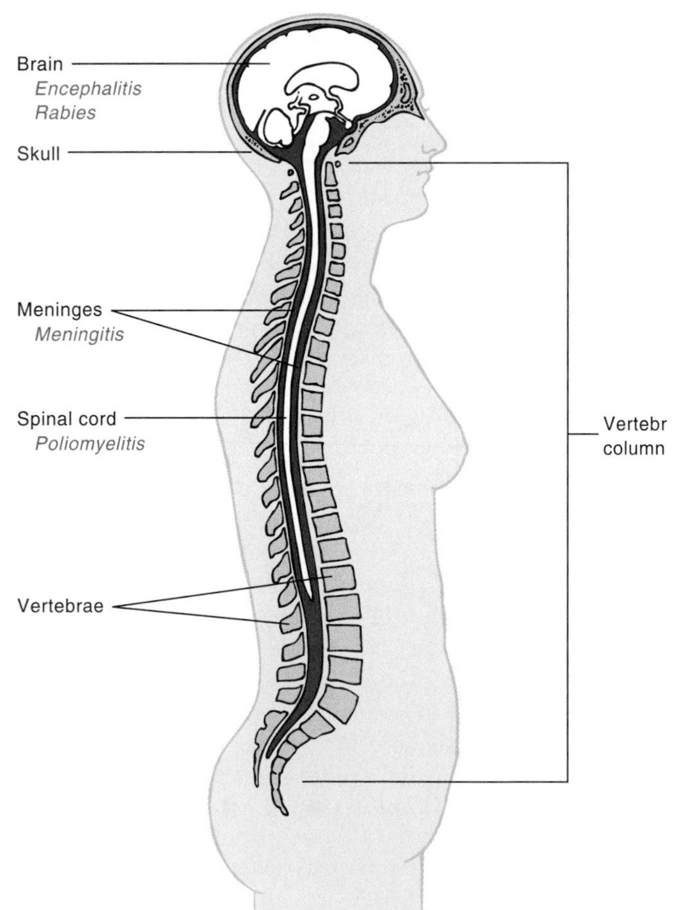

Figure 26.1 The Central Nervous System Terms in color indicate diseases associated with that area.

The brain is a very complex structure, distinct parts having different functions. This means that if an individual has symptoms from a brain abscess, for example, physicians will know in which part of the brain the abscess is located. Generalized inflammation or infection of the brain is termed **encephalitis**. Deep inside the brain are four cavities called **ventricles**, which are filled with a clear fluid called **cerebrospinal fluid** (**CSF**). This fluid is continually derived from the blood by structures in the wall of ventricles, and it flows out through a small opening at the base of the brain. It then spreads over the surface of the brain and spinal cord, and is reabsorbed into the bloodstream through specialized structures between the skull and the brain **(figure 26.2a)**.

Three membranes called **meninges** cover the surface of the brain and spinal cord. Inflammation or infection of these membranes is called **meningitis**. The outer membrane, or **dura**, is tough and fibrous and adheres closely to the skull and vertebrae. It provides a barrier to the spread of infection from bones surrounding the central nervous system. The two inner membranes, the **arachnoid** and **pia**, are separated by a space through which the cerebrospinal fluid flows. Blood vessels and nerves that pass through the meninges and cerebrospinal fluid may become damaged by meningitis. Sometimes both the meninges and the brain are involved by an infection, a condition called **meningoencephalitis**.

To determine if a person has meningitis, a needle can be safety inserted between the vertebrae in the small of the back and a sample of cerebrospinal fluid withdrawn and examined. At this point, the spinal cord has tapered to only a threadlike structure (see figure 26.2b). The causative agent of a central nervous system infection can thus be identified microscopically and cultivated on laboratory media. Once the cause of the disease is known, it can usually be treated effectively with an appropriate antimicrobial medication.

Pathways to the Central Nervous System

Since the nervous system lies entirely within body tissues, it has no normal flora. It is sterile, and no viruses or microorganisms are normally found there. Also, its well-protected environment prevents infectious agents from getting to it readily. When they do, however, it seems that their entry must be accidental, because most microorganisms that infect the nervous system infect other parts of the body much more frequently. The following are routes pathogenic microorganisms and viruses take to reach the brain and spinal cord.

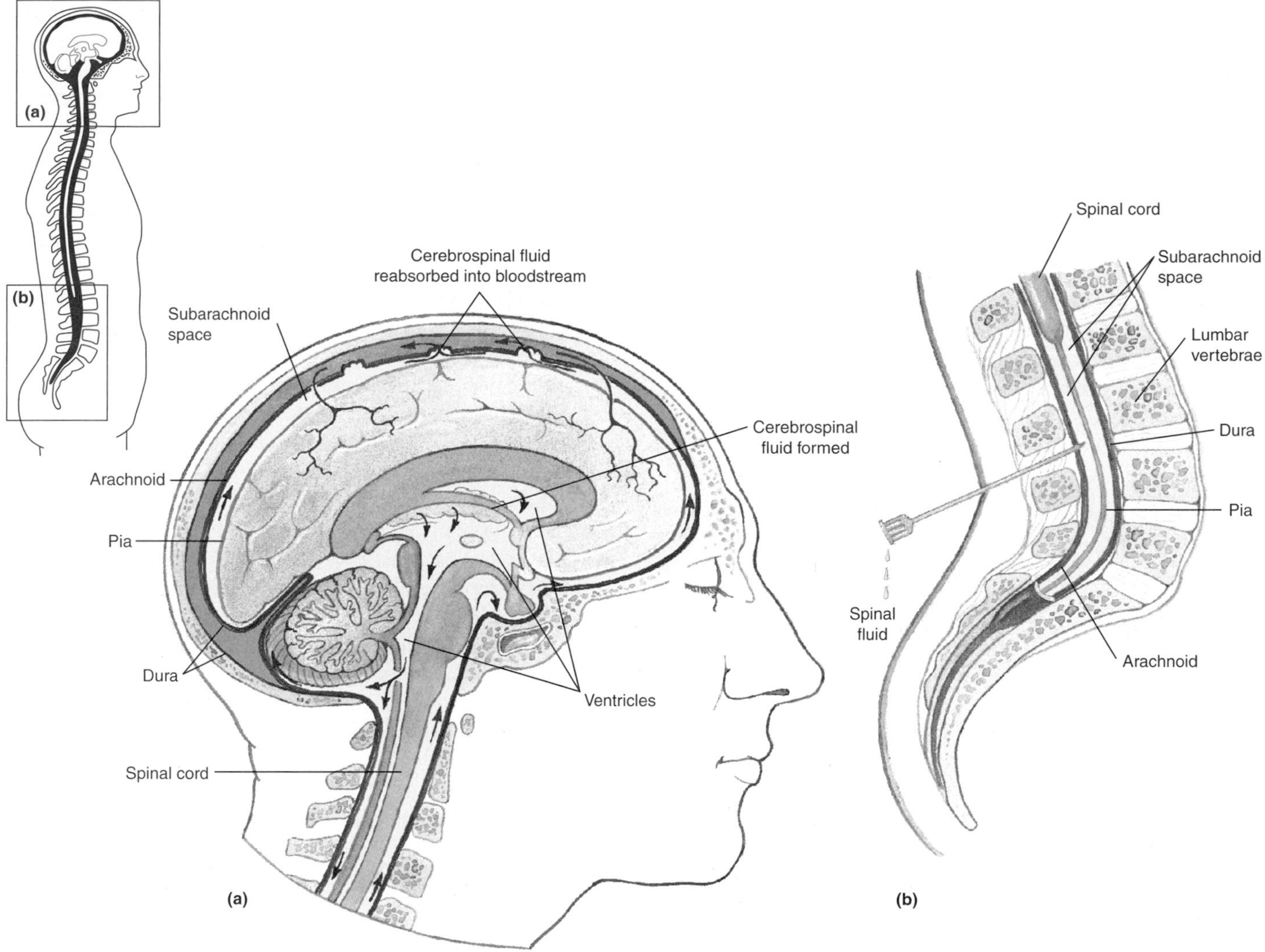

Figure 26.2 Cerebrospinal Fluid (a) The fluid is formed in the ventricles, flows out over the surface of the brain and spinal cord, and is reabsorbed by structures at the top of the brain. **(b)** A sample of cerebrospinal fluid can be withdrawn from the subarachnoid space.

The Bloodstream

The bloodstream is the chief source of CNS infections, although it is very difficult for infectious agents to cross from it to the brain. This barrier to the passage of harmful agents is called the **blood-brain barrier**. It depends on special cells lining the central nervous system capillaries. These cells, only a single layer thick, are so close to each other that except for essential nutrients that are actively taken up, most molecules cannot pass through them. Consequently, many medications, including penicillin, cannot cross the barrier unless their concentrations in the blood are very high; antimicrobial medications must be able to cross the barrier in order to treat CNS infections. The blood-brain barrier generally prevents pathogens from entering nervous tissue except in rare cases when a high concentration of the infectious agent circulates for a long time in the bloodstream.

Infections of the face above the level of the mouth may involve veins that communicate through bone with veins on the brain surface. These communicating veins can provide a direct avenue for spread of infection to the central nervous system. A good rule, therefore, is to never squeeze pimples situated on the upper part of the face.

The Nerves

Some pathogenic agents penetrate the CNS by traveling up the nerve bundles, probably within the axon cytoplasm. Examples of such agents are the rabies and herpes simplex viruses and the tetanus toxin. ■ **rabies, p. 682** ■ **herpes simplex, p. 594** ■ **tetanus, p. 698**

Extensions from Bone

To reach the CNS, microorganisms must penetrate not only the protecting bone of the skull or spine, but also the tough membrane that surrounds all bones, and the dura. Only rarely do infections in the bone surrounding the CNS succeed in reaching the brain or spinal cord. Sometimes, however, they extend through bone from the sinuses, mastoids, or middle ear. Skull fractures commonly produce nonhealing injuries that predispose a person to recurring infection of the CNS. ■ **mastoids, p. 550** ■ **middle ear, p. 550, 558**

MICROCHECK 26.1

The brain and spinal cord make up the central nervous system (CNS). The peripheral nervous system includes sensory nerves, which convey sensations to the CNS, and motor nerves, which carry messages from the CNS to parts of the body and cause them to act. The cerebrospinal fluid fills cavities within the brain and flows out over the brain and spinal cord. Inflammation of the meninges, membranes covering the brain, is called meningitis. Infection can reach the CNS via the bloodstream, via the nerves, or by extension from bone. Infectious diseases of the central nervous system are uncommon compared with infections elsewhere in the body. Antimicrobial medicines are effective in central nervous system infections only if they can cross the blood-brain barrier.

- What are the main entry routes for infections of the central nervous system?
- Why is it uncommon that infectious agents pass from the bloodstream to the CNS?
- Why can an infection in the brain's ventricles usually be detected in spinal fluid obtained from the lower back?

Bacterial Nervous System Infections

Bacteria can infect the brain and spinal cord, causing abscesses. Or in a single example, Hansen's disease, they can infect the peripheral nerves. More commonly, bacteria infect the meninges and cerebrospinal fluid, causing meningitis.

Meningitis provokes an intense inflammatory response, with a marked accumulation of white blood cells in the cerebrospinal fluid and swelling of brain tissue. The incidence of the disease is, on the average, fewer than 10 cases per 100,000 population per year. A large number of bacterial species can infect the meninges. In many cases, these organisms are carried by healthy people, are transmitted by inhalation, and only produce meningitis in a small percentage of people who are infected. Organisms in this category include three important causes of bacterial meningitis, *Haemophilus influenzae*, *Neisseria meningitidis*, and *Streptococcus pneumoniae*. ■ *Haemophilus influenzae*, p. 558 ■ *Streptococcus pneumoniae*, p. 564

Meningitis in infants during the first month of life is usually due to bacteria that colonize the mother's birth canal. *Streptococcus pneumoniae*, *H. influenzae*, and *N. meningitidis* are uncommon causes of meningitis in newborn babies, because most mothers have antibodies against them that cross the placenta and protect the baby. Instead, newborn babies become infected during labor and delivery by Gram-negative rods such as certain strains of *E. coli* originating from the mother's intestinal tract. The reason that the newborn is susceptible to Gram-negative rod infections is that its mother's bactericidal antibodies to the organisms are of the IgM class, which do not cross the placenta and therefore cannot confer immunity to the infant. ■ IgM antibody, p. 394

Streptococcus agalactiae, a Lancefield group B streptococcus, which colonizes the vagina in 15% to 40% of pregnant women, exceeds *E. coli* as a cause of meningitis in newborn infants. Babies also acquire this infection from the mother's genital tract around the time of birth. The American Academy of Pediatrics recommends that cultures selective for Group B streptococci be obtained from the vagina and rectum late in pregnancy. Women with positive cultures can then be treated with an appropriate antibacterial medication before labor begins. Evidence indicates that by using this screening procedure doctors can decrease the incidence of serious Group B streptococcal disease by more than 75%. ■ Lancefield grouping, p. 549, 553

Of the six antigenic types of *Haemophilus influenzae*, labeled **a** through **f**, type **b** is responsible for most cases of serious disease. Starting in the late 1980s, children in the United States have been routinely immunized using a T-cell-dependent antigen consisting of *H. influenzae* b polysaccharide joined covalently to a bacterial protein such as diphtheria toxoid. The result of using these **conjugate vaccines** has been a dramatic decline in meningitis and other serious infections caused by *H. influenzae* type b (**figure 26.3**). The decline in meningitis due to *H. influenzae* has resulted in a marked overall decrease in bacterial meningitis and a shift in the peak incidence to older age groups. ■ T-cell-dependent antigen, p. 400

Streptococcus pneumoniae (**figure 26.4**) is the leading cause of meningitis in adults. The organism is a prominent cause of otitis media, sinusitis, and pneumonia, conditions that often precede pneumococcal meningitis. *Neisseria meningitidis*, the meningococcus, differs from the other causes in that it is often responsible for epidemics of meningitis. ■ otitis media, pp. 558–560

The relationship between age and causative bacterium is shown in **figure 26.5**.

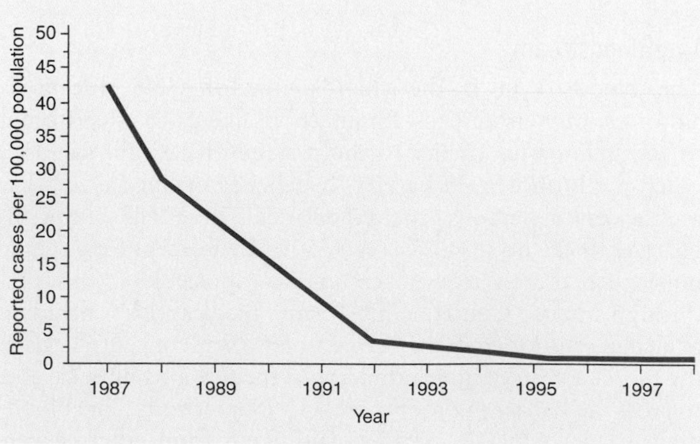

Figure 26.3 Rate of Serious *Haemophilus influenzae* Disease per 100,000 Children Less than Age Five, United States, 1987 through 1998
Before the availability of conjugate vaccines in late 1987, *H. influenzae* type b was the most common cause of bacterial meningitis in preschool children.

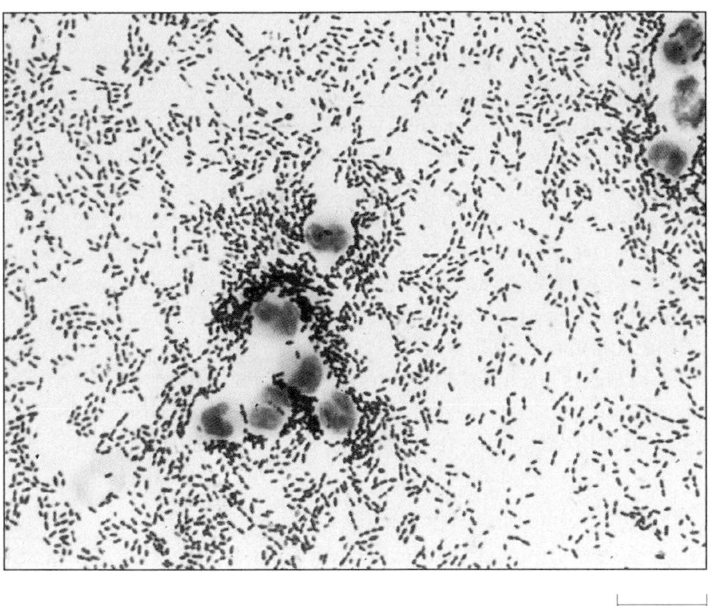

Figure 26.4 ***Streptococcus pneumoniae*** **(Pneumococci) in the Spinal Fluid of a Person with Pneumococcal Meningitis (Gram Stain)** The large red objects are polymorphonuclear leukocytes that have entered the spinal fluid in response to the infection.

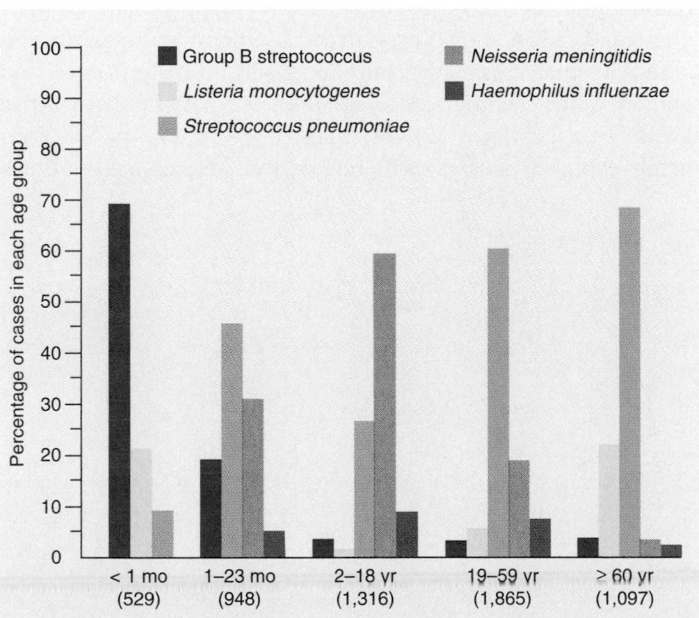

Figure 26.5 The Five Leading Causes of Meningitis in the United States, 1995 Estimates of their relative importance and numbers of cases in different age groups based on surveillance of a population of about 10 million people. *Escherichia coli* and other enterobacteria were not included in the surveillance. The estimated numbers of cases nationwide is given in parentheses.

Meningococcal Meningitis

Meningococcal meningitis occurs most commonly in children aged 6 to 11 months, but it also occurs frequently in older children and adults. It is greatly feared because it can sometimes progress to death within a few hours, but most patients respond well to treatment and recover without permanent nervous system damage.

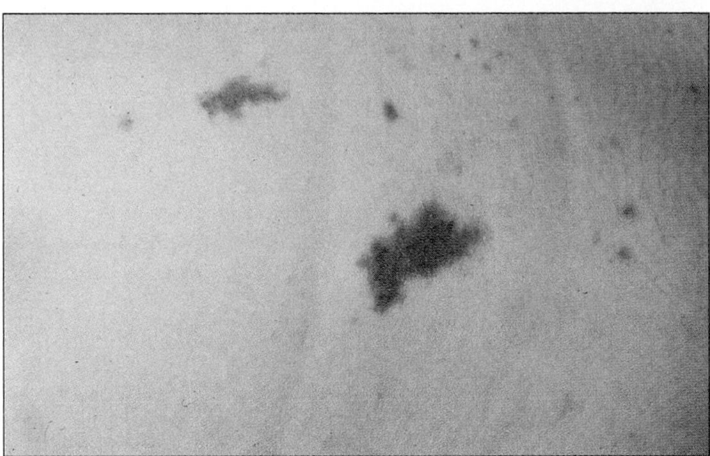

Figure 26.6 Petechiae of Meningococcal Disease *Neisseria meningitidis* is present in the lesions.

Symptoms

The first symptoms of bacterial meningitis are similar regardless of the causative agent. Usually they begin with a mild cold, followed by the sudden onset of a severe throbbing headache, fever, pain and stiffness of the neck and back, nausea, and vomiting. Deafness and alterations in consciousness progressing to coma may develop. Especially likely in meningococcal meningitis, purplish spots called **petechiae**, resulting from small hemorrhages, may appear on the skin **(figure 26.6)**. The infected person may develop shock and die within 24 hours. Usually, though, progression of the illness is slower, allowing time for effective treatment.

Causative Agent

The causative organism of meningococcal meningitis is *Neisseria meningitidis.* Although there are 12 antigenic groups of *N. meningitidis*, most serious infections are due to A, B, C, Y, and W135. The process of identifying the different strains, called **serogrouping**, employs serum containing specific antibody, obtained from the blood of laboratory animals that have been immunized with known strains of meningococci.

Pathogenesis

Infection is acquired by inhaling airborne droplets from the respiratory tract of another person. The meningococci attach to the mucous membrane by pili and multiply. They are then taken in by the respiratory tract epithelial cells, pass through them, and invade the bloodstream. The blood carries the organisms to the meninges and cerebrospinal fluid, where they multiply faster than they can be engulfed and destroyed by the polymorphonuclear neutrophils (PMNs) that enter the fluid in large numbers in response to the infection **(figure 26.7)**. The bacteria and leukocytes degrade glucose and thereby cause a marked drop in the

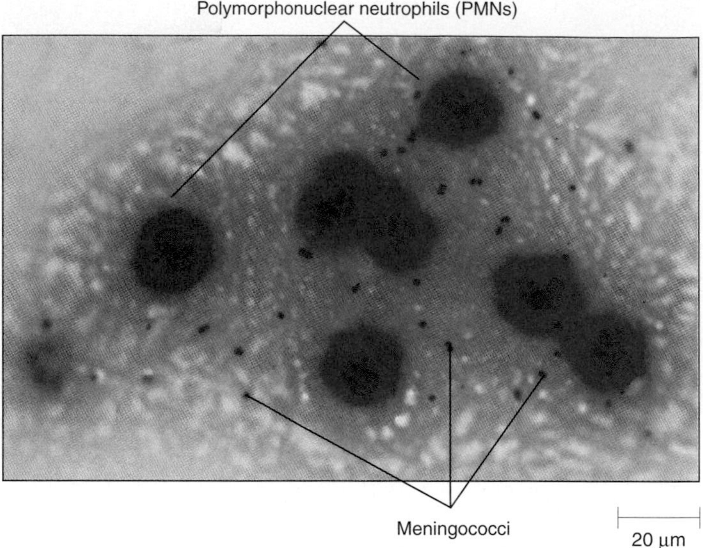

Figure 26.7 Polymorphonuclear Neutrophils and Meningococci in the Spinal Fluid of an Individual with Meningococcal Meningitis This case is somewhat unusual because so many of the meningococci can be seen outside the leukocytes.

glucose concentration normally present in the cerebrospinal fluid. The inflammatory response, with its formation of pus and clots, may cause brain swelling and **infarcts**, meaning death of tissue resulting from loss of blood supply. It can also lead to obstruction of the normal outflow of cerebrospinal fluid, causing the brain to be squeezed against the skull by the buildup of internal pressure. The infection may damage the nerves of hearing or vision or the motor nerves, producing paralysis. Moreover, *N. meningitidis* circulating in the bloodstream releases endotoxin, thereby causing a drop in blood pressure that can lead to shock. Shock is a state in which there is not enough blood pressure to circulate the blood adequately to meet the oxygen requirements of vital body tissues such as the kidneys and central nervous system. The smaller blood vessels of the skin are also damaged by the circulating meningococci, causing the petechiae. ■ pili, p. 69 ■ endotoxin, p. 461

The organism's endotoxin also increases the body's sensitivity to repeated endotoxin exposure. This strikingly increased sensitivity, called the **Shwartzman phenomenon**, can be demonstrated in experimental animals following repeated injections of meningococci. Intermittent "showers" of meningococci entering the blood from infected meninges are probably responsible for the rapidly fatal outcome in some human cases of meningitis. The tendency of meningococci to autolyse, meaning to rupture spontaneously, may enhance their release of endotoxin.

Epidemiology

The reasons are unknown why *N. meningitidis*, unlike *Haemophilus influenzae* and *Streptococcus pneumoniae*, is prone to cause epidemics of meningitis. Meningococcal meningitis can spread rapidly in crowded and stressed populations such as military recruits **(figure 26.8)**. More typically, however, it appears at widely separate locations and occurs throughout the year. Humans are the only source of the infection, and transmission can occur with exposure to a person with the disease or an asymptomatic carrier of *N. meningitidis.* The majority of meningococcal infections are limited to the nose and throat and pass unnoticed. If a selective medium is used to suppress the growth of other nasopharyngeal flora, meningococci can be recovered from the throats of 5% to 15% of apparently healthy people.

Prevention and Treatment

A vaccine composed of purified capsular polysaccharides from the four major disease-causing groups is available. The vaccine is used to control epidemics caused by these groups and to immunize people at high risk for the disease. It is not given routinely because it is not very effective in children less than two years old and its effect is not long lasting, but better vaccines are nearing approval. A number of outbreaks of meningitis due to group B strains have occurred. Group B strains possess a poorly immunogenic polysaccharide, and as yet, no vaccine is available for their control. Mass prophylaxis with an appropriate antibacterial medication, however, can be useful in controlling epidemics in small confined groups as in jails, nursing homes, and schools. Since meningococcal strains from patients with active disease may be more virulent than other strains, people intimately exposed to cases of meningococcal disease are routinely

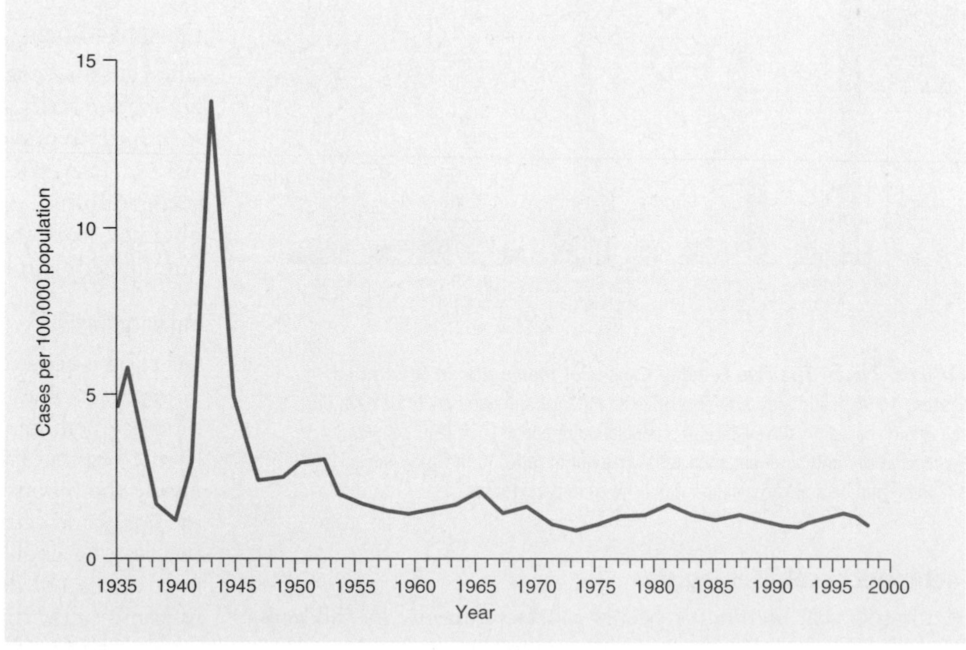

Figure 26.8 Meningococcal Disease in the United States, 1935 to 1998 Notice the high incidence during World War II, when epidemics occurred among military recruits crowded in barracks.

TABLE 26.1 Meningococcal Meningitis

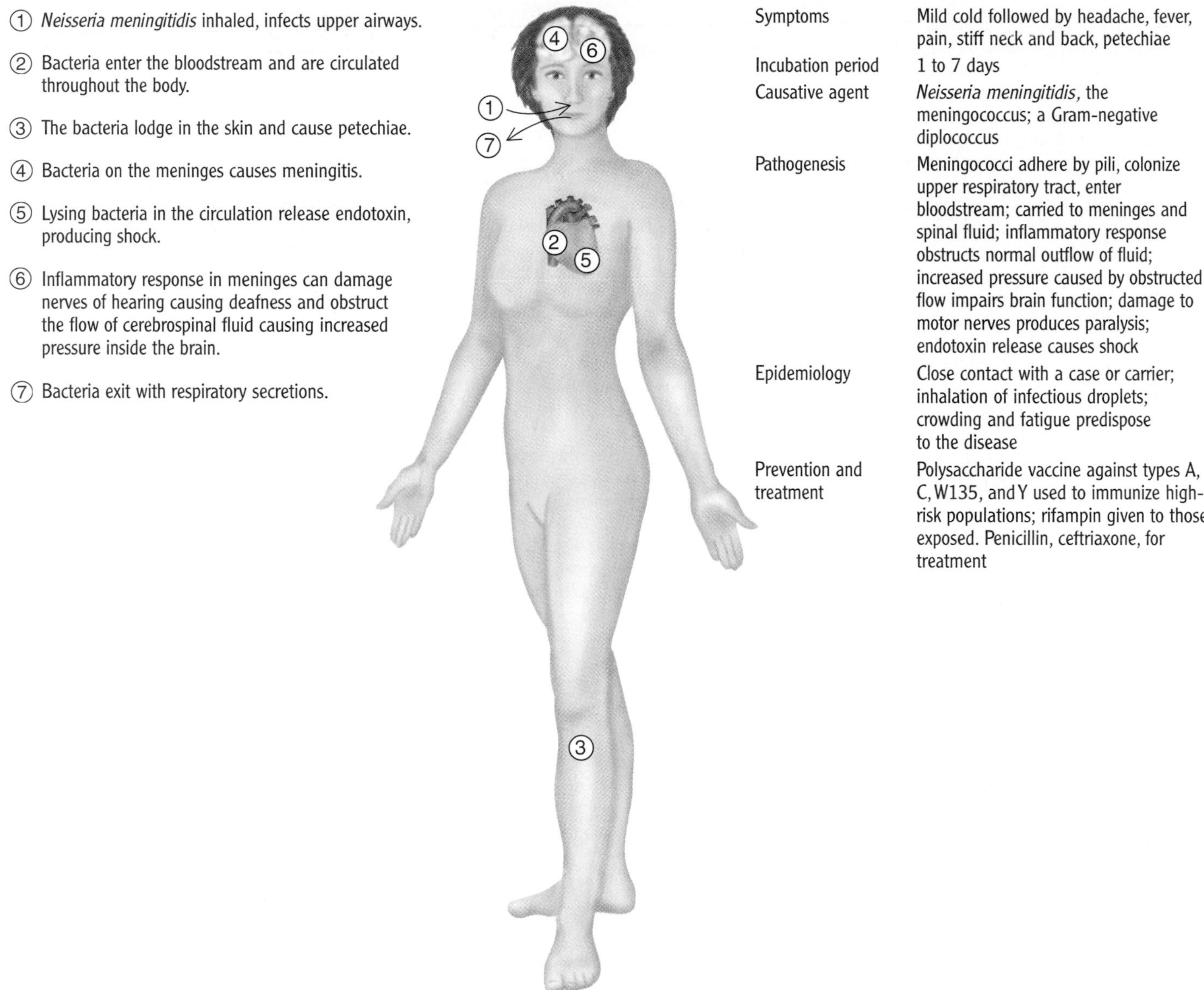

① *Neisseria meningitidis* inhaled, infects upper airways.

② Bacteria enter the bloodstream and are circulated throughout the body.

③ The bacteria lodge in the skin and cause petechiae.

④ Bacteria on the meninges causes meningitis.

⑤ Lysing bacteria in the circulation release endotoxin, producing shock.

⑥ Inflammatory response in meninges can damage nerves of hearing causing deafness and obstruct the flow of cerebrospinal fluid causing increased pressure inside the brain.

⑦ Bacteria exit with respiratory secretions.

Symptoms	Mild cold followed by headache, fever, pain, stiff neck and back, petechiae
Incubation period	1 to 7 days
Causative agent	*Neisseria meningitidis,* the meningococcus; a Gram-negative diplococcus
Pathogenesis	Meningococci adhere by pili, colonize upper respiratory tract, enter bloodstream; carried to meninges and spinal fluid; inflammatory response obstructs normal outflow of fluid; increased pressure caused by obstructed flow impairs brain function; damage to motor nerves produces paralysis; endotoxin release causes shock
Epidemiology	Close contact with a case or carrier; inhalation of infectious droplets; crowding and fatigue predispose to the disease
Prevention and treatment	Polysaccharide vaccine against types A, C, W135, and Y used to immunize high-risk populations; rifampin given to those exposed. Penicillin, ceftriaxone, for treatment

given prophylactic treatment with the antibiotic rifampin. Meningococcal meningitis can usually be cured unless severe brain injury or shock is present. The mortality rate is less than 10% in treated cases.

Table 26.1 gives the main features of this disease.

Listeriosis

Meningitis is a common manifestation of listeriosis, a foodborne disease caused by *Listeria monocytogenes.* Although generally causing only a small percentage of meningitis cases in the United States, epidemics sometimes occur. ■ **foodborne illness, p. 814**

Symptoms

The vast majority of healthy people have asymptomatic infections or trivial symptoms. Symptomatic infections are usually characterized by fever and muscle aches, and sometimes nausea or diarrhea. About three-quarters of the cases coming to medical attention have meningitis, heralded by fever, headache, stiff neck, and vomiting, that may or may not be preceded by other symptoms. Pregnant women who become infected often miscarry, or deliver terminally ill premature or full-term infants.

Causative Agent

Listeria monocytogenes is a motile, non-spore-forming, facultatively anaerobic, Gram-positive rod that can grow at 4°C.

Pathogenesis

The mode of entry in isolated cases is usually obscure, but it is generally via the gastrointestinal tract during epidemics. Gastrointestinal symptoms may or may not occur, but the bacteria

CASE PRESENTATION

The patient was a 31-month-old girl admitted to the hospital because of fever, headache, drowsiness, and vomiting. She had been previously well until 12 hours before admission, when she developed a runny nose, malaise, and loss of appetite.

Her birth and development were normal. There was no history of head trauma. Her routine immunizations had been neglected.

On examination, her temperature was 40°C (104°F), her neck was stiff, and she did not respond to verbal commands.

Her white blood count was elevated and showed a marked increase in the percentage of polymorphonuclear neutrophils (PMNs). Her blood sugar was in the normal range.

A spinal tap was performed, yielding cloudy cerebrospinal fluid under increased pressure. The fluid contained 18,000 white blood cells per microliter (normally, there are few or none), a markedly elevated protein, and a markedly low glucose. Gram stain of the fluid showed many tiny Gram-negative coccobacilli, most of which were outside the white blood cells.

1. What is the diagnosis and what is the causative agent?
2. What is the prognosis in this case?
3. What age group is most susceptible to this illness?
4. Compare the pathogenesis of this disease in children and adults.

Discussion

1. The patient had bacterial meningitis caused by ***Haemophilus influenzae***, serotype b. The fact that she had been healthy prior to her illness makes it highly unlikely that other serotypes could be responsible.
2. With treatment, the fatality rate is approximately 5%. Formerly, ampicillin was effective in most cases, but beginning in 1974, an increasing number of strains possessed a plasmid coding for β-lactamase. So far, however, these strains have been susceptible to the newer cephalosporin-type antibiotics. Unfortunately, about one-third of those who are treated and recover from the infection are left with permanent damage to the nervous system, such as deafness or paralysis of facial nerves. Prompt diagnosis and correct choice of antibacterial treatment minimize the chance of permanent damage.
3. The peak incidence of this disease is in the age range of 6 to 18 months, corresponding to the time when protective levels of transplacental antibody are lacking and acquired active immunity has not yet developed. Since 1987, vaccines consisting of type b capsular antigen conjugated with a protein such as the outer membrane protein of *Neisseria meningitidis* or diphtheria toxin have been used to immunize infants and thus eliminate the immunity gap. As a result, meningitis caused by *H. influenzae* type b is now rare. It is important to know, however, that strains other than type b can sometimes cause meningitis. So far, such strains are uncommon, although they represent an increasing percentage as type b strains decline in importance.
4. *Haemophilus influenzae* type b strains are referred to as "invasive" strains because they establish infection of the upper respiratory tract, pass the epithelium, and enter the lymphatic vessels and bloodstream. In this way, they gain access to the general circulation and are carried to the central nervous system. Most strains of *H. influenzae*, unlike type b, are non-invasive; although they can cause infections of respiratory epithelium, they usually do not enter the circulation. Most adults are immune to type b strains, but some adults develop meningitis from non-invasive *H. influenzae* strains that gain access to the nervous system because of a skull fracture or by direct extension to the meninges from an infected sinus or middle ear.

promptly penetrate the intestinal mucosa and enter the bloodstream. The resulting bacteremia is usually associated with fever and muscle aches, and it is the source of meningeal infection. In pregnant women, *L. monocytogenes* crosses the placenta and produces widespread abscesses in tissues of the fetus. Babies infected at the time of childbirth usually develop meningitis after an incubation period of 1 to 4 weeks.

Epidemiology

Listeria monocytogenes is widespread in natural waters and vegetation, and it can be carried in the intestines of asymptomatic animals and humans. It occurs on all continents except Antarctica. Pregnant women, the elderly, and those with underlying illnesses such as immunodeficiency, diabetes, cancer, and liver disease are especially susceptible to listeriosis. Epidemic disease has resulted from *L. monocytogenes* contamination of foods including coleslaw, milk, pork tongue in jelly, some soft cheeses, and hot dogs. The organisms can grow in food at refrigerator temperatures.

Prevention and Treatment

Preventive measures reflect the epidemiology of listeriosis:

- Thoroughly cook poultry, pork, beef, and other meats, and avoid contaminating countertops and kitchenware with them.
- Wash raw vegetables well before eating them.
- Those who are pregnant or have underlying diseases are advised to avoid soft cheeses such as Brie, Camembert, and Mexican style, and to avoid or reheat cold cuts, hot dogs, and refrigerated leftovers before eating.

So far, most strains of *L. monocytogenes* have remained susceptible to antibacterial medications such as penicillin. Even though the disease is often mild in pregnant women, prompt diagnosis and treatment is important to protect the fetus.

Some of the main features of listeriosis are presented in **table 26.2**.

Hansen's Disease (Leprosy)

Although Hansen's disease, also known as leprosy, is now a relatively minor problem in the Western world, it was once common in Europe and America. Like tuberculosis, the disease began to recede for unknown reasons even before an effective treatment was discovered. Today, it is chiefly a problem in tropical and economically underdeveloped countries, with an estimated worldwide occurrence of under 1 million active cases, most of which are in Asia. In recent years, reported new cases in the United States have averaged about 250 per year, mostly acquired outside the country.

Symptoms

Hansen's disease begins gradually, usually with the onset of increased or decreased sensation in certain areas of skin. These areas typically have either increased or decreased pigmentation. Later, they enlarge and thicken, losing their hair, sweating ability, and all sensation. The nerves of the arms and legs become visibly enlarged with accompanying pain, later changing to numbness, muscle wasting, ulceration, and loss of fingers or toes (**figure 26.9**). In some patients the skin lesions are not sharply defined, but slowly enlarge and spread. Changes are most obvious in the face, with thickening of the nose and ears and deep

TABLE 26.2 Listeriosis

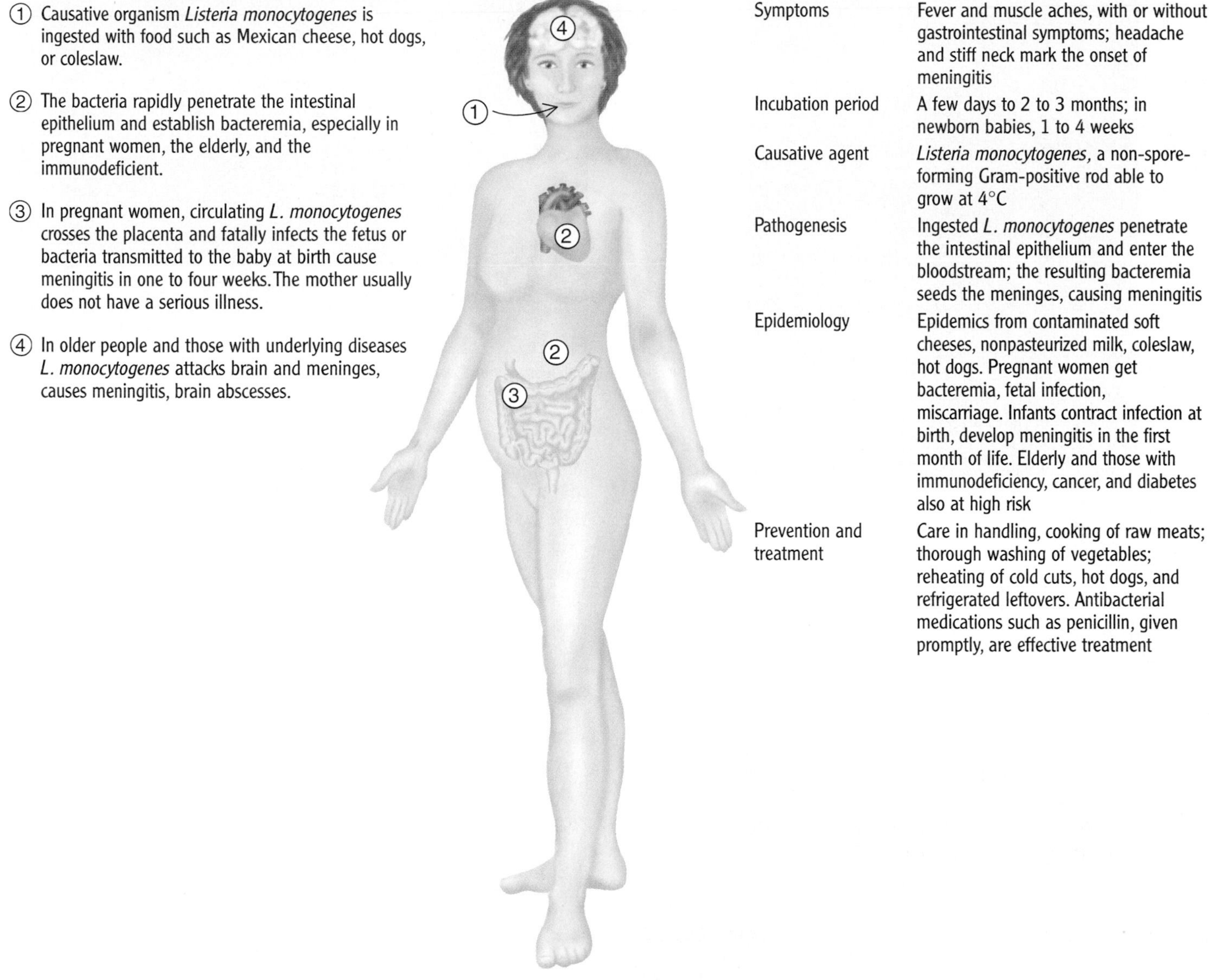

① Causative organism *Listeria monocytogenes* is ingested with food such as Mexican cheese, hot dogs, or coleslaw.
② The bacteria rapidly penetrate the intestinal epithelium and establish bacteremia, especially in pregnant women, the elderly, and the immunodeficient.
③ In pregnant women, circulating *L. monocytogenes* crosses the placenta and fatally infects the fetus or bacteria transmitted to the baby at birth cause meningitis in one to four weeks. The mother usually does not have a serious illness.
④ In older people and those with underlying diseases *L. monocytogenes* attacks brain and meninges, causes meningitis, brain abscesses.

Symptoms	Fever and muscle aches, with or without gastrointestinal symptoms; headache and stiff neck mark the onset of meningitis
Incubation period	A few days to 2 to 3 months; in newborn babies, 1 to 4 weeks
Causative agent	*Listeria monocytogenes,* a non-spore-forming Gram-positive rod able to grow at 4°C
Pathogenesis	Ingested *L. monocytogenes* penetrate the intestinal epithelium and enter the bloodstream; the resulting bacteremia seeds the meninges, causing meningitis
Epidemiology	Epidemics from contaminated soft cheeses, nonpasteurized milk, coleslaw, hot dogs. Pregnant women get bacteremia, fetal infection, miscarriage. Infants contract infection at birth, develop meningitis in the first month of life. Elderly and those with immunodeficiency, cancer, and diabetes also at high risk
Prevention and treatment	Care in handling, cooking of raw meats; thorough washing of vegetables; reheating of cold cuts, hot dogs, and refrigerated leftovers. Antibacterial medications such as penicillin, given promptly, are effective treatment

wrinkling of the facial skin. Collapse of the supporting structure of the nose occurs with accompanying congestion and bleeding.

Causative Agent

Mycobacterium leprae (**figure 26.10**) is identical in appearance to *M. tuberculosis.* It is aerobic, acid-fast, rod-shaped, and typically stains in a beaded manner. Despite many attempts, *M. leprae* has not been grown in the absence of living cells. In the 1960s, it was successfully cultivated to a limited degree in the footpads of laboratory mice, and in 1971, in a major advance, Dr. Eleanor E. Storrs and colleagues demonstrated that the organisms can be grown in armadillos (**figure 26.11**). Later, mangabey monkeys were also found to be susceptible. *Mycobacterium leprae* grows very slowly, with a generation time of about 12 days. A clone bank of the genome of *M. leprae* has been made in *E. coli*, thus making large quantities of the organism's antigens available for study. ■ **acid-fast stain, p. 51** ■ ***Mycobacterium*, p. 281** ■ **cloning, p. 224**

Pathogenesis

The earliest detectable finding in human infection with *M. leprae* is generally the invasion of the small nerves of the skin demonstrated by biopsy of skin lesions. Indeed, *M. leprae* is the only known human pathogen that preferentially attacks the peripheral nerves. The bacterium also initially grows very well within macrophages, before delayed hypersensitivity develops. The course of the infection depends on the immune response of the host. In most cases, cell-mediated immunity and delayed

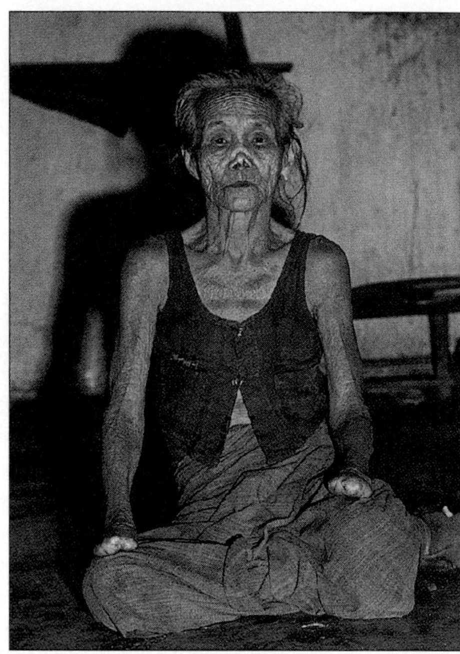

Figure 26.9 A Person with Leprosy Notice the absence of her fingers as a result of the disease.

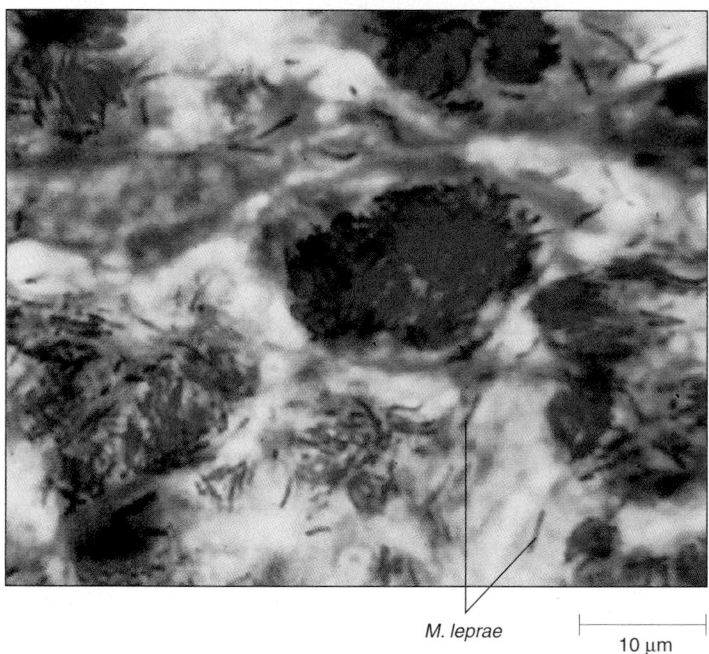

Figure 26.10 Masses of *Mycobacterium leprae* in a Biopsy Specimen from a Person with Lepromatous Leprosy

hypersensitivity develop against the invading bacteria. Immune macrophages limit the growth of *M. leprae*, and the bacteria therefore do not become numerous. The attack of immune cells against chronically infected nerve cells can progressively damage the nerves, however, which leads to disabling deformity, resorption of bone, and skin ulceration. The disease often spontaneously stops progressing, and thereafter the nerve damage, although permanent, does not worsen. This limited type of Hansen's disease in which cell-mediated immunity suppresses proliferation of the bacilli is called **tuberculoid leprosy**. People with tuberculoid leprosy rarely, if ever, transmit the disease to others. In many cases of tuberculoid leprosy, however, the immune system is gradually overwhelmed by the growth of *M. leprae*, leading to the more serious lepromatous form of the disease discussed in further detail next. Early treatment is important to prevent the disease from progressing. ■ macrophage, p. 372 ■ delayed hypersensitivity, p. 441

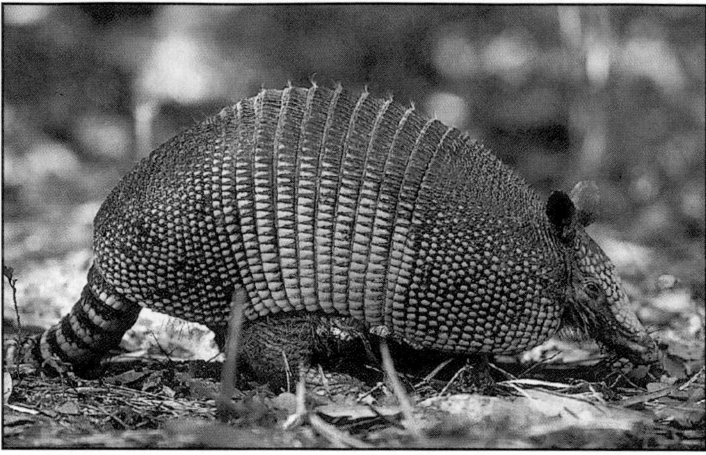

Figure 26.11 Armadillo The discovery that *Mycobacterium leprae* from people with leprosy grows well in the nine-banded armadillo made it possible to obtain large quantities of the bacteria for scientific studies.

When cellular immunity and delayed hypersensitivity to *M. leprae* fail to develop or are suppressed, unrestricted growth of *M. leprae* occurs in the cooler tissues of the body, notably in skin macrophages and peripheral nerves. This relatively uncommon form of Hansen's disease is called **lepromatous leprosy**. The tissues and mucous membranes contain billions of *M. leprae*, but there is almost no inflammatory response to them. The mucus of the nose and throat is loaded with the bacteria, which can readily be transmitted to others. Lymphocytes present in the lesions are mostly T-suppressor cells, and there is little or no evidence of macrophage activation. ■ T-suppressor cells, p. 398

In lepromatous leprosy, cell-mediated immunity to *M. leprae* is absent, although immunity and delayed hypersensitivity to other infectious agents are usually normal. Normal immune function tends to return when the disease is controlled by medication.

The very long generation time of this bacterium most likely accounts for the long incubation period of Hansen's disease, usually more than 2 years and sometimes 10 years or longer. ■ generation time, p. 90

Epidemiology

Transmission of the disease is by direct human-to-human contact. The source of the organisms is mainly nasal secretions of a lepromatous case, which transport *M. leprae* to mucous membranes or skin abrasions of another individual. Although leprosy bacilli readily infect people exposed to a case of lepromatous leprosy, the disease develops in only a tiny minority, being controlled by body defenses in the rest. Hansen's disease is no longer a much feared contagious disease, and the morbid days when lepers were forced to carry a bell or horn to warn others of their presence are fortunately past.

Natural infections with *M. leprae* occur in wild nine-banded armadillos and in mangabey monkeys. An epidemiological study appears to exclude armadillos as an important source of human leprosy, but occasional transmissions to humans have not been ruled out.

Prevention and Treatment

No proven vaccine to control leprosy is yet available. Tuberculoid leprosy can be arrested by treatment with the antimicrobials dapsone and rifampin administered in combination for 6 months. Lepromatous leprosy is generally treated for a minimum of 2 years, with a third drug, clofazimine, included in the treatment regimen. As in tuberculosis, multiple drug therapy is required to minimize the development of strains resistant to the antimicrobials.

Some features of Hansen's disease are summarized in **table 26.3**.

Botulism

Botulism, classically a severe form of food poisoning, is not a nervous system infection. It is considered here because its principal symptom is paralysis. It is one of the most feared of all diseases because it strikes without warning and is often fatal. The name botulism comes from *botulus*, meaning "sausage," chosen because some of the earliest recognized cases occurred in people who had eaten contaminated sausage. The majority of cases worldwide represent **foodborne botulism**, caused by eating food in which the organism has grown and released a poisonous substance. A small percentage of cases occur when the organisms colonize the intestine or a wound, however, causing **intestinal botulism** and **wound botulism**, respectively. ■ foodborne botulism, p. 815

Symptoms

Usually, symptoms begin 12 to 36 hours after ingestion of improperly canned non-acidic food. Most cases begin with dizziness, dry mouth, and blurred or double vision indicating eye muscle weakness. Abdominal symptoms, including abdominal pain, nausea, vomiting, and diarrhea or constipation, are present in 30% to more than 90% of cases, depending on different strains of the causative organism. Progressive paralysis then ensues, generally involving all voluntary muscles, but respiratory paralysis is the most common cause of death. Paralysis distinguishes botulism from most other forms of food poisoning.

Causative Agent

Botulism is caused by the strictly anaerobic, Gram-positive, spore-forming, rod-shaped bacterium, *Clostridium botulinum*. Its endospores, formed near the ends of the cells (**figure 26.12**), generally resist boiling for hours, but they are killed by autoclaving. Different strains of *C. botulinum* vary markedly in their biochemical activity, but all produce toxins that act in a similar way to cause paralysis. Seven antigenically different toxins, designated A, B, C1, D, E, F, and G, are synthesized by different strains of *C. botulinum*. Types A, B, E, and F are responsible for most human cases, while types C and D affect only birds and other animals. In strains C and D, toxin production results from lysogenic conversion. The toxins are released in an inactive form upon lysis of the bacterial cells, and they are activated by proteolytic enzymes. ■ lysogenic conversion, p. 331

Pathogenesis

Like other clostridia, *C. botulinum* produces endospores that are generally highly resistant to heat and therefore can persist in foods such as vegetables, fruit, meat, seafood, and cheese despite cooking and canning processes. These spores germinate if the environment is favorable—such as having a suitable nutrient, anaerobic conditions, a pH above 4.5, and a temperature above 39°F (4°C)—and growth of the bacteria results in the release of exotoxin into the food. When a person eats the food, the exotoxin resists digestion by stomach enzymes and acid, is absorbed by the small intestine, and can circulate in the bloodstream for 3 weeks or more. This exotoxin is a **neurotoxin**, meaning that it acts against the nervous system, and is one of the most powerful

TABLE 26.3 Hansen's Disease (Leprosy)

Symptoms	Anaesthetic skin lesions, ulcerative lesions of mucous membranes, deformed face, loss of fingers or toes
Incubation period	3 months to 20 years; usually 3 years
Causative agent	*Mycobacterium leprae*, an-acid fast, non-culturable rod
Pathogenesis	Invasion of small nerves of skin; multiplication in macrophages; course of disease depends on immune response of host; activated macrophages limit growth of bacterium; attack of immune cells against infected nerve cells produces nerve damage, leading to deformity; immunity sometimes overwhelmed by accumulating bacterial antigen, allowing unrestrained growth of *M. leprae*
Epidemiology	Direct contact with *M. leprae* from mucous membrane or skin lesions
Prevention and treatment	Disinfection of contaminated articles, handwashing; vaccines under evaluation. Treatment: dapsone plus rifampin for months or years; clofazimine added for lepromatous disease

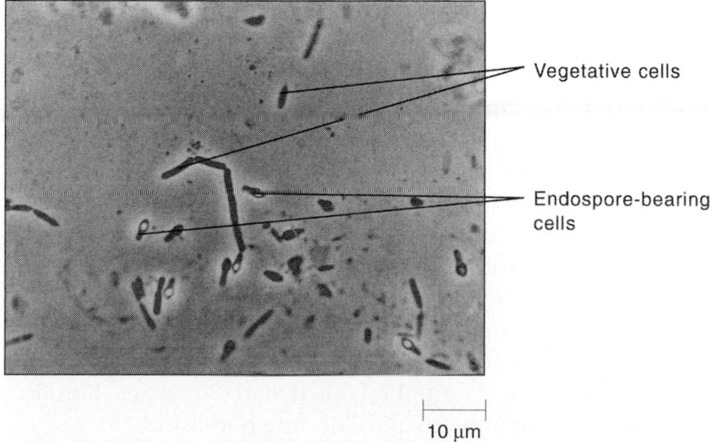

Figure 26.12 ***Clostridium botulinum*** Notice the spores that form near the ends of the bacteria. These spores are not reliably killed by the temperature of boiling water.

poisons known. A few milligrams of the toxin would be sufficient to kill the entire population of a large city. Indeed, cases of botulism have resulted from a person eating a single contaminated string bean, and from licking a finger contaminated with toxin.

The circulating neurotoxin attaches to motor nerves at the point where they meet the muscles, blocking transmission of nerve signals to the muscles, and thereby producing paralysis. The nerve signals are transmitted by a type of chemical known as a **neurotransmitter**, released by the nerve in tiny vesicles. The neurotoxin probably acts by preventing the vesicles from attaching to the muscle cells, likely by destroying a receptor on the vesicles. Like a number of other bacterial exotoxins, botulinal toxin is composed of two portions, A and B. The B portion attaches to specific receptors on motor nerve endings, and the A portion enters the nerve cell. The A portion then becomes an active peptidase enzyme that prevents release of a neurotransmitter, producing the harmful effects of the toxin. ■ **A-B toxins, p. 462**

Intestinal botulism occurs occasionally when *C. botulinum* colonizes the intestine, especially of infants 6 months of age or less, and produces a mild form of the disease characterized by constipation followed by generalized paralysis that can range from mild lethargy to respiratory insufficiency. In infants, *C. botulinum* organisms and toxin are often demonstrable in their feces, but the toxin levels are too low to be detectable in their blood. Most recover without receiving antitoxin treatment, although respirator support and tube feeding may be needed until the organisms are replaced by normal intestinal flora. These infections arise from *C. botulinum* spores, which are commonly present in dust and contaminate foods such as honey. Indeed, ingestion of honey has been implicated in 10% to 30% of infant botulism cases. Intestinal botulism also occurs in adults, particularly in immunodeficient patients whose normal intestinal flora have been suppressed by antibiotic treatment.

Clostridium botulinum can also colonize dirty wounds, especially those containing dead tissue. The bacteria do not invade, but they multiply in dead tissue and their neurotoxin diffuses into the bloodstream. Many cases have been due to wounds caused by abuse of injected drugs.

Clostridium botulinum neurotoxin can be used to treat people with certain chronic spastic conditions. Minute amounts of the toxin injected into the area of spasm give prolonged relief of symptoms. ■ **Botox, p. 463**

Epidemiology

Clostridium botulinum is widely distributed in soils and aquatic sediments around the world. Foodborne botulism, formerly the most common and most severe of the three types of botulism, was first recognized in the late Eighteenth century. In the early part of the Twentieth century, outbreaks of foodborne botulism were common in the United States, often traceable to commercially canned foods. Strict controls were then placed on commercial canners to ensure adequate sterilizing methods. Since then, outbreaks caused by commercially canned foods have been infrequent. Today, in the United States, intestinal botulism is much more common than foodborne botulism.

Prevention and Treatment

Prevention of botulism depends on proper sterilization and sealing of food at the time of canning. Fortunately, the toxin is heat-labile, and heating food to 100°C for 15 minutes just prior to serving generally makes it safe to eat. One cannot rely on a spoiled smell, taste, or appearance to detect contamination, because such changes are not always present. The disease is treated by the intravenous administration of antitoxin as soon as possible after the diagnosis is made. The antitoxin, however, only neutralizes toxin circulating in the bloodstream, and the nerves already affected by it recover slowly, over weeks or months. Enemas, gastric washing, and surgical removal of dirt and dead tissue from infected wounds help remove any unabsorbed toxin. Artificial respiration may be required for prolonged periods. The main features of botulism are presented in **table 26.4**.

TABLE 26.4 Botulism

Symptoms	Blurred or double vision, weakness, nausea, vomiting, diarrhea; generalized paralysis and respiratory insufficiency
Incubation period	Usually 12 to 36 hours
Causative agent	*Clostridium botulinum*, an anaerobic, Gram-positive, spore-forming, rod-shaped bacterium
Pathogenesis	*Clostridium botulinum* multiplies in food and releases neurotoxin. Toxin is ingested, survives stomach acid and enzymes, is absorbed by the small intestine, and is carried by the bloodstream to motor nerves; toxin acts by blocking the transmission of nerve signals to the muscles, producing paralysis; *C. botulinum* can also colonize intestine or wounds, and cause generalized weakness or paralysis
Epidemiology	Ingestion of contaminated, often home-canned, non-acid food that was not heated enough to kill *C. botulinum* spores. Spores widespread in soil, aquatic sediments, and dust. Can result in colonization of the intestine of adults and infants with deficiencies in normal flora, and wounds containing dirt and dead tissue, including those caused by injected-drug abuse
Prevention and treatment	Education in proper home-canning methods; heating food to boiling for 15 minutes just prior to serving. Treatment: enemas and stomach washing to remove toxin, cleaning infected wounds of dirt and dead tissue, intravenous administration of antitoxin, and artificial respiration

MICROCHECK 26.2

Bacterial meningitis is uncommon. Most cases occur in children under 5 years of age. Organisms from the mother's birth canal are responsible in infants less than 1 month old. Bacteria commonly carried in the upper respiratory tract cause most other cases. Meningococcal meningitis sometimes occurs in epidemics and occasionally causes death within 24 hours. Meningitis is a common complication of listeriosis, caused by a Gram-positive rod that can multiply at refrigerator temperatures and is usually foodborne. Hansen's disease (leprosy) is caused by an acid-fast unculturable bacterium; the disease exists in two forms, depending on the immune status of the victim. Botulism is a toxin-mediated disease characterized by generalized paralysis. It is not a nervous system infection.

- Name and describe the organism that was generally the leading cause of bacterial meningitis in the United States until the late 1980s.
- Why are most newborn babies unlikely to contract meningococcal, pneumococcal, or *Haemophilus* meningitis?
- Describe the only bacterial pathogen that preferentially invades peripheral nerves.
- Why would such a high percentage of infant botulism cases be associated with honey?

Viral Diseases of the Nervous System

Many different kinds of viruses can infect the central nervous system, including the Epstein-Barr virus of infectious mononucleosis; the mumps, rubeola, varicella-zoster, and herpes simplex viruses; and more commonly, human enteroviruses and the viruses of certain zoonoses. In most cases, nervous system involvement occurs in only a very small percentage of people infected with the viruses. The following section discusses four kinds of illness resulting from viral central nervous system infections: meningitis, encephalitis, poliomyelitis, and rabies.

Viral Meningitis

Viral meningitis is much more common than bacterial meningitis. It is usually a mild disease that does not require specific treatment, and patients generally recover in 7 to 10 days. Viruses are responsible for most cases of "aseptic meningitis," meaning those with symptoms and signs of meningitis but with negative tests for bacterial pathogens.

Symptoms

The onset of viral meningitis is typically abrupt, with fever and severe headache above or behind the eyes, and a stiff neck with increased pain on forward flexion. Sensitivity of the eyes to light, nausea, and vomiting are common. In addition, depending on the causative agent, there may be a sore throat, chest pain with inspiration, swollen parotid glands, or a skin rash. ■ **parotid glands, p. 587**

Causative Agents

Small, nonenveloped RNA viruses, members of the enterovirus subgroup of picornaviruses, are responsible for at least half of the cases of viral meningitis. Of these, the most common offenders are coxsackie viruses, which can cause throat or chest pain, and echoviruses, which can cause a rash. Formerly common, the mumps virus is now an infrequent cause because of widespread immunization against the disease. ■ **picornaviruses, p. 343** ■ **mumps, p. 596**

Pathogenesis

Enteroviruses characteristically infect the throat and intestinal epithelium and lymphoid tissue, and then seed the bloodstream. The **viremia**, meaning viruses circulating in the bloodstream, results in meningeal infection and sometimes rashes or chest infection. The inflammatory response in the meninges differs from bacterial meningitis in that fewer cells usually enter the cerebrospinal fluid, and a high percentage of them are mononuclear rather than polymorphonuclear neutrophils (PMNs). Typically the cerebrospinal fluid glucose remains normal.

Epidemiology

Enteroviruses are relatively stable in the environment, and they can sometimes even survive in chlorinated swimming pools. Infected individuals, including those with asymptomatic infections, often eliminate the viruses in their feces for weeks. Enteroviral meningitis is transmitted by the fecal-oral route, and the peak incidence occurs in the late summer and early fall in temperate climates. Mumps virus is transmitted by the respiratory route, and mumps meningitis is generally most common in the fall and winter months.

Prevention and Treatment

Handwashing and avoidance of crowded swimming pools are reasonable preventive measures when cases of aseptic meningitis are present in the community. There are no vaccines against coxsackie and echoviruses. Mumps virus disease can be prevented by immunization. No specific treatment is available.

The main features of viral meningitis are presented in **table 26.5**.

Viral Encephalitis

Whereas viral meningitis is usually a benign illness, viral encephalitis is much more likely to cause death or permanent disability. Viral encephalitis can be sporadic, meaning that there are generally a few widely scattered cases occurring all the time, or it can be epidemic, meaning a number of cases appear in a

TABLE 26.5 Viral Meningitis

Symptoms	Abrupt onset, fever, severe headache, stiff neck, often vomiting; sometimes sore throat, large parotid glands, rash, or chest pain
Incubation period	Usually 1 to 2 weeks for enteroviruses, 2 to 4 weeks for mumps
Causative agents	Most cases: small nonenveloped RNA enteroviruses of the picornavirus family, usually coxsackie or echoviruses. Mumps virus common in unimmunized populations
Pathogenesis	Viremia from primary infection seeds the meninges. Fewer leukocytes enter cerebrospinal fluid than with bacterial infections, and many are mononuclear; usually no decrease in CSF glucose
Epidemiology	Enteroviruses transmitted by the fecal-oral route, mumps by respiratory secretions and saliva. Enterovirus transmission mainly summer and early fall; mumps in fall and winter
Prevention and treatment	Handwashing, avoiding crowded swimming pools during enterovirus epidemics; mumps vaccine for mumps prevention. No specific treatment

limited period of time in a given geographical area. Sporadic encephalitis is usually due to herpes simplex virus, but other viruses, including those that cause mumps, measles, and infectious mononucleosis, can occasionally cause encephalitis. Two mechanisms have been proposed to explain herpes simplex encephalitis. Most instances are thought to be due to activation of latent virus, the result of primary infection in the past. In about one-fourth of the cases, however, the encephalitis strain differs from latent virus in the person's ganglia, suggesting that the infection causing the encephalitis is newly acquired, the virus probably entering the CNS via the nerves of smell. Most people recover from the disease but are left with permanent impairment, such as epilepsy, paralysis, deafness, or difficulty thinking. The yearly incidence of herpes simplex encephalitis is estimated to be two to three cases per million people. The disease occurs throughout the year, mostly affecting those under 30 and the elderly. Cases in newborn infants are usually acquired from the mother's birth canal and are due to herpes simplex, type 2, but more than 95% of those in children and adults are caused by herpes simplex virus, type 1. No proven preventive measures for herpes simplex encephalitis exist except for the newborn, for whom cesarean section can be performed to prevent exposure of the baby to a type 2 herpes simplex infection. Medication given promptly can shorten the illness and improve the outcome. ■ **herpes simplex virus, p. 594, 678**

Epidemic viral encephalitis is a bigger problem because of the numbers of people that can be affected. Although symptoms are quite similar to sporadic encephalitis, the causative agents, pathogenesis, epidemiology, prevention, and treatment are different.

Symptoms

The onset is usually abrupt, with fever, headache, vomiting, and one or more nervous system abnormalities such as disorientation, localized paralysis, deafness, seizures, or coma.

Causative Agents

Epidemic viral encephalitis is usually caused by **arboviruses**, meaning arthropod-borne viruses, a diverse group of viruses transmitted by insects, mites, or ticks. The four leading causes of epidemic encephalitis in the United States are all arboviruses transmitted by mosquitoes, as is the West Nile virus, introduced into New York from the Middle East in the summer of 1999. All are enveloped, single-stranded RNA viruses: LaCrosse encephalitis virus, bunyavirus family; St. Louis and West Nile encephalitis viruses, flavivirus family; and the eastern and western equine encephalitis viruses, togavirus family. ■ **mosquitoes, p. 315**

Pathogenesis

Knowledge of the pathogenesis of arboviral encephalitis is incomplete. The viruses multiply at the site of the mosquito bite and in local lymph nodes, producing viremia, meaning viruses circulating in the bloodstream. Levels of viremia are very low and transitory, yet the ratio of overt encephalitis to mild or inapparent infection can be 1:25 or greater. This raises the question of how the virus crosses the blood-brain barrier so readily. The viruses replicate in nerve cells and cause extensive destruction of brain tissue in severe cases. The process is halted with the appearance of neutralizing antibody. Mortality ranges from about 2% with LaCrosse, to 35% to 50% with eastern equine encephalitis. Disabilities occur in those who recover, ranging from 5% to more than 50% of cases, depending largely on the kind of virus and the age of the patient, the very young and the elderly suffering the most. The kinds of disability include emotional instability, mental retardation, epilepsy, blindness, deafness, and paralysis of one side of the body.

Epidemiology

During epidemic viral encephalitis, only a minority of those infected develop encephalitis. Others develop viral meningitis, fever and headache only, or no symptoms at all. These diseases are all zoonoses maintained in nature in birds or rodents, humans being an accidental host. In the United States, LaCrosse encephalitis virus usually causes most of the reported encephalitis cases. In its natural cycle (**figure 26.13**), the LaCrosse virus infects *Aedes* mosquitoes, which pass it directly from one mosquito to another in semen, and it can survive the winter in the mosquito's eggs. These mosquitoes feed on and infect squirrels and chipmunks, which markedly amplify the amount of virus and spread it to uninfected female mosquitoes feeding on the rodents' blood. These mosquitoes are forest dwellers that normally breed in water-containing cavities in hardwood trees or in discarded tires. The natural cycle of other encephalitis-causing viruses involves wild birds rather than rodents, and involves different mosquitoes with different habitats and feeding habits. **Figure 26.14** shows

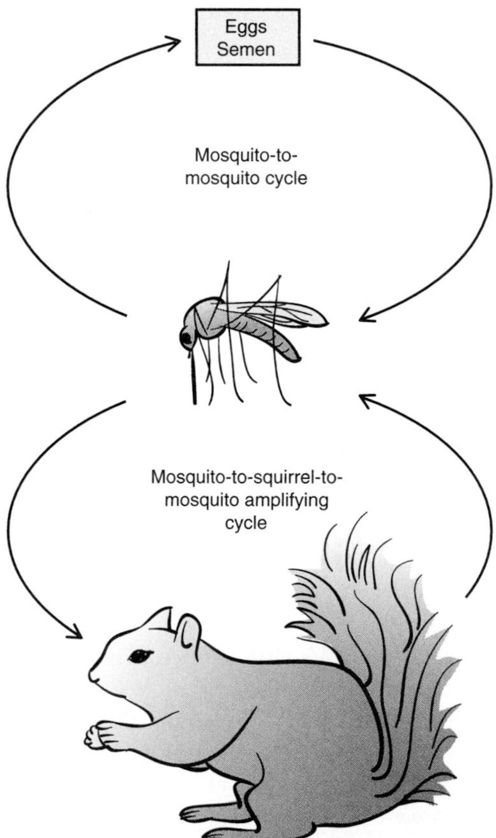

Figure 26.13 LaCrosse Encephalitis Virus, Natural Cycles The virus is maintained both by mosquito-to-mosquito transmission through semen and eggs, and by mosquito-to-squirrel (or chipmunk) -to-mosquito transmission. Infection of the rodent markedly increases the quantity of virus and the potential to infect many mosquitoes.

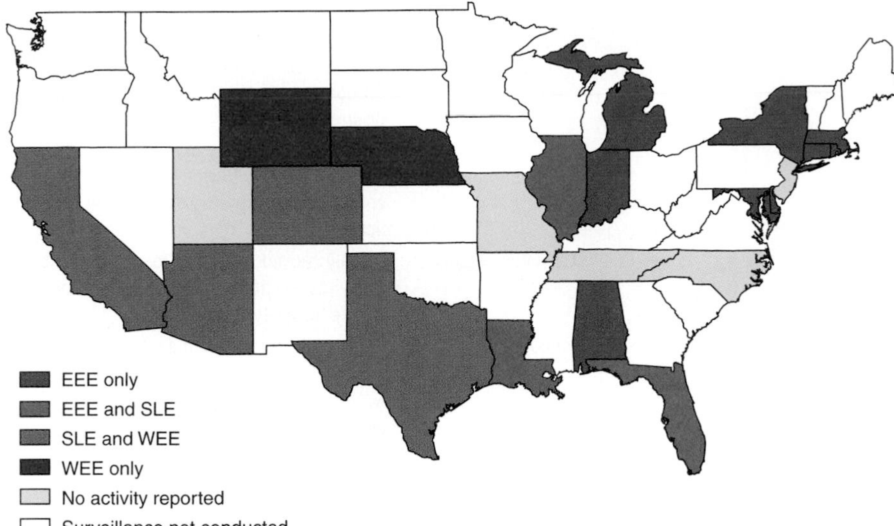

Figure 26.14 Distribution of Encephalitis-Causing Arboviruses Reported presence of St. Louis encephalitis (SLE), eastern equine encephalitis (EEE), and western equine encephalitis (WEE) viruses in mosquitoes, sentinels, or wild birds, United States, 1996 to 1997.

the results of testing for these viruses in mosquitoes, wild birds, and sentinels in states doing surveillance.

There is still much to be learned about the ecological factors responsible for the spread of these viruses to humans, but the proximity of humans to their natural cycles, rainfall, winter and summer temperatures, numbers and range of amplifying hosts, and predators could all be important.

Prevention and Treatment

As equine encephalitis viruses spread from their natural hosts, they generally infect horses 1 or 2 weeks before the first human cases appear. Cases in horses provide a warning to increase protection against mosquitoes. Sentinel chickens serve the same function. They are stationed in cages with free access of mosquitoes, and their blood is tested periodically for evidence of arbovirus infection. A positive test would trigger an encephalitis alert, as for St. Louis encephalitis in Florida:

- Avoid outdoor activity during evening and night, the peak hours of biting for the *Culex* mosquito vector. If outdoors, wear long sleeves and pants.
- Make sure windows and porches are properly screened.
- Use insect repellents and insecticides.

A vaccine against eastern equine encephalitis is approved for use in horses and has also been used to protect emus, a domesticated meat-producing fowl that is highly susceptible to this virus. There is no proven antiviral therapy for arbovirus encephalitis.

The main features of epidemic viral encephalitis are presented in **table 26.6**.

Infantile Paralysis, Polio (Poliomyelitis)

The characteristic feature of poliomyelitis is selective destruction of motor nerve cells, usually of the spinal cord, resulting in permanent paralysis of one particular group of muscles, such as those of an arm or leg. The two individuals most responsible for control of this terrifying disease will not see its imminent elimination from the world; Albert Sabin died in 1992, Jonas Salk in 1995. The two men were bitter rivals, both of whom expected, but did not receive, the Nobel Prize. Each man, one hopes, felt rewarded by knowing how very much they had reduced human suffering, even though failing to accept the enormous contribution made by the other.

Symptoms

Poliomyelitis usually begins with the symptoms of meningitis: headache, fever, stiff neck, and nausea. In addition, pain and spasm of some muscles generally occur, later followed by paralysis. Over the ensuing weeks and months, muscles shrink and bones do not develop normally in the affected area (**figure 26.15**). In the more severe cases, the muscles of respiration are paralyzed, and the victim requires an artificial respirator, a machine to pump air in and out of the lungs. Some recovery of function is the rule if the person survives the acute stage of the illness. The nerves of sensation, touch, pain, and temperature are not affected.

Causative Agent

Poliomyelitis is caused by three types of polioviruses, designated 1, 2, and 3, which are distinguished by using antisera. These small, nonenveloped, single-stranded, positive sense RNA viruses are members of the enterovirus subgroup of the picornavirus family. They can be grown *in vitro* in cell cultures, where they cause cell destruction. With low concentrations of virus, the areas of cell destruction, termed plaques, are separated from one another and are readily seen with the unaided eye (**figure 26.16**).

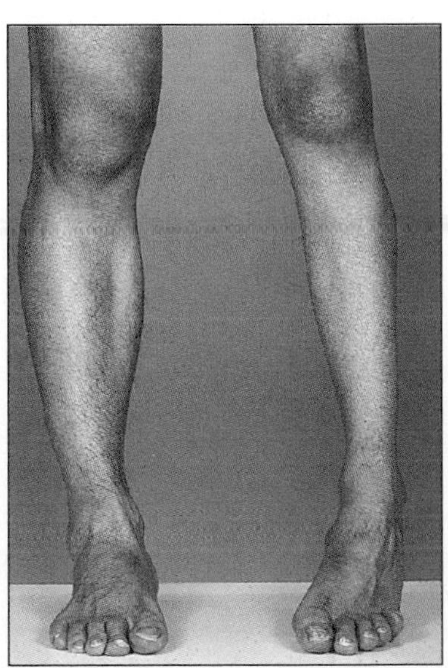

Figure 26.15 Individual with Atrophy of the Left Leg Due to Poliomyelitis

TABLE 26.6 Epidemic Viral Encephalitis

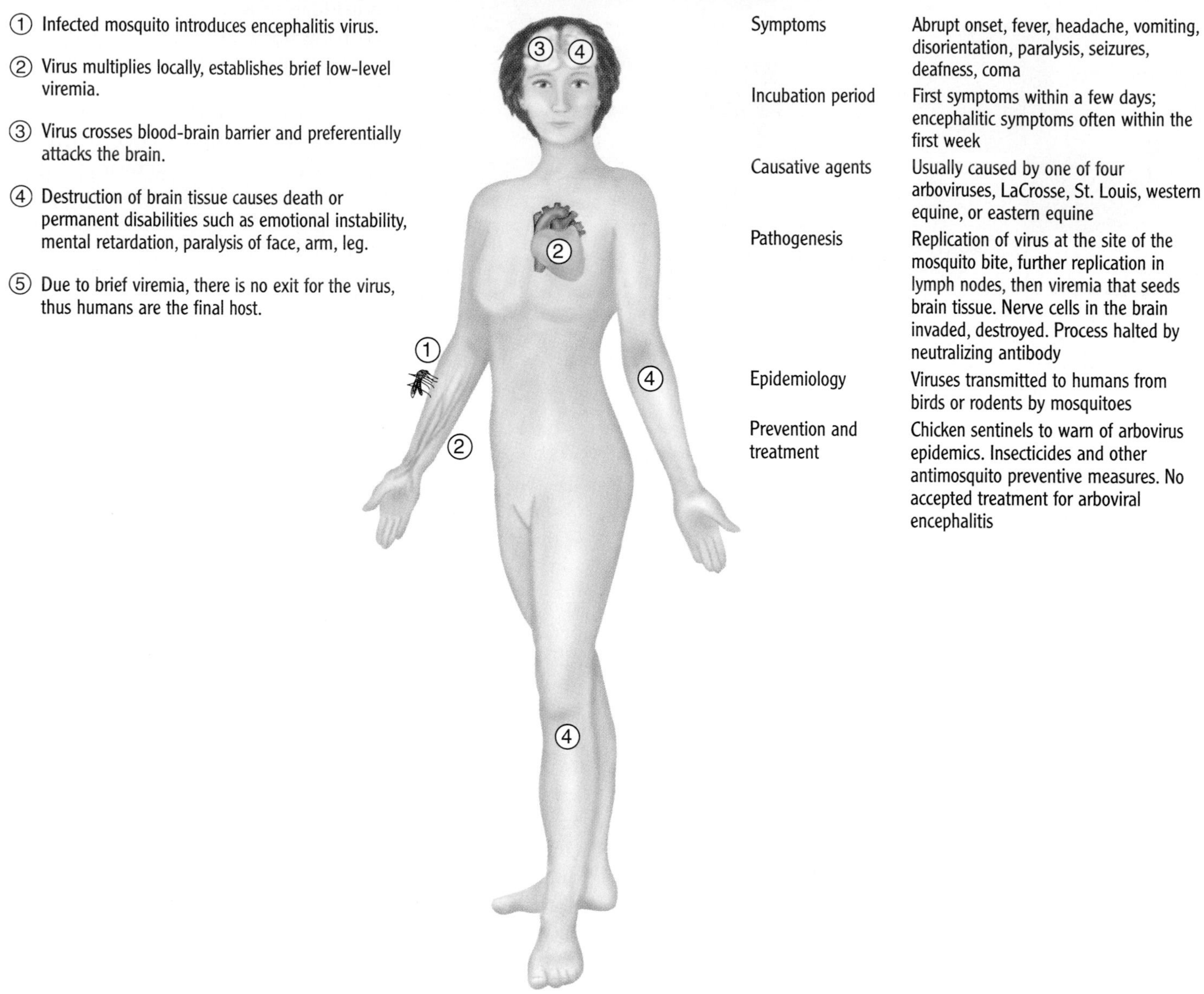
① Infected mosquito introduces encephalitis virus.
② Virus multiplies locally, establishes brief low-level viremia.
③ Virus crosses blood-brain barrier and preferentially attacks the brain.
④ Destruction of brain tissue causes death or permanent disabilities such as emotional instability, mental retardation, paralysis of face, arm, leg.
⑤ Due to brief viremia, there is no exit for the virus, thus humans are the final host.
③ ④
②
①
②
④
④

Symptoms	Abrupt onset, fever, headache, vomiting, disorientation, paralysis, seizures, deafness, coma
Incubation period	First symptoms within a few days; encephalitic symptoms often within the first week
Causative agents	Usually caused by one of four arboviruses, LaCrosse, St. Louis, western equine, or eastern equine
Pathogenesis	Replication of virus at the site of the mosquito bite, further replication in lymph nodes, then viremia that seeds brain tissue. Nerve cells in the brain invaded, destroyed. Process halted by neutralizing antibody
Epidemiology	Viruses transmitted to humans from birds or rodents by mosquitoes
Prevention and treatment	Chicken sentinels to warn of arbovirus epidemics. Insecticides and other antimosquito preventive measures. No accepted treatment for arboviral encephalitis

Pathogenesis

Polioviruses enter the body orally, infect the throat and intestinal tract, and then invade the bloodstream. In most people, the immune system conquers the infection and recovery is complete. Only in a small percentage of people does the virus enter the nervous system and attack motor nerves. The viruses can infect a cell only if the surface of that cell possesses specific receptors to which the virus can attach. This specificity of receptor sites helps explain the fact that polioviruses selectively infect motor nerve cells of the brain and spinal cord, while sparing many other kinds of cells. The infected cell is destroyed when the mature virus is released.

Even though poliomyelitis transmission no longer occurs in the United States, there are about 300,000 survivors of the disease, most of whom recovered years ago. Some of them develop muscle pain, increased weakness, and muscle degeneration 15 to 50 years after they had acute poliomyelitis. This condition, called the **postpolio syndrome**, sometimes involves muscles not obviously affected by the original illness. The late progression of muscle weakness is not due to recurrent multiplication of polio viruses. During recovery from acute poliomyelitis, surviving nerve cells branch out to take over the functions of the killed nerve cells. The late appearance of symptoms is probably due to the death of these nerve cells that have been doing double duty for so many years. The change is not noticeable to individuals who have relatively good muscle function, but those with moderately severe impairment can develop severe disability.

Perspective 26.1 **Bye Bye Polio!**

Poliomyelitis was first recognized as a distinct disease early in the Nineteenth century, although based on ancient Egyptian inscriptions it probably existed long before that. The first polio epidemic was reported in 1887 in Sweden. For the next 30 years, polio epidemics were mostly confined to the Scandinavian countries, Canada, New Zealand, Australia, and the northeastern United States. Polio was a terrifying threat, especially in economically advanced nations. Franklin Delano Roosevelt, the losing vice presidential candidate in 1920, came down with the disease the next year, at age 39. Roosevelt later was elected president of the United States (1933 to 1945) and played an important role in the eventual defeat of the disease.

The National Foundation for Infantile Paralysis was established in 1937, and one of Roosevelt's law partners, Basil O'Connor, became its volunteer president. Eddie Cantor, a famous comedian, developed the idea of a "March of Dimes" to defeat polio, and hundreds of thousands of volunteers collected dimes to support the effort. A major breakthrough occurred in 1949, when Dr. John Enders and his colleagues at Harvard University succeeded in growing polioviruses in monkey kidney cell cultures. John Enders, Thomas Weller, and Frederick Robbins were awarded the Nobel Prize in Medicine in 1954 for this work. Building on their finding, Dr. Jonas Salk perfected a vaccine consisting of formalin-inactivated poliovirus in the same year. His vaccine was quickly and widely employed, and the incidence of paralytic poliomyelitis plummeted. At about this same time, Dr. Albert Sabin, a Russian immigrant and, like Dr. Salk, a graduate of New York University medical school, selected mutant poliovirus strains that did not attack nervous tissue but infected the intestine, could spread from person to person, and stimulated antibody formation against the wild, virulent virus. Orally administered live attenuated vaccines were quickly developed and proved more effective than the killed virus vaccines.

The results of programs using these vaccines was dramatic. In 1952, just before the development of the killed vaccine by Salk, 57,879 cases of polio were reported in the United States. Ten years later, the CDC reported only 910 cases, and the last transmission of wild poliovirus in the United States occurred in 1979. In 1994, the entire Western Hemisphere was declared free of poliomyelitis. Global eradication of poliomyelitis using live attenuated vaccine is currently under way, and if all goes well, the world should be certified free of poliomyelitis by the year 2005. ■ **attenuated virus, p. 425**

Epidemiology

Poliomyelitis has had its greatest impact in countries when the level of sanitation is improved. This characteristic can be explained by the fact that, in areas where sanitation is poor, the polioviruses are widespread, transmitted by the fecal-oral route. Very few people escape childhood without becoming infected and developing immunity to the disease. Therefore, most babies receive antipolio antibodies transplacentally from their mothers. Newborn infants in these nations thus are partially protected against nervous system invasion by poliomyelitis virus for as long as their mothers' antibodies persist in their bodies—usually about 2 or 3 months. During this time, because of exposure to the poliovirus through crowding and unsanitary conditions, infants are likely to develop mild infections of the throat and intestine, thereby achieving lifelong immunity.

In contrast, in areas with efficient sanitation, the poliomyelitis virus sometimes cannot spread to enough susceptible people to sustain itself, and it disappears from the community. When it is reintroduced, people of all ages will lack antibody and will be susceptible, and a high incidence of infection and paralysis will result. This situation occurred in the United States in the 1950s, resulting in many cases of paralysis **(figure 26.17)** and death. With most people now routinely immunized against the disease, however, and the likelihood of imported disease rapidly waning, this scenario should no longer occur.

Prevention and Treatment

Like other enteroviruses, polioviruses are quite stable under natural conditions, often surviving in swimming pools, but are inactivated by pasteurization and properly chlorinated drinking water. Control of poliomyelitis using vaccines represents one of the greatest success stories in the battle against infectious diseases **(figure 26.18)**. Ironically, all cases of paralytic polio acquired in

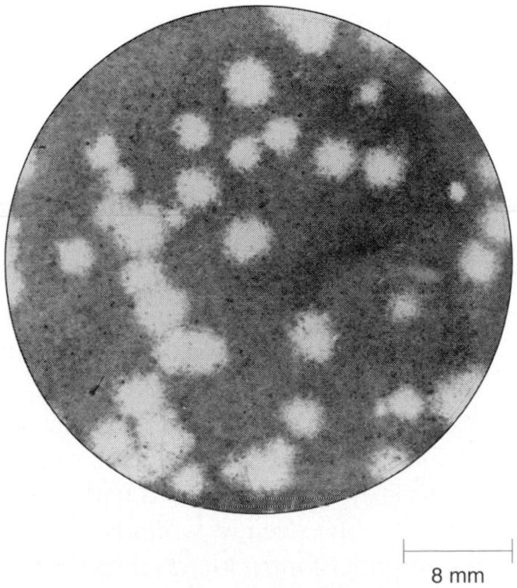

8 mm

Figure 26.16 Plaques Produced by Poliomyelitis Virus in a Cell Culture Monolayer Each plaque represents an area of the monolayer that has been destroyed by replicating poliovirus.

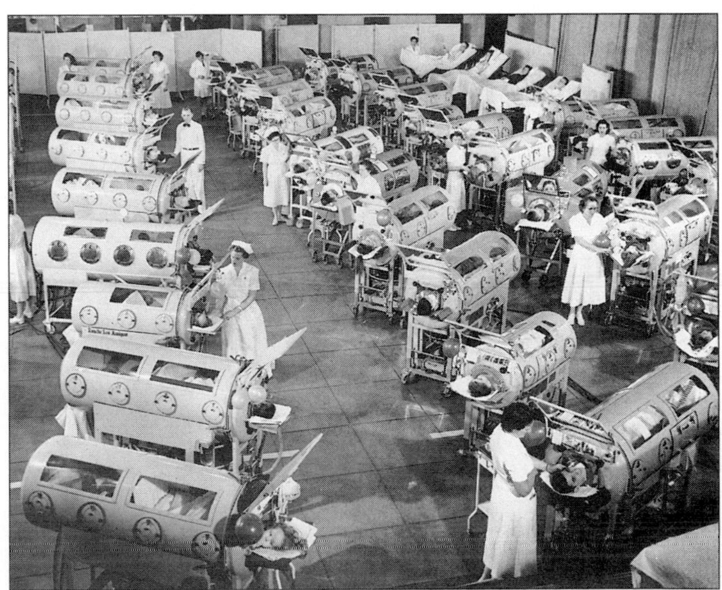

Figure 26.17 The Horror of Poliomyelitis These tanklike respirators ("iron lungs") were used during the 1950s epidemics of poliomyelitis to keep alive people whose respiratory muscles were paralyzed by the disease. Now we can hope to see polio forever banished from the earth!

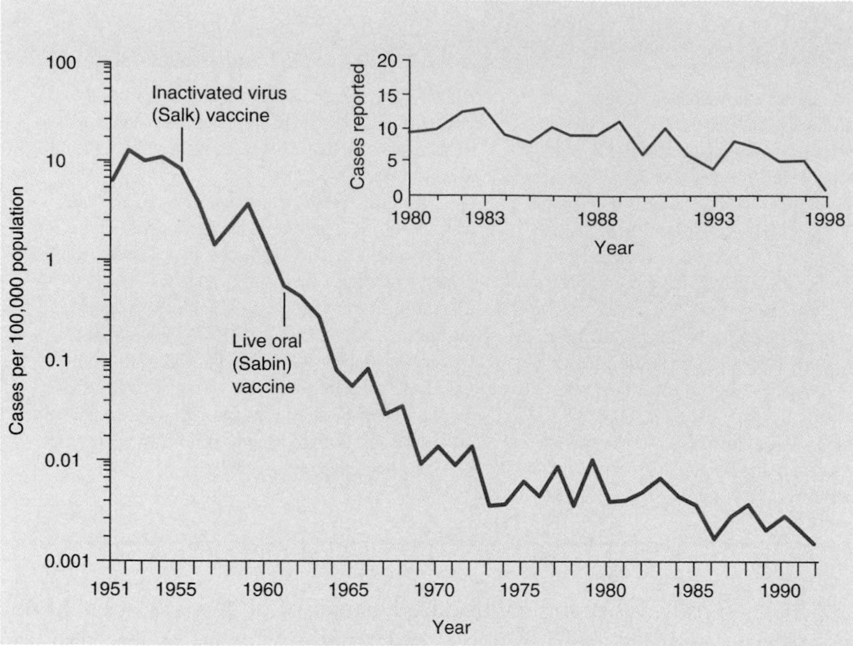

Figure 26.18 Incidence of Poliomyelitis in the United States, 1951 to 1998 Ironically, since 1980, all cases acquired in the United States have been caused by the oral (Sabin) vaccine.

the United States between 1980 and 1999 were caused by Sabin's live oral polio vaccine introduced in 1961. Because of the small risk of developing paralysis from the vaccine virus, approximately one case per 2.4 million doses given, routine use of the live vaccine was discontinued in the United States in mid-1999. The live oral, meaning taken by mouth, vaccine produces better local immunity in the throat and intestine, can spread from person to person, does not require an injection, and is less expensive. It is the preferred vaccine for completing the conquest of the disease in areas of the world where epidemic or endemic poliomyelitis continues to exist.

Poliomyelitis is summarized in **table 26.7**.

TABLE 26.7 Poliomyelitis

Symptoms	Headache, fever, stiff neck, nausea, pain, muscle spasm, followed by paralysis
Incubation period	7 to 14 days
Causative agents	Polioviruses 1, 2, 3, members of the picornavirus family
Pathogenesis	Virus infects the throat and intestine, circulates via the bloodstream, and enters some motor nerve cells of the brain or spinal cord; infected nerve cells lyse upon release of mature virus
Epidemiology	Spreads by the fecal-oral route; asymptomatic and nonparalytic cases common
Prevention and treatment	Prevented by injecting Salk's killed vaccine, or by Sabin's orally administered live vaccine in areas of epidemic or endemic disease. Treatment: artificial ventilation for respiratory paralysis; physical therapy and rehabilitation

Rabies

In the United States, immunization of dogs against rabies has practically eliminated them as a source of human disease. The rabies virus remains rampant among wildlife, however, a constant threat to non-immunized domestic animals and humans. Many questions remain about the pathogenesis of rabies, and no effective treatment exists for the disease.

Symptoms

Rabies is one of the most feared of all diseases because its terrifying symptoms almost invariably end with death. Like many other viral diseases, it begins with fever, head and muscle aches, sore throat, fatigue, and nausea. The characteristic symptom that strongly suggests rabies is a tingling or twitching sensation at the site of viral entry, usually an animal bite. These early symptoms generally begin 1 to 2 months after viral entry and progress rapidly to symptoms of encephalitis, agitation, confusion, hallucinations, seizures, and increased sensitivity to light, sound, and touch. The body temperature then rises steeply, and increased salivation combined with difficulty swallowing result in frothing at the mouth. Hydrophobia, painful spasm of the throat and respiratory muscles provoked by swallowing or even seeing liquids, occurs in half the cases. Coma develops, and about 50% of patients die within 4 days of the first appearance of symptoms, the rest soon after.

Causative Agent

The cause of rabies is the rabies virus (**figure 26.19a**), a member of the rhabdovirus family. This virus is about 180 × 75 nm in size, has a striking bullet shape, is enveloped, and contains single-stranded, negative-sense RNA. It buds from the surface of infected cells.

Pathogenesis

The principal mode of transmission of rabies to humans is via the saliva of a rabid animal (**figure 26.19b**) introduced into bite wounds or abrasions of the skin. Rabies can also be contracted by inhaling aerosols containing the virus, such as from bat feces. Only a few details of the events that follow introduction of the virus into the body are known. During the incubation period of the disease, the virus multiplies in muscle cells and probably other cells at the site of infection. Knoblike projections on the viral surface attach to receptors in the region where the nerve joins the muscle. At some point, the virus enters an axon and is carried along the course of the nerve by the normal flow of the axon's cytoplasm, eventually reaching the brain. The long incubation period, usually 1 to 2 months but sometimes exceeding a year, is partly determined by the length of the journey to the brain. Patients with head wounds into which the virus is introduced tend to have a shorter incubation period than those with extremity wounds. Severe wounds and those with a large amount of introduced rabies virus also generally result in short incubation periods. The virus then multiplies extensively in brain tissue, causing the symptoms of encephalitis. Characteristic inclusion bodies, called **Negri bodies (figure 26.20)**,

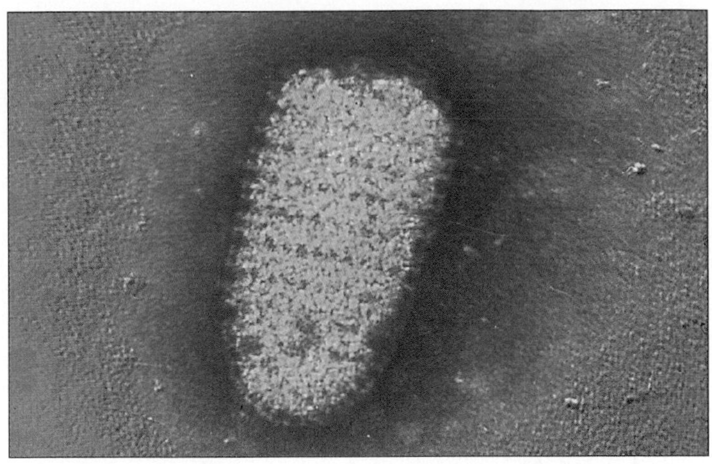

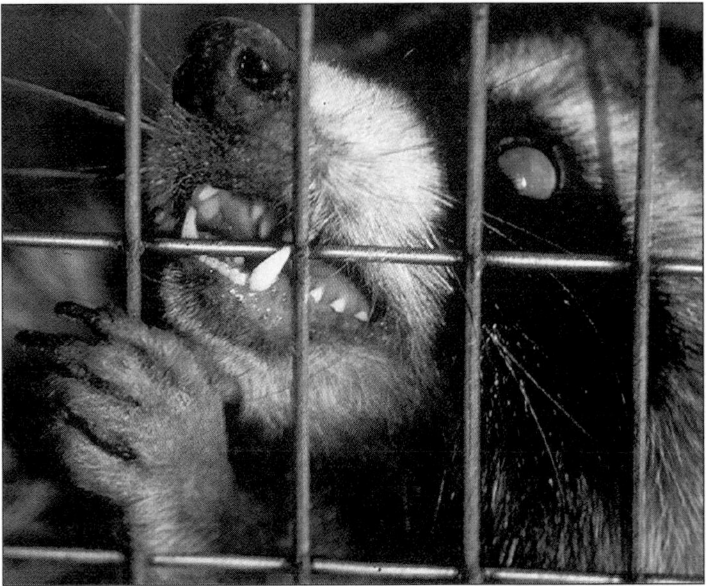

Figure 26.19 **In Most Cases, Rabies Is Transmitted by the Saliva of a Biting Animal** **(a)** Color-enhanced transmission electron micrograph of rabies virus. Notice the bullet shape. **(b)** Rabid raccoon caged in a Virginia animal shelter.

form at the sites of viral replication in the brain, but the cells are not lysed. The virus spreads outward from the brain via the nerves to various body tissues, notably the salivary glands, eyes, and fatty tissue under the skin, as well as to the heart and other vital organs. The presence of the virus in the eyes is of some practical significance, since cases can be diagnosed before death by stained smears made from the surface of the eyes. Moreover, several cases of rabies have occurred in individuals who received corneal transplants from donors who died of an atypical form of rabies that escaped diagnosis (the cornea is the clear middle part of the eye in front of the lens).

Epidemiology

Rabies is widespread in wild animals; about 5,000 wild animal cases are generally reported each year in the United States. This

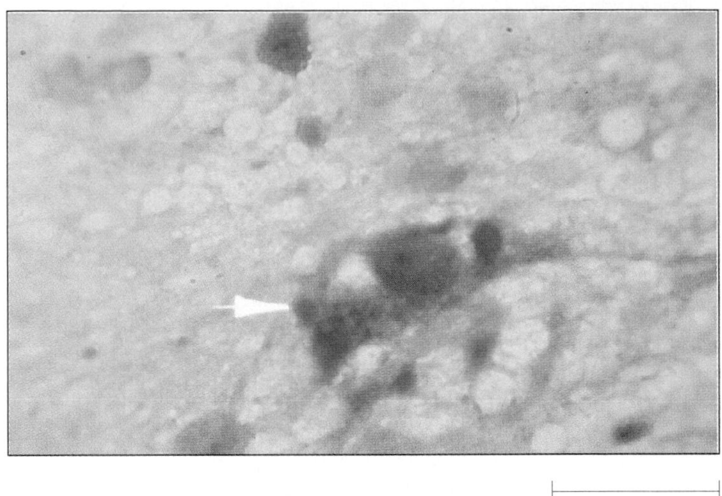

Figure 26.20 **Stained Smear of Brain Tissue from a Rabid Dog** The arrow points to one of several Negri bodies within the triangular-shaped nerve cell. The Negri bodies represent the sites of rabies virus replication.

represents an enormous reservoir from which infection can be transmitted to domestic animals and humans. In the United States, skunks, raccoons, and bats constitute the chief reservoir hosts. The virus can remain latent in bats for long periods, and healthy looking bats can have the virus in their salivary glands.

Until the mid-Twentieth century, most rabies cases in the United States resulted from dog bites, which is the primary mode of transmission in the non-industrialized world. The dog population in the United States is about 25 million, and an estimated 1 million people are bitten each year. Fortunately, since World War II, the incidence of dog rabies has dropped dramatically, and now more than 85% of reported rabies infections are in wild, as opposed to domestic, animals. The incidence of rabies in people has declined because dogs and cats have been immunized against rabies infection, in effect creating a partial barrier to the spread of the rabies virus from wild animal reservoirs to humans. Since pets vary in their tolerance to rabies vaccines depending on their age and species, several kinds of vaccines are available.

About three-quarters of dogs that develop rabies excrete rabies virus in their saliva, and about one-third of these begin excreting it 1 to 3 days before they get sick. Therefore, when a person is bitten by an unvaccinated, apparently healthy dog, the animal should be confined for 10 days to see if symptoms of rabies appear. Some dogs become irritable and hyperactive with the onset of rabies, produce excessive saliva, and attack people, animals, and inanimate objects. Perhaps more common is the "dumb" form of rabies, in which an infected dog simply stops eating, becomes inactive, and suffers paralysis of throat and leg muscles. Obviously, one should not be tempted to try to remove a suspected foreign body from the throat of a sick, choking, nonvaccinated dog! The reported number of rabies cases in humans now generally ranges from zero to four per year in the United States. Only about one-fourth of the cases have a history of animal bites or exposure to sick animals. It is possible that some of them contracted the infection by inhaling dust contaminated by rabies virus. The long incubation period, however, reportedly up to 6 years, and the patients' illness make the animal bite history unreliable.

TABLE 26.8 Rabies

Symptoms	Fever, headache, nausea, vomiting, sore throat, cough at onset; later, spasms of the muscles of mouth and throat, coma, and death
Incubation period	Usually 30 to 60 days; sometimes many months or years
Causative agent	Rabies virus, single-stranded RNA, rhabdovirus family; has an unusual bullet shape
Pathogenesis	During incubation period, virus multiplies at site of bite, then travels via nerves to the central nervous system; here it multiplies and spreads outward via multiple nerves to infect heart and other organs
Epidemiology	Bite of rabid animal, usually dog, cat, skunk, raccoon, or bat. Inhalation is another possible mode
Prevention and treatment	Avoid suspect animals; immunize pets. Effective post-exposure measures: immediately wash wound with soap and water and apply antiseptic; inject rabies vaccine and human rabies antiserum as soon as possible. No effective treatment once symptoms begin

Prevention and Treatment

A person who has been bitten by an animal should wash the wound immediately and thoroughly with soap and water and then apply an antiseptic. In people bitten by dogs having rabies virus in their saliva, the risk of developing rabies is about 30%. Louis Pasteur discovered that this risk can be lowered considerably by administering rabies vaccine as soon as possible after exposure to the virus. Presumably, the vaccine provokes a better immune response than the natural infection, inactivating free virus and killing infected cells during the long incubation period before the virus enters the nerves. Pasteur's vaccine was made from dried spinal cords of rabies-infected rabbits. Unfortunately, the rabbit nervous system tissue in the vaccine sometimes stimulated an immune response against the patient's own brain, causing *allergic encephalitis.* Current vaccines are essentially devoid of nervous tissue and the risk of serious side effects is very low.

If there is a reasonable possibility that the animal is rabid, the bitten individual should then receive a series of five injections of vaccine intramuscularly. Anti-rabies antibody is also injected at the wound site and intramuscularly. The anti-rabies antibody is obtained from humans who have been immunized against rabies. In the United States, about 30,000 people annually receive rabies vaccine to prevent rabies after having been bitten by suspected rabid animals.

There is no effective treatment for rabies, and only three people are known to have ever recovered from the disease. Interestingly, all of them recovered completely.

Some features of rabies are summarized in **table 26.8**.

MICROCHECK 26.3

Many different kinds of viruses can attack the nervous system, but they generally do so in only a small percentage of the people they infect. At least half of viral meningitis cases are caused be the enterovirus subgroup of picornaviruses, which are generally spread by the fecal-oral route. Viral meningitis is usually benign, but viral encephalitis often causes permanent disability. Arboviruses maintained in nature in a mosquito-bird or -rodent cycle are a leading cause of epidemic viral encephalitis. Sporadic viral encephalitis is usually due to herpes simplex virus. Poliomyelitis, characterized by paralysis of one or more muscle groups, is caused by three other enteroviruses. Rabies is a widespread zoonosis, almost uniformly fatal for humans, usually transmitted by animal bites.

- Name one cause of viral meningitis that is preventable by vaccination.
- Discuss the likelihood of person-to-person spread of viral encephalitis.
- Why is rabies now rare in humans when it is so common in wildlife?
- Why is the Sabin vaccine against poliomyelitis the preferred vaccine to use in countries where polio is still prevalent?

Fungal Diseases of the Nervous System

Fungi very rarely invade the central nervous system of healthy people, but patients with cancer, diabetes, and AIDS, as well as those receiving immunosuppressive medications, risk serious disease. Common fungi from the soil and decaying vegetation sometimes infect the nose and sinuses of these patients; from there, they penetrate to the brain and cause the patient's death. These infections are dangerous because they are difficult to treat with antimicrobial medication. Cryptococcal meningoencephalitis differs somewhat from this general pattern, since about half the cases occur in apparently normal people, and many cases are cured by medication.

Cryptococcal Meningoencephalitis

Cryptococcal meningoencephalitis was an uncommon disease until the onset of the AIDS epidemic. Now 2 to 4 cases occur among every 1,000 AIDS patients, as opposed to 0.2 to 0.9 cases per 100,000 in the general population. The disease is among the top four life-threatening infectious complications in AIDS.

Symptoms

In apparently healthy people, symptoms of cryptococcal meningoencephalitis develop very gradually in most cases and generally consist of difficulty in thinking, dizziness, intermittent headache, and slight fever. After weeks or months of slow progression of these symptoms, vomiting, weight loss, stupor, seizures, and impairment of one or more nerves may appear. In people with immunodeficiency, the disease generally progresses much faster; without treatment, death can occur in as little as 2 weeks.

Causative Agent

Cryptococcal meningoencephalitis is an infection of the meninges and brain by the encapsulated yeast form of *Filobasidiella neoformans*, a fungus previously placed in the genus *Cryptococcus.*

In 1975, however, discovery of a sexual form was reported, showing that the fungus was closely related to a group of basidiomycetes pathogenic for plants. Due to this discovery, the organism was renamed *F. neoformans*, but the earlier genus name is still generally used. As can be seen in infected material from patients, the organism is a small, spherical yeast generally 3 to 7 μm in diameter surrounded by a large capsule (**figure 26.21**). ■ **basidiomycetes, p. 309**

Pathogenesis

The *C. neoformans* fungus becomes airborne with dust, enters the body by inhalation, and establishes an infection first in the lung. This infection causes mild or no symptoms in most people and is usually eliminated by body defenses, chiefly lung phagocytes. Phagocytic killing of *C. neoformans* is slow and inefficient in immunocompromised individuals, however. In some cases, the organisms continue to multiply, enter the bloodstream, and are then distributed throughout the body. The capsule is essential to pathogenicity, since nonencapsulated strains do not cause disease. Capsular material inhibits phagocytosis and migration of leukocytes. It also diffuses from the organism and neutralizes opsonins. Progressive infection and dissemination are much more likely to occur when a person's cell-mediated immunity is impaired, particularly in AIDS and certain cancers. Macrophage activation by immune lymphocytes, lacking in many immunodeficiencies, is essential for rapid phagocytic killing of *C. neoformans.* Meningoencephalitis is the most common infection outside of the lung, but organisms spread by the bloodstream can also infect skin, bones, or other body tissues. In meningoencephalitis, the organisms typically cause thickening of the meninges, sometimes impeding the flow of cerebrospinal fluid, thereby increasing the pressure within the brain. They also invade the brain tissue, producing multiple abscesses. ■ **opsonins, p. 377, 394**

Epidemiology

Cryptococcus neoformans is distributed worldwide in soil and vegetation but is especially numerous in soil where pigeon droppings accumulate. For every case of cryptococcal meningoencephalitis, millions of people are infected by the organism without harm. The infection is often the first indication of AIDS. Person-to-person transmission of the disease does not occur.

Figure 26.21 ***Filobasidiella (Cryptococcus) neoformans*** **in the Spinal Fluid of an Individual with Cryptococcal Meningitis** India ink has been added to the fluid to outline the organism's capsule.

TABLE 26.9 Cryptococcal Meningoencephalitis

Symptoms	Headache, vomiting, confusion, and weight loss; fever often absent; symptoms may progress to seizures, paralysis, coma, and death
Incubation period	Widely variable, few to many weeks
Causative agent	*Filobasidiella (Cryptococcus) neoformans*
Pathogenesis	Infection starts in lung; in immunocompromised individuals, encapsulated organisms multiply, enter bloodstream, and are carried to various parts of the body; phagocytosis inhibited and opsonins neutralized; meninges and adjacent brain tissue become infected
Epidemiology	Inhalation of dust containing dried pigeon droppings contaminated with the fungus; most people resistant to the disease
Prevention and treatment	No preventive measures. Treatment: amphotericin B with flucytosine or itraconazole

Prevention and Treatment

There is no vaccine or other preventive measure available. Treatment with the antibiotic amphotericin B is often effective, particularly if given concurrently with flucytosine (5-fluorocytosine) or newer oral antifungal medicines such as itraconazole. Amphotericin B must be given intravenously and the dose carefully regulated to minimize the toxic effects of the antibiotic, mainly against the kidneys. Since amphotericin B does not reliably cross the blood-brain barrier, it is often necessary to administer it through a plastic tube inserted through the skull into a lateral ventricle of the brain. Except in AIDS patients, treatment is successful in about 70% of cases. AIDS patients respond poorly to treatment, most likely because they lack T-cell-dependent killing of *C. neoformans* that normally assists the action of the antifungal medications. Unless their T-cell function can be restored by treatment, AIDS patients are rarely cured of their infection and must remain on antifungal medication for life. ■ **antifungal medicines, p. 514**

The main features of cryptococcal meningoencephalitis are summarized in **table 26.9**.

MICROCHECK 26.4

Fungi rarely invade the CNS of healthy people, but even common soil fungi are a serious threat to individuals with diabetes, cancer, or immunodeficiency. Cryptococcal meningoencephalitis occurs rarely in healthy people, but it is among the leading life-threatening complications in AIDS. The causative organism is a small yeast with a large capsule, often found in soil contaminated with pigeon droppings.

- What underlying condition should one immediately suspect when confronted with a patient with cryptococcal meningoencephalitis?

- Why might it be a good idea for persons with immunodeficiency to avoid soil contaminated with pigeon droppings?
- Why is it so difficult to cure AIDS victims of cryptococcal meningoencephalitis?

Protozoan Diseases of the Central Nervous System

Only a few species of protozoa are important central nervous system pathogens for humans. One interesting example, *Naegleria fowleri*, causes **primary amebic meningoencephalitis** after penetrating the skull along the nerves responsible for the sense of smell. The protozoan occurs worldwide, but less than 200 cases have been reported, almost all rapidly fatal. The disease is usually acquired by swimming in or being splashed with warm fresh water, with or without chlorine, but apparently it can also be contracted by inhaling dust laden with cysts of the protozoa. For every case of *Naegleria* meningoencephalitis, many millions of people are exposed to the organism without harm. *Naegleria fowleri* can exist in three forms: ameba, flagellate, and cyst. It is one of only a few free-living protozoa pathogenic for humans. In contrast, African sleeping sickness is a quite different protozoan disease. ■ protozoa, p. 304

African Sleeping Sickness

African sleeping sickness, also known as **African trypanosomiasis**, is transmitted by its biological vector, the day-biting tsetse fly. The disease is important because it can be contracted by residents and visitors in a wide area across the middle of the African continent. The causative protozoan has an interesting way of protecting itself against the host's immune system.

Symptoms

The first symptoms of African trypanosomiasis appear within a week after a person is bitten by an infected tsetse fly. A tender nodule develops at the site of the bite. The regional lymph nodes may enlarge, but symptoms may all disappear spontaneously. Weeks to several years later, recurrent fevers develop that may continue for months or years. Involvement of the central nervous system is marked by gradual loss of interest in everything, decreased activity, and indifference to food. The eyelids droop, the individuals fall asleep while eating or even standing, the speech becomes slurred, coma develops, and eventually death ensues.

Causative Agent

African sleeping sickness is caused by the flagellated protozoan, *Trypanosoma brucei*. The organisms are slender, with a wavy, undulating membrane and an anteriorly protruding flagellum (**figure 26.22**). There are two subspecies that are morphologically identical, *T. brucei rhodesiense* and *T. brucei gambiense*. The *rhodesiense* subspecies occurs mainly in the cattle-raising areas of East Africa, whereas the *gambiense* subspecies occurs mainly in forested areas of Central and West Africa. Both are transmitted by tsetse flies, a group of biting insects of the genus *Glossina*.

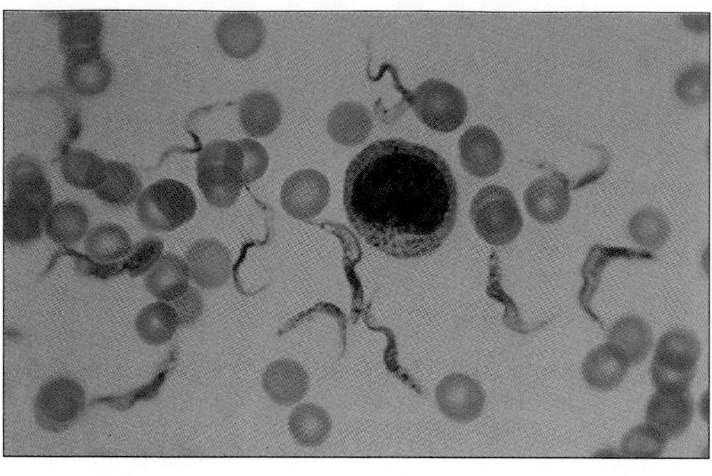

Figure 26.22 ***Trypanosoma brucei*** **in the Blood Smear of an Individual with African Sleeping Sickness (African Trypanosomiasis)**

Pathogenesis

During the bite of an infected tsetse fly, the protozoan enters the bite wound in the fly's saliva. The organism multiplies at the skin site and within a few weeks enters the lymphatics and blood circulation. The patient responds with fever and production of IgM antibody against the protozoa, and symptoms improve. Within about a week and at roughly weekly intervals thereafter, however, there are recurrent increases in the number of parasites in the blood. Each of these bursts of increased **parasitemia**, meaning parasites in the circulating blood, coincides with the appearance of a new glycoprotein on the surface of the trypanosomes. More than a thousand genes, each coding for a different surface glycoprotein, are present in the protozoan chromosome. Only one of these genes is activated at a time, and the patient's immune system must respond to each gene product with production of a new antibody. The recurrent cycles of parasitemia and antibody production continue until the patient is treated or dies.

In *T. brucei rhodesiense* infections, the disease tends to progress rapidly, with the heart and brain invaded within 6 weeks of infection. Irritability, personality changes, and mental dullness result from brain involvement, but the patient usually dies from heart failure within 6 months. With *T. brucei gambiense*, progression of infection is much slower, and years may pass before death occurs, often from secondary infection. Much of the damage to the host is due to immune complexes formed when antibody reacts with complement and high levels of protozoan antigen. ■ immune complexes, p. 440

Epidemiology

African sleeping sickness occurs on the African continent within about 15° of the equator, with 10,000 to 20,000 new cases each year, including a number of American tourists. The occurrence of the disease is determined by the distribution of the tsetse fly vectors. The main reservoirs of the severe Rhodesian form of the disease are wild animals; for the milder Gambian form, humans are the main reservoir, and human-to-human transmission is

TABLE 26.10 African Sleeping Sickness

Symptoms	Fever, rash, enlargement of lymph nodes, liver, and spleen; later, involvement of the central nervous system, uncontrollable sleepiness, headache, poor concentration, unsteadiness, coma, death
Incubation period	3 weeks
Causative agent	*Trypanosoma brucei*, a flagellated protozoan
Pathogenesis	The protozoa multiply at site of a tsetse fly bite, then enter blood and lymphatic circulation; as new cycles of parasites are released, their surface protein changes and the body is required to respond with a new antibody
Epidemiology	Bites of infected tsetse flies transmit the trypanosomes through fly saliva; wild animal reservoir for *T. brucei rhodesiense*
Prevention and treatment	Protective clothing, insecticides, clearing of brush where flies breed. Treatment: suramin; when central nervous system is involved, melarsoprol or eflornithine

more common than animal-to-human. Bites of the infected tsetse fly transmit the disease, but less than 5% of the flies are infected. ■ reservoir, p. 476

Prevention and Treatment

Preventive measures directed against tsetse fly vectors include insect repellents and protective clothing to prevent bites, traps containing bait and insecticides, and clearing of brush that provides breeding habitats for the flies. Populations can be screened for *T. brucei* infection by examining blood specimens. Treatment of infected people helps reduce the protozoan's reservoir. A single intramuscular injection of the medication pentamidine prevents the Gambian form of the illness for a number of months, although it will not necessarily prevent an infection that could progress at a later time. As with other eukaryotic pathogens, treatment is problematic because of toxic side effects of the available medications. Suramin can be used if the disease has not progressed to involve the central nervous system; melarsoprol and eflornithine cross the blood-brain barrier and can be used when the central nervous system is involved. ■ antiprotozoal medications, p. 516

The main characteristics of African sleeping sickness are presented in **table 26.10**.

MICROCHECK 26.5

Only on rare occasions can free-living amebas such as *Naegleria fowleri* cause meningoencephalitis in human beings. Residents and visitors to a wide swath of tropical Africa, however, are at risk of contracting African sleeping sickness, caused by the flagellated protozoan parasite, *Trypanosoma brucei*, and transmitted by a biting insect, the tsetse fly. These protozoa can circulate in the bloodstream for extended periods by changing their surface antigens to escape the host's antibodies. Eventually they are able to penetrate the CNS, causing indifference, sleepiness, coma, and death.

- How likely is it that a person who swims in warm fresh water will contract primary amebic meningoencephalitis?
- How can one explain repeated abrupt increases in *T. brucei* in the blood of African sleeping sickness victims?

FUTURE CHALLENGES

Understanding Transmissible Spongiform Encephalopathies

A mysterious group of chronic degenerative brain diseases involves wild and domestic animals and humans. Microscopically, brain tissue affected by these diseases has a spongy appearance due to the loss of nerve cells and other changes—this is why these diseases are referred to as spongiform encephalopathies. Brain and other tissues from affected animals can transmit the disease to normal animals of the same, and sometimes different, species. The incubation period is often long, measured in months, years, or decades. Some of these encephalopathies can be transmitted to small laboratory animals such as mice, allowing scientific scrutiny. Cow-to-human transmission may possibly occur on rare occasions following consumption of flesh from cattle with bovine spongiform encephalopathy, "mad cow disease," but it is curious that the disease in sheep, ***scrapie****, has apparently not been transmitted to people. Human-to-human transfer of one type of spongiform encephalopathy called Creutzfeldt-Jakob disease has resulted from corneal transplants and via contaminated surgical instruments. Another human spongiform encephalopathy, kuru, was probably transmitted by cannibalism, which was formerly practiced by New Guinea natives. The main signs and symptoms of spongiform encephalopathies are dementia and unsteadiness.* ■ "mad cow disease," p. 364

In 1997, Dr. Stanley Prusiner received a Nobel Prize for his work on the causative agents of these diseases, now widely known as proteinaceous infectious particles, or ***prions****. Prions may prove to be a new class of infectious agents that differ from bacteria, viruses, and viroids. Their nature is still controversial, but present evidence indicates that they are unusual proteins, devoid of nucleic acid and more resistant to proteases than normal proteins are. Prions have an amino acid sequence identical to a normal brain protein but are folded differently. The normal protein contains helical segments, whereas the corresponding segments of the prion protein are pleated sheets.* ■ prions, p. 13, 364 ■ protein folding, p. 30, 40

The main characteristics of prions are are as follows:

- *They increase in quantity during the incubation period of the disease.*
- *They resist inactivation by ultraviolet and ionizing radiation.*
- *They resist inactivation by formaldehyde and heat.*
- *They are not readily destroyed by proteases.*
- *They are not destroyed by nucleases.*
- *They are much smaller than the smallest virus.*
- *They are composed of protein coded by a normal cellular gene but modified after transcription.*

Elegant scientific studies are slowly expanding knowledge of the pathogenesis of these diseases. There are no inflammatory or immune responses to the infections. Replication of prions depends on the presence of the corresponding normal cellular protein. Mice that lack the ability to produce the normal protein are completely resistant to the disease, but they become fully susceptible to it if the gene for the normal protein is introduced into their body. This is strong support for the idea that prions replicate by converting the normal protein into copies of themselves, but it is not known what additional molecules might be involved in the process. In the normal course of infection, replication first occurs in the spleen and other parts of the body, probably in dendritic cells. The prions are then transported to the central nervous system via the nerve axons, by B lymphocytes, and probably by other means. There they aggregate in insoluble masses in the nerve cells, causing malfunction and nerve cell death. But not all prions act the same. They differ in host range, incubation period, and the areas of the nervous system that are attacked.

One potentially important development occurred in 1994 with the first description of a prionlike protein in yeast. This protein, known as sup35, is a normal subunit of a factor responsible for causing ribosomes to terminate translation at nonsense codons. Under certain conditions, sup35 changes form and aggregates into clumps of material that convert all newly produced sup35 to copies of the same aggregating form. The normal translating function of sup35 then ceases. Other yeast prionlike proteins are now known, and these findings are making it easier to study the natural history of prions.

Since there is no treatment for the spongiform encephalopathies, and as shown by the "mad cow episode," they remain a potential threat to humans and other animals, the challenge is to understand better the etiology and pathogenesis of these diseases, and devise methods for their prevention and treatment.

SUMMARY

Anatomy and Physiology (Figure 26.1)

1. The brain and spinal cord make up the central nervous system (CNS); the peripheral nervous system is composed of **motor nerves** and **sensory nerves**.
2. **Cerebrospinal fluid** is produced by structures in cavities inside the brain and flows out over the brain and spinal cord. (Figure 26.2)
3. **Meninges** are the membranes that cover the surface of the brain and spinal cord.

Pathways to the Central Nervous System

1. Infectious agents can reach the CNS by way of the bloodstream, when they penetrate the **blood-brain barrier**; via the cytoplasm of nerve cell **axons**; and by direct extension through bone.

Bacterial Nervous System Infections

1. Bacteria can infect the brain, spinal cord, and peripheral nerves, but more commonly they infect the meninges and cerebrospinal fluid, causing **meningitis**.
2. Bacterial meningitis is uncommon; formerly most victims were children, but childhood immunization has decreased the incidence in these age groups. (Figures 26.5, 26.8)
3. In most but not all of the victims, the causative bacterium is one commonly found among the normal upper respiratory flora of healthy people.
4. *Haemophilus influenzae*, a tiny pleomorphic Gram-negative rod, once the leading cause of childhood bacterial meningitis, is now mostly controlled by a vaccine. (Figure 26.3)

Meningococcal Meningitis (Table 26.1)

1. Greatly feared because it can result in shock and death, meningococcal meningitis can occur in both childhood and adult epidemics.
2. Symptoms are similar to other forms of meningitis: cold symptoms followed by abrupt onset of fever, severe headache, pain and stiffness of the neck and back, nausea, and vomiting. Small hemorrhages into the skin (Figure 26.6), deafness, and coma can occur.
3. The bacteria, *Neisseria meningitidis,* cause a massive PMN response; metabolic activity of the leukocytes and bacteria consumes glucose normally present in the cerebrospinal fluid; shock results from the release of endotoxin into the bloodstream. (Figure 26.7)

Listeriosis (Table 26.2)

1. Listeriosis is caused by *Listeria monocytogenes,* a non-spore-forming Gram-positive rod; it is a foodborne illness often manifest as meningitis in newborn infants and others.
2. The bacterium is widespread, commonly contaminates foods such as unpasteurized milk, cold cuts, and soft cheeses, and can grow in the refrigerator.
3. The bacteria readily penetrate the intestinal epithelium, enter the bloodstream, and infect the meninges.

Hansen's Disease (Leprosy) (Table 26.3, Figure 26.9)

1. Hansen's disease is characterized by invasion of peripheral nerves by the acid-fast bacillus *Mycobacterium leprae,* which has not been cultivated *in vitro.* (Figures 26.10, 26.11)
2. The disease occurs in two main forms, **tuberculoid** and **lepromatous**, depending on the immune status of the patient.

Botulism (Table 26.4)

1. Botulism is not a nervous system infection, but an often fatal type of food poisoning that causes severe generalized paralysis.
2. The causative bacterium, *Clostridium botulinum,* is an anaerobic Gram-positive rod that forms heat-resistant spores (Figure 26.12). Spores that survive canning or other heat treatment of foods germinate, and the bacteria multiply, releasing a powerful toxin into the food.
3. Because of strict controls on food processing, intestinal botulism is now the most common form of the disease in the United States. Relatively mild symptoms result from *C. botulinum* colonization of the intestine of infants and some adults.
4. Wound botulism, caused when *C. botulinum* colonizes dirty wounds containing dead tissue, is rare.

Viral Diseases of the Nervous System

1. Most viral nervous system infections are caused by human enteroviruses or by the viruses of certain zoonoses.
2. Many common viruses of human beings can occasionally infect the nervous system, including the ones that cause infectious mononucleosis, mumps, measles, chickenpox, and herpes simplex ("cold sores," genital herpes).

Viral Meningitis (Table 26.5)

1. Viral meningitis is much more common than bacterial meningitis. It is generally a mild disease for which there is no specific treatment.

Viral Encephalitis (Table 26.6, Figure 26.14)

1. Viral encephalitis has a high fatality rate and often leaves survivors with permanent disabilities. It can be sporadic or epidemic.
2. Herpes simplex virus is the most important cause of sporadic encephalitis; epidemic encephalitis is usually caused by **arboviruses**.
3. LaCrosse encephalitis virus, maintained in *Aedes* mosquitoes and squirrels and chipmunks, is usually the most frequently reported. (Figure 26.13)

Infantile Paralysis, Polio (Poliomyelitis) (Table 26.7, Figure 26.18)

1. Poliomyelitis is the focus of an international effort to certify the earth free of the disease by the year 2005.
2. Destruction of motor nerve cells of the brain and spinal cord leads to paralysis, muscle wasting, and failure of normal bone development. (Figures 26.15, 26.17)
3. **Postpolio syndrome** occurs years after poliomyelitis, and it is probably caused by the death of nerve cells that had taken over for the ones killed by poliomyelitis virus.

Rabies (Table 26.8)

1. Rabies is a widespread zoonosis transmitted to humans mainly through the bite of an infected animal. (Figure 26.19)
2. Once symptoms appear in an infected human being, the disease is almost uniformly fatal.
3. Because of the long incubation period, prompt immunization with inactivated vaccine begun after a rabid animal bite is effective in preventing the disease. Passive immunization given at the same time increases the protection.

Fungal Diseases of the Central Nervous System

1. Fungi rarely invade the nervous system of healthy people, but they can be a threat to the life of individuals with underlying diseases such as diabetes, cancer, and immunodeficiency.
2. Treatment of these infections is usually very difficult.

Cryptococcal Meningoencephalitis (Table 26.9)

1. Infection originates in the lung after a person inhales dust laden with *Filobasidiella* (*Cryptococcus*) *neoformans*, an encapsulated yeast, which resists phagocytosis because of its large capsule. (Figure 26.21)
2. The organism is associated with soil contaminated with pigeon droppings.

Protozoan Diseases of the Central Nervous System

1. Only a few free-living protozoa infect the human nervous system. Certain parasitic protozoa are a much more widespread threat.

African Sleeping Sickness (Table 26.10)

1. African sleeping sickness is a major health problem in a wide area across equatorial Africa, in its late stages marked by indifference, sleepiness, coma, and death.
2. The disease is caused by *Trypanosoma brucei* (Figure 26.22), a flagellated protozoan that lives in the blood of its victims and in its biological vector, the tsetse fly.
3. During infection, the organism shows bursts of growth, each appearing with different surface proteins. The organism has more than a thousand genes coding for these antigens, but only expresses one at a time.
4. Each antigen requires that the body respond with a new antibody.

REVIEW QUESTIONS

Short Answer

1. Name two causes of bacterial meningitis in newborn infants.
2. How can the incidence of meningitis due to Lancefield group B streptococci be reduced?
3. Name and describe the organism that is the leading cause of bacterial meningitis in adults.
4. Of the different antigenic types of *Haemophilus influenzae*, which one is most likely to cause invasive disease?
5. What causes petechiae?
6. What cell wall component of *Neisseria meningitidis* is probably responsible for the shock and death that sometimes occur with infections by this bacterium?
7. Describe the Shwartzman phenomenon.
8. What measures can be undertaken to prevent meningococcal meningitis?
9. What is the usual route of entry for the causative agent in epidemic listeriosis?
10. Why is listeriosis so important to pregnant women even though it usually causes them few symptoms?
11. Give two characteristics of the causative agent of leprosy.
12. Describe the differences between tuberculoid and lepromatous leprosy.
13. How is Hansen's disease transmitted?
14. What is the difference between sporadic encephalitis and epidemic encephalitis? Name one cause of each.
15. What is the least number of different antisera required for treating humans with botulism?
16. Give two ways in which viral meningitis usually differs from bacterial meningitis.
17. Give an example of an enterovirus.
18. What are arboviruses? Give an example.
19. Why does poliomyelitis virus attack motor but not sensory neurons?
20. Can a dog infected with rabies virus transmit the disease while appearing well? Explain.

Multiple Choice

1. Which is the best way to prevent meningococcal meningitis in individuals intimately exposed to the disease?
   A. Vaccinate them against *Neisseria meningitidis.*
   B. Treat them with the antibiotic rifampin.
   C. Culture their throat and hospitalize them for observation.
   D. Withdraw a sample of spinal fluid and begin antibacterial treatment if the cell count is high and the glucose is low.
   E. Have them return to their usual activities, but seek medical evaluation if symptoms of meningitis occur.
2. Which of these statements concerning the causative agent of listeriosis is *false*?
   A. It can cause meningitis during the first month of life.
   B. It is a Gram-positive rod that can grow in refrigerated food.
   C. It is usually transmitted by the respiratory route.
   D. Infection commonly results in bacteremia.
   E. It is widespread in natural waters and vegetation.
3. Of the following statements about Hansen's disease, which one is *false*?
   A. It was once common in the United States.
   B. An early symptom is loss of sensation, sweating, and hair in a localized patch of skin.
   C. The incubation period is usually less than 1 month.
   D. Treatment should include more than one antimicrobial medication given at the same time.
   E. The form the disease takes depends on the immune status of the victim.
4. Pick the one *false* statement about botulism.
   A. It is not a central nervous system infection.
   B. Pathogenicity of some strains of the causative agent is the result of lysogenic conversion.
   C. Food can taste normal but still cause botulism.
   D. Treatment is based on choosing the correct antibiotic.
   E. Control of the disease depends largely on proper food-canning techniques.
5. Which of the following statements about enteroviral meningitis is *true*?
   A. Vaccines are generally available to protect against the disease.
   B. The main symptom is muscle paralysis.
   C. Transmission is usually by the fecal-oral route.
   D. The causative agents do not survive well in the environment.
   E. Recovery is rarely complete.
6. Choose the one *false* statement about arboviral encephalitis.
   A. It is likely to occur in epidemics.
   B. Mosquitoes can be an important vector.
   C. Epilepsy, paralysis, and thinking difficulties are among the possible sequels to the disease.
   D. Use of sentinel chickens helps warn about the disease.
   E. In the United States, the disease is primarily a zoonosis involving cattle.
7. Which one of the following statements about poliomyelitis is *false*?
   A. The nerves of sensation are usually involved.
   B. It can be caused by any of three specific enteroviruses.
   C. Only a small fraction of those infected will develop the disease.
   D. The disease is transmitted via the fecal-oral route.
   E. A postpolio syndrome can develop years after recovery from the original illness.
8. Choose the one *false* statement about rabies.
   A. It is widespread in wildlife.
   B. Vaccines are available to protect humans and domestic animals.
   C. Hydrophobia is a characteristic symptom.
   D. Animals that do not bite do not get the disease.
   E. The incubation period is often a month or more.
9. Which statement concerning cryptococcal meningoencephalitis is *true*?
   A. It is caused by a yeast with a large capsule.
   B. It is a disease of pigeons transmissible to humans.
   C. Typically it attacks the meninges, but spares the brain.
   D. Person-to-person transmission commonly occurs.
   E. Cell-mediated immunity is of little importance in the body's struggle against the disease.
10. Which of the following statements about African sleeping sickness is *true*?
    A. It is transmitted by a species of nighttime-biting mosquito.
    B. It is a threat to visitors to tropical Africa.
    C. The onset of sleepiness is usually within 2 weeks of contracting the disease.
    D. It is caused by free-living protozoa.
    E. Distribution of the disease is determined mainly by the distribution of standing water.

Applications

1. An outbreak of viral meningitis in a small eastern city was linked epidemiologically to swimming in a nonchlorinated pool in an abandoned quarry outside of town. What might public health officials surmise about the probable cause of the outbreak?
2. A group called Concerned Moms of Young Children is petitioning the local health department after a child died of rabies in the area. They argued that health officials were not doing enough to rid the area of rabies. If you represented the health department, how would you respond?
3. Two microbiologists are trying to write a textbook, but they cannot agree where to place botulism. One favored the chapter on nervous system infections, while the other insisted on the chapter covering alimentary system infections. Which microbiologist is correct?

Critical Thinking

1. A pathologist stated that it was much easier to determine the causative agent of meningitis than of an infection of the skin or intestine. Is her statement valid? Why or why not?
2. An epidemiological study concludes that armadillos are not an important source of human leprosy. What findings would lead to that conclusion?
3. Why is it important to learn about rabies when only a few cases occur in the entire United States each year?

Wound Infections

Clostridium tetani, *the cause of lockjaw (tetanus), is found in soil and dust—thus, virtually everywhere. Before its cause and pathogenesis were understood, the disease was widespread, often ending in agonizingly painful death. Dr. Shibasaburo Kitasato (1856–1931), working in Robert Koch's laboratory in Germany, was the first to discover how to cultivate* C. tetani, *and subsequently he made a startling discovery that paved the way for control of the disease. Kitasato was born in a mountainous village in southern Japan in 1856. He was sent by his family to medical schools in Kumamoto and Tokyo. Following his graduation in 1883, Kitasato went to work for the Central Hygienic Bureau in Japan. The Japanese government needed to control epidemics of typhoid, cholera, and blackleg, a clostridial disease of calves, and so in 1885, Kitasato was sent to Germany to study with the famous Dr. Koch.*

In Koch's laboratory, Kitasato was given the tetanus problem to study. In the course of his experiments, he discovered that C. tetani *would only grow under strictly anaerobic conditions. Once he had isolated the organism in pure culture, he was able to show that it produced tetanus in laboratory animals. He was puzzled, however, by a surprise finding: although the animals died of generalized disease, there was no* C. tetani *anyplace other than the site where the organism had been injected. By doing experiments in which he injected the tails of mice and then removed the injected tissue at hourly intervals after injection, he showed that the animals only developed tetanus if the organisms were allowed to remain for more than an hour. He further showed that the organisms remained at the site of inoculation; at no time were they found in the rest of the body. Kitasato reasoned that something other than bacterial invasion was causing the disease.*

About this time, another investigator, Emil von Behring, was busy investigating how Corynebacterium diphtheriae *caused the disease diphtheria. Together, Kitasato and von Behring were able to show that both diseases were caused by poisonous substances, toxins, that were produced by the bacteria. The concept that a bacterial toxin could cause disease in the absence of bacterial invasion was an extremely important advance in the understanding and control of infectious diseases.*

Kitasato published his studies in 1890. In 1892, even though urged to stay in Germany, he returned to Japan. The Japanese government was not prepared to support basic research at that time, and so Kitasato established his own institute for infectious diseases where he worked and trained Japanese scientists for the rest of his life. In 1908, Koch paid a visit to Kitasato in Japan and a Shinto shrine was built in Koch's honor. During his later years, Kitasato was instrumental in establishing laws regulating health practices in Japan. He died in 1931 at age 75. To honor him, a shrine was erected next to Koch's.

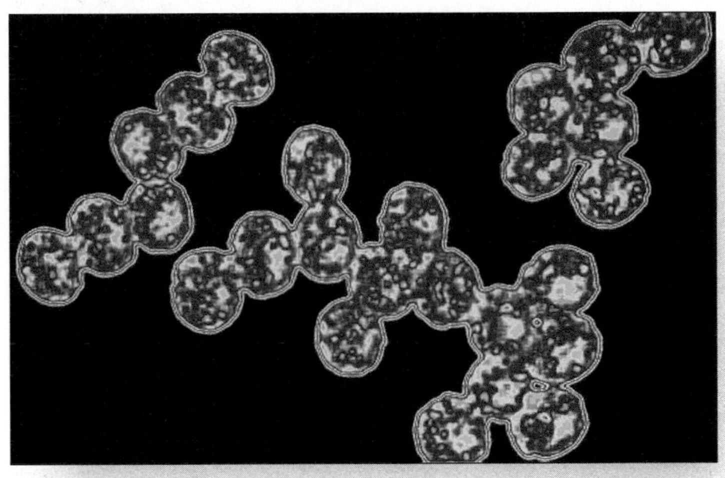

—A Glimpse of History

MOST PEOPLE OCCASIONALLY SUSTAIN WOUNDS that produce breaks in the skin or mucous membranes. Almost always, microorganisms contaminate these wounds from the air, fingers, normal flora, or the object causing the wound. Whether these microorganisms cause disease depends on (1) how virulent they are, (2) how many there are, (3) the status of host defenses, and (4) the nature of the wound, especially whether it contains crushed tissue or foreign material.

Wounds that contain materials such as dirt, leaves, bits of rubber, and cloth usually become infected and do not heal until the foreign material is removed. Often such a wound provides places for microorganisms to multiply and produce injurious substances, out of the reach of phagocytes and other body defenses. Foreign materials may also provide surfaces for biofilms, or the materials reduce available oxygen, thereby impairing phagocytic function and allowing growth of anaerobic pathogens. Clean wounds often heal uneventfully despite microbial colonization, but sometimes even a trivial wound can result in a severe, or even fatal, infection by providing an entryway for infection or microbial toxins to spread throughout the body.

Wounds can be classified as follows:

- Incised, as when produced by a knife or other sharp object, as in surgery
- Puncture, from penetration of a small sharp object, as when an individual steps on a nail
- Lacerated, when the tissue is torn
- Contused, as when caused by a blow that crushes tissue
- Burns

Burns represent an important category of accidental wounds. In the United States each year, more than 2.5 million people sustain thermal burns severe enough for them to seek medical attention. Of these, more than 100,000 require hospitalization, and 12,000 die from their burns. Burns present special problems. Although initially sterile, thermal burns are often extensive, with a large area of weeping devitalized tissue, representing an enormous and nutritious feast for microorganisms.

Intentional wounds inflicted by surgery represent about 23 million cases annually in the United States. From 1% to 9% of them become infected; the percentages vary according to the type of surgery and the status of the patients' host defenses. Costs related to postoperative wound infections amount to approximately $1.5 billion each year, more than half of the total expense for nosocomial infections. They result in about 13,000 deaths each year. ■ **nosocomial infections, p. 488**

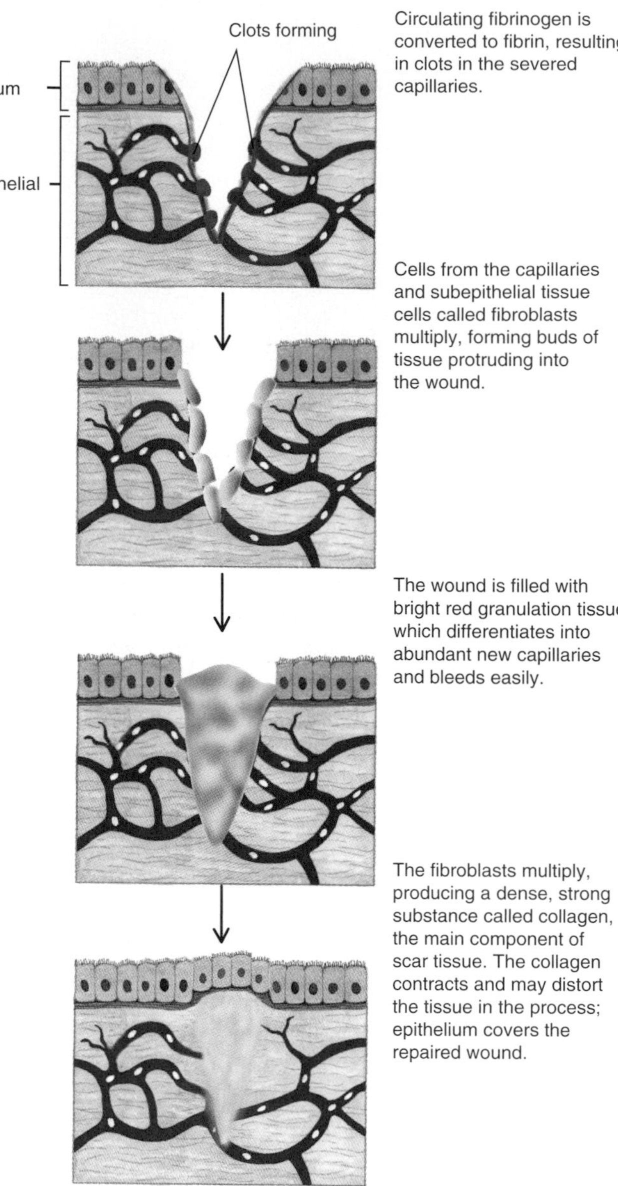

Clots forming
Epithelium
Subepithelial tissue
Circulating fibrinogen is converted to fibrin, resulting in clots in the severed capillaries.
Cells from the capillaries and subepithelial tissue cells called fibroblasts multiply, forming buds of tissue protruding into the wound.
The wound is filled with bright red granulation tissue, which differentiates into abundant new capillaries and bleeds easily.
The fibroblasts multiply, producing a dense, strong substance called collagen, the main component of scar tissue. The collagen contracts and may distort the tissue in the process; epithelium covers the repaired wound.

Figure 27.1 The Process of Wound Repair

Anatomy and Physiology

Wounds expose components of tissue normally protected from the outside world by skin or mucous membranes. These components, including **collagen**, **fibronectin**, **fibrin**, and **fibrinogen**, provide receptors to which potential pathogens specifically attach. Collagen is a fibrous material, the main supportive protein of skin, tendons, scars, and other body structures. Fibronectin is a glycoprotein that occurs both as a circulating form and as a component of tissue, where it ties cells and other tissue substances together. Shortly after the occurrence of a wound, the soluble blood protein fibrinogen is converted to the fibrous material fibrin, thereby forming clots in the damaged vessels. This stops the flow of blood as the first step in the repair process.

Wound healing begins with the outgrowth of connective tissue cells, called fibroblasts, and capillaries from the surfaces of the wound, producing a nodular, red, translucent material called **granulation tissue**. In the absence of dirt or infection, granulation tissue fills the void created by the wound, contracts, and is converted to collagen that composes the scar tissue eventually covered by overlying skin or mucous membrane (**figure 27.1**). Sometimes in the presence of dirt or infection, the granulation tissue overgrows, bulging from the wound to form a **pyogenic granuloma**.

Wound Abscesses

An **abscess** (**figure 27.2**) is a localized collection of pus, composed of living and dead leukocytes, components of tissue breakdown, and infecting organisms surrounded by body tissue. There are no blood vessels in abscesses, because they have been destroyed or pushed aside. A surrounding area of inflammation and clots in adjacent blood vessels separate the abscess from normal tissue. Consequently, abscess formation helps to localize an infection and prevent its spread. Microorganisms in abscesses often are not killed by antimicrobial medications because the microorganisms cease multiplying, and active multiplication is generally required for microbial killing by the medications. In addition, the chemical nature of pus interferes with the action of some antibiotics, and many antimicrobials diffuse poorly into abscesses because of the absence of blood vessels. Microorganisms in abscesses are a potential source of infection of other

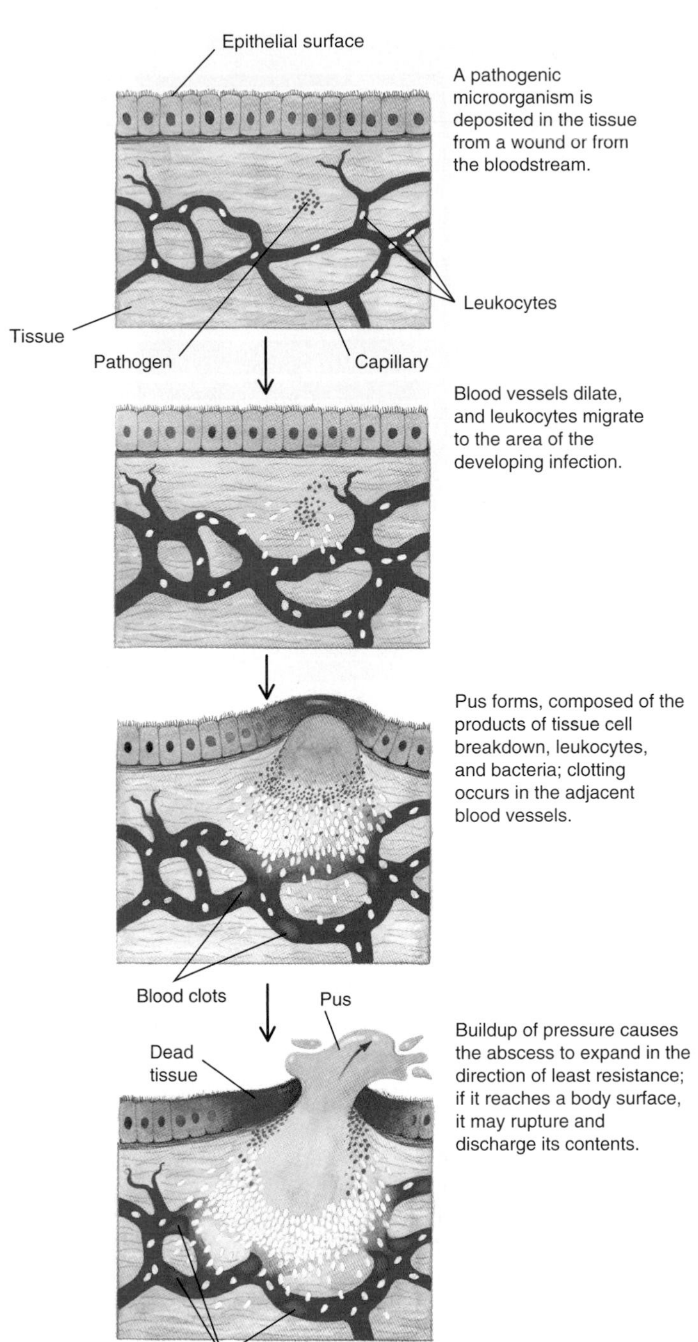

Epithelial surface
A pathogenic microorganism is deposited in the tissue from a wound or from the bloodstream.
Leukocytes
Tissue
Pathogen
Capillary
Blood vessels dilate, and leukocytes migrate to the area of the developing infection.
Pus forms, composed of the products of tissue cell breakdown, leukocytes, and bacteria; clotting occurs in the adjacent blood vessels.
Blood clots
Pus
Dead tissue
Buildup of pressure causes the abscess to expand in the direction of least resistance; if it reaches a body surface, it may rupture and discharge its contents.
Blood clots

Figure 27.2 Abscess Formation

parts of the body if they escape the surrounding area of inflammation and enter the blood or lymph vessels. Generally, abscesses must burst to a body surface or be drained surgically in order to effect a cure. ■ antimicrobial medications, p. 495

Anaerobic Wounds

Another important feature of many wounds is that they are relatively anaerobic, thus allowing colonization by dangerous anaerobic pathogens such as *Clostridium tetani*. Anaerobic conditions are especially likely in dirty wounds, wounds with crushed tissue, and puncture wounds. Puncture wounds caused by nails, thorns, splinters, and other sharp objects can introduce foreign material and microorganisms deep into the body. Bullets and other projectiles can carry fragments of skin or cloth contaminated with microorganisms into the tissues. Projectiles, although causing relatively small breaks in the skin, often produce extensive tissue damage because of the force with which they enter.

MICROCHECK 27.1

Wounds expose components of tissue to which pathogens specifically attach. Wounds can be classified as incised, punctured, lacerated, contused, or burns. Healing involves the outgrowth of fibroblasts and capillaries from the sides of the wound to produce granulation tissue that fills the defect. Abscess formation provides a way of isolating infections enclosed by tissue. Anaerobic conditions in wounds are created by the presence of dead tissue and foreign material.

- Name two substances in wounds to which pathogens specifically attach.
- Give two reasons that an abscessed wound might not respond to antibiotic treatment.
- What kinds of wounds are subject to infection by anaerobic bacteria?
- What is the advantage of the contraction of granulation tissue?

Common Bacterial Wound Infections

The possible consequences of wound infections include (1) delayed healing, (2) formation of abscesses, and (3) extension of the bacteria or their products into adjacent tissues or the bloodstream. Surgical wounds often split open because swelling causes the stitches to pull through tissues softened by the infection. Also, the infection can extend to involve devices such as an artificial hip, often necessitating removal of the device pending control of the infection.

The following sections present some aspects of wound infections caused by staphylococci, streptococci, and *Pseudomonas* sp.

Staphylococcal Wound Infections

Staphylococci lead the causes of wound infections, both surgical (**figure 27.3**) and accidental. Staphylococci are commonly present in the nostrils or on the skin. Of the 30 or more recognized species of staphylococci, only two account for most human wound infections.

Symptoms

Staphylococci are **pyogenic**, meaning that they characteristically cause the production of a purulent discharge, otherwise known as pus. They usually cause an inflammatory reaction, with swelling, redness, and pain. If the infected area is extensive or if the infection has spread to the general circulation, fever is a prominent symptom. Wound infections by some strains produce

Figure 27.3 Surgical Wound Infection Due to *Staphylococcus aureus*
The stitches pull through the infected tissue, causing the wound to open.

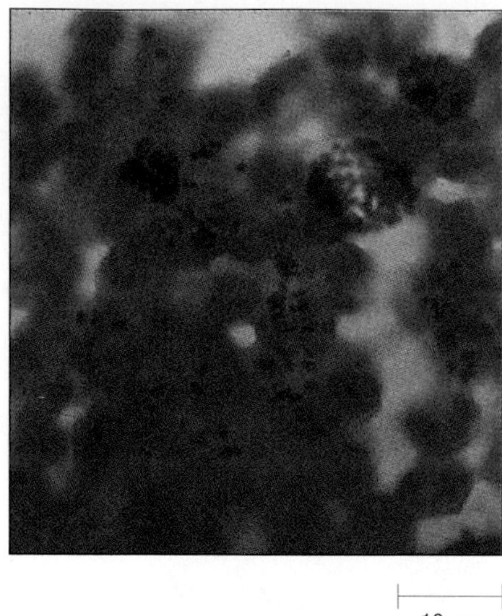

Figure 27.4 *Staphylococcus aureus* in Pus The dark-colored dots are staphylococci; the red objects, leukocytes.

the toxic shock syndrome, with high fever, muscle aches, and life-threatening shock, sometimes accompanied by a rash and diarrhea. ■ **toxic shock syndrome, p. 641**

Causative Agents

Staphylococci are Gram-positive cocci that arrange themselves in clusters (**figure 27.4**). They grow readily aerobically or anaerobically and are salt tolerant, probably because they evolved on the skin where evaporation concentrates the salt in sweat. The most important species are *S. aureus* and *S. epidermidis.* Both species survive well in the environment, making it easy for them to transfer from one person to another.

Staphylococcus aureus One of the most useful identifying characteristics of *S. aureus* is that it produces **coagulase**. Despite its *-ase* ending, coagulase is not an enzyme. This protein product of *S. aureus* is largely extracellular, meaning it is released from the bacterium. It reacts with a substance in blood called prothrombin. The resulting complex, called staphylothrombin, causes blood to clot by converting fibrinogen to fibrin. Conclusive evidence is lacking that coagulase is a virulence factor for *S. aureus.* Some coagulase is tightly bound to the surface of the bacteria, however, and coats their surface with fibrin upon contact with blood. Fibrin-coated staphylococci resist phagocytosis.

Generally, *S. aureus* also possesses **clumping factor**, often called "slide coagulase," because it causes a suspension of the bacteria to clump together when mixed with a drop of blood plasma on a microscope slide. The clumping is caused by a cell-fixed protein that attaches specifically to fibrinogen in the plasma. Clumping factor protein is a virulence factor for *S. aureus* because it attaches to fibrinogen and fibrin present in wounds, thus aiding colonization of wound surfaces. Plastic devices, such as intravenous catheters and heart valves, become coated with fibrinogen shortly after insertion, thus making them, too, a target for colonization. The gene for clumping factor is distinct from the one controlling coagulase.

Other virulence factors possessed by *S. aureus* that aid colonization of wounds include binding proteins for fibronectin, fibrin, fibrinogen, and collagen. **Protein A**, another component of the bacterial surface, also aids virulence. It binds IgG by the Fc portion of the immunoglobulin molecule. Thus, not only does protein A compete with the Fc receptors of phagocytes for antibody molecules, it also coats staphylococci with host protein and therefore hides them from phagocytes and the cells responsible for immunity. Most strains of *S. aureus* also produce **α-toxin**, which kills cells by attaching to specific receptors on host cell membranes and making holes in them. A relatively small percentage of *S. aureus* strains produce one or more additional toxins. ■ **IgG, p. 394** ■ **Fc, p. 394**

Table 27.1 summarizes properties of *S. aureus* that contribute to its virulence. The bacterium also produces a variety of extracellular enzymes that probably do not contribute directly to virulence but supply it with nutrients by degrading blood cells and components of damaged tissue. ■ ***Staphylococcus aureus*, p. 525, 527, 641**

Staphylococcus epidermidis Most strains of *S. epidermidis* have little or no invasive ability for normal people. They commonly cause small abscesses of little consequence around the stitches used in surgery. Most strains bind fibronectin, however, and can therefore colonize the plastic intravenous catheters, heart valves, and other devices employed in modern medicine. Following adherence, the bacteria may produce a kind of slime or glycocalyx that cements the growing colony to the plastic in a biofilm, protecting it from attack by phagocytes and other host defense mechanisms, and antibacterial medications. ■ **glycocalyx, p. 66,** ■ **biofilm, p. 108**

TABLE 27.1 Properties of *Staphylococcus aureus* Implicated in Its Virulence

Virulence Factor	Action Site	Action
Clumping factor	Bacterial surface	Attaches bacterium to fibrin, fibrinogen, plastic devices
Fibronectin-binding protein	Bacterial surface	Attaches bacterium to acellular tissue substance, endothelium, epithelium, clots, indwelling plastic devices
Protein A	Bacterial surface, extracellular	Competes with Fc receptors of phagocytes; coats bacterium with host's immunoglobulin
α-toxin	Extracellular	Makes holes in host cell membranes
Leukocidin	Extracellular	Kills neutrophils or causes them to release their enzymes
Enterotoxins	Extracellular	Superantigens. If systemic, cause toxic shock; cause food poisoning if ingested
Toxic shock syndrome toxin-1	Extracellular	A superantigen. If systemic, causes toxic shock

Pathogenesis

Staphylococcus aureus Many years ago, it was shown that it takes more than 100,000 staphylococci injected into the skin to produce a small abscess. When injected along with a suture, however, only about 100 *S. aureus* are required to produce the same lesion. These studies, which used students as guinea pigs, dramatized the effect of foreign material in the pathogenesis of staphylococcal infections.

Studies cloning the genes controlling suspected staphylococcal virulence factors and then deleting or reinserting the genes have indicated that multiple virulence factors act together to produce the usual wound infection. Clumping factor and the other binding proteins attach the organisms to clots and tissue components, fostering colonization. Clumping factor, coagulase, and protein A serve to coat the organisms with host proteins, giving them a disguise that hides them from attack by phagocytes and the immune system. This may explain why immunity to staphylococcal infection is generally weak or nonexistent. Some protein A is released from the bacterial surface and reacts with circulating immunoglobulin. The resulting complexes activate complement and probably contribute to the intense inflammatory response and accumulation of pus. Colonization of plastics and other foreign materials occurs because they quickly become coated with fibrinogen and fibronectin to which the staphylococci attach. Systemic spread of wound infections can lead to abscesses in other tissues, such as the heart and joints. Staphylococcal toxins that enter the circulation act as superantigens, causing the widespread release of cytokines, thereby producing toxic shock. ■ **superantigens, p. 464** ■ **cytokines, p. 379** ■ **complement, p. 377**

Staphylococcus epidermidis Wound infections by *S. epidermidis* in healthy people are frequently cleared by host defenses alone. Organisms come loose from biofilms on indwelling plastic catheters and are carried by the bloodstream to the heart and other tissues. This can result in subacute bacterial endocarditis or multiple tissue abscesses in people with impaired host defenses as from cancer, diabetes mellitus, or other causes. ■ **bacterial endocarditis, p. 718**

Epidemiology

The epidemiology of *S. aureus* is discussed in the chapter on skin infections. Various studies have shown that in the case of surgical wound infections, nasal carriers have a two to seven times greater risk of infection than do those who are not nasal carriers. From 30% to 100% of the infections in different studies are due to a patient's own staphylococcus strain. Advanced age, poor general health, immunosuppression, prolonged preoperative hospital stay, and infection at a site other than the site of surgery increase the risk of infection. ■ ***S. aureus*, epidemiology, p. 477, 525**

Prevention and Treatment

Cleansing and removal of dirt and devitalized tissue from accidental wounds minimizes the chance of infection, as does prompt closure of clean wounds by sutures. Trying to eliminate the nasal carrier state with antistaphylococcal medications is occasionally successful. Infections in surgical wounds can be reduced by half by administering an effective antistaphylococcal medication immediately before surgery. For unknown reasons, the infection rate is actually increased if the medication is given more than 3 hours before or 2 hours after the surgical incision.

Treatment of staphylococcal infections has been problematic because of the development of resistance to antibacterial medications. When penicillin was first introduced, more than 95% of the strains of *S. aureus* were susceptible to it. These strains soon largely disappeared with the widespread use of the antibiotic. The remaining strains are resistant by virtue of plasmid-encoded β-lactamase. Treatment of these strains became much easier with the development of penicillins and cephalosporins resistant to β-lactamase. Soon thereafter, however, strains of *S. aureus*, referred to as MRSAs, methicillin-resistant *Staphylococcus aureus*, appeared that were resistant because of modified penicillin-binding proteins. In the United States, these strains were reliably treated with the antibiotic vancomycin until 1997, when the first vancomycin resistant strain was identified. Since MRSAs have R plasmids, making them resistant to most antistaphylococcal medications, the appearance of vancomycin-resistance is an alarming development. In late 1999, a new medication active against vancomycin-resistant bacteria, marketed under the trade name Synercid, became available. This new medication is a combination of two substances that act synergistically to block bacterial protein synthesis. It is hoped that other new medications will be developed to keep pace with the growth of resistance, but this will depend on how intelligent human beings are in avoiding overuse of these valuable substances. ■ *β*-lactamase, p. 500 ■ antibacterial resistance, p. 509

Group A Streptococcal "Flesh Eaters"

Streptococcus pyogenes was introduced in the chapters on skin and respiratory infections as the cause of strep throat, scarlet fever, and other conditions. It is also a common cause of wound infections, which have generally been easy to treat since the bacteria are consistently susceptible to penicillin. Occasionally, however, *S. pyogenes* infections can progress rapidly, even leading to death despite antimicrobial treatment. These more severe infections are called invasive and include pneumonia, meningitis, puerperal or childbirth fever, necrotizing fasciitis or "flesh-eating" disease, and worst of all, streptococcal toxic shock, which is similar to staphylococcal toxic shock. Deaths from invasive strep infections have caused widespread popular concern as *S. pyogenes* became the "flesh-eating" bacterium of the tabloid press. This section will focus on necrotizing fasciitis (**figure 27.5**), a rare but dramatic complication of *S. pyogenes* infection. ■ *Streptococcus pyogenes*, p. 528, 553 ■ staphylococcal toxic shock, pp. 641–642

Symptoms

At the site of a surgical wound or accidental trauma, sometimes even without an obvious break in the skin, severe pain develops acutely. Within a short time, swelling becomes apparent, and the injured person develops fever and confusion. The overlying skin becomes tense and discolored because of the swelling. Unless treatment is initiated promptly, shock and death usually follow in a short time.

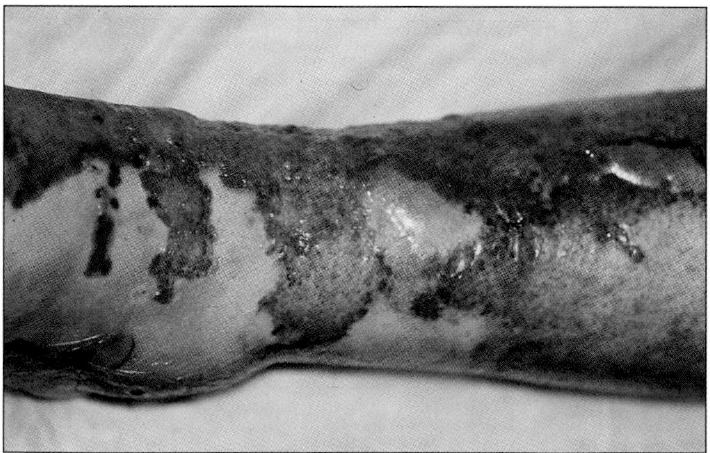

Figure 27.5 Individual with *Streptococcus pyogenes* "Flesh-Eating" Disease (Necrotizing Fasciitis)

Causative Agent

Streptococcus pyogenes is a *β*-hemolytic, Gram-positive, chain-forming coccus with Lancefield group A cell wall polysaccharide. Current evidence indicates that the strains of *S. pyogenes* that cause invasive disease are more virulent than other strains by virtue of at least two extracellular products. The first of these is pyrogenic exotoxin A, which is a superantigen and causes streptococcal toxic shock. The second product, exotoxin B, is a protease that destroys tissue by breaking down protein.

Pathogenesis

Like *Staphylococcus aureus*, *S. pyogenes* has a fibronectin-binding protein that aids colonization of wounds. In necrotizing fasciitis, the subcutaneous fascia and fatty tissue are destroyed. **Fascia** are bands of fibrous tissue that underlie the skin and surround muscle and body organs. In some cases, the fascia surrounding muscle is penetrated, and muscle tissue is also destroyed. Intense swelling occurs as fluid is drawn into the area because of increased osmotic pressure from the breakdown of tissue into small molecules. The organisms continue to multiply and produce toxic products in the mass of dead tissue, using the breakdown products as nutrients. In most cases, toxic products and organisms enter the bloodstream. Superantigens and other streptococcal products cause shock by releasing cytokines, and probably by other as yet unproven mechanisms.

Epidemiology

"Flesh-eating" infections by *S. pyogenes* have probably occurred at least since the Fifth century B.C. based on descriptions of necrotizing fasciitis by Hippocrates. More than 2,000 cases of this condition were reported among soldiers during the Civil War. Cases in this country are generally sporadic, although small epidemics have occurred such as a 1996 outbreak in San Francisco among injected-drug abusers using contaminated "black tar" heroin. The number of cases with *invasive S. pyogenes* in the United States was estimated at 10,000 for 1998, resulting in 1,300 deaths, 60 of which were due to necrotizing fasciitis. There is no firm evidence of a trend toward increasing incidence. Underlying conditions that increase the risk of necrotizing fasciitis and other invasive *S. pyogenes* infections include diabetes, cancer, alcoholism, AIDS, recent surgery, abortion, childbirth, chickenpox, and injected-drug abuse, but such infections have occurred in healthy individuals following minor injuries.

Prevention and Treatment

There are no proven preventive measures. Because of the rapidity with which the toxins spread, urgent surgery is mandatory to relieve the pressure of the swollen tissue and to remove dead tissue. Amputation is sometimes necessary, to promptly rid the patient of the source of toxins. Penicillin is the drug of choice for early infection, but it has little or no effect on streptococci in necrotic tissue and no effect on toxins, and therefore it cannot substitute for surgery.

Pseudomonas aeruginosa Infections

Pseudomonas aeruginosa is an opportunistic pathogen that is widespread in the environment, a major cause of nosocomial infections, and an occasional cause of community-acquired infections, meaning infections acquired outside hospital or other medical facilities. Community-acquired infections include skin rashes from contaminated swimming pools and hot tubs, serious infections of the foot bones from stepping on nails, serious eye infections from contaminated contact lens solutions, heart valve infections in injected-drug abusers, external ear canal infections in swimmers and others who fail to dry their ears properly, and biofilms in the lungs of individuals with the inherited disease **cystic fibrosis**. ■ *Pseudomonas*, p. 281

In hospitals, the bacterium is the leading cause of nosocomial lung infections and a common cause of wound infections, especially of thermal burns. Burns have large exposed areas of dead tissue free of any body defenses and, therefore, are ideal sites for infection by bacteria from the environment or normal flora. Almost any opportunistic pathogen can infect burns, but *Pseudomonas aeruginosa* is among the most common and hardest to treat.

Symptoms

In burns and other wounds, *P. aeruginosa* can often color the tissues green from pigments released from the bacteria (**figure 27.6a, b**). It is especially dreaded because it often invades the bloodstream and causes chills, fever, skin lesions, and shock.

Causative Agent

Pseudomonas aeruginosa is motile by means of a single polar flagellum (figure 27.6c). The organisms grow readily and rapidly; some strains can grow in a variety of aqueous solutions, even in distilled water. The bacterium generally utilizes oxygen as the final electron acceptor in the breakdown of nutrients. Under certain circumstances, however, it can also grow anaerobically, as when nitrate is available. Nitrate substitutes for oxygen as a final electron acceptor, an example of anaerobic respiration. Strains of *P. aeruginosa* often produce one or more water-soluble pigments, which can be red, yellow, blue, or dark brown. Commonly they produce a fluorescent yellowish pigment called pyoverdin, which combines with a blue pigment, pyocyanin, to produce the striking green color characteristically seen in infected wounds and in growth media. ■ anaerobic respiration, p. 153

Pathogenesis

The overall effect of *P. aeruginosa* infection of burns and other wounds is to produce tissue damage, prevent healing, and increase the risk of septic shock. Septic shock results from pathogens circulating in the bloodstream, as opposed to toxic shock caused by the circulation of specific exotoxins. Most strains produce an extracellular enzyme called exoenzyme S that *in vitro* catalyzes the transfer of an ADP-ribose fragment of NAD to G proteins. The signaling mechanism from host cell surface receptors to the appropriate genes is thereby impaired. In other words, exoenzyme S functions *in vitro* in the same way as the toxins of *Vibrio cholerae* and *Bordetella pertussis*. Strains of

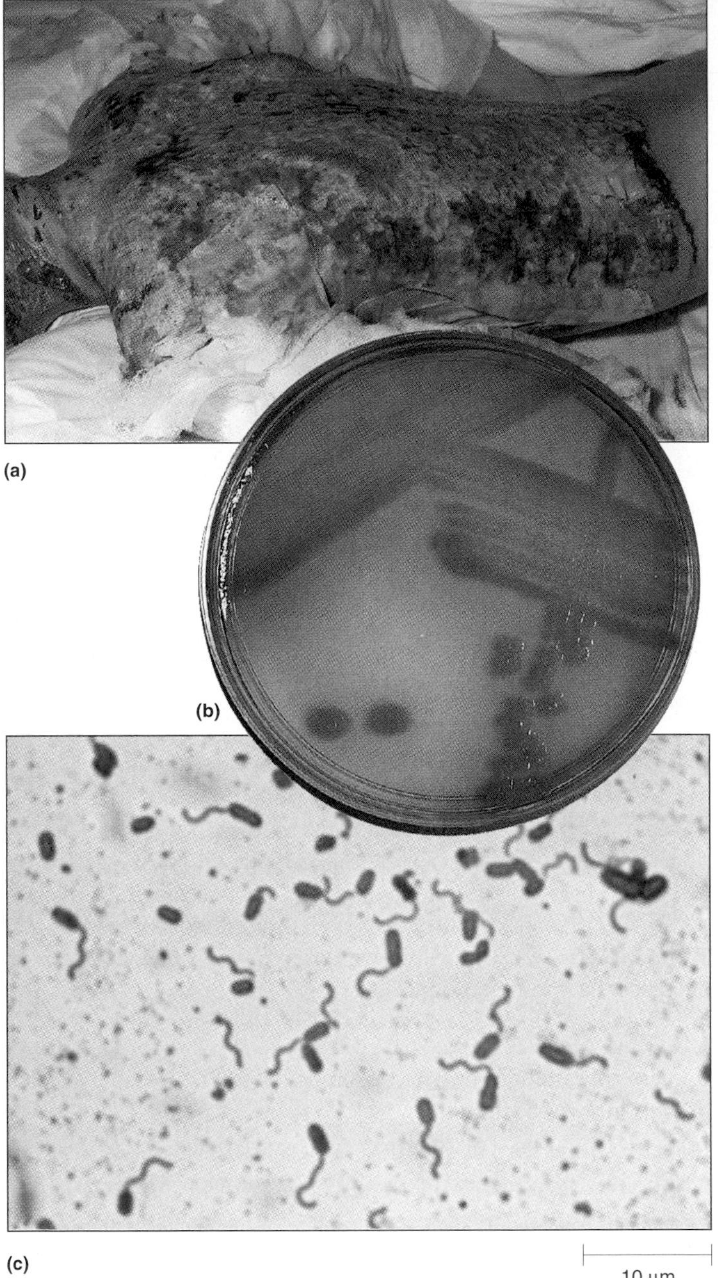

Figure 27.6 ***Pseudomonas aeruginosa*** **(a)** Extensive burn infected with *P. aeruginosa*. Notice the green discoloration. **(b)** Culture. The green discoloration of the medium results from water-soluble pigments diffusing from the *P. aeruginosa* colonies. **(c)** *P. aeruginosa* has a single polar flagellum.

P. aeruginosa having an experimentally impaired gene for the enzyme show marked decrease in virulence. ■ ADP ribosylation, p. 463 ■ signaling, p. 74 ■ *Bordetella pertussis*, pp. 566–567 ■ *Vibrio cholerae*, p. 599

Most strains of *P. aeruginosa* also produce toxin A, an exoenzyme that has a mode of action identical to that of the toxin of *Corynebacterium diphtheriae*, a Gram-positive rod. The two toxins, however, are antigenically distinct and target different cells. These toxins halt protein synthesis by the host cell and are enhanced by lowering the concentration of iron. Toxin A deficient mutants of *P. aeruginosa* show a marked decrease in

TABLE 27.2 Leading Causes of Wound Infections

Causative Organism	Characteristics	Consequences
Staphylococcus aureus	Gram-positive cocci in clusters, coagulase-positive	Delayed healing; abscess formation; extension into tissues, artificial devices, or bloodstream; occasional strains can cause toxic shock syndrome
Streptococcus pyogenes	Gram-positive cocci in chains; Lancefield group A	Same, except occasional strains can cause "flesh-eating" necrotizing fasciitis
Pseudomonas aeruginosa	Aerobic Gram-negative rod, green pigment	Delayed healing; abscess formation; extension into tissues, artificial devices, or bloodstream; septic shock

virulence. Experimentally, most strains *of P. aeruginosa* produce proteases that cause localized hemorrhages and tissue necrosis. Also, many strains produce a heat-labile hemolysin, a phospholipase C, identical in mode of action to the principal toxin of the gas gangrene bacillus to be discussed later in this chapter, *Clostridium perfringens*. Phospholipase C hydrolyses **lecithin**, an important lipid component of cell membranes. ■ ***Corynebacterium diphtheriae*, p. 557** ■ **elongation factor, p. 177**

Epidemiology

Pseudomonas aeruginosa is widespread in nature, in soil and water, and on plants, including fruits and vegetables. It is introduced into hospitals on shoes, on ornamental plants and flowers, and on produce. It can persist in most places where there is dampness or water, and it can contaminate soaps, ointments, eyedrops, contact lens solutions, cosmetics, disinfectants, many kinds of hospital equipment, swimming pools, hot tubs, the inner soles of shoes, and illegal injectable drugs, all of which have been sources of infections.

Prevention and Treatment

Prevention involves elimination of potential sources of the bacterium and prompt care of wounds. Careful removal of dead tissue from burn wounds, followed by application of an antibacterial cream such as silver sulfadiazine, is often effective in preventing infection with *P. aeruginosa*. Established infections are notoriously difficult to treat because *P. aeruginosa* is usually resistant to multiple antibacterial medications. Antibiotic susceptibility tests are done to guide the selection of an effective regimen. Medications must usually be administered intravenously in high doses.

Table 27.2 compares the leading causes of wound infections.

MICROCHECK 27.2

Staphylococcus aureus is the most important cause of wound infections because it is so commonly carried by humans, transfers easily from one person to another, and possesses multiple virulence factors. *Staphylococcus epidermidis* forms biofilms on foreign materials, protecting the bacteria from body defenses and antibacterial medications. "Flesh-eating" strains of *Streptococcus pyogenes* are uncommon, but often life threatening. *Pseudomonas aeruginosa* is a Gram-negative pigment-producing rod, widespread in the environment, and a major cause of nosocomial infections. Bacterial exotoxins from different species of bacteria may have the same mode of action but cause distinctly different diseases because the toxin molecules attach to and enter different body cells.

- Why are antibacterial medications not effective for treating "flesh-eating" disease (necrotizing fasciitis)?
- Why do wound infections due to *Pseudomonas aeruginosa* sometimes produce green pus?
- Why are *P. aeruginosa* infections so hard to treat?
- Why is it not surprising that staphylococci are the leading cause of wound infections?

Diseases Due to Anaerobic Bacterial Wound Infections

Wounds that provide anaerobic conditions allow colonization by certain strictly anaerobic species of bacteria, organisms with their own unique pathogenic abilities. This section describes three distinctive diseases that result from anaerobic bacterial wound infections: lockjaw, gas gangrene, and lumpy jaw.

"Lockjaw" (Tetanus)

Tetanus is frequently fatal; fortunately, it is rare in economically advanced countries. There is no reasonable way to avoid exposure to the causative organism because its spores are widespread in dust and dirt, frequently contaminating clothing, skin, and wounds. Even a trivial wound in a nonimmunized person can result in tetanus if the wound provides conditions sufficiently anaerobic to allow germination of the spores.

Symptoms

Tetanus is characterized by sustained, painful, and uncontrollable cramplike muscle spasms, which are usually generalized but can be limited to one area of the body (**figure 27.7**). The spasms often begin with the jaw muscles, giving the disease the popular name "lockjaw." The early symptoms include

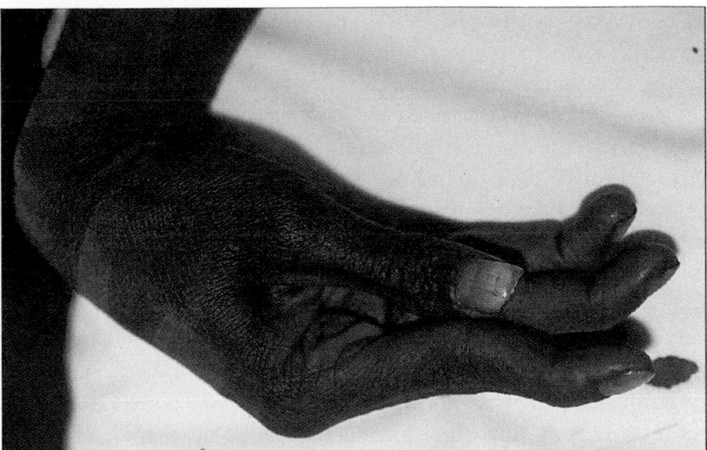

Figure 27.7 Marked Muscular Spasm of the Wrist and Hand of an Individual with Tetanus In most tetanus cases, the spasms are generalized, meaning they involve all the body's muscles.

restlessness, irritability, stiffness of the neck, contraction of the muscles of the jaw, and sometimes convulsions, particularly in children. As more muscles tense, the pain grows more severe and is similar to that of a severe leg cramp. Breathing becomes labored, and after a period of almost unbearable pain, the infected person often dies of pneumonia or from stomach contents regurgitated into the lung.

Causative Agent

The causative agent, *Clostridium tetani*, an anaerobic, spore-forming Gram-positive rod-shaped bacterium **(figure 27.8)**, shows two striking features : (1) a spherical endospore that forms at the end of the bacillus, in contrast to the oval endospore that develops near the center of the rod in other pathogenic species of *Clostridium*, and (2) swarming growth that quickly spreads over the surface of solid media, making it easy to obtain pure cultures. Final identification of *C. tetani* depends on identifying its toxin, which is coded by a plasmid. ■ **bacterial toxins, p. 460** ■ ***Clostridium*, p. 272** ■ **plasmids, p. 70, 208, 210** ■ ***C. botulinum* spores, figure 26.12, p. 675**

Isolation of *C. tetani* from wounds does not prove that a person has tetanus. The reasons are that tetanus endospores may contaminate wounds that are not sufficiently anaerobic to allow germination and toxin production, or the person may simply be immune to the toxin by prior vaccination. Only vegetative cells, not endospores, synthesize toxin. Conversely, failure to find the organism in a person's wound cultures does not eliminate the possibility of tetanus.

Pathogenesis

Clostridium tetani is not invasive, and colonization is generally localized to a wound. Its pathologic effects are entirely the result of a 150,000 molecular weight exotoxin called **tetanospasmin**, released from the organism during the stationary phase of growth. The toxin is composed of two chains joined by a disulfide bond. The heavier chain attaches specifically to receptors on motor neurons, which then take up the toxic, lighter chain by endocytosis. The toxin is carried by the cytoplasm of the neuron's axon to its cell body in the spinal cord. There, the motor neuron is in contact with other neurons that control its action. Some of these other neurons cause stimulation of the motor nerve cell, thereby producing a muscle contraction, and some make the motor neuron resistant to stimulation, thereby inhibiting muscle contraction. Tetanospasmin blocks the action of the inhibitory neurons, so that the muscles continuously contract **(figure 27.9)**. The toxin generally spreads across the spinal cord to the side opposite the wound

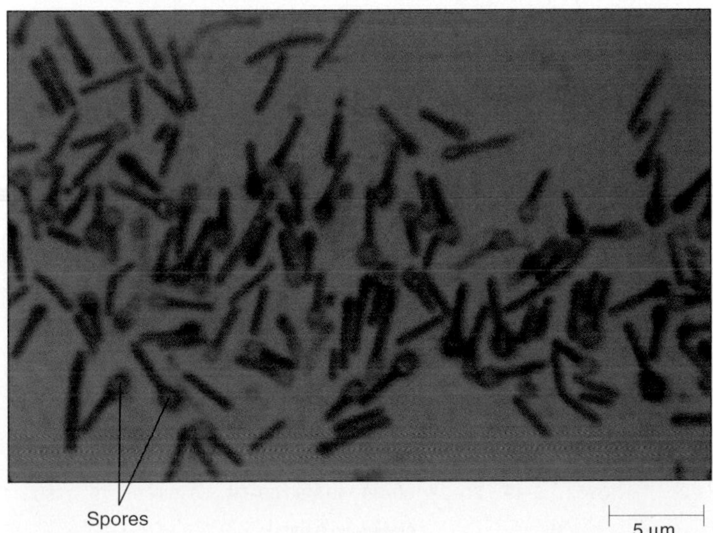
Spores
5 µm

Figure 27.8 *Clostridium tetani* Terminal endospores are characteristic of this species.

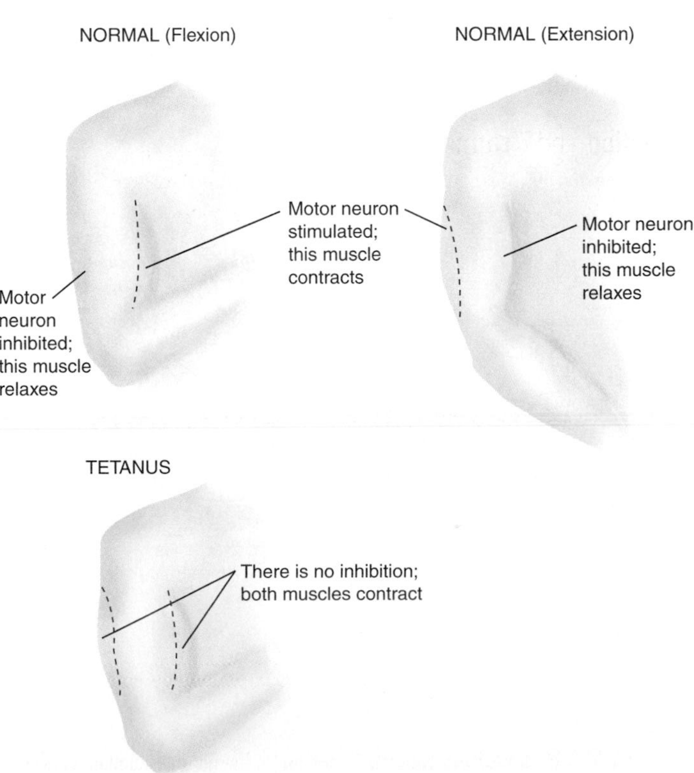
NORMAL (Flexion)
NORMAL (Extension)
Motor neuron stimulated; this muscle contracts
Motor neuron inhibited; this muscle relaxes
Motor neuron inhibited; this muscle relaxes
TETANUS
There is no inhibition; both muscles contract

Figure 27.9 Tetanus The exotoxin tetanospasmin blocks the action of inhibitory neurons, allowing all muscles to contract at the same time.

and then downward. Thus, typically, spastic muscles first appear on the side of the wound, then on the opposite side, and then downward, depending on the amount of toxin. Tetanospasmin released from the infected wound often enters the bloodstream, which carries it to the central nervous system. In these cases, inhibitory neurons of the brain are first affected, and the muscles of the jaw are among the first to become spastic.

Neurons exert their effect on other nerve cells by releasing chemicals called neurotransmitters. Tetanospasmin prevents the release of neurotransmitters from inhibitory neurons. Like other neurotransmitters, glycine and its main counterpart in the brain, γ-aminobutyric acid, exist in the inhibitory neurons in tiny membrane-bound sacs called vesicles. In order for the neurotransmitters to be released to another neuron, the vesicles must first attach specifically to receptors on their cell's membrane to initiate the process of exocytosis. Tetanospasmin is a peptidase, and it acts by removing from the vesicles the ligand responsible for attachment to the cell membrane. Thus, exocytosis cannot occur, and the inhibitory effect of the neuron is blocked **(figure 27.10)**.

Epidemiology

Clostridium tetani occurs not only in dirt and dust, but in the gastrointestinal tract of humans and other animals that have eaten foods contaminated with its spores. About half of the cases result from puncture wounds, including stepping on a nail, body piercing, tattooing, animal bites, splinters, injected-drug abuse, and insect stings. Cases also occur following medical procedures such as hemorrhoid removal or knee surgery. Surface abrasions and burns can result in tetanus if the wound is anaerobic by virtue of dirt, dead tissue, or exclusion of air. Ordinarily, 30 to 60 cases are reported annually in the United States, with a mortality rate around 25%. Age is an important factor **(figure 27.11)**, probably because maintaining immunity is neglected by many older people. Due to lack of immunization and proper care of wounds, tetanus occurs much more frequently on a global basis than it does in the economically advanced countries. In some parts of the world, babies commonly die of **neonatal tetanus** as a result of their umbilical cord being cut with unsterile instruments contaminated with *C. tetani*.

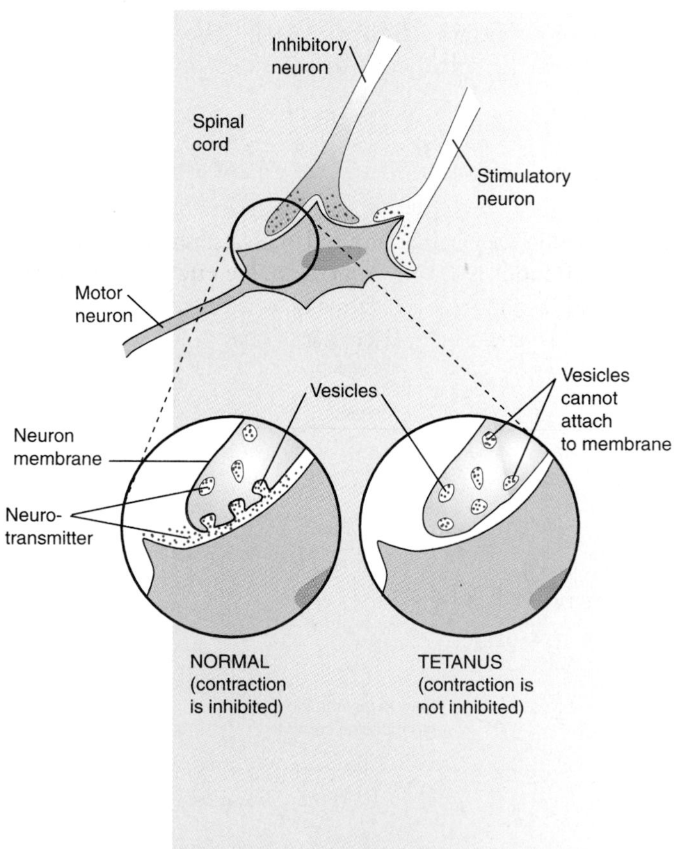

Figure 27.10 Inhibitory Neuron Function In the normal situation, vesicles containing neurotransmitter attach to the inhibitory neuron membrane and discharge their contents by exocytosis. In tetanus, tetanospasmin prevents the vesicles from attaching to the membrane, and therefore exocytosis cannot occur.

Prevention and Treatment

Active immunization with tetanus toxoid, which is inactivated tetanospasmin, is by far the best preventive weapon against tetanus. Immunization is usually begun during the first year of life. For infants and young children, the tetanus toxoid is given in combination with diphtheria toxoid and pertussis vaccine. The three together are commonly known as DPT. A booster dose is given when children enter school. Once immunity has been established by this regimen, additional booster doses of tetanus toxoid are given at 10-year intervals to maintain an adequate level of protection. Any injury, burn, or contemplated surgery is an occasion to make sure immunization is up to date. Updating of booster doses of toxoid more frequently than every 10 years is not recommended because there is danger of an allergic reaction following intensive immunization. This type of reaction can occur when the toxoid is injected into an individual who already has large quantities of antitoxin, meaning antibodies against the toxoid. Individuals who have recovered from tetanus are not immune to the disease and must be immunized. ■ **toxoid, p. 427**

Even though tetanus is easily prevented by immunization with tetanus toxoid, about 97% of the people who developed

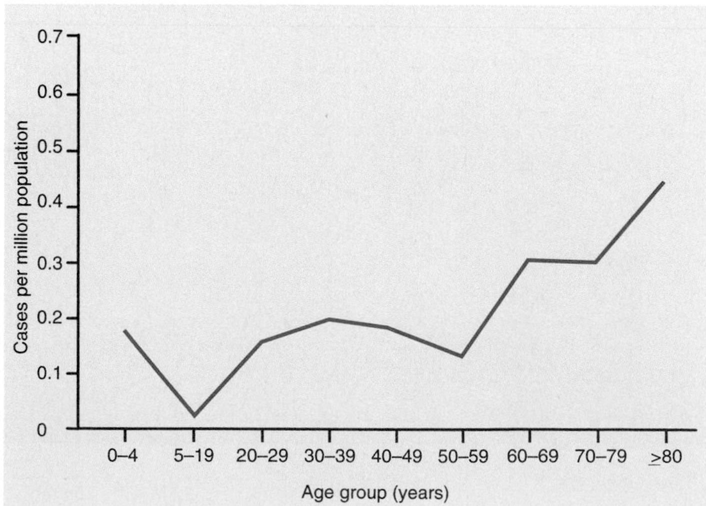

Figure 27.11 Average Annual Incidence of Tetanus for Different Age Groups, United States, 1995 to 1997

TABLE 27.3 Tetanus Prevention in the Management of Wounds

Immunization History	Clean Minor Wounds		All Other Wounds	
	Toxoid	Immune globulin	Toxoid	Immune globulin
Unknown, or fewer than three injections	Yes	No	Yes	Yes
Fully immunized (three or more injections of toxoid)				
(1) 5 years or less since last dose	No	No	No	No
(2) 5 to 10 years since last dose	No	No	Yes	No
(3) more than 10 years since last dose	Yes	No	Yes	No

tetanus in recent years were never immunized. To help prevent tetanus after a wound has been sustained, active immunization with toxoid, and passive immunization with immune globulin containing antitoxin, may be indicated **(table 27.3)**.

Neonatal tetanus can be prevented by immunizing expectant mothers. Their infants are then protected by antitoxin crossing the placenta. Since 1989, the World Health Organization has been attempting to eliminate neonatal tetanus. In 1997, there were an estimated 277,400 neonatal deaths from tetanus, as compared with 490,000 in 1994. The decrease in deaths largely resulted from immunizing millions of expectant mothers in developing countries against tetanus and educating them in the care of the severed umbilical cord.

Tetanus is treated by administering tetanus antitoxin, which is antibody against tetanospasmin, to neutralize any of the toxin not yet attached to motor nerve cells. The antitoxin of choice is **tetanus immune globulin (TIG)**, human gamma globulin with a high titer of tetanus antitoxin, which is prepared from blood of humans immunized with tetanus toxoid. This antitoxin generally does not cause hypersensitivity reactions in humans and is much more effective than the antitoxin formerly used which was derived from the blood of horses immunized with toxoid. TIG, however, cannot neutralize tetanospasmin that is already bound to nerve tissue. This explains why tetanus antitoxin is often ineffective in treating the disease. In addition to antitoxin treatment, the wound is thoroughly cleaned of all dead tissue and foreign material that could provide anaerobic conditions. An antibacterial medication such as metronidazole is given to kill any actively multiplying clostridia and thereby prevent the formation of more tetanospasmin. Antibacterials do not kill endospores or nongrowing bacteria, however. ■ **immune globulin, p. 425**

Table 27.4 describes the main features of tetanus.

Gas Gangrene (Clostridial Myonecrosis)

Almost every sample from soil and dusty surfaces has endospores of the gas gangrene bacillus, and the organism can frequently be recovered from cultures of wounds. Only rarely does contamination of a wound result in gas gangrene, however, because anaerobic conditions are necessary for the disease to develop. Primarily a disease of wartime, gas gangrene occurs mainly in neglected wounds with fragments of bone, foreign material, and extensive tissue damage, as might occur from schrapnel. Gas gangrene is highly unusual in peacetime gunshot wounds, but it occurs occasionally in abdominal and other surgeries, especially in persons with underlying diseases.

Symptoms

Gas gangrene **(figure 27.12)** begins abruptly, with pain rapidly increasing in the infected wound. Increased swelling occurs in the area, and a thin, bloody or brownish fluid leaks from the wound. The fluid may have a frothy appearance due to gas formation by the organism. The overlying skin becomes stretched tight and mottled with blue. The victim appears very ill and apprehensive, but remains quite alert until late in the illness when, near death, he or she becomes delirious and lapses into a coma.

Causative Agent

Several species of *Clostridium* can produce life-threatening gas gangrene when they invade injured muscle, but by far the most common offender is *C. perfringens.* These encapsulated Gram-positive rods are shorter and fatter than *C. tetani* and usually do not exhibit spores in material from wounds or cultures. The

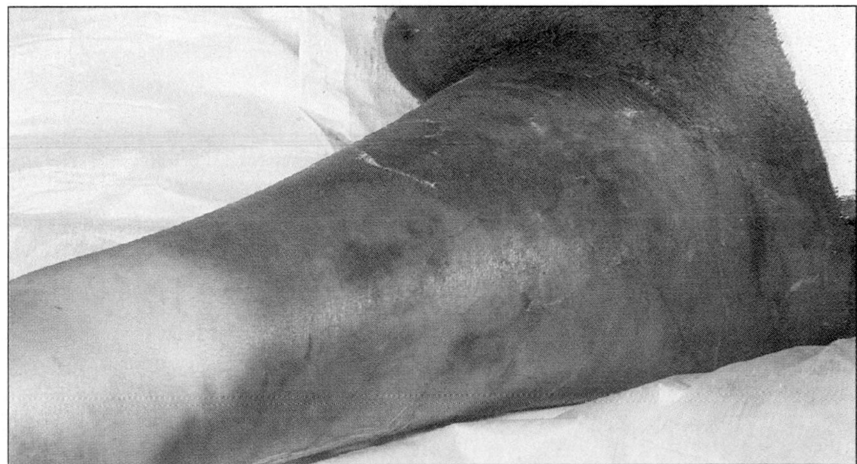

Figure 27.12 Individual with Gas Gangrene (Clostridial Myonecrosis) Fluid seeping from the involved area typically shows bits of muscle digested by *Clostridium perfringens*. Leukocytes are absent because the clostridial toxin kills them.

TABLE 27.4 "Lockjaw" (Tetanus)

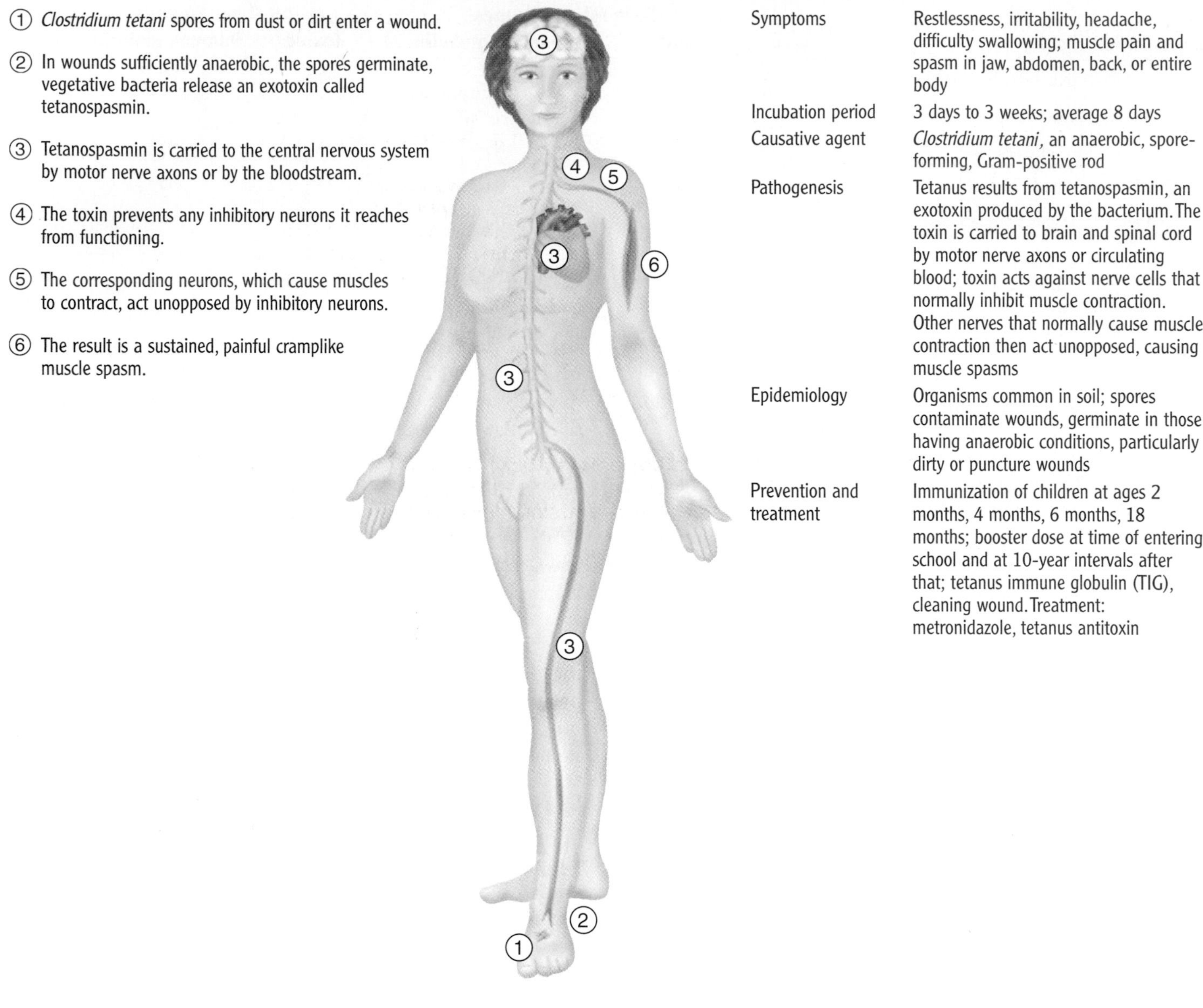

1. *Clostridium tetani* spores from dust or dirt enter a wound.
2. In wounds sufficiently anaerobic, the spores germinate, vegetative bacteria release an exotoxin called tetanospasmin.
3. Tetanospasmin is carried to the central nervous system by motor nerve axons or by the bloodstream.
4. The toxin prevents any inhibitory neurons it reaches from functioning.
5. The corresponding neurons, which cause muscles to contract, act unopposed by inhibitory neurons.
6. The result is a sustained, painful cramplike muscle spasm.

Symptoms	Restlessness, irritability, headache, difficulty swallowing; muscle pain and spasm in jaw, abdomen, back, or entire body
Incubation period	3 days to 3 weeks; average 8 days
Causative agent	*Clostridium tetani,* an anaerobic, spore-forming, Gram-positive rod
Pathogenesis	Tetanus results from tetanospasmin, an exotoxin produced by the bacterium. The toxin is carried to brain and spinal cord by motor nerve axons or circulating blood; toxin acts against nerve cells that normally inhibit muscle contraction. Other nerves that normally cause muscle contraction then act unopposed, causing muscle spasms
Epidemiology	Organisms common in soil; spores contaminate wounds, germinate in those having anaerobic conditions, particularly dirty or puncture wounds
Prevention and treatment	Immunization of children at ages 2 months, 4 months, 6 months, 18 months; booster dose at time of entering school and at 10-year intervals after that; tetanus immune globulin (TIG), cleaning wound. Treatment: metronidazole, tetanus antitoxin

toxin implicated in pathogenicity is α-toxin, an enzyme that attacks a vital component of host cell membranes called lecithin.

Pathogenesis

Two main factors foster the development of gas gangrene: (1) the presence of dirt and dead tissue in the wound and (2) long delays before the wound gets medical attention. *Clostridium perfringens* is unable to infect healthy tissue but grows readily in dead and poorly oxygenated tissue, releasing α-toxin. Growth is fostered by anaerobic conditions, as well as by the presence of growth factors and amino acids in dead tissue. Curiously, these infections are generally not ominous until they invade muscle. The toxin diffuses from the area of infection, killing leukocytes and tissue cells. Several enzymes produced by the pathogen, including collagenase and hyaluronidase, break down macromolecules in the dead tissues to smaller ones, thereby promoting swelling. The organisms grow readily in the fluids of the dead tissue, producing hydrogen and carbon dioxide from fermentation of amino acids and muscle glycogen. These gases accumulate in the tissue and contribute to the rise in pressure, thereby fostering spread of the infection. Without prompt surgical treatment, massive amounts of α-toxin diffuse into the bloodstream and destroy red blood cells, tissue capillaries, and other structures throughout the body and cause death. The reason for the rapid onset of severe toxicity when muscle becomes involved is unclear. It is not reversed by administering antibody to α-toxin.

Epidemiology

Besides being widespread in soil, *C. perfringens* is present in the feces of many animals and humans, and in the vagina in from 1% to 9% of healthy women. Besides neglected battlefield wounds and occasional surgical wounds, gas gangrene of the uterus is fairly common after self-induced abortions, and rarely, it can occur after miscarriages and childbirth. In other cases, impaired oxygenation of tissue from poor blood flow because of arteriosclerosis, otherwise known as hardening of the arteries, and diabetes are predisposing factors. Cancer patients also have increased susceptibility to gas gangrene.

Prevention and Treatment

Neither toxoid nor vaccine is available for immunization against gas gangrene. Prompt cleaning and debridement of wounds, the surgical removal of dead and damaged tissue, dirt, and other foreign material, is highly effective in preventing the disease. Treatment of gas gangrene depends primarily on the prompt surgical removal of all dead and infected tissues, and it may require amputations. Some authorities recommend hyperbaric oxygen treatment because it inhibits growth of the clostridia, thereby stopping release of toxin, and it also improves oxygenation of injured tissues. In this treatment the patient is placed in a special chamber and breathes pure oxygen under three times normal atmospheric pressure. Some people appear to improve dramatically with hyperbaric oxygen, but others show no benefit and develop complications from the treatment. Antibiotics such as penicillin are given to help stop bacterial growth and toxin production, but they are of minor if any value in treating the disease because they do not diffuse well into large areas of dead tissue and they do not inactivate toxin.

Table 27.5 describes the main features of this disease.

TABLE 27.5 Gas Gangrene (Clostridial Myonecrosis)

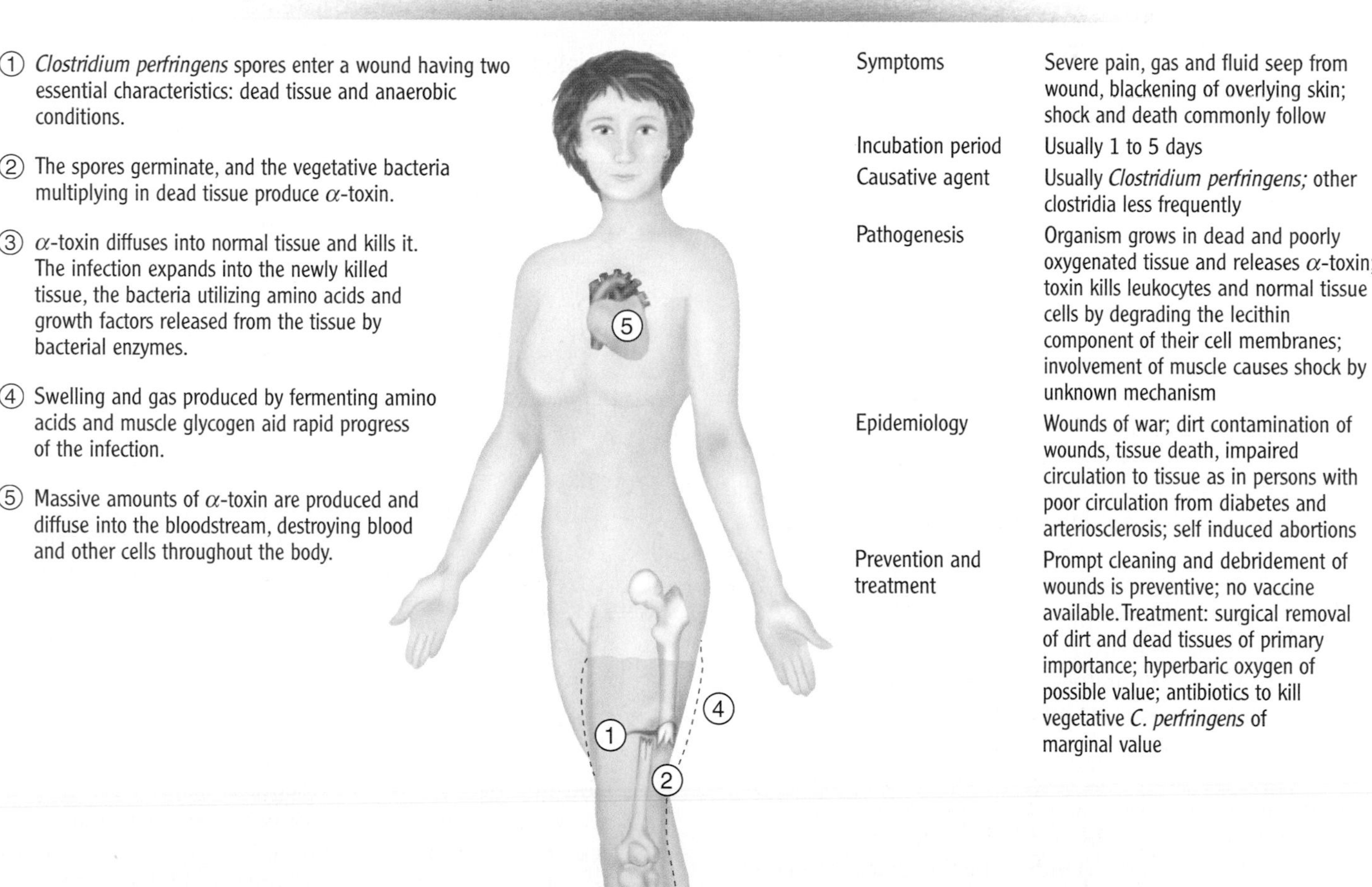
① *Clostridium perfringens* spores enter a wound having two essential characteristics: dead tissue and anaerobic conditions.
② The spores germinate, and the vegetative bacteria multiplying in dead tissue produce α-toxin.
③ α-toxin diffuses into normal tissue and kills it. The infection expands into the newly killed tissue, the bacteria utilizing amino acids and growth factors released from the tissue by bacterial enzymes.
④ Swelling and gas produced by fermenting amino acids and muscle glycogen aid rapid progress of the infection.
⑤ Massive amounts of α-toxin are produced and diffuse into the bloodstream, destroying blood and other cells throughout the body.
5
4
1
2
3

Symptoms	Severe pain, gas and fluid seep from wound, blackening of overlying skin; shock and death commonly follow
Incubation period	Usually 1 to 5 days
Causative agent	Usually *Clostridium perfringens;* other clostridia less frequently
Pathogenesis	Organism grows in dead and poorly oxygenated tissue and releases α-toxin; toxin kills leukocytes and normal tissue cells by degrading the lecithin component of their cell membranes; involvement of muscle causes shock by unknown mechanism
Epidemiology	Wounds of war; dirt contamination of wounds, tissue death, impaired circulation to tissue as in persons with poor circulation from diabetes and arteriosclerosis; self induced abortions
Prevention and treatment	Prompt cleaning and debridement of wounds is preventive; no vaccine available. Treatment: surgical removal of dirt and dead tissues of primary importance; hyperbaric oxygen of possible value; antibiotics to kill vegetative *C. perfringens* of marginal value

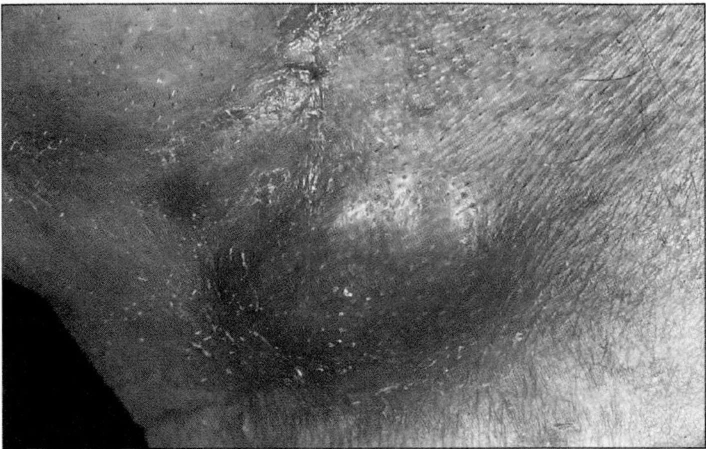

Figure 27.13 **"Lumpy Jaw" (Actinomycosis) of the Face** Abscesses, such as the one shown, form, drain, and heal, and then they recur at the same or different location. Sulfur granules, colonies of *Actinomyces israelii*, can be found in the drainage.

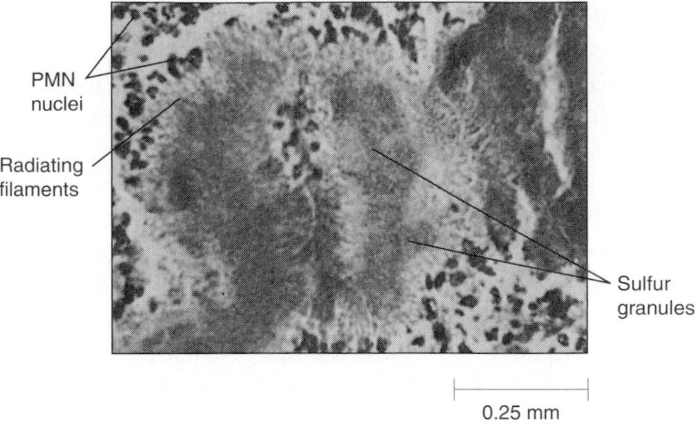

Figure 27.14 ***Actinomyces israelii* Sulfur Granule** Microscopic view of a stained smear of pus from an infected person. Filaments of the bacteria can be seen radiating from the edge of the colony.

"Lumpy Jaw" (Actinomycosis)

Many wound infections are caused by anaerobes other than clostridia. The main offenders are members of the normal flora of the mouth and intestine. Actinomycosis is not a mycosis since it is not caused by a fungus, but its name persists from when it was thought to be a fungal disease.

Symptoms

Actinomycosis is characterized by slowly progressive, sometimes painful swellings under the skin that eventually open and chronically drain pus. The openings usually heal, only to reappear at the same or nearby areas days or weeks later. Most cases involve the area of the jaw and neck; the resulting scars and swellings gave rise to the popular name "lumpy jaw" **(figure 27.13)**. In other cases the recurrent swellings and drainage develops on the chest or abdominal wall, or in the genital tract of women.

Causative Agent

Most cases of actinomycosis are caused by *Actinomyces israelii*, a Gram-positive, filamentous, branching, anaerobic bacterium that grows slowly on laboratory media. A number of similar species can cause the disease in animals and humans.

Pathogenesis

Actinomyces israelii cannot penetrate the normal mucosa, but it can establish an infection in association with other organisms if introduced into tissue by wounds. The infectious process is characterized by cycles of abscess formation, scarring, and formation of sinus tracts, or passageways that generally ignore tissue boundaries. The disease usually progresses to the skin where pus is discharged, but occasionally it penetrates into bone, or into the central nervous system. In the tissue, *A. israelii* grows as dense yellowish colonies **(figure 27.14)** called **sulfur granules** because they are the color and size of particles of elemental sulfur. Since other more rapidly growing bacteria are invariably present, finding sulfur granules in the sinus drainage is a great aid in establishing the diagnosis. Almost half of the cases originate from wounds in the mouth, the remainder from the lung, intestine, or vagina. The detailed mechanisms by which the organism produces the disease have not yet been established.

Epidemiology

Actinomyces israelii can be part of the normal flora of the mucosal surfaces of the mouth, upper respiratory tract, intestine, and sometimes the vagina. It is common in the gingival crevice, particularly with poor dental care. Pelvic actinomycosis can complicate use of intrauterine contraceptive devices (IUDs). Other species of *Actinomyces* responsible for actinomycosis of dogs, cattle, sheep, pigs, and other animals generally do not cause human disease. The disease is sporadic and is not transmitted person to person. ■ **gingival crevice, p. 587**

Prevention and Treatment

There are no proven preventive measures. Actinomycosis responds to treatment with a number of antibacterial medications, including penicillin and tetracycline. To be successful, however, treatment must be given over weeks or months. This need for prolonged therapy probably is due to the slow growth of the organisms and their tendency to grow in dense colonies.

The main features of actinomycosis are presented in **table 27.6**.

MICROCHECK 27.3

Lockjaw (tetanus) is caused by an anaerobic, spore-forming bacterium with essentially no invasive ability, yet it causes death by releasing an exotoxin that is carried to the nervous system. Gas gangrene (clostridial myonecrosis), caused by another anaerobic sporeformer, usually arises from neglected wounds containing dead tissue and foreign material and advances into normal muscle tissue. Lumpy jaw (actinomycosis) is caused by a branching, filamentous, slow-growing, anaerobic bacterium, a member of the normal flora of the mouth, upper respiratory tract, intestine, and vagina. The disease is characterized by recurrent abscesses that drain sulfur granules.

- Why do many tetanus victims fail to respond to tetanus antitoxin?

TABLE 27.6 "Lumpy Jaw" Actinomycosis

Symptoms	Chronic disease; recurrent, sometimes painful swellings open and drain pus, heal with scarring; usually involves face and neck; chest, abdomen, and pelvis other common sites
Causative agent	*Actinomyces israelii*, a filamentous, branching Gram-positive slow-growing anaerobic bacterium
Incubation period	Months (usually indeterminate)
Pathogenesis	Usually begins in a mouth wound, extends without regard to tissue boundaries to the face, neck, or upper chest; sometimes begins in the lung, intestine, or female pelvis. *A. israelii* always accompanied by normal flora. In tissue, grows as dense yellowish colonies called sulfur granules
Epidemiology	No person-to-person spread. *A. israelii* commonly part of normal mouth, upper respiratory, intestine, and vagina flora. Dental procedures, intestinal surgery, insertion of IUDs can initiate infections
Prevention and treatment	No proven preventive measures. Because of its slow growth, *A. israelii* infections require prolonged treatment; a number of antibacterial medications are effective

- What factors favor the development of gas gangrene (clostridial myonecrosis)?
- Why is it necessary to treat lumpy jaw (actinomycosis) for a prolonged period of time?
- Would babies need to be immunized against lockjaw (tetanus) if their mother had been immunized against the disease? Explain.

Bacterial Bite Wound Infections

Each year, more than 3 million animal bites occur in the United States, the most feared result being the viral disease, rabies. The infection risk following an animal bite depends partly on the type of injury (crushing, lacerated, or puncture), and partly on the kinds of infectious agents in the animal's mouth. ■ rabies, p. 682

Pasteurella multocida Bite Wound Infections

Infections caused by bacteria that live in the mouth of the biting animal are much more common and less frightening than rabies. Surprisingly, a single species, *Pasteurella multocida*, is responsible for bite infections from a number of kinds of animals, including dogs, cats, monkeys, and humans, among others.

Symptoms

There are no reliable symptoms or signs that distinguish among most bacterial bite wound infections. Spreading redness and tenderness and swelling of tissues adjacent to the wound, followed by the discharge of pus, are early indications of infection.

Causative Agent

Pasteurella multocida is a Gram-negative, facultatively anaerobic, coccobacillus that is easily cultivated in the laboratory. Most isolates are encapsulated. There are a number of different antigenic types.

Pathogenesis

The details of pathogenesis are not yet known. There is evidence for one or more adhesins. Some strains produce a toxin that kills cells. The capsules of *P. multocida* are antiphagocytic. Abscesses form. When opsonized by specific antibody, the organisms are ingested and killed by phagocytes.

Epidemiology

Pasteurella multocida is best known as the cause of a devastating disease of chickens called **fowl cholera**. This disease is historically significant because while studying it, Pasteur first discovered that the virulence of a pathogen could be attenuated by repeated laboratory culture, and the attenuated organism used as a vaccine. *Pasteurella multocida* also causes diseases in a number of other animal species. Epidemics of fatal pneumonia and bloodstream infection occur in rabbits, cattle, sheep, and mice. Many healthy animals, however, carry the bacterium among their normal oral and upper respiratory flora. Both diseased animals and healthy carriers constitute a reservoir for human infections. Cats are more likely to carry *P. multocida* than dogs, and so cat bites are more likely than dog bites to cause the infection.

Prevention and Treatment

No vaccines are available for use in humans. Immediate cleansing of bite wounds and prompt medical attention usually prevent the development of serious infection and possible permanent impairment of function. Unlike many Gram-negative pathogens, *P. multocida* is susceptible to penicillin. Usually, before the cultural diagnosis is known, amoxicillin plus a β-lactamase inhibitor are administered, available in a single tablet under the trade name Augmentin. This combination is used because amoxicillin is active against *P. multocida*, and with the addition of the β-lactamase inhibitor, the antibiotic is also active against β-lactamase-producing *Staphylococcus aureus*, another common cause of bite wound infections. This and other antibacterial medications are effective if given early in the infection.

Table 27.7 give the main features of *Pasteurella multocida* bite wound infections.

Cat Scratch Disease

In the United States, cat scratch disease is the most common cause of chronic lymph node enlargement at one body site in young children (**figure 27.15**). Despite its name, this disease can be transmitted by bites, and probably other means. Typically the infection is mild and localized, lasting from several weeks to a few months. Of the estimated 24,000 annual cases, about 2,000 require hospitalization.

Symptoms

In about half the cases, the disease begins within a week of a scratch or bite with the appearance of a pus-filled pimple at the

TABLE 27.7 *Pasteurella multocida* Bite Wound Infections

Symptoms	Spreading redness, tenderness, swelling, discharge of pus
Incubation period	24 hours or less
Causative agent	*Pasteurella multocida,* a Gram-negative, facultatively anaerobic, encapsulated coccobacillus
Pathogenesis	Introduced by bite, *P. multocida* attaches to tissue, resists phagocytes because of its capsule; probable cell-destroying toxin. Extensive swelling, abscess formation. Opsonins develop, allow phagocytic killing, limit spread
Epidemiology	Carried by many animals in their mouth or upper respiratory tract
Prevention and treatment	No vaccines to use in humans. Prompt wound care is preventive. Treatment with penicillin, other antibacterials, effective if given promptly

site of injury. Painful enlargement of the lymph nodes of the region develops in 1 to 7 weeks. About one-third of the patients develop fever, and in about half that number the lymph nodes become pus-filled and soften. The disease generally disappears without treatment in 2 to 4 months. Of the remaining 10% of cases, some develop irritation of an eye with local lymph node enlargement, epileptic seizures and coma due to encephalitis, or acute or chronic fever associated with bloodstream or heart valve infection.

Causative Agent

Cat scratch disease is caused by *Bartonella* (formerly *Rochalimaea*) *henselae*, a tiny (0.6 by 1.0 μm) slightly curved Gram-negative rod cultivable on laboratory media.

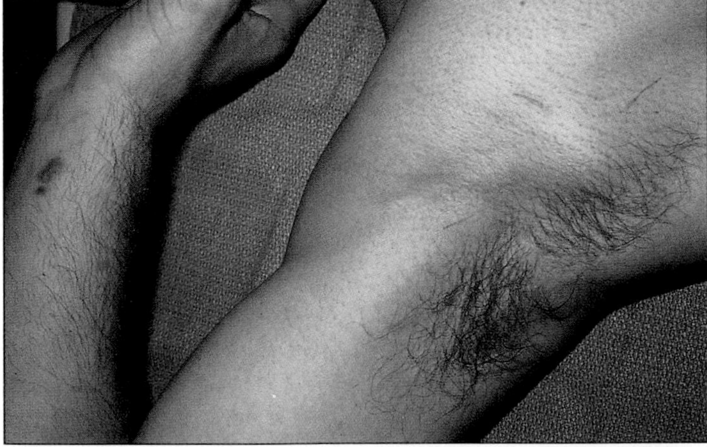

Figure 27.15 Individual with Cat Scratch Disease, Showing Enlarged Lymph Node

Pathogenesis

The virulence factors of *B. henselae* and the process by which it causes disease are not yet known. The organisms enter the body with a cat scratch or bite, are carried to the lymph nodes, and the disease is arrested by the immune system in most cases. Spread by the bloodstream, however, occurs in some individuals. **Peliosis hepatis** and **bacillary angiomatosis** are two complicating conditions seen mostly in people with AIDS. In peliosis hepatis, blood-filled cysts form in the liver. In bacillary angiomatosis, nodules composed of proliferating blood vessels develop in the skin and other parts of the body. *Bartonella henselae* is present in these lesions and those of typical cat scratch disease.

Epidemiology

Cat scratch disease mainly occurs in people less than 18 years old. The disease is a zoonosis of cats, and humans are accidental hosts. Person-to-person spread does not occur. Bites and scratches represent the usual mode of transmission of *B. henselae* from cats to people. Asymptomatic bacteremia is common in cats, however, and transmission from cat to cat occurs by cat fleas. The fleas are biological vectors, and they discharge *B. henselae* in their feces. Cat fleas probably are responsible for transmission in cases where people develop the disease after they handle cats but are not scratched or bitten. The disease occurs worldwide. ■ **biological vectors, p. 315**

Prevention and Treatment

There are no proven preventive measures for cat scratch disease. It is prudent to avoid handling stray cats, however, especially young ones and those with fleas. Any cat-inflicted wound should be promptly cleaned with soap and water and then treated with an antiseptic. If signs of infection develop, or if the cat's immunization status against rabies is uncertain, prompt medical evaluation is indicated. Severely immunodeficient persons should avoid cats if possible and, if not, control of fleas and abstinence from rough play is advisable. Severe *B. henselae* infections can usually be treated with antibacterial medications such as ampicillin, but some strains are resistant.

The main features of cat scratch disease are presented in **table 27.8**.

Streptobacillary Rat Bite Fever

Rat bites are fairly common among the poor people of large cities and among workers who handle laboratory rats. In the past, as many as one out of 10 bites resulted in rat bite fever. Currently, the disease is not reportable in any state and the incidence of the disease in the United States is unknown.

Symptoms

Usually, the bite wound heals promptly without any problem noted. Two to 10 days later, however, chills and fever, head and muscle aches, and vomiting develop. The fever characteristically comes and goes. A rash usually appears after a few days, followed by pain on motion of one or more of the large joints.

Perspective 27.1 Infection Caused by a Human "Bite"

A 26-year-old man injured the knuckle of the long finger of his right hand during a tavern brawl when he punched an assailant in the mouth. Due to his inebriated condition, a night spent in jail, and the insignificant early appearance of his knuckle wound, the man did not seek medical help until more than 36 hours later. At that time, his entire hand was massively swollen, red, and tender. Furthermore, the swelling was spreading to his arm. The surgeon cut open the infected tissues, allowing the discharge of pus. He removed the damaged tissue and washed the wound with sterile fluid. Smears and cultures of the infected material showed aerobic and anaerobic bacteria characteristic of mouth flora, including species of *Bacteroides* and *Streptococcus*. The patient was given antibiotics to combat infection, but the wound did not heal well and continued to drain pus. Several weeks later, X rays revealed that infection had spread to the bone at the base of the finger. To cure the infection, the finger had to be amputated.

Causative Agent

The cause of streptobacillary rat bite fever is *Streptobacillus moniliformis*, a facultatively anaerobic Gram-negative rod. Stained smears show a multiplicity of forms ranging from small coccobacilli to unbranched filaments more that 100 µm long. The organism is unique in that it spontaneously develops **L-forms**. L-forms are cell wall deficient variants, first identified at the famous Lister Institute (hence, the *L* in *L-form*). As might be expected, L-form colonies resemble those of mycoplasmas, bacteria that lack a cell wall. ■ mycoplasmas, p. 291

Pathogenesis

The details of how this bacterium causes streptobacillary rat bite fever are not yet known. The organisms enter the body through a bite or scratch, and sometimes by ingestion. There is typically little enlargement of the local lymph nodes, and *S. moniliformis* quickly enters the bloodstream and spreads throughout the body. Fever often subsides and reappears in an irregular fashion. The majority of cases recover without treatment in about 2 weeks, but others develop serious complications such as brain abscesses or infection of the heart valves. About 13% of untreated cases are fatal.

Epidemiology

From 50% to 100% of wild, and 10% to 100% of healthy laboratory rats, as well as mice and other rodents carry the organism in their nose and throat. Rat bites and scratches are the usual source of human infections and can occur while the victim is sleeping. Epidemics of the disease, however, have arisen from ingesting milk, water, or food contaminated with *S. moniliformis* from rodent droppings. The foodborne disease is called **Haverhill fever** from a 1926 epidemic in Haverhill, Massachusetts. A more recent epidemic occurred in an English boarding school in 1983, when 208 students acquired the disease from drinking raw milk. Cases of rat bite fever have also been associated with exposure to animals that prey on rodents, including cats and dogs.

Prevention and Treatment

Wild rat control and care in handling laboratory rats are reasonable preventive measures. Penicillin, given by injection, is the treatment of choice for cases of rat bite fever. This shows that either *S. moniliformis* L-forms do not occur *in vivo*, or if they do, they are avirulent.

The main features of streptobacillary rat bite fever are presented in **table 27.9**.

TABLE 27.8 Cat Scratch Disease

Symptoms	Pimple appears at the bite or scratch site, followed by local lymph node enlargement, fever; nodes may soften and drain pus; prolonged fever, convulsions, indicate spread to other body parts
Incubation period	Usually less than 1 week
Causative agent	*Bartonella henselae*, a tiny Gram-negative rod
Pathogenesis	The bacteria enter with cat bite or scratch and reach lymph nodes, where the disease is usually arrested. May spread by bloodstream, cause infections of the heart, brain, or other organs. Peliosis hepatis, bacillary angiomatosis mostly in those with AIDS
Epidemiology	A zoonosis of cats, spread from one to another by cat fleas, which are biological vectors and have *B. henselae* in their feces. Infected cats usually asymptomatic, but often bacteremic. Humans are accidental hosts. No person-to-person spread. Mainly a disease of those less than 18 years old
Prevention and treatment	Avoiding rough play. Promptly wash skin breaks, apply antiseptic. Flea control in cats. Most *B. henselae* susceptible to antibacterial treatment

TABLE 27.9 Streptobacillary Rat Bite Fever

Symptoms	Chills, fever, muscle aches, headache, and vomiting; later, rash and pain in one or more of the large joints
Incubation period	Usually 2 to 10 days (range, 1 to 22) after a rat bite
Causative agent	*Streptobacillus moniliformis*, a highly pleomorphic Gram-negative bacterium that spontaneously produces L-forms
Pathogenesis	Bite wound heals without treatment; *S. moniliformis* quickly invades bloodstream. Fevers come and go irregularly. Most victims recover without treatment; in others, Infection established in various body organs, results in death if treatment is not given
Epidemiology	Wild and laboratory rats can carry *S. moniliformis*. Bites of other rodents and animals that prey on them can transmit the disease to humans. Food or drink contaminated with rodent excreta can also transmit the infection; foodborne disease is called Haverhill fever
Prevention and treatment	Control wild rats and mice, care in handling laboratory animals. Effectively treated with penicillin, other antibacterial medications

CASE PRESENTATION

The patient was a 24-year-old woman, a surgical nurse, seen in the clinic for evaluation of a needle puncture wound to the hand. Earlier in the day, while assisting in a frantic attempt to revive a man with cardiac arrest, she sustained a deep puncture wound to her right palm from a needle that had accidentally dropped into the bedclothes. The needle was visibly contaminated with blood. She immediately washed her hand thoroughly with soap and water, applied an antiseptic, and dressed the puncture site with a loose adhesive bandage.

She was married, with one 14-month-old child. There was no history of blood transfusion or injected-drug abuse. She had donated blood the previous month, and it was not rejected. Her tetanus immunization was up to date, but she had not been immunized against hepatitis B.

Two days after the clinic visit, tests for antibody in the cardiac arrest patient's blood revealed that he had a chronic viral infection.

1. What were the main diagnostic considerations?
2. What risk of infection did the patient face?
3. What measures could be taken to reduce the risk? How much time could expire before preventive measures became ineffective?
4. What was the nurse's prognosis?

Discussion

1. The viruses of concern are hepatitis B virus (HBV), human immunodeficiency virus (HIV), and hepatitis C virus (HCV). Each of these could be transmitted to the nurse by a needle stick and cause serious illness.
2. There are an estimated 750,000 to 1 million carriers of HBV in the United States. They typically have large amounts of circulating infectious virus, so that even a tiny amount of their blood can transmit the disease. The risk of infection from a needle puncture wound when the blood originates from a hepatitis B virus carrier is estimated to be 10% to 35%. The AIDS-causing human immunodeficiency virus (HIV) infects approximately 1 million Americans. The blood of these persons is also potentially infectious, but the risk of transmission by a needle stick is considerably lower than the risk for hepatitis B, averaging about 0.4%. The lower risk results from smaller amounts of circulating infectious virus in HIV-infected individuals. The risk is probably higher early in HIV disease, during the acute infection, and later, when AIDS develops, because much higher levels of circulating infectious virus are then present.

   Hepatitis C virus transmission by blood accounts for most cases of post-transfusion hepatitis. Transmission from surgeon to patient has been documented, presumably by the multiple pricks from surgical needles that often penetrate the surgeons' gloves during major surgery. The risk of transmission by needle stick from an HCV-positive individual is about 1.8%. The number of new hepatitis C virus infections in the United States each year has been estimated at between 150,000 and 170,000, but the mode of transmission is unknown in most cases.

   Other viruses, such as cytomegalovirus (CMV) and Epstein-Barr virus (EBV), can be transmitted by blood. The risk from a needle stick injury is unknown but is probably much lower than from the viruses already mentioned. Obviously, all blood should be considered potentially infectious.
3. In the case of needle puncture wounds that expose a person to HBV, hepatitis B immune globulin (HBIG) is given as soon as possible after the wound occurs. HBIG is gamma globulin obtained from individuals that have a high titer of antibody against HBV. At the same time, active immunization is started with hepatitis B vaccine. These measures must be initiated within 7 days of the injury to be effective. This nurse, as with all persons at high risk of blood exposure, should have already been immunized with hepatitis B vaccine; then, no other preventive measures would need to be taken.

   Those exposed to HIV by needle punctures should be given zidovudine (AZT), plus one or more other anti-HIV medications, immediately and for 4 weeks. There is probably little protective effect if therapy is delayed beyond 2 hours.

   There is no proven preventive measure for HCV exposure. Approaches similar to those for HBV may become available in the future.
4. Preventive measures for hepatitis B exposure are highly effective, reducing the risk of infection by 75% or more. Also, the already relatively low risk from needle puncture wound for HIV exposure can probably be reduced by 75% to 80% with preventive medication. The patient's prognosis for remaining free of infection was good. The small chance of becoming infected and the long incubation period of these diseases, however, add up to considerable worry. Every effort should be made to avoid needle puncture wounds in the first place.

Human Bites

Wounds caused from human bites, striking the teeth of another person, or resulting from objects that have been in a person's mouth are common and can result in very serious infections. Rarely, diseases such as syphilis, tuberculosis, and hepatitis B are transmitted this way. Much more commonly, it is the normal mouth flora that cause trouble.

Symptoms

The wound may appear insignificant at first but then becomes painful and swells massively. Discharged pus often has a foul smell. Most of the wounds are on the extensor surface on the hand, and here the swelling may soon involve the palm also, and movement of some or all of the fingers becomes difficult or impossible.

Causative Agents

Stains and cultures usually show members of the normal mouth flora, including anaerobic streptococci, fusiforms, spirochetes, and *Bacteroides* sp., often in association with *Staphylococcus aureus.*

Pathogenesis

The crushing nature of bite wounds provides suitable conditions for anaerobic bacteria to establish infection. Although most members of the mouth flora are harmless alone, together they produce an impressive number of toxins and destructive enzymes. These include leukocidin, collagenase, hyaluronidase, ribonuclease, various proteinases, neuraminidase, and enzymes that destroy complement and antibody. Capsules of some species inhibit phagocytosis. Facultatively anaerobic organisms reduce available oxygen and thus encourage the growth of anaerobes. The result of all these factors is a **synergistic infection**, meaning that the sum effect of all the organisms acting together is greater than the sum of their individual effects. Irreversible destruction of tissues such as tendons and permanent loss of function can be the result. ■ **facultative anaerobes, p. 98**

Epidemiology

Most of the serious human bite infections occur in association with violent confrontations related to alcohol ingestion, or during forcible restraint, as in law enforcement and in mental institutions. The risk is greatly increased when the biting individual has poor mouth care and extensive dental disease. Bites by little children are usually inconsequential.

Prevention and Treatment

Prevention involves avoiding situations that lead to uncivilized behavior such as biting and hitting. Prompt cleansing of wounds followed by application of an antiseptic are advised, and

TABLE 27.10 Human Bite Wound Infections

Symptoms	Rapid onset, pain, massive swelling, drainage of foul-smelling pus
Incubation period	Usually 6 to 24 hours
Causative agent	Mixed mouth flora: anaerobic streptococci, fusiforms, spirochetes, anaerobic Gram-negative rods; sometimes *Staphylococcus aureus*
Pathogenesis	Various mouth bacteria act synergistically to destroy tissue
Epidemiology	Alcohol-related violence; forcible restraint; poor mouth care and extensive dental disease
Prevention and treatment	No proven preventive measures except to avoid altercations. Prompt cleansing of wound and application of antiseptic is advised. Treatment is usually surgical

most important is immediate medical attention if there is any suspicion of developing infection. Treatment of infected wounds consists of opening the infected area widely with a scalpel, washing the wound thoroughly with sterile fluid, and removing dirt and dead tissue. The choice of antibacterial medication includes one effective against anaerobes.

The main features of human bite wound infections are presented in **table 27.10**.

MICROCHECK 27.4

A single species of Gram-negative, encapsulated, facultatively anaerobic rods, *Pasteurella multocida*, can infect bite wounds caused by a number of different animals, notably cats. Cat bites and scratches can also transmit *Bartonella henselae*, cause of cat scratch disease, characterized typically by local lymph node enlargement, but the disease may involve other parts of the body. Streptobacillary rat bite fever, acquired from bites of rats and mice, and animals that prey on them, is marked by fever that comes and goes and a rash. The causative bacterium, *Streptobacillus moniliformis*, spontaneously develops L-forms. Human bite infections can be dangerous because certain members of the mouth flora with little invasive ability when growing alone can invade and destroy tissue when growing together.

- What Gram-negative organism commonly infects wounds caused by animal bites?
- What is the most common cause of chronic localized lymph node enlargement in young children?
- What unusual kind of variant occurs spontaneously in *Streptobacillus moniliformis* cultures?
- Why would normal mouth flora be a more common cause of serious human bite infections than the causative agents of syphilis, tuberculosis, and hepatitis B, which can also be transmitted by human bites?

Fungal Wound Infections

Fungal infections of wounds are unusual in economically developed countries, except that the yeast *Candida albicans* can be troublesome in severe burns and in those with wounds and underlying diseases such as diabetes and cancer. This yeast, commonly present among the normal flora and kept in check by it, becomes pathogenic when the competing microorganisms are eliminated, as in individuals receiving antibacterial therapy. Other fungal wound infections are much more common in impoverished people around the world. For example, Madura foot, a condition caused by various species of fungi, occurs in areas of the world where foot injuries are common, resulting from lack of shoes. Named after the city in India where it was first described, Madura foot is characterized by swellings and draining passageways that spit out yellow or black granules of fungal material. Only a minority of those with foot injuries contract the disease despite exposure to the same fungi, suggesting that other factors such as malnutrition may play a role. Sporotrichosis, another kind of fungal wound infection, occurs worldwide and is not poverty-related.

"Rose Gardener's Disease" (Sporotrichosis)

Sporotrichosis, also known as "rose gardener's disease," is widely distributed around the world, associated with activities that lead to puncture wounds from vegetation. Although many cases are sporadic, the disease can occur in groups of people engaged in the same occupation. Thousands of workers in the warm humid mines of South Africa have contracted the disease from splinters on mine timbers. Epidemics have occurred in the United States among handlers of sphagnum moss from Wisconsin.

Symptoms

In most cases, a hand or arm is involved, but the trunk, legs, and face can also be sites of infection. Typically, a chronic ulcer forms at the wound site, followed by a slowly progressing series of ulcerating nodules that develop sequentially toward the center of the body (**figure 27.16**). Lymph nodes in the region of the wound enlarge, but patients generally do not become ill. If they have AIDS or other immunodeficiency, however, the disease can spread throughout the body, threatening life.

Causative Agent

Sporotrichosis is caused by the dimorphic fungus *Sporothrix schenckii* (**figure 27.17**), which lives in soil and on vegetation.
■ **dimorphic fungus, p. 309**

Pathogenesis

Sporothrix schenckii spores are usually introduced with an injury caused by plant material. After an incubation period that usually ranges from 1 to 3 weeks but can be much longer, the multiplying fungi cause formation of a small nodule or pimple at the site of the injury. This lesion slowly enlarges and ulcerates, producing a red, easily bleeding skin defect. Unless the ulcer becomes secondarily infected with bacteria, there is little or no pus, and the lesion is pain-free. After a week or

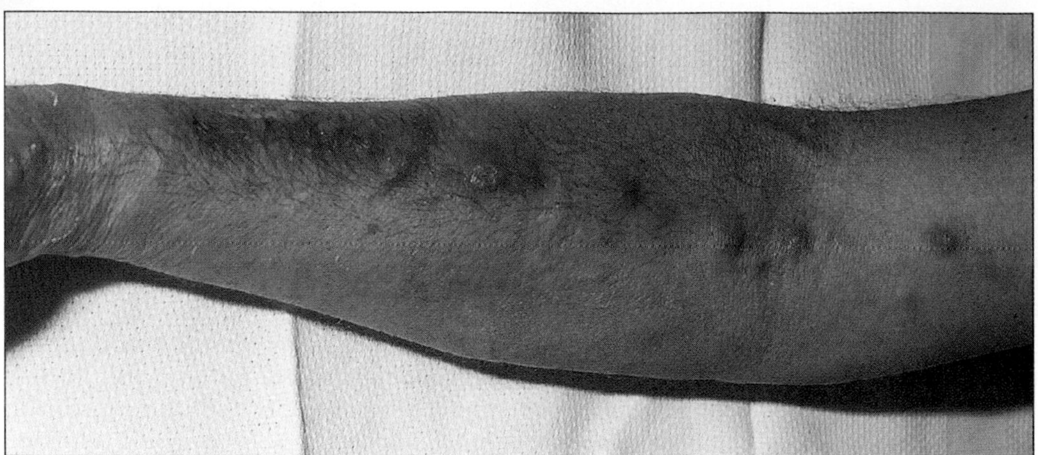

Figure 27.16 Individual with "Rose Gardener's Disease" (Sporotrichosis) Notice the multiple abscesses along the course of the lymphatic drainage from the hand.

longer, the process repeats itself, progression of the disease usually following the flow of a lymphatic vessel. In healthy individuals, the process does not proceed beyond the lymph node. Sometimes, however, satellite lesions appear irrespective of the lymphatic vessels. Without treatment, the disease can go on for years. ■ **lymphatic vessel, p. 716**

Epidemiology

Sporotrichosis is distributed worldwide, mostly in the warmer regions but extending into temperate climates. In the United States, most cases occur in the Mississippi and Missouri river valleys. It is an occupational disease of farmers, carpenters, gardeners, greenhouse workers, and others who deal with plant materials.

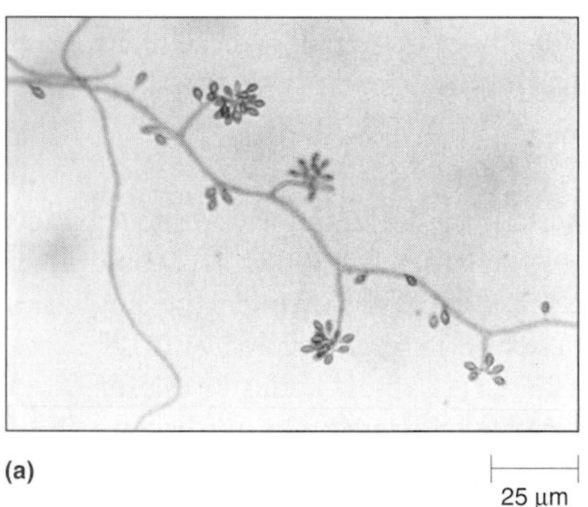

(a) 25 μm

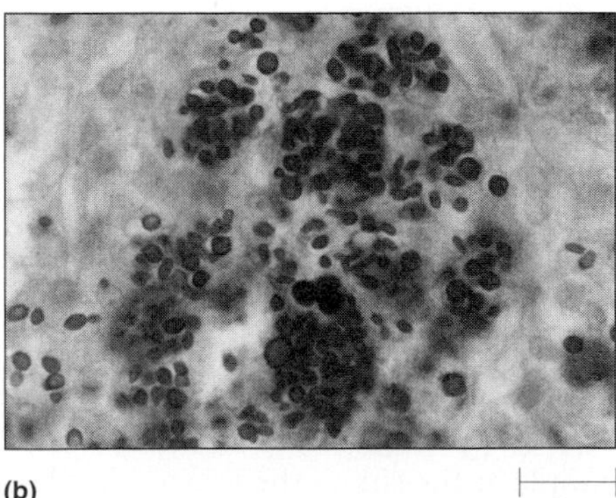

(b) 30 μm

Figure 27.17 ***Sporothrix schenckii*, Showing the Two Forms of This Dimorphic Fungus** (a) Mold form. (b) Yeast form, as seen in infected tissue.

Sporotrichosis is not a reportable disease, and its incidence is unknown. Risk factors for the disease besides occupation include diabetes, immunosuppression, and alcoholism. Children can contract the disease from playing in baled hay. Individuals with chronic lung disease can contract *S. schenckii* lung infections from inhaling dust from hay or cattle feed. Deaths from sporotrichosis are rare.

Prevention and Treatment

Protective gloves and a long-sleeved shirt can help prevent sporotrichosis, especially when handling evergreen seedlings and sphagnum moss. This chronic disease is often misdiagnosed, leading to delayed and inappropriate treatment. Surprisingly, unlike other infections, sporotrichosis can usually be cured by oral treatment with the simple chemical compound potassium iodide (KI). KI is not active against *S. schenckii in vitro*, but somehow enhances the body's ability to reject the fungus. Itraconazole or the antibiotic amphotericin B is used in rare cases when the disease spreads throughout the body.

Table 27.11 presents some of the main features of sporotrichosis.

MICROCHECK 27.5

Candida albicans, often a member of the normal flora, can infect burns and other wounds, especially in individuals with underlying conditions. Madura foot, characterized by draining passageways that discharge colored granules, occurs mainly in areas of the world where shoes are scarce and foot injuries common. Rose gardener's disease (sporotrichosis), caused by the dimorphic fungus *Sporothrix schenckii*, is widely distributed around the world, affecting mainly those associated with occupations that expose them to splinters and sharp vegetation. Unlike most invasive fungal infections, sporotrichosis is usually easy to treat.

- What fungal disease victimizes people who can't afford shoes?
- What fungus infection is likely to result from thorn or splinter injuries? What is unique about the treatment of the disease?
- Why might *Candida albicans* become pathogenic in an individual receiving antibacterial medications?

TABLE 27.11 "Rose Gardener's Disease" Sporotrichosis

Symptoms	Painless, ulcerating nodules appearing in sequence in a linear pattern
Incubation period	Usually 3 days to 3 weeks
Causative agent	*Sporothrix schenckii*, a dimorphic fungus
Pathogenesis	Spores multiply at site of introduction by thorn, splinter, or other plant material, causing a small nodule that ulcerates. Spores carried by lymph flow repeat the process along the course of the lymphatic vessel. May spread beneath the skin irrespective of lymphatic vessels
Epidemiology	Distributed worldwide in tropical and temperate climates. Occupations requiring contact with sharp plant materials at particular risk
Prevention and treatment	Protective gloves and clothing. Most cases effectively treated with potassium iodide. Generalized infections require amphotericin B or itraconazole

FUTURE CHALLENGES

Staying Ahead in the Race with *Staphylococcus aureus*

Over the last half century or so scientists have generally, but not always, kept us at least one step ahead in the race between Staphylococcus aureus *and human health. Today, "staph" seems again poised to challenge us for the lead, but new knowledge from molecular biology promises to help us leave this microscopic menace in the dust.*

In the past, we have relied heavily on developing antibacterial medications to kill or inhibit the growth of S. aureus *and probably will continue to do so. Other options, however, now seem to be within reach. For example, it should be possible to determine the structure of the active portion of toxin molecules and design medications that would bind to that site and inactivate it. Also, working with the staphylococcal genome, it is possible to determine which staphylococcal genes are responsible for virulence. The products of these genes can prove to be unexpected vaccine candidates. For example, in the spring of 1999, a staphylococcal substance was identified that is expressed only during infection, not under laboratory conditions; when used to vaccinate mice, it protected them against lethal injections of* S. aureus.

The race continues!

SUMMARY

Anatomy and Physiology

1. Wounds expose components of tissues to which pathogens specifically attach.
2. Wounds heal by forming **granulation tissue**, which in the absence of dirt and infection, fills the defect, and subsequently contracts to minimize scar tissue. (Figure 27.1)
3. Thermal burns often present large areas of dead tissue devoid of competing organisms and body defenses, ideal conditions for microbial growth.

Wound Abscesses

1. An **abscess** is composed of a collection of pus, which is composed of leukocytes, components of tissue breakdown, and infecting organisms.
2. Abscess formation localizes an infection within tissue to prevent its spread; inflammatory cells and clotted blood vessels separate the abscesses from normal tissue. (Figure 27.2)

Anaerobic Wounds

1. Anaerobic conditions are likely to occur in wounds containing dead tissue or foreign material, and those with a narrow opening to the air.
2. Anaerobic conditions permit infection by particularly dangerous pathogens.

Common Bacterial Wound Infections (Table 27.2)

1. Possible consequences of wound infections include delayed healing, abscess formation, and extension of infection or toxins into adjacent tissue or the bloodstream.
2. Infections can cause surgical wounds to split open, and they can spread to create biofilms on artificial devices.

Staphylococcal Wound Infections (Figures 27.3, 27.4)

1. Staphylococci are the leading cause of wound infections, both surgical and accidental; *Staphylococcus aureus* and *S. epidermidis*, are the most common wound-infecting species.
2. *Staphylococcus aureus* possesses many virulence factors; occasional strains release a toxin that causes toxic shock syndrome (Table 27.1)
3. *Staphylococcus epidermidis* is less virulent but can form biofilms on blood vessel catheters and other devices.

Group A Streptococcal "Flesh Eaters"

1. *Streptococcus pyogenes* (Group A, β-hemolytic streptococcus) causes strep throat, scarlet fever, wound infections, and other conditions.

2. Necrotizing fasciitis–causing strains produce exotoxin B, a protease thought to be responsible for the tissue destruction. (Figure 27.5)

Pseudomonas aeruginosa Infections

1. *Pseudomonas aeruginosa*, an aerobic Gram-negative rod with a single polar flagellum, is an opportunistic pathogen widespread in the environment, and a cause of both nosocomial infections and those acquired outside the hospital.
2. Production of two pigments by the bacterium often colors infected wounds green. (Figure 27.6)
3. Toxin A of *P. aeruginosa* has a mode of action identical to diphtheria toxin, but is it antigenically distinct and attaches to different cells.

Diseases Due to Anaerobic Bacterial Wound Infections

"Lockjaw" (Tetanus) (Table 27.4)

1. The characteristic symptom of tetanus is sustained, painful, cramplike spasms of one or more muscles (Figure 27.7). The disease is often fatal.
2. Tetanus is caused by an exotoxin, **tetanospasmin**, produced by *Clostridium tetani*, a non-invasive, anaerobic Gram-positive rod (Figure 27.8). The toxin renders the nerve cells that normally inhibit muscle contraction inactive by blocking release of their neurotransmitter. (Figures 27.9, 27.10)
3. The spores of *C. tetani* are widespread in dust and dirt, and most wounds are probably contaminated with them.
4. Tetanus can be prevented by active immunization with toxoid (inactivated tetanospasmin), and maintaining immunity throughout life. Any wound, including surgeries, no matter how trivial, is an occasion to make sure immunizations are current. (Table 27.3)

Gas Gangrene (Clostridial Myonecrosis) (Table 27.5)

1. Usually caused by the anaerobe *Clostridium perfringens*, symptoms begin abruptly with pain, swelling, a thin brown bubbly discharge, and dark blue mottling of the tightly stretched overlying skin. (Figure 27.12)
2. The toxin causes tissue necrosis; hydrogen and carbon dioxide gases are produced from fermentation of amino acids and glycogen in the dead tissue.
3. There is no vaccine or toxoid. Prevention depends on prompt medical care of dirty wounds containing dead tissue. Treatment depends on urgent surgical removal of dead and infected tissue, and it may require amputation.

"Lumpy Jaw" (Actinomycosis) (Table 27.6)

1. Actinomycosis is a chronic, slowly progressive disease characterized by repeated swellings, discharge of pus, and scarring, usually of the face and neck. (Figure 27.13)
2. The causative agent is *Actinomyces israelii*, a member of the normal mouth, intestinal, and vaginal flora that enters tissues with wounds such as those with dental and intestinal surgery. (Figure 27.14)
3. The organism is slow growing; treatment must be continued for weeks or months.

Bacterial Bite Wound Infections

1. The kind of bite wound infection depends on the kinds of infectious agents in the mouth of the biting animal, and the nature of the wound, whether puncture, crushed, or torn.

Pasteurella multocida Bite Wound Infections (Table 27.7)

1. *Pasteurella multocida*, a small Gram-negative rod, can infect bite wounds inflicted by a number of animal species.
2. Specific opsonizing antibody permits killing of the bacteria by phagocytes.
3. *P. multocida* causes **fowl cholera** and diseases in other animals, but many animals are asymptomatic carriers.

Cat Scratch Disease (Table 27.8, Figure 27.15)

1. Cat scratch disease is the most common cause of chronic localized lymph node enlargement in children.
2. Caused by *Bartonella henselae*, it begins with a pimple at the site of a bite or scratch, followed by enlargement of local lymph nodes, which often become pus-filled.
3. Most individuals with cat scratch disease recover without treatment, but persons with AIDS are susceptable to two serious conditions caused by *B. henselae*, **peliosis hepatis** and *bacillary angiomatosis.*

Streptobacillary Rat Bite Fever (Table 27.9)

1. Streptobacillary rat bite fever is characterized by relapsing fevers, head and muscle aches, and vomiting, following a rat bite. A rash and joint pains often develop.
2. It is usually caused by *Streptobacillus moniliformis*, a highly pleomorphic Gram-negative rod that characteristically produces cell wall deficient variants called L-forms.
3. Those in rat-infested dwellings and those who handle laboratory rats are at greatest risk of contracting the disease.

Human Bites (Table 27.10)

1. Human bites, injuries from objects that have been in a person's mouth, and those from striking a person in the teeth often result in severe infections, with pain, massive swelling and foul-smelling pus.
2. The infections are usually caused by members of the normal mouth flora acting synergistically, including anaerobic streptococci, fusiforms, spirochetes, and *Bacteroides* sp., often with *Staphylococcus aureus.*
3. The crushing nature of bite wounds causes death of tissue and conditions suitable for growth of anaerobes; prompt cleansing and application of an antiseptic are advised.

Fungal Wound Infections

1. Fungal wound infections are unusual in economically developed countries, except for *Candida albicans* infections of burns and other wounds in individuals receiving antibacterial therapy.
2. Madura foot occurs in many impoverished areas of the world where people do not wear shoes.

"Rose Gardener's Disease" (Sporotrichosis) (Table 27.11)

1. Rose gardener's disease, also called sporotrichosis, is a chronic fungal disease mainly of people who work with wood or vegetation.
2. The usual case is characterized by painless ulcerating nodules that develop one after the other along the course of a lymphatic vessel. (Figure 27.16)
3. The causative organism is the dimorphic fungus, *Sporothrix schenckii* (Figure 27.17), usually introduced into wounds caused by thorns or splinters.
4. The fungus is distributed worldwide in tropical and temperate climates; people at risk include farmers, carpenters, gardeners, and greenhouse workers.

REVIEW QUESTIONS

Short Answer

1. What four things determine whether microorganisms in a wound will cause a problem?
2. Give three undesirable consequences of wound infection.
3. Approximately what proportion of the total cost of nosocomial infections is due to postoperative wound infections?
4. Give an example of a pyogenic coccus and explain why it is called pyogenic.
5. What property of *Staphylococcus epidermidis* aids its ability to colonize plastic materials used in medical practice?
6. What is the role of clumping factor in the pathogenesis of infections due to *Staphylococcus aureus*?
7. What is the relationship between the superantigens of *S. aureus* and the organism's production of toxic shock?
8. Give two ways by which *S. aureus* resists the action of penicillins.
9. Why is there major concern about having only one antibacterial medication effective against *S. aureus*?
10. Name two underlying conditions that predispose to *Streptococcus pyogenes* "flesh-eating" disease.
11. Give two sources of *Pseudomonas aeruginosa*.
12. Outline the pathogenesis of lockjaw.
13. What characteristic of *Clostridium tetani* aids its isolation when other bacteria are present?
14. Explain why *C. tetani* can be cultivated from wounds in the absence of tetanus.
15. What is a neurotransmitter?
16. What characteristics of bite wounds favor anaerobic infections?
17. What is the causative agent of cat scratch disease? Why is it a threat to patients with AIDS?
18. What is a synergistic infection? How might one be acquired?
19. What is Madura foot?
20. Why is sporotrichosis sometimes called rose gardener's disease?

Multiple Choice

1. All of the following are true of *Staphylococcus aureus, except...*
   A. it is generally coagulase-positive.
   B. its infectious dose is increased in the presence of a foreign body.
   C. some strains infecting wounds can cause toxic shock.
   D. nasal carriage increases the risk of surgical wound infection.
   E. it is pyogenic.
2. Which of these statements about *Streptococcus pyogenes* is *false*?
   A. It is a Gram-positive coccus occurring in chains.
   B. Some strains infecting wounds can cause toxic shock.
   C. Some strains infecting wounds can cause necrotizing fasciitis.
   D. It can cause puerperal sepsis.
   E. A vaccine is available for preventing *S. pyogenes* infections.
3. Choose the one *false* statement about *Pseudomonas aeruginosa*.
   A. It is widespread in nature.
   B. Some strains can grow in distilled water.
   C. It is a Gram-positive rod.
   D. It produces a hemolytic toxin having the same mode of action as α-toxin of *Clostridium perfringens.*
   E. Under certain circumstances it can grow anaerobically.
4. Which of these statements concerning tetanus is *true*?
   A. It can originate from a bee sting.
   B. Immunization is carried out using tiny doses of tetanospasmin.
   C. Those who recover from the disease are immune for life.
   D. Tetanus immune globulin is derived from the blood of immunized sheep.
   E. It is easy to avoid exposure to spores of the causative organism.
5. Choose the one *true* statement about gas gangrene.
   A. There are few or no leukocytes in the wound drainage.
   B. It is best to rely on antibacterial medications and avoid disfiguring surgery.
   C. A toxoid is generally used to protect against the disease.
   D. Only one antitoxin is used for treating all cases of the disease.
   E. It is easy to avoid spores of the causative agent.
6. All the following statements about actinomycosis are true, *except...*
   A. it can occur in cattle.
   B. it is caused by a branching filamentous bacterium.
   C. it always appears on the jaw.
   D. it can arise from intestinal surgery.
   E. its abscesses can burrow into bone.
7. Which of the following statements about *Pasteurella multocida* is *false*?
   A. Infections generally respond to a penicillin.
   B. It can cause epidemics of fatal disease in domestic animals.

C. It is commonly found in the mouths of biting animals, including human beings.

D. A vaccine is in general use to prevent *P. multocida* disease in people.

E. Cat bites are more likely to result in *P. multocida* infections than dog bites.

8. Which of these statements about cat scratch disease is *false*?

A. It is a common cause of chronic lymph node enlargement in children.

B. It is a serious threat to individuals with AIDS.

C. Cat scratches are the only mode of transmission to humans.

D. It is a zoonosis of cats transmitted by fleas.

E. It can affect the brain or heart valves in a small percentage of cases.

9. The following statements about *Streptobacillus moniliformis* are all true, *except*...

A. it can be transmitted by food.

B. its colonies can resemble those of mycoplasmas.

C. it can be transmitted by the bites of animals other than rats.

D. human infection is characterized by irregular fevers, rash, and joint pain.

E. it is a Gram-positive spore-forming rod.

10. Pick out the statement concerning sporotrichosis that is *wrong*.

A. It is characterized by ulcerating lesions along the course of a lymphatic vessel.

B. Person-to-person transmission is common.

C. It can occur in epidemics.

D. It can persist for years if not treated.

E. The causative organism is a dimorphic fungus.

Applications

1. Clinicians become concerned when the laboratory reports that organisms capable of digesting collagen and fibronectin are present in a wound culture. What is the basis of their concern?
2. ABC pharmaceutical company produces a drug that inhibits new blood vessel growth into cancers, thus depriving the cancer cells of the nutrients and hormones needed for growth of the tumor. A researcher in the company laboratory has now isolated a compound having the opposite effect. Do you see any possible role for the new compound in the treatment of wounds?
3. An army field nurse performing triage at a mobile surgical hospital asks this question of all the ambulance drivers: "Was the soldier wounded while in a pasture?" Why did he ask this question?

Critical Thinking

1. In what ways would the curve of tetanus incidence in an economically disadvantaged country differ from figure 27.11?
2. Could colonization of a wound by a noninvasive bacterium cause disease? Explain your answer.
3. Surgical wound infections are frequently due to *Staphylococcus aureus*, a bacterium often carried in the nostrils and sometimes on the skin. How would you study the relationship, if any, between carriage of *S. aureus* and the chance of getting a wound infection?

CHAPTER 28

Blood and Lymphatic Infections

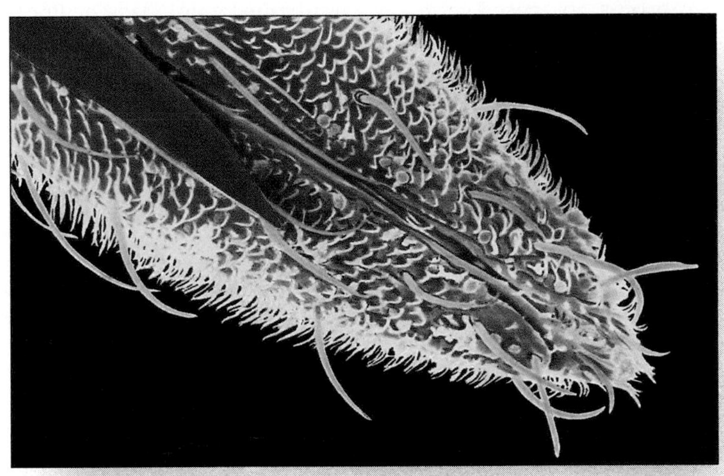

Alexandre Emile John Yersin (1863–1943) is one of the most interesting though relatively unknown contributors to the conquest of infectious diseases. He was born in 1863 in the Swiss village of Lavaux, where his family lived in a gunpowder factory. His father, director of the factory and a self-taught insect expert, died unexpectedly 3 weeks before Yersin's birth. Alexandre developed an interest in science when as a young boy he discovered in an attic trunk the microscope and dissecting instruments of the father he never knew. A local public health physician befriended the family and influenced Yersin to study medicine, and at age 20 he began pre-med studies in Lausanne. Later, while a medical student in Paris, Yersin volunteered to help at the Pasteur Institute, and in his fourth year of medical training, he was hired by Emile Roux, a coworker of Pasteur. Yersin became increasingly recognized for his work at the Institute, but he was bored with research, and he did not want to practice medicine because he also felt it was wrong for physicians to make a living from sickness.

He abruptly left the Pasteur Institute and was employed as a physician on a ship sailing from Marseilles to Saigon. Most of Indochina was under French control and was largely unexplored and undeveloped. Yersin became enchanted with the land and its peoples, exploring crocodile-infested rivers in a dugout canoe and making expeditions on elephant into jungles where tigers roared. For the rest of Yersin's life, Vietnam was his home. His many contributions to the region included the introduction of improved breeds of cattle, cultivation of rubber and quinine-producing trees, and the establishment of a medical school.

Yersin also studied the diseases that affected the Vietnamese people. One of them, plague, was a horrifying recurring problem. The disease had killed millions in Europe in medieval times and had struck France as recently as 1720. This great killer came and went unpredictably, an incurable disease, its cause and transmission completely unknown.

In 1894, Yersin went to study a Hong Kong outbreak of plague because the facilities for studying the disease were better in Hong Kong than in Vietnam. Unfortunately for Yersin, Shibasaburo Kitasato, the famous colleague of Robert Koch and Emil von Behring, had arrived 3 days before he did with a large team of Japanese scientists. British authorities had given them access to patients and laboratory facilities. Since Yersin did not speak English, it was difficult for him to communicate with the authorities. Moreover, although both he and the Japanese scientists spoke fluent German, the Japanese team treated him coolly, perhaps as a reflection of the intense rivalry between Pasteur and Koch. Yersin was reduced to setting up a laboratory in a bamboo shack. He bribed British sailors responsible for disposing of the bodies of plague victims to allow him to get samples of material from their swollen, pus-filled lymph nodes called buboes.

A week after his arrival in Hong Kong in the summer of 1894, Yersin reported to the British authorities his discovery of a bacillus of characteristic appearance, invariably present in the buboes of plague victims. The bacterium could be cultivated, and it caused plaguelike disease when injected into rats. Later, he showed that the disease was transmitted from one rat to another. Yersin's plague bacillus, now known as Yersinia pestis, was used to make an antiplague vaccine, and in 1896, antiserum prepared against the organism provided the world's first cure of a patient with plague. Not long afterward, another Pasteur Institute scientist proved that the rat flea was crucial in transmission. ■ **Robert Koch, p. 4, 451, 585, 691**
■ **Shibasaburo Kitasato, p. 691**

Kitasato, working with blood from the hearts of plague victims, announced shortly after his arrival in Hong Kong that he had isolated the bacterium that caused plague. Although subsequently it proved to be only a laboratory contaminant, Kitasato was named co-discoverer of the plague bacillus because of his great prestige.

—*A Glimpse of History*

THE CIRCULATION OF BLOOD AND LYMPH FLUIDS supplies nutrients and oxygen to the body's tissue cells and carries away the cells' waste products. The circulatory system also heats and cools body tissues to maintain an optimum temperature. Infection of the system can have devastating effects because infectious agents become **systemic**, meaning they can be carried to all parts of the body, producing disease in one or more vital organs or causing the circulatory system itself to stop functioning. Thus, even a small scratch can cause considerable harm if it results in a pathogen entering the bloodstream. The circulation of an agent in the bloodstream is given a name ending in *-emia* that specifies the nature of the agent, as with **bacteremia**, **viremia**, and **fungemia**. In many cases, there are no symptoms associated with the circulating agent. When illness results from a circulating agent or its toxins, the condition is referred to as **septicemia**, or blood poisoning. When, as a result of septicemia, the blood pressure falls to such low levels that blood flow to vital organs is insufficient to maintain their functioning, the condition is called **septic shock**.

Anatomy and Physiology

The Heart

The heart, a muscular double pump enclosed in a fibrous sac called the **pericardium**, supplies the force that moves the blood. As shown schematically in **figure 28.1**, the heart is divided into a right and a left side, separated by a septum, a wall of tissue, through which blood normally does not pass after birth. The right and left sides of the heart are both divided into two chambers, the **atrium**, which receives blood, and the **ventricle**, which discharges it. Blood from the right ventricle flows through the lungs and into the atrium on the left side of the heart. From the atrium, the blood passes into the left ventricle and is then pumped through the aorta to the arteries and capillaries that supply the tissues of the body. Blood that exits the capillaries is returned to the right atrium by the veins. The heart valves are situated at the entrance and exit of each ventricle and ensure that the blood flows in only one direction. Although not common, infections of the heart valves, muscle, and pericardium may be disastrous because they affect the vital function of blood circulation.

Arteries

Arteries have thick muscular walls to withstand the high pressure of the arterial system. The blood in arteries is bright red, the color of oxygenated hemoglobin. Infection of the arteries is unusual, although the aorta can be dangerously weakened by infection in syphilis. Arteries are subject to arteriosclerosis, "hardening of the arteries," a process that begins in teenage years or before and progresses through life, ultimately putting us at risk for heart attacks and strokes. A possible role of bacteria and viruses in arteriosclerosis has been considered for some years.

Veins

Blood becomes depleted of oxygen in the capillaries, causing it to assume the dark color characteristic of venous blood. Since pressure in the veins is low, one-way valves help keep the blood flowing in the right direction. Veins are easily compressed, so the action of muscles aids the flow of venous blood.

Thus, as the blood flows around the circuit, it alternately passes through the lungs and through the tissue capillaries. During each circuit, a portion of the blood passes through organs such as the spleen, liver, and lymph nodes, all of which, like the lung, contain phagocytic cells of the mononuclear phagocyte system. These phagocytes help cleanse the blood of foreign material, including infectious agents, as the blood passes through these tissues.

Lymphatics

The system of lymphatic vessels begins in tissues as tiny tubes that resemble blood capillaries but differ from them in having closed ("blind") ends and in being somewhat larger. Lymph, an almost colorless fluid, is conveyed within these lymphatic vessels. It originates from plasma, the noncellular portion of the blood, that has oozed through the walls of the blood capillaries to become the interstitial fluid that surrounds tissue cells. This fluid

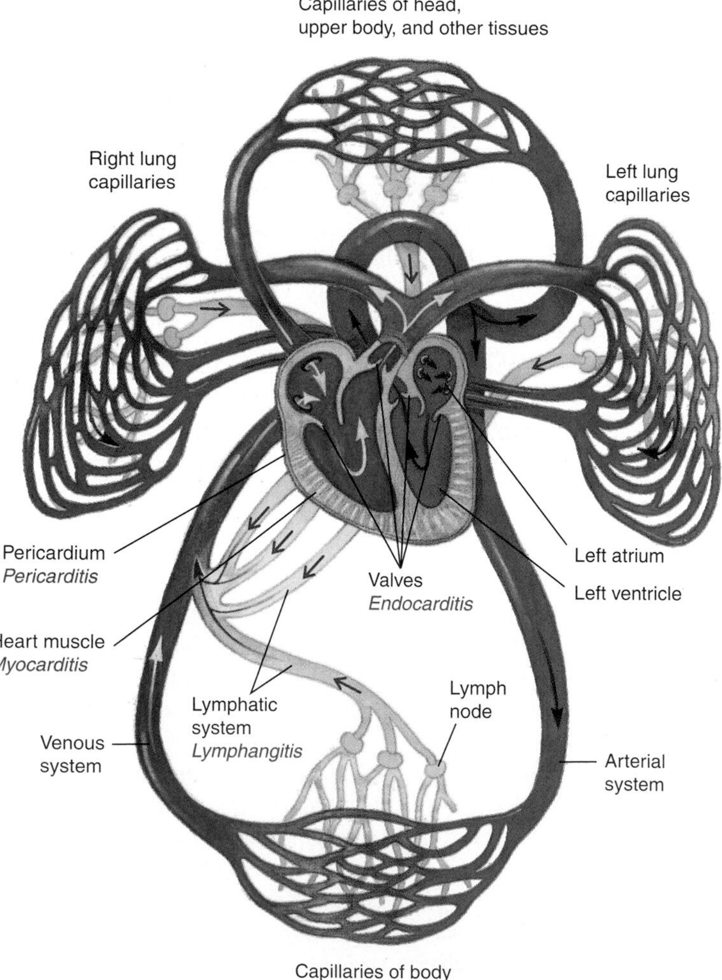

Figure 28.1 The Blood and Lymphatic Systems For simplicity, the spleen is not shown. It is a fist-sized, blood-filled lymphoid organ located high in the left side of the abdomen, behind the stomach. Disease conditions are indicated in red type.

Perspective 28.1 Arteriosclerosis: The Infection Hypothesis

Arteriosclerosis, the main cause of heart attacks and strokes, is characterized by lipid-rich deposits that develop in arteries and can impair or stop the flow of blood.

In 1988, researchers in Helsinki, Finland, reported that patients with coronary artery disease, meaning arteriosclerosis of the arteries that supply the heart muscle, commonly had antibodies against *Chlamydia pneumoniae*. This organism is a tiny obligate intracellular bacterium that is responsible for a variety of upper and lower respiratory infections including sinusitis and pneumonia. The infections are common, and most people have been infected by *C. pneumoniae* by the time they reach adulthood. What was intriguing about the antibody studies was that the higher the titer of antibody, the greater the risk of coronary disease. Then, in 1996, viable *C. pneumoniae* were shown to be present in arteriosclerotic lesions, and other studies conclusively established that there was an association between the bacterium and arteriosclerotic lesions of both heart and brain arteries.

Does *C. pneumoniae* cause arteriosclerosis, does it worsen the effects of arteriosclerosis, or does it merely exist harmlessly in the lesions? How can the hypothesis that *C. pneumoniae* contributes to heart attacks and strokes be tested? Some studies of coronary patients treated with antibiotics effective against the bacterium have suggested, but not proven, a beneficial effect, and so we will have to await further studies for answers to these questions. It would truly be a revolutionary finding if *C. pneumoniae* proves to play an important role in arteriotic heart and brain disease, currently numbers one and two killers of human beings!

bathes and nourishes the tissue cells and then enters the lymphatics. Unlike the blood capillaries, the readily permeable lymphatic vessels take up foreign material such as invading microbes and their products, including toxins and other antigens. The tiny lymphatic capillaries join progressively larger lymphatic vessels. Many one-way valves in the lymphatic vessels keep the flow of lymph moving away from the lymphatic capillaries. Both contraction of the vessel walls and compression by the movements of the body's muscles force the lymph fluid along.

At many points in the system, lymphatic vessels drain into small, bean-shaped bodies called lymph nodes. These nodes are constructed so that foreign materials such as bacteria are trapped in them; the nodes also contain phagocytic cells and antibody-producing cells. Lymph flows out of the nodes through vessels that eventually unite into one or more large tubes that then discharge into a large vein, usually behind the left collarbone, and thus back into the main blood circulation. ■ **lymphoid tissues, p. 374** ■ **phagocytic cells, p. 372**

When a hand or a foot is infected, a visible red streak may spread up the limb from the infection site (**figure 28.2**). This streak represents the course of lymphatic vessels that have become inflamed in response to the infectious agent. This condition is called **lymphangitis**. It may stop abruptly at a swollen and tender lymph node, only to continue later to yet another lymph node. This pause in the progression of lymphangitis demonstrates the ability, even though sometimes temporary, of the lymph nodes to clear the lymph of an infectious agent.

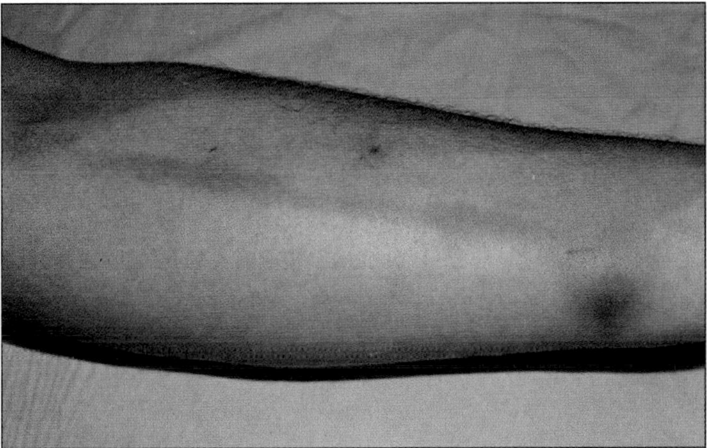

Figure 28.2 Lymphangitis Notice the red streak extending up the arm. The streak represents an inflamed lymphatic vessel.

Blood and lymph carry infection-fighting leukocytes and such antimicrobial proteins as antibodies, complement, lysozyme, β-lysin, and interferon. An inflammatory response may cause lymph and blood to clot in vessels that are close to areas of infection or antibody-antigen reactions, one of the ways the body has to localize an infection from the rest of the body. As with the nervous system, the blood and lymphatic systems lie deep within the body and are normally sterile. ■ **antimicrobial substances, p. 375** ■ **inflammatory response, p. 382**

Spleen

The spleen is a fist-sized organ situated high up on the left side of the abdominal cavity, behind the stomach. It is composed of two kinds of tissue; one consists of multiple blood-filled passageways, and the other, of lymphoid tissue. One of its functions is to cleanse the blood, much as the lymph nodes cleanse the lymph. Large numbers of phagocytes located in the spleen remove aging or damaged erythrocytes, bacteria, and other foreign materials from the blood. Another function, carried on by the lymphoid tissue, is to provide an immune response to microbial invaders. A third function is to produce new blood cells in rare situations where the bone marrow is unable to meet the demand. The spleen enlarges in a number of infectious diseases, such as infectious mononucleosis and malaria, in which its immune and filtering functions are challenged.

MICROCHECK 28.1

The heart can be thought of as two pumps that work together to move the blood around a circular course. The main function of the lymphatic system is to cleanse interstitial fluid and to return it to the bloodstream. Systemic infections threaten the transportation of oxygen and nutrients to body tissues and removal of waste products. The circulatory system can expose all the body's tissues to infectious agents and their toxins.

- What are the functions of the spleen?
- Which side of the heart, right or left, do bacteria in infected lymph reach first? Where do they go from there?
- What is the relationship between easily compressed veins, muscle action, and blood flow?

Bacterial Diseases of the Blood Vascular System

Bacterial infections of the vascular system can be rapidly fatal, or they can smolder for months, causing a gradual decline in health. They are not common, but they are always dangerous. Usually the bacteria are carried into the bloodstream by the flow of lymph from an area of infection in the tissues. Some pathogens multiply in the bloodstream, and they may colonize and form biofilms on structures such as the heart valves or abnormalities in the heart or major arteries.

Endocarditis is the term used for infections of the heart valves or the inner, blood-bathed surfaces of the heart, and it can be acute or subacute. **Acute bacterial endocarditis** starts abruptly with fever, and usually an infection such as pneumonia is present somewhere else in the body or there is evidence of injected-drug abuse. Virulent species such as *Staphylococcus aureus* and *Streptococcus pneumoniae* are usually the cause, and they infect both normal and abnormal heart valves. They can often produce a rapidly progressive disease, often with valve destruction and formation of abscesses in the heart muscle, leading to heart failure. By contrast, subacute bacterial endocarditis is usually caused by organisms with little virulence, and it has a much more protracted course. ■ ***Streptococcus pneumoniae*, p. 564**

Septicemia is caused by both Gram-negative and Gram-positive bacteria, as well as other infectious agents.

Subacute Bacterial Endocarditis

Subacute bacterial endocarditis (SBE) is usually localized to one of the valves on the left side of the heart. It commonly occurs on valves that are deformed as a result of a birth defect, rheumatic fever, or some other disease. ■ **subacute bacterial endocarditis, rheumatic fever, p. 555**

Symptoms

People with subacute bacterial endocarditis usually suffer from marked fatigue and slight fever. They typically become ill very gradually and slowly lose energy over a period of weeks or months. They may abruptly develop a stroke.

Causative Agent

The causative organisms of SBE are usually members of the normal bacterial flora of the mouth or skin, notably α-hemolytic viridans streptococci and *Staphylococcus epidermidis*. The infecting organisms are usually shed from the infected heart valve into the circulation and can be found by culturing samples of blood drawn from an arm vein. In 5% to 15% of cases, however, culturing the blood from an arm vein fails to yield the causative bacteria. This is especially likely to occur when the infection is on the right side of the heart, and blood from the infected site must pass through the lung and spleen, both richly supplied with phagocytes, before reaching the arm vein. Also, some bacteria that cause endocarditis, such as *Coxiella burnetii*, the cause of Q fever, cannot be cultivated on cell-free media. At least one, *Tropheryma whippelii*, the cause of Whipple's disease, can only be identified at the time of valve surgery by using the polymerase chain reaction and a nucleic acid probe. ■ **Q fever, p. 292** ■ **Whipple's disease, p. 255** ■ **polymerase chain reaction, p. 237** ■ **nucleic acid probes, p. 232**

Pathogenesis

The bacteria that cause SBE can gain entrance to the bloodstream during dental procedures, toothbrushing, or other trauma. In an abnormal heart, a thin blood clot may form in areas where there is turbulent blood flow around a deformed valve or other defect. This clot traps circulating organisms; these multiply and may form a biofilm that makes them inaccessible to phagocytic killing, and tends to protect them from antimicrobial treatment. People with bacterial endocarditis often have high levels of antibodies, which are of little value in eliminating the bacteria and may even be harmful because they make the bacteria clump together and adhere to the clot. As the organisms multiply, more clot is deposited around them, gradually building up a fragile mass. Bacteria continually wash off the mass into the circulation, and pieces of infected clot may break off. If large enough, these clots may block important blood vessels and lead to death of the tissue, infarction, supplied by the vessel as occurs in a stroke. They can also cause a vessel to weaken and balloon out, forming an **aneurysm**. Circulating immune complexes may lodge in the skin, eyes, and other body structures. In the kidney, they produce a kind of glomerulonephritis by their inflammatory effects. Even though the organisms normally have little invasive ability, great masses of them growing in the heart are sometimes able to burrow into heart tissue to produce abscesses or damage valve tissue, resulting in a leaky valve. This is a good example of how pathogenicity depends on both host factors and the virulence of the microorganism. ■ **biofilms, p. 108** ■ **immune complexes, p. 440** ■ **virulence, p. 454**

Epidemiology

In recent years, viridans streptococci have accounted for a smaller percentage of SBE cases than previously, partly because most dentists are careful to give antibiotic treatment to patients with prominent heart murmurs. A heart murmur is an abnormal sound the physician hears when listening to the heart, often indicative of a deformed valve or other structural abnormality. Also, more SBE cases occur in injected-drug abusers, in hospitalized patients who have plastic intravenous catheters for long periods, and in those with artificial heart valves. These people are usually infected with *S. epidermidis* or a wide variety of species other than viridans streptococci.

Prevention and Treatment

No scientifically proven preventive methods are available. In people with known or suspected defects in their heart valves, however, the accepted practice is to give an antibacterial medication shortly before dental or other bacteremia-causing procedures. The medication is chosen according to the expected species of bacterium and its likely antimicrobial susceptibility. To prevent nosocomial SBE, rigorous attention to sterile technique when inserting plastic intravenous catheters, changing the catheters to new locations every several days, and discontinuing their use as soon as possible probably helps prevent colonization of the catheters and consequent bacteremia that could lead to heart valve infection. For treatment, antibacterial medications are chosen according to the susceptibility of the causative organism. Only bactericidal medications are effective, and usually two or more antimicrobials are used together. Prolonged treatment over 1 or more months is usually required. Infection of foreign material such as an artificial heart valve is almost invariably associated with biofilm formation and the object must often be replaced to effect a cure. Sometimes, it is necessary to perform surgery to remove the mass of infected clot or to drain abscesses. ■ nosocomial infections, p. 488

The main characteristics of SBE are presented in **table 28.1.**

Gram-Negative Septicemia

Septicemia is a common nosocomial illness, with an estimated 400,000 cases occurring in the United States each year. Approximately 30% of of the cases are caused by Gram-negative bacteria. Endotoxin release by Gram-negative bacteria can lead to shock and death.

Symptoms

The symptoms of septicemia include violent shaking chills and fever, often accompanied by anxiety and rapid breathing. If septic shock develops, urine output drops, the respirations and pulse become more rapid, and the arms and legs become cool and dusky colored.

Causative Agents

Probably because they possess endotoxin, Gram-negative bacteria are more likely to cause fatal septicemias than other infectious agents. Shock is common, and despite treatment, only about half of all people afflicted with this kind of infection survive. Cultures of blood from these patients usually reveal enterobacteria such as *Escherichia coli.* Among the aerobic, Gram-negative rod-shaped bacteria encountered in septicemia are organisms commonly found in the natural environment, such as *Pseudomonas aeruginosa.* Some of the Gram-negative organisms causing septicemia are anaerobes. For example, *Bacteroides* sp., which make up a sizable percentage of the normal flora of the large intestine and the upper respiratory tract, cause many septicemia cases. ■ enterobacteria, p. 282 ■ Pseudomonas, p. 281, 697

Pathogenesis

Septicemia almost always originates from an infection somewhere in the body other than the bloodstream—a kidney infection, for example. Alterations in normal body defenses as the result of medical treatments, such as surgery, placement of catheters, and medications that interfere with the immune response, may allow microorganisms that normally have little invasive ability to infect the blood. Endotoxin is released from the outer cell walls of Gram-negative bacteria growing in a localized infection or in the bloodstream. Unfortunately, antibiotics that act against the bacterial cell wall can also enhance the release of endotoxin from the organisms. These antibiotics are typically used in treating Gram-negative bacterial infections. ■ endotoxin, p. 461

Many of the body's cell types, especially tissue macrophages and circulating leukocytes, respond defensively to endotoxin, just as they do to many other foreign substances, but the response to endotoxin is particularly intense. Quite likely, this exaggerated response to endotoxin is a type of hypersensitivity, which develops at an early age in response to endotoxin that diffuses into the circulation from the normal intestinal flora.

The response of the body cells to endotoxin is appropriate for localizing Gram-negative bacterial infections in tissues and killing the invaders. When localization fails and endotoxin enters the circulation, however, it causes the nearly simultaneous triggering of macrophages throughout the body and a cascade of harmful events **(figure 28.3)**. Although macrophages normally are of central importance in body defense, they also play a key role in septic shock.

The interaction of endotoxin with macrophage cell membrane causes the cell to synthesize and release cytokines such as tumor necrosis factor and interleukin-1, among others. Tumor necrosis factor (TNF) is released from macrophages within minutes of exposure to endotoxin. Levels rise quickly and then fall off. TNF has diverse effects, one of which is a change in the setting of the body's thermostat, causing the temperature to rise. TNF also causes circulating polymorphonuclear neutrophils (PMNs) to adhere to capillary walls, leading to large accumulations of these inflammatory cells in tissues such as the lung, which have large populations of macrophages. Experimentally, antibody against TNF gives substantial protection against endotoxic shock. ■ cytokines, p. 379

Interleukin-1 (IL-1) is another cytokine released from macrophages. Besides acting with TNF to cause fever and the release of leukocytes from bone marrow, it has many other effects. One potentially harmful action is to cause the release of enzymes from polymorphonuclear leukocytes.

TABLE 28.1 Subacute Bacterial Endocarditis

Symptoms	Fever, loss of energy over a period of weeks or months; sometimes, a stroke
Incubation period	Poorly defined, usually weeks
Causative agents	Usually oral α-hemolytic viridans streptococci or *Staphylococcus epidermidis*
Pathogenesis	Normal flora gain entrance to bloodstream through dental procedures, other trauma; in an abnormal heart, turbulent blood flow causes formation of a thin clot that traps circulating organisms; a biofilm forms, makes them inaccessible to phagocytic killing; pieces of clot break off, block important blood vessels, leading to tissue death
Epidemiology	Persons at risk are mainly those with hearts that have congenital defects or are damaged by disease such as rheumatic fever; situations that cause bacteremia
Prevention and treatment	Administration of an antibiotic at time of anticipated bacteremia, such as dental work. Treatment: Bactericidal antibiotics given together, such as penicillin and gentamicin

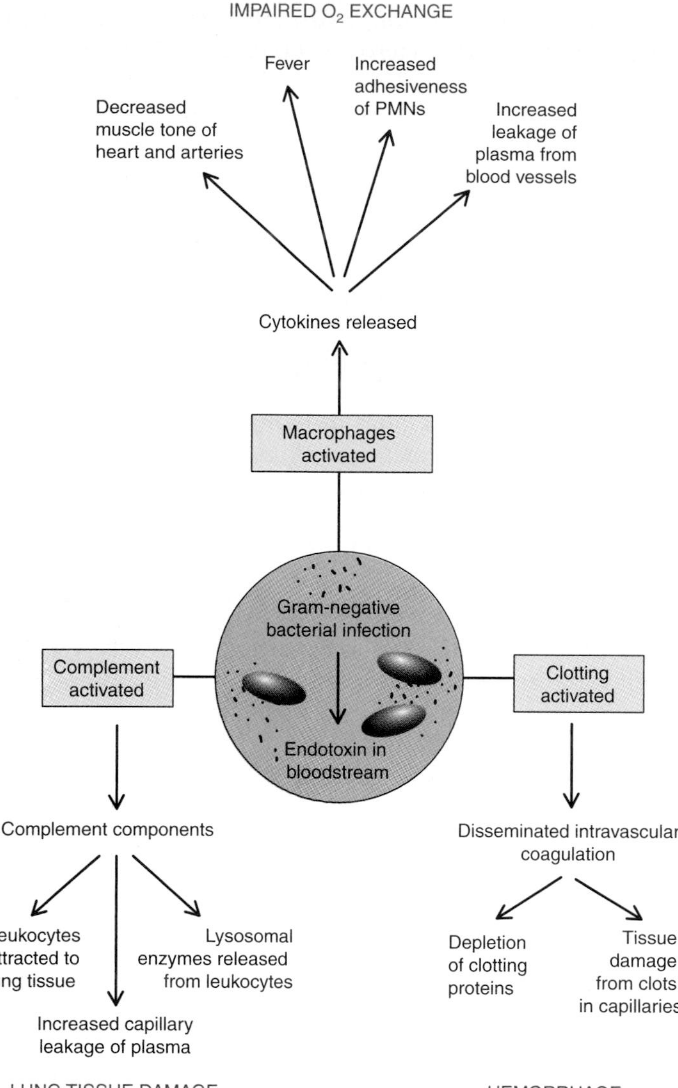

Figure 28.3 Events in Gram-Negative Septicemia

Macrophages also synthesize and secrete complement, which is activated by endotoxin. Components of activated complement attract leukocytes and cause them to release tissue-damaging lysosomal enzymes. Activated complement also causes capillaries to leak excessive amounts of plasma. ■ complement, p. 377

The circulating proteins responsible for blood clotting are also activated by endotoxin. Activation causes small clots to form, which plug capillaries, cutting off blood supply and causing tissue necrosis. Paradoxically, this condition, called **disseminated intravascular coagulation** (**DIC**), is often accompanied by hemorrhage, bleeding. This is because not enough clotting proteins are left to stop blood from flowing out of damaged blood vessels.

These harmful effects of endotoxemia, endotoxin circulating in the bloodstream, are made worse by the hypotension, low blood pressure, that usually is present. The hypotension is caused by decreased muscular tone of the heart and blood vessel walls, and the low blood volume that results from leakage of plasma out of the blood vessels. Current evidence indicates that the release of cytokines and their effects on the heart and blood vessels play a key role in causing hypotension in Gram-negative septicemia. Shock results when the blood pressure falls so low that vital organs are no longer supplied with adequate amounts of blood to maintain their function.

Although multiple organs are affected by endotoxemia, the lung is particularly vulnerable to serious, irreversible damage because of its high concentrations of tissue macrophages. Cytokines released from these macrophages in response to endotoxin attract leukocytes, which in turn discharge tissue-destroying enzymes. The resulting damage to lung tissue often results in death of an individual despite successful cure of their infection and correction of shock.

Epidemiology

Gram-negative septicemia is mainly a nosocomial disease, reflecting the high incidence of Gram-negative bacteremia in hospitalized patients with impaired host defenses. Patients with cancers and other malignancies, diabetes, and organ transplants are particularly vulnerable. There is a general trend toward an increasing incidence of the disease that relates to increasing life span, antibiotic suppression of normal flora, immunosuppressive medications, and medical equipment where biofilms readily develop, such as respirators, and catheters placed in blood vessels and the urinary system.

Prevention and Treatment

Prevention of septicemia depends largely on the prompt identification and effective treatment of localized infections, particularly in people whose host defenses are impaired. Also, conditions such as bedsores and pyelonephritis, which commonly lead to septicemia in patients with cancer and diabetes, can usually be prevented. For treatment, antimicrobial medications directed against the causative organism are given. Measures are taken to correct shock and poor oxygenation. Despite these treatments, the mortality rate remains high, generally 30% to 50%, partly because most of the patients have serious underlying conditions. Treatment with monoclonal antibody directed against endotoxin or TNF has shown some benefit when given early in the illness before shock develops, but no significant decrease in mortality 1 month later. A synthetic protein that blocks the receptors for TNF is more promising. Evaluation of these kinds of treatments is difficult because many of the patients have underlying life-threatening illnesses in addition to their septicemia. ■ pyelonephritis, p. 636 ■ monoclonal antibodies, p. 415

MICROCHECK 28.2

Bacteria of low virulence can cause serious, even fatal, infections in individuals whose only defect is a structural abnormality of the heart. A systemic infection represents failure of the body's mechanisms for keeping infections localized to one area. The inflammatory response, although vitally important in localizing infections, can be life threatening if generalized. People die of Gram-negative septicemia despite antibacterial therapy because of the endotoxin-mediated release of cytokines in the lung and bloodstream.

- What is a "systemic" infection?
- What is the difference between bacteremia and septicemia?
- Why might two or more antimicrobials be used together?
- Why might clots on the heart valves make microorganisms inaccessible to phagocytic killing?

Bacterial Diseases of the Lymph Nodes and Spleen

The enlargement of the lymph nodes and spleen is a prominent feature of diseases that involve the mononuclear phagocyte system. Three examples of these diseases—tularemia, brucellosis, and plague—are discussed next. All three are now uncommon human diseases in the United States. They represent a constant threat, however, because of their widespread existence in animals. Understanding these diseases is the best protection against contracting them. ■ **mononuclear phagocyte system, p. 372**

"Rabbit Fever" (Tularemia)

Tularemia is widespread among wild animals in the United States, involving species as diverse as rabbits, muskrats, and bobcats. Many human cases are acquired when people are skinning animals that appear to be free of disease. The causative organism enters through unnoticed scratches or by penetration of a mucous membrane. The disease can also be acquired from the bites of flies and ticks and by inhalation of the causative organism.

Symptoms

Tularemia is characterized by development of a skin ulceration and enlargement of the regional lymph nodes 2 to 5 days after a person is bitten by a tick or insect or handles a wild animal. The usual symptoms of fever, chills, and achiness that occur in many other infectious diseases are also present in tularemia. Symptoms usually clear in 1 to 4 weeks, but sometimes they last for months.

Causative Agent

Tularemia is caused by *Francisella tularensis*, a nonmotile, aerobic, Gram-negative rod that derives its name from Edward Francis, an American physician who studied tularemia in the early 1900s, and from Tulare County, California, where it was first studied. The organism is unrelated to other common human pathogens and is unusual in that it requires a special medium enriched with the amino acid cysteine in order to grow.

Pathogenesis

Typically, *F. tularensis* causes a steep-walled ulcer where it enters the skin (**figure 28.4**). Lymphatic vessels draining the area carry the organisms to the regional lymph nodes. These nodes then become large and tender, and they may become filled with pus and drain spontaneously. Later, the organisms spread to other parts of the body via the lymphatics and blood vessels. Pneumonia, which occurs in 10% to 15% of the cases, occurs when the organisms infect the lung from the bloodstream or by inhalation. Tularemic pneumonia has a mortality rate as high as 30%, most pneumonias occurring in individuals who work with the organism in laboratories. *Francisella tularensis*, like *Mycobacterium tuberculosis*, is ingested by phagocytic cells and grows within them. This may explain why tularemia persists in some people despite the high titers of antibody in their blood. Cell-mediated immunity is responsible for ridding the host of this infection, as it is with other pathogens that can live intracellularly. Both delayed hypersensitivity and serum antibodies quickly arise during infection, so that even without treatment over 90% of infected people survive. ■ **cell-mediated immunity, p. 404** ■ **delayed hypersensitivity, p. 441**

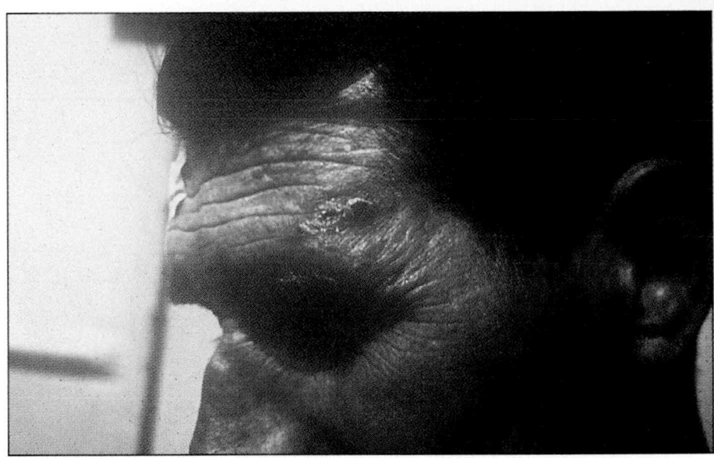

Figure 28.4 Ulceroglandular Form of Tularemia in a Muskrat Trapper The healing ulcer above the patient's left eyebrow is the site of entry of *Francisella tularensis*.

Epidemiology

Tularemia occurs among wild animals in many areas of the Northern Hemisphere, including all the states of the United States except Hawaii. In the eastern United States, human infections usually occur in the winter months, as a result of people skinning rabbits. Hence, the common name, "rabbit fever." Hunters and trappers in various parts of the country, however, have contracted the disease from muskrats, beavers, squirrels, deer, and other wild animals. The animals are generally free of illness. In the West, infections mostly result from the bites of infected ticks and deer flies and, thus, usually occur during the summer. Generally, 150 to 250 cases of tularemia are reported each year from counties across the United States (**figure 28.5**).

Prevention and Treatment

Rubber gloves and goggles or face shields are advisable for people skinning wild animals; remember that the bacteria can enter the body via mucous membranes. Insect repellants and protective clothing help guard against insect and arachnid vectors. It is a good practice to inspect routinely for ticks after exposure to the out-of-doors and to remove them carefully. A vaccine is

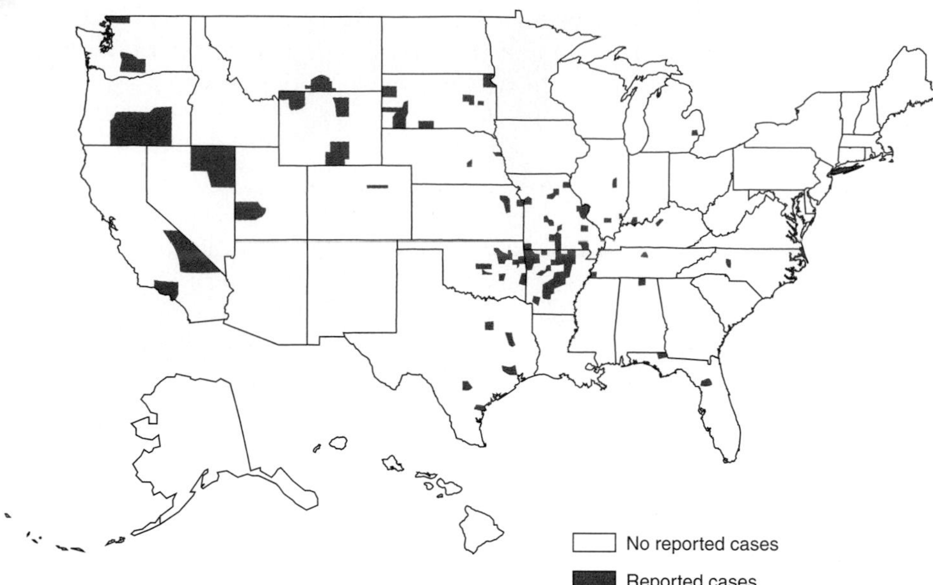

Figure 28.5 Counties Reporting Cases of Tularemia, United States, 1993

available for laboratory workers, veterinarians, trappers, game wardens, and others at high risk for infection. Most cases of tularemia are effectively treated with tetracycline or gentamicin.

The main features of this disease are summarized in **table 28.2**.

"Undulant Fever" (Brucellosis, "Bang's Disease")

Only about 150 cases of human brucellosis are reported each year in the United States. Between 10 and 20 times that number of cases, however, go unreported annually. Brucellosis is often called "undulant fever" or "Bang's disease," after Frederik Bang (1848–1932), a Danish veterinary professor who discovered the cause of cattle brucellosis.

Symptoms

The onset of brucellosis is usually gradual, and the symptoms are vague. Typically, patients complain of mild fever, sweating, weakness, aches and pains, and weight loss. The recurrence in some cases of fevers over weeks or months gave rise to the alternative name "undulant fever." Even without treatment, most cases recover within 2 months, and only 15% will be symptomatic for more than 3 months.

Causative Agent

Four varieties of the genus *Brucella* cause brucellosis in humans. DNA studies show that all members of the genus fall into a single species, *Brucella melitensis*, but traditionally, the different varieties were assigned species names depending largely on their preferred host: *B. abortus* invades cattle; *B. canis*, dogs; *B. melitensis*, goats; and *B. suis*, pigs. *Brucella* sp. are small, aerobic, nonmotile, Gram-negative rods with complex nutritional requirements. The distinctions between the various strains are mainly useful epidemiologically and generally have little pathogenic significance.

Pathogenesis

As with tularemia, the organisms responsible for brucellosis penetrate mucous membranes or breaks in the skin and are disseminated via the lymphatic and blood vessels to the heart, kidneys, and other parts of the body. Like *Francisella tularensis*, *Brucella* sp. are not only resistant to phagocytic killing but also can grow intracellularly in phagocytes, where they are inaccessible to antibody and some antibiotics. Mortality, generally due to endocarditis, is about 2%. Bone infection, **osteomyelitis**, is the most frequent serious complication.

Epidemiology

Brucellosis is typically a chronic infection of domestic animals involving the mammary glands and the uterus, thereby contaminating milk and causing abortions in the affected animals. Abortion is not a feature of human disease. Sixty percent of the cases of brucellosis occur in workers in the meat-packing industry; less than 10% arise from ingesting raw milk or other unpasteurized dairy products. Worldwide, brucellosis is a major problem in animals used for food, causing yearly losses of many millions of dollars. In the United States, infections have been acquired by hunters from elk, moose, bison, caribou, and reindeer. About 20% of the Yellowstone Park bison herd are infected, and more than 1,000 have been killed when they wander outside the park because of fear that they will transmit the disease to cattle.

TABLE 28.2 "Rabbit Fever" (Tularemia)

Symptoms	Ulcer at site of entry, enlarged lymph nodes in area, fever, chills, achiness
Incubation period	1 to 10 days; usually 2 to 5 days
Causative agent	*Francisella tularensis*, an aerobic Gram-negative rod
Pathogenesis	Organisms are ingested by phagocytic cells, grow within these cells, spread throughout body
Epidemiology	Present among wildlife in most states of the United States. Risk mainly to hunters, trappers, game wardens, and others who handle wildlife. Mucous membrane or broken skin penetration of the organism, as with skinning rabbits, for example; bite of infected insect or tick. Occasionally, inhalation
Prevention and treatment	Vaccination for high-risk individuals; avoiding bites of insects and ticks; wearing rubber gloves, goggles, when skinning rabbits; taking safety precautions when working with organisms in laboratory. Treatment: gentamicin or tetracycline

Prevention and Treatment

The most important control measures against brucellosis are pasteurization of dairy products and inspection of domestic animals for evidence of the disease. The use of goggles or face shield and rubber gloves helps protect veterinarians, butchers, and slaughterhouse workers; remember, the bacterium can penetrate mucous membranes. A live attenuated vaccine effectively controls the disease in domestic animals. Treatment using tetracycline with rifampin for 6 weeks is usually effective. ■ pasteurization, p. 117

Table 28.3 gives the main features of brucellosis.

"Black Death" (Plague)

Plague, once known as the "black death," was responsible for the death of approximately one-fourth of the population of Europe between 1346 and 1350. Crowded conditions in the cities and a large rat population undoubtedly played major roles in the spread of the disease.

Symptoms

The symptoms of plague develop abruptly 1 to 6 days after an individual is bitten by an infected flea. The person characteristically develops markedly enlarged and tender lymph nodes called buboes—hence the name **bubonic plague**—in the region that receives lymph drainage from the area of the flea bite. High fever, shock, delirium, and patchy bleeding under the skin quickly develop, and in many cases, there may also be cough and bloody sputum if the lungs become involved, which is called **pneumonic plague**.

TABLE 28.3 "Undulant Fever" (Brucellosis, "Bang's Disease")

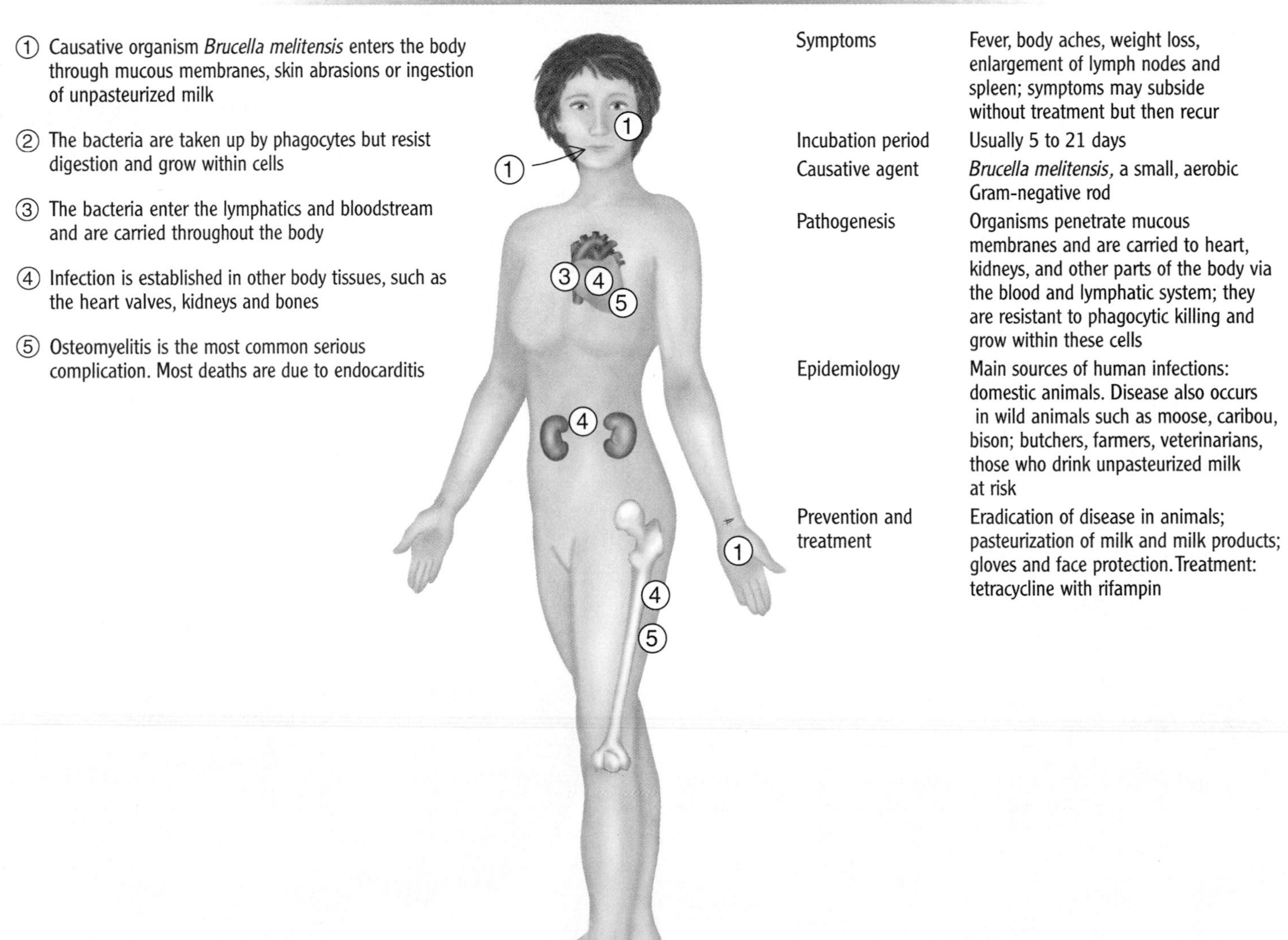

① Causative organism *Brucella melitensis* enters the body through mucous membranes, skin abrasions or ingestion of unpasteurized milk
② The bacteria are taken up by phagocytes but resist digestion and grow within cells
③ The bacteria enter the lymphatics and bloodstream and are carried throughout the body
④ Infection is established in other body tissues, such as the heart valves, kidneys and bones
⑤ Osteomyelitis is the most common serious complication. Most deaths are due to endocarditis

Symptoms	Fever, body aches, weight loss, enlargement of lymph nodes and spleen; symptoms may subside without treatment but then recur
Incubation period	Usually 5 to 21 days
Causative agent	*Brucella melitensis,* a small, aerobic Gram-negative rod
Pathogenesis	Organisms penetrate mucous membranes and are carried to heart, kidneys, and other parts of the body via the blood and lymphatic system; they are resistant to phagocytic killing and grow within these cells
Epidemiology	Main sources of human infections: domestic animals. Disease also occurs in wild animals such as moose, caribou, bison; butchers, farmers, veterinarians, those who drink unpasteurized milk at risk
Prevention and treatment	Eradication of disease in animals; pasteurization of milk and milk products; gloves and face protection. Treatment: tetracycline with rifampin

Causative Agent

Plague is caused by *Yersinia pestis*, a facultatively intracellular enterobacterium that grows best at 28°C. The organism resembles a safety pin in stained preparations of material taken from infected lymph nodes because the ends of the bacterium stain more intensely than does the middle (**figure 28.6**). Extensive study has revealed some of the complex mechanisms by which *Y. pestis* achieves its impressive virulence. The bacterium has three kinds of plasmids distinguishable by their size. The smallest has 9.5 kilobase pairs (9,500 base pairs) and codes for an important protease called **Pla** that causes blood clots to dissolve by activating a substance in the body called plasminogen activator. Another activity of the Pla protease is that it destroys the C3b and C5a components of complement. ■ plasmids, p. 70, 208, 210 ■ complement, p. 377

The middle-sized plasmid has 72 kilobase pairs, and almost its entire genome codes for (1) a group of proteins that interfere with phagocytosis and (2) regulators of the proteins' expression. These proteins are referred to as Yops, for *Yersinia* outer-membrane proteins. The bacterium loses its virulence if this plasmid is lost. Yops are injected into host cells by a type III secretion system and are responsible for several kinds of actions. For example, Yop E destroys microfilaments of host cells, Yop H interferes with the ability of phagocytes to receive environmental signals, Yop J blocks cytokine synthesis and promotes apoptosis, and Yop M prevents the release of cytokines, which would normally call forth an inflammatory response to infection. The sum of the effects of Yops is to interfere with phagocytes that normally would kill *Y. pestis* and initiate an immune response to the bacterium. ■ secretion systems, pp. 598–599 ■ microfilaments, p. 77

The size of the largest plasmid is 110 kilobase pairs, and it codes for **Fra 1**. This protein become part of an antiphagocytic capsule. The stimulus for capsule production is the relatively increased temperature of a mammalian host (37°C for humans) compared with that of a flea (about 26°C). The genes for some other *Y. pestis* virulence factors reside on the bacterial chromosome. Included in these properties are resistance to the lytic action of activated complement, mechanisms for storing hemin and using it as an iron source, and production of a pilus adhesin. These genes are probably activated as a result of intracellular growth. ■ iron requirement, p. 385 ■ adhesins, p. 458

A summary of these potential virulence factors is presented in **table 28.4**.

Pathogenesis

Masses of *Y. pestis* partially obstruct the digestive tract of infected rat fleas (**figure 28.7**). Consequently, not only is the flea ravenously hungry, causing it to bite repeatedly, it also regurgitates infected material into the bite wounds. Chronically infected fleas do not have their digestive tract obstructed but discharge *Y. pestis* in their feces; the organisms can then be introduced into human tissue when a person scratches the flea bite.

The *Y. pestis* protease Pla is essential for the spread of the organisms from the site of entry of the bacteria by clearing the lymphatics and capillaries of clots. The organisms are carried to

5 µm

Figure 28.6 ***Yersinia pestis*** Each pole of the organism is stained intensely by certain dyes, producing a safety pin appearance.

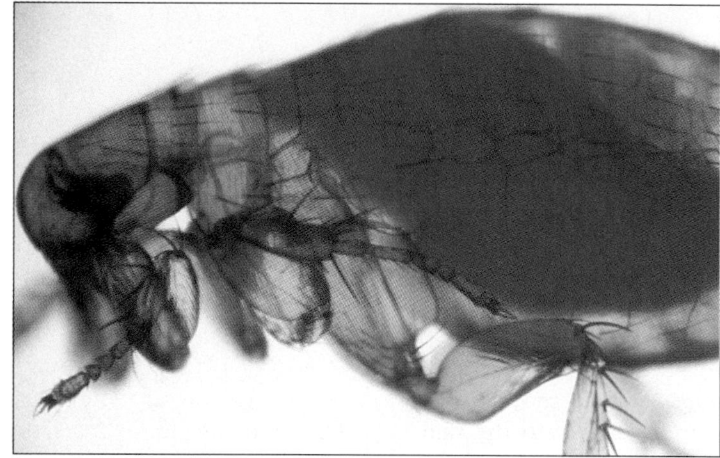

(a)

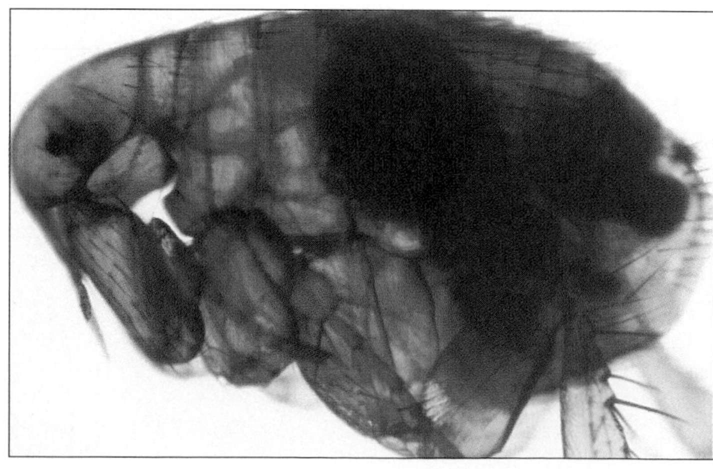

(b)

Figure 28.7 **Fleas Following a Blood Meal** **(a)** Healthy flea; **(b)** flea with obstruction due to *Yersinia pestis* infection.

TABLE 28.4 Virulence Factors of *Yersinia pestis*

Factor	Coded by	Action
Pla (protease)	9.5-kbp plasmid	Activates plasminogen activator; destroys C3b, C5a
Yops (proteins)	72-kbp plasmid	Interferes with phagocytosis and the immune response by differing mechanisms
Fra 1	110-kbp plasmid	Forms antiphagocytic capsule at 37°C
PsaA (adhesin)	Chromosome	Role in attachment to host cells
Complement resistance	Chromosome	Protects against lysis by activated complement
Iron acquisition	Chromosome	Traps hemin and other iron-containing substances; stores iron compounds intracellularly

the regional lymph nodes, where they are taken up by macrophages. The bacteria are not killed by the macrophages and instead multiply and elaborate Fra 1 capsular material and other virulence factors in response to the intracellular growth conditions. The macrophages die and release the bacteria, which now are encapsulated and express Yops, pilus adhesin, complement resistance, and heme storage. After several days, an acute inflammatory reaction develops in the nodes, producing enlargement and marked tenderness. The lymph nodes become necrotic, allowing large numbers of virulent *Y. pestis* to spill into the bloodstream. This stage of the disease is called **septicemic plague**, and endotoxin release results in shock and disseminated intravascular coagulation (DIC). Infection of the lung from the bloodstream occurs in 10% to 20% of the cases, resulting in pneumonic plague. Organisms transmitted to another person from a case of pneumonic plague are already fully virulent and, therefore, especially dangerous. The dark hemorrhages into the skin from DIC and the dusky color of skin and mucous membranes probably inspired the name black death for the plague.

The mortality rate for persons with untreated bubonic plague is between 50% and 80%. Untreated pneumonic plague progresses rapidly and is nearly always fatal within a few days.

Epidemiology

Plague is endemic in rodent populations of all continents except Australia. In the United States, the disease is mostly confined to wild rodents in about 15 states in the western half of the country. Prior to 1974, only a few cases of plague were reported each year. Over the last few decades, however, the number of cases has generally been higher, averaging about 15 reported cases per year, as towns and cities expand into the countryside. Prairie dogs, rock squirrels, and their fleas constitute the main reservoir, but rats, rabbits, dogs, and cats are potential hosts. Hundreds of species of fleas can transmit plague, and the fleas can remain infectious for a year or more in abandoned rodent burrows. Epidemics in humans, initiated by infected rodent fleas, can spread from person to person by household fleas, as well as by aerosols produced by coughing patients with pneumonic plague. *Yersinia pestis* can remain viable for weeks in dried sputum and in flea feces and for months in the soil of rodent burrows. ■ **endemic disease, p. 474** ■ **reservoir, p. 476**

Prevention and Treatment

Plague epidemics can be prevented by rat control measures such as proper garbage disposal, constructing rat-proof buildings, installing guards on the ropes that moor oceangoing ships to keep rats from entering, and rat extermination programs. The latter must be combined with the use of insecticides to prevent the escape of infected fleas from dead rats. A killed vaccine that gives short-term partial protection against plague is available to control epidemics and for those who are at high risk in laboratories or endemic areas. The antibiotic tetracycline can be given as a preventive for someone exposed to plague. The antibiotic is useful in controlling epidemics because of its immediate effect. Treatment with gentamicin or tetracycline is effective, especially if given early in the disease.

The main features of plague are presented in **table 28.5**.

MICROCHECK 28.3

Generally, tularemia is a bacteremic disease of wild animals transmitted to humans by exposure to their blood or tissues, or by biting insects. Brucellosis is most commonly a disease of domestic animals transmitted to humans by handling their flesh or by drinking unpasteurized milk. Plague is endemic in rodents in the western United States and is transmitted to humans by flea bites. Unlike tularemia and brucellosis, plague has the potential to spread rapidly from human to human by coughing, with a high fatality rate.

- Workers in what industry are especially likely to contract brucellosis? How can they protect themselves from the disease?
- How would growth within phagocytes protect *Francisella tularensis* from destruction by antibodies?
- How would crowded conditions in cities favor spread of plague?

TABLE 28.5 "Black Death" (Plague)

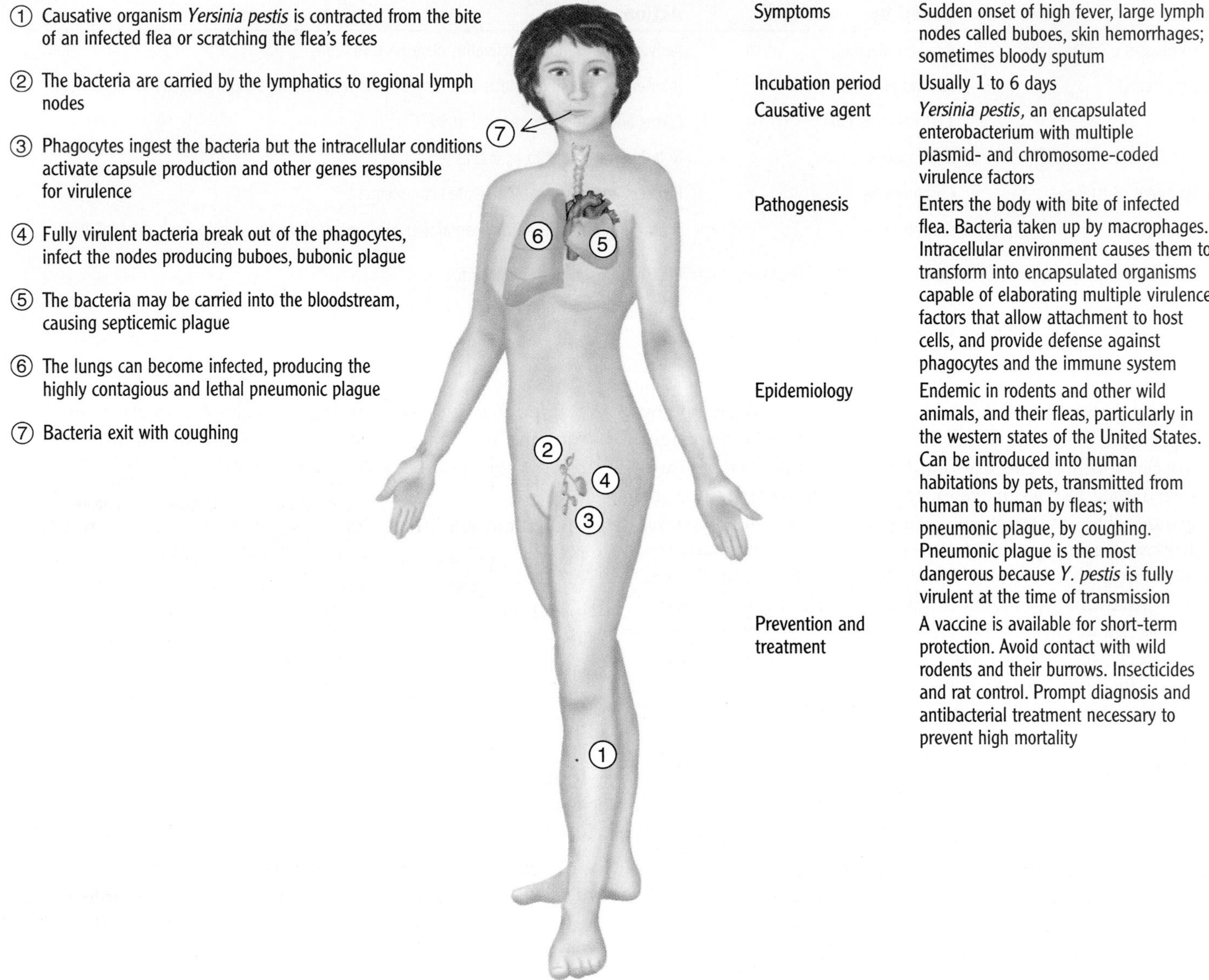

① Causative organism *Yersinia pestis* is contracted from the bite of an infected flea or scratching the flea's feces
② The bacteria are carried by the lymphatics to regional lymph nodes
③ Phagocytes ingest the bacteria but the intracellular conditions activate capsule production and other genes responsible for virulence
④ Fully virulent bacteria break out of the phagocytes, infect the nodes producing buboes, bubonic plague
⑤ The bacteria may be carried into the bloodstream, causing septicemic plague
⑥ The lungs can become infected, producing the highly contagious and lethal pneumonic plague
⑦ Bacteria exit with coughing

Symptoms	Sudden onset of high fever, large lymph nodes called buboes, skin hemorrhages; sometimes bloody sputum
Incubation period	Usually 1 to 6 days
Causative agent	*Yersinia pestis,* an encapsulated enterobacterium with multiple plasmid- and chromosome-coded virulence factors
Pathogenesis	Enters the body with bite of infected flea. Bacteria taken up by macrophages. Intracellular environment causes them to transform into encapsulated organisms capable of elaborating multiple virulence factors that allow attachment to host cells, and provide defense against phagocytes and the immune system
Epidemiology	Endemic in rodents and other wild animals, and their fleas, particularly in the western states of the United States. Can be introduced into human habitations by pets, transmitted from human to human by fleas; with pneumonic plague, by coughing. Pneumonic plague is the most dangerous because *Y. pestis* is fully virulent at the time of transmission
Prevention and treatment	A vaccine is available for short-term protection. Avoid contact with wild rodents and their burrows. Insecticides and rat control. Prompt diagnosis and antibacterial treatment necessary to prevent high mortality

Viral Diseases of the Lymphoid and Blood Vascular Systems

A number of viral illnesses mainly involve the lymphoid system and blood vessels. Two notable examples are the human immunodeficiency virus (HIV), cause of AIDS, and cytomegalovirus, which causes lymph node enlargement in healthy adults and more serious disease in infants and individuals with immunodeficiency. Infectious mononucleosis has similar symptoms, but it is a much more familiar disease. Yellow fever more prominently involves the circulatory system. ■ **human immunodeficiency virus disease, p. 742** ■ **cytomegalovirus disease, p. 759**

"Kissing Disease" (Infectious Mononucleosis, "mono")

Infectious mononucleosis ("mono") is a disease familiar to many students because of its high incidence among people between the ages of 15 and 24 years. The term *mononucleosis* refers to the fact that people afflicted with this condition have an increased number of mononuclear leukocytes in their blood. ■ **leukocytes, p. 370**

Symptoms

Typically, symptoms of infectious mononucleosis appear after a long incubation period, usually 30 to 60 days. They consist of fever, a sore throat covered with pus, marked fatigue, and enlargement of the spleen and lymph nodes. In most cases, the fever and

CASE PRESENTATION

The patients were two American boys, seven and nine years old, living in Thailand, who almost simultaneously developed irritated eyes and slightly runny noses, progressing to fever, headache, and severe muscle pain. It hurt them to move their eyes, and they refused to walk because of pain in their legs. The symptoms subsided after a couple of days, only to recur at lesser intensity. No treatment was given, and they were completely back to normal within a week. Their illness was diagnosed as **dengue** (pronounced DEN-gay), also known as "breakbone fever."

1. What causes dengue and where does it occur?
2. Why is dengue important?
3. What is known about the pathogenesis of dengue?
4. What can be done to prevent dengue?

Discussion

1. Dengue is caused by any of four closely related flaviviruses, dengue 1, 2, 3, and 4. The disease occurs in large areas of the tropics and subtropics around the world. In the Americas, the disease is transmitted mainly by *Aedes aegypti* mosquitoes, which have staged a comeback after being almost eradicated in the 1960s. This vector now occurs year-round along the Gulf of Mexico. Cases of dengue were diagnosed in 18 states and the District of Columbia in 1996. Most of these cases were contracted in the Caribbean region or in Asia.
2. The two individuals presented here had a mild form of the disease that characteristically involves newcomers to an area where the disease commonly occurs. In endemic areas, the disease is characterized by fever, headache, muscle aches, rash, nausea, and vomiting. In a small percentage of cases, however, the effects of dengue infection are much more serious, resulting in dengue hemorrhagic fever. This form of the disease is characterized by bleeding and leakage of fluid from the capillaries. An important result is that the blood pressure drops, and the blood thickens. With expert treatment, the mortality is about 1% to 2%. Dengue shock syndrome is another potential development. It is characterized by profound shock and disseminated intravascular coagulation (DIC) and has a mortality above 40%.
3. About 90% of the cases of hemorrhagic fever or shock occur in subjects who have previously been infected by a dengue virus and have antibody to it. The remaining cases occur largely in infants who still have transplacentally acquired maternal antibody against dengue viruses. These antibodies, whether from earlier infection or from the mother, attach to the infecting dengue virus strain and thereby promote its uptake by macrophages. The virus, instead of being killed, reproduces in the macrophage. This results in the death of the macrophage and release of chemicals that cause leaky capillaries and shock. T lymphocytes may also play a role in this process in older children and adults. Evidence also exists that the virus causes a depression of the bone marrow, accounting for the very low white blood cell and platelet counts seen in dengue.
4. Scientists at Mahidol University in Bangkok and other research centers around the world are working on vaccines designed to bring dengue under control. At present, control efforts are largely directed at killing mosquitoes and their larvae and eliminating water containers around houses where the mosquitoes breed. Scientists at Colorado State University have successfully genetically engineered mosquitoes to render their cells incapable of reproducing dengue virus. This was accomplished by using another virus to introduce an antisense segment of the dengue genome into the mosquito cells. This was a dramatic accomplishment, but years of work remain before it can play a role in the control of dengue. ■ **antisense technology, p. 173**

sore throat are gone in about 2 weeks, the enlarged lymph nodes in 3. Persons can usually return to school or work within 4 weeks, but some suffer severe exhaustion and difficulty concentrating that prohibits return to normal activities for months.

Causative Agent

Infectious mononucleosis is caused by the Epstein-Barr (EB) virus named after its discoverers, M. A. Epstein and Y. M. Barr. It is a double-stranded DNA virus of the herpesvirus family, and although identical in appearance, it is not closely related to any of the other known herpesviruses that cause human disease. This interesting virus was unknown until the early 1960s when it was isolated from **Burkitt's lymphoma**, a malignant tumor derived from B lymphocytes, common in parts of Africa. Subsequent studies showed that EB virus is the cause of infectious mononucleosis. ■ **B lymphocytes, p. 397**

Pathogenesis

Primary infection with EB virus is analogous to throat infections with the herpes simplex virus in that both viruses initially infect the mouth and throat and then become latent in another cell type. The probable sequence of events is shown in **figure 28.8**. Following replication in epithelium of the mouth, saliva-producing glands, and throat, the Epstein-Barr virus is carried by the lymphatics to the lymph nodes. There it infects B lymphocytes, which have specific surface receptors for the virus. During the illness, up to 20% of the circulating B lymphocytes are infected with the virus. The B-cell infection can be (1) productive, in which the virus replicates and kills the B cell, or (2) nonproductive, in which the virus establishes a latent infection, existing as either extrachromosomal circular DNA, or integrated into the host cell chromosome at random sites. For most of these cells, the infection is non-productive, but profound changes in the cells result from the infection. The virus activates the B cells, causing multiple clones of B lymphocytes to proliferate and produce immunoglobulin. The infected cells are also "immortal," meaning that they can reproduce indefinitely in laboratory cultures.

The T lymphocytes respond actively to the infection and destroy any B cells with productive infections since these B cells display viral antigens on their surfaces. The abnormal-appearing lymphocytes characteristically seen in smears of the patient's blood are activated helper T cells (**figure 28.9**).

The proliferating lymphocytes are responsible for the large numbers of mononuclear cells that give the disease its name. Their numbers and appearance sometimes mistakenly suggest the diagnosis of leukemia, which is disproved when the patient spontaneously recovers. In many cases, a consequence of B-cell infection is the appearance of an IgM antibody that will react with an antigen on the red blood cells of certain animal species, notably sheep, horse, and ox. This kind of antibody arising against antigens of another animal is called a **heterophile antibody**. It generally has no pathologic significance, but its presence helps in diagnosing infectious mononucleosis. The antibody does not react specifically with EB virus. Enlargement of the lymph nodes and spleen reflects the active replication of lymphocytes. Hemorrhage from rupture of the enlarged spleen is the chief cause of the rare deaths due to infectious mononucleosis. It is most likely to occur within 3 to 4 weeks of the onset of illness. Patients are instructed to avoid exertion and contact sports, but almost half of the cases of splenic rupture occur in the absence of trauma.

EB virus
Epithelium
EB virus attaches to and infects epithelium of the throat, where it replicates and causes pharyngitis.
Lymphatic vessel
Blood vessel
Lymphatic vessel
Lymph node
Virions enter the lymphatic vessels and are carried to the lymph nodes. Some virions escape the lymph node trap and are carried to the bloodstream.
Virions attach specifically to B lymphocytes and infect them, producing either latent or productive infections.
Productively infected B cell
Viral genome
Latently infected B cell
Activated lymphocyte
The infected B lymphocytes actively replicate and differentiate into cells producing random immunoglobulins, including heterophile antibody.
Plasma cells
Heterophile antibody
Productively infected B cell
T cells respond to infection and destroy the lymphocytes replicating EB virus.
Activated T cell
Lysed B cell
Latently infected B cell
Latently infected B lymphocytes become immortal. They are not attacked by T cells.

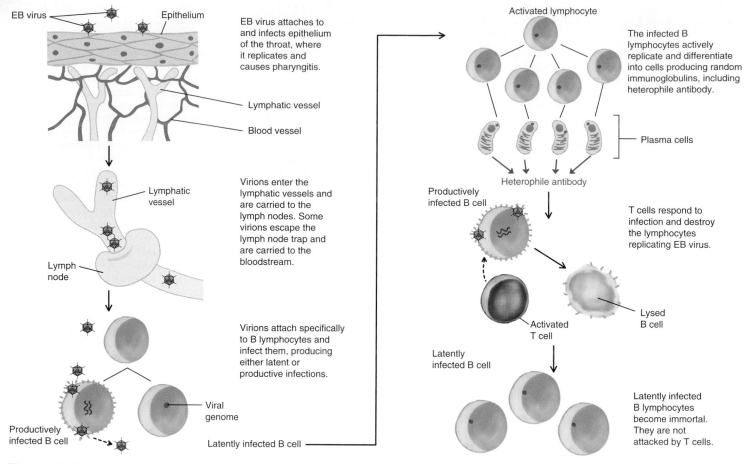

Figure 28.8 **Pathogenesis of Infectious Mononucleosis**

The possibility that EB virus could play a role in causing certain malignant tumors has been intensely investigated. The malignancies most closely related to EB virus infection—Burkitt's lymphoma and nasopharyngeal carcinoma—cluster dramatically in certain populations. This suggests that factors other than simple EB viral infection must be present for these malignancies to develop. EB virus–associated Burkitt's lymphoma cases occur mainly in children in East Africa and New Guinea, whereas nasopharyngeal carcinoma is common in Southeast China. The EB virus genome is detectable in 90% to 100% of these two malignancies. Infectious mononucleosis has not been shown to increase risk of malignant tumors in

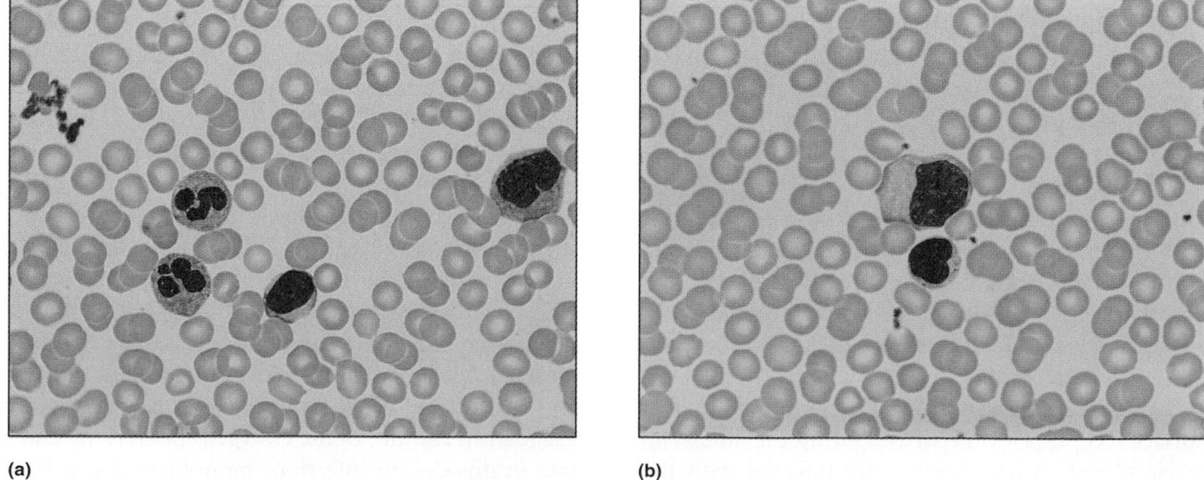

Figure 28.9 **Normal and Infectious Mononucleosis Blood Smears** Photomicrographs of **(a)** a normal blood smear showing two polymorphonuclear neutrophils (PMNs), a lymphocyte and a monocyte and **(b)** an infectious mononucleosis blood smear showing an abnormal lymphocyte.

otherwise healthy American college students. Evidence suggests that EB virus may be a factor, however, in some malignancies in patients with immunodeficiency from AIDS or organ transplantation.

Epidemiology

The EB virus is distributed worldwide and infects individuals in crowded, economically disadvantaged groups at an early age without producing significant illness. In such populations, the characteristic infectious mononucleosis syndrome is quite rare. More affluent populations such as students entering college in the United States often have escaped past infection with the agent and lack immunity to EB virus. In any year, an estimated 1.5 million college students will either have had the disease or will get it during the college year. Infectious mononucleosis occurs almost exclusively in adolescents and adults who lack antibody to the virus. Even in the age group of 15 to 24 years, however, only about half of the EB virus infections produce infectious mononucleosis; the remainder develop few or no symptoms.

EB virus is present in the saliva for up to 18 months after infectious mononucleosis, and thereafter it occurs intermittently for life. Continuous salivary shedding of virus is common in people with AIDS or other immunodeficiencies. Mouth-to-mouth kissing is an important mode of transmission of infectious mononucleosis in young adults, giving rise to the name "kissing disease." The donor of the virus is usually asymptomatic and may have been infected in the past without developing symptoms of infectious mononucleosis. By middle age, most people have demonstrable antibody to the virus, indicating past infection. There is no animal reservoir.

Prevention and Treatment

Prevention is aided by avoiding the saliva of another person, as well as objects such as toothbrushes or drinking glasses possibly contaminated with another's saliva. No vaccine for infectious mononucleosis is available. The antiviral medication acyclovir inhibits productive infection by the virus and is of value in rare serious cases; however, it has no activity against the latent infection.

Table 28.6 gives the main features of infectious mononucleosis.

Yellow Fever

Yellow fever was first recognized in Central America in 1648, probably introduced there from Africa. One of the worst outbreaks of yellow fever in this century occurred in Ethiopia in the 1960s, producing 100,000 cases and 30,000 deaths. In 1989, an epidemic of yellow fever occurred in Bolivia among poor people who moved into the jungle to try to make a living growing coca. There have been no outbreaks in the United States since 1905, but the vector mosquito has reappeared in the Southeast, raising the possibility of outbreaks of the disease if the virus is again introduced.

Symptoms

The symptoms of yellow fever can range from very mild to severe. Symptoms of mild disease, the most common form, may be only fever and a slight headache lasting a day or two. Patients suffering severe disease, however, may experience a high fever, nausea, bleeding from the nose and into the skin, "black vomit" (from gastrointestinal bleeding), and jaundice (hence, the name *yellow fever*). The mortality rate of severe yellow fever cases can reach 50% or more. The reasons for the wide variation in severity of symptoms are unknown, but probably have more to do with the size of the infecting dose and the status of human host defenses than with differences among strains of the causative virus.

Causative Agent

Yellow fever is caused by an enveloped, single-stranded, positive-sense RNA arbovirus of the flavivirus family. The virus multiplies in species of mosquitoes, apparently without harming them, and the mosquitoes transmit the infection to humans.

Pathogenesis

The yellow fever virus is introduced into humans by the bite of an *Aedes* mosquito, the biological vector. It multiplies, enters the bloodstream, and is carried to the liver and other parts of the body. Viral liver damage results in jaundice and decreased production of clotting proteins, and injury to small blood vessels produces petechiae, tiny hemorrhages, throughout the body. The virus affects the circulatory system by directly damaging the heart muscle, by causing bleeding from blood vessel injury in various tissues, and by causing disseminated intravascular coagulation (DIC). Kidney failure is a common consequence of loss of circulating blood and low blood pressure. ■ **biological vector, p. 315**
■ **disseminated intravascular coagulation, p. 440**

Epidemiology

The reservoir of the disease is mainly infected mosquitoes and primates living in the tropical jungles of Central and South

TABLE 28.6 "Kissing Disease" (Infectious Mononucleosis, "Mono")

Symptoms	Fatigue, fever, sore throat, and enlargement of lymph nodes
Incubation period	Usually 1 to 2 months
Causative agent	Epstein-Barr (EB) virus, a DNA virus of the herpesvirus family
Pathogenesis	Productive infection of epithelial cells of throat and salivary ducts; latent infection of B lymphocytes; activation of B and T lymphocytes; hemorrhage from enlarged spleen is a rare but serious complication
Epidemiology	Spread by saliva; lifelong recurrent shedding of virus into saliva of asymptomatic, latently infected individuals
Prevention and treatment	Avoid sharing of articles such as toothbrushes and drinking glasses, which may be contaminated with the virus from saliva. Treatment: usually none needed; acyclovir of benefit in rare cases

Perspective 28.2 Walter Reed and Yellow Fever

Walter Reed was born in Virginia in 1851. After receiving medical degrees, he entered the Army Medical Corps. He was appointed professor of bacteriology at the Army Medical College in 1893, and in 1900, Reed was named president of the Yellow Fever Commission of the U.S. Army. Although improved sanitation had effectively controlled other diseases, the incidence of yellow fever remained high and its cause and transmission were unknown.

Reed suspected that an insect was the vector of yellow fever. As early as 1881, Dr. Carlos Finlay of Havana, Cuba, had suggested that a mosquito might be the carrier of this disease, but he was unable to prove his hypothesis. Beginning with mosquitoes raised from eggs, Reed and the others in his commission proceeded with a series of experiments. First, the laboratory-raised mosquitoes were allowed to feed on patients with yellow fever and, second, on members of the commission.

After initial failures to transmit the disease by mosquitoes, one member of the commission, Dr. James Carroll, came down with classic yellow fever 3 days after an experimental mosquito bite. Dr. Jesse Lazear, another commission member, noted that the yellow fever patient on which the mosquito had originally fed was in his second day of the disease and that 12 days had elapsed before it bit Carroll. This timing was critical for the transmission of the disease because, as we now know, the mosquito can only contract the infection during a brief period when the patient is viremic; the mosquito can only transmit the infection to another individual after the virus has replicated to a high level in the mosquito.

Later, Lazear was bitten by a mosquito while working in a hospital yellow fever ward, developed yellow fever, and died. He was the only yellow fever fatality among commission members, although other volunteers lost their lives before the mysteries of this potentially fatal disease were resolved. The commission had a mosquito-proof testing facility constructed, called Camp Lazear in honor of their deceased colleague, to house volunteers for their experiments.

As a result of their studies, Dr. Reed and his colleagues made the following conclusions: (1) mosquitoes are the vectors of the disease; (2) an interval of about 12 days must elapse between the time the mosquito ingests the blood of an infected person and the time it can transmit the disease to an uninfected person; (3) yellow fever can be transmitted from a person acutely ill with the disease to a person who has never had the disease by injecting a small amount of the ill person's blood; (4) yellow fever is not transmitted by soiled linens, clothing, or other items that have come into contact with infected persons; and (5) yellow fever is caused by an infectious agent so small that it passes through a filter that excludes bacteria.

Armed with these findings, Major William C. Gorgas, the chief sanitary officer for Havana, instituted mosquito control measures that resulted in a dramatic reduction in yellow fever cases. Later, Gorgas used mosquito control in the Panama Canal Zone to allow construction of the Panama Canal. Previously, French workers had tried to build the canal but had failed because of heavy losses from yellow fever and malaria.

America and in Africa (**figure 28.10**). Periodically, the disease spreads from the jungle reservoir to urban areas, where it is transmitted to humans by *Aedes* mosquitoes.

Prevention and Treatment

In urban areas, control of yellow fever is achieved by spraying insecticides and eliminating the breeding sites of its principal vector, *Aedes aegypti*. In the jungle, control of yellow fever is almost impossible because the mosquito vectors live in the forest canopy and transmit the disease among canopy-dwelling monkeys. A highly effective live attenuated vaccine is available to immunize people who might become exposed, including foreign travelers to the endemic areas. There is no proven antiviral treatment. ■ **attenuated vaccine, p. 425**

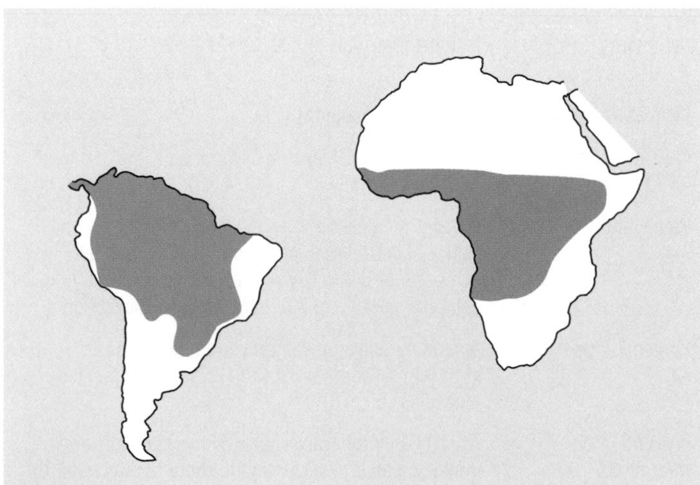

Figure 28.10 Distribution of Yellow Fever Extensions may sometimes also occur into Central America.

The main features of yellow fever are presented in **table 28.7**.

MICROCHECK 28.4

Infectious mononucleosis occurs worldwide and is transmitted from person to person by saliva. Curiously, the causative virus is associated with different malignancies, but only in certain geographic areas or in AIDS patients. Yellow fever virus infects monkeys that live in tropical forests of Africa and South and Central America, and the mosquitoes that bite them. Transmission of the virus among human beings can occur wherever there is a human-biting mosquito species that is susceptible to the virus.

- What characteristic changes occur in the blood of patients with infectious mononucleosis?
- What is the name of the tumor from which EB virus was first isolated?
- Why does it take more than a week before a mosquito just infected with yellow fever virus can transmit the disease?

Protozoan Diseases

Protozoa infect the blood vascular and lymphatic systems of millions of people worldwide. One example, discussed in an earlier chapter, is *Trypanosoma brucei*, fundamentally a bloodstream parasite of African animals and cause of African sleeping sickness in humans. Malaria, another protozoan disease, is much more widespread and is a leading cause of morbidity and mortality, meaning illness and death, worldwide. ■ **protozoa, p. 304** ■ **African sleeping sickness, p. 686**

TABLE 28.7 Yellow Fever

Symptoms	Often only headache and fever. Severe cases characterized by high fever, jaundice, black vomit, and hemorrhages into the skin
Incubation period	Usually 3 to 6 days
Causative agent	Yellow fever virus, an enveloped, single-stranded RNA virus of the flavivirus family
Pathogenesis	Virus multiplies locally at site of introduction by an infected mosquito; spreads to the liver and throughout the body by the bloodstream. Virus destroys liver cells, causing jaundice and decreased production of blood-clotting proteins. Hemorrhages and decreased strength of the heart result in circulatory failure and kidney failure
Epidemiology	Virus persists in forest primates and the mosquitoes that feed on them, in Africa and Central and South America; human epidemics occur when the virus infects household mosquitoes that feed on humans
Prevention and treatment	There is a highly effective live attenuated viral vaccine. No proven antiviral therapy is available

Malaria

Malaria is an ancient scourge, as evidenced by early Chinese and Hindu writings. During the Fourth century B.C., the Greeks noticed its association with exposure to swamps and began drainage projects to control the disease. The Italians gave the disease its name, *mal aria*, which means "bad air," in the Seventeenth century. In early times, malaria ranged as far north as Siberia and as far south as Argentina. In 1902, Ronald Ross received a Nobel Prize for demonstrating the life cycle of the protozoan cause of malaria.

Malaria is the most common serious infectious disease worldwide. In 1955, the World Health Organization (WHO) began a program for the worldwide elimination of malaria. Initially there was great success, as WHO employed insecticides such as DDT against the mosquito vector, detected infected patients by obtaining blood smears, and provided treatment for those who were infected. Fifty-two nations undertook control programs and, by 1960, 10 of them had eradicated the disease. Unfortunately, strains of *Anopheles* mosquitoes resistant to insecticides began to appear, and in cooperation with bureaucracy and complacency, malaria began a rapid resurgence. In 1976, the World Health Organization acknowledged that the eradication program was a failure. Today there are 300 to 500 million people infected annually worldwide, with about 3 million deaths. More people are dying of the disease than when the eradication programs first began. ■ **DDT, p. 782**

Symptoms

The first symptoms of malaria are "flu"-like, with fever, headache, and pain in the joints and muscles. These symptoms generally begin about 2 weeks after the bite of an infected mosquito, but in some cases they can begin many weeks afterward. After 2 or 3 weeks of these symptoms, the pattern changes and symptoms tend to fall into three phases highly suggestive of malaria.

- The patient abruptly feels cold and develops shaking chills that can last for as much as an hour (cold phase).
- Following the chills, the temperature begins to rise steeply, often reaching 40°C (104°F) or more (hot phase).
- After a number of hours of fever, the temperature falls, and drenching sweating occurs (wet phase). Except for fatigue, the patient feels well until 24 or 48 hours later, depending on the causative species, when the pattern of symptoms repeats.

Causative Agent

Human malaria is caused by protozoa of the genus *Plasmodium*. Four species are involved—*P. vivax*, *P. falciparum*, *P. malariae*, and *P. ovale*. These species differ in microscopic appearance and, in some instances, life cycle, type of disease produced, severity, and treatment. In recent years, the majority of patients diagnosed in the United States have been infected with *P. vivax*, but up to 30% have been infected with the more dangerous species, *P. falciparum*.

The *Plasmodium* life cycle is complex (**figure 28.11**), with a number of different forms that differ in microscopic appearance and antigenicity. The parasite grows and divides in the erythrocytes of the host. The earliest form resembles a ring, with a large pale food vacuole in the central area, with the nucleus and cytoplasm being pushed to the periphery. This develops into a larger motile **trophozoite**, which goes on to subdivide, producing a **schizont**. The infected erythrocyte then breaks open, and the offspring of the division, called **merozoites**, are released into the plasma. The merozoites then enter new erythrocytes and multiply, repeating the cycle. Some merozoites that enter erythrocytes develop into **gametocytes**, however, which are specialized sexual forms different from the other circulating plasmodia in both their appearance and susceptibility to antimalarial medicines. These sexual forms do not rupture the red blood cells. They cannot develop further in the human host and are not important in causing the symptoms of malaria. They are, however, infectious for certain species of *Anopheles* mosquitoes and are thus ultimately responsible for the transmission of malaria from one person to another.

When a mosquito dines on a person's blood (see figure 28.11) it digests infected and uninfected erythrocytes, except that it only liberates the gametocytes from the erythrocytes that enclose them. Shortly after entering the intestine of the mosquito and stimulated by the drop in temperature, the male and female gametocytes change in form to become gametes. The male gametocyte transforms into about a half dozen tiny, whiplike **gametes** that swim about until they unite with the female gamete in much the same way as the sperm and ovum unite in higher animals. The resulting **zygote** transforms into a motile form that burrows into the wall of the midgut of the mosquito and forms a cyst. The cyst enlarges as the diploid nucleus undergoes meiosis, dividing asexually into numerous offspring. The cyst then ruptures into the body cavity of the mosquito, and the released parasites, called **sporozoites**, find their way to the

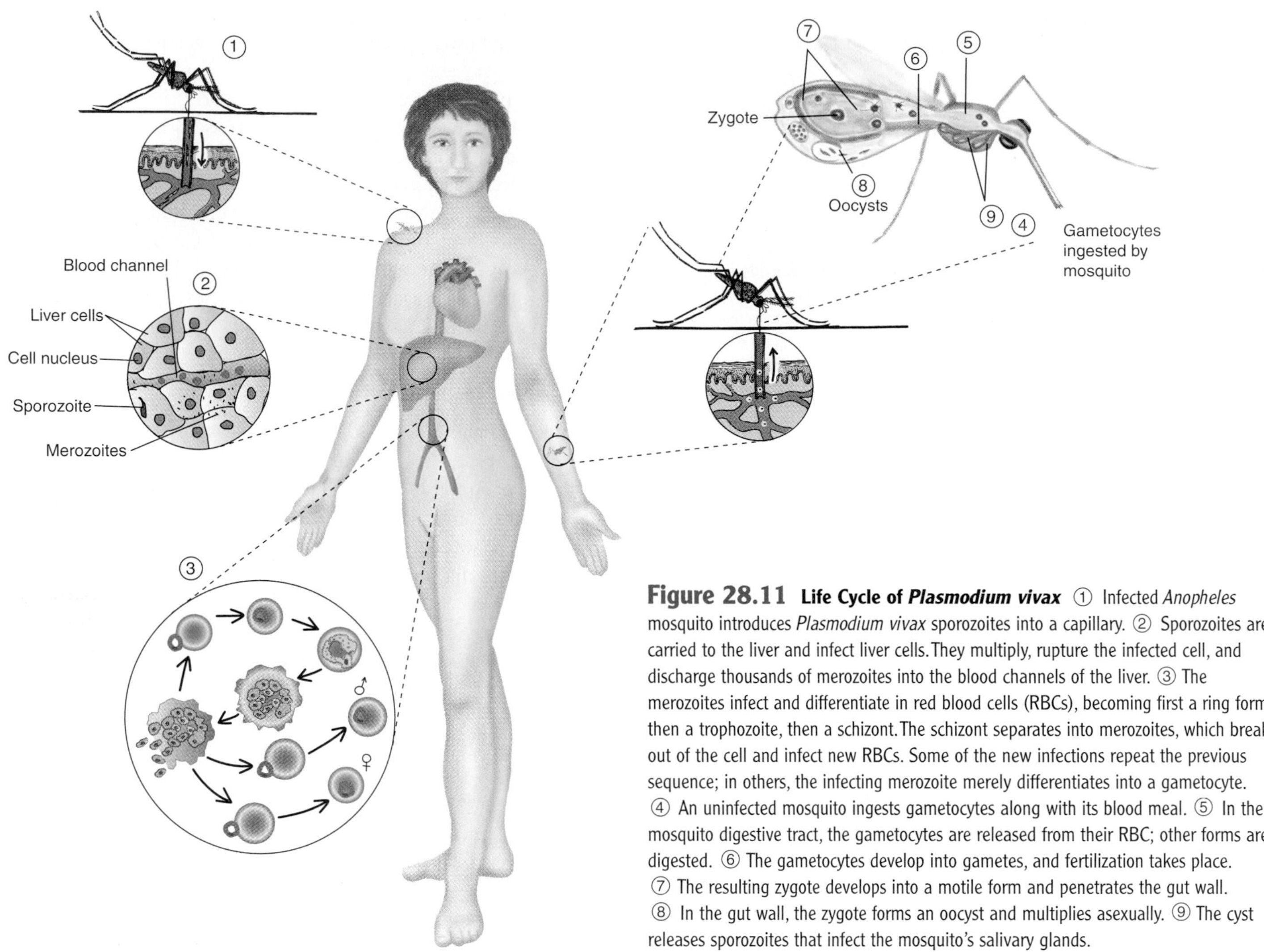

Figure 28.11 Life Cycle of *Plasmodium vivax* ① Infected *Anopheles* mosquito introduces *Plasmodium vivax* sporozoites into a capillary. ② Sporozoites are carried to the liver and infect liver cells. They multiply, rupture the infected cell, and discharge thousands of merozoites into the blood channels of the liver. ③ The merozoites infect and differentiate in red blood cells (RBCs), becoming first a ring form, then a trophozoite, then a schizont. The schizont separates into merozoites, which break out of the cell and infect new RBCs. Some of the new infections repeat the previous sequence; in others, the infecting merozoite merely differentiates into a gametocyte. ④ An uninfected mosquito ingests gametocytes along with its blood meal. ⑤ In the mosquito digestive tract, the gametocytes are released from their RBC; other forms are digested. ⑥ The gametocytes develop into gametes, and fertilization takes place. ⑦ The resulting zygote develops into a motile form and penetrates the gut wall. ⑧ In the gut wall, the zygote forms an oocyst and multiplies asexually. ⑨ The cyst releases sporozoites that infect the mosquito's salivary glands.

mosquito's salivary glands and saliva, from which they may be injected into a new human host.

In the human host, the sporozoites are carried by the bloodstream to the liver, where they infect liver cells. In these cells, each parasite enlarges and subdivides, producing thousands of merozoites. These merozoites are then released into the bloodstream and establish the cycle involving the erythrocytes.

Pathogenesis

The characteristic feature of malaria, recurrent bouts of fever followed by feeling healthy again, results from the erythrocytic cycle of growth and release of merozoites. Interestingly, the infections in all the millions of different red blood cells become nearly synchronous. Thus, cell rupture and release of daughter protozoa occur at roughly the same time for all infected cells, and each release causes a fever. For *P. malariae*, the growth cycle takes 72 hours, so that fever recurs every third day. For the other species, fevers generally occur every other day.

Infections by *P. falciparum* tend to be very severe, probably because all erythrocytes are susceptible to infection, whereas other *Plasmodium* species infect only young or old erythrocytes. Thus, very high levels of parasitemia, meaning parasites in the bloodstream, can develop with *P. falciparum* infections. The infected red blood cells become rigid, in contrast to normal red cells, which are flexible. Also, they adhere to each other and to the walls of capillaries. These tiny blood vessels therefore become plugged, and the affected tissue becomes deprived of oxygen as a result. Involvement of the brain, or cerebral malaria, is particularly devastating, but almost any organ can be severely affected.

Plasmodium vivax and *P. ovale* malaria often relapse after treatment of the blood infection because treatment-resistant forms of the organisms continue to infect the liver, the **exoerythrocytic cycle**; (*exo-* means "outside of," *-erythrocytic* refers to red blood cells). The exoerythrocytic cycle can initiate new erythrocytic cycles of infection after the bloodstream infection has been cured.

The spleen characteristically enlarges in malaria to cope with the large amount of foreign material and abnormal red blood cells, which it removes from the circulation. Malaria is the most common cause of splenic rupture, which can occur with or without trauma.

Especially with *P. falciparum*, the parasites cause anemia by destroying red blood cells and converting the iron in hemoglobin to a form not readily recycled by the body. The large amount of foreign material in the bloodstream strongly stimulates the immune system. In some cases, the overworked immune system fails and immunodeficiency results.

Those who live continuously in areas where malaria is endemic develop some immunity to the lethal effects of the disease, which crosses the placenta and gives partial protection to the newborn. The greatest risk of death from malaria is to children over six months of age as this immunity wanes. Currently, worldwide, a child dies of malaria about every 20 seconds. Others at high risk of death are women with their first pregnancy, and individuals who move into an endemic area.

Epidemiology

Malaria was once common in both temperate and tropical areas of the world, and endemic malaria was only eliminated from the continental United States in the late 1940s. Today, malaria is predominantly a disease of warm climates, but 41% of the world's population live in endemic areas. Certain human-biting mosquito species of the genus *Anopheles* are biological vectors of malaria. Since suitable vectors are abundant in North America, the potential exists for the spread of malaria whenever it is introduced. Infected mosquitoes and humans constitute the reservoir for malaria. Malaria can be transmitted by, besides mosquitoes, blood transfusions or the sharing of syringes among drug users. Malaria contracted in this manner is easier to treat since it involves only red blood cells and not the liver; only sporozoites from mosquitoes can infect the liver. Some people of black African heritage are genetically resistant to *P. vivax* malaria because their red blood cells lack the receptors for the parasite. Also, some genetically determined blood diseases such as sickle cell anemia have survived over the eons, despite their negative effect on health, because they provide partial protection against malaria. ■ **receptors, p. 58, 394, 404**

Prevention and Treatment

Travelers to malarious areas, including 27 million Americans each year, can generally prevent symptomatic malaria, but not necessarily infection by the plasmodia, with weekly doses of chloroquine. Using insect repellants and mosquito netting impregnated with insecticide, avoiding the outdoors from dusk to dawn, and wearing protective clothing all help avoid infection. After leaving malarious areas, people take primaquine to eliminate possible exoerythrocytic infection, which if untreated could cause recurrence of the disease.

Areas of the world with chloroquine-resistant strains of *P. falciparum* present an increasingly vexing problem since the only effective preventive medications can cause serious, even fatal reactions in some people. Travelers to such areas should consult with a health officer familiar with local conditions before deciding on a preventive regimen.

Unfortunately, the malaria problem is likely to worsen unless effective control is achieved soon. The world population is presently 6 billion, with 250 new births occurring every minute. Moreover, various computer models project the climate to warm significantly, resulting in an increase in the areas where malaria is likely to occur. The combination of increasing population, expanding areas where malaria can be easily transmitted, and the development of medication-resistant plasmodia and insecticide-resistant mosquitoes underscores the need for an effective control program.

In 1998, a new initiative called Roll Back Malaria was begun, linking the World Health Organization, the United Nations Childrens Fund (UNICEF), the United Nations Development Program, and the World Bank in the fight against malaria. Their goal is to halve malaria deaths by the year 2010, and halve deaths again by 2015. The initial focus is on detailed mapping of malarious areas using satellite imagery and climate information, and documenting the level of malaria treatment and prevention at the village level. The aim is to organize and fund a sustained effort to improve access to medical care, strengthen local health facilities, and promote the delivery of medications and insecticide impregnated mosquito netting. Besides Roll Back Malaria, other initiatives under way are designed to spur development of new antimalarial medications and to better fund vaccine and other research. In contrast to initiatives in the past, there is now better understanding that malaria control must be part of overall economic development, since it is difficult for one to move forward without the other.

Scientists have been trying to perfect a vaccine against malaria for many years. The first major breakthrough was reported in 1976 from the laboratory of Dr. William Trager at the Rockefeller University. Trager described a method for the continuous *in vitro* cultivation of *P. falciparum*, allowing for the production of antigens from different stages of the organism's life cycle. Later on, other scientists using genetic engineering techniques cloned the genes responsible for these antigens and identified which antigens were potential vaccine candidates. A number of different vaccines are currently under development, including recombinant protein vaccines, and a 15-gene naked DNA vaccine. Current estimates are that it could take another decade before an effective vaccine is widely available. ■ **DNA vaccines, p. 429**

Treatment of malaria is complicated by the fact that different stages in the life cycle of the parasite respond to different medications. Chloroquine, for example, is generally effective against the erythrocytic stages, although it will not cure the liver infection with *P. vivax* or kill the gametocytes of *P. falciparum*. Primaquine is generally effective against the exoerythrocytic stage and the *P. falciparum* gametocytes. Strains of chloroquine-resistant *P. falciparum* are now common in many parts of the world, and resistant strains of *P. vivax* have appeared in some areas (**figure 28.12**). Promising new medications such as a combination, atovaquone and proguanil, may help with this problem. Some patients infected with chloroquine-resistant strains respond to intravenous quinine or quinidine together with an oral medication such as tetracycline or a sulfa drug.

Table 28.8 gives the main features of malaria.

MICROCHECK 28.5

There are a number of different stages in the life cycles of the protozoa that cause malaria, and they have differing microscopic appearance, antigenicity, and susceptibility to antimalarial medications. Malaria is a major hindrance to economic improvement in many countries. It is a leading serious infectious disease worldwide, and unless well-funded sustained control

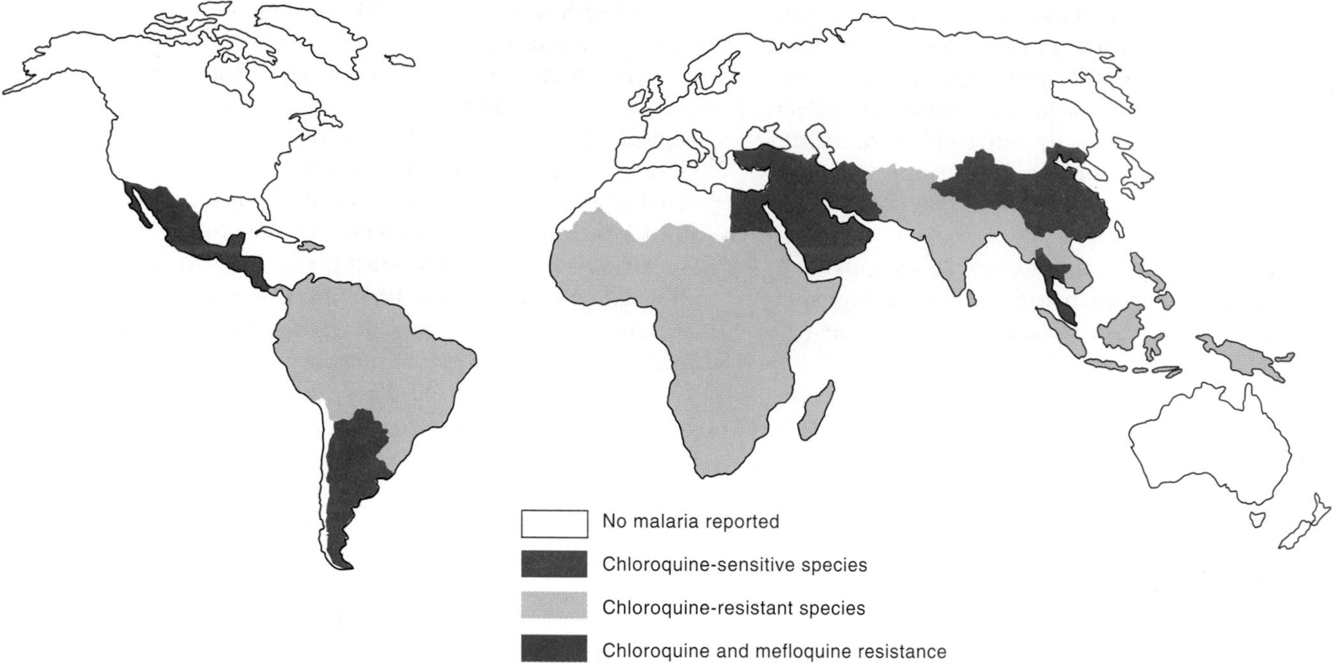

Figure 28.12 Distribution of Malaria in 1996 The malaria range may expand greatly with global warming.

measures are undertaken, it is expected to advance into new geographic areas.

- Why is malaria contracted from a blood transfusion easier to treat than if contracted from a mosquito?
- Why did insecticide-resistant mosquitoes begin to appear?
- Are sporozoites diploid or haploid?

TABLE 28.8 Malaria

Symptoms	Recurrent bouts of violent chills and fever alternating with feeling healthy
Incubation period	Varies with species; 6 to 37 days
Causative agent	Four species of protozoa of the genus *Plasmodium*
Pathogenesis	Cell rupture, release of protozoa cause fever; infected red blood cells adhere to each other and to walls of capillaries; vessels plug up, depriving tissue of oxygen; spleen enlarges in response to removing large amount of foreign material and many abnormal blood cells from the circulation
Epidemiology	Transmitted from person to person by bite of infected anopheline mosquito. Some individuals genetically resistant to infection
Prevention and treatment	Weekly doses of chloroquine while in malarial areas; after leaving, primaquine is given; other medicines for resistant strains; eradication of mosquito vectors; mosquito netting impregnated with insecticide; vaccines under development. Same medicines used in treatment; additional choices available for resistant strains

Multicellular Parasites

A number of species of roundworms can infect the blood and lymphatic vessels of humans. Mostly, they are contracted in tropical countries, transmitted by biting insects. The adult female worms live in the lymphatic vessels and deposit their offspring, tiny microfilaria, directly into the skin, lymphatics, or bloodstream, where they can be identified in biopsies or blood smears to make the diagnosis. The disorder these roundworms produce is termed **filariasis**, and it may be asymptomatic or produce dramatic symptoms. Two examples are **elephantiasis**, marked swelling of a body part due to lymphatic obstruction, caused by *Wuchereria bancrofti* and *Brugia* sp., and **river blindness**, caused by *Onchocerca volvulus*. ■ roundworms, p. 317

Completely different kinds of parasitic worms live in the blood vessels that carry venous blood from the intestines to the liver, or in the veins of the bladder. They are responsible for the disease schistosomiasis.

Schistosomiasis

Schistosomiasis, endemic in countries of Africa, Asia, the Caribbean, and South America, involves more than 200 million people worldwide and causes over 500,000 deaths annually. About 400,000 individuals with the disease now live in the United States, having emigrated from places where schistosomiasis is endemic. Schistosomiasis is caused not by roundworms, but by flukes. Flukes are short, bilaterally symmetrical worms that usually attach by one or more sucking discs. ■ flukes, p. 318

Symptoms

Itching skin may occur at the time of exposure to fresh water containing the worm larvae. The itching subsides, and weeks later, a generalized acute illness occurs, with fever, hives, cough, abdominal, joint and muscle pain, and diarrhea. Some people die during this stage, but usually the symptoms subside and infected individuals are free of symptoms for a number of years. Then a chronic, slowly progressing illness appears, with weakness, accumulation of fluid in the abdominal cavity, and sometimes vomiting blood.

Causative Agent

Most cases of schistosomiasis are caused by three species of flukes in the genus *Schistosoma*. Other genera of flukes that infect humans are hermaphroditic, meaning that each worm has both male and female reproductive organs, but *Schistosoma* species have male and female worms. *Schisto-soma* means "split-body," referring to a deep groove running along the male's body in which he clasps his female partner. *Schistosoma mansoni*, the only species established in the Americas, and the most common cause of schistosomiasis worldwide, is 10 to 20 mm long and lives in the small veins of the human intestine. Because it lives in blood vessels, it is called a blood fluke. The life cycle of *S. mansoni* is shown in **figure 28.13**. The slender female worm deposits ova that rupture through the tiny intestinal veins and wall of the intestine to enter the lumen, ultimately to be eliminated with the feces. The ova hatch in fresh water, releasing ciliated larvae called **miracidia** which can live up to 6 hours. When a miracidium encounters a certain species of snail, it penetrates the snail's body and multiplies asexually. Over about 6 weeks, thousands of elongated fork-tailed larvae called **cercariae** develop and leave their snail intermediate host to enter the water. When they encounter a human being wading in the water, they burrow through his or her skin, leaving their tails behind them. These larvae proceed to enter the circulatory system, are carried by the bloodstream through the heart and lungs, and eventually reach veins of the intestine, where they mature, mate, and begin ova production. ■ intermediate hosts, p. 318

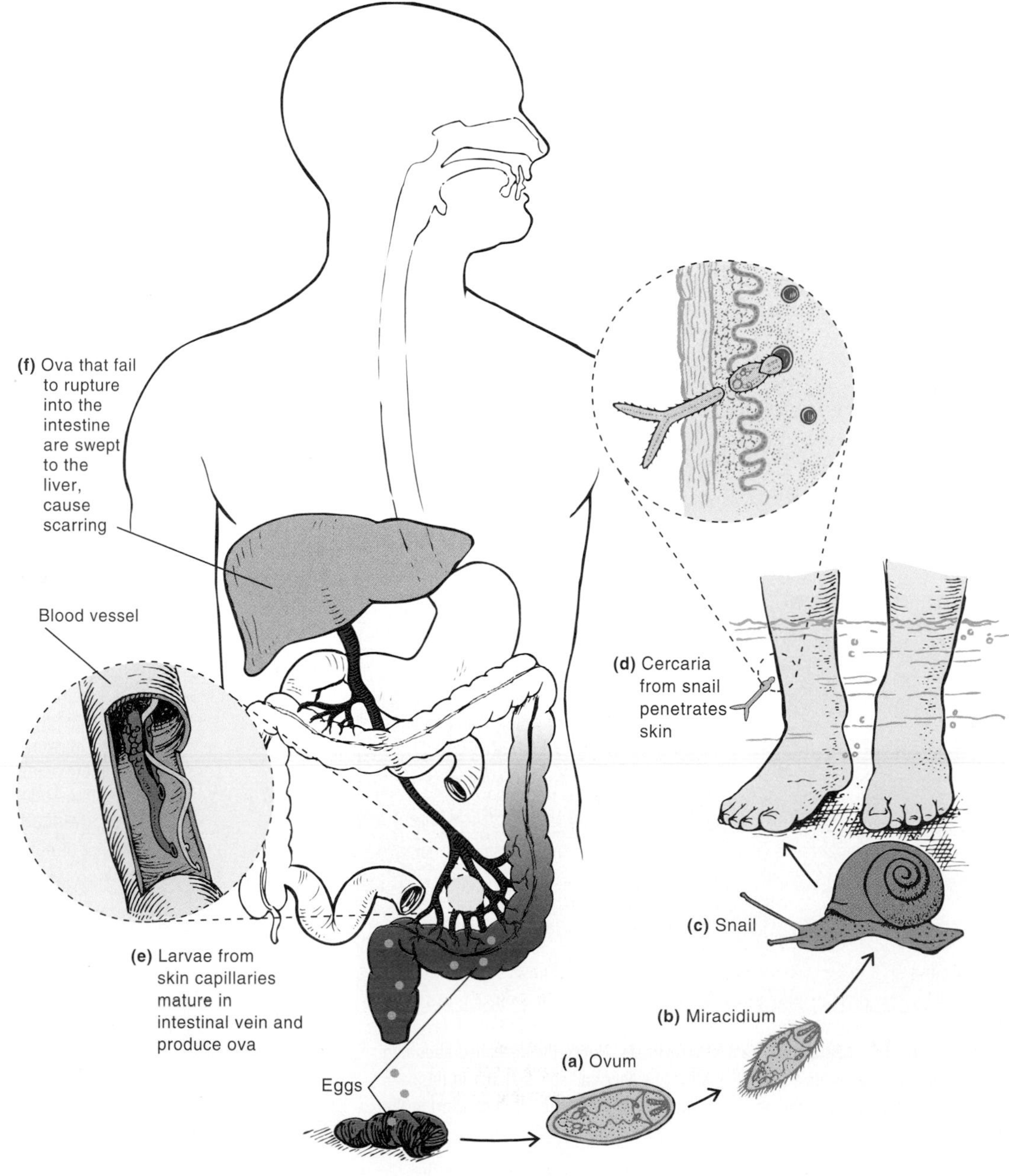
(f) Ova that fail to rupture into the intestine are swept to the liver, cause scarring
Blood vessel
(d) Cercaria from snail penetrates skin
(c) Snail
(e) Larvae from skin capillaries mature in intestinal vein and produce ova
(b) Miracidium
(a) Ovum
Eggs

Figure 28.13 Life Cycle of *Schistosoma mansoni* **(a)** Eggs (ova) from feces reach fresh water. **(b)** First larval form (miracidium) hatches from ovum and **(c)** infects snail. **(d)** Asexual reproduction in body of snail and transformation into another larval form (cercaria). **(e)** Cercariae break out of snail and penetrate human skin. Tail left behind, larva enters capillary, is carried by bloodstream to intestinal veins, and develops into mature fluke. **(f)** Ova deposited, break into the intestine or are swept to the liver by the blood.

Pathogenesis

Many of the cercariae die upon entering the skin, causing an inflammatory and immune response that becomes more and more intense with each exposure to the parasites. Each skin penetration by the schistosomes causes an itchy skin rash that gradually subsides. Weeks later, when the mature worms begin to deposit their ova, a generalized illness occurs, probably due to circulating schistosomal antigens reacting with antibodies. Although some people die from the reaction, usually it subsides within several months. Unfortunately, the worms continue to deposit hundreds of ova per day over a lifetime that can exceed a quarter century. Perpetuation of the species depends on ova staying close to the intestine, where an intense inflammatory response liquefies the tissue and allows them to rupture into the intestinal lumen. The spine on the ovum probably helps hold them in place, but many are swept away by the bloodstream to the liver. The same inflammatory response to the ova occurs in the liver, causing gradual destruction of liver cells and their replacement with scar tissue. The end result is malnutrition and a buildup of pressure in the abdominal veins and connecting veins in the esophagus. Fluid accumulates in the abdominal cavity (**figure 28.14**), and hemorrhage occurs if the engorged esophageal veins rupture.

Swimmer's itch is a schistosomal disease that is common across the United States. The disease is characterized by an itchy rash caused by cercariae of schistosomes of wild birds and other animals. The cercariae penetrate the skin of swimmers and then die. These schistosomes are otherwise harmless, since they are unable to complete their life cycle in humans.

Epidemiology

Transmission occurs in northern and eastern South America, parts of the Caribbean, much of the African continent, parts of the Middle and Far East, Philippines, Southeast Asia, China, and Japan. The distribution depends on abundant fresh water, extensive contamination of water by human feces, and a suitable snail intermediate host. Farmers and people who fish or wade in fresh water are at high risk for the disease. Large irrigation projects to enhance agriculture have extended the range of the disease. Schistosomiasis cannot be contracted in continental United States because there is no appropriate snail intermediate host.

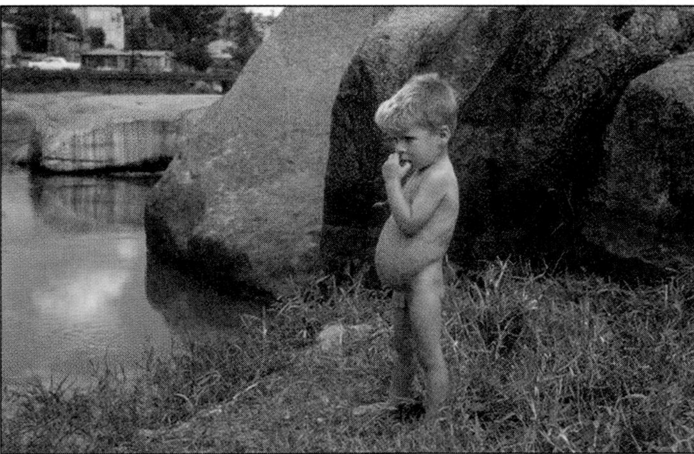

Figure 28.14 Child with Schistosomiasis Notice the distended abdomen, probably due to inflammatory enlargement of the liver and spleen. Later in the development of the disease the abdomen will fill with fluid, the result of scarring that obstructs the flow of blood throught the liver.

Prevention and Treatment

Preventing schistosomiasis depends on avoiding skin exposure to fresh water contaminated with the miracidia, preventing untreated human feces from entering fresh water, treatment of cases of the disease, and measures, both chemical and biological, to control the snail intermediate hosts. Promising synthetic peptide and DNA vaccines are under development. Effective medications are available that kill the parasites in disease victims.

■ vaccines, p. 425

The main features of schistosomiasis are presented in **table 28.9**.

TABLE 28.9 Schistosomiasis

Symptoms	Acutely: fever, hives, cough, abdominal, joint, and muscle pain. These symptoms subside and after years are replaced by generalized weakness and swelling of the abdomen
Incubation period	Usually 2 to 6 weeks
Causative agents	Most cases are caused by three species of *Schistosoma*, the most prevalent being *S. mansoni*. The male schistosome has a longitudinal slit in its body in which he clasps the female. Perpetuation of the life cycle depends on rupture of their eggs into the lumen of the intestine from the tiny intestinal veins where the adults live. The eggs hatch in fresh water, releasing a free-swimming miracidium that penetrates into the flesh of the snail intermediate host. In the snail, the organism multiplies and differentiates into forked-tail cercariae that leave the snail and can infect humans by penetrating their skin
Pathogenesis	Inflammatory and immune response to skin penetration by the cercariae causes intense itching. Allergic reaction to released schistosomal antigens as the worms mature and begin depositing ova causes the acute illness. Chronic malnutrition and swelling of the abdomen result from liver damage caused by ova that fail to penetrate into the intestinal lumen, and are carried to the liver by the bloodstream
Epidemiology	Distribution of the disease is favored by large bodies of shallow fresh water, extensive contamination of the water with human feces, and a suitable snail intermediate host. Farmers, people who fish, and children who wade in fresh water at high risk
Prevention and treatment	Prevention depends on proper disposal of human feces, identifying and treating infected patients, and measures to control the snail intermediate host. Praziquantel is effective treatment

MICROCHECK 28.6

The health and economic burden of multicellular parasite infections challenges that of malaria. The complex life cycle of *Schistosoma mansoni* provides a number of possibilities for attacking schistosomiasis.

- Trace the life cycle of *Schistosoma mansoni.*
- How long would it take *S. mansoni* to become extinct if snail-eating fish eliminated the intermediate host?

FUTURE CHALLENGES

Rethinking Malaria Control

Previous attempts at malaria control have relied heavily on the use of DDT, a non-biodegradable insecticide. Use of this substance was banned in the United States in 1972 because of its accumulation in the environment and its damaging effects on fowl, notably the peregrine falcon and the bald eagle. Many other concerns have been raised about its possible carcinogenicity and potential effects on human fetuses. Although not generally used in agriculture, more than 20 countries still use it for malaria control, and more would use it if they could afford it. Biodegradable insecticides are available, but they are more expensive than DDT, and they are not without toxicity. With several children dying of malaria every minute worldwide, the options for rapid relief from malaria all seem bad, and a choice has to be made of which ones are the least bad. While insecticides may be necessary short term, other options including education, economic development, vaccines, biological control of the Anopheles *vectors, and mosquito habitat alteration may prove more important in achieving sustained relief from the disease.*

The challenge for the future is to better understand the ecology of malaria as it applies in each location, with the aim of minimizing insecticide use and discovering new options for maintaining long-term control.

SUMMARY

Anatomy and Physiology (Figure 28.1)

1. The left side of the heart receives oxygenated blood from the lungs and pumps it into thick-walled arteries; the right side of the heart receives blood depleted of oxygen from the veins and pumps it through the lungs.
2. Arteries have thick muscular walls and begin to develop arteriosclerosis in the teenage years or before. Arteriosclerotic lesions are commonly colonized by *Chlamydia pneumoniae*, and the significance of this for health is under investigation.
3. Lymphatics, lymph vessels, are blind-ended tubes that take up fluid that leaks from capillaries. They also take up bacteria, which are normally trapped by lymph nodes distributed along the course of the lymphatics.
4. The spleen filters unwanted material such as bacteria and damaged erythrocytes from the arterial blood. It becomes enlarged in diseases such as infectious mononucleosis and malaria.

Bacterial Diseases of the Blood Vascular System

1. Bacteria circulating in the bloodstream can colonize the inside of the heart, and they can cause collapse of the circulatory system and death.
2. Infections of the heart valves and lining of the heart are called **endocarditis**; illness resulting from circulating pathogens is called **septicemia**.
3. **Acute bacterial endocarditis** is caused when virulent bacteria enter the bloodstream from a focus of infection elsewhere in the body, or when contaminated material is injected by drug abusers; normal heart valves are commonly infected and destroyed.

Subacute Bacterial Endocarditis (Table 28.1)

1. Subacute bacterial endocarditis (SBE) is commonly caused by organisms of little virulence, including oral streptococci and *Staphylococcus epidermidis*, and infection usually begins on structural abnormalities of the heart.
2. Biofilm formation may complicate treatment.

Gram-Negative Septicemia (Figure 28.3)

1. Gram-negative septicemia is commonly a nosocomial illness; many afflicted individuals have serious underlying illnesses such as cancer and diabetes.
2. Shock precipitated by release of endotoxin from the bacteria is a common complication; it is the result of an exaggerated response to endotoxin in which massive release of cytokines occurs in the circulation.

Bacterial Diseases of the Lymph Nodes and Spleen

1. Tularemia, brucellosis, and plague involve the mononuclear-phagocyte system and are characterized by enlargement of the lymph nodes and spleen.
2. The causative organisms grow within phagocytes, protected from antibody.

"Rabbit Fever" (Tularemia) (Table 28.2, Figure 28.4)

1. Tularemia is usually transmitted from wild animals to humans by exposure to the animals' blood or via insects and ticks.
2. The cause is the Gram-negative aerobe *Francisella tularensis* found throughout the United States except Hawaii. (Figure 28.5)

"Undulant Fever" (Brucellosis, "Bang's disease") (Table 28.3)

1. Brucellosis, caused by *Brucella melitensis*, is usually acquired from cattle or other domestic animals, less often from wild animals such as elk, moose, and bison.
2. Hunters, butchers, and those who drink unpasteurized milk or milk products are at increased risk for the disease.
3. The organisms can infect via mucous membranes and minor skin injuries.

Plague ("Black Death") (Table 28.5)

1. Plague, once pandemic, now persists endemically in rodent populations, including those in many western states of the United States.

2. It is caused by *Yersinia pestis*, an enterobacterium with many virulence factors, chromosomally or plasmid coded, which interfere with phagocytosis and immunity. (Table 28.4, Figure 28.6)
3. **Bubonic plague** is transmitted to humans by fleas. (Figure 28.7) **Pneumonic plague** is transmitted person to person.
4. Untreated, bubonic plague mortality is 50% to 80%; **pneumonic plague**, almost 100%.

Viral Diseases of the Lymphoid and Blood Vascular Systems

"Kissing Disease" (Infectious Mononucleosis, "Mono") (Table 28.6)

1. The incidence is high in 15 to 24-year-olds; most individuals with the disease become lifelong carriers, capable of transmitting the disease by their saliva.
2. The causative agent is Epstein-Barr virus, which establishes a lifelong latent infection of B lymphocytes; the disease is called mononucleosis because victims have an increased number of mononuclear cells in their blood. (Figure 28.9)
3. The disease is confirmed by the presence of a **heterophile antibody**.

Yellow Fever (Table 28.7)

1. Yellow fever is a zoonosis of mosquitoes and monkeys that exists mainly in tropical jungles; it can become epidemic in humans where a suitable *Aedes* mosquito vector is present. (Figure 28.10)
2. The disease involves the heart and blood vessels throughout the body and is characterized by fever, jaundice, and hemorrhaging.
3. A highly effective live attenuated vaccine is available for preventing the disease.

Protozoan Diseases

1. African sleeping sickness is present over much of tropical Africa; of the protozoan blood and lymphatic diseases, however, malaria is the most widespread.

Malaria (Table 28.8)

1. Malaria is caused by four species of *Plasmodium* and is transmitted from person to person by the bite of certain species of *Anopheles* mosquitoes, its biological vectors.
2. Now confined mainly to impoverished warm regions of the world, malaria survives despite massive eradication programs; it is one of the most widespread of all serious infectious diseases. (Figure 28.12)
3. The life cycle is complex; different forms of the organism invade different body cells and have different susceptibility to antimalarial medication. (Figure 28.11)
4. Replication of the organism inside red blood cells results in the almost simultaneous rupture of the infected cells and release of the progeny protozoa to infect new red cells.

Multicellular Parasites

1. Species of roundworms, transmitted in the tropics by biting insects, can infect the blood vessels and lymphatics. **Filariasis**, **elephantiasis**, and **river blindness** are caused by roundworms.

Schistosomiasis (Table 28.9. Figures 28.13, 28.14))

1. Hundreds of thousands of people with Schistosomiasis now live in the United States, having emigrated from endemic areas.
2. The disease is caused by blood flukes of the genus *Schistosoma.*
3. The adult worms live in veins of the intestine or bladder, where the male holds the female in a cleft in his body.
4. Ova must rupture into the intestinal or bladder lumen, escape to the outside world, and continue their life cycle by releasing larvae called **miracidia**.
5. The miracidia must infect a suitable species of snail, reproduce asexually, and differentiate into larvae that leave the snail and penetrate human skin.
6. Ova that fail to reach the outside often lodge in the liver, causing scarring.

REVIEW QUESTIONS

Short Answer

1. Where do the organisms that cause subacute bacterial endocarditis (SBE) originate?
2. How is the bacteriological diagnosis of SBE usually made?
3. What is the significance of immune complex formation in SBE?
4. Give two examples of enterobacteria that can cause septicemia.
5. What is disseminated intravascular coagulation (DIC)?
6. What activities of humans are likely to expose them to tularemia?
7. Why is pasteurization of milk so important in the control of brucellosis?
8. Why is brucellosis a threat to big game hunters?
9. What is the most frequent serious complication of brucellosis?
10. Why might the *Yersinia pestis* bacteria from a patient with pneumonic plague be more dangerous than the ones from fleas?
11. Why might rodent burrows be a source of plague months after they are abandoned?
12. Describe three virulence factors of *Yersinia pestis* that are coded by plasmids.
13. What is the difference between bubonic and pneumonic plague?
14. What type of leukocytes does EB virus infect and immortalize?
15. How is infectious mononucleosis transmitted?
16. Name a viral disease that can be complicated by disseminated intravascular coagulation.
17. Travelers to and from which areas of the world should have certificates of yellow fever vaccination?
18. Which microorganism causes the most dangerous form of malaria?
19. Why did malaria increase despite many years of attempted eradication?
20. What is the current strategy for containing malaria?

Multiple Choice

1. Which of the following infection fighters are found in lymph?
   A. Leukocytes
   B. Antibodies
   C. Complement
   D. Interferon
   E. All of the above
2. All the following are true of the spleen, *except*...
   A. it is located low on the right side of the abdomen.
   B. it cleanses the blood of foreign material and damaged cells.
   C. it provides an immune response to circulating pathogens.
   D. it can help produce new blood cells.
   E. it enlarges in a number of infectious diseases.
3. Which one of the following statements about SBE is *false*?
   A. It is generally a chronic illness characterized by fatigue and slight fever.
   B. Kidney involvement can be one complication.
   C. Infection occurs exclusively on the left side of the heart.
   D. Injected-drug abuse can be responsible for the disease.
   E. It can result in aneurysm formation.
4. Choose the one *true* statement about Gram-negative bacterial septicemia.
   A. It is a rare nosocomial disease.
   B. The output of urine increases if shock develops.
   C. It can only be caused by anaerobic bacteria.
   D. An antibiotic that kills the causative organism can be depended on to cure the disease.
   E. Lung damage is an important cause of death.
5. Which of these statements about tularemia is *false*?
   A. It can be contracted from muskrats and bobcats.
   B. Biting insects and ticks can transmit the disease.
   C. The causative organism is closely related to *E. coli*.
   D. A steep-walled ulcer at the site of entry of the bacteria and enlargement of nearby lymph nodes is characteristic.
   E. Without treatment, 9 out of 10 people can be expected to survive.
6. All of the following statements about brucellosis are true, *except*...
   A. fevers that come and go over a long period of time gave it the name "undulant fever."
   B. the causative agent can infect via mucous membranes.
   C. the causative agent is readily killed by phagocytes.
   D. the disease in cattle is characterized by chronic infection of the mammary glands and uterus.
   E. butchers are advised to wear goggles or a face shield to help protect against the disease.
7. Pick out the one *false* statement about *Yersinia pestis.*
   A. Growth conditions inside human phagocytes activate virulence genes.
   B. The bacterium can multiply in the flea digestive system.
   C. The code for YOPs exists on the bacterial chromosome.
   D. The organism resembles a safety pin in certain stained preparations.
   E. It was responsible for the "black death" in Europe during the 1300s.
8. Which of the following statements about yellow fever is *false*?
   A. There is no animal reservoir.
   B. The name yellow comes from the fact that many victims have jaundice.
   C. Certain mosquitoes are biological hosts for the causative agent.
   D. Outbreaks of the disease could occur in the United States because a suitable vector is present.
   E. A live attenuated vaccine is widely used to prevent the disease.
9. The malarial form infectious for mosquitoes is called a...
   A. gametocyte.
   B. miracidium.
   C. sporozoite.
   D. cercaria.
   E. merozoite.
10. Choose the one *false* statement concerning schistosomiasis.
    A. Transmission cannot occur in the United States.
    B. The disease occurs in the United States.
    C. The disease is caused by a hermaphroditic worm.
    D. A ciliated larval form of the causative agent is infectious for certain snails.
    E. An early symptom of the disease is itching skin.

Applications

1. Some years ago, dentists and doctors began noticing an association between subacute bacterial endocarditis and prior dental work, and they began advising that an antibiotic be administered at the time of dental procedures to those with known or suspected heart defects. What was the rationale for this advice?
2. Several children attending a New Mexico school developed a serious illness characterized by high fever and enlarged tender lymph nodes. Their physicians diagnosed their illnesses as plague, but were mystified because the children all lived and played in town and denied seeing any rodents. Health officials checked and found no evidence of rodents at their homes or school. What other investigations should they carry out to try to establish the source of the disease?
3. A health worker in Honduras is concerned about a potential outbreak of yellow fever in his town. A laborer from a jungle area known to be endemic for the disease had come to the town 2 weeks earlier to work and subsequently developed yellow fever. Several coworkers reported getting mosquito bites while working with him. Why is it important that the health worker determine how long it is since the workers were bitten by the mosquitoes?

Critical Thinking

1. The finding that there is an association between *Chlamydia pneumoniae* infection and arteriosclerotic lesions raised hopes

that new methods to combat arteriosclerosis could be developed. An investigator reviewing this research, however, stated that even a perfect correlation between infection and lesion formation would not prove that infection causes arteriosclerosis. Moreover, even showing that therapeutic antibiotics could prevent infection and lesion formation would not be definitive proof. Is the investigator justified in making this argument? Why or why not?

2. Why do dentists prescribe an antibiotic for patients with damaged heart valves such a short time before performing dental work?
3. Distinguish between EB virus infection and infectious mononucleosis.
4. Even though genetically engineered mosquitoes might be developed that do not allow the reproduction of malaria protozoa, these mosquitoes would have little, if any, immediate effect on the spread of the disease. Why should this be so? What would have to happen for these mosquitoes to significantly affect the spread of malaria?

HIV Disease and Complications of Immunodeficiency

In 1981, five published reports described an illness in previously healthy young homosexual men, characterized by unusual opportunistic infections, certain malignant tumors, and immunodeficiency. The constellation of symptoms and signs associated with this illness came to be known as **AIDS**, *an acronym for acquired immunodeficiency syndrome. Initially, there was wild speculation about what might be the cause of AIDS, but by 1982, the Centers for Disease Control had convincing epidemiological evidence that AIDS was caused by a new infectious agent. Scientists around the world scrambled to identify it.*

Fortunately, the scientists were able to build on some important scientific advances of the 1960s and 1970s—specifically, the discovery of subsets of lymphocytes, the functions of T cells, and the role of cytokines. In 1975, the Nobel Prize was awarded to R. Dulbecco, H. Temin, and D. Baltimore for discovering reverse transcriptase, revealing its function in making a DNA copy of RNA genomes. Also, in early 1978, Dr. Robert Gallo's laboratory at the National Cancer Institute discovered the first human retrovirus, human T-lymphotrophic virus (HTLV). Dr. Gallo developed a technique for cultivating retroviruses in normal lymphocytes activated with interleukin-2 (IL-2). Viral growth in the cultures was detected by an assay for reverse transcriptase and by electron microscopy. ■ **lymphocytes, p. 370** ■ **cytokines, p. 379** ■ **reverse transcriptase, p. 229** ■ **interleukins, p. 380** ■ **retroviruses, p. 355**

In January 1983, an important scientific breakthrough occurred in the laboratory of Dr. Luc Montagnier at the famous Pasteur Institute. Using techniques developed earlier by Gallo for cultivating HTLV, Montagnier recovered a new virus from a patient with lymphadenopathy syndrome, an AIDS-associated condition. Because the IL-2 activated lymphocyte cultures died out quickly, he was only able to obtain small quantities of the new virus, named LAV for "lymphadenopathy virus," but enough to use as antigen in a blood test that showed that AIDS cases were infected with the virus. He sent a sample of this virus to Gallo at the National Cancer Institute, and the two exchanged material from other patients. He also made application to the U.S. Patent Office for a patent on his blood test.

Meanwhile, Gallo's laboratory began recovering a virus from AIDS patients, and reported the finding in the same issue of the journal Science *in which Montagnier reported his finding. A number of these viral isolates were introduced together into continuous cell cultures (as opposed to IL-2 activated normal lymphocytes) to see if a strain of the virus could replicate in the cells. One did replicate well, enabling Gallo to obtain large quantities of*

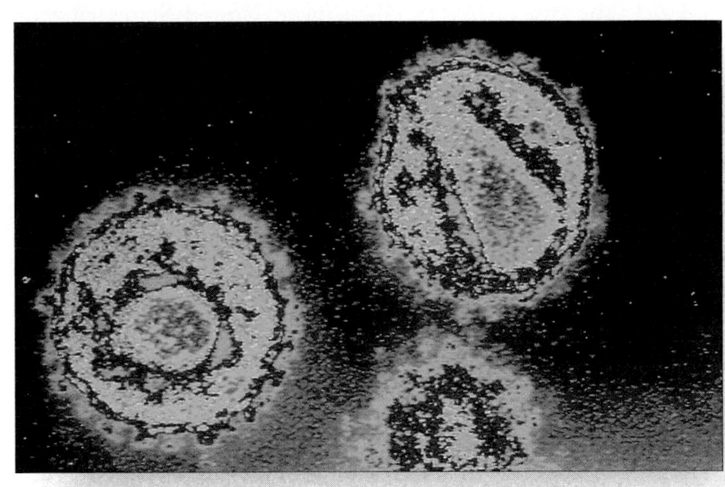

the virus. Gallo named the virus HTLV-III because of a resemblance to two human T-lymphotrophic viruses discovered earlier. It soon became clear, however, that the new virus more closely resembled a virus that caused persistent infections in sheep, a lentivirus. Using "HTLV-III," Gallo perfected a blood test for AIDS and, like Montagnier, applied for a patent. These conflicting patent claims caused a bitter scientific and legal battle that raged for years, the subject of charges and countercharges and investigations by scientific and congressional committees costing hundreds of thousands of dollars.

Under pressure from President Reagan, a settlement was negotiated in 1987 wherein Gallo and Montagnier were named co-discoverers of the test. Eighty percent of the royalties were to go to an AIDS foundation, the remainder to be divided equally between the National Institutes of Health (NIH) and the Pasteur Institute.

The settlement did not end the conflict. Genetic analysis showed that HTLV-III was actually LAV, introduced into Gallo's cultures as described above. Later, the Pasteur Institute suffered its own embarrassment, acknowledging that due to a laboratory mixup of its own, LAV actually came from a different patient than stated in its publications. In 1994, the earlier settlement was modified, NIH admitting that HTLV-III and LAV were the same and giving the French side a larger share of the royalties.

The American share of the royalties has generated many millions of dollars for the NIH over the years, but the value of the blood test far exceeded any monetary figure. It helped prove beyond doubt that HTLV-III/LAV—renamed HIV for human immunodeficiency virus—caused AIDS. It also showed that many asymptomatic people were infected with the virus and could transmit it, and that the epidemic was far more extensive than previously suspected. By 1985, the blood test was generally available for routine testing of donated blood, thus markedly improving the safety of blood transfusion and products prepared from pooled blood. These findings helped generate political support for controlling the disease.

—*A Glimpse of History*

OVER THE FIRST DECADE AFTER ITS RECOGNITION in 1981, the disease AIDS killed more United States citizens than the Korean and Vietnam wars combined. By 1994, it had become the leading cause of death among those 25 to 44 years of age in the United States. The total number of cases there since the epidemic began now exceeds 712,000, of whom more than 420,000 have died of the disease. Currently, the number of people infected with HIV in North America approaches 1 million, and it increases by more than 30,000 each year. Initially introduced into a population of homosexual men, the virus is now increasingly infecting heterosexuals, especially among economically disadvantaged racial and ethnic minorities.

Worldwide, more than 33 million people are infected, and over 16 million have died from the disease, more than were killed by the "black death" of Europe in the Middle Ages. It has become the number one killer in Africa, passing malaria, is devastating Southeast Asia, and spreading rapidly in India and into China. Soon India may have more cases than any other country. The disease is now the fourth leading cause of death in the world, behind heart disease, strokes, and pneumonia. A new infection occurs every 6 seconds, and a person dies of the disease every 5 minutes. ■ black death, p. 723

This is the disease they call AIDS (**figure 29.1**).

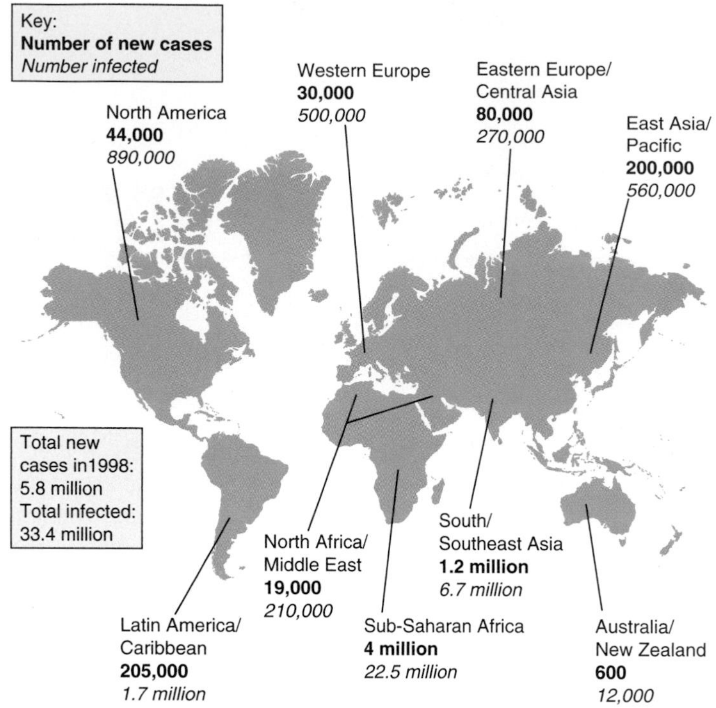

Figure 29.1 The AIDS Epidemic As of 1998 AIDS has killed 16 million people since the beginning of the epidemic, 2.6 million in 1998 alone.

Human Immunodeficiency Virus (HIV) Infection and AIDS

Microorganisms such as *Pneumocystis carinii* are common, but of such low virulence that they only cause disease in individuals with immunodeficiency disorders. When such a disease appears, it strongly suggests that the victim is immunodeficient. In the United States in 1981, a number of cases of *P. carinii* pneumonia in previously healthy homosexual men first led to the recognition of AIDS. The acronym AIDS, for "acquired immunodeficiency syndrome," was first used in 1982 by the Centers for Disease Control for diseases that were at least moderately predictive of a defect in cell-mediated immunity in subjects with no apparent cause for low resistance to the diseases. These AIDS-defining diseases (**table 29.1**) were useful in studying the AIDS epidemic, especially before the cause was known. Immunodeficiency, however, is the end stage of a disease with many other manifestations. AIDS is therefore more appropriately called **HIV disease**, after its causative agent, human immunodeficiency virus (HIV). In this chapter, *HIV disease* implies that replication of the virus is causing symptoms or demonstrable damage to the body, while *HIV infection* means only that the virus has entered the body and is replicating, whether or not disease has occurred. The term *AIDS* refers to the end stage of HIV disease, characterized by unusual tumors and immunodeficiency. ■ immunodeficiency disorders, p. 444

HIV Disease

Almost everyone who becomes infected with HIV develops HIV disease, marked by slow destruction of their immune system, eventually ending in AIDS.

Symptoms

The first symptoms of HIV disease appear after an incubation period of 6 days to 6 weeks and usually consist of fever, headache, sore throat, muscle aches, enlarged lymph nodes, and a generalized rash. Some subjects develop central nervous system symptoms ranging from moodiness and confusion to seizures and paralysis. These symptoms constitute the **acute retroviral syndrome** (**ARS**), and they typically subside within 6 weeks. Many HIV infections are asymptomatic, however, or the symptoms are mild and attributed to the "flu." Following the acute illness, if any, there is an asymptomatic period that typically lasts for years, even though the disease advances in the infected person and can be transmitted to others. The asymptomatic period may end with persistent enlargement of the person's lymph nodes, a condition known as **lymphadenopathy syndrome** (**LAS**). Other symptoms heralding immunodeficiency include fever, weight loss, fatigue, and diarrhea, referred to as the **AIDS-related complex**

TABLE 29.1 AIDS-Defining Conditions*

Cancer of the uterine cervix, invasive	Kaposi's sarcoma
Candidiasis involving the esophagus, trachea, bronchi, or lungs	Lymphomas, such as Burkitt's, or arising in the brain
Coccidioidomycosis, of tissues other than the lung	Mycobacterial diseases, including tuberculosis
Cryptococcosis, of tissues other than the lung	Pneumocystosis (pneumonia due to *Pneumocystis carinii*)
Cryptosporidiosis of duration greater than 1 month	Pneumonias occurring repeatedly
Cytomegalovirus disease of the retina with vision loss or other involvement outside liver, spleen, or lymph nodes	Progressive multifocal leukoencephalopathy (a brain disease caused by the JC polyomavirus)
Encephalopathy (brain involvement with HIV)	*Salmonella* infection of the bloodstream, recurrent
Herpes simplex virus causing ulcerations lasting 1 month or longer or involving the esophagus, bronchi, or lungs	Toxoplasmosis of the brain
Histoplasmosis of tissues other than the lung	Wasting syndrome (weight loss of more than 10% due to HIV); also known as **slim disease**
Isosporiasis (a protozoan disease of the intestine) of more than 1 month's duration	

*Additional conditions are used in surveillance of childhood AIDS

(**ARC**). The first symptoms in many cases are those due to tumors or opportunistic infections resulting from severe immunodeficiency. As one might expect, symptoms at this stage of the disease vary widely according to the kind of infection. For example, a frequently encountered symptom is a fuzzy white patch on the tongue, **hairy leukoplakia (figure 29.2)**, a result of latent Epstein-Barr virus (EBV) reactivation. Severe skin rashes, cough and chest pain, stiff neck, confusion, visual problems, and diarrhea are manifestations of other infections. ■ Epstein-Barr virus, p. 727

Causative Agent

In the United States, and most other parts of the world, AIDS is usually caused by human immunodeficiency virus, type 1 (HIV-1), a single-stranded RNA virus of the retrovirus family. There are many kinds of retroviruses, naturally infecting hosts as diverse as fish and humans. HIV-1 belongs to the lentivirus subgroup of the retroviruses, which characteristically infect mononuclear phagocytes.

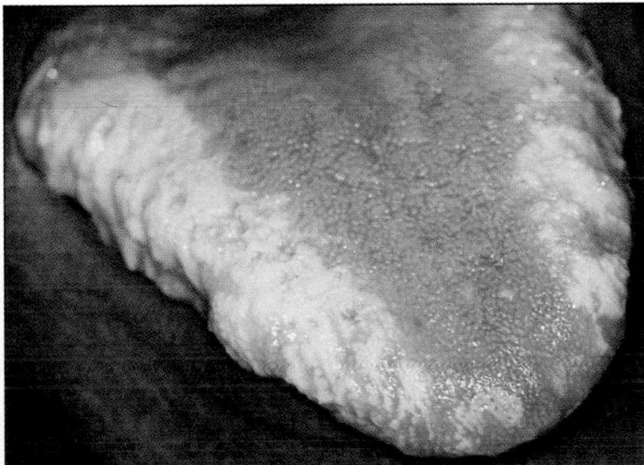

Figure 29.2 Hairy Leukoplakia This AIDS-related condition is probably due to activation of latent Epstein-Barr virus infection.

HIV-1 viruses can be classified into subtypes based on nucleic acid sequences. In the M (for "major") group of HIV-1 viruses, 10 subtypes, clades, have been described, designated A through J. Members of each clade are closely related, as determined by sequencing their genomes. Infection by two different subtypes can give rise to "hybrids" having properties of both. Other HIV-1 groups such as O (for "outlier") and N differ from M group viruses. The importance of being able to recognize all these differing strains of HIV-1 viruses is to aid epidemiological studies and to ensure that the tests for HIV reliably detect infection.

Human immunodeficiency viruses, type 2 (HIV-2), are lentiviruses similar in structure to HIV-1, but antigenically distinct, their genomes differing from HIV-1 by more than 55%. They are a prominent cause of AIDS in parts of West Africa and India, and have appeared in the United States and other countries. HIV-2 transmission has generally been less efficient that HIV-1, and disease progression slower. Otherwise, the biology of HIV-2 is quite similar to HIV-1. For the remainder of this chapter, we will use the term HIV to indicate HIV-1.

The structure of HIV is shown schematically in **figure 29.3a**. The locations of its important antigens are indicated. The numerous knobs projecting from the surface of the virion represent gp120 (SU, for "surface") antigen, partly responsible for attachment to the host cell. The *gp* stands for glycoprotein, indicating that sugar molecules are attached to the protein. The gp41 (TM, for "transmembrane") antigen traverses the viral envelope and is closely associated with SU; it probably plays a role in entry of the virus into the host cell. The p17 (MA) matrix protein, located inside the viral envelope, helps maintain viral structure, transport the viral genome to the host cell nucleus, and assemble new virions. The core of the virion is composed of the p24 (CA) capsid antigen, two copies of the single-stranded

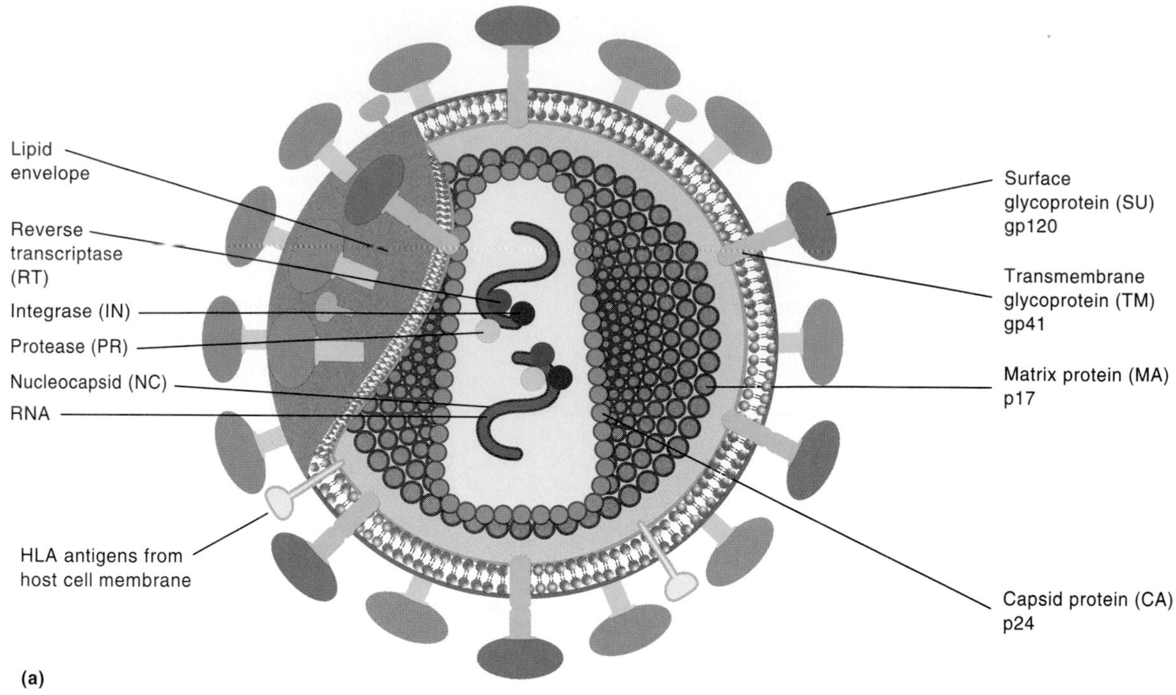

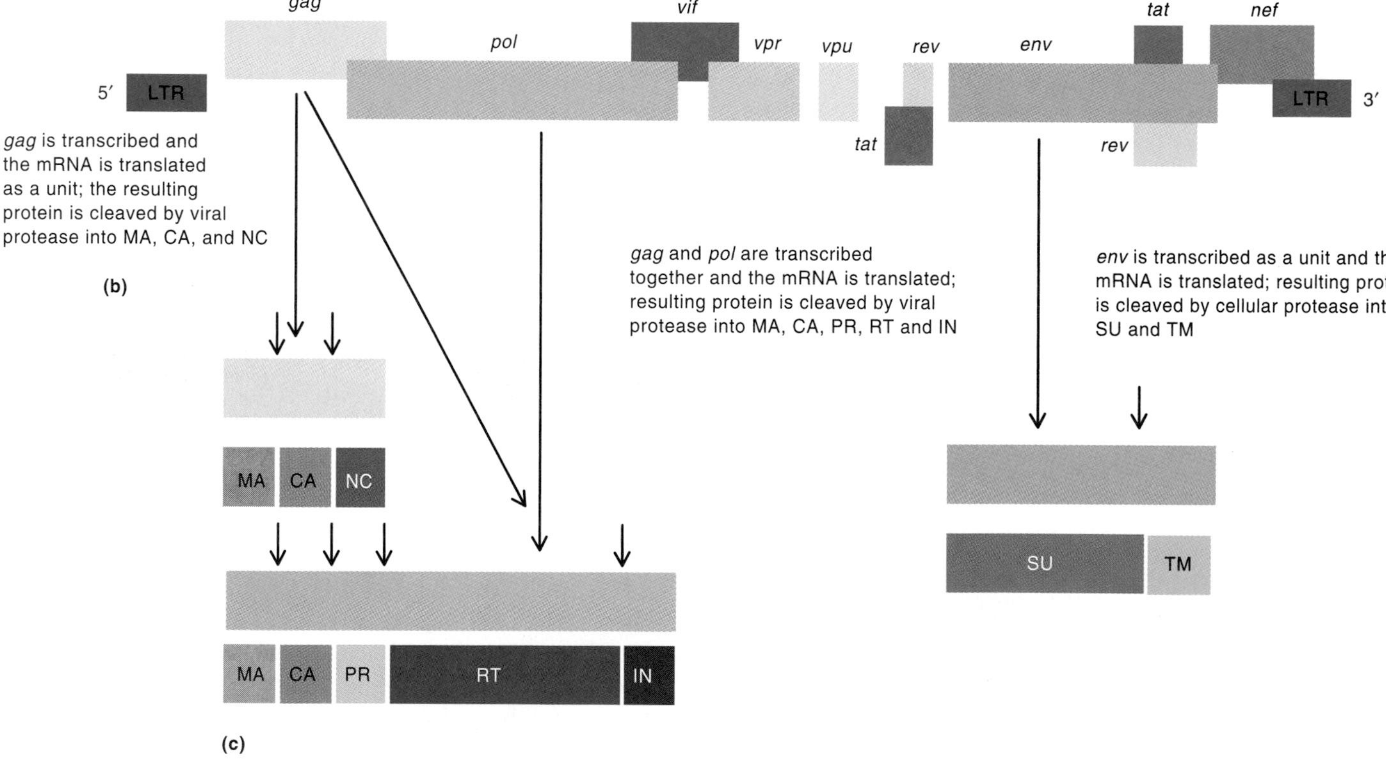

Figure 29.3 Human Immunodeficiency Virus, Type 1 (HIV-1) (a) Diagrammatic representation of the virus showing important antigens. **(b)** Map of the HIV-1 genome showing its nine genes and flanking long terminal repeats (LTRs). The LTRs contain regulatory sequences recognized by the host cell. **(c)** HIV-1 gene products. The accessory gene products are translated into proteins of final size, while *gag* and *pol* products must be cleaved by viral protease, and *env* products by host cell protease. The small arrows indicate the sites of cleavage.

RNA viral genome, and various proteins, including nucleocapsid (NC) and three important viral enzymes, reverse transcriptase (RT), protease (PR), and integrase (IN).

The HIV genome is shown schematically in figure 29.3b. The first two genes, *gag* (from "group antigen") and *pol* (from "polymerase"), can be translated as a unit from the full-length viral messenger RNA (figure 29.3c). The resulting protein is split into four segments by viral protease, yielding three enzymes: protease, reverse transcriptase, and integrase. The fourth segment is split into a number of other proteins, including p24 (CA) capsid and p17 (MA) matrix. The virus, however, has a greater need for structural proteins MA, CA, and NC than for the enzymes PR, RT, and IN, and so *gag* is usually expressed by itself, only rarely joining *pol* through a frameshift. Reverse transcriptase and protease are the targets of the medicines currently available for treating AIDS. The *env* gene is translated from a spliced messenger RNA, yielding a precursor protein that is processed by host cell enzymes to give the gp120 (SU) surface glycoprotein and the gp41 (TM) transmembrane glycoprotein. ■ **RNA splicing, p. 178**

There are six additional genes, known as accessory genes, all translated from spliced messenger RNAs. These genes, *tat*, *rev*, *nef*, *vif*, *vpr*, and *vpu*, code for the proteins Tat, Rev, Nef, Vif, Vpr, and Vpu. These gene products and their known or assumed functions are listed in **table 29.2**.

The functions of these HIV accessory genes are exceedingly complex, as they interact in different ways with host cell substances and vary with the type of cell infected. Developing knowledge of their gene products promises to reveal new ways to attack AIDS.

TABLE 29.2 HIV Accessory Gene Products

Gene Product	Function
Tat (transactivating protein)	Regulates transcription from the integrated DNA form of HIV
Rev (regulator of viral expression)	Binds to unspliced RNA transcripts and targets them for passage out of the host cell nucleus
Nef (negative factor)	Acts to remove CD4 from the host cell surface. Needed for full pathogenicity of the virus
Vif	Aids viral infectivity; regulates virion assembly
Vpr (viral protein R)	Aids transport of uncoated nucleoprotein to the nucleus. Also prevents infected cells from proliferating
Vpu	Blocks transport of CD4 to the cell surface

Pathogenesis

HIV virions enter the body; there, they attach to and infect certain types of cells. Which cells are affected and their response to infection vary with the particular viral strain and the type of host cell. Some of the cell types and the consequences of infection are shown in **figure 29.4**. Infection of intestinal epithelium is

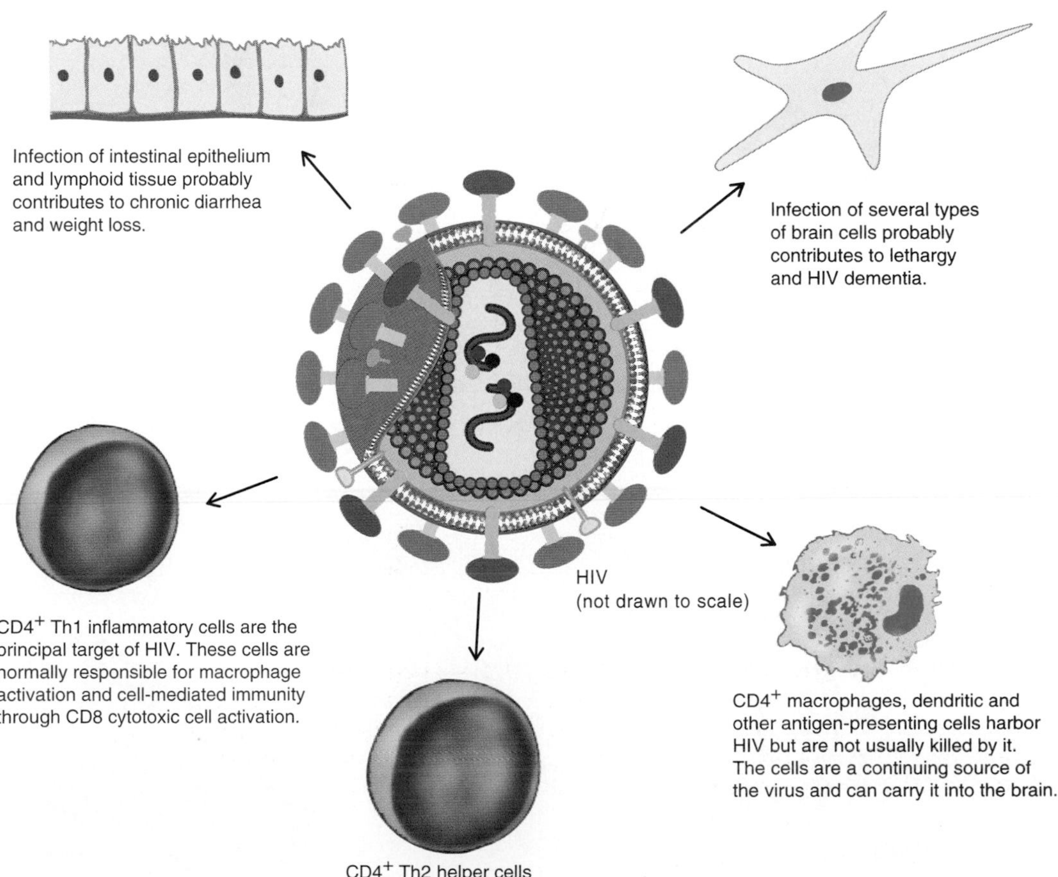

Infection of intestinal epithelium and lymphoid tissue probably contributes to chronic diarrhea and weight loss.
Infection of several types of brain cells probably contributes to lethargy and HIV dementia.
HIV
(not drawn to scale)
CD4+ Th1 inflammatory cells are the principal target of HIV. These cells are normally responsible for macrophage activation and cell-mediated immunity through CD8 cytotoxic cell activation.
CD4+ macrophages, dendritic and other antigen-presenting cells harbor HIV but are not usually killed by it. The cells are a continuing source of the virus and can carry it into the brain.
CD4+ Th2 helper cells normally control antibody production by B cells.

Figure 29.4 Some of HIV's Cellular Targets

partly responsible for the chronic diarrhea and weight loss, and infection of brain cells for HIV dementia. Infected macrophages and other antigen-presenting cells are a continuing source of infectious virus and show impairment of chemotaxis, phagocytosis, and antigen presentation. They and the Th (T-helper) lymphocytes are exceedingly important targets of HIV because of their central role in the body's specific immune response. The cytokines of Th lymphocytes regulate cytotoxic action of Tc cells, immunoglobulin production by B cells, and chemotaxis of antigen-presenting cells such as macrophages.

HIV must attach to and enter the body's cells to establish infection. Like most other cells susceptible to HIV, Th lymphocytes and macrophages are $CD4^+$, meaning they possess the CD4 surface antigen. HIV attaches to CD4 by its surface gp120 (SU) antigen. The presence of the CD4 antigen, however, is by itself not sufficient to cause entry of HIV. In addition, there must be a host cell co-receptor specific for HIV in order for viral entry to occur **(figure 29.5)**. The nature of HIV attachment and entry into the host cell has been a mystery subject to intense study because antibodies and other substances designed to block SU and CD4 have not been very successful in preventing infection. Findings so far indicate that SU is flexible, irregular in shape, and divided into two parts connected by peptide strands. The main binding sites for CD4 and the co-receptor are only

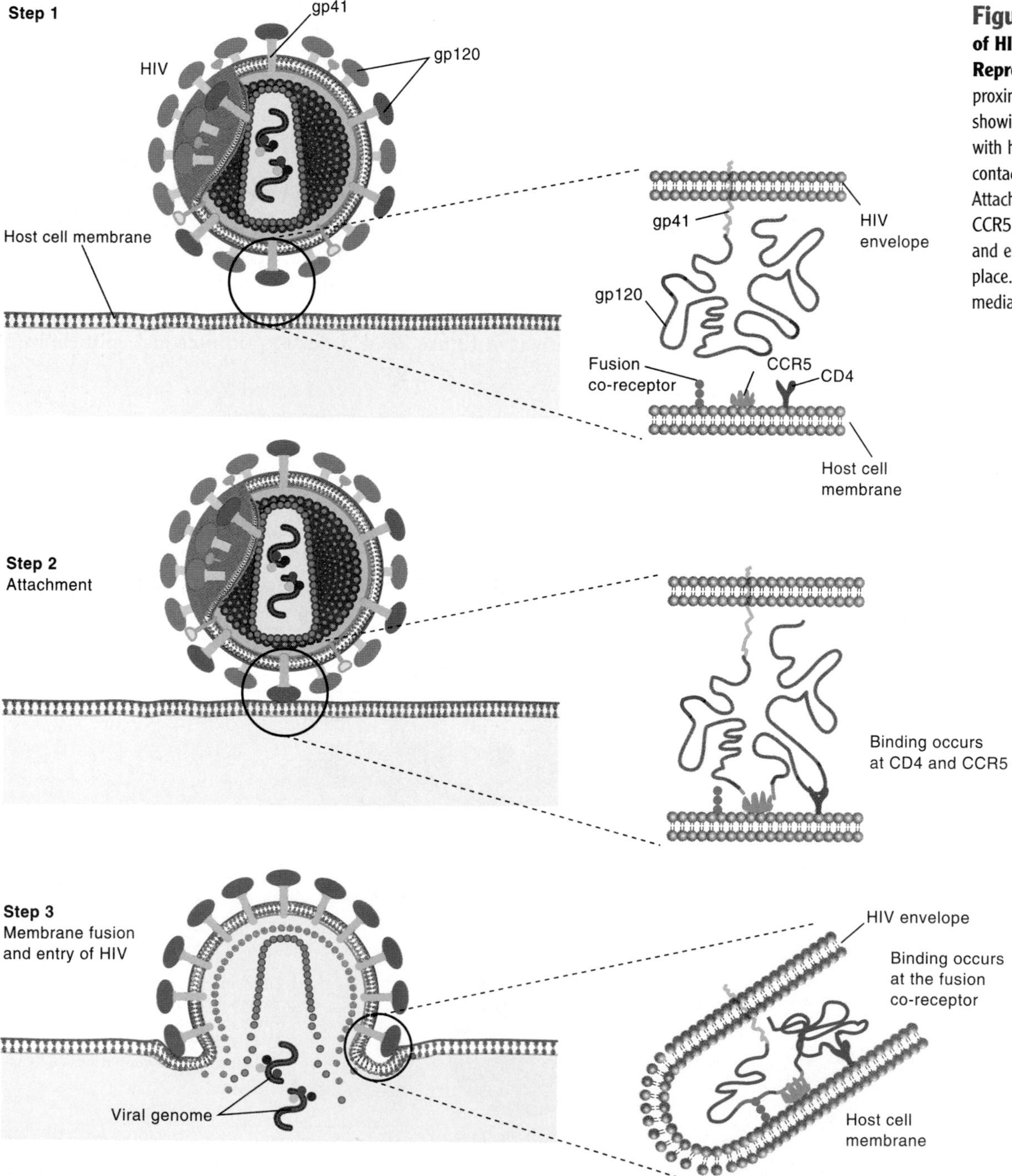

Figure 29.5 Attachment and Entry of HIV into a Host Cell, Schematic Representation Step 1:Virion in close proximity to the cell membrane; blowup showing gp120 protein and sites of reaction with host cell receptors. Step 2: Initial contact of gp120 (SU) is with CD4. Step 3: Attachment to a chemokine receptor such as CCR5 must occur before membrane fusion and entry of the viral genome can take place. Membrane fusion probably is mediated by gp41 (TM).

Perspective 29.1 Origin of AIDS-Causing Viruses

Where did the AIDS-causing viruses come from? Genetic evidence indicates that the HIV-1 virus mutated to its present form fairly recently, between 50 and 150 years ago. Although it first appeared in the United States in the 1970s, serological evidence indicates that it was present in Africa in rare individuals in the 1950s.

Viruses similar to HIV exist in a number of wild and domestic animals, including cats, dairy cattle, and monkeys. A virus closely related to HIV-1 has been found in animals belonging to a single West African subspecies of chimpanzees. The virus does not appear to harm the chimpanzees, suggesting that they have been living together for eons. Another virus, closely related to HIV-2, is present in a species of large monkey, the sooty mangabey. A likely theory is that the AIDS-causing viruses "jumped" to humans from these simian relatives, presumably through contact with their blood. Indeed, in Africa, chimps and mangabeys are often killed for food, exposing humans to their blood, and incidentally, driving some species nearly to extinction. Also, a number of humans were intentionally injected with chimpanzee and mangabey blood in the course of research on malaria carried out between 1922 and 1955.

Genetic comparisons of a large number of simian lentiviruses and AIDS-causing viruses from humans support the idea that the simian viruses can jump to humans. In the case of HIV-1, a jump to humans probably occurred only once, between 1910 and 1950 most likely around 1930. It is unlikely that HIV was transferred to humans by inadequately sterilized Salk polio vaccine grown in simian kidney cell cultures, but the possibility has not been completely ruled out. There is no credible evidence for another popular theory that the viruses resulted from botched biological warfare experiments by Russia or the United States.

It is possible that AIDS-causing viruses have existed for many years in people living in isolated African villages, perhaps even for centuries. According to this idea, population increases and migration to big, crowded cities to find work allowed the viruses to spread rapidly, becoming more virulent in the process.

The answer to the question about the origins of AIDS-causing viruses may never be known precisely, but the question is intriguing and may lead to better understanding of the emergence of new infectious diseases.

exposed after initial contact of the virus with CD4. Penetration into the host cell probably involves gp41 (TM), probably with help from a host cell enzyme.

The co-receptor differs for different cell types. In early 1996, after a 10-year search, identification of a co-receptor for T lymphocytes was accomplished at the National Institutes of Health. Initially called fusin, it was later identified as CXCR-4, one of the cell's chemokine receptors. An HIV co-receptor of macrophages, CCR5, has been identified and is also a chemokine receptor, and other co-receptors are now known. These important findings have opened up several new avenues of research in the battle to control HIV disease.

Once HIV enters the host cell, its reverse transcriptase makes a DNA copy of the viral RNA genome. A complementary DNA strand is then added, and the ends of the resulting double-stranded DNA segment are joined noncovalently. The resulting circular DNA is then moved to the nucleus and is inserted into the host cell chromosome by the viral integrase (IN) enzyme. The viral segment thus becomes a provirus, a step necessary for efficient replication of HIV. No specific location is required for the insertion into the host chromosome. Following integration, spliced mRNA segments appear, mainly transcribed from the regulatory genes, *tat*, *rev*, and *nef*, and are translated into proteins that are probably involved in HIV gene expression. High levels of Tat are associated with replication of infectious virions. Nef protein, despite the name derivation from negative factor, may cause increased virion production and other effects increasing pathogenicity, depending on the viral strain and type of cell infected. In the case of Th lymphocytes, replication of mature virions results in lysis of the cells and release of infectious HIV. Macrophages, on the other hand, usually release infectious virus over long periods of time without death of the cells. The life cycle of HIV is shown schematically in **figure 29.6**, but in an infected person, at any one time, the vast majority of infected cells show neither lysis nor latency. Most of the HIV-infected cells show accumulations of various viral products that impair the normal functioning of the cells. ■ covalent bonds, p. 23

HIV is genetically highly variable. The variant viruses show differences in their preferred host cell, rates of replication, response to host immunity, and other characteristics. The variability is due to a high rate of "error" when reverse transcriptase copies the viral genome. Some regions of the genome are highly conserved, meaning they do not change much from one strain of virus to another. Other regions are highly variable. This is reflected for example in the glycoprotein composing gp120 (SU), which has conserved and variable regions (**figure 29.7**). The V3 variable region is important because it plays a role in the virulence of HIV. This antigenic variability of HIV enormously complicates the task of developing an effective vaccine against the virus. ■ reverse transcriptase, p. 229, 355

Destruction of immune system Th cells by HIV can occur via multiple mechanisms:

- Lysis following HIV replication. Lysis of helper $CD4^+$ T lymphocytes during replication of infectious virus is certainly one mechanism, but by itself it cannot account for the devastation that results from HIV disease.
- Attack by HIV-specific cytotoxic $CD8^+$ T lymphocytes. The earliest detectable immune response to HIV infection involves HIV-specific $CD8^+$ T lymphocytes, which attack and lyse infected cells.
- Natural killer cells. Natural killer cells probably also play a role in cell destruction.
- Antibody-dependent cellular cytotoxicity. Humoral antibody may also play a role through antibody-dependent cytotoxicity.
- Autoimmune process. Autoimmunity could be responsible, since HIV envelope proteins show some homology with MHC II antigen.
- Fusion of infected and uninfected cells. In cell fusion, a large number of uninfected cells fuse with an infected cell, whereupon the resulting syncytium is destroyed.
- Apoptosis, also called programmed cell death, or death due to aging, is accelerated in HIV infections by a number of mechanisms, including induction by Tat, SU interaction with CD4, and certain cytokines.

Accumulation of viral products such as viral RNA and unintegrated viral DNA inside the cytoplasm of the infected cell

Site of action of AZT and other reverse transcriptase inhibitors

(1) Viral genome and reverse transcriptase enter cell.

(2) DNA copy synthesized by reverse transcriptase.

RNA
DNA

(3) RNA degraded; second DNA strand synthesized.

DNA
DNA

(4) DNA circularizes (unintegrated provirus) or integrase functions to incorporate DNA into host cell genome (integrated provirus).

Host cell nucleus

Host cell genome

(5) With host cell activation, viral DNA is transcribed, yielding messenger RNAs and viral genome RNA.

(6) Viral RNAs are translated, yielding viral enzymes (including protease) and structural proteins.

(7) Viral membrane proteins are transported to host cell membrane.

(8) Final viral assembly and budding take place.

Site of action of protease inhibitors

Site of action of antiretroviral drugs under development

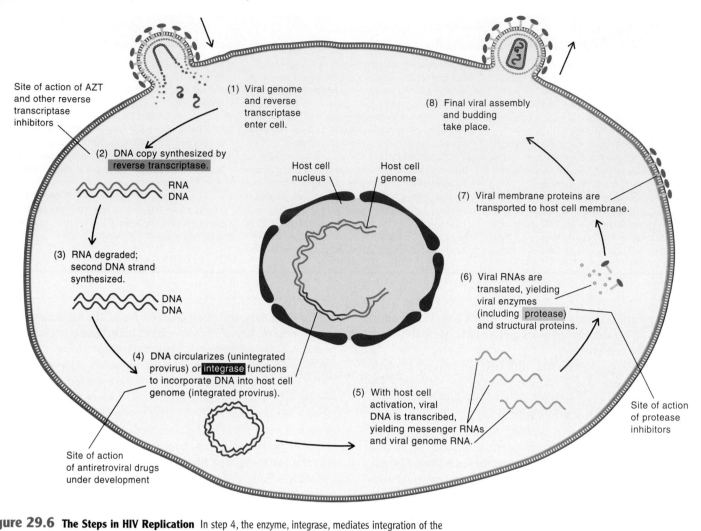

Figure 29.6 The Steps in HIV Replication In step 4, the enzyme, integrase, mediates integration of the provirus into the host cell genome. This enzyme is potentially a target of new anti-HIV medications. In step 7, the accessory gene product Vpr and MA transport viral proteins to the cell membrane.

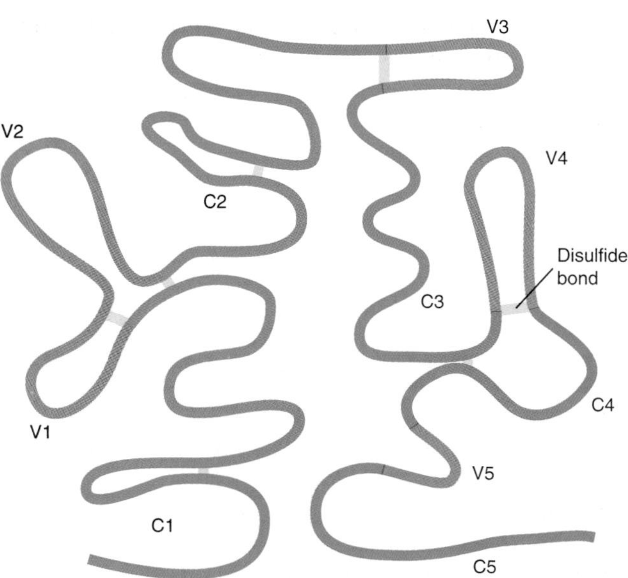

Figure 29.7 Diagram of the SU Glycoprotein There are five "constant" segments (blue) that are conserved in different HIV strains. The segments in orange, labeled V1 to V5, are "variable" and differ from strain to strain, frustrating attempts to develop a vaccine. Note the V3 loop, implicated in pathogenicity.

can also result in its death. ■ **natural killer cells, p. 373** ■ **antibody-dependent cellular cytotoxicity, p. 394** ■ **major histocompatibility complex, type II (MHC-II), p. 399** ■ **apoptosis, p. 407**

An acute retroviral syndrome (ARS) begins in 50% to 70% of HIV-infected subjects after an incubation period of 1 to 6 weeks, with sore throat, fever, muscle and headaches, enlarged lymph nodes, and a rash. These symptoms generally last 1 to 4 weeks and disappear without treatment. Whether or not symptoms occur, the concentration of HIV in the blood rises to high levels as the newly infected cells release their progeny virus (**figure 29.8**). The marked viremia usually subsides as the supply of uninfected cells dwindles and anti-HIV MHC I $CD8^+$ cytotoxic T cells appear. The $CD4^+$ T-cell level falls initially, then slowly rises but does not reach the preinfection level. Following the acute episode, the HIV-infected individual generally enters an asymptomatic period ranging from months to many years. The levels of HIV virus as measured by circulating plasma and lymph node viral RNA are good predictors of the course of the illness, high levels pointing to a more rapid progression to AIDS. Regardless of the lack of symptoms, a silent struggle goes on between HIV and the immune system for the rest of the person's life. Infectious virus, a threat to others, continues to be present in the blood and body secretions.

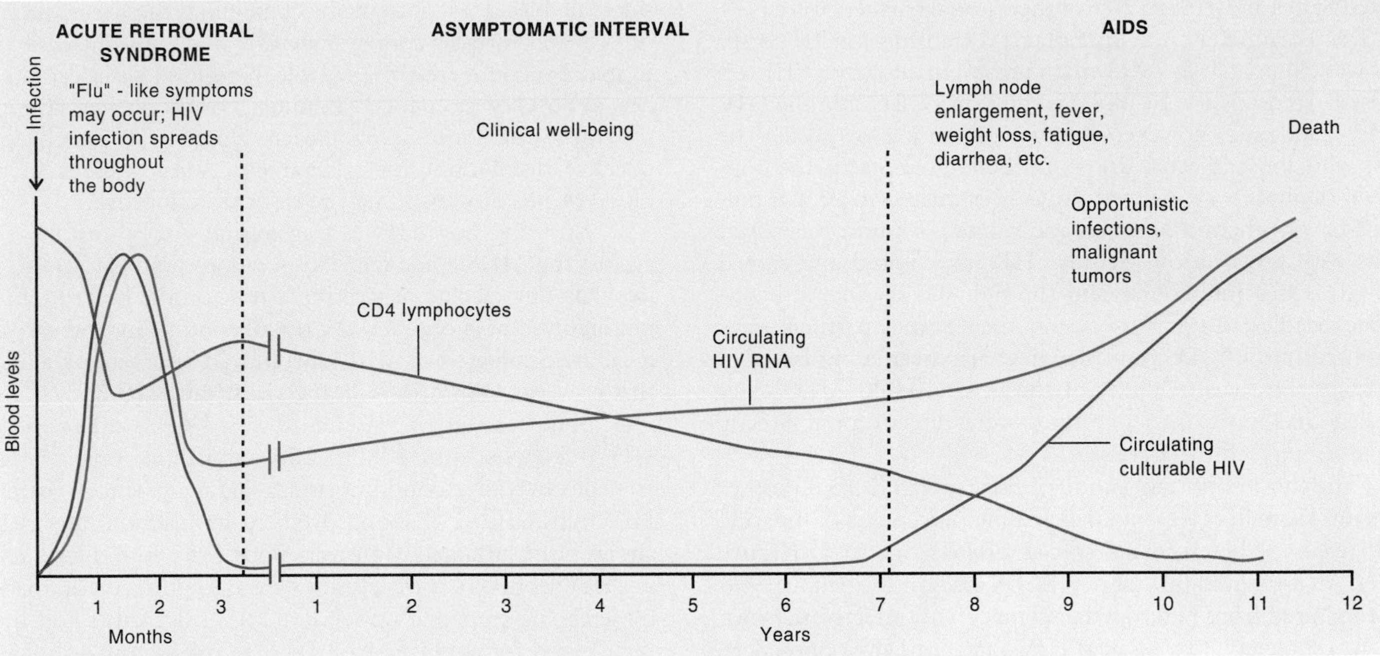

Figure 29.8 Natural History of HIV Disease Blood levels of infectious virus are very high at the beginning, during the acute retroviral syndrome, and at the end of the disease when AIDS ensues. Antibody tests for diagnosing the disease are often negative in the early stage of the disease even though infected people are highly infectious. The disease steadily progresses in the absence of symptoms, as shown by the rising levels of plasma viral RNA and falling $CD4^+$ cell count.

In the typical case, as seen in approximately 80% of HIV-infected people, the immune system slowly loses ground to HIV even though the body can normally replace over a billion $CD4^+$ cells per day. The peripheral blood $CD4^+$ count (normally about 1,000 cells per microliter) steadily falls at a rate of roughly 50 cells/ml/year, symptoms of AIDS usually appearing when $CD4^+$ counts fall below 300 cells per microliter. Levels of infectious virus again rise dramatically. At this point in the illness, the pathology is dominated by malignant neoplasms or opportunistic infections. Half of the patients reach this stage of the illness within 9 to 10 years; the rest take longer. The typical course of HIV disease is summarized in figure 29.8.

Atypical progression to AIDS occurs in about 10% of persons with HIV disease who have high virus levels with the acute infection that do not fall dramatically within a few months. These subjects progress rapidly to AIDS within a few years. Another 5% to 10% of HIV-infected individuals show no fall in $CD4^+$ cells. They maintain high levels of anti-HIV antibody and HIV-specific $CD8^+$ cytotoxic T cells. Estimates suggest that, with these cases and the more slowly progressing typical cases, 10% to 17% of HIV-infected persons will be free of AIDS 20 years after their infection.

Epidemiology

Indiscriminate sexual intercourse with multiple partners without the use of condoms is a major factor in the spread of HIV disease. In the United States, initially, promiscuous homosexual men were the hardest hit by the epidemic. An estimated 1.4% to 10% of American men are active homosexuals. An additional unknown number of men are bisexual, meaning they have sexual intercourse with both men and women. A survey done before the arrival of AIDS found that 33% to 40% of gay men had more than 500 lifetime sexual partners, and another 25% had 100 to 500. By 1978, 4.5% of one group of gay men in San Francisco were infected with HIV; by 1984, two-thirds were infected and almost one-third had developed AIDS. The pace of the epidemic among gay men has been slowing, but still there were more than 22,000 new AIDS cases reported among men who have sex with men in the year ending mid-1999, and these accounted for an estimated 45% of new HIV infections over the same period. These numbers, however, merely reflect how the virus was introduced into Western countries. Heterosexual spread of HIV is increasing and promises to become the dominant mode of transmission, as it is in many African countries. Sex with multiple partners; sex with a person who has had multiple partners; being the receptive partner, especially receptive anal sex; traumatic sex; and any irritation or inflammatory process as from another sexually transmitted disease—all increase the risk of sexual HIV transmission. The disease can be contracted by the insertive partner during vaginal or rectal intercourse with an infected person. Saliva is very unlikely to transmit the disease, but there is evidence that oral-genital contact may be risky.

The next most important mode of transmission of HIV is through blood and blood products. Individuals infected with HIV who donated blood before a screening test became available in 1985 unknowingly infected thousands of transfusion recipients. One of the products from pooled donated blood, clotting factor VIII, was used to treat bleeding episodes among hemophiliacs. By 1984, over half of the hemophiliacs in the

United States and 10% to 20% of their sexual partners were HIV positive. Fortunately, the risk of HIV transmission by factor VIII was eliminated in 1992 when recombinant factor VIII was licensed. Transmission by blood is still a major factor in the HIV disease pandemic, however, because of sharing of needles by those who abuse injected drugs. In the United States, the population of abusers of injected drugs is estimated to be 1.5 million. Many of them, both men and women, support their drug habits with prostitution. Because HIV is acquired and spread through sexual intercourse and through the sharing of hypodermic needles, drug abusers and their sexual partners have been hard hit by AIDS, but they have also been a big factor in spreading the virus. A study of the first 640,000 AIDS cases revealed that more than one-third were directly or indirectly related to injected-drug abuse.

The third important mode of HIV spread is from mother to infant. Women represent an increasing percentage of the total AIDS cases as heterosexual spread of HIV increases **(figure 29.9)**. An estimated 60% of new AIDS cases in women in 1998 were acquired from heterosexual contact with infected persons, and an estimated 70% of new HIV infections (as opposed to AIDS) in women in the year ending mid-1999 were heterosexually acquired. If untreated, about one out of 10 pregnant HIV-positive women will miscarry, and of live-born babies, 15% to 40% will develop AIDS. Remarkably, however, in a study of 219 newborn babies positive for HIV by culture or PCR, almost 3% cleared their infection without any treatment. Breast feeding carries a significant risk of mother-infant transmission, especially if the mother becomes acutely infected with HIV.

Prevention and Treatment

There is no approved vaccine against HIV infection. Infectious HIV persists in samples of blood plasma for at least 1 week after they are taken from AIDS patients. Most people with HIV disease do not know they are infected, and it is advisable to consider all blood as potentially containing the virus. HIV on objects and surfaces contaminated by body fluids is easily inactivated by commercially available high-level disinfectants and heat at 56°C or more for 30 minutes. Freshly opened household sodium hypochlorite 5.25% bleach diluted 1:10 is a cheap and effective disinfectant for general use. Virus present in dried blood or pus, however, may be difficult to inactivate.

Knowing how HIV is transmitted is a powerful weapon against the AIDS epidemic. This weapon can be far more effective than any vaccine or treatment now on the horizon. HIV is not highly contagious, and the risk of contracting and spreading it can be eliminated or markedly reduced by assuming a lifestyle that prevents transmission of the virus **(table 29.3)**.

Some groups of gay men lowered the incidence of new HIV infections from 10% to 20% annually to only 1% to 2% annually by using condoms and avoiding practices that favor HIV transmission. The improvement in infection rates has not always been sustained, however. Since 1985, the risk of acquiring HIV from blood transfusions has been lowered dramatically by screening potential donors for HIV risk factors and testing their blood for antibody to HIV. The risk is now estimated to be less than 1 in 400,000 transfused units. Newer tests for HIV antibody and nucleic acid give positive results within 1 month of infection in most but not all cases. Screening tests used for infection have also markedly reduced the risk of HIV transmission from artificial insemination and organ transplantation.

Education about how HIV is transmitted can be a very effective tool in helping to bring the worldwide epidemic of HIV disease under control. Education of schoolchildren has

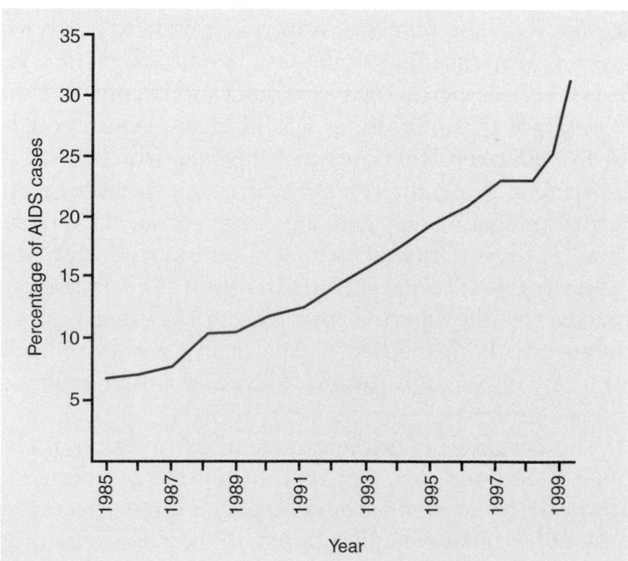

Figure 29.9 In the United States, a Steadily Rising Percentage of AIDS Cases Occur in Women

TABLE 29.3 Lifestyles that Help Control the AIDS Epidemic

1. Not engaging in sexual intercourse.
2. Staying with one faithful sexual partner who you know very well.
3. Avoiding sexual intercourse with persons at risk for HIV infection (see table 29.4).
4. Avoiding trauma to the genitalia and rectum. Small breaks in the skin and mucous membranes allow HIV to infect.
5. Not engaging in sexual intercourse when sores from herpes simplex or other causes are present. They represent sites where HIV can infect.
6. Not engaging in anal intercourse. Receptive anal intercourse carries a high risk of HIV transmission.
7. Using latex condoms from beginning to end of sexual intercourse. Polyurethrane condoms are a reasonable alternative for those allergic to latex. Condoms made from other materials are not reliable for disease prevention, nor are those marketed in many African and other countries outside the United States. Oil-based lubricants are not compatible with latex. Condoms for women are available.
8. Postponing pregnancy indefinitely if you are a woman infected with HIV. If you are not sure of your HIV status, blood tests to rule out HIV disease before considering pregnancy.
9. Using extreme care to avoid needles, razors, toothbrushes, etc., that could be contaminated with someone else's blood.

CASE PRESENTATION

The patient was a young woman from West Africa presenting with the complaint of generalized enlargement of her lymph nodes. She had recently moved to the United States. She had no history of drug abuse or of receiving blood transfusions. She had had three sex partners in her life. Her single pregnancy the year before her arrival was delivered by emergency cesarean section. She and the baby's father had routine tests for HIV at that time, and both tests were negative. The baby and the father remained well. Ten years before her evaluation, she was treated for a fever by scarification, and this was repeated for an unrelated complaint 4 years before her evaluation. The native healer performing the scarification used a razor blade, the sterility of which is unknown to the patient.

Her initial test results included nondiagnostic lymph node biopsies and a negative HIV test (enzyme-linked immunoassay, ELISA). A repeat HIV test some months later was weakly positive by ELISA, and the confirmatory Western blot showed only questionable reactions of the patient's serum with gp41 and two other HIV antigens. A test for HIV-2 was negative. A $CD4^+$ T lymphocyte cell count was very low. A test to detect HIV-1 using the polymerase chain reaction was negative.

1. Could this patient have HIV disease? If so, how could the negative tests be explained?
2. Does the history give any possible ways in which she could have contracted HIV disease?
3. How could the diagnosis be established?
4. Could her baby and the baby's father have the disease?
5. Does this case suggest that any changes should be made in the way HIV disease is diagnosed?

Discussion

1. This patient could well have had HIV disease/AIDS because of her persistent generalized lymphadenopathy and very low $CD4^+$ cell count. Her HIV strain might have differed enough from the pandemic group M HIV-1 strains so that the usual laboratory tests did not detect antibody to it.
2. An AIDS-causing virus could have been contracted during sexual intercourse, or from a blood-contaminated razor blade used for scarification. Contracting such a virus from unsterile instruments during her emergency cesarean is a possibility, but this seems less likely because of the short time span before presenting with severe immunodeficiency.
3. The Centers for Disease Control and Prevention have a global surveillance system designed to detect cases like this because they raise the possibility of new or rare AIDS-causing viruses being introduced into a population. In the present case, samples of the patient's blood were examined by the CDC and she was shown to be infected with a group O HIV-1 virus. The patient's serum reacted with peptides specific to group O strains, and the virus was isolated from her blood. Nucleic acid sequences of the *env*, *gag*, and *pol* genes matched those of previously isolated group O strains. Group O strains of HIV-1 were first found in Cameroon, a country on the West African coast, where they account for 6% of HIV infections.
4. Despite the absence of symptoms, both her baby and the baby's father could have the disease. The baby's risk of infection was reduced but not eliminated by the emergency cesarean section, and infection could have occurred subsequently if it were breast fed. The mother's very low $CD4^+$ cell count suggests a high level of viremia and, therefore, increased risk of transmitting the disease. She could have infected the baby's father during sexual intercourse, or he could have infected her.
5. This patient was the first case of group O HIV disease identified in the United States. Studies indicate that previous introductions, if any, were not accompanied by spread of the virus. Nevertheless, federal agencies worked with manufacturers of HIV tests to increase sensitivity of the tests to group O strains.

Source: Centers for Disease Control and Prevention. 1996. *Morbidity and Mortality Weekly Report* 45(6): 122.

been shown effective in decreasing risky sexual behavior among teenagers. Videotapes and written material developed for one group of people, however, can be completely ineffective and even offensive to another group.

All persons unsure of their HIV status and especially those at increased risk of HIV disease (**table 29.4**) are advised to get tested for HIV. By knowing, those with HIV disease can receive prompt preventive treatment for the infections and cancers that complicate the disease. Antiretroviral treatment of HIV disease before the onset of AIDS shows promise of preventing the complications of immunodeficiency for decades, and it may also reduce transmissibility. Also, HIV-positive individuals who know their status can also do their part in preventing transmission. Federally approved home test kits coupled with counseling are reliable and commercially available. ■ **sexually transmitted diseases, pp. 643–661**

TABLE 29.4 Persons at Increased Risk for HIV Disease

1. Injected-drug abusers who have shared needles.
2. Persons who received blood transfusions or pooled blood products between 1978 and 1985.
3. Sexually promiscuous men and women, especially prostitutes, drug abusers, and homosexual and bisexual men.
4. People with history of hepatitis B, syphilis, gonorrhea, or other sexually transmitted diseases that may be markers for unprotected sexual intercourse with multiple partners.
5. People who have had blood or sexual exposure to any of the people listed.

HIV transmission from infected mother to newborn can safely be prevented in two-thirds of cases by administering zidovudine (AZT) to the mother during pregnancy and to the newborn infant for 12 weeks. Newer, more potent combinations of AZT and other antiretroviral therapies are widely employed and may be more effective, but their long-term safety is still being evaluated. Elective cesarean section significantly reduces the risk of HIV transmission to the newborn baby, perhaps by avoiding HIV-containing fluids in the birth canal.

More than 100 needle and syringe exchange programs, operating in at least 30 states and the District of Columbia, help prevent spread of HIV and other blood-borne diseases among injected-drug abusers and to their sexual partners and children. In these programs, sterile syringes and needles are exchanged for used ones along with drug rehabilitation efforts and education about use of condoms and other safer sexual practices.

Even without a vaccine, better prevention and treatment of opportunistic infections, and better antiviral therapy against HIV have significantly lengthened the asymptomatic stage of the disease and prolonged life once AIDS develops (**figure 29.10**). Advances in antiviral treatment mainly result from the development of new medications with different modes of antiviral action, and use of these medications in "cocktails," combinations of reverse transcriptase and protease inhibitors referred to as **HAART**, "highly active antiretroviral therapy." The effectiveness of HAART probably stems from the fact that each of

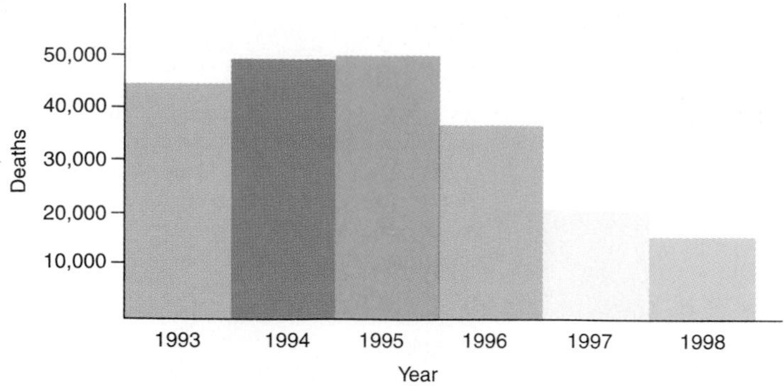

Figure 29.10 Estimated Deaths Due to AIDS, United States, 1993 to 1998 Notice the marked drop in deaths after HAART became available in late 1995.

the medications act on the replicating virus at different parts of the virus life cycle, and despite the high mutation rate of HIV, it is much less likely that any one mutant could develop resistance to all the medications at the same time. Presently available medications fall into two groups: inhibitors of reverse transcriptase and inhibitors of viral protease.

Medications that interfere with reverse transcriptase fall into two categories, nucleoside reverse transcriptase inhibitors (NRTI), and nonnucleoside reverse transcriptase inhibitors (NNRTI). Zidovudine (AZT), stavudine (D4T), and lamivudine (3TC) are among the half dozen NRTIs in wide use, and a number of others are undergoing clinical trials. All are nucleoside analogs, but they differ chemically and in their site of action on the enzyme. These substances owe their effectiveness to their resemblance to the normal purine and pyrimidine building blocks of nucleic acids. During nucleic acid synthesis, the viral reverse transcriptase enzyme incorporates the medication molecule into the growing DNA chain, thereby blocking completion of the DNA strand. This process for AZT is illustrated in **figure 29.11**. ■ nucleosides, p. 513

As for the nonnucleoside inhibitors of reverse transcriptase, nevirapine, efavirenz, and delavirdine are among those in wide use, and at least five newer NNRTIs are in clinical trials. Generally, the NNRTIs are not nucleoside analogs, and they act in a completely different manner from NRTIs, by binding tightly to reverse transcriptase and preventing the enzyme from acting.

Protease inhibitors were a major addition to the anti-HIV arsenal. The first protease inhibitor, saquinavir, was approved for therapy in December 1995. Currently, there at least five protease inhibitors in use, and at least that many more in clinical trials. Unlike the reverse transcriptase inhibitors, the protease inhibitors act late in HIV replication to prevent packaging of viral proteins in the virion. The impact of these inhibitors on the treatment of people with AIDS has in some cases been immediate and dramatic. For example, one 30-year-old woman with AIDS had exhausted all available AIDS therapies. She was a wasted 85 pounds, her hair had fallen out, and she coped with her fatigue by napping every 2 hours. She was given indinavir (a newer protease inhibitor) together with D4T and 3TC, and within a month, her hair grew back, she gained weight, she began skating, and her CD4+ cell count rose from 3 to more than 200 per microliter of blood.

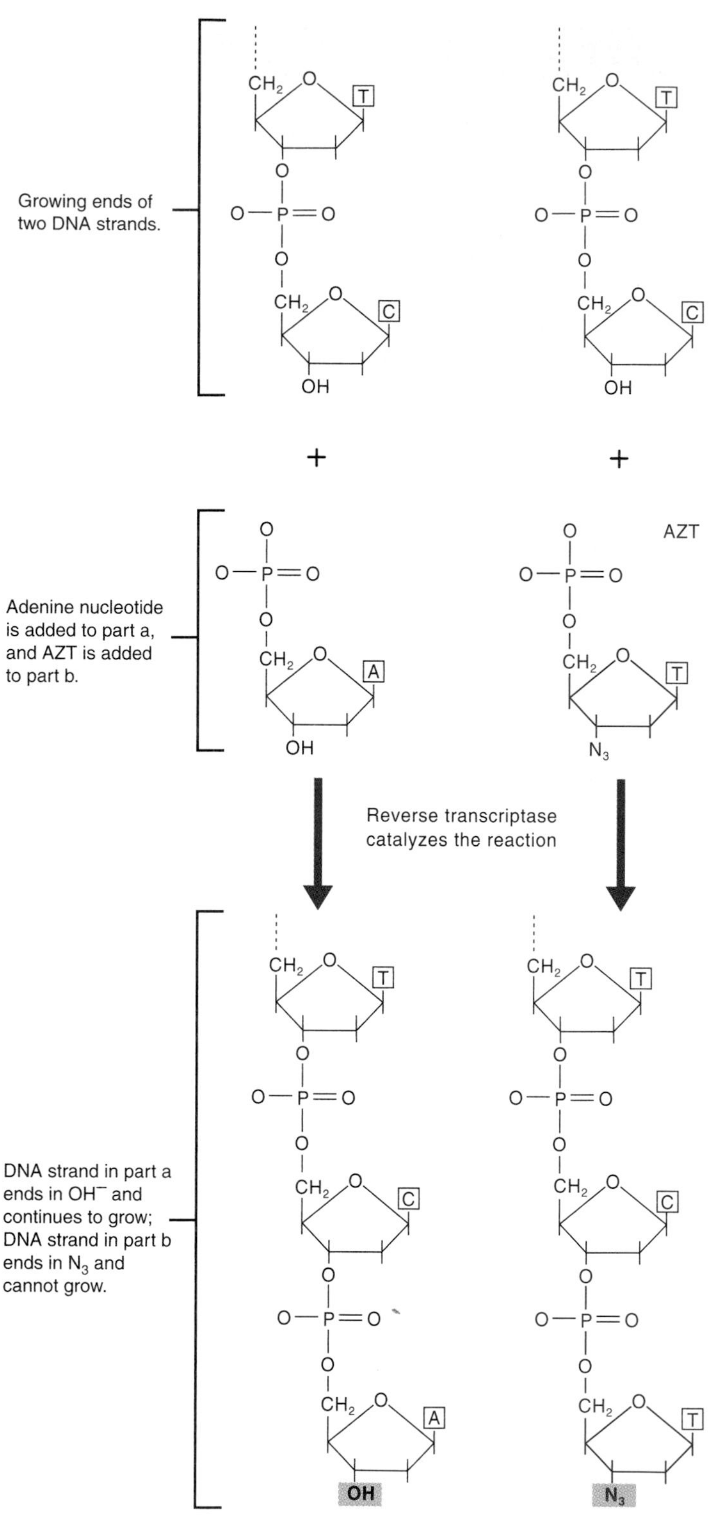

Figure 29.11 Mode of Action of Zidovudine (AZT) **(a)** Normal elongation process of DNA in which reverse transcriptase catalyzes the reaction between the OH group on the chain with the phosphate group of the nucleotide being added. **(b)** Reverse transcriptase catalyzes the reaction of zidovudine with the growing DNA chain. Since zidovudine lacks the reactive OH group, no further additions to the chain can occur, and DNA synthesis is halted. T = thymine, C = cytosine, A = adenine.

HAART does not cure AIDS. Viremia becomes undetectable in only about half the cases, and even then, the viremia returns if the medications are discontinued. While it stops production of virions, it does not eliminate HIV provirus hidden in host cell genomes. In successful HAART, production of infectious virus is largely halted, many fewer $CD4^+$ lymphocytes are killed, and so the $CD4^+$ cell count rises. When the medications are stopped, persistently infected cells release infectious virions that can infect a new population of $CD4^+$ cells, resulting in a quick rise in viremia. Many authorities now feel that the secret to controlling HIV disease lies ultimately in minimizing or eliminating body cells containing the hidden HIV genomes. These authorities are advocating HAART for the acute retroviral syndrome to prevent the buildup in the body of cells containing HIV provirus.

Many strains of HIV fail to respond to HAART because they have become resistant to the medications during past treatment. These resistant strains of HIV are transmissible to other persons. Toxic effects of the medications are another limitation of anti-HIV therapy. For example, AZT can cause anemia, low white blood count, vomiting, fatigue, headache, and muscle and liver damage. Painful peripheral nerve injury, inflammation of the pancreas, rash, mouth and esophagus ulceration, and fever are side effects of other NRTIs. Also, indinavir may be responsible for kidney stone formation, and ritonavir, another protease inhibitor, causes nausea and diarrhea. Diabetes mellitus is another potential side effect of protease inhibitors.

Finally, a big limitation of the use of anti-HIV medications is their cost. The cost of a combination of these medications can easily exceed $1,000 per month for either an asymptomatic or symptomatic individual and is expected to be required for life. While not considered excessive for a serious chronic disease in the United States, the cost of medications puts them completely out of reach for more than 90% of the world's HIV victims.

The main features of HIV disease are presented in **table 29.5**.

TABLE 29.5 HIV Disease

Symptoms	Over half develop fever, sore throat, head and muscle aches, rash, enlarged lymph nodes early in the infection. After an asymptomatic period symptoms from unusual malignant tumors, pneumonia, meningoencephalitis, diarrhea, etc., due to reactivation of latent infections
Incubation period	Usually 1 to 6 weeks for acute symptoms; immunodeficiency symptoms within 10 years in half the infections (10% within 5 years and 90% within 17 years)
Causative agent	Human immunodeficiency virus, type 1 (HIV-1), many subtypes and strains. HIV-2 mainly in West Africa.
Pathogenesis	HIV infects various body cells, notably those vital to specific immunity, $CD4^+$ T lymphocytes and antibody-presenting cells. T cells killed, numbers slowly decline until the immune system can no longer resist infections or development of tumors
Epidemiology	Three main routes of transmission of HIV: intimate sexual contact, via transfer of blood or blood products, and from mother to child around the time of childbirth. Risk of transmission by breast milk, and possibly by oral-genital contact
Prevention and treatment	No vaccine yet available. Medications and vaccines can prevent many infections. Anti-HIV medications and cesarian section decrease mother-to-newborn transmission. Effective preventive measures: sex education of schoolchildren, needle exchange programs for drug addicts, use of condoms. Treatment: HAART therapy, consisting of combinations of several anti-HIV medications, effective for many AIDS victims, and delays progression of HIV disease to AIDS. Not a cure, and too expensive for most of the world's HIV disease victims

HIV Vaccine Prospects

Currently there are no approved vaccines for preventing HIV disease. Development of potential HIV vaccines began soon after the discovery of the causative agent. In theory, a vaccine could be used in either of two ways. One, "preventive vaccine," would be to immunize uninfected individuals against the disease. The other, "therapeutic vaccine," would be to boost the immunity of those already infected with HIV before they became severely immunodeficient. The latter approach would be similar to the post-exposure treatment of rabies infections. A successful vaccine must induce both mucosal and bloodstream immunity, because HIV disease is primarily sexually transmitted. The vaccine must also get around the problem of HIV antigenic variability and stimulate cellular and humoral responses against virulence determinants. Moreover the vaccine has to be safe. A live attenuated agent must not be capable of becoming a disease-causing strain, and it must not be oncogenic, meaning cancer-causing. Additionally, the vaccine must not stimulate an autoimmune response, and it must not cause production of "enhancing antibodies" that could aid the passage of HIV into the body's cells. The vaccine should induce neutralizing antibodies against cell-free virions and also prevent direct spread of HIV from a cell to neighboring cells. ■ rabies, p. 682

Despite enormous difficulties, HIV vaccine research continues to offer the best hope for eventually controlling a worldwide epidemic that every year produces tens of thousands of new HIV infections in the United States alone. Findings that have encouraged vaccine researchers include failure of some African prostitutes to develop AIDS although repeatedly exposed to HIV. Also, HIV-infected pregnant women that have high titers of HIV-neutralizing antibodies are less likely to transmit the infection to their offspring than women with low antibody titers. Finally, infection by some strains of HIV apparently progresses to AIDS very slowly.

Development of vaccines against HIV disease is taking time and costing enormous amounts of money. At present, there is no vaccine in sight that could prevent 80% to 90% of HIV infections, but the epidemic is expanding so rapidly that even a poor vaccine quickly put into use might save more lives than would be saved by waiting for a highly effective vaccine. Candidate vaccines undergo an extensive evaluation for safety and immunogenicity in experimental animals before they are tested in humans. Some of the many substances in this preclinical stage of development are killed HIV, various viruses and bacteria carrying parts of the HIV genome, HIV peptides and proteins, HIV proteins with adjuvants, and naked DNA containing HIV genes. ■ DNA vaccines, p. 429

Vaccine trials in humans have been undertaken for at least 10 experimental vaccines. Vaccine evaluations in humans progress from phase I (testing safety and ability to provoke an immune response), to phase II (determining optimum dose, and characteristics and duration of immune response), to phase III (testing effectiveness in preventing HIV infection or AIDS). Substances that have entered human trials include recombinant gp120 and gp160; vaccinia virus containing HIV envelope genes; canary poxvirus containing HIV genes *env*, *gag*, and *pol*; a synthetic lipopeptide; a DNA vaccine containing HIV envelope genes; a *Salmonella* typhi strain engineered to express HIV gp120; and purified p17 (MA). Currently, only gp120 vaccines are in phase III trials. One trial in Thailand employs gp120 from two HIV strains, one subtype B and the other subtype E; the other trial in North America uses two different subtype B HIV strains.

The "prime-boost" technique employs an initial injection of a live virus such as genetically engineered vaccinia containing HIV surface and core genes, followed later by an injection of recombinant viral surface antigen, gp120. The first injection gives primarily a cytotoxic T lymphocyte response, and the second injection boosts the production of neutralizing antibodies. Vaccines that use vaccinia virus containing HIV genes have generally been discarded in favor of another poxvirus, canary pox, because many people have antibodies to the vaccinia virus from past vaccination against smallpox. Also, the vaccinia virus is potentially pathogenic in people with severe immunodeficiency. The poxviruses have a large genome with multiple sites where foreign genes can be spliced without interfering much with its ability to infect and multiply. ■ vaccinia, p. 413 ■ poxviruses, p. 344

The potential of synthetic peptides in immunization against HIV is also being explored. Since the amino acid sequence of the surface proteins of HIV is known, peptides that exactly mimic immunologically important segments such as the V3 loop (see figure 29.7) can be synthesized. These peptides may prove to be effective as booster injections after primary immunization with a modified virus. ■ peptide, p. 30 ■ peptide vaccines, p. 428

MICROCHECK 29.1

The signs and symptoms of persons with AIDS are mainly due to the opportunistic infections and tumors that complicate HIV disease. HIV is not highly contagious, and the AIDS worldwide epidemic could be stopped by changes in human behavior. Highly active antiretroviral therapy has given miraculous improvement in many patients with AIDS but is unlikely significantly to affect the advance of the AIDS epidemic. The largely silent spread of HIV continues.

- What is hairy leukoplakia?
- How would the ability to recognize differing strains of HIV-1 aid in epidemiological studies?
- If AIDS was present in Africa in the 1950s, why did it not appear in the United States until the 1970s?

Malignant Tumors that Complicate Acquired Immunodeficiencies

Certain malignant tumors are associated with HIV disease, organ transplantation, and other acquired immunodeficiency states. Most of these malignancies fall into one of only three types: Kaposi's sarcoma, lymphomas, and carcinomas arising from anal or cervical epithelium. They tend to metastasize, meaning jump to new areas, and be difficult to treat. Evidence indicates that viruses are a factor in their causation. A popular theory is that certain viral antigens, perhaps with the aid of cytokines, cause rapid multiplication of a host cell type. A mutation to malignancy then occurs among the rapidly dividing cells, the result of insertion of viral DNA into their genome, or other carcinogen. Finally, the malignant cell escapes detection and destruction by immune surveillance because of defective cellular immunity, and it is able to multiply without restraint.

Kaposi's Sarcoma

Kaposi's sarcoma **(figure 29.12)** is an unusual tumor arising from blood or lymphatic vessels in multiple locations. Formerly, it was seen rarely and mainly in older men of Mediterranean and Eastern European origin. It also occurs among all age groups in certain parts of tropical Africa. It is not associated with immunodeficiency in any of these varieties.

Coincident with the spread of HIV, the tumor began to appear in young men with HIV disease, and the incidence rose dramatically. Among a group of never-married San Francisco men, the incidence of Kaposi's sarcoma was 2,000 times as high in 1984 as it was in the period before HIV became widespread. The tumor was so common among AIDS patients that it became an AIDS-defining condition, even though it generally

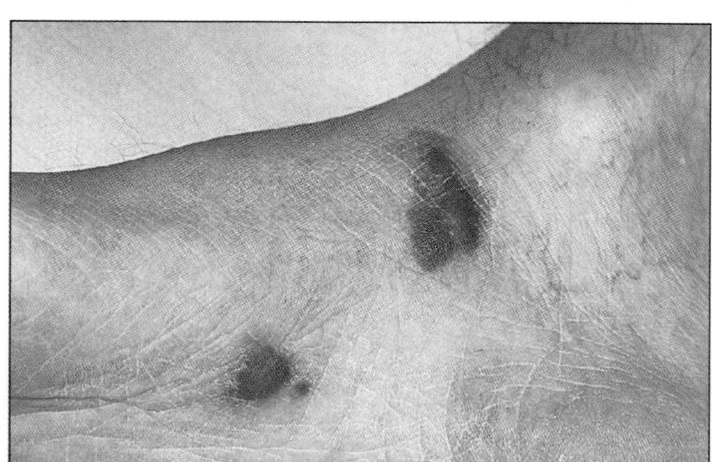

Figure 29.12 **Kaposi's Sarcoma** Two lesions are shown on a person's foot.

appeared before the development of severe immunodeficiency. One of the many unsolved mysteries about this tumor is that its incidence among AIDS patients has fallen dramatically since the adoption of condom use and safer sex practices.

Besides HIV disease, Kaposi's sarcoma is often associated with the use of medications to suppress the immune system in organ transplant patients. Indeed, the incidence of Kaposi's sarcoma is more than 400 times as high in transplant patients as in the general public. Another peculiarity of the tumor is that it can sometimes disappear if a patient's organ graft is rejected, suggesting a strong influence for cellular immunity.

In 1994, scientists at Columbia University detected a previously unknown herpesvirus in Kaposi's sarcomas, named Kaposi's sarcoma associated herpesvirus (KSHV), or human herpesvirus-8 (HHV-8). The virus can be detected in essentially all cases of Kaposi's sarcoma, whether associated with immunodeficiency or not. It is also strongly associated with certain malignant tumors, including multiple myeloma. Serological evidence of infection is present in about 25% of the healthy adult United States population. HHV-8 has been detected in the saliva of patients with HIV disease and Kaposi's sarcoma, and in semen of healthy men. Although there is still much to be learned about the role HHV-8 plays in Kaposi's sarcoma, the virus appears to be necessary for formation of the tumor, but additional factors must be present for the tumor to develop. ■ **multiple myeloma, p. 445**

In Kaposi's sarcoma, HHV-8 infects the endothelial cells that line blood and lymphatic vessels and persists there mostly in a latent form, only a small percentage of the infected cells evidencing a lytic infection at any one time. Presence of the virus is associated with two dramatic changes that result in tumor formation: (1) The cells assume a spindle shape and proliferate; (2) Extensive formation of new blood vessels occurs. These changes are probably due to release from the host cells of the cytokines normally responsible for stimulating cell growth and new vessel formation. This release of cytokines appears to be due largely to a gene product of latent HHV-8. The changes typically seen in cancers and other malignant tumors are usually not present—namely, proliferation of a single clone of microscopically abnormal cells with abnormal chromosomes, and origin in a single location with later distant metastases. When malignant changes occur in Kaposi's sarcoma, as they do occasionally, the abnormal clone of cells does not contain the HHV-8 genome, showing that some other factor than the virus is responsible for the transformation.

Only in a fraction of individuals infected with both HHV-8 and HIV-1 develop Kaposi's sarcoma, but that fraction is enormously greater than with those infected only with HHV-8. Why is HHV-8 infection so much more likely to result in Kaposi's sarcoma in patients with HIV-1 disease? Scientists around the world, hard at work on this question, have discovered some clues. Tat, the protein product of the HIV *tat* gene, is released from infected T lymphocytes and enhances proliferation of endothelial cells. The Tat of HIV-2 does not have the same effect, and there is no increase in Kaposi's sarcoma in patients with HIV-2 disease. Many of the molecular details of the interaction of the two viruses are now known and undoubtedly will lead to new treatment options and better understanding of how malignancy develops.

B-Lymphocytic Tumors of the Brain

Lymphomas are a group of malignant tumors that arise from lymphoid cells. Most of these tumors arise from B lymphocytes, although T-lymphocytic lymphomas also occur. B-cell lymphomas are 60 to 100 times as common in AIDS patients as in the general public, and 60 times as frequent in individuals with organ transplants. Intense, sustained replication of lymphoid cells is a constant feature of HIV. As stated at the beginning of this chapter, HIV was first isolated from a patient with the lymphadenopathy syndrome, a condition in which the lymph nodes are markedly enlarged for 3 months or more, now known to be a prelude to AIDS. The lymph node enlargement reflects proliferation of lymphoid cells in response to high-level unregulated cytokine release and to the HIV gp41 (TM) antigen. Also in HIV disease, sustained replication of T cells occurs as billions of new cells are produced each day to replace those destroyed by HIV. In contrast to Kaposi's sarcoma, with B-lymphocytic tumors there is no epidemiologic evidence of involvement of a sexually transmitted infection.

Epstein-Barr virus (EBV), however, plays a role in many of the B-cell lymphomas associated with AIDS, present in essentially all those that involve the brain. Lymphomas rarely arise in the brain except in AIDS patients. EBV is also present in all the B-cell lymphomas that occur in transplant patients. The EBV-related tumor, Burkitt's lymphoma, is at least 1,000 times as frequent among AIDS patients as in the general public, representing about one-fourth of all the lymphomas associated with AIDS. EBV probably plays an indirect role in B-cell lymphoma formation rather than being the direct cause. HIV infection causes activation of latent EBV infection, with release of the virus to infect new B lymphocytes. This, in turn, causes polyclonal B-cell proliferation and increased life span. Malignant B-cell clones are thought to arise from this population of rapidly dividing cells. ■ **Epstein-Barr virus, Burkitt's lymphoma, p. 727**

Cervical and Anal Carcinoma

Carcinoma (cancer) of the uterine cervix in women and carcinoma of the anus in women and gay men are strongly associated with human papillomaviruses (HPV) types 16 and 18. The cells involved in these cancers are epithelial cells and, therefore, differ from those in Kaposi's sarcomas and lymphomas. HPV is transmitted during sexual activity and infects the cervical and anal epithelium, appearing to cause increased replication of the cells by blocking expression of a cellular gene responsible for controlling cell growth. In HIV disease, organ transplantation, and other immunodeficient conditions, HPV replication increases with the decline of the host's cellular immunity. Precancerous changes can be demonstrated in the anal cells of HIV-positive gay men twice as frequently as in gay men who are HIV negative, and six times as frequently among HIV-positive gay men with low $CD4^+$ T-cell counts as in HIV-positive gay men with high $CD4^+$ T-cell counts. Interestingly, even before the arrival of HIV, the incidence of anal carcinoma in gay men exceeded the incidence of cervical carcinoma in women. One important implication of these findings is that women who engage in anal intercourse should be screened for precancerous lesions of the cervix and anus, and gay men for lesions of the anus, at least twice yearly if they are HIV positive.

MICROCHECK 29.2

Certain DNA viruses are strongly associated with development of malignant tumors in patients with HIV disease. These viruses are not sufficient by themselves to cause malignancy, but require the presence of other conditions. The tumors all arise in a setting of increased cell proliferation caused in part by the viruses. Many of these tumors are preventable by safer sex practices and screening for premalignant lesions.

- What HIV gene product appears to play a role in development of Kaposi's sarcoma?
- What member of the herpesvirus family is associated with almost all of the B-cell lymphomas of the brain in AIDS patients?
- What measure is important to prevent cervical and anal carcinomas in persons with HIV disease?
- Would rapidly dividing cells favor mutation?

Infectious Complications of Acquired Immunodeficiency

Immunodeficient individuals are susceptible to the same infectious diseases as other people. In addition, infections that pose no threat to immunologically normal people can cause severe, even fatal disease in those with impaired immune systems. Prevention of infectious diseases is key to the care of these patients. **Table 29.6** shows some immunizations that are advisable; they should be given as early as possible in HIV disease. Except for measles-mumps-rubella, live vaccines are generally unsafe to use in HIV disease.

Bacteria, viruses, fungi, protozoa, and even parasitic worms such as *Strongyloides stercoralis* can be life threatening. The geographical area may determine which infectious complications are most important. For example, the mold *Penicillium marneffei* is the third most common opportunistic agent among AIDS patients in Thailand but is almost unheard of elsewhere.
■ ***Strongyloides stercoralis*, p. 317, 622**

This section will present some examples of infectious diseases common in patients with immunodeficiency.

TABLE 29.6 Immunizations in Individuals with HIV Disease

Advised	Not Advised
Measles-mumps-rubella	Oral poliomyelitis
Pneumococcal polysaccharide	Oral typhoid
Influenza	BCG (tuberculosis vaccine)
Hepatitis A	Varicella-zoster
Hepatitis B	Yellow fever

Pneumocystosis

Pneumocystosis, a severe infectious lung disease, was recognized just after World War II in Europe when it caused deaths among hospitalized, malnourished, premature infants. Subsequently, scattered cases were recognized among immunodeficient patients until the onset of the AIDS epidemic, when the incidence soared. By 1995, almost 128,000 cases had been reported to the Centers for Disease Control and Prevention.

Symptoms

The symptoms of pneumocystosis typically begin slowly, with gradually increasing shortness of breath and rapid breathing. Fever is usually slight or absent, and only about half of the patients have a cough, which is non-productive. As the disease progresses a dusky coloration of the skin and mucous membranes appears and gradually worsens, a reflection of poor oxygenation of the blood, which can become fatal.

Causative Agent

Pneumocystosis is caused by *Pneumocystis carinii* (**figure 29.13**), a tiny fungus belonging to the phylum Ascomycota, class Archiascomycetes. Formerly considered a protozoan, *P. carinii* has a number of characteristics that support its inclusion among the fungi. It differs from many fungi, however, in the chemical make-up of its cell wall; consequently, it is resistant to the medications often used against fungal pathogens.
■ **ascomycetes, p. 308**

Pathogenesis

The small spores of *P. carinii*, measuring only 1 to 3 μm, are easily inhaled into lung tissue. In experimental infections, the spores attach to the alveolar walls, and the alveoli fill with fluid, mononuclear cells, and *P. carinii* cells in various stages of development.

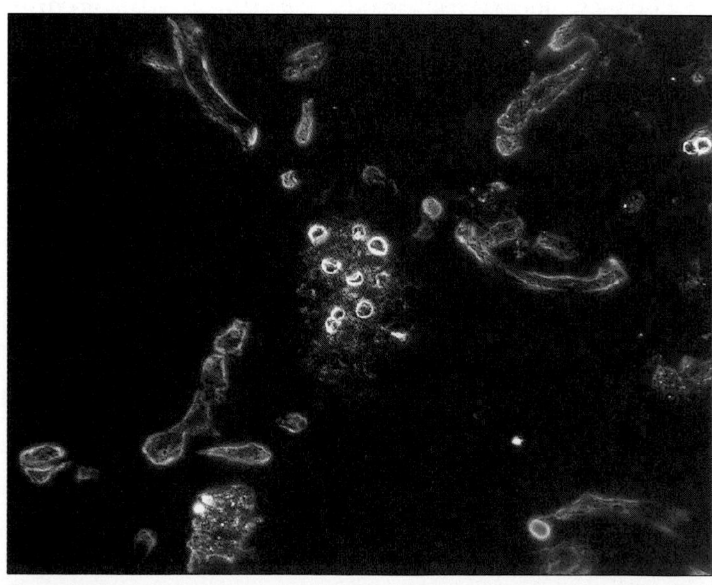

Figure 29.13 Fluorescent Antibody Stain of *Pneumocystis carinii* The yellow circles are *P. carinii* cysts.

Later, the alveolar walls become thickened and scarred, preventing the free passage of oxygen.

Epidemiology

Various strains of *P. carinii* are widespread among animals, including dogs, cats, horses, and rodents, persisting in their lungs as a latent infection. Although identical in appearance to strains infecting humans, those examined have been genetically distinct. Serological tests indicate that most children are infected by age two and a half. The infection is asymptomatic and is generally eliminated within a year. The source and transmission of human infections are unknown. Most cases of pneumocystosis occur in persons with immunodeficiency, but it is uncertain whether their disease is caused by activation of latent infection, or infection newly acquired from inhalation of airborne spores. Epidemics among hospitalized malnourished infants and elderly nursing home residents suggest airborne spread, and *P. carinii* has been detected in indoor and outdoor air by using the polymerase chain reaction.

Prevention and Treatment

Pneumocystosis used to occur in about four-fifths of AIDS patients and was the leading cause of death. The disease is now largely prevented by starting regular doses of a medication such as trimethoprim-sulfamethoxazole for HIV disease as soon as the $CD4^+$ T-cell count falls below 200 per μl, or if other hallmarks of immunodeficiency appear, such as **thrush**, a *Candida* infection of the mouth and throat.

Trimethoprim-sulfamethoxazole is also among the best medications for treating pneumocystosis and, along with oxygen and other measures, can reduce mortality from nearly 100% to about 30%. Alternative medications are available for treating those who cannot tolerate trimethoprim-sulfamethoxazole because of its side effects—mainly rash, nausea, and fever. For unknown reasons, people with HIV disease are more likely to develop these side effects than others. After treatment for pneumocystosis, individuals with HIV disease must receive preventive medication indefinitely, or until they have a sustained rise in $CD4^+$ T-cell count to above 200 cells/μl.

The main features of pneumocystosis are presented in **table 29.7**.

TABLE 29.7 Pneumocystosis

Symptoms	Gradual onset, shortness of breath, rapid breathing, non-productive cough, slight or absent fever, dusky color of skin and mucous membranes
Incubation period	4 to 8 weeks
Causative agent	*Pneumocystis carinii,* a tiny fungus related to the ascomycetes
Pathogenesis	Pneumocystosis can result from reactivation of latent infection or be newly acquired. Spores of *P. carinii* escape body defenses, enter the lungs with inspired air, attach to alveolar walls, multiply. Alveoli fill with fluid, macrophages, and *P. carinii*. The walls thicken, impairing oxygen exchange
Epidemiology	*Pneumocystis carinii* widespread in domestic and wild animals as a latent lung infection, but the source of animal and human infections is unknown. Most humans become infected in early childhood. Disease arises in individuals with immunodeficiency; epidemics can occur in hospitalized premature infants and elderly nursing home residents
Prevention and treatment	Formerly leading cause of death in those with AIDS, now usually prevented by medication, e.g., trimethoprim-sulfamethoxazole, as soon as the $CD4^+$ lymphocyte count drops to 200 cells per μl. Same medication is used for treatment; alternatives available. Medication is continued for life, or until the $CD4^+$ cell count rises and remains above 200 as a result of HAART or other treatment of the underlying immunodeficiency

Toxoplasmosis

Toxoplasmosis is a protozoan disease that rarely develops among healthy people but can be a serious problem for those with malignant tumors, recipients of organ transplantation, people with HIV disease, and the unborn child.

Symptoms

The disease presents itself differently in the three main categories of patients: (1) the immunologically normal, (2) unborn children, and (3) those with immunodeficiency.

Toxoplasmosis can be acquired by immunocompetent individuals who eat raw or undercooked meat or who are exposed to cat feces. Most infections are asymptomatic, but 10% to 20% develop symptoms similar to those of infectious mononucleosis. They usually consist of sore throat, fever, enlarged lymph nodes and spleen, and sometimes a rash. These symptoms subside over weeks or months and do not require treatment. Rarely, a severe, even life-threatening illness develops, due to involvement of the heart or central nervous system. ■ infectious mononucleosis, p. 726

Toxoplasmosis of the unborn results from almost half of maternal infections that occur during pregnancy. Fetal toxoplasmosis during the first trimester of pregnancy is the least common but most severe, often resulting in miscarriage or stillbirth. Babies born live may have severe birth defects including small or enlarged heads, and their lungs and liver may also be damaged by the disease. Later, these babies may develop seizures or manifest mental retardation. Almost two-thirds of fetal toxoplasmosis cases occur during the last trimester of pregnancy, however, and the effects on the fetus are usually less severe. Most of these infants appear normal at birth, although later in life, **retinitis**, infection of the retina, the light-sensitive part of the eye, can be a big problem for them. This manifests itself as recurrent episodes of pain, sensitivity to light, and blurred vision, usually involving only one eye. Less common late consequences of congenital toxoplasmosis include mental retardation and epilepsy.

Toxoplasmosis that complicates immunodeficiency is commonly life threatening. Brain involvement in the form of encephalitis occurs in more than half the cases, manifest by confusion, weakness, impaired coordination, seizures, stiff neck, paralysis, and coma. Involvement of the brain, heart, and other organs often results in death.

Causative Agent

Toxoplasmosis is caused by *Toxoplasma gondii*, a tiny (3 by 7 μm) banana-shaped protozoan **(figure 29.14a)** that has a worldwide distribution and infects many vertebrate hosts, including household pets, pigs, sheep, cows, rodents, and birds. Its species name derives from the fact that the protozoan was first discovered in the gondi, an African desert rodent. The life cycle of *T. gondii* is shown in figure 29.14b. The definitive host, the one in which sexual reproduction occurs, is the cat or other feline (ocelot, puma, bobcat, Bengal tigers, and so on). The organism reproduces in the cat's intestinal lining, resulting in numerous offspring, some of which spread throughout the body and others that differentiate into male and female gametes (sexual forms). Union of gametes results in the formation of a thick-walled oval structure (oocyst) about 12 μm in diameter that is shed in the cat's feces. Millions of the oocysts are shed each day generally over a period of 1 to 3 weeks. In the soil, the oocysts undergo further development over 1 to 5 days into an infectious form containing two sporocysts, each with four sporozoites. These may remain viable for up to a year, contaminating soil and water and, secondarily, hands and food. In general, the cats recover from the acute infection and do not shed oocysts again. The oocysts are infectious for cats and other animals, including humans.

When nonfeline animals ingest the oocysts, usually from contaminated food or water, the sporozoites emerge from the oocysts and invade the cells of the small intestine, especially its lower part, the ileum, but there is no sexual cycle. The intestinal infection spreads by the lymphatics and blood vessels throughout the tissues of the host, infecting cells of the heart, brain, and muscles. As host immunity develops, multiplication slows, and a tough fibrous capsule forms, surrounding large numbers of a smaller form of *T. gondii*. These capsules packed with organisms are called cysts (figure 29.14c), and they remain viable for months or years. The life cycle is completed when a cat becomes infected by eating an animal with cysts in its tissues.

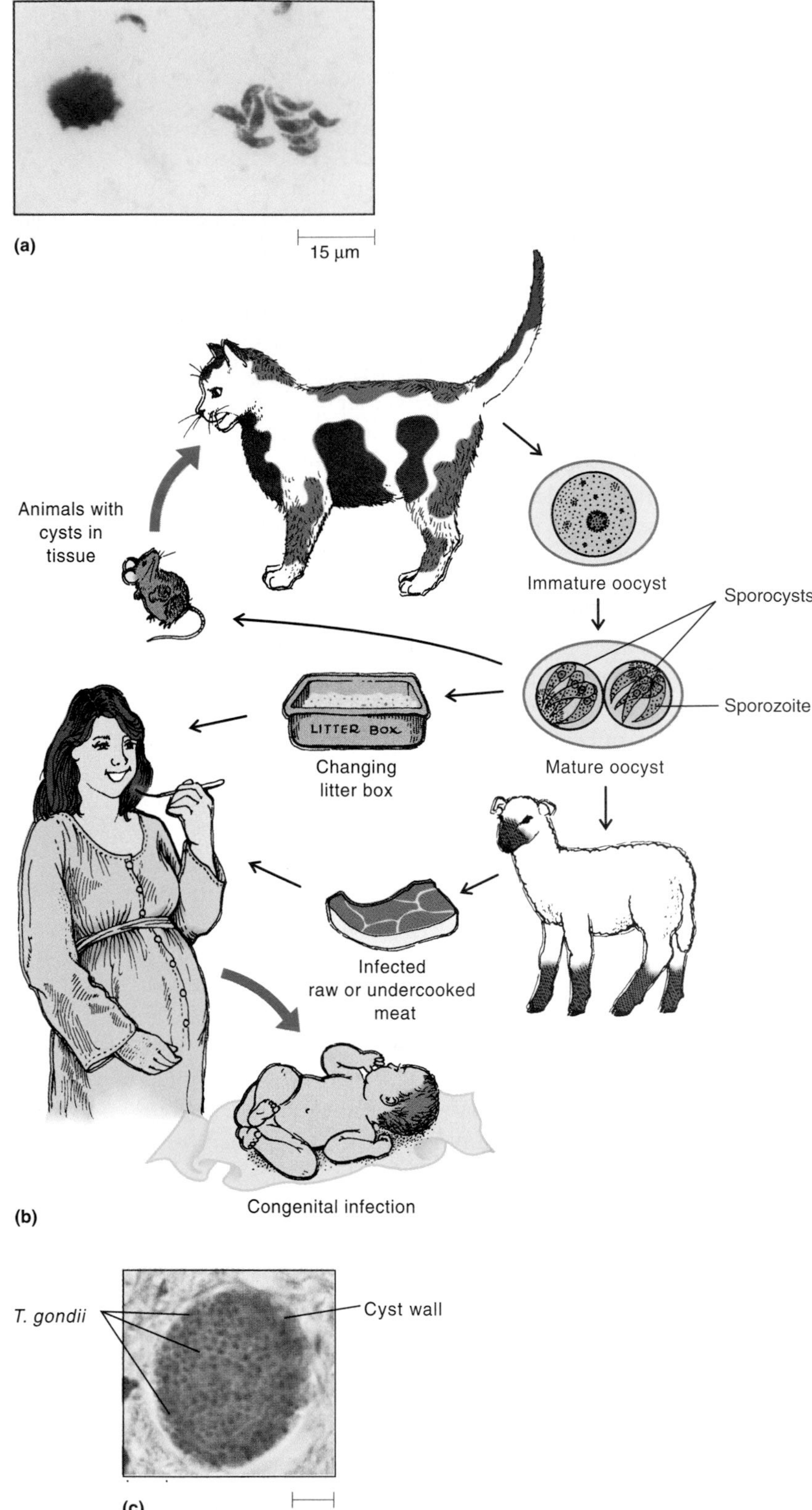

Figure 29.14 ***Toxoplasma gondii*** **(a)** Invasive forms. **(b)** Life cycle. Oocysts from cat feces and cysts from raw or inadequately cooked meat can infect humans and many other animals. **(c)** Cyst in tissue.

Pathogenesis

The organisms enter the body by ingestion of oocysts or inadequately cooked meat containing tissue cysts. *Toxoplasma gondii* is infectious for any kind of animal cell except non-nucleated red blood cells. Entry into the host cell is aided by an enzyme produced by the organism, which alters the host cell membrane. Proliferation of *T. gondii* in cells of the host causes destruction of the cells. Unless the infecting dose of the organism is very high, this process is normally brought under control by the immune response of the host, and infected humans usually show few if any symptoms. In patients with immunodeficiency, however, infection can be widespread and uncontrolled, producing many areas of tissue necrosis. The disease process can result from a newly acquired infection, or declining immunity can allow reactivation of latent infection with escape of *T. gondii* from a person's tissue cysts.

Epidemiology

Toxoplasma gondii is distributed worldwide, but it is less common in cold and in hot, dry climates. Human infection is widespread. Serological surveys show the infection rate increases with age, ranging from 10% to 67% among those over 50. Most infections are acquired from ingesting oocysts that contaminate fingers, food, or drink, and probably from inhaling contaminated dust in an enclosed space such as a barn. Young, homeless cats that commonly eat rodents and birds are likely sources of the organism. Small epidemics have occurred from drinking water contaminated with oocysts. Gardening in areas frequented by cats can result in *T. gondii* contamination of hands or vegetables. Eating rare meat poses a definite risk. Tissue cysts are present in about 25% of pork, in 10% of lamb, and less commonly in beef and chicken.

Prevention and Treatment

General measures for preventing *T. gondii* infection include washing hands after handling raw meat, coming in contact with soil, or changing cat litter. Meat, especially lamb, pork, and venison, should be cooked until the pink color is lost from its interior. Fruits and vegetables should be washed before eating. Cats should not be allowed to hunt birds and rodents, nor should they be fed undercooked or raw meat. Children's sandboxes should be kept tightly covered when not in use. These measures are especially important for pregnant women and persons with immunodeficiency.

HIV-infected patients and those about to receive immunosuppressant medications are tested for antibody to *T. gondii*. A positive test indicates they have latent infection. Therefore, those with positive tests are given prophylactic trimethoprim-sulfamethoxazole if they have fewer than 100 $CD4^+$ T cells per microliter. Treatment of toxoplasmosis employs related medications, pyrimethamine with sulfadiazine. Alternatives are available.

The main features of toxoplasmosis are presented in **table 29.8**.

TABLE 29.8 Toxoplasmosis

Symptoms	In healthy individuals: sore throat, fever, enlarged lymph nodes, rash; with fetal infections: miscarriage, stillbirth, birth defects, epilepsy, mental retardation, retinitis; in immunodeficient individuals: confusion, poor coordination, weakness, paralysis, seizures, coma
Incubation period	Usually indeterminate
Causative agent	*Toxoplasma gondii,* a protozoan infectious for most warm-blooded animals. Sexual reproduction occurs in the intestinal epithelium of cats, the definitive hosts. Infected cats discharge oocysts with their feces. Ingested organisms are released from the oocysts, multiply rapidly, spread throughout the body. As immunity develops, infected cells become filled with the organisms, resulting in tissue cysts, which remain viable and infectious for the lifetime of the animal
Pathogenesis	Organisms penetrate host cells causing necrosis. With development of immunity cell destruction stops, tissue cysts develop. Most healthy individuals have few or no symptoms unless the numbers of ingested organisms is very large. Organisms released from tissue cysts if immunity becomes impaired
Epidemiology	Occurs worldwide, less common in cold or dry locations. Infection acquired by ingesting oocysts from cat feces, or eating inadequately cooked meat
Prevention and treatment	Prevention: avoiding foods potentially contaminated with oocysts from cat feces, and not consuming inadequately cooked meat. Trimethoprim-sulfamethoxazole is given to immunodeficient persons with $CD4^+$ T lymphocyte counts below 100 cells per µl if they have antibodies to *T. gondii* indicating latent infection. Treatment: pyrimethamine with sulfadiazine, or alternative medication

Cytomegalovirus Disease

Cytomegalovirus (CMV) is a member of the herpesvirus family, which includes herpes simplex virus, Epstein-Barr virus, and varicella-zoster virus, any of which can cause troublesome symptoms in patients with immunodeficiency. CMV, like other herpesviruses, is commonly acquired early in life and then remains latent. With impairment of the immune system, the infection activates and can cause severe symptoms.

Symptoms

Symptoms of cytomegalovirus disease follow a pattern similar to that of toxoplasmosis. Acute infections in immunocompetent individuals are usually without symptoms, but adolescents and young adults sometimes develop illness resembling infectious mononucleosis, with fever, fatigue, and enlarged lymph nodes and spleen for weeks or months.

Severe damage can occur to the fetus if the mother develops an acute infection during pregnancy. This condition is known as **congenital cytomegalic inclusion disease** and is characterized by jaundice, large liver, anemia, eye inflammation, and birth defects. The vast majority of infected infants appear normal at birth, but 5% to 25% manifest hearing loss, mental retardation, or other abnormalities later in life.

Blindness is one of the most feared complications of cytomegalovirus disease in immunodeficient individuals. Other symptoms include fever, loss of appetite, painful joints and muscles, rapid, difficult breathing, ulcerations of the gastrointestinal tract with bleeding, lethargy, paralysis, dementia, and coma.

Causative Agent

Human cytomegalovirus is an enveloped, double-stranded DNA virus that looks like other herpesviruses on electron micrographs but has a larger genome. Its name (*cyto* for "cell" and *megalo* for "large") derives from the fact that cells infected by the virus are two or more times the size of uninfected cells. Infected cells show a large intranuclear inclusion body surrounded by a clear halo, inspiring its description as an "owl's eye" **(figure 29.15)**. The envelope is acquired as the virion buds from the nuclear membrane.

There are many different strains of the virus detected by the patterns of endonuclease digests, and antigenic differences

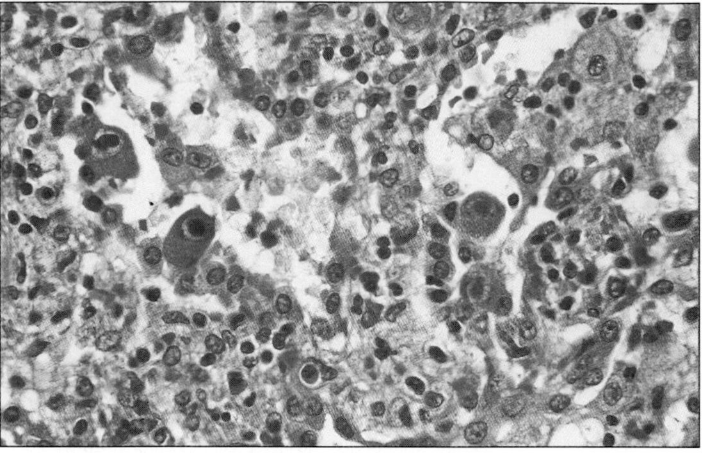

Figure 29.15 Cytomegalovirus Infected cells showing "owl's eye" appearance of intranuclear inclusions.

also occur. Like other herpesviruses, CMV can cause lysis of the infected cell or become latent and subject to later reactivation.

■ **genomic typing, p. *256**

Pathogenesis

In cytomegalovirus disease, a wide variety of tissues, including eye **(figure 29.16)**, central nervous system, and liver, are susceptible to infection. Once cell entry has occurred, the viral genome can exist in a latent non-infectious form, in a slowly replicating form, or as a fully productive infection. Control of viral gene expression depends partly on the type of cell infected. Monocytes allow low levels of infectious virus production. The CMV genome is ordinarily quiescent in T and B lymphocytes, expressing some viral genes but not producing viral DNA. Integration of viral DNA into the host cell genome probably does not occur. If CMV-infected T cells are also infected with HIV, productive CMV infection occurs. Fully productive infection of a wide variety of tissues occurs during acute infections, causing tissue necrosis. Transplanted organs and blood transfusions can transmit the disease, indicating that virus-producing cells are present or that non-producing cells in the blood or organ start producing infectious CMV under conditions of immune suppression. Cell-mediated immunity probably plays a role in suppressing production of infectious virus as well as in lysis of infected cells. In cells infected with CMV, however, transfer of MH antigen to the cell surface is impaired, and therefore, CMV antigens are not recognized as being "foreign." CMV infection is associated with an increase in $CD8^+$ cells and a decrease in $CD4^+$ cells, thus enhancing the effect of HIV infection. Latent infections activate with AIDS, organ transplants, and other immunodeficient states. Also, immunodeficient subjects are highly susceptible to newly acquired infection.

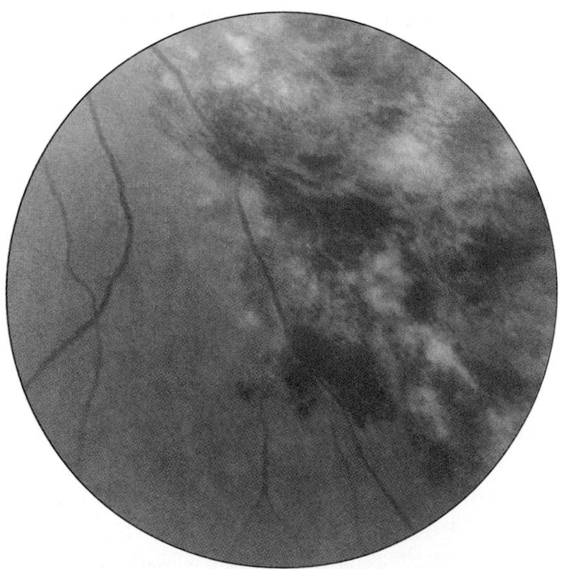

Figure 29.16 Cytomegalovirus (CMV) Retinitis, a Common Cause of Blindness in Persons with AIDS Photograph of a CMV-infected retina as it appears when viewed through the eye's pupil using an ophthalmoscope.

Epidemiology

CMV is found worldwide. One U.S. study found that more than 50% of adults 18 to 25 years old and more than 80% of people over 35 years had been infected. Infection is lifelong. Infants born with CMV infection and those who acquire it shortly after birth excrete the virus in their saliva and urine for months or years. Virus is found in semen and cervical secretions in the absence of symptoms, and sexual intercourse is a common mode of transmission in young adults. Up to 15% of pregnant women secrete the virus, and 1% of newborn infants have CMV in their urine. Breast milk, blood, and tissue transplants may contain CMV and be responsible for transmission. Almost all prostitutes and promiscuous gay men are infected with CMV. CMV spreads readily in day care centers.

Prevention and Treatment

There is no approved vaccine for preventing cytomegalovirus disease. The use of condoms is effective in decreasing the risk of sexual transmission. People with HIV disease or other immunodeficiency syndromes or who lack CMV antibody should avoid contact with day care centers if possible and, in any case, should wash their hands if exposed to saliva, urine, or feces. Tissue and blood donors can be screened for antibody to CMV. Those who have anti-CMV antibody are assumed to be infected and should not donate to those lacking antibody to CMV. The antiherpesviral medication ganciclovir, given orally, halves the incidence of CMV retinitis in HIV disease patients with low $CD4^+$ cell counts and positive tests for CMV antibody. A ganciclovir implant designed to release the medication into the eye over a long period also delays the progression to blindness.

Combination drug therapy with the antiviral drugs ganciclovir and foscarnet can reduce the severity of CMV disease. They inhibit CMV DNA polymerase, the enzyme responsible for assembling viral DNA, at different sites. Both medications have serious side effects, mainly bone marrow suppression with ganciclovir and kidney impairment with foscarnet. ■ **DNA polymerase, p. 171**

The main features of cytomegalovirus disease are presented in **table 29.9**.

Mycobacterial Diseases

Initial exposure to *Mycobacterium tuberculosis*, the cause of tuberculosis, usually causes an asymptomatic infection that is controlled by the immune system and becomes latent. As might be expected, defects in cellular immunity are associated with reactivation of latent tuberculosis, which commonly results in unrestrained disease in AIDS and other immunodeficiencies. *Mycobacterium tuberculosis* is not the only mycobacterium that causes disease in immunodeficient people. Disease has been reported due to *M. kansasii*, *M. scrofulaceum*, *M. xenopi*, *M. szulgai*, *M. ganavense*, *M. haemophilum*, *M. celatum*, and others. This section discusses disease caused by organisms of *Mycobacterium avium* complex, next to *M. tuberculosis* the most common mycobacterial opportunists complicating immunodeficiency diseases.

Symptoms

The vast majority of *Mycobacterium avium* complex (MAC) infections in immunologically normal people are asymptomatic. Elderly people, especially those with underlying lung disease from smoking and those with alcoholism, develop a chronic cough productive of sputum and sometimes lesions in the lungs resembling tuberculosis. Children sometimes develop chronic enlargement of lymph nodes on one side of their neck, easily treated by surgical removal of the affected nodes. Patients with AIDS and other severe immunodeficiencies have slowly progressing symptoms ranging from chronic cough productive of sputum to fever, drenching sweats, marked weight loss, abdominal pain, and diarrhea.

Causative Agent

Mycobacterium avium complex is a group of mycobacteria consisting of two closely related species, *M. avium* and *M. intracellulare*. More than two dozen strains of these acid-fast rods fall into the MAC, distinguishable by serological tests, optimum growth temperature, and host range. Their growth rate is almost as slow as that of *M. tuberculosis*, but they are easily distinguished by using biochemical tests and nucleic acid probes. ■ tuberculosis, p. 569

Pathogenesis

MAC organisms enter the body via the lungs and the gastrointestinal tract. They are taken up by macrophages via phagocytosis, but they resist destruction because they inhibit acid production in the phagosome. Surviving organisms multiply within phagocytes and are carried by the bloodstream to all parts of the body. When cellular immunity is intact, most organisms are destroyed, and disease is localized. With profound immunodeficiency, the disease spreads throughout the body. In these patients, there is persistent bacteremia, occasionally with counts as high as 1 million per milliliter of blood. The small intestine contains macrophages packed with the bacteria, and infected tissues **(figure 29.17)** may resemble leprosy, sometimes containing more than 100 billion of the acid-fast bacteria per gram of tissue. Despite the presence of enormous numbers of the bacteria, there is little or no inflammatory reaction, and the clinical effect is a slow decline in the patient's well-being

TABLE 29.9 Cytomegalovirus Disease

Symptoms	Symptoms rare in immunocompetent individuals, but sometimes an infectious mononucleosis–like illness develops. Infection of the mother during pregnancy can result in disease of the newborn. Immunocompromised individuals may experience blindness, lethargy, dementia, coma, reflecting liver and brain damage
Incubation period	In immunocompetent adults, 20 to 60 days
Causative agent	Cytomegalovirus, a member of the herpesvirus family
Pathogenesis	Many tissues susceptible to infection, damage, especially eyes, brain, and liver. CMV latent infection can reactivate, produce infectious virions, tissue necrosis. $CD4^+$ T-lymphocyte count is depressed and thus can enhance HIV disease
Epidemiology	Common worldwide; lifelong infection. Up to 50% of 18- to 25-year-olds are infected. Infants with congenital infections and those infected shortly after birth shed the virus for months or years. Body fluids, including breast milk, blood, urine, semen, and vaginal secretions, can transmit the disease. Almost all prostitutes and promiscuous homosexual men are infected with CMV
Prevention and treatment	No vaccine available. Condoms decrease transmission. CMV-negative immunodeficient persons advised to avoid day care centers, wash their hands following contact with bodily fluids of infants. Blood and tissue transplants are tested for CMV before being given to CMV-negative individuals. The antiviral medication ganciclovir is considered for immunodeficient individuals who have antibody to CMV and whose $CD4^+$ T-lymphocyte count falls below 50 cells per μl. Treatment: same medication

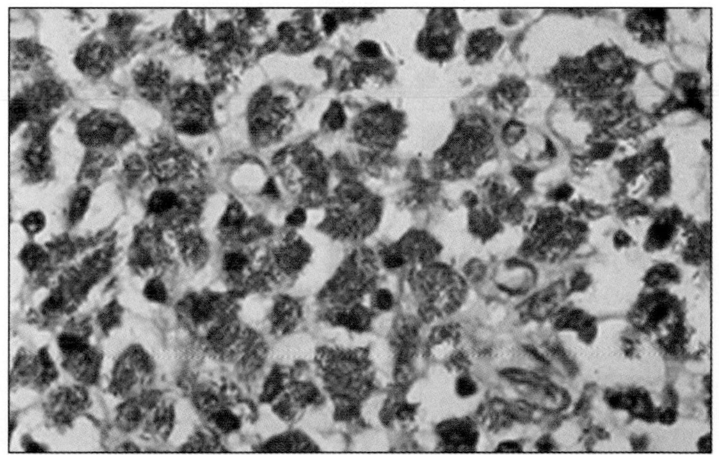

Figure 29.17 **Acid-Fast Stain Showing MAC Bacteria Infecting Cells**

rather than a quickly lethal effect. One cause of deterioration may be activation of HIV replication when HIV-infected macrophages are infected with MAC. ■ leprosy, p. 672

Epidemiology

MAC organisms are widespread in natural surroundings and have been found in food, water, soil, and dust. In the United States, they are most prevalent in the Southeast, parts of the Pacific Coast, and the North Central region. Some strains are important pathogens of chickens and pigs. In AIDS patients, they are the most common bacterial cause of generalized infection. It is not known whether infection in AIDS patients is mostly newly acquired or a reactivation of latent disease. Most infections are from environmental sources rather than from person-to-person spread.

Prevention and Treatment

No effective measures are available to prevent exposure to MAC organisms. For HIV patients whose $CD4^+$ count is below 50 cells/μl, the antibacterial medication clarithromycin is recommended to help prevent MAC disease. If MAC bacteremia develops, patients must be given two or more medications together, such as clarithromycin plus ethambutal.

The main features of MAC disease are presented in **table 29.10**.

TABLE 29.10 MAC Disease

Symptoms	Usually asymptomatic in healthy people. Children can develop chronic localized enlargement of lymph nodes. Can cause chronic cough similar to tuberculosis in the elderly. Cough, fever, sweating, marked weight loss, abdominal pain, and diarrhea in those with severe immunodeficiencies
Incubation period	Usually indeterminate
Causative agents	A group of mycobacterial strains in the species *M. avium* and *M. intracellulare*
Pathogenesis	MAC bacteria enter the body via lungs or gastrointestinal tract, are taken up by macrophages. In immunocompetent individuals, most are destroyed and infection controlled by cellular immunity. In severe immunodeficiency the bacteria multiply without restraint, are carried by macrophages throughout the body, and grow to enormous numbers in tissues but cause little or no inflammatory reaction
Epidemiology	MAC organisms are widespread in food, water, soil, and dust. MAC disease of immunodeficient persons could result from environmental sources, or activation of latent infection
Prevention and treatment	No proven measures to prevent exposure to MAC bacteria. The antibiotic clarithromycin is used to prevent MAC disease in severe immunodeficiency. If MAC bacteremia develops, two or more medications, such as clarithromycin plus ethambutal, are effective

MICROCHECK 29.3

Infectious diseases are a major threat to individuals with immunodeficiency. They get the same infectious diseases as healthy people, and in addition, they are much more susceptible to reactivation of latent infections, and organisms of low virulence generally harmless to others. Prevention of infectious diseases in AIDS patients is key to improving the quality and duration of their lives.

- Bacteria, fungi, protozoa, and viruses can all cause life-threatening diseases in AIDS patients. Name a fifth kind of infectious agent that can do the same.
- The disease pneumocystosis mainly involves which part of the body?
- Why should raw or uncooked meat not be eaten?

FUTURE CHALLENGES

AIDS and Poverty

In the United States and other economically advanced countries, antiviral medications have contributed significantly to the lessening of mother-child transmission of HIV, to delaying progression of HIV disease to AIDS, and to improving the quality and duration of life of AIDS sufferers. Newer drugs have also decreased transmission of HIV somewhat, thereby slowing the progress of the epidemic. The cost in the United States, while high, amounts to less than 1% of the total health care expenditure.

Antiviral medications have provided little benefit to most of the world's AIDS sufferers because they cannot afford them. Although HIV disease is spreading rapidly in India and China, two countries that together contain more than one-third of the world's population, and in the Caribbean, Southeast Asia, and Eastern Europe, the situation is the most desperate in African countries south of the Sahara desert. In this area, 11.5 million people have already died of the disease, and another 22.5 million currently infected with HIV are expected to die over the next decade. Few of them know that they have the disease, because funds for testing are limited. In one city in Zambia, 46% of young pregnant women attending a prenatal clinic were found to be HIV positive. At a hospital in a city of South Africa, 20 children per day are admitted, most with AIDS, many brought in by relatives because their parents have died of the disease. In Uganda, an estimated 250,000 children have lost both parents to AIDS; by the year 2010 an estimated 41 million African children will have lost one or both parents to the disease. Average life expectancy is projected to drop from almost 60 years, early in the last decade, to 45.

The humanitarian, social, and economic implications of these statistics are enormous. Unimaginable suffering, large numbers of starving, uneducated street children, loss of productive work years, spread of other infectious diseases, breakdown of public health, and setbacks for economic development are possible consequences that can affect neighboring countries and the rest of the world.

The challenge is to find effective control measures for the world's poor.

SUMMARY

AIDS

1. Since **AIDS** was first recognized in 1981 in sexually promiscuous homosexual men, it has claimed more than 420,000 lives in the United States, and become the leading cause of death in those 25 to 44 years of age.
2. Worldwide, 33 million people are infected with an AIDS-causing virus; it is estimated that a new infection occurs about every 6 seconds, and a person dies of AIDS every 5 minutes. (Figure 29.1)
3. In Africa, AIDS is now the number one cause of death, surpassing malaria; it is devastating Southeast Asia and spreading rapidly in India and into China.

Human Immunodeficiency Virus (HIV) Infection and AIDS

1. "AIDS-defining conditions" such as unusual tumors or serious infections by agents that normally have little virulence usually reflect immunodeficiency and were especially useful "markers" in early studies of the AIDS epidemic. (Table 29.1)

HIV Disease (Table 29.5, Figure 29.8)

1. Acquired immunodeficiency syndrome (AIDS) is a late manifestation of **human immunodeficiency virus (HIV) disease**.
2. "Flu"-like or infectious mononucleosis–like symptoms representing the **acute retrovirus syndrome** often occur 6 days to 6 weeks after infection by HIV; these symptoms subside without treatment.
3. Despite the absence of symptoms, HIV disease progresses and can be transmitted to others for years.
4. The asymptomatic period ends with the appearance of certain tumors, or the onset of immunodeficiency (AIDS) as marked by **LAV**, **ARC**, and opportunistic or reactivated infections.
5. Human immunodeficiency virus, type 1 (HIV-1), possesses single-stranded RNA. It belongs to the lentivirus genus of the retrovirus family, and is the main cause of AIDS. (Figure 29.3)
6. The disease is spread mainly by sexual intercourse, by blood or contaminated hypodermic needles, and from mother to fetus or newborn. (Tables 29.3, 29.4)
7. Different kinds of cells can be infected by HIV, notably two kinds vital to the immune system, T-helper (Th) lymphocytes and macrophages. (Figures 29.4, 29.5)
8. Inside the host cell, a DNA copy of the viral genome is made through the action of reverse transcriptase, a complementary DNA strand is made, and the resultant double-stranded DNA is inserted into the host genome as a provirus by viral integrase (IN). (Figure 29.6)
9. Antimicrobial medications are used to prevent and treat opportunistic infections, and combinations of antiretroviral medications (**HAART**) are used to treat HIV disease. (Figure 29.11)

HIV Vaccine Prospects

1. No approved vaccine is currently available, but a number of prospects have moved from animal to human trials.

Malignant Tumors that Complicate Acquired Immunodeficiencies

Kaposi's Sarcoma (Figure 29.12)

1. Kaposi's sarcoma is a tumor arising from blood or lymphatic vessels.
2. The incidence of the tumor is also markedly increased among other immunodeficient individuals.
3. Infection by human herpesvirus 8 appears to be required for the tumor to develop.

B-Lymphocytic Tumors of the Brain

1. Lymphomas are malignant tumors that arise from lymphoid cells.
2. Both B and T lymphocytes can give rise to lymphomas in individuals with immunodeficiencies, but B-cell lymphomas are more common. In contrast to nonimmunodeficient cases, B-cell tumors often arise in the brain.
3. Strong evidence exists that Epstein-Barr virus plays a causative role in these tumors.

Cervical and Anal Carcinoma

1. There is an increased rate of anal, genital, and cervical carcinoma in people with HIV disease. These tumors arise from squamous epithelial cells, thus differing from Kaposi's sarcoma and lymphomas.
2. These cancers are strongly associated with human papillomaviruses (HPV), transmitted by sexual intercourse.

Infectious Complications of Acquired Immunodeficiency

1. Infections that occur in healthy individuals also occur and produce more severe disease in those with immunodeficiency.
2. Latent infections such as those by *Mycobacterium tuberculosis* and herpesviruses commonly activate, and organisms rarely capable of causing disease in healthy individuals can be life threatening.

Pneumocystosis (Table 29.7)

1. Before effective preventive regimens were developed, the disease was the most common cause of death among AIDS sufferers.
2. Symptoms develop slowly, with gradually increasing shortness of breath and rapid breathing; patients can die from lack of oxygen.
3. The causative agent, *Pneumocystis carinii*, a tiny fungus, is widespread in humans and other animals, generally causing asymptomatic infections that become latent.

Toxoplasmosis (Table 29.8)

1. Toxoplasmosis is rare among healthy people but can be a serious problem for those with cancer, organ transplants, and HIV disease. The disease can also be congenital.
2. Although infection is common in healthy people, only about 10% develop symptoms of sore throat, fever, and enlarged lymph nodes and spleen, sometimes with a rash. Symptoms usually disappear without treatment.

3. With infections early in pregnancy, miscarriage can occur due to toxoplasmosis or babies can be born with birth defects or damage. Infections later in pregnancy are usually milder but can result in epilepsy, mental retardation, or recurrent **retinitis** in the child.
4. More than half of the cases with toxoplasmosis and AIDS develop encephalitis, and death often occurs because of involvement of the brain, heart, and other organs.
5. *Toxoplasma gondii* is a tiny banana-shaped protozoan that undergoes sexual reproduction in the intestinal epithelium of cats but can infect humans and many other vertebrates. Oocysts discharged in the feces of acutely infected cats become infectious in soil and contaminate food, water, and fingers. (Figure 29.14)
6. *T. gondii* is present worldwide, and most people become infected from oocysts. Epidemics have resulted from contaminated drinking water. Eating rare meat can also cause infection.

Cytomegalovirus Disease (Table 29.9)

1. Cytomegalovirus is a common cause of impaired vision in people with AIDS. (Figure 29.16)
2. Normal individuals generally have asymptomatic infections; fetuses may develop **cytomegalic inclusion disease** or appear normal at birth but show mental retardation or hearing loss later; immunodeficient individuals develop fever, gastrointestinal bleeding, mental dullness, and blindness.
3. Cytomegalovirus (CMV) is an enveloped, double-stranded DNA virus; infected cells are enlarged and have an "owl's eye" appearance. (Figure 29.15)
4. The virus can exist in a latent form, a slowly replicating form, or a fully replicating form. Co-infection with HIV results in fully productive infection and tissue death.
5. The virus occurs worldwide in breast milk, semen, and cervical secretions, and in saliva and urine of infected infants.
6. No vaccine is available. Condoms decrease the risk of sexual transmission. The antiviral medication ganciclovir can be given to prevent CMV retinitis.

Mycobacterial Diseases (Table 29.10)

1. *Mycobacterium tuberculosis* and *Mycobacterium avium* complex (MAC) organisms are most commonly responsible for infection.
2. Normal people usually get asymptomatic or mild infections with MAC organisms, but immunodeficient patients may have fever, drenching sweats, severe weight loss, diarrhea, and abdominal pain.
3. MAC organisms enter the body via the lungs and gastrointestinal tract, are taken up by macrophages but resist destruction, and are carried to all parts of the body. In immunodeficiency, MAC organisms multiply without restriction, producing massive numbers of the organisms in blood, intestinal epithelium, and other tissues. (Figure 29.17)
4. No generally effective measures are available for preventing exposure to MAC bacteria. Prophylactic medication is advised for severely immunodeficient patients, but it can fail to prevent infection.

REVIEW QUESTIONS

Short Answer

1. What was the purpose of "AIDS-defining conditions"?
2. What is the main symptom of patients with lymphadenopathy syndrome (LAS)?
3. Give one function of HIV accessory genes.
4. Which cells of the immune system are prime targets of HIV?
5. What role do asymptomatic people with HIV disease play in the epidemiology of AIDS?
6. How was the risk of contracting HIV disease from clotting factor VIII eliminated?
7. Why might the infant son of a hemophiliac man develop AIDS when the son's parents were strictly monogamous nonabusers of drugs?
8. What are two consequences of the rising percentage of HIV disease cases in women?
9. Name a disinfecting agent active against human immunodeficiency virus (HIV).
10. List five groups of people at increased risk for HIV infection.
11. Give two reasons it is a good idea to know whether you are infected with HIV.
12. What are four requirements of an acceptable HIV vaccine?
13. Why is HIV protease so important to replication of HIV?
14. What is a phase III vaccine trial?
15. Why would you expect gp120 vaccines to be the most advanced?
16. What are the three main types of malignant tumors that complicate HIV disease?
17. How do physicians prevent pneumocystosis in AIDS patients?
18. In AIDS patients with toxoplasmosis, which part of the body is affected in more than half the cases?
19. Name a feared complication of cytomegalovirus infection in AIDS patients.
20. Where in an AIDS patient's surroundings might MAC organisms be found?

Multiple Choice

1. About how many people have died of AIDS in the United States since the epidemic of HIV disease began?
   A. 4,000,000
   B. 4,100
   C. 42,000

D. 450,000

E. 120,000

2. All of the following symptoms are characteristic of the AIDS related complex (ARC), *except...*

   A. fever.

   B. fatigue.

   C. diarrhea.

   D. blindness.

   E. weight loss.

3. Which one of the following is true of Kaposi's sarcoma?

   A. HHV-8 is necessary for development of the tumor.

   B. HIV-1 is necessary for development of the tumor.

   C. Both HHV-8 and HIV-1 are necessary for development of the tumor.

   D. HHV-8 alone is sufficient for development of the tumor.

   E. Both HHV-8 and HIV-1 together are sufficient for the tumor to develop.

4. All of the following are HIV accessory genes, *except...*

   A. *tat.*

   B. *env.*

   C. *vpr.*

   D. *rev.*

   E. *vpu.*

5. When was AIDS first recognized as representing a new disease?

   A. 1973

   B. 1959

   C. 1981

   D. 1989

   E. 1999

6. All of the following are AIDS-defining conditions, *except...*

   A. influenza.

   B. herpes simplex of the esophagus.

   C. *Pneumocystis carinii* pneumonia.

   D. invasive cancer of the uterine cervix.

   E. Kaposi's sarcoma.

7. Which of the following types of cells can be infected by HIV?

   A. Th cells

   B. Intestinal epithelium

   C. Antigen-presenting cells

   D. Brain cells

   E. All of the above

8. All of the following are HIV antigens, *except...*

   A. CD4.

   B. TM.

   C. RT.

   D. MA.

   E. CA.

9. Which of the following is a cause of Th cell death in HIV disease?

   A. Replication of HIV lyses the cell.

   B. Infected cells are destroyed by cytotoxic T cells (Tc).

   C. Infected cells are attacked by natural killer cells.

   D. Cells are killed by fusion and syncytium formation.

   E. All of the above.

10. Highly active antiretroviral therapy (HAART) is less than ideal because...

   A. it does not eliminate latent HIV infection.

   B. its cost is too great for 90% of AIDS sufferers.

   C. it often has severe side effects.

   D. some HIV strains are resistant to it.

   E. All of the above.

Applications

1. An epidemiologist from the Centers for Disease Control and Prevention was presenting a report on the status of AIDS to a congressional committee. In concluding her remarks, she noted that from an epidemiological perspective it was more important to focus on HIV infection than on AIDS, and urged that the Congress consider redirecting funding of AIDS research to reflect this fact. What was the rationale for her request?
2. A historian researching the influence of society on the spread of communicable disease began to speculate on what it would be like if AIDS had appeared at a different time. What differences might one expect, for example, if AIDS had appeared in 1928 instead of 1978?
3. A newly emerging virus lethal for all felines is rapidly killing off household cats and related zoo animals. The CDC urgently appeals for funds to develop a vaccine against the virus, but a scientific adviser to Congress states that it would be very expensive and may not be possible. Moreover, she states that getting rid of cats would have the side benefit of ridding the world of toxoplasmosis. Is she correct? If so, how long would it take?

Critical Thinking

1. Vaccines have effectively prevented many viral diseases, witness smallpox and poliomyelitis. Attempts over many years to develop an effective vaccine against HIV disease and AIDS, however, have so far met with little success. Why might this be so?
2. A basic problem in searching for a cure of HIV disease is that the viral genome is "hidden" as a provirus in many host cells. These cells do not express viral antigens on their surface and are not detected by cellular immunity. If the cell becomes activated, however, the viral genome leaves the host cell chromosome, it begins to replicate, and viral antigens appear on the host cell surface. Also, antiviral medications are active against the

replicating virus. Why not try to find a way deliberately to activate all the infected cells?

3. The graph to the right shows the survival of individuals with HIV disease and different quantities of circulating virus (copies of HIV RNA per ml of blood plasma) early in their infection. How do you interpret these data? What are the implications for treatment of people with HIV disease?
4. Why is reverse transcriptase needed in order for HIV to become a provirus?

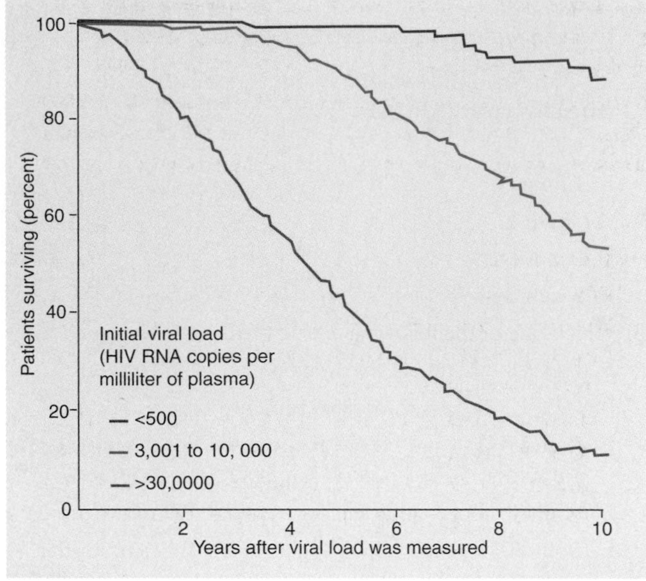

CHAPTER 30

Environmental Microbiology

For centuries, farmers have understood that they could not continue growing the same crop on the same piece of land year after year without reducing the crop's yield. They knew that allowing a field to lie unplanted for one or more seasons enables it to recover its productivity. The wild plants that grow on the field for a year or two appear to rejuvenate the soil. It was not until the late Nineteenth century that scientists began to discover why this was so. They isolated soil microorganisms that could take nitrogen from the air and chemically transform it into forms of nitrogen that plants could use. This process is called **nitrogen fixation**. *Although many scientists have worked for years to understand how microorganisms fix nitrogen, one scientist from the Netherlands, Martinus Willem Beijerinck, stands out as an early contributor in these studies.*

In 1887 Beijerinck reported on the properties of the root nodule bacterium, which he called Bacterium radicola. *(The genus name of this bacterium was later changed to* Rhizobium.*) He showed in 1890 that root nodules were formed when* B. radicola *was incubated with legume seedlings. Russian microbiologist Sergei Winogradsky then showed that the bacteria formed a symbiotic relationship with the roots of the legumes. The bacteria-containing nodules could fix nitrogen.*

Beijerinck also made major contributions to other areas of microbiology. He worked on yeasts, plant viruses such as tobacco mosaic virus, and plant galls. Beijerinckia, *a group of Gram-negative aerobic rods, is named for him. The genera* Azotobacter *and* Beijerinckia *include bacteria that can fix nitrogen under aerobic conditions in the absence of plants.*

Beijerinck was described as a "keen observer," a person who was able "to fuse results of remarkable observations with a profound and extensive knowledge of biology and the underlying sciences." This ability was undoubtedly responsible for the great success of his work.

—A Glimpse of History

MICROORGANISMS ARE VITAL TO THE CYCLING OF elements, the maintenance of fertile soil and usable water supplies, and the decomposition of wastes and other pollutants. Without microbial activities, life on earth could not survive. People would quickly become buried by the tons of waste products they generate, and nutrients would be depleted, limiting growth and reproduction.

The study of microorganisms in their natural environments is much more complex than laboratory studies. Instead of growing as pure cultures under controlled conditions, organisms in nature exist as members of heterogeneous communities under poorly defined and often changing environmental conditions. In the laboratory, the investigator provides the nutrients necessary to ensure the optimal growth of the culture. In nature, organisms usually exist in low-nutrient environments where nutrient supplies are limiting to microbial growth. In the laboratory, organisms are often grown in rich liquid culture, whereas in nature, most microorganisms grow in association with solid surfaces, on plants or animals and in biofilms or biomats, from which they can gain nutrients.

The growth and development of organisms in different environments as well as the role of microorganisms in the cycling of key biological components are discussed in this chapter. Bioremediation, the employment of microorganisms in the cleanup of pollutants such as pesticides, oil spills, and metals, will also be discussed.

Principles of Microbial Ecology

Ecology is the study of the relationship of organisms to each other and to their environment. Living organisms interact with one another in various ways, such as symbiosis, mutualism, or parasitism. Living organisms in a given area, the **community**, interacting with the nonliving environment form an ecological system, or **ecosystem**. Major ecosystems include the sea, rivers and lakes, deserts, marshes, grasslands, forests, and tundras. The human body also provides the environment for growth of many communities of microorganisms. The rumens of cattle and sheep contain ecosystems with protozoa and anaerobic bacteria that can ferment plant materials. Each of these ecosystems possesses its own particular physical environment and spectrum of living organisms. Microorganisms play a major role in most ecosystems, and many ecosystems host unique microorganisms. The role that an organism plays in its ecosystem is called its ecological niche. ■ **microbial interactions, symbiosis, mutualism, parasitism p. 452, normal flora of humans, p. 451, 452**

The region of the earth inhabited by living organisms is called the **biosphere**. Within the biosphere, ecosystems vary, both in the number of species of organisms inhabiting them, constituting **biodiversity**, and in the weight of all organisms there, called the **biomass**. The fundamental requisite for all organisms in any environment is that they be capable of multiplying, at least from time to time. In an environment conducive to life, such as the top few inches of fertile soil, there is likely to be great biodiversity, as well as a large biomass. The top 6 inches of fertile soil may contain more than 2 tons of bacteria and fungi per acre. When a sample of such fertile soil is cultured, millions of organisms per gram of soil are isolated. The number of organisms cultured is extremely misleading, however, since less than 0.1% of the organisms in many environments can be grown in the laboratory. Most of the other organisms are metabolically active, as shown by many different tests, but cannot be cultured. Nucleic acid probes, measurement of ATP in the sample to estimate the numbers of living cells present, and staining with a dye that stains all living cells but not dead cells are a few of the techniques used to show that the number and the diversity of organisms are much greater than would be expected based solely on culture results.

In harsh environments, exposed, for example, to extremes of temperature, pressure, or other environmental influences, relatively few species are capable of multiplying and, therefore, of existing for a significant period. Microorganisms that survive environmental extremes are called **extremophiles**. Other microbes may survive adverse conditions for a time by producing resting forms such as **endospores**.

The environment immediately surrounding an individual microorganism, the **microenvironment**, is most relevant, but because organisms are so small the microenvironment is difficult to identify and measure. Remember that the usual size of a bacterium is about one-millionth that of a human being. The more readily measured gross environment, or **macroenvironment**, may be very different from the microenvironment. To a bacterial cell living, for example, within a biofilm (**figure 30.1**), the microenvironment inside this film is most important. Growth of aerobic organisms there can deplete oxygen and create anaerobic microzones that will then support the survival of obligate anaerobes. Fermenters can produce organic acids that are metabolized by other organisms in the film. Other growth factors can be transferred directly within the microenvironment of the biofilm.

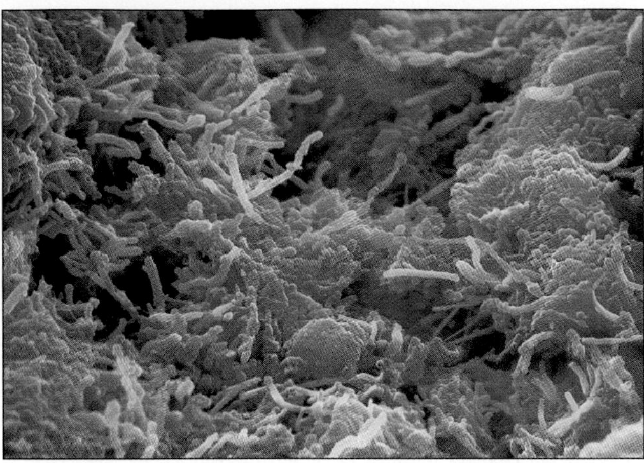

Figure 30.1 Bacteria Within a Biofilm This scanning electron micrograph image of a human tooth surface shows a variety of bacteria within the biofilm of plaque. Growth of aerobic organisms depletes oxygen, creating anaerobic microzones where obligate anaerobes can grow. On teeth, these anaerobes ferment sugars from foods, producing acids. The acids corrode tooth enamel, resulting in cavities.

Bacteria in Low-Nutrient Environments

Since low-nutrient environments are common in nature, microorganisms capable of growth in dilute aqueous solutions are also common. Most microbial growth in these environments is in biofilms, and the organisms are shed from the film into the aqueous solution. These organisms are by no means restricted to lakes, rivers, and streams. Indeed, microorganisms even grow in distilled-water reservoirs such as those found in research laboratories and pulmonary mist therapy units used in hospitals. In these environments, the microbes can extract trace amounts of nutrients absorbed by the water from the air or adsorbed onto the biofilm. Although the organisms grow slowly, they can reach concentrations as high as 10^7 per milliliter. Since this cell concentration is not high enough to result in a cloudy solution, the growth usually goes unnoticed. This can have serious consequences for the health of hospitalized patients and for the success of laboratory experiments that depend on water purity. Organisms that grow in dilute environments contain highly efficient transport systems for moving nutrients inside the cell.

Microbial Competition and Antagonism

Although many different species may be capable of living in a given microenvironment, only one or a few species actually occupy it. The species best adapted to live in a particular microenvironment is the one that inhabits it to the exclusion of all others. A large number of different species are found, for example, within a particle of soil. If, however, the particle is carefully dissected into squares of about 70 µm per side, only a single species will be found living as a microcolony within any one of these squares.

Perhaps nowhere in the living world is competition more fierce and the results of competition more quickly evident than among microorganisms. The ability of an organism to compete successfully for its ecological niche is generally related to the rate at which the organism multiplies, as well as to its ability to withstand adverse environmental conditions. The complete takeover of one species by another is especially likely when two species have similar nutrient requirements and one of these nutrients is in limited supply. The organism that multiplies faster will yield the larger population, utilizing more of the limited nutrient supply. Because bacteria often can divide at least once every several hours, any small differences in their generation times will result in a very large difference in the total number of cells of each species after a relatively short time. For example, if 10 cells each of a rod and a coccus are growing in the same natural environment, under conditions in which the generation time of the rod is 100 minutes and that of the coccus is 99 minutes, and if the number of cells reaches about 10^9, there will be 1.35 times as many cocci as rods. If the cells are then placed in conditions where they can continue to multiply, the ratio of cocci to rods will increase to the point where there are very few rods left **(figure 30.2)**.

Antagonism between groups of organisms also helps determine the make-up of a community. In the soil, for example, some microbes produce antibiotics that not only deter the growth of other organisms but protect plants as well. ■ **antibiotics, p. 496**

Competition and antagonism both probably contribute to the observation that the community of bacteria living in the human intestine tends to remain quite stable with time. Although humans swallow large numbers of bacteria, including some closely related to those already present, the new strains have great difficulty becoming established. The resident community not only competes successfully for a limited food supply, it also produces toxic substances that inhibit the growth of newcomers. The resident microorganisms thus provide a natural defense mechanism against intestinal pathogens.

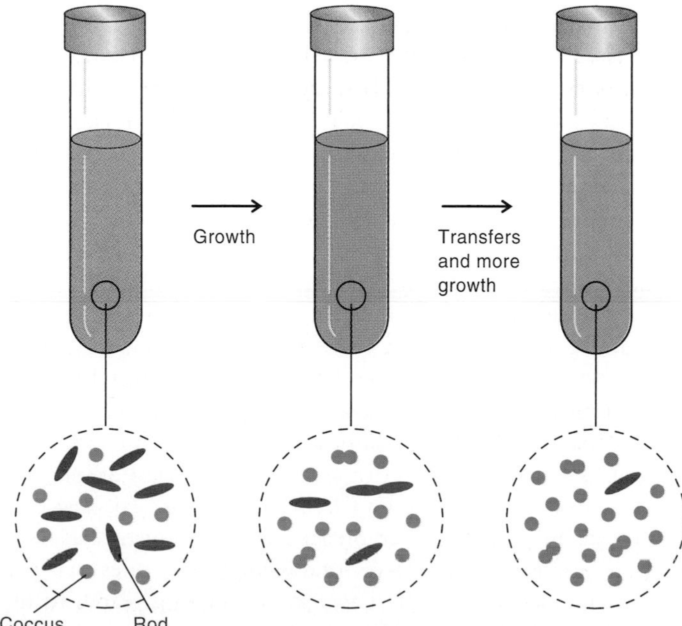

Figure 30.2 Competition Between Two Bacteria The coccus has a generation time of 99 minutes, while the rod divides every 100 minutes.

Microorganisms and Environmental Changes

Environmental changes often result in changes in a community. Those organisms that have adapted to live several inches beneath the surface of an untilled field will probably not be well adapted to that field if it is plowed, fertilized, and irrigated. If the organisms are to survive under these new conditions, they must adapt to the altered environment. Cells may be able to synthesize different enzymes that help them cope with a new environment. Generally, cells synthesize only the enzymes they absolutely require for growth. Under changed environmental conditions, however, additional or different enzymes may become important. In response to the new stimuli, the microbes are induced to synthesize these enzymes and stop producing others that are no longer useful. For example, some bacteria can produce an enzyme that inactivates mercury, a toxic metal, but the enzyme is only formed when mercury is present. ■ **enzyme induction, p. 179**

A mutant within a population may be ill adapted to its environment and remain in the minority. It may, however, be especially well adapted to a new changed environment. For example, bacteria that mutate to become resistant to tetracycline are at a distinct advantage when tetracycline is added to the environment. As the antibiotic inhibits growth of competing organisms, the tetracycline-resistant bacteria then become the dominant organisms **(figure 30.3)**.

In addition to external sources of environmental change, the growth and metabolism of organisms themselves may alter the environment dramatically. Nutrients may become depleted, and a variety of waste products, many of which are toxic, may accumulate. In some environments, the changing conditions bring about a highly ordered and predictable succession of bacterial species. First one species becomes dominant, then another, and then a third. An example of such a microbial succession is the one that occurs in unpasteurized milk. Most of the microorganisms involved are destroyed by pasteurization.

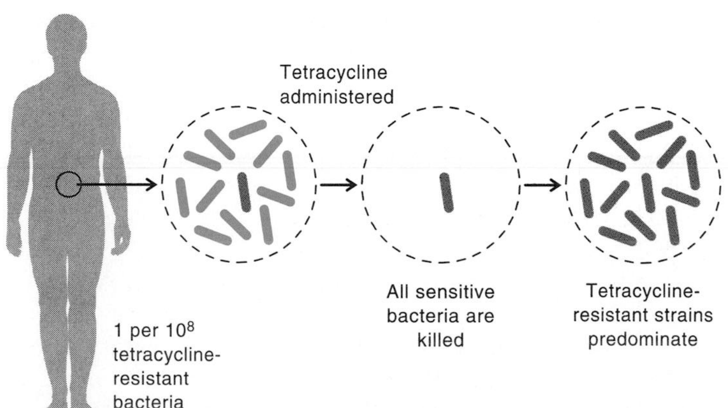

Figure 30.3 Selection of an Antibiotic-Resistant Mutant in a Population by the Application of Antibiotic Therapy The mutant is in the minority in the original environment, but it can adapt to a new environment containing tetracycline. Bacteria not adapted to the new environment are inhibited, and the mutant strain becomes predominant.

Unpasteurized milk usually contains various species of bacteria, yeasts, and molds, which are derived mainly from the immediate environment around the cow. Initially, the dominant organism is the bacterium *Streptococcus lactis*, which breaks down the milk sugar lactose to lactic acid (**figure 30.4**). The resulting acid inhibits the growth of most other organisms in the milk, and eventually enough acid is produced to prevent even the further growth of the *S. lactis*. The acid sours the milk and also curdles it, a result of denaturation of the milk proteins. Bacterial species such as *Lactobacillus casei* and *L. bulgaricus* can multiply in this highly acidic environment. These species metabolize any remaining lactose, forming more acid until their growth is also inhibited. Yeasts and molds, which grow very well in this highly acidic environment, then become the dominant group and convert the lactic acid into non-acidic products. Because most of the milk sugar has already been used, the streptococci and lactobacilli cannot resume multiplication since they require this substrate. Milk protein (casein) is still available and can be utilized for energy by bacteria of the spore-forming genus *Bacillus*, and some other bacteria, all of which secrete proteolytic enzymes that digest the protein. This breakdown of protein, known as **putrefaction**, yields a completely clear and very odorous product. The milk thus goes through a succession of changes with time, first souring and finally putrefying.

MICROCHECK 30.1

Microorganisms are found throughout the biosphere. The best adapted organism takes over its environment. Microorganisms can change their environments and can adapt to environmental changes.

- Why do antibiotic-resistant bacteria become common in a hospital environment?
- Why do people not very often observe the souring and putrefaction of milk?
- Why might it not be possible to culture any more than 0.1% of the organisms from some environments?

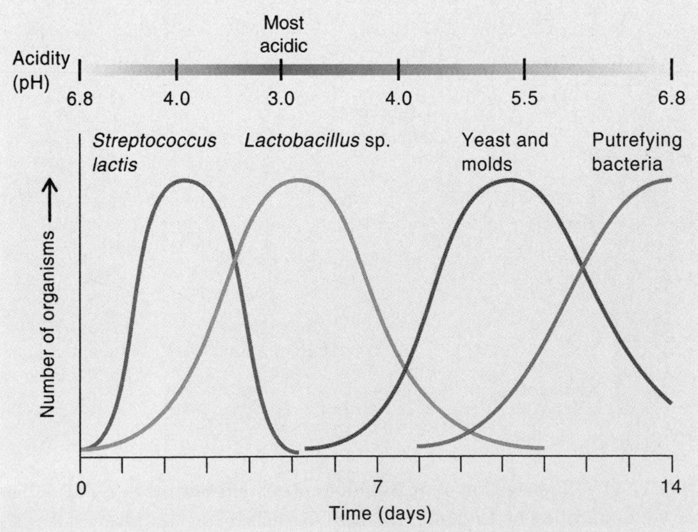

Figure 30.4 Growth of Microbial Populations in Unpasteurized Raw Milk at Room Temperature Production of acid causes souring and encourages growth of fungi. Eventually bacteria digest the proteins and fat, causing putrefaction.

Terrestrial Environments

Soil is the loose material on the outer portion of the earth's surface, largely resulting from the weathering of rocks and other inorganic materials. As rock weathers, rain dissolves water-soluble components. This, together with temperature changes, leads to cracking and fragmentation of the rock. Wind-blown particles are abrasive and aid in disintegration. Photosynthetic microbes such as lichens, algae, and mosses living on exposed rocks also play a role in soil formation. They synthesize organic materials used by nonphotosynthetic heterotrophic bacteria and fungi. These heterotrophs produce acids, which dissolve limestone rocks, and other chemicals contributing to the weathering of rocks. As the soil forms, plants begin to appear. When these plants die and decay, their organic material is added to the soil. During the course of weathering and soil formation, the spaces between individual fragments or particles of soil become smaller. The surface layer of soil is darkened by the organic matter known as **humus**. The next layer, the subsoil, is mineral soil in which microorganisms have decomposed the organic materials into inorganic substances. The lowest layer is bedrock, only slightly weathered.

Microorganisms and Soil

Soil teems with life, from the tiniest viruses to larger bacteria, fungi, algae, and protozoa, up to the macroscopic worms and insects. One estimate is that soil communities contain as many as 4,000 different species per gram of soil. Many microorganisms that live in the soil play an indispensable role in maintaining life on this planet by degrading or chemically modifying molecules. One of the most important functions of microbes in soil is their degradation of dead and waste materials. Their role in recycling matter is discussed in a later section.

Soil also contains organisms that transform chemical substances not readily utilized into some that serve as nutrients for a variety of other organisms. For example, although we live in a virtual ocean of air that is 79% nitrogen, this element cannot be used by humans unless it is first converted into amino acids. Certain soil microorganisms such as those of the genus *Rhizobium* can reduce atmospheric nitrogen to ammonia (NH_3). This ammonia is then available in the dissolved form of ammonium ions (NH_4^+) to the plants and animals that use it to synthesize amino acids and eventually protein.

Considerable human interest in soil organisms stems from their ability to synthesize a variety of useful chemicals. Many species produce antibiotics that function in the natural environment in microbial antagonism and persistence, and also in protecting plant life. For example, 500 different antibiotic substances are produced by *Streptomyces* sp., at least 50 of which have proved useful for human beings and other species, or in agriculture and industry. The pharmaceutical industry has tested many thousands of soil microorganisms to find ones that produce useful antibiotics. There is also an ongoing search to find organisms capable of synthesizing other useful compounds. Probably in no other habitat can one hope to find a greater range of synthetic capabilities than are represented in the soil. ■ **antibiotics from *Streptomyces* sp. p. 496**

Not all soil organisms are beneficial. For example, soil is also the habitat of a number of fungi and bacteria that are potentially pathogenic for plants and animals. Among those pathogenic for humans are members of the bacterial genera *Clostridium* and *Nocardia* and the fungal genera *Coccidioides*, *Histoplasma*, and *Blastomyces*. For example, *Clostridium tetani* produces a toxin that causes tetanus, and the other organisms named cause chronic infections of the lung and other organs. ■ **diseases caused by clostridia and soil fungi, p. 579, 581, 709**

Bacteria

The physiological diversity of bacteria allows them to colonize all types of soil. Since soil contains aerobic and anaerobic microhabitats, both aerobes and anaerobes are present in it. In acid or alkaline soils, specifically adapted organisms will be present. In desert or arctic soils, bacteria with resistant forms, such as endospores or cysts, will be present, along with extremophiles.

In general, the bacteria most commonly found in the soil are some of the **actinomycetes**. Actinomycetes are often resistant to dry conditions and can survive even in desert soils. Soil actinomycetes are aerobes and use organic compounds obtained by decomposing dead plant and animal matter. The most common genera are *Nocardia* and *Streptomyces*. Bacteria of the genus *Streptomyces*, in addition to producing antibiotics, produce metabolites called **geosmins**, which give soil its characteristic musty odor. Members of the genus *Arthrobacter* are Gram-positive club-shaped rods that are also abundant in soil. ■ **actinomycetes, p. 285, 704**

Another common group of heterotrophic soil bacteria is the **myxobacteria**. Many of these are pigmented and form brightly colored masses that can sometimes be seen on forest debris. Under starvation conditions, myxobacteria will swarm together to form complex macroscopic fruiting bodies (**figure 30.5**). These fruiting bodies contain spores that can survive nutrient-poor conditions. ■ **myxobacteria, p. 284**

Soil bacteria are important in cycling the biologically important elements nitrogen, oxygen, carbon, sulfur, and phosphorus. Each of these cycles is discussed later in this chapter. Bacteria are also involved in controlling the availability of some key metals such as iron, manganese, and selenium that are necessary trace elements for living cells.

Fungi

Many types of fungi can be found in soils, but the specific genera will vary with the type of soil and with its physical and chemical properties. Although bacteria are the most numerous soil inhabitants, the biomass of fungi is greater. Some soil fungi are free-living, and some occur in symbiotic relationships. Species of the genera *Candida*, *Aspergillus*, *Penicillium*, *Rhizopus*, and *Mucor* are common. Some fungi are capable of degrading and using components of soil that cannot be attacked by other microorganisms. Other fungi are important plant pathogens, attacking wheat and corn among others.

The symbiotic relationships of fungi in soil are especialy important to many plants. **Mycorrhizae** are fungi that grow in a symbiotic relationship with certain plant roots. They help plants to take up phosphorus and other substances from the soil, while gaining nutrients for their own growth from the area around the root hairs. Many trees depend on mycorrhizal roots. The expensive delicacy truffles are mycorrhizae found in association with oak or beech trees. Indian pipe and some other mycorrhizal plants have lost the ability to produce chlorophyll and thus are totally dependent on their fungal partners. **Lichens** represent another mutually beneficial symbiotic relationship, consisting of algae and fungi. The fungi protect the algae from the environment and supply some nutrients, while the algae may utilize nitrogen from the air, making it available to its fungal partner, or use CO_2 and sunlight to make organic materials for the pair. Cyanobacteria may also associate with fungi in a symbiotic relationship to form lichens. In barren soils, the symbiotic lichens are common. ■ **mycorrhizae, p. 312** ■ **lichens, p. 311, 312**

Figure 30.5 Myxobacteria Fruiting Bodies Fruiting bodies are about 150 to 400 μm tall.

Fungi in soil generally exhibit active metabolism during periods when environmental conditions are favorable and become dormant when conditions worsen or when nutrient supplies diminish. Since they are aerobes, they are usually found in the top 10 cm of soil. The soil fungi are crucial in the decomposition of plant matter, degrading and utilizing complex macromolecules such as cellulose and lignin, the major component of cell walls of woody plants.

Algae

A number of algal types may be found in the soil. Because they depend on sunlight and photosynthesis to provide their energy needs, they mostly live on or near the soil surface. Algal populations are quite sensitive to environmental conditions, declining during periods of drought or low temperature. Algae are a major nutrient source for a number of soil inhabitants, including protozoa, fungi, earthworms, and nematodes.

Protozoa

Protozoa are found in most soils at a density of about 10^4 to 10^5 organisms per gram of soil. They generally occur near the surface, since most require oxygen. Protozoa are consumers of soil bacteria and algae. An important relationship exists between termites and the protozoa in their stomachs. Termites are notorious for destroying wood, but they cannot do it by themselves.

The termite bites off wood, but it must be digested by the activities of large numbers of protozoa that live in the termite gut. Neither the termites nor the protozoa can live without the other, and they both live in soil.

Environmental Influences in Soil

Environmental conditions affect the density and composition of the flora of the soil. The primary environmental influences include moisture, acidity, temperature, and nutrient supply.

Moisture

Wet soils are unfavorable for aerobic bacteria simply because the spaces in the soil fill up with water, thus diminishing the amount of air in the soil. Water is essential for life, however, so when the moisture content of soil drops to a low level, as in a desert environment, the metabolic activity and number of soil bacteria are markedly reduced. Many species of soil bacteria, both those that form spores and those that do not, are markedly resistant to drying. Representatives of these groups can survive drought for years.

Because the filamentous fungi, the molds, are strict aerobes, they are found close to the surface of the soil. In waterlogged soils, the numbers of vegetative cells of these fungi are also markedly reduced, but their spores often persist. Once the water in the soil diminishes and the level of soil oxygen increases, the molds resume multiplication and become an important part of the community.

Acidity

Highly acid or alkaline conditions generally inhibit the growth of common bacteria, but many species of fungi can multiply over a broad pH range, from alkaline above pH 9 to acid almost as low as pH 2. The fungi, however, generally are more successful in acidic environments. Adding lime to a soil reduces the abundance of fungi and treating the soil with acid-forming fertilizers, such as those containing ammonium, increases the fungal abundance. This is why mushrooms may appear in profusion in a lawn fertilized with an acid-producing fertilizer such as ammonium chloride. The increased abundance of fungi at low pH results primarily from suppression of bacteria, which thus do not compete with the fungi for nutrients.

Temperature

Temperature governs the rates of biochemical processes through its effect on the rates of enzyme reactions. A warming trend favors the biochemical changes in soil that are brought about by its microorganisms. The bulk of the soil bacteria are mesophiles that grow best at temperatures between 20°C and 50°C. True psychrophiles, bacteria that cannot grow above 20°C, are rare or absent in most soil, being found mainly in polar ice or water. In winter, the bacteria in soil are usually cold-tolerant mesophiles rather than psychrophiles. Thermophiles that can grow above 50°C are ubiquitous, especially thriving in such environments as compost heaps, which generate a great amount of heat.

Nutrient Availability

The great majority of soil bacteria and fungi are heterotrophic. The dead and decaying organic matter that serves as their source of energy is produced largely by the photosynthetic activity of higher plants. The size of the microbial community is limited by the organic matter available. The addition of organic material, such as occurs when manure or crop residues are plowed into the soil, dramatically increases the number of bacteria and also fungi in the soil.

In soils in which plants are growing, a variety of organic materials exude from the plant roots. The zone that surrounds the roots and contains these exudates is termed the **rhizosphere**. The exudates provide an abundant energy source for bacteria, fungi, and other heterotrophic organisms that tend to concentrate on or near the root surfaces of plants.

MICROCHECK 30.2

The soil teems with a broad diversity of organisms that are essential for modifying, degrading, and producing biologically important substances. Environmental influences such as moisture, pH, temperature, and nutrient supply affect the numbers and kinds of organisms in soil.

- What is the largest group of bacteria found in most soils?
- How could temperature affect the growth of microorganisms in the soil?
- How could the biomass of fungi in soil be greater if the number of bacteria is greater?
- What do the protozoa, but not termites, possess that allows the termite to live by consuming wood?

Aquatic Environments

Virtually all aquatic environments contain living microorganisms. Water has unique properties that make it a necessary part of the environment for many microorganisms. For example, water can absorb light of certain wavelengths, an important property for photosynthetic aquatic microbes. Also, it is an extremely efficient solvent, a property that is of paramount importance for all organisms, terrestrial or aquatic. The numbers and varieties of organisms vary greatly in different aquatic environments.

Fresh water may be found in either surface waters, such as rivers and lakes, or underground water, called **groundwater**. Groundwater close to the surface may break through to form springs, while deeper groundwater is accessed by digging or drilling wells. Water underground gains dissolved nutrients and dissolved gases. The temperature of natural waters can vary from nearly freezing to nearly boiling at surface atmospheric conditions. The temperature increases about 1°C for every 30 meters of depth of underground water. Hot springs occur as a result of groundwater passing through deep strata of the earth's crust.

As water follows its course to the ocean, it becomes progressively more enriched with dissolved salts. Seawater contains about 3.5% salt, compared with about 0.05% for fresh water.

Phosphate and nitrate are scarce in seawater. Thus, seawater usually contains fewer organisms than fresh water, but halophilic organisms, those that prefer or require high salt concentrations, thrive in it. Barometric pressure increases with depth, whereas temperature generally decreases with depth to about 2°C in the deep sea, except around hot vents. Thus, the oceans serve as habitats for a variety of extremophiles. ■ halophiles, p. 99

Specialized aquatic habitats include **salt lakes**, such as the Great Salt Lake in Utah, which have no outlet. Water in the lake evaporates, leaving concentrations of salt much higher than in seawater. Only extremophile halophilic organisms live in this environment. Other specialized habitats include iron springs that contain large quantities of ferrous ions; these springs are the habitat for species of *Gallionella* and *Sphaerotilus*. Also, sulfur springs support the growth of both photosynthetic and non-photosynthetic sulfur bacteria.

Organisms pathogenic for humans are often found in water. For example, water is the natural habitat for some members of the genera *Vibrio* and *Aeromonas*. Many other pathogens are transmitted through water. ■ waterborne infections, p. 789

Any body of water represents a complex ecosystem. Energy enters the system largely in the form of light and is converted to chemical energy by photosynthetic organisms, either algae in aerobic conditions or bacteria in anaerobic conditions. These are the **primary producers**, organisms that use CO_2 as a carbon source and convert it to organic materials. Primary producers are the first step in the food chain, providing food for protozoa and small invertebrates, which in turn feed the fish. About two-thirds of the earth's surface is covered by oceans, in addition to all the lakes, rivers, and other bodies of water, making aquatic environments a major source of primary producers. The diverse microorganisms in marine environments are being actively studied as potential sources of new compounds, such as pharmaceuticals.

MICROCHECK 30.3

Virtually all water contains living organisms, which vary greatly with different aquatic environments, from fresh water to seawater and some extreme habitats, such as salt lakes. Primary producers in aquatic environments convert inorganic substances such as CO_2 to organic matter by photosynthesis.

- Compare the salt content of seawater with that of fresh water and salt lakes.
- Name at least one specialized aquatic habitat and explain how organisms survive there.

Energy Sources for Ecosystems

All ecosystems must have access to energy. The energy trapped in chemical bonds cannot be totally recycled, since a portion is always lost as heat when bonds are broken. Thus, energy is continually lost from biological systems. To compensate for this outflow, energy is continually added to most ecosystems in the form of solar or light energy. Before this energy is useful, however, it must be converted into chemical bond energy. The conversion process, called **photosynthesis**, is carried out by chlorophyll-containing plants and microorganisms. These organisms, the producers, make the light energy available to heterotrophs. ■ chemical bonds, p. 21, photosynthesis, p. 158

The requirement for solar energy has traditionally been used to explain why life is not equally abundant everywhere in and on the earth. Solar energy is confined primarily to a thin shell in the upper levels of the soil and water and to the lower levels of the atmosphere. It does not penetrate to deep soil, rocks, or the depths of the oceans. Until recently, it was thought that any organism present in soil or water must depend for its food on organisms that have access to light. The discovery of different types of ecosystems, however, has dramatically changed these ideas.

First, scientists found large communities of extremophiles in the hot water around vents on the ocean floor near the Galapagos Islands, in the Juan de Fuca vent off the northwestern coast of the United States, and in other areas. Hundreds of meters below the surface, a number of undersea hot springs, or vents from the interior of the earth, spew out vast quantities of minerals and gases, including hydrogen sulfide (**figure 30.6**). These are called **warm vents**, because the cold ocean water mixes with the hot spring water to give temperatures of 6°C to 23°C, compared with seawater temperatures of about 2°C in surrounding areas. The communities that grow on the dark ocean floor near these relatively hot undersea geysers are supported not by photosynthesis, but by **chemoautotrophy**. Large numbers of sulfur-oxidizing *Thiomicrospira* sp. have been found in the area, both free-living and in association with the large tube worms and clams discovered there. The chemoautotrophs obtain energy from oxidation of hydrogen sulfide, and they reduce CO_2 to synthesize organic compounds, on which the

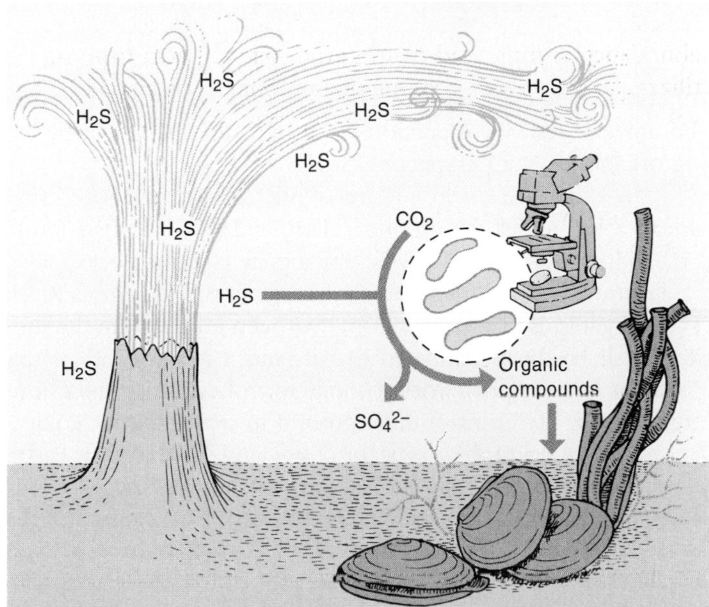

Figure 30.6 A Hydrothermal Vent Community In this ecosystem, the producers are chemoautotrophic bacteria. Water escaping from the vent is rich in minerals and dissolved gases, including hydrogen sulfide.

Perspective 30.1 Life on Mars?

The world listened with extreme interest in 1996 when NASA scientists reported finding what could be life forms in a meteorite from Mars. The meteorite, called ALH 84001, is a 4-pound potato-shaped rock found in Antarctica in the 1980s. It is estimated to have formed on the surface of the red planet 4.5 billion years ago and to have been launched into space by an asteroid collision with Mars about 15 million years ago, finally falling on Antarctica 13,000 years before it was found there. Chemical analysis has established the Martian origin of the rock. The evidence for living forms within the rock is not so conclusive.

Four types of evidence were put forward to support the thesis that life once existed within the rock. Carbonate globules, mixtures of calcium and magnesium carbonates, were present along fractures and pores in the rock. These could have been deposited by water flowing along the rock, water being a prerequisite for life. Magnetite and iron sulfide particles were present. These are often formed by bacteria and other living organisms. Polycyclic aromatic hydrocarbons (PAHs) were present. These are compounds left behind when living organisms decay. Very tiny microscopic fossils, resembling bacteria but smaller, were found inside the carbonate globules. It is proposed that these were tiny bacteria or bacteria-like organisms that lived on Mars long ago, died, and left their remains within the rock.

In 1997, at the next large gathering of planetary scientists, those who made the first report were more confident that they were right, but other researchers doubted their conclusions. The latter suggested that the carbonate globules were formed by high-temperature impacts on the surface of Mars, rather than by water. Some thought the magnetite and iron sulfide particles looked like the trail left by gases from volcanic eruptions. Still others felt that the PAHs were contaminants from the Antarctic ice and snow. The supposed bacterial fossils were about 100 times smaller than any known bacteria on earth, bolstering the skeptics.

The opinions and data accumulate. One reason for disbelief has been the extreme conditions under which these purported organisms must have survived. The recent discoveries of extremophiles thriving under previously unbelievable conditions on earth has made the reports of Martian microbes easier to accept or at least to imagine. It was hoped that the explorer sent to Mars in late 1999 would at least settle the question whether there is water on the red planet. That mission, however, was lost, setting the program back a bit. The series of spacecraft being sent to Mars during the coming years may bring back more evidence to enable scientists to answer the question, "Life on Mars?"

tube worms and clams then feed. This type of symbiosis is now thought to be widespread among species that live in environments where oxygen, carbon dioxide, and hydrogen sulfide are present. ■ **chemoautotrophy, p. 100**

Subsequently, another type of seafloor vent has been studied, the conical mounds up to 9 meters high, called **black smokers**. The smokers spout superheated waters at temperatures up to 300°C or even higher. Water remains liquid at this temperature because the pressure at these depths reaches 265 atmospheres. The large numbers of microbes isolated from the smokers were found to be anaerobic hyperthermophiles as well as barophiles that thrive at very high pressure. The finding of large numbers of organisms living under extreme conditions suggested that the growth of microorganisms is limited not by high temperatures, but by a requirement for liquid water, as long as other conditions are suitable for growth. If this is so, it follows that conditions suitable for life exist in many environments on the earth and in the universe that were previously thought uninhabitable by living organisms (see **Perspective 30.1**).

In 1994, there were reports of microbial populations living almost 3 km under the ground. Then, in 1995, scientists found large populations of microbes thriving in volcanic rocks called basalts from nearly 1 km below the surface of the Columbia River. These organisms gain energy from hydrogen produced in the subsurface. It has been estimated that if (and it is a big "if" at this point) most appropriate rocks contain microbes, there could be as much as 2×10^{14} tons of underground microorganisms, equivalent to a layer 1.5 m thick over the entire land surface of the earth.

Extremophiles found in superheated smokers require very high temperatures in order to multiply. For example, the archaen *Pyrolobus fumarii* can multiply at temperatures as high as 106°C, but it cannot grow below 91°C. Other extremophiles isolated from Antarctic sea ice require opposite extremes of cold. The bacterium *Polaromonas vacuolata* grows best at about 4°C and does not multiply at temperatures above 10°C. The extremophile archaea and bacteria have some enzymes that function at extremes of environmental conditions, and these enzymes are proving to be very useful in laboratory and industrial applications. For example, the heat-stable polymerase of *Thermus aquaticus* (*Taq* polymerase) has made possible the polymerase chain reaction used to amplify tiny quantities of DNA for analysis. ■ **polymerase chain reaction, p. 237**

MICROCHECK 30.4

Energy for ecosystems can come from sunlight via photosynthesis or from chemical synthesis of inorganic and organic materials by chemoautotrophic microorganisms.

- How is light energy converted to chemical bond energy?
- How do communities of microorganisms living near warm vents on the dark ocean floor gain energy?

Biogeochemical Cycling

A fixed and limited amount of the elements that make up living cells exists on the earth and in the atmosphere. For an ecosystem to sustain its characteristic life forms, these chemical elements must eventually be recycled. Only a relatively small percentage of the total amount of these elements is found in available biomass; the rest is unavailable to the biological community. It is either buried too deep in the earth's crust or present in forms that are not usable. Therefore, elements such as oxygen, carbon, nitrogen, phosphorus, and sulfur must be continuously recycled from nonutilizable to utilizable forms if the growth of new organisms is to continue.

All living organisms are involved in the important process of cycling elements. Plants are ingested by animals, which in turn die and are decomposed through the action of numerous microorganisms. A number of microorganisms can degrade complex cell constituents to inorganic substances such as ammonia, sulfates, phosphates, and carbon dioxide. Microorganisms play a major role in those processes owing to their ubiquity, their unique metabolic capabilities, and their high rates of conversion.

Recycling processes require the activities of organisms in three ecological roles: **production**, **consumption**, and **decomposition**. On the basis of these roles, organisms can be divided into three groups: (1) the **producers**, which convert carbon dioxide or other inorganic materials into organic material; (2) the **consumers**, which metabolize the organic matter synthesized by the producers; and (3) the **decomposers**, which digest and convert dead plant and animal material into small molecules that can be used by both the consumers and producers (**figure 30.7**). For each chemical element used in cell structures, recycling will occur only if members of all three functional groups are present; the loss of any group breaks the cycle.

Oxygen Cycle

Oxygen is cycled by the processes of respiration and photosynthesis (**figure 30.8**). During photosynthesis, cyanobacteria, algae, and higher plants produce oxygen from water. It is believed that the earth's early atmosphere contained little or no oxygen and that, as cyanobacteria developed, with chlorophyll and other compounds that permitted them to carry out photosynthesis, they increased the level of oxygen in the atmosphere. This made possible the development of organisms that respire using oxygen. The current level of oxygen in the atmosphere is kept in balance by chemical reactions in the upper atmosphere, aerobic respiration, and photosynthesis.

Carbon Cycle

All organisms are composed of organic molecules built from carbon "skeletons." The carbon cycle revolves around carbon dioxide, its fixation into organic compounds by primary producers, and its regeneration, mostly by microorganisms (**figure 30.9**). Consumers utilize the organic compounds and release the carbon as carbon dioxide through respiration.

The decomposition of organic matter is carried out by "decomposers"—chiefly, bacteria, fungi, protozoa, and small animals. The decomposition of organic debris involves the combined activity of a large number of organisms. Fungi and many bacteria utilize the more readily decomposable organic substances, such as sugars, amino acids, and proteins. Members of one group of bacteria, the Cytophaga, can degrade cellulose,

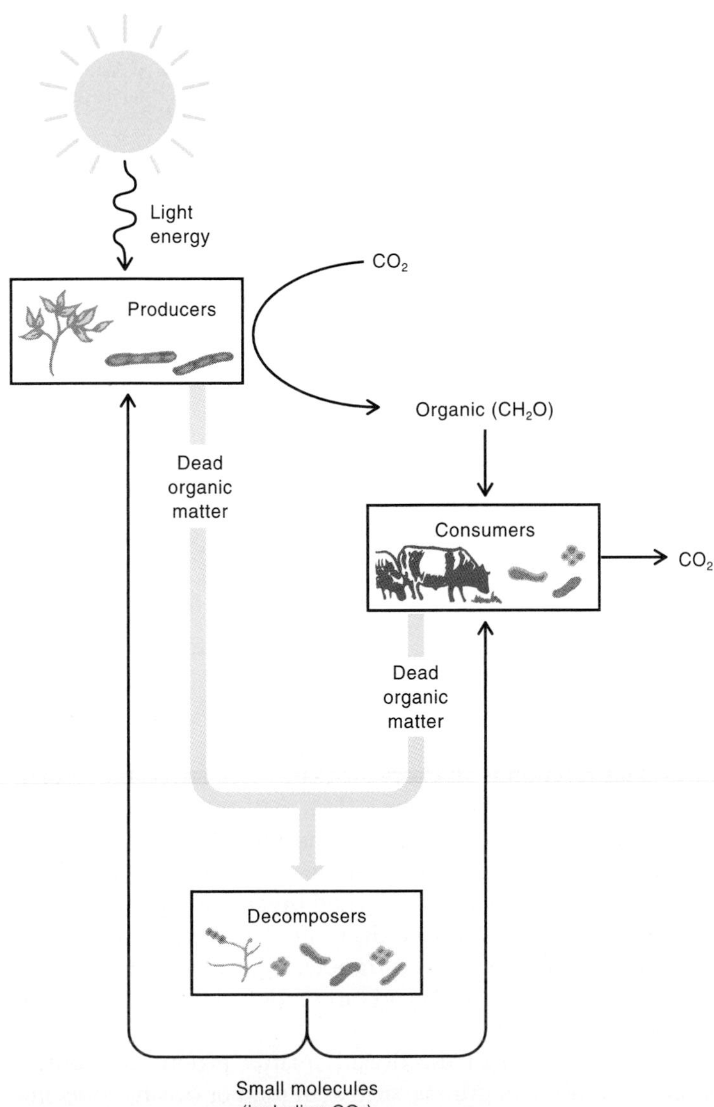

Figure 30.7 Relationship Between Producers, Consumers, and Decomposers in an Ecosystem

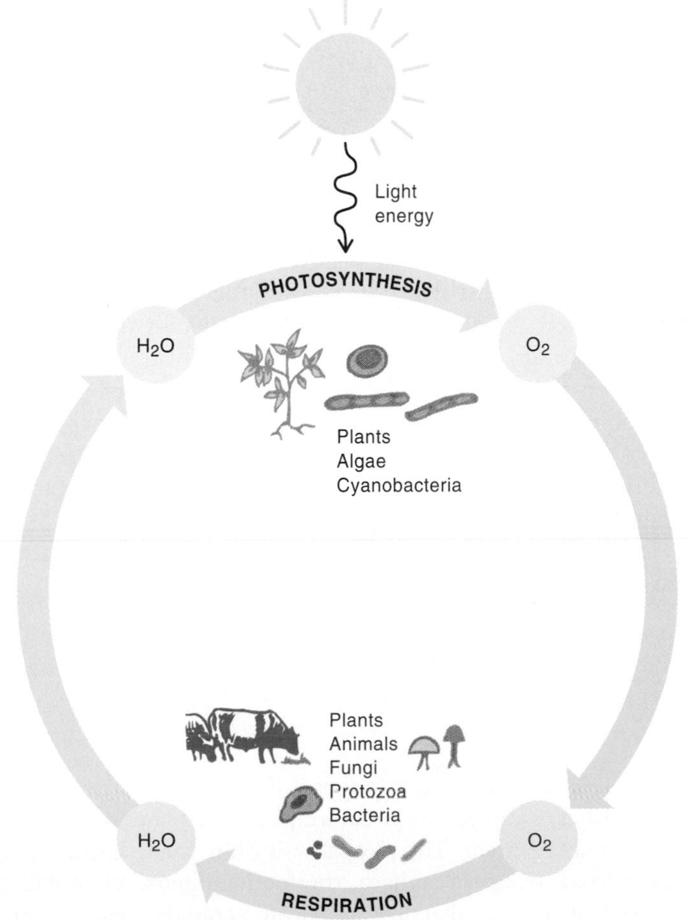

Figure 30.8 Oxygen Cycle

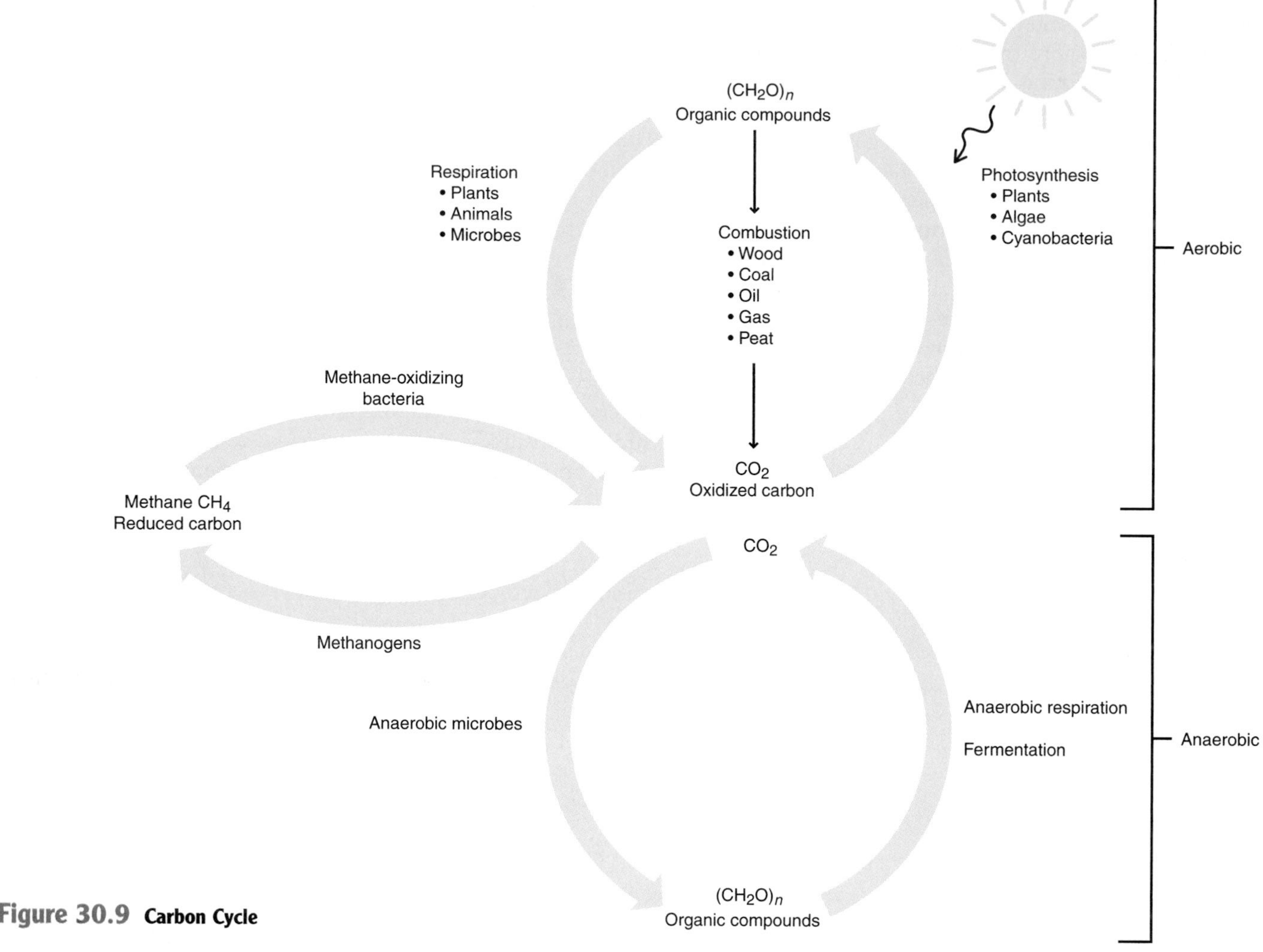

Figure 30.9 Carbon Cycle

a glucose polymer found in plants that is highly resistant to degradation by most other bacteria. Other major plant constituents, such as lignin and pectins, are generally decomposed initially by fungi. For example, lignin is resistant to degradation but can be degraded aerobically by some of the fungal group of basidiomycetes (**figure 30.10**). Some of the breakdown products of these constituents can then be further metabolized by bacteria. As a general rule, bacteria appear to play the dominant role in the decomposition of animal flesh.

During the aerobic decomposition of organic matter, gases are formed; the main gas evolved is carbon dioxide:

$$\underset{\text{Organic matter}}{(CH_2O)_n} + \underset{\text{Oxygen}}{(O_2)_n} \longrightarrow \underset{\text{Carbon dioxide}}{CO_2} + \underset{\text{Water}}{H_2O}$$

When the level of oxygen is low, as is the case in wet rice-paddy soil, marshes, swamps, and manure piles, methane (CH_4) production may be great. The biosynthesis of methane is limited to a specialized group of archaea, the methanogens, all of which are strict anaerobes that live in poorly aerated environments. These organisms have the ability to gain energy from the oxidation of hydrogen gas and the reduction of carbon dioxide, according to the following reaction:

$$\underset{\text{Hydrogen}}{4\,H_2} + \underset{\text{Carbon dioxide}}{CO_2} \longrightarrow \underset{\text{Methane}}{CH_4} + \underset{\text{Water}}{2H_2O}$$

This reaction is an excellent example of anaerobic respiration in which CO_2 is reduced. The methanogens cannot use common organic compounds, as can most heterotrophs. The carbon cycle is completed by the methane oxidizers, a group able to oxidize methane to CO_2 and organic compounds under aerobic conditions. Methane that enters the atmosphere is oxidized photochemically (by ultraviolet light and chemical ions) to carbon monoxide (CO) and carbon dioxide (CO_2). ■ **oxidation-reduction reactions, p. 138**

A proportion of the organic matter produced is buried under anaerobic conditions, such as in mud or ocean sediments. Under these conditions, it is not decomposed and eventually forms fossil fuels such as coal, oil, and peat. Fossil fuels, the

Figure 30.10 Wood-Destroying Fungus Growing on a Dead Tree These fungi thrive in wet conditions. They digest lignin, the major cell wall component of woody plants, and consequently destroy the wood.

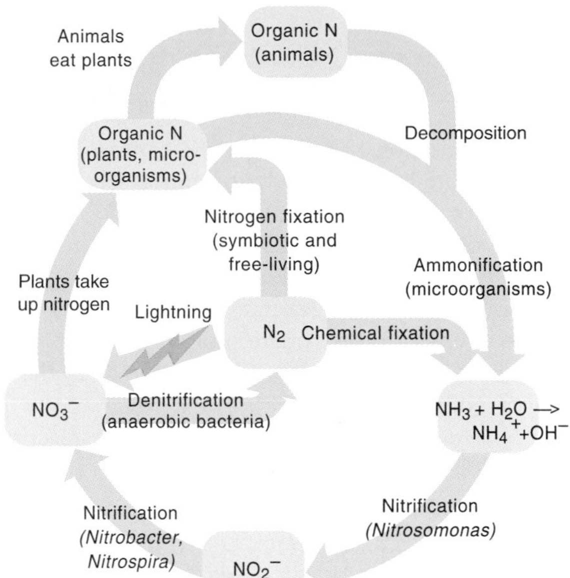

Figure 30.11 Nitrogen Cycle

biomass, and CO_2 dissolved in the ocean are examples of carbon reservoirs that keep CO_2 from the atmosphere. Burning fossil fuels and trees releases this carbon from the reservoirs and increases the amount of CO_2. Currently, more CO_2 is being released than can be used, and methane in the atmosphere is also increasing. Normal amounts of methane and CO_2 in the atmosphere help to maintain normal temperatures by absorbing infrared radiation and reflecting it back to earth. But increased burning of fossil fuels and trees and increased production of methane are contributing to a greenhouse effect, with gradual global warming that could have dire results.

Nitrogen Cycle

Next to carbon and oxygen, nitrogen is the element that organisms require in greatest quantity. Nitrogen is an important constituent of proteins and nucleic acids. Although the atmosphere consists of 79% nitrogen, relatively few organisms can use this gaseous form of the element. In most microorganisms and plants, inorganic nitrogen is taken up as nitrate (NO_3^-) or ammonium (NH_4^+) ions. These compounds of nitrogen are called **fixed nitrogen** because the nitrogen is combined with hydrogen (NH_4^+) or oxygen (NO_3^-). Usually any ammonia (NH_3) formed reacts with water to give NH_4^+ (ammonium ions) **(figure 30.11)**.

Fixed nitrogen is provided to soils and water mainly by the action of the few microorganisms that can convert nitrogen gas into these forms. Only prokaryotic cells have developed the ability to fix nitrogen. Alternatively, the fixation of molecular nitrogen can be carried out industrially by an energy-expensive chemical process that is used to make nitrogen fertilizers. Fertilizers produced chemically by humans are assuming an increasingly larger role in the nitrogen cycle. On a global scale, however, the bulk of the nitrogen added to soil is in the form of organic materials, largely as crop waste.

To be used by plants and many microorganisms, the organic nitrogen in these materials must be converted first to ammonium ions, a process known as **ammonification**, and then to nitrate, by **nitrification** (see figure 30.11). Nitrogen can be lost from the soil because some species of bacteria convert nitrate to gaseous nitrogen by using nitrate as a metabolic electron acceptor in place of oxygen. This process, another example of anaerobic respiration, is called **denitrification**. The result is a loss of soil fertility, since gaseous nitrogen is only available after it is fixed.

All forms of nitrogen undergo a number of transformations. The entire series of reactions involving nitrogen makes up the nitrogen cycle, and microorganisms are essential at each stage of this cycle. The cycle consists of several individual transformations of nitrogen, in which the nitrogen becomes progressively oxidized from its highly reduced state (NH_4^+) to its highly oxidized state (NO_3^-).

Ammonification

Almost all of the nitrogen found in the upper surface of the soil and in aquatic sediments exists in organic molecules—primarily purines and pyrimidines, the building blocks of nucleic acids, and amino acids that make up proteins.

A wide variety of microorganisms, including aerobic and anaerobic bacteria as well as fungi, can decompose protein. They do this initially through the action of extracellular proteolytic enzymes that break down proteins into chains of amino acids of varying lengths. Other enzymes then degrade these long chains into smaller chains, ammonium and sulfate ions (derived from the sulfur-containing amino acids), and water. During this process, called ammonification, some nitrogen may be released into the atmosphere as ammonia (NH_3, a gas). This is likely to happen only in alkaline environments, such as heavily limed soils, in which large amounts of organic nitrogen are metabolized.

Nitrification

The process that converts NH_4^+ to NO_3^- is called nitrification. During this process, the ammonium ion is oxidized to nitrite (NO_2^-) and then to nitrate (NO_3^-).

It might be helpful to consider the conversion of ammonium ion to nitrate in terms of the oxidation levels of the various elements involved. The ammonium ion (NH_4^+) is the most reduced form of nitrogen. As such, it can be oxidized by bacteria with the release of energy. Nitrification can be carried out only by members of a few genera. Most often, this oxidation occurs in two steps: (1) the oxidation of NH_4^+ to NO_2^-, carried out by members of the genus *Nitrosomonas*; and (2) the further oxidation of NO_2^- to NO_3^-, performed by a second group, including *Nitrospira* sp. and *Nitrobacter* sp. The reactions can be written as follows:

$$NH_4^+ + \tfrac{1}{2}O_2 \xrightarrow{\textit{Nitrosomonas}} NO_2^- + H_2O + 2H^+ + \text{energy}$$

$$NO_2^- + \tfrac{1}{2}O_2 \xrightarrow[\textit{Nitrobacter}]{\textit{Nitrospira}} NO_3^- + \text{energy}$$

As might be expected from looking at these reactions, they occur only under aerobic conditions, and therefore no nitrification takes place in waterlogged soils or in anaerobic regions of aquatic environments. The nitrifiers are generally autotrophs, utilizing CO_2 as their sole source of carbon. The energy required for the reduction of CO_2 to the constituents of cytoplasm is obtained from the oxidation of NH_4^+ to NO_2^- to NO_3^-.

Nitrite ion (NO_2^-) is highly toxic to plants, and it is therefore important that the organisms oxidizing NO_2^- to NO_3^- be in the same environment as those converting NH_4^+ to NO_2^-. Nitrate is soluble and can be leached from soil by rainwater. If it accumulates in groundwater, it can be dangerous to humans who drink the water. Nitrate itself is not highly toxic, but it can be converted to nitrite by intestinal bacteria. Nitrite is toxic for humans because it combines with hemoglobin of the blood, causing a reduction in its oxygen-carrying capacity.

Denitrification

Nitrate ions (NO_3^-) represent fully oxidized nitrogen. Certain facultatively anaerobic bacteria, such as some *Pseudomonas* sp., can use nitrate as a final electron acceptor under anaerobic conditions (anaerobic respiration). The net result of this process is the reduction of nitrate with the liberation of nitrogen gas and the net loss of nitrogen from the ecosystem. Under anaerobic conditions in wet soils, denitrifying bacteria may use the nitrates of expensive fertilizers, resulting in the loss of nitrogen as a gas into the atmosphere and consequent economic loss to the farmer. ■ **anaerobic respiration, p. 153**

Nitrogen Fixation

Biologically available nitrogen is continually being removed from the soil through denitrification; through the leaching action of water, which removes nitrate; and through the incorporation of nitrogen into plants, which are subsequently harvested. Usable nitrogen must, therefore, be continually added to the soil to maintain its nitrogen status. Growth in many ecosystems, including agricultural ecosystems, is limited by the availability of usable nitrogen.

The natural supply of nitrogen fixed in nongaseous compounds is limited, in turn limiting the capacity of world agriculture. None of the processes described thus far explains how nitrogen is added to the soil. One source is the small amounts of nitrogen, in the form of ammonium and nitrate ions, that occur in rainwater. Fertilization of soil also adds nitrogen. The combination of these two processes, however, usually does not compensate for the amount of nitrogen lost from the soil by denitrification. Nitrogen must, therefore, be gained in other ways, and nitrogen-fixing bacteria represent the most important means for this. No eukaryotic organisms have been shown to fix nitrogen without the aid of prokaryotes.

Nitrogen-fixing bacteria may be free-living, or they may live in symbiotic association with various plants. A variety of these bacteria can reduce nitrogen gas to the amino group ($—NH_2$) of the amino acids that make up proteins. This process, carried out by the enzyme **nitrogenase**, occurs in several steps. The nitrogen gas is first reduced to the ammonium ion. Since nitrogen gas (N_2) has a very stable triple bond, the overall reaction requires a tremendous expenditure of energy, about 15 molecules of ATP for every molecule of nitrogen fixed. The ammonium formed inside the cell is incorporated into amino acids by an enzymatic reaction.

Free-Living Nitrogen Fixers A variety of free-living nonsymbiotic microorganisms can fix nitrogen. Among these free-living forms of bacteria are members of the genus *Azotobacter*. These heterotrophic, aerobic, Gram-negative rods may be the chief suppliers of fixed nitrogen in grasslands and other ecosystems lacking plants with nitrogen-fixing symbionts. Azotobacters are rare in soils with a pH below 6.0. Members of the genus *Beijerinckia*, another genus containing heterotrophic nitrogen-fixing species, are found in highly acid soil. For reasons that are not clear, members of the genus *Beijerinckia* occur principally in the tropics and rarely in temperate zones. The dominant free-living, anaerobic, nitrogen-fixing organisms of the soil are members of the genus *Clostridium*, which have widespread distribution. Other nitrogen-fixing bacteria are found in genera that include photosynthetic, methane-oxidizing, sulfate-reducing, and sulfur-oxidizing organisms.

Symbiotic Nitrogen Fixers Although free-living bacteria are potentially capable of adding a considerable amount of fixed nitrogen to the soil, symbiotic nitrogen-fixing organisms are far more significant in benefiting plant growth and crop production. In aquatic environments, the most significant nitrogen fixers are cyanobacteria. They are especially important in flooded soils. Rice has been cultivated successfully for centuries without the addition of nitrogen-containing fertilizer because of the cyanobacterium *Anabaena* that grows on the aquatic fern *Azolla* in the water of paddy fields. Before planting rice, the farmer allows the flooded rice paddy to overgrow with *Azolla* ferns. As the rice grows, it eventually crowds out the fern, which dies and releases the nitrogen into the water. Sufficient fixed nitrogen is produced in this way to fertilize the rice plants, and the farmer does not need to add any chemical fertilizers.

The rhizobia, found in the three genera *Rhizobium*, *Bradyrhizobium*, and *Azorhizobium*, are the most agriculturally important symbiotic nitrogen-fixing bacteria. The plants with which these organisms associate are all leguminous plants and include alfalfa, clover, peas, beans, and peanuts. The input of soil nitrogen from the microbial symbionts of leguminous plants such as alfalfa may be roughly 10 times the annual rate of nitrogen fixation attainable by nonsymbiotic organisms in a natural ecosystem. The amount of nitrogen fixed by rhizobia depends primarily on two factors, the organism and the habitat (**table 30.1**).

Both the leguminous plants and the bacteria can be cultivated separately without each other; however, in soils lacking leguminous plants, rhizobia compete poorly with other bacteria and tend to disappear from the soil if leguminous plants are not grown in them for long periods of time. To encourage plant growth, people often plant the seeds of certain legumes together with inocula of the appropriate rhizobia. The symbiotic association of the bacteria with the plant results in the formation of nodules on the roots of the plant. Generally speaking, rhizobia fix nitrogen only when they are growing in the root nodules of the plants (**figure 30.12**). The plant sends chemical signals to *Rhizobium* bacteria, which then synthesize factors that activate plant genes necessary for the invasion of the root by the bacteria (see **Perspective 30.2**). Communication occurs between plants and bacteria, just as it does between bacteria and animal cells. ■ **bacterial invasion of human cells, p. 458**

In addition to legumes, several genera of nonleguminous trees, including alder and gingko, possess nitrogen-fixing root nodules at some stages in their life cycle. The microorganisms involved in symbiosis with the alder tree are actinomycetes of the genus *Frankia*.

TABLE 30.1 Quantities of Nitrogen Fixed by Microorganisms

Group	Species or Habitat	N_2 Fixed per Acre per Year (lbs)
Nodulated legumes	Alfalfa	113–297
	Soybean	57–105
	Red clover	75–171
Nodulated nonlegumes	Alder tree	200
Cyanobacteria	Arid soil in Australia	3
	Paddy field in India	30
Free-living heterotrophs	Soil under wheat	14
	Soil under grass	22
	Rain forest in Nigeria	65

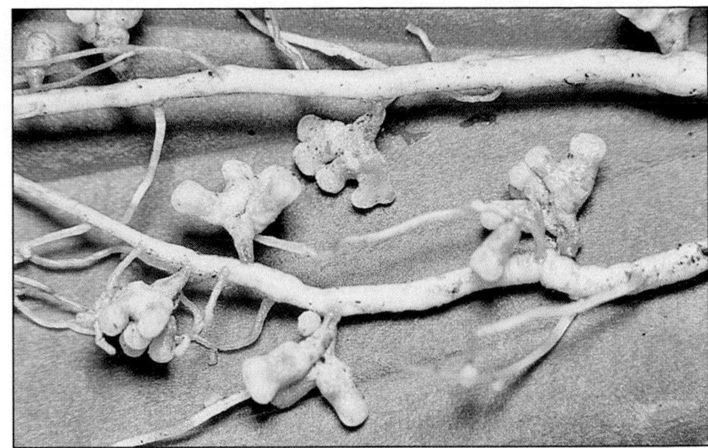

Figure 30.12 **Nitrogen-Fixing Bacterium *Rhizobium* on Clover Roots**

Sulfur Cycle

Sulfur also occurs in all living matter, chiefly as a component of the amino acids methionine, cystine, and cysteine. The sulfur cycle bears many similarities to the nitrogen cycle (**figure 30.13**).

Like nitrogen, sulfur is present in the soil chiefly as a part of proteins, and it is taken up by plants in its oxidized form, sulfate (SO_4^{2-}). The sulfur-containing proteins are first degraded into their constituent amino acids by proteolytic enzymes secreted by a large variety of soil organisms. The sulfur of the amino acids is generally converted to hydrogen sulfide (H_2S) by a variety of soil microorganisms, a process analogous to NH_4^+ formation. The H_2S formed would be highly toxic to many biological systems if it were allowed to accumulate. Under aerobic conditions, however, H_2S oxidizes spontaneously to sulfur and is then converted to its most readily utilized form, sulfate, by

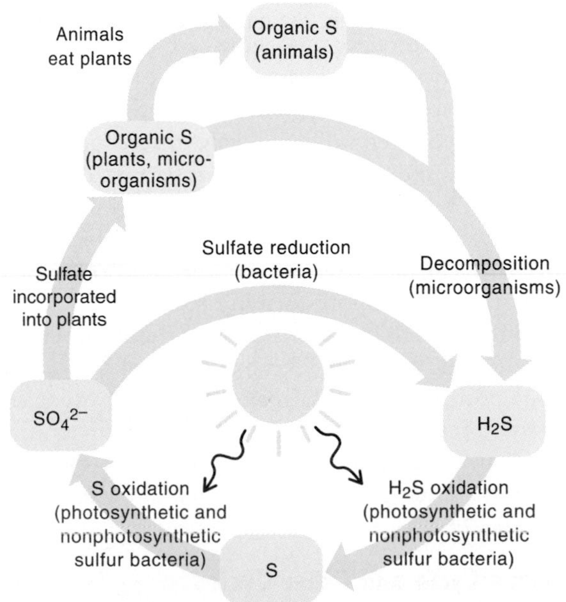

Figure 30.13 **Sulfur Cycle**

Perspective 30.2 How Roots Are Infected with Rhizobia

The infection process of legumes by rhizobia consists of several stages. First, the bacteria recognize and attach to specific molecules on the root hairs of the legume. Some of these bacteria have a broad host range, and others can nodulate only one or a few host plant species. Next, the root hair curls and forms an infection thread through which the bacteria invade. Once inside the plant, the bacteria change in appearance to a form called **bacteroids** and begin to fix nitrogen. Finally, the repeated division of both plant cells and bacteroids leads to the formation of complete root nodules **(figure 1)**.

The genes that code for these activities are distributed among the plant, the bacterial chromosome, and a large bacterial plasmid called Sym. The genes for nodulation, the **nod genes**, are largely on the Sym plasmids. Products of the nod genes are called **NOD factors**. Also on the plasmid are some genes that account for the host specificity. Some NOD factors, secreted by the bacteria in the vicinity of the root hairs, signal the plant to switch on the nodulation process. NOD factors cause changes in the host root hairs, with branching and curling of the hairs and development of the cellulose infection threads. A special oxygen carrier called **leghemoglobin** is coded for by plant genes.

Nitrogen fixation is carried out by the rhizobia in the microaerobic environment of the nodule. Leghemoglobin regulates oxygen concentration in the nodule by binding to oxygen to keep free oxygen at a very low level and maintain anaerobic conditions. The fixed nitrogen then diffuses out of the cells into the root environment, where it is used by the plant for the synthesis of proteins.

Nodules are not found on all species of legumes, and not all species of rhizobia can nodulate leguminous plants. Moreover, the ability to develop a symbiotic relationship resulting in nitrogen fixation varies markedly among strains of rhizobia. Transfer of the Sym plasmid can sometimes transfer host specificity.

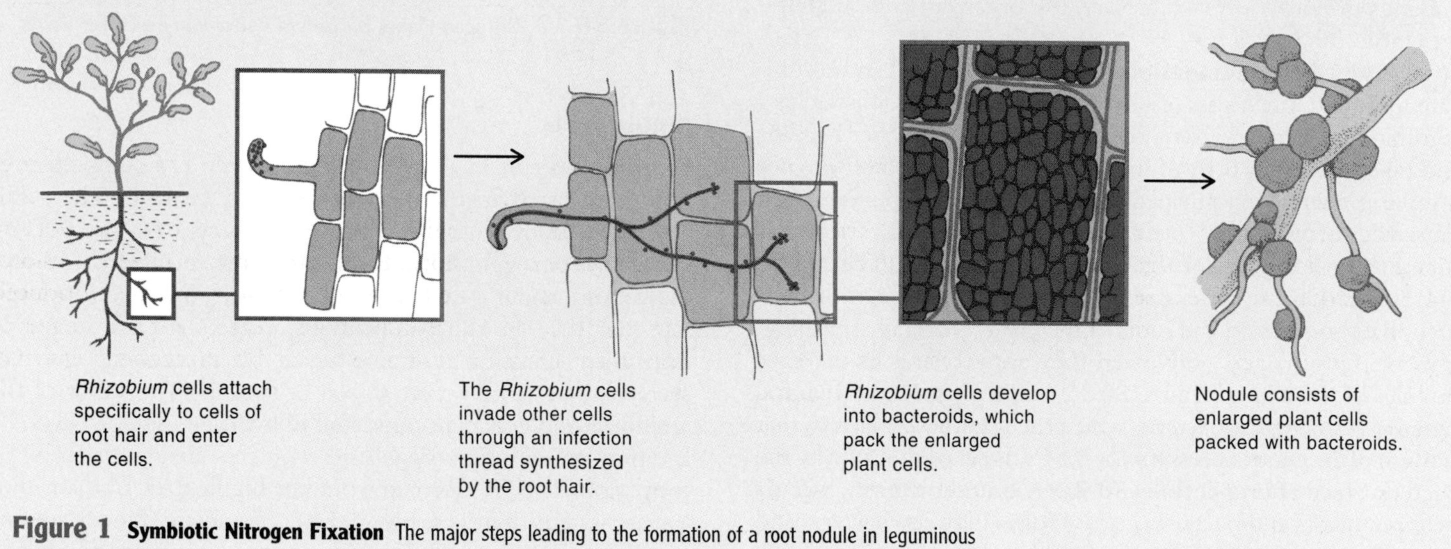

Rhizobium cells attach specifically to cells of root hair and enter the cells.
The *Rhizobium* cells invade other cells through an infection thread synthesized by the root hair.
Rhizobium cells develop into bacteroids, which pack the enlarged plant cells.
Nodule consists of enlarged plant cells packed with bacteroids.

Figure 1 **Symbiotic Nitrogen Fixation** The major steps leading to the formation of a root nodule in leguminous plants by *Rhizobium*.

sulfur bacteria, a process analogous to nitrification. Under anaerobic conditions, the counterparts of the denitrifying organisms operate—these sulfate-reducing bacteria reduce SO_4^{2-} to H_2S. Hydrogen sulfide is likely to accumulate in waterlogged soils. The reduction of sulfate is carried out by a small group of anaerobic bacteria that are capable of utilizing sulfate as the final electron acceptor in their anaerobic respiration. These anaerobes include organisms of the genus *Desulfovibrio* as well as a few anaerobic spore-forming rods.

The microbial flora concerned with the sulfur cycle have a number of unique features. The oxidation of H_2S to SO_4^{2-} is carried out principally by two major groups of organisms, the nonphotosynthetic autotrophs, members of the genera *Thiobacillus* and *Beggiatoa*, and less commonly, by the photosynthetic autotrophic green and purple sulfur bacteria. ■ **sulfur bacteria, p. 272, 276, 279, 293**

Phosphorus Cycle and Other Cycles

Phosphorus occurs in the soil in both organic and inorganic forms. Organic phosphorus-containing compounds occur almost exclusively near the surface. They are derived mostly from surface vegetation and are accordingly found as constituents of plant tissue, animals, and microorganisms, for example, in nucleic acids. The inorganic forms of phosphorus occur largely as insoluble salts of calcium, iron, and aluminum. Plants generally cannot use these inorganic and organic forms readily, but orthophosphate (PO_4^{3-}) is readily used by most plants and microorganisms, being part of such biologically important molecules as ATP and nucleic acid.

In many aquatic habitats, growth of algae and cyanobacteria, the primary producers, is often limited because concentrations of phosphorus are too low. Addition of phosphates to such habitats from agricultural fields or overuse of phosphate-containing detergents can result in algal blooms and deterioration of water quality, called **eutrophication**. **Figure 30.14** shows a polluted river in which eutrophication has occurred. Fertilizers rich in nitrates and phosphates were washed out of farmland by rain and carried into the river, where they stimulated overgrowth of algae and weeds. The overabundant vegetation produced by eutrophication blocks out light and, as it decays, exhausts the oxygen supply, allowing growth of anaerobic bacteria that produce obnoxious odors.

Figure 30.14 **Eutrophication in a Polluted River**

Microorganisms play three major roles in phosphorus transformation: They convert organic forms of phosphorus to inorganic forms; they convert inorganic to organic forms; and they convert insoluble inorganic phosphates in minerals to soluble forms that can be utilized.

Bacteria and fungi are largely responsible for making organic phosphorus available to microorganisms as well as to plants. Most microorganisms do this by synthesizing the enzyme **phosphatase**, which separates the organic phosphate from a wide variety of compounds in which it is found. The breakdown of phosphate-containing nucleic acids is initiated by the action of nucleases, which degrade DNA and RNA into individual nucleotides. The phosphate is then separated from these nucleotides by phosphatase to form nucleosides:

$$\text{nucleotide} \xrightarrow{\text{Phosphatase}} \text{nucleoside} + PO_4^{3-}$$

Phosphate, like nitrogen, is converted by soil organisms from organic to the inorganic forms, which can than be changed by bacteria into their own organic forms of phosphorus. Also, some minerals containing phosphate may be converted to orthophosphate by microorganisms. A wide variety of commonly occurring soil organisms can dissolve calcium phosphate salts. The amount of phosphate that dissolves depends primarily on the amount of acid these organisms produce in their metabolic processes.

Important elements, including iron, calcium, zinc, manganese, cobalt, and mercury, are also recycled by microorganisms. Many bacteria contain plasmids coding for enzymes that carry out oxidation of metallic ions. Bacteria also produce special iron transport compounds to bring iron into their cells, where it can be recycled.

MICROCHECK 30.5

Microorganisms are essential in biogeochemical cycling of biologically important elements such as oxygen, carbon, nitrogen, sulfur, and phosphorus, among others. A vital facet of the nitrogen cycle is the ability of a number of both free-living and symbiotic organisms to convert atmospheric nitrogen to biologically useful forms.

- What are the roles of producers, consumers, and decomposers, and which of these groups is nonessential?
- Compare the sulfur cycle with the nitrogen cycle.
- Could recycling occur in populations of bacteria living deep underground? How?
- Why is hydrogen sulfide likely to accumulate in waterlogged soils?

Bioremediation: The Biological Cleanup of Pollutants

Bioremediation is the biological cleanup of **pollutants**, substances that are harmful or injurious, in a given environment. In bioremediation, biological systems are used to degrade toxic substances or to transform them into harmless materials. Bioremediation may involve the use of special organisms introduced into the polluted environment, or it may take advantage of organisms already present, possibly adding nutrients to encourage their growth. It is essential to remove the harmful materials without producing new injurious substances.

Pollutants

Traditionally, pollutants from domestic and industrial wastes have been dumped into the environment, with the hope that they will be degraded and removed through natural recycling. Most organic compounds of natural origin can be degraded by one or more species of soil or aquatic organisms, although some natural materials can cause huge problems, as when oil spills occur in the ocean. The degradation of most naturally occurring compounds by microorganisms led some people to believe that organic compounds would not persist in the environment. This, however, is not the case with a large number of chemicals synthesized industrially. People have increased the number of slowly degraded or non-biodegradable compounds in the soil and in aquatic systems of many parts of the world by adding thousands of tons of pesticides, herbicides, nonbiodegradable detergents, and plastics. Although some synthetic compounds disappear from an ecosystem within the reasonably short time of a few days or weeks, many remain undegraded for years. This is especially true when large amounts of pollutants, as from a pulp mill or oil spill, are concentrated in a new environment. Chemically synthesized compounds are most likely to be biodegradable if they have a chemical composition similar to that of naturally occurring compounds. If a synthetic chemical compound is totally different from any that occurs in nature, organisms are less likely to have the enzymes necessary for degrading it rapidly, and the compound is likely to persist for long periods.

Many factors influence the rate of degradation of compounds by the soil population. As a general rule, any practice that favors the multiplication of microorganisms will increase the rate of degradation. Thus, raising the temperature, maintaining the pH near neutrality, and providing an optimal amount of moisture are all likely to increase the rate of degradation of most materials added to the soil.

Relatively slight molecular changes markedly alter the biodegradability of a compound. Specific alterations of a compound, such as the addition of a chlorine or other halogen atom, often result in the need for totally new mechanisms to degrade it. Perhaps the best-studied example involves the herbicides 2,4-dichlorophenoxyacetic acid (2,4-D) and 2,4,5-trichlorophenoxyacetic acid (2,4,5-T). The only difference between these two compounds is the additional chlorine atom on 2,4,5-T. When 2,4-D is applied to the soil, it completely disappears within a period of several weeks, as a result of its degradation by a segment of the microbial population in the soil. When 2,4,5-T is applied, however, it is often still present more than a year later **(figure 30.15)**. Its persistence is apparently due to the third chlorine atom, which blocks the enzyme that makes the initial attack on 2,4-D.

Most herbicides and insecticides not only are toxic to their target weeds or insects, but also have far-ranging deleterious effects on fish, birds, and other animals. DDT is an example of such a compound. This non-biodegradable pesticide accumulates in the fat of predatory birds. **Table 30.2** illustrates the phenomenon of biological magnification that occurs. Small amounts of the pollutant in water are concentrated in minute plankton, which are eaten by minnows, accumulating even more in the fish. When large birds, such as herons or ducks, in turn eat the fish, the amount of chemical in the tissues is tremendously magnified. The continuing ingestion of DDT, which accumulates in fat, results in an ever greater concentration of the DDT as it is passed upward through the food chain. Ducks have large amounts of fat in which to accumulate the chemical.

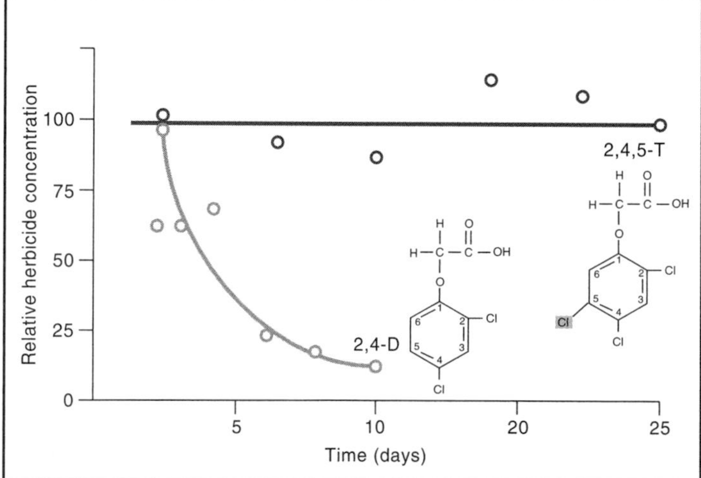

Figure 30.15 Comparison of the Rates of Disappearance of Two Structurally Related Herbicides, 2,4-D, and 2,4,5-T The addition of a third chlorine atom in the 2,4,5-T molecule blocks the enzyme that degrades the substance, so that compound remains in the environment.

TABLE 30.2 Biological Magnification of DDT

Parts per Million DDT	Source
0.00005	Water
0.04	Plankton
0.23	Minnow
3.57	Heron
22.8	Merganser (fish-eating duck)

DDT, as well as other chlorinated hydrocarbon insecticides, also interferes with the reproductive process of birds, leading to the production of fragile eggs, which break before the young can hatch. Furthermore, DDT and other chlorinated hydrocarbons are carcinogenic. Although banned in the United States, DDT is still heavily used in other countries, where its effects can be far-reaching. For example, DDT use in South America caused a depletion of predatory birds that normally summer in Greenland. The result was an overpopulation of Greenland's rodents. Thus, although a biodegradable compound must be applied repeatedly to maintain its effectiveness, it is preferred over a toxic, non-biodegradable one that may have widespread and long-term effects.

Another class of dangerous compounds is the polychlorinated biphenyls (PCBs), which have been used in many manufacturing operations. These molecules are not readily biodegradable, and their concentration in the environment increased steadily for several years. In 1968, cooking oil contaminated with PCBs poisoned 1,000 people in Japan. PCBs have also been blamed for reproductive failure in birds and in mink and are carcinogenic. The use of PCBs was banned in the United States in 1978, but even so, one-third of the United States population is estimated to contain PCBs in concentrations greater than 1 part per million in their tissues.

It seems unlikely that organic chemical compounds can be created that will selectively affect only certain undesirable groups of plants or insects. Therefore, greater emphasis is being placed on developing and using synthetic materials, including pesticides, detergents, and plastics, that can be broken down in the natural environment.

Means of Bioremediation

The problem of toxic chemicals in natural environments has prompted a great deal of research into the development of rapid means to degrade the polluting compounds. One goal of such research is developing strains of microorganisms that can be used in biological pollution control, either by large-scale inoculation of polluted areas or by introduction of the necessary degradative capabilities into the natural population of a given area. Pseudomonad strains capable of growth on 2,4,5-T have been used successfully to remove this chemical from soil samples in the laboratory, and dual cultures of an *Acinetobacter* and a pseudomonad have been shown to degrade PCBs.

Figure 30.16 Oil Spill Bioremediation Bioremediation was used to clean the shoreline contaminated by oil spilled from the *Exxon Valdez*. The growth of indigenous oil-degrading bacteria was stimulated by addition of nutrients. Comparing the uncleaned rocks on the left to the rocks cleaned by bioremediation on right shows the dramatic efficiency of bioremediation.

Bioremediation in soil and aquatic environments is usually carried out by organisms attached to particles. Petroleum hydrocarbons and many other pollutants adsorb firmly to the particles, allowing contact and degradation by the attached organisms (**figure 30.16**). All of the petroleum-degrading bacteria are limited in their growth by the low concentrations of nitrogen and phosphorus in seawater. The development of a nitrogen- and phosphorus-containing fertilizer that adheres to the spilled oil has greatly improved the results of bacterial treatment of oil spills. Use of the fertilizer has led to better growth of the bacteria and thereby increased the effectiveness of degradation by the bacteria at least threefold. Bioremediation is becoming ever more important in maintaining a healthy environment, and many areas of research are devoted to its development.

MICROCHECK 30.6

Bioremediation is the biological cleanup of pollutants. It may involve the use of specially selected organisms introduced into the polluted habitat, or it may use organisms already present, perhaps with added nutrients to encourage their activities. Much research is directed toward developing new methods of bioremediation.

- Give several reasons that pollutants may persist in the environment.
- What is meant by biological magnification?
- How would adding a third chlorine atom to 2,4-D (see figure 30.15) block the enzyme attack? (Hint: Explain in terms of enzyme and substrate structure.)

FUTURE CHALLENGES

Healing the Past, Designing the Future

Microorganisms have kept the world going for eons by recycling essential materials, yet we still have much to learn about how to take advantage of some of these myriad microbial capacities. Many of the bacteria that have been isolated by virtue of their ability to grow on toxic chemicals contain plasmids with the genes required for degrading these compounds. Therefore, it should be possible either by selective breeding or by using recombinant DNA techniques to combine several characteristics on one plasmid and produce a strain with multiple degradative capacities. With this method, a pseudomonad that rapidly degrades certain fractions of petroleum has been bred. Although naturally occurring microorganisms are responsible for removing much of the oil that is regularly spilled into the oceans, the use of genetically selected organisms can help speed the recovery from ecological accidents and disasters. Currently, federal laws generally prohibit the use of genetically engineered microorganisms in the environment. The many ethical and legal issues surrounding release of bioengineered organisms must be addressed.

Much of what we know about microorganisms has been learned in laboratory settings, where conditions can be controlled and competition between organisms is not the same as in natural environments. Particularly in the field of bioremediation, efforts are under way to carry out field trials, studying populations and the persistence of organisms that can remediate. If populations of these organisms are increased by fertilization of the site, will they persist or become dispersed? Are there unwanted consequences of this increase of one portion of the microbial community? How can these populations be measured accurately? What genes are involved in the desired microbial activities, and can these genes be amplified in the microorganisms? If they are, can the organisms still compete successfully? These and many other questions offer great challenges, but hold out the hope that their answers will make bioremediation an increasingly efficient way to deal with pollution of the environment.

SUMMARY

Principles of Microbial Ecology

1. Microorganisms are found throughout the biosphere. The best adapted organism takes over its environment.
2. Within the **biosphere**, **ecosystems** vary in their **biodiversity** and **biomass**.
3. The **microenvironment** immediately surrounding a microorganism is most relevant to its survival and growth.
4. Microorganisms can change their environments and can adapt to environmental change.

Bacteria in Low-Nutrient Environments

1. Low-nutrient environments are common in nature, and most organisms in such environments grow in biofilms. **(Figure 30.1)**
2. Organisms in low-nutrient environments transport nutrients into their cells very efficiently.

Microbial Competition and Antagonism

1. Microbial competition demands rapid reproduction and efficient nutrient use. **(Figure 30.2)**

2. Antagonism helps determine the make-up of a community.

Microorganisms and Environmental Changes

1. Microbial populations both cause and adapt to environmental changes. (Figure 30.4)
2. Mutants may be selected or enzymes may be induced to allow microorganisms to adapt to a new environment. (Figure 30.3)
3. Growth and metabolism of organisms may change environmental conditions significantly.

Terrestrial Environments

1. The soil teems with a broad diversity of organisms that are essential for modifying, degrading, and producing biologically important substances.
2. Environmental influences such as moisture, pH, temperature, and nutrient supply affect the numbers and kinds of organisms in soil.
3. Some soil microorganisms are pathogens of plants, animals, or people.

Microorganisms and Soil

1. Bacteria called **actinomycetes** are the most common soil bacteria. They can produce antibiotics and **geosmins**, and are essential in biogeochemical cycling.
2. Soil fungi may be free-living or they may be symbiotic in **mycorrhizae** or **lichens**. They are important in decomposing plant matter.
3. The algae in soil serve as nutrients for fungi, protozoa, and worms.
4. Protozoa are consumers of soil bacteria and algae. Together with termites, they decompose wood.

Environmental Influences in Soil

1. Important environmental influences in soil include moisture, acidity, temperature, and nutrient availability.

Aquatic Environments

1. Virtually all water contains living organisms, which vary greatly with different aquatic environments, from fresh water to seawater and some extreme habitats, such as salt lakes.
2. **Primary producers** in aquatic environments convert inorganic substances such as CO_2 to organic matter by photosynthesis.

Energy Sources for Ecosystems

1. Energy for ecosystems can come from sunlight via **photosynthesis** or from chemical synthesis of inorganic and organic materials by **chemoautotrophic** microorganisms.

Biogeochemical Cycling

1. Microorganisms are essential in biogeochemical cycling of biologically important elements such as oxygen, carbon, nitrogen, sulfur, and phosphorus, among others. (Figure 30.7)
2. Recycling processes require the activities of **producers**, **consumers**, and **decomposers**.

Oxygen Cycle (Figure 30.8)

1. Oxygen is cycled by the processes of photosynthesis and respiration.

Carbon Cycle (Figure 30.9)

1. The carbon cycle revolves around CO_2, its fixation into organic compounds by primary producers, and its regeneration, mostly by microorganisms.

Nitrogen Cycle (Figure 30.11)

1. Processes fueling the nitrogen cycle include **ammonification**, **nitrification**, **denitrification**, and **nitrogen fixation** by free-living and symbiotic nitrogen fixers. (Table 30.1)
2. A vital facet of the nitrogen cycle is the ability of microorganisms to convert atmospheric nitrogen to biologically useful forms.

Sulfur Cycle (Figure 30.13)

1. The sulfur cycle bears many resemblances to the nitrogen cycle.

Phosphorus Cycle and Other Cycles

1. Bacteria and fungi are largely responsible for making organic phosphorus available through the action of the enzyme phosphatase.

Bioremediation: The Biological Cleanup of Pollutants

1. Bioremediation is the biological cleanup of pollutants. It may involve the use of specially selected organisms introduced into the polluted habitat, or it may use organisms already present, perhaps with added nutrients to encourage their activities.

Pollutants

1. Biodegradable pollutants are removed within a relatively short time, but many synthetic compounds remain in the environment for long periods of time, or indefinitely. (Figure 30.15)
2. Biological magnification occurs when compounds are taken up by sequential members of the food chain, thereby concentrating large amounts of the polluting compound in higher levels in the food chain. (Table 30.2)

Means of Bioremediation

1. Organisms already in the environment or specially selected organisms are used to degrade and/or recycle materials in pollutants. Sometimes, nutrients are added to encourage growth of the organisms. (Figure 30.16)

REVIEW QUESTIONS

Short Answer

1. Why are microorganisms well suited to recycle elements?
2. How is the decomposition of organic matter achieved?
3. List several functions of fungi in soil.
4. How are proteins decomposed in natural environments?
5. What is the importance of nitrogen fixation?
6. List at least four genera of soil microorganisms that are pathogenic for humans and note whether each genus is bacterial, fungal, or something else.
7. Why is there a high concentration of microbes in the rhizosphere of plants?
8. Contrast ecosystems supported by photosynthesis with those that depend on chemoautotrophy.
9. What are some differences between warm sea vents and black smokers, and how do these differences affect the microbial flora found in each location?
10. How can apparently barren basalt rocks deep under the surface of the earth support microbial growth?
11. Give examples of free-living and symbiotic nitrogen-fixing microorganisms. Are these prokaryotic or eukaryotic?
12. Outline the symbiotic relationship between rhizobia and leguminous plants.
13. Describe the use of bioremediation in the cleanup of oil spills.

Multiple Choice

1. Atmospheric nitrogen can be used…
   A. directly by all living organisms.
   B. only by aerobic bacteria.
   C. only by anaerobic bacteria.
   D. in symbiotic relationships between rhizobia and plants.
   E. in photosynthesis.
2. Extremophiles…
   A. are found in the polar ice caps.
   B. exist under great atmospheric pressure in the ocean floor.
   C. live in salt lakes.
   D. live within basalt rocks.
   E. All of the above
3. Mycorrhizae represent associations between plant roots and microorganisms that…
   A. are antagonistic.
   B. help plants take up phosphorus and other nutrients from soil.
   C. involve algae in the association with plant roots.
   D. form nodules on the plant's leaves.
   E. lead to the production of antibiotics.
4. The decomposition of organic matter…
   A. is carried out by only a few bacterial species.
   B. causes the production of oxygen.
   C. involves all the biogeochemical cycles discussed.
   D. involves photosynthesis.
   E. is largely symbiotic.
5. In symbiotic nitrogen fixation by rhizobia and legumes…
   A. the amount of nitrogen fixed is much greater than by nonsymbiotic organisms.
   B. neither the bacteria nor the legume can exist independently.
   C. the bacteria enter the leaves of the legumes.
   D. bacteroids are found in the leaves of the legumes.
   E. the bacteria operate independently of the legume.
6. To compete successfully, microorganisms must…
   A. produce endospores.
   B. reproduce more rapidly than their competitors.
   C. be aerobic.
   D. be anaerobic.
   E. be symbiotic.
7. Individual microorganisms are most affected by…
   A. the amount of oxygen in the region.
   B. sunlight.
   C. their microenvironment.
   D. the temperature.
   E. the pH.
8. Energy for ecosystems can come from…
   A. sunlight via photosynthesis.
   B. chemical synthesis by chemoautotrophs.
   C. Both A and B
   D. Only A
   E. Only B
9. Chemically synthesized compounds are most likely to be biodegradable if they…
   A. are totally different from anything found in nature.
   B. have three chlorine atoms per molecule.
   C. are plastics.
   D. are present in very large amounts.
   E. are chemically similar to naturally occurring substances.
10. Numbers of living organisms in an environment can be estimated by all of the following techniques, *except*…
    A. measurement of ATP in the sample.
    B. nucleic acid probes.
    C. Gram staining.

D. staining with dyes that only stain living cells.

E. A and D

Applications

1. A farmer who was growing soybeans, a type of legume, saw an Internet site advertising an agricultural product for safely killing soil bacteria. The ad claimed that soil bacteria were responsible for most crop losses. The farmer called the agricultural extension office at a local university for advice. Explain what the extension office crop adviser most likely told the farmer about the usefulness of the product.
2. Recent reports suggest that human activities, such as the generous use of nitrogen fertilizers, have doubled the rate at which elemental nitrogen is fixed, raising concerns of environmental overload of nitrogen. What problems could arise from too much fixed nitrogen, and what could be done about this situation?
3. Nearly every summer, a huge area of oxygen-depleted, almost lifeless ocean spreads off the coast of Louisiana into the Gulf of Mexico. Scientists claim that this "dead zone" is the result of nitrogen- and phosphorus-containing fertilizer used in farmland along the Mississippi River. Fertilizer washes down the river, into the Gulf of Mexico, and nourishes the increased growth of algae and tiny organisms that feed on the algae. Two questions can be asked about this effect:
   a. How can increased algae and other organisms deplete oxygen from the surface layers of the water?
   b. Algae live only near the surface, where sunlight is available. Why would oxygen be depleted far below the surface?

Critical Thinking

1. A student argued that if soil particles were all the same size and all the same composition, then only one kind of microorganism would be found in the soil. What was the student's probable reasoning? How would you criticize the student's argument?
2. Large populations of bacteria are found living almost 3 km underground. Most of these are autotrophs and derive their nutrients and energy from inorganic chemicals in the immediate environment. Surprisingly, many other bacterial species are found that require organic material as a nutrient and energy source. Since organic material will not be carried from the surface to such depths, where can these bacteria obtain the necessary organic material?
3. Each colony of microorganisms growing on an agar plate arises from a single cell (see photo). Colonies growing close together are much smaller than those that are well separated. Why would this be so?

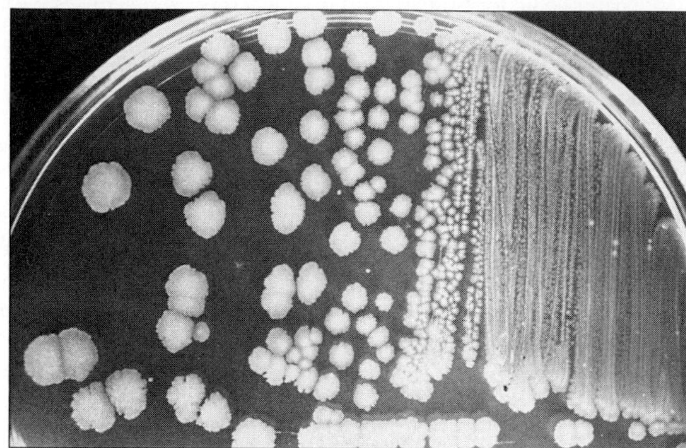

4. An entrepreneur found an economically feasible way of collecting large amounts of sulfur from underwater hot vents in the Pacific Ocean. The sulfur will be harvested from the microorganisms found in the vent areas. A group of ecologists argued that the project would destroy the fragile ecosystem by depleting it of usable sulfur. The entrepreneur argued that the environment would not be harmed because the vents produce an unlimited source of sulfur for the clams and tube worms in the area. Explain who is correct in their assessment.
5. Describe the kind of life forms one would expect to find on Mars or the moon, given what we know about conditions there.

Water and Waste Treatment

Delivering fresh water to urban areas and removing human wastes have been practiced at least since Roman times. Ruins of aqueducts used to deliver fresh water long distances can be seen today in many parts of Europe. Ridding cities of human wastes has been more difficult, and the sewers that were used until the mid-Nineteenth century were not much more than large, open cesspools.

Long before the discovery of the microbial world, it was recognized that some diseases are associated with water supplies. As early as 330 B.C., Alexander the Great had his armies boil their drinking water, a habit that probably contributed to his huge successes. Certainly, many battles have been lost over the years as a result of waterborne diseases that decimated the combatants. Years before the cholera-causing Vibrio cholerae *was identified, it was obvious that cholera epidemics were associated with drinking water. The desire for clean, clear water led to the use of a sand filtration system in London and elsewhere in the early Nineteenth century. Late in that century, Robert Koch showed that not only did this kind of filtration yield clear water, it also removed more than 98% of bacteria from the water.*

As early as the 1840s, Edwin Chadwick, an English activist, championed a new idea on how wastes could be removed. His idea was to construct a system of narrow, smooth ceramic pipes through which water could be flushed along with solid waste materials. This system would carry the waste materials away from the inhabited part of the city to a distant collection site. There, he hoped to collect the waste materials and turn them into fertilizer to sell to farmers. The system he envisioned required the installation of new water and sewer pipes along with pumps to deliver water under pressure to houses. With the water under pressure and smooth narrow pipes, the system could be kept well flushed.

In 1848, with the threat of a cholera epidemic imminent, the Board of Health in England instituted widespread reforms and began the installation of a sewage system along the lines envisioned by Chadwick. New York City did not establish its Board of Health and a proper sewage disposal system until 1866, again in response to a threatened cholera epidemic. By the end of the Nineteenth century, most large European and United States cities had established water-sewer systems to deliver safe drinking water and remove and treat waste materials. Cholera in the industrialized nations of Europe and North America virtually disappeared.

—***A Glimpse of History***

IN THIS CHAPTER WE WILL IDENTIFY SOME OF THE microorganisms and chemicals that cause water pollution, and discuss some ways that water can be purified and tested for safety. Even the clearest water is not completely "pure," because it contains dissolved substances and likely some microorganisms. The term **potable water** refers to water that is safe to drink but often far from pure in the chemical or microbiological sense. Also, we will consider ways in which microorganisms are essential in breaking down and recycling wastes.

Water Pollution by Microorganisms and Chemicals

In 1998 the Centers for Disease Control and Prevention (CDC) reported a disease outbreak among the hundreds of athletes that had participated in triathlons held in Wisconsin and Illinois. The triathlon competition included biking, running, and a 1.5-mile lake swim. Days after competing, three of the athletes were hospitalized with fever and acute illness. Leptospirosis, a waterborne disease, was suspected. Extensive investigation of these and other athletes who had participated in the event identified 74 (12%) of the participants with evidence of leptospirosis, and as a result the lake in Illinois where the event was held was closed to swimming and other water sports for a period of time. Episodes such as this, resulting from water pollution, are common. ■ leptospirosis, p. 636

Pollution of water occurs from a variety of sources (**figure 31.1**). Contamination of water with pathogenic organisms remains a major cause of disease epidemics. The CDC monitors outbreaks of waterborne diseases, including those associated with both drinking water and recreational water, such as swimming pools, whirlpools, hot tubs, water parks, rivers, ponds, and lakes. Relatively few cases of contamination are associated with chemical contaminants, such as lead, copper, fluoride, or nitrate. The remainder are caused by known organisms or by microbial infection with unidentified organisms. **Table 31.1** lists a large variety of waterborne organisms that cause infections or intoxications of humans, along with their acute and chronic health effects.

During 1995 and 1996 alone almost 12,000 cases of waterborne diseases were reported in 59 outbreaks in the United States. Most of these cases (9,129) were associated with recreational water; the remainder were attributed to drinking water. The largest outbreak from drinking water, 1,449 cases, was caused by the flagellate *Giardia lamblia*. Other organisms causing drinking water outbreaks were *Shigella sonnei*, *Escherichia coli* O157:H7, and a small round virus that was not identified. Well over a third of the drinking water outbreaks (comprising 27% of the cases) were acute gastrointestinal illness of unknown cause (AGI), with characteristics of an infectious disease (**figure 31.2**). In illness associated with recreational water, the most common causative agent was *Cryptosporidium parvum*, accounting for 8,512 cases of gastroenteritis. Almost all of these cases were from two outbreaks associated with water parks. Several small outbreaks of gastroenteritis were caused by *E. coli* O157:H7 in lake water and a swimming pool (figure 31.2). There were 48 cases of dermatitis caused by *Pseudomonas aeruginosa* from hot tubs, and 121 cases of dermatitis caused by schistosomes from lake water. The only reported fatalities from waterborne diseases within this 2-year period were six separate

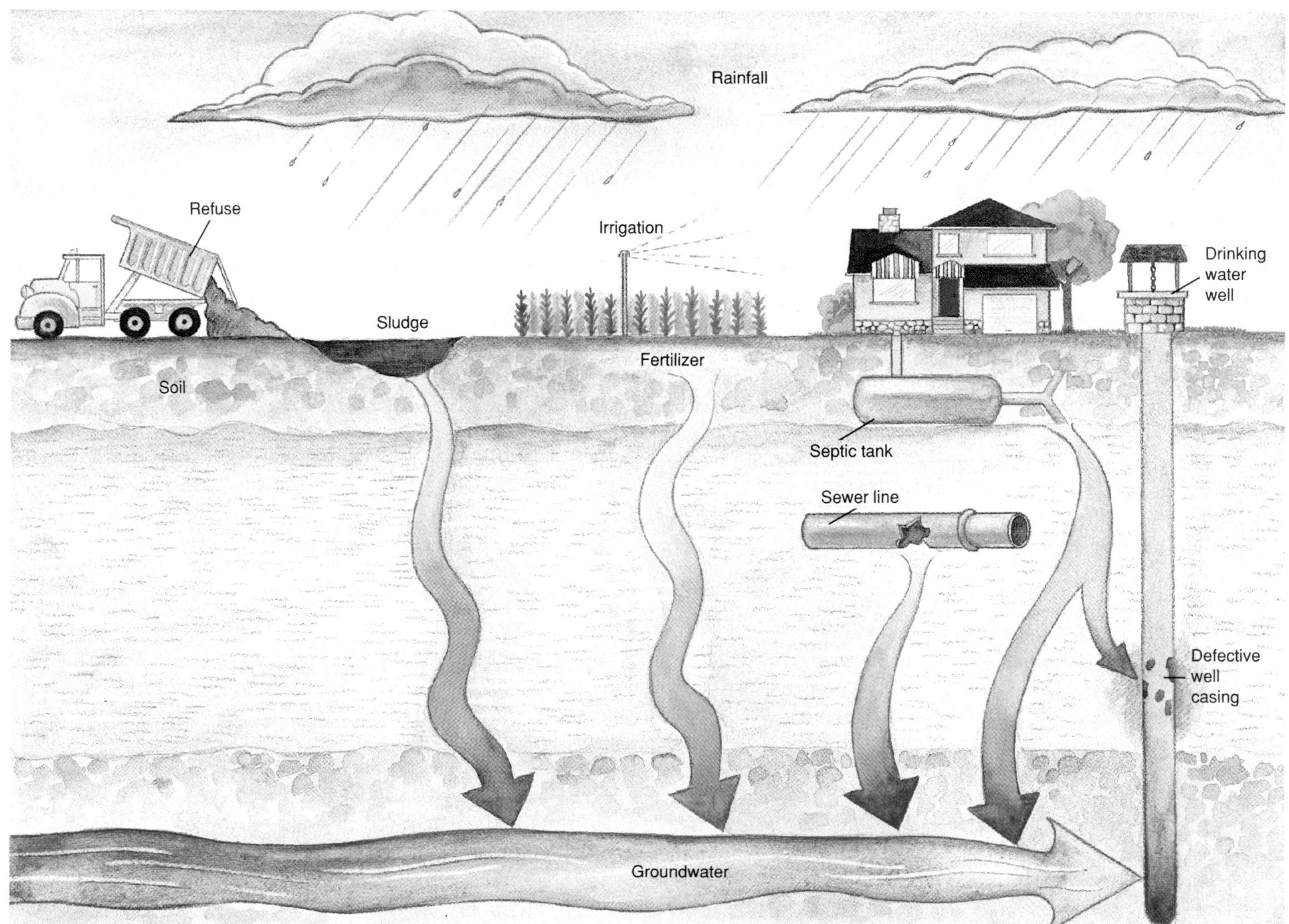

Figure 31.1 Groundwater Environment and Relationships Between Pollution Sources and Drinking Water Quality Some important soil factors influencing the fate of chemical and microbiological contaminants in water are texture, organics (humic acids), ionic strength, adsorption, structure, pH, and permeability.

TABLE 31.1 Acute and Chronic Health Effects Associated with Waterborne Microorganisms

	Agent	Acute Effects	Chronic or Ultimate Effects
Bacteria	*E. coli* O157:H7	Diarrhea	Adults: death (thrombocytopenia) Children: death (kidney failure)
	Legionella pneumoniae	Fever, pneumonia	Elderly: death
	Helicobacter pylori	Gastritis	Ulcers and stomach cancer
	Vibrio cholerae	Diarrhea	Death
	Vibrio vulnificus	Skin and tissue infection	Death in those with liver disorders or problems
	Campylobacter	Diarrhea	Death: Guillain-Barré syndrome
	Salmonella	Diarrhea	Reactive arthritis
	Yersinia	Diarrhea	Reactive arthritis
	Shigella	Diarrhea	Reactive arthritis
	Cyanobacteria	Diarrhea	Potential cancer
	Leptospirosis	Fever, headache, chills, muscle aches, vomiting	Weil's disease, death (not common)
	Aeromonas hydrophila	Diarrhea	
Parasites	*Giardia lamblia*	Diarrhea	Failure to thrive Severe hypothyroidism Lactose intolerance Chronic joint pain
	Cryptosporidium	Diarrhea	Death in immunocompromised host
	Toxoplasma gondii	Newborn syndrome Hearing and visual loss Mental retardation Diarrhea	Dementia and/or seizures
	Acanthamoeba	Eye infections	
	Microsporidia (Enterocytozoon and Septata)	Diarrhea	
Viruses	Hepatitis viruses	Liver infection	Liver failure
	Adenoviruses	Eye infections, diarrhea	
	Caliciviruses small round structured viruses, Norwalk virus	Diarrhea	
	Coxsackieviruses	Encephalitis Aseptic meningitis Diarrhea Respiratory disease	Heart disease (myocarditis), reactive insulin-dependent diabetes
	Echoviruses	Aseptic meningitis	

Source: CDC Emerging Infectious Diseases, vol. 3, no. 4, Oct–Dec 1997.

cases of meningoencephalitis, inflammation of the brain and meninges, caused by the protozoan *Naegleria fowleri*. The CDC reports that every year, despite federal and state standards for water systems, many outbreaks of disease occur as a result of water pollution. **Figure 31.3** shows that the number of outbreaks from drinking water varies from year to year; a total of 674 outbreaks were reported between 1971 and the end of 1996. There was a large increase in the number of cases reported in 1992, largely due to a single outbreak of cryptosporidiosis in Milwaukee, where about 403,000 cases were reported.

Chemicals contaminating water supplies caused almost a third of the reported outbreaks from drinking water in 1995 and 1996 (see figure 31.2a); however, the numbers affected were low, a total of only 90 cases. Copper leaching from plumbing systems caused some outbreaks. A number of cases of disease caused by nitrate contamination of water have been reported in recent years. In fecally contaminated water, breakdown of large amounts of organic matter in the feces leads to accumulation of nitrates in the water. Intestinal bacteria convert nitrate to nitrite, which combines with hemoglobin, reducing the oxygen-carrying

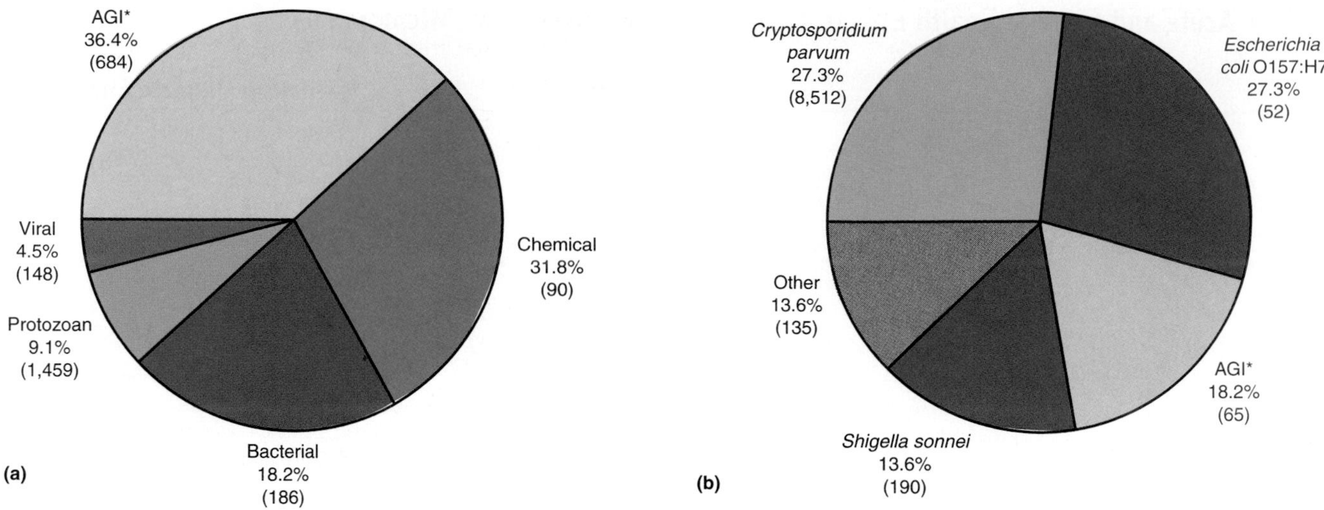

Figure 31.2 Causative Agents of Waterborne Diseases in the United States During 1995 and 1996 Percent of outbreaks is shown with number in parentheses. **(a)** Outbreaks associated with drinking water (22 outbreaks). **(b)** Outbreaks of gastroenteritis associated with recreational water (22 outbreaks).

capacity of the blood. Infants and children, more likely to be affected than are adults, may develop respiratory distress and a bluish skin color (cyanosis) from their poorly oxygenated blood. The CDC reported outbreaks of miscarriages of pregnancies in a small number of women in Indiana during 1993 and 1994; all drank water from private wells that were subsequently shown to have high levels of nitrate. Some of the wells tapped into the same aquifer, which was apparently contaminated by a leaking waste pit on a hog farm in the area. In 1996, nitrite contamination of drinking water caused 9 cases of acute cyanosis. The nitrite entered the water through faulty backflow check valves from air conditioning and boiler conditioning units.

Another example of chemical contamination of water that has proved fatal in the past is mercury contamination from industrial wastes. In this type of contamination, metallic mercury, thought to be harmless, was discharged into water. In this environment, the mercury was converted by microbes into methyl mercury, a powerful nerve toxin for humans. Fish concentrated the methyl mercury in their tissues. Many humans who ate contaminated fish died, and many more suffered severe mercury poisoning.

MICROCHECK 31.1

Waterborne diseases result from contamination of water with a variety of microorganisms and chemicals.

- Name five microbial causes of waterborne diseases.
- What kind of microorganism has caused a majority of cases of waterborne disease in recent years; bacteria, protozoa, fungi, or viruses?
- How can potable water be safe to drink and yet be far from pure?

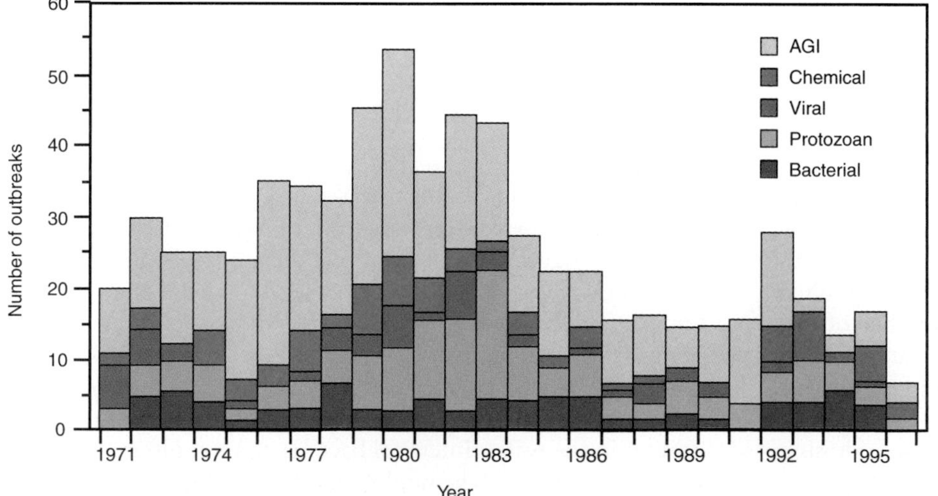

Figure 31.3 Number of Waterborne Disease Outbreaks Associated with Drinking Water, United States, 1971 to 1996 A total of 674 outbreaks were reported to the Centers for Disease Control and Prevention (CDC) during this period. The causes are indicated by the colors: **pink**—acute gastrointestinal illness of unknown cause (AGI); **orange**—chemical; **blue**—viral; **green**—protozoan; **purple**—bacterial. Because they are acute and short-lived diseases, the causes of many of these illnesses are never established with certainty. Many more episodes of disease occur but are not reported.

Water Treatment and Testing

To ensure its quality, water is treated in various ways. The principal concern is detecting and removing contamination, especially by pathogenic microorganisms. A typical treatment procedure is diagrammed in **figure 31.4**.

First, the water is allowed to stand long enough for particulate matter to settle. The water is then removed from the sediment and placed in a tank where it is mixed with a flocculent chemical, such as aluminum potassium phosphate, or alum. Alum causes a flocculation (fluffy clumping) of materials still suspended in the liquid. The clumps settle, thereby removing unwanted materials from the water, including some bacteria and viruses. Following the flocculation, the water is filtered through sand or diatomaceous earth. Sometimes, filtration through an activated charcoal filter is also used, especially when toxic chemicals may remain. Organisms grow in biofilms on filter materials, including activated charcoal. This is advantageous because these organisms use carbon in the water, resulting in lower organic carbon, and as a consequence there is less microbial growth in pipes delivering the water and less disinfectant demand.

As a last step, the water is treated with chlorine or other disinfectants to kill any harmful organisms that might remain. Ultraviolet irradiation has been used for disinfection, but a small amount of chlorination usually has to be added to prevent post treatment contamination or growth. In Europe, ozone has often been used to disinfect water supplies, but it has some disadvantages, especially its high cost. ■ **diatomaceous earth, p. 302**

Regular testing must be done to be sure that the water is safe. It is not feasible to test for pathogens in the water on a regular basis, so some accepted methods test for **coliform bacteria**. Coliforms, including *Escherichia coli* , are Gram-negative, rod-shaped, non-spore-forming bacteria that ferment lactose, forming acid and gas. They are commonly found in soil and in the gut and feces of warm-blooded animals. Their presence in water may indicate contamination with human and/or animal feces. A subset, the **fecal coliforms**, are better indicators of fecal contamination. These bacteria are called **thermotolerant**, because they can grow at a higher temperature. For municipal water supplies in the United States, zero coliform organisms per 100 milliliters is considered safe for treated potable water. Higher numbers are acceptable in water used for other purposes, such as recreational waters. The U.S. Environmental Protection Agency criterion is no more than 200 organisms per 100 ml of recreational water, but this requirement varies from state to state.

Two procedures used to test for coliforms in water are described in **figure 31.5**. The test shown in **(a)** requires more time and materials than the membrane filtration test shown in **(b)**. Aside from being quicker and cheaper, the membrane filtration test allows identification of colonies that grow on the filter. Also helpful in identification, fecal coliforms produce the enzyme β-galactosidase, which can easily be detected by its ability to change a colorless galactoside compound into a colored product, or by acting on a different galactoside compound to give a fluorescent end product. In addition to the coliforms, other bacteria that have been used as indicators of fecal contamination are some of the clostridia and enterococci, and bacteriophages.

Work is under way to develop gene probe methods to detect pathogens, including viruses, in water. The polymerase chain reaction (PCR) method can be applied to microorganisms and viruses concentrated on filtration membranes. The technique is fast and sensitive. For example, by combining gene probes for several groups of viruses, many viruses can be detected simultaneously. The methods must be improved to make them quantitative and to distinguish between living and dead cells. ■ **genetic probe methods, PCR, p. 232, 254**

Public water systems in the United States are regulated under the Safe Drinking Water Act of 1974, amended in 1986 and again in 1996. In 1996, a bill was enacted to help Americans know what is actually in the water they drink. The law requires local water agencies to issue annual reports disclosing the chemical and bacterial components of the tap water. This information is required to be sent directly

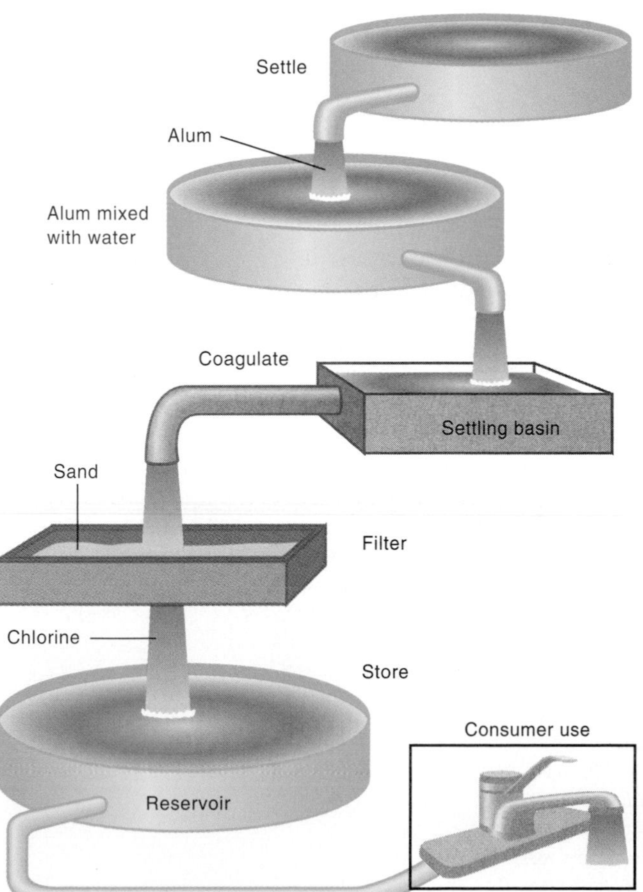

Step 1 Water is held in a series of reservoirs, where large materials sediment. Aluminum potassium sulfate (alum) may be added to cause flocculation of organic matter, which then settles out.

Step 2 The water is then filtered through beds of sand, which removes almost all of the bacteria. Filtration through activated charcoal may be used to remove toxic or objectionable organic materials.

Step 3 Finally, chlorination is used to disinfect the water, killing any pathogens that might remain. It is possible to treat drinking water with ultraviolet irradiation or with ozone so that it does not have to be chlorinated. Some cities also add fluoride to protect against dental cavities.

Figure 31.4 Steps in the Treatment of Metropolitan Water Supplies

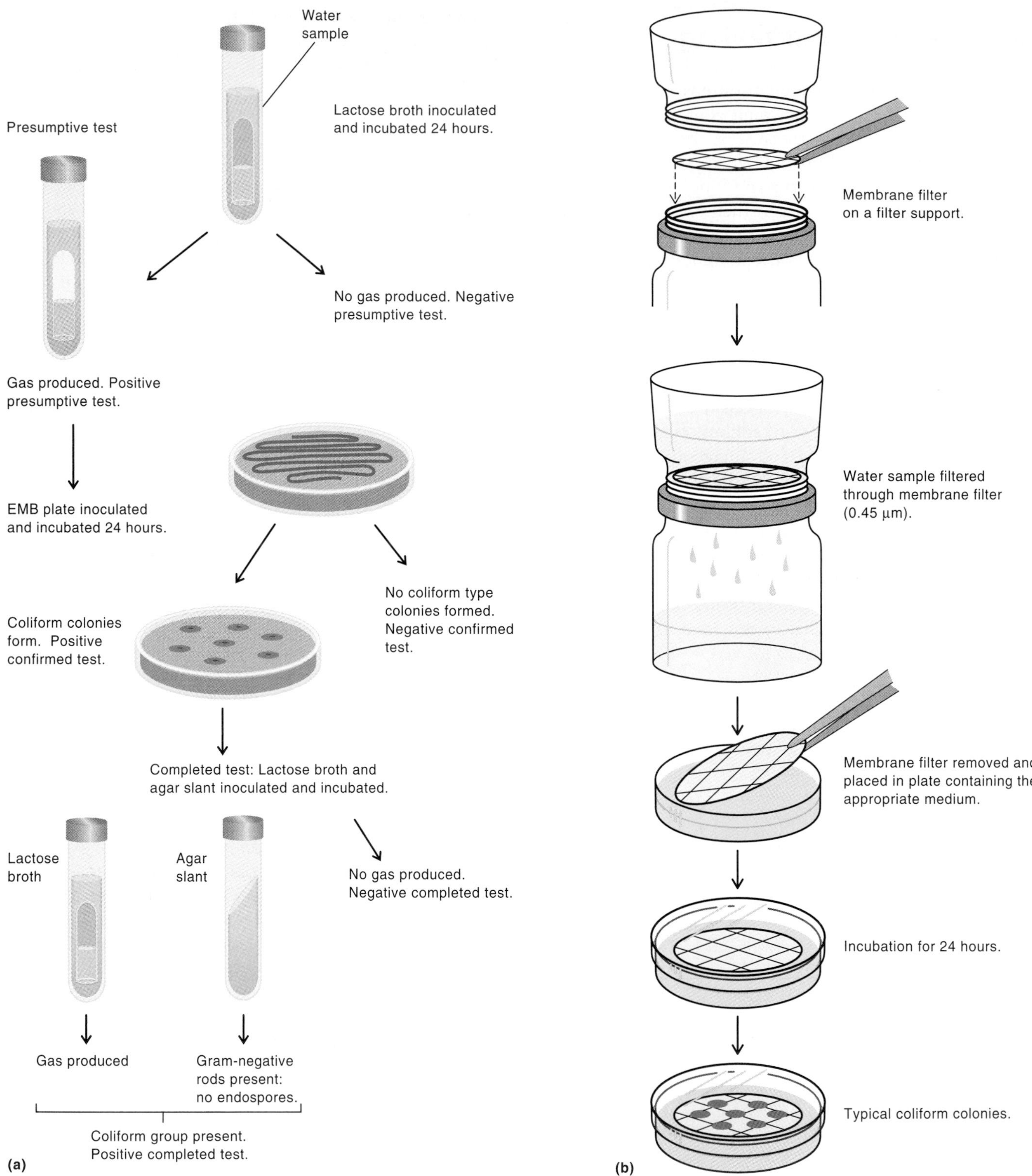

Figure 31.5 Methods Used for Testing Water Fecal contamination of water is indicated by the presence of the coliform *Escherichia coli*. **(a)** Traditional method for detecting *E. coli*. *E. coli* ferments lactose, producing acid and gas. In the first step, the water sample is added to lactose broth and incubated for 24 hours. The production of gas is a positive presumptive test. The next step is to confirm that the gas-forming organism in the sample is *E. coli*. To do this, the 24-hour culture is streaked onto an EMB (eosin–methylene blue) plate and incubated for 24 hours. *E. coli* grown on EMB gives characteristic colonies with a metallic sheen. The presence of such colonies is taken as a positive confirmed test. To complete the investigation, lactose broth and an agar slant are inoculated. The presence of Gram-negative rods and the absence of endospores on the agar slant and the presence of gas production in the lactose broth constitute a positive completed test. **(b)** The membrane filter procedure is used for direct recovery of indicator bacteria from water. Which method would allow easier detection of more dilute populations of microorganisms?

to water users, along with their water bills. Also, water suppliers are required to provide 24-hour public notification when a contaminant poses a significant risk. It is hoped that public awareness will offer even more protection against waterborne diseases.

MICROCHECK 31.2

Adequate water treatment and regular testing help to assure safe drinking water and recreational waters.

- Does chlorination of water supplies guarantee safe drinking water?
- Describe two methods of water testing.
- If ultraviolet radiation kills the microorganisms in drinking water, why would post-treatment chlorination be needed to control growth?

Microbiology of Sewage Treatment

A large majority of the population of the United States is concentrated in cities that cover a very small percentage of the land. To understand what this means in terms of waste treatment, consider that every day the average American uses about 150 gallons of water, 4 pounds of food, and 19 pounds of fossil fuel, which are converted into some 120 gallons of sewage, 5 pounds of trash, and almost 2 pounds of air pollutants! This means that a city with only 1 million inhabitants is faced with the disposal of 120 million gallons of domestic sewage daily, to say nothing of industrial and other wastes. Microbial activity is extremely important in the recycling of all these waste materials.

Reduction of Biochemical Oxygen Demand

The **biochemical oxygen demand** (**BOD**) represents the amount of oxygen required for the microbial decomposition of organic matter in a wastewater sample; it is roughly proportional to the amount of degradable organic material present in the water sample. To measure the BOD, a well-aerated sample of water is incubated in a sealed container under standard conditions of time and temperature (usually 5 days at 20°C), and the amount of oxygen consumed by the growth of microorganisms naturally present in the wastewater sample is determined. The higher the BOD, the more oxygen is being used in biological degradation processes during the incubation. High BOD values reflect large amounts of degradable organic materials in a sample of wastewater or other material. For example, rich nutrient media used for laboratory cultivation of bacteria have BOD values of from approximately 2,000 to 7,000 milligrams of oxygen per liter of solution, as compared with 100 to 300 milligrams per liter for unpolluted natural waters. The decrease in BOD of wastewater during treatment reflects the effectiveness of the treatment in converting organic wastes into inorganic materials.

Effective treatment decreases the BOD of sewage as much as possible and removes toxic materials and other objectionable matter from the sewage. The methods of sewage treatment chosen depend on a number of factors, including the amount of sewage material, the BOD, the presence of toxic materials, and the nature of the **receiving waters**, that is, the bodies of water into which the sewage treatment products are emptied.

Organic Matter Degraded to Inorganic Compounds

During the aerobic treatment of sewage, microbial oxidations of organic compounds yield carbon dioxide and inorganic nitrogen-containing nutrients for plants. During the anaerobic decomposition of sewage, similar changes occur, except that anaerobic microbes ferment organic compounds. The products of this fermentation are subsequently utilized through aerobic respiration. The methanogens are important in anaerobic degradation, transforming the small breakdown products formed by other bacteria from organic carbon compounds into CO_2 and CH_4 (methane). In the presence of hydrogen, methanogens can further convert CO_2 into methane as well. The remaining CO_2 can be metabolized by photosynthetic organisms and plants, and the CH_4 generated by the methanogens can either be discarded, conserved for fuel, or oxidized to CO_2 by the methane-oxidizing bacteria (see **Perspective 31.1**). The conversion of organic to inorganic matter is called **stabilization**. ■ fermentation, p. 153 ■ aerobic respiration, p. 140

Large-Scale Sewage Treatment Methods

In general, the primary treatment of sewage involves the removal of large objects and much of the particulate matter through the physical processes of screening and sedimentation. Secondary treatment entails the biological processes of converting the remaining materials in sewage into odorless inorganic substances that can be reused. In tertiary treatment, phosphates and nitrogen compounds are removed, mostly by chemical means. These compounds in water can cause nutrient enrichment leading to the overproduction of algae and other organisms, **eutrophication**. **Figure 31.6** outlines the steps in large-scale sewage treatment.

Primary Treatment

Primary treatment of sewage is designed to remove materials that will settle or sediment out. During this step, sewage is passed through a series of screens to remove large objects such as sticks, rags, and trash and then allowed to settle for a period of from 90 minutes to 2 hours. Sometimes, aluminum sulfate, ferrous sulfate, or other chemicals are added to clump particles so they settle more rapidly. After the settling period, the remaining fluid is given secondary treatment, and the sedimented material from the primary tanks is usually either sent to a large tank called a digester for further treatment or is incinerated.

Secondary Treatment

Secondary treatment of sewage is designed to convert most of the organic materials to inorganic and reduce the BOD of the sewage. The Clean Water Act requires that all municipal treatment plants provide for secondary treatment.

The mechanisms by which communities of aerobic organisms stabilize sewage at most secondary treatment plants is called the **activated sludge method**. In this treatment, the sewage serves as a nutrient source for mixed aerobic organisms adapted to grow in it. They grow in flocculent masses. Although these organisms are often naturally present, large numbers of them are inoculated into the wastes by leaving a small portion of leftover sludge from the previous load of treated wastes. An abundance of oxygen is supplied by mixing the sewage in an aerator. Most of the biologically degradable organic material is then converted into gases or oxidized products, and a very small percentage is incorporated

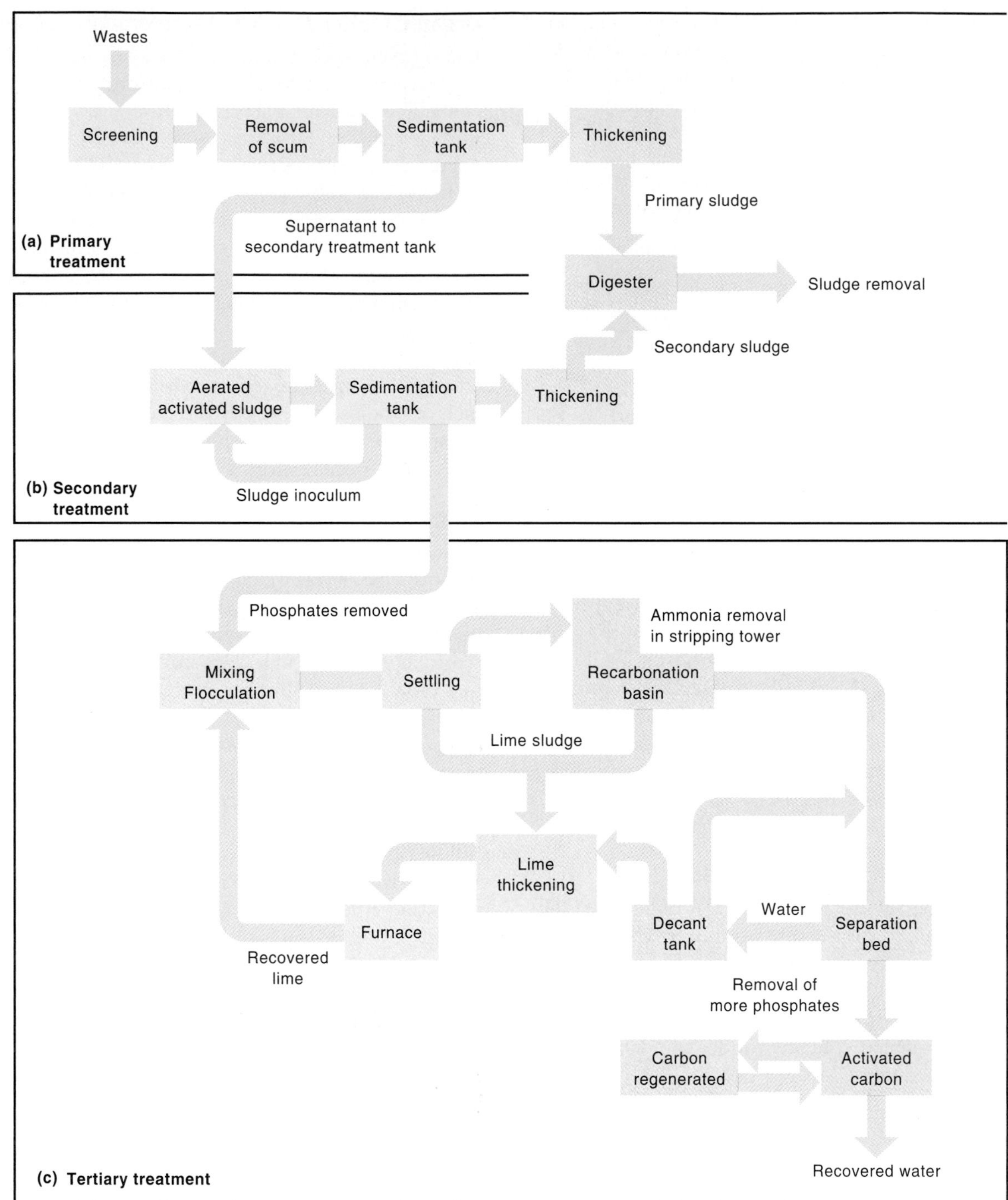

Figure 31.6 Metropolitan Waste Treatment The most advanced waste treatment schemes utilize tertiary as well as primary and secondary treatments. **(a)** Primary (physical) treatment; **(b)** secondary (biological) treatment; and **(c)** tertiary (largely chemical) treatment.

into the cell material of the organisms growing in the sewage. This process is often the most satisfactory way to treat domestic sewage, but it can be rendered totally ineffective by the presence of toxic industrial wastes, which poison the microbial population.

After the bacteria and fungi in sewage treatment have used certain nutrients, they serve as the food for ciliates, protozoa, nematodes, and other forms of life. The end result is a small increase in the mass of organisms present in a given amount of treated sewage and a large decrease in the amount of biodegradable organic materials. The increased microbial mass in the sewage is then removed to the digester; a portion is left behind as the inoculum to act on a new load of waste materials.

Within the sewage digester, anaerobic organisms act on the solids remaining in sewage after its aerobic treatment. The digester provides anaerobic conversion of organic to inorganic matter and removes water from the sewage so that a minimum of solid matter remains in it. Various populations act sequentially. In the sewage digester, the anaerobic methane-forming

Perspective 31.1 Now They're Cooking with Gas

Farmers in rural China have historically been very frugal when it came to using any waste materials. In rural areas, however, fuel for cooking and heating is often in short supply; as a consequence, up to 80% of the hay that should go for animal feed must be used for fuel. To alleviate this problem, many of the farmers build methane-producing tanks on their farms **(figure 1)**. Near the farmhouse is an underground cement tank that is connected to the latrine and pigpen. Human and animal wastes along with water and some other organic materials such as straw are added to the tanks. As the natural process of fermentation occurs, methane gas (CH_4) is produced. This gas rises to the top of the tank and is connected to the house with a hose. Enough gas is produced to provide lights and cooking fuel for the farm family. By producing its own gas, a family also saves money because it does not need to buy coal for cooking.

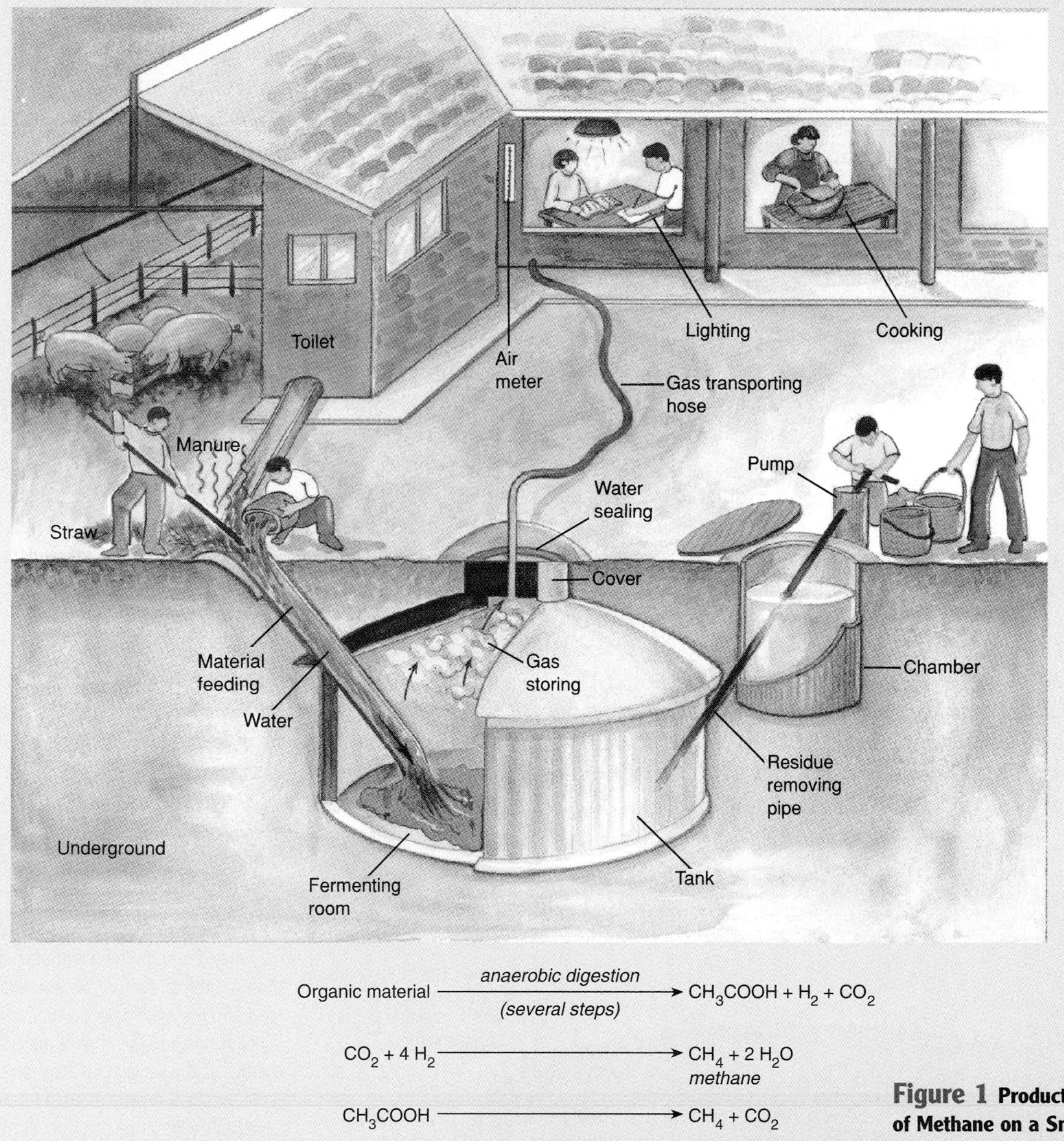

Figure 1 Production and Use of Methane on a Small Scale

organisms can perform their role of converting the simple organic acids in sewage into the useful end product methane (CH_4). Many sewage treatment plants are equipped to use their methane, thereby avoiding the cost of other sources of energy to run their equipment.

Pathogenic bacteria are generally eliminated from sewage during secondary treatment, but disease-producing viruses may survive. Pathogens account for only a small proportion of the total number of bacteria in feces, and they are greatly diluted by the water in sewage. During secondary treatment of sewage, these pathogens must compete for nutrients with the huge mass of bacteria that have been adapted to grow best at the temperature and conditions provided. As a result, most pathogenic bacteria are rapidly overgrown and eliminated by their competitors. Animal viruses lack appropriate host cells in sewage and cannot replicate there, although they may survive for long periods. If large quantities of virus particles are present in raw sewage, some may be recovered after secondary treatment Although sewage effluents are often chlorinated before being discharged into receiving waters, chlorine treatment at this stage does not inactivate viruses, because virus particles are commonly enclosed within small clumps of effluent materials where they are protected from the chlorine. Because the viruses do adhere to larger particles, they can be removed along with the other solid materials.

Tertiary Treatment

At present, many sewage treatment plants discard the effluent from secondary treatment into lakes, rivers, and other bodies of water, often causing eutrophication. The large quantities of phosphates or nitrates remaining in sewage after secondary treatment may increase the growth of microorganisms, which gradually deplete the oxygen and thus threaten other forms of aquatic life. The tertiary treatment of sewage, to remove nitrates and phosphates, can greatly alleviate this problem.

In some designs for the tertiary treatment of sewage, chemical precipitation of phosphates has been combined with biological removal of nitrates. Certain bacteria (particularly species of *Pseudomonas* and *Bacillus*) can completely reduce nitrates (NO_3^-) to N_2 (denitrification). The N_2 gas is inert, nontoxic, and easily removed.

Use of Treated Waste Residues

One area of waste treatment in which research and development are greatly needed is in the utilization of treated waste residues. Clearly, receiving waters deteriorate following the addition of treated wastes as a result of the nutrients added to them, as well as from toxic materials and changes in their pH and temperature. Even when the best available methods of tertiary sewage treatment are used, the quality of receiving water decreases. For instance, although the recovered water at some treatment plants is pure enough to drink, it still contains some microbial nutrients. For this reason, the purified water is collected in reservoirs rather than being rechanneled into a lake.

The use of sewage sludge, the solid residue of waste treatment, is also a difficult problem. When this waste is burned, polluting gases are often formed and must be removed. In some areas, the treated sludge is used for fertilizer, but this is possible only if toxic compounds are not present in the sludge (see **Perspective 31.2**).

The treatment of wastes and utilization of waste effluents and residues currently constitute one of the most challenging areas of research for microbiologists and other scientists working in this area.

Small-Scale Sewage Treatment

In recent years, considerable progress has been made in developing better methods for the disposal of urban wastes. At present, however, these methods are not practical for small communities or for isolated dwellings. In such settings, other means of waste disposal that employ much the same basic principles of microbial degradation, but in different ways, must be used.

Small towns sometimes depend for their waste disposal on a process called **lagooning**, in which sewage is channeled into shallow ponds, or lagoons, where it remains for several days to a month or more, depending on the design of the lagoon. During this period, settling occurs and sewage materials are stabilized by anaerobic or aerobic organisms or both. Pathogenic bacteria are usually eliminated by competition, as described previously.

Trickling filters (figure 31.7) are frequently used for smaller sewage treatment plants. They are also sometimes used in place of activated sludge methods for secondary treatment. The rotating arm of the trickling filter sprays controlled amounts of sewage over a bed of coarse gravel and rocks. The pebbles and rocks in the bed then become coated with a biofilm of organisms that aerobically degrade the sewage. The film, about 2 mm thick, consists of an outer layer in which fungi predominate; a middle layer of fungi, algae, and cyanobacteria; and an inner layer composed largely of bacteria, fungi, and algae. Protozoa, especially

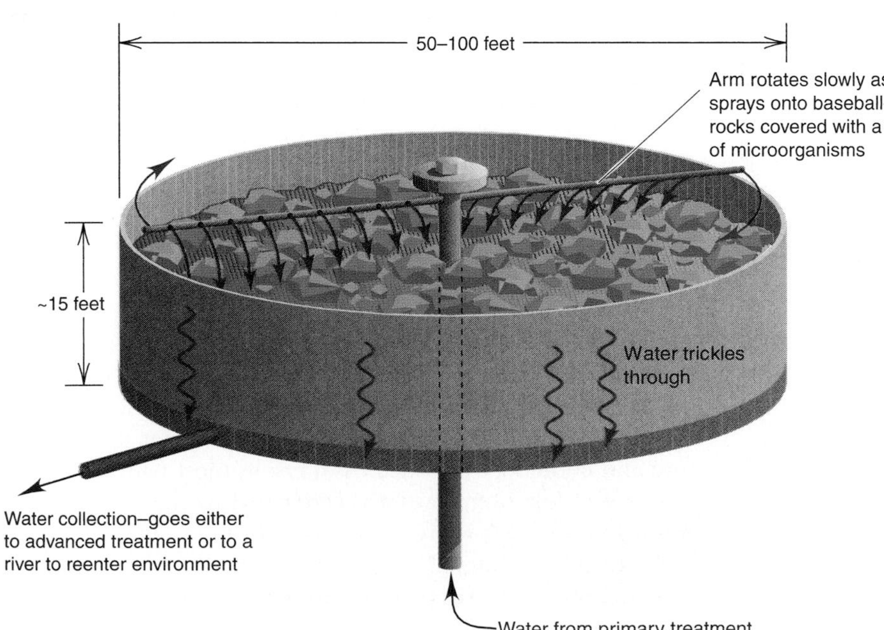

50–100 feet
Arm rotates slowly as water sprays onto baseball-sized rocks covered with a variety of microorganisms
~15 feet
Water trickles through
Water collection–goes either to advanced treatment or to a river to reenter environment
Water from primary treatment

Figure 31.7 Trickling Filter Sewage wastes are channeled into the revolving arm, and they trickle through holes in the bottom of the arm onto a gravel and rock bed. The rocks are coated with microorganisms, which convert much of the organic materials into inorganic matter as the sewage waste trickles through the bed. The effluent has a greatly reduced load of degradable organic materials when it reenters the environment.

Perspective 31.2 Sludge for Trees

Sludge is a by-product of sewage treatment that can be difficult to dispose of. If trucked to landfills, it takes up valuable space. If discharged into waterways or incinerated, it can become a major source of pollution.

Scientists from the Forestry Department at the University of Washington in Seattle are experimenting with at least one strategy for dealing with the problem of sludge disposal. They have taken the sludge from the city's sewage treatment plant to their experimental forest and used it as fertilizer for Douglas-fir trees. They have found that the sludge boosts the growth of these trees by over 400%.

There are two major problems with using sludge as fertilizer. The first problem is that heavy metals and other similar pollutants remain in the sludge following digestion of the sewage. The second is that pathogenic microorganisms and viruses also may be left in the residues. Because of these two problems, sludge cannot be used as fertilizer on food crops or in other situations in which humans might be contaminated. The sludge can be used, however, to grow trees that will be processed for pulp or lumber. The sites where sludge is used must be carefully monitored to ensure that heavy metals, organic pollutants, large amounts of nitrogen, and pathogenic microorganisms do not present public health problems in the runoff or in groundwater leaching. Whether large-scale use of sludge for forest production is an economical or practical solution to the disposal of sludge is still open to debate.

some of the ciliates, are also present. The rate of sewage flow over the rocks can be adjusted so that waste materials are maximally degraded. As is the case with the activated sludge method, different populations act in turn to degrade various compounds. Nematodes, rotifers, ciliates, bacteria, and other organisms cooperate during sewage stabilization with trickling filters and in lagoons.

Isolated dwellings or very small communities customarily rely on **septic tanks** for sewage disposal. In theory, the septic tank makes sense; in practice, however, it often does not work correctly. **Figure 31.8** illustrates the design of a septic tank. Sewage is collected in a large tank in which much of the solid material settles and is degraded by anaerobic microorganisms. The fluid overflow from the tank has a high BOD and must be passed through a drainage field of sand and gravel. Theoretically, stabilization should occur in the drainage field in the same manner described for the trickling filter. Stabilization, however, depends on adequate aeration and sufficient action by aerobic organisms associated with sand and gravel in the drainage field, conditions that are often not met. For example, clay soil under a drainage field may prevent adequate drainage, allowing anaerobic conditions to develop, or toxic materials may inhibit microbial activity in the drainage field. The possibility exists that the drainage from a septic tank contains pathogens; therefore, the tank must never be allowed to drain where it can contaminate water supplies.

Better ways of dealing with small quantities of sewage are constantly being sought. One method that has proven satisfactory in some small communities is the use of **artificial wetlands**. With an artificial wetlands system, individual septic tanks and drainage fields are not needed. Instead, land can be subdivided into more lots of smaller size, with a common portion of the land set aside for a series of ponds (**figure 31.9**). Sewage is channeled into successive ponds, which are carefully designed to carry out both aerobic and anaerobic stabilization. By the time the water reaches the final pond, it is suitable for decorative or recreational use. This type of system depends on proper design, adequate maintenance, and careful monitoring. In communities where it has been used, it has provided safe and efficient waste disposal with the added bonus of a park area around the pond system.

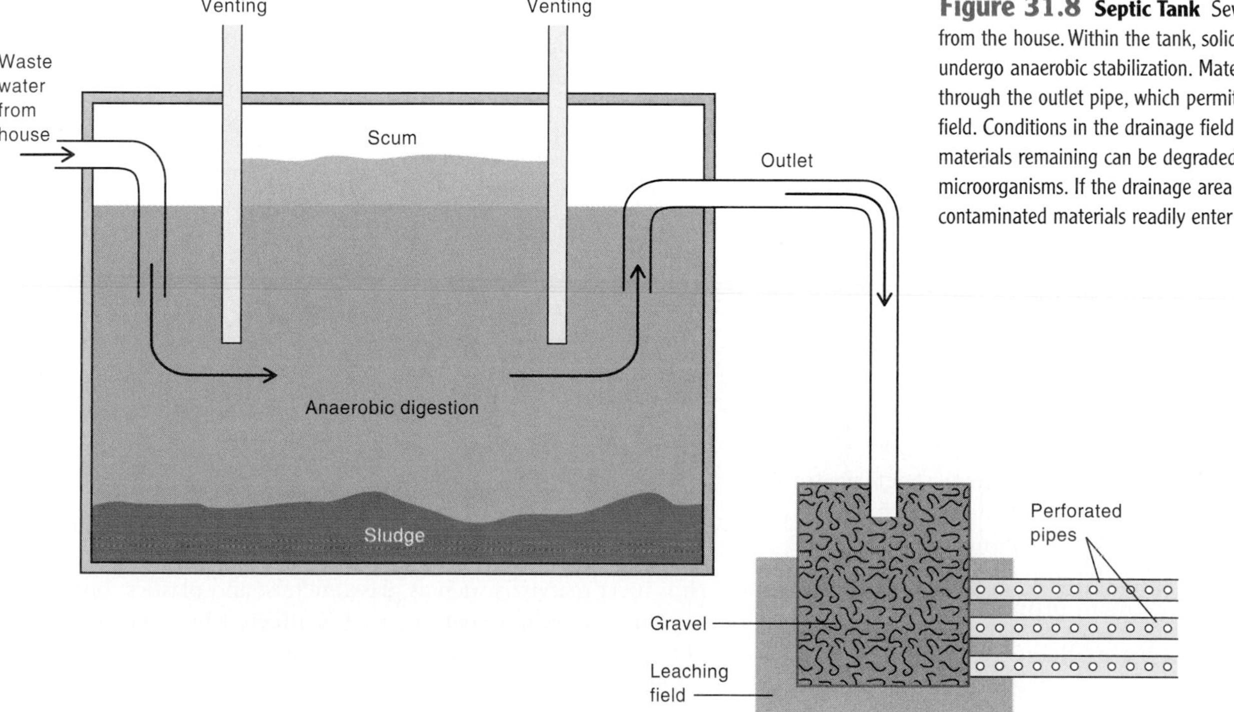

Figure 31.8 Septic Tank Sewage wastes enter the tank from the house. Within the tank, solid materials settle and undergo anaerobic stabilization. Materials that do not settle exit through the outlet pipe, which permits seepage into the drainage field. Conditions in the drainage field must be aerobic so that the materials remaining can be degraded by the activities of aerobic microorganisms. If the drainage area is not properly designed, contaminated materials readily enter adjacent waters.

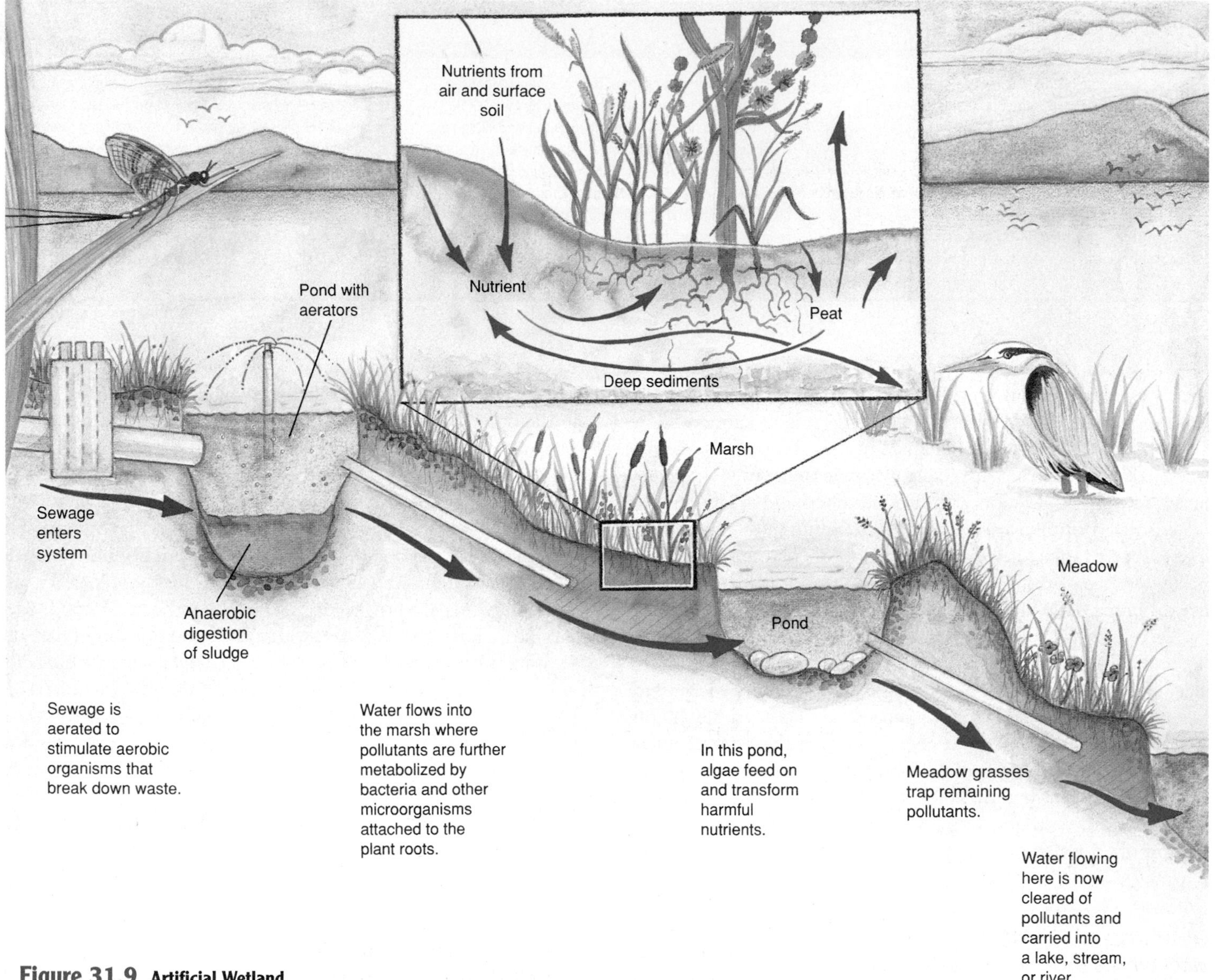
Nutrients from
air and surface
soil
Nutrient
Peat
Deep sediments
Pond with
aerators
Marsh
Sewage
enters
system
Anaerobic
digestion
of sludge
Pond
Meadow
Sewage is
aerated to
stimulate aerobic
organisms that
break down waste.
Water flows into
the marsh where
pollutants are further
metabolized by
bacteria and other
microorganisms
attached to the
plant roots.
In this pond,
algae feed on
and transform
harmful
nutrients.
Meadow grasses
trap remaining
pollutants.
Water flowing
here is now
cleared of
pollutants and
carried into
a lake, stream,
or river.

Figure 31.9 Artificial Wetland

Progress is also being made in many other areas of waste treatment. For example, soil filters are being developed to remove gases, such as H_2S, that are formed during sewage treatment. If the area is kept moist so that bacteria can grow, the foul-smelling and toxic hydrogen sulfide is removed during passage through the soil and converted by bacteria (such as *Thiobacillus* sp.) to sulfate ions.

MICROCHECK 31.3

Proper sewage treatment, necessary to ensure the health of a community, depends on the stabilization of wastes by microorganisms. Pathogenic bacteria are usually eliminated by secondary sewage treatment. Methods such as septic tanks, trickling filters, and the production of artificial wetlands are good solutions to small-scale sewage treatment problems.

- What is the origin of the methane gas produced during sewage treatment?
- What is meant by BOD?
- What would be the consequences if the BOD of sewage were *not* decreased?

Microbiology of Solid Waste Treatment

In addition to ridding our environment of wastes in water, we must take care of the solid wastes that are generated each day. In the United States, more than 150 million tons of solid waste are produced each year by industries and households. Some of it is inert materials such as glass, metals, and plastics, but a large amount is organic material that is affected by microorganisms. Eliminating waste products from the environment has become an increasingly complex problem.

Use of Landfills for Disposal of Solid Wastes

Landfills are used to dispose of solid wastes near towns and cities. Usually, the area of land chosen for the landfill is not particularly valuable. Because just piling the solid wastes up on the ground attracts insects and rodents, causing both aesthetic and public health problems, sanitary landfill methods are most often used. In this method, a site is excavated. The lowest part is typically a layer of clay soil, to minimize leaching of contaminants from the wastes into the surrounding area. A plastic liner may be placed above this layer. Next comes a layer of sand with drainage pipes, and above this the wastes are compacted and covered with a layer of soil every day. When a landfill is finished, it is covered with soil and plants, and it can be used for recreation and eventually as a site for construction. Methane and other gases are vented, and the methane is burned or recovered for use.

There are several disadvantages to this type of waste management. First, there are only a limited number of sites that are available for use near urban and suburban areas. Second, the organic content of landfills anaerobically decomposes very slowly, over a period of at least 50 years. During this time, the methane gas that is produced must be removed. If buildings are constructed before most of the methane is removed, disastrous gas explosions can occur. Pollutants such as heavy metals and pesticides can leach from landfill sites into the underground aquifers. It is very difficult to purify these aquifers once they have become contaminated.

Sanitary landfills have traditionally been a low-cost method of handling large quantities of solid waste. Because of increased costs and decreased availability of land, however, many cities are looking for ways to decrease the amount of solid waste dumped in landfills. In Seattle, the fees charged to people for garbage collection are based on the size of the container collected. The smaller the can, the lower the cost. Waste disposal companies are also considering weighing garbage and charging people accordingly to further reduce the amount of solid waste generated. Both of these methods are intended to raise people's awareness of how much solid waste they are generating as well as to offer an incentive to recycle. Companies in some areas offer lower costs for garbage collection to people who recycle. Programs to recycle paper, plastics, glass, and metal are being implemented in many cities and counties with great success. Through these programs, landfill areas can be expected to be available for a longer period of time.

Commercial and Backyard Composting—Alternative to Landfills

Commercial and home composting programs are becoming popular in many areas and are succeeding in reducing the amount of solid waste generated. Composting is the natural decomposition of organic solid material.

Backyard composting involves mixing garden debris with kitchen organic waste, excluding meats and fats (**figure 31.10**). If proper amounts of sunlight, water, and air are provided to the mix, in a matter of months the bulk of the waste is reduced by two-thirds. The black organic material generated by composting can then be used as fertilizer in garden beds. This material is rich in organic compounds and improves the soil for growing plants. **Figure 31.11** shows the steps involved in composting.

Composting on a large scale offers cities a way to reduce the amount of garbage sent to their landfills. In Seattle, yard wastes are collected separately from the main garbage. These wastes are then composted and used in various ways, including improving soils in city parks. Special machinery is used to compost on a large scale, and the composting can be accomplished in a very short time (see figure 31.11).

A backyard compost pile usually starts with a supply of organic material such as leaves and grass clippings, as well as kitchen wastes. Often, some soil and perhaps water are added to it. In a few days, the inside of the pile heats up. At 38°C to 66°C , pathogens are killed but other organisms that are thermophilic are not affected. When the pile cools down, it is aerated by physically stirring it up and turning it. This stirring adds oxygen and allows the bacteria to resume growing and breaking down the organic materials. If a compost pile is turned frequently and other conditions are good for aerobic digestion, the composting can be completed in as little as 6 weeks. ■ **thermophile, p. 96**

MICROCHECK 31.4

Both commercial and backyard composting reduce the need for large landfills to dispose of solid waste.

- What are the advantages and disadvantages of landfills?
- List the steps that must take place in successful composting of organic material.
- Developing plants that can take up and concentrate heavy metal pollutants in soil water is an active field of research. Would this approach effectively remove polluting metals from the environment?

Figure 31.10 A Backyard Compost Heap

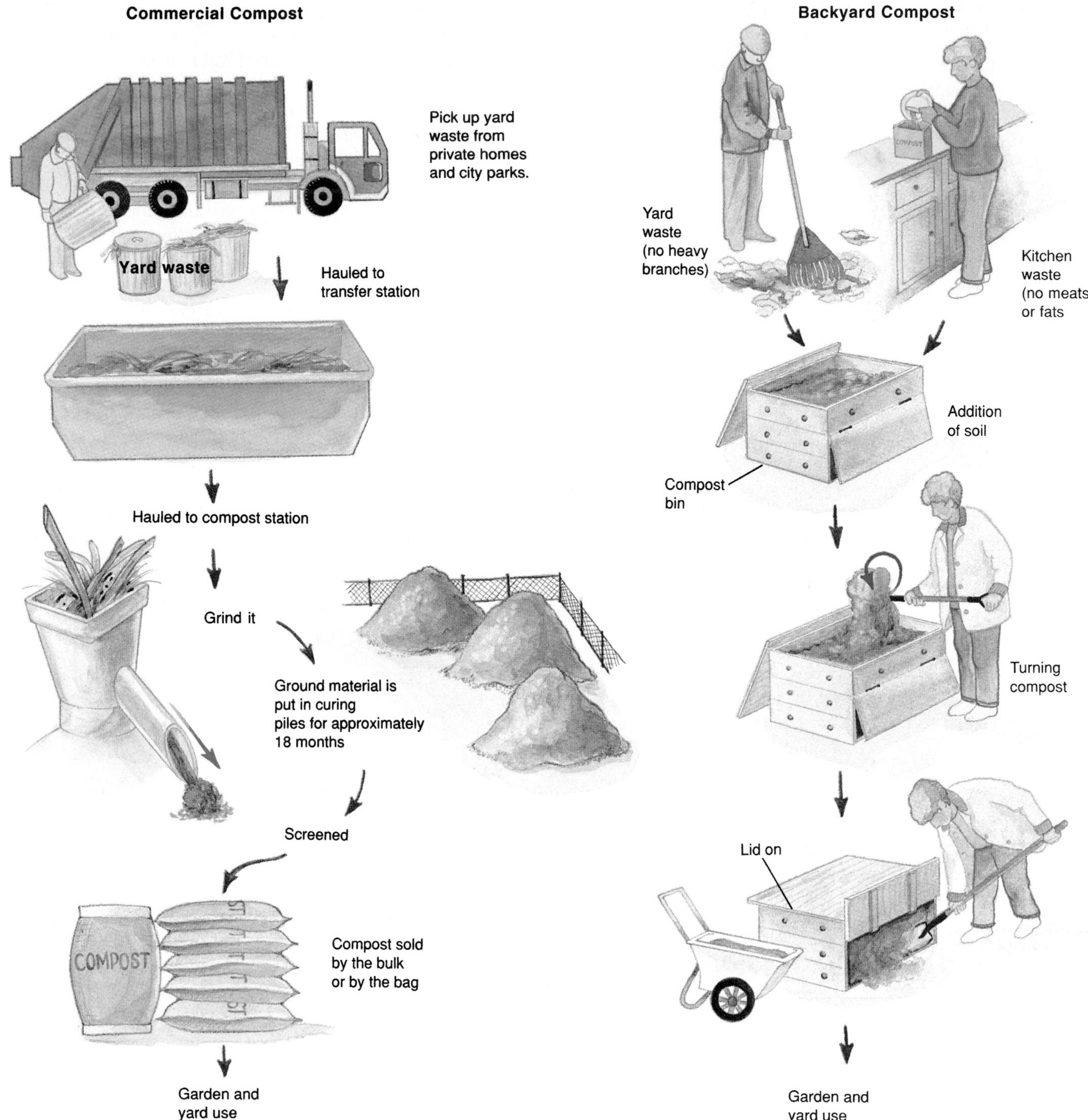

Figure 31.11 **Industrial and Backyard Composting**

FUTURE CHALLENGES

Better Identification of Pathogens in Water and Wastes

One of the most important challenges in the field of water and waste treatment is the development of new and better methods to detect waterborne contaminants in both drinking water and environmental water samples. This would make it possible to follow the occurrence and persistence of pathogens in water supplies with greater accuracy, and aid in better reporting of waterborne illnesses. Methods for detecting protozoan pathogens are not very accurate or dependable, and there is a need to find new approaches to this problem. Methods being studied, but not yet perfected for use in this field, include the polymerase chain reaction (PCR) and a variety of fluorescence techniques and radioactivity labeling methods.

Cysts of Giardia *and* Cryptosporidium *have been detected by amplifying specific regions of their DNA by PCR. In fact, using this method, it is possible to detect a single cyst of*

Giardia and to distinguish between species that are pathogenic for humans and those that are not. But problems arise in using these techniques with environmental samples, which contain substances that inhibit the reaction. In addition, the PCR is not quantitative, and it detects DNA from dead organisms as well as living. Studies are needed to make these techniques feasible for use in water testing.

Viruses can also be detected by PCR. The viruses are concentrated by filtration onto membranes, and many viruses can be detected simultaneously by combining gene probes from different groups of viruses. A problem is that viruses inactivated by disinfection procedures are still detected by PCR. To overcome this, viruses may be put into cell cultures to allow them to replicate, indicating that they are not inactivated, and the PCR is done on the infected cell cultures.

Many scientists are trying to improve PCR and other techniques so that they are quantitative and can detect living microorganisms and active viruses in water.

SUMMARY

Water Pollution by Microorganisms and Chemicals (Figure 31.1)

1. Waterborne diseases result from contamination of water with a variety of microorganisms. (Table 31.1, Figures 31.2, 31.3)
2. Chemicals contaminating water supplies include copper, nitrate, and mercury, among others.

Water Treatment and Testing (Figure 31.4)

1. Metropolitan water supplies are treated by (1) removing large and particulate matter, (2) filtering the water through sand and activated charcoal or other filters, and (3) disinfecting, usually with chlorination.
2. Regular testing of the water uses **coliform** organisms as indicator organisms of contamination. (Figure 31.5)

Mircrobiology of Sewage Treatment (Figure 31.6)

Reduction of Biochemical Oxygen Demand (BOD)

1. The decrease of **BOD** in wastewater reflects the effectiveness of treatment.

Organic Matter Degraded to Inorganic Compounds

1. During aerobic treatment of sewage, microbial oxidation converts organic wastes to inorganic matter.
2. During anaerobic treatment, microbes ferment organic materials and methanogens convert them into carbon dioxide and methane.

Large-Scale Sewage Treatment Methods

1. Primary treatment removes materials that will settle or sediment out.
2. Secondary treatment uses microbes in an activated sludge to convert organic materials to inorganic, reducing the BOD, followed by anaerobic conversion in a digester. Pathogenic bacteria are usually removed by secondary treatment.
3. Tertiary treatment removes phosphates and nitrates that could cause **eutrophication** if discharged into **receiving waters**.

Use of Treated Waste Residues

1. Treated waste residues present a problem because thay may contain toxic materials and unwanted nutrients.

Small-Scale Sewage Treatment (Figures 31.7, 31.8, 31.9)

1. **Lagooning**, **trickling filters**, **septic tanks**, and **artificial wetlands** are useful methods for dealing with small-scale sewage treatment.

Microbiology of Solid Waste Treatment

Use of Landfills for Disposal of Solid Wastes

1. Disadvantages of landfills include limited available sites and slow decomposition of wastes.

Commercial and Backyard Composting—Alternative to Landfills

1. Both commercial and backyard composting reduce the need for large landfills for disposal of solid waste. (Figures 31.10, 31.11)

REVIEW QUESTIONS

Short Answer

1. Discuss the possible sources of several chemical contaminants found in water supplies.
2. Why do water-testing procedures look for coliform organisms rather than other pathogens?
3. How is water treated to make it safe to drink?
4. Outline the steps in large-scale metropolitan sewage treatment.
5. Explain why primary, secondary, and tertiary treatments of waste are each important in processing sewage.
6. How effective are the different stages of sewage treatment in removing pathogens from the materials?
7. How does a septic tank system work?
8. In what situation might lagooning be a good choice for sewage treatment?
9. What are some advantages of using artificial wetlands for dealing with small amounts of sewage?
10. Describe a trickling filter.

Multiple Choice

1. Outbreaks of infectious diseases are often associated with contaminated drinking water, but they may also be associated with…

   A. hot tubs.

   B. whirlpools.

C. swimming pools.

D. water parks.

E. All of the above

2. Although *E. coli* O157:H7 is usually transmitted in foods, several outbreaks have occurred due to contaminated...

A. whirlpools.

B. hot tubs.

C. lake swimming areas.

D. ocean swimming areas.

E. river swimming areas.

3. Diarrhea often results from drinking water contaminated with...

A. *Staphylococcus aureus.*

B. *Streptococci.*

C. *Clostridium botulinum.*

D. *Listeria monocytogenes.*

E. *Cryptosporidium parvum.*

4. A marked decrease in BOD during secondary sewage treatment indicates...

A. lack of oxidation during treatment.

B. effective aerobic decomposition during treatment.

C. effective anaerobic decomposition during treatment.

D. removal of all pathogenic bacteria.

E. removal of all toxic chemicals.

5. Tertiary treatment of sewage is designed to remove...

A. BOD.

B. nitrates and phosphates.

C. bacteria.

D. protozoa.

E. methane.

6. Septic tanks should be placed...

A. as close to the well as possible.

B. at least 500 feet from the house.

C. under the house.

D. in deep clay soil.

E. where overflow cannot contaminate any water supply.

7. Activated sludge is a nutrient source for...

A. anaerobic bacteria.

B. anaerobic fungi.

C. mixed communities of aerobic organisms.

D. mostly pathogenic bacteria.

E. viruses.

8. Overgrowth of algae and plants in receiving waters, eutrophication, occurs when sewage effluent is rich in...

A. nitrates and phosphates.

B. carbon dioxide.

C. methane.

D. organic matter.

E. toxic chemicals.

9. Landfills are often used to dispose of...

A. household sewage.

B. commercial sewage.

C. solid wastes.

D. petroleum wastes.

E. sewage effluent.

10. Backyard composting is an excellent way to dispose of...

A. cooking fats.

B. garden debris and most organic kitchen wastes.

C. spoiled meats.

D. insecticides.

E. cleaning supplies.

Applications

1. A developer is interested in building vacation homes on 150 acres of oceanfront property. A priority is to retain as much as possible the natural beauty of the area. Safe and effective sewage treatment must be part of the plan, yet there is no nearby sewage treatment facility. An environmental consultant came up with three options for sewage disposal. What advantages and disadvantages of each of the following proposals must the developer consider before selecting one?

   a. Individual septic tanks for each home

   b. Trickling filter central treatment system

   c. Artificial wetlands

2. A public health official is investigating waterborne diseases in Illinois. She notes that over half of the actual cases of waterborne diseases originating from drinking water were caused by *Giardia lamblia.* Other data showed that most cases of gastroenteritis attributed to exposure to recreational waters were caused by *Cryptosporidium parvum.* What does this suggest about controlling waterborne diseases?

Critical Thinking

1. It has been said that "the solution to pollution is dilution." Discuss the reasons this is not a valid statement.

2. The figure below shows the effects of different treatments of drinking water on the incidence of typhoid fever in Philadelphia, 1890–1935. If filtration of drinking water caused such a dramatic decrease in the disease incidence, was it necessary to introduce chlorination a few years later? Why or why not?

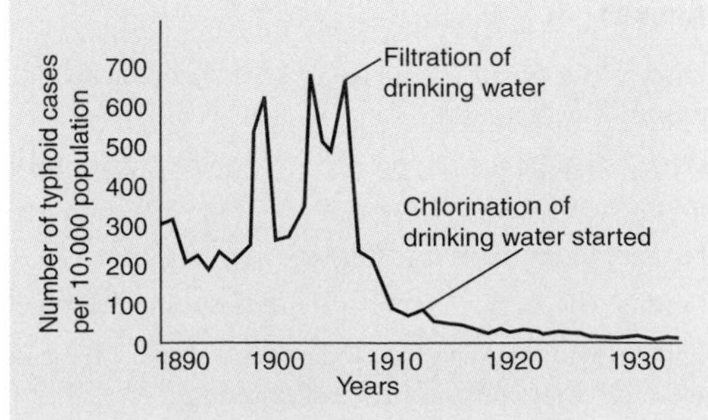

CHAPTER 32

Food Microbiology

Pasteurization of milk and milk products is so common today that it is hard to fathom that only during this century have federal, state, and local laws mandated it for all milk products. Although most large cites have required milk to be pasteurized since 1900, as late as 1930 most milk sold in rural areas was not routinely pasteurized. As a result, human diseases such as brucellosis and tuberculosis were fairly common.

Alice Catherine Evans, the first woman president of the Society of American Bacteriology (now the American Society for Microbiology), helped establish the connection between unpasteurized milk and brucellosis in humans. A graduate of both Cornell University and the University of Wisconsin, Evans worked for the U.S. Department of Agriculture, seeking out the sources of microbial contamination of dairy products. In 1917, Evans reported that cases of human brucellosis were related to the finding of Brucella abortus *in cows' milk. Her conclusion that* B. abortus *could be transmitted from cows to humans through milk conflicted with the prevailing view of a number of prominent scientists, including Robert Koch. In 1900, Koch had declared that bovine tuberculosis and brucellosis could not be transmitted to humans. As a result, at least 30 years elapsed before many scientists and dairy workers would accept the increasing evidence that diseases could be transmitted from cows to humans through the milk supply.*

In the 1920s, dairy herds were inspected and vaccinated for tuberculosis. Herds that passed inspection and were vaccinated were called certified herds, and their milk could be sold commercially without pasteurization. In the late 1930s, after a number of the children of dairy workers had died of brucellosis even though they had drunk only certified milk, the problem of milk-borne disease was finally acknowledged. Today, milk is routinely pasteurized, and only very small amounts of unpasteurized milk are sold in the United States.

—*A Glimpse of History*

WHEN YOU PREPARE A MEAL, YOU ALSO INVITE A host of microorganisms to dinner. Practically all the food we purchase or grow, be it fruit, vegetable, meat, or dairy product, harbors a variety of microorganisms. This is not surprising when one considers that bacteria and fungi are ubiquitous and are especially plentiful in soil and around animals. From the microbial perspective, food could be viewed as a fertile ecosystem in which these organisms vie for the rich nutrients. The successful microorganisms earn the right to multiply and predominate.

Principles of Food Microbiology

A variety of microbes find their way onto foods, introduced from the soil in which they were grown, and during harvest, packaging, storage, and handling. Those that are most suited for growth in the environment flourish, making end products such as acids, alcohols, and gas.

Microorganisms on foods are not always undesirable. Sometimes, their growth results in a more pleasant taste or texture. Foods that have been intentionally altered by carefully controlling the activity of bacteria, yeasts, or molds are called **fermented (figure 32.1)**. For example, food manufacturers purposely encourage some microorganisms to flourish in milk to produce foods such as sour cream and blue cheese. Alcoholic beverages, such as beer and wine, and many Asian food condiments, such as soy sauce and miso, also rely on microbial metabolism for their production. Strictly speaking, the term fermentation is used to describe only those metabolic activities that utilize pyruvate or another organic compound as an electron acceptor, with the result that alcohols and acids are produced. Food scientists, however, use the term more generally, to encompass any desirable change that a microorganism imparts to food. ■ **fermentation, p. 153**

Figure 32.1 Fermented Foods Fermented foods have been intentionally altered in their production by carefully controlling the growth and activity of microorganisms.

Biochemical changes in foods, when perceived as undesirable, are called **spoilage (figure 32.2)**. The processes that cause spoilage are often the same ones involved in fermentation of foods. In fact, a food product considered by one cultural population to be fermented may be considered spoiled by another. Sour milk and moldy bread are examples of foods considered spoiled as a result of microbial growth. The souring of milk, however, involves microbial processes analogous to those that cause the agreeable acidic flavor of sour cream, and the mold growing on bread may be related to the one that causes the blue veining in Gorgonzola cheese.

Growth of pathogens such as *Clostridium botulinum*, *Staphylococcus aureus*, *Salmonella*, and *E. coli* O157:H7 can result in foodborne illness but generally does not result in perceptible changes in quality of a food. Depending on the type of pathogen, the illness may be the result of consumption of the living organisms or caused by toxins they have produced during growth.

Limiting microbial growth can preserve the quality of foods and prevent foodborne illnesses. Foods can be canned, pasteurized, or irradiated to eliminate or decrease the numbers of microorganisms. Alternatively, the multiplication of microorganisms can be suppressed by storing food at cold temperatures, or by adding growth-inhibiting ingredients, called **preservatives**. The end products of some fermentation processes also inhibit the growth of many microorganisms.

MICROCHECK 32.1

A variety of microorganisms can use food as a growth medium. End products they produce can be desirable (fermented foods), undesirable (spoilage), or harmful (foodborne illness).

- Name three foods that rely on microbial metabolism for their production.
- Differentiate between fermentation and spoilage.
- Which end products of fermentation might inhibit other microorganisms?

Factors Influencing the Growth of Microorganisms in Foods

Select microorganisms have a competitive advantage in those foods with characteristics that enable them to multiply most rapidly. Understanding the factors that influence microbial growth is essential to maintaining food quality, whether utilizing microorganisms to produce fermented foods, or suppressing organisms to prolong the shelf life of perishable foods.

The conditions naturally present in the food, such as moisture, acidity, and nutrients, are called **intrinsic factors**. Environmental conditions, such as the temperature and atmosphere of storage, are called **extrinsic factors**. All of these factors

(a)

(b)

(c)

Figure 32.2 Examples of Spoiled Food **(a)** Apples infected by the fungus *Venturia inaequalis*, cause of apple scab. **(b)** Orange affected by a *Penicillium* mold. **(c)** Ear of corn and a lemon found after several weeks in the author's refrigerator.

combine to determine which microorganisms can grow in a particular food product and the rate of that growth. For example, bacteria are well suited for growth in the environment found in fresh meat and other moist, pH-neutral, nutrient-rich foods. Although other microorganisms, such as yeasts and molds, can also grow under these conditions, the more rapid increase of bacteria overwhelms these competitors. Competition also occurs between different types of bacteria. On moist, pH-neutral foods, members of the genus *Pseudomonas* tend to flourish. Slightly acidic foods inhibit *Pseudomonas* species, however, and the slower-growing lactic acid bacteria become dominant. When the growth of most common bacteria is restricted by conditions such as lack of moisture or high acidity, fungi predominate despite their relatively slow growth.

Intrinsic Factors

The multiplication of microorganisms in a food is greatly influenced by the inherent characteristics of that food. In general, microorganisms multiply more rapidly in moist, nutritionally rich, pH-neutral foods.

Water Availability

Food products vary dramatically in terms of how much water is accessible to the organisms that might grow in them. Fresh meats and milk, for example, have ample water and will support growth of many microorganisms. Bread, nuts, and dried foods, on the other hand, provide a relatively arid environment. Some sugar-rich foods, such as jams and jellies, are seemingly moist, but most of that water is chemically interacting with the sugar, making it unavailable for use by microbes. Highly salted foods, for similar reasons, have little available moisture.

The term **water activity** (a_w) is used to designate the amount of water available in foods. By definition, pure water has an a_w of 1.0. Most fresh foods have an a_w above 0.98, whereas ham has an a_w of 0.91, jam has an a_w of 0.85, and some cakes have an a_w of 0.70.

Most bacteria require an a_w above 0.90 for growth, which explains why fresh moisture-rich foods spoil more quickly than dried, sugary, or salted foods. Fungi can grow at an a_w as low as 0.80, which explains why forgotten bread, cheese, jam, and dried foods often become moldy. *Staphylococcus* species, which are adapted to grow on the dry, salty surfaces of human skin, can grow at an a_w of 0.86, which is lower than the minimum required by most common spoilage bacteria. *Staphylococcus* species normally do not compete well with other bacteria, but on salty products, such as ham and other cured meats, they can multiply with little competition. Ham, when improperly handled, is a common vehicle for *S. aureus* food poisoning. ■ ***Staphylococcus aureus*** **foodborne illness, p. 814**

pH

The pH of a food is also important in determining which organisms can survive and thrive on it. Many species of bacteria, including most pathogens, are inhibited by acidic conditions and cannot grow at a pH below 4.5. A commercially valuable exception is the lactic acid bacteria, which can grow at a pH as low as 3.5 and are used in the production of fermented foods such as yogurt and sauerkraut. This group of bacteria produces lactic acid as a result of fermentative metabolism. Although they are useful in food production, their growth and accompanying acid production is also a prime cause of spoilage of unpasteurized milk and other foods. Fungi can grow at a lower pH than most spoilage bacteria, leading to some foods, such as fruit, eventually becoming moldy. For example, the pH of lemons is approximately 2.2, which inhibits the growth of bacteria, including the lactic acid group, but some fungi can grow on them. ■ **pH, p. 25** ■ **lactic acid bacteria, p. 275** ■ **pasteurization, p. 117**

The pH of a food product may also determine whether toxins can be produced. For example, *Clostridium botulinum*, the causative agent of botulism, does not grow or produce toxin below pH 4.5, and so it is not considered a danger in highly acidic foods (**figure 32.3**). This is why the canning process for acidic fruits and pickles is less stringent than that for foods with a higher pH. There are cases, however, such as some newer varieties of tomatoes that are less acidic than older types, that require the addition of acid if they are to be safely canned using these less stringent procedures.

Nutrients

The nutrients present in a food as well as other intrinsic factors determine the kinds of organisms that can grow in it. An organism requiring a particular vitamin cannot grow in a food lacking that vitamin. A microbe capable of synthesizing that vitamin, however, can grow if other conditions are favorable. Members of the genus *Pseudomonas* often spoil foods because they can synthesize essential nutrients and can multiply in various environments, including refrigeration.

Biological Barriers

Rinds, shells, and other coverings aid in protecting some foods from the invasion of microorganisms. Eggs, for example, retain their quality much longer with intact shells. Whole lemons keep longer than slices. Even so, a microorganism that can degrade

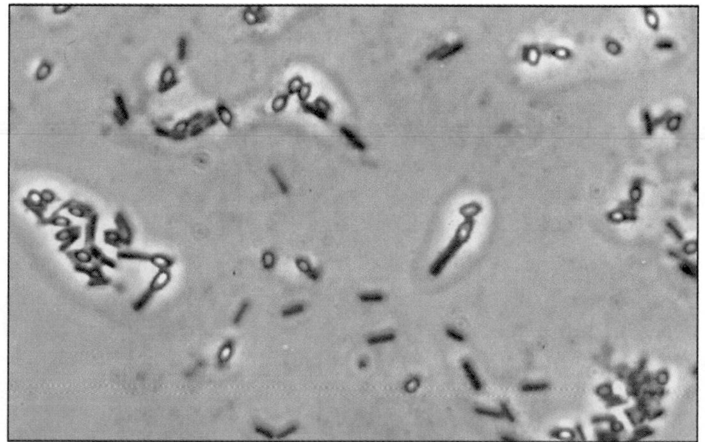

Figure 32.3 Phase-Contrast Micrograph of the Foodborne Pathogen *Clostridium botulinum* This pathogen does not grow or produce toxin in highly acidic foods.

these protective materials will eventually break down the coverings and cause spoilage.

Antimicrobial Chemicals

Some foods contain natural antimicrobial chemicals that may help prevent spoilage. Egg white, for instance, is rich in lysozyme. If lysozyme-susceptible bacteria breach the protective shell of an egg, they are destroyed by lysozyme before they can cause spoilage. Other examples of naturally occurring antimicrobial chemicals are benzoic acid, which is found in cranberries, allicin in garlic, and an antibacterial peroxidase system, similar to that found in phagocytes, which is in raw milk. ■ **lysozyme, p. 65** ■ **peroxidase enzymes, p. 375**

Extrinsic Factors

The extent of microbial growth varies greatly depending on the conditions under which a food is stored. Microorganisms multiply rapidly in warm, oxygen-rich environments such as the surface of meat stored at room temperature.

Storage Temperature

The temperature of storage affects the rate of growth of microorganisms in food. Below its freezing point, water becomes crystalline and inaccessible, effectively halting microbial growth that requires moisture. At low temperatures above freezing, many enzymatic reactions are either very slow or nonexistent, with the result that some microorganisms are unable to grow. Those that can do so at a reduced rate. Microorganisms that grow on refrigerated foods are most likely psychrophiles such as some members of the genus *Pseudomonas.* ■ **psychrophiles, p. 96**

Atmosphere

The presence or absence of oxygen affects the type of microbial population able to grow in food. For example, members of the genus *Pseudomonas* are obligate aerobes, and they cannot grow in foods stored under conditions that exclude all of their required oxygen. Excluding oxygen from a food, however, may enable the growth of other bacteria, including the obligate anaerobe *Clostridium botulinum.* A case of botulism was traced to the consumption of a thick homemade stew that had been slowly cooked and then left at room temperature overnight. The cooking process did not destroy the endospores of *C. botulinum* and had created anaerobic conditions in which the organism thrived and produced toxin. ■ **oxygen requirements for growth, p. 97**

MICROCHECK 32.2

Intrinsic factors (such as available moisture, pH, and the presence of antimicrobial chemicals), and extrinsic factors (including storage temperature and atmosphere) can influence the type of microorganisms that grow and predominate in a food product.

- Why is *Staphylococcus aureus* more likely to be found in high numbers on ham than on fresh meat?
- Which is more important to refrigerate: homemade stew or bread? Why?
- Why would the cooking process create anaerobic conditions?

Microorganisms in Food and Beverage Production

Microorganisms have been used to produce food for thousands of years. Yogurt, cheese, pickled vegetables, and other fermented foods are not only perceived as pleasant tasting, but the acids produced as a by-product of microbial metabolism inhibit the growth of many spoilage organisms as well as foodborne pathogens. Thus, fermentation historically has been, and continues to be today, an important method of food preservation, particularly when modern conveniences such as refrigeration are lacking.

The development of methods for isolating and culturing microorganisms has enabled the scientific study of the microbial processes involved in the production of fermented foods. In turn, this has resulted in changes, such as the careful selection of microbial strains, that have increased the efficiency of production and/or the quality of fermented food products. Nevertheless, some ancient processes, including the production of some fine cheeses and wines, are not fully understood, and still remain somewhat of an art.

Lactic Acid Fermentations by the Lactic Acid Bacteria

The tart taste of yogurt, pickles, sharp cheeses, some sausages, and other foods is due to the production of lactic acid by one or more members of a group of bacteria known as the lactic acid bacteria **(table 32.1)**. This group of organisms, including members of the genera *Lactobacillus, Lactococcus, Streptococcus, Leuconostoc,* and *Pediococcus,* are obligate fermenters that characteristically produce lactic acid as an end product of their metabolism. Some also produce flavorful and aromatic compounds that contribute to the overall quality of fermented foods. ■ **lactic acid bacteria, p. 275** ■ **obligate fermenters, p. 98**

Cheese, Yogurt, and Other Fermented Milk Products

In a cow's udder, milk is sterile, but it rapidly becomes contaminated with a variety of microorganisms during milking and handling. Various species of lactic acid bacteria commonly reside on the udder and are inevitably introduced into the milk. If the milk is not refrigerated, these bacteria readily ferment lactose, the predominant sugar in milk, producing lactic acid. The removal of the primary carbohydrate as a nutrient source, combined with the production of lactic acid, inhibits the growth of many other microorganisms in the milk. Aesthetic features of the milk change as well because lactic acid lowers the pH, which in turn causes the milk proteins to coagulate or curdle, and sours the flavor. Historically, people most likely acquired a taste for sour curdled milk out of necessity and, later, began experimenting to intentionally produce more desirable fermented milk products. ■ **milk spoilage, p. 770**

Today, with high quality control standards and the use of pasteurized milk, which has been heated to kill most bacteria, the commercial production of fermented milk products does not rely on naturally introduced lactic acid bacteria. Instead, **starter cultures** containing one or more strains of lactic acid bacteria are added to the milk. These strains are carefully selected to produce the most desirable flavors and textures to enhance the quality of

TABLE 32.1 Foods Produced Using Lactic Acid Bacteria

Food	Characteristics
Milk products	
Cheese (unripened)	Employs a starter culture usually containing *Lactococcus cremoris* and *L. lactis*.
Cheese (ripened)	Employs rennin and a starter culture containing *Lactococcus cremoris* and *L. lactis*; ripened for weeks to years; other bacteria and/or fungi may be added to enhance flavor development.
Yogurt	Employs a starter culture containing *Streptococcus thermophilus* and *Lactobacillus bulgaricus*.
Sweet acidophilus milk	*Lactobacillus acidophilus* added for purported health benefits.
Vegetables	
Sauerkraut	Cabbage; succession of naturally occurring bacteria including *Leuconostoc mesenteroides, Lactobacillus brevis,* and *Lactobacillus plantarum*.
Pickles	Cucumbers; naturally occurring bacteria.
Poi	Taro root; naturally occurring bacteria; Hawaii.
Olives	Green olives
Kimchee	Cabbage and other vegetables; Korea.
Meats	
Dry and semidry sausages	Employs a starter culture containing species of *Lactobacillus* and *Pediococcus*; meat is stuffed into casings, incubated, heated, and then dried.

the fermented milk product. Precious starter cultures must be carefully maintained and protected against contamination, particularly by bacteriophages, which can damage or destroy them. Efforts are currently being made to select or develop bacteriophage-resistant strains of lactic acid bacteria. ■ bacteriophage, p. 323

Cheese Cheese-making probably originated in Asia more than 8,000 years ago, yet even with today's modern production methods, it has remained an art requiring experience, timing, and patience. Cheese can be made from the milk of a wide variety of animals, but most common cheeses are made with cow's milk. Sheep or goat's milk may be used to give characteristic flavors. Cheeses are classified as very hard, hard, semisoft, and soft, according to their percentage of water.

Cottage cheese is one of the simplest cheeses to make. Pasteurized milk is inoculated with a starter culture, usually containing *Lactococcus* (*Streptococcus*) *cremoris* and *L. lactis*, and then incubated until fermentation of the lactose produces enough lactic acid to cause the proteins in milk to coagulate. The coagulated proteins, or **curd**, are heated and cut into small pieces to facilitate drainage and removal of the liquid waste portion, or **whey**. Unlike most cheeses that undergo further microbial processes called **ripening** or **curing**, cottage cheese is unripened.

The initial steps in the production of ripened cheese are the same as those of cottage cheese, except the enzyme **rennin** is added to the fermenting milk to hasten protein coagulation (**figure 32.4**). Rennin is an enzyme naturally found in the stomach of young calves, where it aids in digestion of the mother's milk. Today, rennin is commercially produced using genetically engineered microorganisms. After the whey is separated from the curds, the curds are salted and pressed, and then they are molded into the traditional forms of cheese, usually bricks or wheels. The molded cheese is then ripened to encourage characteristic changes in texture and flavor.

Depending on the type of cheese, the ripening process can take from several weeks to several years. Changes imparted during this time are generally due to the metabolic activities of naturally occurring or starter lactic acid bacteria. Longer ripening processes give rise to more acidic, sharper cheeses. Some cheeses are inoculated with other bacteria or fungi that impart characteristics particular to the kind of cheese. For example, the bacterium *Propionibacterium shermanii* ripens Swiss cheese and gives it the characteristic holes, or eyes, and a nutty flavor. This bacterium ferments organic compounds to produce propionic acid and CO_2. The CO_2 gas causes the holes in the cheese, while the propionic acid imparts the typical flavor. Propionic acid also inhibits spoilage organisms. Roquefort, Gorgonzola, and Stilton cheeses are ripened by the fungus *Penicillium roquefortii*. Growth of the fungus along cracks in the cheese gives these cheeses the distinctive bluish-green veins. Brie and Camembert are ripened by a white fungus such as *P. candidum* or *P. camemberti* inoculated on the surface of the cheese. The mycelia of the fungus produce enzymes that alter texture and flavor as they gradually work their way into the cheese. Limburger cheese is made in a similar manner, but with the bacterium *Brevibacterium linens*.

The fermentation and the ripening process must both be carefully controlled to guard against growth of contaminating microorganisms that can produce undesired flavors, textures, and appearances. Fortunately, salting, low oxygen levels, acidity,

Lactic acid production and rennin activity cause the milk proteins to coagulate. The coagulated mixture is cut to facilitate the separation of the solid curds and liquid whey.

The curds are heated and cut into small pieces. The liquid whey is removed by draining.

Curds are salted and pressed into blocks or wheels for aging.

Figure 32.4 Commercial Production of Cheese Why does a longer ripening process give rise to a sharper cheese?

and presence of microbial metabolites such as propionic acid inhibit the growth of most pathogens in cheese.

From a microbiological standpoint, an interesting challenge in cheese-making is the disposal of copious volumes of whey produced during cheese production. Roughly 9 liters of protein and carbohydrate-rich whey are generated for every liter of curd. If large volumes of untreated whey are dumped into a body of water, aerobic microorganisms metabolize the organic matter, depleting the water's dissolved oxygen during the process, which results in the death of fish and other aquatic animals that require oxygen. Therefore, before disposal, whey must be treated to decrease its organic content. Alternatively, whey is sometimes spread on agricultural land. This disposal method poses environmental problems when the whey has a high salt or acid content. Commercial uses for whey, such as animal feed and protein food additives, are being developed in response to environmental and economic concerns associated with its disposal. ■ **biochemical oxygen demand, p. 793**

Yogurt To produce yogurt, pasteurized milk is concentrated slightly and then inoculated with a starter culture containing equal amounts of *Streptococcus thermophilus* and *Lactobacillus bulgaricus.* The mixture is incubated at 40°C to 45°C for several hours, during which time these thermophilic bacteria grow rapidly and produce lactic acid and other end products, such as acetaldehyde, that contribute to the flavor. Carefully controlled incubation conditions favoring the balanced growth of *S. thermophilus* and *L. bulgaricus* ensure the proper portions of acid and flavor compounds.

Acidophilus Milk Traditional acidophilus milk is the product of fermentation by *Lactobacillus acidophilus.* The more readily available **sweet acidophilus milk** retains the flavor of fresh milk because it is not fermented. Instead, a culture of *L. acidophilus* is added immediately before packaging. The bacteria are simply included for their purported health benefits. Some evidence suggests they may aid in the digestion of lactose as well as prevent and reduce the severity of some diarrheal illnesses, but the role they play in the complex interactions of the human intestinal tract is not clear. Unlike most lactic acid bacteria used as starter cultures, *L. acidophilus* can potentially colonize the intestinal tract.

Pickled Vegetables

Another fermentation process known as **pickling** originated as a way to preserve vegetables such as cucumbers and cabbage. Today, pickled products such as sauerkraut (cabbage), pickles (cucumbers), and olives are valued for their flavor. Unlike the fermentation of milk products that rely on the use of starter cultures in the manufacturing process, fermentation of most vegetables utilizes naturally occurring lactic acid bacteria residing on the vegetables.

One of the most well-studied natural fermentations is the production of sauerkraut. The cabbage is first shredded and layered with salt. The layers are firmly packed and pressed down to provide an anaerobic environment. The salt draws water and nutrients from the cabbage cells, creating a salty brine that inhibits the growth of many bacteria but permits the growth of the naturally occurring lactic acid bacteria. Controlled conditions are important to ensure that only the desired lactic acid bacteria grow. Too much salt inhibits the lactic acid bacteria, enabling the growth of other organisms that spoil the product. Under controlled conditions, natural successions of lactic acid bacteria grow. These bacteria, *Leuconostoc mesenteroides*, *Lactobacillus brevis*, and *Lactobacillus plantarum*, produce lactic acid, which lowers the pH, further inhibiting undesired bacteria. The lactic acid and other end products of the fermentation give sauerkraut its characteristic tangy taste. When the desired flavor has been attained, usually after 2 to 4 weeks at room temperature, the sauerkraut is often canned. Similar processes are used to make some pickles, olives, and other vegetable products.

Fermented Meat Products

Traditionally, fermented meat products, such as salami, pepperoni, and summer sausage, were produced by enabling the initially small numbers of lactic acid bacteria naturally present to multiply to the point of dominance. Relying on the natural fermentation of meat is inherently risky, however, because the incubation conditions used to initiate fermentation can potentially support the growth and toxin production of pathogens such as *Staphylococcus aureus* and *Clostridium botulinum*. The development and use of reliable starter cultures assures a more rapid production of lactic acid, inhibiting the growth of pathogens and enhancing flavor development. Starter cultures used by U.S. sausage-makers typically contain species of *Lactobacillus* and/or *Pediococcus*, depending on the type of sausage.

To make fermented sausages, meat is ground and combined with a starter culture and other ingredients including sugar, salt, and nitrite. The sugar serves as a substrate for fermentation, because meat does not naturally contain enough fermentable carbohydrate to produce sufficient amounts of lactic acid. Salt and nitrite contribute to the flavor of sausage; they also inhibit the growth of natural microflora and, most importantly, *Clostridium botulinum*. After thorough blending, the mixture is stuffed into a casing and incubated from one to several days. When the desired amount of acid has been produced or the fermentation is complete, the product may be smoked or otherwise heated to kill bacteria, and it is then dried.

Alcoholic Fermentations by Yeast

Some yeasts, such as members of the genus *Saccharomyces*, ferment simple sugars to produce ethanol and carbon dioxide. These yeasts are found naturally on the skin of many fruits, and so it is not surprising that early humans learned to make alcoholic drinks. As many as 10 million yeast cells may be found on the dull waxy film or "bloom" that covers a single grape. Juice from any of a variety of fruits, such as grapes, pears, plums, or apples, sitting at room temperature for several days will usually form bubbles on the surface as the sugars in the fruits are fermented, producing carbon dioxide and ethanol. Commercial production of wine, beer, and other alcoholic beverages is really no more than a variation on this process, although the source and amount of fermentable sugar, pH, and strain of yeast are strictly controlled.

Yeast is also used to make bread, but in this case, the alcohol is lost to evaporation during baking. Instead, the production of carbon dioxide is critical to bread-making. The expansion of the gas in the dough causes bread to rise.

Table 32.2 describes some of the products made utilizing alcoholic fermentation by yeast.

Wine

Wine is the product of the alcoholic fermentation of naturally occurring sugars in the juices of fruit, most commonly grapes. An amazing variety of microbial transformations occurs during both the fermentation and aging of wines, giving rise to complex flavors and aromas. Enzymes of yeast and other microorganisms catalyze the conversion of sugars, organic acids, amino acids, pigments, and many other components of fruit into numerous different substances including alcohol. Only a few of these reactions are completely understood.

One of the most important variables in the quality of wine is the variety and quality of grapes used. The growing conditions and ripeness as well as other factors affect the grapes' content of sugar, acids, and various organic compounds, which in turn critically influences the final product. Commercially, wine is made by crushing carefully selected grapes in a machine that removes the stems and collects the resulting solids and juices, or **must** (**figure 32.5**). For red wine, the entire must, including skins and pulp, of red grapes is put into the fermentation vat. The red color and complex flavors contributed by the chemical **tannin** of red wines are derived from components of the grape skin and seeds. The solids are removed during fermentation by a wine press that separates them from the partially fermented juice once the desired amount of color and tannin have been extracted. For the production of white wines, the solids are removed immediately and only the clear juice of white, or occasionally red, grapes is fermented. Rose wines obtain their light pink color from the

TABLE 32.2 Foods Produced Using Alcoholic Fermentation by Yeast

Food	Characteristics
Alcoholic Beverages	
Wine	Sugars in grape juice are fermented.
Sake	Amylase from mold (*Aspergillus oryzae*) converts the starch in rice to sugar, which is then fermented by *Saccharomyces*.
Beer	Enzymes in germinated barley convert starches of barley and other grains to sugar, which is then fermented by *Saccharomyces*.
Distilled spirits	Sugars, or starches that are converted to sugars, are fermented by *Saccharomyces*; distillation purifies the alcohol.
Vinegar	Alcohol produced by fermentation is oxidized to acetic acid by species of *Gluconobacter* or *Acetobacter*.
Breads	*Saccharomyces* ferments sugar; expansion of CO_2 causes the bread to rise; alcohol is lost to evaporation.

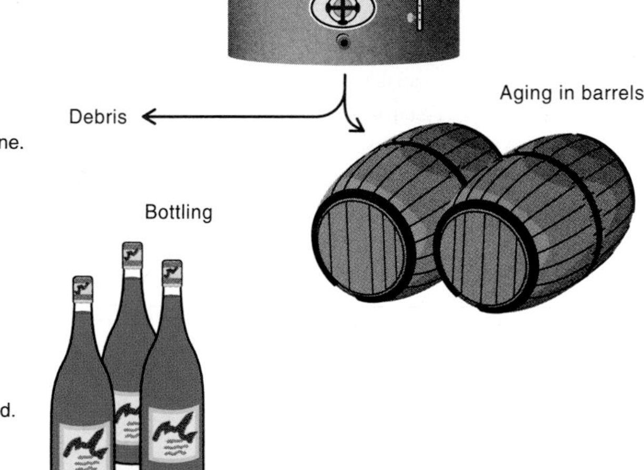
Stemmer-crusher
Fruit, most commonly grapes, is crushed to yield must, composed of solids and juices. Stems are removed. Red wines are made from the entire must of grapes. White wines are made from the clear juice of white or red grapes.
Stems
Fermenting vat
Sulfur dioxide is added to inhibit wild yeasts and spoilage bacteria. Specially selected strains of Saccharomyces are added. The fermentation process begins.
Settling vat
The wine is siphoned several times to separate the fermented juice from the particulate debris.
Aging in barrels
During the aging process, chemical and microbial changes occur that contribute to the complex flavors of wine.
Debris
Bottling
Wine is clarified by filtration and bottled.

Figure 32.5 Commercial Production of Wine

entire crushed red grape fermented for about 1 day, after which the juice is removed and fermented alone.

The fermentation must be carefully controlled to ensure that desired reactions occur. Sulfur dioxide (SO_2) is generally added to inhibit the growth of the natural microbial population of the grape, including acetic acid bacteria. These bacteria can convert alcohol to acetic acid (vinegar) and are the most prevalent cause of wine spoilage.

The fermentation process is initiated by the addition of specially selected strains of *Saccharomyces cerevisiae*. Species of *Saccharomyces* are more resistant to the antimicrobial action of SO_2 and produce a higher alcohol content than naturally occurring yeasts. Fermentation is carried out at a carefully controlled temperature, which varies with the type of wine, for a period ranging from a few days to several weeks. During fermentation most of the sugar is converted to ethanol and CO_2, generally resulting in a final alcohol content of less than 14%. Dry wines result from the complete fermentation of the sugar, whereas sweet wines contain residual sugar.

In addition to the alcoholic fermentation of the grape sugars, a distinctly different type of fermentation, called **malolactic fermentation**, may occur during wine production. Lactic acid bacteria, primarily species of *Leuconostoc*, convert malic acid to the less acidic lactic acid. Red wines made from grapes grown in cool regions tend to have high levels of malic acid, and their flavor is mellowed by this fermentation.

After fermentation, the wine is siphoned several times to remove the clear juice from the sediment of yeast and particulate debris. Most red wines and some white wines are then aged in oak barrels, contributing to the complexity of the flavor. Red wines are generally aged for 1 to 2 years. Wine is clarified by filtration and bottled. The CO_2 produced during fermentation is usually released before the wine is bottled, resulting in a "still" (noncarbonated) wine. Other processes are used to prepare carbonated wines such as champagne.

The Japanese wine, sake, depends on several microbial fermentation reactions. First, cooked rice is inoculated with the fungus *Aspergillus oryzae*. The fungus produces the enzyme amylase, which degrades the rice starch to sugar. Then, a strain of *Saccharomyces* that converts the sugar to alcohol and CO_2 is added. Lactic acid bacteria add to the flavor by producing lactic acid and other fermentation end products.

Beer

The production of beer is a multistep process designed to break down the starches of grains such as barley to produce simple sugars, which can then serve as a substrate for alcoholic fermentation by yeast. Yeast alone cannot convert grain to alcohol because they lack the enzymes that degrade starch, the primary carbohydrate of grain. Sprouted or germinated barley, however, known as

malted barley or **malt**, naturally contains these and other important enzymes.

Dried, roasted malt is ground, mixed with **adjuncts** (starches, sugars, or whole grains such as rice, corn, or sorghum), and then soaked in warm water in a process called **mashing (figure 32.6)**. During this process enzymes of the malt act on the starches, converting them to fermentable sugars. The final characteristics of the beer such as color, flavor, and foam are derived entirely from compounds in the roasted malt. The adjuncts simply serve as readily available, less expensive sources of carbohydrates for alcohol production.

After mashing, the residual solids or **spent grains** are removed to yield the sugary liquid called **wort. Hops**, the flowers of the vinelike hop plant, are added to the wort to impart a desirable bitter flavor to the beer and contribute antibacterial substances. The mixture is boiled for several hours to extract the flavor components of hops, concentrate the wort, inactivate enzymes, kill most microorganisms, and precipitate proteins, facilitating their removal. The wort is then centrifuged to remove the solids, including hops and precipitated proteins, and cooled before being transferred to the fermentation tank.

Special strains of **brewer's yeasts** (as opposed to baker's yeasts) are commonly used in beer-making. **Bottom yeasts** such as *Saccharomyces carlsbergensis*, which tend to form clumps that sink to the bottom of the fermentation vat, are used to make lager beers. These yeasts ferment best at temperatures between 6°C and 12°C and usually take 8 to 14 days to complete fermentation. In contrast, top-fermenting yeasts, such as *Saccharomyces cerevisiae*, are used to make ales. **Top-fermenting yeasts** are distributed throughout the wort but are carried to the top of the vat by the rising CO_2. They also ferment at higher temperatures (14°C to 23°C) and over a shorter period (5 to 7 days). Most American beers are lager beers produced by the bottom yeasts. Porter and stout beers are made using top-fermenting yeasts.

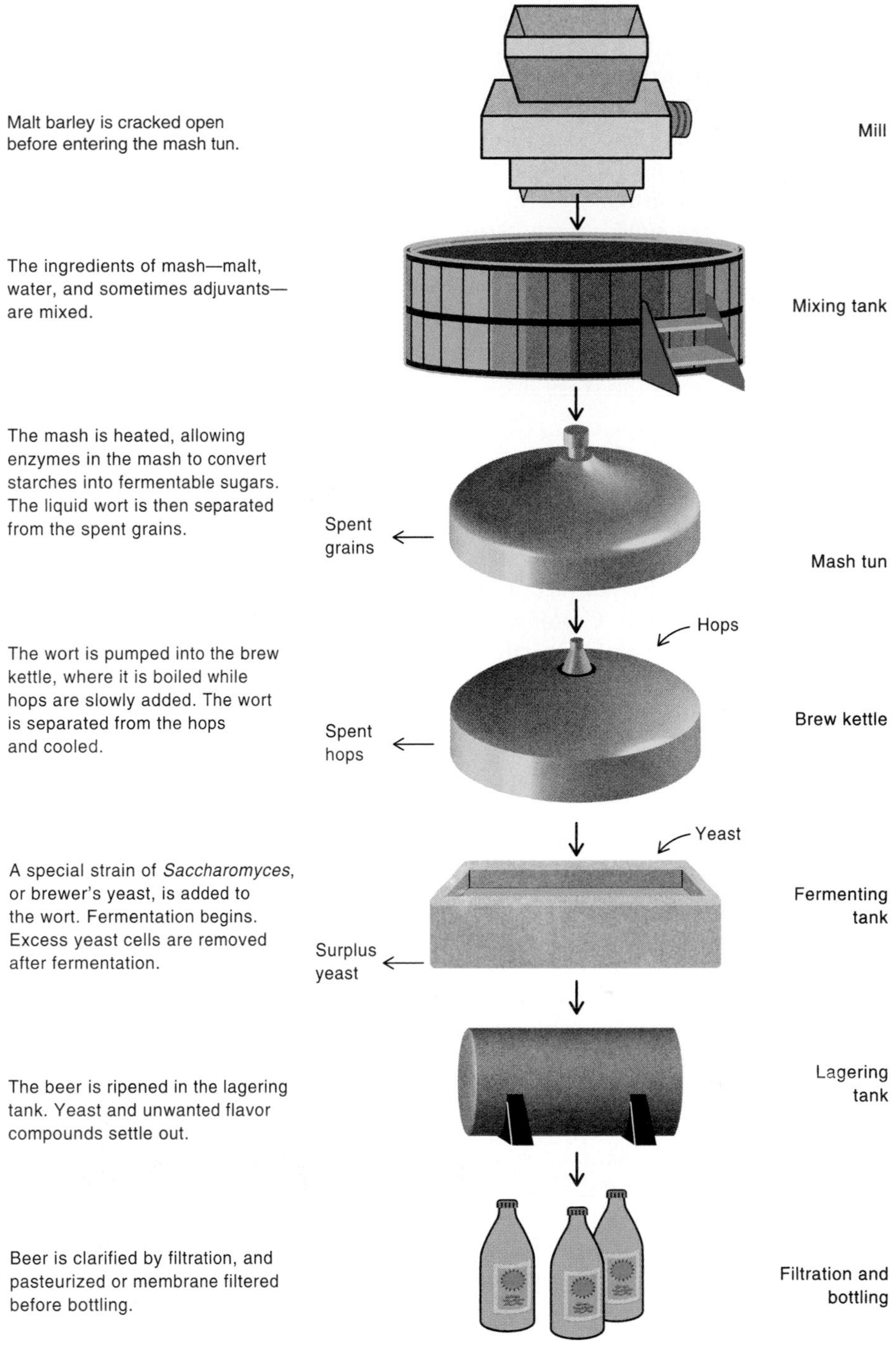

Malt barley is cracked open before entering the mash tun.
Mill
The ingredients of mash—malt, water, and sometimes adjuvants—are mixed.
Mixing tank
The mash is heated, allowing enzymes in the mash to convert starches into fermentable sugars. The liquid wort is then separated from the spent grains.
Spent grains
Mash tun
Hops
The wort is pumped into the brew kettle, where it is boiled while hops are slowly added. The wort is separated from the hops and cooled.
Spent hops
Brew kettle
Yeast
A special strain of *Saccharomyces*, or brewer's yeast, is added to the wort. Fermentation begins. Excess yeast cells are removed after fermentation.
Surplus yeast
Fermenting tank
The beer is ripened in the lagering tank. Yeast and unwanted flavor compounds settle out.
Lagering tank
Beer is clarified by filtration, and pasteurized or membrane filtered before bottling.
Filtration and bottling

Figure 32.6 Commercial Production of Beer What is the purpose of the membrane filtering in the last step?

The fermentation process generates beer with an alcohol content ranging from 3.4% to 6%. Most of the yeasts settle out following fermentation and are removed. Excess yeast can be sold as flavor and dietary supplements. The beer is then aged, during which time residual unwanted flavor compounds are metabolized by remaining yeast cells or settle out. Cask-conditioned beer undergoes a second fermentation, which generates CO_2 in the cask. Other beers must be carbonated to replace the CO_2 allowed to escape during fermentation. After aging, beer is clarified by filtration, microorganisms are removed or killed using membrane filtration or pasteurization, and the product is packaged.

Distilled Spirits

The manufacturing process of distilled spirits such as scotch, whiskey, and gin is initially similar to that of beer, except the wort is not boiled. Consequently, degradation of starch by the enzymes in the wort continues during the fermentation. When the fermentation is complete, the ethanol is purified by distillation.

Different types of spirits are made in different ways. For example, rum is made by fermenting sugar cane or molasses. Malt scotch whiskey is the product of the fermentation of barley that is then aged for several years in oak sherry casks. The wood and the residual sherry contribute both flavor and color to the whiskey as it ages. Lactic acid bacteria are used to produce lactic acid in grain mash for making sour-mash whiskey. The yeast *S. cerevisiae* subsequently ferments the sour mash to form alcohol. The distilled spirit tequila is traditionally made from the fermentation of juices from the agave plant using the bacterium *Zymomonas mobilis*. This bacterium ferments sugars to ethanol and CO_2 via a pathway similar to the yeast alcoholic fermentation pathway.

Vinegar

Vinegar, which is an aqueous solution of at least 4% acetic acid, is the product of the oxidation of ethanol by the acetic acid bacteria, *Acetobacter* and *Gluconobacter* species. Acetic acid bacteria are strictly aerobic, Gram-negative rods, characterized by their ability to carry out a number of oxidations. They can tolerate high concentrations of acid as they oxidize alcohol to acetic acid (acetification).

Alcohol is commercially converted to vinegar using processes that provide readily available oxygen to hasten the oxidation reaction. The **vinegar generator** sprays alcohol onto loosely packed wood shavings that harbor a biofilm of acetic acid bacteria. As the alcohol trickles through the bacteria-coated shavings, it is oxidized to acetic acid. In principle, the vinegar generator operates much like the trickling filter used in wastewater treatment by providing a large surface area for aerobic metabolism. The **submerged culture reactor** is an enclosed system that continuously pumps small air bubbles into alcohol that has been inoculated with acetic acid bacteria. ■ **trickling filter, p. 796** ■ **biofilm, p. 108**

Bread

Yeast bread rises through the action of **baker's yeast**, strains of *Saccharomyces cerevisiae* carefully selected for the commercial baking industry. The CO_2 produced during fermentation causes the bread to rise, producing the honeycombed texture characteristic of yeast breads. The alcohol evaporates during baking. Modern microbiological techniques are used to produce a high-quality yeast inoculum that gives the reliable performance demanded by commercial and home bakers alike.

Yeast bread is made from a mixture of flour, sugar, salt, milk or water, yeast, and sometimes butter or oil (**figure 32.7**). Packaged baker's yeast that can be reconstituted in warm water is readily available as pressed cakes or dried granules. An excess of yeast is added to enable adequate production of CO_2 in a time period too short to permit multiplication of spoilage bacteria.

Sourdough bread is made with a combination of yeast and lactic acid bacteria. Alcohol and CO_2 are produced as well as lactic acid, giving the bread its sour flavor.

Ingredients (mix)

Ingredients of bread include yeast, milk or water, oil, flour, salt, and sugar.

Knead

The dough is thoroughly mixed and kneaded.

Rise (fermentation)

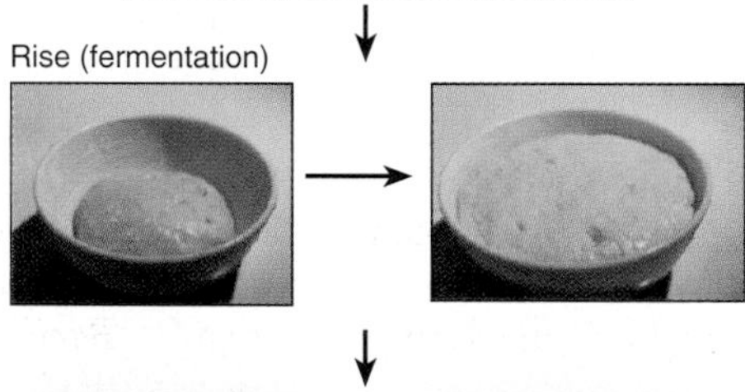

The dough is put in a bowl and allowed to rise. During this time, the yeast produces ethanol and CO_2. The dough rises to approximately double the original volume. The gas generated creates air pockets in the dough, producing the texture we see in the finished bread.

Following the period of rising, the dough is shaped into loaves and then goes through a second rising. The bread dough is then baked.

Baking

Note the holes caused by the production of CO_2 and the evaporation of ethanol in the finished loaf of bread.

Figure 32.7 Bread Production

Changes Due to Mold Growth

Molds contribute to the flavor and texture of some cheeses, as already discussed. In addition, many traditional dishes and condiments used throughout the world are produced by encouraging the growth of molds on food (**table 32.3**). Successions of naturally occurring microorganisms are often involved. Like other fermented foods not produced commercially, the microbiological and chemical aspects of many of these traditional foods have not been extensively studied.

Soy sauce

Soy sauce is made by inoculating equal parts of cooked soybeans and roasted cracked wheat with a culture of either *Aspergillus oryzae* or *A. sojae*. The mixture, called **koji**, is allowed to stand for several days, during which time carbohydrates and proteins in the soybeans are broken down, producing a yellow-green liquid containing fermentable sugars, peptides, and amino acids. After this initial step, the mixture is put into a large container with an 18% NaCl solution, or brine. Salt-tolerant microorganisms then grow and impart changes over an extended period. Microorganisms involved in this stage of fermentation include lactobacilli and pediococci, as well as the yeasts *Zygosaccharomyces rouxii* and a species of *Torulopsis*. The brine mixture is allowed to ferment for 8 to 12 months, after which the liquid soy sauce is removed. The residual solids are used as animal feed.

MICROCHECK 32.3

Lactic acid bacteria are used to produce a variety of foods including cheese, yogurt, sauerkraut, and some sausages. Yeast is used to produce alcoholic beverages and breads. Some cheeses and many traditional foods owe their characteristics to changes caused by molds.

- Describe how the metabolism of lactic acid bacteria differs from that of most other microorganisms that can grow aerobically.
- How does the use of starter cultures improve the safety of fermented meat products?
- How could cottage cheese be produced without bacteria?

Food Spoilage

Food spoilage encompasses any undesirable changes in food. Spoilage microorganisms and their microbial metabolites that cause repugnant tastes and odors, although aesthetically disagreeable, generally are not harmful. This is not surprising when their growth requirements are considered. Most human pathogens grow best at temperatures near 37°C, whereas most foods are usually stored at temperatures well below the normal body temperature. Similarly, the nutrients available in fruits, vegetables, and other foods are often not suitable for the optimum growth of human pathogens. In addition, the pH of foods is often more acid or alkaline than the pH of body fluids. As a result, the nonpathogens can easily outgrow the pathogens when competing for the same nutrients. Spoiled foods are considered unsafe to eat, however, because the presence of spoilage organisms growing to high numbers indicates that foodborne pathogens may have had the opportunity to grow as well.

Common Spoilage Bacteria

A wide range of bacteria is important in food spoilage. *Pseudomonas* species can metabolize a wide variety of compounds, and they can grow on and spoil many different kinds of foods, including meats and vegetables. Psychrophilic pseudomonads can multiply at refrigeration temperatures and are notorious for spoiling meats as well as other foods. Members of the genus *Erwinia* can produce enzymes that degrade pectin, and so they commonly cause soft rot of fruits and vegetables. *Acetobacter* species can transform ethanol to acetic acid, the principal acid of vinegar. Although this property is very beneficial to commercial producers of vinegar, it presents a great problem to wine producers. Milk products are sometimes spoiled by members of the genus *Alcaligenes* that form a glycocalyx, causing strings of slime, or "ropiness," in raw milk. The lactic acid bacteria, including species of *Streptococcus*, *Leuconostoc*, and *Lactobacillus*, all produce lactic acid. Anyone who has unexpectedly consumed sour milk, knows that this can be disagreeable. The genera *Bacillus* and *Clostridium* are particularly troublesome because their heat-resistant endospores survive cooking and, in some cases, canning. *Bacillus coagulans* and *B. stearothermophilus* cause spoilage of some canned foods. ■ glycocalyx, p. 66

TABLE 32.3 Foods Produced Using Molds

Food	Characteristics
Soy sauce	Koji is produced by inoculating soybeans and cracked wheat with a starter culture of *Aspergillus oryzae* or *A. sojae*; the mixture is then added to a brine and incubated for many months.
Tempeh	Soybeans are fermented by lactic acid bacteria and then inoculated with the mold *Rhizopus*; Indonesia.
Miso	Rice, soybeans, or barley are inoculated with *Aspergillus oryzae*; Asia.
Cheeses	
Roquefort, Gorgonzola, and Stilton	Curd is inoculated with *Penicillium roquefortii*.
Brie and Camembert	Wheels of cheese are inoculated with selected species of *Penicillium*.

Common Spoilage Fungi

A wide variety of fungi, including species of *Rhizopus*, *Alternaria*, *Penicillium*, *Aspergillus*, and *Botrytis*, spoil foods. Since fungi grow readily in acidic as well as low-moisture environments, fruits and breads are more likely to be spoiled by fungi than by bacteria. *Aspergillus flavus* infects peanuts and other grains, producing **aflatoxin**, a potent carcinogen monitored by the Food and Drug Administration.

MICROCHECK 32.4

The metabolic products of microorganisms can spoil foods by imparting undesirable flavors, odors, and textures.

- What characteristics of members of the genus *Pseudomonas* enable them to spoil such a wide variety of foods?
- Why do fungi most commonly spoil breads and fruits?

Foodborne Illness

Foodborne illness occurs when a pathogen, or a toxin it has produced, is consumed in a food product. Food production is carefully regulated in the United States to prevent foodborne illness. Federal, state, and local agencies cooperate in inspections to help enforce protective laws. The Federal Food, Drug, and Cosmetic Acts set standards for foods that are shipped between states or territories, and they require accurate labeling of the foods. Other federal laws require meats and poultry produced for interstate shipment to be inspected. State and local agencies enforce many of the laws that protect food consumers. The United States and other countries, however, also rely on food manufacturers to act as their own watchdogs. In these days of instant news media, it is very much in the interest of these companies to maintain quality control, as even a small outbreak of food poisoning can lead to a nationwide lack of consumer confidence and the resulting heavy financial loss.

Generally, the microbiological quality of most foods sold in the United States is very good compared with that found in some other parts of the world, but consumers must employ sound preserving, preparation, and cooking techniques to avoid hazards from both home-prepared and purchased food products. In spite of strict controls on the quality of the food supply, the United States is far from free of foodborne illness. Millions of cases are estimated to occur each year from foods prepared either commercially, at home, or in institutions such as hospitals and schools. The vast majority of these cases could have been prevented with proper storage, sanitation, and preparation. **Table 32.4** lists some bacteria that cause foodborne illness in the United States.

One of the challenges in preventing foodborne illness is developing reliable methods to detect pathogens before a contaminated product is marketed and consumed. Not only are culture methods time-consuming, they often lack sensitivity because of the large number of competing organisms generally found on foods. Rapid methods utilizing antibody tests, such as ELISA, and DNA probes are increasingly being developed to detect many pathogens. ■ **DNA probes, p. 232, ELISA, p. 422**

Foodborne Intoxication

Foodborne intoxication is an illness that results from the consumption of a toxin produced by a microorganism growing in a food product. When such a food product is ingested, it is the toxin that causes illness, not the living organisms (**figure 32.8**). The symptoms of the illness, which in some cases appear within a few hours of ingestion of the food, may indicate the type of toxin ingested. *Staphylococcus aureus* and *Clostridium botulinum* are two examples of organisms that cause foodborne intoxication.

Staphylococcus aureus

Many strains of *Staphylococcus aureus* produce a toxin that, when ingested, causes nausea and vomiting. Although *S. aureus*

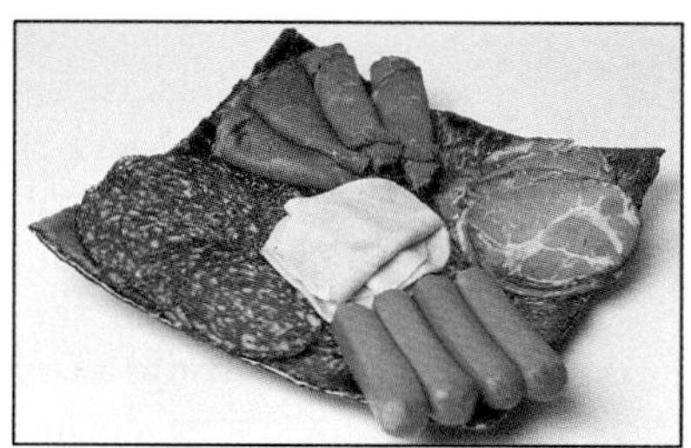

- Most bacteria that normally compete with *Staphylococcus aureus* are either killed by cooking or inhibited by high salt conditions.
- Food handler inoculates *S. aureus* onto food.
- *S. aureus* grows and produces toxin when food is allowed to slowly cool or is stored at room temperature.
- A person ingests the toxin-containing food. Symptoms of staph food poisoning, including nausea, abdominal cramping and vomiting, begin after 4–6 hours.

- *Clostridium botulinum* endospores, common in soil and marine sediments, contaminate many different foods.
- Endospores survive inadequate canning processes. Canned foods are anaerobic.
- Surviving *C. botulinum* endospores germinate, grow, and produce toxin in low-acid canned foods.
- A person ingests the toxin-containing food. Symptoms of botulism, including weakness, double vision, and progressive inability to speak, swallow, and breathe, begin in 12–36 hours.

Figure 32.8 **Typical Events Leading to *Staphylococcus aureus* and *Clostridium botulinum* Foodborne Intoxications**

TABLE 32.4 Common Foodborne Illnesses

Organism	Page	Symptoms	Foods Commonly Implicated
Intoxication			
Clostridium botulinum	p. 675	Weakness; double vision; progressive inability to speak, swallow, and breathe	Low-acid canned foods such as vegetables and meats
Staphylococcus aureus	p. 525	Nausea, vomiting, abdominal cramping	Cured meats, creamy salads, cream-filled pastries
Infection			
Bacillus cereus		Two disease types: nausea and vomiting, or watery diarrhea and abdominal cramps	Meats, milk, shellfish, rice products
Campylobacter species	p. 606	Diarrhea, fever, abdominal pain, nausea, headache	Poultry, raw milk
Clostridium perfringens	p. 701	Intense abdominal cramps, watery diarrhea	Meats, meat products
Escherichia coli 0157:H7	p. 603	Severe abdominal pain, bloody diarrhea; may progress to hemolytic uremic syndrome	Ground beef
Listeria monocytogenes	p. 671	Influenza-like symptoms, fever; may progress to septicemia, meningitis	Raw milk, cheese, meats, vegetables
Salmonella species	p. 604	Nausea, vomiting, abdominal cramps, diarrhea, fever	Poultry, eggs, milk, meat
Shigella species	p. 602	Abdominal cramps; diarrhea with blood, pus, or mucus; fever; vomiting	Salads, raw vegetables
Vibrio parahaemolyticus	p. 289	Diarrhea, abdominal cramps, nausea, vomiting, headache, fever, chills	Fish and shellfish
Vibrio vulnificus	p. 789	Diarrhea, nausea; may progress to septicemia	Oysters, clams, crabs
Yersinia enterocolitica		Abdominal pain, fever, diarrhea, vomiting; may mimic appendicitis	Meats, oysters, fish, raw milk

does not compete well with most spoilage organisms, it thrives in moist, rich foods in which other organisms have been killed or their growth has been inhibited. For example, *S. aureus* can grow with little competition on unrefrigerated salty products such as ham (a_w of 0.91). Creamy pastries and starchy salads stored at room temperature also offer ideal conditions for the growth of this pathogen, because the cooking of the ingredients kills most other competing organisms. The source of *S. aureus* is usually a human carrier who has not followed adequate hygiene procedures, such as handwashing, before preparing the food. If the organism is inoculated into a food that can support its growth, and the food is left at room temperature for several hours, *S. aureus* can grow and produce the toxin. Unlike most exotoxins, *S. aureus* toxin is heat-stable, so that cooking the food will not destroy it. ■ ***Staphylococcus aureus*, p. 525**

Botulism

Botulism is caused by ingestion of a neurotoxin produced by the anaerobic, spore-forming, Gram-positive rod *Clostridium botulinum*. Unfortunately, growth of the organism and production of the deadly toxin may not result in any noticeable changes in the taste or appearance of the food. Canning processes are specifically designed to destroy the endospores of this organism, but processing errors or post-processing contamination can allow the germination of the endospores and growth of the vegetative cells that produce toxin. Such errors are extremely rare in commercially canned foods, and most cases of botulism are due to processing errors in home-canned foods. As an added safety measure, home-canned foods should be boiled for 10 to 15 minutes immediately before consumption. The toxin is heat-labile (sensitive), and so the heat treatment will destroy any toxin that may have been produced. Cans that are damaged should be discarded, because they might have small holes through which *C. botulinum* could enter. Likewise, bulging cans indicate gas production, which could indicate microbial growth, and should be discarded.

Symptoms of botulism typically begin 12 to 36 hours after a person has eaten toxin-containing food. Blurred or double vision occurs initially, indicating eye muscle weakness. Progressive paralysis then ensues, generally involving all voluntary muscles, but respiratory paralysis is the most common cause of death. Botulism is treated by administering an antitoxin. The antitoxin, however, only neutralizes toxin circulating in the bloodstream, and the nerves already affected by the toxin

Perspective 32.1 Changing Habits

As the eating habits of our society change, so does the pattern of foodborne illnesses. Twenty years ago, foodborne disease outbreaks were primarily local occurrences that followed a family gathering or church picnic at which undercooked meat or eggs, or improperly stored potato salad, were served. Today, a greater variety of foods are being implicated, including more fresh fruits and vegetables, and the outbreaks are often widespread, sometimes involving hundreds or thousands of people.

In the past, food was primarily grown and prepared at home. Fresh fruits and vegetables, harvested immediately before use from the garden patch, were only available during the growing season. Today, thanks to modern transportation and technology, we can readily purchase an amazing array of foods year-round at the grocery store. These foods, often produced on a large scale, may originate from any of a number of states or countries. The risks that these foods can carry, however, are changing the scenario of foodborne illness. For example, when we import produce from other countries, the adequacy of the sanitation standards under which those foods were grown and harvested cannot be ensured. This was illustrated by widespread outbreaks of diarrheal disease in 1996 and 1997 caused by the protozoan parasite *Cyclospora cayetanensis* that were traced to raspberries imported from Guatemala. At least 1,400 individuals in more than 20 states and two Canadian provinces were affected.

Even foods grown in this country can pose a risk if they are not washed and stored properly. Lettuce and other field grown vegetables may be contaminated with *Listeria monocytogenes*, a common environmental bacterium. Careless washing, coupled with prolonged storage particularly at room temperature, can allow this organism to multiply to numbers sufficient to cause disease. *Listeria monocytogenes* poses the greatest threat to those with a compromised or poorly developed immune system, including AIDS patients and the elderly. In these patients, gastrointestinal infection is more likely to progress to an invasive, and often fatal, disease such as meningitis or septicemia.

Another factor influencing the spectrum of foodborne disease is the growing popularity of unpasteurized juices. While their wholesome and natural image may be appealing, some of these juices have been implicated in large-scale disease outbreaks. In 1999, an outbreak of salmonellosis was traced to unpasteurized orange juice. Several years earlier, unpasteurized apple juice was implicated in an outbreak of *E. coli* O157:H7.

While the overall quality of our food production and storage have improved, the widespread distribution of a food from a single source, where quality controls have broken down, now has the potential of creating far-reaching disease outbreaks.

recover slowly, over weeks or months. Despite treatment, about one-fourth of victims of botulism die. ■ **antitoxin, p. 389**

Foodborne Infection

Unlike foodborne intoxication, foodborne infection requires the consumption of living organisms. The symptoms of the illness, which usually do not appear for at least 1 day after ingestion of the contaminated food, usually include diarrhea, but vary according to the type of organism ingested. Thorough cooking of food immediately before consumption will kill the organisms, thereby preventing foodborne infection. *Escherichia coli* O157:H7 and species of *Salmonella* and *Campylobacter* are examples of organisms that cause foodborne infection (**figure 32.9**).

Salmonella and *Campylobacter*

Salmonella and *Campylobacter* are two different genera of pathogens that are commonly associated with poultry products, such as chicken, turkey, and eggs. Inadequate cooking of these products can result in foodborne infection. **Cross-contamination** of other foods can result in the transfer of pathogens to those foods. For example, if a cutting board on which raw chicken was cut is then immediately used to cut up vegetables for a salad, the salad can become contaminated with *Salmonella* or *Campylobacter* species. ■ ***Salmonella*, p. 604** ■ ***Campylobacter*, p. 606**

Escherichia coli O157:H7

The newly emerging strain of *Escherichia coli*, O157:H7, causes bloody diarrhea that sometimes develops into hemolytic uremic syndrome (HUS), which is life-threatening. In the past decade, *E. coli* O157:H7 has caused several large food poisoning outbreaks, including an epidemic in Japan in 1996 that sickened more than 9,000 people and a 1993 outbreak in the United States that spread over four states. The latter epidemic was traced to undercooked contaminated hamburger patties served at a fast-food restaurant. Ground meats are a troublesome source of foodborne infection. While the initial contamination by bacteria such as *E. coli* O157:H7 normally occurs on the surface and is easily destroyed by searing the exterior of whole meats such as steaks, grinding the meat distributes those same bacteria throughout the food. Prevention then involves more thorough cooking so that enough heat reaches the center to kill all the organisms. ■ **hemolytic uremic syndrome, p. 603**

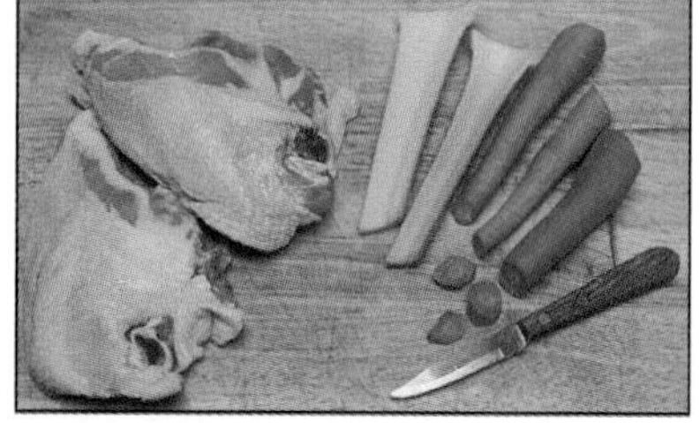

- Incomplete cooking fails to kill all pathogens. Surviving *Salmonella* and/or *Campylobacter* can multiply as food is cooled slowly or stored at room temperature.
- Live organisms are ingested. They multiply in the intestinal tract and cause disease. Symptoms include diarrhea, abdominal pain, and nausea.

- Incomplete cooking fails to kill all pathogens. Even low numbers of surviving *E. coli* O157:H7 can cause illness.
- Live organisms are ingested. They multiply in the intestinal tract and cause disease. Symptoms include severe abdominal pain and bloody diarrhea.

Figure 32.9 Typical Events Leading to *Salmonella*, *Campylobacter*, and *E. coli* O157:H7 Foodborne Infections

MICROCHECK 32.5

Foodborne intoxication results from the consumption of toxins produced by organisms growing in a food. Foodborne infection results from consumption of living organisms.

- How does cooking a home-canned food immediately prior to consumption prevent botulism?
- Which foodborne pathogen can cause hemolytic uremic syndrome?
- Why would a large number of competing microorganisms in a food sample result in lack of sensitivity of culture methods for detecting pathogens?

Food Preservation

Preventing the growth and concurrent metabolic activities of microorganisms that cause spoilage and foodborne illness preserves the quality of food. Some methods of food preservation, such as drying and salting, have been known throughout the ages, whereas others have been discovered or developed more recently. The major methods of preserving foods—high-temperature treatment, low-temperature storage, addition of antimicrobial chemicals and irradiation—are briefly summarized below and described in more detail in chapter 5.

- *Canning.* The canning process destroys all spoilage and pathogenic organisms capable of growth at normal storage temperatures. Low-acid foods are processed using steam under pressure (autoclaving) in order to reach temperatures high enough to destroy the endospores of *Clostridium botulinum*. Acidic foods are not subjected to as stringent heating conditions because *Clostridium botulinum* cannot grow and produce toxin in those foods. ■ **canning, p. 119**
- *Pasteurization.* Heating foods under carefully controlled conditions at high temperatures for short periods of time destroys non-spore-forming pathogens and substantially reduces the numbers of spoilage organisms without significantly altering the flavor. ■ **pasteurization, p. 117**
- *Cooking.* Cooking, like pasteurization, can destroy non-spore-forming pathogens and spoilage organisms. Cooking obviously alters the characteristics of food, however, and heat distribution may be uneven, resulting in survival of organisms in inadequately heated regions.
- *Refrigeration.* Refrigeration preserves food by slowing the rate of microbial growth. Many organisms, including most pathogens, are unable to multiply at low temperatures. ■ **temperature of growth, p. 96**
- *Freezing.* Freezing stops microbial growth because water in the form of ice is unavailable for biological reactions. A portion of the organisms will actually be killed by damage caused by ice crystals, but remaining organisms can grow and spoil food once it is thawed. Formation of ice crystals can also damage the cellular structure of the food product.
- *Drying/reducing the* a_w. Drying foods or adding high concentrations of sugars or salts inhibits the growth of microorganisms by decreasing the available moisture. Given enough time, however, molds may eventually grow on these foods. ■ **drying, p. 128**
- *Lowering the pH.* Lowering the pH, either by the addition of acids or by encouraging fermentation by lactic acid bacteria, inhibits a wide range of spoilage organisms and pathogens.
- *Adding antimicrobial chemicals.* Numerous chemicals inhibit the growth of microorganisms. Organic acids such as propionic acid, benzoic acid, and sorbic acid are naturally occurring antimicrobial chemicals that are added to a wide variety of foods primarily to inhibit fungal growth. Nitrates are added to cured meats to inhibit growth of *Clostridium botulinum* and other organisms. Wine, fruit juices, and other products are preserved by the addition of sulfur dioxide. ■ **chemical preservatives, p. 127**
- *Irradiation.* Gamma irradiation destroys microorganisms without significantly altering the flavor of some foods; however, its use is regulated. It has been approved by the Food and Drug Administration for use in controlling insects in fruits, vegetables, and grains, and in destroying pathogens such as *E. coli* O157:H7 and *Salmonella* species in meats. ■ **irradiation, p. 126**

MICROCHECK 32.6

Food spoilage can be eliminated or delayed by destroying microorganisms or altering conditions to inhibit their growth.

- Why does the canning process for low-acid foods require higher temperatures than the canning process for acidic foods?
- Why are nitrates added to cured meats?
- Microorganisms are often grouped according to their optimum growth temperatures. Which one of these groups is most likely to spoil refrigerated foods?

FUTURE CHALLENGES

Using Microorganisms to Nourish the World

The world's steadily increasing population mandates the efficient use of finite natural resources and the development of new protein supplies to nourish that growing populace. One potential solution that addresses both of these needs is to cultivate microorganisms as a protein source, employing industrial by-products currently considered wastes as the growth medium.

*The term **single-cell protein** or **SCP** was coined in the 1960s to describe the use of unicellular organisms such as yeast and bacteria as a protein source. Today, the term generally encompasses the use of multicellular microorganisms as well, and might more accurately be called microbial biomass. While whole cells can be consumed, a more palatable alternative is to extract proteins from those cells and use them to form textured protein products.*

Yeasts are considered the most promising large-scale source of single-cell protein. They multiply rapidly, are larger than bacteria, and are more readily acceptable as a potential food. The most suitable type of yeast depends on the growth medium employed. The various genera of yeasts utilize differing carbohydrate sources and conditions for growth. For example, Kluyveromyces marxianus *can be grown on whey, a by-product of cheese-making;* Saccharomyces cerevisiae *will grow on molasses, a by-product of the sugar industry; and* Candida utilis *can multiply on cellulose-containing by-products of the pulp and paper industry. Most of these wastes must be supplemented with a nitrogen source, as well as various vitamins and minerals, in order to support the growth of yeasts or other microorganisms.*

To a lesser extent, the use of bacteria as source of SCP is also being explored. The cyanobacterium Spirulina maxima *can be cultivated in alkaline lakes and then harvested and dried for use as a food source. Unfortunately, this requires adequate sunlight and warmth, making large-scale, year-round production impractical in many parts of the world.*

One of the chief concerns regarding the consumption of microorganisms as a protein source is their high concentration of nucleic acid, primarily RNA. Yeast and bacteria contain approximately three to eight times as much RNA per gram as does meat. High levels of nucleic acid in the diet cause an increase in uric acid in the blood, which can lead to gout and kidney stones. Chemical and enzymatic methods are being developed to decrease the nucleic acid in SCP without altering the nutritional value of the protein.

SUMMARY

Principles of Food Microbiology

1. Food is an ecosystem in which microorganisms compete to metabolize the nutrients, making end products such as acids, alcohols, and gas.
2. Foods that have been intentionally altered by carefully controlling the activity of bacteria, yeasts, or molds are called **fermented**. (Figure 32.1)
3. Biochemical changes in foods, when perceived as unpleasant, are called **spoilage**. (Figure 32.2)
4. Growth of pathogens generally does not result in perceptible changes in quality of food, but it can result in foodborne disease.
5. Spoilage can be delayed or prevented and foodborne illness can be avoided by slowing the growth of microorganisms, or by reducing or eliminating the initial numbers of them in food.

Factors Influencing the Growth of Microorganisms in Foods

1. **Intrinsic** and **extrinsic factors** together determine which microorganisms can grow and eventually predominate in a food product.

Intrinsic Factors

1. Bacteria require a high a_w. They grow quickly in fresh, moisture-rich foods, but not in dry, sugary, or salted foods. Fungi can grow in foods that have an a_w too low to support the growth of bacteria.
2. The pH of a food is important in determining which organisms can survive and thrive in it; many species of bacteria, including most pathogens, are inhibited by acidic conditions.
3. Rinds, shells, and other coverings aid in protecting some foods from the invasion of microorganisms.
4. Some foods contain natural antimicrobial chemicals that may help prevent spoilage.

Extrinsic Factors

1. Low temperatures halt or inhibit the growth of most foodborne microorganisms. Psychrophiles, however, are able to grow at refrigeration temperatures.
2. The presence or absence of oxygen affects the type of microbial population able to grow in a food. Excluding oxygen from a food prevents the growth of obligate aerobes but enables the growth of obligate anaerobes, including *Clostridium botulinum.*

Microorganisms in Food and Beverage Production

1. Not only are fermented foods perceived as pleasant tasting, the acids produced inhibit the growth of many spoilage organisms as well as foodborne pathogens.

Lactic Acid Fermentations by the Lactic Acid Bacteria (Table 32.1)

1. The tart taste of yogurt, pickles, sharp cheese, and some sausages is due to the production of lactic acid by species of *Lactobacillus, Lactococcus, Streptococcus, Leuconostoc,* and/or *Pediococcus.*
2. Commercial production of fermented milk products relies on the use of starter cultures. The lactic acid produced by the bacteria causes milk proteins to coagulate or curdle, and sours the flavor.
3. Starter cultures, and sometimes rennin, are added to pasteurized milk to make cheese. Acid production and enzyme activity in milk results in formation of **curds** and **whey**. The curds are used directly to make **unripened** cheeses, or they are **ripened** to make other cheeses. Other bacteria or fungi are sometimes added to ripened cheese to impart characteristic flavors or textures. (Figure 32.4)
4. Whey is a waste product of cheese-making.
5. Fermentation of vegetables, **pickling**, relies on naturally occurring lactic acid bacteria.
6. Commercial sausage production utilizes starter cultures to rapidly decrease the pH and prevent the growth of pathogens.

Alcoholic Fermentations by Yeast (Table 32.2)

1. Species of *Saccharomyces* ferment sugar to produce ethanol and carbon dioxide.
2. Wine is the product of the alcoholic fermentation of naturally occurring sugars in the juices of fruit, most commonly grapes, by specially selected strains of *Saccharomyces cerevisiae.* Sulfur dioxide is added to inhibit the naturally occurring microbial population of the grape. (Figure 32.5)

3. The production of beer is a multistep process designed to break down the starches of grains such as barley to produce simple sugars, which can then serve as a substrate for alcoholic fermentation by yeast. During **mashing**, enzymes in malted barley convert starch to sugar. Spent grains are removed from the liquid, **hops** are added, and the **wort** is boiled. Specially selected strains of *Saccharomyces* are then added to ferment the sugars in the wort. (Figure 32.6)
4. Distilled spirits are produced using distillation to purify the alcohol generated during fermentation.
5. Vinegar is the product of the oxidation of alcohol by the acetic acid bacteria, *Acetobacter* and *Gluconobacter*.
6. In bread-making, the CO_2 produced by yeast causes bread to rise, and the alcohol is lost to evaporation. (Figure 32.7)

Changes Due to Mold Growth (Table 32.3)

1. Some cheeses and many traditional dishes used throughout the world are produced by encouraging the growth of molds on foods.
2. Soy sauce is made by allowing species of *Aspergillus* to degrade a mixture of soybeans and wheat, which is then fermented in brine.

Food Spoilage

1. Food spoilage is most often due to the metabolic activities of microorganisms as they grow and utilize the nutrients in the food.

Common Spoilage Bacteria

1. Psychrophilic species of *Pseudomonas* can multiply at refrigeration temperatures and metabolize a wide variety of compounds; therefore, they cause spoilage of many different kinds of foods, including meats and vegetables.
2. *Erwinia*, *Acetobacter*, *Alcaligenes*, lactic acid bacteria, and endosporeformers are important causes of food spoilage.

Common Spoilage Fungi

1. Fungi grow readily in acidic as well as low-moisture environments; therefore, fruits and breads are more likely to be spoiled by fungi than by bacteria.

Foodborne Illness

Foodborne Intoxication

1. Foodborne intoxication is any illness that results from consuming of a toxin produced by a microorganism growing in a food product. (Figure 32.8)
2. Many strains of *Staphylococcus aureus* produce a toxin that, when ingested, causes nausea and vomiting.
3. Botulism is caused by ingestion of a neurotoxin produced by the anaerobic, spore-forming, Gram-positive rod *Clostridium botulinum*.
4. As an added safety measure, home-canned foods should be boiled for 10 to 15 minutes immediately before consumption to ensure destruction of potential botulinum toxin.

Foodborne Infection

1. Foodborne infection requires the consumption of living organisms. (Figure 32.9)
2. Thorough cooking of food immediately before consumption will kill bacteria, thereby preventing foodborne infection.
3. *Salmonella* and *Campylobacter* species are commonly associated with poultry products.
4. Some epidemics of *E. coli* O157:H7 have been traced to undercooked contaminated hamburger patties.

Food Preservation

1. Food spoilage can be eliminated or delayed by destroying microorganisms or altering conditions to inhibit their growth.
2. Methods used to preserve foods include canning, pasteurization, cooking, freezing, refrigeration, reducing the a_w, lowering the pH, adding antimicrobial chemicals, and irradiation.

REVIEW QUESTIONS

Short Answer

1. What is the purpose of rennin in cheese-making?
2. What causes the bluish-green veins to form in blue cheese?
3. What causes the holes to form in Swiss cheese?
4. What is the difference between traditional acidophilus milk and sweet acidophilus milk?
5. What is the purpose of the mashing step in beer-making?
6. Explain how *Alcaligenes* species cause "ropiness" in raw milk.
7. Explain the significance of *Aspergillus flavus* growing in grain products.
8. Explain the typical sequence of events that lead to botulism.
9. Explain the typical sequence of events that lead to staphylococcal food poisoning.
10. How does canning differ from pasteurization?

Multiple Choice

1. Benzoic acid is an antimicrobial chemical that is naturally found in which of the following foods?
   A. Apples
   B. Cranberries
   C. Eggs
   D. Milk
   E. Yogurt
2. The a_w of a food product reflects which of the following?
   A. Acidity of the food
   B. Presence of antimicrobial constituents such as lysozyme
   C. Amount of water available
   D. Storage atmosphere
   E. Nutrient content

3. What is a generally minimum pH for growth and toxin production by *Clostridium botulinum* and other foodborne pathogens?
   A. 8.5
   B. 7.0
   C. 6.5
   D. 4.5
   E. 2.0
4. Which of the following organisms cause foodborne intoxication?
   A. *E. coli* O157:H7
   B. *Campylobacter* species
   C. *Lactobacillus* species
   D. *Salmonella* species
   E. *Staphylococcus aureus*
5. Which group of organisms most commonly spoils breads, fruits, and dried foods?
   A. *Acetobacter*
   B. Fungi
   C. Lactic acid bacteria
   D. *Pseudomonas*
   E. *Saccharomyces*
6. Canned pickles require less stringent heat processing than canned beans, because pickles…
   A. contain fewer nutrients.
   B. are more acidic.
   C. have a lower a_w.
   D. contain antimicrobial chemicals.
   E. are less likely to be contaminated with endospores.
7. Which of the following genera is used in bread, wine, and beer production?
   A. *Lactobacillus*
   B. *Pseudomonas*
   C. *Saccharomyces*
   D. *Streptococcus*
   E. *Staphylococcus*
8. Most spoilage bacteria cannot grow below an a_w of…
   A. 0.3.
   B. 0.5.
   C. 0.7.
   D. 0.9.
   E. 1.0.
9. Which of the following is often added to wine to inhibit growth of the natural microbial population of grapes?
   A. Benzoic acid
   B. Lactic acid
   C. Carbon dioxide
   D. Sulfur dioxide
   E. Oxygen
10. In the brewing process, the sugar and nutrient extract obtained by soaking germinated grain in warm water is called…
    A. baker's yeast.
    B. hops.
    C. malt.
    D. must.
    E. wort.

Applications

1. A small cheese-manufacturing company in Wisconsin is looking for ways to reduce the costs of disposing of the cheese by-product whey. A company meeting was held to consider what to do with the thousands of liters of whey being produced per month. As a food microbiologist working at the company, what would you suggest that the company do with the whey so they can actually profit from it?
2. A microbiologist is troubleshooting a batch of home-brewed ale that did not ferment properly. She noticed that the alcohol content was only 2%, which is well below the desired level. Microscopic examination showed that the yeast were healthy and numerous. Chemical analysis of the beer showed low levels of sugar, high levels of CO_2, and large amounts of protein in the liquid. What did the microbiologist conclude as the probable cause of the beer not coming out properly?

Critical Thinking

1. It has been argued that the nature of the growth of fungi in Roquefort cheese, indicated by the appearance of bluish-green veins, is evidence that these fungi require oxygen for growth. How does this evidence lead to the conclusion?
2. In the production of sauerkraut, a natural succession of lactic acid bacteria is observed growing in the product. What causes the succession? What does this tell you about the optimal growth conditions of the different species of lactic acid bacteria?

Microbial Mathematics

Because prokaryotes are very tiny and can multiply to very large numbers of cells in short time periods, convenient and simple ways are used to designate their numbers without resorting to many zeros before or after the number. In the study of microbiology, it is important to gain an understanding of the metric system, which is used in scientific measurements.

The basic unit of measure is the meter, which is equal to about 39 inches. All other units are fractions of a meter:

1 decimeter is one tenth = 0.1 meter

1 centimeter is one hundredth = 0.01 meter

1 millimeter is one thousandth = 0.001 meter

Because prokaryotes are much smaller than a millimeter, even smaller units of measure are used. A millionth of a meter is a micrometer = 0.000001 meter, and is abbreviated µm. This is the most frequently used size measurement in microbiology, since bacteria are in this size range. For comparison, a human hair is about 75 µm wide.

Since it is inconvenient to write so many zeros in front of the 1, an easier way of denoting the same number is through the use of superscript, or exponential, numbers (exponents). One hundred dollars can be written 10^2 dollars. The 10 is called the base number and the 2 is the exponent. Conversely, one hundredth of a dollar is 10^{-2} dollars; thus the exponent is negative. The base most commonly used in biology is 10 (which is designated as $\log_{10}$). The above information can be summarized as follows:

1 millimeter = 1 mm = 0.001 meter = 10^{-3} meter

1 micrometer = 1 µm = 0.000001 meter = 10^{-6} meter

1 nanometer = 1 nm = 0.000000001 meter = 10^{-9} meter

The same prefix designations can be used for weights. The basic unit of weight is the gram, abbreviated g. Approximately 450 grams are in a pound.

1 milligram = 1 mg = 0.001 gram = 10^{-3} g

1 microgram = 1 µg = 0.000001 gram = 10^{-6} g

1 nanogram = 1 ng = 0.000000001 gram = 10^{-9} g

1 picogram = 1 pg = 0.000000000001 gram = 10^{-12} g

Note that the number of zeros before the 1 is one less than the exponent.

The value of the number is obtained by multiplying the base by itself the number of times indicated by the exponent.

Thus, $10^1 = 10 \times 1 = 10$

$10^2 = 10 \times 10 = 100$

$10^3 = 10 \times 10 \times 10 = 1{,}000$

When the exponent is negative, the base and exponent are divided into 1.

For example, $10^{-2} = 1/10 \times 1/10 = 1/100 = 0.01$

When multiplying numbers having exponents to the same base, the exponents are added.

For example, $10^3 \times 10^2 = 10^5$ (not 10^6)

When dividing numbers having exponents to the same base, the exponents are subtracted.

For example, $10^5 \div 10^2 = 10^3$

In both cases, only if the bases are the same can the exponents be added or subtracted.

Metabolic Pathways

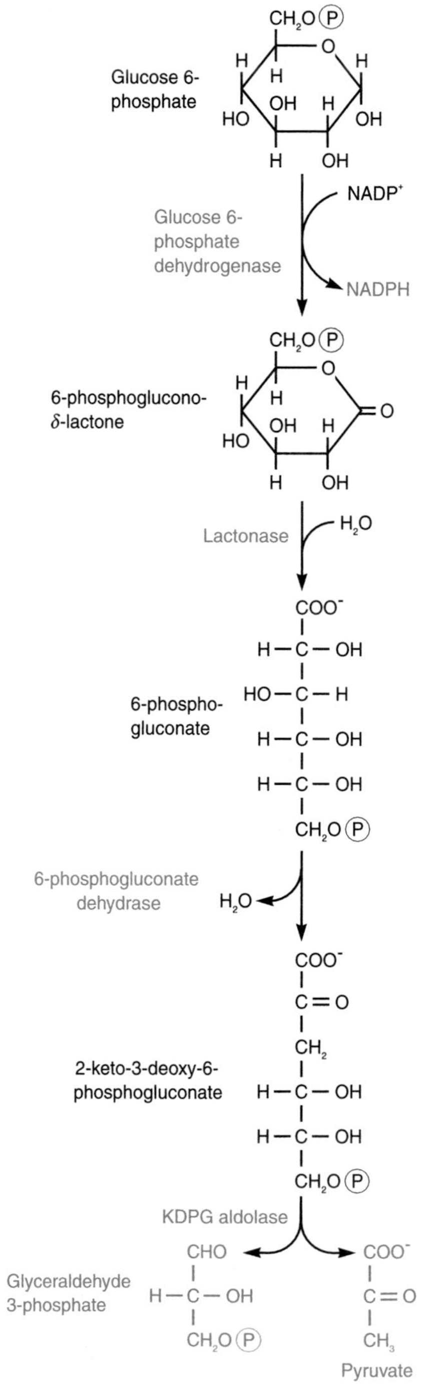

Glucose 6-phosphate

Glucose 6-phosphate dehydrogenase

$NADP^+$

NADPH

6-phosphoglucono-δ-lactone

Lactonase

H_2O

6-phosphogluconate

6-phosphogluconate dehydrase

H_2O

2-keto-3-deoxy-6-phosphogluconate

KDPG aldolase

Glyceraldehyde 3-phosphate

Pyruvate

Figure II-1 **The Entner-Doudoroff Pathway**

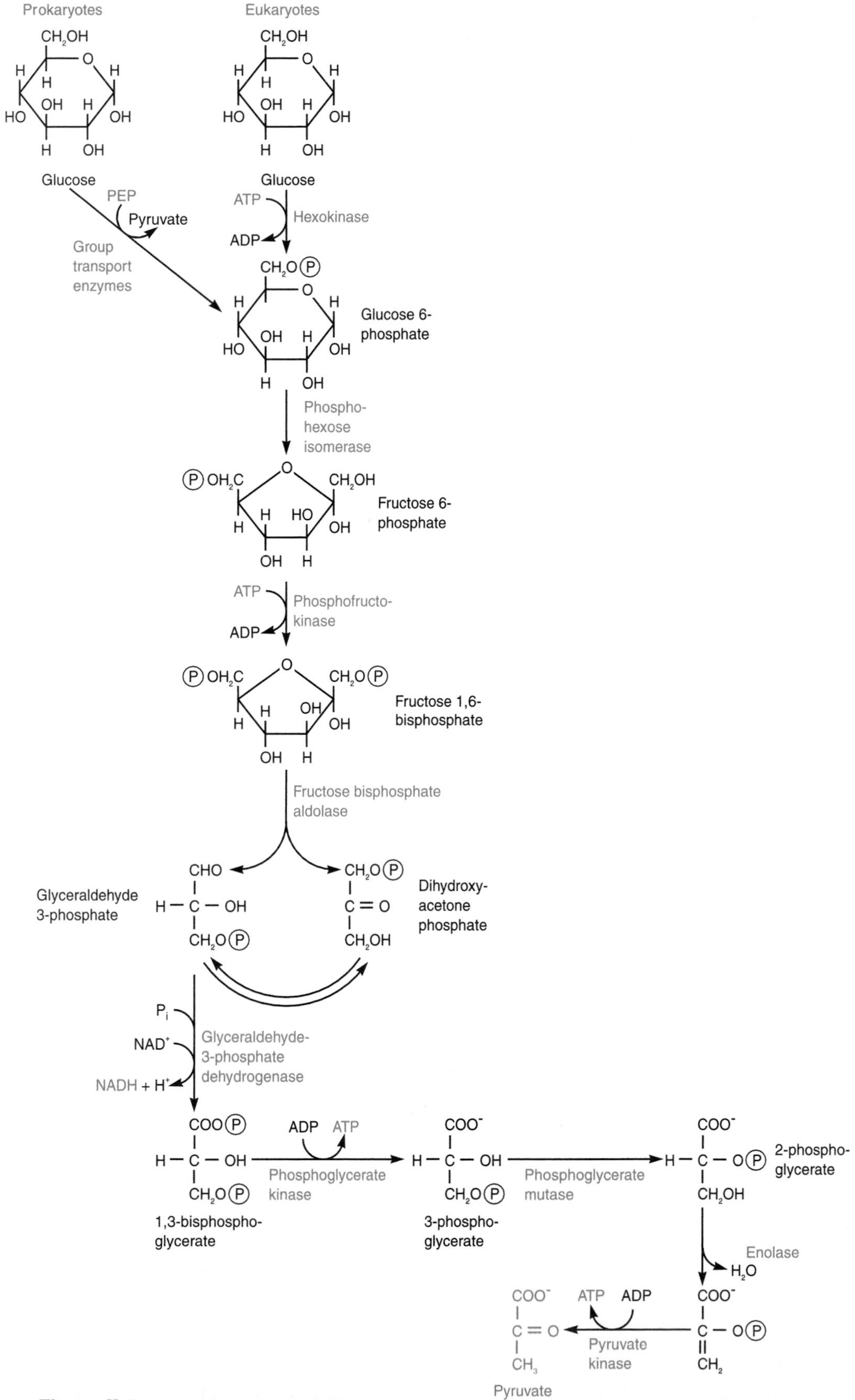

Figure II-2 The Embden-Meyerhof Pathway

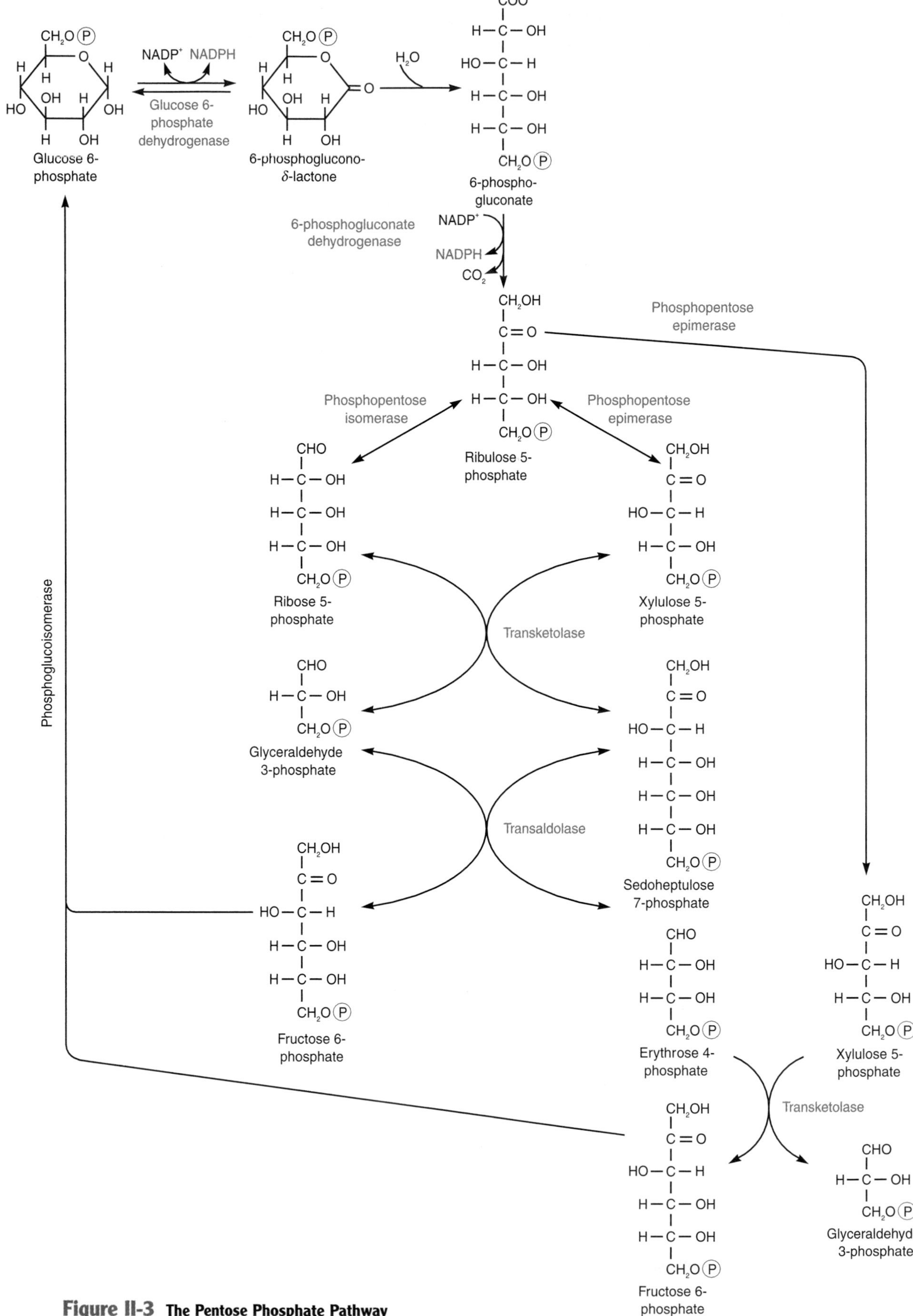

Figure II-3 The Pentose Phosphate Pathway

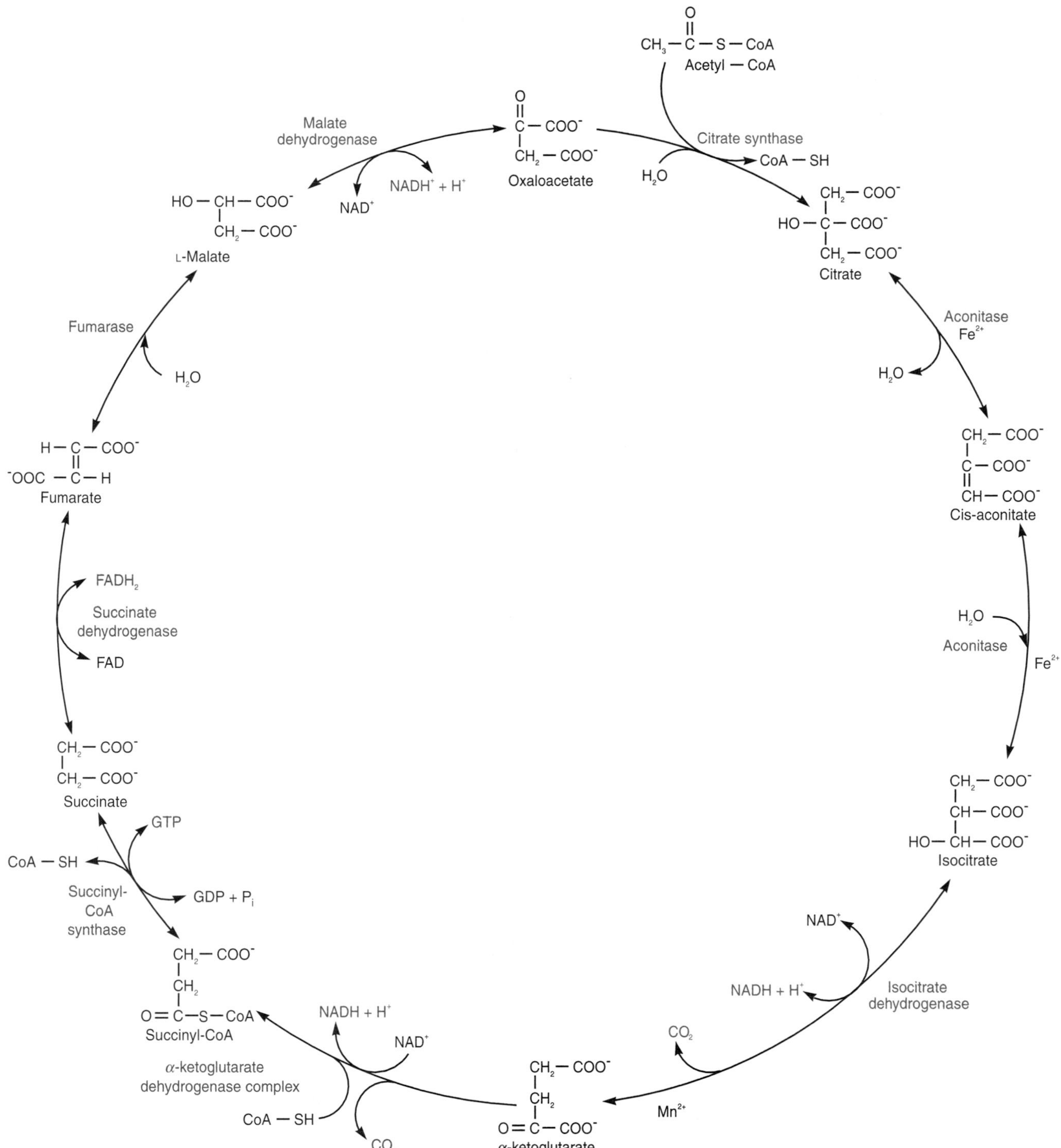

Figure II-4 The Tricarboxylic Acid Cycle (TCA cycle)

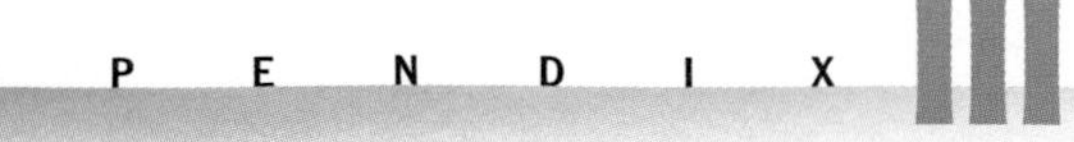

Answers to Multiple Choice Questions

Chapter 1
1. D
2. A
3. B
4. E
5. B
6. B
7. C
8. A
9. E
10. A

Chapter 2
1. C
2. A
3. E
4. A
5. B
6. C
7. B
8. D
9. E
10. A

Chapter 3
1. A
2. A
3. E
4. B
5. B
6. C
7. E
8. C
9. D
10. A

Chapter 4
1. B
2. B
3. C
4. B
5. D
6. B
7. E
8. A
9. D
10. B

Chapter 5
1. C
2. D
3. A
4. B
5. B
6. C
7. E
8. D
9. C
10. A

Chapter 6
1. D
2. C
3. D
4. E
5. B
6. A
7. D
8. D
9. D
10. E

Chapter 7
1. B
2. A
3. A
4. A
5. C
6. C
7. B
8. C
9. D
10. D

Chapter 8
1. C
2. A
3. D
4. A
5. B
6. D
7. A
8. B
9. A
10. B

Chapter 9
1. C
2. B
3. B
4. C
5. C
6. E
7. B
8. B
9. B
10. D

Chapter 10
1. E
2. B
3. E
4. B
5. D
6. B
7. C
8. D
9. B
10. E

Chapter 11
1. E
2. B
3. E
4. B
5. D
6. A
7. D
8. B
9. C
10. D

Chapter 12
1. C
2. A
3. B
4. A, C
5. A
6. B
7. A, B, E
8. A, B
9. C
10. A, B

Chapter 13
1. C
2. B
3. E
4. A
5. C
6. E
7. D
8. C
9. A
10. A

Chapter 14
1. D
2. A
3. B
4. B
5. D
6. B
7. A
8. D
9. E
10. A

Chapter 15
1. C
2. E
3. D
4. B
5. A
6. C
7. D
8. B
9. A
10. C

Chapter 16
1. C
2. E
3. E
4. A
5. E
6. E
7. E
8. E
9. C
10. A

Chapter 17

1. D
2. C
3. E
4. D
5. D
6. C
7. C
8. A
9. A
10. C

Chapter 18

1. A
2. E
3. A
4. D
5. E
6. A
7. A
8. C
9. A
10. C

Chapter 19

1. A
2. A
3. D
4. E
5. D
6. D
7. A
8. C
9. D
10. C

Chapter 20

1. A
2. B
3. A
4. B
5. E
6. B
7. B
8. E
9. C
10. E

Chapter 21

1. D
2. C
3. A
4. B
5. A
6. B
7. D
8. D
9. A
10. C

Chapter 22

1. E
2. E
3. A
4. E
5. B
6. A
7. C
8. A
9. D
10. D

Chapter 23

1. E
2. C
3. E
4. A
5. B
6. A
7. A
8. C
9. D
10. D

Chapter 24

1. C
2. A
3. B
4. D
5. A
6. D
7. D
8. E
9. A
10. C

Chapter 25

1. D
2. D
3. C
4. B
5. C
6. B
7. A
8. B
9. D
10. C

Chapter 26

1. B
2. C
3. C
4. D
5. C
6. E
7. A
8. D
9. A
10. B

Chapter 27

1. B
2. E
3. C
4. A
5. A
6. C
7. D
8. C
9. E
10. B

Chapter 28

1. E
2. A
3. C
4. E
5. C
6. C
7. C
8. A
9. A
10. C

Chapter 29

1. D
2. D
3. A
4. B
5. C
6. A
7. E
8. A
9. E
10. E

Chapter 30

1. D
2. E
3. B
4. C
5. A
6. B
7. C
8. C
9. E
10. C

Chapter 31

1. E
2. C
3. E
4. B
5. B
6. E
7. C
8. A
9. C
10. B

Chapter 32

1. B
2. C
3. D
4. E
5. B
6. B
7. C
8. D
9. D
10. E

Classification of Prokaryotes according to *Bergey's Manual of Systematic Bacteriology*, second edition

Domain Archaea

Phylum AI Crenarchaeota

Class I. Thermoprotei
  Order I. Thermoproteales
    Family I. Thermoproteaceae
      Genus I. Thermoproteus
      Genus II. Caldivirga
      Genus III. Pyrobaculum
      Genus IV. Thermocladium
    Family II. Thermofilaceae
      Genus I. Thermofilum
  Order II. Desulfurococcales
    Family I. Desulfurococcaceae
      Genus I. Desulfurococcus
      Genus II. Aeropyrum
      Genus III. Ignicoccus
      Genus IV. Staphylothermus
      Genus V. Stetteria
      Genus VI. Sulfophobococcus
      Genus VII. Thermodiscus
      Genus VIII. Thermosphaera
    Family II. Pyrodictiaceae
      Genus I. Pyrodictium
      Genus II. Hyperthermus
      Genus III. Pyrolobus
  Order lil. Sulfolobales
    Family I. Sulfolobaceae
      Genus I. Sulfolobus
      Genus II. Acidianus
      Genus III. Metallosphaera
      Genus IV. Stygiolobus
      Genus V. Sulfurisphaera
      Genus VI. Sulfurococcus

Phylum AII. Euryarchaeota

Class I. Methanobacteria
  Order I. Methanobacteriales
    Family I. Methanobacteriaceae
      Genus I. Methanobacterium
      Genus II. Methanobrevibacter
      Genus III. Methanosphaera
      Genus IV. Methanothermobacter
    Family II. Methanothermaceae
      Genus I. Methanothermus
Class II. Methanococci
  Order I. Methanococcales
    Family I. Methanococcaceae
      Genus I. Methanococcus
      Genus II. Methanothermococcus
    Family II. Methanocaldococcaceae
      Genus I. Methanocaldococcus
      Genus II. Methanotorris
  Order II. Methanomicrobiales
    Family I. Methanomicrobiaceae
      Genus I. Methanomicrobium
      Genus II. Methanoculleus
      Genus III. Methanofollis
      Genus IV. Methanogenium
      Genus V. Methanolacinia
      Genus VI. Methanoplanus
    Family II. Methanocorpusculaceae
      Genus I. Methanocorpusculum
    Family III. Methanospirillaceae
      Genus I. Methanospirillum
  Order III. Methanosarcinales
    Family I. Methanosarcinaceae
      Genus I. Methanosarcina
      Genus II. Methanococcoides
      Genus III. Methanohalobium
      Genus IV. Methanohalophilus
      Genus V. Methanolobus
      Genus VI. Methanosalsum
    Family II. Methanosaetaceae
      Genus I. Methanosaeta
Class III. Halobacteria
  Order I. Halobacteriales
    Family I. Halobacteriaceae
      Genus I. Halobacterium
      Genus II. Haloarcula
      Genus III. Halobaculum
      Genus IV. Halococcus
      Genus V. Haloferax
      Genus VI. Halogeometricum
      Genus VII. Halorhabdus
      Genus VIII. Halorubrum
      Genus IX. Haloterrigena
      Genus X. Natrialba
      Genus XI. Natrinema
      Genus XII. Natronobacterium
      Genus XIII. Natronococcus
      Genus XIV. Natronomonas
      Genus XV. Natronorubrum
Class IV. Thermoplasmata
  Order I. Thermoplasmatales
    Family I. Thermoplasmataceae
      Genus I. Thermoplasma
    Family II. Picrophilaceae
      Genus I. Picrophilus
Class V. Thermococci
  Order I. Thermococcales
    Family I. Thermococcaceae
      Genus I. Thermococcus
      Genus II. Pyrococcus
Class VI. Archaeoglobi
  Order I. Archaeoglobales
    Family I. Archaeoglobaceae
      Genus I. Archaeoglobus
      Genus II. Ferroglobus
Class VII. Methonopyri
  Order I. Methanopyrales
    Family I. Methanopyraceae
      Genus I. Methanopyrus

Domain Bacteria

Phylum BI. Aquificae

Class I. Aquificae
  Order I. Aquificales
    Family I. Aquificaceae
      Genus I. Aquifex
      Genus II. Calderobacterium
      Genus III. Hydrogenobacter
      Genus IV. Thermocrinis
    Family II. Incertae sedis
      Genus I. Desulfurobacterium

Phylum BII. Thermotogae

Class I. Thermotogae
  Order I. Thermotogales
    Family I. Thermotogaceae
      Genus I. Thermotoga
      Genus II. Fervidobacterium
      Genus III. Geotoga
      Genus IV. Petrotoga
      Genus V. Thermosipho

Phylum BIII. Thermodesulfobacteria

Class I. Thermodesulfobacteria
  Order I. Thermodesulfobacteriales
    Family I. Thermodesulfobacteriaceae
      Genus I. Thermodesulfobacterium

Phylum BIV. "Deinococcus-Thermus"

Class I. Deinococci
  Order I. Deinococcales
    Family I. Deinococcaceae
      Genus I. Deinococcus
  Order II. Thermales
    Family I. Thermaceae
      Genus I. Thermus
      Genus II. Meiothermus

Phylum BV. Chrysiogenetes

Class I. Chrysiogenetes
  Order I. Chrysiogenales
    Family I. Chrysiogenaceae
      Genus I. Chrysiogenes

Bergey's Manual of Systematic Bacteriology, second editon, 5 volumes (2000)

Phylum BVI. Chloroflexi
Class I. "Chloroflexi"
Order I. "Chloroflexales"
Family I. Chloroflexaceae
Genus I. Chloroflexus
Genus II. Chloronema
Genus III. Heliothrix
Genus IV. Oscillochloris
Order II. Herpetosiphonales
Family I. Herpetosiphonaceae
Genus I. Herpetosiphon

Phylum BVII. Thermomicrobia
Class I. Thermomicrobia
Order I. Thermomicrobiales
Family I. Thermomicrobiaceae
Genus I. Thermomicrobium

Phylum BVIII. Nitrospira
Class I. Nitrospira
Order I. Nitrospirales
Family I. Nitrospiraceae
Genus I. Nitrospira
Genus II. Leptospirillum
Genus III. Magnetobacterium
Genus IV. Thermodesulfovibrio

Phylum BIX. Deferribacteres
Class I. Deferribacteres
Order I. Deferribacterales
Family I. Deferribacteraceae
Genus I. Deferribacter
Genus II. "Flexistipes"
Genus III. Geovibrio
Genera incertae sedis
Genus I. Synergistes

Phylum BX. "Cyanobacteria"
Class I. 'Cyanobacteria'
Subsection I.
Genus I. Chamaesiphon
Genus II. Chroococcus
Genus III. Cyanobacterium
Genus IV. Cyanobium
Genus V. Cyanothece
Genus VI. Dactylococcopsis
Genus VII. Gloeobacter
Genus VIII. Gloeocapsa
Genus IX. Gloeothece
Genus X. Microcystis
Genus XI. Prochlorococcus
Genus XII. Prochloron
Genus XIII. Synechococcus
Genus XIV. Synechocystis
Subsection II.
Subgroup I.
Genus I. Cyanocystis
Genus II. Dermocarpa
Genus III. Stanieria
Genus IV. Xenococcus
Subgroup II.
Genus I. Chroococcidiopsis
Genus II. Myxosarcina
Genus III. Pleurocapsa
Subsection III.
Genus I. Arthrospira
Genus II. Borzia
Genus III. Crinalium
Genus IV. Geitlerinema
Genus V. Leptolyngbya
Genus VI. Limnothrix
Genus VII. Lyngbya
Genus VIII. Microcoleus
Genus IX. Oscillatoria
Genus X. Planktothrix
Genus XI. Prochlorothrix
Genus XII. Pseudanabaena
Genus XIII. Spirulina
Genus XIV. Starria
Genus XV. Symploca
Genus XVI. Trichodesmium
Genus XVII. Tychonema
Subsection IV.
Subgroup I.
Genus I. Anabaena
Genus II. Anabaenopsis
Genus III. Aphanizomenon
Genus IV. Cyanospira
Genus V. Cylindrospermopsis
Genus VI. Cylindrospermum
Genus VII. Nodularia
Genus VIII. Nostoc
Genus IX. Scytonema
Subgroup II.
Genus I. Calothrix
Genus II. Rivularia
Genus III. Tolypothrix
Subsection V.
Family I.
Genus I. Chlorogloeopsis
Genus II. Fischerella
Genus III. Geitleria
Genus IV. Iyengariella
Genus V. Nostochopsis
Genus VI. Stigonema

Phylum BXI. 'Chlorobi'
Class I. 'Chlorobia'
Order I. Chlorobiales
Family I. Chlorobiaceae
Genus I. Chlorobium
Genus II. Ancalochloris
Genus III. Chloroherpeton
Genus IV. Pelodictyon
Genus V. Prosthecochloris

Phylum BXII. Proteobacteria
Class I. "Alphaproteobacteria"
Order I. Rhodospirillales
Family I. Rhodospirillaceae
Genus I. Rhodospirillum
Genus II. Azospirillum
Genus III. Magnetospirillum
Genus IV. Phaeospirillum
Genus V. Rhodocista
Genus VI. Rhodospira
Genus VII. 'Rhodothalassium'
Genus VIII. 'Rhodovibrio'
Genus IX. Roseospira
Genus X. Skermanella
Family II. Acetobacteraceae
Genus I. Acetobacter
Genus II. Acidiphilium
Genus III. Acidocella
Genus IV. Acidomonas
Genus V. Craurococcus
Genus VI. Gluconacetobacter
Genus VII. Gluconobacter
Genus VIII. Paracraurococcus
Genus IX. Rhodopila
Genus X. Roseococcus
Genus XI. Stella
Genus XII. Zavarzinia
Order II. Rickettsiales
Family I. Rickettsiaceae
Genus I. Rickettsia
Genus II. Orientia
Genus III. Wolbachia
Family II. Ehrlichiaceae
Genus I. Ehrlichia
Genus II. Aegyptianella
Genus III. Anaplasma
Genus IV. Cowdria
Genus V. Neorickettsia
Family III. Holosporaceae'
Genus I. Holospora
Genus II. Caedibacter
Genus III. Lyticum
Genus IV. Polynucleobacter
Genus V. Pseudocaedibacter
Genus VI. Symbiotes
Genus VII. Tectibacter
Order III. "Rhodobacterales"
Family I. Rhodobacteraceae
Genus I. Rhodobacter
Genus II. Ahrensia
Genus III. Amaricoccus
Genus IV. Antarctobacter
Genus V. Gemmobacter
Genus VI. Hirschia
Genus VII. Hyphomonas
Genus VIII. Maricaulis
Genus IX. Octadecabacter
Genus X. Paracoccus
Genus XI. Rhodovulum
Genus XII. Roseivivax
Genus XIII. Roseobacter
Genus XIV. Roseovarius
Genus XV. Rubrimonas
Genus XVI. Ruegeria
Genus XVII. Sagittula
Genus XVIII. Staleya
Genus XIX. Stappia
Genus XX. Sulfitobacter
Order IV. "Sphingomonadales"
Family I. "Sphingomonadaceae"
Genus I. Sphingomonas
Genus II. Blastomonas
Genus III. Erythrobacter
Genus IV. Erythromicrobium
Genus V. Erythromonas
Genus VI. Porphyrobacter
Genus VII. Rhizomonas
Genus VIII. Sandaracinobacter
Genus IX. Zymomonas
Order V. Caulobacterales
Family I. Caulobacteraceae
Genus I. Caulobacter
Genus II. Asticcacaulis
Genus III. Brevundimonas
Genus IV. Phenylobacterium
Order VI. "Rhizobiales"
Family I. Rhizobiaceae
Genus I. Rhizobium
Genus II. Agrobacterium
Genus III. Carbophilus
Genus IV. Chelatobacter
Genus V. Ensifer
Genus VI. Sinorhizobium
Family II. Bartonellaceae
Genus I. Bartonella
Family III. Brucellaceae
Genus I. Brucella
Genus II. Mycoplana
Genus III. Ochrobactrum
Family IV. "Phyllobacteriaceae"
Genus I. Phyllobacterium
Genus II. Allorhizobium
Genus III. Aminobacter

      Genus IV. Aquamicrobium
      Genus V. Defluvibacter
      Genus VI. Mesorhizobium
      Genus VII. Pseudaminobacter
    Family V. "Methylocystaceae"
      Genus I. Methylocystis
      Genus II. Methylopila
      Genus III. Methylosinus
    Family VI. "Beijerinckiaceae"
      Genus I. Beijerinckia
      Genus II. Chelatococcus
      Genus III. Derxia
    Family VII. "Bradyrhizobiaceae"
      Genus I. Bradyrhizobium
      Genus II. Afipia
      Genus III. Agromonas
      Genus IV. Blastobacter
      Genus V. Bosea
      Genus VI. Nitrobacter
      Genus VII. Oligotropha
      Genus VIII. Rhodopseudomonas
    Family VIII. Hyphomicrobiaceae
      Genus I. Hyphomicrobium
      Genus II. Ancalomicrobium
      Genus III. Ancylobacter
      Genus IV. Angulomicrobium
      Genus V. Aquabacter
      Genus VI. Azorhizobium
      Genus VII. Blastochloris
      Genus VIII. Devosia
      Genus IX. Dichotomicrobium
      Genus X. Filomicrobium
      Genus XI. Gemmiger
      Genus XII. Labrys
      Genus XIII. Methylorhabdus
      Genus XIV. Pedomicrobium
      Genus XV. Prosthecomicrobium
      Genus XVI. Rhodomicrobium
      Genus XVII. Rhodoplanes
      Genus XVIII. Seliberia
      Genus XIX. Xanthobacter
    Family IX. "Methylobacteriaceae"
      Genus I. Methylobacterium
      Genus II. Protomonas
      Genus III. Roseomonas
    Family X. "Rhodobiaceae"
      Genus I. Rhodobium
Class II. "Betaproteobacteria"
  Order I. "Burkholderiales"
    Family I. "Burkholderiaceae"
      Genus I. Burkholderia
      Genus II. Cupriavidus
      Genus III. Lautropia
      Genus IV. Thermothrix
    Family II. "Ralstoniaceae"
      Genus I. Ralstonia
    Family III. "Oxalobacteraceae"
      Genus I. Oxalobacter
      Genus II. Duganella
      Genus III. Herbaspirillum
      Genus IV. Janthinobacterium
      Genus V. Telluria
    Family IV. Alcaligenaceae
      Genus I. Alcaligenes
      Genus II. Achromobacter
      Genus III. Bordetella
      Genus IV. Pelistega
      Genus V. Sutterella
      Genus VI. Taylorella
    Family V. Comamonadaceae
      Genus I. Comamonas
      Genus II. Acidovorax
      Genus III. Aquabacterium
      Genus IV. Brachymonas
      Genus V. Delftia
      Genus VI. Hydrogenophaga
      Genus VII. Ideonella
      Genus VIII. Leptothrix
      Genus IX. Polaromonas
      Genus X. Roseateles
      Genus XI. Rhodoferax
      Genus XII. Rubrivivax
      Genus XIII. Sphaerotilus
      Genus XIV. Thiomonas
      Genus XV. Variovorax
  Order II. "Hydrogenophilales"
    Family I. "Hydrogenophilaceae"
      Genus I. Hydrogenophilus
      Genus II. Thiobacillus
  Order III. "Methylophilales"
    Family I. "Methylophilaceae"
      Genus I. Methylophilus
      Genus II. Methylobacillus
      Genus III. Methylovorus
  Order IV. "Neisseriales"
    Family I. Neisseriaceae
      Genus I. Neisseria
      Genus II. Alysiella
      Genus III. Aquaspirillum
      Genus IV. Catenococcus
      Genus V. Chromobacterium
      Genus VI. Eikenella
      Genus VII. Formivibrio
      Genus VIII. Iodobacter
      Genus IX. Kingella
      Genus X. Microvirgula
      Genus XI. Prolinoborus
      Genus XII. Simonsiella
      Genus XIII. Vitreoscilla
      Genus XIV. Vogesella
  Order V. "Nitrosomonadales"
    Family I. "Nitrosomonadaceae"
      Genus I. Nitrosomonas
      Genus II. Nitrosospira
    Family II. Spirillaceae
      Genus I. Spirillum
    Family III. Gallionellacea
      Genus I. Gallionella
  Order VI. "Rhodocyclales"
    Family I. "Rhodocyclaceae"
      Genus I. Rhodocyclus
      Genus II. Azoarcus
      Genus III. Propionibacter
      Genus IV. Propionivibrio
      Genus V Thauera
      Genus VI. Zoogloea
Class III. "Gammaproteobacteria"
  Order I. "Chromatiales"
    Family I. Chromatiaceae
      Genus I. Chromatium
      Genus II. Allochromatium
      Genus III. Amoebobacter
      Genus IV. Halochromatium
      Genus V. Isochromatium
      Genus VI. Lamprobacter
      Genus VII. Lamprocystis
      Genus VIII. Marichromatium
      Genus IX. Nitrosococcus
      Genus X. Pfennigia
      Genus XI. Rhabdochromatium
      Genus XII. Thermochromatium
      Genus XIII. Thiocapsa
      Genus XIV. Thiococcus
      Genus XV. Thiocystis
      Genus XVI. Thiodictyon
      Genus XVII. Thiohalocapsa
      Genus XVIII. Thiolamprovum
      Genus XIX. Thiopedia
      Genus XX. Thiorhodococcus
      Genus XXI. Thiorhodovibrio
      Genus XXII. Thiospirillum
    Family II. Ectothiorhodospiraceae
      Genus I. Ectothiorhodospira
      Genus II. Arhodomonas
      Genus III. Halorhodospira
      Genus IV. Nitrococcus
      Genus V. Thiorhodospira
  Order II. "Xanthomonadales"
    Family I. "Xanthomonadaceae"
      Genus I. Xanthomonas
      Genus II. Frateuria
      Genus III. Lysobacter
      Genus IV. Nevskia
      Genus V. Pseudoxanthomonas
      Genus VI. Rhodanobacter
      Genus VII. Stenotrophomonas
      Genus VIII. Xylella
  Order III. "Cardiobacteriales"
    Family I. Cardiobacteriaceae
      Genus I. Cardiobacterium
      Genus II. Dichelobacter
      Genus III. Suttonella
  Order IV. "Thiotrichales"
    Family I. "Thiotrichaceae"
      Genus I. Thiothrix
      Genus II. Achromatium
      Genus III. Beggiatoa
      Genus IV. Leucothrix
      Genus V. Macromonas
      Genus VI. Thiobacterium
      Genus VII. Thiomargarita
      Genus VIII. Thioploca
      Genus IX. Thiospira
    Family II. "Piscirickettsiaceae"
      Genus I. Piscirickettsia
      Genus II. Cycloclasticus
      Genus III. Hydrogenovibrio
      Genus IV. Methylophaga
      Genus V. Thiomicrospira
    Family III. Francisellaceae
      Genus I. Francisella
  Order V. "Legionellales"
    Family I. Legionellaceae
      Genus I. Legionella
    Family II. Coxiellaceae
      Genus I. Coxiella
      Genus II. Rickettsiella
  Order VI. "Methylococcales"
    Family I. Methylococcaceae
      Genus I. Methylococcus
      Genus II. Methylobacter
      Genus III. Methylocaldum
      Genus IV. Methylomicrobium
      Genus V. Methylomonas
      Genus VI. Methylosphaera
  Order VII. "Oceanospirillales"
    Family I. "Oceanospirillaceae"
      Genus I. Oceanospirillum
      Genus II. Balneatrix
      Genus III. Fundibacter
      Genus IV. Marinomonas
      Genus V. Marinospirillum
      Genus VI. Neptunomonas
    Family II. Halomonadaceae
      Genus I. Halomonas
      Genus II. Alcanivorax

Genus III. Carnimonas
Genus IV. Chromohalobacter
Genus V. Deleya
Genus VI. Zymobacter
Order VIII. Pseudomonadales
Family I. Pseudomonadaceae
Genus I. Pseudomonas
Genus II. Azomonas
Genus III. Azotobacter
Genus IV. Cellvibrio
Genus V. Chryseomonas
Genus VI. Flavimonas
Genus VII. Lampropedia
Genus VIII. Mesophilobacter
Genus IX. Morococcus
Genus X. Oligella
Genus XI. Rhizobacter
Genus XII. Rugamonas
Genus XIII. Serpens
Genus XIV. Thermoleophilum
Genus XV. Xylophilus
Family II. Moraxellaceae
Genus I. Moraxella
Genus II. Acinetobacter
Genus III. Psychrobacter
Order IX. "Alteromonadales"
Family I. "Alteromonadaceae"
Genus I. Alteromonas
Genus II. Colwellia
Genus III. Ferrimonas
Genus IV. Glaciecola
Genus V. Marinobacter
Genus VI. Marinobacterium
Genus VII. Microbulbifer
Genus VIII. Moritella
Genus IX. Pseudoalteromonas
Genus X. Shewanella
Order X."Vibrionales"
Family I. Vibrionaceae
Genus I. Vibrio
Genus II. Allomonas
Genus III. Enhydrobacter
Genus IV. Listonella
Genus V. Photobacterium
Genus VI. Salinivibrio
Order XI. "Aeromonadales"
Family I. Aeromonadaceae
Genus I. Aeromonas
Genus II. Tolumonas
Family II. *Succinivibrionaceae*
Genus I. Succinivibrio
Genus II. Anaerobiospirillum
Genus III. Ruminobacter
Genus IV. Succinimonas
Order XII. "Enterobacteriales"
Family I. Enterobacteriaceae
Genus I. Enterobacter
Genus II. Alterococcus
Genus III. Arsenophonus
Genus IV. Brenneria
Genus V. Buchnera
Genus VI. Budvicia
Genus VII. Buttiauxella
Genus VIII. Calymmatobacterium
Genus IX. Cedecea
Genus X. Citrobacter
Genus XI. Edwardsiella
Genus XII. Erwinia
Genus XIII. Escherichia
Genus XIV. Ewingella
Genus XV. Hafnia
Genus XVI. Klebsiella
Genus XVII. Kluyvera
Genus XVIII. Leclercia
Genus XIX. Leminorella
Genus XX. Moellerella
Genus XXI. Morganella
Genus XXII. Obesumbacterium
Genus XXIII. Pantoea
Genus XXIV. Pectobacterium
Genus XXV. Photorhabdus
Genus XXVI. Plesiomonas
Genus XXVII. Pragia
Genus XXVIII. Proteus
Genus XXIX. Providencia
Genus XXX. Rahnella
Genus XXXI. Saccharobacter
Genus XXXII. Salmonella
Genus XXXIII. Serratia
Genus XXXIV. Shigella
Genus XXXV. Sodalis
Genus XXXVI. Tatumella
Genus XXXVII. Trabulsiella
Genus XXXVIII. Wigglesworthia
Genus XXXIX. Xenorhabdus
Genus XL. Yersinia
Genus XLI. Yokenella
Order XIII. "Pasteurellales"
Family I. Pasteurellaceae
Genus I. Pasteurella
Genus II. Actinobacillus
Genus III. Haemophilus
Genus IV. Lonepinella
Genus V. Mannheimia
Genus VI. Phocoenobacter
Class IV. "Deltaproteobacteria"
Order I. "Desulfurellales"
Family I. "Desulfurellaceae"
Genus I. Desulfurella
Genus II. Hippea
Order II. "Desulfovibrionales"
Family I. Desulfovibrionaceae
Genus I. Desulfovibrio
Genus II. Bilophila
Genus III. Lawsonia
Family II. "Desulfomicrobiaceae"
Genus I. Desulfomicrobium
Family III. "Desulfohalobiaceae"
Genus I. Desulfohalobium
Genus II. Desulfomonas
Genus III. Desulfonatronovibrio
Order III. "Desulfobacterales"
Family I. "Desulfobacteraceae"
Genus I. Desulfobacter
Genus II. Desulfobacterium
Genus III. "Desulfobacula"
Genus IV. "Desulfobotulus"
Genus V. Desulfocella
Genus VI. Desulfococcus
Genus VII. Desulfofaba
Genus VIII. Desulfofrigus
Genus IX. Desulfonema
Genus X. Desulfosarcina
Genus XI. Desulfospira
Genus XII. Desulfotalea
Family II. "Desulfobulbaceae"
Genus I. Desulfobulbus
Genus II. Desulfocapsa
Genus III. Desulfofustis
Genus IV. Desulforhopalus
Family III. "Desulfoarculaceae"
Genus I. "Desulfoarculus"
Genus II. Nitrospina
Genus III. Desulfobacca
Genus IV. Desulfomonile
Order IV. "Desulfuromonadales"
Family I. "Desulfuromonadaceae"
Genus I. Desulfuromonas
Genus II. Desulfuromusa
Family II. "Geobacteraceae"
Genus I. Geobacter
Family III. "Pelobacteraceae"
Genus I. Pelobacter
Genus II. Malonomonas
Order V. "Syntrophobacteriales"
Family I. "Syntrophobacteriaceae"
Genus I. Syntrophobacter
Genus II. Desulfacinum
Genus III. Desulforhabdus
Genus IV. Thermodesulforhabdus
Family II. "Syntrophaceae"
Genus I. Syntrophus
Genus II. Smithella
Order VI. "Bdellovibrionales"
Family I. "Bdellovibrionaceae"
Genus I. Bdellovibrio
Genus II. Micavibrio
Genus III. Vampirovibrio
Order VII. Myxococcales
Family I. Myxococcaceae
Genus I. Myxococcus
Genus II. Angiococcus
Family II. Archangiaceae
Genus I. Archangium
Family III. Cystobacteraceae
Genus I. Cystobacter
Genus II. Melittangium
Genus III. Stigmatella
Family IV. Polyangiaceae
Genus I. Polyangium
Genus II. Chondromyces
Genus III. Nannocystis
Class V. "Epsilonproteobacteria"
Order I. "Campylobacterales"
Family I. Campylobacteraceae
Genus I. Campylobacter
Genus II. Arcobacter
Genus III. Sulfurospirillum
Genus IV. Thiovulum
Family II. "Helicobacteraceae"
Genus I. Helicobacter
Genus II. Wolinella

Phylum BXIII. Firmicutes

Class I. "Clostridia"
Order I. Clostridiales
Family I. Clostridiaceae
Genus I. Clostridium
Genus II. Acetivibrio
Genus III. Acidaminobacter
Genus IV. Anaerobacter
Genus V. Caloramator
Genus VI. Natronincola
Genus VII. Oxobacter
Genus VIII. Sarcina
Genus IX. Sporobacter
Genus X. Thermobrachium
Genus XI. Tindallia
Family II. "Lachnospiraceae"
Genus I. Lachnospira
Genus II. Acetitomaculum
Genus III. Anaerofilum
Genus III. Butyrivibrio
Genus IV. Catonella
Genus V. Coprococcus
Genus VI. Johnsonella

Genus VII. Pseudobutyrivibrio
Genus VIII. Roseburia
Genus IX. Ruminococcus
Genus X. Sporobacterium
Family III. "Peptostreptococcaceae"
Genus I. Peptostreptococcus
Genus II. Filifactor
Genus III. Fusibacter
Genus IV. Helcococcus
Genus V. Tissierella
Family IV. "Eubacteriaceae"
Genus I. Eubacterium
Genus II. Acetobacterium
Genus III. Pseudoramibacter
Family V. Peptococcaceae
Genus I. Peptococcus
Genus II. Anaeroarcus
Genus III. Anaerosinus
Genus IV. Anaerovibrio
Genus V. Carboxydothermus
Genus VI. Centipeda
Genus VII. Dehalobacter
Genus VIII. Dendrosporobacter
Genus IX. Desulfitobacterium
Genus X. Desulfonispora
Genus XI. Desulfosporosinus
Genus XII. Desulfotomaculum
Genus XIII. Mitsuokella
Genus XIV. Propionispira
Genus XV. Succinispira
Genus XVI. Syntrophobotulus
Genus XVII. Thermoterrabacterium
Family VI. "Heliobacteriaceae"
Genus I. Heliobacterium
Genus II. Heliobacillus
Genus III. Heliophilum
Family VII. "Acidiminococcaceae"
Genus I. Acidaminococcus
Genus II. Acetonema
Genus III. Anaeromusa
Genus IV. Dialister
Genus V. Megasphaera
Genus VI. Pectinatus
Genus VII. Phascolarctobacterium
Genus VIII. Quinella
Genus IX. Schwartzia
Genus X. Selenomonas
Genus XI. Sporomusa
Genus XII. Succiniclasticum
Genus XIII. Veillonella
Genus XIV. Zymophilus
Family VIII. Syntrophomonadaceae
Genus I. Syntrophomonas
Genus II. Acetogenium
Genus III. Aminobacterium
Genus IV. Aminomonas
Genus V. Anaerobaculum
Genus VI. Anaerobranca
Genus VII. Caldicellulosiruptor
Genus VIII. Dethiosulfovibrio
Genus IX. Syntrophospora
Genus X. Thermaerobacter
Genus XI. Thermanaerovibrio
Genus XII. Thermohydrogenium
Genus XIII. Thermosyntropha
Order II. "Thermoanaerobacteriales"
Family I. "Thermoanaerobacteriaceae"
Genus I. Thermoanaerobacterium
Genus II. Ammonifex
Genus III. Coprothermobacter
Genus IV. Moorella
Genus V. Sporotomaculum
Genus VI. Thermoanaerobacter
Genus VII. Thermoanaerobium
Order III. Haloanaerobiales
Family I. Haloanaerobiaceae
Genus I. Haloanaerobium
Genus II. Halocella
Genus III. Halothermothrix
Genus IV. Natroniella
Family II. Halobacteroidaceae
Genus I. Halobacteroides
Genus II. Acetohalobium
Genus III. Haloanaerobacter
Genus IV. Orenia
Genus V. Sporohalobacter
Class II. Mollicutes
Order I. Mycoplasmatales
Family I. Mycoplasmataceae
Genus I. Mycoplasma
Genus II. Eperythrozoon
Genus III. Haemobartonella
Genus IV. Ureaplasma
Order II. Entomoplasmatales
Family I. Entomoplasmataceae
Genus I. Entomoplasma
Genus II. Mesoplasma
Family II. Spiroplasmataceae
Genus I. Spiroplasma
Order III. Acholeplasmatales
Family I. Acholeplasmataceae
Genus I. Acholeplasma
Order IV. Anaeroplasmatales
Family I. Anaeroplasmataceae
Genus I. Anaeroplasma
Genus II. Asteroleplasma
Class III. "Bacilli"
Order I. Bacillales
Family I. Bacillaceae
Genus I. Bacillus
Genus II. Amphibacillus
Genus III. Exiguobacterium
Genus IV. Gracilibacillus
Genus V. Halobacillus
Genus VI. Saccharococcus
Genus VII. Salibacillus
Genus VIII. Virgibacillus
Family II. Planococcaceae
Genus I. Planococcus
Genus II. Filibacter
Genus III. Kurthia
Genus IV. Sporosarcina
Family III. Caryophanaceae
Genus I. Caryophanon
Family IV. "Listeriaceae"
Genus I. Listeria
Genus II. Brochothrix
Family V. "Staphylococcaceae"
Genus I. Staphylococcus
Genus II. Gemella
Genus III. Macrococcus
Genus IV. Salinicoccus
Family VI. "Sporolactobacillaceae"
Genus I. Sporolactobacillus
Genus II. Marinococcus
Family VII. "Paenibacillaceae"
Genus I. Paenibacillus
Genus II. Ammoniphilus
Genus III. Aneurinibacillus
Genus IV. Brevibacillus
Genus V. Oxalophagus
Genus VI. Thermobacillus
Family VIII. "Alicyclobacillaceae"
Genus I. Alicyclobacillus
Genus II. Pasteuria
Genus III. Sulfobacillus
Family IX. "Thermoactinomycetaceae"
Genus I. Thermoactinomyces
Order II. "Lactobacillales"
Family I. Lactobacillaceae
Genus I. Lactobacillus
Genus II. Pediococcus
Family II. "Aerococcaceae"
Genus I. Aerococcus
Genus II. Abiotrophia
Genus III. Dolosicoccus
Genus IV. Eremococcus
Genus V. Facklamia
Genus VI. Globicatella
Genus VII. Ignavigranum
Family III. "Carnobacteriaceae"
Genus I. Carnobacterium
Genus II. Agitococcus
Genus III. Alloiococcus
Genus IV. Desemzia
Genus V. Dolosigranulum
Genus VI. Lactosphaera
Genus VII. Trichococcus
Family IV. "Enterococcaceae"
Genus I. Enterococcus
Genus II. Melissococcus
Genus III. Tetragenococcus
Genus IV. Vagococcus
Family V. "Leuconostocaceae"
Genus I. Leuconostoc
Genus II. Oenococcus
Genus III. Weisella
Family VI. Streptococcaceae
Genus I. Streptoccus
Genus II. Lactococcus
Family VII. Incertae sedis
Genus I. Acetoanaerobium
Genus II. Oscillospira
Genus III. Syntrophococcus

Phylum BXIV. Actinobacteria

Class I. Actinobacteria
Subclass I. Acidimicrobidae
Order I. Acidimicrobiales
Suborder I. "Acidimicrobineae"
Family I. Acidimicrobiaceae
Genus I. Acidimicrobium
Subclass II. Rubrobacteridae
Order I. Rubrobacterales
Suborder I. "Rubrobacterineae"
Family I. Rubrobacteraceae
Genus I. Rubrobacter
Subclass III. Coriobacteridae
Order I. Coriobacteriales
Suborder I. "Coriobacterineae"
Family I. Coriobacteriaceae
Genus I. Coriobacterium
Genus II. Atopobium
Genus III. Collinsella
Genus IV. Cryptobacterium
Genus V. Eggerthella
Genus VI. Slackia
Subclass IV. Sphaerobacteridae
Order I. Sphaerobacterales
Suborder IV. "Sphaerobacterineae"
Family I. Sphaerobacteraceae
Genus I. Sphaerobacter
SubclassV. Actinobacteridae

Order I. Actinomycetales
SuborderV. Actinomycineae
Family I. Actinomycetaceae
Genus I. Actinomyces
Genus II. Actinobaculum
Genus III. Arcanobacterium
Genus IV. Mobiluncus
SuborderVI. Micrococcineae
Family I. Micrococcaceae
Genus I. Micrococcus
Genus II. Arthrobacter
Genus III. Bogoriella
Genus IV. Demetria
Genus V. Kocuria
Genus VI. Leucobacter
Genus VII. Nesterenkonia
Genus VIII. Renibacterium
Genus IX. Rothia
Genus X. Stomatococcus
Genus XI. Terracoccus
Family II. Brevibacteriaceae
Genus I. Brevibacterium
Family III. Cellulomonadaceae
Genus I. Cellulomonas
Genus II. Oerskovia
Genus III. Rarobacter
Family IV. Dermabacteraceae
Genus I. Dermabacter
Genus II. Brachybacterium
Family V. Dermatophilaceae
Genus I. Dermatophilus
Genus II. Dermacoccus
Genus III. Kytococcus
Family VI. Intrasporangiaceae
Genus I. Intrasporangium
Genus II. Janibacter
Genus III. Ornithinicoccus
Genus IV. Sanguibacter
Genus V. Terrabacter
Family VII. Jonesiaceae
Genus I. Jonesia
Family VIII. Microbacteriaceae
Genus I. Microbacterium
Genus II. Agrococcus
Genus III. Agromyces
Genus IV. Aureobacterium
Genus V. Clavibacter
Genus VI. Cryobacterium
Genus VII. Curtobacterium
Genus VIII. Frigoribacterium
Genus IX. Leifsonia
Genus X. Rathayibacter
Family IX. "Beutenbergiaceae"
Genus I. Beutenbergia
Family X. Promicromonosporaceae
Genus I. Promicromonospora
SuborderVII. Corynebacterineae
Family I. Corynebacteriaceae
Genus I. Corynebacterium
Family II. Dietziaceae
Genus I. Dietzia
Family III. Gordoniaceae
Genus I. Gordonia
Genus II. Skermania
Family IV. Mycobacteriaceae
Genus I. Mycobacterium
Family V. Nocardiaceae
Genus I. Nocardia
Genus II. Rhodococcus
Family VI. Tsukamurellaceae
Genus I. Tsukamurella
Family VII. "Williamsiaceae"
Genus I. Williamsia
SuborderVIII. Micromonosporineae
Family I. Micromonosporaceae
Genus I. Micromonospora
Genus II. Actinoplanes
Genus III. Catellatospora
Genus IV. Catenuloplanes
Genus V. Couchioplanes
Genus VI. Dactylosporangium
Genus VII. Pilimelia
Genus VIII. Spirilliplanes
Genus IX. Verrucosispora
Suborder IX. Propionibacterineae
Family I. Propionibacteriaceae
Genus I. Propionibacterium
Genus II. Luteococcus
Genus III. Microlunatus
Genus IV. Propioniferax
Genus V. Tessaracoccus
Family II. Nocardioidaceae
Genus I. Nocardioides
Genus II. Aeromicrobium
Genus III. Friedmanniella
Genus IV. Kribbella
Genus V. Micropruina
Suborder X. Pseudonocardineae
Family I. Pseudonocardiaceae
Genus I. Pseudonocardia
Genus II. Actinopolyspora
Genus III. Amycolatopsis
Genus IV. Kibdelosporangium
Genus V. Kutzneria
Genus VI. Prauserella
Genus VII. Saccharomonospora
Genus VIII. Saccharopolyspora
Genus IX. Streptoalloteichus
Genus X. Thermobispora
Genus XI. Thermocrispum
Family II. Actinosynnemataceae
Genus I. Actinosynnema
Genus II. Actinokineospora
Genus III. Lentzea
Genus IV. Saccharothrix
Suborder XI. Streptomycineae
Family I. Streptomycetaceae
Genus I. Streptomyces
Genus II. Kitasatospora
Genus III. Streptoverticillium
Suborder XII. Streptosporangineae
Family I. Streptosporangiaceae
Genus I. Streptosporangium
Genus II. Herbidospora
Genus III. Microbispora
Genus IV. Microtetraspora
Genus V. Nonomuraea
Genus VI. Planobispora
Genus VII. Planomonospora
Genus VIII. Planopolyspora
Genus IX. Planotetraspora
Family II. Nocardiopsaceae
Genus I. Nocardiopsis
Genus II. Thermobifida
Family III. Thermomonosporaceae
Genus I. Thermomonospora
Genus II. Actinomadura
Genus III. Spirillospora
Suborder XIII. Frankineae
Family I. Frankiaceae
Genus I. Frankia
Family II. Geodermatophilaceae
Genus I. Geodermatophilus
Genus II. Blastococcus
Genus III. Modestobacter
Family III. Microsphaeraceae
Genus I. Microsphaera
Family IV. Sporichthyaceae
Genus I. Sporichthya
Family V. Acidothermaceae
Genus I. Acidothermus
Family VI. "Kineosporiaceae"
Genus I. Kineosporia
Genus II. Cryptosporangium
Genus III. Kineococcus
Suborder IV. Glycomycineae
Family I. Glycomycetaceae
Genus I. Glycomyces
Order II. Bifidobacteriales
Family I. Bifidobacteriaceae
Genus I. Bifidobacterium
Genus II. Falcivibrio
Genus III. Gardnerella
Family II. Unknown Affiliation
Genus I. Actinobispora
Genus II. Actinocorallia
Genus III. Excellospora
Genus IV. Pelczaria
Genus V. Turicella

Phylum BXV. Planctomycetes

Class I. "Planctomycetacia"
Order I. 'Planctomycetales
Family I. Planctomycetaceae
Genus I. Planctomyces
Genus II. Gemmata
Genus III. Isosphaera
Genus IV. Pirellula

Phylum BXVI. Chlamydiae

Class I. "Chlamydiae"
Order I. Chlamydiales
Family I. Chlamydiaceae
Genus I. Chlamydia
Genus II. Chlamydophila
Family II. Parachlamydiaceae
Genus I. Parachlamydia
Family III. Simkaniaceae
Genus I. Simkania
Family IV. Waddliaceae
Genus I. Waddlia

Phylum BXVII. Spirochaetes

Class I. "Spirochaetes"
Order I. 'Spirochaetales'
Family I. Spirochaetaceae
Genus I. Spirochaeta
Genus II. Borrelia
Genus III. Brevinema
Genus IV. Clevelandina
Genus V. Cristispira
Genus VI. Diplocalyx
Genus VII. Hollandina
Genus VIII. Pillotina
Genus IX. Treponema
Family II. "Serpulinaceae"
Genus I. Brachyspira
Genus II. Serpulina
Family III. Leptospiraceae
Genus I. Leptonema
Genus II. Leptospira

Phylum BXVIII. Fibrobacteres

Class I. "Fibrobacteres"
Order I. "Fibrobacterales"

Family I. 'Fibrobacteraceae'
Genus I. Fibrobacter

Phylum BXIX. Acidobacteria

Class I. "Acidobacteria"
Order I. 'Acidobacteriales'
Family I. 'Acidobacteriaceae'
Genus I. Acidobacterium
Genus II. Geothrix
Genus III. Holophaga

Phylum BXX. Bacteroidetes

Class I. "Bacteroides"
Order I. "Bacteroidales"
Family I. "Bacteroidaceae"
Genus I. Bacteroides
Genus II. Acetofilamentum
Genus III. Acetomicrobium
Genus IV. Acetothermus
Genus V. Anaerorhabdus
Genus VI. Megamonas
Family II. "Rikenellaceae"
Genus I. Rikenella
Genus II. Marinilabilia
Family III. "Porphyromonadaceae"
Genus I. Porphyromonas
Family IV. "Prevotellaceae"
Genus I. Prevotella

Class II. "Flavobacteria"
Order I. "Flavobacteriales"
Family I. Flavobacteriaceae
Genus I. Flavobacterium
Genus II. Bergeyella
Genus III. Capnocytophaga
Genus IV. Cellulophaga
Genux V. Chryseobacterium
Genus VI. Coenonia
Genus VII. Empedobacter
Genus VIII. Gelidibacter
Genus IX. Ornithobacterium
Genus X. Polaribacter
Genus XI. Psychroflexus
Genus XII. Psychroserpens
Genus XIII. Riemerella
Genus XIV. Weeksella
Family II. 'Myroideaceae'
Genus I. Myroides
Genus II. Psychromonas
Family III. "Blattabacteriaceae"
Genus I. Blattabacterium

Class III. "Sphingobacteria"
Order I. "Sphingobacteriales"
Family I. Sphingobacteriaceae
Genus I. Sphingobacterium
Genus II. Pedobacter
Family II. "Saprospiraceae"
Genus I. Saprospira
Genus II. Haliscomenobacter
Genus III. Lewinella
Family III. "Flexibacteraceae"
Genus I. Flexibacter
Genus II. Cyclobacterium
Genus III. Cytophaga
Genus IV. Flectobacillus
Genus V. Hymenobacter
Genus VI. Meniscus
Genus VII. Microscilla
Genus VIII. Runella
Genus IX. Spirosoma
Genus X. Sporocytophaga
Family IV. "Flammeovirgaceae"
Genus I. Flammeovirga
Genus II. Flexithrix
Genus III. Persicobacter
Genus IV. Thermonema
Family V. 'Crenotrichaceae'
Genus I. Crenothrix
Genus II. Chitinophaga
Genus III. Rhodothermus
Genus IV. Toxothrix

Phylum BXXI. Fusobacteria

Class I. "Fusobacteria"
Order I. "Fusobacteriales"
Family I. "Fusobacteriaceae"
Genus I. Fusobacterium
Genus II. Ilyobacter
Genus III. Leptotrichia
Genus IV. Propionigenium
Genus V. Sebaldella
Genus VI. Streptobacillus
Family II. Incertae sedis
Genus I. Cetobacterium

Phylum BXXII. Verrucomicrobia

Class I. Verrucomicrobiae
Order I. Verrucomicrobiales
Family I. Verrucomicrobiaceae
Genus I. Verrucomicrobium
Genus II. Prosthecobacter

Phylum BXXIII. Dictyoglomi

Class I. "Dictyoglomi"
Order I. "Dictyoglomales"
Family I. "Dictyoglomaceae"
Genus I. Dictyoglomus

Pronunciation Key For Bacterial, Fungal, Protozoan, and Viral Names

A

Acetobacter (a-see´-toe-back-ter)
Achromobacter (a-krome´-oh-back-ter)
Acinetobacter calcoaceticus (a-sin-et´-oh-back-ter kal-koh-ah-see´-ti-kus)
Actinomyces israelii (ak-tin-oh-my´-seez iz-ray´-lee-ee)
Actinomycetes (ak-tin-oh-my´-seats)
Adenovirus (ad´-eh-no-vi-rus)
Agrobacterium tumefaciens (ag-rho-bak-teer´-ee-um too-meh-faysh´-ee-enz)
Alcaligenes (al-ka-li´-jen-ease)
Amoeba (ah-mee´-bah)
Arbovirus (are´-bow-vi-rus)
Archezoa (ar-key-zoe´-ah)
Aspergillus niger (ass-per-jill´-us nye´-jer)
Aspergillus oryzae (ass-per-jill´-us or-eye´-zee)
Azolla (aye-zol´-lah)
Azotobacter (ay-zoh´-toe-back-ter)

B

Bacillus anthracis (bah-sill´-us an-thra´-siss)
Bacillus cereus (bah-sill´-us seer´-ee-us)
Bacillus coagulans (bah-sill´-us coh-ag´-you-lans)
Bacillus fastidiosus (bah-sill´-us fas-tid-ee-oh´-sus)
Bacillus stearothermophilus (bah-sill´-us steer-oh-ther-maw´-fill-us)
Bacillus subtilis (bah-sill´-us sut´-ill-us)
Bacillus thuringiensis (bah-sill´-us thur´-in-jee-en-sis)
Bacteroides (back´-ter-oid´-eez)
Baculovirus (back´-you-low-vi-rus)
Beggiatoa (beg-gee-ah-toe´-ah)
Beijerinckia (by-yer-ink´-ee-ah)
Bordetella pertussis (bor-deh-tell´-ah per-tuss´-iss)
Borrelia burgdorferi (bor-real´-ee-ah berg-dor´-fir-ee)
Bradyrhizobium (bray-dee-rye-zoe´-bee-um)
Branhamella (bran-ham-el´-lah)
Brucella abortus (bru-sell´-ah ah-bore´-tus)

C

Campylobacter jejuni (kam´-peh-low-back-ter je-june´-ee)
Candida albicans (kan´-did-ah al´-bi-kanz)
Caulobacter (caw´-loh-back-ter)
Ceratocystis ulmi (see´-rah-toe-sis-tis ul´-mee)
Chlamydia trachomatis (klah-mid´-ee-ah trah-ko-ma´-tiss)
Claviceps purpurea (kla´-vi-seps purr-purr´-ee-ah)
Clostridium acetobutylicum (kloss-trid´-ee-um a-seat-tow-bu-til´-i-kum)
Clostridium botulinum (kloss-trid´-ee-um bot-you-line´-um)
Clostridium difficile (kloss-trid´-ee-um dif´-fi-seal)
Clostridium perfringens (kloss-trid´-ee-um per-frin´-gens)
Clostridium tetani (kloss-trid´-ee-um tet´-an-ee)
Coccidioides immitis (cock-sid-ee-oid´-eez im´-mi-tiss)
Coronavirus (kor-oh´-nah-vi-rus)
Corynebacterium diphtheriae (koh-ryne´-nee-bak-teer-ee-um dif-theer´-ee-ee)
Coxsackievirus (cock-sack-ee´-vi-rus)
Cryptococcus neoformans (krip-toe-cock´-us knee-oh-for´-manz)
Cytophaga (sigh-taw´-fa-ga)

D

Desulfovibrio (dee-sul-foh-vib´-ree-oh)

E

Eikenella corrodens (eye-keh-nell´-ah kor-roh´-denz)
Entamoeba histolytica (en-ta-mee´-bah his-toh-lit´-ik-ah)
Enterobacter aerogenes (en´-ter-oh-back-ter)
Enterococcus faecalis (en´-ter-oh-kock´-us fee-ka´-liss)
Enterovirus (en´-ter-oh-vi-rus)
Epidermophyton (eh-pee-der´-moh-fy-ton)
Escherichia coli (esh-er-ee´-she-ah koh´-lee)

F

Filobasidiella neoformans (fee-loh-bah-si-dee-ell´-ah knee-oh-for´-manz)
Flavivirus (flay´-vih-vi-rus)
Flavobacterium (flay-vo-back-teer´-ee-um)
Francisella tularensis (fran-siss-sell´-ah tu-lah-ren´-siss)
Frankia (frank´-ee-ah)
Fusobacterium (fu´-zoh-back-teer-ee-um)

G

Gallionella (gal-ee-oh-nell´-ah)
Gardnerella vaginalis (gard-nee-rel´-lah va-jin-al´-is)
Giardia intestinalis (jee-are´-dee-ah in-test´-tin-al-is)
Giardia lamblia (jee-are´-dee-ah lamb´-lee-ah)
Gluconobacter (glue-kon-oh-back´-ter)
Gonyaulax (gon-ee-ow´-lax)
Gymnodinium breve (jim-no-din´-i-um brev-eh)

H

Haemophilus influenzae (hee-moff´-ill-us in-flew-en´-zee)
Helicobacter pylori (he´-lih-koh-back-ter pie-lore´-ee)
Hepadnavirus (hep-ad´-nah-vi-rus)
Hepatitis virus (hep-ah-ti´-tis vi-rus)
Herpes simplex (her´-peas sim´-plex)
Herpes zoster (her´-peas zoh´-ster)
Histoplasma capsulatum (his-toh-plaz´-mah cap-su-lah´-tum)
Hyphomicrobium (high-foh-my-krow´-bee-um)

I

Influenza virus (in-flew-en´-za vi-rus)

K

Klebsiella pneumoniae (kleb-see-ell´-ah new-moan´-ee-ee)

L

Lactobacillus brevis (lack-toe-ba-sil´-lus bre´-vis)
Lactobacillus bulgaricus (lack-toe-ba-sil´-lus bull-gair´-i-kus)
Lactobacillus casei (lack-toe-ba-sil´-us kay´-see-ee)
Lactobacillus plantarum (lack-toe-ba-sil´-us plan-tar´-um)

Lactobacillus thermophilus (lack-toe-ba-sil′-us ther-mo′-fil-us)
Lactococcus lactis (lack-toe-kock′-us lak′-tiss)
Legionella pneumophila (lee-jon-ell′-ah new-moh′-fill-ah)
Leptospira interrogans (lep-toe-spire′-ah in-ter-roh′-ganz)
Leuconostoc citrovorum (lew-kow-nos′-tok sit-ro-vor′-um)
Listeria monocytogenes (lis-tear′-ee-ah mon′-oh-sigh-to-jen′-eze)

M

Malassezia (mal-as-seez′-e-ah)
Methanobacterium (me-than′-oh-bak-teer-ee-um)
Methanococcus (me-than-oh-ko′-kus)
Microsporum (my-kroh-spore′-um)
Mobiluncus (moh-bi-lun′-kus)
Moraxella catarrhalis (more-ax-ell′-ah kah-tah-rah′-liss)
Moraxella lacunata (more-ax-ell′-ah lak-u-nah′-tah)
Mucor (mu′-kor)
Mycobacterium leprae (my-koh-bak-teer′-ee-um lep-ree)
Mycobacterium tuberculosis (my-koh-bak-teer′-ee-um too-ber-kew-loh′-siss)
Mycoplasma pneumoniae (my-koh-plaz′-mah new-moan′-ee-ee)

N

Neisseria gonorrhoeae (nye-seer′-ee-ah gahn-oh-ree′-ee)
Neisseria meningitidis (nye-seer′-ee-ah men-in-jit′-id-iss)
Neurospora sitophila (new-rah′-spor-ah sit-oh-phil′-ah)

O

Orthomyxovirus (or-thoe-mix′-oh-vi-rus)
Oscillatoria (os-sil-la-tor′-ee-ah)

P

Papillomavirus (pap-il-oh′-ma-vi-rus)
Parainfluenza virus (par-ah-in-flew-en′-zah vi-rus)
Paramecium (pair′-ah-mee-see-um)
Paramyxovirus (par-ah-mix′-oh-vi-rus)
Parvovirus (par′-vo-vi-rus)
Pasteurella multocida (pass-ture-ell′-ah mul-toe-sid′-ah)
Pasteurella piscicida (pass-ture-ell′-ah pis-si-si′-dah)
Pediococcus soyae (ped-ih-ko′-kus soy′-ee)
Penicillium camemberti (pen-eh-sill′-ee-um cam-em-bare′-tee)
Penicillium roqueforti (pen-eh-sill′-ee-um rok-e-for′-tee)
Peptostreptococcus (pep′-to-strep-to-ko-kus)
Phytophythora infestans (fy′-toe-fy-thor-ah in-fes′-tanz)
Picornavirus (pi-kor′-na-vi-rus)
Plasmodium falciparum (plaz-moh′-dee-um fall-sip′-air-um)
Plasmodium malariae (plaz-moh′-dee-um ma-lair′-ee-ee)
Plasmodium ovale (plaz-moh′-dee-um oh-vah′-lee)
Plasmodium vivax (plaz-moh′-dee-um vye′-vax)
Pneumocystis carinii (new-mo-sis′-tis car′-nee-ee)
Poliovirus (poe′-lee-oh-vi-rus)
Polyoma virus (po-lee-oh′-mah vi-rus)
Propionibacterium acnes (proh-pee-ah-nee-bak-teer′-ee-um ak′-neez)
Propionibacterium shermanii (proh-pee-ah-nee-bak-teer′-ee-um sher-man′-ee-ee)
Proteus mirabilis (proh′-tee-us mee-rab′-il-us)
Pseudomonas aeruginosa (sue-dough-moan′-ass aye-rue-gin-o′-sa)

R

Rabies virus (ray′-bees vi-rus)
Retrovirus (re′-trow-vi-rus)
Rhabdovirus (rab′-doh-vi-rus)
Rhinovirus (rye′-no-vi-rus)
Rhizobium (rye-zoh′-bee-um)
Rhizopus nigricans (rise′-oh-pus nye′-gri-kanz)
Rhizopus stolon (rise′-oh-pus stoh′-lon)
Rhodococcus (roh-doh-koh′-kus)
Rickettsia rickettsii (rik-kett′-see-ah rik-kett′-see-ee)
Rotavirus (row′-tah-vi-rus)
Rothia (roth′-ee-ah)
Rubella virus (rue-bell′-ah vi-rus)
Rubeola virus (rue-bee-oh′-la vi-rus)

S

Saccharomyces carlsbergensis (sack-ah-row-my′-sees karls-berg-en′-siss)
Saccharomyces cerevisiae (sack-ah-row-my′-sees sara-vis′-ee-ee)
Saccharomyces rouxii (sack-ah-row-my′-sees roos′-ee-ee)
Salmonella choleraesuis (sall-moh-nell′-ah koh′-er-ay-sues)
Salmonella enteritidis (sall-moh-nell′-ah en-ter-it′-id-iss)
Salmonella typhi (sall-moh-nell′-ah tye′-fee)
Salmonella typhimurium (sall-moh-nell′-ah tye-fe-mur′-ee-um)
Serratia marcescens (ser-ray′-sha mar-sess-sens)
Shigella dysenteriae (shig-ell′-ah diss-en-tair′-ee-ee)
Spirillum minus (spy-rill′-um my′-nus)
Sporothrix schenckii (spore′-oh-thrix shenk-ee-ee)
Staphylococcus aureus (staff-ill-oh-kok′-us aw′-ree-us)
Staphylococcus epidermidis (staff-ill-oh-kok′-us epi-der′-mid-iss)
Streptobacillus moniliformis (strep-tow-bah-sill′-us mon-ill-i-form′-is)
Streptococcus agalactiae (strep-toe-kock′-us a-ga-lac′-tee-ee)
Streptococcus cremoris (strep-toe-kock′-us kre-more′-iss)
Streptococcus durans (strep-toe-kock′-us dur′-anz)
Streptococcus mutans (strep-toe-kock′-us mew′-tanz)
Streptococcus pneumoniae (strep-toe-kock′-us new-moan′-ee-ee)
Streptococcus pyogenes (strep-toe-kock′-us pie-ah-gen-ease)
Streptococcus salivarius (strep-toe-kock′-us sal-ih-vair′-ee-us)
Streptococcus sanguis (strep-toe-kock′-us san′-gwis)
Streptococcus thermophilis (strep-toe-kock′-us ther-moh′-fill-us)
Streptomyces griseus (strep-toe-my′-seez gree′-see-us)

T

Thiobacillus (thigh-oh-bah-sill′-us)
Torulopsis (tore-you-lop′-siss)
Treponema pallidum (tre-poh-nee′-mah pal′-ih-dum)
Trichomonas vaginalis (trick-oh-moan′-as vag-in-al′-iss)
Trichophyton (trick-oh-phye′-ton)
Trypanosoma brucei (tri-pan′-oh-soh-mah bru′-see-ee)

V

Varicella-zoster virus (var-ih-sell′-ah zoh′-ster vi-rus)
Veillonella (veye-yon-ell′-ah)
Vibrio anguillarum (vib′-ree-oh an-gwil-air′-um)
Vibrio cholerae (vib′-ree-oh kahl′-er-ee)

Y

Yersinia enterocolitica (yer-sin′-ee-ah en-ter-koh-lih′-tih-kah)
Yersinia pestis (yer-sin′-ee-ah pess′-tiss)

ABC transport systems A type of active transport system that requires ATP as an energy source (***A***TP ***B***inding-***C***assette).

abscess A localized collection of pus within a tissue.

acellular Not composed of cells, therefore not living.

acetyl-CoA Product of the transition reaction; a precursor metabolite used in fatty acid synthesis.

acid-fast staining A procedure used to stain certain microorganisms, particularly members of the genus *Mycobacterium* that do not readily take up dyes used in microbiology.

acidic amino acids Amino acids with more acid (—COOH) than amino (—NH_2) groups.

acidophiles Organisms that grow optimally at a pH below 5.5.

acquired resistance Development of antimicrobial resistance in a previously sensitive organism; occurs through spontaneous mutation or acquisition of new genetic information.

acridine orange A fluorescent dye that binds DNA and can be used to determine the total number of microorganisms in a sample.

actin Polymer that makes up microfilaments of eukaryotic cells; it can rapidly assemble and subsequently disassemble to cause motion.

actinomycetes Filamentous bacteria; many are valuable in the production of antibiotics.

activated macrophages Macrophages stimulated by cytokines to enlarge and become metabolically active, with greatly increased capability to kill and degrade intracellular organisms and materials.

activated sludge method A method of sewage treatment in which wastes are degraded by complex populations of aerobic microorganisms.

activated (or **immune**) **T lymphocytes** T cells activated by exposure to antigen.

activation energy Initial energy required to break a chemical bond.

activator-binding site Sequence of DNA that precedes an ineffective promoter; binding of an activator to this site enhances the ability of RNA polymerase to initiate transcription at that promoter.

active immunity Protective immunity produced by an individual in response to an antigenic stimulus.

active site Site on an enzyme molecule to which substrate binds; also known as the catalytic site.

active transport Energy-requiring process by which molecules are carried across cell boundaries; can accumulate compounds against a concentration gradient.

acute infections Infections in which the symptoms and signs have a rapid onset and are usually severe, often with fever, but short-lived.

acute phase response Changes in the blood that occur early during an infection, with the production of acute phase proteins and cells that contribute to the inflammatory response.

acylated homoserine lactone (AHL) Small molecules that can move freely in and out of a cell; provides cells with a mechanism of assessing cell density (quorum sensing).

ADCC (antibody-dependent cellular cytotoxicity) Nonspecific killing of target cells by cells, such as macrophages, granulocytes, or natural killer cells, that contact the target via their Fc receptors binding to Fc of antibodies on the target.

adenosine diphosphate (ADP) The acceptor of free energy in a cell; that energy is used to add a inorganic phosphate (P_i) to ADP, generating ATP.

adenosine triphosphate (ATP) The energy currency of a cell, serving as the ready and immediate donor of free energy.

adherence A necessary first step in colonization and infection, in which the pathogen attaches to host cells to avoid being removed from the body.

adjuvant Substance that increases the immune response to antigen.

ADP Abbreviation for adenosine diphosphate.

adsorption Attachment of one substance to the surface of another.

aerobic respiration Metabolic process in which electrons are transferred from the electron transport chain to molecular oxygen (O_2).

aerosol Material dispersed into the air as a fine mist.

aerotaxis Movement toward or away from molecular oxygen.

aerotolerant anaerobes Organisms that can grow in the presence of O_2 but do not use it to generate energy; also called obligate fermenters.

aflatoxin Potent toxin made by species of *Aspergillus*; may contaminate peanuts and other grains.

agar Polysaccharide extracted from marine algae; used to solidify microbiological media.

agarose Highly purified form of agar used in gel electrophoresis.

agar slant Microbiological medium that has been solidified with agar and stored in a tube that was held at a shallow angle as the medium solidified, creating a larger surface area.

agglutination Clumping together of cells or particles.

AIDS Acquired immunodeficiency syndrome.

AIDS-related complex (ARC) A group of symptoms, fever, fatigue, diarrhea, and weight loss, that herald the onset of AIDS.

alga (pl. **algae**) A primitive photosynthetic eukaryotic organism.

alkaliphiles Organisms that grow optimally at a pH above 8.5.

alkylating agent Chemical that adds alkyl groups, short chains of carbon atoms, to purines and pyrimidines; promotes mutations.

alkyl group Short chain or single carbon atom such as a methyl group (—CH_3).

allele One form of a gene.

allergen Antigen that causes an allergy.

allergic rhinitis Hay fever; sneezing, runny nose, teary eyes resulting from exposure of a sensitized person to inhaled antigen; an IgE-mediated allergic reaction.

allergy Hypersensitivity, especially of the IgE-mediated type.

allograft Organ or tissue graft transplanted between genetically nonidentical members of the same species.

allosteric site Site on an allosteric enzyme that binds an effector molecule; binding alters the activity of the enzyme.

alpha (α) hemolysis Type of hemolysis observed on blood agar, characterized by zone of greenish clearing around the colonies.

alternative pathway Pathway of complement activation nonspecifically initiated by bacterial substances such as endotoxin and polysaccharides.

amalgam Mixture of mercury with other metals to form a paste that hardens; used to fill cavities in teeth.

amino acids Subunits of a protein molecule.

aminoglycosides Group of antimicrobial medications that interferes with protein synthesis.

amino terminal (or **N terminal**) The end of the protein molecule that has an unbonded —NH_2 group.

ammonification The reactions that result in the release of ammonia (NH_3) from organic nitrogen-containing molecules.

amphibolic pathways Metabolic pathways that play roles in both catabolism and anabolism.

amylases Enzymes that digest starches.

anabolism Cellular processes that use the energy stored in ATP to synthesize and assemble subunits such as amino acids; synonymous with biosynthesis.

anaerobic Containing no molecular oxygen (O_2).

anaerobic respiration Metabolic process in which electrons are transferred from the electron transport chain to an inorganic terminal electron acceptor other than O_2.

analytical study An epidemiological study done to identify specific risk factors associated with developing a certain disease.

anamnestic response (or **memory response, secondary response**) Enhanced immunological response to a second or subsequent dose of antigen.

anaphylaxis Allergic reaction caused by IgE; generalized hypersensitive reaction to an allergen that can cause a profound drop in blood pressure.

anion Negatively charged ion.

anoxic Devoid of O_2.

anoxygenic phototrophs Photosynthetic bacteria that use hydrogen sulfide or organic compounds rather than water as a source of electrons for reducing power; they do not generate O_2.

antagonistic In antimicrobial therapy, a combination of antimicrobial medications in which the action of one interferes with the action of the other.

antenna complex Complex in photosynthetic organisms composed of hundreds of light-gathering pigments; acts as a funnel, capturing light energy and transferring it to reaction-center chlorophyll.

antibacterial drug Chemical used to treat bacterial infections.

antibiogram Antibiotic susceptibility pattern; used to distinguish different strains.

antibiotic Chemical produced by certain molds and bacteria that kills or inhibits the growth of other microorganisms.

antibiotic-associated colitis Intestinal disease caused by overgrowth of toxin-producing strains of *Clostridium difficile*; typically, occurs when a person is taking antimicrobial medications.

antibody Immunoglobulin protein produced by the body in response to a substance and that reacts specifically with that substance.

anticodon Sequence of three nucleotides in a tRNA molecule that is complementary to a particular codon in mRNA.

antigen Molecule that reacts specifically with an antibody or immune lymphocyte.

antigen-binding sites Hypervariable regions at the ends of the two arms of an antibody molecule that recognize specific antigen; there are two identical antigen-binding sites on each monomer of antibody.

antigenic determinant Part of an antigen molecule that binds the specific antibody; an epitope.

antigenic drift Slight changes that occur in the antigens of a virus; specific antibodies made to the antigen before the change occurred are only partially effective.

antigenic shift Major changes that can occur in the antigens of a virus.

antigenic variation Routine alteration by an organism in the characteristics of certain of its surface proteins.

antigen presentation Process in which macrophages, dendritic cells, or B cells ingest and partially degrade antigens, which then appear on the cell surface in a form that can react with CD4 T lymphocytes.

antimicrobial drug Chemical used to treat microbial infections; also called an antimicrobial.

antiparallel Term used to describe opposing orientations of the two strands of DNA in the double helix; one strand is oriented in the 5′ to 3′ direction and its complement is oriented in the 3′ to 5′ direction.

antiporter Transport system that exchanges the location of one molecule or ion for another.

antisense strand Complement to the sense (or plus) strand of RNA; also called the minus (–) strand.

antiseptic A disinfectant that is nontoxic enough to be used on skin.

aplastic anemia Potentially lethal condition in which the body is unable to make blood cells.

apoptosis Programmed cell death, with enzymatic degradation of DNA.

arbovirus Arthropod-borne virus. One of a large group of RNA viruses carried by insects and mites that act as biological vectors.

Archaea One of the two domains of prokaryotes; most, but not all, archaea grow in extreme environments.

arteriosclerosis Condition characterized by thickening and loss of elasticity of the walls of arteries; "hardening of the arteries."

arthropod Classification grouping of invertebrate animals that includes insects, ticks, lice, and mites.

Arthus reaction Hypersensitivity reaction caused by immune complexes and neutrophils.

artificial wetland method Method of sewage treatment in which sewage is channeled into successive ponds where both aerobic and anaerobic stabilization occurs.

aseptic Microorganisms and viruses; sterile.

aseptic technique Use of specific methods and sterile materials to exclude contaminating microorganisms from an environment.

asexual Reproduction not preceded by the union of cells or genetic exchange.

A-site (or **aminoacyl site**) Site on the ribosome to which tRNAs enter to donate their amino acid; acceptor site.

asthma Immediate respiratory allergy resulting from mediator release from mast cells in the lower airways.

astrobiology Study of life in the universe.

atomic force microscope Type of scanning probe microscope that has a tip mounted so it can bend in response to the slightest force between the tip and the sample.

ATP Abbreviation for adenosine triphosphate.

ATP synthase Protein complex that harvests the energy of a proton motive force to synthesize ATP.

attack rate Proportional number of cases developing in a population exposed to an infectious agent.

auramine A fluorescent dye that can be used to selectively stain members of the genus *Mycobacterium*.

autoclave Device employing steam under pressure used for sterilizing materials that are stable to heat and moisture.

autoimmune disease Disease produced as a result of an immune reaction against one's own tissues.

autolyze To spontaneously disintegrate as a result of enzymes within the cell.

autoradiography The use of film to detect a radioactive molecule.

autotroph Organism that can use CO_2 as its main source of carbon.

auxotroph Mutant of a microorganism that requires an organic growth factor.

avirulent Lacking disease-causing attributes, such as a capsule.

a_w Abbreviation for water activity.

axial filaments Charactersitic structure of motility found in spirochetes.

azoles Large family of chemically synthesized medications, some of which have antifungal activity.

---

bacillus (pl. **bacilli**) Cylindrical-shaped bacterium; also referred to as a rod.

bacitracin Antimicrobial medication that inhibits cell wall biosynthesis by interfering with the transport of peptidoglycan precursors across the cytoplasmic membrane.

bacteremia Bacteria circulating in the bloodstream.

Bacteria One of the two domains of prokaryotes; all medically important prokaryotes are in the domain Bacteria.

bacterial artificial chromosome (BAC) Derivative of the F-plasmid used as a vector to clone DNA fragments as large as 300,000 base pairs in length into bacteria.

bactericidal Able to kill bacteria.

bacteriochlorophyll Type of chlorophyll used by purple and green bacteria; absorbs wavelengths of light that penetrate to greater depths and are not used by other photosynthetic organisms.

bacteriophage A virus that infects bacteria; often abbreviated to phage.

bacteriorhodopsin Pigment of some extreme halophiles that absorbs energy from sunlight and uses it to expel protons from the cell, generating a proton gradient.

bacteriostatic Able to inhibit the growth of bacteria.

balanced pathogenicity Host parasite relationship in which the parasite persists in the host without causing obvious harm.

barophiles Bacteria that can grow under high pressure.

basal body Structure that anchors the flagella to the cell wall and cytoplasmic membrane.

base Refers to the purine or pyrimidine ring structure found in nucleic acids.

base analog Compound that resembles a purine or pyrimidine base closely enough to be incorporated into DNA in place of a natural base.

base pairing The hydrogen bonding of adenine (A) to thymine (T) and cytosine (C) to guanine (G); occurs between two complementary strands of DNA.

basic amino acids Amino acids with more basic ($—NH_3^+$) groups than acid ($—COO^-$) groups.

basophil Leukocyte with large dark-staining granules that contain histamine and other mediators of inflammation; receptors on cell surfaces bind monomers of IgE.

B cells Lymphocytes that produce immunoglobulin.

beta- (β) hemolysis Type of hemolysis observed on blood agar that is characterized by a clear zone around a colony.

beta- (β) lactam drugs Group of antimicrobial medications that inhibit peptidoglycan synthesis and have a shared chemical structure called a β-lactam ring.

bilayer membrane (or **unit membrane**) Double layer of phospholipid molecules that forms the major structure of the cytoplasmic (plasma) membrane.

binary fission Asexual process of reproduction in which one cell divides into two independent daughter cells.

binding protein Protein that functions in the ABC transport system; resides immediately outside of the cytoplasmic membrane to deliver a given molecule to a specific transport complex within the membrane.

binomial system System of naming each species of organism with two Latin words.

biochemical oxygen demand (BOD) Measure of the amount of biologically degradable organic material in water.

biocide Compound such as a disinfectant that is toxic to many forms of life, including microorganisms.

biodiversity Diversity in the number of species of organisms inhabiting an ecosystem.

biofilm Polysaccharide-encased community of microorganisms attached to a surface.

bioinformatics Developing and using computer technology to store, retrieve, and analyze nucleotide sequence data.

bioleaching Conversion of metals to a soluble form due to the metabolic oxidation of insoluble metal sulfides by microorganisms.

biological vector Organism that acts as a host for a disease organism before it is transmitted to another organism; the pathogen can multiply to high numbers within it.

bioluminescence Biological production of light.

biomass Total weight of all organisms in any particular environment.

bioremediation Process that uses microorganisms to degrade harmful chemicals.

biosphere The sum of all the regions of the earth where life exists.

biotechnology The use of microbiological and biochemical techniques to solve practical problems and produce more useful products.

biotype A strain that has a characteristic biochemical pattern, which differs from other strains; also called a biovar.

blood agar Type of agar medium that contains red blood cells; a rich medium that can be used to detect hemolysis.

blood-brain barrier Property of the central nervous system blood vessels that restricts passage of infectious agents and certain molecules (such as medications) into the brain and spinal cord.

blunt end The type of DNA ends generated by a restriction enzyme that cuts directly in the middle of the recognition sequence.

BOD Abbreviation for biochemical oxygen demand.

boil Painful localized collection of pus within the skin and subcutaneous tissue; a furuncle.

bone marrow Soft material that fills bone cavities and contains stem cells for all blood cells.

botulinum toxin Toxin produced by *Clostridium botulinum* that can cause a fatal paralysis in people who consume it.

bright-field microscope Type of light microscope that illuminates the field of view evenly.

brine Salty water; used to cure fish and meats.

broad-spectrum antimicrobials Antimicrobials that inhibit or kill a wide range of microorganisms, often including both Gram-positive and Gram-negative bacteria.

Bt-toxin Protein crystal naturally produced by the bacterium *Bacillus thuringiensis* as it forms endospores; toxic to insect larvae that consume it.

bubo Enlarged, tender lymph node characteristic of plague and some venereal diseases.

bubonic plague Form of plague that typically develops when *Yersinia pestis* is injected via the bite of an infected flea.

budding Asexual reproductive technique that involves a pushing out of a part of the parent cell that eventually gives rise to a new daughter cell.

buffer Substance in a solution that acts to prevent changes in pH.

bulking Overgrowth of filamentous microorganisms in sewage at treatment facilities; interferes with the separation of the solid sludge from the liquid effluent.

burst size Number of newly formed virus particles released from a single cell.

calcofluor white Fluorescent dye that binds to a component of fungal cells, causing them to fluoresce bright blue.

Calvin cycle Metabolic pathway used by many autotrophs to incorporate CO_2 into an organic form; also called the Calvin-Benson cycle.

cAMP Abbreviation for cyclic AMP.

cancer Abnormally growing cells that can spread from their site of origin; malignant tumors.

candidiasis Fungal diseases caused by *Candida albicans*.

candle jar Closed jar in which a lit candle converts some of the O_2 in air to CO_2 and water vapor; used to cultivate capnophiles.

CAP Abbreviation for cyclic AMP–activating protein.

cap Methylated guanine derivative added to the 5′ end of eukaryotic mRNA before transcription is complete.

capnophiles Organisms that require increased concentrations of CO_2 (5% to 10%) and approximately 15% oxygen.

capsid Protein coat that surrounds the nucleic acid of a virus.

capsule Glycocalyx that is distinct and gelatinous; sometimes correlated with an organism's ability to cause disease.

carbapenems Group of antimicrobial medications that interferes with peptidoglycan synthesis; very resistant to β-lactamases.

carbohydrate Compounds containing principally carbon, hydrogen, and oxygen atoms in a ratio of 1:2:1.

carbon fixation Process of converting inorganic carbon (CO_2) to an organic form; in photosynthetic organisms, the dark or light-independent reactions.

carboxy terminal (or **C terminal**) The end of the protein molecule that has an unbonded —COOH group.

carbuncle Painful infection of the skin and subcutaneous tissues; manifests as a cluster of boils.

carrier (1) Type of protein found in cell membranes that transports certain compounds across the membrane; may also be called a permease or transporter protein. (2) A human or other animal that harbors a pathogen without noticeable ill effects.

carrier state State of infection in which the agent can be detected in body fluids without causing disease symptoms.

cascade In biology, a series of reactions that, once started, continues to the final step by each step triggering the next in a special order; activation of complement is an example.

caseous necrosis Type of localized tissue death having a cheeselike consistency, characteristic of tuberculosis and certain other chronic infectious diseases.

catabolism Cellular processes that harvest the energy released during the breakdown of compounds such as glucose and use that energy to synthesize ATP, the energy currency of all cells.

catabolite Product of catabolism.

catabolite repression The mechanism by which cells decrease the expression of genes that encode certain degradative enzymes in the presence of a compound such as glucose.

catalase Enzyme that breaks down hydrogen peroxide (H_2O_2) to produce water (H_2O) and oxygen gas (O_2).

catalyst Substance that speeds up the rate of a chemical reaction without being altered or depleted in the process.

cations Positively charged ions.

CD antigens Abbreviation for cluster of differentiation antigens.

CD4 lymphocytes T lymphocytes bearing the CD4 (cluster of differentiation) surface molecules; T-helper cells are CD4.

CD8 cells T lymphocytes bearing the CD8 cluster of differentiation surface molecules; T-cytotoxic cells and some T-suppressor cells are CD8 cells.

cDNA DNA obtained by using reverse transcriptase to synthesize DNA from an RNA template *in vitro*; lacks introns that characterize eukaryotic DNA.

cell culture (or **tissue culture**) Cultivation of animal or plant cells in the laboratory.

cell wall Rigid barrier that surrounds a cell, keeping the contents from bursting out; in prokaryotes, peptidoglycan provides rigidity to the cell wall.

cell-mediated immunity (CMI) Immune responses mediated by T lymphocytes.

cellulose Polymer of glucose subunits; principal structural component of plant cell walls.

central metabolic pathways The amphibolic pathways—glycolysis, the TCA cycle, and the pentose phosphate pathway—that are used by most chemoorganoheterotrophs.

cephalosporins Group of antimicrobial medications that interfere with peptidoglycan synthesis.

cestode Tapeworm.

challenge In immunology, to give an antigen to provoke an immunologic response in a subject previously sensitized to the antigen.

chancre Sore resulting from an ulcerating infection; the "hard chancre" of primary syphilis is typically firm and painless.

chaperones Proteins that help other proteins fold properly.

chemical bond Force that holds atoms together to form molecules.

chemically defined media Bacteriological media composed of ingredients of known chemical composition; generally used for specific experiments when nutrients must be precisely controlled.

chemiosmotic gradient Accumulation of protons on one side of a membrane due to expulsion of protons by the electron transport chain; used to power the synthesis of ATP, fuel certain transport processes, and drive the rotation of flagella; also called the proton motive force.

chemiosmotic theory The theory that a proton gradient is formed by the electron transport chain and is then used to power the synthesis of ATP.

chemoautotrophs Organisms that use inorganic chemicals as a source of energy and CO_2 as the major source of carbon.

chemoheterotrophs Organisms that use chemical energy and an organic source of carbon.

chemolithoautotrophs Organisms that obtain energy by degrading reduced inorganic compounds such as hydrogen gas (H_2), and use CO_2 as a source of carbon.

chemolithotrophs Organisms that obtain energy by degrading reduced inorganic chemicals such as hydrogen gas (H_2); in general, chemolithotrophs are chemolithoautotrophs.

chemoorganoheterotrophs Organisms that obtain both energy and carbon from organic compounds.

chemoorganotrophs Organisms that obtain energy by degrading organic compounds such as glucose; in general, chemoorganotrophs are chemoorganoheterotrophs.

chemostat Device used to grow bacteria in the laboratory that allows nutrients to be added and waste products to be removed continuously.

chemotaxis Directed movement of an organism in response to a certain chemical in the environment.

chemotherapeutic agent Chemical used as a therapeutic medication to treat a disease.

chemotrophs Organisms that obtain energy by degrading chemical compounds.

chickenpox Disease caused by the herpesvirus, varicella.

chloramphenicol Antimicrobial medication that interferes with protein synthesis.

chlorophylls The primary light-absorbing pigments used in photosynthesis.

chloroplasts Organelles in photosynthetic eukaryotic cells that harvest the energy of sunlight and use it to synthesize ATP.

chlorosomes Structures of green bacteria in which the light-harvesting pigments are located.

chocolate agar Type of agar medium that contains red blood cells that have been heated under controlled conditions to lyse them, releasing their nutrients; used to culture fastidious bacteria.

cholesterol Sterol found in animal cell membranes; provides rigidity to eukaryotic membranes.

chorea Constant complex, rapid, jerky involuntary movements; an occasional sequel to untreated *Streptococcus pyogenes* infections.

chromatin Complex of histones and DNA that make up the chromosomes of eukaryotic cells.

chromosome Array of genes responsible for the determination and transmission of hereditary characteristics.

chronic infections Infections that develop slowly and persist for months or years.

cilium (pl. **cilia**) Short, projecting hairlike organelle of locomotion, similar to a flagellum.

circulative transmission Transmission of viruses to plants by insects within which the virus circulates but does not multiply.

citric acid cycle Metabolic pathway that incorporates acetyl-CoA, ultimately generating CO_2 and reducing power; also known as the tricarboxylic acid (TCA) cycle and the Krebs cycle.

clade Subtype of a virus such as human immunodeficiency virus (HIV), defined by similar amino acid sequences of their envelope proteins.

class Collection of similar orders; a collection of several classes makes up a phylum.

classical pathway Pathway of complement activation initiated by specific antigen-antibody interaction.

classification Process of arranging organisms into similar or related groups, primarily to provide easy identification and study.

clathrin Proteins that line regions of eukaryotic cell membranes that are internalized during the process of receptor-mediated endocytosis.

clonal anergy Inability of a clone of potentially responsive cells to respond immunologically.

clonal deletion Elimination of immature lymphocytes upon binding to self-antigens to produce tolerance to self.

clonal selection and expansion Selection and activation of a lymphocyte by interaction of antigen and specific antigen receptor on the lymphocyte surface, causing the lymphocyte to proliferate to form an expanded clone.

clone Group of cells derived from a single cell.

closed system Batch system (such as a tube or flask of broth, or an agar plate) used for growing microorganisms; nutrients are not replenished and wastes are not removed.

clusters of differentiation (CD) antigens Cell surface antigens identified by clusters or groups of monoclonal antibodies; used to distinguish subgroups of white blood cells.

CMI Abbreviation for cell-mediated immunity.

CO_2 fixation Process of converting inorganic carbon (CO_2) to an organic form.

coagulase Non-enzymatic product of *Staphylococcus aureus* that clots plasma.

coccus (pl. **cocci**) Spherical-shaped bacterial cell.

codon Set of three nucleotides.

coenzyme Non-protein organic compounds that assist some enzymes, acting as a loosely bound carrier of small molecules or electrons.

coenzyme A Coenzyme involved in decarboxylation.

cofactor Non-protein component required for the activity of some enzymes.

cohesive ends Single-stranded overhangs generated when DNA is digested with a restriction enzyme that cuts asymmetrically within the recognition sequence; sticky ends.

cohort group Population with a known exposure to a specific risk factor that is followed over time in a prospective study.

coliforms Facultative, non-spore-forming, Gram-negative rods that ferment lactose, producing acid and gas within 48 hours at 35°C; because most typically reside in the intestine, they are used as indicators of fecal pollution.

colonization Establishment of a site of reproduction of microbes on a material, animal, or person without necessarily resulting in tissue invasion or damage.

colony Population of bacterial cells arising from a single cell.

colony blot Technique that uses a probe to detect a given DNA sequence in colony.

colony-forming unit A unit that gives rise to a single colony; may be a single cell or multiple cells attached to one another.

combination therapy Administration of two antibiotics simultaneously to prevent growth of mutants that might be resistant to one of the antibiotics.

commensalism Relationship between two organisms in which one partner benefits from the association and the other is unaffected.

commercially sterile Free of all microorganisms capable of growing under normal storage conditions; the endospores of some thermophiles may remain.

common-source epidemic Outbreak of disease due to contaminated food, water, or other single source of infectious agent.

communicable diseases Diseases that are spread from an infected animal or person to another animal or person.

communities All of the living organisms in a given area.

competent Condition in which a bacterial cell is capable of taking up and integrating longer fragments of DNA into its chromosome.

competitive inhibition Type of enzyme inhibition that occurs when the inhibitor competes with the normal substrate for binding to the active site.

complement System of serum proteins that act in sequence, producing biological effects concerned with inflammation, the immune response, and the lysis of cells.

complementary Describes bases in nucleic acid which hydrogen bond to one another; A (adenine) is complementary to T (thymine), and G (guanine) is complementary to C (cytosine).

complement fixation test Serological method in which antibody-antigen reactions are detected by the consumption (fixation) of complement.

complex medium Medium for growing bacteria that has some ingredients of unknown chemical composition.

compound microscope Microscope that employs two magnifying lenses—an objective lens and an ocular lens; the lenses in combination visually enlarge an object by a factor equal to the product of each lens's magnification.

condenser lens Lens of a microscope that is used to focus the illumination; positioned between the light source and the specimen and does not affect the magnification.

conditional lethal mutant Mutant that under some environmental conditions will grow, with lethal results, but under other conditions will not grow.

confocal scanning laser microscope Type of microscope that focuses a laser beam to illuminate a given point on one vertical plane of a specimen; after successive regions and planes have been scanned, a computer can construct a three-dimensional image of a thick structure.

congenital A condition existing from the time of birth.

conidia Asexual spores borne on hyphae; produced by fungi and bacteria of the genus *Streptomyces*.

conjugation Mechanism of gene transfer in bacteria that involves cell-to-cell contact.

conjugative plasmid Plasmid that carries the genes for sex pili and can transfer copies of itself to other bacteria during conjugation.

consensus sequence Common theme of nucleotides that characterize a stretch of DNA that has a certain function.

constant region That part of the antibody molecule that does not vary in amino acid sequences among molecules of the same immunoglobulin class.

consumption Stage of biogeochemical recycling in which consumer organisms metabolize the organic materials made by producers.

contact dermatitis AT-cell-mediated inflammation of the skin occurring in sensitized individuals as a result of contact with the particular antigen; a form of delayed hypersensitivity.

contagious diseases Diseases that are highly communicable and are spread from one host to another very readily.

continuous culture Method used to maintain cells in a state of uninterrupted growth by continuously adding nutrients and removing waste products; a type of open system.

convalescence Period of recuperation and recovery from an illness.

convergent evolution Process of evolution when two genetically different organisms develop similar environmental adaptations.

corepressor Molecule that binds to an inactive repressor and, as a consequence, enables it to function as a repressor.

cortex Layer of the endospore that helps maintain the core in a dehydrated state, protecting it from the effects of heat.

counterstain In a differential staining procedure, the stain applied to impart a contrasting color to bacteria that do not retain the primary stain.

covalent bond Strong chemical bond formed by the sharing of electrons between atoms.

critical items Medical instruments such as needles and scalpels that come into direct contact with body tissue.

cross-contamination Transfer of pathogens from one item to another.

cross-sectional study Study that surveys a range of people to determine the prevalence of characteristics including disease, risk factors associated with disease, or previous exposure to a disease-causing agent.

croup Acute obstruction of the larynx occurring mainly in infants and young children, often resulting from respiratory syncytial or other viral infection.

crown gall tumor A tumor on a plant caused by *Agrobacterium tumefaciens*.

CSF Abbreviation for colony-stimulating factor.

curd Coagulated milk proteins, produced during cheese-making.

cyclic AMP–activating protein (CAP) Protein that binds to cAMP to promote gene transcription.

cyclic photophosphorylation Type of photophosphorylation in which electrons are returned directly to the chlorophyll; used to synthesize ATP without generating reducing power.

cyst Dormant resting protozoan cell characterized by a thickened cell wall.

cysticercus (pl. **cysticerci**) Cystlike larval form of tapeworms.

cystitis Inflammation of the urinary bladder.

cytochromes Proteins that carry electrons, usually as members of electron transport chains.

cytokines Low molecular weight regulatory proteins made by cells that affect the behavior of other cells; they attach to specific cytokine receptors and are essential for communication between cells.

cytopathic effect Observable change in a cell *in vitro* produced by viral action such as lysis of the cell.

cytoplasmic membrane Thin, fluid, lipid bilayer that surrounds the cytoplasm and defines the boundary of a cell.

cytoskeleton Dynamic filamentous network that provides structure and shape to eukaryotic cells.

cytotoxic Kills cells.

---

dark-field microscope Type of microscope that directs light toward the specimen at an angle, so that only light scattered by the specimen enters the objective lens; material in the specimen stands out as bright objects against a dark background.

dark reactions Process of carbon fixation in photosynthetic organisms; the ATP used to drive the process is obtained in the light reactions; the dark reactions are called the light-independent reactions.

dark repair Enzymes of DNA repair that do not depend on visible light.

death phase Stage in which the number of viable bacteria in a population decreases at an exponential rate.

decarboxylation Removal of carbon dioxide from a chemical.

decimal reduction time Time required for 90% of the organisms to be killed under specific conditions; D value.

decomposition Stage of biogeochemical recycling in which decomposer organisms digest and convert dead plant and animal material into small molecules that can be used by both consumers and producers.

decontamination Treatment to reduce the number of disease-causing organisms to a level that it is considered safe.

degranulation Release of mediators from granules in the cell, as histamine is released from mast cells.

dehydration synthesis Chemical reaction in which H_2O is removed with the result that two molecules are joined together.

dehydrogenation Oxidation reaction in which both an electron and an accompanying proton are removed.

delayed hypersensitivity Hypersensitivity caused by cytokines released from sensitized T lymphocytes; reactions occur within 48 to 72 hours after exposure of a sensitized individual to antigen.

denaturation (1) Disruption of the three-dimensional structure of a protein molecule. (2) The separation of the complementary strands of DNA.

dendritic cells Cells with a branched morphology, some of which are related to macrophages; some are antigen-presenting cells.

denitrification Bacterial conversion of nitrate to gaseous nitrogen by anaerobic respiration.

dental plaque A biofilm on teeth.

deoxyribonucleic acid (DNA) Macromolecule in the cell that carries the genetic information.

depth filter Type of filter with complex, tortuous passages that allow the suspending fluid pass through while retaining microorganisms.

dermatophytes Certain moldlike fungi that live on the skin and can be responsible for disease of the hair, nails, and skin.

descriptive study Type of study that seeks to characterize a disease outbreak by determining the characteristics of the persons involved and the place and time of the outbreak.

diapedesis Movement of leukocytes from blood vessels into tissues in response to a chemotactic stimulus during inflammation.

diatomaceous earth Sedimentary soil composed largely of the skeletons of diatoms; contains large amounts of silicon.

diauxic growth Two-step growth frequency observed when bacteria are growing in media containing two carbon sources.

dichotomous key Flowchart of tests used for identifying an organism; each test gives either a positive or negative result.

dideoxynucleotide (ddNTP) Nucleotide that lacks the 3′ OH group, the portion required for the addition of subsequent nucleotides during DNA synthesis.
differential media Culture media that contain ingredients such as sugars and pH indicators used to distinguish among organisms based on their metabolic traits.
differential staining Type of staining procedure used to distinguish one group of bacteria from another by taking advantage of the fact that certain bacteria have distinctly different chemical structures in some of their components.
diffusion Movement of substances from a region of high concentration to a region of low concentration.
diluent Sterile solution used to make dilutions.
dimorphic Able to assume two forms, as the yeast and mold forms of pathogenic fungi.
diphtheroids Gram-positive cells that are club-shaped and arranged to form V-shapes and palisades; the typical microscopic morphology of *Corynebacterium*.
diplococci Cocci that typically occur in pairs.
directly observed therapy Method used to ensure that patients comply with their antimicrobial therapy; health care workers routinely visit patients in the community and watch them take their medications.
direct microscopic count Method of determining the number of bacteria in a measured volume of liquid by counting them microscopically using special glass slides.
direct selection Technique of selecting mutants by plating organisms on a medium on which the desired mutants but not the parent will grow.
disaccharide Carbohydrate molecule consisting of two monosaccharide molecules.
disease Process resulting in tissue damage or alteration of function, producing body changes noticeable by physical examination or laboratory tests.
disinfection Process of reducing or eliminating pathogenic microorganisms or viruses in or on a material so that they are no longer a hazard.
disseminated intravascular coagulation Devastating condition in which clots form in small blood vessels, leading to failure of vital organs.
division Taxonomic rank that groups similar classes; also called a phylum. A collection of similar divisions makes up a kingdom.
DNA Abbreviation for deoxyribonucleic acid.
DNA chip Silicon chip of less than an inch in diameter that can carry an array of hundreds of thousands of oligonucleotides.
DNA fingerprinting The use of characteristic patterns in the nucleotide sequence of DNA to match a specimen to a probable source.
DNA gyrase Enzyme that helps relieve the tension in DNA caused by the unwinding of the two strands of the DNA helix.
DNA library Population of cells that together contain the entire cloned genome of an organism of interest; each cell can be viewed as containing one "book" of the total genetic information of the organism of interest.
DNA ligase Enzyme that joins short fragments of DNA.
DNA-mediated transformation Process of gene transfer in which DNA is transferred as a "naked" molecule.
DNA polymerases Enzymes that synthesize DNA; they use one strand as a template to generate the complementary strand.
DNA probe A piece of DNA, labeled in some manner, that is used to identify the presence of homologous DNA by hybridizing to its complement.
DNA replication Duplication of a DNA molecule.
domain Level of taxonomic classification above the kingdom level; there are three domains—Bacteria, Archaea, and Eukarya.
donor Refers to the cell that donates DNA in DNA transfer.
double-blind Type of study where neither the physicians nor the patients know who is receiving the actual treatment.
double diffusion in gel Qualitative precipitation reaction in which antigen and antibody diffuse toward each other, making a line of precipitation in the area of optimal proportions.
doubling time Time it takes for the number of cells in population to double; the generation time.
downstream Direction toward the 3′ end of an RNA molecule or the analogous (+) strand of DNA.
droplet transmission Transmission of infectious agents through inhalation of respiratory droplets.
Durham tube Small inverted tube placed in a broth of sugar-containing media that is used to detect gas production by a microorganism.
D value Abbreviation for the decimal reduction time.
dysentery Condition characterized by crampy abdominal pain and bloody diarrhea.

---

eclipse period Time during which viruses exist within the host cell separated into their protein and nucleic acid components.
ecosystem An environment and the organisms that inhabit it.
eczema Condition characterized by a blistery skin rash, with weeping of fluid and formation of crusts, usually due to an allergy.
edema Swelling of tissues caused by accumulation of fluid.
effector Regulatory molecule that binds to the allosteric site of an enzyme; the binding alters the shape of the enzyme, altering its affinity for the substrate.
electrochemical gradient A separation of charged ions across the membrane.
electron Negatively charged component of an atom that orbits the nucleus.
electron microscope Microscope that uses electrons instead of light and can magnify images in excess of 100,000×.
electron transport chain Series of electron carriers that transfer electrons from donors such as NADH to acceptors such as oxygen.
electrophoresis Technique that uses an electric current to separate either DNA fragments or proteins.
electroporation Process of treating cells with an electric current to introduce DNA into them.
elementary body Small dense-appearing infectious form of *Chlamydia* species that is released upon death and rupture of the host cell.
elephantiasis Massive enlargement of the legs and/or external genitalia due to lymphatic obstruction, caused by the inflammatory response to larval roundworms such as *Wuchereria bancrofti*.
ELISA Abbreviation for enzyme-linked immunosorbent assay.
Embden-Meyerhof pathway Metabolic pathway that oxidizes glucose to pyruvate, generating ATP and reducing power; also known as glycolysis and the glycolytic pathway.
emerging diseases Diseases that have increased in incidence in the past two decades.
encephalitis Inflammation of the brain.
endemic Constantly present in a population.
endergonic Chemical reaction that requires a net input of energy because the products have more free energy than the reactants.
endocarditis Inflammation of the heart valves or lining of the heart chambers.
endocytosis Process through which cells take up particles by enclosing them in a vesicle pinched off from the cell membrane.
endoplasmic reticulum Organelle of eukaryotes where macromolecules destined for the external environment of other organelles are synthesized.
endosome Vesicle formed when a cell takes up material from the surrounding environment using the process of endocytosis.
endospore A kind of resting cell, characteristic of a limited number of bacterial species; highly resistant to heat, radiation, and disinfectants.
endosymbiont Microorganism that resides within another cell, providing a benefit to the host cell.
endosymbiont theory Theory that the ancestors of mitochondria and chloroplasts were bacteria that had been residing within other cells in a mutually beneficial partnership.
endotoxin A poisonous compound (Lipid A) within the outer membrane of Gram-negative that can elicit symptoms such as fever and shock.
end product repression Inhibition of gene activity by the end product of a biosynthetic pathway.
energy The capacity to do work.
energy source Compound that is oxidized by a cell to release energy; also called an electron donor.
enrichment culture Culture method that provides conditions to enhance the growth of one particular organism in a mixed population.
enterics A common name for members of the family *Enterobacteriaceae*.
enterotoxin Poisonous substance, usually of bacterial origin, that acts on the intestinal lining cells to cause diarrhea and vomiting.
Entner-Doudoroff pathway Pathway that converts glucose to pyruvate and glyceraldehyde-3-phosphate by producing 6-phosphogluconate and then dehydrating it.
entropy The degree of disorder in a system.
enveloped viruses Viruses that have a double layer of lipid surrounding their nucleocapsid.
enzyme A protein that functions as a catalyst.
enzyme-linked immunosorbent assay (ELISA) Technique used for detecting and quantifying specific antigens or antibodies by using an antibody labeled with an enzyme.
enzyme-substrate complex Transient form that occurs in an enzyme-mediated reaction, as the enzyme converts a substrate into a product.
EPA Abbreviation for Environmental Protection Agency, a federal agency.
epidemic A disease or other occurrence whose incidence is higher than expected within a region or population.
epidemiology The study of factors influencing the frequency and distribution of diseases.
epitope Area of an antigen molecule that stimulates the production of, and combines with, specific antibodies.
ergosterol Sterol found in fungal cell membranes; the target of many antifungal drugs.
ergot Poisonous substance produced by the fungus that causes rye smut.
erythrocytes Red blood cells.
E-site (or **exit site**) Site on the ribosome from which tRNAs exit after donating their amino acid to the adjacent tRNA.
ester bond Covalent bond formed between a —COOH group and an —OH group with the removal of H_2O.
E test Modification of the disc diffusion test that utilizes a strip impregnated with a gradient of concentrations of an antimicrobial drug.
ethambutol Antimycobacterial drug that inhibits enzymes required for synthesis of mycobacterial cell wall components.
ethidium bromide Mutagenic dye that binds to nucleic acid by intercalating between the bases; ethidium bromide–stained DNA is fluorescent when viewed with UV light.
eubacteria Term formerly used to describe those prokaryotes that are now separated into the domain bacteria.

Eukarya Name of the domain comprising eukaryotic organisms.

eukaryote Complex cell type differing from a prokaryote mainly in having a nuclear membrane.

Euryarchaeota Phylogenetic group of the domain Archaea.

eutrophication Nutrient enrichment leading to the overproduction of algae.

exanthema A skin rash.

excision repair Mechanism of DNA repair in which a fragment of single-stranded DNA containing mismatched bases is cut out.

exergonic Describes a chemical reaction that releases energy because the starting compounds have more free energy than the products.

exocytosis Process by which eukaryotic cells expel material; membrane-bound vesicles inside the cell fuse with the plasma membrane, releasing their contents to the external medium.

exoenzyme Enzyme that acts outside the cell that produces it.

exoerythrocytic Occurring outside the red blood cells, as the developmental cycle in malaria that occurs in the liver.

exons Portions of eukaryotic genes that are expressed; interrupted by introns.

exotoxin Soluble poisonous protein substance released by a microorganism.

experimental study Type of study done to assess the effectiveness of measures to prevent or treat disease.

exponential phase Stage of growth of a bacterial culture in which cells are multiplying exponentially; log phase.

expression vectors Vectors that facilitate transcription and translation of cloned DNA.

external node Point on a phylogenetic tree that represents a named species that still exists.

external transmission (or **temporary transmission**) Refers to transmission of viruses to plants by insects in which the virus is associated with the external mouthparts of the insect.

extremophiles Organisms that live under extremes of temperature, barometric pressure, or other environmental conditions.

extrinsic factors In food microbiology, environmental conditions, such as the temperature and atmosphere, that influence the rate of microbial growth.

Fab (fragment antigen-binding) region Portion of an antibody molecule that binds to the antigen.

facilitated diffusion Transport process that enables movement of impermeable compounds from one side of the membrane to the other by exploiting a concentration gradient; does not require expenditure of energy by the cell.

facultative Flexible with respect to growth conditions; for example, able to live with or without O_2.

facultative anaerobe Organism that grows best in the presence of oxygen (O_2), but can grow in its absence.

FAD Abbreviation for flavin adenine dinucleotide.

family Taxonomic group between order and genus.

fastidious Exacting; refers to organisms that require growth factors.

F^- cell Recipient bacterial cell in conjugation.

F^+ cell Donor bacterial cell in conjugation.

Fc portion of antibody Crystallizable end of the constant region of an immunoglobulin molecule; responsible for binding to Fc receptors on cells, for initiating the classical pathway of complement activation, and for other biological functions.

fecal coliforms Thermotolerant coliform bacteria.

fecal-oral transmission Transmitting organisms that colonize the intestine by ingesting fecally contaminated material.

feedback inhibition Inhibition of the first enzyme of a biosynthetic pathway by the end product of the pathway; also called allosteric or end product inhibition.

feeding tolerance Lack of immune response to a specific antigen resulting from introducing the antigen orally.

fermentation Metabolic process in which the final electron acceptor is an organic compound.

fertility plasmid (or **F-plasmid**) Plasmid found in donor cells of *E. coli* which codes for the sex pilus and makes the cell F^+.

fever An increase in internal body temperature to 37.8°C or higher.

fibronectin Glycoprotein occurring on the surface of cells and also in a circulating form that adheres tightly to medical devices; certain pathogens attach to it to initiate colonization.

filterable viruses The old terminology for viruses.

fimbria (pl. **fimbriae**) Type of pilus that enables cells to attach to a specific surface.

first-line antimicrobials In antimycobacterial drug therapy, the group of antimicrobials that is preferred because they that are most effective as well as least toxic.

flagellin Protein subunits that make up the filament of flagella.

flagellum (pl. **flagella**) (1) In prokaryotic cells, a long protein appendage composed of subunits of flagellin that provides a mechanism of motility. (2) In eukaryotic cells, a long whiplike appendage composed of microtubules in a 9 + 2 arrangement that provides a mechanism of locomotion.

flavin adenine dinucleotide (FAD) A derivative of the vitamin riboflavin.

flow cytometer Instrument that counts cells in a suspension by measuring the scattering of light by individual cells as they pass by a laser.

fluid mosaic model Model that describes the dynamic nature of the cytoplasmic membrane.

fluke Short, nonsegmented, bilaterally symmetrical flatworm.

fluorescence-activated cell sorter (FACS) Machine that sorts fluorescent-labeled cells in a mixture by passing single cells in a stream past photodetectors.

fluorescence microscope Special type of microscope used to observe cells that have been stained or tagged with fluorescent dyes.

fluoroquinolones Group of antimicrobial drugs that interferes with nucleic acid synthesis.

fomites Inanimate objects such as books, tools, or towels that can act as transmitters of pathogenic microorganisms or viruses.

foraminifera Protozoa that have silicon or calcium in their cell walls.

forespore Portion of the endospore formed during the process of sporulation that will ultimately become the core of the endospore.

fowl cholera Worldwide septicemic illness of wild and domestic fowl, caused by *Pasteurella multocida*; focus of the discovery by Pasteur that an attenuated organism could be used as a vaccine.

F-plasmid Plasmid found in donor cells of *E. coli* that codes for sex pilus biosynthesis.

fragmentation Form of asexual reproduction in which a filament composed of a string of cells breaks apart, forming multiple reproductive units.

frameshift mutation Mutation resulting from the addition or deletion of a number of nucleotides not divisible by three.

free energy Amount of energy that can be gained by breaking the bonds of a chemical; does not include the energy that is always lost as heat.

freeze-etching Process used to prepare specimens for transmission electron microscopy that allows the shape of underlying regions within structures of a cell to be viewed.

fruiting body With respect to myxobacteria, a complex aggregate of cells, visible to the naked eye, produced when nutrients or water are depleted.

fungemia Fungi circulating in the bloodstream.

fungicide Kills fungi; used to describe the effects of some antimicrobial chemicals.

fungistatic Able to inhibit the growth of fungi.

fungus (pl. **fungi**) A nonphotosynthetic eukaryotic heterotroph.

furuncle A boil; a localized skin infection that penetrates into the subcutaneous tissue, usually caused by *Staphylococcus aureus*.

GALT Abbreviation for gut-associated lymphoid tissue.

gametes Haploid cells that fuse with other gametes to form the diploid zygote in sexual reproduction.

gamma globulin Portion of blood serum proteins that is separated by electrophoresis and contains most of the immunoglobulins of the blood.

gas chromatography Technique of separating and identifying gaseous components of a substance.

gastroenteritis Acute inflammation of the stomach and intestines; often applied to the syndrome of nausea, vomiting, diarrhea, and abdominal pain.

gas vesicles Small rigid compartments produced by some aquatic bacteria that provide buoyancy to the cell; gases, but not water, flow freely into the vesicles, thereby decreasing the density of the cell.

G + C content Percentage of guanine plus cytosine in double-stranded DNA; also called the GC content.

gel electrophoresis Technique that uses electric current to separate either DNA fragments or proteins according to size by drawing them through a slab of gel, which has the consistency of very firm gelatin.

gene The functional unit of a genome.

gene cloning Procedure by which genes are inserted into a replicon such as a plasmid or bacteriophage, which is then introduced into cells where the replicon can replicate.

gene library Sum total of all of the genes of an organism that have been inserted into cloning vectors.

generalized transducing phage Bacteriophage that is capable of transferring any part of the bacterial chromosome from one cell to another. (By contrast, a specialized transducing phage transfers only specific parts of the genome.)

generalized transduction Transfer of any bacterial gene to other bacteria by phage.

general paresis Group of symptoms arising from nervous system damage, usually occurring 10 to 20 years after contracting syphilis; often manifest by emotional instability, memory loss, hallucinations, abnormalities of the eyes, and paralysis.

general secretory pathway Primary mechanism bacterial cells use to secrete proteins; those destined for secretion are recognized by their characteristic sequence of amino acids that make up the amino terminal end.

generation time Time it takes for the number of cells in a population to double; doubling time.

genetic engineering Process of deliberately altering an organism's genetic information by changing its nucleic acid sequences.

genetic reassortment Exchange of genetic information following two different segmented viruses infecting the same cell.

genetic recombination The joining together of genes from different organisms.

genetics The study of the function and transfer of genes.

genome Complete set of genetic information in a cell.

genome mining Searching genomic databases; for example, companies might search genomic databases to locate ORFs that may encode proteins of medical value.

genomics Study and analysis of the nucleotide sequence of DNA.

genus (pl. **genera**) Category of related organisms, usually containing several species. The first name of an organism in the Binomial System of Nomenclature.

germicide Agent that kills microorganisms and inactivates viruses.

germination Sum total of the biochemical and morphological changes that an endospore or other resting cell undergoes before becoming a vegetative cell.

giant cell Very large cell with many nuclei, formed by the fusion of many macrophages during a chronic cell-mediated response; found in granulomas.

gingivitis Inflammation of the gums.

global regulation Simultaneous regulation of numerous unrelated genes; global control.

global repressor Regulatory protein involved in the control of several different biosynthetic or degradative pathways.

glucose-salts Type of chemically defined medium that contains only glucose and certain inorganic salts; supports the growth of *E. coli*.

glycan chain High molecular weight linear polymer of alternating subunits of *N*-acetylglucosamine and *N*-acetylmuramic acid that serves as the backbone of the peptidoglycan molecule.

glycocalyx Gel-like layer that surrounds some cells and generally functions as a mechanism of either protection or attachment.

glycogen Polysaccharide composed of glucose molecules.

glycolipids Lipids that have various sugars attached.

glycolysis Metabolic pathway that oxidizes glucose to pyruvate, generating ATP and reducing power; also called the Embden-Meyerhoff pathway and the glycolytic pathway.

glycoproteins Proteins with covalently bonded sugar molecules.

goblet cells Mucus-secreting epithelial cells.

Golgi apparatus Series of membrane-bound flattened sacs within eukaryotic cells that serve as the site where macromolecules synthesized in the endoplasmic reticulum are modified before they are transported to other destinations.

Gram-negative Bacteria that lose the crystal violet in the Gram stain procedure and therefore stain pink; the cell wall of these organisms is composed of a thin layer of peptidoglycan surrounded by an outer membrane.

Gram-positive Bacteria that retain the crystal violet stain in the Gram stain procedure and therefore stain purple; the cell wall of these organisms is composed of a thick layer of peptidoglycan.

Gram stain Staining technique that divides bacteria into one of two groups, Gram-positive or Gram-negative, on the basis of color; among bacteria, the staining reaction correlates well with cell wall structure.

granulation tissue New tissue formed during healing of an injury, consisting of small, red, translucent nodules containing abundant blood vessels.

granulocytes White blood cells characterized by the presence of prominent granules; basophil granules stain dark with basophilic dyes, eosinophils stain bright red with eosinophilic dyes, and neutrophils do not take up either stain.

granuloma Found in a chronic cell-mediated response, collections of lymphocytes, and stages of macrophages; an attempt by the body to wall off and contain persistent organisms and antigens.

griseofulvin Antifungal medication that appears to interfere with the action of tubulin, a necessary factor in nuclear division.

group translocation Type of transport process that chemically alters a molecule during its passage through the cytoplasmic membrane.

growth curve Growth pattern observed when cells are grown in a closed system; consists of four stages—lag phase, log phase (or exponential phase), stationary phase, and death phase.

growth factors Compounds that a particular bacterium cannot synthesize and therefore must be included in a medium that supports the growth of that organism.

gumma Localized area of chronic inflammation and necrosis in tertiary syphilis, often manifest as a swelling.

gut-associated lymphoid tissue (GALT) Part of the mucosa-associated lymphoid tissue found in the intestines, including Peyer's patches, the appendix, and some lymph nodes.

HAART Highly active antiretroviral therapy; a cocktail of medications that act at different sites during replication of human immunodeficiency virus.

hairy leukoplakia Whitish patch, usually appearing on the tongue of individuals with severe immunodeficiency, thought to be caused by reactivation of latent Epstein-Barr virus (EBV) infection.

half-life Time it takes for one-half of the original number of molecules of a compound to be eliminated or degraded.

halophile Organism that prefers or requires a high salt (NaCl) medium.

hapten Substance that can combine with specific antibodies but cannot incite the production of those antibodies unless it is attached to a large carrier molecule.

haustoria Specialized hypha of parasitic fungi that can penetrate plant or animal cell walls.

helicase Enzyme that unwind the DNA helix ahead of the replication fork.

helminth A parasitic worm.

helper T lymphocytes $CD4^+$ lymphocytes that provide assistance needed for other cells such as B lymphocytes and macrophages to carry out their functions; T-helper cells usually act by producing cytokines.

hemagglutination Clumping of red blood cells.

hematopoietic stem cells Bone marrow cells that give rise to all blood cells.

hemolytic disease of the newborn (HDN) Disease of the fetus or newborn caused by transplacental passage of maternal antibodies against the baby's red blood cells, resulting in red cell destruction; usually anti-Rhesus (Rh) antibodies are involved and the disease is called Rh disease; also called erythroblastosis fetalis.

hemolytic uremic syndrome Serious condition characterized by red cell breakdown and kidney failure; a sequel to infection by certain Shiga toxin-producing strains of *Shigella dysenteriae* and *Escherichia coli*.

hepatitis B virus An enveloped DNA virus with an unusual mode of replication involving reverse transcriptase; cause of hepatitis B.

herd immunity Phenomenon that occurs when a critical concentration of immune hosts prevents the spread of an infectious agent.

hermaphroditic Having both male and female reproductive structures in the same organism.

herpes zoster Another name for shingles; a disease that results from reactivation of the herpesvirus causing chickenpox.

heterocyst Specialized nonphotosynthetic cells of cyanobacteria within which nitrogen fixation occurs.

heterophile antibody Antibody that reacts with the red blood cells of another animal.

heterotroph Organism that obtains carbon from an organic compound such as glucose.

Hfr cells (high frequency of recombination cells) Rare cells in the F^+ population that can transfer their chromosome to an F^- cell.

high-copy-number plasmid Plasmid whose numbers in the cell range from 50 to 500.

high efficiency particulate air (HEPA) filters Special filters that remove from air nearly all particles, including microorganisms, that have a diameter greater than 0.3 μm.

high-energy phosphate bond Bond that joins a phosphate group to a molecule and releases a relatively high amount of energy when hydrolyzed; denoted by the symbol ~.

high-level disinfectant Chemical used to destroy all viruses and vegetative cells, but not endospores.

high-temperature-short-time (HTST) method Most common pasteurization protocol; using this method, milk is pasteurized by holding it at 72°C for 15 seconds.

histamine A substance found in basophil and mast cell granules that upon release can cause dilation and increased permeability of blood vessel walls and other effects; a mediator of inflammation.

histocompatibility antigens Cell surface molecules involved in the immunological rejection of transplanted tissues and organs and in recognition between cells during the immune response; also called major histocompatibility (MHC) antigens, and in people, the human leukocyte antigens (HLA).

HIV disease The illness caused by human immunodeficiency virus, marked by gradual impairment of the immune system, ending in AIDS.

HLA Abbreviation for human leukocyte antigen.

homologous With respect to DNA, stretches that have similar or identical nucleotide sequences and probably encode similar characteristics.

homologous recombination Genetic recombination between stretches of similar or identical nucleotide sequences.

homoserine lactone (HSL) Freely diffusible molecule that is used by certain types of bacteria to sense the density of cells within their population.

hook Curved structure that connects the filament of the flagella to the cell surface.

hops Flowers of the vinelike hop plant; they are added to wort to impart a desirable bitter flavor to beer and contribute antibacterial substances.

horizontal evolution With respect to antimicrobial resistance, the acquisition of resistance through gene transfer.

horizontal gene transfer Transmission of DNA from one species to another; also called lateral gene transfer.

horizontal transmission Transfer of a pathogen from one person to another through contact, ingestion of food or water, or via a living agent such as an insect.

host Organism on or in which smaller organisms or viruses live, feed, and reproduce; a definitive host is an animal in which the sexually mature form of a parasite occurs; an intermediate host is an animal in which the asexual developmental stages of a parasite occur.

host range The range of cell types that a pathogen can infect.

HSV-1 (herpes simplex virus-1) Member of the herpes family of viruses that causes cold sores and other types of infection.

HSV-2 (herpes simplex virus-2) Member of the herpes family of viruses; principal cause of genital herpes.

HTST Abbreviation for high-temperature-short-time pasteurization.

human leukocyte antigen (HLA) Group of human cell surface antigens.

humoral immune response Antibody response.

hybridization The annealing of two complementary strands of DNA from different sources to create a hybrid double-stranded molecule.

hybridoma Cell made by fusing a lymphocyte, such as an antibody-producing B cell, with a cancer cell.

hydrogenation Reduction reaction in which an electron and an accompanying proton is added to a molecule.

hydrogen bond Weak attraction between a positively-charged hydrogen atom of one compound and a negatively charged atom of another compound; the charges of the two atoms are due to polar covalent bonds.

hydrolysis Chemical reaction in which a molecule is broken down as H_2O is added.

hydrophilic Water loving; soluble in water.

hydrophobic bonds Weak bonds formed between molecules as a result of their mutual repulsion of water molecules.

hyperimmune globulin Immunoglobulin prepared from the sera of donors with large amounts of antibodies to certain diseases, such as tetanus; used to prevent or treat the disease.

hypersensitivity Also termed allergy; heightened immune response to antigen.

hyperthermophiles Organisms that have an optimum growth temperature between 70°C and 110°C.

hypervariable regions Small areas in the Fab portion of the immunoglobulin light and heavy polypeptide chains that bind the antigenic epitope.

hypha (pl. **hyphae**) Threadlike structure that characterizes the growth of most fungi and some bacteria such as members of the genus *Streptomyces*.

hyposensitization (or **desensitization**) Form of therapy for immediate IgE-mediated allergies in which extremely small but increasing amounts of antigen are injected regularly over a period of months, directing the response from IgE to IgG.

IFN Abbreviation for interferons.

illness Period of time during which symptoms and signs of disease occur.

immune complex Complex of antigen and antibody bound together, often with some complement components included.

immune serum globulin Immunoglobulin G portion of pooled plasma from many donors, containing a wide variety of antibodies; used to provide passive protection.

immunity Protection against infectious agents and other substances.

immunoassay Tests using immunological reagents such as antigens and antibodies.

immunodeficiency Inability to produce a normal immune response to antigen.

immunodiffusion tests Precipitation reactions carried out in agarose or other gels.

immunoelectrophoresis Technique for separating proteins by subjecting the mixture to an electric current followed by diffusion and precipitation in gels using antibodies against the separated proteins.

immunofluorescence Technique used to identify particular antigens microscopically in cells by the binding of a fluorescent antibody to the antigen.

immunogen Antigen that can induce an immune response.

immunoglobulin Glycoprotein molecules that react specifically with the substance that induced their formation; antibodies.

immunological tolerance State of specific unresponsiveness; failure to respond to a specific antigen, such as self-antigen.

immunology The study of immunity, or protection against infectious and other agents, and conditions arising from the mechanisms involved in immunity, such as hypersensitivities.

immunosuppression Nonspecific suppression of acquired immune responses.

inapparent (or **subclinical**) **infections** Infections in which symptoms do not occur or are mild enough to go unnoticed.

incidence rate Number of new cases of a disease within a specific time period in a given population.

inclusion body Microscopically visible structure within a cell representing the site at which an infecting virus replicates; can occur within the nucleus or the cytoplasm.

incubation period Interval between entrance of a pathogen into a susceptible host and the onset of illness caused by that pathogen.

index case First identified case of a disease in an epidemic.

indirect contact Means of transmitting infectious disease by coughing or sneezing, through food and water, insect bites, and other indirect contact.

indirect selection Technique for isolating mutants and identifying organisms unable to grow on a medium on which the parents do grow; often involves replica plating.

induced With respect to gene expression, a gene product that is synthesized only under certain conditions.

induced mutation Mutation that results from the organism being treated with an agent that alters its DNA.

inducer Substance that activates transcription of certain genes.

inducible enzyme Enzyme synthesized only when a substrate on which it can act is present.

induction Process by which a prophage is excised from the host cell DNA; activation of gene transcription.

infection Growth and multiplication of a parasitic organism or virus in or on the body of the host with or without the production of disease.

infectious disease Disease caused by a microbial or viral infection.

infectious dose Number of microorganisms or viruses sufficient to establish an infection.

infective dose (ID) Concentration of infective virions; often expressed as ID_{50} in which 50% of the hosts are infected.

inflammation Nonspecific response to injury characterized by swelling, heat, redness, and pain in the affected area.

initiation complex Complex of a 30S ribosomal subunit, a tRNA that carries f-Met, and elongation factors that comes together at a start codon on mRNA and begins the process of translation.

innate immunity Nonspecific immunity that is not acquired or affected by prior contact with the infectious agent or other material involved and is not mediated by lymphocytes.

innate resistance Resistance of an organism to an antimicrobial medication due to the inherent characteristics of that type organism; also called intrinsic resistance.

inner membrane (1) In prokaryotic cells, the cytoplasmic membrane of Gram-negative bacteria. (2) In eukaryotic cells, the membrane on the interior side of an organelle that has a double membrane.

insert DNA that is (or will be) joined to a vector to create a recombinant DNA molecule.

insertion mutation Mutation resulting from the integration of a transposon into a gene.

insertion sequence (IS) Short piece of DNA that has the ability to move from one site on a DNA molecule to another; simplest type of transposable element.

insulin-dependent diabetes mellitus (IDDM) Diabetes caused by autoimmune destruction of pancreatic cells by cytotoxic T cells.

intercalating agents Agents that insert themselves between two nucleotides in opposite strands of a DNA double helix.

interference microscope Type of light microscope that employs special optical devices to cause the specimen to appear as a three-dimensional image; an example is the Nomarski differential interference contrast microscope.

interferons Cytokines that induce cells to resist viral replication.

interleukins Cytokines produced by leukocytes.

intermediate filaments Component of the eukaryotic cell cytoskeleton.

intermediate-level disinfectant Type of chemical used to destroy all vegetative bacteria including mycobacteria, fungi, and most, but not all, viruses.

internal node Branch point on a phylogenetic tree that represents an ancestor to modern organisms.

intranuclear inclusion body Structure found within the nucleus of cells infected with certain viruses such as the cytomegalovirus.

intrinsic resistance Resistance of an organism to an antimicrobial medication due to inherent characteristics of that type of organism; also called innate resistance.

intron Part of the eukaryotic chromosome that does not code for a protein; removed from the RNA transcript before the mRNA is translated.

inverted repeat Sequence of nucleotides on one strand of DNA that is identical to DNA on another strand when both are read in the same direction, that is, 5′ to 3′; associated with transposable elements.

in vitro In a test tube or other container as opposed to inside a living plant or animal.

in vivo Inside a living plant or animal as opposed to a test tube or other container.

ion Charged atom or molecule.

ionic bond Bond formed by the attraction of positively charged atoms or molecules to negatively charged ones.

IS Abbreviation for insertion sequence.

isomer Molecule with the same number and types of atoms as another but differing in its structure.

isotope Form of an element that differs in atomic weight from the form most common in nature.

Jarisch-Herxheimer reaction Abrupt but transitory worsening of symptoms after starting effective antibacterial treatment, thought to be caused by substances released by the death of the bacteria.

---

kinetic energy Energy of motion.

kingdom Taxonomic rank that groups several phyla or divisions; a collection of similar kingdoms makes up a domain.

Kirby-Bauer disc diffusion test Procedure used to determine whether a bacterium is susceptible to concentrations of an antimicrobial usually present in the bloodstream of an individual receiving the antimicrobial.

Koch's postulates Group of criteria used to determine the cause of an infectious disease by culturing the agent and reproducing the disease.

Koplik's spots Lesions of the oral cavity caused by measles virus that resemble a grain of salt on a red base.

Krebs cycle Metabolic pathway that incorporates acetyl-CoA, and generates CO_2 and reducing power; also called the tricarboxylic acid (TCA) cycle and the citric acid cycle.

Kupffer cells Macrophages of the liver.

***lac* operon** Operon that encodes the proteins required for the degradation of lactose; it has served as one of the most important models for studying gene regulation.

lactic acid bacteria Group of Gram-positive bacteria that generate lactic acid as a major end product of their fermentative metabolism.

lactoferrin Iron-binding protein found in leukocytes, saliva, mucus, milk, and other substances; helps defend the body by depriving microorganisms of iron.

lactose Disaccharide consisting of one molecule of glucose and one of galactose.

***lacZ′* gene** Gene used to visually determine whether or not a vector contains a fragment inserted into the multiple-cloning site.

lagging strand Strand of double-stranded DNA that must be synthesized as a series of discontinuous fragments because of its 5′ to 3′ orientation with respect to the replication fork.

lagooning Sewage treatment method in which sewage is channeled into shallow lagoons, during which time it is stabilized by anaerobic and/or aerobic organisms.

lag phase Stage in the growth of a bacterial culture characterized by extensive macromolecule and ATP synthesis but no increase in the number of viable cells.

laminar flow hood Biological safety cabinet in which laboratory personnel work with potentially dangerous airborne pathogens; a continuous flow of incoming and outgoing air is filtered through HEPA filters to contain microorganisms within the cabinet.

Lancefield grouping Classification of β-hemolytic streptococci based on serological identification.

Langerhan's cells Antigen-presenting dendritic cells of the skin, similar in some respects to macrophages.

latent infection Infection in which the infectious agent is present but not active.

lateral gene transfer Transfer of DNA from one prokaryotic species to another; also called horizontal gene transfer.

laws of thermodynamics The fundamental principles of energy relationships. The first of these laws states that the energy in the universe can never be created or destroyed. The second states that entropy always increases.

leading strand Strand of double-stranded DNA that, because of its 5′ to 3′ orientation with respect to the replication fork, is synthesized continuously.

leaky Refers to a mutation in which the mutant gene codes for a protein that is partially functional.

lecithin Component of mammalian cell walls; attacked by the α-toxin of *Clostridium perfringens* and other lecithinases.

lectin pathway Pathway of complement activation non-specifically initiated by a host protein binding to mannose present in many bacterial cell walls.

leghemoglobin Protein synthesized by leguminous plants that carries O_2 within a *Rhizobium*-harboring nodule.

lethal dose (LD) Concentration of an infectious agent that causes death; often expressed as LD_{50}, the concentration of substances in which 50% of the hosts are killed by the agent.

leukemia Cancer of the leukocytes (white blood cells).

leukocidins Substances that kill white blood cells.

leukocytes White blood cells.

leukotrienes Substances active in inflammation, leading to chemotaxis and increased vascular permeability; produced by mast cells, basophils, and macrophages.

L-forms Bacterial variants that have lost the ability to synthesize the peptidoglycan portion of their cell wall.

lichen Organism composed of a fungus in a symbiotic association with either a green alga or a cyanobacterium.

light-dependent reactions Processes used by phototrophs to harvest energy from sunlight; the energy-gathering component of photosynthesis.

light-independent reactions Stage of photosynthesis in which the ATP generated in the light-dependent reactions is used to fix CO_2; also called dark reactions.

light microscope Microscope that uses visible light to illuminate objects.

light reactions Processes used by phototrophs to harvest energy from sunlight; the energy-gathering component of photosynthesis; also called light-dependent reactions.

light repair Process by which bacteria repair UV damage to their DNA only in the presence of light.

lincosamides Group of antimicrobials that interferes with protein synthesis.

lipid One of a diverse group of organic substances all of which are relatively insoluble in water, but soluble in alcohol, ether, chloroform, or other fat solvents.

lipid A Portion of lipopolysaccharide (LPS) that forms the outer leaflet in the lipid bilayer of the outer membrane of Gram-negative cells; it plays an important role in the body's ability to recognize the presence of invading bacteria, but is also responsible for the toxic effects of LPS.

lipopolysaccharide (LPS) Molecule formed by bonding of lipid to polysaccharide; a part of the outer membrane of Gram-negative bacteria.

lipoprotein Macromolecule formed by the bonding of lipid to protein.

lipoteichoic acids Component of the Gram-positive cell wall that is linked to the cytoplasmic membrane.

localized infections Infections limited to one site in or on the body, as a furuncle.

log phase Stage of growth of a bacterial culture in which the cells are multiplying exponentially.

low-copy-number plasmid Plasmid whose numbers in the cell are one or two copies.

low-level disinfectants Type of chemical used to destroy fungi, enveloped viruses, and vegetative bacteria except mycobacteria.

LPS Abbreviation for lipopolysaccharide.

luciferase Enzyme that catalyzes the chemical reactions that produce bioluminescence.

lymph Clear yellow liquid that flows within lymphatic vessels; generally contains lymphocytes and may contain globules of fat.

lymphadenopathy syndrome (LAS) Marked generalized enlargement of lymph nodes that often occurs at the end of the period of clinical well-being in HIV disease.

lymphocyte Small, round, or oval white blood cell with a large nucleus and a small amount of cytoplasm; involved in specific immunity.

lymphoid tissues and organs Collections of lymphocytes and related cells involved in immune responses.

lymphokines Cytokines secreted by lymphocytes.

lysate Remains of cells and virions that are released after lysis of cells.

lyse To burst.

lysogenic conversion Change in the properties of bacteria as a result of carrying a prophage.

lysogens Bacteria that carry a prophage integrated into their chromosome.

lysosome Membrane-bound structure in eukaryotic cells that contains powerful degradative enzymes.

lysozyme Enzyme that degrades the peptidoglycan layer of the bacterial cell wall.

MacConkey agar Type of selective and differential bacteriological medium used to isolate Gram-negative rods such as those that typically reside in the intestine.

macroenvironment Overall environment in which an organism lives.

macrolides Group of antimicrobial medications that interfere with protein synthesis.

macromolecule Very large molecule composed of repeating subunits.

macrophages Large mononuclear phagocytes of the tissues; professional phagocytes of the mononuclear phagocyte system that can engulf and destroy microorganisms and other extraneous materials, function as antigen-presenting cells, and carry out ADCC (antibody-dependent cellular cytotoxicity).

magnetotaxis Movement by bacterial cells containing magnetite crystals in response to a magnetic field.

major histocompatibility complex (MHC) Cluster of genes coding for key cell surface proteins important in cell-to-cell recognition. (See *histocompatibility antigens*.)

malaise Vague feeling of uneasiness or discomfort.

malignant tumor Abnormal growth of cells no longer under normal control that have the potential to spread to other parts of the body.

MALT Abbreviation for mucosal-associated lymphoid tissue.

mammalian artificial chromosomes Vectors that can be used to clone large fragments of DNA into animal cells.

mast cells Granule-containing tissue cells similar in appearance and function to the basophils of the blood, with receptors for the Fc portion of IgE; important in the inflammatory response and immediate allergic reactions.

MBC Abbreviation for minimum bactericidal concentration.

mechanical vector Organism such as a fly that physically moves contaminated material from one location to another.

medium (pl. **media**) Any material used for growing organisms.

meiosis Process in eukaryotic cells by which the chromosome number is reduced from diploid (2*N*) to haploid (1*N*).

melting Denaturating of double-stranded DNA.

membrane attack complex (MAC) Complex of the later components of complement that inserts through the cell membrane, resulting in lysis of the cell.

membrane filtration A technique that employs a membrane filter to determine the number of bacteria in a liquid sample that has a relatively low number of organisms.

membrane proteins Specialized proteins embedded in the membrane bilayer; some function as receptors and others function as transport proteins.

memory cells Lymphocytes specific for an antigen that persist in the body after an immune response to that antigen; upon subsequent exposure to the same antigen, they must differentiate and usually proliferate to become effector cells.

memory response (or **anamnestic response**) Enhanced immunological response to a second or subsequent dose of antigen.

meninges Membranes covering the brain and spinal cord.

meningitis Inflammation of the meninges.

merozoite Stage in the life cycle of certain protozoa, such as the malaria-causing *Plasmodium* species.

mesophiles Bacteria that grow most rapidly at temperatures between 20°C and 45°C.

messenger RNA (mRNA) Single-stranded RNA synthesized during transcription from DNA that binds to ribosomes and directs the synthesis of protein.

metabolism Sum total of all the chemical reactions in a cell.

metabolite Any product of metabolism.

metachromatic granules Polyphosphate granules found in the cytoplasm of some bacteria that appear as different colors when stained with a basic dye.

methanogens Group of Archaea that generate energy by oxidizing hydrogen gas, using CO_2 as a terminal electron acceptor; this process generates methane (CH_4).

MHC Abbreviation for major histocompatibility complex.

MIC Abbreviation for minimum inhibitory concentration.

microaerophiles Organisms that require small amounts of oxygen (2% to 10%) for growth, but are inhibited by higher concentrations.

microenvironment Environment immediately surrounding an individual microorganism.

microfilaments Cytoskeleton structures of eukaryotic cells that enable the cell cytoplasm to move.

microtubules Cytoskeleton structures of a eukaryotic cell that form mitotic spindles, cilia, and flagella; long hollow cylinders composed of tubulin.

microvillus (pl. **microvilli**) Tiny cylindrical process from luminal surfaces of cells such as those lining the intestine; increases surface area of the cell.

mineralization Conversion from organic to inorganic form; stabilization.

minimum bactericidal concentration (MBC) Lowest concentration of a specific antimicrobial medication that kills 99.9% of a given strain of bacteria.

minimum inhibitory concentration (MIC) Lowest concentration of a specific antimicrobial medication that prevents the growth of an organism *in vitro*.

minus (−) strand (1) The DNA strand that is used as a template for RNA synthesis. (2) The complement to the plus (or sense) stand of RNA. Also called the antisense strand.
miracidium First larval form of a fluke, hatching from the ovum as a ciliated organism.
mismatch repair Repair mechanism in which a repair enzyme recognizes improperly hydrogen-bonded bases and excises a short stretch of nucleotides containing these bases.
mitochondrion Organelle in eukaryotic cells in which the majority of ATP synthesis occurs.
mitogen Substance that induces mitosis; causes proliferation of cells.
mitosis Nuclear division process in eukaryotic cells that ensures the daughter cells receive the same number of chromosomes as the original parent.
MMWR Abbreviation for *Morbidity and Mortality Weekly Report*, published by the Centers for Disease Control and Prevention (CDC).
mold A filamentous fungus.
mole Amount of a chemical in grams that contains 6.023×10^{23} molecules; it is equal to the molecular weight of the chemical, or the sum of the atomic weight of all the atoms in a molecule of that chemical.
molecular postulates Group of criteria used to determine the cause of an infectious disease by using genetic and other molecular techniques.
molecular weight Relative weight of an atom or molecule based on a scale in which the H atom is assigned the weight of 1.0.
molecule Smallest part of a compound that retains all the properties of the compound.
monobactams Group of antimicrobial medications that interferes with peptidoglycan synthesis; very resistant to β-lactamases.
monocistronic RNA transcript that encodes one gene.
monoclonal antibodies Antibodies with a single specificity produced *in vitro* by lymphocytes that have been fused with a type of malignant myeloma cell.
monocytes Mononuclear phagocytes of the blood; part of the mononuclear phagocyte system of professional phagocytes.
monomer Repeating subunit of a polymer.
mononuclear phagocyte system (MPS) System of mononuclear cells (monocytes and macrophages) scattered throughout the body that are highly efficient at phagocytosis; formerly known as the reticuloendothelial system.
monosaccharide A sugar; a simple carbohydrate generally having the formula $C_nH_{2n}O_n$, where n can vary in number from three to eight.
morbidity Illness; most often expressed as the rate of illness in a given population at risk.
morbidity rate Number of cases of a specific disease per given population at risk.
mordant Substance that increases the affinity of cellular components for a dye.
morphology Form or shape of a particular organism or structure.
mortality Death; most often expressed as a rate of death in a given population at risk.
mortality rate Fraction of people who die from a given disease.
most probable number (MPN) method Statistical estimate of cell numbers based on the theory of probability; a sample is successively diluted to determine the point at which subsequent dilutions receive no cells.
M protein Heat- and acid-resistant protein found in the cell walls of Group A streptococci.
mRNA Messenger RNA.
mucociliary escalator Moving layer of mucus and cilia lining the respiratory tract that traps bacteria and other particles and moves them into the throat.
mucosal-associated lymphoid tissue (MALT) Lymphoid tissue present in the mucosa of the respiratory, gastrointestinal, and genitourinary tracts.
multiple-cloning site Small sequence of DNA that contains several unique restriction enzyme recognition sites into which foreign DNA can be cloned.
mushroom Filamentous multicelled fungus with macroscopic fruiting bodies.
mutagen Any agent that increases the frequency at which DNA is altered (mutated).
mutant Organism that has a changed nucleotide sequence or arrangement of nucleotides in its DNA, resulting in properties that make the organism different from the parent strain.
mutation Modification in the base sequence of DNA in a gene resulting in an alteration in the protein encoded by the gene.
mutualism Association in which both partners benefit.
myasthenia gravis Autoimmune disease characterized by muscle weakness, caused by autoantibodies.
mycelium (pl. **mycelia**) Tangled, matlike mass of fungal hyphae.
mycology The study of fungi.
mycorrhiza Symbiotic relationship between certain fungi and the roots of plants.
mycosis (pl. **mycoses**) Disease caused by a fungus.

***N*-acetylglucosamine (NAG)** One of the two alternating subunits of the glycan chains that make up peptidoglycan.
NAD/NADH Abbreviations for the oxidized/reduced forms of nicotinamide adenine dinucleotide, a diffusible electron carrier.
NADP/NADPH Abbreviations for the oxidized/reduced forms of nicotinamide adenine dinucleotide, a diffusible electron carrier.
narrow host range plasmid Plasmid that only replicates in one or a few closely related species of bacteria.
narrow-spectrum antimicrobials Antimicrobial medications that inhibit or kill a limited range of bacteria.
National Molecular Subtyping Network for Foodborne Disease Surveillance Surveillance network established by the Centers for Disease Control to facilitate the tracking of foodborne disease outbreaks.
natural killer cell See *NK cell*.
natural selection Selection by the environment of those cells best able to grow in that environment.
necrotic Dead; refers to dead cells or tissues in contact with living cells, as necrotic tissue in wounds.
negative (−) sense strand Strand of RNA that does not function as mRNA; strand of DNA that is not transcribed into mRNA.
negative staining Staining technique that employs an acidic dye to stain the background against which colorless cells can be seen.
Negri body Viral inclusion body characteristic of rabies.
nematodes Roundworms.
neurotransmitter Any of a group of substances released from the terminations of nerve cells when they are stimulated; they cross to the adjacent cell and cause it to be excited or inhibited.
neutralization tests Tests in which antibodies neutralize viruses by preventing them from infecting cells or neutralize toxins by binding to them and making them nontoxic.
neutron Uncharged component of an atom found in the nucleus.
neutrophiles Organisms that can live and multiply within the range of pH 5 (acidic) to pH 8 (basic) and have a pH optimum near neutral (pH 7).
neutrophils (or **PMNs**) Actively phagocytic leukocytes with granules containing antimicrobial and degradative substances that destroy engulfed organisms and materials; professional phagocytes. See *polymorphonuclear neutrophils*.
nitrification Conversion of NH_3 to nitrate (NO_3^-).
nitrogen fixation Conversion of nitrogen gas to ammonia.
NK cell Large granular, non-T, non-B lymphocyte that can recognize and destroy certain cells that it senses as foreign in a nonspecific fashion.
nodule Swelling on the root of a leguminous plant caused by *Rhizobium* species; the site of nitrogen fixation.
Nomarski differential interference contrast microscope Type of microscope that has a device for separating light into two beams that pass through the specimen and then recombine; light waves are out of phase when they recombine, resulting in the three-dimensional appearance of material in the specimen.
nomenclature System of assigning names to organisms; a component of taxonomy.
non-communicable diseases Disease that cannot be transmitted from one individual to another.
noncompetitive inhibition Type of enzyme inhibition that results from a molecule binding to the enzyme at a site other than the active site.
nonconjugative plasmid Plasmid that lacks some of the genetic information required for its transfer to other bacteria by conjugation.
noncritical items Medical instruments and surfaces such as stethoscopes and countertops that only come into contact with unbroken skin.
noncyclic photophosphorylation Type of photophosphorylation in which high-energy electrons are drawn off to generate reducing power; electrons must still be returned to chlorophyll, but they must come from a source such as water.
nonpolar covalent bond Bond formed by sharing electrons between atoms that have equal attraction for the electrons.
nonspecific immunity Immune responses that are nonspecific with respect to the infectious agent or other material involved, and that are not affected by prior exposures. See *innate immunity*.
normal microbial flora That group of microorganisms that colonizes the body surfaces but does not usually cause disease; also called normal flora.
Northern blot Procedure that is similar in principle to a Southern blot, except that it uses a nucleic acid probe to detect sequences of RNA.
nosocomial infection Infection acquired during hospitalization.
notifiable diseases Group of diseases that are reported to the CDC by individual states; typically these diseases are of relatively high incidence or otherwise a potential danger to public health.
nuclear membrane Membrane that separates the nucleus from the cytoplasm in eukaryotic cells.
nucleic acid hybridization Technique in which single strands of DNA are mixed together and allowed to reassociate. The degree to which the single strands reassociate indicates how complementary the strands are to one another.
nucleic acids Ribonucleic acid (RNA) and deoxyribonucleic acid (DNA).
nucleocapsid Viral nucleic acid and its protein coat.
nucleoid Region of a prokaryotic cell containing the DNA.
nucleolus Region within the nucleus where ribosomal RNAs are synthesized.
nucleosome Unit of the chromatin of eukaryotic cells that consists of a complex of histones around which the linear DNA wraps twice.
nucleotide array technology Use of a solid support to which a two-dimensional arrangement of numerous different single-stranded DNA fragments of known sequences has been attached.
nucleotides Basic subunits of ribonucleic or deoxyribonucleic acid consisting of a purine or pyrimidine covalently bonded to ribose or deoxyribose, which is covalently bound to a phosphate molecule.

nucleus Membrane-bound organelle in a eukaryotic cell that contains chromosomes and the nucleolus.

nuisance bloom Odiferous scum caused when buoyant cyanobacteria cells float to the surface of a body of stagnant water, and then lyse and decay.

numerical taxonomy Method of classification based on the phenotypes of prokaryotes; determines the relatedness of different organisms based on the percentage of characteristics that two groups have in common.

O antigen Antigenic polysaccharide portion of lipopolysaccharide, the molecule that makes up the outer leaflet of the outer membrane of Gram-negative bacteria.

objective lens Lens of a compound microscope that is closest to the specimen.

obligate aerobes Organisms with an absolute requirement for oxygen.

obligate anaerobes Organisms that cannot multiply if O_2 is present; they are often killed by traces of O_2 because of its toxic derivatives.

obligate fermenters Organisms that can grow in the presence of O_2 but never use it as a terminal electron acceptor; also called aerotolerant anaerobes.

obligate intracellular parasites Organisms that grow only inside living cells.

occlusion bodies Masses of viruses inside or outside cells.

ocular lens Lens of a compound microscope that is closest to the eye.

Okazaki fragment Nucleic acid fragment synthesized as a result of the discontinuous replication of the lagging strand of DNA.

oligonucleotide Short chain of nucleotides.

oligosaccharide Short chain of monosaccharide subunits joined together by covalent bonds; shorter than a polysaccharide.

oligotrophic environment An environment that is deficient in nutrients.

oligotrophs Organisms that can grow in a nutrient-deficient environment.

oncogene Gene whose activity is involved in turning a normal cell into a cancer cell.

open reading frames (ORFs) Stretches of DNA, generally longer than 300 base pairs, that begin with a start codon and end with a stop codon; they suggest that the region encodes a protein.

open system Method used to maintain cells in a state of continuous growth by continuously adding nutrients and removing waste products; also called a continuous culture.

operator Region located immediately downstream of a promoter to which a repressor can bind; binding of the repressor to the operator effectively prevents RNA polymerase from progressing past that region and blocks transcription.

operon Group of linked genes that are controlled as a single unit.

opine Unusual amino acid derivative; the portion of the Ti plasmid of *Agrobacterium tumefaciens* that is transferred to plant cells directs the recipient cells to synthesize this compound.

opportunist Organism that causes disease only in hosts with impaired defense mechanisms or when introduced into an unusual location; also called opportunistic pathogens.

opsonization Enhanced phagocytosis, usually caused by coating of the particle to be ingested with either antibody or complement components.

opthalmia neonatorum Eye infection of newborns usually caused by *Neisseria gonorrhoeae* or *Chlamydia trachomatis*, acquired from infected mothers during the birth process.

optical isomer (or **stereoisomer**) Mirror image of a compound.

optimum growth temperature Temperature at which a microorganism multiplies most rapidly.

order Taxonomic classification between class and family.

organ A structure composed of different tissues coordinated to perform a specific function.

organelle A structure within a cell that performs a specific function.

organic matter Material that contains carbon atoms bonded to other carbon atoms.

origin of replication Distinct region of a DNA molecule at which replication is initiated.

origin of transfer Short stretch of nucleotides, a part of which is transferred first when a plasmid is transferred to a recipient cell; necessary for plasmid transfer.

osmosis Movement of water across a membrane from a dilute solution to a more concentrated solution.

osmotic pressure Pressure exerted by water on a membrane due to a difference in the concentration of molecules on each side of the membrane.

osmotolerant Describes organisms that can tolerate relatively high salt concentrations, up to approximately 10% NaCl.

O-specific polysaccharide side chain The portion of LPS that is directed away from the membrane, at the end opposite of Lipid A; because its composition varies, it can be used to identify species or strains.

outbreak Cluster of cases occurring during a brief time interval and affecting a specific population; may herald the onset of an epidemic.

outer membrane (1) In prokaryotic cells, the unique lipid bilayer of Gram-negative cells that surrounds the peptidoglycan layer. (2) In eukaryotic cells, the membrane on the cytoplasmic side of organelles that have double membranes.

oxazolidinones Group of antimicrobial drugs that interferes with protein synthesis.

oxidase test Rapid biochemical test used to detect cytochrome c.

oxidation Removal of an electron.

oxidation-reduction reactions Chemical reactions in which one or more electrons is transferred from one molecule to another; the compound that loses electrons becomes oxidized and the chemical that gains electrons becomes reduced.

oxidative phosphorylation Cellular process that synthesizes the energy of ATP, using the proton motive force that is established as electrons are passed along the electron transport chain to a terminal electron acceptor.

oxygenic photosynthesis Photosynthetic reaction that synthesizes carbohydrate from CO_2 and H_2O with the release of oxygen.

---

palindrome Two stretches of DNA on opposite strands that are identical when oriented in the same direction, that is, 5′ to 3′ or 3′ to 5′.

pandemic A worldwide epidemic.

para-aminobenzoic acid (PABA) Intermediate in the pathway for folic acid synthesis in bacteria; sulfa drugs have a similar structure to PABA.

parasitism Association in which one organism, the parasite, benefits at the expense of the other organism, the host.

parent strain Refers to the original strain of a bacterium used in an experiment; often used in place of wild-type strain.

passive diffusion Process in which molecules flow freely into and out of a cell so that the concentration of any particular molecule is the same on the inside as it is on the outside of the cell.

passive immunity Protective immunity resulting from the transfer of immune serum or immune cells produced by other individuals or animals.

pasteurization Process of heating food or other substances under controlled conditions of time and temperature to kill pathogens and reduce the total number of microorganisms without damaging the substance.

pathogen Organism or virus causing a disease.

pathogenesis Process by which disease develops.

pathogenicity islands Short stretches of DNA in bacteria that code for factors that cause disease and have been transferred from disease-causing organisms.

peliosis hepatis Serious condition characterized by formation of blood-filled cysts in the liver, caused by *Bartonella henselae*; usually a complication of AIDS or other severe immunodeficiency.

penicillin Antibiotic that interferes with the synthesis of the peptidoglycan portion of bacterial cell walls.

penicillin-binding proteins (PBPs) Target of β-lactam antimicrobial drugs; their role in bacteria is peptidoglycan synthesis.

penicillin enrichment Method for increasing the relative proportion of auxotrophic mutants in a population by killing off the growing prototrophic cells with penicillin.

pentamer Polymer composed of five monomeric structural units.

pentose phosphate pathway Metabolic pathway that converts glucose to pyruvate, generating reducing power in the form of NADPH, and two precursor metabolites.

peptide bond Covalent bond formed between the —COOH group of one amino acid and the $—NH_2$ group of another amino acid; characteristic of proteins.

peptide interbridge Component of the peptidoglycan layer of Gram-positive bacteria; the short chain of amino acids that links the peptide side chains of adjacent *N*-acetylmuramic acid molecules.

peptidyl site (or **P-site**) Site on the ribosome to which the tRNA that temporarily carries the elongating amino acid chain residues.

peptidoglycan Macromolecule found only in bacteria that provides rigidity to the bacterial cell wall. The basic structure of peptidoglycan is an alternating series of two major subunits *N*-acetylmuramic acid (NAM) and *N*-acetylglucosamine (NAG); chains of these alternating sub units are cross-linked by peptide chains.

peptone Common component of bacteriological media; consists of proteins originating from any of a variety of sources that have been hydrolyzed to amino acids and short peptides by treatment with enzymes, acids, or alkali.

perforin Molecule produced by T-cytotoxic cells and NK cells; functions in killing target cells by forming a pore through the target cell membrane.

periplasm (or **periplasmic gel**) Very narrow gel that fills the region between the outer membrane and the cytoplasmic membrane in Gram-negative bacteria.

peritrichous flagella Distribution of flagella over the entire surface of a cell.

peroxidase enzymes Enzymes found in neutrophil granules, saliva, and milk that together with hydrogen peroxide and halide ions make up an effective antimicrobial system.

persistent Refers to infection in which the causative agent remains in the body for long periods of time, often without causing symptoms of disease.

petechia (pl. **petechiae**) Small purplish spot on the skin or mucous membrane caused by hemorrhage.

petri dish Two-part dish of glass or plastic often used to contain medium solidified with agar, on which bacteria are grown.

Peyer's patches Collections of lymphoid cells in the gastrointestinal tract; part of the gut-associated lymphoid tissue (GALT) and mucosal-associated lymphoid tissue (MALT).

pH Scale of 0 to 14 that expresses the acidity or alkalinity of a solution.

phage Shortened term for bacteriophage.
phage induction Process by which phage DNA is excised from bacterial DNA.
phagocytosis (v. **phagocytize**) The process by which certain cells ingest particulate matter by surrounding and enveloping those materials, bring them into the cell in a membrane-bound vesicle.
phase-contrast microscope Type of light microscope that employs special optical devices to amplify the difference in the refractive indexes of cells and the surrounding medium, increasing the contrast of the image.
phase variation The reversible and random alteration of expression of certain bacterial structures such as fimbriae by switching on and off the genes that encode those structures.
phenotypic mixing (or **transcapsidation**) Exchange of protein coats by two viruses when two virions infect the same cell.
phospholipid Lipid that has a phosphate molecule as part of its structure.
phosphotransferase system Type of group translocation in which the transported molecule is phosphorylated as it passes through the cytoplasmic membrane.
photoautotrophs Organisms that use light as the energy source and CO_2 as the major carbon source.
photoheterotrophs Organisms that use light as the energy source and organic compounds as the carbon source.
photooxidation Chemical reaction occurring as a result of absorption of light energy in the presence of oxygen.
photophosphorylation Processes that utilize light as an energy source to drive the synthesis of ATP.
photoreactivation (or **light repair**) Breakage of the covalent bonds joining thymine dimers in the light, thereby restoring the DNA to its original state.
photosynthesis Reactions used to harvest the energy of light to synthesize ATP, and then use of that energy to power CO_2 fixation.
photosystems Protein complexes within which chlorophyll and other light-gathering pigments are organized; located in special photosynthetic membranes.
phototaxis Directed movement in response to variations in light.
phototrophs Organisms that use light as a source of energy.
phycobiliproteins Light-harvesting pigments of cyanobacteria; they absorb energy from wavelengths of light that are not well absorbed by chlorophyll.
phylogenetic tree Type of diagram that depicts the evolutionary heritage of organisms.
phylogeny Evolutionary relatedness of organisms.
phylum (pl. **phyla**) Collection of similar classes; a collection of similar phyla makes up a kingdom; a phylum may also be called a division.
phytoplankton Floating and swimming algae and photosynthetic prokaryotic organisms of lakes and oceans.
pilus (pl. **pili**) Hairlike appendages on many Gram-negative bacteria that function in conjugation and for attachment.
pinocytosis Process by which eukaryotic cells take in liquid and small particles from the surrounding environment by internalizing and pinching off small pieces of their own membrane, bringing along a small volume of liquid and any material attached to the membrane.
plankton Primarily microscopic organisms floating freely in most waters.
plaque (1) Clear area in a monolayer of cells. (2) In dentistry, a collection of bacteria that adhere to a tooth surface.
plasma Fluid portion of nonclotted blood.
plasma cell End cell of the B-cell series, fully differentiated to produce and secrete large amounts of antibody.
plasma membrane Semipermeable membrane that surrounds the cytoplasm in a cell; cytoplasmic membrane.
plasmid Small extrachromosomal circular DNA molecule that replicates independently of the chromosome; often codes for antibiotic resistance.
plasmolysis Process in which water diffuses out of a cell, causing the cytoplasm to dehydrate and shrink from the cell wall.
plate count Method used to determine the number of viable cells in a specimen by determining the number of colonies that arise when the specimen is added to an agar medium.
platelets (or **thrombocytes**) Small cell fragments in the blood that are essential for blood clotting; arise from large bone marrow cells called megakaryocytes.
pleomorphic Bacteria that characteristically vary in shape.
pleurisy Inflammation of the pleura, membranes that line the lung and chest cavity; often marked by a sharp pain associated with breathing.
plus (+) strand (1) The DNA strand that is complementary to the strand used as a template for RNA synthesis. (2) Of the two RNA molecules that can theoretically be transcribed from double-stranded DNA, the one that can be translated to make a protein; also called the sense strand.
PMN Abbreviation for polymorphonuclear neutrophil.
pneumonia Inflammation of the lungs accompanied by filling of the air sacs with fluids such as pus and blood.
pneumonic plague Disease that develops when *Yersinia pestis* infects the lungs.
point mutation Mutation in which only a single base pair is involved.
polar covalent bond Bond formed by sharing electrons between atoms that have unequal attraction for the electrons.
polarity (1) The degree of affinity that an atom has for electrons; this results in positive or negative charges on atoms in a molecule. (2) The 5′ to 3′ directionality of a nucleic acid fragment.
poly A tail Series of approximately 200 adenine derivatives that are added to the 3′ end of an mRNA transcript in eukaryotic cells; thought to stabilize the transcript and enhance translation.
polycistronic (or **polygenic**) An mRNA molecule that carries more than one gene.
polymer Large molecules formed by the joining together of repeating small molecules (subunits).
polymerase chain reaction (PCR) Method used to create millions of copies of a given region of DNA in only a matter of hours.
polymorphic Having different distinct forms at various stages of the life cycle.
polymorphonuclear neutrophils (PMNs) Phagocytic cells that together with the macrophages are known as "professional phagocytes"; the nuclei of these cells are segmented and composed of several lobes.
polymyxin B Type of antimicrobial medication that damages cytoplasmic membranes.
polypeptide Chain of amino acids joined by peptide bonds; also called a protein.
polyribosome Assembly of multiple ribosomes attached to a single mRNA molecule; also called a polysome.
polysaccharide Long chains of monosaccharide subunits.
polyunsaturated fatty acid Fatty acid that contains numerous double bonds.
porins Proteins in the outer membrane of Gram-negative bacteria that form channels through which small molecules can pass.
portal of entry Place of entry of microorganisms into the host.
portal of exit Place where infectious agents leave the host to find a new host.
positive (+) sense strand Strand of RNA which functions as mRNA; the strand of DNA that gives rise to mRNA.
potential energy Stored energy; it can exist in a variety of forms including chemical bonds, a rock on the top of a hill, and water behind a dam.
pour plate method Method of inoculating an agar medium with bacteria while the agar is liquid and then pouring it into a petri dish, where the agar hardens; the colonies grow both on the surface and within the medium.
precipitation reaction Reaction of an antibody with a soluble antigen to form an insoluble substance.
precursor metabolites Metabolic intermediates that are produced in catabolic pathways but can be siphoned off for use in anabolic pathways.
prevalence Total number cases, both old and new, in a given population at risk at a point in time.
primary culture Cells taken and grown directly from the tissues of an animal.
primary immune response Immune response that occurs upon first exposure to an antigen.
primary infection Infection in a previously healthy individual, such as measles in a child who has not had measles before.
primary lymphoid organs Organs in which lymphoid stem cells mature, including the thymus and bone marrow.
primary metabolites Compounds synthesized by a cell during the log phase.
primary producers Organisms that harvest the energy of sunlight, and use it to convert CO_2 into organic compounds; by doing so, they sustain other life forms, including humans.
primary stain First dye applied in a multistep differential staining procedures; generally stains all cells.
primary structure Refers to the sequence of amino acids in a protein.
primase Enzyme that synthesizes small fragments of RNA to serve as primers for DNA synthesis during DNA replication.
primer RNA molecule that initiates the synthesis of DNA.
prion Infectious protein that has no nucleic acid.
production Stage of biogeochemical recycling in which producer organisms convert carbon dioxide or other inorganic compounds into organic materials.
productive infection Virus infection in which more virions are produced.
proglottid One of the segments that make up most of the body of a tapeworm.
prokaryote Cell characterized by lack of a nuclear membrane and the absence of membrane-bound organelles.
promoter Nucleotide sequence to which RNA polymerase binds to initiate transcription.
propagated epidemic Outbreak of contagious disease in which the infectious agent is transmitted to others, resulting in steadily increasing numbers of people becoming ill.
prophage Latent form of a temperate phage whose DNA has been inserted into the host's DNA.
prophylaxis Prevention of disease.
prospective study Study that looks ahead to see if the risk factors identified by a retrospective study predict a tendency to develop the disease.
protease Enzyme that degrades protein; the protease that is encoded by HIV is the target of several anti-HIV medications.
protein Macromolecule containing one or more polypeptide chains.
Protein A Protein produced by *Staphylococcus aureus* that inhibits phagocytosis of the organism by binding to the Fc portion of antibodies.
protist Designation for eukaryotic organisms other than plants, animals, and fungi; may be unicellular or multicellular.
proton Positively charged component of an atom found in the nucleus.

proton motive force Form of energy generated by the electron transport chain, which expels protons to create a chemiosmotic gradient.

proto-oncogene Genes in the bodies of animals that code for proteins that activate transcription.

protoplast Gram-positive cell from which the rigid cell wall has been removed.

prototroph Organism that has no organic growth requirements other than a source of carbon and energy.

protozoa Group of single-celled eukaryotic organisms.

provirus Latent form of a virus in which the viral DNA is incorporated into the chromosome of the host.

pseudopods Transient armlike extensions formed by phagocytes and protozoa; they surround and enclose extracellular material, including bacteria, during the process of phagocytosis.

P-site (or **peptidyl site**) Site on the ribosome where the tRNA that temporarily carries the elongating amino acid chain resides.

psychrophile Microorganism that grows best between –5°C and 15°C.

psychrotroph Organism that can grow well at low temperatures but has an optimum temperature above 15°C.

pulsed-field gel electrophoresis Type of gel electrophoresis that is used to separate very large fragments of DNA.

PulseNet Surveillance network established by the Centers for Disease Control to facilitate the tracking of foodborne disease outbreaks; catalogues the RFLPs of certain pathogenic organisms. Also called the National Molecular Subtyping Network for Foodborne Disease Surveillance.

pure culture Culture that contains only a single strain of an organism.

purine Component of RNA and DNA; the two major purines are adenine and guanine.

pus Thick, opaque, often yellowish material that forms at the site of infection, made up of dead neutrophils and tissue debris.

putrefaction Digestion of proteins by enzymes to yield foul-smelling products.

pyoderma Any skin disease characterized by production of pus.

pyogenic Pus-producing.

pyrimidine Component of RNA and DNA; the three major pyrimidines are thymidine, cytosine, and uracil.

pyrogens Fever-inducing substances.

pyruvate End product of glycolysis; a precursor metabolite used in the synthesis of amino acids.

---

quaternary ammonium compounds Cationic (positively charged) detergents that are nontoxic enough to be used to disinfect food preparation surfaces; also called quats.

quaternary structure Level of structure of a protein molecule resulting from the interaction of one or more protein chains.

quorum sensing Communication between bacteria by means of small molecules, permitting the bacteria to sense when there is an adequate quorum or number of organisms present to activate virulence genes.

---

radial immunodiffusion test Quantitative antigen-antibody precipitation in gel test in which one reactant is distributed throughout the gel and the other reactant diffuses into the gel, producing a ring of precipitation.

radioallergosorbent test (RAST) Radioimmunoassay that measures the IgE antibody reacting with a specific antigen, used in the diagnosis of immediate allergies.

radioimmunoassay (RIA) Competitive inhibition assay using a radioactively labeled reactant to measure antigens or antibodies.

rDNA DNA that encodes ribosomal RNA (rRNA).

reading frames Grouping of a stretch of nucleotides into sequential triplets; an mRNA molecule has three reading frames, but only one is typically used in translation.

receptor Type of membrane protein that binds to specific molecules in the environment, providing a mechanism for the cell to sense and adjust to its surroundings.

receptor-mediated endocytosis Type of pinocytosis that allows cells to internalize extracellular ligands that bind to the cell's receptors.

recombinant DNA molecule DNA molecule created by joining DNA from two different sources *in vitro*; a vector-insert chimera is a recombinant DNA molecule.

recombinant vaccines Subunit vaccines produced by genetic engineering.

redox reactions Transfer of electrons from one compound to another; one compound becomes reduced and the other becomes oxidized.

reducing agents Compounds that readily donate electrons to another compound, thereby reducing the other compound.

reducing power Reduced electron carriers such as NADH, NADPH, and $FADH_2$; their bonds contain a form of usable energy.

reduction Process of adding electrons and hydrogen atoms to a molecule.

refraction Bending of light rays that occurs when light passes from one medium to another.

regulatory gene Gene that functions in the control of the rate of synthesis of other gene products.

regulatory protein Protein that binds to DNA, either blocking or enhancing the function of RNA polymerase.

regulon Set of related genes that are transcribed as separate units but are controlled by the same regulatory protein.

replica plating Technique for the simultaneous transfer of organisms in separated colonies from one medium to another medium.

replication fork In DNA synthesis, the site at which double helix is being unwound to expose the single strands that can function as templates.

replicon Piece of DNA that is capable of replicating; contains an origin of replication.

reporter gene Gene that has a detectable phenotype and can be fused to a gene of interest, providing a mechanism by which to monitor the expression of the gene of interest.

repressible pathway Pathway in which the enzymes are not synthesized when the end product of the pathway is present.

repressor Protein that binds to the operator site and prevents transcription.

reservoir Source of a disease-producing organism.

resistance factor (or **R factor**) Plasmid that carries genetic information for resistance to one or more chemotherapeutic agents.

resistance plasmid (or **R plasmid**) Plasmid that carries genetic information for resistance to one or more antibiotics and heavy metals.

resolve To clearly separate.

respiration Sum total of metabolic steps in the degradation of foodstuffs when the electron acceptor is an inorganic compound.

respire To use the processes of respiration.

response regulator Regulatory protein of a two-component regulatory system; receives a phosphoryl group from the membrane-spanning sensor.

restriction endonuclease Enzyme that recognizes and cuts DNA at specific purine and pyrimidine sites.

restriction enzyme Type of enzyme that recognizes and cleaves a specific sequence of DNA.

restriction fragment length polymorphism (RFLP) Pattern of fragment sizes obtained by digesting DNA with one or more restriction enzymes.

reticulate body Fragile, replicating, non-infectious intracellular form of *Chlamydia* species.

retrospective study Type of study done following a disease outbreak; compares the actions and events surrounding clinical cases with those of controls.

retroviruses Group of viruses that carry their genetic information as single-stranded RNA; they have the enzyme reverse transcriptase, which forms a DNA copy that is then integrated into the host cell chromosome.

reverse transcriptase Enzyme that synthesizes double-stranded DNA complementary to an RNA template.

reversion Process by which a second mutation corrects a defect caused by an earlier mutation.

Reye's syndrome Often fatal condition characterized by vomiting, coma, and brain and liver damage, mostly occurring in children treated with aspirin for influenza or chickenpox.

RFLP (restriction fragment length polymorphism) Pattern of fragment sizes obtained by digesting DNA with one or more restriction enzymes.

rheumatoid arthritis Severe crippling autoimmune disease of the joints, caused by cytokines from inflammatory Th1 cells and immune complexes.

rhizosphere Zone around plant roots containing organic materials exuded by the roots.

rhodamine Fluorescent dye that binds to a compound found only in the cell walls of members of the genus *Mycobacterium*.

rhuMab (recombinant human monoclonal antibody) Hybrid recombinant anti-IgE molecule being tested in the treatment of asthma.

RIA Abbreviation for radioimmunoassay.

ribonucleic acid (RNA) Macromolecules in a cell that play a role in converting the information coded by the DNA into amino acid sequences in protein.

ribose A 5-carbon sugar found in RNA.

ribosomal RNA (rRNA) Type of RNA present in ribosomes; the nucleotide sequences of these are increasingly being used to classify and, in some cases, identify microorganisms.

ribosome Structure that facilitates the joining of amino acids during the process of translation; composed of protein and ribosomal RNA.

ribosome-binding site Sequence of nucleotides on mRNA to which a ribosome binds; the first time the codon for methionine (AUG) appears after that site, translation generally starts.

ribotyping Technique used to distinguish among related strains; detects RFLPs in ribosomal RNA genes.

ribozymes RNA molecules that have a catalytic function.

rifamycins Group of antimicrobial medications that block transcription.

risk factors Specific conditions associated with high frequencies of disease.

RNA Abbreviation for ribonucleic acid.

RNA polymerase Enzyme that catalyzes the synthesis of RNA using a DNA template.

RNases Enzymes that degrade RNA.

rod Cylindrical-shaped bacterium; also called a bacillus.

rolling circle replication Mechanism of DNA replication in which a single strand of DNA is synthesized.

rough endoplasmic reticulum Organelle where proteins that are not located in the cytoplasm are synthesized.

roundworm (or **nematode**) Helminth that is often parasitic and causes disease.

R plasmids Plasmids that encode resistance to one or more antimicrobial medications.

rRNA Ribosomal RNA.

RTF Abbreviation for resistance transfer factor.

rubisco Enzyme that initiates the Calvin cycle by joining CO_2 to the 5-carbon compound ribulose-1,5-bisphosphate.

---

salinity Amount of salt in a solution.

SALT Skin-associated lymphoid tissues.

sanitization Process of substantially reducing the microbial populations on objects to achieve acceptably safe public health levels.

saprophyte Organism that takes in nutrients from dead and decaying matter.

saturated Refers to a fatty acid that contains no double bonds.

scanning electron microscope (SEM) Type of electron microscope that scans a beam of electrons scans back and forth over the surface of a specimen; used for observing surface details, but not internal structures of cells.

scanning probe microscope Microscope that makes it possible to view images at an atomic scale; produces a map showing the bumps and valleys of atoms on a surface.

scanning tunneling microscope Type of scanning probe microscope with a sharp metallic probe that causes electrons to tunnel between the probe and a conductive surface.

schizogony Process of multiple fission in which the nucleus divides a number of times before individual daughter cells are produced.

schizont Multinucleate stage in the development of certain protozoa, such as the ones that cause malaria.

scolex Attachment organ of a tapeworm, the head end.

scrapie Common name for a neurological disease of sheep thought to be caused by a prion.

sebum Oily secretion of the sebaceous glands of the skin.

secondary response (or **memory response**) Enhanced immune response that occurs upon second or subsequent exposure to specific antigen, caused by the rapid activation of long-lived memory cells; anamnestic response.

secondary infection Infection that occurs along with or immediately following another infection, usually as a result of the first infection.

secondary lymphoid organs Peripheral lymphoid organs throughout the body where mature lymphocytes function in immune responses; their locations include the adenoids, tonsils, spleen, appendix, and lymph nodes, among others.

secondary metabolites Metabolic products synthesized during late-log and stationary phase.

secondary structure Refers to the arrangement of amino acids in a protein; the two major arrangements are helices and sheets.

secretion system Mechanism by which bacterial pathogens transfer gene products into host cells to cause endocytosis or other effects.

segmented virus Virus with more than one RNA molecule enclosed in the capsid.

selectable marker Gene that encodes a selectable phenotype such as antibiotic resistance.

selective enrichment Method of increasing the relative proportion of one particular organism in a mixed population by providing conditions in a broth that enhance the growth of the desired organism and including a selective agent that inhibits the growth of other organisms.

selectively permeable membrane Membrane that allows some but not other molecules to pass through freely.

selective medium Culture medium that inhibits the growth of certain microorganisms and therefore favors the growth of desired microorganisms.

selective toxicity Causing greater harm to a pathogen than to the host.

self-assembly Spontaneous formation of a complex structure from its component molecules without the aid of enzymes.

self-transmissible plasmid Plasmid that codes for all of the information necessary for its own transfer.

semiconservative replication Type of nucleic acid replication that results in each of the two double-stranded molecules containing one of the original strands (the template strand) and one newly synthesized strand.

semicritical items Medical instruments such as endoscopes that come into contact with mucous membranes, but do not penetrate body tissue.

semipermeable Describes material that allows the passage of some but not other molecules.

sense strand Of the two RNA molecules that can theoretically be transcribed from double-stranded DNA, the one that can be translated to make a protein; also called the plus (+) strand.

sensitization Prior immunization; allergic reactions to an antigen occur only in sensitized individuals who have been exposed to that particular antigen.

sepsis A bloodstream infection.

septicemia Acute illness caused by infectious agents or their products circulating in the bloodstream; blood poisoning.

septic tank method Method of treating sewage in which wastes are collected in a large tank and degraded by anaerobic organisms, with the resulting fluid stabilized in a drainage field by aerobic organisms.

serial dilutions Series of dilutions, usually twofold or tenfold, used to determine the titer or concentration of a substance in solution.

seroconversion Change from negative serum without specific antibodies to serum positive for specific antibodies.

serogroup Microorganisms within a species that are the same antigenically as determined by specific antisera.

serology Use of serum antibodies to detect and measure antigens, or conversely, the use of antigens to detect serum antibodies.

serotonin A basic amine, stored in mast cell granules, that upon release can cause contraction of some smooth muscles and increased vascular permeability; a mediator of inflammation.

serotype A strain that has a characteristic antigenic structure that differs from other strains; also called a serovar.

serum Fluid portion of blood that remains after blood clots.

sex pilus Thin protein appendage required for attachment of one bacterium to another prior to DNA transfer by conjugation.

shake tube Tube of agar medium that has been uniformly inoculated with a bacterial culture in order to determine the oxygen requirement of that organism.

sheath (1) Tube that surrounds and holds a linear chain of cells; characteristic of some aquatic bacteria. (2) Refers to a rodlike protein component found on many phages; often called a "tail."

shingles (or **herpes zoster**) Condition resulting from the reactivation of the varicella-zoster virus.

shock Condition with multiple causes characterized by low blood pressure and circulation of the blood inadequate to sustain normal function of vital organs; septic shock results from growth of microorganisms in the body; toxic shock results from a circulating exotoxin.

sigma (σ) factor Component of RNA polymerase that recognizes the promoter.

signal sequence Characteristic series of hydrophobic amino acids at the amino terminal end of proteins that are destined for secretion; functions as a tag, directing transport of the protein through the membrane.

signal transduction Process that transmits information from outside of a cell to the inside, allowing that cell to respond to changing environmental conditions.

signature sequences Characteristic sequences in the genes that encode ribosomal RNA that can be used to classify or identify certain organisms.

signs Effects of a disease observed by examining the patient.

similarity coefficient Numerical value that can be used to classify prokaryotes based on their phenotypic characteristics.

simple diffusion Movement of molecules or ions from a region of high concentration to a region of low concentration; does not involve transport proteins.

simple staining Staining technique that employs a basic dye to impart color to cells.

single-cell protein (SCP) Use of microorganisms such as yeast and bacteria as a protein source.

site-specific recombination Mechanism by which a piece of DNA becomes part of a larger piece of DNA; involves identical sequences on each piece of DNA.

skin-associated lymphoid tissue (SALT) Secondary lymphoid tissue consisting of collections of lymphoid cells under the skin.

slime layer Type of glycocalyx that is diffuse and irregular.

slime mold Terrestrial organism that is similar to the fungi but not related genetically.

smear In a staining procedure, the film obtained by placing a drop of a liquid containing a microbe on a glass microscope slide and allowing it to air dry.

smooth endoplasmic reticulum Organelle of eukaryotic cells that is the site of lipid synthesis and degradation and calcium ion storage.

solute Dissolved molecules.

SOS repair Complex, inducible repair process used to repair highly damaged DNA.

Southern blot Technique to detect a given nucleotide sequence in DNA fragments that have been separated by gel electrophoresis and transferred to a membrane filter.

specialized transduction Transfer of only specific bacterial genes by phage from one bacterium to another.

species Group of related isolates or strains; the basic unit of taxonomy. In the binomial nomenclature scheme, the second name given to an organism.

specific immunity Immune response that depends on the recognition and elimination of foreign antigens by specialized lymphocytes.

spheroplast Gram-negative cell from which the peptidoglycan component of the cell wall has been removed; retains some portions of the outer membrane.

spikes (or **attachment proteins**) Structures on the outside of the virion that bind to host cell receptors.

spirillum (pl. **spirilla**) Curved rod long enough to form spirals.

spirochete Type of long helical cell with a flexible cell wall that is characterized by an axial filament.

splicing Process that removes introns from eukaryotic precursor RNA to generate mRNA.

spoilage Biochemical changes in foods that are perceived as undesirable.

spontaneous generation Discredited theory that organisms can arise from nonliving matter.

spontaneous mutation Mutation that occurs naturally without the addition of mutagenic agents.

sporadic Occurring in a population widely spaced in time and place.

spore Type of differentiated, specialized cell formed by certain organisms; includes some types of dormant cells that are resistant to adverse conditions and the reproductive structures formed by fungi.

sporogenesis In bacteria, a complex, highly ordered sequence of morphological changes during which a bacterial vegetative cell produces a specialized cell greatly resistant to environmental adversity; eukaryotes can also experience sporogenesis.

sporozoite Elongated infectious form of certain protozoa; for example, in malaria, the form entering the body from a mosquito bite, infectious for liver cells.

spread plate Technique used to cultivate bacteria by uniformly spreading a suspension of cells onto the surface of an agar plate.

sputum Material coughed from the lungs.

stabilization Conversion of organic to inorganic matter.

start codon Codon at which translation is initiated; in prokaryotes, typically the first AUG after a ribosome-binding site.

starter cultures Strains of microorganisms added to a food to initiate the fermentation process.

stationary phase Stage of growth of a culture in which the number of viable cells remains constant.

stereoisomer (or **optical isomer**) Mirror image of a compound.

sterile Completely free of all microorganisms and viruses; an absolute term.

steroid Type of lipid with a specific four-membered ring structure.

sticky ends Single-stranded overhangs generated when DNA is digested with a restriction enzyme that cuts asymmetrically within the recognition sequence; also called cohesive ends.

stock culture Culture stored for use as an inoculum in later procedures.

stop codon Codon that does not code for an amino acid and is not recognized by a tRNA; signals the end of the protein.

strain Population of organisms that has characteristics differing from others within the species.

streak plate Simplest and most commonly used technique for isolating bacteria; a series of successive streak patterns is used to sequentially dilute an inoculum on the surface of an agar plate.

streptococcal pyrogenic exotoxins (SPEs) Family of genetically related poisonous proteins produced by certain strains of *Streptococcus pyogenes,* responsible for scarlet fever, toxic shock, and "flesh-eating" necrotizing fasciitis.

stromatolite Coral-like mat of filamentous microorganisms.

structural isomers Molecules that contain the same elements but in different arrangements that not mirror images; structural isomers have different names.

substrate (1) Substance on which an enzyme acts to form products. (2) Surface on which an organism will grow.

substrate-level phosphorylation Transfer of the high-energy phosphate from a phosphorylated compound to ADP to form ATP.

subunit vaccines Vaccines made of products or portions of an agent, bacteria, or viruses.

sucrose Disaccharide consisting of a molecule of glucose bonded to fructose; common table sugar.

sugar-phosphate backbone Series of alternating sugar and phosphates moieties of a DNA molecule.

sulfa drugs Group of antimicrobial drugs that inhibit folic acid synthesis.

sulfanilamide Antimicrobial drug that inhibits folic acid synthesis; one of the sulfa drugs.

sulfate-reducers Group of obligate anaerobes that use sulfate (SO_4^{2-}) as a terminal electron acceptor, producing hydrogen sulfide as an end product.

S unit Unit of measurement that expresses the sedimentation rate of a compound; reflects the mass and density of the compound; "S" stands for Svedberg.

superantigens Molecules that stimulate T lymphocytes by binding to MHC class II molecules and to part of the T-cell receptor distinct from the antigen-binding site, resulting in activation of many T cells, overproduction of cytokines, severe reactions, and sometimes fatal shock.

superficial mycoses Fungal infections that affect the hair, skin, or nails.

superoxide (O_2^-) Toxic derivative of O_2.

superoxide dismutase Enzyme that degrades superoxide to produce hydrogen peroxide.

swarmer cells Motile cells of sheathed bacteria that disperse to new locations.

symbiosis The living together of two dissimilar organisms or symbionts.

symporters Transport systems in which the passage of one type of molecule or ion facilitates the transport of another.

symptoms Effects of a disease experienced by the patient.

syncytium (pl. **syncytia**) Multinucleate body formed by the fusion of cells.

synergistic Describes the acting together of agents to produce an effect greater than the sum of the effects of the individual agents.

synthetic medium Medium in which the chemical composition and quantity of every component is known.

systemic infection Infection in which the infectious agent spreads throughout the body.

systemic mycoses Fungal infections that affect the tissues deep in the body.

---

tandem repeat Repetitive core sequence located in regions between human genes; probes that bind to tandem repeats are used in DNA fingerprinting.

tapeworm (or **cestode**) Helminth that is often parasitic and causes disease.

***Taq* polymerase** Heat-stable DNA polymerase of the thermophilic bacterium *Thermus aquaticus*.

target DNA In the PCR procedure, the region to be amplified.

tautomeric shift Movement of H atoms from one site on a nitrogenous base to another on DNA; alters the hydrogen-bonding properties and therefore, may result in mutations.

taxa Groups into which organisms are classified.

taxonomy The science that studies organisms in order to arrange them into groups; those organisms with similar properties are grouped together and separated from those that are different. Taxonomy encompasses identification, classification, and nomenclature.

T-cell-dependent antigens Proteins and other antigens that require the cooperation of CD4 Th2 and B cells for the production of antibodies.

T-cell-independent antigens Antigens that can stimulate the production of IgM antibody without the aid of Th cells.

T cells Lymphocytes that mature in the thymus; they originate from stem cells in bone marrow, differentiate in the thymus, and move through the bloodstream to secondary lymphoid organs where they are responsible for cellular immune responses and function as helper cells in the antibody response.

T-cytotoxic cells $CD8^+$ lymphocytes that recognize antigen on the surface of target cells and kill the target cells.

T-DNA Portion of the Ti plasmid of *Agrobacterium tumefaciens* that is typically transferred into a plant cell.

teichoic acids Component of the Gram-positive cell wall, composed of chains of a common subunit, either ribitol-phosphate or glycerol-phosphate, to which various sugars and D-alanine are usually attached.

temperate phage Bacteriophage that can either become integrated into the host cell DNA as a prophage or replicate outside the host chromosome leading to cell lysis.

terminal electron acceptor Chemical that is ultimately reduced as a consequence of chemotrophic metabolism.

tertiary structure Level of structure of a protein described by its three-dimensional nature; two major shapes exist, globular and fibrous.

tetracyclines Group of antimicrobial medications that interfere with protein synthesis.

Th1 cells CD4 T lymphocytes called inflammatory T cells because they produce cytokines that drive the development of cytotoxic T cells and that activate macrophages.

Th2 cells CD4 T lymphocytes that produce cytokines that stimulate B cells to produce antibodies.

T-helper (Th) cells $CD4^+$ T lymphocytes that produce cytokines, influencing and helping the specific immune response; two subgroups are Th1 and Th2.

therapeutic index Ratio of minimum toxic dose to minimum effective dose of a medication.

thermophile Organism with an optimum growth temperature between 45°C and 70°C.

thrush Infection of the mouth by *Candida albicans*.

thylakoids Membrane-bound disclike structures within the stroma of chloroplasts; they contain chlorophyll.

thymine dimer Two adjacent thymine molecules on the same strand of DNA joined together through covalent bonds.

thymus Primary lymphoid organ, located in the upper chest, in which T lymphocytes mature.

Ti plasmid Plasmid of *Agrobacterium tumefaciens* that enables it to cause tumors in plants; a derivative of the plasmid is used a vector by scientists to introduce DNA into plants; "Ti" stands for tumor inducing.

tissue culture Culture of plant or animal cells that grows in an enriched medium outside the plant or animal.

titer Measure of the concentration of a substance in solution; for example, the amount of a specific antibody in serum, usually measured as the highest dilution of serum that will test positive for antibody.

toxemia Circulation of toxins in the bloodstream.

toxin Poisonous chemical substance.

toxoid Modified form of a toxin that is no longer toxic but is able to stimulate the production of antibodies that will neutralize the toxin.

trace elements Elements that are required in very minute amounts by all cells; they include cobalt, zinc, copper, molybdenum, and manganese.

trachoma Potentially serious chronic eye disease caused by certain strains of *Chlamydia trachomatis.*

transamination Transfer of an amino group from an amino acid to another organic compound; converts the recipient compound to an amino acid.

transcapsidation (or **phenotypic mixing**) Exchange of protein coats in viral assembly when two virions infect the same cell.

transcript Fragment of RNA, synthesized using one of the two strands of DNA as a template.

transcription Process of transferring genetic information coded in DNA into messenger RNA (mRNA).

transduction Mechanism of gene transfer between bacteria in which bacterial DNA is transferred inside a phage.

transfer RNA (tRNA) Type of RNA that delivers the appropriate amino acid to the ribosome during translation.

transformed cells Bacterial or animal cells containing inheritable changes.

transfusion reaction Reaction characterized by fever, low blood pressure, pain, nausea, and vomiting, resulting from the transfusion of immunologically incompatible blood.

transgenic Plants and animals into which new DNA has been introduced.

transition step Step in metabolism that links glycolysis to the TCA cycle; converts pyruvate to acetyl-CoA.

translation Process by which genetic information in the messenger RNA directs the order of amino acids in protein.

translocation Advancement of a ribosome a distance of one codon during translation.

transmissible spongiform encephalopathies Group of fatal neurodegenerative diseases of humans and animals in which brain tissue develops spongelike holes.

transmission electron microscope (TEM) Type of microscope that directs a beam of electrons at a specimen; used to observe fine details of cell structure.

transport protein Type of protein found in cell membranes that functions in the transport of certain compounds across the membrane; may be called a permease or a carrier.

transposable element (or **transposon**) Gene that moves from one DNA molecule to another within the same cell or from one site on a DNA molecule to another site on the same molecule.

transposition Movement of a piece of DNA from one site in a molecule to another site in the same cell.

trematodes Flatworms known as flukes.

tricarboxylic acid (TCA) cycle Metabolic pathway that incorporates acetyl-CoA, ultimately generating CO_2 and reducing power; also called the Krebs cycle and the citric acid cycle.

trichomes Filamentous multicellular associations of cyanobacteria that may or may not be enclosed within a sheath.

trickling filter method Treatment method for small sewage plants in which a rotating arm of the filter sprays sewage onto a bed of rocks coated with a biofilm of organisms that aerobically degrades the wastes.

triglyceride Molecule consisting of three molecules of the same or different fatty acids bonded to glycerol.

trimethoprim Antimicrobial medication that interferes with folic acid synthesis.

tRNA Transfer RNA.

trophozoite Vegetative feeding form of some protozoa.

***trp* operon** Operon that encodes the five enzymes of the pathway for tryptophan biosynthesis.

T-suppressor (Ts) cells T lymphocytes that specifically suppress immune responses, often $CD8^+$ T lymphocytes.

tubercle Granuloma formed in tuberculosis.

tumble Rolling motion of a motile cell that is caused by an abrupt change in the direction of rotation of flagella.

tumor-inducing (Ti) plasmid Plasmid of *Agrobacterium tumefaciens* that enables the organism to cause tumors in plants; a derivative of the plasmid is used a vector by scientists to introduce DNA into plants.

tumor necrosis factors (TNFs) Cytokines with many functions in immune responses; some kill tumor cells.

turbidity Cloudiness; the turbidity of a bacterial suspension is proportional to the number of cells in that suspension.

tyndallization Repeated cycles of heating and incubation to kill spore-forming bacteria.

---

ubiquity Widespread prevalence.

ultra-high-temperature (UHT) method A method that uses heat to render a product free of all microorganisms that can grow under normal storage conditions.

ultraviolet (UV) light Electromagnetic radiation with wavelengths between 175 and 350 nm; invisible.

uniporters Transport systems that translocate a single molecule or ion.

unsaturated Refers to a fatty acid with one or more double bonds.

upstream Direction toward the 5′ end of either an RNA molecule or the analogous (+) strand of DNA.

urticaria Hives; an allergic skin reaction characterized by the formation of itchy red swellings.

UV Abbreviation for ultraviolet light.

---

vaccine Preparation of living or dead microorganisms or viruses or their components used to immunize a person or animal against a particular disease.

vancomycin Antimicrobial medication that interferes with peptidoglycan synthesis.

vector (1) In molecular biology, a piece of DNA that acts as a carrier of a cloned fragment of DNA. (2) In epidemiology, any living organism that can carry a disease-causing microbe; most commonly arthropods such as mosquitoes and ticks.

vegetative cell Typical, actively multiplying cell.

vehicle Inanimate carrier of an infectious agent from one host to another.

vertical evolution Acquisition of antimicrobial resistance through spontaneous mutation.

vertical transmission Transfer of a pathogen from a pregnant woman to the fetus, or from a mother to her infant during childbirth.

vibrio (pl. **vibrios**) Short, curved rod-shaped bacterial cell.

villus (pl. **villi**) Narrow protrusion from a membrane such as the intestinal lining.

viremia Viruses circulating in the bloodstream.

virion Viral particle in its inert extracellular form.

viroid Piece of RNA that does not have a protein coat but does replicate within living cells.

virulence Relative ability of a pathogen to overcome body defenses and cause disease; properties of microorganisms that assist pathogenicity.

virus Acellular or nonliving agent composed of nucleic acid surrounded by a protein coat.

vitamin One of a group of organic compounds found in small quantities in natural foodstuffs that are necessary for the growth and reproduction of an organism; usually converted into coenzymes.

volutin Storage form of phosphate found inside certain bacterial cells; because granules of volutin exhibit characteristic staining with the dye methylene blue, they are called metachromatic granules.

---

water activity (a_w) Quantitative measure of the water available.

water molds Nonphotosynthetic members of the heterokons; similar to the fungi but not related genetically.

Western blot Procedure that uses a labeled antibody to detect specific proteins; similar in principle to a Southern blot.

whey Liquid portion that remains after milk proteins coagulate during cheese-making.

wide host range plasmid Plasmid that can replicate in unrelated bacteria.

wild type Form of an organism that is isolated from nature.

---

yeast Unicellular fungus.

yeast artificial chromosome (YAC) Vector that can be used to clone segments of DNA up to a million nucleotides in length and used to introduce foreign DNA into the yeast *Saccharomyces cerevisiae*.

---

zone of inhibition Region around a chemical saturated disc where bacteria are unable to grow due to adverse effects of the compound in the disc.

zoonosis (pl. **zoonoses**) Disease of animals that can be transmitted to humans.

zooplankton Floating and swimming small animals and protozoa found in marine environments, usually in association with the phytoplankton.

zygote Diploid cell formed by the sexual fusion of two haploid cells.

Photographs

Table of Contents, Part Openers

1: E. coli: © Dr. Gopal Murti/SPL/Custom Medical Stock Photo; 2: T phage: © Oliver Meckes/Photo Researchers, Inc.; 3: Ragweed: © Ralph Eagle/Photo Researchers, Inc.; 4: Staphylococcus: © SPL/Photo Researchers, Inc.; 5: Farm: © Grant Heilman Photography

Chapter One

Chapter Opener 1: © Science VU/Visuals Unlimited; Fig. 1.1: Courtesy J. William Schopf; Fig. 1.4: Courtesy of J.P. Dalmasso; Fig. 1.5a: Courtesy of L. Santo; L. Santo, H. Hohl, and H. Frank, *J. Bacteriol.* 99:824, 1969. American Society for Microbiology; Fig. 1.5b: © Terry J. Beveridge/Visuals Unlimited; Fig. 1.6: Courtesy of Dr. Thomas Fritsche; Fig. 1.7a, Fig. 1.7b: © David M. Phillips/Visuals Unlimited; Fig. 1.7c: © Cabisco/Visuals Unlimited; Fig. 1.8a: © Alex Rakosy/Custom Medical Stock Photos; Fig. 1.8b: © Cabisco/Visuals Unlimited; Fig. 1.8c: © V. Burmeister/Visuals Unlimited; Fig. 1.9a: © A.M. Siegelman/Visuals Unlimited; Fig. 1.9b: © C. Shih & R. Kessel/Visuals Unlimited; Fig. 1.9c: © Robert Simpson/Tom Stack & Associates; Fig. 1.10a: Courtesy Gregory Antipa, San Francisco State University, Eugene Small, Donald Marszalek; Fig. 1.10b: © Manfred Kage/Peter Arnold; Fig. 1.11a: Original photo courtesy G.W. Kelley, from *A Pictorial Presentation of Parasites*, edited by H.Zaiman; Fig. 1.11b: © Manfred Kage/Peter Arnold, Inc.; Fig. 1.11c: © E.J.Cable/Tom Stack & Associates; Fig. 1.12a: © K.G. Murti/Visuals Unlimited; Fig. 1.12b: © Thomas Broker/Phototake; Fig. 1.12c: © K. G. Murti/Visuals Unlimited; Fig. 1.13: Dr. Diemer, U.S. Department of Agriculture; Fig. 1.14: © Stanley B. Prusiner/Visuals Unlimited; Perspective 1.1, Fig. 1: Courtesy of Esther R. Angert and Norman R. Pace, Indiana University; Perspective 1.1, Fig. 2: Courtesy J. N. Schulz/Max Planck Institute for Marine Microbiology; Perspective 1.1, Fig. 3: Courtesy Dr. Philippa J.R. Uwins.

Chapter Two

Chapter Opener 2: © Scott Camazine/Photo Researchers, Inc.; Perspective 2.1, Fig. 1a: Courtesy of David M. Prescott; Perspective 2.1, Fig. 1b: Courtesy of David M. Prescott.

Chapter Three

Chapter Opener 3: © Terry Beveridge/BPS/Tony Stone Images; Fig. 3.1: Photograph courtesy of Leica, Inc., Deerfield, FL; Fig. 3.2 top, Fig. 3.2 bottom: Courtesy of W.A. Jensen; Fig. 3.3a: Courtesy of Dr. Edwin R. Jones; Fig. 3.4: Courtesy of Roger Burks, University of California Riverside, and Mark A. Schneegurt, University of Notre Dame, and Cyanosite (http://www-cyanosite.bio.purdue.edu/index.html); Fig. 3.5: © P. Johnson/Photo Researchers, Inc.; Fig. 3.6: © T.E. Adams/Visuals Unlimited; Fig. 3.7: Evans Roberts; Fig. 3.8: Courtesy Center for Biofilm Engineering, Montana State University-Bozeman, Image by Wendy Cochran.; Fig. 3.10a: © R. Kessel & C. Shih/Visuals Unlimited; Fig. 3.10b: © H. Aldrich/Visuals Unlimited; Fig. 3.11: © David M. Phillips/Visuals Unlimited; Fig. 3.12: © Science VU/Visuals Unlimited; Fig. 3.14b: © Leon J. Lebeau/Biological Photo Service; Fig. 3.15: © John D. Cunningham/Visuals Unlimited; Fig. 3.16: © Arthur M. Siegelman/Visuals Unlimited; Fig. 3.17: © Jack M. Bostrack/Visuals Unlimited; Fig. 3.18: © E. Chan/Visuals Unlimited; Fig. 3.19a: Courtesy Molecular Probes; Fig. 3.19b: (Richard L. Moore/Biological Photo Service; Fig. 3.19c: (Evans Roberts; Fig. 3.20a, Fig. 3.20b: © David M. Phillips/Visuals Unlimited; Fig. 3.20c: © Dennis Kunkel/Phototake Inc.; Fig. 3.20d: © Veronika Burmeister/Visuals Unlimited; Fig. 3.20e, Fig. 3.20f: © David M. Phillips/Visuals Unlimited; Fig. 3.21a: Courtesy William Stoeckenius; Fig. 3.21b: Courtesy James T. Staley; Fig. 3.22a-1: © George Musil/Visuals Unlimited; Fig. 3.22a-2: © David M. Phillips/Visuals Unlimited; Fig. 3.22b: © R. Kessel/G. Shih/Visuals Unlimited; Fig. 3.22c: © Oliver Meckes/Photo Researchers, Inc.; Fig. 3.31: Courtesy of Dale C. Birdsell; Fig. 3.33c: © S.C. Holt/Biological Photo Service; Fig. 3.34c: © John J. Cardamone, Jr./Biological Photo Service; Fig. 3.36: Courtesy of E.S. Boatman; Fig. 3.37a: Courtesy of K.J. Cheng and J.W. Costerton; Fig. 3.37b: Courtesy of A. Progulske and S.C. Holt, *Journal of Bacteriology* 143:1003-1018, 1980; Fig. 3.38: © Science VU/EPA/Visuals Unlimited; Fig. 3.39a: © Fred Hossler/Visuals Unlimited; Fig. 3.39b: © Science VU/Visuals Unlimited; Fig. 3.42: © D. Balkwill and D. Maratea/Visuals Unlimited; Fig. 3.43a: Courtesy of C. Brinton, Jr.; Fig. 3.43b: Courtesy Harley W. Moon, U.S. Department of Agriculture; Fig. 3.44a: © CNRI/SPL/Photo Researchers, Inc.; Fig. 3.44b: ; Fig. 3.46: Courtesy of E.S. Boatman; Fig. 3.47: Courtesy of J.F.M. Hoeniger, *J. Bacteriol.* 96:1835, 1968. American Society for Microbiology.; Fig. 3.53: © M. Schliwa/Visuals Unlimited; Fig. 3.54 right: © Dr. Gopal Murti/Photo Researchers, Inc.; Fig. 3.55a: © Biophoto Associates/Photo Researchers, Inc.; Fig. 3.55b: © Garry T. Cole, Ph.D., Dept. of Botany, Univ. of Texas/Biological Photo Service; Fig. 3.56a: Courtesy of A.L. Olins; A.L. Olins and D.E. Olins, *Science* 183:330, 1974. Copyright © 1974 American Association for the Advancement of Science; Fig. 3.57b: © Keith Porter/Photo Researchers, Inc.; Fig. 3.59: © B.F. King/Biological Photo Service; Fig. 3.60: © M. Powell/Visuals Unlimited.

Chapter Four

Chapter Opener 4: © A. Brooks/Tony Stone Images; Fig. 4.1: © William H. Mullins/Photo Researchers, Inc.; Fig. 4.2: © Michael Gabridge/Visuals Unlimited; Fig. 4.3: © Fred E. Hossler/Visuals Unlimited; Fig. 4.8a: © Dennis Kunkel/Phototake Inc.; Fig. 4.8b: © K. Talaro/Visuals Unlimited; Fig. 4.13: © WHOI-D. Foster/Visuals Unlimited; Fig. 4.14a: © Fred Hossler/Visuals Unlimited; Fig. 4.14b: © G. W. Willis, M.D./Biological Photo Service; Fig. 4.15: © Michael Gabridge/Visuals Unlimited; Fig. 4.17: Supplied by Forma Scientific, Inc.; Fig. 4.21 a, Fig. 4.21 b: Courtesy of Dr. Eshel Ben-Jacob; Fig. 4.23: Courtesy Center for Biofilm Engineering, Montana State University-Bozeman, Image by P. Stoodley.

Chapter Five

Chapter Opener 5: © Southern Illinois University/Photo Researchers, Inc.; Fig. 5.1a: © Yoav Levy/Phototake Inc.; Fig. 5.1b, Fig. 5.1c: © Rosenfeld Images/LTD/SPL/Photo Researchers, Inc.; Fig. 5.1d: © Nieto/Herrican/Photo Researchers, Inc.; Fig. 5.5a, Fig. 5.5b: Evans Roberts; Fig. 5.7: © Dennis Kunkel/Phototake Inc.; Fig. 5.8: © Pall/Visuals Unlimited; Fig. 5.9: Evans Roberts.

Chapter Six

Chapter Opener 6: © Brock May/Photo Researchers, Inc.; Fig. 6.2: © Farrell Grehan/Photo Researchers, Inc.; Fig. 6.3b: Photodisc Vol Series 74, photo by Robert Glusie; Fig. 6.3c: © Jane Burton/Bruce Coleman Inc.

Chapter Seven

Chapter Opener 7: © D. Struthers/Tony Stone Images; Fig. 7.4: From J. Cairns, "The Chromosome of E. coli" in *Cold Spring Harbor Symposia on Quantitative Biology*, 27. © 1963 by Cold Spring Harbor Laboratory Press.; Fig. 7.22: From Blattner F.R., Plunkett G. 3rd, Block C.A., Perna N.T., Burland V., Riley M., Collado-Vides J., Glasner J.D., Rode C.K., Mayhew G.F., Gregor J., Davis N.W., Kirkpatrick H.A., Goeden M.A., Rose D.J., Mau B., Shao Y., The Complete Genome Sequence of Escherichia coli K-1", *Science*, 1997 Sep. 5;277 (5331):1453-74.

Chapter Eight

Chapter Opener 8: © Dr. Gopal Murti/ SPL/Custom Medical Stock Photo; Fig. 8.5: © Matt Meadows/Peter Arnold, Inc.; Fig. 8.17: Courtesy Jerome Perry; Fig. 8.23: Courtesy of C. Brinton, Jr. and J. Carnahan.

Chapter Nine

Chapter Opener 9: © Jean Claude Levy/ Phototake NYC; Fig. 9.1 right: © James Holmes/CellMark Diagnostic/SPL/Photo Researchers, Inc.; Fig. 9.1 left: © Leonard Lessin/Peter Arnold, Inc.; Fig. 9.1 center: Courtesy Dr. Ingo Potrkus; Fig. 9.3: Courtesy Milton P. Gordon and Syed Tanveer Haider; Fig. 9.4: Courtesy Pamela Silver and Jason Kahana, Harvard Medical School, Dana Farber Cancer Institute; Fig. 9.8: ; Fig. 9.11: Reprinted with permission from Edvotek, Inc.; Fig. 9.14: Courtesy Bio-Rad Laboratories; Fig. 9.18: © Richard T. Nowitz/Photo Researchers, Inc.; Fig. 9.22: © Simon Fraser/ SPL/Photo Researchers, Inc.; Fig. 9.23: © Jean Claude Levy/Phototake.

Chapter Ten

Chapter Opener 10: © Meckes/Ottawa/Photo Researchers, Inc.; Fig. 10.3a: © George Wilder/ Visuals Unlimited; Fig. 10.3b, Fig. 10.3c: Courtesy of Dr. Thomas R. Fritsche, M.D., Ph.D., Clinical Microbiology Division, University of Washington, Seattle; Fig. 10.4a: © E. Koneman/Visuals Unlimited; Fig. 10.4b: © Biophoto Associates/Photo Researchers, Inc.; Fig. 10.5a, Fig. 10.5b, Fig. 10.5c: © Raymond B. Otero/Visuals Unlimited; Fig. 10.7a: © Analytab Products, A division of Sherwood Medical; Fig. 10.7b: Courtesy Becton Dickinson; Fig. 10.11: Courtesy Patricia Ezekiel Moor; Fig. 10.13c, Fig. 10.14: Evans Roberts; Fig. 10.15a: © Tom E. Adams/ Visuals Unlimited; Fig. 10.15b: J. William Schopf. Reprinted with permission from *Science* 260, April 30, 1993, p. 643, Figure 4A. American Association for the Advancment of Science.

Chapter Eleven

Chapter Opener 11: © F. Widdel/Visuals Unlimited; Fig. 11.1: USGS Photo by David E. Wieprocht; Fig. 11.3a: © F. Widdel/Visuals Unlimited; Fig. 11.3b: © Ralph Robinson/ Visuals Unlimited; Fig. 11.4a: © Thomas Tottleben/Tottleben Scientific Company; Fig. 11.4b: © David M. Phillips/Visuals Unlimited; Fig. 11.5: © John Walsh/Photo Researchers, Inc.; Fig. 11.6a: From *ASM News* 53(2): cover, 187, American Society for Microbiology. Photo by H. Kaltwasser; Fig. 11.6b: From M. P. Starr et al (Eds.), *The Prokaryotes.* Springer-Vergla; Fig. 11.7: From J. G. Holt (Ed.), *The Shorter Bergey's Manual of Determiniative Bacteriology*, 8/e, 1977. Williams and Wilkins Col, Baltimore; Fig. 11.8a: Courtesy Isao Inouye, University of Tsukuba, Mark A. Schneegurt, Wichita State University, and Cyanosite (http://www-cyanosite.bio.purdue.edu/index.html); Fig. 11.8b: © M.I.Walker/Photo Researchers, Inc.; Fig. 11.9: Courtesy Wayne Carmichael, Wright State University, Mark A. Schneegurt, Wichita State University, and Cyanosite (http://www-cyanosite.bio.purdue.edu/index.html); Fig. 11.10: Courtesy Roger Burks, University of California Riverside, Mark A. Schneegurt, Wichita State University and Cyanosite (http://www-cyanosite.bio.purdue.edu/index.html); Fig. 11.11a: Courtesy of J.T. Staley and J.P. Dalmasso; Fig. 11.11b: Courtesy of Dr. James Staley; Fig. 11.12: © John D. Cunningham/Visuals Unlimited; Fig. 11.13: © Dennis Kunkel/Phototake NYC; Fig. 11.14: Evans Roberts; Fig. 11.15: © CNRI/ Phototake NYC; Fig. 11.17a: © Eric Grave/ Photo Researchers, Inc.; Fig. 11.17b: © Soad Tabaqchali/Visuals Unlimited; Fig. 11.18a/b: Courtesy Harold L. Sadoff; Fig. 11.19a: Courtesy of Mary F. Lampe; Fig. 11.19b: © Patricia L. Grilione/Phototake; Fig. 11.20: From S.T. Williams, M.E. Sharpe, and J.G. Holt (Eds.), *Bergey's Manual of Systematic Bacteriology*, Vol. 4. © 1989 Williams and Wilkins Col, Baltimore. Micrograph from T. Cross, U. of Bradford, Bradford, U.K.; Fig. 11.21: © John D. Cunningham/Visuals Unlimited; Fig. 11.22a: © John Cunningham/ Visuals Unlimited; Fig. 11.22b: From N. R. Krieg and J.G. Holt (Eds.), *Bergey's Manual of Systematic Bacteriology*, Vol. 1, 1984. Williams and Wilkins Col, Baltimore; Fig. 11.23, 11.24a: Courtesy of J.T. Staley and J.P. Dalmasso; Fig. 11.25a: From J.T. Staley, M.P. Bryant, N. Pfennig, and J.G. Holt (Eds.) *Bergey's Manual of Systematic Bacteriology*, Vol. 3. © 1989 Williams and Wilkins Col, Baltimore; Fig. 11.26a: © Alfred Pasieka/Peter Arnold, Inc.; Fig. 11.27a: Courtesy of Amy Cheng Vollmer, Swarthmore College; Fig. 11.27b: © Kenneth Lucas, Steinhart Aquarium/Biological Photo Service; Fig. 11.28: Courtesy of Betsy L. Williams; Fig. 11.29: © Michael Gabridge/Visuals Unlimited; Fig. 11.30: © Alfred Pasieka/Peter Arnold, Inc.; Fig. 11.31: Courtesy of R. R. Friis, *J. Bacteriol.* 110:706, 1972. American Society for Microbiology; Fig. 11.32: © Wayne P. Armstrong, Life Sciences Department, Palomar College; Fig. 11.33: Courtesy Lansing Prescott.

Chapter Twelve

Chapter Opener 12: © John D. Cunningham/ Visuals Unlimited; Fig. 12.2: © Cabisco/ Visuals Unlimited; Fig. 12.2b: © John D. Cunningham/Visuals Unlimited; Fig. 12.4: Martha Nester; Fig. 12.6: © David M. Phillips/ Visuals Unlimited; Fig. 12.8: © Arthur M. Siegelman/Visuals Unlimited; Fig. 12.9a, Fig. 12.9b, Fig. 12.9c: Courtesy of Fritz Schoenknecht; Fig. 12.12a: © E. Chan/Visuals Unlimited; Fig. 12.12b: © Inga Spence/Tom Stack & Associates; Fig. 12.12c: © Bob Pool/ Tom Stack & Associates; Fig. 12.13: © A.B. Dowsett/Photo Researchers, Inc.; Fig. 12.16a: Martha Nester; Fig. 12.16b: © Bruce Iverson; Fig. 12.18b: Martha Nester; Fig. 12.18c: Martha Nester; Fig. 12.18d: Martha Nester; Fig. 12.20b: © Carolina Biological Supply/ Phototake; Fig. 12.22, Fig. 12.24, Fig. 12.25: Courtesy of S. Eng and F. Schoenknecht.

Chapter Thirteen

Chapter Opener 13: © Oliver Meckes/Photo Researchers, Inc.; Fig. 13.1d: Courtesy of Harold Fisher, University of Rhode Island and Robley Williams, University of California at Berkeley; Fig. 13.1e: © Dennis Kunkel/ Phototake; Fig. 13.1f: © Tomas Broker/ Phototake; Fig. 13.9a: Courtesy Joachim Messing, from Messing, J., Gronenbom, B., Muller-Hill, B. and Hofschneider P.H. (1977). Filamentous coliphage M13 as a cloning vehicle: Insertion of a Hind II fragment of the lac regulatory region in the M13 replicative form in vitro. Proc. Natl. Acad. Sci. USA 74, 3642-3646.; Fig. 13.14a: © Prof. L. Caro/ SPL/Photo Researchers, Inc.; Fig. 13.14b: Courtesy of Dr. Rafael Martinez, University of California, Los Angeles; Fig. 13.14c: © Oliver Meckes/Photo Researchers, Inc.

Chapter Fourteen

Chapter Opener 14: © Scott Camazine/ Photo Researchers, Inc.; Fig. 14.1a: © CDC/ Science Source/Photo Researchers, Inc.; Fig. 14.1b: © K. G. Murti/Visuals Unlimited; Fig. 14.1c: Courtesy Dr. Frederick A. Murphy, Centers for Disease Control and Prevention; Fig. 14.2: © Michael Gabridge/Custom Medical Stock Photo; Fig. 14.3a, Fig. 14.3b, 14.3c: Courtesy of M. Cooney; Fig. 14.4a: © Terry Hazen/Visuals Unlimited; Fig. 14.4b: © E. Chan/Visuals Unlimited; Fig. 14.5: Courtesy of Fred Williams, National Exposure Research Laboratory/U.S. EPA; Fig. 14.6b: Courtesy of M. Cooney; Fig. 14.11b: © Dr. Jack Griffith; Fig. 14.19a: © Dean A. Glawe/Biological Photo Services; Fig. 14.19b: © Bruce Iverson; Fig. 14.19c: © Dean Glawe/ Biological Photo Service; Perspective 14.1 Fig. 1: From W. Archer and M.J. Fraser, Dept. of Biological Sciences, The University of Notre Dame; Fig. 14.21: © Science VU/Visuals Unlimited; Fig. 14.21: © Barbara J. Miller/ Biological Photo Service.

Chapter Fifteen

Chapter Opener 15: © David M. Phillips/ Photo Researchers, Inc.; Fig. 15.4b: ; Fig. 15.8: © Reprinted from Wendell F. Rosse et al., "Immune Lysis of Normal Human and Parozysmal Nocturnal Hemogloginuria (PNH) Red Blood Cells," *Journal of Experimental Medicine*, 123:969, 1966. Rockefeller University Press.

Chapter Sixteen

Chapter Opener 16: Boehringer Ingelheim International, Lennart Nilsson; Fig. 16.10: Courtesy of Dr. James W. Vardiman, University of Chicago; Fig. 16.10b: © NIBSCS/ SPL/Photo Researchers, Inc.; Fig. 16.10c: © D. Fawcett/Visuals Unlimited; Perspective 16.1: © *The Seattle Times*; Fig. 16.19 a/b: From D. Zagury, J. Bernard, N. Thierness, M. Feldman, and G. Berke, Eur. J. Immunol. 5:818-822, 1975.

Chapter Seventeen

Chapter Opener 17: © Kevin Noran/Tony Stone Images; Fig. 17.1: © Terry C. Haze/ Visuals Unlimited; Fig. 17.6: © Ed Reschke/ Peter Arnold, Inc.; Fig. 17.7c: Centers for Disease Control; Fig. 17.10: © The McGraw-Hill Companies, Inc./ Bob Coyle, photographer;

Fig. 17.11: Courtesy of Philip Mackham, Advanced Bioscience Laboratories, Inc.

Chapter Eighteen

Chapter Opener 18: © Ralph Eagle/Photo Researchers, Inc.; Fig. 18.2: © STU/Custom Medial Stock Photo; Fig. 18.6: Courtesy Kathy Parks; Fig. 18.8: © Kenneth E. Greer/Visuals Unlimited; Fig. 18.9: © SIU/Visuals Unlimited.

Chapter Nineteen

Chapter Opener 19: © SPL/Photo Researchers, Inc.; Fig. 19.6: © A.M. Siegleman/Visuals Unlimited; Fig. 19.12: Courtesy Pascale F. Cossart, From Finlay, B. and Cossart, P.F., *Science* 276:720, 4 April, 1997.

Chapter Twenty

Chapter Opener 20: © Gianni Tortoli/Photo Researchers, Inc.; Fig. 20.1a-1: Couple: © PhotoDisc, Inc.; Fig. 20.1a-2: Mouse: © Corbis Digital Stock CD; Fig. 20.1b-1: Shaking hands: © PhotoDisc, Inc.; Fig. 20.1b-2: Tick: Courtesy University of Nebraska/ Department of Entomology; Fig. 20.1b-3: Phone: © PhotoDisc, Inc.; Fig. 20.1b-4: Sneeze: © Kent Wood/Photo Researchers, Inc.; Fig. 20.1b-5: Foodgroups: © PhotoDisc, Inc.; Fig. 20.3a: © Camermann International; Fig. 20.3b: © Camermann International; Fig. 20.9: © WHOI, Geneva Switzerland; Fig. 20.13: © B. Seitz/Photo Researchers, Inc.

Chapter Twenty-One

Chapter Opener 21: © SPL/Photo Researchers, Inc.; Fig. 21.5: Reprinted with permission of Robert L. Perkins from A.S. Klainer & R.L. Perkins, *Journal of Infectious Diseases*, 12:4, 1970 © University of Chicago Press; Fig. 21.10: © Fred Marsik/Visuals Unlimited; Fig. 21.11: Courtesy bioMerieux, St. Louis, MO.; Fig. 21.12: Courtesy AB Biodisk, Solna Sweden

Chapter Twenty-Two

Chapter Opener 22: (© SPL/Photo Researchers, Inc.; Fig. 22.2: Courtesy Karen A. Holbrook; Fig. 22.3a, 22.3b: © Carroll H. Weiss/Camera M.D.; Fig. 22.3c: Courtesy Glenn D. Roberts, Mayo Clinic; Fig. 22.5: Courtesy National Institute Slide Bank/The Wellcome Centre for Medical Sciences; Fig. 22.6: © SPL/Photo Researchers, Inc.; Fig. 22.7: Evans Roberts; Fig. 22.9: © Ken Greer/Visuals Unlimited; Fig. 22.10: Rocky Mountain Laboratory, Hamilton, Montana, Centers for Disease Control; Fig. 22.12: © Ed Reschke/Peter Arnold, Inc.; Fig. 22.13: © C.C. Duncan/ Medical Images, Inc.; Fig. 22.14: © Charles W. Stratton/Visuals Unlimited; Fig. 22.16: © C.C. Duncan/Medical Images, Inc.; Fig. 22.18: © John D. Cunningham/Visuals Unlimited; Fig. 22.19: © Carroll H. Weiss/ Camera M.D. Studios; Fig. 22.20: © Oliver Meckes/Gelderblom/Photo Researchers, Inc.; Fig. 22.21: © Lowell Georgia/Photo Researchers, Inc.; Fig. 22.22: © Dr. P. Marazzi/ SPL/Photo Researchers, Inc.; Fig. 22.24: © Carroll H. Weiss/Camera M.D.; Fig. 22.26: © CNRI/SPL/Photo Researchers, Inc.; Fig. 22.27a: © Dr. P. Marazzi/SPL/Photo Researchers, Inc.; Fig. 22.27b: Evans Roberts; Fig. 22.28a: © Dr. P. Marazzi/SPL/Photo Researchers, Inc.; Fig. 22.28b: © A.M. Siegelman/Visuals Unlimited.

Chapter Twenty-Three

Chapter Opener 23: © Omikron/Photo Researchers, Inc.; Fig. 23.2, Fig. 23.5: Evans Roberts; Fig. 23.7: Courtesy Smith & Nephew Ent.; Fig. 23.8: © A.B. Dowsett/SPL/Photo Researchers, Inc.; Fig. 23.9: Evans Roberts; Fig. 23.10a: © St. Francis Hospital/ Cameramann Int'l; Fig. 23.10b: © St. Francis Hospital/Cameramann Int'l; Fig. 23.11a: Evans Roberts; Fig. 23.11b: © Larry Jensen/ Visuals Unlimited; Fig. 23.12: © Michael Gabridge; Fig. 23.13: Courtesy of J.B. Baseman; from P.C. Hu, A.M. Collier, and J.B. Baseman, *J. Exp. Med.* 145:1328, 1977. Rockefeller University Press; Fig. 23.14: Courtesy Dr. Kenneth Bromberg; Fig. 23.18: © BioPhoto Assoc./Photo Researchers, Inc.; Fig. 23.19: © John D. Cunningham/Visuals Unlimited; Fig. 23.19 b: © CNRI/Phototake; Fig. 23.20: © SPL/Photo Researchers, Inc.; Fig. 23.21: © Science VU/Visuals Unlimited; Fig. 23.25a: Courtesy of the Centers for Disease Control; Fig. 23.25b: © E. Chan/Visuals Unlimited; Fig. 23.27a, Fig. 23.27b: Evans Roberts.

Chapter Twenty-Four

Chapter Opener 24: © F. Hossler/Visuals Unlimited; Fig. 24.3: Courtesy of W. Fischischweiger and D.O. Birdsell; Fig. 24.5a, Fig. 24.5b, Fig. 24.6a: Courtesy of R.C. Page; from S. Schluger, R.A. Yaodel's and R.C. Page, *Peridontal Disease*, 1977. Lea & Febiger, Philadelphia; Fig. 24.6b: Evans Roberts; Fig. 24.9: © Frederick C. Skvara, M.D.; Fig. 24.10: © Science VU/Visuals Unlimited; Fig. 24.13: © Veronika Burmeister/Visuals Unlimited; Fig. 24.15: Courtesy Philippe J. Sansonetti, M.D., Professeur Institut Pasteur; Fig. 24.16: © David M. Rollins/Visuals Unlimited; Fig. 24.17: © Moredun Animal Health/SPL/Photo Researchers, Inc.; Fig. 24.18: © Science Source/Photo Researchers, Inc.; Fig. 24.23: © P.M. Motta & F.M. Magliocca/SPL/Photo Reserchers, Inc.; Fig. 24.24: © Michael Abbey/Photo Researchers, Inc.; Fig. 24.25: © Oliver Meckes/Photo Researchers, Inc.; Fig. 24.26a,b, 24.27b, 24.28 lower, 24.31, 24.33a, 24.34: Courtesy of S. Eng and F. Schoenknecht; Fig. 24.27a: © Arthur M. Siegelman/Visuals Unlimited; Fig. 24.28 top: © Bruce Iverson/Visuals Unlimited; Fig. 24.29d: © Bruce Iverson; Fig. 24.30: © Carroll Weiss/Camera MD Studios.

Chapter Twenty-Five

Chapter Opener 25: © Kwangshin Kim/Photo Researchers, Inc.; Fig. 25.3: Science VU/Fred Marsik/Visuals Unlimited; Fig. 25.4: © Mary Stallone/Medical Images, Inc.; Fig. 25.5: Courtesy of the Centers for Disease Control; Fig. 25.8: © Biophoto Assoc./Photo Researchers, Inc.; Fig. 25.9: Courtesy of Stanley Falkow; Fig. 25.10: Courtesy of Morris D. Cooper, Ph.D., Professor of Medical Microbiology, Southern Illinois University School of Medicine, Springfield, IL.; Fig. 25.11: Courtesy National Medical Slide Bank/ The Wellcome Centre for Medical Sciences; Fig. 25.12: Courtesy of S. Eng and F. Schoenknecht; Fig. 25.13: Courtesy Thomas Diepgen, University Erlangen; Fig. 25.14: © Carroll Weiss/Camera MD Studios; Fig. 25.15: © L. Winograd/Biological Photo Service; Fig. 25.16a: ©Carroll H. Weiss/ Camera MD Studios; Fig. 25.16b: © Harvey Blank, M.D./Camera MD Studios; Fig. 25.17: Courtesy National Medical Slide Bank/ The Wellcome Centre for Medical Sciences; Fig. 25.18: © SPL/Photo Researchers, Inc.; Fig. 25.19: © Biophoto Associates/Photo Researchers, Inc.; Fig. 25.21: © David M. Phillips/Visuals Unlimited; Fig. 25.22, 25.23a: Courtesy of S. Eng and F. Schoenknecht.

Chapter Twenty-Six

Chapter Opener 26: © Dr. Linda Stannard, UCT/Science Photo Library/Photo Researchers, Inc.; Fig. 26.4: © John Durham/ SPL/Photo Researchers, Inc.; Fig. 26.6: © Kenneth Greer/Visuals Unlimited; Fig. 26.7: © Mary K. Stallone/The Stock Shop, Medi-Chrome; Fig. 26.9: © Paolo Koch/Photo Researchers, Inc.; Fig. 26.10: © John D. Cunninham/Visuals Unlimited; Fig. 26.11: © Joe McDonald/Visuals Unlimited; Fig. 26.12: Evans Roberts; Fig. 26.15: Courtesy National Medical Slide Bank/The Wellcome Centre for Medical Sciences; Fig. 26.16: Evans Roberts; Fig. 26.17: © Bettmann/Corbis; Fig. 26.19a: © Tektoff-BM/CNRI/SPL/Photo Researchers, Inc.; Fig. 26.19b: © Tobey Sanford; Fig. 26.20: Evans Roberts; Fig. 26.21: © Mary K. Stallone/ The Stock Shop, MediChrome; Fig. 26.22: © Cabisco/Visuals Unlimited.

Chapter Twenty-Seven

Chapter Opener 27: © Alfred Pasieka/Peter Arnold, Inc.; Fig. 27.3: © Michael English/ Medical Images, Inc.; Fig. 27.4: Evans Roberts; Fig. 27.5: © Kenneth E. Greer/Visuals Unlimited; Fig. 27.6a: © Leonard Morse, M.D./Medical Images, Inc.; Fig. 27.6b: © Carroll H. Weiss/Camera M.D.; Fig. 27.6c: © A.M. Siegelman/Visuals Unlimited; Fig. 27.7: Courtesy National Medical Slide Bank/ The Wellcome Centre for Medical Sciences; Fig. 27.8: © A.M. Siegelman/Visuals Unlimited; Fig. 27.12, 27.13: Courtesy National Medical Slide Bank/The Wellcome Centre for Medical Sciences; Fig. 27.14: Courtesy of Dr. John Slack; Fig. 27.15: © Kenneth E. Greer/Visuals Unlimited; Fig. 27.16: © Leonard Morse, M.D./Medical Images, Inc.; Fig. 27.17a, b: Evans Roberts.

Chapter Twenty-Eight

Chapter Opener 28: © Dr. Tony Brain/Science Photo Library/Photo Researchers, Inc.; Fig. 28.2: © Ken E. Greer/Visuals Unlimited; Fig. 28.4: Centers for Disease Control, Atlanta; Fig. 28.6: © A.M. Siegelman/Visuals Unlimited; Fig. 28.7a, b: Courtesy Joe Hinnebusch, From Hinnebusch, B.J., Perry, R.D., and Schwan, T.G., *Science* 273:367, 19 July 1996.; Fig. 28.9a: © Ed Reschke; Fig. 28.9b: © Jacquin Carrillo-Farga/Science Source/Photo Researchers, Inc.; Fig. 28.14: Courtesy Centers for Disease Control.

Chapter Twenty-Nine

Chapter Opener 29: © Manfred Kage/Peter Arnold, Inc.; Fig. 29.2: Courtesy Stuart Fischman, D.M.D.; Fig. 29.12: © St. Mary's Hospital Medical School/SPL/Photo Researchers, Inc.; Fig. 29.13: © Cecil H. Fox/Photo Researchers, Inc.; Fig. 29.14a: © E. Koneman/Visuals Unlimited; Fig. 29.14c: Courtesy of Dr. Thomas R. Fritsche, M.D., Ph.D., Clinical Microbiology Division, University of Washington, Seattle; Fig. 29.15: © Dr. F.C. Skvara/Peter Arnold, Inc.; Fig. 29.16: © Paula Inhad/Custom Medical Stock Photo; Fig. 29.17: Courtesy Dr. Edwin P.Ewing/ Centers for Disease Control.

Chapter Thirty

Chapter Opener 30: © Grant Heilman Photography; Fig. 30.1: © Merckes/ Ottawa/Photo Researchers, Inc.; Fig. 30.5: © M. Dworkin-H. Reichenbach/Phototake; Fig. 30.10: © Nancy Pearsall; Fig. 30.12: © John D. Cunningham/Visuals Unlimited; Fig. 30.14: © Manfred Kage/Peter Arnold Inc.; Fig. 30.16: © Science VU/Visuals Unlimited; Critical Thinking page 786: Courtesy J.P. Dalmasso.

Chapter Thirty-One

Chapter Opener 31: © Myrleen Cate/Photo Edit; Fig. 31.10: © Alan L. Detrick/Photo Researchers, Inc.

Chapter Thirty-Two

Chapter Opener 32: © Michael Newman/ PhotoEdit, Inc.; Fig. 32.1: © Paul Webster/ Tony Stone Images; Fig. 32.2, Fig. 32.2b, Fig. 32.2c: Martha Nester; Fig. 32.3: © Robert Calentine/Visuals Unlimited; Fig. 32.4a, Fig. 32.4b, Fig. 32.4c: Courtesy of the Wisconsin Milk Marketing Board; Fig. 32.7a, Fig. 32.7b, Fig. 32.7c, Fig. 32.7d, Fig. 32.7e, Fig. 32.7f: Martha Nester; Fig. 32.8a: © Rachel Epstein/Photo Edit, Inc.; Fig. 32.8b: © William J. Weber/Visuals Unlimited; Fig. 32.9a, Fig. 32.9b: © Felicia Martinez/ PhotoEdit, Inc.

Text and Line Art Credits

Chapter 1

1.3: Adapted from WHO, 1993. *Ten Years of Progress 1984-1993.* The Programme for Vaccine Development.

Chapter 2

2.1, 2.15: Adapted from Raymond Chang *Chemistry 6/e* 1998 The McGraw-Hill Companies. Reprinted by permission. All rights reserved; 2.6, 2.10: adapted from Martin Silberberg Chemistry: The molecular Nature of Matter and Change 1996 The McGraw-Hill Companies. Reprinted by permission. All rights reserved; 2.9 Burton Guttman *Biology* 1999 The McGraw-Hill Companies. Reprinted by permission. All rights reserved; 2.28b, 2.29: adapted from Robert Weaver *Molecular Biology* 1999 The McGraw-Hill Companies. Reprinted by permission. All rights reserved.

Chapter 3

3.3b, 3.9, 3.45, 3.58: Lansing Prescott et. al. *Microbiology 4/e* 1999 The McGraw-Hill Companies. Reprinted by permission. All rights reserved; 3.56b: adapted from Leland Hartwell, et. al. *Genetics: From Genes to Genomes* 2000 The McGraw-Hill Companies. Reprinted by permission. All rights reserved; TA5.1: adapted from Robert Allen *Biology: A Critical Thinking Approach* 1995 The McGraw-Hill Companies. Reprinted by permission. All rights reserved.

Chapter 4

4.9: Table excerpt from Standard Methods for the Examination of Water and Wastewater, *18th edition,* The American Water Works Association, Denver CO; 4.6: Talaro/Talaro *Foundations in Microbiology 2/e* 1996 The McGraw-Hill Companies. Reprinted by permission. All rights reserved; 4.11, 4.22: Lansing Prescott et. al. *Microbiology 4/e* 1999 The McGraw-Hill Companies. Reprinted by permission. All rights reserved.

Chapter 6

6.10: Adapted from Lansing Prescott et. al. *Microbiology* 1996 The McGraw-Hill Companies. Reprinted by permission. All rights reserved; 6.17b, 6.24a: adapted from Raven/Johnson *Biology 5/e* 1999 The McGraw-Hill Companies. Reprinted by permission. All rights reserved.

Chapter 10

10.1: Adapted from G. J. Olsen and C. R. Woese "Ribosomal RNA: A Key to Phylogeny" FASEB Journal, 7: 113-123, 1993; 10.8: adapted from Sigma-Aldrich //www.sigma-aldrich.com; 10.16: J. Mandelstam and K. McQuillen *Biochemistry of Bacterial Growth,* 1968, Blackwell Scientific Publications, Ltd.; 10.2, 10.19: Lansing Prescott et. al. *Microbiology* 1999 The McGraw-Hill Companies. Reprinted by permission. All rights reserved.

Chapter 11

11.2: Adapted from a figure by Ken Nealson, Jet Propulsion Laboratory, California Institute of Technology, Pasadena CA; 11.25b: adapted from Lansing Prescott et. al. *Microbiology* 1999 The McGraw-Hill Companies. Reprinted by permission. All rights reserved.

Chapter 13

13.8, 13.12: Adapted from Lansing Prescott et. al. *Microbiology* 1999 The McGraw-Hill Companies. Reprinted by permission. All rights reserved.

Chapter 15

15.1: Adapted from Sylvia Mader *Inquiry Into Life, 9/e* 2000 The McGraw-Hill Companies. Reprinted by permission. All rights reserved; 15.8a: adapted from Lansing Prescott et. al. *Microbiology* 1999 The McGraw-Hill Companies. Reprinted by permission. All rights reserved.

Chapter 16

16.4: Robert Weaver *Molecular Biology* 1999 The McGraw-Hill Companies. Reprinted by permission. All rights reserved; 16.11: Burton Guttman *Biology* 1999 The McGraw-Hill Companies. Reprinted by permission. All rights reserved.

Chapter 17

Table 17.3: Table adapted from CDC, *MMWR* Vol. 49 (2), 36-37, January 21, 2000; 17.1b: adapted from Talaro/Talaro *Foundations in Microbiology* 1996 The McGraw-Hill Companies. Reprinted by permission. All rights reserved.

Chapter 19

19.10: Adapted from Arousing the Fury of the Immune System, 1998 Howard Hughes Medical Institute.

Chapter 20

20.2: Source: CDC, *MMWR* 45 (6): 135, 1996; 20.4: Source: Summary of Notifiable Diseases, United States, 1997. *MMWR* 46 (No. 54-001); 20.6: Source: CDC, *MMWR* 44 (48): 901, 1995. 20.7: Source: Incidence of Foodborne Illness: Preliminary Data from the Foodborne Disease Active Surveillance Network (FoodNet), United States, 1998 *MMWR* 48 (09): 189-195; 20.8: CDC, *MMWR* (49) 4, Feruary 2000. 20.10: Source: WHO, World Health Report 1999, figure 2.7 20.11: Source: http://www.whitehouse.gov/WH/EOP/OSTP/CISET/html/ciset.html; 20.12: Source: National Nosocomial Infections Surveillance, CDC.

Chapter 21

21.1: From Rolinson, George N., (1998) *Journal of Antimicrobial Chemotherapy* 41, 589-603; 21.7: adapted from Lansing Prescott et. al. *Microbiology* 1999 The McGraw-Hill Companies. Reprinted by permission. All rights reserved.

Chapter 22

22.11: Source: CDC, *MMWR* 42 (53): 48, October 21, 1994. Summary of Notifiable Diseases, United States, 1993; 22.15: Source: CDC, *MMWR* Summary of Notifiable Diseases, United States 1997. 46 (54): 42, November 20, 1998; 22.12: Source: CDC, 1998. *MMWR* 47 (RR-11): 3 and *MMWR* 48 (49): 1127, December 17, 1999; 22.25: Source: CDC. Summary of Notifiable Diseases, United States, 1997. *MMWR* 46 (54): 54, November 20, 1998; 22.8: adapted from Lansing Prescott et. al. *Microbiology* 1996 The McGraw-Hill Companies. Reprinted by permission. All rights reserved.

Chapter 23

23.16: Source: CDC. 1999. Summary of Notifiable Diseases, United States, 1998. *MMWR* http://www.cdc.gov/epo/mmwr/

preview/mmwrhtml/mm.4753al.htm; 23.24: Source: National Center for Infectious Diseases http://www.cdc.gov/ncidod/diseases/hanta/hps/index.htm; *A Field Guide to the Mammals, 3rd edition,* New York, NY, Houghton-Mifflin Co., 1987.

Chapter 24

24.11: Source: CDC 1999. Summary of Notifiable Diseases, United States, 1998. *MMWR* 47 (53): 1-93; 24.19: Source: CDC 1999. Summary of Notifiable Diseases, United States, 1998. *MMWR* 47 (53): 1-93; 24.22: Source: CDC 1999. Summary of Notifiable Diseases, United States, 1998. *MMWR* 47 (53): 1-93.

Chapter 25

25.6: Source: CDC 1994. Summary of Notifiable Diseases, United States 1993. *MMWR* 42 (53): 57; 25.20: Adapted from *Science* 4 December 1998, Vol. 282.

Chapter 26

26.3: Source: CDC 1998, 1999. *MMWR* 47 (46): 993, and Summary of Notifiable Diseases, United States, 1998. *MMWR* 47 (53): viii, 38; 26.5: Source: Schuchat, Anne et. al. Bacterial Meningitis in the United States in 1995, 1997. *The New England Journal of Medicine* 337 (14): 970; 26.8: Source: CDC 1993, 1999. Summary of Notifiable Diseases, United States, 1992, 1998. *MMWR* 41 (55): 41 and 47 (53): 49; 26.14: Source: CDC 1998. *MMWR* 47 (25):517-522; 26.18: Source: CDC 1993, 1999. Summary of Notifiable Diseases, United States, 1992, 1998. *MMWR* 41 (55): 46, 47 (53): 54.

Chapter 27

27.11: Source: CDC, CDC Surveillance Summaries. 1998. *MMWR* 47 (NO. SS-2); 5.

Chapter 28

28.5: Source: CDC, 1994. Summary of Notifiable Diseases, 1993. *MMWR*, 42 (53): 62; 28.12: Adapted from CDC Health Information for International Travel, 1996-1997 and WHO Malaria Recommendations, 1997.

Chapter 29

29.1: Adapted from United Nations HIV/AIDS Program © Knight-Ridder / Tribune, Tuesday, November 24, 1998; 29.9: Adapted from CDC. HIV/AIDS Surveillance in Women, L264 slides series, slide #1, and *HIV/AIDS Surveillance Report* 1999; 11 (No. 1): 3; 29.10: Source: CDC. *HIV/AIDS Surveillance Report* 1999; 11 (No. 1): 36.

Chapter 31

31.1: CDC, Emerging Infectious Diseases, Vol. 3 (4), Oct-Dec. 1997; 31.2: Adapted from *MMWR,* Vol. 47/No. SS-5, 24, December 11, 1998; 31.3: Adapted from *MMWR,* Vol. 47/No. SS-5, 25, December 11, 1998; 31.7: Adapted from Paul Kelter, et. al. *Chemistry: A World of Choices* 1999 The McGraw-Hill Companies. Reprinted by permission. All rights reserved.

Appendix 2

Adapted from Lansing Prescott et. al. *Microbiology* 1999 courtesy of Lansing Prescott, The McGraw-Hill Companies. Reprinted by permission. All rights reserved.

INDEX

Page numbers followed by *t* and *f* indicate tables and figures, respectively.

A

B

C

D

E

F

G

I

J

K

L

M

N

O

P

Q

R

S

T

U

V

W

X

Y

Z

SINGULARS AND PLURALS

Singular	Plural	Singular	Plural
alg*a*	alg*ae*	fung*us*	fung*i*
ameb*a*	ameb*ae*	hydroly*sis*	hydroly*ses*
bacill*us*	bacill*i*	hyph*a*	hyph*ae*
bacteri*um*	bacteri*a*	inocul*um*	inocul*a*
cili*um*	cili*a*	medi*um*	medi*a*
clostridi*um*	clostridi*a*	mucos*a*	mucos*ae*
cocc*us*	cocc*i*	myceli*um*	myceli*a*
conidi*um*	conidi*a*	myco*sis*	myco*ses*
dat*um*	dat*a*	phyl*um*	phyl*a*
diagno*sis*	diagno*ses*	pil*us*	pil*i*
fimbri*a*	fimbri*ae*	nucle*us*	nucle*i*
flagell*um*	flagell*a*	sept*um*	sept*a*
foc*us*	foc*i*	synthe*sis*	synthe*ses*

MEANINGS OF PREFIXES AND SUFFIXES

Prefix or Suffix	Meaning	Example
a-, an-	not, without	avirulent (lacking virulence) anaerobic (without air)
aer-	air	aerobic
anti-	against	antiseptic
-ase	enzyme	penicillinase
chlor-	green	chlorophyll
-chrom-	color	metachromatic (staining differently with the same dye)
-cide	causing death	germicide
co-, com-, con-	together	coenzyme
cyan	blue	pyocyanin (a blue bacterial pigment)
de-	down, from	dehydrate (remove water)
-dem-	people, district	epidemic
end-	within	endospore (spore within a cell)
-enter-	intestine	enteritis (inflammation of the intestine)
epi-	upon	epidermis
erythro-	red	erythocyte (red blood cell)
eu-	well, normal	*Eubacteriales* (the "true" bacteria) eukaryotic (true nucleus)
exo-	outside	exotoxin (toxin found outside a cell)
extra-	outside of	extracellular
flav-	yellow	flavoprotein (a protein containing a yellow enzyme)
-gen	produce, originate	antigen (a substance that induces a production of antibodies)
glyc-	sweet	glycemia (the presence of sugar in the blood)